HVAC and Refrigeration Systems

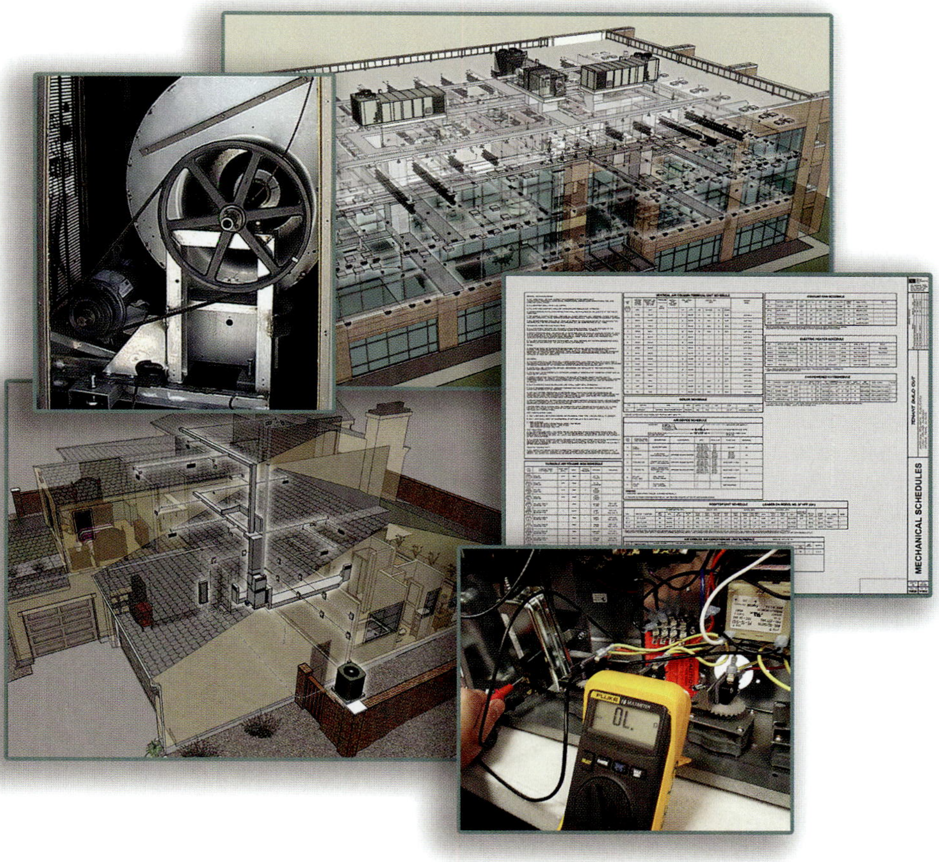

AMERICAN TECHNICAL PUBLISHERS
Orland Park, Illinois 60467-5756

Ronnie J. Auvil

American Technical Publishers, Inc., Editorial Staff

Editor in Chief:
Jonathan F. Gosse
Vice President—Production:
Peter A. Zurlis
Art Manager:
Jennifer M. Hines
Multimedia Manager:
Carl R. Hansen
Technical Editors:
Russell G. Burris
Richard A. Scheffrahn
Copy Editors:
Talia J. Lambarki
Cathy A. Mini
Jeana M. Platz

Cover Design:
Jennifer M. Hines
Illustration/Layout:
Nicholas W. Basham
Melanie G. Doornbos
Robert M. McCarthy
Joshua P. Hugo
Mark S. Maxwell
Jennifer M. Hines
Thomas E. Zabinski
Bethany J. Fisher
CD-ROM Development:
Robert E. Stickley
Daniel Kundrat
Nicole S. Polak
Kathleen A. Moster
Cory S. Butler

1 2 3 4 5 6 7 8 9 – 14 – 9 8 7 6 5 4 3 2

Printed in the United States of America

ISBN 978-0-8269-0785-1

This book is printed on recycled paper.

Acknowledgments

The author and publisher are grateful for the technical information and assistance provided by the following individuals.

Shane McCarthy
Training Director
The Refrigeration Workers of Ontario
United Association Local 787

Jim Coates
Training Director
International Union of Operating Engineers
Local 399—Chicago, Illinois

Arthur P. Abbey
Service Apprentice Instructor
United Association Industry Training Center (UA)
Troy, Michigan

The author and publisher would like to thank the following companies, organizations, and individuals for providing images.

Air Conditioning Contractors of America
Allied Air Enterprises Inc.
American Society of Mechanical Engineers
AMTROL, Inc.
Andersen Windows, Inc.
ASHRAE Handbook-Fundamentals
ATOFINA Chemicals, Inc.
Bacharach, Inc.
Baldor Electric Company
R.W. Beckett Corp
Carlo Gavazzi, Inc.
Carrier Corporation
Cleaver-Brooks
Copeland Corporation
Cooper Instruments
Crane Valves
Crown Boiler Co.
Cutler-Hammer
Danfoss Drives
Detroit Radiant Products Company
Ductsox
Dunham-Bush
Dunkirk
DuPont Company
Dwyer Instruments, Inc.
Electrical Apparatus Service Association, Inc.
Empire Comfort Systems Inc.
FAMCO
FirstAidStore.com
Fluke Corporation
Fluke Thermography
The Fulton Companies
Furnas Electric Company
The Gates Rubber Company
GE Panametrics
Goodman Global Group, Inc.
Greenheck Fan Corp
Greenless Textron Inc.
Heat Wagon, Inc.
Henry Valve Co.

Acknowledgments

Honeywell Chemicals
Honeywell International, Inc.
Ideal Industries, Inc.
INDEECO
International Ground Source Heat Pump Association
International Rectifier
ITT Bell & Gossett
Katy Instrument Sales
Kebco Power Transmission
Klein Tools, Inc.
Lau, a Division of Tomkins Industries
Lab Safety Supply, Inc.
LEESON Electric Corporation
Lennox Industries, Inc.
Loctite Corporation
L.S. Starrett Company
Lunkenheimer Co.
Manchester Tank & Equipment Co.
March Manufacturing, Inc.
McDonnell & Miller
McQuay International
Michigan Precision Fabricators, Inc.
Miller Equipment
Mill-Rose Company
Milwaukee Electric Tool Corporation
Mueller Co.
The New York Blower Company
Noritz America Corp.
Omron Electronics LLC
OSHA Office of Training and Education
Parker Boiler Co.
Parker Hannifin–Pneumatic Division North America
Peerless Boilers
Plan Ahead, Inc.
Ranco Inc.
Reed Manufacturing Co
Rheem Air Conditioning Division
Riello Burners
RIDGID®
Rockwell Automation, Allen-Bradley Company, Inc.
Rockwell Automation/Reliance Electric
Rothenberger USA Inc.
R. W. Beckett Corp.
Siemens
The Sinco Group, Inc.
The Stanley Works
Spirax-Sarco Inc.
Sporlan Valve Company
SPX Robinair
The Stanley Works
Sterling
Super Radiator Coils
Taco, Inc.
TSI Incorporated
Trane
Turbotec Products, Inc.
Unico, Inc.U.S. Boiler Company
U.S. Department of Energy
U.S. Energy Information Administration
Vertiflo Pump Company
Watts Regulator Company
Weil-McLain
White-Rodgers Div., Emerson Electric Co.
Yellow Jacket Div., Ritchie Engineering Company, Inc.
York International Corp.

Contents

Contents

SECTION FOUR: HEATING SYSTEMS

Contents

SECTION FIVE: REFRIGERATION SYSTEMS

Contents

SECTION SEVEN: HVAC AND REFRIGERATION SYSTEMS

Contents

SECTION EIGHT: HVAC SYSTEM LOAD CALCULATIONS, DESIGN, AND BALANCING

Contents

SECTION TEN: ENERGY EFFICIENCY TECHNOLOGY

Digital Resources

- Quick Quizzes®
- Illustrated Glossary
- Forms and Documents
- Interactive Animations
- Library
- Step-by-Step Procedures
- Flash Cards
- Media Library
- Review Questions
- ATPeResources.com

Features

HVAC and Refrigeration Systems covers all aspects of residential and light commercial heating, ventilation, and air conditioning systems, focusing specifically on the operation, installation, service, maintenance, and troubleshooting of these systems. The textbook covers heating and refrigeration fundamentals, psychrometrics, building mechanical systems, and electrical and electronic devices and controls. The textbook also covers air- and water-source heat pump systems and chiller systems and includes 100 installation and 65 step-by-step service procedures. Energy efficiency practices, energy auditing, building commissioning, and retrofitting are covered as part of Energy Star® and LEED® certifications.

HVAC and Refrigeration Systems offers proven and thorough content in a format organized by sections and units. Section and unit color designations help learners find the desired pages in the textbook. The appendix contains many useful tables and charts, as well as refrigerant information.

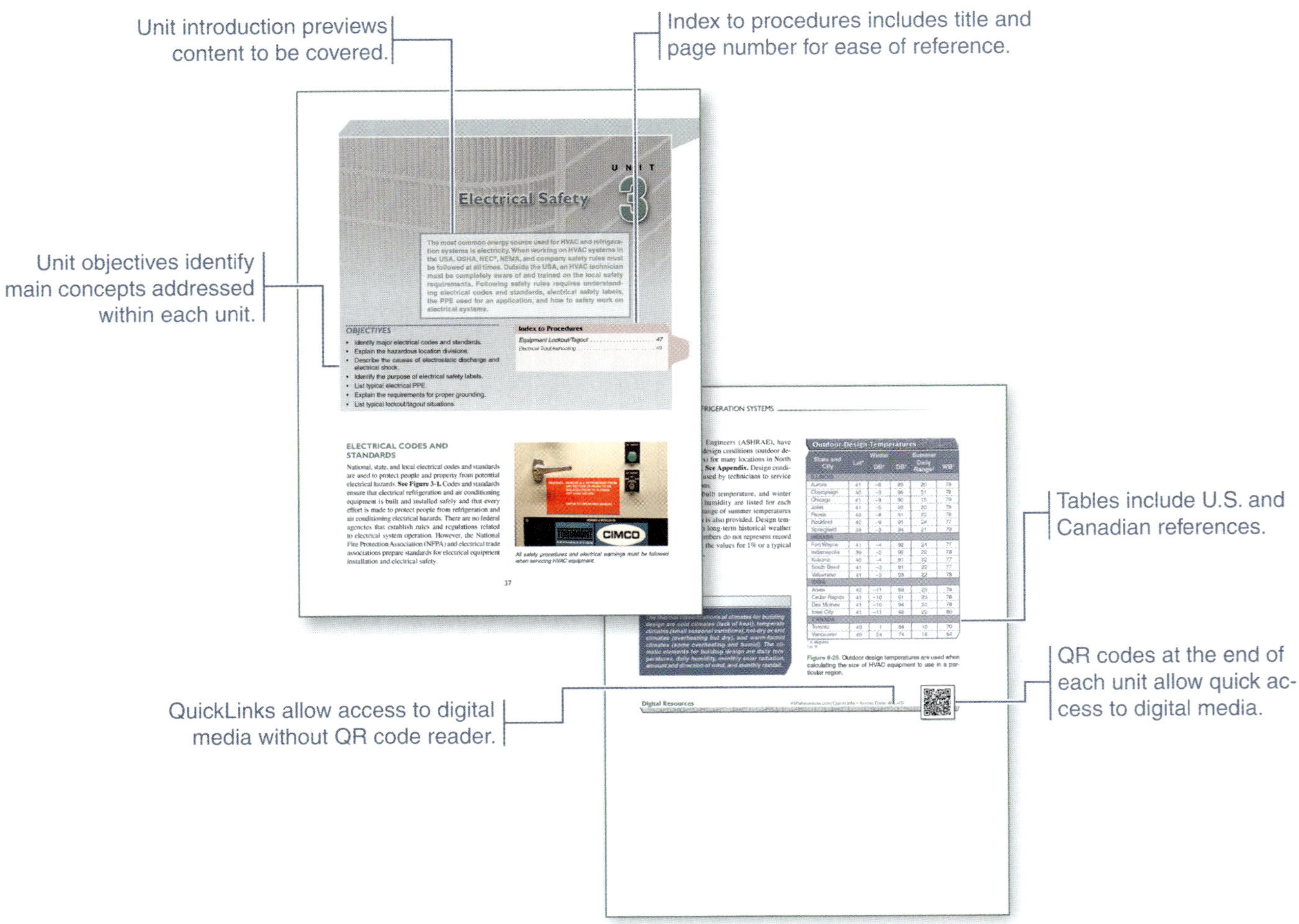

Color designations help learners find specific sections.

Detailed illustrations depict fundamental concepts.

HVAC industry equipment and procedures shown in typical real-world troubleshooting applications.

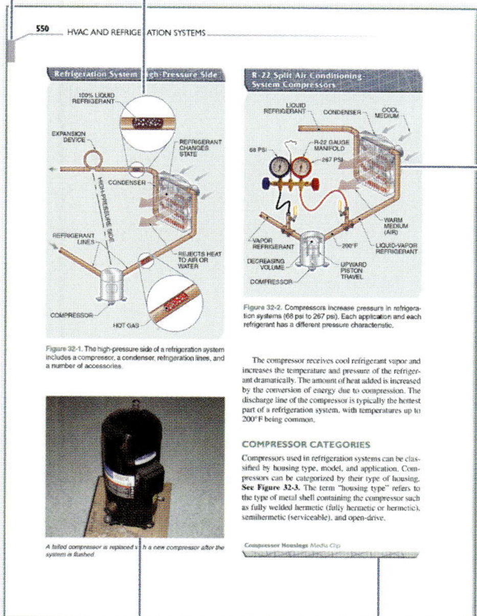

550 HVAC AND REFRIGERATION SYSTEMS

Refrigeration System High-Pressure Side

Figure 32-1. The high-pressure side of a refrigeration system includes a compressor, a condenser, refrigeration lines, and a number of accessories.

A failed compressor is replaced with a new compressor after the system is flushed.

R-22 Split Air Conditioning System Compressors

Figure 32-2. Compressors increase pressure in refrigeration systems (68 psi to 267 psi). Each application and each refrigerant has a different pressure characteristic.

The compressor receives cool refrigerant vapor and increases the temperature and pressure of the refrigerant dramatically. The amount of heat added is increased by the conversion of energy due to compression. The discharge line of the compressor is typically the hottest part of a refrigeration system, with temperatures up to 200°F being common.

COMPRESSOR CATEGORIES

Compressors used in refrigeration systems can be classified by housing type, model, and application. Compressors can be categorized by their type of housing. **See Figure 32-3.** The term "housing type" refers to the type of metal shell containing the compressor such as fully welded hermetic (fully hermetic or hermetic), semihermetic (serviceable), and open-drive.

Compressor Housings Media Clip

3D modeling used to develop illustrations depicting installation, servicing, and troubleshooting fundamentals.

Residential and commercial application photographs supplement text and illustrations.

Icons indicate media clips and interactive animations, available in the Digital Resources.

UNIT 41—Servicing and Troubleshooting Residential Forced-Air HVAC Systems 751

Cooling System Troubleshooting Scenario #4—Loss of Charge

Figure 41-31. When brazing or soldering the hole in the line set (depending on location) precautions must be taken so that the heat caused by brazing or soldering does not damage any valves or remote bulbs in the system.

Cooling System Troubleshooting Scenario #5—Thermostat Turned Too Low (Ice Buildup)

Figure 41-32. During technician inspection of the air conditioning system, ice is allowed to melt off of the evaporator.

Cooling System Troubleshooting Scenario #5—Thermostat Turned Too Low (Ice Buildup)

A "too hot" complaint is received on a summer morning. The technician arrives to find that the thermostat was turned down as far as possible by an occupant the previous afternoon. The air conditioning system has accumulated an ice buildup on the evaporator. **See Figure 41-32.**

Service Work Performed: The air conditioning system is shut down and the ice is allowed to melt off of the evaporator. The system is turned back on and the thermostat setting changed to normal. A lockable thermostat guard is used to prevent future tampering with the setting of the thermostat.

Cooling System Troubleshooting Scenario #6—Dirty Condenser

A customer calls in and complains that their air conditioning system is not cooling sufficiently. It is the first really warm day of spring. The technician notices that the high-side pressure is higher than normal. The customer states that the unit has not been serviced for sometime. **See Figure 41-33.**

Service Work Performed: The condenser (an outdoor unit) is inspected and found to be very dirty. It is cleaned, and the unit is put back into operation. The head pressures are now lower, capacity is increased, and the system cools properly. As a result, the customer is signed to a maintenance contract.

Clear, concise text is used throughout the textbook.

Features

Enlarged, detailed drawings simplify complex equipment and systems.

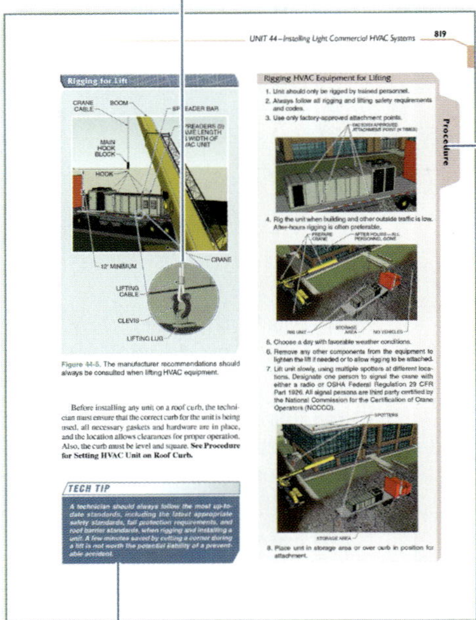

Step-by-step procedures with illustrations detail specific tasks.

Tech tips provide supplemental information related to topics presented.

Illustrated sequences of operation for residential and light commercial equipment facilitate understanding.

Troubleshooting principles and procedures are covered throughout

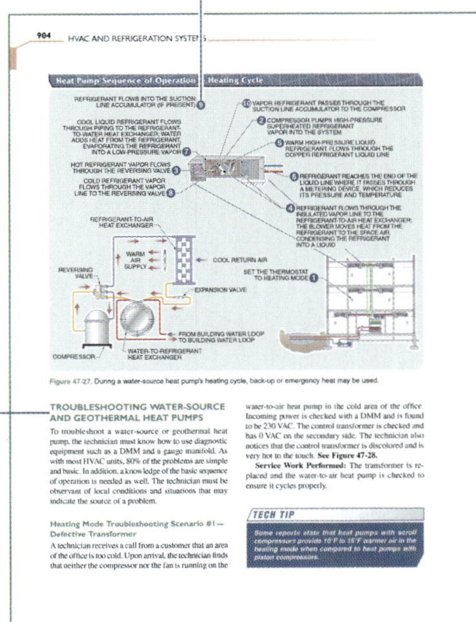

3D Features

3D Residential House

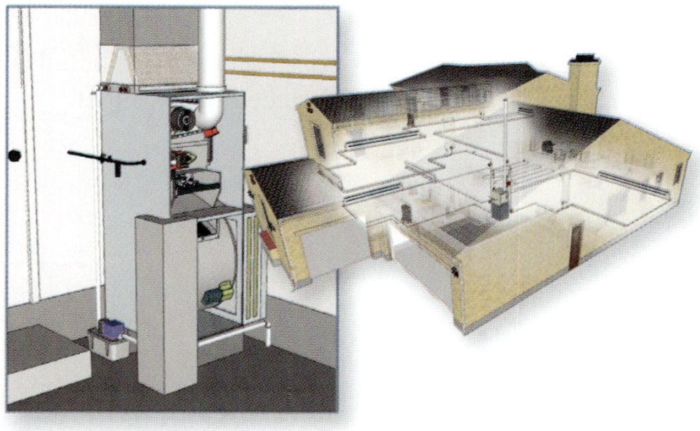

The residential house 3D model used in the installation, servicing, and troubleshooting units is a typical residence for the United States and Canada. The house is designed with a forced-air heating system with central air conditioning. The ductwork, dampers, filters, chimney, registers, and grills are clearly depicted. The house has also been designed with a hydronic heating system. The piping, control valves, safety valves, gauges, air eliminators, circulating pump, and expansion tank are all in place.

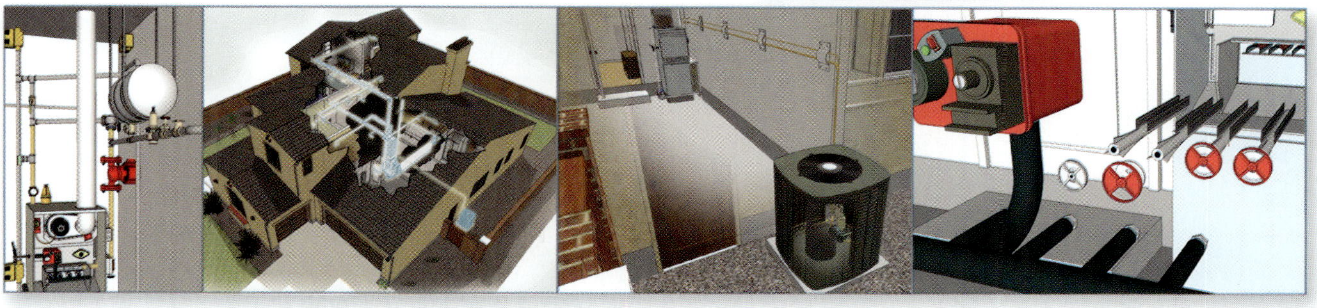

3D Light Commercial Building

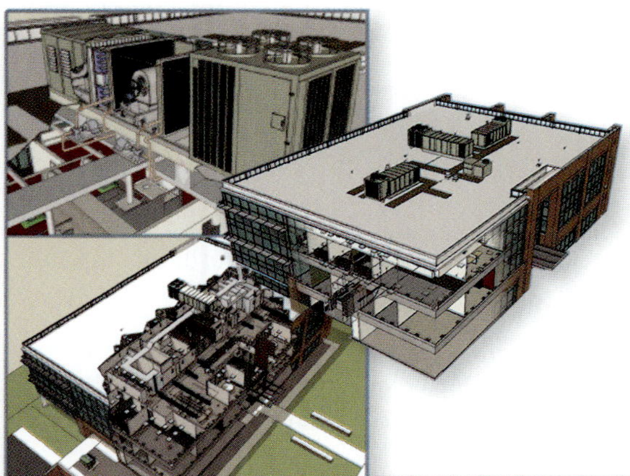

The light commercial office building 3D model used in the installation, servicing, and troubleshooting units is a typical light commercial building for the United States and Canada. The building is designed with forced-air and air conditioning rooftop units. The supply and return ductwork, filters, dampers, registers, grills, and vertical air handlers are clearly depicted. The building has also been designed with a pressurized floor with no supply ductwork, a hydronic heating system with heat pumps, and a hydronic heating system with geothermal ground loops, heat pumps, and complete control systems.

Digital Resources

HVAC and Refrigeration Systems also includes access to digital resources that reinforce and enhance learning. These resources can be accessed by either:

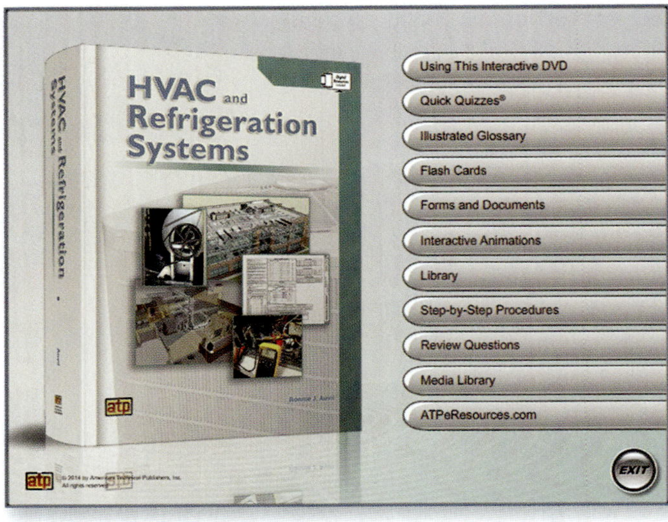

The Digital Resources include the following:

- **Quick Quizzes®** that provide 20 interactive questions per section, with embedded links to textbook content and to the Illustrated Glossary
- An **Illustrated Glossary** that provides a helpful reference to commonly used terms, with selected terms linked to illustrations and media clips
- **Flash Cards** that provide a self-study/review tool for identifying common HVAC terms, symbols, tool types, measurement devices, and fuels
- **Forms and Documents**, which are a collection of various HVAC equipment and maintenance forms; company safety, invoice, and related documents; and IAQ and system design documents and data sheets
- **Interactive Animations** that provide an interactive experience with fundamental HVAC topics, such as identifying components of an HVAC system, using standard HVAC procedures to perform a task, troubleshooting burner and flame problems, recovering refrigerant from Type I systems, using a psychrometric chart and virtual test meters, and assembling and wiring a motor control enclosure
- A **Library** containing illustrations from the textbook that can be enlarged and manipulated for greater clarity, such as building prints, electrical schematics, and manufacturer equipment manuals
- **Step-by-Step Procedures** covering 100 common installation procedures and 65 service procedures with illustrations that detail specific task steps
- **Review Questions** that allow learners to demonstrate comprehension of unit content
- A **Media Library** consisting of animated illustrations and video clips that reinforce textbook content
- **ATPeResources.com**, which links to online instructional resources that support continued learning

QR Codes and AR Icons

Quick Response (QR) codes located at the end of each chapter offer easy access to related information on the Internet via mobile device technology. To access content using a QR code, follow these steps:

1. Download a QR code reader app to a mobile device. (Visit atplearning.com/QR for more information.)
2. Open the app and scan the QR code on the book page.
3. Instantly access the digital content.

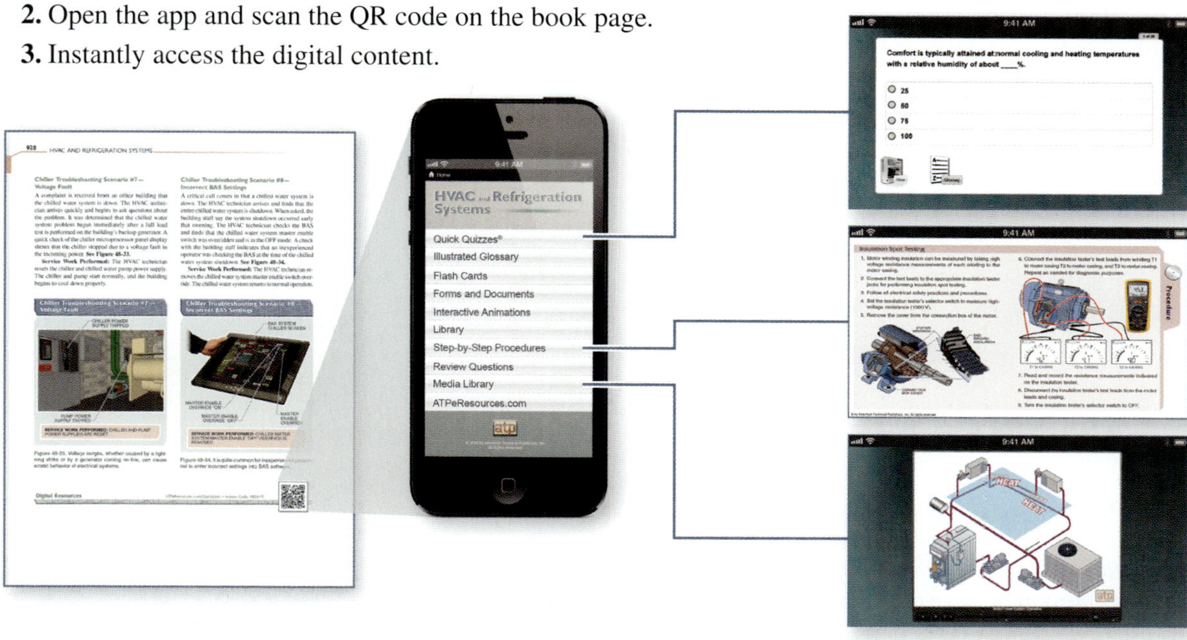

Augmented Reality (AR) icons located next to specific figures offer easy access to interactive 3D models on the Internet via mobile device technology. To access AR content, download the HVAC AR app from the iTunes App Store™ or Google Play™ and use the following steps:

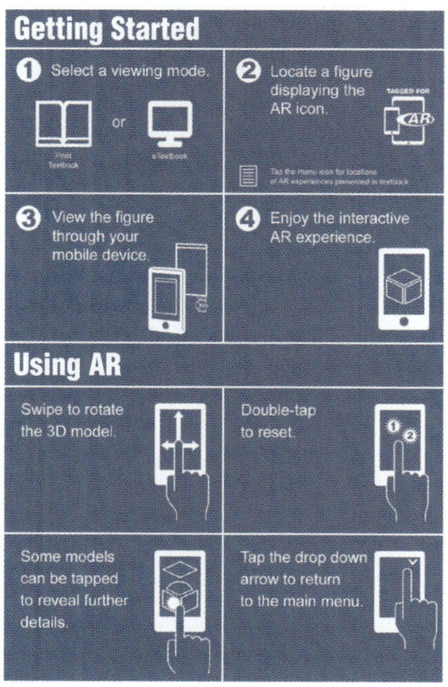

Available on the App Store™ and Google Play™

Resources

Study Guide

The *HVAC and Refrigeration Systems Study Guide* includes review questions and activities that reinforce and expand upon the information presented in the textbook.

Review Questions for each unit include true-false, multiple choice, and completion questions that reinforce comprehension of chapter concepts.

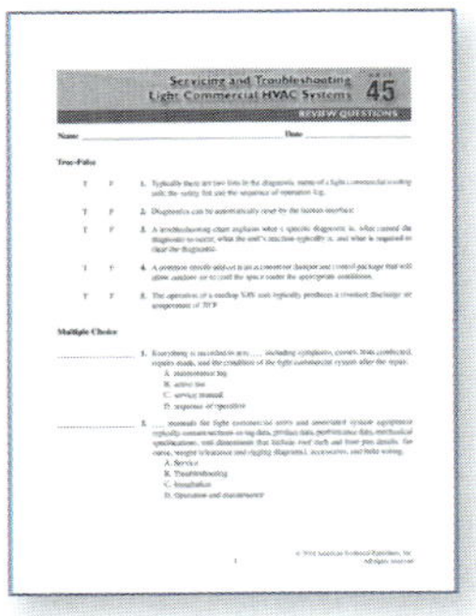

Activities include identification questions and activities based on real-world service calls to help learners understand and apply unit concepts.

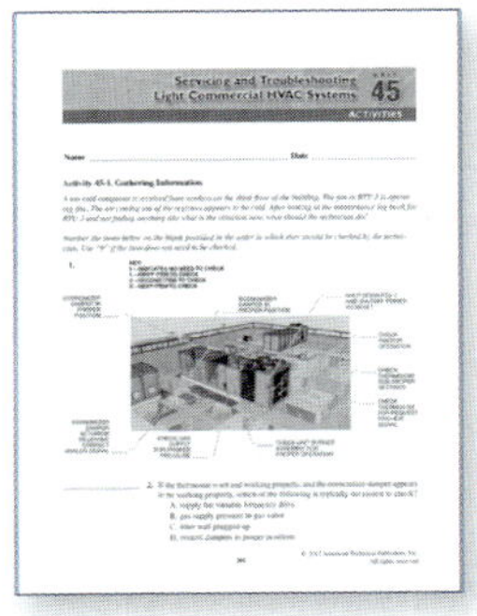

Instructor's Resource Guide

The *HVAC and Refrigeration Systems Instructor's Resource Guide* provides a comprehensive teaching resource that includes a detailed Instructional Guide, editable PowerPoint® Presentations, an interactive Image Library, Assessments, and Answer Keys as well as access to the digital resources used with the textbook.

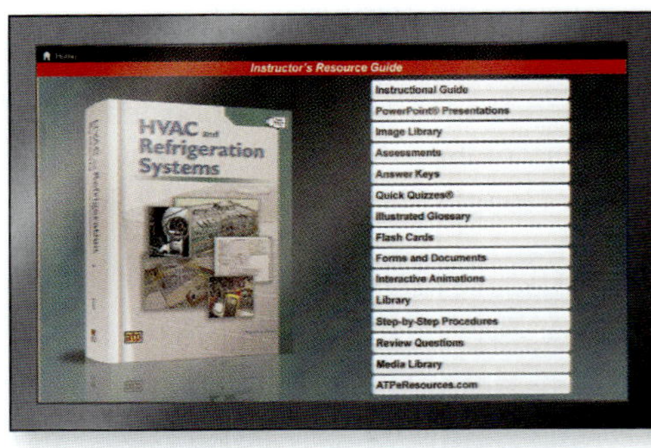

- The **Instructional Guide** explains how to best use the learning resources provided and includes Instructional Plans for each chapter that identify images and media pieces for an enhanced learning experience.
- **PowerPoint® Presentations** provide a review of key concepts from each chapter of *HVAC and Refrigeration Systems* and include illustrations, chapter objectives, and review questions. Instructor notes are also provided.
- The **Image Library** provides all the numbered illustrations in a format that can be manipulated for maximum instructional use.
- **Assessments** include sets of questions based on objectives and key concepts from each unit and consist of a pretest, posttest, and test banks. The test banks can be used with popular test development software packages and learning management systems.
- **Answer Keys** list answers to pretest, posttest, and study guide questions.

To obtain information on related products, visit the American Technical Publishers website at atplearning.com.

The Publisher

UNIT 1

HVAC and Refrigeration Systems

Heating, ventilating, and air conditioning (HVAC) systems used in residential and light commercial buildings are designed and operated to provide comfort for occupants. A feeling of comfort results when temperature, humidity, circulation, filtration, and ventilation of the air are controlled within a comfortable range. When operating properly, HVAC systems are designed to fulfill comfort requirements efficiently.

OBJECTIVES

- Describe the five factors that affect comfort.
- Identify environmental factors that may affect the perceived level of comfort.
- Explain how to control indoor air quality (IAQ).
- Describe common HVAC systems and refrigeration processes.

COMFORT

Comfort is the condition that occurs when a person cannot sense a difference between themselves and the surrounding air. *Discomfort* is the condition that occurs when a person can sense a difference between themselves and the surrounding air.

Heating, ventilating, and air conditioning (HVAC) systems are designed to provide comfort using mechanical, pneumatic, electrical, electronic, and chemical components and equipment. **See Figure 1-1.** HVAC system components include fans, compressors, valves, dampers, actuators, fuels, refrigerants, and controls. The five environmental conditions for comfort are proper temperature, humidity, circulation, filtration, and ventilation. Pressure is a consideration in some new buildings. Discomfort can occur when any of the five requirements for comfort are not met.

Comfort Requirements Media Clip

Temperature

Controlling the temperature of the human body is an important physiological function. A *physiological function* is the natural physical and chemical function of an organism. The body produces energy by metabolizing food. Energy is used for normal living processes, such as producing thermal energy (heat). Physiological systems regulate body temperature to maintain comfort.

People typically have a body temperature of 98.6°F. The body has natural heating and cooling systems to maintain this temperature. The body responds to the conditions of the air by controlling blood flow at the surface of the skin, radiating heat from body surfaces, or using evaporation of perspiration from the skin. *Evaporation* is the process that occurs when a liquid changes to a vapor by absorbing heat.

When the body is clothed, the body's temperature control system provides comfort at an air temperature of approximately 72°F. When the air temperature varies above 76°F or below 68°F, the body begins to feel uncomfortably warm or uncomfortably cool.

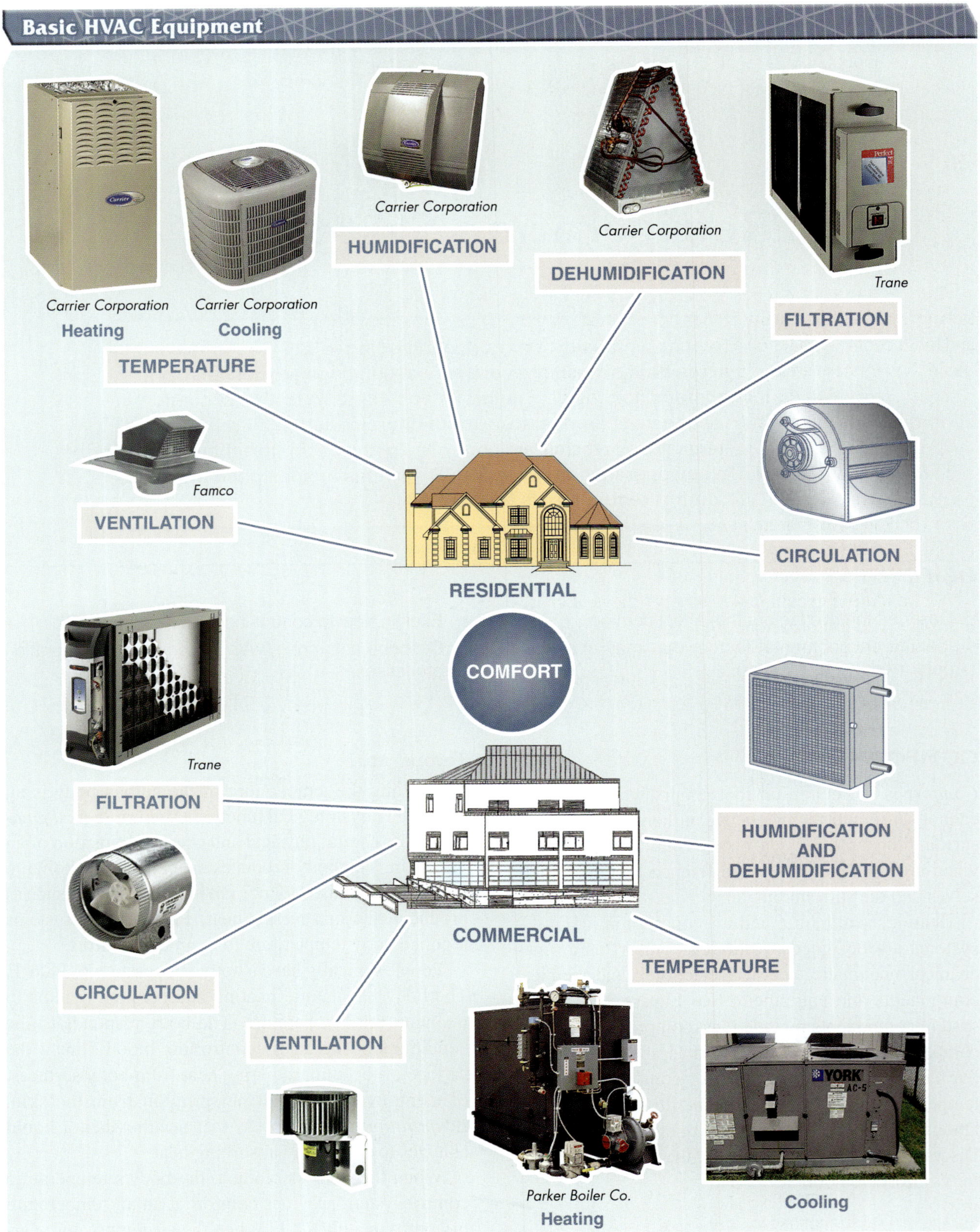

Figure 1-1. Residential and light commercial HVAC equipment work as a system to provide the proper temperature, humidity, circulation, filtration, and ventilation required for comfort.

Different points of the body send signals through a network of nerves to the hypothalamus, a gland in the brain. The hypothalamus regulates body temperature by controlling blood flow to capillaries (tiny blood vessels) located in the skin. Capillaries regulate blood flow to the surface of the skin. If the body temperature rises, blood flow to the skin increases.

Blood carries heat to the skin, and the heat is rejected to the cooler surrounding air. A person who becomes overheated also becomes flushed because the blood flow to the skin increases. When signals indicate the body is cooling off, the system reduces the cooling effect by allowing less blood flow to the skin. In cold weather, fingers and toes become cold before the rest of the body because blood flow to these areas is reduced. Cold areas also appear pale because of the reduced blood flow.

Temperature control in building spaces is provided by warm air from heating equipment or cool air from air conditioning equipment. Heating equipment supplies the proper amount of heat in cold weather to offset heat loss from a building. Heating equipment is rated in Btu per hour (Btu/hr). A *British thermal unit (Btu)* is the quantity of heat required to change the temperature of 1 lb of water 1°F.

Air conditioning equipment removes the heat gained into the space in hot weather, providing a cooler more comfortable indoor space. Air conditioning equipment is rated in Btu per hour or ton of cooling. A *ton of cooling* is the amount of heat required to melt one ton of ice (2000 lb) over 24 hr. One ton of cooling equals 12,000 Btu/hr or 288,000 Btu/24 hr (12,000 × 24 = 288,000).

The HVAC system equipment and components are selected based on ability to control the temperature in buildings. **See Figure 1-2.** For maximum comfort, temperature control equipment is typically set to maintain air temperatures in buildings within 1°F or 2°F of the temperature necessary for comfort.

TECH TIP

Correct humidity levels must be maintained for human comfort and to ensure the quality of wood, paper, and electronic products.

Humidity

Humidity is the amount of moisture, or water vapor, in air. Humidity is always present in air. A low humidity level indicates dry air that contains little moisture. A high humidity level indicates damp air that contains a significant amount of moisture. Humidity is important in determining comfort.

Heating and Air Conditioning Equipment

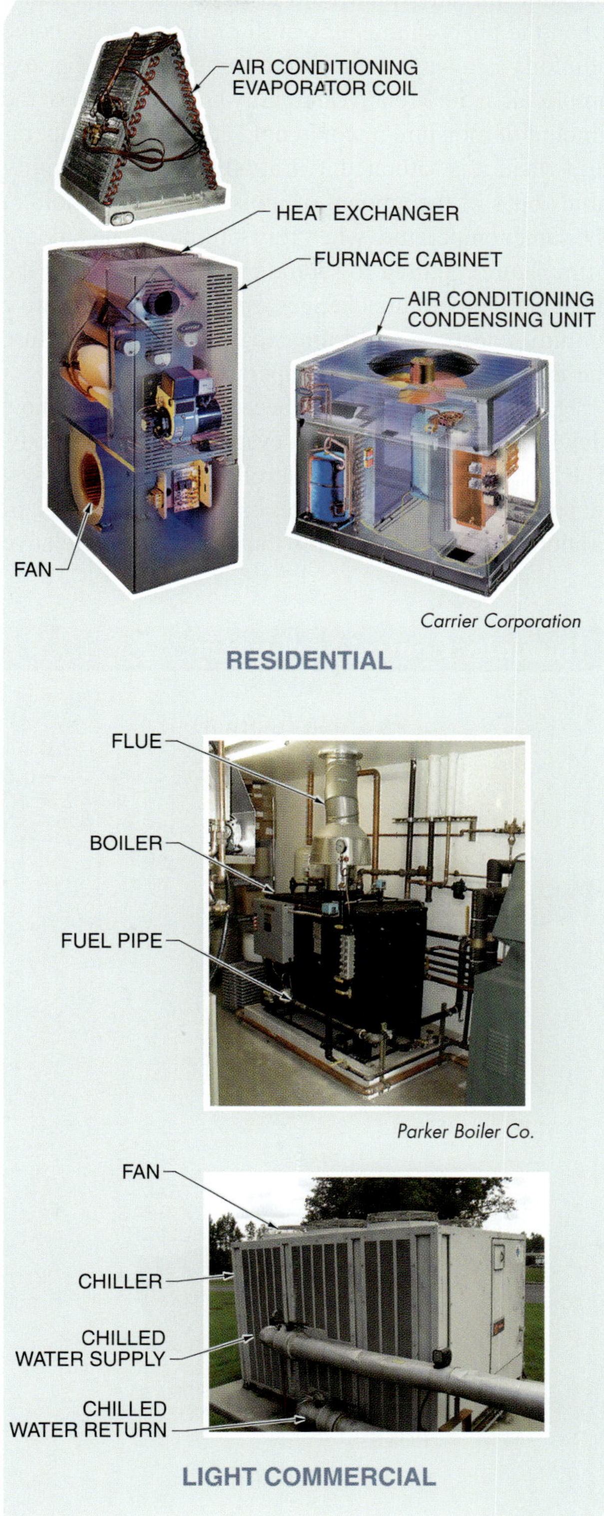

Figure 1-2. Heating and air conditioning equipment is used to maintain the temperature in building spaces.

Relative Humidity. *Relative humidity (RH)* is the amount of moisture in air compared to the amount of moisture the air would hold at the same temperature when the air is saturated, or full of moisture. Relative humidity is always expressed as a percentage. For example, air at 50% relative humidity holds one-half of the amount of moisture it would hold at the same temperature when it is saturated. Air at 60% relative humidity holds 60% of the amount of moisture it would hold at the same temperature when it is saturated. The amount of moisture required to saturate the air changes as the dry-bulb temperature changes. In addition, the relative humidity decreases and the capacity to hold moisture increases as the dry-bulb temperature increases.

Humidity affects comfort because it determines how slowly or rapidly perspiration evaporates from the body. The flow of perspiration is controlled by the cooling system of the body, which regulates body temperature. Evaporation of perspiration cools the body. Higher relative humidity lowers the evaporation rate. Lower relative humidity raises the evaporation rate. For example, with no temperature change and an increase in relative humidity, a person feels warmer because of the slower evaporation rate. With no temperature change and a decrease in relative humidity, a person feels cooler because of the faster evaporation rate. A steam room is an example of an area with high temperature and high relative humidity. These conditions cause the body to have a high perspiration rate with a slow evaporation rate.

Controlling Relative Humidity. Comfort is typically attained at normal cooling and heating temperatures with a relative humidity of about 50%. When the humidity is too low, a higher air temperature is required to feel comfortable. When the humidity is too high, a lower air temperature is required for the same feeling of comfort. In some cases the humidity in a building may be too low or too high for comfort. Humidity level is controlled by humidifiers and dehumidifiers. **See Figure 1-3.**

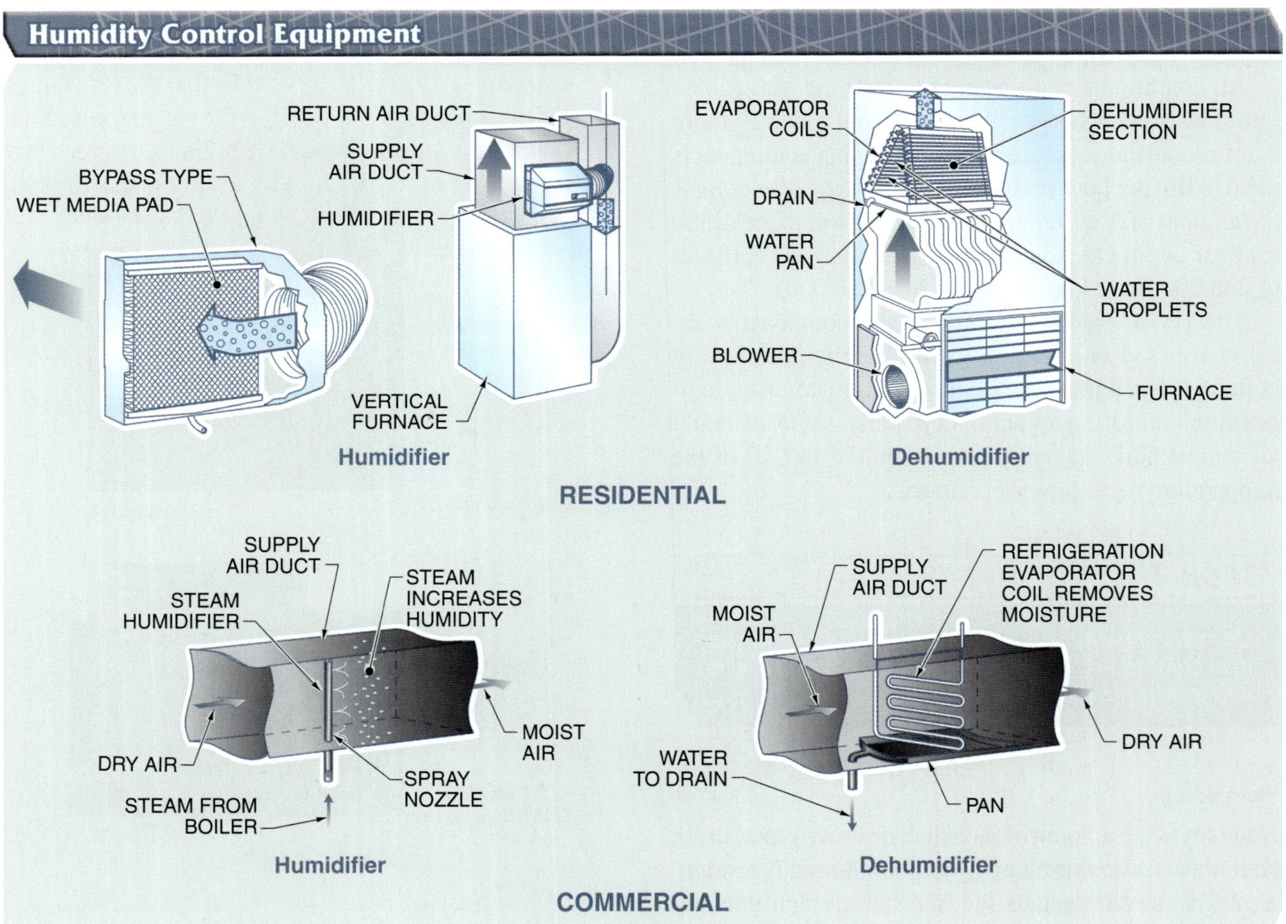

Figure 1-3. Humidifiers and dehumidifiers are used to control the levels of humidity in building spaces.

A *humidifier* is a device that adds moisture to air by causing water to evaporate into the air. In cold climates and dry climates, the humidity in buildings may be too low for comfort. For example, in cold climate areas, such as Chicago in the winter, and dry climate areas, such as southern Arizona, the relative humidity is low. In these cases, humidifiers are used in building HVAC systems to add moisture to the air and to maintain a comfortable humidity level.

A *dehumidifier* is a device that removes moisture from air by causing the moisture to condense or be trapped. *Condensation* is the formation of liquid (condensate) as moisture or other vapor cools below its dew point. *Dew point* is the temperature below which moisture begins to condense. In buildings with swimming pools or a large number of potted plants, the humidity level may be too high for comfort. HVAC and refrigeration systems remove moisture from the air as the air is cooled by the evaporator.

Air Circulation

Air circulation is the movement of air through distribution systems and building spaces. Air in buildings must be circulated continuously to provide occupant comfort. Total comfort systems always include gently circulating air that is clean and fresh. In buildings where there is improper circulation, air rapidly becomes stagnant and occupants become uncomfortable. *Stagnant air* is air that is not circulating, typically contains an excess of impurities such as carbon dioxide, and lacks sufficient levels of oxygen required to provide comfort.

When air is improperly circulated, temperature stratification occurs. *Temperature stratification* is the variation of air temperature in a building space that occurs when warm air rises to the ceiling and cold air drops to the floor. Temperature stratification in building spaces causes discomfort. **See Figure 1-4.**

Air velocity is the speed air moves from one point to another through ductwork, a register, or a grill and is measured in feet per minute (fpm). When air is not moving, heat cannot be carried away from skin surfaces. When heat collects at the surface of the skin, the body overheats. Air circulation is important because it helps cool the body by causing perspiration to evaporate. An increase in air velocity increases the rate of evaporation of perspiration from the skin, which causes a person to feel cooler. A decrease in air velocity or no air velocity reduces the evaporation rate, which causes a person to feel warmer. Air movement is used to supply clean, fresh air and to remove stagnant air. Air velocity of 40 fpm in building spaces is considered ideal for comfort.

Temperature Stratification

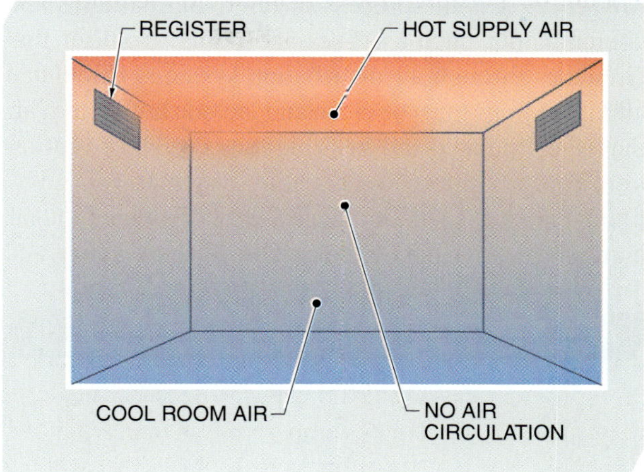

Figure 1-4. Temperature stratification occurs when there is improper air circulation in building spaces.

Forced-air HVAC systems circulate air throughout the building. Supply air ductwork and registers are used to distribute air to building spaces. Supply air ductwork is sized and located for efficient air distribution. The type, location, and size of registers determine the amount of supply air introduced into each building space and the distribution pattern of the air. Return air ductwork and grills are used to move air from building spaces back to the fan. Return air ductwork and grills are sized and located for efficient air collection. **See Figure 1-5.**

Air Circulation

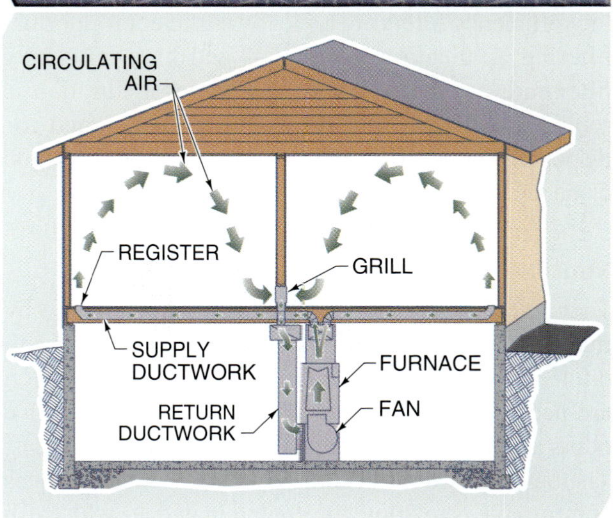

Figure 1-5. Registers and grills are located to efficiently circulate air in building spaces. Air is circulated by the furnace fan through supply and return ductwork.

Air Filtration

Air filtration is the process of removing particles and contaminants, such as gases and odors, from air that circulates through air distribution systems. Circulated air in buildings must be clean and include both return air that is recirculated and fresh air that is used for ventilation. The air in a forced-air heating system is cleaned by filters that are placed in the air ductwork where the air enters the heating or cooling equipment. **See Figure 1-6.**

Air Filtration

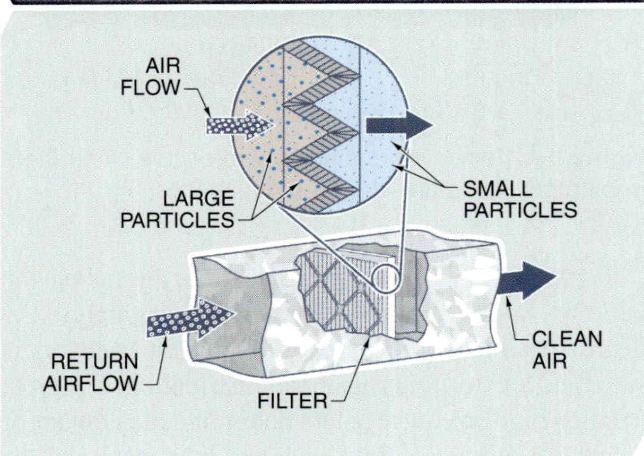

Figure 1-6. Circulating air is cleaned by filters in the return ductwork or in the heating and cooling equipment.

The primary consideration in air filter selection is the level of clean air desired. Low-, medium-, and high-efficiency filters are commercially available. In one-family dwellings or small commercial buildings, low-efficiency filters are typically sufficient for normal filtering applications. Medium- and high-efficiency filters are used when a higher degree of filtration is required. Most commercial filters are designed for disposal after use. However, electrostatic filters may be cleaned and reused.

Ventilation

Ventilation is the process of reducing air pollution by introducing enough fresh outdoor air into a building to dilute indoor pollutants. Ventilation and air circulation are necessary for comfort inside buildings. Offensive odors and vapors are created in closed building spaces. The best method for keeping the air in building spaces comfortable is to dilute the contaminated air with fresh air from outdoors.

Some odors that originate in buildings are simply offensive, while others may be dangerous. Offensive odors are reduced by bringing fresh air into a system from outdoors and introducing it into the return air portion of the HVAC system. **See Figure 1-7.** The amount of ventilation necessary for comfort is determined by the number of occupants and the kind of activity taking place inside the building. When odors or contaminants originating in a building are offensive or potentially toxic, the air in the building may require continuous exhaust with no recirculation. The building would require 100% makeup air for ventilation. *Makeup air* is air that is used to replace air that is discharged as exhaust.

Ventilation

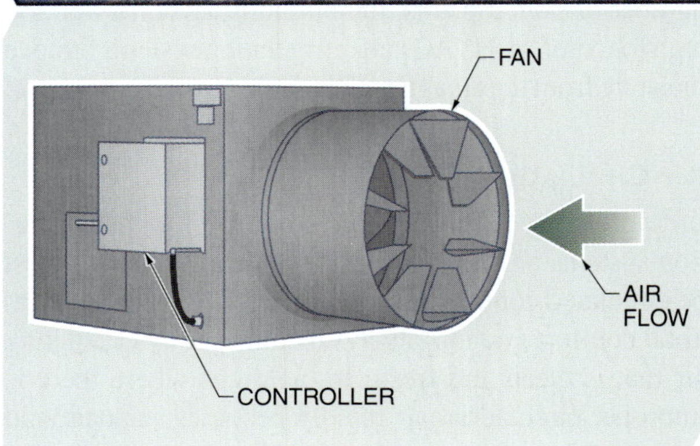

Figure 1-7. Fresh air intake fans are used to draw fresh outdoor air into a building to dilute contaminants in the indoor air.

Control of Comfort Factors

While comfort is defined as meeting specific criteria, at its core, comfort is still somewhat subjective. The described level of comfort varies with each individual and is affected by comfort factors in the general environment. **See Figure 1-8.** HVAC technicians should be prepared to recognize some environmental factors that may affect the perceived level of comfort. When HVAC technicians respond to complaints, environmental factors should be considered.

Sounds. Buildings or building spaces that produce large amounts of sound, or noise, may be distracting and perceived as uncomfortable. The source or transmission of building noise typically includes the HVAC system. For HVAC technicians, the most common complaint is that a building space is too hot or cold. The second most common complaint is that a building space is too noisy.

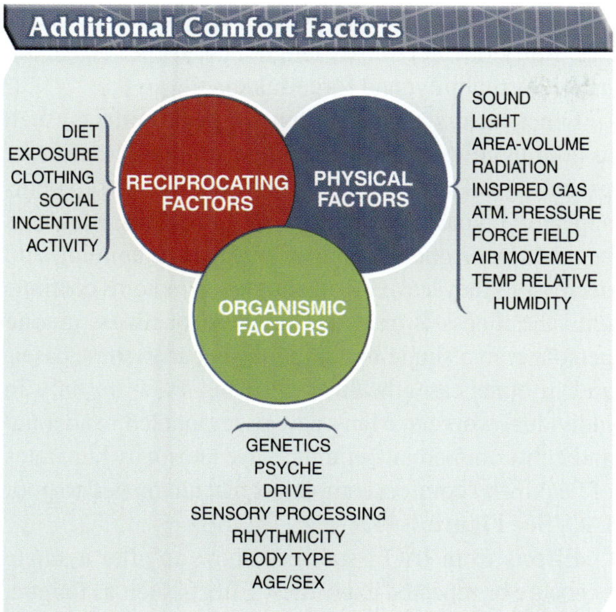

Figure 1-8. Many additional factors affect the perceived level of occupant comfort in building spaces.

Lights. Incorrect levels of indoor lighting can add to perceived discomfort. Building spaces that use inconsistent or insufficient levels of light to reduce energy use may be perceived as uncomfortable. Localized high levels of lighting use, such as at displays, typically produce hot spots within building spaces. Air circulation for affected building spaces needs to be adjusted to maintain building temperatures.

Size of Building Space Areas. Building spaces are typically divided into working areas that create personal space for the occupants. The partitions used may affect the building space air circulation. The space allowed may not be sufficient for the work done, creating a crowded feeling. Odors may accumulate creating an unpleasant work environment. The feeling of excessive closeness lowers the level of comfort. Typically this condition requires increased air circulation.

Activity. The physical activity level of building space occupants affects their perceived level of comfort. Building spaces with sedentary occupants working at desks decreases the demands on HVAC systems but can make people more sensitive to perceived comfort issues. Building spaces where occupants are constantly moving, such as when moving product and restocking shelves, increases the demands on HVAC systems. However, the occupants during these activities are considered to be less sensitive to perceived comfort issues.

Use of Building Spaces. The use of building spaces determines the expected level of comfort. Corporate offices and meeting rooms where formal attire is expected will create a high demand for comfort. Commercial buildings used for sales, where the level of use varies significantly, need HVAC systems that can adjust to the varying loads and maintain comfort levels. Factory floors and warehouse storage areas typically have a low expectation of comfort due to cost concerns.

Regional, Cultural, and Personal Differences. People from different regions and various cultures can perceive comfort differently. For example, people who have lived for years in very cold climates and then move to a warmer climate perceive comfort differently. People with different body types vary in the amount of surface exposed to the ambient conditions in the building. Dietary norms in specific regions can affect how the human body perceives the temperature around it and impact the feeling of comfort. Age and health conditions of occupants can play a role in perceived comfort. Older individuals and those having health problems will perceive comfort differently.

INDOOR AIR QUALITY

Indoor air quality (IAQ) is the condition of indoor air as it relates to contaminant levels and air properties that affect the health and comfort of building occupants. IAQ is an important concern. ASHRAE 62, *Ventilation for Acceptable Indoor Air Quality*, is a standard that specifies ventilation rates based on chemical, physical, and biological contaminants. Energy conservation measures reduce the amount of outside air used in buildings, which contributes to the buildup of indoor air contaminants. **See Figure 1-9.**

Indoor Air Quality (IAQ) Guidelines

Temperature*	Humidity†	Carbon Dioxide‡
Winter—70 to 74	Min—30	Well ventilated—600 to 1000
Summer—72 to 76	Max—60	Average—800
Combustible Gases	**Formaldehyde‡**	**Hydrogen Sulfide‡**
Max 10% LEL (lower explosive limit)	Max—0.1	Max 8 hr—10 ppm Max 15 min—15 ppm

* in °F
† in %
‡ in parts per million (ppm)

Figure 1-9. IAQ guidelines are standards for fresh air circulation that limit the buildup of indoor air contaminants.

Typically IAQ issues are the result of many sources and causes. Sensitivities vary from individual to individual because of previous exposures, preexisting medical conditions, and age. Complaints about IAQ range from simple complaints, such as dry, scratchy throat and eyes, to more complex complaints where the air quality may cause illness and loss of work.

In some cases there may be one pollutant, while other spaces may have a combination of pollutants. Some effects are made worse by an inadequate supply of outside (ventilating) air. Some effects are from the building heating, cooling, or humidifying equipment. **See Figure 1-10.**

Determining Causes of Pollutants

Fluke Corporation

Figure 1-10. Building spaces may be affected by a combination of several pollutants, including pollutants brought in with ventilating air.

Health Effects

Two terms that are associated with the health effects of buildings are "building-related illness" and "sick building syndrome." A building-related illness is an illness that is traced to a building space and attributed to a definite cause. Building-related illnesses include Legionnaires' disease, asthma, hypersensitivity pneumonitis, and humidifier fever. These diseases pose serious risks but can be treated. Sick building syndrome is an illness that can be traced to a building space but not to a specific cause. Building personnel may complain of dry or burning mucous membranes in the nose, eyes, and throat; sneezing; stuffy or runny nose; fatigue; headache; dizziness; nausea; irritability; and forgetfulness.

Other factors may contribute to health effects, such as poor lighting, noise, vibration, thermal discomfort, or psychological stress. Building personnel health problems appear in no particular pattern. In some cases, problems begin as personnel enter work areas of a building and decrease as they leave. At other times, symptoms continue until the illness is treated. Outbreaks of illness among personnel in a single building may occur in some cases, and in other cases health symptoms show up only in individual workers. Many new or remodeled residential and light commercial buildings have unusually high rates of health and comfort complaints that are related to poor IAQ. **See Figure 1-11.**

Effects from IAQ issues can show up after a single exposure or repeated exposures. Effects such as fatigue, dizziness, headaches, and irritation of the eyes, nose, and throat are usually short-term and treatable. The treatment may be as simple as eliminating the exposure of an individual to the source of the pollution.

Symptoms of some diseases, including asthma, hypersensitivity pneumonitis, and humidifier fever, may also show up soon after exposure to certain pollutants. Common effects of these diseases are similar to those from colds or other viral diseases, so it is often difficult to determine when the symptoms are a result of exposure to indoor air pollution. When the symptoms decrease or go away while the person is away from the building space and return when the person is in a building space, an effort must be made to identify the source.

Causes of Poor IAQ

There are three primary reasons for poor IAQ in building spaces. These reasons include the presence of indoor air pollution sources; poorly designed, maintained, or operated ventilation systems; and using the building in a way that was not anticipated or poorly planned for when the building was designed or renovated.

Indoor Air Pollution Sources. Probably the most important factor influencing IAQ is the presence of pollutant sources. **See Figure 1-12.** In the past, environmental tobacco smoke was a major source of pollution in building spaces. With the increase of local, state, and federal laws that limit or abolish smoking indoors, the potential effects of tobacco smoke have been dramatically reduced. Reducing the sources of pollution within the building reduces the distribution of pollution within the building by the HVAC system.

Building-Related Illnesses

Health and Comfort Complaints	Sources	Pollutant
Headache, fatigue, queasiness, poor vision, poor concentration, heart pains, possibly death	Malfunctioning gas appliances, fireplaces, kerosene heaters, grills, running cars in attached garages, contaminated outside air	Carbon monoxide
Difficult breathing, lung damage, respiratory infections	Malfunctioning gas appliances, fireplaces, kerosene heaters, grills, running cars in attached garages, contaminated outside air	Nitrogen dioxide
Eye, nose, and throat irritation, cancer of the bladder, stomach, skin, and lungs	Tobacco smoke, fireplaces, kerosene heaters, grills, self-cleaning ovens, hobbies, contaminated outside air	Polycyclic aromatic hydrocarbons
Eye, nose, and throat irritation, emphysema, bronchitis, allergies, asthma, respiratory and ear infections	Tobacco smoke, fireplaces, kerosene heaters, grills, incense burning, hobbies, contaminated outside air	Particles

Figure 1-11. Many health and comfort complaints are related to contaminants that are recirculated in building spaces.

Building Space Pollutant Sources

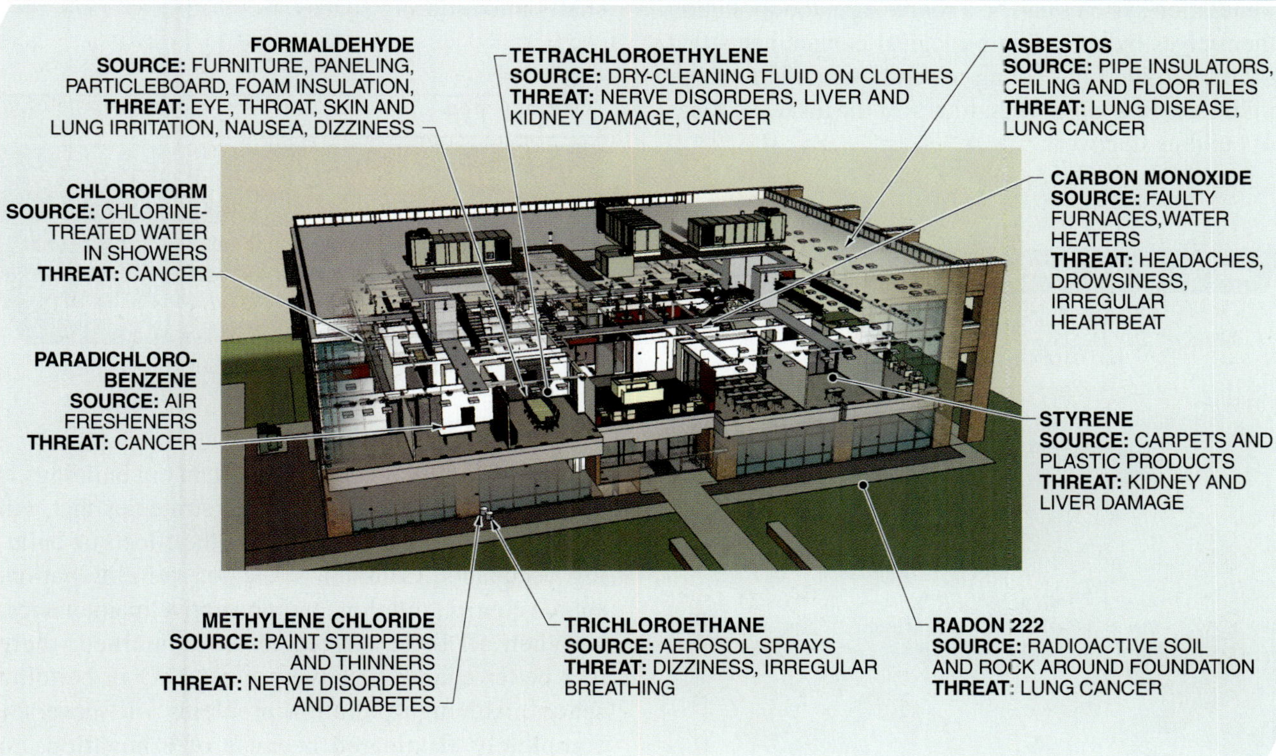

Figure 1-12. Pollutant sources contribute to reducing indoor air quality in building spaces.

Volatile organic compounds (VOCs), such as formaldehyde, are a source of pollution that is found in glues used in many building materials. VOCs are found in many products, including construction materials, carpets, furnishings, cleaning materials, air fresheners, paints, adhesives, and copying machines, or develop due to company processes and activities.

Biological contaminants from dirty ventilation systems or water-damaged walls, ceilings, and carpets add to the source of pollutants found in all building spaces. Biological contaminants include hair, skin, animal dander, mold, and mildew. Pesticides used for pest management in building spaces have the potential to be very serious pollutant sources if not used correctly.

Ventilation Systems. The ventilation systems in buildings are designed to operate during heating and cooling demand times or to bring fresh outdoor air into building spaces. Poorly designed, installed, operated, or maintained ventilation systems contribute to indoor air problems in several ways.

Problems may arise when, in an effort to save energy costs, ventilation systems are not used to bring in the required amounts of outdoor air. Inadequate ventilation also occurs when air supply registers and air return grills within each room are blocked or placed in such a way that proper air circulation does not occur in the building space. Improperly located outdoor air intakes can also bring in air contaminated with automobile or truck exhaust, combustion emissions from furnaces and boilers, fumes from waste containers, or air from lowlands or refuse collection areas. **See Figure 1-13.** Ventilation systems can be a source of indoor pollution themselves by spreading biological contaminants that have multiplied in cooling towers, humidifiers, dehumidifiers, air conditioners, filters, or the inside surfaces of building ductwork.

Building Air Intake

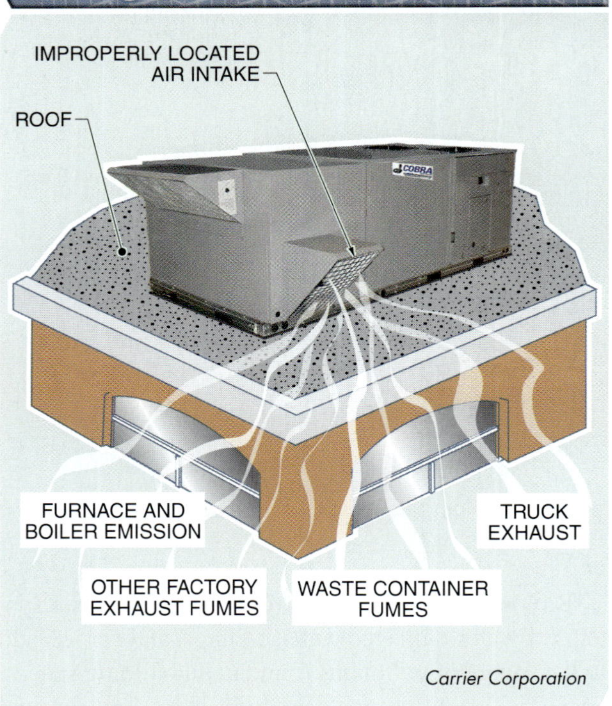

Figure 1-13. Improperly located outdoor air intakes can bring in air contaminated with automobile exhaust and other fumes.

Building Use. Buildings originally designed for a specific use may be converted for other uses. When building renovations are not planned and executed properly, room partitions and ventilation systems can contribute to IAQ problems by restricting air circulation or providing an inadequate supply of outdoor air. Ventilation systems are typically overlooked during renovation projects because it is difficult to modify distribution systems.

Indoor air pollutants can also be circulated from portions of a building used for specialized purposes such as utility rooms, restaurant kitchens, shops, and stores. Pollutants travel through the HVAC system to other occupied spaces of the building. Carbon monoxide and other components of automobile exhaust can be brought in from garages and loading docks and travel through hallways, stairwells, and elevator shafts into building spaces.

TECH TIP

Ventilation requirements can be very strict for residential and light commercial buildings. Since buildings that do not have enough ventilation air can suffer from poor indoor air quality, following current standards and using the proper equipment can help ensure that the ventilation requirements are met.

Control of Pollutants

The control of all indoor air pollutants in buildings is very difficult, but there are some strategies that will control most pollutants and reduce the effects on building occupants. Common strategies are elimination, source control, dilution, and preventive maintenance.

When all these strategies are combined, there is a better chance of having good IAQ in building spaces. Although pollution problems will never be completely eliminated because of renovations to building spaces, new products being introduced into the work environment, and the ever-changing environments surrounding buildings, pollution can be kept at acceptable levels.

Elimination. The best way to control indoor air pollution in a building space is to eliminate the product or material of contamination from the building. For example, the elimination of tobacco smoke from public buildings has dramatically improved the indoor air quality in those buildings.

Source Control. Pollution emissions can be reduced by controlling the source of the pollution. Pollutants can be controlled at the source by venting the emissions to a location where they do not create a problem. For example, by placing a fume hood over a printing press, the emissions from the printing process can be vented outdoors. Combustion equipment and appliances can also be adjusted to decrease emissions. **See Figure 1-14.**

Figure 1-14. Properly calibrated equipment reduces pollution emissions from heating equipment.

Dilution. Controlling indoor air pollution by dilution is common and has been done for many years. A past method used to control the escalating use and cost of energy was to decrease the fresh air supplied to building spaces for ventilation. Reducing the amount of fresh air used allowed pollutants to build up in building spaces.

To counteract the problem, ASHRAE recommended that the ventilation requirements be increased in all residential and light commercial buildings. The normal office building space requirement for fresh outside air (ventilation) was increased to approximately 20 cfm per person. The increased ventilation requirement did not eliminate indoor air pollution problems, but it did eliminate the buildup of pollutants and keep the pollutants at acceptable levels. **See Figure 1-15.**

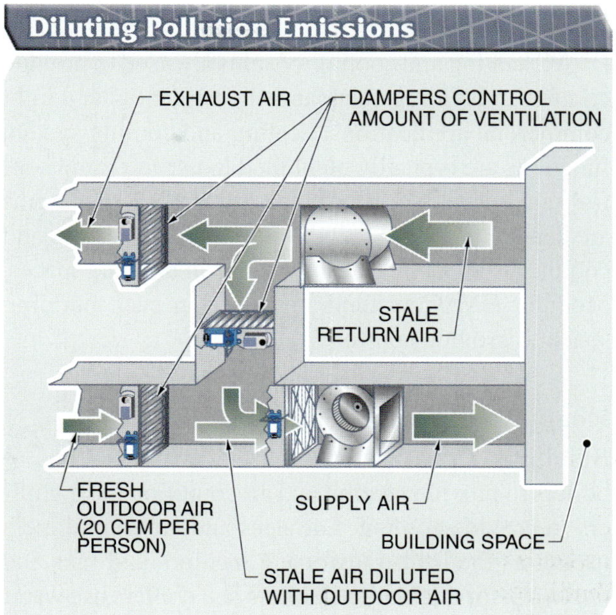

Figure 1-15. Normal office building spaces require 20 cfm per person of fresh outside air.

Preventive Maintenance. All building owners should have preventive maintenance systems in place to keep HVAC equipment operating at design conditions. Preventive maintenance programs ensure that the proper quality of air enters building spaces in the proper quantity. Preventive maintenance involves cleaning all surfaces in the building including registers and grills, heating and cooling coils, ventilation intake areas, and ductwork. Areas to be cleaned also include related areas such as stairwells, mechanical spaces, and closets.

A good maintenance program ensures that air filters are changed before system performance and air quality are affected.

HVAC SYSTEMS

HVAC heating and cooling systems are used to provide comfort to building occupants. In residential and light commercial applications, heating and cooling system functions are typically combined to use the same controls and distribution systems. Central HVAC systems are centrally located in buildings with the heating and cooling transported to terminal units in building spaces. Modular HVAC systems use units in each building space to provide comfort.

Heating and Cooling

Building HVAC systems typically use furnaces or boilers to provide heating and air conditioners or chillers to provide cooling. Furnaces and air conditioners use air as the heat-transferring medium and fans and ductwork for circulation. Boilers and chillers use water as the heat-transferring medium and pumps and piping for circulation.

Split Systems. A split HVAC system has separate cabinets for the evaporator coil and the condenser coil. **See Figure 1-16.** Split systems are typically used for heat pumps and small air conditioning systems. The inside coil is typically installed in the supply plenum of the furnace distribution system. The outdoor coil is typically located adjacent to or on the roof of the building to discharge heat.

Packaged Units. A packaged HVAC unit is a self-contained building cooling unit that has the components contained in one cabinet. **See Figure 1-17.** Controls and components are installed by the manufacturer. The electrical power, control connections, and heating/cooling distribution systems need to be connected to the building systems.

Combination Units. A combination HVAC unit is a self-contained heating and cooling unit that has all of the HVAC components contained in one cabinet. The cabinet is divided into a heating unit, evaporator, and fan in one section and a condenser and compressor in the other section. **See Figure 1-18.** Controls and components are installed by the manufacturer. Most combination units are air-cooled and located outdoors adjacent to or on the roof of the building.

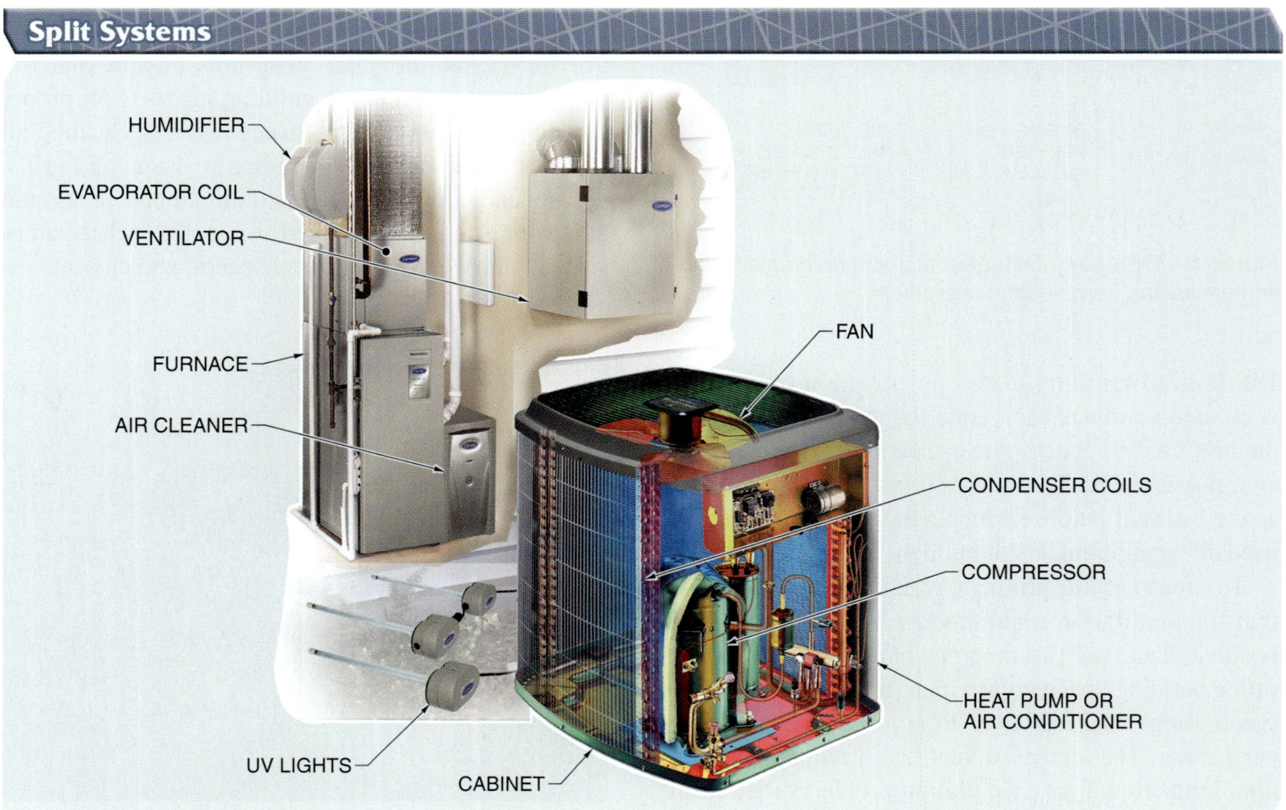

Figure 1-16. Split systems separate the evaporator from the condensing unit. The evaporator is placed in the furnace airflow.

Packaged HVAC Units

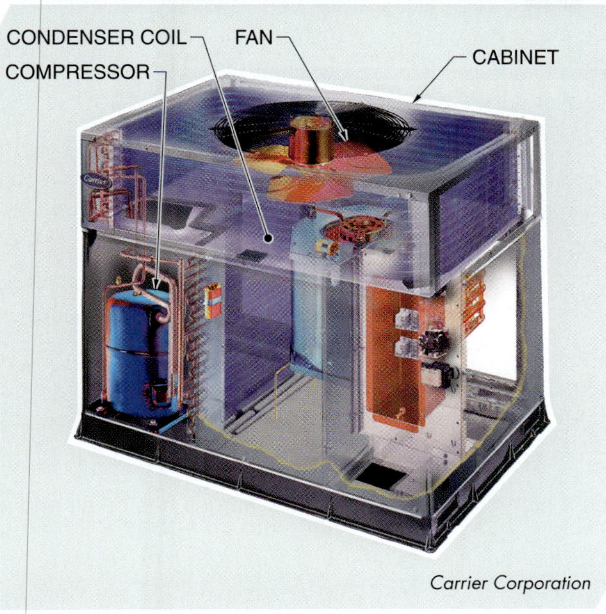

Figure 1-17. Packaged HVAC units are assembled by the manufacturer for easy installation.

Combination HVAC Units

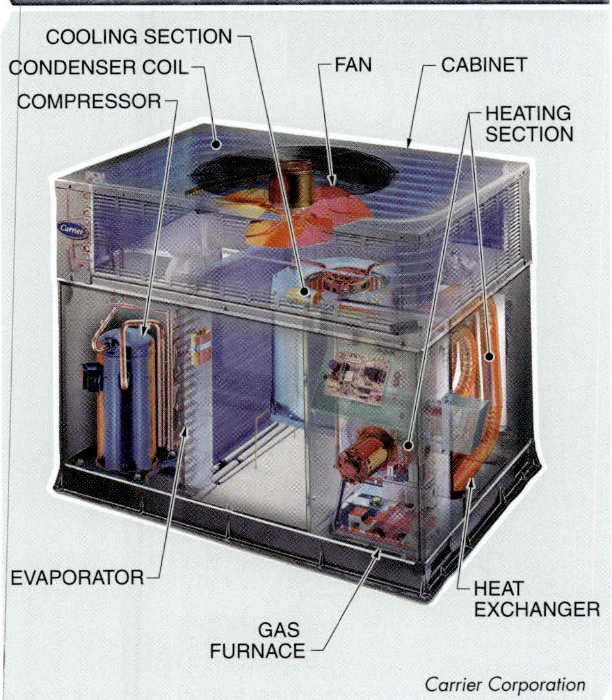

Figure 1-18. Combination units include a gas furnace and heat exchanger with the air conditioning system to provide for system heating and cooling.

TECH TIP

Most residential and light commercial HVAC units are designed to produce a specific amount of heating and cooling.

Refrigeration Processes

Air conditioners use either the mechanical compression refrigeration process or the absorption refrigeration process. Air conditioning systems are designed to produce a refrigeration effect at a fixed rate. Controls cycle the components of an air conditioning system on and off to produce the required refrigeration effect.

Mechanical Compression Refrigeration. In the mechanical compression refrigeration process, a compressor increases the pressure and temperature of a refrigerant so the refrigerant can reject heat to the condensing medium. A mechanical compression refrigeration system also includes an expansion device, evaporator, and condenser. **See Figure 1-19.** The expansion device causes pressure to decrease in the refrigerant. The evaporator transfers heat from the evaporating medium to the refrigerant in the system. The condenser transfers heat from the refrigerant to the condensing medium.

Refrigeration System Component Identification
Interactive Animation

Absorption Refrigeration. In the absorption refrigeration process, an absorbant controls the pressure and temperature of a refrigerant. An absorption refrigeration system includes an absorber, generator, condenser, orifice, and evaporator. **See Figure 1-20.** In the absorber, the absorbant combines with the refrigerant to form a solution. In the generator, the solution absorbs heat and separates into refrigerant and absorbant. In the condenser, the refrigerant rejects heat and condenses to a high-pressure liquid. In the orifice, the pressure decreases. In the evaporator the low-pressure refrigerant evaporates. The refrigerant absorbs heat from the evaporating medium that flows through the evaporator. This cools the evaporating medium, which is used for cooling product or air.

Liquid Chiller Media Clip

Mechanical Compression Refrigeration

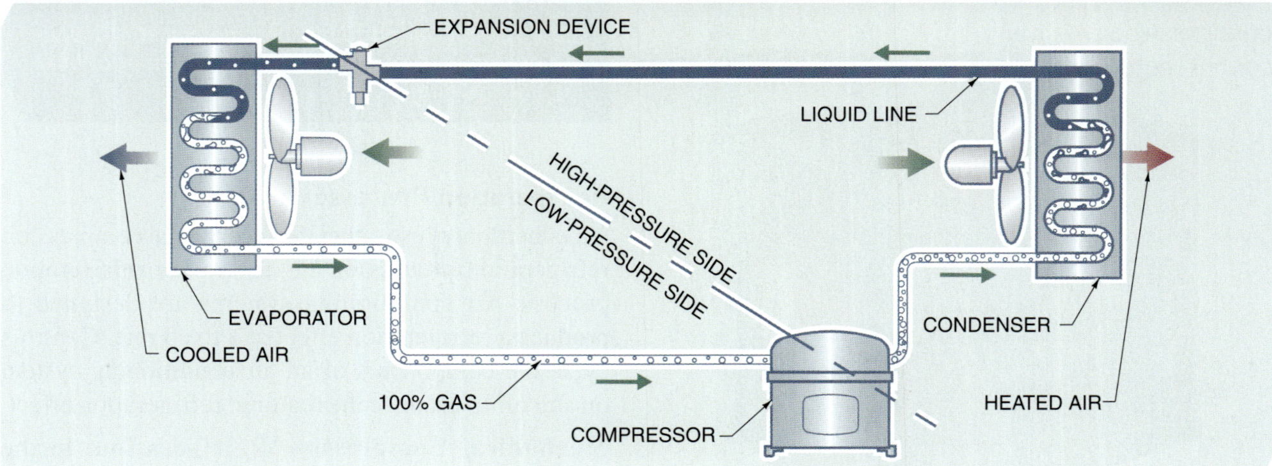

Figure 1-19. A mechanical compression refrigeration system has a high-pressure side that discharges heated air and a low-pressure side that discharges cooled air.

Absorption Refrigeration

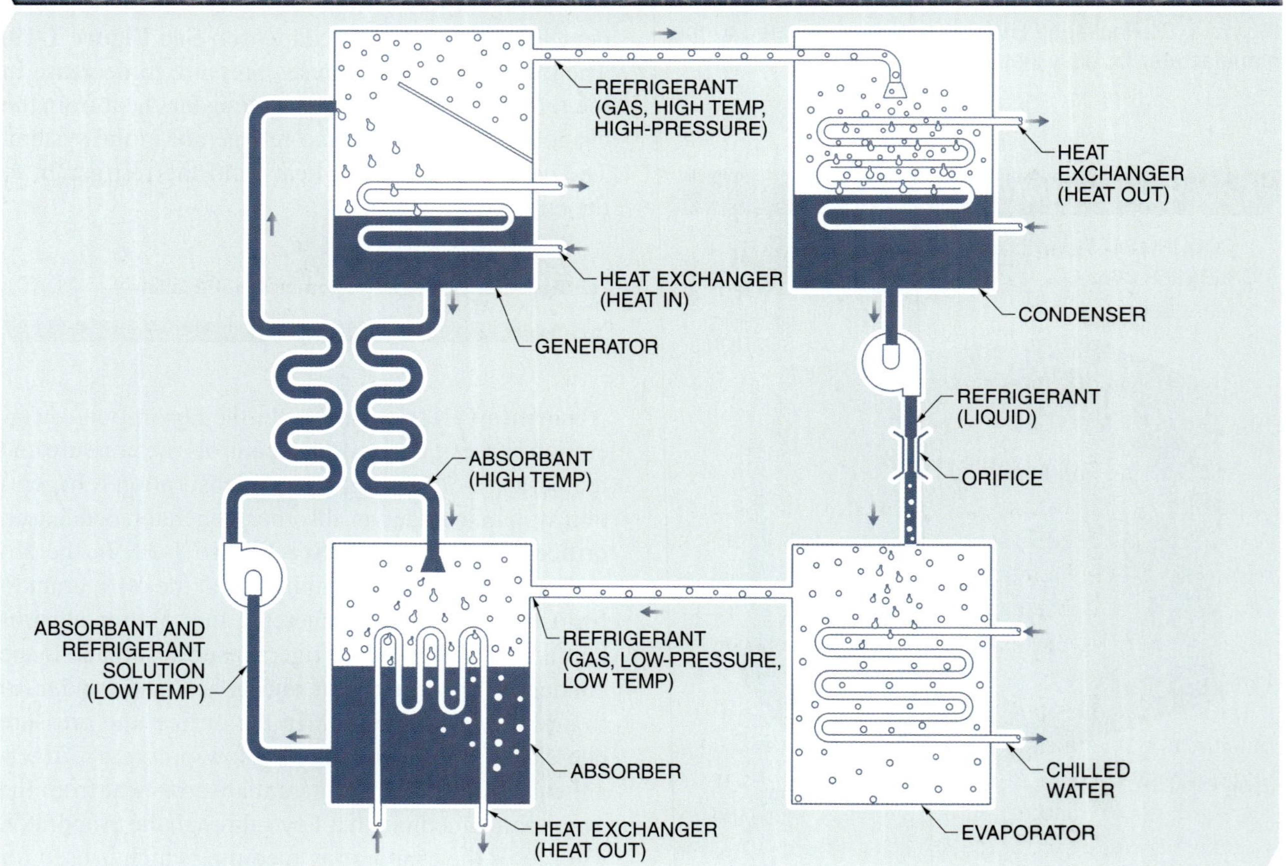

Figure 1-20. Absorption refrigeration systems produce the refrigeration effect when one chemical is absorbed by another and heat is then removed.

UNIT 2

HVAC and Refrigeration System Safety

Working on HVAC equipment and systems requires following safety rules. Two important safety rules are wearing approved protective clothing and using protective equipment. An additional requirement for personnel safety includes understanding the hazards and how to protect against them by identifying danger, warning, and caution labels used with HVAC equipment and systems.

OBJECTIVES

- Define the differences between codes and standards.
- List common regulatory agencies, trade associations, and standards and testing organizations.
- Describe personal safety methods and procedures.
- Identify key requirements of height and environmental safety.
- Describe mechanical and chemical safety.
- Describe key factors of safety programs and training.

CODES AND STANDARDS ORGANIZATIONS

National, state, and local codes and standards are used to protect people and property from potential hazards. A *code* is a set of regulations or minimum requirements related to a particular trade or environment. A *regulation* is a rule that has been made mandatory by a federal, state, or local government. A *standard* is an accepted reference or practice.

Following codes and standards ensures that HVAC equipment is manufactured, installed, and operated safely and efficiently. Codes and standards detail requirements for worker safety and protecting people from hazards.

Guidelines are used to reduce costs, standardize practices, and improve building operations. A *guideline* is a completely voluntary recommendation that is related to a particular trade or environment. Guidelines are not intended to be codes or requirements, and their use does not guarantee any specific outcome. **See Figure 2-1.**

Industry Regulations

CODES

- Developed by organization or government entity
- Usually broad in scope addressing many trades
- Enforceable when adopted by an authority having jurisdiction

STANDARDS

- Developed by trade or industry organization
- Usually narrow in scope addressing specific requirements
- Enforceable only when specified in a code, otherwise voluntary

GUIDELINES

- Developed by trade or industry organization
- Scope varies
- Never enforceable

Figure 2-1. Codes, standards, and guidelines are used to manufacture, install, and safely operate HVAC equipment and to protect people from HVAC hazards.

Government Regulatory Agencies

A *government regulatory agency* is a federal, state, or local government organization that establishes rules and regulations related to safety, health, equipment installation, and equipment operation. **See Figure 2-2.** Technicians must be familiar with the regulatory agencies affecting HVAC system procedures. The Occupational Safety and Health Administration (OSHA) ensures that all technicians are working in safe conditions while the Environmental Protection Agency (EPA) protects the environment from HVAC contaminants. Standards organizations, state government agencies, and local government agencies may also have specific requirements.

Government Regulations

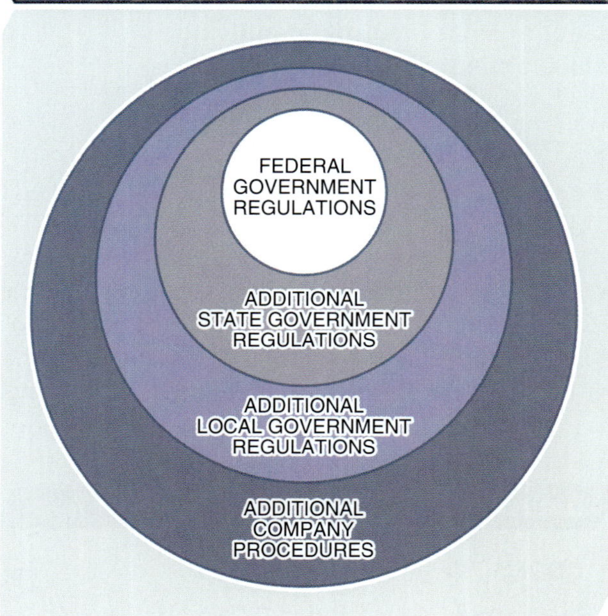

Figure 2-2. Company procedures must comply with the minimum requirements of federal, state, and local regulations.

With the exception of federal regulations, codes are not enforceable unless they are adopted by an authority having jurisdiction. An *authority having jurisdiction (AHJ)* is an organization, office, or individual designated by a government agency with legal powers to administer, interpret, and enforce building codes. An AHJ may choose to develop its own codes, but most jurisdictions adopt national codes with small modifications to suit their locale.

OSHA. The *Occupational Safety and Health Administration (OSHA)* is a federal government regulatory agency that requires all employers to provide a safe environment for their employees. OSHA was established under the Occupational Safety and Health Act of 1970. The Occupational Safety and Health Act requires that all employers provide work areas free from recognized hazards that are likely to cause serious harm. Under OSHA guidance, states may develop and administer state occupational safety and health plans. State plans may include the private and/or public sector and must be revised as necessary to comply with minimum federal OSHA standards.

OSHA Media Clip

EPA. The *Environmental Protection Agency (EPA)* is a federal government regulatory agency established in 1970 to control and abate air, water, and land pollution in the form of solid waste, radiation, pesticides, and various other toxic substances. The primary responsibilities of the EPA include creating and establishing standards affecting public health and the environment. In addition, the EPA coordinates the efforts of federal, state, and local governmental agencies.

EPA Regulations. In the 1970s, scientists began to realize that chemicals in the atmosphere were destroying the ozone layer that protects the Earth from ultraviolet radiation. Chlorofluorocarbon (CFC) and other refrigerants that are used in many HVAC and refrigeration systems were determined to be among the causes of ozone depletion.

On September 16, 1987, the United States signed an international agreement (Montreal Protocol on Substances that Deplete the Ozone Layer) that established phase-out dates for CFC refrigerants. This protocol was implemented in the United States through Title VI of the Clean Air Act (CAA) of 1990. Under CAA Section 608, the EPA established regulations that do the following:

- require service practices that maximize the recycling of ozone-depleting substances such as chlorofluorocarbon (CFC) and hydrochlorofluorocarbon (HCFC) refrigerants during the servicing and disposal of air conditioning and refrigeration equipment
- set certification requirements for technicians and recovery, recycling, and reclaiming equipment
- restrict the sale of refrigerants to certified technicians
- require technicians servicing or disposing of air conditioning and refrigeration equipment to prove to the EPA that recycling and recovery equipment being used is in compliance with EPA rules

- require the immediate repair of substantial leaks in air conditioning and refrigeration equipment with a charge of greater than 50 lb
- establish safe disposal requirements to prevent refrigerants from entering the waste stream by ensuring their removal from appliances

Regulations have also been implemented that prohibit individuals from knowingly venting CFCs and HCFCs into the atmosphere while servicing or repairing air conditioning and refrigeration equipment. The following four types of releases are permitted:

- "de minimis" (minimal) quantities released in good faith efforts to recover, recycle, or safely dispose of refrigerants
- refrigerant releases during normal operation such as those resulting from purging and leaks (the EPA requires leak repairs in equipment above a certain size)
- releases of CFCs and HCFCs that are not being used as refrigerants
- small releases caused by connecting, disconnecting, and purging hoses (recovery and recycling equipment made after November 15, 1993 must be equipped with low-loss fittings)

EPA Certification Program. The EPA has established a technician certification program for those who perform maintenance on, service, repair, or dispose of equipment that is likely to release refrigerants into the atmosphere. The duties of a certified technician include and exclude certain activities. Apprentices are exempt from certification requirements as long as a certified technician closely supervises them at all times.

Activities requiring a certified technician include the following:

- attaching and detaching hoses
- adding or removing refrigerants
- activities that violate the integrity of the device, such as opening the device

Activities that do not require a certified technician include the following:

- activities that do not violate the integrity of the refrigeration circuit, such as painting, rewiring, replacing insulation, or tightening nuts and bolts
- maintenance, service, and repair of devices that have already been evacuated properly
- motor vehicle air conditioning (MVAC) servicing and disposal (MVACs have their own set of rules and certifications under Section 609 of the CAA)
- most operator activities of air conditioning and refrigeration systems, such as taking regular readings of conditions and observing equipment operation (however, many employers require certification for any air conditioning and refrigeration task)

Section 608 of the CAA establishes technician certification for the following four types of certification:

- Type 1: servicing small appliances (i.e., residential refrigerators)
- Type 2: servicing or disposing of medium, high, or very-high pressure appliances (i.e., rooftop units)
- Type 3: servicing or disposing of low-pressure appliances (i.e., centrifugal chillers)
- Universal: servicing all types of equipment, including Type 1, Type 2, and Type 3

Standards Organizations

A *standards organization* is a nongovernmental organization that develops standards based on an accumulation of knowledge and experience with materials, methods, and practices. Standards organizations are typically trade groups or industrial organizations that develop a set of rules through the consensus of members in their organization. Standards organizations often impose stricter standards than government organizations. Examples of standards organizations specializing in certain safety-related areas of HVAC include the American National Standards Institute (ANSI), the American Society of Heating, Refrigerating and Air-Conditioning Engineers (ASHRAE), the National Fire Protection Association (NFPA), and the American Society of Mechanical Engineers (ASME). **See Figure 2-3.**

ANSI. The *American National Standards Institute (ANSI)* is a national standards organization that helps identify industrial and public need for national standards. ANSI standards are typically produced and published jointly with technical societies, trade associations, and U.S. government departments. ANSI acts as a national coordinator of voluntary standards and activities and as an approval organization and clearinghouse for consensus standards.

For example, ANSI provides standards for the manufacture of hard hats, respirators, safety glasses, and other personal protective equipment (PPE). Standards are submitted and approved through democratic procedures. The approval process ensures impartial consensus as well as eliminates any conflict with, or duplication of, other national standards.

Codes and Standards Organizations

Organization	Role
OSHA Occupational Safety and Health Administration www.osha.gov	Requires employers to provide a safe working environment Regulations are mandatory Enforces compliance through inspections
EPA Environmental Protection Agency www.epa.org	Protects human health and environment Regulations are mandatory Enforces compliance through inspections
NFPA National Fire Protection Association www.nfpa.org	Provides guidance in assessing hazards of products of combustion Publishes the National Electrical Code® Develops hazardous materials information
ANSI American National Standards Institute www.ansi.org	Coordinates and encourages activities in national standards development Identifies industrial and public needs for standards Acts as national coordinator and clearinghouse for consensus standards
ASHRAE American Society of Heating, Refrigerating and Air-Conditioning Engineers www.ashrae.org	Develops standards and guidelines for the HVAC and refrigeration industry
NEMA National Electrical Manufacturers Association www.nema.org	Assists with information and standards concerning proper selection, ratings, construction, testing, and performance of electrical equipment
NIOSH The National Institute for Occupational Safety and Health www.cdc.gov/niosh/	Acts in conjunction with OSHA to develop recommended exposure limits for hazardous substances or conditions located in the workplace Recommends preventive measures to reduce or eliminate adverse health and safety effects
UL Underwriters Laboratories Inc.® www.ul.com	Tests equipment and products to verify conformance with national codes and standards
CSA Canadian Standards Association www.csa.ca	Tests equipment and products to verify conformance with Canadian national standards
ASME American Society of Mechanical Engineers www.asme.org	Develops standards and guidelines for the HVAC and refrigeration industry and other industries

Figure 2-3. Codes and standards organizations develop codes and standards to ensure technician and public safety.

ANSI also manages U.S. participation in international standards activities. The *International Organization for Standardization (ISO)* is a nongovernmental, international organization that is comprised of national standards institutions from over 160 countries and includes over 2700 technical committees. The ISO provides a worldwide forum for the standards-developing process. ANSI is the U.S. representative to the ISO.

ASHRAE. The *American Society of Heating, Refrigerating and Air-Conditioning Engineers (ASHRAE)* is an international organization established in 1894 whose mission is to advance the arts and sciences of heating, ventilation, air conditioning, and refrigeration to serve humanity and promote a sustainable world. This mission is accomplished through research, standards writing, publishing, and continued education. ASHRAE produces many standards that have been adopted for manufacturing and installation specifications and local codes. ASHRAE standards define minimum values and acceptable levels of performance for HVAC equipment and process.

ASHRAE 90.1 Media Clip

NFPA. The *National Fire Protection Association (NFPA)* is a national organization that provides guidance in assessing hazards of the products of combustion. The NFPA publishes more than 300 consensus codes and standards intended to minimize the possibility and effects of fire and other risks. NFPA codes and standards are widely adopted because they are developed using an open, consensus-based process. Most buildings, processes, and installations are affected by NFPA codes and standards.

Joint Commission. The *Joint Commission* is a major nonprofit organization that provides accreditation services for health care organizations and sets most of the standards by which health care quality is measured in the U.S. The Joint Commission is also referred to as the Joint Commission on the Accreditation of Healthcare Organizations (JCAHO). The Joint Commission is governed by a board that includes physicians, nurses, and consumer advocates. Since HVAC installations provide comfort to personnel in health care facilities, safety programs and precautions are of interest to the Joint Commission.

Trade Associations

A *trade association* is an organization that represents the manufacturers or contractors that make, install, and service specific products. The Air-Conditioning, Heating, and Refrigeration Institute (AHRI) is a national trade association representing more than 90% of manufacturers of central air conditioning and commercial refrigeration equipment produced in North America. The Air Conditioning Contractors of America (ACCA) is an organization that represents contractors who install and service refrigeration and air conditioning equipment.

United Association of Journeymen and Apprentices of the Plumbing and Pipe Fitting Industry of the United States and Canada (UA). The *United Association of Journeymen and Apprentices of the Plumbing and Pipe Fitting Industry of the United States and Canada (UA)* is a multicraft union. UA journeymen and apprentices are engaged in the fabrication, installation, and servicing of piping systems. UA members are employed by mechanical contractors and plumbing, heating, and cooling contractors. The UA is committed to providing hands-on training programs for journeymen and apprentices to improve the quality of craftsmanship and provide highly skilled craft workers to the HVAC industry.

International Union of Operating Engineers (IUOE). The *International Union of Operating Engineers (IUOE)* is a diversified trade union established in 1896 that primarily represents stationary engineers, operating engineers, and members in the service industry. Stationary engineers work in the operations and maintenance of buildings, including HVAC equipment and processes. The IUOE provides training and apprenticeship programs that ensure its members are safe, productive, and highly skilled craft workers.

HVAC Excellence. *HVAC Excellence* is a not-for-profit organization that provides improved competency through validation of technical education for the HVAC and refrigeration industry. HVAC Excellence developed HVAC teaching standards and a program to verify that they have been met. HVAC Excellence provides an accreditation program and certification program for highly skilled workers in the HVAC and refrigeration industry.

Professional Organizations

A professional organization is a group of individuals with the same profession that collaborate to advance their particular field of study. Professional organizations help develop and advance individuals, while promoting their profession and educating the public on issues related to their industry. Most professional organizations cooperate with other associations to network new ideas and practices to set standards in their industry. Professional organizations include the Mechanical Contractors Association of America (MCAA) and the Mechanical Service Contractors of America (MSCA).

Mechanical Contractors Association of America. The Mechanical Contractors Association of America (MCAA) is an organization that serves the needs of companies in the heating, air conditioning, refrigeration, plumbing, piping, and mechanical services trades. These companies are provided quality training programs and materials to further develop knowledge needed in these trades. The MCAA also strives to help workers attain high levels of managerial and technical knowledge.

Mechanical Service Contractors of America. The Mechanical Service Contractors of America (MSCA) is a subsidiary of the MCAA and helps contractors with management and marketing skills, training and educational programs, and government and labor relations. The MSCA contributes time and resources to benefit individuals in the fields of heating, ventilation, air conditioning, and refrigeration.

Testing Organizations

Underwriters Laboratories Inc. (UL) is an independent organization that tests equipment and products to verify conformance to national codes and standards. Equipment tested and approved by UL carries the UL label. UL-approved equipment and products are listed in UL's data base for verification. *CSA International* is a Canadian government agency that tests electrical equipment and products to verify their conformance to Canadian national standards. CSA functions similarly to UL.

Insurance Agencies

The insurance agency insuring the facility and/or refrigeration and air conditioning equipment often determines the procedural requirements for refrigeration and air conditioning equipment installation and operation. For example, commercial and industrial refrigeration and air conditioning manufacturers may be required to provide special trim and accessories to meet insurance agency requirements. The requirements typically pertain to added safety features.

PERSONAL SAFETY

To prevent an accident or damage to equipment, HVAC and refrigeration system components must be installed, programmed, and serviced by technicians who have received training in OSHA 1926 guidelines. This training enables technicians to understand and manage the risks that may be encountered while servicing, maintaining, and installing HVAC equipment. OSHA 1926 also includes guidelines requiring a competent person to be present on a jobsite to review and manage site specific risks. A competent person is a person who is capable of identifying existing and predictable hazards in the surroundings or working conditions that are unsanitary, hazardous, or dangerous to employees and who has authorization to take prompt, corrective measures to eliminate them.

A competent HVAC person must meet the following requirements:

- safely work with refrigeration and air conditioning equipment and follow all OSHA and company safety procedures and practices
- perform the appropriate task required during an accident or emergency situation
- understand the operation of refrigeration test equipment and follow all manufacturer procedures
- know how to find help from equipment manuals and manufacturer representatives
- inform other technicians and facility operators of tasks being performed
- maintain all required records in computerized and written form

Safety Labels

A *safety label* is a label that indicates areas or tasks that can pose a hazard to personnel and/or HVAC equipment. Safety labels provide an instant reminder for personnel to wear the appropriate PPE. Safety labels appear in several ways on equipment and in equipment manuals. Safety labels use signal words to communicate the severity of a potential problem. The three most common signal words are danger, warning, and caution. **See Figure 2-4.**

Other symbols and signal words may appear with the danger, warning, and caution safety words used by manufacturers. Additional information concerning safety labels is detailed in ANSI/NEMA Standard Z535.4-2011, *Product Safety Signs and Labels.* Other signal words may be used on safety labels alone or in varying combinations. Additional symbols and signal words used by refrigeration and air conditioning equipment manufacturers include "electrical warning," "hazardous gases," and "flammable vapors."

Danger. The *danger signal word* is a word that indicates an imminently hazardous situation that, if not prevented, will result in death or serious injury. A danger signal word indicates the most extreme type of potential situation and, like the other signal words, accompanying instructions must be followed. The danger symbol is an exclamation mark enclosed in a triangle followed by the word "danger" written in bold print in a red box.

Warning. The *warning signal word* is a word that indicates a potentially hazardous situation that, if not prevented, could result in death or serious injury. The warning symbol is an exclamation mark enclosed in a triangle followed by the word "warning" written in bold print in an orange box.

Caution. The *caution signal word* is a word that indicates a potentially hazardous situation that, if not prevented, may result in minor or moderate injury or damage to equipment. In addition, the caution signal word warns of problems due to unsafe work practices. The caution symbol is an exclamation mark enclosed in a triangle followed by the word "caution" written in bold print in a yellow box.

Electrical Warning. The *electrical warning symbol and signal word* is a combination signal and symbol that indicates a high-voltage location and condition that could result in death or serious personal injury from electrical shock if proper precautions and procedures are not taken. The electrical warning safety label is typically placed where there is the potential of coming in contact with live wires, terminals, or parts. The electrical warning symbol is a lightning bolt enclosed in a triangle. The safety label may be shown with no words or may include the word "warning" written in bold print.

Explosion Warning. The *explosion warning symbol and signal word* is a combination signal and symbol that indicates locations and conditions where exploding parts may cause death or serious personal injury if proper precautions and procedures are not followed. The explosion warning symbol is an explosion drawing enclosed in a triangle. The safety label may be shown with the symbol only and no words or may include the word "warning" written in bold print.

PPE

Personal protective equipment (PPE) is gear worn by technicians to reduce the possibility of injury in the work area. **See Figure 2-5.** Appropriate PPE should be worn when installing or maintaining any HVAC equipment. All PPE must meet OSHA 29 CFR 1926-Personal Protective Equipment.

The PPE required for an application depends on the level of exposure, type of contamination, and type of work performed. PPE includes protective clothing, special clothing, head protection, eye and face protection, ear protection, hand protection, foot protection, back protection, and knee protection.

Protective Clothing. *Protective clothing* is clothing made of durable material, such as denim, that provides protection from contact with sharp objects, hot equipment, and harmful materials. Protective clothing should be snug yet allow ample movement. Pockets should allow convenient access but should not snag on tools or equipment. Soiled protective clothing should be washed to reduce the risk of flammability.

Safety Labels and Words

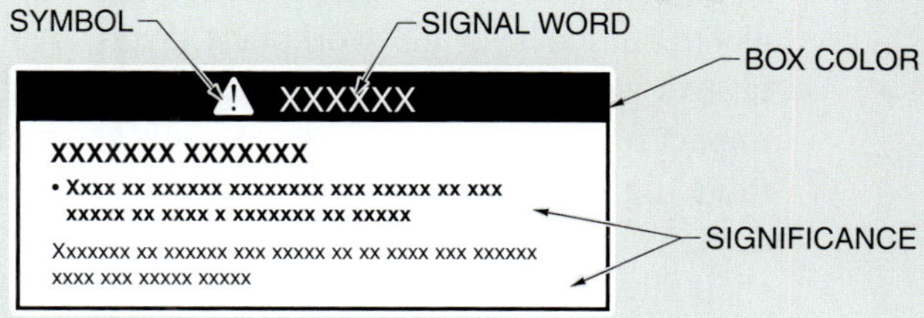

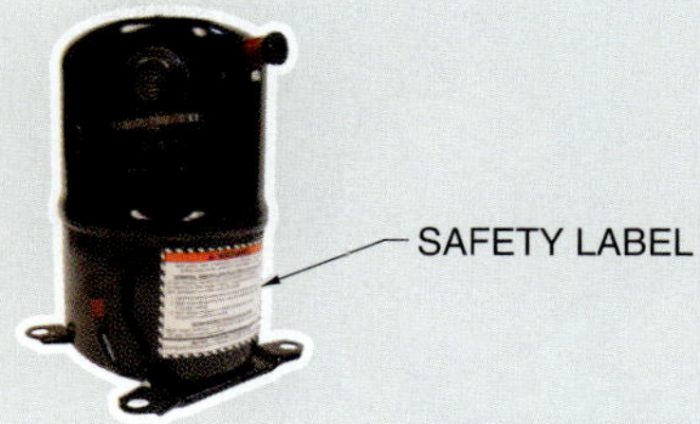

Safety Label	Box Color	Symbol	Significance
DANGER **HAZARDOUS VOLTAGE** **• Ground equipment using screw provided. Electric motor drive panel must be properly grounded before applying power.** **• Do not use metallic conduits as a ground conductor.** Failure to observe these precautions will cause shock or burn, resulting in severe personal injury or death!	RED		DANGER—INDICATES AN IMMINENTLY HAZARDOUS SITUATION THAT, IF NOT PREVENTED, WILL RESULT IN DEATH OR SERIOUS INJURY
WARNING **MEASUREMENT HAZARD** When taking measurements inside the electric motor drive, make sure that only the test lead tips touch internal metal parts. Keep hands behind the protective finger guard provided on the test leads.	ORANGE		WARNING—INDICATES A POTENTIALLY HAZARDOUS SITUATION THAT, IF NOT PREVENTED, COULD RESULT IN DEATH OR SERIOUS INJURY
CAUTION **MOTOR OVERHEATING** **This drive controller does not provide direct thermal protection for the motor. Use of a thermal sensor in the motor may be required for protection at all speeds and loading conditions. Consult motor manufacturer for thermal capability of motor when operated over desired speed range.** Failure to observe this precaution can result in equipment damage.	YELLOW		CAUTION—INDICATES A POTENTIALLY HAZARDOUS SITUATION THAT, IF NOT PREVENTED, MAY RESULT IN MINOR OR MODERATE INJURY OR DAMAGE TO EQUIPMENT; MAY ALSO BE USED TO ALERT AGAINST UNSAFE WORK PRACTICES
WARNING **Disconnect electrical supply before working on this equipment.**	ORANGE		ELECTRICAL WARNING—INDICATES HIGH VOLTAGE LOCATIONS AND CONDITIONS THAT COULD RESULT IN DEATH OR SERIOUS INJURY FROM AN ELECTRICAL SHOCK
WARNING **Do not operate the meter around explosive gas, vapor, or dust.**	ORANGE	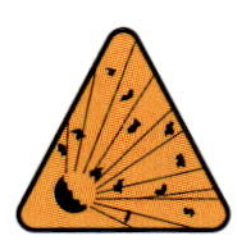	EXPLOSION WARNING—INDICATES LOCATIONS AND CONDITIONS WHERE EXPLODING ELECTRICAL PARTS MAY CAUSE DEATH OR SERIOUS INJURY

Figure 2-4. Safety labels are used to indicate the degree of likelihood of death or injury to personnel in different situations.

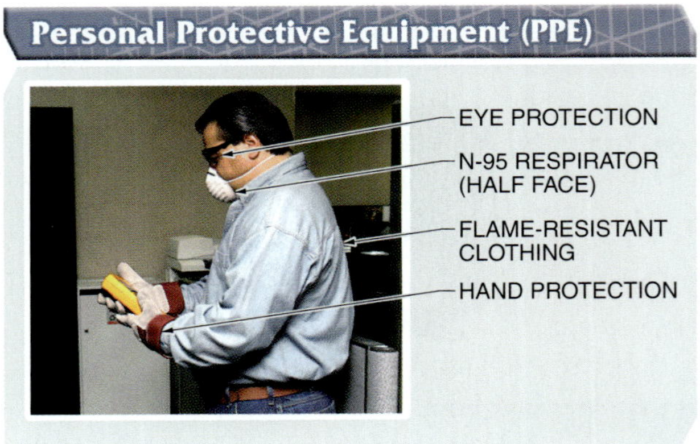

Figure 2-5. PPE is used to reduce the possibility or severity of an injury.

Special Clothing. *Special clothing* is a coverall or bodysuit made of a specific material that is treated to be impervious to gases, chemicals, or specific materials. Many bodysuits also include hoods and shoe covers that are either separate or part of one-piece bodysuits. **See Figure 2-6.** Loose-fitting clothing and long hair must be secured to prevent clothing and hair from getting caught in moving HVAC equipment.

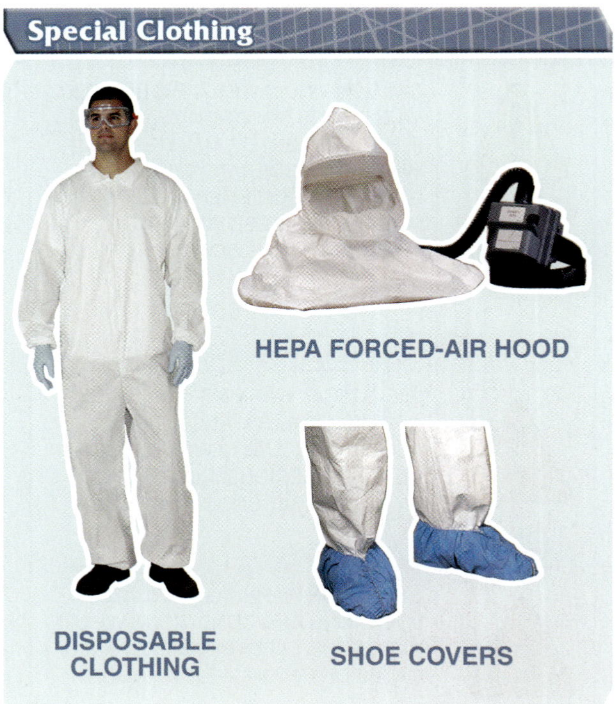

Lab Safety Supply, Inc.

Figure 2-6. Special clothing is needed when encountering hazardous substances. The protection chosen depends on the situation.

Head Protection. A *protective helmet,* or hard hat, is a specially designed hat used to prevent injury from impacts to the head, falling and flying objects, and electrical shock in the workplace. Protective helmets protect workers by resisting penetration and absorbing impact forces. Protective helmet shells are made of durable, lightweight materials. A shock-absorbing lining consists of crown straps and a headband that keeps the shell away from the head and allows ventilation.

Protective helmets are identified by their class of protection against specific hazardous conditions. **See Figure 2-7.** Type I helmets protect against impacts from the top, and Type II helmets protect against top and side impacts. Class G, Class E, and Class C helmets are used for construction and industrial applications.

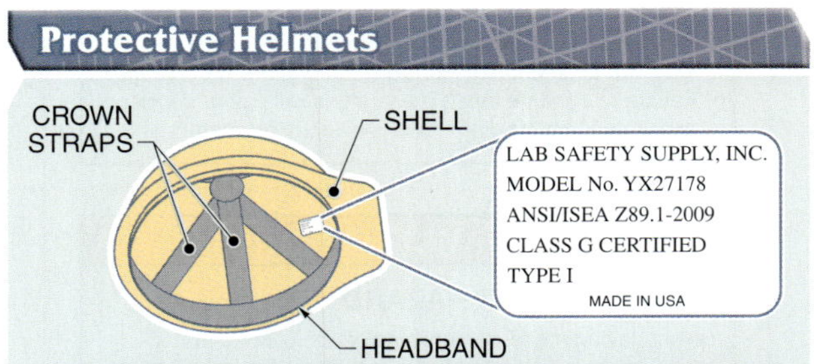

Type	Impact Protection
I	Impacts from top
II	Impacts from top and side
Class	**Use**
G	General service, voltage protection up to 2200 V
E	Utility service, voltage protection up to 20,000 V
C	Special service, no voltage protection

Figure 2-7. Protective helmets are identified by type and class for protection against specific hazards.

Class G (general) protective helmets are used for general service and provide protection up to 2200 V. Class E protective helmets are used for utility service and provide protection up to 20,000 V. Class C protective helmets are used for special service and provide no voltage protection.

Eye and Face Protection. Eye protection must be worn to prevent eye or face injuries caused by flying particles. Eye protection must comply with OSHA 29 CFR 1926—Eye and Face Protection. Eye protection standards are specified in ANSI/ISEA Standard Z87.1-2010, *American National Standard for Occupational and Educational Eye and Face Protection Devices.* Eye protection includes safety glasses, goggles, face shields, and respirator masks. **See Figure 2-8.**

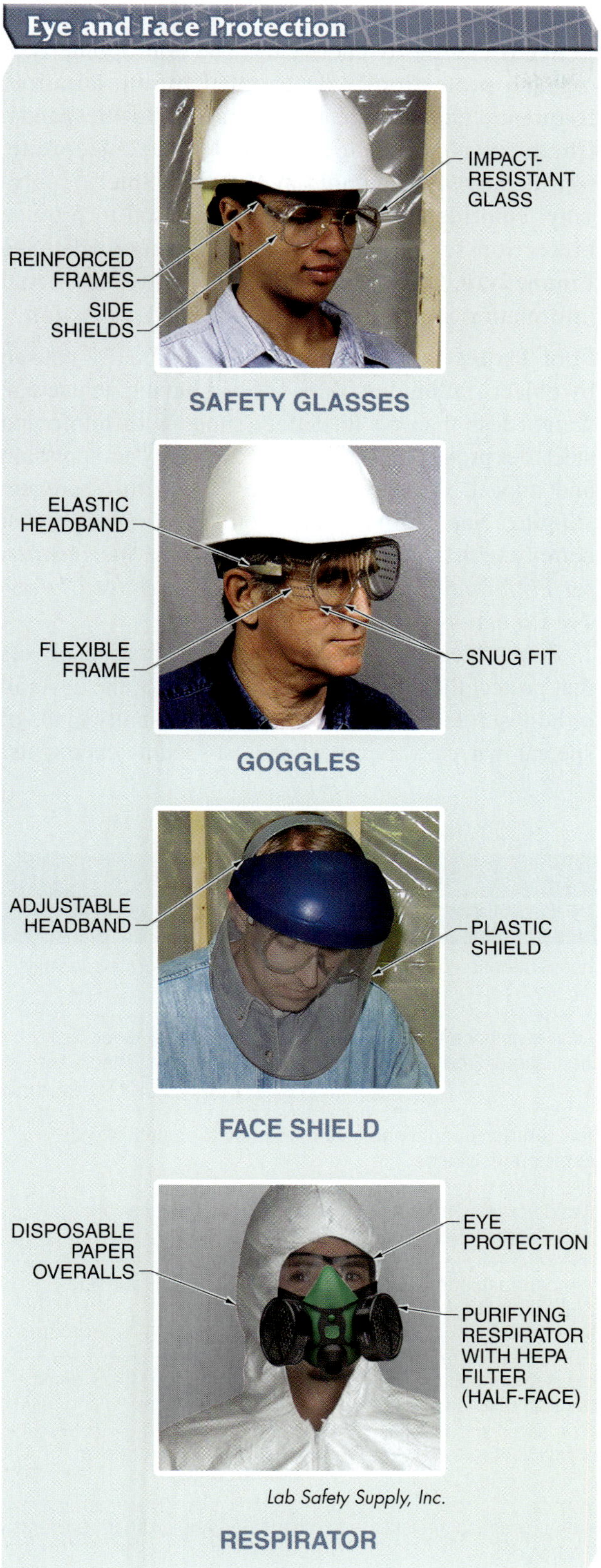

Figure 2-8. Eye and face protection is worn to prevent injuries caused by flying particles.

Safety glasses are an eye protection device with special impact-resistant glass or plastic lenses, reinforced frames, and side shields. The plastic frames are designed to keep the lenses secured in the frame when an impact occurs. Side shields provide additional protection from flying objects. *Goggles* are a type of eye protection with a flexible frame that is secured on the face with an elastic headband. Goggles fit snugly against the face to seal the areas around the eyes. A *face shield* is an eye and face protection device that covers the entire face with a plastic shield and is used for protection from flying objects.

A *full-face respirator mask* is a type of respiratory protection that covers the entire face, including the eyes, and has filter cartridges that protect against dusts, fumes, and particulates. A *half-face respirator mask* is a type of respiratory protection that covers the nose and mouth and has filter cartridges that protect against dusts, fumes, and particulates. Half-face respirator masks are typically worn with safety glasses or goggles.

Safety glasses and face shields must be properly maintained to provide the proper protection and clear visibility. Lens cleaners are available that clean without risk of lens damage. Crazing is a defect caused by exposure to aggressive solvents, chemicals, or heat that leaves microscopic cracks within the lenses. Pitted, scratched, and crazed lenses reduce vision and may cause lenses to fail on impact.

Ear Protection. Power tools, refrigeration, and air conditioning equipment can all produce excessive noise levels. Technicians subjected to excessive noise levels may develop hearing loss over time. The severity of hearing loss depends on the intensity and duration of exposure. **See Figure 2-9.**

Permissible Noise Exposures*

Duration†	Sound Level‡	Approximate Equivalent
8.0	90	Heavy truck at 3′
6.0	92	Power lawn mower at 3′
4.0	95	Newspaper press
3.0	97	Major road at 7′
2.0	100	Jackhammer at 7′
1.5	102	Jet flyover at 1000′
1.0	105	Bulldozer
0.5	110	Chainsaw at 3′
0.25 or less	115	Pneumatic chipper at 3′

* OSHA 29 CFR 1926.52(D)(1)
† in hours per day
‡ in dBA slow response

Figure 2-9. Hearing protection must be used when noise levels exceed permissible exposures.

Ear protection is worn to prevent hearing loss. Ear protection includes earplugs and earmuffs. An *earplug* is an ear protection device made of moldable rubber, foam, or plastic and is inserted into the ear canal. An *earmuff* is an ear protection device worn over the ears. A tight seal around an earmuff is required for proper protection.

Ear protection devices are assigned a noise reduction rating (NRR) based on the amount of noise they reduce. For example, a device with an NRR of 27 will reduce a noise level by 27 dB when tested at the factory. To determine approximate noise reduction in the field, 7 dB is subtracted from the NRR. For example, an NRR of 27 provides a noise reduction of approximately 20 dB in the field.

A *decibel (dB)* is a unit of measure used to express the relative intensity of sound. A typical human ear is able to hear sounds with frequencies from 20 Hz to 20,000 Hz. A particular sound has different strengths in the frequency range, with a peak response of around 3000 Hz. An A-weighted scale, or dBA, was developed to correlate with the frequency response of the human ear.

Hand Protection. *Hand protection* is a set of hand coverings (gloves) that is worn to prevent cuts or injuries to hands caused by chemicals. The appropriate type of hand protection is determined by the duration, frequency, and degree of the hazard to the hands. Chemical gloves made of butyl, latex, neoprene, nitrile, polyvinyl alcohol (PVA), Silver Shield®, polyvinyl chloride (PVC), or Viton® are used to provide protection for handling hazardous chemicals. **See Figure 2-10.** OSHA 29 CFR 1926 provides additional information regarding the need for hand protection.

Foot Protection. Foot injuries are typically caused by objects falling less than 4 ft and having an average weight less than 65 lb. Safety shoes with reinforced steel toes protect against injuries caused by compression and impact. Rubber-soled shoes are worn to prevent slipping. **See Figure 2-11.** Protective footwear must comply with ASTM F2413-11, *Standard Specification for Performance Requirements for Protective (Safety) Toe Cap Footwear.*

Foot protection also includes boots, shoes, or covers that protect the foot and seal the bottom of the coverall or bodysuit leg. Footwear covers are typically made of special materials to protect against specific chemicals.

Safety Gloves

Glove Material	Applications
Butyl	A synthetic rubber material that offers the highest permeation resistance to gas and water vapors, especially suited for use with esters and ketones
Latex	A natural rubber material that offers good chemical resistance to acids, alkalis, and detergents. Good tear resistance. **WARNING:** Presents a risk of irritation, allergic reaction, or significant skin reaction. The use of latex gloves should be considered carefully
Neoprene	A synthetic rubber material that provides excellent tensile strength and heat resistance. Compatible with some acids and caustics and has a moderate abrasion resistance
Nitrile	A synthetic rubber material that offers chemical and abrasion resistance. Provides protection from oils, greases, petroleum products, and some acids and caustics. Very good general-duty glove
PVA (polyvinyl alcohol)	A water-soluble synthetic material that is highly impermeable to gases. Excellent chemical resistance to aromatic and chlorinated solvents. **WARNING:** Glove material cannot be used in water or water-based solutions
Silver Shield®	A lightweight, flexible, laminated material that resists permeation from a wide range of toxic and hazardous chemicals. Offers the highest level of overall chemical resistance. **WARNING:** No cut resistance
PVC (polyvinyl chloride)	A synthetic thermoplastic polymer that provides excellent resistance to most acids, fats, and petroleum hydrocarbons. Good abrasion resistance
Viton®	A fluoroelastomer material that provides exceptional chemical resistance to chlorinated and aromatic solvents. Flexible but has minimal resistance to cuts and abrasions

Figure 2-10. Gloves protect hands from harmful substances. Glove types and materials are selected depending on the substance that is handled.

Figure 2-11. Foot protection is required to protect feet from injury.

Back Protection. Back injury is one of the most common injuries resulting in lost time in the workplace. Back injuries are often the result of improper lifting procedures but can be prevented through proper planning and by following established procedures. **See Figure 2-12.** Assistance should be sought when moving heavy objects. Before lifting objects from the ground, the worker should ensure that the path is clear of obstacles and free of hazards.

Long objects such as piping may not be heavy, but the weight might not be balanced. Therefore, two or more people should carry long objects when possible. When carried on the shoulder by one person, piping should be transported with the front end pointing downward to minimize the possibility of injury to others when walking around corners or through doorways. **See Figure 2-13.**

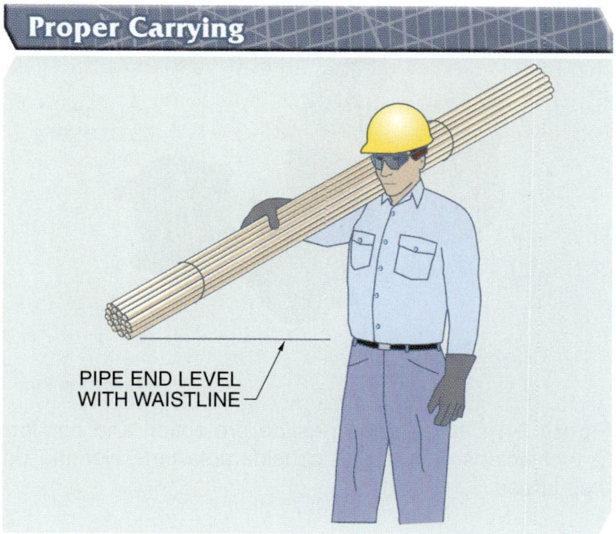

Figure 2-13. When carried on the shoulder by one person, pipe should be transported with the front end down.

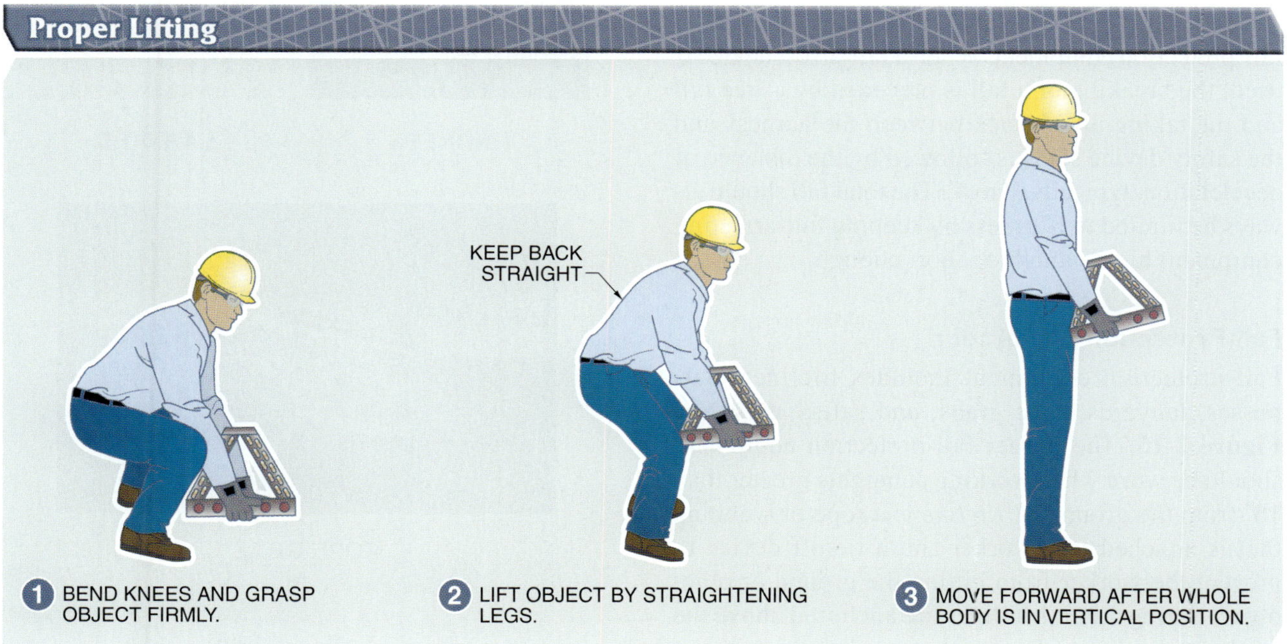

Figure 2-12. Lifting an object using the muscles in the legs reduces the possibility of a back injury.

Knee Protection. A *knee pad* is a rubber, leather, or plastic pad strapped onto the knees for protection. Technicians may require knee pads when they spend considerable time working on their knees or work in close areas and must kneel for proper access to equipment. Buckle straps or Velcro® closures secure knee pads in position. **See Figure 2-14.**

Knee Protection

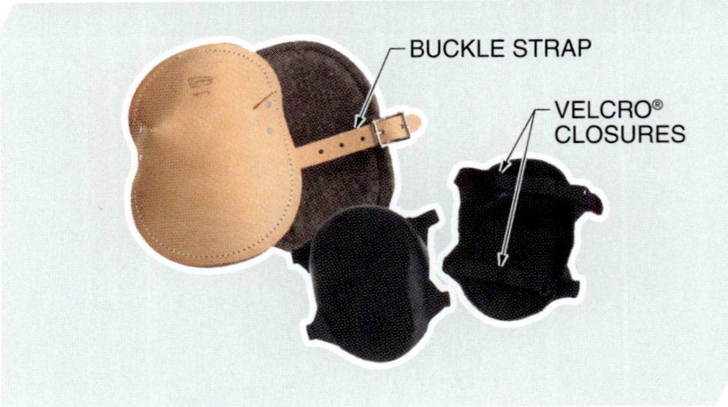

The Stanley Works

Figure 2-14. Knee pads provide protection and comfort to technicians who spend considerable time working on their knees.

HEIGHT SAFETY

Ladders and lifts allow HVAC technicians to reach the proper working heights required for some work. The safe use of ladders and lifts includes the use of fall-protection equipment. When a fall-arrest device is used, the breaking of a fall is preceded by a free fall and the taking up of slack between the harness and the safety device. This is followed by the distance of deceleration, typically 3′ to 4′. The total fall should always be limited to 6′ or less by keeping fall-arresting equipment high enough or short enough.

Fall-Protection Equipment

Fall-protection equipment includes lifelines, harnesses, lanyards, rope grabs, and safety nets. **See Figure 2-15.** The proper fall-protection equipment should be worn when working at heights greater than 10′ from the ground. A *lifeline* is a rope or webbing that is attached to a worker and a tie-off device to prevent the worker from hitting the ground or other object during a fall. Lifelines are anchored above the work area and must be strong enough to support the force of a fall.

Fall-Protection Equipment

Miller Equipment

Figure 2-15. The appropriate fall-protection equipment should be worn when working in elevated spaces.

Vertical lifelines must never have more than one person attached per line and must be long enough to reach the ground or landing below the work area when released. The lifeline must be terminated, or tied up, to prevent the safety sleeve from sliding off its end. When anchoring a lifeline to an anchor system, the path of a fall must be visualized so that there are no obstructions to the fall.

Harnesses protect internal body organs, the spine, and other bones in a fall. Harnesses and body belts must fit snugly and be securely attached to a lanyard. A *lanyard* is a rope or webbing device that connects a harness or body belt to a lifeline. A *rope grab* is a device that clamps securely to a rope and contains a ring to which a lifeline can be attached. Rope grabs provide protection from falls while allowing freedom of movement.

A *safety net* is a net made of rope or webbing designed to catch and protect a falling worker. A safety net is used when a worker is at least 10′ above the ground and is not otherwise protected by a body belt, lifeline, or scaffold. **See Figure 2-16.**

Safety Nets

The Sinco Group, Inc.

Figure 2-16. A safety net is made of rope or webbing for catching and protecting a falling technician.

Ladders

Ladders are devices used to reach otherwise inaccessible areas, such as rooftops. Many injuries are caused by falls from ladders due to improper maintenance or use. A ladder must be rated to support the weight of the technician and any necessary tools and supplies. Weight restrictions are found on the label for the ladder.

The base of an extension ladder must be 1′ away from the wall for every 4′ of height. **See Figure 2-17.** The upper end of the ladder is placed against a secure surface capable of supporting the weight on the ladder. When accessing an elevated area, such as a rooftop, the ladder should extend 3′ above the uppermost support point. The technician should never climb higher than the fourth rung from the top of an extension ladder. The technician's body must remain in the center of the ladder and not reach out too far to the side.

Positioning Ladders

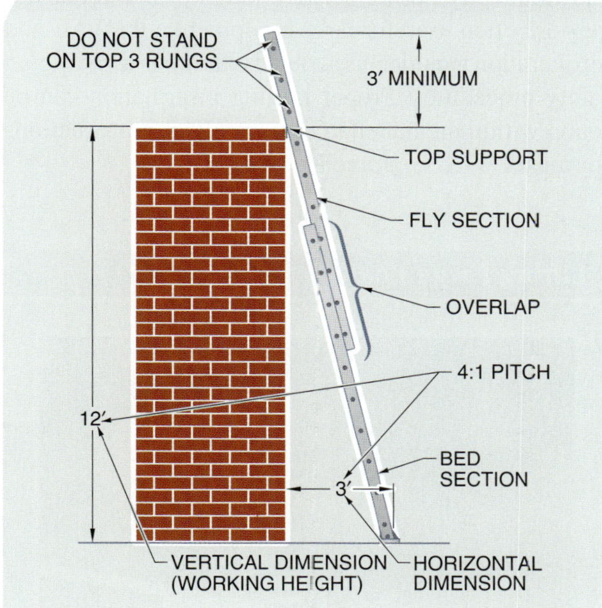

Figure 2-17. Ladders must be set up with the base 1′ from the wall for every 4′ of height up to the top support.

Stepladders are either self-supporting or fold-out ladders. The top of the ladder must be kept clear to prevent items from falling. Only one person may use a step ladder at a time, and the back support of the ladder must never be used for climbing. The person on a stepladder must not stand above the second rung from the top of the ladder.

Personnel Lifts

Personnel lifts are mechanized devices that lift a technician beyond the typical reach of a ladder. Personnel lifts can become unstable as the technician and tools rise, so stabilizing outriggers are typically required. The manufacturer instructions and information on the height, weight, and extension limits must be followed. Technicians in the bucket of a personnel lift must wear fall-protection equipment. When raising a personnel lift, particular attention must be paid to overhead obstructions. The required minimum distances to overhead electrical wires must be maintained.

Rigging

Some residential and light commercial refrigeration and air conditioning installations require the lifting of equipment. *Rigging* is securing equipment in preparation for lifting. Rigging is accomplished by means of wire ropes, fiber ropes, chains, round slings, or webbing. **See Figure 2-18.** For safety, the loads must be balanced and load weights calculated for load lifting. All rigging components should be inspected initially and frequently during use. Although professional riggers are often used for large equipment, all HVAC and refrigeration technicians should be familiar with rigging safety procedures. Proper rigging information can be found within the manufacturer installation instructions for most HVAC equipment.

Lifting Large Loads

Figure 2-18. Large loads, such as roof top units, are installed using slings placed by experienced riggers.

ENVIRONMENTAL SAFETY

In addition to awareness of individual and group safety, it is important to recognize that the environment must be safeguarded as well. As HVAC and refrigeration equipment is installed or serviced, the release of minute amounts of harmful substances into the environment may occur. The use of oils and refrigerants is regulated by governing agencies such as the EPA. Local jurisdictions are often involved as well. When these regulations are violated, heavy fines or even imprisonment may result. Whenever these materials are disposed of or consumed, all appropriate laws and standards must be followed. Proper procedures must be established and followed to eliminate the damage to the environment.

Radiological Safety

Possible exposure to radioactive materials is another risk to technicians. While HVAC and refrigeration systems typically do not contain radioactive elements, many facilities they serve do. Hospitals, research labs, and industrial facilities often have radioactive materials that are used in a variety of ways. HVAC and refrigeration equipment may be used in medical treatment facilities or where trace elements for diagnostic equipment, such as X-ray generators, are created. Special care must be taken when the radiological hazard symbol is observed. **See Figure 2-19.**

Radiological Hazard Symbols

Figure 2-19. Radiological materials are commonly found in hospitals, research labs, and industrial facilities.

HVAC technicians may service equipment that brings them in contact with radioactive elements. Technicians must be familiar with standard procedures developed by qualified personnel. In many cases, HVAC and refrigeration technicians wear a dosimeter badge that indicates the amount of exposure to radioactive elements.

Biohazard Safety

In addition to radiological safety, HVAC technicians must be aware of biohazard safety. Some biohazards include materials such as blood, fluids, treatment serums, and DNA research compounds. Hospitals, research labs, and industrial facilities often use biohazardous materials in a variety of ways, such as to develop new medical treatments or calibrate diagnostic equipment.

Most HVAC and refrigeration systems do not contain biohazardous materials. However, some building ductwork systems in certain circumstances generate biohazards, such as the bacteria linked to Legionnaires' disease. Also, when fume hoods are used during experiments, technicians may be called to service the airflow system of the hood. Exhaust fan ductwork may also come in contact with these biohazardous materials. Equipment and systems that serve biohazard areas should be clearly marked. Special care must be taken when the biohazard symbol is observed. **See Figure 2-20.**

Biohazard Symbols

Figure 2-20. Biohazard materials include materials such as blood, fluids, and treatment serums. Special care must be taken when the biohazard symbol is observed.

Technicians must be familiar with standard procedures developed by experienced, trained personnel. When exposure to a biohazardous material is suspected, immediate reporting and disinfecting are required.

Emergency Showers

Some facilities have emergency showers or decontamination stations located at marked points in the building. Emergency showers, also called drench showers, must comply with the ANSI Standard Z358.1-2009, *American National Standard for Emergency Eyewash and Shower Equipment*. When exposure to a dangerous substance occurs, a chain can be pulled or a button can be pushed to turn on a water shower to wash off the contaminant. Full-body showers or mobile decontamination areas are available as well. In addition to emergency showers being clearly marked, their proper use should also be a part of a comprehensive safety training program.

TECH TIP

A technician must perform a thorough review and check of the environment for safety hazards before troubleshooting or servicing equipment in uncommon or unfamiliar locations. For example, a technician called to service equipment at a medical facility may be faced with a biohazard situation or other uncommon safety issues.

Confined Spaces

A *confined space* is a space large enough that a technician can enter and perform assigned work but has limited or restricted means of entry and exit and is not designed for continuous occupancy. Confined spaces are subject to the accumulation of toxic or flammable contaminants or an oxygen-deficient atmosphere. Confined spaces include storage tanks, process vessels, boilers, chiller evaporators and condensers, some maintenance rooms, utility vaults, tunnels, pipelines, trenches, and any space a refrigerant can fill when released.

Confined Space Hazards. Confined spaces expose technicians to unsafe conditions such as excessive heat or cold, oxygen deficiency, combustible gases, and/or toxic gases. Oxygen deficiency in a space can be caused by the displacement of leaking refrigerant gases or vapors. Oxygen deficiency can result in serious injury or death. Proper requirements for entering a confined space must be followed.

Confined Space Entry Permit. A *confined space entry permit* is a document issued by an employer to allow and control entry into a confined space. **See Figure 2-21.** Confined space entry permits must be posted at the entry point to a confined space and kept as a permanent record.

Confined Space Entry Permits

CONFINED SPACE ENTRY PERMIT

NOTE: This form must be completed prior to entry of the confined space and posted near the entry opening. Upon completion of the project or expiration of a permit, a copy must be forwarded to EHSS.

Date of Issue:	Expiration Date:
Time:	Time:
Location:	
Description of Space:	
Job Task:	
Hazard Assessment:	
Attendant(s):	
Signature of Employees to Enter:	

TIME/DATE	%LEL (10%)	$\%O_2$ (19.5-23.5)	H_2S 10 ppm	CO 35 ppm	INITIALS	INSTRUMENT

SUPERVISOR AUTHORIZING ENTRY (PRINT)	SIGNATURE

Figure 2-21. A confined space entry permit is posted at the entry of a confined space prior to entering the space.

SYSTEM SAFETY

HVAC and refrigeration systems use equipment consisting of mechanical and chemical components. Each of these components has characteristics that could be hazardous when proper safety precautions are not followed. Mechanical hazards are related to mass, motion, or the condition of the heating and air conditioning equipment. Chemical hazards relate to the flammability, caustic action, or high temperature of chemicals used in refrigeration and air conditioning systems.

Mechanical Safety

Mechanical components of HVAC systems include moving parts such as motors, compressors, fans, pulleys, and belts. Loose clothing, neckties, and jewelry must not be worn near moving equipment. Clothing can become tangled in rotating shafts, resulting in moderate to serious injuries. Caution must be taken when working around moving parts.

Installation and Safeguards. All HVAC equipment must be installed according to the specifications of the manufacturer. A complete inspection must be conducted on all equipment after installation or servicing and before turning on the system. Access to HVAC system components must allow for maintenance tasks to be carried out with the necessary tools and test instruments in accordance with the operating methods specified by manufacturers.

Mechanical safety rules include the following:

- Do not operate any machinery or HVAC equipment unless all guards are in place.
- Replace any machine guards removed for maintenance or repairs as soon as maintenance or repairs have been completed.
- Exercise care when working with or near pneumatically driven HVAC equipment. Sudden or unexpected motion can inflict serious injury.
- Do not lubricate or service HVAC equipment unless locked out. It is possible to adjust, maintain, repair, clean, and service equipment while machinery is at a standstill.

Nonelectrical Lockout Devices. *Nonelectrical lockout devices* are lightweight enclosures that allow the lockout of standard control devices. Lockout devices are available in various shapes and sizes that allow for the lockout of ball valves, gate valves, and other nonelectrical devices.

Lockout devices resist chemicals, cracking, abrasion, and temperature changes and are available in colors to match ANSI pipe colors. Lockout devices are sized to fit standard industry control device sizes. **See Figure 2-22.** Locks used to lockout a device may be color-coded and individually keyed. Locks are rust resistant and available with various size shackles.

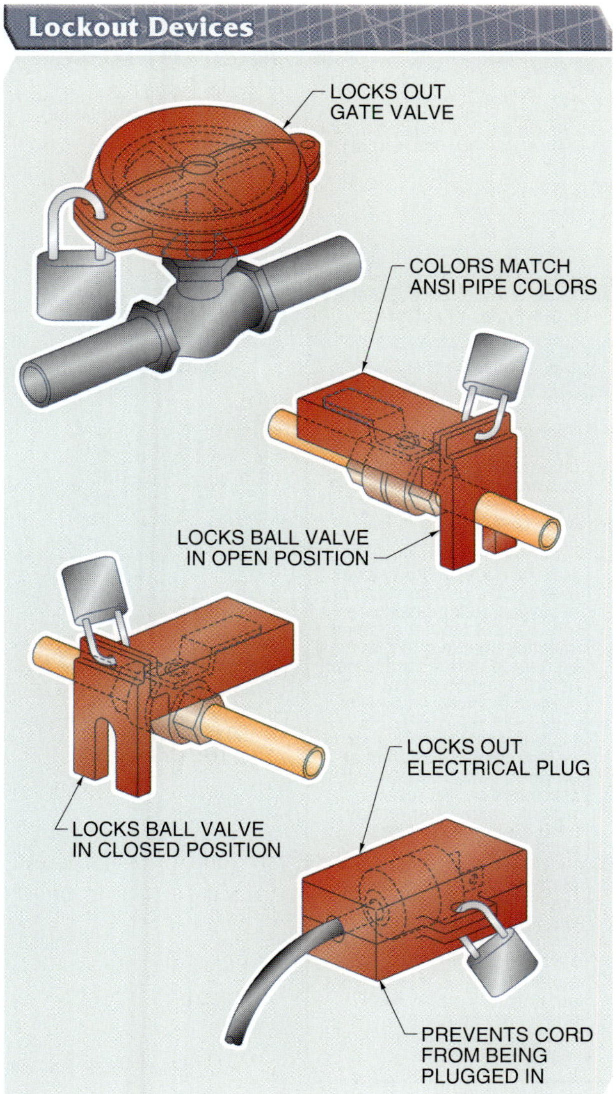

Figure 2-22. Lockout devices are sized to fit standard industry control device sizes.

A lockout/tagout kit contains items required to comply with OSHA lockout/tagout standards. Lockout/tagout kits contain reusable danger tags, tag ties, multiple lockouts, locks, magnetic signs, and information on lockout/tagout procedures. **See Figure 2-23.**

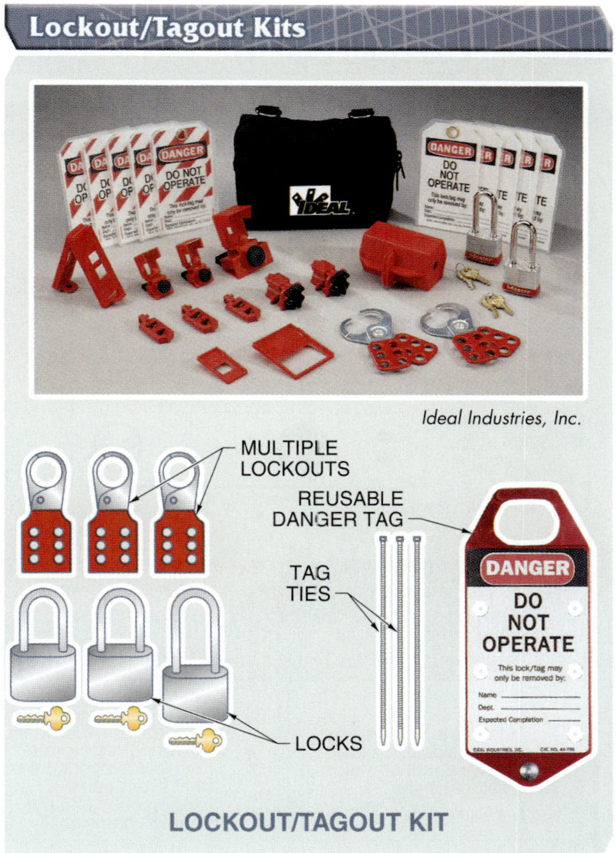

Figure 2-23. Lockout/tagout kits contain reusable danger tags, tag ties, multiple lockouts, locks, magnetic signs, and information on lockout/tagout procedures.

Nonelectrical Lockout Media Clip

Chemical Safety

Chemicals present in HVAC systems include fuels, refrigerants, and chemicals used for operation, cleaning, and water treatment. When fuels are properly contained in fuel lines, tanks, or other vessels, hazards are kept to a minimum. Danger occurs when fuels leak out of contained areas and mix with air. When this happens, the fuel can ignite or explode. All fuels, regardless of quantity, must be handled with care.

Refrigerants typically used in HVAC and refrigeration systems can cause injuries when handled improperly. Ammonia, which is used in large systems, is dangerous. Other refrigerants are dangerous when exposed to open flames or high temperatures. When this occurs, the refrigerant compounds change chemically and toxic gases form. Most refrigerants boil at low temperatures. Refrigerants coming in contact with skin cause severe frostbite. Protective clothing must always be worn to prevent injury from refrigerant leaks. Technicians should always refer to the manufacturer-recommended safety procedures when handling chemicals used in HVAC and refrigeration systems. **See Figure 2-24.**

Labeling Hazardous Material Containers

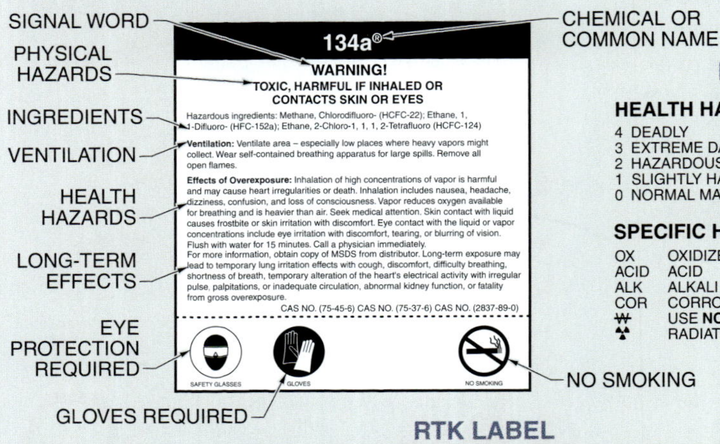

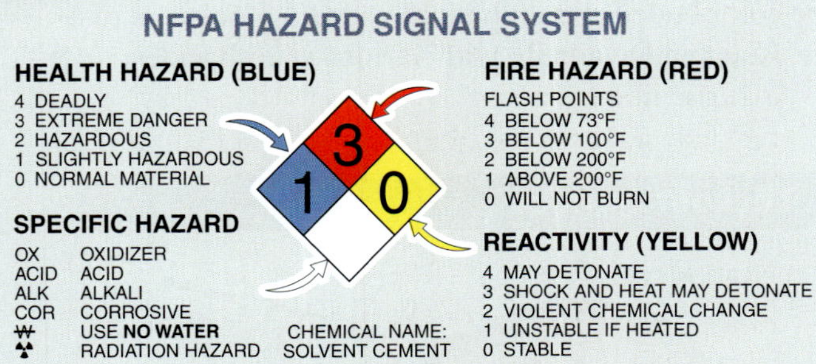

Identification of Health Hazard Color Code: BLUE		Identification of Flammability Color Code: RED		Identification of Reactivity (Stability) Color Code: YELLOW	
Signal	Type of Possible Injury	Signal	Susceptibility of Materials to Burning	Signal	Susceptibility to Release of Energy
4	Materials that on very short exposure could cause death or major residual injury	4	Materials that will rapidly or completely vaporize at atmospheric pressure and normal ambient temperature or that are readily dispersed in air and that will burn readily	4	Materials that in themselves are readily capable of detonation or of explosive decomposition or reaction at normal temperatures and pressures
3	Materials that on short exposure could cause death or major residual injury	3	Liquids and solids that can be ignited under almost all ambient temperature conditions	3	Materials that in themselves are capable of detonation or explosive decomposition or reaction but require a strong initiating source and that must be heated under confinement before initiation or that react explosively with water
2	Materials that on intense or continued but not chronic exposure could cause temporary incapacitation or possible residual injury	2	Materials that must be moderately heated or exposed to relatively high ambient temperatures before ignition can occur	2	Materials that readily undergo violent chemical change at elevated temperatures and pressures or that react violently with water or may form explosive mixtures with water
1	Materials that on exposure would cause irritation but only minor residual injury	1	Materials that must be preheated before ignition can occur	1	Materials that in themselves are normally stable but can become unstable at elevated temperatures and pressures
0	Materials that on exposure under fire conditions would offer no hazard beyond that of ordinary combustible material	0	Materials that will not burn	0	Materials that in themselves are normally stable even under fire exposure conditions and are not reactive with water

REFRIGERANT CYLINDERS ARE COLOR CODED ACCORDING TO THE REFRIGERANT IN THE CYLINDER

Figure 2-24. The NFPA Hazard Signal System and the Hazardous Material Information Guide (HMIG) provide quick reference information regarding hazardous materials.

Chemical safety rules include the following:

- Treat every chemical and refrigerant as hazardous.
- Make sure all chemicals and refrigerants are clearly and currently labeled with the substance name, concentration, date, and name of the individual responsible.
- Never return chemicals or refrigerants to their original containers.
- Comply with fire regulations concerning storage quantities, types of approved containers and cabinets, and proper labeling.
- Use volatile and flammable compounds only under a fume hood. Procedures that produce aerosols must be performed under a hood to prevent inhalation of hazardous materials.
- Never allow a solvent to come in contact with skin. Always use gloves.
- Read the labels on solvent bottles to identify contents. Do not sniff solvent.
- Clean up all spills immediately.

Safety Data Sheets. A *safety data sheet (SDS)* is a printed document used to relay hazardous material information from the manufacturer, importer, or distributor to the technician. **See Figure 2-25.** SDSs were previously known as material safety data sheets (MSDSs). OSHA's Hazard Communication Standard (HCS) specifies that certain information must be included on SDS forms. An SDS provides precautionary information regarding proper handling of a material. An SDS also lists emergency and first aid procedures as well as appropriate PPE to use when performing a task.

Chemical (refrigerant) manufacturers, distributors, and importers must offer an SDS for each hazardous material. When an SDS is not provided, the employer must contact the manufacturer, distributor, or importer to obtain the missing SDS. SDS files must be kept up-to-date and be readily available to all building or company personnel.

SDSs may be filed according to product name, manufacturer, or a company-assigned number. When two or more SDSs are found for the same hazardous material, the latest version is used. There are multiple potential SDS formats. The formats detailed in ANSI Standard Z400.1/Z129.1-2006, *Hazardous Workplace Chemicals – Hazard Evaluation and Safety Data Sheet Precautionary Labeling Preparation,* are typically used.

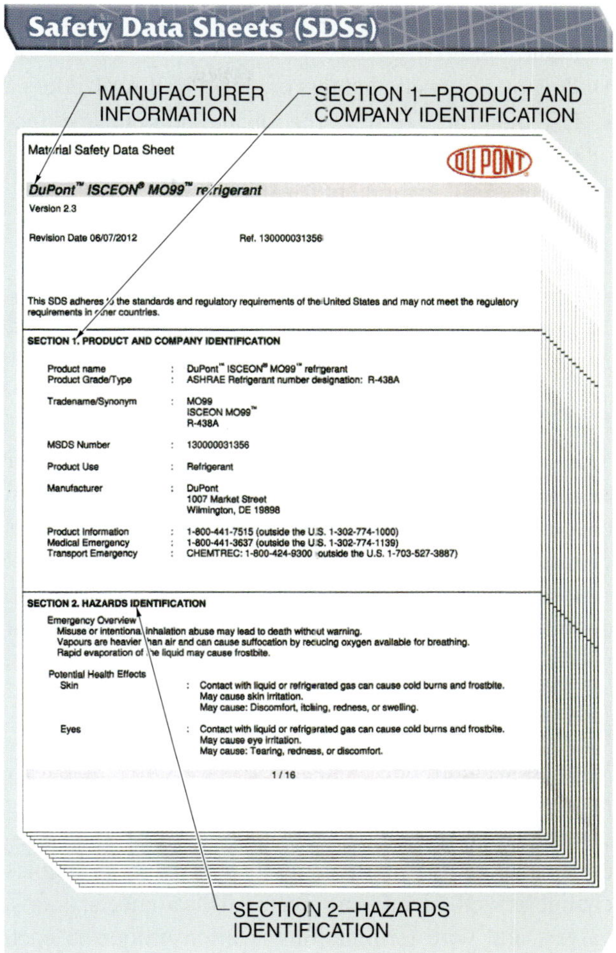

DuPont Company

Figure 2-25. An SDS, previously known as a material safety data sheet (MSDS), provides chemical hazard information regarding proper handling, emergency, and first aid procedures.

Safety Precautions

A technician must develop safety habits to prevent personal injury, injury to others, and damage to equipment. Safety procedures vary depending on the type and size of the HVAC equipment. The following are basic safety rules that are common for working with all HVAC and refrigeration system equipment:

- Wear approved PPE at all times.
- Wear gloves when handling energized lines and refrigerant system charging equipment.
- Wear appropriate respirators.
- Wear a protective helmet (hard hat) when there is a possibility of head injury.
- Do not use hands to stop moving equipment.

- Use approved safety cylinders to store combustible liquids and refrigerants.
- Do not use unsafe ladders or substitutes for ladders.
- Use the proper tool for the job and do not use defective tools.
- Precheck all equipment for starting hazards.
- Ensure that only qualified personnel are allowed to repair electrical equipment.
- Never enter an enclosed area that has toxic or flammable gas without adequate ventilation.

SAFETY PROGRAMS AND TRAINING

Many facilities implement ongoing safety training for maintenance personnel. Ongoing safety training is often encouraged or mandated by governing agency regulations or insurance requirements. Safety training may consist of a variety of presentations. Vendors may be used to demonstrate hands-on safety procedures with tools or power equipment.

Equipment Safety Training

When new equipment is installed in a facility, all safety procedures associated with the new equipment must be distributed. As technology and equipment operating principles change, the implementations of safety principles change as well. Identifying electrical disconnects, drains, valves, and wire terminations is often unique to each installation. As new HVAC equipment is installed, the operation and maintenance information must be obtained from the manufacturer. Standard operating procedures are developed or updated as new equipment is installed. **See Figure 2-26.** Equipment safety training is part of a comprehensive safety-training program.

Standard Operating Procedures

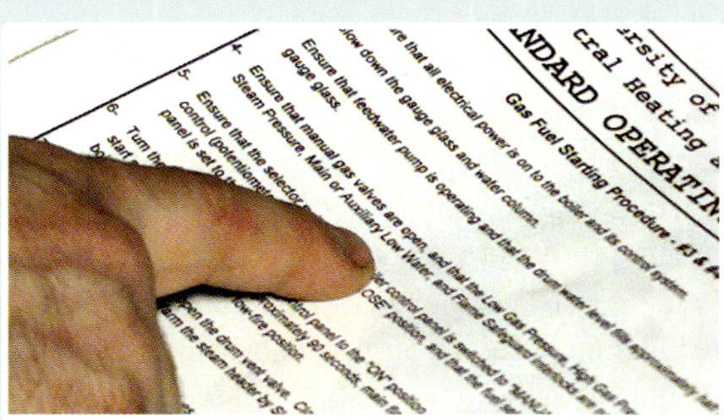

Figure 2-26. Standard operating procedures are used to ensure safe and consistent operation of a process.

First Aid

In the event of injury, it is important to obtain proper skilled medical help as quickly as possible. Depending on the type and severity of the injury, a fellow team member or employee is typically the first on the scene. Having designated, trained employees for first aid is often encouraged or mandated by governing agency regulations or insurance requirements. Typically, some employees are trained to respond to specific medical emergencies such as lacerations and burns. In some facilities, individuals may be trained in cardiopulmonary resuscitation (CPR) as well. These designated employees are trained to respond to certain public address system alerts or pages in the facility.

In order to treat injuries as quickly as possible, first aid kits are located at strategic spots in a facility. It is common to have a first aid kit in maintenance areas, mechanical rooms, and equipment penthouse areas. Also, many facilities have bloodborne pathogen protection kits. First aid kits and bloodborne pathogen kits can be purchased from any industrial supply house.

A typical first aid kit consists of bandages, antiseptic salve, and other items. Bloodborne pathogen kits typically contain a disposable gown, shoe covers, biohazard bags, a fluid solidifier, and a biohazard scoop. **See Figure 2-27.** The location of each kit should be noted and announced at safety meetings. Signs should be used to indicate the locations of first aid kits and bloodborne pathogen kits.

Bloodborne Pathogen Protection Kits

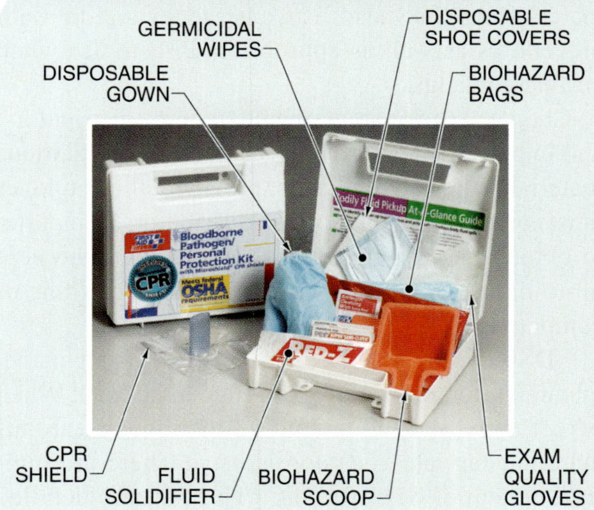

FirstAidStore.com

Figure 2-27. Bloodborne pathogen protection kits allow building personnel to work safely when remedying a blood or bodily fluid incident.

All kits must be checked periodically to ensure that they have a complete set of supplies. Missing supplies must be replenished or the entire kit should be replaced. First aid kits are mainly used to treat minor injuries quickly until proper medical care is received. Use of a first aid kit does not replace the need for proper medical care.

Fire Safety

Fuel, heat, and oxygen are required to start and sustain a fire. A fire is extinguished when any one of the three is taken away. Guidelines in assessing hazards of any products of combustion are provided by the NFPA.

Fire Prevention. HVAC technicians must prevent conditions that could result in a fire. Fire prevention includes the proper handling of fuels, immediate cleanup of combustible materials, and proper use and storage of lubricants, oily rags, and solvents. For example, oily rags must be stored in properly sealed containers to prevent fires caused by spontaneous combustion. *Spontaneous combustion* is combustion, or fire, caused by self-ignition through chemical action. A technician must know the locations of all fire alarm boxes and stations in or near the HVAC and refrigeration equipment and the procedure for contacting the fire department.

In the event of a fire, the technician must act quickly to minimize injury and damage. All facilities must have a fire safety plan. A fire safety plan establishes procedures that must be followed when a fire occurs. Locations of fire alarms, fire extinguishers, the main electrical breaker, the fire main, and exits are listed for each area of the facility on a fire safety plan.

Fire Extinguishers. Fire extinguishers must be readily available in a residential or light commercial facility in case of a fire. The NFPA classifies fires and fire extinguishers as Class A, B, C, D, and K based on the material burning in a fire. **See Figure 2-28.** The fire extinguisher designed for a specific combustible material is the only extinguisher that will safely and quickly extinguish that material fire.

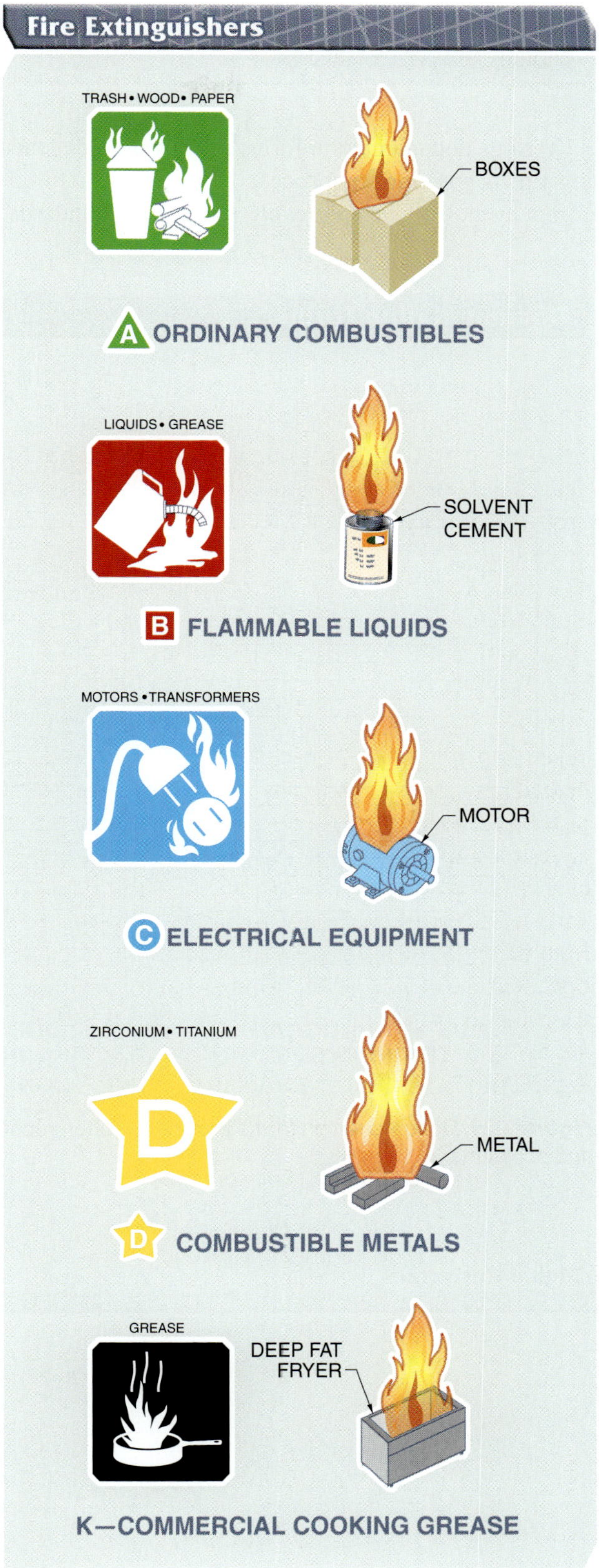

Figure 2-28. Fire classes are based on the combustible material and type of fire extinguisher needed.

Documentation Forms

Documentation forms provide a written record of employee safety training, compliance with safety standards, and accidents. Setting safety goals and rules and providing safe tools, safe working conditions, and training are all important factors in reducing accidents in the workplace. Safety steps must be documented to ensure proper compliance with industry safety requirements and to protect employees and employers.

Documentation Forms Media Clip

Various documentation forms may be used depending on the product or process being manufactured, company policy, or applicable codes and standards. Documentation generally requires several different steps and forms for recordkeeping. Documentation form titles include Safety Program Goals, Danger/Equipment Identification, Protection Device Location Identification, Protection Device Log, Safety Program Training Log, Job Site Safety Inspection, and Accident Report. **See Figure 2-29.**

Documentation Forms

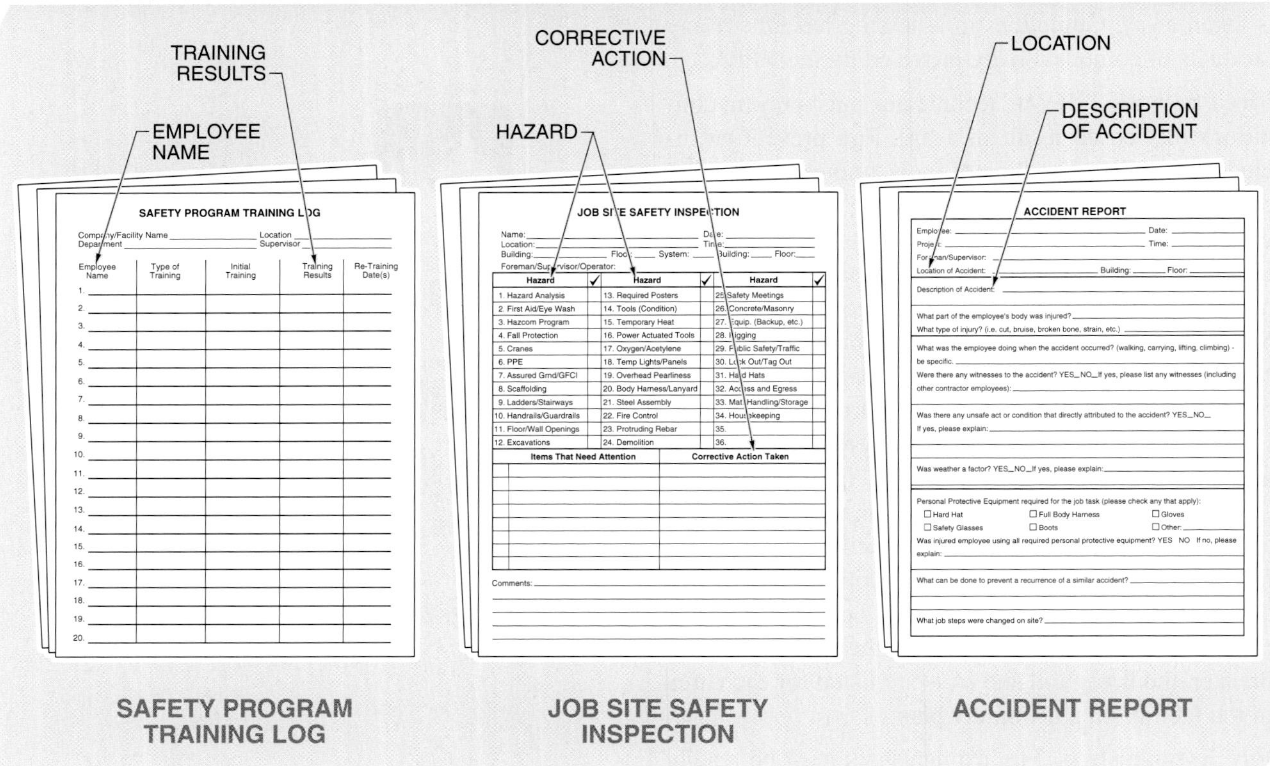

Figure 2-29. Documentation forms provide a written record of employee safety training, compliance with safety standards, and accidents.

Digital Resources ATPeResources.com/QuickLinks • Access Code: 482610

UNIT 3

Electrical Safety

The most common energy source used for HVAC and refrigeration systems is electricity. When working on HVAC systems in the USA, OSHA, NEC®, NEMA, and company safety rules must be followed at all times. Outside the USA, an HVAC technician must be completely aware of and trained on the local safety requirements. Following safety rules requires understanding electrical codes and standards, electrical safety labels, the PPE used for an application, and how to safely work on electrical systems.

OBJECTIVES

- Identify major electrical codes and standards.
- Explain the hazardous location divisions.
- Describe the causes of electrostatic discharge and electrical shock.
- Identify the purpose of electrical safety labels.
- List typical electrical PPE.
- Explain the requirements for proper grounding.
- List typical lockout/tagout situations.

Index to Procedures

ELECTRICAL CODES AND STANDARDS

National, state, and local electrical codes and standards are used to protect people and property from potential electrical hazards. **See Figure 3-1.** Codes and standards ensure that electrical refrigeration and air conditioning equipment is built and installed safely and that every effort is made to protect people from refrigeration and air conditioning electrical hazards. There are no federal agencies that establish rules and regulations related to electrical system operation. However, the National Fire Protection Association (NFPA) and electrical trade associations prepare standards for electrical equipment installation and electrical safety.

All safety procedures and electrical warnings must be followed when servicing HVAC equipment.

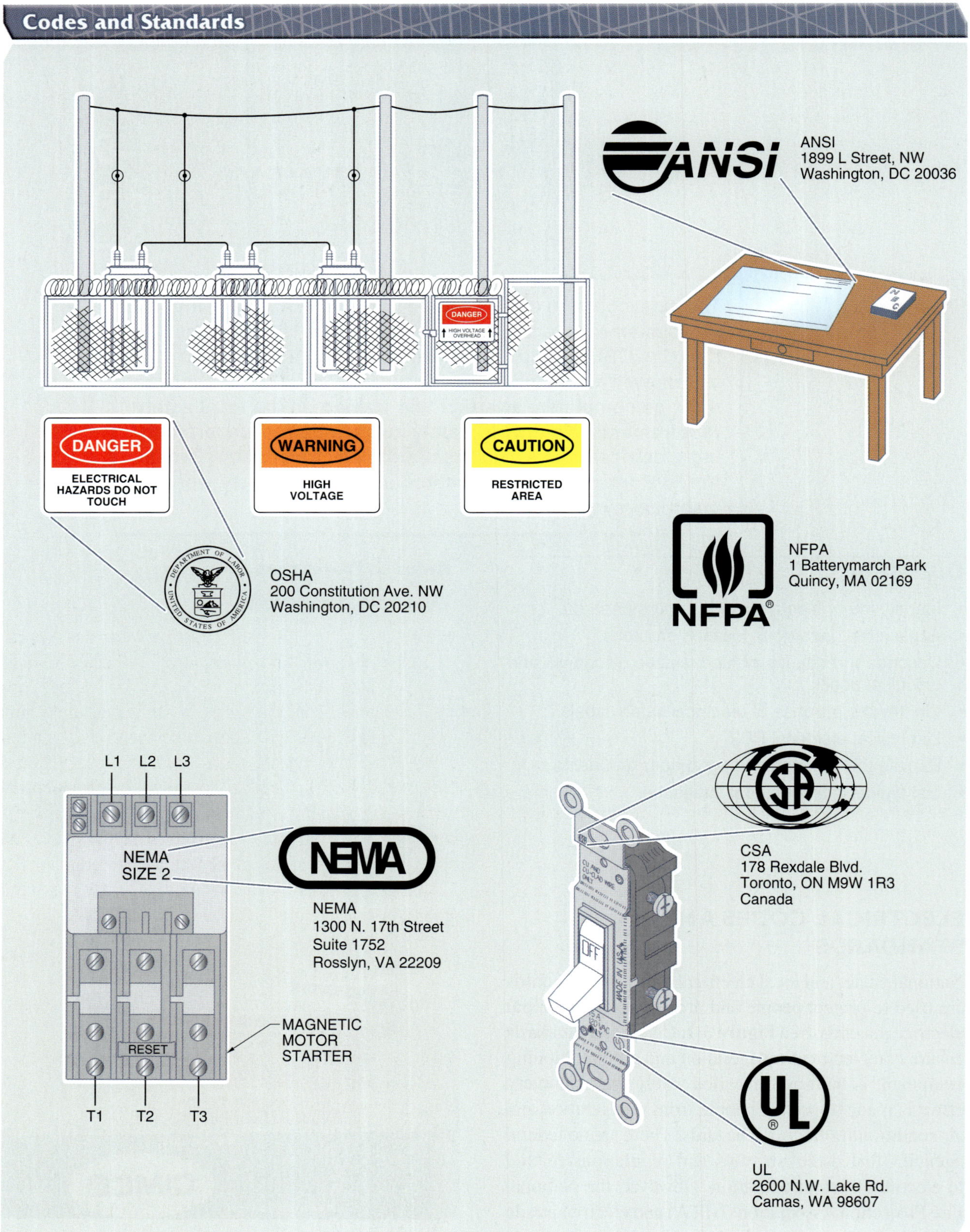

Figure 3-1. Codes and standards ensure that people are protected from electrical hazards.

NFPA

The NFPA publishes NFPA 70®, National Electrical Code® (NEC®). **See Figure 3-2.** The purpose of the NEC® is the practical safeguarding of persons and property from hazards arising from the use of electricity and electrical equipment. The NEC® is updated every three years. While the NEC® is not a regulation, many federal, state, and local agencies use the NEC® to set requirements for electrical installations.

National Electrical Code® (NEC®)

Figure 3-2. The NFPA publishes NFPA 70®, National Electrical Code® (NEC®), to safeguard people and property from hazards arising from the use of electricity and electrical equipment.

NFPA Media Clip

The NFPA also publishes NFPA 70E, *Standard for Electrical Safety in the Workplace®*. The purpose of NFPA 70E® is to convey the safety requirements for persons in the workplace installing electrical conductors, electrical equipment, signaling and communications conductors, communications equipment, and raceways. Many federal, state, and local agencies use NFPA 70E® to set safety requirements for electrical equipment installations. Although NFPA 70E® is not a regulatory standard, it provides information regarding the safe installation of electrical conductors and equipment.

Electrical Trade Associations

An *electrical trade association* is an organization that represents the manufacturers of electrical equipment. The National Electrical Manufacturers Association (NEMA) is the largest trade association representing manufacturers of products used in the generation, transmission, distribution, control, and end use of electricity. The Independent Electrical Contractors (IEC) is a trade association for electrical systems contractors.

ELECTRICAL HAZARDS

In order to practice electrical safety, personnel must be able to recognize hazards. Existing hazardous conditions must be identified and isolated. Electrical circuits must be deenergized prior to working on equipment. Common electrical hazards include hazardous locations, electrostatic discharge, and electrical shock.

Disconnect Media Clip

NEC® Hazardous Locations

NEC® Article 500 classifies hazardous locations according to the properties and quantities of the hazardous material that may be present in the air. Hazardous locations are divided into three classes, two divisions, and seven groups. **See Figure 3-3.** *Class* is the hazardous location category that refers to the generic hazardous material present. *Division* is the hazardous location category that refers to the probability that a hazardous material is present. A *group* is a designation used to classify gases and dusts. Gases and vapors are classified into Groups A, B, C, and D for Class I locations, and dusts are classified into Groups E, F, and G for Class II locations.

Electrostatic Discharge

Electrostatic discharge (ESD) is the movement of electrons from a source to an object. *Static electricity* is an electrical charge at rest. The most common way to build up static electricity is by friction. Friction causes electrons from one source to flow to another source, causing an electron buildup (negative static charge). When a person (negatively charged) contacts a grounded object (positively charged), the excess electrons flow (jump) to that object. Electrostatic discharge can be 35,000 V or more. People typically do not feel electrostatic discharges until the discharge reaches 3000 V or more.

Hazardous Locations

Hazardous Location – A location where there is an increased risk of fire or explosion due to the presence of flammable gases, vapors, liquids, combustible dusts, or easily ignitable fibers or flyings

Location – A position or site

Flammable – Capable of being easily ignited and burning quickly

Gas – A fluid (such as air) that has no independent shape or volume but tends to expand indefinitely

Vapor – A substance in the gaseous state as distinguished from the solid or liquid state

Liquid – A fluid (such as water) that has no independent shape but has a definite volume; a liquid does not expand indefinitely and is only slightly compressible

Combustible – Capable of burning

Ignitable – Capable of being set on fire

Fiber – A thread or piece of material

Flyings – Small particles of material

Dust – Fine particles of matter

Classes	Likelihood that a flammable or combustible concentration is present
I	Sufficient quantities of flammable gases and vapors present in air to cause an explosion or ignite hazardous materials
II	Sufficient quantities of combustible dust are present in air to cause an explosion or ignite hazardous materials
III	Easily ignitable fibers or flyings are present in air but not in a sufficient quantity to cause an explosion or ignite hazardous materials

Divisions	Location containing hazardous substances
1	Hazardous location in which hazardous substance is normally present in air in sufficient quantities to cause an explosion or ignite hazardous materials
2	Hazardous location in which hazardous substance is not normally present in air in sufficient quantities to cause an explosion or ignite hazardous materials

Class I Division I:
Spray booth interiors

Areas adjacent to spraying or painting operations using volatile flammable solvents

Open tanks or vats of volatile flammable liquids

Drying or evaporation rooms for flammable vents

Areas where fat and oil extraction equipment using flammable solvents is operated

Cleaning and dyeing plant rooms that use flammable liquids that do not contain adequate ventilation

Refrigeration or freezer interiors that store flammable materials

All other locations where sufficient ignitable quantities of flammable gases or vapors are likely to occur during routine operations

Class II Division I:
Grain and grain products

Pulverized sugar and cocoa

Dried egg and milk powders

Pulverized spices

Starch and pastes

Potato and wood flour

Oil meal from beans and seeds

Dried hay

Any other organic materials that may produce combustible dusts during their use or handling

Class III Division I:
Portions of rayon, cotton, or other textile mills

Manufacturing and processing plants for combustible fibers, cotton gins, and cotton seed mills

Flax processing plants

Clothing manufacturing plants

Woodworking plants

Other establishments involving similar hazardous processes or conditions

Hazardous Locations		
Class	**Group**	**Material**
I	A	Acetylene
	B	Hydrogen, butadiene, ethylene oxide, propylene oxide
	C	Carbon monoxide, ether, ethylene, hydrogen sulfide, morpholine, cyclopropane
	D	Gasoline, benzene, butane, propane, alcohol, acetone, ammonia, vinyl chloride
II	E	Metal dusts
	F	Carbon black, coke dust, coal
	G	Grain dust, flour, starch, sugar, plastics
III	No Groups	Wood chips, cotton, flax, nylon

Figure 3-3. NEC® Article 500 classifies hazardous locations according to hazardous material in the air.

Solid-state devices and circuits may be damaged or destroyed by a 10 V electrostatic discharge. HVAC technicians must discharge any static charge buildup to ground before working on any solid-state devices or circuits. The safety rules that are effective in preventing electrostatic discharge damage include the following:

- Always keep work areas clean and clear of unnecessary materials, especially common plastics.
- Never handle electronic devices by the device leads.
- Test grounding devices daily to ensure that they have not become loose.
- Never work on an ESD-sensitive object without a properly working grounding device.
- Always handle printed circuit (PC) boards by the outside corners.
- Always transport PC boards in antistatic trays.
- Never stack PC boards.
- When transporting single electronic devices, always keep them sealed in conductive static shielded containers.

Electrical Shock

Improper electrical wiring or misuse of electricity can cause destruction of HVAC equipment and electrical shock to personnel. An *electrical shock* is the result of a body becoming part of an electrical circuit. **See Figure 3-4.** Electrical shock effects vary from a mild sensation to paralysis or death. The severity of an electrical shock depends on the amount of electric current (in milliamps) that flows through the body, the length of time the body is exposed to the current flow, the path the current takes through the body, the physical size and condition of the body through which the current passes, and the amount of body area in contact with the electricity.

The amount of current that passes through a body or circuit depends on the voltage and resistance of the completed circuit. **See Figure 3-5.** The resistance a person's body offers to the flow of current varies. A person with sweaty hands has less resistance than a person with dry hands. A person standing on a wet floor has less resistance than a person standing on a dry floor. When the resistance is lowered, the current flow is greater and the severity of the shock increases.

Electrical Shock Media Clip

Electrical Shock

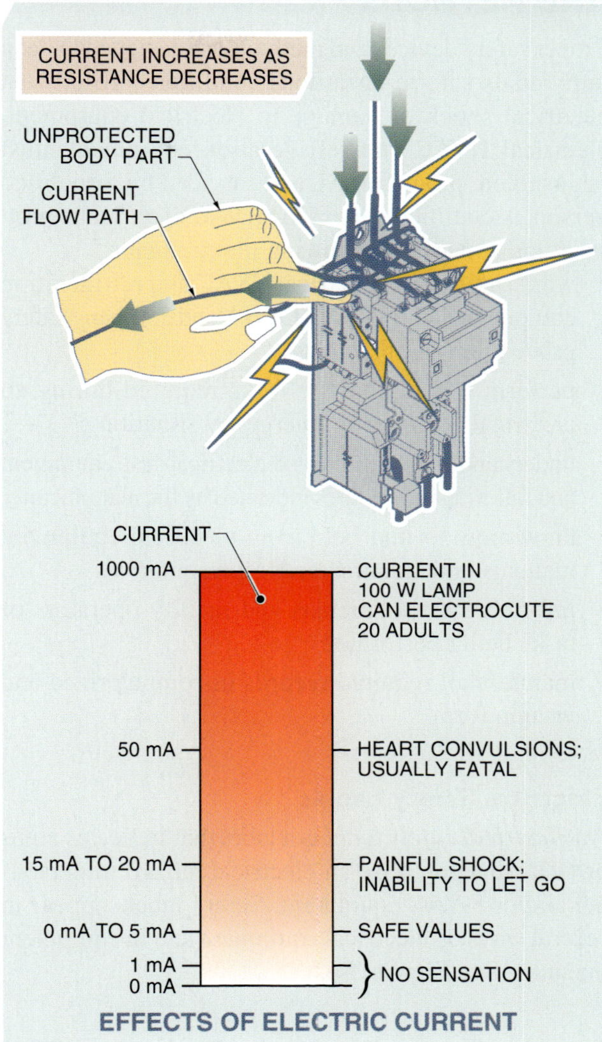

Figure 3-4. An electrical shock results when a body becomes part of an electrical circuit.

Effects of Electrical Current

Current*	Effect on Body
less than 8	Sensation of shock but probably not painful
8 to 15	Painful shock Removal from contact point by natural reflexes
16 to 20	Painful shock May be frozen or locked to point of electrical contact until circuit is deenergized
over 20	Causes severe muscular contractions, paralysis of breathing, heart convulsions

* in mA

Figure 3-5. An increase in current flow increases the severity of an electrical shock and its effects on a body.

ELECTRICAL SAFETY DEVICES AND METHODS

Proper safety devices and methods are designed to keep individuals safe in hazardous conditions. To prevent electrical shock or damage to electrical equipment, electrical HVAC and refrigeration equipment must be installed, programmed, and serviced by a qualified person. A qualified person for HVAC electrical equipment must meet the following requirements:

- work safely with electrical HVAC and refrigeration equipment and follow all OSHA and company safety procedures and practices
- perform the appropriate tasks required during an electrical accident or emergency situation
- understand the operation of electrical test equipment and follow all procedures indicated by the manufacturer
- know how to find help from electrical equipment manuals and manufacturer representatives
- inform other technicians and facility operators of tasks being performed
- maintain all required records in computerized and written form

Electrical Safety Labels

An *electrical safety label* is a label that indicates areas or tasks that can pose an electrical hazard to personnel and/or HVAC equipment. Safety labels appear in several ways on electrical equipment and in equipment manuals. **See Figure 3-6.**

Electrical Personal Protective Equipment (PPE)

Electrical personal protective equipment (PPE) is gear worn by electricians and HVAC technicians to reduce the possibility of injury in the work area. PPE must be worn when installing or maintaining any electrical equipment or motors. All PPE must meet OSHA 29 CFR 1910 Subpart I—Personal Protective Equipment, applicable ANSI standards, and other safety mandates. Electrical PPE includes arc-rated clothing, head protection, eye protection, hand protection, and foot protection. **See Figure 3-7.**

Electrical Personal Protective Equipment (PPE)

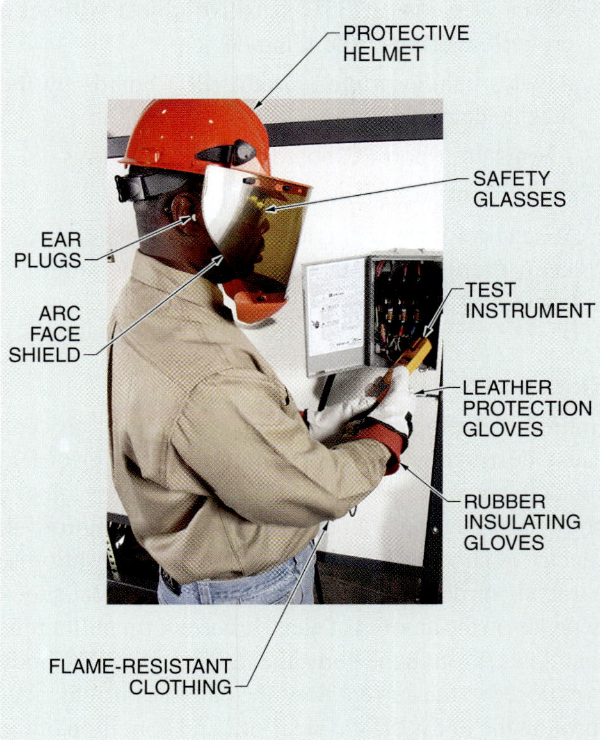

Figure 3-7. Electrical PPE is used by all personnel to reduce the possibility or severity of injury from an electrical shock.

Electrical Safety Labels

Figure 3-6. Electrical safety labels are used to warn personnel about electrical hazards that may be present.

Arc-Rated Clothing. Arc-rated clothing must be used when working with live electrical circuits. Arc-rated clothing is made of materials such as Nomex®, Basofil®, and Kevlar® fibers. **See Figure 3-8.** The arc-rated fibers can be coated with polyvinyl chloride (PVC) to offer weather resistance and to increase arc resistance. Arc-rated clothing must meet the following three requirements:

- must not ignite and continue to burn
- must provide an insulative value to the wearer that dissipates heat through the clothing away from the skin
- must provide resistance to the break-open forces generated by the shock wave of an arc

PPE Required for Electrical Work

Hazard/Risk Category	Clothing	Other PPE
0	Nonmelting fabrics	Safety glasses Hearing protection Hand protection*
1	Arc-rated clothing with minimum 4 cal/cm² arc rating	Hard hat Safety glasses Hearing protection Hand protection* Leather safety shoes
2	Arc-rated clothing with minimum 8 cal/cm² arc rating	Hard hat Safety glasses Hearing protection Hand protection* Leather safety shoes
3	Arc-rated clothing and flash suit with minimum 25 cal/cm² arc rating	Hard hat Safety glasses Hearing protection Hand protection* Leather safety shoes
4	Arc-rated clothing and flash suit with minimum 40 cal/cm² arc rating	Hard hat Safety glasses Hearing protection Hand protection* Leather safety shoes

* Hand protection may consist of rubber insulating gloves and leather protectors, depending on voltage.

Figure 3-8. PPE required for electrical work is based on hazard/risk categories for various tasks listed in NFPA 70E®.

The NFPA specifies live circuit approach boundaries where arc protection is required. **See Figure 3-9.** All personnel working within these specified boundary distances are required to wear arc-rated clothing and equipment. Approach boundary distances vary depending on the voltage involved.

Limits of Approach

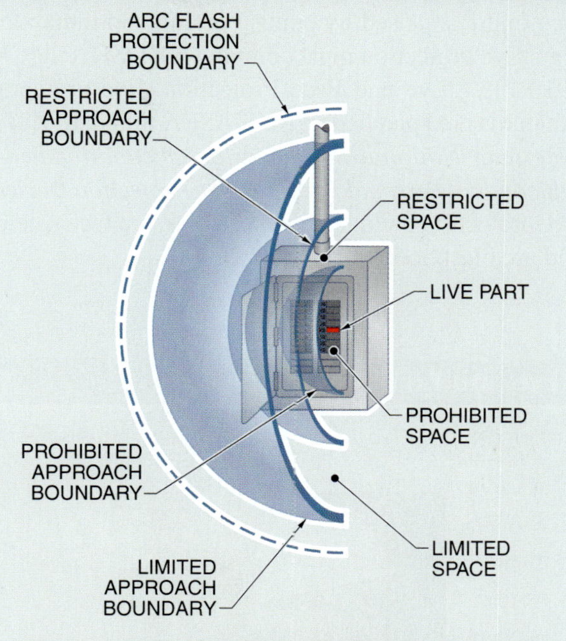

Working Voltage*	Limited Approach Boundary Distance†
Up to 750	3
750 to 2000	4
2000 to 15,000	16
15,000 to 36,000	19
Over 36,000	Must wear protection

* in V
† in ft

Figure 3-9. NFPA 70E® defines arc flash protection boundaries and shock protection boundaries to specify the required PPE and procedures based on the distance from live parts.

Limits of Approach Media Clip

Electrical Head Protection. Protective helmets are identified by their class of protection against electrical hazards. Class G (general) protective helmets are used for general service and provide protection up to 2200 V. Class E (electrical) protective helmets are used for utility service and provide protection up to 20,000 V. Class C (conductive) protective helmets are used for special service and provide no voltage protection. The proper class of protective helmet must be worn.

Electrical Eye and Face Protection. Electrical eye and face protection must be worn to prevent eye or face injuries caused by contact arcing and radiant energy. Eye protection must comply with OSHA 29 CFR 1910.133—Eye and Face Protection. Eye protection standards are specified in ANSI/ISEA Standard Z87.1, *American National Standard for Occupational and Educational Personal Eye and Face Protection Devices*. Eye protection includes safety glasses, arc face shields, and arc blast hoods. **See Figure 3-10.**

Eye and Face Protection

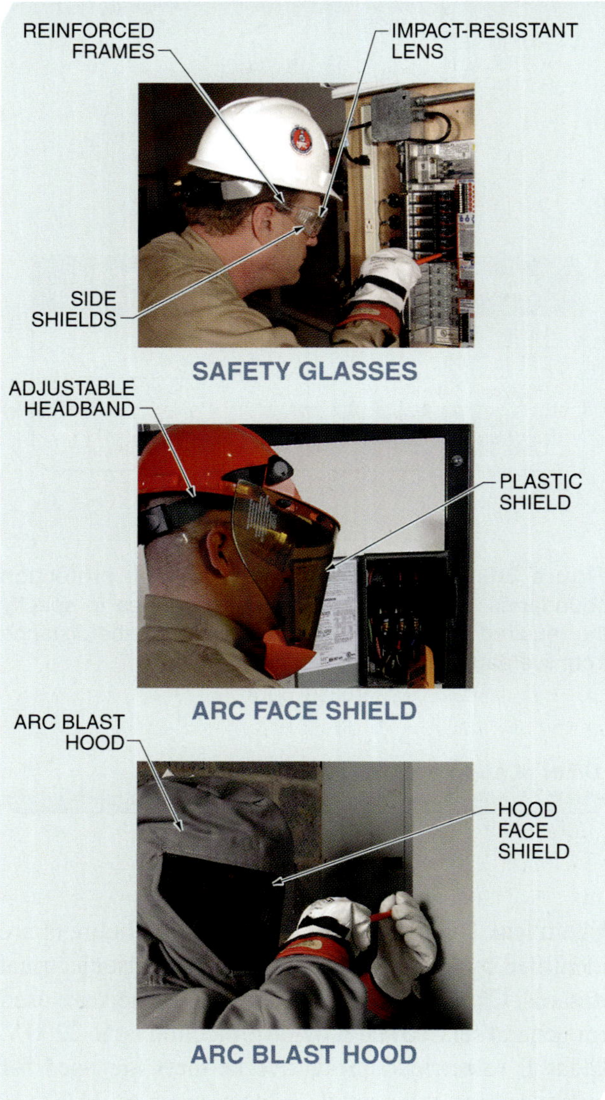

Figure 3-10. Eye and face protection must be worn to prevent injuries caused by flying particles, contact arcing, or radiant energy.

Electrical safety glasses are eye protection devices with plastic frames designed to minimize the shock hazard when working with electrical equipment. Tinted-lens safety glasses protect against low-voltage arc hazards. Non-tinted-lens safety glasses with side shields can be worn when working on low-voltage electrical circuits of 50 V or less.

An *arc face shield* is an eye and face protection device that protects against low-voltage arc hazards by covering the entire face with a tinted face shield. A tinted face shield should be worn when working on any electrical circuit that has voltages over 50 V or in any environment where a chance of a low-voltage arc is possible.

An *arc blast hood* is an eye and face protection device that consists of a flame-resistant hood and face shield. Arc blast hoods are used to protect against high-voltage arc blasts. Electricians working with energized high-voltage equipment must wear arc blast protection.

Electrical Hand Protection. *Electrical hand protection* is a set of hand coverings (gloves) that are worn to prevent injuries to hands caused by electrical shock. The hand protection required is determined by the duration, frequency, and degree of the hazard to the hands. *Rubber insulating gloves* are gloves made of latex rubber and are used to provide maximum insulation from electrical shock. Rubber insulating gloves are stamped with a working voltage range, such as 500 V to 26,500 V. **See Figure 3-11.**

Rubber Insulating Glove Ratings

Class	Maximum Use*	Label Color
00	500 V	Beige
0	1kV (1000 V)	Red
1	7.5 kV (7500 V)	White
2	17 kV (17,000 V)	Yellow
3	26.5 kV (26,500 V)	Green
4	36 kV (36,000 V)	Orange

* Refer to ASTM D120-09 Standard Specification for Rubber Insulating Gloves

Figure 3-11. Rubber insulating gloves use a color-coded label to represent their voltage rating.

Leather protector gloves are gloves worn over rubber insulating gloves to prevent penetration of the rubber insulating gloves and provide added protection against electrical shock. Safety procedures for wearing rubber insulating gloves and leather protector gloves must be followed at all times.

The primary purpose of rubber insulating gloves and leather protectors is to protect hands and lower arms from contact with live conductors. Rubber insulating gloves offer a high resistance to current flow to help prevent electrical shock. Leather protector gloves help protect rubber insulating gloves and provide additional insulation. Electrically rated rubber insulating gloves and leather protector gloves must be worn when working on any electrical system that includes voltages over 50 V. Cotton liners can be used to improve comfort.

The proper care of leather protector gloves is essential to user safety. Leather protector gloves must be inspected when inspecting rubber insulating gloves. Metal particles or any substance that could physically damage rubber insulating gloves must be removed from leather protector gloves before use. Leather protector gloves are checked for cuts, tears, holes, abrasions, defective or worn stitching, oil contamination, and any other condition that might prevent the leather gloves from adequately protecting rubber insulating gloves.

The entire surface of rubber insulating gloves must be field tested (visually inspected and air tested) before each use. Every six months rubber insulating gloves must be laboratory tested by an approved laboratory. Visual inspection of rubber insulating gloves is performed by stretching a small area (particularly the fingertips) and checking for defects such as punctures, pinholes, embedded or foreign material, deep scratches, cracks, cuts, snags, and deterioration caused by oil, heat, grease, insulating compounds, or any other substance that harms rubber.

Rubber insulating gloves must also be air tested when damage is suspected. The entire surface of an insulating glove is inspected by rolling the cuff tightly toward the palm in such a manner that air is trapped inside the glove or by using a mechanical inflation device. When using a mechanical inflation device, care must be taken to not overinflate the glove. **See Figure 3-12.** Puncture detection is made easier by listening for escaping air while holding the glove to the face or ear.

Rubber insulating gloves or leather protector gloves found to be defective in any way should not be discarded or destroyed in the field. They must be tagged as unsafe and returned to a supervisor.

Electrical Foot Protection. *Electrical foot protection* is protection for feet in the form of insulated, rubber-soled shoes and is typically worn during electrical work to prevent electrical shock. Protective electrical footwear must comply with ASTM F2413-11, *Standard Specification for Performance Requirements for Protective (Safety) Toe Cap Footwear.* Insulated, rubber-soled boots are used when working in damp locations.

Inspecting Rubber Insulating Gloves

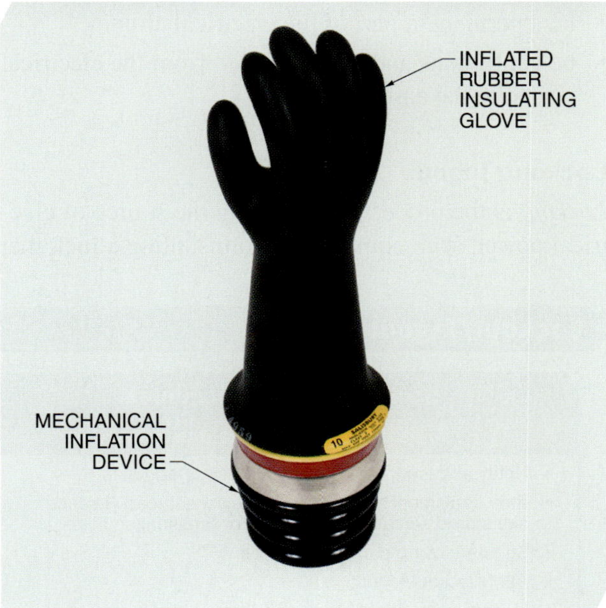

Figure 3-12. Mechanical inflation devices can be used to inflate rubber insulating gloves to inspect them for punctures or other defects.

Grounding

Grounding is the connection of all exposed, non-current-carrying metal parts to the earth. Grounding provides a direct path to the earth for unwanted fault current without causing harm to persons or equipment. Electrical circuits are grounded to safeguard equipment and personnel against the hazards of electrical shock and electrostatic discharge. Proper grounding of electrical tools, motors, refrigeration and air conditioning equipment, enclosures, PLCs, and refrigeration system control circuitry helps prevent hazardous conditions. Improper electrical wiring or misuse of electricity can cause destruction of equipment, fire damage to property, and personal injury.

HVAC and refrigeration equipment must not be operated unless properly grounded. HVAC and refrigeration equipment must be grounded per NEC® Article 250 and applicable state and local codes. Equipment is grounded by connecting the equipment ground terminals to the electrical system ground. The ground screw is green and marked with the symbol for ground and/or the abbreviation for ground (GND).

To prevent problems, a grounding path must have the following characteristics:

- be as short as possible and of sufficient size as recommended by the manufacturer (minimum 14 AWG copper)
- never be fused or switched
- be a permanent part of the electrical circuit
- be continuous and uninterrupted from the electrical circuit to the earth

Lockout/Tagout

Lockout is the process of removing the source of electrical power from equipment and installing a lock that prevents the power from being turned on. To ensure the safety of personnel performing installation or preventive maintenance on HVAC and refrigeration equipment, all electrical power is removed and the equipment must be locked out and tagged out. **See Figure 3-13.** *Tagout* is the process of placing a danger tag on the source of electrical power, which indicates that the equipment may not be operated until the danger tag is removed. OSHA standards specify that all electrical equipment must be locked out and tagged out before any installation or preventive maintenance is performed.

Electrical Lockout Media Clip

Lockout/Tagout Devices

Lockout/Tagout Procedures

Before Working on Equipment

- Notify all personnel affected by lockout/tagout.
- Stop equipment and remove all power sources. Release any stored electrical energy and/or pressurized fluid.
- Secure moving parts as required.
- Apply lockout/tagout.
- Measure for voltage using voltmeter and test all controls to ensure power is off.
- Verify all controls are in OFF position.

After Working on Equipment

- Remove tools, release secured moving parts, and prepare equipment for operation.
- Make sure all affected personnel are clear of equipment.
- Replace all safety guards and controls.
- Remove lockout/tagout.
- Energize and test equipment.

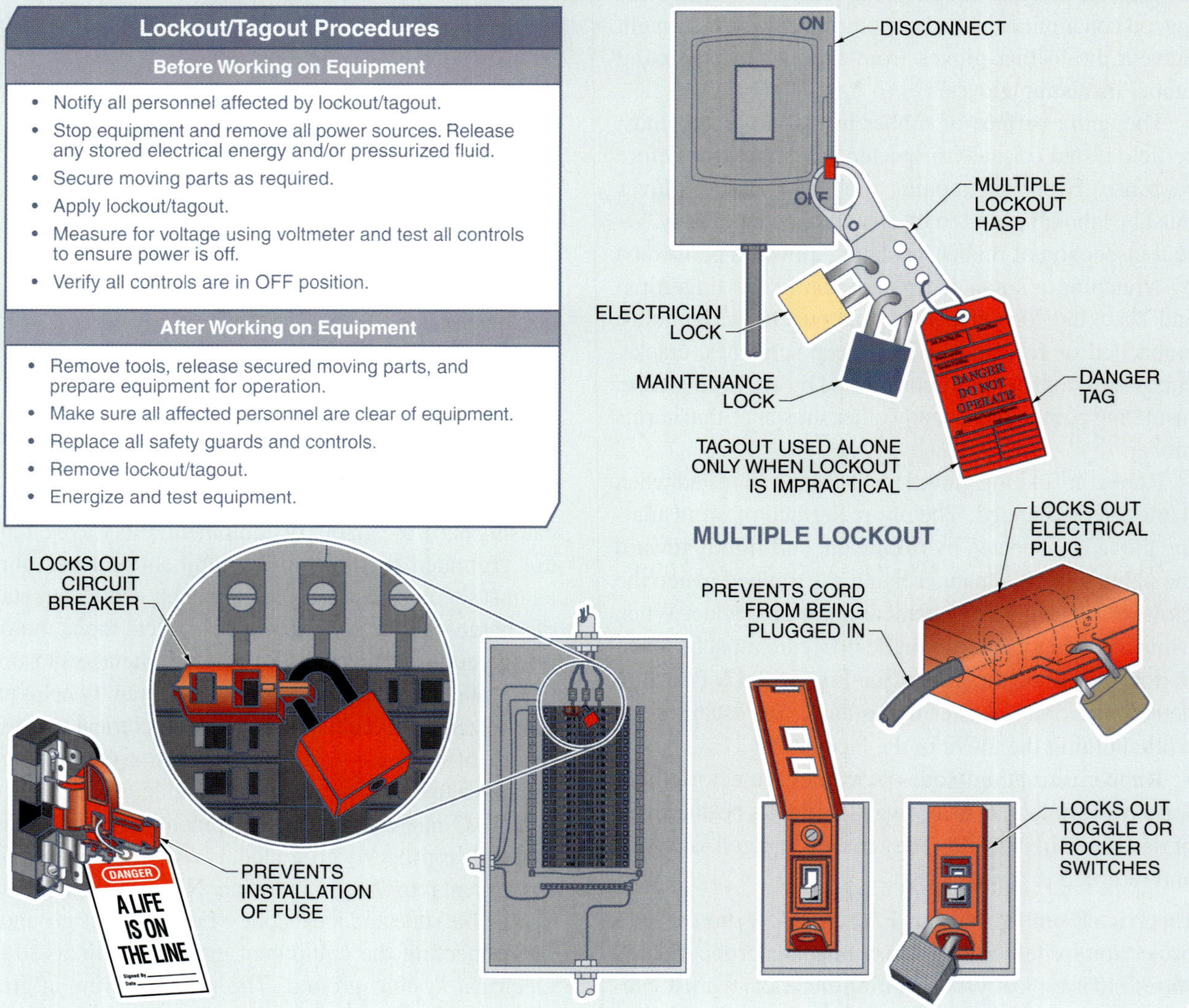

Figure 3-13. Lockout devices are used to ensure equipment is deenergized while it is serviced.

A danger tag has the same importance and purpose as a lock. A danger tag is used alone only when a lock does not fit the disconnect device. The danger tag must be attached at the disconnect device with a tag tie or equivalent device and must have space for the name of the technician, trade, and other company-required information. A danger tag also must withstand the elements and expected atmosphere for the maximum period of time that exposure is expected.

Servicing HVAC units requires that they be locked out and tagged out before service to eliminate the possibility of electrical shock.

When more than one technician is required to perform a task on a piece of equipment, each technician must place a lockout and/or tagout on the energy-isolating device or devices. A *hasp* is a multiple lockout/tagout device. A hasp must be used because energy-isolating devices typically cannot accept more than one lockout/tagout.

Lockout/tagout is used in the following situations:

- when power to a piece of equipment is not required to be on to perform a task
- when removing or bypassing machine guards or other safety devices
- when the possibility exists of being injured or caught in moving machinery
- when clearing jammed equipment
- when the danger of being injured exists when the equipment's power is turned on

Lockouts and tagouts do not by themselves remove electrical power from a machine or its circuitry. OSHA provides a standard procedure for equipment lockout/tagout. Lockout is performed and tagouts are attached only after the equipment is turned off and tested. **See Procedure for Equipment Lockout/Tagout.**

Equipment Lockout/Tagout

Procedure

1. Notify all affected persons that a lockout/tagout is required. Notification should include the reason for the lockout/tagout and the expected duration.
2. When the equipment is operating, shut down the equipment using normal procedures.
3. Apply the energy-isolating device or devices so that the equipment is isolated from all energy sources. Stored energy, such as in springs, elevated machine members, capacitors, and pneumatic lines, must be dissipated or restrained by blocking, discharging, or other appropriate methods.

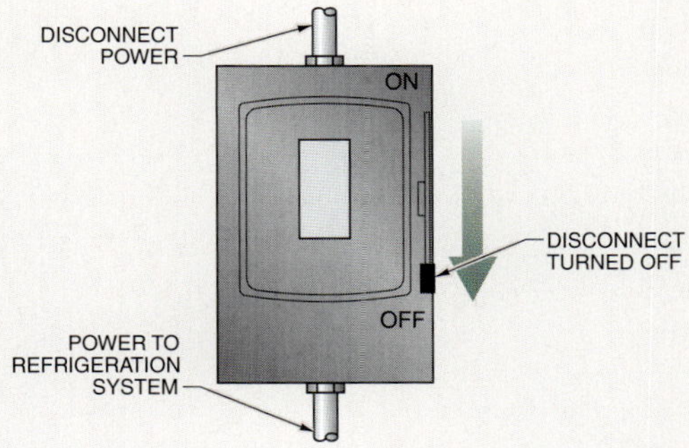

4. Apply the lockout and/or tagout to the energy-isolating device or devices with assigned locks and/or danger tags.

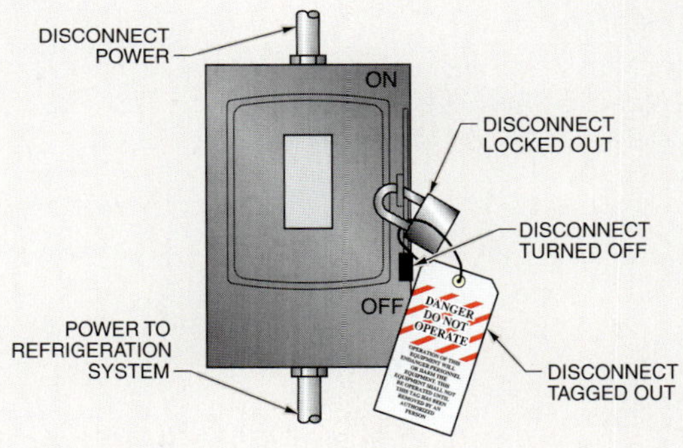

5. After ensuring that personnel are not exposed, operate the normal operating controls to verify that the equipment is inoperable and that all energy sources have been isolated.
6. Inspect and test the equipment with the appropriate test instruments to verify that all energy sources are disconnected. Multiphase electrical power requires that each phase be tested. The equipment is now locked out and/or tagged out.

Note: A lockout/tagout must not be removed by any person other than the authorized person who installed the lockout/tagout, except in an emergency. In an emergency, only authorized supervisory personnel may remove a lockout/tagout and only upon notification of the authorized person who installed the lockout/tagout.

Electrical Safety Programs and Training

Many facilities implement ongoing electrical safety training as part of a general safety training program. Electrical equipment safety training must be part of a comprehensive safety training program. Identifying electrical disconnects and wire terminations is often unique to each installation. When new equipment is installed in a facility, all electrical safety procedures associated with the new equipment installation must be distributed to maintenance technicians.

Digital Resources

ATPeResources.com/QuickLinks • Access Code: 482610

UNIT 4

Tools and Equipment

In order to successfully service HVAC and refrigeration equipment and systems, technicians must master the use of basic tools and equipment. The correct tools and equipment must be selected for the job and used for the tasks for which they were designed. Tools must be organized and ready for use. The operator's manual should be consulted for correct operation and use before using any new tool or equipment. Injury and damage to tools, equipment, or workpieces can be prevented through the use of proper operating and safety procedures.

OBJECTIVES

- Identify the purposes of common hand tools used in the HVAC and refrigeration industry.
- Explain the functions of common power tools used when working on HVAC equipment.
- Describe methods used to cut and join copper tubing.
- Describe methods used to cut and join steel and plastic pipe.

Index to Procedures

HAND TOOLS

A *hand tool* is a tool that uses no external power other than that supplied by the individual using the tool. Safety must be the primary concern when using any tool in order to prevent death or injuries due to improper tool use. Tool manufacturer safety information must be read and followed. The proper tool for the job must be used and the condition of the tool must be checked before use. Clamps and/or tables should be used to support the material on which work is being done.

While many hand tools are typically used by other trades, some are used mainly within the HVAC and refrigeration industry. Hand tools typically used in the HVAC and refrigeration industry include screwdrivers, wrenches, pliers, hammers, metalworking tools, tubing and piping tools, and other common tools.

Screwdrivers

A *screwdriver* is a hand tool with a tip designed to fit into a screw head for fastening operations. HVAC technicians use screwdrivers in most installation and troubleshooting activities to secure and remove various threaded fasteners. Many types of screwdrivers are available, but the two types of screwdrivers typically used are the Phillips and flat head. Phillips and flat head screwdrivers are available as standard and offset.

Offset screwdrivers are also available as a combination with a Phillips head and a flat head. A screw-holding attachment can be used with standard screwdrivers. **See Figure 4-1.**

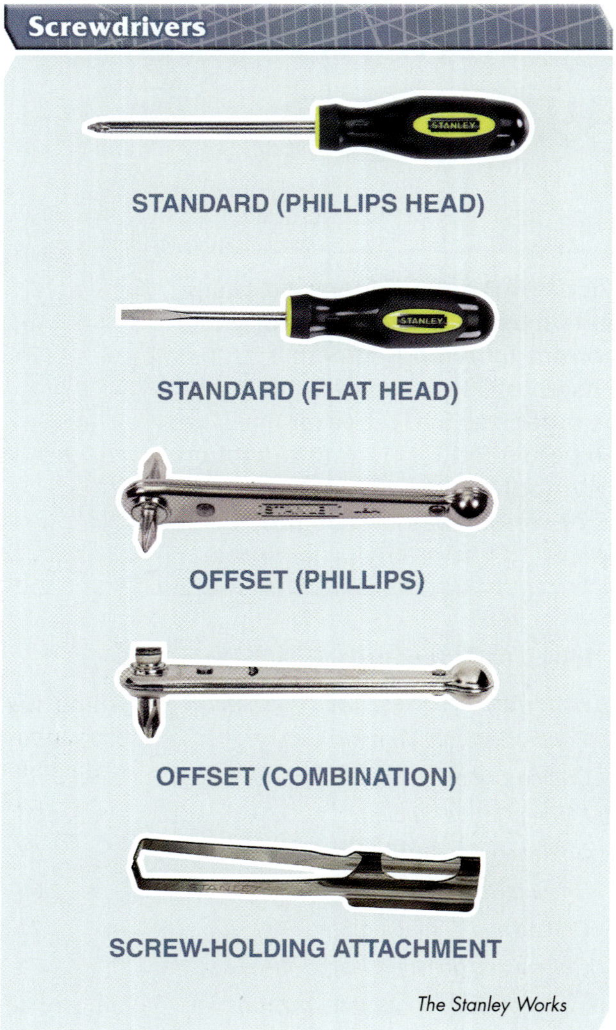

Figure 4-1. Different types of screwdrivers are used for various turning operations.

Standard screwdrivers are used for the installation and removal of threaded fasteners. Offset screwdrivers provide a means for reaching screws in difficult locations. A screw-holding attachment is used with standard screwdrivers to hold screws in place when working in tight spots. Once started, the screw is released and tightened with a standard screwdriver. Screwdrivers are available with square shanks that a wrench can be applied to for the removal of stubborn screws. Screwdrivers can also have a long shank to reach and drive screws in deep, counterbored holes.

When using a screwdriver, the tip must fit the slot of the screw snugly and not project beyond the screw head. Screwdrivers should never be used as a cold chisel or punch. Screwdrivers should not be used near energized electrical wires and should never be exposed to excessive heat. A worn tip should be redressed with a file to regain a good, straight edge. A screwdriver that has a worn or broken handle should be discarded.

Special types of screws are used in select locations to prevent occupants from tampering with or adjusting HVAC controls. Specially designed screwdrivers may be used for tamperproof screws that will not allow a standard screwdriver to tighten or loosen the screw. Tamperproof screws deny unauthorized access to the inside of devices. Special-head tamperproof screws are commonly used to attach thermostat covers in areas where people may try to adjust the thermostat.

Jeweler screwdrivers are typically used to adjust controls. Jeweler screwdrivers are small in size to allow access to tight locations and fit on the small heads of control adjustment screws. The small size discourages tampering with controls.

TECH TIP

Screwdriver accidents can cause puncture wounds in the hand. To avoid hand injuries, hold the screwdriver properly and secure the tip before applying force.

Wrenches

A *wrench* is a hand tool with jaws at one or both ends that is designed to turn bolts, nuts, or pipes. Common wrenches used for HVAC applications include socket, adjustable, hex key (Allen), combination, and pipe. **See Figure 4-2.**

Socket wrenches are used to tighten a variety of items, such as hex-head lag screws, bolts, and various electrical connectors. Socket wrenches typically include a ratchet with a set of sockets available in a series of sizes. Adjustable wrenches are used to tighten items such as various-size hex-head lag screws, bolts, and large conduit couplings.

Hex wrenches are used for tightening hex-head bolts. Combination wrenches have an open-end wrench on one end and a closed-end box wrench on the other. Pipe wrenches can be straight, offset, strap, or chain. Pipe wrenches are used to tighten and loosen pipes and large conduit.

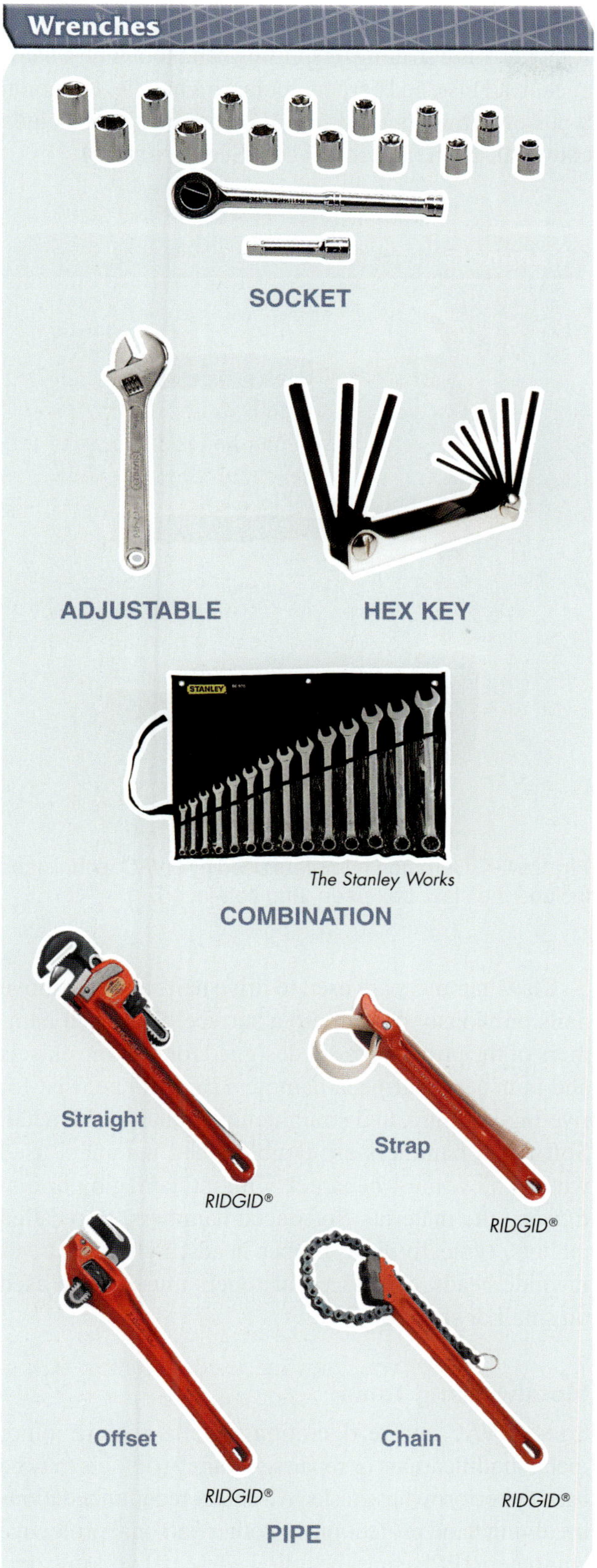

Figure 4-2. Wrenches have jaws or fixed ends and are designed to turn bolts, nuts, or pipes.

TECH TIP

A wrench should be used only for its intended purpose. Using the wrong type of wrench or improperly using a wrench can cause scraped knuckles or injured muscles if the wrench slips.

When using a wrench, a pipe extension or other form of "cheater" should never be used to increase the leverage of the wrench. Wrenches are selected with an opening that corresponds to the size of the nut to be turned. A wrench with too large an opening for the workpiece can spread the jaws of an open-end wrench and batter the points of a box or socket wrench. Care must be taken in selecting wrenches sized in inches for fasteners sized in inches and metric wrenches for metric fasteners. It is a good practice to pull on a wrench handle instead of pushing. The safest wrenches that can be used are box or socket wrenches because these types of wrenches cannot slip and cause injury to a technician. A straight handle is better than an offset handle when conditions permit.

Pliers

Pliers are a hand tool with opposing jaws for gripping and/or cutting. HVAC technicians use pliers for various gripping, turning, cutting, positioning, and bending operations. Common pliers include slip-joint, tongue-and-groove, long nose, locking, diagonal-cutting, side-cutting, end-cutting, and self-adjusting pliers. **See Figure 4-3.**

Slip-joint pliers are used to tighten box connectors, lock nuts, and small-size conduit couplings. Tongue-and-groove pliers are used for a wide range of applications involving gripping, turning, and bending. The adjustable jaws of tongue-and-groove pliers enable adjustment to a wide range of sizes.

Long nose pliers are used for bending and cutting thin wires and positioning small components. Locking pliers, such as Vise-Grip™ pliers, are used to lock onto a workpiece for positioning or holding. Locking pliers can be adjusted to lock at any size with any desired amount of pressure.

Diagonal-cutting pliers are used for cutting cables and wires that are too difficult to cut with side-cutting pliers. Side-cutting pliers are used for cutting cable, removing knockouts, twisting wire, and deburring metal pipe. End-cutting pliers are used for cutting wire, nails, rivets, etc., close to the workpiece. Self-adjusting pliers work in the same manner as slip-joint pliers but automatically adjust to the size of the workpiece and lock in place.

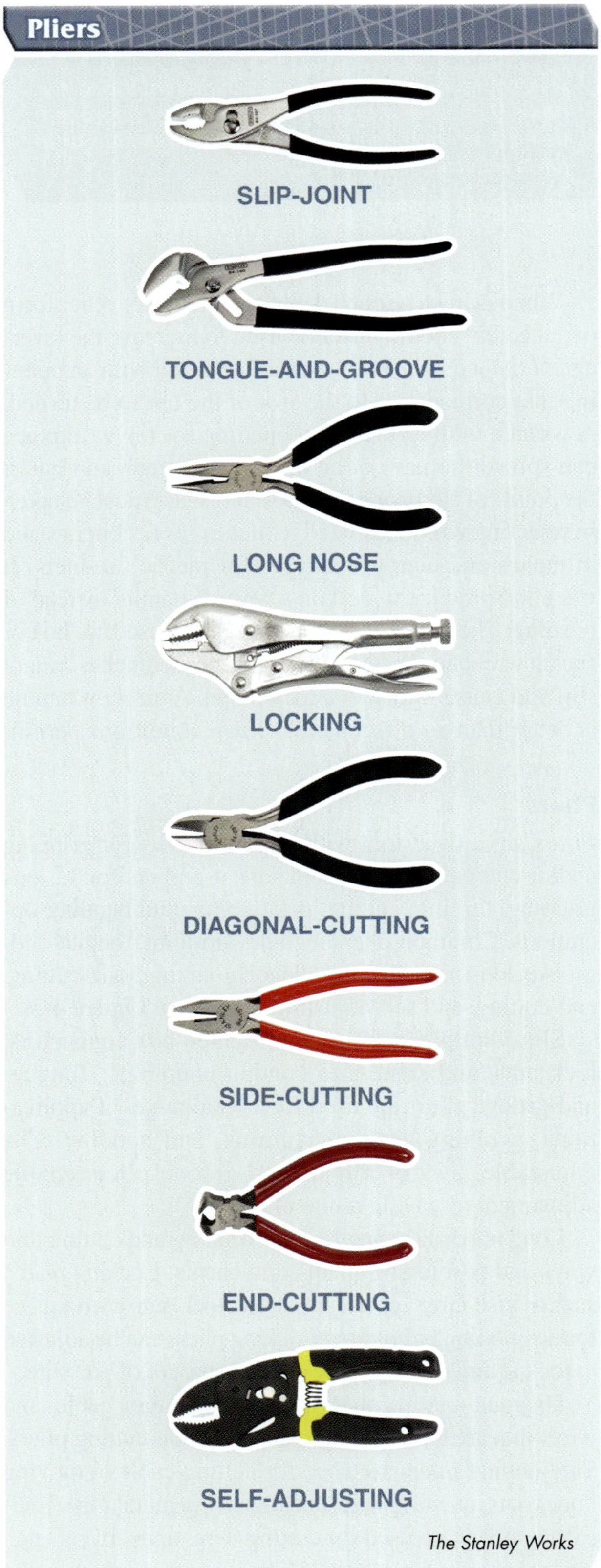

Figure 4-3. Pliers are used by HVAC technicians for various gripping, turning, cutting, positioning, and bending operations.

Hammers

A *hammer* is a striking or splitting hand tool with a hardened head fastened perpendicular to a handle. Common types of hammers used in HVAC applications include claw, ball-peen, and soft-faced. **See Figure 4-4.**

Figure 4-4. Common hammers used by HVAC technicians include the claw, ball-peen, and soft-faced.

Claw hammers are used to drive nails and to remove nails previously driven into a surface. Ball-peen hammers of the proper size are designed for striking chisels and punches. Ball-peen hammers can also be used for riveting, shaping, and straightening unhardened metal. Soft-faced hammers are double-faced hammers used when a blow must be struck without damaging or disfiguring the material. Soft-faced hammers, also called mallets, typically have rubber heads, nylon heads, or rawhide heads. Soft-faced hammers must not be used on small or sharp objects.

Metalworking Tools

Some HVAC and refrigeration systems require minor metal modifications or repairs. A variety of tools may be used to perform these tasks. All safety recommendations involving hand protection and other personal protective equipment (PPE) must be followed. Metalworking tools used typically include metal snips, hacksaws, files, taps and dies, and hand breaks.

Metal Snips. A *metal snip* is a hand tool resembling pliers that is specifically used for cutting or shearing sheet metal. A sheet metal snip, also called tin snip, is a snip used to make straight cuts in light-gauge sheet metal. An aviation snip is a snip used to make straight or angled cuts in light-gauge sheet metal and small metal parts. **See Figure 4-5.** The compound action of aviation snips results in greater cutting power with less effort. For maximum cutting effectiveness, snips must be in good condition and the sheet metal secured.

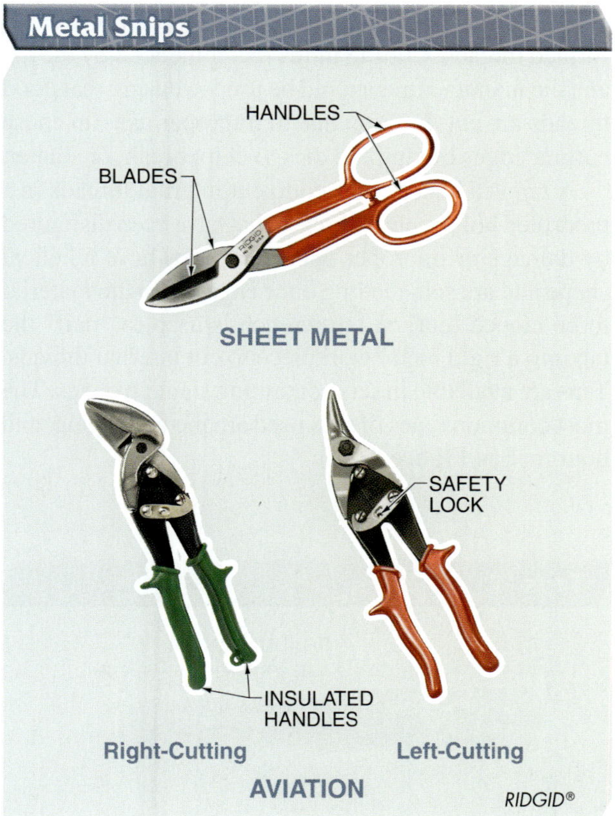

Figure 4-5. Sheet metal snips and aviation snips are used to cut light-gauge sheet metal.

Hacksaws. A *hacksaw* is a metal-cutting hand tool with an adjustable steel frame for holding various lengths and types of blades. Blades are inserted with the teeth pointing away from the handle. Hacksaws can be straight handle or pistol-grip handle. **See Figure 4-6.**

A straight-handle hacksaw is usually preferred for fine work. Either type of frame is adjustable for various blade lengths. Tension is applied to a blade to make it taut by using a wing nut on the pistol-grip frame or by turning a threaded handle on the straight-handle hacksaw. Hacksaws are used to cut workpieces such as metal pipe, PVC pipe, conduit, small-diameter metal rods, rigid plastics, bolts, and nails.

Figure 4-6. Common hacksaw designs are straight handle and pistol-grip handle.

A miter box may be used with a hacksaw or backsaw to guide the saw for accurate angled or straight cuts. Miter boxes are particularly useful for making square cuts when cutting larger pipes. A miter box can be set to the desired angle and locked in position.

Files. A *file* is an abrasive hand tool with single or double rows of fine teeth cut into the surface. Files are used to remove burrs from sheet metal, to square the ends of band iron, to straighten uneven edges, and for various other operations that require removal of small amounts of metal. File parts include the point, edge, face, heel, and tang. The most common types of files used for HVAC projects include flat, mill, three-square, round, half-round, square, and knife. **See Figure 4-7.**

File cutting faces can be single-cut or double-cut. A single-cut file has a single set of teeth cut at an angle of 65° to 85°. A double-cut file has two sets of teeth crossing each other. Double-cut files are used for rough filing since they remove material faster than single-cut files. File handles are usually made of wood and are designed to fit the hollow of the hand. A metal ferrule on the end of the handle prevents it from splitting.

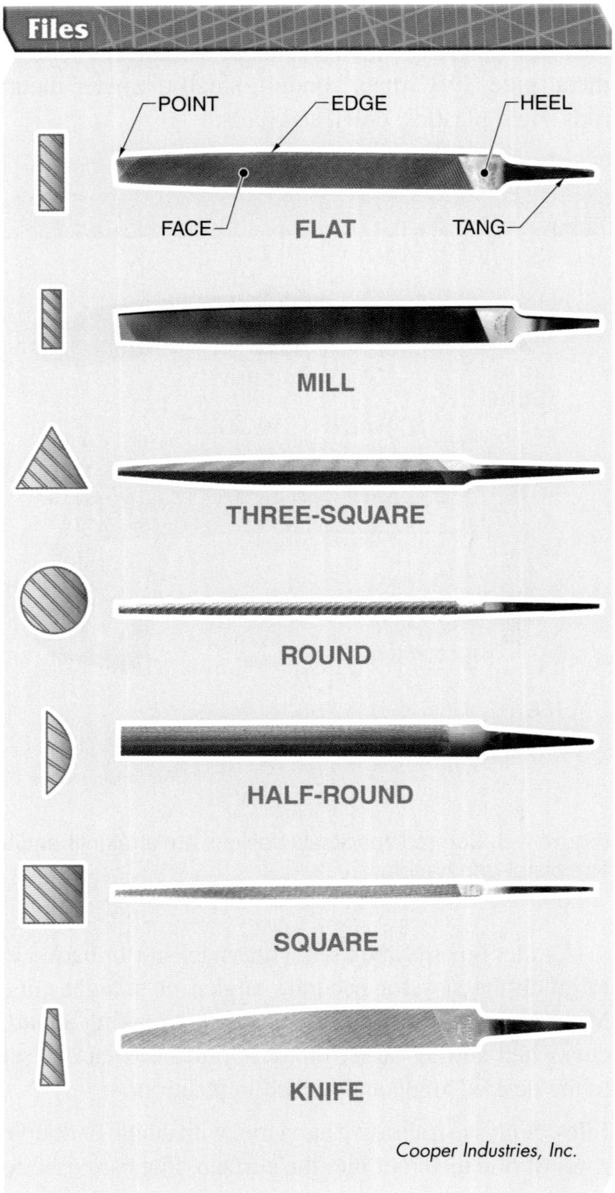

Figure 4-7. Different files are used for various sheet metal working applications.

A *flat file* is a file that is used to file flat surfaces as well as for other operations that require a fast-cutting file. A *mill file* is an all-purpose, single-cut file especially adapted for finish filing. A *three-square file,* also known as a three-cornered file, is a file that has angles of 60° and is used for filing internal angles and clearing out corners and threads. Round (rat tail) files are used for filing burrs out of holes and off of bevels and can come in small sizes. Half-round files do the same and come in larger sizes. Square files are used in the same manner as flat or mill files, with four different teeth styles.

A *knife file* is a file with a blade cross-section tapering from a square to a pointed edge. A knife file is used to smooth areas enclosed by an acute angle. It is suited for finishing sharp corners of grooves and slots where other files cannot fit. A file card or brush with a scorer is used to remove particles that clog files.

Taps and Dies. Taps and dies are used to cut threads into a workpiece or on a round rod. Taps and dies are typically used to recut disfigured threads. Taps and dies are manufactured in sets of various standard diameters and thread spacing.

Cutting metal with taps and dies requires the proper cutting fluids to reduce friction and to achieve cleanly formed threads. Cutting fluids recommended by the tap and die manufacturer should be used to ensure that good threads are cut. Friction due to improper use can cause cutting edges of taps and dies to chip, break, or shatter.

A *tap* is a hand tool used to cut internal threads in a predrilled hole or clean threads that have been disfigured by damage or improper operation. Taps have a helical shape and are self-feeding once engaged in the material to be tapped (cut). A tap wrench is used to "pull" the tap into a rigid metal workpiece to cut internal threads. Taps are available in sets containing standard sizes. The most common types of taps used are the taper, plug, and bottom. **See Figure 4-8.**

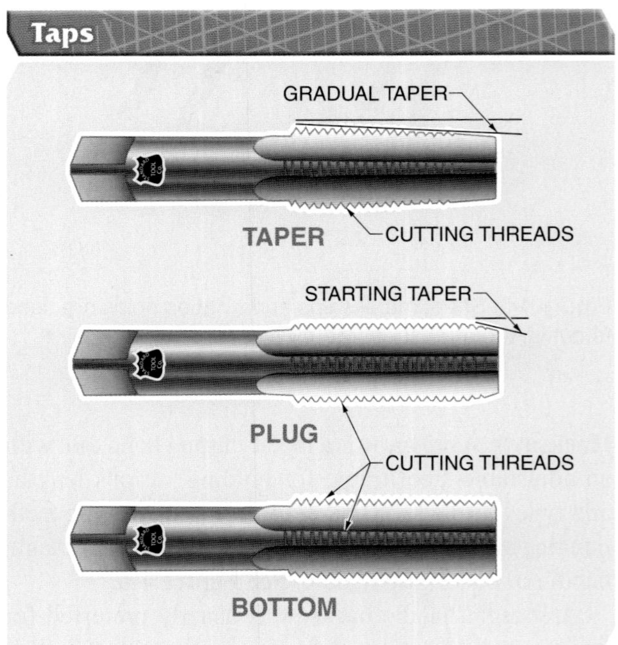

Figure 4-8. Taps have a helical design and are self-feeding once engaged in the material to be tapped.

TECH TIP

All HVAC technicians should be able to drill and tap a hole. A common field occurrence is that a bolt breaks off, must be removed, and the metal tapped for a larger size bolt.

A *taper tap* is a tap used to start a true and straight thread and when a tap is required completely through a workpiece. Taper taps are used for threading small-diameter holes and alloy steels because they cause less breakage. Taper taps have a long, gradual taper that allows the tap to start easily. The taper also allows for a gradual cutting action, which keeps the tap square with a drilled hole.

A *plug tap,* also known as an intermediate tap, is a tap used for through tapping and as the second tap in bottom tapping. Plug taps are used after a taper tap is used to start a true and straight thread.

A *bottom tap* is a tap designed to cut threads at the bottom portion of a hole and should not be used for cutting threads in an unthreaded hole. A lack of a taper at the cutting edge greatly lessens the starting capability of a bottom tap.

A properly tapped hole begins with a properly drilled hole and requires determining the number of threads per inch and diameter of the finished tapped hole. A tap-drill size chart can be used to make this determination. **See Appendix.** When a tap-drill size chart is not available, tap-drill size is determined by applying the following formula:

$$TD = MD - \frac{1}{N}$$

where

TD = tap-drill size

MD = major diameter

N = number of threads per inch

For example, find the tap-drill size for a ¼″-20 threaded hole.

$$TD = MD - \frac{1}{N}$$

$$TD = 0.250 - \frac{1}{20}$$

$$TD = 0.250 - 0.050$$

$$TD = \mathbf{0.200''}$$

Checking the drill sizes on a drill index indicates that the tap drill for a ¼″-20 hole is a number 7 drill (0.201″). **See Procedure for Hand-Tapping Holes.**

Hand-Tapping Holes

Procedure

1. Secure workpiece and drill a hole perpendicular to work surface.

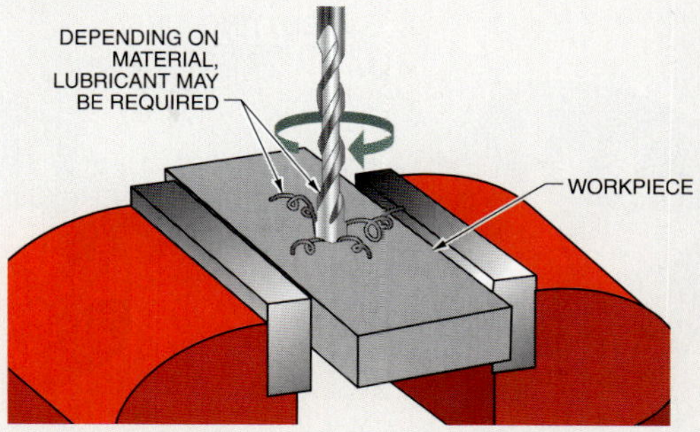

2. Place a taper tap in a tap wrench ("T" handle wrench).
3. Hold tap wrench square to surface and start cutting action while applying cutting fluid and turning tap clockwise. *Note:* Cutting fluid is used to disperse heat and assist in formation and removal of metal chips.

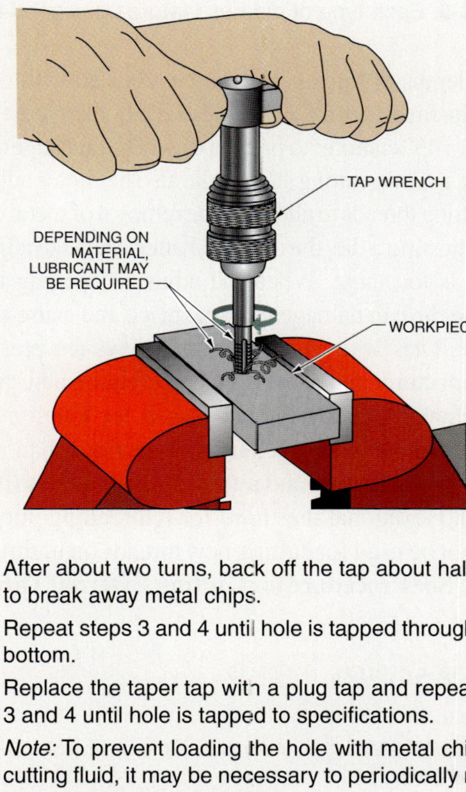

4. After about two turns, back off the tap about half a turn to break away metal chips.
5. Repeat steps 3 and 4 until hole is tapped through to the bottom.
6. Replace the taper tap with a plug tap and repeat steps 3 and 4 until hole is tapped to specifications.

   Note: To prevent loading the hole with metal chips and cutting fluid, it may be necessary to periodically remove the tap to clean out the chips.

A *die* is a hand tool used to cut external threads on round rods or clean threads that have been disfigured by damage or improper operation. Dies are available in a variety of designs, typically in round or hexagonal shapes. Dies are placed in a diestock (handle) for hand threading. Dies are available in sets containing standard sizes. **See Figure 4-9.**

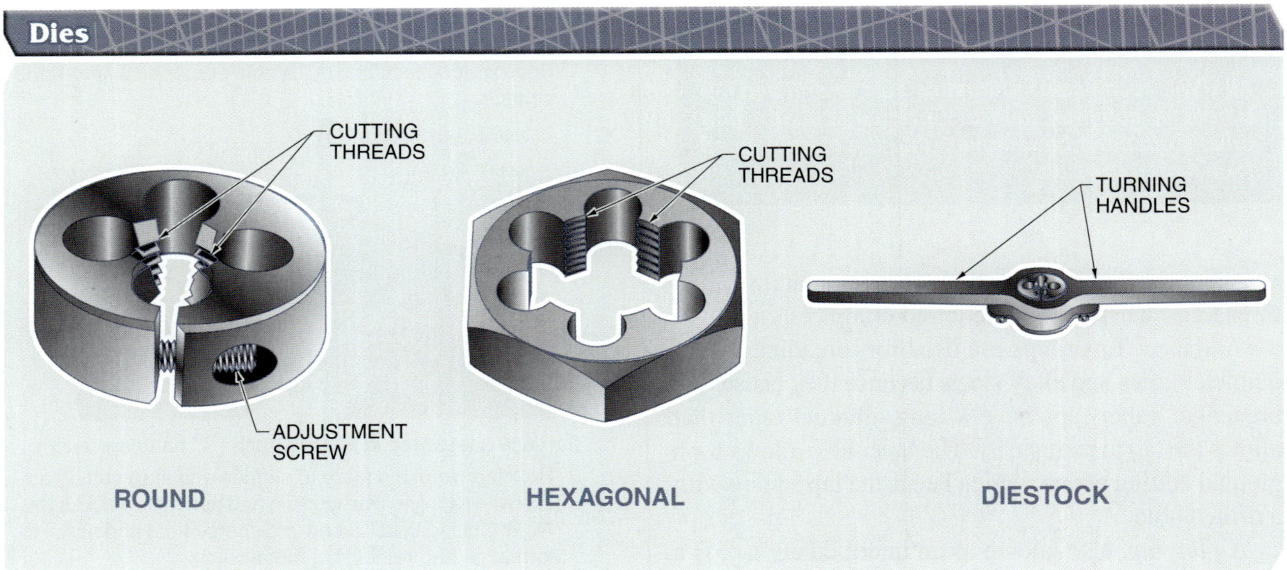

Figure 4-9. Each type of die has features designed for specific threading applications.

The depths of thread cuts can be adjusted with one or more adjustment screws in the die or die holder. A die has a side with a 45° chamfer to prevent die-tooth breakage and to allow for a gentle cutting start. A die also has holes adjacent to the cutting threads to allow for the removal of metal chips. When adjusting a die, the die can shatter when the adjusted opening is too large. When the adjusted opening is too small, the die can damage the workpiece and cause thread breakage. Dies without adjustment screws are preset by the manufacturer and generally do not require adjustment.

Hexagonal dies are typically used for tough cutting and are designed with thicker cross-sections than round dies to permit cleaning and rethreading of threads (thread chasing). Hexagonal dies used for rethreading purposes should not be used for cutting new threads on unthreaded material. **See Procedure for Cutting External Threads.**

Loctite Corporation

A thread anti-seize lubricant is applied to a threaded fastener to prevent the assembled parts from locking together over time.

Procedure

Cutting External Threads

1. Place die in diestock.
2. Place 45° starting chamfer side of die on end of workpiece to be threaded.
3. Align die squarely with starting point of workpiece.
4. Rotate die clockwise (counterclockwise for left-hand threads) while applying cutting fluid. *Note:* Cutting fluid is used to disperse heat and assist in formation and removal of metal chips.
5. Apply equal pressure to both sides of die to verify a straight cut.
6. Repeat steps 4 and 5 until workpiece is threaded to specifications. *Note:* To prevent chipped die threads, do not turn the die completely against the shoulder (bottom) of the workpiece.

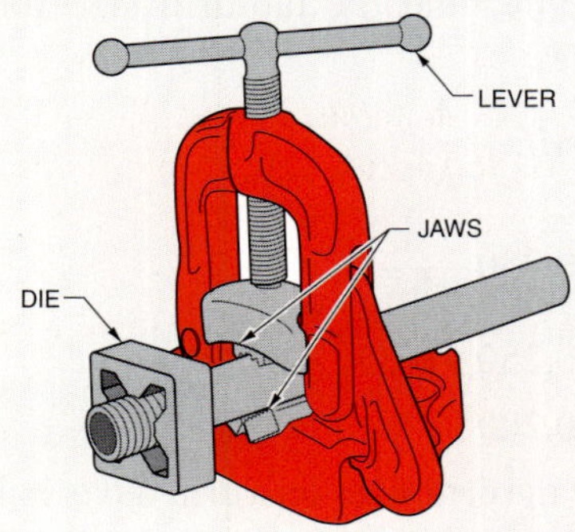

RIDGID®

Hand Brakes. A *hand brake* is a hand tool used to make accurate bends in sheet metal. Hand brakes are folding tools with adjustable stops used for varying the depth of bends. Hand brakes are typically 6″, 12″, 18″, or 24″ long and made of stainless steel. Adjustable stops are set to the desired depth of the bend, and the hand brake is placed over the edge of sheet metal. The sheet metal edge is bent to the desired position. Hand brakes are typically used to bend tabs needed in joining ductwork in custom fit sheet metal sections.

Tubing and Piping Tools

Most tubing and piping used with HVAC and refrigeration equipment are standard sizes with ends prepared for connections. Although tubing and piping are available in standard lengths, some custom fitting typically is necessary. Copper tubing is typically installed using flared, soldered, or brazed connections to HVAC equipment. Steel pipe is typically installed using threaded connections. Plastic pipe is typically installed using solvent cemented connections.

Tubing Cutters. Copper tubing is cut with a tubing cutter and deburred before it is joined to other tubing or fittings. A *tubing cutter* is a hand tool that consists of a cutting wheel and guide wheels attached to a rigid frame used to make square cuts on tubing. **See Figure 4-10.** A cutting wheel is attached to the feed screw. When the feed screw knob is turned, the guide wheels and cutter wheel contact to the tubing tighten. The tubing cutter is rotated several times around the tubing. The feed screw knob tightening and the tubing cutter rotations continue until the tubing is cut through.

Tubing cutters do not remove any metal. The cutter wheel squeezes the metal and forces it ahead of the cutter wheel until the tubing is cut through. The tubing cutter may have an attached reamer that is used to remove any burrs on the cut tubing ends.

Copper Tubing Media Clip

Deburring Tools. A *deburring tool* is a hand tool used to remove burrs left inside the tubing by tubing cutters. Deburring tools clean burrs, which would restrict flow, from the internal edge of tubing. Ridges on the outside of the tubing may not allow the tubing to be easily inserted into a fitting. A deburring tool or reamer is used to remove burrs and ridges, preparing the tubing for joining and assembly. **See Figure 4-11.** A deburring tool or reamer is rotated around the inside edge of the tubing to remove burrs without flaring or thinning the tubing edge.

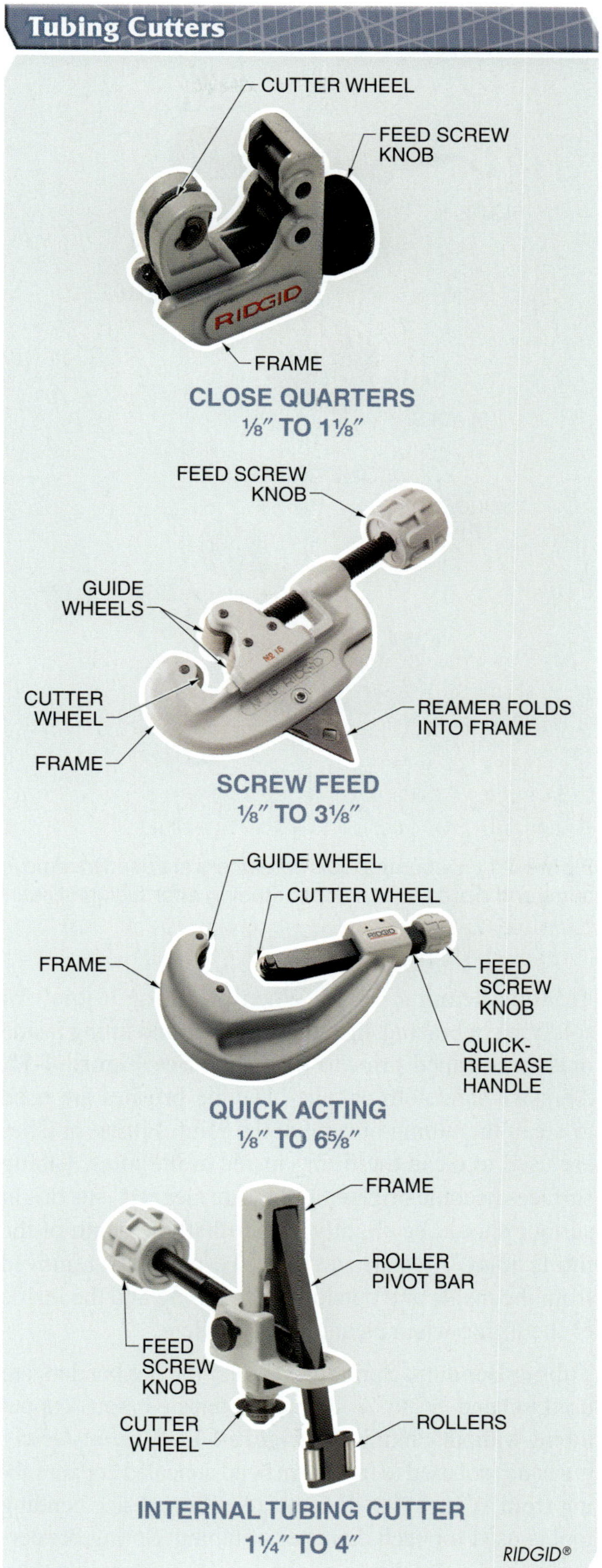

Figure 4-10. Tubing cutters are used to make square cuts on tubing.

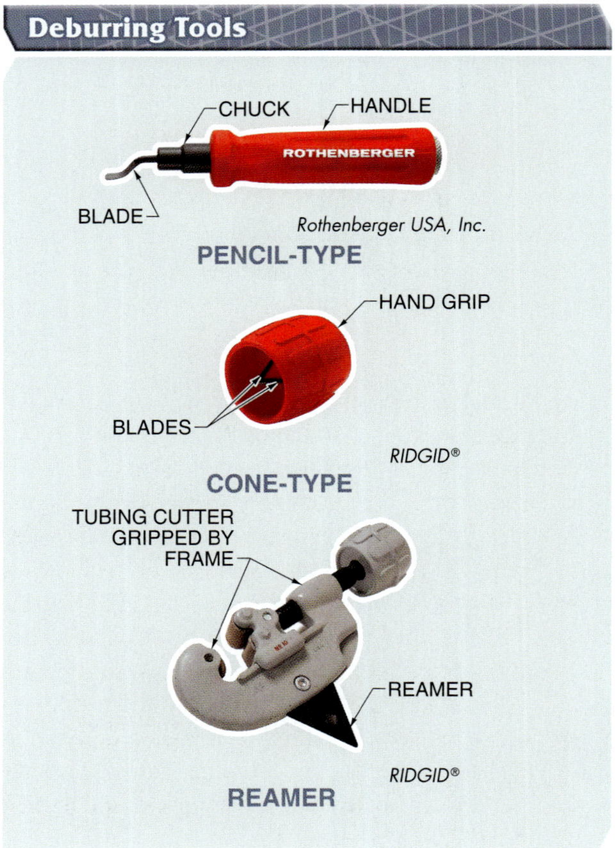

Figure 4-11. Deburring tools or reamers are used to remove burrs and ridges left on copper tubing after tubing is cut.

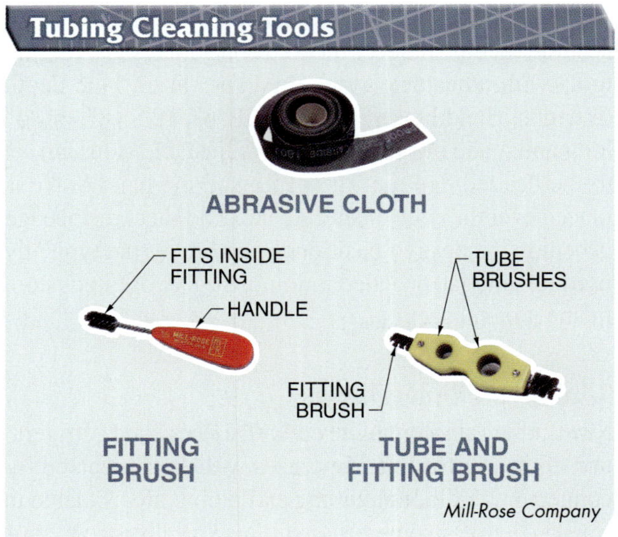

Figure 4-12. Abrasive sandcloth and brushes are used to clean copper tubing and fittings prior to assembly.

Tubing Cleaning Tools. When preparing a joint for soldering or brazing, the tubing outside and fitting inside must be cleaned prior to assembly. **See Figure 4-12.** Abrasive sandcloth or tube-cleaning brushes are used to clean the tubing outside at the joint. Fitting brushes are used to clean the fitting inside at the joint. Tubing surfaces become brighter as they are cleaned. The bright surface should be slightly longer than the depth of the fitting joint. All particles must be carefully removed from the inside and outside of the tubing and the inside of the fitting when cleaning is complete.

Tubing Benders. Spring, lever, and ratchet benders are used to bend ¼″ to ¾″ diameter annealed (soft) copper tubing without kinking. **See Figure 4-13.** A *spring bender* is a hand tool used to free-form bend annealed copper tubing from ¼″ to ¾″ in diameter. A different size bending tool is used for each diameter of tubing. Spring benders are slipped over tubing ends and slid to the bending area. The spring bender and tubing are bent to the desired shape, and the spring bender is slid off the tubing.

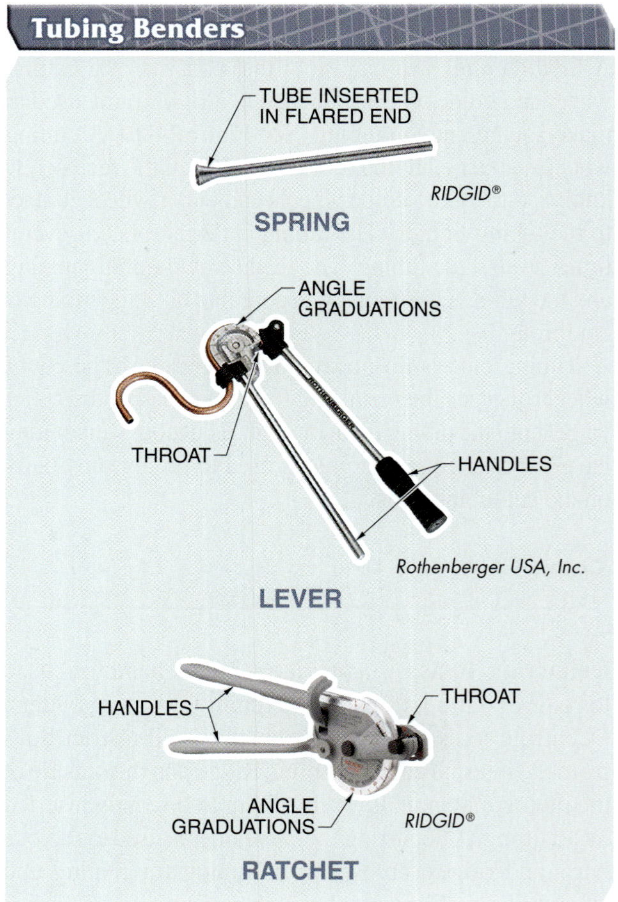

Figure 4-13. Spring, lever, and ratchet benders are used to bend small diameter copper tubing.

A *lever bender* is a hand tool with a hinge and lever used to bend ¼″ and ⅜″ annealed copper tubing. A *ratchet bender* is a hand tool with gears used to bend ½″ to ¾″ annealed copper tubing. Ratchet benders contain a series of gears that make bending large sizes of drawn copper tubing easy. Lever and ratchet benders operate similarly to each other. Copper tubing is inserted into the throat of the bender and properly positioned. The handles are then pulled together to form the desired angle. Graduations on the benders indicate the approximate angle of bend.

Tubing Flaring Tools. Annealed (soft) copper tubing is typically joined together using flared fittings. A *flaring tool* is a tool applied to the end of tubing used to flare the end of the tubing. **See Figure 4-14.** A flare is created by forcing the end of the tubing out at an angle. An impact flaring tool of the correct size is inserted into the end of the tubing. A hammer is used to drive the flaring tool against the end of the tubing to form an outward angle, or flare. Yoke-and-screw flaring tools have a clamp that can hold different tubing diameters. A screw assembly is tightened against the end of the tubing using a hand screw, forming an outward angle or flare.

Tubing Flaring Tools

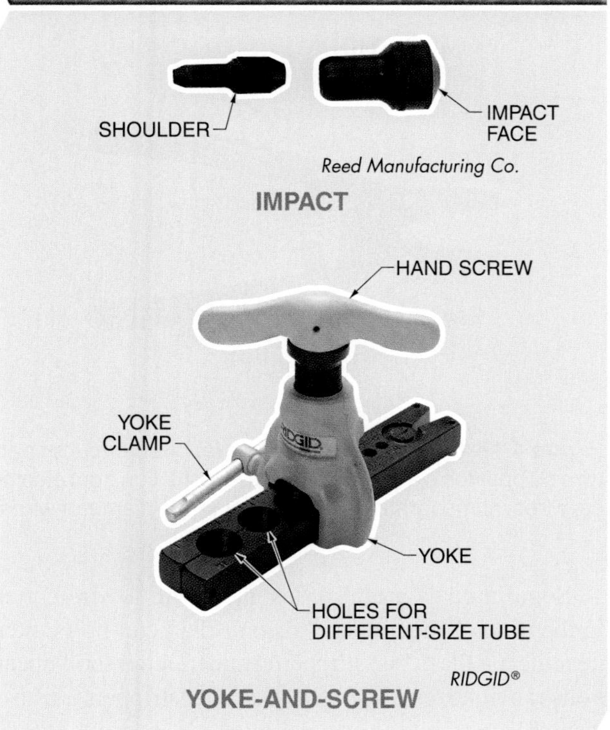

Figure 4-14. Impact or yoke-and-screw flaring tools are used to flare the ends of copper tubing.

Hand Flaring Media Clip

Soldering and Brazing Torch Assemblies. Hard copper tubing is typically joined by soldering or brazing. These processes are also used to join annealed (soft) copper tubing. *Soldering* is a copper tubing joining method that uses nonferrous filler metal with a melting temperature less than 840°F. Soldered joints are typically used in HVAC piping systems. *Brazing* is a copper tubing joining method in which nonferrous filler metal with a melting temperature of 840°F or higher is used. Brazing joints is typically done where greater joint strength is required, such as some refrigeration piping systems.

Joints (fittings and tubing) must be heated to the filler metal melting point to ensure that the filler metal flows properly into the fitting around the tubing. HVAC technicians use several types of torch assemblies to solder and braze copper tubing. **See Figure 4-15.** A torch assembly that uses MAPP (methyl acetylene propadiene) stabilized gas is typically used as a heat source when soldering small diameter copper tubing. Larger diameter copper tubing is typically soldered using either liquefied propane (LP) gas or acetylene gas.

Brazing requires more heat than soldering. Smaller diameter copper tubing is typically brazed with an acetylene gas torch. Oxyfuel equipment may be needed when brazing larger diameter tubing.

Various sizes of torch tips are available for soldering and brazing copper tubing and fittings. Larger diameter copper tubing and fittings require larger torch tips to ensure that adequate heat is delivered to the joint. A #1 torch tip (pencil tip) is used for ½″ to 1″ diameter tubing, a #2 tip is used for 1¼″ to 2″ diameter tubing, a #3 tip is used for 3″ to 4″ diameter tubing, and a #4 multi-orifice tip, or swirl tip, is used for copper tubing larger than 4″ in diameter.

WARNING: Soldering and brazing involve the use of fire and molten metals. Swirl combustion torches emit a loud whistling noise when they are being used. Proper PPE, including appropriate gloves, must be worn when soldering or brazing.

TECH TIP

Improperly brazed and soldered lines are a major source of leaks. All line joints should be thoroughly leak-tested to ensure a leak-free system.

Soldering and Brazing Torch Assemblies

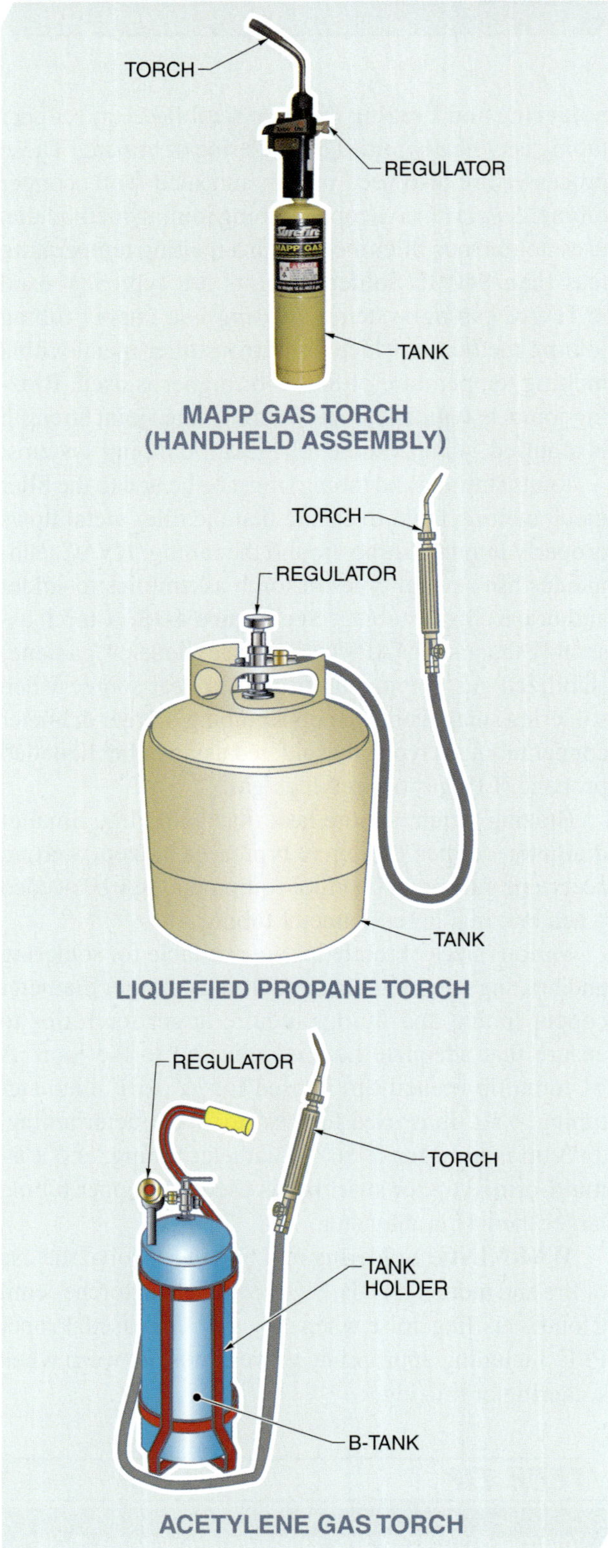

Figure 4-15. HVAC technicians use a variety of torch assemblies to solder and braze copper tubing.

Other Common Hand Tools

Other common hand tools are used by HVAC technicians to install and troubleshoot HVAC equipment. Typically, hand tools are designed for specific uses. Hand tools typically used include wire strippers, hand levels, tape rules, fish tape, flashlights, inspection mirrors, fin combs, nut drivers, staple guns and heat guns, spray bottles, and cleaning tools.

Wire Strippers. A *wire stripper/crimper/cutter* is a hand tool used for the removal of insulation from small-diameter wire. **See Figure 4-16.** Most wire strippers strip stranded wire from American Wire Gauge (AWG) size 22 to AWG size 10 and solid wire from AWG size 18 to AWG size 8. A wire stripper/crimper/cutter is also used to crimp terminals from AWG size 22 to AWG size 10.

Wire Strippers

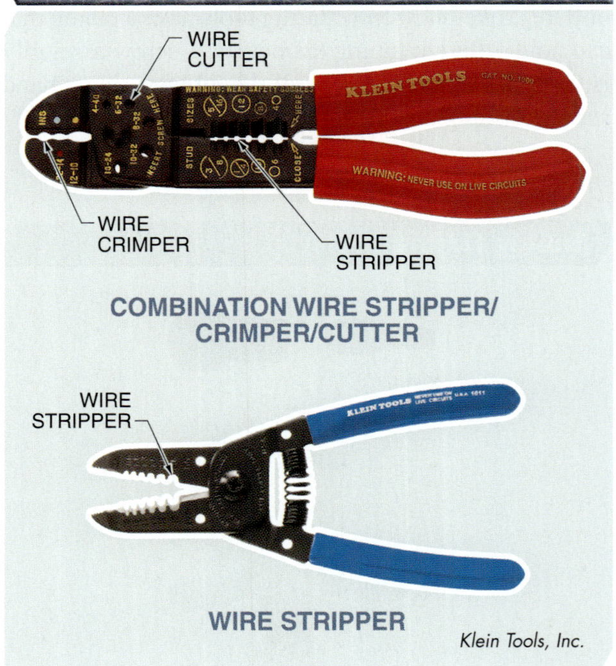

Figure 4-16. Combination wire stripper/crimper/cutter and wire stripper tools are used to remove insulation from electrical wires, crimp terminals, and shear small-diameter wires.

Some models are also designed with a wire cutter in the nose of the tool and a small-diameter bolt cutter near the handle. Most stripper/crimper/cutter tools shear bolts ranging from 4-40 to 10-32. Wire strippers are also available without the crimper and bolt-shear functions. Wire strippers without these functions are smaller and easier to handle than a stripper/crimper/cutter tool.

Hand Levels. A *hand level* is a hand tool used to establish accurate horizontal and vertical orientations. Hand levels are short levels that use an air bubble inside liquid-filled transparent vials. **See Figure 4-17.** Levels ensure that pipes, electrical conduit, or control panels are mounted in a horizontal or vertical orientation.

Hand Levels

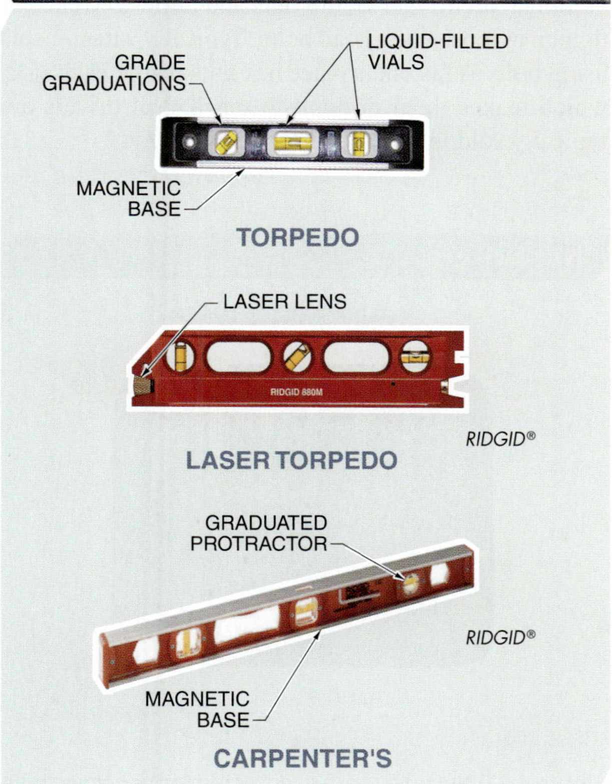

Figure 4-17. Hand levels are used to establish accurate horizontal and vertical reference lines.

Levels are also used to install drains so water will drain away from an evaporator coil and toward a floor drain. The vials on some levels are graduated to indicate ⅛″ per foot grades. Some levels have a magnetic base used to attach the level to equipment. Other levels have a laser light attached.

Tape Rules. A *tape rule* is a hand tool used for measuring lengths and consists of a long, continuous strip of steel that is graduated in regular increments. The steel strip is coiled inside a case and can be retracted manually or with a spring. Tape rules with steel strip lengths of 10′, 12′, 16′, 25′, 30′, and 50′ are typically used for taking general length measurements. **See Figure 4-18.** Once the steel tape is extracted from the case, it can be locked in place using a locking mechanism for reading measurements.

Tape Rules

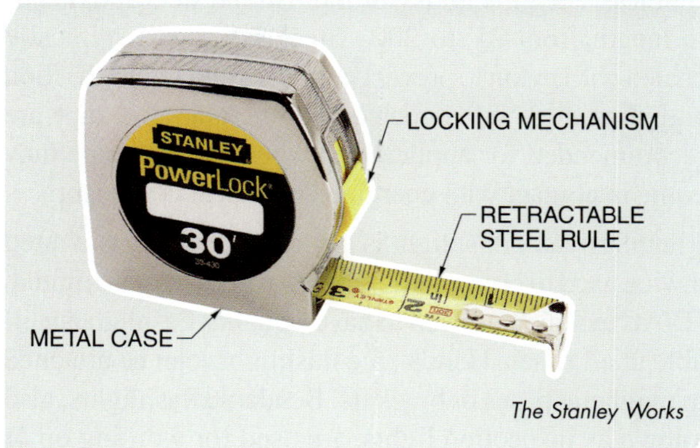

Figure 4-18. Tape rules are typ cally used for taking general measurements.

Fish Tape. *Fish tape* is a retractable tape, usually with a rectangular cross section, that is pushed through an inaccessible space, such as a conduit or partition, in order to draw in electrical conductors. Fish tape is used when running electrical conductors through conduit or walls. **See Figure 4-19.** Conductors are attached to the end of the fish tape and the fish tape is pulled back through the conduit or wall with the conductors attached.

Fish Tape

Figure 4-19. Fish tape is a retractable tape used to pull wires through conduit and other inaccessible spaces.

Fish tape is constructed of nylon, tempered steel, stainless steel, fiberglass, or multistrand steel and ranges in length from 25′ to 200′. Steel fish tape is typically coated with nylon to prevent corrosion and reduce friction and jamming. Nylon-coated and fiberglass fish tape are recommended for applications where the fish tape may come in contact with energized electrical conductors.

Flashlights. A flashlight is a portable, battery-powered device used to create light in areas where light is limited. HVAC technicians need to have a quality flashlight available at all times. Hands-free flashlights can be attached to magnetic bases or brackets. Bendable flashlights, also known as inspection lights, are used for working on or inspecting equipment in hard-to-reach areas.

Inspection Mirrors. HVAC and refrigeration components are not always located in easily accessible and viewable spots. Components may be mounted on a wall or behind other components, making viewing difficult. Inspection mirrors are used to allow viewing around corners and at awkward angles. Inspection mirrors are also used to inspect the interior of small storage tanks and pressure vessels.

Fin Combs. A *fin comb* is a hand tool used to straighten the edges of thin metal fins that extend from condenser or evaporator coils. Fin combs, also known as fin straighteners, are used to restore fins to their original orientation. HVAC and refrigeration systems typically have finned heat exchangers with air flowing across them. When the fins are bent during operation, heat exchange capability is decreased. Fin combs often have multiple sides that provide different fin spacing. Fin spacing is measured by the number of fins per inch. **See Figure 4-20.**

Fin Combs

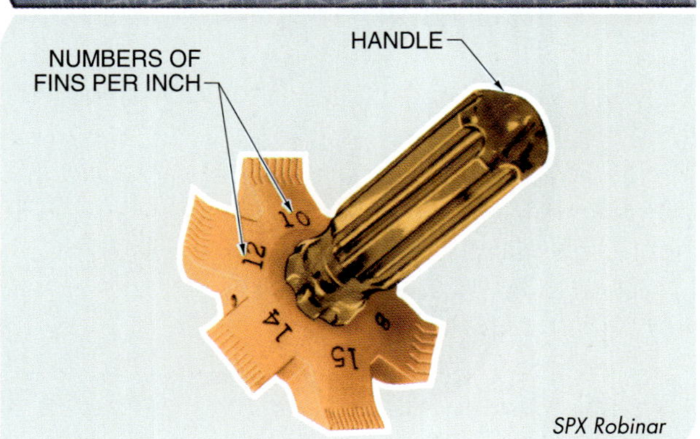

Figure 4-20. A fin comb is used for condenser maintenance to straighten damaged or bent fins, which limit airflow and reduce condenser efficiency.

Fin combs must be used to straighten bent fins. Fins have sharp edges and could harbor contaminants. Great care must be used when the fins are straightened to prevent damage or breakage. It is important that all safety procedures are followed and gloves and other appropriate PPE are used.

Nut Drivers. A *nut driver* is a hand tool with a tip designed to fit over a hex nut and used for turning or holding. Nut drivers secure nuts and allow a wrench to tighten or loosen hex-head bolts. Typically, attachments using bolts as fasteners place hex nuts on the back side, which makes them difficult to reach. Nut drivers are typically sold in sets. **See Figure 4-21.**

Nut Drivers

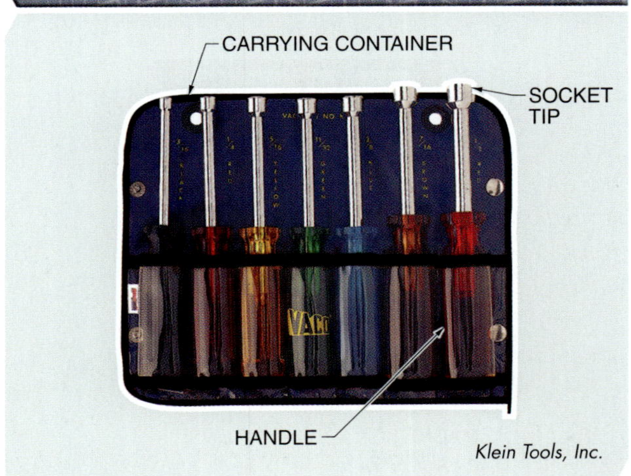

Figure 4-21. Nut drivers are used for turning or holding nuts.

Staple Guns and Heat Guns. Staples are often used to attach insulation to fixed materials such as wood and drywall. Staple guns drive the staples into the fixed material. Heat guns are used to direct heat at a surface. The heat may serve to dry an adhesive or shrink electrical tubing to make connections. It is important to follow all safety procedures, including using gloves and other appropriate PPE, while using staple and heat guns.

Spray Bottles. Small plastic bottles containing a water and soap solution are typically used for detecting leaks. Once the tubing or piping system is pressurized, the soapy solution is sprayed on joints and connections. Instruments and controls must be protected from the water and soap solution. When soap bubbles form, the joint or connection is leaking.

Leaking joints or connections must be tightened, reconnected, and retested. When checking refrigerant tubing, the tubing is typically pressurized with nitrogen to check for leaks prior to filing the system with a refrigerant. All water and soap solution must be cleaned up immediately after testing.

Cleaning Tools. Soft-bristle and hard-bristle brushes are typically used to remove dirt and debris from work areas. Soft-bristle brushes lightly touch surfaces to remove loose dirt without damaging sensitive equipment and finished surfaces. Hard-bristle brushes are typically used to remove dirt and debris from work areas. Cleaning cloths, typically made of cotton, are used to remove remaining dirt and grime from workpieces. New or clean cloths must be used when cleaning instruments and finished surfaces to prevent damage. A shop vacuum is used to remove dirt and debris from inside work areas.

POWER TOOLS

A *power tool* is a hand tool that uses externally supplied power from an electrical outlet or battery. Power tools are used extensively in the HVAC and refrigeration industry. Power tools with cords typically use 115 VAC power. Portable battery-powered tools are convenient and are used in locations where electrical outlets are not available.

When using any power tool, all safety information provided by the manufacturer must be read and followed in order to prevent death and injuries due to improper tool use. The proper tools for the work must be used, and the condition of each tool must be checked before use. Clamps and tables must be used to support workpieces when necessary. All electrical safety codes regarding tool power supply and grounding must be followed.

Extension cords may be used to provide power to corded power tools at a distance. Using an extension cord with a ground-fault circuit interrupter (GFCI) reduces the risk of electrocution in wet work areas. Extension cords must have a wire gauge that is the appropriate size for the power tool used. Extension cords must be placed where they will not cause a tripping hazard. In general, joining two or more cords together to form a longer power cord is not recommended.

Power Drills

A *power drill* is a power-driven rotary tool used with a drill bit with cutting edges for boring holes in materials such as wood, metal, and plastic. HVAC technicians use power drills to create holes for fasteners, such as screws and bolts, and larger holes for electrical conduit and tubing. Power drills can also be used with attachments for screwing, grinding, and buffing. Power drills are available as cordless (battery-powered) or corded. **See Figure 4-22.**

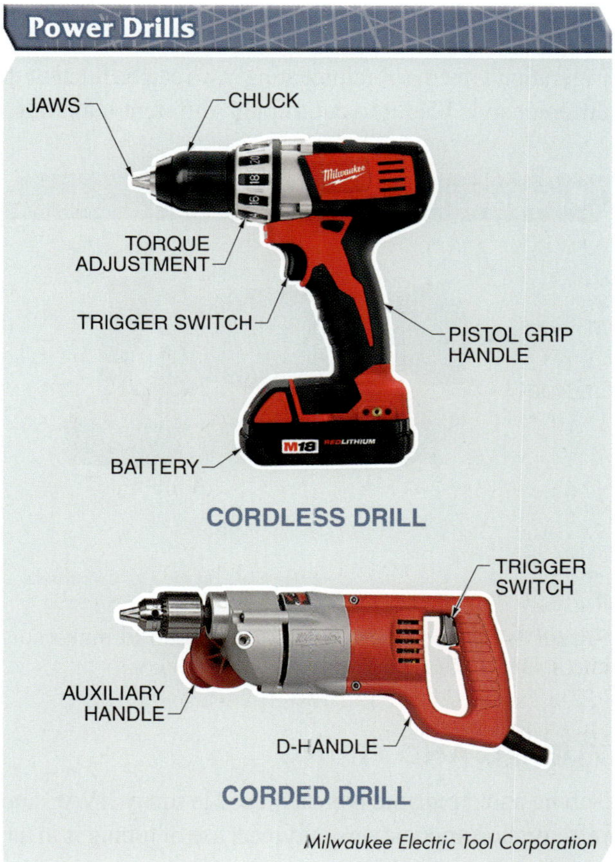

Figure 4-22. Power drills can be fitted with different types of attachments for operations such as drilling, screwing, grinding, and buffing.

Drill bits are available in various sizes and types. Drills bits are selected based on the size of the hole and the material to be drilled. In most cases, drill bits are packaged as sets that include typical bit sizes. Drill bits must be sharpened or replaced when necessary. The material being drilled must be secured to allow both hands to operate the drill.

When using a power drill in a control cabinet, all electronic control equipment must be turned off or removed to prevent damage due to the electromagnetic interference created by the drill. All chips created during the drilling process must be removed properly so the chips do not short out electrical components.

Reciprocating Saws

A *reciprocating saw* is a multipurpose power tool with a blade that reciprocates (quickly moves back and forth) to create a cutting action. Reciprocating saws are typically used to cut through existing structures, such as wood, pipe, or other materials. **See Figure 4-23.** Reciprocating saws are also used to create openings during installation of new diffusers, ductwork, and piping for HVAC and refrigeration systems. Reciprocating saws can be fitted with different style blades to cut through different materials.

Reciprocating Saws

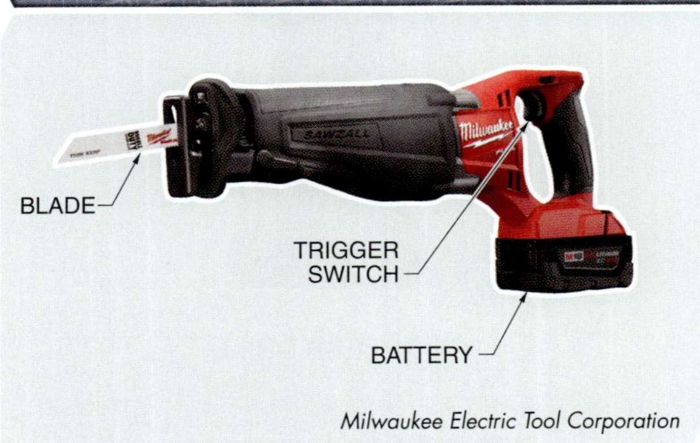

Figure 4-23. Reciprocating saws are used to make cuts into finished walls, ceilings, and floors.

TUBING AND PIPING

Tubing and piping are typically used in many HVAC and refrigeration applications. A typical use of tubing is in the transfer of a refrigerant to an evaporator, a compressor, a condenser, and a metering device. Tubing also returns oil to a compressor. Tubing and piping are also used in condensate drains and for carrying chilled and condensed water, natural gas, and fuel oil.

Working with Copper Tubing

It is important for HVAC technicians to be proficient when working on tubing with different types of tools. Proficiency means using the correct tools and techniques for the proper application. When the wrong tubing, tools, or techniques are used, the installation is incorrect and possibly dangerous.

Tubing Materials. HVAC and refrigeration system components are typically connected with copper tubing. Copper tubing may be annealed (soft) or rigid (hard). Annealed tubing is pliable and useful when making bends or connections to equipment. Annealed tubing is typically used to connect components of residential and light commercial HVAC systems. Rigid tubing cannot be easily bent and requires the use of fittings to change tubing directions and make connections. Rigid tubing is typically used to connect components of refrigeration and commercial/industrial HVAC systems.

Air conditioning and refrigeration (ACR) tubing is copper tubing used in HVAC and refrigeration applications. One of the distinguishing features of ACR tubing is that it is sized by outside diameter (OD) instead of inside diameter (ID). With non-ACR tubing, measuring the OD adds as much as ⅛″ to the listed diameter. For example, ½″ (non-ACR) copper tubing is the same size as ⅝″ ACR tubing.

Four different thicknesses of tubing are available: K, L, M, and DWV. K tubing has a thicker wall and is considered heavy-duty tubing. L tubing has a thinner wall and is considered the standard tubing. M tubing and DWV tubing have thinner walls and are rarely used in HVAC and refrigeration systems. Residential and light commercial HVAC units typically use ¼″ diameter (DIA) type L ACR tubing. Annealed copper tubing can be bought in rolls up to 100′ long, depending on the tubing diameter and thickness. Commercial/industrial HVAC and refrigeration systems, due to their large capacity, typically use large tubing of up to 6″ in diameter or greater.

ACR tubing typically is delivered with caps on both ends to ensure it is clean and dry. Some ACR tubing is charged with nitrogen to prevent the formation of internal moisture as a result of humidity in the air. When using only part of the tubing at a time, the unused tubing should be capped immediately after cutting. When handling rolled tubing, the tubing is unrolled carefully on a flat surface without twisting it. Repetitive bending of the tubing must not be done in order to prevent the tubing from hardening.

HVAC tubing is typically insulated to reduce heat transfer to the cold refrigerant and improve operating efficiency. Insulation also prevents condensation from forming on the cool temperature refrigeration lines that operate below the dew point temperature of the air. Refrigerant tubing that is pre-charged and pre-insulated is also available in line sets that have quick-connect fittings. Line sets are typically used on small-capacity residential split-system units and packaged units.

Cutting Tubing. Tubing cutters are used to make square cuts on tubing. Hacksaws are typically used for cutting large diameter tubing. Tubing must be properly marked and supported during the cutting process. The cut must be made at a 90° angle to the length of tubing. When the cutting wheel becomes loose, a different cutter must be used. The wobble of a loose cutting wheel will cause a spiral, out-of-round cut instead of a sharp, single cut. **See Procedure for Cutting Copper Tubing.**

Cutting Copper Tubing

1. Measure and mark cut location on tubing.
2. Align cutting wheel with mark and rotate feed screw knob until wheels fully contact tubing.

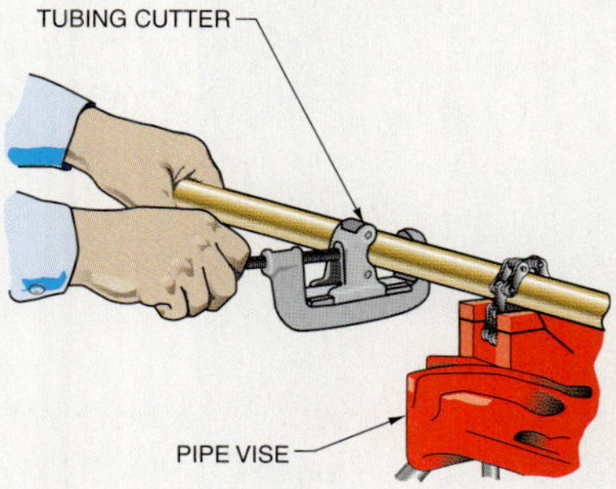

3. Rotate cutter around tubing.
4. Apply moderate pressure to tubing by tightening hand screw. Continue rotation of cutter around tubing. Apply additional pressure as tubing is cut. **CAUTION:** If excessive pressure is used, end of tubing may be pinched and the cut may be ruined.
5. Ream tubing ends to remove burrs remaining on inside of tubing using a deburring tool or reamer.

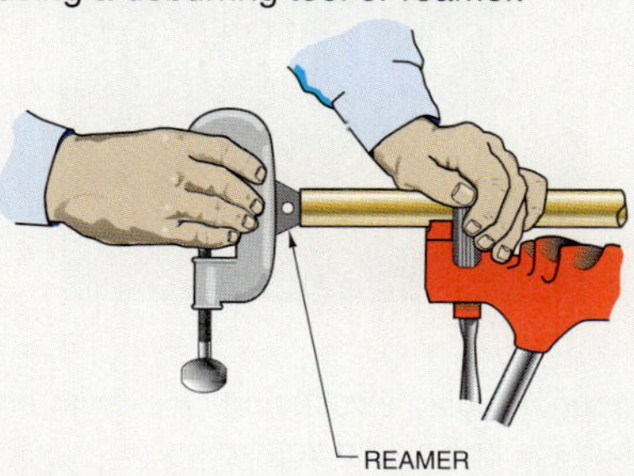

6. Deburr outside of tubing end to remove burrs remaining on outside using a deburring tool or file.

Procedure

Tubing Connections. Copper tubing is typically connected using flared or soldered joints. Flaring is a method used to join annealed tubing to fittings. The tubing end is expanded so that the coupling nut secures the tubing to the fitting or component. Soldering and brazing are methods used to join annealed or rigid tubing to couplings or fittings. Soldering is typically used for HVAC systems, while brazing is typically used for refrigeration systems. Both methods involve using a filler material to join tubing to fittings. **See Figure 4-24.** When heated, the filler material is drawn by capillary action to fill the space between the base metals (tubing and fitting).

Tubing Connections

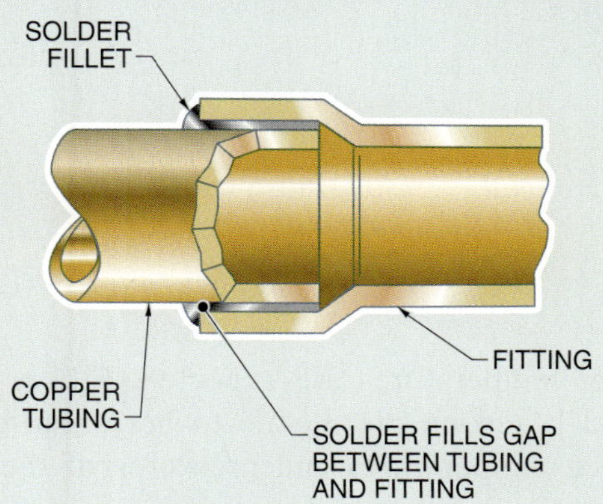

Figure 4-24. Capillary attraction draws molten solder into the gap between the copper tubing and fitting.

Copper tubing must be installed correctly for a refrigeration compressor to operate properly.

Flaring and Joining Tubing. Flared joints are used with annealed copper tubing when connecting it to equipment or when soldering is not practical. A coupling nut, with the threads facing the end of the tubing, must be placed on the tubing before flaring the end. Yoke-and-screw flaring tools are typically used to flare copper tubing. The coupling nut will be used to secure the end of the tubing to the fitting or equipment. **See Procedure for Flaring Copper Tubing.**

Procedure

Flaring Copper Tubing

1. Check length of tubing and condition of tubing ends.
2. Slip coupling nut over end of tubing. Ensure that nut threads face end of tubing being flared.

3. Insert copper tubing into flaring block with end flush with top of surface and clamp tightly.

4. Place yoke of flaring tool on block so that beveled end of flare cone is directly over tubing end.

5. Turn hand screw to force flare cone into end of tubing. Continue to turn hand screw to force tubing against block to form flare.

6. Remove tubing from flaring tool when flaring operation is complete. Check joint and attaching tubing to fitting.

Soldering. Soldering is typically used in systems operation in a temperature range of 350°F to 550°F. Soldering is typically done with a simple MAPP or propane torch assembly. Also, 95/5 tin-antimony fillers are used to add strength to the joint when higher temperatures are expected. Fillers are selected based on the metals to be joined and the service temperatures and pressures of the joints. The soldering process provides a strong joint. **See Procedure for Soldering Copper Tubing.**

Soldering Copper Tubing

1. Check length of tubing and condition of tubing ends.
2. Clean outside of tubing end thoroughly to depth of joint with abrasive cloth. Wipe tubing end clean using a cloth. *Note:* Parts that look clean may contain small amounts of contaminants. Any contaminant present may cause joint to fail. Contaminants may also be introduced into refrigeration system and cause motor or compressor failure.

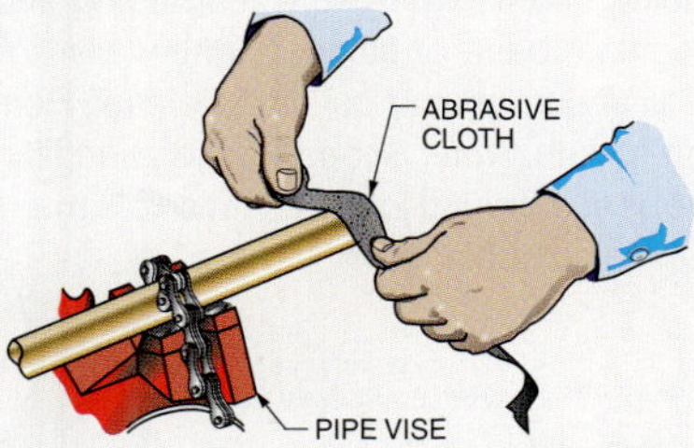

3. Clean fitting inside thoroughly to depth of joint with wire brush. Wipe fitting clean using cloth.

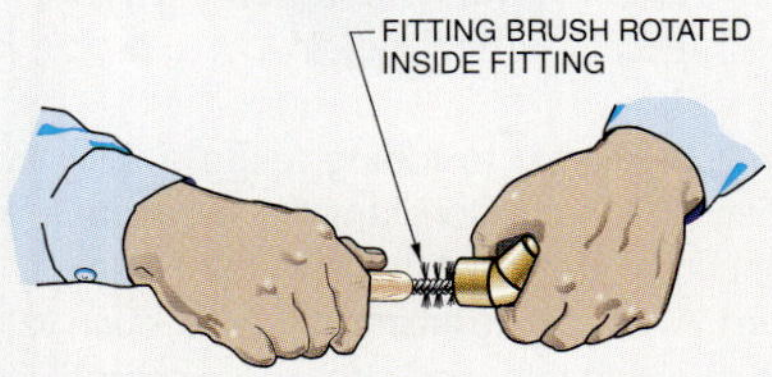

4. Apply flux to outside of tubing end to depth of joint soon after cleaning using only enough flux to lightly cover surface. *Note:* Flux is used to remove oxidation from joint. Flux should be applied only on joint area and not be allowed to get into tubing. Only clean flux should be used. Do not apply flux with fingers.

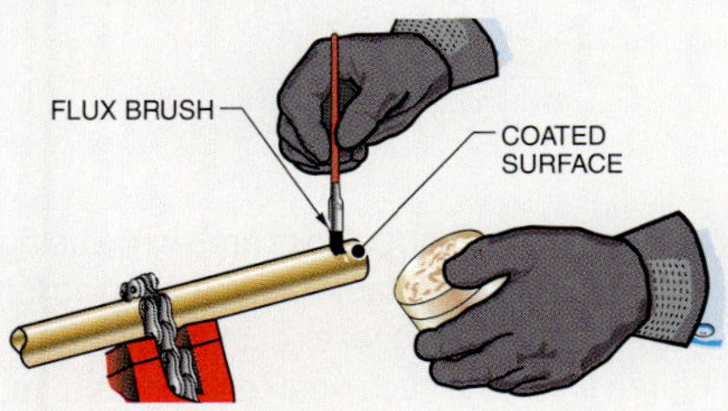

5. Apply flux to depth of cap soon after cleaning. Use only enough flux to lightly coat surface, avoiding excess flux to enter (get inside) pipe when assembling joints. Never apply flux to inside of fitting to avoid system contamination.
6. Insert tubing into fitting. Ensure that tubing end is properly seated against base of fitting. Slightly twist tube as it is inserted to ensure proper coverage of flux. Remove excess flux using a cloth. *Note:* Support tubing and fitting to ensure adequate space around joint. When necessary, a heat shield must be placed to protect anything behind joint.

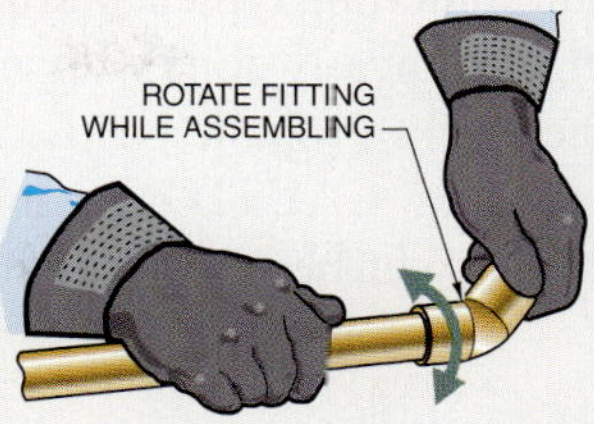

7. Apply heat evenly to tubing with flame directed perpendicular to tubing. Constantly move flame around tubing. The amount of heat applied to tubing depends on size of tubing and fitting. Do not directly heat flux or it will burn. When joining tubing to a valve, open valve completely before applying heat and apply heat primarily to tubing.

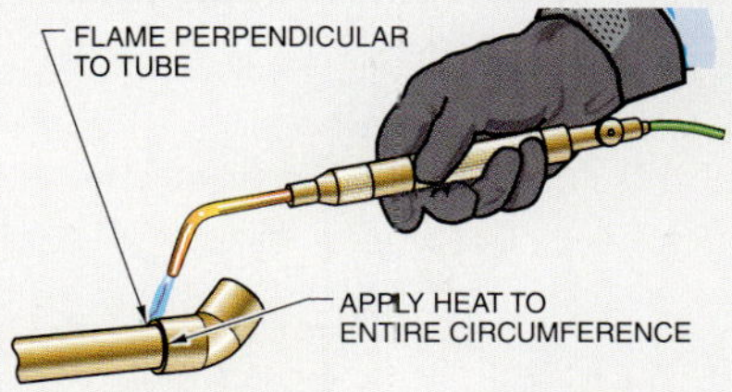

8. Apply heat evenly to fitting at joint area while being careful not to overheat fitting. Move flame back and forth between fitting and tubing. Do not over heat joint or apply flame directly to fitting.

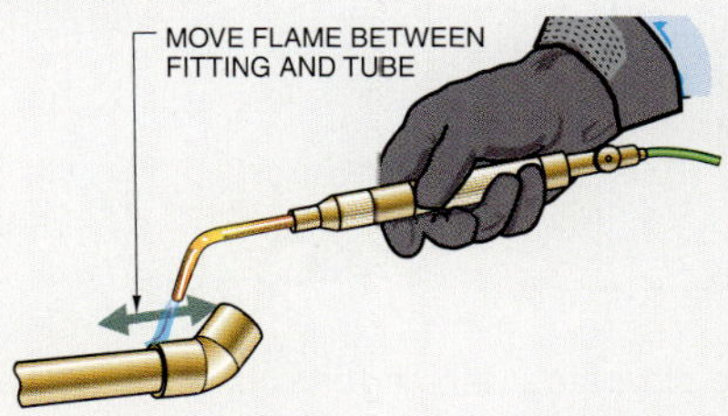

9. Remove flame from tubing and fitting and immediately feed solder to joint at one or two points until a ring of solder appears around fitting end. *Note:* The proper amount of solder is about equal to diameter of fitting. For example, use ¼″ of solder for a ¼″ fitting. The filler must not be heated directly.

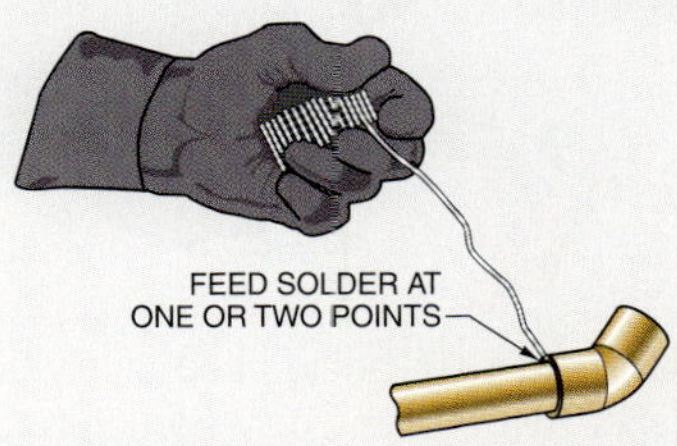

10. Allow joint to cool naturally to prevent stresses from forming in the joint.

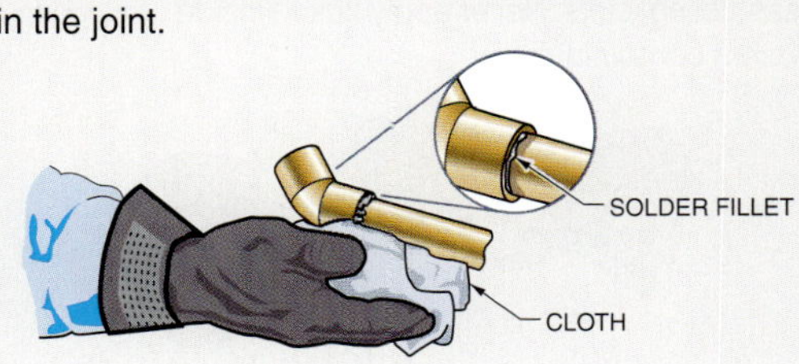

Brazing. Brazing is typically done at a much higher temperature than soldering, typically around 1100°F to 1500°F. Brazing is typically done with an air acetylene or oxyacetylene torch. Propane gas is also used with small diameter tubing. **CAUTION:** Air acetylene and oxyacetylene torches can be dangerous and should only be used after proper instruction and in the presence of an experienced technician.

Brazing fillers are often alloys that contain silver or phosphorous. The fillers are selected based on the metals to be joined and the service temperatures and pressures of the joints. **See Procedure for Brazing Copper Tubing.**

Brazing Copper Tubing

Procedure

1. Check length of tubing and condition of tubing ends.
2. Clean outside of tubing end thoroughly to depth of joint with abrasive cloth. Wipe tubing end clean using a cloth. *Note:* Parts that look clean may contain small amounts of contaminants. Any contaminant present may cause joint to fail. Contaminants may also be introduced into refrigeration system and cause motor or compressor failure.

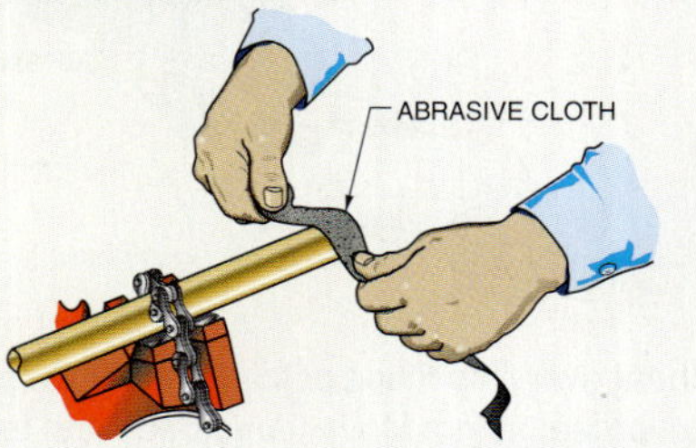

3. Clean fitting inside thoroughly to depth of joint with a wire brush. Wipe fitting clean using cloth.

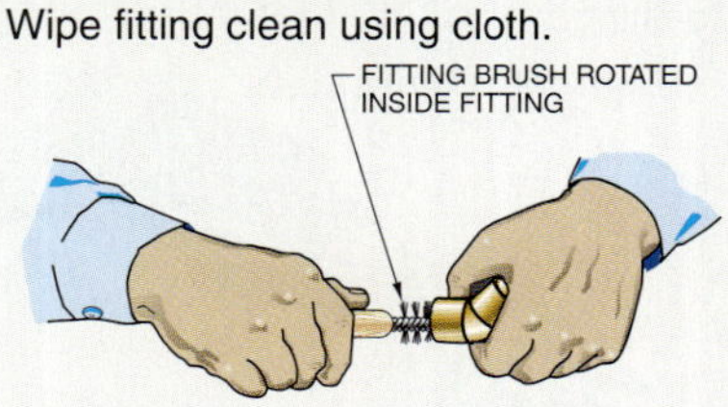

4. Brazing alloys are self-fluxing on a copper tube to copper fitting joint. When copper is brazed to a dissimilar metal, flux can be applied. Always purge with nitrogen while heat is applied to the joint.

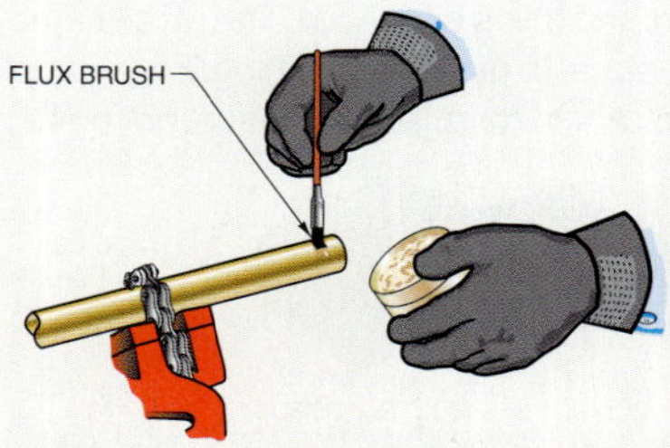

5. If dissimilar metals are being used, apply flux to depth of cap soon after cleaning. Use only enough flux to lightly coat surface, avoiding excess flux to enter (get inside) pipe when assembling joints. Never apply flux to inside of fitting to avoid system contamination.

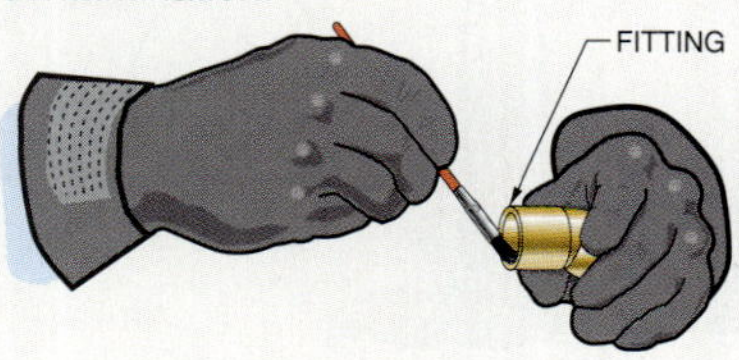

6. Insert tubing into fitting. Ensure that tubing end is properly seated against base of fitting. Slightly twist tubing as it is inserted to ensure proper coverage of flux. Remove excess flux using a cloth. *Note:* Support tubing and fitting to ensure adequate space around joint. A heat shield may be placed to protect anything behind joint.

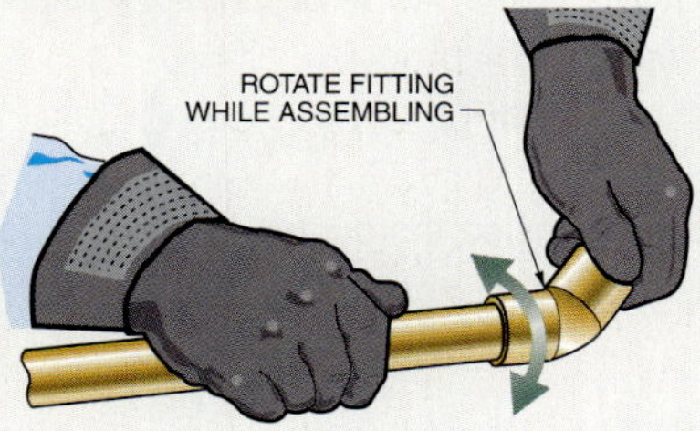

7. Apply heat evenly to tubing using a swirl combustion or oxyacetylene torch. Constantly move flame around tubing to prevent burning through tubing. Do not directly heat flux or it will burn. When joining tubing to a valve, open valve completely before applying heat and apply heat primarily to tubing.

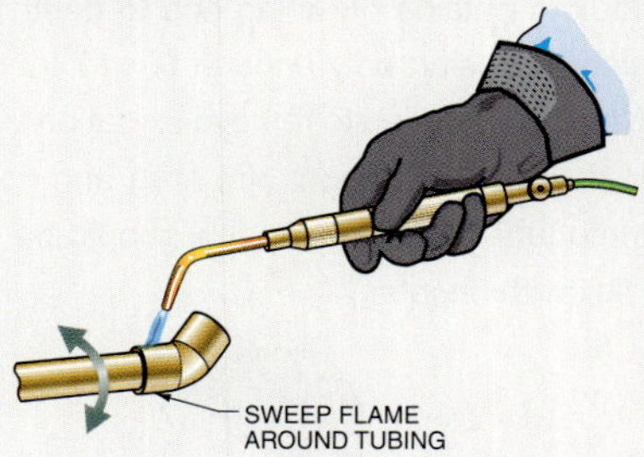

8. Apply heat evenly to fitting at joint area while being careful not to overheat fitting. Move flame slowly back and forth between fitting and tubing. Apply heat all of the way around joint. Apply about three times as much heat to fitting as to tubing.
9. As heating progresses, touch end of filler rod to joint periodically. *Note:* Rod will melt when proper temperature is attained. Filler rod must not be heated directly.

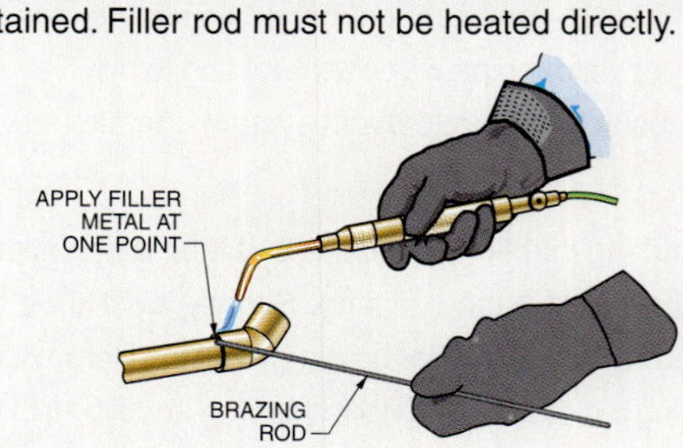

continued on next page

Brazing Copper Tubing *(continued)*

10. When proper temperature is reached, apply brazing filler metal at one point along fitting edge. Filler metal will readily flow into space between tubing and fitting. Move torch around fitting and feed filler rod into joint with light pressure. Do not direct flame at filler material as it is fed into joint. Remove flame and stop feeding filler rod as soon as a fillet appears completely around joint.
11. Allow joint to cool naturally to prevent stresses.

Procedure

Working with Steel Pipe

Steel pipe is often used in hydronic heating systems and for gas piping. Steel pipe may be black or zinc-coated to resist rusting. The pipe may be formed by welding or drawing the steel through an automatic forming machine.

Steel Pipe Threads. Steel pipe is typically joined with threaded fittings. Two types of American National Standard Pipe Threads that are typically used are straight threads and taper threads. Straight threads follow the industry standard American Standard Straight Pipe Thread (NPS) and are used for mechanical joints. An O-ring or metal seal may be required. Tapered threads follow the industry standard National Pipe Thread Taper (NPT) and are used for leak-free joints. A sealant compound or Teflon® tape must be used. Both tapered threads and straight threads are used in the HVAC industry.

NPT pipe threads are standardized to taper at a specific rate, such as 1⁄16″ per inch of pipe. A properly threaded pipe end will have six to eight perfect threads and then two to four imperfect threads. **See Figure 4-25.** Care must be taken not to nick the perfect threads, as damage may cause leaks.

National Pipe Thread Taper (NPT)

National Pipe Thread Taper Characteristics				
Nominal Pipe Size	Threads Per Inch	Approximate Thread Length*	Approximate Number of Threads	Approximate Total Thread Makeup*
1⁄8	27	3⁄8	10	1⁄4
1⁄4	18	5⁄8	11	3⁄8
3⁄8	18	5⁄8	11	3⁄8
1⁄2	14	3⁄4	10	7⁄16
3⁄4	14	3⁄4	10	1⁄2
1	11½	7⁄8	10	9⁄16
1¼	11½	1	11	9⁄16
1½	11½	1	11	9⁄16
2	11½	1	11	5⁄8
2½	8	1½	12	7⁄8
3	8	1½	12	1
3½	8	1⅝	13	1 1⁄16
4	8	1⅝	13	1 1⁄16
5	8	1¾	14	1 3⁄16
6	8	1¾	14	1 3⁄16
8	8	1⅞	15	1 5⁄16
10	8	2	16	1½
12	8	2⅛	17	1⅝

*in in.

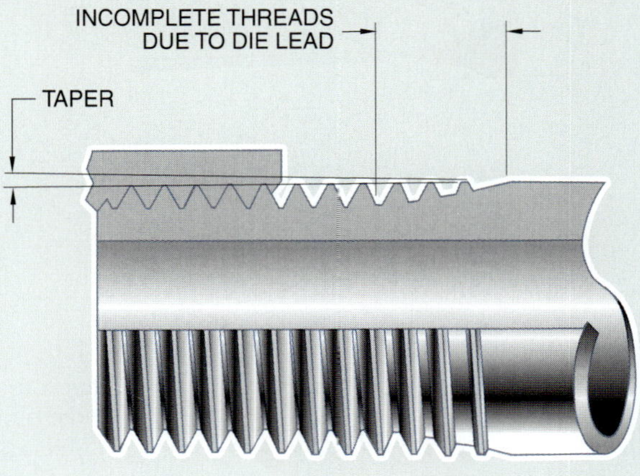

Figure 4-25. Threads are tapered so that the pipe and fittings will seal tightly to form a leakproof joint.

Thread Sealant Media Clip

Cutting and Joining Steel Pipe. When working with steel pipe, a hacksaw or pipe cutter is used to cut the pipe, and a reamer is used to remove burrs from the pipe. A threader or die is used to cut threads into the pipe. **See Figure 4-26.** Threaders or dies are either manual or power operated. Cutting fluid is applied during the cutting process.

Pipe Threaders

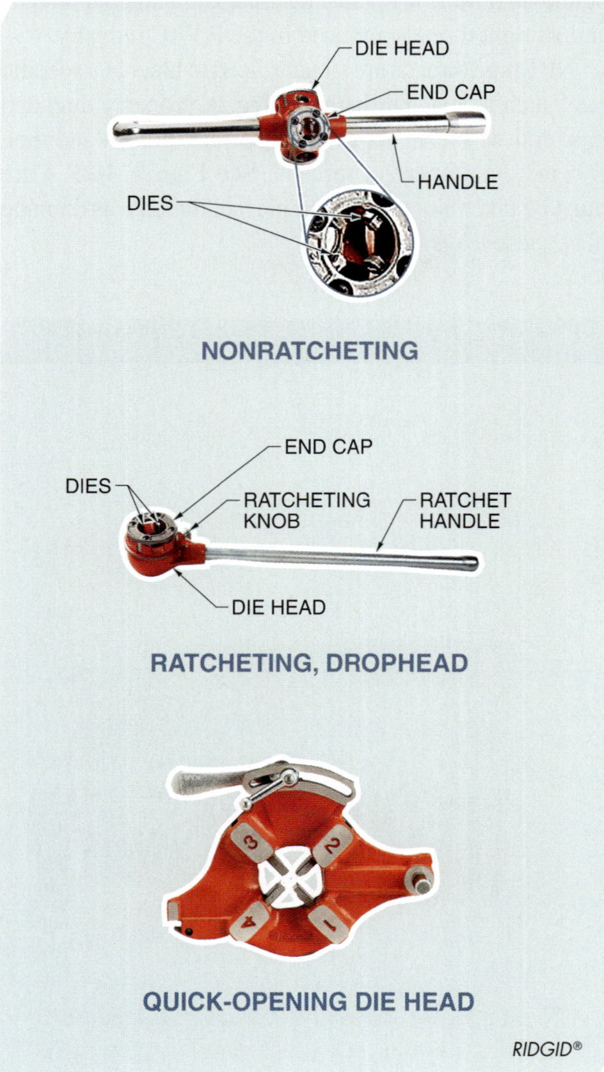

Figure 4-26. Pipe threaders are used to cut threads in steel pipe.

A vise or wrench is used to hold the pipe and stabilize it so it can be worked with or threaded. A pipe-joint compound, or thread sealant, must be applied to the pipe threads at the joint. To prevent pipe-joint compound from getting into the system, it should not be applied to the first few threads at the end of the pipe. Pipe wrenches are used to turn pipe into the fittings. **See Procedure for Cutting and Threading Steel Pipe.**

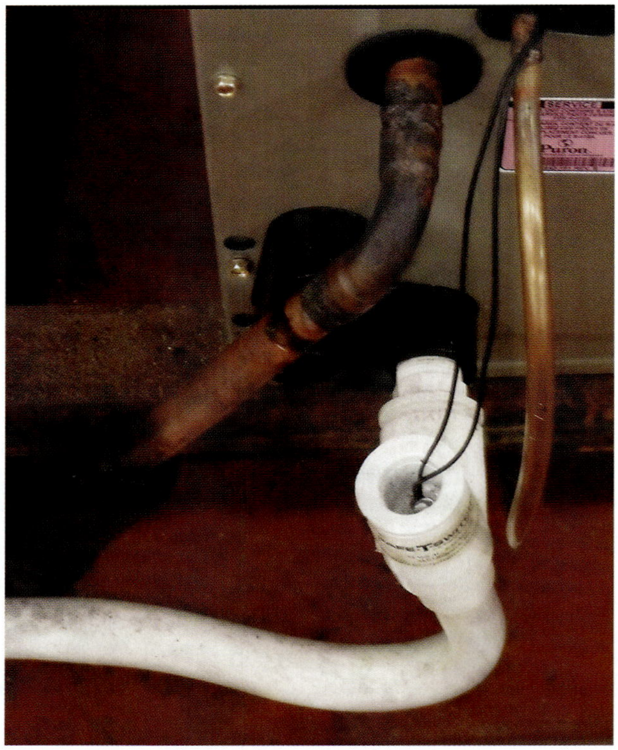

Condensate lines often use plastic pipe to remove water after it has condensed on an air conditioning coil.

Working with Plastic Pipe

Plastic pipe is used for many HVAC and refrigeration applications from condensate drains to venting equipment. Polyvinyl chloride (PVC) piping is used for high-pressure applications at low temperatures.

Plastic Pipe Material. PVC and chlorinated polyvinyl chloride (CPVC) plastic pipe are typically used in HVAC and refrigeration for water supply and drainage. PVC is a plastic material typically used for piping water supply, drainage, and venting. CPVC is similar to PVC and is useful with temperatures up to 180°F.

Cutting and Threading Steel Pipe

1. Check length of steel pipe and condition of ends.
2. Measure and mark location of cut on steel pipe.
3. Properly secure pipe in a pipe vice or pipe machine, and cut pipe to required length using a pipe cutter.

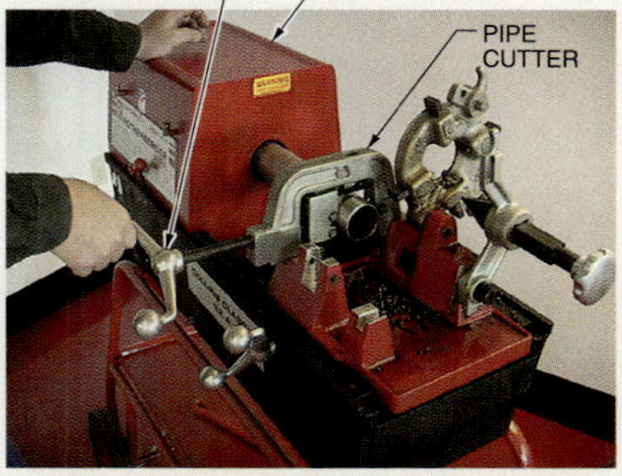

4. Ream pipe ends to remove burr on inside of pipe using a pipe reamer or half-round file, depending on pipe size.

5. Thread pipe using die while applying cutting fluid.

6. Remove pipe from pipe vice or pipe machine. Wipe pipe threads and inside of pipe clean using cloth.

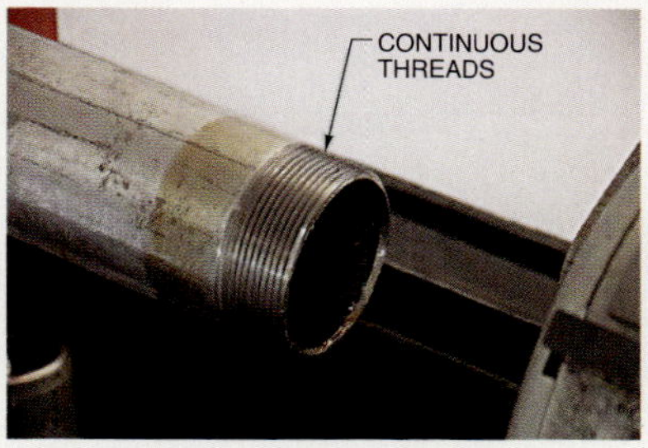

7. Apply thread sealant to pipe threads.

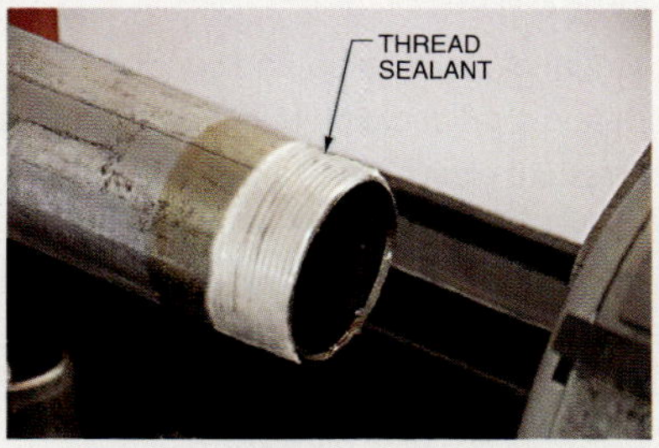

8. Start fitting onto pipe thread by hand while turning pipe or fitting two or three rotations.

9. Tighten fitting with a pipe wrench about two to three rotations. Do not overtighten fitting.

Cutting and Joining Plastic Pipe. PVC and CPVC are typically cut with a hacksaw or backsaw. A miter box secures the pipe and ensures a square cut. Solvent cement is used to fuse plastic pipe and fittings together by softening the adjoining surfaces with a chemical agent. **See Figure 4-27.** Softened surfaces are forced together to form a solid joint that is as strong as the pipe itself. Threaded PVC fittings that join with threaded pipe are also available. **See Procedure for Cutting and Joining Plastic Pipe.**

PVC Connections

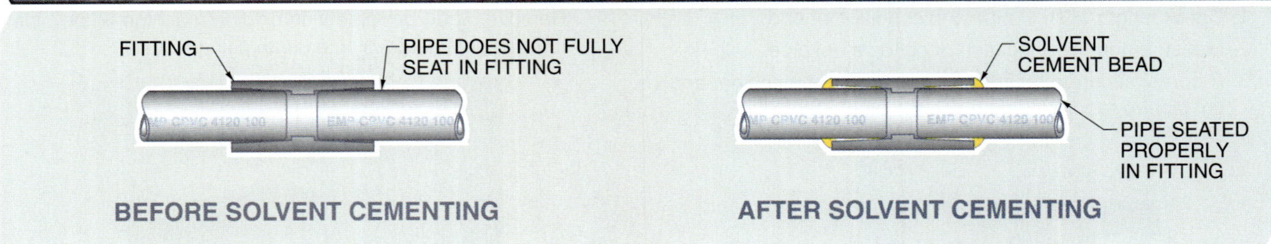

Figure 4-27. Solvent cement softens plastic pipe and fittings, ensuring a tight joint when the pipe is forced all of the way into the fitting.

Cutting and Joining Plastic Pipe

Procedure

1. Check length of plastic pipe and condition of ends.
2. Measure and mark location of cut on plastic pipe.
3. Properly secure plastic pipe, and cut pipe to required length using a hacksaw or backsaw.

4. Smooth pipe ends using a deburring tool for small pipe or a half-round file for large pipe.

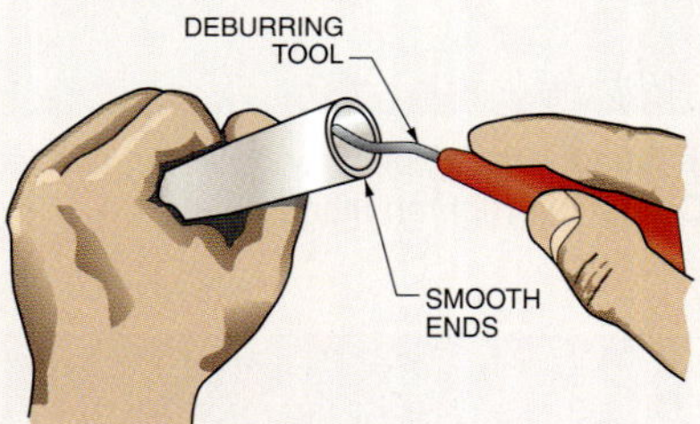

5. Dry fit joint to check for proper fit. Pipe should fit tightly into fitting. Ensure pipe and fitting surfaces are clean and dry.

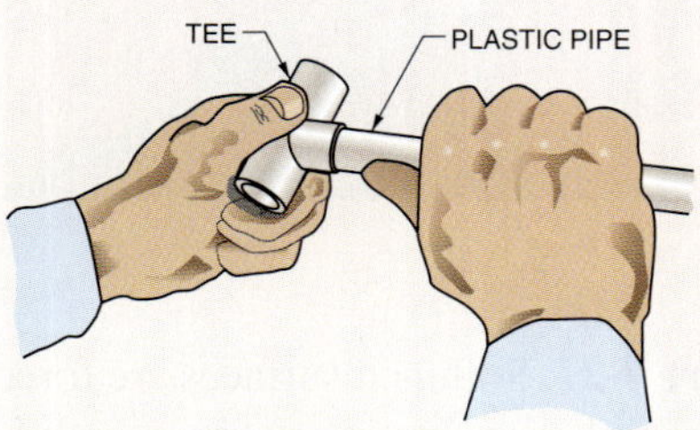

6. Select appropriate primer based on pipe material. Apply primer to fitting inside, ensuring a puddle does not remain in bottom of fitting. Apply primer to pipe outside to depth of fitting. Wait about 10 sec to 15 sec before applying solvent cement. Ensure that primer is still wet.

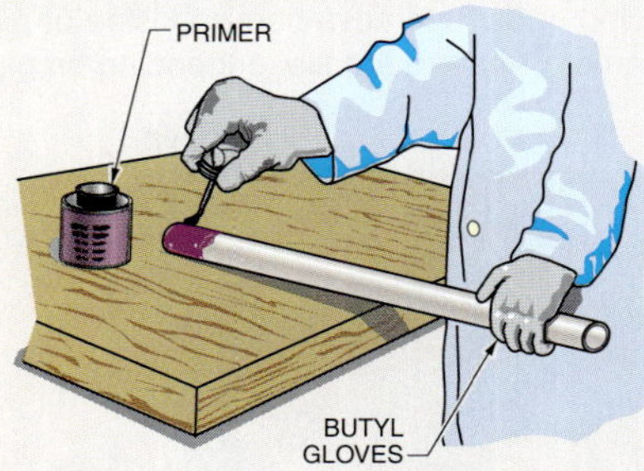

7. Apply full and even coat of solvent cement to fitting inside, ensuring a puddle does not remain in bottom of fitting. Apply solvent cement to pipe to same depth as that of primer.

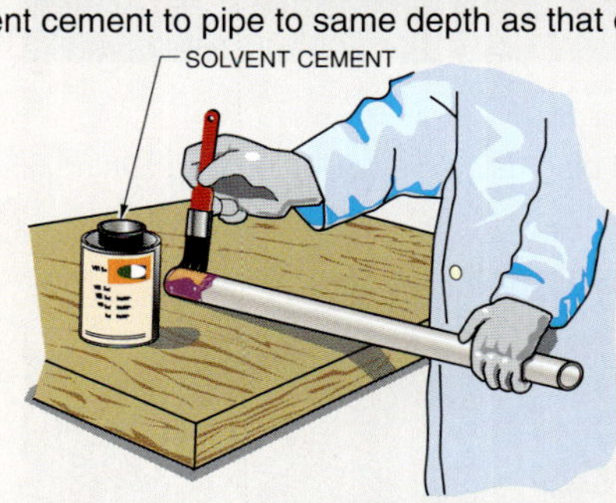

8. Fit and position pipe and fitting before solvent cement evaporates. Use enough force to ensure that pipe is fully inserted and seated in fitting. Rotate pipe ¼ turn to evenly spread solvent. Hold in position for 30 sec to allow solvent cement to start bonding two surfaces together.

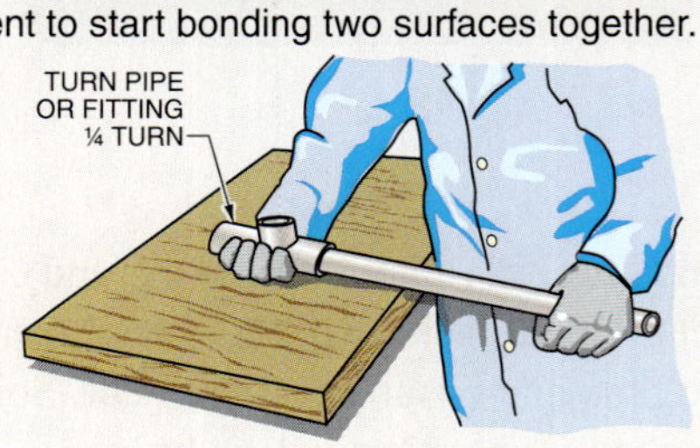

continued on next page

Cutting and Joining Plastic Pipe *(continued)*

9. Check for proper bead of solvent cement entirely around joint.

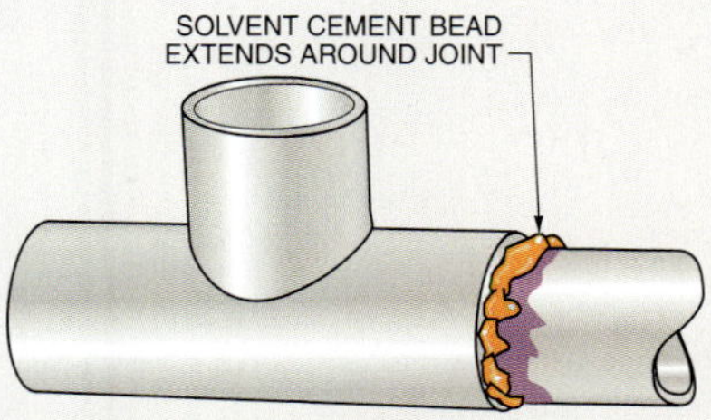

10. When bead looks complete, wipe off excess cement with a clean rag, leaving an even fillet all of the way around joint. Allow solvent to dry for a few minutes.

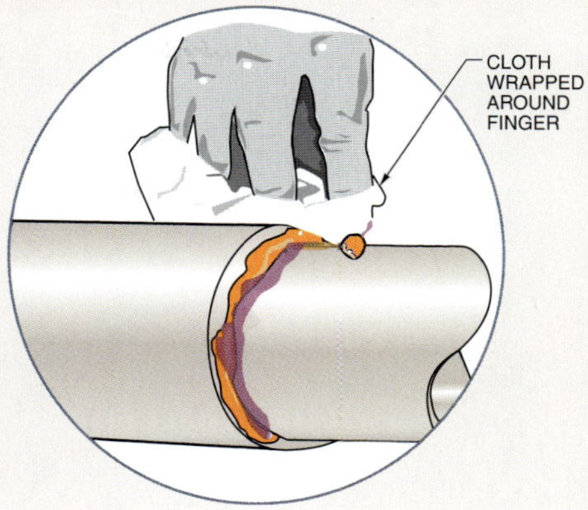

UNIT 5

Measurement Devices

Accurate measurements of HVAC and refrigeration components and functions are necessary during systems installation and troubleshooting. HVAC and refrigeration technicians must properly use the correct measuring devices. HVAC and refrigeration technicians typically measure linear dimensions, temperatures, humidity, pressures, airflow, rotation, and electrical circuits.

OBJECTIVES

- List and explain the use of linear measurement devices.
- List and explain the use of temperature, humidity, and pressure measurement devices.
- List and explain the use of airflow measurement devices.
- List and explain the use of electrical measurement devices.
- List and explain the use of air contaminant measurement devices.
- Describe the use of recovery, recycling, and reclaiming equipment.

Index to Procedures

TAKING MEASUREMENTS

HVAC and refrigeration technicians must be able to take measurements accurately. Proper handling, calibration, and storage of instruments help consistently attain the most accurate measurements. Proper care will prolong the working life of the instruments.

Accuracy

Accuracy is the degree to which a measurement conforms to a specific standard. *Precision* is the level of accuracy or mechanical exactness. Precision measurement allows technicians to perform tasks that require small dimensional tolerances.

Precision measuring instrument handling requires proper cleaning and lubrication of the instrument before and after each use. Prior to precision measuring instrument use, the measuring surfaces should be cleaned with a lint-free cloth to remove dust and oil deposits. Depending on the accuracy of a desired measurement, dirt and other foreign debris particles can cause inaccurate measurements. After instrument use, a drop of light machine oil or all-purpose oil should be added to any screws or moving parts. For best results, a clean cloth containing light oil should be used to apply a light coat of oil to the instrument's surface.

Calibration

Calibration is the verification of graduations and incremental values of a precision measuring instrument for accuracy and adjustments. Precision measuring instruments must be properly calibrated to ensure accurate readings and measurements. A *calibration standard* is a finely ground, precisely sized object that is used as a basis of dimensional comparison. Calibration standards are measured and used to adjust a measuring instrument to the proper calibration. Precision measuring instruments must be calibrated with calibration standards annually to ensure their accuracy.

LINEAR MEASUREMENT DEVICES

HVAC and refrigeration technicians must be able to take measurements through the use of rules, calipers, and micrometers. Accurate measurements are necessary for proper adjustment of HVAC and refrigeration components and systems. Proper handling, calibration, and storage of rules, calipers, and micrometers help consistently attain accurate measurements.

Rules and Scales

A *rule* is a semiprecision measuring tool used for measuring length. Typically not regarded as a precision instrument, a rule is used by experienced technicians to read measurements within 0.062″ (1⁄16″). Most rules consist of a rigid strip of steel, plastic, or wood with a straight edge that is marked in increments of inches or centimeters. Rules are typically available in lengths of 6″, 12″, 18″, and 24″. **See Figure 5-1.**

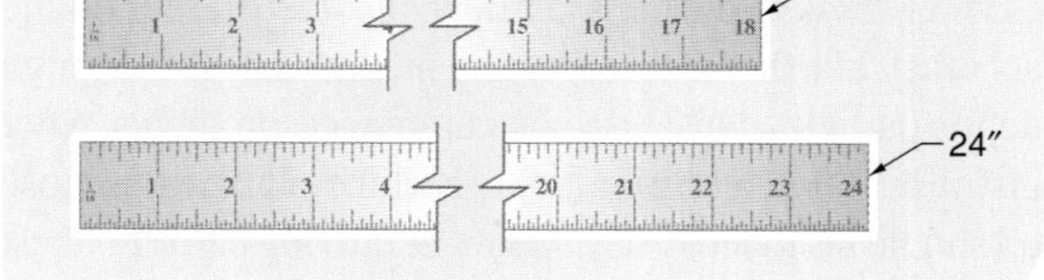

Figure 5-1. Rules used for semiprecision measuring are composed of aluminum or stainless steel, with the most common lengths being 6″, 12″, 18″, and 24″.

Rule markings consist of incremental vertical lines of various lengths. Each incremental vertical line indicates a unit of measure. The longer lengths of incremental vertical lines represent larger units of measurement. For example, on a rule with inch markings, the 1″ increment line is longer than the 1⁄2″ increment line, and the 1⁄4″ increment line is longer than the 1⁄8″ increment line. **See Figure 5-2.**

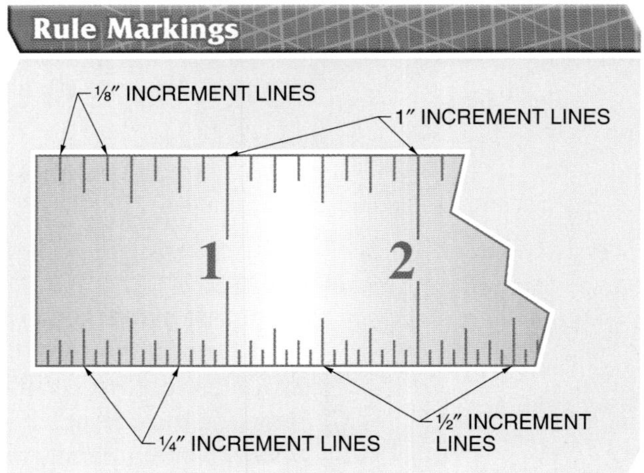

Figure 5-2. Incremental vertical lines indicate each unit of measure on a rule.

Rules used for measurements are typically manufactured to offer the user a choice of scales. For example, one edge of a rule can be calibrated in increments of 1⁄10″ (0.100″) and the opposite edge calibrated in increments of 1⁄100″ (0.010″). The other side of the rule can have an edge calibrated in increments of 1⁄32″ (0.032″) and the opposite edge calibrated in increments of 1⁄64″ (0.015″). Some rules are available with U.S. Customary units on one edge and metric units on the opposite edge. **See Figure 5-3.**

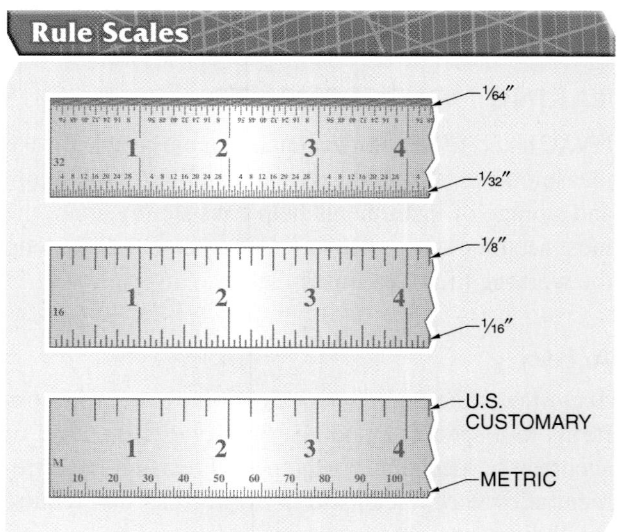

Figure 5-3. Rules used for semiprecision measurements are generally manufactured to offer the user a choice of scales.

Calipers

A *caliper* is a hand tool with one fixed and one adjustable jaw used for making precision measurements. Markings for taking precision measurements are graduated on the body of the tool. Calipers are used to make precision outside diameter (OD), inside diameter (ID), and length measurements. Some calipers are available with a depth measurement blade for taking depth measurements of hollow or tubular workpieces. Common calipers used by HVAC technicians are vernier, dial, and digital. **See Figure 5-4.**

Vernier Calipers. The ability to measure to a greater accuracy than a rule allows began with the invention of the vernier scale. A *vernier scale* is a short auxiliary scale placed along the main scale of an instrument to provide accurate fractional readings of the smallest division on the main scale. A *vernier caliper* is a caliper with a main scale and a parallel sliding vernier scale. **See Figure 5-5.** The main scale is fixed and has each unit normally divided into 10 sub units or decimal units. The vernier scale is also laid out with 10 units, except they total one-tenth shorter in length than the main scale. Each vernier space is 1⁄100 smaller than the main scale space.

U.S. Customary vernier caliper main scales have one inch increments with each inch divided into tenths (0.100″). The tenths are further divided into either 0.025″ section scales or 0.050″ section scales, which allows a caliper to dependably measure lengths as precise as 0.001″.

Calipers

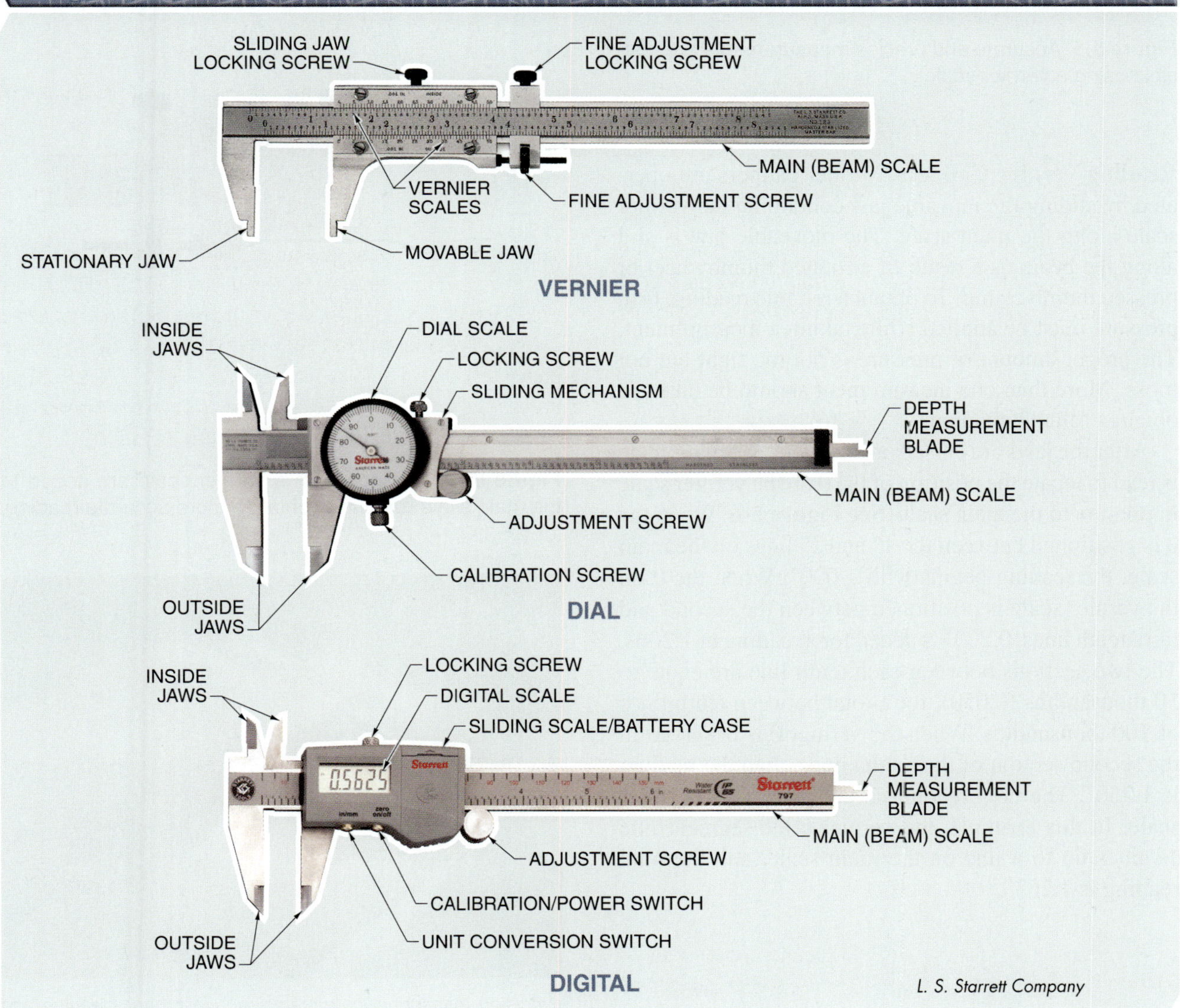

Figure 5-4. The most common types of calipers used for HVAC applications are vernier, dial and digital.

Vernier Scales

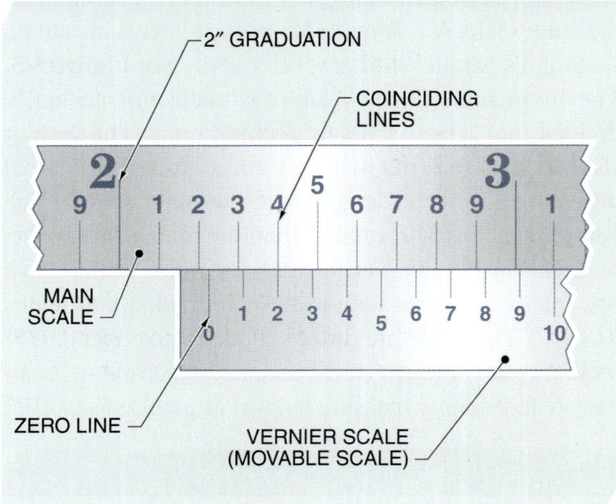

Figure 5-5. Accurate and precise measurements are possible using a vernier scale.

Reading Vernier Calipers. Vernier calipers are operated by sliding the movable jaw containing the vernier scale along the main scale. The moveable jaw is slid along the beam as a result of a rotated thumbwheel or pressed thumb clamp. To obtain a reliable reading, firm pressure must be applied while taking a measurement. The proper amount of pressure is not too tight but not loose. More than one measurement should be taken to obtain a reliable reading.

After the jaws are closed on an object, the main scale is read to locate the position of the 0 on the vernier scale in relation to the main scale. **See Figure 5-6.** When the 0 is positioned between the 1″ and 2″ lines on the main scale, the reading begins with 1.000″. When the 0 on the vernier scale is positioned between the second and third tenth lines, 0.200″ is added for a reading of 1.200″. The two sections between each tenth line are equal to 50 thousandths (0.050), for a total between tenth lines of 100 thousandths. When the vernier 0 is observed in the second section of the tenths lines, then the reading is 1.250″. The next reading is taken from the vernier scale. In this example, the vernier scale segment line 13 lines up to a line on the main scale, and the final reading is 1.263″.

Outside Calipers Media Clip

Inside Calipers Media Clip

Reading Vernier Calipers

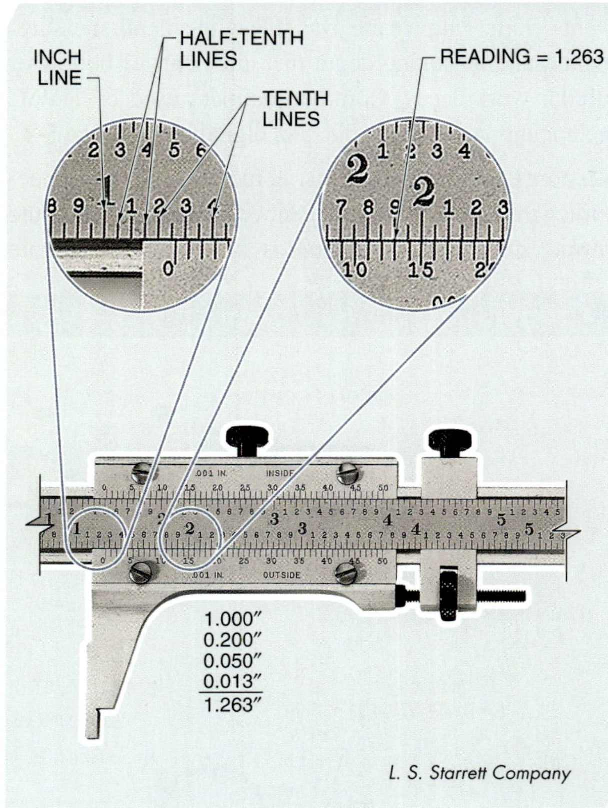

Figure 5-6. Readings from a vernier scale are added to the main scale reading to obtain a more accurate reading.

Precision measuring instruments require proper cleaning and lubrication before and after each use. Light machine oil is applied to a cloth, which is used to lightly coat the instrument.

Dial Calipers. A *dial caliper* is a caliper with a dial indicating gauge. Dial calipers are easier and quicker to read than vernier calipers and are commonly used by HVAC and refrigeration technicians because of ease of use and durability. However, dial calipers must be calibrated prior to each use. U.S. Customary dial calipers show inches and tenths (millimeters on the metric scale) on a main scale similar to a rule.

Dial calipers can take only one type of unit measurement. For example, metric dial calipers can only take measurements in metric units. A separate U.S. Customary dial caliper is used to take measurements in U.S. Customary units. The dial on a U.S. Customary dial caliper indicates measurements in thousandths of an inch (0.001″). **See Procedure for Measuring with Dial Calipers.**

Dial Calipers Media Clip

Digital Calipers. A *digital caliper* is a caliper that indicates measurements with a digital display. Digital calipers are more precise and easier to read than vernier or dial calipers, but they require a DC power source (battery) for operation. Digital calipers are used for taking many precision measurements over a short period. Digital calipers have components similar to those of a dial caliper but with a digital display instead of a dial. Most digital calipers can take measurements in either U.S. Customary or metric units. Digital calipers indicate digital readings in thousandths of an inch (0.001″) or fiftieths of a millimeter (0.02 mm). **See Procedure for Measuring with Digital Calipers.**

TECH TIP

Most digital calipers and micrometers are available with an electronic interface that transmits measurement data directly to a PC with applicable software.

Measuring with Dial Calipers

Procedure

1. Close caliper jaws and calibrate by using calibration screw, or rotate dial to set indicating needle on dial to 0.

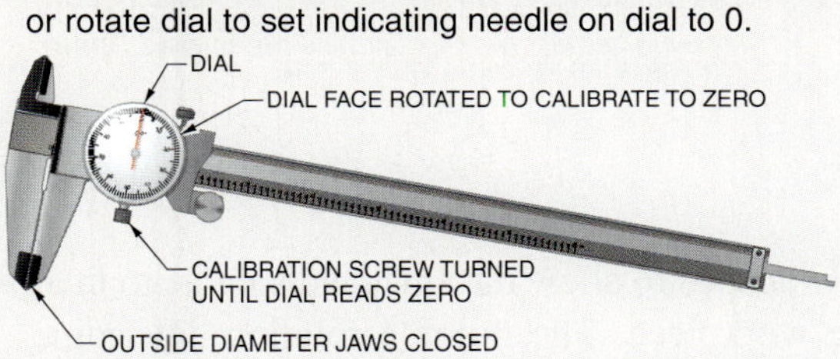

2. Close caliper jaws gently on workpiece to be measured by moving adjustment screw. Do not overtighten caliper jaws on workpiece.

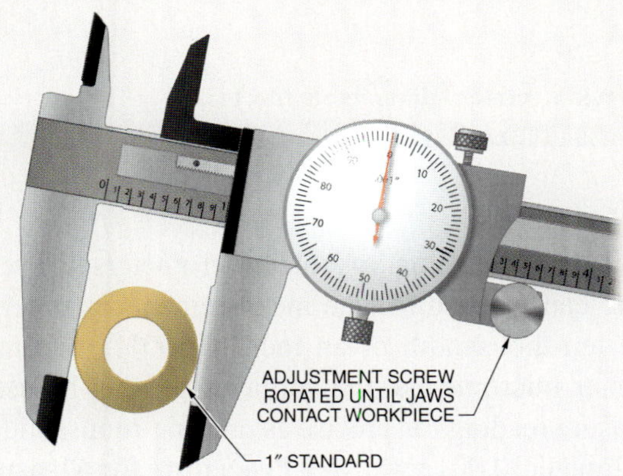

3. Read measurement by reading value on main scale and adding it to value on dial scale. When required, tighten locking screw to prevent losing measurement reading.

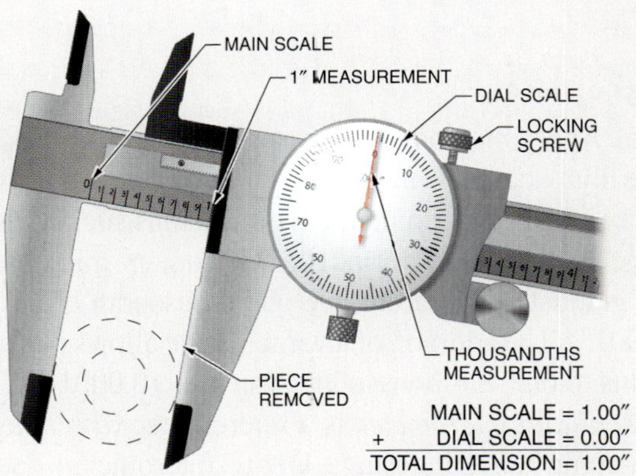

4. Record the measurement.

Procedure

Measuring with Digital Calipers

1. Turn caliper ON.

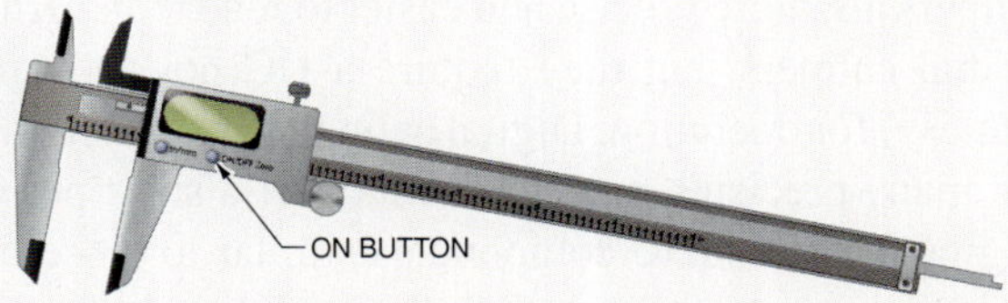

2. Select a unit of measure by depressing "in/mm" button until desired unit of measure is displayed.

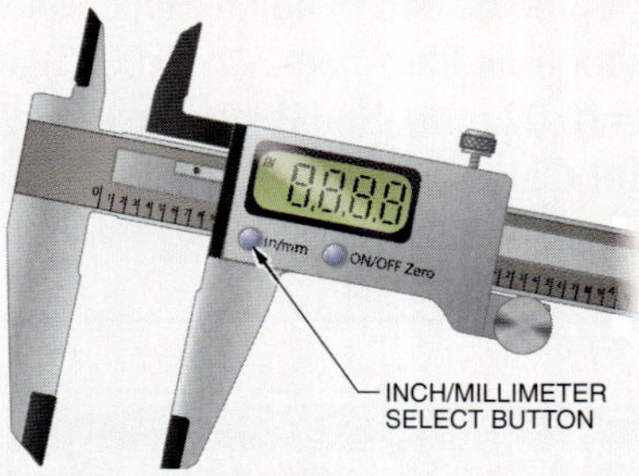

3. Close caliper jaws and calibrate by depressing "ZERO," "ON/ZERO," or "CAL" switch, depending on design of caliper. Digital reading of calibrated caliper displays 0.000.

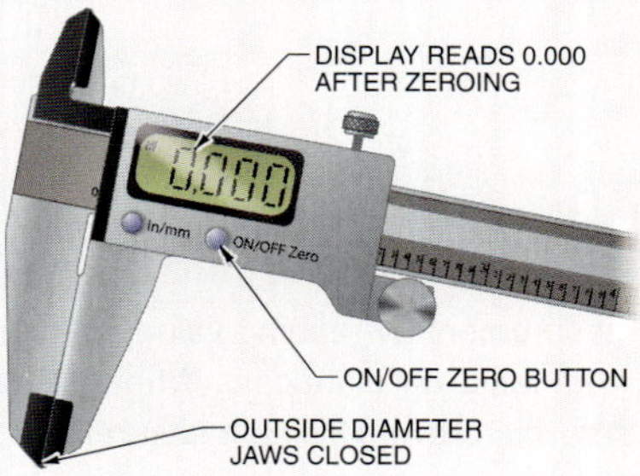

4. Close caliper jaws gently on workpiece to be measured by moving adjustment screw. Do not overtighten caliper jaws on workpiece. When required, tighten locking screw to prevent losing measurement reading.

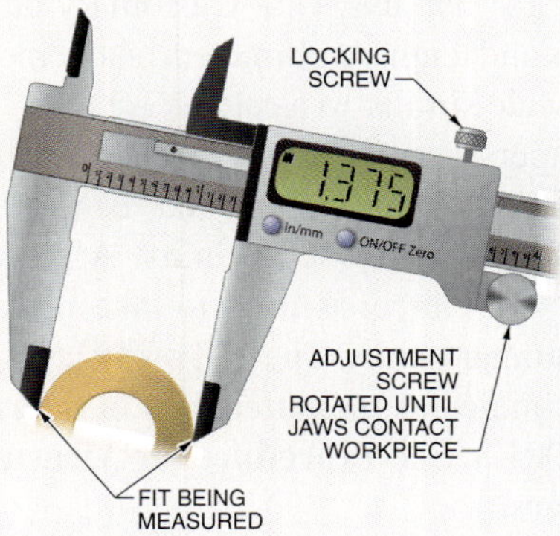

5. Read measurement on digital display.

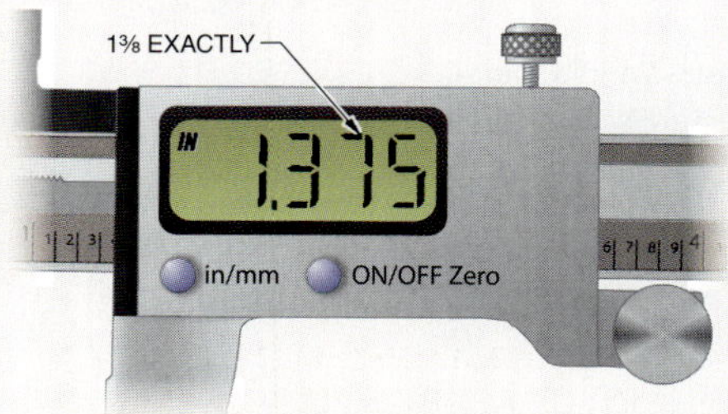

NOTE: WHEN USING CALIPERS, SEVERAL READINGS CAN BE TAKEN TO VERIFY ACCURACY OF MEASUREMENTS. WHEN THE FIRST AND SECOND READINGS ARE THE SAME, THE MEASUREMENT IS PROBABLY CORRECT. WHEN THE FIRST AND SECOND READINGS ARE NOT THE SAME, RETAKE THE READINGS UNTIL TWO OR MORE READINGS ARE THE SAME.

Micrometers

A *micrometer* is a precision measuring instrument used to take high-accuracy measurements. Micrometers are used for verification of component dimensions such as thickness, diameter, and depth. U.S. Customary micrometers are graduated to read to one one-thousandth of an inch (0.001″). The addition of a vernier scale allows measurements to one ten-thousandth of an inch (0.0001″).

A *digital micrometer* is a micrometer with a digital electronic indicating gauge. Digital micrometers operate similarly to standard micrometers but are battery powered with a digital display. The accuracy of micrometer measurements is dependent on the quality of the tool, care the tool receives, and skill of the HVAC and refrigeration technician using the tool.

Outside Micrometers. An *outside micrometer* is a micrometer used for measuring the outside diameter and thickness of parts. The frame of an outside micrometer is U-shaped to allow the anvil and spindle to fit around the workpiece. The thimble rotation determines the movement of the spindle. In addition to the barrel and thimble scales, many outside micrometers have a vernier scale located on top of the barrel. **See Figure 5-7.**

Outside Vernier Micrometers Media Clip

On U.S. Customary vernier micrometers, the vernier scale can be used to take measurements as precise as one ten-thousandth of an inch (0.0001″). On metric vernier micrometers, the vernier scale can be used to measure readings as precise as one one-thousandth of a millimeter (0.001 mm). **See Procedure for Measuring with Outside Micrometers.**

Outside Micrometers

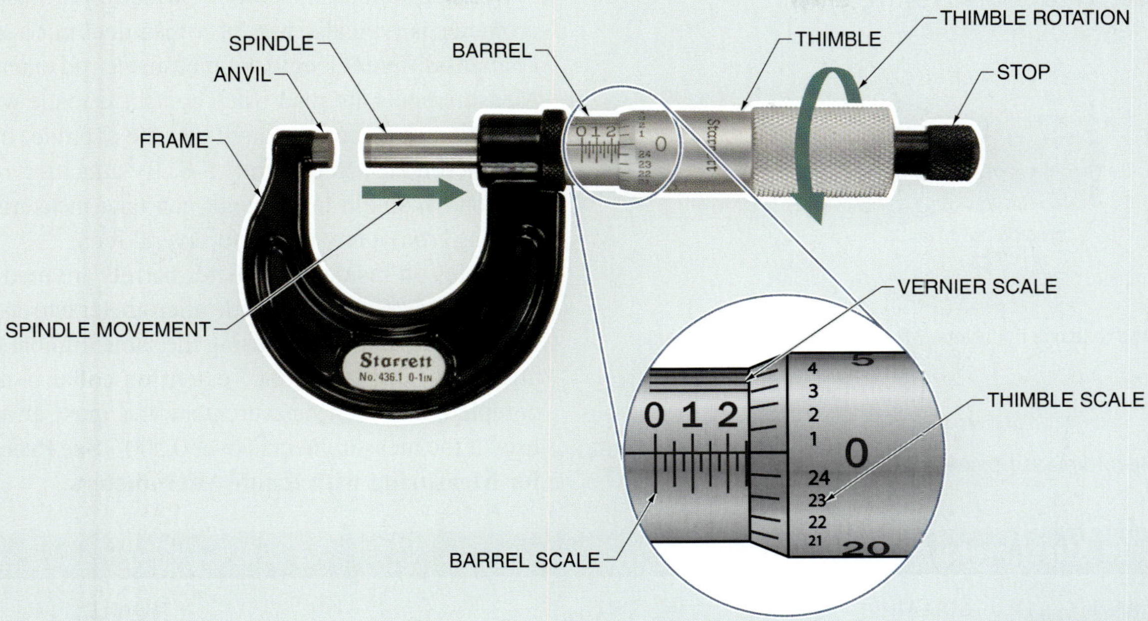

L. S. Starrett Company

Figure 5-7. An outside micrometer is used for measuring outside diameters and thicknesses of parts.

Measuring with Outside Micrometers

1. Check micrometer for proper calibration.

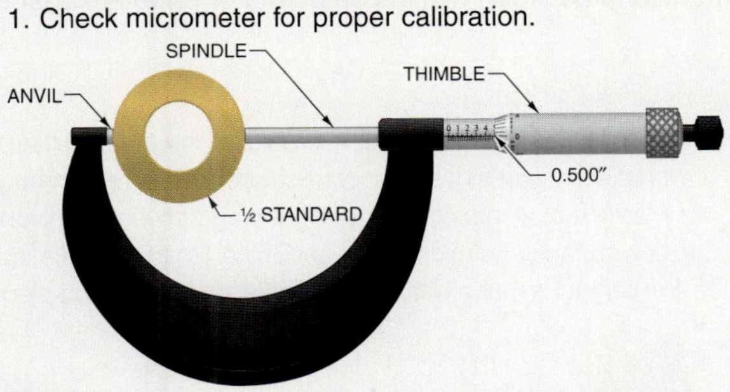

2. Close micrometer anvil and spindle gently on workpiece to be measured by moving thimble. Do not overtighten micrometer on workpiece.

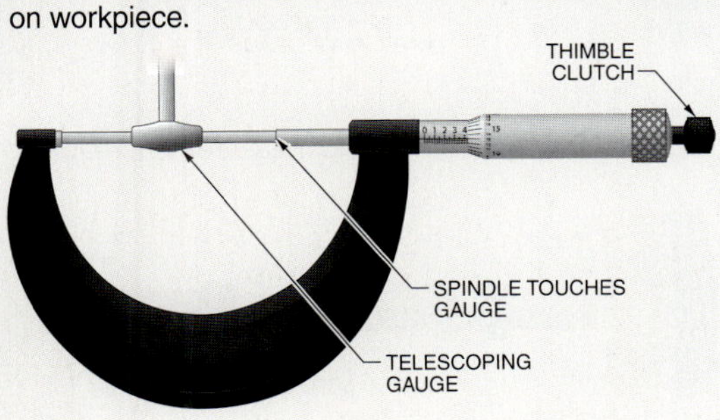

3. Read last uncovered division number on barrel closest to thimble.

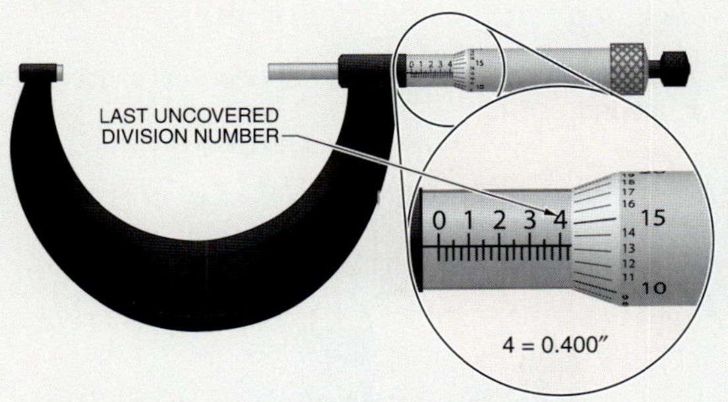

4. Read number of graduations on barrel after uncovered division number.

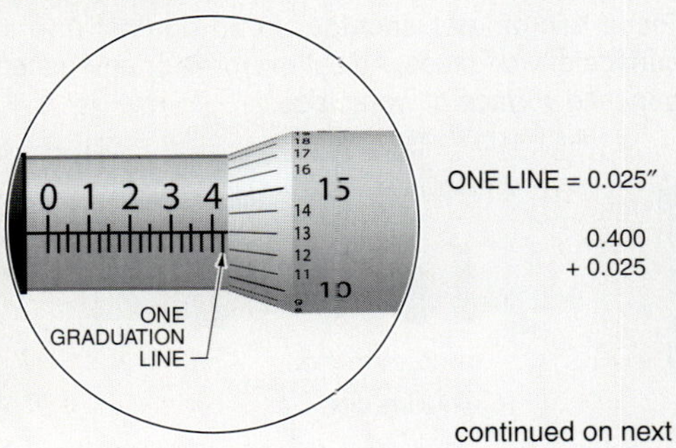

continued on next page

Procedure

Procedure

Measuring with Outside Micrometers *(continued)*

5. Read division on thimble that is most closely aligned with horizontal line on barrel.

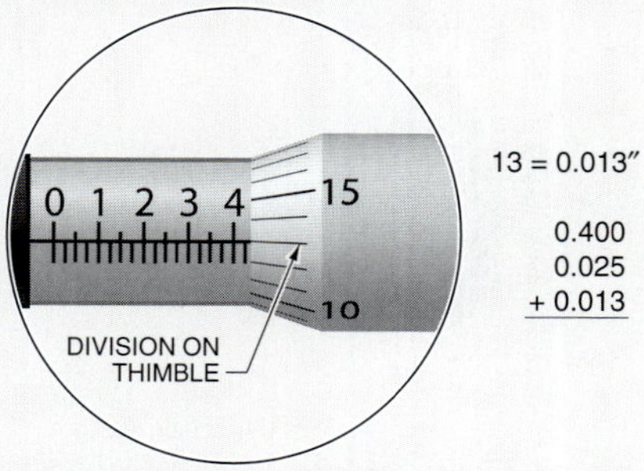

6. Add readings from steps 3 through 5 for total measurement.

0.400
0.025
+ 0.013
TOTAL MEASUREMENT = 0.438″

7. Record measurement.

Inside Micrometers. An *inside micrometer* is a micrometer used to measure linear dimensions between two inside points or parallel surfaces. An inside micrometer is typically part of a manufacturer's kit that contains different lengths of measuring rod extensions. Measurements are read when contact is made with the workpiece between the anvil and the shoulder head of the micrometer. **See Figure 5-8.** By attaching rod extensions, a single micrometer can take measurements ranging from a few inches to several feet.

Scales on inside micrometer barrels are read in the same way as scales on outside micrometer barrels (from left to right) and operate using the same thimble graduations of 0.001″. A 0.500″ extension collar is used to complete any inch measurement and must always be used if the measurement is over 0.500″. **See Procedure for Measuring with Inside Micrometers.**

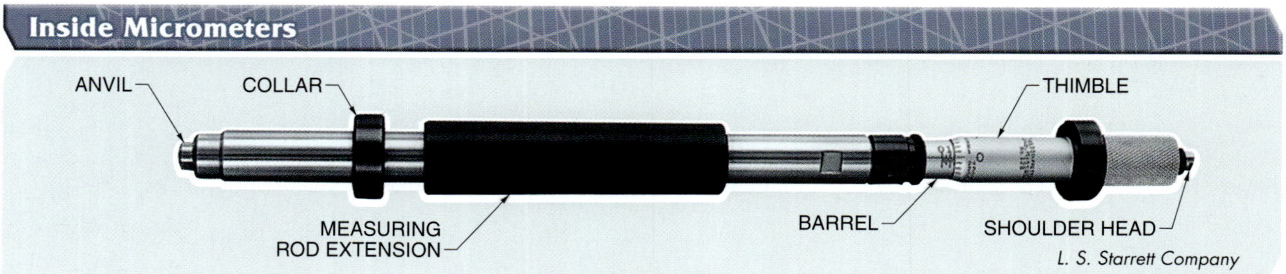

Figure 5-8. An inside micrometer is used for measuring linear dimensions between two inside points or parallel surfaces.

Procedure

Measuring with Inside Micrometers

1. Attach appropriate measuring rod extension to micrometer, if needed.

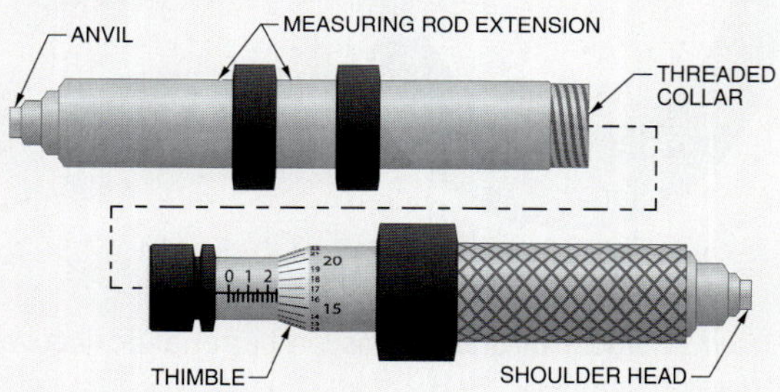

2. Place micrometer shoulder head against one internal surface of workpiece. Adjust micrometer anvil face toward opposite surface of workpiece.

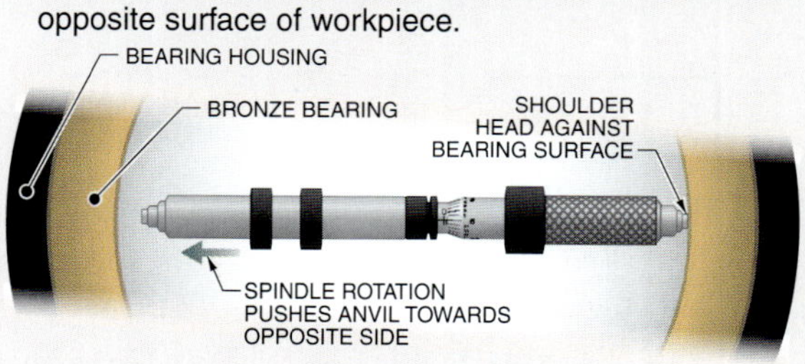

3. Use a gentle rocking movement to adjust micrometer to best fit workpiece dimensions. Fit is correct when movement through centerline of workpiece has a slight drag between the workpiece surface and micrometer anvil face. Rotate spindle until firm contact is made with anvil, shoulder head, and workpiece.

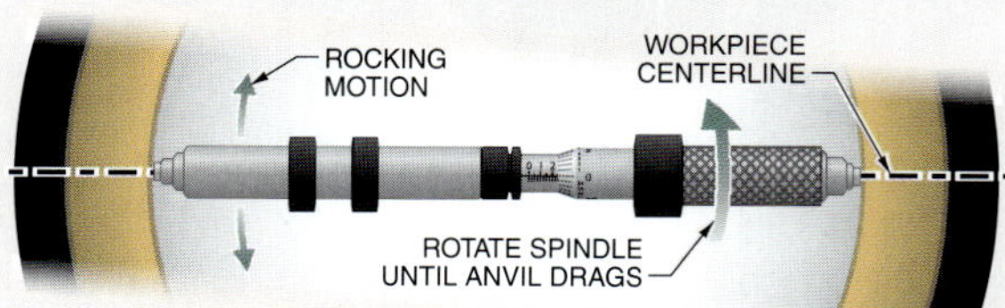

4. Tighten locking screw to hold measurement reading.

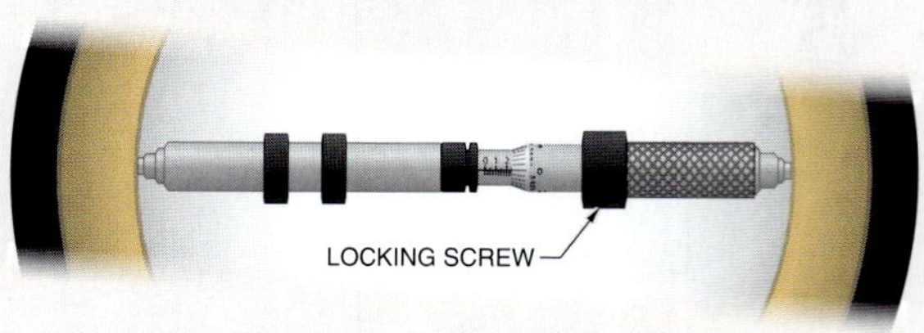

continued on next page

Measuring with Inside Micrometers *(continued)*

5. Read measurement on inside micrometer.

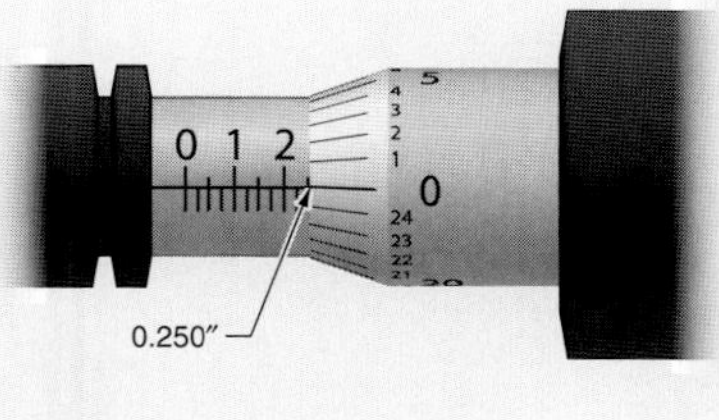

6. Add length of rod, collar, and micrometer reading for total measurement.

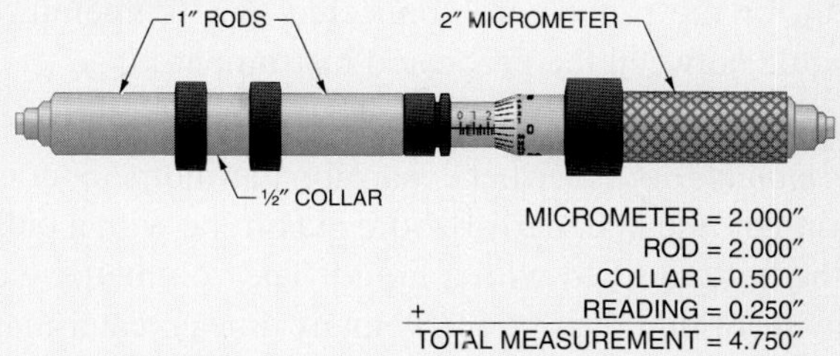

7. Verify total measurement using an outside micrometer.

Procedure

Depth Micrometer. A *depth micrometer* is a micrometer that measures component heights and slot depths. Depth micrometers are designed with standard micrometer heads (thimble and barrel) attached to a flat "tee" base. The base design allows the micrometer to be positioned on the workpiece perpendicular to the surface from which the depth is measured (reference surface). **See Figure 5-9.** Depth micrometers are used to measure distances of workpieces that have critical inside length dimensions, holes, slots, shoulders, and projections such as splines, filter housings, sleeve bearings, and bushings.

Depth Micrometers

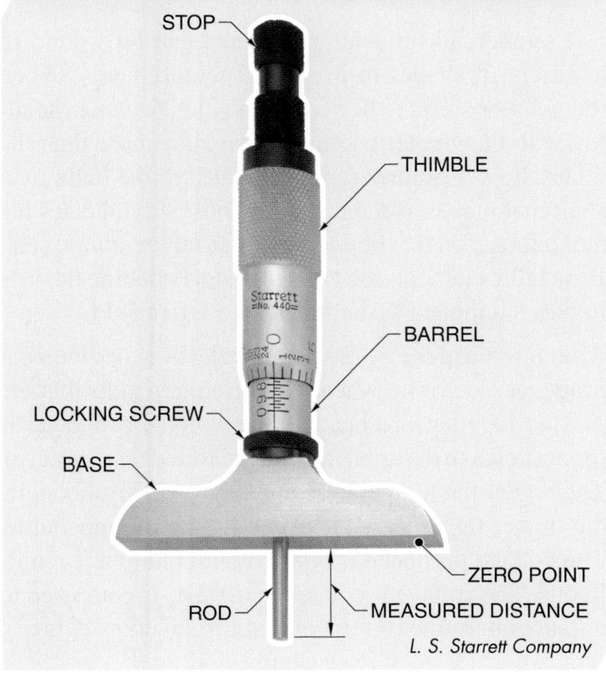

Figure 5-9. The base of a depth micrometer must be positioned parallel (square) to the reference surface of the workpiece.

Depth Micrometers Media Clip

TEMPERATURE MEASUREMENT DEVICES

Temperature is the measurement of the intensity of heat. The amount of increase in temperature depends upon the amount of heat produced, the mass of the object being heated, and the material of the heated object. The higher the amount of produced heat, the faster the temperature increases in the mass of the object being heated. An object being heated with a large mass has a slow temperature rise. The better the heated material is at conducting heat, the faster the rate of heat transfer. The rate at which a material conducts heat is dependent upon the thermal conductivity rating of the material.

Most temperature measuring instruments use a temperature probe to take the temperature of the test point. A *temperature probe* is the part of a temperature measuring instrument that measures the temperature of liquids, gases, surfaces, and pipes. The temperature probe required depends on the material being measured, the temperature measurement range, and the accuracy required. Temperatures are measured using contact temperature probes or noncontact temperature probes.

Temperature Measurements Media Clip

Contact Temperature Probes

A *contact temperature probe* is a device used for taking temperature measurements at a single point by direct contact with the area being measured. Contact temperature probes are used to measure the temperature of various solids, liquids, and gases, depending on the type of probe used. Contact temperature probes are connected directly to a temperature measuring instrument, or the probes are connected to a temperature module between the probe and instrument.

Liquid-in-Glass Thermometers. A *liquid-in-glass thermometer* is a thermal expansion thermometer consisting of a sealed, narrow-bore glass tube with a bulb at the bottom filled with a liquid. The volumetric expansion of liquids is typically many times greater than that of glass. Since the volume of the liquid changes more than the change in the glass, the liquid moves up or down in the tube with changes in temperature. **See Figure 5-10.**

Liquid-in-Glass Thermometers

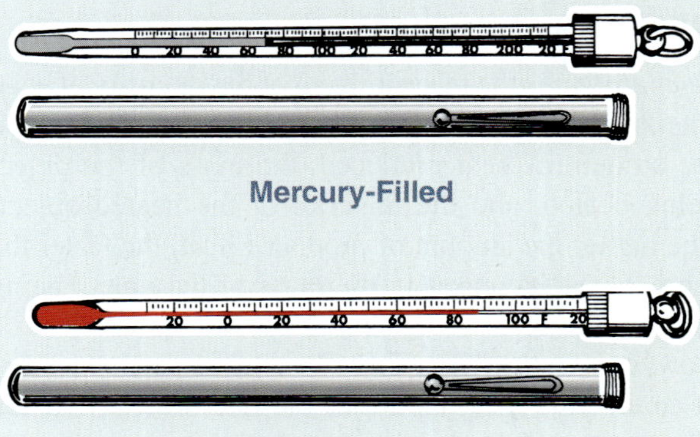

Figure 5-10. The simplest thermometer is a common liquid-in-glass expansion device.

Reference marks are made on the thermometer with the spaces between the reference marks evenly subdivided. A larger number of reference marks allows for more accurate readings. Some liquid-in-glass thermometers are calibrated to be completely immersed and others are calibrated for partial immersion. The thermometer should be immersed as recommended by the manufacturer to obtain accurate readings.

Bimetallic Thermometers. A *bimetallic thermometer* is a thermal expansion thermometer that uses a strip consisting of two alloys with different coefficients of thermal expansion that are fused together and formed into a single strip and a pointer or indicating mechanism calibrated for temperature reading. A *bimetallic element* is an element consisting of a bimetallic strip that is usually wound into a spiral, helix, or coil to allow the element to be placed in a smaller space than the straight element requires.

All thermometers need to be checked for accuracy.

TECH TIP

Temperature is the most important variable measured and controlled in an HVAC system. The two substances that are most often measured are air and water.

Commercial bimetallic elements can be wound in a variety of shapes to fit special requirements. When the helix or spiral is heated, it unwinds because the alloy with the greater coefficient expands more than the other alloy. A pointer can be attached to the helix by a shaft that moves as the helix unwinds and indicates the temperature on a calibrated circular temperature scale. Bimetallic elements are connected to switching devices to turn equipment on and off. **See Figure 5-11.**

Thermocouples. A *thermocouple* is a device that produces electricity when two different metals that are twisted together are heated. The voltage produced is proportional to the applied heat (measured temperature). The higher the temperature applied to a thermocouple, the higher the voltage produced by the thermocouple. The voltage produced is typically limited to a few millivolts. The voltage produced in mV DC is converted to a temperature measurement and displayed as degrees, typically in °F or °C. **See Figure 5-12.**

CAUTION: When taking a temperature measurement, always use a temperature probe that is rated higher than the highest possible temperature to be measured.

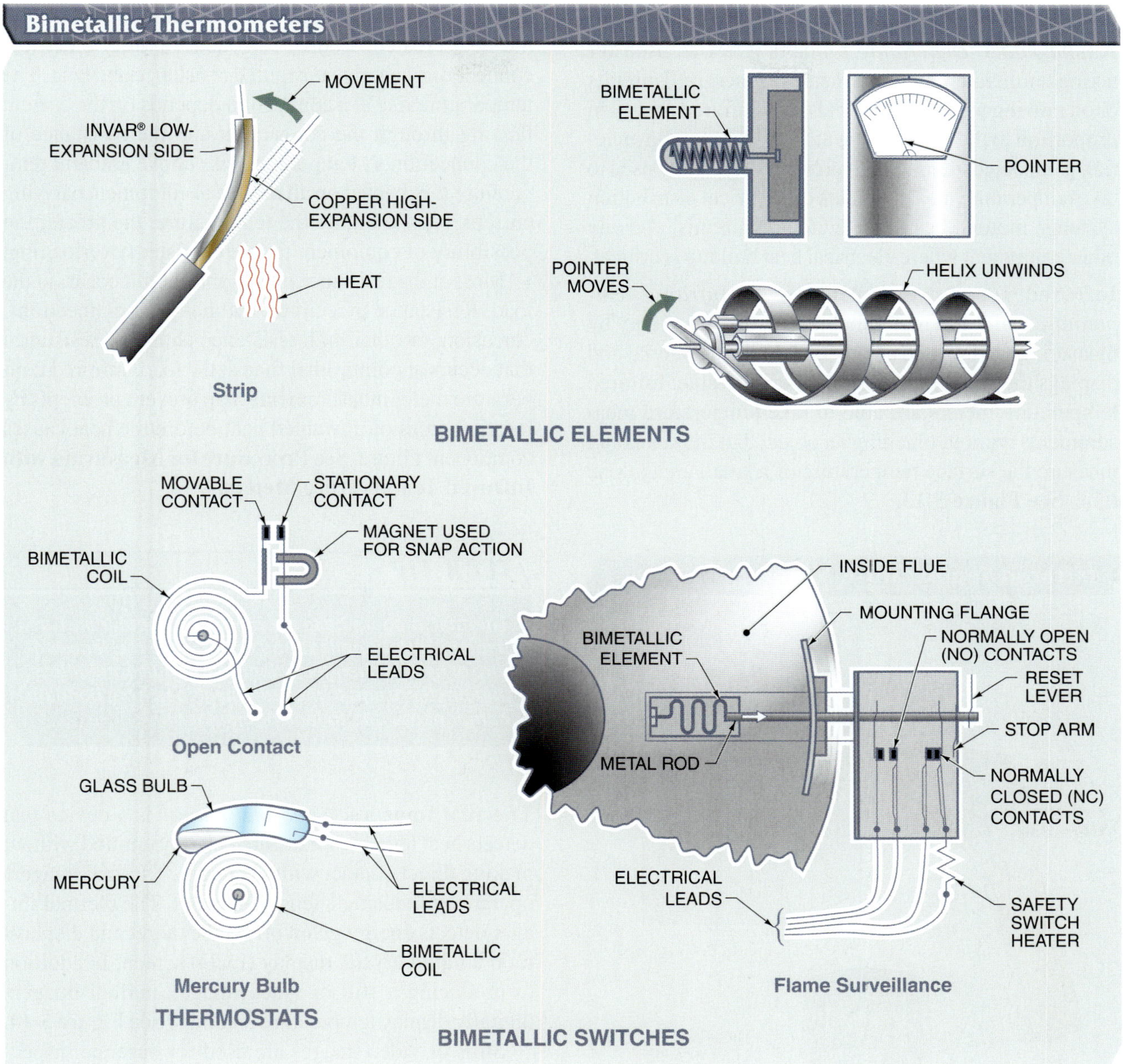

Figure 5-11. Bimetallic thermometers work through differential thermal expansion and can be used with switches to activate circuits.

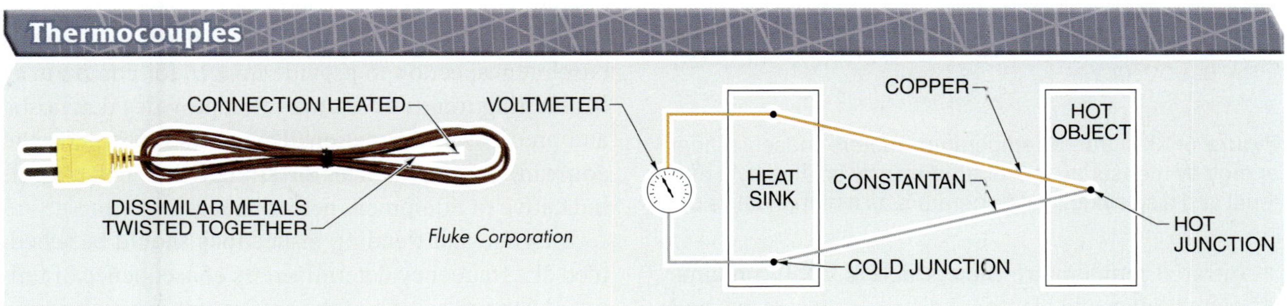

Figure 5-12. A thermocouple creates an electrical potential when the junction is at an elevated temperature, which can be converted to a temperature measurement.

Noncontact Temperature Probes

A *noncontact temperature probe* is a device used for taking temperature measurements on energized circuits or on moving parts. All materials emit infrared energy in proportion to the temperature at the surface of the material. Noncontact temperature probes are typically used to take temperature measurements of electrical distribution systems, motors, bearings, switching circuits, and any other equipment where electrical heat buildup is critical.

Infrared Temperature Meters. An *infrared temperature meter* is a meter that measures heat energy by measuring the infrared energy emitted by a material and displays the temperature as a numerical value. Infrared temperature meters are able to take temperature measurements without touching an object, but they can only measure the surface temperature of a small area at one time. **See Figure 5-13.**

Infrared Temperature Meters

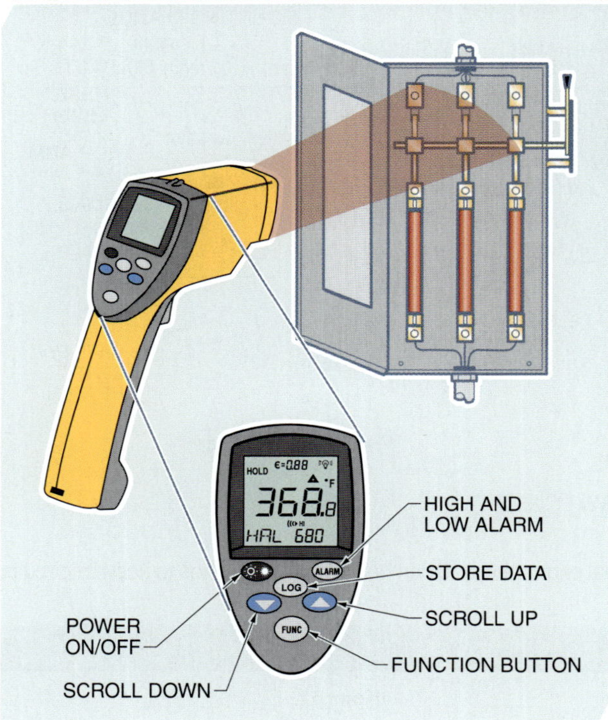

Figure 5-13. Infrared temperature meters measure heat energy by measuring the infrared energy emitted by a material and displaying the temperature as a numerical value.

Infrared temperature meters detect small amounts of heat variances in electrical distribution systems and loads. A poor electrical connection is a typical problem that causes a small to moderate heat increase. The worse the connection, the higher the resistance and the higher the heat created. Loose, corroded, or dirty electrical connections generate unwanted resistance and heat. The temperature rise at a connection depends on the current flowing through the connection and the resistance of the connection. A temperature rise above ambient temperature is expected on all electrical equipment carrying current, but the higher the temperature, the greater the possibility of equipment failure or system overloading.

Most of the resistance in a normal circuit occurs at the load. Resistance in a circuit that has poor connections, corrosion, or other high resistance paths has resistance that occurs at points other than at the load. Infrared temperature meter measurements help prevent problems by locating areas of unwanted heat before the heat causes component failure. **See Procedure for Measuring with Infrared Temperature Meters.**

TECH TIP

Thermal imagers have become very popular and are used in HVAC applications to measure the temperature of heat exchangers, electrical circuit conductors, and breakers and to measure heat loss from building components due to lack of insulation.

Thermal Imagers. A *thermal imager* is a device that detects heat levels from infrared energy emitted without making direct contact with the target. Thermal imagers operate by focusing a lens on a target. The thermal imager detects energy given off by the target and displays it on a liquid crystal display (LCD) screen. In addition to producing a still or video image, thermal imagers digitally display temperature readings. **See Figure 5-14.**

Stills or video images are used for baseline inspections and trending inspections. A baseline inspection is used to establish a reference point using equipment operating under normal conditions and in good working order. A trending inspection is performed after a baseline inspection to provide images for comparison. Monitoring trends over time often provides diagnostic and predictive information. This allows technicians to compare any differences or similarities that may be indicative of equipment performance.

Baseline and trending inspections should be scheduled at a frequency determined by consequence of failure and the condition of the equipment. By monitoring trends, preventive maintenance can be performed to reduce unplanned downtime and costly failures.

Procedure

Measuring with Infrared Temperature Meters

1. Turn meter ON.

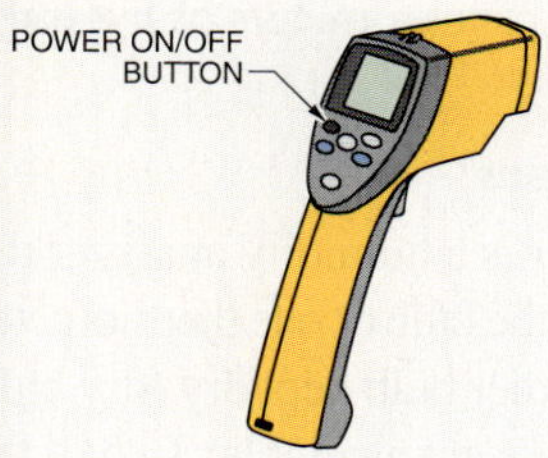

2. Take an ambient temperature reading for reference.

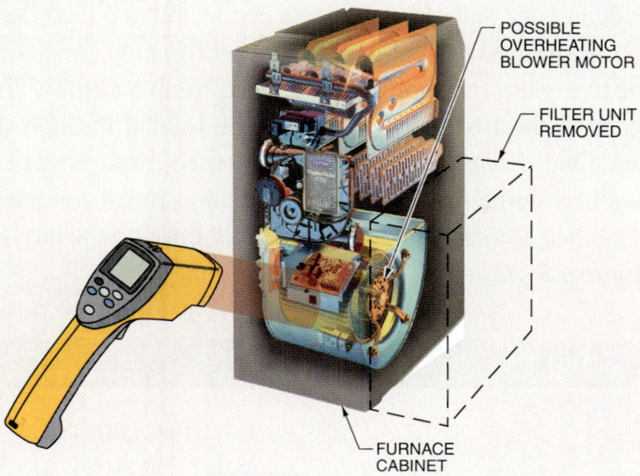

3. Aim infrared temperature meter at area being measured. Meter is focused based on distance between object and meter.

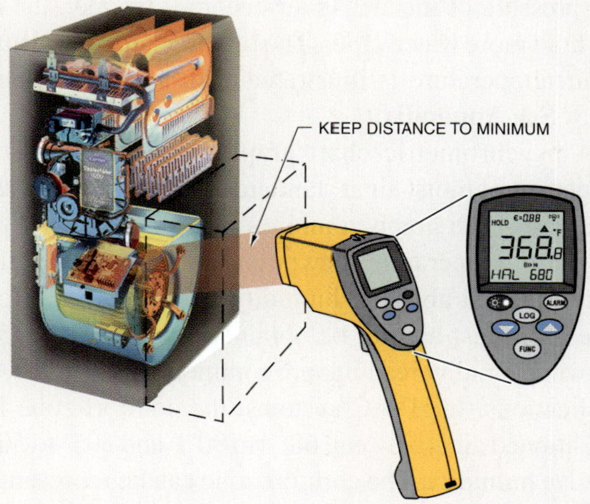

4. Take temperature readings of any areas suspected of having temperatures above ambient temperature.
5. Turn meter OFF.

NOTE: TO DETERMINE TEMPERATURE READINGS IN SUCH AREAS, SUBTRACT THE AMBIENT TEMPERATURE FROM THE READING OBTAINED ON THE INFRARED TEMPERATURE METER. THE DIFFERENCE IN TEMPERATURE IS THE AMOUNT OF HEAT PRODUCED BY THE ELECTRICAL COMPONENT, CIRCUIT, OR EQUIPMENT BEING TESTED.

Thermal Imagers

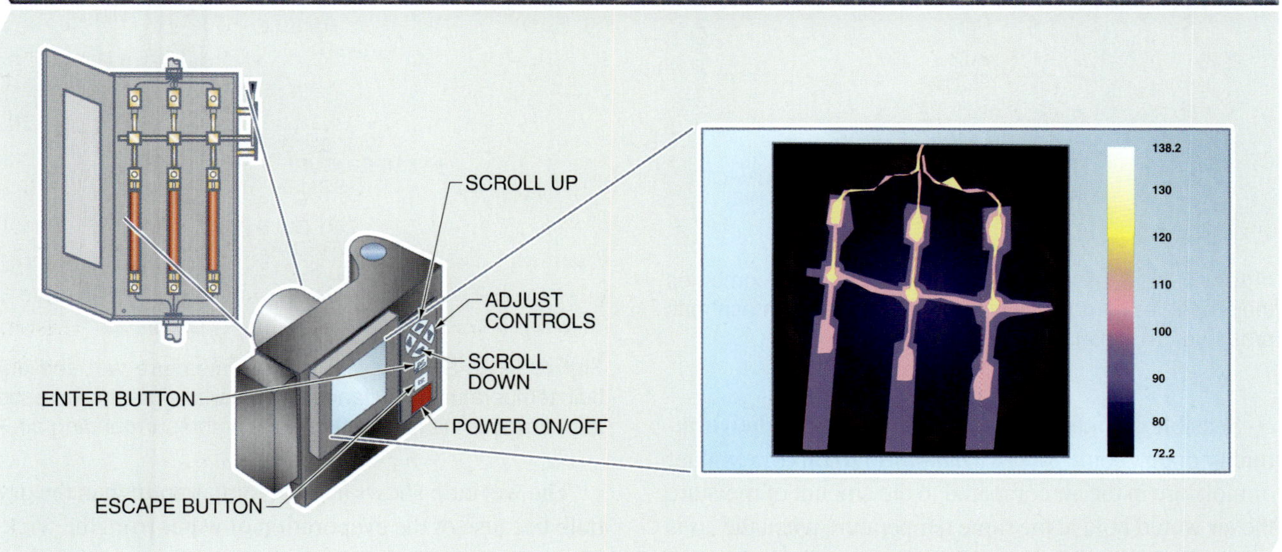

Figure 5-14. Thermal imagers detect heat energy by measuring the infrared energy emitted by materials in the image area and displaying the temperature as an image.

HUMIDITY MEASUREMENT DEVICES

Humidity is the amount of moisture (water vapor) in the air as a gas. Humidity varies according to the temperature and pressure of the air. As air becomes warmer, the air can hold more water vapor. The relationship of moisture to air temperature is illustrated with a psychrometric chart. **See Appendix.**

A psychrometric chart graphically combines the properties of moist air at standard atmospheric pressure (14.7 psi). The graph combines dry bulb temperature, wet bulb temperature, dewpoint temperature, relative humidity, and absolute humidity so that all the properties of moist air can be obtained when any two are known. Humidity readings are commonly used in HVAC applications. In HVAC systems, the comfort zone for conditioned air is between 68°F to 80°F and 30% to 50% relative humidity. The comfort zone can be represented on a psychrometric chart. **See Figure 5-15.**

Psychrometric Charts

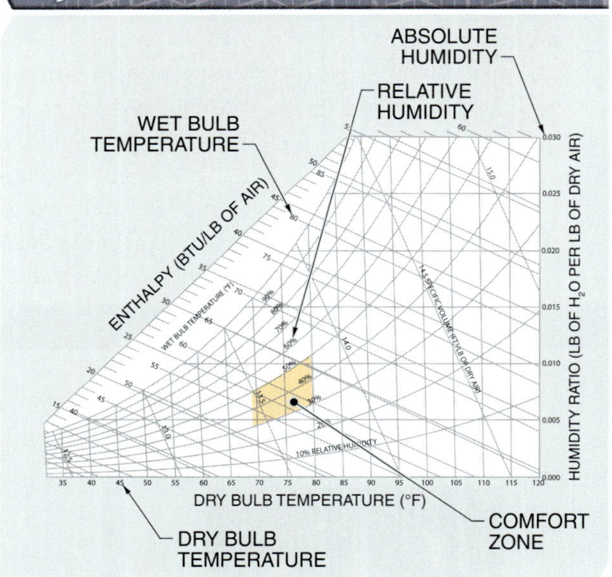

Figure 5-15. A psychrometric chart graphically combines the properties of moisture in the air at standard atmospheric pressure (14.7 psi).

Sensible humidity is typically expressed as relative humidity or dewpoint. *Relative humidity (RH)* is the amount of moisture in the air compared to the amount of moisture the air would hold at the same temperature when the air is saturated, or full of moisture. Relative humidity is the most common method of expressing the amount of water vapor contained in air. For example, when air contains only half of the water vapor the air can hold, the relative humidity is 50%. *Saturated air* is a mixture of water vapor and air where the relative humidity is 100%. Humidity is typically measured with psychrometers or hygrometers.

Psychrometers

A *psychrometer* is a humidity analyzer that uses two thermometers with the bulb of one thermometer kept moist (wet bulb) and the other bulb kept dry (dry bulb). Several types of psychrometers are available. Two of the most common are the recording psychrometer and the sling psychrometer. A *recording psychrometer* is a psychrometer that has the wet and dry bulb thermometers connected to a recorder.

A *sling psychrometer* is a psychrometer consisting of two glass thermometers attached to an assembly that permits the two thermometers to be rotated through the air. One thermometer has a moistened cloth cover (or wick) over the sensing bulb to maintain a moist condition. The thermometers are rotated rapidly through the air. **See Figure 5-16.**

Sling Psychrometers

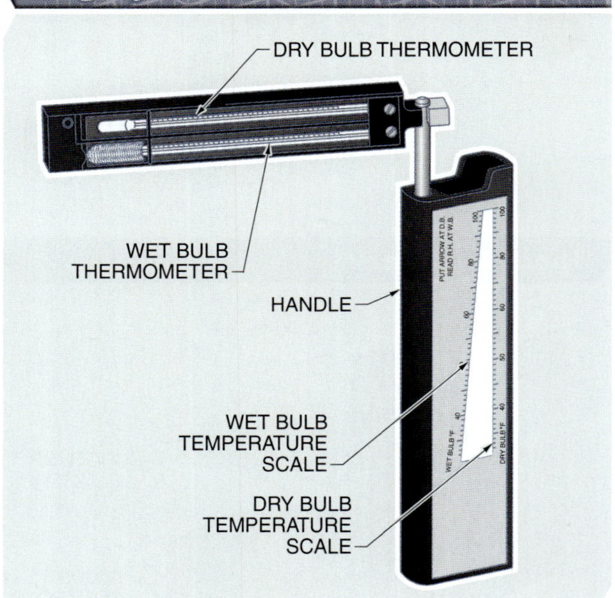

Figure 5-16. Sling psychrometers measure wet and dry bulb temperature simultaneously. Charts or graphs are used to determine the relative humidity of the surrounding air.

The wet bulb shows a lower temperature than the dry bulb because of the evaporation of water from the wick. These temperature readings, along with a psychrometric chart, are used to determine the relative or absolute humidity of the surrounding air. The hygrometer and thermohygrometer family of instruments is replacing the sling psychrometer because they provide direct readouts in relative humidity values.

Hygrometers

A *hygrometer* is a humidity analyzer that measures the physical or electrical changes that occur in various materials as they absorb or release moisture. Some hygrometers employ human hair, animal membrane, or other materials that lengthen or stretch when they absorb water.

A *hygroscopic element* is a material that changes its physical size or electrical characteristics as the humidity changes. A hygroscopic element undergoes changes in physical properties as the amount of moisture changes. In some applications, the element changes physical dimensions and a strain gauge is used to measure the change in tension of the fiber of the material. **See Figure 5-17.**

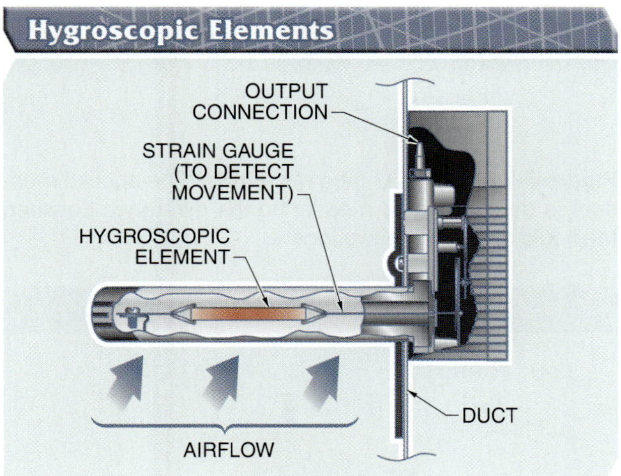

Figure 5-17. A hygroscopic element changes length with changes in humidity. The change in length is measured with a strain gauge and converted to a humidity measurement.

Another type of electrical hygrometer uses a capacitance probe as its primary element. A *capacitance probe* is an anodized aluminum strip with a porous aluminum oxide layer used as a dielectric. A thin film of gold is deposited on the oxide layer. The gold and the aluminum serve as the plates and the oxide layer as the dielectric of an electrical capacitor. An alternative design uses a polymer layer instead of the aluminum oxide layer as the dielectric. The output of the cell is a voltage that varies with the relative humidity. **See Figure 5-18.**

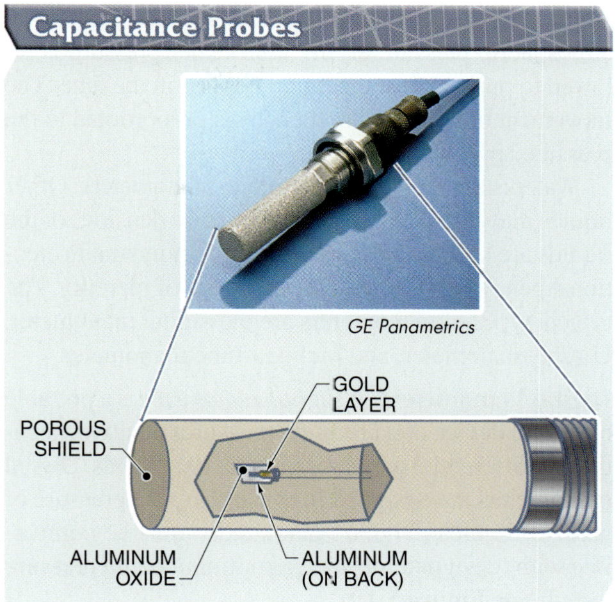

Figure 5-18. Capacitance hygrometers measure the change in capacitance as the sensor absorbs moisture. The change in capacitance is converted to a humidity measurement.

When capacitance probes are installed in ducts containing water vapor, the capacitor's dielectric constant changes with humidity levels. This alters the electrical capacitance of the probe. The instrument can be calibrated for digital readouts in relative humidity units. This type of hygrometer is typically used because of its small size, rapid response, and wide range of humidity and dewpoint measurement capabilities.

PRESSURE MEASUREMENT DEVICES

Pressure is force, in pounds, exerted over a surface divided by the area of the surface. Exerted force always produces a deflection or change in the volume or dimension on the area to which it is applied. Pressure readings are taken by technicians and engineers to ensure safe and proper operation of equipment and systems. Pressure is expressed in pounds per square inch (psi). Low pressures are typically expressed in inches of water column (in. WC). One psi equals 27.68″ WC. Low pressures are typically measured using manometers. High pressures are typically measured using mechanical pressure gauges.

Manometers

A *manometer* is a device used for taking precise pressure measurements using a portable, electrically powered device or using a liquid-filled tube. A manometer is the simplest device for measuring pressure. In a manometer, atmospheric pressure or any fluid under pressure is allowed to push against the liquid trapped in the tube. The movement of the liquid in the tube is proportional to the pressure applied.

Water is the typical liquid used in manometers. Other liquids may be used as long as the precise densities of the liquids are known. However, the Environmental Protection Agency (EPA) discourages the use of mercury. The typical types of manometers are the digital manometer, U-tube manometer, and inclined-tube manometer.

Digital Manometers. A *digital manometer* is a portable electrical device that precisely measures positive, negative, or differential pressures in small increments. Digital manometers are designed for measuring the pressure of gases or liquids. Digital manometers must be compatible with (designed for) the gas or liquid being pressure tested. **See Figure 5-19.**

U-Tube Manometers. A *U-tube manometer* is a U-shaped section of glass or plastic tubing that is marked in increments and partially filled with water or another liquid used to measure pressure. The liquid in the two legs of a U-tube manometer remains level due to gravity when the pressures on the two legs are equal. When more pressure is exerted on the air in one leg, the liquid in that leg will be pushed down and the liquid in the other leg rises. The difference in the liquid levels is a measurement of the pressure. **See Figure 5-20.**

U-Tube Manometers

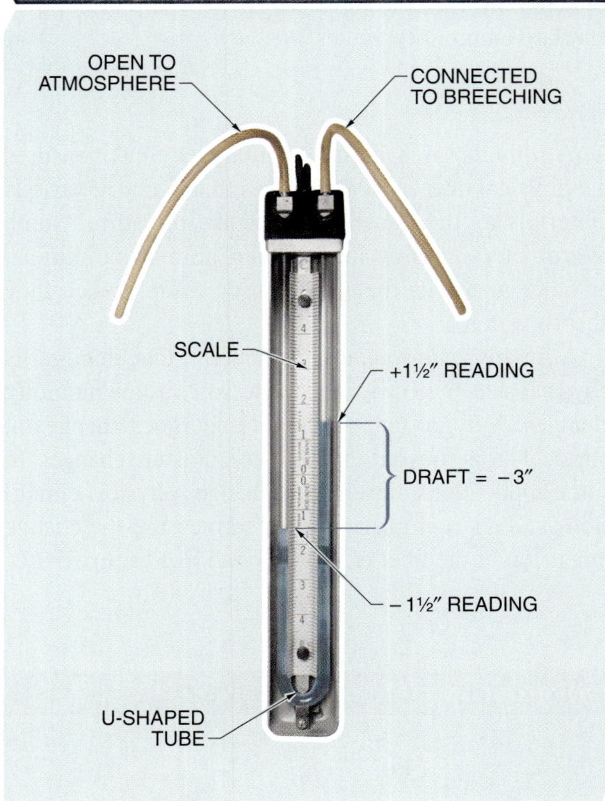

Figure 5-20. Using a U-tube manometer, the applied pressure is determined by measuring the difference between the liquid levels in the two legs.

Digital Manometers

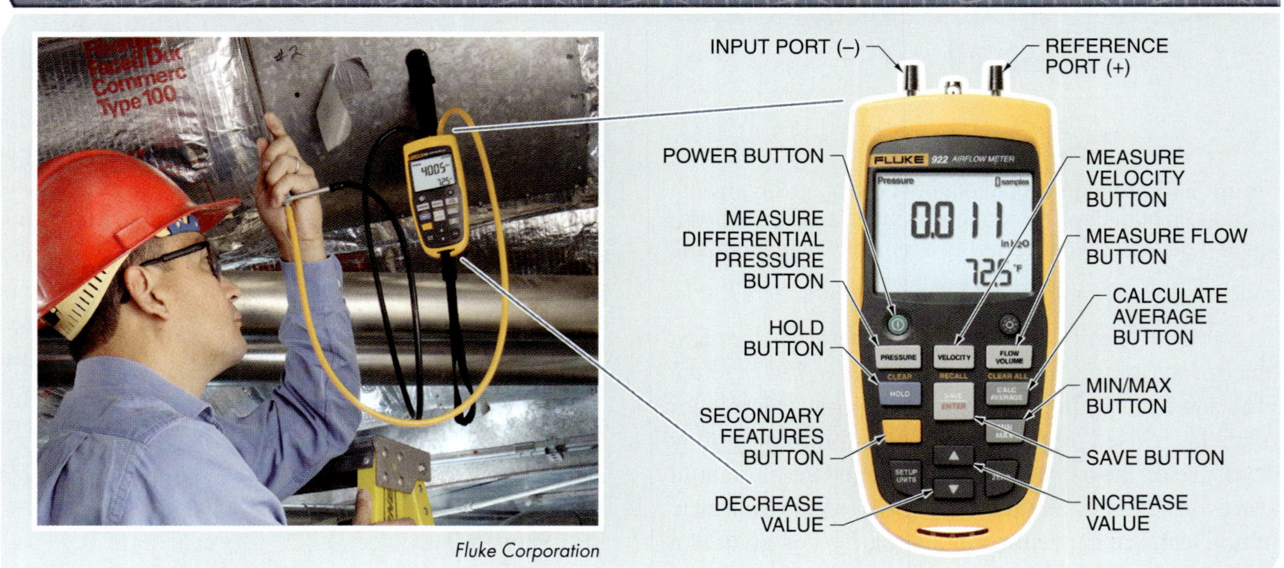

Figure 5-19. Digital manometers are capable of positive, negative, and differential pressure measurements and storing the measurements for future reference.

Inclined-Tube Manometers. An *inclined-tube manometer* is a very precise manometer designed so a long inclined section of glass or plastic tubing has more graduations per inch than a U-tube manometer, allowing for greater accuracy. The liquid in an inclined manometer fills part of the inclined tube and part of the vertical leg. A change in air pressure causes the liquid to move along the inclined tube. Small changes in pressure move the liquid a long distance along the incline. **See Figure 5-21.**

Inclined-Tube Manometers

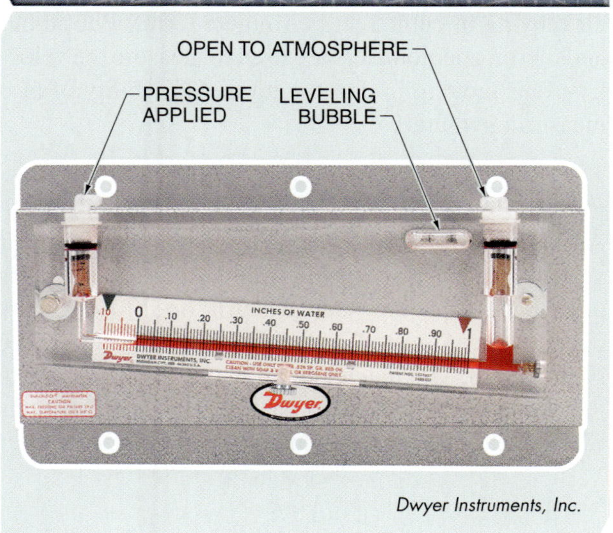

Dwyer Instruments, Inc.

Figure 5-21. An inclined-tube manometer has one column at an angle to lengthen the scale, improving accuracy and readability.

Mechanical Pressure Gauges

Typical mechanical pressure gauges use an elastic deformation element. An *elastic deformation element* is a device consisting of flexible components such as pressure springs or pistons that flex, expand, or contract in proportion to the pressure applied within or against them. Most gauge elements are attached to mechanical linkages. The movement of the pressure element is transferred through a mechanical linkage that moves a pointer on the face of the gauge. **See Figure 5-22.**

Bourdon Tube Pressure Gauges. A *Bourdon tube pressure gauge* is a mechanical pressure gauge with a curved tube made of thin metal that acts as a pressure-sensing element. To construct a Bourdon tube, a tube made from flattened, seamless metal tubing is formed into the desired shape. Beryllium copper, steel, and stainless steel are commonly used materials. The tube is welded, brazed, threaded, or flanged to metal tubing that admits the pressurized fluid from the measured process. Bourdon tubes are available to measure pressures up to 20,000 psi and are accurate to within 5% of full face value. These gauges can be accurate to within 2½% of full face value when filled with glycerin. **See Figure 5-23.**

Compound Pressure Gauges

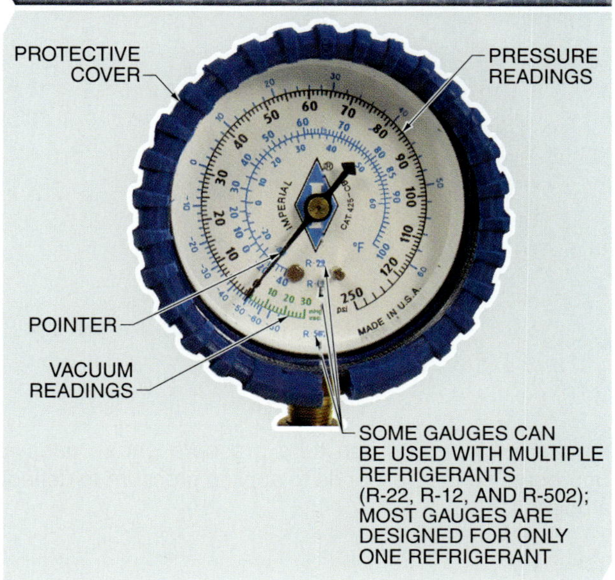

Figure 5-22. A compound refrigeration gauge reads positive pressures in psi and negative pressures in inches of vacuum. Most refrigeration compound gauges will include pressure/temperature readings for the gauge rated refrigerants.

Schrader Pressure Gauges. A *Schrader (spring-loaded) pressure gauge* is a mechanical pressure gauge that has a small, single-acting spring return cylinder that acts as a pressure-sensing element. The amount of movement of the piston in the cylinder is dependent on the diameter of the piston and the strength of the biasing spring. Schrader pressure gauges are only accurate to within 15% of full face value. Schrader pressure gauges are, however, more tolerant of sudden pressure variations or high pressure surges.

Magnehelic® Pressure Gauges. A *Magnehelic® pressure gauge* is a mechanical pressure gauge consisting of two pressure-tight compartments, a flexible diaphragm, a biasing spring, and a magnetic pointer. Most of these gauges are used in applications where maximum pressure is 15 psi or 30 psi. Magnehelic® pressure gauges have a rubber plug installed, which acts as a relief valve when needed. In most applications, these gauges are used as differential pressure gauges (two pressurized compartments). Magnehelic® pressure gauges are accurate to within 2% of full face value. **See Figure 5-24.**

Bourdon Tube Gauges

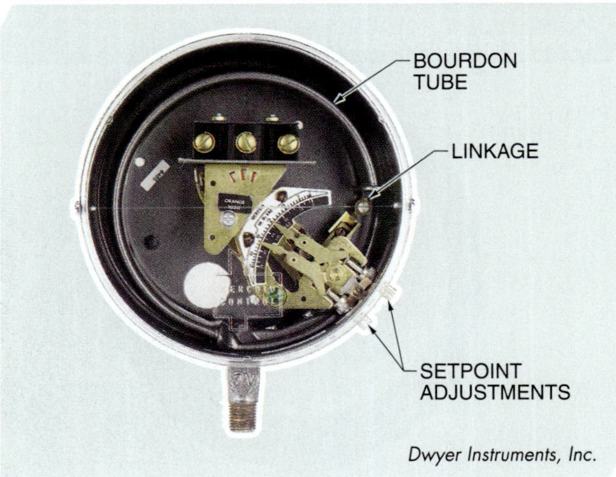

Figure 5-23. A Bourdon tube pressure gauge uses a Bourdon tube that responds to applied pressure to deflect a pointer.

Magnehelic® Pressure Gauges

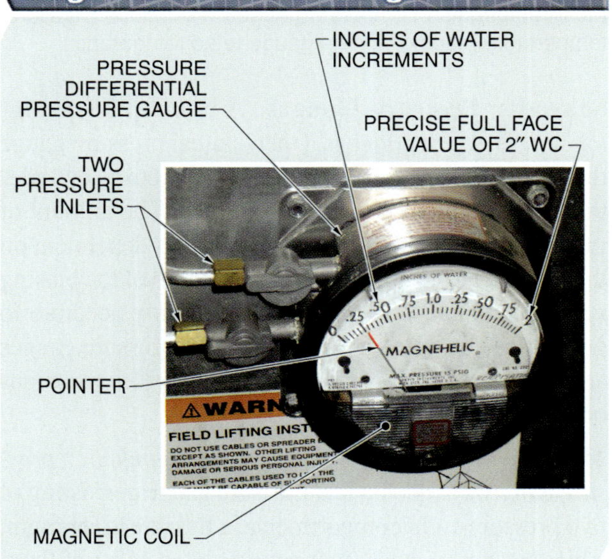

Figure 5-24. A Magnehelic® pressure gauge has a small full face value (inches of water column) but is the most accurate of all pressure gauges.

AIRFLOW MEASUREMENT DEVICES

Velocity is the speed at which gases, liquids, or solids travel through a system. The movement of air over heating elements and coils and cooling coils provides heating and cooling by HVAC systems. When air moves at an excessive velocity, air does not gain or lose the appropriate amount of heat from heat exchangers in the system. Air velocity instruments are used to measure the velocity of air within ductwork.

Airflow measurements are taken to confirm that the volume and/or velocity of airflow meets the design specifications. Anemometers measure the velocity of air, and flow hoods are used to measure the volume of air moving in cubic feet per minute (cfm). Pitot tubes and thermoanemometers are used to measure the velocity of air moving in ducts, with the possibility of also measuring volume.

Anemometers

An *anemometer* is a device that measures average air velocity or average airflow rate. An anemometer is placed at an opening, such as a grill or diffuser, and is operated according to the manufacturer instructions. **See Figure 5-25.** The manufacturer specifications indicate the recommended number of readings and locations where measurements should be taken.

Anemometers

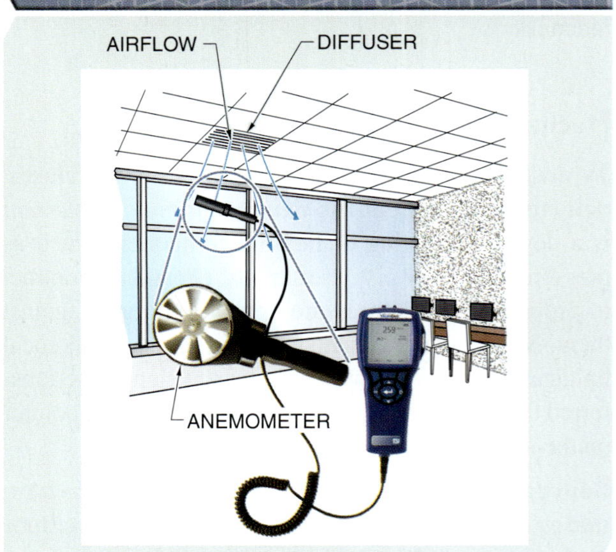

Figure 5-25. An anemometer is used to measure the airflow rate in cfm from diffusers and into grills.

Velocity measurements are averaged and then multiplied by the K factor for the specific type of diffuser in use. The *K factor* is a number that represents the actual open area and design characteristics of a specific grill or diffuser. The result of the K factor equation is the airflow rate in cfm out of the diffuser.

Flow Hoods

A *flow hood* is a device that is placed over a grill or diffuser to measure the rate of airflow. **See Figure 5-26.** Flow hoods are easier to use than anemometers because flow hoods provide a direct reading without the need for calculations. Flow hoods also save a considerable amount of time when performing multiple field measurements.

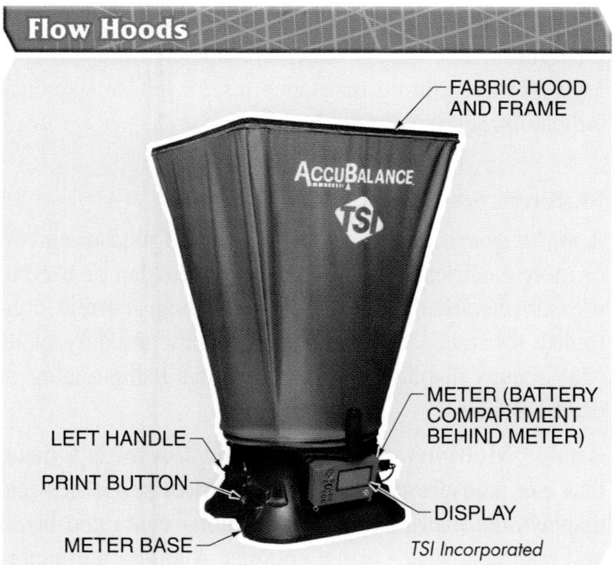

Figure 5-26. A flow hood directly measures airflow in cfm into grills and from diffusers.

Pitot Tubes

Duct pressures are measured to determine the airflow inside an air-handling system. The total pressure and static pressure in a duct are measured with an insertion pitot tube. A pitot tube is used to measure velocity pressures in a duct. A pitot tube is inserted into the direction of the flow and measures both static and total pressures.

When two pitot tubes are being used, one is the static pressure probe, and the second, which is upstream from the first, is the total pressure probe. **See Figure 5-27.** The total pressure probe is bent or scooped to force moving air into the tube. The static pressure tube is connected to the low-pressure port of a differential pressure transmitter, and the total pressure tube is connected to the high-pressure port of the transmitter. The velocity pressure can be found by subtracting the static pressure from total pressure. Once the velocity pressure is found, the value can be converted into air velocity in cfm.

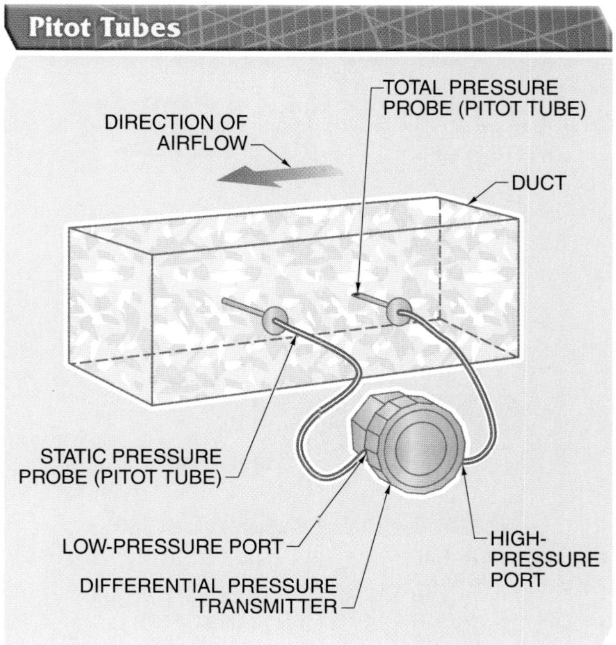

Figure 5-27. A pitot tube measures static pressure and total pressure in ducts.

Thermoanemometers

A *thermoanemometer* is a permanently installed air-handling system device used to measure air velocity in a duct. Thermoanemometers are also known as hot-wire anemometers because they use the principle of heat transfer to determine air velocity. Thermoanemometers can accurately measure a wide range of air velocities, from 10 fpm to 10,000 fpm.

The probe portion of a thermoanemometer is mounted within a duct. The external portion of the device passes a current through a very fine wire located in the probe. The probe is heated by the wire's resistance to current flow, but it is cooled by the air passing over it. **See Figure 5-28.** The amount of airflow affects the amount of cooling of the probe. The temperature of the probe affects the current flow (since temperature affects the resistance of most metals), which is used to determine the air velocity.

Thermoanemometers

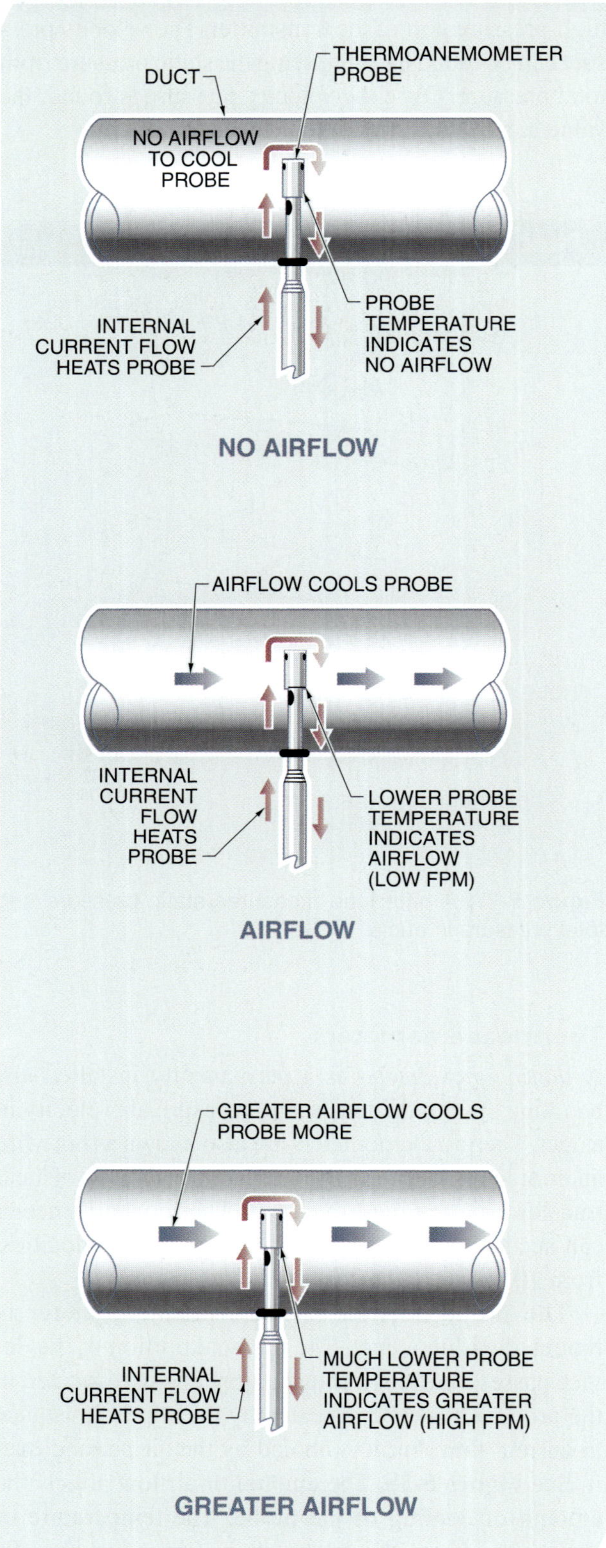

Figure 5-28. Thermoanemometers are permanently installed in ducts and are used to measure air velocities.

ELECTRICAL MEASUREMENT DEVICES

Electrical measurement devices are used by HVAC and refrigeration technicians to take various electrical measurements. Care must be taken when using electrical measurement devices because damage to the instrument or personal injury may result from improper or unsafe use. The owner's manual must be consulted and all functions of a measurement instrument fully understood before it is used. Multimeters must be designed to take measurements on the circuit being tested. Required PPE must always be worn and safety rules must always be followed when taking measurements.

Many types of measurement devices are used by HVAC technicians. Most measurement instruments are available in either analog or digital versions and are designed to perform specific functions. Electrical measurement devices include multimeters, clamp-on ammeters, continuity testers, megohmmeters, voltage testers, receptacle testers, digital logic probes, ground resistance testers, phase sequence indicators, and tachometers.

Multimeters

A *multimeter* is a meter that is capable of measuring two or more electrical quantities. Multimeters can be used to measure electrical functions such as voltage, current, continuity, resistance, capacitance, frequency, and duty cycle. Multimeters display measured readings using analog or digital displays.

Analog Multimeters. An *analog multimeter* is a meter that can measure two or more electrical properties and displays the measured properties along calibrated linear and nonlinear scales using a pointer. Analog multimeters use electromechanical components to display measured values. Most analog multimeters have several calibrated scales that correspond to the different selector switch settings (AC, DC, and R) and placement of the test leads (mA jack and 10 A jack). **See Figure 5-29.**

When reading a measurement on an analog multimeter, the correct scale must be used. The most common measurements made with analog multimeters are voltage, resistance, and current. Analog multimeters may also include scales for measuring decibels (dB) and checking batteries.

Reading Analog Displays. An *analog display* is an electromechanical device that indicates a value by the position of a pointer on a scale. Analog scales may be linear or nonlinear. A *linear scale* is a scale that is divided into equally spaced segments. A *nonlinear scale* is a scale that is divided into unequally spaced segments. **See Figure 5-30.**

Analog Multimeters

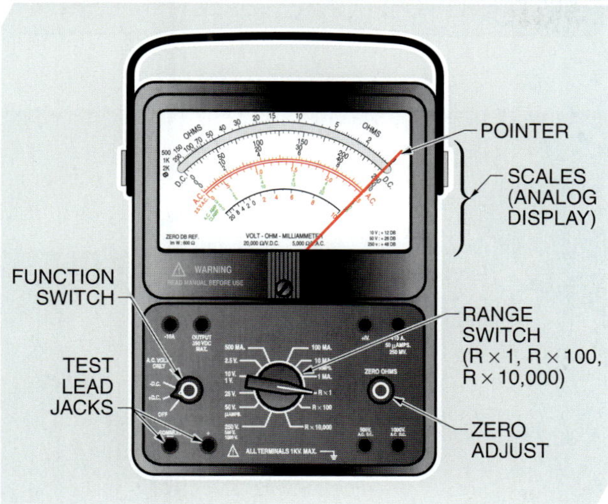

Figure 5-29. Analog multimeters are capable of measuring two or more electrical quantities and use an analog display.

Analog Displays

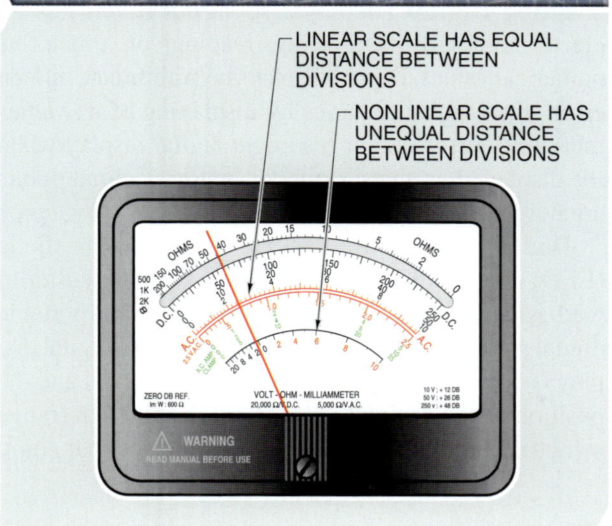

Figure 5-30. An analog display is an electromechanical device that indicates a value by the position of a pointer on a scale.

Analog scales are divided using primary divisions, secondary divisions, and subdivisions. A *primary division* is a division with a listed value. A *secondary division* is a division that divides primary divisions in halves, thirds, fourths, fifths, etc. A *subdivision* is a division that divides secondary divisions in halves, thirds, fourths, fifths, etc. Secondary divisions and subdivisions do not have listed numerical values. When reading an analog scale, the primary, secondary, and subdivision readings are added. **See Procedure for Reading Analog Multimeter Scales.**

Reading Analog Multimeter Scales

Procedure

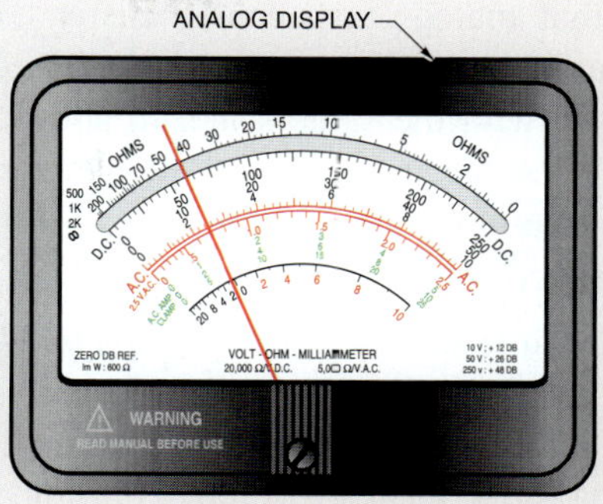

1. Read primary division.

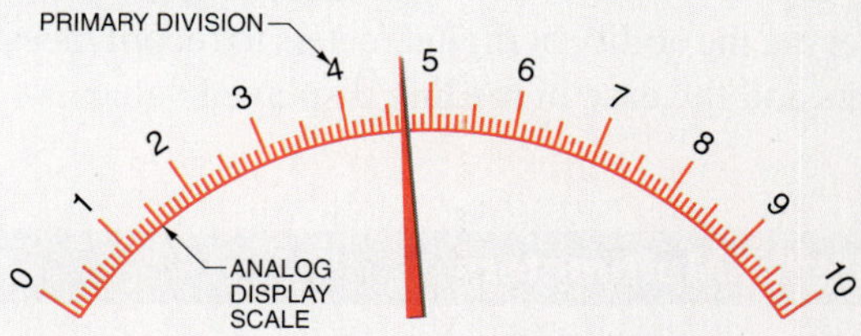

2. Read secondary division if pointer moves past a secondary division. *Note:* This may not occur with very low readings.

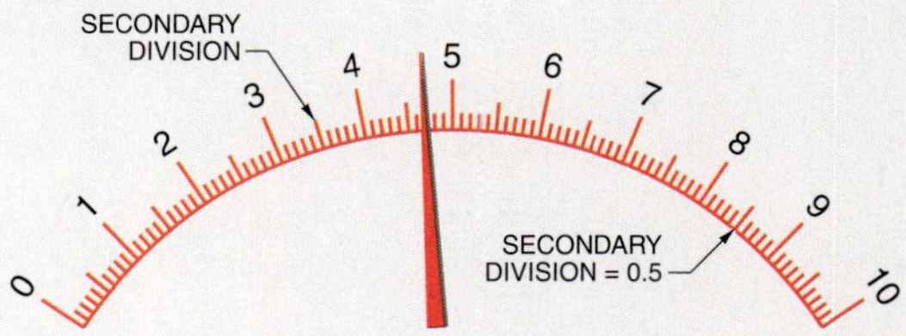

3. Read subdivision if pointer is not directly on a primary or secondary division. *Note:* Round the reading to the nearest subdivision if the pointer is not directly on a subdivision. Round the reading to the next highest subdivision if rounding to the nearest subdivision is unclear.

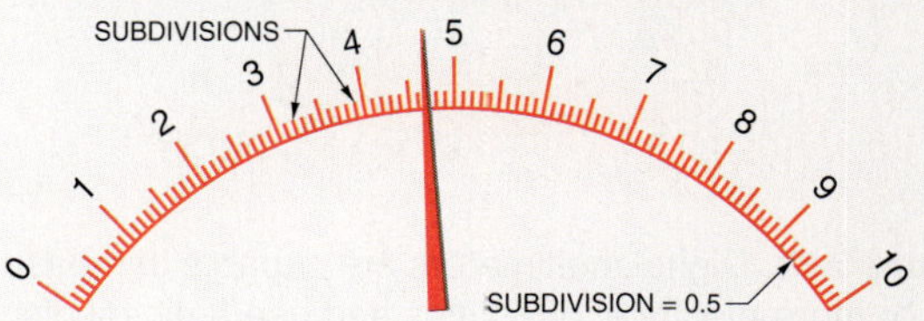

4. Add primary division, secondary division, and subdivision readings to obtain analog reading.

PRIMARY DIVISION = 4.0
SECONDARY DIVISION = 0.5
\+ SUBDIVISION = 0.2
ANALOG READING = 4.7

Analog meters are less susceptible to electrical noise than digital meters. A disadvantage of analog meters is that analog scales can be misread, especially if the meter has multiple settings (1 V, 10 V, 25 V, 50 V, 250 V) used with the same scale. In addition, analog meters are less accurate than digital meters (typical voltage measurements are within 1% to 5%, depending on meter specifications).

Digital Multimeters. A *digital multimeter (DMM)* is a meter that can measure two or more electrical properties and displays the measured properties as numerical values. Basic DMMs can measure voltage, current, and resistance. Advanced DMMs include functions such as measuring capacitance and/or temperature. **See Figure 5-31.** The main advantages of a digital meter over an analog meter are the ability of digital meters to record measurements and the ease in reading displayed values.

Digital Multimeters

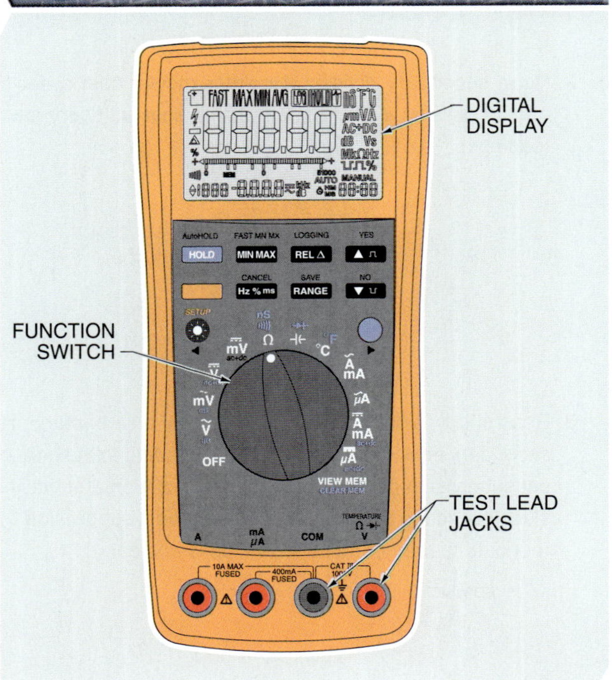

Figure 5-31. Digital multimeters are capable of measuring two or more electrical quantities and use a digital display.

Wireless DMMs provide displays that are detachable from the sensing probes. Wireless DMMs allow connecting sensing probes to circuits in hazardous areas before they are energized. The circuit values are read from a safe distance while the system is in operation.

Digital Multimeters Media Clip

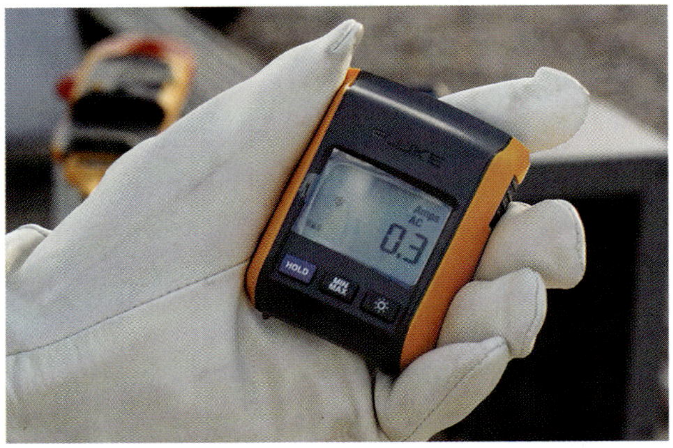

Wireless meters have a display that can be detached and read from a distance for improved safety.

Reading Digital Displays. A *digital display* is an electronic device that displays readings on a meter as numerical values. Digital displays help eliminate human error when taking readings by displaying exact values measured. Errors occur reading a digital display when the displayed prefixes, symbols, and/or decimal points are not properly applied.

Digital displays use either a light-emitting diode (LED) display or an LCD. LED displays are easier to read than LCDs, but they use more power. Most portable digital meters use LCDs. The exact value on a digital display is determined from the numbers displayed and the position of the decimal point. A selector switch (range switch) determines the placement of the decimal point.

TECH TIP

Multimeters can be used to measure electrical functions such as voltage, current, continuity, resistance, capacitance, frequency, and duty cycle. Most multimeters have several calibrated scales that correspond to different selector switch settings.

Ohmmeters, typically a function of a DMM, are used to measure current and resistance. Ohmmeters must be periodically tested and adjusted. **See Procedure for Testing Multimeter Ohmmeters.**

Testing Multimeter Ohmmeters

Procedure

1. Find three new resistors of known value. Select resistors with resistance values that vary by a factor of 10, such as 50 Ω, 500 Ω, and 5000 Ω.

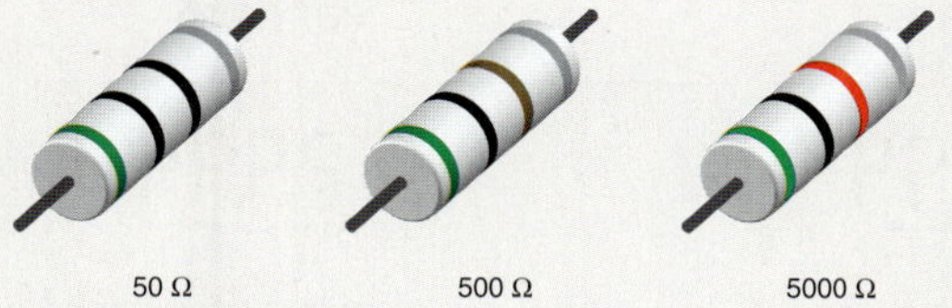

2. Turn multimeter ON.

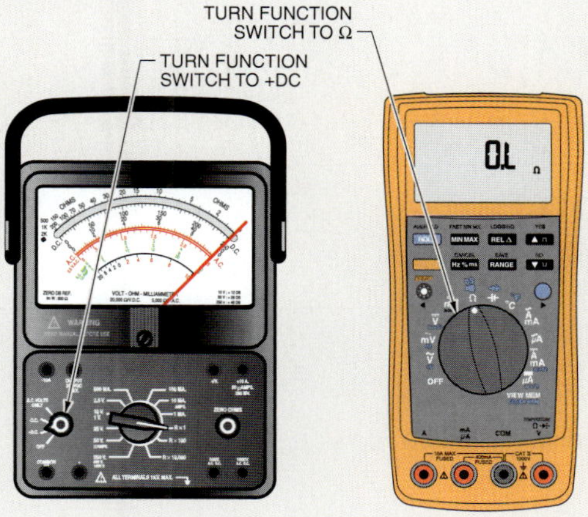

3. Select ohmmeter function setting on multimeter and select 10 Ω range. Touch multimeter red test lead to black test lead. Meter should read 0 Ω of resistance. On analog multimeters, use "zero" adjustment knob to zero meter.

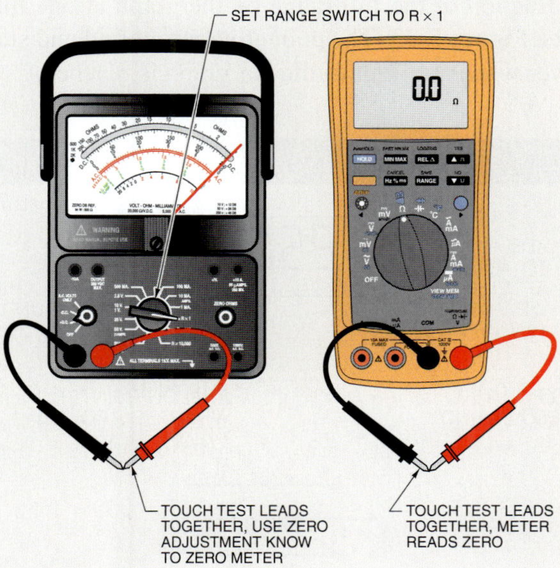

4. Place 50 Ω resistor on a nonconducting surface and touch one test lead to one side of resistor and the other test lead to the other side. Meter should read a value close to 50 Ω.

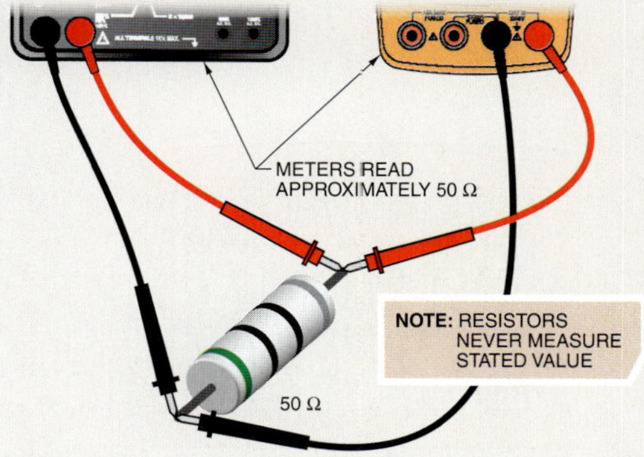

5. Place 500 Ω resistor on a nonconducting surface, change analog multimeter to 100 Ω range, and connect test leads to each side of resistor. Meter should read a value close to 500 Ω.

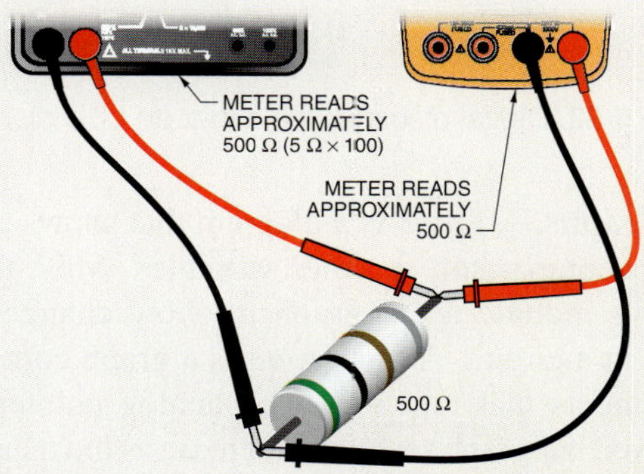

6. Place 5000 Ω resistor on a nonconducting surface, leave analog multimeter on 100 Ω range, and connect test leads to each side of resistor. Meter should read a value close to 5000 Ω.

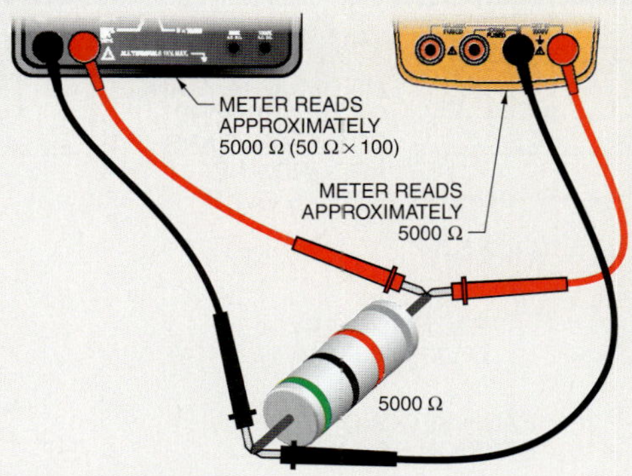

7. Turn multimeter OFF.

Typical voltage ranges on a digital display are 3 V, 30 V, and 300 V. The highest possible reading with the range switch on 3 V is 2.999 V. The highest possible reading with the range switch on 30 V is 29.99 V. The highest possible reading with the range switch on 300 V is 299.9 V. Accurate readings are obtained by using the range that gives the best resolution without overloading. **See Figure 5-32.**

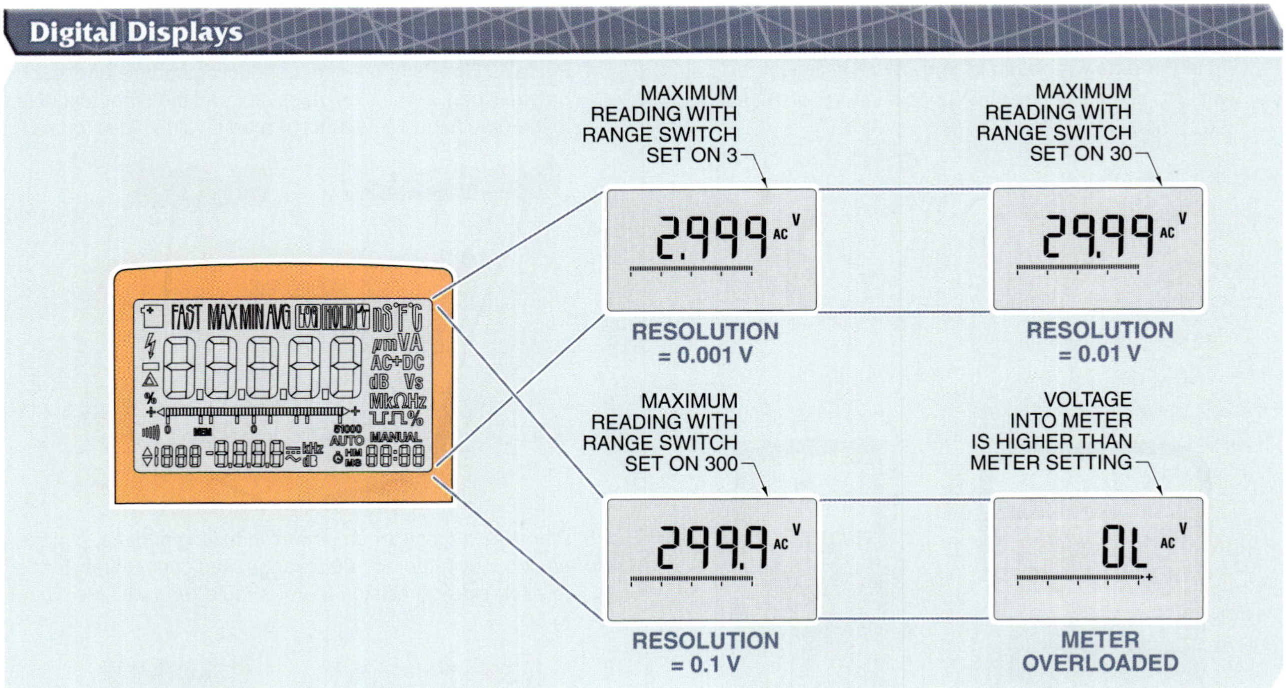

Figure 5-32. Digital displays use either an LED display or an LCD to display values.

Bar Graphs. A *graph* is a diagram that shows a variable in comparison to other variables. Most digital displays include a bar graph to show changes and trends in a circuit. A *bar graph* is a graph composed of segments that function as an analog pointer. The displayed bar graph segments increase as the measured value increases and decrease as the measured value decreases. The polarity of test leads should be reversed when a negative sign is displayed at the beginning of a bar graph.

A *wraparound bar graph* is a bar graph that displays a fraction of the full range on the graph at one time. **See Figure 5-33.** The pointer wraps around and starts over when the limit of the bar graph is reached.

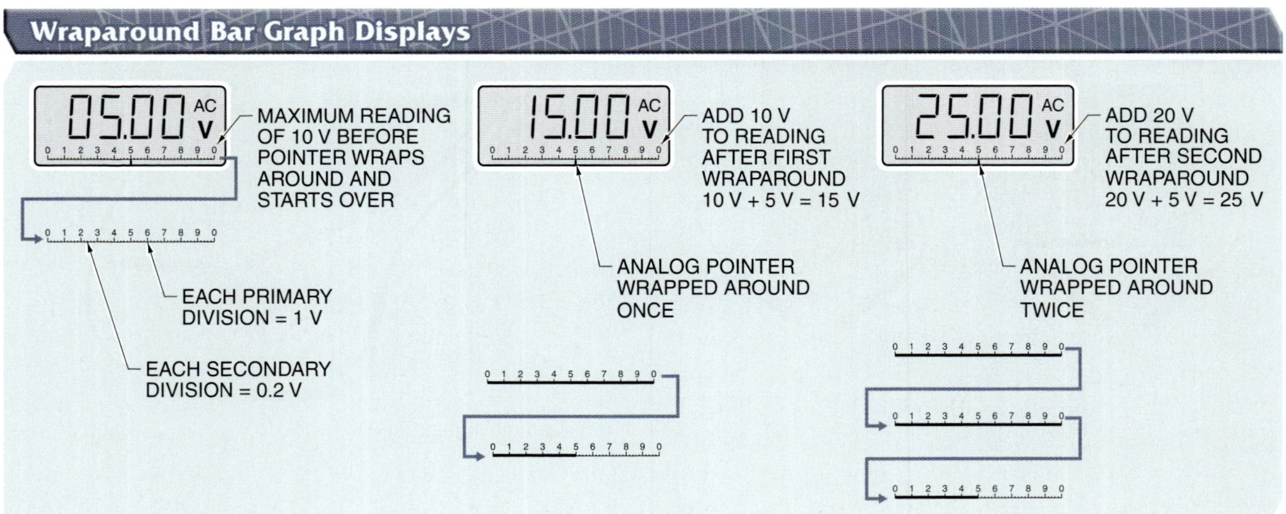

Figure 5-33. A wraparound bar graph is composed of segments that function as an analog pointer. Wraparound bar graphs quickly respond to changes in signals.

A bar graph reading is typically updated 30 times per second. A digital reading is typically updated four times per second. The bar graph is used when quickly changing signals cause the digital display to flash or when there is a change in the circuit that is too rapid for the digital display to detect. For example, mechanical relay contacts may bounce open when exposed to vibration. Contact bounce causes intermittent problems in electrical equipment. Frequency and severity of contact bounce increase as a relay ages.

A contact's resistance changes momentarily from zero to infinity and back when a contact bounces open. *Infinity* is an unlimited number or amount. Digital displays cannot indicate contact bounce because most digital displays typically require more than 250 ms (milliseconds) to update their reading. The quick response of bar graphs enables detection of most contact bounce problems. The contact bounce is displayed by the movement of one or more segments the moment the contact opens.

Most digital multimeters are autoranging, meaning that once a measuring function (such as VAC) is selected, the meter automatically selects the best meter range for taking the measurement (400 mV, 4 V, 40 V, 400 V, or 1000 V). In addition, digital multimeters have a greater accuracy than analog meters. Typical voltage measurements with digital multimeters are within 0.01% to 1.5%, depending on meter specifications.

Ghost Voltages. A meter set to measure voltage may display a reading before the meter is connected to a powered circuit. The displayed voltage is a ghost voltage that appears as changing numbers on a digital display or as a vibrating analog display. A *ghost voltage* is a voltage that appears on a meter that is not connected to a circuit. Ghost voltages are produced by the magnetic fields generated by current-carrying conductors, fluorescent lighting, and operating electrical equipment. Ghost voltages enter a meter through the test leads because test leads not connected to a circuit act as antennae for stray voltages. **See Figure 5-34.**

Ghost voltages do not damage a meter. Ghost voltages may be misread as circuit voltages when a meter is connected to a circuit that is believed to be powered. A circuit that is not powered can also act as an antenna for stray voltages. To ensure true circuit voltage readings, a meter is connected to a circuit for a long enough time that the meter displays a constant reading.

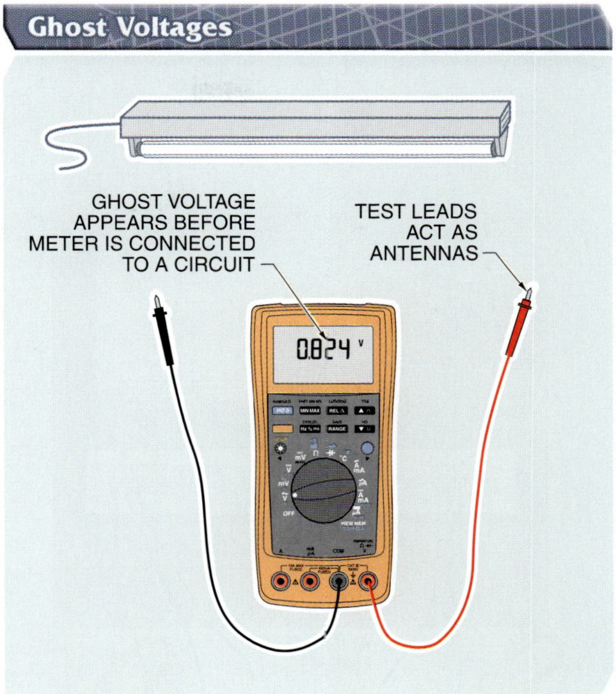

Figure 5-34. Ghost voltages are produced by the magnetic fields generated by current-carrying conductors, fluorescent lighting, and operating electrical equipment.

Clamp-On Ammeters

A *clamp-on ammeter* is a meter that measures the current in a circuit by measuring the strength of the magnetic field around a single conductor. Clamp-on ammeters measure currents from 0.01 A or less to 1000 A or more. A clamp-on ammeter is normally used to measure current in a circuit with over 1 A of current and in applications in which current can be measured by easily placing the jaws of the ammeter around one of the conductors. Most clamp-on ammeters can also measure voltage and resistance. To measure voltage and resistance, the clamp-on ammeter must include test leads, a voltage mode, and a resistance mode. **See Figure 5-35.**

Current is a typical troubleshooting measurement because only a current measurement can be used by an HVAC technician to determine how much a circuit is loaded or when a circuit is operating correctly. Current measurements vary because current can vary at different points in parallel or series/parallel circuits. Current in series circuits is constant throughout the circuit. The largest amount of current in a parallel circuit is at the point closest to the power source. Current decreases as the system distributes current to each of the parallel loads.

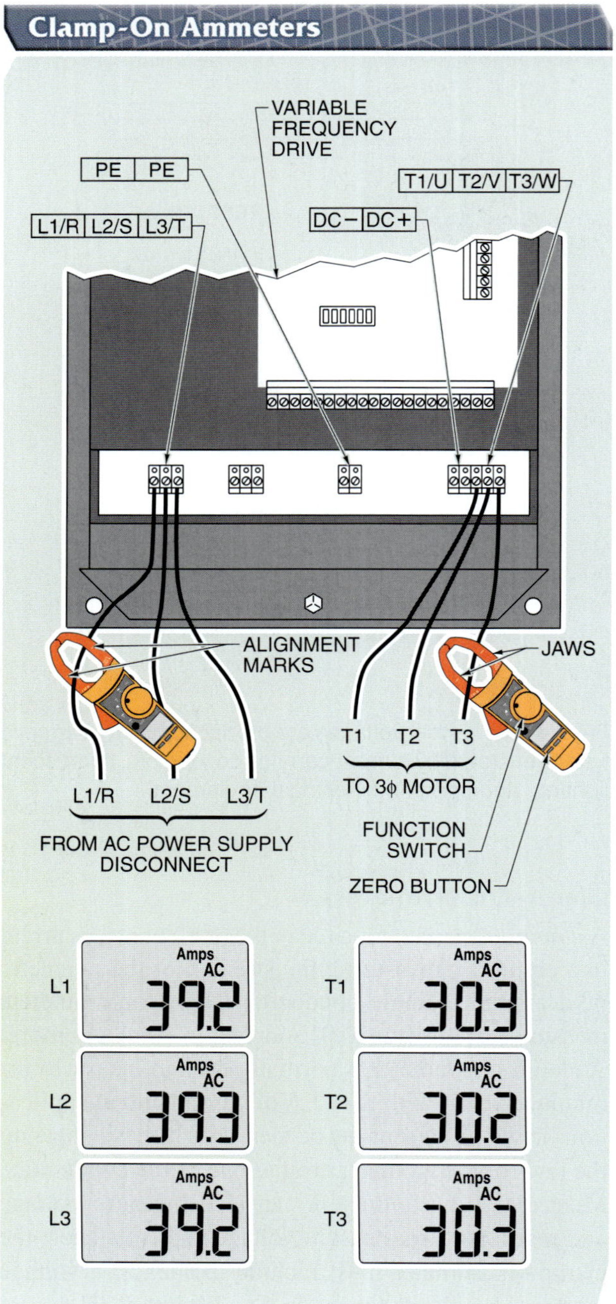

Figure 5-35. A clamp-on ammeter measures the current in a circuit by measuring the strength of the magnetic field around a single conductor.

Any variation that is excessively high must be investigated because the current measurement may indicate that a partial short exists on one of the lines and a small amount of current is flowing to ground. Technicians must ensure that clamp-on ammeters do not pick up stray magnetic fields during testing by separating conductors being tested from other conductors as much as possible. When stray magnetic fields are possibly affecting a measurement, several measurements at different locations along the same conductor must be taken.

Continuity Testers

A *continuity tester* is an instrument that tests for a complete path for current to flow. Continuity testers are economical instruments used to test switches, fuses, and grounds. Continuity testers are also used for identifying individual conductors in a multiwire cable. Most DMMs have a built-in continuity test mode. **See Figure 5-36.** In the continuity test mode, DMMs emit an audible response (beep) when there is a complete path. Indication of a complete path can be used to determine the condition of a component as open or closed.

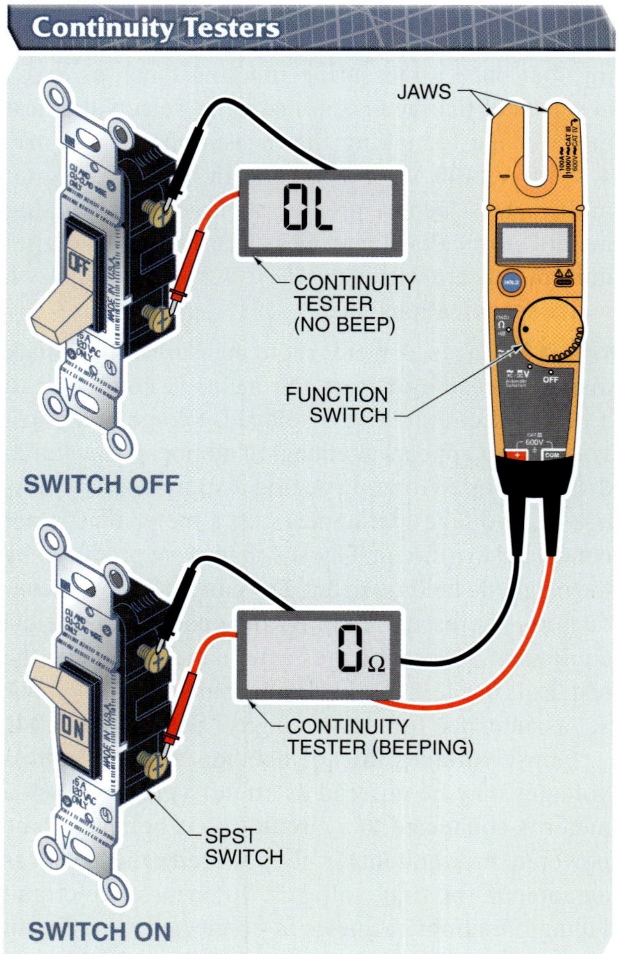

Figure 5-36. A continuity tester tests for a complete path for current to flow.

For example, a good fuse has continuity, whereas a bad fuse does not. The main advantage of using the continuity test mode of a DMM is that an audible response is often more desirable than reading a resistance measurement. The problem is that the continuity mode of a meter can be fooled, providing false information. Circuits should always be checked for voltage before taking a continuity test.

WARNING: Continuity testers must only be used on de-energized circuits or components. Voltage applied to a continuity tester can cause damage to the instrument and/or harm the technician.

Megohmmeters

A *megohmmeter* is a meter that detects insulation deterioration by measuring high resistance values under high-voltage test conditions. Megohmmeter test voltages range from 50 V to 5000 V. Megohmmeters detect insulation failure or potential failure of insulation caused by excessive moisture, dirt, heat, cold, corrosive vapors or solids, vibration, and aging. **See Figure 5-37.**

Some insulation, such as that found on conductors used to wire branch circuits, is thick and harder to damage or break down. Other insulation, such as the insulation used on motor windings, is very thin (to save weight and space) and breaks down much more easily. Megohmmeters are used to test for insulation breakdown in long wire runs or motor windings.

Insulation allows conductors to stay separated from each other and from earth ground. Insulation must have a high resistance to prevent current from leaking through the insulation. All insulation has a resistance value that is less than infinity, which allows some current leakage to occur. *Leakage current* is current that flows through insulation. Under normal operating conditions, the amount of leakage current is so small (only a few microamperes) that the leaking current has no effect on the operation or safety of a circuit.

Voltage Testers

A *voltage tester* is an instrument that indicates approximate voltage level and type (AC or DC) through the movement and vibration of a pointer on a scale. Voltage testers contain a scale marked 120 VAC, 240 VAC, 480 VAC, 600 VAC, 120 VDC, 240 VDC, and 600 VDC. Some voltage testers include a colored plunger to indicate the polarity of the test leads. When the red indicator is up, the red test lead is connected to the positive DC voltage. When the black indicator is up, the black test lead is connected to the positive DC voltage. **See Figure 5-38.**

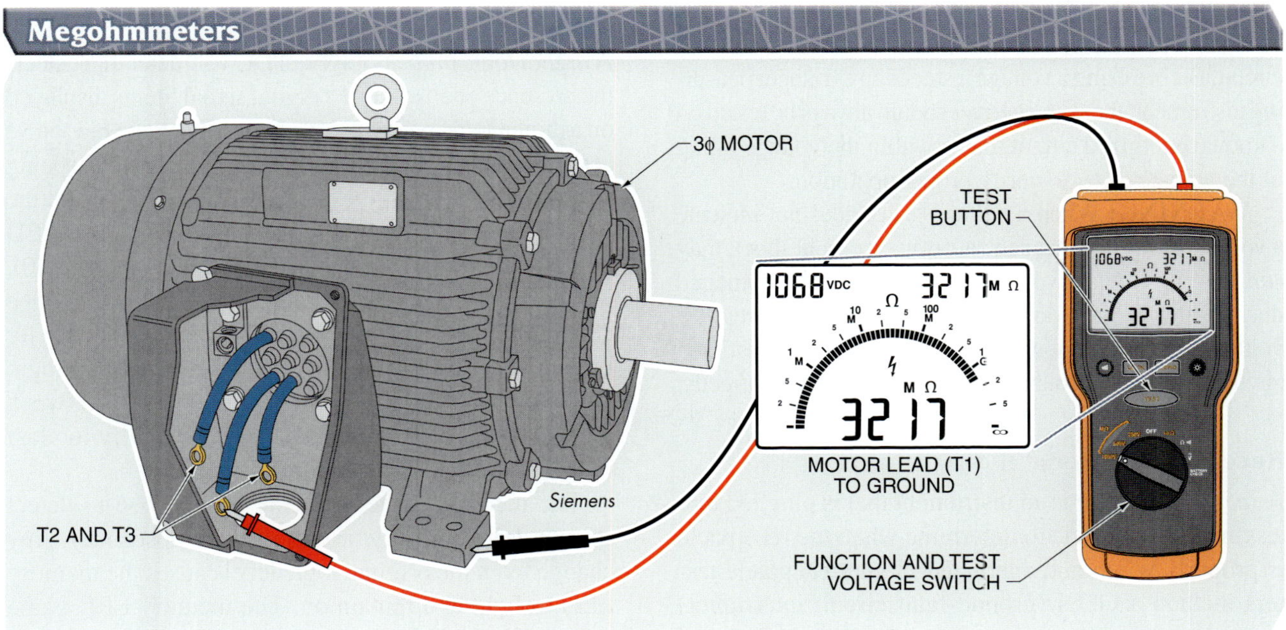

Figure 5-37. A megohmmeter detects insulation deterioration by measuring high resistance values under high-voltage test conditions.

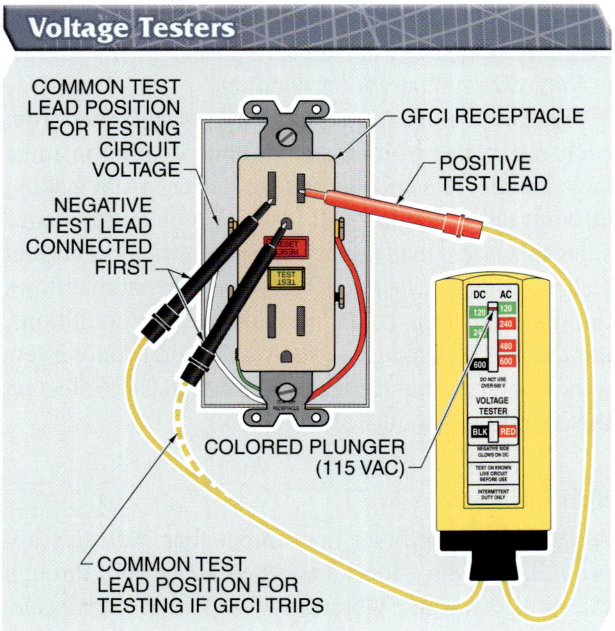

Figure 5-38. A voltage tester indicates approximate voltage level and type (AC or DC) through the movement and vibration of a pointer on a scale.

Voltage testers are used to take voltage measurements when the voltage of a circuit is within the rating of the tester and an exact voltage measurement is not required. Exact voltage measurements are not required when testing to determine which receptacles are energized, when a system is grounded, which fuses or circuit breakers are good or bad, and when a circuit is a 115 VAC, 230 VAC, or 460 VAC circuit. Before using a voltage tester or any voltage measuring instrument, the voltage tester should always be tested on a known energized circuit that is within the voltage rating of the voltage tester to verify proper operation.

WARNING: When a voltage tester does not indicate a voltage, a voltage that can cause an electrical shock may still be present. The voltage tester can also be damaged during a test (from too high a voltage). Always retest a voltage tester on a known energized circuit after testing a circuit to ensure the voltage tester is still operating correctly.

Receptacle Testers

A *receptacle tester* is an instrument that is plugged into a standard receptacle to determine when the receptacle is properly wired and energized. Some receptacle testers include a GFCI (ground-fault circuit interrupter) test button that allows the receptacle tester to be used on GFCI receptacles. When testing a receptacle, the indicator light code indicates whether the receptacle is wired correctly. When energized and neutral wires are reversed it is a safety hazard and must be corrected. Improper grounds are also a safety hazard and must be corrected. **See Figure 5-39.**

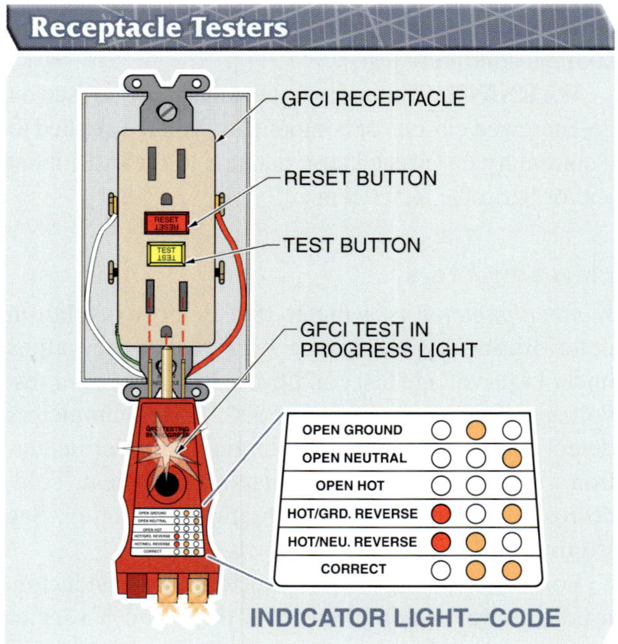

Figure 5-39. Receptacle testers use a sequence of lights to indicate the condition of receptacles and GFCI receptacles.

Digital Logic Probes

A *digital logic probe* is a special DC voltmeter that detects the presence or absence of a digital signal. Items displayed on a digital logic probe include logic high, logic low, pulse activity, memory, and type of family (TTL/CMOS) being tested. **See Figure 5-40.** The high LED lights when the logic probe detects a high logic level (1). The low LED lights when the logic probe detects a low logic level (0). The pulse LED flashes relatively slowly when the probe detects logic activity present in a circuit. Logic activity indicates that the circuit is changing between logic levels. The pulse light displays the changes between logic levels because the changes are typically too fast for the high and low LEDs to display.

The memory switch sets the logic probe to detect short pulses, typically lasting a few nanoseconds. Any change from the original logic level causes the memory LED to light and remain on. The memory LED is the pulse LED switch in the memory position. The memory switch is manually moved to the pulse position and back to the memory position to reset the logic probe.

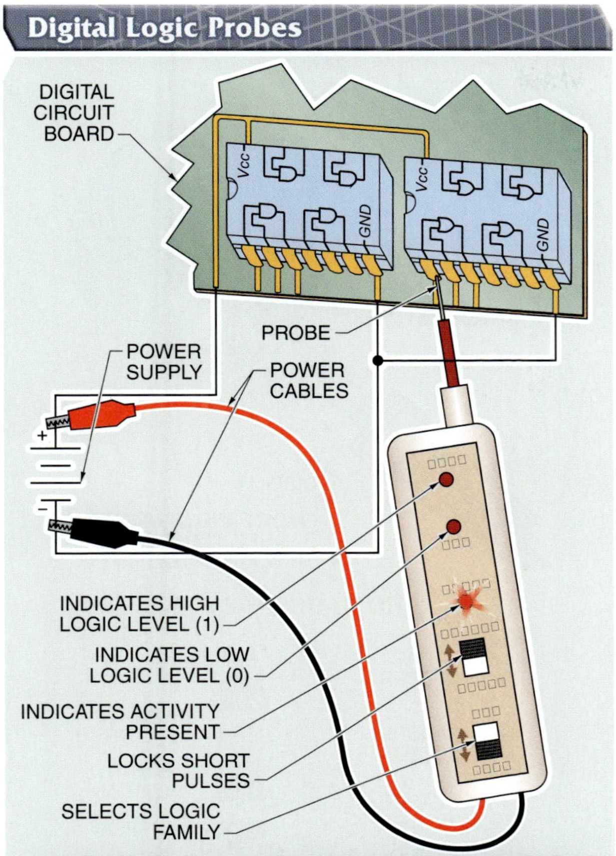

Logic Probe Functions

LED Condition	Meaning
No LEDs lit	Open circuit
High LED lit	Logic 1
Low LED lit	Logic 0
Pulse LED lit	Circuit activity

Figure 5-40. A logic probe is a digital test instrument that can only test a single signal at a time. When many logic signals need to be observed or recorded at the same time, logic analyzers are used.

Ground Resistance Testers

A *ground resistance tester* is an instrument used to measure ground resistance for grounding rod installations. Ground resistance tests are required, as specified by NEC® Article 230.95(C). **See Figure 5-41.** *Grounding* is the connection of all exposed non-current-carrying metal parts to the earth. Grounding provides a direct path for unwanted fault current to the earth without causing harm to persons or equipment. Electrical circuits are grounded to safeguard personnel and equipment against the hazards of electrical shock and electrostatic discharge.

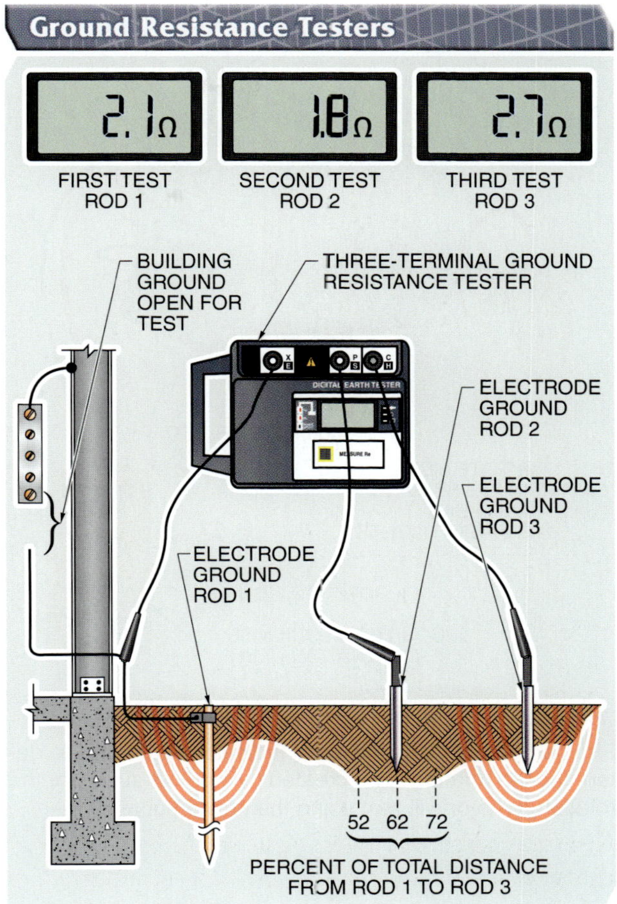

Figure 5-41. Resistance to current through an earth electrode actually has three components: resistance of the electrode, contact resistance between the electrode and the soil, and resistance of the surrounding soil.

Phase Sequence Indicators

A *phase sequence indicator* is a device used to determine phase sequence and open phases. Phase sequence indicators are used to test electrical connections to three-phase equipment. Phase sequence indicators help protect motors, generators, and other equipment from damage due to incorrect motor rotation. Phase sequence indicators are available with either a rotating disk with a colored dot or with two or three LED lamps to indicate phase. **See Figure 5-42.**

TECH TIP

Out-of-phase conditions can easily damage large motors and compressors and can be caused by the electrical distribution system or by the utility.

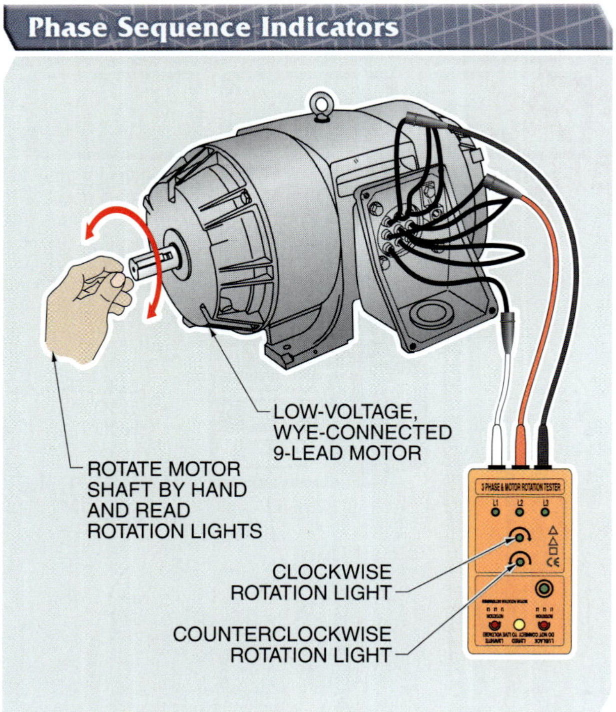

Figure 5-42. Phase sequence indicators are used to determine the rotation to be created by a power supply or the rotation a motor will turn using the power supply.

Tachometers

A *tachometer* is an instrument used to measure the speed of rotation of a shaft or wheel. Rotating speeds are displayed in revolutions per minute (rpm), and linear speeds are displayed in feet per minute (fpm). Electronic tachometers optically measure a mark of tape or chalk as the mark revolves or travels with a motor or pump shaft. Some electronic tachometers also contain a contact tachometer on one end that can be applied directly to a shaft to measure rpm. Typical types of tachometers used are photo-contact, laser-contact, and strobe. **See Figure 5-43.**

CONTAMINANT MEASUREMENT DEVICES

Air sampling helps identify contaminants in building spaces. HVAC technicians should perform building air sample testing on a regular basis as a preventive measure in order to limit occupants' exposure to contaminants. Air-sampling devices are used to collect samples of building air, which are analyzed for the level of contamination.

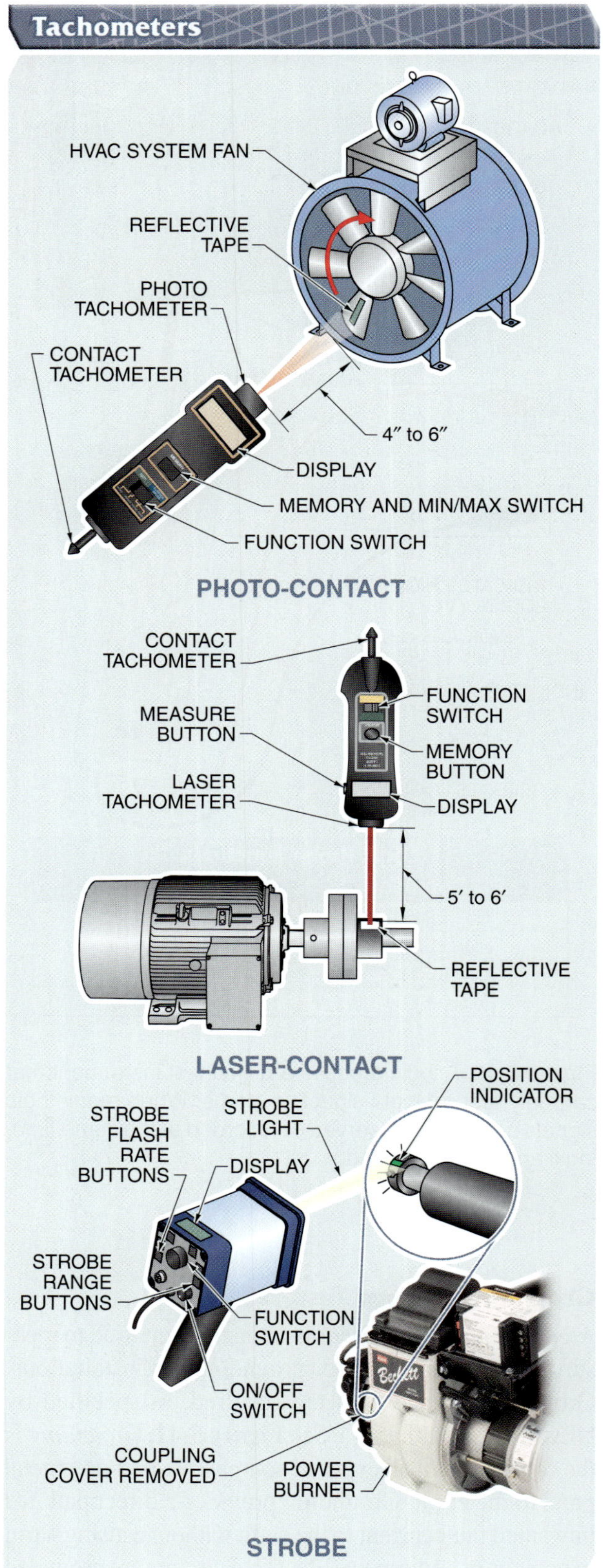

Figure 5-43. Tachometers measure the speed of a moving object.

Hazardous Substance Air Monitoring

A wide variety of contaminants are found in air, some of which are potentially hazardous. Exposure to hazardous substances is regulated by federal, state, or local enforcement agencies. Air monitoring and testing must be performed because of the potential risks involved from exposure to airborne contaminants such as asbestos and lead.

Asbestos is a mineral fiber that can cause cancer or asbestosis when inhaled. Asbestos has been used in a variety of building construction materials for insulation and as a fire-retardant. The EPA and the Consumer Product Safety Commission (CPSC) have banned several asbestos products. Asbestos is often found in mechanical rooms, pipe shafts, pipe insulation, fireproofing materials, acoustical materials, shingles, siding, and ceiling and floor tiles. Testing for the presence of asbestos is accomplished with personal air-sampling devices. **See Figure 5-44.**

Lead is a heavy metal that is a health hazard when inhaled or swallowed. Occupants are exposed to lead through air, drinking water, food, contaminated soil, deteriorating paint, and dust. Lead becomes an indoor air problem only when lead-based particulates become airborne.

Test instruments can be used to indicate whether an indoor air quality issue is present.

Air Quality Monitoring

Maintaining the desired level of oxygen (O_2) in building space air is controlled by the amount of ventilation air provided. Testing for the presence of gases, such as nitrogen dioxide (NO_2), carbon monoxide (CO), carbon dioxide (CO_2), and ozone (O_3), is easily accomplished using handheld air test meters. **See Figure 5-45.**

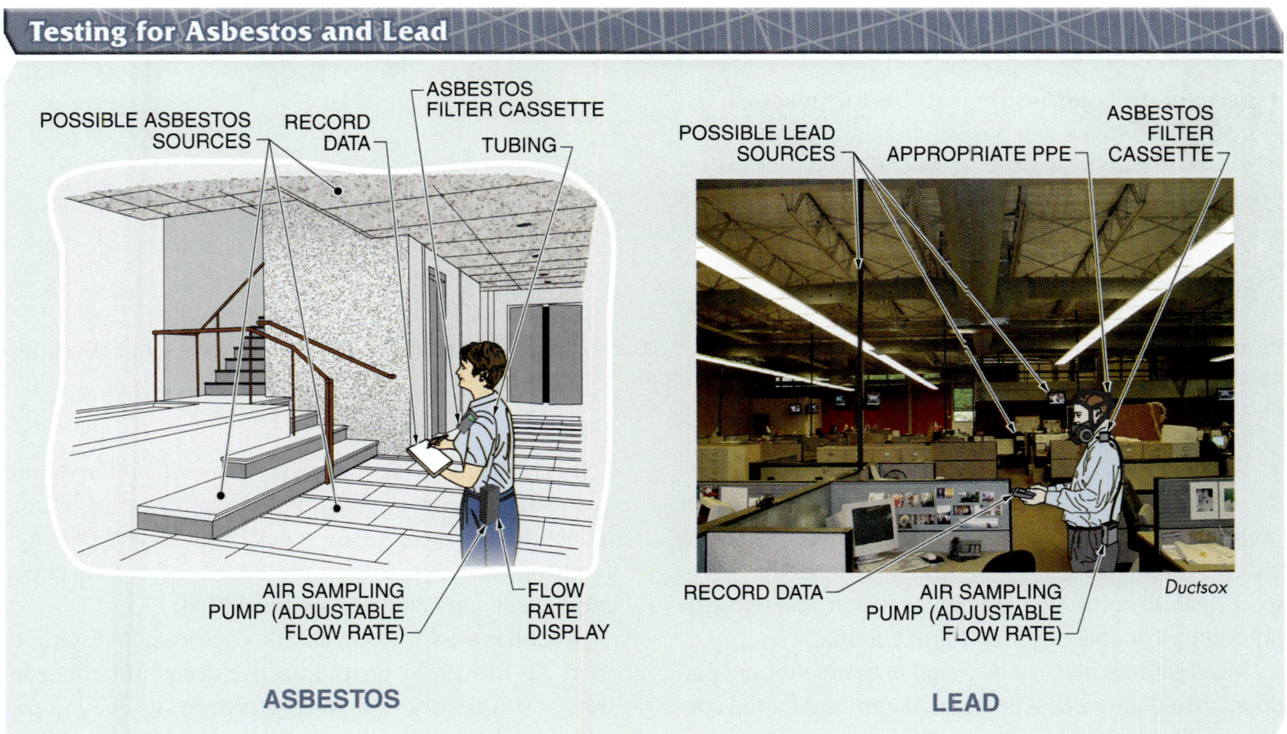

Figure 5-44. Asbestos testing and lead testing require special cassette filters that only work correctly when the proper airflow is passed through them for the correct amount of time.

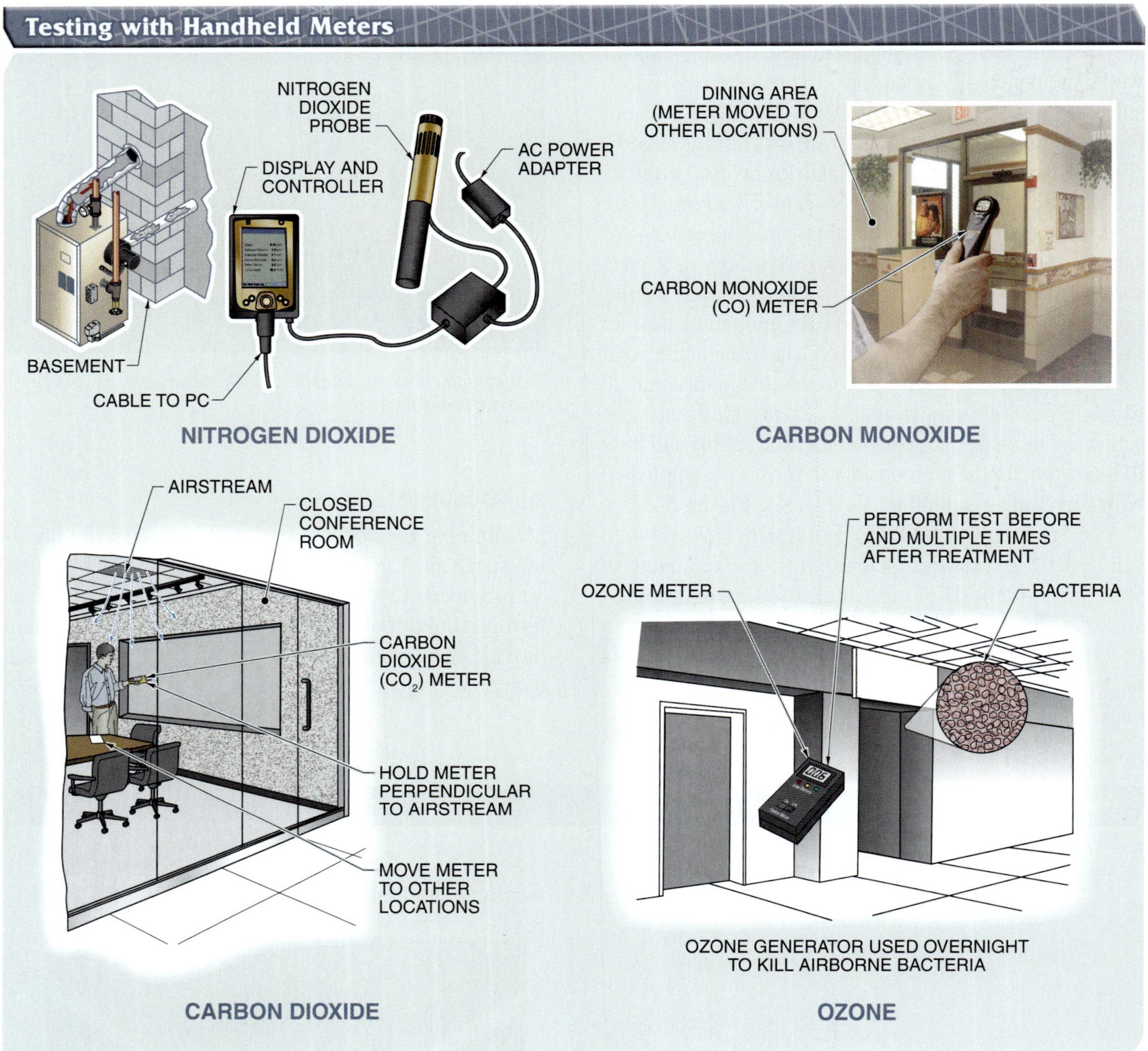

Figure 5-45. Nitrogen dioxide, carbon monoxide, carbon dioxide, and ozone each have a special meter that is used to measure the contaminants in the air in parts per million (ppm).

Aeroallergens, such as mold and pollen, are a growing environmental concern. Formaldehyde and radon are organic compounds that are typically found in building spaces. Indoor air quality sampler kits are an easy way to collect airborne contaminants, which are sent to a laboratory for analysis. **See Figure 5-46.**

Mold is fungi that can be found both indoors and outdoors. Mold grows best in warm, damp, and humid conditions. Mold spreads and reproduces by making spores. Mold spores can survive harsh, dry climates, which are not typically considered conducive to mold growth.

Formaldehyde is a colorless, pungent, and irritating gas used as a preservative in building materials. Formaldehyde is used in many types of glues and coatings for products such as particle board, plywood, paper products, paints, and certain insulation.

Radon is a colorless, naturally occurring, radioactive, inert gas formed by the radioactive decay of radium atoms in soil or rocks. Radon is a proven carcinogen and a form of ionizing radiation. Radon is found in outdoor air and indoor air of all kinds of buildings. Exposure to radon may result in lung cancer.

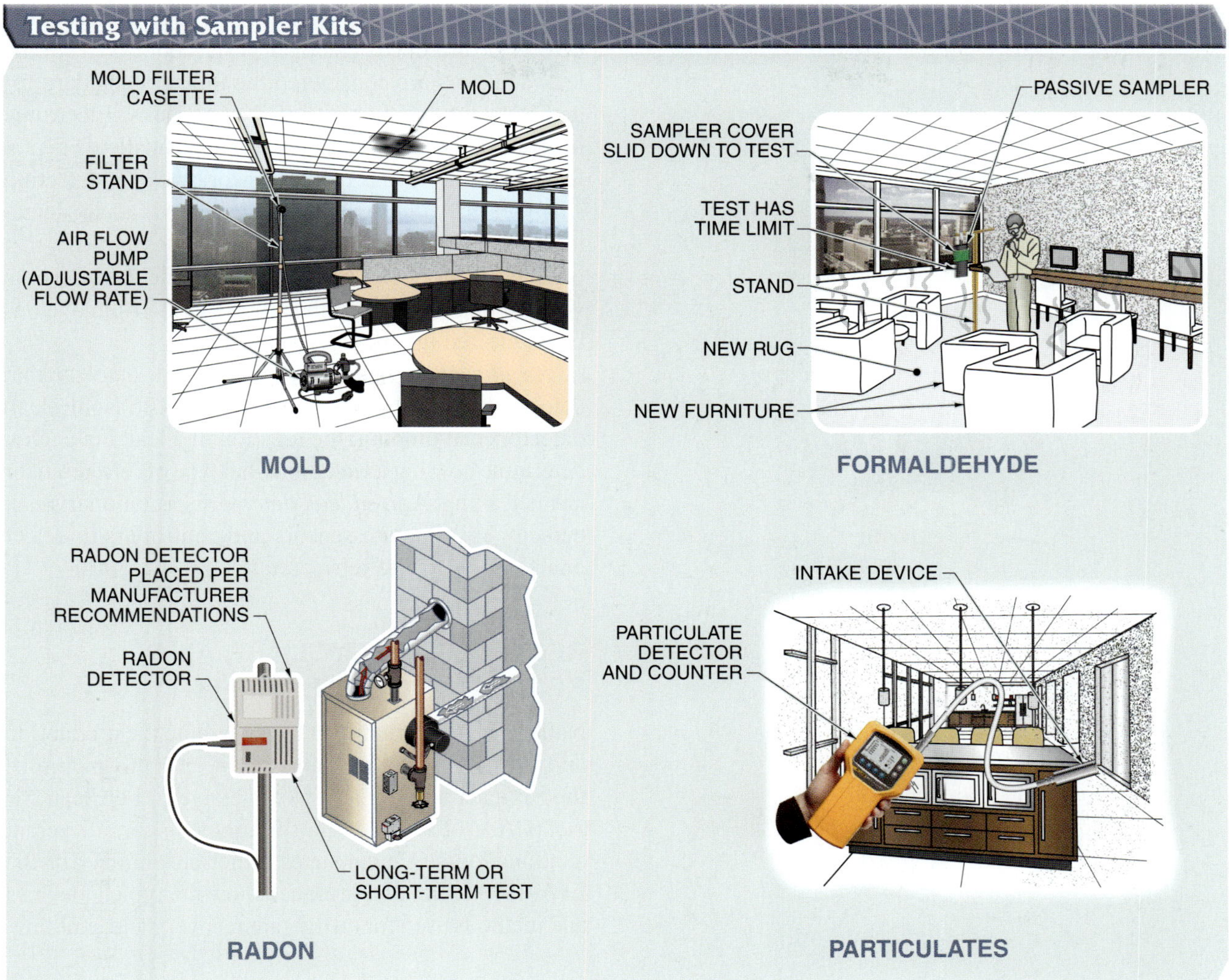

Figure 5-46. Once a filter cassette or passive sampler has been exposed to air for the proper amount of time, it is sent to a laboratory for analysis.

Particulates in the air are categorized by type and size. The concentration levels of small particulates, such as those from combustion, are measured in micrograms per cubic meter (μg/m^3). The acceptable level for small particulates is 15 μg/m^3. Large particulates are measured in milligrams per cubic meter (mg/m^3). The limit for large particulates, such as dust and smoke, is 50 mg/m^3.

Flue gas is the exhaust gas that exits a furnace or boiler to the atmosphere using a flue. A flue-gas analyzer is typically used to measure the percentage of carbon dioxide, oxygen, and other gases in flue gas. The temperature of the flue-gas sample must also be measured using a stack thermometer. The carbon dioxide reading, oxygen reading, and flue gas temperature is used to determine the actual combustion efficiency.

Leak Detectors

Refrigerant gases and other gases are kept under pressure in storage tanks when not being used in a system. With any system under pressure, leaks can develop. Large leaks are soon detected by smell or loss of processes, such as cooling. Small leaks can be much harder to detect. Refrigerant leak detection instruments and meters are used to detect large and small leaks. **See Figure 5-47.**

A *leak detector* is a device that is used to detect refrigerant or other gas leaks in pressurized air conditioning or refrigeration systems. Leak detection is performed with soap bubbles or one of many different types of leak detectors available such as electronic, fluorescent, ultrasonic, or fixed leak detectors.

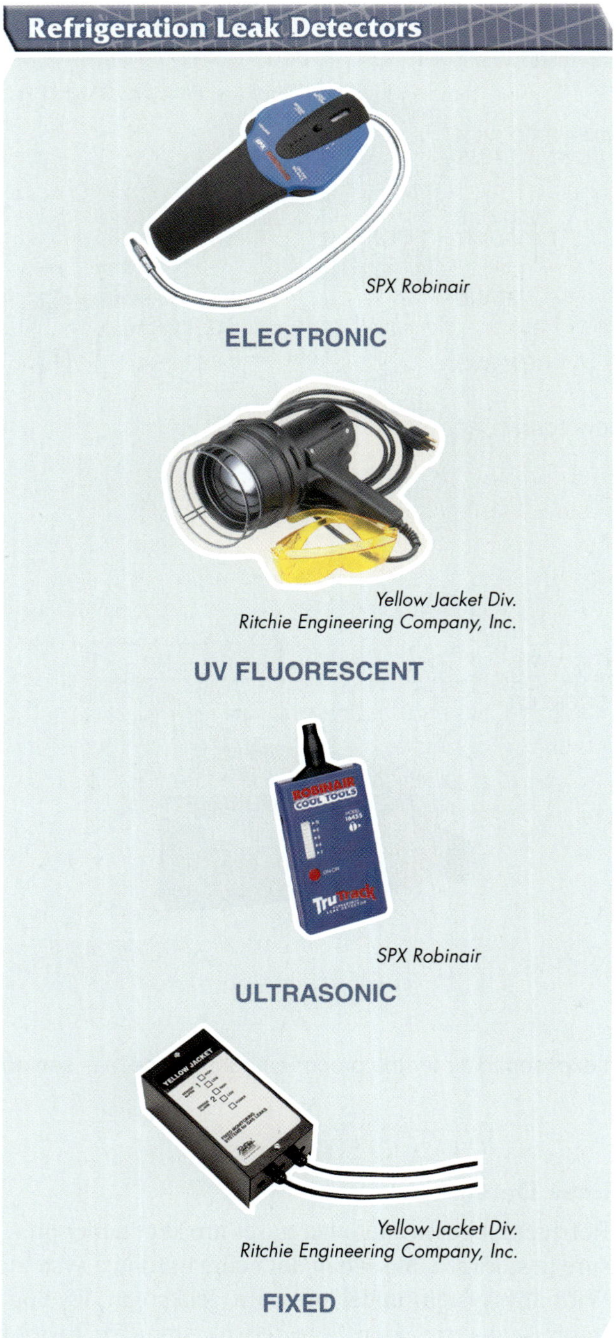

Figure 5-47. Leak detection devices detect refrigerant leaks in HVAC or refrigeration systems.

Soap bubbles are used to pinpoint leaks in a system by covering the piping with a soapy solution. As the refrigerant leaks, bubbles are formed at the point of the leak. Soap bubbles are a safe way to detect leaks in a system in a hazardous environment. Soap bubbles can only be used in a system that is fully pressurized.

An *electronic leak detector* is a leak detector that detects the presence of halogen gas. Handheld electronic leak detectors are considered the industry standard for detecting the location of refrigerant leaks. Electronic leak detectors detect the general area of small refrigerant leaks. Electronic leak detectors work well for detecting small leaks. A *fluorescent leak detector* is a leak detector that uses a UV light to detect fluorescent dye that is added to a system. Fluorescent leak detection works well as long as the system can be run after the dye has been injected into the system.

An *ultrasonic leak detector* is a leak detector that detects the sounds created by a leak. Ultrasonic leak detectors can pinpoint the location of a leak but cannot determine how big a leak is or what type of refrigerant or gas is leaking. A *fixed leak detector* is a stationary leak detector system with sensors and controllers to detect one specific type of refrigerant.

RECOVERY, RECYCLING, AND RECLAIMING EQUIPMENT

National and international laws regulate the production and handling of refrigerants in an effort to minimize their impact on the Earth's ozone layer. The Clean Air Act (CAA) of 1990 sets the requirements for refrigerant handling for the United States, which are enforced by the EPA. HVAC technicians must know the typical devices and methods used for refrigerant recovery and handling.

Service Practice Requirements

EPA regulations affect the way HVAC equipment is serviced. Refrigerant leak testing, recovery, evacuation, and charging must be performed following the most recent regulations. All procedures in place for refrigerant recovery, evacuation, and charging have the intent of minimizing any system contamination and refrigerant loss.

Evacuation Requirements. HVAC technicians are required to evacuate refrigeration equipment to established vacuum levels. Deeper vacuum levels ensure that the maximum amount of refrigerant is recovered. The required vacuum levels are based on the date the evacuation equipment was manufactured. Newer equipment must achieve deeper vacuum levels. An EPA-approved equipment testing organization must certify all recovery and recycling equipment. HVAC technicians who are adding refrigerants to top off appliances or systems are not required to evacuate the systems.

Reclamation Requirements. Refrigerant must be recovered when removed from a system. *Refrigerant recovery* is a process that allows refrigerant to be removed from a system in any condition and stored in an external container without necessarily testing or processing it in any way. Recovery is accomplished using a recovery machine. **See Figure 5-48.** Refrigerant that is lightly contaminated is recycled using a filter in a recovery machine that is capable of recycling refrigerant.

Refrigerant Recovery Equipment

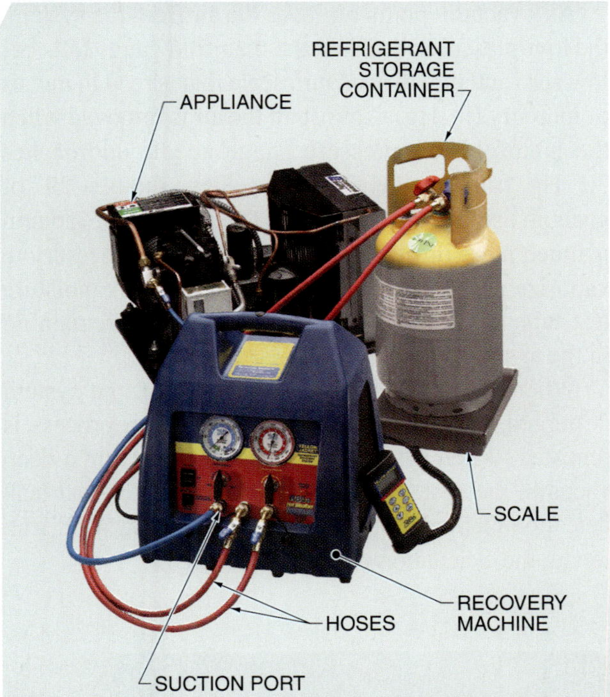

Yellow Jacket Div. Ritchie Engineering Company, Inc.

Figure 5-48. Recovery machines are used to recover refrigerant from appliances for reuse.

The recovery machine suction hose is connected to the appliance service valve for refrigerant removal. The discharge hoses are connected to a refrigerant storage container placed on a scale to receive the removed refrigerant. The removed refrigerant is carefully measured and recorded.

Refrigerant recycling is the process of removing dirt, oil, and moisture from lightly contaminated refrigerant that has been removed from a refrigeration system. The EPA has established that refrigerant recovered or recycled can be returned to the same system or other systems owned by the same individual without restriction. When the refrigerant changes ownership, the refrigerant must be reclaimed unless the refrigerant was used only in a motor vehicle air conditioner or similar appliance and will be used in the same type of appliance.

Refrigerant reclaiming is the process of cleaning heavily contaminated refrigerant. Although recovery and recycling can be accomplished using special equipment, reclaiming must be accomplished at specialized facilities.

Gauge Manifolds

A *gauge manifold* is a device that has two gauges, a manifold with valves, and connecting hoses to control the transfer of a specific refrigerant. Like recovery machines, gauge sets can only be used with one type of refrigerant. A gauge manifold is an important tool when servicing air conditioning or refrigeration systems. A gauge manifold may have two valves or four valves, depending on whether the manifold has separate valves for vacuum, low-pressure, high-pressure, and refrigerant cylinder or recovery machine connections. **See Figure 5-49.**

Refrigerant Gauge Manifolds

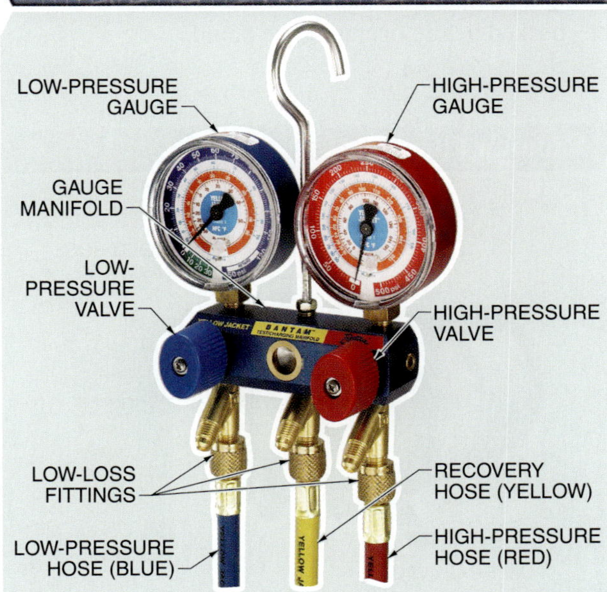

Yellow Jacket Div. Ritchie Engineering Company, Inc.

Figure 5-49. Refrigerant gauge manifolds are used to control the transfer of refrigerants.

Gauges are used when charging a system with refrigerant or evacuating a system. The low-pressure (compound) gauge and connecting hose are typically blue. The low-pressure gauge provides pressure readings above (psi) and below (0 in. Hg) atmospheric pressure. The high-pressure gauge and connecting hose are typically color-coded red. The high-pressure gauge provides pressure readings from 0 psi to 1000 psi. The center port on a three-port gauge manifold is used during the recovery, evacuation, and charging processes. A yellow hose is typically used with the center port.

Gauge manifolds connect the low-pressure hose (blue) to a suction line access port. The high-pressure hose (red) connects to a liquid line access port. The center hose (yellow) connects to the vacuum pump or recovery machine.

System Evacuation

Evacuation of a refrigeration system involves removing all the air from inside the system. Air enters the tubing and components when a system is opened for repairs or after new construction. All the air and moisture contained in a system must be evacuated from the system before charging the system with refrigerant. This is accomplished with a vacuum pump that creates a deep low pressure while it draws all the air from inside the system. **See Figure 5-50.** Before beginning evacuation, the vacuum pump oil must be changed according to manufacturer instructions to prevent the formation of sludge in the pump. Only vacuum pump oil should be used, and the oil level must be checked before operating the pump.

The vacuum pump is connected to the system service valves using a gauge manifold and hoses. **See Figure 5-51.** As the air is removed by the vacuum pump, the pressure inside the system falls below atmospheric pressure and any water in the system vaporizes and is removed. The vaporization point of water is controlled by the pressure exerted on it. Water vaporizes (boils) at 212°F at sea level. At higher elevations, the pressure is lower because there is less air pressing on the surface of the water. As pressure is reduced, water vaporizes at a lower temperature. In a complete vacuum, water vaporizes at a temperature above –90°F. **See Appendix.**

As a vacuum pump removes the air from the system, the low-pressure needle on the manifold gauge falls below zero and into the vacuum scale that is read in inches of mercury (in. Hg). Almost all the air is removed when the manifold low-pressure gauge needle approaches 30″ Hg. It is relatively easy to achieve the first 29″ of vacuum using a vacuum pump. The last inch of vacuum is much more difficult to achieve, but it is necessary to vaporize all the moisture in some systems. A moisture test tube is used to check for any moisture remaining in the system.

When moisture remains in the system, the system is purged with nitrogen and the evacuation process is repeated. When system leaks prevent achieving a deep vacuum, the system is charged with nitrogen and leak detection methods are used to locate the leak. After leak repair, the evacuation process is repeated.

Vacuum Pumps

Figure 5-50. A vacuum pump is used to remove all of the air from a refrigeration system.

Test instruments can be used to detect vibration in cooling system equipment.

Evacuating Systems

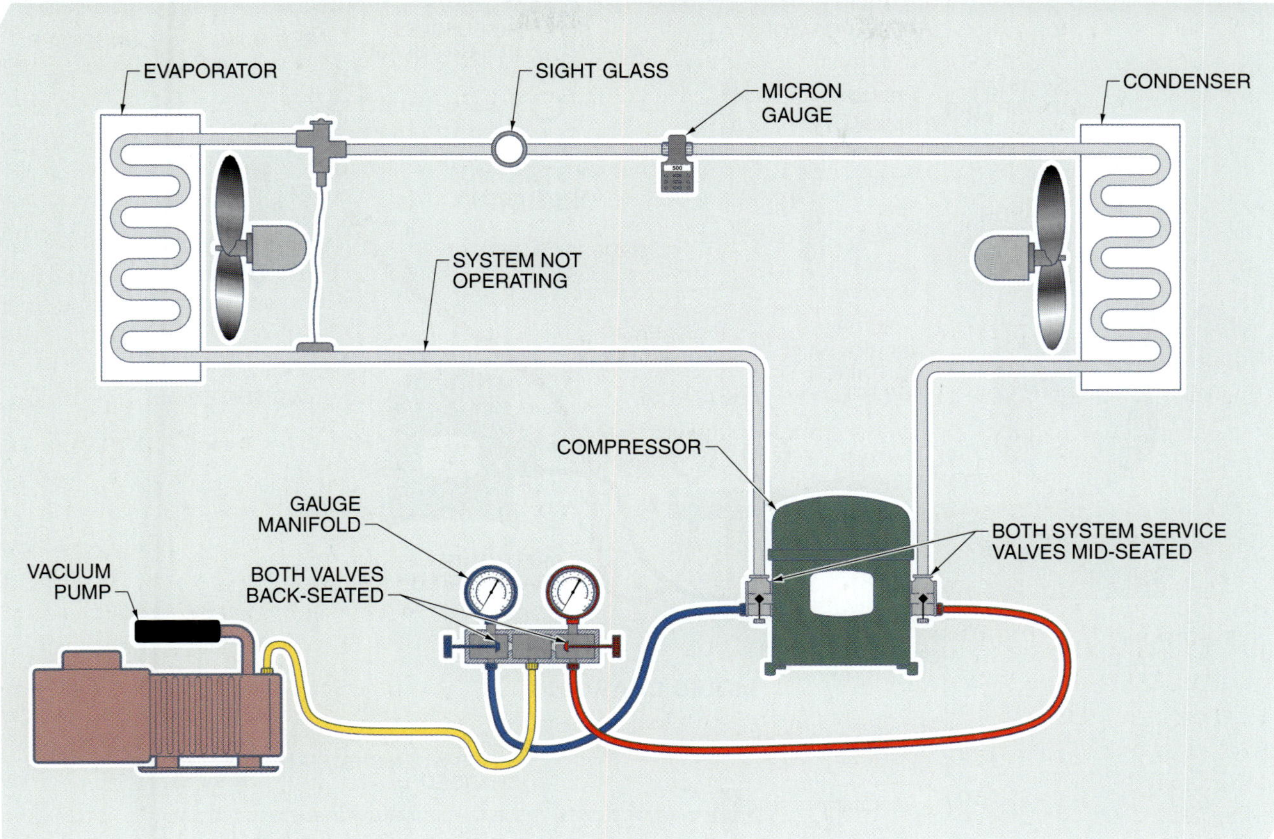

Figure 5-51. A vacuum pump is connected to the system using a gauge manifold and hoses.

System Charging

A system is charged with refrigerant after it has been completely evacuated. Liquid refrigerant is typically charged into the high-pressure side of the system when the system is completely empty. Smaller systems and some refrigerants require that refrigerant be added through the low-pressure side. A special fitting is attached to the red charging hose or the refrigerant container. This fitting contains a small opening that causes the liquid refrigerant to flash into vapor so that only vapor is added to the system even though it is being charged from the liquid side of the refrigerant container. The system is not operated during liquid charging and the refrigerant must be weighed into the system. **See Figure 5-52.**

The system is operated when the correct amount of refrigerant has been added or when the refrigerant stops flowing because pressure in the system equals pressure in the refrigerant container. When additional refrigerant is required to fill the system, it is charged into the low-pressure side as a vapor while the system is operating. An operating system that is low on refrigerant is also charged at the low-pressure side as a vapor.

The low pressure of the system's low-pressure side allows the high-pressure refrigerant from the container to flow into the system. Charging is complete when the correct amount is added by weight or when the bubbles disappear from the sight glass and system pressures and temperatures are within normal operating specifications.

The refrigerant container is placed upside down to charge with liquid and right side up to charge with vapor. Four-hose gauge manifolds do not require the removal of hoses during evacuation and charging and do not require purging. This prevents refrigerant release and prevents air from entering the system because no air enters the hoses once the evacuation is complete. Low-pressure chillers and ammonia systems require special evacuating and charging procedures.

Charging Systems

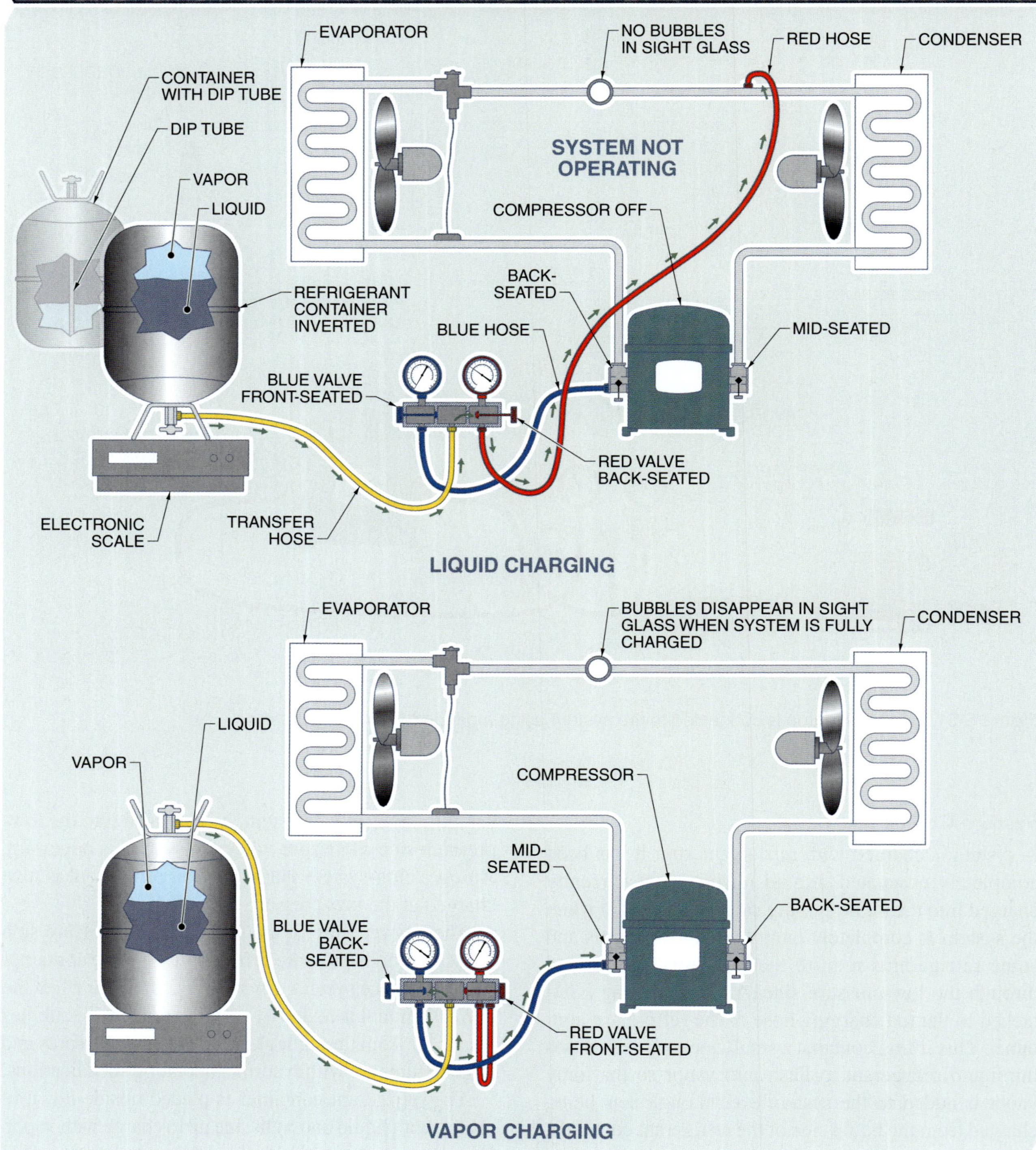

Figure 5-52. The refrigerant container is placed upside down (or right side up for a container with a dip tube) to charge with liquid and right side up to charge with vapor.

UNIT 6

Temperature, Pressure, and Thermodynamics

HVAC systems involve changing the physical property of substances such as air, water, and refrigerants. These properties are defined and should be understood in order to make calculations for purchasing equipment.

OBJECTIVES

- Describe the use of work, power, and horsepower formulas.
- Describe the basic principles and measurements of matter.
- Explain the various ways to measure heat.
- Define conduction, convection, and radiation.
- Explain pressure, atmospheric pressure, gauge pressure, and absolute pressure.
- Demonstrate the use of Charles' law, Boyle's law, Dalton's law, and the combined gas laws.
- Identify the components of refrigerant temperature-pressure charts.

ENERGY

In physics, the term "energy" describes the amount of work that can be performed by a force. There are different forms of energy such as heat (thermal), gravity, sound, light, chemical, electrical, and mechanical energy. Each form of energy exists in either a potential or kinetic state. *Potential energy* is stored energy that is not in motion. *Kinetic energy* is energy that is in motion and, therefore, doing work. Work, power, and horsepower are used to define exactly what kinetic energy is accomplishing.

TECH TIP

Work, power, and horsepower are found as ratings for HVAC equipment. For example, a steam boiler may have a boiler horsepower (BHP) rating. The equivalent of 1 BHP is 34.5 lb/hr of water evaporated at 212°F, or steam.

Work

Work is the amount of energy transferred by a force (in lb) over a specific distance (in ft). Work is performed when a force overcomes resistance. When a force does not cause movement, work has not been produced. Work is found by applying the following formula:

$W = F \times d$

where

W = work (in ft-lb)

F = force (in lb)

d = distance (in ft)

Example: Calculating Work

What is the work produced when a 1220-lb load (force) is raised a distance of 36′? **See Figure 6-1.**

$W = F \times d$

$W = 1220 \times 36$

$W =$ **43,920 ft-lb**

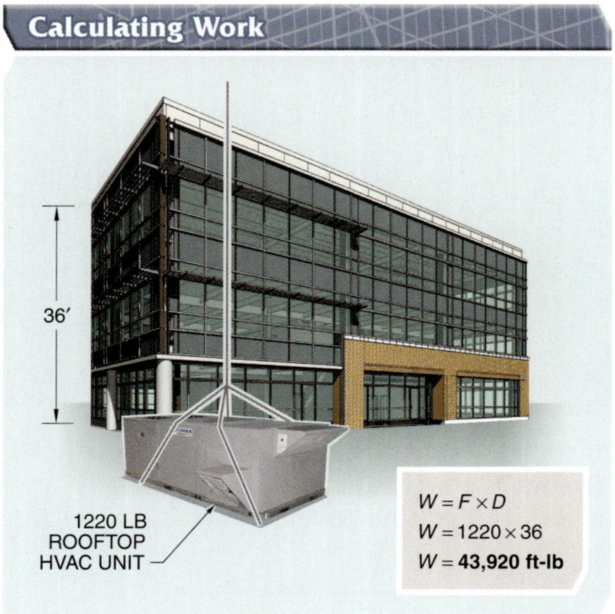

Figure 6-1. Work is determined by multiplying force (in lb) by distance (in ft).

Power

Power is the rate at which work (in ft-lb) is produced per unit of time (in sec). It is equal to work, or the product of force and distance, divided by the time it takes to do the work. Power is found by applying the following formula:

$$P = \frac{W}{t}$$

where
P = power (in ft-lb/sec)
W = work (in ft-lb)
t = time (in sec)

Example: Calculating Power

What is the power produced when a 1220-lb load (force) is moved a distance of 36′ in 40 sec? **See Figure 6-2.**

$$P = \frac{W}{t}$$

$$P = \frac{43,920}{40}$$

$$P = \mathbf{1098\ ft\text{-}lb/sec}$$

Horsepower

Horsepower is a unit of power equal to 550 ft-lb/sec or 745.7 W of electricity. The measurement equivalent of horsepower is derived from the power used by horses to pull a 550-lb load 1′ in 1 sec. It is used less frequently in modern applications. Horsepower is found by applying the following formula:

$$HP = \frac{P}{550}$$

where
HP = horsepower (in ft-lb/sec)
P = power (in ft-lb/sec)
550 = constant

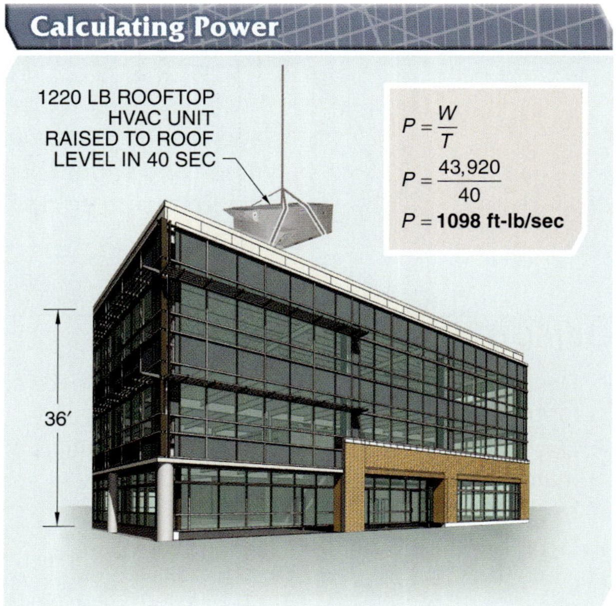

Figure 6-2. Power is determined by dividing work (ft-lb) by time (in sec).

Example: Calculating Horsepower

What is the horsepower developed when a 1220-lb load is moved a distance of 36′ in 40 sec? **See Figure 6-3.**

$$HP = \frac{P}{550}$$

$$HP = \frac{1098}{550}$$

$$HP = \mathbf{2\ HP}$$

ENERGY UNITS

In physics, energy is defined as the ability to do work. Different types of measurements are used for different forms of energy. The forms of energy used in refrigeration and HVAC systems include mechanical, electrical, and heat energy. **See Figure 6-4.**

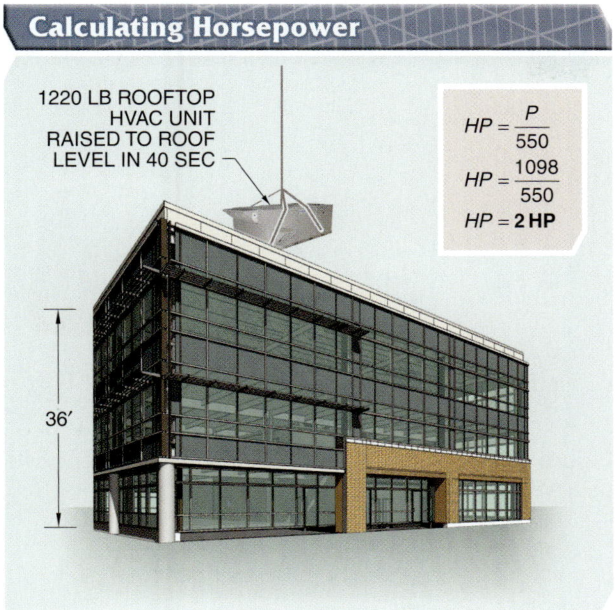

Figure 6-3. Horsepower is determined by dividing power (ft-lb/sec) by 550 (a constant number).

Energy Units

Energy Unit	Description
Horsepower (HP)	Equal to 550 lb being moved 1 ft in 1 sec
Joule (J)	Equal to the force of 1 newton acting on a distance of 1 meter in 1 sec
Watt (W)	Equal to the release of 1 A of electricity at 1 V; 1 W = 3.413 Btu
British thermal unit (Btu)	Equal to the amount of heat required to raise 1 lb of water 1°F
Calorie (cal)	Equal to the amount of heat required to raise 1 g of water 1°C
Therm	Equal to 100,000 Btu

Figure 6-4. Each form of energy has its own specific measurement system and abbreviation.

Mechanical

Mechanical energy is the energy of work. It is measured in foot-pounds (ft-lb). Mechanical energy can be in the form of either potential energy, which is stored energy, or kinetic energy, which is energy in motion. The sum of potential and kinetic energy is total mechanical energy. When heat energy is used to do work, it is also called mechanical energy.

Electrical

Electrical energy typically is measured in watts or kilowatts (1000 W). A watt is a measurement of the amount of work performed by electrical energy. A watt is equal to current (electrical flow in amperes) multiplied by voltage (in V).

Heat

Heat is a form of energy identified by a temperature change or change in the state of matter. All matter exists as a solid, liquid, or gas. In refrigeration and HVAC systems, a change of state occurs when enough heat has been added or removed from matter to change the matter from one physical state to another.

Heat Measurement. The U.S. customary system of measurement uses the British thermal unit (Btu) as the unit of measure for heat. A Btu is the quantity of heat required to raise the temperature of 1 lb of water 1°F. Large quantities of heat are measured in thousands of Btu per hour (Mbh). A *therm* is a measurement of heat that equals 100,000 Btu.

MATTER AND MEASUREMENTS

Matter is any substance that has weight and occupies space. All matter is composed of elements and atoms. An *element* is a substance that cannot be broken down into a simpler form of matter. An *atom* is the smallest unit of an element that holds the properties of the element. **See Figure 6-5.**

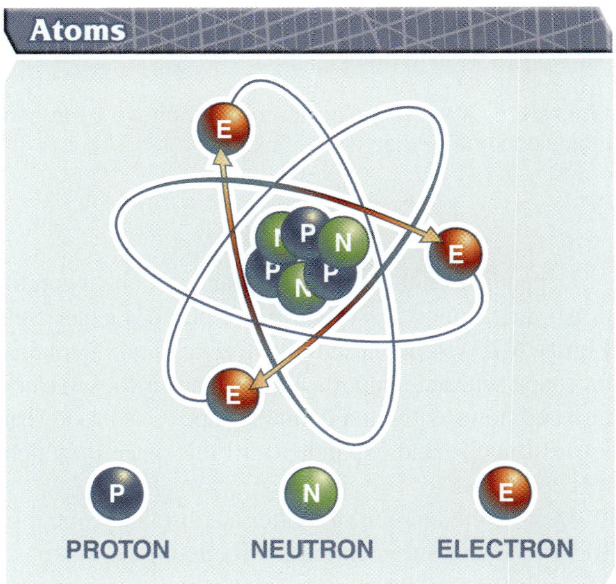

Figure 6-5. An atom consists of a nucleus, which includes protons and neutrons, and a cloud of electrons.

An atom consists of a nucleus, which is composed of particles called protons and neutrons. A *proton* is a positively charged particle in the nucleus of an atom. A *neutron* is an uncharged particle in the nucleus of an atom. A cloud of particles, called electrons, surrounds the nucleus. An *electron* is a negatively charged particle that makes up the cloud of an atom.

Atoms for each element have a certain number of neutrons, protons, and electrons. Common examples of elements are oxygen and hydrogen. When elements are joined together, they form molecules. A molecule of water, for example, consists of two hydrogen atoms and one oxygen atom. **See Figure 6-6.**

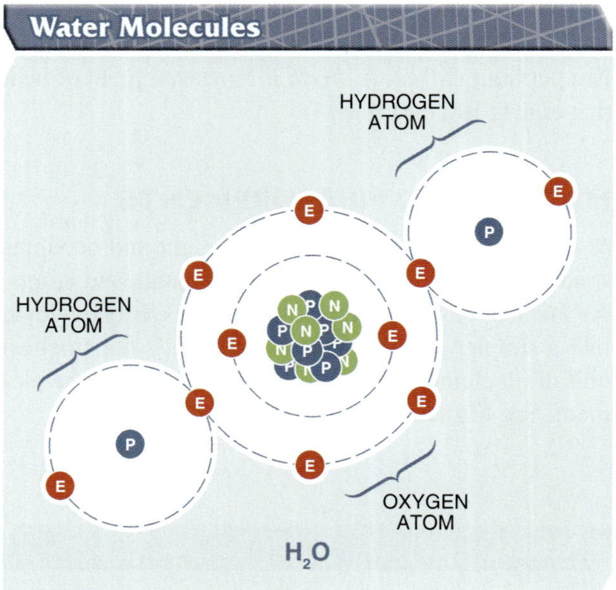

Figure 6-6. A water molecule consists of two hydrogen atoms and one oxygen atom.

Depending on the type of substance and heat conditions, matter may exist as a solid, liquid, or gas. **See Figure 6-7.** A solid has a tight structure and maintains its shape without support. A liquid has a loose structure and flows without a defined shape. Gas has a very loose structure and expands to fill the space in which it is contained.

Mass is the amount of matter a substance contains. *Weight* is a measurement of gravity acting on the mass of a substance. All three states of matter have weight and can be measured or compared by their density and specific gravity.

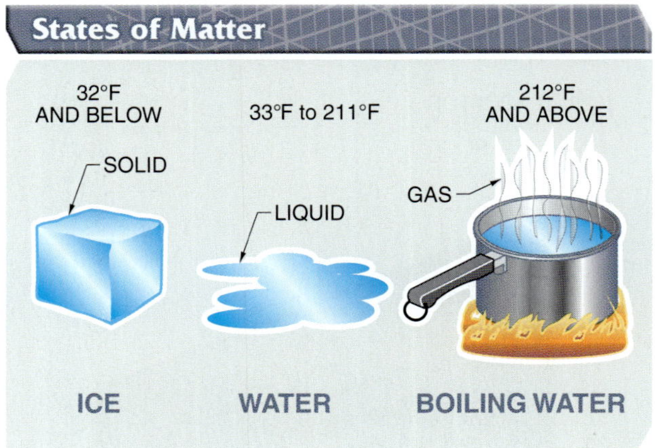

Figure 6-7. Depending on the temperature, matter exists as a solid, liquid, or gas at 0 psi.

Density

Density is the mass of a substance per unit of volume. In the U.S. customary system of measurement, density is expressed in pounds per cubic feet (lb/ft^3). Density can be used to compare substances. **See Figure 6-8.** For example, the density of gold is 1204 lb/ft^3, while the density of copper is 555 lb/ft^3. This shows that gold is more than two times heavier than copper.

Density and Specific Gravity

Substance	Density*	Specific Gravity
Aluminum	168	2.7
Brass	536	8.7
Copper	555	8.92
Gold	1204	19.3
Ice	57	0.92
Iron	490	7.86
Lithium	33	0.53
Tungsten	1186	19.0
Mercury	845	13.54
Water	62.4	1.0

* in lb/ft^3

Figure 6-8. All substances have a unique density and specific gravity that can be used as identification.

Specific Gravity

Specific gravity is the density of a solid, liquid, or gas based on the density of water. The specific gravity of water is 1. Since iron has a specific gravity of 7.86, a cubic foot of iron weighs 7.86 times a cubic foot of water.

The density of gas is also measured by its specific volume. *Specific volume* is the volume that one pound of a gas substance occupies at a particular temperature and pressure condition and is measured in cubic feet per pound (ft^3/lb). Specific volume is measured in cubic feet per pound (ft^3/lb). For example, at a temperature of 68°F and pressure of 29.92 in. Hg absolute, the specific volume of air is 13.33 ft^3/lb and the specific volume of hydrogen is 179 ft^3/lb. When air and hydrogen mix together, the hydrogen rises because it has a much lighter specific volume.

TECH TIP

Most thermometers used in the United States typically include the Fahrenheit (°F) scale. The Fahrenheit scale is used for weather, industrial, cooking, and medical purposes. However, the Celsius scale (°C) is used for scientific work in laboratories. Outside the United States, the Celsius scale is used for all purposes.

HEAT

Heat is the energy of a substance resulting from the internal motion of its molecular structure. *Temperature* is the measurement of the intensity of heat. Temperature is measured with a thermometer. A number of units can be used to measure temperature. In the U.S. customary system, temperature is measured in degrees Fahrenheit (°F). **See Figure 6-9.** Water freezes at 32°F and boils at 212°F. Refrigeration and HVAC technicians in the United States primarily use Fahrenheit units to measure temperature.

Fahrenheit and Celsius Temperature Comparisons

°F*	°C†	°F	°C
212	100	60	16
200	93	40	4
180	82	32	0
160	71	20	-7
140	60	0	-18
120	49	-20	-29
100	38	-40	-40
80	27	-60	-51

* in degrees Fahrenheit
† in degrees Celsius

Figure 6-9. The U.S. customary temperature scale is the Fahrenheit scale, while the metric system temperature scale is the Celsius scale.

The vast majority of nations outside the United States use the metric system. In the metric system, temperature is measured in degrees Celsius (°C). Water freezes at 0°C and boils at 100°C. Conversion formulas are used to convert degrees Fahrenheit to degrees Celsius and degrees Celsius to degrees Fahrenheit. **See Figure 6-10.** To convert degrees Fahrenheit to degrees Celsius, apply the following formula:

$$°C = \frac{°F - 32}{1.8}$$

where
$°C$ = metric system temperature (in °C)
$°F$ = U.S. customary system temperature (in °F)
32 = constant
1.8 = constant

Fahrenheit To Celsius Temperature Conversions
Interactive Animation

Temperature System Conversion

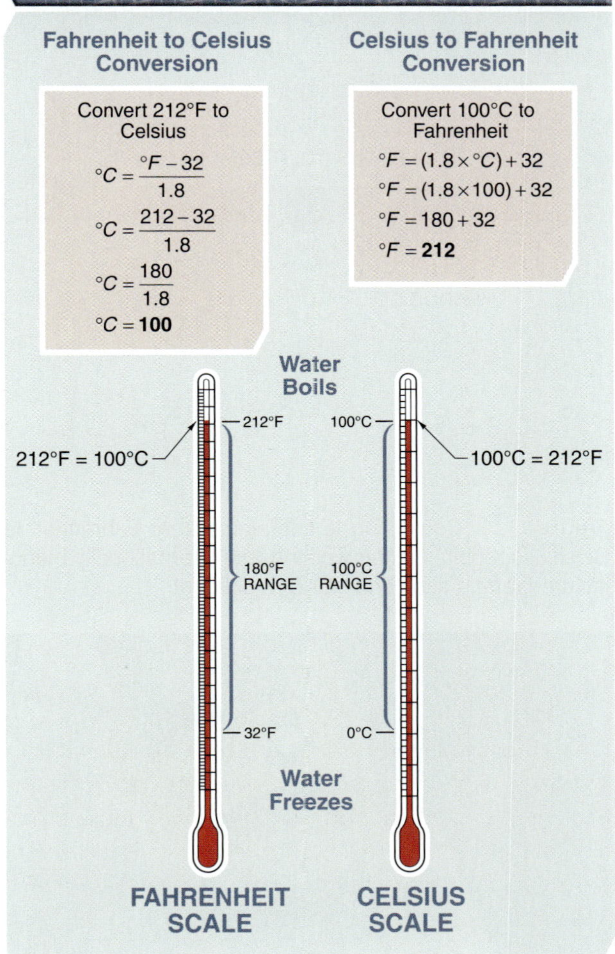

Figure 6-10. Temperatures in refrigeration and HVAC systems are commonly expressed in degrees Fahrenheit or Celsius and can be converted from one system to the other using formulas.

Example: Converting Fahrenheit to Celsius

An air conditioning system delivers air of 55°F to a building space. What is the temperature in degrees Celsius? **See Figure 6-11.**

$$°C = \frac{°F - 32}{1.8}$$

$$°C = \frac{55 - 32}{1.8}$$

$$°C = \frac{23}{1.8}$$

$$°C = \mathbf{12.8}$$

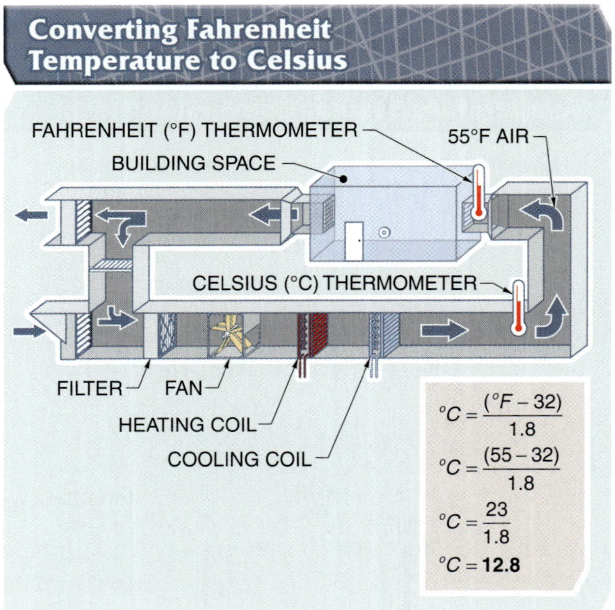

Figure 6-11. Converting temperatures from Fahrenheit to Celsius is required when working with internationally manufactured refrigeration and HVAC equipment.

Fluke Corporation

All HVAC meters can switch between Fahrenheit and Celsius scales with the push of a button.

To convert degrees Celsius to degrees Fahrenheit, apply the following formula:

$$°F = (1.8 \times °C) + 32$$

where

$°F$ = U.S. customary system temperature (in °F)

1.8 = constant

$°C$ = metric system temperature (in °C)

32 = constant

Example: Converting Celsius to Fahrenheit

The air temperature in a room is 22°C. What is the temperature in degrees Fahrenheit? **See Figure 6-12.**

$$°F = (1.8 \times °C) + 32$$

$$°F = (1.8 \times 22) + 32$$

$$°F = 39.6 + 32$$

$$°F = \mathbf{71.6}$$

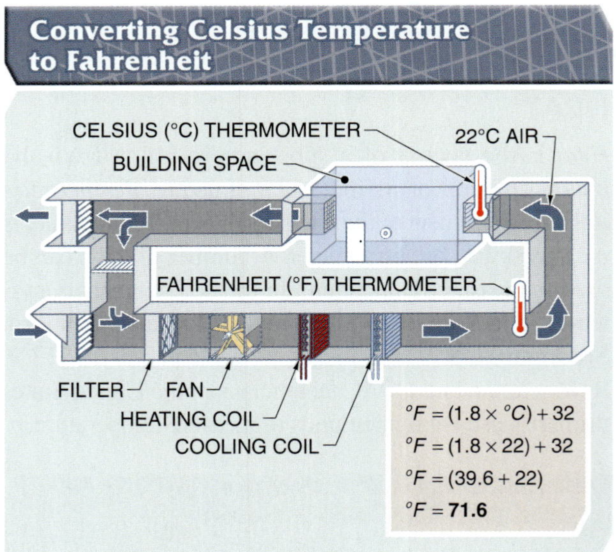

Figure 6-12. Converting temperatures from Celsius to Fahrenheit is a common task performed for refrigeration system devices and components that are purchased from other countries to allow rating comparison.

Temperature Scales

Both the Fahrenheit scale and Celsius scale are based on human observations of water freezing and boiling points. A human observation of water uses a narrow, limited range of temperatures. When heat is a measurement of molecular activity, 0°C or 32°F does not imply that molecular activity is absent or at a low level.

In fact, molecular activity itself continues well below human observations of water. *Absolute zero* is a theoretical condition in which a substance has no molecular movement. U.S. customary and metric temperature scales were developed for human experiences with naturally occurring substances, such as water. They were not developed from a temperature scale based on absolute zero.

The Rankine scale indicates absolute zero using U.S. customary measurements. **See Figure 6-13.** Fahrenheit is easily converted to degrees Rankine (°R) by adding 460 to the base Fahrenheit value. For example, by adding 460 to the freezing point of water in Fahrenheit (32°F), the corresponding absolute temperature is 492°R on the Rankine scale. To obtain degrees Kelvin (K), a value of 273 is added to the base Celsius value. For example, the boiling point of water in degrees Celsius (100°C) is 373 K on the Kelvin scale.

Absolute Temperature Conversions
Interactive Animation

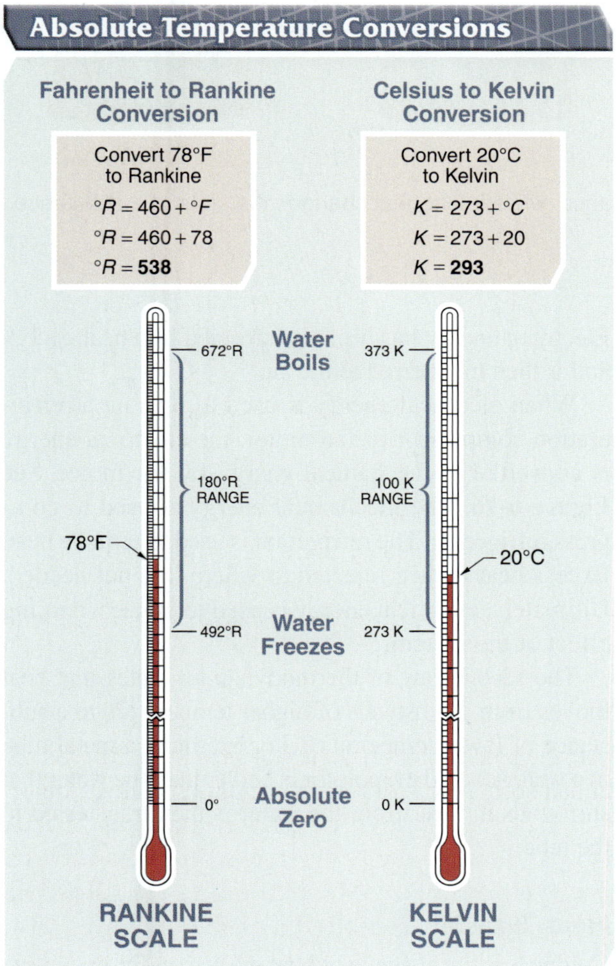

Figure 6-13. The base of both the Rankine and Kelvin scales is absolute zero. The Rankine scale is used with Fahrenheit temperatures and the Kelvin scale is used with Celsius.

British Thermal Units

It is important to measure the intensity of heat. However, it is also important to measure the amount of heat contained in a substance. Heat content is measured in British thermal units on the U.S. customary scale. A *British thermal unit (Btu)* is the quantity of heat required to change the temperature of 1 lb of water 1°F. **See Figure 6-14.** Water has a specific heat of 1, so it takes 1 Btu to raise 1 lb of water 1°F. To raise 500 lb of water from 50°F to 150°F would require adding 50,000 Btu (500 lb × 100°F = 50,000 Btu).

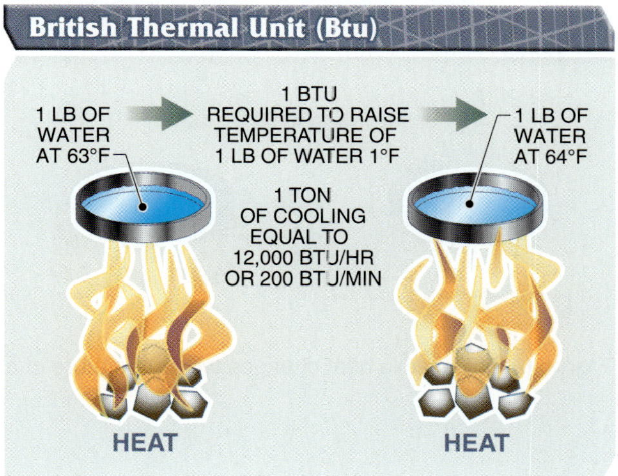

Figure 6-14. One British thermal unit (Btu) is the quantity of heat required to raise 1 lb of water 1°F at sea level.

The phrase "ton of cooling" is often used in refrigeration. Ton of cooling originally referred to the cooling available in a physical ton of ice. Today, ton of cooling equals 12,000 Btu of heat removed in 1 hr or 200 Btu removed in 1 min.

Heat can also be categorized by its effect on a substance. The corresponding metric unit to the Btu is the calorie. However, HVAC-R technicians often do not use calorie measurements. A common energy conversion used for HVAC is 1 ton of refrigeration = 3.5168 kW of refrigeration.

Sensible Heat

Sensible heat is the heat applied to or removed from a substance that results in a temperature change. For example, there is a transfer of sensible heat when water temperature changes from 100°F to 150°F. Sensible heat can be measured with a thermometer. **See Figure 6-15.**

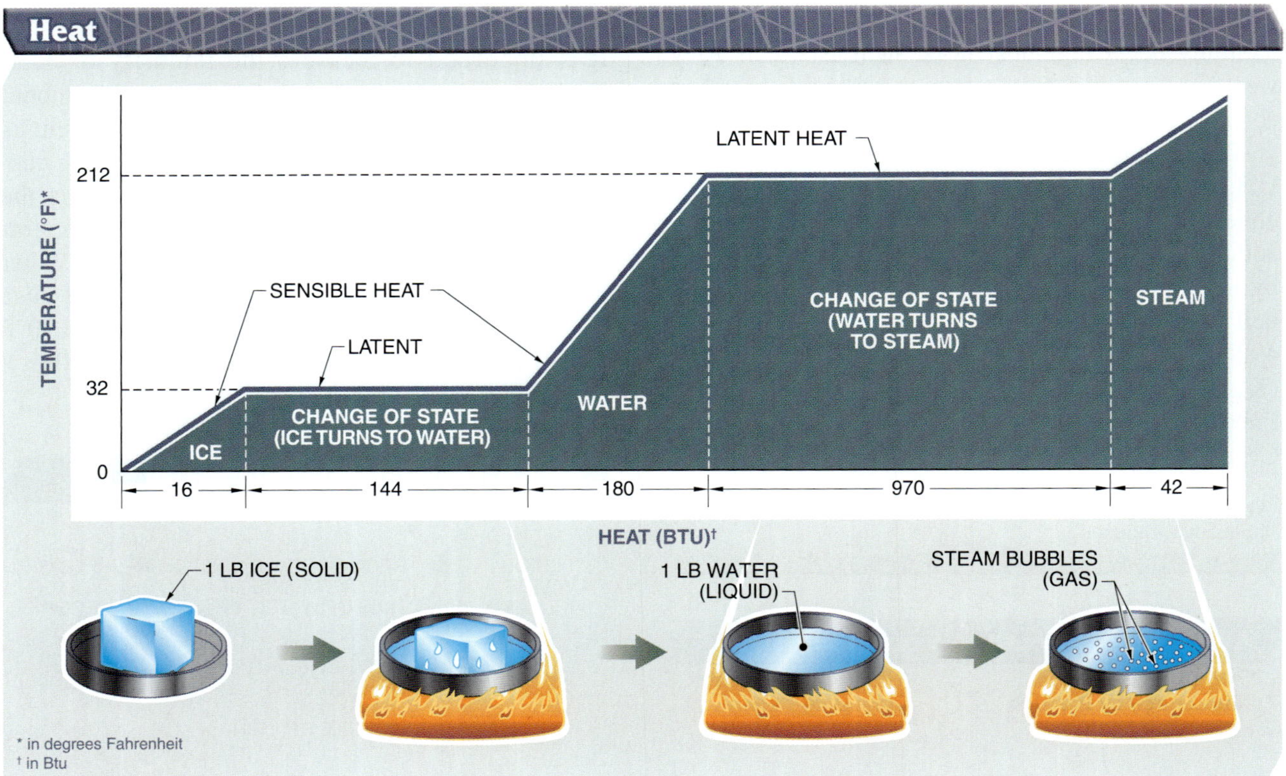

Figure 6-15. Sensible heat changes the temperature of a substance, while latent heat changes the state of a substance.

Latent Heat

Latent heat is the heat applied to or removed from a substance that results in a change of state. While a substance is changing state, its sensible heat remains constant. When latent heat is applied to ice at 32°F, the ice changes state to become water at 32°F. Latent heat has been added even though the sensible heat is the same.

Once ice melts into water, any additional heat results in a sensible heat change, such as when water is heated from 32°F to 212°F. When more heat is added, the water begins to boil into steam. Latent heat also causes this change of state from a liquid to a gas.

THERMODYNAMICS

Thermodynamics is the scientific study of the behavior of heat. Two major laws of thermodynamics affect refrigeration and HVAC systems. The first law of thermodynamics states that energy cannot be created or destroyed but can only be converted from one form to another. For example, when electric current flows through resistance heating elements located in an air duct, the elements become hot. Electrical energy has been transformed into heat energy and is then transferred to the air.

When electrical energy is used to operate a refrigeration compressor or fan motor, the electrical energy is converted to mechanical energy by the motor. **See Figure 6-16.** The mechanical energy is used to compress refrigerant. The refrigerant is used to remove heat. Excess heat is then rejected to where it is not needed. Ultimately, electrical energy is used to create a cooling effect of the system.

The second law of thermodynamics states that heat moves from a substance of higher temperature to a substance of lower temperature. For example, a metal tube in a water-cooled evaporator is cooler than the water that surrounds it. Heat from the water is then transferred to the tube.

Heat Transfer

Coldness is the absence of heat or the condition that exists after heat has been removed. Heat can be transferred from one material to another by means of conduction, convection, or radiation. **See Figure 6-17.**

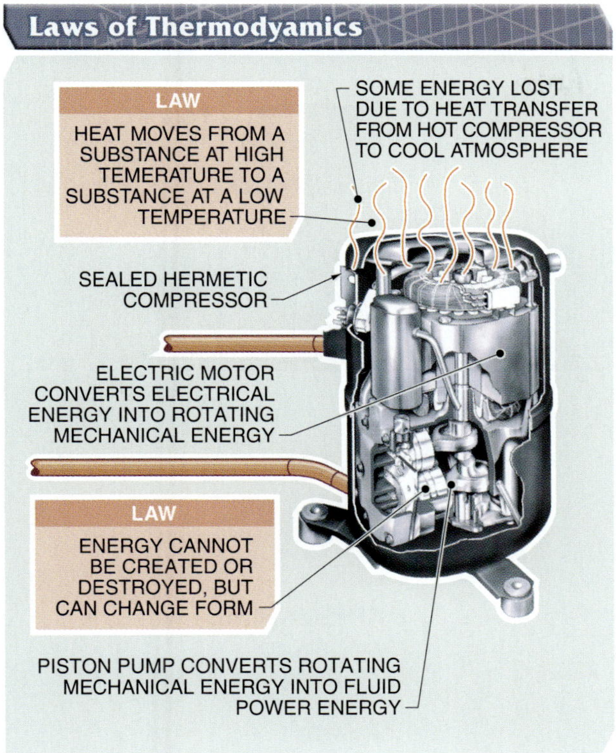

Figure 6-16. The laws of thermodynamics state that energy can neither be created nor destroyed but can only change state, and that heat always travels from a substance of high temperature to a substance of low temperature.

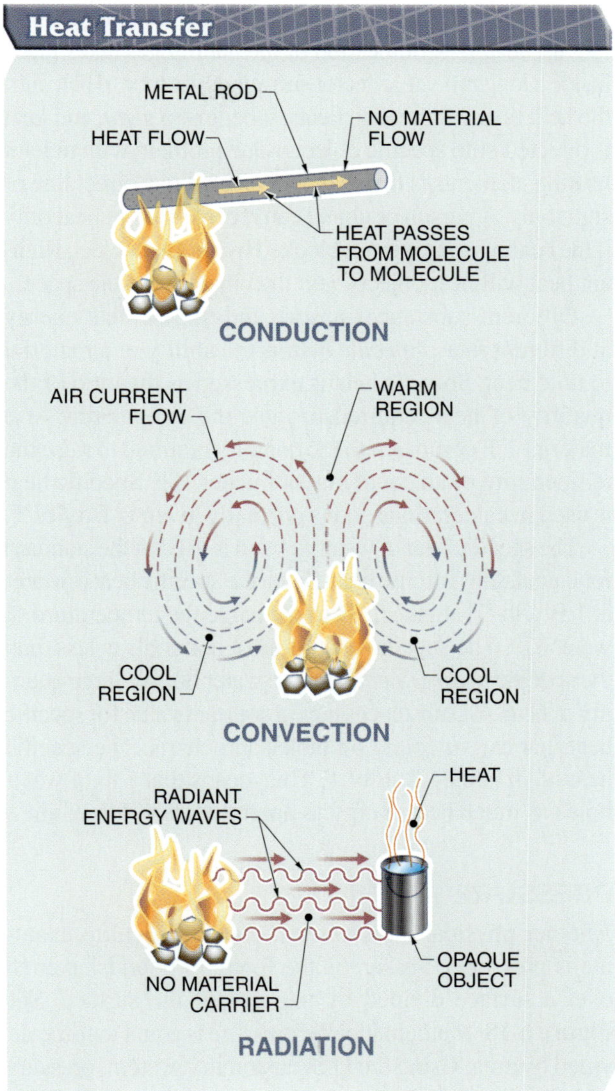

Figure 6-17. Heat transfer takes place by conduction, convection, or radiation.

Conduction. *Conduction* is a method of heat transfer caused by the direct physical contact of substances. In conduction, heat travels from molecule to molecule until it reaches the end of a substance. For example, in a water-cooled evaporator, metal tubes physically touch the water. The direct contact of the tubes with the water allows the conduction of heat.

Convection. *Convection* is a method of heat transfer in which heat is moved from one location to another through the circulation of a substance. The most common substances for circulation used in the convection method are air and water.

For example, hot water baseboard heaters transfer heat from piping to room air by convection and the room air rises as it is heated. Heat is then transferred to an object, such as a person. As heat is transferred to an object, the air cools and falls back to the floor. A natural circulation pattern is created that causes heat to be transferred repeatedly. In some systems, natural convection works slowly, so fans are used to force air movement.

Radiation. *Radiation* is a method of heat transfer in which invisible heat rays move heat. Hot objects emit infrared heat rays, most of which are invisible to the naked eye. When an object is heated to a temperature high enough, it begins to glow and the radiation can be seen as visible light.

Radiation heat transfer is used in applications such as radiant heat panels and natural-gas-fired radiant heaters. A radiant heat panel has electric heating elements embedded in it and may be disguised as a typical ceiling tile. As the heating elements heat the panel, it emits heating rays onto objects and people.

Many garages and loading docks use natural-gas-fired radiant heaters. Combustion of the natural gas takes place inside a long tube. A reflector surrounds the tube to help aim the heat rays. As the tube heats, it begins to glow, and heat is directed onto specific objects. One problem with radiant heating elements is that heat is projected in a direct line of sight only. A person or object can receive radiant heat only if the heating source is not blocked by another object. Radiant heat will heat objects, and the objects heat the space.

Different substances absorb and release heat energy at different rates. *Specific heat* is the ability of a material to hold heat. Specific heat is expressed as the ratio of the quantity of heat required to raise the temperature of a material 1°F compared to the quantity required to raise the temperature of an equal mass of water 1°F. Specific heat is used in calculations and is normally given in Btu/lb/°F.

The specific heat of water is 1 and is used as the standard for calculating specific heat. Since the specific heat of water is 1 Btu/lb/°F, it takes 1 Btu to change the temperature of water 1°F. The specific heat of most materials is less than the specific heat of water because water holds a large quantity of heat. All substances have a constant value for specific heat that can be found on tables and charts. The specific heat of air is 0.24 Btu/lb/°F. This means that 1 lb of water holds as much heat energy as approximately 4 lb of air.

PRESSURE

Another physical measurement that is critical to examine is pressure. *Pressure* is the force, in pounds, exerted over a surface divided by the area of the surface. **See Figure 6-18.** Mathematically, pressure is equal to force divided by area. Using the U.S. customary system, pressure is defined as the weight of a substance pushing against a square inch (lb/in^2 or psi). For refrigeration and fluid power system purposes, pressure is defined as resistance to flow.

Atmospheric Pressure

Atmospheric pressure is the force exerted by the weight of the atmosphere (gas molecules) on the earth's surface. Pressure due to the weight of molecules may be stated as 14.696 lb/in^2 or 14.7 psia at sea level. **See Figure 6-19.**

Gauge Pressure

Gauge pressure is a pressure measurement above atmospheric pressure that is used to express pressure derived from the operation of a closed system. Since atmospheric pressure does not contribute to the operation of a system, it is not counted in measurements. Gauge pressure is widely used and expressed as pounds per square inch gauge (psig or psi). Since gauge pressure does not include atmospheric pressure, 0 psi = 14.7 psia (pounds per square inch absolute).

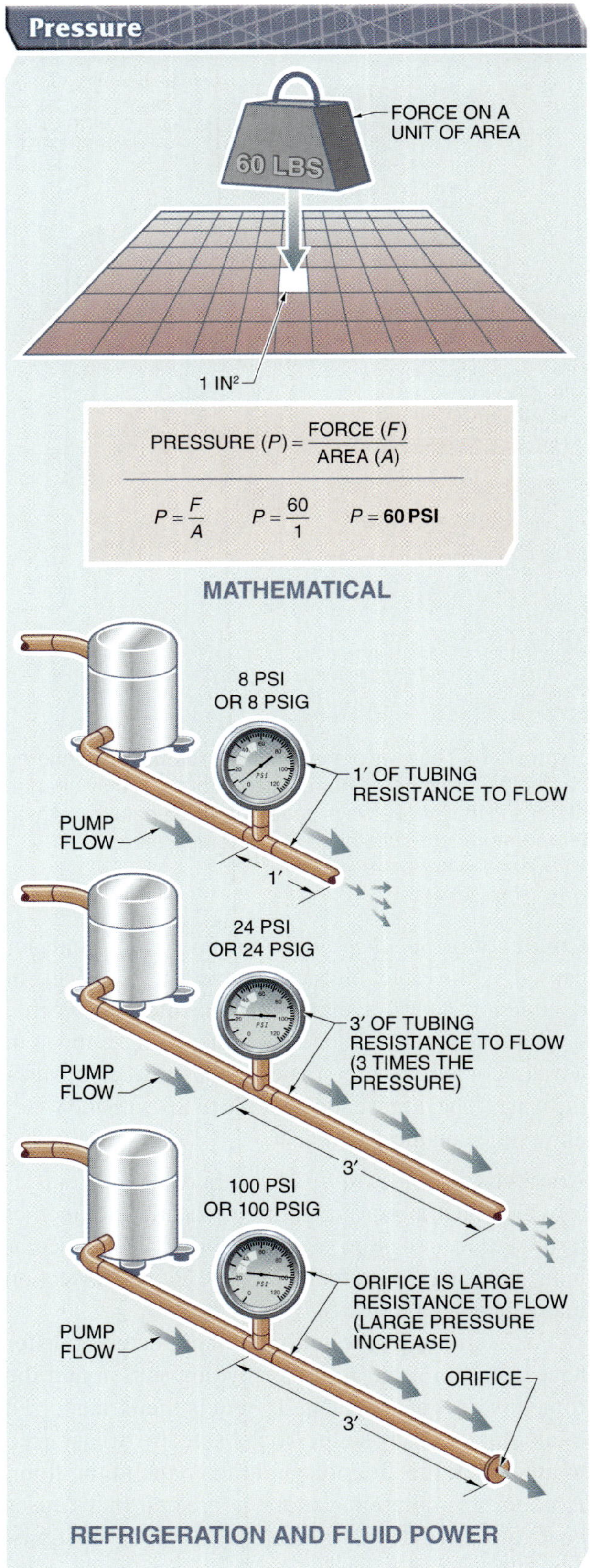

Figure 6-18. For refrigeration and HVAC system usage, pressure is defined as resistance to flow.

Atmospheric Pressure

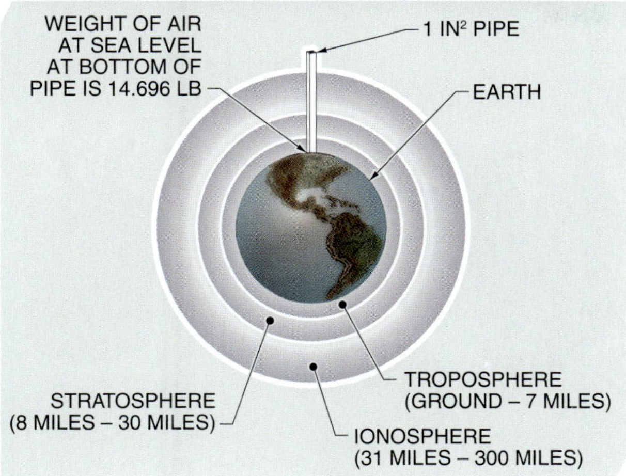

Figure 6-19. Atmospheric pressure is the weight of the atmosphere pushing against 1 in^2 of area at sea level.

Absolute Pressure

Absolute pressure is any pressure above a perfect vacuum (0 psia). Absolute pressure is the sum of gauge pressure plus atmospheric pressure. Absolute pressure is measured in pounds per square inch absolute (psia). Gauge pressure can be easily converted to absolute pressure by adding atmospheric pressure (14.7 psi). Absolute pressure is expressed mathematically as the following:

$$psia = psi + 14.7$$

where

$psia$ = pounds per square inch absolute

psi = pounds per square inch gauge

14.7 = constant

Pressure and Absolute Pressure Conversions
Interactive Animation

Example: Calculating Absolute Pressure

If a refrigeration system gauge reads 260 psi, what is the absolute pressure? **See Figure 6-20.**

$$psia = psi + 14.7$$
$$psia = 260 + 14.7$$
$$psia = \mathbf{274.7}$$

Absolute Pressure Media Clip

Sections of refrigeration systems operate at pressures less than atmospheric pressure. *Vacuum* is any pressure that is less than atmospheric pressure. Vacuum is not measured on a standard psi scale.

A *mercury barometer* is a test instrument used to measure atmospheric pressure. It is a glass tube that is closed on one end and filled with mercury. The tube is inverted with the open end submerged in a dish of mercury. **See Figure 6-21.** Atmospheric pressure on the mercury in the open dish prevents the mercury in the tube from running out. A vacuum is created at the top of the tube as the mercury tries to run out. The height of mercury in the tube corresponds to the atmospheric pressure acting on the mercury in the open dish.

When the altitude of the mercury-filled dish increases, the amount of atmosphere and weight of air above the dish decrease. This causes the pressure on the surface of the mercury to decrease.

Vacuum is measured in inches of mercury (in. Hg). The highest vacuum possible is 29.92 in. Hg. When air is removed from a sealed container, pressure decreases and a partial vacuum is formed. When all air is removed, a perfect vacuum exists. *Note:* It is impossible to remove 100% of the air from any container or space.

Refrigeration system gauges and large fan gauges use the inches of mercury scale to measure the vacuum in a system. **See Figure 6-22.** Atmospheric pressure has no vacuum, so atmospheric pressure (0 psi or 14.7 psia) is equal to 0 in. Hg. A perfect vacuum is 29.92 in. Hg and is often rounded to 30 in. Hg.

Inches of mercury measurements are not precise enough for all vacuum measurements. For precise vacuum measurements, the micron scale is used. A micron is based on metric millimeters. For example, 1000 microns = 1 mm, and 1 micron = 1⁄1000 mm absolute pressure.

Fluke Corporation

Some HVAC meters can measure barometric pressure and the pressure differentials of air or liquid systems to check flow.

Calculating Absolute Pressure

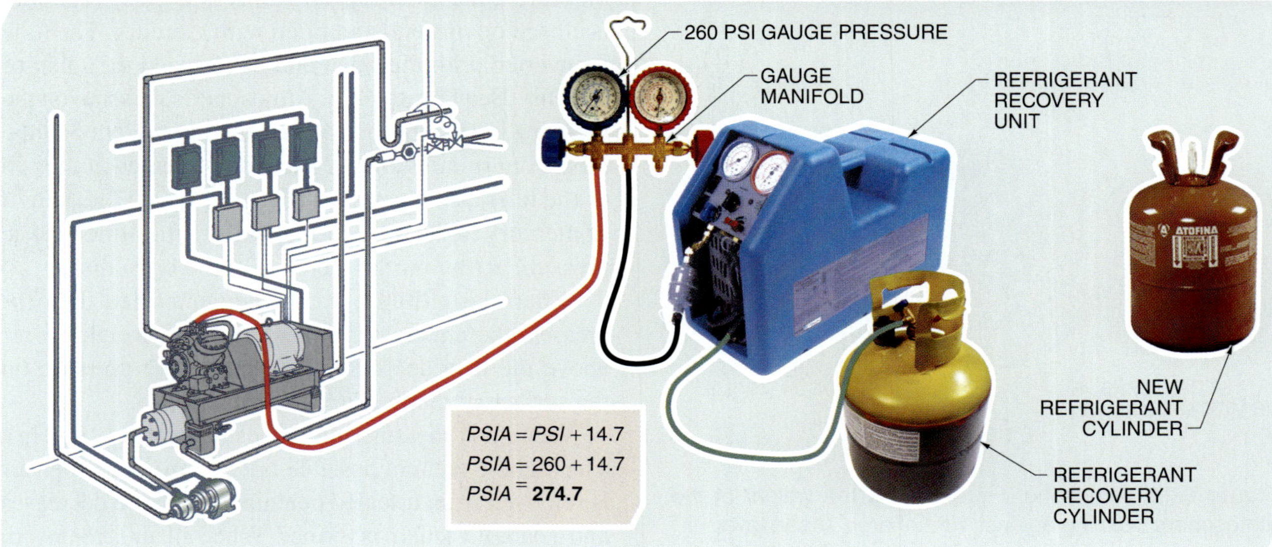

Figure 6-20. All gas law formulas require that absolute pressure be used when making calculations.

Mercury Barometers

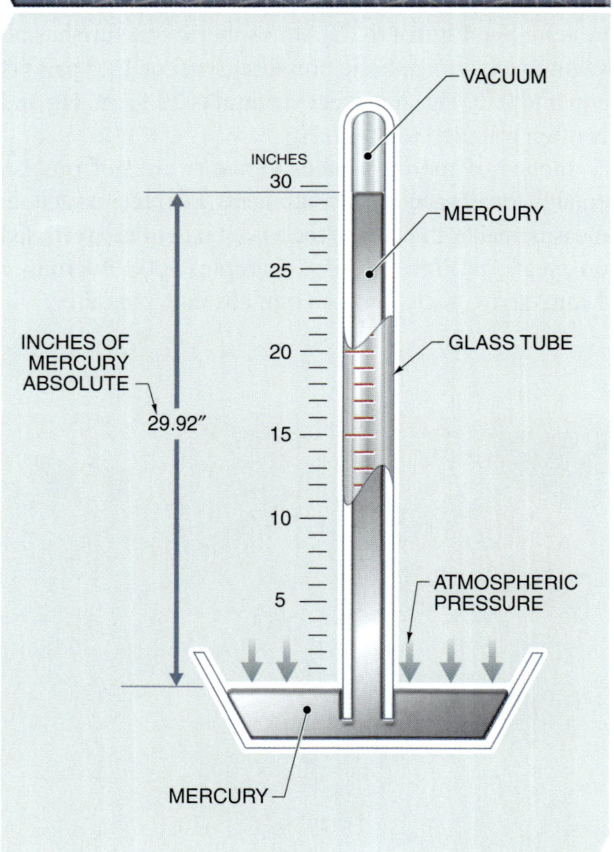

Figure 6-21. Atmospheric pressure measurements correspond with the height of mercury (in in. Hg abs) in a mercury barometer.

Pressure and Vacuum Equivalents

	PSI or PSIG	PSIA	Hg ABS*	Vacuum	
				Hg*	Micron
2 Atmosphere	14.696	29.392	59.84	—	—
1 Atmosphere	0.0	14.696	29.92	0.0	760,000
	—	12.24	24.92	5.0	632,968
	—	4.912	10.0	19.92	254,000
	—	2.456	5.0	24.92	127,000
	—	1.0	2.036	27.884	51,715
	—	0.452	0.92	29.00	23,368
	—	0.099	0.2	29.72	5,000
	—	0.019	0.1	29.82	1,000
	—	0.010	0.019	29.901	500
	—	0.002	0.003	29.917	100
	—	0.001	0.001	29.919	50
Perfect vacuum	—	0	0	29.92	0

* in in.

Figure 6-22. Vacuum is measured in inches of mercury and works below atmospheric pressure. Extremely accurate vacuums are measured in microns.

Pressure and Vacuum Conversions
Interactive Animation

Pressure Gauges

Bourdon tube and Schrader pressure gauges are used to indicate pressures in refrigeration and HVAC systems. A *Bourdon tube pressure gauge* is a mechanical pressure gauge with a curved tube made of thin metal that acts as a pressure-sensing element. **See Figure 6-23.** The tube straightens as pressure is applied in a system. Typically, the tube is made of brass.

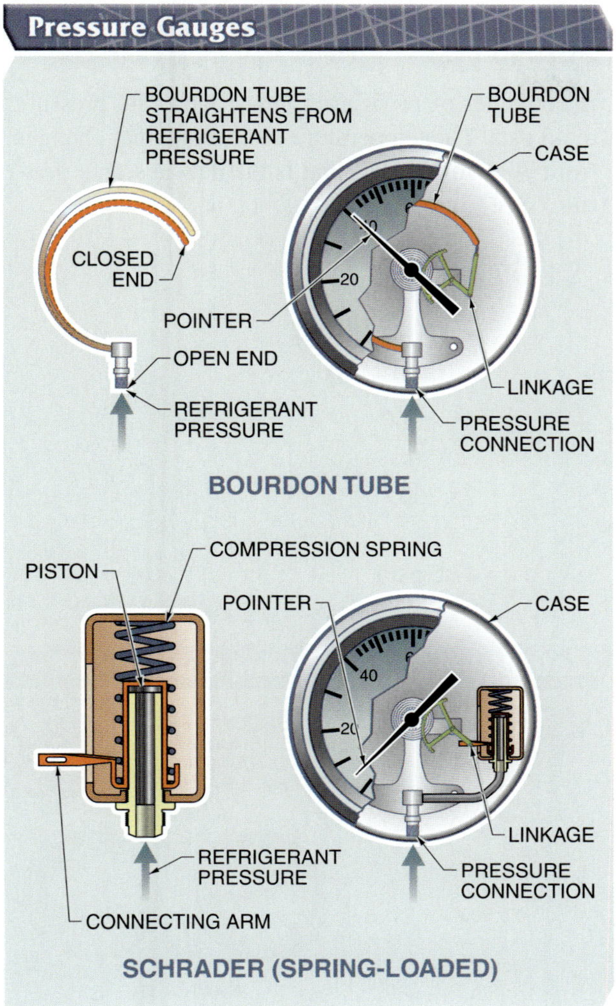

Figure 6-23. Bourdon tube and Schrader (spring-loaded) pressure gauges are used to determine pressure in refrigeration and HVAC systems.

The opposite end of a Bourdon tube is connected to a gear and pointer mechanism that translates the straightening movement of the tube into an indicated number on the gauge. The gauge indicates how much pressure (in psi) has been applied. Bourdon tube gauges are the most accurate type of mechanical pressure gauge.

A *Schrader (spring-loaded) pressure gauge* is a mechanical pressure gauge that has a small, single-acting spring return cylinder that acts as a pressure-sensing element. A piston is connected to a gear and pointer mechanism that translates piston movement into an indicated number on the gauge. Schrader gauges are not as accurate as Bourdon tube gauges but they receive less damage from pressure surges.

GAS LAWS

Scientists around the world have observed the behavior of gases at different conditions and created gas laws. These gas laws are Charles' law, Boyle's law, Dalton's law of partial pressures, and the combined gas law.

Charles' Law

Charles' Law is named after the French scientist Jacques Charles. *Charles' law* is a gas law stating that when the pressure on a gas is constant, the volume of the gas is directly proportional to absolute temperature, and when the volume of a gas is constant, the pressure on the gas is directly proportional to absolute temperature. The volume of a gas increases as heat is added to the gas. Absolute temperature is based on the Rankine scale. The volume-temperature relationship of a gas can be expressed using the following formula:

$$\frac{V_1}{T_1} = \frac{V_2}{T_2}$$

where

V_1 = volume of the first condition (in ft^3)

T_1 = absolute temperature of the first condition (in °R)

V_2 = volume of the second condition (in ft^3)

T_2 = absolute temperature of the second condition (in °R)

Charles' law can be rearranged mathematically to find any desired value when the other three values are known. For example, the formula can be rearranged as:

$$V_2 = \frac{V_1 \times T_2}{T_1}$$

Example: Calculating Volume Using Charles' Law

If 500 ft^3 of air is cooled from 105°F to 70°F, what is the new volume of air? **See Figure 6-24.**

$$V_2 = \frac{V_1 \times T_2}{T_1}$$

$$V_2 = \frac{500 \times (460 + 70)}{(460 + 105)}$$

$$V_2 = \frac{500 \times 530}{565}$$

$$V_2 = \frac{265{,}000}{565}$$

$$V_2 = \mathbf{469\ ft^3}$$

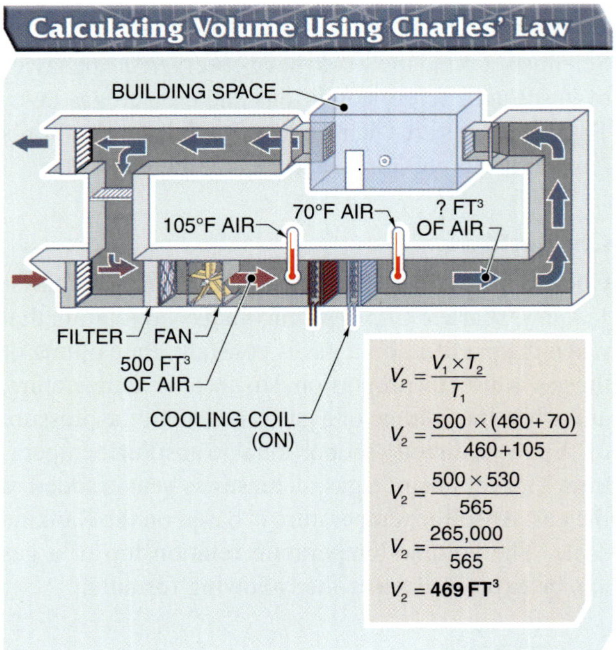

Figure 6-24. Air decreases in volume as it cools.

The pressure-temperature relationship of a gas can be expressed using the following formula:

$$\frac{P_1}{T_1} = \frac{P_2}{T_2}$$

where

P_1 = absolute pressure of the first condition (in psia)

T_1 = absolute temperature of the first condition (in °R)

P_2 = absolute pressure of the second condition (in psia)

T_2 = absolute temperature of the second condition (in °R)

Charles' law can be rearranged mathematically to find the desired value when the other three values are known. For example, the formula is rearranged as:

$$P_2 = \frac{P_1 \times T_2}{T_1}$$

TECH TIP

In a steam boiler, water is vaporized (changed into steam). As the water is changed to steam, it increases in volume and pressure. Due to the pressure, water boils at a temperature above 212°F.

Example: Calculating Pressure Using Charles' Law

A container of refrigerant has an absolute pressure of 45 psia. The temperature of the container changes from 70°F to 100°F. What is the new absolute pressure? **See Figure 6-25.**

$$P_2 = \frac{P_1 \times T_2}{T_1}$$

$$P_2 = \frac{45 \times (460 + 100)}{(460 + 70)}$$

$$P_2 = \frac{45 \times 560}{530}$$

$$P_2 = \frac{25{,}200}{530}$$

$$P_2 = \mathbf{47.5\ psia}$$

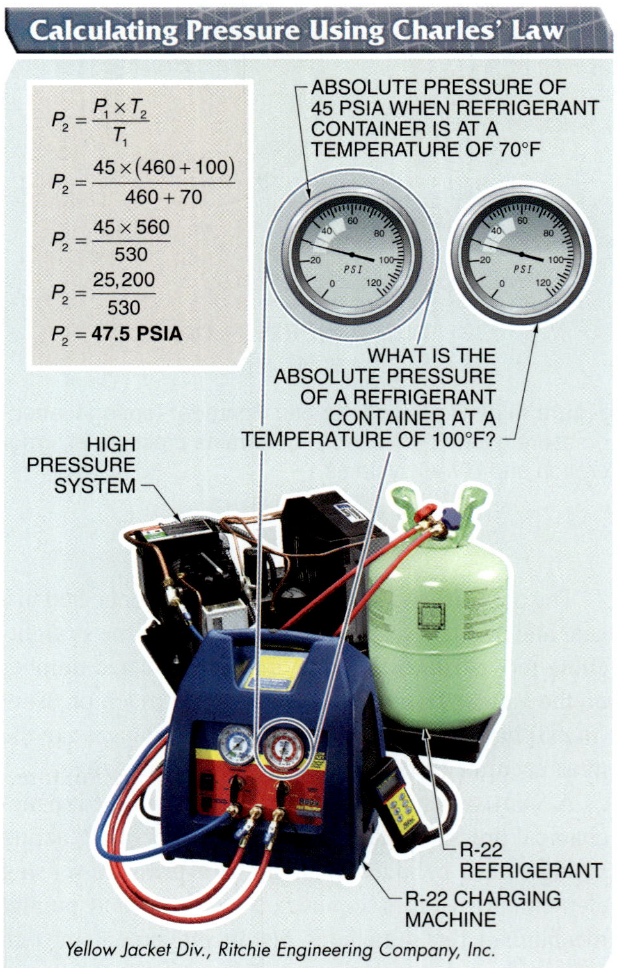

Figure 6-25. As the temperature of a gas increases, pressure increases.

Boyle's Law

Boyle's law is a gas law stating that the volume of a gas decreases as pressure is applied. Boyle's law can be expressed by the formula $P_1 \times V_1 = P_2 \times V_2$. It can be rearranged to find any unknown value when the other three values are known. For example, the volume-pressure relationship of a gas can be stated with the following formula:

$$V_1 = \frac{P_2 \times V_2}{P_1}$$

where
V_1 = volume of the first condition (in in^3)
P_2 = absolute pressure of the second condition (in psia)
V_2 = volume of the second condition (in in^3)
P_1 = absolute pressure of the first condition (in psia)

Example: Calculating Volume Using Boyle's Law

The pressure on a gas is increased from 5 psi to 35 psi. The new volume is found to be 120 in^3 (0.833 ft^3). What was the original volume? **See Figure 6-26.**

$$V_1 = \frac{P_2 \times V_2}{P_1}$$

$$V_1 = \frac{(35 + 14.7) \times 120}{(5 + 14.7)}$$

$$V_1 = \frac{49.7 \times 120}{19.7}$$

$$V_1 = \frac{5964}{19.7}$$

$$V_1 = \mathbf{302.7\ in^3}$$

Dalton's Law

Dalton's law is derived from the observation that the earth's atmosphere is a mixture of gases. *Dalton's law* is a gas law stating that the total pressure of a gas mixture is the sum of the pressures of each gas in the mixture. For example, if two gases, such as oxygen and nitrogen, each have an absolute pressure of 5 psia, then mixed together, the total absolute pressure is 10 psia.

Combined Gas Law

The combined gas law is a combination of both Charles' and Boyle's laws. The *combined gas law* is a gas law stating that the volume of gas is directly proportional to absolute temperature and inversely proportional to pressure. It is very useful because it combines temperature, pressure, and volume. The combined gas law is expressed as the following:

$$\frac{P_1 \times V_1}{T_1} = \frac{P_2 \times V_2}{T_2}$$

where
P_1 = absolute pressure of the first condition (in psia)
V_1 = volume of the first condition (in in^3)
T_1 = absolute temperature of the first condition (in °R)
P_2 = absolute pressure of the second condition (in psia)
V_2 = volume of the second condition (in in^3)
T_2 = absolute temperature of the second condition (in °R)

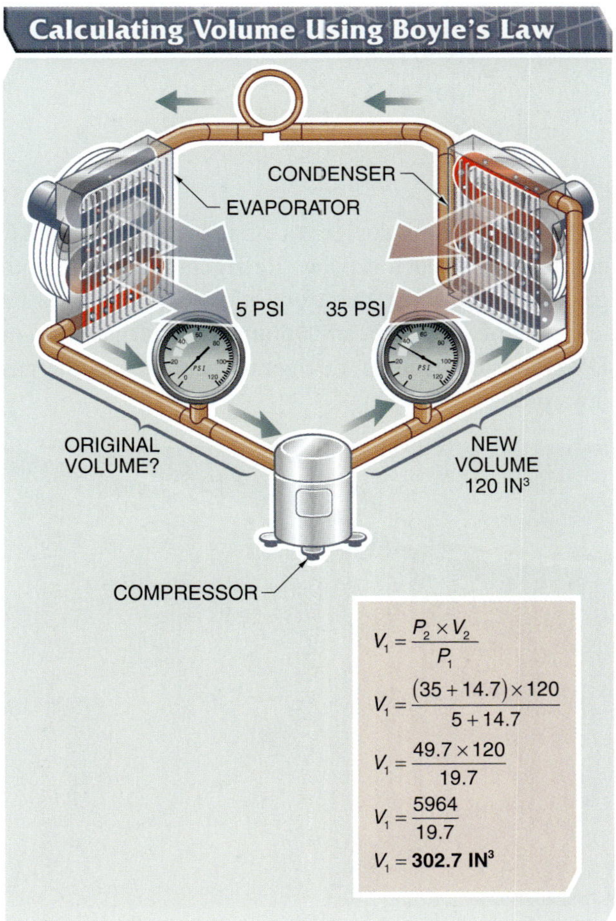

Figure 6-26. As the pressure of a gas increases, volume decreases.

As with the other gas laws, an unknown value may be found by rearranging the formula to isolate an unknown variable and then substituting known values. For example, the law can be rewritten as the following:

$$V_1 = \frac{P_2 \times V_2 \times T_1}{T_2 \times P_1}$$

Example: Calculating Volume Using the Combined Gas Law

A refrigeration system compressor increases the pressure on a gas from 50 psi to 85 psi. The temperature also increases from 115°F to 162°F. The new volume is 185 in^3. What was the original volume? **See Figure 6-27.**

$$V_1 = \frac{P_2 \times V_2 \times T_1}{T_2 \times P_1}$$

$$V_1 = \frac{(85 + 14.7) \times 185 \times (115 + 460)}{(162 + 460) \times (50 + 14.7)}$$

$$V_1 = \frac{99.7 \times 185 \times 575}{622 \times 64.7}$$

$$V_1 = \frac{10{,}605{,}587}{40{,}243.4}$$

$$V_1 = \mathbf{263.5\ in^3}$$

Engineers and designers of refrigeration HVAC equipment use each gas law. Refrigerant exists in both gas and liquid form, therefore, it obeys these laws. As temperature, pressure, and volume in a refrigeration or HVAC system change, refrigerant performs as the gas laws predict.

Refrigeration and HVAC systems come in many different sizes and shapes. The equipment design and performance are determined by how the gas laws are applied.

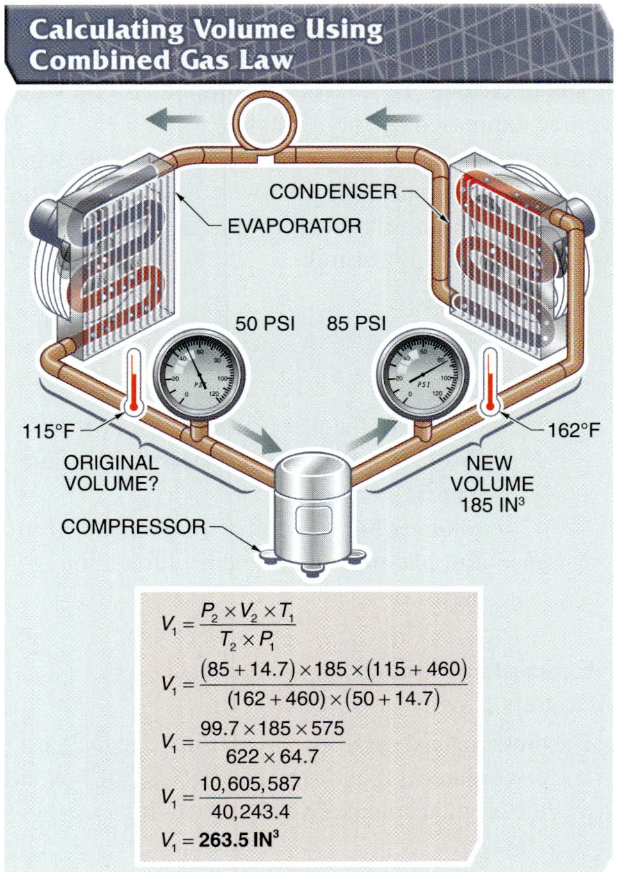

Figure 6-27. The combined gas law is a universal law that can be used for all applications involving gas.

TEMPERATURE-PRESSURE RELATIONSHIPS

Understanding the temperature-pressure relationship of a fluid is critical for understanding the operation of refrigeration systems. **See Figure 6-28.** Water begins to boil at 212°F and begins to freeze at 32°F as long as the pressure exerted on the water is 14.7 psia. Pressure and temperature decrease with increasing altitude. When pressure changes, the temperature at which boiling or freezing occurs also changes. For example, when the pressure of water is reduced to 9.3 psia, it boils at 40°F. This new boiling temperature is well below normal room temperature.

In a vapor-compression refrigeration system, a refrigerant is pressurized to a high pressure, which raises its temperature higher than the outdoor air. This creates a temperature difference to allow heat transfer. As the refrigerant moves through the system, its pressure is reduced in the evaporator, resulting in a lower pressure-temperature relationship to absorb heat from the medium (air or water). **See Figure 6-29.**

Temperature-Pressure Boiling Points

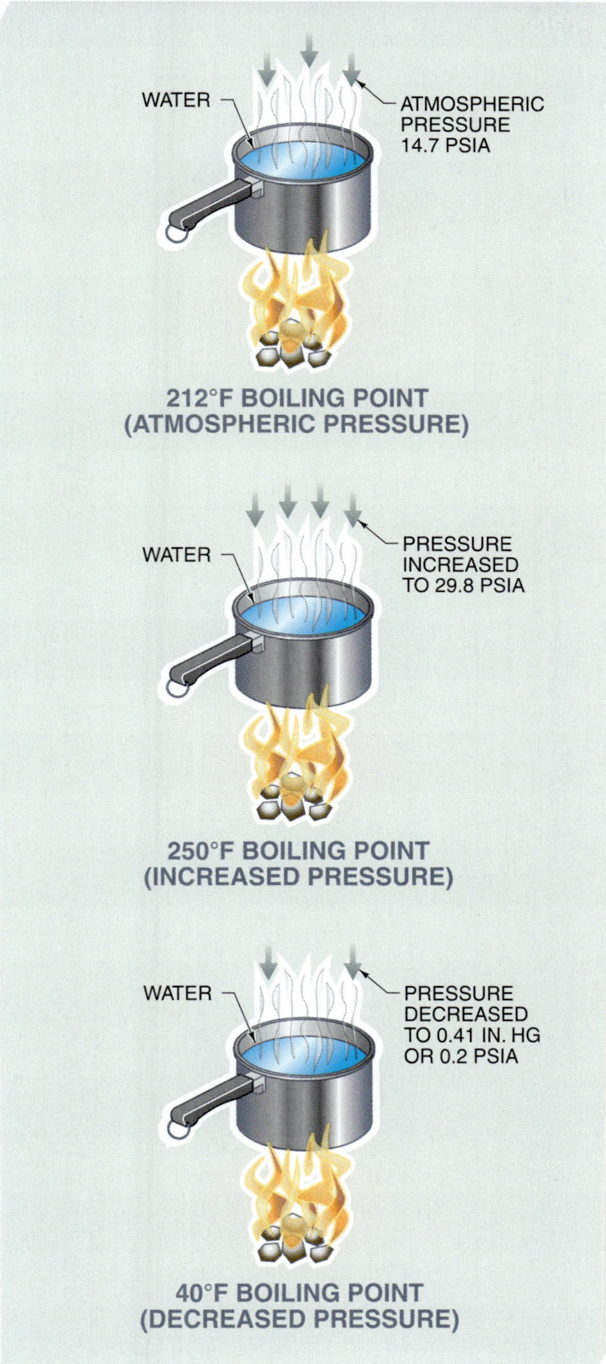

Figure 6-28. As altitude increases, the atmospheric pressure and boiling temperature of a liquid decrease.

A *refrigerant* is a substance that moves heat by changing state from a liquid to a gas when absorbing heat and from a gas to a liquid when rejecting heat. Refrigeration systems are designed to use one of the many different types of refrigerants available.

Vapor-Compression Refrigeration Systems

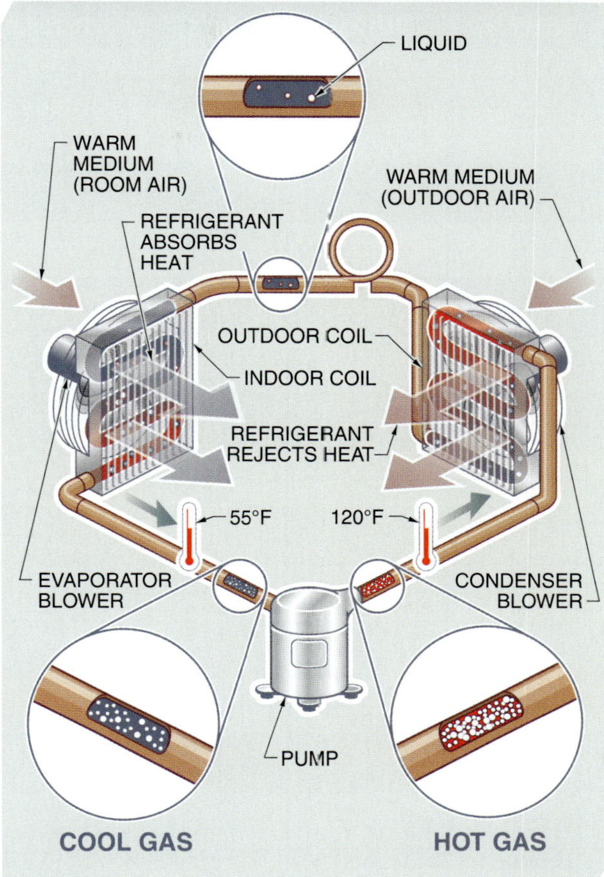

Figure 6-29. The refrigerant at a high pressure condenses to liquid as it rejects heat and the liquid boils as it absorbs heat at a lower pressure.

Temperature-Pressure Charts

Temperature-pressure relationship charts are available for each refrigerant. **See Figure 6-30.** For example, at 120°F, refrigerant R-134a has a corresponding pressure of 171.1 psi. At 65°F, refrigerant R-134a has a corresponding pressure of 64.0 psi. Temperature-pressure charts are available from refrigerant manufacturers, wholesalers, and distributors. Usually, they are pocket-size for easy reference in the field.

TECH TIP

Temperature is a measure of the energy in a substance. As temperature increases and molecules move faster, pressure increases as more molecules collide. As temperature decreases and molecules move slower, pressure decreases as fewer molecules collide.

Temperature-Pressure Charts

TEMP °F	R-123	R-11	R-134a	R-500	R-22	R-407c Liquid	R-407c Vapor	R-410a
-150	–	–		*29.5	*29.4	–	–	–
-140	–	–	*29.6	*29.2	*29.1	–	–	–
-130	–	–	*29.4	*28.8	*28.5	–	–	–
-120	–	–	*29.1	*28.3	*27.7	–	–	–
-110	–	–	*28.7	*27.5	*26.6	–	–	–
-100	–	*29.7	*28.0	*26.9	*25.1	–	–	–
-90	–	*29.6	*27.1	*24.9	*23.0	–	–	–
-80	*29.7	*29.5	*25.7	*22.9	*20.2	–	–	*14
-70	*29.6	*29.4	*24.0	*20.3	*16.6	–	–	*7.0
-60	*29.4	*29.1	*21.6	*17.0	*11.9	–	–	0.0
-50	*29.2	*28.8	*18.6	*12.8	*6.1	–	–	6.0
-40	*28.8	*28.3	*14.7	*7.6	0.6	3.0	*4.4	11
-35	*28.6	*28.0	*12.3	*4.6	2.6	5.4	*0.6	14
-30	*28.3	*27.7	*9.7	*1.2	4.9	8.0	1.8	18
-25	*28.1	*27.4	*6.8	1.2	7.5	10.9	4.1	23
-20	*27.7	*26.9	*3.6	3.2	10.2	14.1	6.6	26
-15	*27.3	*26.5	0.0	5.4	13.2	17.6	9.4	31
-10	*26.9	*25.9	2.0	7.8	16.5	21.3	12.5	36
-5	*26.4	*25.3	4.1	10.4	20.1	25.4	15.9	42
0	*25.8	*24.6	6.5	13.3	24.0	29.9	19.6	48
5	*25.2	*23.9	9.1	16.4	28.3	34.7	23.6	54
10	*24.5	*23.0	12.0	19.7	32.8	39.9	28.0	62
15	*23.7	*22.1	15.1	23.3	37.8	45.6	32.8	70
20	*22.8	*21.0	18.4	27.2	43.1	51.6	38.0	78
25	*21.8	*19.8	22.1	31.4	48.8	58.2	43.6	86
30	*20.7	*18.5	26.1	36.0	54.9	65.2	49.6	96
35	*19.5	*17.1	30.4	40.8	61.5	72.6	56.1	106
40	*18.1	*15.5	35.0	46.0	68.5	80.7	63.1	171
45	*16.6	*13.8	40.0	51.6	76.1	89.2	70.6	129
50	*15.0	*12.0	45.4	57.5	84.1	98.3	78.7	141
55	*13.1	*9.9	51.2	63.8	92.6	108	87.3	154
60	*11.2	*7.7	57.4	70.6	101.6	118	96.8	170
65	*9.0	*5.3	64.0	77.7	111.3	129	106	183
70	*6.6	*2.7	71.1	85.3	121.4	141	117	200
75	*4.1	0.1	78.6	93.4	132.2	153	128	218
80	*1.3	1.6	86.7	101.9	143.7	166	140	236
85	0.9	3.2	95.2	110.9	155.7	180	153	255
90	2.5	4.9	104.3	120.5	168.4	195	166	275
95	4.2	6.8	113.9	130.5	181.8	210	181	287
100	6.1	8.8	124.1	141.1	196.0	226	196	318
105	8.1	10.9	134.9	152.2	210.8	243	211	340
110	10.2	13.2	146.3	163.9	226.4	261	229	364
115	12.6	15.7	158.4	176.3	242.8	280	247	390
120	15.0	18.3	171.1	189.2	260.0	300	266	416
125	17.7	21.1	184.5	202.7	278.1	321	286	445
130	20.5	24.0	198.7	216.9	297.0	342	307	474
135	23.5	27.1	213.5	231.8	316.7	365	329	505
140	26.7	30.5	229.2	247.4	337.4	389	353	537
145	30.2	34.0	245.6	263.7	359.1	–	–	572
150	33.8	37.7	262.8	280.7	381.7	–	–	608

* Pressures are shown as PSIG except that (*) indicates inches of mercury vacuum.

Figure 6-30. A temperature-pressure chart is used as reference to diagnose refrigeration system performance.

UNIT 7

Combustion and Fuels

The quantity of heat produced by burning fuel is directly related to the elements in the fuel. Combustion and electrical resistance are the primary methods of producing heat for buildings. Combustion is a chemical reaction between fuel and oxygen. During a reaction, chemical energy is converted into thermal energy. When enough oxygen is available for complete combustion, the products of combustion are acceptable. When there is not enough oxygen available for complete combustion, the products of combustion, such as CO and CO_2, can be dangerous.

OBJECTIVES

- Explain combustion.
- Identify the types of fuels used in residential and light commercial heating systems.
- Explain the importance of combustion air.
- Describe the sections of a flame.
- Explain complete combustion and incomplete combustion.
- Describe how efficient combustion is accomplished.

COMBUSTION

Combustion is a chemical reaction in fuel that occurs when oxygen reacts with hydrogen (H) and carbon (C) at an ignition temperature. A *chemical reaction* is a chemical change of one or more substances that produces another substance. The three requirements for combustion are fuel, oxygen, and ignition temperature. **See Figure 7-1.** In combustion, fuel provides chemical energy, oxygen reacts with the fuel, and the ignition temperature starts the reaction.

Combustion produces heat and light as oxygen and elements in fuel recombine to form different substances. Combustion takes place in a furnace or boiler to heat a building. It must be controlled for the heat to be used efficiently.

Fuels and Heating Values

A *fuel* is any substance that is burned to produce heat. Fuels are classified by physical characteristics or chemical composition. The physical characteristics of a fuel identify the fuel and determine the type of equipment required for transporting, storing, and burning. The most important physical characteristic of fuel is its state.

Fuel is available as a solid, liquid, or gas. A *chemical composition* is the constitution of all elements that make up a compound. A *compound* is a substance that can be broken down into elements. Each compound has a unique chemical composition. Most fuels are compounds. The unit of measure used for a fuel depends on the state of the fuel. Solid fuels are measured in pounds (lb), liquid fuels in gallons (gal.), and gas fuels in cubic feet (ft^3).

Combustion Requirements

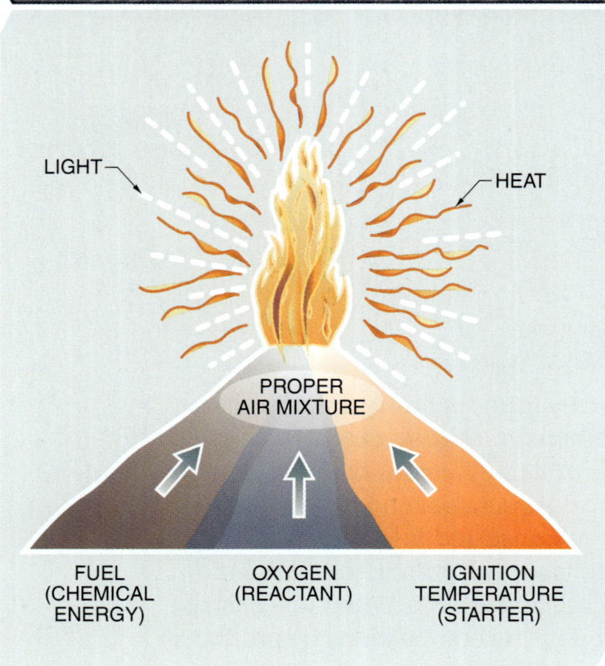

Figure 7-1. There must be fuel, oxygen, and proper ignition temperature for combustion to occur.

Gross Heating Values of Elements and Compounds

Substance	Molecular Symbol	Heating Value*
Acetylene	C_2H_2	2355
Butane	C_4H_{10}	3271
Carbon (to CO)	C	308
Carbon (to CO_2)	C	1733
Carbon monoxide	CO	339
Ethylene	C_2H_4	1641
Hydrogen	H_2	315
Methane	CH_4	1016
Propane	C_3H_8	2524
Propylene	C_3H_6	2477

* in Btu/ft³

Figure 7-2. A heating value is the quantity of heat (in Btu) that is produced when one unit of fuel is burned.

Commercial Fuel Gross Heating Values

Liquids (Btu/gal.)	Gases (Btu/ft³)	Solids (Btu/lb)
#2 Fuel Oil 140,000	Natural Gas 1030	Wood 8600
	Propane 2524	Coal 14,000
	Butane 3271	

Figure 7-3. The unit of measure used for a fuel depends on the state of the fuel.

All fuels contain hydrocarbons. A *hydrocarbon* is a compound that contains hydrogen and carbon. Some fuels also include small quantities of other elements. Each element burns at a different rate. Hydrogen burns quickly, and carbon, the major element in most fuels, burns slowly.

Each element has a specific heating value. *Heating value* is the amount of heat produced per unit of fuel per hour; expressed in British thermal units per hour (Btu/hr). When a given quantity of an element is burned, a fixed amount of heat is produced. When the elements in a fuel are known, the heating value of the fuel can be calculated. It is possible to calculate the quantity of heat that will be produced when a fuel is burned because each element has a unique heating value. **See Figure 7-2.**

The quantity of heat produced by burning fuel is determined by the heating value and amount of fuel used in a given time. Heating values are measured in Btu per pound (Btu/lb) for solid fuels, Btu per gallon (Btu/gal.) for liquid fuels, and Btu per cubic foot (Btu/ft³) for gas fuels. **See Figure 7-3.** Because fuels are compounds, different fuels have different heating values. However, the heating value for a fuel can be determined when a chemical analysis is performed and the amount of oxygen combined with the fuel is known. **See Appendix.**

For example, when 1 lb of hydrogen is burned, 51,000 Btu is produced. When 1 lb of carbon is burned, 14,093 Btu is produced. When a sample of fuel containing 1 lb of hydrogen and 1 lb of carbon is burned, the heat output would be 14,093 Btu for the carbon plus 51,000 Btu for the hydrogen, or 65,093 total Btu. This assumes that combustion is 100% efficient.

Fuels are used depending on regional availability and cost. Because the prices of fuel and electricity vary, selecting a fuel for a specific job is a difficult and important task. Petroleum-base fuels are popular but subject to availability. Electricity is used for heating but the cost is high. The availability of fuel oil and gas fuels make them popular fuels for heating residential homes and commercial facilities. Coal is also used in boilers to make steam for heating or electricity in industrial facilities and areas where it is inexpensive and readily available.

Fuel Oils. Fuel oil is a petroleum-base product made from crude oil. Different grades of fuel oil are available. Crude oil is a mixture of semisolids, liquids, and gases formed from the remains of organic materials that have been changed by pressure and heat over millions of years. Crude oil deposits are found under the surface of the earth in various parts of the world.

Geologists use various devices and techniques to locate structures underground that could hold oil. One method of finding these structures is to detonate an explosive charge on the surface of the earth. A seismograph is used to listen to the echoes that bounce back from the various layers of rock underground.

A *seismograph* is a device that measures and records vibrations in the earth. Seismograph readings are used to determine if an underground structure may hold oil. If oil is suspected to be present, a pilot well is then drilled at the site. When crude oil is found, more wells are drilled to collect the oil. The crude oil either flows from a well under natural pressure or it is pumped from the ground. **See Figure 7-4.**

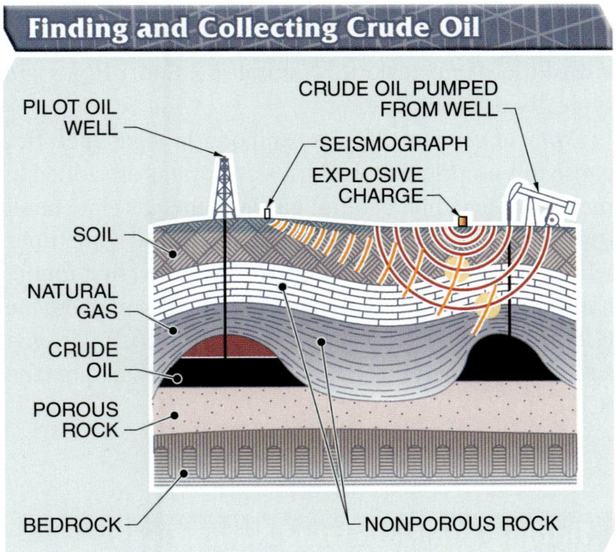

Figure 7-4. Crude oil flows from deposits deep underground by natural pressure or pumping.

Once crude oil is collected, it is transported to a refinery. A *refinery* is a plant where crude oil is processed into petroleum products. Heat and pressure are used to vaporize (change liquid to a gas) crude oil in one part of the refinery. Various petroleum products boil out of the vapor at different temperatures. The products are collected and stored separately in holding and pressure tanks. **See Figure 7-5.**

Petroleum products are classified as either distillates (light oils) or residuals (heavy oils). Distillates are refined into fuels such as liquefied petroleum gas, jet and automobile gasoline, kerosene, light heating oils, and diesel fuels. Residuals are refined into industrial fuel oils and other commercial and industrial products.

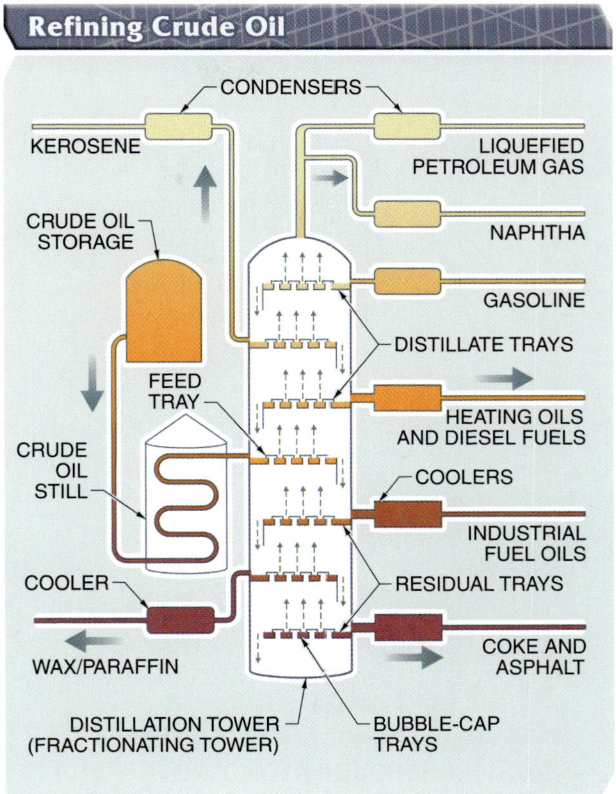

Figure 7-5. Petroleum products boil at different temperatures and are collected separately during the refining process.

The grade of fuel oil is based on the weight and viscosity of the oil. *Viscosity* is the measurement of a liquid's resistance to flow (thickness). As numbers of fuel oil grades increase, so does the viscosity. Each grade of oil has properties required for specific applications.

Grade No. 1 fuel oil is a light distillate fuel oil used in open pot burners. It is almost as light as diesel oil, which is burned in automobiles and trucks. An *open pot burner* is a burner used in kerosene heaters. Oil flows into an open pot and begins to vaporize. Combustion air is introduced around the pot and combustion begins when oil vapors ignite at the top of the pot. **See Figure 7-6.**

Grade No. 2 fuel oil is slightly heavier than Grade No. 1. Grade No. 2 is typically used for heating with pressure atomizing burners. A *pressure atomizing burner* is a burner that sprays oil into a combustion chamber through an atomizing nozzle. *Atomization* is the process of breaking a liquid into small droplets. Air mixes with the oil droplets and an electric spark ignites the air-oil mixture.

Fuel Oil Burners

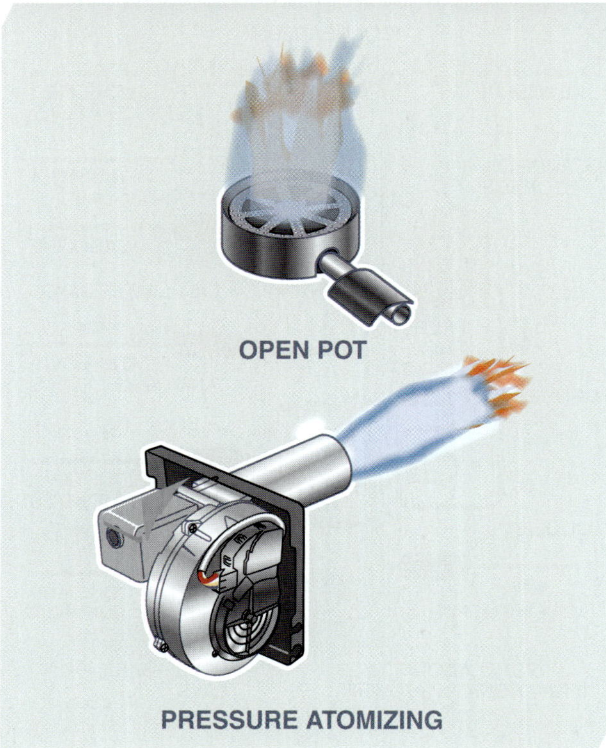

Figure 7-6. Fuel oil burners fall into two general categories of open pot and pressure atomizing.

Grade No. 4 fuel oil is an intermediate fuel oil used for commercial and industrial burners. Grade No. 5 (light) is a residual fuel oil that is heavier than Grade No. 4. Grade No. 5 (heavy) is a residual fuel oil heavier than Grade No. 5 (light). Preheating is typically required when using Grade No. 5 and heavier oils in cold climates. Fuel tanks and fuel lines are heated when using heavier fuel oils to lower the viscosity. Grade No. 6, sometimes called bunker C, is the heaviest fuel oil and is used for heating. It requires heating in a tank, in lines, and before it enters a burner.

Fuel oil is transported from a refinery to a distributor by ship, barge, railroad tank car, or truck. A tank truck transports fuel oil from a distributor to the consumer. The fuel oil is deposited in a storage tank that is piped to the furnace or boiler inside a building. A pump is used to create a flow of oil from the storage tank to the burner.

The heating values of fuel oil range from 132,900 Btu/gal. to 155,900 Btu/gal. **See Figure 7-7.** Grade No. 2 is used for residential heating. The standard heating value for Grade No. 2 is 140,000 Btu/gal. Higher grades of fuel oil have higher heating values.

Fuel Oil Heating Values

Grade	Heating Value*
No. 1	132,900 – 137,000
No. 2	137,000 – 141,800
No. 4	143,100 – 148,100
No. 5 (light)	146,800 – 150,000
No. 5 (heavy)	149,400 – 152,000
No. 6	151,300 – 155,900

* in Btu/gal.

Figure 7-7. The various grades of fuel oils have different heating values.

Gas Fuels. The gas fuels used for heating are natural gas, liquefied petroleum (LP) gas, and manufactured gas. These three fuels are stored and transported differently but burn in the gas state. This ensures clean combustion with little or no residue. Natural gas and LP gas are typically used for heating.

Natural gas is a colorless and odorless gas fuel. Because it is colorless and odorless, odorants are added to make it noticeable. Natural gas is in the gas state at all times. It is found near crude oil deposits and sometimes where there is no oil. **See Figure 7-8.** When first found, natural gas may contain varying amounts of methane (CH_4), ethane (C_2H_6), water vapor (H_2O), hydrogen sulfide (H_2S), helium (He), propane (C_3H_8), and butane (C_4H_{10}).

Finding and Collecting Natural Gas

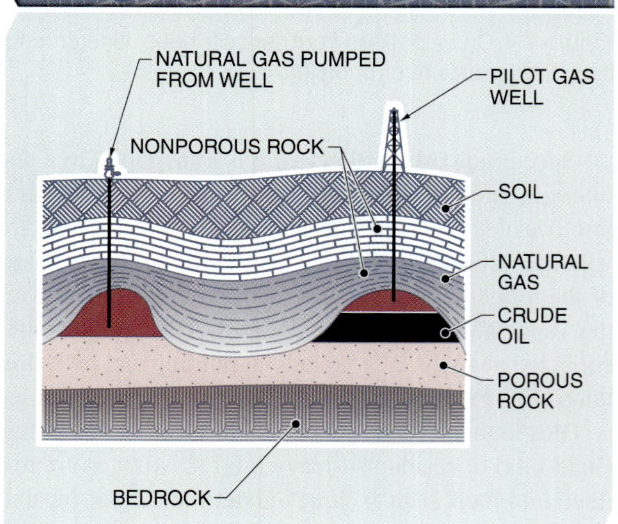

Figure 7-8. Natural gas is found near crude oil deposits and sometimes where there is no oil.

Before being used as a fuel, natural gas is scrubbed (cleaned) to remove LP gas, gasoline, and noncombustible materials. The product that is distributed as natural gas is composed of 70% to 96% methane and 1% to 14% ethane. Small percentages (0.4% to 17%) of pentane, hexane, carbon dioxide, oxygen, and nitrogen are also found in natural gas.

Natural gas for distribution has a specific gravity of 0.660 to 0.708. The specific gravity of gas is the weight of 1 ft^3 of gas compared to the weight of 1 ft^3 of dry air at 29.92 in. Hg at 60°F. Because natural gas is lighter than air, it will dissipate into the atmosphere when released. Natural gas is available for heating in most urban areas.

The heating value for natural gas varies with its chemical composition. The heating value for gas delivered by a gas utility is about 950 Btu/ft^3 to 1050 Btu/ft^3. The natural gas supplied by a utility is composed of about 25% hydrogen and 75% carbon, which is an efficiently burning fuel. **See Figure 7-9.** For general use, the standard heating value of natural gas is 1000 Btu/ft^3. Gas is sold to consumers by the therm. A *therm* is a unit for quantity of heat that equals 100,000 British thermal units.

Gas Heating Values

Fuel	Heating Value*
Natural gas	950 – 1050
Manufactured gas	500 – 600
Liquid petroleum	
Propane-butane mixture (depends on mixtures)	2500 – 3200
Propane	2500
Butane	3200

* in Btu/ft^3

Figure 7-9. Heating values of gas fuels range between 500 Btu/ft^3 and 3200 Btu/ft^3.

Liquefied petroleum (LP) gas is a fuel refined from crude oil or natural gas. It is the second most popular gas fuel used for heating. LP gas is used in rural areas where natural gas pipelines are not available. It is collected as a gas but reduced to liquid form by pressure and cooling for transportation in tank trucks. LP gas has higher heating values than natural gas. When the fuel of LP gas is used, the liquid vaporizes upon leaving the storage tank. LP gas consists of propane-butane mixtures.

Propane is a gas fuel composed of 90% to 95% propane gas and 5% to 10% propylene (C_3H_6). Propane gas has a specific gravity of 1.5, which means it is heavier than air. Because propane is heavier than air, care must be taken to prevent leaks in storage and piping systems. When propane leaks, it accumulates in low areas and creates the potential for a fire or explosion. Propane can be used as a fuel in most climates because it has a low boiling point. The boiling point of propane is –43.7°F.

Butane is a gas fuel composed of 95% to 99% butane gas and up to 5% butylene (C_4H_8). Butane has a specific gravity of 2.0, which is heavier than propane and air. It will settle in low areas and must be handled with care. Butane cannot be used where storage temperature would drop below freezing because it has a boiling point of 32°F.

A *propane-butane mixture* is a fuel mixture of propane and butane that is used when the properties of each gas alone are not sufficient for an application. Butane is stored as a liquid. When the temperature falls below 32°F, butane will not vaporize. By mixing butane with propane, which has a boiling point of –43.7°F, the mixture will vaporize at winter temperatures in most climates. The heating value of propane is 2500 Btu/ft^3, and the heating value of butane is 3200 Btu/ft^3. The properties of propane-butane mixtures vary as the ratio of the two gases vary.

Manufactured gas is any gas produced from coal, oil, liquefied petroleum, or natural gas. Most of these gas fuels are by-products of manufacturing processes and are used for industrial operations or as specialty fuels. At one time, many cities had gas plants where gas was manufactured for lighting and heating. Today, most manufactured gas systems are specialty refineries that do not typically provide fuel for gas heating systems.

Flame Troubleshooting Interactive Animation

Manchester Tank & Equipment Co.

Liquefied petroleum gas, propane, butane, and propane-butane mixtures can be purchased in various sizes of cylinders and tanks including 250 gal., 500 gal., 1000 gal., 1500 gal., and 2500 gal. sizes.

Coal. The standard heating value of coal is 12,000 Btu/lb, which is slightly less than the heating value of both hydrogen and carbon. The heating value of coal varies with its chemical content. Heating values range from 6900 Btu/lb to 14,350 Btu/lb. **See Appendix.**

Alternative Fuels. Alternative fuels are combustible materials other than oil, gas, and coal. Alternative fuels include wood pellets, sawdust, refuse, biomass, and waste gas from manufacturing processes. Most alternative fuels are not used as often as primary fuels and therefore, are not readily available in large quantities for commercial use. Even when they are available, the unusual form of the fuel may make it difficult to use for heating. Biomass in standard form is converted into energy in three ways: thermal conversion, chemical conversion, and biochemical conversion.

Wood is the alternative fuel used most often for residential heating. Wood burns most efficiently in airtight fireplaces or stoves. Because the heating values of various woods and the efficiencies of different stoves vary, it is difficult to evaluate the savings of using wood as a fuel. However, wood is cost-effective when it can be obtained locally at low cost. **See Figure 7-10.**

Alternative Fuel Heating Values

Fuel	Btu/lb
Sawdust - Green	4500
- Kiln dry	7500
Wood pellets	8000
Corn - Cob	8300
- Shelled	8100
Rubber pellets	16,000
Manure	12,000
Refuse - Type 0	8500
- Type 1	4300
- Type 2	1100
Waste Gas	20*

* in Btu/ft^3

Figure 7-10. Heating values of alternative fuels are about 20 Btu/ft^3 for gases and up to 16,000 Btu/lb for solids.

Determining the heating values of different alternative fuels is difficult. Firing methods, which affect combustion efficiency, densities, and moisture contents, vary among alternative fuels and affect heating values. For example, the heating values of wood vary depending on the density and moisture content of the wood. Heating values are about 5980 Btu/lb for softwoods and about 9500 Btu/lb for hardwoods. **See Appendix.**

Combustion Air

Combustion air is air provided to a burner for the proper combustion of fuel. The precise amount of air to achieve proper combustion must be available. The actual amount of needed air depends on the fuel used. When enough air is not available, dangerous by-products of combustion are generated.

Each burner in a heating system must have the proper amount of combustion air. Combustion air can be taken from inside a building where air can get to the burner without any obstruction. **See Figure 7-11.** When heating equipment is isolated from the rest of the building, louvers may be installed to provide a passage for combustion air. When a building area is small or combustion air cannot be acquired from inside the building, outside air must be ducted directly to the heating equipment area.

Combustion Air

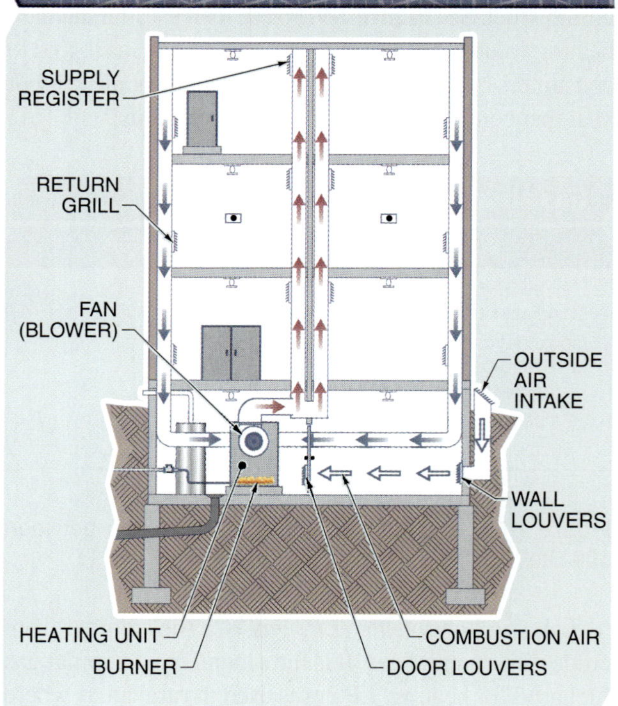

Figure 7-11. Heating equipment that is isolated from the rest of a building must have louvers installed or have direct access to outside air by ductwork.

TECH TIP

Preheated combustion air is typically used for large furnaces and boilers. It is now being considered for usage in residential and light commercial systems, because the efficiency of these systems can be increased by 1% for every 40°F increase in temperature.

Oxygen. In a typical heating unit, oxygen for combustion comes from atmospheric air. Air consists of approximately 78% nitrogen and 21% oxygen. During combustion, oxygen reacts with the various elements in fuel. Nitrogen goes through the combustion process unchanged. Various fuels require different air-fuel ratios for combustion. When too little or too much air mixes with the fuel, the mixture will not burn properly.

For example, the airflow to a gasoline engine must mix with fuel for the engine to fire. When the air mixture is set too lean or too rich, the engine will not run. An air mixture that is too lean or too rich can be adjusted by changing the amount of air in the mixture. The air-fuel ratio for burning natural gas is 86% to 96% air to 4% to 14% gas. At 4% gas, an air-fuel mixture is lean. At 14% gas, an air-fuel mixture is rich. Combustion takes place readily when the mixture is near the center of the range. **See Figure 7-12.**

Combustion Air-Fuel Mixture

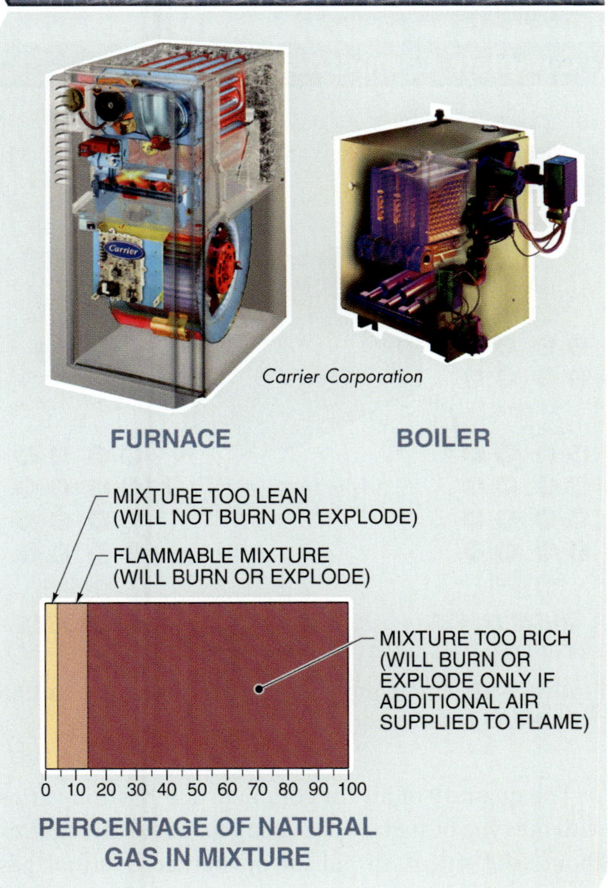

Figure 7-12. The air-fuel ratio for natural gas must be between 4% and 14% for combustion to occur.

Ignition Temperature. *Ignition temperature* is the heat required to start a chemical reaction between oxygen and the elements in a fuel. An igniter, such as a pilot light, electric spark, or hot surface igniter, provides ignition temperature to a fuel and oxygen mixture. Once combustion has begun, the temperature must remain at or above the ignition temperature or combustion will stop. The heat produced by combustion maintains temperature above the required ignition temperature.

Flames

Flames are generated as fuel burns. Flames are the result of the chemical reaction between fuel and oxygen. Combustion is a chemical reaction where compounds break apart into elements and release chemical energy in the form of heat and light. The elements recombine into the products of combustion.

Flame Sections. Flames consist of either two or three sections. Each section of a flame relates to the chemical reaction taking place. **See Figure 7-13.** An *inner mantle* is the center section of a flame that is produced as hydrogen reacts with oxygen. It is dark to light blue in color. An *outer mantle* is the light blue area of a flame that surrounds the inner mantle and is produced as carbon reacts with oxygen. The carbon reaction takes place farther away from the burner face because it occurs after the hydrogen reaction. Carbon does not burn as quickly as hydrogen. The outermost, yellow, tip of a flame consists of incandescent particles of carbon that give off heat but are not yet consumed by the reaction.

Flame Sections

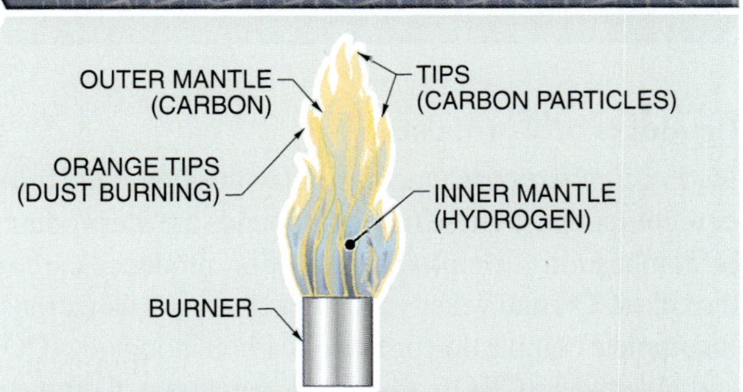

Section	Color
Tip	Yellow
Outer Mantle	Light Blue
Inner Mantle	Dark Blue

Figure 7-13. The different burning rates of elements in a fuel create two or three sections in a flame.

Flame Types. The size and color of a flame and the products of combustion depend on the kind of fuel burned, how combustion air mixes with the fuel, and the kind of burner used. When combustion air mixes with fuel before the mixture reaches the burner face, a blue flame is produced. A *burner face* is the point at which a flame is established. **See Figure 7-14.** A blue flame is typically produced on a gas fuel burner. Fuel in a gas state mixes with air before the air-fuel mixture enters the burner. Flames produced by natural gas are blue flames.

Flame Analysis

Indicator	Possible Problem
Stack Smoke	
- Black	Too much fuel in mixture
- White	Too much air in mixture
Flame Color	
- Light	Too much air in mixture
- Dark (dull)	Too much fuel
- Yellow Tipped	Too much natural gas in mixture

* in Btu/ft³

Figure 7-14. A blue flame is produced by gas fuel that mixes with air before the air-fuel mixture enters a burner. A yellow flame is produced when combustion air mixes with fuel at the burner face.

A yellow flame is produced when combustion air mixes with fuel at the burner face. Solid or liquid fuels cannot mix efficiently with combustion air before they burn. Burning liquid and solid fuels is less efficient than burning gas fuels. Coal and fuel oil fires produce yellow flames for this reason. Modern oil burners are built for high efficiency, but the fuel oil flames are still yellow.

Understanding Burner Flames Media Clip

Products of Combustion

During combustion, elements react with oxygen and form new compounds. These new compounds are the products of combustion. Complete combustion produces carbon dioxide (CO^2) and water vapor, which are not dangerous. Incomplete combustion produces carbon monoxide (CO) and aldehydes (CHO), which are dangerous. Complete combustion requires enough combustion air to provide enough oxygen for the process.

Stoichiometric Combustion. *Stoichiometric combustion* is a theoretical condition of combustion where the air-fuel ratio is perfect and there is complete combustion with 100% efficiency. All flame heating equipment strive for stoichiometric combustion, but it cannot be met in reality.

TECH TIP

Furnaces and boilers typically incorporate 5% to 10% more air (excess air) for gas and 10% to 20% more air for fuel oil than what is needed to burn the fuel completely. When a burner is not supplied enough air, the unburned fuel, smoke, and carbon monoxide are exhausted and build up on surfaces. This results in lower combustion efficiency, flame instability, and a possibility for explosion.

Complete Combustion. *Complete combustion* is a type of combustion that occurs when enough oxygen is supplied by combustion air to combine with all elements in a fuel. In complete combustion, hydrogen, carbon, and oxygen recombine to form carbon dioxide and water vapor. Nitrogen passes through the combustion process unchanged. Chemical energy is changed to thermal energy, with no loss or gain of energy. **See Figure 7-15.** With complete combustion, there is the same number of chemical molecules in the products of combustion as there was in the air and fuel.

Complete Combustion

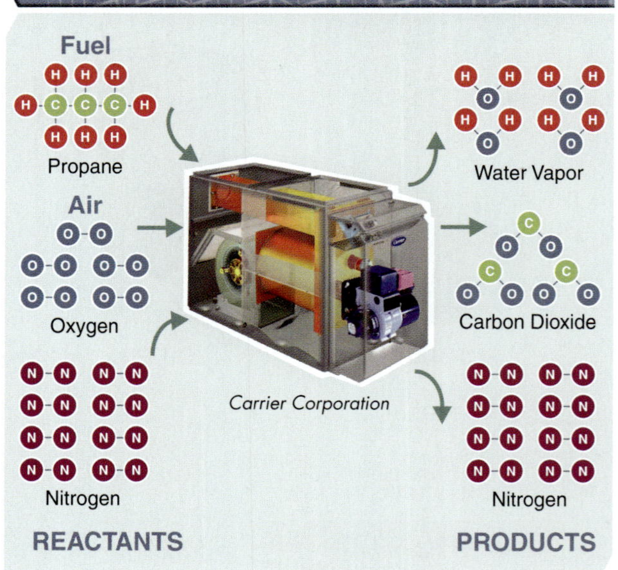

Figure 7-15. Complete combustion generates carbon dioxide and water vapor.

The quantity of air for complete combustion varies with the type of fuel used. Grade No. 2 fuel oil requires about 1400 ft³ of air/gal. of oil burned. Natural gas requires 10 ft³ of air/ft³ of gas burned, butane requires 30 ft³ of air, and propane requires 24 ft³ of air. Coal requires approximately 135 ft³ of air/lb of coal burned.

Incomplete Combustion. *Incomplete combustion* is a type of combustion that occurs when not enough oxygen is supplied by combustion air to combine with all the elements in a fuel. Incomplete combustion produces carbon monoxide, aldehydes, and water vapor. Carbon monoxide is a dangerous gas. An *aldehyde* is a toxic chemical compound that contains hydrogen, carbon, and oxygen. It has a harsh odor that burns the eyes and nose. Exposure to the gases created by incomplete combustion can result in CO poisoning and death. Incomplete combustion is caused by a lack of combustion air and rapid cooling of the flame. **See Figure 7-16.**

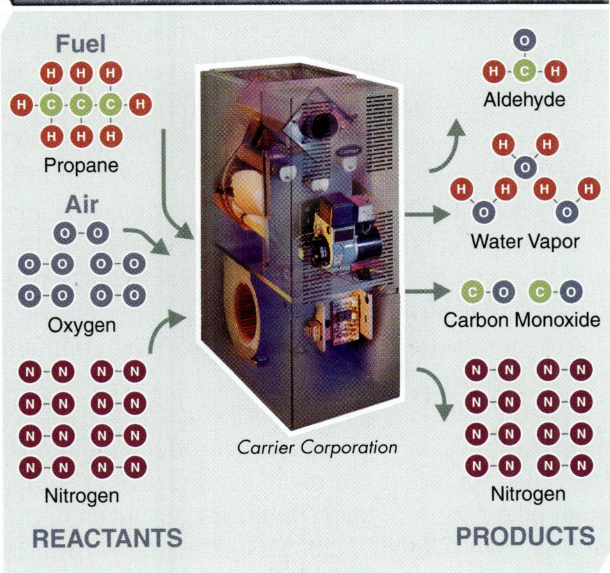

Figure 7-16. Incomplete combustion generates carbon monoxide, aldehydes, and water vapor.

Rate of Combustion. The rate of combustion of a system is measured by the amount of fuel burned per hour. Fuel can be measured in pounds, gallons, or cubic feet. Solid fuel is measured in pounds. Liquid fuel is measured in gallons. Gas fuel is measured in cubic feet.

Combustion Efficiency

Combustion efficiency is a measure of how well chemical energy in a fuel is converted to thermal energy during combustion. **See Figure 7-17.** When combustion efficiency is high, most chemical energy is converted to thermal energy. When combustion efficiency is low, most chemical energy is wasted and dangerous products of combustion form.

Common Fuel Chemical Composition

Fuel Components	Fuel Oil*	Natural Gas*
Carbon	84	74
Hydrogen	12	25
Sulfur	3	–
Oxygen	1	Trace
Nitrogen	Trace	1
Ash	Trace	–
Water	Trace	–

* in percent (%)

Figure 7-17. Most fuels are made up of mostly carbon and hydrogen.

When a chemical analysis of a fuel is available, the total heating value can be calculated by adding the heating values of each element and the atom bonding energy of the fuel. The main elements of all fuels are hydrogen and oxygen. Hydrogen and bonding energy has a heating value of 61,500 Btu/lb, and carbon and bonding energy has a heating value of 14,550 Btu/lb. These numbers make it possible to calculate the heating value of a fuel containing 1 lb of hydrogen and 1 lb of carbon. **See Figure 7-18.**

For 1 lb of hydrogen to burn completely, 8 lb of oxygen is required. This reaction produces 9 lb of water vapor and 61,500 Btu. For 1 lb of carbon to burn completely, 2.66 lb of oxygen is required. This reaction produces 3.66 lb of carbon dioxide and 14,550 Btu.

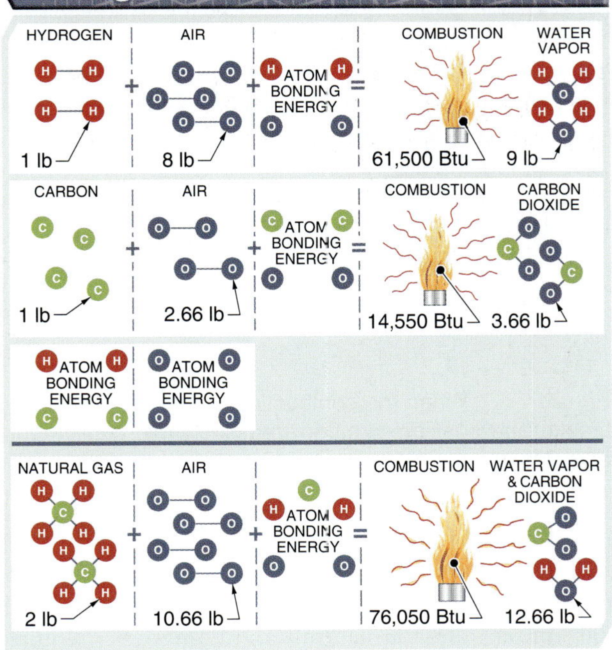

Figure 7-18. The heating values of individual fuel components can be analyzed and added together to find the heating value and combustion results of the original fuel.

If 100% combustion efficiency were achieved, 76,050 Btu would be produced by burning a fuel that contains 1 lb of hydrogen and 1 lb of carbon in the presence of 10.66 lb of oxygen. The combustion products would total 12⅔ lb. However, combustion is never 100% efficient. Combustion efficiency is directly related to the amount of excess air present during combustion.

Combustion Inefficiency

The two main causes of combustion inefficiency are incomplete combustion and an excess amount of air. Hydrogen in fuel burns rapidly and therefore, cleanly. Carbon burns slowly. When there is not enough oxygen for complete combustion, some carbon particles will leave the process unburned. When this happens, smoke and soot form. **See Figure 7-19.**

Combustion Inefficiencies

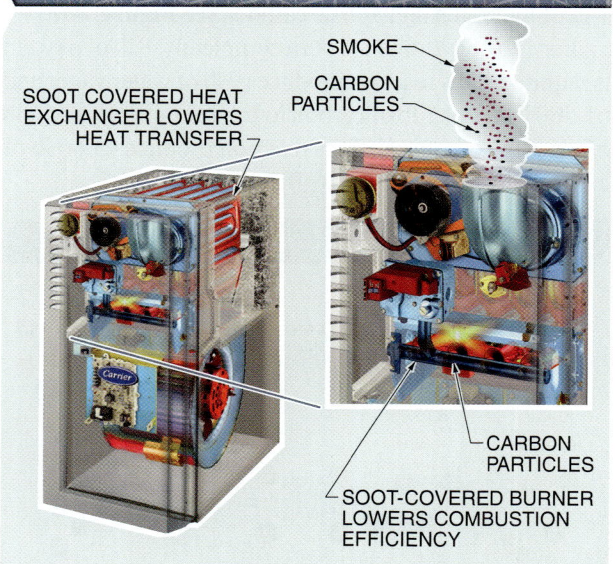

Carrier Corporation

Figure 7-19. When the combustion process is inefficient, smoke and soot are created from the unburned carbon particles of fuel.

Smoke is a collection of unburned carbon particles that is carried away by the convection currents generated by a flame. *Soot* is a collection of unburned carbon particles on heating and combustion equipment that forms when the surface temperature is lower than the ignition temperature of the carbon. The lower temperature prevents the carbon particles from burning. Soot causes a smoky fire and acts as an insulator when it collects on the inside of a furnace. A coat of soot on the surface of a heat exchanger reduces efficiency by slowing heat transfer through exchanger materials.

Burners and combustion air blowers must be adjusted to provide more air than required for complete combustion. For maximum efficiency, 5% to 20% excess air is required for combustion, depending on the system. However, too much excess air results in combustion inefficiency.

When too much combustion air is introduced into a burner, air that is not required for combustion passes through the process unchanged, except that it is heated. Excess air removes heat from combustion. This reduces combustion efficiency because fuel is used to heat the air that goes up a flue.

Annual Fuel Utilization Efficiency (AFUE). HVAC combustion equipment, as used in the HVAC industry, is rated by its annual fuel utilization efficiency (AFUE). The *annual fuel utilization efficiency (AFUE) rating* is a measure of the thermal efficiency of HVAC combustion equipment, expressed as a percent. The Department of Energy (DOE) establishes the efficiency standard for furnaces. The minimum AFUE for furnaces is 78%.

Testing for AFUE is not an instant, steady-state test performed in a lab under controlled conditions. Instead, it represents the average efficiency of equipment under varying load conditions during an entire season. **See Figure 7-20.** For example, a furnace with a peak thermal efficiency of 88% may have an AFUE of only 74% under normal conditions. A furnace with an AFUE of 80% transfers 80% of the heat energy to room air and the remaining 20% goes up the chimney. ASHRAE Standard 103 determines the standard AFUE.

Efficiency Tests. Combustion efficiency is determined by measuring flue-gas temperature and the amount of oxygen in the flue gas. A theoretical maximum percentage of carbon dioxide (ultimate carbon dioxide) is produced when precisely enough air is supplied to burn all of the fuel. When more air is supplied than required for combustion, the excess air will dilute the carbon dioxide.

Measuring the percentage of carbon dioxide in flue gas is a way to determine the amount of excess air in combustion. A proper amount of oxygen will ensure clean complete combustion and a safety margin for variations in fuel, draft, and burner conditions.

Annual Fuel Utilization Efficiency (AFUE) Rating

Fuel	Furnace/Boiler	AFUE
Natural Gas	Conventional	55 – 65
	Midefficiency	78 – 84
	Condensing	90 – 97
Propane	Conventional	55 – 65
	Midefficiency	79 – 85
	Condensing	88 – 95
Oil	Cast iron	60
	Retention head burner	70 – 78
	Midefficiency	83 – 89
Electric	Central or baseboard	100
	Geothermal heat pump	–
	Air-source heat pump	–
Firewood	Conventional	45 – 55
	Advanced	55 – 65

Figure 7-20. An annual fuel utilization efficiency (AFUE) rating is a measure of the thermal efficiency of combustion equipment used in HVAC systems.

Fuel Identification Flash Cards

It is recommended that natural gas burners contain between 8% and 9.5% carbon dioxide. Grade No. 2 fuel oil burners should contain between 10% and 12.5% carbon dioxide. Bituminous coal furnaces should contain between 13% and 15% carbon dioxide. These values are approximate percentages for heating equipment in good operating condition. Manufacturer recommendations must be followed at all times.

Flue Gases. *Flue gas* is the carbon dioxide and other vapors that exit a furnace or boiler to the atmosphere using a flue, which is a pipe or channel for conveying exhaust gases. A flue-gas analyzer is typically used to measure the percentage of carbon dioxide and oxygen in flue gas. **See Figure 7-21.** The temperature of the flue-gas sample must also be measured. Flue-gas temperature is measured with a stack thermometer. A stack thermometer is used to measure high flue-gas temperatures. Tables are used along with the carbon dioxide reading, oxygen reading, and net flue-gas temperature to determine the actual combustion efficiency of the flue gas.

Carbon dioxide and flue-gas temperature readings are used to adjust a burner so that it will operate as efficiently as possible. When a carbon dioxide reading is too low and oxygen is too high, the amount of combustion air should be reduced. The combustion air is adjusted based on the carbon dioxide, oxygen level, and flue-gas temperature.

Analyzing Flue Gases

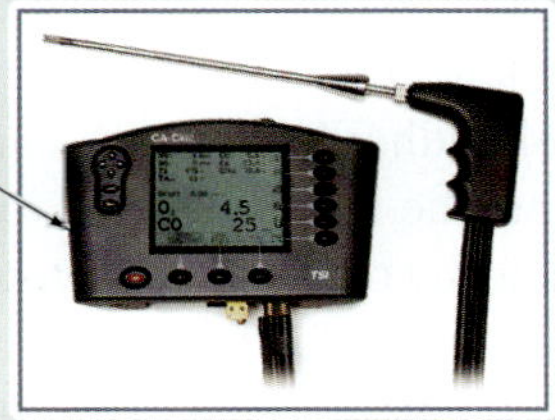

Gas*	Sensor/Probe	Resolution
CO-carbon monoxide	IR	1 PPM-10 PPM
CO2-carbon dioxide	IR	0.01%-0.1%
NO-nitric oxide	IR	1 PPM-10 PPM
SO2-sulphur dioxide	IR	1 PPM-10 PPM
CH4-methane	IR	1 PPM-10 PPM
O2-oxygen	Electromechanical	0.01%-0.1%
HO2-nitrogen dioxide	Electromechanical	0.01%
H2S-hydrogen sulphide	Electromechanical	1 PPM
C12-chlorine	Electromechanical	1 PPM
H2-hydrogen	Electromechanical	1 PPM

* volumetric concentration

FLUE-GAS ANALYZER SENSOR PROBES

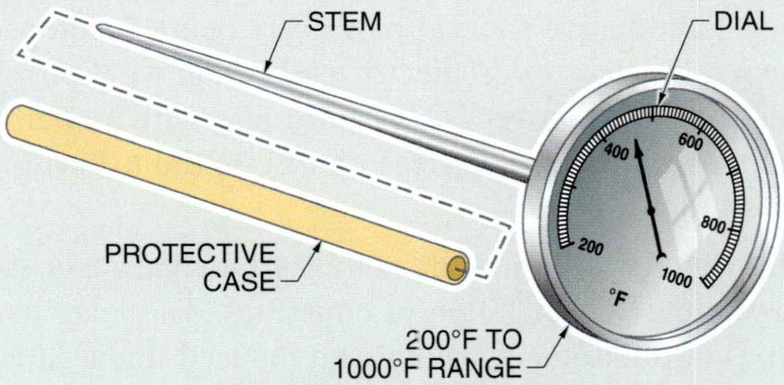

STACK THERMOMETER

Figure 7-21. Flue-gas analyzers and stack thermometers are used to analyze flue gas.

Fuel Combustion Efficiency and Emissions

When testing for the combustion efficiency of fuels, oxygen and carbon dioxide levels must be analyzed. Oxygen and carbon dioxide analyzers are similar to flue-gas analyzers. These analyzers are available in conventional readout or electronic readout models.

TECH TIP

The typical combustion efficiency for a space heater is 65%. It is 76% for a residential gas boiler with an atmospheric burner, 82% for a residential gas boiler with a power vent, 78% for an induced draft furnace, and 93% for a condensing furnace.

Conventional analyzers require several steps and multiple samples to achieve results. Electronic analyzers may require only one sample. Some models are permanently attached to a heating unit while others are handheld. Oxygen, carbon dioxide, and carbon monoxide levels are used to determine the combustion efficiency of a fuel. To determine proper oxygen levels, burner manufacturer recommendations should be followed. There should never be more than a trace amount of carbon monoxide present in flue gas.

Combustion has been used throughout history to produce heat. Carbon dioxide is a product of combustion. Today, there are concerns about the increasing amounts of carbon dioxide in the atmosphere. Industrial plants and combustion engines that burn fossil fuel emit most of the carbon dioxide in the atmosphere. A *fossil fuel* is a fuel derived from organic matter that has decayed over millions of years beneath the earth's surface.

The United States is responsible for 24% of the total carbon emissions from fossil fuels. As carbon dioxide emissions continue to increase, the earth's atmosphere will become warmer. To reduce carbon dioxide emissions, alternative energy sources are being developed such as nuclear, hydroelectric, solar, and wind power energy. **See Figure 7-22.** These alternative energy sources produce electrical energy without burning fossil fuels.

Dilution Air. Dilution air combines with flue gases and lowers the concentration of emissions. There are two types of dilution air: natural and induced. Induced air is artificially created diluted air.

Energy

Energy can be classified into renewable energy and nonrenewable energy. *Renewable energy* is energy that is generated from rivers, the sun, oceans, and from beneath the surface of the planet to generate energy over and over as needed. Renewable energy sources are environmentally friendly to the planet. *Nonrenewable energy* is energy that is generated from fossil fuels and cannot be generated over and over as needed. Nonrenewable energy sources are not environmentally friendly. **See Figure 7-23.**

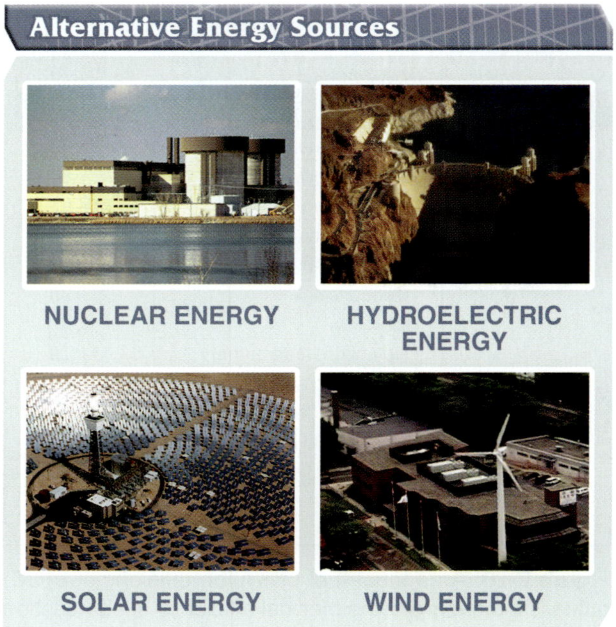

Figure 7-22. Alternative energy sources, such as nuclear, hydroelectric, solar, and wind power energy, produce electrical energy without burning fossil fuels, which generate carbon dioxide emissions.

Energy

Renewable Energy	
Hydropower	Biomass power
Wind power	Nuclear
Solar power	Wood
Geothermal	Ocean energy
Nonrenewable Energy	
Natural Gas	Coal
Oil	

Figure 7-23. Renewable energy technology has not developed as fast as most people would like, but wind, geothermal, and ocean energies have made significant advancements.

TECH TIP

In addition to the 5% to 20% excess air provided to burners for combustion, another 10 ft^3 of air and 50% excess air must be provided as dilution air. This allows draft diverters to carry the products of combustion up the flue.

UNIT 8

Psychrometrics

To maintain proper comfort levels for building occupants, every worker in the HVAC industry must be familiar with the principles of comfort and psychometrics. First, human comfort is defined, and then the principles of air (psychometrics) that determine comfort are discussed. The principles addressed in this unit can be used to properly size, install, and maintain HVAC systems.

OBJECTIVES

- Describe human comfort.
- List the uses for ANSI/ASHRAE Standard 55-2004.
- Define psychrometrics.
- List the air properties that can be found using a psychrometric chart.
- Explain how to plot a heating or cooling process on a psychrometric chart.
- Perform common calculations using a psychrometric chart.

Index to Procedures

COMFORT

The heating, ventilating, and air conditioning (HVAC) industry is based on the pursuit of comfort. The idea of comfort is an integral part of all HVAC system designs. *Human comfort* is a condition that occurs when people cannot sense a difference between themselves and the surrounding air. Human comfort is subjective. One person may be comfortable in an environment while another person experiences discomfort. A broad comfort range acceptable to most people can be illustrated with a psychrometric chart. **See Figure 8-1.**

The HVAC industry developed as people demanded greater levels of comfort in their homes and at work. As the idea of comfort evolved, it became necessary to scientifically describe the conditions that make up comfort. The heating, ventilating, and air conditioning processes needed to produce a desired comfort level require mathematical descriptions.

A mathematical description makes it possible to size HVAC equipment for various situations. In addition to human comfort, the representation of these applications is used for controlling temperature and air for product creation, storage, and quality.

ANSI/ASHRAE Standard 55-2004

ANSI/ASHRAE Standard 55-2004, *Thermal Environmental Conditions for Human Occupancy,* is a revision of Standard 55-1992. Standard 55-2004 specifies the combinations of indoor thermal factors and personal factors that produce acceptable environmental conditions for most occupants. **See Figure 8-2.** Factors that affect the environment include temperature, thermal radiation, humidity, and air speed. Personal factors controlled by individuals include occupant activity and type of clothing worn.

Comfort

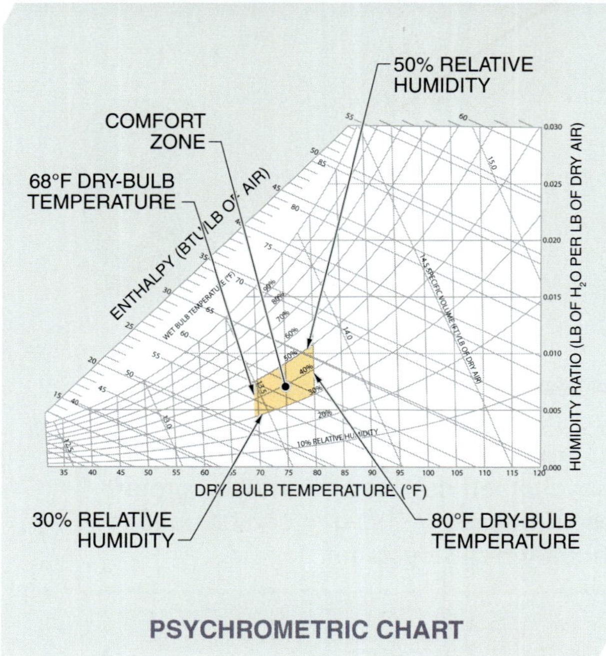

Figure 8-1. A broad comfort range that is acceptable to most people can be illustrated on a psychrometric chart.

ANSI/ASHRAE Standard 55-2004

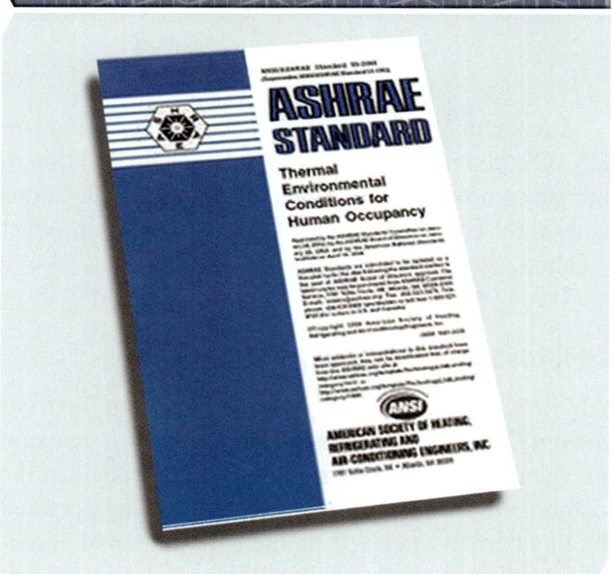

Figure 8-2. ANSI/ASHRAE Standard 55-2004 specifies the combinations of indoor thermal environmental factors that affect human comfort, such as temperature, thermal radiation, humidity, and air speed.

Standard 55-2004 is used in the design, commissioning, and testing of HVAC systems. It is also used to evaluate thermal environments. The standard contains a predicted mean vote (PMV) calculation, which is used where various levels of comfort may be required in the same building. Use of the PMV can improve occupant comfort by allowing comfort analysis to be performed on occupants in various areas in a building. Based on human physical parameters, it is possible to evaluate comfort conditions using PMV software to complete the individual calculations.

PSYCHROMETRICS

Psychrometrics is the mathematical and graphical study of the properties of air. Although air is composed of many substances, the properties of air analyzed in psychrometrics are humidity, dry air, and other compounds. The category for other compounds consists of items such as pollution, airborne pollen, and bacteria.

Water Vapor

Water vapor is generated from the activity of plant life, oceans, lakes, rivers, streams, and pools. **See Figure 8-3.** It can be compared to steam at a low temperature and pressure. Similar to the heat of steam, water vapor adds heat to the air, which makes the environment feel warmer than usual.

Water Vapor Sources

Figure 8-3. Water vapor enters the air from plant life, oceans, lakes, rivers, streams, and pools.

Dry Air

Dry air is different from pure air. Pure air is a mixture of oxygen and nitrogen only. Compounds, such as pollution and bacteria, change the composition of pure air. Dry air is a mixture of gases that consists of 21% oxygen, 78% nitrogen, and 1% argon, carbon dioxide, and other gases. **See Figure 8-4.**

Air Composition

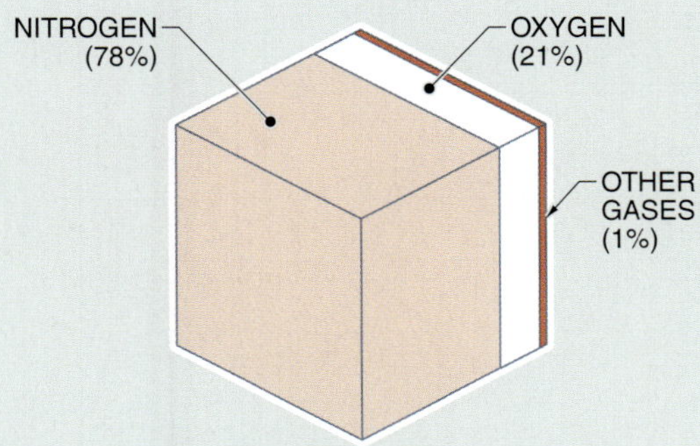

ONE CUBIC FOOT OF AIR

Gas	Ratio Compared to Dry Air *		Chemical Symbol
	By Volume	By Weight	
Oxygen	20.95	23.20	O_2
Nitrogen	78.09	75.47	N_2
Carbon dioxide	0.03	0.046	CO_2
Hydrogen	0.00005	≈0.0	H_2
Argon	0.933	1.28	Ar
Neon	0.0018	0.0012	Ne
Helium	0.0005	0.00007	He
Krypton	0.0001	0.0003	Kr
Xenon	9×10^{-6}	0.00004	Xe

* in %

Figure 8-4. Air is a mixture of gases that consists of 21% oxygen, 78% nitrogen, and 1% argon, carbon dioxide, and other gases.

TECH TIP

A standard cubic foot of air is defined as air at sea level, or 14.7 psia (1 atmosphere), at a temperature of 60°F, and with a relative humidity of 0%. Other gases that can be found in a standard cubic foot of air are methane (2.0 ppm), sulfur dioxide (1.0 ppm), nitrous oxide (0.5 ppm), ozone (0.07 ppm), nitrogen dioxide (0.02 ppm), iodine (0.01 ppm), carbon monoxide (trace amount), and ammonia (trace amount).

Indoor air such as in restaurants can contain many of the same odorous contaminants and compounds (auto pollution, pollen, and bacteria) that are found in outdoor air.

Other Compounds

The variety of compounds present in outdoor air include automobile pollution, pollen, and bacteria. When an automobile is running, the natural process of the automobile creates pollution and mixes with the surrounding air. Also, the pollen produced by plants mixes with the air.

PSYCHROMETRIC CHARTS

In the early 1900s, a team led by Willis Carrier, considered the father of modern air conditioning, introduced a method to represent the different properties of air in a graphical format. This format is known as a psychrometric chart. **See Figure 8-5.** A *psychrometric chart* is a graph that defines the properties of air at various conditions. The chart is used primarily to show comfort levels and the effects HVAC processes have on comfort.

The ideal storage conditions for various products, such as textiles, paper, chocolate, and pharmaceuticals, can be found by using a psychrometric chart. The chart may also be useful for selecting the proper capacity of HVAC equipment and aiding in troubleshooting. Psychrometric chart elements include relative humidity, humidity ratio, wet-bulb temperature, specific volume, enthalpy, dewpoint temperature, and dry-bulb temperature. **See Figure 8-6.**

Psychrometric Charts

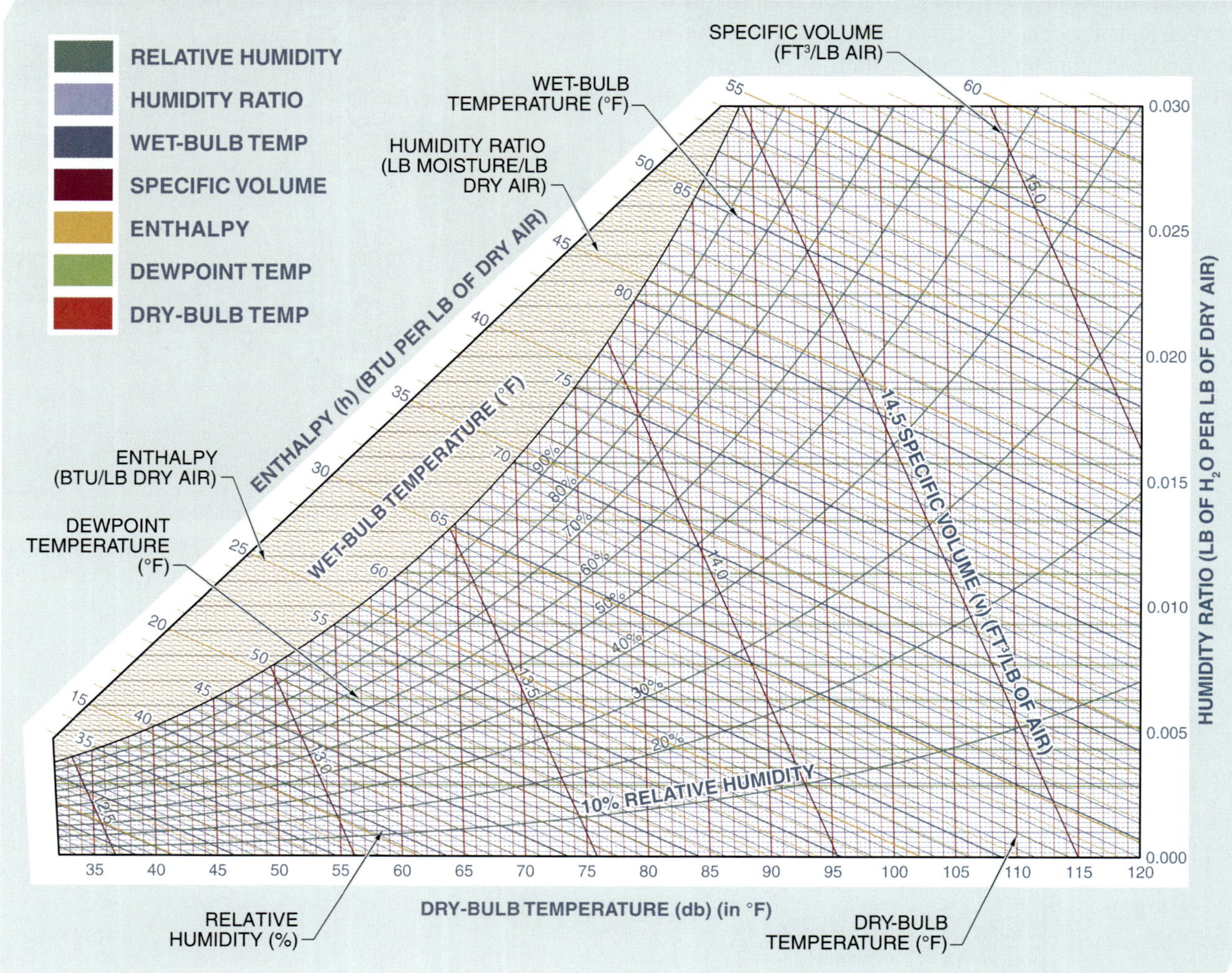

Figure 8-5. A psychrometric chart is a graph that defines the properties of air at various conditions and is used to analyze human comfort and the effects of HVAC processes.

Dry-Bulb Temperatures

A *dry-bulb temperature* is the temperature of air as taken with a standard thermometer. Since standard thermometers are readily available, dry-bulb temperature (db) is one of the most commonly used elements of a psychrometric chart. The dry-bulb temperature scale is represented by vertical lines across the bottom of the chart. Standard psychrometric charts start with a temperature of about 30°F on the left side and end with approximately 115°F to 120°F on the right. **See Figure 8-7.**

Dry-Bulb Temperature Applications. Correct dry-bulb temperature is required to prevent an environment from becoming too hot or too cold. Dry-bulb temperature is also a primary consideration in storing perishable items such as ice cream, milk, and frozen foods. When a technician monitors the temperature sensor in a room, duct, or outside a building, the technician is typically monitoring a dry-bulb temperature.

TECH TIP

Wet bulb sling psychrometers that measure the amount of water (humidity) in the air have a wick (sock) that is wetted and whirled around to acquire the wet bulb temperature. It is recommended that distilled water be used to wet the wick.

Psychrometric Chart Elements

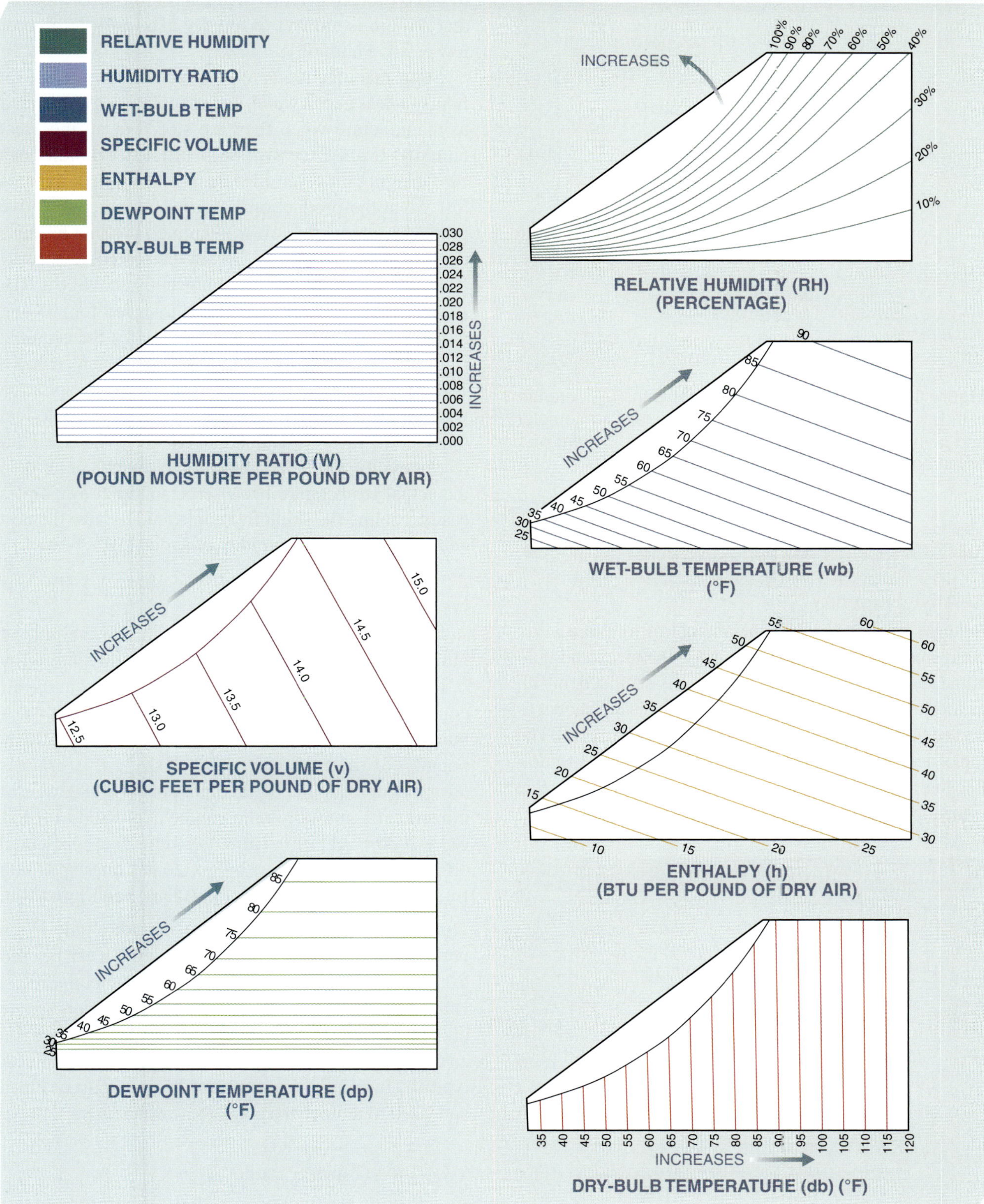

Figure 8-6. The elements of a psychrometric chart include dry-bulb temperature, relative humidity, humidity ratio, wet-bulb temperature, specific volume, enthalpy, and dewpoint temperature.

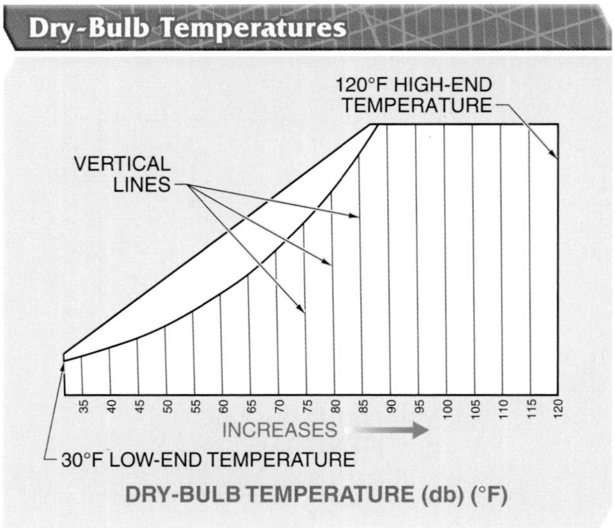

Figure 8-7. Dry-bulb temperature (db) is the temperature of the air as taken with a standard Fahrenheit thermometer and is indicated by vertical lines across the bottom of a psychrometric chart.

Psychrometric Chart Interactive Animation

Relative Humidity

Relative humidity (RH) is the amount of moisture in air compared to the amount of moisture the air would hold at the same temperature when the air is saturated, or full of moisture. Relative humidity percentages are shown as curved lines on a psychrometric chart. **See Figure 8-8.** The top of a relative humidity curve is referred to as the saturation line and represents 100% humidity. The saturation line is the top line because the air cannot hold more moisture.

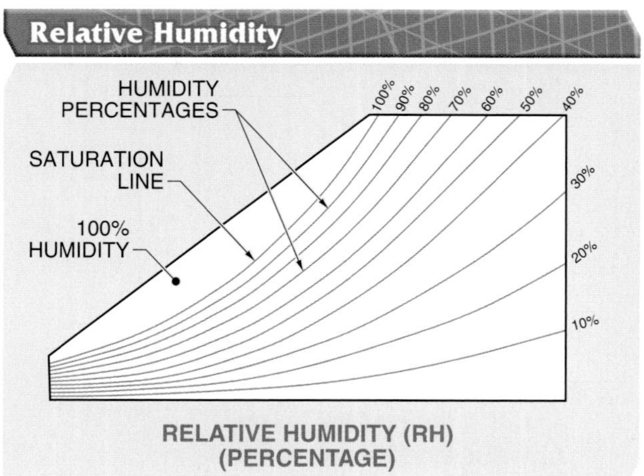

Figure 8-8. Relative humidity (RH) is shown as curved lines from the bottom to the top of a psychrometric chart.

Relative Humidity Applications. Relative humidity is often considered in terms of human comfort. The feeling that a room is too wet or too dry is a result of high or low relative humidity.

Some manufactured products are humidity sensitive. Items such as paper, wood, and chocolate can be difficult to manufacture when they are stored at an incorrect humidity level. Excessive or insufficient humidity can be damaging on several levels. For example, money is lost when the production line of a humidity-sensitive product is stopped to clear a jammed product because the humidity of the area is not at the correct level.

Mold growth accelerates significantly above 60% RH. An RH value below 60% is recommended for slowing mold growth. In many areas of the United States, mold is an indoor air quality (IAQ) problem. When air has a 100% relative humidity (air is saturated), perspiration will not evaporate into the air. As a result, people feel much hotter than the actual temperature. When the relative humidity is low, people can feel much cooler than the actual temperature because perspiration evaporates easily, cooling the skin off. People tend to have the best comfort at a relative humidity of about 45%.

Humidity Ratio

A humidity ratio is a measurement, by weight, of moisture in the air. Unlike relative humidity, humidity ratio represents the actual weight of moisture in the air compared to the weight of moisture and air together. A humidity ratio is expressed in grains (or alternatively pounds) of moisture per pound of dry air. A grain is 1/7000 lb. For example, if 1 lb of air contains 78 grains of moisture, its humidity ratio can be expressed as 0.111 (78 ÷ 7000 = 0.111 lb). Humidity ratios are represented on a psychrometric chart by horizontal lines spanning from the bottom to the top of the chart. **See Figure 8-9.**

Humidity Ratio Applications. A humidity ratio is not used as often as relative humidity, however, it can be used to size equipment such as humidifiers and dehumidifiers. Humidity ratios can also be used to help size condensate drainage systems. As air is cooled, moisture condenses on an air conditioning coil. This condensate must be drained, typically by using a drip pan. The drip pan must be piped and sized to remove the condensate efficiently.

Wet-Bulb Temperature

A *wet-bulb temperature* is the temperature of air if it was cooled to 100% relative humidity. A sling psychrometer (two thermometers) is used to measure wet-bulb temperature. The sock on one thermometer is wet and then whirled through the air. Some of the water in the sock evaporates, cooling the thermometer bulb, and results in a lower temperature. This lower temperature is the wet-bulb temperature. Wet-bulb temperature is represented on a psychrometric chart by angled lines that begin at the saturation line and angle down to the right. **See Figure 8-10.** The values are read at the saturation line.

Humidity Ratio

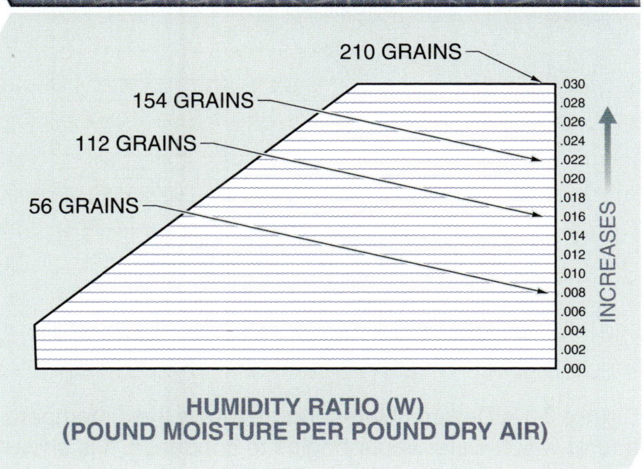

Figure 8-9. A humidity ratio (W) is a measurement of the actual weight of air moisture and is indicated by horizontal lines spanning from the bottom to the top of the chart.

Wet-Bulb Temperatures

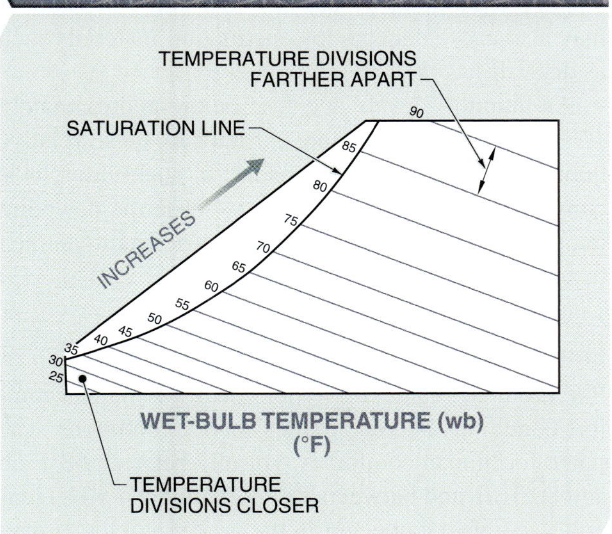

Figure 8-10. Wet-bulb temperature (wb) is the temperature of air if it was cooled to 100% relative humidity and is represented by angled lines that begin at the saturation line and angle down to the right of the chart.

Wet-Bulb Temperature Applications. Wet-bulb temperature is often used to express a combination of dry-bulb and humidity measurements. It is often used to determine the appropriate humidity settings that will achieve human comfort. This information is often used in the design of air conditioning systems.

Wet-bulb temperature also affects the efficiency of cooling towers. Cooling towers operate by rejecting heat into the atmosphere. They work more efficiently when outdoor air has a lower wet-bulb temperature.

Specific Volume

Air expands when heated and contracts when cooled. *Specific volume* is the volume that one pound of a gas substance occupies at a particular temperature and pressure condition; expressed in cubic feet per pound (ft^3/lb). The specific volume of air varies as the temperature and humidity of the air changes. Specific volume on a psychrometric chart is represented by steep lines from top to bottom or lines that intersect the right side of the chart. **See Figure 8-11.**

Specific Volume

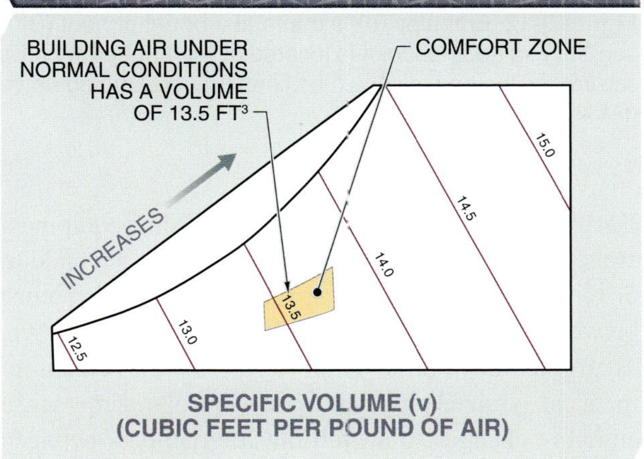

Figure 8-11. Specific volume (v) is the amount of volume (in cubic feet per pound) that 1 lb of air occupies at a specific temperature and humidity condition. It is indicated by steep lines from the top to bottom right of the chart.

Specific Volume Applications. Specific volume is often used to size fans. Air handling units move air to provide HVAC capability to a space or product. The amount of air moved depends on the specific volume of the air. Once the condition of the air is known, the fan for a system can be sized appropriately.

Enthalpy

Water vapor in the air represents heat content, or enthalpy. *Enthalpy* is the total amount of heat contained in a substance. The British thermal unit (Btu) is used to express enthalpy. An enthalpy scale is located at the upper left of the saturation line and is shown as lines angled down and to the right. **See Figure 8-12.**

Enthalpy

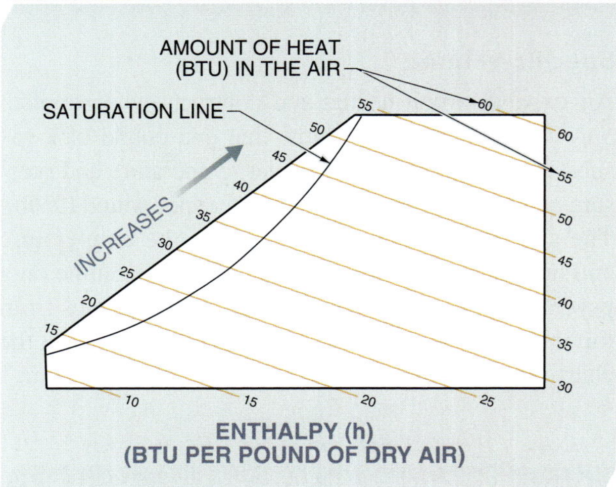

Figure 8-12. Enthalpy (h) is the total amount of heat (Btu) content in a substance. It is located at the upper left of the saturation line and is shown as lines that are angled down to the right.

Enthalpy Applications. Heating and cooling equipment are sized based on the number of Btus they can add to or remove from a desired space. When the difference between the beginning and ending condition is known, as well as the volume of air used, then HVAC equipment can be sized properly. Many control systems sense enthalpy changes to switch outside dampers between free cooling mode and minimum air mode. Switching between modes occurs when the outdoor air heat content is excessively high and would impose a large load on the cooling equipment of a system.

Dewpoint Temperature

Dewpoint temperature (dp) is the temperature at which water vapor in the air begins to condense. Similar to the humidity ratio, dewpoint temperature lines are shown horizontally on the chart. **See Figure 8-13.** Unlike humidity ratio lines, dewpoint temperature lines get closer together at the bottom of the scale. Some charts list the numerical values directly on the chart. Other charts require the user to find the intersection point of the dewpoint temperature line and the saturation line, where the numerical value is obtained.

Dewpoint Temperature

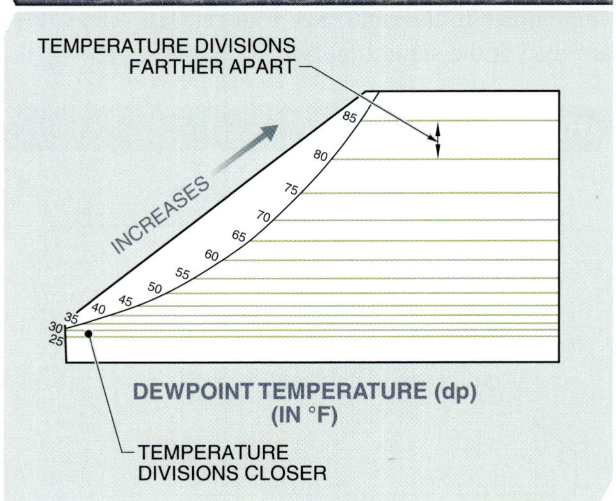

Figure 8-13. Dewpoint temperature (dp) is the air temperature at which water vapor begins to condense. It is shown as horizontal lines that get closer together at the bottom of the psychrometric chart.

Dewpoint Applications. Condensation forms when the dewpoint temperature is reached. Condensation may damage equipment by causing rust and corrosion. It may also cause damage to construction materials such as drywall.

As humidity levels decrease below approximately 10% RH, it becomes more difficult to sense relative humidity using standard sensors. In such cases, it is common to use optical sensors to sense the dewpoint temperature of the air. A psychrometric chart can then be used to find the relative humidity.

Human Comfort vs. Product Quality

Psychrometric chart applications deal with human comfort conditions and/or product quality conditions. The range for human comfort is typically between 68°F db and 80°F db and between 30% RH and 60% RH. Generally, comfort can occur in the winter at a lower dry-bulb temperature when the humidity is high enough. **See Figure 8-14.** Comfort in the summer can occur at a higher dry-bulb temperature when the humidity is low enough.

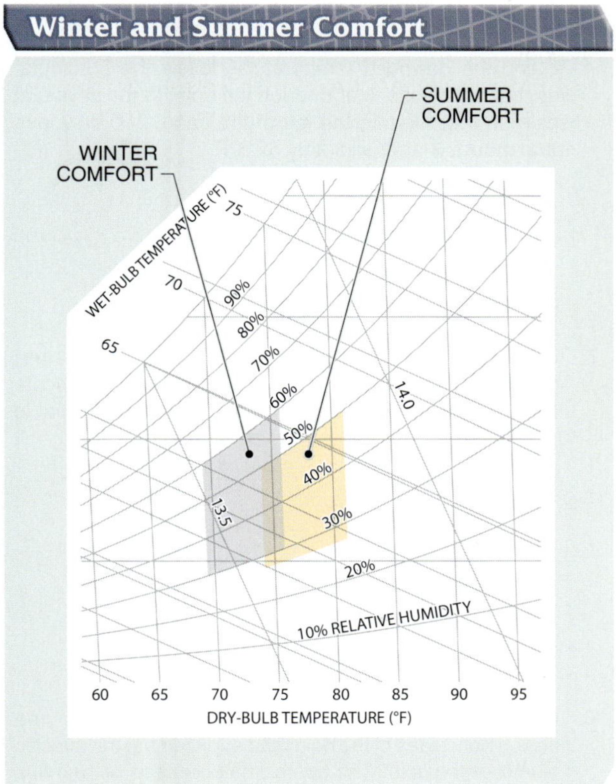

Figure 8-14. The range for human comfort in the summer is between 73°F and 82°F dry-bulb temperature with 28% to 50% relative humidity. In the winter, the range is between 68°F and 76°F dry-bulb temperature with 30% to 60% relative humidity.

In certain applications, the requirements for maintaining product-quality conditions are more stringent than the requirements for maintaining human comfort, and may take precedence over human comfort conditions at work. Examples of product quality conditions include the following:

- Museum art and antiquities are displayed at 72°F at 50% RH, ±5% RH.
- The relative humidity in a food processing plant must be less than 65%.
- The moisture content (humidity) of stored peanuts must be 9% or less to slow the growth of mold and mildew.
- A relative humidity of 20% or less is required to retard the corrosion of electronics in stored aircraft.
- Textile factories are kept as close as possible to 100% RH to facilitate the proper operation of cutting machines.

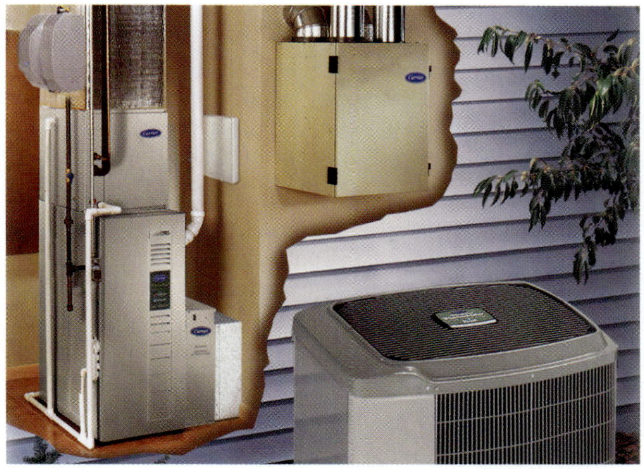

Carrier Corporation

Due to various environmental conditions, humans require furnaces, ventilation fans, humidifiers, air cleaners, and air conditioners to achieve air conditions suitable for human comfort.

PROCESS PLOTTING

When any two conditions of air are known, a psychrometric chart can be used to find the other five conditions. For example, two known conditions are found on the chart, and their intersection forms a single point that is plotted (drawn) on the chart. **See Figure 8-15.** The other five air conditions are found using that point. The following procedure is used to find the five air conditions for a single point. **See Procedure for Plotting Air Properties.**

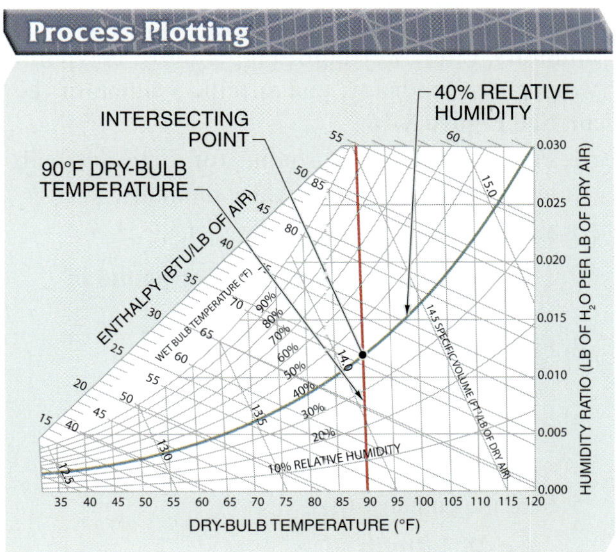

Figure 8-15. When any two properties of air are known (air condition intersection), the remaining five properties can be found using a psychrometric chart.

Procedure

Plotting Air Properties

1. To find the properties of air at 90°F db temperature and 40% RH, locate the 90°F dry-bulb temperature line on the scale across the bottom of the chart. Follow this line upward until it crosses the 40% relative humidity line. The point where the two lines intersect defines the condition of the air.

2. To find the humidity ratio at this point, follow the horizontal line that runs through the point to the humidity ratio scale on the right side of the chart. The humidity ratio is 0.0122 lb of moisture per pound of dry air.

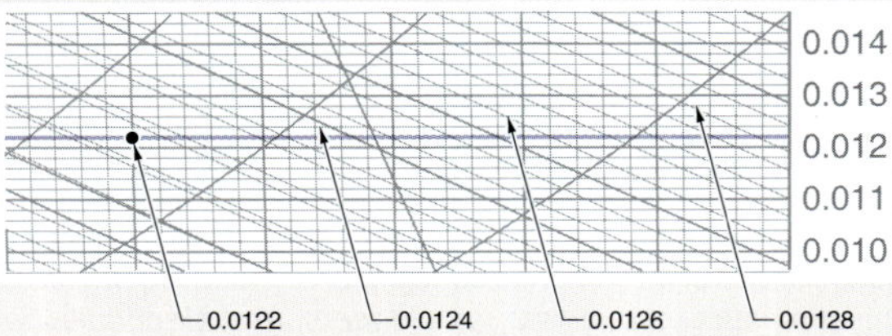

3. To find the wet-bulb temperature, follow the wet-bulb temperature line that runs through the point to the wet-bulb temperature scale on the curve to the left of the chart. The wet-bulb temperature is 71.2°F.

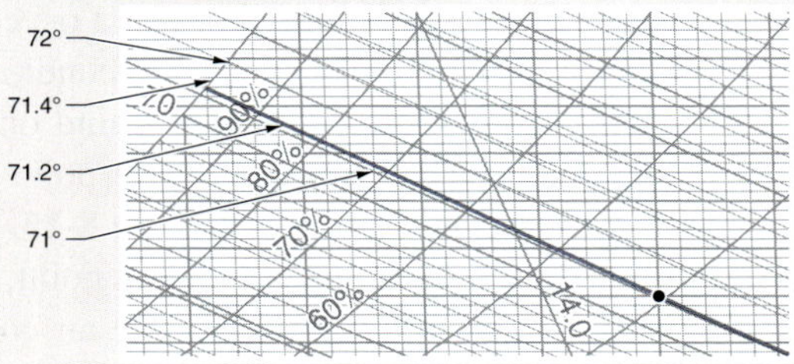

4. To find the dewpoint temperature, follow the horizontal line that runs close to or through the point to the dewpoint scale on the curve to the left of the chart. The dewpoint temperature is approximately 62.5°F.

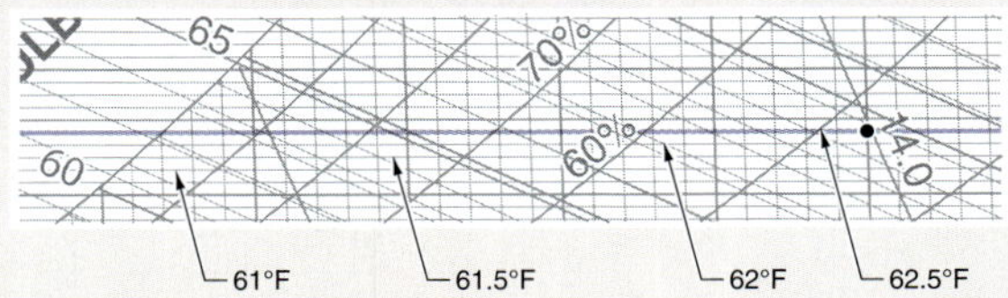

5. To find the enthalpy, extend the wet-bulb temperature line of the point through the curve to the enthalpy scale on the left side of the chart. The enthalpy for the point is 35.2 Btu/lb.

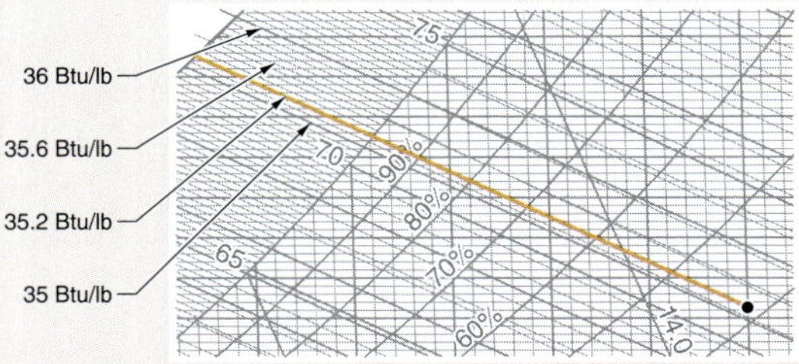

6. To find the specific volume, locate the specific volume lines on both sides of the point and approximate the specific volume for the point. The approximate value is 14.13 ft³/lb.

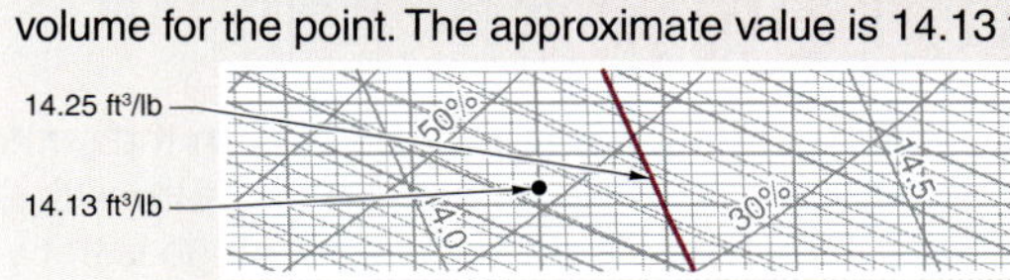

Example: Finding Air Properties

The air in a building is at 75°F db temperature and 50% RH. Using the proper procedure, find the humidity ratio, wet-bulb temperature, dewpoint temperature, enthalpy, and specific volume of the air. **See Figure 8-16.**

1. Find the intersection point for 75°F dry-bulb temperature and 50% relative humidity.
2. Find the humidity ratio (*W*).

   ***W* = 0.0093 lb of moisture per pound of dry air**

3. Find the wet-bulb temperature (*wb*).

   ***wb* = 62.4°F**

4. Find the dewpoint temperature (*dp*).

   ***dp* ≅ 55.1°F**

5. Find the enthalpy (*h*).

   ***h* = 28.2 Btu/lb**

6. Find the specific volume (*v*).

   ***v* = 13.65 ft³/lb**

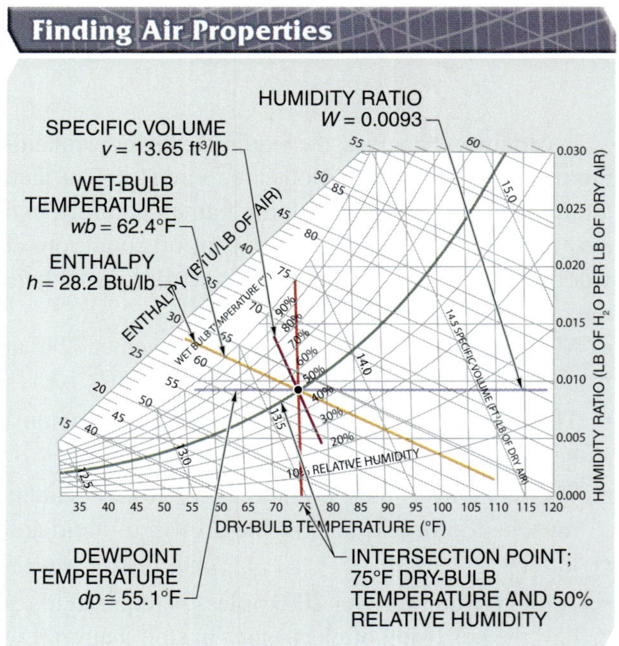

Figure 8-16. Using the intersection point of the 75°F dry-bulb temperature line and the 50% relative humidity line makes it possible to find other properties of air.

Plotting Two Points

As air is heated or cooled, its position on a psychrometric chart will change. When the beginning and end conditions are known, both points can be plotted.

Example: Finding Process Points—Heating Air

Air enters a furnace at 75°F db temperature and 50% RH and leaves at 95°F db. No humidity is added or removed from the air. Plot both points. **See Figure 8-17.**

1. Locate the point at the intersection of the 75°F dry-bulb temperature line and the 50% relative humidity line. This point represents the initial condition of the air.
2. Locate the point at the intersection of the humidity ratio line of the initial point with the 95°F dry-bulb temperature line. This point represents the final condition of the air.

Finding Process Points–Heating Air

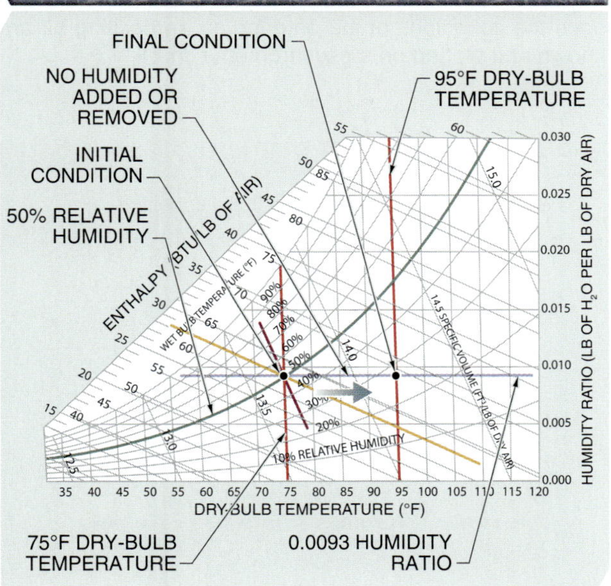

Figure 8-17. Having two intersection points (initial condition and final condition) allows heating processes to be analyzed.

TECH TIP

Standard HVAC-R psychrometric charts plot humidity ratio against dry-bulb temperatures. Other psychrometric charts include different barometric pressures with extreme temperature ranges for specific applications.

Example: Finding Process Points—Cooling Air

Air enters a cooling coil at 75°F db and 71°F wb, and it exits at 60°F db and 50°F wb. Plot both points. **See Figure 8-18.**

1. Locate the point at the intersection of the 75°F dry-bulb temperature line and the 71°F wet-bulb line. This point represents the initial condition of the air.
2. Locate the point at the intersection of the 60°F dry-bulb temperature line and the 50°F wet-bulb line. This point represents the final condition of the air.

Finding Process Points–Cooling Air

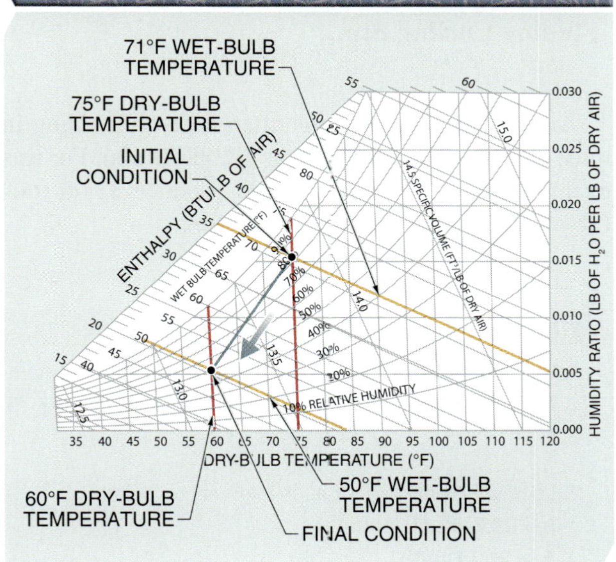

Figure 8-18. Having two intersection points (initial condition and final condition) allows cooling processes to be analyzed.

Once two points are plotted on a psychrometric chart, the points can be connected with a straight line. The line denotes changes as the air is heated, cooled, humidified, and dehumidified. When the line is drawn, it is customary to add an arrowhead to the line pointing from the initial condition toward the final condition.

On a psychrometric chart, the direction of the line represents whether the air is heated, cooled, humidified, or dehumidified. **See Appendix.** The length of the line denotes the amount of process change. A small line indicates a small process change, and a long line indicates a large process change.

CALCULATIONS USING THE PSYCHROMETRIC CHART

In addition to plotting points and process lines, a psychrometric chart may be used to perform calculations. Psychrometric chart calculations can be used to identify oversized or undersized equipment, identify equipment that is not operating correctly, and verify the amount of heating or cooling produced by equipment.

Example: Finding Cooling Btu

A cooling coil cools 12,500 cubic feet per minute (cfm) of air. Air enters the coil at 74°F db and 63.2°F wb. Air exits the coil at 55°F db and 52.5°F wb. **See Figure 8-19.** The following procedure is used to calculate the total cooling performed as air passes over the cooling coil. **See Procedure for Finding Cooling Btu.**

Air conditioning systems often measure cooling in tons. One ton of cooling equals 12,000 Btu/hr. For this example, the cooling coil would produce 32.00 tons (384,052 ÷ 12,000 = 32.00) of cooling.

Finding Cooling Btu

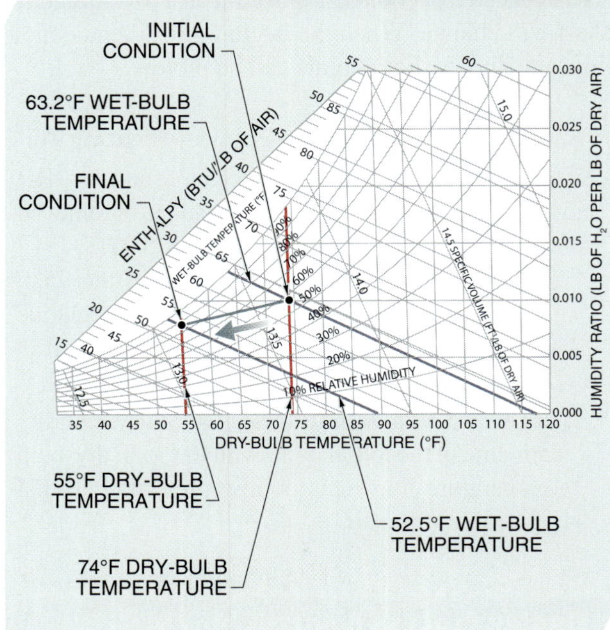

Figure 8-19. The total cooling Btu of a coil can be found once the conditions of the entering air and exiting air are known and plotted on a psychrometric chart.

Finding Cooling Btu

Procedure

1. Plot the initial condition of 74°F db and 63.2°F wb temperatures and the final condition of 55°F db and 52.5°F wb temperatures on the psychrometric chart.

2. Locate the enthalpy for each intersection point.
   h_I = **28.7 Btu/lb**
   h_F = **21.7 Btu/lb**

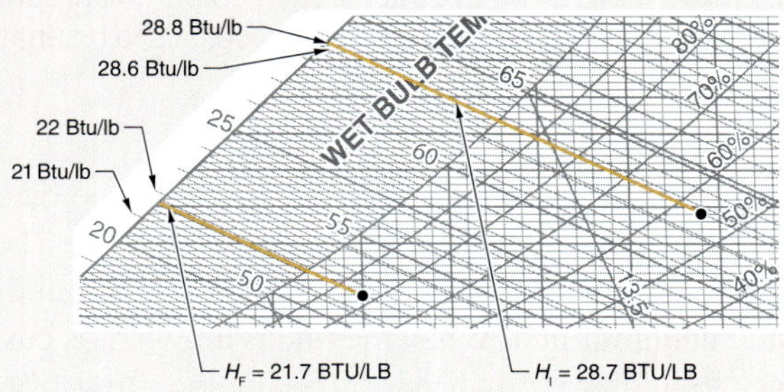

3. Subtract the final condition (h_F) from the initial condition (h_I) to find the amount of heat removed.
   heat removed = $h_I - h_F$
   heat removed = 28.7 Btu/lb − 21.7 Btu/lb
   heat removed = **7.0 Btu/lb**

4. Find the specific volume of the initial condition.
   v = **13.67 ft³/lb**

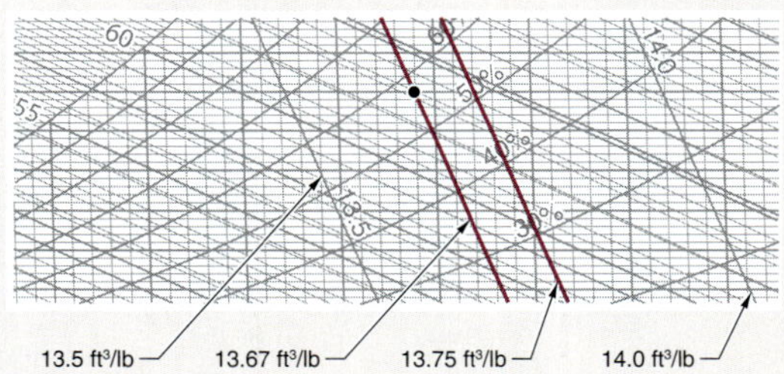

5. Calculate the rate of airflow.

$$\textit{mass flow} = \frac{\textit{quantity of air}}{v}$$

$$\textit{mass flow} = \frac{12{,}500}{13.67}$$

$$\textit{mass flow} = \mathbf{914.41\ lb / min}$$

6. Calculate the total cooling per hour.
   cooling = *mass flow* × *heat removed* × 60
   cooling = 914.41 × 7 × 60
   cooling = **384,052.2 Btu/hr**

Many furnace designs can be used to heat air for human comfort or product stability.

Example: Finding Heating Btu

A furnace is heating 5000 cfm of air. Air enters the coil at 70°F db and 40% RH. Air exits the coil at 95°F db with no humidity added or subtracted. **See Figure 8-20.** The following procedure is used to calculate the total heating performed as air passes over the heating coil. **See Procedure for Finding Heating Btu.**

Finding Heating Btu

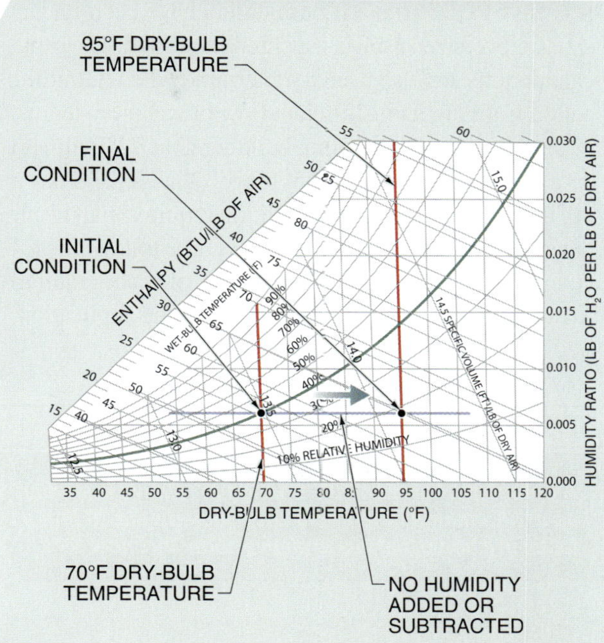

Figure 8-20. The total Btu of a furnace can be found once the conditions of the entering air and exiting air are known and plotted on a psychrometric chart.

Finding Heating Btu

1. Plot the initial condition of 70°F db temperature and 40% RH and the final condition of 95°F db temperature (no humidity added or subtracted) on the psychrometric chart.

2. Locate the enthalpy for each intersection point.

   h_I = **23.7 Btu/lb**

   h_F = **29.7 Btu/lb**

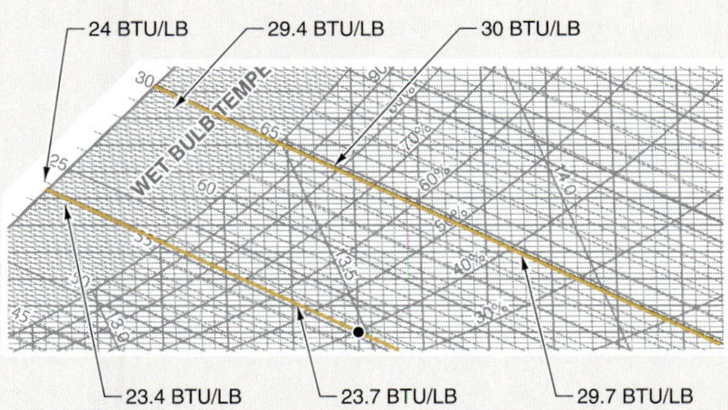

3. Subtract the initial condition (h_I) from the final condition (h_F) to find the amount of heat added.

   heat added = $h_F - h_I$

   heat added = 29.7 Btu/lb − 23.7 Btu/lb

   heat added = **6.0 Btu/lb**

4. Find the specific volume of the initial condition.

   v = **13.48 ft³/lb**

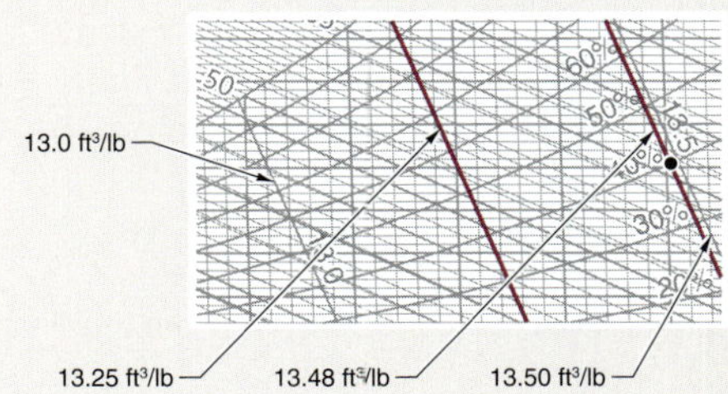

5. Calculate the rate of airflow.

   $$mass\ flow = \frac{quantity\ of\ air}{v}$$

   $$mass\ flow = \frac{5000}{13.48}$$

   mass flow = **370.92 lb / min**

6. Calculate the total heating per hour.

   heating = *mass flow* × *heat added* × 60

   heating = 370.92 × 6.0 × 60

   heating = **133,531.2 Btu/hr**

Procedure

Example: Finding Humidifier Size

It is easy to determine if humidifiers have been undersized or oversized using a psychrometric chart. The same calculations are also used to determine when humidification is giving an insufficient flow of water or steam.

For example, the air in a building is at 72°F db and 40% RH. The building is 45′ long, 37′ wide, and has a 10′ high ceiling. The final condition requires that the air have an RH of 55%. **See Figure 8-21.** The following procedure is used to calculate the quantity of water required to raise the relative humidity to 55% for the application. **See Procedure for Finding Humidifier Size.**

TECH TIP

A person perceives a 10% change in relative humidity as a 1°F change in temperature. Bacteria die 20 times faster in a home that has a relative humidity under 55%.

Finding Humidifier Size

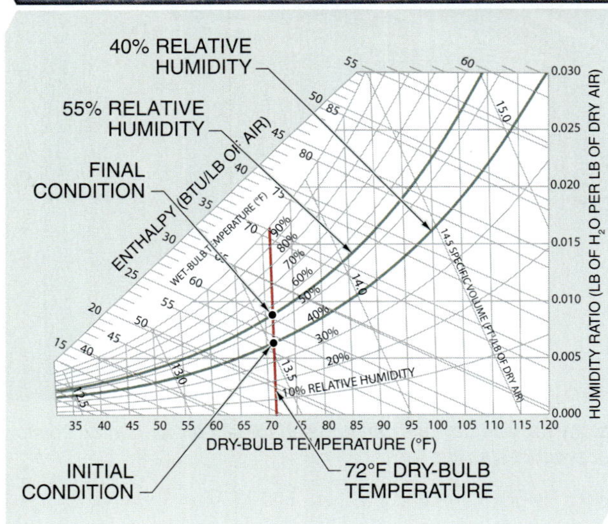

Figure 8-21. A psychrometric chart can be used to determine if humidifiers have been undersized or oversized.

Procedure

Finding Humidifier Size

1. Plot the initial condition of 72°F db temperature and 40% RH and the final condition of 55% RH.
2. Locate the humidity ratio for each intersection point.

   W_I = **0.0064 lb of moisture/lb of air**

   W_F = **0.0089 lb of moisture/lb of air**

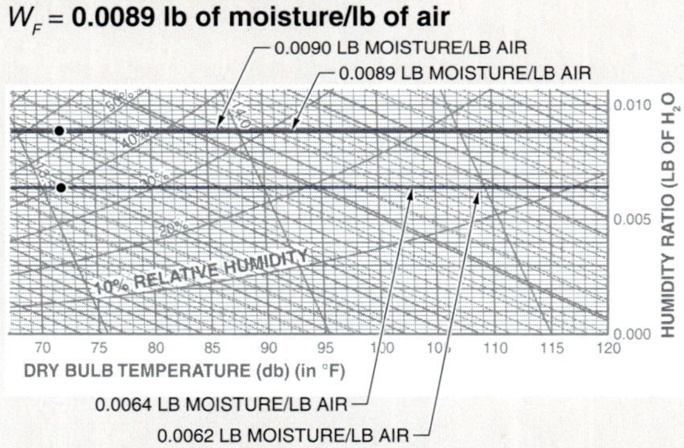

3. Subtract the initial condition (W_I) from the final condition (W_F).

   $W = W_F - W_I$

   $W = 0.0089 - 0.0064$

   W = **0.0025 lb moisture/lb of air**

4. Find the specific volume of the initial condition.

   v = **13.48 ft³/lb**

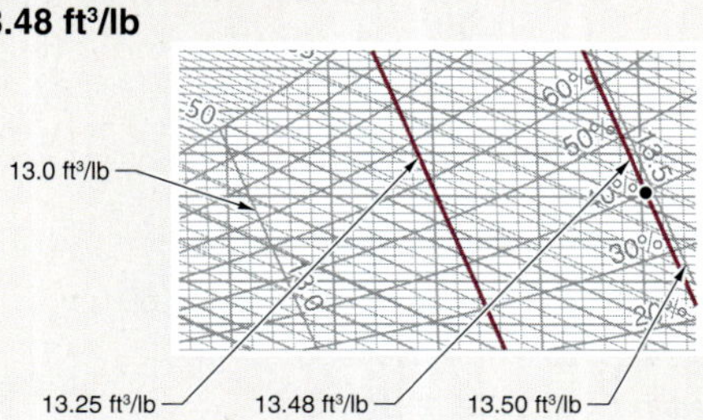

5. Calculate the volume of the building space (v_B).

   where

   l = length

   w = width

   h = height

   $v_B = l \times w \times h$

   $v_B = 45 \times 37 \times 10$

   $v_B = \mathbf{16{,}650\ ft^3}$

6. Calculate the total weight of air in the building space.

   $$weight_{total} = \frac{v_B}{v}$$

   $$weight_{total} = \frac{16{,}650\ \text{ft}^3}{13.48\ \text{ft}^3/\text{lb}}$$

   $$weight_{total} = \mathbf{1235.16\ lb}$$

7. Calculate the weight of water to be added.

   $weight_{water} = weight_{total} \times W$

   $weight_{water} = 1235.16 \times 0.0025$

   $weight_{water} = \mathbf{3.09\ lb\ of\ water}$

8. Calculate the amount of water (in gallons) that must be added.

   $$amount_{water} = \frac{weight_{water}}{8.345\ \text{lb/gal.}}$$

   $$amount_{water} = \frac{3.09}{8.345}$$

   $$amount_{water} = \mathbf{0.37\ gal.\ of\ water}$$

TECH TIP

A minimum amount of outdoor airflow must be introduced into most air-handling units to ensure good IAQ and to pressurize a building. The flow rate must match or exceed the amount of exhaust air taken from the building or specific area unless the building or area is required to operate at a negative pressure.

Mixed Air Calculations

Air-handling units use outdoor air for ventilation purposes. Outdoor (makeup) air is mixed with return air and then heated or cooled before a supply fan sends it to the required space. The final condition of the mixed air can be found by using a psychrometric chart.

Example: Determining Mixed Air Conditions

An air-handling unit requires 20% outdoor air for ventilation purposes. The outdoor air is at 90°F db and 70% RH. This air is mixed at the air handler with 80% return air at 75°F db and 65°F wb. **See Figure 8-22.** The following procedure is used to calculate the final condition of the air mixture. **See Procedure for Determining Mixed Air Conditions.**

Determining Mixed Air Conditions

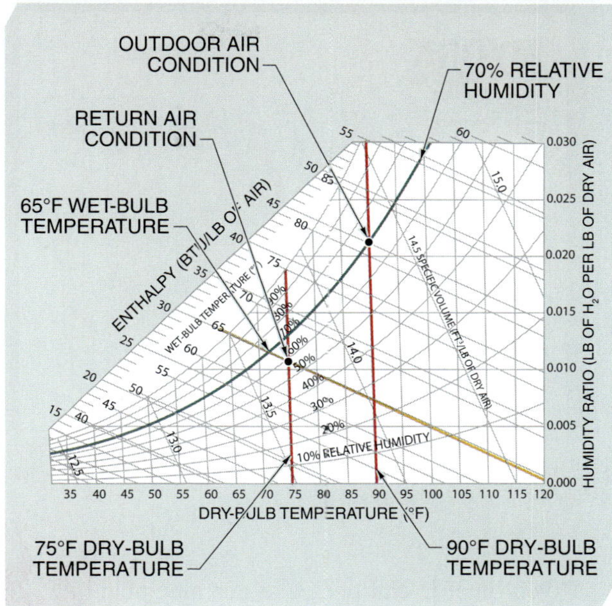

Figure 8-22. When outside air (ventilation or makeup air) is mixed with return air and then heated or cooled, the final condition of mixed air can be found by using a psychrometric chart.

Determining Mixed Air Conditions

1. Plot the outdoor air condition of 90°F db and 70% RH and the return air condition of 75°F db and 65°F wb on the psychrometric chart.
2. Draw a line on the psychrometric chart that connects the outdoor air condition and the return air condition.

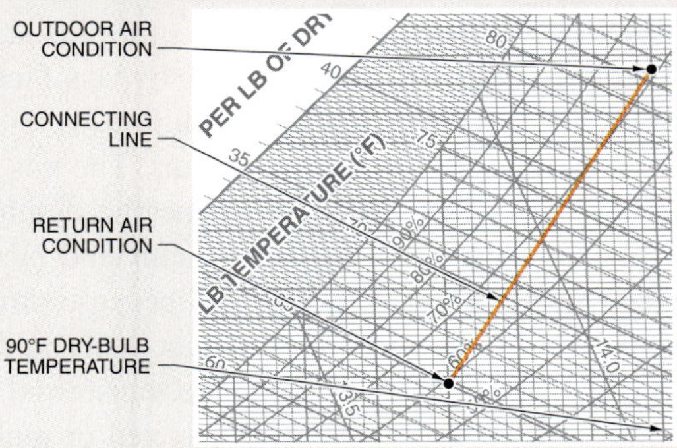

3. Find the temperature ratio for the outdoor air and return air.

For outdoor air:

temp ratio = *db* × *percentage of outdoor air*

temp ratio = 90 × 0.2

temp ratio = **18°F**

For return air:

temp ratio = *db* × *percentage of return air*

temp ratio = 75 × 0.8

temp ratio = **60°F**

4. Find the dry-bulb temperature of the air mixture.

$\textit{air mixture}_{db}$ = *outdoor air temp ratio* + *return air temp ratio*

$\textit{air mixture}_{db}$ = 18 + 60

$\textit{air mixture}_{db}$ = **78°F**

5. The point that represents the condition of the mixed air (outdoor air and return air) is found at the intersection of the 78°F dry-bulb temperature line and the connecting line.

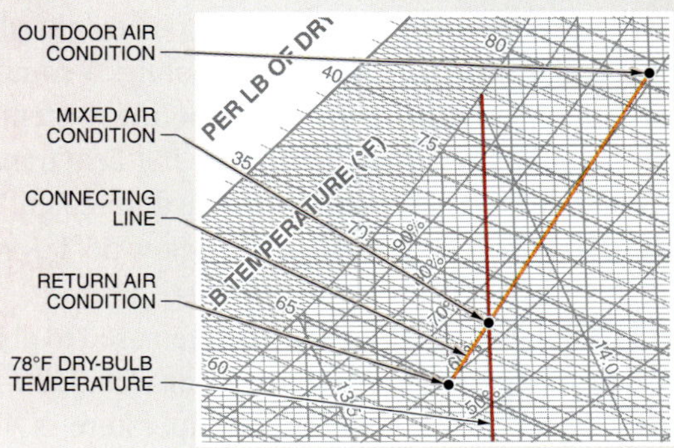

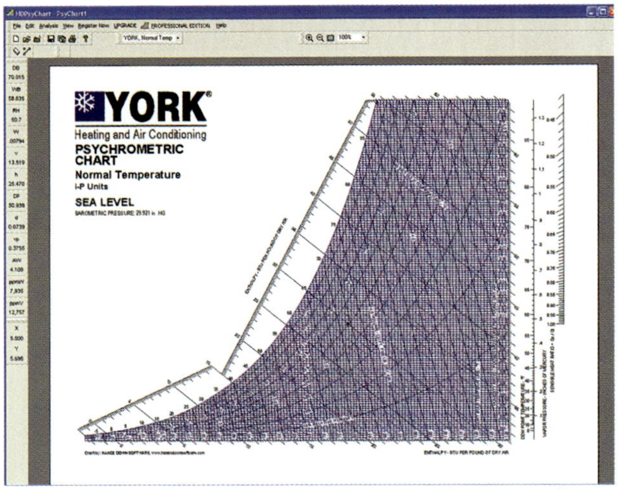

York International Corp.

Electronic psychrometric charts are quite useful when troubleshooting refrigeration and HVAC systems.

Psychrometric chart calculations may help indicate when there is an insufficient amount of outdoor air. Insufficient amounts of outdoor air may be the result of improper controls or damper actuation. When calculations show that too much outdoor air is being introduced, the excess air may be the result of missing damper edge seals or dampers stuck in the open position.

ADDITIONAL APPLICATIONS

A psychrometric chart has uses beyond those derived from the elements listed. For example, the chart is often used by engineers to perform apparatus dewpoint, bypass factor, and sensible heat factor calculations.

Apparatus Dewpoint and Bypass Factor

An *apparatus dewpoint (ADP)* is the temperature in which a cooling coil is set to achieve proper cooling. A common discharge temperature setpoint for cooling systems is 55°F. Humidity is set around 90% RH. For heat transfer to take place, there must be a temperature differential. The cooling coil must be at a temperature below 55°F for the discharge air temperature to be cooled.

An ADP range of 45°F to 48°F is often used to obtain the 55°F discharge air temperature. When a chilled water system is used, the chilled water temperature is 42°F to 44°F. **See Figure 8-23.** A cooling coil with an ADP of approximately 45°F will both cool and dehumidify the air. Air cooling represents the removal of sensible heat from the air, while dehumidification represents the removal of latent heat.

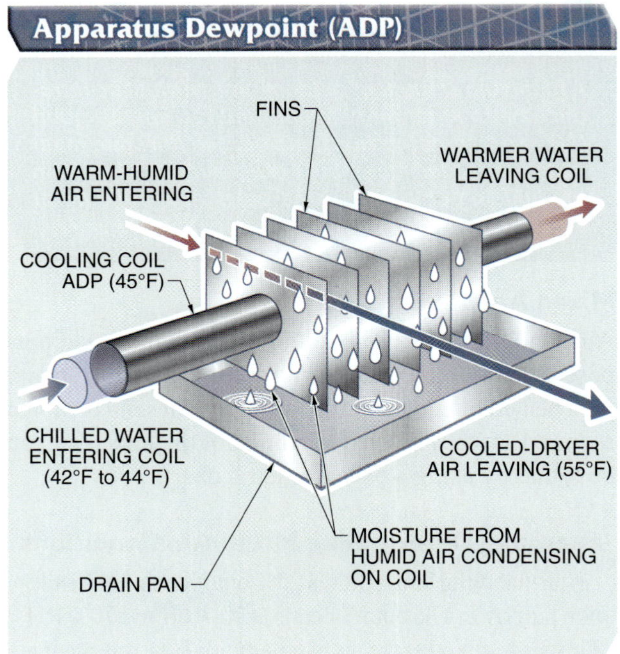

Figure 8-23. An apparatus dewpoint (ADP) is the temperature in which a cooling coil is set to achieve proper cooling. A cooling coil with an ADP of approximately 45°F is typically used to achieve a discharge temperature of 55°F.

There are factors that affect cooling coil efficiency. Not all air comes into contact with the coil. Some air bypasses the coil surface. A *coil bypass factor* is a number that represents the amount of air that bypasses the surface of a coil unchanged. As the fin density of a coil increases, there is more resistance, and the bypass factor is reduced. A bypass factor is a function of coil design and installation.

Coils are designed with rows of fins. The fins are a specific distance apart and with a specific number of fins per inch to minimize bypass factor and increase efficiency. Coils must be kept clean because dirt will prevent air from coming into contact with cold coil surfaces. As a result, less heat will be transferred from the air. When a dirty coil is fully cleaned or replaced with a new comparable coil, the cleaned or new coil will typically provide a 30% to 50% increase in performance.

Sensible Heat Factor

When cooling air, two processes are involved. The first process is cooling the air. The second process is dehumidification. Cooling air is a sensible heat removal process and dehumidification is a latent heat removal process.

The two processes are represented by the different loads present in the process being performed. The ratio between sensible heat removal and total heat removal is known as the sensible heat factor (SHF). The SHF can be calculated using the following formula:

$$SHF = \frac{Q_S}{(Q_S + Q_L)}$$

where

SHF = sensible heat factor

Q_S = sensible heat removed (in Btu/hr)

Q_L = latent heat removed (in Btu/hr)

Example: Finding Sensible Heat Factors

Consider a chilled water coil that removes 6400 Btu/hr of sensible heat and 3600 Btu/hr of latent heat. Calculate the sensible heat factor of the water coil.

$$SHF = \frac{Q_S}{(Q_S + Q_L)}$$

$$SHF = \frac{6400}{(6400 + 3600)}$$

$$SHF = \frac{6400}{10{,}000}$$

$$SHF = \mathbf{0.64\ Btu/hr}$$

Some psychrometric charts include a sensible heat factor indicator. **See Figure 8-24.** When a line is drawn from the design point to the saturation line using the slope of the sensible heat factor line, the required ADP of the equipment will be indicated.

The climatic design conditions for a building's location must be known to properly size the components of the building's HVAC system.

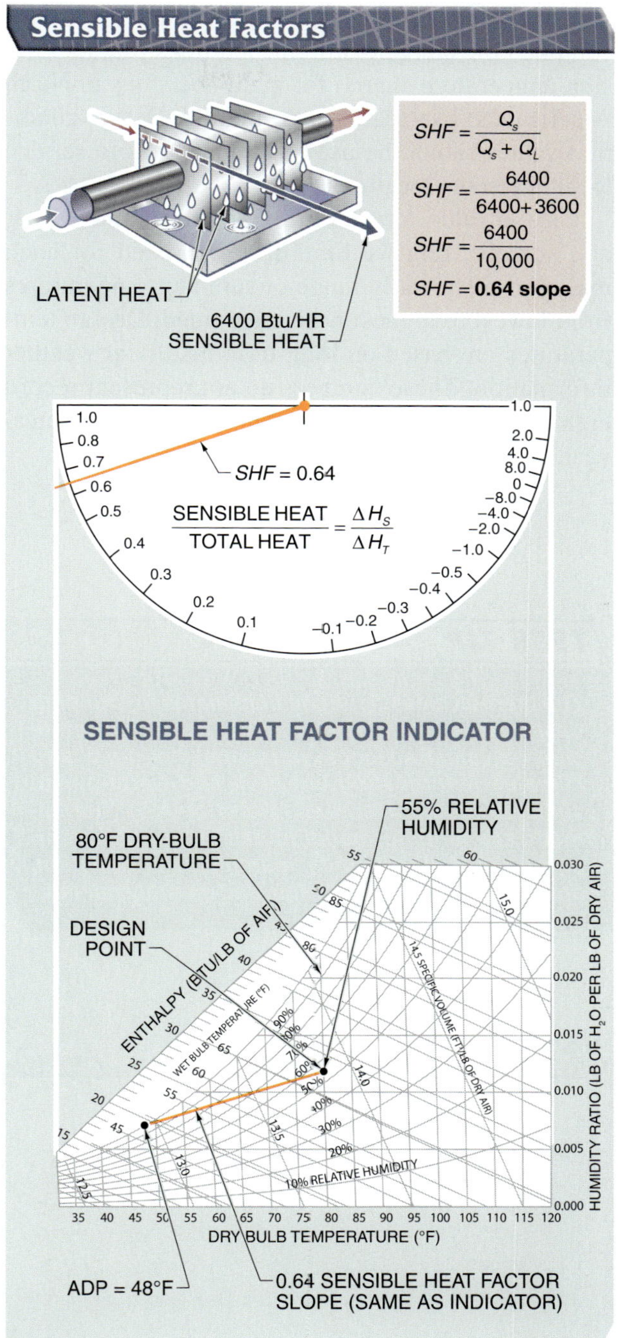

Figure 8-24. A sensible heat factor indicator is used to determine the slope of a line that is drawn from a design point to the ADP of a coil on a psychrometric chart.

Local Outdoor Conditions

Equipment should be sized and selected based on the local outdoor climatic conditions. Organizations, such as the American Society of Heating, Refrigerating, and Air-Conditioning Engineers (ASHRAE), have compiled the climatic design conditions (outdoor design temperature charts) for many locations in North America and overseas. **See Appendix.** Design condition values should be used by technicians to service local or remote locations.

City latitude, dry-bulb temperature, and winter and summer relative humidity are listed for each major city. The daily range of summer temperatures from lowest to highest is also provided. Design temperatures are based on long-term historical weather information. These numbers do not represent record high or low values but the values for 1% or a typical year. **See Figure 8-25.**

TECH TIP

The thermal classifications of climates for building design are cold climates (lack of heat), temperate climates (small seasonal variations), hot-dry or arid climates (overheating but dry), and warm-humid climates (some overheating and humid). The climatic elements for building design are daily temperatures, daily humidity, monthly solar radiation, amount and direction of wind, and monthly rainfall.

Outdoor Design Temperatures

State and City	Lat*	Winter	Summer		
		DB†	DB†	Daily Range†	WB†
ILLINOIS					
Aurora	41	−6	93	20	79
Champaign	40	−3	95	21	78
Chicago	41	−9	90	15	79
Joliet	41	−5	93	20	78
Peoria	40	−8	91	22	78
Rockford	42	−9	91	24	77
Springfield	39	−3	94	21	79
INDIANA					
Fort Wayne	41	−4	92	24	77
Indianapolis	39	−2	92	22	78
Kokomo	40	−4	91	22	77
South Bend	41	−3	91	22	77
Valparaiso	41	−3	93	22	78
IOWA					
Ames	42	−11	93	23	78
Cedar Rapids	41	−10	91	23	78
Des Moines	41	−10	94	23	78
Iowa City	41	−11	92	22	80
CANADA					
Toronto	43	1	84	18	70
Vancouver	49	24	74	18	64

* in degrees
† in °F

Figure 8-25. Outdoor design temperatures are used when calculating the size of HVAC equipment to use in a particular region.

UNIT

9

Indoor Air Quality

One of the responsibilities of a good HVAC and refrigeration system technician is to ensure the health of building occupants. Therefore, it is important to be aware of the health effects of the building environment. Companies that market services for HVAC systems are experiencing rapid growth. While these services very often involve the installation, maintenance, and troubleshooting of mechanical equipment, issues such as indoor air quality must be considered.

OBJECTIVES

- Define indoor air quality.
- Explain acute and chronic health effects.
- Identify an HVAC technician's role when responding to an IAQ complaint.
- List the information that HVAC technicians and building personnel must release to building tenants.
- Discuss sick building syndrome.
- Identify the typical contaminant sources in outdoor and indoor air.
- List the categories of contaminants found in indoor air.

Index to Procedures

INDOOR AIR QUALITY

Indoor air quality (IAQ) is the condition of indoor air as it relates to contaminant levels and air properties that affect the health and comfort of building occupants. *Good IAQ* is indoor air that is free of harmful particle levels and has a comfortable temperature and humidity level. *Poor IAQ* is indoor air that contains harmful particle levels and/or has an uncomfortable temperature or humidity level. Poor IAQ affects the health of building occupants and can lead to serious health problems.

Recently, scientific evidence has indicated that the air within some buildings can be more contaminated than the outdoor air of the largest industrialized cities. **See Figure 9-1.** Research also indicates that many people spend approximately 90% of their time indoors. People who are exposed to indoor air pollutants for the longest periods are most susceptible to the effects of indoor air pollution. Within this group, the most highly susceptible populations to poor IAQ include those suffering from respiratory or cardiovascular diseases.

TECH TIP

The Indoor Environments Division (IED), located within the Office of Radiation and Indoor Air (ORIA), under the Office of Air and Radiation (OAR), is responsible for implementing the Environmental Protection Agency's Indoor Environments Program (www.epa.gov).

State Outdoor Air Pollution Episode Levels

Pollutant	Advisory	Yellow Alert	Red Alert	Emergency
Particulate matter*	2-hour 420	24-hour 350	24-hour 420	24-hour 500
Sulfur dioxide†	2-hour 0.30	4-hour 0.30	4-hour 0.35	4-hour 0.40
Carbon dioxide†	2-hour 30	8-hour 15	8-hour 30	8-hour 40
Nitrogen dioxide†	2-hour 0.40	1-hour 0.60	1-hour 1.20	1-hour 1.60
Ozone†	1-hour 0.12	1-hour 0.20	1-hour 0.30	1-hour 0.50

* in μg/m³
† in ppm

Figure 9-1. Television, radio, and the Internet are used to notify people locally that the outdoor air is poor, but the quality of indoor air in a building is unknown at a local level.

IAQ FACTORS

Factors that affect IAQ include pollutant and odor sources; moisture and humidity; occupant perceptions and susceptibilities; and the design, maintenance, and operation of building HVAC systems. The quality of indoor air varies by building, season, date, and time of day.

Contaminants come from cleaning products, building materials, carpet, furniture, combustion sources, office equipment, and certain products such as perfumes and deodorants. **See Figure 9-2.** Indoor air contains dust, dirt, pollen, mold, bacteria, pesticides, and many other types of particles generated indoors or brought in from the outdoors. Volatile organic compounds (VOCs) are organic, carbon-based compounds that evaporate at room temperature.

Health Effects

The health effects of poor IAQ are serious. The Environmental Protection Agency (EPA) ranks poor IAQ among the top five environmental causes of poor health faced in modern societies. In the United States, asthma leads to 2 million emergency room visits per year and 5000 deaths per year. IAQ is considered to be one of the many causes of asthma. Rates of asthma have risen sharply over the past 30 years, particularly among children ages 5 to 14.

Schools must be especially concerned about IAQ. Twenty percent of the U.S. population spends weekdays in school. In the 1990s, studies showed that one in five schools reported unsatisfactory indoor air quality, and one in four schools reported unsatisfactory ventilation. **See Figure 9-3.** Children are at greater risk for IAQ-related illnesses because they are still growing and have less effective disease-fighting systems than adults.

Indoor Air Contaminants

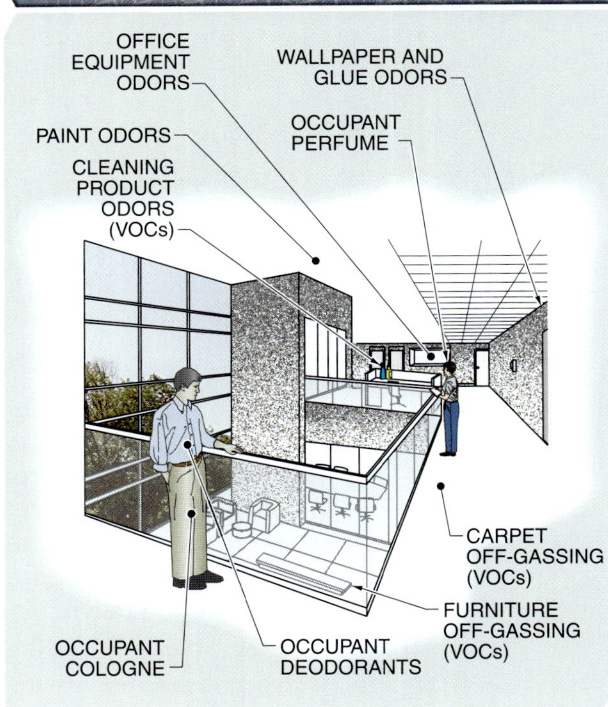

Figure 9-2. Contaminants generated indoors come from a variety of sources such as furniture, building materials, and equipment.

Educational Building IAQ Recommendations

School Space	Winter Temp*	Summer Temp*	Ventilation Rate†
Classrooms, auditoriums, libraries, and administrative areas	72	78	15
Corridors	68	80	5
Laboratories	72	78	–
Locker rooms and shower areas	75	–	5
Mechanical rooms	60	–	5
Industrial technology areas	72	78	5
Storage areas	65	–	5
Restrooms	72	–	20

* in degrees Fahrenheit
† flow rate in cubic feet per minute per person

NOTE: Corridors, mechanical rooms, industrial technology shops, and storage areas are not usually air conditioned. Shops, laboratories, and restrooms usually have additional exhaust system requirements.

Figure 9-3. The ventilation rates for specific sections of a school are important when providing a positive learning environment for students.

Indoor air contaminants may be unnoticed, annoying, or life threatening. Some contaminants have highly variable effects because people's susceptibility can differ. In many cases, the harm that each contaminant has on the body increases as its concentration in the air increases. Reactions to indoor air contaminants are either acute or chronic. **See Figure 9-4.**

Acute vs. Chronic Health Effects

Effect	Time Frame	Related Symptons	Sources
Acute	Short-term exposure	Headaches, dizziness, and fatigue	VOCs and fiberglass
Chronic	Long-term exposure	Asthma, heart disease, and cancer	Environmental tobacco smoke and asbestos

Figure 9-4. Contaminants cause acute health effects (symptoms experienced immediately) and chronic health effects (symptoms experienced only after a long period of time).

Acute Health Effects. An *acute health effect* is a health effect that occurs immediately after exposure to a substance and typically disappears soon after the exposure ends. Acute health effects develop within 24 hr after exposure to substances like formaldehyde, ammonia, cleaning agents, fiberglass, and pesticides.

Acute health effects may appear after a single exposure or repeated exposures and are typically short-term and treatable. Acute health effects include headaches, dizziness, and fatigue. Sometimes the only treatment necessary is to eliminate the person's exposure to the source of pollution. However, some acute effects can lead to death. Even short-term exposure to high concentrations of some substances, such as carbon monoxide, can be life threatening.

Chronic Health Effects. A *chronic health effect* is a health effect that appears either years after exposure or only after long or repeated periods of exposure to a substance has occurred. Chronic health effects, which include some respiratory diseases, heart disease, and cancer, can be severely debilitating or fatal. Contaminants that cause chronic effects include environmental tobacco smoke, radon, and asbestos. Some contaminants, such as formaldehyde, can have both acute and chronic health effects.

Communication and documentation are important in diagnosing IAQ-related effects. For this reason, it is important for HVAC technicians and/or building personnel to pay attention to the time and place symptoms occur. Some effects may worsen due to an inadequate supply of outdoor air (ventilation) for the heating, cooling, or humidity conditions prevalent in the building.

HVAC Technician and Building Personnel Roles

When IAQ problems arise, HVAC technicians are often responsible for identifying and responding to IAQ complaints. Therefore, the HVAC technician's role must include ongoing training that also incorporates the building's IAQ as an everyday concern with the aid of building personnel.

In addition to solving HVAC system and IAQ issues as they arise, building personnel and HVAC technicians must work together to develop an effective preventive maintenance program to prevent IAQ problems. HVAC systems and controls must be maintained and operated correctly to prevent the development of IAQ issues and serious illnesses.

People Skills. Many HVAC technicians do not understand how building occupants or business customers can be affected by IAQ problems. At times, the causes of many IAQ complaints are not completely understood by experts in the field. HVAC and maintenance technicians sometimes work with chemicals on a daily basis, so trying to understand that someone can become sick by what is a normal day-to-day circumstance can be difficult.

Some people can suffer acute and/or chronic health effects from exposure to substances such as inks and toners from copy machines, printers, and fax machines.

A building occupant who complains about IAQ may be more sensitive than the people near the source of air pollutants. **See Figure 9-5.** This person might serve as the first warning of IAQ problems that could affect others later. Therefore, every complaint must be treated seriously. Just as not everyone gets the measles, mumps, flu, or the cold strain circulating in a workspace, not all individuals react the same way to indoor contaminants. The location, amount of contaminant, and occupant susceptibility all affect how an occupant feels.

Good Manners. HVAC technicians and building personnel who receive complaints should always be polite and patient. The anger that an occupant would like to direct to a building manager or coworker might be directed to whoever is available. It is vital to never respond rudely to a building occupant for any reason. The goodwill of all individuals will help in the investigation and mitigation of an IAQ problem. Sometimes just paying attention to someone's complaints and making a few small changes will reduce or stop IAQ complaints.

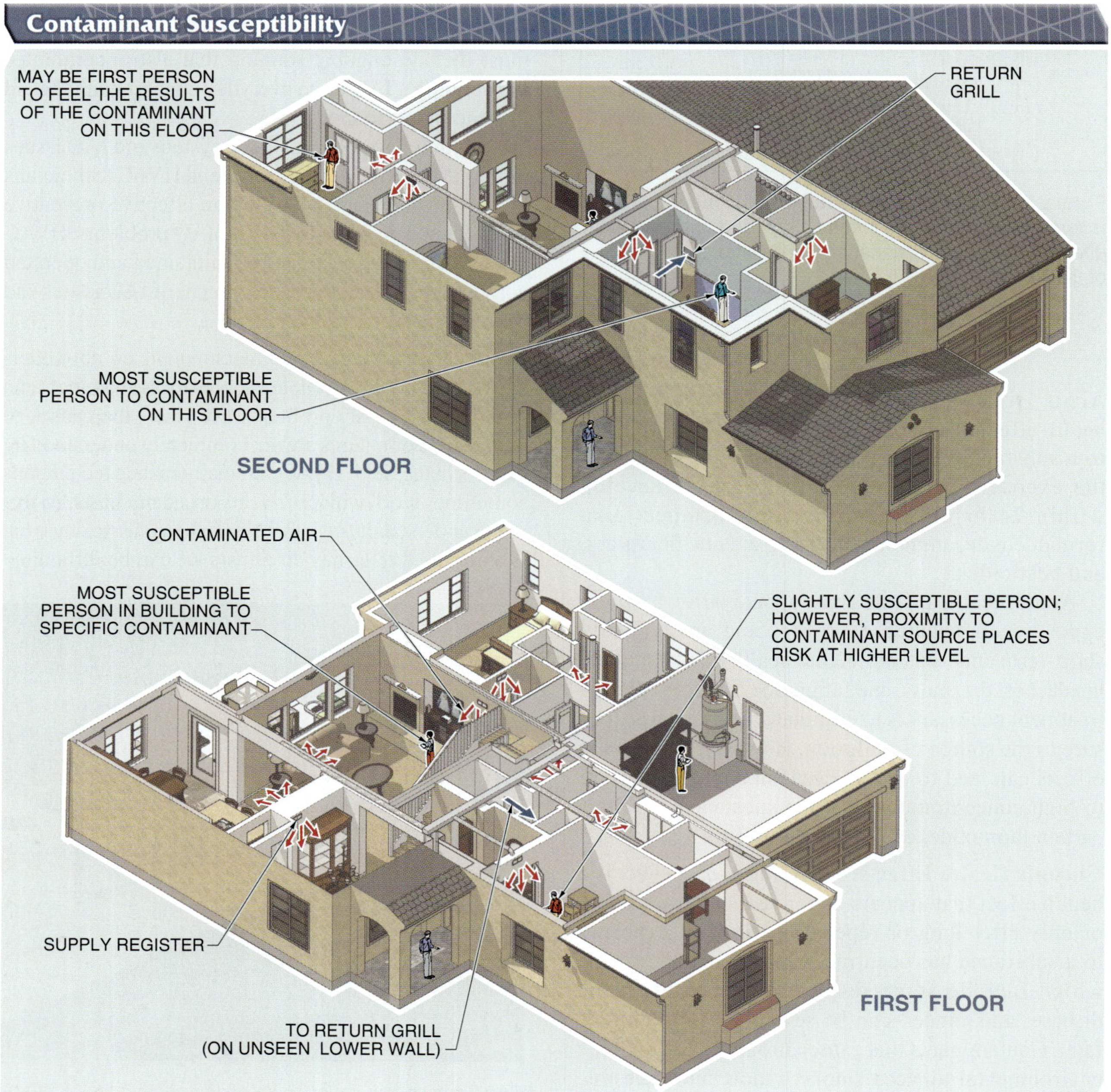

Figure 9-5. A person close to a contamination source is more likely to experience symptoms than someone at a distance.

Communication. One important task of an HVAC technician is to participate in the IAQ communication process with building personnel. A good IAQ communication program encourages building occupants to improve their work environment through specific actions. **See Figure 9-6.** The following guidelines are important when communicating IAQ issues with building occupants:

- Provide accurate information about factors that affect indoor air quality.
- Clarify the responsibilities of the parties involved such as building management, HVAC technicians, building personnel, and tenants.
- Establish an effective system for logging and responding to IAQ complaints.

Honeywell International, Inc.

Good communication can help to prevent IAQ problems and eliminate unnecessary fears. In addition, building owners must respond promptly and effectively to any IAQ issues that may arise.

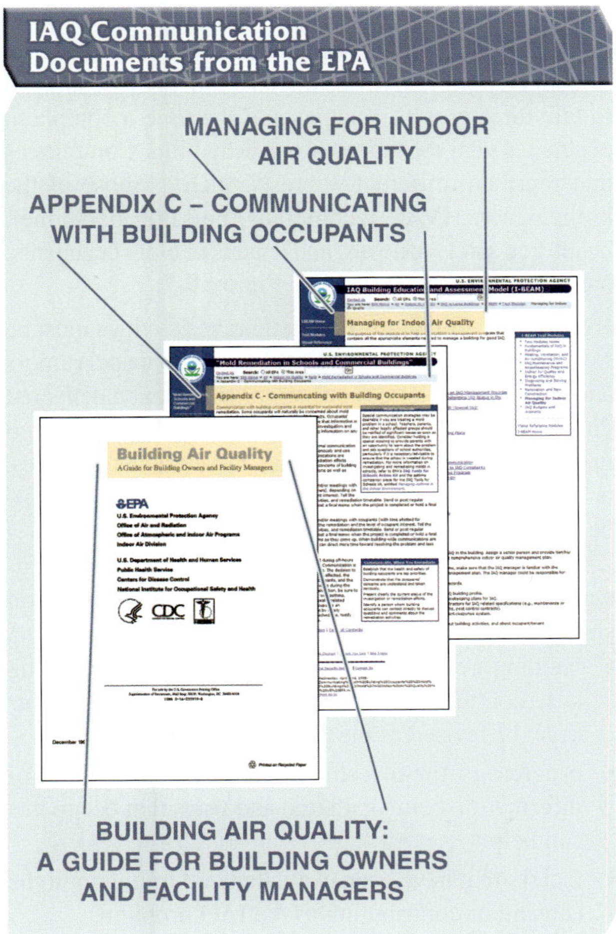

Figure 9-6. The Environmental Protection Agency (EPA) sets standards and has resources for homeowners, building owners, managers, HVAC technicians, and occupants concerning indoor air quality.

Establishing and maintaining a communication system can prevent or resolve IAQ problems. Communication becomes more critical when there are delays in identifying and resolving IAQ problems and when serious health concerns are involved. An HVAC technician responding to an IAQ problem must convey that building personnel are dedicated to providing a healthy and safe building, that good IAQ is an essential component of a healthy indoor environment, and that all complaints about IAQ are taken seriously.

Important information to communicate includes the proper use of building spaces, design occupancy rates, building modifications, and planned activities such as construction projects. The responsibilities of building management, staff, and occupants must be well defined. These responsibilities can be formalized by incorporating them into documents such as employee handbooks or tenant lease agreements.

Complaints. In many cases, occupants may be the first to know and alert building personnel about potential IAQ problems. Occupant observations about patterns of symptoms or building conditions provide valuable information. However, occupant complaints are often vague. For example, someone may complain of feeling sick or may have noticed an unusual odor. Even when a complaint is vague, the person is typically reacting to a real problem. Therefore, all complaints must be taken seriously.

Many building owners have established procedures for responding to occupant complaints, including indoor air quality concerns. **See Complaint Processing Procedures.**

Procedure

Complaint Processing

1. Building personnel receives occupant complaint.

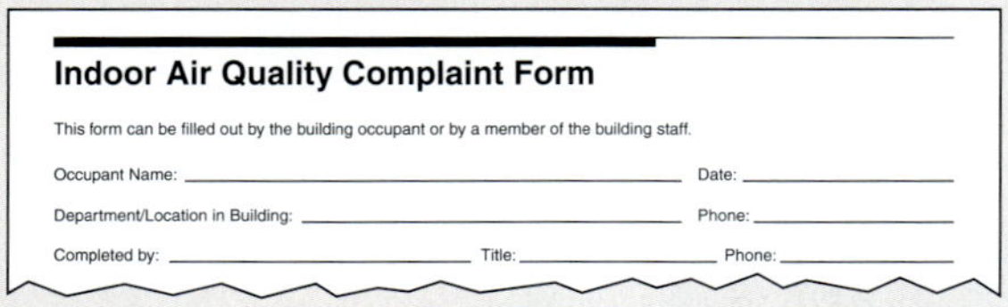

2. Building personnel documents complaint and creates work order for HVAC technician.

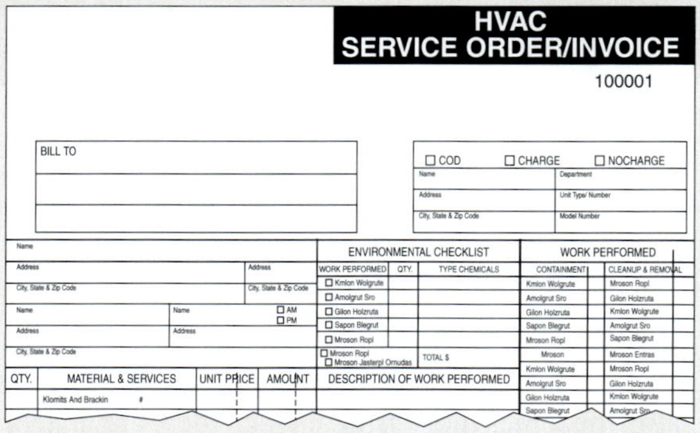

3. HVAC technician receives work order ticket with all known information.
4. HVAC technician investigates IAQ complaint.

5. HVAC technician records the date and time of the response and the final resolution of the IAQ complaint.

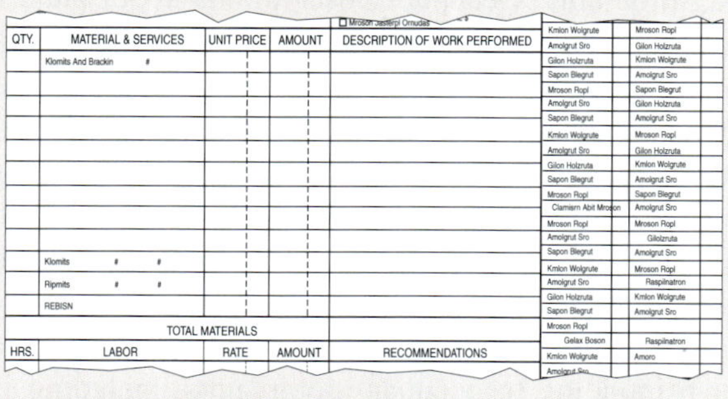

6. Building personnel processes IAQ complaint form.
7. All information is entered into the building's database.

Building occupants must know to contact the person(s) responsible for receiving and responding to IAQ complaints. Complaints may be delivered over the phone, by posted information, or by complaint forms.

Complaints must be handled promptly and every incident must be given serious attention. Some indoor air quality complaints can be resolved quickly such as harmless odors coming from an easily identified source. Complaints that involve small numbers of building occupants can be handled with little confusion and disruption. Many light commercial buildings and facilities have tracking and recordkeeping systems that can be helpful in identifying specific patterns related to IAQ complaints.

Health and Safety Committees

Building owners and facility managers can establish health and safety committees that work to promote good working conditions and prevent many IAQ problems. Health and safety committees are useful in disseminating information to building occupants and promoting a sense of cooperation. HVAC technicians must be encouraged to talk directly with occupants at the time a complaint occurs, as well as during the investigation. Committees that represent different interests, such as those of the homeowner, HVAC technicians, building personnel, health and safety officials, and tenants or other occupants, have the greatest success. **See Figure 9-7.**

Release of Information. Investigative activities must be explained with the known facts in order to dispel rumors and suspicions. Notices or memoranda can be delivered directly to selected occupants through mailings or posted in general-use areas. **See Figure 9-8.** Newsletters, articles, or other established communication channels could also be used to keep building occupants up-to-date.

HVAC technicians and building managers must always be factual and direct when presenting information such as the following:

- building areas where the IAQ complaints are occurring based on the location and distribution of complaints (may be revised as the investigation progresses)
- progress of the investigation, including the kind of information being gathered and ways that occupants can help
- factors that have been evaluated and found not to be causing or contributing to the IAQ problem
- projected duration of the investigation
- attempts made to improve indoor air quality
- remaining work required and the schedule for its completion

Figure 9-7. The larger a building is, the more impact a health and safety committee will have on disseminating factual information acquired through an IAQ complaint investigation.

Problems can arise from either too much or too little information. A premature release of information before all data has been gathered can produce confusion, frustration, and distrust. Similar problems can result when no information is released at all and also lead to occupants assuming the worst-case scenario. When progress reports are not given, people may assume that no progress is being made. The best practice is to obtain the consent of the homeowner, facility manager, and/or legal counsel before releasing information.

TECH TIP

Developing an IAQ policy should be an open process. A health and safety committee could provide a forum for developing consensus recommendations. In the absence of an existing committee, an ad hoc committee could be established comprised of homeowners, support personnel, local health officers, building maintenance personnel, and community leaders. Interested police and fire department members may also wish to serve on the committee.

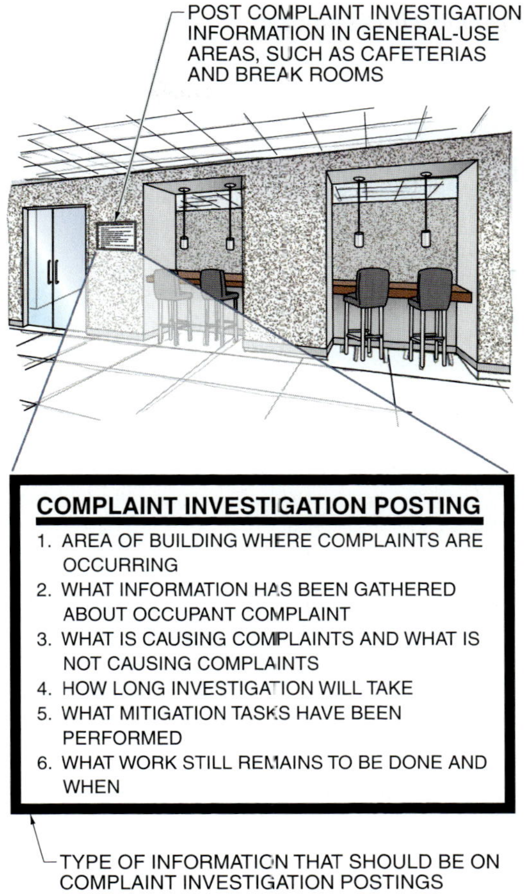

Figure 9-8. Complaint investigation materials, such as notices or memoranda, work well when displayed in the general-use areas of a building.

IAQ Response. HVAC technicians should work with building personnel to implement a quality preventive maintenance program. Preventive maintenance programs include activities such as performing regularly scheduled random air sampling. When problems do occur, these programs may demonstrate due diligence to limit any legal liability.

IAQ problems are often difficult to investigate because of simultaneous factors such as intermittent symptoms, vague discomfort, and the complex interactions between job stress and environmental issues. These factors may also hinder the effects of IAQ remediation efforts. **See Complaint Investigation Procedures.**

Procedure

Complaint Investigation

1. Respond to the complaint immediately.
2. Speak to the person who generated complaint.

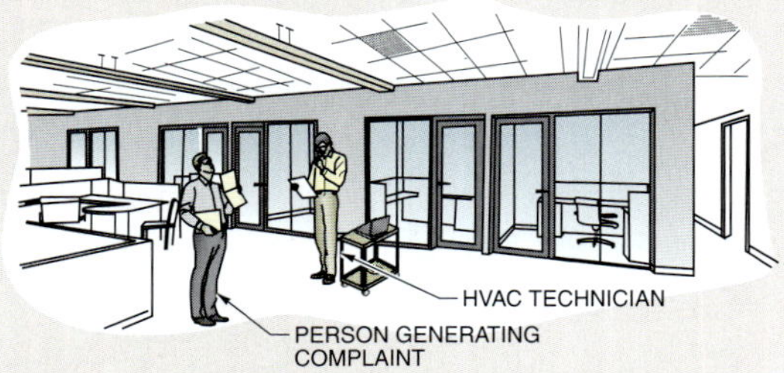

3. Document all work and record all information.
4. Visually inspect the area of the complaint.
5. Check the temperature and humidity of the complaint area.

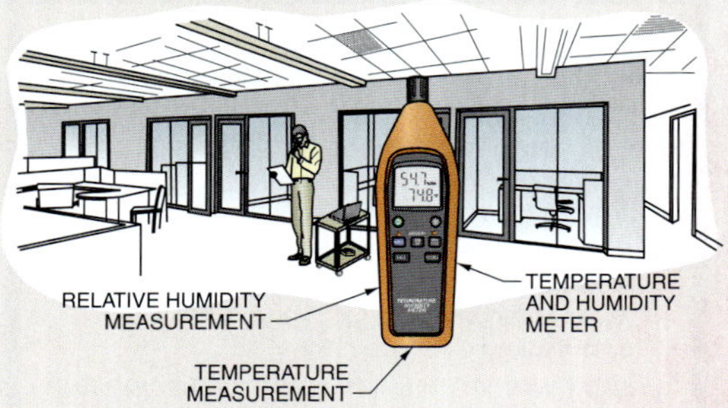

6. Check the airflow from space diffusers.

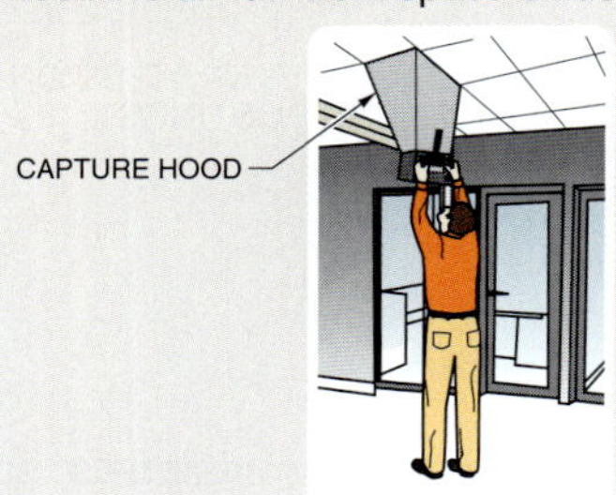

7. Investigate the complaint area for contaminants.

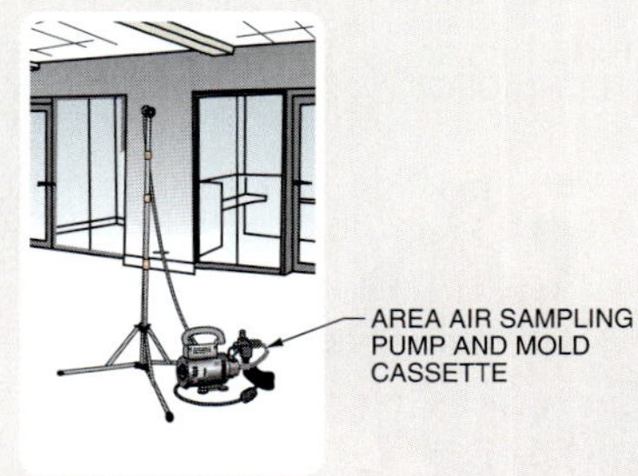

8. Investigate personnel usage of the complaint area.
9. Investigate the equipment and furniture usage.

10. Review all relevant repair and work orders.
11. Disclose information to building occupants.
12. Complete and file all complaint investigation paperwork.

Even after the proper remediation strategy is in place, it may take days or weeks for contaminants to dissipate and symptoms to disappear.

LEGAL LIABILITIES

A large variety of legal issues can arise from IAQ problems. Three of the most common issues include negligence, constructive eviction, and workers' compensation.

Negligence is a legal issue that occurs when there is an injury due to a lack of performance by building owners. Building occupants might accuse building owners of negligence when management fails to remedy a problem in a reasonable fashion or time frame. *Constructive eviction* is a legal issue that occurs when a tenant is unable to use a building space due to special circumstances and results in termination of the landlord-tenant relationship. *Workers' compensation* is compensation offered to an employee who is temporarily unable to work because of a job-related injury.

Confidentiality of records is important to homeowners who are concerned that IAQ complaints will lead to negative reactions from their neighbors. Investigators are more likely to obtain honest and complete information by reassuring homeowners that their privacy is respected. Violating confidentiality of medical records often carries its own legal penalties.

Future IAQ Regulations

Concerns about IAQ have increased since the 1970s when energy conservation measures started to be taken in commercial buildings. Energy conservation measures minimized the infiltration of outdoor air and contaminant build-up in residential and commercial buildings.

Due to the health problems associated with IAQ and the lawsuits that have already been filed in courts, federal and state officials are considering enacting laws to govern IAQ. Comprehensive indoor air quality bills have been introduced in both the U.S. House of Representatives and the Senate in recent years. However, none of these bills have ever been passed. Still, the Occupational Safety and Health Administration (OSHA) and the Federal Register have standards and notices that apply to IAQ. **See Figure 9-9.** The various bills included the following considerations:

- specific ventilation standards
- guidelines to identify and prevent indoor air hazards
- funds for research into IAQ contaminants
- requirement of OSHA inspections of buildings and facilities with repeated IAQ complaints

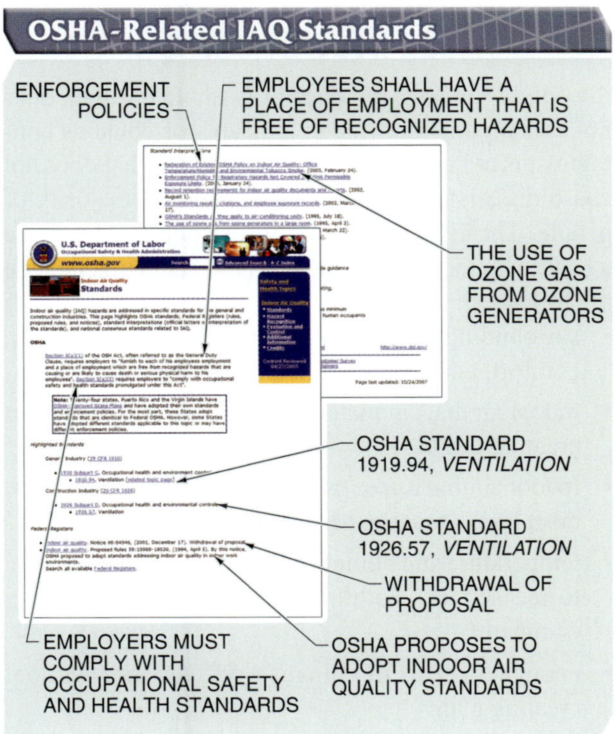

Figure 9-9. Twenty-four states have IAQ standards that have been passed into law. The Occupational Safety and Health Administration (OSHA) and the Federal Register maintain notices of these laws for all HVAC technicians.

Twenty-four states, Puerto Rico, and the Virgin Islands have OSHA-approved state plans. In these plans, the states have adopted standards and enforcement policies. In general, these states adopt standards that are identical to those set by federal agencies. However, some states have adopted different standards applicable to this topic or may have different enforcement policies.

Other associations, such as the American National Standards Institute (ANSI) and the American Society of Heating, Refrigerating, and Air Conditioning Engineers (ASHRAE), have also enacted their own standards to protect members and building occupants against possible IAQ problems. ANSI/ASHRAE Standard 62.1-2007, *Ventilation for Acceptable Indoor Air Quality,* specifies minimum rates of ventilation and IAQ considered acceptable for occupants to avoid health risks.

BUILDING-RELATED ILLNESSES

A *building-related illness (BRI)* is an identifiable illness, disease, or infirmity caused by exposure to contaminants in the air of a building. Symptoms may vary depending on the illness. Victims may require prolonged recovery time after leaving the building. A BRI is traceable to a specific cause in a building. Building-related illnesses include Legionnaires' disease, hypersensitivity pneumonitis, humidifier fever, and asthma.

SICK BUILDING SYNDROME

Sick building syndrome (SBS) is a situation in which common symptoms experienced by building occupants can be associated with their presence in a specific building but cannot be linked to a specific illness or cause. With SBS, people often feel worse as the day or week progresses and typically recover when they are away from the building.

Complaints may come from one area or the whole building. Typically, a broad spectrum of specific and nonspecific complaints are involved. The key factors of SBS are a commonality of symptoms among building occupants and an absence of symptoms among building occupants when the individuals are not in the building.

Sick building syndrome should be suspected when a substantial proportion of those spending extended time in a building experience acute on-site discomfort. It is important, however, to distinguish sick building syndrome from a building-related illness. **See Figure 9-10.** With a BRI, the symptoms of a diagnosable illness are identified and can be attributed directly to a specific airborne contaminant, which is not possible with SBS.

Building-Related Illness vs. Sick Building Syndrome

Illness	Symptoms	Causes	Time Period
Building-related Illness	Cough; chest tightness; fever; chills; muscle aches	Identifiable	Prolonged recovery
Sick building syndrome	Headache; eye, nose, and throat irritation; dry cough; dry or itchy skin; dizziness and nausea; difficulty concentrating; fatigue; sensitivity to odors	Unknown	Immediate recovery

Figure 9-10. A building-related illness has identifiable causes, whereas the causes of sick building syndrome cannot be identified.

Possible causes of SBS may include poor design, maintenance, and/or operation of the structure's ventilation system. Interior redesign, such as the rearrangement of offices or installation of partitions, may also interfere with the efficient functioning of such systems. Humidity may be another factor. While high relative humidity may contribute to biological pollutant problems, an unusually low relative humidity (below 30%) may heighten the effects of mucosal irritants that also prove bothersome.

Multiple Chemical Sensitivity

Multiple chemical sensitivity (MCS) is an allergy-like illness resulting from contact with, or proximity to, a spectrum of substances, including airborne agents. These substances may include recognizable contaminants and materials that would typically be considered harmless. MCS is also known as chemical hypersensitivity or environmental illness.

The cause and existence of multiple chemical sensitivity have not been proven. Doctors disagree if symptoms are physiologically or psychologically generated or a combination of both.

Mass Psychogenic Illness

Mass psychogenic illness is a condition that occurs when groups of people start feeling sick at the same time even though there is no proven physical or environmental reason. Mass psychogenic illness has occurred all around the world in many different social settings. Outbreaks of mass psychogenic illness typically start with an environmental trigger. The trigger can be a bad smell, a suspicious-looking substance, or something else that makes people in a group believe they have been exposed to a germ or poison. Signs indicating a mass psychogenic illness include the following:

- surroundings that contribute to human productivity, comfort, and a sense of health and well-being
- indoor air that circulates to prevent stuffiness without creating drafts and is free from significant levels of odors, dust, and contaminants
- a large group of people sick simultaneously
- individual medical exams and tests showing no abnormalities
- no substances found in the group's environment, such as a poison, that would cause sickness

ACCEPTABLE INDOOR AIR QUALITY

Acceptable indoor air quality is air that either does not contain any known contaminants or contains contaminants at harmless concentrations and that does not cause dissatisfaction among a substantial majority of people. The EPA applies several standards to declare a healthy indoor environment. These standards include the following:

- surroundings that contribute to human productivity, comfort, and a sense of health and well-being
- indoor air that circulates to prevent stuffiness without creating drafts
- indoor air that is free from significant levels of odors, dust, and contaminants
- temperature and humidity levels that are appropriate to the season, clothing, and activities of building occupants
- enough light to illuminate work surfaces without creating glare
- noise levels that do not interfere with activities
- properly managed sanitation, drinking water, fire protection, and other systems that affect health and safety

POOR INDOOR AIR QUALITY

Poor indoor air quality can lead to maintenance problems, building damage, and health issues. Most residential and light commercial buildings contain a multitude of IAQ problems. Contaminants may be produced from a variety of activities and/or environmental sources, which include the following:

- building design
- system and filter installation
- system operation
- building maintenance
- renovation
- occupant activities

Contaminants may originate outdoors and be drawn into the building or may originate inside the building itself. **See Figure 9-11.**

Common Airflow Pathways for Contaminants

Common Pathway	Comment
Outdoors to Indoors	
Outdoor air intake	Polluted outdoor air or exhaust air enters building through air intake
Windows, doors, Cracks and crevices	Negatively pressurized building draws air and outside pollutants into building through any available opening
Substructures and slab penetrations	Radon and other soil gases and moisture-laden air or microbial contaminated air travel through crawlspaces and other substructures into the building
Indoors	
Stairwell, Elevator shaft, and Vertical electrical or plumbing shafts	Stack effect causes airflow by drawing air toward shafts on lower floors and away from shafts on higher floors, affecting flow of contaminants
Wall openings and penetrations	Contaminants enter and exit building cavities and move from space to space
Duct or plenum	Contaminants commonly carried by HVAC system throughout occupied spaces
Duct or plenum leakage	Duct leakage accounts for significant unplanned airflow and energy loss in buildings
Flue or exhaust leakage	Leaks from sanitary exhausts or combustion flues cause serious health problems
Room spaces	Air and contaminants move within room or through doors and corridors to adjoining spaces

Figure 9-11. Contaminants may be produced from a variety of outdoor and indoor activities but take typical pathways through a building.

Outdoor Sources

Outdoor air can be contaminated by many different sources including vehicle exhaust, trash and refuse, fertilizers, and tobacco smoke. Environmental tobacco smoke was once the largest source of pollution within buildings. Now, with the advent of local, state, and federal laws that limit or abolish indoor smoking, the presence of tobacco smoke has been moved outdoors.

Potential sources of contamination must be considered when determining the location for outdoor air intakes. Outdoor air intakes located near contaminant sources may draw contaminants indoors and increase the chance of IAQ problems. Contaminated outdoor air must be cleaned for use indoors. This typically means improved filtration.

Indoor Sources

Contaminants are typically introduced into indoor air by means of an improper building design or by products and equipment used within a building. Some sources of indoor contaminants include the following:

- VOCS used in glues and sealants found in building materials and furniture
- special-use and mixed-use areas such as smoking lounges, laboratories, exercise rooms, beauty salons, dry cleaners, and restaurants
- redecorating, remodeling, and repair work
- areas that collect dust such as open shelving and old or deteriorated furnishings
- locations where fibers, such as carpeting and textiles, are produced
- fax and photocopy machines
- insufficient air management systems

Building damage that can produce indoor contaminants include defective ventilation systems and water-damaged walls, ceilings, and carpets. Indoor moisture problems, such as excessive moisture, standing water, and water-damaged furnishings, may also promote the growth of indoor contaminants.

Effects of Poor Indoor Air Quality

Health and comfort complaints are common problems resulting from poor IAQ. When the concentration of substances increases to the point of causing health-related symptoms (even minor symptoms), the number of occupant complaints typically increases. The increased number of illnesses, symptoms, and complaints are serious consequences of poor IAQ.

When contaminated air is inhaled, the contaminants affect the respiratory system (nose, airways, lungs, etc.) and, in some cases, may be absorbed into the bloodstream. The resulting illnesses can be mild, such as general fatigue, or life threatening such as Legionnaires' disease. The economic impact is serious. In buildings with poor IAQ, employees become less productive, costing employers time and money. When people become seriously ill, they may miss work and require expensive medical care.

Contaminant Categories

While contaminants typically found in indoor air are responsible for many harmful effects, there is a level of uncertainty regarding what concentration levels or periods of exposure produce specific health problems. Indoor air contaminants can be categorized by the symptoms and health effects they cause. Indoor air contaminant categories include irritants, asphyxiates, allergens, neurotoxins, pathogens, and carcinogens. **See Figure 9-12.** Contaminants are found as particles, gases, and as biological (microbial) contaminants.

Contaminant Categories

Category	Health Effects	Examples
Irritants	Inflammation, redness, and discoloration	Fiberglass, pesticides, cleaning agents, and perfumes
Asphyxiates	Drowsiness, headaches, and slowed breathing	Carbon monoxide and refrigerants
Allergens	Extreme fever and asthma	Pollen, molds, and animal dander
Neurotoxins	Cardiac problems, convulsions, and loss of memory	Lead and chemicals in paint and glue
Pathogens	Fever, chills, and chest congestion	Some molds and bacteria
Carcinogens	Cancer	Radon gas, asbestos fibers, and environmental tobacco smoke

Figure 9-12. Contaminants are classified by the type of health effect they produce.

An *irritant* is a contaminant that can cause redness and/or inflammation of the skin, eyes, nose, throat, or upper-respiratory tract. An *asphyxiate* is a contaminant that prevents oxygen from entering the body, which causes drowsiness, headaches, and depressed respiration. HVAC technicians and building maintenance personnel should be aware of the lack of oxygen and possibility of asphyxiation when conducting confined space entry procedures.

An *allergen* is a contaminant that can trigger an allergic reaction. A *neurotoxin* is a contaminant that can damage the central nervous system. A *pathogen* is a microorganism that can cause fever or disease. A *carcinogen* is a contaminant that can cause cancer. Cancer may develop years after exposure to carcinogens.

Particles

A *particle* is a solid or liquid that is light enough to be suspended in air. Each particle may be one material, such as asbestos, or a cluster of many materials, such as dust. Particles can land on surfaces and become airborne again when disturbed.

Large particles that are inhaled are captured in the upper respiratory tract. Soluble particles, which can be dissolved and then absorbed through the mucus lining, can also irritate the upper respiratory tract. Insoluble particles (and some soluble materials) are swept out of the upper lungs and nasal passages in mucus and enter the stomach. Particulate contaminants include asbestos, lead, mold, and fiberglass.

Asbestos. *Asbestos* is a collection of impure magnesium silicate mineral fibers that can pollute air and water and cause cancer or asbestosis when inhaled. It has been used commonly in a variety of building construction materials for insulation and as a fire-retardant. The EPA and the Consumer Product Safety Commission (CPSC) have banned most asbestos products. Asbestos is a dangerous contaminant, and steps must be taken to prevent it from becoming airborne in the indoor air environment. **See Figure 9-13.**

Lead. *Lead* is a toxic metallic element. It has long been recognized as a harmful environmental pollutant. Humans may be exposed to lead through air, drinking water, food, contaminated soil, deteriorating paint, and dust. **See Figure 9-14.** Lead becomes an indoor air quality problem only when lead-based particles become airborne. Airborne lead enters the body when an individual inhales or swallows lead particles. Before the harmful effects of lead were known, it was used in many products including paint, gasoline, and water pipes.

Asbestos Exposure Limits

Classification	10 Minutes	30 Minutes	1 Hour	4 Hours	8 Hours*
Chrysotile	0.9*	0.7	0.5	0.3	0.2
Nonchrysotile	0.6	0.5	0.4	0.2	0.1

Exposure	Symptoms	Prevention	First Aid Treatment
Inhalation	May cause asbestosis, pulmonary fibrosis, and/or other lung disorders	Exhaust ventilation and full-face, self-contained breathing apparatus	Rest in well ventilated area; oxygen possibly needed
Skin	Irritation	Special full-body suit, hood, gloves, and boots	Wash with plenty of water and disinfectant soap
Eyes	Scratching of eyes	Splash goggles–do not wear contacts	Remove contacts; do not use ointments; flush eyes with water
Ingestion	May cause gastrointestinal irritation or cancer	Handling of materials, food, or drink prior to washing is prohibited	Do not induce vomiting; rinse mouth with water

* in fiber per cubic centimeter

Figure 9-13. Asbestos exposure limits vary according to the type of fiber found in the air.

Lead Exposure Limits

Classification	10 Minutes	30 Minutes	1 Hour	4 Hours	8 Hours*
Nondisabling	—	—	—	—	0.03
Disabling	—	—	—	—	0.05

Exposure	Symptoms	Prevention	First Aid Treatment
Inhalation	Irritation of upper respiratory tract and lungs; possible chest and abdominal pain	Half-face, high-efficiency particulate respirator or full-face, positive-pressure air supplied respirator	Move to fresh air and provide oxygen; if not breathing, give artificial respiration
Skin	Short exposure may cause local irritation, redness, and pain; prolonged exposure; absorbed through skin; lead posoning	Use impervious protective clothing that includes boot, gloves, lab coat, apron, or coveralls	Flush skin with soapy water for 15 min; remove contaminated clothing and shoes
Eyes	Local irritation, blurred vision, and possibly severe tissue burns	Ventless safety goggles or full-face shield; eye wash station	Flush eyes with water for at least 15 min; move lower and upper eyelids while flushing
Ingestion	Poisoning that includes abdominal pain, spasms, nausea, vomiting, and headache; acute poisoning causes muscle weakness, insomnia, dizziness, loss of appetite, coma, and/or death	—	Induce vomiting immediately

* in mg/m^3

Figure 9-14. Exposure to lead typically occurs during building renovations when materials containing lead, such as paint, are disturbed.

Mold. *Mold* is fungal growth produced on damp or decaying matter. Mold is nature's recycling mechanism. There are hundreds of different types of mold. The reproductive cells of mold, known as microscopic spores, float in the air. When something dies, mold spores germinate after landing on the deceased object. **See Figure 9-15.** Mold then consumes the object and recycles its organic materials.

Molds range in toxicity from nontoxic to highly toxic. Mold spores can be found both indoors and outdoors. In buildings, sources of mold growth include water-damaged ceiling tiles, walls, and floors. Mold spores grow in areas where there is excessive moisture or accumulated water.

Fiberglass. *Fiberglass* is a material composed of thin strands of glass. Fiberglass is the most common insulating material used in walls, ceilings, and ventilation systems. Large quantities of fiberglass particles can be released into the air during remodeling. Fiberglass is often introduced into the air stream of ventilation systems, and eventually fiberglass particles blow out of the system into the indoor air. **See Figure 9-16.**

Laboratory Mold Densities

Density of Growing Mold Spores	
Level <1	Very few mold spores and mold parts observed; spores cover less than 10% of surface; up to 20,000 spores/cm²
Level 1	Few spores and mold parts observed; spores cover 10% to 25% of surface; up to 200,000 spores/cm²
Level 2	A number of spores and mold parts observed; spores cover 25% to 50% of surface; up to 2 million spores/cm²
Level 3	Many spores and mold parts observed; spores cover 50% to 75% of surface; up to 4 million spores/cm²
Level 4	Many spores and mold parts observed; spores cover more than 75% of surface; at least 4 million spores/cm²
Density of Non-Growing Mold Spores	
Very Few	Very few mold spores detected; densities range up to 10 spores/cm²
Few	Few mold spores detected; densities range up to 25 spores/cm²
Moderate	A number of mold spores detected; densities range up to 50 spores/cm²
Many	Many mold spores detected; at least 50 spores/cm²

Figure 9-15. Laboratory test reports state the type of mold in a building and the amount present.

TECH TIP

Mold spores are much larger than one micron in size and can be removed by applying air filtration according to ASHRAE. A MERV 6 pleated-panel filter has an efficiency rating of over 80% for most molds. However, the contaminated building space air must flow through the filter for the mold spores to be removed. When contamination from elevated mold levels is a problem, increasing the air changes to the space will increase contaminant removal efficiency.

Gases

Gas spreads evenly through air and cannot be collected by the ordinary particle filters used in a building's air management system. Certain gases are odorless, such as carbon monoxide, whereas other gases produce odors, such as formaldehyde. Some gases are emitted from liquids such as solvents or paints.

Water-soluble gases, such as formaldehyde and nitrogen dioxide, are absorbed by the upper or lower respiratory system. Other gases, such as carbon monoxide, dissolve into the bloodstream and are carried through the body. The more hazardous a gas is, the greater the probability that the gas is being monitored by a system designed to warn of the presence of the gas. Gases that often contribute to IAQ issues include carbon dioxide, carbon monoxide, nitrogen dioxide, formaldehyde, and ozone.

Carbon Dioxide. *Carbon dioxide (CO_2)* is an odorless, colorless gas that is typically present in indoor air. Carbon dioxide is used in some industrial processes and in restaurants for carbonated beverages. One source of CO_2 is the exhaled breath of building occupants. About 4% to 5% of exhaled breath is CO_2. Carbon dioxide is also a by-product of automobiles, so parking garages and busy traffic areas may impact a building's IAQ as well. Environmental tobacco smoke is another source of CO_2.

Indoor CO_2 concentrations can, under some test conditions, provide a good indication of the adequacy of ventilation. The level of CO_2 in the air indicates how much outdoor air is entering a space. Carbon dioxide readings are useful in determining when a problem is caused or aggravated by an inadequate supply of outdoor air. A peak CO_2 concentration over 1000 parts per million (ppm) is a general indicator of inadequate ventilation. **See Figure 9-17.**

Fiberglass Exposure Limits

Classification	10 Minutes	30 Minutes	1 Hour	4 Hours	8 Hours*
Respirable	—	—	—	—	5.0
Total Dust	—	—	—	—	15.0

Exposure	Symptoms	Prevention	First Aid Treatment
Inhalation	Sore throat, mouth, and nose	Half-face mask (N-95)	Move to fresh air
Skin	Redness and itchy, sometimes bumpy rash	Protective clothing that minimises or eliminates skin contact	Gently wash area with soap and water; do not rub or scratch
Eyes	Blurred vision and slight burning	Safty glasses with side shields or goggles	Flush eyes with water for at least 15 min
Ingestion	—	Handling of materials, food, or drink prior to washing is prohibited	Typically not required

* in mg per cubic meter (mg/m³)

Figure 9-16. Fiberglass is found in respirable form (can be inhaled) and nonrespirable form (cannot be inhaled) and is used in many different applications.

Carbon Dioxide Exposure Limits

Classification	10 Minutes*	30 Minutes*	1 Hour*	4 Hours*	8 Hours*
Nondisabling	5000	—	2000	—	1000
Disabling	30,000	—	20,000	—	10,000

Exposure	Symptoms	Prevention	First Aid Treatment
Inhalation	Asphyxiation; headache, sweating, shortness of breath, dizziness, and convulsions	Ventilation-area exhaust; supplied-air respirator	Artificial respiration and cardiopulmonary resuscitation
Skin	Possible frostbite	Appropriate gloves (not rubber or neoprene)	Rinse with warm water
Eyes	Mechanical injury; freezing of tissue	Nonvented eye protection or face shield	Gently flush with warm water
Ingestion	—	—	—

* in mg per cubic meter (mg/m³)

Figure 9-17. Testing for the concentration of carbon dioxide (CO_2) in a space can determine if there is a proper amount of ventilation air.

Carbon Monoxide. *Carbon monoxide (CO)* is an odorless, colorless, and toxic gas. Carbon monoxide can cause harm before occupants are aware of it because the toxic fumes are impossible to see, taste, or smell. Carbon monoxide is the product of incomplete combustion. Improperly vented heaters, gasoline engine-powered equipment, and worn or poorly adjusted combustion devices, such as those found in furnaces, boilers, and other natural gas heating systems, may emit CO.

Carbon monoxide problems may also occur when the discharge flues of fossil-fueled equipment are leaking, blocked, disconnected, or improperly sized. Cars, buses, and trucks are major sources of CO. Therefore, unventilated attached garages, parking lots, or heavy traffic areas may affect building spaces. Environmental tobacco smoke is another major source of CO. **See Figure 9-18.**

TECH TIP

The U.S. Consumer Product Safety Commission (CPSC) recommends that every home have at least one carbon monoxide (CO) alarm. The CPSC also urges consumers to have a professional inspection conducted of all fuel-burning appliances, such as furnaces, stoves, fireplaces, clothes dryers, water heaters, and space heaters, for the detection of deadly CO leaks.

Nitrogen Dioxide. *Nitrogen dioxide (NO_2)* is a toxic gas and corrosive that is a highly reactive oxidant. Sources of NO_2 include truck and bus emissions, kerosene heaters, unvented gas stoves and heaters, power plant emissions, gasoline engines, and environmental tobacco smoke. **See Figure 9-19.**

Carbon Monoxide Exposure Limits

Classification	10 Minutes*	30 Minutes*	1 Hour*	4 Hours*	8 Hours*
Nondisabling	2500	1500	600	150	35-50
Disabling	6000	3000	800	400	80

Exposure	Symptoms	Prevention	First Aid Treatment
Inhalation	Asphyxiation	Ventilation area exhaust; supplied-air respirator	Artificial respiration and cardiopulmonary resuscitation
Skin	Possible frostbite	Appropriate gloves (not rubber or neoprene)	Rinse with warm water
Eyes	Mechanical Injury	Nonvented eye protection or face shield	Force eyes open and rinse with lukewarm water
Ingestion	—	—	—

* in ppm

Figure 9-18. Several severe health effects result when the exposure limits to carbon monoxide (CO) are exceeded.

Nitrogen Dioxide Exposure Limits

Classification	10 Minutes*	30 Minutes*	1 Hour*	4 Hours*	8 Hours*
Nondisabling	0.50	0.50	0.50	0.50	0.50
Disabling	20	15	12	8.2	6.7

Exposure	Symptoms	Prevention	First Aid Treatment
Inhalation	Burning sensation, cough, headache, dizziness, sore throat, nausea, and vomiting	Ventilation area exhaust; supplied air respirator	Half-upright position with rest and fresh air; artificial respiration may be required
Skin	Burns, redness, and pain	Protective clothing and neoprene gloves	Rinse with water for 15 min; remove clothes (discard); rinse again for 15 min
Eyes	Severe burns, redness, and pain	Nonvented eye protection or face shield	Rinse eyes with water for 15 min
Ingestion	—	Do not drink or eat in nitrogen dioxide environment	Rinse mouth for 15 min

* in ppm

Figure 9-19. Exposure limits vary from industry to industry for nitrogen dioxide (NO_2) because no federal limits exist.

Formaldehyde. *Formaldehyde* is a colorless gas with a strong, pungent odor used as a preservative. Formaldehyde is used in many types of glues, paints, cosmetics, permanent press fabrics, and insulation. **See Figure 9-20.** It is also used to preserve tissue specimens in medical and research facilities. Formaldehyde is released through auto emissions and tobacco smoke and by burning wood, kerosene, and natural gas.

The off-gassing of formaldehyde is what causes IAQ issues. *Off-gassing* is the release of a chemical compound from a material into the air. Off-gassing is higher when products are new. The rate of off-gassing increases with higher temperatures and humidity.

Ozone. *Ozone* is an unstable gas with a characteristic pungent odor. An ozone molecule is composed of three atoms of oxygen. The third oxygen atom can detach from the ozone molecule and reattach to molecules of other substances, thereby altering their chemical composition. This is why ozone is used in air cleaning devices. However, this same property can also make ozone hazardous.

Ozone is produced by high-voltage electrical equipment such as laser printers, photocopy and fax machines, welding equipment, ion generators, and some air cleaners. It is the major component of smog and can be brought indoors by an HVAC system during high outdoor pollution episodes. Several U.S. government agencies have different standards regarding ozone concentration. **See Figure 9-21.**

Mixtures

A *mixture* is a substance that contains various combinations of solids, liquids, and/or gases. A mixture may contain thousands of separate components. Mixtures include radon, environmental tobacco smoke, wood smoke, pesticides, and volatile organic compounds.

Radon. *Radon* is an odorless, cancer-causing gas that is a form of ionizing radiation. It is found in outdoor and indoor air. Radon is produced by the decay of naturally occurring uranium in soil and water. It may enter a building or facility from naturally occurring formations or any underground structure. **See Figure 9-22.**

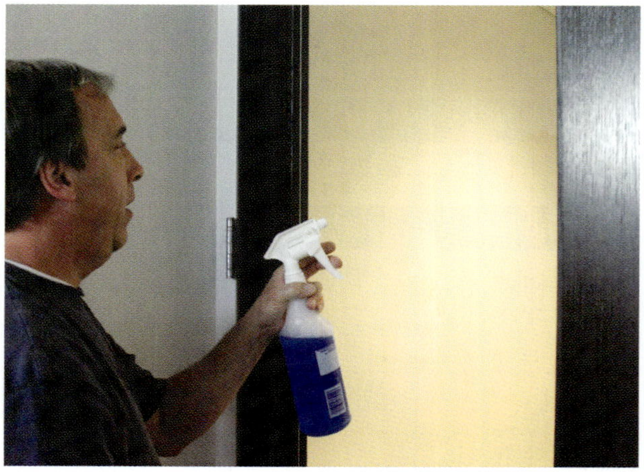

Volatile organic compounds (VOCs) are found in many products, such as cleaners, paints, and glues.

Formaldehyde Exposure Limits

Classification	10 Minutes*	30 Minutes*	1 Hour*	4 Hours*	8 Hours*
Nondisabling	2	1.5	1.25	1	0.5
Disabling	10	5	4	2	1

Exposure	Symptoms	Prevention	First Aid Treatment
Inhalation	Respiratory tract irritation; short exposure causes headache, vomiting, dizziness, and drowsiness; prolonged exposure causes sensory impairment	Full-face, self-contained breathing apparatus	Remove to fresh air; if not breathing, provide artificial respiration; oxygen may be required
Skin	Short exposure may cause burns; prolonged exposure allows harmful amounts to be absorbed into body	Plastic or rubber gloves; coated protective clothing and footwear	Remove clothing, gloves, and shoes; flush skin with warm water
Eyes	Severe irritation, tearing, burns, and permanent eye injury	Ventless safety glasses and full-face shield	Flush eyes with water for at least 15 min; lift eyelids to ensure all surfaces are flushed
Ingestion	Severe irritation and inflammation of mouth, throat, and stomach; nausea, blindness, and may cause death	—	Do not induce vomiting; if conscious, provide water or milk

* in ppm
Sources: Occupational Safety & Health Administration (OSHA), Office of Environmental Health & Safety (EH&S), and Praxair, Inc.

Figure 9-20. Formaldehyde must be tested for a period of time in a new building due to off-gassing from furniture and construction materials.

Ozone Exposure Limits

Classification	10 Minutes*	30 Minutes*	1 Hour*	4 Hours*	8 Hours*
Nondisabling	0.3	0.2	0.12	0.11	0.1
Disabling	25	10	2	1	0.5

Exposure	Symptoms	Prevention	First Aid Treatment
Inhalation	Nose and throat irritation, chest pain, nausea, drowsiness, and may cause death	Full-face supplied-air respirator or self-contained breathing apparatus	Artificial respiration using oxygen – (do not breathe air from victim)
Skin	Irritation and possible chemical burns	Plastic clothes and gloves	Remove clothing and shoes; rinse with warm water
Eyes	Stinging	Face shield	Force eyes open and rinse with lukewarm water
Ingestion	—	—	—

* in ppm
Sources: Environmental Protection Agency (EPA), Occupational Safety & Health Administration (OSHA), Illinois Department of Public Health (IDPH), and Praxair, Inc.

Figure 9-21. Ozone is used to kill bacteria in building spaces, however, a concentration too high is harmful to humans and animals.

Environmental Tobacco Smoke. *Environmental tobacco smoke (ETS)* is a mixture of smoke from the burning end of a cigarette, pipe, or cigar and smoke exhaled by a smoker. Environmental tobacco smoke, also referred to as secondhand smoke, contains a few thousand substances, most of which are harmful to humans and can cause cancer.

Wood Smoke. *Wood smoke* is a smoke mixture of carbon monoxide, cancer-causing polycyclic aromatic hydrocarbons, and various small particles derived from burning wood. Two sources of wood smoke are wood-burning stoves and fireplaces. Cooking and food preparation should be restricted to areas that are under negative pressure and vented to the outdoors. **See Figure 9-23.**

Exposure to Radon

Radon Level*	If 1000 people who smoked were exposed to this level over a lifetime†...	The risk of cancer from radon exposure compares to‡...
20	About 260 people could get lung cancer	250 times the amout of drowning
10	About 150 people could get lung cancer	200 times the amout of dying in a home fire
8	About 120 people could get lung cancer	30 times the risk of dying from a fall
4	About 62 people could get lung cancer	5 times the risk of dying in a car crash
2	About 32 people could get lung cancer	6 times the risk of dying from poison
1.3	About 20 people could get lung cancer	(Average indoor radon level)
0.4	About 3 people could get lung cancer	(Average outdoor radon level)

* in pCi/L
† Lifetime risk of lung cancer deaths from EPA Assessment of Risks from Radon in Homes (EPA 402-R-03-003)
‡ Comparison data calculated using the Centers for Disease Control and Prevention's 1999-2001 National Center for Injury Prevention and Control Reports

Source: Environmental Protection Agency (EPA)

Figure 9-22. The Environmental Protection Agency (EPA) provides information regarding radon exposure limits.

Wood Smoke Exposure Limits

Classification	10 Minutes*	30 Minutes*	1 Hour*	4 Hours*	8 Hours*
Nondisabling	301	201	101	51	0
Disabling	500	300	200	100	50

Exposure	Symptoms	Prevention	First Aid Treatment
Inhalation	Coughing, scratchy throat, headaches; possible chest pain, and palpitations	—	Artificial respiration and cardiopulmonary resuscitation
Skin	—	—	—
Eyes	Irritation or stinging	—	—
Ingestion	—	—	—

* in ppm
Sources: Environmental Protection Agency (EPA), Occupational Safety & Health Administration (OSHA), Illinois Department of Public Health (IDPH), and Praxair, Inc.

Figure 9-23. The EPA does not have exposure limits for wood smoke.

Pesticides. A *pesticide* is a product used to kill household pests and insects. Pesticides, rodent poisons, and herbicides applied indoors or outdoors can lead to high indoor levels of harmful gases. Pesticide containers should not be reused. Leftover spray mixtures and rinse water should be disposed of by an approved waste disposal facility. **See Figure 9-24.**

Volatile Organic Compounds. A *volatile organic compound (VOC)* is a carbon-containing compound that easily evaporates at room temperature. VOCs are emitted as gases from certain solids and liquids. They include a wide variety of chemicals, some of which may have short- and long-term adverse health effects. The concentrations of many VOCs are higher indoors than outdoors (up to ten times).

The presence and concentration of VOCs change as conditions and activities change. VOCs are used for many products that are designed to be applied wet and then dry out. **See Figure 9-25.** Examples of VOCs include acetone, alcohol, ammonia, benzene, hydrocarbon, and phenol. Hundreds of products contain VOCs such as deodorizers, pesticides, solvents, paints, glues, marking pens, copy machine fluid, cosmetics, perfumes, and newly dry-cleaned fabrics. VOCs are often released during and after renovations.

TECH TIP

According to the EPA, 75% of U.S. households have used at least one pesticide product indoors during the past year. Studies suggest that 80% of pesticide exposure occurs indoors, and up to a dozen different pesticides have been found in the air inside of homes. It is important to remember that the "-cide" in pesticides means "to kill."

Pesticide (Permethrin) Exposure Limits

Classification	10 Minutes*	30 Minutes*	1 Hour*	4 Hours*	8 Hours*
Nondisabling	—	—	—	—	—
Disabling	2000	—	1000	750	375

Exposure	Symptoms	Prevention	First Aid Treatment
Inhalation	Irregular breathing, dizziness, drowsiness, and respiratory irritation	Good ventilation; half-face or full-face air respirator	Remove to fresh air; artificial respiration may be required
Skin	Numbing, burning, or tingling	Rubber full-body suit, rubber gloves (nitrile or neoprene), and rubber boots	Remove contaminated clothing; wash with soap for 15 to 20 min
Eyes	Irritation	Chemical goggles or face shield	Rinse eyes with water for 15 min
Ingestion	Vomiting; may cause pulmonary edema or death	—	Do not induce vomiting; call poison control center

* in mg/Kg

Figure 9-24. Permethrin is a common chemical used in the manufacture of many pesticides.

Volatile Organic Compound Exposure Limits (Acetone)

Classification	10 Minutes*	30 Minutes*	1 Hour*	4 Hours*	8 Hours*
Nondisabling	—	1000	750	—	500
Disabling	—	2400	1000	—	750

Exposure	Symptoms	Prevention	First Aid Treatment
Inhalation	Drowsiness; dizziness, nausea, headache, paralysis, unconsciousness, and death	Low levels: half-face mask air purifier with organic vapor cartridge. High levels: full-face supplied-air respirator or self-contained breathing apparatus (SCBA)	Remove to fresh air; artificial respiration or oxygen may be required
Skin	Moderate irritation, drying of skin	Chemical resistant clothes (neoprene), gloves, and footwear	Remove clothing; flush with water for 15 min to 20 min; cover skin with antibacterial cream
Eyes	Severe irritation	Chemical goggles or full-face shield	Flush with cool water for 15 min
Ingestion	Pulmonary aspiration, coma, and death	Wash hands and face before eating; wash all clothing before reuse	Do not induce vomiting; give glass of water or milk; call poison control center

* in ppm

Figure 9-25. Volatile organic compounds (VOCs), such as acetone, are found in many daily household products.

Biological Contaminants

A biological contaminant is a mass of particulate living organisms that can cause adverse health effects. Biological contaminants include molds, mildew, viruses, mites, pollen, and animal dander. **See Figure 9-26.** Major sources of biological contaminants include wet or damp materials, standing water, cooling towers, and contaminated outdoor air. Contaminated central air handling systems can distribute contaminants through a building. Many biological contaminants are small enough to be inhaled.

Microbial Contaminants. A *microbial contaminant* is a contaminant invisible to the naked eye that is either a living organism or derived from a living organism and can cause adverse health effects. Examples of microbial contaminants include viruses, bacteria, fungi, yeasts, and animal antigens. Microbial contaminants can be inhaled and cause many types of health problems.

Excessive concentrations of bacteria, viruses, fungi (including molds), dust mites, animal dander, and pollen may result from inadequate maintenance and housekeeping. Inadequate housekeeping includes water spills, poor humidity control, or allowing microbial contaminants to be brought into the building by occupants, infiltration, or ventilation air. Microbial contamination occurs anywhere accumulated carbon-containing material, such as dirt, plastic, wood, or paper, becomes moist or wet.

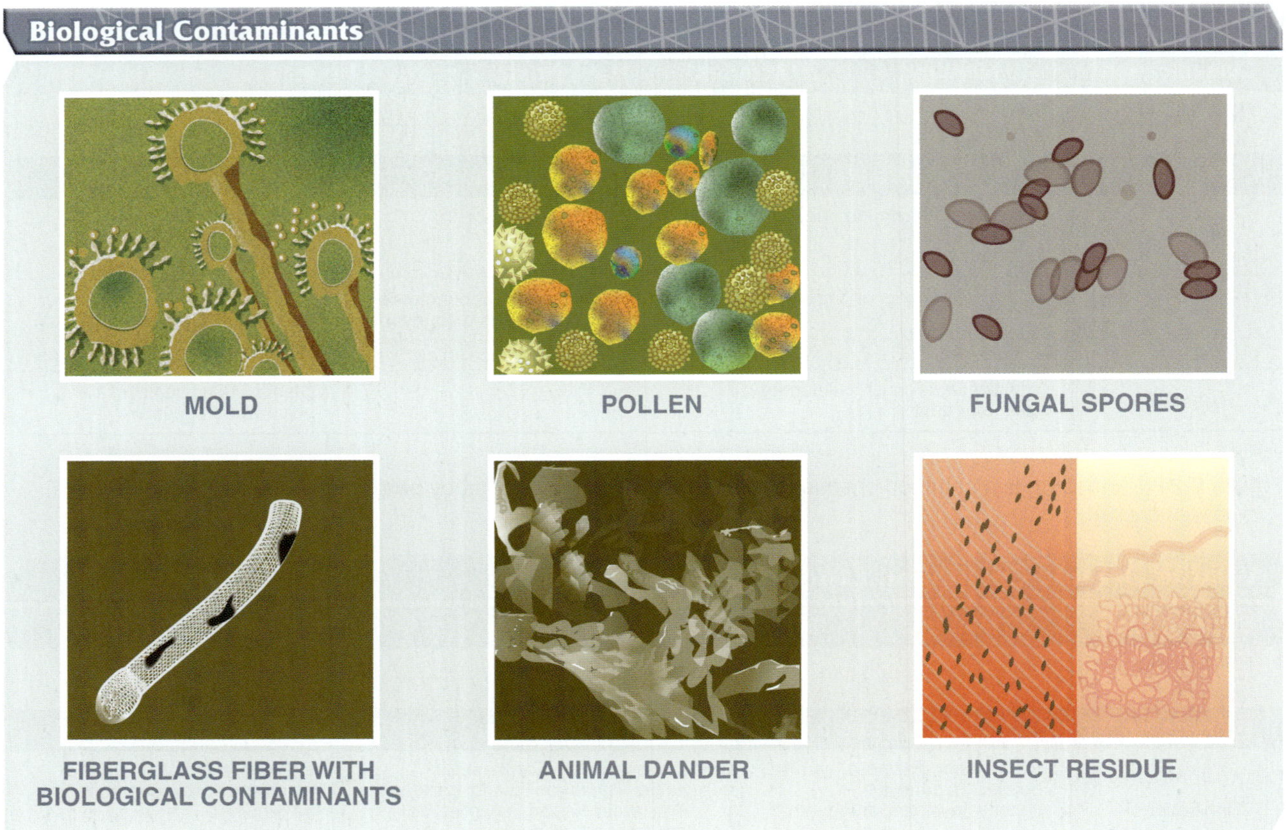

Figure 9-26. Biological contaminants must be viewed under a microscope in order to be identified.

Digital Resources

ATPeResources.com/QuickLinks • Access Code: 482610

UNIT 10

HVAC and Refrigeration System Printreading

In order to be effective in the industry, HVAC technicians must be competent in the reading and usage of HVAC mechanical system prints. These prints are used to properly install, service, and troubleshoot HVAC and refrigeration systems.

OBJECTIVES

- Identify the subsystems that may be integrated together and included on HVAC equipment prints.
- Discuss the abbreviations used on HVAC system prints.
- Sketch the various symbols used on HVAC system prints.
- Explain the content in a set of HVAC prints.
- Explain the purposes of HVAC wiring diagrams and detail drawings.
- Describe how bills of materials and sequences of operations are used with HVAC system drawings.

CSI MASTERFORMAT™

The CSI MasterFormat™ is a master list of titles and numbers used to organize specifications and other project information for most building design and construction projects in North America. HVAC systems are a separate category in the 2004 through 2012 editions of the CSI MasterFormat™. Division 23, *Heating, Ventilating, and Air Conditioning (HVAC),* covers all aspects of HVAC system equipment. Division 25, *Integrated Automation,* covers HVAC control systems. **See Figure 10-1.**

A wide variety of prints are used to install and maintain HVAC systems and controls. HVAC system and control system prints consist of abbreviations and symbols, wiring diagrams, detail drawings, and sequence-of-operation narratives. **See Figure 10-2.** For small projects, a single print may contain more than one type of HVAC control system. Prints for large (light commercial) projects include separate prints for each section or floor of the HVAC system and controls.

HVAC SYSTEM PRINTS

HVAC system prints include information on many different types of HVAC equipment and automated systems. Large HVAC equipment manufacturers design and engineer automated systems for integration into a building. These automated systems incorporate several subsystems that have different functionalities. Subsystems, which are sometimes integrated together, include HVAC control, fire alarm and life safety, access control, closed-circuit TV, and energy management systems.

TECH TIP

The Construction Specifications Institute (CSI) develops and maintains the CSI MasterFormat™. The CSI MasterFormat™ contains construction and materials specifications used in building contracts. It is found in outline form with 16 divisions. The divisions are subdivided into sections, which contain three parts: general, products, and execution. Each part consists of articles and paragraphs.

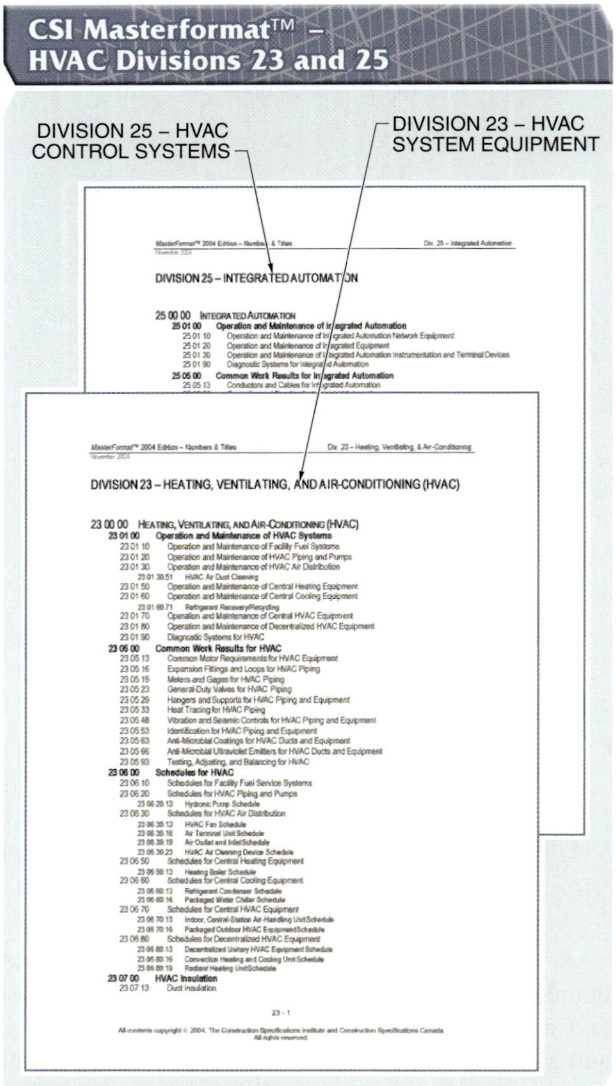

Figure 10-1. Division 23 of the 2004 CSI MasterFormat™ covers all HVAC equipment and Division 25 covers HVAC control systems and building integration.

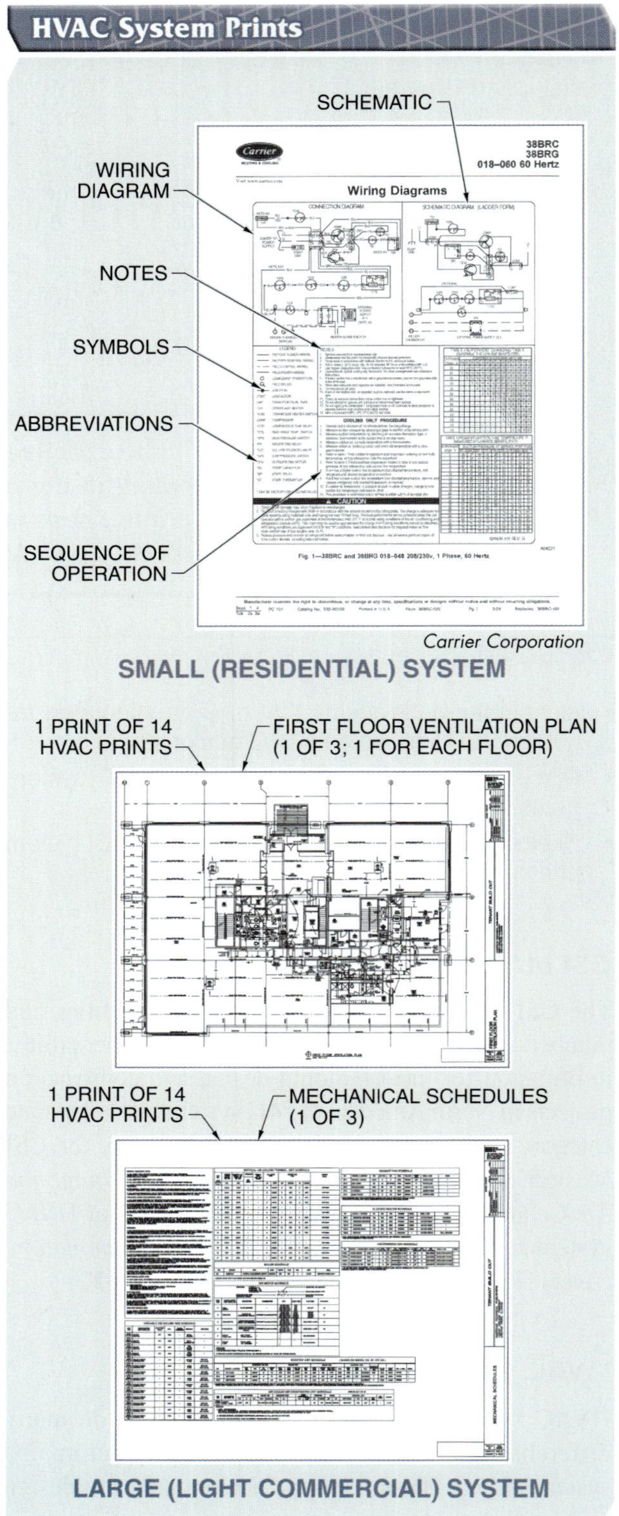

Figure 10-2. HVAC system and control system prints consist of abbreviations and symbols, wiring diagrams, detail drawings, and sequence-of-operation narratives that can be on a single sheet (print) or the prints for each section or floor of the HVAC system and controls.

Print Abbreviations and Symbols

HVAC system prints include a list of abbreviations and symbols. Abbreviations and symbols provide detailed information in a small amount of space on a print.

HVAC Abbreviations. HVAC system and control system prints include a unique set of abbreviations. **See Figure 10-3.** Abbreviations are used for both electrical and mechanical components. Some of the abbreviations used on HVAC system prints are also used on electrical prints. For example, the abbreviation for valve is either "V" or "VLV." HVAC abbreviations and definitions for a project are found in the legend or on the abbreviation sheet of a set of mechanical prints.

HVAC Abbreviations

Abbreviations	Description
AHU	air-handling unit
ASD	adjustable-speed drive
BLR	boiler
C	common
CC	cooling coil
CHWR	chilled water return
CHWS	chilled water supply
CLG	cooling
CWR	condenser water return
CWS	condenser water supply
DA-T	discharge air temperature sensor
DMPR	damper
DP	differential pressure
DPS	differential pressure sensor
EA	exhaust air
EP	electric pneumatic switch
HC	heating coil
HWR	hot water return
HWS	hot water supply
LL	low limit
MA	mixed air
MA-LL	mixed air low limit
NC	normally closed
NO	normally open
OA	outside air
P	pump
PE	pneumatic electric switch
PI	pressure indicator
R	relay
RA	return air
RF	return fan
RTU	roof top unit
SA	supply air
SF	supply fan
SP	static pressure
S-SP	sensor static pressure
TSP	shielded twisted pair cable
T	temperature
VFD	variable frequency drive
VLV	valve

Figure 10-3. A set of HVAC prints includes abbreviations for both mechanical and electrical terms.

Some abbreviations have multiple meanings and their interpretation depends on the application. For example, "NO" (normally open) and "NC" (normally closed) typically refer to electrical contacts but are also used for mechanical dampers and valves. The abbreviation "NO" refers to de-energized electrical contacts that do not allow the flow of electricity, but it also refers to mechanical dampers and valves that allow the flow of fluid or gas. The interpretation of NO or NC depends on whether the abbreviation refers to an electrical or mechanical component. **See Figure 10-4.**

Significance of Normally Open (NO) and Normally Closed (NC)

	Normally Open (NO)
Electrical (relay, motor starter)	Contacts are open: no flow of electricity
Mechanical (damper, valve)	Mechanical device is open: flow of fluid or gas
	Normally Closed (NC)
Electrical (relay, motor starter)	Contacts are closed: flow of electricity
Mechanical (damper, valve)	Mechanical device is closed: no flow of fluid or gas

Figure 10-4. The abbreviations "NO" (normally open) and "NC" (normally closed) have different meanings for electrical and mechanical equipment.

HVAC Symbols. A set of symbols is used to represent HVAC system devices and control components. Symbols represent the electrical, electronic, mechanical, and pneumatic devices and components of HVAC equipment and control system. **See Figure 10-5.**

Although some symbols are from ANSI standards, many are not. The use of nonstandard symbols is the result of HVAC equipment manufacturers developing their own unique symbols. Many manufacturer-developed symbols resemble the items they represent. **See Figure 10-6.** Symbols and their definitions are found in legends and on abbreviation sheets.

Every HVAC system device and component has a symbol that represents it on an HVAC print.

HVAC Symbols

Device	Type	Symbol
Combination 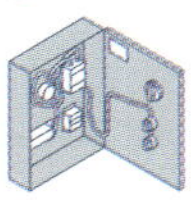	Magnetic Motor Starter and Disconnect	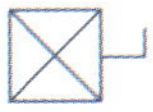
Adjustable-Speed Drive 	Abbr = ASD	ASD
Automatic Temperature Control Panel 	Abbr = ATC	ATC
Sensors 	Aquastat Abbr = AST	(A)
	Firestat Abbr = ATC	(F)
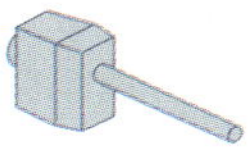	Humidistat Abbr = ATC	(H)
Thermostat	Abbr = TMST	(T)
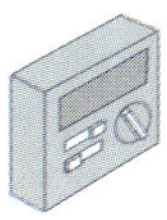	Line Voltage Abbr = LVT	(T) L
	Low Voltage Abbr = LWVT	(T) LV
Time Switch 	Abbr = TSW	TS
Pipe Trace Heater 	Abbr = PTHR	

Figure 10-5. Symbols for electrical, electronic, mechanical, and pneumatic devices and components are used on HVAC system prints.

Manufacturer-Developed HVAC Symbols

Device	Type	Symbol
Sheet-Metal Duct 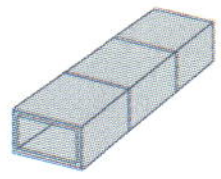	Abbr = SMD	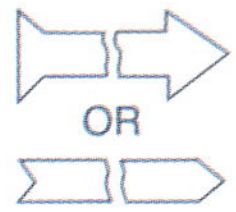
Automatic Valves	2-Way	
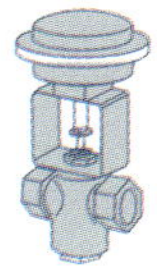	3-Way	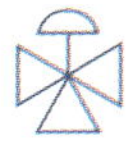
	Solenoid-Actuated Abbr = SOL	
Dampers 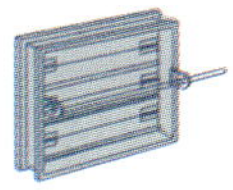	Pneumatic Abbr = PNDM	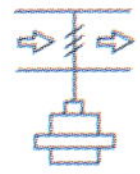
	Electric Abbr = ELDM	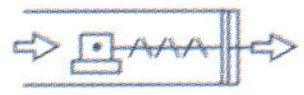
Fan 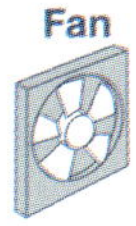	Abbr = FAN	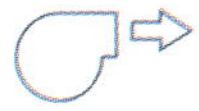
Pump 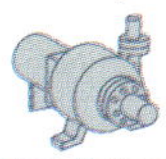	Abbr = PMP	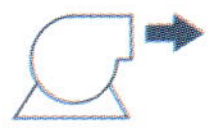
Electric Pneumatic Switch 	Electric Controlling Pneumatic Abbr = EPSW	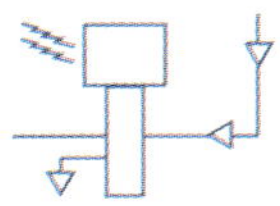
Pneumatic Electric Switch 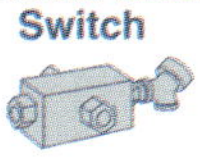	Pneumatic Controlling Electric Abbr = PESW	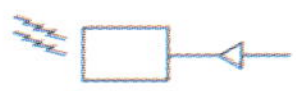
Shielded Twisted Pair Cable	Abbr = STPC	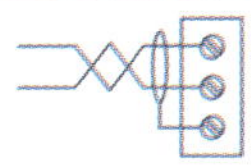
Heating Coil 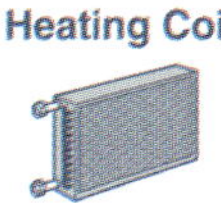	Abbr = HC	
Cooling Coil 	Abbr = CC	
Filter 	Abbr = FLT	

Figure 10-6. The use of manufacturer-developed symbols is the result of HVAC equipment manufacturers developing their own set of symbols for the products they developed.

HVAC SYSTEM PRINT SETS

A separate set of prints are provided for each subsystem of an HVAC system in a building. For example, separate prints are used for ventilation plans, roof plans, mechanical schedules, ceiling plans, and flooring plans. **See Figure 10-7.** A set of prints can include more than 30 prints. Information provided on a set of HVAC system prints includes the following:

- sheet metal ducts, including the direction of airflow in each duct
- mechanical piping system, including the direction of liquid flow within the piping of the system
- general location of HVAC control system devices and components in relation to the mechanical equipment, sheet metal ducts, and mechanical piping systems
- addresses of the sensors and actuators that are part of the networked HVAC control system
- general location of control system component terminations
- notes covering contractor installation responsibilities

HVAC system prints do not provide the exact location of, or routing and mounting information for, electrical equipment such as control panels, disconnects, motor starters, or variable frequency drives. Also, HVAC system prints do not provide specifics on the sizing of conduit and wire, motor starters, VFDs, or control and power wiring. Electrical prints must be consulted for this information.

HVAC system prints do not provide the exact location of, or routing and mounting information for, mechanical equipment such as furnaces, boilers, air-handling units, sheet metal ducts, or mechanical piping. Also, HVAC system prints do not provide specifics on duct size, pipe size, air-handler output, or furnace output. Mechanical schedules must be consulted for this information.

HVAC Wiring Diagrams

An *HVAC wiring diagram* is a diagram that shows the wiring connections between HVAC control panels and system components such as sensors and actuators. **See Figure 10-8.** Some HVAC wiring diagrams show only the terminal strip of a control panel, while other wiring diagrams show the entire control panel. The abbreviations, symbols, and addresses that appear on plans also appear on wiring diagrams.

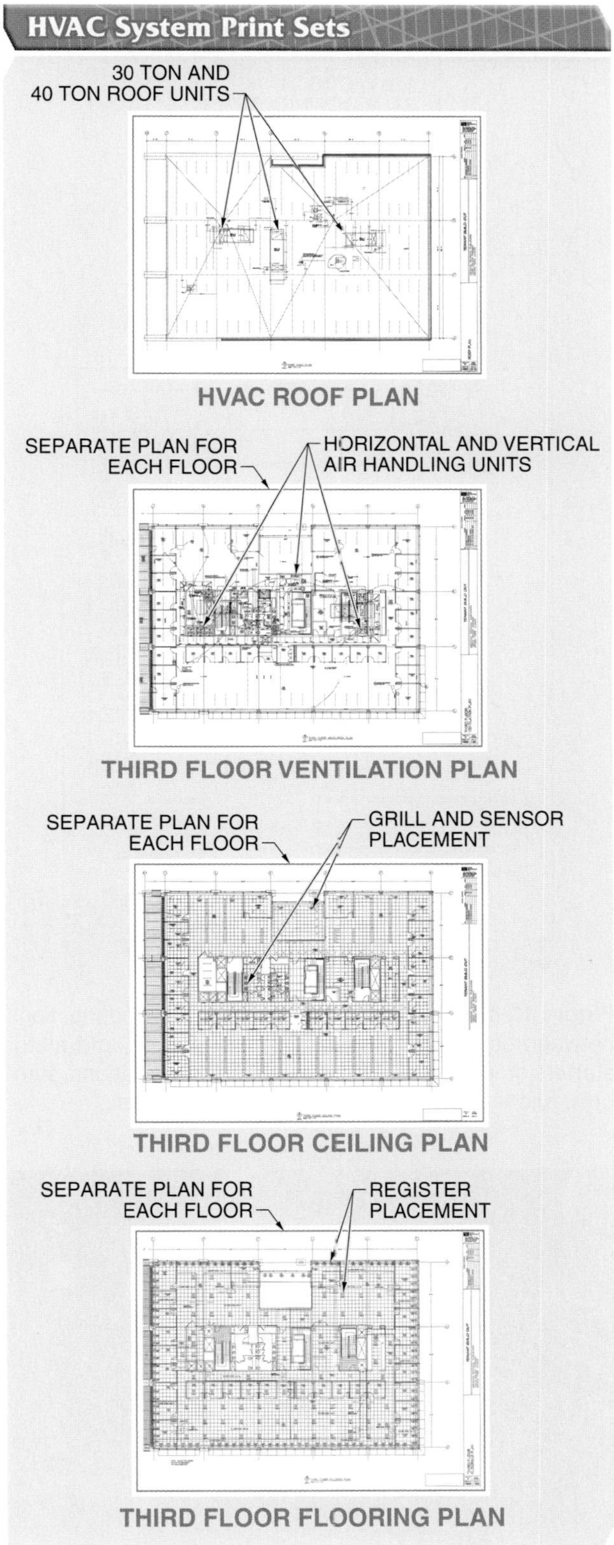

Figure 10-7. A set of HVAC system prints can include HVAC roof plans, per floor ventilation plans, per floor ceiling plans, per floor flooring plans, and wiring diagrams that indicate the control system connections.

HVAC Wiring Diagrams

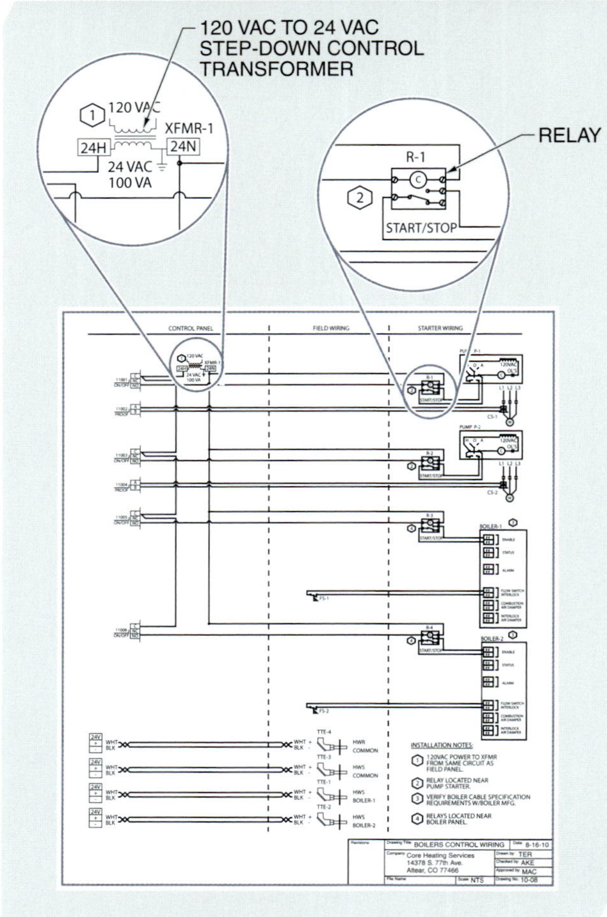

Figure 10-8. HVAC wiring diagrams show the connections between equipment, sensors, control panels, and motor starters or VFDs while using the same abbreviations, symbols, and addresses that appear on HVAC plans.

Wiring diagrams and plans are required to wire HVAC panels and system components together.

HVAC control systems interface with mechanical equipment and other systems that operate at different voltage levels. Consequently, various voltage levels, such as 24 VDC, 24 VAC, 120 VAC, and 460 VAC, are found on HVAC wiring diagrams. Many HVAC control systems operate at 24 VAC. The actuators and sensors that connect directly to the control panel also operate at 24 VAC. When a control system must control a component operating at a different voltage, a relay is used. Typical information found on an HVAC wiring diagram includes the following:

- the location of the terminal strip for connecting conductors to HVAC control panels or related equipment such as furnaces or boiler boards, interposing relays, or duct smoke detectors
- the type and size of conductors used to connect control devices and components
- the addresses of the sensors and actuators that are part of the networked HVAC control system
- voltage information for the HVAC control system and related equipment
- any notes covering contractor installation responsibilities and locations of the control components
- any section of an HVAC system print needed for clarification

HVAC panel wiring diagrams are drawn to indicate the terminal strips and board connections for components. Terminal strips and board connections are located inside HVAC control panels and mechanical equipment panels. Panel wiring diagrams indicate how all devices and components are wired together. HVAC wiring diagrams typically include the terminal strips for actuators and sensors located in the field. **See Figure 10-9.**

Installation of an HVAC control system requires the use of both plans and wiring diagrams to determine what devices and components must be wired together. HVAC wiring diagrams, like most HVAC plans, do not provide locations or specifics on mounting and cable routing. Electrical and mechanical prints must be consulted for this information.

HVAC Detail Drawings

Detail drawings provide additional information that cannot be shown on HVAC system prints or wiring diagrams due to space limitations. Although HVAC detail drawings are not the most frequently used prints, they contain essential installation information. Common types of HVAC detail drawings include panel details, wiring details, and bills of materials.

HVAC Panel Wiring Diagrams

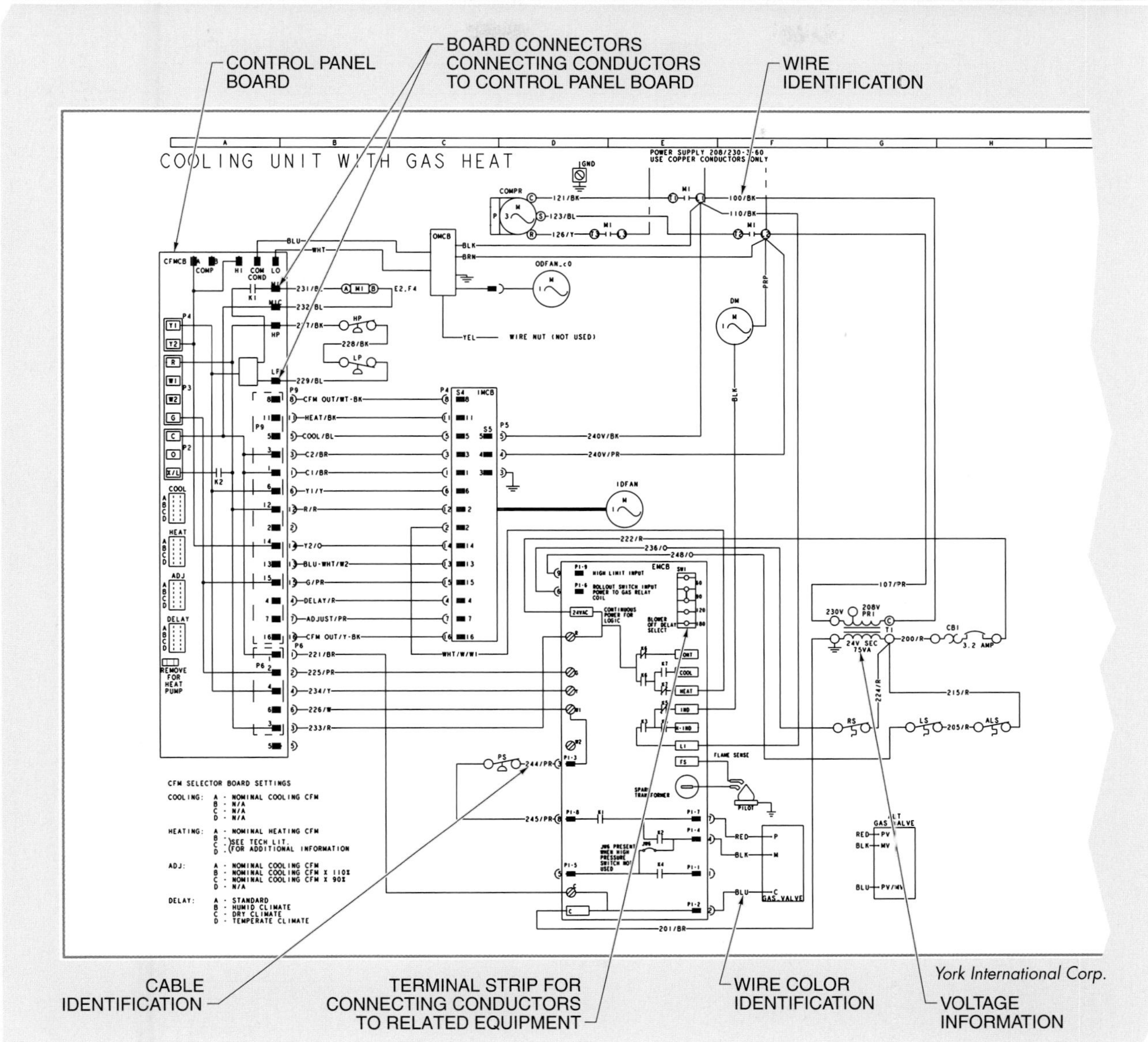

Figure 10-9. An HVAC panel wiring diagram must indicate how all panel devices and field components are connected.

Panel Detail Drawings. There are a variety of panel detail drawings used for HVAC systems. A *panel detail drawing* is a detail drawing that provides specific information about the mountings and connections found in an HVAC control panel. Panel detail drawings include panel layout details, panel enclosure details, and panel mounting details. **See Figure 10-10.**

A *panel layout detail* is a detail drawing that provides an elevation view of the interior of an HVAC control panel. Controllers, relays, terminal strips, transformers, and wireways are shown in their proper locations, attached to the back panel of the HVAC control panel. The dimensions of the back panel are usually provided.

A *panel enclosure detail* is a detail drawing that provides an elevation view of the front and side of an HVAC control system panel. The dimensions of the enclosure are usually provided.

A *panel mounting detail* is a detail drawing that provides an elevation view of HVAC control panels and mounting support. Information on the mounting height, mounting support structure, panel dimensions, and interconnecting raceway or gutter is also provided.

Panel Detail Drawings

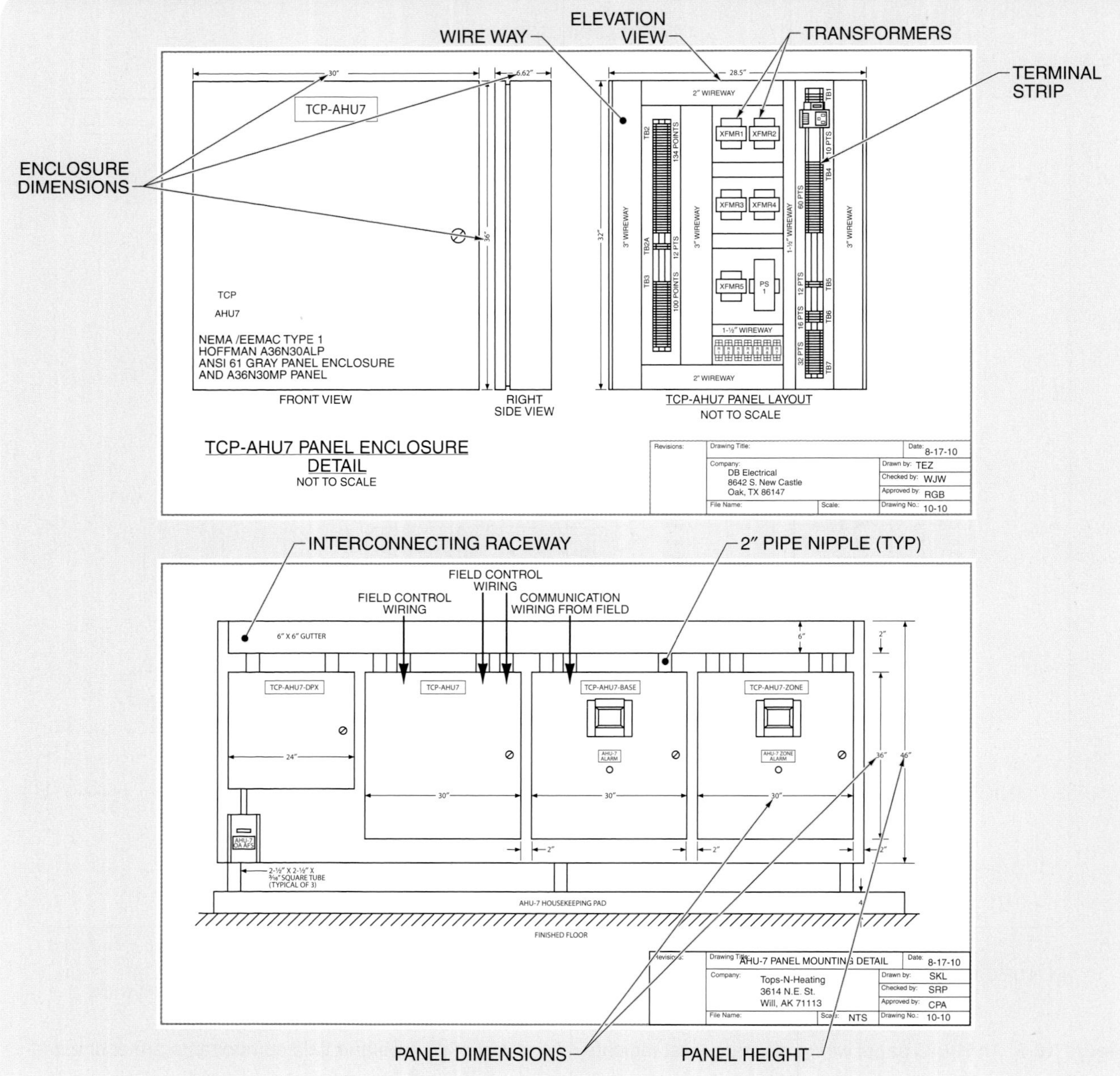

Figure 10-10. There are a variety of panel detail drawings used for HVAC systems. Panel detail drawings include panel layout details, panel enclosure details, and panel mounting details.

Wiring Detail Drawings. Wiring detail drawings are provided for most HVAC systems. A *wiring detail drawing* is a detail drawing that provides wiring termination information for sensors, actuators, or pieces of equipment that have several termination possibilities or that connect to multiple voltage sources. The sensor or actuator terminal strip, HVAC control panel terminal strip, and conductors connecting the terminal strips are shown on wiring detail drawings. **See Figure 10-11.**

Bills of Materials. Most HVAC system prints include a bill of materials. Typically, each HVAC control system panel has a specific bill of materials that is organized like a spreadsheet. **See Figure 10-12.** A bill of materials provides the abbreviation and/or symbol, quantity, and a description of each item on a print. Model information can also be obtained from the bill of materials. All HVAC CAD softwares produce bills of materials from HVAC drawings.

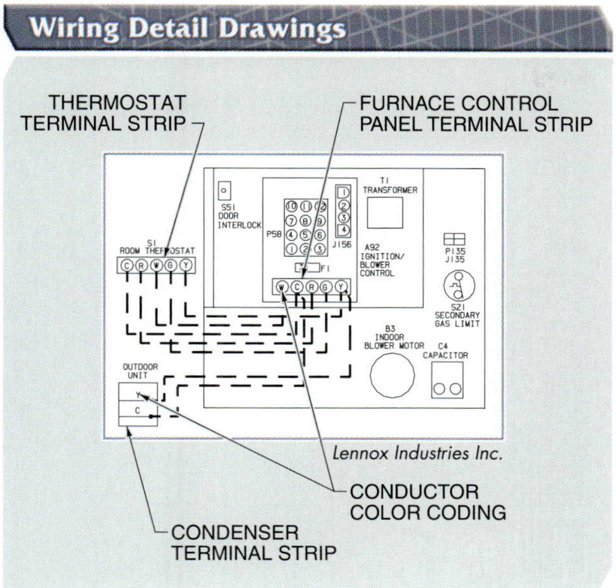

Figure 10-11. Wiring detail drawings provide wiring information on terminating conductors at sensors, actuators, or pieces of equipment that have several termination possibilities or that connect to multiple voltage sources.

A bill of materials lists the major items that make up a control panel along with the sensors and actuators that are mounted in the field. Items that typically make up a control panel include the enclosure, control transformers, and relays. Sensors mounted in the field include temperature, pressure, and smoke sensors. The actuators mounted in the field are damper actuators. Information about miscellaneous hardware, such as screws, washers, and nuts, is not included on a bill of materials.

SEQUENCES OF OPERATION

Most HVAC control system prints include a sequence of operation. A sequence-of-operation narrative is also provided for each subsystem of an HVAC system. It may appear on the same sheet as an HVAC system print or on a separate sheet. **See Figure 10-13.** Typically, it is provided for air handlers, boilers, chillers, and cooling towers. When a building contains multiple units with different control strategies, separate sequence-of-operation narratives are provided for each unit.

During the startup of an HVAC control system, a sequence of operation is used to verify that the newly installed system functions per system prints. When it is necessary to troubleshoot an existing HVAC control system, the sequence of operation functions as a guide for how the system should work.

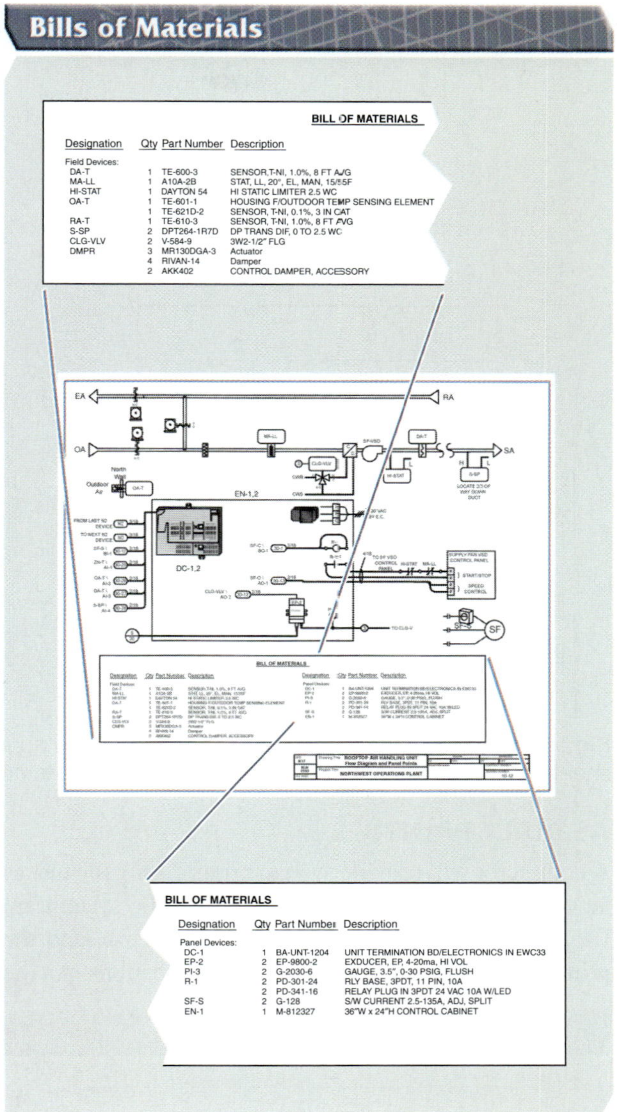

Figure 10-12. Typically, every HVAC control panel has a specific bill of materials that is organized like a spreadsheet, which lists the major items such as the enclosure, control transformer, and the relays that make up the panel along with all field mounted sensors and equipment actuators.

Installation jobs contain job site specifications in addition to the prints. Depending on the size of the installation, the sequence of cperation, installation wiring diagrams, and other specifications will be included in the job site prints or may be provided in separate specification documents in addition to blueprints. The job site specifications may contain specific procedures, or sequence to follow, job safety requirements, specific materials to use, etc.

HVAC Sequences of Operation

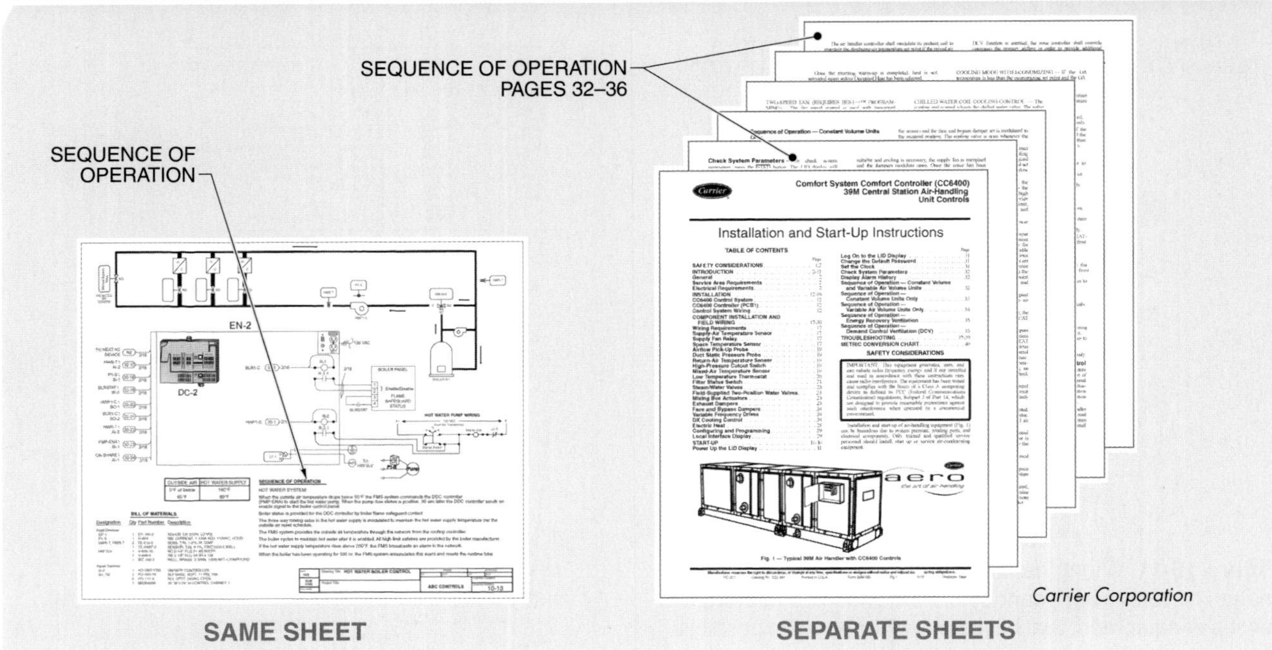

Figure 10-13. A sequence-of-operation narrative may appear on the same sheet as an HVAC system print or on a separate sheet. It explains the proper function of an HVAC system or subsystem.

AS BUILT PRINTS

A contractor will provide system prints during the installation of an HVAC system. Changes in the system are very common during the installation process, and the prints are updated to reflect these modifications. As the project is completed, prints are created and labeled "as built" to reflect the final project. These "as built" prints are extremely important for the future service of the equipment. Care must be taken to secure them safely. **See Figure 10-14.**

As Built Prints

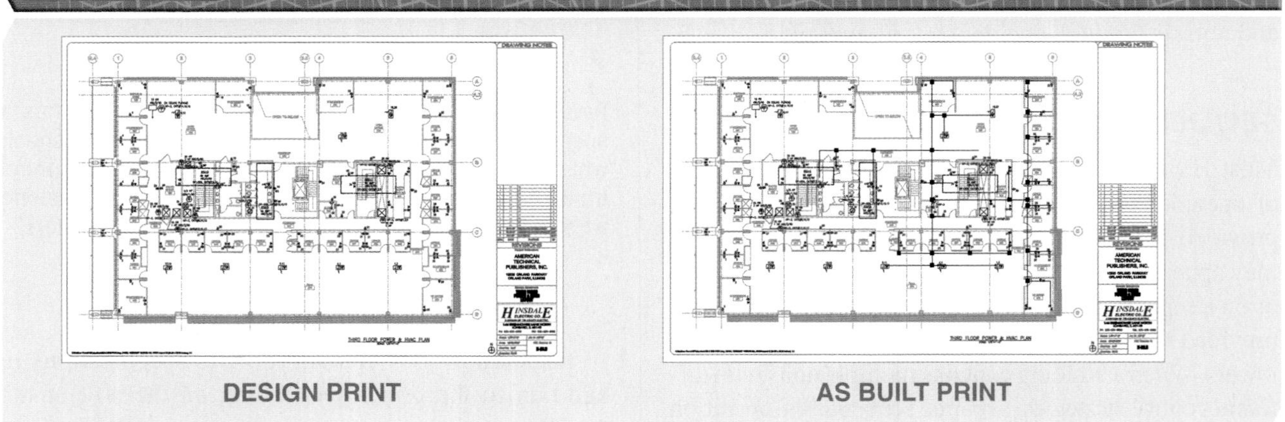

Figure 10-14. As built prints are not available immediately after a building is constructed. It takes time to incorporate all the changes to the prints according to the work performed by the contractor.

UNIT 11

Electrical System Principles

A solid foundation in electricity and electrical systems is vital for refrigeration and HVAC technicians. The technician must be able to recognize and troubleshoot the electrical circuits of refrigeration and HVAC systems. Most problems, are related to the electrical components of these systems.

OBJECTIVES

- Explain how electricity is produced.
- Define electricity.
- Demonstrate the use of Ohm's law.
- Compare direct and alternating current.
- Explain series and parallel circuits.
- Find the load voltage drops in series and parallel circuits.
- Describe how induced electricity works.
- Identify single-phase and three-phase circuits.
- Describe common power quality problems.

Index to Procedures

ELECTRICAL ENERGY

Electrical energy is used to produce heating and cooling for residential, commercial, and industrial use. Electrical energy is the most popular form of energy used in noncombustion heating units. Other forms of energy, such as solar energy, are used in limited applications.

Electrical Energy Production

Many different sources of energy are used to produce electricity. Water is used for producing electricity in hydroelectric-power generation plants. A *hydroelectric-power generation plant* is a power plant that changes energy from moving water into electrical energy. **See Figure 11-1.** A *cogeneration power plant* is a power plant that generates both electricity and heat for use.

Water is piped from a reservoir that is at a higher elevation than the generator. The reservoir is fed from a river or lake. The force of the water falling through large pipes (penstocks) creates energy, which turns turbines in the power generation plant. The mechanical energy of the turbines turns the generators, which changes the mechanical energy into electrical energy. Because of the high costs of building dams and hydroelectric-power generation plants, these facilities are constructed by the central government in most countries.

Combustion-power generation plants use steam to produce electricity. When steam is used to produce electricity, water must first be changed into steam. The most common method of producing steam is to use boilers. Coal, natural gas, or fuel oil is burned to heat water to produce steam. The steam is then piped at high pressures to steam turbines, which turn generators, and electricity is produced. **See Figure 11-2.** Steam from other sources can also be used to operate a turbine such as steam geysers with high enough temperatures and pressures. Nuclear energy also generates heat for steam production.

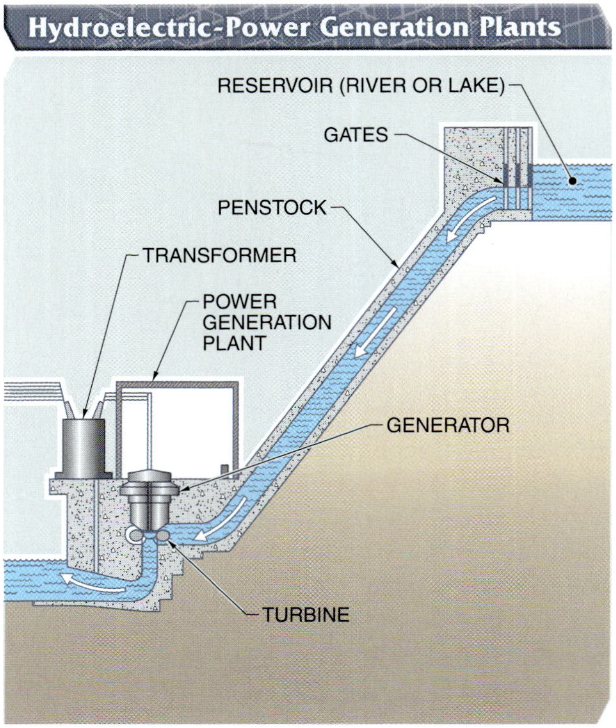

Figure 11-1. Hydroelectric-power generation plants use the energy of falling water to turn turbines. The turbines then turn generators that produce electricity.

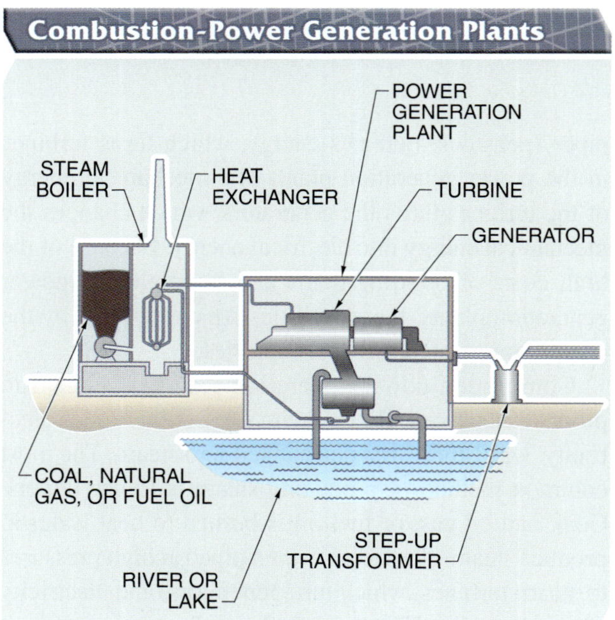

Figure 11-2. Steam used to turn steam turbines in coal-, natural gas-, or fuel oil-fired power generation plants is produced by combustion.

Electrical energy produced by nuclear power plants, hydroelectric power plants, solar power, and wind power are cleaner forms of energy because they do not emit carbon dioxide or other pollutants to the atmosphere. Hydroelectric power plants, solar power, and wind power are renewable sources of energy. Nuclear power plants produce large amounts of energy efficiently, but the hazardous waste produced must be handled properly.

ELECTRICITY

Electricity is the energy produced from the flow of electrons through a conductor (wire). Electrons are particles that are negatively charged and orbit the nucleus of an atom. A *nucleus* is the heavy, dense center of an atom. The nucleus of an atom contains protons and neutrons. Protons are particles with a positive charge and neutrons are particles with no electrical charge. **See Figure 11-3.**

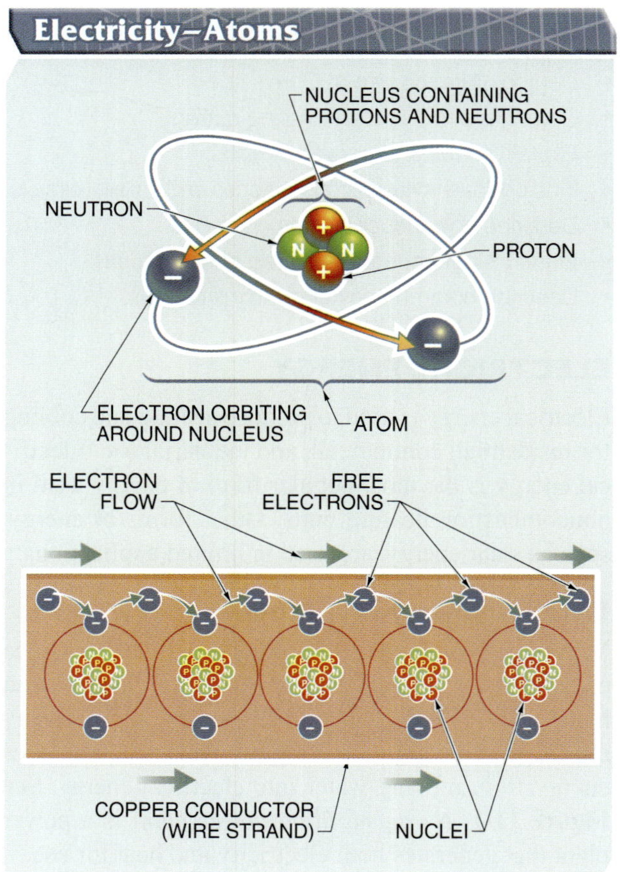

Figure 11-3. Electricity is the movement of electrons from atom to atom.

Electrons flow from a source that has an excess of electrons. Electrons flow easily through substances with little resistance, which are known as conductors. A *conductor* is a material with low resistance that permits electrons to flow through it easily. Materials that are good conductors include copper, silver, and aluminum. A metal wire is an example of a conductor.

The most common type of wire used today is copper. Aluminum wire is used, but has limited applications. Conductors are insulated and rated according to the American Wire Gauge system (AWG). **See Figure 11-4.** The smaller the wire number, the larger the wire diameter and the more current the wire can safely carry. Control circuits use wire in the 22 gauge to 18 gauge range, while line voltage systems use wire in the 12 gauge to 8 gauge range or larger. The wire size used for each application is closely regulated by the National Electrical Code® (NEC®) in the U.S. and the Canadian Electrical Code (CEC®) in Canada.

Electrons do not flow easily through substances with high resistance, which are called insulators. An *insulator* is a material that resists the flow of electrons. Materials that are good insulators include plastic, rubber, and glass.

Current is the amount of electrons flowing through a conductor. Current is measured in amperes (A). *Voltage* is the amount of electrical pressure in a conductor. Voltage is measured in volts (V). *Resistance* is the amount of opposition a material has to the flow of electrons. Resistance is measured in ohms (Ω).

Ohm's Law

Ohm's law expresses the relationship between voltage, current, and resistance in a circuit. **See Figure 11-5.** It states that current in a circuit is proportional to voltage and inversely proportional to resistance. When using Ohm's law, any value in the relationship can be found when the other two values are known. Any value of Ohm's law can be determined by using the following formula:

$E = I \times R$

where

E = voltage (in V)

I = current (in A)

R = resistance (in Ω)

Ohm's Law Media Clip

TECH TIP

Benjamin Franklin made the first attempt to identify the direction of electrical flow. Today, conventional flow designations indicate that electron holes move from positive (+) to negative (−). Electron flow designations indicate that electrons move from negative (−) to positive (+). Electrical circuits use conventional flow designations, and electronic circuits use electron flow designations.

Standard Conductor Ratings

AWG Gauge	Conductor Diameter*	Ohms per 1000′	Maximum Amps-Chassis Wiring	Maximum Amps-Power Transmission
0000	0.460	0.049	380	230
000	0.409	0.062	328	200
00	0.365	0.078	283	175
0	0.325	0.098	245	150
1	0.289	0.124	211	130
2	0.257	0.156	181	115
3	0.229	0.197	158	100
4	0.204	0.248	135	70
5	0.181	0.313	118	60
6	0.162	0.395	101	55
7	0.144	0.498	89	50
8	0.128	0.628	73	40
10	0.102	0.999	55	30
12	0.080	1.588	41	25
14	0.064	2.525	32	20
16	0.051	4.016	22	15
18	0.040	6.385	16	10
20	0.032	10.15	11	5
22	0.025	16.14	7	0.92
24	0.020	25.67	3.5	0.577
26	0.016	40.81	2.2	0.361
28	0.013	64.9	1.4	0.226
30	0.010	130.2	0.86	0.142
32	0.008	164.1	0.53	0.091

* in in.

Figure 11-4. Tables 310.15(B)(16) through 310.15(B)(21) of the NEC® are used to determine the ampacities of specific materials and types of wires for specific applications.

Example: Finding Voltage Using Ohm's Law

A circuit has a power source delivering 0.5 A of current flow through a pressure switch and a motorized actuator that has 48 Ω of resistance. Find the supply voltage for the circuit. **See Figure 11-6.**

$E = I \times R$

$E = 0.5 \times 48$

$E = \mathbf{24\ V}$

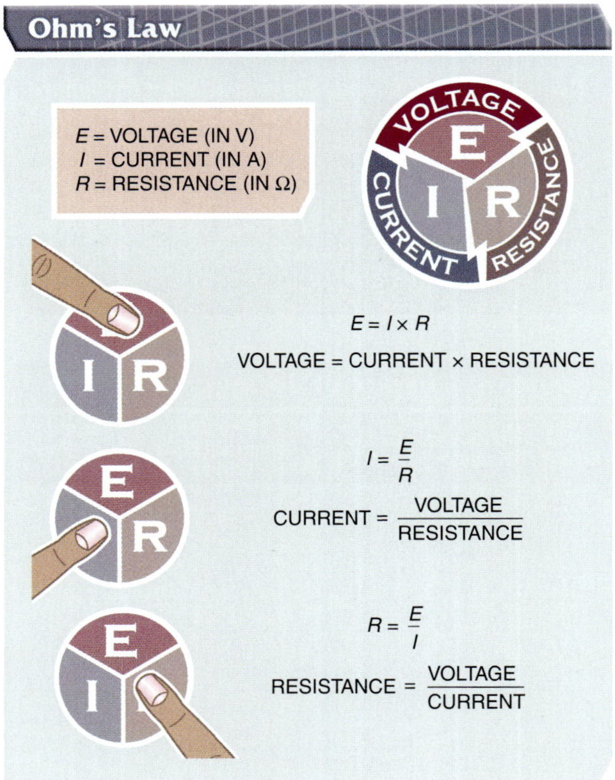

Figure 11-5. Any missing value in Ohm's law can be found when the other two values are known.

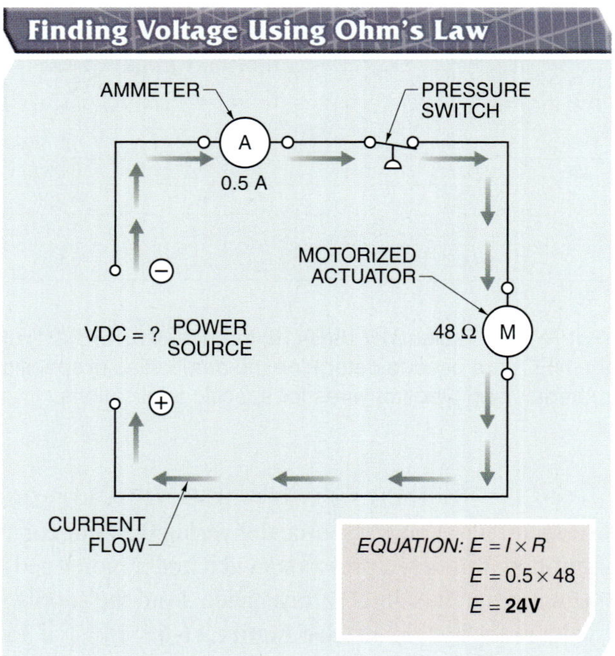

Figure 11-6. The supply voltage for a circuit can be found by multiplying the total circuit resistance by the total current of the circuit.

Example: Finding Current Using Ohm's Law

A circuit has a power source delivering 24 V to a pressure switch and a motorized actuator that has 48 Ω of resistance. Find the current in the circuit. **See Figure 11-7.**

$$I = \frac{E}{R}$$

$$I = \frac{24}{48}$$

$$I = \mathbf{0.5\ A}$$

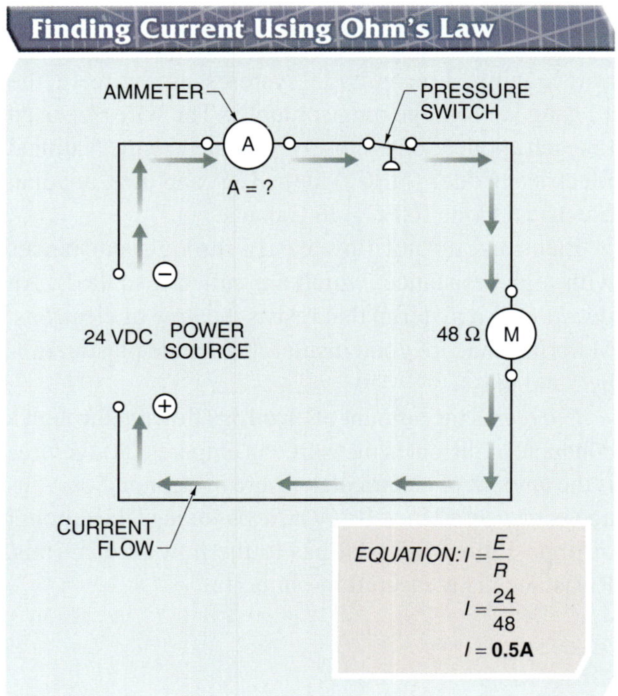

Figure 11-7. The current in a circuit can be found by dividing the supply voltage to a circuit by the total resistance of the circuit.

Example: Finding Resistance Using Ohm's Law

A circuit has a power source delivering 24 V with 0.5 A of current flow through a pressure switch and a motorized actuator. Find the resistance of the motorized actuator in the circuit. **See Figure 11-8.**

$$R = \frac{E}{I}$$

$$R = \frac{24}{0.5}$$

$$R = \mathbf{48\ \Omega}$$

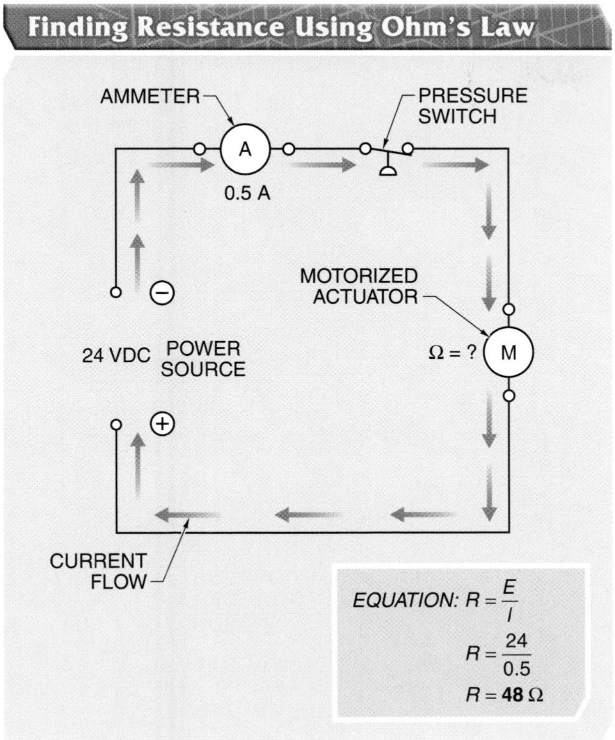

Figure 11-8. The total resistance of a circuit can be found by dividing the supply voltage to a circuit by the current in the circuit.

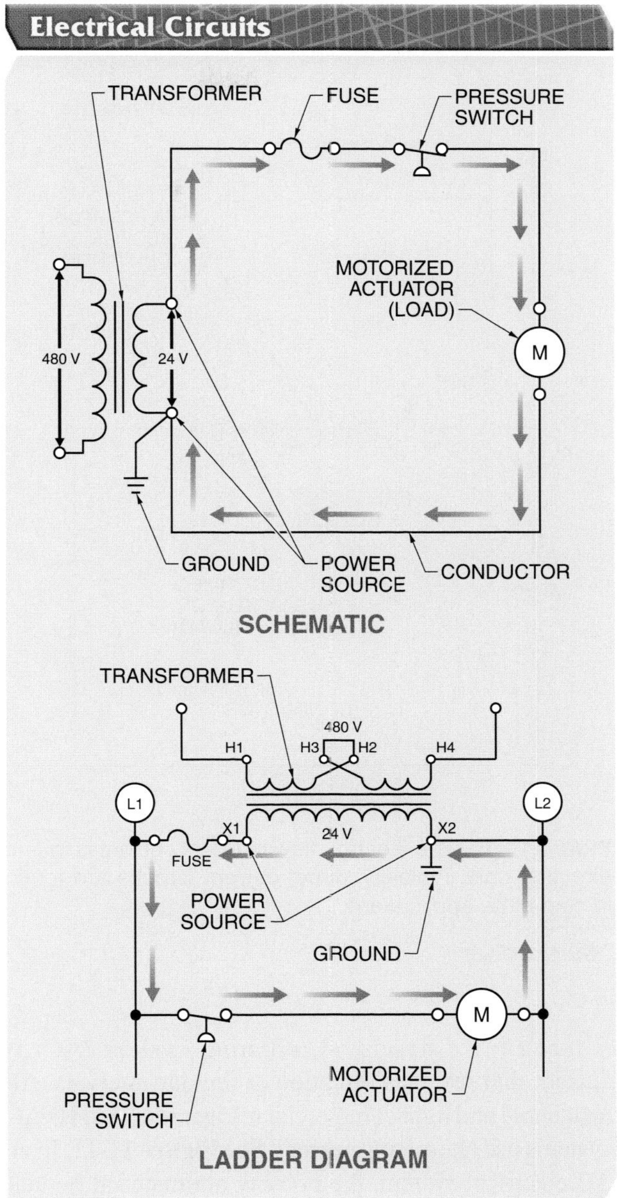

Figure 11-9. HVACR diagrams are typically represented in component or pictorial view and in schematic form, called "ladder diagrams."

Electrical Circuits

An *electrical circuit* is the interconnection of conductors and electrical elements through which current is designed to flow. An electrical circuit consists of a power source, conductor, switch, and load. A *load* is a device that performs work. **See Figure 11-9.** For a control circuit to operate properly, each element must perform its task. Problems occur when any part of the circuit does not function properly. A technician must be able to identify each component and its function in an electrical circuit.

Control System Current

When current is applied to a control circuit, current flow performs work. In general, the higher the current flow, the higher the amount of work performed by the circuit. Using control drawings, a refrigeration, heating, and air conditioning technician must be able to trace current flow through the complete circuit, from its source to its return. Ammeters are used to measure the current flow in a circuit. Current may be direct or alternating.

Direct Current. *Direct current (DC)* is current that flows in one direction only. All DC voltage sources have positive and negative terminals, which establish polarity in a circuit. **See Figure 11-10.** Polarity is the positive (+) or negative (–) state of an object. Many electronic devices, such as personal computers, PLCs, and controllers, operate using direct current. Today, HVAC systems can include DC motors in some of the circuits.

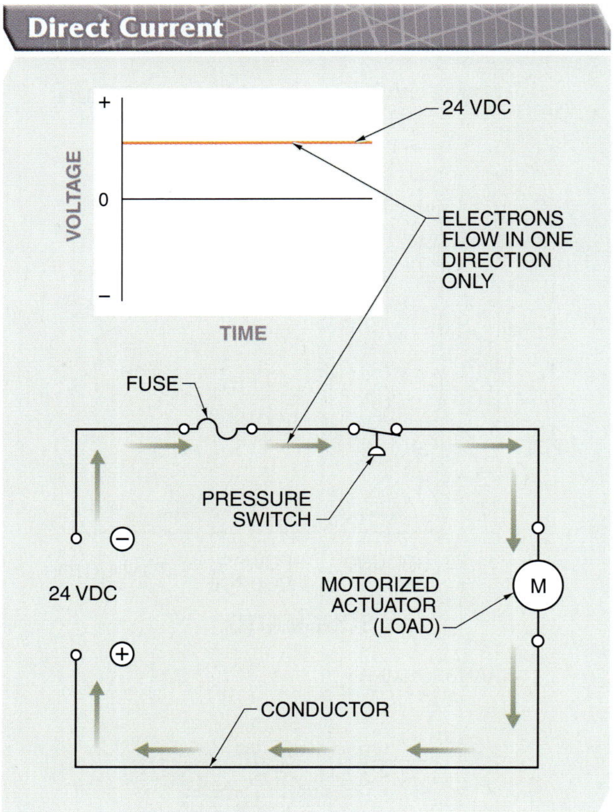

Figure 11-10. The electron flow of direct current is in one direction only. Typically, direct current circuits are found in sold-state applications.

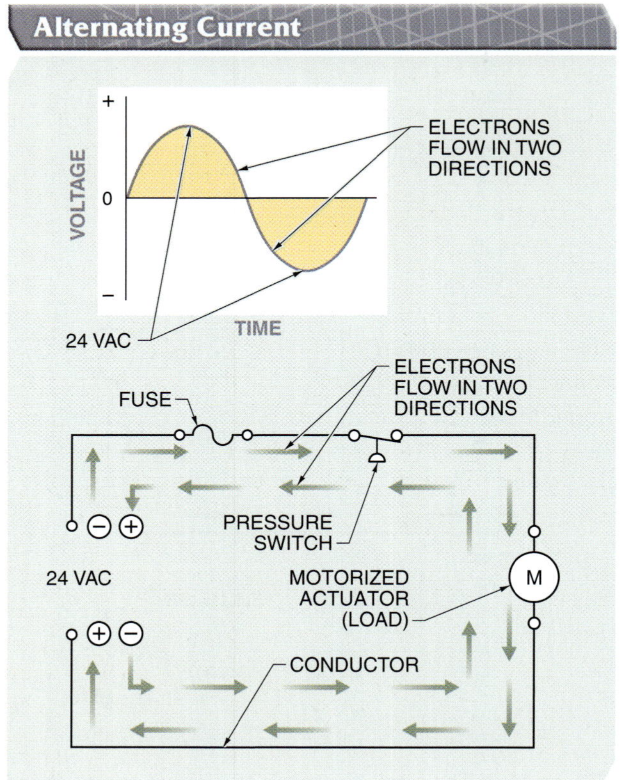

Figure 11-11. The electron flow of alternating current changes direction at given intervals. Typically, alternating current circuits are found in most refrigeration and HVAC systems.

Alternating Current. *Alternating current (AC)* is current that reverses direction at regular intervals. All residential and light commercial refrigeration and HVAC systems use alternating current. **See Figure 11-11.** In an AC electrical system, the polarity alternates at regular intervals. The number of times (frequency) that a circuit alternates is measured in cycles per second. *Hertz (Hz)* is the international unit of frequency equal to one cycle per second. The United States, Canada, and Mexico use AC power with a frequency of 60 Hz.

Circuit Connections

Residential and light commercial refrigeration and HVAC system electrical circuits have components such as switches, conductors, and loads. These components are connected in series, parallel, or series-parallel combinations. A *series circuit* is a circuit that has two or more components connected so that there is one path for current to flow. **See Figure 11-12.**

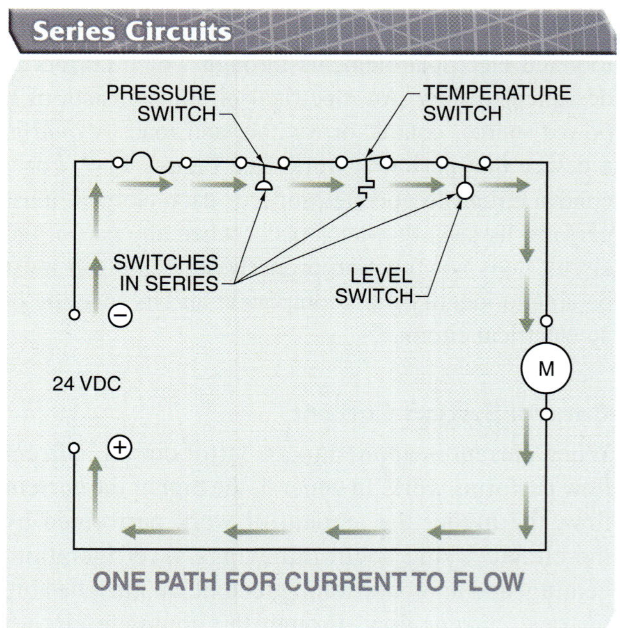

Figure 11-12. A series circuit can have as many components as required, but there is only one flow path for current to flow.

In a series circuit, when any component is not connected, or if any switch is open, there is not a complete path for current flow and circuit operation stops. Without a complete path for current to flow, the load, such as a compressor motor, is not energized. Series connections are often used for devices that provide a safety function in a system. In this application, when any safety device is open, the load does not operate.

A *parallel circuit* is a circuit that has two or more components connected so that there is more than one path for current to flow. **See Figure 11-13.** Parallel connections are often used in applications that contain multiple switching devices that are able to operate the same electrical load. For example, some systems can use either a manual switch or a thermostat to cause a pump motor to run.

Parallel Circuits

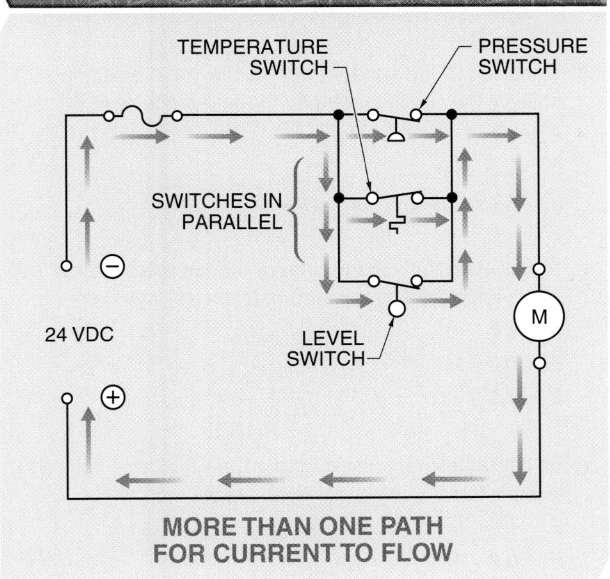

Figure 11-13. A parallel circuit can have as many components as required, and there are multiple flow paths for current to flow.

A *series-parallel circuit* is a combination of series- and parallel-connected components in a circuit. **See Figure 11-14.** Many refrigeration and HVAC system control circuits use series- and parallel-connected switches to control a load.

Load Voltage Drops in Series Circuits

In a series circuit, the voltage applied across individual loads is divided across the loads. The voltage drop across each load depends on the resistance of the load. The load with the lowest resistance has the lowest voltage drop, and the load with the highest resistance has the highest voltage drop. The sum of all voltage drops equals the supply voltage. This is because all of the voltage must be consumed by the load(s) of a circuit.

Series-Parallel Circuits

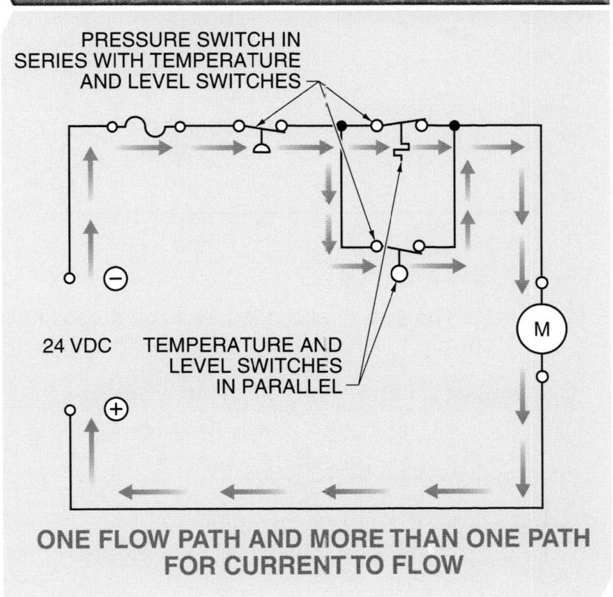

Figure 11-14. A series-parallel circuit is a combination of series- and parallel-circuit components and is found in most refrigeration and HVAC system control circuits.

TECH TIP

Due to wiring, the National Electrical Code® (NEC®) recommends no more than a 5% voltage drop between the electrical panel and an outlet. High current levels and long runs cause wire to have large voltage drops.

Example: Finding Load Voltage Drop in a Series Circuit

Find the voltage drop created by each load of a 24 VAC series circuit. The loads include a 40 Ω valve solenoid, a 25 Ω fan motor starter coil, and a 15 Ω indicator light. In an energized circuit, the voltage drop of the loads can be measured using a DMM. **See Figure 11-15.** The voltage drops created by each of the series connected loads can also be calculated by using Ohm's law. A procedure is used to find the voltage drops of the loads in a series circuit. **See Procedure for Calculating Load Voltage Drop in a Series Circuit.**

Measuring Voltage Drop of Loads in Series Circuits

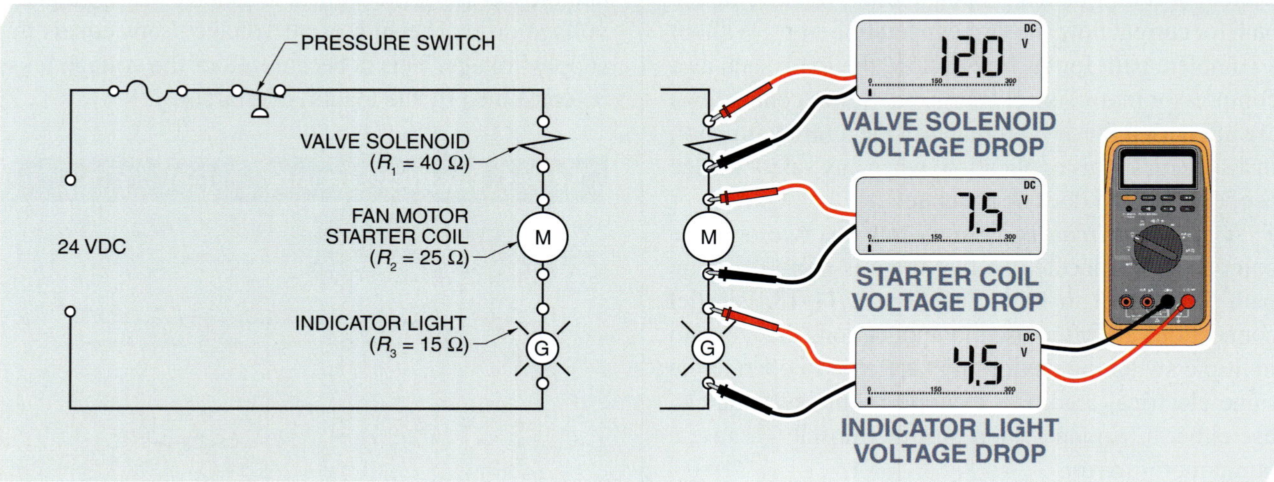

Figure 11-15. The sum of all the load voltage drops in a series circuit will equal the supply voltage to the circuit.

Calculating Load Voltage Drop in a Series Circuit

Procedure

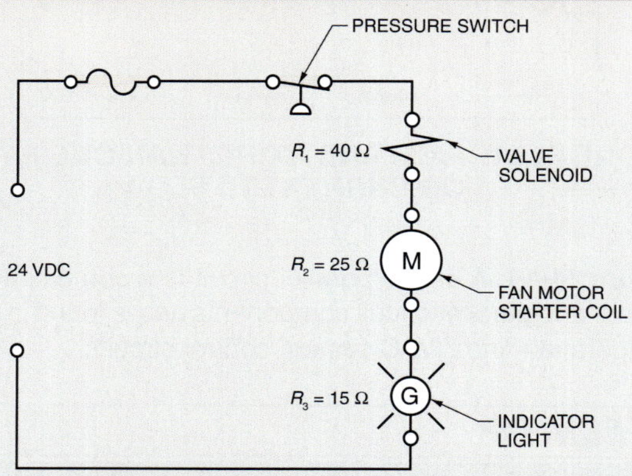

1. To calculate the total resistance in a series circuit, add all of the resistances together.

$R_T = R_1 + R_2 + R_3$

$R_T = 40 + 25 + 15$

$R_T = \mathbf{80\ \Omega}$

2. To calculate the current in a series circuit, divide the supply voltage by the total resistance.

$I = \frac{E}{R_T}$

$I = \frac{24}{80}$

$I = \mathbf{0.3\ A}$

3a. To calculate the voltage drop of the valve solenoid (R_1), multiply the circuit current by the resistance of R_1.

$E_1 = IR_1$

$E_1 = 0.3 \times 40$

$E_1 = \mathbf{12\ V}$

3b. To calculate the voltage drop of the fan motor starter coil (R_2), multiply the circuit current by the resistance of R_2.

$E_2 = IR_2$

$E_2 = 0.3 \times 25$

$E_2 = \mathbf{7.5\ V}$

3c. To calculate the voltage drop of the indicator light (R_3), multiply the circuit current by the resistance of R_3.

$E_3 = IR_3$

$E_3 = 0.3 \times 15$

$E_3 = \mathbf{4.5\ V}$

4. To calculate the total voltage drop created by the circuit loads, add the individual load voltage drops together.

$E_T = E_1 + E_2 + E_3$

$E_T = 12 + 7.5 + 4.5$

$E_T = \mathbf{24\ V}$

Load Voltage Drops in Parallel Circuits

The voltage drop across each load is the same when loads are connected in parallel. When loads are added or removed in parallel circuits, the voltage drop across each load remains the same. Parallel circuits are often used for control circuits where multiple controls have the ability to energize or de-energize a particular load. One part of a circuit may be energized even though a control has shut off power to another part.

Example: Finding Load Voltage Drop in a Parallel Circuit

Find the voltage drop created by each load of a 24 V parallel circuit. The circuit contains a 40 Ω valve solenoid, a 25 Ω fan motor starter coil, and a 15 Ω indicator light connected in parallel. In an energized circuit, the voltage drop of the loads can be measured by using a digital multimeter (DMM). **See Figure 11-16.** The voltage drops created by the parallel loads can also be calculated by using Ohm's law. A procedure is used to find the voltage drops of the loads in a parallel circuit. **See Procedure for Calculating Load Voltage Drop in a Parallel Circuit.**

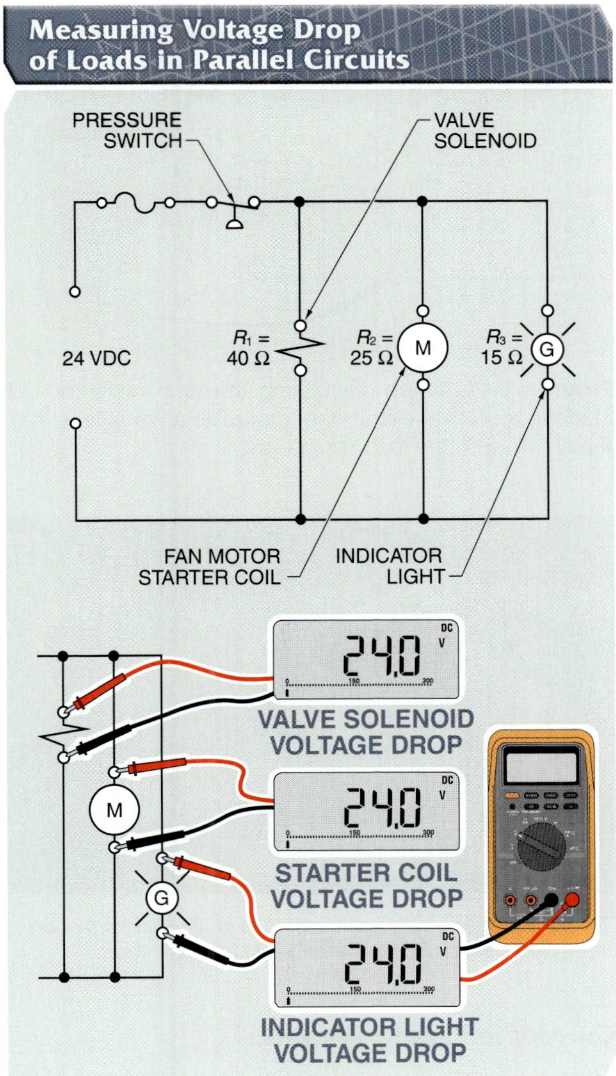

Figure 11-16. All loads in a parallel circuit will have the same standard voltage drop (supply voltage).

Calculating Load Voltage Drop in a Parallel Circuit

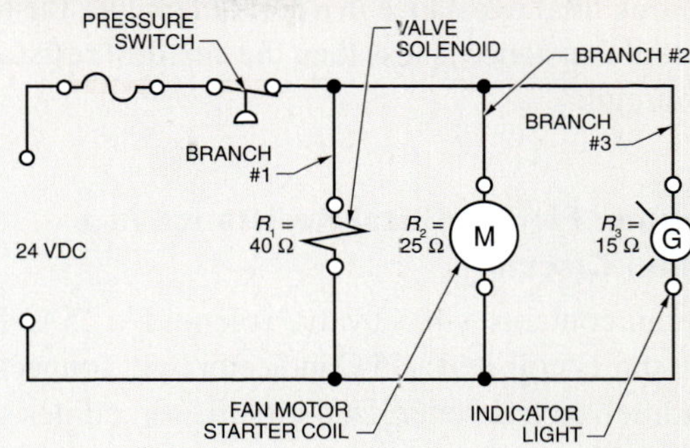

Procedure

1a. To calculate the current through Branch 1 (I_1), divide the supply voltage by the resistance of R_1.

$$I_1 = \frac{E}{R_1}$$
$$I_1 = \frac{24}{40}$$
$$I_1 = \mathbf{0.6\ A}$$

1b. To calculate the current through Branch 2 (I_2), divide the supply voltage by the resistance of R_2.

$$I_2 = \frac{E}{R_2}$$
$$I_2 = \frac{24}{25}$$
$$I_2 = \mathbf{0.96\ A}$$

1c. To calculate the current through Branch 3 (I_3), divide the supply voltage by the resistance of R_3.

$$I_3 = \frac{E}{R_3}$$
$$I_3 = \frac{24}{15}$$
$$I_3 = \mathbf{1.6\ A}$$

2a. To calculate the voltage drop of resistance R_1, multiply the branch current (I_1) by the resistance of R_1.

$$E_1 = I_1 \times R_1$$
$$E_1 = 0.6 \times 40$$
$$E_1 = \mathbf{24\ V}$$

2b. To calculate the voltage drop of resistance R_2, multiply the branch current (I_2) by the resistance of R_2.

$$E_2 = I_2 \times R_2$$
$$E_2 = 0.96 \times 25$$
$$E_2 = \mathbf{24\ V}$$

2c. To calculate the voltage drop of resistance R_3, multiply the branch current (I_3) by the resistance of R_3.

$$E_3 = I_3 \times R_3$$
$$E_3 = 1.6 \times 15$$
$$E_3 = \mathbf{24\ V}$$

Resistance in Parallel Circuits

The opposition to current flow is called resistance. When calculating total resistance in a parallel circuit, the total connected resistance is less than the smallest resistance in the circuit.

Example: Finding Total Resistance in a Parallel Circuit

A circuit contains a 40 Ω valve solenoid, a 25 Ω fan motor starter coil, and a 15 Ω indicator light connected in parallel. In a de-energized circuit, sometimes the total resistance of the loads in a parallel circuit can be measured by using a DMM. **See Figure 11-17.** The total resistance created by the parallel loads can also be calculated by using Ohm's law. A procedure is used to find the total resistance of the loads in a parallel circuit. **See Procedure for Calculating Total Resistance of Loads in a Parallel Circuit.**

Procedure

Calculating Total Resistance of Loads in a Parallel Circuit

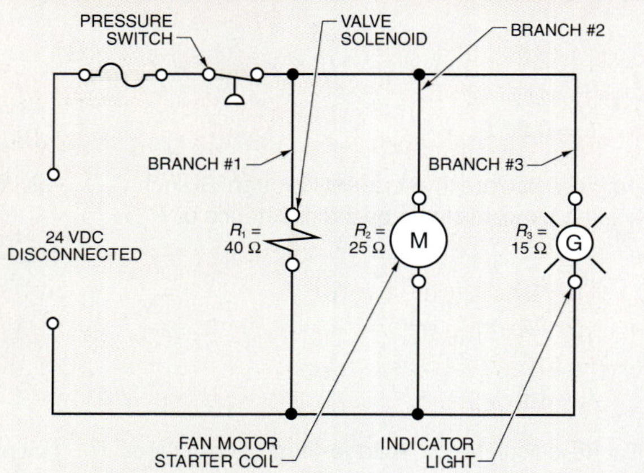

1. To calculate the total parallel circuit resistance, divide the number 1 by the sum of the resistance reciprocals of R_1, R_2, and R_3.

$$R_T = \frac{1}{\frac{1}{R_1} + \frac{1}{R_2} + \frac{1}{R_3}}$$

$$R_T = \frac{1}{\frac{1}{40} + \frac{1}{25} + \frac{1}{15}}$$

$$R_T = \frac{1}{0.025 + 0.04 + 0.067}$$

$$R_T = \frac{1}{0.132}$$

$$R_T = \mathbf{7.576}\ \Omega$$

Measuring Total Resistance of Loads in Parallel Circuits

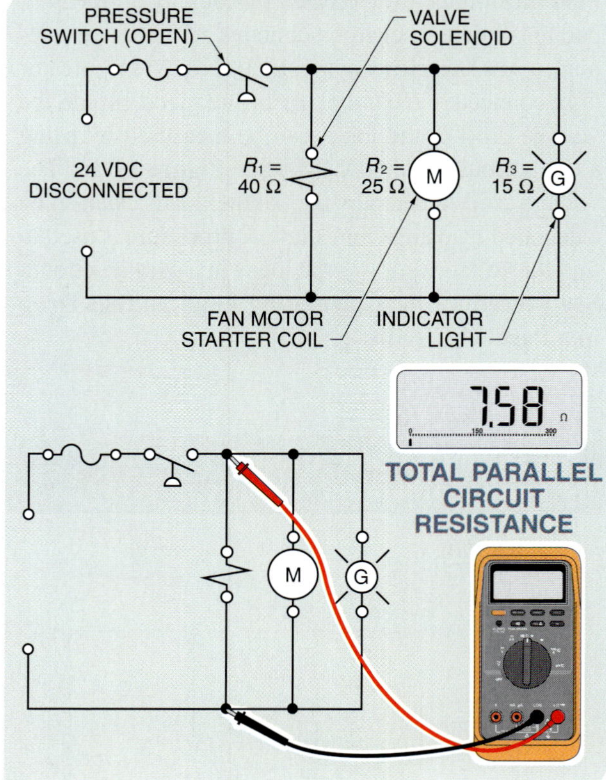

Figure 11-17. When calculating the total resistance of loads in a parallel circuit, the total resistance is less than the smallest single load resistance.

To take a proper resistance measurement, the component being measured must be isolated or removed from its circuit.

Current in Parallel Circuits

The total current flow in a parallel circuit is the sum of the current through all the loads. Total current increases when loads are added and decreases when loads are removed.

Example: Finding Total Current in a Parallel Circuit

A 24 V circuit contains a 40 Ω valve solenoid, a 25 Ω fan motor starter coil, and a 15 Ω indicator light connected in parallel. Find the total current supplying the parallel loads when the total circuit resistance is 7.576 Ω. In an energized circuit, the total current of a parallel circuit can be measured by using a DMM. **See Figure 11-18.** The total current of a parallel circuit can also be calculated by using Ohm's law. A procedure is used to find the total current of a parallel circuit. **See Procedure for Calculating Total Current in a Parallel Circuit.**

Measuring Total Current Flow in Parallel Circuits

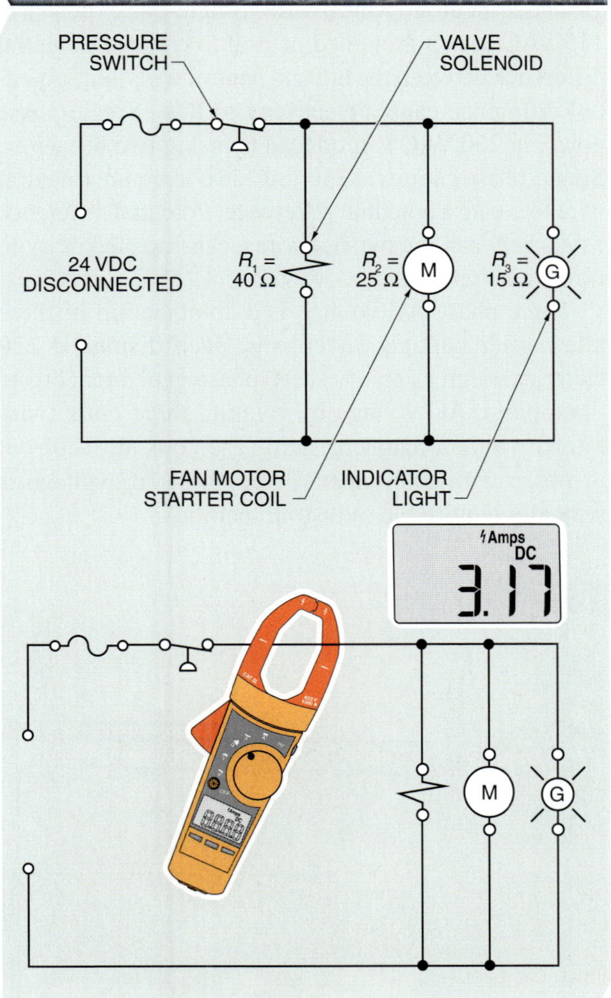

Figure 11-18. The total current flow in a parallel circuit is the sum of the current through each branch. Total current increases as loads are added and decreases as loads are removed.

Calculating Total Current in a Parallel Circuit

Procedure

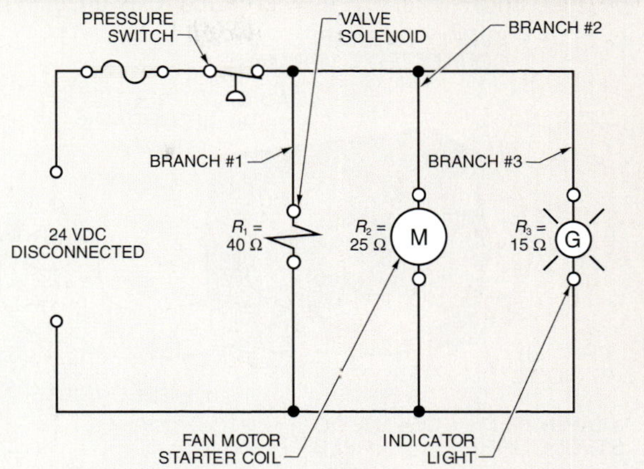

1. To calculate the total current in a parallel circuit, divide the supply voltage by the total circuit resistance.

$$I = \frac{E_S}{R_T}$$

$$I = \frac{24}{7.576}$$

$$I = \mathbf{3.17\ A}$$

Electricity Produced from Magnetism

When a conductor passes through a magnetic field, current is produced. Conversely, when current passes through a conductor, a magnetic field is produced. *Inductance* is the production of current flow due to a magnetic field on a coil.

Transformers use inductive flow to change a voltage level. A transformer consists of two sets of windings wrapped around an iron or laminated steel core. When current flows through one set of windings, the magnetic field induces flow in the other set. The first set of windings (supply) are known as the primary windings, and the second set is known as the secondary windings (load). The ratio of turns between the primary and secondary windings determines the voltage output of the transformer.

Transformers. Transformers are classified as step-up, step-down, or as isolation transformers. **See Figure 11-19.** A step-up transformer has a higher output voltage than input voltage, while a step-down transformer has a lower output voltage than input voltage.

An isolation transformer has the same number of primary and secondary turns in the windings. An isolation transformer is not used to change voltage but to electrically isolate the secondary circuit. This isolation reduces the chances of problems on the primary side of the circuit affecting the secondary side.

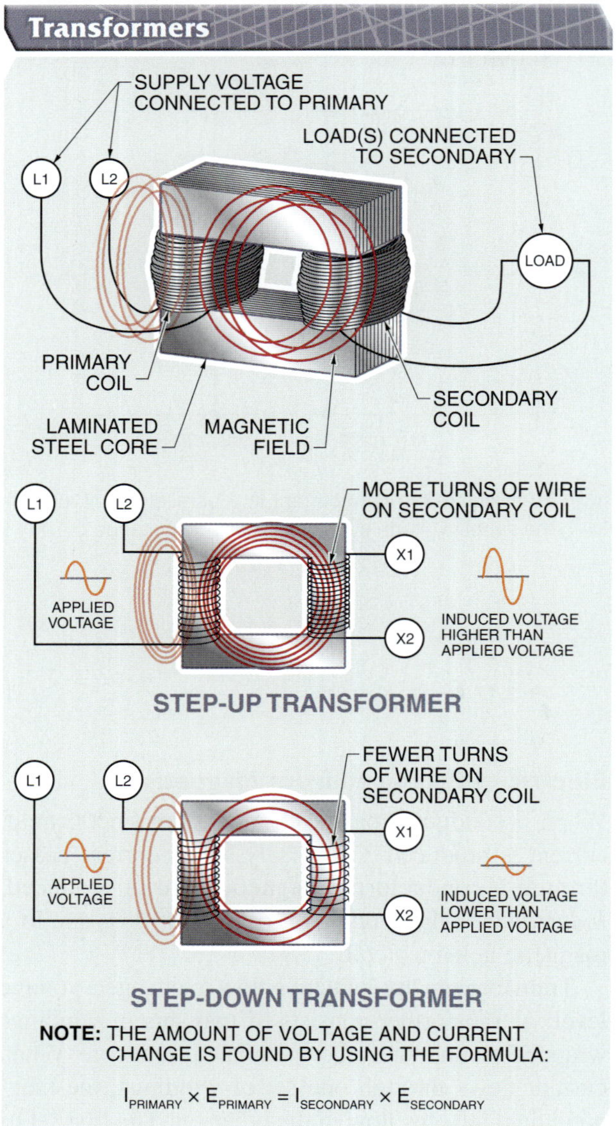

Figure 11-19. A transformer is an alternating current device that steps up or steps down voltage.

Step-up transformers are often used to increase voltage for use in spark ignition systems. They are also used to increase AC voltages for long-distance electrical transmission. Step-down transformers are often used in refrigeration and air-conditioning control systems to reduce line voltage to low voltage for safe use in the system.

GENERATION AND TRANSMISSION OF ELECTRICAL POWER

Electricity is produced by converting one form of energy into another form of energy. In almost all cases, potential energy from fossil fuels, water, and nuclear sources is converted into electrical energy. The potential energy is used to create steam that is then used to drive a turbine connected to a generator. In the generator, a wire coil (rotor) is turned inside a set of powerful fixed electromagnets (stator). **See Figure 11-20.** As the wires on the rotor cut the lines of magnetic flux, current flow is induced in the wires of the rotor. The voltage produced by this process depends on the strength of the magnetic field and the speed of the rotor.

Generating an AC Sine Wave Media Clip

AC voltage is either single phase (1ϕ) or three phase (3ϕ). **See Figure 11-21.** Single-phase AC voltage contains one alternating voltage waveform. Single-phase power at 115 VAC is produced with one hot wire and a grounded neutral (white) wire. Single-phase power at 115 VAC uses a grounded neutral to create a potential difference between the hot and neutral wire. This potential difference causes electricity to flow. Single-phase power at 230 VAC is produced by using two hot wires. Single-phase circuits at 230 VAC do not require a neutral wire to create a potential difference. Potential difference exists between the two hot wires, causing electricity to flow in the circuit.

Three-phase AC voltage is a combination of three alternating voltage waveforms, each displaced 120 electrical degrees apart. Three-phase generators create three-phase AC voltage by rotating three coils (wire loops) within a magnetic field. The coils are 120° out of phase with each other. Three-phase AC voltage is typically required in industrial facilities.

Three-phase (3ϕ) AC power is transmitted by high-voltage power lines and towers. It is then stepped down in switching yards by transformers for local use.

Three-Phase AC Generators

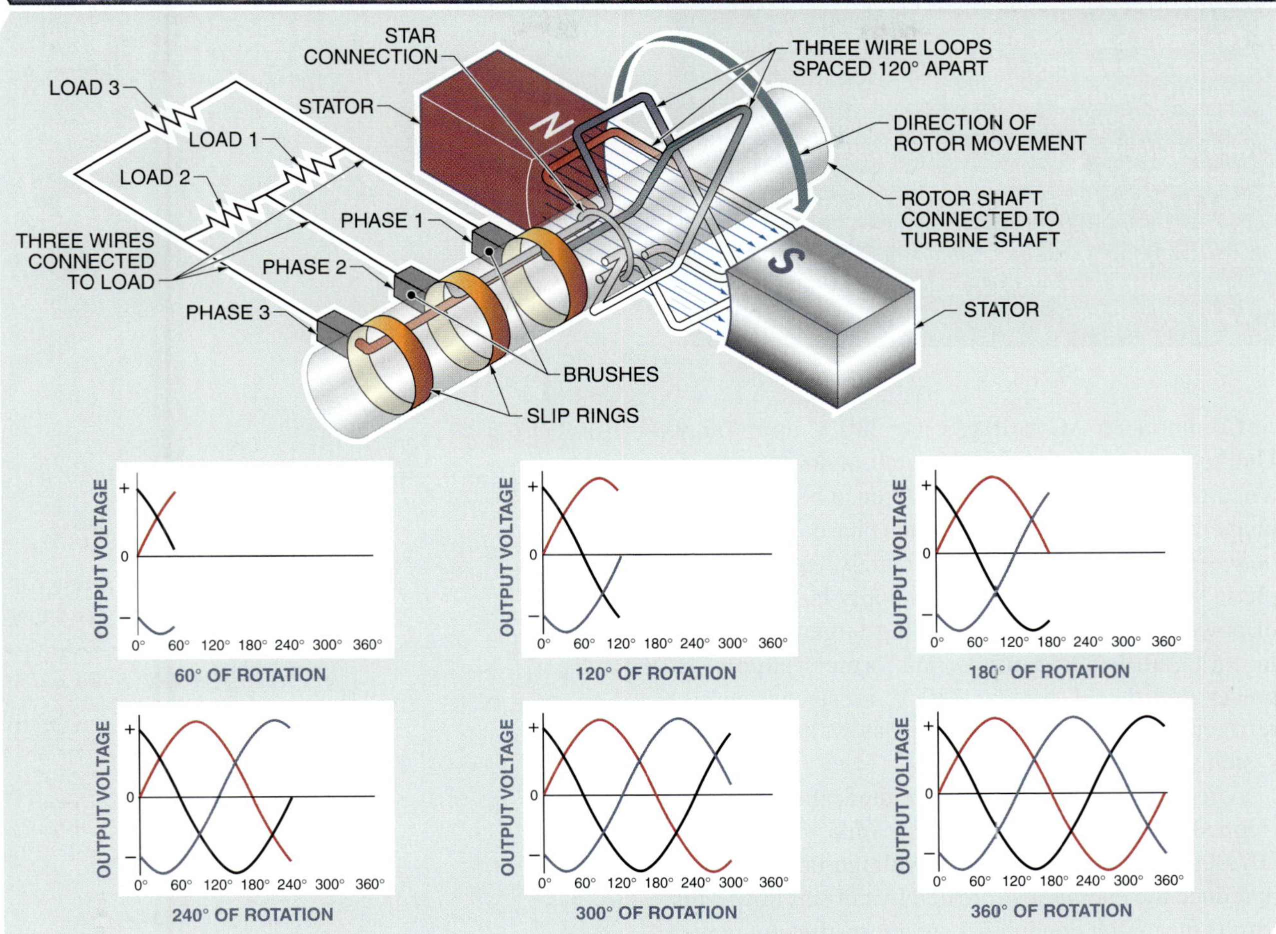

Figure 11-20. Three-phase generators create three-phase AC voltage by rotating three coils that are connected to a rotor within a magnetic field.

AC Voltages

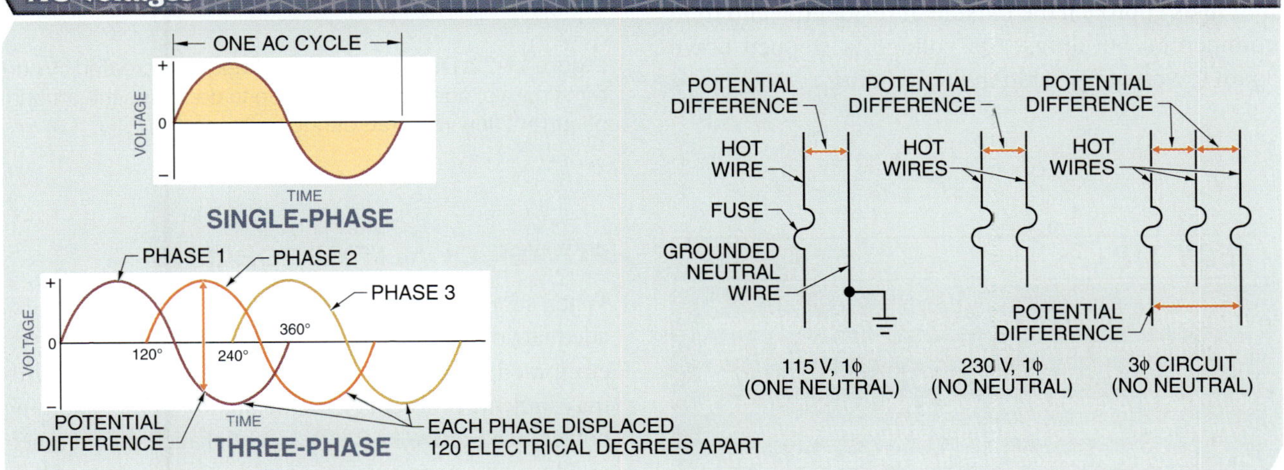

Figure 11-21. Single-phase power is used in residential and light commercial refrigeration and HVAC applications while three-phase power is used in heavy commercial and industrial applications.

TECH TIP

The NEC® does not require color-coded identification of 3ϕ conductors other than that of the neutral (white), ground (green), and the high leg of a delta system (orange). However, on diagrams and for system balancing purposes in the United States, the conductors of low-voltage systems are marked black for L1, red for L2, and blue for L3. The conductors of high-voltage systems (277/480 V) are marked brown for L1, orange for L2, and yellow for L3. In Canada, the standard is red for L1, black for L2, and blue for L3 for all voltages.

Common 3ϕ AC voltages are 230 V and 460 V. Unlike 1ϕ, 115 V circuits, 3ϕ circuits do not require a neutral wire. The potential difference between each phase causes current to flow in each phase. The voltage between each phase is approximately equal. Three-phase voltage is widely used to supply large amounts of power and is ideal for operating large compressor motors. While residential transformers supply 1ϕ power at either 115 V or 230 V, most commercial refrigeration and HVAC systems, as well as lighting systems, require 3ϕ power.

After electrical power is generated, the voltage is stepped up for long distance transmission to about 207,000 V. **See Figure 11-22.** Voltage is stepped up to reduce the energy loss created by current flow. This results in smaller conductors and more efficient power transmission.

As power lines approach the residential or light commercial point of use, utility substations step down the voltage from 207,000 V to 23,000 V. As distribution lines approach the outside of residential and light commercial buildings, the voltage is stepped down again to voltages usable in the building.

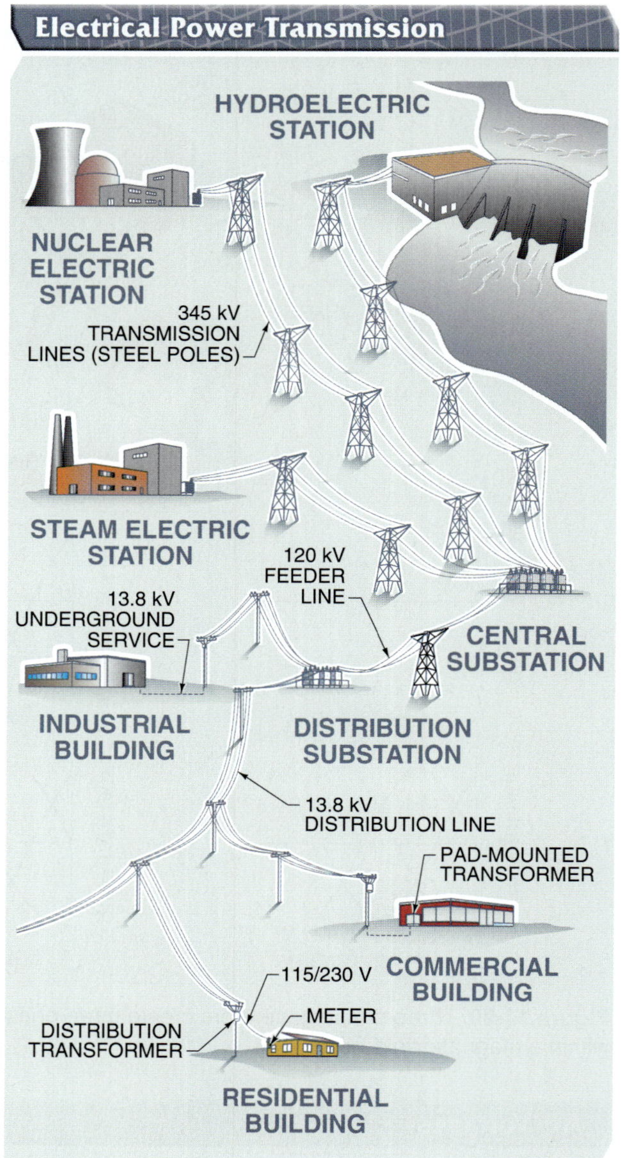

Figure 11-22. Due to the energy (heat) loss created by current flow, voltages are stepped up to decrease the amount of current flow required during transmission.

TECH TIP

The neutral conductor of a 3ϕ electrical system allows high-voltage systems to operate 1ϕ circuits. A 3-wire, 1ϕ system that has a single center-tapped transformer with two live conductors and a neutral conductor is typical for residential and light commercial buildings. A 3ϕ industrial system that has no neutral conductor uses three live conductors to connect the loads phase to phase.

POWER QUALITY PROBLEMS

When serviced properly, electrical equipment provides adequate operation during its normal lifetime. However, equipment that has continual problems and failures may indicate problems with the quality of the incoming power. **See Figure 11-23.** Residential and light commercial building power quality problems vary but include voltage interruptions, voltage fluctuations, transient voltages, harmonics, and noise.

Voltage interruptions may be momentary, temporary, or sustained. Interruptions are often caused by storms, circuit breakers performing their function, or damaged equipment. Voltage fluctuations vary from small changes of less than 10% for normal line voltage to voltage sags and swells of more than 10% that can last up to a minute. Undervoltage and overvoltage are conditions where a 10% or greater change from normal voltage lasts more than one minute.

A *transient voltage* is a temporary, unwanted voltage that ranges from a few volts to several thousand volts and lasts from a few microseconds to a few milliseconds. **See Figure 11-24.** Transient voltages are often caused by lightning or stored voltages being released from capacitors. Transient voltages are often called transients.

Harmonics are power frequencies that are whole integer multiples of the base frequency (60 Hz). Harmonics may cause overheating of motors and power distribution wiring. Overheating of a neutral conductor is especially dangerous. Harmonic distortion often occurs when using motor drives, electronic lighting ballasts, personal computers, or medical equipment.

Noise is sine wave distortion from motors and electrical devices that enters a power distribution system directly on the wires/grounds or through the magnetic coupling of adjacent wires. Noise in a system produces false signals in electrical circuits. This leads to PLC errors and incorrect data being transferred to devices. All motor power problems can be monitored and checked by trained personnel with the use of high-quality diagnostic equipment.

Kebco Power Transmission

Power quality problems, such as transients and harmonics, can cause devastation to the power supply sections of HVAC devices.

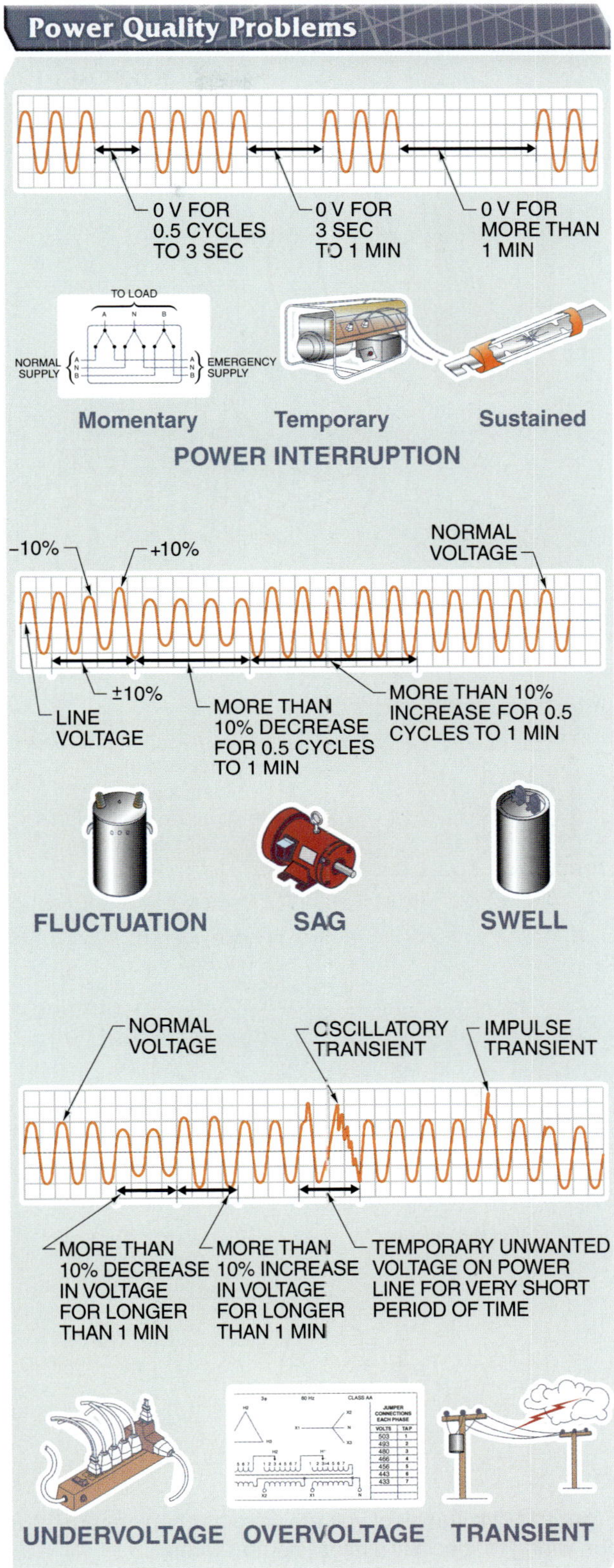

Figure 11-23. Residential and light commercial power quality problems vary but include voltage interruptions, voltage fluctuations, transient voltages, harmonics, and noise.

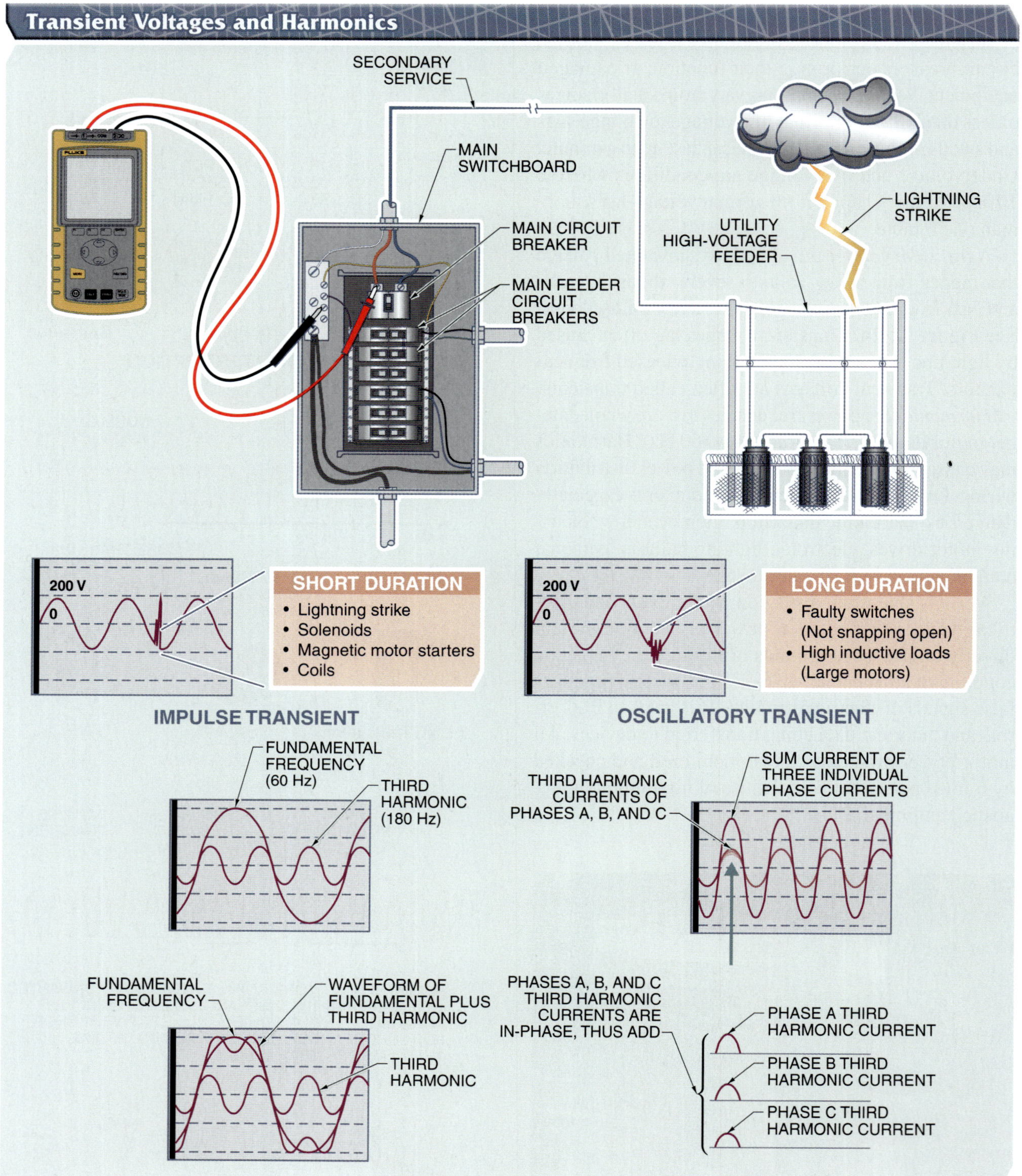

Figure 11-24. Transient voltages and harmonics cause overheating of motors, solid-state components, and power distribution wiring (especially system neutral wires).

UNIT 12

Electrical and Electronic Control Devices

To be effective in the field, an HVAC technician must be knowledgeable of electrical and electronic control systems. A large percentage of the problems found with refrigeration and HVAC mechanical systems concerns control systems that are not operating correctly.

OBJECTIVES

- List the devices used in electrical control systems.
- Identify the purpose of capacitors.
- Explain low-voltage, line-voltage, and high-voltage AC voltage supplies.
- Describe features of disconnects, overcurrent protective devices, relays, contactors, and motor starters.
- Explain the basic function of electronic control systems.
- Describe the process of semiconductor doping.
- Describe the use of diodes.
- Explain the operation of transistors and thyristors.
- Describe integrated circuits.
- Identify common electronic control systems applications.

Index to Procedures

ELECTRICAL CONTROL SYSTEMS

Electrical control systems use AC voltage. Most residential package units, such as furnaces, heat pumps, and split system air conditioning, fall into this category of control. **See Figure 12-1.** The two most common voltages for electrical control systems are 24 VAC and 115 VAC.

ELECTRICAL CONTROL SYSTEM DEVICES

Electrical control systems consist of a number of devices such as capacitors, power supplies, disconnects, overcurrent protection, relays, contactors, and motor starters. These devices are used in combination to cycle refrigeration and HVAC equipment in response to a thermostat or other controller.

The large control panels for light commercial HVAC rooftop units include starters for multiple motors and relays and receive input from multiple system sensors and switches.

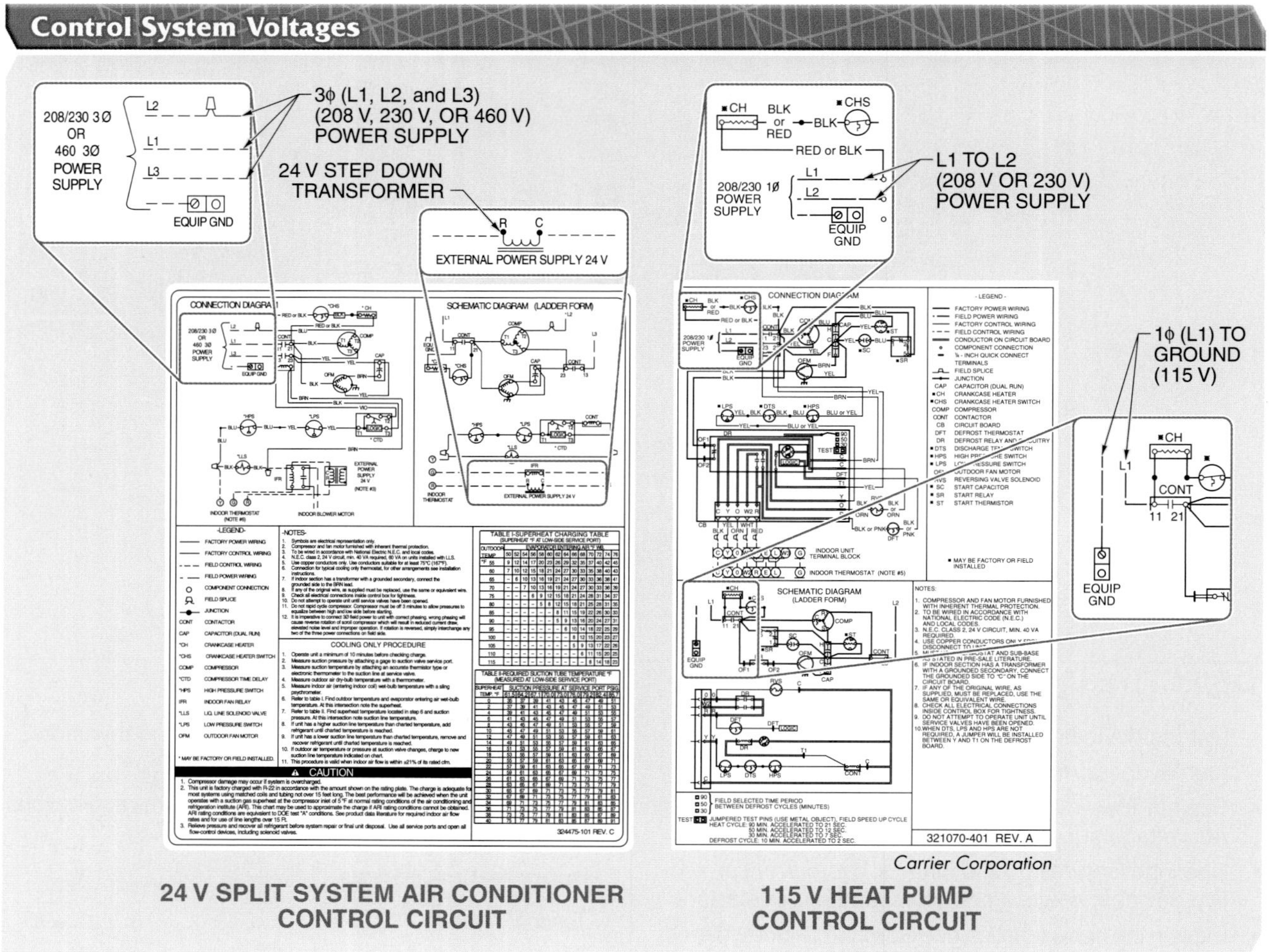

Figure 12-1. Refrigeration and HVAC systems use 24 V or 115 V as control circuit voltages.

Capacitors

A *capacitor* is an electrical device that consists of two thin metal plates with an insulating dielectric material separating them. When a charge is placed on one of the plates, electrons on the other plate are forced to flow through a conductor attached to that plate. **See Figure 12-2.** Capacitors are often used to increase the starting torque of motors. Capacitor banks are also used to counteract the effect of large inductive loads on the power supply of a facility. Capacitors are rated in farads (F) or microfarads (μF). It is important to use the correct capacitor rating to match the application. Small capacitors can be checked with a digital multimeter (DMM) that includes a capacitor check function.

Power Supplies

The power supply for a refrigeration HVAC system controls is typically located between the building power supply and the refrigeration and HVAC equipment. Many electrical control systems for refrigeration and HVAC systems operate at 24 VAC. Twenty-four volts is used because loads that operate at low voltage are much safer for technicians to work on. A transformer is used to step down the line voltage or high voltage to the required 24 VAC for use in refrigeration and HVAC control system circuits.

AC Voltage Supplies

AC voltage is constantly changing in polarity and may change in magnitude. AC voltage supplied to a refrigeration or HVAC system may be at a low voltage, line voltage, or high voltage.

Low Voltage. In the United States, *low voltage* is voltage of 50 VAC or less (typically 24 VAC). Low-voltage control systems are typically used in refrigeration and HVAC control systems. **See Figure 12-3.** At the low-voltage level, there is little danger of accidental shock or fire. Low-voltage control system conductors can be relatively small and do not require metal conduit. In a low-voltage control system, the low voltage is used to open and close relay contacts that switch line- or high-voltage loads in the system ON and OFF.

Capacitor Operation

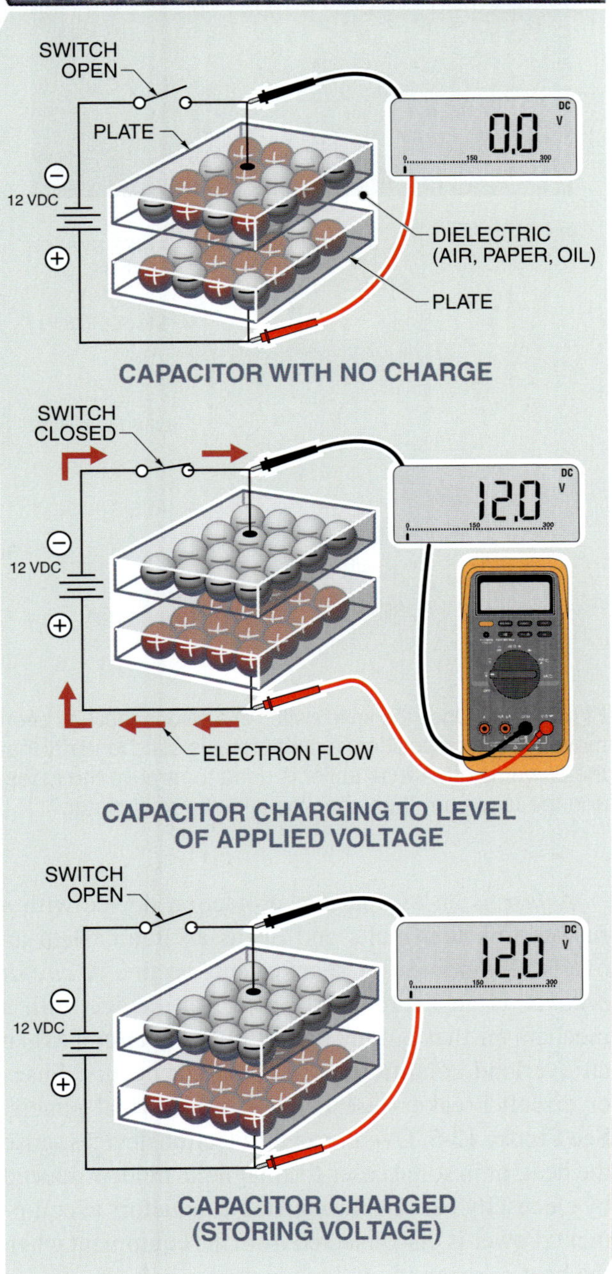

Figure 12-2. Capacitors allow an electrical circuit the ability to store energy in the form of an electrical charge.

Low-Voltage Power Supplies

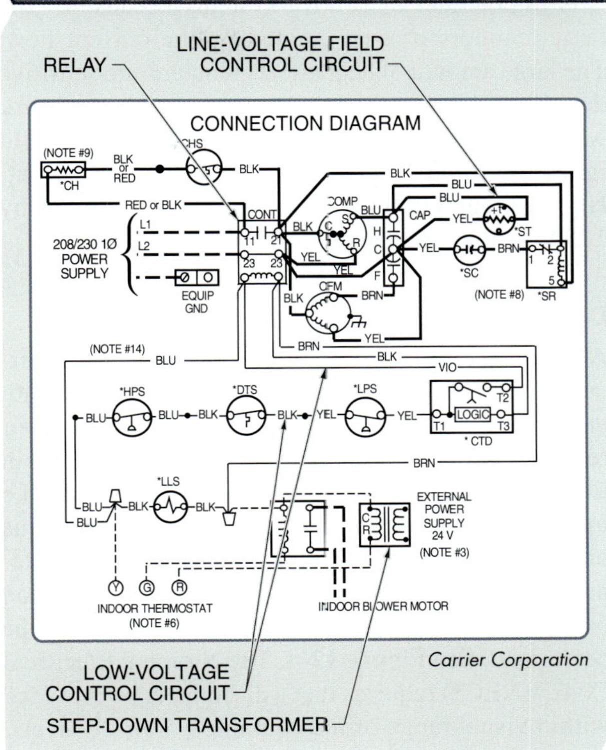

Figure 12-3. Low voltage (typically 24 VAC) is used for most control circuits in residential and light commercial refrigeration and HVAC systems.

Line Voltage. In the United States, *line voltage* is voltage at 115 VAC to 600 VAC. The most common line-voltage level for control systems is 115 VAC. Line-voltage control systems are used in some light commercial systems, but are typically used in heavy commercial and industrial areas and systems where access can be restricted to authorized personnel. Specific measures are taken to reduce the possibility of shock and fire in heavy commercial and industrial applications.

High Voltage. In the United States, *high voltage* is voltage over 600 VAC. High voltage is used in light commercial, heavy commercial, and industrial applications. In a high-voltage system, power is supplied from a utility at a high voltage level to power fan motors, compressors, and pumps. The high voltage allows refrigeration and HVAC equipment to come up to speed rapidly and generate large amounts of starting torque for large loads. High voltage is always supplied as 3ϕ power.

While high voltage is needed to start large loads, it is impractical in control systems because of the large amounts of heat created by the current flow. The large amount of heat affects temperature sensitive devices. High voltage levels may also cause a spark when contacts are opened or closed, which could ignite flammable substances in the environment. High-voltage equipment is typically controlled by 115 V or low-voltage control circuits.

Disconnects

A *disconnect* is a switch that isolates electrical circuits from a voltage source to allow safe access into an electrical panel or equipment for maintenance or repair. All refrigeration and HVAC systems contain a disconnect. Disconnects usually have a lever on the upper right-hand side of a disconnect panel that opens or closes electrical contacts. The lever enables maintenance personnel to shut off the power to refrigeration or HVAC equipment so that maintenance tasks can be performed. **See Figure 12-4.** The National Electrical Code® (NEC®) requires that a disconnect be provided within visual range of the equipment it disconnects.

The National Electrical Code® (NEC®) requires that a disconnect be within visual range of the HVAC equipment it disconnects.

Disconnects

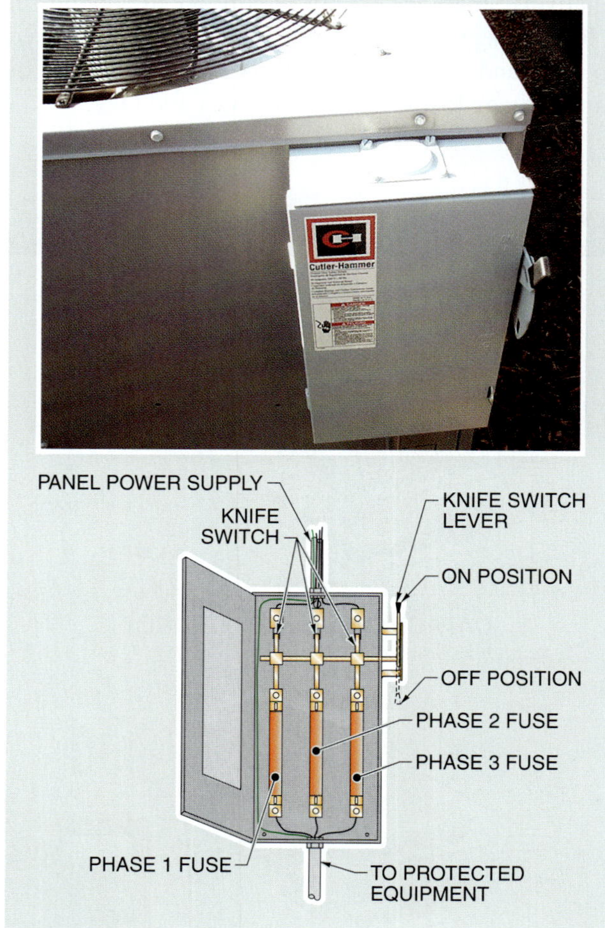

Figure 12-4. Once the knife switch of a disconnect has been moved to the off position, a multimeter is used to verify that the power is off. A fuse puller is used to remove the fuses, and the fuses are then checked with the multimeter.

Overcurrent Protection Devices

All electrical equipment must be protected from excessive current flow. An *overcurrent protection device* is a device that shuts off the power supply to a circuit when current flow is excessive. Overcurrent protection devices include fuses and circuit breakers.

A *fuse* is an overcurrent protection device with a fusible link that melts and opens a circuit when an overload condition or short circuit occurs. A *circuit breaker* is an overcurrent protection device with a mechanism that automatically opens a circuit when an overload condition or short circuit occurs. Fuses or circuit breakers are located in electrical panels. **See Figure 12-5.** Overcurrent protection devices sense the heat, or in some cases the magnetic field, produced by electricity passing through the conductors to equipment. Power is disconnected from the equipment when the heat or magnetic field is excessively high.

Internal motor overload protection can also be provided. To accomplish internal motor overload protection, a bimetallic thermodisc is embedded inside motor windings where it directly senses the temperature. **See Figure 12-6.** The bimetallic thermodisc warps and opens the motor starter coil circuit when the motor is overloaded.

Overcurrent Protection Devices

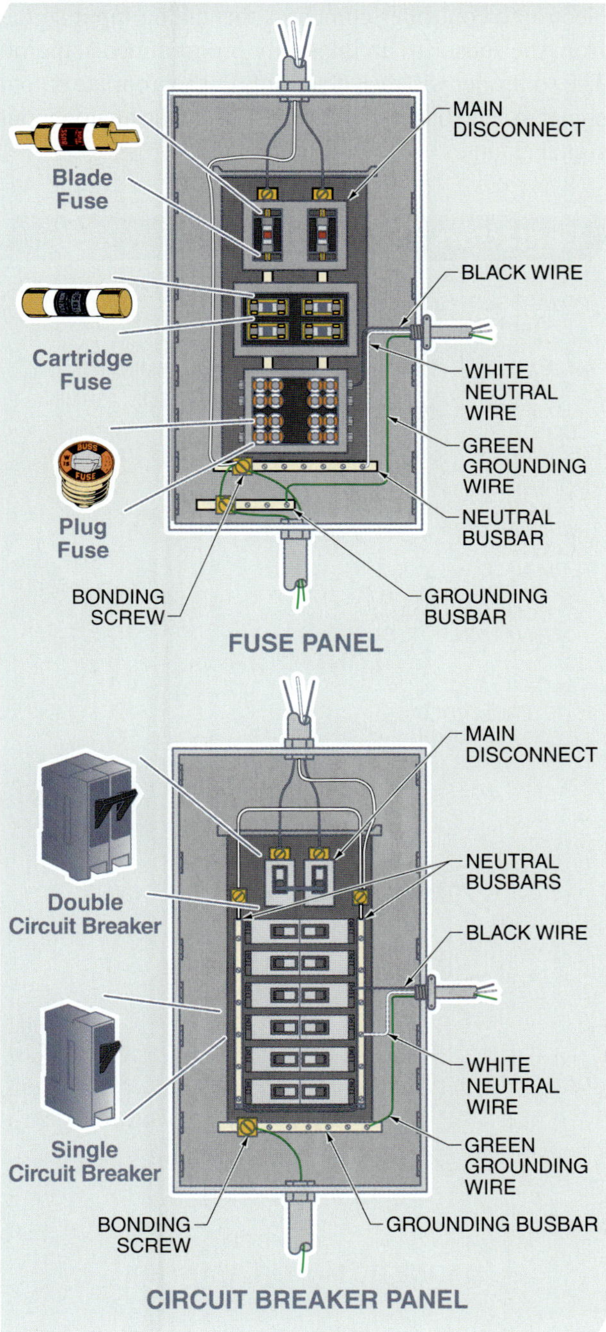

Figure 12-5. Overcurrent protection devices include current and voltage ratings.

Internal Motor Overload Protection

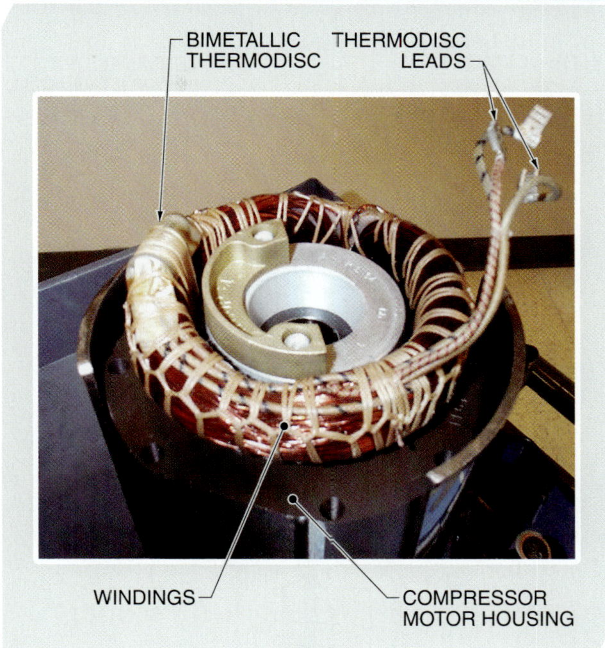

Figure 12-6. Internal thermal overload circuits in motors use semiconductor temperature switches for protection. These internal overload switches, depending on design, can have a reset button on the outside of the motor housing.

Relays

A *relay* is a device that uses a low voltage coil to close one or more sets of contacts to switch high-voltage loads on and off. Relays may be solid-state or electromechanical. A *solid-state relay* is a switching device that uses semiconductor materials to change on and off characteristics. An *electromechanical relay* is a switching device that has sets of contacts that are closed by a magnetic force. An electromechanical relay consists of a coil, fixed contact, and movable contact. **See Figure 12-7.** When current passes through the coil (low voltage), a magnetic field is developed that pulls the movable contact to touch the stationary contact. This closes the load circuit (high voltage), turning on the device. Relays can consist of a variety of contact configurations for use in various applications.

Contactors and Motor Starters

A *contactor* is a large heavy-duty, relay-like switching device that uses low voltage to energize or de-energize the high-current, nonmotor load connected to it. **See Figure 12-8.** Contactors are used in applications that have high current flow requirements such as with lighting or heating elements. Contactors use the same principle as a relay but in a heavy-duty package.

Relays

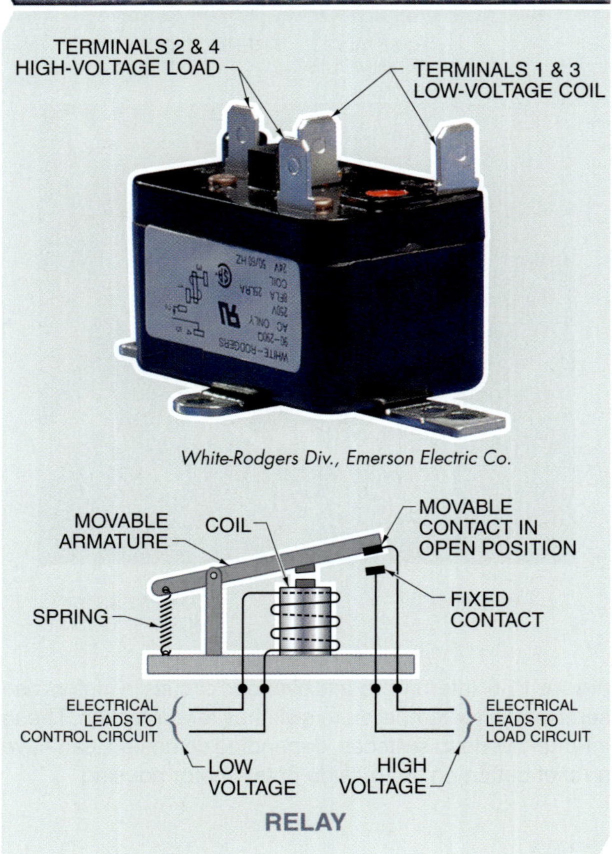

Figure 12-7. The two basic types of relays are electromechanical and solid-state relays. Electromechanical relays use magnetic force to close contacts, while solid-state relays use electronic switching.

A *motor starter* is a heavy-duty switching device that includes motor overload protection and typically uses a low voltage coil to energize or de-energize the high-voltage motor connected to it. The overloads in a motor starter open the thermal overload circuit when high or excess temperatures are being generated, due to a motor's high current draw.

Wiring Electrical Systems Media Clip

ELECTRONIC CONTROL SYSTEMS

Electronic control systems have similar power supplies to electrical control systems that use 30 VDC or less. The solid-state devices used in electronic control systems are also used to construct automated control system controllers.

An *electronic control system* is a type of control system that uses solid-state devices to control refrigeration and HVAC systems. **See Figure 12-9.** Electronic control systems use analog sensors to measure the temperature, humidity, or pressure in refrigeration and HVAC systems. These variables are represented by resistance values of 0 Ω to over 1000 Ω, voltage values of 0 VDC to 10 VDC, or current values of 0 mA to 20 mA. The electronic controller compares the analog input signal from the sensor to an internally programmed setpoint. The controller will then determine the appropriate output signal to send based on the amount that the analog input signal changes.

Contactors and Motor Starters

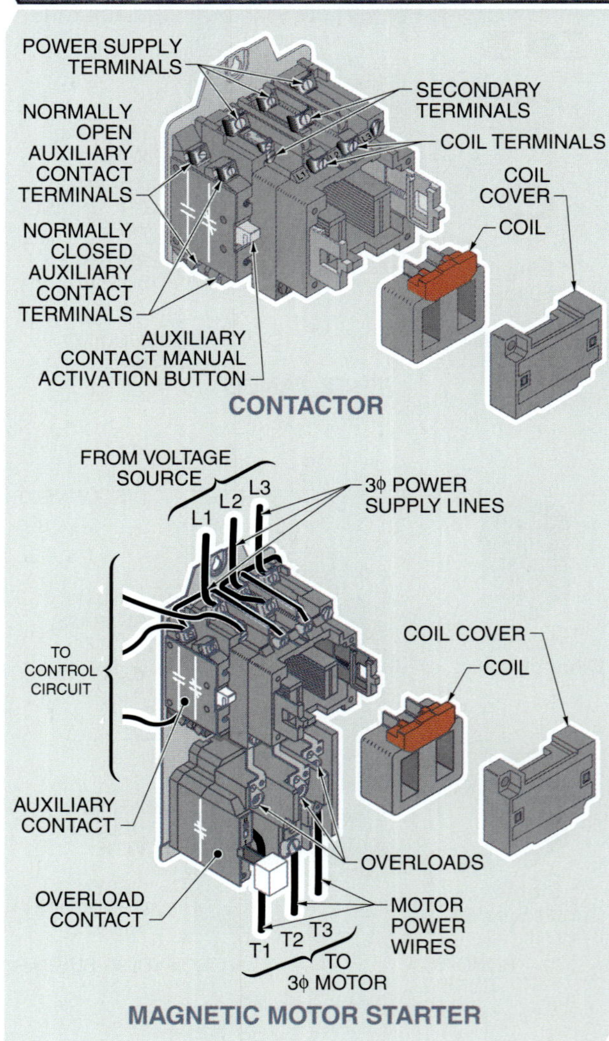

Figure 12-8. Contactors are used in high current draw applications such as with banks of lights, while motor starters are used exclusively with 1ϕ and 3ϕ motors.

Electronic Control Systems

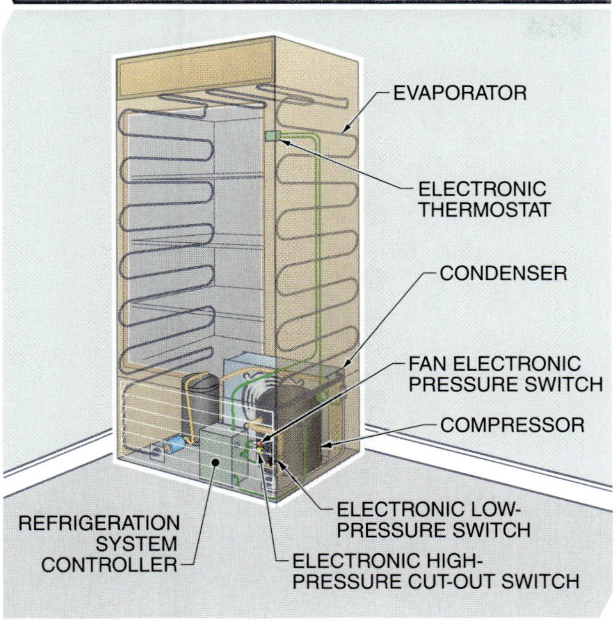

Figure 12-9. Electronic control systems use solid-state devices that measure temperature, humidity, and pressure and a controller to control refrigeration and HVAC equipment.

The output signals of an electronic controller are sent to motor- or solenoid-driven dampers and valve actuators or to electric motor drives. Electronic control systems may also include switching, averaging, or other electronic controls that complete the sequence of operation for the specific refrigeration and HVAC equipment.

Electronic control systems are often referred to as analog electronic systems because they provide variable (analog) switching. With the advent of digital automated systems, analog electronic control systems have lost some of their popularity. Modern electronic devices use analog control systems as well as digital automated systems.

SEMICONDUCTORS

A *semiconductor* is a material in which the electrical conductivity is between that of a conductor (high conductivity) and that of an insulator (low conductivity). **See Figure 12-10.** Applying voltages to semiconductor materials causes the materials to change electrical characteristics.

Semiconductors use small amounts of power and produce little heat. The small amount of heat produced increases the reliability of the devices. The systems used for device cooling can also be reduced or eliminated from the system. Semiconductors (solid-state devices) are continually being manufactured in smaller and smaller packages, allowing increased capability in a smaller housing.

Semiconductors

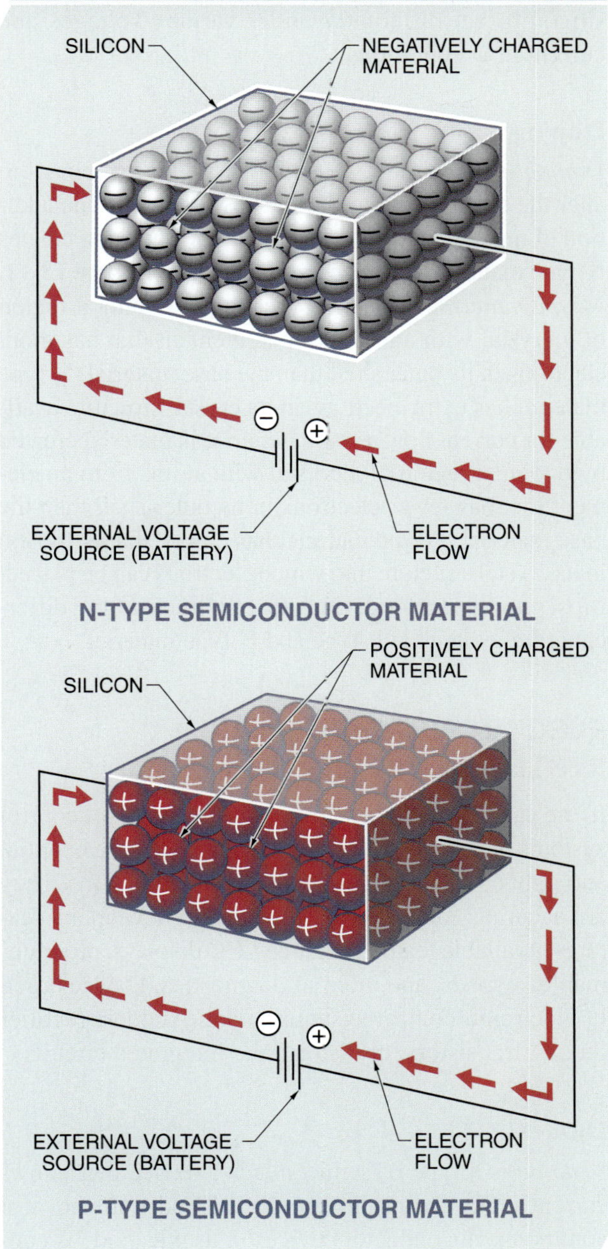

Figure 12-10. The most common material used for semiconductor devices is silicon, which has electrical conductivity between that of a conductor (high conductivity) and that of an insulator (low conductivity).

The most common material used for a semiconductor device is silicon. Silicon is one of the most common naturally occurring elements, which makes devices made from silicon inexpensive. The main costs involved in the manufacturing of semiconductors are the engineering and design of the solid-state device, not in the raw materials. Other elements and/or materials are added to silicon to create semiconductor materials with various characteristics. The various materials cause the silicon to act differently under various voltages and current flow conditions.

Doping

Doping is the addition of material to a base element to alter the crystal structure of the base element. The addition of material to the crystal structure of silicon creates N-type materials or P-type materials. **See Figure 12-11.** *N-type material* is material created by doping a region of a crystal with atoms from an element that has more electrons in its outer shell than the base material. N-type material has extra electrons in its crystal structure available for current flow. *P-type material* is material created by doping a region of a crystal with atoms from an element that has fewer electrons in its outer shell than the base material. P-type material has empty spaces (holes) in its crystal structure into which electrons can be placed. Most semiconductor devices are constructed from different arrangements of P-type and N-type material.

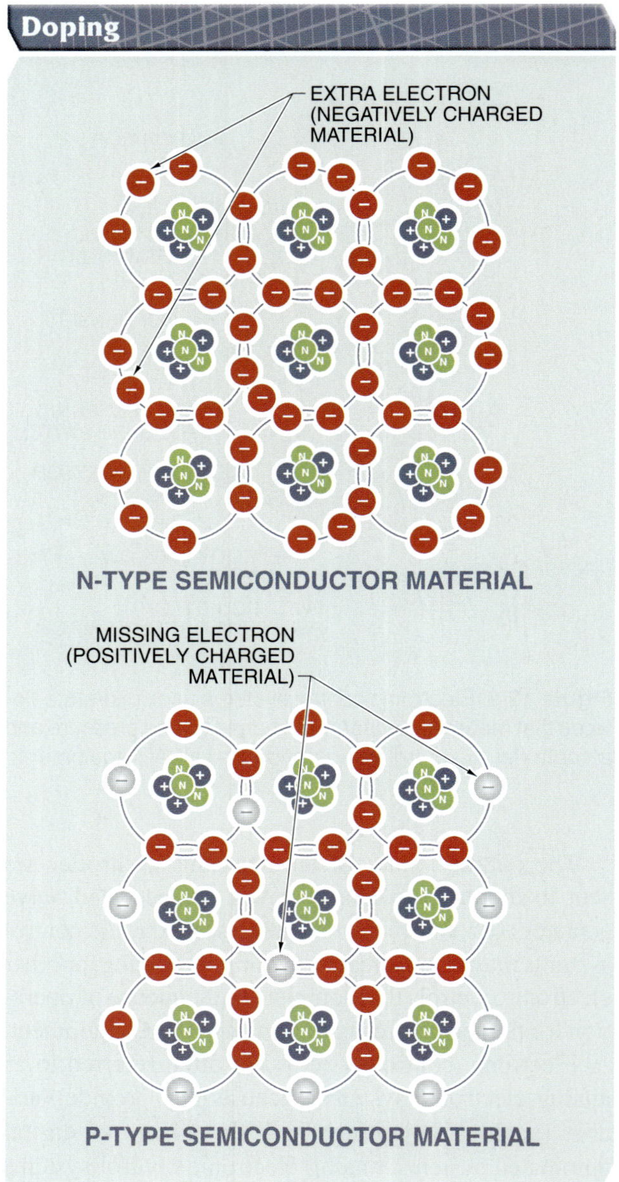

Figure 12-11. Semiconductor devices are constructed from different arrangements of P-type and N-type materials created by the doping process.

ELECTRONIC CONTROL SYSTEM DEVICES

In the past few years, electronic microprocessor control systems have become very common in both residential and light commercial HVAC system applications. They are accurate, inexpensive, rugged, and have many features. Available features include LCD displays, programmable keypads, and internal diagnostics. Devices used in electronic control systems include diodes, rectifier diodes, transistors, thyristors, and integrated circuits.

Diodes

A *diode* is a type of semiconductor device that allows current to flow in one direction only. Diodes are the most common semiconductor devices in use. A diode is constructed by placing one layer of P-type material and one layer of N-type material back to back. The area where the materials come into contact with each other is referred to as the PN junction (PN region). **See Figure 12-12.**

TECH TIP

When doping, either phosphorous or arsenic is added to silicon in small amounts to create N-type semiconductor material. Phosphorous and arsenic each have five electrons in their outer shell. The fifth electron is free to move and create electrical flow. Either boron or gallium is added to silicon to create P-type semiconductor material. Boron and gallium have an absence of electrons, which create holes. An excess of holes also allows electrons to flow.

Diodes

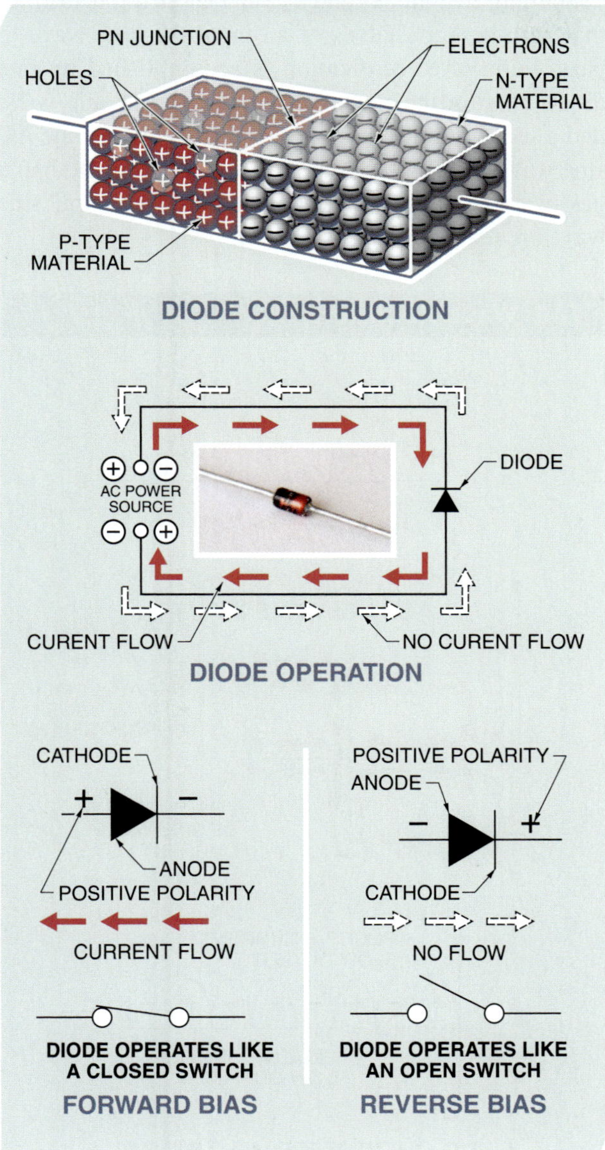

Figure 12-12. Diodes are semiconductor devices that allow current to flow in only one direction.

Diode Operation Media Clip

The P-type material of a diode has holes that are available to receive electrons, and the N-type material has extra electrons available. When a voltage source of 0.6 VDC or more is connected to the N-type material, the extra electrons of the N-type material jump across the PN junction (current flow) and fill the holes of the P-type material. The 0.6 VDC is used to forward bias the diode (allow current flow). **See Procedure for Testing Diodes.**

Testing Diodes

Procedure

1. To test a diode using an analog multimeter, disconnect power from the circuit, and isolate the diode by removing it from the circuit.

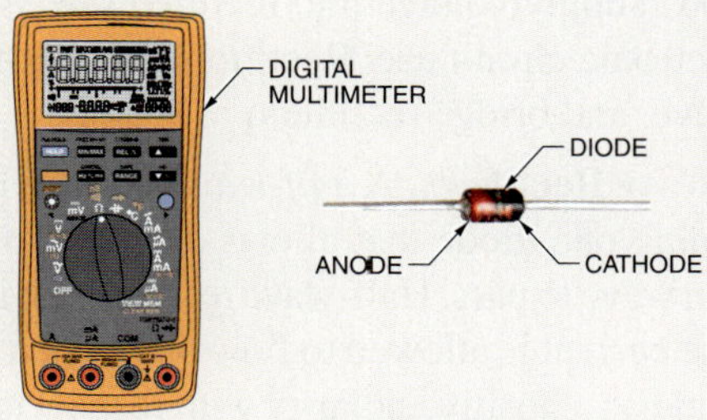

2. Connect the test leads to the appropriate analog multimeter jacks for taking resistance measurements.
3. Follow all electrical safety practices and procedures.
4. Set the analog multimeter range switch to measure 10 Ω or 100 Ω of resistance (R × 1 or R × 100). Set the function switch to positive DC voltage (+DC).
5. Zero out the meter by holding the test leads together and rotating the zero ohms switch until the meter reads 0 Ω.
6. Connect the red (+) test lead to the anode and the black (–) test lead to the cathode of the diode. The diode should conduct and the analog meter should display a low resistance value (the exact value is not critical).

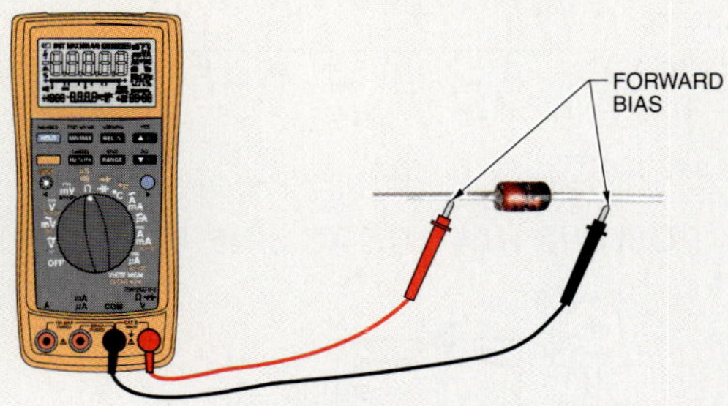

7. Reverse the test lead connections. Connect the black (–) test lead to the anode and the red (+) test lead to the cathode. The diode should not conduct (infinite resistance).

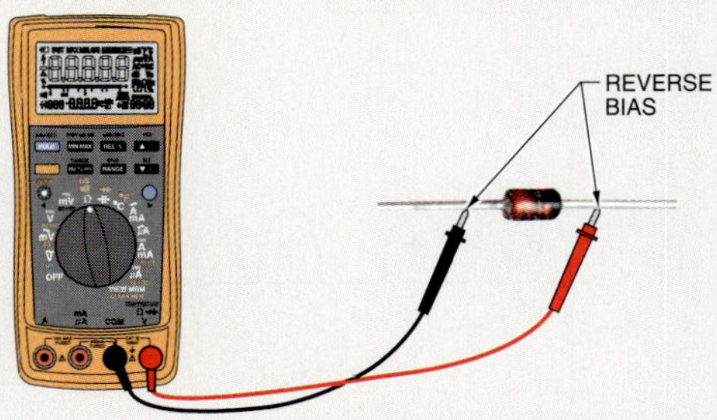

8. When in forward bias, the diode has low resistance. When in reverse bias, the diode has infinite resistance (the diode is good). If the diode has low or zero resistance in both directions, the diode is shorted. If the diode has infinite resistance in both directions, the diode is open.

Rectifier Diodes

Diodes are used as electronic check valves in rectifiers. A *rectifier* is a device that changes AC voltage into DC voltage. Rectifiers are used in power supplies to change 115 VAC supply voltage into rectified low-voltage DC for electronic circuit use. Rectifiers include half-wave, full-wave, and bridge rectifiers.

Half-Wave Rectifiers. A *half-wave rectifier* is a circuit containing one diode that allows only half of an input AC sine wave to pass. Half-wave rectification is possible because current is allowed to flow only when the anode terminal has a positive polarity with respect to the cathode. Current is not allowed to flow through the rectifier when the cathode has a positive polarity with respect to the anode. Half-wave rectification is inefficient for most applications because half of the input sine wave is not used. **See Figure 12-13.**

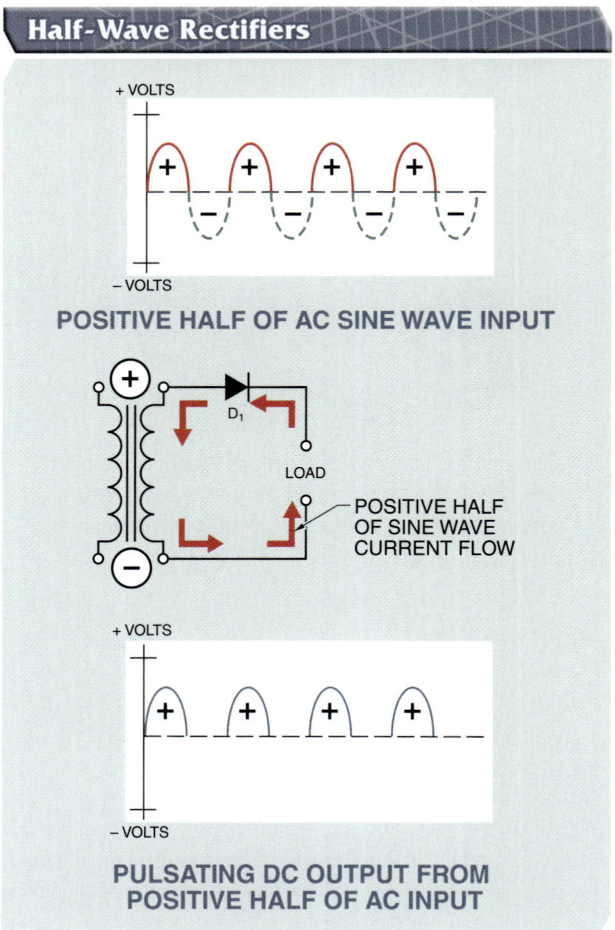

Figure 12-13. Half-wave rectification is obtained with one diode and converts AC voltage into pulsating DC voltage.

Full-Wave Rectifiers. A *full-wave rectifier* is a circuit containing two diodes and a center-tapped transformer that permits both halves of an input AC sine wave to pass. Full-wave rectification is accomplished by one diode passing the positive half of the AC sine wave and a second diode passing the negative half of the AC sine wave. A full-wave rectifier is more efficient than a half-wave rectifier because both halves of the input sine wave are used. **See Figure 12-14.**

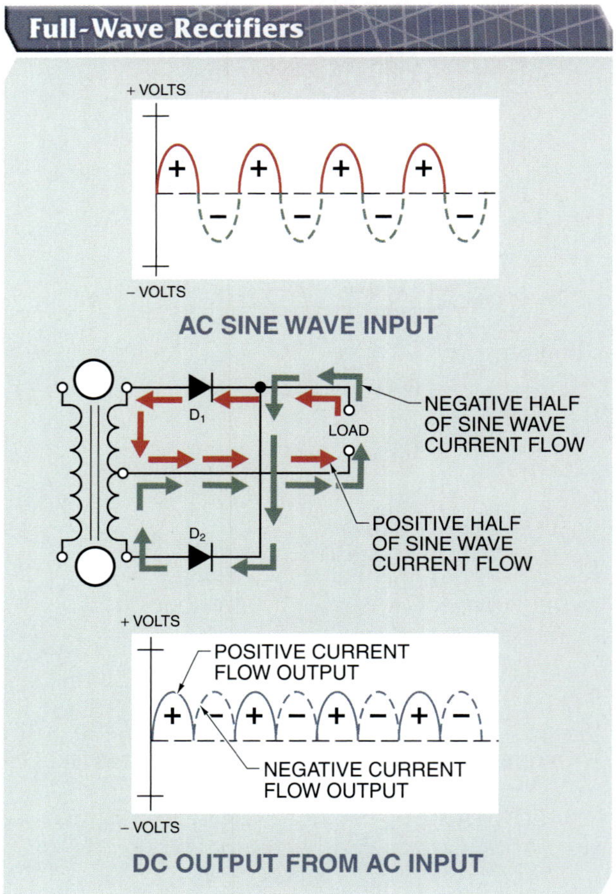

Figure 12-14. Full-wave rectification is obtained with two diodes and a center-tapped transformer. It converts AC voltage into DC voltage with more pulsations than half-wave rectification but with less intensity.

Bridge Rectifiers. A *bridge rectifier* is a circuit containing four diodes that permits both halves of an input AC sine wave to pass. A bridge rectifier is more efficient than a half-wave and full-wave rectifier, and it is the most common rectifier circuit used in rectification circuits. The output of a bridge rectifier is pulsating DC voltage that must be filtered (smoothed) before it can be used in most electronic equipment. **See Figure 12-15.**

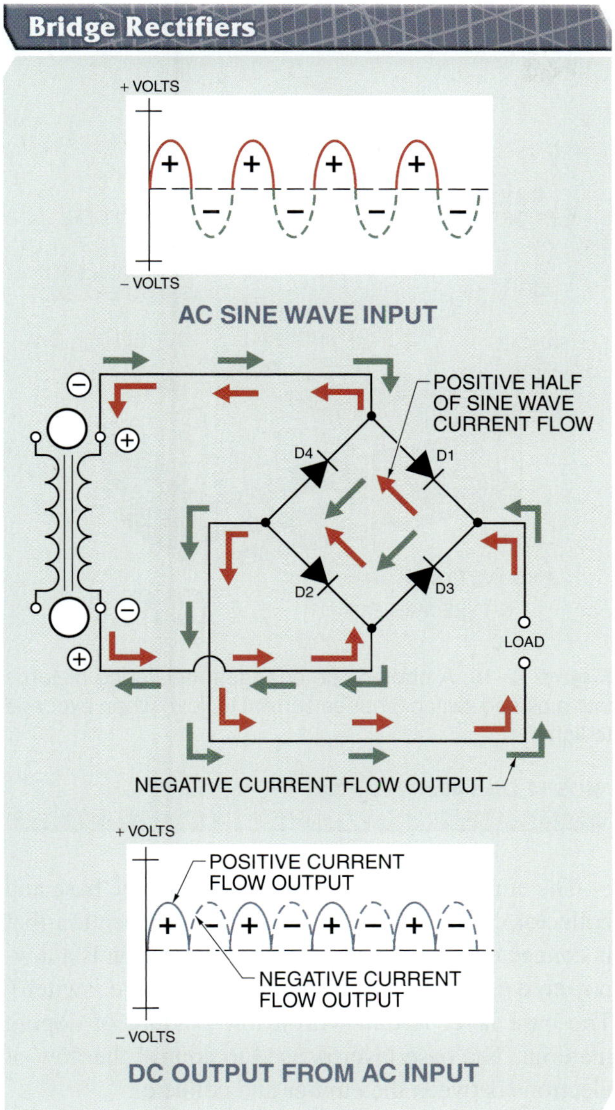

Figure 12-15. Full-wave bridge rectification is obtained with four diodes and is the most energy-efficient method of rectification.

Full-Wave Bridge Rectifiers Media Clip

Diode Types

Diodes are used for rectification and to block current flow. Some diodes are designed with certain characteristics that are used to perform specific tasks. These diodes include zener diodes, light-emitting diodes, and photodiodes.

Zener Diodes. A *zener diode* is a type of diode designed to operate in a reverse-biased mode without being damaged. **See Figure 12-16.** Different types of zener diodes are available that use various voltages to trigger (turn on) the diode. Zener diodes are typically used as voltage shunts or electronic safety valves. When a power-supply voltage to an electronic circuit increases due to a malfunction, a zener diode passes the unwanted voltage to ground. This protects the sensitive electronic circuit from overvoltage.

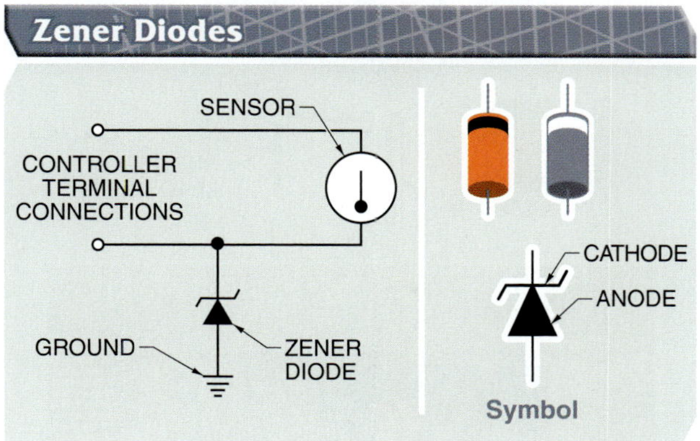

Figure 12-16. Zener diodes operate in reverse-biased mode without being damaged and are used to protect electronic devices from overvoltage.

Light-Emitting Diodes (LEDs). A *light-emitting diode (LED)* is a type of diode designed to produce light when forward-biased. The light emitted from most diodes is small and not visible, but LEDs are specifically designed to create the maximum amount of light. An LED consists of a small diode chip enclosed in an epoxy housing that contains light diffusing particles. **See Figure 12-17.**

When an LED is forward-biased, light is released and diffused in different directions by the epoxy housing. The amount of light produced is directly proportional to the current flow through the LED. The light wavelength determines the color of the LED, with the most common colors being red and green. To prevent LEDs from being damaged by overcurrent, a resistor is normally placed in series with the LED to limit current flow.

Photodiodes. A *photodiode* is a type of diode that changes resistance or switches on when exposed to light. In many ways, photodiodes are the opposite of LEDs. **See Figure 12-18.** An adequate light source striking the surface of the PN junction (window) of a photodiode chip causes electrons to jump the PN junction and current to flow. Photodiodes can be designed to switch only when exposed to specific amounts of light or certain wavelengths of light. Due to their sensitivity to light, photodiodes are typically used as light detectors.

LED Operation Media Clip

Light-Emitting Diodes (LEDs)

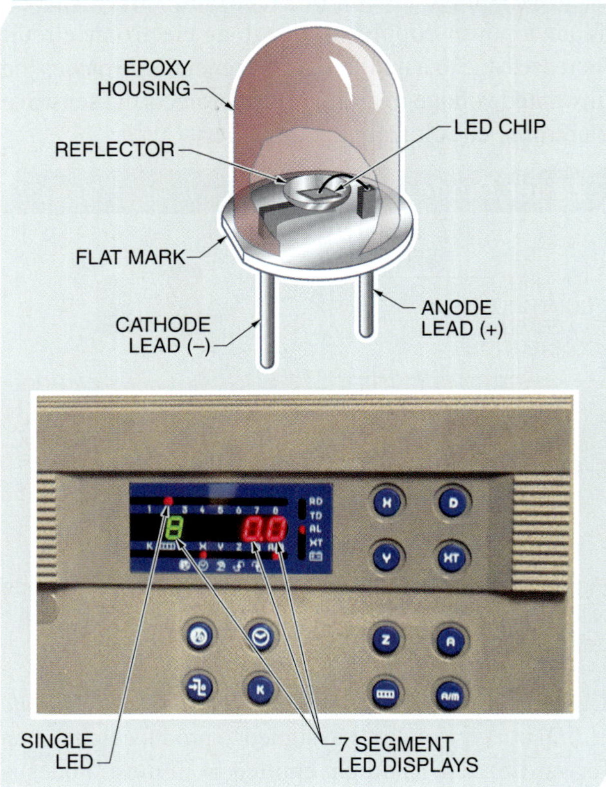

Figure 12-17. LEDs produce a small amount of light when properly biased by voltage. A blue light has a wavelength of 490 nm, a green light has a wavelength of 565 nm, a yellow light has a wavelength of 590 nm, and a red light has a wavelength of 640 nm to 690 nm.

Photodiodes

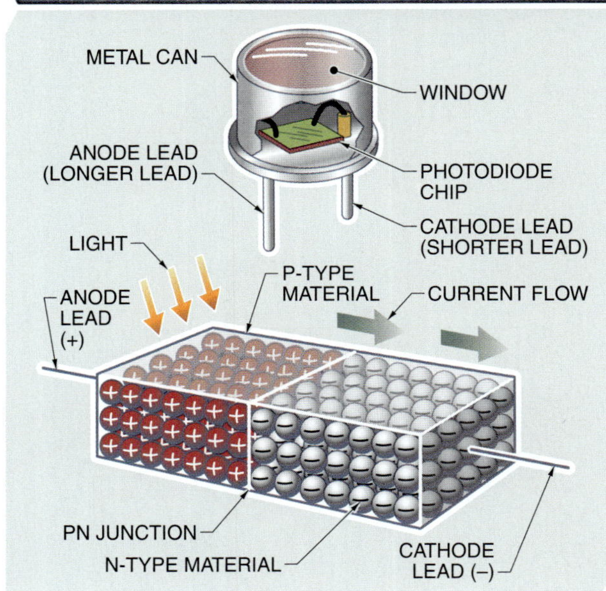

Figure 12-18. A photodiode changes resistance or turns into a closed switch (causes current to flow) when exposed to light.

Transistors

A *transistor* is a three-terminal semiconductor, switch-like device that controls current according to the amount of voltage applied to the base. Transistors are used as amplifiers or DC switches, with some transistors being able to be used as either one. Transistors include bipolar junction transistors, field-effect transistors, phototransistors, and photodarlington transistors.

Bipolar Junction Transistors. A *bipolar junction transistor (BJT)* is a transistor that controls the flow of current through the emitter (E) and collector (C) when the base (B) is properly biased. Bipolar junction transistors have three separate layers of P-type and N-type material and can be layered in a PNP configuration or an NPN configuration. **See Figure 12-19.** The layers of a bipolar junction transistor are in physical contact with each other and have terminals connected to the layers. The three terminals are the emitter, collector, and base.

MOSFET Operation Media Clip

The emitter is the common contact for the base and collector circuits. The collector is the connection that is connected to the load. The base connection is a low-powered connection that functions as a valve (switch). The base layer is thin with a low amount of doping material. The base layer is used to control the flow of electrons between the emitter and collector.

A small flow of electrons between the base and emitter can cause a much larger flow of electrons between the emitter and collector. Therefore, the BJT functions as an amplifier. When a variable resistor is placed in the base circuit, the collector output signal can be changed. **See Figure 12-20.** BJTs are typically used to amplify small control signals into large signals that are used to drive large-capacity devices such as electronically controlled valve and damper actuators.

Field-Effect Transistors (FETs). A *field-effect transistor (FET)* is a type of transistor that controls the flow of current through the drain (D) and source (S) when the gate (G) is properly biased. Field-effect transistors were developed to be better switching devices than bipolar junction transistors. Field-effect transistors include junction field-effect transistors (JFETs) and metal-oxide semiconductor field-effect transistors (MOSFETs).

Bipolar Junction Transistor Operation Media Clip

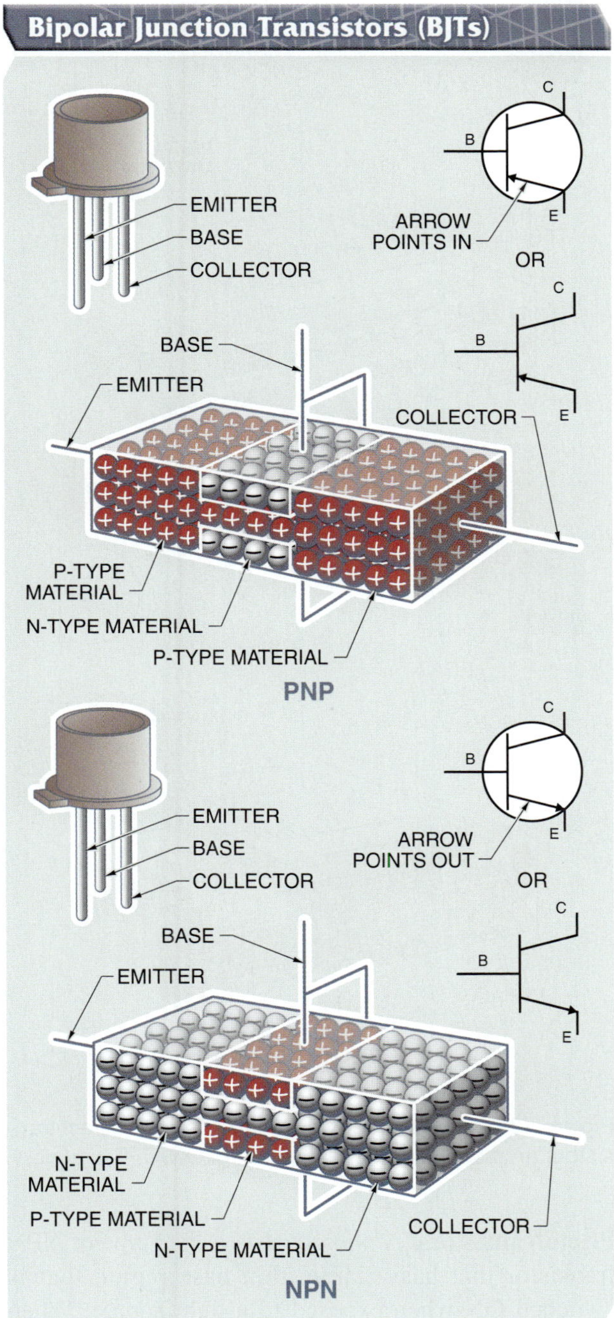

Figure 12-19. Each layer that forms a bipolar transistor has either an N-type or P-type material and is provided with a wire contact for connection to a circuit.

JFET Operation Media Clip

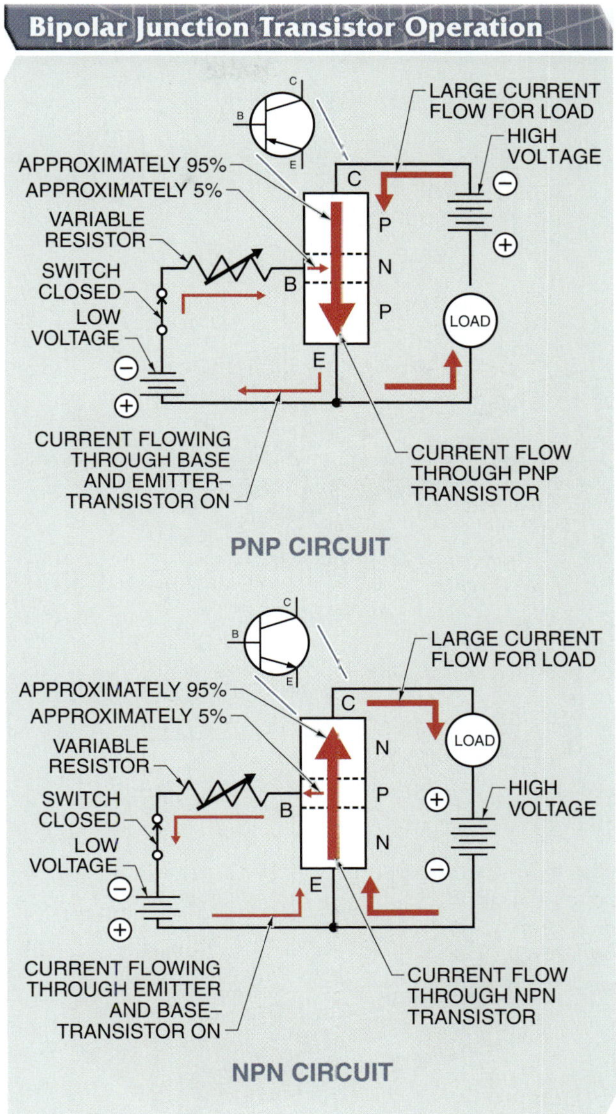

Figure 12-20. Bipolar junction transistors have a small flow of electrons between the base and emitter that is used to create a much larger flow of electrons between the emitter and collector.

Junction field-effect transistors may be N-channel or P-channel JFETs. **See Figure 12-21.** When voltage is applied to the material that surrounds the channel, the effect of the electrons being present causes the channel to have various states of conductivity. The channel acts like a valve or switch. The action takes place with a low current and voltage at the gate. An advantage of JFETs over bipolar junction transistors is that current and voltage requirements are much lower in JFETs. The low–power consumption of JFETs makes them ideal for use in a wide variety of circuits.

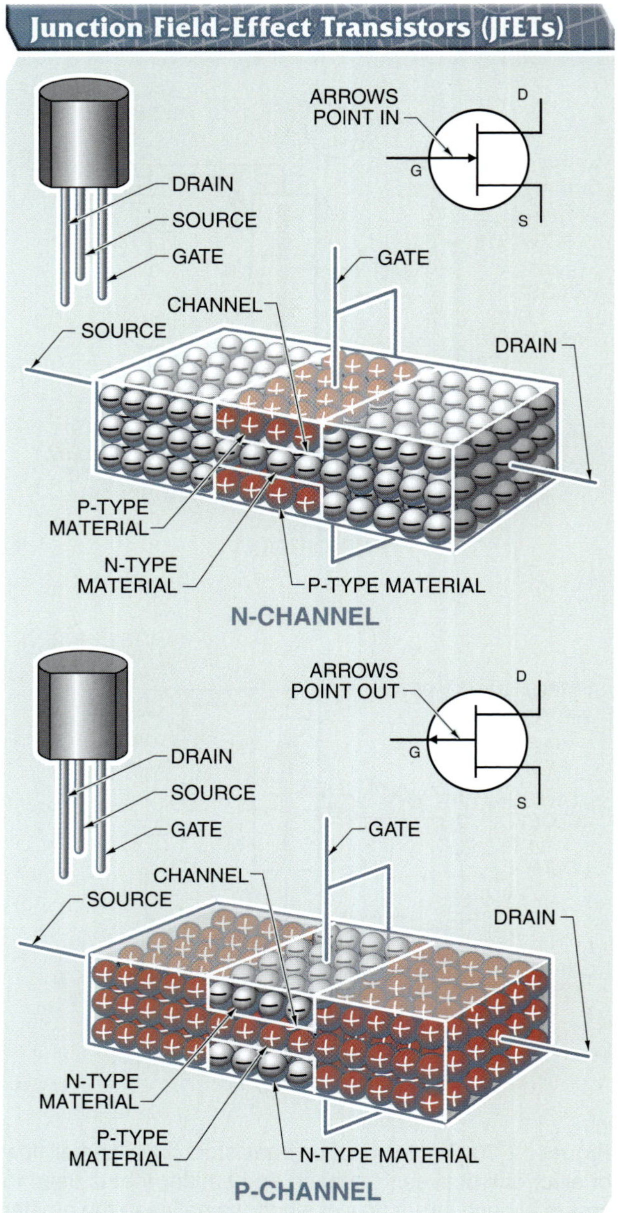

Figure 12-21. The output current of a JFET is controlled by the electrical field within the channel of the JFET.

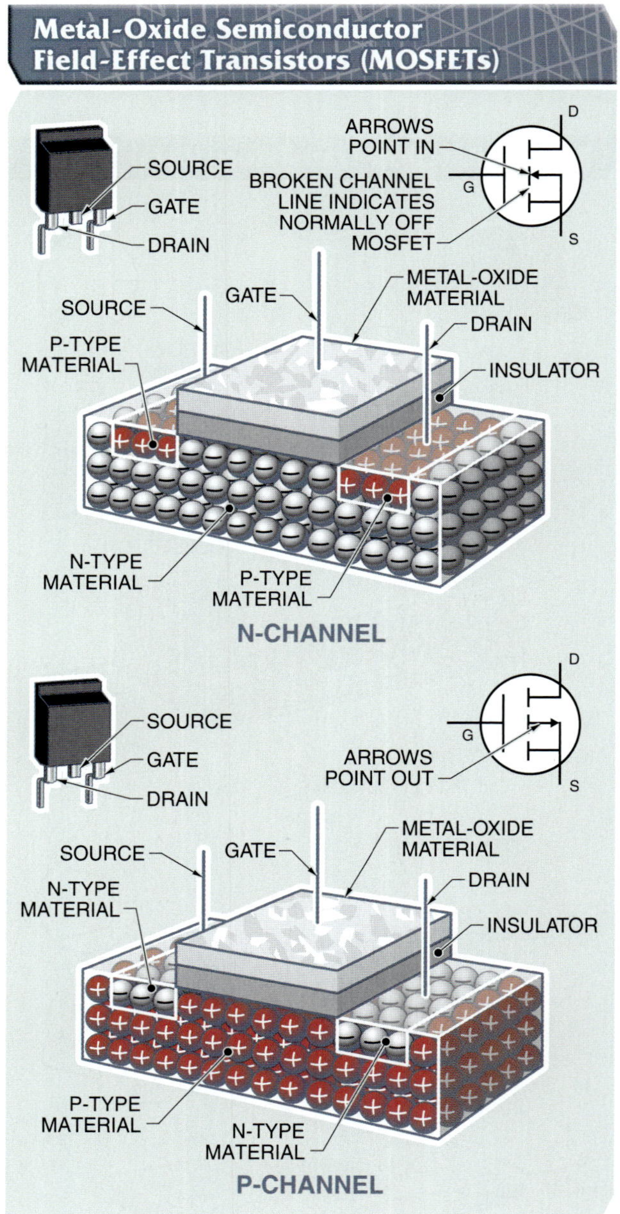

Figure 12-22. Metal-oxide semiconductor field-effect transistors are used as switches to control ON/OFF signals.

The most common transistors used today are metal-oxide semiconductor field-effect transistors (MOSFETs). MOSFETs are constructed from two separate sections of either P-type or N-type material at opposite ends of the transistor. **See Figure 12-22.** Between the two sections is a gate lead that is connected to the transistor by a metal-oxide material and insulator. The insulator allows no physical contact between the gate and the rest of the transistor. A voltage present at the gate of a MOSFET causes conductivity between the source and drain leads.

Phototransistors. A *phototransistor* is a type of NPN transistor that has a large, thin base region that is switched ON when exposed to a light source. When light strikes the base region, current flow is induced that causes the emitter-collector current to switch on. **See Figure 12-23.** The drawback of a phototransistor is that the large, thin base region causes the current-carrying ability of the transistor to be reduced. In many cases, a device must have a large current-carrying ability to function properly in a circuit.

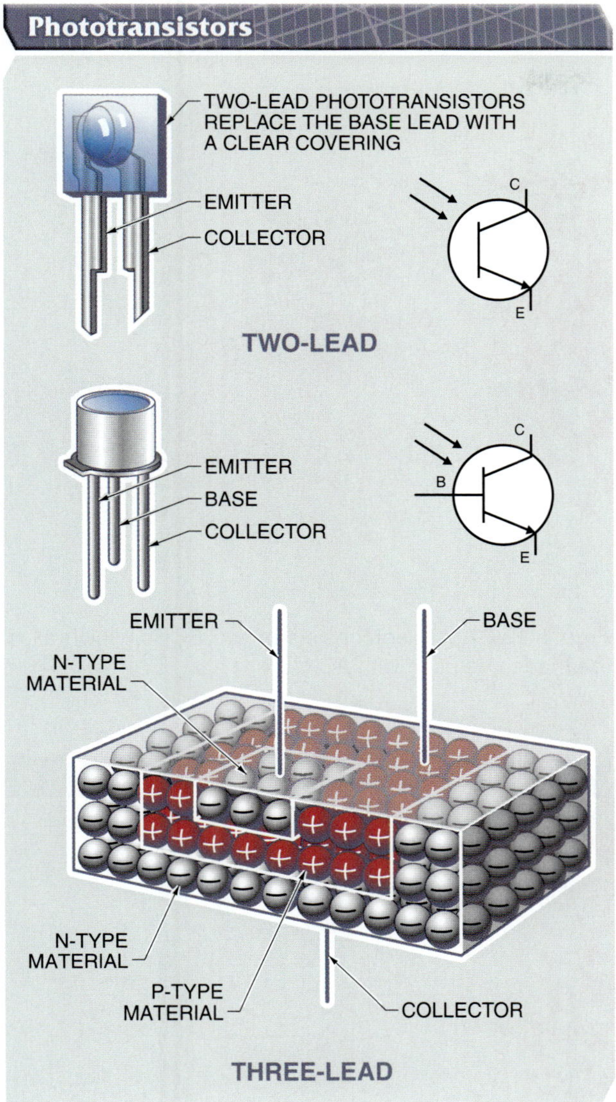

Figure 12-23. A phototransistor requires light to strike the base region of the transistor and induce current flow between the emitter and collector.

Photodarlington Transistors. A *photodarlington transistor* is a type of transistor that consists of a phototransistor and a standard NPN transistor in a single package. In a photodarlington transistor, the sensitive phototransistor is used to switch the standard NPN transistor. **See Figure 12-24.** The phototransistor provides high sensitivity while the standard NPN transistor provides the high current flow necessary for specific applications such as energizing relay coils.

SCR Operation Media Clip

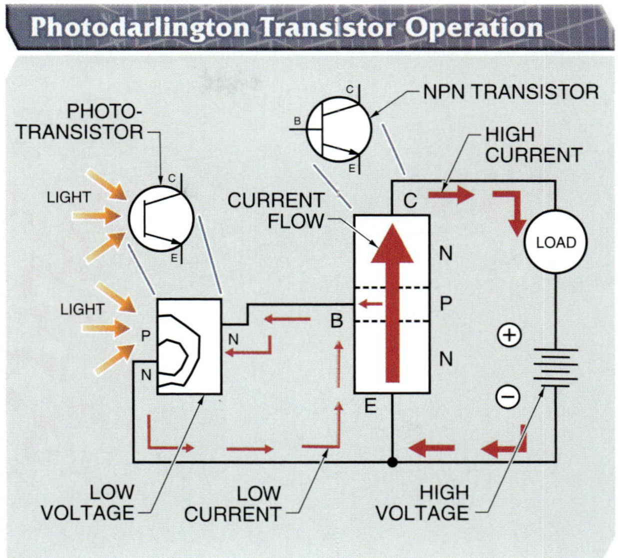

Figure 12-24. Photodarlington transistors use a phototransistor to switch a standard NPN transistor ON and OFF.

Thyristors

Another major category of semiconductor devices is thyristors. A *thyristor* is a solid-state switching device that switches current ON by a quick pulse of control current. Thyristors act as solid-state relays that can switch devices, such as heaters, compressors, motors, and relays, ON and OFF. The two major categories of thyristors are silicon-controlled rectifiers (SCRs) and triacs.

Silicon-Controlled Rectifiers (SCRs). A *silicon-controlled rectifier (SCR)* is a type of thyristor that is capable of only switching direct current. **See Figure 12-25.** Silicon-controlled rectifiers have four layers of P-type and N-type material. One layer is referred to as the gate.

When a small voltage is introduced at the gate, the middle layer switches on, and the SCR allows current flow in one direction only. The SCR continues to conduct current even when the gate voltage is removed. **See Figure 12-26.** To stop an SCR from conducting, the system power must be disconnected. Silicon-controlled rectifiers have a high current-carrying capability that causes them to create high amounts of heat. The heat must be removed by heat sinks or fans.

Triacs. A *triac* is a solid-state switching device used to switch alternating current. A triac is triggered into conduction in either direction by a small voltage to its gate. **See Figure 12-27.** Triacs are used as solid-state relays. Unlike electromechanical relays, triacs have no moving parts and are capable of cycling on/off millions of times without failure.

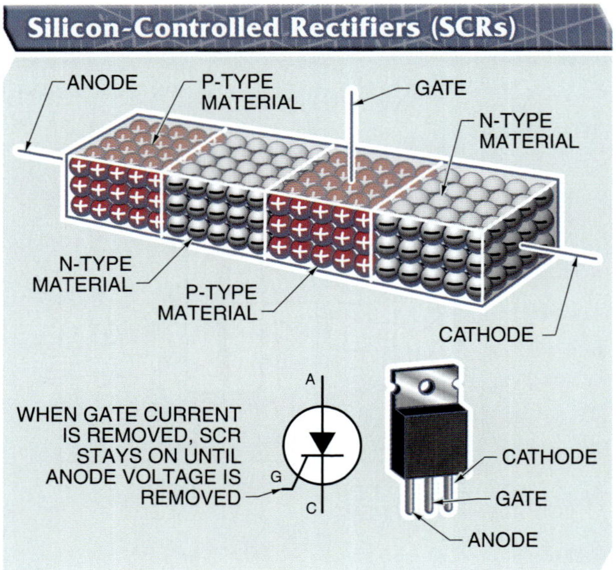

Figure 12-25. Silicon-controlled rectifiers have very high DC current-carrying capability that allows them to create high amounts of heat.

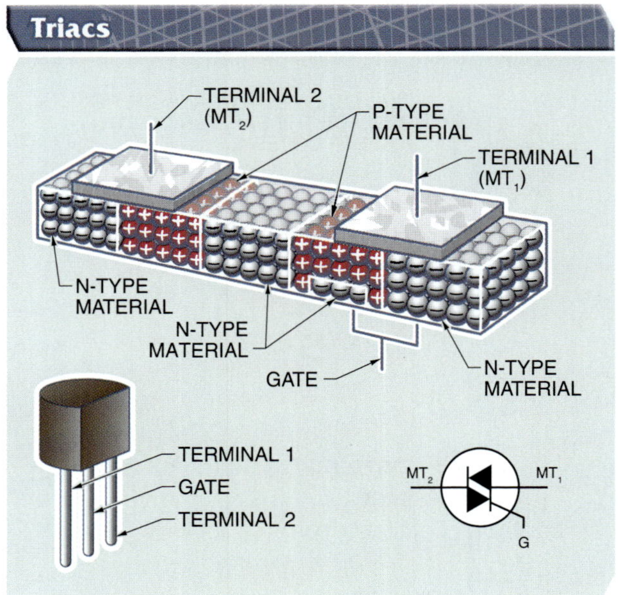

Figure 12-27. A triac consists of two SCRs, which allow the triac to be used with AC current.

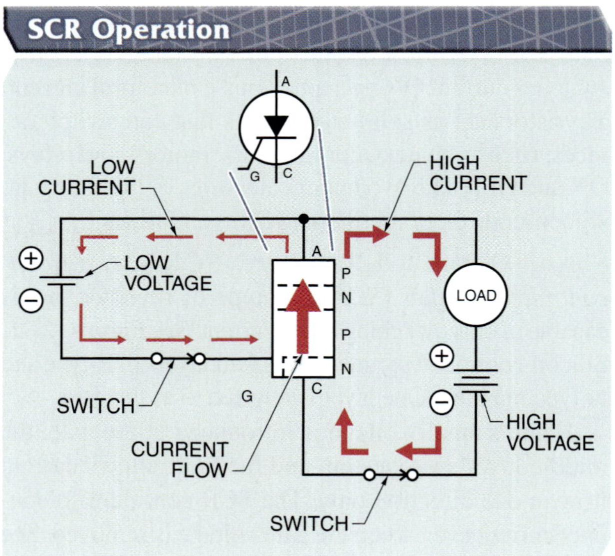

Figure 12-26. When a small voltage is introduced at the gate of an SCR, current flows in only one direction, even after the gate voltage is removed.

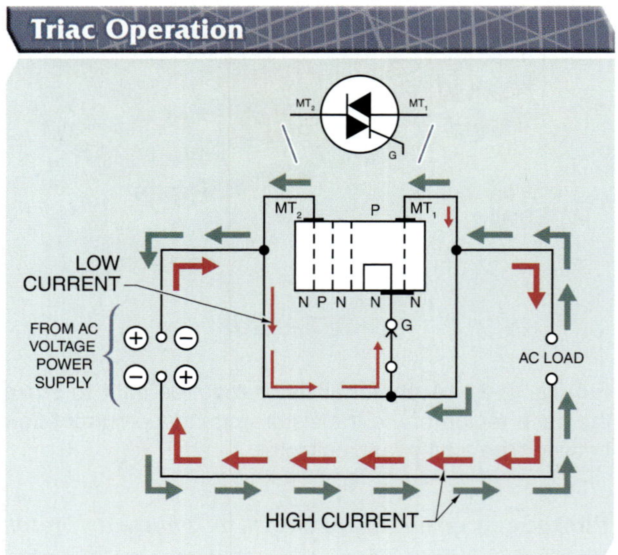

Figure 12-28. One SCR conducts in one direction of current flow, while the other SCR conducts current in the opposite direction of current flow.

A triac consists of two SCRs connected in a reverse-parallel configuration. **See Figure 12-28.** One SCR conducts current in one direction, while the other SCR conducts current in the opposite direction. Triacs are used in automated systems to turn electrical devices ON and OFF and are limited only by the amount of current they can carry.

TECH TIP

Snubber circuits are typically used between T1 and T2 of a triac to assist in turn-off. Snubber circuits are also used to prevent premature triac triggering caused by voltage spikes in the supply power.

Integrated Circuits

An *integrated circuit* is an electronic device in which all devices, such as transistors, diodes, and resistors, are contained in a single package or chip as components. Integrated circuits were developed because as electronic devices were being used to perform more and more functions, it was discovered that many transistors, diodes, and other devices could be deposited microscopically on a small chip of silicon. **See Figure 12-29.**

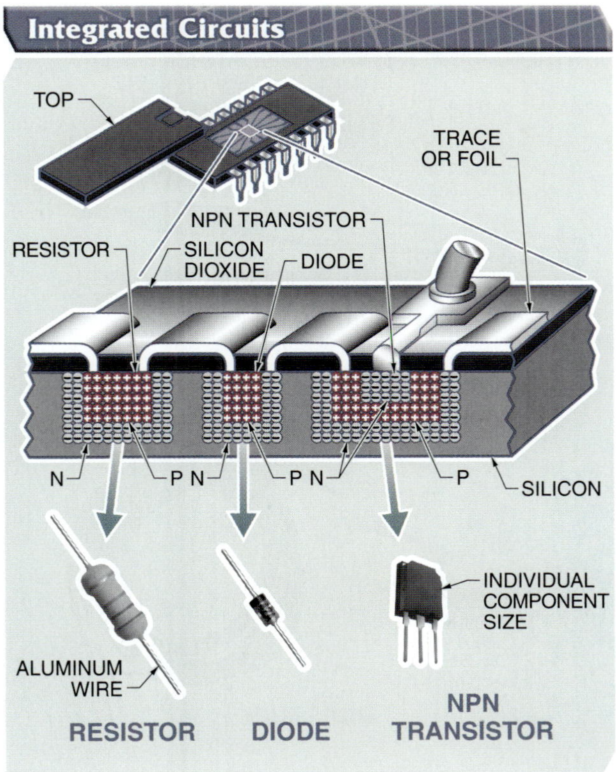

Figure 12-29. Integrated circuits can contain thousands or millions of microscopic transistors, diodes, and resistors in one small package.

Microscopically depositing many semiconductor devices on a small chip of silicon enables many functions to be performed by a single semiconductor chip. One function is to represent ON/OFF (digital) decisions. The ON/OFF decision devices are referred to as logic gates and follow a set of logic rules known as Boolean logic. Logic gates include AND, OR, NOT, NOR, and NAND gates. **See Figure 12-30. See Appendix.** Combining logic gates onto a single chip in the minimum amount of space so that they use the minimum amount of power enables a ¼″ square chip of silicon to contain hundreds of thousands or millions of individual semiconductor devices.

AND Logic Gates

Logic Functions	
Logic function	AND
Circuit description	ENERGIZED Output energized if all inputs are activated DE-ENERGIZED Output de-energized if any one input is deactivated
Ladder diagram	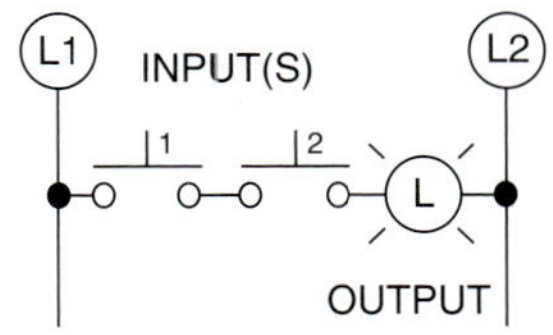
Programming diagram	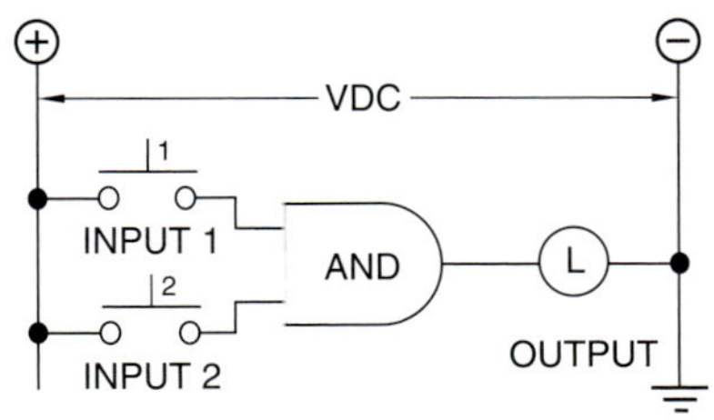

Figure 12-30. All electronic control circuits use logic gates to control loads by using the logic function to control various damper and valve actuators.

International Rectifier

Integrated circuits (ICs) are the principal components of the HVAC original equipment manufacturer (OEM) control boards found in furnaces, boilers, air handlers, and building automation systems.

Troubleshooting Semiconductor Devices

Refrigeration and HVAC technicians require an understanding of electronic component troubleshooting procedures and methods. Before troubleshooting any semiconductor device, it is vital to always verify that the power is disconnected. When power is connected, serious injury, death, and equipment damage can result. Even low-voltage power can damage electronic circuit components.

Additionally, it is important to always verify adequate personal grounding. An inexpensive wrist strap can be connected to ground to protect against static electricity discharge to electronic components. Electronic components are damaged or destroyed by even small amounts of static electricity discharges. When the cover is removed from the packaging of an electronic device, the device is unprotected and vulnerable to static electricity discharges. HVAC technicians do not replace board levels components, but will replace an entire control board.

ELECTRONIC CONTROL SYSTEM APPLICATIONS

Electronic control systems use various semiconductor devices to control refrigeration and HVAC systems. Electronic control system devices include sensors, controllers, switches, and damper and valve actuators. These input and output devices provide temperature, pressure, and humidity control in a building. Electronic control system input devices include sensors and electronic thermostats that are used in applications such as multizone unit control, and boiler controls.

Electronic Sensors

Many types of sensors use semiconductor materials as the electronic components. **See Figure 12-31.** Electronic temperature sensors (thermistors) use semiconductor materials that change resistance characteristics as the temperature around the sensor changes. Humidity sensors change capacitance as humidity is absorbed or adsorbed by the semiconductor materials. A pressure sensor typically uses a piezoelectric crystal, which increases output voltage as pressure is exerted on the crystal.

Electronic Thermostats. Many residential and light commercial thermostats are electronic. Applications such as air conditioning, heat pumps, and other packaged equipment typically use electronic thermostats. **See Figure 12-32.** Electronic thermostats use electronic sensors to sample the building space temperature and use triacs to energize and de-energize heating, cooling, and fan functions. Some electronic thermostats include LEDs to indicate system operation or trouble conditions.

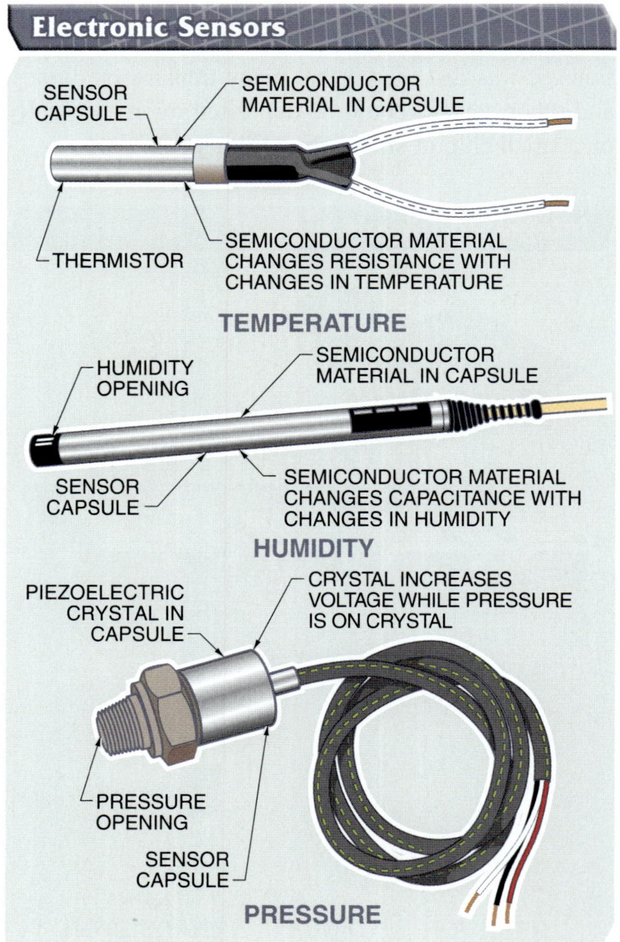

Figure 12-31. Electronic sensors made from semiconductor materials are used in refrigeration and HVAC systems for temperature, humidity, pressure, and sometimes flow measurements.

Multizone Unit Controls. In multizone unit control, a zone thermostat controls each zone damper or valve. Typically, the system's air-handling unit is controlled by several electronic controllers. **See Figure 12-33.** In this application, the supply fan runs continuously. The hot deck controller controls the hot deck electronic valve to maintain the hot deck setpoint. The setpoint of the hot deck controller is reset from outside air. As the outside air temperature varies from –10°F to 65°F, the hot deck setpoint varies from 70°F to 120°F. The cold deck controller maintains a constant temperature of approximately 55°F.

Electronic Thermostats

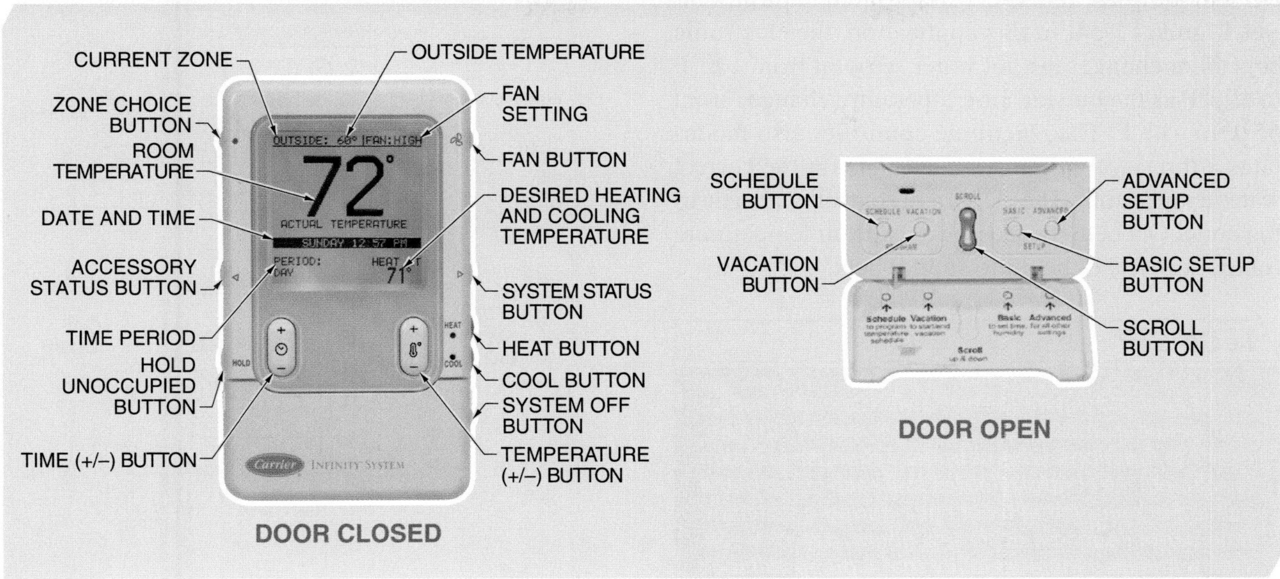

Carrier Corporation

Figure 12-32. Electronic thermostats are typically programmed to change the temperature setpoint for specific times and conditions of a day, week, or month.

Multizone Unit Controls

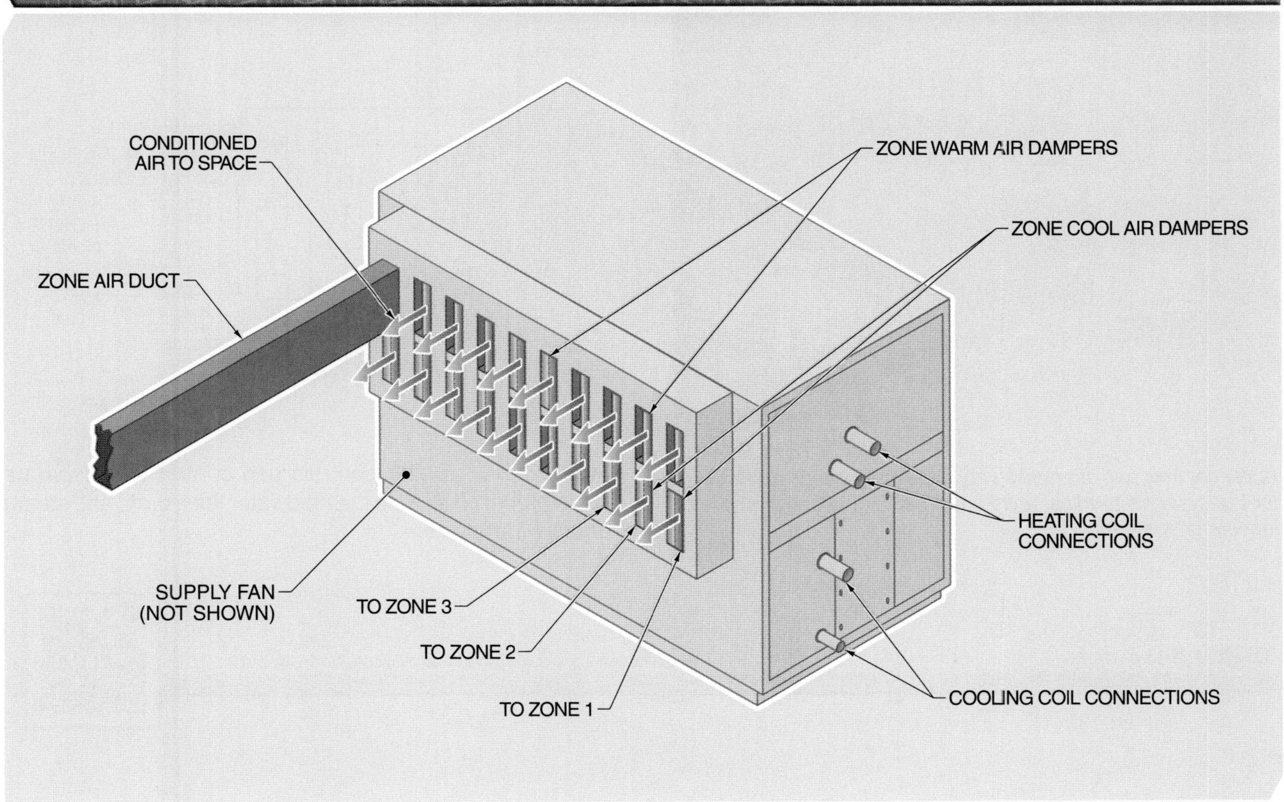

Figure 12-33. Multizone HVAC units use a number of dampers to provide the proper air to building spaces and equipment.

Boiler Controls. Electronic controls are used to control boilers and their associated equipment. **See Figure 12-34.** In this application, the electronic controller changes the hot water setpoint from 100°F to 190°F as the outside air temperature changes from 65°F to –10°F. The electronic controller also modulates a three-way mixing valve to obtain the correct water temperature. The electronic controller also turns the boiler ON below a specific outside air temperature (outdoor reset) to begin to supply heat.

TECH TIP

Boiler control systems, as well as furnace, air handler, and building automation control systems, include exercise functions. When a boiler's exercise function is enabled, the control ensures that each pump of the system is operated at least once during a specified time period, such as every few hours, day, or week. This minimizes the possibility that a pump will lock up due to inactivity.

Parker Boiler Co.

Boiler controls and strategies vary depending on the applications for the system's terminal units and whether the boiler is located indoors or outdoors.

Boiler Controls

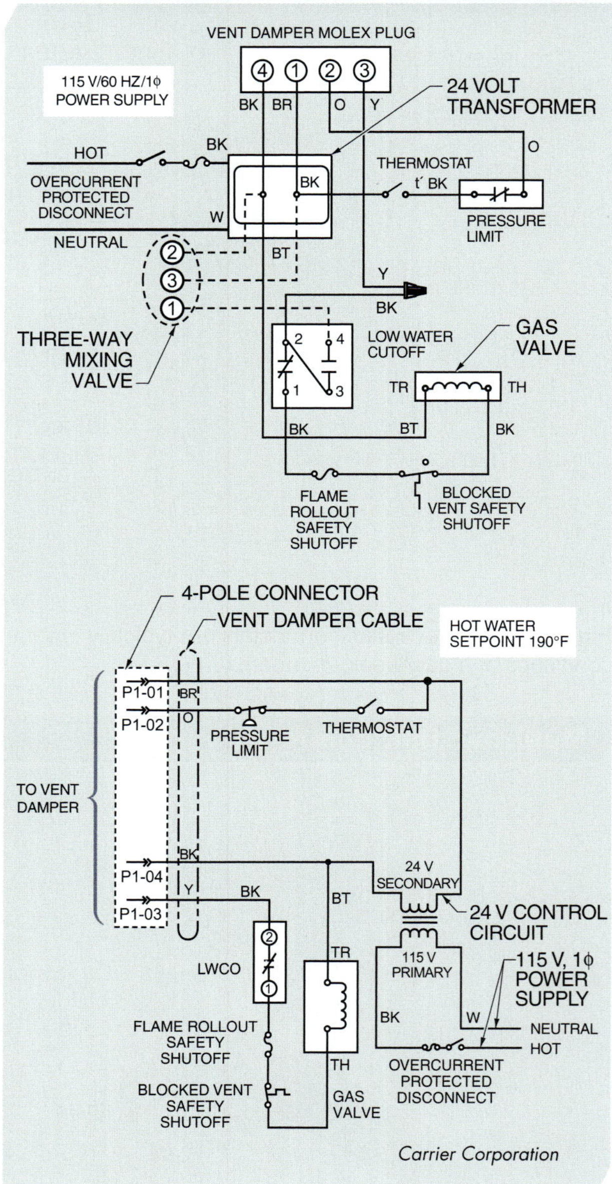

Figure 12-34. Boiler controls can control four separate systems, which include the feedwater, fuel, draft, and steam or hot water system.

UNIT 13

Electric Motors and Solenoids

The electrical systems that operate refrigeration and HVAC systems use many different types of electrical devices. Two of the most common devices are electric motors and solenoids. Recent technological advances in electric motor construction require that technicians have more knowledge than ever on how motors operate. Solenoids are traditionally used to control flow valves in HVAC systems and, therefore, knowledge on how they operate is important in understanding how an HVAC system will function.

OBJECTIVES

- Describe the construction of a motor.
- Describe the information found on a motor nameplate.
- Explain synchronous speed.
- Describe the features of motor mountings and enclosures.
- Identify the types of couplings used in direct drive and flexible drive applications.
- List the designations for the various classifications of V-belts.
- Compare types of shaft misalignment.
- Describe how 1ϕ and 3ϕ motors operate.
- Explain how solenoids operate.
- Explain how to troubleshoot motors.

Index to Procedures

ELECTRIC MOTORS

Electric motors are used in almost all refrigeration and HVAC applications. Motors convert electrical energy into rotating mechanical motion. **See Figure 13-1.** The rotating mechanical motion created by motors is used to rotate fans, pumps, and compressor shafts. Electric motors come in a variety of styles, ratings, and applications. A refrigeration and HVAC technician must be proficient in troubleshooting common electric motors used in various applications.

Motor Construction

Electric motors vary in style, size, and application, but the basic construction among all motors is similar. The two main parts of a motor are the stator and rotor. **See Figure 13-2.** A *stator* is the stationary part of a motor that produces a rotating magnetic field. A *rotor* is the moving part of a motor that spins in response to current flow. Most motors have metal rotor bars that use induced magnetic fields to cause rotor rotation. Other important parts of typical motors include the motor shaft, frame, endbells, bearings, and fan.

Electric Motors

Figure 13-1. Electric motors can be single-phase (1ϕ) or three-phase (3ϕ), come in fractional horsepower to thousands of horsepower, and are available in frames for any application that requires an electric motor.

Motor Construction

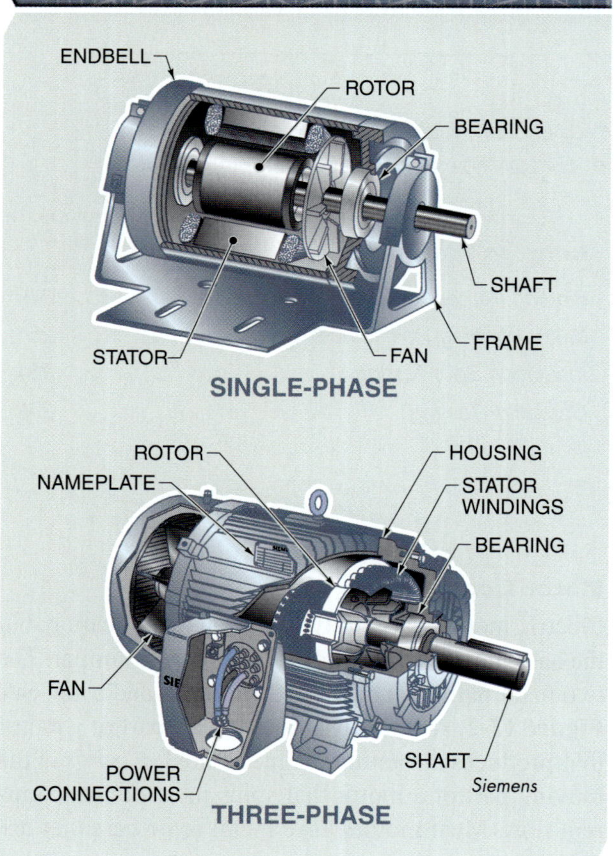

Figure 13-2. The standard components of a motor include the stator, rotor, shaft, frame, endbells, bearings, and fan.

Motors and Magnetism

Electric motors use electricity to create rotating mechanical motion. Electrical current flowing through the stator windings creates a rotating magnetic field. **See Figure 13-3.** A magnetic field of opposite polarity is induced in the bars of the rotor. This magnetic field is opposite in polarity of that in the stator windings. As the magnetic field changes and moves through the stator windings, the rotor moves in response. The motor shaft then rotates.

Three-Phase Motor Rotation

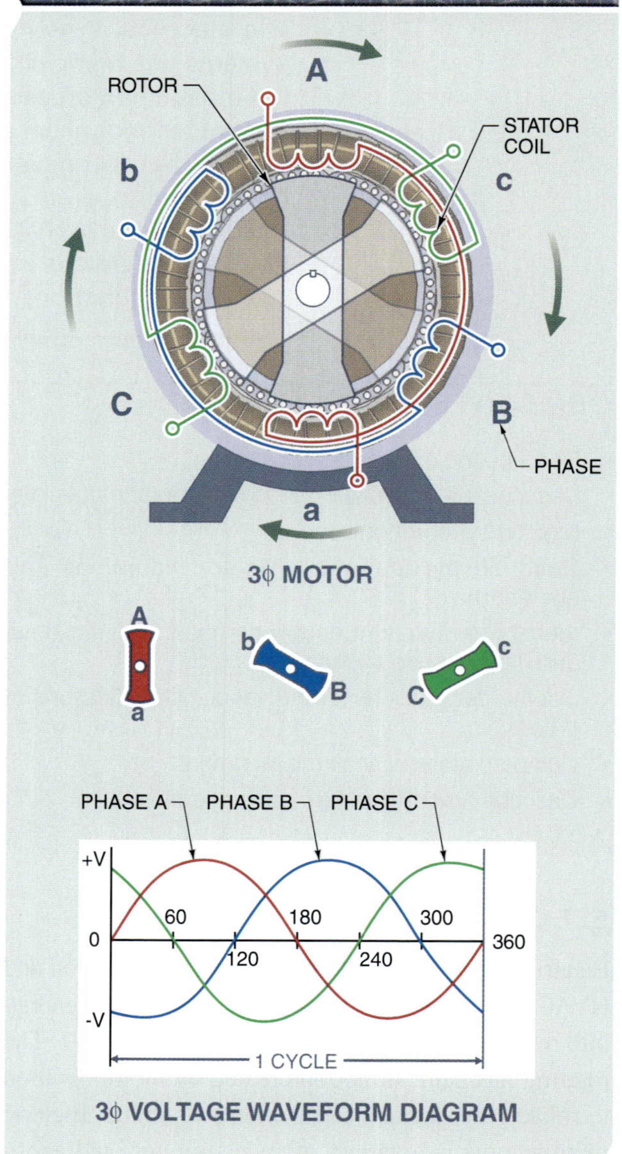

Figure 13-3. Electricity flowing through the stator windings of a 3ϕ motor creates the rotating magnetic fields and induces the current flow that cause the rotor to rotate.

Most of the motors used in refrigeration and HVAC applications are known as induction motors. Induction motors use an induced magnetic field to cause shaft rotation. No direct physical contact exists between the rotor and stator.

Motor Nameplate Data

Virtually, all electric motors have an attached nameplate. Exceptions would be a motor mounted inside a hermetically sealed compressor and a motor where the nameplate has been destroyed. A *motor nameplate* is a plate attached to a motor that lists pertinent motor and manufacturer information. The information on a motor nameplate includes, but is not limited to, the manufacturer's name, motor serial number, voltage rating, and current rating.

Manufacturer Name and Serial Number. The motor manufacturer's name and the serial number are used when replacing or repairing a motor. **See Figure 13-4.** A cross-reference motor chart may indicate a possible replacement motor from another manufacturer. The serial number of a motor is useful when a rare motor is used. The serial number aids in finding information about a specific motor to use from another manufacturer's catalog.

Manufacturer Name and Motor Serial Number

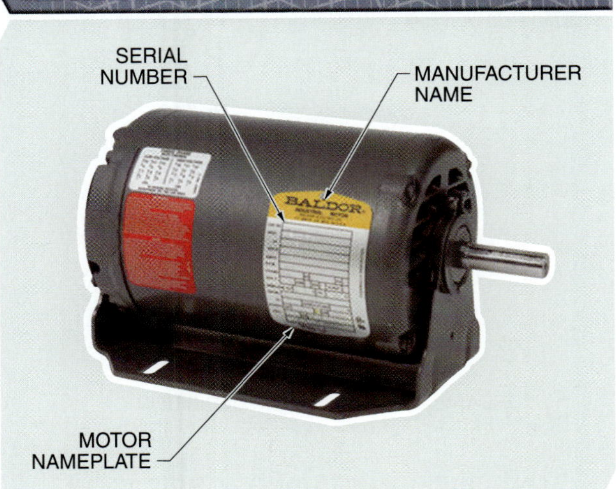

Baldor Electric Company

Figure 13-4. The metal nameplate on a motor always displays the manufacturer name and includes the serial number for future reference.

Voltage Ratings. A *voltage rating* is the voltage that a motor uses to operate properly. Common voltage ratings include 24 V, 120 V, 208 V, or 230/460 V. **See Figure 13-5.** Typically, a voltage variation of ±10% is allowed. When voltage is outside of this safe range, motor damage will result.

Motor Nameplate Ratings

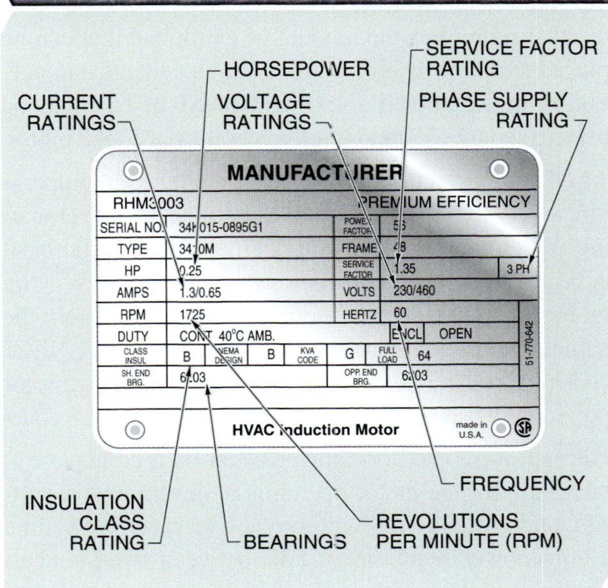

Figure 13-5. The ratings on a motor nameplate include the voltage rating, current rating, phase supply, service factor rating, type of thermal protection, horsepower, and rpm.

Current Ratings. Motors typically have two current ratings, which are full-load and locked-rotor amperage. The *full-load amperage (FLA)* is the amperage drawn when a motor is at full load. Full-load amperage is often called run-load amperage (RLA). Technicians use the FLA to diagnose proper motor operation. Another amperage listed is the locked-rotor amperage. The *locked-rotor amperage (LRA)* or startup amperage is the amount of current drawn on a motor if a rotor is unable to turn or is locked. Locked-rotor amperage is typically six times the amount of full-load amperage. For instance, when a motor has a listed FLA of 10 A, its LRA would be listed as 60 A.

Phase Supply Rating. Motors are designed to use either single-phase (1ϕ) or three-phase (3ϕ) electrical power. The motor nameplate indicates the correct phase of power required.

Frequency Rating. In the United States, 60 hertz (Hz) frequency is used, while in other countries 50 Hz is common. Some 1ϕ, small HP motors can be used with either frequency. Motors that can use either frequency list the relevant information for voltage and amperage for both frequencies. Using any frequency other than the nameplate frequency stated for a motor will change the operating characteristics (speed and torque) of the motor and can cause damage.

Service Factor Rating. A *service factor (SF)* is a number that represents the amount of extra load that can be placed temporarily on a motor without causing damage. For example, a 20 HP motor with an SF of 1.15 may be overloaded by 15% and briefly operate as a 23 HP motor.

Thermal Protection. *Thermal protection* is a temperature safety method that protects a motor from overheating. When thermal protection is included in a motor, it is typically indicated on the motor nameplate. When a motor is equipped with internal thermal protection, the nameplate will read "thermally protected," if all NEC® codes are covered, or "over temp. prot.," when the motor is thermally protected according to part of the NEC® code.

Horsepower. The horsepower listed on a nameplate is the rating for the motor operating at design phase, voltage, and current (full-load torque and speed). Anytime a horsepower requirement falls between two standard horsepower ratings, the larger size is selected for use.

Revolutions per Minute. *Revolutions per minute (rpm)* is the rated motor rotation speed per minute at full-load condition. Typically, full-load speed is 96% to 99% of no-load speed.

Bearings. A *bearing* is a mechanical device that allows free shaft movement by reducing the friction between moving surfaces. The two main types of bearings are plain and antifriction bearings. **See Figure 13-6.** The use of bearings depends on the motor application, orientation, and other factors.

TECH TIP

Plain bearings are used in many electric motors, but the most common type of bearing used for electric motors is the AFBMA-7 C-3 rated antifriction bearing. This bearing is manufactured in all sizes with internal clearances of 0.003" to 0.005". This amount of clearance allows lubrication to flow through the bearing to reduce the friction and wear of machine surfaces.

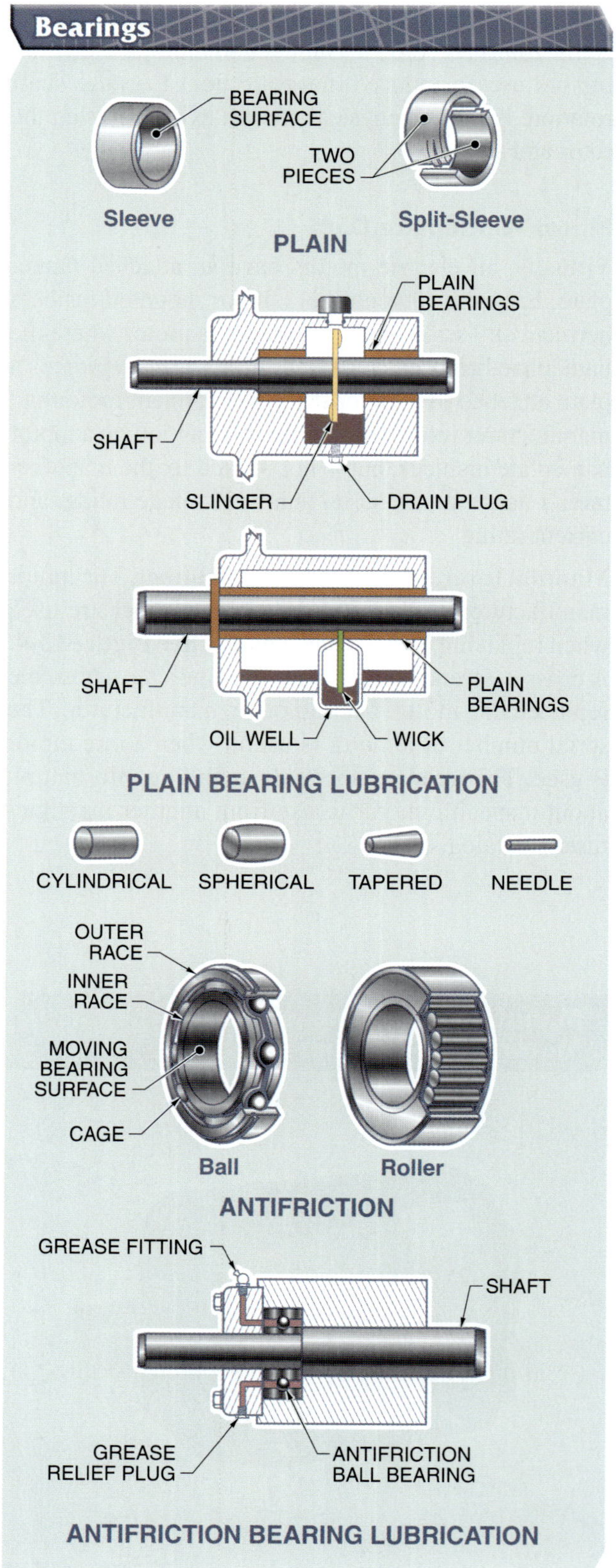

Figure 13-6. The model numbers of the bearings used in a motor are typically listed on the nameplate as reference for future replacement.

A motor nameplate will provide drawing numbers for plain bearings or catalog numbers for antifriction bearings. The motor shaft and plain bearings do not come into actual contact but are separated by a thin layer of oil. Motors of low horsepower with antifriction bearings may be permanently lubricated, while larger motors will have oil ports or grease fittings for lubrication to be performed on a regular basis.

Insulation Class Rating. An *insulation class rating* is a rating that describes how hot a motor can safely operate in a specific ambient environment. Some motors are rated by an allowable temperature rise on the motor.

A typical motor has an allowable ambient temperature of 104°F (40°C). When the ambient temperature is 104°F (40°C), Class B motor windings may be at a maximum temperature of 130°C = [40°C (ambient) + 80°C (temperature rise) + 10°C (hot spot allowance)]. However, conversion from Fahrenheit to Celsius and back again may be required. Instead of calculating maximum temperature, insulation classes are used to determine the rated temperature of motors. **See Figure 13-7.**

NEMA Insulation Class Ratings

NEMA Insulation Classes*	Maximum Operating Temperature	Allowable Temperature Rise — SF 1.0	Maximum Temperature Rise — SF 1.15
A	221°F (105°C)	140°F (60°C)	158°F (70°C)
B	266°F (130°C)	176°F (80°C)	194°F (90°C)
F	311°F (155°C)	221°F (105°C)	239°F (115°C)
H	356°F (180°C)	257°F (125°C)	—

*104°F (40°C) Ambient Temperature

Figure 13-7. Insulation class is a code letter that is used to signify the maximum operating temperature for the wire insulation used in a motor.

Synchronous Speed

The frequency of the supply voltage to a motor and the number of poles in the motor windings determine the synchronous speed. **See Figure 13-8.** The following formula is used to find the synchronous speed of a motor:

$$RPM_{syn} = \frac{120 \times f}{N_P}$$

where

RPM_{syn} = synchronous speed (in rpm)

120 = a constant

f = supply voltage frequency (in Hz)

N_P = number of motor poles

Synchronous Speed

Rockwell Automation/Reliance Electric

Number of Poles	Synchronous Speed*	With a 4% Slip*	Actual Speed*
2	3600	144	3456
4	1800	72	1728
6	1200	48	1152
8	900	36	864

* in rpm

Figure 13-8. Synchronous speed is the mathematical speed of a motor that does not include slip. The actual operating speed of a motor is listed on the nameplate.

Motors do not operate at the calculated synchronous speed because they have a small percentage of slip. For instance, the calculated synchronous speed of a motor can be 1200 rpm. The nameplate rpm is 1152, which means that the motor has a slip of 4%.

Synchronous Motor Operation Media Clip

Example: Motor Synchronous Speed

What is the synchronous speed of a six-pole motor operating at 60 Hz?

$$RPM_{syn} = \frac{120 \times f}{N_P}$$

$$RPM_{syn} = \frac{120 \times 60}{6}$$

$$RPM_{syn} = \frac{7200}{6}$$

$$RPM_{syn} = \mathbf{1200\ rpm}$$

LEESON Electric Corporation

Electric motor rigid mounts include cradle, rigid-base, end flange, and strap mounts.

The number of poles used by the motor generally determines motor speed. However, motor speed can be changed by substituting a motor with a different number of poles or by using a variable-speed drive.

Motor Mountings

Motor mounts are used to secure motors in position. A motor can be mounted to a refrigeration unit in a number of ways. The two most common motor mounts are the rigid mount and resilient mount. Rigid-mount motors are attached using metal-to-metal surface contact. With resilient-mount motors, rubber grommets are placed between the metal surface being connected to and the motor. **See Figure 13-9.** The four subcategories of motor mounts include cradle, rigid-base, end flange, and strap mounts. Motors use standard dimensions called frame numbers, which are established by the National Electrical Manufacturer's Association (NEMA).

Motor Enclosures

Motor enclosures promote safety, reduce maintenance, and protect internal motor components from external environmental conditions. The two major categories of motor enclosures are open and totally enclosed. **See Figure 13-10. See Appendix.**

Many motors used in refrigeration and air conditioning applications have open motor enclosures. The enclosure openings allow ambient air to circulate through the motor to cool the windings and components. The fan creates air movement and is connected to the motor shaft. Subclassifications of open motor enclosures include general, dripproof, splashproof, guarded, semiguarded, and dripproof fully guarded.

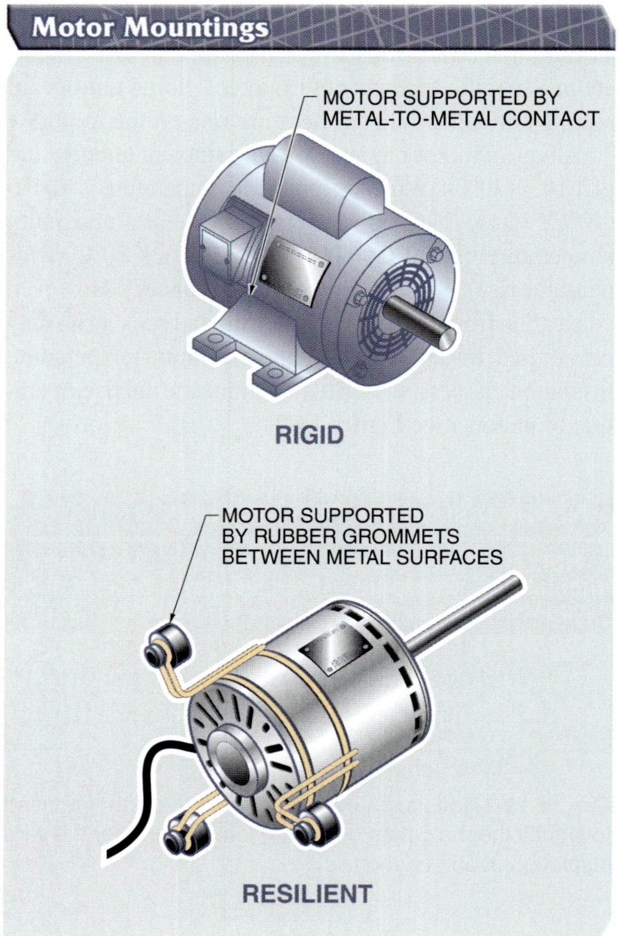

Figure 13-9. A motor mounting style (rigid or resilient) is dependent on the application and frame size of the motor.

Totally enclosed motor enclosures are rarely used except with compressors and in conditions where ambient air cannot be used for cooling. Locations suitable for totally enclosed motors include labs, production areas, or chemical plants with explosive or corrosive gases present in the atmosphere. Further subclassifications of totally enclosed motor enclosures include fan-enclosed, nonventilated, pipe-ventilated, water-cooled, dust-ignitionproof, explosionproof, and waterproof enclosures.

Motor Enclosures

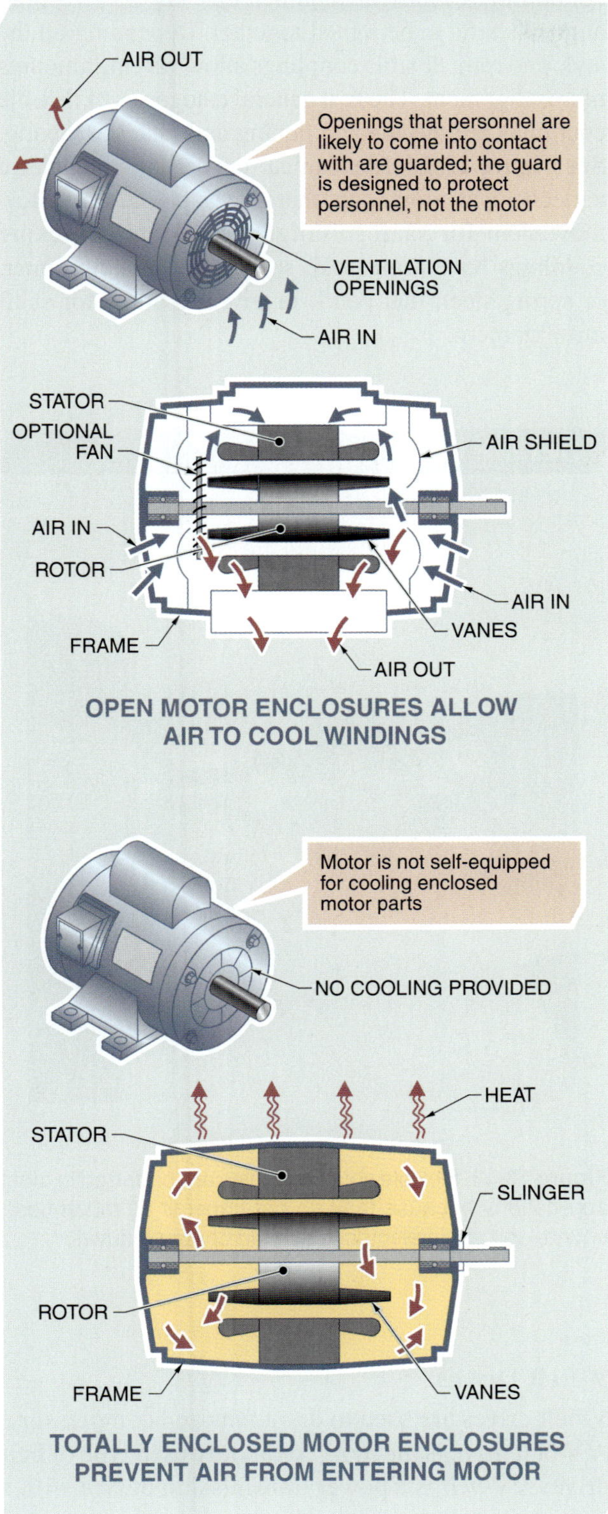

Figure 13-10. A motor enclosure protects a motor from the outside environment, and it protects people from the electrical and mechanical parts of the motor.

MOTOR DRIVE SYSTEMS

Motors must be connected to a load in order to produce work. Motors are connected to loads as direct, flexible, or belt drive systems. Direct drives have the load, such as a fan, mounted to the motor shaft, or they have a machined fitting between the motor and load housings to align the rigid coupling. Flexible drives have a coupling that contains a flexible element to allow a small amount of shaft misalignment. Belt drives allow a load to operate at a different speed from the motor.

Direct Drives

Many motors are rigidly connected to a load via a shaft. The load may be a fan, pump, or compressor crankshaft. Direct drive systems must have sufficient torque to begin movement of the load on startup. As the motor housing and load housing are connected, alignment of the shafts and rigid coupling is ensured. Before the motor is replaced, it must be aligned and the coupling hubs must be properly loaded. **See Figure 13-11.**

Direct Drives

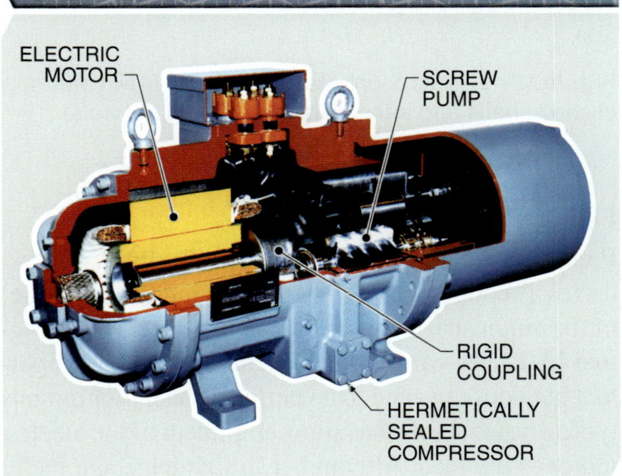

Figure 13-11. A direct drive has a load mounted or coupled to the shaft of a motor with a rigid coupling.

Rigid Couplings. A *rigid coupling* is an inflexible coupling that is used to join two precisely aligned shafts. Rigid couplings do not allow any type of shaft misalignment. The two general categories of rigid couplings are flange and sleeve. Flange couplings have two hubs, male and female, each with keyways and turned up faces. The hubs are machine fitted with the hubs bolted together by the turned up flanges. A *sleeve coupling* is a hollow cylinder that is machined to match the size of the shafts. Compression clamping or multiple setscrews are used to hold a sleeve coupling in place. **See Figure 13-12.**

Rigid Couplings

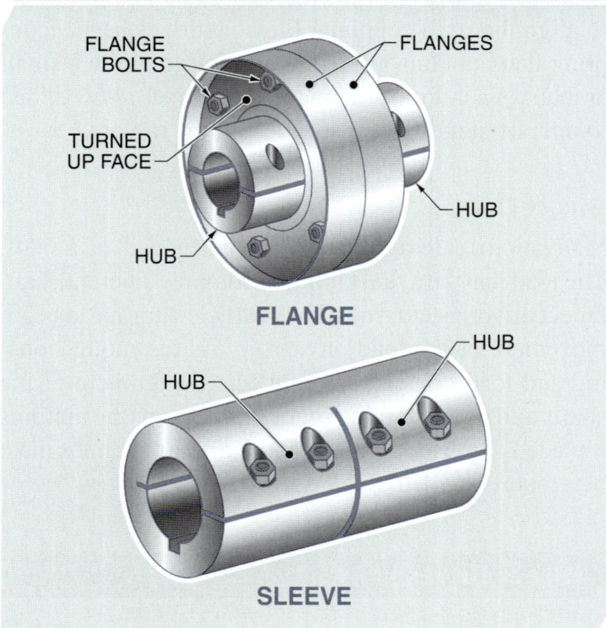

Figure 13-12. Rigid couplings are used to join two precisely aligned shafts and transmit large amounts of torque.

Flexible Drives

Pumps and compressors often use flexible coupling drives. A coupling transmits the moving force (torque) of the motor shaft to the pump or compressor. **See Figure 13-13.** As with all couplings, alignment is important but does not need to be perfect. The motor mounts typically use shims that allow alignment of the electric motor to the load. Although a straightedge and feeler gauges are commonly used to align flexible couplings, laser alignment devices are becoming more popular.

TECH TIP

The service factor (SF) of a motor is important to consider when selecting a coupling for an application. Applications with infrequent stops require devices of 1.00 SF to 1.50 SF. The standard is typically 1.50 SF to 2.50 SF for cycling loads and conveyors, 1.75 SF to 2.50 SF for internal combustion engines, 2.00 SF for machine tools, and 3.00 SF for heavier loads.

Flexible Couplings. A *flexible coupling* is a coupling containing a resilient member that allows two misaligned shafts to be joined together. Depending on the style and rpm, flexible couplings allow varying amounts of misalignment. The two general categories of flexible couplings are mechanical flexing and material flexing. **See Figure 13-14.** Mechanical-flexing couplings have a mechanical component that moves or slides to allow movement for shaft misalignment. Material-flexing couplings have a material, such as rubber, elastomer, or spring steel, that bends or gives to allow for shaft misalignment.

Flexible Coupling Drives

Figure 13-13. Flexible coupling drives are preferred in most applications because they do not transmit as much heat, sound, and static electricity as rigid coupling drives.

V-Belt Drives

V-belt drives are used to drive fans and compressors. A V-belt drive is the most commonly used type of belt drive. A *V-belt* is a power transmission device with a trapezoidal cross section. V-belts are made of molded rubber and contain fiber or steel reinforcement cords as tension members. **See Figure 13-15.** A *tension cord* is the load-carrying element of a V-belt.

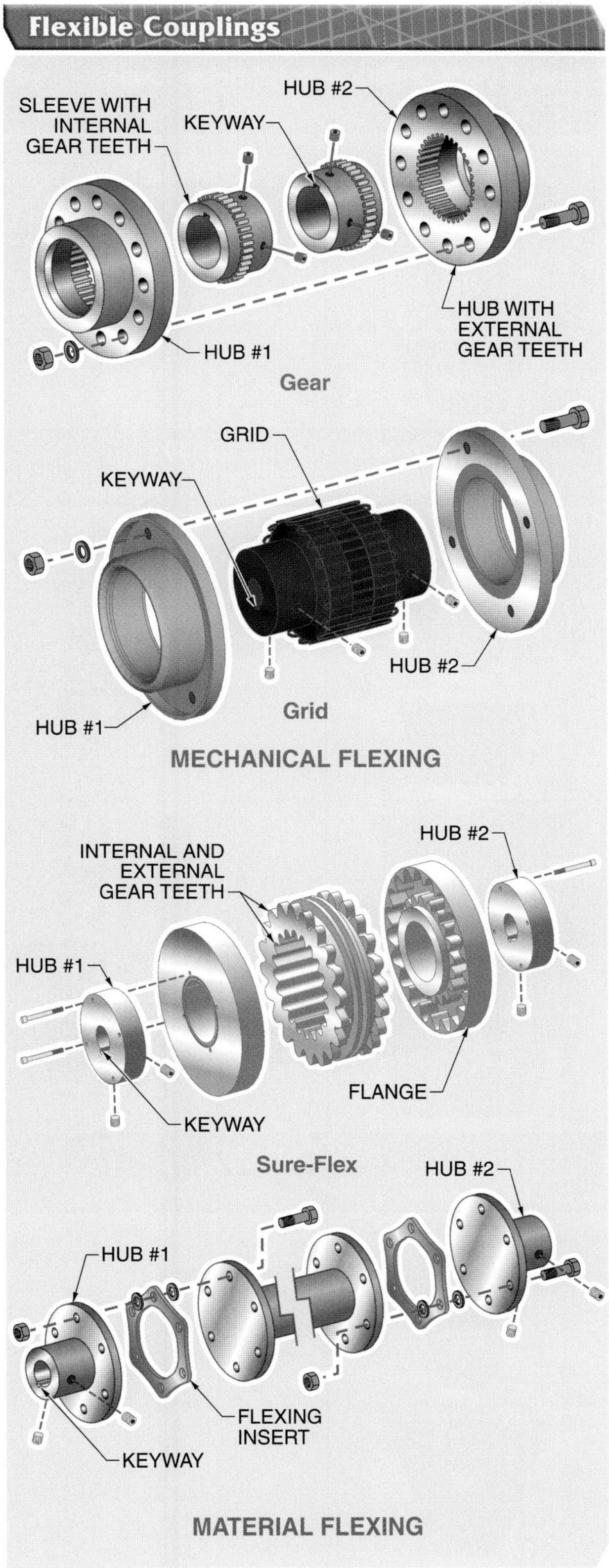

Figure 13-14. Flexible couplings contain a resilient member that allows the shafts being joined to be misaligned and still function properly.

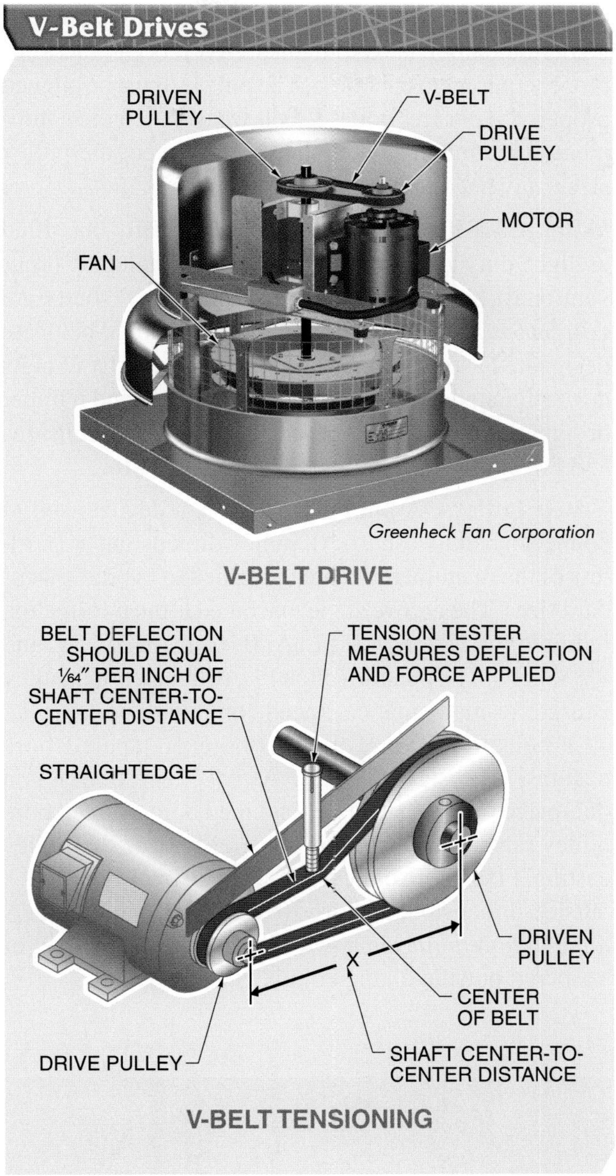

Figure 13-15. V-belts are commonly used to drive fans and compressors because of their smooth operation and ability to allow a load to be driven at any rotational speed.

A V-belt is attached to a pulley on one side of the motor shaft. Another pulley is attached to the fan or compressor shaft on the other side. The pulleys are sized to provide the load with any rotational speed required. Some motors provide easy adjustment bases to facilitate V-belt changes and tensioning.

On small equipment, one V-belt is used. On large systems, multiple V-belts are used. V-belts must be checked for wear and replaced when needed. When multiple V-belt sets are used, all belts must be replaced at the same time with a matching set. The tension must be periodically checked to ensure correct tightness for proper equipment operation. The pulleys must be aligned properly to ensure proper V-belt wear. When the equipment operating characteristics are changed, the pulley sizes may need to be changed as well.

V-Belt Classifications. V-belts are typically classified as light duty (fractional horsepower), standard (conventional), and high capacity. Light-duty V-belts are designated as 2L, 3L, 4L, and 5L. Standard V-belts are designated as A, B, C, D, and E. High-capacity V-belts are designated as 3V, 5V, and 8V. The letter and number designations represent the cross-sectional dimensions and thicknesses of V-belts. **See Figure 13-16.**

V-Belt Pulleys. A *pulley (sheave)* is a device used to connect a belt to a shaft. All pulley dimensions, regardless of the manufacturer, are machined to the same standard sizes. The groove angle machined is the information required for 2L, 3L, 4L, 5L, A, B, C, D, E, 3V, 5V, and 8V V-belts. The outside (pitch line) diameter of a pulley is used for mathematical speed computations.

V-belt pulleys are of fixed bore or tapered bore designs. **See Figure 13-17.** A *fixed bore pulley* is a one-piece pulley that has a machined bore. A fixed bore pulley has a hub that is an integral part of the pulley casting. Fixed bore pulleys are selected by the bore size, outside dimension, and groove dimensions required. A *tapered bore pulley* is a two-piece pulley consisting of a tapered outside diameter hub.

Baldor Electric Company

When used in groups, all V-belts must be exactly the same length.

V-Belt Classifications

Light-Duty (Fractional Horsepower)		
Classification	Width	Height
2L	1/4″	1/8″
3L	3/8″	7/32″
4L	1/2″	5/16″
5L	21/32″	3/8″
Standard (Conventional)		
Classification	Width	Height
A	1/2″	5/16″
B	21/32″	13/32″
C	7/8″	17/32″
D	1 1/4″	3/4″
E	1 1/2″	29/32″
High-Capacity		
Classification	Width	Height
3V	3/8″	21/64″
5V	5/8″	35/64″
8V	1″	7/8″

Figure 13-16. V-belts are classified as light-duty (2L, 3L, 4L, and 5L), standard (A, B, C, D, and E), and high-capacity (3V, 5V, and 8V).

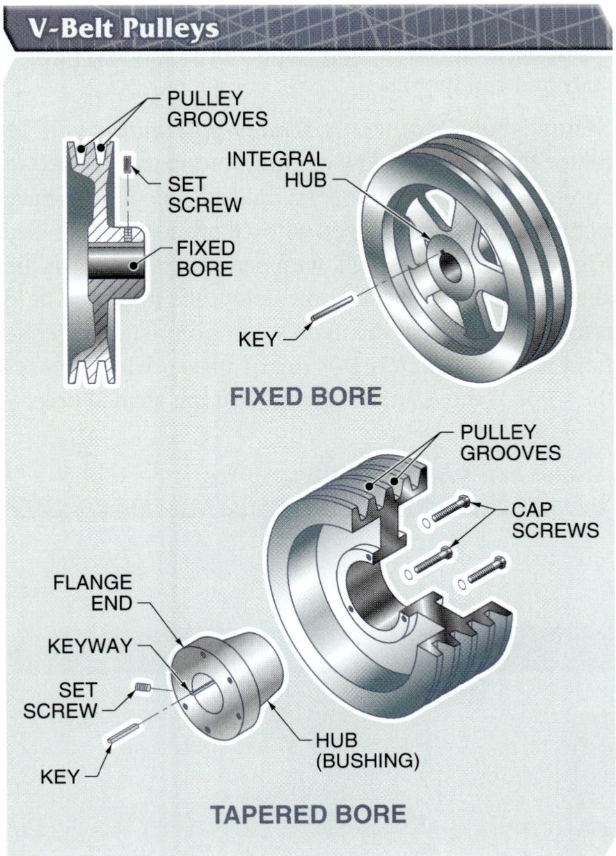

Figure 13-17. V-belt pulleys are single grooved or multiple grooved and are of fixed bore or tapered bore designs.

Shaft Alignment

Shafts that are not aligned properly destroy bearings, couplings, and belts quite quickly. Shaft misalignment is categorized as offset, nonparallel, or angular. **See Figure 13-18.**

Offset Misalignment. *Offset misalignment* is a condition in which coupling hubs or pulleys are not on the same axis. Typically, a string or straightedge across the coupling or pulley surfaces can be used to correct offset misalignment.

Nonparallel Misalignment. *Nonparallel misalignment* is a condition in which shafts are not parallel with one another. Typically, a string or straightedge across the pulley faces or a straightedge on the shafts of couplings can be used to correct nonparallel misalignment.

Angular Misalignment. *Angular misalignment* is a condition in which shafts are parallel in one plane but at different angles in another. Typically, a level on the surface of a coupling or pulley can be used to correct angular misalignment.

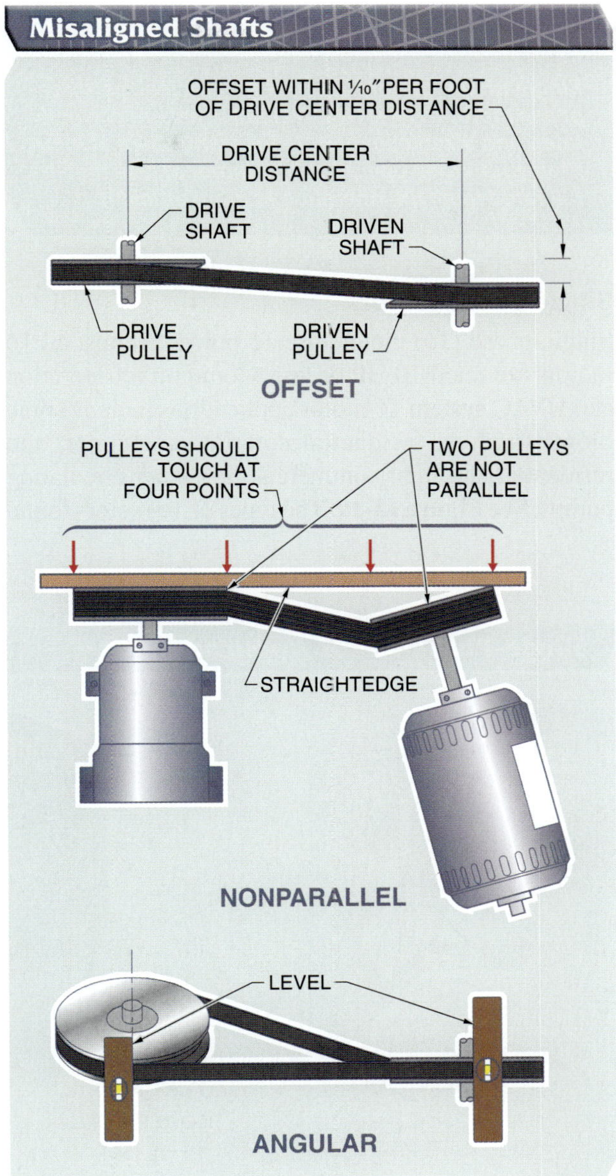

Figure 13-18. Shaft misalignment is the condition where the centerlines of two shafts are not aligned within specified tolerances.

SINGLE-PHASE AND THREE-PHASE MOTORS

While motors used in refrigeration and HVAC systems can be categorized in many ways, it is common to break them down into single-phase (1ϕ) and three-phase (3ϕ) types. In general, 1ϕ motors are used in smaller residential and light commercial applications. Three-phase motors are used in light commercial, heavy commercial, and industrial refrigeration and HVAC applications.

TECH TIP

Single-phase motors are not as efficient as three-phase motors. However, 1ϕ motors that are correctly rated and sized can operate for many years without problems. When operating a 1ϕ motor at less than its full rated load, the operation is more efficient.

Single-Phase Motors

Single-phase (1ϕ) motors use 1ϕ power. Almost all 1ϕ motors are rated 10 HP or less. Common refrigeration and HVAC system 1ϕ motor applications include some blower motors, residential compressor motors, and residential and light commercial hot water circulating pumps. **See Figure 13-19.** The types of 1ϕ motors found in refrigeration and HVAC systems are shaded-pole, split-phase, capacitor-start, capacitor-run, and capacitor start-and-run motors.

Shaded-Pole Motors. A *shaded-pole motor* is a 1ϕ motor that develops low starting torque and is used in applications of 1⁄20 HP or less. Shaded-pole motors have a small pole that is wrapped with a thick copper wire. **See Figure 13-20.** This small, wrapped pole is known as the shaded pole. The shaded pole delays the magnetic field effect at that part of the main stator pole. The magnetic field is approximately 90° out of phase, which causes the rotor to move in the direction of the shaded pole.

Shaded-Pole Motor Operation Media Clip

Single-Phase Motor Applications

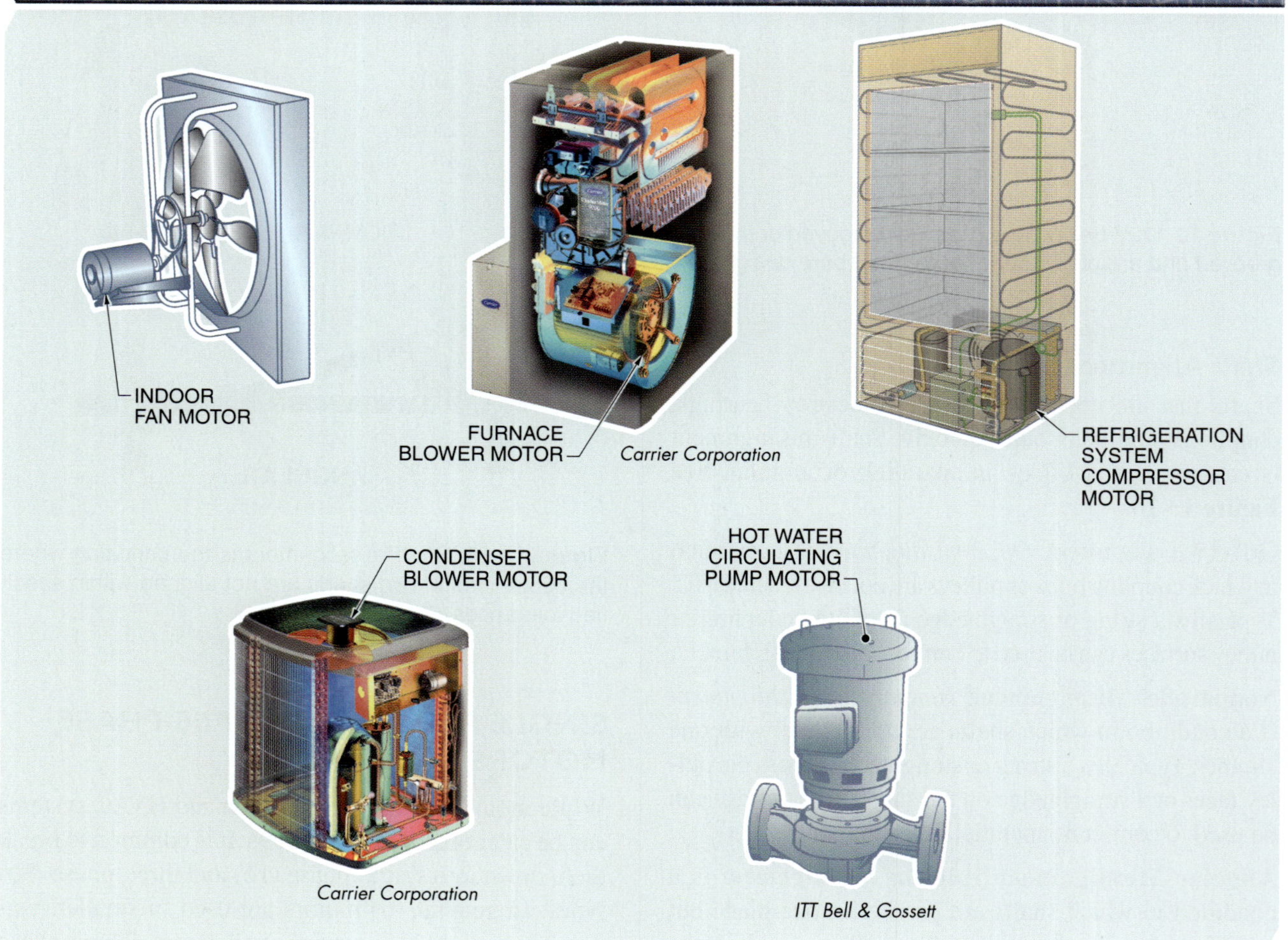

Figure 13-19. Single-phase motors are commonly used for fans, furnaces, refrigeration and HVAC compressors, condenser units, and hot water circulating pumps in residential and light commercial applications.

Shaded-Pole Motors

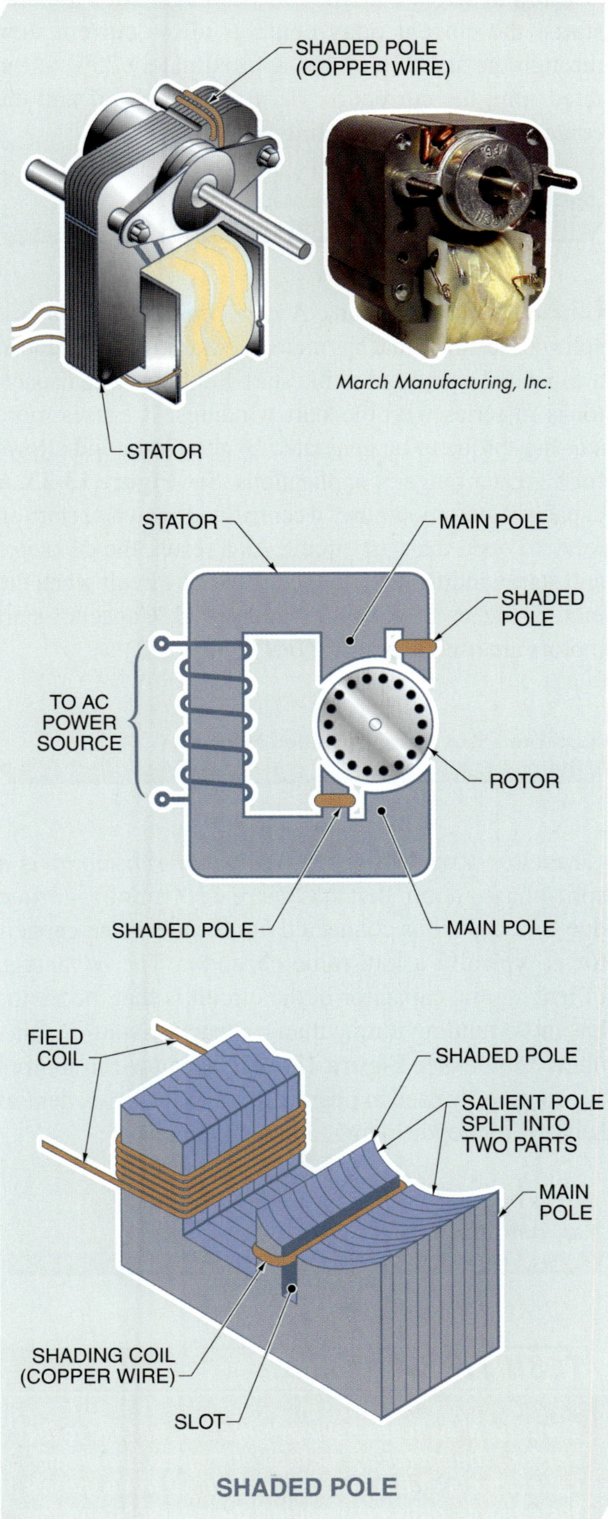

Figure 13-20. A shaded-pole motor is a 1ϕ motor that develops low starting torque and is used in applications of 1⁄10 HP or less.

Split-Phase Motors. A *split-phase motor* is a 1ϕ motor that has separate start windings and run windings. **See Figure 13-21.** The start windings have smaller diameter wire than the run windings. The start and run windings have magnetic fields out of phase with each other. The start winding provides enough starting torque to overcome resistance to movement of the load. It also brings the motor up to speed quickly.

Split-Phase Motors

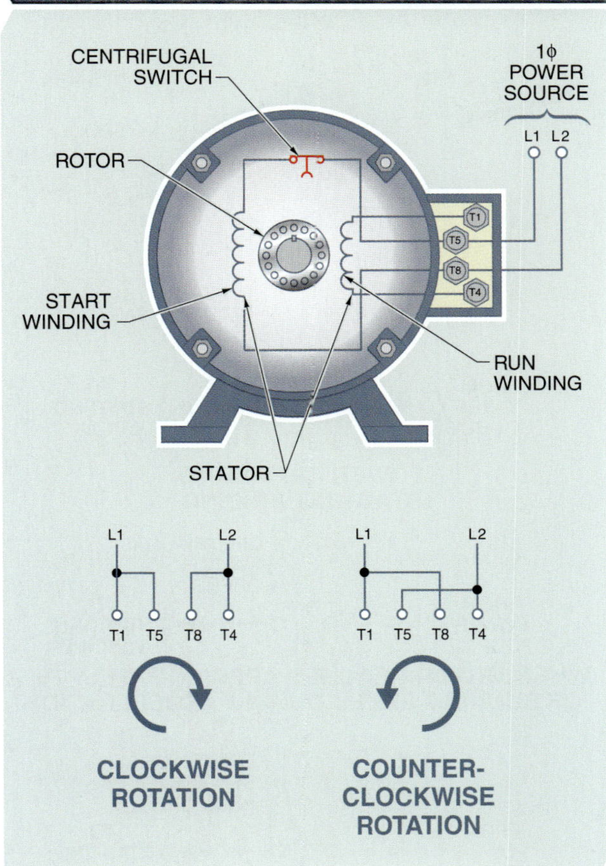

Figure 13-21. A split-phase motor has separate start windings and run windings that provide extra starting torque and allow the motor to get up to speed quickly.

The run winding continues to rotate the motor shaft and keep it at the desired speed. Most motors with start and run windings disconnect the start winding from the circuit after the motor is up to speed. The start winding is disconnected because of its high resistance, high current draw, and heat generation. The start and run windings are in parallel and energized upon motor startup.

Split-phase motors have centrifugal switches attached to the motor shaft. **See Figure 13-22.** When the motor shaft reaches approximately 75% of full speed, the centrifugal switch opens and disconnects the start winding from the circuit. Even with both start and run windings energized, most split-phase motors have a relatively low starting torque and are limited to applications of ⅓ HP or less.

Centrifugal Switches

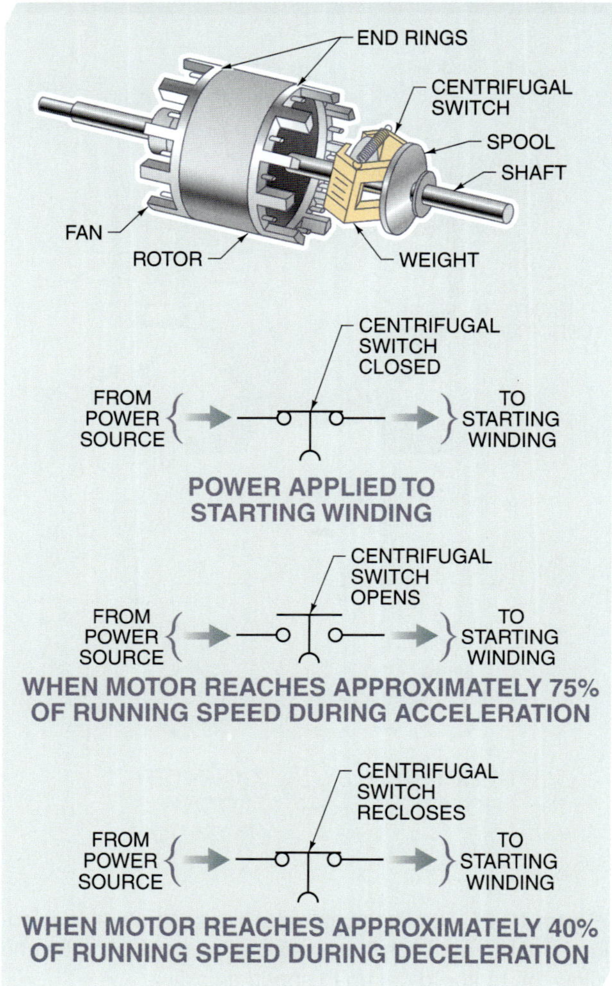

Figure 13-22. A split-phase motor has a centrifugal switch that opens and disconnects the start winding when the motor shaft reaches approximately 75% of full speed.

Besides centrifugal switches, potential relays and current relays open the start circuit when the motor reaches approximately 75% of its rated speed. A potential relay consists of a coil that senses the counter electromotive force (CEMF) of the start circuit. A current relay coil is wired in series with the run winding. When a motor starts, the current relay contacts allow current flow through the start circuit. At approximately 75% of the rated rpm, the current to the motor drops off and the contacts open, disconnecting the start circuit.

Split-Phase Motor Operation Media Clip

Capacitor-Start Motors. A *capacitor-start motor* is a split-phase motor that has increased starting torque due to a capacitor connected to the start windings. The capacitor is in series with the start windings. It causes more starting torque to be generated by the motor and allows for a greater range of applications. **See Figure 13-23.** A capacitor-start motor uses a centrifugal switch or starting relay to open the start circuit. As a result, the capacitor and start winding are removed from the circuit when the motor reaches 60% to 80% full speed. Capacitor-start motors are used in applications up to 10 HP.

Capacitor-Start Motor Operation Media Clip

Capacitor-Run Motors. A *capacitor-run motor* is a split-phase motor that has increased running torque due to a capacitor connected at all times. The capacitor is typically a low-value capacitor. The advantage of leaving the capacitor in the circuit is that the motor has more running torque than a capacitor-start or split-phase motor. **See Figure 13-24.** Capacitor-run motors are commonly used to operate indoor fans in residential heating, air conditioning, and heat pumps.

Capacitor-Run Motor Operation Media Clip

TECH TIP

Centrifugal switches have weights that are held to the shaft of the motor by springs and close the electrical contacts of the switch. Centrifugal force throws the weights outward, opening the contacts and the electrical circuit. Centrifugal switches deactivate or activate motor control circuits when the motor reaches 75% to 80% of its operating speed.

Capacitor-Start Motors

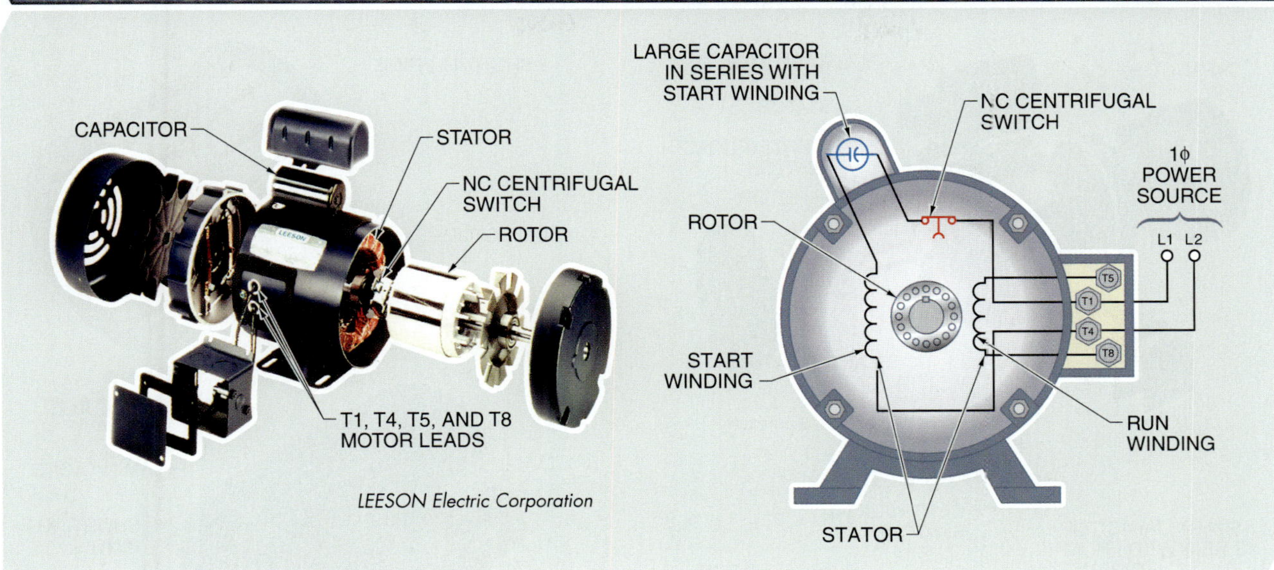

Figure 13-23. A capacitor-start motor is a split-phase motor that has increased starting torque due to a capacitor connected to the start windings.

Capacitor-Run Motors

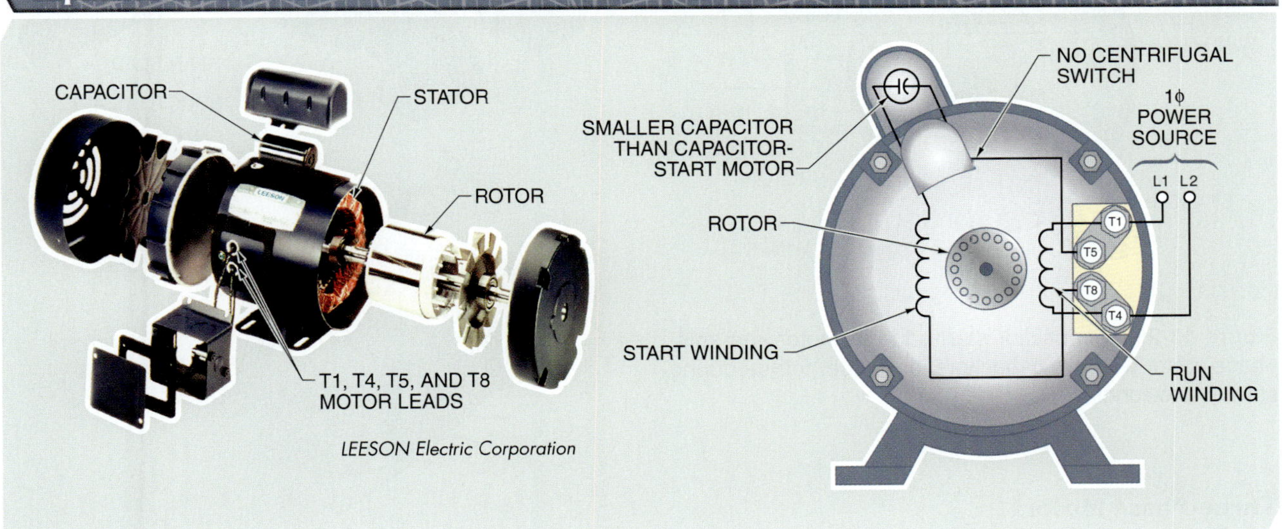

Figure 13-24. A capacitor-run motor is a split-phase motor that has increased running torque due to a capacitor connected at all times.

Capacitor Start-and-Run Motors. A *capacitor start-and-run motor* is a split-phase motor with two capacitors to allow greater torque during startup and running. **See Figure 13-25.** One capacitor is in the start circuit and the other is in the run circuit. The start capacitor generates additional starting torque and is removed from the circuit when the motor comes up to speed. The run capacitor has a low capacitance value and remains in the circuit while the motor is operating.

Capacitor Start-and-Run Motor Operation
Media Clip

Capacitor Start-and-Run Motors

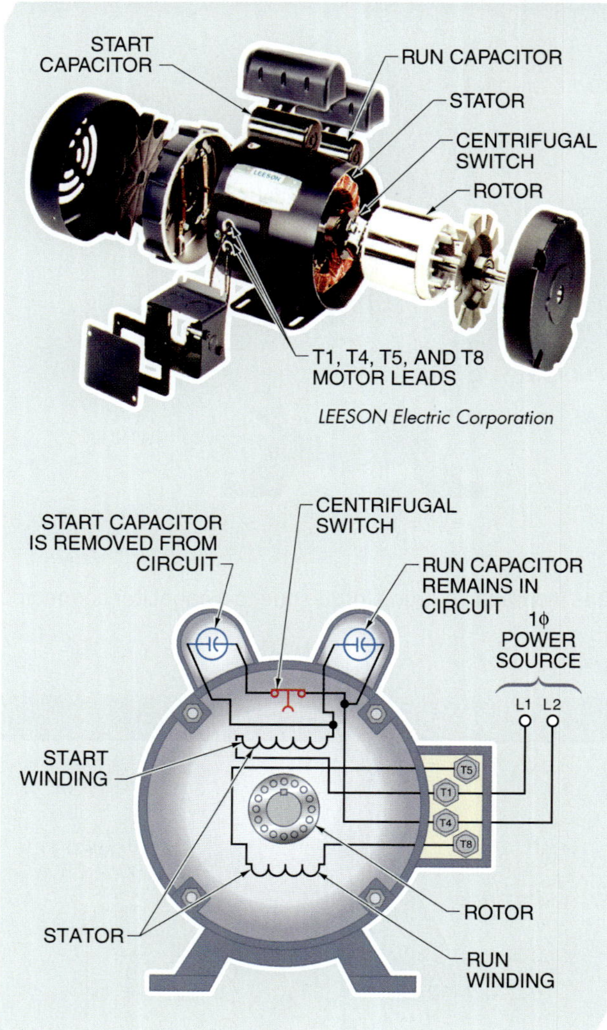

Figure 13-25. A capacitor start-and-run motor is a split-phase motor with two capacitors for greater torque during startup and running.

Three-Phase Motors

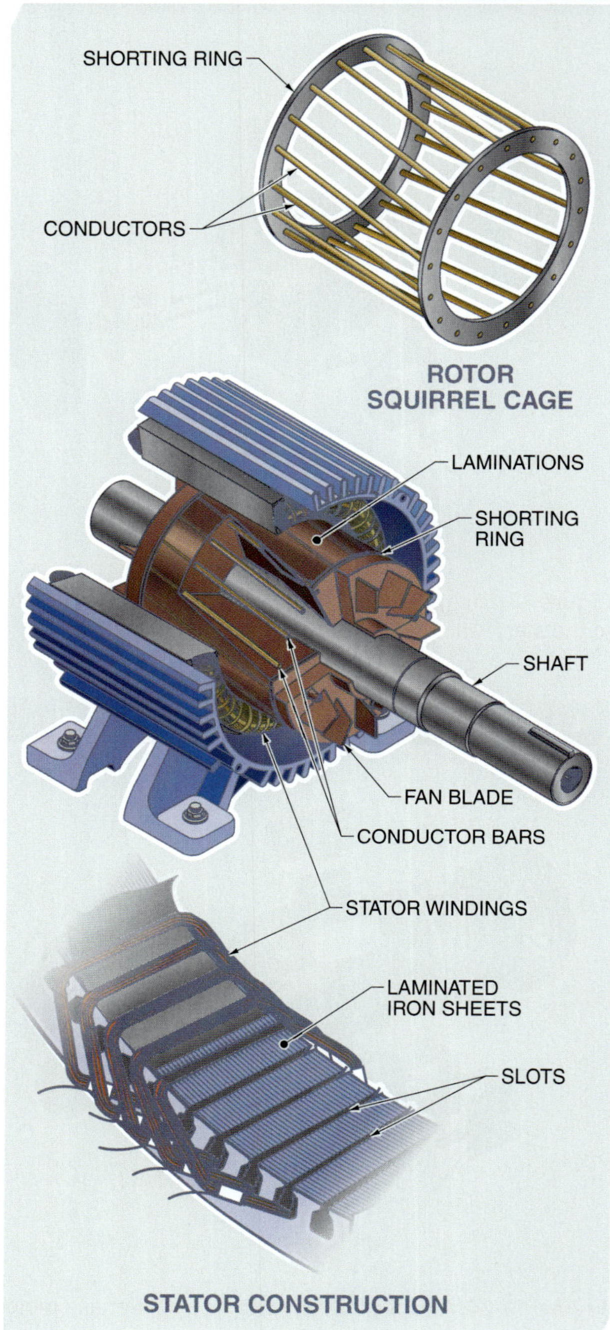

Figure 13-26. A 3ϕ motor requires little maintenance, is simple in construction, creates large amounts of torque, and costs less to operate than a 1ϕ motor.

Three-Phase Motors

A *three-phase (3ϕ) motor* is a motor that uses three phases of electricity to create a rotating magnetic field. A 3ϕ motor is considered to be heavy-duty. They are used in most industrial and commercial applications. The advantages of 3ϕ motors are that they require little maintenance, are simple in construction, create large amounts of torque, and cost less to operate than 1ϕ motors. While 1ϕ motors are generally used in applications of 3 HP or less, 3ϕ motors are often used in applications of very high horsepower, such as 1000 HP or more. **See Figure 13-26.**

Most 3ϕ motors are induction motors. Induction motors create a rotating magnetic field in the stator windings, which induces opposite polarity magnetic fields in the rotor bars. **See Figure 13-27.** Three-phase motors use a 3ϕ power supply that is wired to three separate windings in the motor stator. Since each phase is 120° out of phase with the others, large amounts of starting and running torque are guaranteed.

Induction Motor Operation Media Clip

Three-Phase Induction Motors

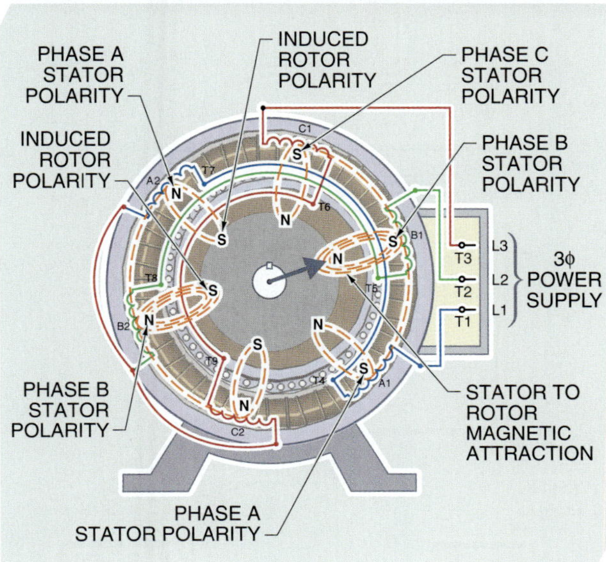

Figure 13-27. Three-phase induction motors consist of a stator and rotor inside a frame that have no physical connections between each other.

Three-phase motors are available in different varieties. Some 3ϕ motors are single-voltage motors that use only a specific 3ϕ voltage, such as 230 VAC. Others are dual-voltage motors that can be used with multiple voltages. Single-voltage, 3ϕ motors are either wye- or delta-connected.

Wye- and Delta-Connected Motors. Each winding in a wye-connected motor is joined at the center to the other two windings. **See Figure 13-28.** The windings in a delta motor form a continuous closed-loop circuit. In either connection type, the main motor leads are labeled T1, T2, and T3. These motor leads correspond to the power supply leads labeled L1, L2, and L3.

Dual-Voltage, 3ϕ Motors. The availability of dual-voltage motors increases the flexibility of motor application and reduces the motor inventory needed for a facility. For any 3ϕ motor to operate properly, the voltage and frequency of the power supplied must match that of the motor rating.

Wye- and Delta-Connected Motors

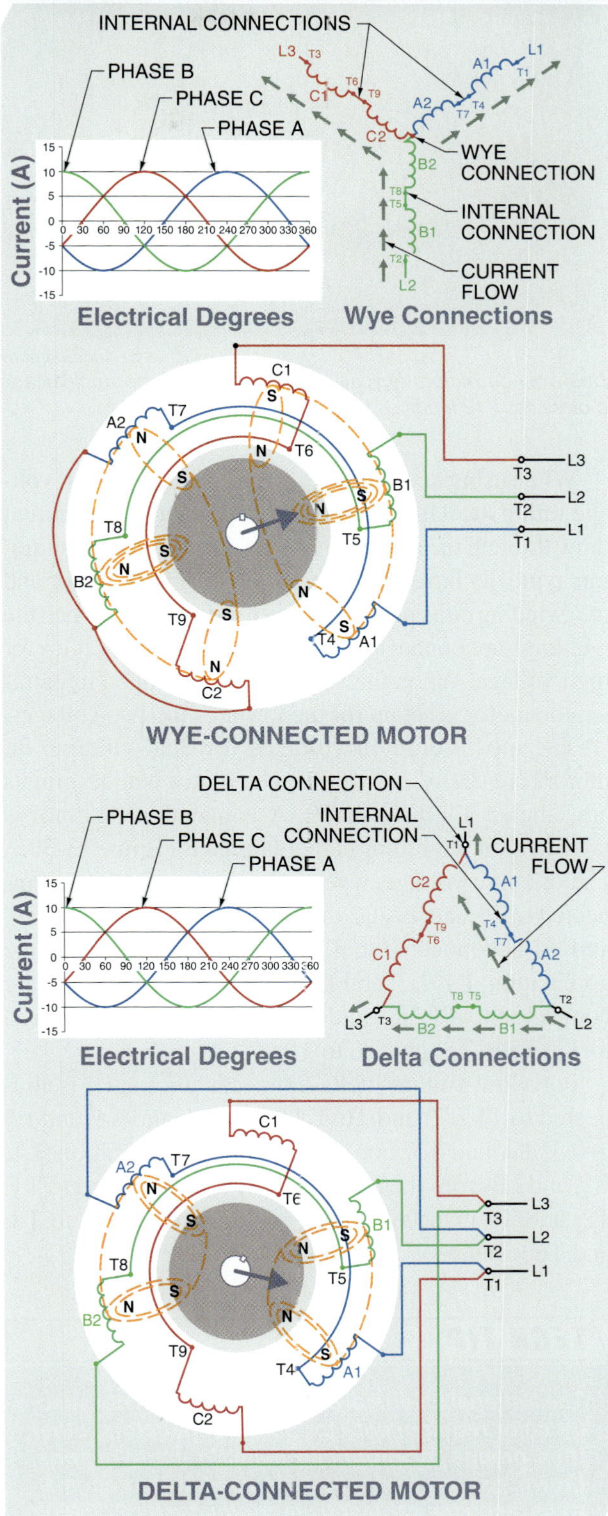

Figure 13-28. Wye- and delta-connected motors develop different torque and horsepower values for continuous run and start-stop applications.

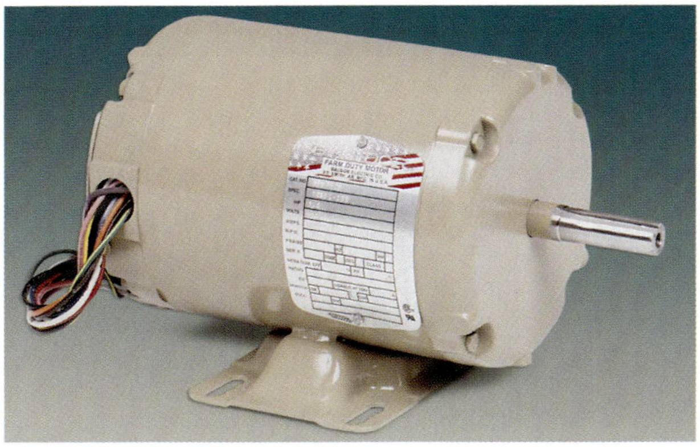

Baldor Electric Company

Dual-voltage motors typically have 9 leads. Some dual-voltage motors have 12 leads.

When using dual-voltage, 3ϕ motors, the highest voltage rating available should be used because the current draw through the motor is reduced. Dual-voltage, 3ϕ motors typically have nine wiring leads from the motor and two winding coils per phase. **See Figure 13-29.** When the windings are connected in series, the motor is set up for high voltage. When the windings are arranged in parallel, the motor is set up for the lower of the two voltages.

The most common voltages for dual-voltage, 3ϕ motors are 230 V and 460 V. The motor lead terminals are labeled T1 through T9. A connection diagram is provided on the motor nameplate. **See Figure 13-30.**

For a low-voltage, wye-connected motor, L1 is connected to T1 and T7, L2 to T2 and T8, L3 to T3 and T9, and T4 is connected to T5 and T6. For a high-voltage application, L1, L2, and L3 from the power supply are connected to T1, T2, and T3 respectively. T4 is wired to T7, T5 to T8, and T6 to T9.

For a low-voltage, delta-connected motor, L1 is connected to T1, T7, and T6; L2 to T2, T8, and T4; and L3 to T3, T9, and T5. For a high-voltage application, L1, L2, and L3 from the power supply are connected to T1, T2, and T3 respectively. T4 is wired to T7, T5 to T8, and T6 to T9.

TECH TIP

The National Electrical Code® (NEC®) does not specify color requirements for phase conductors, other than orange for grounded delta. In many regions, brown, orange, and yellow are used, with orange always being the center phase conductor for 277/480Y systems. Per NEC® 517.160(A)(5), conductor colors are to be used for isolated power systems in health care facilities.

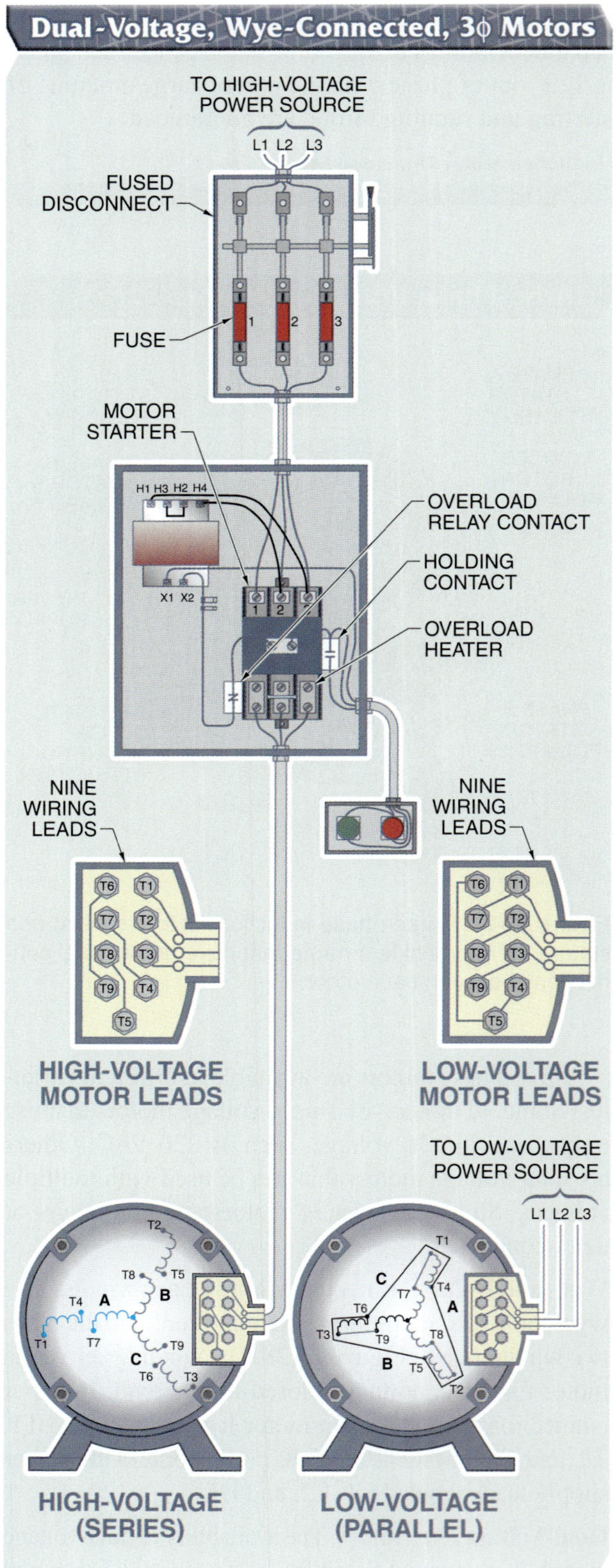

Figure 13-29. Dual-voltage motors increase the flexibility of motor applications and reduce the motor inventory of a facility.

Motor Wiring Diagrams

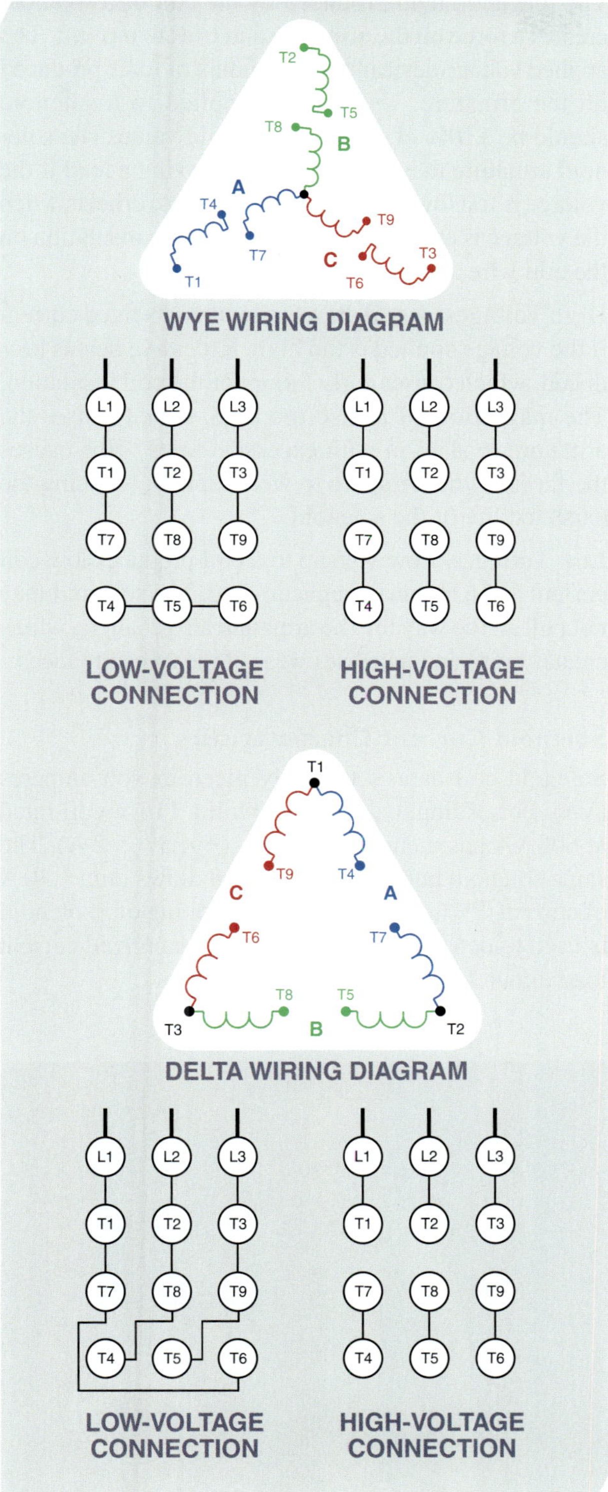

Figure 13-30. The most common voltages for wye- and delta-connected, dual-voltage, 3ϕ motors are 230 V and 460 V. Each motor and voltage requires special connection wiring.

SOLENOIDS

A *solenoid* is an electric output device that converts electrical energy into linear mechanical force. The magnetic attraction of the solenoid coil is used to create force. The coil is combined with an armature that transmits the force into useful mechanical work. An *armature* is the movable part of a solenoid.

Solenoid Configurations

Solenoids are configured in various ways for different applications and operating characteristics. The solenoid configurations are clapper, bell-crank, horizontal action, vertical action, and plunger. **See Figure 13-31.**

Solenoid Configurations

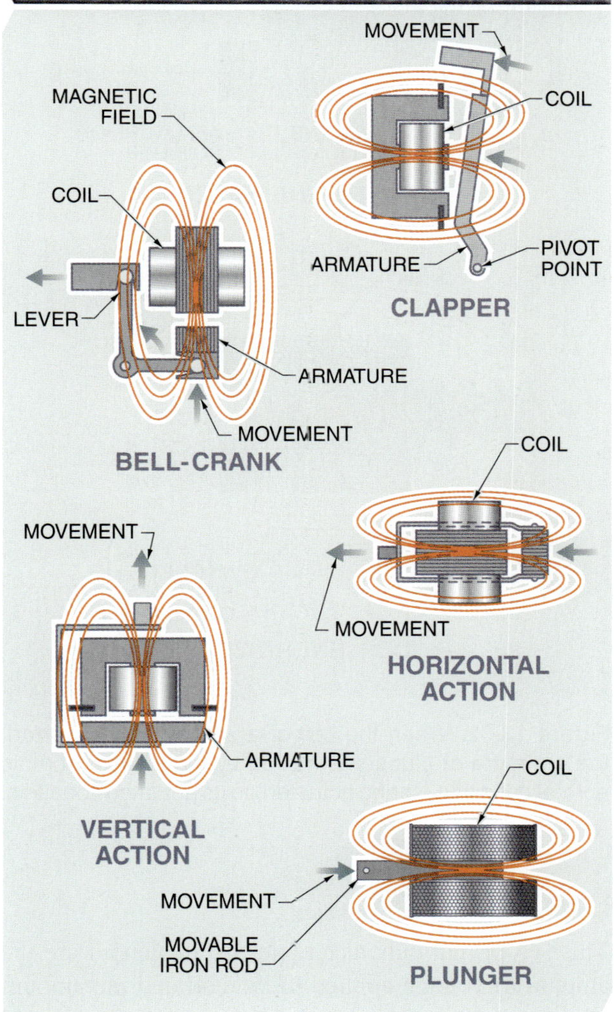

Figure 13-31. Solenoids use magnetic attraction to transmit mechanical force regardless of the solenoid configuration.

Solenoid Operation

Solenoids are constructed of many turns of wire wrapped around an iron frame. The iron frame increases the magnetic force produced by the coil. Passing electric current through the coil causes the armature or plunger to be pulled to the center of the coil. Devices are attached to the armature or plunger to accomplish tasks like opening and closing contacts. **See Figure 13-32.**

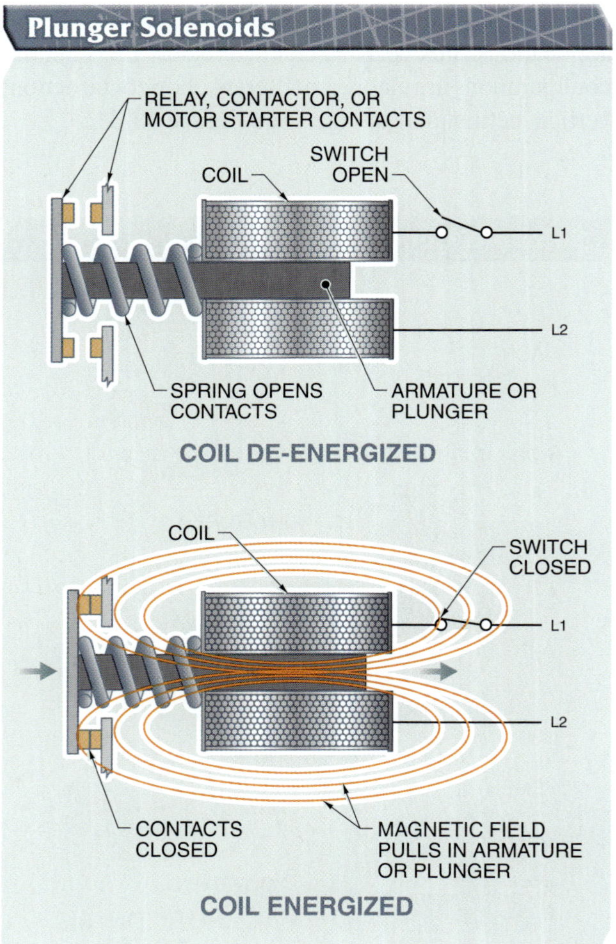

Figure 13-32. When the coil of a solenoid is energized, the armature or plunger is moved by magnetism from its normal condition (spring actuated) to its actuated condition.

Solenoid Characteristics

The two primary characteristics of a solenoid are the amount of voltage applied to the coil and the amount of current allowed to pass through the coil. A solenoid develops a magnetic field in the coil when voltage and current are applied.

Solenoid Voltage Characteristics

The magnetic field produced by the coil of a solenoid creates a force on the armature and tries to move it. The applied voltage determines the amount of force produced on the armature. The voltage applied to a solenoid should be ±10% of the rated solenoid value. The solenoid armature may have difficulty moving a load if the voltage is too low. However, a solenoid overheats when the voltage is excessive. Heat destroys the insulation on the coil wire and burns out the solenoid.

High Voltages. A coil draws more than its rated current if the voltage applied is too high. Excessive heat is produced, which causes early failure of the coil insulation. The magnetic pull is also too high, which causes the armature to slam-in with excessive force. This causes the faces of the armature to wear rapidly, reducing the expected life of the solenoid.

Low Voltages. Low voltage to a coil produces low coil current and a reduced magnetic field. The solenoid may not pull all the way in. The armature also chatters, which creates noise and increases wear of the armature faces.

Solenoid Current Characteristics

Solenoid coil data is typically given in volt-amperes (VA). For example, a solenoid with a 115 V coil rated at 600 VA has a current of 5.2 A ($^{600}/_{115}$ = 5.2 A). The same solenoid but with a 460 V coil draws only 1.30 A of current ($^{600}/_{460}$ = 1.30 A). The VA rating of a solenoid is used to determine the starting and energized current load drawn from the power supply.

Carrier Corporation

In refrigeration and HVAC systems, solenoids are commonly used to shift valves that control water, air, or refrigerant flow.

Solenoid Applications

Solenoids are found in a wide range of equipment. In residential equipment, solenoids are used in items such as doorbells, washing machines, and kitchen appliances. Solenoids are also used in light commercial control-circuit applications such as refrigeration systems, combustion systems, and general-purpose HVAC controls.

Refrigeration. Direct-acting, two-way valves are found in many refrigeration and HVAC systems. **See Figure 13-33.** Two-way (shutoff) solenoid-actuated valves have one inlet and one outlet pipe connection. These valves are constructed as normally open (NO) or normally closed (NC). In NO valve construction, the valve is open when the solenoid is de-energized and closed when energized. In NC valve construction, the valve is closed when the solenoid is de-energized and open when energized.

Solenoid-Actuated Valves

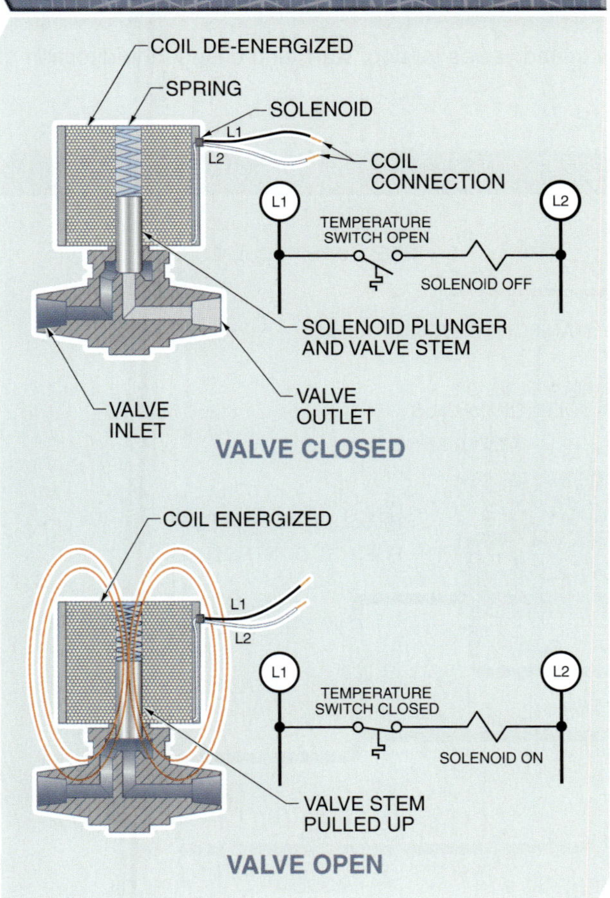

Figure 13-33. All two-way and three-way valves used in refrigeration and HVAC systems have one solenoid actuator.

A number of different solenoids may be used in a typical refrigeration or HVAC system. For example, a hot gas solenoid valve remains closed until the defrost cycle, and then it feeds the evaporator with hot gas for the defrost operation. **See Figure 13-34.** Liquid-line solenoid valves may be operated by 2-wire or 3-wire thermostats.

Combustion. Solenoids are also used in oil-fired, single-sequence burner systems. Solenoids are required in the startup and in the normal operation of oil-fired burner systems. **See Figure 13-35.**

TROUBLESHOOTING MOTORS

Refrigeration and HVAC technicians must be able to troubleshoot common motor problems encountered in the field. Most motor problems involve mechanical, incoming power, or electrical control issues.

Mechanical Problems

Mechanical problems that lead to motor failure are typically due to issues with the load connected to the motor or the motor-to-load mechanical connection. Mechanical problems are typically found by examining the load or the connection mechanism of the shafts. **See Figure 13-36.** When a motor protection device has opened and stopped the motor, or a circuit overcurrent protection has tripped the circuit breaker, the load must be checked for the following problems:

- misalignment between the motor and load
- improperly fastened motor and/or load
- defective bearings in motor or load
- object that prevents the load from turning
- defective, incorrectly installed, or improperly sized motor and/or connection mechanism (belt or coupling)

Incoming Power Problems

Incoming power problems mainly consist of voltage imbalance, current imbalance, or phases out of angle. Any phase-to-phase voltage imbalance greater than 2% causes motor windings to overheat and damage to occur. Voltage imbalance can be determined by taking voltage measurements of each phase supplying power to a system. **See Procedure for Finding Percent of Voltage Imbalance.** Loose conductors, improper grounding, or internal motor problems may cause unbalanced current flow in the conductors. Three-phase motors require that all three phases be ±10% of the rated voltage.

Refrigeration and HVAC System Solenoids

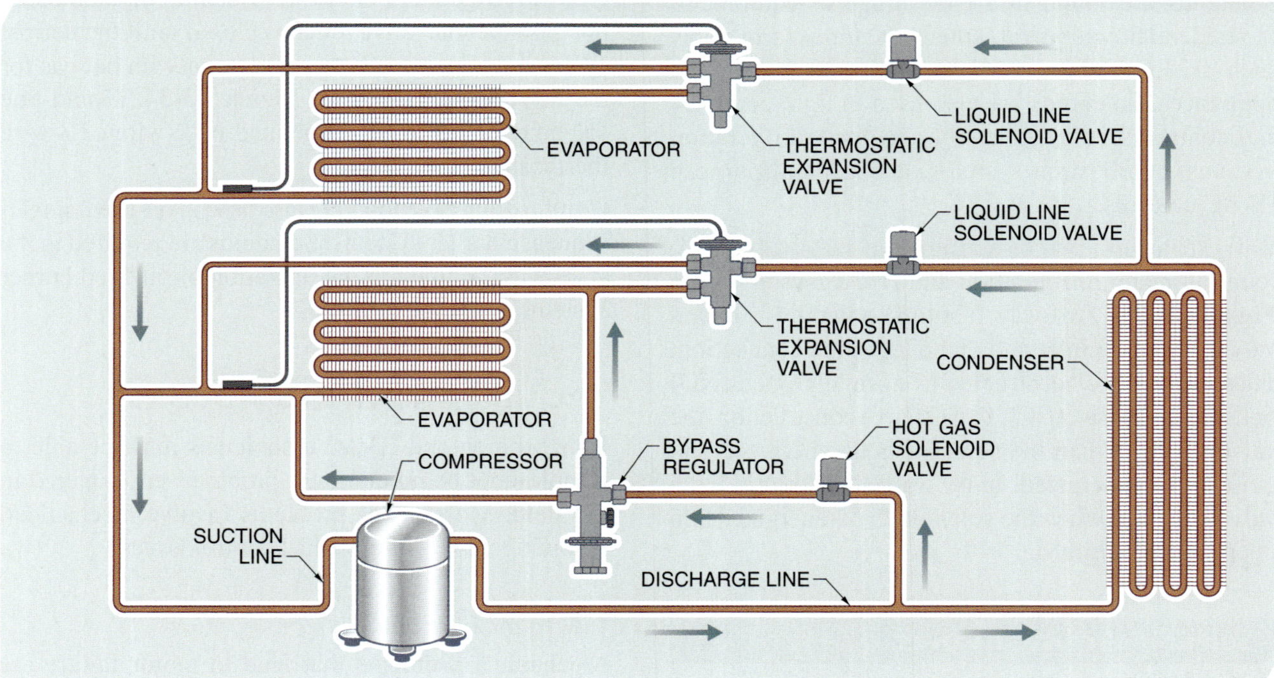

Figure 13-34. Refrigeration and HVAC systems use solenoid-actuated valves to stop, start, and change the direction of refrigerant flow.

Oil-Fired Burner Solenoid Valves

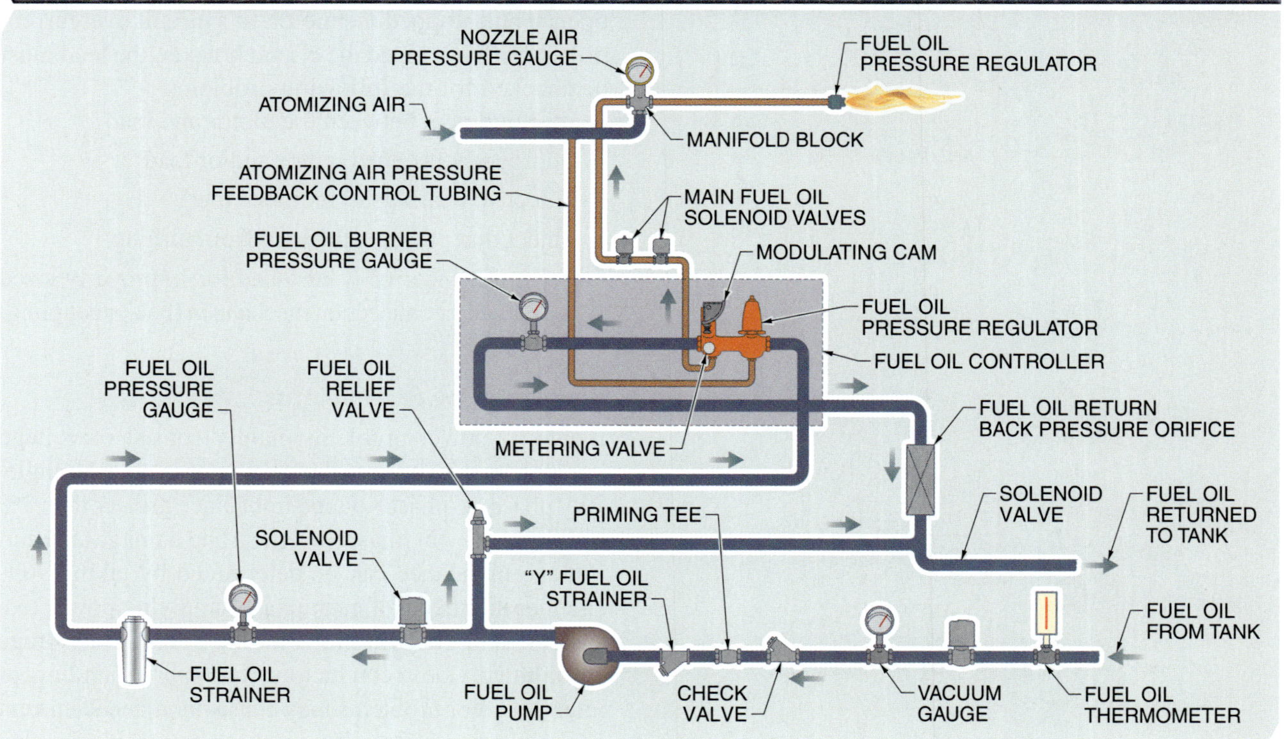

Figure 13-35. Many different solenoid-actuated valves are used to safely control the flow of oil to furnace and boiler burners.

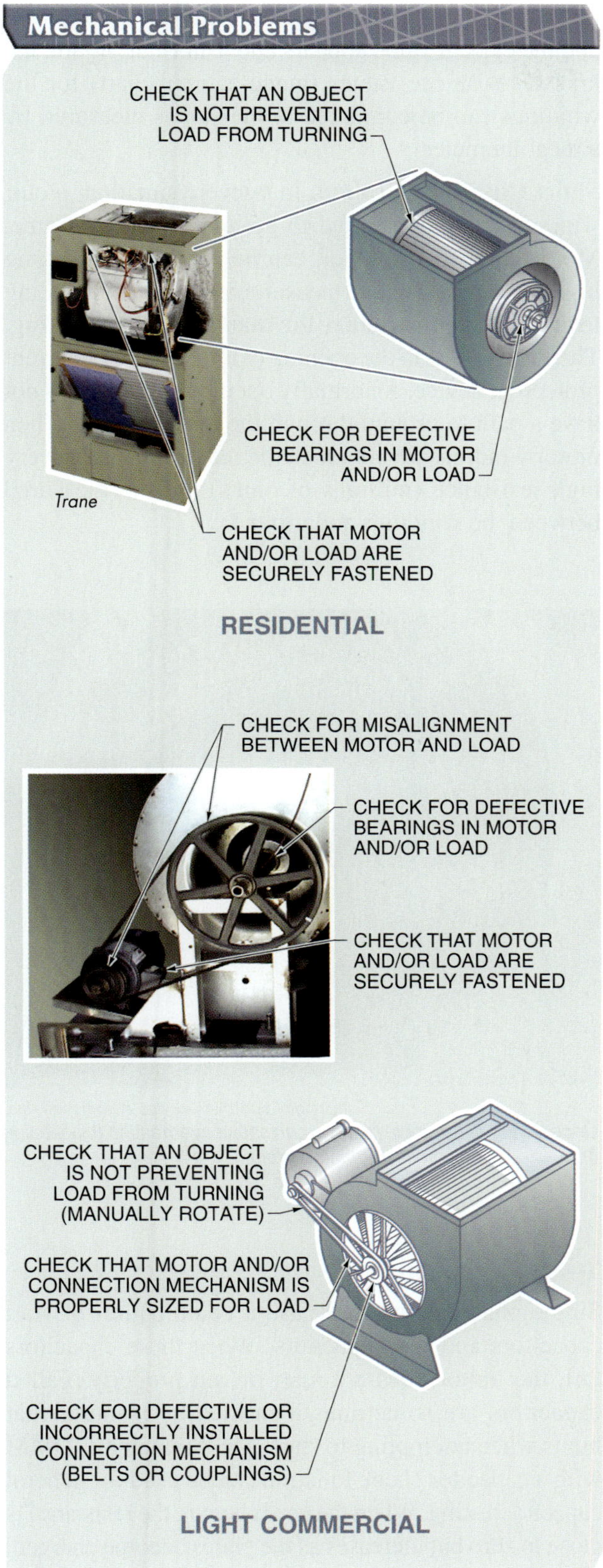

Figure 13-36. Many mechanical problems can lead to motor electrical overloads and thermal problems.

Finding Percent of Voltage Imbalance

Procedure

1. Voltage imbalance can be determined by taking voltage measurements of each phase supplying power to a system and using these measurements in calculations.
2. Connect the test leads to the appropriate multimeter meter jacks for taking AC voltage measurements.
3. Follow all electrical safety practices and procedures.
4. Set the multimeter selector switch to measure AC voltage ($\tilde{V}$).
5. Open the access door to the light commercial HVAC rooftop unit condenser fan motor starter.

6. Connect the voltmeter's test leads from phase 1 to phase 2. Repeat for phase 1 to phase 3 and phase 2 to phase 3.

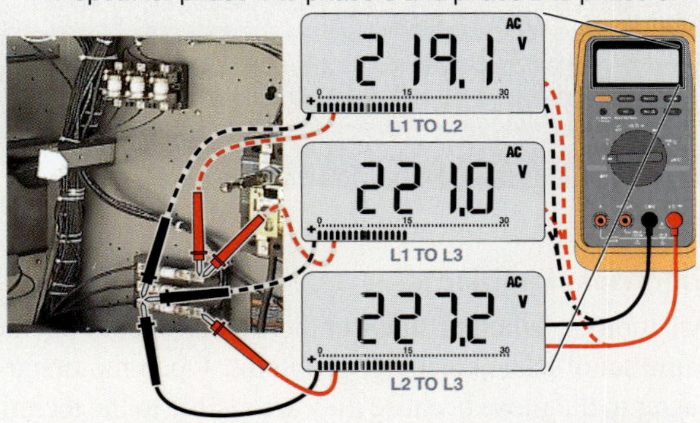

7. Read and record the voltage measurements indicated on the voltmeter.
8. Disconnect the voltmeter's test leads from the motor starter contacts.
9. Turn the multimeter's selector switch to OFF.
10. Add the measured incoming voltages together.

$$\begin{array}{r} 219 \\ 221 \\ +227 \\ \hline \mathbf{667\ V} \end{array}$$

11. Calculate the average of the measured voltages.

$$V_a = \frac{V}{3}$$

$$V_a = \frac{667}{3}$$

$$V_a = \mathbf{222\,V}$$

12. Calculate the largest voltage deviation from the average voltage.

$$V_d = V - V_a$$

$$V_d = 227 - 222$$

$$V_a = \mathbf{5\ V}$$

13. Calculate the voltage imbalance for the system.

$$V_a = \frac{V_d}{V_a} \times 100$$

$$V_a = \frac{5}{222} \times 100$$

$$V_a = 0.0225 \times 100$$

$$V_a = \mathbf{2.25\%}$$

When the phases of the incoming power are out of adjustment, a motor will overheat and become damaged. Most overheating problems are caused by poor incoming power from the utility company. Many firms with multiple, large critical motors install monitoring and alarm equipment at the incoming power supply that alerts the facility maintenance personnel when the incoming power supply is outside normal parameters. The power monitoring devices may automatically shut down the affected motors to prevent damage. The information the devices generate may also be used to contact the utility company and alert them of the condition.

TECH TIP

When incoming power has lower than the required voltage, two problems with HVAC equipment occur. First, motor life is shortened as motors draw excess current to produce the needed horsepower. Second, electronics that operate at 5 V or 24 V will not function properly because they have insufficient voltage to charge capacitors. Generally, this is not a problem when operating on utility power. However, HVAC technicians report numerous problems with voltage and frequency when operating on standby generators.

Electrical Problems

All motors, whether open or hermetic, encounter the same set of basic electrical problems. Open motors are easier to diagnose because they are visible to the technician. However, electrical diagnostic equipment such as a DMM must be used to understand the condition of a motor inside the hermetically sealed compressor can of a refrigeration system.

Open Windings. A common 3ϕ motor problem is that at least one motor winding is open. This means that the winding has a break that opens the electrical circuit inside the motor. Open windings can be found by using a quality DMM set to measure resistance or continuity. **See Procedure for Motor Winding Resistance Testing.** The run windings of a 1ϕ motor have a lower resistance than the start windings due to the difference in conductor size. Manufacturers provide winding resistance charts for the motors of hermetic compressors.

Shorted Motor Windings. In addition to the windings of a motor being open, a short circuit can also exist. The conductors that make up a motor winding are separated by an insulating material. When the insulator is worn, part of the winding may be bypassed. During operation, the winding resistance value appears below normal and the current value appears above normal. Again, the known resistance values (manufacturer chart) for the windings can be compared to the values measured by a megohmmeter.

Short Circuit to Ground. In a severe situation, motor windings may be shorted to ground through a frame. Motor winding insulation can be measured by taking high-voltage resistance measurements of each winding to ground. **See Procedure for Insulation Spot Testing.** The short results in the opening of the circuit overcurrent protection device. A normally operating motor does not have a path to ground through the motor frame. When motor windings do not touch the motor housing, a very high resistance (millions of ohms) will be measured between the windings and ground.

Electrical Apparatus Service Association, Inc.

Undervoltage and single phasing can cause overheating (blackening or burning) of the coils carrying current.

Testing Capacitors

Single-phase motors may use a combination of start capacitors and run capacitors. When these capacitors fail, the motor does not start or run properly. Failed capacitors are sometimes identified by a motor that hums when attempting to start. An ohmmeter or DMM with a capacitor check function can be used for general capacitor testing. When the test is begun, the resistance is close to zero but increases as the plates become charged. **See Procedure for Capacitor Testing.** If the resistance does not change or only changes slightly, the capacitor is probably bad. *Note:* This is a general test only.

Motor Winding Resistance Testing

1. Motor winding resistance can be measured by taking resistance measurements of each winding in a motor (9 or 12 leads must be present in motor connection box).
2. Ensure all power is OFF to the circuit and remove the motor being tested from the circuit.
3. Connect the test leads to the appropriate DMM jacks for taking resistance measurements.
4. Follow all electrical safety practices and procedures.
5. Set the DMM function switch to the resistance mode.
6. Remove the cover from the connection box of the motor.

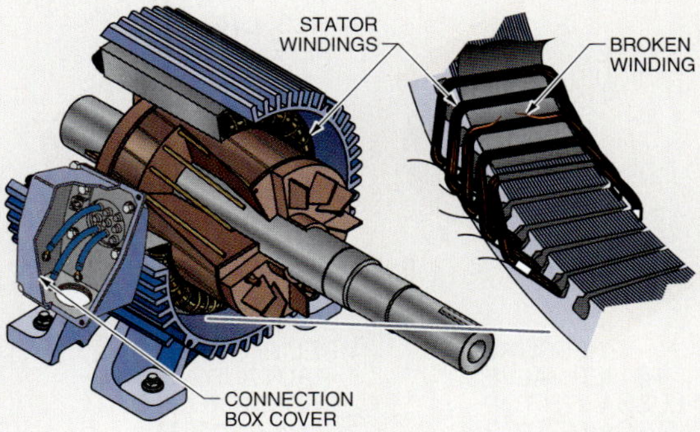

7. Connect the DMM test leads from windings T1 to T2. Repeat for windings T3 to T4, T5 to T6, T7 to T8, and T1 to T9. Repeat as need for diagnostics.

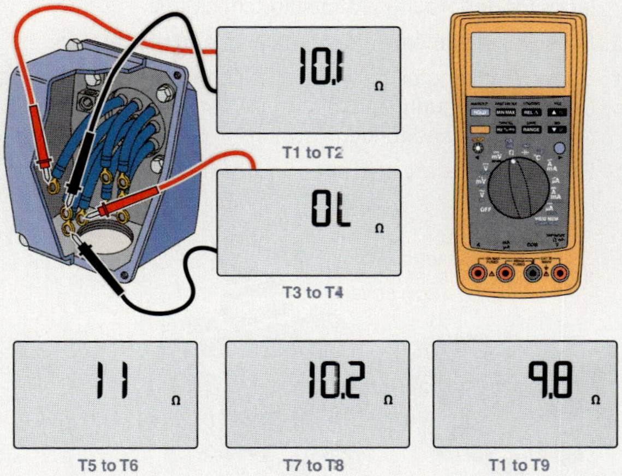

8. Read and record the resistance measurements indicated on the DMM.
9. Disconnect the DMM test leads from the motor leads.
10. Turn OFF the DMM.

Procedure

Insulation Spot Testing

1. Motor winding insulation can be measured by taking high voltage resistance measurements of each winding to the motor casing.
2. Connect the test leads to the appropriate insulation tester jacks for performing insulation spot testing.
3. Follow all electrical safety practices and procedures.
4. Set the insulation tester's selector switch to measure high-voltage resistance (1000 V).
5. Remove the cover from the connection box of the motor.

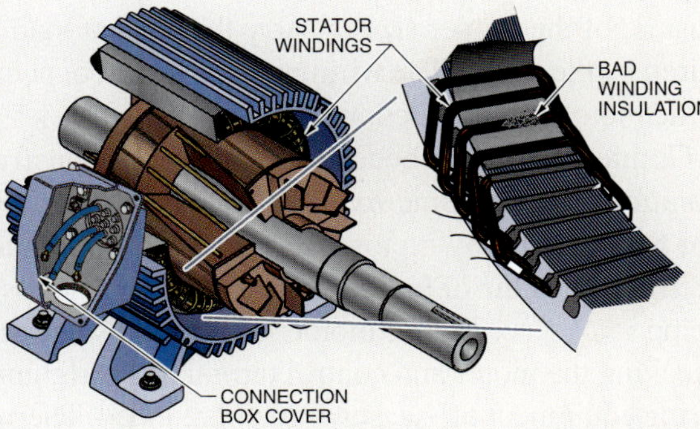

6. Connect the insulation tester's test leads from winding T1 to motor casing T2 to motor casing, and T3 to motor casing. Repeat as needed for diagnostic purposes.

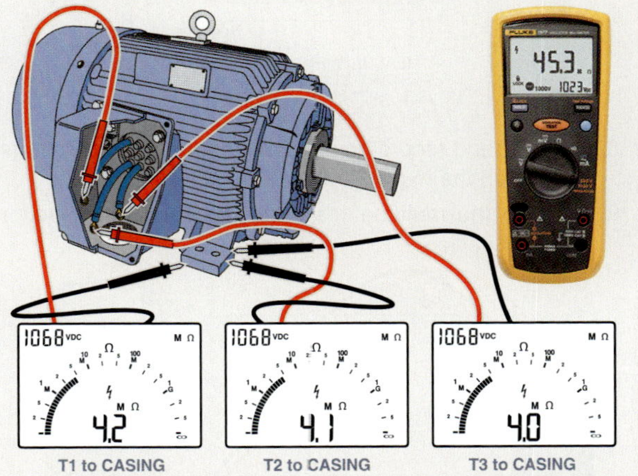

7. Read and record the resistance measurements indicated on the insulation tester.
8. Disconnect the insulation tester's test leads from the motor leads and casing.
9. Turn the insulation tester's selector switch to OFF.

Procedure

Procedure

Capacitor Testing

1. Motor run and/or start capacitors can be tested by using capacitor check measurements.
2. Connect the test leads to the appropriate multimeter jacks for performing a capacitor check test.
3. Follow all electrical safety practices and procedures.
4. Set the multimeter's selection switch to capacitor check (⊣⊢) for measuring microfarads. *Note:* It may be required to push a second function switch.
5. Remove the start or run capacitor from the motor and circuit.

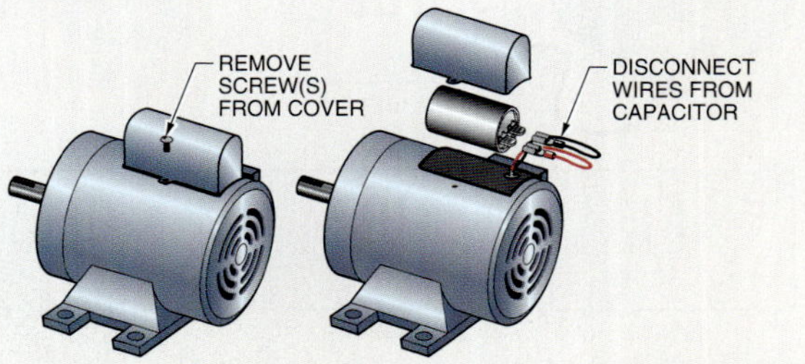

6. Discharge the capacitor by using a discharge resistor across the terminals to short the capacitor.

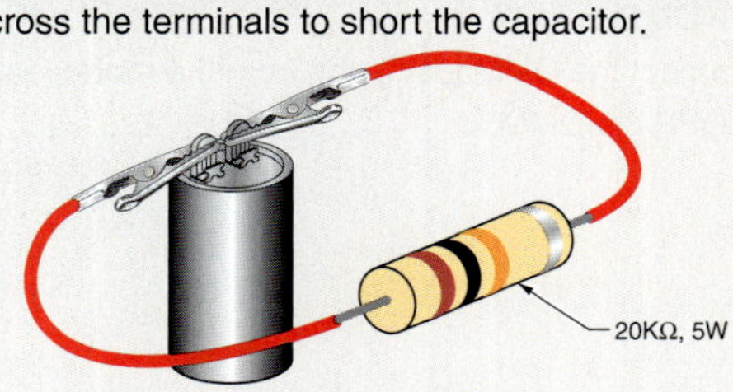

7. Connect the multimeter's test leads across the two terminals of the capacitor.

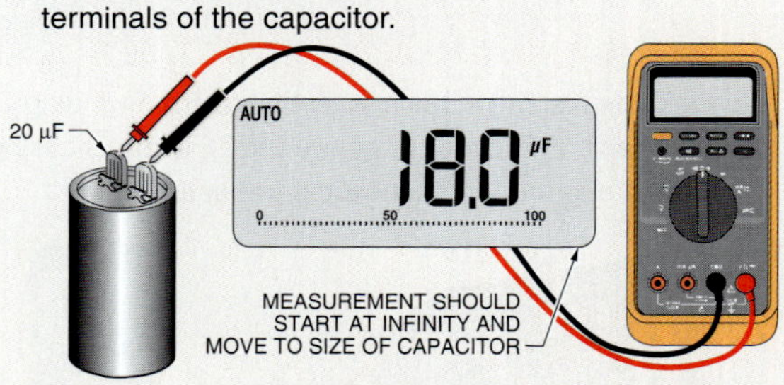

8. Observe and record how the microfarad measurements change on the multimeter.
9. Reverse the multimeter's test leads on the capacitor terminals.

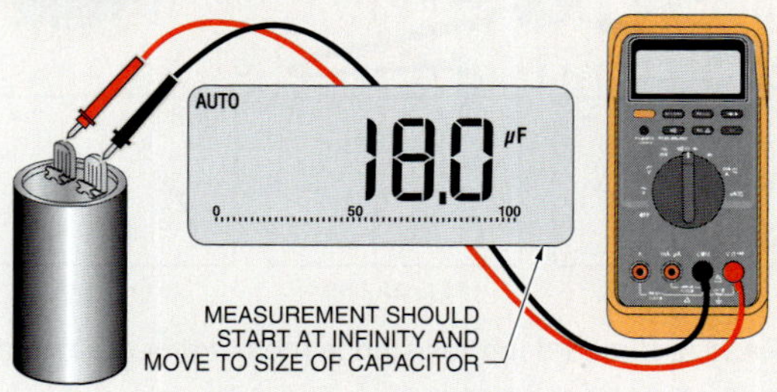

10. Observe and record how the microfarad measurements change on the multimeter.
11. Disconnect the multimeter's test leads from the capacitor.
12. Turn the multimeter's selection switch to OFF.

A capacitor analyzer can provide the exact capacitance of a capacitor in microfarads. The capacitor can then be compared to the nameplate value. Start capacitors have a cylindrical plastic shell. Run capacitors are typically filled with oil and are physically larger than start capacitors. The signs of a bad capacitor include bulging, leaking, discoloration, or excessive heating. **See Figure 13-37.**

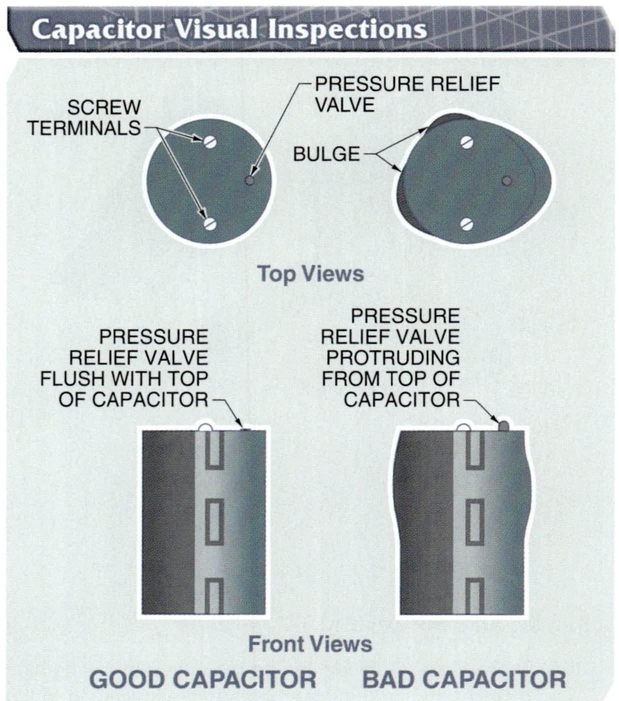

Figure 13-37. A visual inspection of a capacitor can identify deficiencies.

Wiring Connections

Another potential problem with motor operation is the condition of the conductors and terminations. The wiring must be of the proper size to carry the current load required by the motor. The wiring must also be connected to the motor leads and controls properly.

Connections that become loose over time from vibrations may become oxidized due to the high current flow created by motors. Terminal discoloration is often a visual clue of loose connections. In some cases, arcing is observed upon motor startup and shutdown. Checking the motor and control terminals for tightness during equipment off-season is recommended. Thermal imaging devices can also be used to check wiring connections. **See Figure 13-38.**

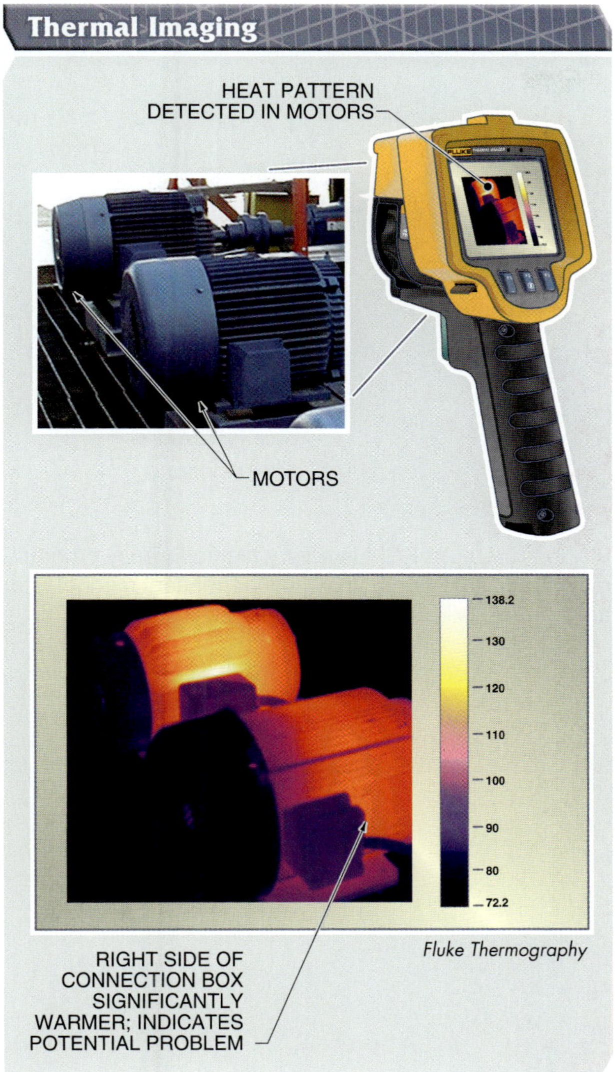

Figure 13-38. New technology, such as thermal imaging, is quite helpful when inspecting motors for problems while the motors are still operating.

Hermetic Motors

In a hermetically sealed compressor, the motor and compressor are sealed in the same housing. The motor is cooled by system refrigerant. The only access to the motor is through the shell of the compressor using the motor leads. The terminals protrude through the shell with a ceramic or plastic grommet. **See Figure 13-39.** While the motor itself can only be checked indirectly, any external devices, such as capacitors, current relays, potential relays, or wiring connections, can be easily seen and diagnosed.

Hermetic motors have the same problems as other motors such as open windings, shorted windings, or shorts to the motor frame. However, these problems are more difficult to detect in a hermetic system because there is no way to observe the motor in operation. When a motor fails in a hermetic compressor, the only solution is total replacement. Since the motor and compressor are sealed together, anything that affects the motor also affects the compressor and vice versa.

Figure 13-39. The only accessible electrical components of a hermetically sealed compressor are the terminal connections for supplying power.

TECH TIP

A serious and costly event encountered by HVAC technicians and system owners is hermetic motor burnout. Hermetic motor burnout sends contaminants, such as acids and other particulates, throughout an HVAC system.

ECM™ Motors

HVAC equipment manufacturers have used ECM™ motors in residential and light commercial furnaces and air conditioners since the 1980s. Originally, these motors were only available in voltages below 230 V, but they are now available in all voltages. ECM motors allow SEER 18 ratings for furnaces and air conditioners to be possible.

ECM Motor Operation

An *ECM (electronically commutated motor)* is a very high efficiency programmable brushless DC motor with a permanent magnet rotor and a built-in variable-frequency drive (VFD). **See Figure 13-40.** Because DC motor speed is easier to control and DC motor applications are more energy efficient than most AC motor applications, ECM motors are finding their way into more HVAC systems. Today, many ECM applications run the ECM motor at a fixed speed but achieve very high efficiency ratings.

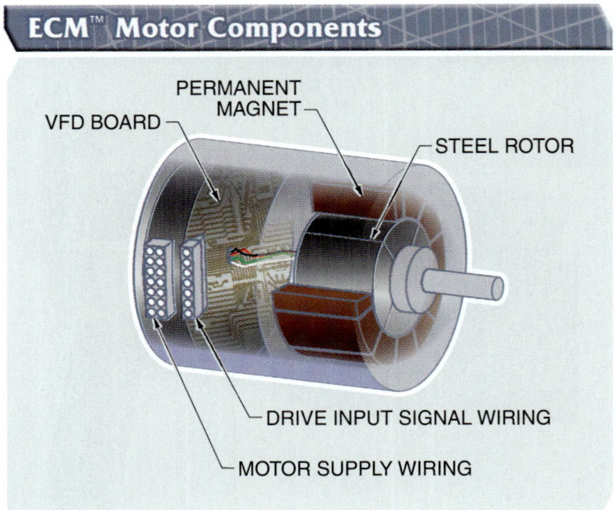

Figure 13-40. An ECM is a very high efficiency programmable brushless DC motor with a permanent magnet rotor and a built-in VFD.

Digital Resources ATPeResources.com/QuickLinks • Access Code: 482610

UNIT 14

HVAC and Refrigeration Control Systems

It is very important for HVAC technicians to get equipment up and running quickly after a failure. Over 50% of all refrigeration and HVAC system failures are electrical. It is critical that HVAC technicians are able to troubleshoot common control systems quickly and accurately.

OBJECTIVES

- Identify common HVAC and refrigeration control applications.
- Explain how a digital multimeter is used to troubleshoot common HVAC and refrigeration control system problems.
- Identify common HVAC and refrigeration control system problems.

Index to Procedures

ELECTRICAL CONTROL SYSTEMS

Most devices used in the refrigeration and HVAC industry are either powered or controlled by electricity. Refrigeration, heating, and air conditioning devices powered by electricity include compressors, fans, electric heaters, analog pressure and temperature switches, and solenoid valves. Electrical control systems are the most common control systems used in the refrigeration and HVAC industry.

An *electrical control system* is a control system that uses electricity to operate. **See Figure 14-1. See Appendix.** An *electromechanical control system* is a control system that uses electricity in combination with a mechanical component to operate. The mechanical component may be a pivot, mechanical bellows, or other device.

Electrical Control System Applications

Refrigeration systems require controls so that equipment can be cycled ON and OFF to maintain a specific temperature. Systems also require safety controls to stop operation when unsafe conditions occur. There are many varieties of refrigeration controls. Electrical control systems are commonly used for heating; cooling; and packaged unit, refrigeration, humidity, hot water, and heat pump control.

Heating Controls. In a heating control application, a room thermostat energizes a gas valve that allows gas to flow to a burner. **See Figure 14-2. See Appendix.** The burner ignites, producing heat for the building spaces. When the plenum temperature rises enough, a fan switch turns on the fan motor. When a fan failure occurs, the plenum temperature rises excessively. As a result, the fan switch and a hot-surface safety switch de-energize the solenoid gas valve. The hot-surface safety switch also de-energizes the fan motor.

TECH TIP

HVAC systems can represent up to 70% of the energy consumed by a facility. By programming temperature set points and space ventilation according to occupancy, 15% to 20% of HVAC energy can be saved.

Electrical Control Systems

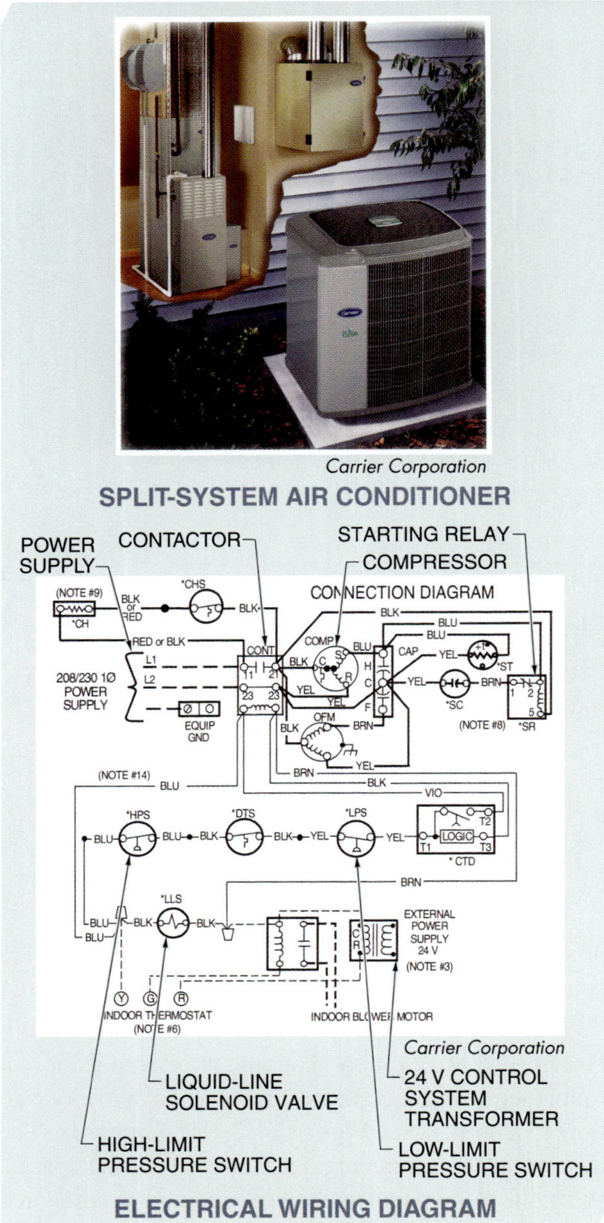

Figure 14-1. Electrical control systems use 24 VAC or higher to operate systems such as furnaces, boilers, humidifiers, split systems, rooftop units, and horizontal and vertical air handlers.

Cooling Controls. In a cooling control application, a thermostat energizes a compressor, condenser fan motor, and indoor fan motor when the thermostat is in the cool position and a call for cooling is received. **See Figure 14-3. See Appendix.** The high-limit and low-limit pressure switches open and stop the compressor and condenser fan motors. This occurs when the suction pressure becomes excessively low or the discharge pressure becomes excessively high. When the thermostat is in the fan position, the indoor fan is turned ON. When the thermostat is in the OFF position, the compressor, condenser fan, and indoor fan are all OFF.

Heating Controls

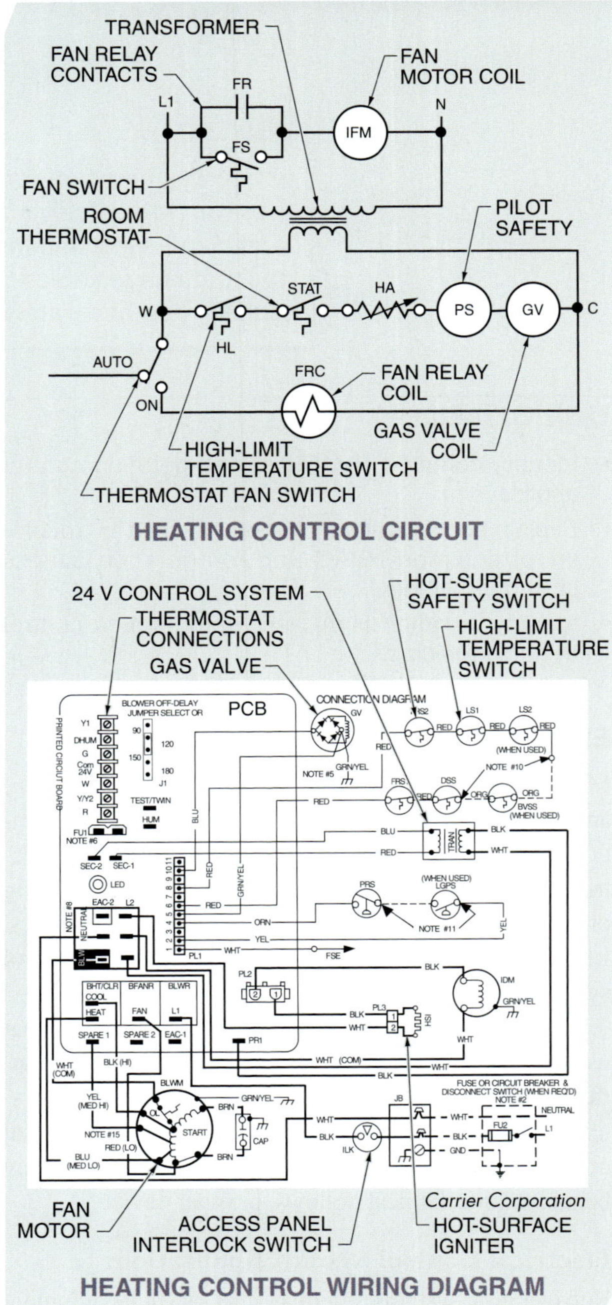

Figure 14-2. In most heating control applications, electrical controls are used to control a gas valve, burner, fan or pump motor, and sometimes an induction fan motor.

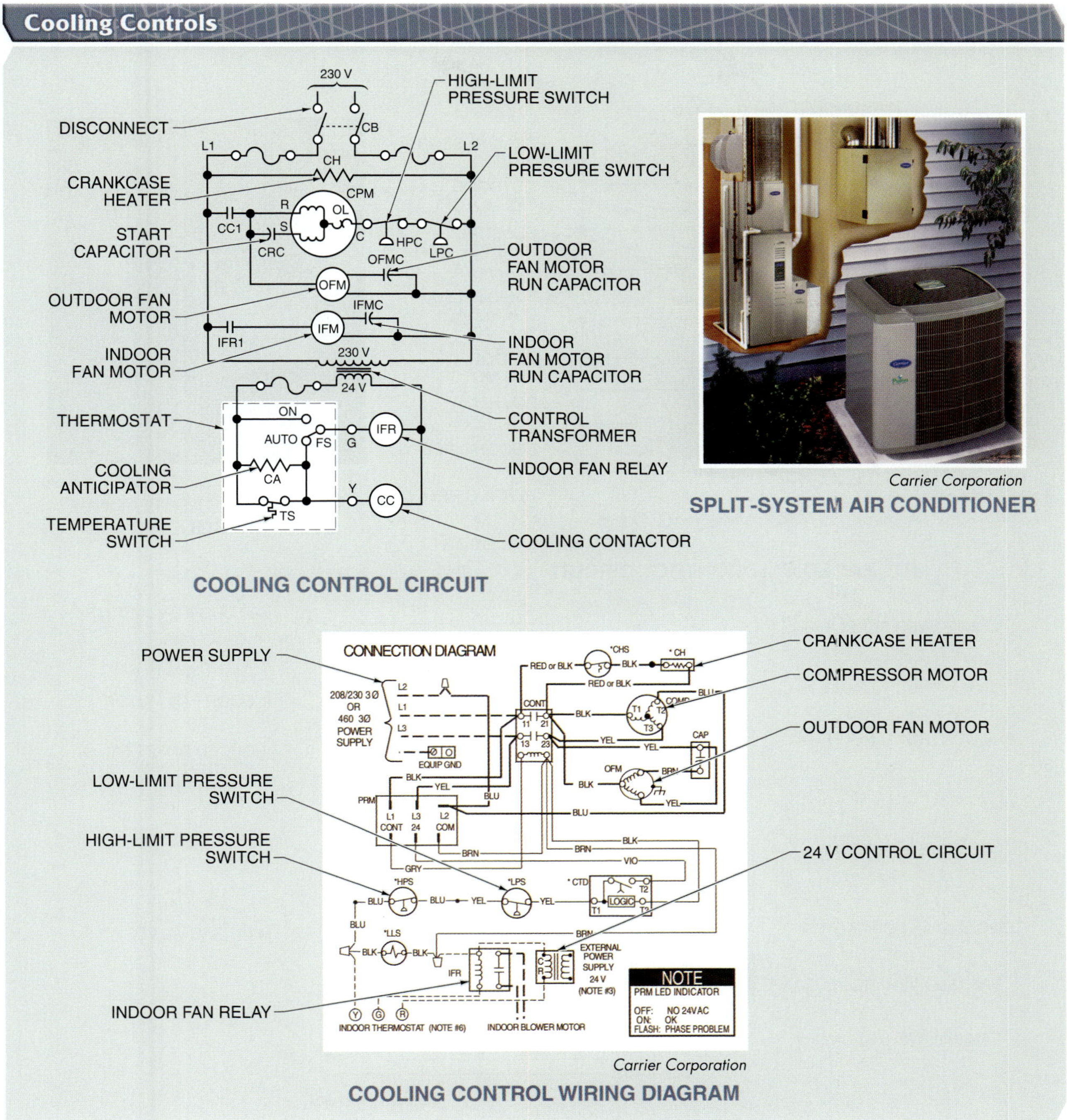

Figure 14-3. In most cooling control applications, electrical controls are used to control the compressor motor, condenser and indoor fan motors, and valves.

Packaged Unit Controls. In a packaged unit control application during a cooling demand, a room thermostat causes a compressor, outdoor fan motor, and indoor fan motor to run. **See Figure 14-4. See Appendix.** After the cooling need is satisfied and the compressor is OFF, the cooling anticipator adds false heat to a thermostat element. This causes the compressor to energize slightly before the setpoint. The indoor fan runs continuously when the thermostat is in the ON position. To provide heat, the thermostat energizes an electric heating element and/or gas valve and burner controls on some units.

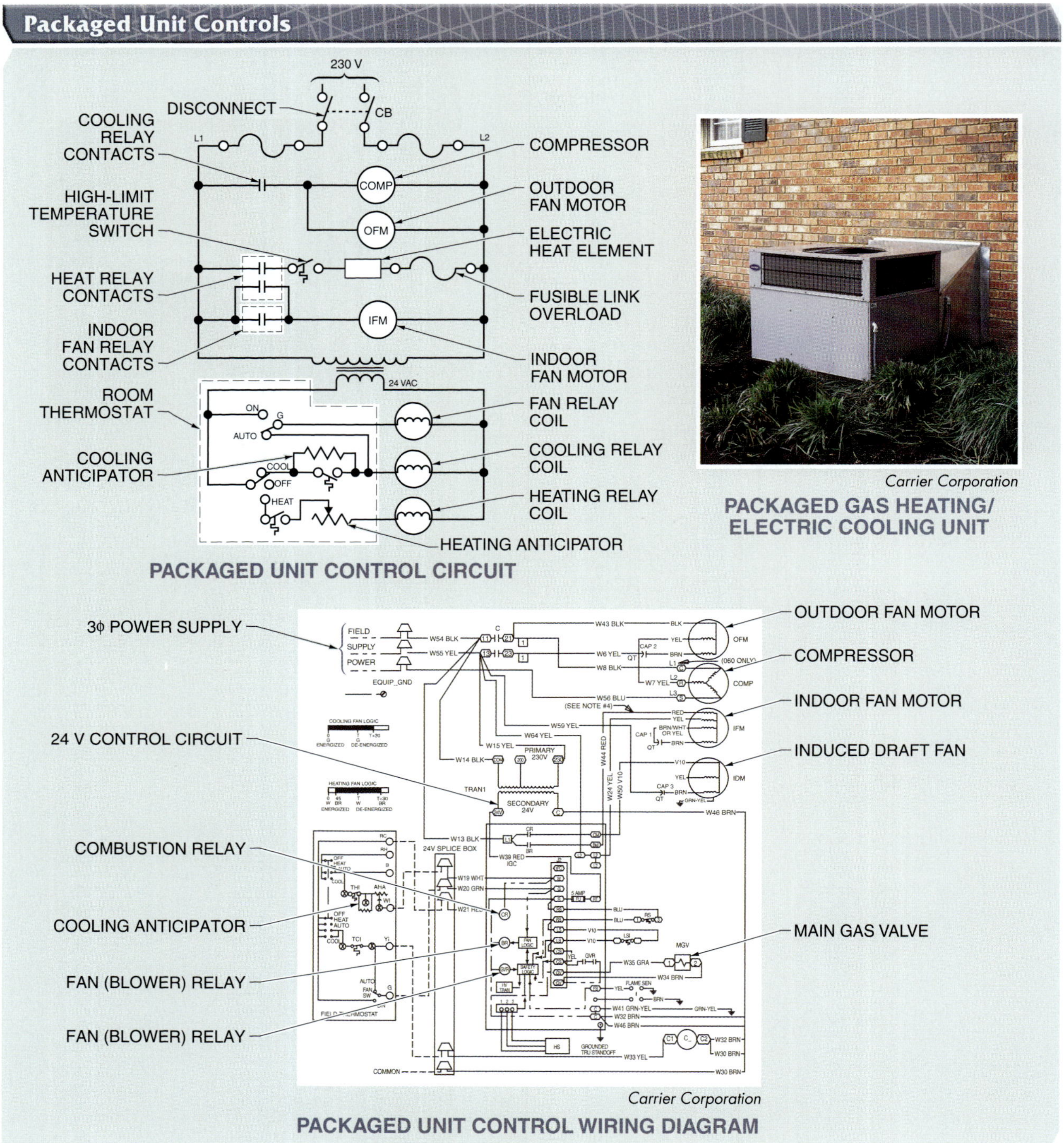

Figure 14-4. In most packaged unit control applications, electrical controls are used to control the outdoor fan motor, indoor fan motor, and heating elements.

Refrigeration Controls. In a refrigeration control application, a compressor crankcase heater is energized when power is applied to the unit. **See Figure 14-5.** During a cooling demand, the compressor, evaporator fan, and liquid-line solenoid valve are all energized. When the temperature drops below the setpoint, the liquid-line solenoid valve closes, the evaporator pressure drops, and the low-limit pressure switch de-energizes the compressor. During a high-pressure condition, the high-limit pressure switch opens, de-energizing the compressor.

Refrigeration Controls

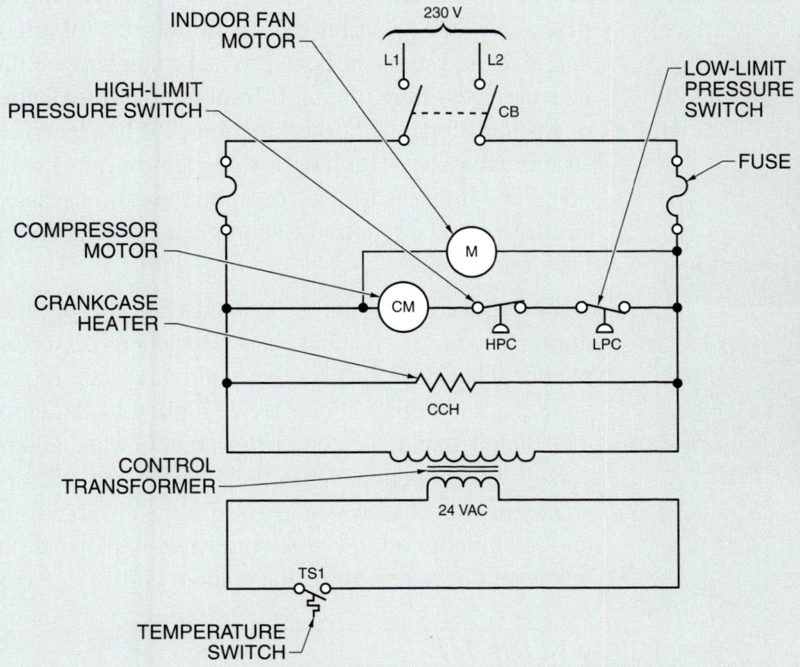

REFRIGERATION CONTROL CIRCUIT

WALK-IN STORAGE/ DISPLAY COOLERS

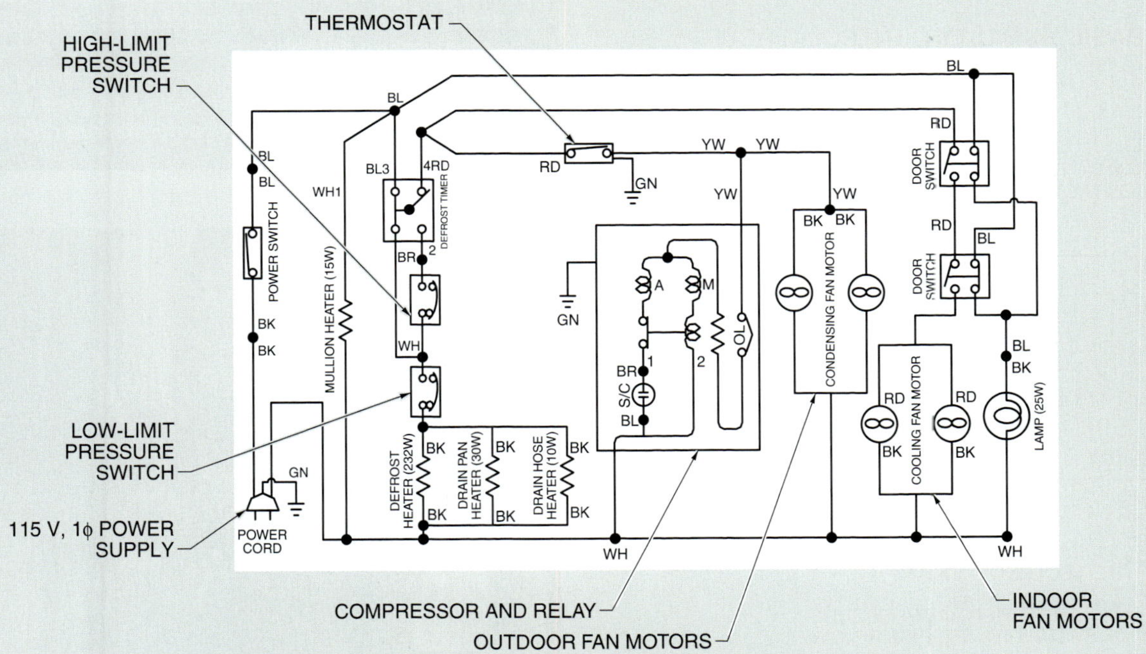

REFRIGERATION CONTROL WIRING DIAGRAM

Figure 14-5. In most refrigeration control applications, electrical controls are used to control the compressor, evaporator fan, and liquid-line solenoid valve.

Humidity Controls. In a humidity control application, a humidistat energizes a humidifier water solenoid valve when humidity drops. **See Figure 14-6.** The humidifier adds moisture to the air flowing to the building space. When the discharge humidity is excessively high or airflow is lost, the humidifier water solenoid valve is de-energized by the high humidity switch or differential pressure switch in the ductwork.

Humidity Controls

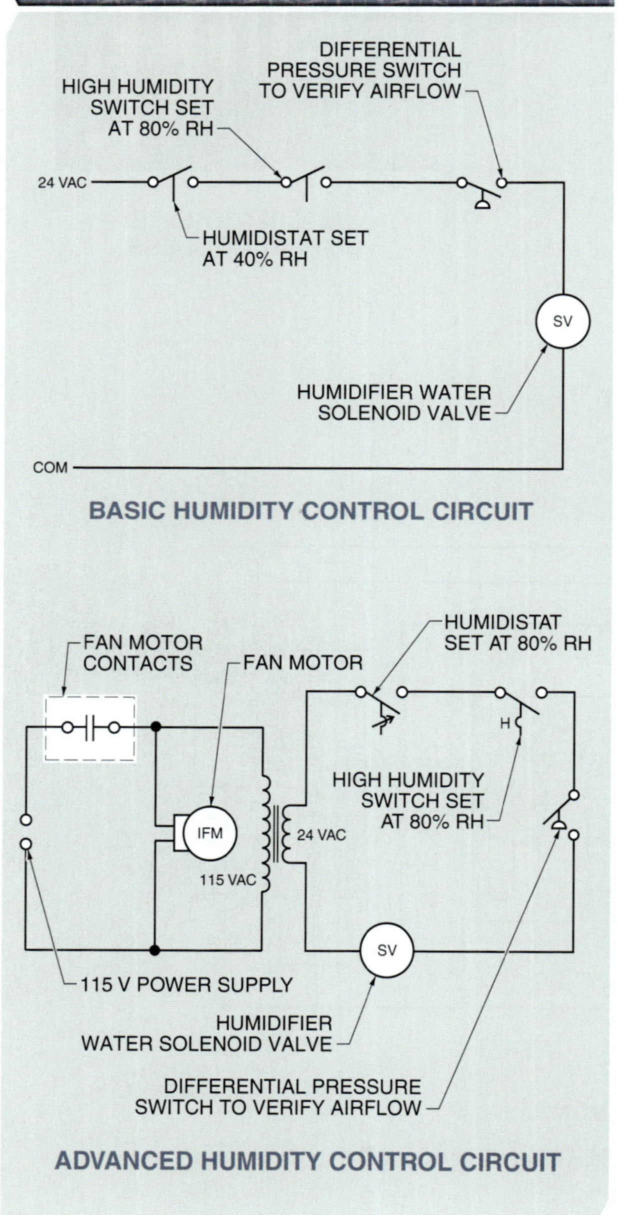

Figure 14-6. In most humidity control applications, electrical controls are used to control water solenoid valves.

Hot Water Controls. In a hot water control application, an aquastat closes when the supply water temperature drops below a setpoint. **See Figure 14-7. See Appendix.** When the aquastat closes, the burner relay coil energizes and starts the heater. When water temperature is excessively high, the high-limit temperature switch opens and stops the burner operation. When gas pressure is excessively high or low, the appropriate switch opens and stops the heater's operation. Additional safety controls may be required to comply with all state and local codes.

Heat Pump Controls. In a heat pump control application during a cooling demand, a thermostat energizes a compressor relay, outdoor fan relay, and indoor fan through an indoor unit terminal block. **See Figure 14-8.** During a heating demand, the compressor, outdoor fan, and reversing valve operate. When the air temperature drops, an auxiliary heat relay is energized. When ice builds up on the outdoor coil, the reversing valve will switch the refrigerant flow, defrosting the outdoor coil.

TECH TIP

The outdoor unit of an air-to-air heat pump operates below dewpoint temperature. For this reason, air-to-air heat pumps require a defrost mode. A thermostat is used to shut off the heat pump outdoor unit when the temperature outdoors is below 38°F. The defrost mode turns the outdoor coil into a heater, melting the frost or ice. Once the defrost cycle has ended, the outdoor unit is placed back into normal operation.

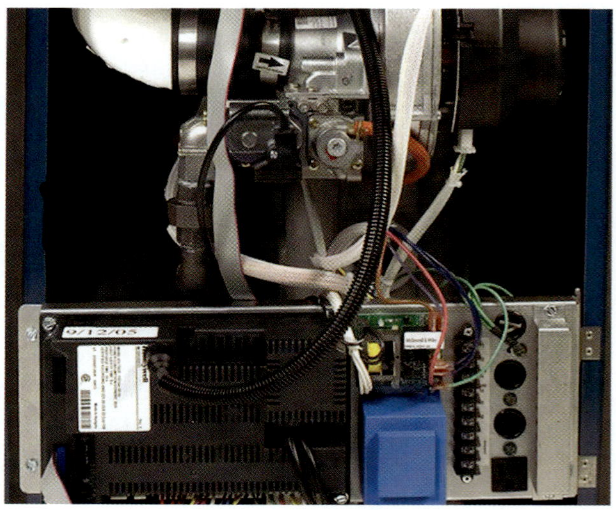

Dunkirk

Most HVAC equipment include controllers that can work with the information provided by analog humidity, level, and temperature sensors.

Hot Water Controls

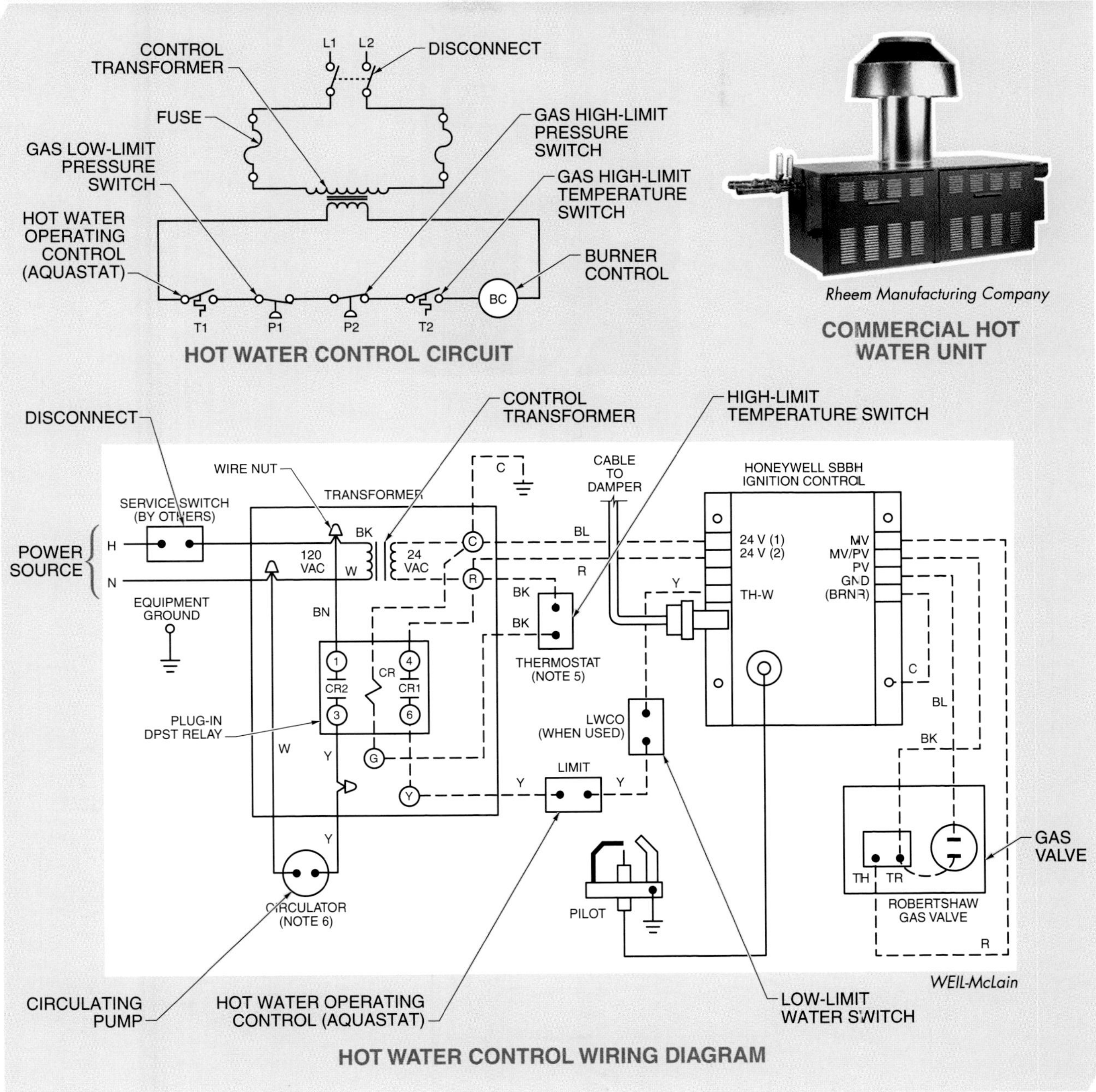

Figure 14-7. In most hot water control applications, electrical controls are used to control a burner and circulating pump.

Troubleshooting Electrical Control Systems

A technician must have knowledge of basic control system troubleshooting. The standard tool used for troubleshooting an electrical control system is a digital multimeter (DMM). The most common measurements taken with a DMM are voltage, current, and resistance.

TECH TIP

HVAC equipment has emergency cut-out switches that are activated when a door or service panel is removed. Service doors that are not replaced properly prevent the equipment from starting. It should always be verified that access doors are properly fitted.

Heat Pump Controls

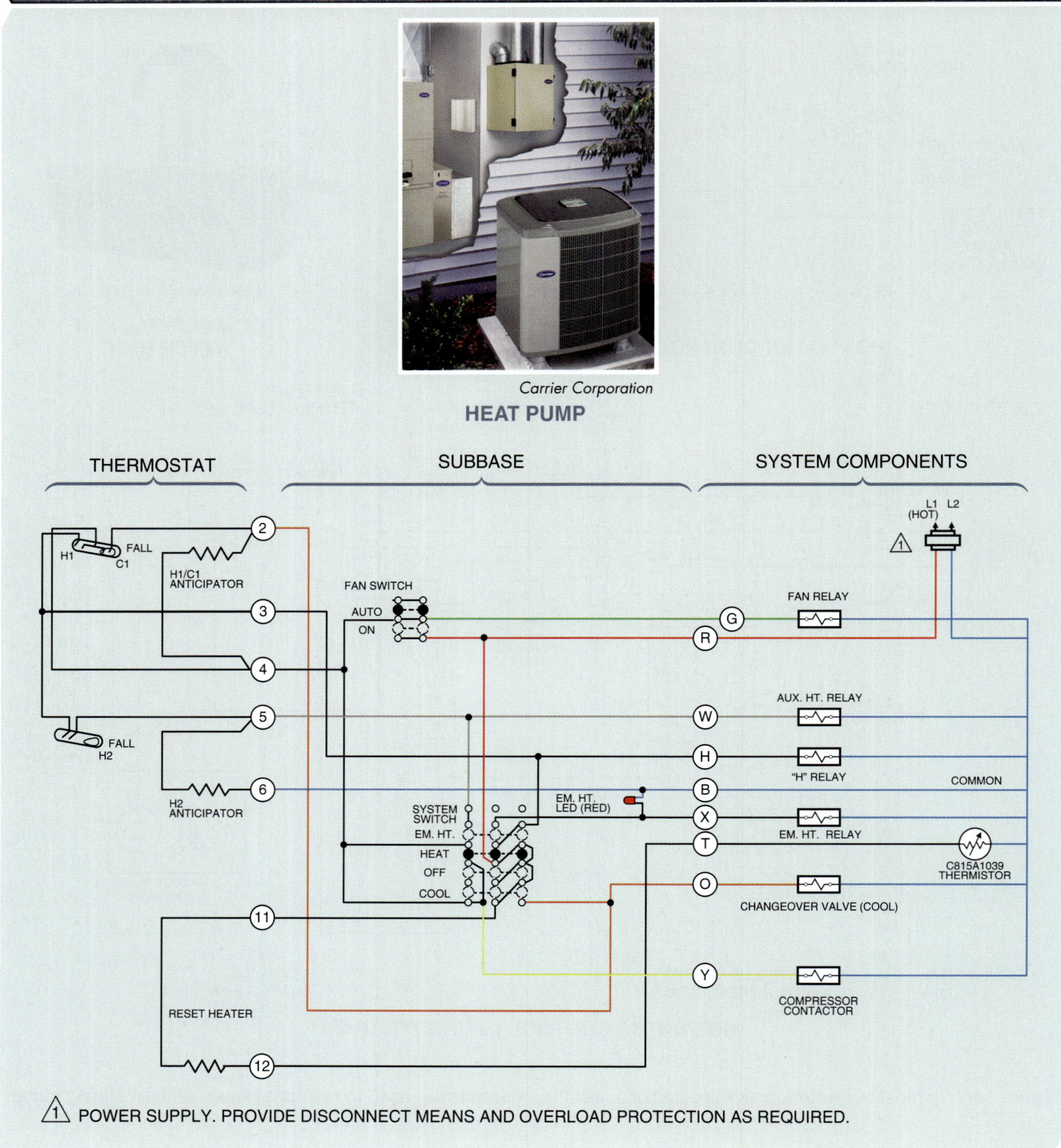

Figure 14-8. In most heat pump control applications, electrical controls are used to control the compressor, crankcase heater, outdoor fan, indoor fan, and reversing valve.

Voltage Measurements. Voltage is measured by connecting the DMM leads in parallel across the device being tested. **See Figure 14-9.** A device, such as a switch, is open when the DMM measures a voltage level. A switch is closed when no voltage (0 V) is measured. By connecting a DMM from point to point in a circuit, an open switch can be found. An open switch prevents a unit from operating.

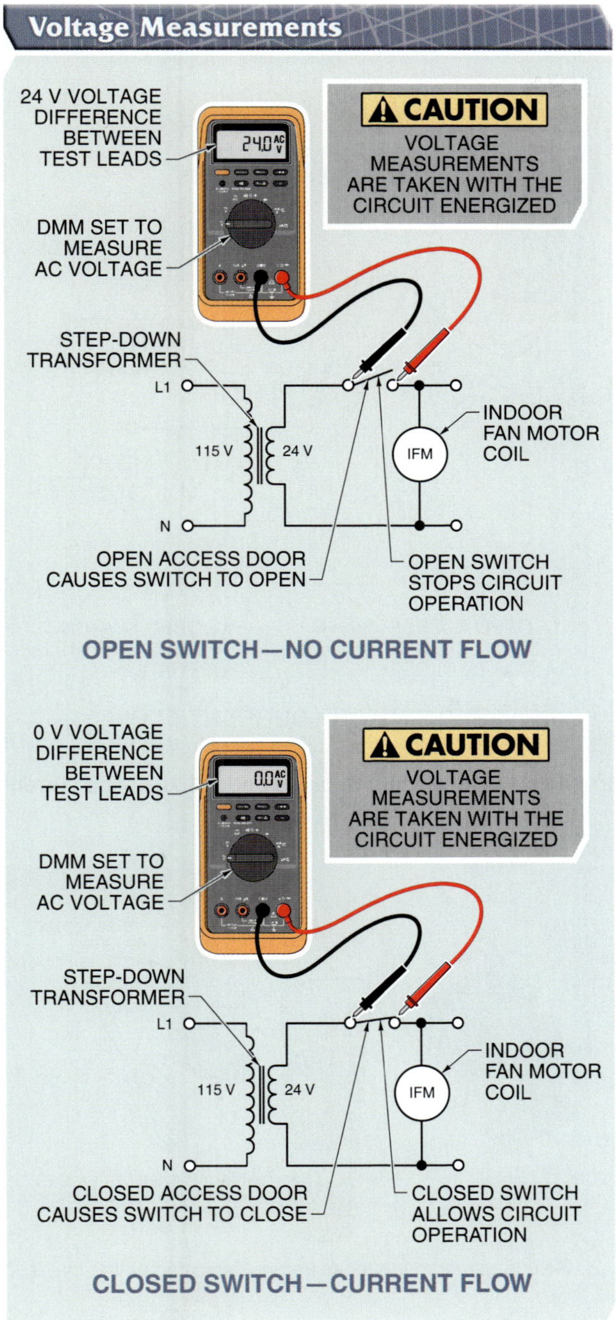

Figure 14-9. When taking voltage measurements, a voltmeter measures the difference in voltage between two test leads.

Voltage testing is also performed at the power supply to determine if the correct amount of voltage is being provided to the control circuit. A meter can be used to find the voltage drop of a circuit component. **See Procedure for Voltage Measurement.**

Voltage Measurement

Procedure

1. Connect the test leads to the appropriate meter jacks.
2. Follow all electrical safety practices and procedures.
3. Rotate the meter function switch to measure AC voltage ($\tilde{V}$).

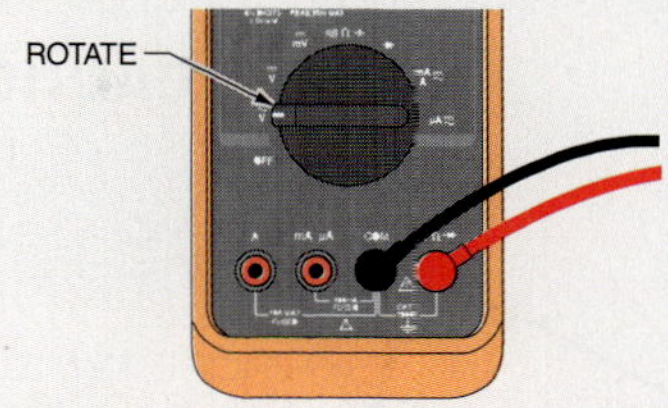

4. Connect the test leads to a circuit to measure the supply voltage to a control circuit or the voltage drop of a device.

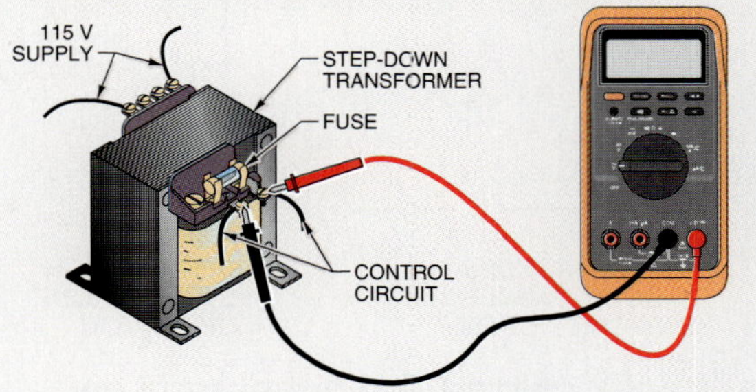

5. Read and record the measurement indicated on the meter.

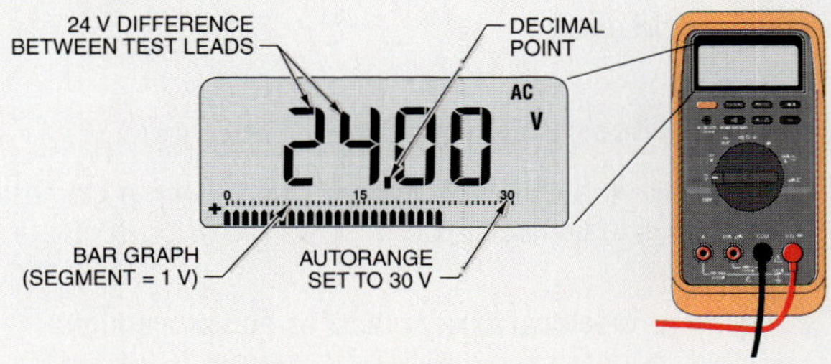

6. Disconnect the test leads from the power supply or device.
7. Turn the meter function switch to OFF.

Current Measurements. Current measurements of a circuit are valuable because the reading indicates the amount of work being performed. Current can be measured in a variety of ways. For example, a DMM can be inserted in series with the circuit under test as an in-line current meter. **See Figure 14-10.**

A clamp-on meter is the recommended test tool for taking current measurements. A clamp-on meter is an electrical test instrument that measures the strength of the magnetic field around a conductor. A DMM with a clamp-on adapter can also be used to measure current flow. A clamp-on meter or a DMM with a clamp-on adapter does not require that a circuit be open to measure the current flow. **See Procedure for Current Measurement.**

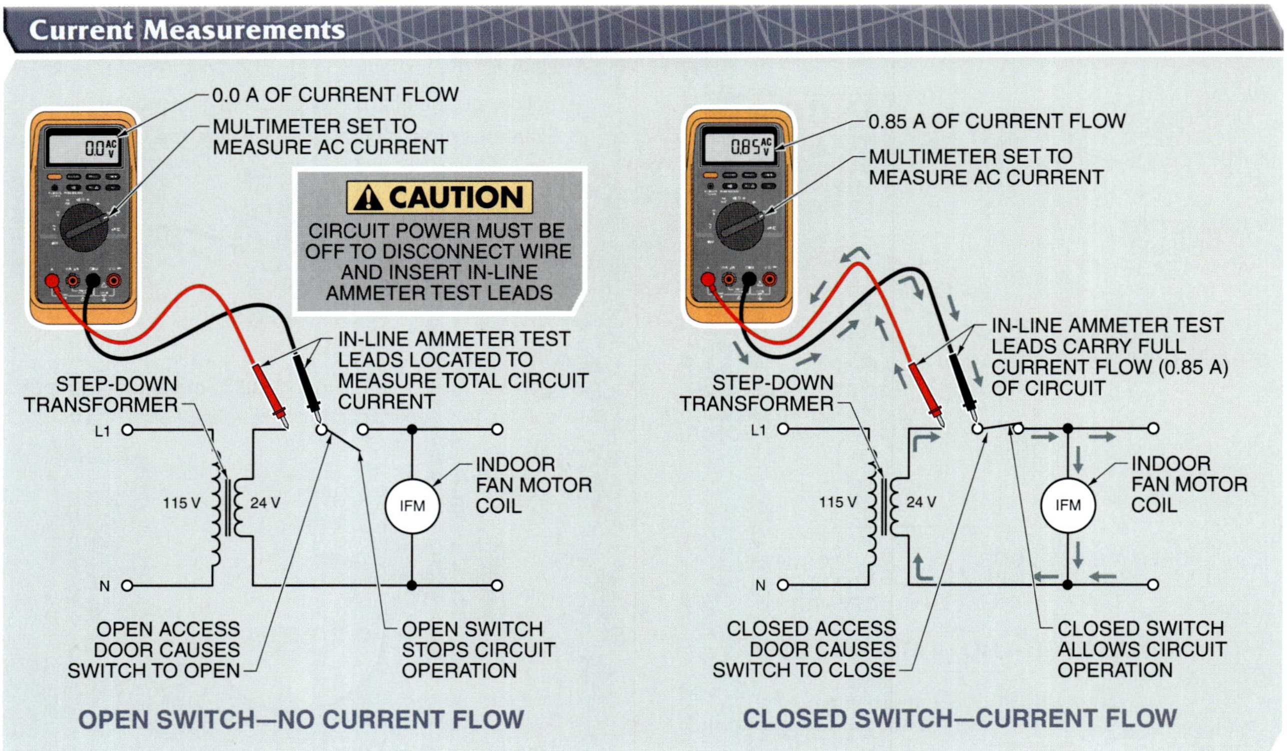

Figure 14-10. For most current measurement applications, if current flow is suspected of being under 1 A, in-line current meters can be used.

Current Measurement

Procedure

1. Determine if a clamp-on meter or DMM with a clamp-on adapter is to be used.
2. Follow all electrical safety practices and procedures.
3. Set the meter function switch to measure AC current (A).

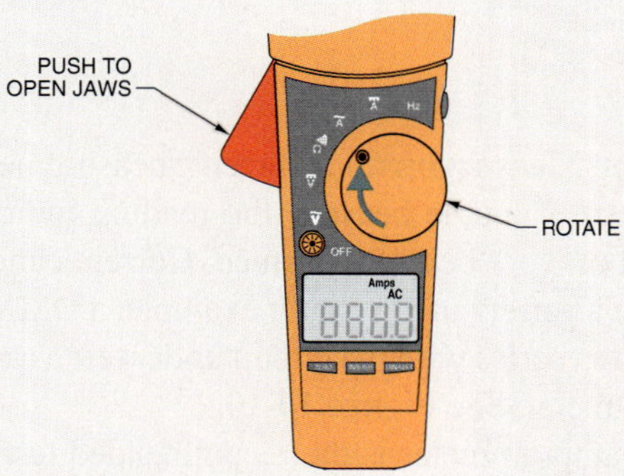

4. Attach the meter to the conductor in the circuit to measure the total current flow of the circuit.

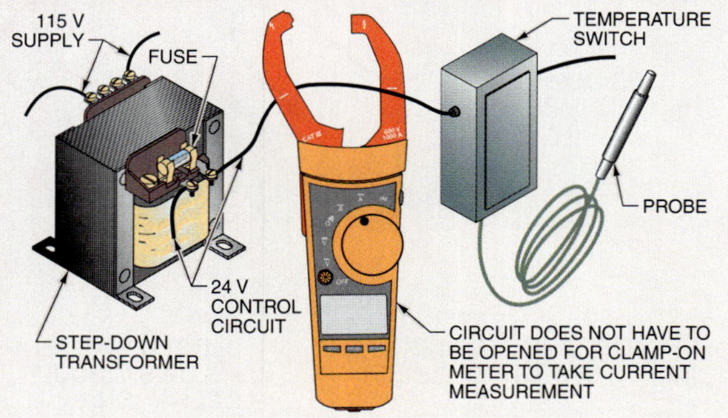

5. Read and record the measurement indicated on the meter.

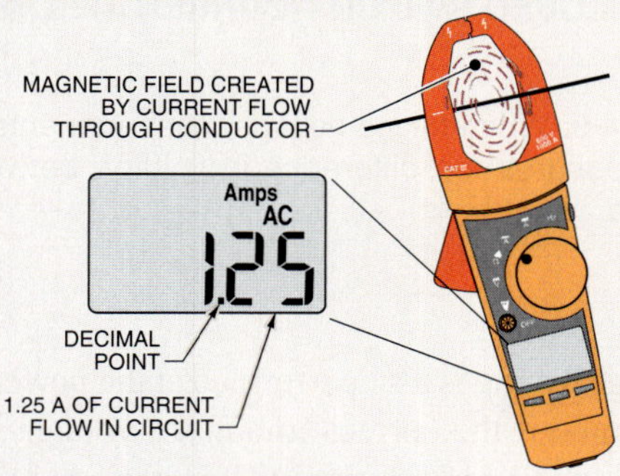

6. Remove the meter from the circuit conductor.
7. Turn the meter function switch to OFF.

TECH TIP

For motors that are not hermetically sealed, there is no pass/fail standard for interpreting insulation-resistance values. However, insulation-resistance trending is a clear indicator of the overall condition of a motor. A rule of thumb for the pass/fail measurement is to use 1 MΩ plus 1 MΩ/kV of motor operating voltage.

Resistance Measurements. Motor windings and switches are often measured for resistance to determine their condition. Resistance measurements must be taken with all power disconnected from the circuit. The Continuity Test mode of a DMM may be used to check a small amount of resistance. **See Figure 14-11.** A *Continuity Test mode* is a digital multimeter mode that detects a complete path for current to flow.

DMMs typically offer a Continuity Test mode with the Resistance mode. DMMs set to continuity typically provide an audible signal when there is very little resistance between two points in a circuit. This signal indicates a complete electrical path or electrical continuity between the points tested.

Continuity checking is helpful when performing large numbers of point-to-point tests to isolate a fault in a control panel. A technician can concentrate on the correct placement of the leads and not be concerned with specific meter readings. **See Procedure for Resistance Measurement.**

Resistance measurements are often used to determine the condition of flow switches and contacts in relays and HVAC pressure and temperature.

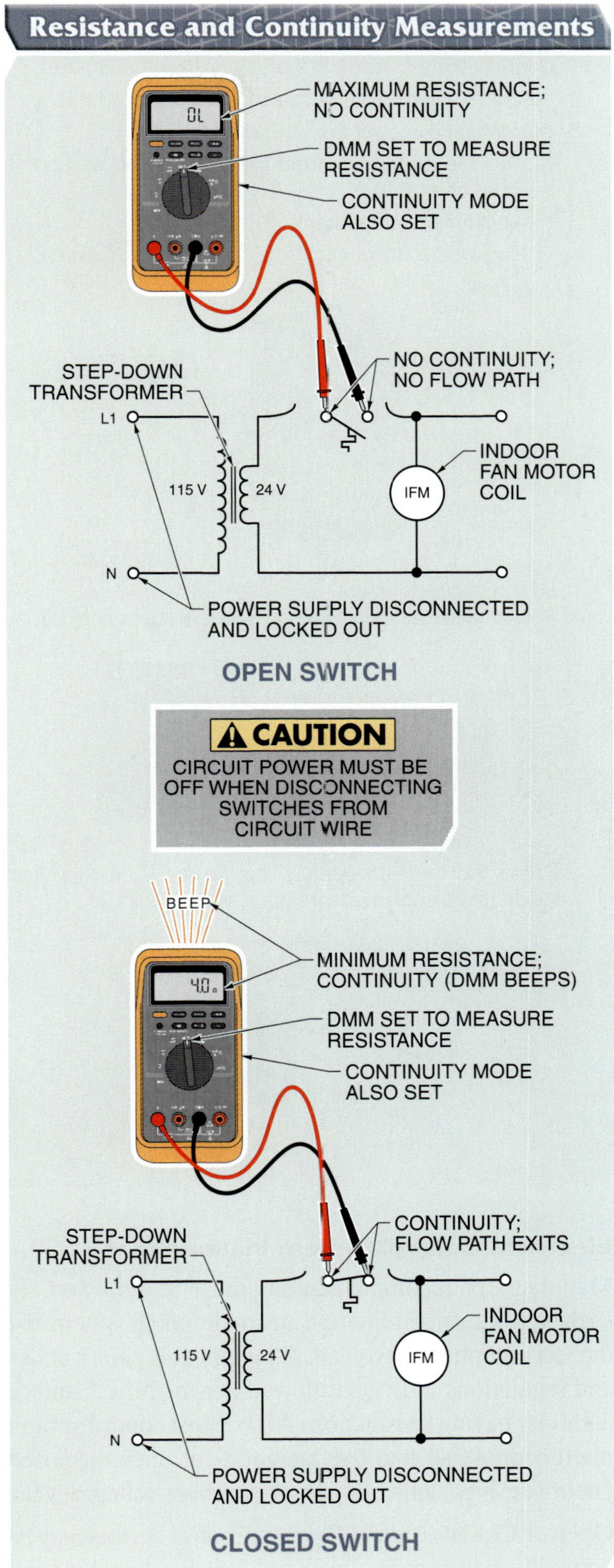

Figure 14-11. To check the resistance or continuity of a device or conductor, it must be isolated from the rest of the electrical circuit.

Procedure

Resistance Measurement

1. Connect the test leads to the appropriate meter jacks.
2. Follow all electrical safety practices and procedures.
3. Disconnect all power from the circuit.
   a. Pull the disconnect from the circuit to be worked on (turn off power).
   b. Lockout the disconnect.
   c. Tagout the disconnect.

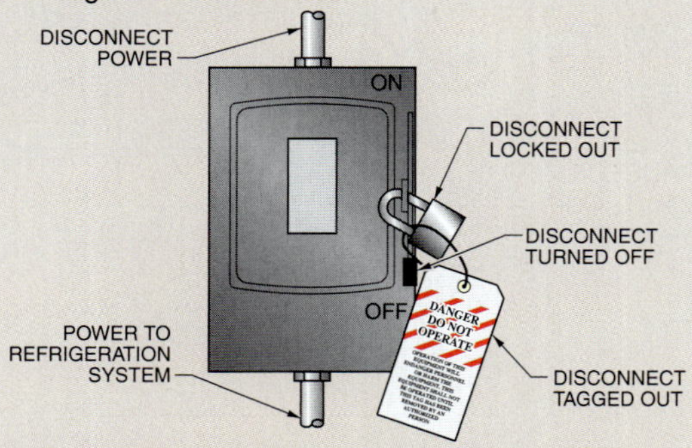

4. Set the meter function switch to measure resistance (Ω).

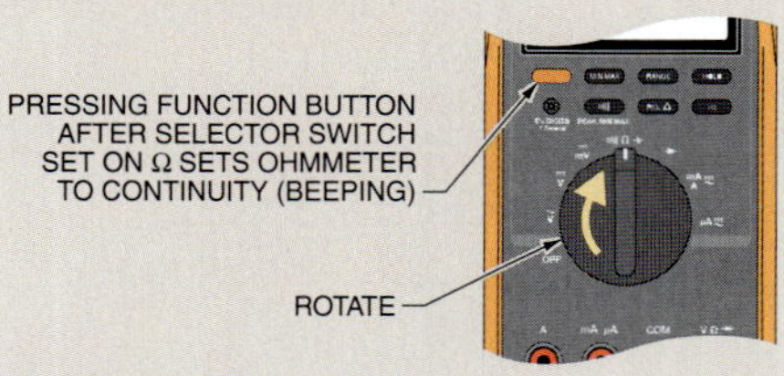

5. Press the function (yellow) button for Continuity Test mode (resistance measurements with beeping).

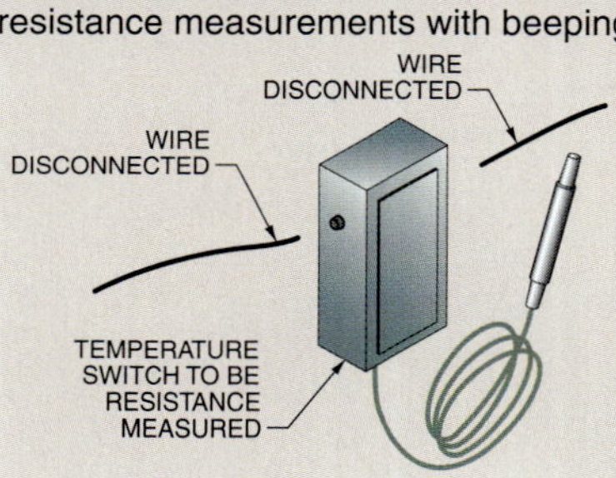

6. Isolate the device to be measured from the rest of the circuit by disconnecting all wires and connections.

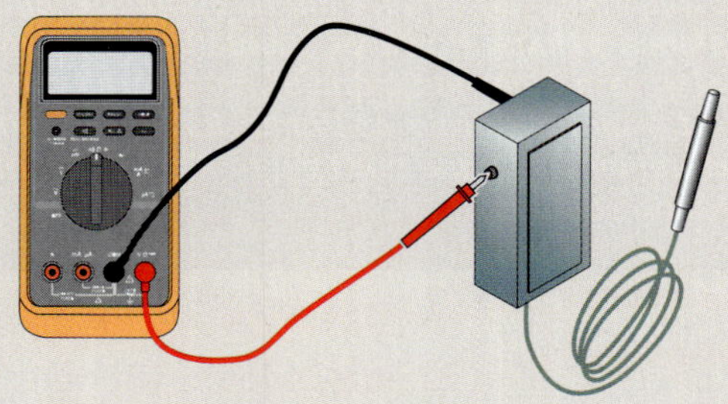

7. Connect the test leads to the electrical device to be measured.

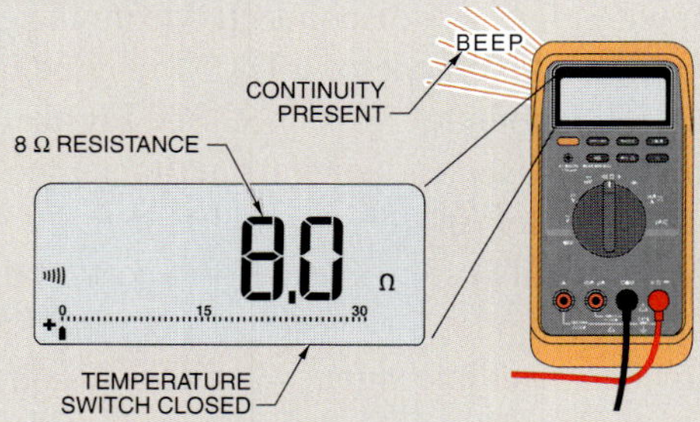

8. Read and record the measurement indicated on the meter.
9. Disconnect the test leads from the device.
10. Turn the function switch to OFF.

Electrical Control System Malfunctions

Manufacturer recommendations must be followed for each type of refrigeration and air conditioning system and the test equipment involved. All applicable safety codes and regulations must be followed per the NEC® and/or authority having jurisdiction (AHJ). Most control system malfunctions fall into the categories of open or closed control devices, short circuits, or improper voltage levels.

Open or Closed Control Devices. Control devices may be normally open or normally closed. When a device is open that is normally closed, it prevents a unit from operating. When a device is closed that is normally open, it allows a unit to operate. Electrical control system malfunctions are typically found by using a DMM. The DMM is set to measure voltage in order to check whether the device functions (opens or closes) as designed. **See Figure 14-12.** For example, an HVAC unit may have an access door that prevents the unit from operating when the door is open.

Short Circuits. A *short circuit* is a condition in which extremely high current leaves its normal current-carrying path and goes around a load and back to the power source or to ground. A control device and/or load is bypassed by the current flow when a short circuit occurs. A *dead short* is a short circuit that opens an overcurrent protection device as soon as the circuit is energized. A *partial short* is a short circuit that occurs only in sections of a machine.

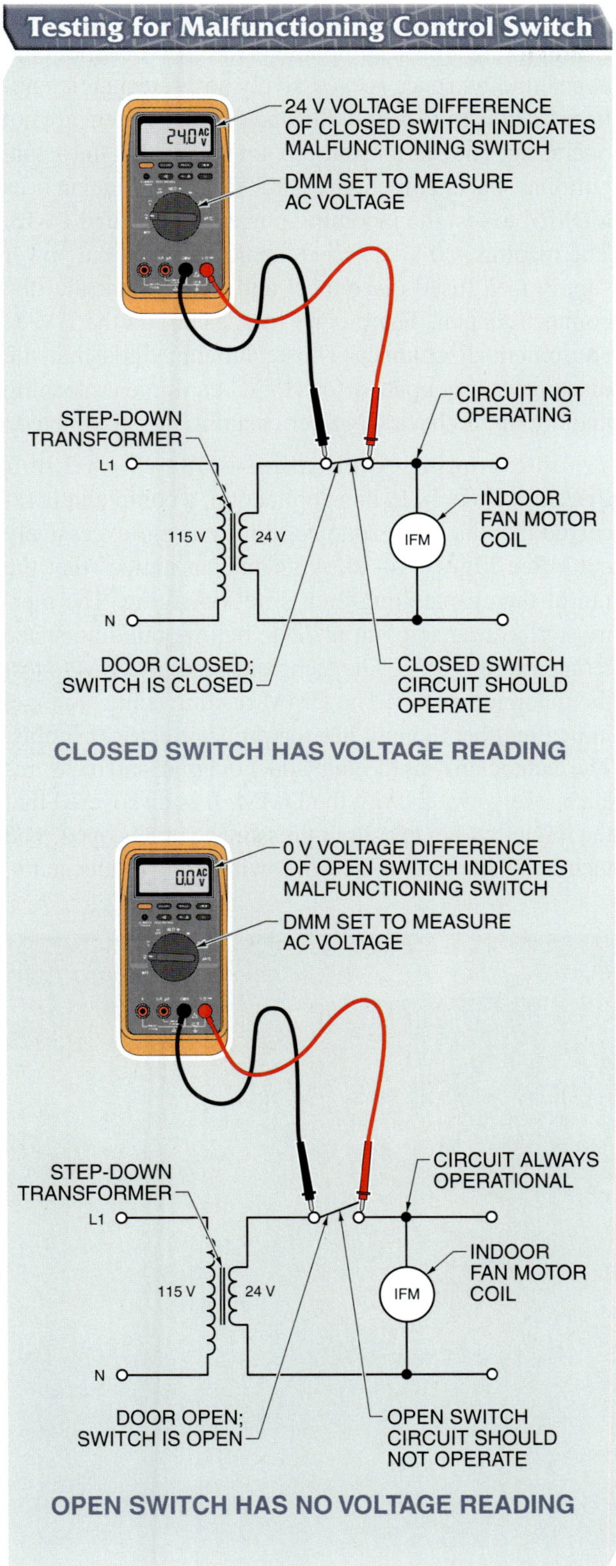

Figure 14-12. Typically, when a closed switch has a measured voltage drop, such as 24 V, the switch is internally open and does not allow a system to operate properly.

A short circuit can be determined by measuring the resistance between a device and ground. **See Figure 14-13.** For example, a reading of 0 Ω between the leads of a motor and ground indicate that a motor winding is touching the motor housing.

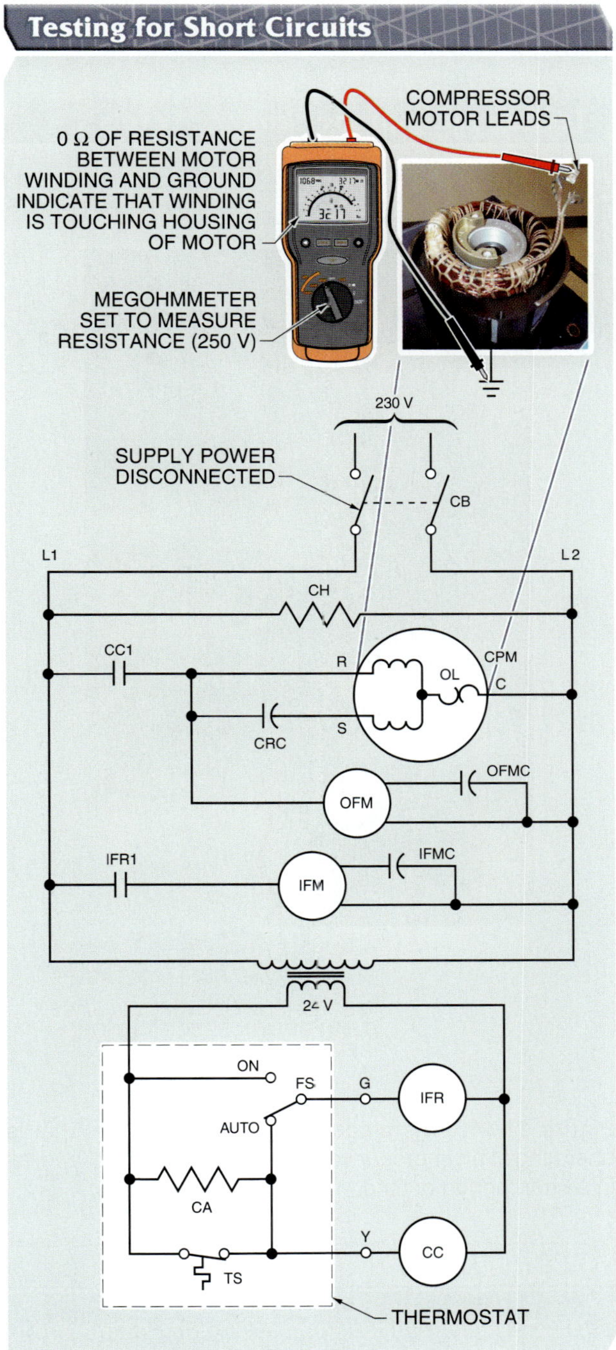

Figure 14-13. Low-resistance shorts can be detected by using an ohmmeter. High-resistance shorts require a megohmmeter for taking resistance measurements.

Improper Voltage Levels. An *improper voltage level* is a voltage level in a circuit that is at least 10% higher or lower than the designed level. **See Figure 14-14.** When there is an improper voltage level, either excessively high or low, a control system will not operate properly. For example, electrical interference due to lightning can affect the power supply. The MIN/MAX recording mode of a DMM can be used to show fluctuations in a power supply.

Figure 14-14. Any refrigeration or HVAC device that is operating at an improper voltage level will overheat and fail in a short period of time.

TECH TIP

HVAC systems create harmonic currents. An average responding meter will not provide accurate readings if harmonics are present. Therefore, a true-rms meter should always be used when taking voltage and current measurements in an HVAC system.

Troubleshooting Application—Open Incoming Power. In this application, a complaint is received that the south zone building space is excessively hot. A technician notices that the rooftop unit fan and cooling system are not operating. The thermostat setpoint is 76°F, but the actual building space temperature is 85°F. The technician uses a DMM to test the incoming power. **See Figure 14-15.** The reading is 0 VAC. The technician traces the power supply to a fused disconnect and notices that the disconnect is open. There is no lock or tag on the HVAC equipment disconnect. The technician finds that the disconnect was opened for HVAC equipment cleaning and not turned back ON after cleaning was completed.

Troubleshooting Application—Open Low-Limit Pressure Switch. In this application, a complaint is received that the north zone building space is excessively cold. **See Figure 14-16.** A technician notices that the fan of the air-handling unit is not operating. The thermostat heating setpoint is 72°F, but the building space temperature is 65°F. The technician uses a DMM to test the incoming power. The DMM reading is line voltage, indicating that the unit has the proper electrical supply. The technician uses the unit ladder diagram and measures each safety switch with the DMM. It is discovered that the induction fan low-limit pressure switch is open. The technician manually resets the switch and the unit starts.

Fluke Corporation

Amperage measurements are typically used to troubleshoot flame sensors that send either a 1 mA signal or a signal between 2 mA and 5 mA to a controller.

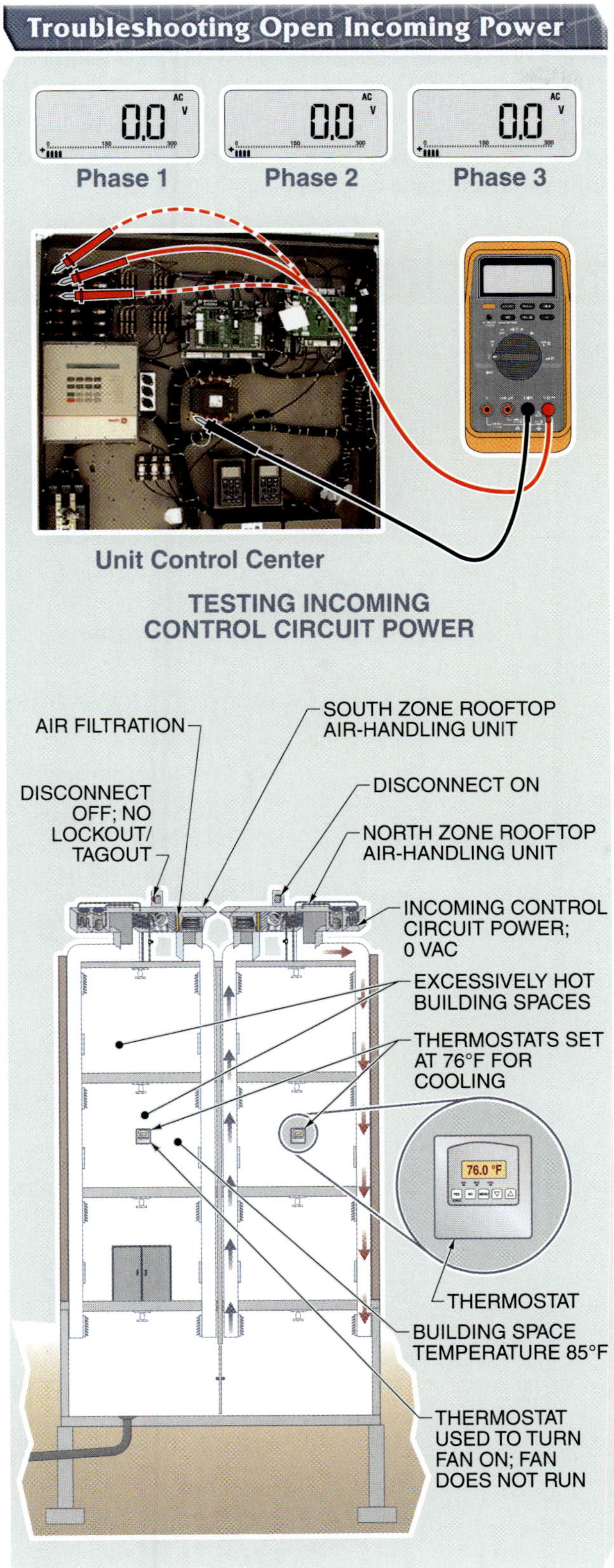

Figure 14-15. When taking voltage measurements of 3ϕ power, a good ground for the common test lead should always be used.

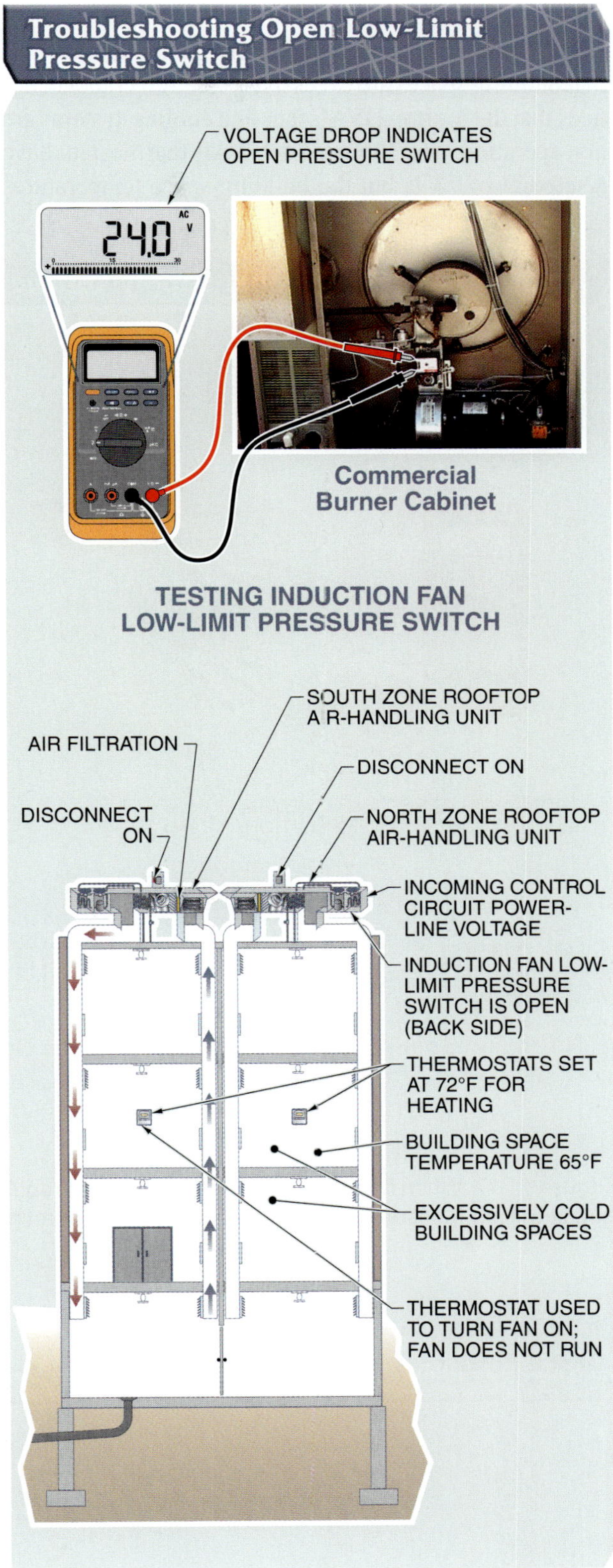

Figure 14-16. When a pressure switch is closed and properly operating, there is no voltage drop. A 24 V voltage drop indicates that the switch is open.

Troubleshooting Application—Improper Voltage Level. In this application, multiple complaints are received that building spaces are excessively hot. A technician notices that the rooftop HVAC fans and cooling systems are not operating. **See Figure 14-17.** All thermostats have a setpoint of 76°F, but the building space temperatures range from 85°F to 90°F. The north and south rooftop fans do not run. The technician uses a DMM to test the supply power to the HVAC units. The reading is more than 10% less than the line voltage required for the HVAC units to operate. The technician or owner must contact the power utility to resolve the electrical supply problem.

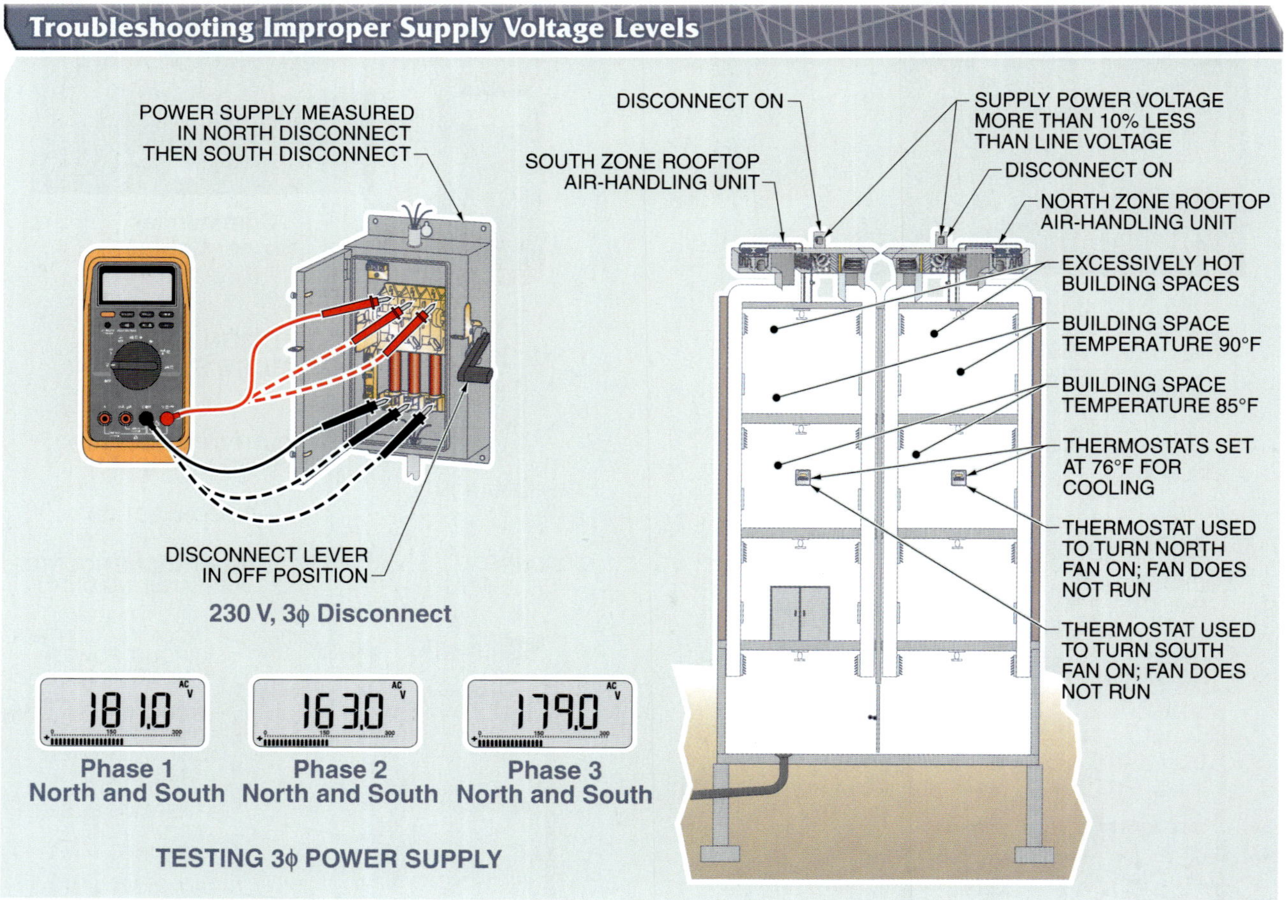

Figure 14-17. When line voltage is 10% less than the required line voltage of 230 V, motors will not operate. Also, the control circuit will not function, stopping the system from operating.

UNIT 15

Motor Starters and Variable-Frequency Drives

HVAC devices, such as compressors, electric heaters, and fans, must be started and stopped electrically. A motor starter can start and stop a motor. A variable-frequency drive can change the speed and control the starting and stopping of a motor. These devices must be understood in order to properly troubleshoot them.

OBJECTIVES

- Describe the operation of relays, contactors, and motor starters.
- Identify motor thermal protection methods.
- Describe the operation of variable-frequency drives.
- Explain general troubleshooting procedures for motor starters and variable-frequency drives.

Index to Procedures

MOTOR CONTROL DEVICES

Several different types of devices are used to connect and disconnect the power supply to electrical loads such as motors. These devices include relays, contactors, motor starters, and variable-frequency drives. There are many variations of each type of device.

Relays

A *relay* is a device that uses a low voltage coil to close one or more sets of contacts to switch high-voltage loads on and off. Relays are classified as low-current switching devices. They are sealed devices that are replaced when they fail. Relays have many different types of contact configurations. **See Figure 15-1.** Common relay configurations are single-pole single-throw (SPST) and double-pole single-throw (DPST). Many other configurations of relay contacts are available. Some relays are pilot-duty rated, which means that they are light-duty devices used to energize large-capacity contactor and starter coils.

Contactors

A *contactor* is a large heavy-duty, switching device that uses low voltage to energize or de-energize the high-current load connected to it. Contactors come in many configurations of physical size, amperage, and number and order of switched poles. **See Figure 15-2.** A contactor consists of a coil, movable and stationary contacts, contact springs, and a shell. Contactors can be disassembled and rebuilt.

Motor Starters

A *motor starter* is a heavy-duty switching device that includes motor overload protection and typically uses a low-voltage coil to energize or de-energize the high-voltage motor connected to it. **See Figure 15-3.** Motor starters are available in a variety of sizes and can be disassembled and rebuilt. The contacts of a motor starter continually open and close against full-load current, which can cause pitting and arcing.

Relays

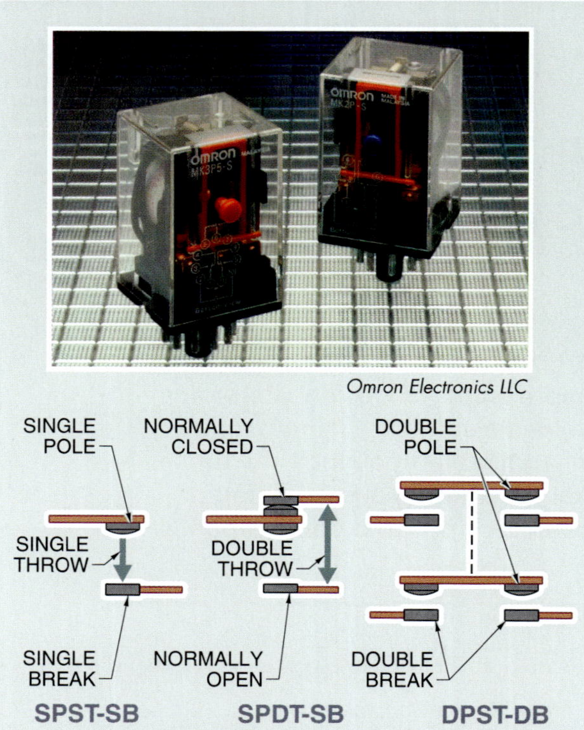

Relay Contact Abbreviations	
Abbreviation	**Meaning**
SP	SINGLE POLE
DP	DOUBLE POLE
3P	THREE POLE
ST	SINGLE THROW
DT	DOUBLE THROW
NO	NORMALLY OPEN
NC	NORMALLY CLOSED
SB	SINGLE BREAK
DB	DOUBLE BREAK

Figure 15-1. Relays have many different types of contact configurations that permit a low current to switch a high current.

Contactors

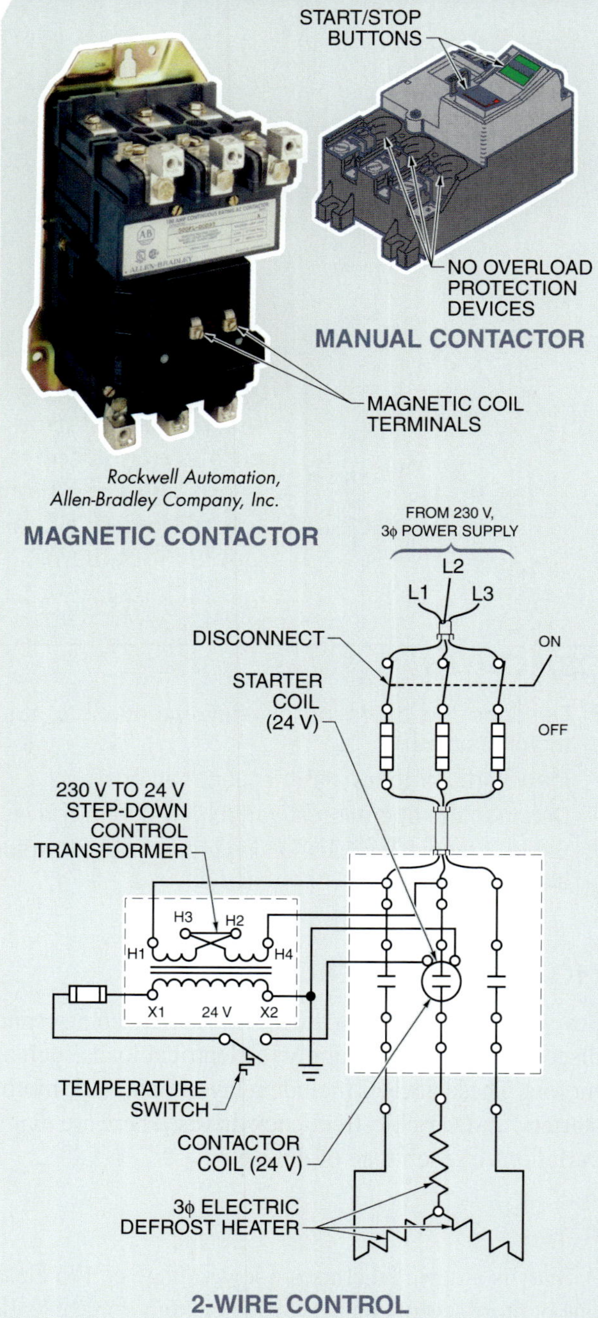

Figure 15-2. A contactor turns on and off high-current loads that do not require overload protection such as lights, heaters, and transformers.

Motor starter contacts can be tested by taking voltage drop measurements across each contact when a motor is running. They can also be tested by taking resistance measurements of each contact when a system is de-energized and the contacts are isolated. **See Procedure for Motor Starter Contact Testing.** When the contacts are closed and are operating properly under full load, no voltage drop should be present. When a voltage drop is present across contacts, it is due to the resistance caused by pitting and wear. Worn contacts on a motor starter are typically discolored and must be replaced.

Motor Control Enclosure Interactive Animation

Motor Starters

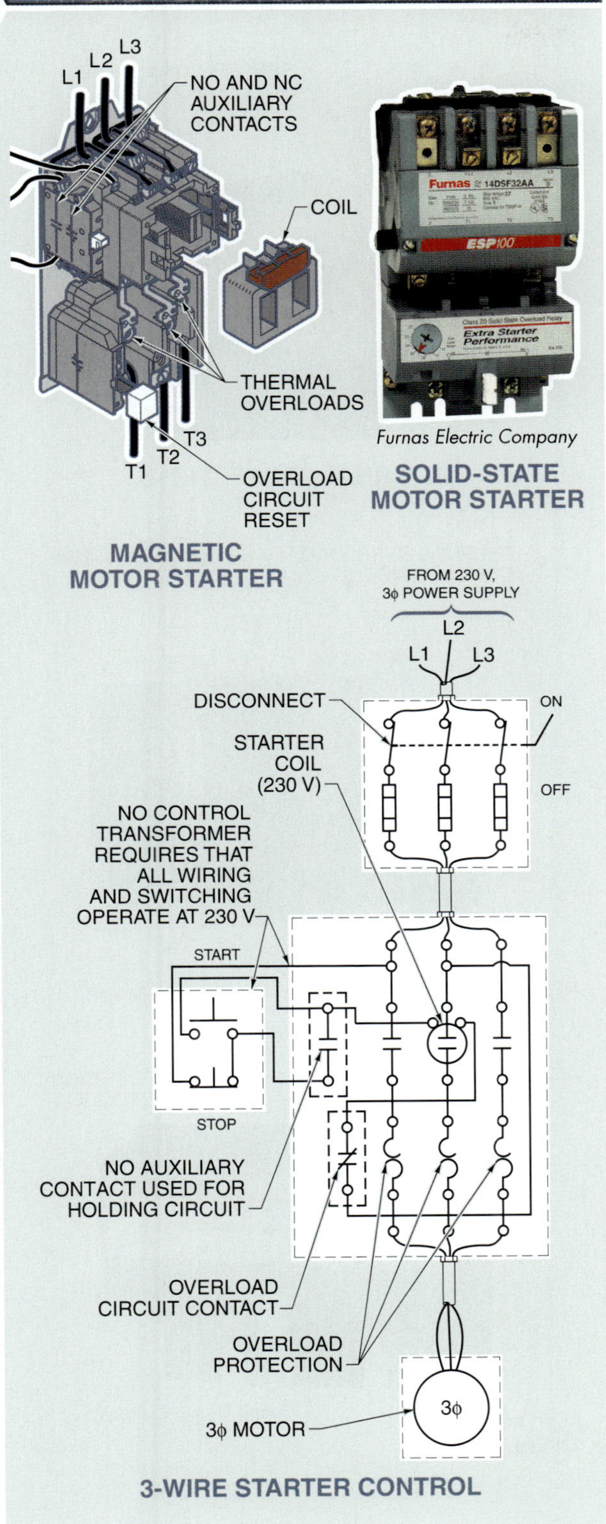

Figure 15-3. A motor starter is an electrically operated switch like a contactor, but it includes overload protection for motor applications.

Motor Starter Contact Testing

Procedure

1. Connect the test leads to the appropriate meter jacks for taking AC voltage measurements.
2. Follow all electrical safety practices and procedures.
3. Set the meter function switch to measure AC voltage ($\tilde{V}$).
4. Open the access panel to the residential HVAC condenser fan motor starter.

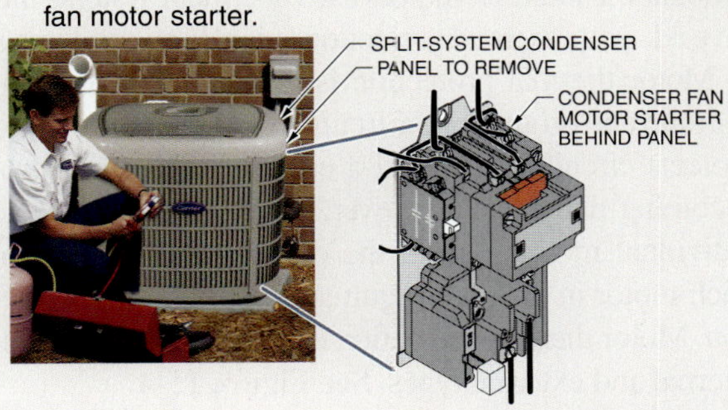

5. Start the HVAC cooling system. Wait until the system starts the condenser fan motor, because the contacts of the condenser fan motor starter will be closed.

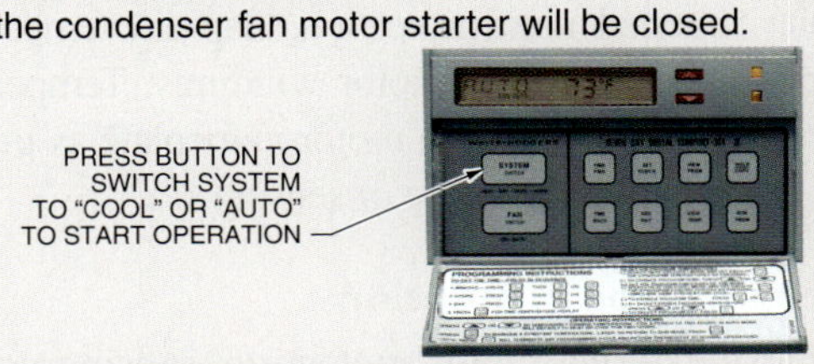

6. Connect the meter test leads across the starter L1 contacts to measure the voltage drop.

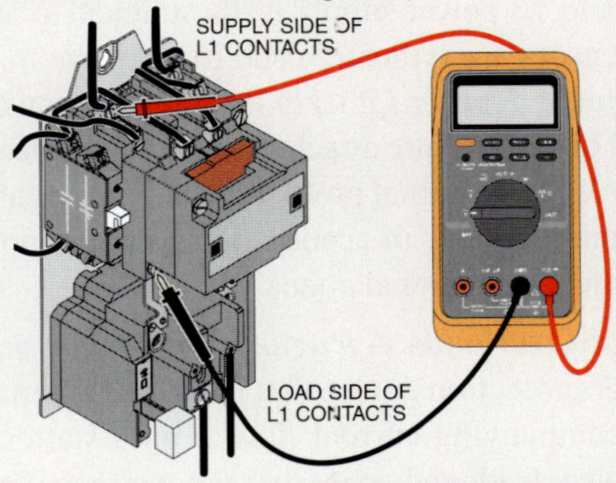

7. Connect the meter test leads across the starter L2 or N contacts to measure the voltage drop.
8. Read and record the voltage drop measurements indicated on the meter.

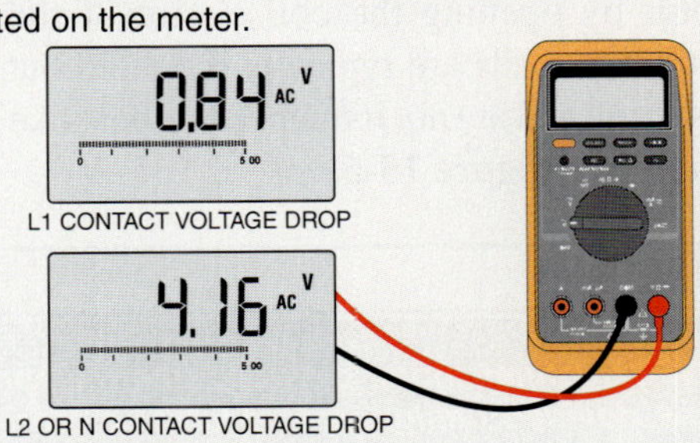

9. Disconnect the meter test leads from the motor starter contacts.
10. Turn the meter function switch to OFF.

MOTOR THERMAL PROTECTION

The motors used in residential and light commercial refrigeration and HVAC systems can be complex and expensive. The shutdown of a cooler or freezer by a motor failure can lead to large financial losses in the thousands of dollars. Therefore, motors used in critical systems must have adequate thermal protection against serious damage.

Motor thermal protection is not the same as circuit overcurrent protection. Circuit overcurrent protection protects circuit wiring and a cabinet or structure from becoming damaged. However, it does not protect each individual motor load in the circuit from burning up. Each motor in a circuit requires its own thermal protection. Motor thermal protection is typically separated into internal and external types. **See Figure 15-4.**

Internal Motor Protection

To achieve internal motor protection, snap-disc temperature sensors are embedded in motor windings. Temperature sensors are often listed on a motor nameplate as internal motor protection and by type of sensor.

External Motor Protection

To achieve external motor protection, sensing devices (magnetic overloads) are connected to the T1 and N or T1, T2, and T3 power supply leads attached to a motor. These sensors are typically used to open the magnetic coil circuit of a starter. Other external protection devices (thermal overloads) are attached to the starter itself. The trip point of any external protection device is established by the manufacturer in accordance with all applicable local, state, and national codes.

Magnetic Overloads. A *magnetic overload* is an external motor protection device that measures the magnetic field accompanying current flow from a starter to the motor power leads and opens the starter circuit if needed. As current flow increases, the magnetic field increases. The magnetic overload circuit contact opens and stops the motor by opening the coil circuit of the starter. Magnetic overloads are typically mounted outside the starter section and are not influenced by heat like thermal overloads. **See Figure 15-5.**

TECH TIP

The NEC® defines motor overload (thermal) protection as devices designed to protect motors and circuits from excessive heating. However, motor overload protection is not intended to protect motors from short-circuits or ground faults.

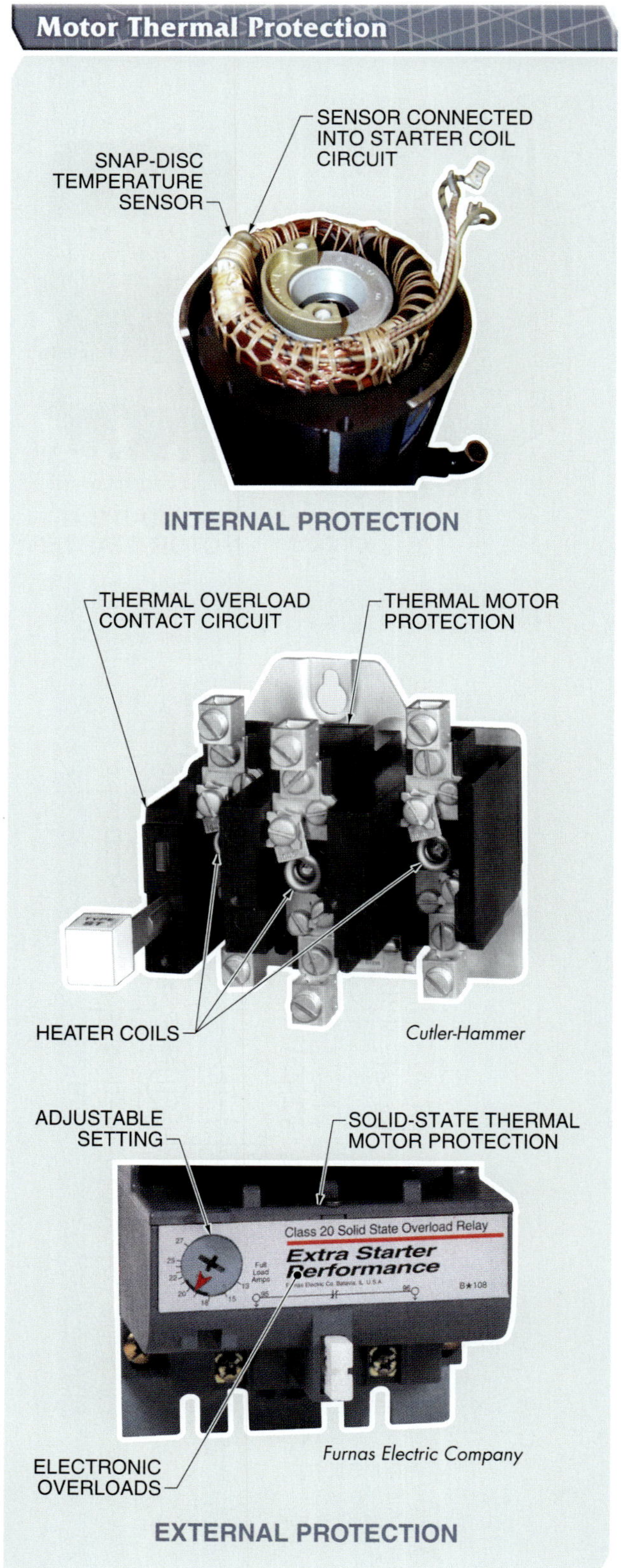

Figure 15-4. Internal motor snap discs and external thermal overload protectors are connected to the thermal contact circuits of motor starters to protect motors from overheating.

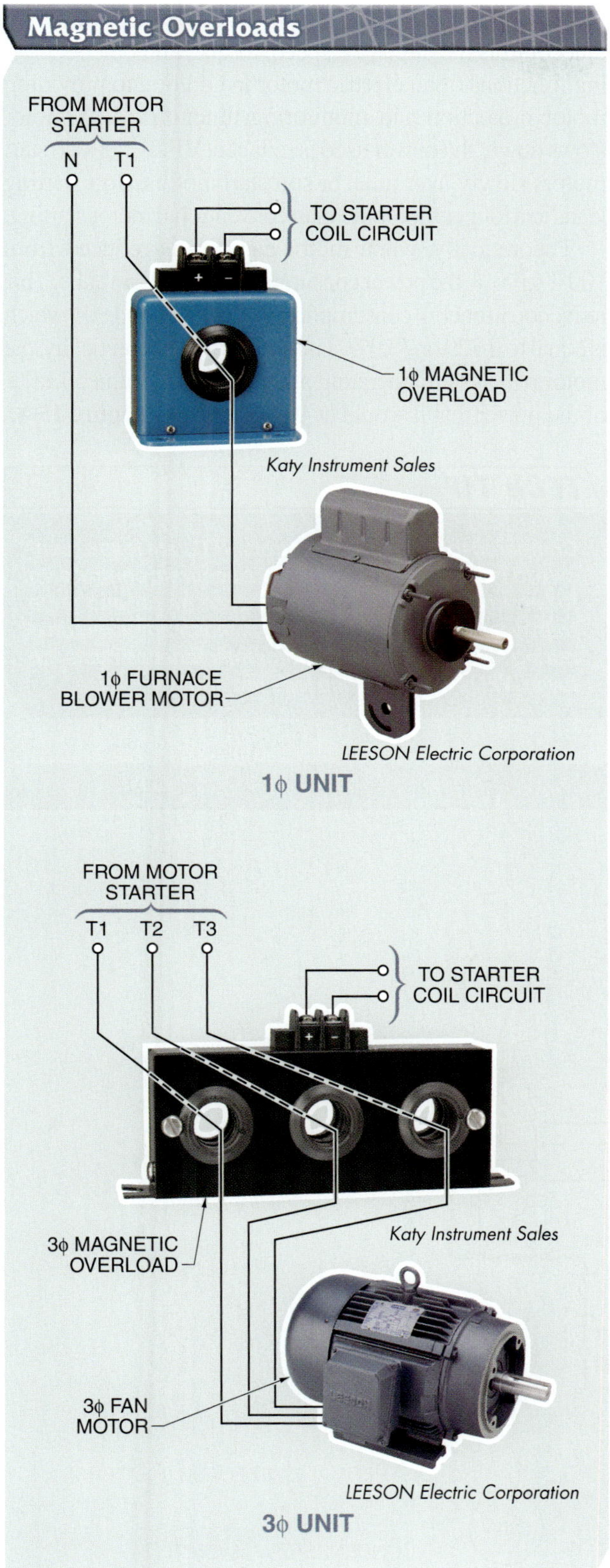

Figure 15-5. Magnetic overloads have an advantage over thermal overloads in that they do not require a cooling-off period before being reset.

Thermal Overloads. A *thermal overload (heater)* is an external motor protection device that uses the high temperatures caused by overcurrent conditions to open a motor circuit. **See Figure 15-6.** Current flows through the conductor and the metal coil in the thermal overload. As current flow increases, the heat in the thermal overload and conductor increases. The metal begins to warp and, in some cases, melt as the heat builds up. Thermal overloads are snap-acting, meaning they open suddenly. This reduces the possibility of arcing across the starter contacts.

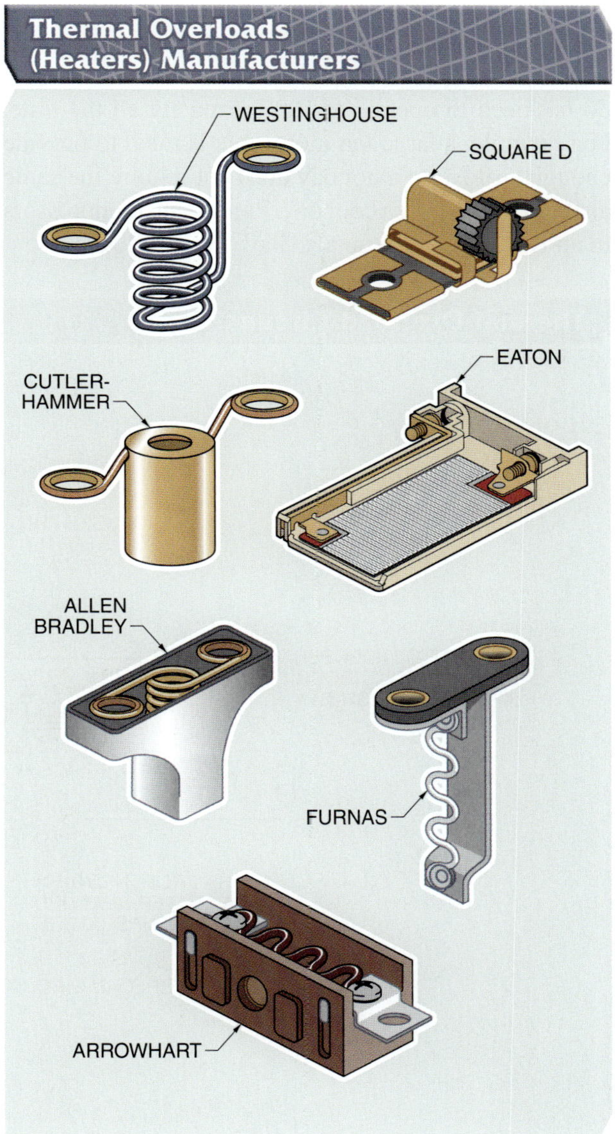

Figure 15-6. Each manufacturer of thermal overloads has a unique design for their motor starters.

When a thermal overload circuit contact opens in a motor starter, the coil circuit of the motor starter opens, and the motor stops. **See Figure 15-7.** When the temperature of the heaters decreases, the thermal overload circuit contact can be reset, or closed, once again. Since thermal overloads sense temperature, any external heat source may influence the opening and closing of the overload circuit. For example, the temperature of a hot boiler room may affect the motor overload circuit for the draft fan motor of a boiler.

VARIABLE-FREQUENCY DRIVES

Variable-frequency drives (VFDs) have become extremely common in the refrigeration and HVAC industry. VFDs provide a number of benefits to refrigeration and HVAC systems. Electrical loads in most systems do not need to operate at 100% capacity all the time. For example, a large fan motor that is rated to provide enough airflow on a hot day does not require the same airflow capacity on a cool day. Pumps and compressors in an HVAC system can also have varying loads.

A *variable-frequency drive (VFD)* is an electronic device that controls the direction, speed, torque, and other operating functions of an electric motor in addition to providing motor protection and monitoring functions. As a motor slows down, the power used is reduced. VFDs can also start motors slowly over time, or soft start motors. Soft starting leads to a longer motor life than conventional motor starting.

Theoretically, when motor capacity is reduced from 100% to 90%, the power consumed is reduced as a cube. This reduction in energy consumption is expressed as 0.90^3, which is equal to 0.729 or 72.9%. This means that, theoretically, the motor of a system operating at 90% flow consumes 72.9% of the power that it would at 100% flow. **See Figure 15-8.**

TECH TIP

Compared to mechanical controls, VFDs are less expensive to maintain and provide more accurate control with fewer adjustments. They also reduce motor and drive train wear and tear by reducing the motor amps required during starting. Therefore, motor maintenance costs are lower. VFDs also produce less fan and duct noise.

Thermal Overload Circuits

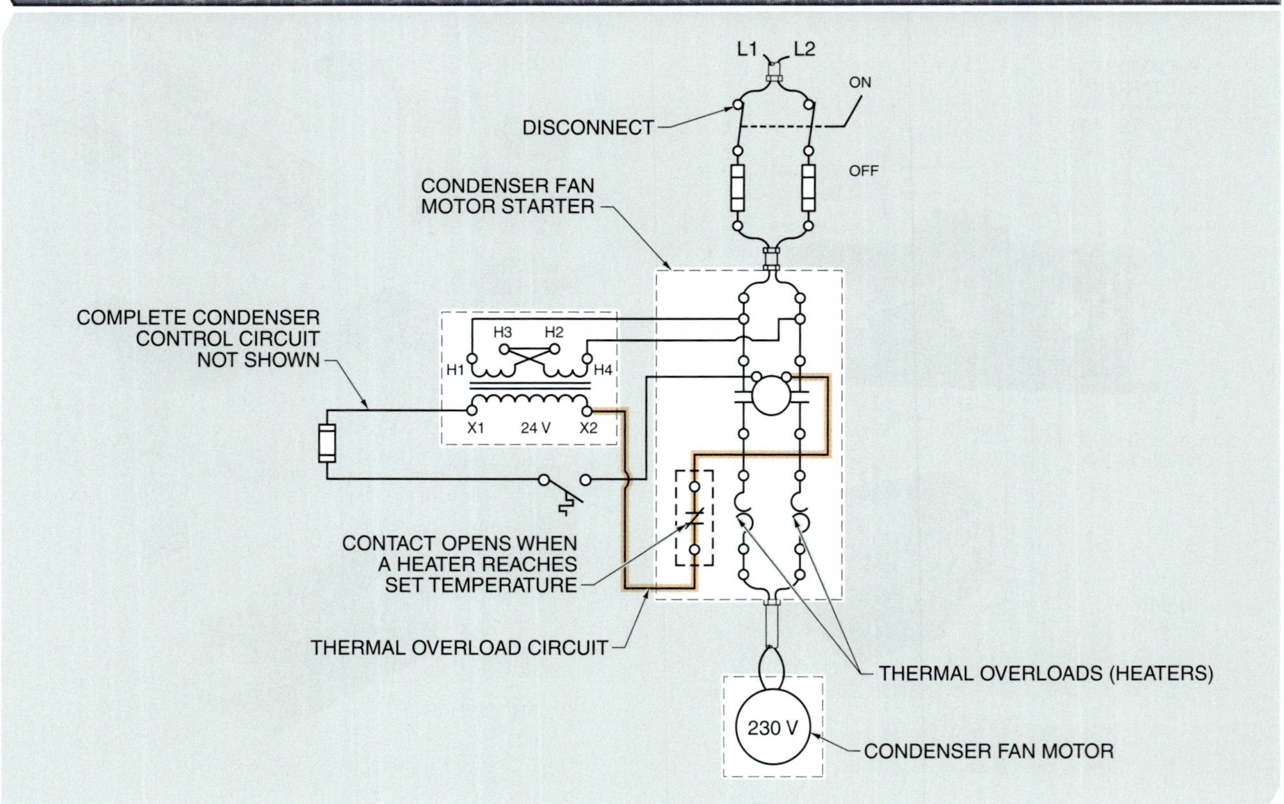

Figure 15-7. All thermal overloads mechanically open the normally closed contact of a thermal overload circuit to stop the operation of a motor.

Variable-Frequency Drives (VFDs)

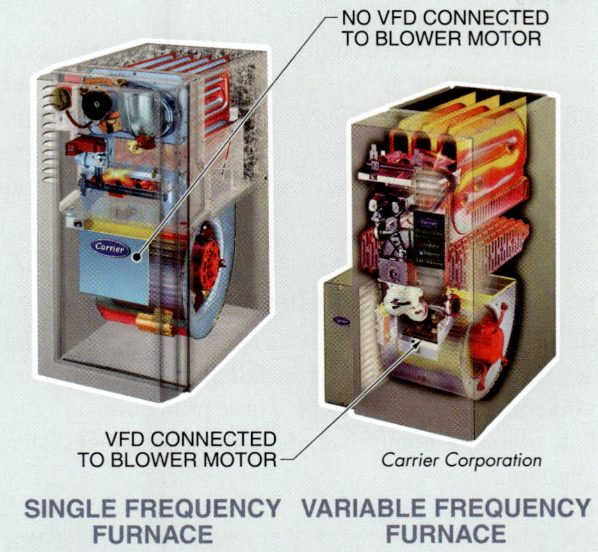

Motor System Capacity*	Motor Power Consumption with VFD†
95	85.7
90	72.9
85	61.4
80	51.2
75	42.2
70	34.3
65	27.5
60	21.6
55	16.6
50	12.5

* percent of maximum system capacity
† percent of power consumed by motor

Figure 15-8. Besides saving energy and money, variable-frequency drives (VFDs) also use 1ϕ, 115 V or 230 V power supplies to operate 3ϕ motors.

Variable-frequency drives can be combined with the latest building automation system (BAS) technologies to provide precise control for energy savings. Most VFDs have a built-in microprocessor panel that provides access to information such as power consumption and faults. The microprocessor panel also allows a programmer to change VFD settings through a keypad menu. VFDs are used in many refrigeration and HVAC applications including furnaces, air handler fans, hot water and chilled water pumps, and refrigeration compressors. **See Figure 15-9.**

HVAC Drive Applications

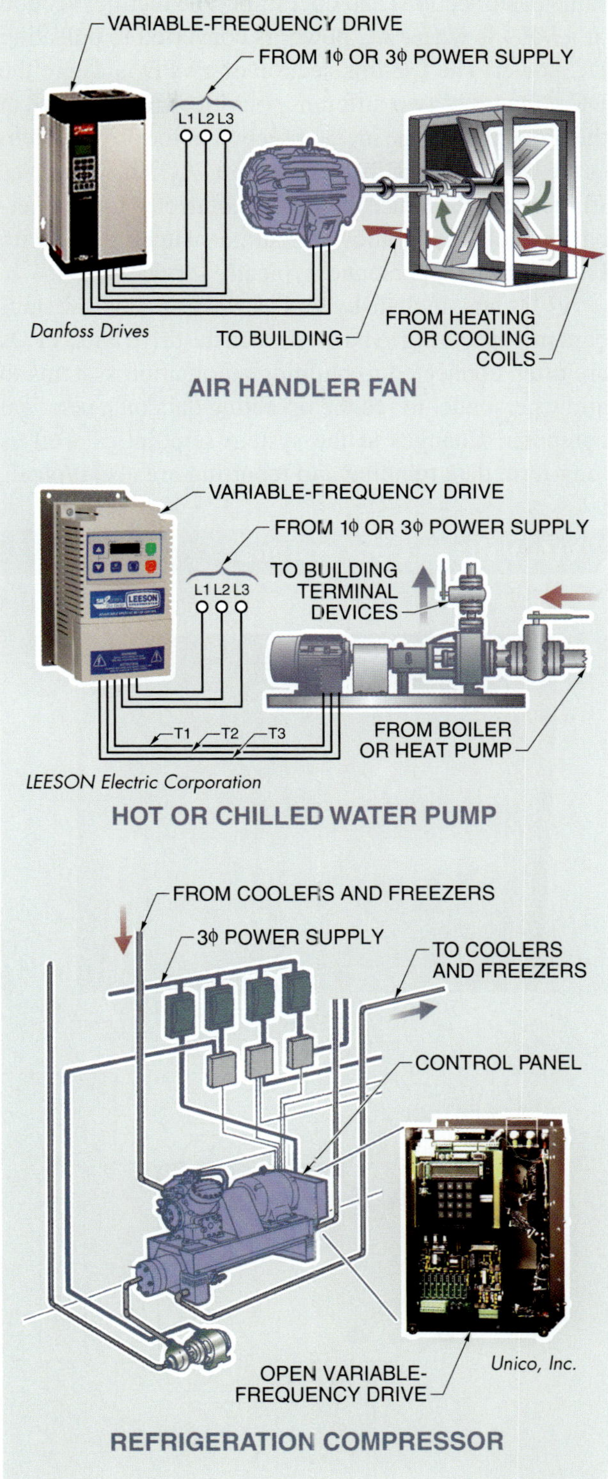

Figure 15-9. Variable-frequency drives are also used in HVAC systems to reduce the mechanical and electrical wear associated with the starting and stopping of system motors.

Variable-Frequency Drive Principles

Various types of VFDs are available but most have the same basic sections and operation. The rectifier section of a VFD is where AC power is converted to pulsating DC power. The DC link section of a VFD is where the DC voltages of two different polarities are generated at the proper level. The inverter section of the drive switches the DC buses rapidly to simulate an AC sine wave. **See Figure 15-10.** A microprocessor connected to the internal control system allows accurate switching to occur. Factory-trained personnel typically service VFDs.

VFDs may include bypass packages that allow standard motor starting to be used when the drive fails. VFDs are often connected to building automation systems to allow personnel to see the operating data on a personal computer. Changes to the system setpoint as well as long-term data trending and reporting are also typical.

TROUBLESHOOTING MOTOR STARTERS AND VFDs

Motor starters and VFDs are the first devices checked when motors are not running properly. They are checked first because they are where the incoming power, load, and control circuit wiring come together. Voltage and current measurements are taken at the motor starter or VFD to determine which section of the motor circuit has the problem.

Motor Starter Troubleshooting

Motor starter problems may be identified as application and selection, wiring and installation, or component problems. **See Figure 15-11.** These problems can cause motor damage, compressor damage, or incorrect system operation.

VFD Voltage Conversions

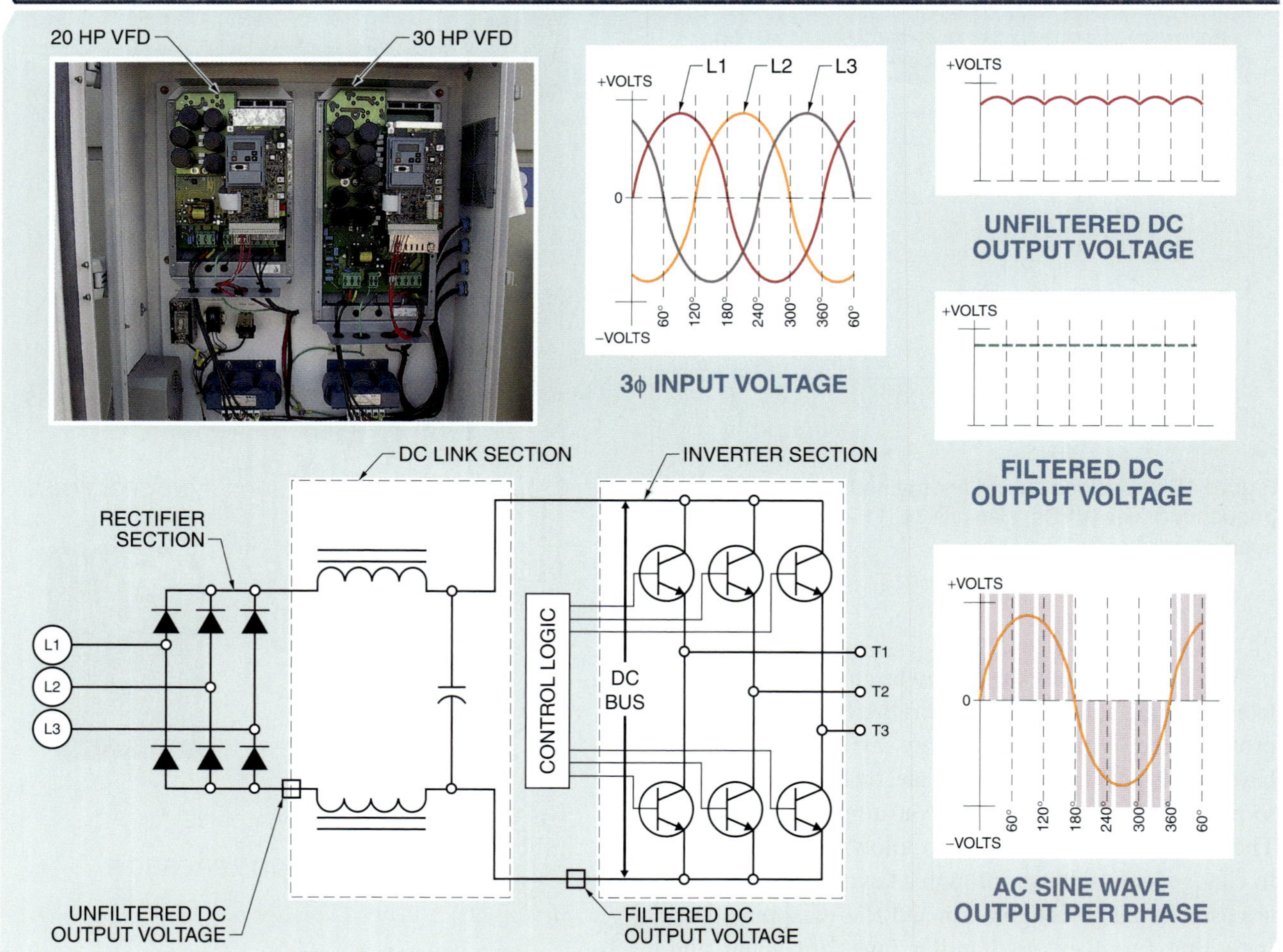

Figure 15-10. The rectifier section of a VFD converts AC voltage into DC voltage. The DC link section changes the two DC voltage polarities into one polarity. The inverter section converts DC voltage into either 1ϕ or 3ϕ AC voltage.

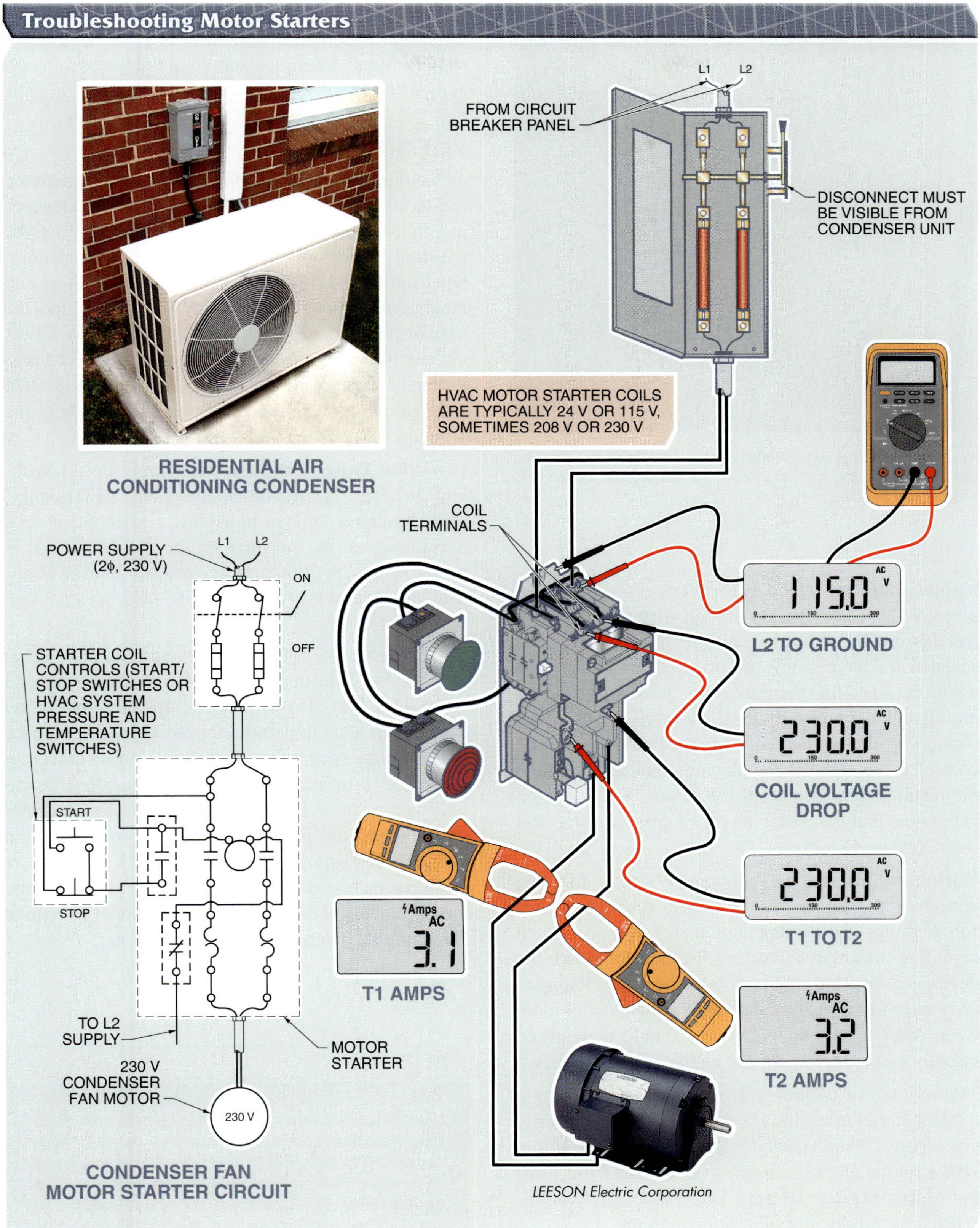

Figure 15-11. Voltage and current measurements are taken at the motor starter to determine in which section of the motor circuit a problem is occurring.

VFD troubleshooting requires that the input bridges, output devices, bus capacitors, and internal power supplies as well as the VFD program be checked.

Application and Selection Problems. In some cases, a motor does not operate properly from the date of installation. The wrong capacity device may have been selected for the application, or the device itself may not be designed for the application. Selection problems are often found upon equipment startup and commissioning by the technician. Documentation should be consulted regarding the application and, if necessary, the manufacturer contacted for advice. While rare, it is possible that the wrong motor or drive package was sent from the factory.

Wiring and Installation Problems. Wiring or installing a motor starter improperly may also cause problems. Job drawings and documentation should be consulted regarding the environment and location for which the device is rated. Motor starters operating in high temperatures or in frequent start-stop environment suffer more wear. After installation, all wires should be checked visually for proper routing and tightness of connections.

Motor Starter Component Problems. Troubleshooting a motor starter assembly is easy because there are only a few parts. The contactor coil can be checked with a DMM for the proper resistance values. **See Procedure for Motor Starter Testing.** The contacts may be inspected for pitting, discoloration, or arcing.

Springs can be checked visually to determine their condition. The springs that return the contacts to their normal position may break or wear down over time. Replacement components are available to rebuild motor starters. Small-size motor starters are replaced immediately when found to be operating improperly.

VFD Troubleshooting

To troubleshoot a VFD, information must be gathered and an inspection of the VFD must be made. Depending on the model, a factory-trained technician may be required to perform the troubleshooting procedure. **See Figure 15-12.** It must be kept in mind that a large number of manufacturers make VFDs, and the specific troubleshooting procedures vary widely. Always follow the manufacturer's recommendations and procedures. Problems with a VFD and/or circuit can be categorized as incoming power problems or as problems with the VFD itself.

Incoming Power Problems. Incoming power problems manifest as overvoltage faults on a VFD display, phase-to-phase voltage unbalances greater than 2%, harmonics on the power supply, and out-of-phase power supply. A difference between no-load and full-load voltage that is greater than 3% is also an incoming power problem.

VFD Problems. VFDs may have overvoltage or overcurrent faults due to poor drive programming and setup. They may also have high temperature faults due to high ambient temperatures, defective drive cooling fans, or dirty heat sinks. They may operate correctly when set to original default parameters but not to field parameters. This could be caused by incorrect or changed field parameters, motor problems, load changes, or improper external control systems. A VFD can be tested when energized by taking voltage measurements of the various sections. **See Procedure for Variable-Frequency Drive Testing.**

TECH TIP

When selecting test instruments for taking VFD measurements, it is important to remember that AC and DC voltages are inside the VFD. DC voltages will be equal to the peak AC input voltage of the drive. The DC voltage is stored in capacitors capable of holding charges for extended periods of time. Due to safety concerns, voltage measurements of the VFD capacitors should always be taken first.

Motor Starter Testing

1. Connect the test leads to the appropriate meter jacks for taking resistance measurements.
2. Follow all electrical safety practices and procedures.
3. Lockout/tagout the disconnect to the residential condenser unit.
4. Open the access panel to the residential HVAC condenser unit.
5. Set the meter function switch to measure AC voltage ($\tilde{V}$).
6. Check the incoming power lines to the condenser for voltage (0 V).

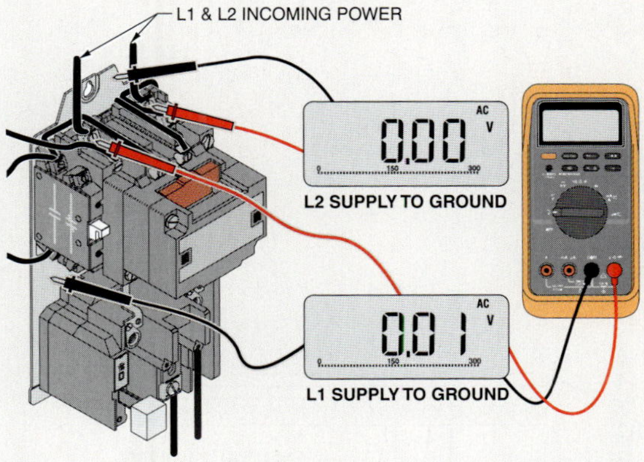

7. Set the meter function switch to measure resistance (Ω).
8. Connect the meter test leads across the coil terminals of the motor starter to measure the resistance or continuity of the coil.

   Note: The coil must be isolated by disconnecting the two circuit wires from the coil terminals.

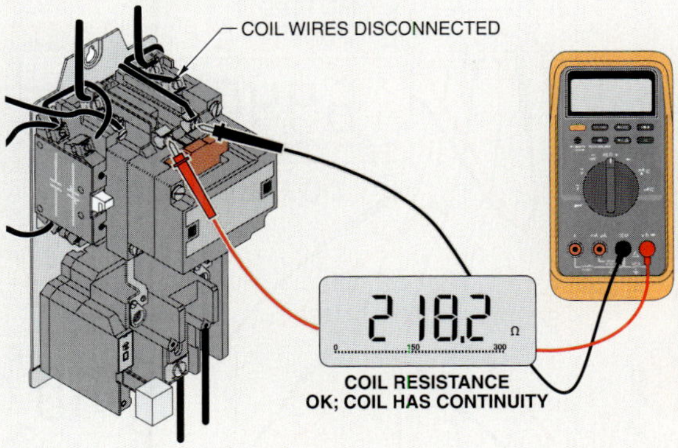

9. Read and record the resistance measurement indicated on the meter.
10. Connect the meter test leads from the input of phase 1 to the top of the phase 1 heater to measure the resistance or continuity of the contacts. Repeat for phase 2.

    Note: Contacts must be manually closed.

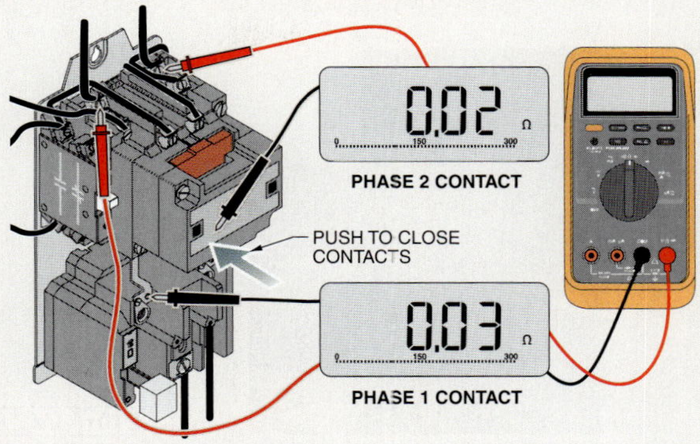

11. Read and record the resistance measurements (phase 1 and phase 2) indicated.
12. Connect the meter test leads across the phase 1 heater to measure the resistance or continuity. Repeat for phase 2.

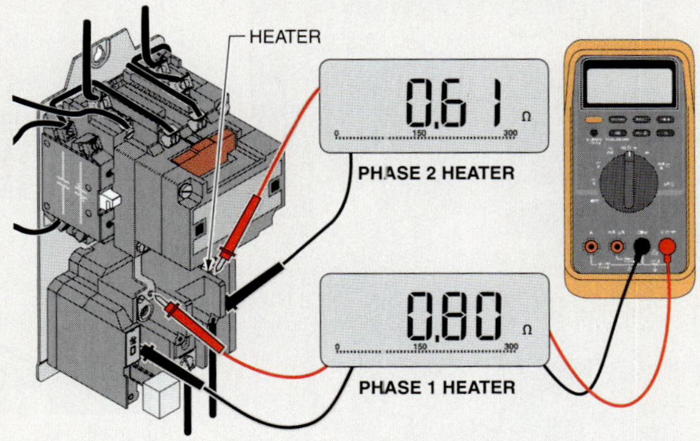

13. Read and record the resistance measurements (phase 1 heater and phase 2 heater) indicated.
14. Connect the meter test leads across the normally open auxiliary contacts. Repeat for all other normally open and normally closed auxiliary contacts.

    Note: Contactors must be manually closed.

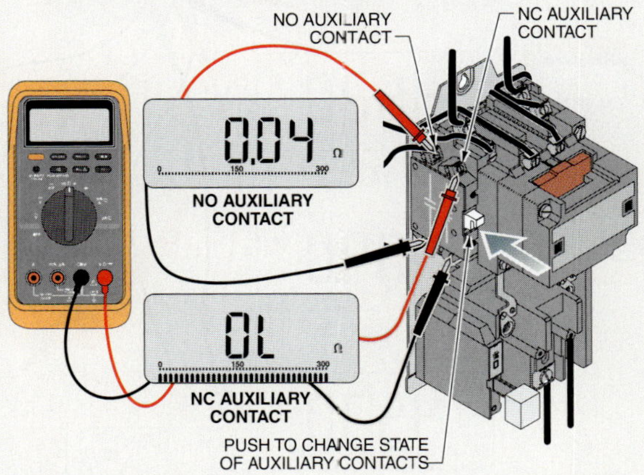

15. Read and record the resistance measurements (normally open and normally closed contacts) indicated.
16. Disconnect the meter test leads from the motor starter.
17. Turn the meter function switch to OFF.

Procedure

Troubleshooting VFDs

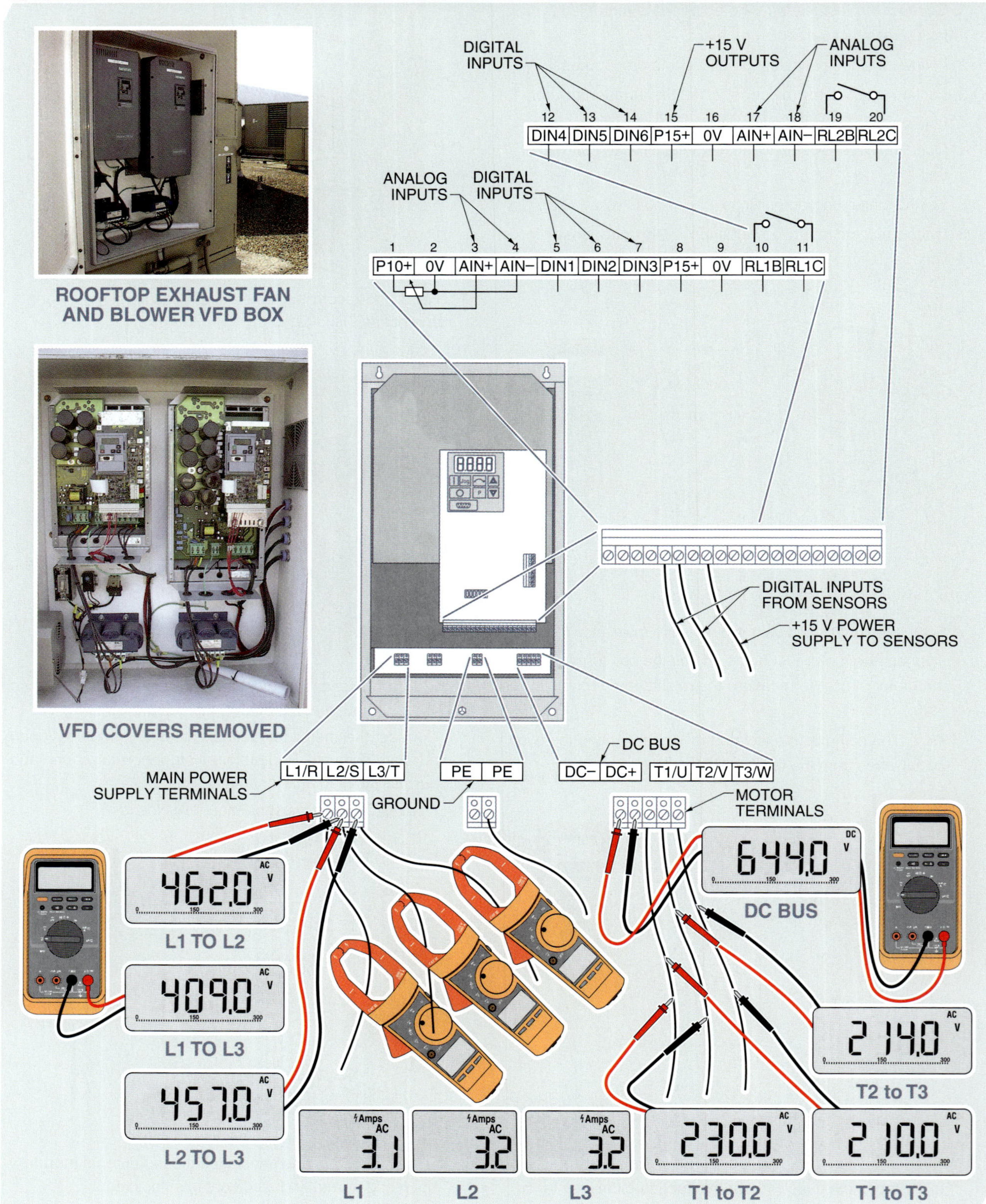

Figure 15-12. Voltage and current measurements can be taken at a VFD to determine where a problem occurs.

Variable-Frequency Drive Testing

1. Connect the test leads to the appropriate meter jacks for taking voltage measurements.
2. Follow all electrical safety practices and procedures.
3. Open the access panel to the residential HVAC furnace.
4. Inspect the (varible-frequency drive) application for any physical mechanical damage.

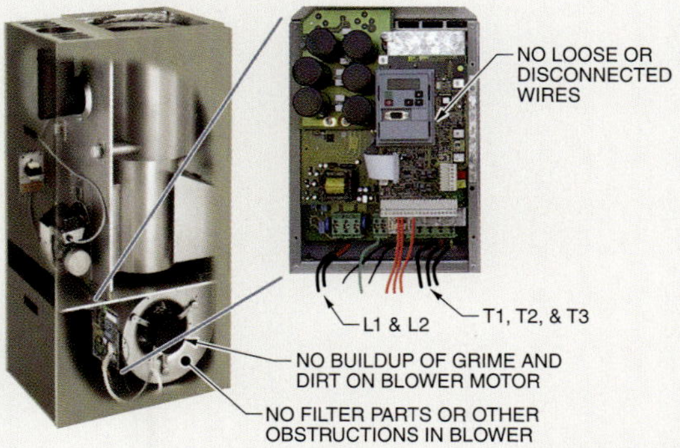

5. Set the meter function switch to measure AC voltage ($\tilde{V}$).
6. Test the incoming power lines L1, L2, and L3 to the VFD.

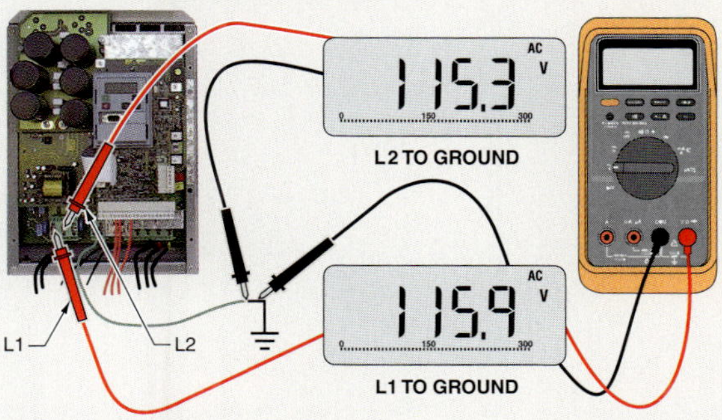

7. Read and record the voltage measurements indicated on the meter. Repeat for L2 and L3.
8. Test the output voltage T1, T2, and T3 to the motor.

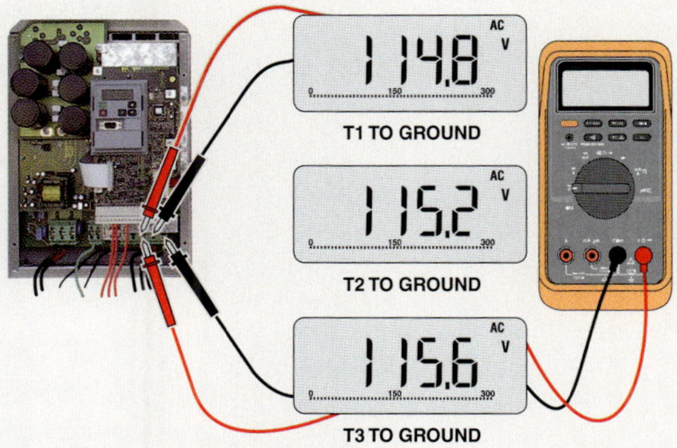

9. Read and record the voltage measurements indicated on the meter. Repeat for T2 and T3.
10. Set the meter function switch to measure AC current ($\tilde{A}$).
11. Test the output current on T1, T2, and T3 to the 3ϕ blower motor.

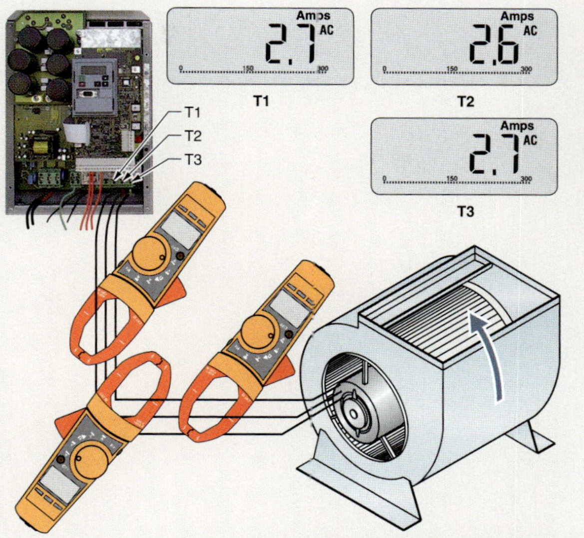

12. Read and record the current measurements indicated on the meter. Repeat for T2 and T3.
13. Test the voltage of the individual VFD input signals. Repeat for all digital and analog input signals.

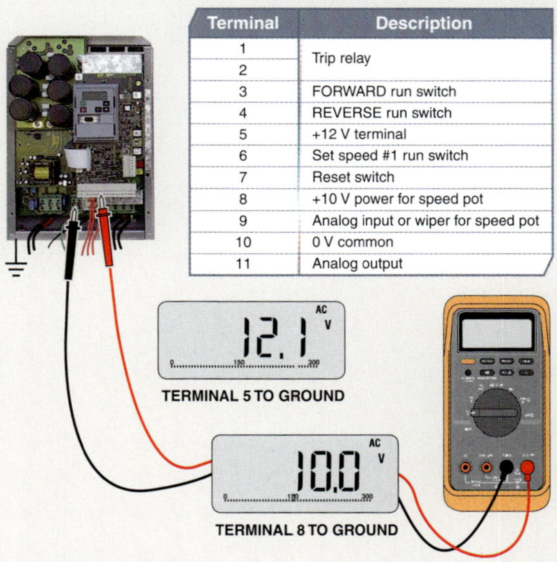

Terminal	Description
1	Trip relay
2	
3	FORWARD run switch
4	REVERSE run switch
5	+12 V terminal
6	Set speed #1 run switch
7	Reset switch
8	+10 V power for speed pot
9	Analog input or wiper for speed pot
10	0 V common
11	Analog output

14. Read and record the voltage measurements indicated on the meter. Repeat for T2 and T3.
15. Disconnect the meter test leads from the VFD.
16. Turn the meter function switch to OFF.

UNIT 16

Forced-Air Furnaces

Forced-air heating systems consist of a heating unit, a distribution system, and numerous controls. Supply air is heated by a furnace and distributed to building spaces through supply air ductwork. Air is returned to the furnace through return air ductwork. Forced-air heating systems are used in residential buildings and light commercial buildings where ductwork is not found in long runs. However, some special-purpose heating units do not require ductwork. The air in a special-purpose heating unit is heated and distributed throughout a building space such as a garage or warehouse.

OBJECTIVES

- Describe the difference between the two types of forced-air heating systems.
- Discuss the different styles of furnaces.
- Explain the ratings of furnaces.
- Describe the various types of special-purpose heating systems.

FORCED-AIR HEATING SYSTEMS

A *forced-air heating system* is a heating system that uses a fan to force air across a heat source to heat a building. The heated air is distributed throughout building spaces to change the temperature of the spaces. A forced-air heating system includes a heating unit (furnace), a distribution system, and a number of controls. Different types of forced-air heating systems are used for various applications. The type of system used for a particular application depends on the amount of heat required and the layout of the building. The two main types of forced-air heating systems are the central and modular forced-air heating systems. **See Figure 16-1.**

Central Forced-Air Heating Systems

A *central forced-air heating system* is a heating system that uses a centrally located heating unit to produce heat for a building. Supply air ductwork runs from the heating unit to the building spaces. Return air ductwork runs from the building spaces back to the heating unit.

Modular Forced-Air Heating Systems

A *modular forced-air heating system* is a heating system that uses multiple heating units to produce heat for a building. Each heating unit produces heat for a building floor, module, or zone. A *zone* is the section of a building that is controlled by one HVAC controlling device, such as a thermostat. The heating unit of a modular forced-air heating system is typically located on a roof.

TECH TIP

Forced-air furnaces offer a wide variety of options and accessories, such as air cleaners, filters, humidifiers, and ultraviolet lights, that can be used to improve indoor air quality.

Forced-Air Heating Systems

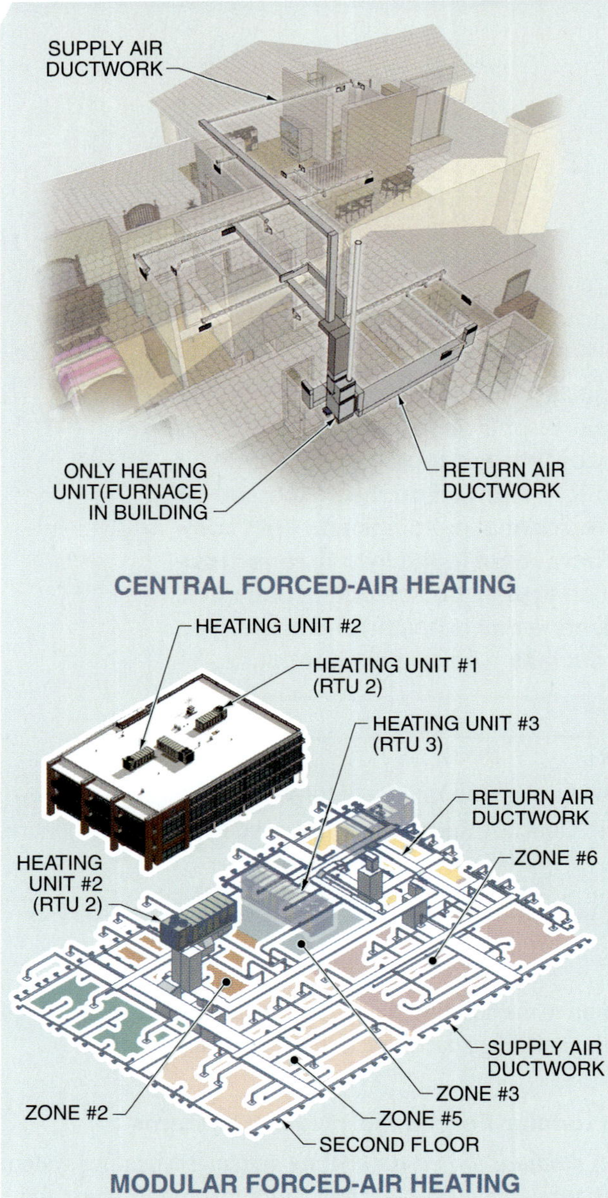

Figure 16-1. A forced air heating system uses either one centrally located heating unit or multiple heating units to produce heat for a building.

FURNACES

A furnace is the central element in a forced-air heating system. A *furnace* is a self-contained heating unit that includes a blower, at least one burner, a heat exchanger, and a number of controls. **See Figure 16-2.** Some furnaces use electric heating elements in place of the burner and heat exchanger. Furnaces are available in different sizes and styles for a variety of applications. Forced-air furnaces are classified by the style (direction of airflow), energy consumption, dimensions, and heating capacity.

Furnaces

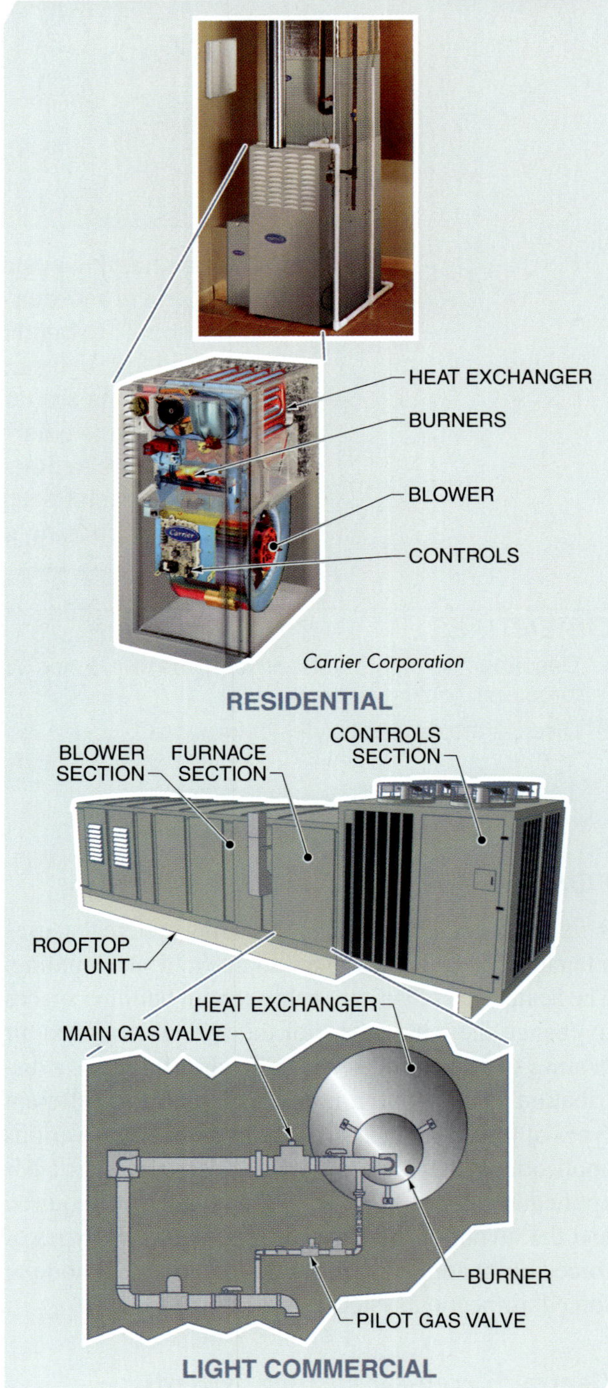

Figure 16-2. Regardless of the manufacturer, all furnaces include a blower, at least one burner, a heat exchanger, and a number of controls.

Furnace Styles

Furnaces are classified by the direction of airflow out of the furnace. The three most common styles of furnaces are upflow, horizontal, and downflow. **See Figure 16-3.**

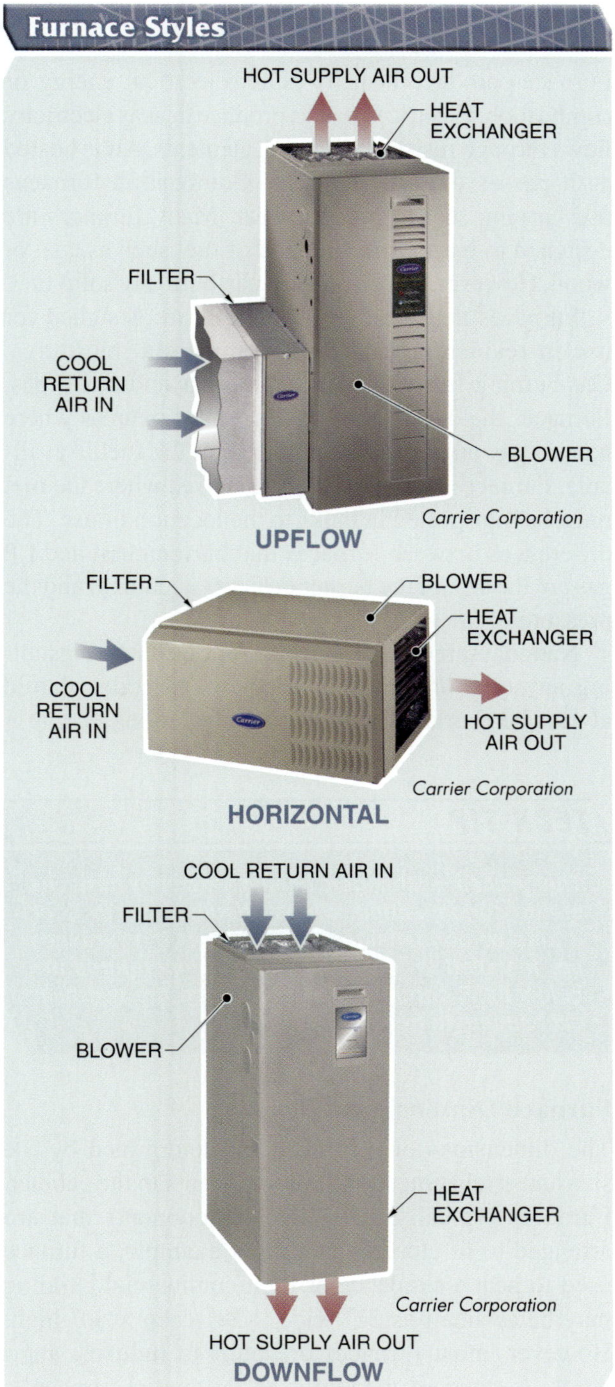

Figure 16-3. Furnaces are classified by the direction of airflow out of the furnace. The common furnace styles are upflow, horizontal, and downflow.

Upflow Furnaces. An *upflow (highboy) furnace* is a furnace that heats air and distributes the heated air upward through ductwork. Return air enters from or near the bottom of the furnace and exits from the top. Upflow furnaces are used in areas such as basements where ductwork is located above the furnace.

Horizontal Furnaces. A *horizontal furnace* is a furnace that heats air and distributes the heated air horizontally through ductwork. Return air enters horizontally from one end of the furnace, and supply air exits horizontally from the other end. Horizontal furnaces are used where headroom is limited, such as in attics or crawl spaces. The ductwork of a horizontal furnace is typically located on the same level as the furnace.

Downflow Furnaces. A *downflow (counterflow) furnace* is a furnace that heats air and distributes the heated air downward through ductwork. Return air enters from the top of the furnace, and supply air exits from the bottom. The supply ductwork of a downflow furnace is located below the furnace, such as in the crawl space below the furnace.

Some furnace designs allow furnaces to be installed vertically or horizontally to the left or right.

Electric Furnaces

An *electric furnace* is a furnace that uses resistance heating elements to warm passing air. A *resistance heating element* is a device consisting of a grid of electrical resistance wires that are attached to a support frame with ceramic insulators. Nichrome, a nickel-chromium alloy, is the material typically used for the electrical resistance wire. Nichrome wire resists the flow of electricity and becomes red-hot when connected to an electrical circuit. Resistance heating elements are installed in a furnace so that air is heated as it flows over the hot elements while passing through the furnace. **See Figure 16-4.** The electrical circuit is protected from overcurrent and overheating by fuses and temperature-sensing limit switches.

Electric Furnaces

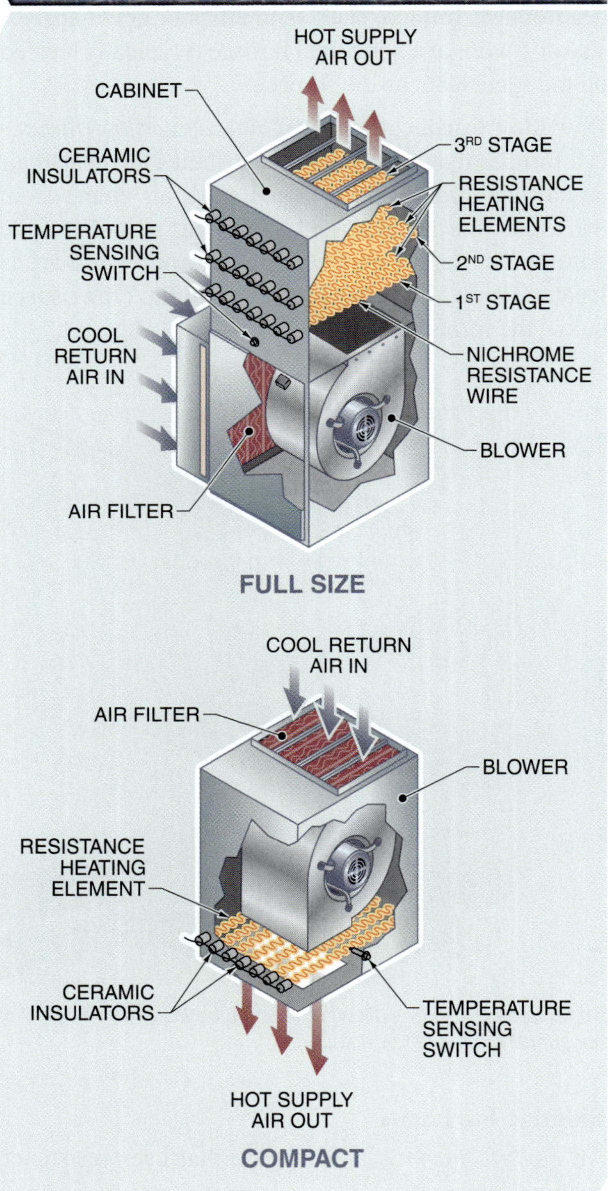

Figure 16-4. Electric furnaces use resistance heating elements made of Nichrome to warm the air passing through the furnace.

Most electric furnaces can be used in any position: upflow, horizontal, or downflow. Electric furnaces require 230 V or higher to power the resistance heating elements. However, like conventional combustion furnaces, the controls use 24 V.

Furnace Energy Consumption

Furnaces produce heat by either electrical energy or combustion. Electric furnaces produce heat as electricity flows through resistance heating elements. Air is heated as it passes the hot elements. Combustion furnaces use burning fuel to produce heat. Many furnaces are designed to burn a specific kind of fuel such as coal or wood. However, some furnaces can burn any solid fuel.

Furnaces that burn No. 2 fuel oil are designed for use in residential and light commercial buildings. Gas burning furnaces burn natural gas and/or LP gas. Furnaces that burn natural gas are used in areas where natural gas pipelines are in place and the fuel is available. Furnaces that burn LP gas are used where the fuel must be transported in tanks to the location of use. The differences between furnaces that burn natural and LP gas are the size of the burner orifices (openings) and the pressure of the gas.

National safety codes must be followed when installing any type of furnace. Jurisdictional codes should always be checked for additional regulations.

TECH TIP

The two heat exchangers of high-efficiency furnaces create condensation, which should drain to floor drains or condensate pumps. When the furnaces are installed in locations where temperatures are below 32°F, condensation can cause the two heat exchangers to freeze.

Furnace Dimensions

The dimensions of a furnace are determined by the size and arrangement of the components in the cabinet. Furnaces typically have small components that are arranged to fit close together. For example, a furnace used to heat a residence or light commercial building may be as small as 21″ wide × 28″ deep × 40″ high. However, most furnaces are built to industry standards. A standard medium-size residential furnace is 32″ wide × 28″ deep × 54″ high. Dimensional standards allow for the replacement of furnaces without the need to modify the building or structure. **See Figure 16-5.**

Common Upflow Gas Furnace Dimensions

Model	Dimensions*		
	W	D	H
GMH8	17.5	28	39
58CVA	21	28	32
6MG05	21	29.75	31.5
DF-48C-090	21	28.5	40
TDCIC9481	21	28	40
58MEB	21	28.5	40
G61MPV	21	28.5	40
INF96	21	28	40

* in in.

Figure 16-5. Furnaces are built to industry standards, which depend on the size and arrangement of the components in the cabinet.

Heating Capacities

Manufacturer specification sheets are used to find the heating characteristics of a furnace, which include the input rating, output rating, and efficiency rating. Input ratings of 50,000 Btu to 150,000 Btu are common. Most furnaces can be mounted in different configurations and can use multiple fuels.

Input Ratings. An *input rating* is the amount of heat produced by a furnace or boiler as fuel is burned; expressed in Btu/hr. The input rating is found by multiplying the heating value of the fuel by the flow rate of the furnace. A *heating value* is the amount of heat produced per unit of fuel per hour; expressed in Btu/ft^3. The heating value of a fuel depends on the chemical composition of the fuel. *Flow rate* is the rate at which a furnace burns fuel. A furnace produces a given amount of heat for each unit of fuel burned. The input rating of a furnace is found by applying the following formula:

$$IR = HV \times Q$$

where

IR = input rating (in Btu/hr)

HV = heating value (in Btu/ft^3)

Q = flow rate (in ft^3/hr for natural gas)

Example: Finding Furnace Input Rating

A natural gas-fired furnace burns natural gas at 100 ft^3/hr. What is the input rating of the furnace? **See Figure 16-6.** *Note:* The heating value for natural gas is 1000 Btu/ft^3.

$$IR = HV \times Q$$
$$IR = 1000 \times 100$$
$$IR = \mathbf{100{,}000\ Btu/hr}$$

Finding Furnace Input Ratings

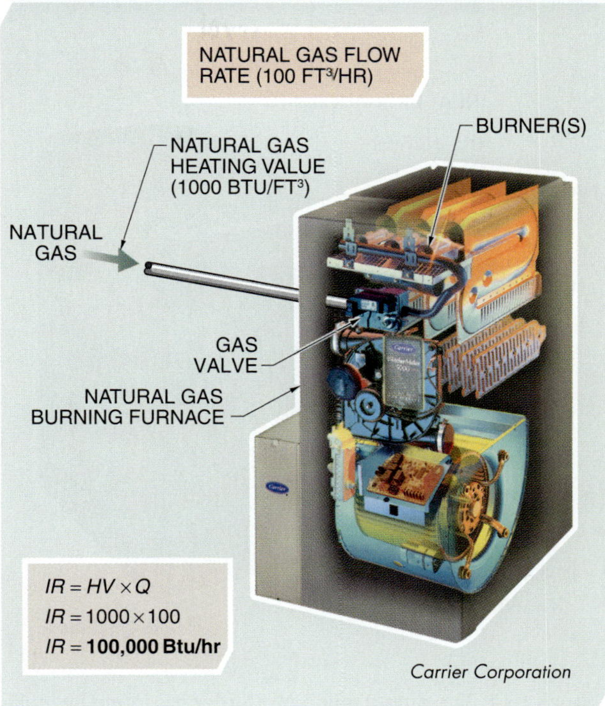

Figure 16-6. The input rating of a furnace is found by multiplying the heating value of the fuel by the fuel flow rate being used by the furnace.

Output Ratings. An *output rating* is the total amount of heat a furnace or boiler will produce in one hour; expressed in Btu/hr. The output rating of a furnace is also the heating capacity. The output rating is found by multiplying the input rating by the efficiency rating of the furnace. An *efficiency rating* is a number provided by a manufacturer that represents the percentage of heat that is delivered to the space versus the total input. The output rating is found by applying the following formula:

$$OR = IR \times ER$$

where

OR = output rating (in Btu/hr)

IR = input rating (in Btu/hr)

ER = efficiency rating (in percent)

Example: Finding Furnace Output Ratings

A furnace with an input rating of 100,000 Btu/hr has an efficiency rating of 80%. What is the output rating of the unit? **See Figure 16-7.**

$$OR = IR \times ER$$
$$OR = 100{,}000 \times 0.80$$
$$OR = \mathbf{80{,}000\ Btu/hr}$$

Finding Furnace Output Ratings

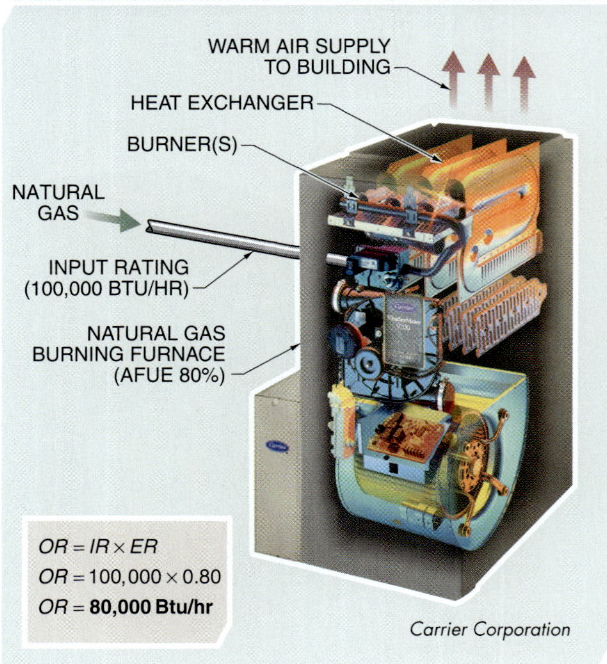

Figure 16-7. The output rating of a furnace is found by multiplying the input rating by the efficiency rating of the furnace.

Annual Fuel Utilization Efficiencies. The *annual fuel utilization efficiency (AFUE) rating* is the measure of the thermal efficiency of HVAC combustion equipment; expressed as a percent. The AFUE for most furnaces is typically between 80% and 95%. Some furnaces can have an AFUE as low as 72%, while others can reach 96.6%. **See Figure 16-8.** A furnace with an AFUE of 80% has 20% of the heat produced rising up through the flue. This rising heat produces draft in the stack.

Furnace Heating Capacities and Efficiency Ratings

Model	Input Rating*	Output Rating*	AFUE
58CVA	192,000	154,000	80
6MG05	75,000	60,000	80
DF-48C-090	75,000	60,000	80
TDCIC9481	103,250	94,000	91
58MEB	137,500	128,000	93
G61MPV	75,000	60,000	80
INF96	124,200	120,000	96.6

* in Btu/hr

Figure 16-8. Some furnaces have an AFUE as low as 72%, while others can reach 96.6%. A furnace with an AFUE of 80% has 20% of the heat produced rising up through the flue.

A certain amount of stack draft is necessary for proper furnace operation. *Draft* is the movement of air across fire and through a heat exchanger. Because modern condensing furnaces allow less heat to be used for producing draft, condensing furnaces operate with higher efficiency ratings.

SPECIAL-PURPOSE HEATING UNITS

Special-purpose heating units are furnaces built for special applications. Special-purpose heating units include unit heaters, duct heaters, wall units, and rooftop units. The following spaces require special-purpose heating units:

- large open spaces
- spaces with a common air supply but not a central heating unit
- spaces that receive 100% of the air required to replace the air exhausted from a given space through a flue
- spaces that do not require heat but have occupants requiring heat

Unit Heaters

A *unit heater* is a self-contained forced convection heater that is not connected to any type of ductwork and contains a fan, a hot water coil, and controls in one cabinet. Unit heaters are also available that burn natural gas or fuel oil for combustion. **See Figure 16-9.** Small unit heaters may have resistance heating elements instead of burners. When a large amount of hot water or steam is available, unit heaters include hot water or steam coils. Unit heaters are used as space heaters in large areas, such as garages and storage spaces.

A blower in the unit heater draws air in through the back of the unit and blows heated air out through louvers in the front. A unit heater is typically suspended from a ceiling or the roof frame of a building. It is never connected to any type of ductwork. A fuel line, electric conductors, and possibly control circuit wiring are run to the heater.

Duct Heaters

A *duct heater* is a unit heater that is installed in a duct and supplied with air from a remote blower. The blower for a duct heater is typically controlled by a central control system, such as a furnace. A duct heater has an air-proving limit switch that shuts off the unit when no air is moving through the duct. In residential and light commercial applications, duct heaters typically use resistance heating elements.

Unit Heaters

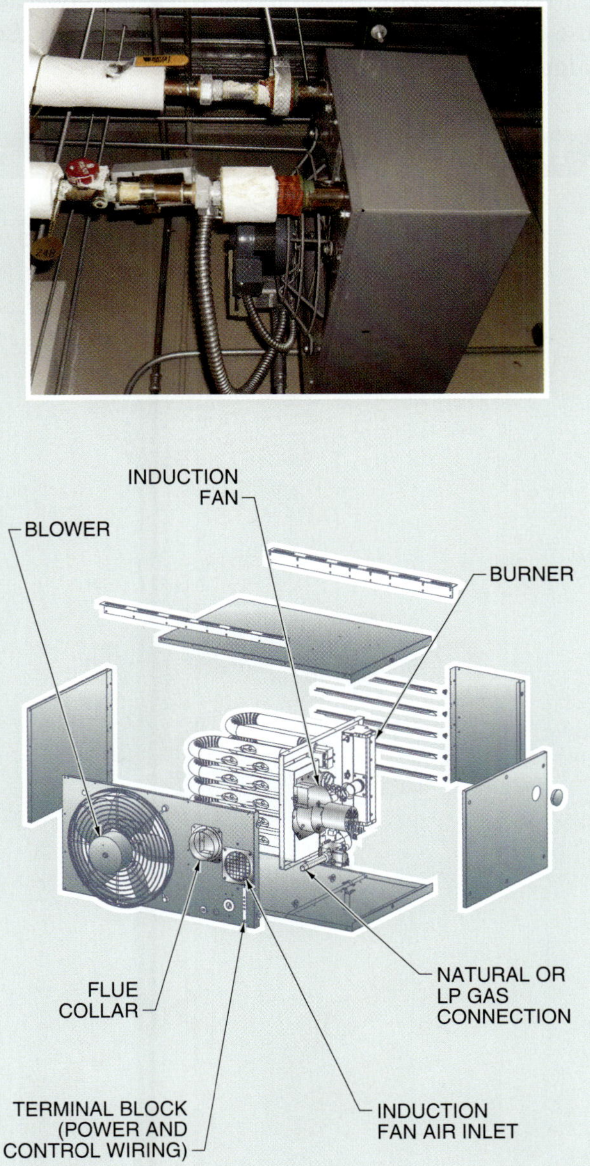

Figure 16-9. Unit heaters heat air by using resistance heating elements, steam or hot water coils, or by fuel combustion. They are typically suspended from ceilings.

Electric Duct Heaters. Electric duct heaters are used in large duct systems that have central blowers for pushing supply air through the duct heater for reheating. **See Figure 16-10.** *Reheat* is heat added to supply air, at the point of use, when the air traveling through a duct system has dropped in temperature. The electric heating elements can be used as either primary or backup heat. The electric heating elements have at least one stage. The stages are measured in kilowatts (kW).

Electric Duct Heaters

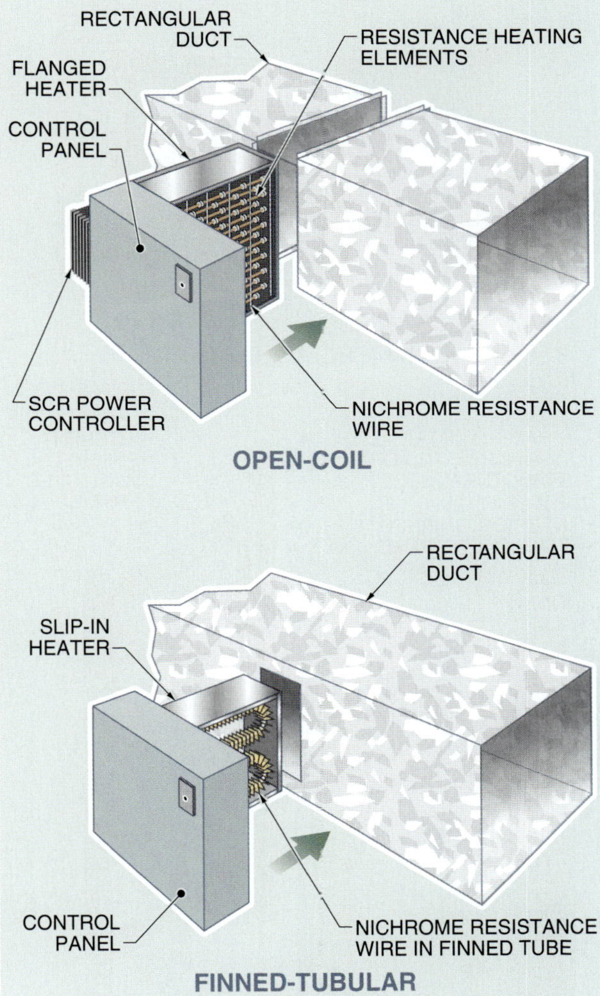

Figure 16-10. Electric duct heaters are used as either primary or backup heat in large duct systems to heat or reheat air at the point of use.

Wall Units

A *wall unit* is a packaged HVAC unit that heats a building space by using a fan and heat exchanger mounted in a metal enclosure, which is installed in or against an outside wall. **See Figure 16-11.** Wall units provide a convenient way to heat a building space. These units are commonly used in additions to residences and light commercial buildings. They may be attached to a chassis, which can be easily moved into and out of the enclosure opening in the structure. Wall units are typically hardwired to a building.

Electric Wall Units. Small wall units, such as packaged terminal air conditioner (PTAC) and packaged terminal heat pump (PTHP) units, are powered by cords that are plugged into electrical receptacles. PTHP units typically use 25% to 40% less electricity than PTAC units. **See Figure 16-12.**

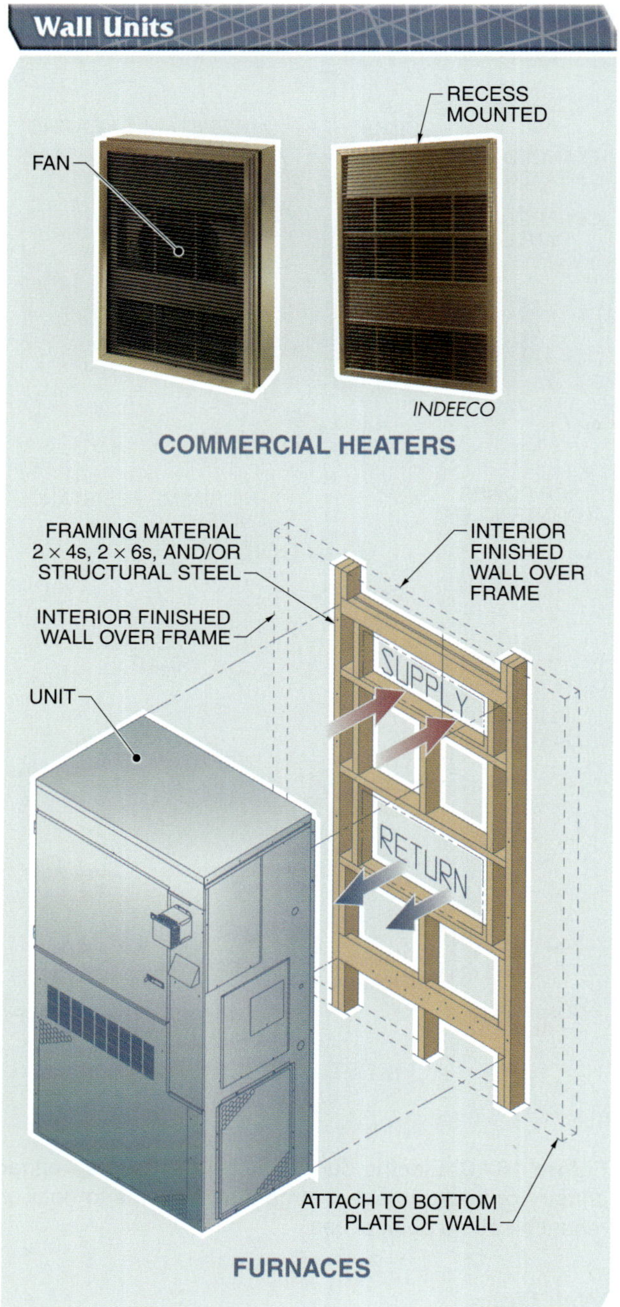

Figure 16-11. The fan and heat exchanger of a wall unit are mounted in a metal enclosure that is installed in or against an outside wall.

Figure 16-12. Small wall units, such as packaged terminal heat pump (PTHP) units, are powered by cords that are plugged into ordinary 115 V electrical receptacles.

Direct-Fired Heaters. A *direct-fired heater* is a unit heater that does not have a heat exchanger. Direct-fired heaters used in garages and construction applications are commonly called "torpedo heaters." These temporary heaters blow air directly through the combustion chamber and out the supply end of the unit.

Direct-fired heaters are used in applications where 100% makeup air is required for ventilation and 100% of the air is constantly exhausted outside. Only portable direct-fired heaters are used in garages and construction applications. **See Figure 16-13.** All manufacturers of direct-fired heaters must comply with ANSI standards. Per ANSI Standard Z83.4a-2001/CSA 3.7a-2001, *Gas Unit Heaters and Direct-Fired Duct Furnaces,* a maximum of 5 ppm of carbon monoxide (CO) and 0.5 ppm of nitrogen dioxide (NO_2) may be generated by direct-fired heating equipment.

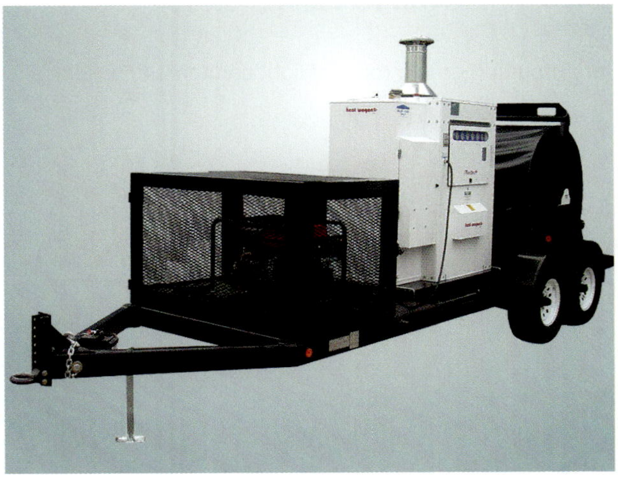

Heat Wagon, Inc.

Some direct-fired heaters are capable of thawing the ground during the winter, allowing excavation work to take place for construction projects.

Rooftop Units

A *rooftop unit* is a packaged HVAC unit mounted on the roof of a light commercial building. **See Figure 16-14.** Rooftop units typically provide heating and cooling. Mounting a unit on a roof creates valuable free space inside a building and may make maintenance access easier. However, this ease of access may be offset by technician exposure to the elements during service and the possibility of carrying heavy tools and equipment up to the roof. When unit replacement is necessary, it is usually simple to bring up a new unit using a crane. A rooftop unit typically has an electromechanical or electronic programmable thermostat in the building.

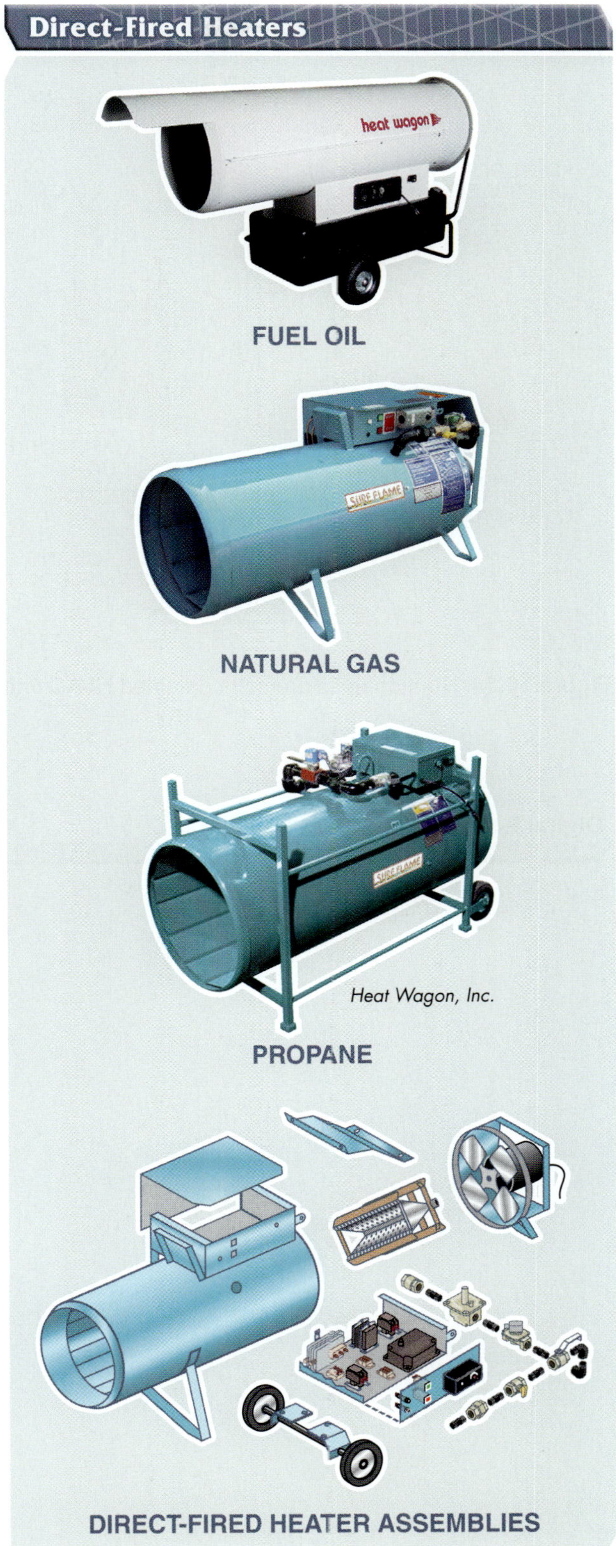

Figure 16-13. Direct-fired heaters do not have heat exchangers and can be used only in applications where 100% makeup air is available for ventilation and 100% of the heated air is constantly exhausted outside.

Rooftop Units

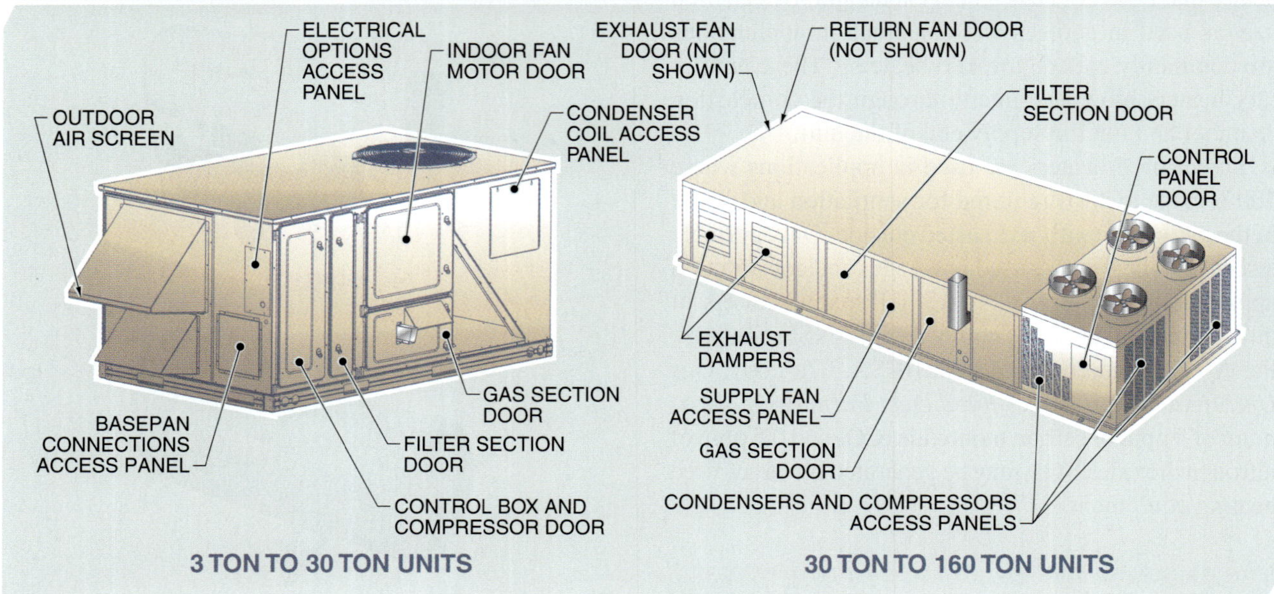

Figure 16-14. Rooftop units are self-contained HVAC units that are mounted on the roofs of light commercial buildings.

Digital Resources ATPeResources.com/QuickLinks • Access Code: 482610

UNIT

17

Forced-Air Furnace Components and Controls

A forced-air furnace consists of a number of components that must be properly connected together. For the furnace to operate properly, each component must perform its task correctly and at the proper time. Also, each component must be matched to the other components in the furnace. In some cases, these individual components must be calibrated for the furnace to operate.

OBJECTIVES

- List the various forced-air furnace components.
- Describe the operation of the various types of furnace burners and fuel igniters.
- Explain the types of gas control valves and their uses.
- Describe how the various types of furnace heat exchangers operate.
- Describe the operation of common furnace controls.

Index to Procedures

FURNACE COMPONENTS

A *furnace component* is a major piece of equipment that makes up a furnace. Combustion furnace components include a blower, fuel valve, burner(s), heat exchanger, and controls. Electric furnace components include a blower, resistance heating element(s), and controls. These components are arranged differently, but they are all found in most furnace cabinets. **See Figure 17-1.**

Cabinets

A *cabinet* is a sheet-metal enclosure that completely covers and provides support for the components of an HVAC device. Cabinets are painted with corrosion-resistant paint to make them attractive and protect them from rust and deterioration. Cabinets for small furnaces are made of heavy-gauge sheet metal, which provides all the support for other components. Larger furnaces have a frame within the sheet-metal cabinet, which supports heavier components and adds strength to the cabinet.

Furnace cabinets are rectangular in shape, come in many standard sizes, have louvered and unlouvered access doors, and support components for mounting.

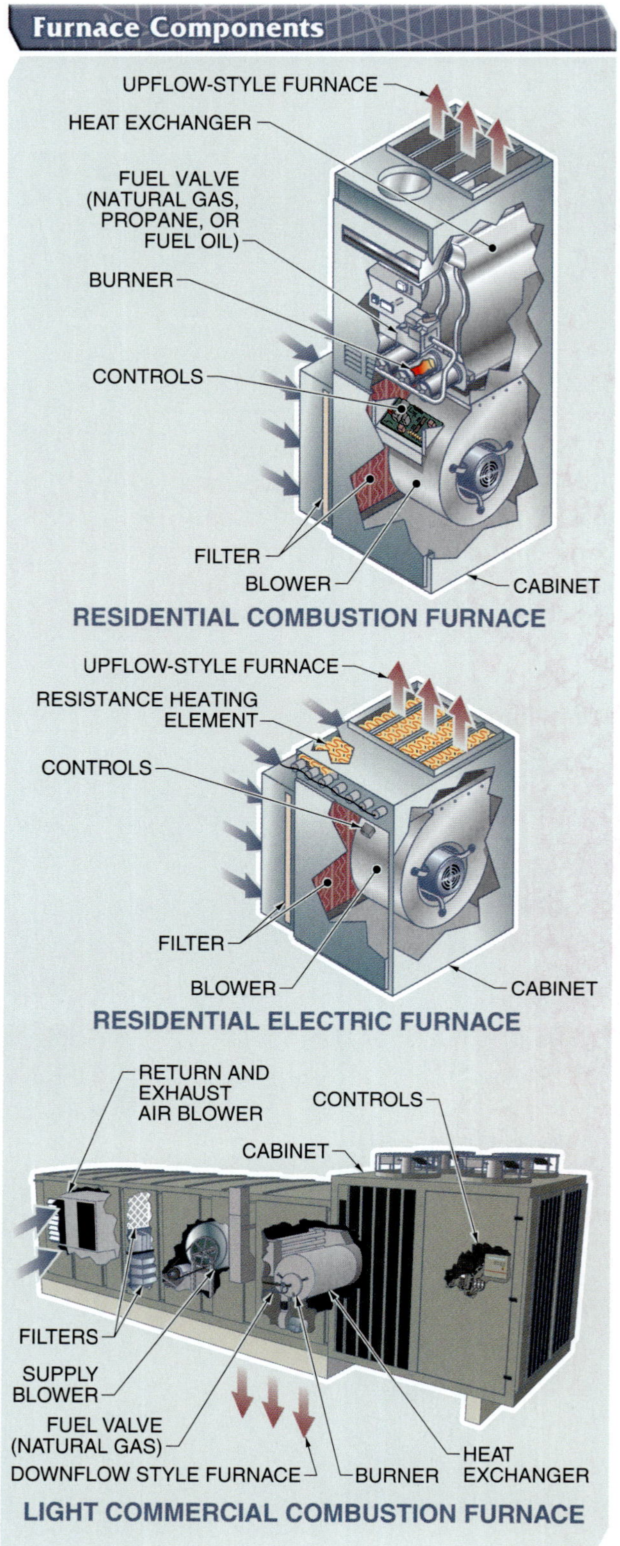

Figure 17-1. Combustion furnaces include components such as a cabinet, blower, fuel valve, heat exchanger, burner, and controls. Electric furnaces include a cabinet, blower, resistance heating element, and controls.

TECH TIP

Many furnaces are constructed of 22-gauge sheetmetal. However, large residential, light commercial, and the bases of small residential furnaces are often constructed of 13-gauge to 16-gauge sheetmetal for increased strength.

Furnace cabinets completely enclose the other components except where the return air and supply air duct connections are made. Louvered access panels on the front of the cabinet allow combustion air to the burner(s) and enclose the burner vestibule, controls, and blower compartment. **See Figure 17-2.** A *burner vestibule* is the area in a furnace where the burner and controls are located.

Blowers

A *blower* is a mechanical device that consists of moving blades or vanes that force air to travel through ductwork. **See Figure 17-3.** Any opening or restriction to airflow in the ductwork causes change as air moves through it. The blower in a forced-air heating system is the component that moves air over the heat exchanger or the resistance-heating element and through the ductwork to building spaces. Supply air blowers and return air blowers are both used to move air in large systems with a large amount of ductwork.

A blower may be a part of the furnace or a separate part within a forced-air heating system. Blowers are available in different sizes for different applications. Blowers used in forced-air systems include propeller fans, centrifugal blowers, and axial flow blowers.

Trane

Residential furnace blowers typically use PSC motors ranging from 0.25 HP to 1.0 HP in size.

Cabinets

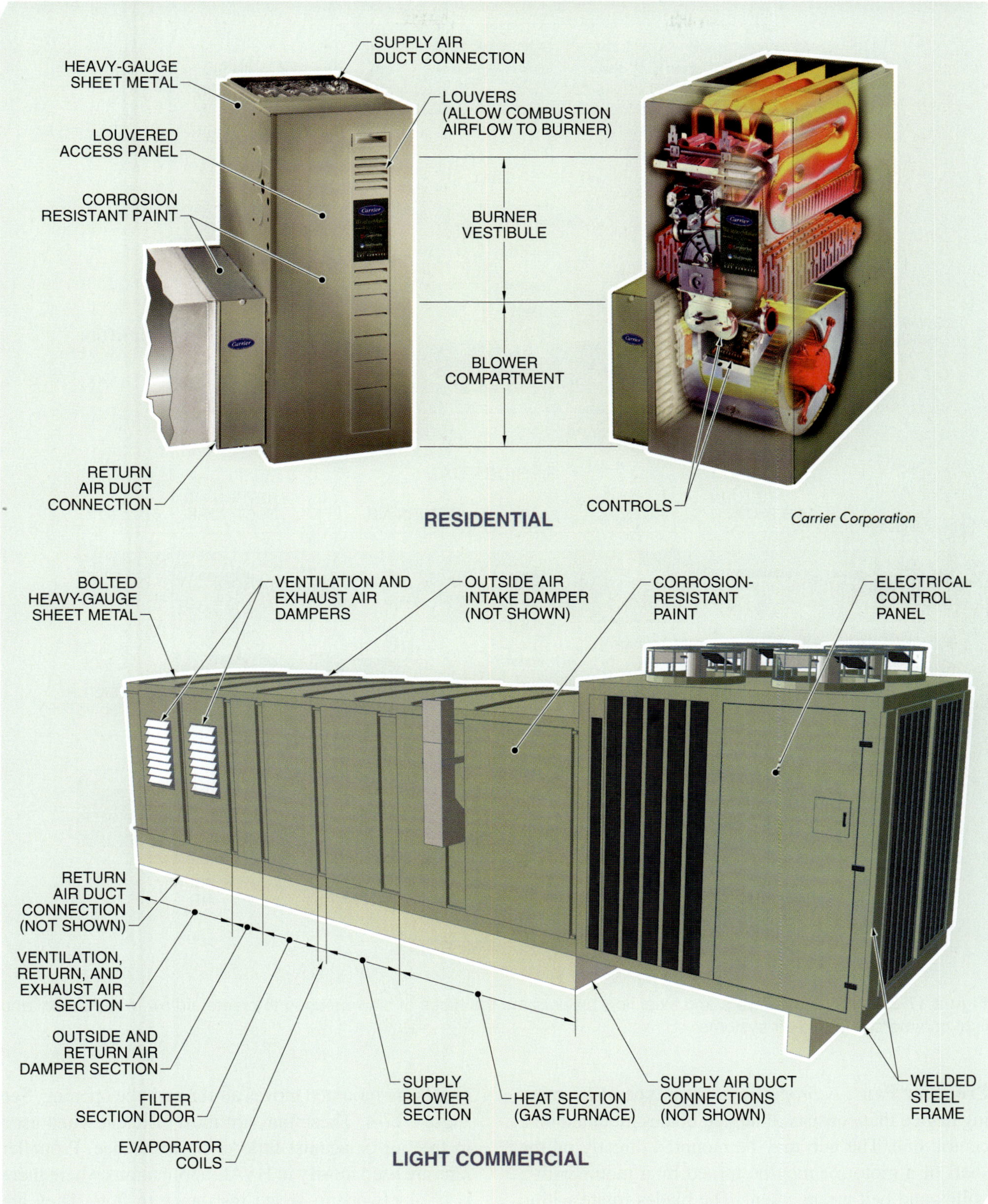

Figure 17-2. Louvered access panels allow a burner vestibule and blower compartment to be enclosed separately but let combustion air in to the burners.

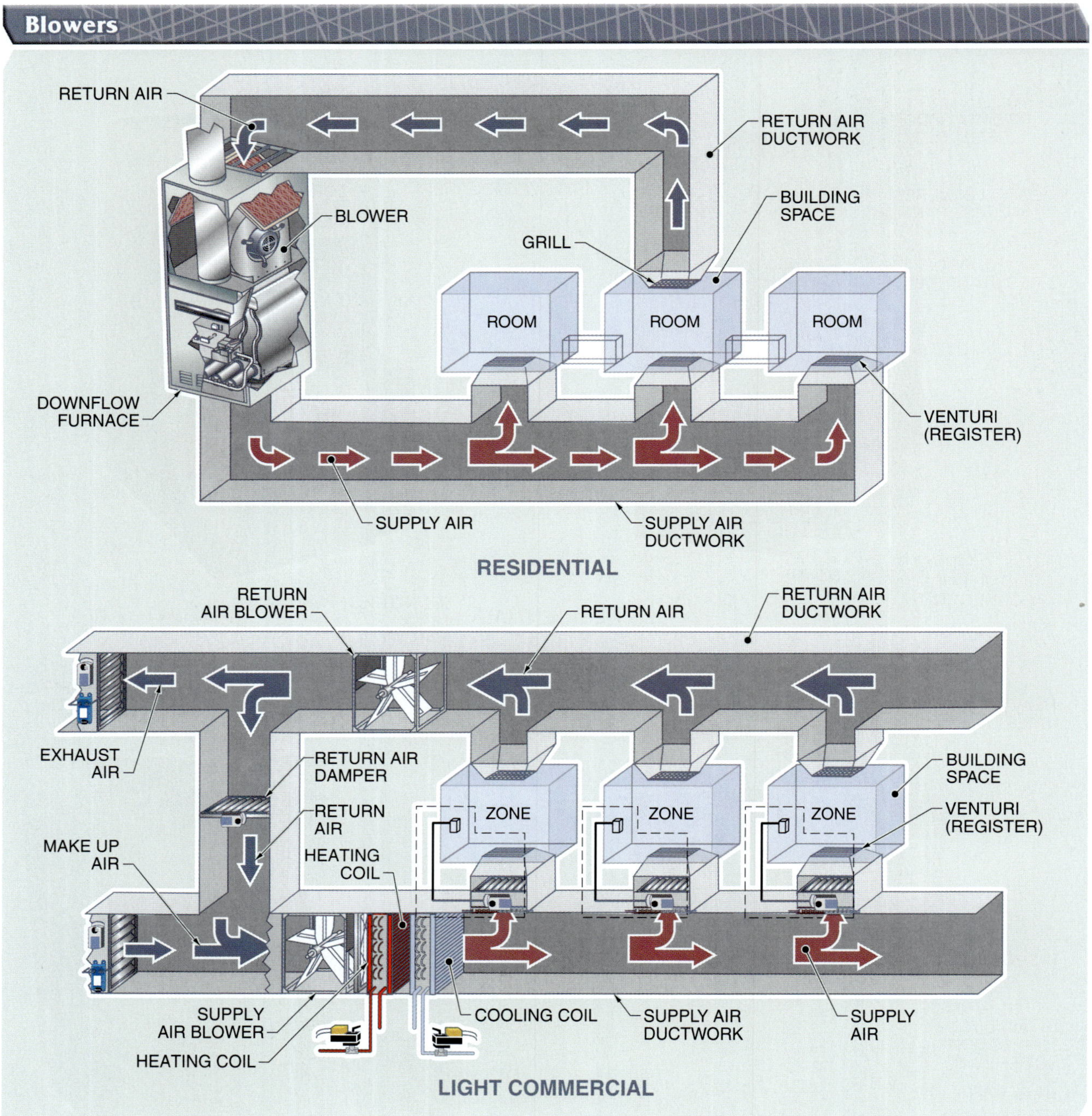

Figure 17-3. Centrifugal blowers and axial flow blowers are the types of blowers used to create airflow in residential and light commercial forced-air systems.

Propeller Fans. A *propeller fan* is a type of air moving device that consists of angled blades mounted on a central hub. The hub may be mounted directly on the shaft of a motor or may be turned by a motor with a pulley and belt arrangement. The blades rotate within an opening in a plate. As the fan hub rotates, the blades move rapidly through the air. The angle at which the blades are mounted moves air through the opening. **See Figure 17-4.** These fans are most efficient when used in free air or against little or no resistance. Propeller fans are used mostly in HVAC applications where there is no ductwork to create resistance to flow. They are also used outdoors as the coil fan on air conditioning condenser units.

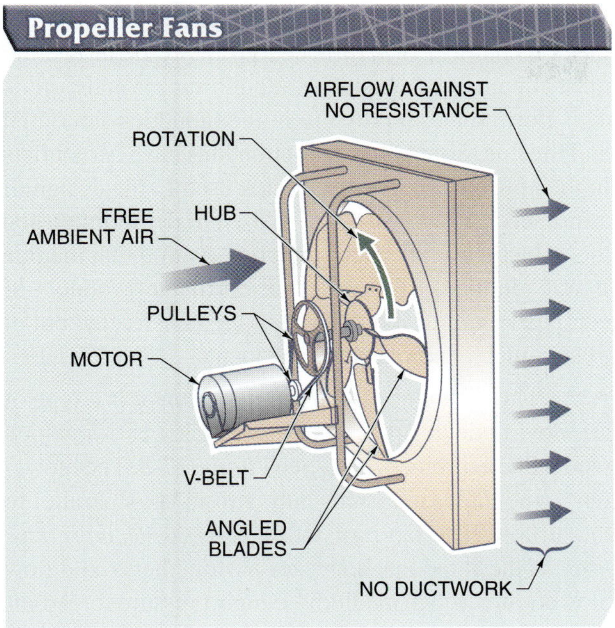

Figure 17-4. A propeller fan has angled blades mounted on a central hub. Propeller fans are efficient when used in free ambient air or applications where there is no ductwork.

Centrifugal Blowers. A *centrifugal blower* is a type of air moving device that consists of a scroll, blower wheel, shaft, and inlet vanes. **See Figure 17-5.** A *scroll* is a sheet-metal enclosure that surrounds a blower wheel. A *blower wheel* is a sheet-metal cylinder with curved vanes along its perimeter. The blower wheel rotates on the blower shaft.

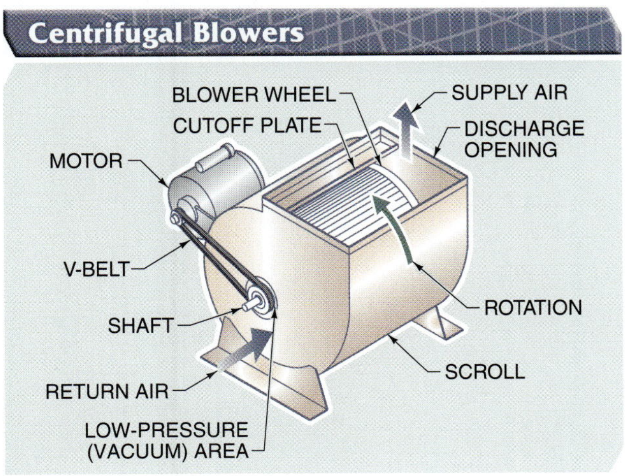

Figure 17-5. A centrifugal blower has a scroll, blower wheel, and shaft that rotate to create a low pressure (vacuum) area at the center of the wheel and a cutoff plate that directs air out of the blower.

As the blower wheel rotates on its shaft, a low-pressure (vacuum) area is created at the center of the wheel. Return air is pushed into the air inlet by the atmosphere and to the center of the wheel. Air passes through the vanes of the blower wheel and is thrown off by centrifugal force through the discharge opening in the scroll. *Centrifugal force* is the force that pushes a body outward when it is spinning around a center. A cutoff plate directs the air out of the discharge opening after the air passes through the vanes of the blower wheel.

Most residential centrifugal blowers have forward-curved blower wheels. Forward-curved blower wheels have blades that are curved in the direction of the wheel rotation, which is the direction of the airflow. Heavy commercial and industrial blower wheels have blades inclined in the direction opposite to the rotation and are classified as backward-inclined blower wheels. Backward-inclined blower wheels are capable of greater pressure ratings. **See Figure 17-6.**

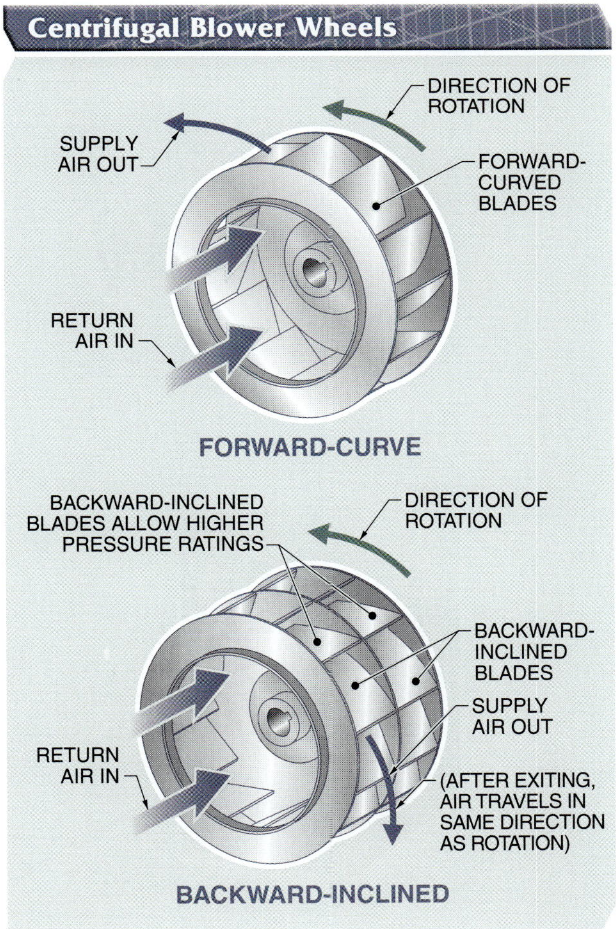

Figure 17-6. Centrifugal blower wheels may be forward-curved or backward-inclined blower wheels. Backward-inclined blower wheels have a slightly higher pressure rating.

Centrifugal blowers are used in low-pressure and many medium- to high-pressure forced-air heating systems. Centrifugal blowers operate with high efficiency at pressures common in most small to medium-size duct systems. The pressure inside ducts is typically expressed in inches of water column. *Inches water column (in. WC)* is the pressure exerted by a column of water one inch in height. **See Appendix.** Low-pressure forced-air systems have pressures between –0.5″ WC and 2″ WC. Medium- to high-pressure forced-air systems have pressures between 3″ WC and 10″ WC.

Centrifugal blowers are also used as draft induction fans in furnaces. A *draft induction fan* is a fan designed to create the correct amount of flow through a combustion chamber and to handle the products of combustion. **See Figure 17-7.** A *combustion chamber* is the area in a heating unit where combustion takes place.

Centrifugal Blower Draft Induction Fans

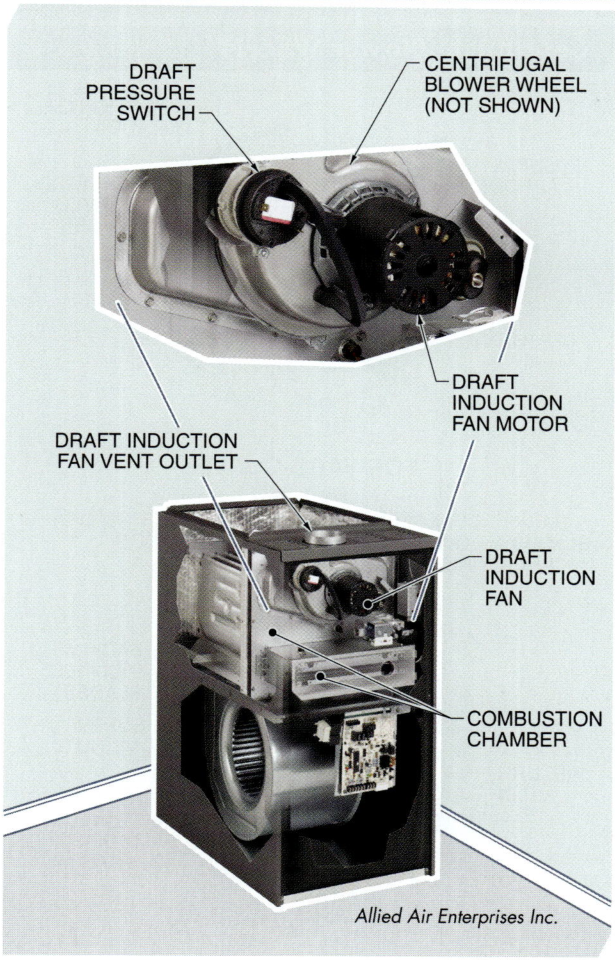

Figure 17-7. Draft induction fans create the correct amount of flow through a combustion chamber to create a specific flue pressure.

The products of combustion travel through the combustion chamber of a furnace, or other type of heating unit, and are sent outside of a building through a flue. This flow creates a pressure in the induction fan outlet and the flue. Some draft induction fans have two outlets in the housing. Only one outlet is used in the design of a furnace. A pressure switch is used to determine when the induction fan is operating properly and that the flue is not plugged. To handle the corrosive products of combustion, draft induction fans are constructed out of treated metal and plastic components.

Axial Flow Blowers. An *axial flow blower* is a type of air moving device that consists of a blower fan, which works like a turbine wheel. **See Figure 17-8.** The blower fan is mounted on a shaft hub with its axis parallel to the airflow. The fan turns at high speed. *Compressive force* is the force that squeezes air together. Axial flow blowers are used in medium- to high-pressure forced-air heating systems. Axial flow blowers are more efficient in high-pressure applications than other kinds of blowers. High-rise buildings often have high-pressure duct systems that use multiple axial flow blowers.

Axial Flow Blowers

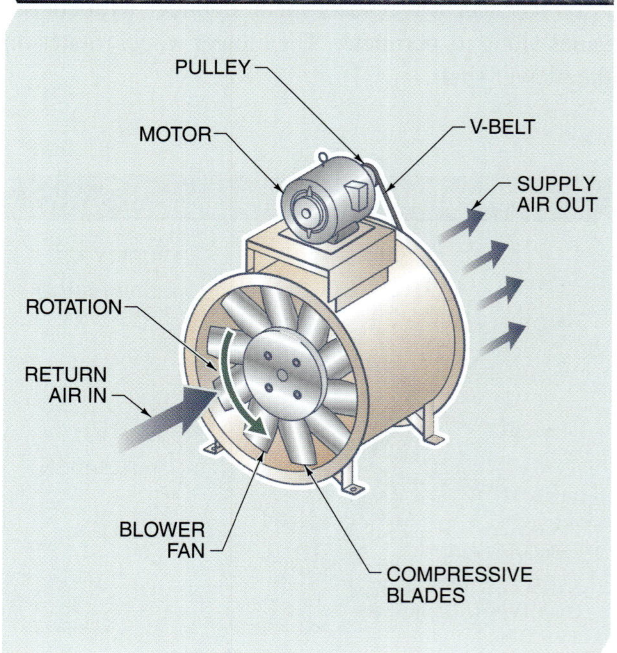

Figure 17-8. An axial flow blower has a turbine-like blower fan that is mounted on a hub and shaft and is used in medium- to high-pressure forced-air heating systems.

Furnace Blower Drive Systems

Electric motors provide the power to turn the blower wheel of a furnace blower. A *furnace blower drive system* is a type of furnace connection used between the electric motor and the blower wheel, or motor-to-wheel connection, of a furnace. The two kinds of electric motor-to-wheel connections used with furnace blowers are the direct drive system and the belt drive system. **See Figure 17-9.**

Direct Drive Systems. A *direct drive system* is a motor-to-wheel connection that has a blower wheel mounted directly on a motor shaft. Most furnaces use a direct drive system, which has the motor mounted in the center of the return air inlet of the blower. Many direct drive blowers are connected to VFD-controlled motors. Changing the speed of the motor changes the speed of the blower. Large blowers use four- or five-speed motors and small blowers use two- or three-speed motors.

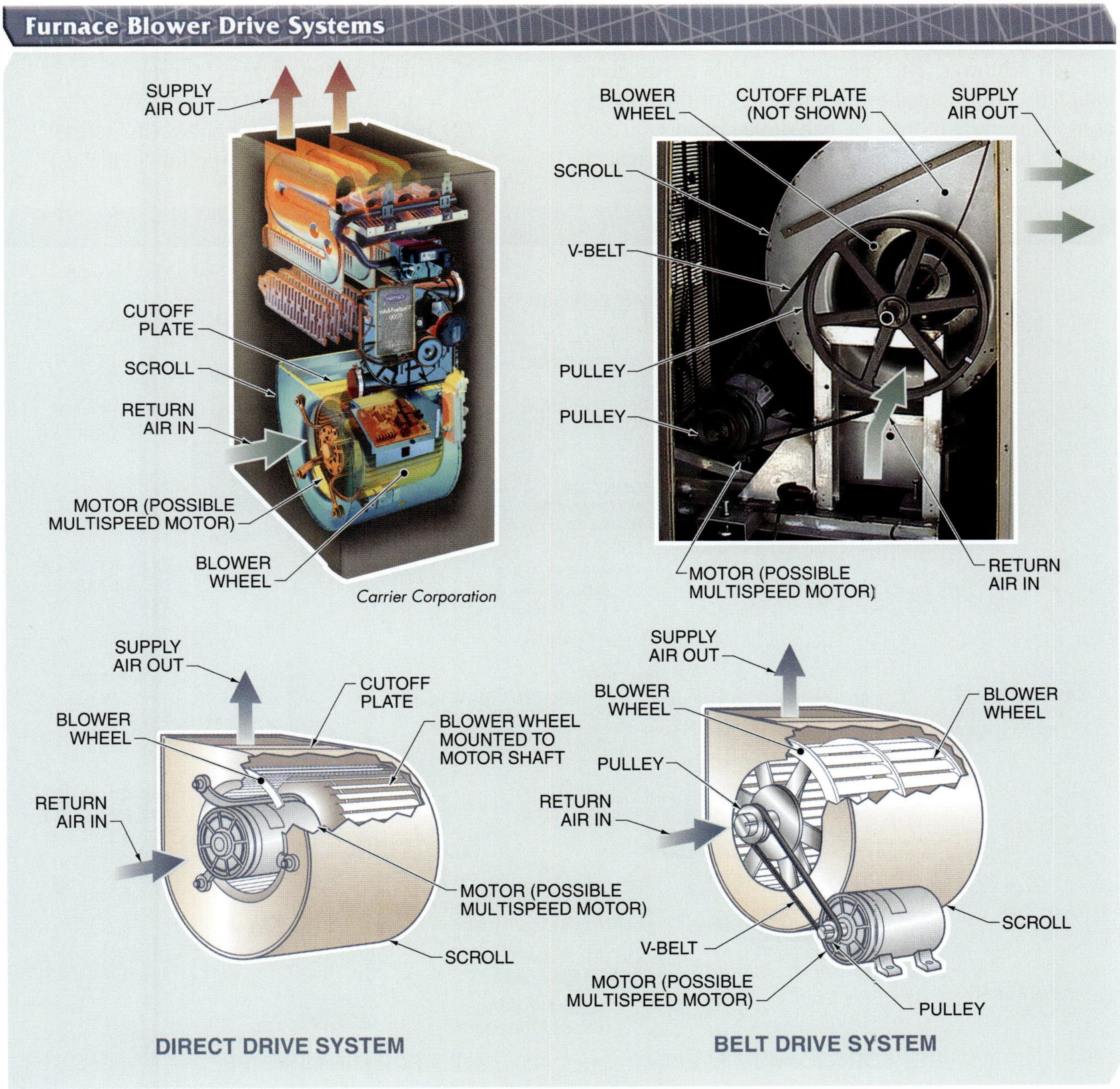

Figure 17-9. The two types of electric motor to wheel connections typically used with furnace blowers today are the direct drive system and the belt drive system.

Belt Drive Systems. A *belt drive system* is a motor-to-wheel connection that has a motor mounted to a frame and the motor is connected to the blower wheel through a belt and pulley arrangement. Pulleys with one or more V-belt grooves are mounted on the motor shaft and the blower wheel shaft. V-belts made of rubber, nylon, polyester, or rayon are used to transmit power from the motor shaft to the blower wheel shaft.

Blower Wheel Speed

The speed of a blower wheel, in revolutions per minute (rpm), determines the volume of air that will flow through a blower. The speed of the blower wheel is indirectly proportional to the diameters of the pulleys used in the system. In some belt drive systems, an adjustable blower pulley is used to make blower wheel speed adjustments. **See Figure 17-10.** The outer flange of the pulley slides onto the sleeve of the inner flange.

With this arrangement, the distance between the two side flanges of the pulley can be changed. A variable-speed belt rides the outer part of the pulley when the flanges are close together and the inner part of the pulley when the flanges are far apart. The speed of a blower wheel that is belt driven is found by applying the following formula:

$$N_b = \frac{N_m \times PD_m}{PD_b}$$

where

N_b = speed of blower wheel (in rpm)
N_m = speed of motor (in rpm)
PD_m = diameter of motor pulley (in inches)
PD_b = diameter of blower pulley (in inches)

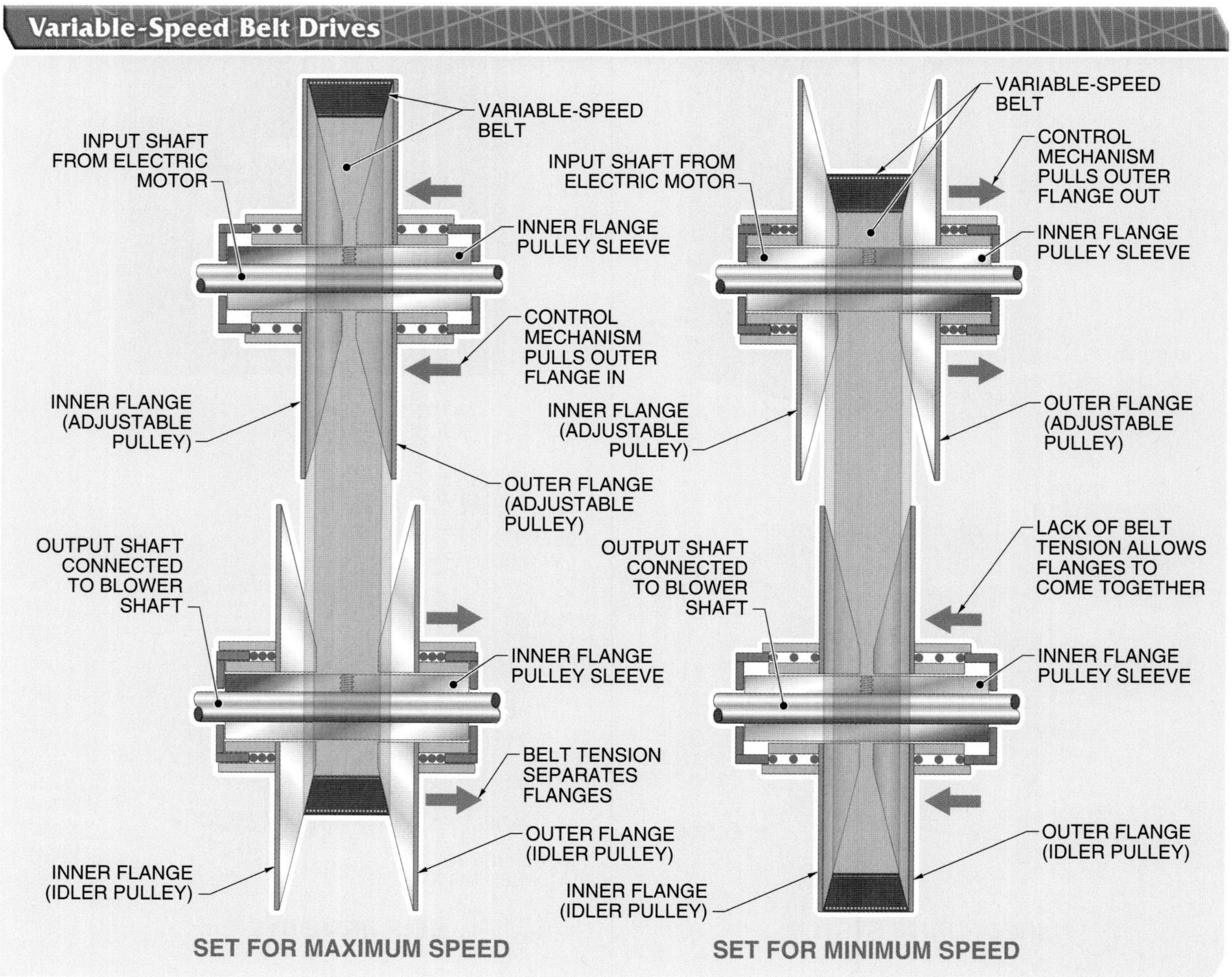

Figure 17-10. A variable-speed belt drive system has a motor-to-wheel connection using a variable-speed belt and an adjustable sheave diameter arrangement.

Example: Finding Belt Drive Blower Wheel Speed

A furnace that has a V-belt drive blower with a motor speed of 1725 rpm has a 3″ diameter motor sheave and a 7″ diameter blower wheel sheave. Find the speed of the blower wheel. **See Figure 17-11.**

$$N_b = \frac{N_m \times PD_m}{PD_b}$$

$$N_b = \frac{1725 \times 3}{7}$$

$$N_b = \frac{5175}{7}$$

$$N_b = \mathbf{739.29\ rpm}$$

Finding Belt Drive Blower Wheel Speed

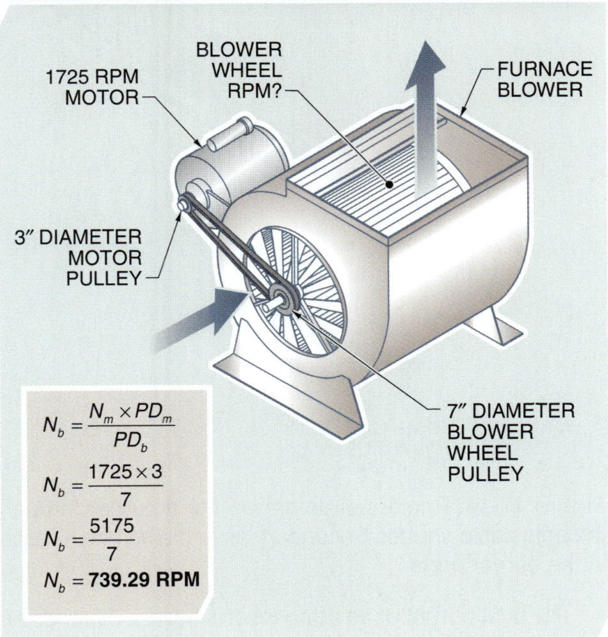

Figure 17-11. Blower wheel speed is indirectly proportional to the speed of a motor and the diameter of the pulleys used on the motor shaft and blower wheel.

Blower Performance

The performance characteristics of a blower are horsepower, speed, volume, and static pressure. These characteristics are found on blower performance charts. *Static pressure* is the force that nonmoving air exerts against surfaces such as ducts; measured in inches of water column (in. WC).

Blower performance charts are developed for individual blowers or class of blowers. These charts illustrate the performance characteristics of a particular type and size of blower in cubic feet per minute, static pressure, revolutions per minute, and horsepower (HP). For example, a furnace blower that has a 1 HP motor, a speed of 1733 rpm, and moves 1350 cfm of air with 2.5″ WC static pressure can be plotted on a chart. **See Figure 17-12.**

Blower Performance Charts

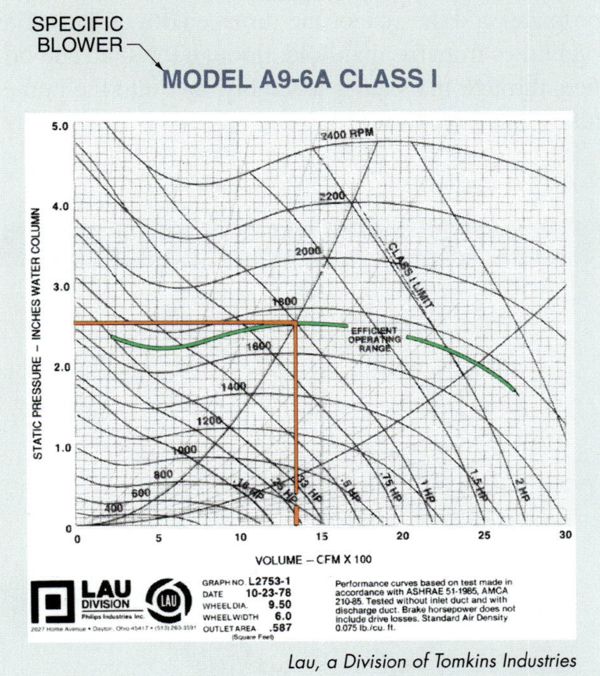

Lau, a Division of Tomkins Industries

Figure 17-12. Blower performance charts show the horsepower, speed, volume, and static pressure performance characteristics of a specific blower during operation.

Burners

A *burner* is the heat-producing component of a combustion furnace. Some furnaces use one large burner, while others use many smaller burners. A burner mixes air and fuel to provide a combustible mixture. While a burner supplies the air-fuel mixture to the burner face (place of combustion), it also meters fuel to maintain a controlled firing rate. The three types of burners found in furnaces are atmospheric, power, and low excess air burners.

Atmospheric Burners. An *atmospheric burner* is a type of burner that uses ambient air as combustion air. *Ambient air* is unconditioned atmospheric air at normal atmospheric pressure. Atmospheric burners typically burn natural gas as a fuel and sometimes fuel oil to produce heat.

A *gas-fired atmospheric burner* is a burner that mixes ambient air with a gas fuel to create a flame. The gas fuel is directed through a manifold to the burner tubes. A *manifold* is a pipe that has outlets for connecting other pipes. A *burner tube* is an open or closed tube that has an opening on one end and burner ports located along the top. A spud (fitting) that contains a small orifice is located in the outlets of the manifold. An *orifice* is a small precisely sized hole through which gas fuel flows. Orifice size is determined by the type of gas (natural or propane) and the size of the furnace (Btu output). Gas fuel flows from the manifold, through the spud and orifice, through the shutter assembly, and into the burner tubes. **See Figure 17-13.**

Gas-Fired Atmospheric Burners

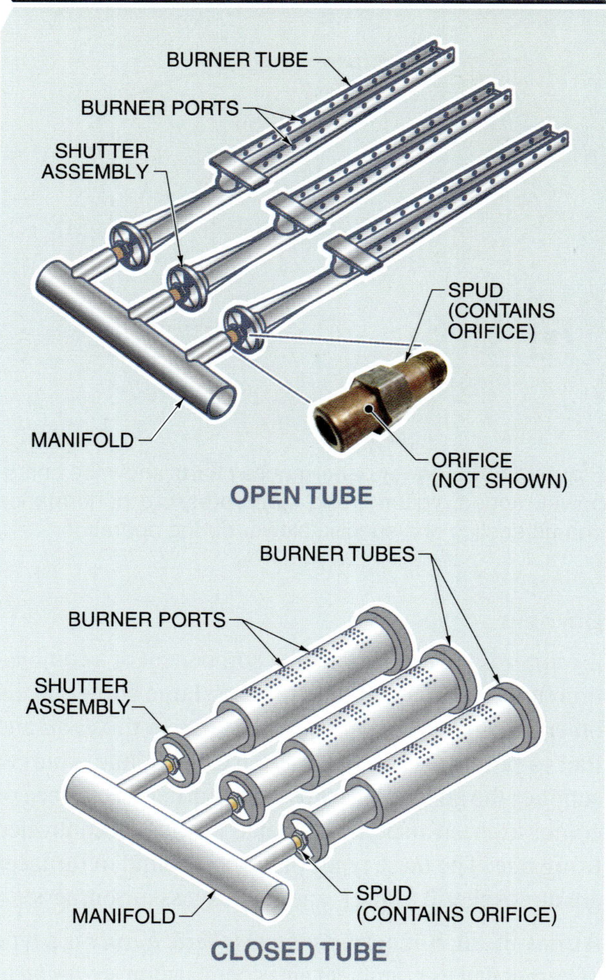

Figure 17-13. Gas-fired atmospheric burners use orifices to mix ambient air with natural gas or propane to create a flame of proper size for a furnace according to its Btu output rating.

A burner tube has an adjustable shutter, which is the primary air inlet for the burner. The gas fuel enters the burner tube at a high velocity. The burner tube draws air in through the openings in the adjustable shutter. **See Figure 17-14.** The primary air mixes with the gas and produces a combustible air-fuel mixture. The air-fuel mixture passes through the burner tube and out the burner ports where the air-fuel mixture is ignited. Secondary air is drawn into the flame at the burner ports to maintain flame integrity.

Flame Troubleshooting Interactive Animation

Burner Tubes

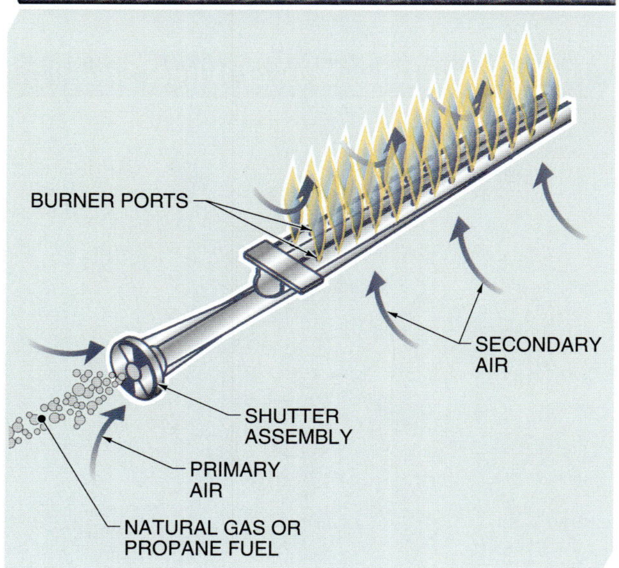

Figure 17-14. Primary air is drawn into a burner through the adjustable shutter. Secondary air is drawn into a flame at the burner ports.

The heat output of an atmospheric burner depends on the size of the burner, size of the orifice, and pressure of the fuel. Most natural gas burners are designed to operate with a pressure of 3.5″ WC. Liquid petroleum gas burners are designed to operate with a pressure of 10.5″ WC. The orifices in the spud are sized to provide the amount of fuel that will produce the rated heat output for that particular burner.

An *oil-fired atmospheric burner* is a type of burner that consists of an open pot into which fuel oil flows at a controlled rate. A metering device (carburetor) controls the flow rate of fuel oil from a reservoir into the pot burner. A *float valve* is a valve consisting of a hollow ball that floats on the liquid in an internal reservoir to restrict flow through the valve when the liquid reaches a certain level. The float valve is adjusted to maintain a small quantity of oil in the reservoir. The oil flows from the reservoir into the pot at a controlled rate.

Most oil-fired atmospheric burners must be ignited manually. After a flame is established, heat from the fire vaporizes oil in the pot and the vapors burn at the burner face locations in the pot. **See Figure 17-15.** These types of burners are used in space heaters (kerosene heaters) and small manually controlled heating units. Oil-fired atmospheric burners are also known as pot burners.

Oil-Fired Atmospheric Burners

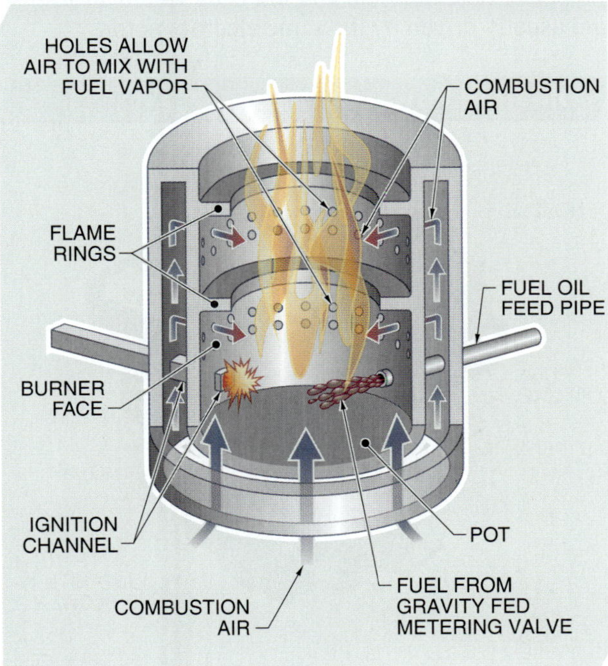

Figure 17-15. After an oil-fired atmospheric burner is ignited manually, heat from the fire vaporizes oil in the pot and causes the vapors to burn at the burner face locations.

Power Burners

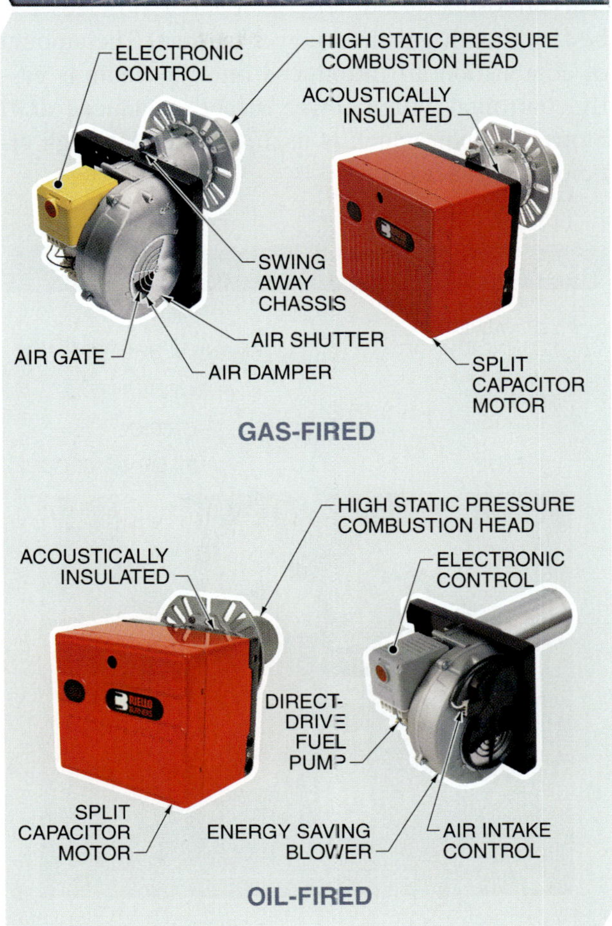

Riello Burners

Figure 17-16. Gas-fired and oil-fired power burners use blowers to control the pressure, amount of fuel, and combustion air used by the burners.

Power Burners. A *power burner* is a type of burner that uses a fan or blower to supply and control combustion air. Air and fuel are introduced under pressure at the burner face. The two basic kinds of power burners are gas-fired and oil-fired power burners. **See Figure 17-16.**

A *gas-fired power burner* is a type of burner that uses natural or LP gas and contains a fan or blower. Gas-fired power burners are located on the outside of the combustion chamber. The blower provides combustion air, which is blown into the combustion chamber under pressure through a perforated bulkhead.

Gas fuel of a gas-fired power burner flows from a fuel line through an orifice at a regulated pressure (manifold pressure) to the burner head. The burner head is located inside the perforated bulkhead. Gas fuel flows through the burner head as air flows through the holes in the perforated bulkhead. Combustion air and the fuel mix to provide a combustible air-fuel mixture. **See Figure 17-17.** The proper air-fuel mixture is attained by using a properly sized orifice in the gas fuel line and by adjusting the combustion air blower. An electric spark igniter is used to start combustion.

Gas-fired power burners are typically used in light commercial natural gas furnaces with outputs from 60,000 Btu/hr to 240,000 Btu/hr. However, because power burners are considered to be 30% more efficient than atmospheric burners, residential furnaces can be found with gas-fired power burners. The amount of combustion air introduced into the flame is easily controlled with a power burner or induced-draft blower system, which is useful in creating high efficiency furnaces.

Gas-Fired Power Burners

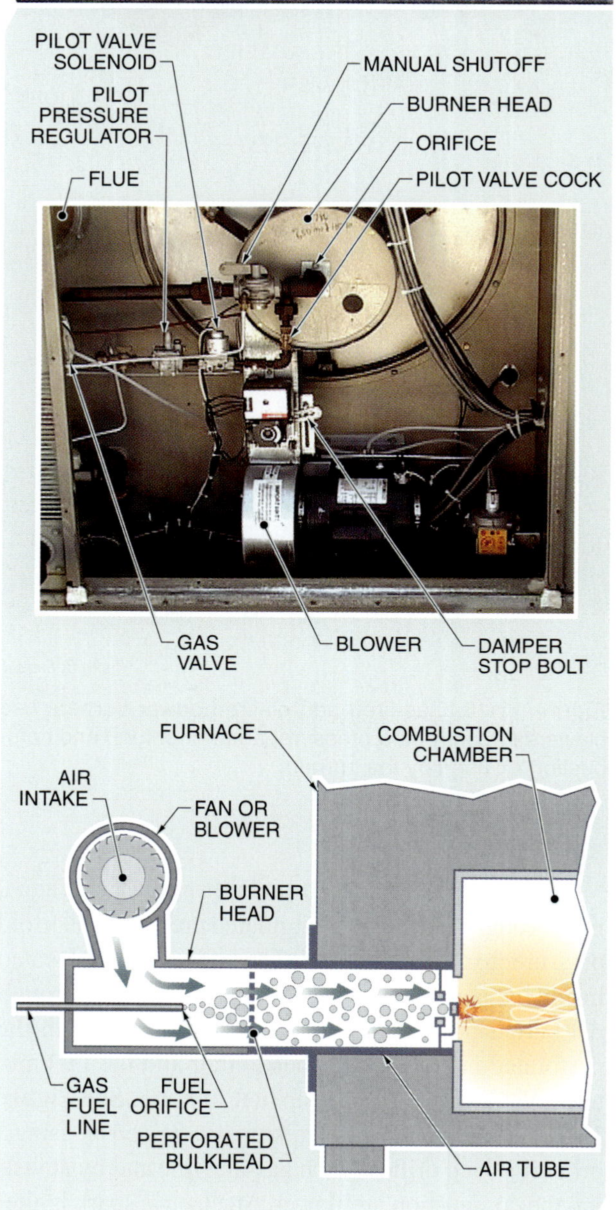

Figure 17-17. A gas-fired power burner creates the proper air gas mixture by using a properly sized orifice in the gas fuel line and by adjusting the amount of combustion air from the blower.

An *oil-fired power burner* is a type of power burner that atomizes fuel oil with the air. An oil-fired power burner, which is the most popular type of power burner, consists of a fuel oil pump, burner assembly, and combustion air blower. **See Figure 17-18.** The fuel oil pump receives oil from a storage tank and pumps it through the burner assembly. The burner assembly consists of a tube that holds a fuel oil line, nozzle, and electrodes. The combustion air blower forces combustion air through the tube. The fuel oil line carries oil from the pump to the nozzle. The nozzle releases atomized fuel oil, which mixes with combustion air at the burner face. Electrodes produce an electric spark that ignites the air-fuel mixture. The fuel oil pump and the combustion air blower are usually driven by the same electric motor.

Oil-Fired Power Burners

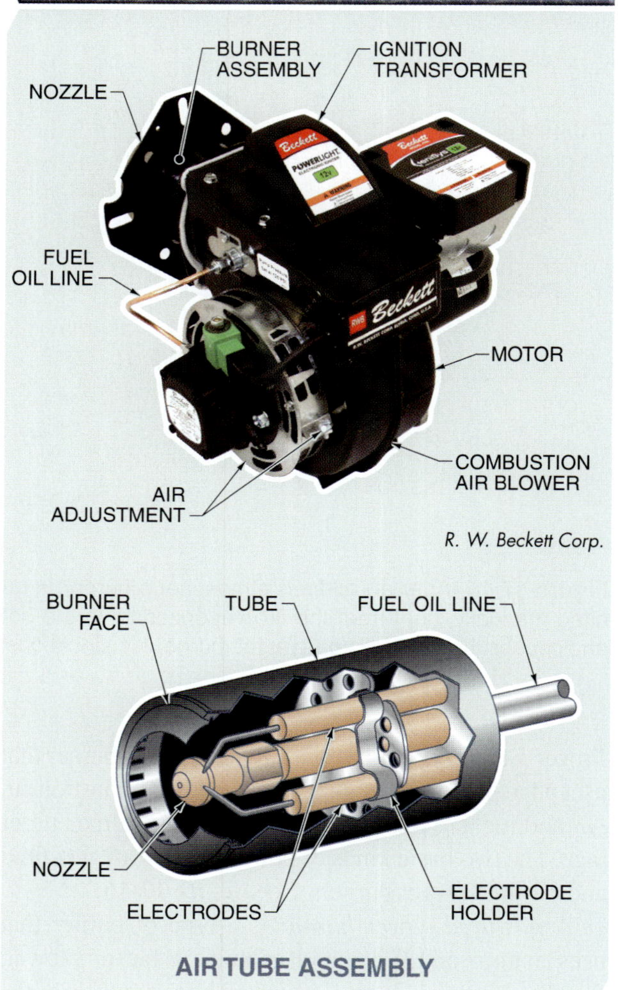

Figure 17-18. The oil in an oil-fired power burner is atomized as it flows through the nozzle. Combustion air mixes with the oil at the burner face.

Fuel oil power burners are available in many sizes. Different burners burn different grades of fuel oil. The smallest burners have an output of 50,000 Btu/hr. The heating capacity of an oil-fired power burner depends on the fuel oil pressure and the size of the nozzle. Nozzles are sized in gallons per hour (gph) of fuel oil flow.

Low Excess Air Burners. Low excess air burners are used in many new furnace designs. A *low excess air burner* is a burner that uses only the amount of air necessary for complete combustion in a furnace. This ensures that the combustion will be as efficient as possible.

A *pulse burner* is a type of low excess air burner that introduces an air-fuel mixture to a burner face in small amounts or pulses. Pulse burners are used in both forced-air and hydronic heating systems. **See Figure 17-19.** The pulses of the mixture are controlled by the pressure in the firing chamber. The pulses are introduced to the firing chamber through a flapper valve. A *flapper valve* is a valve that opens to let an air-fuel mixture in but closes when a firing cycle begins because of backpressure. *Backpressure* is the pressure produced by the ignition of an air-fuel mixture against the normal pressure of a gas flow. The cycles are short but continuous.

Pulse Burners

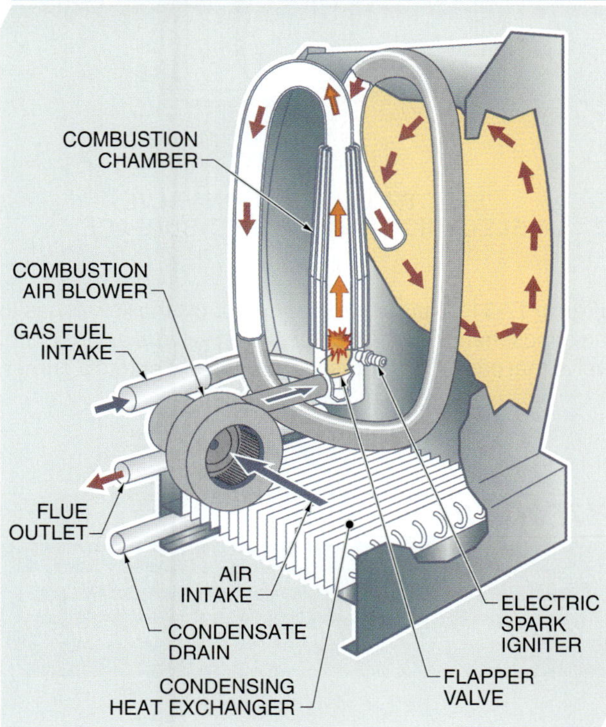

Figure 17-19. A pulse burner is more efficient than a conventional burner because it uses low excess air.

A pulse burner is more efficient than a conventional burner because no excess air is introduced. This ensures a more complete combustion, which produces cleaner products of combustion. Draft over a fire is produced within the burner so that no heat rises up the stack. Most pulse burners are more than 90% efficient. Conventional burners are seldom more than 80% efficient.

Burner Efficiencies. Standard atmospheric burners used in furnaces typically provide efficiencies of 65% to 80%. Power burners in furnaces provide efficiencies up to 86%. Furnaces using condensing heat exchangers may approach efficiencies of 97%. The products of combustion are then vented outside using PVC pipe.

Fuel Igniters

A fuel igniter may be a standing flame that burns constantly, such as the pilot light of a pilot burner, or a device that electrically ignites a gas on each call for heat. **See Figure 17-20.** The two types of electric igniters are the electric spark (flame) igniter and the hot surface igniter.

Pilot Burners. A *pilot burner* is a small burner located near the burner tubes that ignites the fuel and air mixture released into the burner tubes. A pilot burner produces a pilot light, which is a small standing flame. The pilot light ignites an air-fuel mixture when a gas fuel valve opens. The gas control valve will have a pilot light gas line to maintain the standing flame.

Electric Spark Igniters. An *electric spark (flame) igniter* is a device that is located at burners that produces a spark to ignite fuel. On a call for heat, an electrode is energized near the burner face. When a flame is established, the electrode igniter is de-energized until the next call for heat. Electric spark igniters are controlled by combustion safety control systems.

Hot Surface Igniters. A *hot surface igniter* is a device located at burners that uses a small piece of silicon carbide, which glows when electric current passes through it, to produce extremely high temperatures. The current is usually 120 VAC. The silicon carbide material is very durable, lasts a long time, and produces few problems.

TECH TIP

When the metal electrode of a spark igniter or the ceramic of a hot surface igniter is touched by a bare hand, a hot spot develops on the igniter. The hot spot will cause the igniter to burn out much faster than normal.

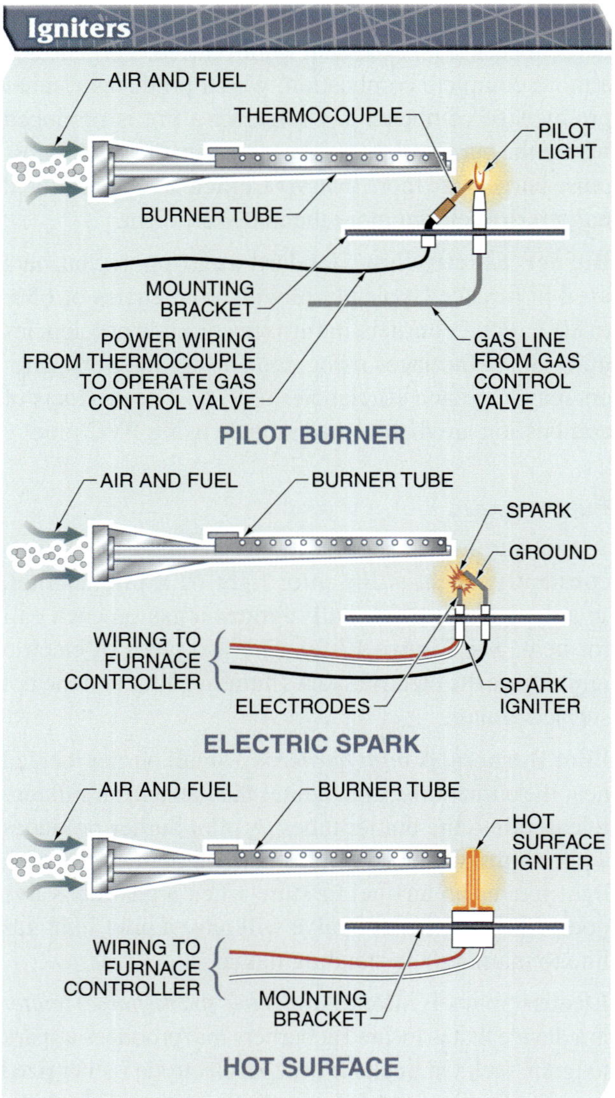

Figure 17-20. Fuel igniters are standing flame pilot lights or devices that electrically ignite gas on each call for heat. The two types of electric igniters are the electric spark igniter and the hot surface igniter.

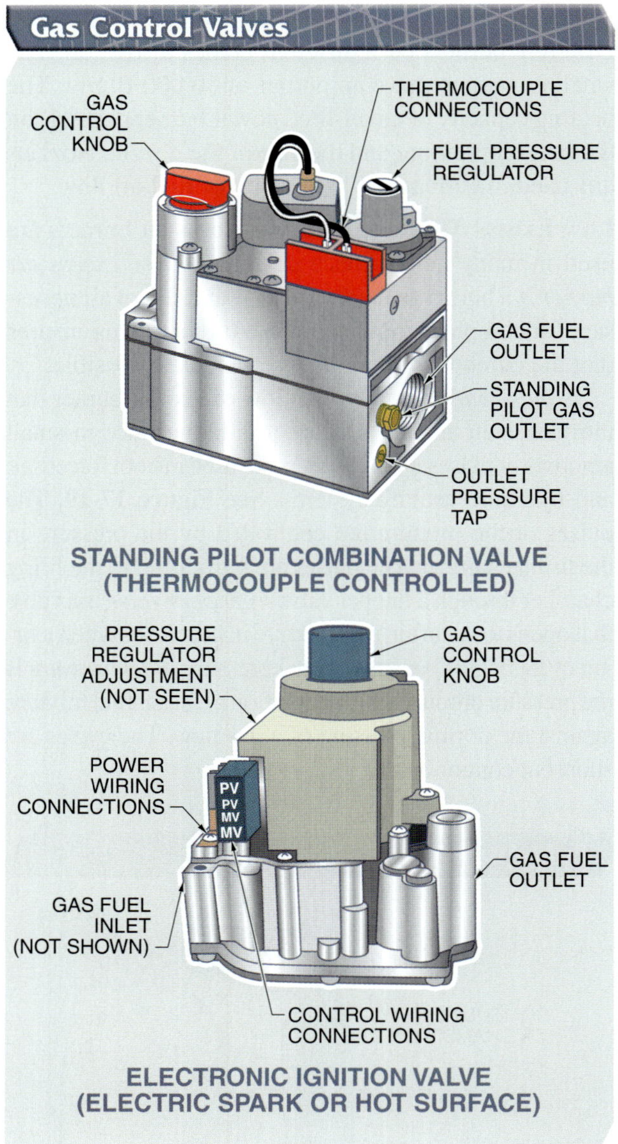

Figure 17-21. A gas control valve controls the flow of gas to an atmospheric burner. A combination valve is a gas control valve that controls the pressure of gas as it enters a burner.

Gas Control Valves

A gas control valve controls the flow of fuel to a gas-fired atmospheric burner. A *gas control valve* is a 100% shutoff safety valve that controls the flow of fuel to a main burner and a pilot burner, if used. When combustion does not occur on a call for heat, the valve will close and not allow gas to flow to the main burner. Also, some valves will shut off the flow to the pilot burner. A combination valve is a control valve that has a built-in fuel pressure regulator, which controls the pressure of the fuel gas as it enters the burner. **See Figure 17-21. See Procedure for Gas Valve Pressure Measurement and Adjustment.**

TECH TIP

To reset most gas control valves, the control knob or control switch on the gas valve must be turned to the OFF position and then, after 5 min, to the ON position.

Gas Control Valves Media Clip

Gas Valve Pressure Measurement and Adjustment

1. Turn OFF all electrical power to the furnace and system at the main fuse or circuit breaker panel.
2. Follow all electrical and hazardous environment safety practices and procedures.
3. Close the main gas cock at the furnace.

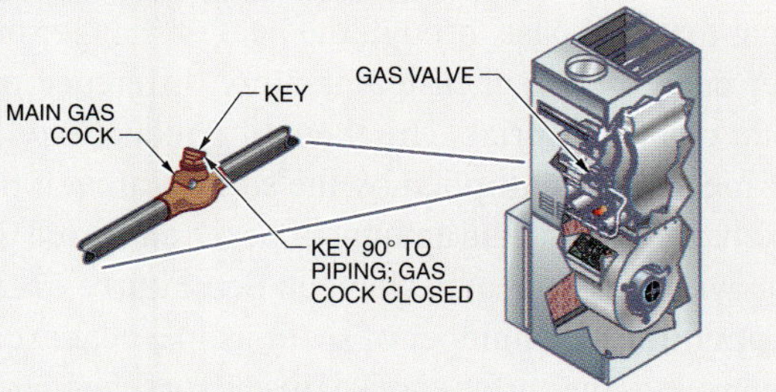

4. Connect a manometer to the outlet pressure tap of the gas valve.
   a. Remove the plug from the outlet pressure tap opening.

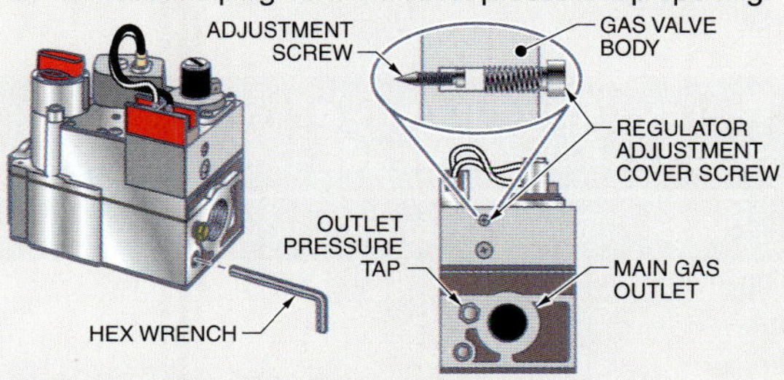

UNIVERSAL GAS VALVE **SIDE VIEW**

   b. Screw the manometer barb (hose-to-thread adapter) into the outlet pressure tap opening.

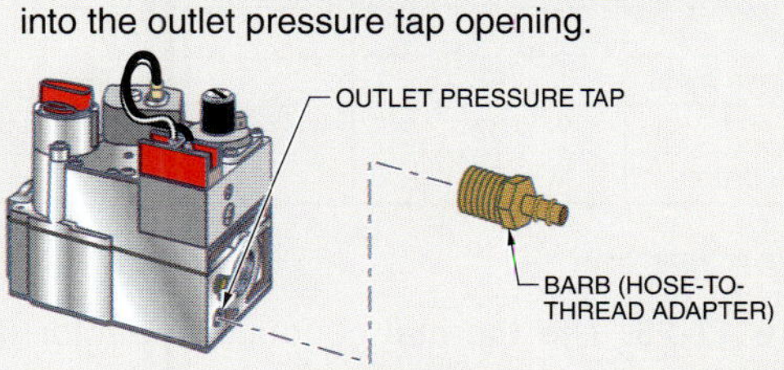

   c. Use an appropriate size hose to connect the barb in the outlet pressure tap to the input (+) port of the manometer. The reference (–) port is left unconnected (left open to ambient conditions).

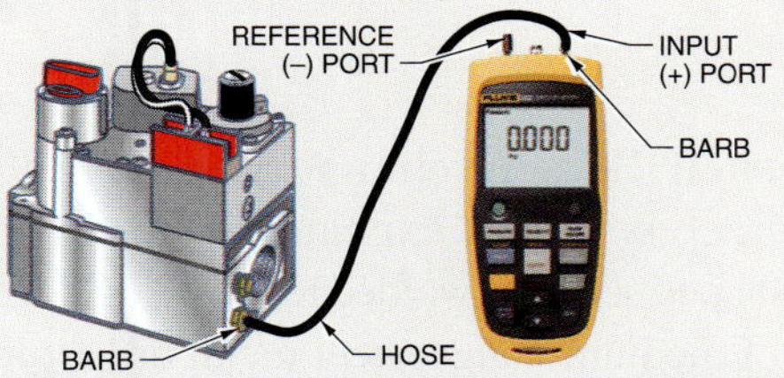

NATURAL GAS COMPATIBLE MANOMETER

5. Turn ON the manometer.

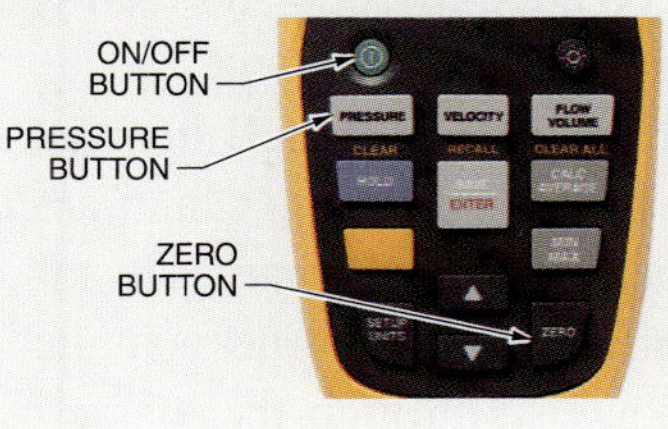

6. Set the manometer to measure pressure by pressing the PRESSURE button.
7. Zero the manometer by pressing and holding the ZERO button for 2 sec.
8. Turn ON the furnace and system.
9. Slowly open the main gas cock on the furnace.
10. Set the system thermostat to ignite the main burner(s).

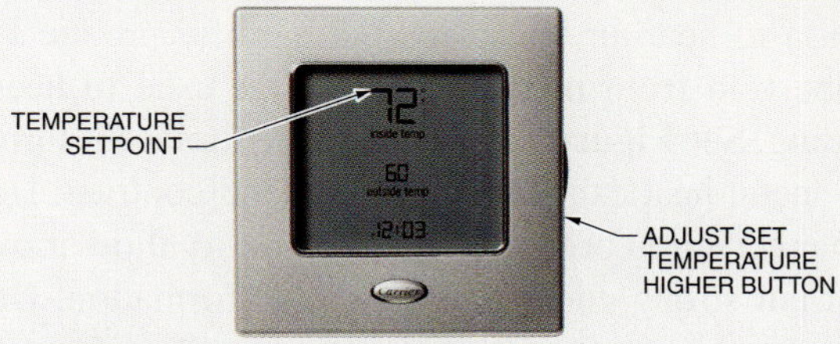

11. Read and record the differential pressure reading on the manometer.

12. If needed, adjust the setting of the gas valve pressure regulator.
    a. Remove the gas valve's regulator adjustment cover screw.
    b. To increase outlet pressure of the gas valve, turn the adjustment screw clockwise. To decrease outlet pressure, turn the adjustment screw counterclockwise. Adjust regulator until pressure indicated on manometer matches the specified pressure on the furnace data plate (typically 3.5).
    c. Replace the regulator adjustment cover screw.

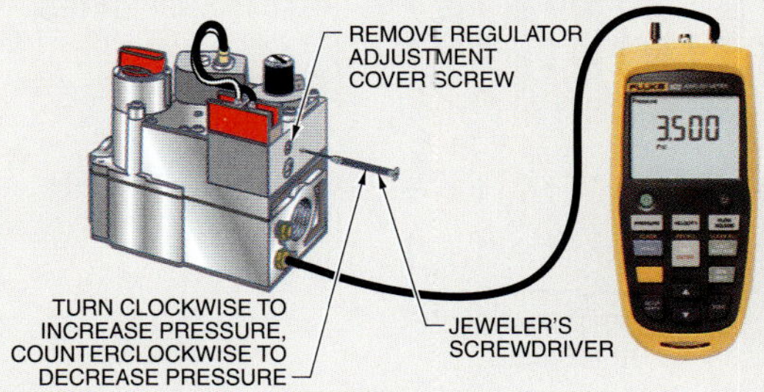

13. Disconnect the manometer and adapter from the gas valve and replace the outlet pressure tap opening plug.

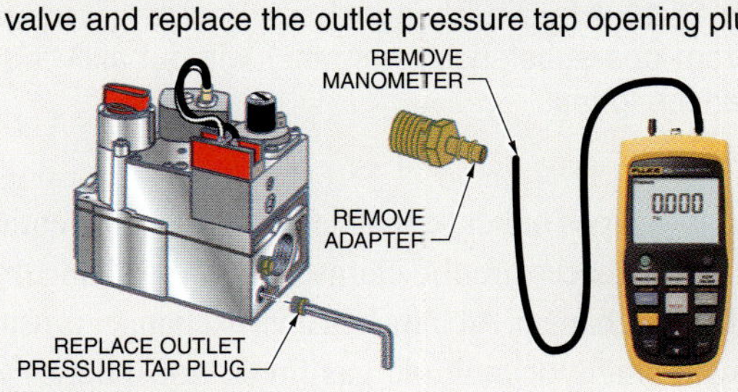

14. Turn OFF the manometer.
15. Cycle the furnace several times to verify proper burner operation.

Heat Exchangers

A *heat exchanger* is any device that transfers heat from one substance to another without allowing the substances to mix. The heat exchanger in a forced-air furnace transfers heat from the hot products of combustion to the cool air returning from the building. The products of combustion often contain toxic chemical compounds.

A heat exchanger uses the heat generated by combustion to heat air but keeps the harmful products of combustion from mixing with the air used to heat a building. **See Figure 17-22.** While the furnace is turned ON, metal heat exchangers expand and contract. Heat exchangers are corrugated and curved to allow movement but strong enough to prevent deformation. Heat exchangers are often coated with corrosion-resistant ceramic glazing.

Heat Exchangers

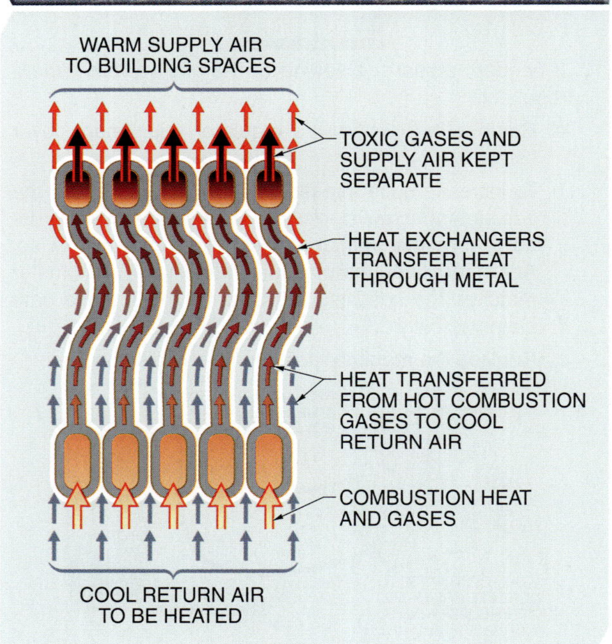

Figure 17-22. A heat exchanger is any device that transfers heat from one substance to another without allowing the substances to mix.

The two types of heat exchangers used in conventional forced-air furnaces are the clamshell heat exchanger and drum heat exchanger. A clamshell heat exchanger is usually used with single or multiple gas-fired atmospheric burners. The drum heat exchanger is used with oil-fired atmospheric burners and power burners. Many high-efficiency furnaces use secondary heat exchangers (condensing heat exchangers), which extract additional heat from the products of combustion.

The heating capacity of a heat exchanger depends on the thickness of the metal, thermal conductivity of the metal, and surface area of the heat exchanger. The thermal conductivity factor (k) represents the amount of heat that will pass through the heat exchanger material per degree Fahrenheit temperature difference on each side of the material. The thermal conductivity factors of metals are multiplied by the surface area of the heat exchanger to find the amount of heat transferred through the walls of the heat exchanger. **See Figure 17-23. See Appendix.** The number of sections in a heat exchanger determines the surface area of the heat exchanger. A heat exchanger with a given surface area produces a given heating capacity.

Thermal Conductivity of Heat Exchanger Metals

Metal	Temperature*	Thermal Conductivity (k)†
Aluminum	68	118
Brass	73	125
Bronze	62	42
Copper	68	380
Cast iron (gray)	70	27–46
Iron	68	42
Stainless steel	68	7–26
Tin	32	36–39
Yellow brass	68	67

* in °F
† in Btu/hr°F difference

Figure 17-23. The thermal conductivity factor (k) of a heat exchanger represents the amount of heat that will pass through the heat exchanger material per degree Fahrenheit temperature difference that exists on each side of the material.

Clamshell Heat Exchangers. A *clamshell heat exchanger* is a heat exchanger that has multiple clam-shaped sections welded together. The sections are made of medium-gauge sheet metal or cast iron. Each section consists of two clam-shaped pieces of metal that are placed edge-to-edge and then welded together to produce an airtight fit.

Combustion in a clamshell heat exchanger takes place at the burner openings, which are located at the bottom of each section. A burner is installed inside each burner opening. When the burner is ignited, hot products of combustion rise through the inside of the sections of the heat exchanger and out the flue openings at the top of each section. **See Figure 17-24.** Air is blown between the clamshell sections by a blower. Heat from the products of combustion is transferred through the metal to the return air that flows across the outside of the sections.

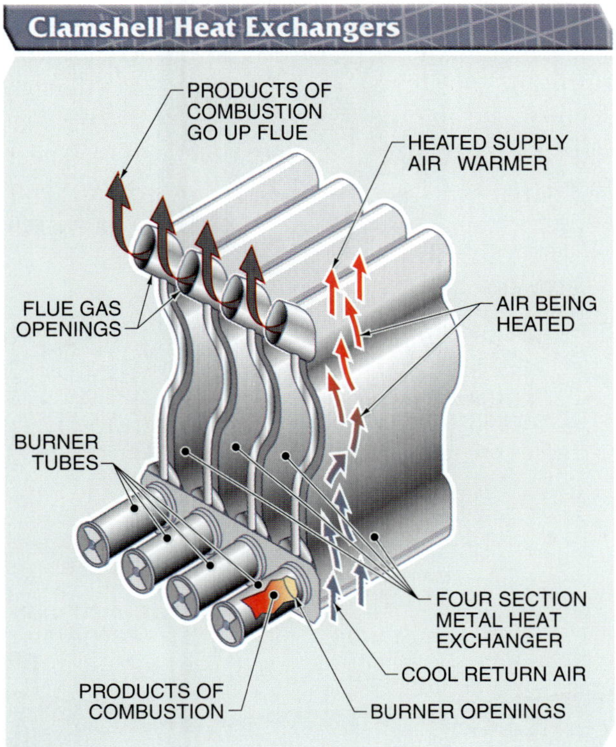

Figure 17-24. The products of combustion flow through a clamshell heat exchanger from top to bottom while cool return air is warmed by flowing over the hot metal.

Heat transfer occurs because of the temperature difference between the hot products of combustion and the cool return air. The ignition temperature for natural gas is approximately 1200°F. The temperature of the flue gas when burning natural gas is about 400°F. The difference between these temperatures, which is 800°F (1200 – 400 = 800), is a result of the heat being transferred through the walls of the heat exchanger. Return air temperature increases in the range of 80°F to 100°F as the air flows over the heat exchanger. Heat exchangers used with combustion burners are designed to provide a 100°F temperature rise between the initial temperature of the return air and the final temperature of the supply air.

Hot flue gas that leaves the flue openings is collected in the draft diverter before the gas rises up the flue. A *draft diverter* is a box made of sheet metal that runs the width of a clamshell heat exchanger to trap flue gases and then travels up a flue with dilution air. The draft diverter is open across the bottom, which allows dilution air to mix with the flue gases that leave the clamshell heat exchanger. **See Figure 17-25.** *Dilution air* is atmospheric air that mixes with, dilutes, and cools the products of combustion. A draft diverter assures a constant updraft and eliminates downdrafts in the flue that would affect burner operation.

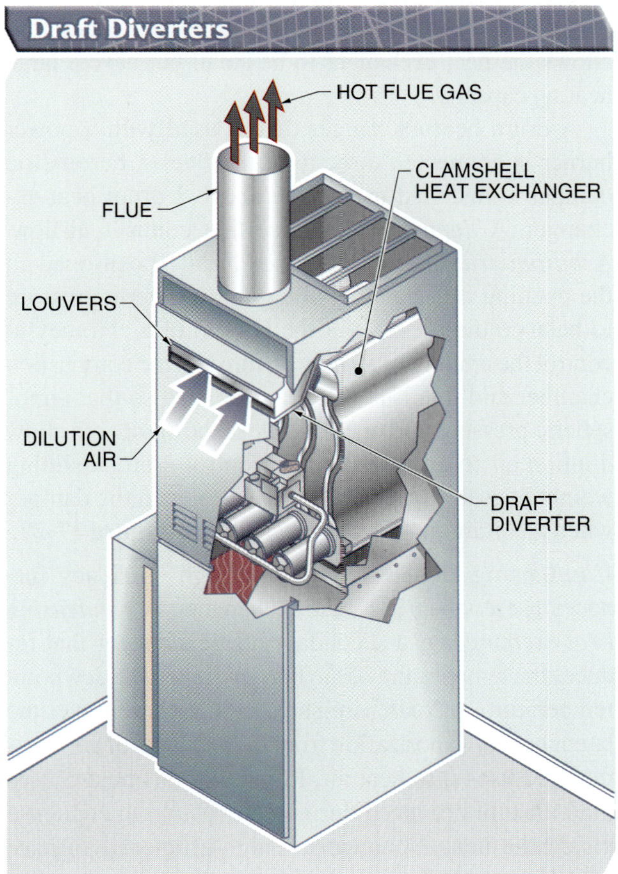

Figure 17-25. A draft diverter is a sheet-metal box that runs the width of a clamshell heat exchanger to make flue gases travel up the flue and to eliminate downdrafts.

Drum Heat Exchangers. A *drum heat exchanger* is a round drum or tube that is located on a combustion chamber to make the products of combustion flow through it. Combustion chambers are designed to retain heat, which helps ignite the fuel. The flame from a power burner is usually contained by a drum heat exchanger. A drum heat exchanger may be positioned horizontally or vertically, depending on the unit in which it is used. A flue connection at the top of the drum carries the products of combustion away from the heat exchanger.

As the hot products of combustion flow through a drum heat exchanger, air is blown around and across the outside surface of the heat exchanger. Heat is transferred through the walls of the heat exchanger because of the temperature difference between the hot products of combustion inside and the cool air outside. **See Figure 17-26.** Drum heat exchangers are made from heavy-gauge sheet metal and may be coated with ceramic glazing to resist corrosion. Drum heat exchangers have secondary or multiple tubes for multiple-pass operation. Multiple-pass operation allows the heat exchanger to be small but have a large heating capacity.

A drum heat exchanger that is used with a power burner is connected directly to the flue. A barometric damper is installed in the flue above a drum heat exchanger. A *damper* is a device that controls airflow. A *barometric damper* is a metal plate positioned in the opening of a flue that allows atmospheric pressure to balance the pressure at the bottom of a chimney to control the amount of airflow up through the combustion chamber and flue. The plate is balanced so that atmospheric pressure will open and close the damper to allow dilution air to enter the flue to control draft. A sliding weight is used to balance the plate to open the damper when a certain pressure is reached. **See Figure 17-27.**

Condensing Heat Exchangers. High-efficiency furnaces use condensing heat exchangers. A *condensing heat exchanger* is a secondary heat exchanger that reduces the temperature of the flue gas below the dewpoint temperature. A condensing heat exchanger removes the latent heat of vaporization from the products of combustion and uses it to heat air. Removing this heat causes the moisture produced during combustion to condense inside the heat exchanger. A high-efficiency furnace must be provided with a drain to remove the condensate. **See Figure 17-28.**

TECH TIP

Many condensing furnaces have two drain connections. One drain connection carries condensation (water) from the vent pipe to a trap. The other drain connection carries condensation from the secondary heat exchanger to a trap.

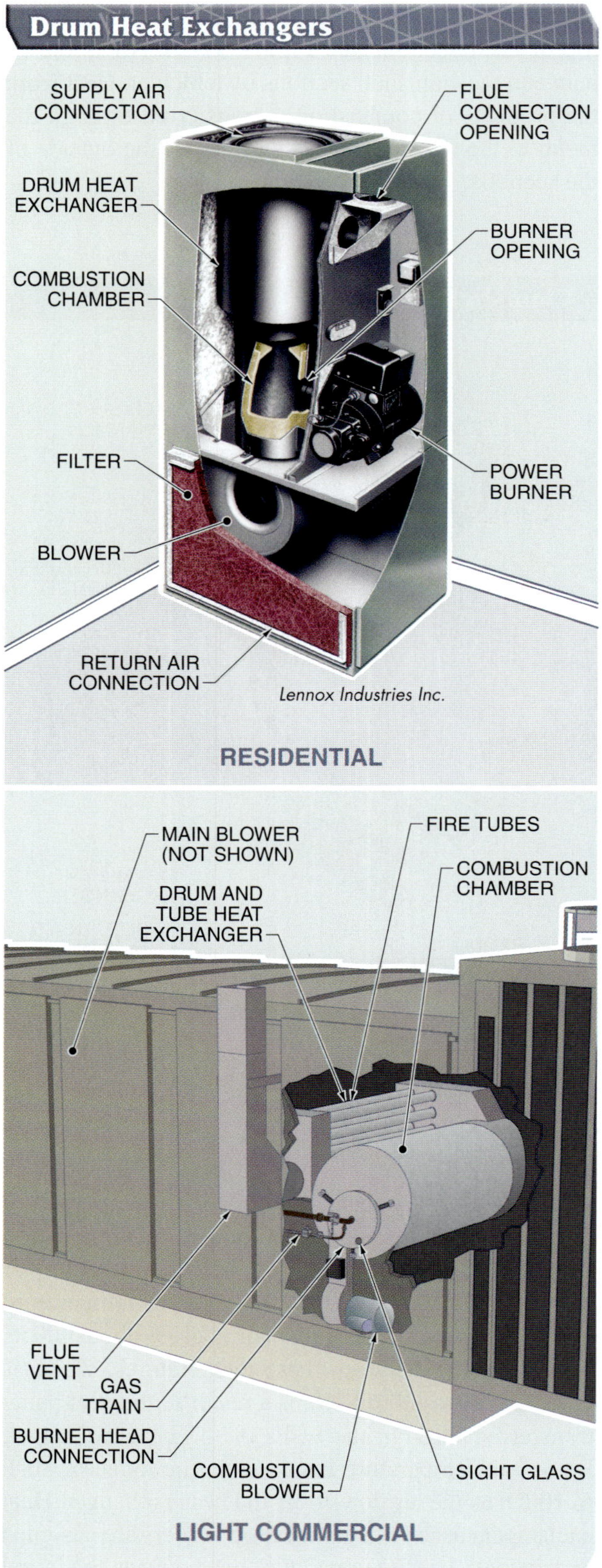

Figure 17-26. Drum heat exchangers typically have secondary or multiple tubes for multiple pass operation.

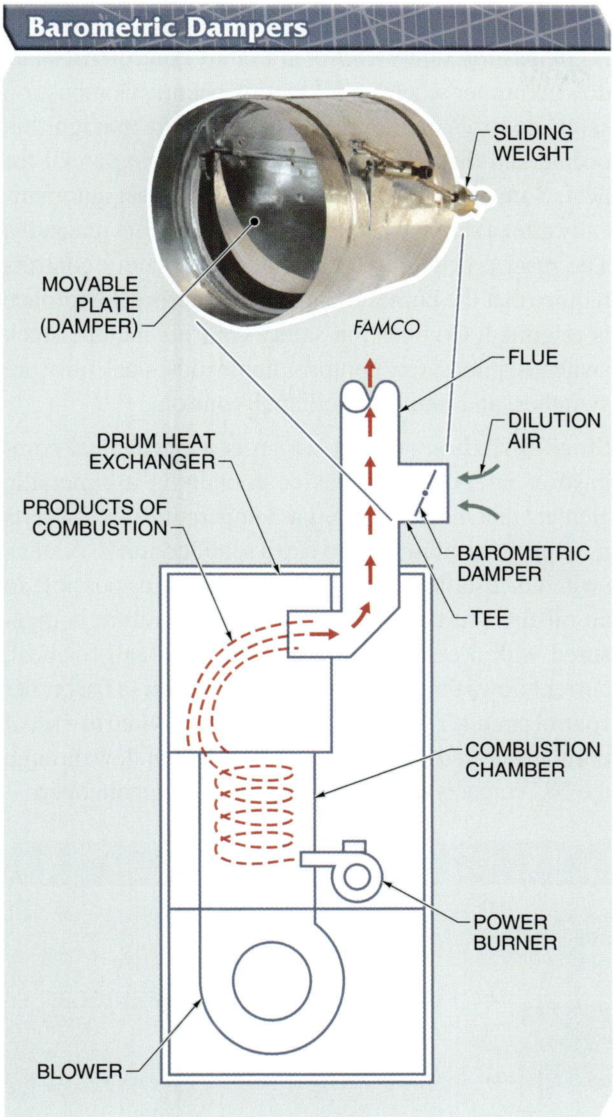

Figure 17-27. Barometric dampers are used when power burners and drum heat exchangers are found in a heating unit.

High-efficiency furnaces reduce the temperature of the flue gas to the dewpoint temperature of the condensing heat exchanger. The lower temperature provides less energy to carry the flue gas up the flue and out the stack. Because of less heat in the flue, a combustion air blower or draft inducer is required to provide positive pressure in the flue. *Positive pressure* is pressure greater than atmospheric pressure. A *combustion air blower* is a blower used to provide positive pressure combustion air at a burner face. A *draft inducer* is a fan that overcomes the resistance to flow in the combustion chamber and provides positive pressure in the flue and vacuum at the burners, which allows the flame to travel through the heat exchanger. Draft inducers are typically attached to the heat exchanger and are connected to the vent to guarantee airflow. The flue carries the products of combustion up the stack. A smaller flue such as a 2½″ diameter plastic pipe can be used when a draft inducer blower is used.

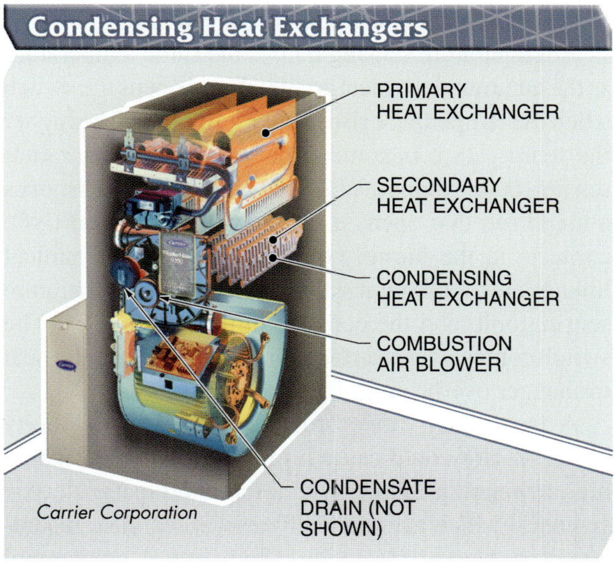

Figure 17-28. A high-efficiency furnace has a condensing heat exchanger that removes the latent heat of vaporization, or condensate, from combustion air and uses it to heat the building air.

CONTROLS

Controls operate the individual components of a furnace. Controls are installed on a furnace by the manufacturer or replaced by a technician to maintain safe, efficient furnace operation. Power controls, safety controls, combustion safety controls, and operating controls are the various kinds of controls found with a furnace.

Power Controls

A *power control* is a control system that controls the flow of electricity to equipment, such as a furnace. Power controls are installed in the electrical circuit between the power source and the furnace. Power controls include disconnects, fuses, and circuit breakers. Power controls should be installed by a licensed electrician.

Safety Controls

A *safety control* is a control that monitors the operation of a furnace. When a furnace causes a hazard to personnel or equipment, safety controls will shut off the furnace. A safety control that is found on all furnaces is a high temperature limit switch.

High Temperature Limit Switches. A *high temperature limit switch* is an electric switch that shuts off a furnace when it begins to overheat. When the temperature of a furnace rises above a safe temperature, the high temperature limit switch shuts off the burner(s) or electric heating element. **See Figure 17-29.**

A temperature sensing bulb is the same temperature as the air around the electric switch and opens the switch when the temperature rises above a setpoint. The higher the temperature becomes, the more the temperature sensing bulb expands. When the bulb expands, it moves a metal rod that opens a set of normally closed (NC) contacts in the burner control circuit. A mechanical linkage and diaphragm system connect the temperature sensing bulb and the contacts in the control box. The switch contacts open as the bulb expands due to the heat exchanger overheating.

A faulty fuel valve, broken blower belt, or faulty blower motor could cause a furnace to overheat. Most high temperature limit switches are automatic electric switches, which automatically reset when the temperature drops below the setpoint. Manual-reset temperature limit switches are typically used for some light commercial applications.

Combustion Safety Controls

A *combustion safety control* is a safety control that shuts down a burner when a malfunction occurs. Combustion safety controls monitor firing to make sure that ignition occurs and that the flame remains on during a call for heat. Some combustion safety controls reset automatically after a shutdown, but others must be reset manually. The reset procedures for manual-reset safety controls require that the burner(s) be checked before the furnace is reignited. Combustion safety controls include stack switches, pilot safety controls, flame rods, stack pressure switches, and flame surveillance controls.

Stack Switches. A *stack switch* is a mechanical combustion safety control device containing a bimetallic element that measures flue-gas temperature and converts it to mechanical motion to trip a set of contacts. A stack switch is installed in the flue and as close as possible to an oil-fired furnace. The flue-gas temperature is measured with a piston or diaphragm. On a call for heat, current flows through a safety switch heater in the burner control circuit. The safety switch heater is wired to a set of normally closed (NC) contacts. The current flow through the NC contacts allows the fuel valve to remain open.

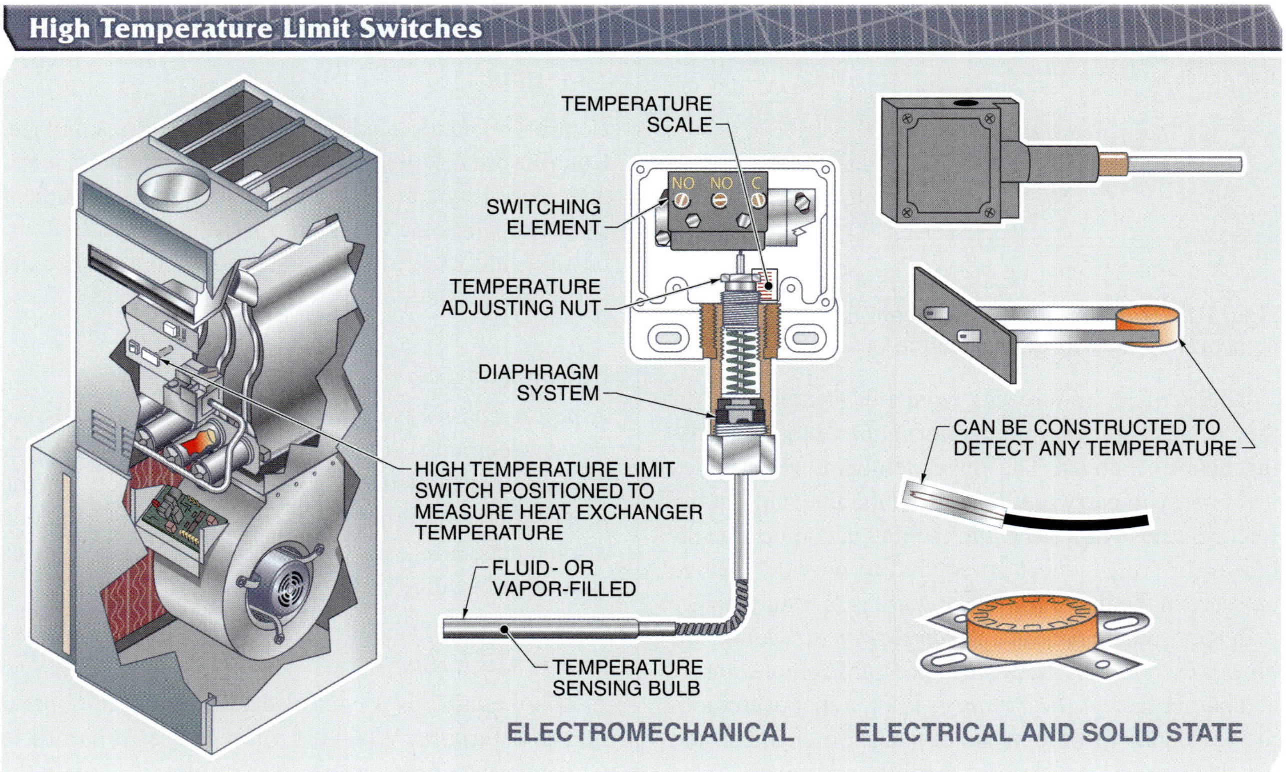

Figure 17-29. When the temperature of a furnace rises above a safe temperature, a high temperature limit switch shuts off the burners or electric heating element.

When the flue-gas temperature does not rise in the allotted time (approximately 90 seconds), the safety switch heater opens the NC contacts. When ignition does occur, the bimetallic element expands. This moves a metal rod that closes a set of normally open (NO) contacts that are part of the burner control circuit. This allows the fuel valve to remain open. **See Figure 17-30.** A mechanical linkage or rod connects the bimetallic element and the contacts in the control box of the switch. The contacts open or close as the bimetallic element expands and contracts.

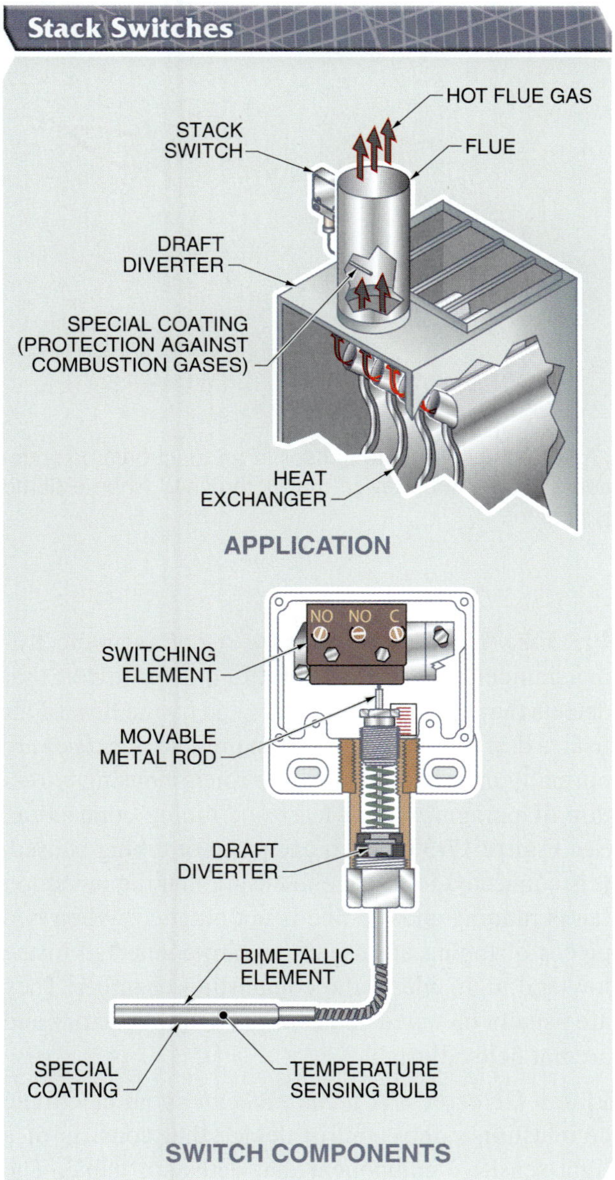

Figure 17-30. A stack switch is installed in a flue as close as possible to the furnace to measure the temperature of combustion gases in the flue.

Pilot Safety Controls. A *pilot safety control* is a safety control that determines when a pilot light is burning. Burners that use a pilot light for initial combustion have pilot safety controls. A pilot safety control contains a thermocouple. A *thermocouple* is a device that produces electricity when two different metals that are twisted together are heated.

A thermocouple is used to detect the presence of a pilot flame in older gas furnaces. **See Appendix.** The thermocouple is installed in the pilot flame, which heats the hot junction. When the hot junction is heated, a low-voltage electric signal is produced. The free ends of the wires (the cold junction) are connected to an electromagnetic coil in a safety valve ahead of the gas fuel valve or to a special safety valve section within the gas fuel valve.

When the pilot light is burning, the low-voltage electric signal generated by the thermocouple produces a magnetic field in the coil. The magnetic field holds the safety valve open. **See Figure 17-31.** When the pilot light goes out, there is no magnetic field and the safety valve closes. The pilot safety control monitors the pilot light in a burner and does not allow the gas fuel valve to open or a burner to ignite unless the pilot flame is established. Pilot safety controls are used mainly on fuel-burning equipment that has a standing pilot.

These older furnaces are called "standing pilot furnaces" because the pilot light is always lit. These types of furnaces are lower efficiency furnaces, typically 70% or lower. Many of these types of furnaces are still being used in buildings today. Universal replacement kits are available for thermocouples for easy replacement.

A thermopile is a variation of a thermocouple. A *thermopile* is a device that generates milli-voltage that is used with a special thermostat designed to operate with the low voltage levels. The thermopile generates all of the current used in the heating control circuit and system. Thermopiles are used with very old furnaces called "gravity furnaces" that do not have a blower but rely on the fact that heated air will rise into a space.

Flame Sensors. A *flame sensor* is an electronic combustion safety control device that determines when a flame is present by using the burner flame to conduct electricity. Flame sensors are used on residential and light commercial furnaces. While a furnace is firing, a combustion control device sends out a low-voltage electric signal to the metal rod of the sensor located in the flame. When a flame is established on a call for heat, the safety circuit is closed by the flame sensor. When a flame is not established within a reasonable length of time, the furnace shuts off. **See Figure 17-32.**

Pilot Safety Controls

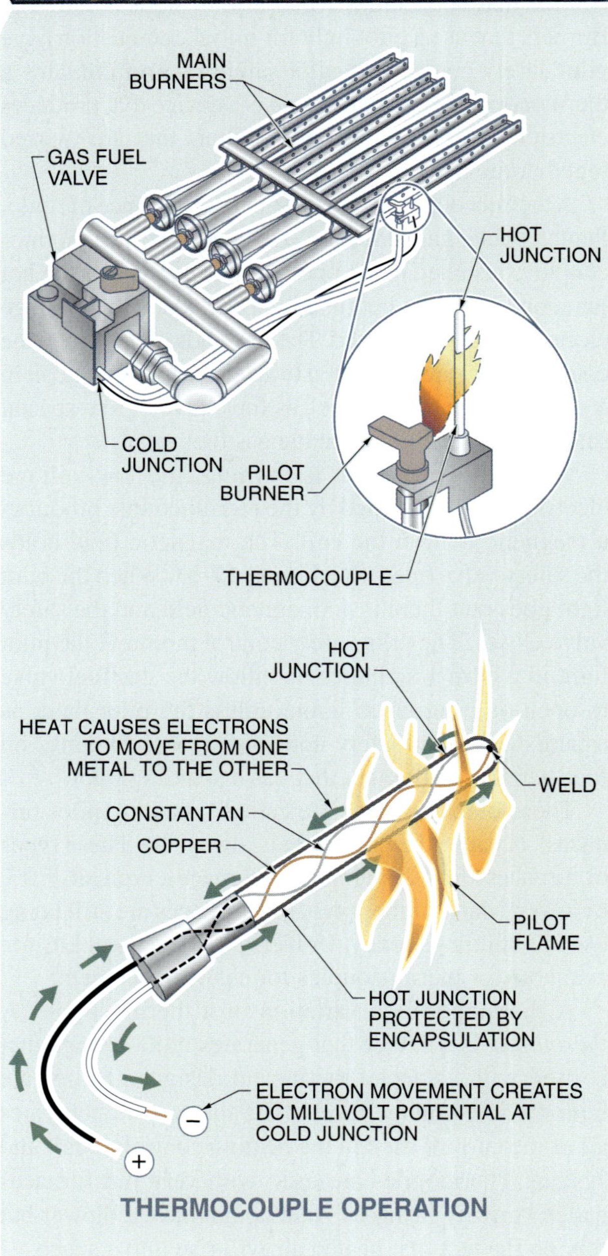

Figure 17-31. Pilot safety controls use a thermocouple to determine when a pilot light is burning.

Flame Sensors

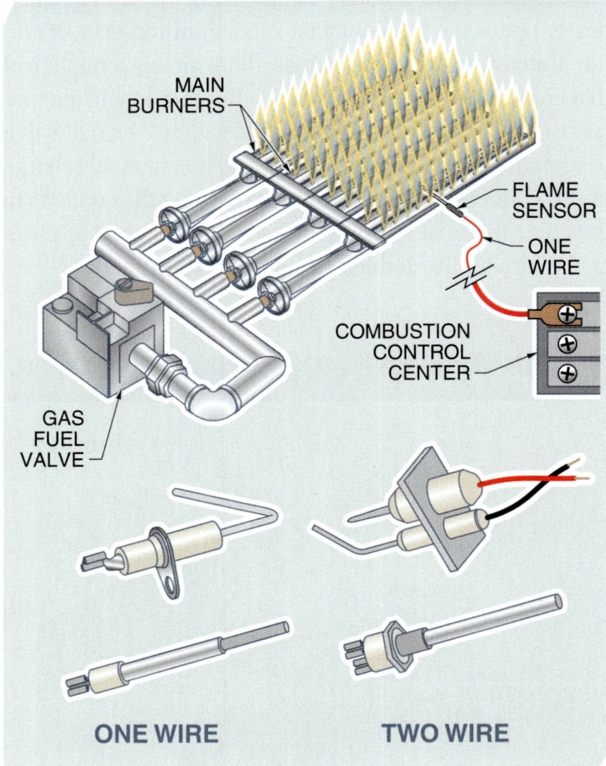

Figure 17-32. A flame sensor is an electronic combustion safety control device that determines when a flame is present.

Flame sensors are smaller and more durable than stack switches or thermocouples. Flame sensors are used in systems that have hot surface igniters or intermittent pilots. Intermittent pilots establish a pilot flame when the thermostat calls for heat. Intermittent pilots extinguish a pilot flame when the burner flame is extinguished.

Pressure Switches. A *pressure switch* is an electromechanical combustion safety control device that detects the pressure in a flue created by the flow of air from a draft induction fan. Most pressure switches are normally open and consist of a round housing with a thin diaphragm with at least one tubing connection. **See Figure 17-33.** When one piece of tubing is used, it is connected to the flue to ensure that the induction fan is running and the flue is not plugged. When two pieces of tubing are used, they are connected to the low and high side of the combustion chamber. They are usually mounted and calibrated at the factory and are not field adjustable.

Flame Detectors. A *flame detector* is an electronic combustion safety control device that consists of a light-sensitive component that detects a flame. The three types of flame detectors in use are the cadmium sulfide (cad) cell, infrared (IR), and ultraviolet (UV) flame detectors.

Pressure Switches

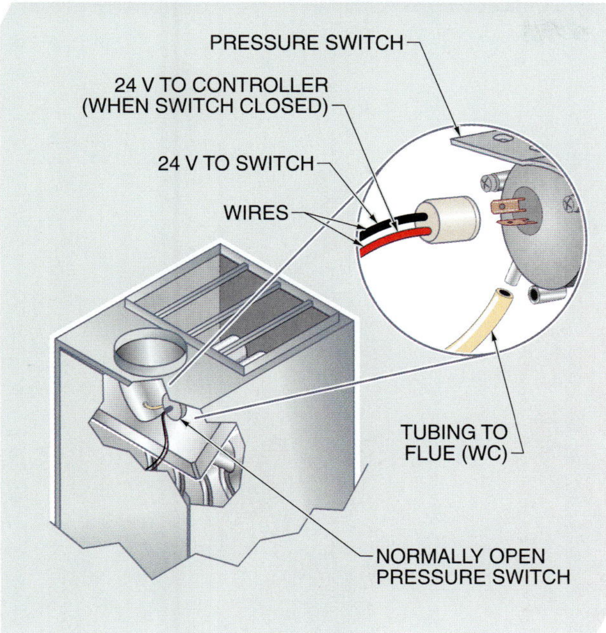

Figure 17-33. A pressure switch is a combustion safety device that detects the pressure in a flue created by the flow of air from a draft induction fan.

Cad Cell Optical Flame Detectors

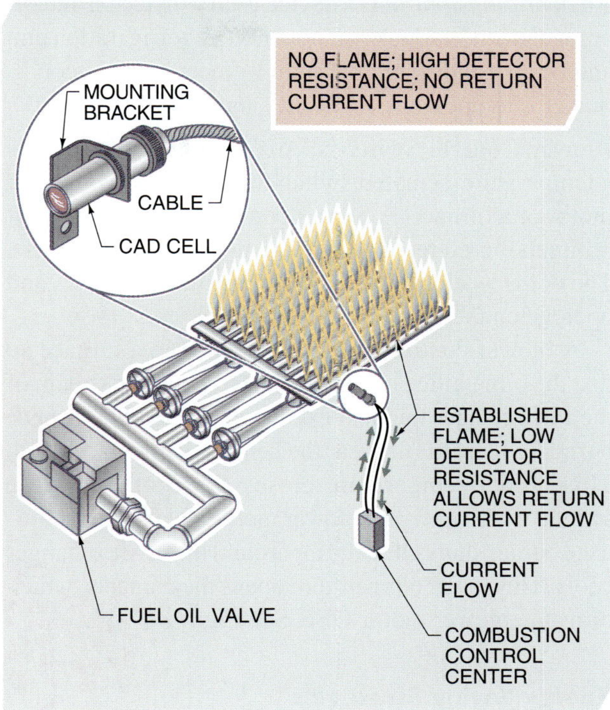

Figure 17-34. A cad cell optical flame detector is a light-sensitive device that uses cadmium sulfide to detect a flame.

Cadmium sulfide is the light-sensitive substance used in cad cell flame detectors. Cad cell optical flame detectors are used in the circuits of oil burning furnaces. The resistance of cadmium sulfide to electric current depends on the intensity of the light that strikes the material. When a cad cell is exposed to light, the resistance to the flow of electricity through the cell is low. When the cad cell is in darkness, resistance through the cell is high. Electrical leads are connected to each side of the cell. Current flow through the cell is monitored to determine when the cell detects light.

A cad cell is mounted on an oil burner in such a way that it is in a direct line of sight of the burner flame. When the oil burner ignites and a flame is established on a call for heat, the resistance through the cell is low. Low resistance allows the electric signal to pass to the control center, which actuates the furnace. When the burner does not ignite, the resistance through the cell is high, which prevents the electric signal from reaching the control center. After a reasonable time, the control center shuts the furnace OFF. Devices for cad cell combustion safety control are used on furnaces that contain fuel oil burners. **See Figure 17-34.**

Infrared (IR) flame detectors respond to the infrared spectrum of light, which most gases emit. Ultraviolet (UV) flame detectors respond to the ultraviolet spectrum of light. Both IR and UV flame detectors can be used in furnaces that are not burning fuel oil.

Ignition Modules. Most modern furnaces are equipped with a solid state ignition module. These modules may also be retrofitted to older furnaces. These solid state ignition modules include wiring termination blocks and wiring harnesses for typical applications.

Typically, the purpose of an ignition module is to attempt to ignite a pilot and prove pilot ignition before allowing main burner ignition. The induction fan on the burner will operate for a period of time, such as 30 sec, to ensure that no fuel from the pilot has accumulated in the combustion chamber. Specific flame detectors are used with certain models. A cross-reference chart is provided for replacement between manufacturers. Usually, the modules are nonadjustable by the technician. Manufacturer literature should always be referenced and followed.

Operating Controls

An *operating control* is a device used to cycle on and off or regulate the various components of a refrigeration unit and/or air conditioning system to meet specific conditions for temperature, pressure, and humidity.

Blower Control Switches. A *blower control switch* is a temperature-actuated switch that controls the blower motor of a furnace. A blower control switch consists of a bimetallic element that operates electrical contacts. The contacts close when the temperature increases and open when the temperature decreases.

A blower control switch is installed in a furnace so that the bimetallic element is located in the air stream of the furnace near the heat exchanger. When the furnace burner ignites, the air near the heat exchanger is heated. This higher temperature closes the contacts of the switch, which turns the blower motor ON. When the furnace burner shuts OFF, the air around the heat exchanger cools. This lower temperature opens the contacts, which turns the blower motor OFF. **See Figure 17-35.**

Blower Control Switches

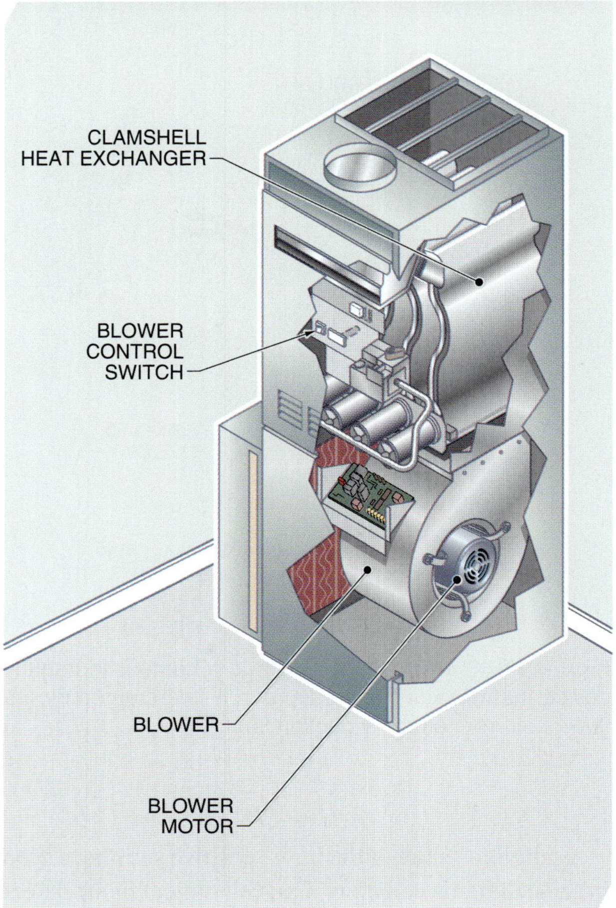

Figure 17-35. The bimetallic element of a blower control switch is installed in the air stream near the heat exchanger so that the nearby air and switch can be heated.

Energy Saving Strategies

New strategies have been developed to save increasing amounts of energy and help to ensure comfort. One method is to use a multispeed fan instead of a single-speed fan. With a multispeed fan system, the fan has more than one set speed. When only a small amount of heat is needed, the fan runs at a low set speed. When the demand for heat increases, the fan speed will increase to a higher set speed. This reduces the energy consumption by the fan and helps to maintain accurate temperature control in building spaces.

UNIT 18

Forced-Air Heating Distribution Systems

To achieve comfort, forced-air HVAC units use the flow of air to heat and cool building spaces. When HVAC systems are operating properly, building occupants are unaware of the HVAC equipment. When the equipment is not maintained or there is a component failure, system noise, and airflow or temperature changes, the attention of the building occupants will be on the HVAC system.

OBJECTIVES

- Explain basic duct design and system terminology.
- Identify the different types of duct designs and layouts.
- Describe the various types of duct materials used.
- Describe the features of balancing dampers, insulation, registers, and grills.
- Describe conditioned air supply and return duct systems.
- Explain how to lay out and size ducts and related components.
- Identify considerations for duct installations and retrofits.

Index to Procedures

FORCED-AIR DISTRIBUTION SYSTEMS

While many variations are possible, forced-air distribution systems typically consist of blowers, supply ductwork, return ductwork, registers, and grills. **See Figure 18-1.** The blower circulates air across heating coils, cooling coils, humidifiers, and air filtration elements. The air moves through the ductwork into the room, and then returns to the air handling unit. Outside air ventilation requirements can be satisfied by using a forced-air distribution unit with an outside air intake duct or energy recovery ventilator (ERV). When these units are operating properly, building occupants will be unaware of equipment noise or airflow unless one of these units fails and the space temperature changes dramatically.

The distribution system of a forced-air heating system must function correctly to achieve proper HVAC system operation.

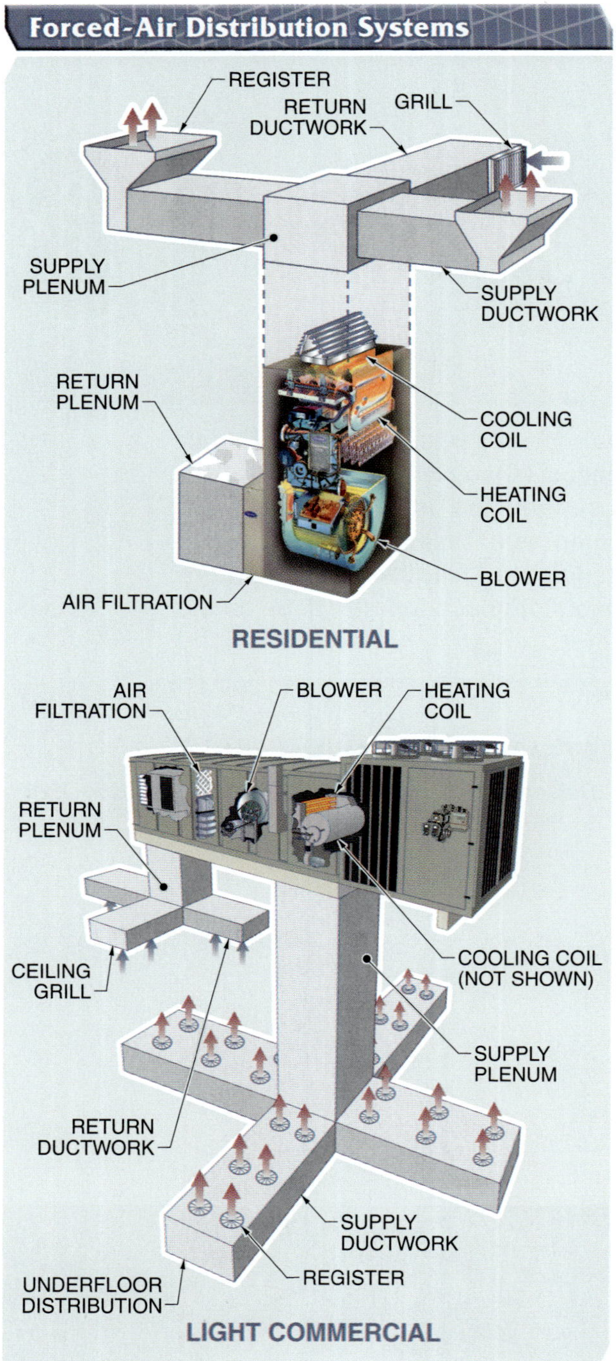

Figure 18-1. Forced-air distribution systems can be assembled with various components, but all distribution systems consist of blowers, supply ductwork, return ductwork, registers, and grills.

Airflow Principles and Measurements

To maintain and troubleshoot forced-air distribution units properly, an HVAC technician must understand the principles and measurements of airflow. When airflow principles are not understood or airflow is not properly measured, a misdiagnosis of a problem or incorrect installation may take place. The characteristics of airflow and their principles include airflow rate, air velocity, and system pressures.

Airflow Rates

The most important function of the air distribution system is to deliver the correct volume of air, which is the airflow rate, to building spaces. *Airflow rate* is the measure of the volume (in cubic feet) of air that passes a given point in one minute (cfm). HVAC forced-air distribution systems must deliver the correct volume of heated or cooled air. **See Figure 18-2.**

In a residential or commercial building, each individual room has specific airflow requirements (rates). These airflow rates are determined by the load calculations performed prior to system installation. Obviously, the heating and cooling loads of a conference room are different from those of a small office. A corner office with outside exposure on two sides will require more airflow than a space with no outside exposure. In a residential system, a kitchen and a family room will have different airflow requirements.

Forced-air distribution systems are required to deliver the correct volume of air to each space in order to satisfy the individual load demand in that space. The air delivered to the space will blend with the room air and provide occupant comfort.

Air Velocity

Air velocity is the speed air moves from one point to another through ductwork, a register, or a grill; measured in feet per minute (fpm). **See Figure 18-3.** Maintaining the proper air velocity through the system is very important. Typically, air velocity should be about 600 fpm in branch ducts and 900 fpm in trunk ducts. Higher velocities create noise in the ducts and decrease efficiency. Lower velocities in air conditioning systems allows for the premature warming of the air, which also decreases system efficiency.

TECH TIP

HVAC units operate most efficiently with the correct amount of airflow across the heating and cooling coils. The primary issue with most HVAC systems is airflow. Having the correct airflow in the air conditioning systems is critical.

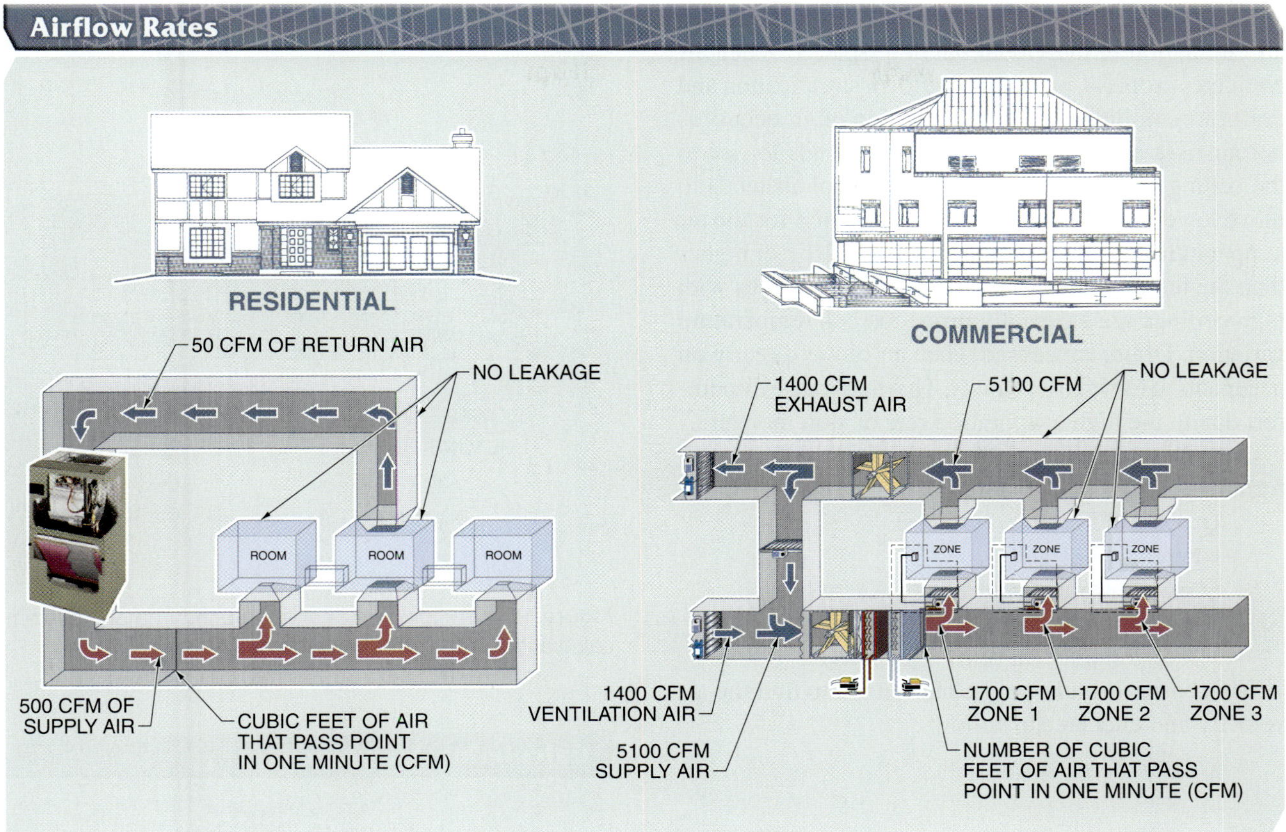

Figure 18-2. The most important function of the forced-air distribution system is to deliver the correct amount of air, which is the volume of air in cubic feet that passes a point in one minute, to building spaces.

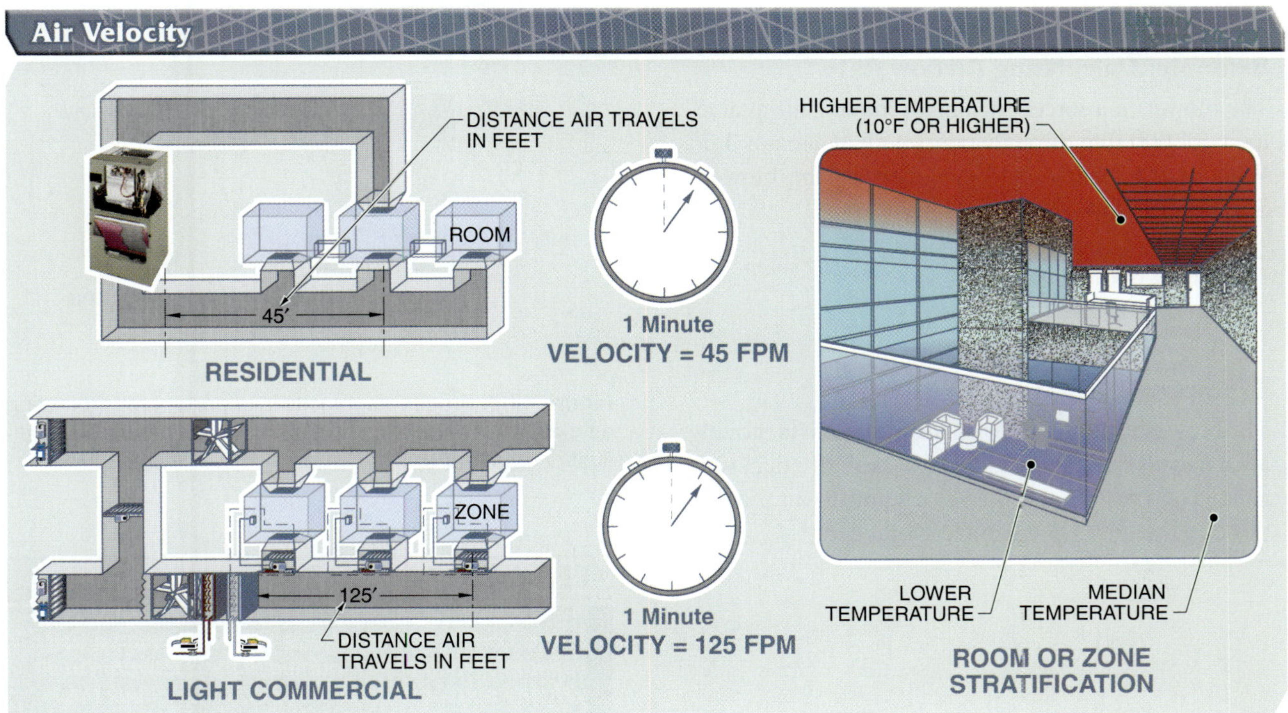

Figure 18-3. Air velocity is the speed (in feet per minute) of air as it moves through ductwork, registers, and grills and should be about 600 fpm in branch ducts and 900 fpm in trunk ducts to keep noise to a minimum.

The proper air velocity is important for promoting air circulation in individual building spaces. When air circulates properly, problems due to air stratification and drafts are minimized. The stratification of air occurs as hot air rises and cold air falls. Hot air tends to rise to the ceiling of a building space, while cool air tends to move lower to the floor. It is not uncommon for the air temperature at the ceiling of a room to be 10°F or higher than the temperature at the floor. Building spaces with high ceilings are especially prone to such temperature variation. Drafts are caused when air blows directly on occupants at too high a velocity. This may cause discomfort due to the high evaporation rate of skin moisture.

Airflow rate can be calculated using the following formula:

$$Q = A \times V$$

where

Q = airflow rate (in cfm)

A = area of duct (in ft^2)

V = air velocity (in fpm)

This basic formula can be rearranged to find the air velocity and duct area as follows:

$$V = \frac{Q}{A}$$

or

$$A = \frac{Q}{V}$$

Example: Calculating Airflow Rate

The blower in a forced-air unit is moving air at a velocity of 600 fpm through a duct with an area of 3 ft^2. What is the airflow rate provided by the blower? **See Figure 18-4.**

$$Q = A \times V$$
$$Q = 3 \times 600$$
$$Q = \mathbf{1800\ cfm}$$

Example: Calculating Air Velocity

The blower in a forced-air distribution unit is moving air at a measured rate of 2400 cfm through a duct with an area of 4 ft^2. What is the velocity of the air moving through the duct? **See Figure 18-5.**

$$V = \frac{Q}{A}$$
$$V = \frac{2400}{4}$$
$$V = \mathbf{600\ fpm}$$

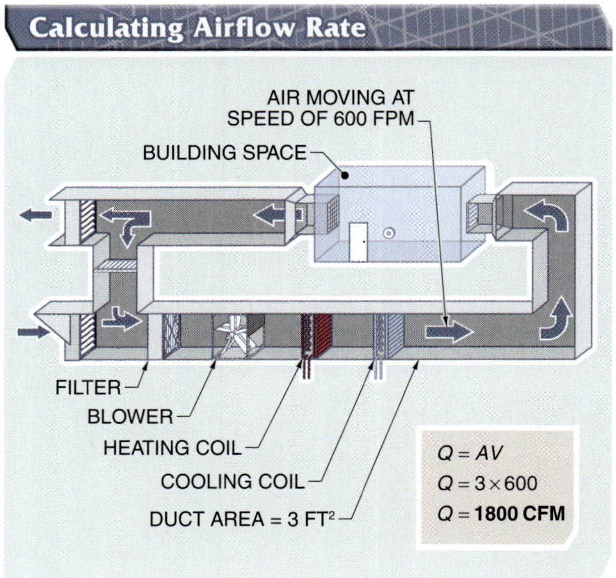

Figure 18-4. Airflow rate is a value used in many system calculations and in troubleshooting.

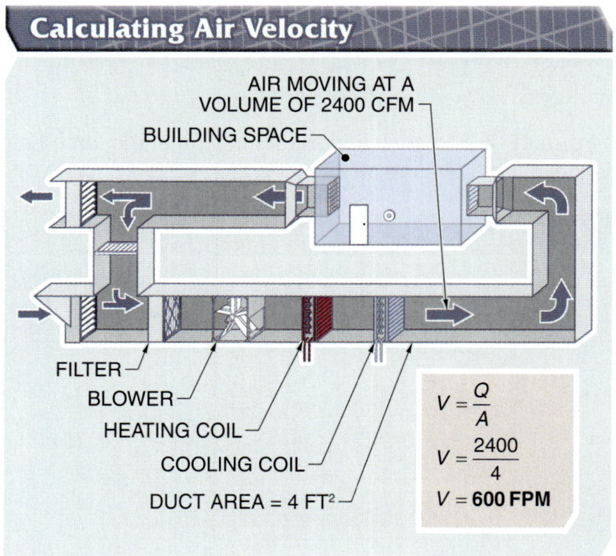

Figure 18-5. Proper air velocity is important for promoting air circulation in building spaces and to minimize air stratification and drafts.

TECH TIP

Large commercial systems may have air velocity sensors permanently mounted in the ductwork. An air velocity sensor enables an operator to check the air velocity on a building automation system computer.

Example: Calculating Duct Area

The blower in a forced-air unit is moving 2400 cfm of air at a velocity of 800 fpm. What is the area of the duct? **See Figure 18-6.**

$$A = \frac{Q}{V}$$

$$A = \frac{2400}{800}$$

$$A = \mathbf{3\ ft^2}$$

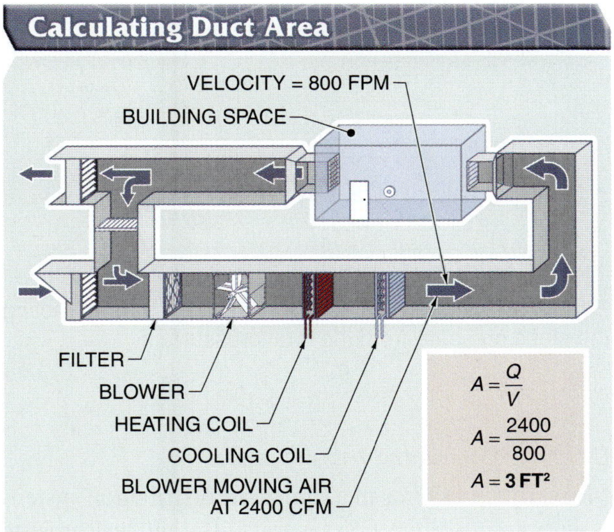

Figure 18-6. Duct sizes are provided in square inches to be used with round duct or rectangular duct tables.

Duct System Pressures

Air exerts pressure as it moves through a duct. It is important to know the difference between the various types of duct system pressures and understand how they are measured in forced-air distribution systems. The types of duct system pressures include static pressure, velocity pressure, and total pressure. Duct system pressure is measured in inches of water column (in. WC). The water column unit of measurement is very precise, as 1″ WC is equal to 0.036 psi (1⁄27.7 psi). **See Appendix.**

Static Pressure. *Static pressure* is the force that nonmoving air exerts against surfaces such as ducts; measured in inches of water column (in. WC). It is typically measured by installing a pressure gauge, pressure sensor, or pressure tap into the surface of the duct. Static pressure at a given point is the same in all directions. **See Figure 18-7.**

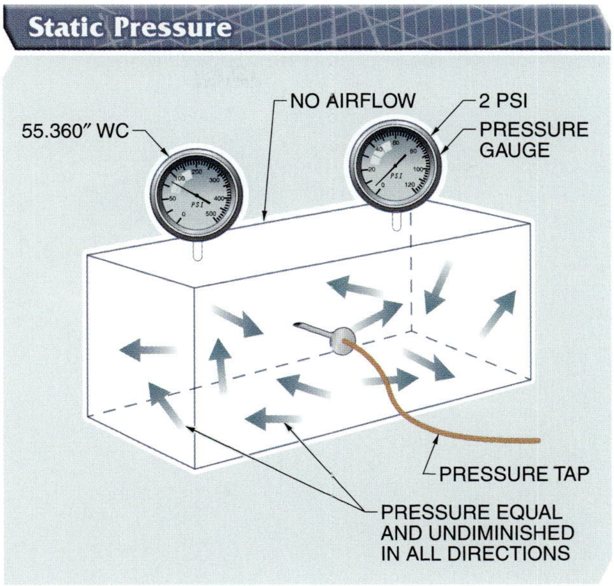

Figure 18-7. Static pressure is the pressure (in. WC) of nonmoving air as it creates a pressure against the surfaces of a duct in all directions. It is typically measured by installing a pressure gauge, pressure sensor, or pressure tap into the surface of the duct.

Velocity Pressure. *Velocity pressure* is the pressure created by air as it travels, such as through a duct. Velocity pressure is expressed in inches of water column (in. WC). It is measured inside the duct and parallel to the airflow. Air velocity in a duct is related to velocity pressure. This relationship can be expressed as the following:

$$P_V = \left(\frac{V}{4005}\right)^2$$

where

P_V = velocity pressure (in in. WC)

V = air velocity (in fpm)

4005 = air properties constant

Example: Calculating Velocity Pressure Using Air Velocity

What is the velocity pressure in a duct with an air velocity of 1000 fpm? **See Figure 18-8.**

$$P_V = \left(\frac{V}{4005}\right)^2$$

$$P_V = \left(\frac{1000}{4005}\right)^2$$

$$P_V = 0.249^2$$

$$P_V = \mathbf{0.062''\ WC}$$

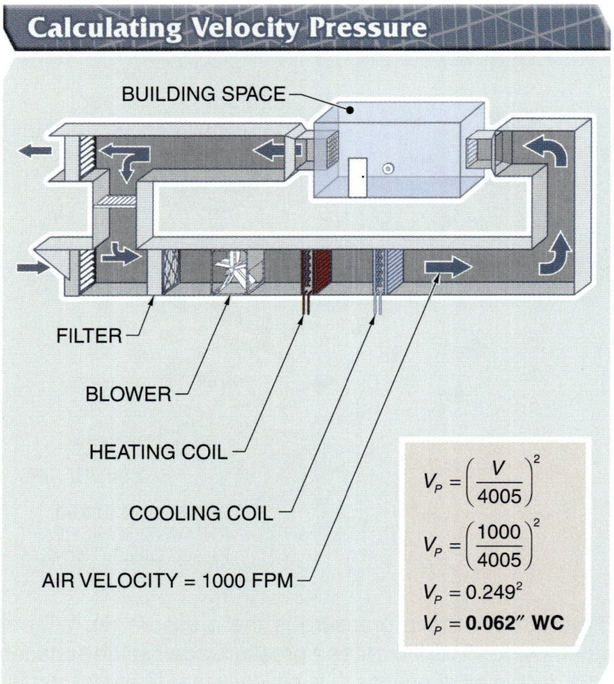

Figure 18-8. Velocity pressure is the pressure (in. WC) created by moving air as it travels through a duct and is measured parallel to the airflow.

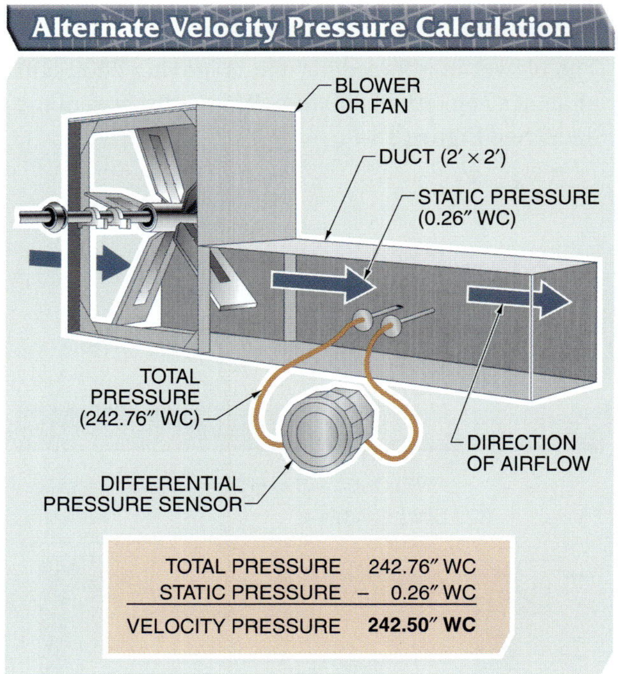

Figure 18-9. Velocity pressure can be found by subtracting the static pressure from the total pressure.

Total Pressure. *Total pressure* is the sum of the static and velocity pressures in a duct system. Total pressure is not a measurement that is typically used alone, but it is used in formulas to find other values. For instance, when the total and static pressures are known, the velocity pressure can be easily determined. The relationships between total, static, and velocity pressure can be expressed as follows:

$$P_V = P_T - P_S$$

where

P_V = velocity pressure (in in. WC)

P_T = total pressure (in in. WC)

P_S = static pressure (in in. WC)

Example: Calculating Velocity Pressure Using Total and Static Pressures

What is the velocity pressure in a duct with 242.76″ WC of total pressure and 0.26″ WC of static pressure? **See Figure 18-9.**

$$P_V = P_T - P_S$$

$$P_V = 242.76 - 0.26$$

$$P_V = \mathbf{242.50'' \ WC}$$

Duct Air Pressure Instruments

Measuring pressures in a forced-air distribution system is important. There are several different instruments available for measuring the air pressure, but all are classified as manometers. In the past, U-tube manometers were the instrument of choice. A U-tube manometer is a U-shaped piece of clear plastic that is filled with water. The pressure from the duct displaces a certain height of a column of the water. **See Figure 18-10.** Today, inclined manometers, Magnehelic® differential pressure gauges, and digital electronic manometers are commonly used.

Many forced-air distribution systems have permanently mounted taps that allow easy access for the attachment of manometers. A manometer is also very handy for evaluating the pressure drop across a filter element in a system. A large pressure drop indicates that the filter needs to be changed.

TECH TIP

Part of the unit checkout process consists of taking readings of the air pressure in the ductwork. As systems age or building layouts change, airflow in the ductwork may change, resulting in a loss of comfort.

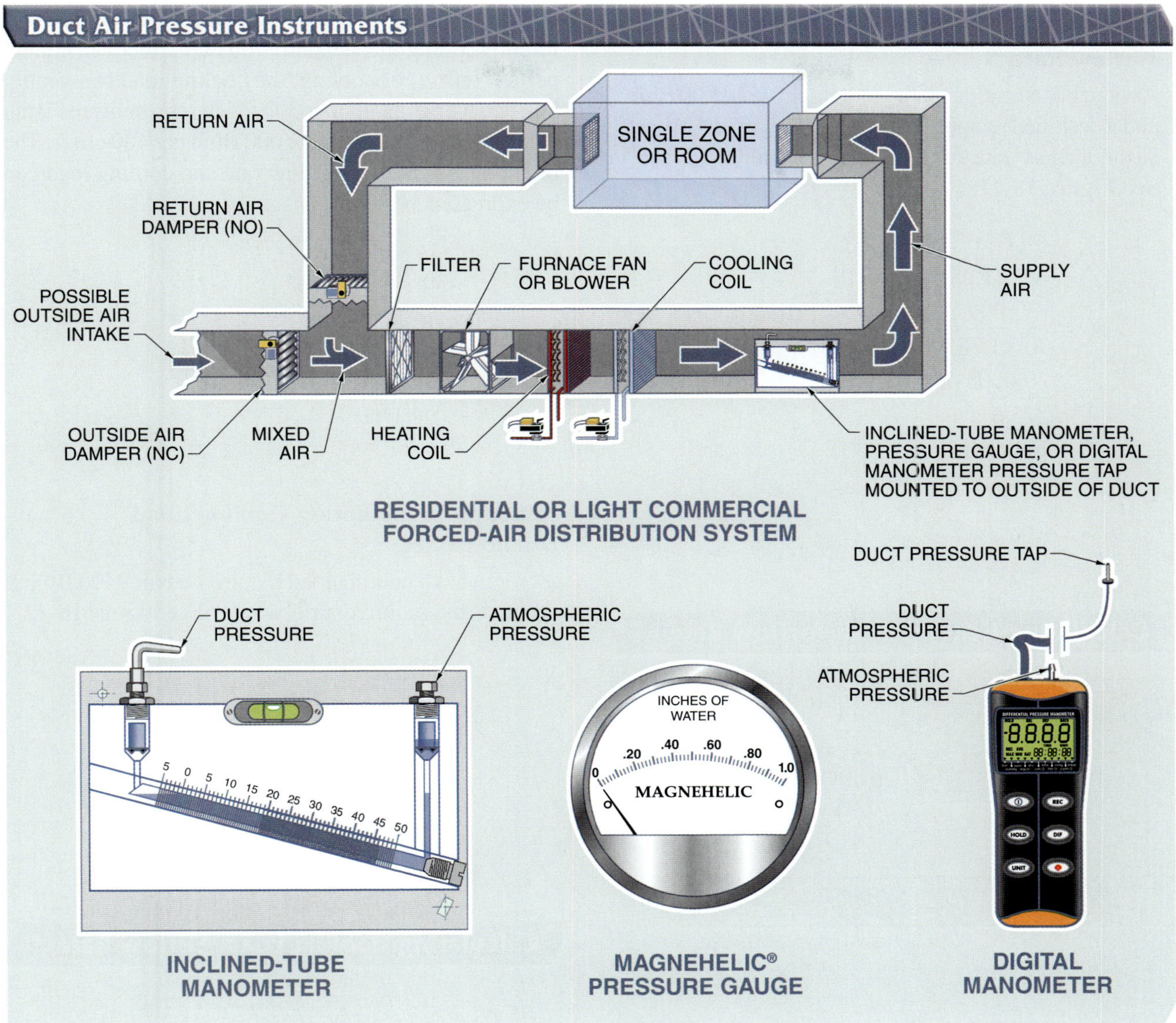

Figure 18-10. Many forced-air distribution systems have permanently mounted taps that allow easy access for attaching inclined manometers, Magnehelic® differential pressure gauges, or digital manometers.

Forced-Air System Blowers

Blowers force air through ductwork to building spaces. The vast majority of blowers are motor driven, centrifugal, squirrel-cage-type fans. It takes approximately 400 cfm of air to deliver 1 ton (t) of cooling to a building space. Blower motors often have multiple leads to permit various operating speeds. Typically, blowers and fans run at higher speeds in the cooling mode than in the heating mode. A higher volume of air is required to cool a space than to heat a space.

Load calculations are used to find the correct airflow rate for each building space. The airflow rate for heating can be found using the following formula:

$$Q = \frac{L_H}{\Delta T \times 1.08}$$

where
Q = airflow rate (in cfm)
L_H = heating load (in Btuh)
ΔT = temperature differential (in °F)
1.08 = constant

Example: Calculating Heating Load Airflow Rate

A room has a heating load requirement of 20,000 Btuh and a designed temperature differential of 70°F. What is the airflow rate required for the building space? **See Figure 18-11.**

$$Q = \frac{L_H}{\Delta T \times 1.08}$$

$$Q = \frac{20{,}000}{70 \times 1.08}$$

$$Q = \frac{20{,}000}{75.6}$$

$$Q = \mathbf{264.6\ cfm}$$

The airflow rate needed for cooling is calculated differently. Using the standard 400 cubic feet per minute per ton (cfm/t) of cooling, the cooling load can simply be divided by the number 30, which represents the Btuh value per cfm/t value, or 12,000 Btuh per 400 cfm/t. The relationship between airflow rate and cooling load can be expressed as follows:

$$Q = \frac{L_C}{30}$$

where
Q = airflow rate (in cfm)
L_C = cooling load (in Btuh)
30 = constant

Example: Calculating Cooling Load Airflow Rate

A room has a cooling load requirement of 7500 Btuh. What is the required airflow rate? **See Figure 18-12.**

$$Q = \frac{L_C}{30}$$

$$Q = \frac{7500}{30}$$

$$Q = \mathbf{250\ cfm}$$

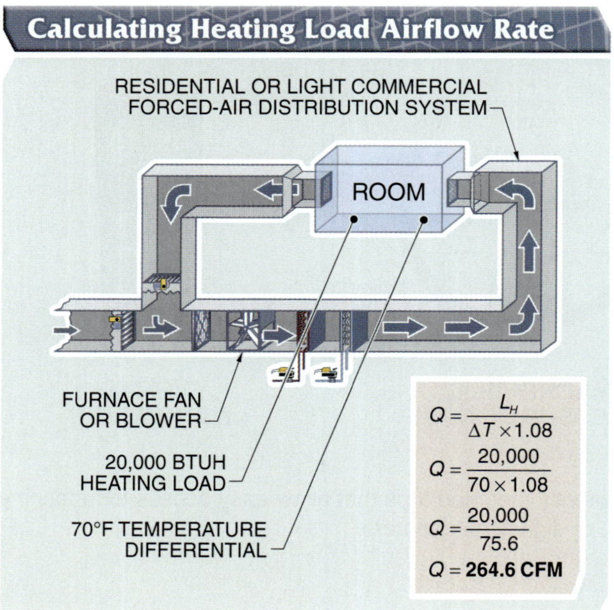

Figure 18-11. Heating mode typically requires 250 cfm of airflow to heat a building space.

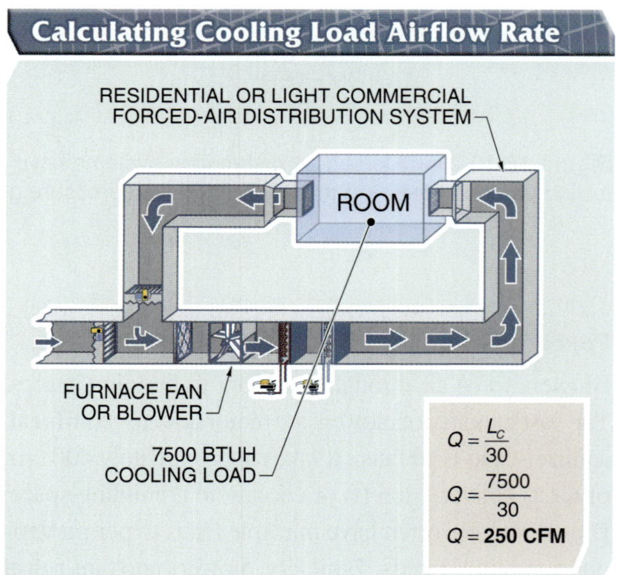

Figure 18-12. Cooling mode requires approximately 400 cfm of airflow per ton of cooling to cool building spaces, because it takes higher flow rates to cool a space than to heat a space.

TECH TIP

Residential split systems, furnaces, and heat pumps, are available with adjustable-speed supply fans. The speed of the fan can be adjusted by connecting the motor wire leads to the correct terminals on the electronic control board of the unit.

Duct Dimensions

The ductwork of a forced-air distribution system directs airflow from the blower to the individual registers in rooms or building spaces. Duct size is expressed in inches. For example, the diameter of round ductwork and the width and height of rectangular ductwork are measured and expressed in inches. By convention, the width measurement is given first. For example, a duct measurement of 24″ × 12″ means that the duct has a width of 24″ and a height of 12″. **See Figure 18-13.**

Duct Dimensions

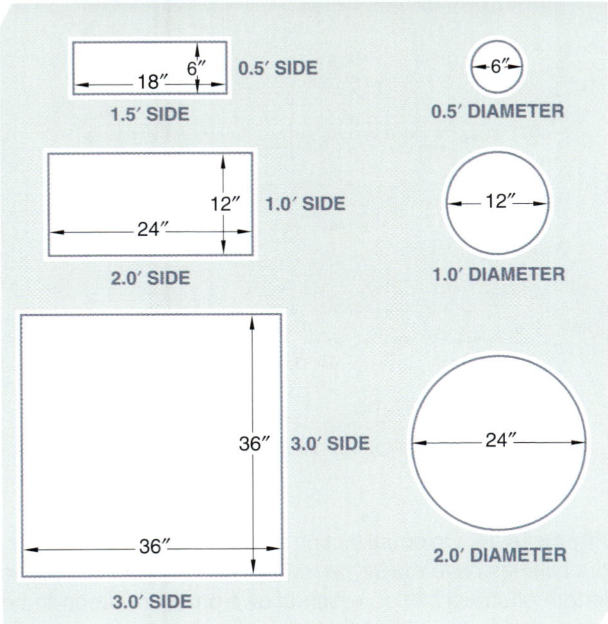

Figure 18-13. The duct size of round duct is provided in inches of diameter. Width and height inch measurements are used for rectangular ducts, with the width measurement given first.

While widely used, flex duct has a high static pressure drop.

Friction Losses

As air moves through a duct, it comes into contact with the duct walls. This contact causes a loss of energy (pressure drop) due to friction. This pressure drop reduces the pressure of the air as it moves through the duct. *Friction loss* is the decrease in air pressure due to the friction created when air moves through a duct. Friction loss is represented as static pressure drop. This drop in pressure indicates the friction loss that has occurred in the duct due to the contact. **See Figure 18-14.**

Duct Friction Losses

	Round Duct Size†						
Flow Rate*	**4**	**5**	**6**	**8**	**10**	**12**	**16**
100	0.65‡	0.21‡	0.09‡	0.02‡	0.01‡		
200		0.08	0.32	0.08	0.02	0.01	
400			1.19	0.28	0.09	0.04	0.01
800					0.34	0.14	0.03
1600							0.12

* flow rate (in cfm)
† round duct diameter (in in.)
‡ pressure loss per 100′ (in in. WC)

	Rectangular Duct Area†						
Flow Rate*	**12.5**	**20**	**28**	**50**	**79**	**115**	**200**
100	0.65‡	0.21‡	0.09‡	0.02‡	0.01‡		
200		0.08	0.32	0.08	0.02	0.01	
400			1.19	0.28	0.09	0.04	0.01
800					0.34	0.14	0.03
1600							0.12

* flow rate (in cfm)
† rectangular duct (in in²)
‡ pressure loss per 100′ (in in. WC)

Figure 18-14. Friction loss is the difference in static pressure between two points in a duct and is based on a standard pressure drop per 100′ of ductwork.

Design static pressure drop is the pressure drop per unit length of duct for a given size of duct at a particular flow rate. Ductwork is designed based on a standard pressure drop per 100′ of ductwork. Since most lengths of ductwork are not exactly 100′, the length may be divided by 100′. To find static pressure drop using the design static pressure drop, apply the following formula:

$$P_D = \frac{L}{100} \times D_{PD}$$

where

P_D = static pressure drop (in in. WC)
L = length of duct section (in ft)
D_{PD} = design static pressure drop per 100″ (in in. WC)
100 = constant

Example: Calculating Duct Static Pressure Drop

A duct 150′ long has a design static pressure drop of 0.08″ WC. What is the static pressure drop of the indicated duct segment? **See Figure 18-15.**

$$P_D = \frac{L}{100} \times D_{PD}$$

$$P_D = \frac{150}{100} \times 0.08$$

$$P_D = 1.5 \times 0.08$$

$$P_D = \mathbf{0.12'' \ WC}$$

Equal Friction and Duct Friction Loss Charts

Manufacturers and distributors provide equal friction and duct friction loss charts to HVAC technicians. Equal friction and duct friction loss charts are used to size ducts when two variables are known. **See Figure 18-16.** Equal friction charts are often used to estimate the sizes and prices of distribution systems for quotes. They can also be used by field technicians to troubleshoot system problems due to inadequate sizing or poor installation. In addition to the straight sections of ductwork, there are pressure drops in individual fittings and coils in the ducts. These pressure drops are converted to equivalent feet of ductwork. In these equal friction charts, duct fittings have the same resistance to flow as specific lengths of straight duct. **See Procedure for Determining Duct Size.**

Duct Friction Loss Charts

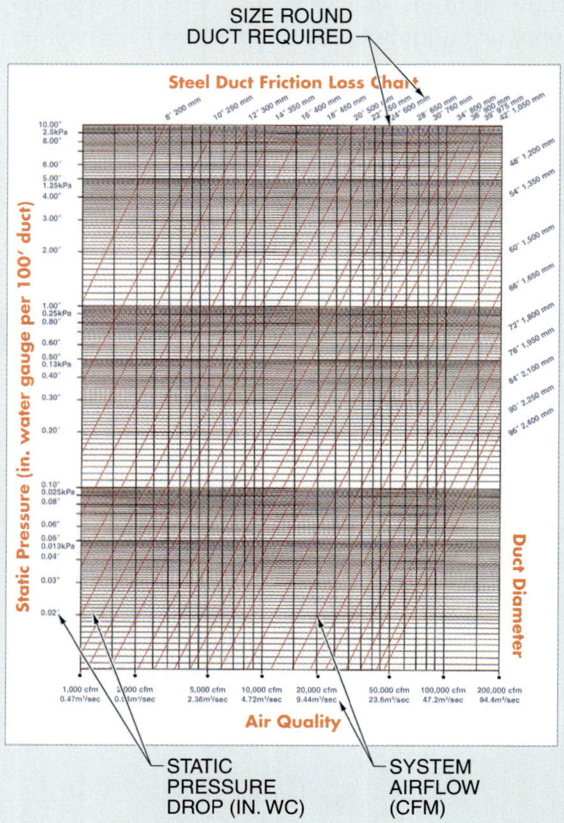

Figure 18-16. On equal friction and duct friction loss charts, duct fittings have the same resistance to flow as a specific length of straight duct, which allows pressure drops to be converted to equivalent feet of ductwork.

Calculating Duct Segment Static Pressure Drop

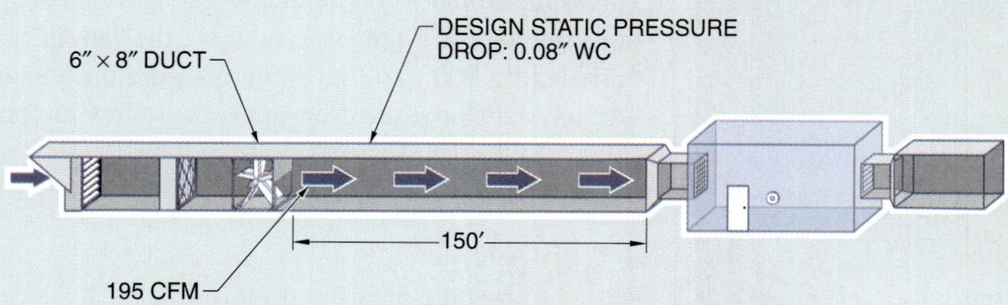

$$P_D = \frac{L}{100} \times D_{PD}$$

$$P_D = \frac{150}{100} \times 0.08$$

$$P_D = 1.5 \times 0.08$$

$$P_D = \mathbf{0.12'' \ WC}$$

Figure 18-15. As air moves through a duct, it comes into contact with the walls of the duct, which causes a loss of energy (pressure drop) due to the friction.

Procedure

Determining Duct Sizes

1. Turn OFF all electrical power to the furnace and air conditioning system at the main fuse or circuit breaker panel.

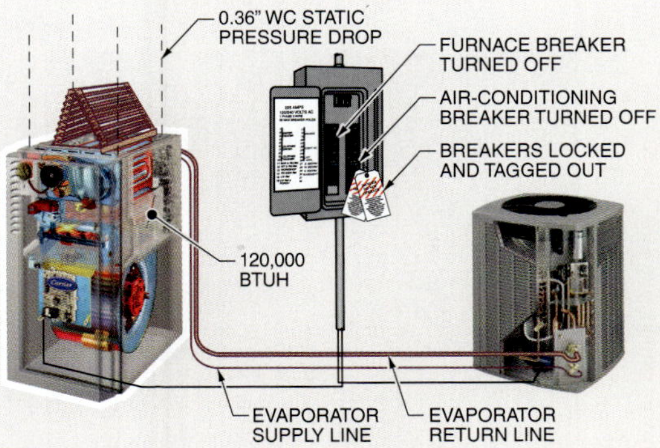

2. Follow all electrical and hazardous environment safety practices and procedures.
3. Close the main gas cock or gas valve at the furnace.

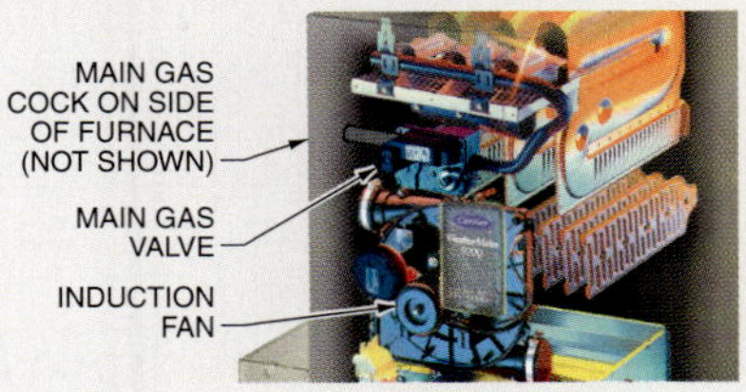

4. Find the Btu rating of the furnace using the furnace specifications (manual).

Specifications: Model: 58MXB Furnace

Performance	
Efficiency	From 92% up to 95.5% AFUE
Limited Warranty	10-year parts limited warranty when properly registered; Lifetime limited warranty on heat exchanger
ENERGY STAR®	
Details	As an ENERGY STAR partner, Carrier Corporation has determined that the Comfort™ 92 gas furnace meets the ENERGY STAR guidelines for energy efficiency.
Controls	
Controls Supported	Thermidistat™, Thermostat
Diagnostics	On Board
Technology	
Capacity	60,000 - 120,000 Btu
Blower Motor	Single Speed
Primary Heat Exchanger	Serpentuff™ polypropylene-coated
Fuel	Natural Gas; convertible to liquid propane (LP)
Aesthetics	
Profile	2 Stage Multipoise
Construction	Galvanized steel
Paint	Baked on urethane base coat and polyester top coat

Back to Top

5. Divide the Btu rating by 10,000.

$$\frac{120{,}000}{10{,}000} = 12$$

6. Determine whether the furnace is a natural draft, induced draft, or condensing gas furnace. Multiply the result from Step 5 by 100 for a natural draft furnace, 130 for an induced draft furnace, or 150 for a condensing furnace to determine the airflow rate.

$12 \times 130 = 1560$ cfm

7. Locate the calculated airflow (in cfm) on the bottom of the chart, and trace a line from that point up.

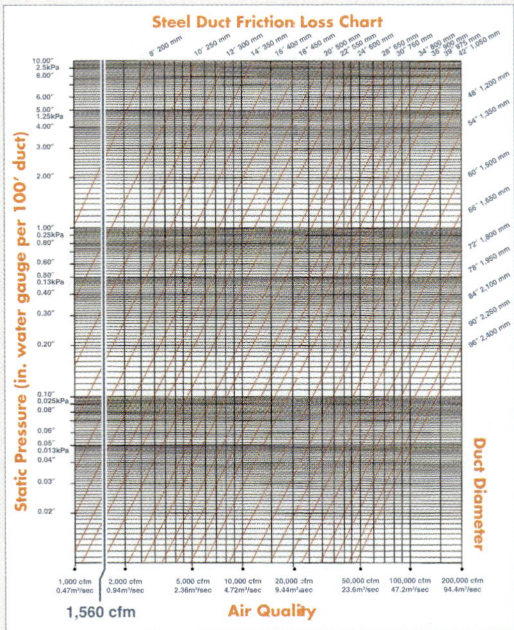

8. Locate the static pressure drop on the left side of the chart. Then, draw a line from that point to the right until the vertical cfm line is met.

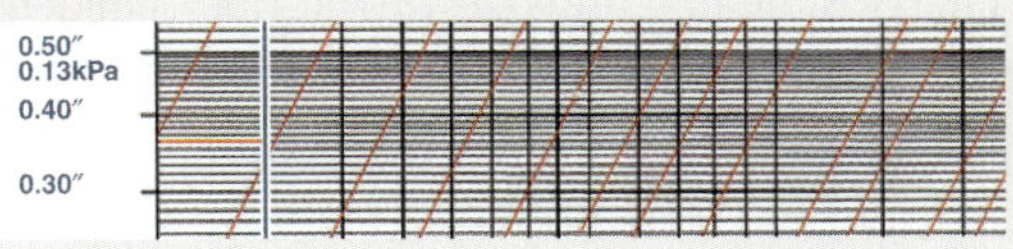

9. From the intersection point, follow the red diagonal line up and to the right, to the edge of the chart, to determine the size duct required for the condition.

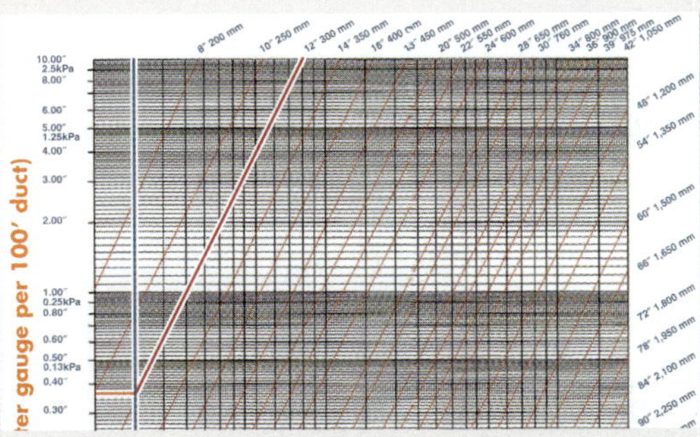

Note: Duct size is provided as a diameter for round ducts. Rectangular duct size must be determined by using additional area tables that are typically supplied by a distributor of rectangular duct parts and fittings.

Example: Calculating Duct Size

A 154,000 Btuh condensing gas furnace is being installed in a residence. There is one main duct connected to the supply plenum to distribute the air. What size should the main duct be when designed with a 0.42″ WC static pressure drop? **See Figure 18-17.**

1. Find the Btu rating of the furnace.

   154,000 Btuh

2. Divide the Btu rating by 10,000.

   154,000 ÷ 10,000 = **15.4 factor**

3. Determine whether the furnace is a natural draft, induced draft, or condensing gas furnace. Multiply the result in step 2 by 100 for a natural draft furnace, 130 for an induced draft furnace, or 150 for a condensing furnace to determine the airflow rate (in cfm).

   condensing furnace: 15.4 × 150 = **2310 cfm**

4. Locate the furnace airflow (in cfm) on the bottom of the chart, and trace a line from that point up.

   2310 cfm ≈ **2500**

5. Locate the design static pressure drop on the left-hand side of the chart and draw another line from that point to the right.

   0.40 ≈ **0.42**

6. The intersection (upper right-hand side of the chart) indicates the size of the duct needed for these conditions.

   14″ diameter – round duct

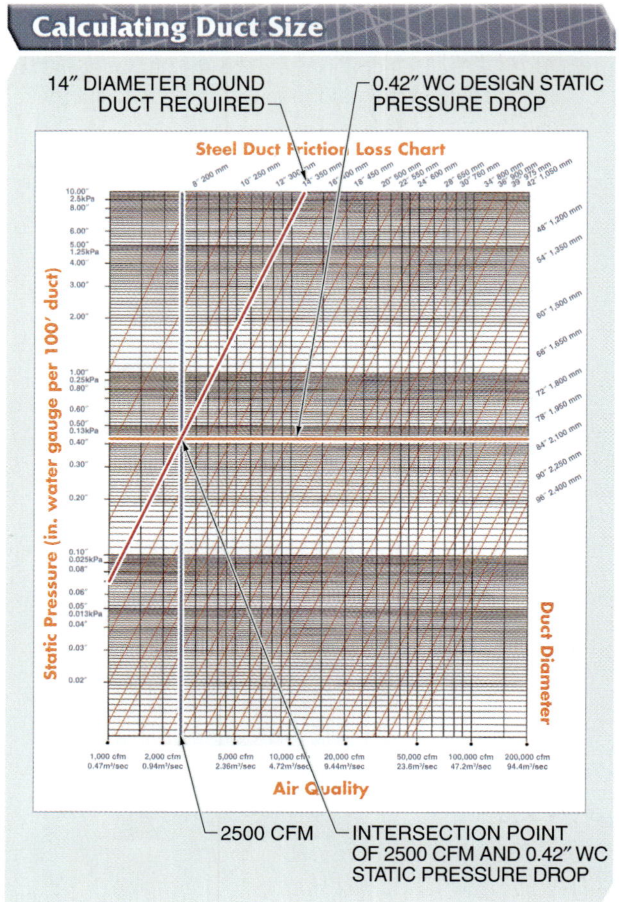

Figure 18-17. To find the size of duct required for an application, the trunk and/or branch airflow rates must be known.

System Pressure Drops

The system pressure drop in a forced-air HVAC system is known as the total static pressure drop. The total static pressure drop is the sum of all pressures associated with the HVAC unit and distribution system. This includes all the ductwork, fittings, transitions, filters, and the HVAC unit.

TECH TIP

System pressures and ductwork design are often checked as part of the commissioning process. When an existing building is commissioned, it is common to find a number of problems. These problems may include low airflow, temperature variations, excessive noise, or indoor air quality issues. These problems can significantly affect building occupants.

DUCT SYSTEM DESIGN

A forced-air distribution system is originally designed using a floor plan of the structure or building. **See Figure 18-18.** In the vast majority of cases, the location of the forced-air heating and distribution unit is fixed by the floor plans, and there is very little that the installation company can do about the design. In the case of retrofits, there may be more flexibility, especially when a little creativity is used.

Another factor involved in the design of duct systems is the local codes that restrict the duct location, type of unit, and location that the heating unit must be placed. After the location is determined, the duct system can be designed using individual heating and cooling loads for the spaces. Also, duct system installation is often controlled by local codes and national standards, which vary by geographic location. It should be ensured that furnace and duct installations follow all applicable state and local codes.

Forced-Air Distribution System Locations

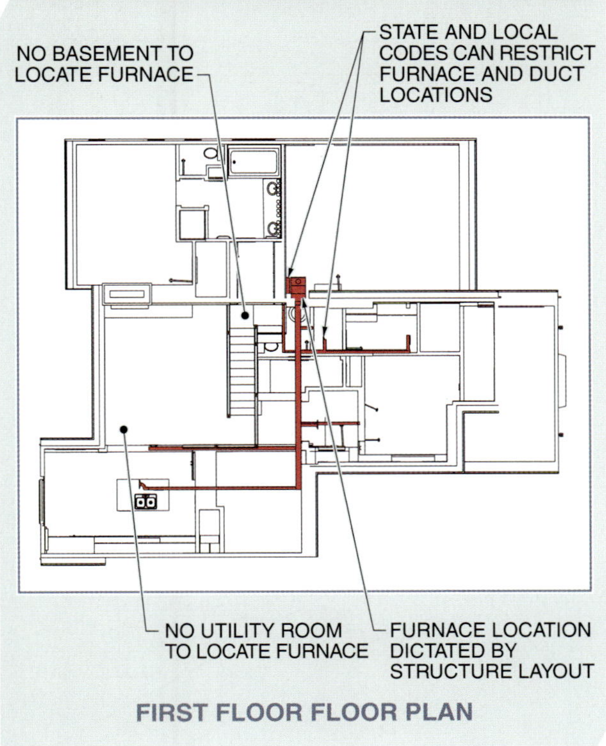

Figure 18-18. A forced-air distribution system is designed by using a floor plan of the structure or building. The installation of the system is often regulated by local codes and national standards.

Duct sections are often labeled to ensure their proper installation in a building.

Supply Duct Systems

The purpose of the supply duct system is to direct air from the forced-air heating unit to the individual building spaces. At the heating unit, the supply duct system may begin with a vibration eliminator. A *vibration eliminator* is a piece of flexible material placed between sheet metal collars used to isolate rigid ductwork from the vibrations of an HVAC unit. **See Figure 18-19.** Vibration eliminators are used to reduce the noise transmitted from heating units, through the ductwork, and to the rooms. They are especially useful in units within close proximity to living areas such as bedrooms where noise is a concern. However, the flexible member can fail, allowing leakage. Vibration eliminators are not used in every heating unit.

Duct Vibration Eliminators

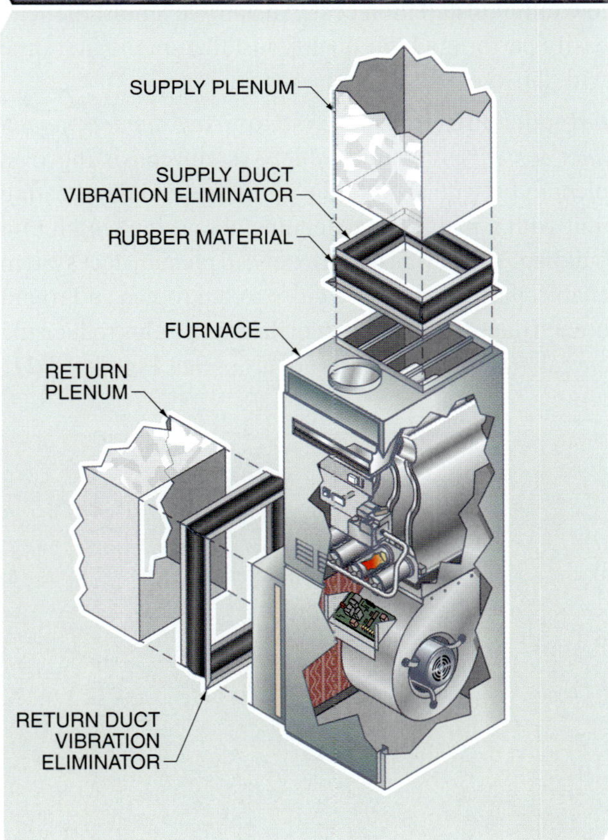

Figure 18-19. A duct vibration eliminator is a piece of flexible material between sheet metal collars that isolates the rigid ductwork from the vibrations of the HVAC unit. It reduces the noise level transmitted from the heating unit, through the ductwork, to building spaces.

Overall, there are a number of duct system designs available. Duct system designs include close plenum, extended plenum, reducing plenum, and perimeter loop system designs. The proper duct system design for a given application depends on various factors such as the building layout, available space, accessibility, and room construction.

Close Plenum Duct Systems. The easiest type of duct system to lay out and install is the close plenum duct system. A *close plenum duct system* is a supply duct system consisting of a plenum box that is attached to the discharge end of a heating unit with individual small ducts connected from the plenum box to each zone.

A close plenum duct system is ideal for a centrally located HVAC unit, with relatively short duct runs to each outlet. **See Figure 18-20.** This type of system does not work well when longer runs are needed for multiple zones, as the pressure drops are too high in the smaller ducts. This system is also good for installations with a low initial installation cost. An entry level installer can easily perform a high-quality installation the first time with this type of system.

Extended Plenum Duct Systems. An *extended plenum duct system* is a supply duct system consisting of a plenum box attached to the discharge end of a heating unit with a main duct extending across the length of a building. The design of the extended plenum duct system enables large ducts to be close to the rooms and round branch ducts to be small, which in turn helps reduce the overall system static pressure drop. **See Figure 18-21.**

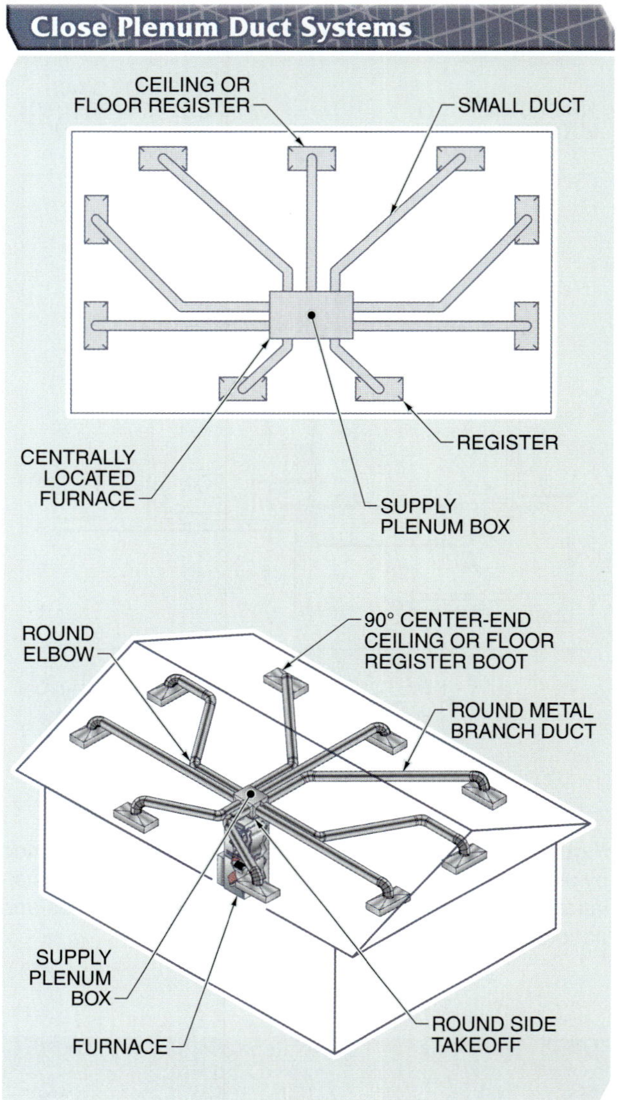

Figure 18-20. The close plenum duct system has a plenum box attached to the discharge end of the heating unit. Individual small ducts are connected directly from the plenum box to each zone.

Reducers are used to transition from one size of duct to another.

Reducing Plenum Duct Systems. The location of an HVAC unit is typically determined by the floor plan of the building. When the HVAC unit is located at one end of a building, a reducing plenum duct system is often used. A *reducing plenum duct system* is a supply duct system consisting of a plenum box attached to the discharge end of a heating unit with a plenum that gets progressively smaller and ducts that branch to individual rooms extending along the length of the main duct. **See Figure 18-22.**

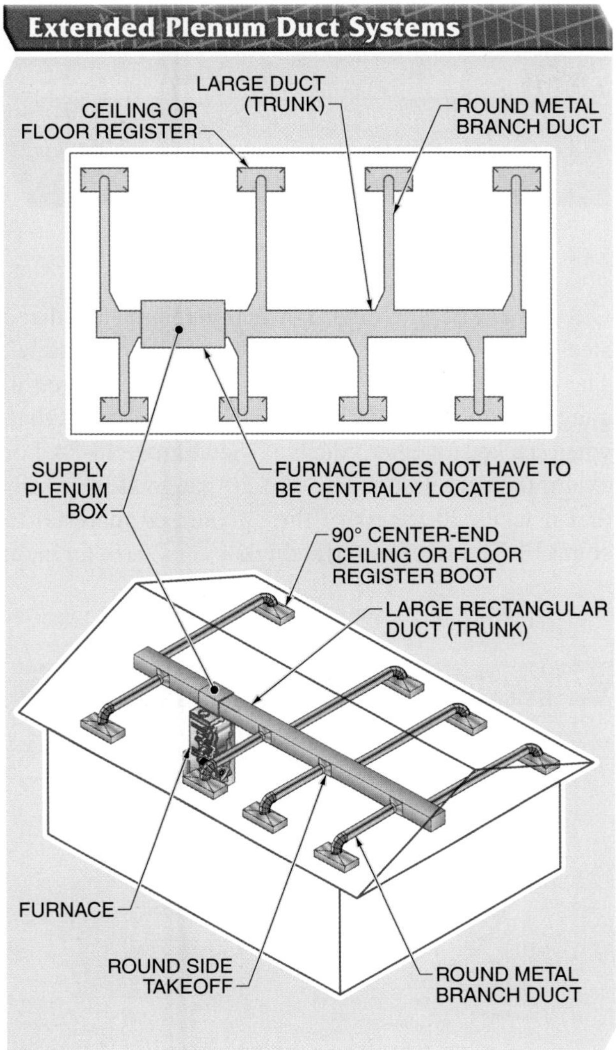

Figure 18-21. The extended plenum duct system has a plenum box attached to the discharge end of the heating unit. A section of main duct (trunk) extends across the length of the building to reduce overall system static pressure drop.

The reducing plenum duct system is designed with a duct that progressively reduces in size because air is removed progressively from the plenum. Each section of duct is sized to keep pressure the same throughout the duct system.

Perimeter Loop Duct Systems. For some types of construction, perimeter loop duct systems are appropriate. This is especially true for residential heating systems with slab construction. A *perimeter loop duct system* is a supply duct system consisting of a single duct loop that connects all outlets (registers) together and includes supply ducts. The perimeter loop duct system has multiple supply ducts that feed the loop at various spots. This helps to ensure that the duct pressure is the same everywhere in the distribution system and that the registers have enough air. Also, the most heat transfer takes place at the exterior of the building, so this is where the registers are located in most cases. **See Figure 18-23.**

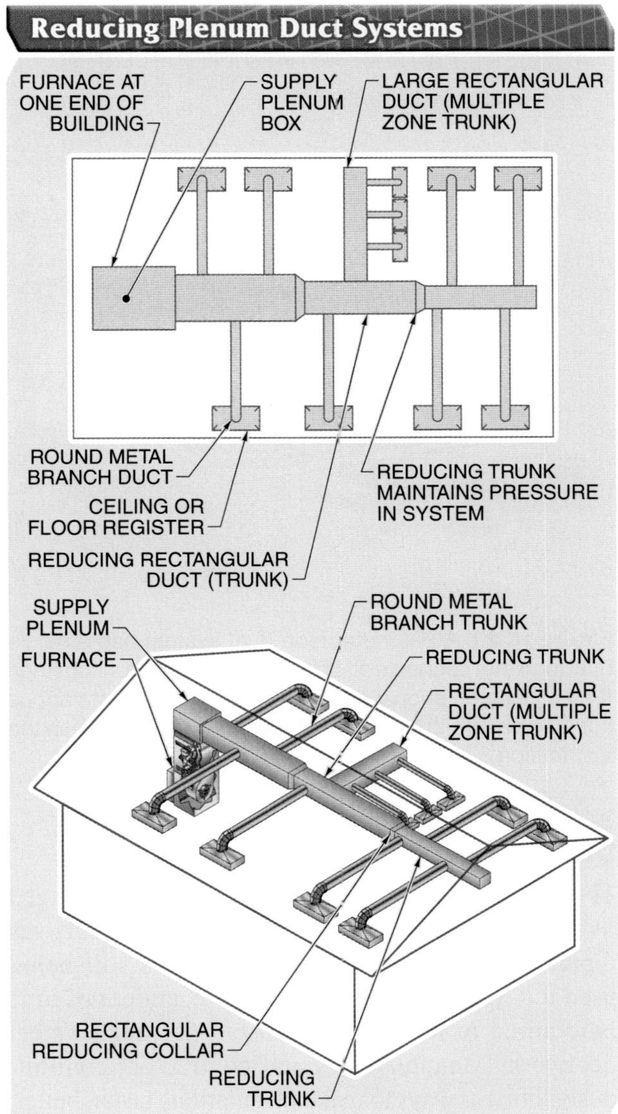

Figure 18-22. The reducing plenum duct system has a large plenum box attached to the discharge end of the heating unit. The plenum gets smaller as the trunk extends away from the heating unit and smaller ducts branch to rooms or zones.

Perimeter Loop Duct Systems

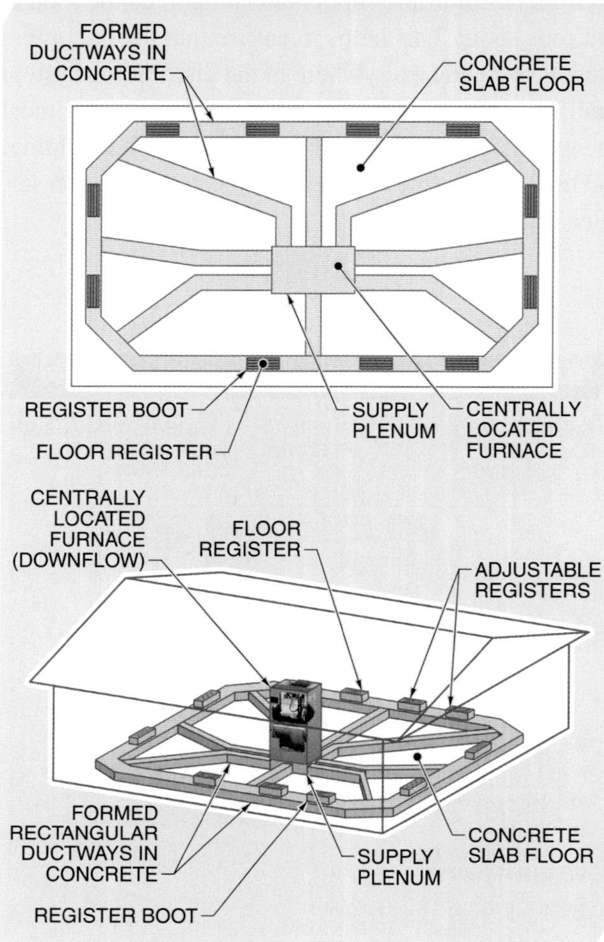

Figure 18-23. A perimeter loop duct system has a single duct loop that connects all of the outlets (registers) together. It also includes supply ducts that feed the loop at various spots to ensure the same duct pressure everywhere in the distribution system.

Duct Materials

The materials used to construct ducts have changed over the years. In the past, galvanized steel was typically used to construct ducts and is still being used today. However, a considerable amount of time is required to fabricate and install galvanized steel ductwork. Manufacturers and contractors continuously look for ways to make installations faster, better, and more economical. One way to improve ductwork installation is to reduce the time needed to fabricate the ductwork. Today, many installation contractors use fiberglass duct board and flexible (flex) duct as substitutes for galvanized steel.

TECH TIP

One consideration in ductwork design and selection is the existing layout of a building. For example, the ease of access to required building areas may determine the location of the ductwork.

Galvanized Steel Ducts. For many years, galvanized steel duct was preferred when choosing duct material. The thickness of galvanized steel duct is measured in gauge. *Gauge* is the number of sheet metal sheets that, when stacked together, equals 1″. **See Figure 18-24.** For example, a steel duct made of 20 gauge steel means that it takes 20 sheets of the galvanized material to equal 1″. The steel material thickness is 1/20 of an inch.

Duct Gauge

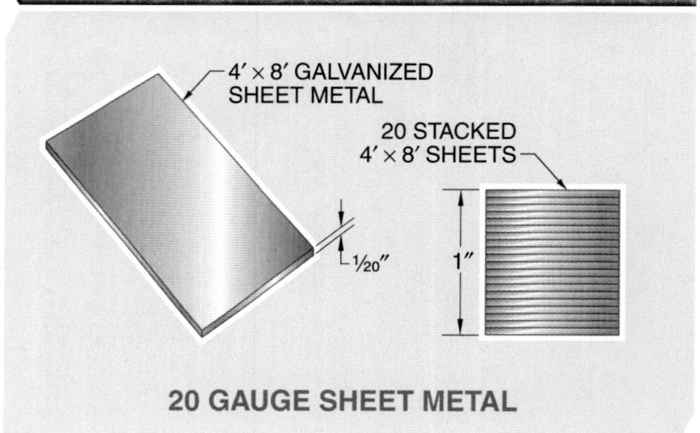

Figure 18-24. Gauge is the number of sheet metal sheets that equal 1″ when stacked together.

Galvanized steel duct can be purchased at any HVAC supply house, with 4′ straight lengths being the most common in the U.S. and 5′ most common in Canada. In addition to the main supply ducts, there are elbows, tees, and other fittings available in various sizes for the attachment of branch ducts, diffusers, and registers to main duct plenums. **See Figure 18-25. See Appendix.** When measurements are provided, many supply houses will fabricate the duct sections, which can then be transported to the job site.

Duct Fittings and Clips Media Clip

Common Galvanized Steel Duct Fittings

Figure 18-25. Galvanized steel duct can be purchased at any HVAC supply house. In addition, elbows, tees, and other fittings are available in various sizes to attach branch ducts, diffusers, and registers to main duct plenums.

Rectangular sheet metal ducts are connected using sheet metal screws, S-clips, and drive clips. **See Figure 18-26.** To prevent air leakage, fittings may be taped with duct or aluminum foil tape after the fasteners are attached. Although the use of galvanized steel duct has decreased in many areas due to the time needed to fabricate and install the duct system, galvanized steel duct is still being installed.

TECH TIP

Ductwork may separate and thus supply conditioned air into undesirable areas. This air leakage can reduce system efficiency and cause mechanical problems by forcing the unit to run longer to meet the same load. Ductwork must be connected and taped securely since it is often located in difficult to access areas.

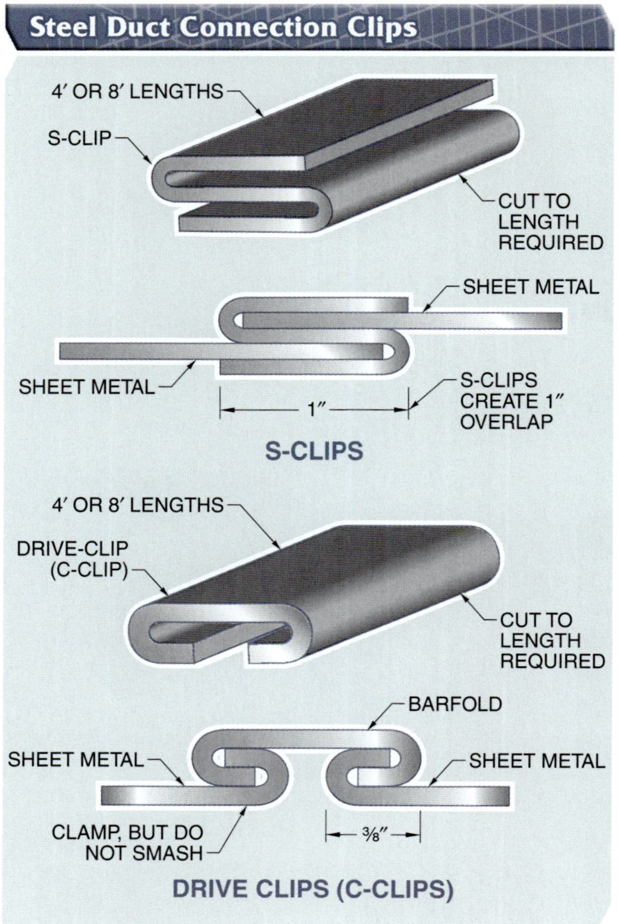

Figure 18-26. Rectangular sheet metal ducts are connected using sheet metal screws, S-clips, and drive clips to prevent air leakage. The fittings may be taped with duct or aluminum foil tape after the fasteners are attached.

Duct board can be used when angles and transitions from one size to another are required.

Fiberglass Ducts. Fiberglass duct (duct board) was introduced to reduce the time and effort required for duct fabrication while still allowing installations to look professional. Duct board is typically available as a flat sheet but is also available as round or rectangular prefabricated duct. The material has a foil backing and a thickness of 1″. The material is easily cut with a knife. Special tools are provided by the manufacturer to work with and shape the duct in different ways as needed. The foil backing is left after cutting so that the sections can be easily joined together. Some large contractors have machines that can quickly cut and shape duct board and in large sections. Therefore, the installation cost of duct board is reduced at the job site. **See Figure 18-27.**

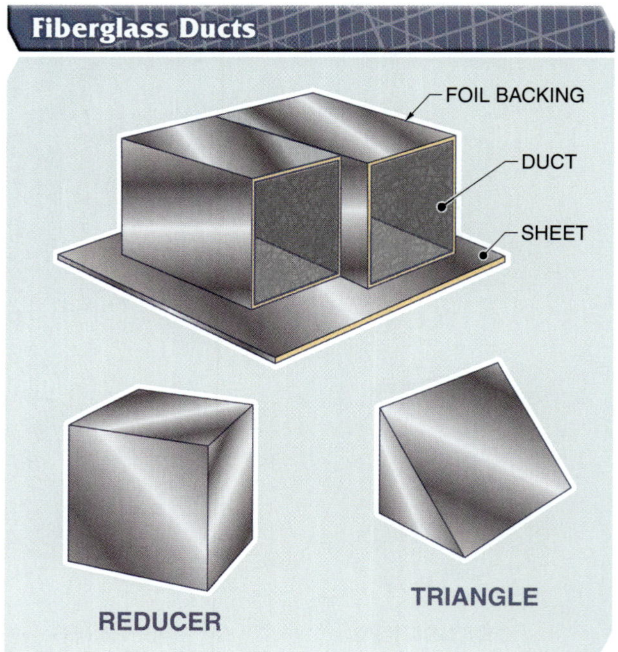

Figure 18-27. Fiberglass duct (duct board) comes in flat sheets but can also be found in round and rectangular prefabricated duct that is 1″ thick with a foil backing. The foil backing is left after cutting so that the sections can be easily joined together.

Spiral Metal Ducts. Spiral metal duct is used in large commercial systems. Flat metal from rolls are fed into a special forming machine, which turns the metal into a spiral form. Installation of this type of ductwork is performed by specially trained contractor personnel. One advantage of spiral metal duct is that it fits in well with many types of interior decor. Spiral metal duct may be exposed to view in restaurants, atriums, and other venues with open ceiling construction. **See Figure 18-28.**

Spiral Metal Ducts and Fittings

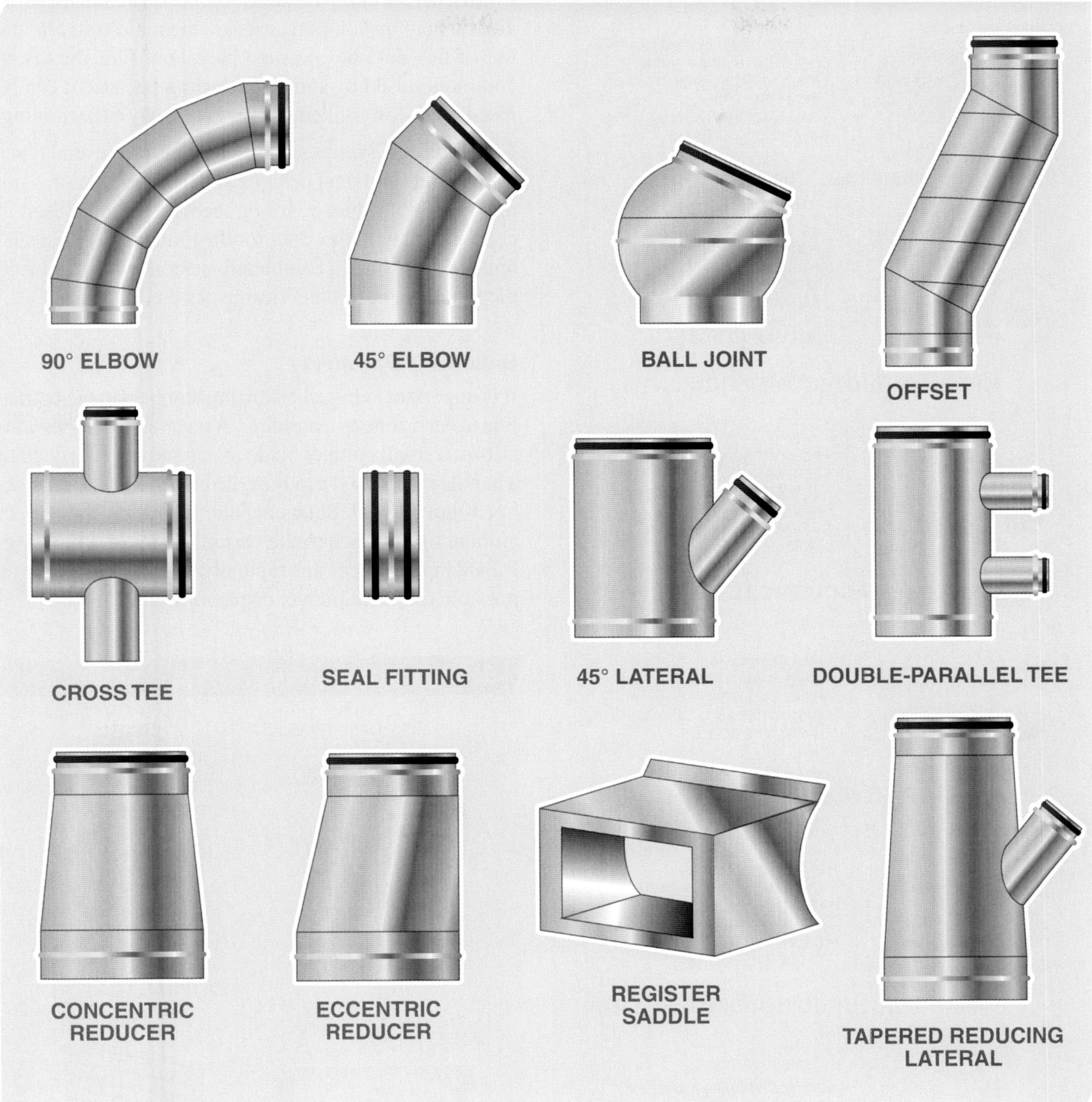

Figure 18-28. Spiral metal duct is ductwork that is formed by special forming machines and contracting personnel. It is typically exposed to view in restaurants, atriums, and other venues with an open ceiling construction.

Flexible Ducts. Flexible (flex) duct is available from the manufacturer compressed in a box. Flex duct is typically insulated and comes in different diameters and lengths, with 25′ and 50′ being the most common lengths. Flex duct can be routed easily around obstructions. One disadvantage of flex duct is that it has a very high resistance to flow. It must be kept to the minimum length possible for the application. **See Figure 18-29.**

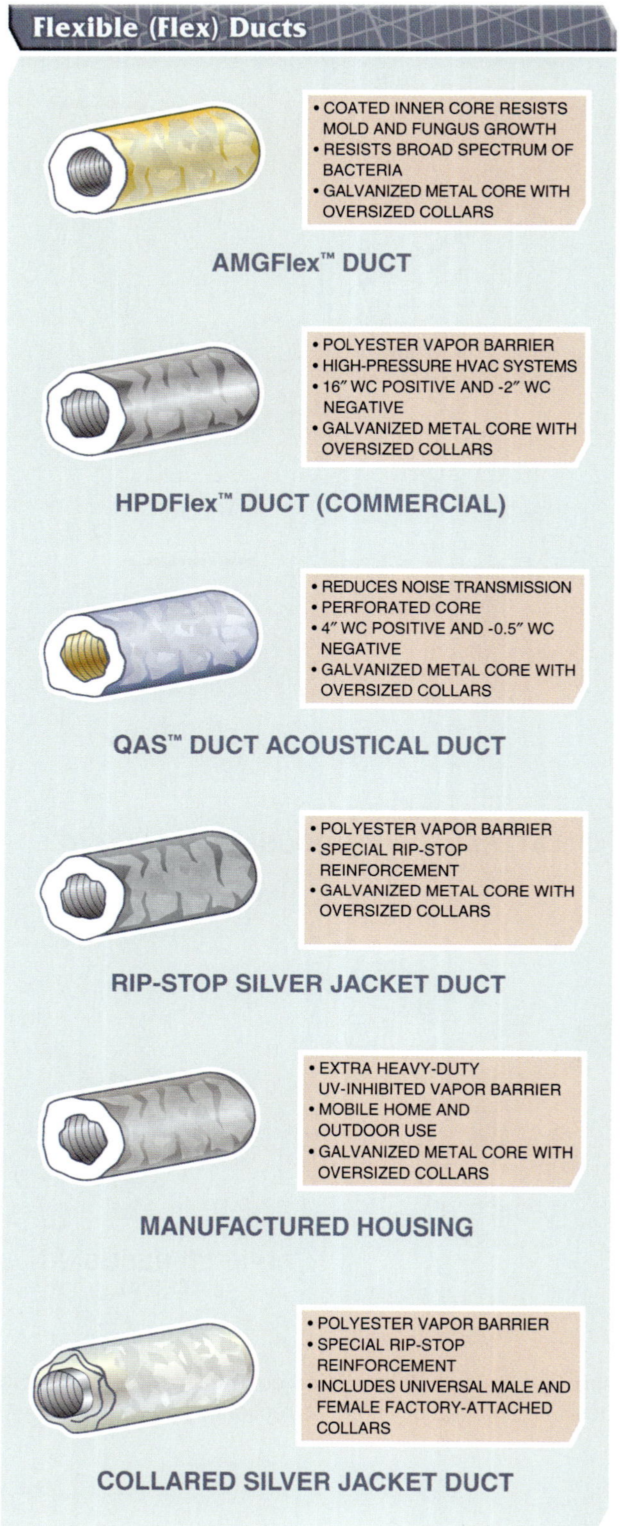

Figure 18-29. Flexible duct is typically insulated, comes in various diameters and lengths, and is easily routed around obstructions. The main disadvantage is that it has a very high resistance to flow and must be used in the shortest lengths possible.

While flex duct can be useful, unfortunately it is overused, especially in new home construction. In some areas where installation costs are the main concern, the use of flex duct has become prevalent. Flex duct is so commonly used by some contractors because it can be installed by low-skilled workers with very little training.

Combination Systems. Many systems use several types of materials and duct designs at the same time. It is not uncommon for duct board or sheet metal to be used in conjunction with flex duct for the branches. A duct system may also have a combination extended and reduced plenum design, or other designs as well.

Balancing Dampers

It is important to install balancing dampers in the branch line to each zone of a building. A balancing damper consists of a small damper blade inside the duct attached to a handle that is used to adjust the blade from the outside. **See Figure 18-30.** Balancing dampers allow the correct amount of airflow into the various zones of a building. Balancing dampers are typically installed as close as possible to the main duct or plenum.

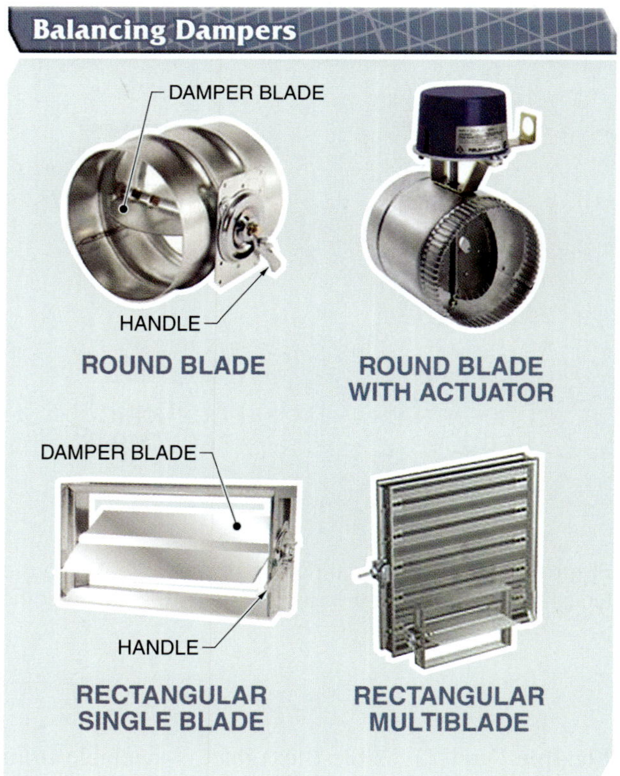

Figure 18-30. A balancing damper installed in the branch line to each zone consists of a small damper blade attached to a handle that is used to adjust the blade.

It is important that balancing dampers be oriented in such a manner that the position of the handle can be easily seen and the damper easily reached. Some technicians mark the final position of the handle with a permanent marker so that if the position of the damper is accidentally changed, it can be returned to the correct position at a later time.

Duct Insulation

As air moves through a duct, heat transfers between the duct and the surrounding environment. Ductwork is often routed through an unconditioned space such as an attic, crawl space, or garage. **See Figure 18-31.** In such locations, the duct and dampers should be insulated using recommended "R" values. Insulation is included with duct board and flex duct. With galvanized steel duct, insulation must be added to the duct and then attached with tape and/or staples.

Registers and Grills

Registers and grills are part of the forced-air distribution system in a building. Registers are located on the supply side of the system and distribute heated air to building spaces. Grills are located on the return side of the system and return air from building spaces back to the furnace.

A *register* is a device that covers the opening of supply air ductwork and consists of a panel with vanes, a frame for mounting, and a damper for controlling airflow. **See Figure 18-32.** The airflow pattern from the register is controlled by the vanes on the panel. Registers throw air straight out from the register and/or spread air evenly in all directions. Registers are sized according to the quantity and velocity of air required in each building space. Registers are located for efficient distribution of air to each building space.

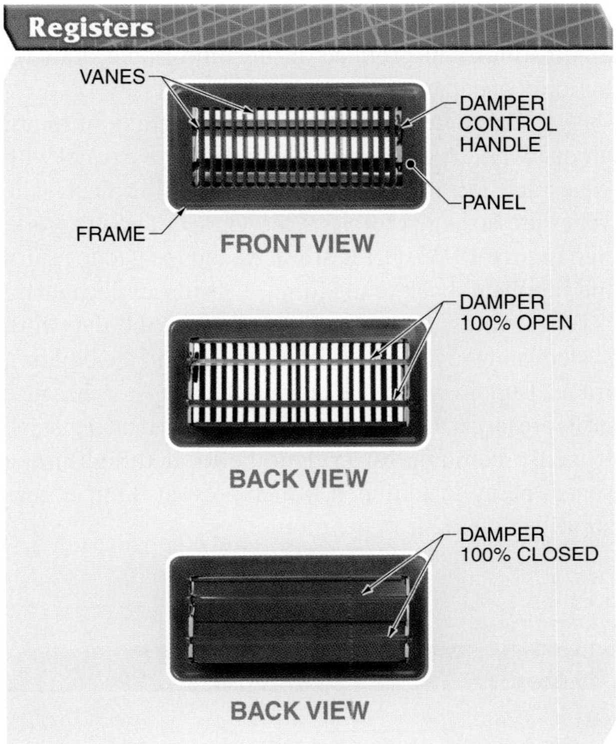

Figure 18-32. A register covers a supply ductwork opening and consists of a panel with vanes, frame for mounting, and damper for controlling the amount of airflow.

Recommended Duct Insulation "R" Values

Climate	Heating System*	R-Values for Unconditioned Attics	R-Values for Unconditioned Basements or Crawl spaces
Warm—cooling and minimal heating (FL, HI, coastal inland CA, southeast TX, and southern LA, AR, MS, AL, and GA)	Gas/oil, electric resistance, or heat pump	R-4 to R-8	None to R-4
Mixed—moderate cooling and heating (VA, WV, KY, MO, NE, OK, OR, WA, ID, southern IN, KS, NM, AZ, northern LA, AR, MS, AL, GA, and inland CA and NV)	Gas/oil, electric resistance, or heat pump	R-4 to R-8	R-2 to R-8
Cold—PA, NY, New England, northern Midwest, Great Lakes area, and mountainous CO, WY, and UT	Gas/oil, electric resistance, or heat pump	R-6 to R-11	R-2 to R-11

* assumes electric air conditioning

Figure 18-31. Ductwork routed through an unconditioned space, such as an attic, crawl space, or garage, should be insulated using recommended "R" values for the region.

The airflow pattern determines the area of influence of the air from the register. Registers with unusual airflow patterns are known as diffusers. An *area of influence* is the area from the front of a register or diffuser to the point where air velocity drops below 50 fpm. Air velocity above 40 fpm for residential applications and 160 fpm to 1200 fpm for various light commercial applications can produce an uncomfortable draft on building occupants.

A *grill* is a device that covers the opening of return air ductwork. A grill consists of a decorative panel with vanes and a frame that holds the panel in place. The vanes are arranged to block the view to the ductwork. **See Figure 18-33.** Grills are sized and located to return the supply air back to the furnace. Grills are located on ceilings, walls, or floors, depending on the ductwork system and the location of the ductwork in the building frame. Buildings have fewer grills than registers because grills are larger and more centrally located than registers. It is also common for grills to be located high in the spaces being conditioned, which prevents temperature stratification of air in the rooms.

Grills

Figure 18-33. Grills, consisting of a decorative panel, vanes, and frame, are located on ceilings, walls, or floors and cover the openings of return air ductwork.

Blending Conditioned Air with Room Air

Supply duct systems send conditioned air into rooms. When the air arrives, it must be distributed properly throughout the space. Hot or cold spots should be avoided. Also, air should not be discharged directly onto the occupants in the rooms. Ideally, the discharge air from the registers will blend (mix) with the room air in a consistent and predictable pattern. Typically, it is desired that the air be discharged close to the outside walls, where the load is the greatest. **See Figure 18-34.**

Blending Conditioned Air with Room Air

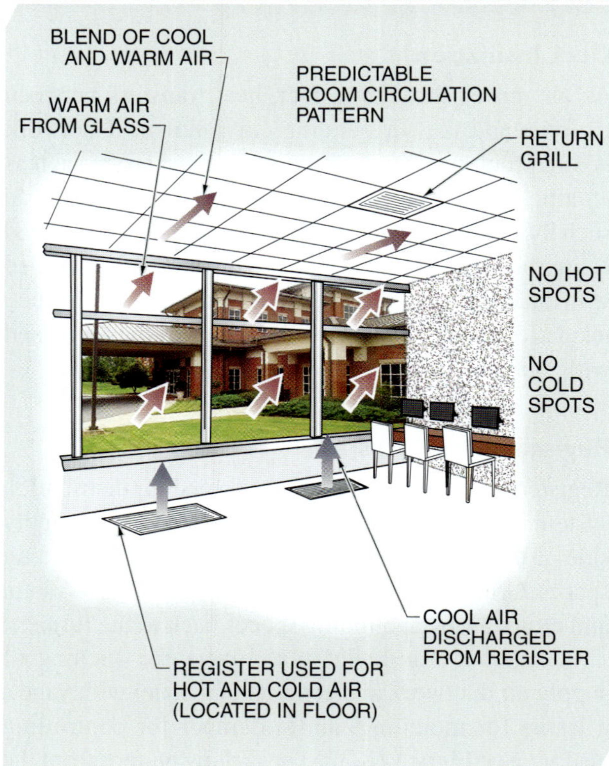

Figure 18-34. When air arrives in a building space, it must be distributed in a consistent, predictable pattern throughout the space to avoid hot or cold spots or from discharging directly onto occupants.

The amount of air discharged from a register or diffuser depends on the air pressure at the discharge and the throw of the diffuser. *Throw* is the distance discharged air flows from a register or diffuser. The throw of a diffuser is the distance from the diffuser where the air velocity decreases below 50 fpm. Different register and diffuser designs are available with various throw characteristics. **See Figure 18-35.** Registers and diffusers also deliver air in specific patterns. These patterns are known as spread. *Spread* is the left to right pattern and distance from a unit where air velocity decreases below design fpm. The spread can be created by using fixed or movable vanes that direct the airflow pattern.

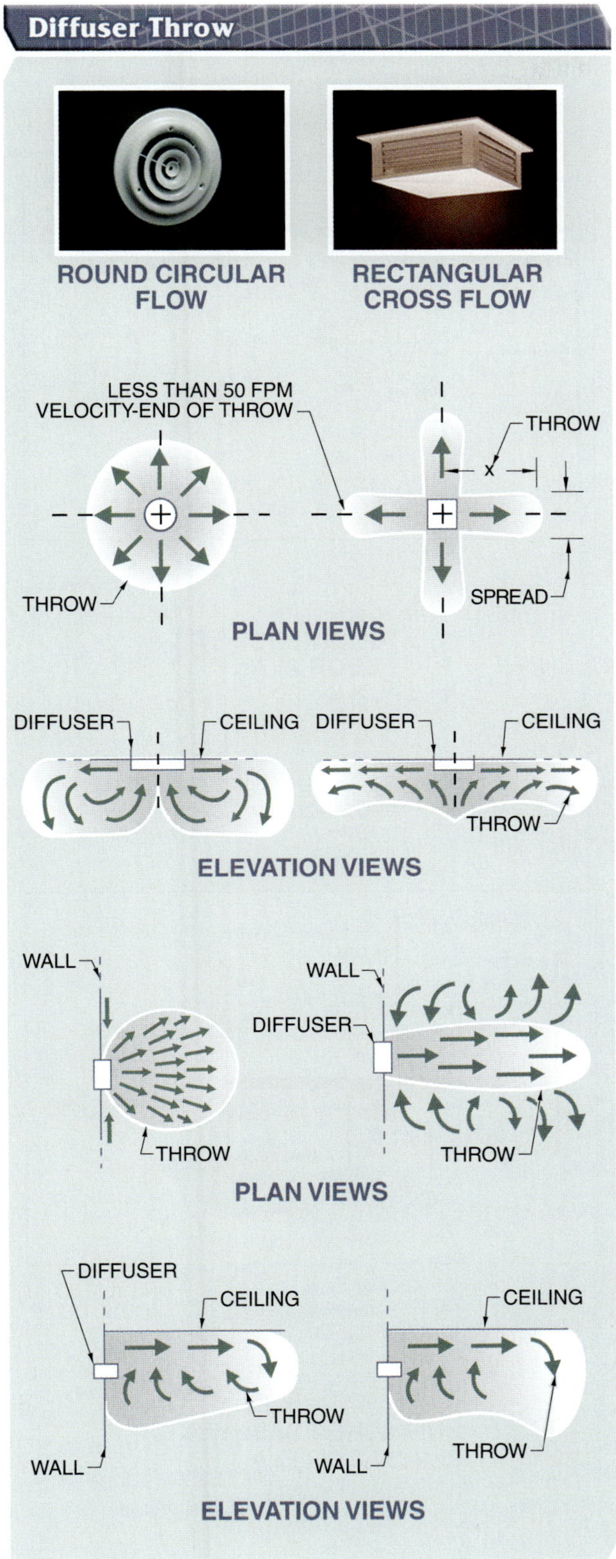

Figure 18-35. Register and diffuser throw is the distance from the unit discharged air flows while keeping air velocity above 50 fpm.

Return Duct Systems

Once supply air is discharged into a room, it mixes with room air and provides a comfortable environment for the occupants. The air is then returned to the furnace to be heated again. A return duct system directs the air back to the heating unit.

The most common type of return duct system is the open return system. **See Figure 18-36.** In an open return system, the return air opening is not connected to a duct but is simply left open to a common space or area. Large return grills are located at different areas in the building. Room air is then pulled through the return grills and into the heating unit.

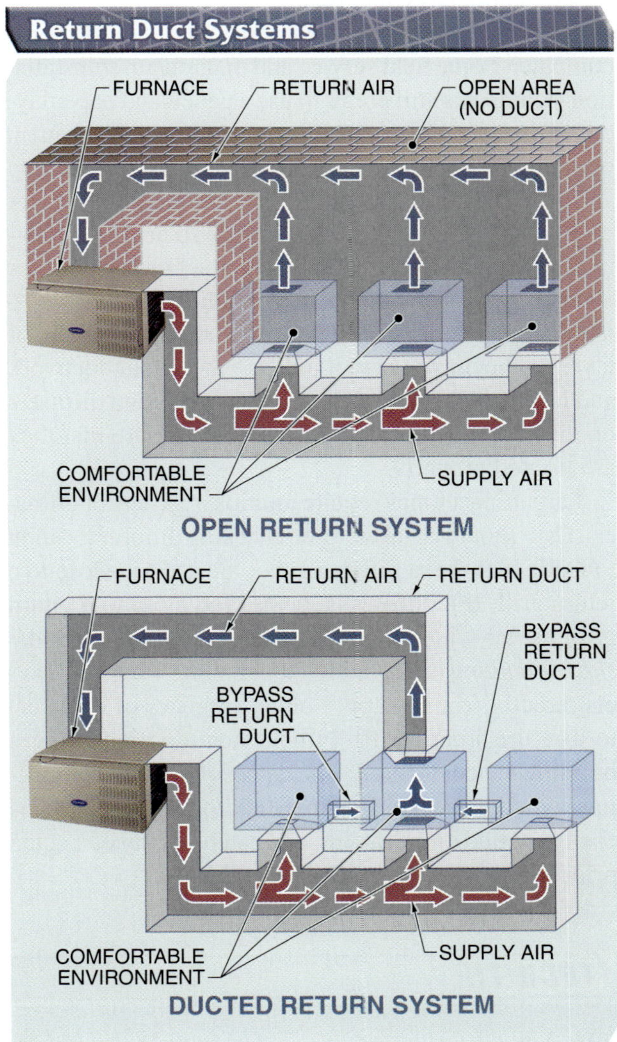

Figure 18-36. The two types of return duct systems are the open return system, where the return air opening is not connected to a duct but is simply left open, and the fully ducted return system, which is similar to the supply duct system.

Some systems have fully ducted return systems. In this case, an entire duct system is built that is similar to the supply duct. This doubles the costs of the duct system. However, fully ducted return systems are used, especially in residential and light commercial high-efficiency systems.

Duct System Layout

Duct system layout begins with a floor plan of the building. The location of the HVAC equipment is typically predetermined, but when it is not, it is typically located in a spot that allows for easy installation of the duct system with short runs. This keeps the installation cost of the duct system low and allows for easy access for future maintenance and repair.

Duct layout is most often performed on a personal computer. Some field service and manufacturer installation technicians still prefer to use a pencil and paper layout of the duct system laid over an unfinished basement or floor plan. The drawing will also indicate the heating and cooling desired and the airflow rate requirements for each space. **See Figure 18-37.** This will help in achieving the proper duct layout. Another set of drawings or another layer on the drawing will indicate floors, walls, doors, and other fixtures. The next set of drawings or layer on the drawings will indicate the actual ductwork and fittings. The placement of return grills and diffusers on the drawings should be checked to ensure that they are located properly.

Larger spaces may require multiple registers or diffusers. Care should be taken when checking diffuser location to make sure that a diffuser is not located too close to a return grill. If a diffuser is located too close to a return grill, the air will "short circuit." **See Figure 18-38.** *Air short circuiting* is a condition that results when air travels directly from a supply outlet (register or diffuser) to a return inlet (grill) without heating or cooling a building space or room. The cause of air short circuiting, which is incorrect diffuser and/or grill location, is easily spotted on a drawing and can be easily corrected prior to installation.

TECH TIP

The actual location of registers and diffusers may be different from those on the layout drawings. Often, changes are made because of ease of construction or unplanned obstructions.

Duct System Layout

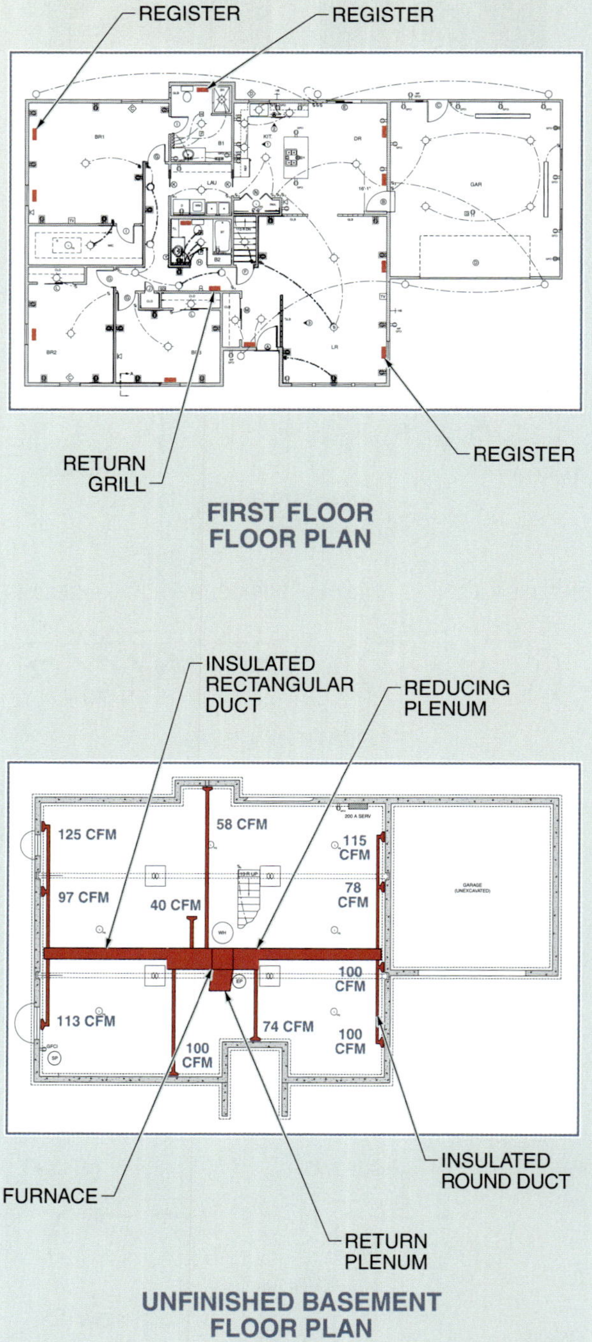

Figure 18-37. A duct system drawing indicates the building floors, walls, and doors as well as the distribution system ductwork, fittings, heating and cooling desired, and airflow requirements for each space.

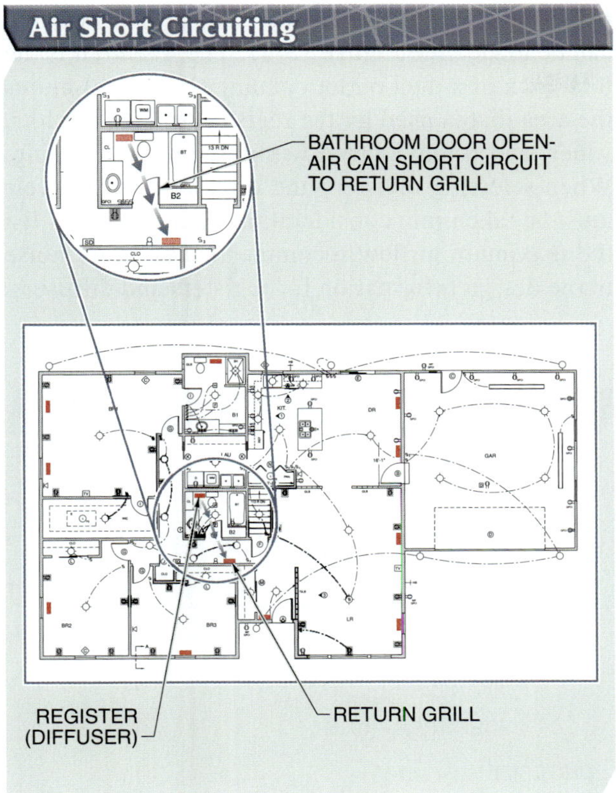

Figure 18-38. Air short circuiting is a condition in which air travels directly from the supply outlet (register or diffuser) to the return inlet (grill) without heating or cooling the building space or room.

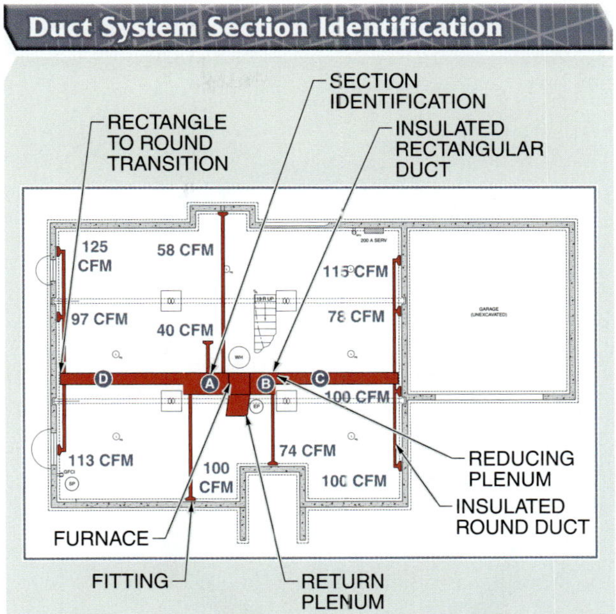

Section	Size Duct*
A	8 × 12
B	8 × 11
C	8 × 10
D	8 × 9

* in in.

Figure 18-39. Each section of a duct distribution system must be shown on an HVAC drawing and sized with a common convention to show a section as "A" and then another section as "B" etc.

After the duct design is chosen, the ductwork is sized and each section is shown on the drawing, with the sizes provided. Fittings and transitions are shown on the drawing as well. A common convention is to show the closest section to the heating unit as "A". The next section would be labeled "B". **See Figure 18-39.** Especially important on the drawings are where walls, ceilings, and floors are penetrated. One of the most important walls to note are the load bearing walls, as they cannot be penetrated by ductwork. Fire code requirements may also apply to the ductwork installation since smoke and fire functions are affected by the location of the ductwork.

Duct Sizing

Duct design must provide enough airflow to satisfy the heating and cooling loads in a space. The design and layout of the duct system determines the airflow into the space or room, and therefore the comfort in that area.

In the past, duct sizing was performed with paper and pencil or entered on a form. Later, a slide rule device called a Ductulator™ enabled quick accurate calculations. **See Figure 18-40.** Even now, many technicians use the Ductulator™ as a trusted tool for quick field measurements. However, personal computers with duct sizing applications have taken precedence. The main duct sizing methods include equal friction, velocity reduction, and static regain sizing.

Equal Friction. One of the most widely used methods for sizing residential and small commercial duct systems is equal friction sizing. The *equal friction method* is a method of sizing ductwork that sets static pressure approximately equal at each branch takeoff in a distribution system. An equal friction chart is used to calculate each branch based on the fittings and duct design. A reliable static pressure drop to be used for small residential and light commercial systems is 0.06″ WC to 0.1″ WC.

Ductulators™

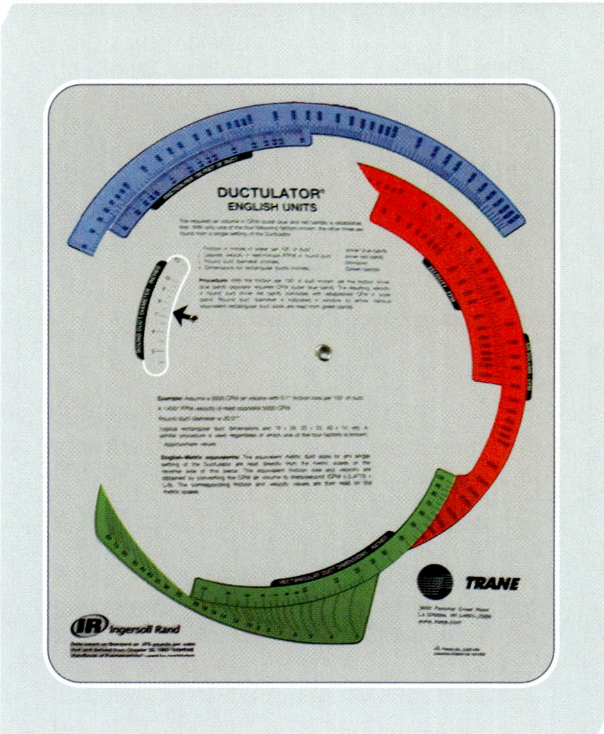

Figure 18-40. For years, the Ductulator™ enabled quick and accurate calculations of duct sizing by technicians, but the personal computer with a duct sizing application has now taken precedence.

Velocity Reduction. This method uses the velocity of the air instead of the static pressure drop. The *velocity reduction method* is a method of sizing ducts that causes air velocity to be reduced at each branch takeoff. Each velocity drop is calculated separately.

Static Regain. The static regain method is used for larger commercial duct systems. The *static regain method* is a method of duct sizing in which each duct section is sized so that the static pressure increase at each takeoff offsets the friction loss in the preceding duct section. This method will cause the same approximate static pressure at each takeoff.

Register, Diffuser, and Grill Sizing

Register, diffuser, and grill sizing for a forced-air distribution system is vital for allowing the proper airflow to be distributed to each building space. To size a register or diffuser, the required airflow and velocity of air must be known. When determining whether to use a register or diffuser, it is important to know the free area of both components so a proper decision can be made. **See Figure 18-41.** The *free area* is the total area of a duct outlet opening (hole size) minus the area that is used by the register or diffuser slots, vanes, and frame located within the outlet opening. When selecting registers and diffusers, noise levels must be taken into consideration. Manufacturers list the maximum airflow recommended to avoid noise in the design information for registers and diffusers.

Free Areas

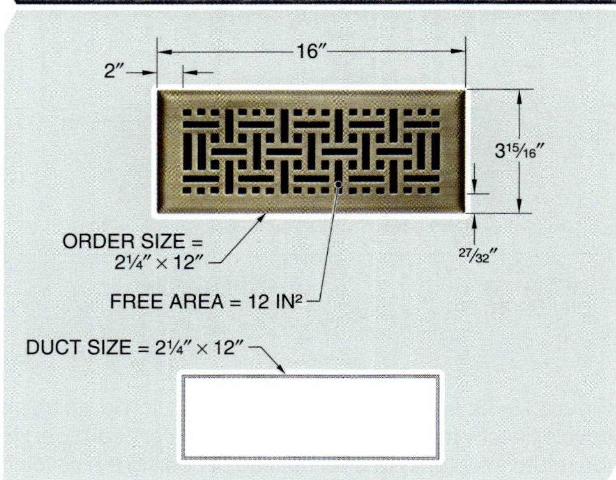

Figure 18-41. The free area of a register or diffuser is the total area of the duct outlet opening (hole size) minus the area that is used by the various slots, vanes, and frame located within the outlet opening.

Return grills must also be sized properly. It is desired that an airflow velocity of 500 cfm to 750 cfm be used as a selection and design parameter for return airflow through a grill. When the free area of a return grill must be found for a specific application, standard formulas are used. The standard formula for finding the free area of a grill is as follows:

$$A_S = \frac{Q}{V} \times 1.3 \times 144$$

where

A_S = free area of the return grill (in in^2)

Q = airflow rate (in cfm)

V = air velocity (in fpm)

1.3 = constant

144 = constant converts square feet to square inches

Example: Calculating Grill Size (Free Area)

A return grill must handle 1800 cfm of air at a maximum face velocity of 600 fpm. Find the required free area needed for the grill. **See Figure 18-42.**

$$A_S = \frac{Q}{V} \times 1.3 \times 144$$

$$A_S = \frac{1800}{600} \times 1.3 \times 144$$

$$A_S = 3 \times 1.3 \times 144$$

$$A_S = \mathbf{561.6\ in^2}$$

Calculating Grill Size (Free Area)

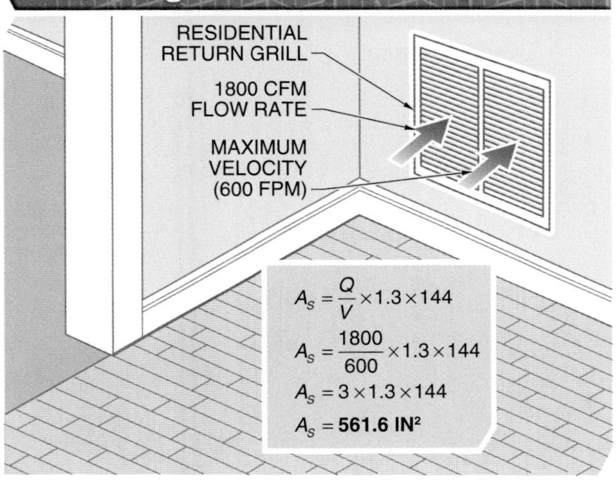

Figure 18-42. When return grills are properly sized, the grills have a face velocity of between 500 cfm and 750 cfm.

Duct Installation Considerations

HVAC forced-air systems must be installed by qualified personnel. The trades involved are mechanical/refrigeration, electrical, and duct. Some individuals perform more than one of the three trades. Typically, certification is required for each trade.

Sheet metal duct is typically fabricated in a shop and then taken to the job site for final assembly and installation. For smaller residential systems, prefabricated ductwork pieces at various lengths are often available at local HVAC supply houses. This allows the opportunity for smaller HVAC shops to simply piece a job together quickly without the need for fabrication machines such as breaks and shears.

Residential Duct Systems Media Clip

Ductwork is connected together using drive clips. Ductwork may need to be supported using straps. Vibration isolation is also important for reducing blower-transmitted noise through the duct system. Ductwork insulation is always needed, especially when the duct is routed through unconditioned spaces. When steel duct is internally insulated, it is installed in the shop. When steel duct is externally insulated, it may be field installed but must be attached with a combination of glue and fasteners. It is important to attach duct insulation securely. If duct is difficult to access, it will be very difficult to access the insulation for repairs at a later time.

Round metal duct systems come in standard diameters and lengths that can be purchased over the counter at many supply houses. Duct diameters typically range from 4″ to 12″ but can be larger. Standard fittings, such as adapters, are available for each size. The simplicity of the duct system and lower installation cost of round metal duct should not be minimized, especially when installation jobs are so cost competitive. Self-tapping sheet metal screws are used with cordless drills to attach the sections together and fittings together quickly and securely. **See Figure 18-43.** Often the connections must be taped to eliminate or dramatically reduce air losses due to leakage.

Duct board systems are also popular today. A special set of tools are used for forming duct board into specific shapes and sections. The sections are taped together and include foil backing that is used to seal the sections together.

Flex duct is also used on many systems. One application of flex duct is to connect the final section from the main trunk to a diffuser boot. Plastic tie wraps are often used to clamp the flex duct to the diffuser boot.

Retrofits

In addition to the considerations for the layout and installation of new duct systems, there are considerations that must be taken into account for duct system retrofits. When an old heating unit is replaced with a new one, the air return and/or air supply openings of the new unit may be of a different orientation and/or size. Assuming that only the heating unit is being replaced and the duct system is remaining, an adapter or new plenum must be constructed.

Connecting Round Duct Sections and Fittings

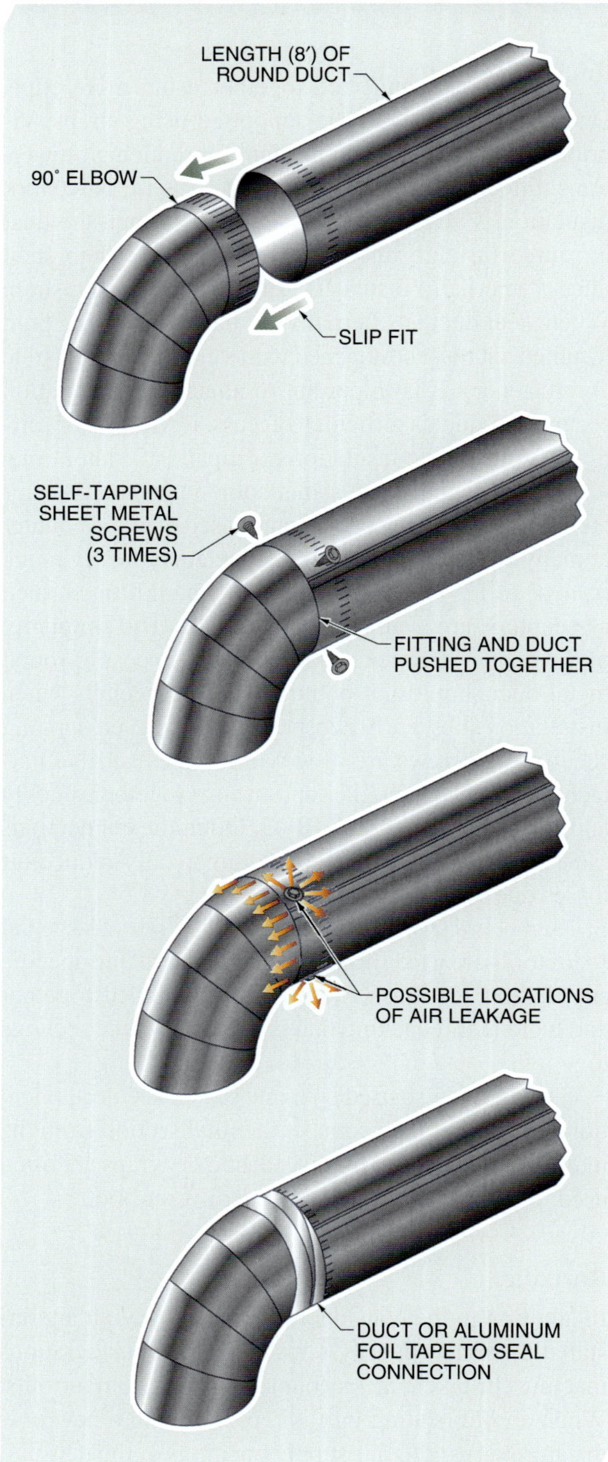

Figure 18-43. Self-tapping sheet metal screws are used with cordless drills to attach duct sections and fittings together. The connections must be taped to eliminate or dramatically reduce air losses due to leakage.

A *new unit adapter* is a section of duct that has the new unit discharge size on one side and the old plenum size opening on the other. **See Figure 18-44.** It is best if the difference in system size and orientation is kept to a minimum. These adapters are typically designed in the field as needed. A small fabricated transition (adapter) can be very useful for connecting the new heating unit to old ductwork. A rough sketch of the needed adapter can be taken to a local shop or supply house where the adapter can be built and then delivered in a short period of time.

New Heating Unit Adapters

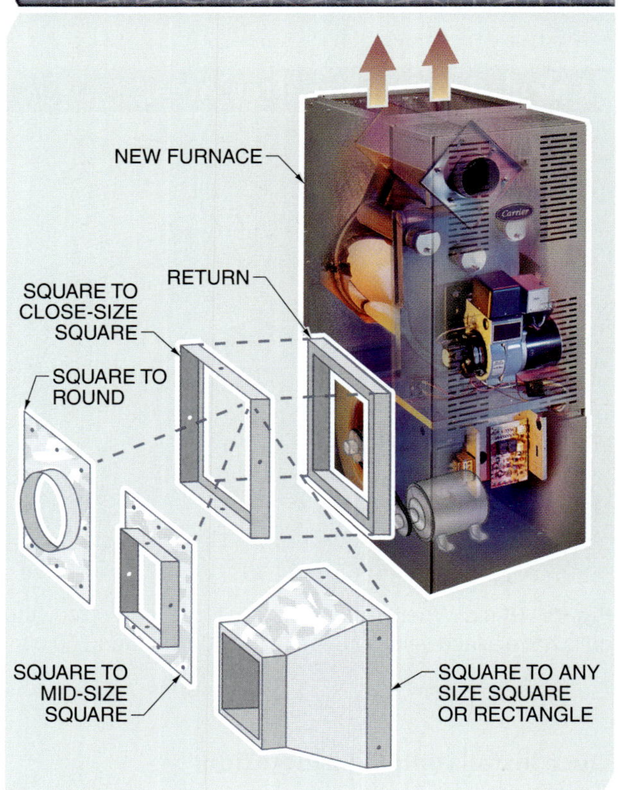

Figure 18-44. When an old heating unit is replaced with a new one, the air return and/or air supply openings of the unit may be of a different orientation and/or size. Therefore, a new unit adapter must be used that has the new unit discharge size on one side and the old plenum size opening on the other.

Zoning Retrofits

Another type of retrofit is the addition of zoning to an HVAC system. A standard HVAC system has one thermostat that cycles the heating unit. While the thermostat is located in a central area that should represent the appropriate temperature, individual rooms are often at a different temperature. **See Figure 18-45.** Zoning systems have become very popular because they allow zone temperature setpoints in each zone. A zoned heating system consists of the following basic components:

- zone damper: Zone dampers are located in the individual branch runs to each space, which has its own thermostat. Zone dampers open and close in response to calls for heating or cooling in the space from its thermostat.
- central controller: A central controller evaluates all the space temperatures and demands for heating. The central unit then decides, based on setup parameters, whether to cycle the unit to heating or cooling mode.
- bypass damper: Since the dampers in multiple zones can be closed, static pressure can build up. A bypass damper is installed with a static pressure sensor and controller, which can open the damper to bypass air from supply to return.

Example: Sizing Duct Systems

A one story home has been sized for a standard 3 t unit for cooling. The HVAC heating unit (blower) location has already been determined to be in the center of the house. It is decided that the reducing plenum duct system will be used. Heating is provided by a natural gas 95% efficiency furnace. The heat input is 90,000 Btuh for an output rating of 85,500 Btuh. A 3 t blower unit uses 400 cfm/t for a total of a 1200 cfm flow rate. This represents high speed on the fan. Low speed is used for heating, which requires less air. The unit fan has a factory pressure rating of 0.4″ WC at 1400 cfm, so there is plenty of air.

The drawings show that 12 registers are needed: 2 registers at 150 cfm each, 8 at 100 cfm, and 2 at 50 cfm. There are also two return grills in a common central hallway. One grill is in the middle of the return duct, and the other is at the end of the return duct. Filters are installed in each return grill.

TECH TIP

Zoning can provide increased comfort in building spaces. However, if zones are sized, installed, or applied incorrectly, reduced comfort and high energy bills may be the result.

Zoning Retrofits

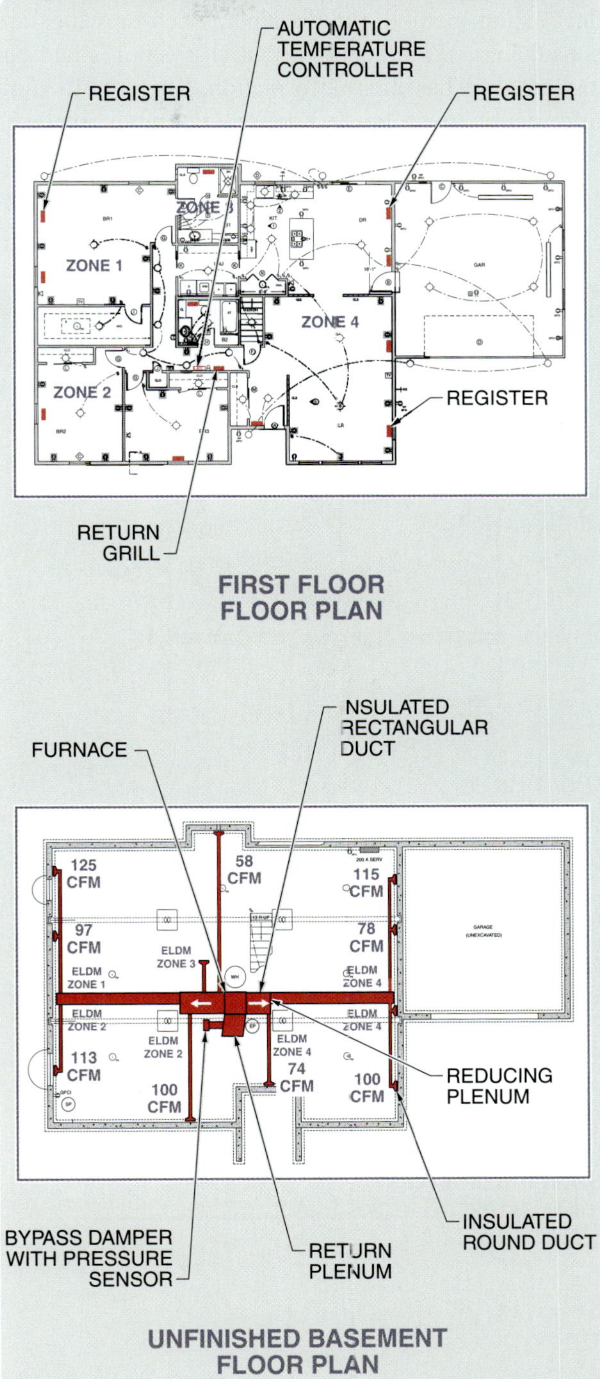

Figure 18-45. A zoned retrofit of a heating system consists of zone dampers located in the individual branch runs to each space, which have their own thermostat, a central controller that evaluates all the space temperatures and demand for heating, and a bypass damper, which opens to bypass air from the supply to return ductwork.

It was decided to use the equal friction method for duct sizing. **See Figure 18-46.** After determining the duct sizing method, the floor plans and drawings are consulted, and the duct distribution system is laid out based on the basement information. Based on the unit information, the following determinations are made:

- The fan outlet pressure rating is 0.4″ WC.
- Discharge air after the heating and cooling coils is 0.1″ WC.
- Based on the branch ducts, section A will carry 300 cfm, section B will carry 500 cfm, section C will carry 700 cfm, section D will carry 600 cfm, and section E will carry 100 cfm.

The airflow rate and duct information determined for each section is entered into a computer. The size of each duct is determined using a computerized or manual equal friction chart. Based on the drawings, the diffusers, grills, and filters are selected.

Sizing Duct Systems

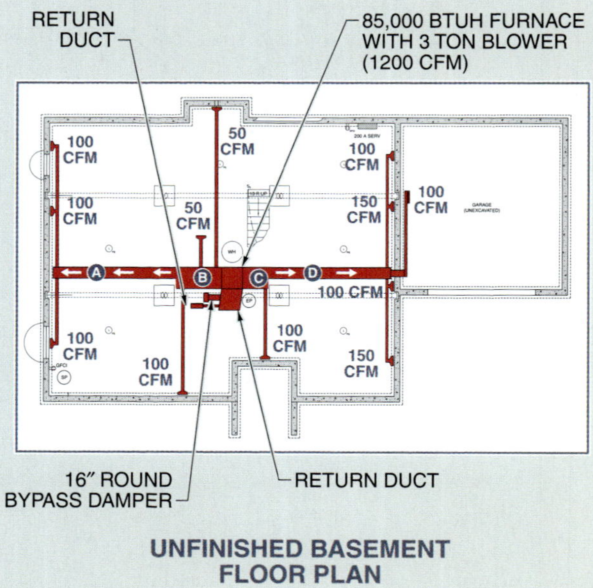

Rectanular Duct

Section	Flow Rate*	Duct Size†
A	300	8 × 8
B	500	8 × 12
C	700	8 × 16
D	600	8 × 14
E	100	4 × 8

* in cfm
† in in.

Round Duct

Flow Rate*	Duct Size†
50	5
100	6
150	7
200	8
250	8
1200	16

* in cfm
† in in.

Figure 18-46. Various rectangular and round duct sizes are required to properly install a forced-air distribution system.

UNIT 19

Hydronic Heating System Boilers

Many residential and light commercial buildings have a hydronic heating system or a forced-air furnace. Hydronic heating systems use water that is heated in a boiler to carry heat to individual rooms or zones. Water has relatively high specific weight and specific heat. These properties make water an excellent medium for carrying heat.

OBJECTIVES

- Explain the purpose of using hydronic heating systems.
- Describe combination heating systems.
- Identify and describe the various types of boilers.
- Explain how boilers are rated.

HYDRONIC HEATING SYSTEMS

A *hydronic heating system* is a heating system that uses water or other fluid to carry heat from the point of generation to the point of use. Hydronic heating systems can be used where the heat generating equipment (boiler) is located at a distance from the building spaces that require the heat. In a light commercial application, hydronic heating systems are used where the boiler is in a mechanical room and the boiler delivers heat to several businesses in a building. A typical hydronic heating system consists of a boiler, fittings, accessories, supply piping and valves, circulating pump, terminal devices, controls, and return piping. **See Figure 19-1.**

A hydronic heating system uses a boiler to store heat and add it to water. Fittings are components attached to a boiler that maintain the safe and efficient operation of the boiler. An accessory is a piece of equipment not directly attached to a boiler but is required for the operation of the boiler and its associated systems.

The piping system of a hydronic heating system distributes hot water to building spaces. Terminal devices transfer the heat from hot water to the air in building spaces that require heat. The circulating pump moves hot water from the boiler through the supply piping system to the terminal devices and back to the boiler. Boiler controls regulate the temperature of the water in the boiler and the operation of the circulating pump. System controls regulate the temperature of the terminal units heating the air in building spaces.

Many residential and light commercial buildings use hydronic heating systems to supply the required heat to building spaces. A boiler located in a utility room, basement, or mechanical room supplies the hot water. A terminal unit that uses hot water to heat air allows easier control of the air temperature than a steam system. In a hot water heating system, heated water is distributed by a circulating pump through supply piping to terminal units and is then returned to the boiler by return piping. **See Figure 19-2.** A *hydronic heating system Btu* is a unit that represents the amount of energy it takes to raise 1 pound of water by 1 degree Fahrenheit.

Hydronic Heating Systems Media Clip

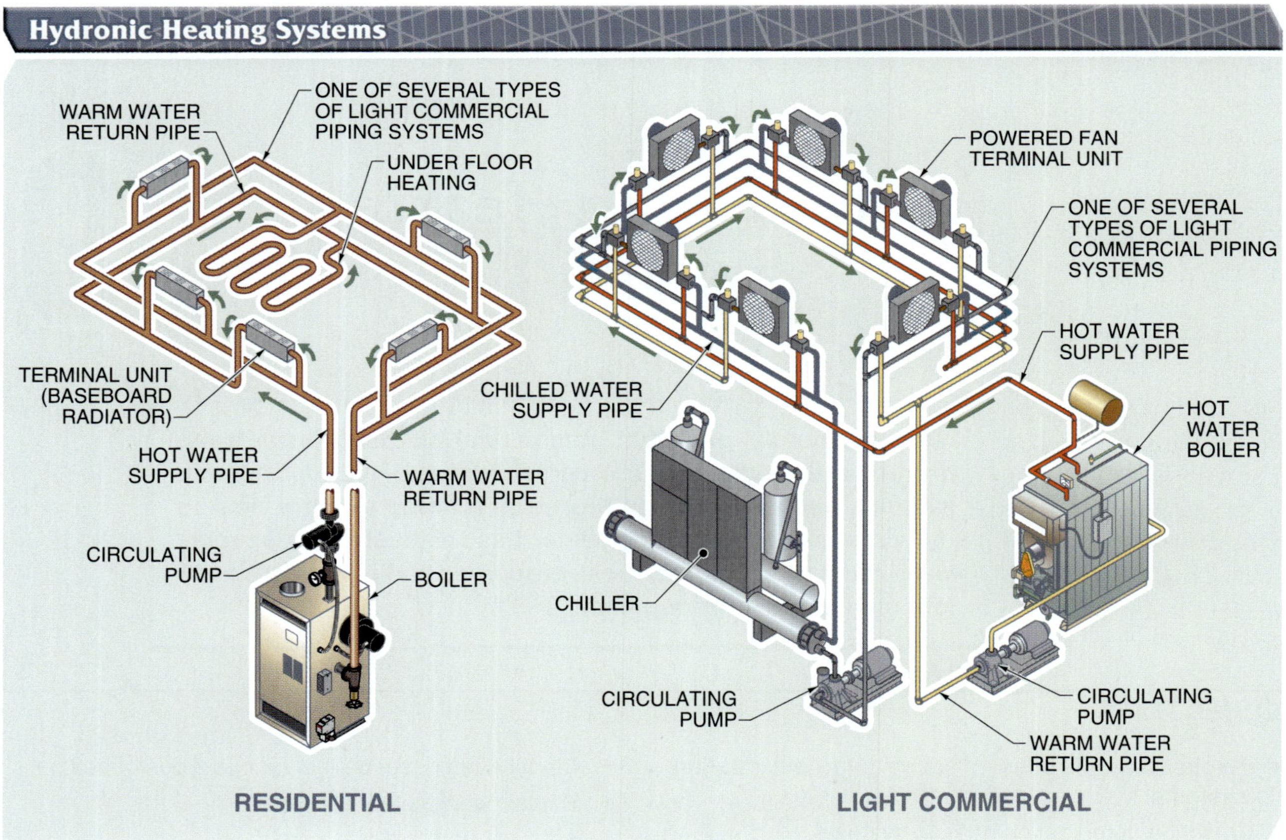

Figure 19-1. Most hydronic heating systems consist of a boiler, fittings, accessories, supply piping and valves, circulating pump, terminal devices, controls, and return piping.

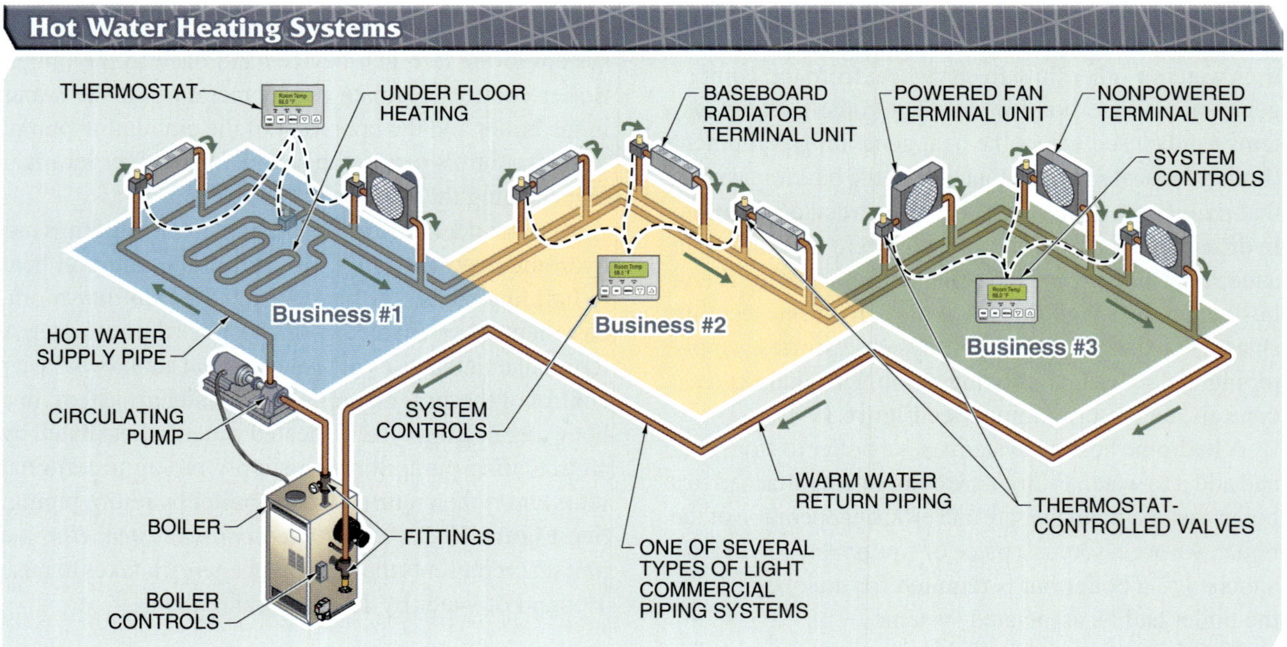

Figure 19-2. In a hot water heating system, heated water is distributed by a circulating pump through supply piping to terminal units and is then returned to the boiler through return piping.

COMBINATION HEATING SYSTEMS

A *combination heating system* is a type of heating system that uses hot water and steam to control building space temperature within a small range. Steam from a boiler is typically used to heat the water for a hydronic heating system using a tube and shell heat exchanger. This tube and shell heat exchanger is typically called a converter. A combination heating system uses all of the components in hydronic heating systems and steam heating systems.

In some light commercial buildings, combination heating systems are used to provide steam for all the heating and processing needs of the building. **See Figure 19-3.** In this type of system, a boiler is used to produce the steam required for building processes. Some of the steam required is piped to a heat exchanger to heat water. The hot water is then circulated through the terminal units by supply and return piping to heat the air in building spaces.

Rust and corrosion may occur if boiler discharge gases are allowed to condense on the boiler heat exchanger.

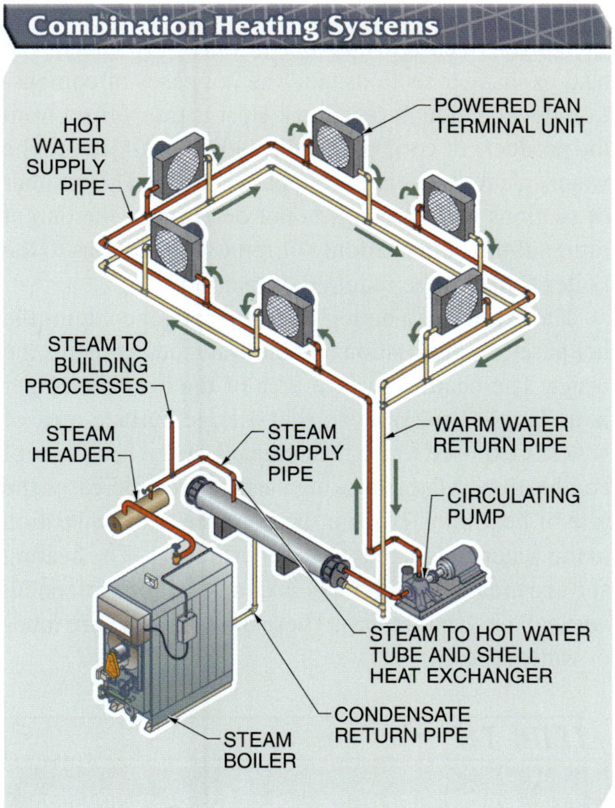

Figure 19-3. In some light commercial applications, combination heating systems use a boiler to furnish steam. Some steam is sent to building processes, and the rest is used to heat the building.

Boilers

The boiler is the central device of any hydronic heating system. A *boiler* is a pressure vessel that transfers heat to water. A *pressure vessel* is a tank or container that operates at a pressure greater than atmospheric pressure. Boilers use hot water or steam to carry heat to residential and light commercial building spaces. The heat is transferred from the hot water or steam to the air or opaque objects in building spaces by terminal devices.

Hydronic boilers are classified by working pressure, temperature, method of heat production, and materials of construction. Residential and light commercial hydronic boilers are typically classified as low-pressure hot water heating boilers and low-pressure steam heating boilers. **See Figure 19-4.**

Low-Pressure Hot Water Heating Boilers. A *low-pressure hot water heating boiler* is a hydronic boiler in which water is heated for the purpose of supplying heat at pressures not to exceed 160 psi and temperatures not to exceed 250°F. The piping systems of hot water boilers are completely filled with water.

Low-Pressure Steam Heating Boilers. A *low-pressure steam heating boiler* is a boiler operated at pressures not to exceed 15 psi of steam. This type of boiler is used for heating purposes only and is typically used in residential and light commercial buildings.

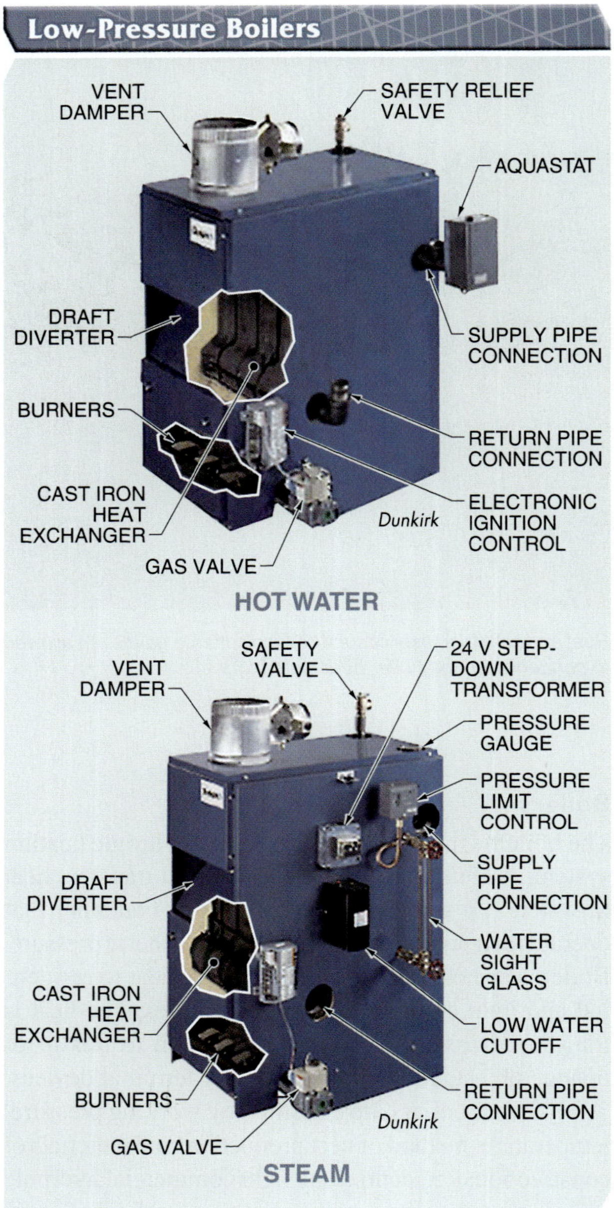

Figure 19-4. Low-pressure residential and light commercial heating boilers have pressures below 160 psi for hot water boilers and 15 psi for steam boilers.

Heat Production

Boilers burn fuel or use electricity to generate the heat that is added to water. Combustion boilers burn fuel, such as natural gas, LP gas, or fuel oil, to produce heat. Combustion boilers are used in most residential and light commercial buildings. Electric boilers use resistance heating elements to produce heat. Electric boilers are used mainly in residential buildings.

Combustion Boiler Heat Exchangers. Boilers that produce heat by combustion have steel, stainless steel, cast aluminum, or cast iron heat exchangers. Steel heat exchangers can be welded together into a single unit to provide a specific output rating. Steel heat exchangers in boilers typically provide AFUEs of 82% or less. They have removable manhole covers for maintenance and service. **See Figure 19-5.**

Stainless steel and cast aluminum heat exchangers are used on condensing boilers. Boilers with stainless steel heat exchangers are designed to extract the maximum amount of heat from the combustion process. Stainless steel heat exchangers in boilers typically provide AFUEs as high as 95%. Cast aluminum heat exchangers are one piece, durable, and designed to extract as much Btu of heat as possible from the combustion process. Cast aluminum heat exchangers in boilers provide the highest AFUEs, as high as 96% or 97%.

Cast iron (sectional) heat exchangers for boilers are made of hollow sections and bolted together. Cast iron heat exchangers provide AFUEs up to 86%. Openings within the sections allow water connections and the hot gases of combustion to travel up through the heat exchanger. **See Figure 19-6.** Hot water flows inside the heat exchanger sections, and the hot gases of combustion flow outside the sections. Heat is transferred from the products of combustion through the cast iron to the water inside the heat exchanger sections. The number of sections in a cast iron boiler determines the output rating of the boiler. Adding or removing sections to the boiler achieves the required output rating.

The fireside of a boiler's heat exchanger contains the hot gases of combustion and the waterside contains the water. The heating surface area of the heat exchanger is as large as possible so that a large surface area of water is exposed to the heat created by the hot gases of combustion. A large heating surface area increases the rate of heat transfer from the hot gases of combustion to the water and increases boiler efficiency. The heating surface area of steel boilers is composed of horizontal, vertical, or slanted tubes. These tubes may be fire tubes or water tubes.

TECH TIP

If the temperature of the hot gases of combustion drops too low, the gases condense and acid is formed. The acid may corrode and damage the boiler heat exchanger. Today, many high-efficiency boilers operate in the condensing mode.

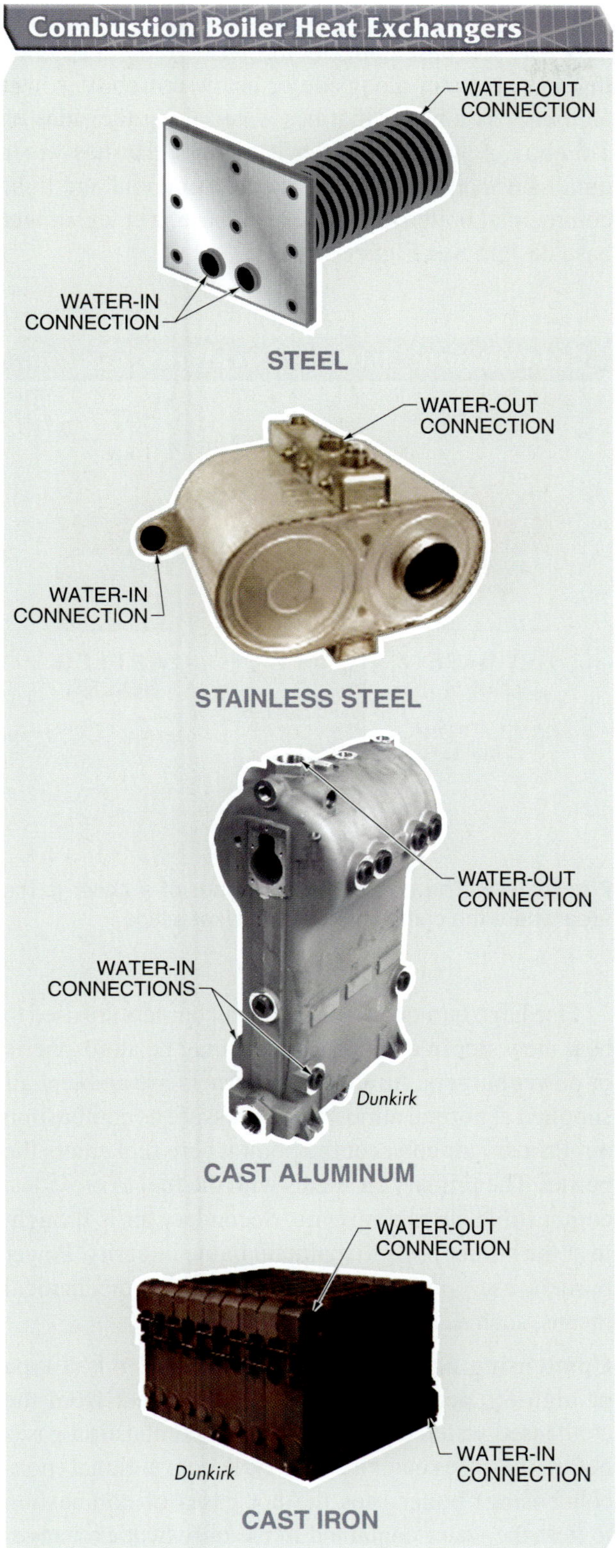

Figure 19-5. Steel heat exchangers are typically welded together to form a single unit. Cast aluminum heat exchangers are available only as one piece. Cast iron heat exchangers consist of hollow sections that are bolted together.

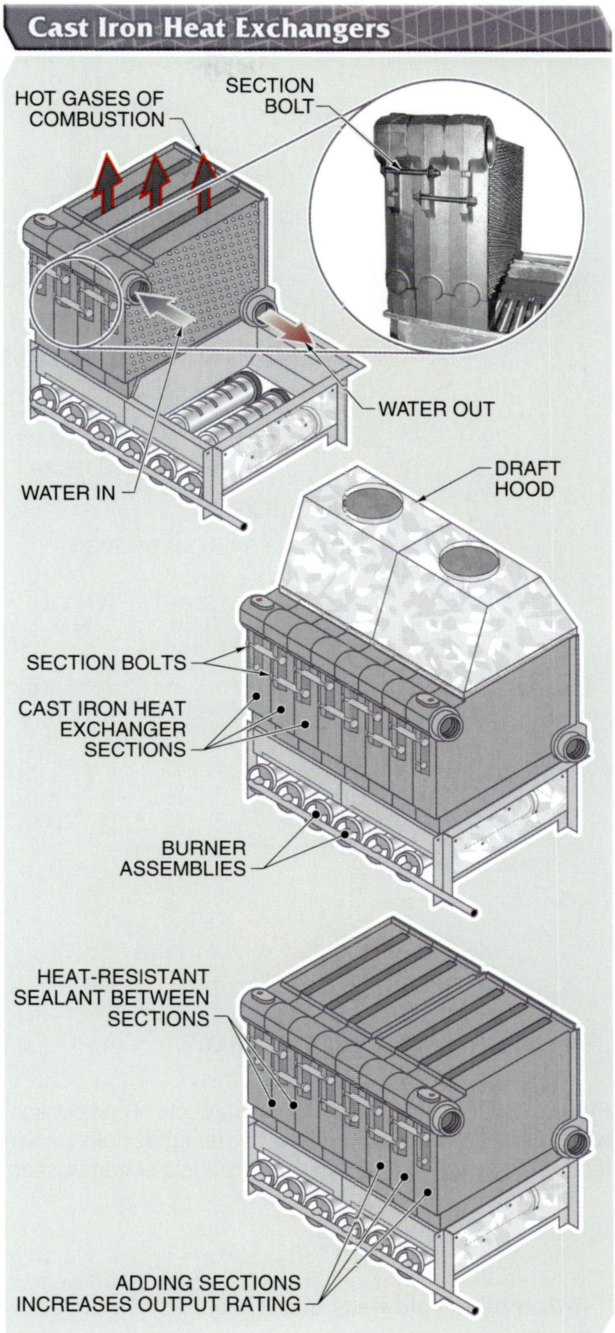

Figure 19-6. Cast iron heat exchangers with controls and draft systems in place can provide AFUEs up to 86%.

Fire tubes hold the hot gases of combustion. A *fire-tube boiler* is a boiler that heats the water surrounding fire tubes as hot gases of combustion pass through the tubes. The combustion chamber is located below or at one end of the boiler. A breeching connected to the flue stack collects the hot gases of combustion and directs the gases up the stack. **See Figure 19-7.**

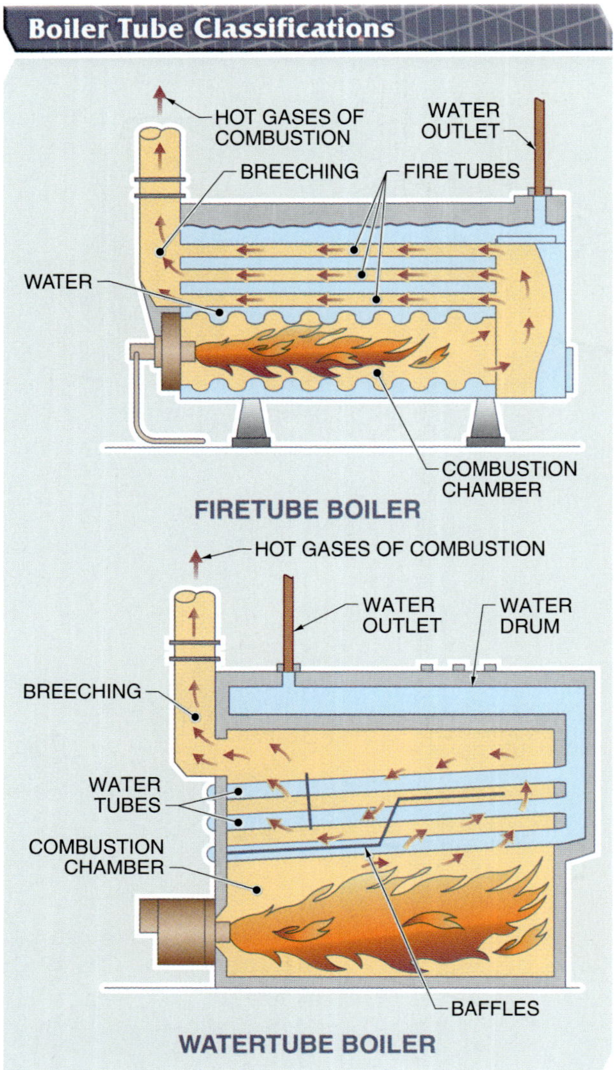

Figure 19-7. Fire tubes hold the hot products of combustion, which heat the surrounding water. Water tubes hold the water heated by the surrounding hot products of combustion.

Water tubes hold water that is heated by the hot gases of combustion. A *watertube boiler* is a boiler that passes hot gases of combustion over the outside surfaces of water-filled tubes to heat the water. The water tubes run to each end of the boiler through the combustion chamber (firebox). The combustion chamber is located below or at one end of the boiler. A breeching connected to the flue stack collects the hot products of combustion.

Boiler Bases. The combustion chamber of a boiler is the area where the combustion of fuel takes place. The relationship between the water in a boiler and the firebox determines whether a boiler is a dry base boiler, wet leg boiler, or wet base boiler. A *dry base boiler* is a boiler that has no water alongside or below a firebox. A *wet leg boiler* is a boiler that has water along the sides of a firebox. A *wet base boiler* is a boiler that has water on all sides and below a firebox. Residential and light commercial boilers can have a dry base, wet leg, or wet base design. **See Figure 19-8.**

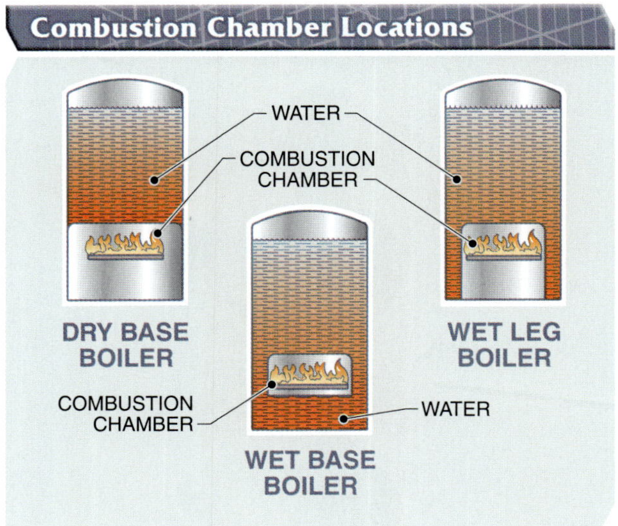

Figure 19-8. The combustion chamber of a boiler is the area where the combustion of fuel takes place.

One large burner or multiple small burners are used to heat the water in a boiler. Burners may be atmospheric or power burners. Atmospheric burners use ambient air supplied at normal atmospheric pressure as combustion air. Primary air enters at the point where fuel enters the burner. The primary air mixes with the fuel to provide a combustible air-fuel mixture. Secondary air is brought in at the burner ports to maintain flame integrity. Power burners use combustion air provided by mechanical means, such as a blower.

Condensing Boilers. A *condensing boiler* is a type of high-efficiency boiler that extracts heat from the condensed water vapor found in hot combustion gases before the gases enter a chimney. A conventional (noncondensing) boiler uses the hot gases of combustion to heat the water contained in the only heat exchanger of the boiler. The gases stay hot (355°F to 400°F) for some time. When the waste gases go up the chimney, a significant amount of heat is lost to the atmosphere. **See Figure 19-9.**

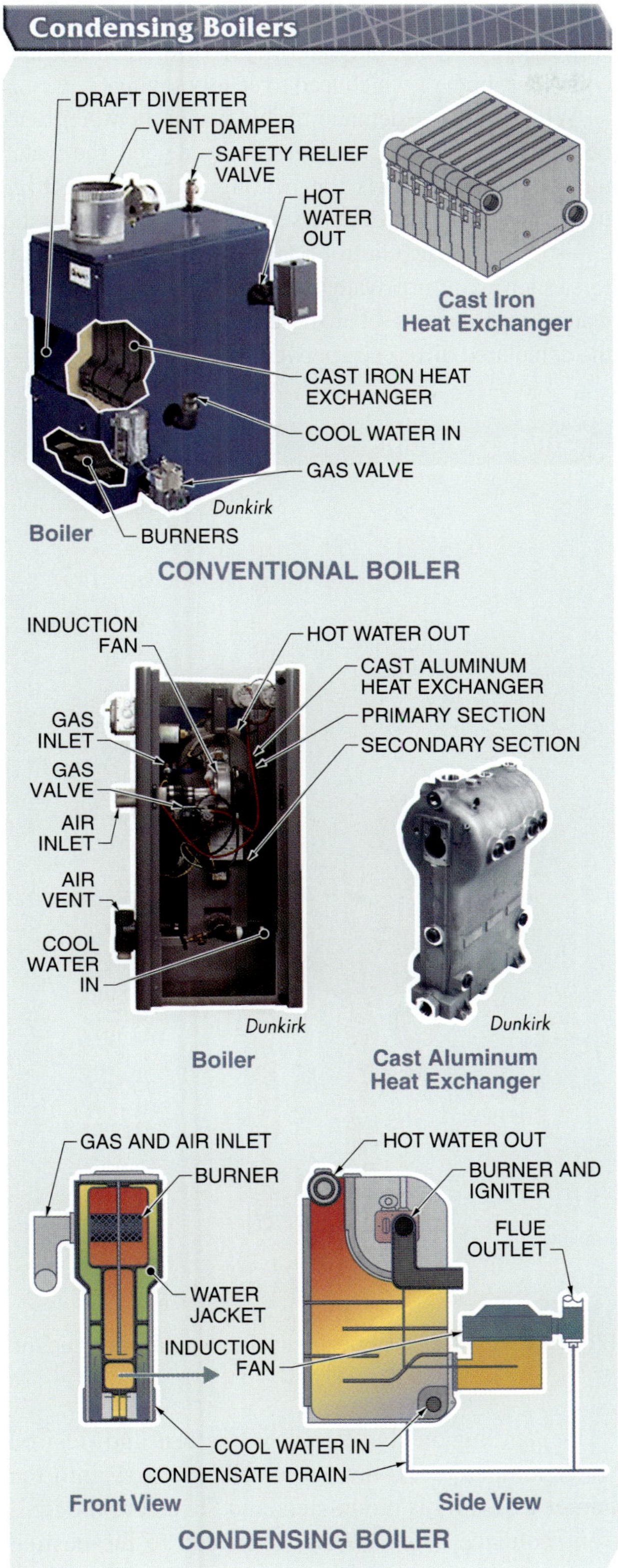

Figure 19-9. In a condensing boiler, the water vapor and heat that normally would go up the chimney are condensed back into liquid water.

In a condensing boiler, the water vapor and heat that normally would go up the chimney with the waste gases is condensed back into liquid water. During the process of condensing the water vapor, the secondary section of the heat exchanger recovers the heat from the water vapor. Nearly the same amount of heat energy is recovered from the condensation of the water vapor in the secondary section of the heat exchanger as in the primary section.

According to manufacturers, the operating efficiency of condensing boilers can be as high as 98%. The actual operating efficiency depends on the temperature of the condensing water stream. When the stream is too warm, little or no condensation takes place, very little heat energy is recovered, and the boiler efficiency drops.

Many of the new condensing boilers include microprocessor-controlled combustion chambers that modulate the air-fuel mixture. Controlling the ratio of air to fuel to the burner(s) approximately every 50 milliseconds ensures maximum heat recovery. **See Figure 19-10.** The microprocessor program uses the outdoor air temperature, temperature of the hot water supply, and the temperature of the return water, to control the air-fuel mixture. Modulating the air-fuel mixture allows the boiler to supply only the amount of heat to the building that is being lost.

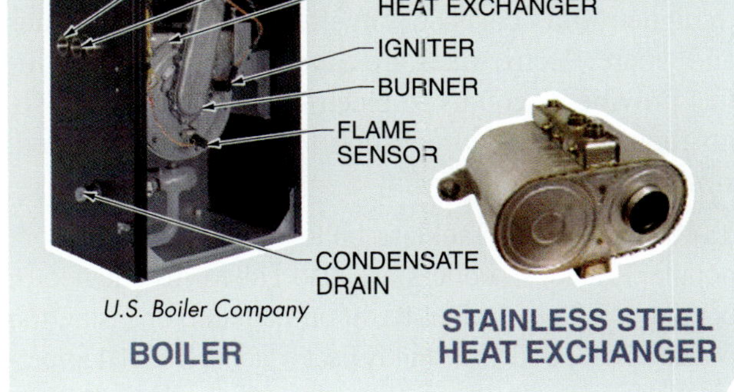

Figure 19-10. Controlling the ratio of air to fuel approximately every 50 milliseconds ensures maximum heat recovery and allows the boiler to supply only the amount of heat being lost.

Electric Boilers. An *electric boiler* is a type of boiler that uses heat produced by resistance heating elements to heat water. Electric boilers are classified by water volume and wattage. The resistance heating element used in an electric boiler is an immersion heater. An *immersion heater* is a series of copper rods enclosed in waterproof tubes that are installed in a boiler and surrounded by water to provide heating by electricity. No heat exchanger is required because heat is transferred directly from the immersion heaters to the water. **See Figure 19-11.**

Electric Boilers

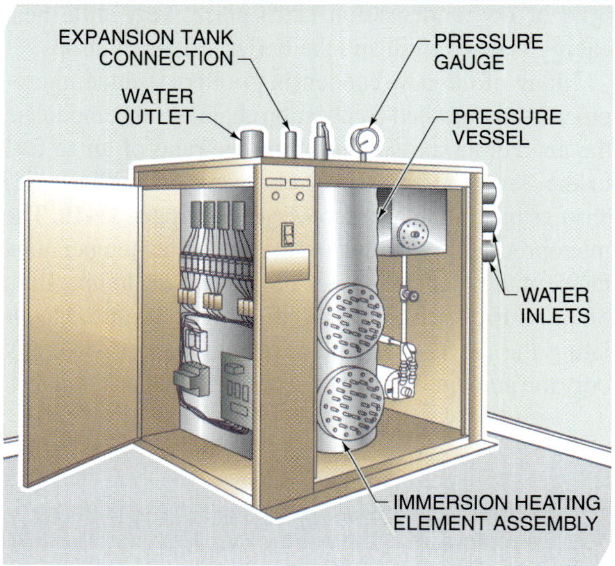

Figure 19-11. An electric boiler is a boiler without a heat exchanger that uses heat produced by submerged resistance heating elements to heat water.

Electric boilers may be more expensive to operate due to higher utility costs but are effective in mild climates and where electric rates are advantageous. Some utilities provide discounts to electric boiler owners. Also, no fuel piping or combustion gas venting is needed, simplifying installation.

Tankless Boilers. Tankless boilers are becoming more popular. Tankless boilers are also called instantaneous, continuous flow, in-line, flash, on-demand, or instant-on boilers. These boilers instantly heat water to 8 or 10 temperature settings as the water flows through the device. Tankless boilers do not retain any water internally except for what is in the heat exchanger coil. These boilers are often installed to also provide all the hot water requirements for an entire house. The main advantages of tankless boilers, in comparison to conventional boilers, is that they do not waste energy storing water, and they have lower and cleaner emissions, both of which lead to energy savings.

When there is a demand for hot water, the water temperature sensor of the tankless boiler detects the water temperature and starts the heating process. **See Figure 19-12.** The temperature sensor sends a signal to the control board. The control board analyzes multiple factors, such as return water temperature, desired supply water temperature as set on the temperature controller, and the calculated difference between the two temperatures.

Tankless Boilers

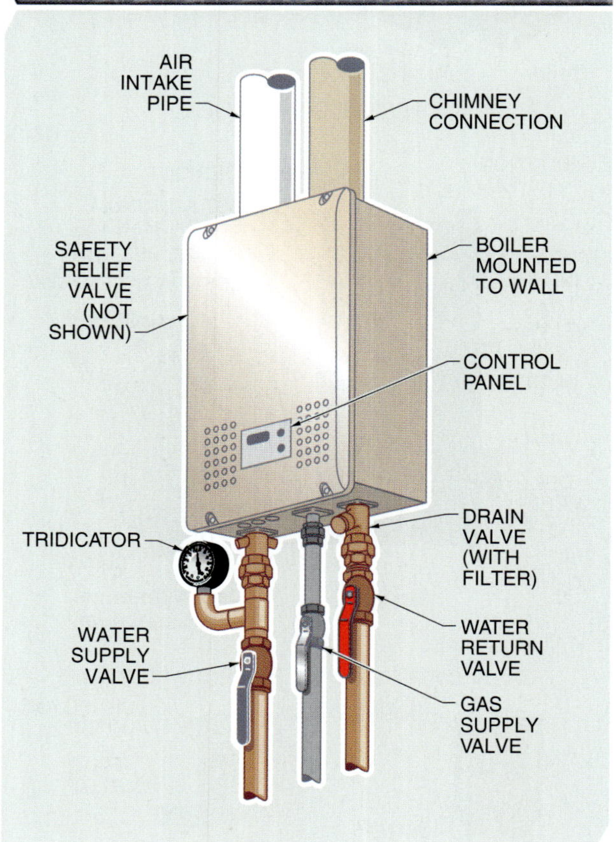

Figure 19-12. Tankless boilers instantly heat water as the water flows through the boiler.

Depending on the calculated return and desired water temperatures, the gas or electric flow into the burner assembly is modulated, and the electronic ignition sequence begins. Water is heated to the desired temperature as it is circulated through the copper heat exchanger. When the temperature in the building space is adequate, the tankless boiler shuts down and is placed in standby mode, pending the next call for heat.

Boiler Ratings

An *input rating* is the amount of heat produced by a furnace or boiler as fuel is burned; expressed in Btu/hr. An *output rating* is the total amount of heat a furnace or boiler will produce in 1 hr; expressed in Btu/hr. The output rating is typically 75% to 80% of the input rating for combustion boilers.

Gross unit output is the heat output of a hydronic or steam boiler when it is fired continuously. *Net unit output* is the gross output of a water or steam boiler multiplied by the percentage of loss due to pickup. *Pickup* is additional heat that is needed to warm the water in a hydronic or steam heating system once a period of off-time, such as overnight, has passed. The pickup factor of a boiler can be as high as 30% to 50%. Electric boilers have output ratings of 100% because all the energy used to produce heat is transferred to the water. **See Figure 19-13.**

The amount of heat produced by a combustion boiler depends on the size of the boiler and the type of fuel used. Fuel gas- and fuel oil-fired boilers have output ratings from 60,000 Btu/hr to 300,000 Btu/hr for residential applications to millions of Btu/hr for light commercial applications. Coal-fired boilers are typically used for industrial applications and have output ratings of hundreds of millions of Btu/hr. The amount of heat produced by an electric boiler depends on the size of the boiler and the number of amps used. Electric boilers are rated in kilowatts per hour (kW/h). One kilowatt per hour equals 3414 Btu.

Hot water from a boiler is often used in reheat coils to provide heat in commercial buildings.

TECH TIP

Care must be taken when sizing a boiler to meet the building heat load. In some buildings, zones are added or the physical layout is changed after the boiler is installed. In these buildings, the boiler becomes undersized for the amount of heat required. Often, an area is set aside in the mechanical room for an additional boiler to be installed at a later date.

Representative Boiler Ratings

Boiler	Gas Input MBH	I=B=R Gross Output MBH		Net I=B=R Ratings			Combustion Efficiency		Water Content	
		Water	Steam	Steam ft²	Steam* MBH	Water† MBH	Water	Steam	Water	Steam
xx-03	320	256	256	800	192	223	80.0	80.0	21.12	11.28
xx-04	360	288	288	900	216	250	80.0	80.0	23.18	12.36
xx-05	400	320	320	1000	240	278	80.0	80.0	25.34	13.44
xx-06	440	352	352	1100	264	306	80.0	80.0	27.50	14.52
xx-07	480	384	384	1200	288	334	80.0	80.0	29.66	15.60
xx-08	520	416	416	1300	312	362	80.0	80.0	31.82	16.88
xx-09	560	448	448	1400	336	390	80.0	80.0	33.98	17.76
xx-10	630	504	504	1575	378	438	80.0	80.0	44.12	31.64
xx-11	840	672	672	2100	504	584	80.0	80.0	55.15	39.55
xx-12	1050	840	840	2625	630	730	80.0	80.0	66.18	47.46
xx-13	1260	1008	1008	3150	756	877	80.0	80.0	77.21	55.37

* net I = B = R steam ratings based on an allowance of 4 to 8 = 1.333, 9 (nat) = 1.324, 9 (LP) = 1.327, 10 (nat) = 1.310, 10 (LP) = 1.312, 11 (nat) = 1.298, 11 (LP) = 1.300, 12 (nat) = 1.290, 12 (LP) = 1.291, and 13 to 19 = 1.288

† net I = B = R water ratings based on an allowance of 1.15

Figure 19-13. The output rating of a combustion boiler is typically 75% to 80% of the input rating.

UNIT 20

Hydronic Heating Boiler Components and Controls

Hydronic heating systems use boilers to supply hot water to terminal units. Boilers have components and controls that allow them to heat water safely and efficiently. Hydronic boilers typically operate at pressures between 12 psi and 25 psi.

OBJECTIVES

- Identify the functions of typical fittings found on a hot water boiler.
- Describe the features of common hot water boiler accessories.
- Describe common hydronic system and boiler control devices.

Index to Procedures

BOILER FITTINGS AND ACCESSORIES

Hot water (hydronic) boilers have multiple types of fittings and many accessories that permit safe and efficient operation. The technician must be familiar with the purpose of, operation of, and testing procedures for these fittings and accessories.

HOT WATER BOILER FITTINGS

A *fitting* is a device directly attached to a boiler that is required for proper operation of the boiler and system. An *accessory* is a component not directly attached to the boiler but is needed for proper boiler operation. Many accessories are of the same type for all residential and light commercial hot water boilers. **See Figure 20-1.** The fittings of residential hot water boilers are also similar to those of light commercial hot water boilers. For example, instead of the pressure gauge that is present on a residential boiler, a light commercial boiler has a pressure-temperature gauge (tridicator).

Since hot water boilers are completely filled with water, there is no need for the water column-type devices that are found on steam boilers. Fittings, such as Aquastats®, are used because hot water boilers are used to maintain water temperature. The major differences between boilers are their temperature and pressure ratings.

Safety Relief Valves

A *safety relief valve* is a valve used only on a hot water boiler that prevents pressure from exceeding the maximum allowable working pressure (MAWP) of the boiler during a malfunction. A safety relief valve is mounted at the top of the boiler with no other valves between it and the boiler. When the pressure in the boiler exceeds the MAWP, the safety relief valve opens and discharges hot water to a drain.

TAGGED FOR AR

Boiler Fittings and Accessories

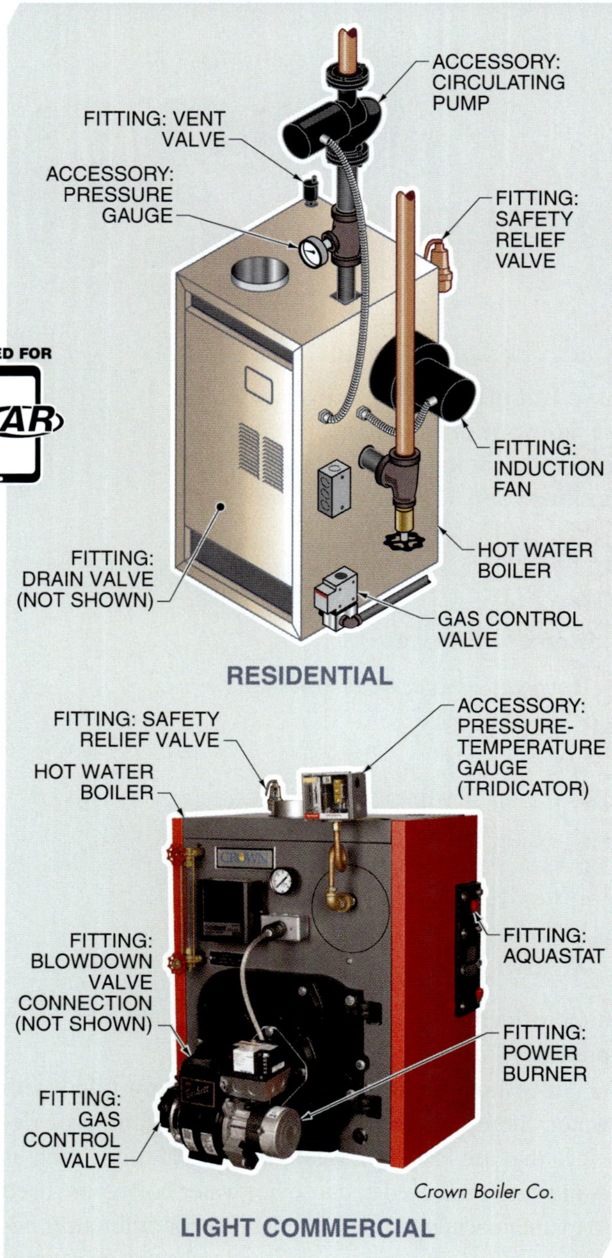

Figure 20-1. Residential and light commercial hot water boilers typically have the same types of fittings and accessories.

Safety Relief Valves

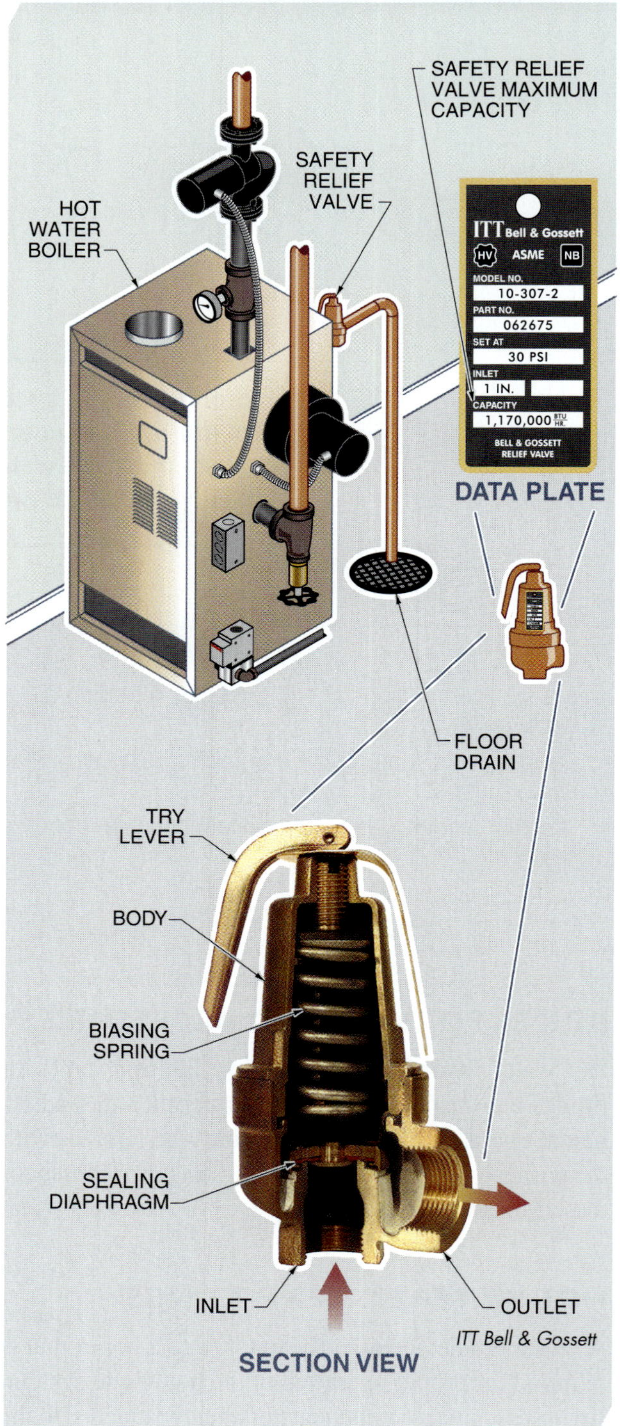

Figure 20-2. A safety relief valve is a hot water boiler valve that prevents excessive pressure from building up during a malfunction.

Safety Relief Valves Media Clip

According to the 2009 ASME International Boiler and Pressure Vessel Code, all hot water boilers must have at least one pressure relief valve. The pressure relief valve (safety relief valve) is rated in Btu relieved per hour. It consists of an inlet, outlet, body, sealing diaphragm, biasing spring, and try lever. **See Figure 20-2.**

Pressure Gauges

A *pressure gauge* is a boiler fitting that displays the amount of pressure in pounds per square inch (psi) inside a boiler. A pressure gauge allows an operator to monitor boiler pressure changes and maintain pressure within a safe operating range. A pressure gauge is connected to the highest part of the load side of the boiler and must be positioned to allow easy viewing. A *compound gauge* is a pressure gauge that indicates vacuum in inches of mercury (in. Hg) and pressure in pounds per square inch (psi). **See Figure 20-3.**

Pressure gauges are susceptible to pressure surges. To protect pressure gauges from the damage caused by pressure surges, snubbers are used. A *snubber* is a pipe-coupling-like device that protects pressure gauges from surges. It has an orifice that slows down the transmission of a pressure surge so that the gauge does not feel the surge. The smaller the orifice, the longer it takes for a surge to travel through the snubber. **See Figure 20-4.**

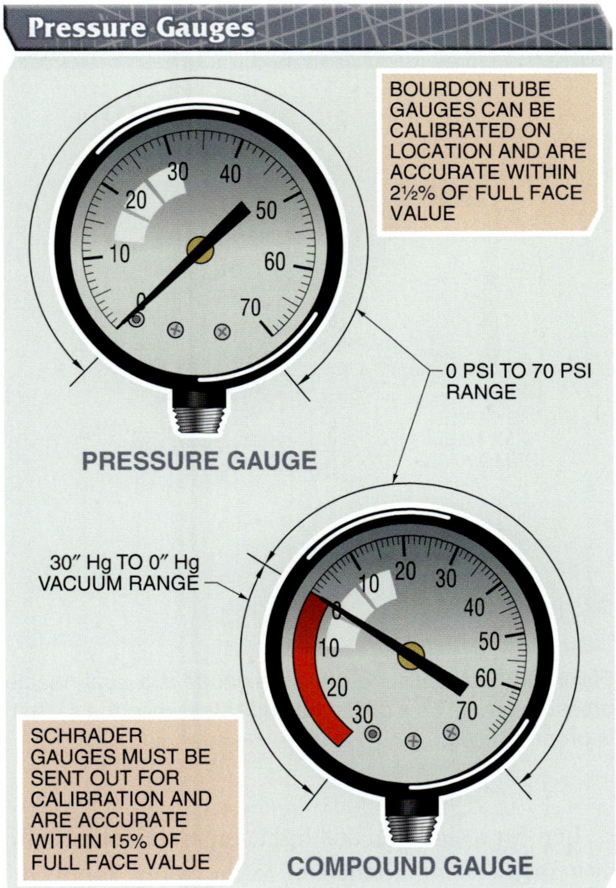

Figure 20-3. Pressure gauges indicate the amount of pressure above or below atmospheric pressure in a boiler.

TECH TIP

A gauge that has a cracked or discolored faceplate should be replaced immediately. Damaged gauges cannot be trusted to accurately measure the boiler pressure and can result in a dangerous condition.

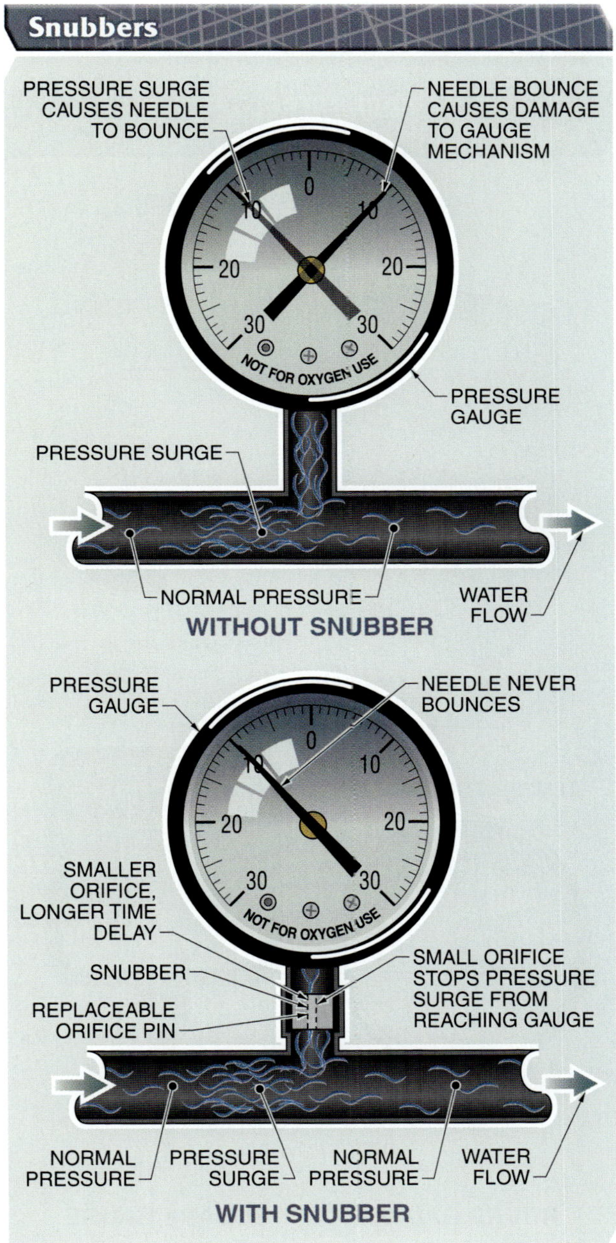

Figure 20-4. Snubbers use orifices to prevent pressure surges from entering a pressure gauge and destroying its pressure-sensing mechanism.

Pressure-Temperature Gauges

A pressure-temperature gauge (tridicator) is used on a hot water boiler to ensure safe and efficient boiler operation. A *pressure-temperature gauge (tridicator)* is a boiler gauge that measures water pressure and temperature at the point where the gauge is located. A pressure-temperature gauge is typically installed at the top and front of the boiler so that the gauge can easily be seen. **See Figure 20-5.**

Pressure-Temperature Gauges (Tridicators)

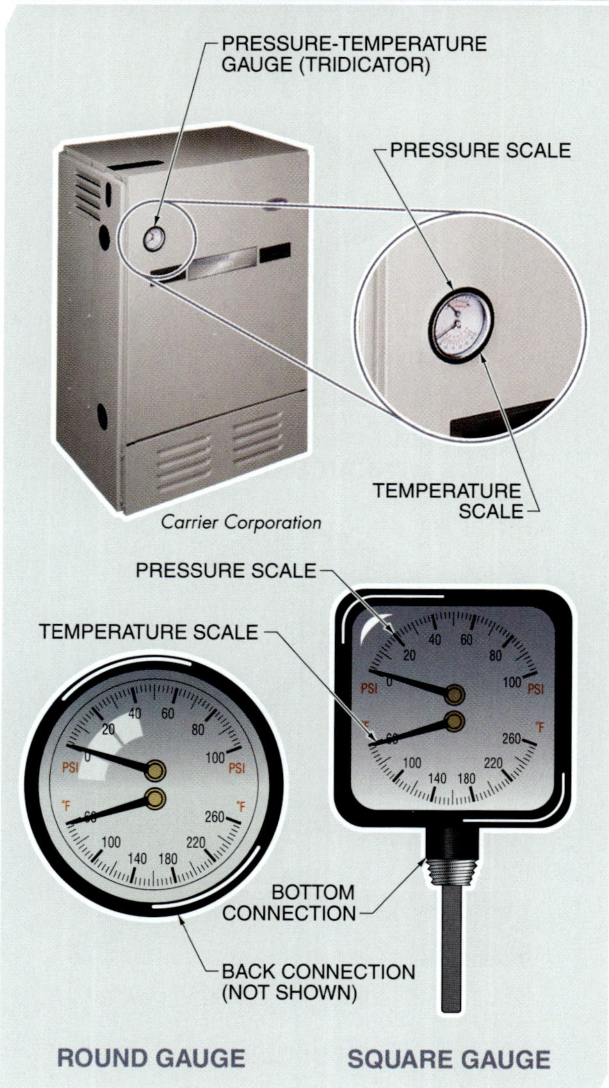

Figure 20-5. A pressure-temperature gauge (tridicator) measures the pressure and temperature of boiler water at the point on the boiler where the gauge is located.

Altitude Pressure Gauges

An *altitude pressure gauge* is a needle gauge on a boiler that indicates when the system is properly filled. Altitude pressure gauges have one black needle and another needle that is typically red. The red needle is manually set at the proper water level for the system. The black needle indicates the true water level and varies as the water level changes. When the black needle is over the red, the system is properly filled with water. The red needle can be reset to conform to the water level in the system or expansion tank by manual positioning of the needle. **See Figure 20-6.**

Altitude Pressure Gauges

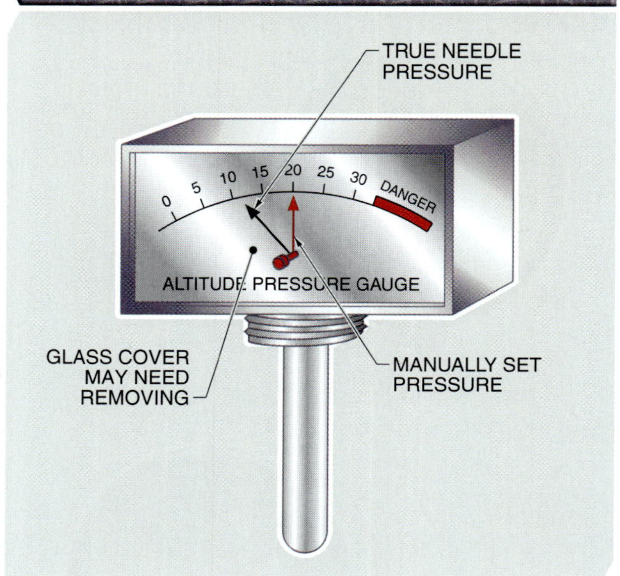

Figure 20-6. An altitude pressure gauge is a dual-needle pressure gauge on a boiler that indicates when the system is properly filled.

In most residential and light commercial hydronic systems, a valve connected to the water supply of the structure automatically controls the amount of water in the boiler. When the water level and pressure drop, the water supply valve opens and allows water to enter the boiler. Reading the altitude gauge is a way to verify that the water supply valve is operating properly. If the hydronic system has no pressure, the boiler could be damaged. When an altitude gauge reads higher than normal, the water supply valve should be examined and any malfunctions corrected before turning on the boiler.

Bottom Blowdown and Drain Valves

A *bottom blowdown valve* is a valve located on the blowdown line at the lowest point of the water side of a steam boiler that is used to release water and sediment from the bottom of the boiler. Residential boilers use drain valves instead of bottom blowdown valves. All boilers have bottom blowdown valves or drain valves. **See Figure 20-7.** The size, capacity, and type of boiler determine the type of valves allowed. Some boilers can have two valves on the bottom blowdown or drain line. One valve is a quick-opening valve that is lever-operated, and the other valve is a slow-opening valve that is screw-operated. Some manufacturers place two slow-opening valves on the bottom blowdown line. When present, the quick-opening valve is always located closest to the boiler shell.

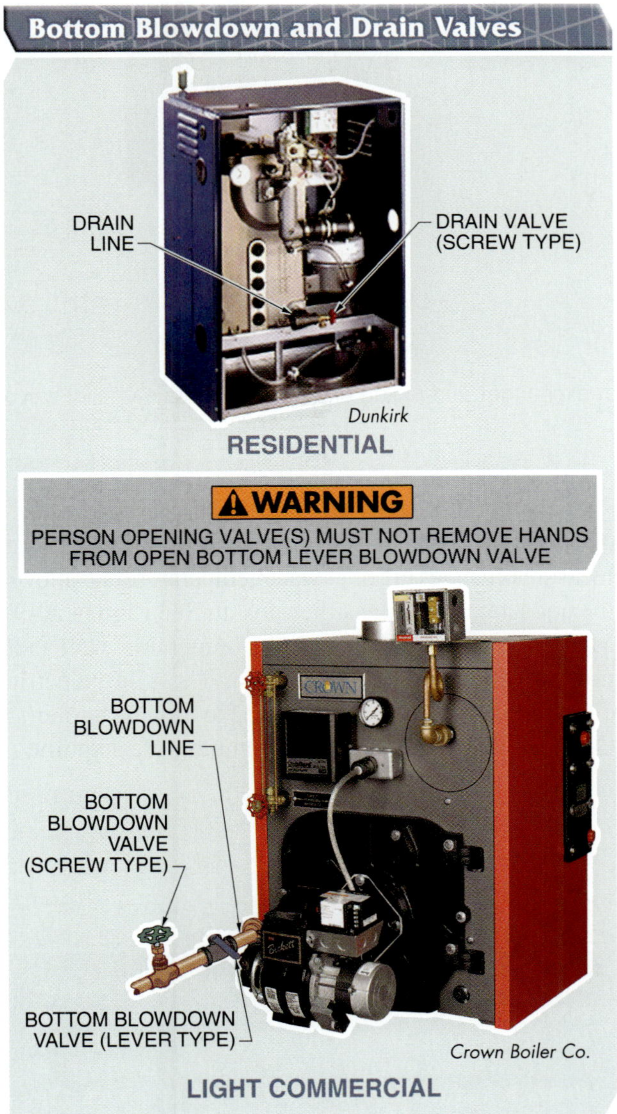

Figure 20-7. Bottom blowdown and drain valves drain boiler water and sludge from the bottom of the heat exchanger.

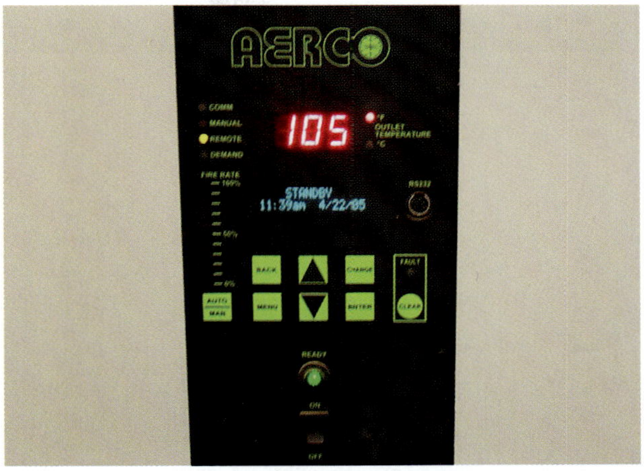

Most new boilers have microprocessor controls that display setpoints and system faults.

A *quick-opening valve* is a lever-operated boiler fitting valve that functions as a sealing valve. The slow-opening valve is used as the blowing valve and takes all the wear during blowdown. When performing a bottom blowdown, the quick-opening valve is always opened first and closed last. Bottom blowdown is performed for water level control, sludge and sediment removal, chemical concentration control, or boiler dumping.

Vent Valves

A *vent valve* (boiler vent or air cock) is a boiler fitting used to vent air from inside a boiler. The vent valve is opened when filling the boiler with water to prevent air from being trapped inside the boiler. The vent valve may be installed on a residential boiler vent line (1/8″ or 1/4″ piping) and light commercial boiler vent line (1/2″ or 3/4″ piping) that is connected to the highest part of the boiler. **See Figure 20-8.** Not all boilers are equipped with a vent valve. However, the safety relief valve of a boiler must never be used to vent air from a boiler. Damage to the safety relief valve can occur, resulting in a possible safety valve malfunction.

TECH TIP

When using a bottom blowdown valve, a hand must always be in contact with the valve as a reminder that the valve should not remain open too long. If the valve is open too long, the boiler water level could drop excessively. If the safety controls fail, a catastrophic condition could result.

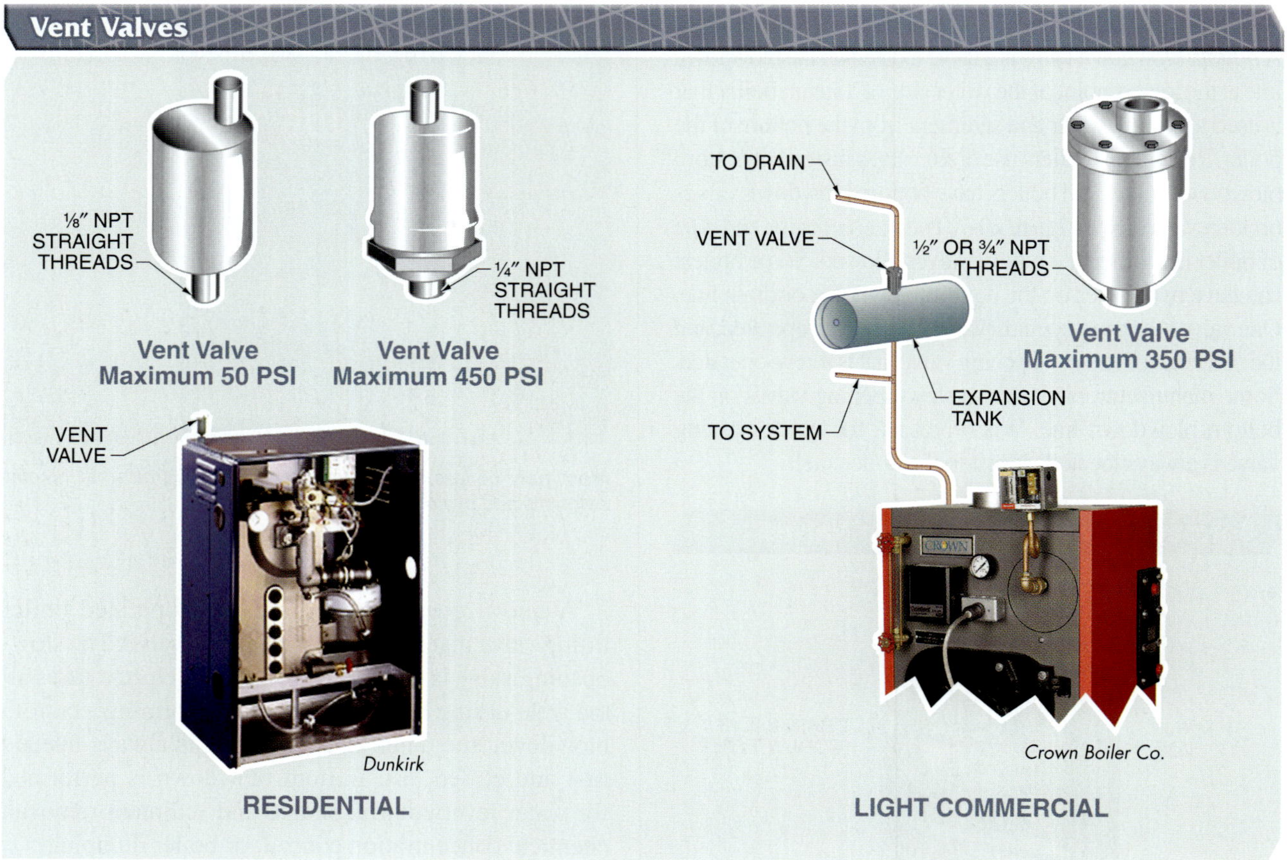

Figure 20-8. A vent valve is used to vent air from inside the heat exchanger and piping of a boiler. The safety relief valve must never be used to vent air from a boiler.

A vent valve must be open to allow air to escape when warming up a boiler before putting it in service. When a boiler is not vented, air trapped above the water is heated and expands, causing a pressure buildup in the boiler. The trapped air affects gauge pressure readings. A pressure gauge will typically indicate a false pressure reading in these conditions.

When the water is drained from a boiler for servicing, the vent valve must be opened to prevent a vacuum from forming. The vent valve is also used to prevent a dangerous vacuum from forming inside a boiler when it is taken out of service. By opening the vent valve when the boiler pressure drops to 1 psi or 2 psi, a vacuum condition is prevented.

Aquastats

An *aquastat* is a device that controls the starting, stopping, and firing rate of a hot water boiler burner by measuring the temperature of the water in the boiler. An aquastat consists of a bulb or remote bulb sensor that measures changes in boiler water temperature and allows the aquastat to act like a thermostat. **See Figure 20-9.** Many aquastats include a bulb immersion well that protects the bulb from corrosion. The relay electric circuit is energized or de-energized to turn the burner(s) ON or OFF, depending on the temperature setpoint.

TECH TIP

An aquastat should be checked for proper operation regularly. Age and aquastat location may cause the aquastat to be off by 10°F to 20°F. This may lead to wasted energy and/or a safety problem. A thermal imager can be used to easily check the water temperature and compare it with the setpoint on the aquastat.

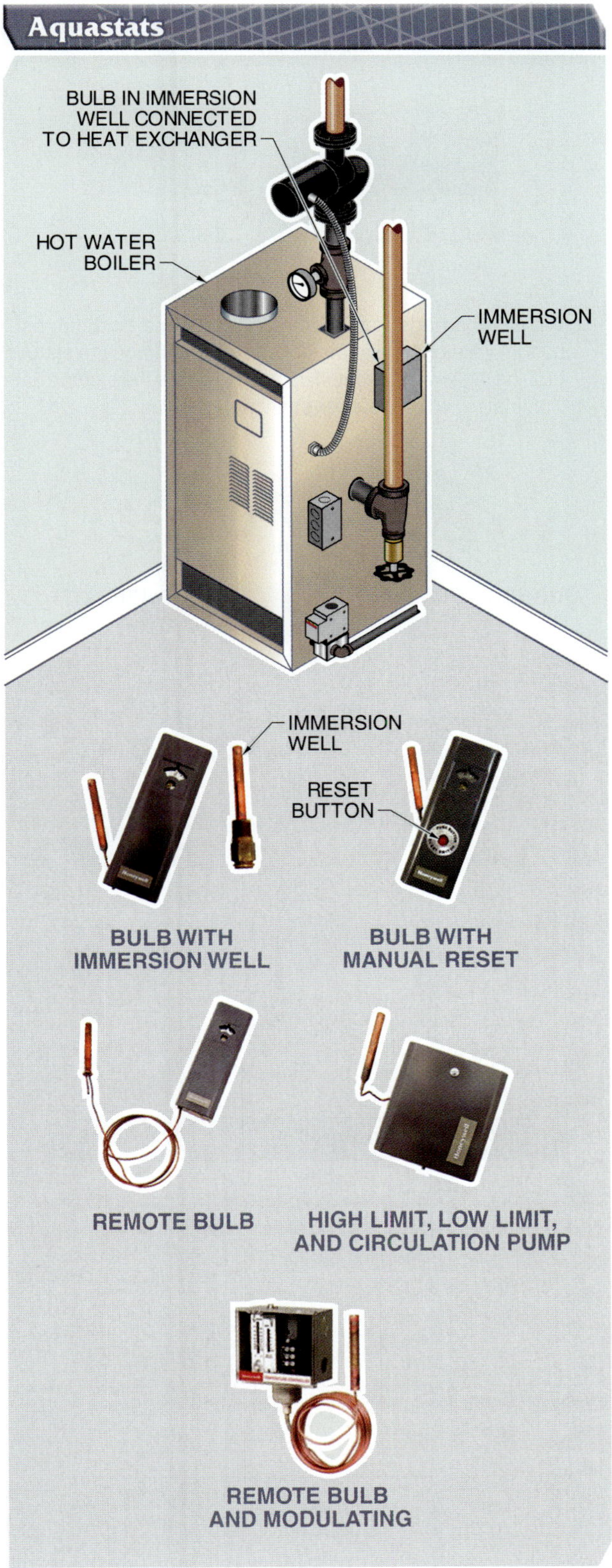

Figure 20-9. An aquastat constantly measures the temperature of boiler water and cycles the burners as necessary to maintain the temperature of the boiler water at the setpoint.

Most hot water boilers are equipped with a second aquastat. It is typically a manual-reset type that is set at a higher temperature to provide a safety backup to the primary aquastat. Many light commercial hot water boilers may have a modulating aquastat to control the boiler firing rate and limit the number of times the boiler cycles ON and OFF.

Cleaver-Brooks

A low water cutoff stops operation of a boiler if the water level drops too low.

Low Water Cutoffs

A *low water cutoff (LWCO)* is a sensor and switch or probe that opens the combustion circuit of a boiler in the event of a low water condition. An LWCO is a very important boiler fitting. An LWCO consists of a small drum or container that includes a float or probe. It is attached to the side of a boiler, boiler return, or supply piping above the minimum safe operating water level. Boiler burners will stop operating when the water level drops below the level of the LWCO.

A light commercial boiler may have an LWCO that includes a blowdown line that can be opened. Low water cutoffs must be checked at least once a year or when local ordinances require. Any applicable codes and manufacturer recommendations should always be followed when testing a LWCO.

In order to ensure the safe operation of a boiler, the LWCO must be tested. Since there are a number of design types and installation methods, exact procedures vary. The manufacturer should be consulted for the procedure on a specific device. **See Procedure for Testing Low Water Cutoff.**

Procedure

Testing Low Water Cutoff (LWCO)

1. On a new installation, do not fill the boiler with water.

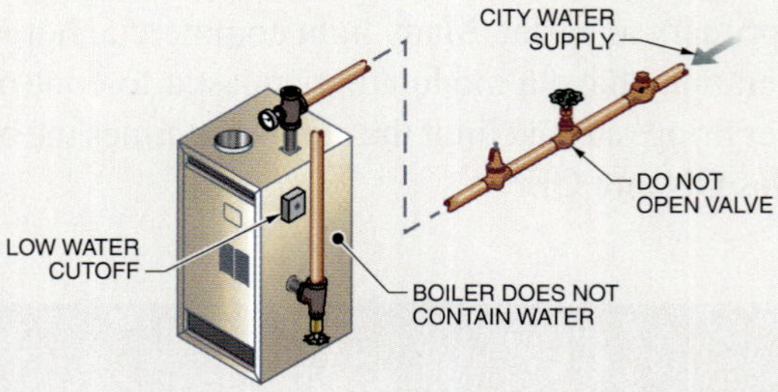

2. Follow all electrical and hazardous environment safety practices and procedures.
3. Set the system thermostat to its lowest possible setting and turn on the electric power to the boiler. On the initial startup, the low water cutoff will indicate a low water condition, which indicates proper function. The status LED on the switch control unit should illuminate red. The burner must not operate without a proper level of water in the boiler.

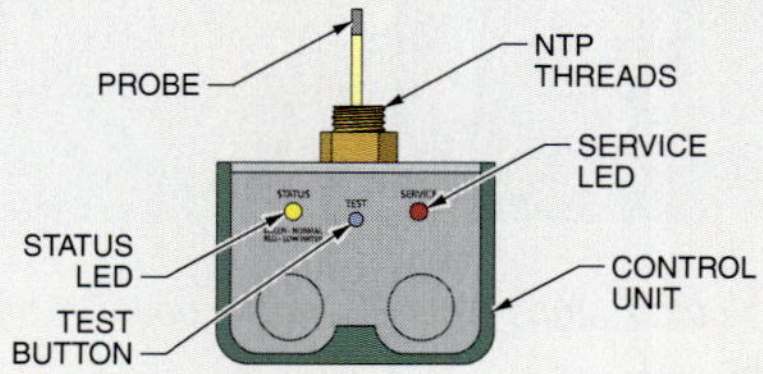

LOW WATER CUTOFF (TOP VIEW)

4. Fill the boiler with water. Once water covers the probe of the low water cutoff, the status LED on the control unit will turn from red to green.

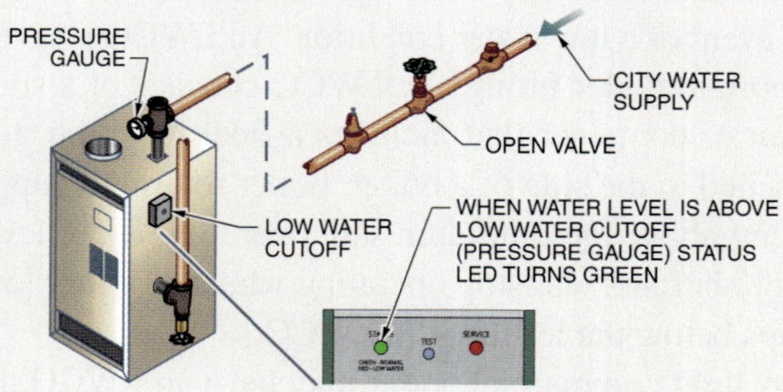

5. Set the thermostat to call for heat and verify that the burner ignites.
   a. Push "View Temperature" button to see space temperature.
   b. Push the "Adjust Temperature Higher" button until display indicates a minimum of 68°F.

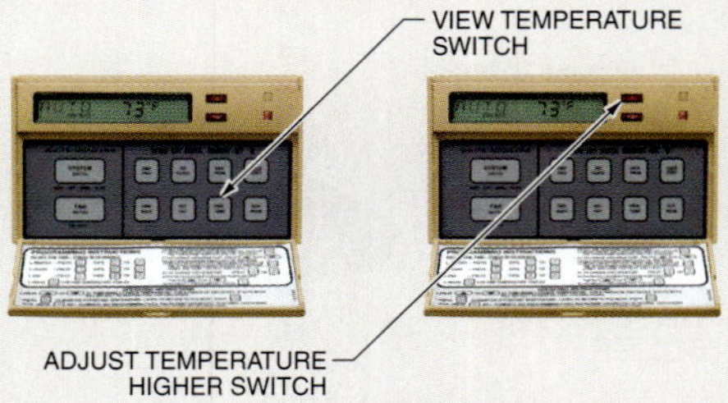

6. With the burner(s) on (aflame), depress and hold the test button while observing the boiler's burner(s) and the status LED on the low water cutoff control unit. After the status LED illuminates red, the burner(s) should go out.

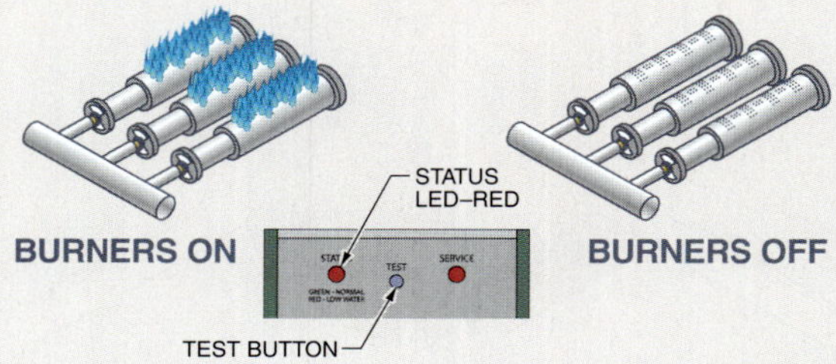

7. Release the test button. The status LED on the control unit should turn from red to green. Verify that the burner ignites (providing water covers the LWCO probe).

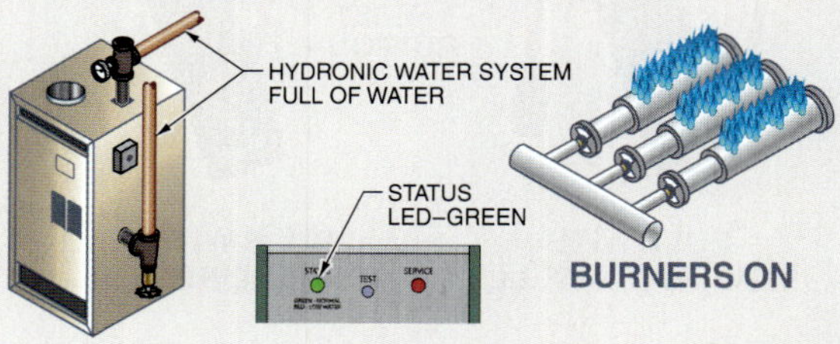

8. If burner fails to cutoff or relight, see the troubleshooting section for details. If all looks good, set the thermostat to its normal setting.

Symptom	Possible Cause	Remedy
Boiler will not ignite burner	No power to LWCO	Verify that the LWCO wiring is correct. Ensure power to boiler is on.
	Low water level	Make sure that the water level is above the probe of the LWCO.
Boiler will not shut down	Shorted probe	Turn off power to the boiler and remove the LWCO. Verify that there is at least ¼″ clearance from all metal surfaces to the metal probe on the LWCO.
Amber service LED is on	Dirty probe	Remove the LWCO and thoroughly wipe it with a clean rag. Refill the system with water and test LWCO for proper operation.

Gas Control Valves

A gas control valve controls the flow of fuel to a gas-fired atmospheric burner. A *gas control valve* is a 100% shutoff safety valve that controls the flow of fuel to a main burner and a pilot burner, if used. When combustion does not occur on a call for heat, the valve will close and not allow gas to flow to the main burner or to the pilot burner, if used. Boiler gas control valves can be thermocouple and standing pilot, automatic, combination, or modulating. **See Figure 20-10.**

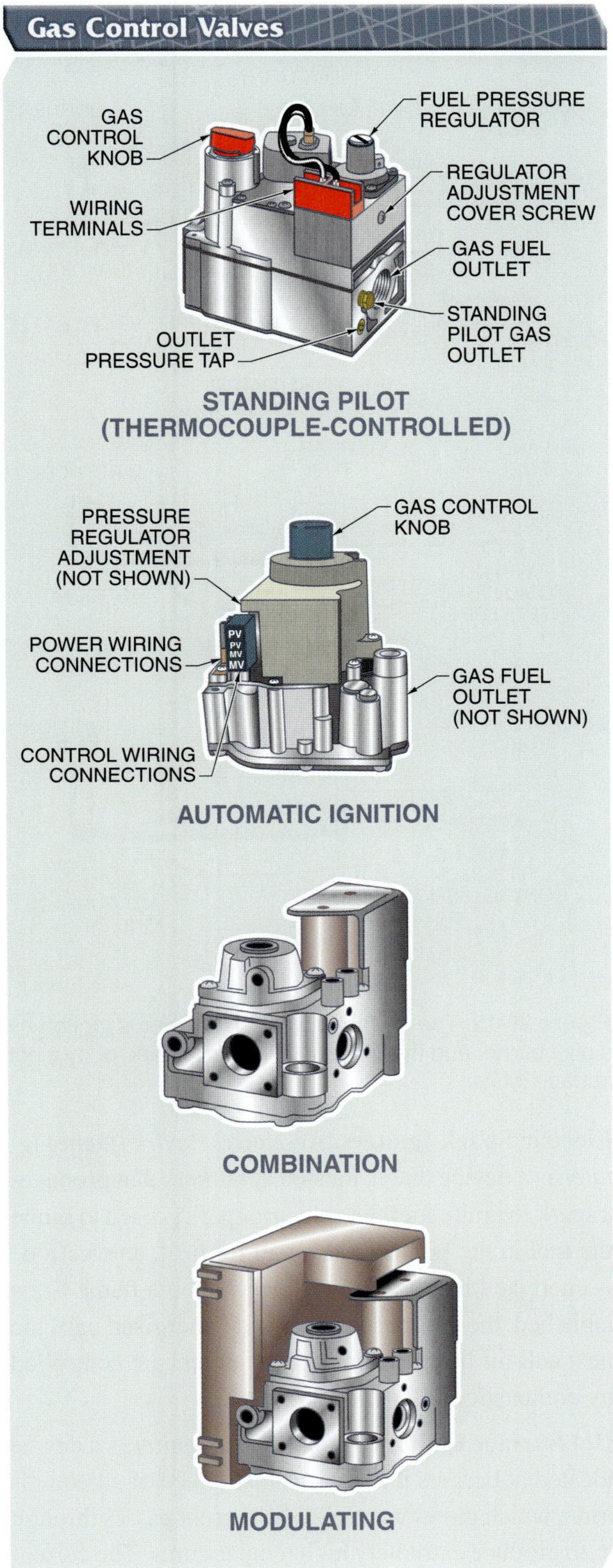

Figure 20-10. A gas control valve controls the flow of fuel to a gas-fired atmospheric burner. When combustion does not occur on a call for heat, the valve will close.

A *thermocouple and standing pilot gas valve* is a gas control valve that is used to control the burners of a boiler and uses the power generated by the thermocouple. An *automatic gas valve* is a gas control valve used with direct spark ignition (DSI) systems or hot surface ignition systems. A *combination valve* is a gas control valve that has a built-in fuel pressure regulator, which controls the pressure of fuel gas as it enters a burner. All residential boilers use combination valves. A *modulating gas valve* is a gas control valve that regulates the high-low flow of gas to the burners of a boiler and typically has an attached ignition control system.

Burners

A *burner* is the heat-producing component of a combustion boiler. Most residential boilers typically use many small burners, while others and light commercial boilers use power burners. A burner mixes air and fuel to provide a combustible mixture. A burner supplies an air-fuel mixture to the burner face (place of combustion) while metering the fuel to maintain a controlled firing rate. Residential boilers typically use tubular burners, but many larger residential boilers use power burners. **See Figure 20-11.**

Tubular Burners. Tubular burners are mounted in an array side by side with each burner being fed fuel and combustion air. Each burner has a percentage of the total heating capacity. They may be removed for cleaning and maintenance as needed.

Power Burners. A power burner consists of a forced air fan that is used to bring air under pressure into close contact with the fuel. A power burner is a single, extremely large burner compared to other burners that are found on a particular boiler.

Boiler power burners force air and fuel together under pressure.

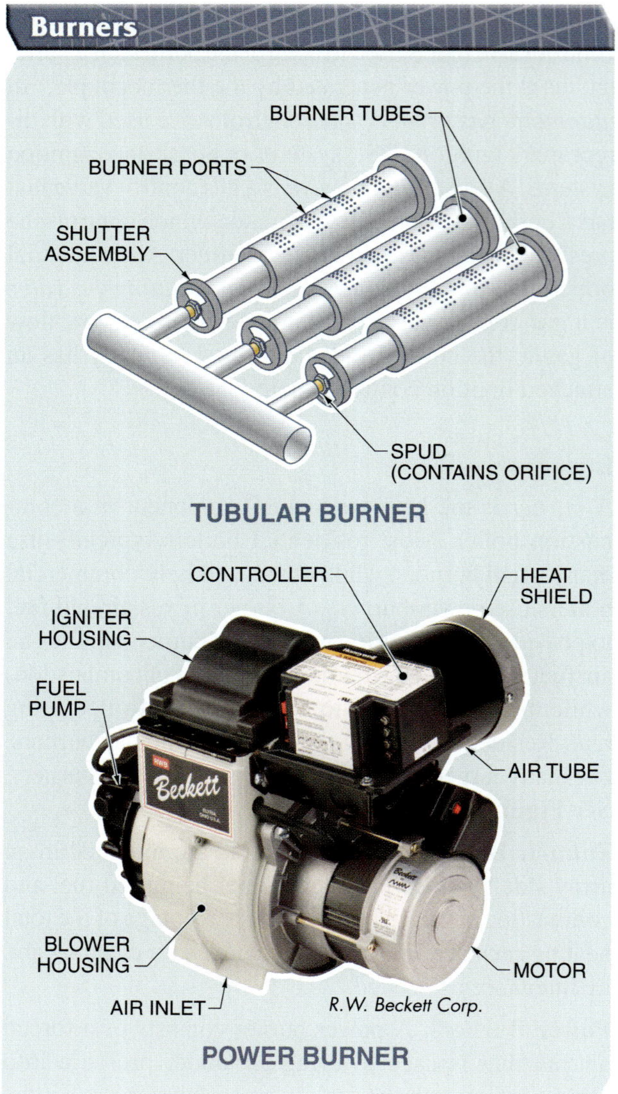

Figure 20-11. A burner produces heat by supplying an air fuel mixture to the burner face and maintaining a controlled firing rate.

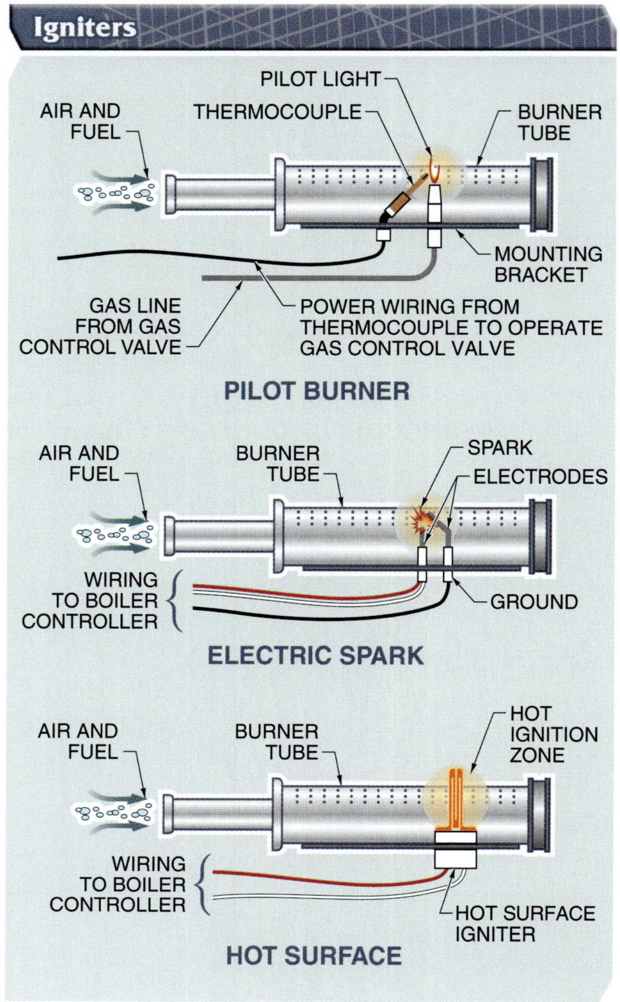

Figure 20-12. The air-fuel mixture in a boiler is ignited by a pilot burner and thermocouple, electric spark, or by a hot surface igniter.

Fuel Igniters

A fuel igniter may be a standing flame that burns constantly, such as the pilot light of a pilot burner, or a device that electrically ignites a gas on each call for heat. **See Figure 20-12.** The two types of electric igniters are the electric spark (flame) igniter and the hot surface igniter.

Pilot Burners. A *pilot burner* is a small burner located near burner tubes that ignites the fuel and air mixture released into the burners. A pilot burner produces a pilot light, which is a small standing flame. The pilot light ignites an air-fuel mixture when the gas fuel valve opens. The gas control valve will have a pilot light gas line to maintain the standing flame.

Electric Spark Igniters. An *electric spark (flame) igniter* is a device that is located at burners that produces a spark to ignite fuel. The electric spark is used to ignite the fuel in the burners. On a call for heat, an electrode is energized near the burner face. When a flame is established, the electrode igniter is de-energized until the next call for heat. Electric spark igniters are controlled by combustion safety control systems.

Hot Surface Igniters. A *hot surface igniter* is a device located at burners that uses a small piece of silicon nitride, which glows when electric current passes through it, to produce extremely high temperatures. The current is usually 115 VAC. The silicon nitride material is very durable and produces few problems. Hot surface igniters are suscepable to hot spots if mishandled.

ACCESSORIES FOR HOT WATER BOILERS

An accessory is a component not directly attached to a system but is needed for proper system operation. Residential and light commercial hot water boilers do not have feedwater systems. Accessories used in the hydronic heating system include pressure-reducing valves, flow control valves, and air separators.

Pressure-Reducing Valves

Hot water boiler hydronic systems that are working properly do not lose or require more water. When the boiler, piping, and terminal devices are filled with water, the water stays in the system virtually forever. The water heated by the boiler is constantly circulated throughout the entire heating system. When problems occur, water can leak out of the system and this leaking water must be replaced. The makeup water typically comes from the city water supply.

A *pressure-reducing valve* is a valve that reduces the pressure of makeup water from a water supply line. The pressure is reduced from approximately 40 psi down to 12 psi. The pressure-reducing valve can be set to higher pressures for residences and commercial buildings that are two or three stories tall to accommodate for the weight of the water in these structures. Typically, pressure-reducing valves are not set higher than 18 psi. A pressure-reducing valve is also known as an automatic water feed valve. Pressure-reducing valves are found as automatic, automatic with backflow preventer, and automatic with manual quick fill. **See Figure 20-13.**

Automatic pressure-reducing valves will automatically fill a hydronic system when pressure drops below the setpoint because water has leaked out and the system is not operating. Some pressure-reducing valves have backflow preventers, or check valves, to stop system water from flowing back into the water supply lines when the system is hot and at higher pressures. Other pressure-reducing valves have handles that allow the valve seats to be manually lifted and allow full flow into the system for quick fill.

TECH TIP

Automatic pressure-reducing valves may be plugged by particulate in the incoming water supply, allowing system pressure to drop. This may cause the low water cutoff to stop boiler operation.

Pressure-Reducing Valves (Automatic Water Feed Valves)

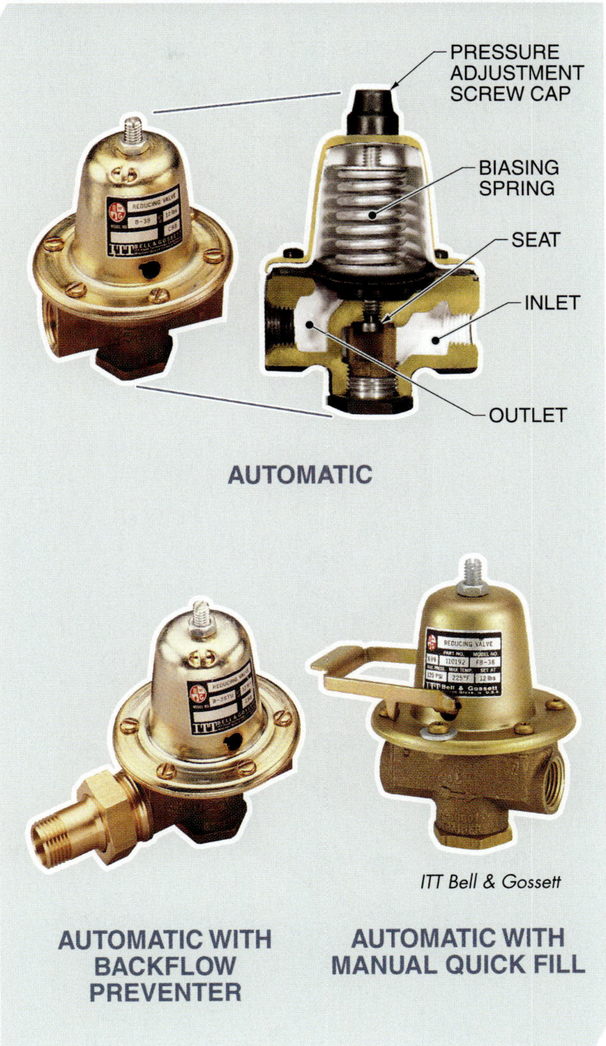

Figure 20-13. A pressure-reducing valve (automatic water feed valve) reduces the pressure of makeup water from the water supply line and automatically fills a hydronic system when pressure drops below 12 psi.

A pressure-reducing valve is located in the water supply piping that leads to the boiler and hydronic system. **See Figure 20-14.** Backflow preventers do not need to be installed with the pressure-reducing valve. Backflow preventers are installed to local, state, and provincial codes as individual devices between the water supply line and hydronic system.

Pressure-Reducing Valves Media Clip

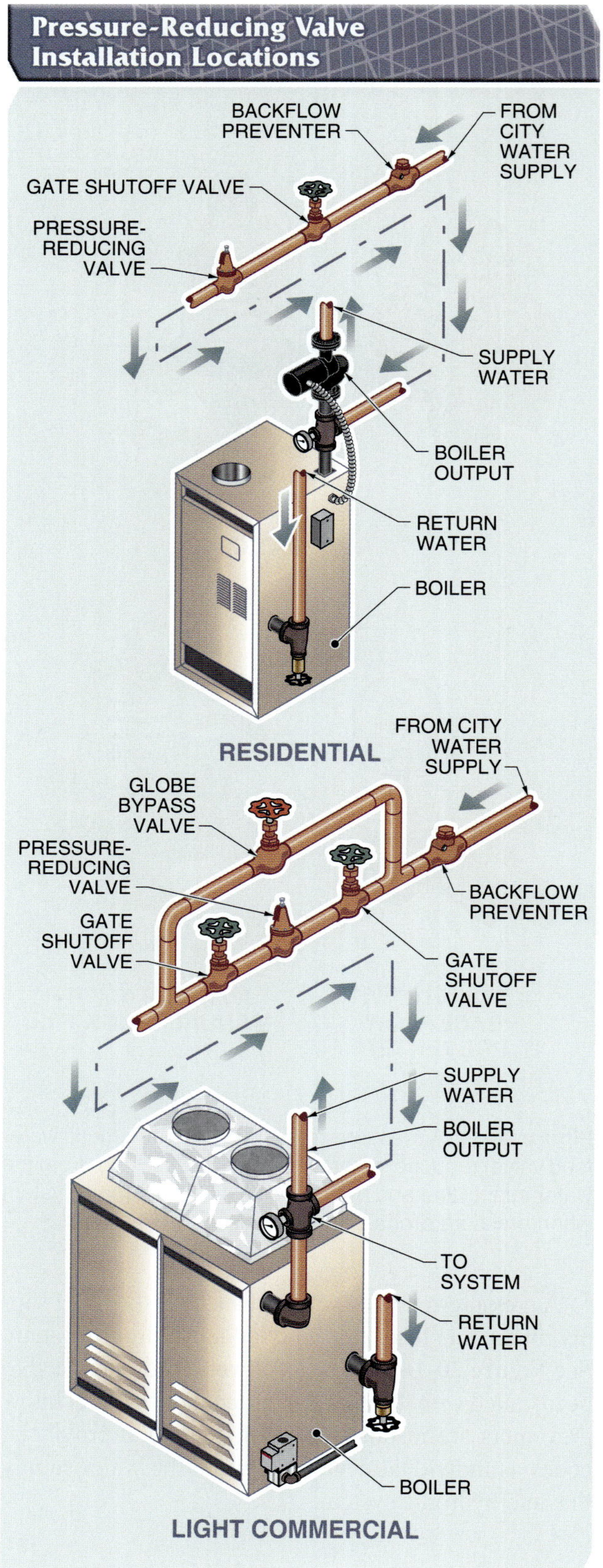

Figure 20-14. Pressure-reducing valves are typically located with shutoff valves and backflow preventers.

Flow Control Valves

A *flow control valve* is a valve that limits the flow of water (in gpm) in a hydronic heating system. A flow control valve may be manual or automatic. A *manual flow control valve* is a globe valve that manually limits the flow of water in a hydronic heating system. An *automatic flow control valve* is a check valve that opens and closes automatically. When a circulating pump is ON, the flow control valve opens to allow water flow. When the circulating pump is OFF, the flow control valve closes and prevents water from circulating. A flow control valve is located on the outlet piping of a boiler. **See Figure 20-15.**

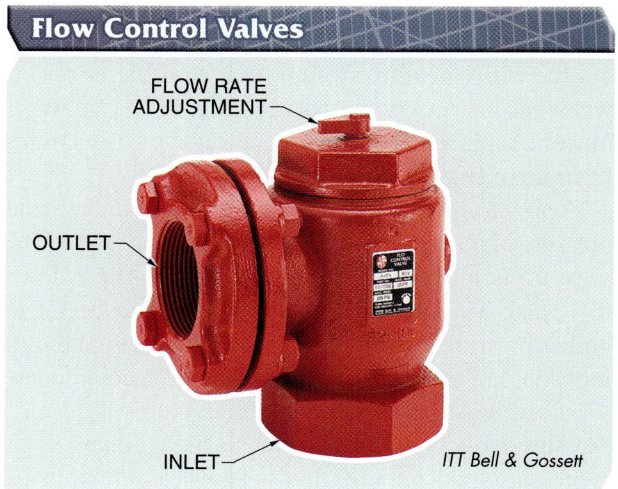

Figure 20-15. Flow control valves limit the amount of water flow throughout a hydronic heating system.

Air Separators

Free air is present in the hot water that is being circulated throughout a hydronic heating system. When air is not removed, heat transfer is reduced, the system becomes noisy, and the air accumulates in the terminal devices of the system. Air separators of different designs are used to remove the air from the system, allowing trouble-free operation. **See Figure 20-16.**

HYDRONIC SYSTEM AND BOILER CONTROL DEVICES

A number of different control devices and schemes are used for the accurate temperature control and energy-efficient operation of hot water heating systems. Depending on the building design and location, these schemes can vary widely.

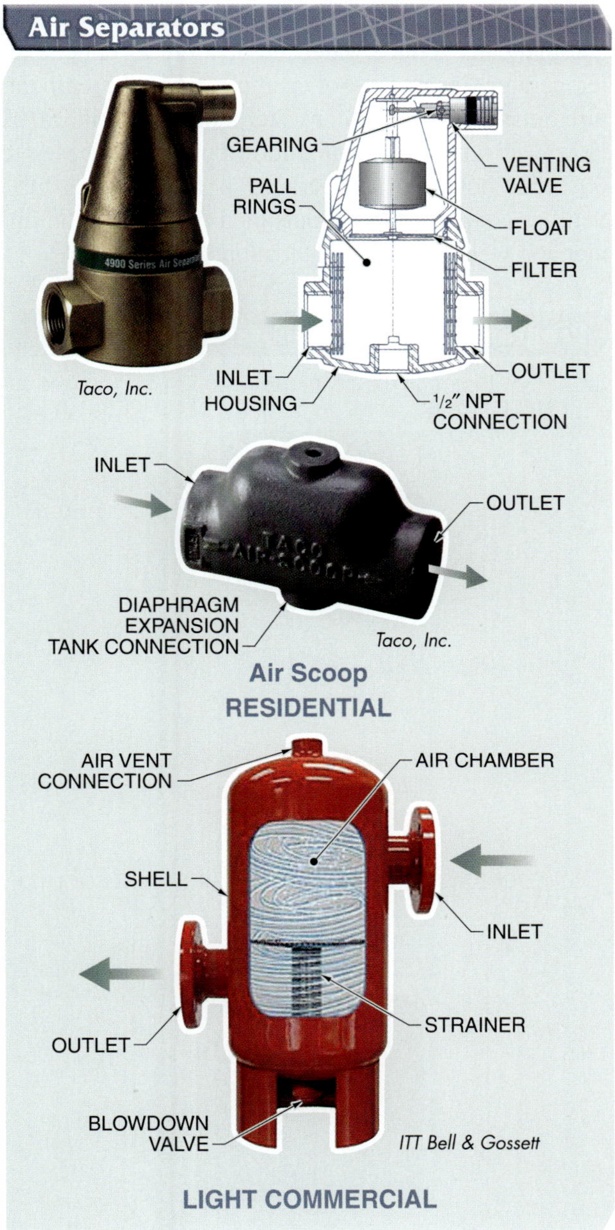

Figure 20-16. Air separators are used to remove the air from a hydronic system to prevent noise and a reduction in heat transfer to terminal units.

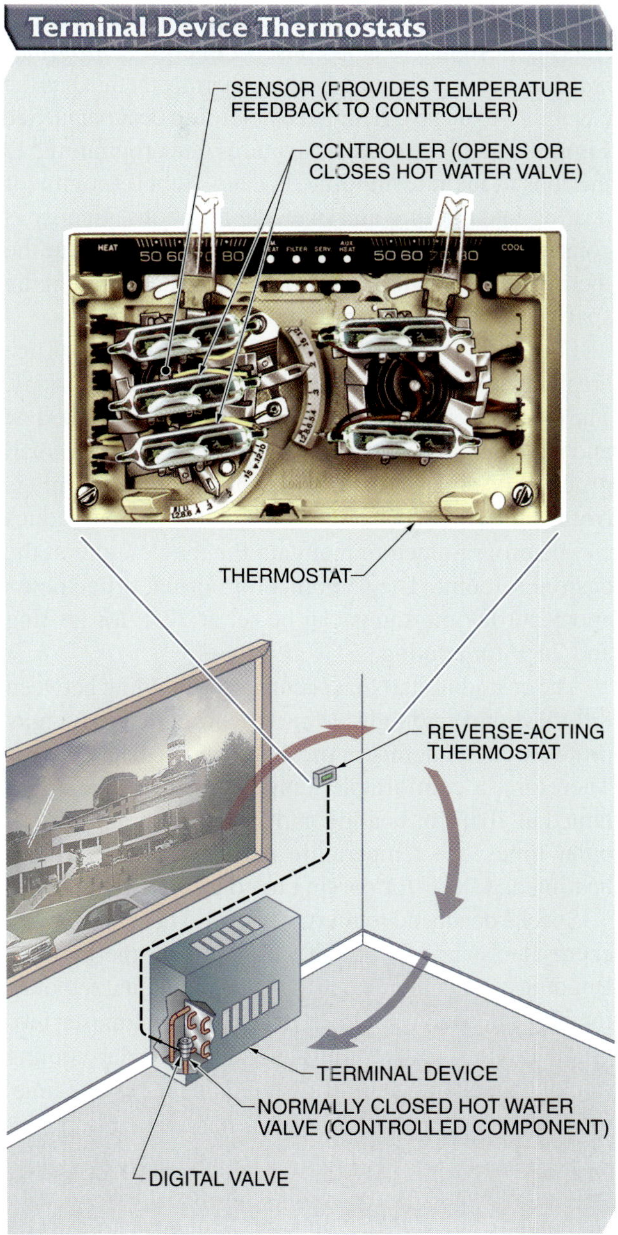

Figure 20-17. Terminal device thermostats regulate water and airflow by controlling circulating pumps and by opening and closing zone valves.

Thermostats

Thermostats used with terminal devices regulate water and airflow by controlling circulating pumps, zone valves, and/or terminal device blowers. A digital thermostat is used to control digital valves. On a call for heat, the thermostat opens the valve. When the temperature reaches the thermostat setpoint, the thermostat closes the valve. **See Figure 20-17.**

A proportional thermostat is used to control modulating valves. A *proportional thermostat* is a thermostat containing a potentiometer that sends out an electric signal that varies as the temperature varies. A potentiometer is a variable-resistance electric device that divides voltage proportionally between two circuits. An actuator receives the signal and converts it to mechanical action in the modulating valve.

Programmable Thermostats. Programmable thermostats are designed to help conserve energy. A *programmable thermostat* is a stand-alone thermostat that has the capability to be programmed by building occupants. **See Figure 20-18.** The standard features of a programmable thermostat include multiple day and night setpoints for heating and cooling and override functions. Energy is conserved by programming separate setpoints into the thermostat for the times throughout the day or night the space is occupied and unoccupied.

For example, the normal working hours for an office are from 8:00 AM to 5:00 PM, Monday through Friday. The custodial staff cleans between the hours of 8:00 PM and midnight, Monday through Friday. There is a small group of occupants that come in on Saturday mornings from 8:00 AM until noon. All of these time frames can be programmed to maintain the temperature at the desired setpoint. The setpoints for normal office hours and Saturday mornings can be set at 72°F for heating and 74°F for cooling.

The custodial staff that occupies the building between 8:00 PM and midnight, Monday through Friday, may prefer a cooler temperature because of their activity. Therefore, a comfortable temperature could be maintained at 70°F for heating and 72°F for cooling. At all other times, the temperature is set down to 65°F for heating and 78°F for cooling to conserve energy.

Special deadband settings, which also help to conserve energy, are standard on most programmable thermostats. Building occupants are able to set the individual setpoints for the desired temperatures. However, the thermostat program will not allow heating setpoints to overlap cooling setpoints to avoid heating and cooling at the same time. In many cases, there is a minimum of a 2°F differential between the heating setpoint and the cooling setpoint. *Deadband* is the temperature range between setpoints in which no mechanical heating or cooling occurs.

Programmed setpoints can be overridden. For example, the Saturday staff is required to work until 4:00 PM instead of noon to complete a large project. The thermostat would normally set the temperature setpoint back to 65°F at noon. The occupants can override the setting and increase the temperature setpoint to 72°F and continue to work. At 5:00 PM, the override is removed, and the thermostat sets the setpoint back to the programmed 65°F temperature setting.

Outside air reset controls are sometimes used to automatically reduce water temperature as outside air temperature increases. For example, when the temperature outside is 0°F, the hot water temperature is 180°F. However, when the temperature outside increases to 60°F, the hot water temperature is reduced to 100°F. Sensors are placed in the outside air and inside an immersion well in the hot water supply piping.

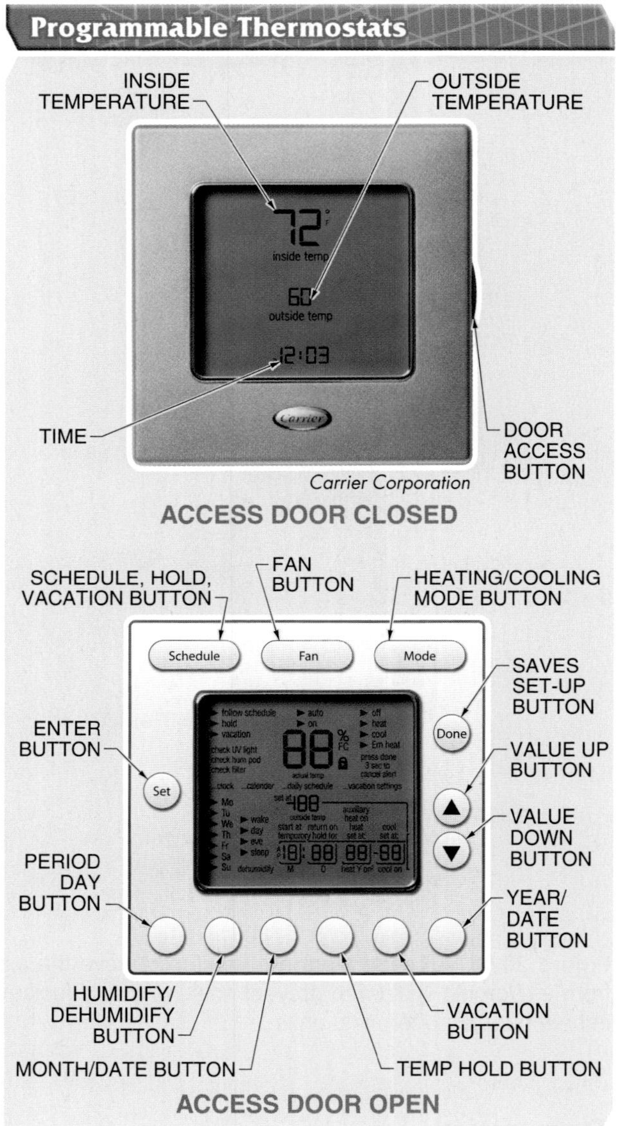

Figure 20-18. The standard features of a programmable thermostat include multiple day and night temperature setpoints and special override functions.

UNIT

21 Hydronic Heating Distribution Systems

A hydronic heating system heats water in a boiler at a central location. The heated water is distributed through supply piping to building spaces. In the building spaces where the heat is required, the hot water flows through terminal devices. The terminal devices transfer heat from the hot water to the air of the building spaces. The water returns to the boiler from the terminal devices through return piping and accessories.

OBJECTIVES

- Identify common hydronic distribution system components.
- Describe common hydronic piping systems.
- List types of hydronic system terminal devices and terminal device controls.

Index to Procedures

HYDRONIC DISTRIBUTION SYSTEMS

Hydronic distribution systems use water to transport heat from the point of generation to the point of use. Residential and light commercial hot water boilers do not have feedwater or steam systems. Most residential distribution systems constantly circulate boiler-heated water throughout the entire heating system. Light commercial distribution systems typically circulate boiler-heated water through one, two, or more zones of the system at any given point in time. The boiler and system are capable of having flow through the entire system. **See Figure 21-1.** Accessories used in a hydronic distribution system include circulating pumps, expansion tanks, valves, and terminal devices.

Circulating Pumps

A *circulating pump* is a type of pump that creates a flow of water through the boiler, system piping, and terminal devices of a hydronic heating system. A circulating pump may be connected on either the supply or return piping but is typically connected close to the boiler on the return piping. The required number and size of circulating pumps are determined by the volume of water to be moved and the resistance to flow in the system. Hydronic heating systems that have many piping loops that feed different zones may have a pump located in each loop.

TECH TIP

Hydronic system loops must be sized, installed, tested, and balanced properly. Failure to do so may result in system problems such as parts of the system being too cold, wasted energy, or physical damage due to frozen coils. Any ongoing hydronic heating problems should be checked using the installation prints, balancing report, or if possible, the actual engineering calculations that were used for system sizing.

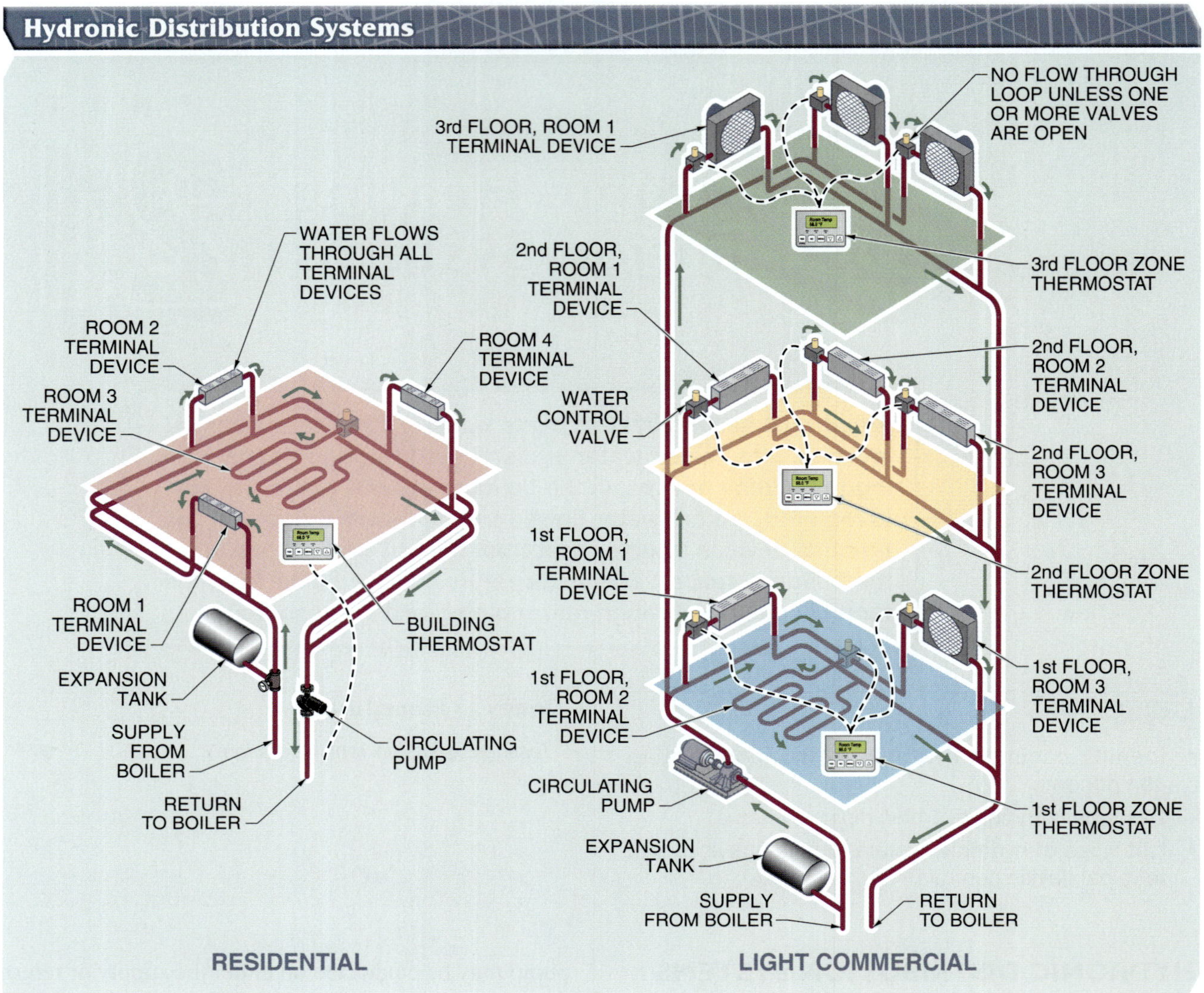

Figure 21-1. Most residential hydronic distribution systems constantly circulate water throughout the entire heating system, while light commercial distribution systems circulate water through one or more zones at any given point in time.

Almost all circulating pumps that are used in hydronic heating systems are centrifugal pumps. A *centrifugal pump* is a pump that has a rotating impeller inside a cast iron or steel housing (volute). An *impeller* is a plate with vanes that radiate from a central hub (eye). The water inlet is an opening in the pump housing that is at the center of the impeller. The water outlet is an opening in the pump housing that is located on the outer perimeter of the pump housing. **See Figure 21-2.**

Return piping is connected to the water inlet connection of a centrifugal pump. Supply piping is connected to the water outlet connection of a centrifugal pump. The impeller of a circulating pump rotates at speeds between 1150 rpm and 3550 rpm. As the impeller rotates, a low-pressure area is created at the eye of the impeller to allow fluid to enter the pump. The vanes of the impeller throw water outward (centrifugal force) to the outlet of the volute.

Circulating pumps are typically driven by electric motors. The motor is often part of the pump housing on residential circulating pumps that have a pumping rate of 0 gpm to 100 gpm. The motor on a circulating pump that has a pumping rate above 100 gpm is typically connected by an in-line connection on the shafts of the pump and motor. Pipe-mounted circulating pumps are small pumps that are installed in the piping of a hydronic system. Floor-mounted circulating pumps are large pumps that must have piping run to the pump. **See Figure 21-3.**

Centrifugal Pumps

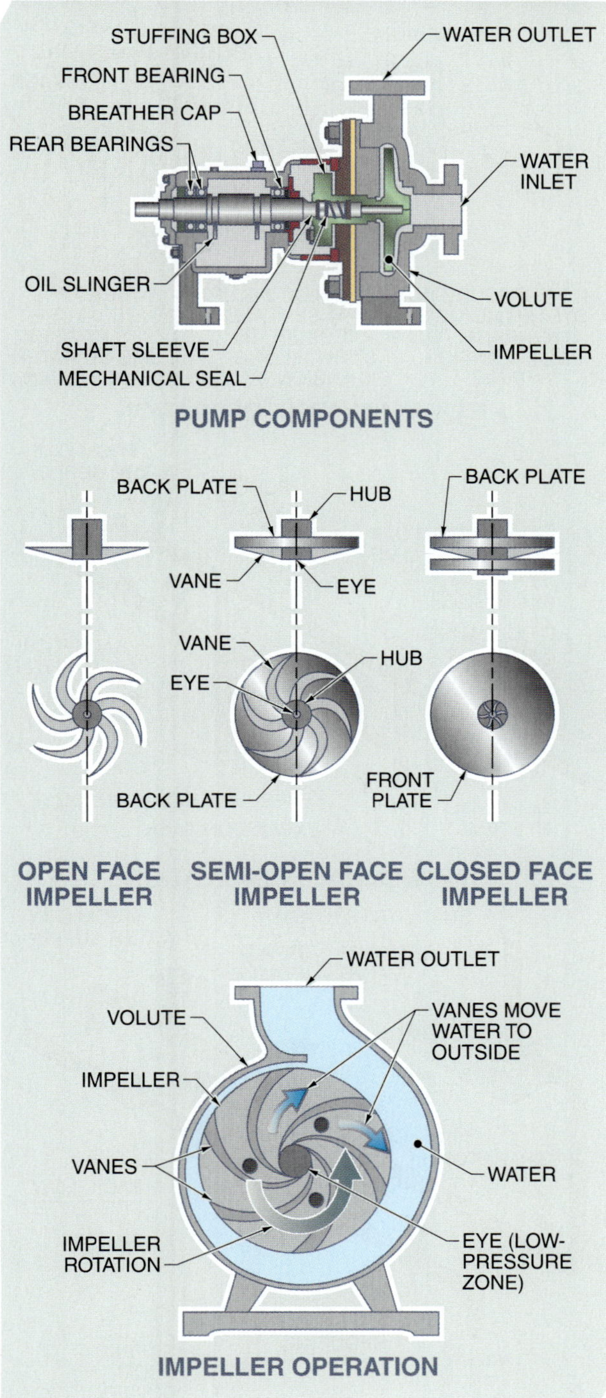

Figure 21-2. The discharge flow rate of a centrifugal pump is dependent on impeller size and rpm. The pressure rating of the pump is dependent on the style of impeller and number of stages.

Circulating Pump Operation Media Clip

Circulating Pumps

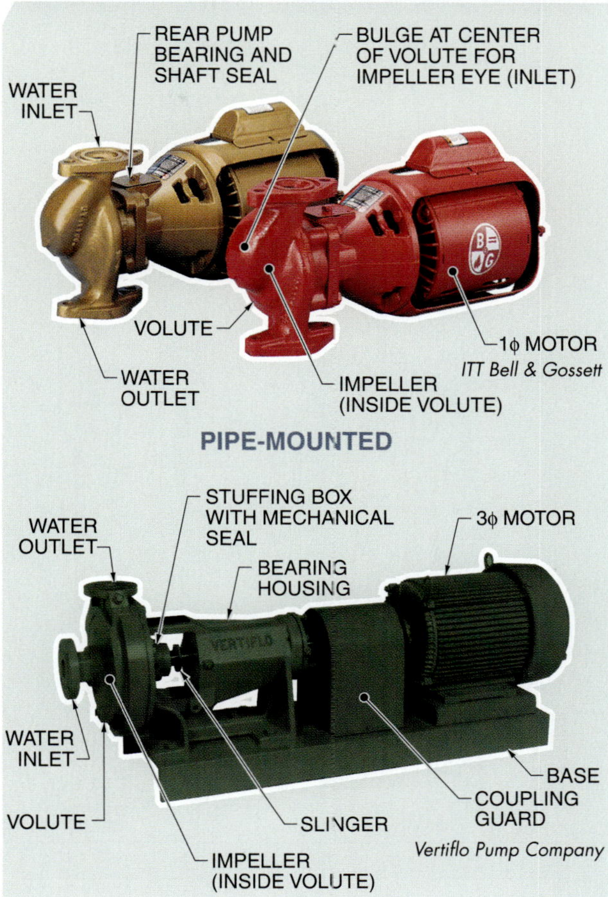

Figure 21-3. Pipe-mounted circulating pumps are small with low flow rates, while floor-mounted circulating pumps have medium flow rates.

Variable-Frequency Drives (VFD). A *variable-frequency drive (VFD)* is an electronic device that controls the direction, speed, torque, and other operating functions of an electric motor in addition to providing motor protection and monitoring functions. To save energy, many motor starters for circulating pump motors are being converted to VFDs. These motor starters are being replaced with electronic devices that change the frequency of electricity to the motor from the normal 60 Hz to any hertz, depending on the application. **See Figure 21-4.**

Changing the frequency to 30 Hz slows the motor to half the original speed. Typically, the speed can be modulated between 20% and 100%. By increasing or decreasing the motor rpm, the discharge flow rate of a pump can be controlled to maintain desired system settings.

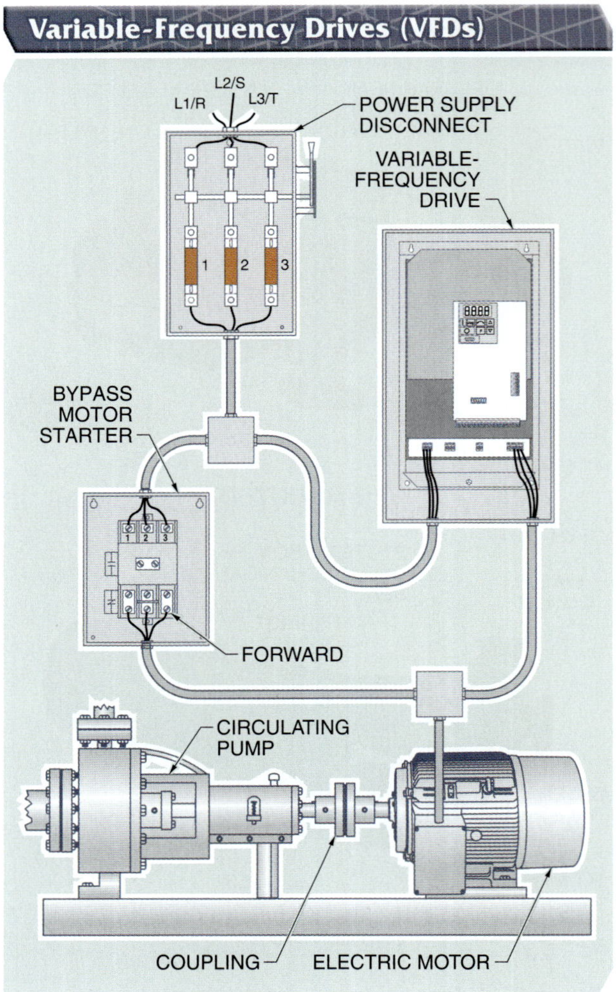

Figure 21-4. A variable-frequency drive (VFD) is an electronic device that controls the functions of an electric motor, such as direction, speed, and torque, in addition to providing motor protection and monitoring of motor operation.

Expansion Tanks

All components, piping, and terminal devices of a hydronic heating system operate while full of water. Water expands as it is heated. An *expansion tank* is a vessel that allows the water in a hydronic heating system to expand without raising the system water pressure to dangerous levels.

There are two types of expansion tanks. There is the older, large horizontal steel tank and the newer, small bladder tank. The newer expansion tank has a rubber diaphragm that is filled with a precharge of air (residential—12 psi) when ready for normal system operating conditions. As water in the system expands, the water flows into the tank, compressing the air and diaphragm. **See Figure 21-5.** The precharge of air in the tank diaphragm acts as a cushion. **See Procedure for Testing Expansion Tank Operation.**

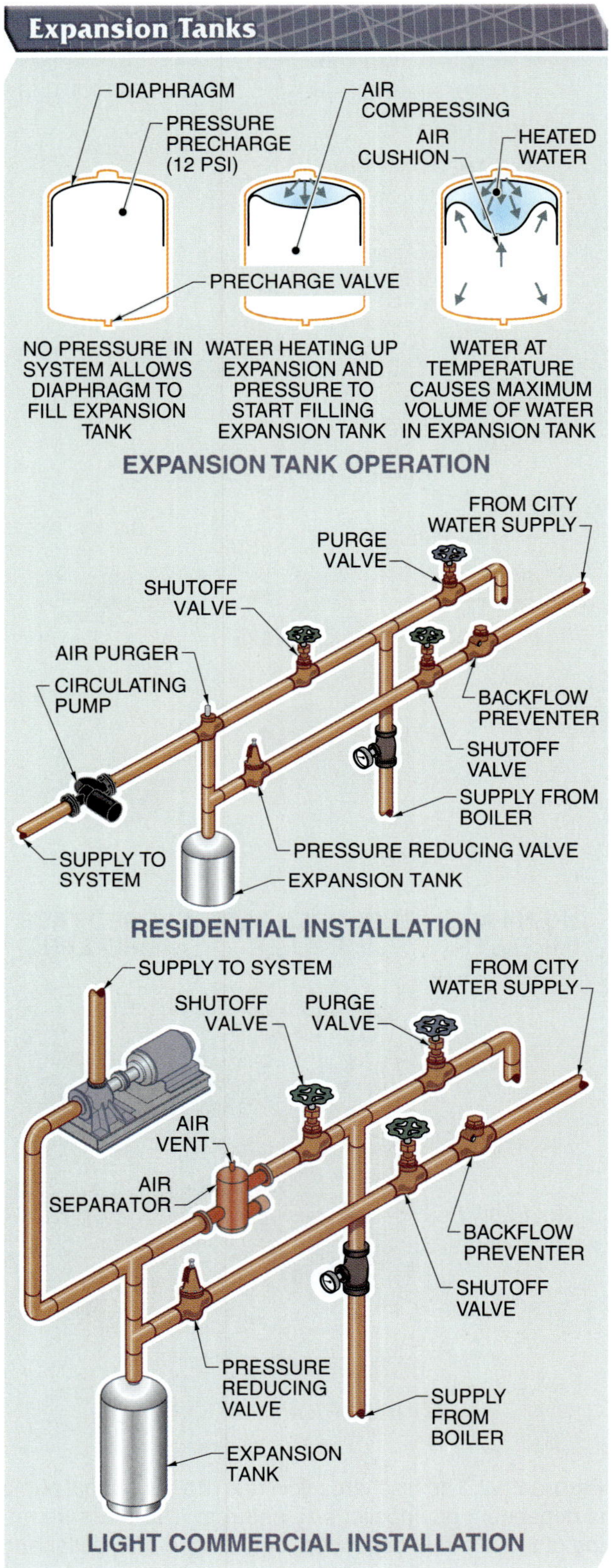

Figure 21-5. An expansion tank is partially filled with an air precharge of 12 psi that acts as a cushion as water in a hydronic system expands due to heating.

Testing Expansion Tank Operation

1. If the hydronic system has been shut down for a period of time and is cold, verify that the precharge pressure of the expansion tank is 12 psi.
   a. Ensure hydronic system pressure is 0 psi. Close the city water supply shutoff valve, and open the purge valve of the hydronic system.

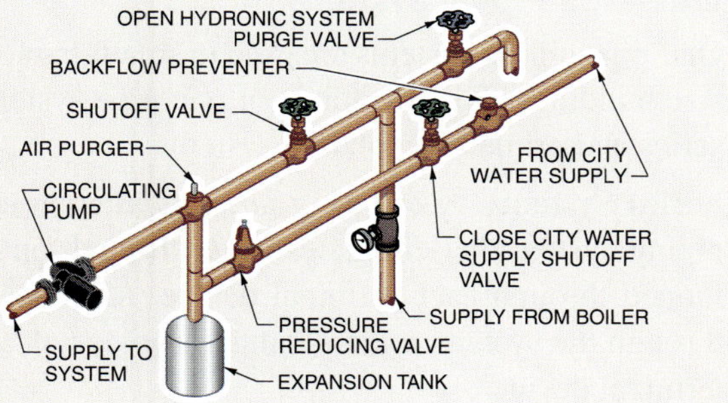

   b. Connect the gauge to the expansion tank precharging valve.

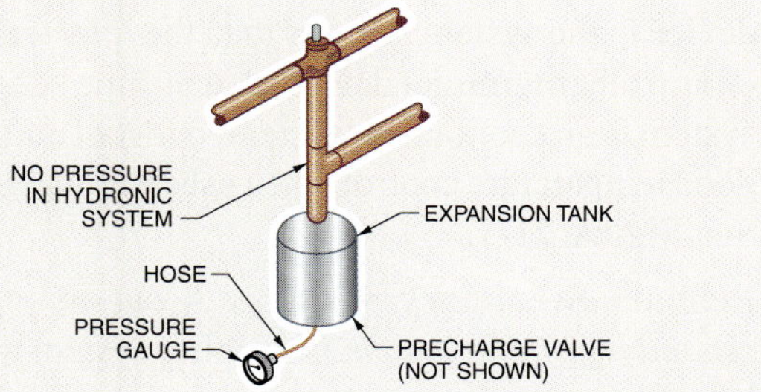

   c. Measure the diaphragm precharge pressure. If below 12 psi, pump more air into the diaphragm. If above 12 psi, discharge some air.

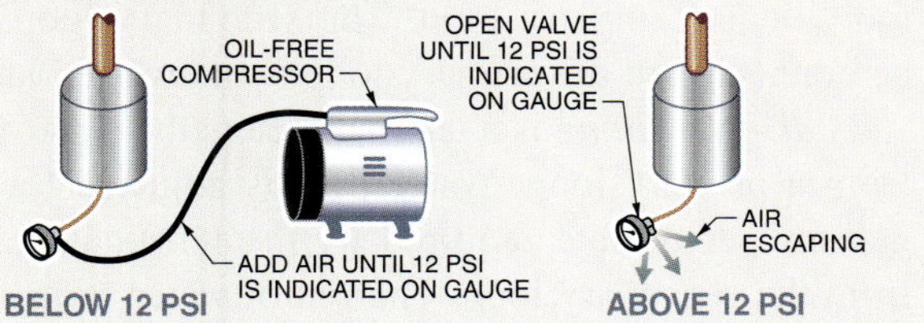

2. With the hydronic system shutdown and cold, verify that the cold fill pressure of the hydronic system is 12 psi.
   a. If the cold fill pressure is below 12 psi, close the purge valve and open the city water supply shutoff valve to add water to the hydronic system until pressure reaches 12 psi.

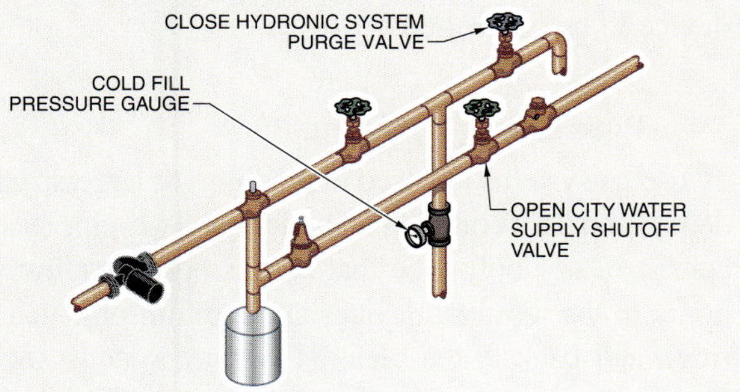

   b. If cold fill pressure is above 12 psi, close the city water supply shutoff valve and open the purge valve to drain some water from the hydronic system until pressure drops to 12 psi.

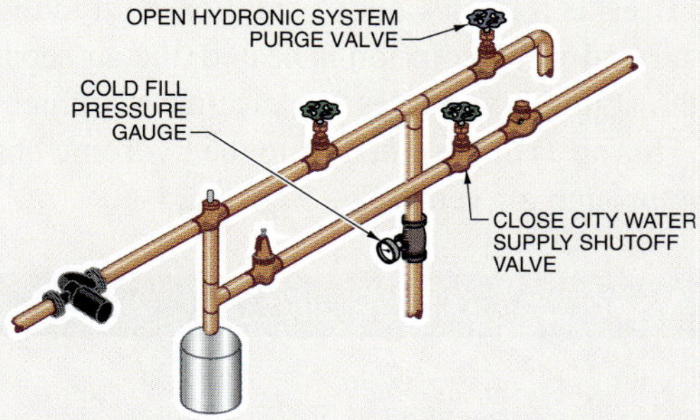

3. When the system is operating and pressure exceeds normal operating pressures of 18 psi to 20 psi, check the safety relief valve for leakage. Leakage from the safety relief valve can be caused by expansion tank overfilling (water logging), allowing pressures to reach the safety relief valve setting of 30 psi. Check the relief valve for proper operation.

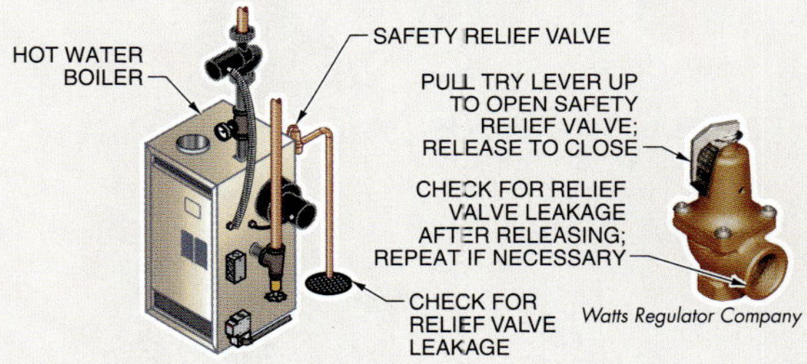

Watts Regulator Company

4. If during normal burner operation, water pressure rises very fast as temperature increases:
   a. Verify diaphragm has not failed and that the expansion tank is not full of water (water logged) by tapping the expansion tank. The bottom should sound hollow (air-filled), and the top should have a thud sound (water-filled).

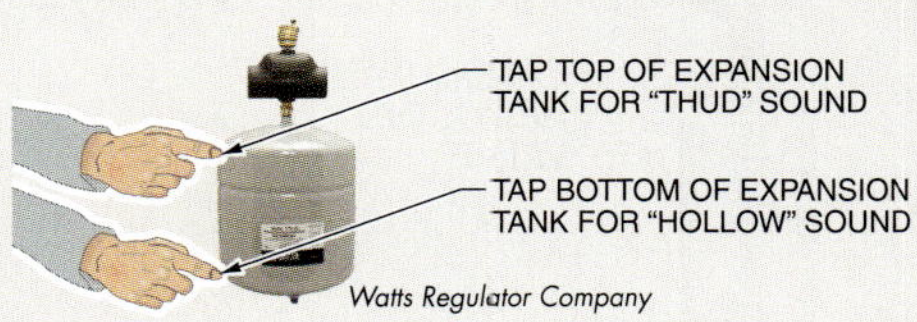

Watts Regulator Company

   b. Verify that the expansion tank is the correct size for the hydronic system.

EXPANSION TANK SIZING*

Boiler Output†	Copper Baseboard & Radiant	Type of Radiation‡		
		Convectors and Unit Heaters	Cast Iron	
			Radiators	Baseboard
25	2.25	2.25	2.25	2.25
50	2.25	5.00	5.00	5.00
75	5.00	5.00	7.50	7.50
100	5.00	5.00	7.50	7.50
125	5.00	7.50	7.50	7.50
150	5.00	7.50	15.00	15.00

* system water temperature of 200°F
† in 1000s Btuh
‡ approximate gallons in capacity

Air Vents

An *air vent* is an automatic or manual device that vents air from a hydronic heating system without allowing water to escape or air to enter the system. Air vents are installed at the highest parts of the piping system where the air rises as it comes out of solution with the water. As an air and water solution is heated, the air separates from the solution and rises because air is lighter than water. This air is then vented from the hydronic heating system through air vents. **See Figure 21-6.**

Air Vents

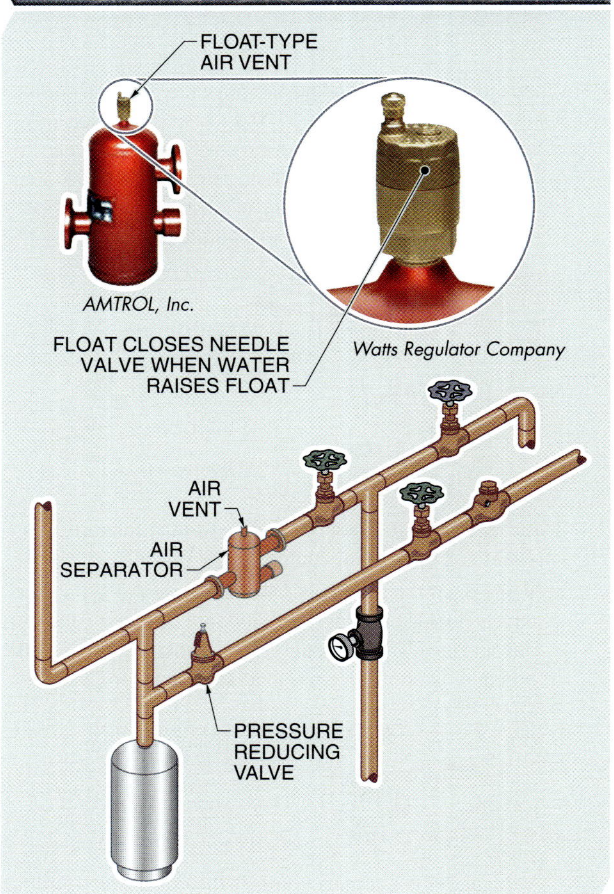

Figure 21-6. Air vents are installed at the highest part of the piping system so that air can be vented from a hydronic heating system without water escaping or air entering the system.

PIPING SYSTEMS

Piping systems for hydronic heating systems are used to distribute heated water from a boiler into building spaces and return the water to the boiler for reheating. Piping systems vary based on the degree of temperature control desired and the specific application. In most systems, the pipe is made of steel or plastic. In small systems, the pipe may be copper. Piping systems used for hydronic heating systems include one-, two-, three-, and four-pipe systems. Because of variations in buildings, each piping system is designed for a specific kind of building.

One-Pipe Piping Systems

One-pipe piping systems are used in small- to medium-size buildings. The two kinds of one-pipe systems are series and primary-secondary systems.

Series Systems. A *one-pipe series piping system* is a piping system that circulates water through one pipe, looped through each terminal device. As water flows through the system, a temperature drop occurs at each terminal device.

The inlet temperature at each terminal device is lower than the inlet temperature at the preceding terminal device. The inlet temperature of the last terminal device in the system is lower than the inlet temperature of any other terminal device. A one-pipe series piping system is an economical system to install and provides good temperature control when used in small buildings. **See Figure 21-7.**

Primary-Secondary Systems. A *one-pipe primary-secondary piping system* is a piping system that circulates boiler water through a primary loop. The terminal devices are connected in parallel to the primary loop by a secondary loop. Each terminal device has an individual secondary loop. The secondary loop can be connected to the primary loop by a diverter valve.

A *diverter fitting* is a tee that meters the water flow from a primary loop. Water flow is adjusted for the gallons per minute required by the terminal device from the secondary loop. The return water from each terminal device returns to the primary loop. Therefore, the temperature drop of each device affects the next terminal device in the system, but the flow rate in each terminal device is controlled within its own secondary loop. Flow control allows the heat output of the terminal device to be controlled.

Two-Pipe Piping Systems

Two-pipe systems are used in medium- to large-size residential and light commercial buildings. Two-pipe systems consist of a supply pipe that carries hot water from the boiler to the terminal devices and a return pipe that carries water back to the boiler. Two-pipe systems include direct-return and reverse-return systems. **See Figure 21-8.**

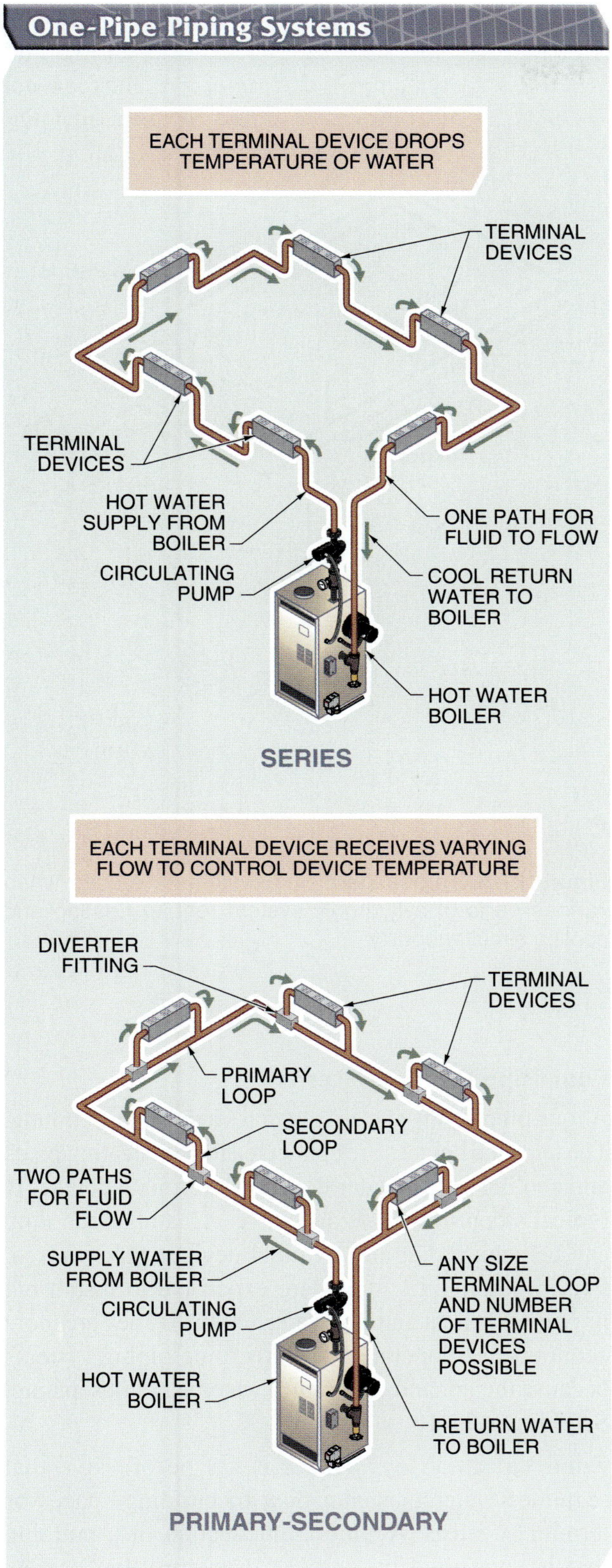

Figure 21-7. Each terminal device of a one-pipe piping system drops the temperature of the supply water to the next terminal device.

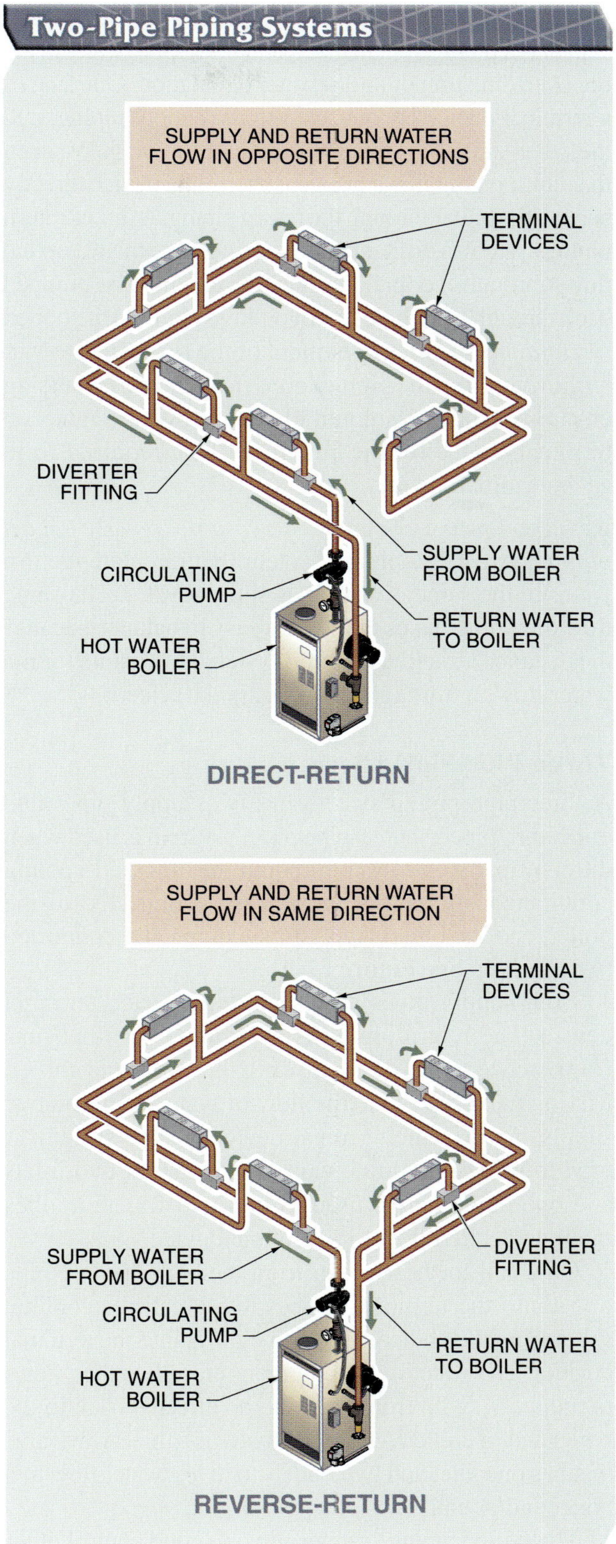

Figure 21-8. The supply and return water of two-pipe, direct-return piping systems travel in opposite directions. The supply and return water of two-pipe, reverse-return piping systems travel in the same direction.

Direct-Return Systems. A *two-pipe, direct-return piping system* is a piping system that circulates supply water in the opposite direction of return water circulation. One side of a terminal device is connected to the supply piping, and the other side is connected to the return piping. Water in the return piping flows from each terminal device directly back to the boiler through the return piping. A direct-return piping system is difficult to balance because the water that flows through the final terminal device must be pumped back through the whole return piping system to the boiler.

Balancing is the adjustment of the resistance to water flow through the piping loops in a system. Balancing ensures the proper amount of water flow through each terminal device for the amount of heat required from each terminal device.

Reverse-Return Systems. A *two-pipe, reverse-return piping system* is a piping system that circulates return water in the same direction as supply water. The same flow direction makes the system easy to balance because the distance around the piping system is the same for the water flowing through each terminal device.

Three-Pipe Piping Systems

A three-pipe piping system has two supply pipes and one return pipe. A three-pipe piping system is used when different parts of a system require heating and cooling simultaneously. One supply pipe is connected to the boiler, and the other supply pipe is typically connected to the chiller. **See Figure 21-9.**

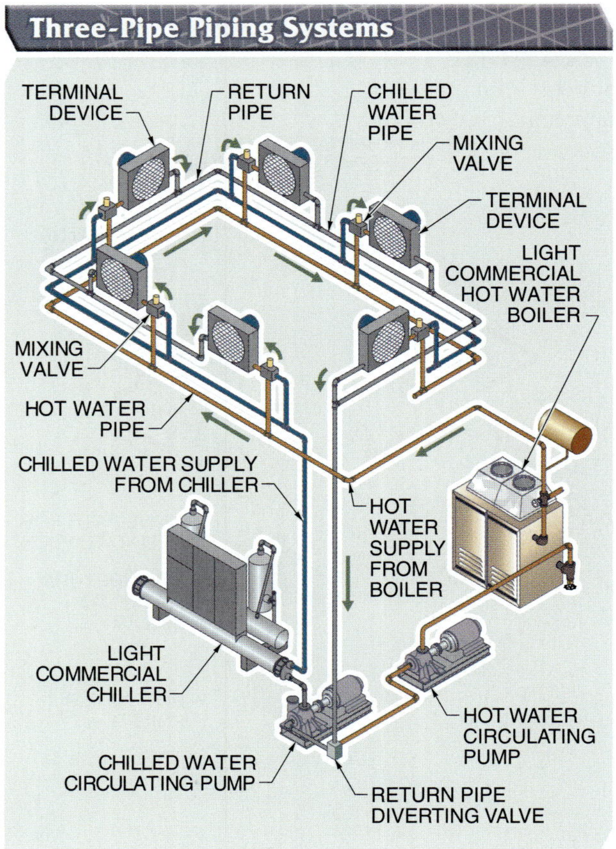

Figure 21-9. Three-pipe piping systems are used when various parts of a hydronic system require heating and cooling simultaneously.

Both supply pipes are connected to each terminal device. The return piping is a common line that runs from each terminal device back to the boiler and chiller. Mixing valves regulate the flow of water from the two supply pipes to the terminal devices in the system. A *mixing valve* is a three-way valve that has two inlets and one outlet and is used to mix two water supplies into one desired flow to a terminal device.

On a call for heat, water from the boiler supply pipe flows into the terminal devices. On a call for cooling, water from the chiller supply pipe flows into the terminal devices. The water in the common return line then flows to a three-way diverting valve to be directed back to the boiler and/or chiller. A *diverting valve* is a three-way valve that has one inlet and two outlets to divert water flow to a boiler and/or chiller.

Three-pipe piping systems provide good control of air temperature but are more expensive to install than two-pipe piping systems. Three-pipe piping systems are also more expensive to operate because the hot and cold return water are mixed.

Four-Pipe Piping Systems

A four-pipe piping system separates heating and cooling. The terminal devices are connected to both heating pipes and cooling pipes. Water flow to the terminal devices is typically controlled by zone valves. Heating or cooling can be achieved at any terminal device when desired. Four-pipe piping systems are expensive to install but provide excellent air temperature control. They are more economical to operate than three-pipe piping systems because the hot and cold return water are kept separate. **See Figure 21-10.**

Zone Valves. A *zone valve* is a type of valve that regulates water or steam flow to building spaces or terminal devices. A *zone* is the section of a building that is controlled by one HVAC controlling device, such as a thermostat. Zone valves may be manual or automatic. A *manual zone valve* is a valve that is set by hand to regulate the flow of water in a piping loop.

An *automatic zone valve* is a valve that is opened or closed automatically by an actuator, such as a motor, solenoid, or pneumatic device. Automatic zone valve actuators are controlled by electric or electronic control systems or by a pneumatic control system. The control systems typically receive a signal from a zone thermostat. **See Figure 21-11.**

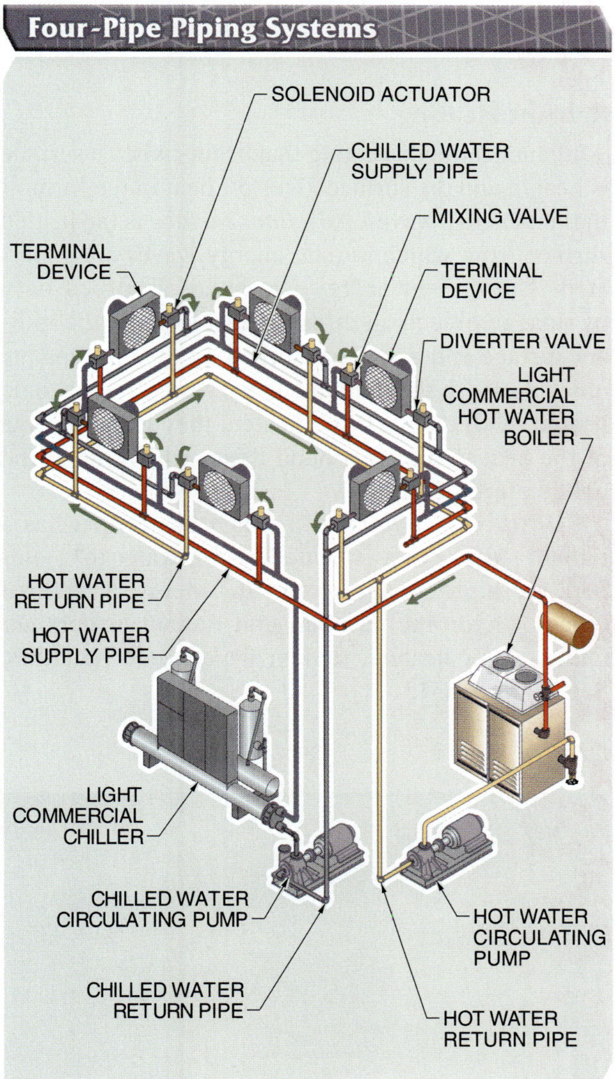

Figure 21-10. Four-pipe piping systems allow heating or cooling to occur at a terminal device at any time.

Valves used for controlling water flow are either digital or modulating valves. A *digital valve* is a two-position (ON/OFF) valve. The valve is either completely open or completely closed. A *modulating valve* is an infinite-position valve. The valve may be completely open, completely closed, or at any intermediate position in response to the control signal the valve receives.

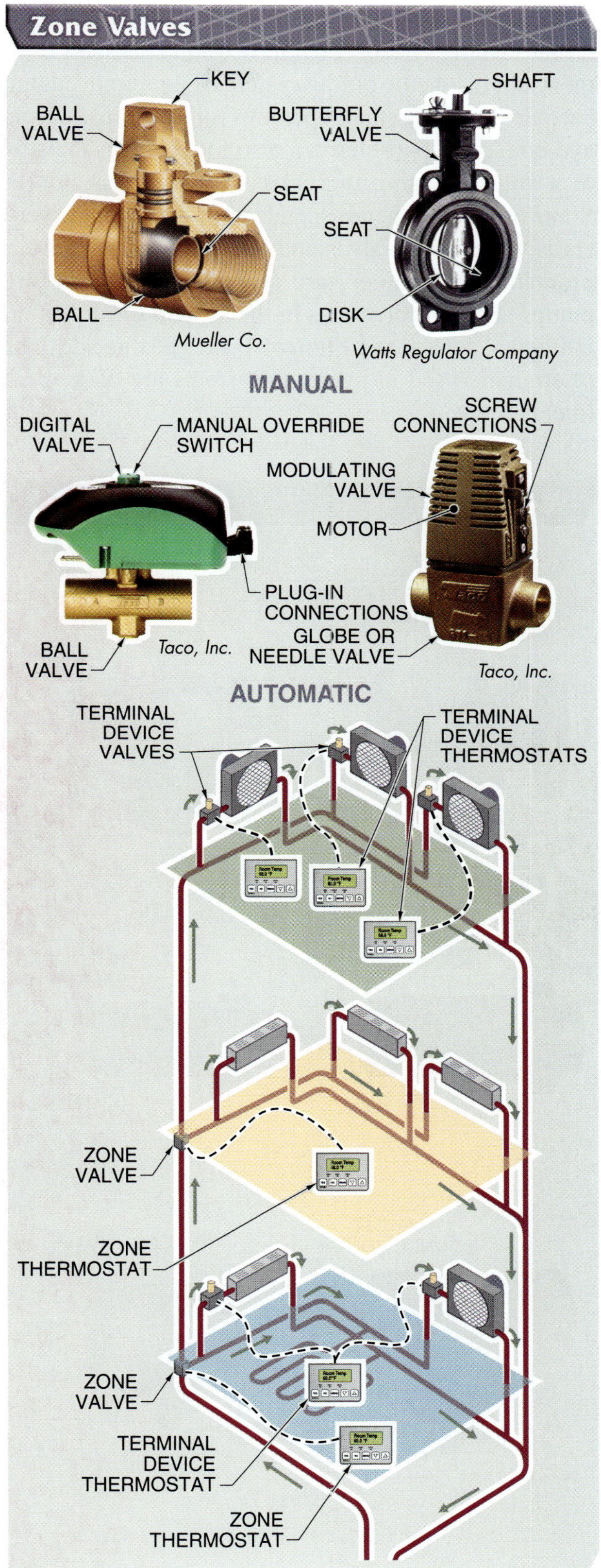

Figure 21-11. Zone valves regulate the flow of water in a zone (building space) or terminal device of a building.

Water Strainers. A *water strainer* is a device in a hydronic heating system that removes solid particles from the water inside the piping system. If not removed, the particles can plug small openings (orifices) in controls and coils. The particles can also lodge under the seat of a valve, causing the valve to leak when closed. Some particles, when suspended in water become very abrasive, causing valves and pump impellers to wear prematurely. Strainers are typically located before pumps and control valves in the system to protect an individual device. **See Figure 21-12.** The three designs of strainers used in hydronic systems are basket, in-line, and wye.

Water Strainers

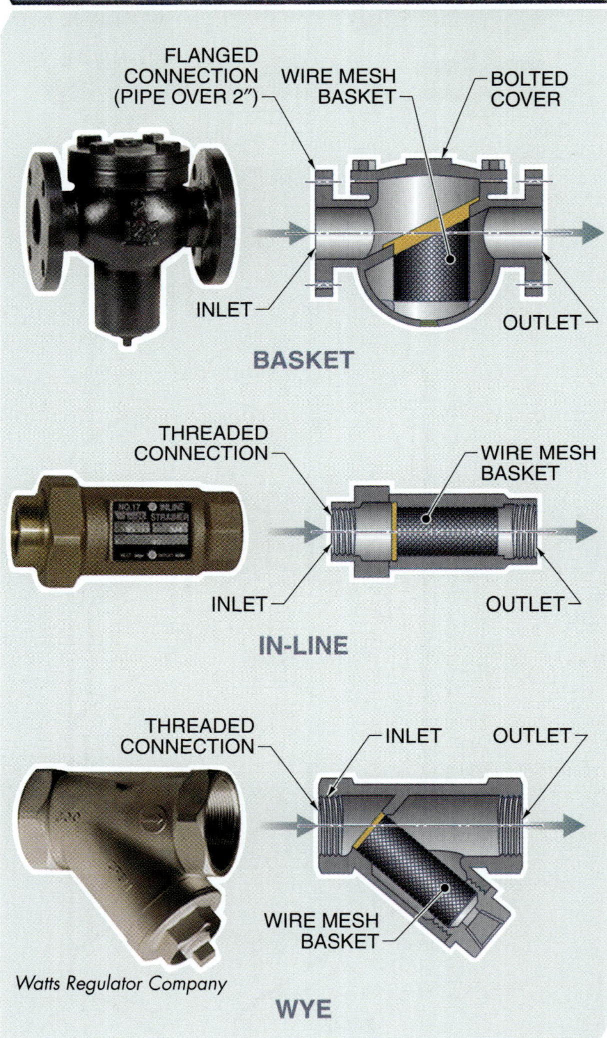

Watts Regulator Company

Figure 21-12. Water strainers must be routinely cleaned to ensure efficient removal of particles from the circulating water.

TERMINAL DEVICES

A *terminal device* is a component at the end of a heating system that transfers heat from the hot water in a hydronic heating system to the air in building spaces. A supply pipe runs from the boiler to each terminal device. A return pipe runs from each terminal device to the boiler. Radiant heating and convection heating terminal devices are the two types of terminal devices found in hydronic distribution systems.

Radiant Heating

Radiant heating is heating that occurs when a surface is heated and the surface gives off heat in the form of radiant energy waves. A *radiant surface* is the heated surface from which radiant energy waves are generated. The radiant energy waves are absorbed only by opaque objects in direct line of sight of the radiant surface and are not absorbed by the air. Opaque objects absorb the radiant energy waves and become heated. As the objects are heated, they give up some of the heat to the air around them, which raises the temperature of the air.

A *hydronic radiant heater* is a heater that has a radiant surface that is heated by hot water to a temperature high enough to radiate energy. The three types of hydronic terminal unit radiant heaters are radiant floor heaters, radiant panels, and radiators. **See Figure 21-13.**

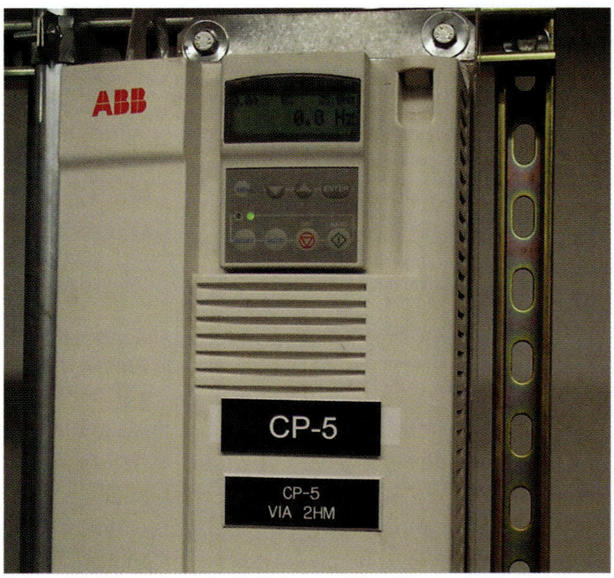

Many commercial hydronic systems use a variable-frequency drive at the pump to control system pressure.

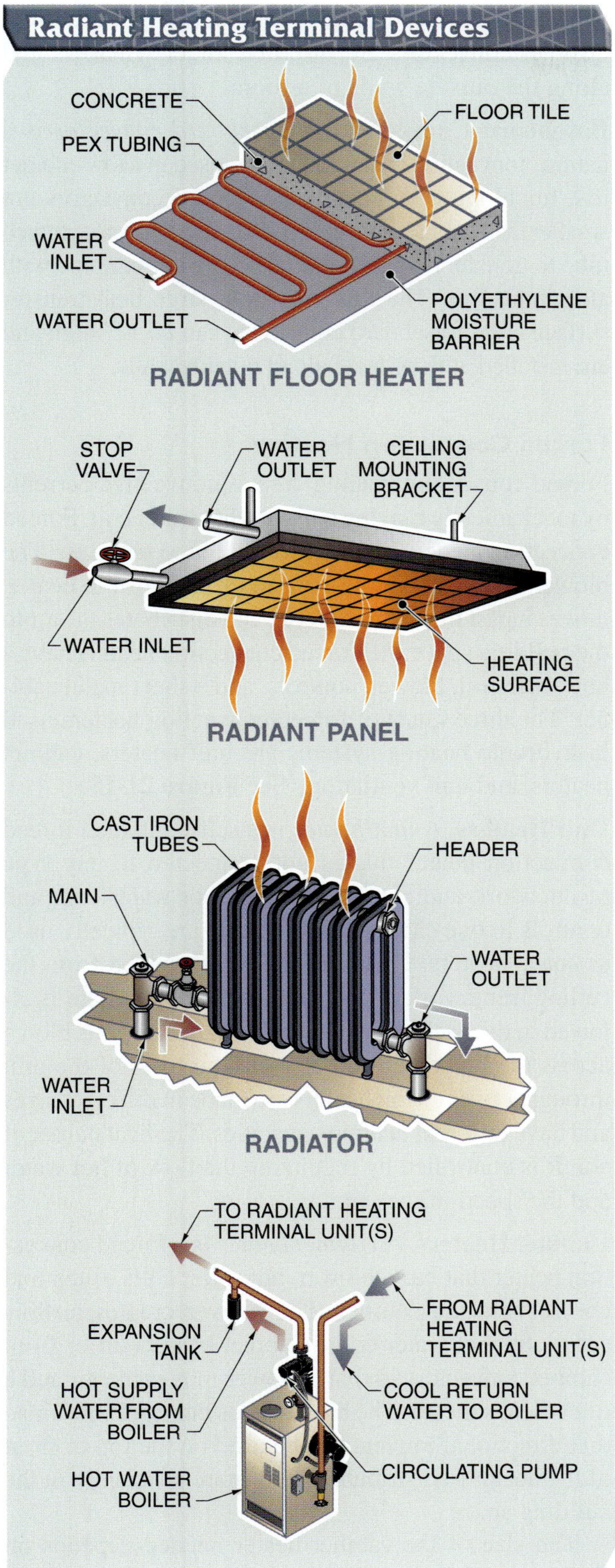

Figure 21-13. Radiant heating terminal devices use hot water to heat surfaces to a temperature high enough for heat to flow from the surfaces.

Radiant Floor Heaters. A *radiant floor heater* is a heating system that uses the interior surface of a floor as a radiant surface. A piping coil or grid is built into the construction material of a floor. Insulation is installed behind the piping, and a thin layer of surfacing material is installed over the insulation.

The radiant floor surface is heated by hot water that flows through the pipes. When the temperature of the radiant surface (floor) is higher than the temperature of objects in the room, heat flows from the radiant surface to the objects. The amount of radiant heat produced is regulated by controlling the temperature of the radiant surface.

Radiant Panels. A *radiant panel* is a factory-built panel with a radiant surface that is heated by a piping coil or grid built into it. Radiant panels are individual heaters that are not an integral part of a building. Radiant panels are enclosed in an insulated cabinet and used for spot heating. *Spot heating* is a heating method in which heat is provided for a particular part of a building space. Entire buildings are not heated by spot heating from radiant panels alone.

Radiators. A *radiator* is a terminal device that consists of hollow metal coils or tubes. Hot water passes through the coils or tubes. Heat from the hot water is transferred to the tubes, which transfer the heat to the air. The air in a building space is heated by radiation and convection when using radiators.

A *standing radiator* is a radiator that has a section of vertical tubes that are connected with headers at each end. Hot water from a boiler flows through the tubes. Heat is transferred to the air by radiation and convection when air flows over the tubes. Standing radiators are usually made of cast iron. Because radiators are considered to be unattractive, baseboard radiators are typically used. A *baseboard radiator* is a radiator that has a section of finned horizontal tubes that are connected with headers at each end, which are concealed by an attractive housing.

TECH TIP

In cold climates, radiant floor heat has become popular in residential applications. Some homeowners prefer the gentle, even heat of a hydronic system to the heat produced by a forced air fan.

Convection Heating

Convection heating is based on heat transfer by convection. Convection heating is heat transfer that occurs when currents circulate between warm and cool regions of a fluid. The two kinds of convection heaters are cabinet convectors and baseboard convectors. **See Figure 21-14.**

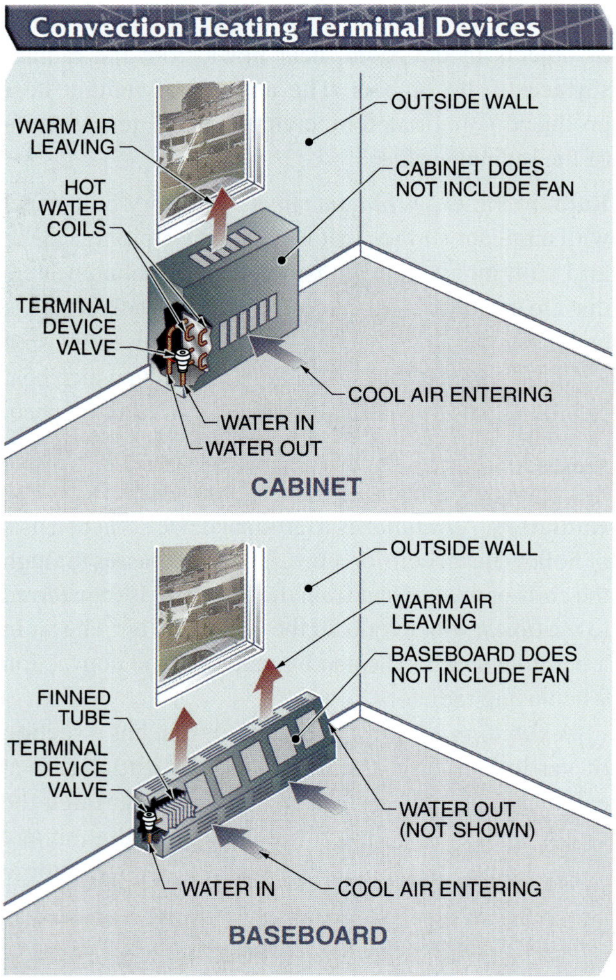

Figure 21-14. Convection heating terminal devices contain hot water coils that heat air as the air flows over the coils, without the assistance of a fan.

Cabinet Convectors. A *cabinet convector* is a convection heater that is enclosed in a cabinet. The hot water coil or tubes and the front of the heater are covered by a sheet metal cabinet. The cabinet has a grill in the bottom for cool air to enter and a grill on the top for warm air to exit. Cabinet convectors can be assembled in sections to produce the amount of heat required to heat a building space. Cabinet convectors are available in sizes between 12″ and 32″ high, are 7″ deep, and vary in length according to the heat output required. Cabinet convectors are installed under windows and along the outside walls of a room.

Baseboard Convectors. A *baseboard convector* is a natural convection heater that is enclosed in a low cabinet that fits along a baseboard. Baseboard convectors are smaller than cabinet convectors and use a section of finned tube to transfer heat. A *finned tube* is a copper pipe with aluminum fins. The fins provide a large heat transfer surface area. Baseboard convectors can be 12″ high and are installed at floor level along outside walls.

Forced Convection Heating

Forced convection heating creates convective currents by mechanically moving air past a hot water coil. Forced convection heaters use a blower or fan to move air. The blower allows air in the bottom or back of the heater, moves it past the hot water coil, and distributes the air to the building space. All forced convection heaters have a hot water coil, blower, controls, and a sheet metal cabinet. The three kinds of forced convection heaters used in hydronic heating systems are unit heaters, cabinet heaters, and unit ventilators. **See Figure 21-15.**

Unit Heaters. A *unit heater* is a self-contained forced convection heater that is not connected to any type of ductwork and contains a fan, a hot water coil, and controls in one cabinet. Unit heaters are typically used as space heaters. Unit heaters are suspended from the ceiling framework of a building space. Return air is allowed in the back of the unit by the fan. The air is blown across the hot water coil and out the front of the unit through louvers. Unit heaters are made in different sizes and have different heating capacities. The heat output of a unit is controlled by regulating the flow of hot water and the speed of the fan.

Cabinet Heaters. A *cabinet heater* is a forced convection heater that has a blower, hot water coils, filter, and controls in one cabinet. The blower creates airflow across the hot water coils. Cabinet heaters differ from cabinet convectors in that a blower moves the air and a filter cleans the air. The blower adds energy to the air so that directional louvers can be used on the top or front of the heater to distribute the air more efficiently in the building space.

The size of the cabinet heater needed depends on the heat output required. Most cabinet heaters are from 30″ to 36″ high, 6″ to 7″ deep, and of different lengths. Cabinet heaters are typically installed along the outside wall of a room under a window.

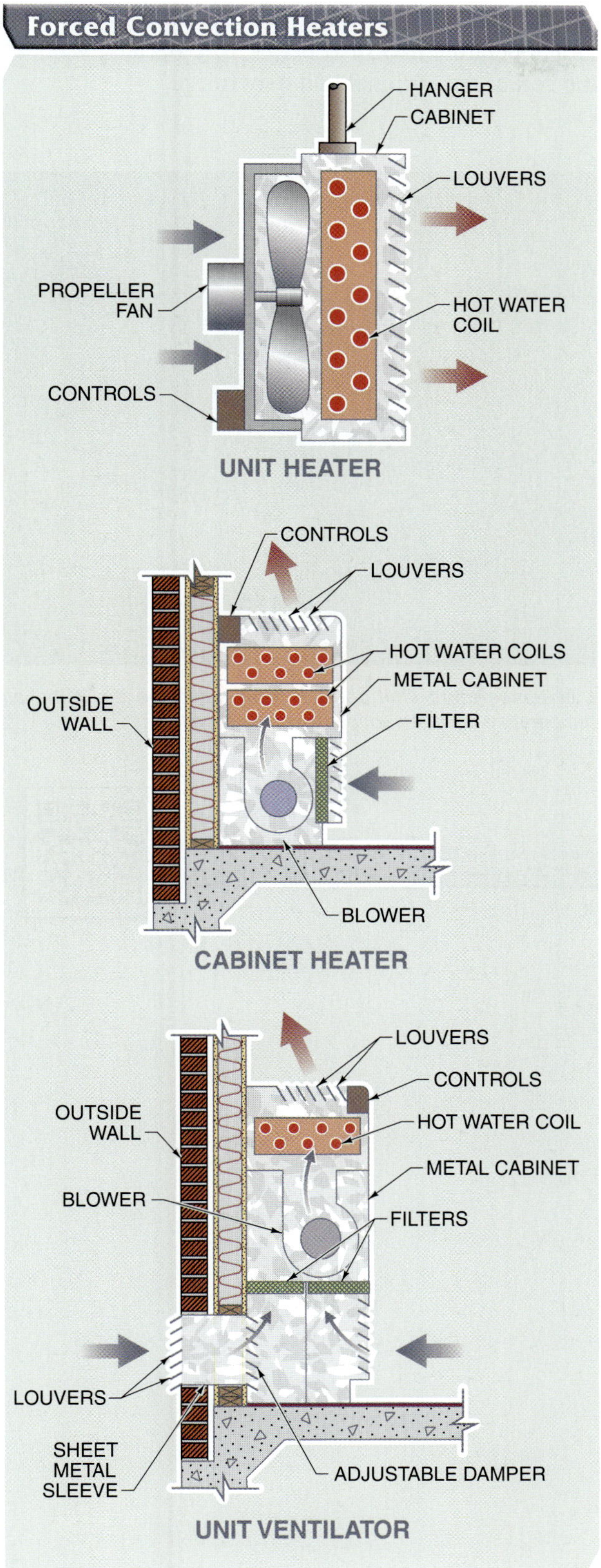

Figure 21-15. Forced convection heaters contain a blower that moves air over the hot water coils.

Unit Ventilators. A *unit ventilator* is a forced convection heater that has a blower, a hot water coil, filters, controls, and a cabinet that has an opening for outdoor ventilation. The blower creates airflow of mixed outside air and return air (mixed air) across the hot water coil.

Unit ventilators differ from cabinet heaters in that they have an outside opening to bring in air from outdoors for ventilation. The cabinet fits on an outside wall so that the connection to the outdoors can be made with an opening in the back of the cabinet for an outdoor air inlet. A sheet metal sleeve extends from the inlet through the wall. A louver covers the opening to the outdoors. The outdoor air inlet has an adjustable damper on the inside for regulating the amount of air admitted to the unit for ventilation. Unit ventilators are often used in buildings that have a large number of people and a need for constant ventilation.

Terminal Device Blowers. The blowers in terminal devices are controlled by relays, which cycle the blower ON or OFF. The relays are the same as those used in forced-air heating systems. A signal from the thermostat closes the relay contacts, which supplies electricity to the blower motor. Thermostats control the blower in a unit heater or unit ventilator. Thermostats used to control terminal devices may be located inside the cabinet of the unit they control or located remotely in the space being heated. **See Figure 21-16.**

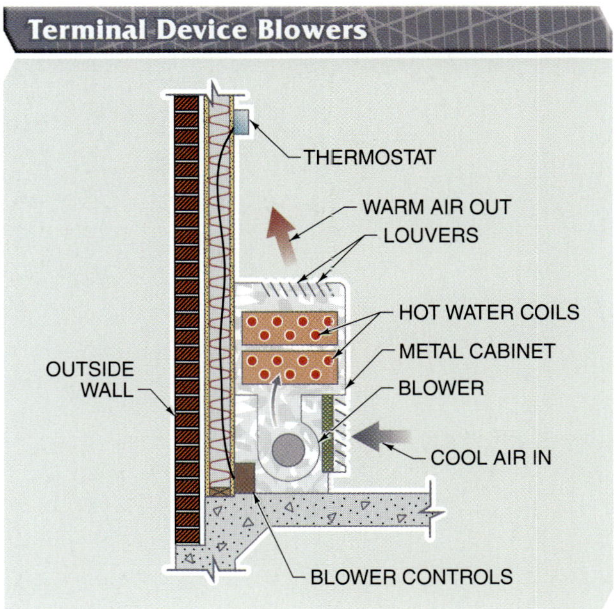

Figure 21-16. The blowers in terminal devices are controlled by relays that are energized by zone thermostats.

Terminal Device Controls

A *terminal device control* is a thermostat and valve combination that controls the temperature of the air in a building space by regulating the air and water flow through a terminal device. Circulating pumps and zone valves regulate water flow, and blowers regulate airflow. Circulating pumps are typically switched ON or OFF by a signal from the thermostat, which produces the flow of hot water in a hydronic piping system. A circulating pump control provides hot water for a complete hydronic heating system or secondary piping loop in a system.

Low-Temperature Limit Controls. A low-temperature limit control (freeze stat) is used on terminal devices that use outdoor air for ventilation. A *low-temperature limit control* is a temperature-actuated electric or pneumatic switch that can energize a damper actuator to close the damper and open a water or steam valve if ventilation air drops below 35°F. The terminal unit fan will continue to run. In other applications, the low-temperature limit control shuts down the terminal unit, completely closes the damper, turns the fan OFF, and opens the water valve. Freeze-protection controls are required when outdoor air temperature may drop below 32°F because water in the coil of the terminal unit may freeze.

Low-temperature limit controls shut down the fan if the air temperature downstream of the coil falls too low.

Steam Heating—Boilers

UNIT 22

Steam heating systems use steam to carry heat. The relatively high specific weight and heat of water make steam an excellent medium for carrying heat. Steam heating systems boil water in a boiler at a central location to produce steam. The steam is distributed through supply piping to terminal units in the building spaces where the heat is required. The terminal units transfer heat from the steam to the air of the building spaces. Steam and condensate return to the boiler from the terminal units through return piping and accessories.

OBJECTIVES

- List steam heating system components.
- Describe steam boiler types and applications.
- Explain steam efficiency ratings.
- Calculate boiler horsepower.
- Calculate boiler capacity in terms of pounds of steam per hour.

Index to Procedures

STEAM HEATING

Some residential and light commercial buildings use steam to supply heat to building spaces. Steam heating carries more British thermal units per pound (Btu/lb) of steam than water. Steam carries 970 Btu/lb of steam, which condenses back into a liquid. Water carries one Btu per degree temperature change per pound of water. A residential cast iron radiator or light commercial terminal unit that uses steam to heat air will have a very quick heat-up time. However, it is more difficult to control building space air temperature with a steam heating system than with a hot water system. One advantage of a steam heating system is that it can be used where there is a limited amount of space for terminal units. **See Figure 22-1.**

Baseboard heating units are commonly used with steam heating systems.

Steam Heating Advantages and Disadvantages

Steam Heating Advantages	Steam Heating Disadvantages
Short air heating times	Not as good as hydronic air temperature control
Terminal units have small surface areas	Increased installation costs
Shorter boiler heating times than hydronic systems	Water requires treatment
System distributes warmth even after burners have been turned off	Lag time from burners turning on to heat reaching radiators
Each zone or room can have independent temperature control	
Choice of various piping systems	
Piping has a long life expectancy	
System operates quietly	

Figure 22-1. Steam heating systems have many advantages. However, they cost more to install than hot water systems.

STEAM HEATING SYSTEMS

A *steam heating system* is a heating system that uses steam to carry heat from the point of generation to the point of use. Steam heating systems do not need to pump steam to the point of use. Steam flows anytime there is a difference in pressure between two points in the system. For example, steam flows from the high-pressure area of a boiler to the low-pressure area of a terminal unit.

Steam Heating Systems Media Clip

A steam boiler is typically located in a central mechanical room. Steam is distributed to the individual zones or rooms of the building where it releases heat, then condenses back into a liquid (condensate). **See Figure 22-2.** The condensate is pumped or uses gravity to travel back to the boiler through piping and steam accessories for reheating.

In a residential or light commercial building, the steam boiler is typically located on the lowest floor of the building, such as a basement or first floor utility room. When terminal units are located on several floors throughout a building, the boiler is likely to be located in the basement. From the basement, steam is allowed to flow up through the floors and supply piping without the assistance of a pump. Gravity typically causes the condensate to flow back to the boiler through return piping.

When all terminal units are located on a single floor, the boiler is most likely to be located on the same floor. Having all terminal units and the boiler on the same floor minimizes the length of the piping runs required, which minimizes heat loss.

Steam Heating Systems

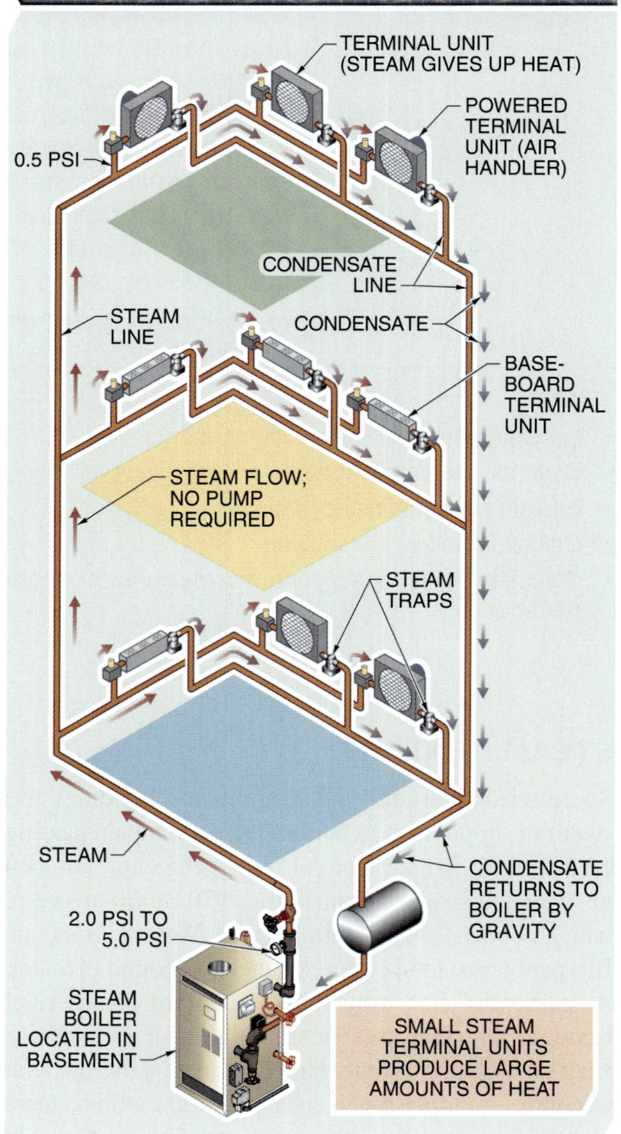

Figure 22-2. A steam heating system uses steam to carry heat from a boiler in a central location to the individual terminal units in various zones or rooms of a building.

A steam heating system consists of a steam boiler, a steam supply system, a condensate return system, and a number of fittings, accessories, terminal units, and controls. **See Figure 22-3.** The boiler is used to heat water above the boiling point to produce steam. The fittings on a steam boiler allow the boiler to operate safely and efficiently. Steam heating system accessories are required system components that support boiler operation.

Steam Heating System Components

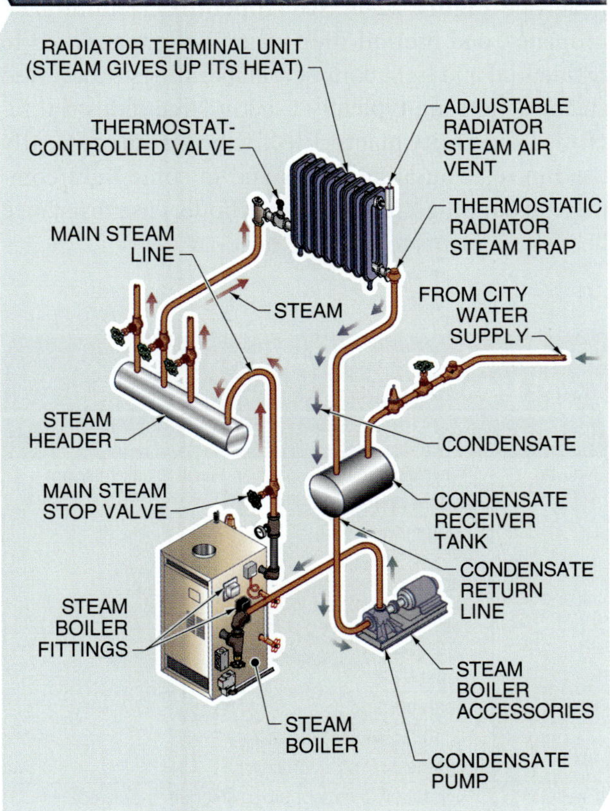

Figure 22-3. The main components of a steam heating system include a steam boiler, piping, and one or more terminal units.

A *steam supply system* is a system of pipes, valves, traps, and strainers that provide a path for steam to flow from a boiler to the point of use. Heat exchangers in the terminal units transfer heat from steam to the air in building spaces. Controls are used to regulate air temperature, such as the temperature of the air discharged from the heating coil of the terminal unit. A *condensate return system* is a system used to return condensate to a boiler after all useful heat has been removed.

Steam Boilers

The steam boiler is the central component of a steam heating system. A *steam boiler* is a pressure vessel that transfers large amounts of heat to water to generate steam. A *pressure vessel* is a tank or container that operates at a pressure greater than atmospheric pressure. Steam boilers superheat water above the 212°F boiling point, which changes the water to vapor (steam). **See Figure 22-4. See Appendix.**

Steam Properties

Saturated Condition		Enthalpy‡	Specific Volume§	
Pressure*	Temperature†	Vapor	Water	Vapor
0	212.0	1150.3	0.0167	29.799
5	227.2	1156.1	0.0168	20.830
10	239.4	1160.6	0.0169	16.491
15	249.8	1164.2	0.0170	13.878
20	258.8	1167.3	0.0171	11.966
30	274.0	1172.2	0.0172	9.460
40	286.7	1176.0	0.0173	7.826
50	297.7	1179.2	0.0174	6.683
60	307.3	1181.9	0.0175	5.837
70	316.0	1184.1	0.0176	5.185
80	323.9	1186.1	0.0177	4.666
90	331.2	1187.8	0.0178	4.233
100	337.2	1189.4	0.0179	3.891

* in psi
† in °F
‡ in Btu/lb
§ in ft^3/lb

Figure 22-4. Residential and light commercial steam boilers can superheat steam up to 274°F. Heavy commercial and industrial steam boilers can superheat steam up to 448°F.

Steam boilers and distribution piping systems are filled with steam. Steam heating systems are classified by the following:

- working pressure (low pressure or high pressure)
- temperature
- method of heat production (fire-tube, water-tube, or angled tube)
- method of condensate return (gravity or mechanical)
- piping system (one-pipe or two-pipe)

For residential and light commercial applications, ASME International classifies steam boilers as low-pressure steam heating boilers and power steam boilers. **See Figure 22-5.**

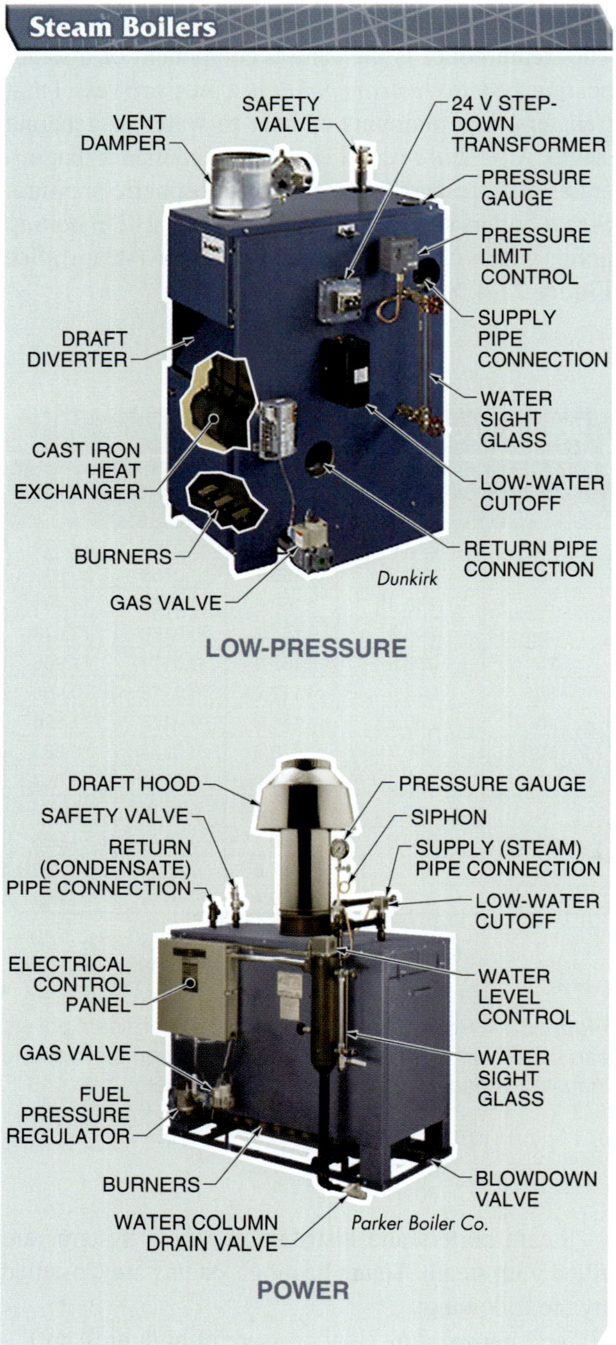

Figure 22-5. Residential and light commercial steam boilers are classified as low-pressure steam heating boilers and power steam boilers.

Low-Pressure Steam Heating Boilers. A low-pressure steam heating boiler is a steam boiler that operates at pressures that do not exceed 15 psi. Residential and light commercial steam boilers typically operate at pressures less than 5 psi and are used for heating purposes only.

Power Steam Boilers. A power steam boiler is a steam boiler that operates at pressures above 15 psi. Small power steam boilers are used in light commercial buildings for heating purposes and/or other processes. Large power steam boilers are used in the utility industry for the production of electricity.

Heat Production

Steam boilers burn fuel or use electricity to create steam. Combustion steam boilers burn fuel, such as natural gas, propane fuel oil, or coal, to produce heat. **See Figure 22-6. See Appendix.** Natural-gas-, propane-, and fuel-oil-fired steam boilers are used in residential and light commercial buildings. Coal-fired steam boilers are typically used in large industrial facilities and utility plants. Electric boilers are typically used in residential buildings and in some light commercial buildings. Electric steam boilers use resistance heating elements to produce heat.

Heating Source Advantages and Disadvantages

Sources	Advantages	Disadvantages
Natural Gas	Available in cities	Rural areas may not have access
	Boilers operate quietly	
	Equipment is very dependable	
	Very little electricity consumed, except for possible circulating pump	
	Less than half the cost of electricity	
Propane	Clean burning	Requires delivery
	High efficiencies possible	
	Half the cost of electricity	
Fuel Oil	Highest efficiency of all boiler fuels	Requires delivery
	Considerably less expensive than other boiler fuels	Requires more maintenance than gas or electric fuel systems
	Can be easily supplied to areas	Older system equipment can be noisy, dirty, and difficult to maintain
		Requires electricity to operate oil burner
Electricity	Plentiful supply	Very high expense to operate
	Requires minimal routine maintenance	
	Does not require the ventilation needed to burn a fuel	

Figure 22-6. Residential and light commercial steam boilers burn fuel, such as natural gas, propane, or fuel oil, or use electricity to create steam.

Combustion Steam Boilers. Steam boilers that produce heat by combustion are made of steel or cast iron. The steel tubes of light commercial steam boilers are welded together to form a tube bundle that has a specific Btu output rating. Steel steam boilers have removable manhole covers for maintenance and service. Hollow cast iron heat exchangers are used in many residential steam boilers. The sections of cast iron heat exchangers are typically bolted together.

Openings in the heat exchanger allow steam to flow inside the sections of the heat exchanger and flue-gas to flow outside the sections. Heat is transferred from the products of combustion through the metal walls of the heat exchanger tubes to the water. **See Figure 22-7.** The number of metal tubes or cast iron sections determines the output rating of the steam boiler. Adding or removing tubes or sections achieves the required Btu output rating desired.

The fireside of a boiler contains the hot products of combustion, and the waterside contains water. The heating surface must be as large as possible so that a large surface area of water can be exposed to the hot products of combustion. A large heating surface increases boiler efficiency and the rate of heat transfer from the hot products of combustion to the water.

The heating surface of a steel-boiler tube bundle consists of circular, horizontal, vertical, or slanted tubes. The tubes may be fire tubes or water tubes. Multiple small burners, called atmospheric burners, are used to heat the water in a steam boiler. Alternatively, one large burner, called a power burner, may be used.

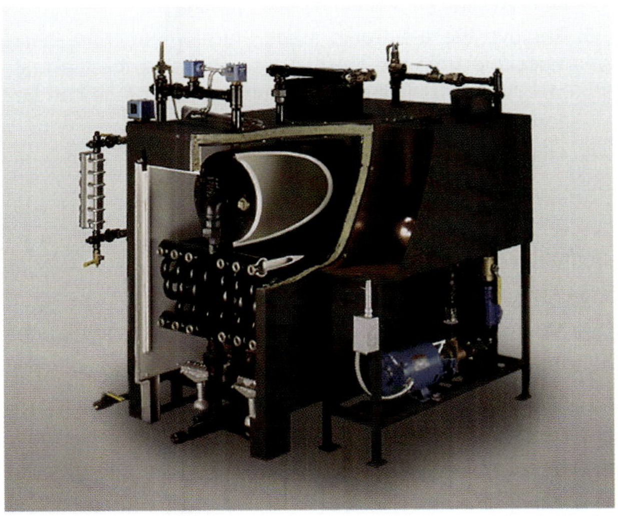

Parker Boiler Co.

Steam boilers are always constructed of heavy gauge steel and cast iron components

Figure 22-7. Heat is transferred from the products of combustion to the water inside the steel tubes of a heat exchanger.

Atmospheric burners use ambient air supplied at normal atmospheric pressure as combustion air. **See Figure 22-8.** Primary air enters the burner where the fuel enters. *Primary air* is air that mixes with fuel, such as in a boiler, to produce a combustible air-fuel mixture that controls the rate of combustion. Secondary air enters the flame at the burner face. *Secondary air* is air that is added to a combusting air-fuel mixture, such as in a boiler, to ensure complete combustion. Power burners use large volumes of air provided by mechanical devices, such as blowers, to efficiently burn fuel. **See Procedure for Taking Stack Temperature Measurements.**

Steam Boiler Burners

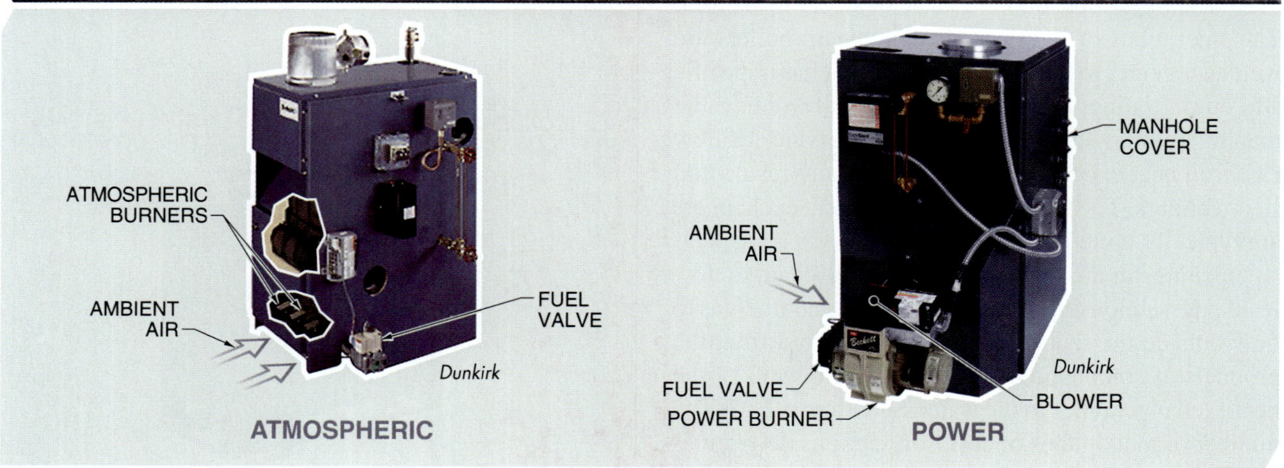

Figure 22-8. Steam boilers may use multiple atmospheric burners or one power burner.

Procedure

Taking Stack Temperature Measurements

1. Determine if the flue (stack) has a thermal well for a stack thermometer thermocouple or a drilled hole for a flue-gas thermometer and probe.

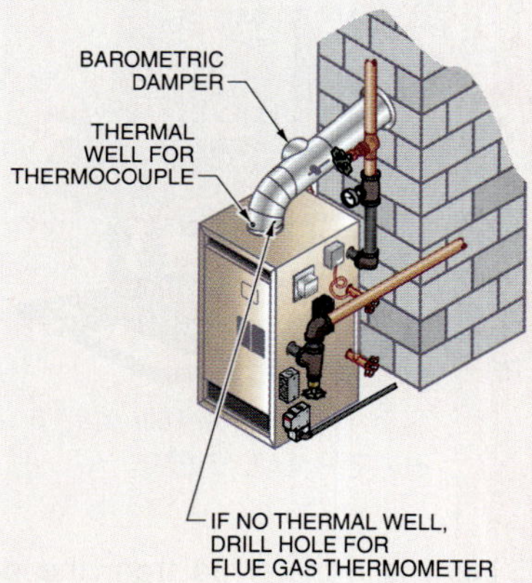

2. Follow all electrical and hazardous environment safety practices and procedures.
3. Screw the stack thermometer thermocouple into the thermal well, or insert the flue-gas thermometer and probe into the drilled hole and seal.

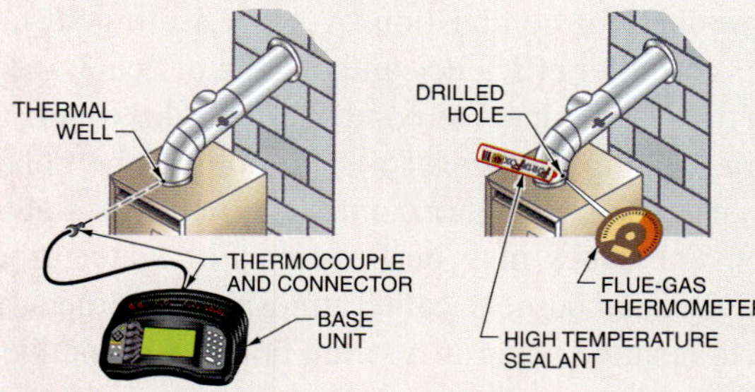

4. Set the electronic stack thermometer to record readings for 8 hr to 24 hr, or read and record the flue-gas thermometer readings for at least 4 boiler cycles.
5. Remove the electronic stack thermometer and thermocouple from the thermal well and plug the thermal well. *Note:* The flue-gas thermometer and probe can be permanently installed in the flue.
6. Analyze the data collected.

Fuel Storage Tanks. Fuel storage tanks are used to store fuel oil for combustion steam boilers. They are available in sizes ranging from several hundred gallons to over 50,000 gal. They can be made of steel or fiberglass and come in single or multiple wall construction. The location and installation of the fuel oil tank must comply with Environmental Protection Agency (EPA) regulations. Although fuel storage tanks can be underground or outside, it is recommended that they be inside in a basement.

Electric Steam Boilers. An *electric steam boiler* is a steam boiler that uses steam produced by resistance heating elements to generate heat. Electric boilers are classified separately from combustion boilers. The resistance heating element used in a steam boiler is an immersion heater. An *immersion heater* is a series of copper rods enclosed in waterproof tubes that are installed in a boiler and surrounded by water to provide heat by electricity. An electric steam boiler does not require a heat exchanger because heat is transferred directly from the immersion heater(s) to the water. **See Figure 22-9.**

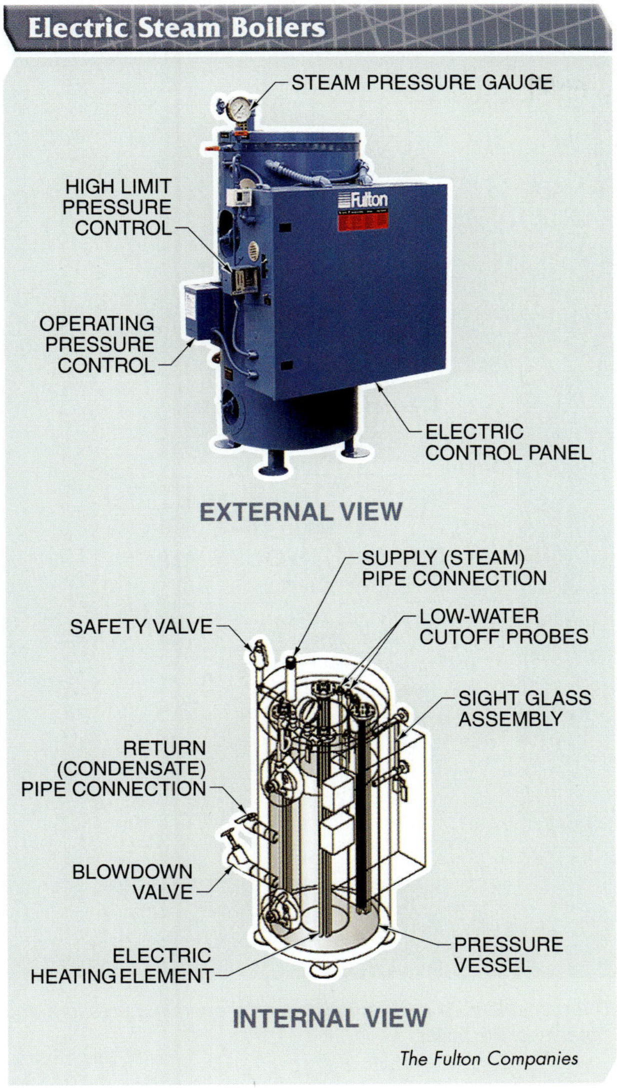

Figure 22-9. Electric steam boilers use resistance heating elements made of copper rods enclosed in waterproof tubes to produce heat.

Electric steam boilers are more expensive to operate than gas-fired and oil-fired boilers. However, they are found in residential and light commercial applications because of their simplicity. Some small electric steam boilers can operate on single-phase power supplies. Most small electric steam boilers operate on two or three phases because of the large amount of current required. Electric steam boilers convert electrical energy into thermal energy with almost 100% efficiency.

Steam Boiler Ratings

The input rating of a combustion steam boiler represents the amount of heat produced in British thermal units per hour (Btuh) per unit of fuel burned. The output rating of a steam boiler represents the actual heat output (in Btuh) after heat losses from draft. The output rating of a combustion steam boiler is typically 75% to 86% of the input rating. Some combustion steam boilers have attained output ratings of 96%. Electric steam boilers have output ratings of 100% because all the energy used to produce heat is transferred to the water.

Parker Boiler Co.

Although fuel storage tanks can be underground or outside, it is recommended that they be inside in a basement.

Gross unit output is the heat output of a hydronic or steam boiler when it is fired continuously. *Net unit output* is the gross output of a water or steam boiler multiplied by the percentage of loss due to pickup. *Pickup* is the additional heat that is needed to warm the water in a hydronic or steam heating system once a period of off-time, such as overnight, has passed. The pickup factor of a boiler can be as high as 50%.

The output ratings of residential gas-fired and oil-fired steam boilers are between 80,000 Btuh and 160,000 Btuh. The output ratings of light commercial steam boilers are between 180,000 Btuh and 1,000,000 Btuh. The output ratings of heavy commercial steam boilers are in the millions of Btuh. **See Figure 22-10.** Coal-fired boilers that are used for industrial applications have output ratings in the hundred millions of Btuh. Electric boilers are rated in kilowatts per hour (kW/h). One kilowatt-hour equals 3414 Btu.

TECH TIP

Electric boilers can be very useful in some applications. For example, electric boilers may be used when fossil fuel is not available or would be very costly to bring to a building site. In addition, electric boilers may be used in conjunction with fossil fuel boilers to take advantage of low utility rates during off-peak periods, such as nighttime hours.

Figure 22-10. Steamer boiler output ratings vary based on the size of the boiler and the boiler's intended application.

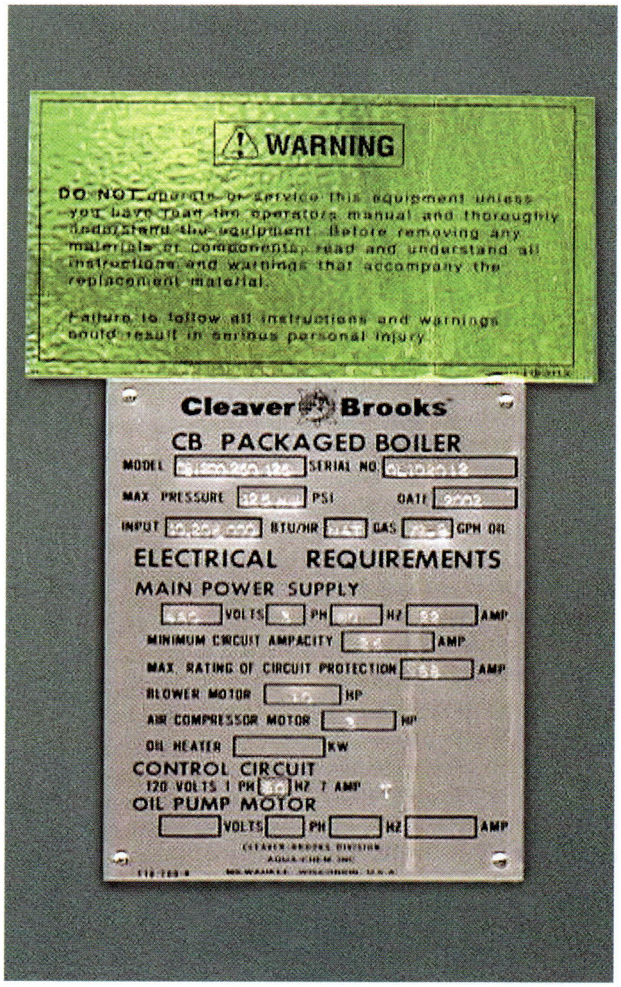

The nameplate on a boiler indicates the operating pressure and capacity of the boiler.

Steam Boiler Horsepower Ratings. The oldest method of rating steam boilers involves the use of horsepower (HP). One boiler horsepower (BHP) is the amount of energy required to evaporate 34.5 lb/hr of water into steam at 212°F. To convert pounds of steam into horsepower, the amount of steam (in lb/hr) is divided by 34.5. In some countries, boiler horsepower is a function of the heat transfer area in a boiler, and 1 BHP is approximately 17 ft² of heat exchanger heating surface area. To find boiler horsepower, apply the following formula:

$$BHP = \frac{S}{34.5}$$

where

BHP = boiler horsepower

S = steam (in lb/hr)

34.5 = constant

Example: Calculating Boiler Horsepower

What is the boiler horsepower of a steam boiler that generates 862.5 lb/hr of steam? **See Figure 22-11.**

Example: Calculating Pounds of Steam per Hour

What is the amount of steam generated (in lb/hr) by a 25.0 HP steam boiler? **See Figure 22-12.**

Figure 22-11. Horsepower (HP) is used for steam boiler ratings. One boiler horsepower (BHP) is the energy required to evaporate 34.5 lb/hr of water at 212°F.

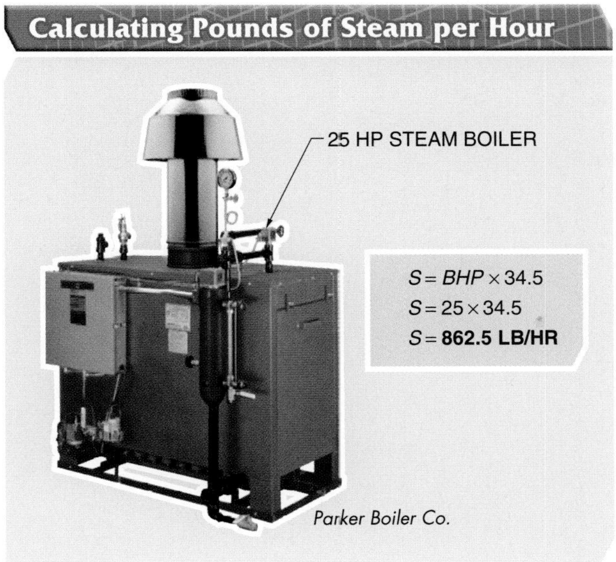

Figure 22-12. Boiler capacity is typically a calculation of the pounds per hour (lb/hr) of steam evaporated and maintained for 24 hr under specified conditions.

Large boiler capacity is typically provided in pounds per hour of steam evaporated, under specified conditions. The *maximum continuous rating* is the amount per hour of evaporation that can be maintained by a steam boiler for 24 hr. To find the pounds per hour of steam, apply the following formula:

$S = BHP \times 34.5$

where

S = steam (in lb/hr)

BHP = boiler horsepower

34.5 = constant

Boiler Loads. The units of measure used to indicate the maximum capacity of a boiler are the British thermal unit and horsepower (in lb/hr of steam). When a boiler operates at its maximum rated capacity, it is operating at maximum load. When the load of a boiler varies each hour, the boiler is operating at a varying load. Loads and load variations can affect the processes of steam heating system applications.

One load horsepower is the consumption of 34.5 lb/hr of steam at a pressure of 0 psi and temperature of 212°F. One boiler horsepower is approximately 33,479 Btuh (about 9810 W or 8430 Kcal/hr).

UNIT 23

Steam Heating Boiler Components and Controls

Residential and light commercial buildings use steam boilers for heating purposes. These boilers include various types of fittings and accessories that allow for the safe and efficient operation of the steam heating system. HVAC service technicians must know how these steam boiler fittings and accessories operate and be familiar with applicable testing procedures to respond to service calls safely.

OBJECTIVES

- List the functions of common steam boiler fittings and accessories.
- Identify gas valve train components.
- Describe the operation of common steam boiler controls.

Index to Procedures

STEAM BOILER FITTINGS

A *steam boiler fitting* is any device directly attached to a steam boiler and required for the proper operation of the steam boiler and system. The fittings for steam boilers include safety valves, steam pressure gauges, water columns, low water cutoffs, bottom blowdown valves, surface blowdown valves, fusible plugs, steam vent valves, and pressure controls. **See Figure 23-1.** Steam boiler fittings have higher temperature and pressure ratings than hydronic boiler fittings.

Steam Boiler Fittings Media Clip

Safety Valves

A *safety valve* is a steam boiler fitting that prevents a boiler from exceeding its maximum allowable working pressure (MAWP). The safety valve is the most important valve on a steam boiler. When the pressure in a steam boiler exceeds the safety valve setting, the safety valve opens. Failure of the safety valve to open can lead to a boiler explosion.

Safety valves are tested to ensure proper operation.

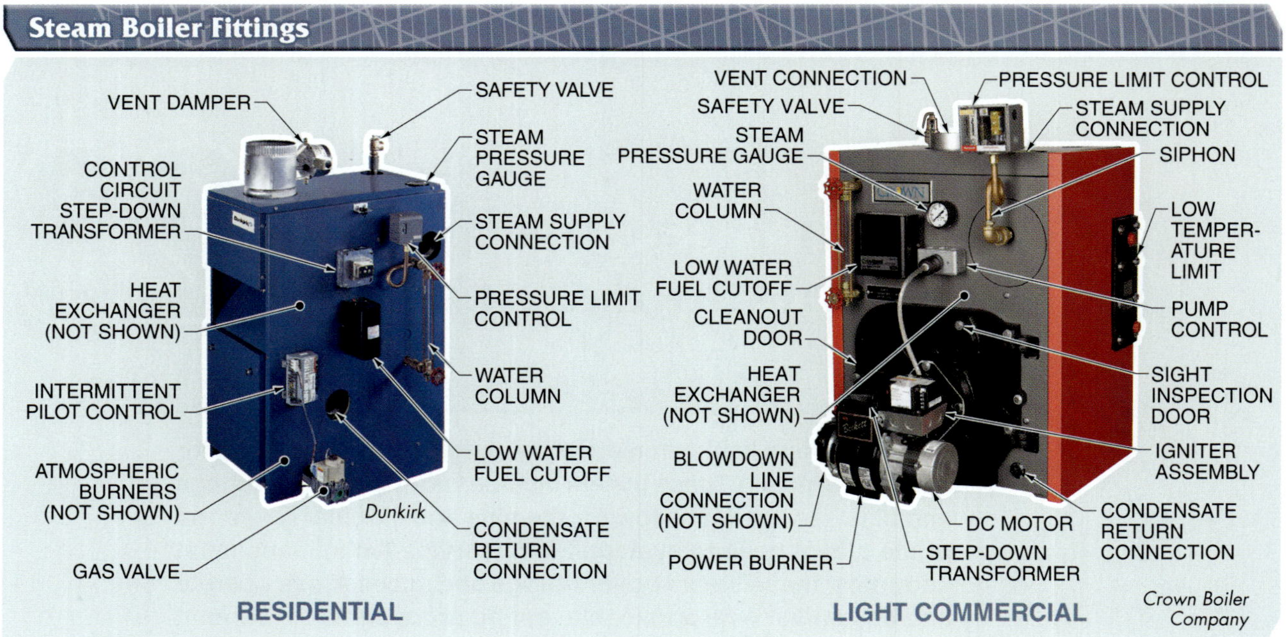

Figure 23-1. Fittings are directly attached to a steam boiler and the devices required to operate the boiler.

The safety valve must be located at the highest part of the steam side of the boiler and connected directly to the boiler shell. There must not be any valves located between the safety valve and the heat exchanger or the point of valve discharge. **See Figure 23-2.**

Safety valves are designed to open rapidly when boiler pressure exceeds a setpoint. The setpoint for residential and light commercial steam boilers is typically from 0.5 psi to 2 psi. Maximum working pressure of a low pressure steam boiler is 15 psi. The safety valve stays open until a sufficient amount of steam is released out of the system to drop the pressure inside the boiler. When the desired decrease in pressure is achieved, the safety valve closes.

According to the ASME International Boiler and Pressure Vessel Code, at least one safety valve (pressure relief valve) must be mounted on each steam boiler. When the heat exchanger surface of a steam boiler exceeds 500 ft^2, it must have two or more safety valves. Also, as required by ASME International, all valves manufactured for steam service must have a lift lever handle.

The capacity of a safety valve is the amount of steam in pounds per hour (lb/hr) the valve can vent at rated pressure. Safety valve capacity is listed on the data plate attached to the safety valve. **See Figure 23-3.**

Safety valves do not have operator adjustments. When a safety valve requires testing, a technician trained and certified by the safety valve manufacturer must perform the work.

Figure 23-2. A safety valve prevents a boiler from exceeding the maximum allowable working pressure (MAWP). It must be located at the highest part of the steam side of the boiler.

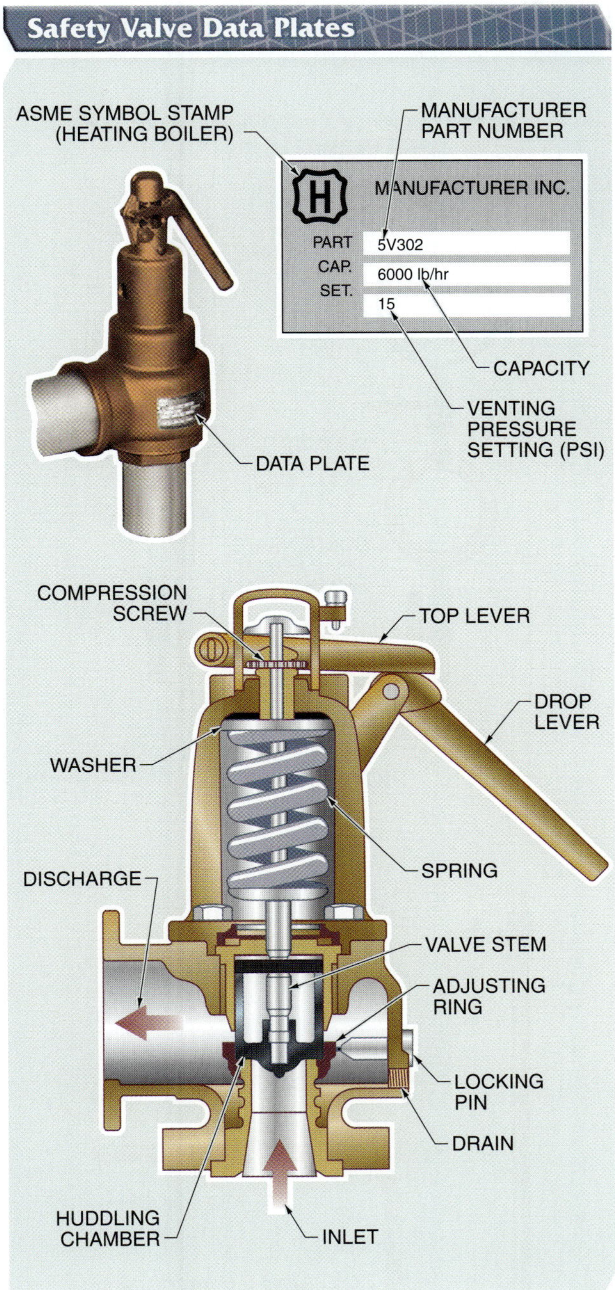

Figure 23-3. The capacity of a safety valve is the amount of steam in pounds per hour (lb/hr) the valve is capable of venting at rated pressure.

TECH TIP

Safety valves are tested manually to clear debris from the valve seat and ensure that the valves open when needed. Care must be taken to ensure that steam does not discharge where it may cause damage.

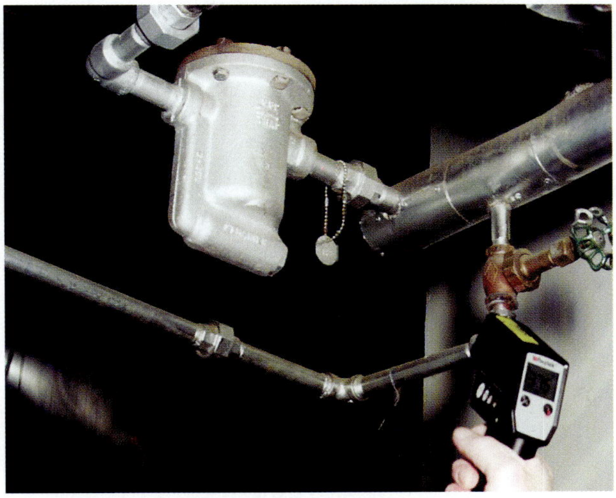

Diagnostic equipment, such as an infrared thermometer, can be used to check boiler components for efficiency and proper operation.

Steam Pressure Gauges

A *steam pressure gauge* is a steam boiler fitting that indicates the amount of pressure in a steam boiler; expressed in pounds per square inch (psi). A steam pressure gauge allows an HVAC technician to monitor steam pressure changes in the boiler and verify that steam pressure stays within the safe operating range. A steam pressure gauge is connected to the highest point on the steam side of the boiler. It must be positioned to allow easy viewing from the front of the steam boiler.

On many steam boilers, a compound gauge is used. A *compound gauge* is a pressure gauge that indicates vacuum in inches of mercury (in. Hg) and pressure in pounds per square inch (psi).

The most common type of gauge used in steam systems is the Bourdon tube pressure gauge. A Bourdon tube is either C-shaped or coil-shaped. The tube is typically made of brass or bronze. However, for low-pressure residential and light commercial applications, the tube is made of stainless steel. **See Figure 23-4.** The internal mechanism of a Bourdon tube pressure gauge can be damaged by vibrations and ambient corrosives. To help protect the steam pressure gauge from harmful conditions, the tube is typically filled with glycerin to dampen vibration.

Steam is prevented from entering the steam pressure gauge by a siphon. A *siphon* is a bent tube installed between a steam boiler and a boiler fitting that provides a water seal to prevent the fitting from coming into direct contact with steam. The water seal in the siphon

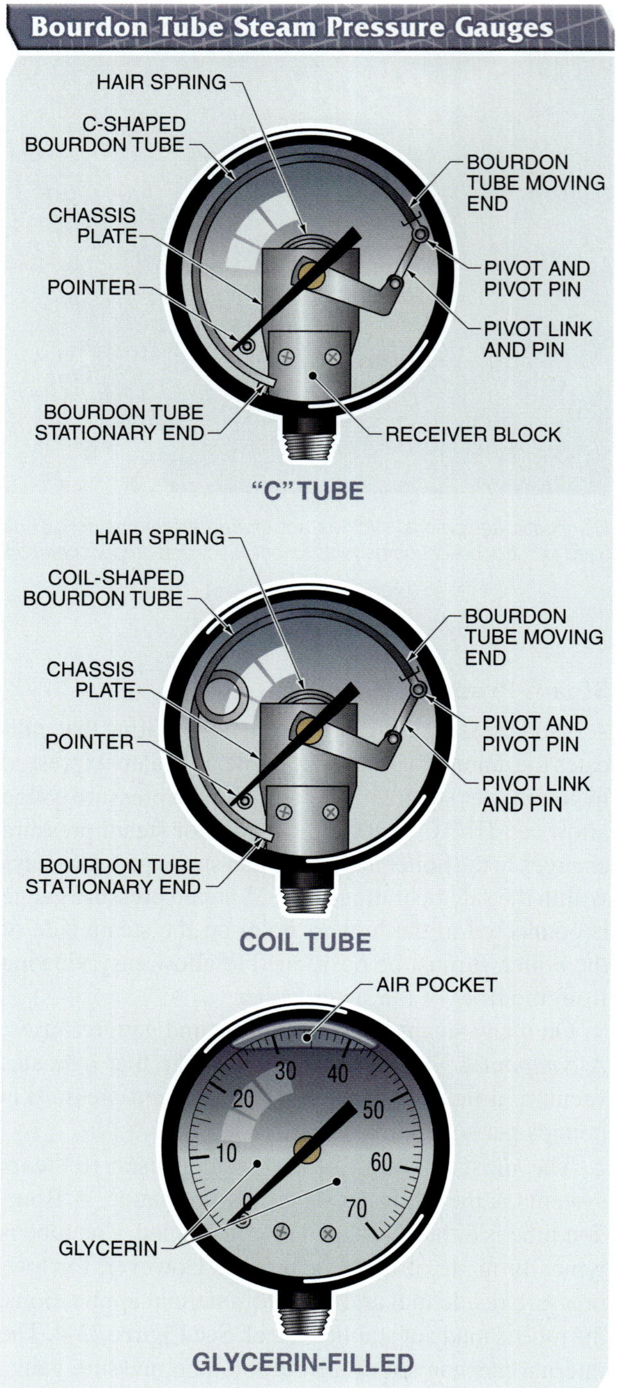

Figure 23-4. A Bourdon tube steam pressure gauge is the most accurate type of pressure gauge that can be used on a steam boiler. It is typically filled with glycerin.

is formed by the collection of condensate in the lower area of the tube. The common types of siphons are the pigtail siphon and the U-tube siphon. Both siphons are designed to trap condensate. **See Figure 23-5.**

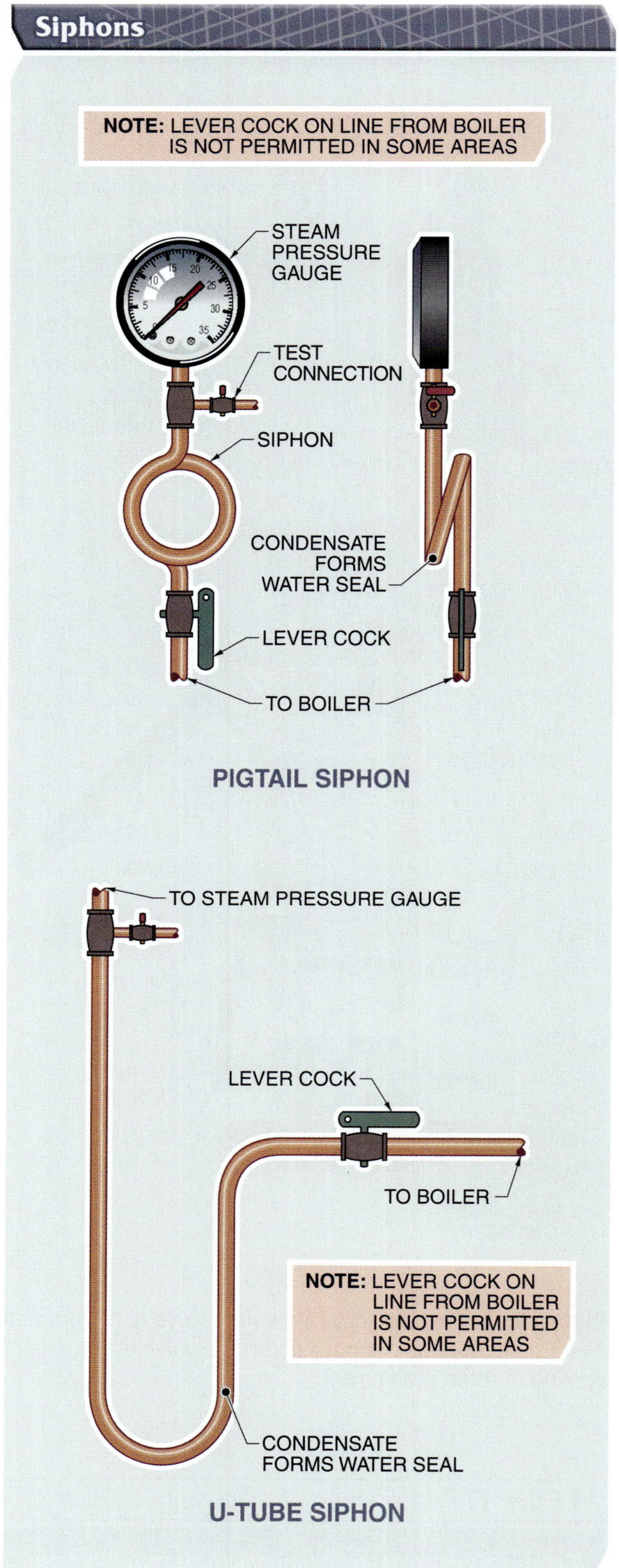

Figure 23-5. A water seal, which is created by the collection of condensate in the lower section of a siphon, prevents steam from entering a pressure gauge.

A lever cock installed on a pigtail siphon allows the steam pressure gauge to be changed for maintenance and service. A test connection allows for the connection of test equipment when checking the steam pressure gauge for accuracy. *Note:* Some municipalities do not allow lever cocks to be installed between the boiler and steam pressure gauges. Check local codes and regulations for proper installation procedures.

Water Columns

A *water column* is a steam boiler fitting that reduces the movement of boiler water in a gauge glass to provide an accurate boiler water level reading. A *gauge glass* is a tubular glass column that indicates water level in a steam boiler. A water column is fitted with isolation valves to facilitate the repair and replacement of the gauge glass. The 2009 ASME International Boiler and Pressure Vessel Code does not require water columns on hot water boilers. However, most steam boilers have a water column. A gauge glass is typically mounted directly on the boiler shell of small boilers that do not encounter rapid changes in load.

Steam boilers for residential and light commercial buildings must have two methods for determining water level. Using a gauge glass is the easiest method for determining boiler water level. The other method involves the use of a try cock. A *try cock* is a valve located on the water column of a boiler and is used to determine boiler water level when the gauge glass is not functional. **See Figure 23-6.**

Crown Boiler Co.

A gauge glass attached to the side of a residential steam boiler.

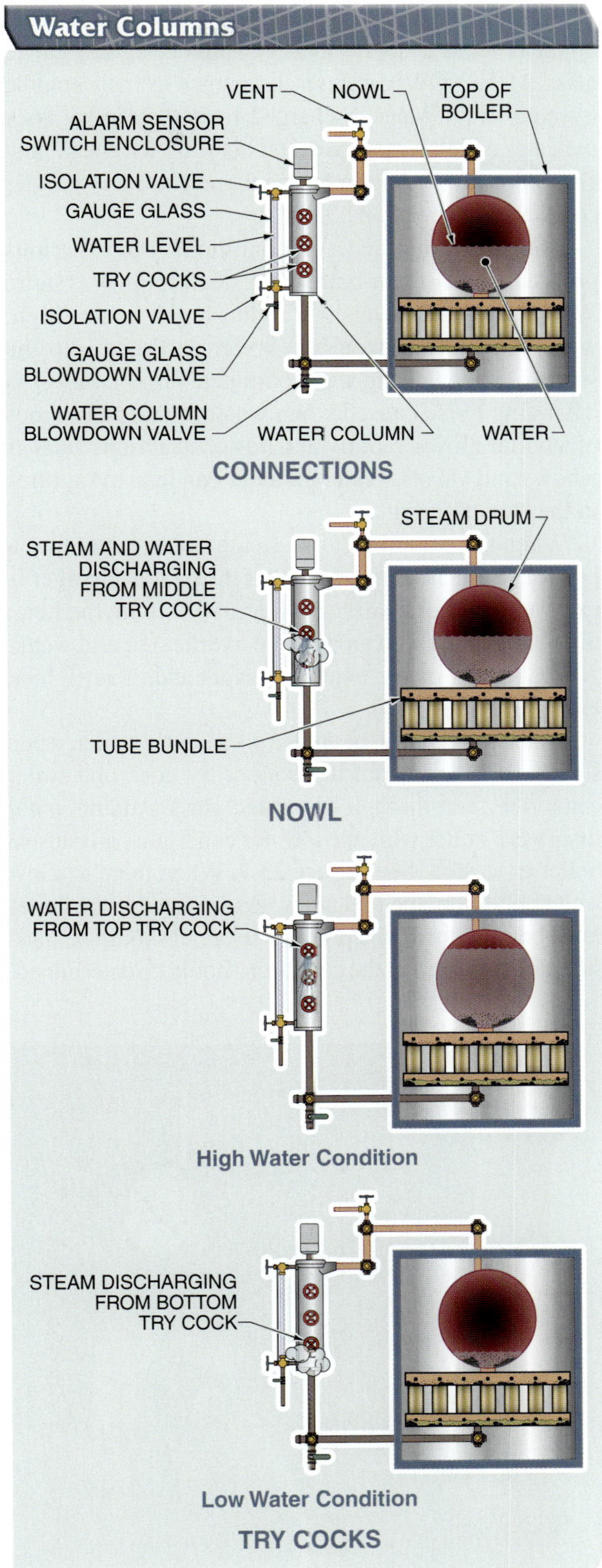

Figure 23-6. Try cocks and gauge glasses are used to determine the water level in steam boilers.

The normal operating water level (NOWL) of a steam boiler is located approximately halfway up the gauge glass. At the NOWL, water is discharged from the middle try cock only. Water discharged from the top try cock indicates a high water condition in the boiler. Steam discharged from the bottom try cock indicates a low water condition in the boiler.

Both high and low water conditions lead to serious problems in a steam boiler and steam piping system. When the water level in a steam boiler is too high, water may escape from the boiler and seep into the steam piping, causing water hammer. *Water hammer* is a banging noise caused when water in the steam lines of a boiler moves rapidly and hits obstructions such as elbows and valves. Water hammer can lead to ruptures in the steam piping.

When the water level in a steam boiler is too low, a portion of the heating surface of the heat exchanger is exposed. Without water present to dissipate the heat, the heating surface can become overheated and weak. A weakening of the heating surface can lead to heat exchanger meltdown.

Water must never be added to a steam boiler when steam discharges from the bottom try cock and water cannot be seen through the gauge glass. Adding water to a steam boiler with a low water condition can cause a boiler explosion. **See Figure 23-7.** When there is a low water condition, the fuel and water systems to the boiler must be shut down, the proper personnel should be notified, and the cause of the condition should be determined.

Crown Boiler Co.

A low water cutoff stops fuel to the burner if the water level is too low.

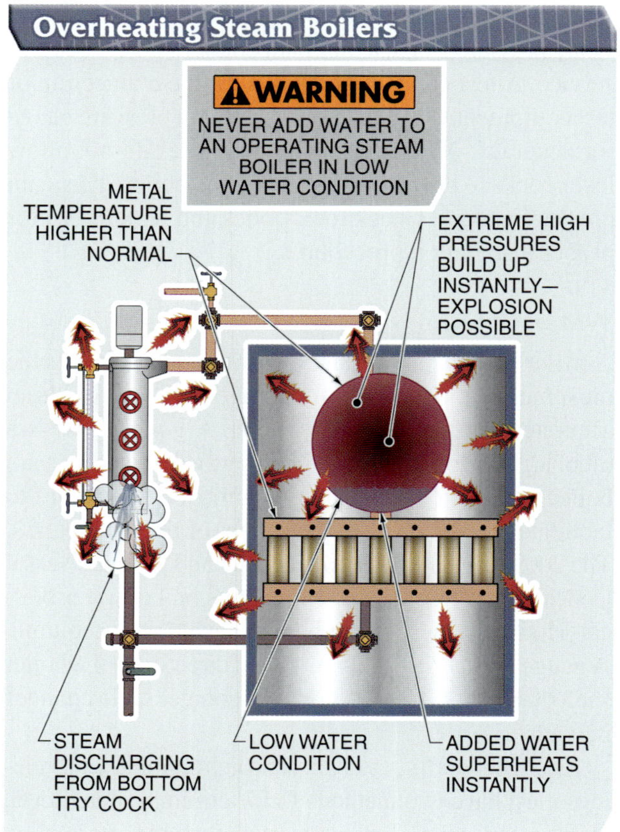

Figure 23-7. Adding water to a steam boiler with a low water condition can cause a boiler explosion.

Water Columns Media Clip

Low Water Cutoffs

A *low water cutoff (LWCO)* is a sensor and switch or probe that opens the combustion circuit of a boiler in the event of a low water condition. ASME CSD-1 requires that all steam boilers have at least one LWCO. Automatic makeup water feeder circuits can malfunction or have an interruption to the water supply, requiring that a LWCO be present. When only one LWCO is present, it must be manually reset. **See Figure 23-8.**

LWCO Operation. When the water level in a boiler is at the NOWL, the burner continues to fire. As the water level drops to an unsafe level, the float or sensor of the LWCO opens the electric circuit to shut off the burner. The burner shuts down, preventing damage to the steam boiler caused by overheating.

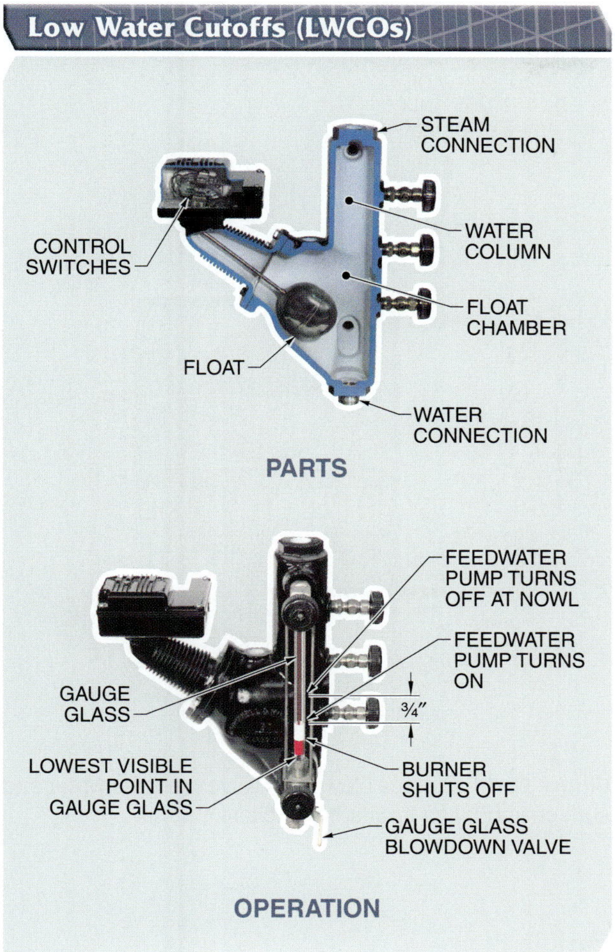

Figure 23-8. LWCOs shut off burners in the event of low water conditions.

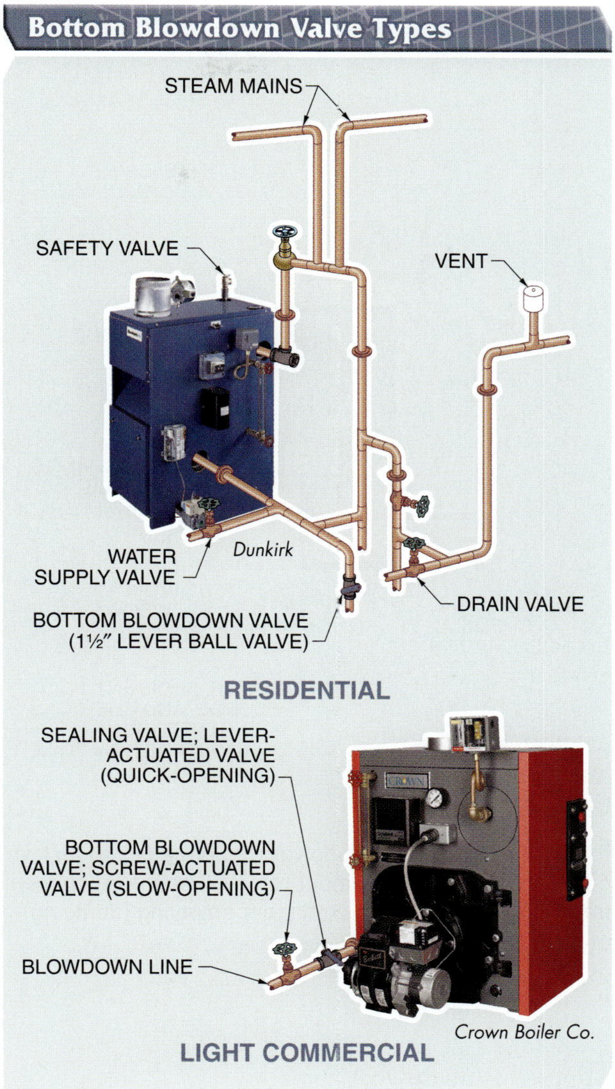

Figure 23-9. A residential steam boiler typically has one blowdown valve. A light commercial steam boiler typically has a quick-opening and slow-opening blowdown valve.

Bottom Blowdown Valves

A *bottom blowdown valve* is a valve located on the blowdown line at the lowest point of the water side of a steam boiler that is used to release water and sediment from the bottom of the boiler. All steam boilers have bottom blowdown valves. **See Figure 23-9.** The steam boiler size, capacity, and type determine the number and type of bottom blowdown valves to be used. Residential steam boilers typically have one bottom blowdown valve.

Light commercial steam boilers typically have two valves on the bottom blowdown line. One valve is a quick-opening (lever) valve, and the other valve is a slow-opening (screw) valve. Some manufacturers place two slow-opening valves on the bottom blowdown line.

The quick-opening valve is a lever-operated valve that functions as a sealing valve. It is always located closest to the boiler shell. The slow-opening valve is used as the blowing valve and takes all the wear during blowdown. When performing a bottom blowdown, the quick-opening valve is always opened first and closed last. During blowdown, an HVAC technician must not let go of a bottom blowdown valve that is open. Bottom blowdown is performed for water level control, sludge and sediment removal, chemical concentration control, and heat exchanger emptying (dumping). **See Figure 23-10.**

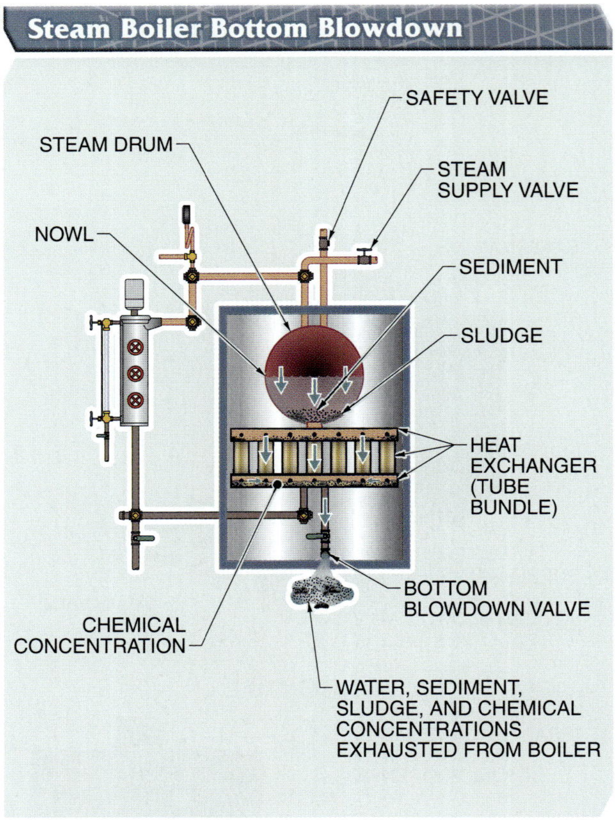

Figure 23-10. Bottom blowdown is performed for water level control, sludge and sediment removal, chemical concentration control, and heat exchanger emptying (dumping).

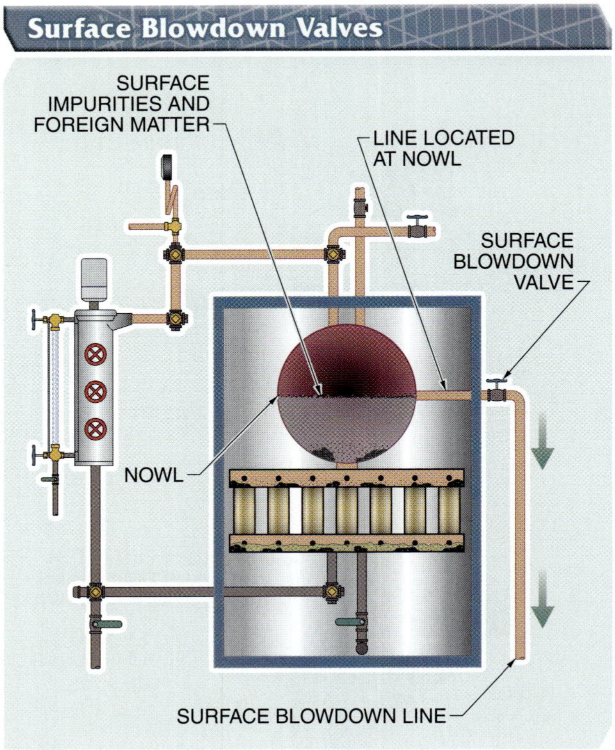

Figure 23-11. Surface blowdown valves release water and surface impurities from steam boilers.

Surface Blowdown Valves

Some light commercial and all heavy commercial steam boilers have surface blowdown lines and valves. A *surface blowdown valve* is a valve located on the surface blowdown line of a steam boiler that is used to release water and surface impurities from the boiler. **See Figure 23-11.** Residential steam boilers are not equipped with surface blowdown lines and valves.

Surface blowdown line piping is located at the normal operating water level (NOWL) of the steam boiler. The film produced by surface impurities increases surface tension by preventing steam bubbles from breaking through the boiler water surface. Increased water surface tension can lead to foaming, which causes a rapid fluctuation of the water level. Foaming can also lead to a false water level reading, which can result in water hammer due to water escaping from the boiler into the steam piping.

Fusible Plugs

When a boiler overheats, boiler tubes may melt. A *fusible plug* is a warning device used to indicate the overheating of a steam boiler from a low water condition. A fusible plug is typically used on coal-fired steam boilers and serves as a last warning to the boiler operator of a low water condition. The fusible plug is a brass or bronze plug with a tapered core that contains a tin alloy. The core melts at approximately 450°F.

When water is not covering the fusible plug to dissipate heat, the heat from the gases of combustion will melt the plug core. Once the core melts, steam blows through the tapered hole and makes a whistling sound, which warns personnel that a low water condition exists. **See Figure 23-12.**

TECH TIP

Residential boilers do not have a surface blowdown valve. However, they are commonly filled, operated for a period of time, and then drained to remove oil or debris that entered the boiler during installation.

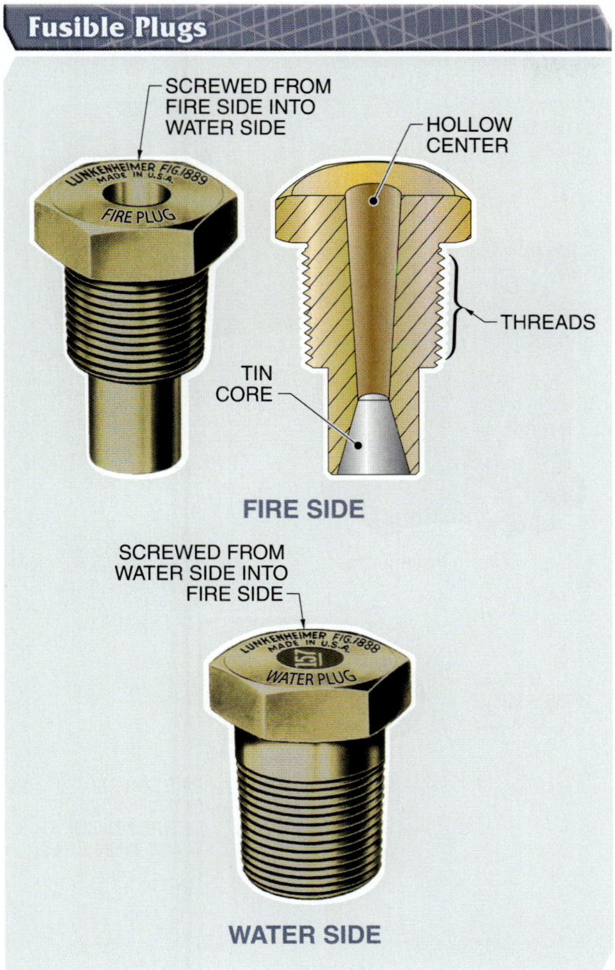

Figure 23-12. During an extreme condition, the tin in a fusible plug melts, allowing steam to exit the plug with a whistling sound to alert personnel.

Typically, one fusible plug is installed on a boiler. Fusible plugs are designed for use on either the fire side or water side of a boiler. A fireside fusible plug is screwed from the fire side into the water side of the boiler. A waterside fusible plug is screwed from the water side into the fire side. A fusible plug is located 1″ to 2″ above the highest heating surface in the direct path of the gases of combustion.

Steam Vent Valves

A *steam vent valve* (steam boiler vent or air cock) is a steam boiler fitting used to vent air from inside a boiler. The steam vent valve is opened when filling the steam boiler with water to prevent air from being trapped inside the boiler. To prevent water hammer or steam vent valves from sputtering water, minimum elevation dimensions must be followed. The steam vent valve must be installed on a vent line (½″ or ¾″ piping) that is connected to the highest part of the boiler. **See Figure 23-13.**

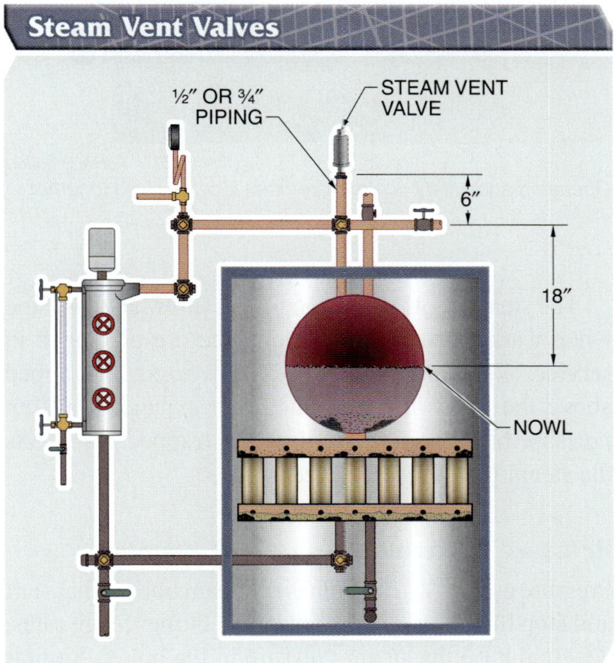

Figure 23-13. To prevent water hammer or steam vent valves from sputtering water, minimum elevation dimensions must be followed.

Steam boilers are not all equipped with vent valves. When a steam boiler does not have a vent valve, the top try cock must be used to vent the boiler. A safety valve must never be used to vent air from a steam boiler. Damage to the safety valve may occur, resulting in a possible safety valve malfunction.

When water is drained from a steam boiler for servicing, the air vent valve must be opened to prevent a vacuum from forming in the heat exchanger. The steam vent valve is also used to prevent a dangerous vacuum from forming inside a steam boiler when it is taken out of service. Steam occupies space as water vapor in a semigaseous condition. When a boiler that has been producing steam is taken off-line, the main steam stop valve is closed. The steam in the boiler cools and condenses, creating a vacuum. A vacuum condition is prevented by opening the vent valve when the boiler drops to 1 psi or 2 psi.

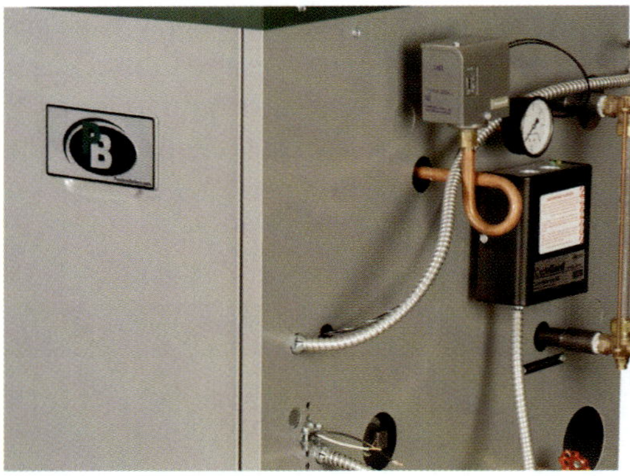

Peerless Boilers
Operating and safety controls are often located close together.

The steam vent valve is opened to allow air to escape when warming up a steam boiler before placing it in service. When a steam boiler is not vented, air trapped above the water line is heated and expands, causing pressure buildup in the boiler. The trapped air affects the steam pressure gauge readings.

Pressure Controls

Pressure controls are switches on steam boilers that start and stop the burners and control the burner firing rates, depending on the steam pressure in the boiler. Similar to steam pressure gauges, pressure controls must be protected from live steam using siphons.

The pressure control must be installed in a vertical position to function properly. In addition, the pressure control face must be mounted perpendicular (90°) to the siphon bend. When a pressure control is mounted incorrectly, expansion and contraction of the siphon may cause the pressure control to move, resulting in improper pressure of the steam boiler. **See Figure 23-14.**

The pressure controls used to control the pressure in a steam boiler include the operating pressure control, modulating pressure control, and high-pressure limit control. The operating pressure control turns the burner ON at a specified steam pressure and OFF at the desired operating pressure. Most high-pressure steam boilers use modulating pressure controls that control the firing rate of the burners, typically between 30% and 100%, depending on the steam load. The high-pressure limit control is required on a steam boiler as a safety device in the event of an operating pressure control failure. The high-pressure limit control is a manually reset control. **See Procedure for Testing Steam Boiler Efficiency—Gas-Fired Boiler.**

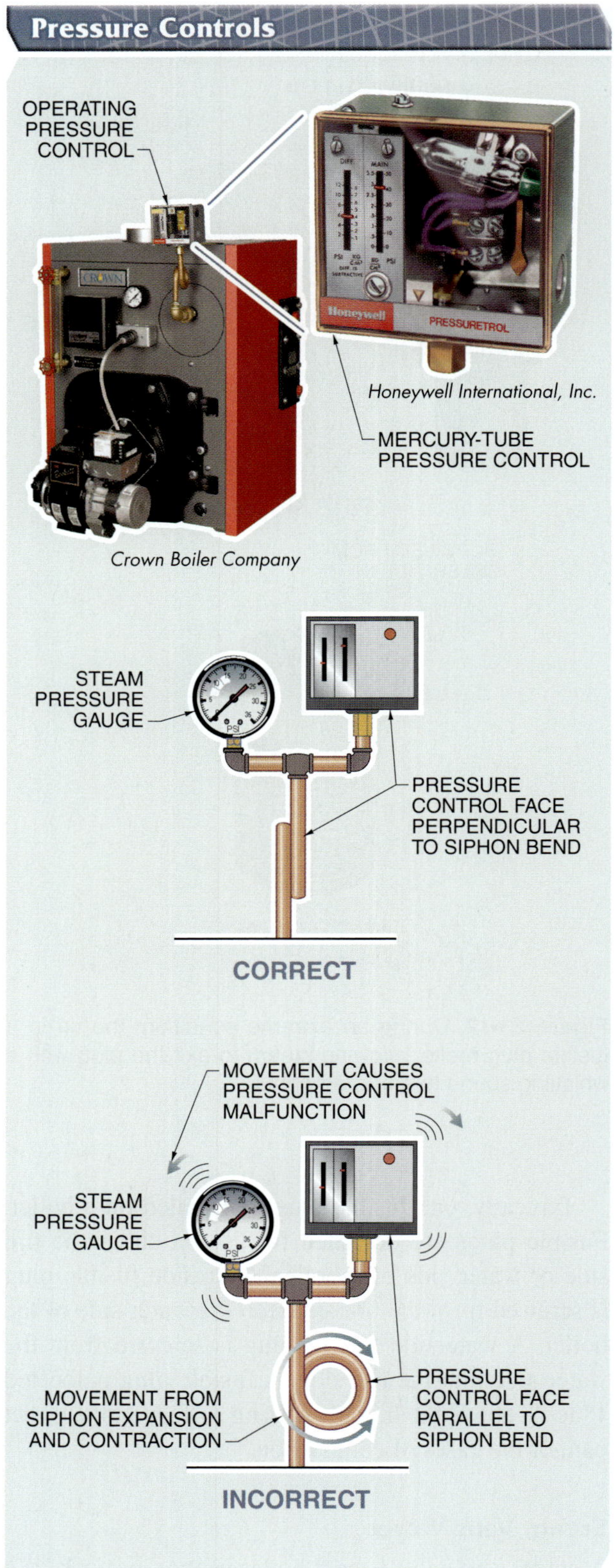

Figure 23-14. Operating pressure controls, modulating pressure controls, and high-pressure limit controls must be properly mounted so that they do not move.

Testing Steam Boiler Efficiency—Gas-Fired Boiler

1. Determine if stack temperature measurements and flue-gas analyses have been performed in the past. If not, shut down the boiler for the preparations required to install an electronic stack thermometer and portable flue-gas analyzer.

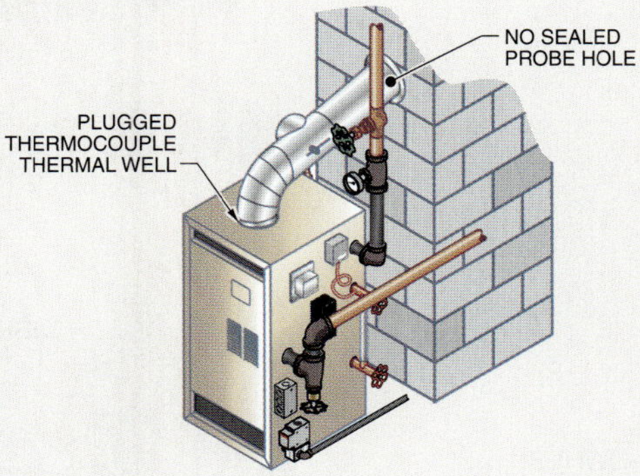

2. Ensure that the portable flue-gas analyzer is properly calibrated and fitted with sensors and filters for the combustion of natural gas. Follow all manufacturer recommendations.
   a. Select the fuel for the combustion system to be measured.
   b. Ensure that the gas outlets are free, so that the gases can escape without obstruction. Measurements could be inaccurate if the outlets are not free.

   c. There must always be oxygen in the flue-gas to prevent damage to the sensor analyzers.
3. Follow all boiler operation and hazardous area safety practices and procedures.
4. Install the electronic stack thermometer. **See Procedure for Taking Stack Temperature Measurements.**
5. Install the probe of the portable flue-gas analyzer into the flow of flue-gases.
   a. Determine a location for the flue-gas analyzer probe to take proper gas samples.

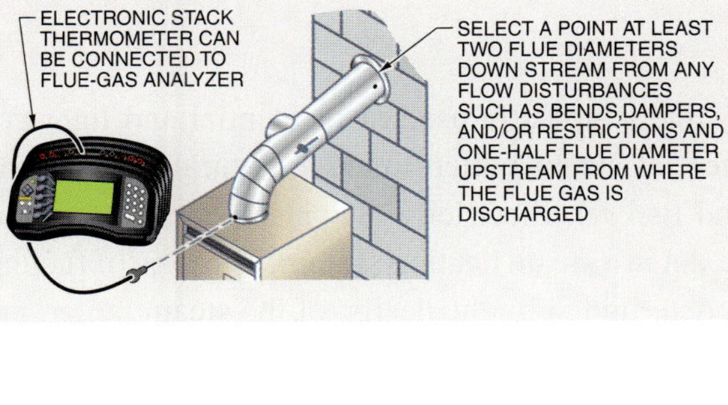

   b. Drill a small hole for the probe.
   c. Firmly insert the probe into the hole until the taper on the probe shaft seals in the escaping flue-gases.

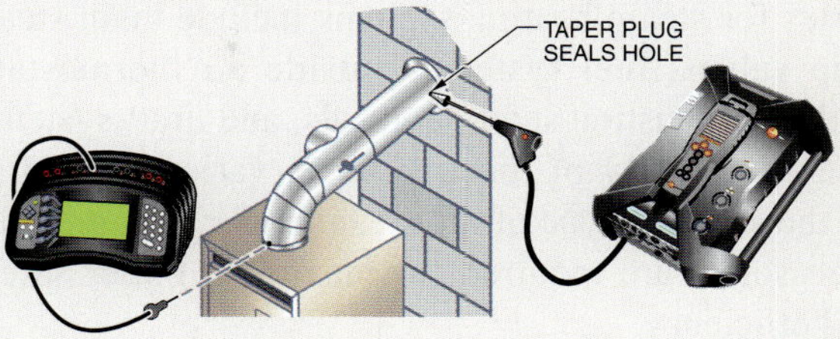

6. Start up the boiler to normal firing conditions.
7. Continuously measure and record the temperature of the flue-gases. **See Procedure for Taking Stack Temperature Measurements.**
8. During the heat-up phase of the boiler, set the flue-gas analyzer to sample the flue-gases and record measurements every 30 sec for a 5-min period.
9. Set the flue-gas analyzer to sample the flue-gases and record measurements every 15 sec for a 10-min to 90-min period.
10. When the measurements are complete, remove the electronic stack thermometer from the thermal well and the flue-gas analyzer probe from the drilled hole.
11. Plug the thermal well and use a high-temperature sealant to seal hole.

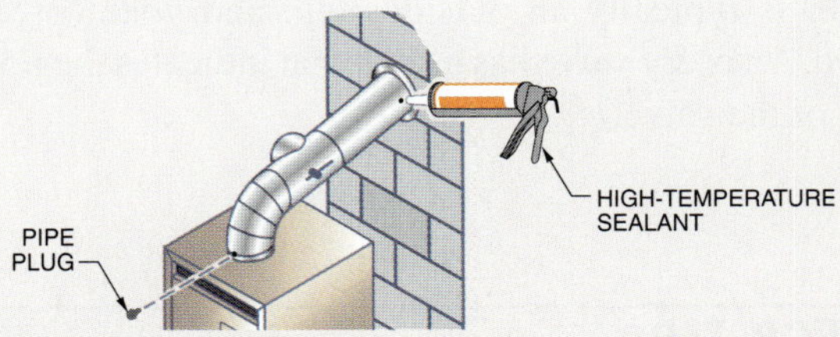

12. Clean the portable flue-gas analyzer and sensors.
    a. With the probe in ambient air, turn the portable flue-gas analyzer ON. Clean the analyzer for 5 min to 90 min, depending on the concentration of gas. Follow the specific manufacturer recommendations.
13. Analyze the data recorded.
    a. Use the data collected to calculate an average concentration for each gas tested.
    b. Take extra note of the nitrogen oxide, carbon monoxide, hydrocarbon, hydrogen, and excess air readings. These gases indicate boiler combustion efficiency.

STEAM BOILER ACCESSORIES

A *steam boiler accessory* is a component not directly attached to a steam boiler but is required for the proper operation of the heating system and/or boiler. Accessories for steam heating systems include main steam stop valves, fuel systems, outside air thermostats, boiler combustion safety controls, and draft systems. The complexity of these systems varies depending on the size and type of the steam boiler. Steam boiler accessories are required for maximum boiler safety and efficiency.

Main Steam Stop Valves

A *main steam stop valve* is a valve that is used to isolate a steam boiler from the main distribution header. It is used in typical residential applications. Angle valves provide less resistance to flow and reduce the number of fittings required. Main steam stop valves must not be used as throttling valves and should only be completely open or completely closed. **See Figure 23-15.**

Main steam stop valves must be opened slowly to prevent water hammer and sudden pressure increases or decreases. The main steam stop valve on a large steam boiler is typically an outside stem and yoke (os&y) valve. An os&y valve has a stem that indicates the position of the valve.

Main Steam Stop Valves

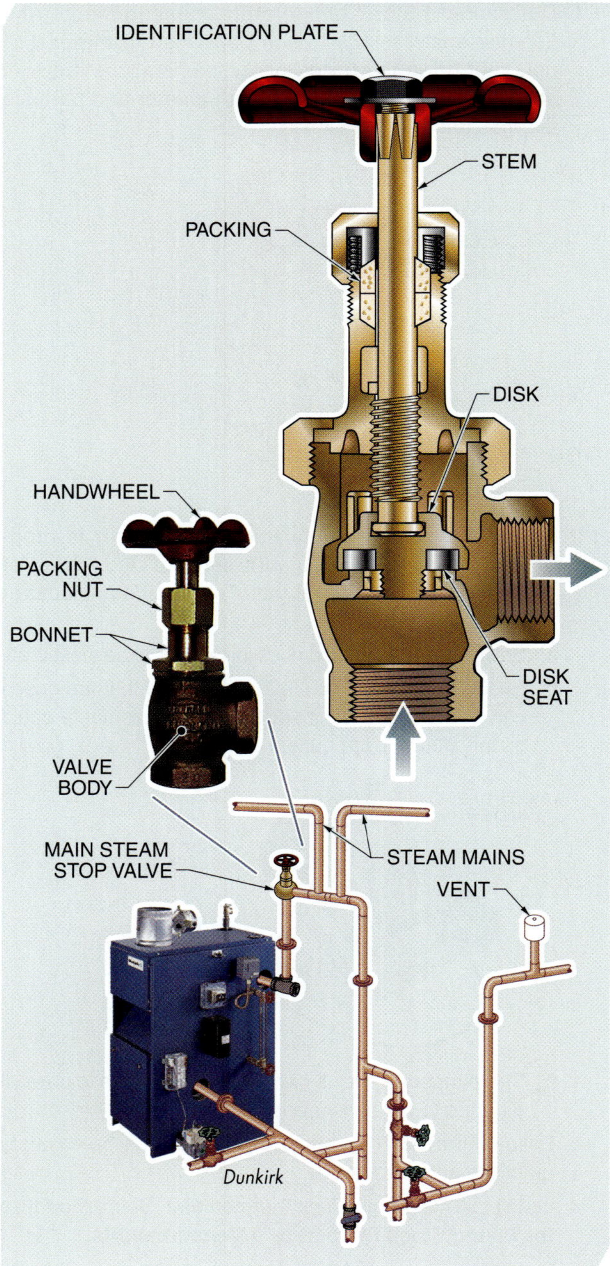

Figure 23-15. Main steam stop valves are used to isolate steam boilers from main distribution headers and should be completely open or completely closed.

TECH TIP

Over the lifetime of a boiler, the availability and price of the different fuels may vary significantly. As a number of different fuels may be used to create heat in a boiler, savings can be realized and efficiencies gained by using boilers that can be adapted for multiple fuel types.

Fuel Systems

A fuel system for a steam boiler is used to supply, store, and control the fuel supplied to the boiler. During the combustion process, an air-fuel mixture is burned to produce the heat necessary to operate the boiler. Heat from the gases of combustion is transferred throughout the heat exchanger of the steam boiler for maximum efficiency.

Fuels typically used in residential and light commercial steam boilers include natural gas, propane, and fuel oil. Electricity may also be used to produce steam in a steam heating system. The type of fuel used is determined by the design of the steam boiler, price and availability of the fuel, and air pollution regulations. The type of fuel used determines the fuel system accessories that can be used. All fuels are dangerous when safety precautions are not followed.

The systems that use natural gas as fuel for steam boilers include low-pressure gas systems and high-pressure gas systems. A *low-pressure gas system* is a type of fuel system that mixes gas with air in a mixing chamber at approximately atmospheric pressure (0 psi) before the gas enters the burner. **See Figure 23-16.**

A *high-pressure gas system* is a fuel system that mixes gas with air at pressures up to 5 psi. **See Figure 23-17.** The main components of fuel systems (valve trains) include gas shutoff cocks, gas pressure regulators, main gas valves, gas butterfly valves, low- and high-pressure gas switches, and gas vent valves.

Low-Pressure Gas Systems

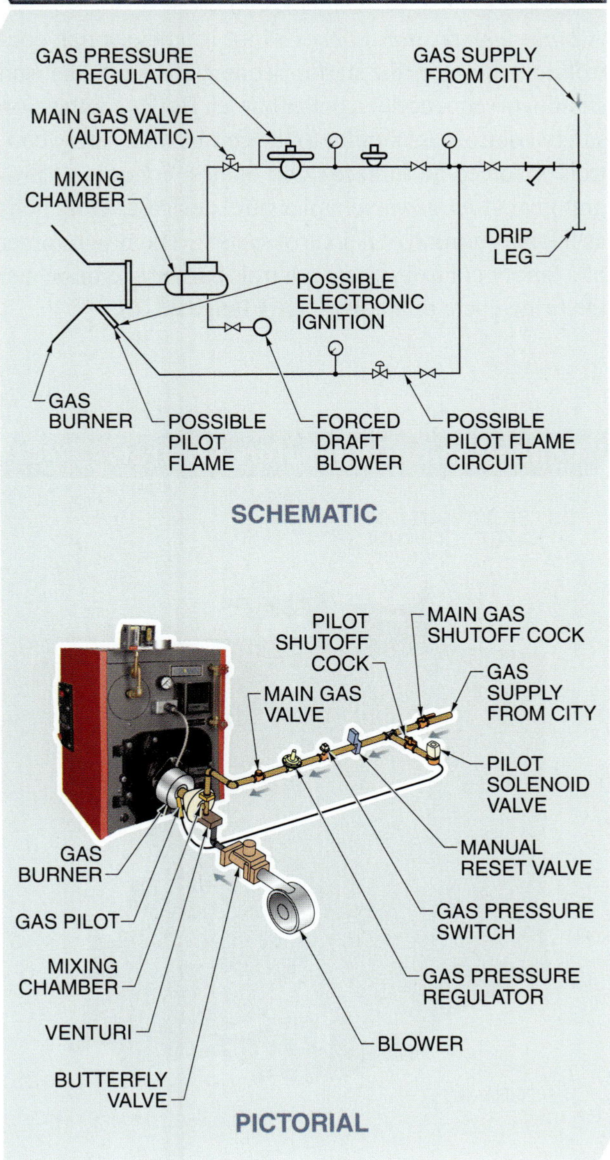

Figure 23-16. Low-pressure gas systems mix fuel with air at approximately atmospheric pressure (0 psi).

High-Pressure Gas Systems

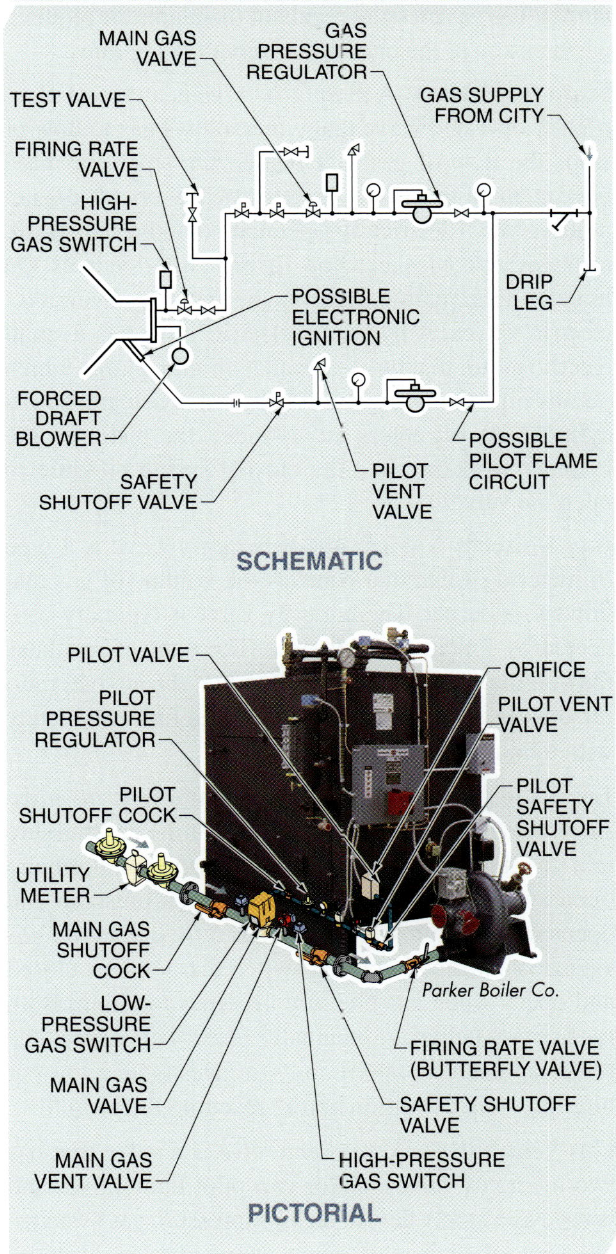

Figure 23-17. High-pressure gas systems mix fuel with air at pressures up to 5 psi.

Gas Shutoff Cocks. A *gas shutoff cock* is a type of valve used on a natural-gas fuel system to manually isolate the system or parts of the system. A gas shutoff cock is normally located on the pilot light gas line between the gas fuel train and the main gas valve. Gas shutoff cocks are normally quick opening and closing valves that have special packing and grease to prevent leaks around the valve stem.

Gas Pressure Regulators. A *gas pressure regulator* is a type of valve that adjusts the supplied gas pressure to a burner. The gas pressure regulator maintains the required gas pressure at the burner through all firing rates.

Main Gas Valves. A *main gas valve* is a type of electrically operated valve that either allows gas to flow or stops the flow of gas to a burner. On small gas-fired boilers, main gas valves are electrically operated solenoid valves. Electrically operated solenoid valves are also used to control gas flow for pilot light systems. On large boilers, main gas valves are typically hydraulic-electric valves. A hydraulic-electric valve has a small electric motor that turns a small hydraulic pump, which pumps oil from a reservoir in the valve actuator into a cylinder. As oil enters the cylinder, the piston in the cylinder works against the closing spring pressure to open the valve.

Gas Butterfly Valves. A *gas butterfly valve* is a type of metering valve that controls the volume of gas that flows to a burner. The butterfly valve is typically connected by linkage to a damper. The damper regulates the volume of air to the burner so that the air-fuel ratio remains at the desired mix to burn the fuel completely with a minimal amount of air.

Low- and High-Pressure Gas Switches. A *gas pressure switch* is a switch used to monitor the gas pressure in a gas line. A *low-pressure gas switch* is a normally open switch that is held closed by gas pressure and opens when gas pressure is too low. A *high-pressure gas switch* is a normally closed switch that remains closed and opens when gas pressure becomes too high. Both types of switches are manually reset. This allows the boiler operator an opportunity to investigate a low- or high-pressure condition before resetting the switch.

Gas Vent Valves. A *gas vent valve* is a valve that has two main gas valves and/or two pilot light valves and is used as a safety device on high-pressure gas systems. Gas vent valves are electrically operated solenoid valves that are open when the main gas valves are closed. When a main gas valve leaks, the gas will be vented out of the building into the atmosphere. When the main gas valve opens, the vent valve will close and the steam boiler will operate as normal.

Outside Air Thermostats

Many steam boilers have an outside air thermostat. The thermostat or sensor shuts down the boiler above a setpoint, when heat is no longer needed. A typical setpoint is between 50°F and 60°F. On a steam system, an outside air thermostat operates the condensate pump as well as the boiler.

Boiler Combustion Safety Controls

A *boiler combustion safety control* is a device that controls and monitors the startup sequence, main flame, and shutdown sequence of a boiler burner. Boiler combustion safety controls are similar to the combustion safety controls on forced-air furnaces. One of these controls is a programmer. A *programmer* is a control device that functions as the mastermind of a control system. The programmer of a burner control system controls burner operation and the firing cycle of a boiler. **See Figure 23-18.**

Boiler Combustion Safety Controls

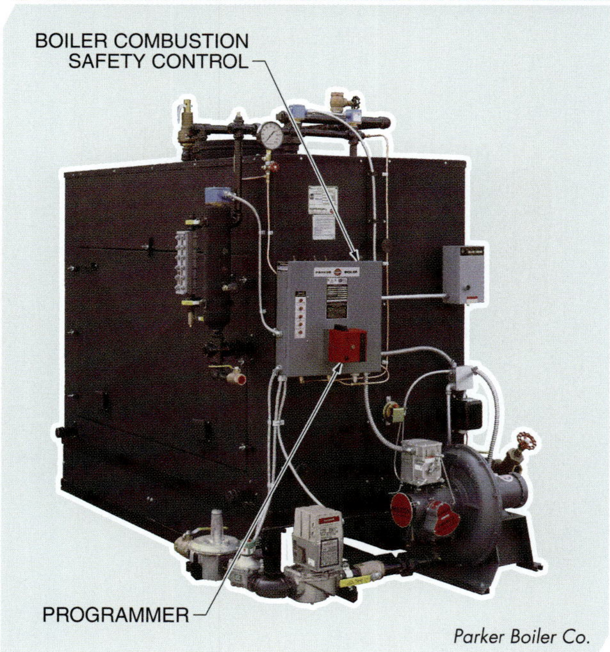

Figure 23-18. Boiler combustion safety controls monitor and control the startup sequence, main flame, and shutdown sequence of burners.

The programmer monitors the safety circuit and the flame safeguard circuit. The safety circuit includes temperature, pressure, and water level switches to ensure that all conditions are normal. The flame safeguard circuit detects the presence of a flame for the proper operation of the pilot light and main flame. Depending on the conditions of the safety and flame safeguard circuits and the timing of the sequences, the programmer determines whether the burner should operate.

Draft Systems

A boiler draft system is used to ensure the adequate flow of air for the combustion process and remove the by-products of combustion from the combustion chamber. The draft system consists of induced draft fans and exhaust breeching and/or a vent system. The vent system must be constructed of Code-approved materials.

TECH TIP

Almost all new boilers are manufactured and shipped with a microprocessor programmer as part of their standard equipment. A common retrofit application is to replace an old electromechanical programmer with a new built-in microprocessor programmer. Fortunately, most manufacturers make this easy by supplying detailed instructions for the retrofit. The new programmer often provides greater reliability, diagnostics, and energy savings.

The controls of the draft system use airflow sensors to start and stop the fans. On small residential steam boilers, barometric dampers are often used to maintain proper draft pressure in the systems. A manometer is often used to check the draft pressure in the breeching and/or vent system.

Digital Resources

ATPeResources.com/QuickLinks • Access Code: 432610

U N I T 24

Steam Heating Distribution Systems

Steam heating distribution systems consist of steam controls, accessories, and supply piping; terminal devices; and condensate return piping. During operation, a steam boiler and distribution piping system are filled with steam. Steam heating systems are classified by working pressure, temperature, method of heat production, and method of condensate return.

OBJECTIVES

- Identify common steam heating applications and associated components.
- Explain the operation of steam system accessories.
- Explain the operation of various steam heating piping systems.
- Identify common steam terminal devices.
- Describe the operation of common steam terminal device controls.

Index to Procedures

STEAM DISTRIBUTION SYSTEMS

Steam boilers are not as popular as hydronic boilers, but an HVAC technician performing field service work will encounter steam boilers in residential and light commercial applications. Steam boilers are typically located in central mechanical rooms, which supply steam to distribution systems. Steam distribution systems rely on the difference in pressure to move steam (vapor) from the boiler through piping to individual terminal units where the steam gives up its heat as it condenses back into a liquid. **See Figure 24-1.** In individual residential and light commercial buildings the steam boiler is typically located on the lowest floor of the building, such as in a basement or first floor mechanical room.

Steam pipes are well insulated to reduce the amount of condensation occurring in the piping.

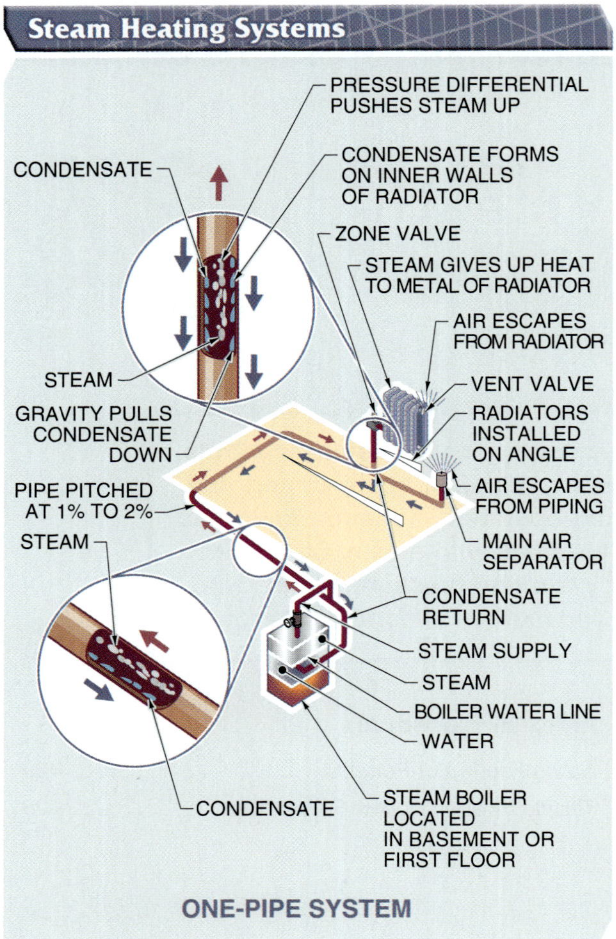

Figure 24-1. Steam boilers are centrally located in mechanical rooms. Steam relies on a difference in pressure to move from the boiler through piping to individual terminal units to give up its heat as it condenses back into a liquid.

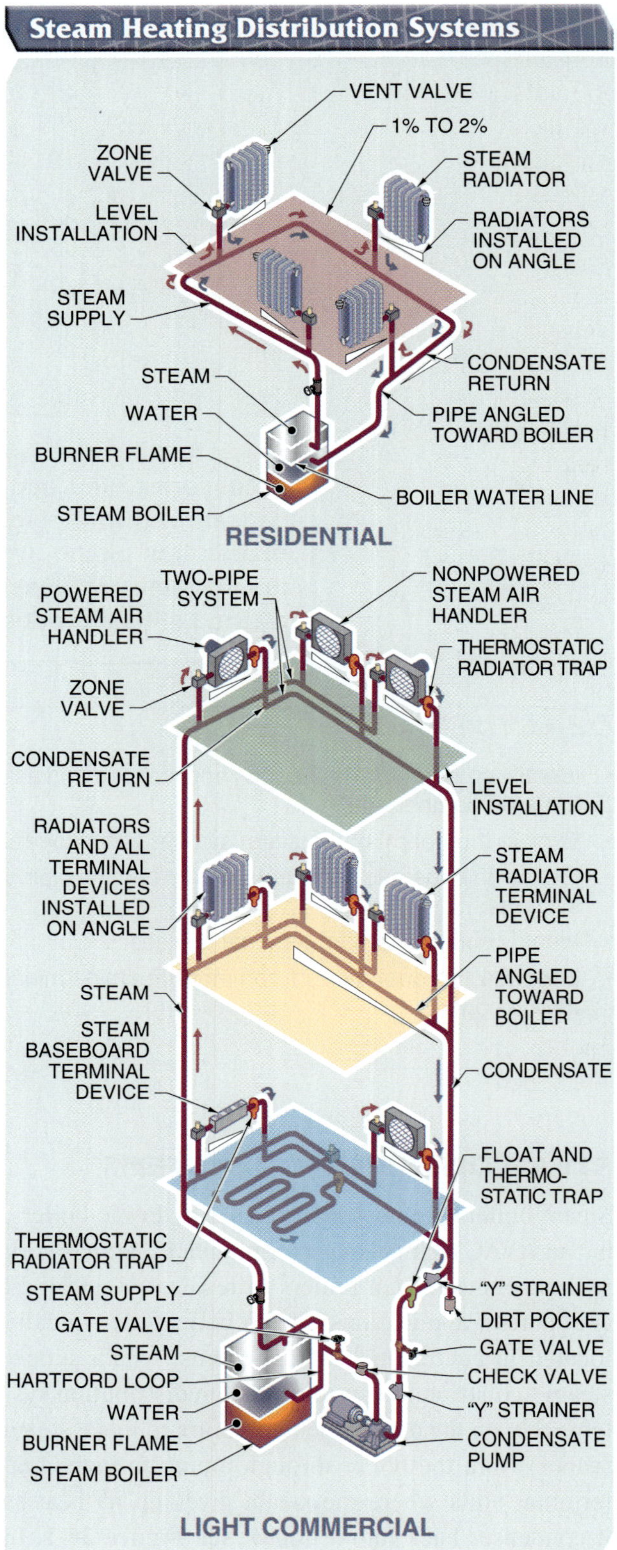

Figure 24-2. When radiators or pieces of terminal air-handling equipment are located on one floor or several floors of a building, the steam boiler is likely to be located in the basement.

One of the factors in locating steam boilers and mechanical rooms is the location of the radiators or terminal air-handling equipment. When the radiators or pieces of terminal air-handling equipment are located on several floors throughout a building, the steam boiler is likely to be located in the basement. Steam flows up through the floors and through the supply piping without the assistance of a pump. The condensate flows back to the boiler by gravity through the return piping. **See Figure 24-2.** When all radiators and air-handling equipment are located on a single floor, the steam boiler is likely to be located near the equipment to minimize the length of piping runs, which minimizes the amount of heat loss.

A steam heating system uses heat from the steam boiler to provide comfort in building spaces. Water in the boiler is heated and turned into steam. Steam leaves the boiler through the main steam line (boiler outlet) and enters the main steam header. From the main steam header, branch lines direct the steam to the terminal units. Heat is released to the air in building spaces as steam travels through the terminal units. Steam in the terminal units cools and turns into condensate. The condensate is separated from the steam by steam traps that allow condensate, but not steam, to pass. **See Figure 24-3.** The condensate is directed through the condensate return line to the condensate return tank. The condensate is pumped (optional on some systems) from the condensate return tank to the boiler as feedwater. The condensate and/or makeup water flow back to the boiler through check valves and stop valves on the feedwater line. The feedwater enters the boiler and is turned into steam to repeat the process.

Steam Distribution System Operation Media Clip

STEAM SYSTEM COMPONENTS

A feedwater system supplies the proper amount of makeup water to a boiler. The fuel system provides fuel for combustion to produce the necessary heat in the boiler. The draft system regulates the flow of air to and from the burners. The steam system piping controls and distributes the steam produced and collects the condensate for reuse by the steam boiler.

Feedwater Systems

Feedwater is boiler-ready water that is a combination of steam condensate and makeup water. A feedwater system requires accessories to supply the correct amount of water in the proper condition to the steam boiler. A *feedwater system accessory* is equipment that is not directly attached to a steam boiler but controls the amount, pressure, and/or temperature of water supplied to the boiler.

Maintaining the correct water level in a steam boiler is critical for safety and efficiency. In residential and light commercial applications, feedwater is rarely treated, but pressure is always regulated automatically. Valves are installed in feedwater lines to permit access for maintenance and repair. The feedwater system and accessories must be capable of supplying water to the steam boiler under any circumstance. **See Figure 24-4.**

Steam Heating Distribution System Operation

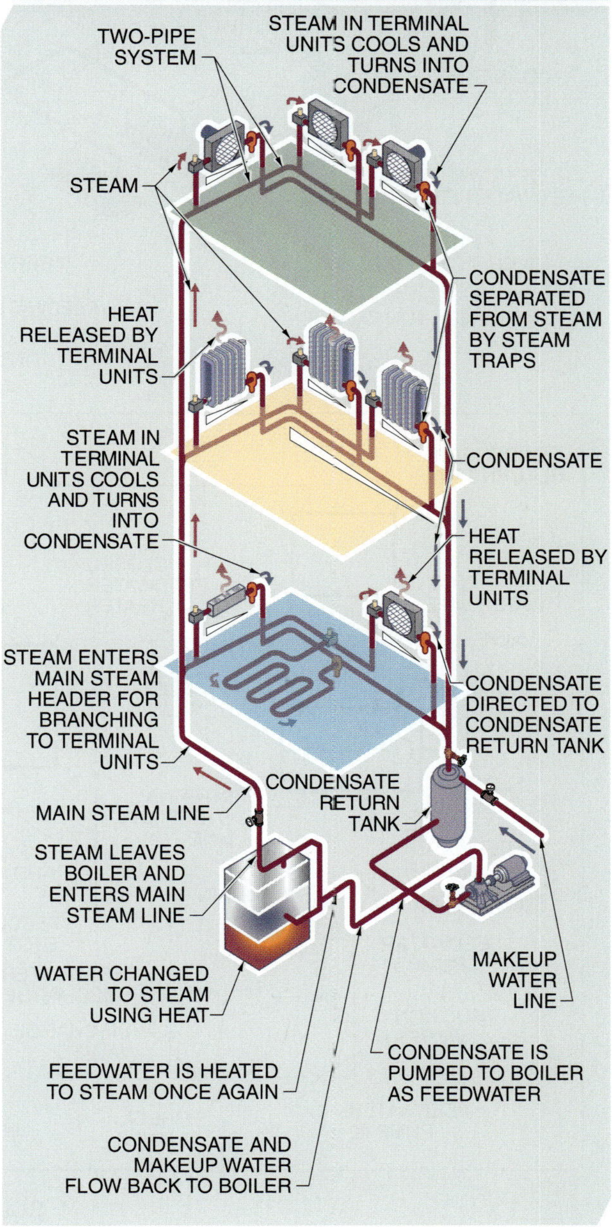

Figure 24-3. Steam heating distribution systems transmit heat from steam boilers to terminal units located throughout a building to provide comfort.

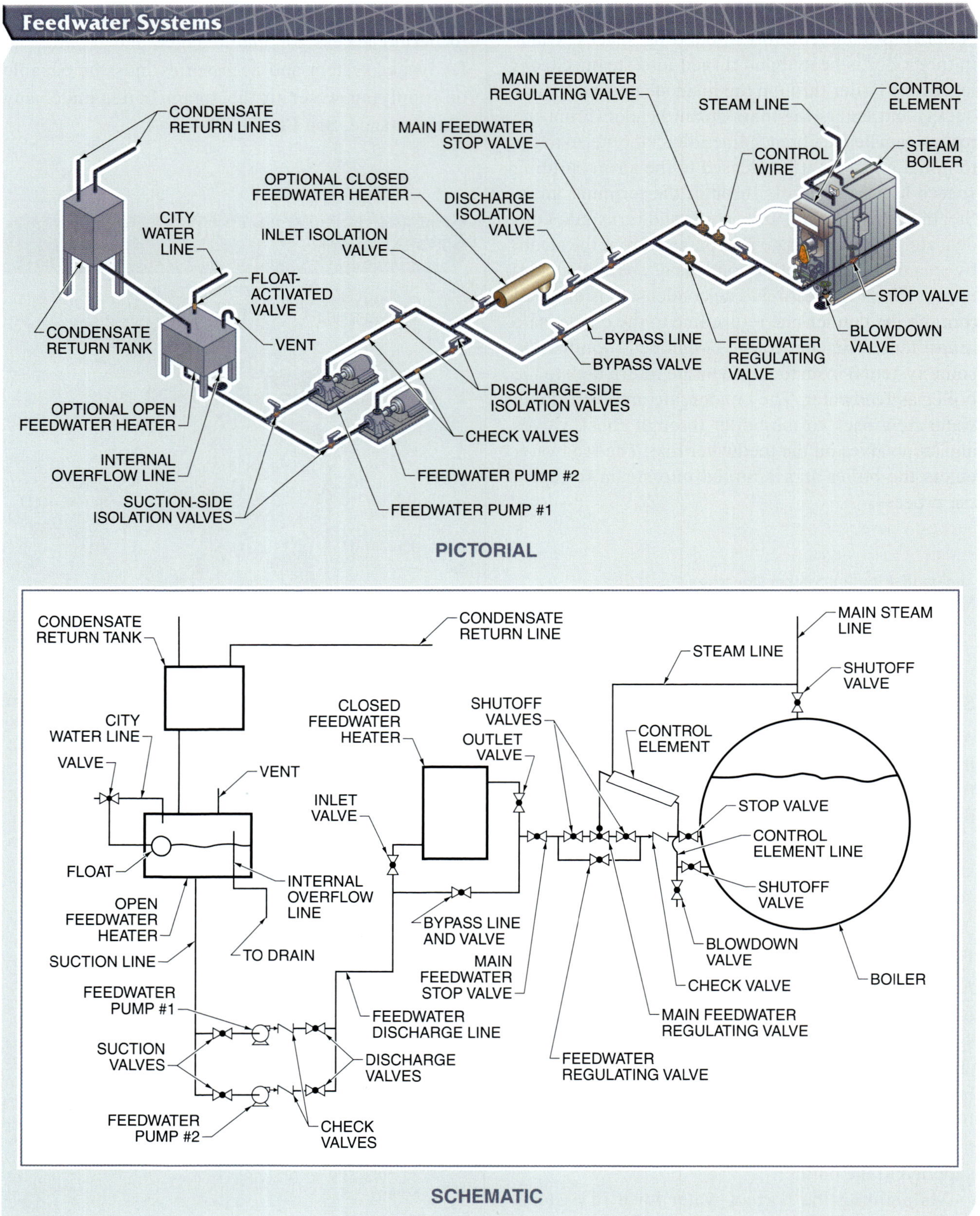

Figure 24-4. A feedwater system must supply sufficient water to a steam boiler at the proper temperature and pressure to maintain the NOWL in the boiler.

TECH TIP

It is vital to maintain the proper water level in a boiler. A loss of water can lead to the burning out of tubes and/or a boiler explosion. The feedwater regulator maintains the proper water level in a boiler. The check valve does not allow water to flow from the boiler in the event of stop valve malfunciton.

Feedwater Valves. A *feedwater valve* is a valve that controls the flow of makeup water into a steam boiler to compensate for water that is lost. The makeup water piping has a check valve and stop valve. **See Figure 24-5.** A *feedwater check valve (backflow preventer)* is a valve that controls the direction of city water flow by allowing flow in only one direction through the valve. Check valves prevent boiler water from draining back into the city water supply piping. Verify the application of proper backflow devices with local building codes and authority having jurisdiction.

A *check valve* is a valve that allows flow in only one direction. Many different types of check valves are available. The ball, poppet, or disc in a check valve opens when fluid pushes against it in one direction but closes when the direction of flow reverses. Ball- and poppet-style check valves are typically found in high-pressure, low-flow applications, such as pilot lines, and as bypass valves inside other valves. Disc-style check valves (swing and lift) are typically found in low-pressure, high-flow applications. **See Figure 24-6.**

Backflow preventers are installed on the water supply line to stop flow back to the city water supply.

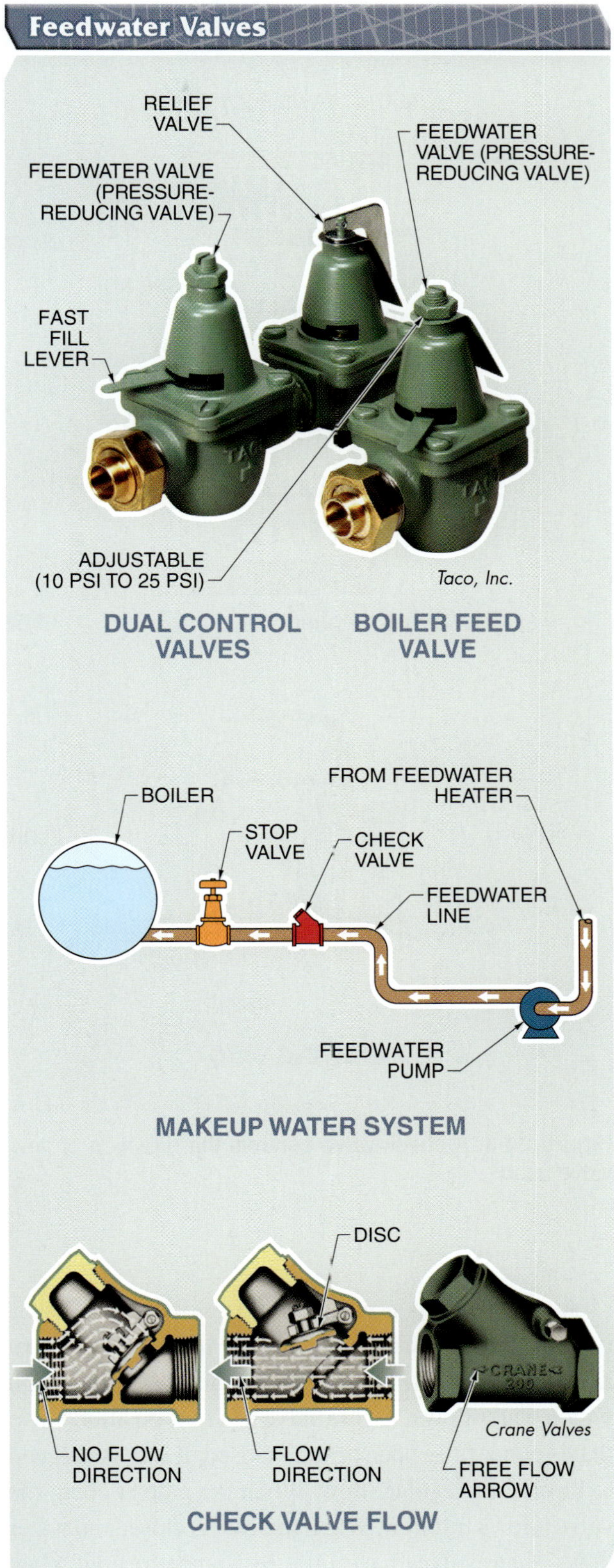

Figure 24-5. A feedwater valve controls the flow of makeup water into a steam boiler to compensate for the water that is lost by the distribution system.

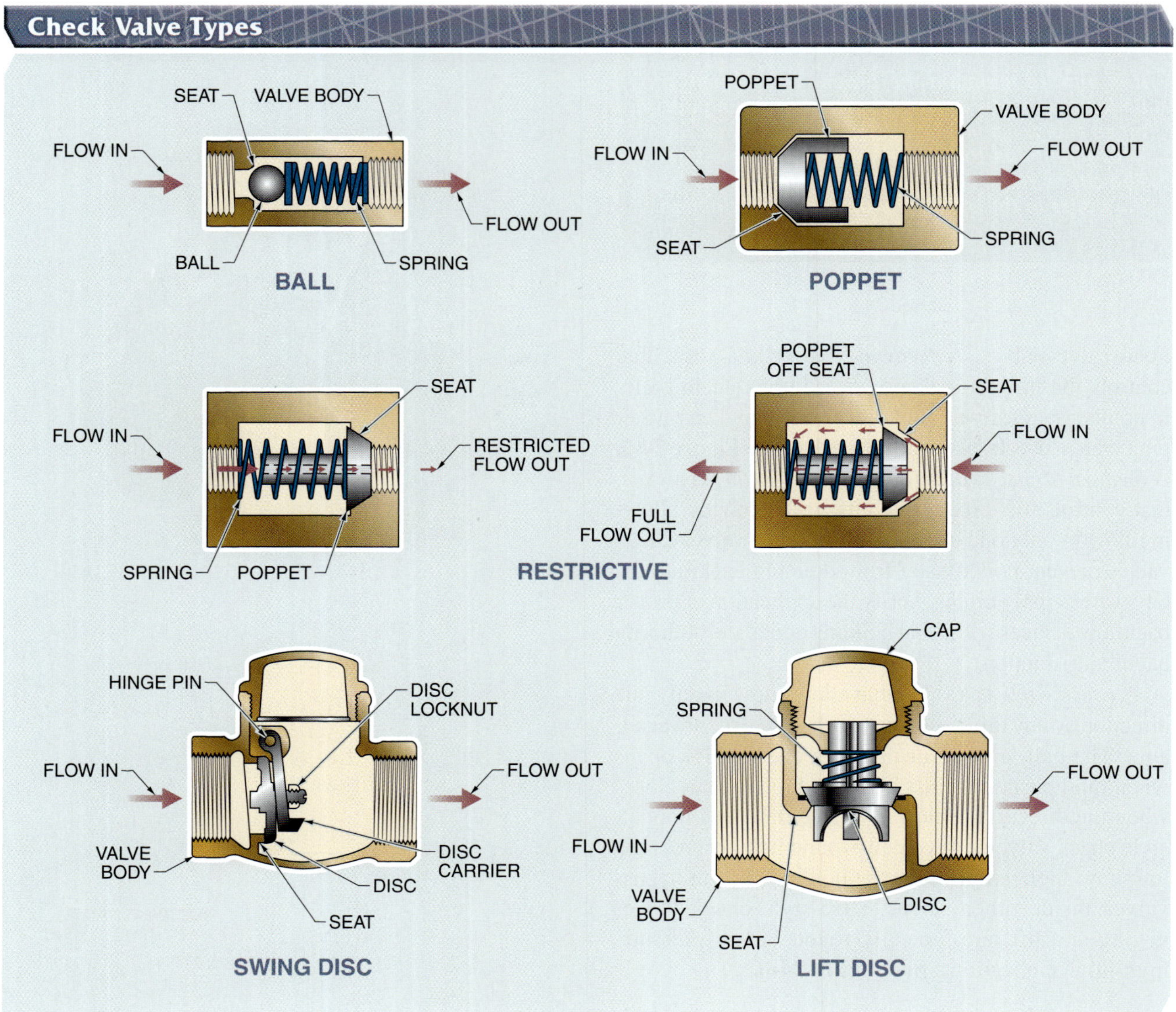

Figure 24-6. A check valve controls the direction of flow in a system, and the application determines the type of check valve used.

A *stop valve* is a gate valve or globe valve that is used to shut off steam flow. A *gate valve* is a valve that has an internal gate that slides over an opening through which water flows. A gate valve is a two-position valve that is either fully open or fully closed. It is not designed to throttle or regulate flow. When the gate is open, the valve allows full flow. When the gate is closed, no water flows. Gate valves have little resistance to flow when open and have a tight seal when closed. **See Figure 24-7.**

A *globe valve* is a valve that has a disc (globe) that rises or lowers over a seat through which fluid flows. The seat is at right angles to the flow of steam. The fluid must make two 90° turns to travel through the valve. A globe valve is an infinite position valve that is used to regulate steam flow. The resistance and pressure drop created by the valve are controlled by the amount the valve is opened. **See Figure 24-8.**

Feedwater Pumps

A *feedwater pump* is a steam boiler accessory used to pump feedwater into a steam boiler. Water flow produced by the feedwater pump must overcome boiler water pressure and system resistance to maintain the normal operating water level (NOWL) in the boiler. Feedwater pumps vary in size depending on system requirements. Feedwater pumps must have adequate capacity to maintain the required boiler water level under all operating conditions.

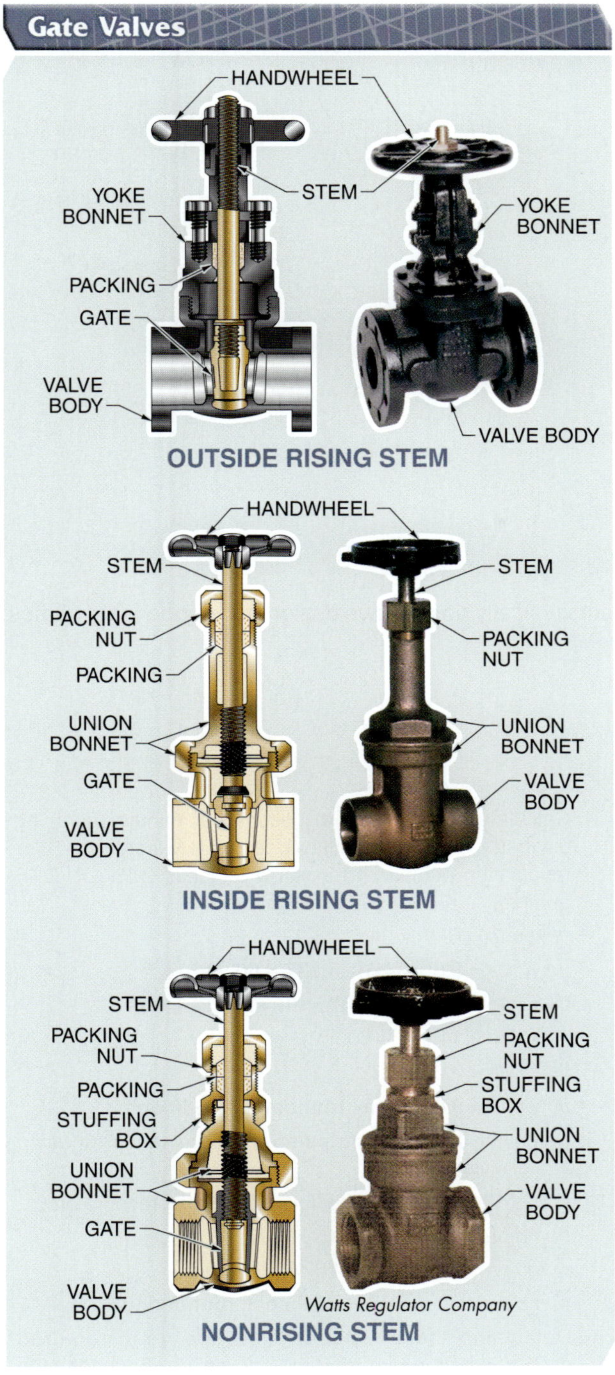

Figure 24-7. A gate valve is an ON/OFF valve that allows nonturbulent flow when fully open and no flow when fully closed. It is not designed to throttle or regulate the flow of steam.

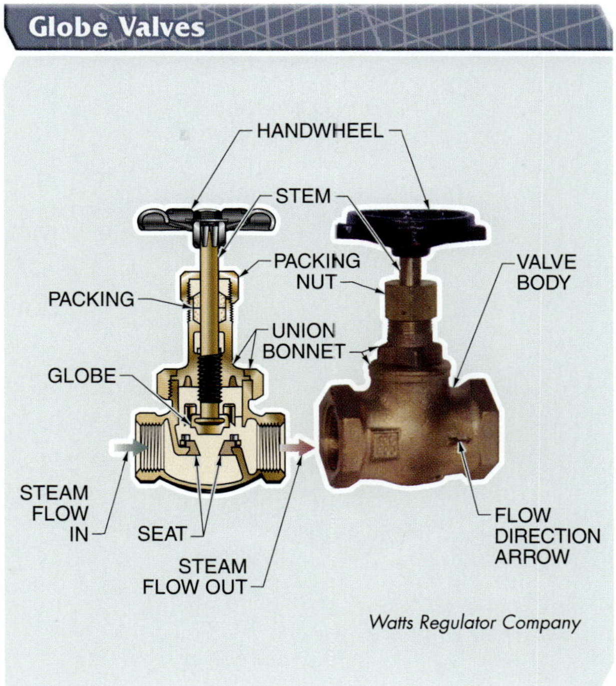

Figure 24-8. A globe valve can be used to regulate flow and to stop flow, but it must be used with the port below the seat as the inlet and the port above the seat as the outlet.

Condensate Pumps

A *condensate pump* is a pump that transports condensate from anywhere in a system back to the steam boiler or condensate return tank for reheating and reuse. A condensate return tank collects condensate from large steam distribution systems. **See Figure 24-9.**

Most residential and light commercial steam boilers do not have condensate or feedwater pumps. A commonly used type of condensate or feedwater pump is the centrifugal pump driven by an electric motor.

A centrifugal pump is a type of pump that converts centrifugal force to flow for transporting water or condensate. Centrifugal force is the outward force produced by a rotating impeller. The feedwater (condensate and makeup water mixture) is thrown out from the impeller against the pump housing (volute). The pump housing directs the feedwater to the discharge piping connected to the boiler. The pressure rating of a centrifugal pump is dependent on the number of stages. The discharge flow rate of a centrifugal pump is dependent on the impeller size and speed. **See Procedure for Sizing Feedwater Pumps.**

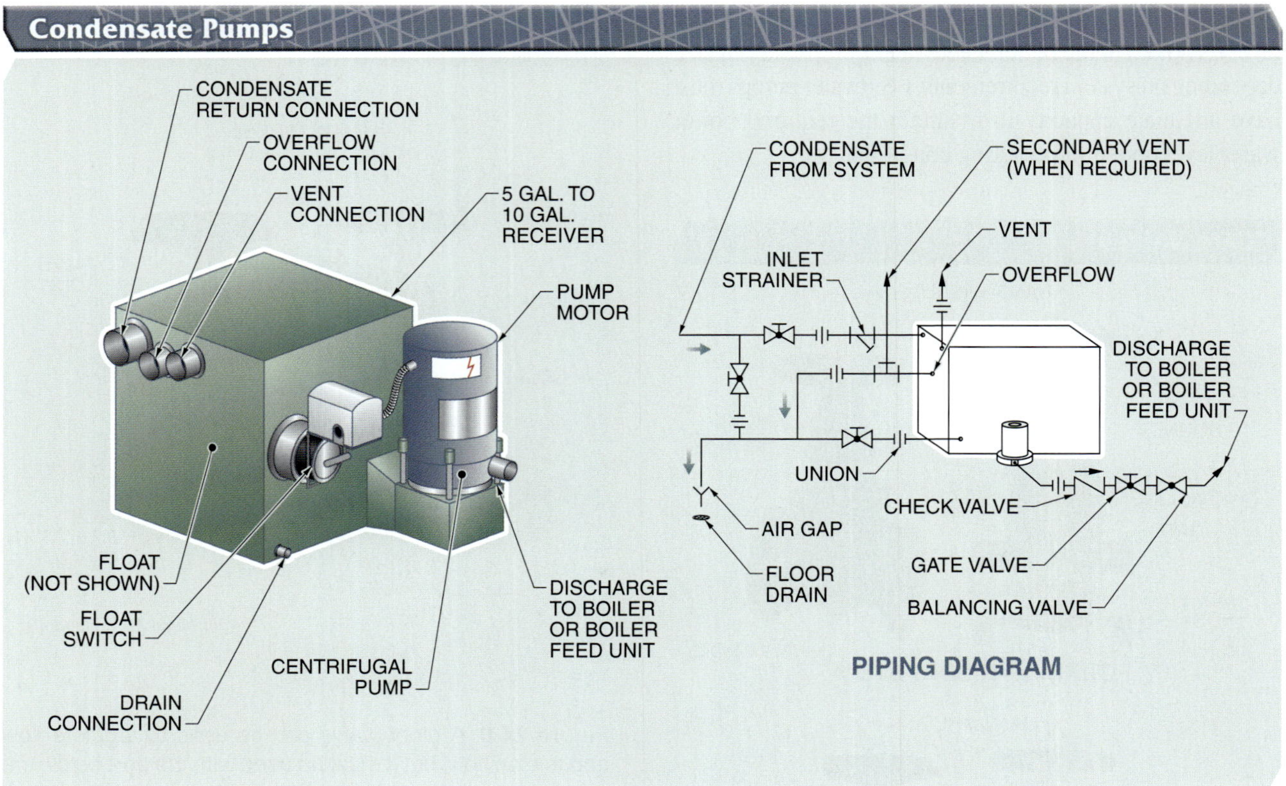

Figure 24-9. Small residential and light commercial steam systems typically do not have designated condensate pumps.

Procedure: Sizing Feedwater Pumps

1. One boiler horsepower (BHP) = 34.5 lb/hr of steam at 212°F or 33,475 Btu/hr. Ten minutes of water should always be available to a steam boiler with a relief valve set at 15 psi. One gallon of water weighs 8.337 lb.
2. Follow all steam system and hazardous environment safety practices and procedures.
3. Calculate the feedwater tank size required for a residential steam boiler producing 122,000 Btu/hr (3.6 BHP). Most residential steam boilers do not have feedwater tanks.

   For a boiler that has a tank, calculate the feedwater tank size required by using the following formula:

   feedwater tank size = *BHP* × 34.5 ÷ 8.337 lb ÷ 60 min × 10

   feedwater tank size = 3.6 × 34.5 ÷ 8.337 ÷ 60 × 10

   feedwater tank size = 124.2 ÷ 8.337 ÷ 60 × 10

   feedwater tank size = 14.897 ÷ 60 × 10

   feedwater tank size = 0.248 × 10

   feedwater tank size = 2.48 gal. × 1.5 (safety factor)

   feedwater tank size = **3.72 gal., rounded to 5 gal.**
4. Calculate the flow rate in gallons per minute (gpm) by using the following formula: (Use a safety factor of 1.5 for modulating pumps and a safety factor of 2 for ON/OFF pumps.)

   flow rate = *BHP* × 34.5 ÷ 8.337 ÷ 60 × 2

   flow rate = 3.6 × 34.5 ÷ 8.337 ÷ 60 × 2

   flow rate = 124.2 ÷ 8.337 ÷ 60 × 2

   flow rate = 14.897 ÷ 60 × 2

   flow rate = 0.248 × 2

   flow rate = **0.496 gpm, rounded to 0.500 gpm**
5. Calculate the minimum pump pressure rating by using the following formula:

   pump pressure rating = psi × 1.03 + 5 (lb pressure drop)

   pump pressure rating = 15 × 1.03 + 5

   pump pressure rating = 15.45 + 5

   pump pressure rating = **20.45 psi, rounded to 25 psi**
6. Determine the net positive suction head required (NPSHR) by the pump for the application. To choose the correct net positive suction head (NPSH), refer to the pump selection tables and application. Check the pump curve for the NPSHR, and then add 1′.

Feedwater Heaters

A *feedwater heater* is a light commercial, commercial, and industrial steam boiler accessory that heats feedwater before the feedwater enters the boiler. Feedwater heaters are used to increase overall boiler efficiency. Feedwater heater systems are typically required on large steam boilers but can also be found on smaller steam boilers to collect and remove oxygen and other gases that may cause boiler corrosion. Maintaining the proper feedwater temperature also reduces the amount of scale-causing material that settles on the heating surfaces of the boiler steam drum. **See Figure 24-10.**

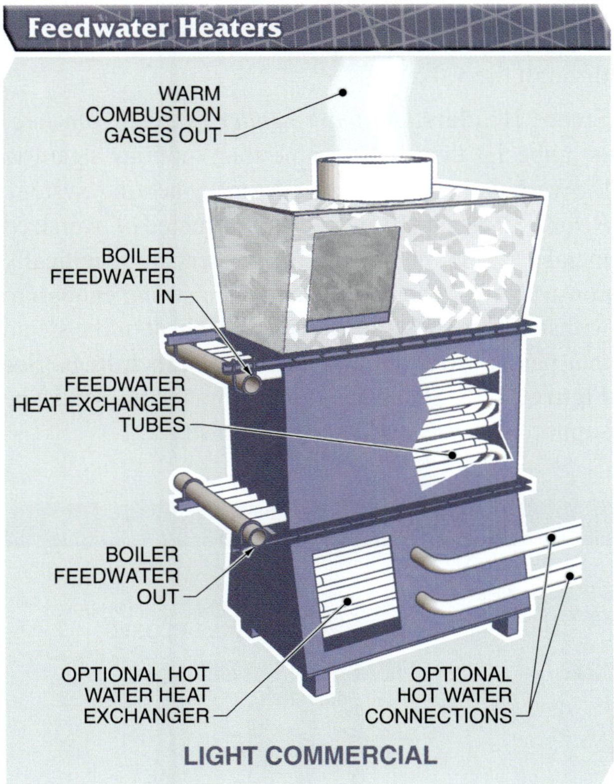

Figure 24-10. Feedwater heaters are used on large steam boilers to increase overall system efficiency and to reduce boiler corrosion.

Feedwater Regulators

A *feedwater regulator* is a steam boiler accessory that maintains the normal operating water level (NOWL) in a boiler. The most common way to maintain the NOWL on residential or light commercial steam boilers used for heating purposes is to control the feedwater valve using a float feedwater regulator or electrode assembly. **See Figure 24-11.**

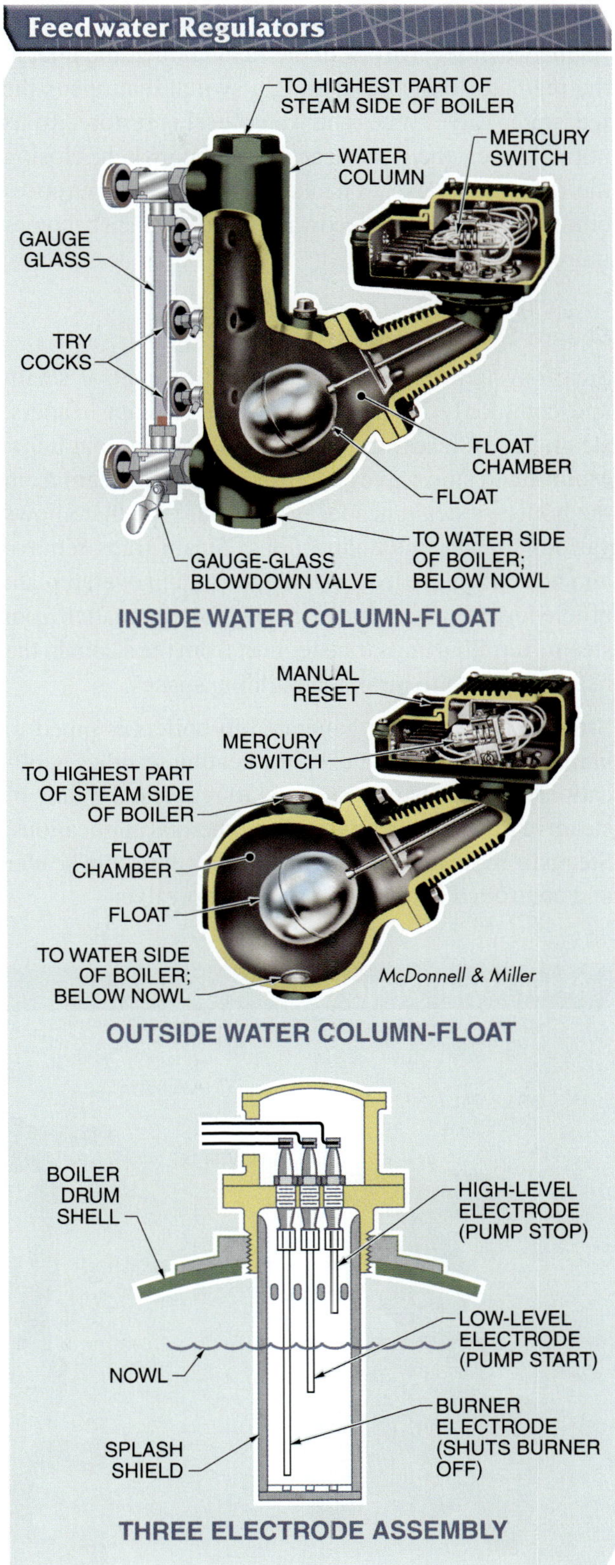

Figure 24-11. Feedwater regulators can be installed attached to or separated from the water column, depending on their specific design.

A bowl containing a float is mounted at the NOWL. As the water level drops in the steam boiler, the float in the regulator drops, activating a switch that opens the feedwater valve. When the water level is restored to its normal level, the float deactivates the switch, closing the feedwater valve. When steam is used for purposes other than heating, the feedwater regulator then becomes more sophisticated.

Steam System Accessories

Steam system accessories are used to control steam and commonly include steam valves, steam headers, steam traps, steam strainers, and steam terminal units. Main steam stop valves control the flow of steam from the boiler. A steam header is a special pipe that allows various branch line connections. Steam traps remove air and condensate from steam to maintain overall plant efficiency. Steam strainers remove foreign matter from steam. Terminal units transfer heat from the steam in the heating system to air in the building spaces.

Steam Valves. Steam generated in boilers is piped to various locations to be used for heating or other applications. Various valves are used to regulate the flow of steam in light commercial and commercial applications. Steam from steam boilers is directed by the steam header and controlled by the main steam stop valve.

A *main steam stop valve* is a valve that is used to isolate a steam boiler from the main distribution header. Typically, the main steam stop valves used in light commercial and commercial steam boilers are rising-stem outside stem and yoke (os&y) valves. **See Figure 24-12.**

A *rising-stem outside stem and yoke (os&y) valve* is a gate valve that indicates whether it is open or closed by the position of the stem. An os&y valve allows steam to flow when the gate is lifted to the open position, which places the stem in the up position. It stops steam flow when the gate is lowered to close the valve and the stem is in the down position. An os&y valve is always used in either the fully open position or fully closed position to prevent damage to the valve. When wide open, steam flows unrestricted through the valve.

Steam Headers. A *steam header (main steam line)* is a special distribution pipe that supplies steam to the various branch lines of a steam heating system. Residential steam headers are constructed of oversized metal pipe. Commercial steam headers are specifically constructed for applications and are strong enough to withstand the pressure, velocity, and heat of the steam that passes from the boilers to the branch lines. **See Figure 24-13.** Steam headers and branch lines must be supported to prevent damage from strain.

Main Steam Stop Valves

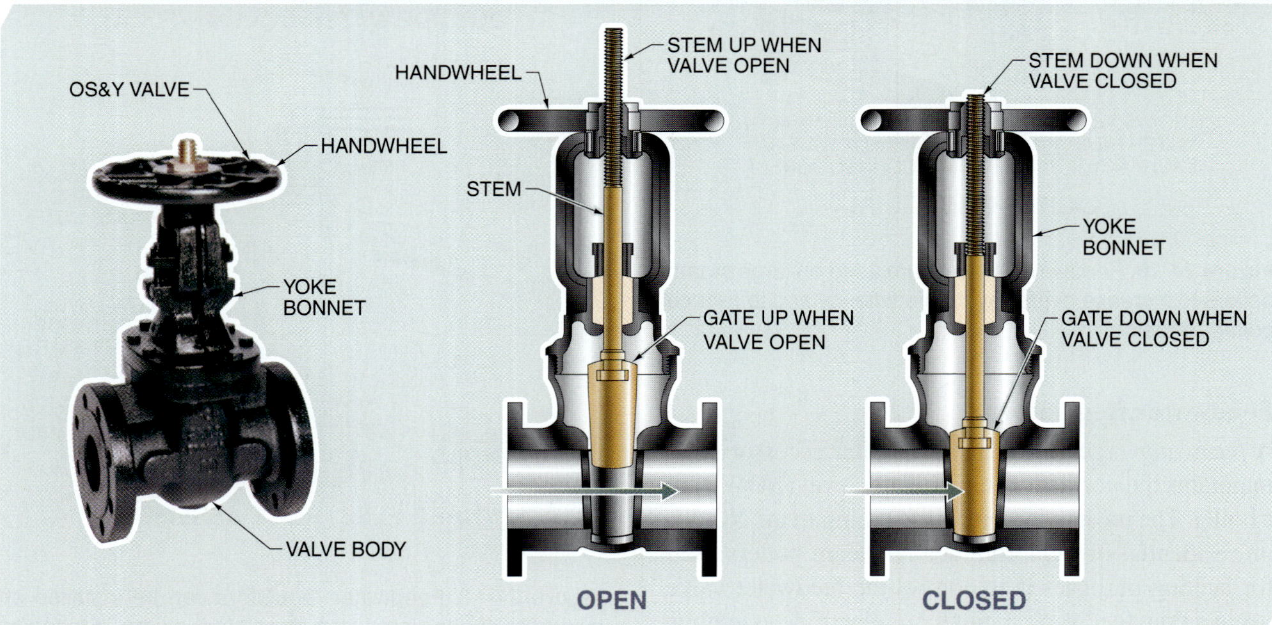

Figure 24-12. An os&y valve is used as the main steam stop valve for a boiler and does not restrict the flow of steam when open.

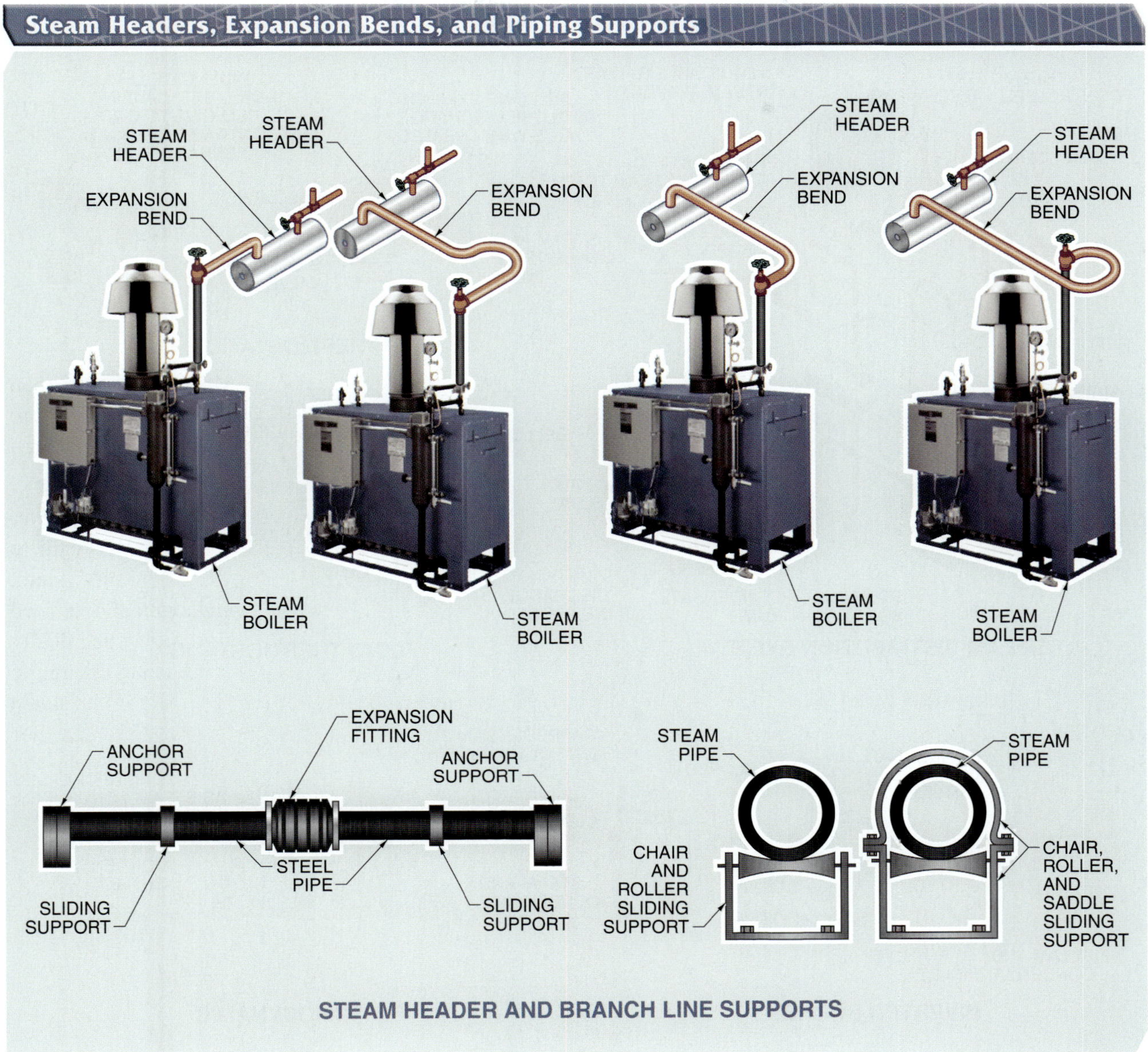

Figure 24-13. Steam headers supply steam to the various lines of steam heating systems, while expansion bends allow for pipe movement due to piping expansion and contraction. All steam distribution system piping must be supported to prevent damage from strain.

In addition, the heating and cooling of steam lines cause expansion and contraction during normal operation and shutdown. Different methods are used to compensate for the expansion and contraction that takes place.

Steam Traps. As steam gives up its heat, it condenses back into a liquid. This condensate must be removed and brought back to the boiler. A *steam trap* is a commercial and industrial steam distribution system accessory that removes air and condensate from steam lines and terminal units and returns the condensate back to the boiler for reuse as feedwater.

Air must be removed from the steam lines to maintain the supply of steam for maximum efficiency. Steam is not permitted to bypass a steam trap, and condensate is always diverted back to the condensate return tank. Steam traps operate automatically and increase overall steam system efficiency by removing air and condensate without the loss of steam. **See Figure 24-14.**

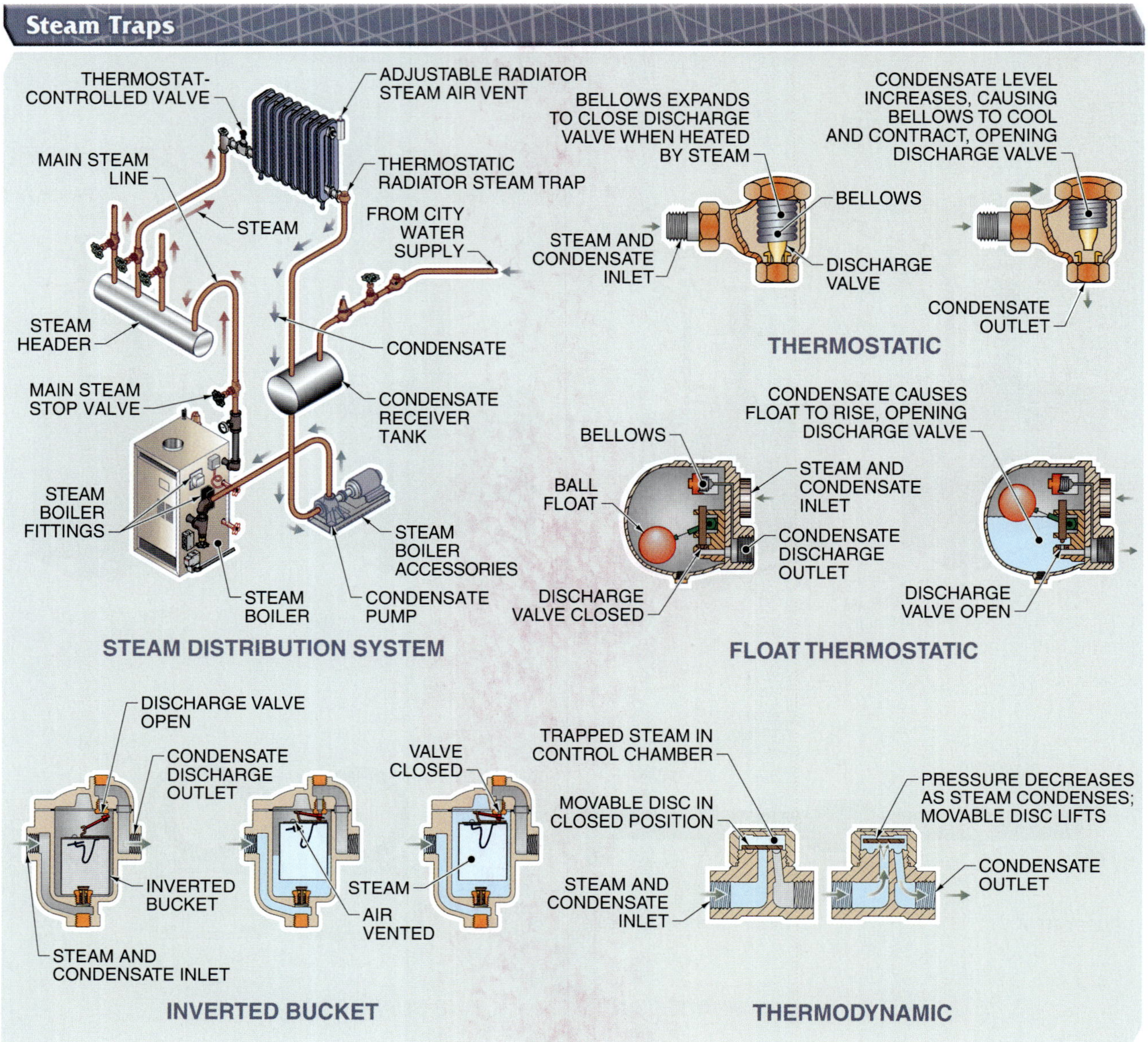

Figure 24-14. Steam traps operate automatically and are located throughout a steam system after each device, where steam is used or heat is released.

Steam temperature drops slightly as it travels from the boiler through the steam header and from the steam header to the terminal units. The drop in temperature causes some of the steam to change into condensate. Condensate that is formed in a system must be removed. Typical steam traps include thermostatic expansion, ball float, inverted bucket, and thermodynamic steam traps. Manufacturers are best at determining the size and type of steam trap that should be chosen for an application.

TECH TIP

One of the main problems in steam heating systems is water hammer. Water hammer is a condition where a hammering or banging sound occurs when condensate is carried along in steam lines at the same velocity as the steam. When the condensate strikes an obstruction such as an elbow, tee, or valve, it causes vibration and noise from the impact. To prevent water hammer, condensate must be trapped and properly removed.

A common way to test steam trap operation is to use a temperature-measuring device to see if the trap opens and closes.

It is important to ensure that steam traps operate properly and do not allow the loss of live steam. Water hammer results when condensate is picked up with the steam as it travels through the steam lines.

To keep steam traps operating efficiently, traps must be tested periodically. There is no single test method that provides usable test results for all the varieties of steam traps.

The three methods of steam trap testing are visual, temperature, and sound or ultrasound testing. Visual testing uses special valve arrangements or in-line sight glasses to view steam trap operations (opening and closing). Temperature measurements taken before and after a steam trap are useful, but the information has limited usefulness. Sound tests provide the most usable and accurate information for testing steam traps. **See Figure 24-15.**

When testing a steam trap in an area of multiple traps, all of the traps should be tested. A good trap can transmit the sound waves of a bad trap. The trap with the loudest sounds is probably the bad trap. Low-frequency listening devices, such as stethoscopes, screwdrivers, and metal rods, are quite useful for detecting steam trap mechanical sounds. High-frequency listening devices, such as ultrasonic leak detectors, are the best devices for detecting some flow sounds. However, these devices do not detect well the mechanical sounds created by traps. Before purchasing a high-frequency listening device, it should be tested on a known system to see if it provides useful information. **See Procedure for Testing Inverted Bucket Steam Traps.**

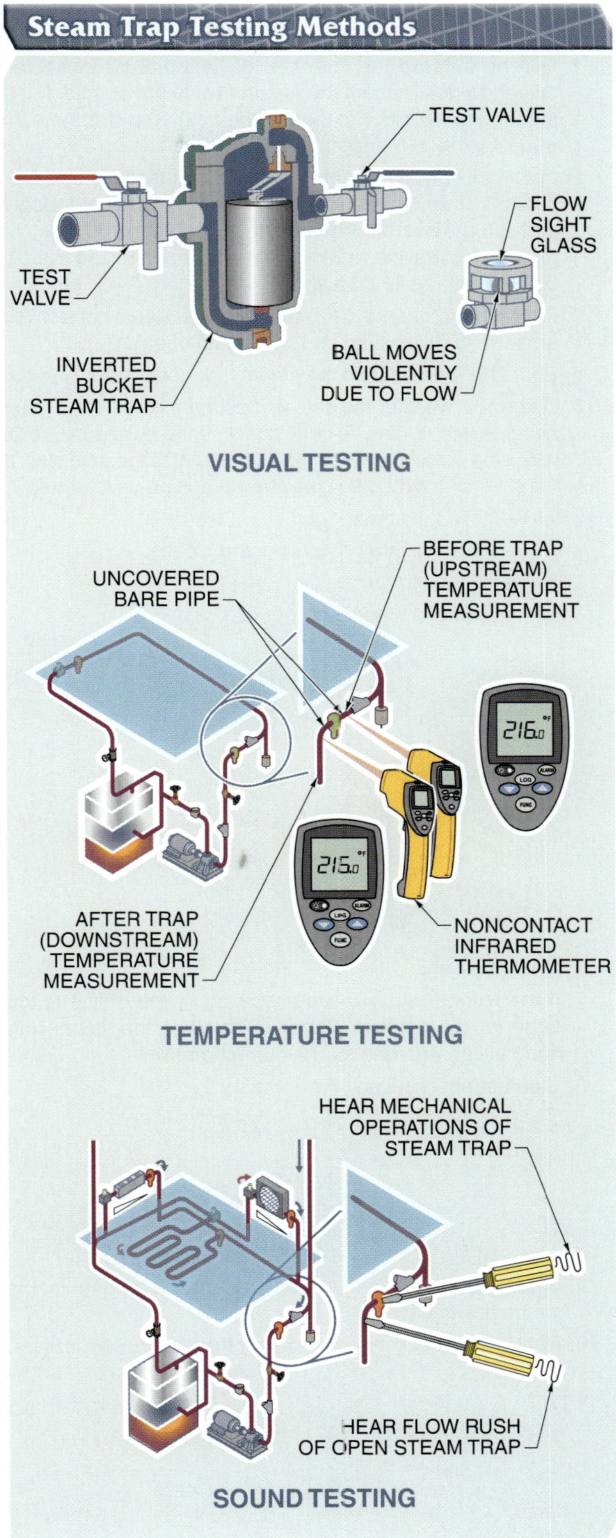

Figure 24-15. Of the three steam trap testing methods, which are visual, temperature, and sound testing, sound testing provides the most useful information when troubleshooting steam traps.

Procedure

Testing Inverted Bucket Steam Traps

1. Use an infrared thermometer or temperature indicating crayon to determine if the steam trap is hot or cold. If the steam trap is cold, proceed to the next step. If the steam trap is warm or hot (operating), proceed to step 11.
2. Follow all specific steam system safety practices and procedures. Isolate the steam trap from the pressurized supply and return lines before opening for inspection, repair, or testing. Isolate the strainer from the pressurized supply before opening for cleaning.
3. If the steam is shut OFF, turn ON the steam to test the steam trap (allow time for the purging of air).
4. Shut OFF the steam if the steam trap is still cold.
5. Determine whether the strainer upstream of the steam trap being tested is dirty. A dirty strainer may be the cause of problems. If the strainer is clean, proceed to the next step. If the strainer is dirty, clean the strainer and proceed to step 7.
   a. Remove the strainer cap.
   b. Remove the strainer basket and clean.
   c. Install the strainer basket.
   d. Install the strainer cap.

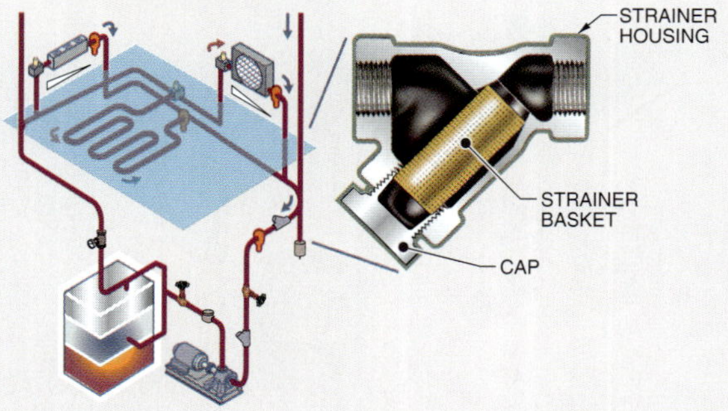

6. Open the trap.
7. If the steam trap is overpressurized or the orifice is too large for the applied differential, replace the steam trap mechanism with one for the correct pressure.
8. Clean the bucket vent, if necessary.

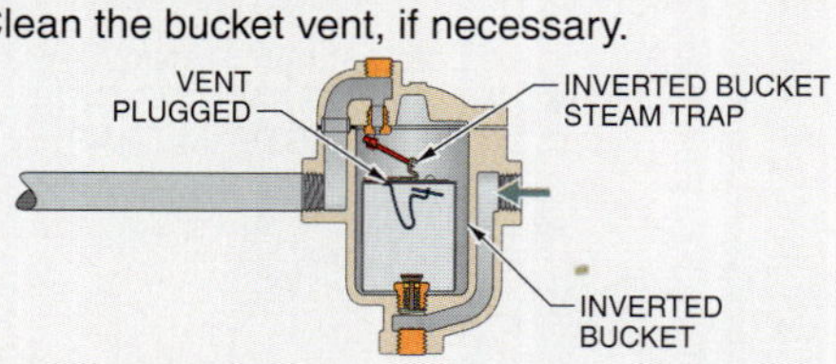

9. Clean the inlet or outlet if plugged or the mechanism is jammed with dirt.
10. Replace the bucket if unhooked or the mechanism is worn.

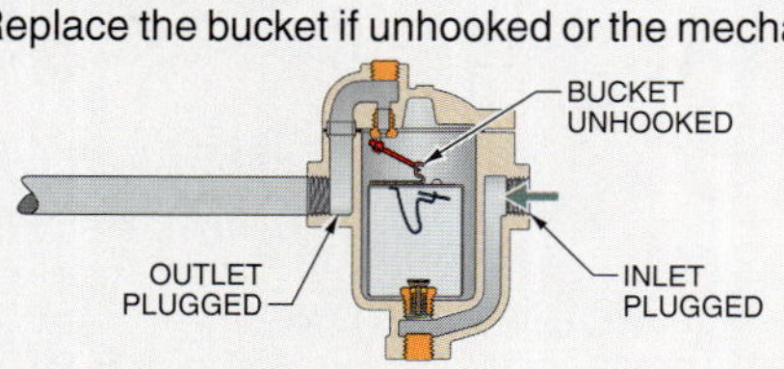

11. Listen to the steam trap in operation.

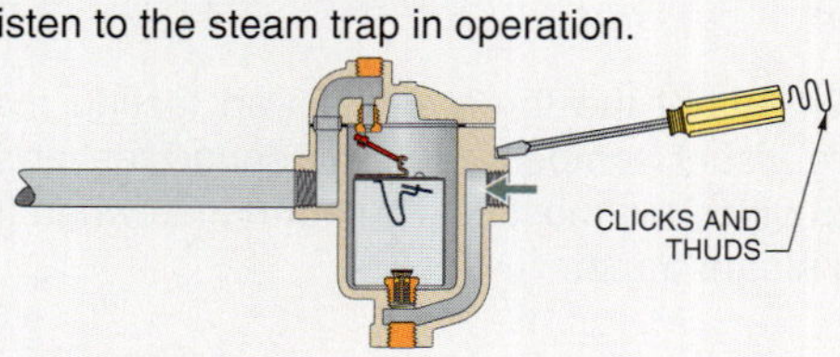

12. Determine whether the steam trap is discharging intermittently. If the steam trap is not discharging intermittently, proceed to the next step. Otherwise, the trap is good.

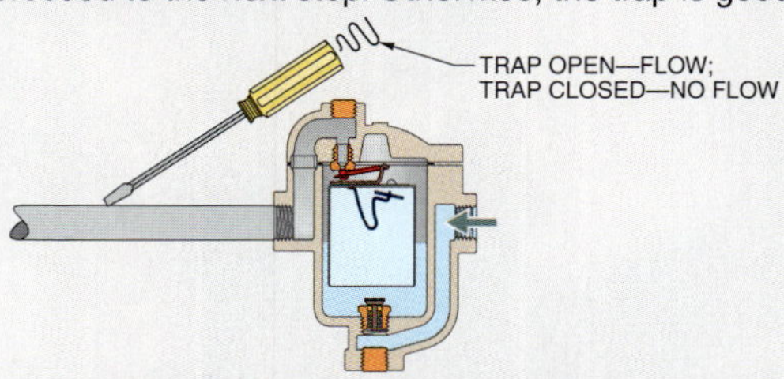

13. Determine whether the steam trap is relatively quiet, but a steady "bubbling" sound can be heard through the bucket vent. If a bubbling sound is heard, the steam trap is good (there is a light load).

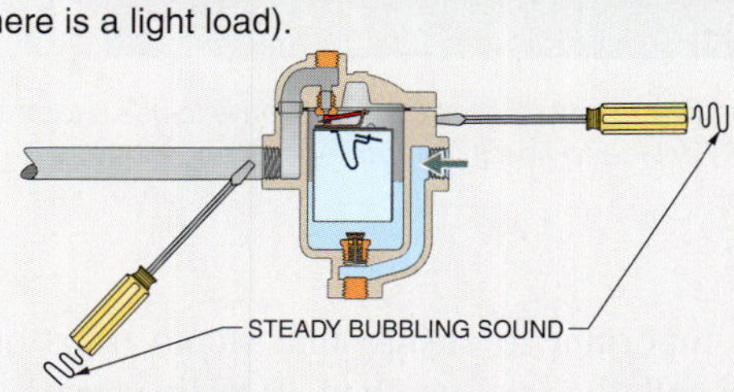

14. Determine whether the bubbling sound increases and decreases in a "rolling" sound. If so, the steam trap is handling air. Check the trap in 1 hr. If the steam trap is still handling air after 1 hr, replace the bucket.
15. If the steam trap steadily discharges with no bucket sound, the trap is too small and must be replaced with a larger trap.
16. If the steam trap steadily discharges with the bucket moving up and down, the trap has lost its prime. Close the valve downstream of the trap for a few minutes and then reopen. If the steam trap is not good, the mechanism is worn or the guide assembly is misaligned.

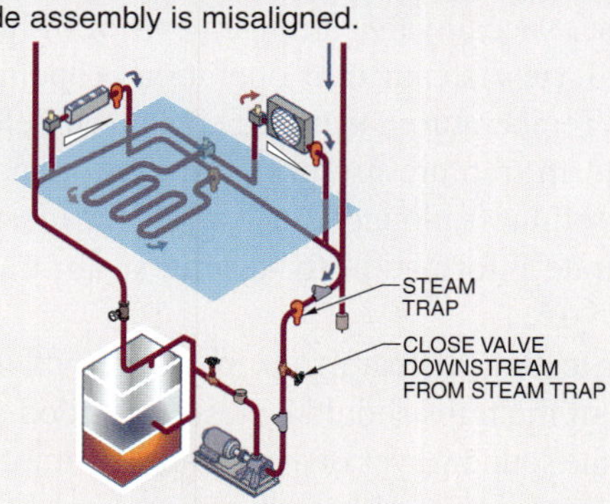

17. For an irreparable steam trap, remove the trap from the line, in case of any malfunction, and apply compressed air or water to the steam trap outlet and then its inlet piping. If the problem is not corrected, replace the trap.

Steam Trap Discharge Modes

Trap Type	Mode of Operation				
	No Load	Light Load	Normal Load	Full or Overload	Usual Failure Mode
Float and thermostatic	No action	Usually continuous but may cycle at high pressures		Continuous	Closed, A.V. open
Inverted bucket	Small dribble	Intermittent	Intermittent	Continuous	Open
Balanced pressure thermostatic	No action	May dribble		Continuous	
Bimetallic thermostatic	No action	Usually dribble action	May blast at high pressure	Continuous	Open
Impulse	Small dribble	Usually continuous with blast at high loads		Continuous	Open
Disc thermo-dynamic	No action	Intermittent	Intermittent	Continuous	Open

Steam Strainers. A *steam strainer* is a commercial and industrial steam distribution system accessory that is located in steam lines before steam traps to remove foreign matter from the steam that could cause a steam trap or valve to malfunction. **See Figure 24-16.** Steam trap function is greatly affected by steam strainer function.

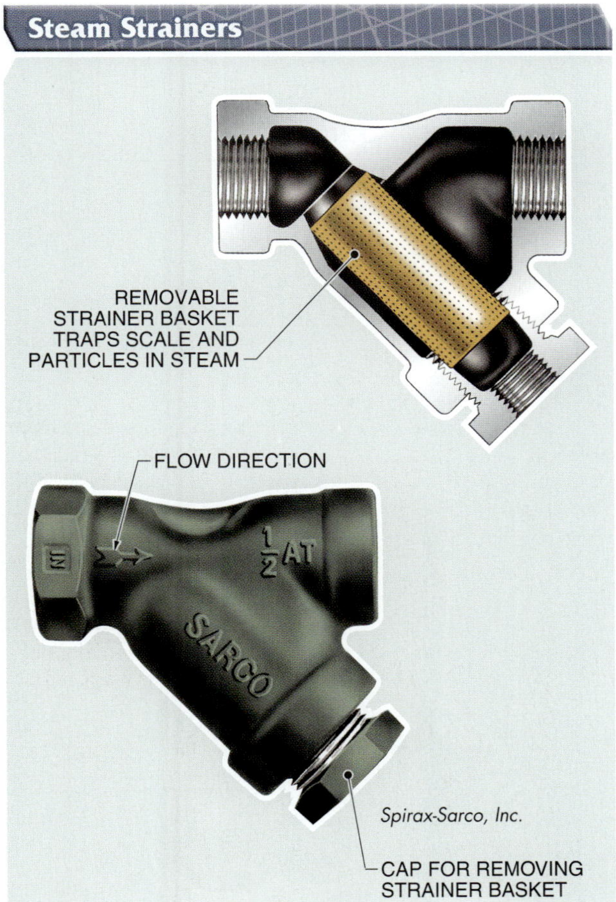

Figure 24-16. Steam strainers are located before steam traps and other devices and must be routinely cleaned to ensure efficient removal of foreign matter from the flow of steam.

Dirt, rust, scale, and/or other particles in steam lines can lead to steam trap failure. Small particles of foreign matter clog orifices (openings) in the steam traps and prevent the proper discharge of condensate. The particles can also lodge under the seat of a steam trap, causing the trap to leak when closed. Various types of steam strainers are typically located before pumps and control valves in the steam system to protect the individual devices.

Steam Terminal Units. Steam terminal units transfer heat from the steam in a heating system to the air in building spaces. Steam terminal units may include simple controls but typically use system controls that are not part of the unit. Steam terminal units are located in open building spaces for maximum efficiency and are designed to maximize heat transfer from the steam to the building spaces. When steam terminal units release heat, steam traps remove air and condensate from the remaining steam.

Condensate Return Tanks

A *condensate return tank* is a steam system device that collects condensate returned from terminal units. Condensate that is not contaminated is pumped back to the steam boiler for reheating and reuse. **See Figure 24-17.** The condensate return tank and pump are installed at the lowest point in the building where steam is used. Condensate is directed to the condensate return tank through condensate return lines. The condensate return tank is vented to atmosphere to prevent pressure from building up in the condensate return system. Small residential and light commercial systems will have an expansion tank instead of a condensate return tank.

The condensate return tanks or feedwater heaters are equipped with automatic makeup water systems to maintain a minimum water level in the tank. Feedwater is pumped to a boiler during a demand for more water by the boiler. When there is an automatic makeup feedwater pump on the condensate return tank, the feedwater pump or the feedwater regulator is controlled by the condensate tank water level. When there is an automatic makeup water valve on the steam boiler, the valve is controlled by the level in the boiler.

STEAM PIPING SYSTEMS

Steam piping systems are used to distribute steam from a boiler throughout a building and return the condensate to the boiler. Piping systems vary depending on the degree of temperature control desired and the specific application. In most systems, the pipe is made of steel. In small systems, the pipe may be made of copper. Piping systems used for residential and light commercial steam heating systems include one- and two-pipe systems. Because of variations in buildings, each piping system is designed for a specific type of building.

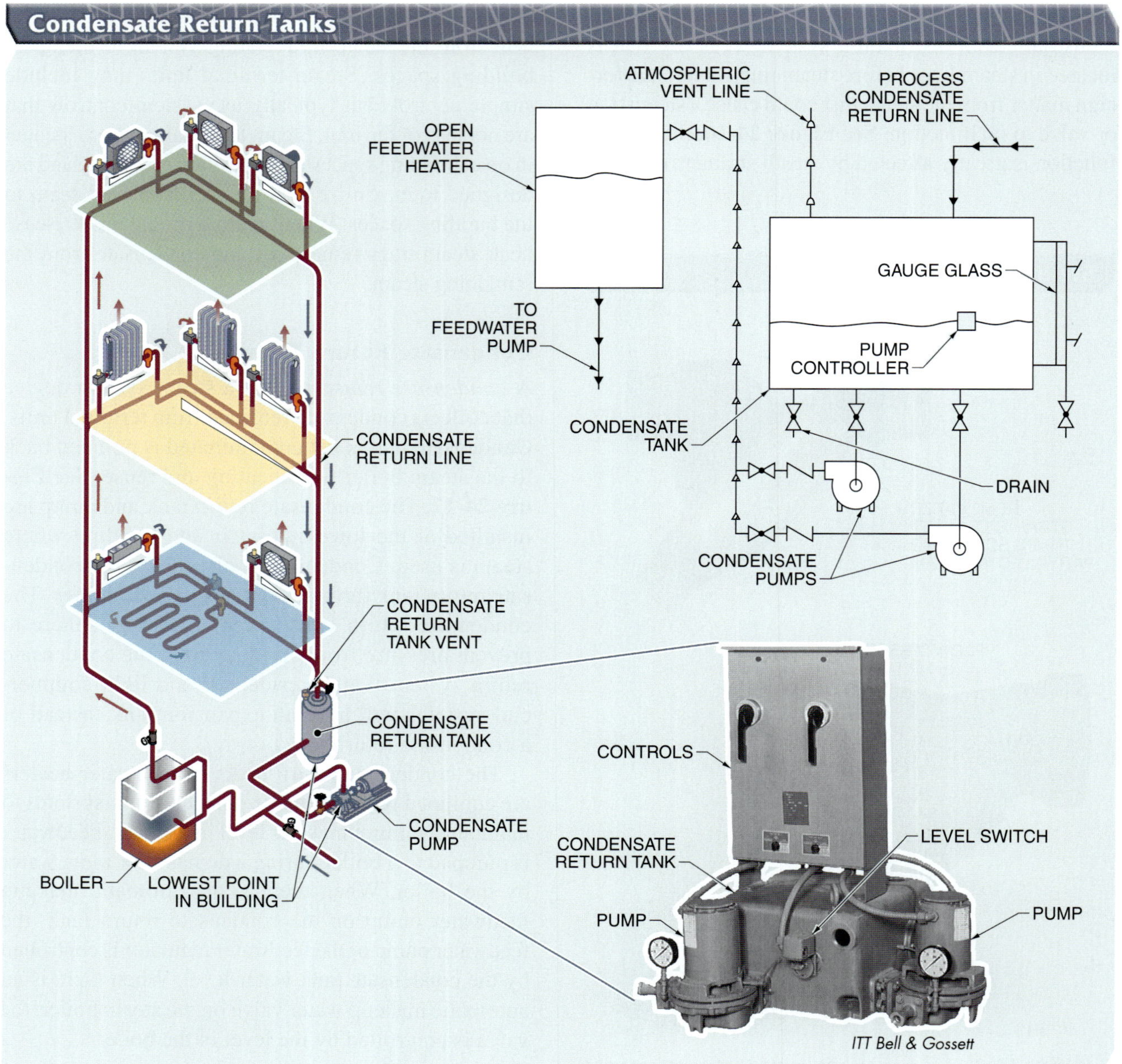

Figure 24-17. Small residential and light commercial steam systems typically do not have condensate return tanks. The condensate return tank and pump are installed at the lowest point in the building where steam is used.

One-Pipe Steam Piping Systems

One-pipe (gravity-flow) steam piping systems are used in small- to some medium-size buildings. The steam and condensate use the same piping simultaneously. The steam flows up through the pipe to the terminal devices. As the steam condenses, the condensate flows down the pipe in the opposite direction of the steam back to the boiler. **See Figure 24-18.** All terminal devices must be above the water line of the boiler so that gravity will make the condensate flow back to the boiler. The two types of one-pipe systems are parallel-flow and counterflow systems.

Parallel-Flow. Parallel-flow steam heating systems typically have a wet return. In a one-pipe, gravity-return parallel-flow system, the lowest steam carrying pipe in the system must be 28" above the boiler to avoid water hammer. **See Figure 24-19.**

One-Pipe Steam Piping Systems

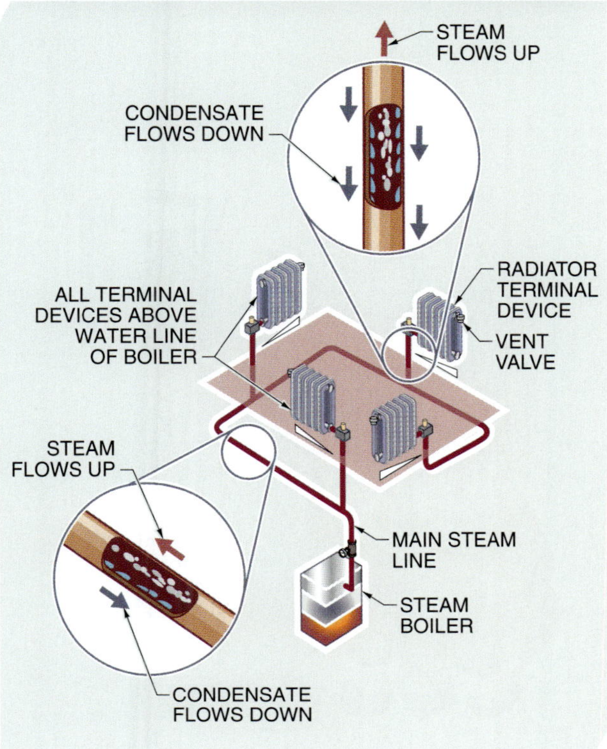

Figure 24-18. One pipe is used to supply steam to a system and to return condensate back to the boiler.

One-Pipe Steam Piping System Types

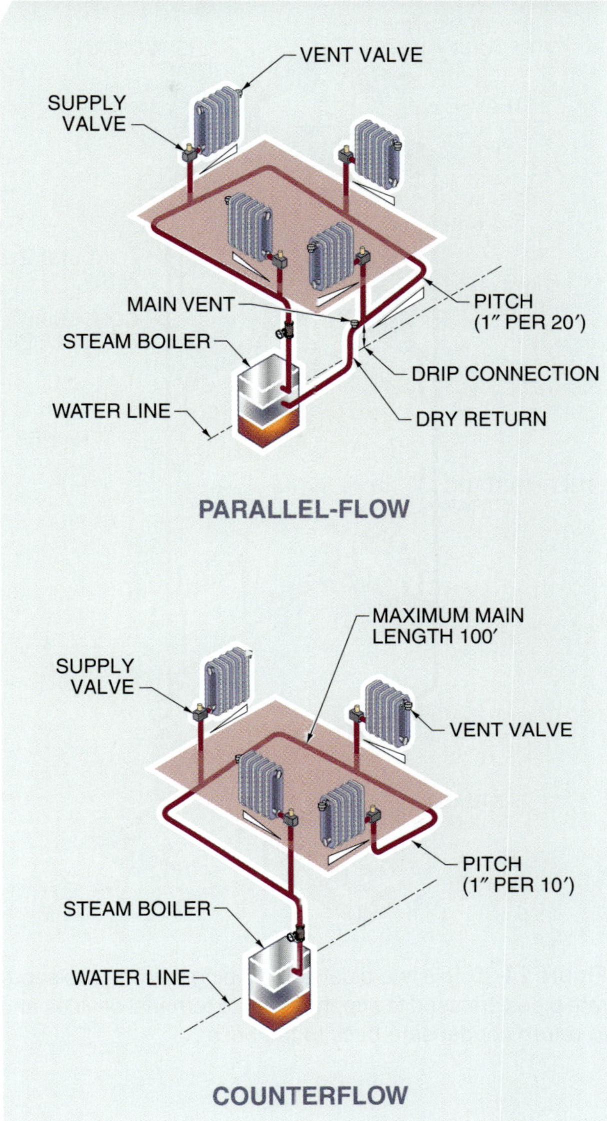

Figure 24-19. The two types of one-pipe steam piping systems are parallel-flow and counterflow systems.

Counterflow. A *counterflow steam heating system* is a type of one-pipe piping system that has steam and condensate flowing in the same pipe but in opposite directions. The steam flows away from the steam boiler and the condensate returns in the opposite direction in the same pipe. When not sized, sloped, or drained properly, a counterflow system can be very noisy due to excessive water hammering. Typically, the main steam lines are sized larger than normal to help with condensate removal and reduce water hammer. Drip lines are often used to ensure adequate condensate removal.

Two-Pipe Steam Piping Systems

Two-pipe steam piping systems are used in medium- to large-size residential and light commercial buildings. A two-pipe piping system consists of a supply pipe that carries steam from the boiler to the terminal devices and a separate return pipe that carries condensate back to the boiler. **See Figure 24-20.** Two-pipe systems require the addition of steam traps and strainers in the system.

STEAM TERMINAL DEVICES

A *steam terminal device* is a heating system component that transfers heat from the steam in a heating system to the air in building spaces and includes controls. A steam supply pipe runs from the steam boiler to each terminal device, and typically a condensate return pipe leaves each terminal device and runs back to the boiler.

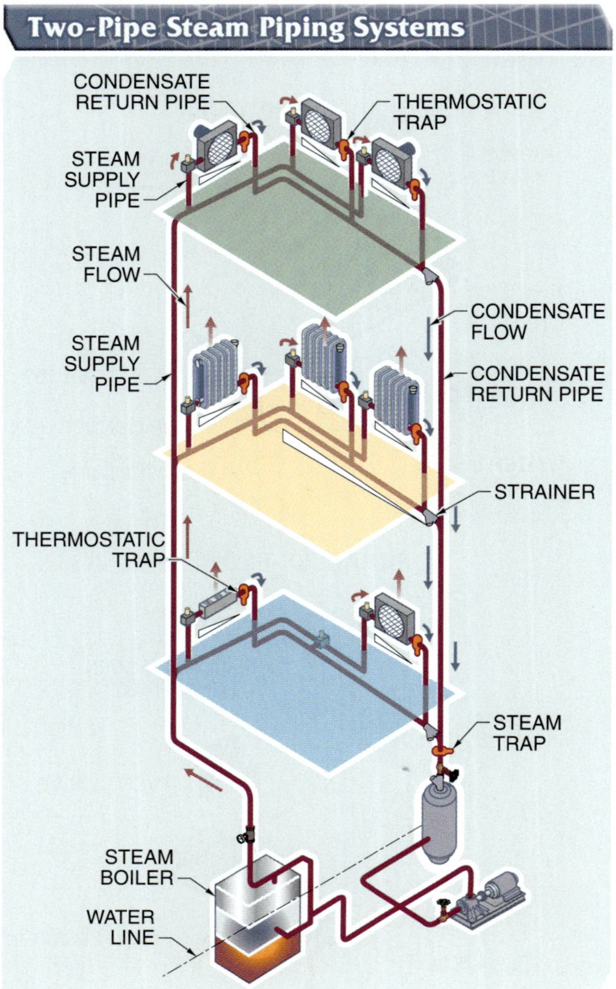

Figure 24-20. In a two-pipe steam piping system, two separate pipes are used to supply steam to terminal devices and to return condensate back to the boiler.

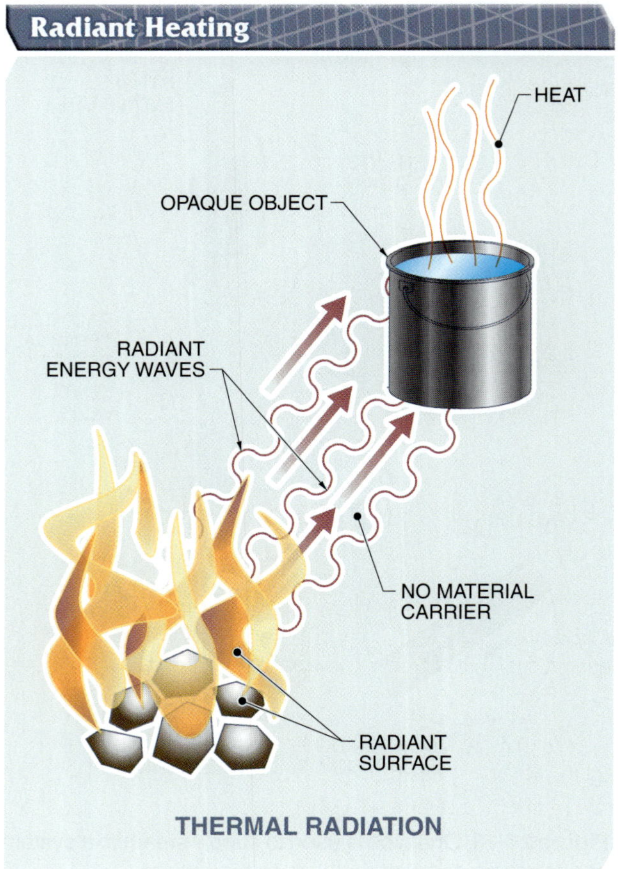

Figure 24-21. Radiant heating occurs when a surface is heated and then gives off heat in the form of radiant energy waves.

Terminal devices are located in building spaces for maximum efficiency and are designed to maximize heat transfer from the steam to the building spaces. When steam terminal devices release heat, steam traps remove air and condensate from the remaining steam. The two types of steam terminal devices are radiant heating and convection heating terminal devices.

Steam Radiant Heating

Thermal radiation is the heat transfer that occurs as radiant energy waves transport energy from one object to another. Radiant heating systems are based on thermal radiation. **See Figure 24-21.** *Steam radiant heating* is the heating that occurs when a surface is heated by steam and gives off heat in the form of radiant energy waves.

A *radiant surface* is the heated surface from which radiant energy waves are generated. Radiant energy waves are absorbed only by opaque objects in the direct line of sight of the radiant surface and are not absorbed by the air. Opaque objects absorb the radiant energy waves and become heated. As the objects are heated, they give up some of the heat to the air around them, which raises the temperature of the air. The three types of steam radiant heaters are surface radiant systems, radiant panels, and radiators.

Steam Surface Radiant Systems. A *steam surface radiant system* is a type of radiant heating system terminal device that uses steam to heat the interior surfaces of a room for use as radiant surfaces. A piping coil or grid is built into the construction material of the ceiling, walls, or floor of a room. **See Figure 24-22.** Insulation is installed behind the piping and a thin layer of surfacing material is installed over the insulation.

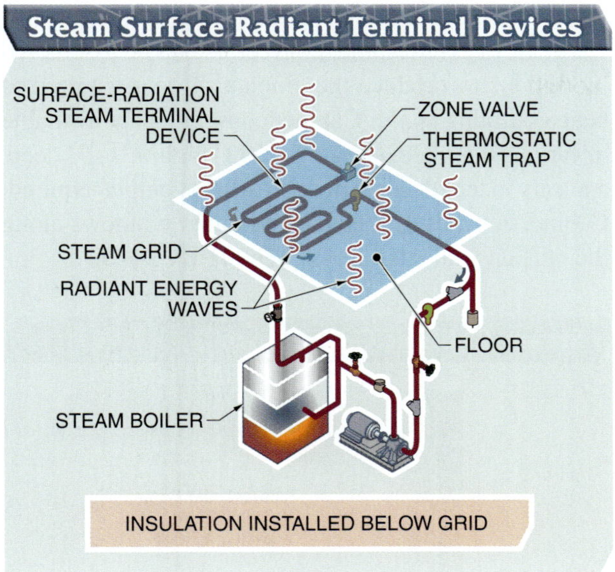

Figure 24-22. A steam surface radiant system is a type of radiant-heating-system terminal device that uses steam to heat the interior surfaces of a room for use as radiant surfaces.

The radiant surface is heated by steam that flows through the pipes or tubing. When the temperature of the radiant surface is higher than the temperature of objects in the room, heat flows from the radiant surface to the objects. The amount of radiant heat produced is regulated by controlling the temperature of the radiant surface.

Steam Radiant Panels. A *steam radiant panel* is a type of radiant heating system terminal device that is factory-built using a panel with a radiant surface that is heated by steam through a piping coil or grid built into it. Steam heated radiant panels are individual heaters that are not integrated into a building. The radiant panels are enclosed in an insulated cabinet and are only used for spot heating. *Spot heating* is a heating method in which heat is provided for a particular part of a building space. Entire buildings are not heated by spot heating from steam heated radiant panels alone. **See Figure 24-23.**

Steam Radiators. *A steam radiator* is a type of radiant heating system terminal device that consists of hollow metal coils or tubes through which steam flows. Steam passes through the coils or tubes. Heat from the steam is transferred to the tubes, which transfer the heat to the air. The air in a building space is heated by radiation and convection when using radiators. **See Figure 24-24.**

A *standing steam radiator* is a type of steam radiator with vertical tubes that are connected by headers all in an upright vertical position with legs and is installed on outside walls. Steam from a boiler flows through the tubes. Heat is transferred to the air by radiation and convection when air flows over the tubes. Standing radiators are typically made of cast iron. Because standing radiator units are for the most part unattractive, baseboard radiators are typically used today. A *steam baseboard radiator* is a section of horizontal tubes that are connected by headers at each end and is installed at the base of outside walls.

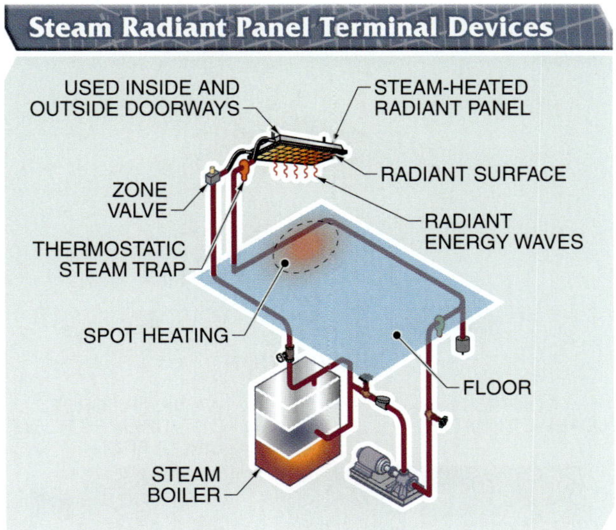

Figure 24-23. Steam radiant panels are individual heaters that are not an integral part of a building but are used for spot heating and are enclosed in an insulated cabinet.

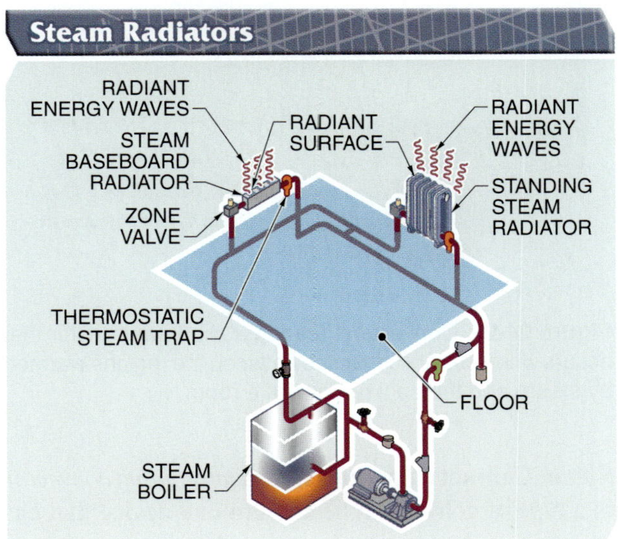

Figure 24-24. Steam radiators can be of a standing or baseboard design.

Steam Convection Heating

Convection heating is based on heat transfer by convection. *Steam convection heating* is the heat transfer that occurs when currents circulate between the regions warmed by steam and the cool regions of a room. **See Figure 24-25.** The two types of steam convection heaters are steam cabinet convectors and steam baseboard convectors.

Convection Heating

Figure 24-25. Convection heating is the heat transfer that occurs when current circulates between the regions warmed by steam and the cool regions of a room.

Steam Cabinet Convectors. A *steam cabinet convector* is a type of convection heater terminal device that has its steam coil and controls enclosed in a metal cabinet. The steam coil or tubes and the front of the heater are covered by a sheet metal cabinet. The cabinet has a grill in the bottom for cool air to enter and a grill on the top for warm air to exit. Cabinet convectors can be assembled in sections to produce the amount of heat required to heat a building space. Cabinet convectors are available in various sizes between 20″ and 32″ high, 7″ to 9″ deep, and vary in length depending on the heat output required. Cabinet convectors are installed under windows along the outside walls of a room. **See Figure 24-26.**

Steam Cabinet Convectors

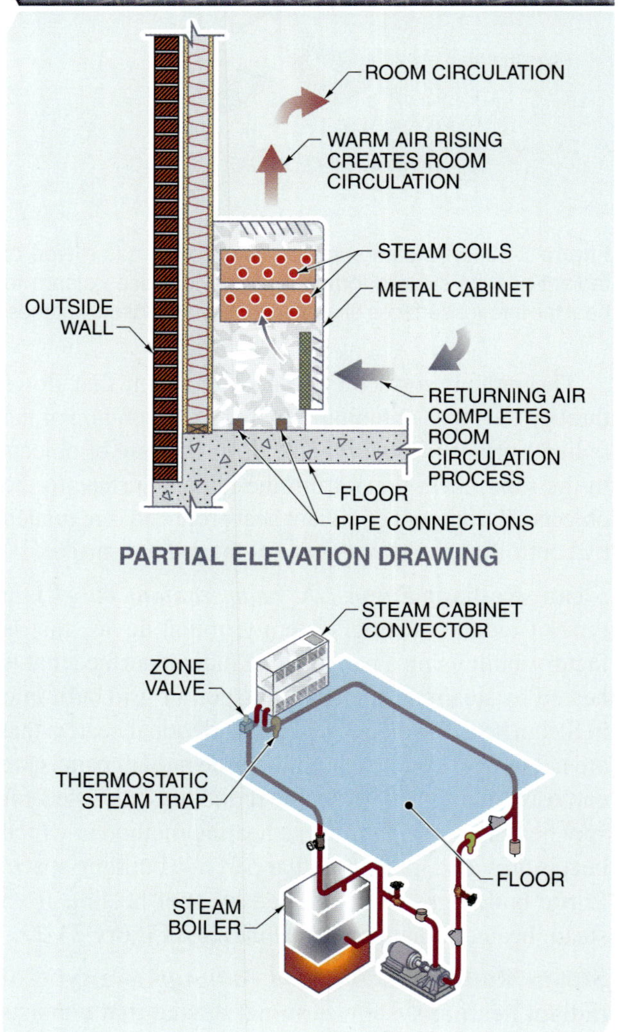

Figure 24-26. A steam cabinet convector is a type of convection heater terminal device that has its steam coil and controls enclosed in a metal cabinet with no blower.

Steam Baseboard Convectors. A *steam baseboard convector* is a type of convection heater terminal device that is enclosed in a 12″ high or lower cabinet and is installed at floor level along the outside wall of a building. Baseboard convectors are smaller than cabinet convectors and typically use one section of finned tubes to transfer heat. A *finned tube* is a copper pipe with aluminum fins. The fins provide a large surface area for heat transfer. **See Figure 24-27.**

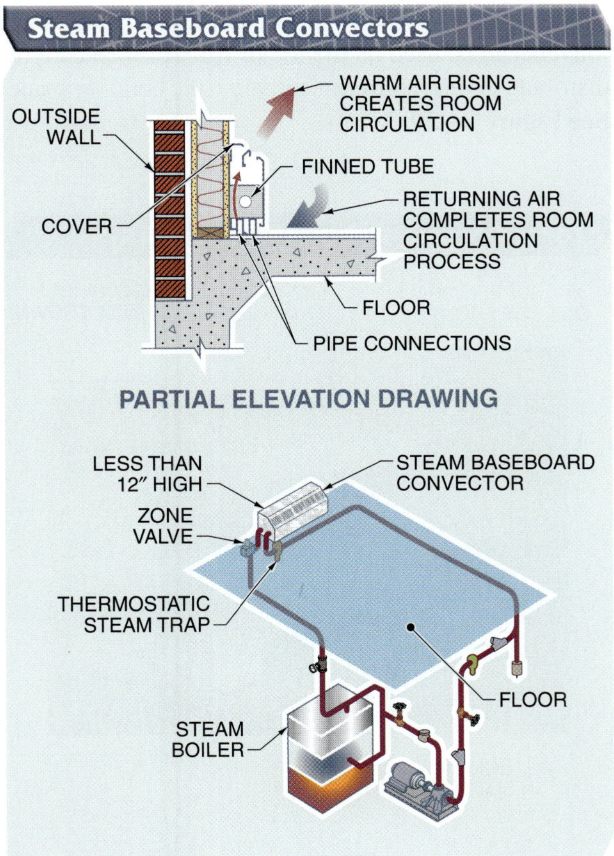

Figure 24-27. A steam baseboard convector is a convection heater terminal device that is enclosed in a 12″ high or lower cabinet and installed at floor level along the outside wall of a building.

Forced Convection Heating

Forced convection heating is a form of heating in which convective currents are mechanically moved over a steam coil. Forced convection heaters use blowers or fans to move air. The blower draws air in the bottom or back of the heater, moves it past the steam coil, and distributes the air to the building space. All forced convection steam heaters have a steam coil, blower, sheet metal cabinet, and number of controls. The three types of forced convection heaters used in steam heating systems are forced steam unit convectors, forced steam cabinet convectors, and forced steam cabinet-ventilator convectors. **See Figure 24-28.**

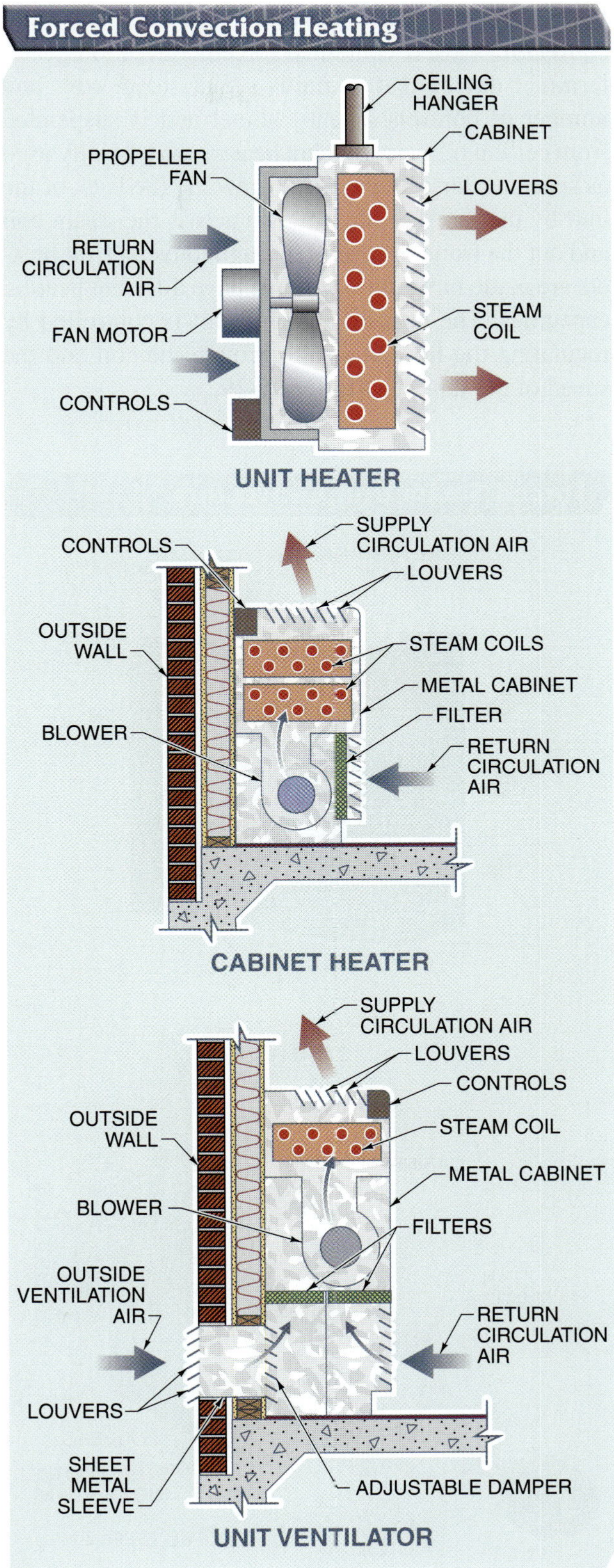

Figure 24-28. Forced convection heating is a type of heating in which convective currents are mechanically moved over a steam coil using a blower.

Forced Steam Unit Convectors. A *forced steam unit convector* is a self-contained forced convection heater terminal device that contains a fan, steam coil, and number of controls in one cabinet and is suspended from ceiling framework. Unit heaters are typically used as space heaters. Return air is drawn in the back of the unit by the fan. The air is blown across the steam coil and out the front of the unit through louvers. Unit heaters are made in various sizes and have different heating capacities. The heat output of a unit is controlled by regulating the flow of steam through the coil and the speed of the fan. **See Figure 24-29.**

Forced Steam Cabinet Convectors. A *forced steam cabinet convector* is a self-contained forced convection heater terminal device that contains a blower, steam coil, filter, and number of controls in one cabinet. The blower forces air across the steam coils. A cabinet convector uses a blower to move air through a filter for cleaning. The blower adds energy to the air so that directional louvers can be used on the top or front of the heater to distribute the air more efficiently into the building space. **See Figure 24-30.**

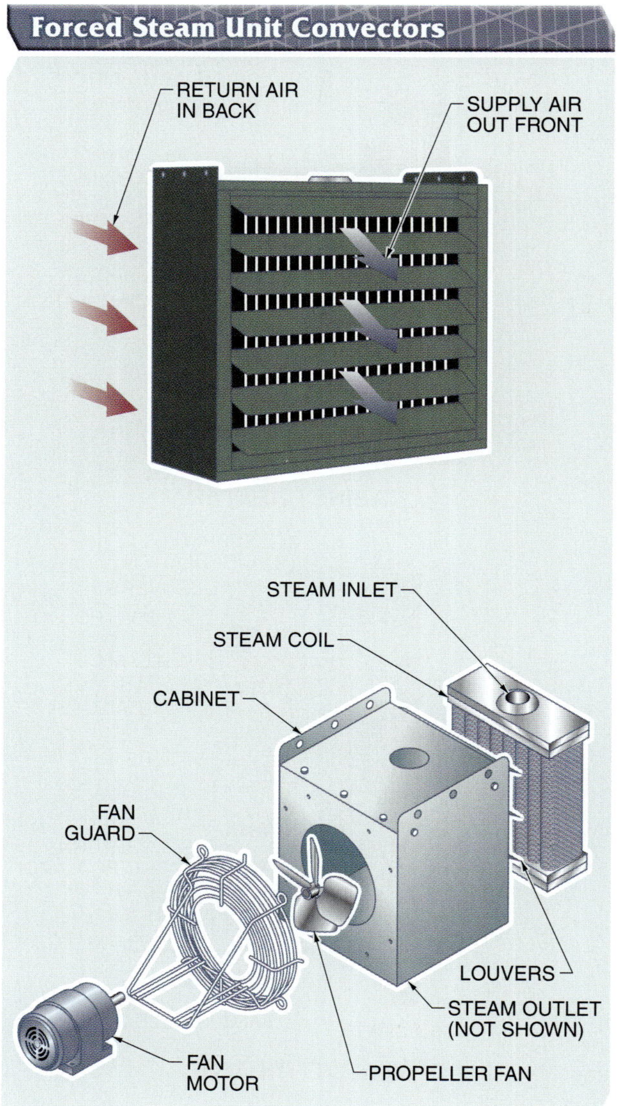

Figure 24-29. A forced steam unit convector contains a fan, a steam coil, and possibly a number of controls in one unit and is suspended from ceiling framework.

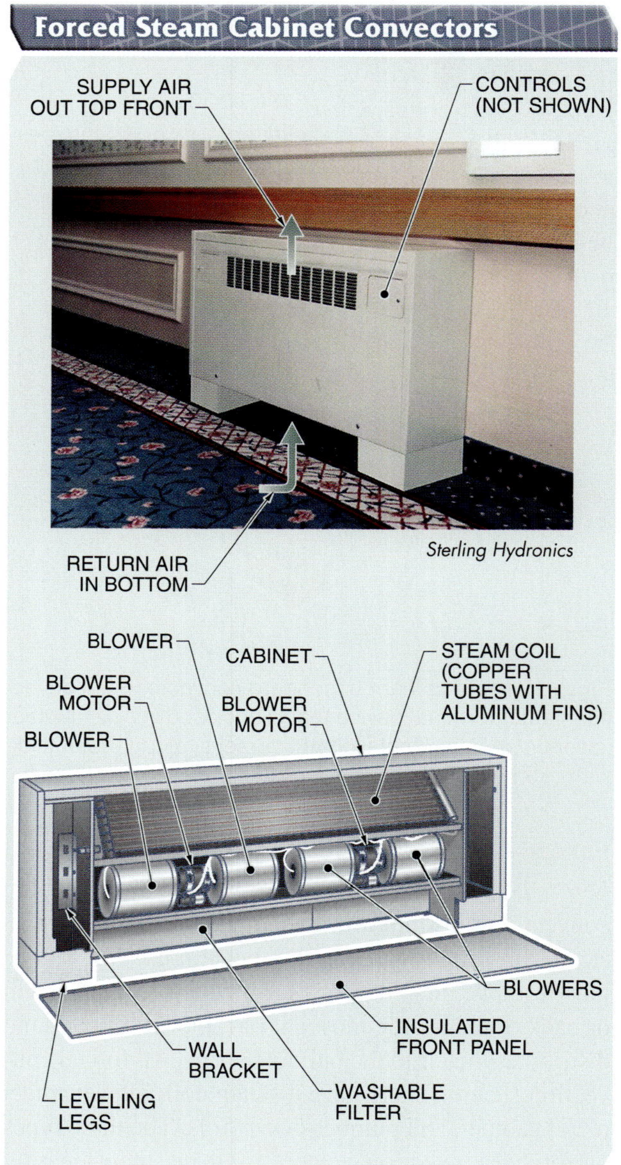

Figure 24-30. A forced steam cabinet convector contains a blower, steam coil, filter, and number of sensors and controls in one cabinet.

The size requirements of cabinet heaters depend on the heat output required. Most cabinet heaters are from 30″ to 36″ high, 6″ to 7″ deep, and of different lengths. Cabinet heaters are typically installed along the outside wall of a room under a window.

Forced Steam Cabinet-Ventilator Convectors. A *forced steam cabinet-ventilator convector* is a forced convection heater terminal device that contains filters and controls as well as a blower, steam coil, and a cabinet with an opening for outdoor ventilation. The blower forces an airflow of mixed outside air and return air across the steam coil.

Cabinet ventilators differ from cabinet heaters in that ventilators have an outside opening to bring in air from the outdoors for ventilation. The cabinet fits on an outside wall so that the connection to the outdoors can be made with an opening in the back of the cabinet for an outdoor air inlet. A sheet metal sleeve extends from the inlet through the wall. A louver covers the opening to the outdoors. The outdoor air inlet has an adjustable damper on the inside for regulating the amount of air admitted to the ventilator for ventilation. Cabinet ventilators are often used in buildings that have a large number of people and a need for constant ventilation. **See Figure 24-31.**

Blower Controls. The blowers in steam terminal devices are controlled by relays, which cycle the blowers in the terminal devices ON or OFF. The relays are the same as those used in forced-air heating systems. A signal from the space thermostat closes the relay contacts, which supply electricity to the blower motor. Thermostats control the blowers in unit convectors or cabinet ventilator convectors. The thermostats used to control terminal devices may be located inside the cabinets of the units they control or remotely in the space being heated. **See Figure 24-32.**

Steam Terminal Device Controls

A *steam terminal device control* is a steam distribution component that controls the air temperature in a building space by regulating the flow of air and steam through a terminal device in the space. Zone valves regulate steam flow and blowers regulate the airflow through terminal devices. Blowers, which are typically switched ON and OFF by signals from the thermostats, force air through the steam terminal devices. Diverter valves control the amount of steam flow being sent to terminal devices.

TECH TIP

A unit ventilator's (steam or other) control sequencing is classified as ASHRAE Cycle I, II, or III. ASHRAE cycles describe how to regulate the outdoor air damper in relation to the heating control valve.

Forced Steam Cabinet-Ventilator Convectors

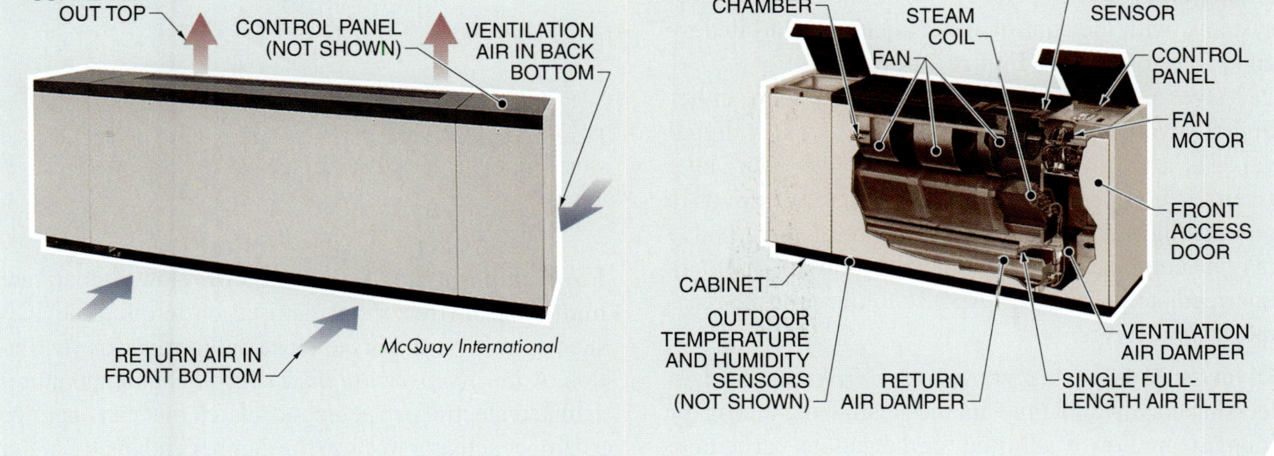

Figure 24-31. A forced steam cabinet-ventilator convector contains filters and controls as well as a blower, steam coil, and cabinet with an opening for outdoor ventilation.

Steam Terminal Device Blower Controls

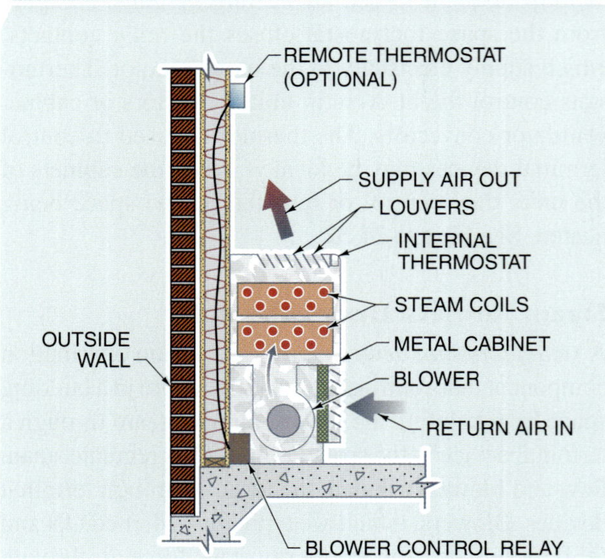

Figure 24-32. The blowers in steam terminal devices are powered by relays, which cycle the blowers ON or OFF. Signals from the space thermostats close the relay contacts, which supply electricity to the blower motors.

Zone Valves. A *zone valve* is a type of valve that regulates water or steam flow to building spaces or terminal devices. A *zone* is the section of a building that is controlled by one HVAC controlling device, such as a thermostat. Zone valves may be manual or automatic. A *manual zone valve* is a valve that is set by hand to regulate the flow of steam in a piping loop. An *automatic zone valve* is a valve that is opened or closed automatically by an actuator such as a motor, solenoid, or pneumatic device. Automatic zone valve actuators are driven by electric, electronic, or pneumatic control systems, with the control systems being controlled by zone thermostats. **See Figure 24-33.**

Valves used for controlling steam flow are either digital (two-position) or modulating valves. A digital valve is a two-position (ON/OFF) valve. The valve is either completely open or completely closed. A modulating valve is an infinite position valve. The valve may be completely open, completely closed, or at any intermediate position in response to the control signal the valve receives.

Diverter Valves. A *diverter valve (zone valve)* is a tee-shaped valve that meters the steam from a primary loop. Steam flow is adjusted for the gallons per minute required by the terminal device. The condensate from each terminal device returns to the same main. Flow control by the diverter valve allows the heat output of each steam distribution section or terminal device to be somewhat controlled.

Zone Valves

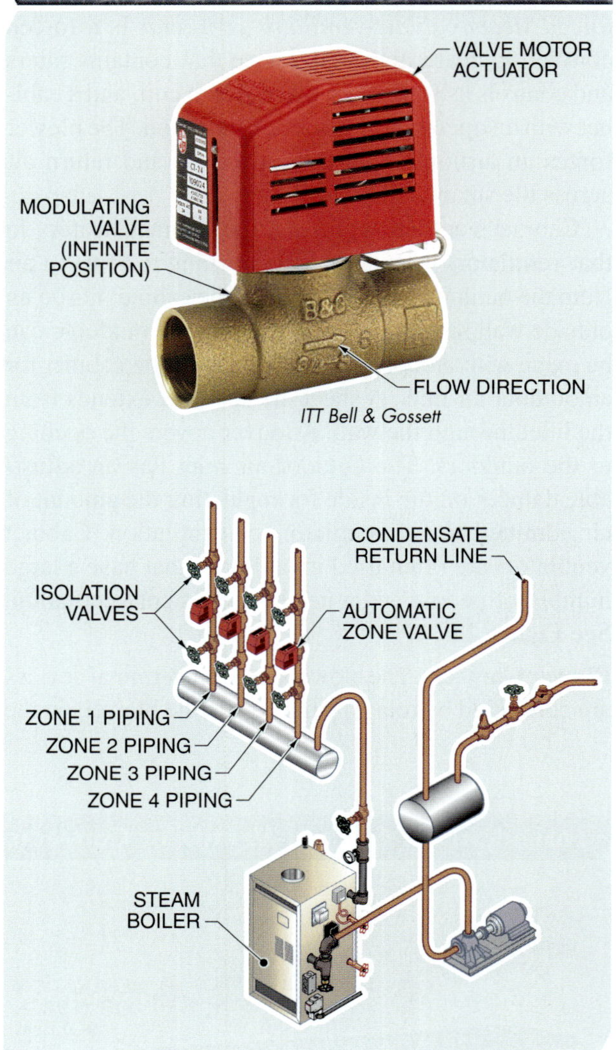

Figure 24-33. A zone valve controls where in a zone or to which terminal device steam travels.

Low Temperature Limit Controls. A low temperature limit control (freeze stat) is used on terminal devices such as unit ventilators that use outdoor air for ventilation. A *low temperature limit control* is a temperature-actuated electric or pneumatic switch that can energize a damper actuator to close the damper and open a water or steam valve if ventilation drops below 35°F. The terminal unit fan will continue to run.

In other applications, the low temperature limit control shuts down the terminal unit, completely closes the damper, turns the fan OFF, and opens the steam valve. Freeze-protection controls are required when outdoor air temperature drops below 32°F, because condensate in the steam coil of the terminal unit may freeze. **See Figure 24-34.** Some low temperature limit controls are not suitable for monitoring the face of a coil for freeze protection.

Thermostats. Thermostats used with steam terminal devices regulate steam and airflow by controlling the zone valves, diverter valves, and terminal device blower relays. A thermostat is used to control digital valves (two-position). On a call for heat, the thermostat opens the valve. When the temperature reaches the thermostat setpoint, the thermostat closes the valve.

TECH TIP

Low temperature limit controls should be tested occasionally. In some installations, the only way that a failure becomes evident is when a coil freezes and bursts. Low temperature limit control testing is especially required on older installations that have not been tested in years. Low temperature limit controls have adjustable setpoints that can be raised to the current temperature, at which point the low temperature limit control opens the circuit.

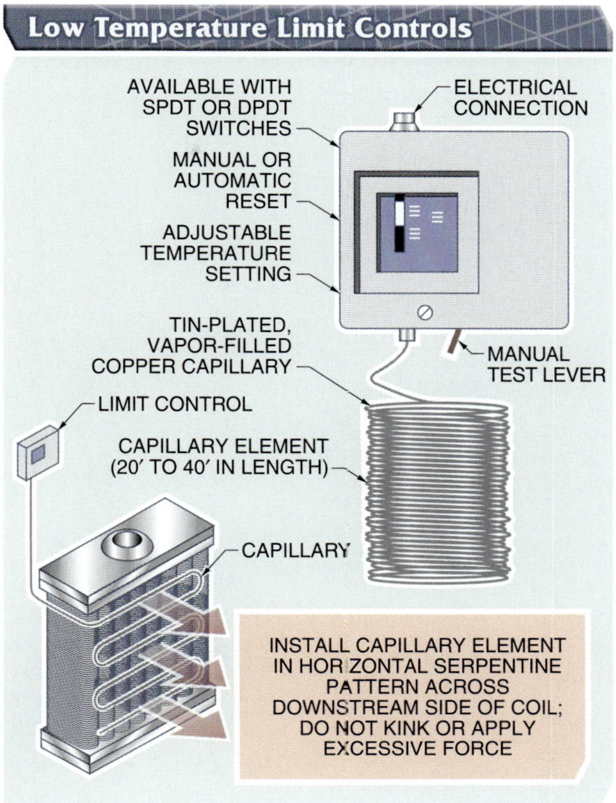

Figure 24-34. A low temperature limit control is a temperature-actuated electric or pneumatic switch that, in some cases, energizes a damper motor to shut a damper and open a steam valve when ventilation air temperature drops below a setpoint.

UNIT 25

Air-Source Heat Pump Systems

An air-source heat pump is a mechanical compression refrigeration system that moves heat from one area to another area using air as a heat source. When an air-source heat pump is in cooling mode, it moves heat from inside a building to outside the building. When an air-source heat pump is in heating mode, it moves heat from outside a building to inside the building. Changing the mode of the heat pump from cooling to heating reverses the flow of heat energy between the condenser and the evaporator.

OBJECTIVES

- Identify the basic components of an air-source heat pump system.
- List auxiliary heat sources.
- Describe the functions of common heat pump controls.
- Explain the modes of heat pump operation.
- Identify factors to consider when selecting a heat pump.

Index to Procedures

AIR-SOURCE HEAT PUMPS

A *heat pump* is a mechanical compression refrigeration system that reverses the flow of refrigerant through a system to move heat from one area to another. A reversing valve is used to reverse the flow of refrigerant through the system. Reversing the flow of refrigerant switches the relative position of the evaporator and condenser in a heat pump system. A heat pump can absorb heat from inside a building and reject the heat outdoors. A heat pump also can absorb heat from the outdoors and reject the heat to the inside of a building.

Heat Pump Modes

A heat pump can operate in either cooling or heating mode. The operation of a heat pump in cooling mode is similar to the operation of an air conditioning system. Air conditioning systems and heat pumps in cooling mode move heat from the inside to the outside of a building. When a heat pump is in cooling mode, the indoor unit is the evaporator and the outdoor unit is the condenser. A heat pump in heating mode moves heat from the outside to the inside of a building. When a heat pump is in heating mode, the indoor unit is the condenser and the outdoor unit is the evaporator. **See Figure 25-1.**

During heat pump operation, refrigerant flows from a compressor to a reversing valve. The refrigerant flows from the reversing valve to either the coil in the indoor unit or the coil in the outdoor unit. The direction of refrigerant flow depends on whether the system is in the cooling or heating mode. **See Figure 25-2.**

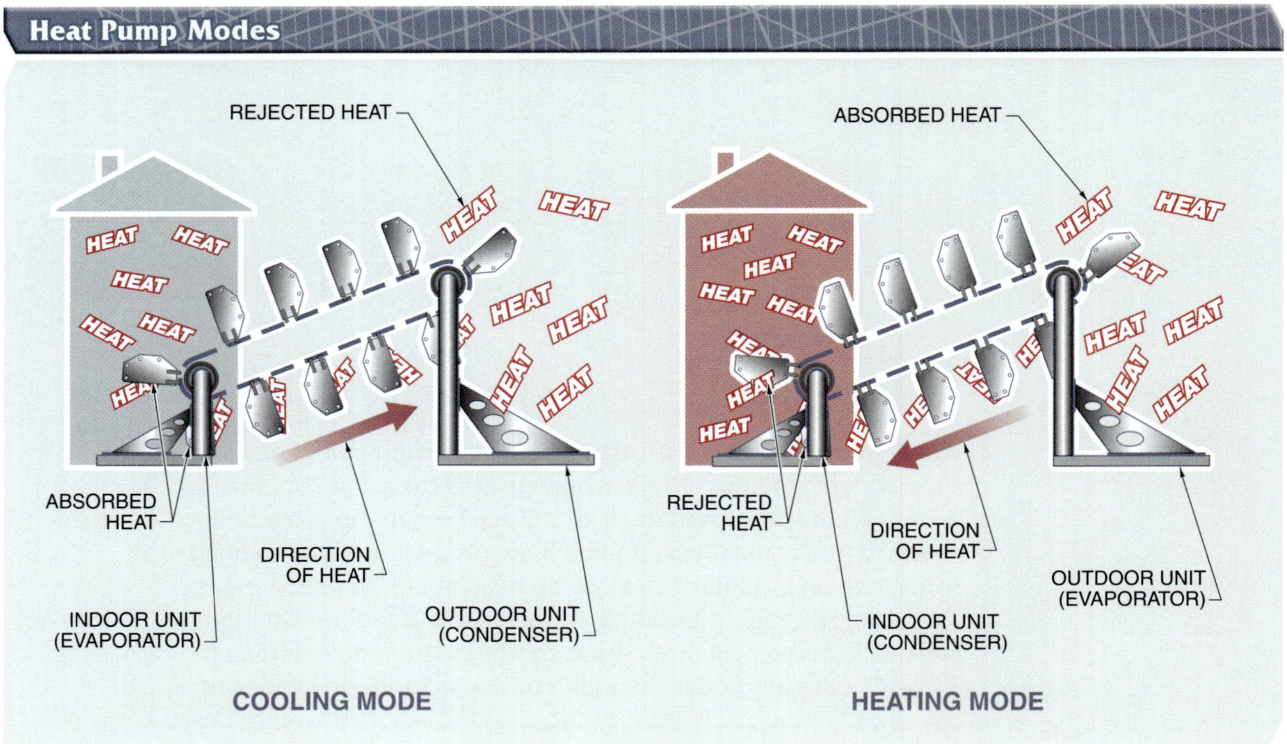

Figure 25-1. Heat pumps operate in either the cooling or heating mode.

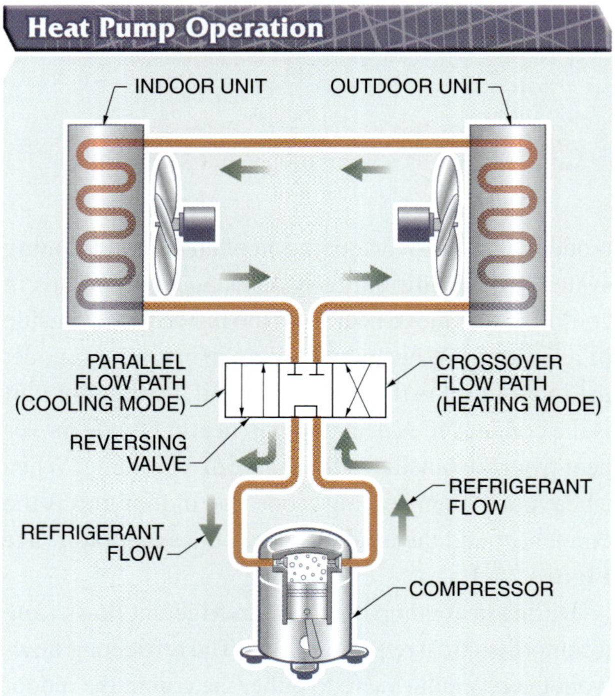

Figure 25-2. During heat pump operation, the refrigerant flows from the reversing valve to either the coil in the indoor unit or the coil in the outdoor unit.

TECH TIP

Air-source heat pumps are popular in the southern and western United States where the primary need is for air conditioning and where there is a relatively low amount of heat needed. In general, heat pumps are more expensive to operate in the heating mode than fossil-fuel-fired units such as natural gas furnaces.

Heat Sinks

Unlike a furnace or boiler, a heat pump does not produce heat. A heat pump transfers heat from one area to another area. A heat sink must always be available for a heat pump. A *heat sink* is a device with a relatively warm or cool surface that is capable of absorbing heat from a substance such as air or water. The heat of a heat sink is moved by a heat pump and used for heating or cooling building spaces. **See Figure 25-3.**

The temperature of a substance is the measurement of the heat intensity of the substance. Any substance that has a temperature above absolute zero (–460°F) contains heat. The higher the temperature of a substance, the more heat energy it contains. As the heat energy contained in a substance is increased, the better the substance is as a heat sink, or source of heat. While air and water are typically used as heat sink sources, most residential and light commercial heat pump applications use air as the heat sink source.

Heat Sinks

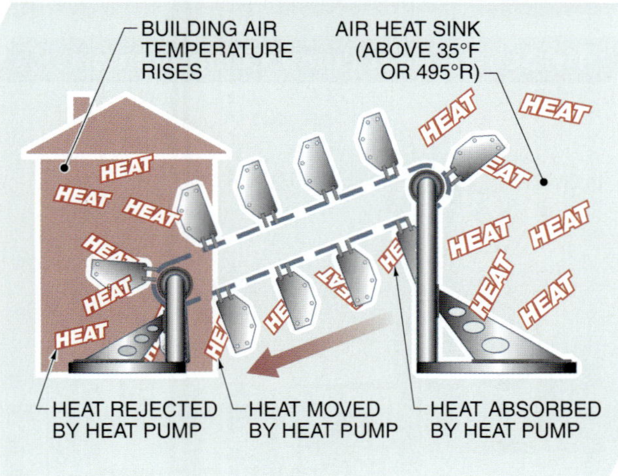

Figure 25-3. The heat of a heat sink is moved by a heat pump and used for heating building spaces.

Scroll compressors are widely used for heat pumps.

Because the temperature of outdoor air varies during different times of the day and year, the amount of heat available from outdoor air varies. During heating mode, air temperature above 35°F works well for a heat sink.

Heat Sinks Media Clip

AIR-SOURCE HEAT PUMP COMPONENTS

An air-source heat pump consists of a compressor, outdoor unit, indoor unit, reversing valve, and a number of fixed expansion devices with bypass circuits. Residential air-source heat pumps are available as split systems and as packaged systems, with cooling capacities from 1 ton to 5 tons.

An air-source heat pump uses outdoor air as the heat source when the system operates in heating mode. Return air from the building is blown across the indoor coil to absorb heat for heating the building spaces. An *outdoor unit* is a component that contains a heat exchanger and a fan. Depending on the direction of refrigerant flow, the outdoor unit acts as the evaporator or condenser for the heat pump. An *indoor unit* is a component that contains a heat exchanger and a blower. Depending on the direction of refrigerant flow, the indoor unit acts as the evaporator or condenser for the heat pump.

A reversing valve reverses the flow of refrigerant in the refrigerant lines depending on whether the heat pump is in heating or cooling mode. The valve is required because the compressor creates a flow of refrigerant in one direction only.

Bypass circuits allow refrigerant to expand regardless of the direction of flow. **See Figure 25-4.** A *bypass circuit* is a refrigerant line that contains a check valve to allow refrigerant to flow in the opposite direction of an expansion device. The bypass circuits run parallel to the expansion devices. The check valve in the bypass circuit allows refrigerant to pass around the expansion device when refrigerant flow is opposite the direction of normal flow through the expansion device. **See Figure 25-5.** Only capillary tube expansion devices produce a pressure decrease when refrigerant flows in either direction.

TECH TIP

Variable-speed blowers compensate for dirty ducts, filters, and coils. Copper tubing with grooves on the bore to increase surface area and more precisely designed TXV valves allow air-source heat pumps to be 1½ to 2 times more efficient than the heat pumps of 25 years ago.

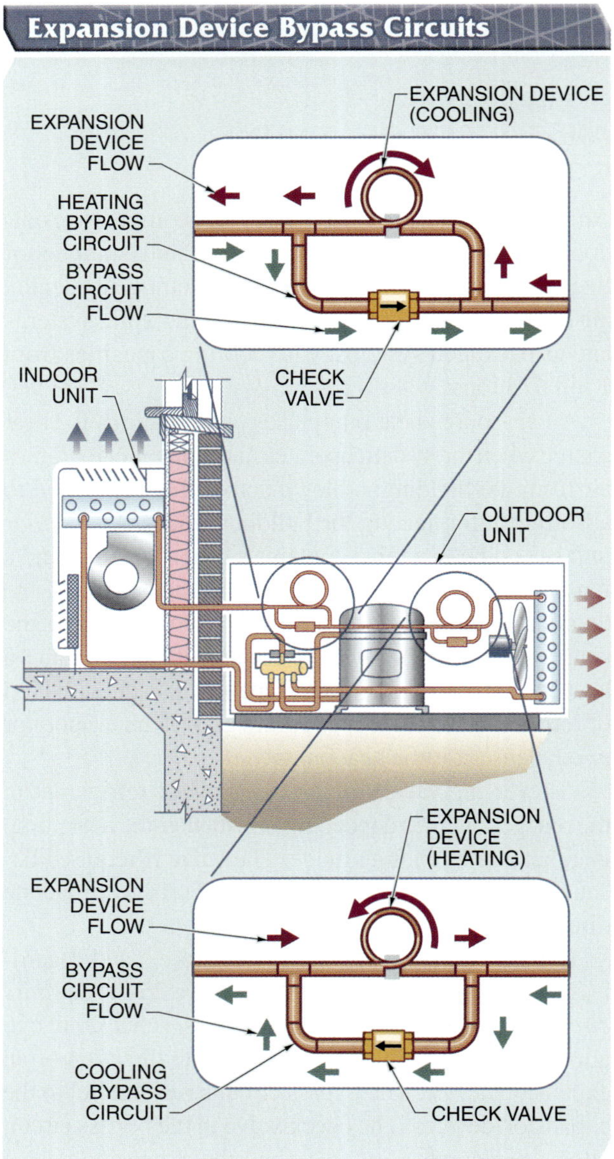

Figure 25-4. A bypass circuit is a refrigerant line that runs parallel to an expansion device. It contains a check valve to allow flow in the opposite direction of the expansion device.

Reversing Valves

A *reversing valve* is a four-way directional valve that reverses the flow of refrigerant through a heat pump system. A reversing valve consists of a piston and valve body (cylinder) that has four refrigerant line connections. The piston moves from one end of the valve body to the other, changing the direction of refrigerant flow by opening and closing outlets. **See Figure 25-6.**

When the piston is at one end of the cylinder, refrigerant flows into the valve through the hot gas discharge line and is directed out of one of the outlet ports to the opposing unit. When the piston moves to the other end of the cylinder, refrigerant flows into the valve through the same inlet port and is discharged to another outlet port. The refrigerant is then directed to the opposite unit of the heat pump. In either position, the suction line to the compressor returns the refrigerant from the reversing valve to the compressor. **See Figure 25-7.**

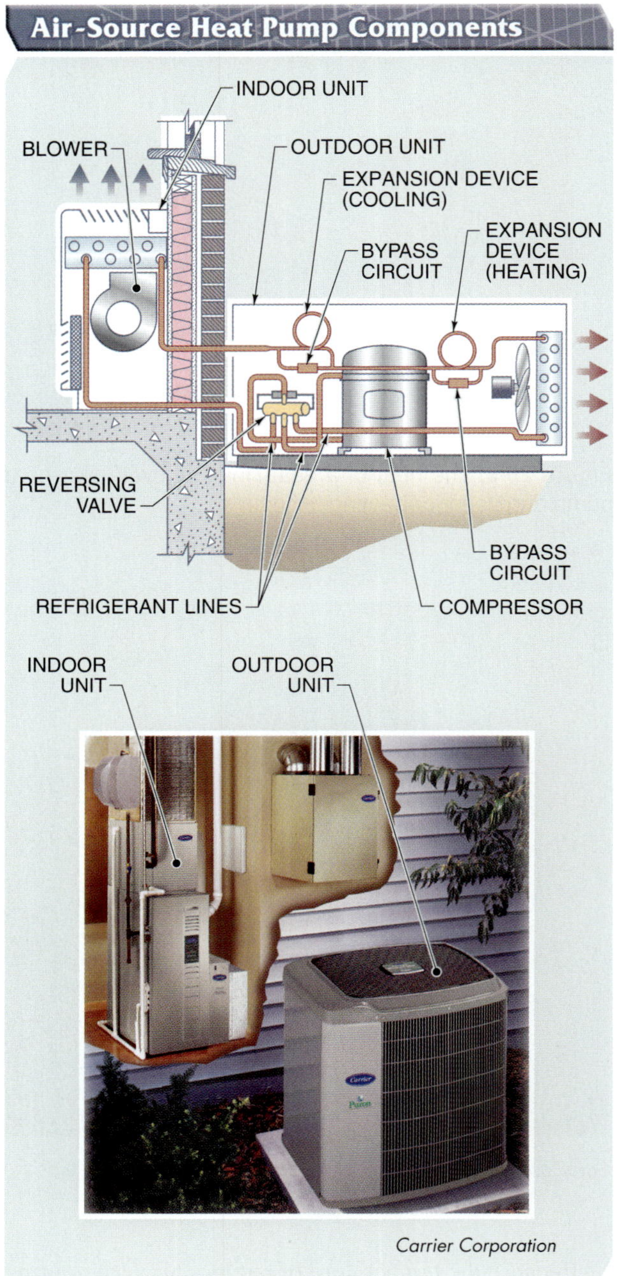

Figure 25-5. An air-source heat pump consists of a compressor, outdoor unit, indoor unit, reversing valve, and a number of expansion devices with bypass circuits.

Reversing Valves

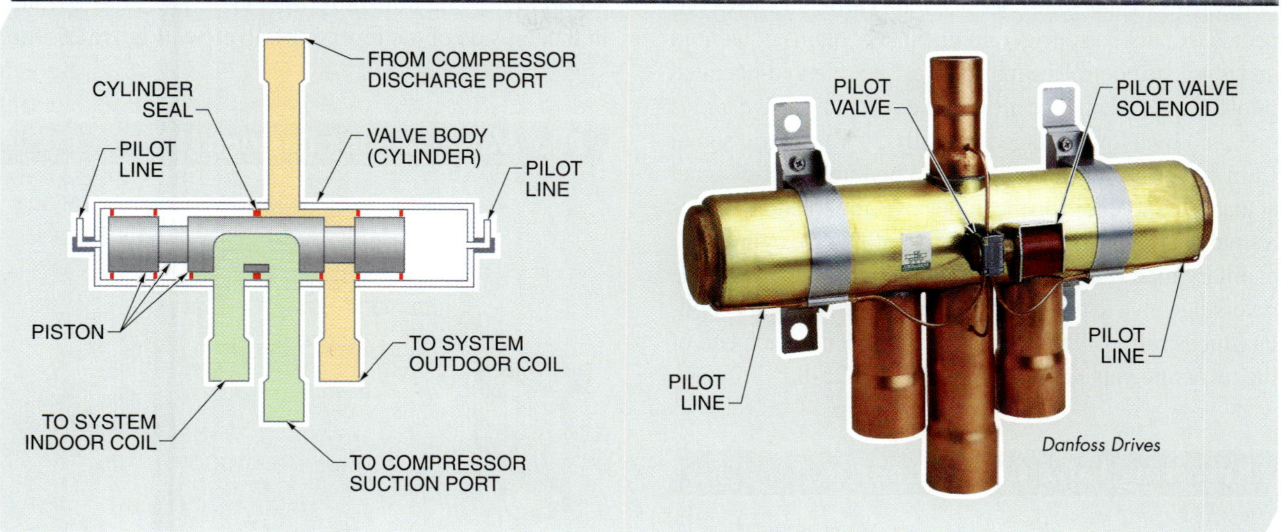

Figure 25-6. A reversing valve is a piston and valve body (cylinder) with four refrigerant line connections. It reverses the flow of refrigerant through a heat pump system, except for the compressor.

Reversing Valve Piston Positions

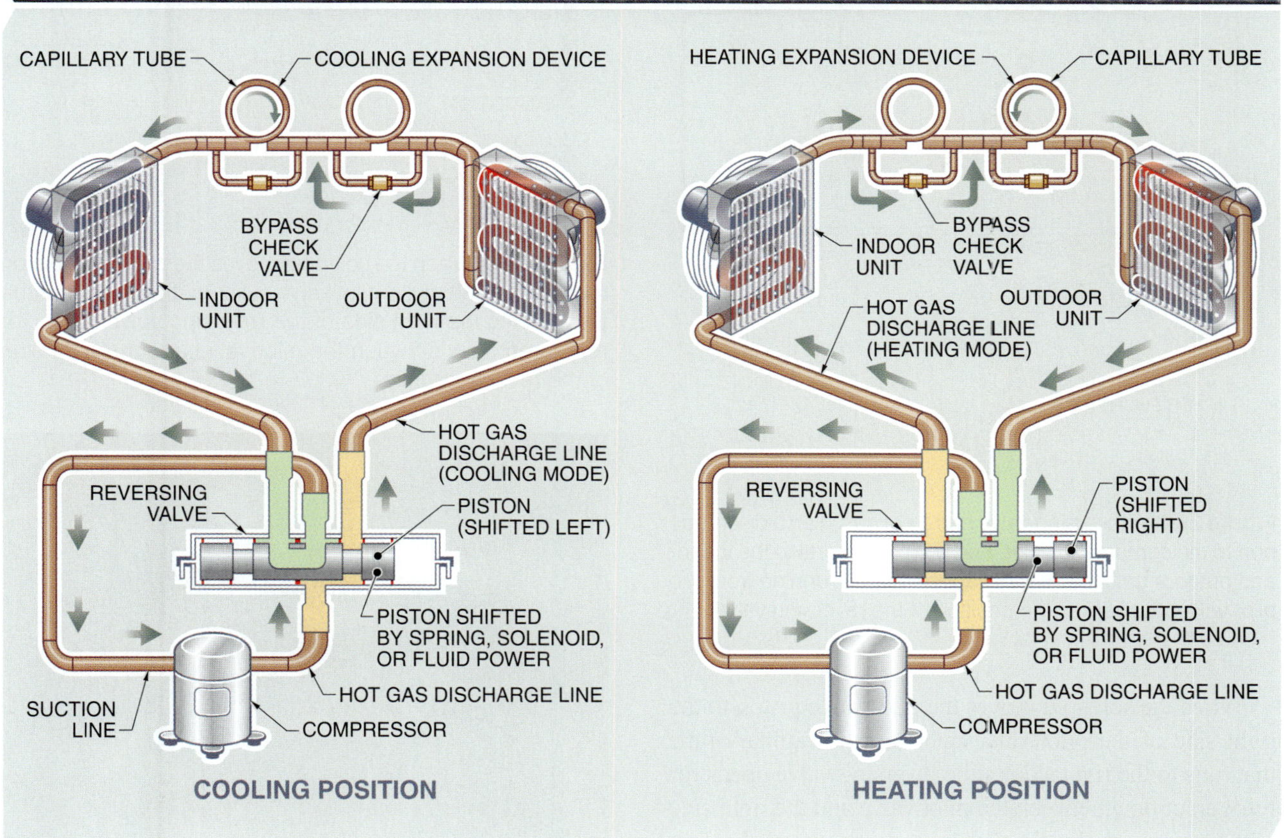

Figure 25-7. Refrigerant flows into a reversing valve through the hot gas discharge line and is directed out one of the two outlet ports to either the indoor or outdoor unit.

Reversing valves are controlled by solenoid-operated pilot valves. A *solenoid-operated pilot valve* is a valve connected to a hot gas refrigerant line that contains a solenoid and is used to control the main valve. The solenoid-operated pilot valve is located near or on the reversing valve and is operated by a low-voltage (typically 24 V) electric signal from a thermostat or controller. When an electric signal actuates the solenoid on the pilot valve, the solenoid moves a small piston back and forth in the pilot valve cylinder. A capillary tube runs from the hot gas discharge line on the reversing valve to the pilot valve cylinder. Two other capillary tubes run from the pilot valve cylinder to each end of the reversing valve cylinder. **See Figure 25-8.**

Reversing Valve Connections

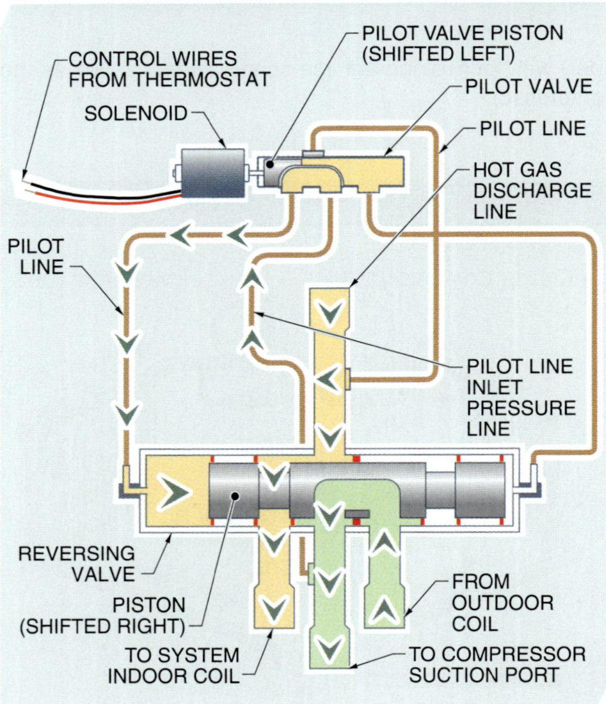

Figure 25-8. A reversing valve has a hot gas inlet, a discharge port to the compressor inlet, and two outlet ports to the indoor and outdoor unit. It also has capillary tubes running from the pilot valve that control the position of the reversing valve.

When the solenoid moves the pilot valve piston to the right side of the pilot valve cylinder, the capillary tube that runs to the right side of the reversing valve opens for refrigerant flow between the pilot valve and the right side of the reversing valve. The refrigerant pressure on the right side of the reversing valve pushes the reversing valve piston to the left side of the cylinder. The refrigerant also flows from the indoor unit, to the pilot valve, to the compressor suction line. The flow of refrigerant reduces the pressure in the left side of the reversing valve. **See Figure 25-9.**

Pilot Valve Cooling Position

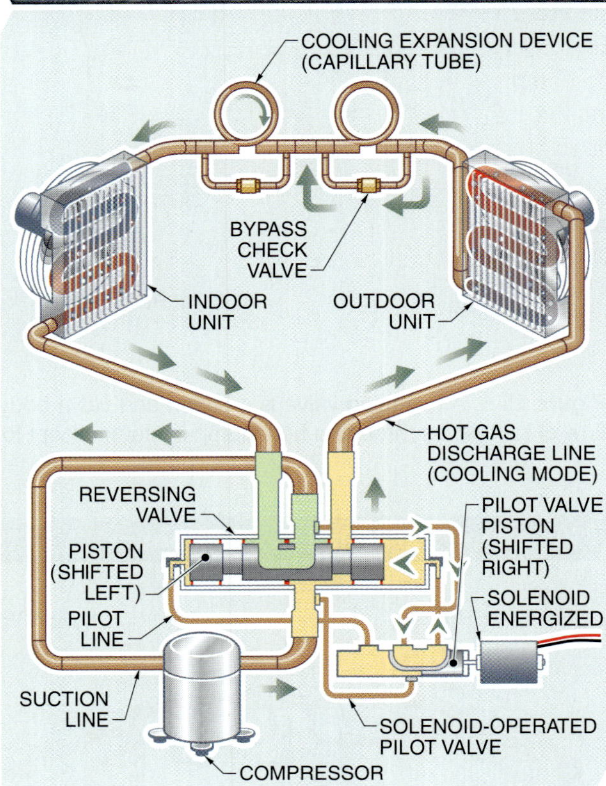

Figure 25-9. When the solenoid moves the pilot valve piston to the right side of the pilot valve cylinder, the capillary tube that runs to the right side of the reversing valve opens for refrigerant flow between the pilot valve and the right side of the reversing valve.

Reversing valves are often located in difficult-to-reach areas.

Refrigerant pressure on the left side of the reversing valve pushes the reversing valve piston to the right side of the cylinder. When the solenoid moves the pilot valve piston to the left side of the pilot valve cylinder, the capillary tube that runs to the left side of the reversing valve opens for refrigerant flow between the pilot valve and the left side of the reversing valve. The refrigerant also flows from the outdoor unit, to the pilot valve, to the compressor suction line. The flow of refrigerant into the compressor suction line reduces the pressure in the right side of the reversing valve. **See Figure 25-10.**

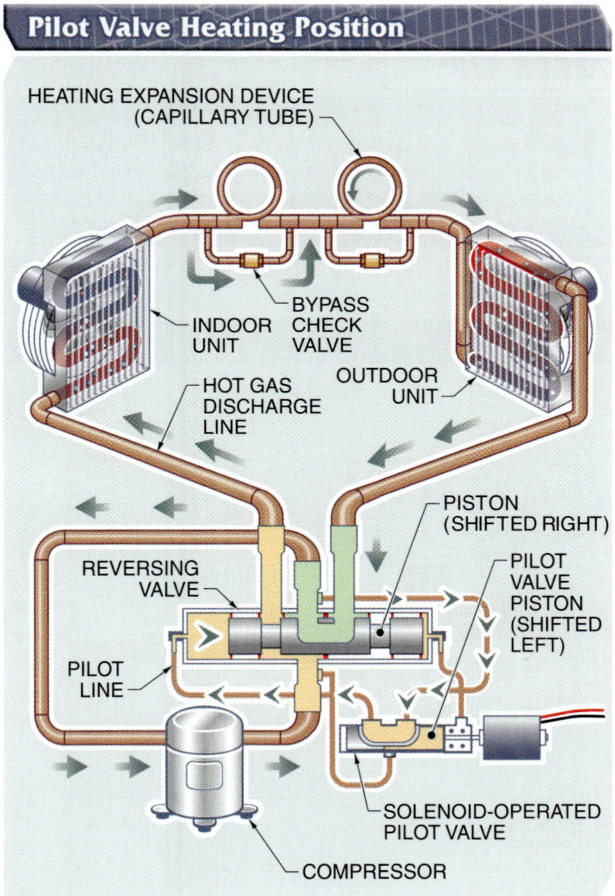

Figure 25-10. When the solenoid moves the pilot valve piston to the left side of the pilot valve cylinder, the capillary tube that runs to the left side of the reversing valve opens for refrigerant flow between the pilot valve and the left side of the reversing valve.

Expansion Devices

The coil in an indoor unit and the coil in an outdoor unit each require an expansion device because the refrigerant in a heat pump can flow in either direction. In small packaged heat pumps where the coils are close together, a capillary tube may be used as the expansion device. Large heat pumps, especially split systems, use thermostatic expansion valves as expansion devices.

Capillary Tubes. A *capillary tube expansion device* is a long length of copper tubing that produces a pressure decrease regardless of the direction of refrigerant flow through it. Filters are located in the capillary tube bypass because the refrigerant must be clean. The filters remove small contaminant particles but allow liquid refrigerant to pass.

When a capillary tube expansion device is used in a heat pump system, a bypass circuit is typically installed to allow the reverse flow of the refrigerant. One capillary tube can be used on small packaged or split heat pump systems that have cooling capacities of up to 5 tons. **See Figure 25-11.** Two capillary tubes are used on larger heat pump systems.

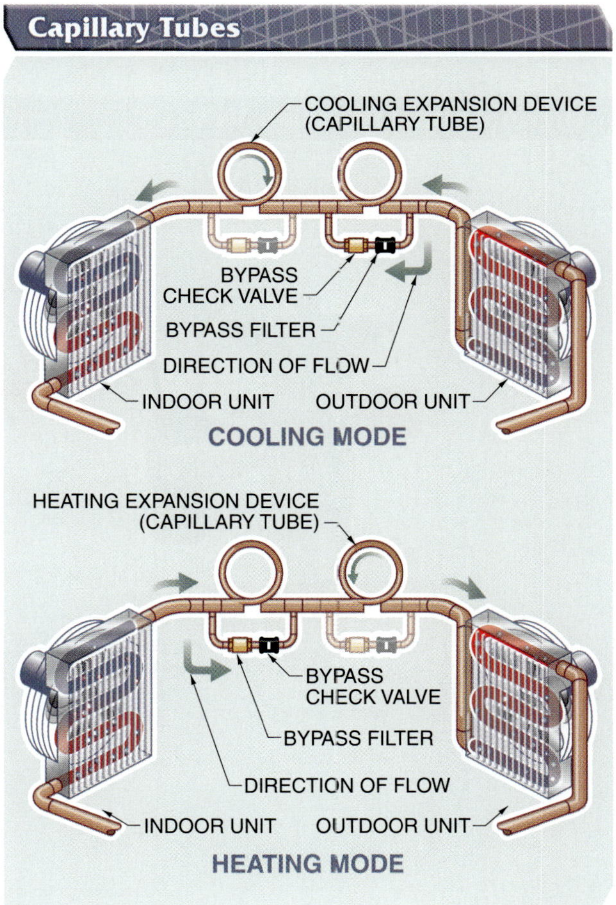

Figure 25-11. One capillary tube (expansion device) can be used on heat pump systems that have cooling capacities of up to 5 tons. Two capillary tubes are used on systems that have greater cooling capacities.

Thermostatic Expansion Valves. Large heat pumps and air conditioning systems typically have thermostatic expansion valves as expansion devices. A *thermostatic expansion valve (TXV)* is an expansion device that uses pressure and temperature to control the amount of refrigerant entering the coil. On a heat pump either the indoor coil or the outdoor coil works as an evaporator depending if its in cooling or heating mode.

In a thermostatic expansion valve, the pressure transmitted from the remote bulb to the diaphragm inside the valve determines the maximum flow through the valve, which determines the size of the valve outlet. A pushrod connects the diaphragm at the top of the valve to a needle inside the valve body. When pressure on the diaphragm decreases, the needle closes the valve. When pressure on the diaphragm increases, the needle opens the valve. The pressure is exerted by a refrigerant-filled remote bulb. The top of the TXV is commonly called the powerhead which includes the diaphragm assembly and the sensing bulb connected by a capillary tube. **See Figure 25-12.**

When the remote bulb senses a temperature rise from the refrigerant leaving the coil, the refrigerant in the bulb increases in pressure. This increase in pressure causes the thermostatic expansion valve to open. The open thermostatic expansion valve increases the flow of refrigerant into the coil (evaporator). When the temperature of the refrigerant leaving the coil decreases, the refrigerant in the bulb condenses and decreases the pressure in the remote bulb. This decrease in pressure causes the thermostatic expansion valve to close. The more closed a thermostatic expansion valve is, the more the flow of refrigerant into the coil decreases.

TECH TIP

Heat pump TXV piping is often installed in difficult to reach areas of a unit. If installed incorrectly, TXVs may be the cause of numerous small leaks. Always check these systems for leaks when performing any service work.

Thermostatic Expansion Valves

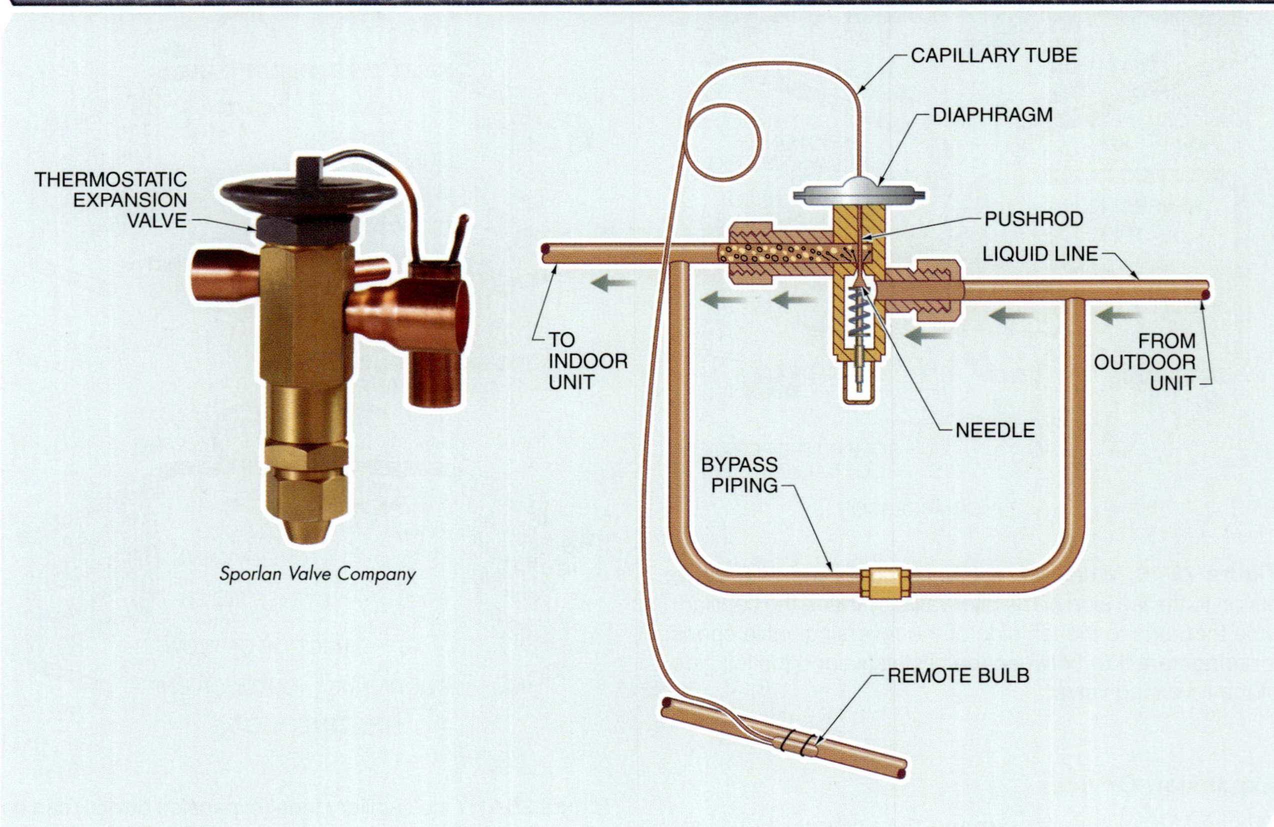

Figure 25-12. A thermostatic expansion valve increases the flow of refrigerant to an evaporator due to pressure increases in a capillary tube.

Thermostatic Expansion Valve Locations. In large heat pumps, a thermostatic expansion valve is located on each coil. The valves are necessary because the indoor coil is the evaporator in the cooling mode and the outdoor coil is the evaporator in the heating mode. A thermostatic expansion valve is located near each coil at the end of the liquid line. **See Figure 25-13.**

Because refrigerant cannot be reversed through a thermostatic expansion valve, bypass piping is installed around each expansion device. The bypass piping has a check valve that prevents liquid refrigerant from flowing back through the expansion valve by allowing the refrigerant to flow through the bypass. The thermostatic expansion valve of the outdoor coil is bypassed when the system operates in cooling mode. The thermostatic expansion valve of the indoor coil is bypassed when the system operates in heating mode.

AUXILIARY HEAT

An auxiliary heat source is used during cold weather because the heat output of a heat pump sized for cooling a building in the summer may not be adequate for heating the building in the winter. The heat output of a heat pump is directly related to the amount of heat available from the heat sink.

Auxiliary Heat Sources

In cold weather, the temperature of air falls, and the amount of heat needed to heat a building increases. When a heat pump is sized for the heat requirements of a building at the coldest outside temperatures, it is oversized for cooling. To prevent this problem, the heat pump is sized to provide the heat required to heat a building with an outdoor air temperature of about 35°F. An auxiliary heat source increases the heat output of the heat pump at outdoor temperatures below 35°F. Sources of auxiliary heat for heat pumps are resistance heating elements and gas- or fuel-fired heaters.

Resistance Heating Elements. A resistance heating element is the most common auxiliary heat source for a heat pump. Electricity is a convenient energy source for auxiliary heat because heat pumps are already supplied with electricity. **See Figure 25-14.**

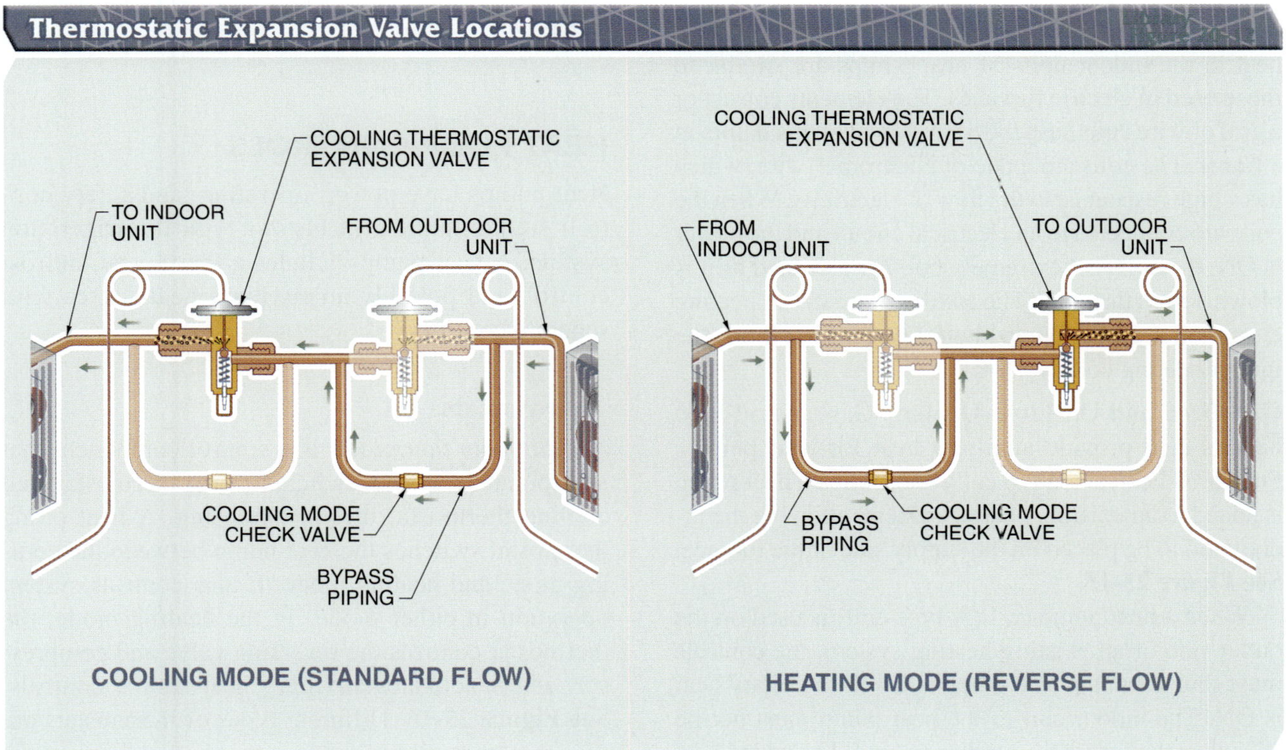

Figure 25-13. Heat pumps typically have two expansion devices located near each coil. One is used when the indoor coil is the evaporator (cooling mode), and the other is used when the outdoor coil is the evaporator (heating mode).

Resistance Heating Element Auxiliary Heat

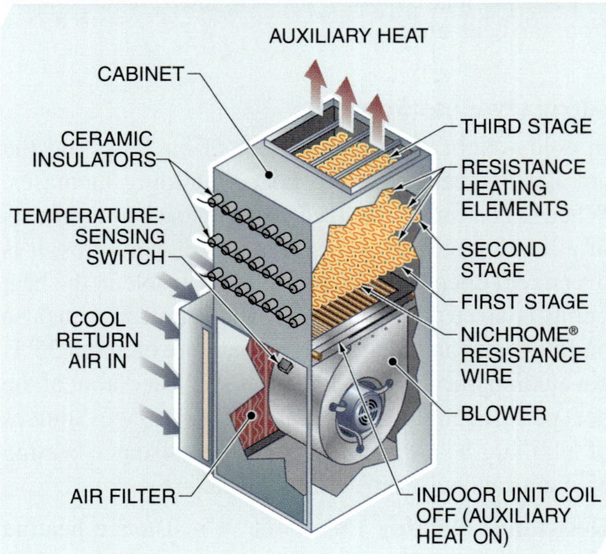

Figure 25-14. A resistance heating element is the most common type of auxiliary heat source for a heat pump with resistance wires made of Nichrome®, which has a high resistance to the flow of electricity.

Gas- and Oil-Fired Auxiliary Heat

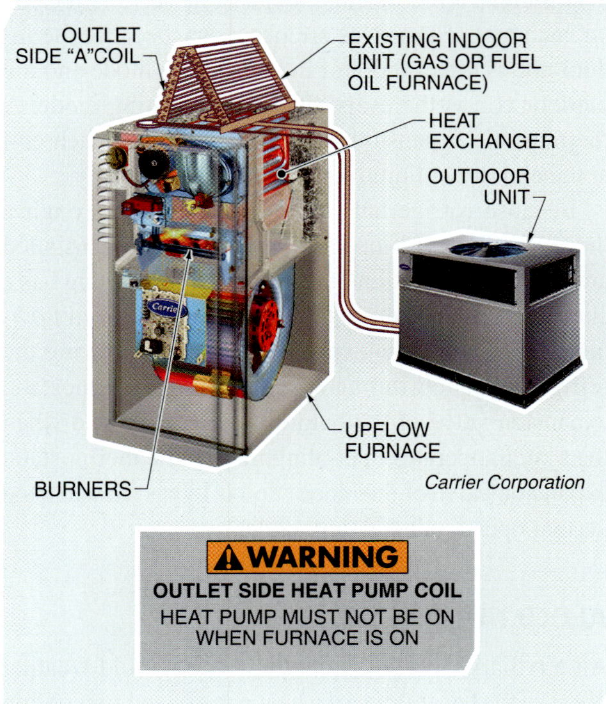

Figure 25-15. When a heat pump coil (A-type coil) is used on the outlet side of an existing heating system, the controls must shut down the heat pump when the auxiliary heat is ON.

The resistance heating elements used for auxiliary heat in the indoor units of heat pumps are similar to those used in electric furnaces. The elements consist of a grid of wire coils suspended by porcelain insulators in a frame. The coils are made of Nichrome® wire, which has a high resistance to the flow of electricity. When the coils are connected to an electrical circuit and the power is ON, the coils get extremely hot. Air is heated as it is blown across the coils. The auxiliary resistance heating elements of a heat pump should be installed after the indoor heating coil.

Gas-Fired and Oil-Fired Heaters. Gas- and oil-fired heaters also provide auxiliary heat for heat pumps. Fuel-fired heaters are typically used when a heat pump is added to an existing heating system, allowing the indoor coil to be placed on the supply side of the furnace. **See Figure 25-15.**

When a heat pump coil (A-type coil) is used on the outlet side of an existing heating system, the controls must shut down the heat pump when the auxiliary heat is ON. The indoor coil of the heat pump must not be ON when the furnace is producing heat. The added heat (super heat) from the furnace will cause excessive pressure in the heat pump system.

HEAT PUMP CONTROLS

Heat pumps have power, operating, and safety controls similar to the controls of a typical refrigeration system. A heat pump includes a thermostat, defrost control, and possibly an auxiliary heat source with separate controls.

Thermostats

A *heat pump thermostat* is a control component that incorporates a system switch, heating thermostat, and cooling thermostat into one housing. A heat pump thermostat switches the heat pump between the cooling mode and heating mode. It also controls system operation in either mode. In the heating mode, the thermostat controls the reversing valve and compressor, and it activates auxiliary heat source controls. **See Figure 25-16.** Different types of thermostats are used for controlling heat pumps. Some thermostats are switched manually from cooling mode to heating mode while others switch automatically.

Heat Pump Thermostats

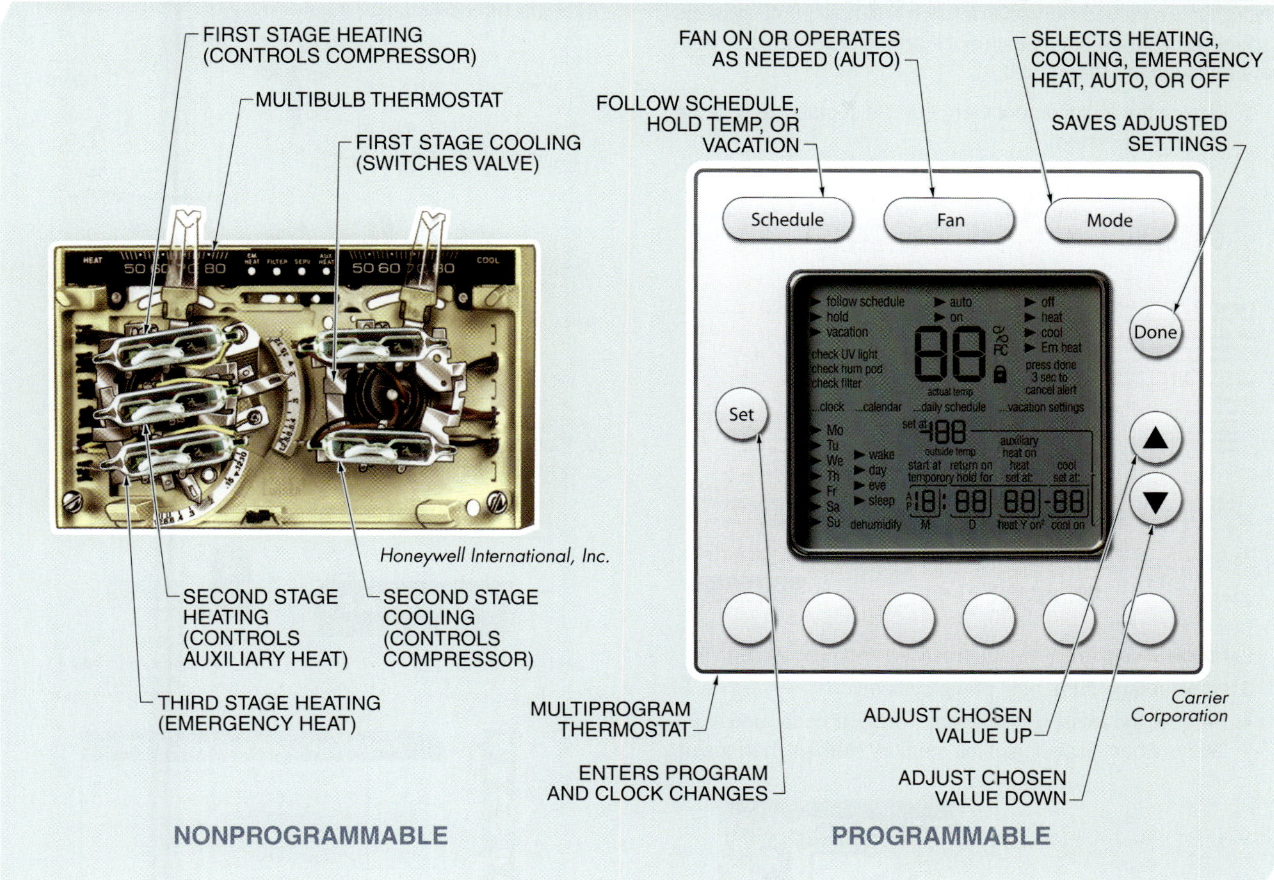

Figure 25-16. A heat pump thermostat switches the heat pump between the cooling mode and heating mode, and it controls system operation in either mode.

Most heat pumps require a multibulb, multistage thermostat. A *multibulb thermostat* is a thermostat that contains more than one mercury bulb switch. A *multistage thermostat* is a thermostat that contains several mercury bulb switches that make and break contacts in stages. When a multibulb, multistage thermostat is connected to two different devices, the devices are energized at different times, depending on temperature changes.

For example, a thermostat used to control a heat pump may have two stages. In the cooling mode, the first-stage mercury bulb controls the reversing valve, and the second-stage mercury bulb controls the compressor. When the temperature increases, the first-stage contact switches the system to the cooling mode (reversing valve), and the second-stage mercury contact energizes the compressor to provide cooling.

When using the same heat pump system in the heating mode, the first-stage contact controls the compressor, and the second-stage contact controls the auxiliary heat. As the temperature falls, the first-stage contact switches ON the compressor to provide heat. As the temperature continues to fall, the second-stage contact switches ON the auxiliary heat. **See Heat Pump Thermostat Replacement Procedure.**

TECH TIP

Almost all modern electronic heat pump thermostats use solid-state contacts in place of mercury bulb contacts because mercury is a known toxic substance. The new thermostats are extremely accurate when combined with electronic sensing capabilities.

Procedure

Heat Pump Thermostat Replacement

Note: Only a trained technician familiar with heat pump systems should replace a mercury element thermostat with an appropriate replacement thermostat.

1. Ensure the replacement thermostat is compatible with the heat pump system.

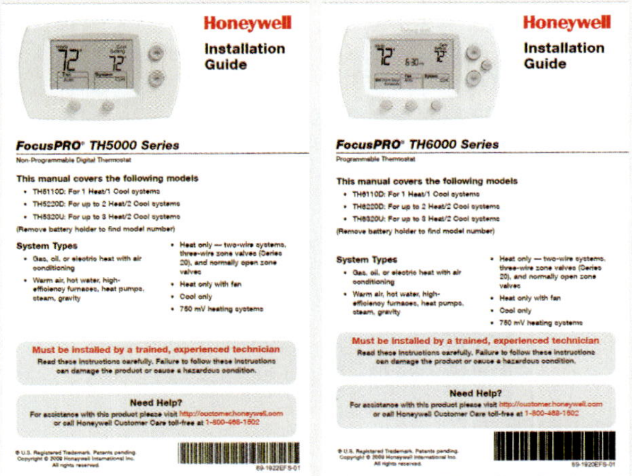

2. Follow all electrical safety practices and procedures.
3. Lockout/tagout the heat pump system.
4. Remove the existing thermostat. Follow all codes and regulations when disposing of the mercury element thermostat.

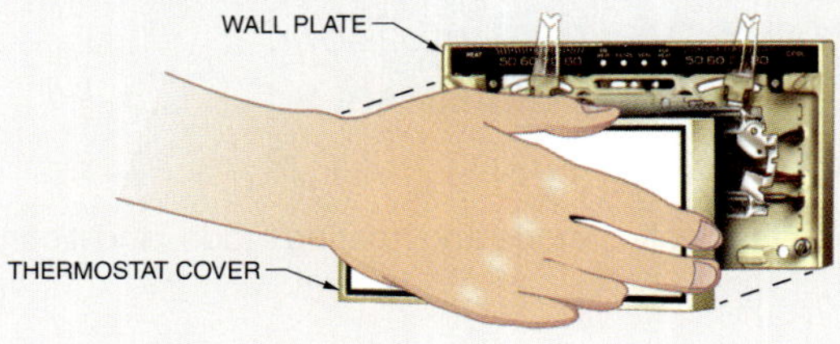

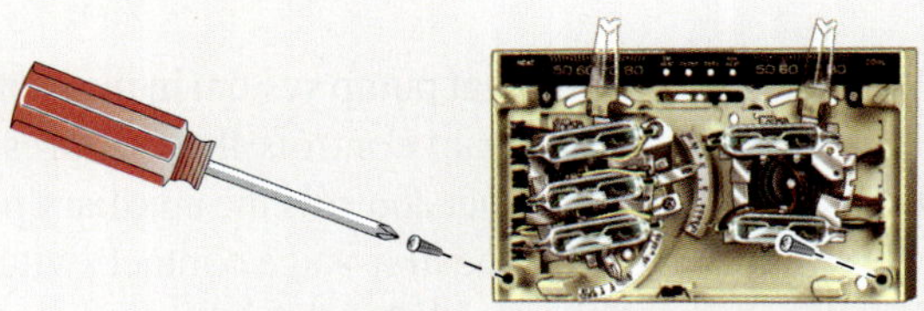

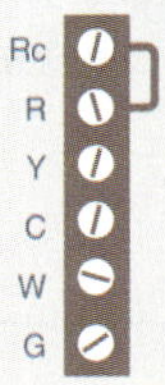

Conventional Terminal Designations	
Rc	24 V AC Power from cooling transformer
R	24 V AC Power from heating transformer
W	Heat Relay (stage 1)
W2	Heat Relay (stage 2)
Y	Compressor Starter (stage 1)
Y2	Compressor Starter (stage 2)
G	Fan Relay
C	24 V AC Common

IT IS ILLEGAL TO DISPOSE OF MERCURY THERMOSTATS IN SOLID WASTE IN 9 STATES; HVAC CONTRACTORS MUST RECYCLE ALL MERCURY THERMOSTATS THEY REMOVE FROM SERVICE; IT IS ILLEGAL TO LEAVE THEM AT THE CUSTOMER'S PREMISES

5. Install the wall plate of the thermostat.

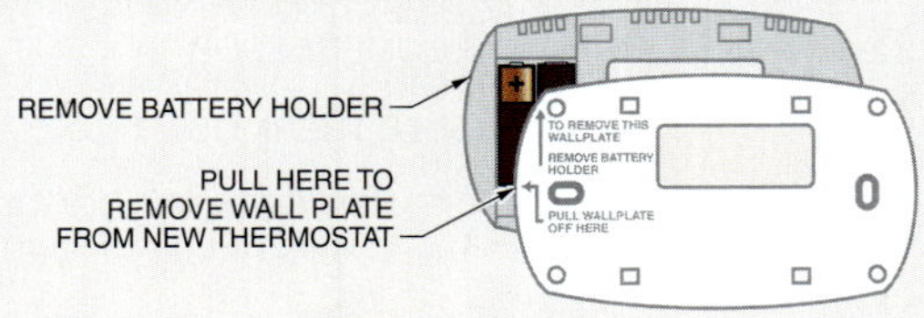

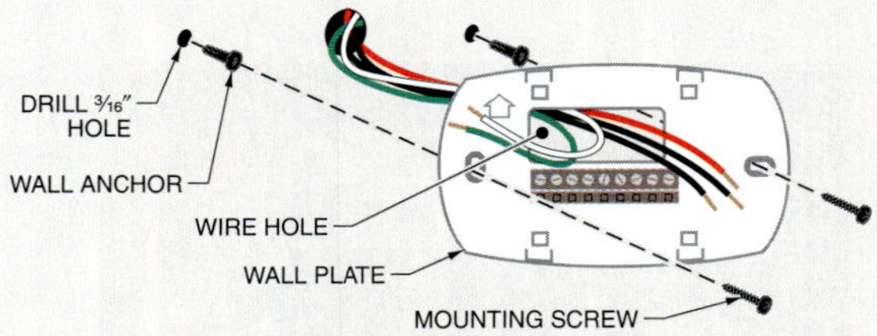

6. Connect the wires to the terminals of the thermostat.

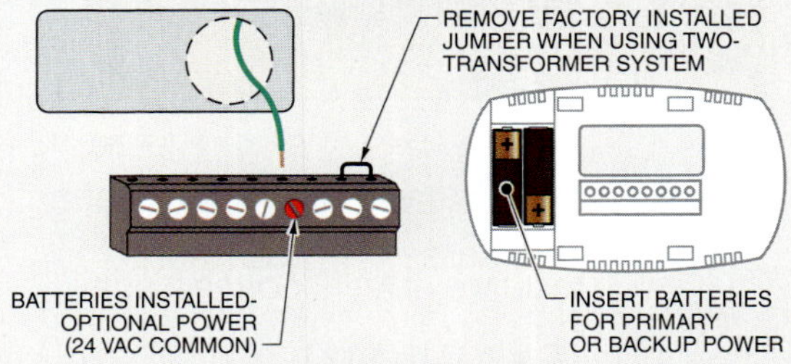

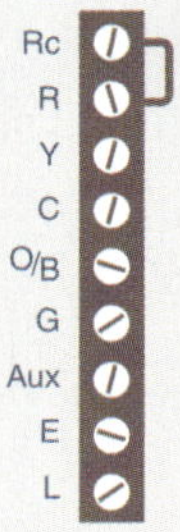

Heat Pump Terminal Designations	
L	Output when set to Emergency Heat
E	Emergency Heat Relay
Aux	Auxiliary Heat Relay
G	Fan Relay
O/B	Reversing Valve
C	24V AC common
Y	Compressor Starter (stage 1)
Y2	Compressor Starter (stage 2)
R	24 VAC power from heating transformer
Rc	24 VAC power from cooling transformer

7. Mount the thermostat.

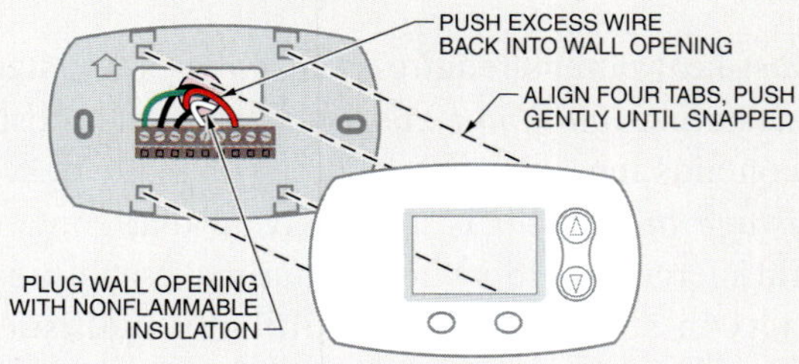

8. Turn the power ON.
9. Program the system using the keypad.

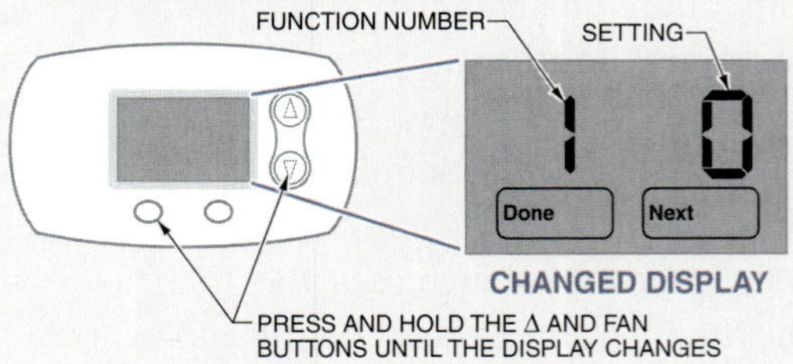

10. Check the operation of the heat pump system.

Defrost Controls

Defrost controls regulate the defrost cycles of a heat pump. A *defrost cycle* is a mechanical procedure used to temporarily reverse the refrigerant flow in a heat pump to melt frost or ice on an outdoor unit coil. When an air-source heat pump operates in heating mode, the outdoor unit functions as an evaporator. The temperature of the outdoor unit coil is typically below 32°F. At this temperature, moisture in the outdoor air freezes and forms frost or ice on the surface of the outdoor coil. The frozen moisture blocks airflow across the outdoor coil. **See Figure 25-17.**

The flow of refrigerant in the heat pump system is reversed by temporarily switching the system to the cooling mode. Hot refrigerant gas flows through the outdoor coil, heating the coil. The defrost cycle (cooling mode) is activated when the position of the reversing valve is changed. The outdoor fans are shut OFF, and the auxiliary heat is turned ON. This raises the temperature of the hot gas, which flows through the outdoor coil and melts the frost or ice. The air blown into the building is cooled slightly during the defrost cycle.

Most defrost control systems are based on time cycles. A timer in the defrost control system calls for defrost cycles at regular time intervals. When defrost is needed, an initiation sensor signals a need for defrost. Initiation sensors sense the temperature of either the air or the refrigerant leaving the outdoor coil to control the initiation of the defrost cycle. Initiation sensors can also use the pressure of the refrigerant leaving the outdoor coil to control the defrost cycle. When defrost is not needed when the timer signals, the initiation sensor prevents the system from entering into a defrost cycle. **See Figure 25-18.**

TECH TIP

One of the more common failures of an air-source heat pump is failure of the defrost system during the heating season. An outdoor unit covered with ice is a clear indication that the defrost system has failed.

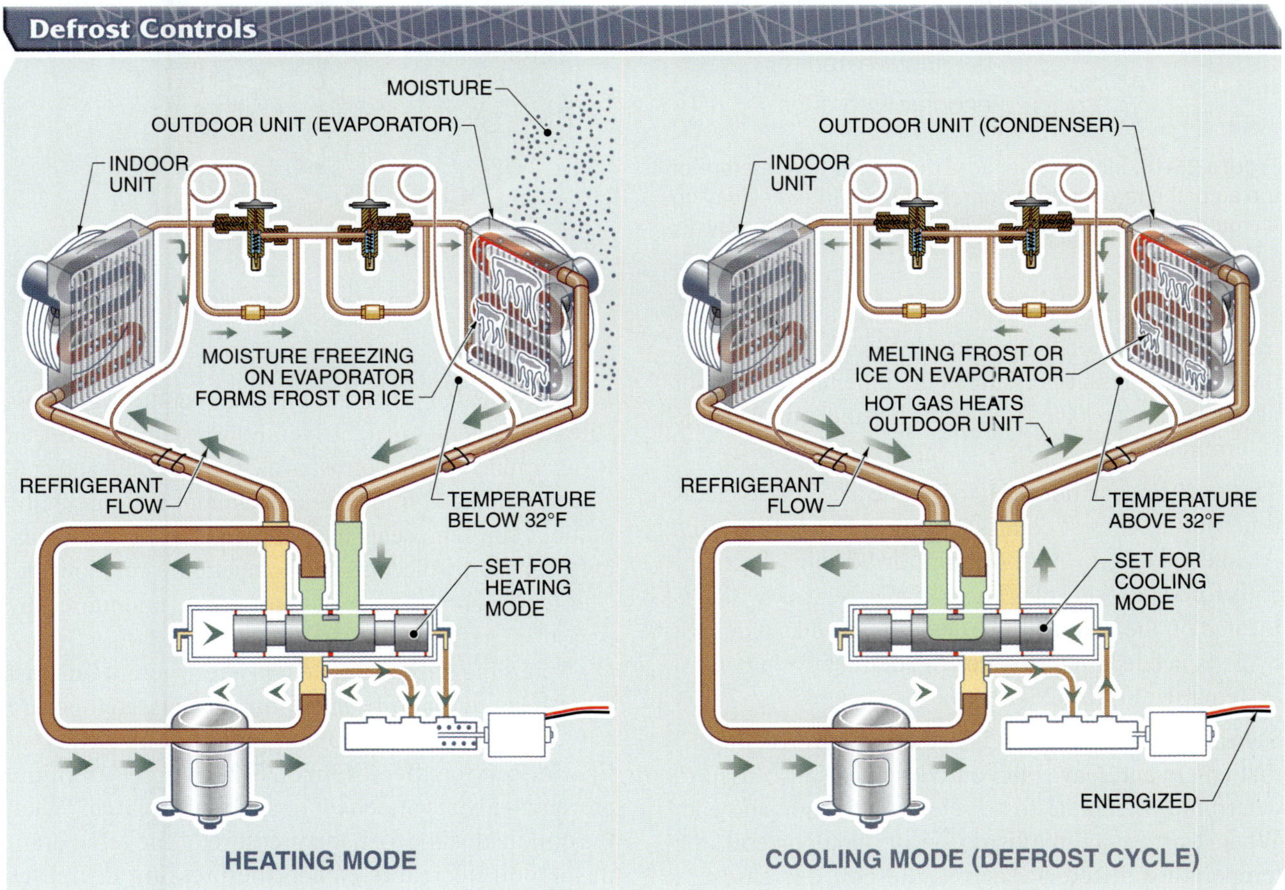

Figure 25-17. When a heat pump is in heating mode and the temperature of the outdoor coil is below 32°F, the outdoor air freezes and forms frost or ice on the surface of the outdoor coil.

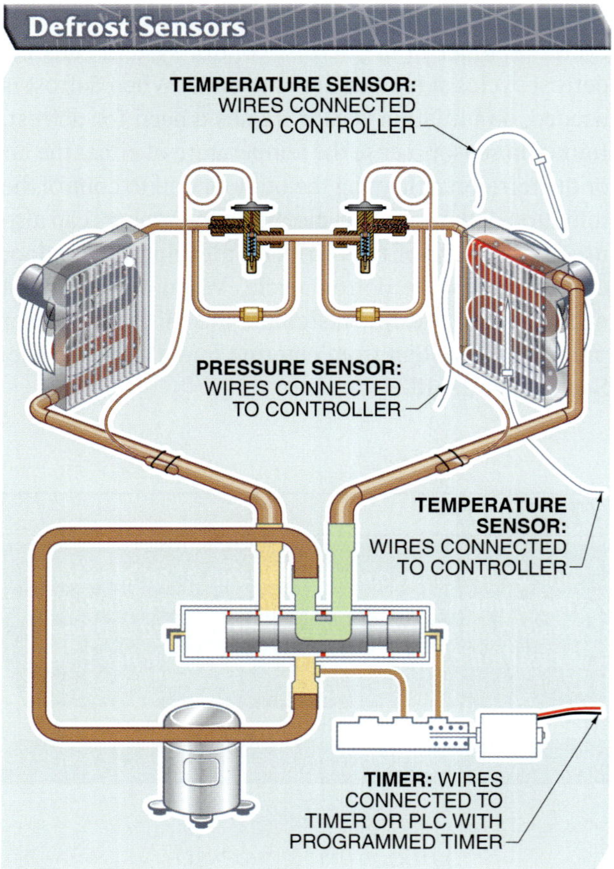

Figure 25-18. Initiation sensors detect the temperature of the air or the temperature or pressure of refrigerant leaving the outdoor coil to determine when to initiate a defrost cycle.

During a defrost cycle, a termination sensor senses the temperature of the outdoor coil either directly or through refrigerant pressure. When the temperature or pressure indicates the frost or ice has melted, the cycle is terminated.

Timers. Time is often used to initiate a defrost cycle. A timer in the defrost control system runs continuously. At intervals of 45 min, 60 min, or 90 min, the timer calls for a defrost cycle. The initiation sensor, which is located on the outdoor coil, determines when a defrost cycle is needed and activates the defrost cycle control system. **See Figure 25-19.**

Temperature Sensors. The temperature of the air that enters and leaves the outdoor coil is also used to determine when a defrost cycle should begin and end. When frost or ice builds up on the outdoor coil, the temperature difference across the coil decreases. A temperature-actuated defrost control is mounted on the discharge side of the coil.

A *temperature-actuated defrost control* is a control device that consists of a remote bulb thermostat and an electric control switch. The switch is wired into the defrost control. The temperature-actuated defrost control sends a signal to the defrost control, which monitors temperature. When the temperature indicates a need for defrost, the defrost cycle is initiated. As the frost or ice on the coil is melted by hot discharge gases, the temperature of the refrigerant leaving the coil increases. This temperature increase is sensed by the thermostat and the temperature-actuated defrost sensor. The sensor sends a signal to the defrost control to terminate the defrost cycle. **See Figure 25-20.**

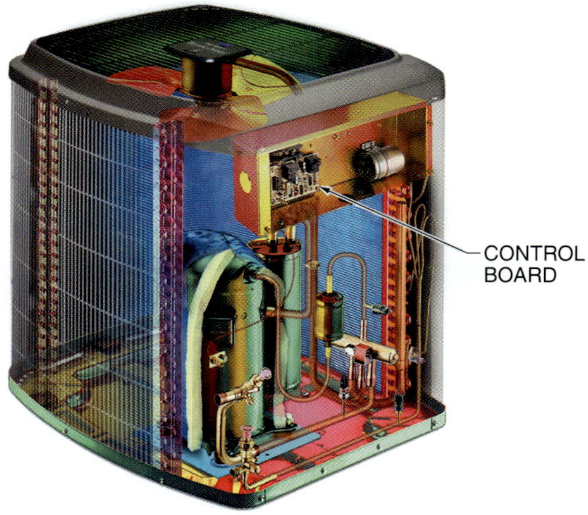

Carrier Corporation

An electronic control board on an exterior heat pump controls the defrost cycle.

Pressure Sensors. In addition to temperature, the pressure of refrigerant in a refrigeration system can be used to determine when a defrost cycle should begin and end. At saturation conditions, the refrigerant pressure in the outdoor coil represents the temperature of the refrigerant because of the pressure-temperature relationship. The temperature of the outdoor coil is monitored by measuring refrigerant pressure.

A piercing valve is placed on a refrigerant line just after the outdoor coil. A *piercing valve* is a temporary valve that is attached to and taps into a refrigerant line for service. **See Figure 25-21.** Tubing from the piercing valve is attached to a pressure switch. When the coil begins to frost, the pressure of the refrigerant in the coil decreases. When the pressure decreases enough to indicate a freezing temperature, the pressure switch activates the defrost control.

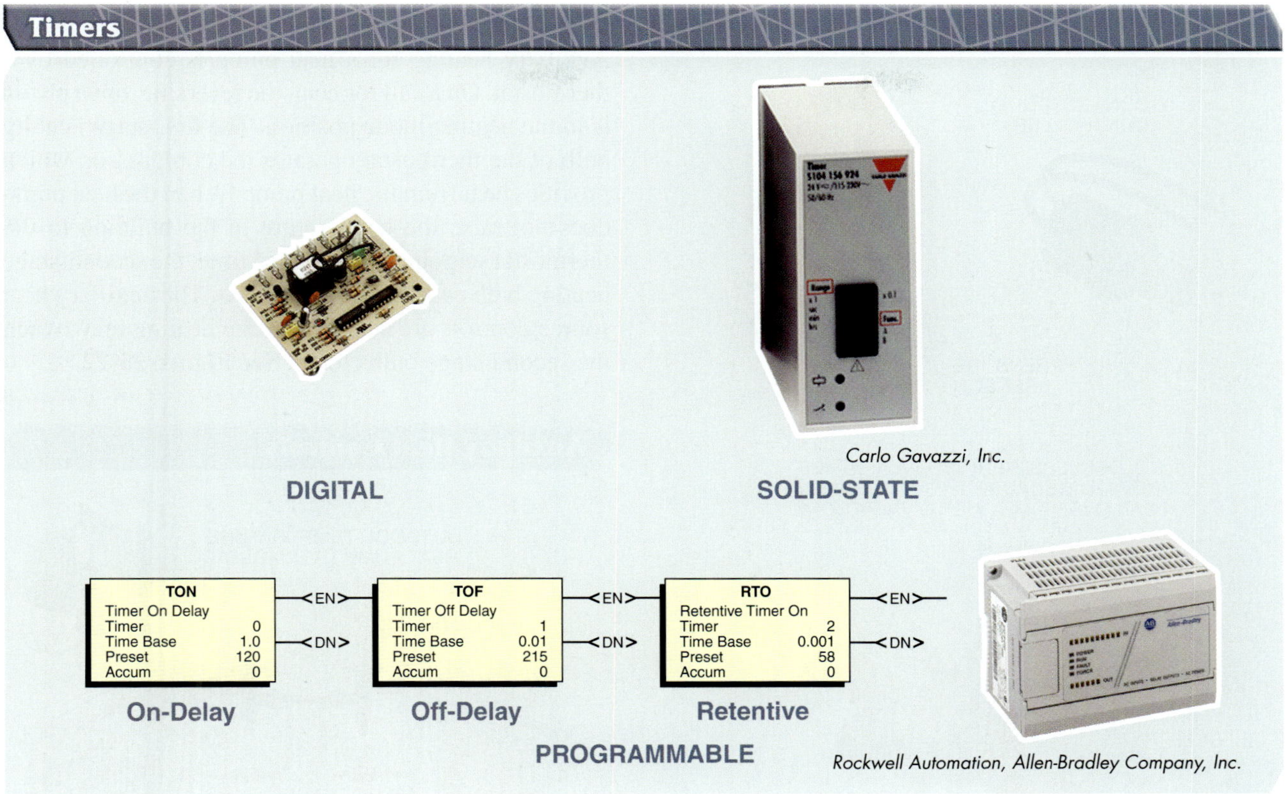

Figure 25-19. The timer in a defrost control system runs continuously. It is typically set to call for a defrost cycle at intervals of 45 min, 60 min, or 90 min.

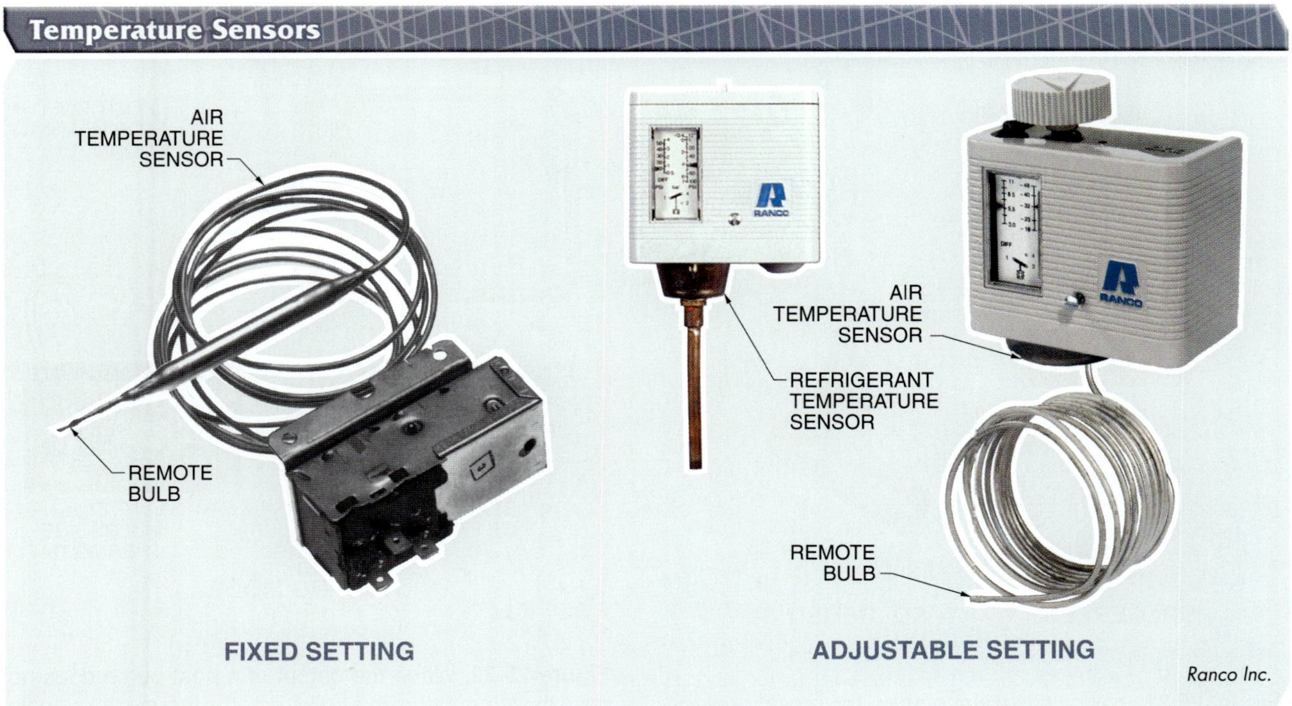

Figure 25-20. The temperature of refrigerant leaving a coil is sensed by a thermostat and temperature-actuated defrost sensor. When the temperature has increased enough, the sensor sends a signal to the defrost control to terminate the defrost cycle.

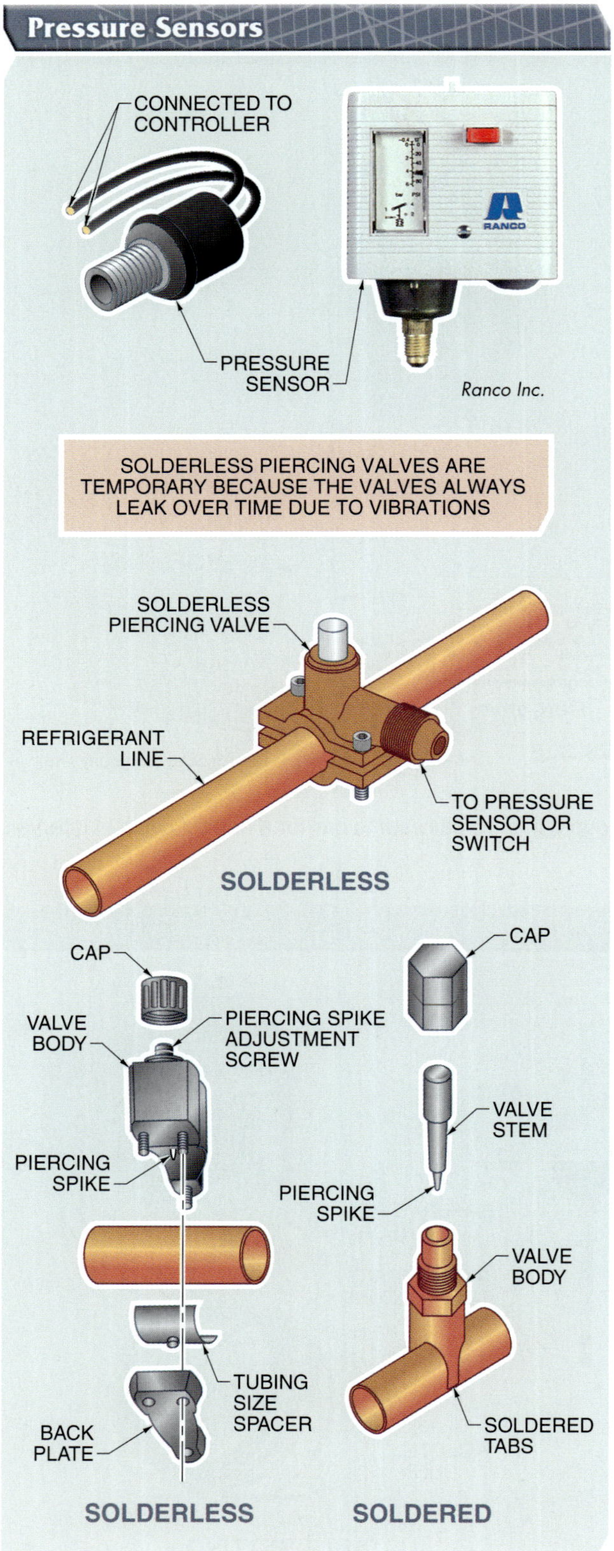

Figure 25-21. In addition to temperature, the pressure of refrigerant in a refrigeration system can be used to determine when a defrost cycle should begin and end.

Auxiliary Heat Controls

Auxiliary heating for a heat pump is controlled by a thermostat. On a call for heat, the reversing valve piston is in the heating mode position. The first-stage heating bulb of the thermostat operates the compressor, which provides heat from the heat pump. When the heat pump does not raise the temperature in the building to the thermostat setpoint in a specified time, the second-stage heating bulb calls for auxiliary heat. The auxiliary heat source controls are actuated by the heating relay when the second-stage bulb closes. **See Figure 25-22.**

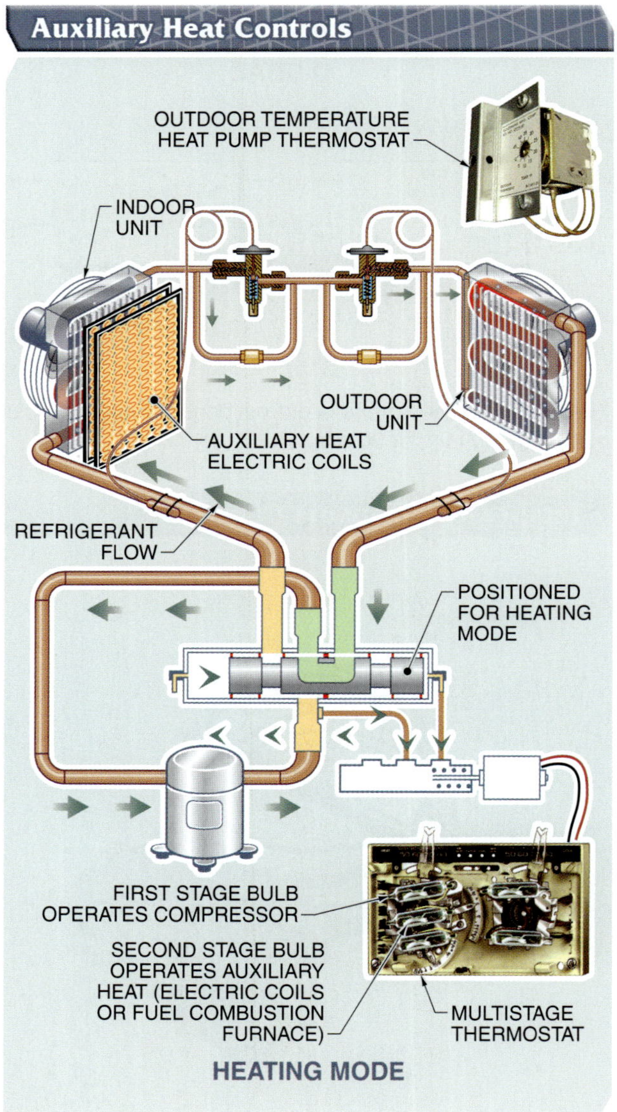

Figure 25-22. When the output of a heat pump does not raise the temperature in a building to the thermostat setpoint in a certain amount of time, auxiliary heat source controls energize the electric heating elements or burners.

To ensure the heat pump heats as much as possible and auxiliary heat is used only when necessary, an outdoor thermostat is often used with the regular thermostat to control the amount of auxiliary heat produced. The outdoor thermostat has a set of contacts that close when the temperature falls below the outdoor temperature setpoint. When the outdoor temperature is above the setpoint temperature, the auxiliary heat will not turn ON.

The auxiliary heat is also controlled by the defrost control system. When the heat pump switches to the defrost cycle, the auxiliary heating is turned ON so that the air blown into the building will be heated. Typically, this air would have been cooled by the refrigerant that changed direction.

OPERATION

When a heat pump operates in cooling mode, refrigerant vapor is compressed in the compressor. **See Figure 25-23.** The compressed refrigerant flows through the hot gas discharge line to the reversing valve. From the reversing valve, the refrigerant flows to the outdoor coil. In the outdoor coil, the refrigerant condenses to a liquid as it rejects heat to the condensing medium (air). The pressure of the refrigerant decreases as the refrigerant flows through the liquid line and then flows to the indoor coil.

In the indoor coil, the refrigerant vaporizes as it absorbs heat from the evaporating medium. The vaporized refrigerant leaves the indoor coil to the compressor suction line and returns to the reversing valve. The refrigerant then flows through the reversing valve into the compressor.

Air-Source Heat Pump Operation Media Clip

Cooling Mode Operation

Until recently, heat pumps typically used R-22 refrigerant. In the past few years, R-410a refrigerant has been used for heat pumps. For an R-410a system that is operating in cooling mode, the pressure of the high-pressure side of the heat pump is approximately 411 psi, and the low-pressure side is approximately 163 psi. In the cooling mode, the outdoor coil (condenser) is the high-pressure side and the indoor coil (evaporator) is the low-pressure side. **See Figure 25-24.** *Note:* All listings for pressure, temperature, and enthalpy are approximate. A heat pump in cooling mode operates in the following manner:

1. Refrigerant leaves the compressor with a pressure of 411 psi, temperature of 175°F, and enthalpy of 142 Btu/lb.

Air-Source Heat Pump Operation

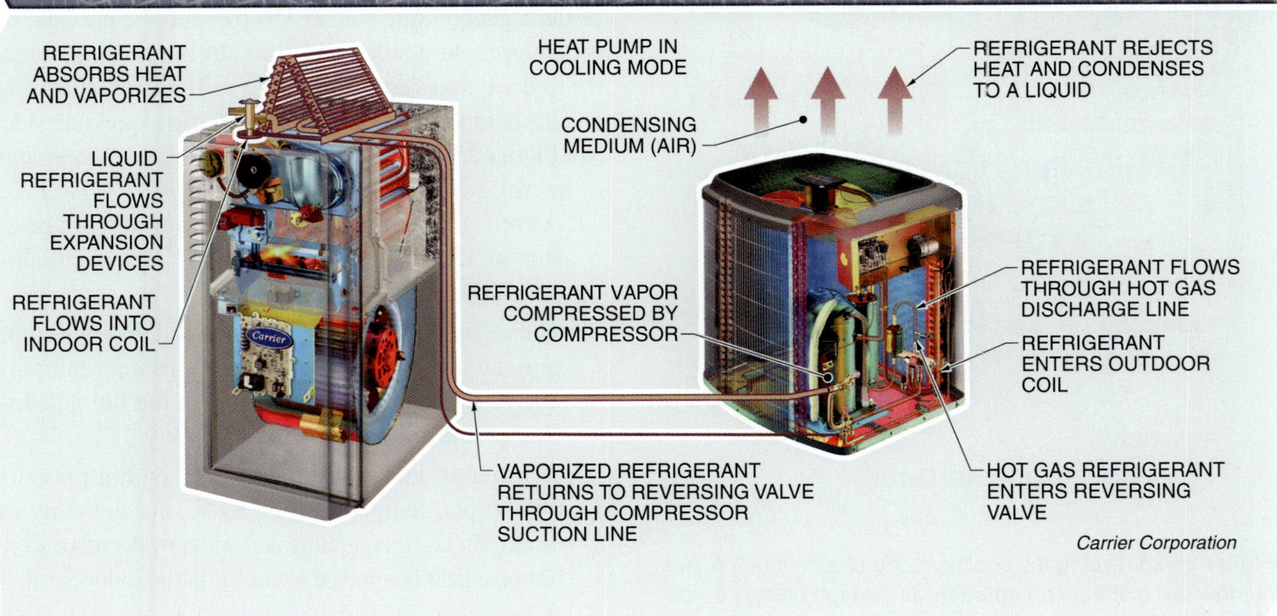

Figure 25-23. Air-source, water-source, and geothermal heat pumps operate in the same manner.

2. The refrigerant remains at a pressure of 411 psi as it enters the outdoor coil. The temperature and enthalpy have both decreased due to the heat loss in the hot gas line between the compressor and the outdoor coil.
3. The refrigerant leaves the outdoor coil with a pressure of 411 psi, temperature of 101°F, and enthalpy of 51 Btu/lb. The temperature decreases because the refrigerant is subcooled before it leaves the coil in the outdoor unit. Enthalpy decreases because the refrigerant rejects heat to the air as the refrigerant flows through the outdoor coil.
4. The refrigerant enters the expansion device from the outdoor coil with the same properties.

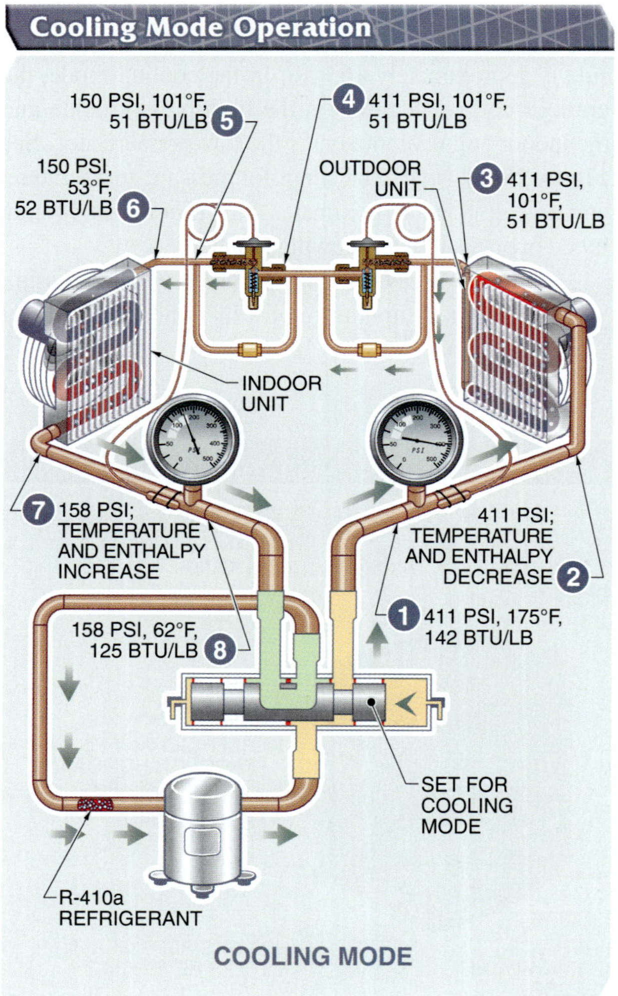

Figure 25-24. During the cooling mode of a heat pump, the outdoor coil (condenser) operates as the high-pressure side (411 psi) and the indoor coil (evaporator) operates as the low-pressure side (163 psi).

5. After the refrigerant flows through the expansion device, the pressure decreases to 150 psi. The pressure decrease is caused by the refrigerant flashing from a liquid to a gas.
6. The refrigerant enters the indoor coil. The temperature is now 53°F and enthalpy is 52 Btu/lb. The temperature decreases because the refrigerant vaporizes at the lower pressure as the refrigerant enters the indoor coil.
7. The refrigerant leaves the indoor coil with a pressure of 158 psi, temperature of 62°F, and enthalpy of 125 Btu/lb. The temperature and enthalpy increase because the refrigerant in the coil of the indoor unit absorbed heat from the hot building air.
8. The refrigerant enters the reversing valve and compressor with the same properties. As the refrigerant flows through the compressor, the pressure, temperature, and enthalpy increase. The temperature and enthalpy increase because of the heat of compression and possibly heat from the compressor motor.

Heating Mode Operation

When a heat pump operates in the heating mode, the refrigerant flows from the reversing valve to the indoor coil. It flows through the indoor unit, liquid line, and expansion device. From the expansion device, the refrigerant flows through the outdoor unit and returns to the reversing valve.

For example, when a heat pump that contains R-410a refrigerant operates in heating mode, the pressure of the high-pressure side is 350 psi, and the pressure of the low-pressure side is 114 psi. In the heating mode, the indoor unit (condenser) is the high-pressure side and the outdoor unit (evaporator) is the low-pressure side. **See Figure 25-25.** A heat pump in heating mode operates in the following manner:

1. As refrigerant leaves the compressor, it has a pressure of 350 psi, temperature of 167°F, and enthalpy of 142 Btu/lb.
2. The refrigerant enters the indoor coil with the same pressure. However, the temperature and enthalpy decrease because of heat losses in the hot gas discharge line and the reversing valve.
3. Refrigerant leaves the indoor coil with a pressure of 350 psi, temperature of 89°F, and enthalpy of 46 Btu/lb. The temperature and enthalpy decrease more because heat is rejected to the air at the indoor coil.
4. Refrigerant enters the expansion device at the outdoor coil with approximately the same properties.

5. The refrigerant leaves the expansion device and enters the outdoor coil with a pressure of 110 psi, temperature of 36°F, and enthalpy of 46 Btu/lb. The pressure decreases because of the restriction in the expansion device. The temperature decreases because some of the liquid refrigerant vaporizes due to the lower pressure.
6. The refrigerant leaves the outdoor coil at a pressure of 114 psi, temperature of 43°F, and enthalpy of 123 Btu/lb. The increase in temperature and enthalpy is due to heat absorbed from the heat sink at the outdoor coil.
7. The refrigerant enters the compressor with approximately the same properties. As the refrigerant flows through the compressor, the pressure, temperature, and enthalpy increase. The temperature and enthalpy increase because of the heat of compression and possibly from the heat of the compressor motor.
8. A timer in the defrost control system calls for defrost cycles at regular time intervals. When defrost is needed, an initiation sensor signals a need for defrost. If defrost is not needed, when the timer signals, the initiation sensor prevents the system from entering into a defrost cycle.

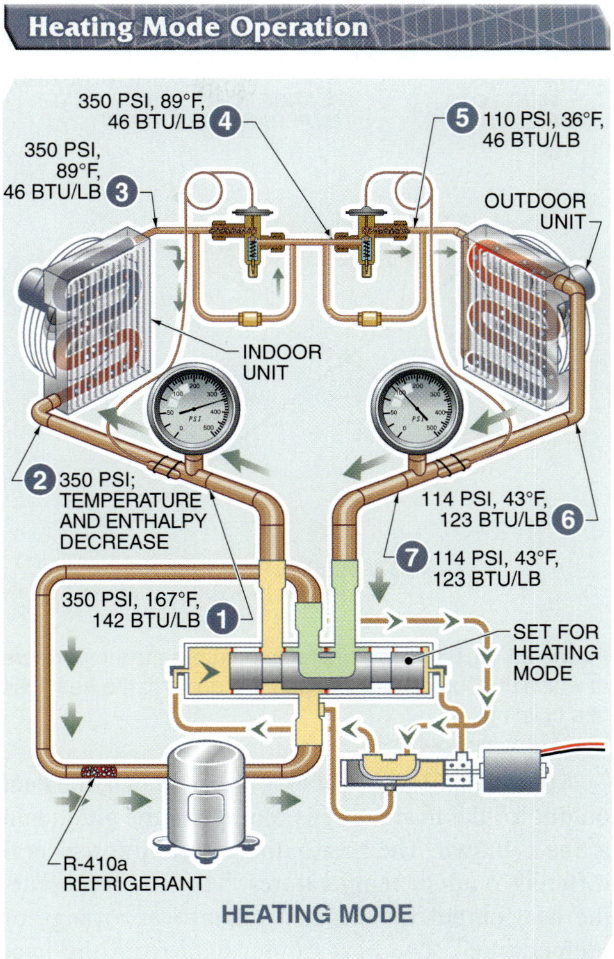

Figure 25-25. During the heating mode of a heat pump, the indoor unit (condenser) operates as the high-pressure side (350 psi) and the outdoor unit (evaporator) operates as the low-pressure side (114 psi).

HEAT PUMP SELECTION

A heat pump is selected for a given application based on the available heat source, type of application, and cooling and heating capacities of the heat pump. A heat pump selected to cool and heat a building located in a warm climate is sized to accommodate the cooling load of the building. Because of the warm climate, the heating capacity of the heat pump may be large enough to handle the heating load. However, a heat pump selected to cool and heat a building located in a cold climate typically requires auxiliary heat to meet the required heating capacity. To determine the amount of auxiliary heat required, the heating capacity of the unit at the design temperature for the area should be checked against the heat loss of the building at the same design temperature.

TECH TIP

In some locations, special low rates are offered by utility companies for all-electric homes. These special rates are designed to encourage the use of electricity. In general, electricity is more expensive than natural gas when used as a heat source. However, the special rates make electricity a less expensive option.

Balance Point Temperatures

Balance point temperature is the outdoor temperature at which the output of a heat pump equals the heat loss of a building. The balance point temperature for a heat pump used to heat and cool a particular building space is found on a graph. On the graph, the heat loss of the building space and the heat output of the heat pump are plotted at different outdoor temperatures. **See Figure 25-26.**

The heat loss of the building is plotted on the point that represents the total heat loss of the building at the outdoor design temperature. A line is drawn from this point to the base line of the graph at which the heat loss is 0 Btu/hr. The heat loss is typically about 5°F less than the indoor design temperature for the building.

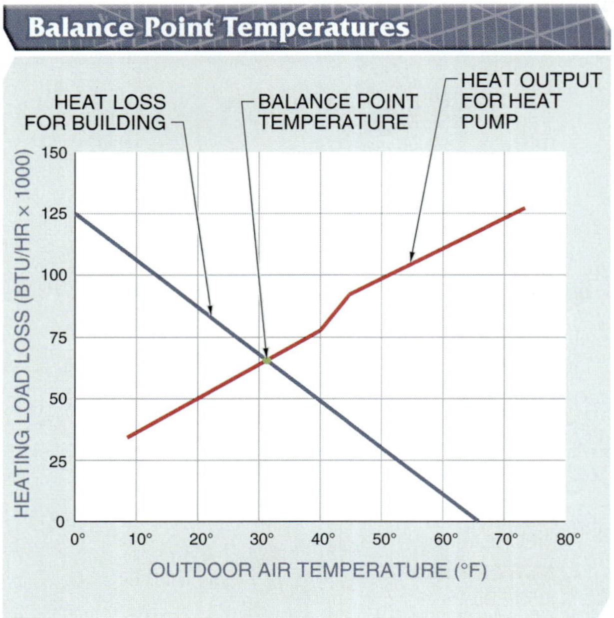

Figure 25-26. Balance point temperature is the temperature at which the output of a heat pump is equal to the heat loss of a building.

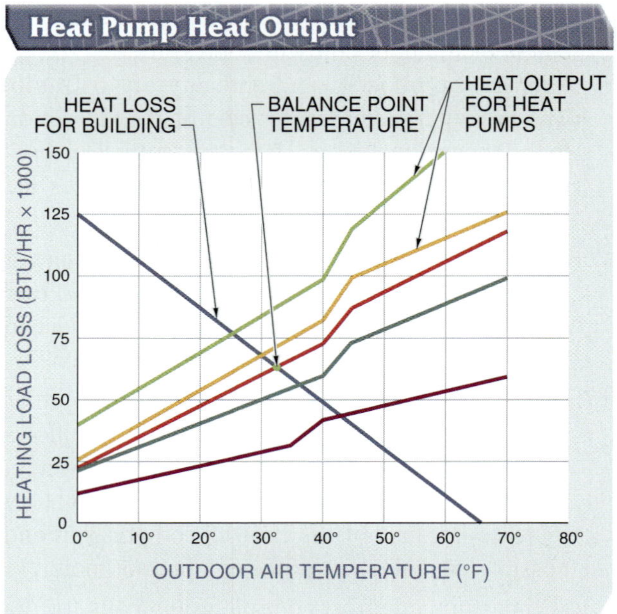

Figure 25-27. Data sheets reveal that the heat output for an air-source heat pump varies with outdoor air temperature.

After the building heat loss line is drawn, the heat output for the heat pump is plotted on the graph and a line is drawn. The heat output is found for several different outdoor temperatures. The line represents the heat output for the heat pump over a range of temperatures. The heat output data from the heat pump data sheet provided by the manufacturer is used to plot the heat output. The data sheet shows that the heat output varies with outdoor temperature. Lines can be drawn for several sizes of heat pumps on the same graph. Some heat pump manufacturers provide completed output graphs for their heat pumps. **See Figure 25-27.**

The point on the graph where the heat loss line for the building crosses the heat output line for the heat pump is the balance point temperature for the heat pump. At this temperature, the heat output of the heat pump balances the heat loss of the building.

At temperatures above the balance point temperature, a heat pump heats the building without requiring auxiliary heat. At temperatures below the balance point temperature, auxiliary heat must be used with the heat pump to heat the building. The amount of heat that must be provided by the auxiliary heat source is found by calculating the difference between the two lines at that temperature (in Btu per hour).

UNIT 26

Water-Source and Geothermal Heat Pump Systems

Water-source heat pumps are available in a variety of designs. Residential heat pump systems are typically air-source heat pumps. Light commercial water-source heat pump systems use ponds or wells as heat sinks. Many light commercial heat pump systems use water as part of a geothermal system.

OBJECTIVES

- Explain water-source heat sinks and heat pumps.
- Describe the components of boiler/tower systems.
- Describe the seasonal operations of heat pumps.
- Discuss how heat pump system controls contribute to occupant comfort and building energy efficiency.
- Describe water-source heat pump modes.
- Describe the operation of common types of geothermal ground loops.
- Identify considerations for geothermal loop design.

Index to Procedures

WATER AND AIR HEAT SINKS

Heat sinks are needed for water-source and geothermal heat pump systems. Although residential heat pump systems typically use air heat sinks, water heat sinks provide several advantages for residential and light commercial systems. **See Figure 26-1.** Water heat sinks provide the following advantages:

- Heat can be moved quickly due to the high specific heat of water.
- Water can be easily piped to different parts of a building, allowing heat pumps to be installed some distance from the heat sinks.
- Water temperature can be maintained at a close tolerance. Therefore, the daily temperature of water heat sinks fluctuates much less than that of air heat sinks.
- The use of moderate water temperatures may prevent the need for emergency heat.
- Heat rejected to a loop by a heat pump in the cooling mode can be used by another heat pump in the heating mode, allowing for higher system efficiency.
- They allow heat pumps to be located anywhere inside a building.
- They do not require outdoor units for each heat pump.
- Due to the moderate temperatures of water, the need for outdoor units to have defrost cycles is eliminated.
- They allow valuable floor space to be used for other purposes since the mechanical rooms for water-source systems are small.

TECH TIP

Care must be exercised when installing any heat pump. Always ensure that enough clearance is allowed for maintenance and replacement. Clearance distance is listed in the manufacturer literature.

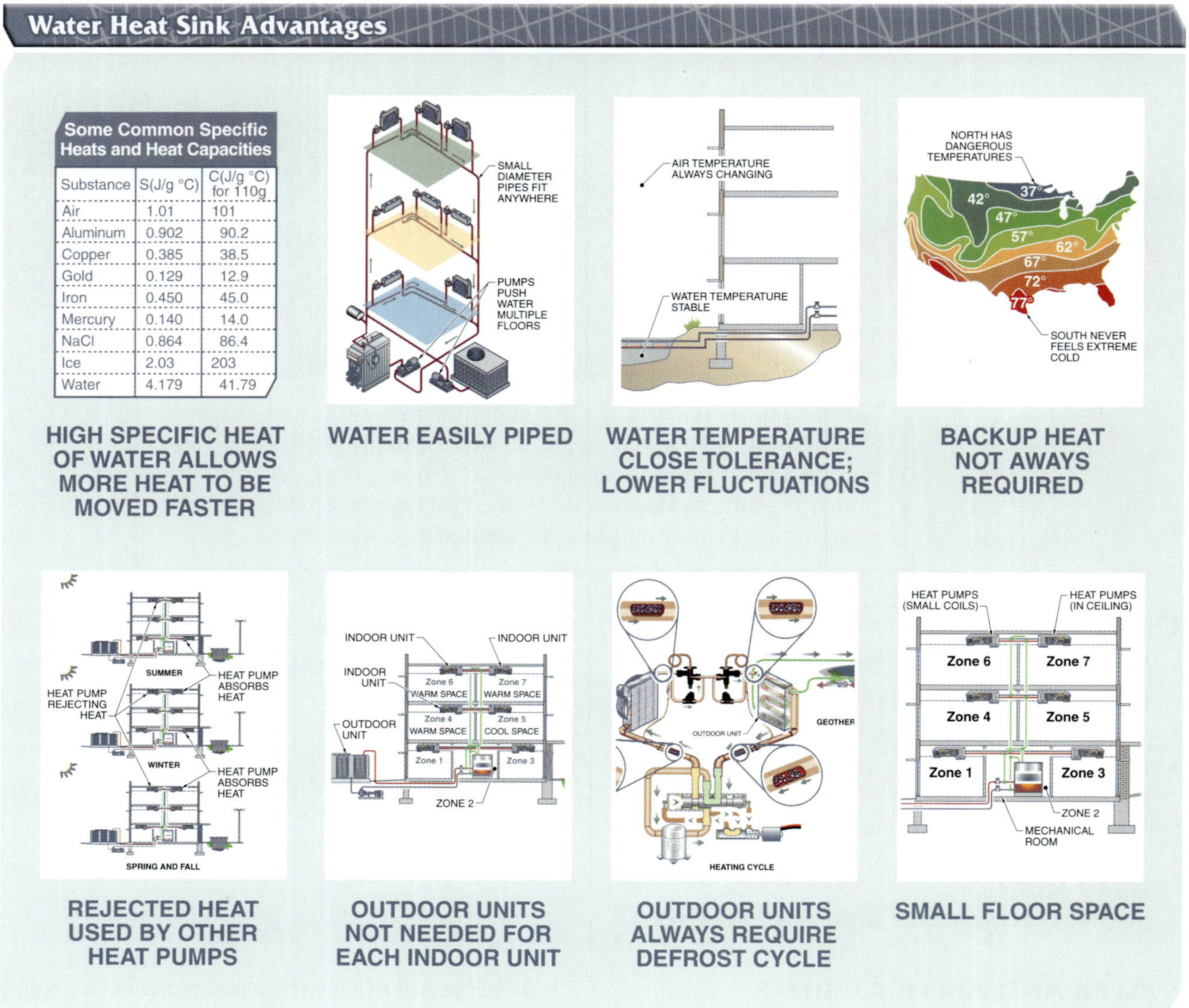

Some Common Specific Heats and Heat Capacities

Substance	S(J/g °C)	C(J/g °C) for 110g
Air	1.01	101
Aluminum	0.902	90.2
Copper	0.385	38.5
Gold	0.129	12.9
Iron	0.450	45.0
Mercury	0.140	14.0
NaCl	0.864	86.4
Ice	2.03	203
Water	4.179	41.79

Figure 26-1. Water heat sinks have many advantages over air heat sinks such as higher specific heat, less daily temperature fluctuations, and higher system efficiency.

Heat Pump Comparison

There are many similarities between water-source and air-source heat pumps. Each type of heat pump contains a compressor, outdoor heat exchanger, indoor unit, and a number of refrigerant lines. A heat pump also contains expansion devices and reversing valves. The expansion devices with bypass circuits allow refrigerant to expand, regardless of the direction of fluid flow. The direction of refrigerant flow depends on whether the heat pump is in the heating or cooling mode. The reversing valves reverse the flow of refrigerant in the refrigerant lines. The controls for water-source and air-source heat pumps are also similar in operation.

The major difference between water-source and air-source heat pumps is that water is used in the refrigerant heat exchangers of water-source heat pumps. **See Figure 26-2.** The heat exchangers are typically of the tube-in-tube design, which have a spiral appearance.

TECH TIP

The installation of water and geothermal heat pumps differs from other HVAC equipment. Some manufacturers provide installation training so personnel can gain experience with these systems.

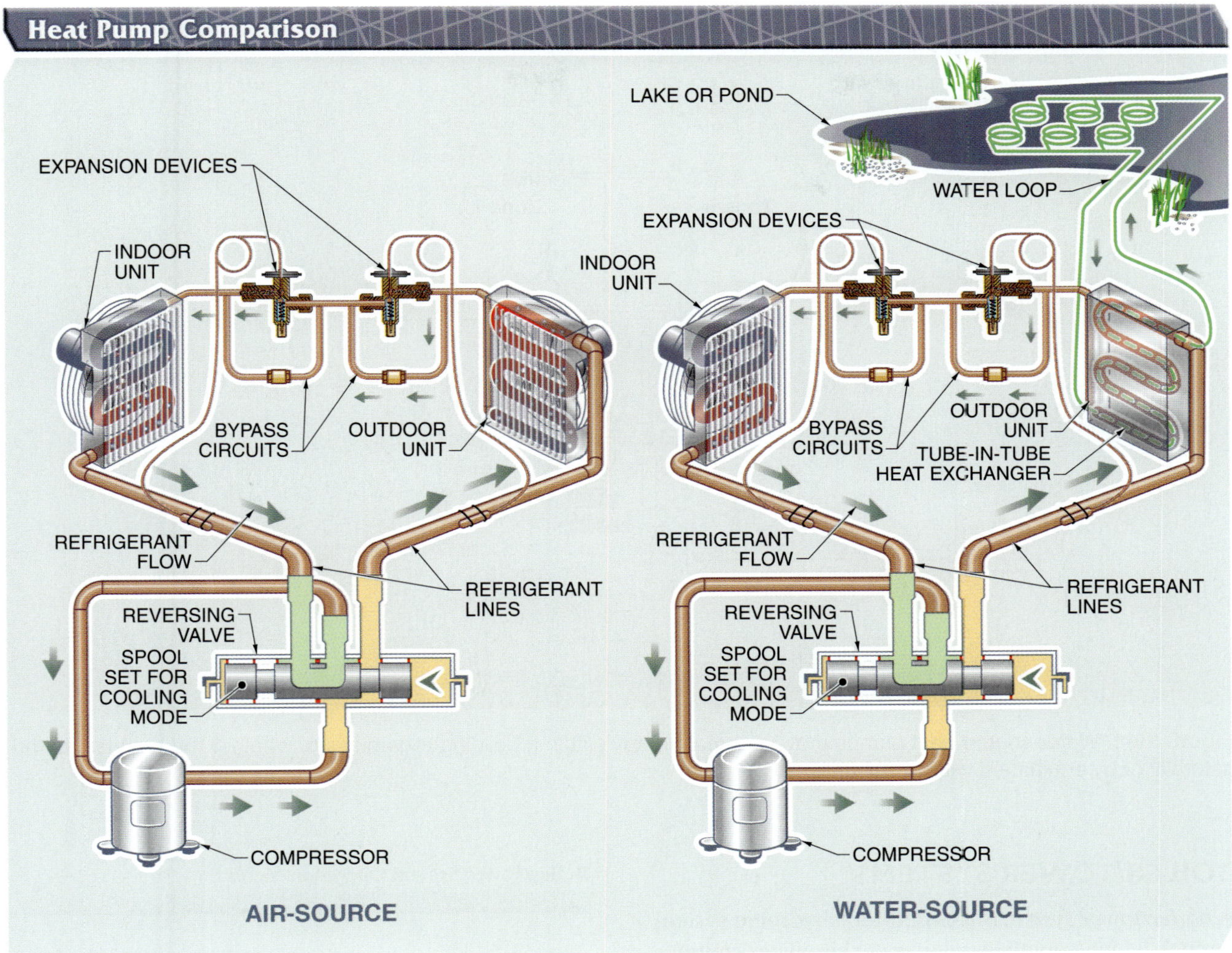

Figure 26-2. The difference between air- and water-source heat pump systems is the refrigerant-to-water, tube-in-tube heat exchanger used in water-source heat pumps.

Water-Source Heat Pump Advantages

In light commercial applications, water-source heat pumps allow HVAC heating systems to be decentralized. Most light commercial buildings are under partial load a majority of the time. The term "diversity" is used to describe the fact that most building zones are not in the cooling mode at the same time. There is a slight possibility that all occupants will be present in the building at the same time or that all lights and heat-producing equipment will operate simultaneously at peak design load. **See Figure 26-3.**

Nearly all central HVAC systems for light commercial buildings have poor partial-load efficiencies. Ideally, a terminal unit in each zone or room provides inherent energy conservation. Each unit heats or cools as required and only to the extent necessary, thus allowing diversity in heating, cooling, and electrical use. By making the terminal units water-source heat pumps and interconnecting them with a closed hydronic loop, energy is transferred from satisfied building spaces to areas lacking sufficient energy.

WATER-SOURCE HEAT PUMP SYSTEMS

Since water is such an efficient heat transfer medium, two types of systems have been developed to obtain water at the right temperature for a hydronic loop. The two types of water-source heat pump systems are boiler/tower systems and geothermal systems.

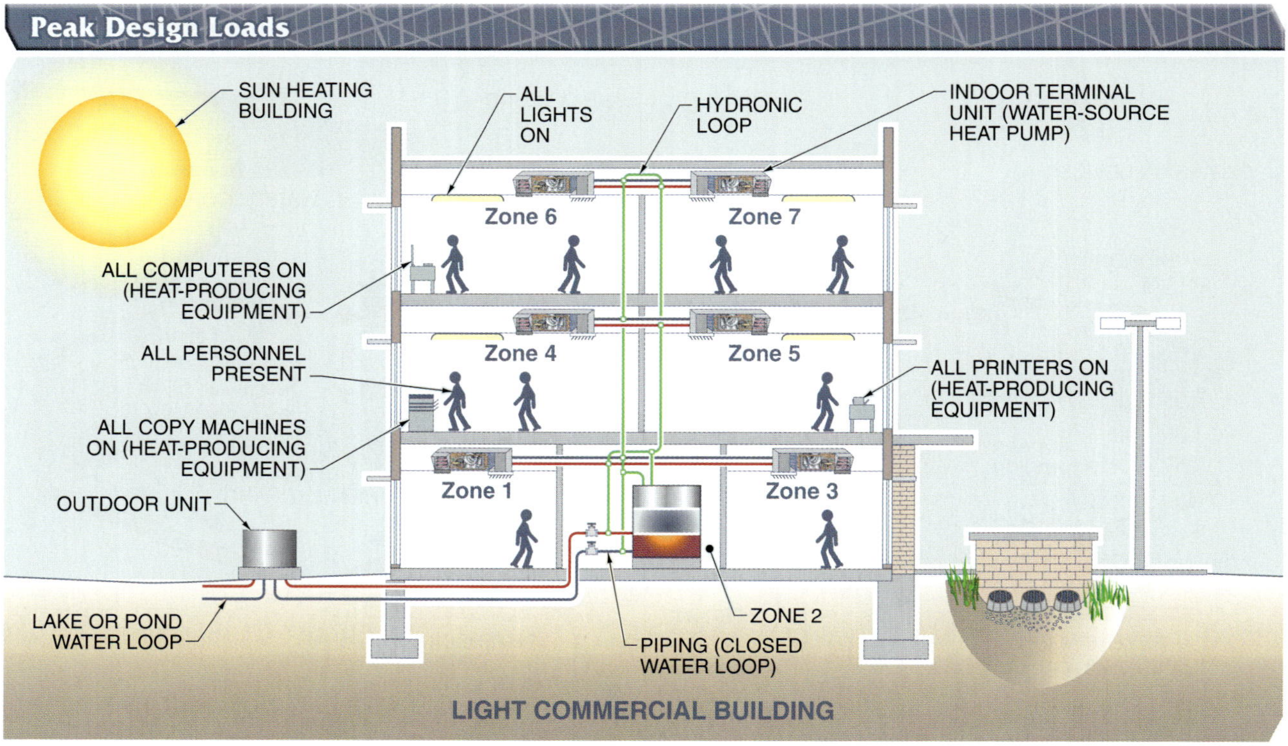

Figure 26-3. Water-source heat pumps move thermal energy within a building regardless of whether the heat is created naturally or by equipment within the building.

BOILER/TOWER SYSTEMS

A *boiler/tower system* is a closed-loop hydronic system through which nonrefrigerated water circulates continuously through a building. Because the piping is within the building where the air temperature is controlled, there is no need for piping insulation. A boiler, used for supplemental heating, adds heat to the hydronic loop at the lower end of the temperature range. A cooling tower, used for supplemental cooling, capable of removing heat at the high end of the range, maintains the hydronic loop temperature throughout the year in an approximate range of 65°F to 95°F (18.3°C to 35°C). **See Figure 26-4.**

This hydronic loop system provides both a heat sink and source of thermal energy. The system conserves energy by pumping heat from warm to cool spaces whenever they coexist in the building. When a space requires heating, the heat pump absorbs heat from the hydronic loop. When a space requires cooling, the heat pump rejects heat to the hydronic loop. **See Figure 26-5.**

This type of system offers the choice of heating or cooling. The occupant may select heating or cooling or choose to shut off the heat pump that serves an individual space without affecting the conditions of other spaces.

Boiler/Tower System Operation Media Clip

Even though most heat pumps have a small capacity, systems with many heat pumps have large cooling towers that include multiple fans.

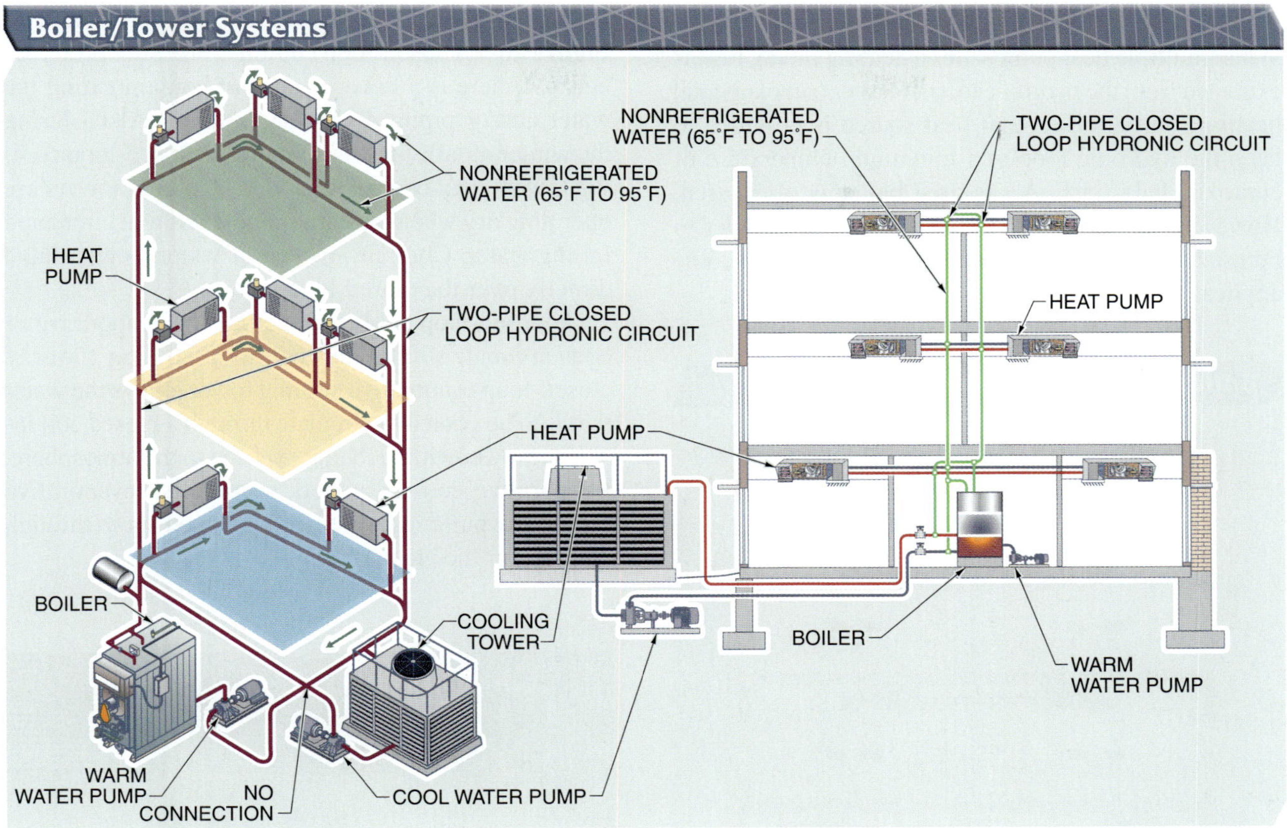

Figure 26-4. Boiler/tower hydronic loop systems use boilers to add heat to the loop and cooling towers to reject heat from the loop as needed.

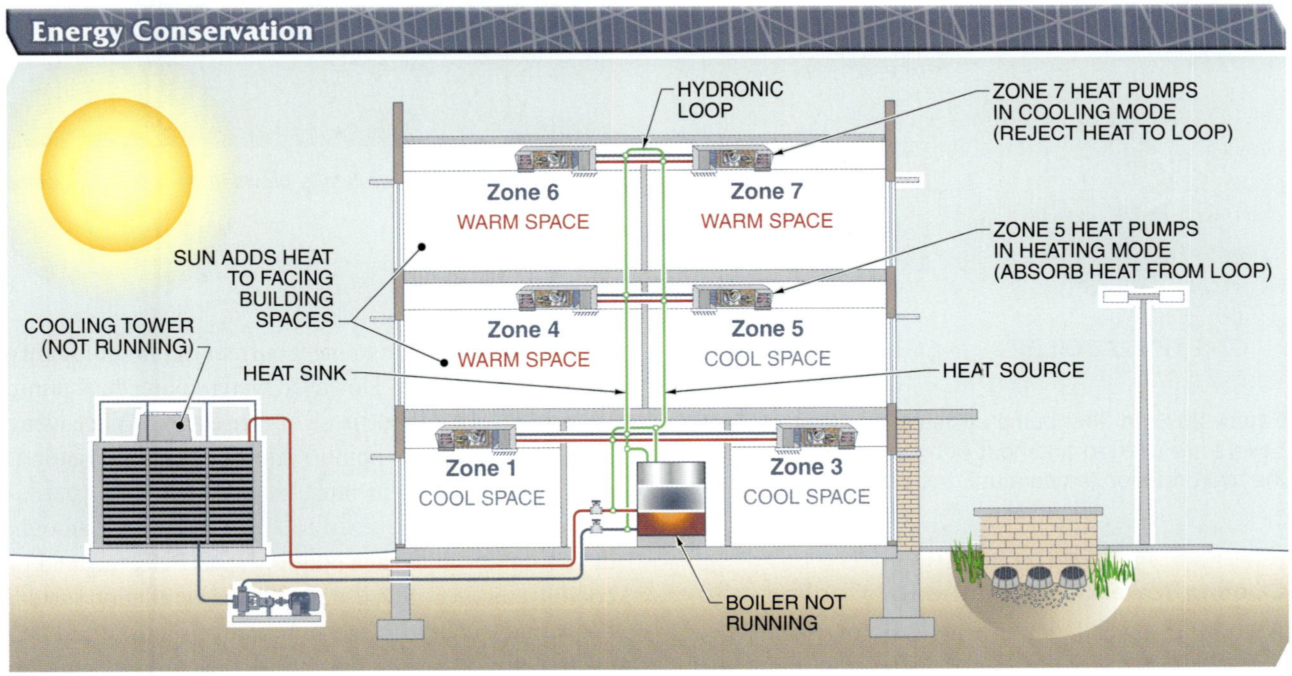

Figure 26-5. Water-source heat pump systems conserve energy by moving thermal energy from warm building spaces and to cool other building spaces by absorbing heat from the hydronic loop to warm building spaces.

Boilers—Supplemental Heating

When multiple heat pumps are in heating mode, heat is extracted from the hydronic loop. A boiler (supplemental heating source) or another heat source is provided to keep the hydronic loop at a minimum temperature of approximately 65°F. A gas-fired boiler is often used. However, it is not uncommon for an electric boiler to be used, since they are more energy-efficient in some applications. **See Figure 26-6.**

Boilers—Supplemental Heat

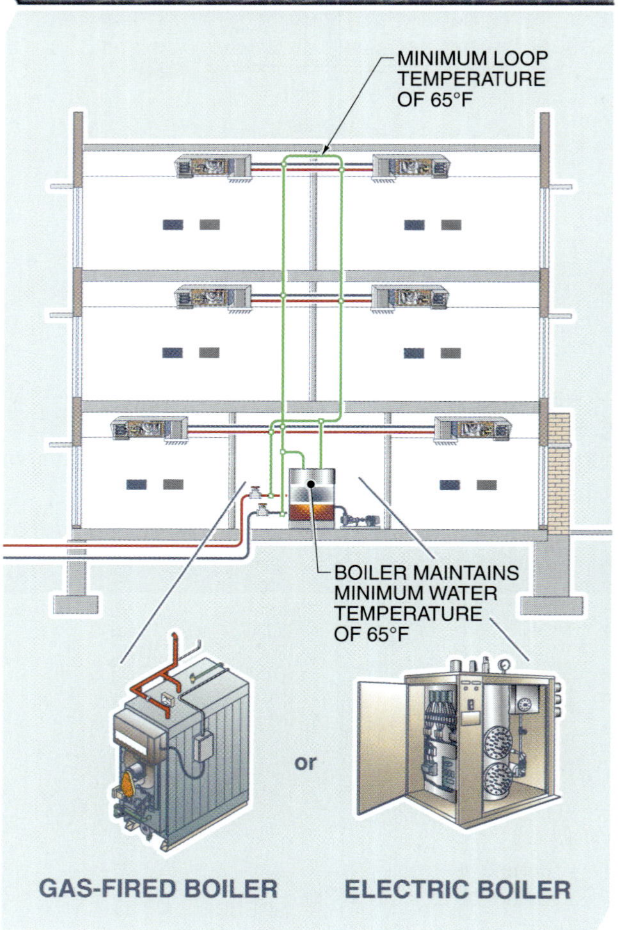

Figure 26-6. In heat pump systems, gas-fired or electric boilers are used to add heat to hydronic loops whenever the hydronic loop temperature drops below 65°F.

Cooling Towers—Supplemental Cooling

A cooling tower (supplemental cooling source) is a type of nonrefrigerated heat rejecter that maintains a hydronic loop temperature of 65°F to 95°F. Typically, the type of rejecter used is a cooling tower. The cooling tower can be of the open- or closed-loop type. **See Figure 26-7.** Closed-loop cooling towers are preferred because there is a lower chance of contaminating the water and/or piping to the heat pumps. Also, during the winter months in some locations, closed-loop cooling towers may be operated "dry." Cooling towers are operating dry when they are running without circulated falling water. Cool air, instead of water, is circulated directly over the closed loop coils.

A closed-loop cooling tower can be evaporative or conventional. Unlike an open-loop cooling tower, a closed-loop evaporative cooling tower allows the water that is to be cooled to circulate through a closed coil inside the tower without being exposed to the atmosphere. Evaporative cooling towers accomplish evaporative cooling by pumping water from an open sump through sprays over the closed coil.

A separate ventilation system is provided on many heat pump installations.

Ventilation

Ventilation is needed to meet current indoor air quality standards and codes. However, water-source heat pump systems do not provide fresh air ventilation. When using a water-source heat pump system, a separate outdoor air ventilation system must be installed. The ventilation system must be sized to allow the proper amount of ventilation air to each building space. Heating and/or cooling coils are often installed in the ventilation system ductwork to condition the air before it is introduced into the building. Controls must be provided to start and stop system fans and the flow of water through the heating and/or cooling coils as needed. **See Figure 26-8.**

Cooling Towers—Supplemental Cooling

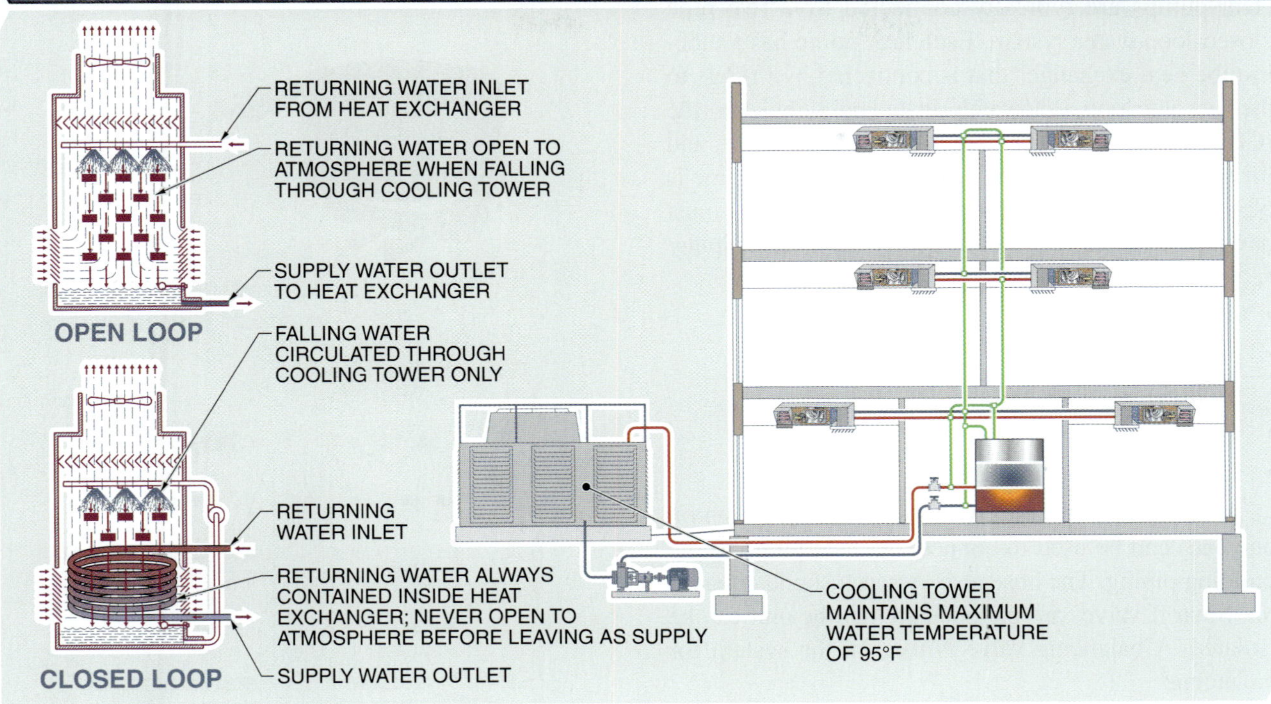

Figure 26-7. In heat pump systems, cooling towers are used to reject heat to the atmosphere whenever the hydronic loop temperature is above 95°F.

Ventilation for Buildings with Heat Pumps

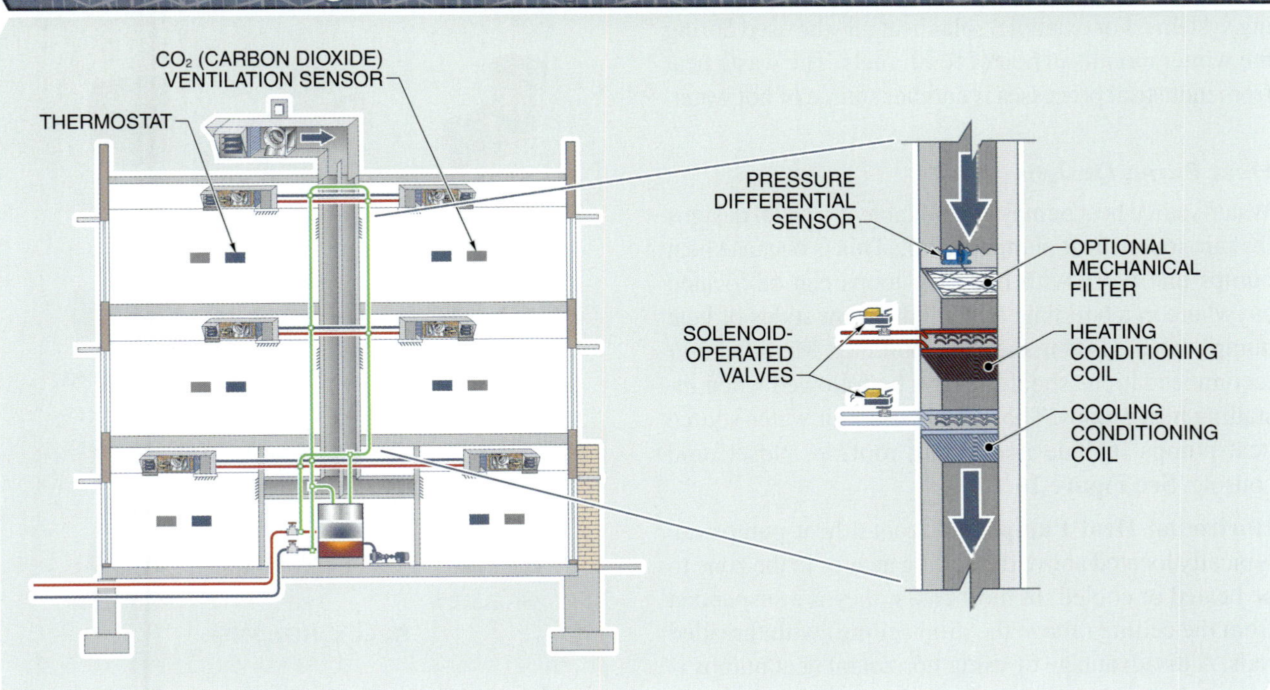

Figure 26-8. Per ASHRAE Standard 62-1999, *Ventilation for Acceptable Indoor Air Quality,* a separate ventilation system must be installed to allow proper airflow into a building when using water-source heat pumps.

Piping

Heat pumps are typically connected by a two-pipe closed-loop water system. Each heat pump has a tube-in-tube heat exchanger that is connected by fittings to the hydronic loop. Heavy chlorinated polyvinyl chloride (CPVC) piping is acceptable in some jurisdictions and for some applications, but copper or black iron pipe is typically required. **See Figure 26-9.** The following must be considered before installing heat pump system piping:

- piping layout
- required flow rates
- pipe size of each unit
- friction loss of each unit
- pump size
- accessories to be used

A short, flexible hose (with a swivel connection on one end) can be used to connect the heat pump to the building piping. The hose also acts as a sound attenuator. Shutoff valves are included so that the unit can be isolated. A balancing valve is used in the system for balancing.

An advantage of water-source heat pumps is that the piping system may be equipped with stubs for future expansion. In most cases, the determinations for piping have already been made by an engineering firm.

Variations. There are many variations of heat pump piping systems. For example, solar heat may be used during the winter months in lieu of fossil fuels. The waste heat from industrial processes is another source of hot water.

Heat Pump Designs

Water-source heat pumps are available in several designs that air-source heat pumps are not. This is because heat pumps that work with hydronic loops can be located anywhere in a building. Several different styles of heat pumps can be used in the same building. Manufacturer recommendations should always be followed when installing heat pumps. The designs used for water-source heat pumps include horizontal, rooftop, closet, and console. **See Figure 26-10.**

Horizontal Heat Pumps. Horizontal heat pumps are typically located above the ceiling near or in the zone to be heated or cooled. In most cases, they are suspended from the ceiling (above the drop ceiling) with threaded rods. A disadvantage of using horizontal heat pumps is that performing maintenance tasks, such as replacing compressors, may be difficult. Such maintenance tasks require the use of a lift.

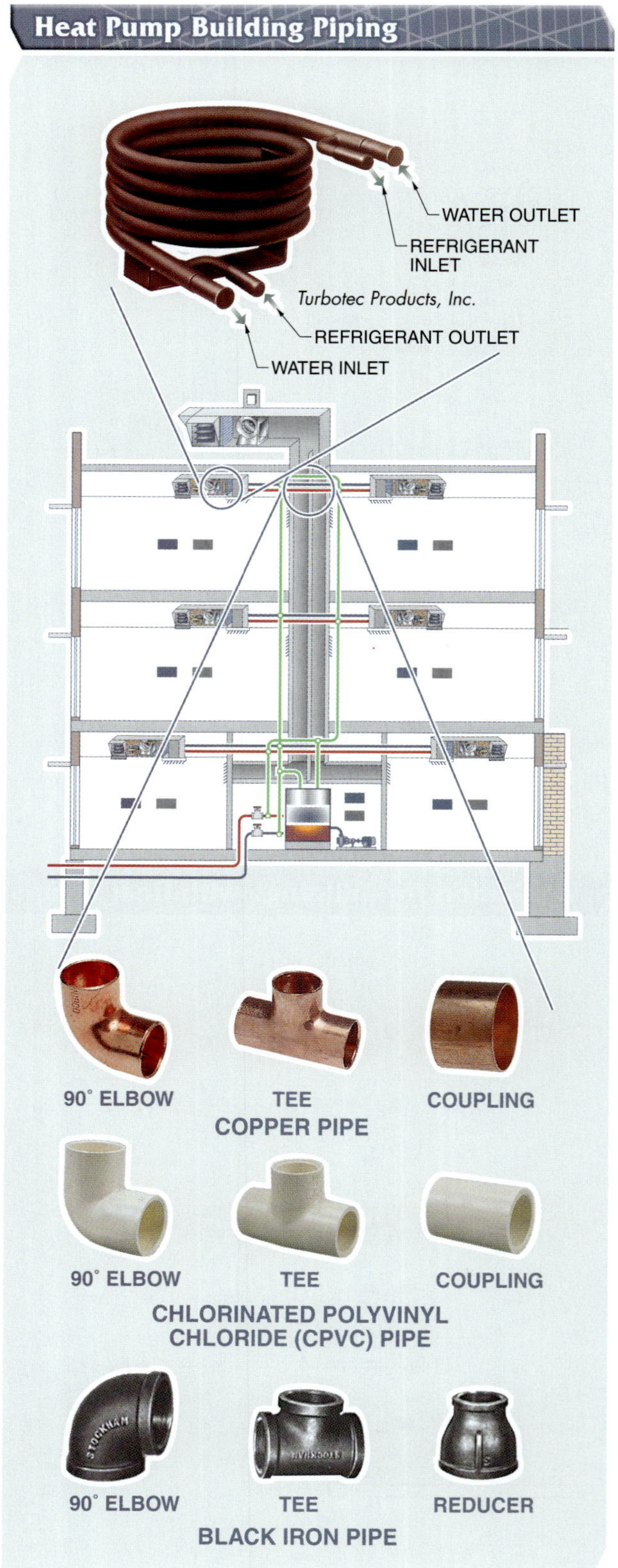

Figure 26-9. In most residential buildings, copper is used for the heat pump system piping, but CPVC or black iron pipe is used in many light commercial buildings.

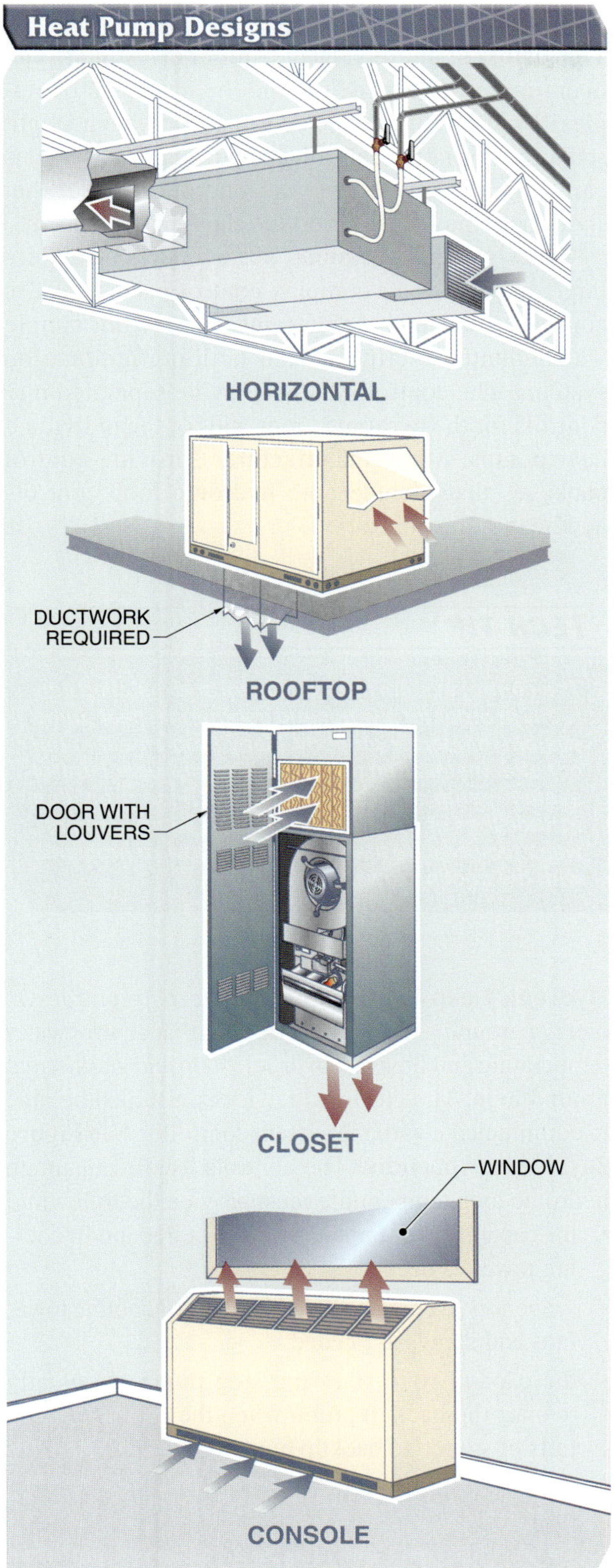

Figure 26-10. Various designs of heat pump systems are available to fit a particular application for increased system efficiency.

Rooftop Heat Pumps. Large heat pumps are available as rooftop systems. Rooftop heat pumps are useful for retrofitting conventional rooftop air-conditioning units. Rooftop heat pumps are used for retrofits because the heat pumps may be placed in the same locations as the old rooftop units.

Closet Heat Pumps. For closet heat pumps, small closets are designed into the building floor plans. Consideration must be given to adequate sizing and noise issues. One advantage of closet heat pumps is that they are typically easy to access for service, assuming adequate clearance is given for maintenance.

Console Heat Pumps. Console heat pumps resemble the units used in hotel rooms. Console heat pumps are used where unit ventilators or fan coils are retrofitted into older buildings. As with the rooftop design, it may be possible to fit a new water-source heat pump in the opening made by the removal of the old unit.

HEAT PUMP SEASONAL OPERATIONS

Water-source heat pumps have operating modes for the summer, winter, spring, and fall. With most units operating in cooling mode during hot weather, the heat removed from the air is transferred to the hydronic loop. The cooling tower rejects the excess heat to the outdoor air to maintain a maximum water loop temperature of 95°F.

With most units operating in heating mode during cold weather, it is necessary to add heat to the water loop with some type of water heater (boiler). This is done when the temperature of the water loop falls to 65°F. The amount of heat in the hydronic loop is reduced any time a unit operates in the heating mode.

The source of heating the water is never sized larger than two-thirds the size normally required because heat pumps in the core of the building continuously operate in the cooling mode. Heat pumps in cooling mode are constantly rejecting heat to the hydronic loop. **See Figure 26-11.**

During the spring and fall, the water-source heat pumps serving the shady side of a building are often in heating mode, while those serving the sunny side of the building are in cooling mode. When approximately one-third of the units in operation are in cooling mode, a sufficient amount of heat is added to the hydronic loop. Neither the addition of heat to or the rejection of heat from the water is required. Therefore, a significant amount of energy can be saved during the spring and fall.

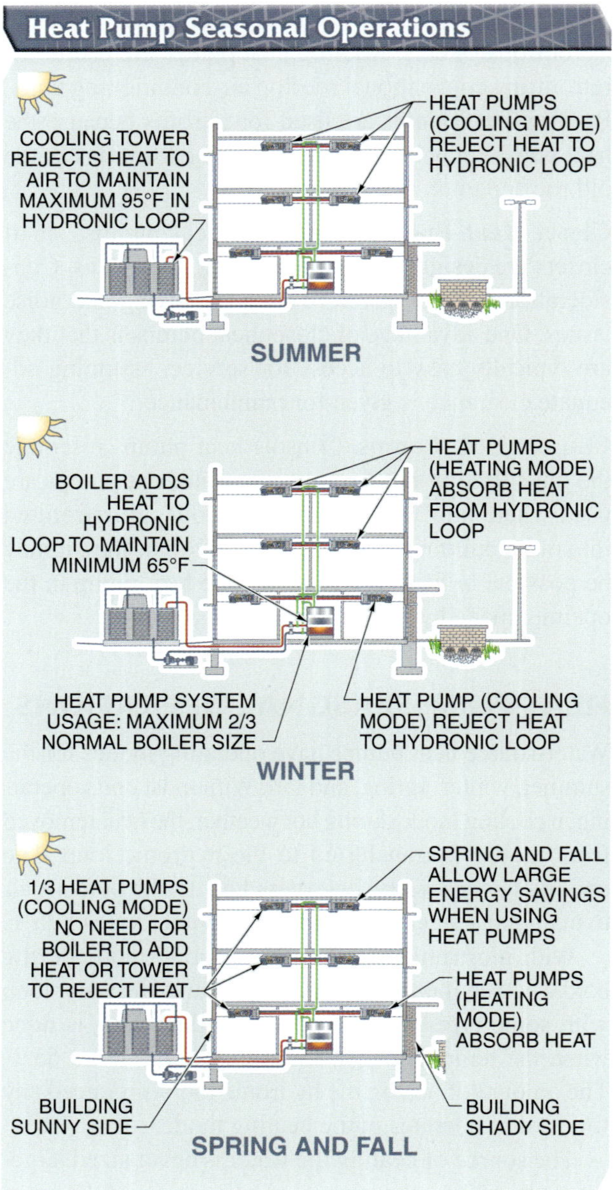

Figure 26-11. Heat pump systems are the most energy efficient in the spring and fall when thermal energy is moved around within a building and there is no need to add to or reject heat from the hydronic loop.

Office buildings with high heat gain from lights, occupants, or equipment in interior areas may require year-round cooling of certain spaces. Heat taken from hot areas is rejected to the hydronic loop, providing enough heat for the building perimeter any time at least one-third of all heat pumps operate in cooling mode. Energy savings are the greatest during the spring and fall since heat is circulated through the building by the water loop and heat pumps.

Controls

The factors of building usage, initial investment, and operating efficiency determine the nature and complexity of the temperature controls of a system. Night setback, ventilation control, and other control systems can dramatically reduce the operating costs that shorten the payback period of a heat pump system.

Water-source heat pump systems do not require large, expensive, or complex control packages. The controls used in these systems range from simple wall-mounted thermostats to building automation systems. The control systems may be separated into controls for the hydronic loops and for the individual heat pumps. Many manufacturers provide control packages that include both hydronic loop controls and heat pump controls.

> **TECH TIP**
>
> ***The control system must be set up and installed correctly in order for a water and geothermal heat pump system to operate properly. Many problems that occur after installation are caused by improper control system setup and/or installation. These problems include loss of water flow due to the pump being off, water temperature that is too high or low, or problems with the cooling tower and/or boiler.***

Hydronic Loop Controls. A *hydronic loop control system* is a grouping of electrical devices that control water temperature and flow in a hydronic loop and geothermal ground loop. The electrical devices must also be able to communicate with the system controller. **See Figure 26-12. See Appendix.** The controls used to maintain hydronic loop temperature must provide the following:

- the capability to start and stop the boiler and/or cooling tower as needed
- sequences to energize several boilers or cooling tower fans and sprays as needed
- the capability to start and stop pumps (typically to start the backup pump when the primary pump fails or switch to backup pump operation when the primary pump has operated for a specified number of hours)
- the ability to open and close heat exchanger and other system valves
- alarms to alert personnel when there is a problem with the water loop

Water Loop Controls

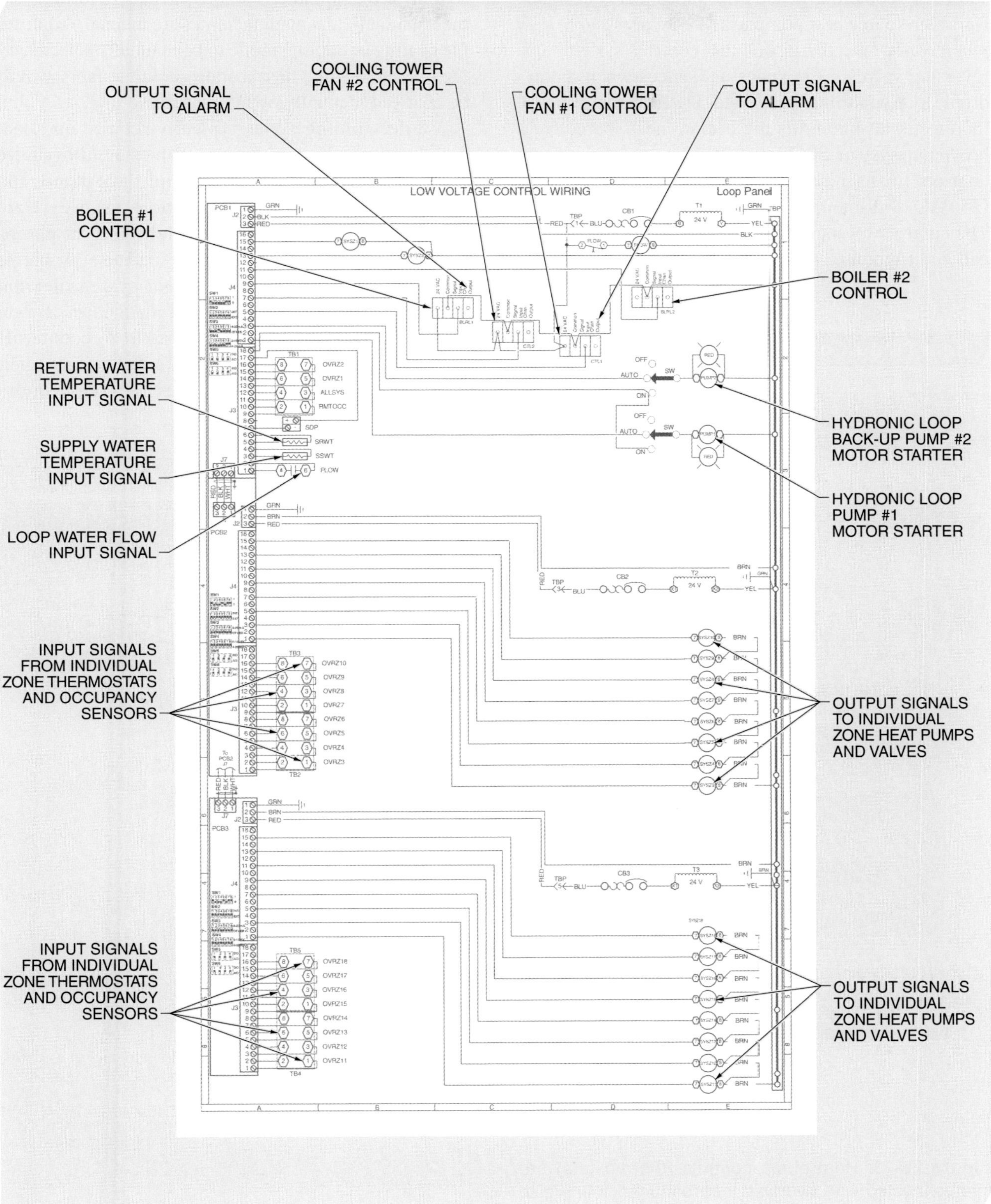

Figure 26-12. Hydronic loop controls receive temperature, pressure, and gas input signals to energize or de-energize boilers, cooling towers, pumps, fans, valve solenoids, and alarms.

Heat Pump Controls. The control system for a water-source or geothermal heat pump is specifically designed for water-source heat pump usage. A *water-source heat pump control* is a thermostat that controls system fans, reversing valves, and expansion devices when the hydronic loop controls are activated. The heat pump control thermostat also controls the backup heat source for a heat pump system. **See Figure 26-13.** Space temperature controls for terminal units are typically low-voltage (24 VAC) wall-mounted special heat pump thermostats. The controls for under-window console units are typically unit-mounted thermostats.

Typically, the thermostats used in office buildings and schools are automatic. The thermostats used in apartments, motels, and nursing homes are manual, requiring the heating or cooling mode to be manually selected by the operator. Some thermostats allow the fan speed to be changed manually or automatically.

While building zones typically require one heat pump with one thermostat, sometimes buildings have at least one large zone where multiple heat pumps and thermostats are required to condition the space. Care must be taken when installing multiple heat pumps and thermostats in a single zone because problems will result from improper control system design. Multiple automatic thermostats in a large single zone must be spaced far enough apart to counteract the tendency of the heat pumps to fight each other, which wastes energy and produces poor temperature control. **See Figure 26-14.**

Heat Pump Controls

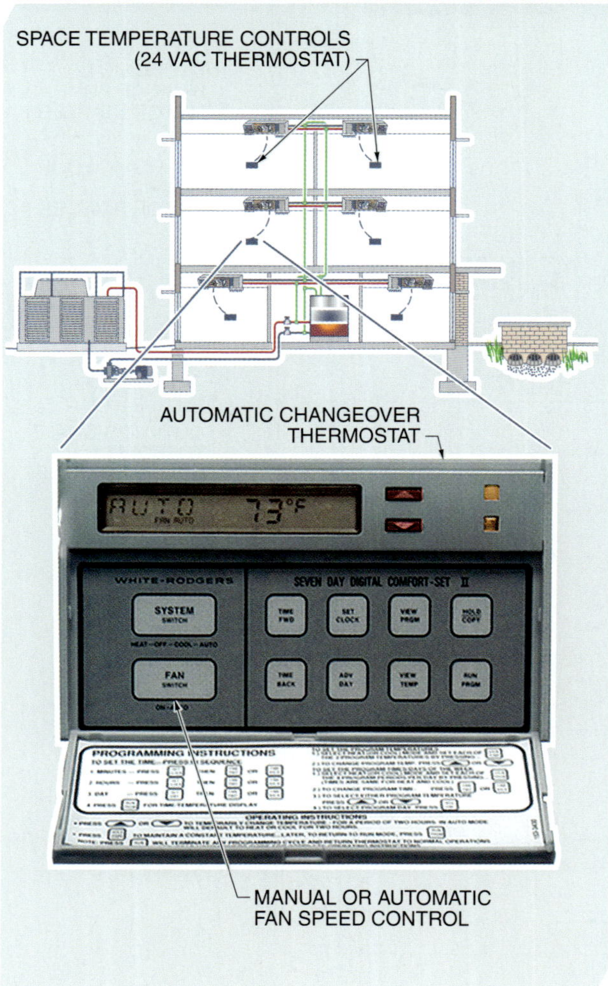

Figure 26-13. Heat pump controls (thermostats) are interconnected with hydronic loop controls to energize and de-energize heat pumps, control internal fan speeds, and allow automatic changeover between heating and cooling modes.

Large Building Zones

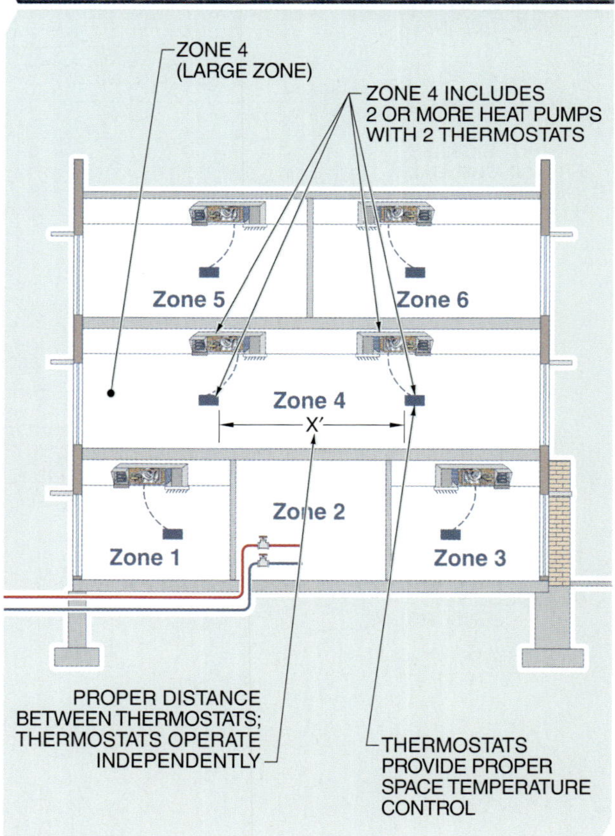

Figure 26-14. Maintaining a proper distance between heat pumps and between thermostats is critical for maintaining proper temperatures in large building spaces.

Heat Pump Reset Controls. Thermostat cycling should never be used to reset heat pumps after an operational fault. The disconnect switch or circuit breaker provides the only acceptable method of protecting equipment and the building from possible damage due to resetting HVAC equipment that is not in operating condition. The power disconnect method of reset always involves maintenance or service personnel.

When a thermostat is routinely reset by unqualified persons, changes in the perceived load requirements of the conditioned space can occur, which could make minor service problems into major repairs. To prevent the problems associated with thermostat resetting, some public area thermostats include locked guards to prevent tampering.

Heat Pump Control Functions

A heat pump is a part of an HVAC system, which typically includes a central control system. **See Figure 26-15.** Central control systems control all HVAC system functions and heat pump modes, including the following tasks:

- restarting all heat pumps from a central location after a general shutdown
- stopping all heat pumps from the same central location
- randomly restarting heat pumps to limit instantaneous current demand to a reasonable minimum
- real time overrides that keep shut off heat pumps available to occupants working after hours so their respective units can be restarted as desired
- overriding the signal to maintain continuous occupied (normal) or unoccupied control by permitting the control of each zone to occur from a central location or computer
- scheduling individual heat pumps to include a morning warm-up or cool-down period

TECH TIP

For maximum efficiency, the equipment in a water and geothermal heat pump system must be sequenced properly. This is especially true when starting equipment after a weekend, when the building was unoccupied. In this case, the heat pumps may all be started at the same time and call for maximum heating or cooling. This causes the temperature of the water loop to spike until the cooling tower or boiler can catch up. Often, a time delay is imposed between heat pump startups.

Heat Pump Control Functions

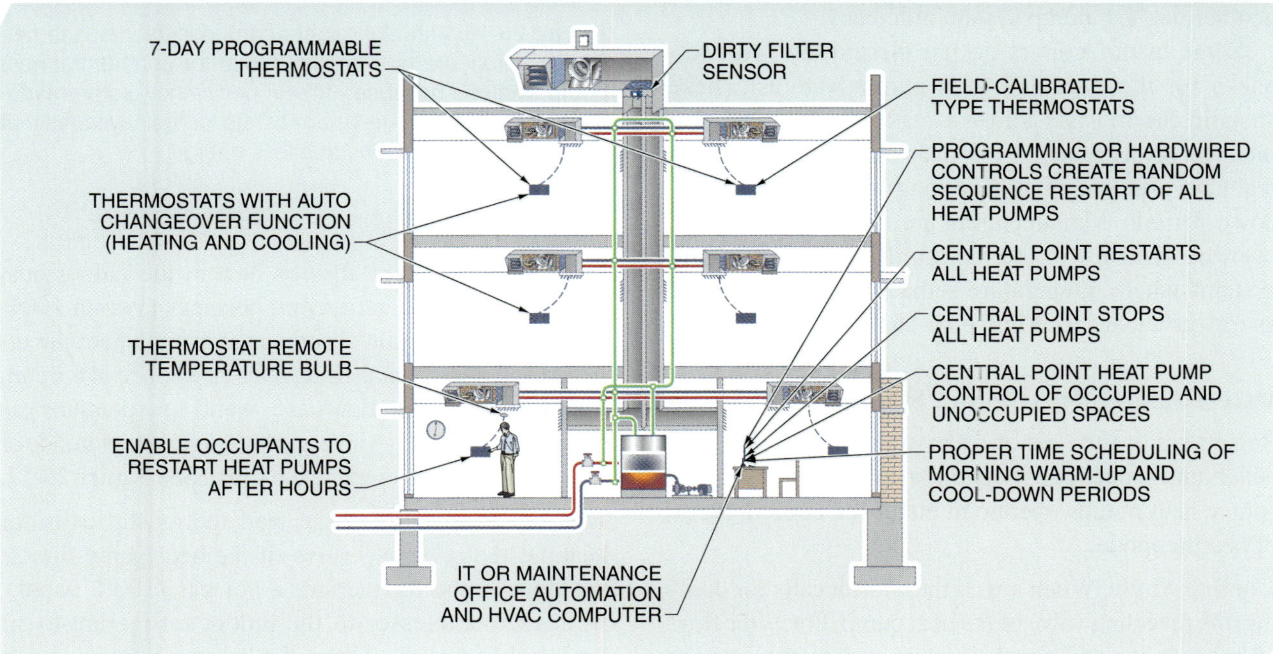

Figure 26-15. Thermostats, hydronic loop controls, and building automation systems (central control) are all connected to control heat pumps for proper indoor air quality and energy efficiency.

Seven-Day Programmable Thermostats. For heat pumps to be centrally controlled, seven-day thermostats must be used. A seven-day thermostat has a seven-day programmable schedule, an automatic changeover function between heating and cooling, and the capability to be calibrated in the field. Some thermostats may require remote temperature sensing capabilities, which measure the temperature of discharge air from registers. Some thermostats are equipped with filter sensors that indicate the need for filter maintenance.

Night Setback Control. Night setback and setup typically reduces energy consumption more than any other single HVAC feature. During setback, building space temperatures are allowed to decrease at night during the heating mode of the fall and winter seasons. During setup, building space temperatures are allowed to increase at night during the cooling mode of the spring and summer seasons. Up to 30% energy can be saved using this strategy versus simply allowing heating and cooling to take place at occupied set points during unoccupied periods.

However, night setback for geothermal heat pump systems is not recommended. Geothermal systems use the temperature of the earth to heat or cool the water of buildings. Recent studies show that a night setback of 2°F reduces the overall efficiency of a geothermal heat pump system by 25%. This is due to the use of back-up heat to get the building space back to occupied temperatures. Essentially, the use of back-up heat reduces geothermal heat pump system efficiency.

Some manufacturers design thermostats with setbacks for their specific heat pump systems. These specific thermostats allow a maximum temperature increase or decrease of 1°F over time. These systems maintain efficiency but have long warm-up and cool-down periods. Manufacturers are also adding setback override switches to thermostats for their heat pump systems where temperature setback is not considered energy efficient. **See Figure 26-16.**

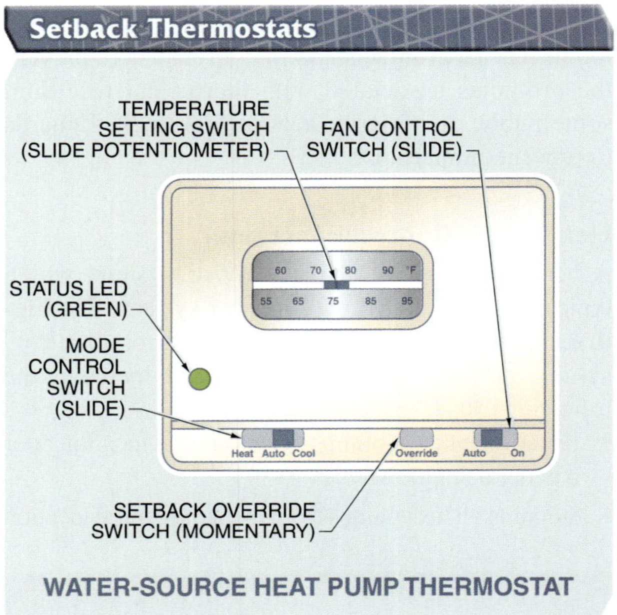

Figure 26-16. Although night and unoccupied-space setback controls are not recommended for geothermal heat pump systems because of their inefficiency, some manufacturers have designed thermostats for their systems that allow the system to consume less energy.

Water-Source Heat Pump Modes

Regardless of the type of water-source heat pump, the heat pump sequences of operation are the same. Water-source heat pumps operate in either the cooling mode or heating mode.

Cooling Mode. When a wall thermostat calls for cooling, the reversing valve of the heat pump directs the flow of hot refrigerant from the compressor to the outdoor refrigerant-to-water heat exchanger coil. The hot refrigerant, often referred to as "hot gas," is 100% vapor when leaving the compressor. The heat is removed by the geothermal water circuit, and the hot refrigerant condenses to become a cool refrigerant that is 100% liquid. The liquid refrigerant then flows through a thermal expansion device or valve to the indoor air-to-refrigerant heat exchanger coil. Refrigerant flow through the expansion device or valve causes the refrigerant to enter a low-pressure area and become 80% liquid.

As the refrigerant absorbs heat in the indoor heat exchanger coil, the refrigerant becomes a warm 100% gas. As the refrigerant absorbs heat from the air, the refrigerant cools the air passing over the surface of the coil. The refrigerant then flows as a warm low-pressure gas through the reversing valve and back to the suction side of the compressor to complete the cycle. **See Figure 26-17.**

Heating Mode. When the wall thermostat calls for heating, the reversing valve of the heat pump directs the flow of the refrigerant, a hot gas (100% vapor), from the compressor to the indoor refrigerant-to-air heat exchanger coil. There, the heat is removed by the air passing over the surface of the coil, and the hot gas condenses and becomes 100% liquid.

Cooling Mode Operation

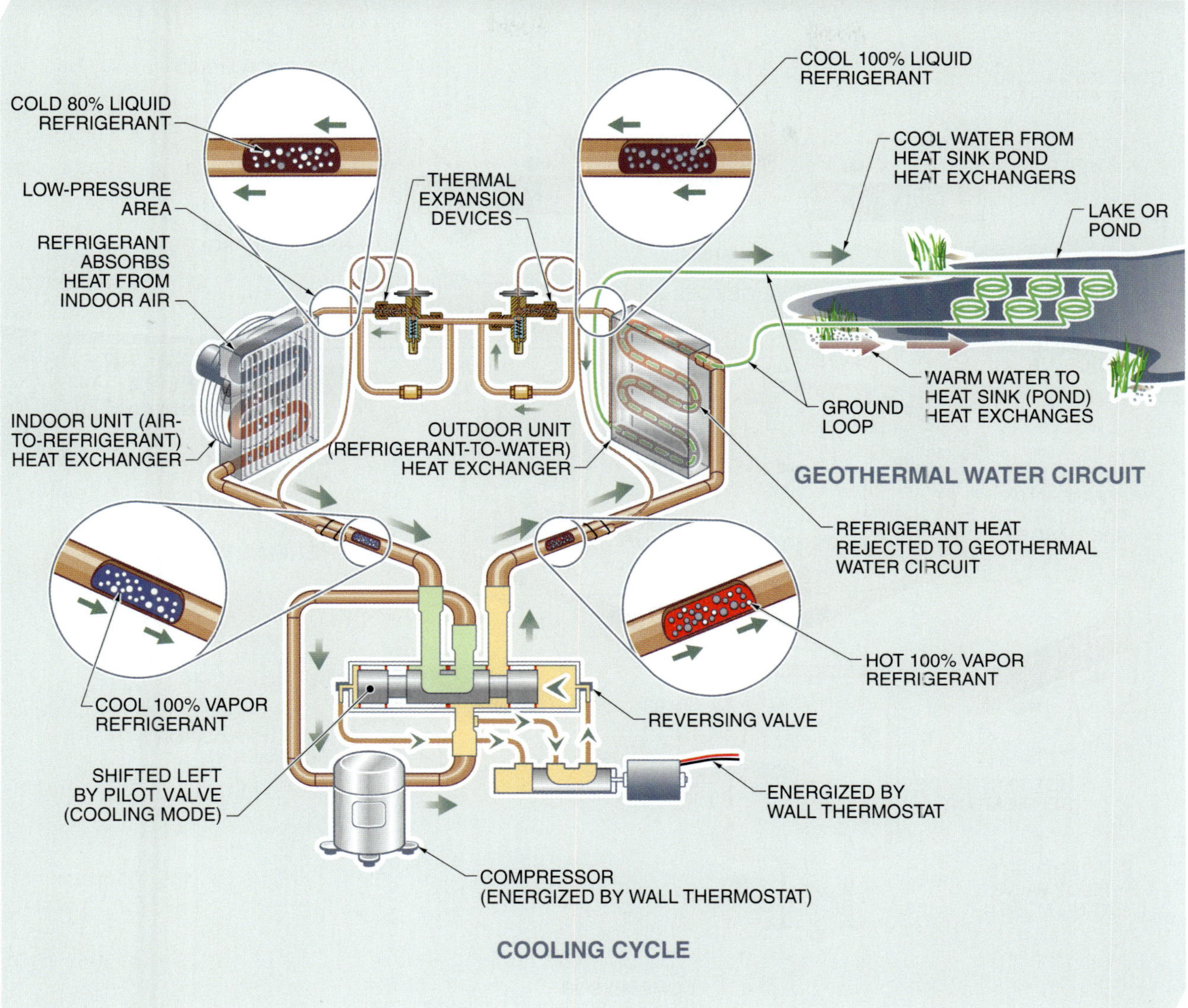

Figure 26-17. During cooling mode, the refrigerant of a water-source heat pump system absorbs heat from building spaces and moves the heat to the water loop and then to the ground loop to be rejected to the earth.

The liquid refrigerant then flows through a thermal expansion device or valve to the outdoor water-to-refrigerant heat exchanger. The 80% liquid then absorbs heat from the geothermal water loop, becomes a 100% gas, and cools the geothermal water loop. The refrigerant then flows as a low-pressure gas through the reversing valve and back to the suction side of the compressor to complete the cycle. **See Figure 26-18.**

Water-Source Heat Pump Operation Media Clip

GEOTHERMAL SYSTEMS

A *geothermal system* is a water system that uses the temperature of the earth to heat or cool the water loop of a building. The term "geothermal" means heat of the earth. Geothermal systems use water loops and ground loops to heat buildings. A water loop is located inside the building, while a ground loop is located outside the building in the earth (ground or groundwater). The water loop of a geothermal heat pump directs water through the ground loop where heat exchangers are buried in the earth.

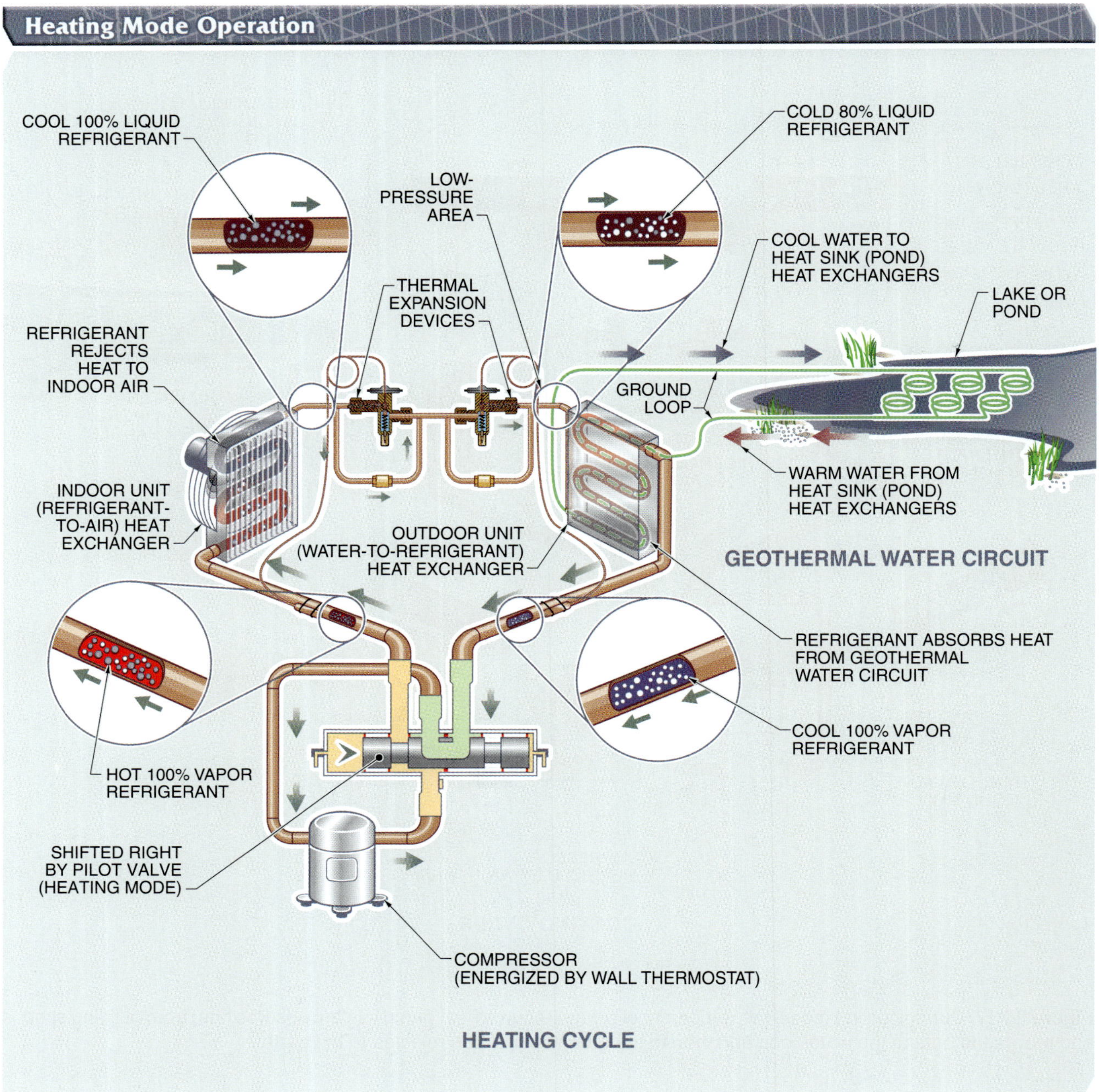

Figure 26-18. During the heating mode, the ground loop absorbs heat from the earth and moves the heat to the water loop and then to the refrigerant to be rejected into building spaces.

GEOTHERMAL GROUND LOOPS

A *geothermal ground loop* is piping that is used to transfer heat either to or from the earth. The earth alone is not a perfect heat sink/source because the energy added to the earth by the ground loop can change the temperature of the sink/source over time. Geothermal ground loop systems are available as open loop and closed loop systems.

TECH TIP

Geothermal heat pumps are popular in parts of the country that experience mild weather conditions. Geothermal heat pump manufacturers provide training and technical support for contractors installing these systems. These systems must be installed in accordance with all local codes.

Geothermal Ground Loop Systems Media Clip

Open Loop Systems

An *open loop system* is a type of geothermal ground loop system that draws groundwater from a well directly into a building where it is used by heat pumps to heat or cool the building spaces and returns the water to the earth through a second well. The second well is not connected by any means to the first well. Open loop systems require sufficient groundwater from the supply well (approximately 3 gpm/ton) to meet the needs of the building. The temperature of the well water entering an open loop system should be above 40°F to prevent freezing. Optimum water temperature is between 50°F and 60°F. **See Figure 26-19.** Open loop systems are most common in residential and light commercial applications.

Open Loops

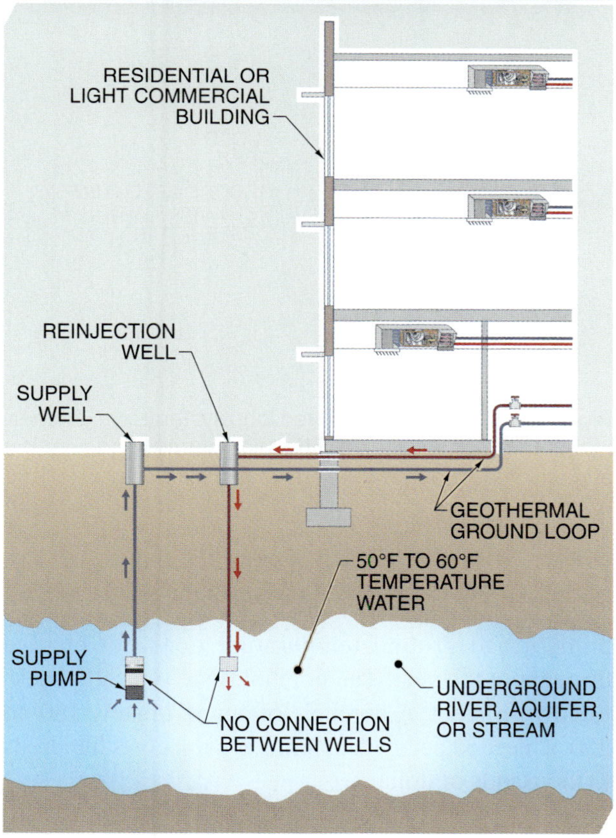

Figure 26-19. Open loop systems are geothermal ground loop systems that draw groundwater from wells into buildings to be used for heating and/or cooling. These systems must also allow for the increased pumphead from the well water level to the height of the heat pump.

Groundwater often contains minerals and other contaminants that create problems for the HVAC equipment. Filtration equipment or secondary heat exchangers are required to filter out the minerals and contaminants in the well water used by open loop systems.

For the open loop system to operate properly, an engineer must be consulted to design the piping system for both the supply well and reinjection well. Also, the system must comply with all applicable federal, state, and local codes as well as regulations from the Environmental Protection Agency (EPA) regarding groundwater use.

Closed Loop Systems

A *closed loop system* is a type of geothermal ground loop system that circulates water through piping in the ground to a water loop in order to exchange thermal energy. Ground loop water and building loop water do not mix. The ground loop is a heat exchanger similar to a cooling coil or an evaporator in a chiller. The goal of a closed loop system is to transfer thermal energy from the ground loop water to or from the ground.

The process of loop design involves estimating the required loop length. Computer software is typically used to estimate loop length, but a basic understanding of the design process is helpful. The heating and cooling loads provide a designer with the thermal energy transfer rates for sizing the loop. The supply fluid temperatures must be estimated. The larger the loop for a known load, the cooler the supply fluid temperature will be. The types of closed loop systems include horizontal, vertical, and surface-water closed loop systems.

Circulating pumps provide water flow through the hydronic loop.

Horizontal Closed Loop Systems. A *horizontal closed loop system* is a type of geothermal ground loop system with piping parallel and close to the surface of the ground that circulates water to exchange thermal energy. The undisturbed ground temperature often changes seasonally depending upon where the loops are installed. Horizontal closed loop systems are easier to install but require significantly more area (approximately 2500 ft²/ton) than other loop systems. Areas such as parking lots must be avoided.

Horizontal closed loop systems should be designed and installed by certified contractors. Loops are typically installed in trenches 3′ to 5′ deep. Deeper trenches may require sidewall supports, which increase the total installation cost. In some applications, several circuits can be installed one on top of the other in the same trench. **See Figure 26-20.** The trenches range from 150′ in length/ton to 220′ in length/ton.

Placing loops in an undesirable location lowers system efficiency and increases the total pipe length required. Loops should not be placed in shallow depths where the undisturbed ground temperature changes with the seasons. The typical loop temperature operating range is 35°F to 95°F for northern locations and 40°F to 100°F for southern locations. Horizontal closed loop systems are typically easier to install than vertical and surface-water closed loop systems.

TECH TIP

When installing a geothermal ground loop, a contractor or personnel that have experience in installation should be used. Otherwise, training regarding the installation of these systems should be factored in. The additional cost can ensure a quality installation and can help prevent problems arising from improper installation and any legal problems.

Horizontal Closed Loops

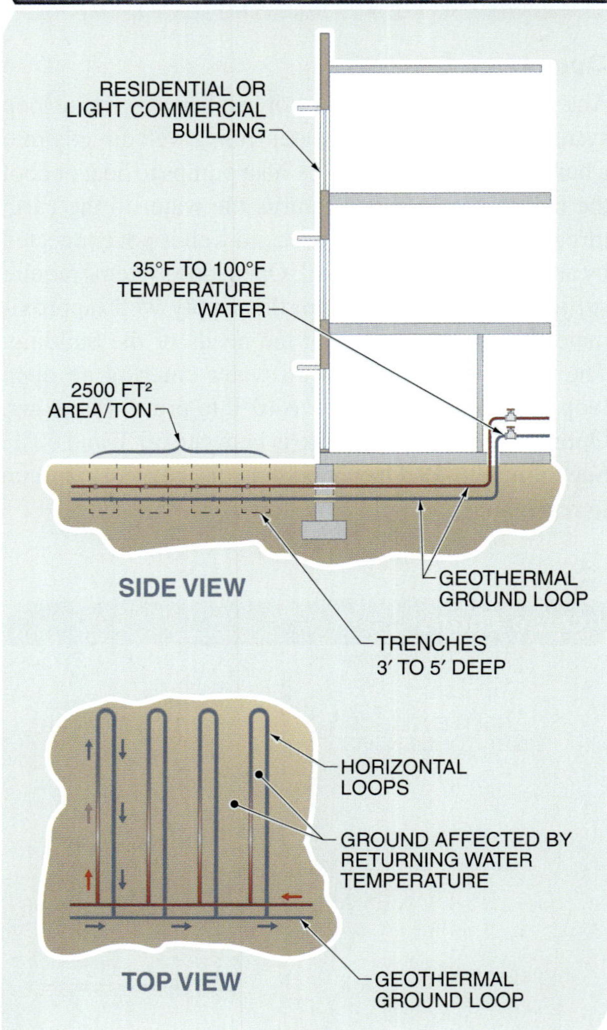

Figure 26-20. Horizontal closed loop systems are geothermal ground loop systems with piping parallel and near the surface of the ground. They are used to circulate water and exchange thermal energy.

Vertical Closed Loop Systems. A *vertical closed loop system* is a type of geothermal ground loop system in which U-tube ground loops that circulate water are inserted into holes drilled perpendicular to the surface in order to exchange thermal energy. Vertical closed loop systems use less land than horizontal closed loop systems. Vertical closed loop systems require approximately 250 ft² of area/ton to 300 ft² of area/ton. A certified contractor should be used to install a vertical closed loop system, and all applicable codes and regulations must be followed.

Vertical closed loops are installed in holes, known as boreholes, 200′ to 400′ deep. At these depths, the undisturbed ground temperature does not change throughout the year. The boreholes are typically backfilled with grout. A borehole of approximately 180′ to 250′ is needed per system ton. Each ton costs hundreds to thousands of dollars.

The undisturbed ground temperatures at the depth in which vertical systems operate remain constant throughout the year. Typical loop temperature operating range is 35°F to 90°F in northern climates. In southern climates, loop temperatures can reach over 100°F. **See Figure 26-21.**

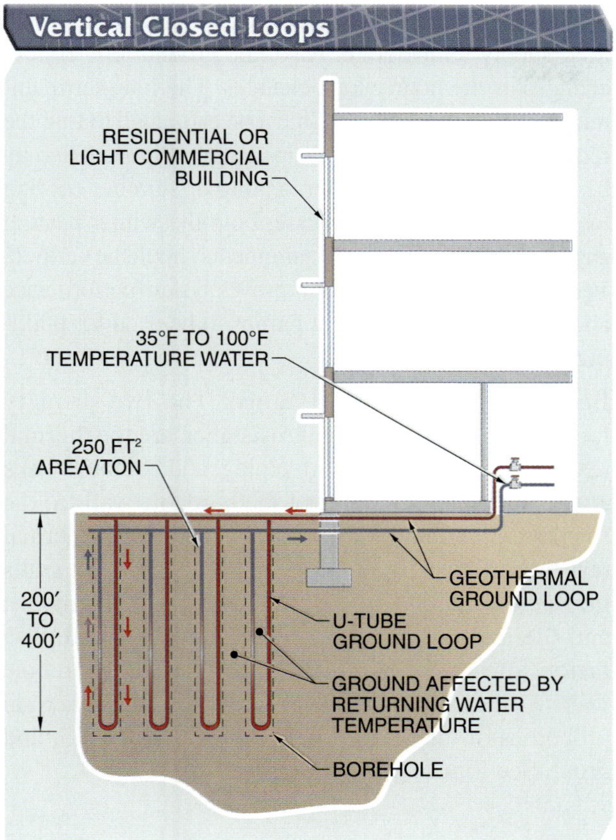

Figure 26-21. Vertical closed loop systems are geothermal ground loop systems in which deep holes are drilled perpendicular to the surface and U-tube ground loops are inserted unto the holes that circulate water to exchange thermal energy.

Ground Loop Operations Media Clip

Surface-Water Closed Loop Systems. A *surface-water closed loop system* is a type of geothermal loop system that circulates water through a loop heat exchanger in a lake or pond in order to exchange thermal energy. Heat escapes the water through surface evaporation, so the process is closely connected to pond temperature and ambient wet bulb temperature. In the winter, when the pond may be frozen, heat transfer is dominated by contact between the loops, the bottom water, and the surface soil at the bottom of the pond. Professionals must be consulted for proper design and water temperature considerations. Loop temperatures are typically 35°F to 90°F. **See Figure 26-22.**

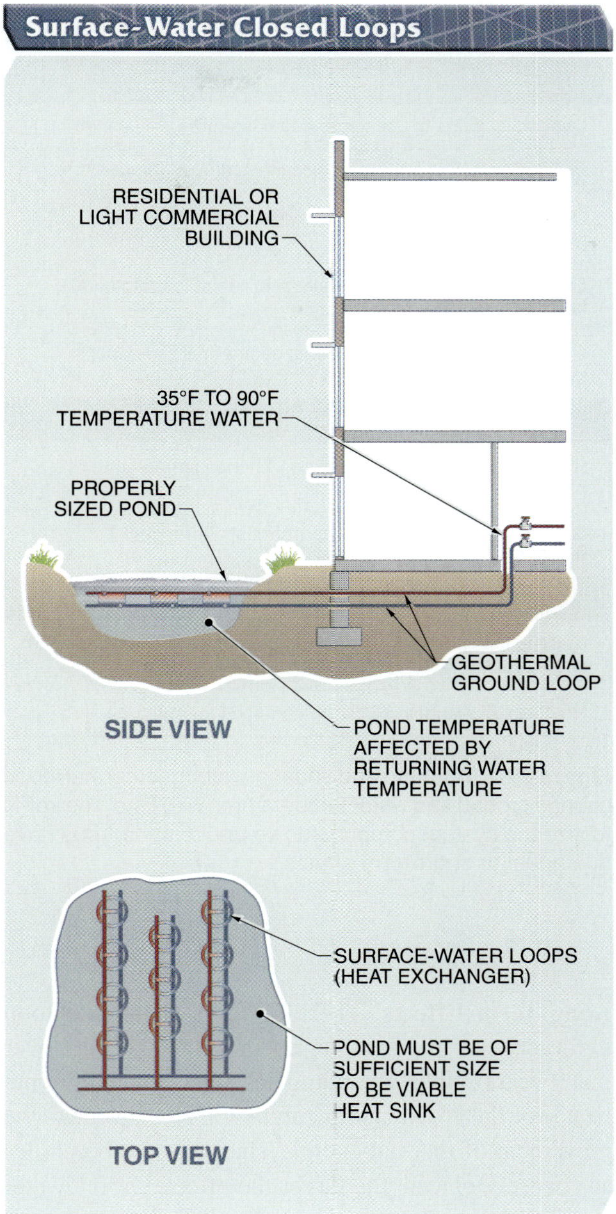

Figure 26-22. Surface-water closed loop systems are geothermal loop systems that circulate water through loop heat exchangers found in lakes or ponds to exchange thermal energy.

Geothermal Loop Considerations

Because ground temperature does not stay constant, it is a challenge to design loops. With horizontal loops, in which the pipe is near the surface, the ground temperature can change with the weather. The loop itself affects ground temperature. There are long-term, annual, and short-term loop effects to ground temperature. **See Figure 26-23.**

Geothermal Loop Considerations

Long-Term Effects
Over many years of use, ground temperatures will change
Building heat gain or heat loss changes ground temperature
Densely placed boreholes have huge impact on ground temperature changes
Groundwater moving in borehole field can stabilize ground temperatures
Light commercial and commercial geothermal water loops increase ground temperatures
Annual Effects
Building heat and cooling loads will change temperature of a borehole field on monthly basis
Ground temperature will be the warmest after a peak cooling demand
Short-Term Effects
Building load and environment temperatures constantly change temperature of the water loop
Hourly load changes will change ground temperature

Figure 26-23. Ground and surface-water geothermal loops change ground and water temperatures with time. The key is to size the system components so that temperatures have no annual or short-term effects.

Long-Term Effects. The long-term effect of a loop is the change in ground temperature that occurs over many years. When a building has a net heat gain or net heat loss, the ground temperature will change due to the movement of thermal energy. The more the boreholes are densely placed, the larger the effect. Groundwater moving through the borehole field can help remove energy and limit the long-term temperature change. For light commercial applications, the ground temperature typically climbs over time.

An example of a long-term effect is a 6°F average ground temperature rise over 10 years due to the heat added to the borehole field. The penalty will not be present during the first year, but the heat buildup will change the performance of the system as years pass.

Annual Effects. Over the course of a year, the heat load on a borehole field changes. The change affects the ground temperature on a monthly basis. This may cause the warmest ground loop temperature to occur after the peak load has occurred.

Short-Term Effects. The actual load on a loop affects the fluid supply temperature. Also, the ground temperature changes as the hourly load changes. The long-term, annual, and short-term effects must be estimated to find the required heat sink areas and pipe length. Pipe length may be established by the winter heating or summer cooling load requirements. In any case, both the winter heating and summer cooling load requirements should be verified. Verifying load requirements improves system performance and may allow smaller heat pumps to be used for building spaces.

Borehole Thermal Resistance. The two primary factors of borehole thermal resistance are the thermal resistance between the fluid, piping, and borehole walls and the thermal resistance of the borehole walls to the surrounding ground. The factors that affect the thermal resistance between the fluid, piping, and borehole walls include the fluid flow rate, design of the flow channels, and thermal properties of the piping and fluid. The factors affecting the thermal resistance of the borehole walls to the surrounding ground include the thermal properties of the borehole walls, ground, backfill, and grout. **See Figure 26-24.**

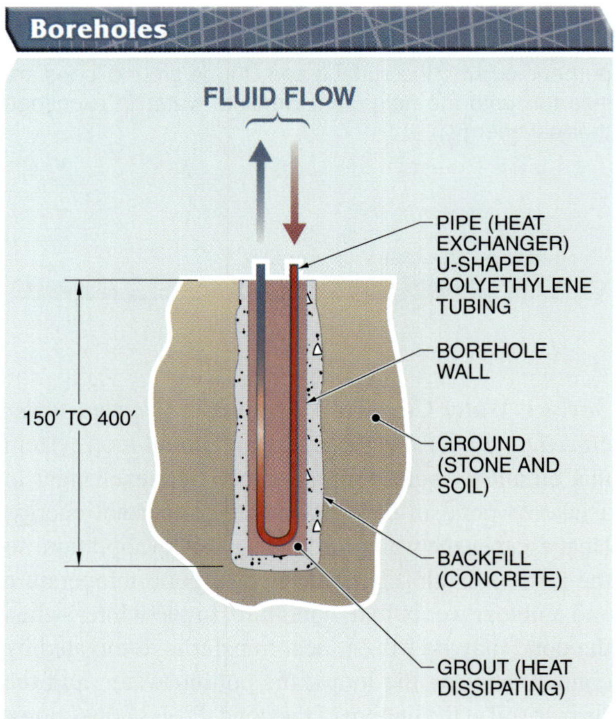

Figure 26-24. Borehole thermal resistance can be large if the wrong materials are used for the fluid, piping, grout, backfill, and borehole walls.

Loop Flow Rates and Temperatures. Heat pumps operating in cooling mode reject the heat collected from the building space plus an additional 25% of heat from the compressor motor and from the act of compression. Most geothermal systems are designed with low flow rates. Lower flow rates reduce the pump and piping size, but they also derate the performance of the heat pumps. **See Figure 26-25.**

Also, in heating mode, the heat from the compressor is added to the heat absorbed from the loop to meet the space heating requirements. It is important to note that when a low flow rate is used as a design flow rate and all of the heat pumps are in heating mode, the loop temperature will drop by 5°F to 8°F. Since freezing must be avoided, the supply water temperature must not drop below 42°F. When temperatures below 42°F are likely, antifreeze should be added to the loop water. The antifreeze may change the design flow rate requirement. **See Taking Water and Air Delta-T Measurements Procedure.**

Console heat pump controls energize the heat pump compressor for either the heating or cooling mode as needed.

Loop Flow Rates and Temperatures

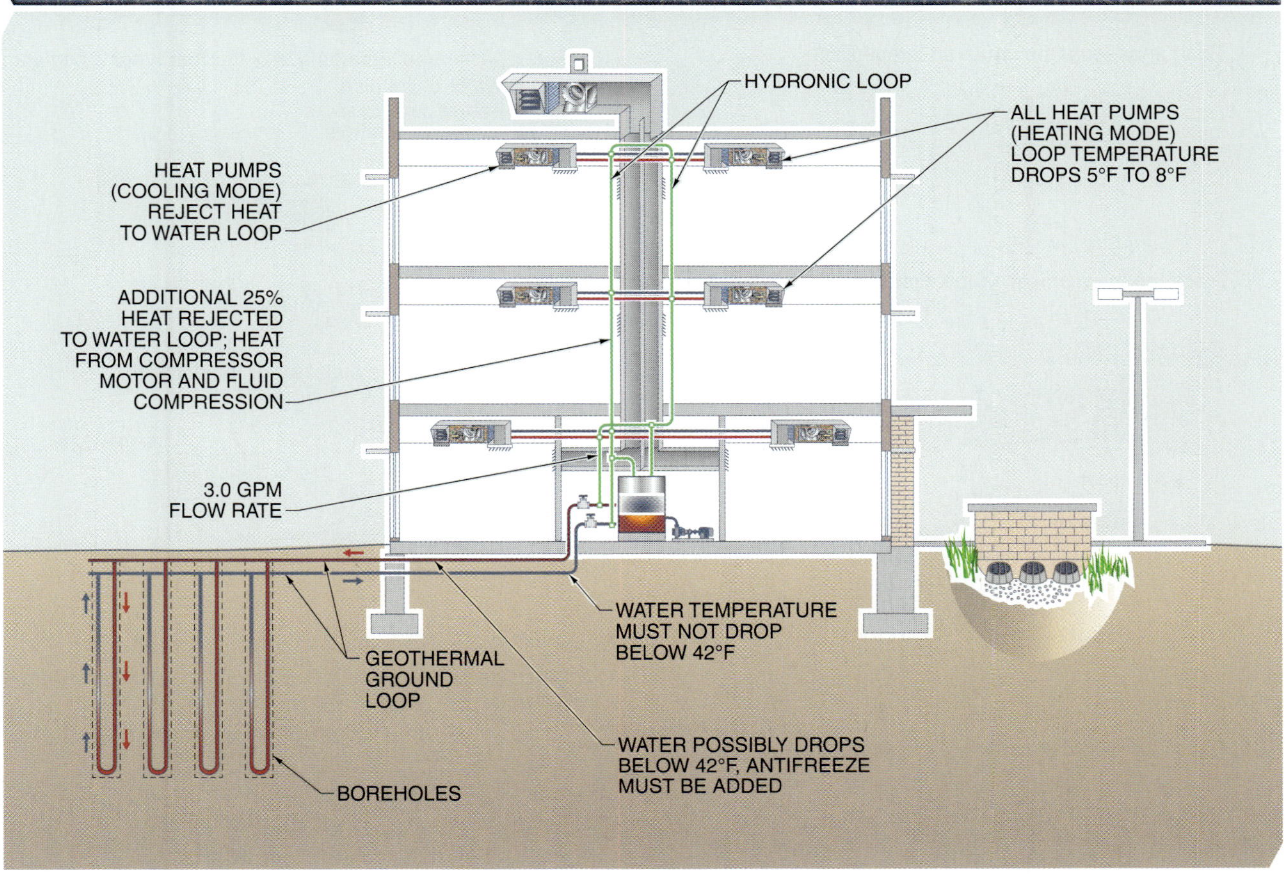

Figure 26-25. Typically, loop flow rates should not be below 3 gpm and loop temperature should not fall below 42°F.

Procedure

Taking Water and Air Delta-T Measurements

1. Set the heat pump system on cooling mode.

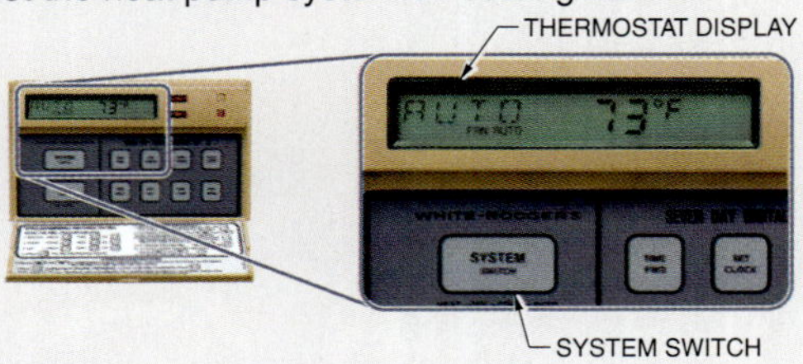

2. Follow all electrical and thermography safety practices and procedures.
3. Place a thermometer in the airstream of the return air.

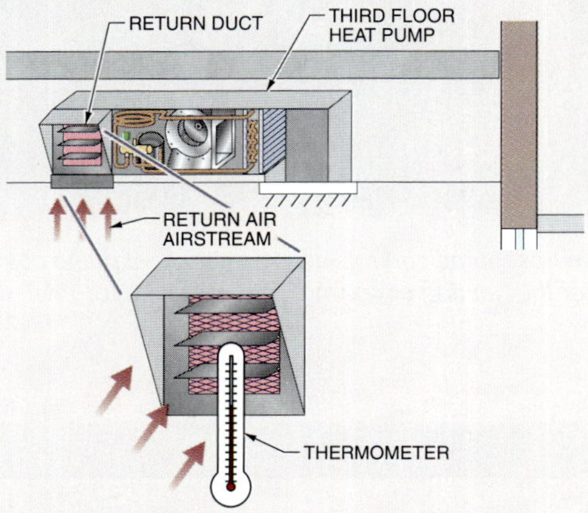

4. Read and record the return air temperature.

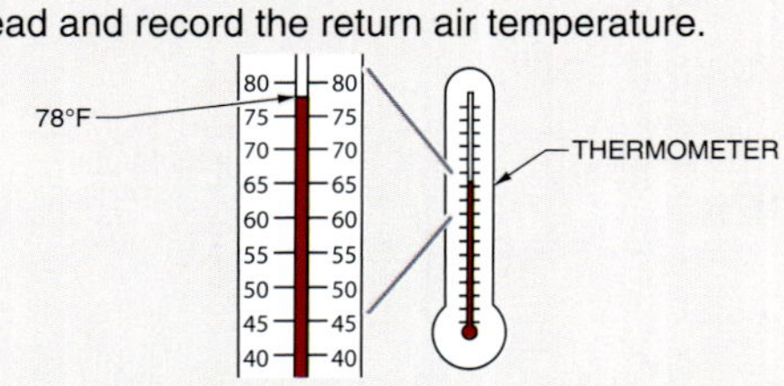

5. Place the thermometer in the airstream of the supply air.

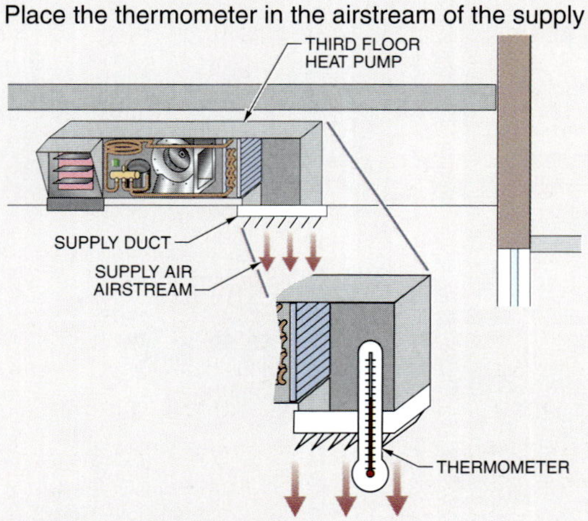

6. Read and record the supply air temperature.

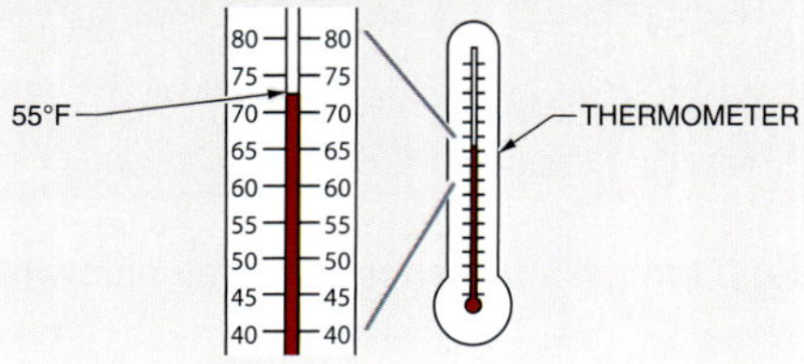

7. Calculate the air delta-T, which is the change in temperature (ΔT) of the air.

   78°F — Return Air Temperature
   − 55°F — Supply Air Temperature
   23 Air Delta-T (during cooling)

8. Focus an infrared thermometer or thermal imaging camera on the water inlet pipe.

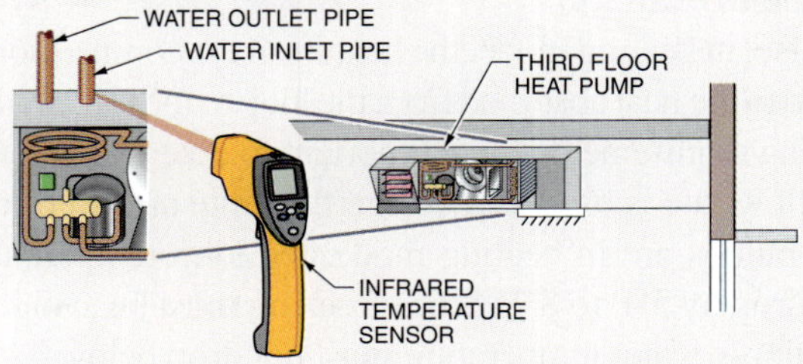

9. Read and record the inlet water temperature.

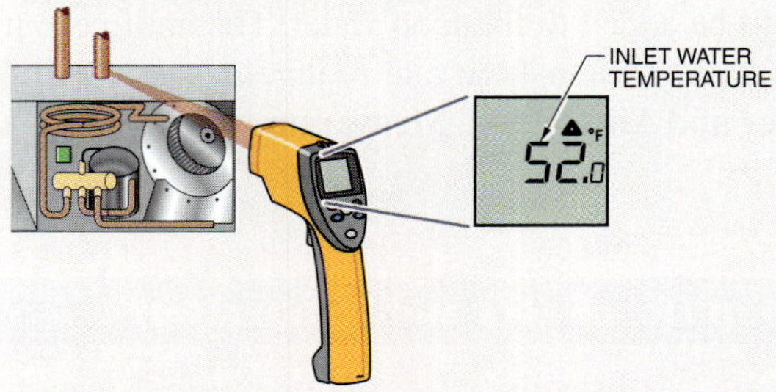

10. Focus an infrared thermometer or thermal imaging camera on the water outlet pipe.

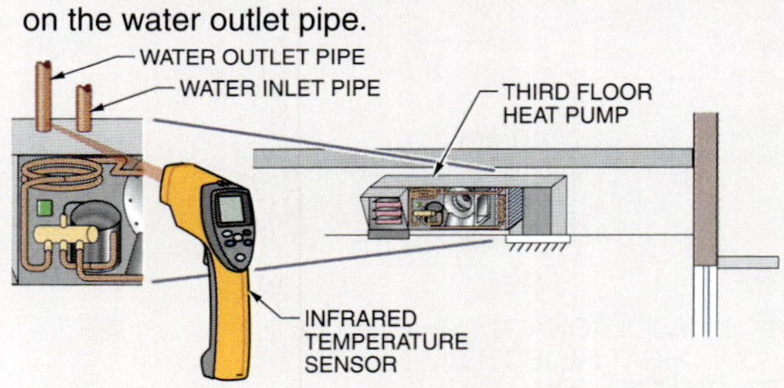

11. Read and record the outlet water temperature.

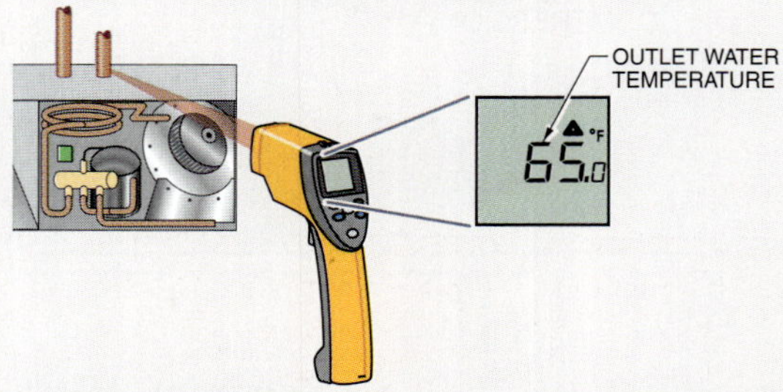

12. Calculate the water delta-T, which is the change in temperature (ΔT) of the water.

   65.0°F — Outlet Water Temperature
   − 52.0°F — Inlet Water Temperature
   13 Water Delta-T (during cooling)

13. Set the heat pump system on heating mode.
14. Repeat Steps 2 through 12 in heating mode.

   Note: The temperature readings of systems upon startup will be different from those of systems that have been operating for some time.

Inlet Water Temperature Selection. Selection of the inlet (supply) water temperature is typically the result of trial and error. The key is to find an economical combination between the inlet water temperature and ground loop size that will meet heating and cooling requirements. For heating, the inlet water should be 20°F to 30°F warmer than the undisturbed ground. For cooling, the inlet water should be 10°F to 20°F cooler than the undisturbed ground.

Once the inlet water temperature is selected, sizing software can then be used to calculate the required loop size. Common software used for loop sizing is GLHEPRO 4.0, which is distributed by the International Ground Source Heat Pump Association. **See Figure 26-26.**

Sizing Ground Loops and Equipment

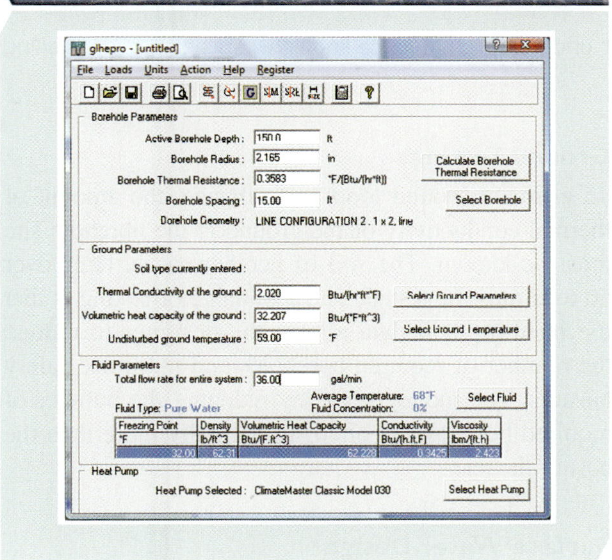

International Ground Source Heat Pump Association

Figure 26-26. Companies have knowledge of sizing heat pumps and loop systems due to trial and error. However, using *The Professional Ground Loop Heat Exchanger Software* (GLHEPRO 4.0) can make any technician knowledgeable on heat pump and ground loop sizing.

Piping. Geothermal system piping within a building is similar to boiler/tower system piping. The difference is that the piping of the geothermal system must be insulated when the minimum temperature is expected to be 50°F or lower. Ground loops use thermally fused high-density polyethylene (HDPE) piping. This is typically the same piping used by utilities for natural gas pipelines. **See Figure 26-27.**

Ground Loop Piping

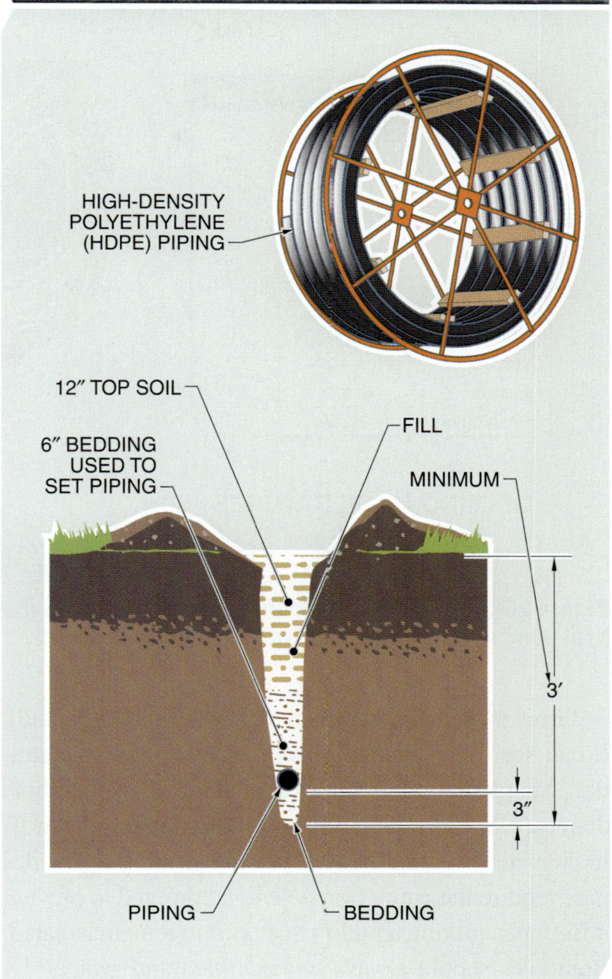

Figure 26-27. Most contractors recommend the use of high-density polyethylene piping (HDPE) for building ground loop systems.

Ground Temperatures. The data for undisturbed ground temperature is best obtained from local water well logs and geological surveys. Data is available that shows the approximate groundwater temperatures in the United States and Canada. Below 30′ underground, the undisturbed ground temperature remains constant throughout the year. Above 30′, the ground temperature changes with the season. **See Figure 26-28. See Appendix.**

Backfill and Grout. Backfill plays a major role in the performance of geothermal ground loops. Backfill prevents the development of air gaps, which should be avoided because air is a natural insulator that restricts the transfer of heat.

Ground and Groundwater Temperatures

State	City	Temp*
Arizona	Phoenix	73
California	Los Angeles	64
Canada	Ontario	37
Canada	Toronto	42
Colorado	Denver	52
Florida	Miami	78
Illinois	Chicago	51
Louisiana	New Orleans	70
New York	New York	52
Virginia	Norfolk	61
Washington	Seattle	53

* in °F

GROUND TEMPERATURES

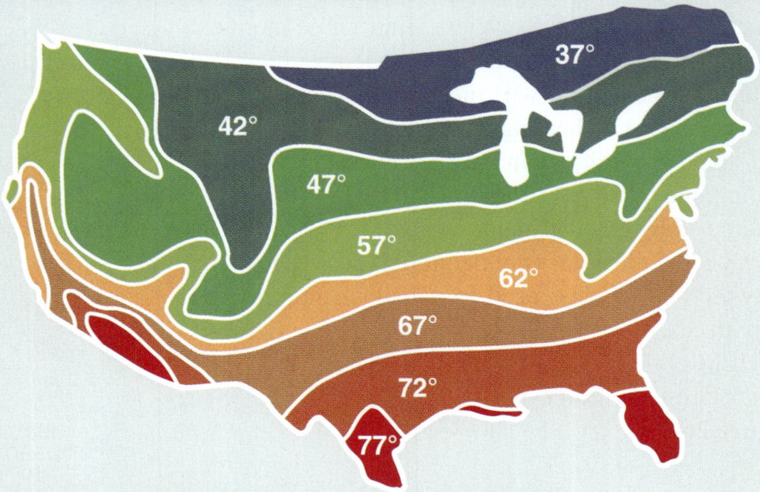

GROUNDWATER TEMPERATURES

Figure 26-28. Ground and groundwater temperatures (below 30′ underground) must be known when sizing and installing heat pumps, water loops, and ground loop systems.

Grout is the most common material used for backfill. It can seal off a borehole from surface water penetration. Standard grout is a poor conductor, so the borehole diameter should be minimized (approximately 5″ to 6″ in diameter) to limit the effect of the grout. Local codes may require the entire borehole to be grouted to prevent it from penetrating aquifers that need to remain isolated. Boreholes should always be backfilled and grouted by an experienced geothermal system installer, and all applicable codes should be followed.

Equipment and Equipment Ratings

Geothermal heat pumps are slightly different from boiler/tower heat pumps. Geothermal heat pumps typically require additional insulation of the internal coaxial coils, thermal expansion (TX) valves, and optimized tube-in-tube heat exchangers. A geothermal heat pump can be safely operated in a boiler/tower system, but the reverse is not true. It is important to select equipment for a specific geothermal project that can operate as required under the expected conditions. Details of the equipment are provided by the manufacturer.

Ground Testing

To estimate ground loop performance, the amount of thermal conductivity of the ground at the borehole site must be known. The soil of geothermal systems over 50 tons must be tested. The advantage of testing is that the resulting soil data allows the designer to reduce the number of required boreholes and minimize safety hazards. The money saved by reducing the number of required boreholes is often substantially more than the cost of the test.

Surface Water Design

Surface water, such as a pond or lake, requires heat transfer mechanisms other than vertical and horizontal loop systems. Surface water absorbs heat through solar radiation, air convection (when the air is warmer than the water), and ground convection. Ground convection, particularly from a frozen lake, is the dominant source of heat transfer in the winter. Cooling is mainly accomplished by the evaporation of warm surface water, but radiant and convective heat transfer also contribute to the cooling process. Evaporation is dependent on the surface temperature of the water, wind speed, and ambient wet bulb temperature.

Fuel Oil Heating Systems

UNIT 27

Fuel oil is used in many furnaces and boilers to heat homes and businesses. Approximately 10 million of the 110 million homes in the United States use fuel oil. The vast majority of these homes, approximately 80%, are in the northeast region of the United States. The other 20% are scattered throughout the northern and mid-Atlantic states. Fuel oil demand and prices are seasonal, with spikes and drops being quite common. These spikes and drops are often caused by abnormal weather conditions. While an HVAC technician may not live in a part of the country where fuel oil heating is common, it is still possible for technicians to run across fuel oil heating systems.

OBJECTIVES

- Identify the different grades of fuel oil.
- Name the various components of a fuel oil unit.
- Describe the combustion process of fuel oil.
- Describe the components of fuel oil burners.
- Describe a fuel oil supply system.

FUEL OIL CHARACTERISTICS

Fuel oil is a crude oil distillate that is burned in a furnace or boiler to create heat or used by engines to create power. The crude oil is separated into different products using heat and pressure. After the crude oil is separated into various distillates and residues, these products are then stored separately.

Classifications of the products include distillates (light oils) and residuals (heavy oils). Distillates are refined into various fuels such as liquefied petroleum gas, jet and automobile gasoline, kerosene, light heating oil, and diesel fuel. About 11 gal. of distillates are produced from a barrel (42 gal.) of crude oil. Less than 2 gal. of the distillates are #2 grade fuel oil. **See Figure 27-1.** Residuals are refined into industrial fuel oils and other commercial and industrial products. The grade of the fuel oil is based on the weight and viscosity of the oil. Viscosity is the ability of a liquid to resist flow.

Crude Oil Distillates and Residues

Products	Gallons per Barrel
Gasoline	19.5
Distillate fuel oil (No. 1, No. 2, & No. 4)	9.2
Jet fuel	4.1
Residual fuel oil (No. 4, No. 5, & No. 6)	2.3
Liquefied refinery gases	1.9
Still gas	1.9
Coke	1.8
Asphalt and road oil	1.3
Petrochemical feedstocks	1.2
Lubricants	0.5
Kerosene	0.2
Other	0.3

Total of 44.2 gal. The total volume of products made is 2.2 gal. greater (processing gain) than the original 42 gal. of crude oil.

Figure 27-1. A 42 gal. barrel of oil creates 44.2 gal. of distillate and residue products.

As the number designation for fuel oil increases, so does the viscosity. Fuel oil grade No. 1 is a very light distillate and not often used for HVAC applications. A slightly higher viscosity fuel oil, fuel oil No. 2, is used for most HVAC heating applications. **See Figure 27-2.** Fuel oil grades No. 4 through No. 6 are not often used in HVAC applications, but are often used in commercial and industrial boilers.

As fuel oil gets heavier, the oil must be heated more and more while in storage, pumped through piping, and as the oil enters the burner. Heating values of fuel oil range from 132,000 Btu/gal. to 156,000 Btu/gal. for the different grades. The standard heating value for No. 2 fuel oil is 140,000 Btu/gal.

Fuel oil is transported from refineries to distributors in ships, barges, railroad tank cars, or trucks. Tank trucks then transport the oil from the distributors to consumers. The fuel oil is stored in storage tanks from where it is pumped to the HVAC heating units.

TECH TIP

Prices often fluctuate, and fuel oil can be more expensive as a heating fuel than natural gas. Fuel oil is commonly used where natural gas infrastructure is not present, such as in rural areas.

OIL-FIRED HEATING UNITS

Oil-fired heating units are available in various styles and layouts. Oil-fired furnaces and boilers are quite similar to gas-fired furnaces and boilers that use natural gas. Common types of oil-fired furnaces are upflow, downflow, and horizontal. Common types of oil-fired boilers are hydronic and steam. **See Figure 27-3.**

Oil-Fired Heating Units

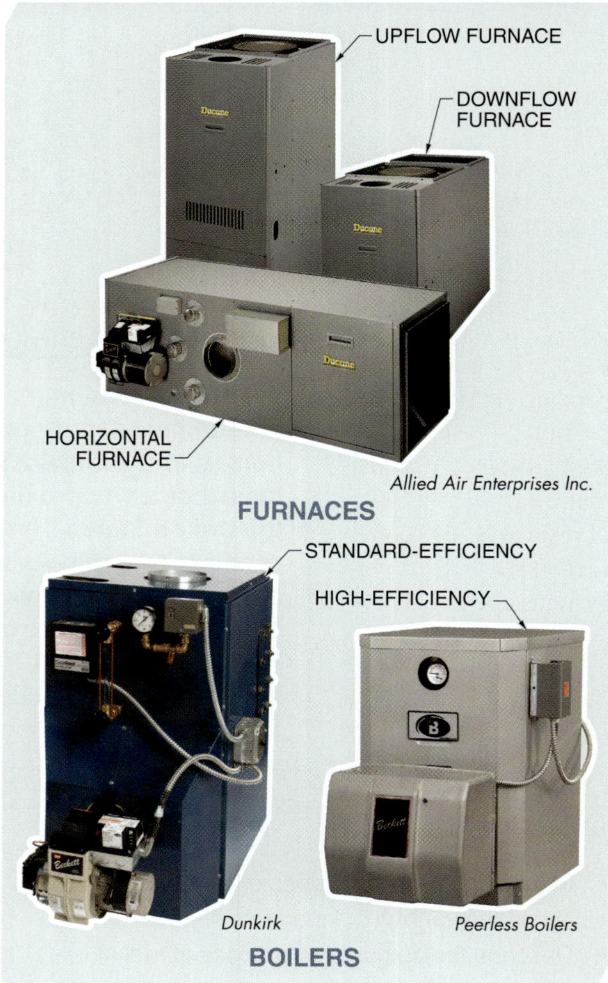

Figure 27-3. Oil-fired furnaces and boilers are quite similar to natural-gas-fired heating units, with standard-efficiency models being rated above 80% and high-efficiency models being rated above 90%.

Fuel Oils

Grade	Category	Usage	Burner Type	Btu/Gal.
No. 1	Light distillate (Kerosine)	Small space heaters	Vaporizing	132,900 to 137,000
No. 2	Distillate (No. 2 heating oil)	Residential heating	Pressure-atomizing (gun)	137,000 to 141,800
No. 3 ASTM merged w/No. 2	Distillate	ASTM merged w/No. 2	ASTM merged w/No. 2	ASTM merged w/No. 2
No. 4 (Blend of No. 2 & No. 6)	Heavy distillate or light residual	Commercial and industrial burners	Burners requiring no preheaters	143,100 to 148,100
No. 5 (light)	Residual	Ships	Preheating sometimes	146,800 to 150,000
No. 5 (heavy)	Residual	Ships	Handling and use preheating	149,400 to 152,000
No. 6 (bunker C)	Residual	Asphalt (road paving)	Atomizing	151,300 to 155,900

Figure 27-2. Fuel oil No. 2 is used for most residential and light commercial heating applications. Fuel oils No. 4 through No. 6 are used in commercial and industrial boilers.

Fuel Oil Combustion Overview

Fuel oil of any grade cannot be just dumped into the combustion chamber of a furnace or boiler. The fuel oil must first be converted into a mist by being forced through an atomizing burner under pressure. The atomizing burner prepares the fuel oil for combustion. The fuel oil mist combines with oxygen, and when a heat source is present, combustion takes place. An ignition transformer is used to generate a spark to ignite the fuel oil mist. **See Figure 27-4.**

There are three forms of fossil fuel combustion: perfect, complete, and incomplete. Since perfect combustion is not practical or possible and incomplete combustion leads to carbon monoxide formation, excess air must be used to allow for the complete and safe combustion of fuel oil. Fuel oil at a pressure of approximately 100 psi is forced into the combustion chamber of residential and light commercial heating units by the fuel oil pump. Commercial and industrial fuel oil supply systems operate at pressures of up to 300 psi.

The oil burner assembly is mounted inside the furnace or boiler cabinet.

Fuel Oil Burners

Fuel oil is prepared for combustion by vaporization or atomization through the use of fuel oil burners. Fuel oil burners vary depending on the grade of fuel oil and the method used to vaporize, spray, or atomize the fuel oil. The three types of fuel oil burners are pot, rotary, and gun. **See Figure 27-5.** A *pot-type fuel oil burner* is a fuel oil burner that vaporizes fuel oil by heating it from below. A *rotary-type fuel oil burner* is a fuel oil burner that sprays fuel oil by rotating a nozzle assembly at high speed in a circular pattern. A *gun-type fuel oil burner* is a fuel oil burner that provides atomized fuel oil to a combustion chamber to increase mixability with combustion air for maximum combustion efficiency. Gun-type fuel oil burners are the most commonly used.

Fuel Oil Combustion

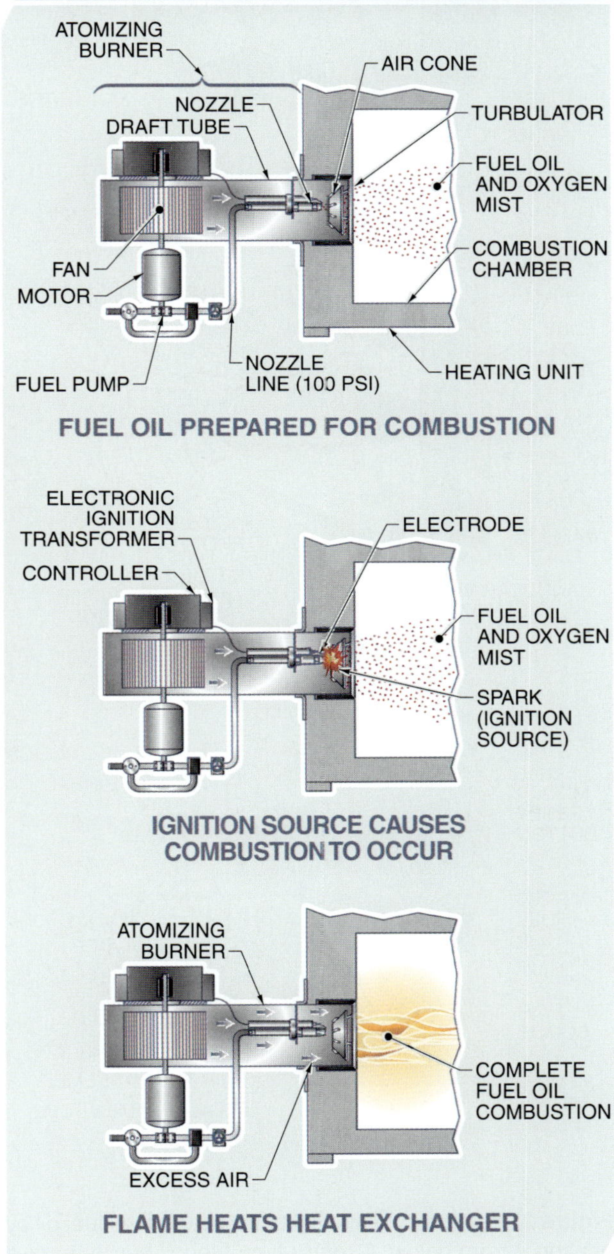

Figure 27-4. Atomizing burners convert fuel oil into a pressurized mist and mix the mist with oxygen for combustion.

Fuel Oil Burners

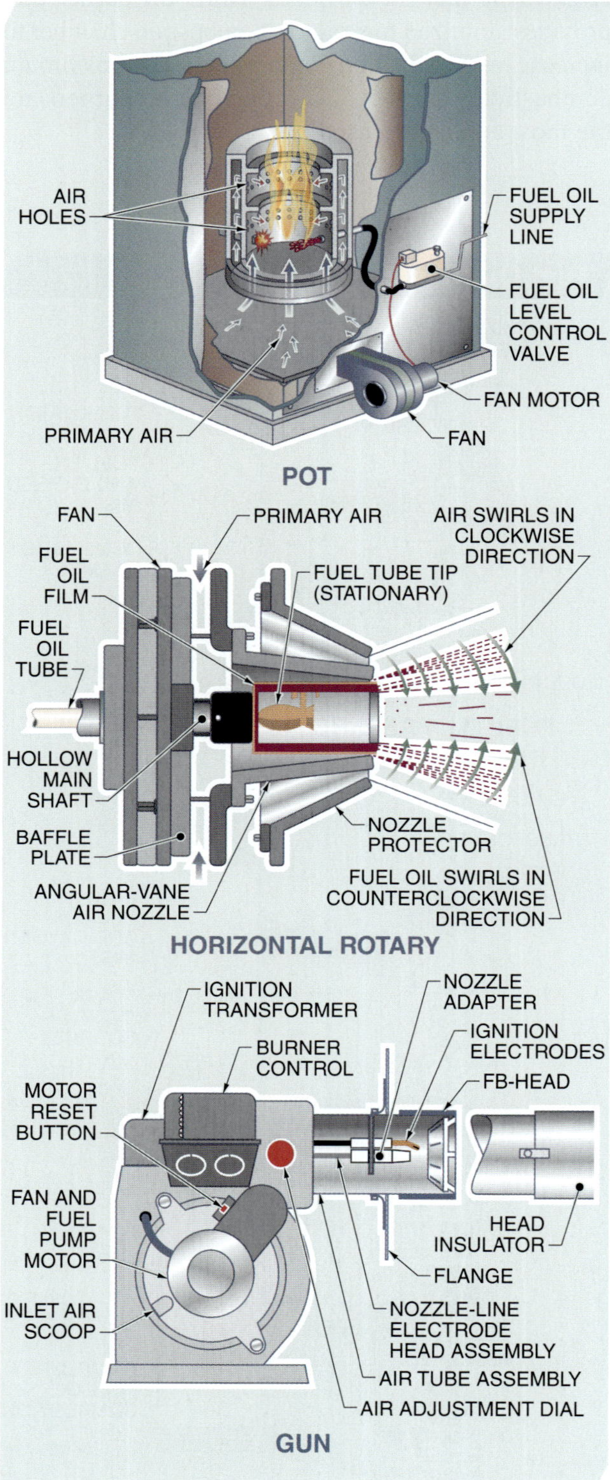

Figure 27-5. A pot-type fuel oil burner vaporizes fuel oil by heating it from below. A rotary-type fuel oil burner sprays fuel oil by rotating a nozzle. A gun-type fuel oil burner atomizes fuel oil to increase air mixability for maximum combustion efficiency.

Gun-type fuel oil burners for residential and light commercial applications are typically designed to be low pressure. A gun-type fuel oil burner consists of a fuel oil pump, fan, motor, fuel oil nozzle, air tube, ignition electrode, ignition transformer, and a primary control unit.

Fuel Oil Burner Operation Media Clip

Fuel Oil Pumps. A *fuel oil pump* is a gun-type fuel oil burner component that forces fuel oil into the combustion chamber of furnaces and boilers under pressure. Fuel oil pumps are typically internal gear pumps. There are different pump styles and piping systems available, depending on the burner. A relief valve bypass is provided at the output of the pump to return any excess oil back to the supply tank. **See Figure 27-6.**

When residential and light commercial fuel oil pumps have problems, they are simply replaced rather than repaired. Care must be taken not to introduce air into the system, as any air in the system must be bled out. An adjustable pressure reducing valve (PRV) is provided in the nozzle line. PRVs are sometimes adjustable, but are typically set at the factory for an application.

Fan and Fuel Pump Motors. The fan of a gun-type fuel oil burner pushes air into the combustion chamber under pressure. The motor and fan shaft can be attached to the shaft of the fuel pump by another flexible coupling. The motor is typically a fractional-horsepower permanent split-capacitor motor. The fan is a squirrel-cage-type fan with adjustable air inlets for combustion control adjustment. **See Figure 27-7.**

Fuel Oil Nozzles. One of the most important parts of a fuel oil burner is the fuel oil nozzle. A *fuel oil nozzle* is a gun-type fuel oil burner component that atomizes fuel oil by converting it from a solid stream into very small droplets (mist). It also controls the amount and shape of the droplets as they are forced into the combustion chamber. The three main types of spray nozzles used in gun-type fuel oil burners are hollow spray (type A), solid spray (type B), and semihollow spray (type W).

TECH TIP

Many HVAC technicians replace fuel oil burner nozzles on a yearly basis. This can be done as part of the routine cleaning and maintenance of a unit at the beginning of the heating season.

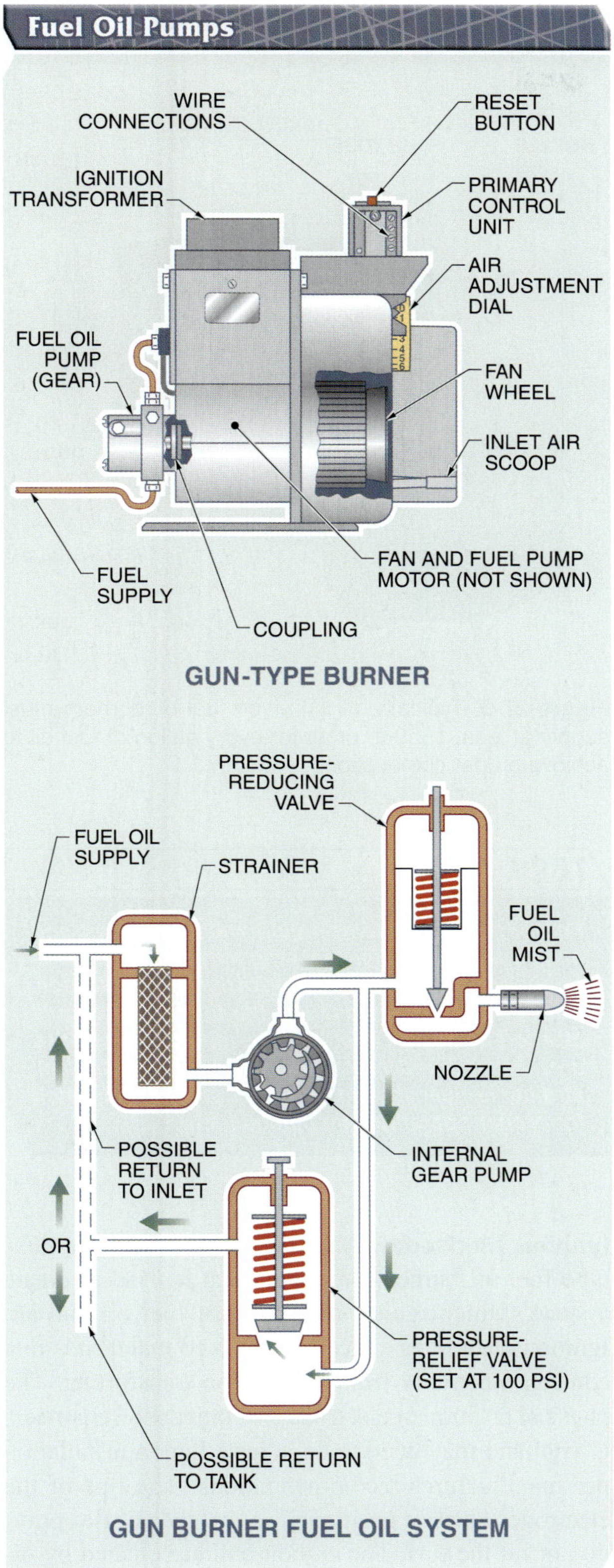

Figure 27-6. Gun-type fuel oil burners have built-in fuel pumps, while other types of burners require that fuel pumps be designed into the systems.

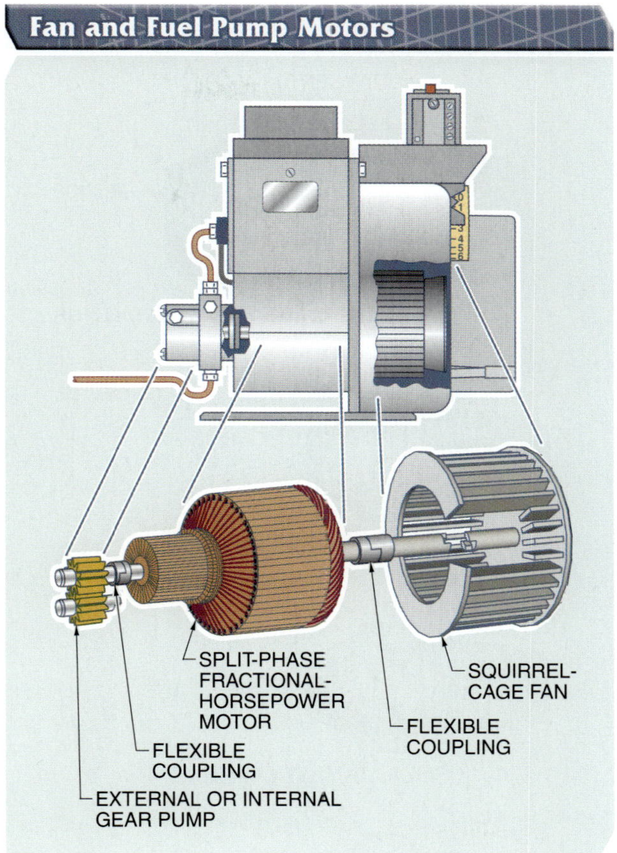

Figure 27-7. Dual-use single-phase motors (fan and pump) are typically split-phase motors because of the high starting torque the motors have available.

The nozzle is stamped with a number, which signifies the spray angle and the design flow of that nozzle in gallons per hour (gph) at a standard pressure of 100 psi. For example, a nozzle that is stamped 60 and 0.8 has a spray angle of 60° and will deliver 0.8 gph of fuel oil. **See Figure 27-8.** It is absolutely vital that a fuel oil nozzle only be replaced with a nozzle that is in the allowable range of the rating plate. Nozzles cannot be field repaired but only replaced. Technicians that work on fuel oil systems often carry a few common nozzle sizes and types in their parts inventory. Typically, a special tool is required to change a nozzle.

Air Tubes. An *air tube,* also known as a draft tube, is a gun-type fuel oil burner component that is used to change the angle of air introduced into the combustion chamber. The air tube can cause a swirling, or any other, effect on the air as it comes into contact with the fuel oil mist. The amount of air needed for proper combustion is approximately 1500 cubic feet per gallon (ft^3/gal.) of fuel oil. **See Figure 27-9.**

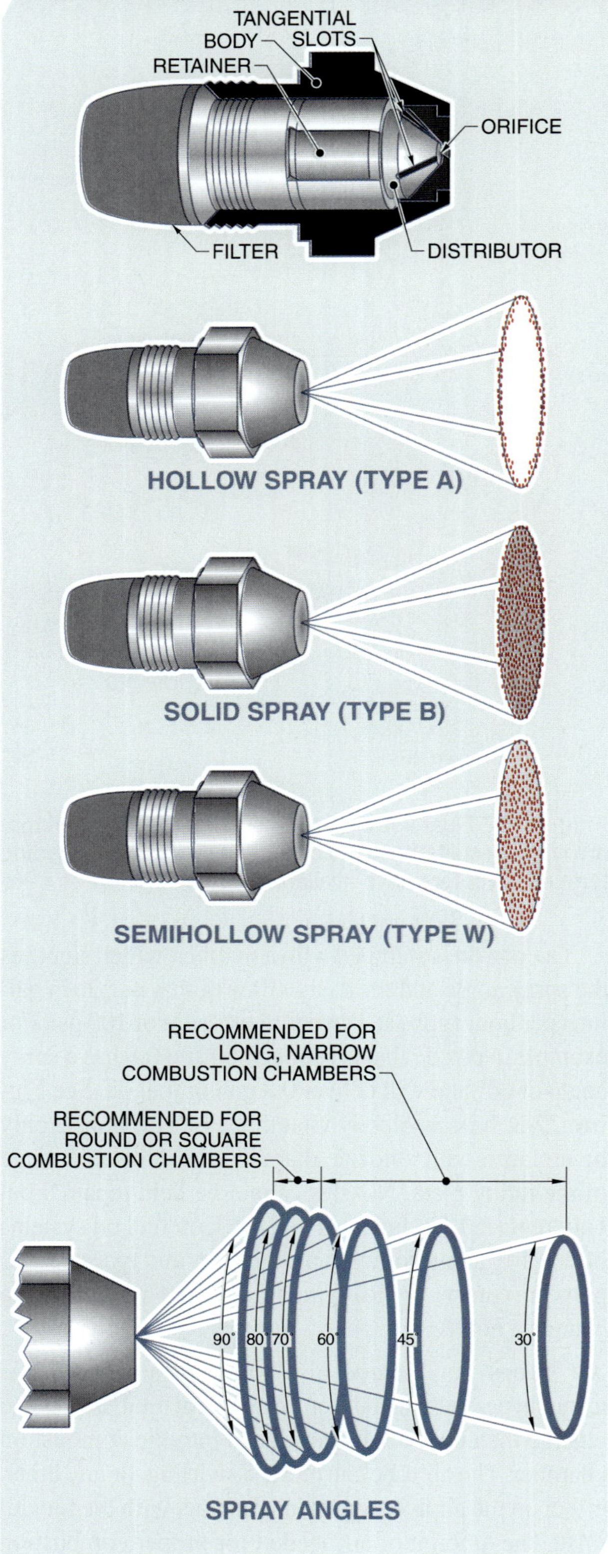

Figure 27-8. A fuel oil nozzle that is stamped 60 and 0.8 has a spray angle of 60° and will deliver 0.8 gph of fuel oil.

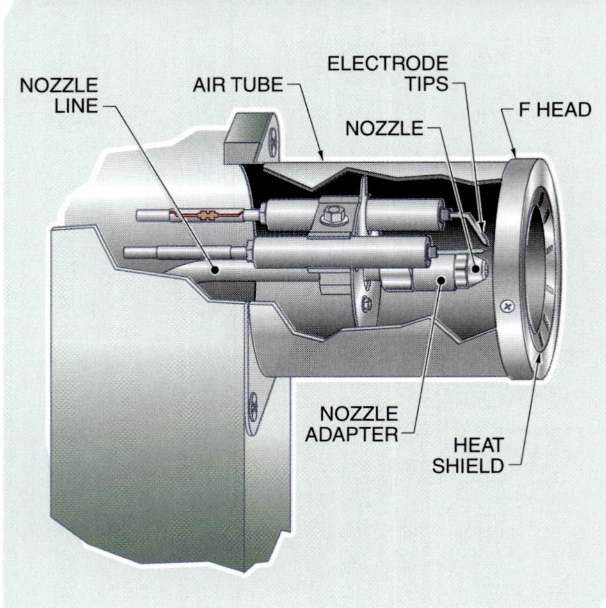

Figure 27-9. Typically, all gun-type fuel oil burners must supply at least 1500 ft^3 of air for every gallon of fuel oil to achieve proper combustion.

TECH TIP

Fuel oil heating systems require routine cleaning and maintenance. Routine cleaning and maintenance is normally done at the beginning of the heating season. During routine maintenance, the unit is thoroughly cleaned of any oil residue, the ignition electrodes are checked and replaced as needed, the fuel oil strainer is cleaned, and the unit is started and tested for proper operation and combustion.

Ignition Electrodes. An *ignition electrode* is a gun-type fuel oil burner component that is used to create a spark, which ignites a mixture of fuel oil and air. Ignition electrodes are ceramic-coated metal rods that conduct electricity from an ignition transformer. The physical position of electrodes in combustion chambers is vital and may need to be adjusted upon installation per manufacturer recommendations. The tips of the electrodes must be positioned so that the nozzle sprays do not hit the tips. The carbon buildup created by direct spray will short circuit the electrodes over time. In most fuel oil heating systems, ignition current from the transformers is connected continuously to the electrodes. **See Figure 27-10.**

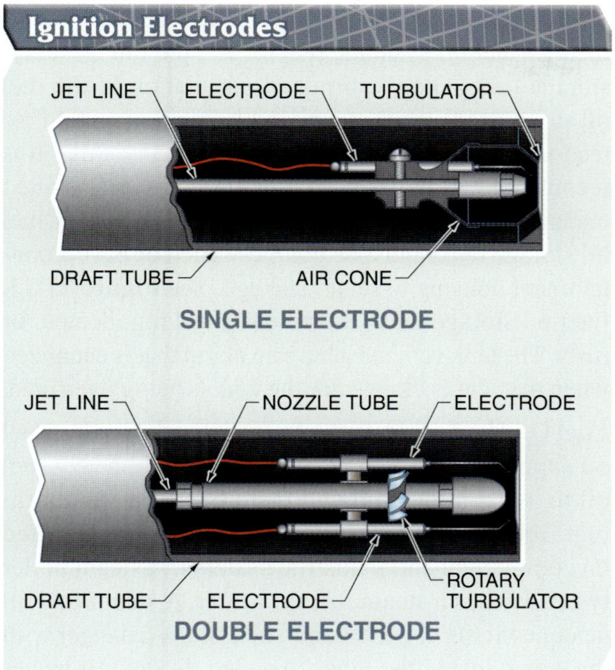

Figure 27-10. The physical position of electrodes in combustion chambers is vital because the nozzle sprays must not hit the tips of the electrodes.

Ignition Transformers. An *ignition transformer* is a step up transformer on a gun-type fuel oil burner that steps up voltage from 120 VAC on the primary side to 10,000 VAC to 17,500 VAC on the secondary side. The transformer is typically attached to a hinge that allows easy access to ignition electrodes for service. **See Figure 27-11.**

Primary Control Units. A *primary control unit* is a special control device for gun-type fuel oil burners that allows a low-voltage (24 VAC) thermostat to control a burner and provides a safety feature when no flame is detected. It cycles the burner on and off through the use of a relay to allow the 24 VAC transformer signal to energize the line voltage burner. **See Figure 27-12.** When the flame is lost or does not ignite, the primary control unit will shut down the burner.

The primary control unit senses the flame through either a bimetallic thermostat in the stack or with a flame-sensing element. A cadmium-cell flame-sensing element is typically used. The cadmium (cad) cell has a low resistance when light is detected. When light is not present, the cad cell greatly increases in resistance, shutting down the burner.

Fuel Oil Supply Systems

A *fuel oil supply system* is a system of tanks, valves, piping, strainers, and pumps that transport and condition fuel oil for combustion. Fuel oil storage tanks are commonly located below ground in cold climates to keep the fuel oil at the correct viscosity. When the fuel oil heating unit is located high above the storage tanks, a booster pump is typically required.

In each fuel oil supply system, a strainer must be installed in the supply line to protect the burner from sediment in the storage tanks and lines. Care must be taken to not allow air to enter the supply line, as it may cut off flow to the fuel oil pump in the burner. All fuel oil lines must be enclosed in a protective covering and run where people will not find them as an obstacle.

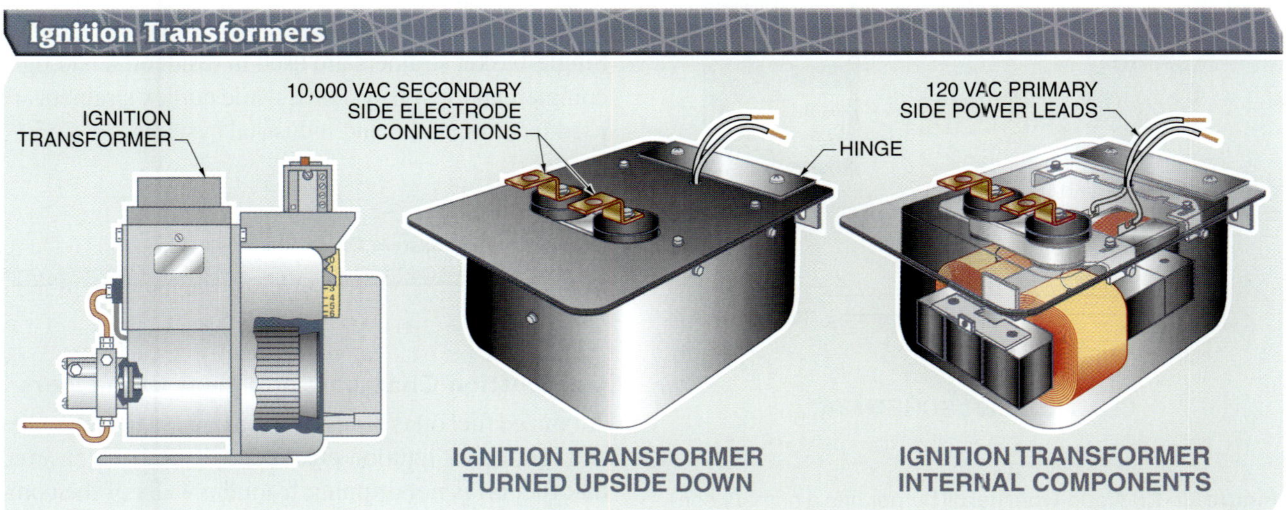

Figure 27-11. The secondary side contacts of an ignition transformer must be checked for proper alignment with the electrodes before being energized for the first time.

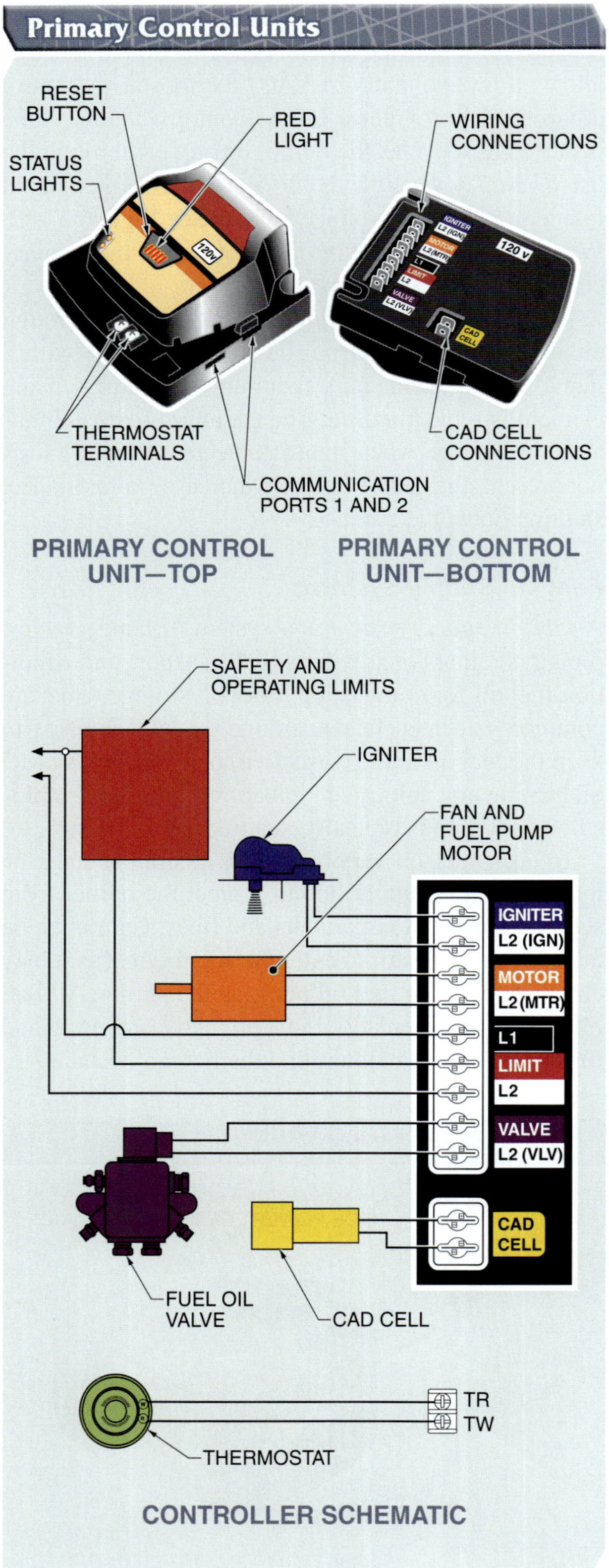

Figure 27-12. A gun-type fuel oil burner has a primary control unit that cycles the burner on and off and shuts down the burner when no flame is detected by a cad cell light sensor.

Fuel Oil Storage Tanks. In residential and light commercial fuel oil systems, fuel oil is stored in a fuel oil storage tank. The location and installation of the fuel oil storage tank must comply with Environmental Protection Agency (EPA) regulations. In many areas, it is recommended that fuel oil storage tanks be installed underground to keep the fuel oil warm. Also, all fuel oil storage tanks must be double-walled or have a containment housing to retain leakage. **See Figure 27-13.** Fuel oil storage tanks must not be leaking, dented, or rusty. The tank vent must have an alarm that is energized when the vent is clogged or the tank is being overfilled.

Fuel Oil Heaters. A *fuel oil heater* is a device in a fuel oil supply system that is used to heat cold heavy fuel oil to assist in pumping and allow for efficient burning. Both steam and electric fuel oil heaters are used on commercial and industrial systems. A steam boiler typically uses a steam fuel oil heater. A steam fuel oil heater consists of a tube-and-shell heat exchanger with the steam inside the tubes. An electric fuel oil heater is used on small heating units in residential and light commercial applications, and as a backup on large heating units that have a steam fuel oil heater. **See Figure 27-14.**

Fuel Oil Strainers. A *fuel oil strainer* is a device that removes solid foreign matter from a fuel oil system. Fuel oil strainers use fine mesh screens to trap foreign matter. **See Figure 27-15.** On a heavy fuel oil system, strainers are located on the discharge line of the fuel oil heater and on the suction line before the fuel oil pump. On a light fuel oil system, a strainer is typically located on the discharge line of the pump. The type of fuel oil used determines how coarse the mesh is for the screen and whether it is a single-basket strainer or a duplex strainer. Single-basket strainers are used in residential and light commercial fuel oil systems, while duplex strainers are used in commercial and industrial systems.

Fuel Oil Supply System Operation Media Clip

Combustion Chambers and Heat Exchangers

Atomized fuel oil is brought into a combustion chamber and ignited by ignition electrodes. It must be ensured that fuel oil is not dripping from the walls of the combustion chamber when being ignited. The fuel oil mist is ignited when mixed with air.

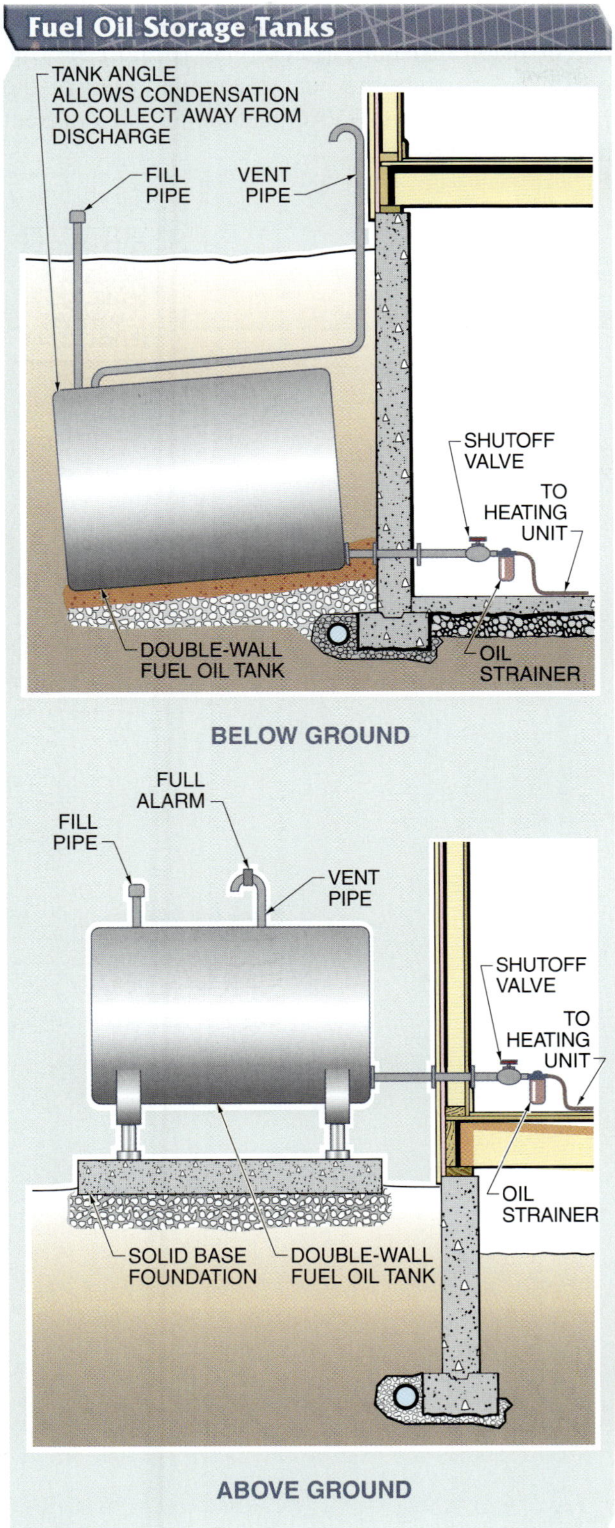

Figure 27-13. Most states require that below ground fuel oil storage tanks of any size be registered. Some states require that fuel oil storage tanks be registered if they contain more than 250 gal. of fuel oil.

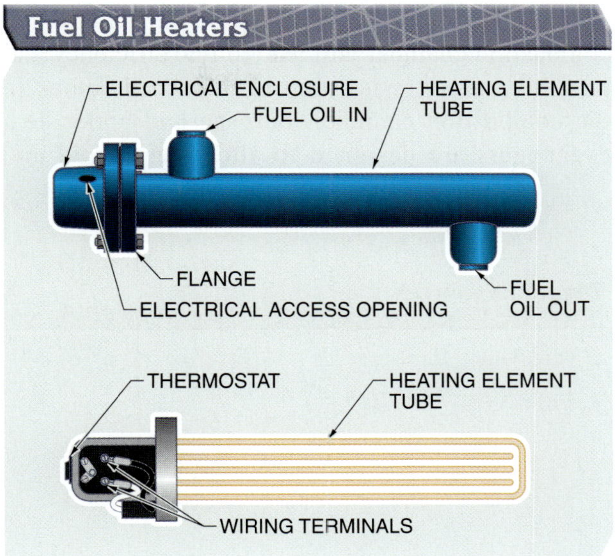

Figure 27-14. Fuel oil heaters used for residential and light commercial applications are electrically powered.

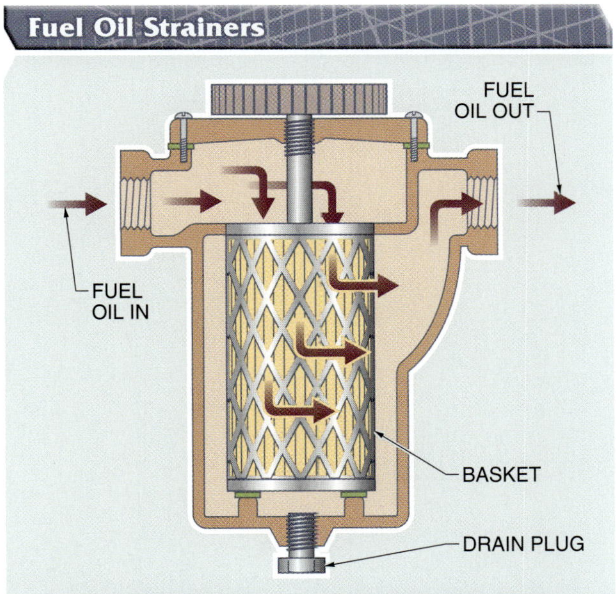

Fuel Oil	Temperature Range	Strainer Mesh	Strainer Pressure Drop Allowed
No. 2	-20°F to 100°F	100 mesh (0.00055″) to 1/16″ openings	1″ Hg to 2″ Hg Clean at 5″ Hg
No. 6	100°F and up	100 mesh (0.074″) to 1/16″ perforations	1″ Hg to 2″ Hg Clean at 5″ Hg

Figure 27-15. Residential fuel oil systems use single-basket strainers to remove solids from the systems.

Fuel oil that is burning while on the walls of the combustion chamber will lead to low efficiency and possibly safety issues due to unburned fumes in the combustion chamber. Furnace and boiler heat exchangers are designed to allow very rapid and efficient heat transfer. They are also designed to be corrosion resistant, as acid may form in the flue gases and corrode the heat exchanger. One of the primary tasks of an HVAC technician is to inspect heat exchangers.

Digital Resources

ATPeResources.com/QuickLinks • Access Code: 482610

UNIT 28

Miscellaneous Manufactured and Field Installed Systems

For many years, there were not a lot of choices available for residential HVAC systems. Fortunately, due to economic pressures, advances in technology, and other factors, this is no longer true. Today, a wide variety of heating systems are available to suit many situations and needs. These range from solar heating and cooling to ductless mini-split systems and more.

OBJECTIVES

- Describe the different types of solar heating systems.
- List the advantages of the various types of electric heat.
- Identify the components of a ductless mini-split air conditioning system.
- List the types of draft systems and the standards for combustion air and vent piping.
- Identify the major components of fuel oil, natural gas, and propane fuel systems.

Index to Procedures

SOLAR HEATING AND COOLING

Over the last number of years, the future of energy has been uncertain. This includes the cost of energy, sources of energy, and pollution restrictions associated with different energy sources. The cost of energy has trended upward over time, but short-term fluctuations have been up and down. The source of energy has been an issue, with foreign sources of energy being predominant and undesirable.

The pollution due to energy sources, such as fossil fuels, has also raised concerns. There has been a new emphasis on renewable energy sources such as solar energy. Solar heating and cooling technologies collect thermal energy from the sun and use this heat to provide hot water, space heating, cooling, and the heating of pool water. Solar heating and cooling is found in residential, light commercial, heavy commercial, and industrial applications. **See Figure 28-1.**

Solar heating and cooling systems can reduce the need for electricity or natural gas. Recently, solar heating systems have shown steady growth. Some of this growth is due to federal and state financial incentives such as the residential solar investment tax credit. Typically, residential solar hot water systems range in price from $2000 to $8000 and have payback periods of three to six years.

Heating and Cooling Energy

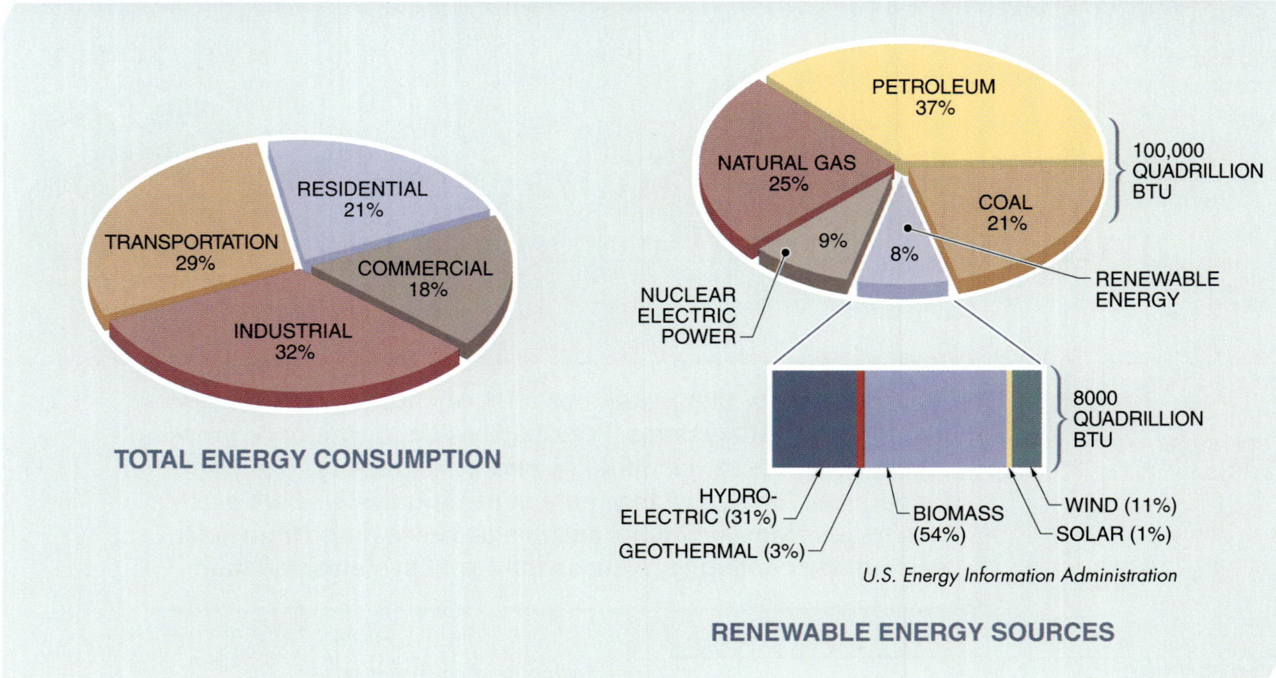

Figure 28-1. Solar energy is a small percentage of renewable energy sources, but the percentage of worldwide solar energy market share is increasing.

Solar Hot Water Heating Systems

Solar hot water heating systems can fulfill most residential and light commercial hot water requirements. This includes water for shower, kitchen, and bathroom use. Solar hot water heating systems should not be confused with solar photovoltaic (PV) systems, which produce electricity. Solar hot water heating systems may be combined with PV systems. It should be noted that solar hot water heating systems can be more cost effective and have a much shorter payback period than solar PV systems.

TECH TIP

Solar hot water heating systems can be effectively combined with solar photovoltaic systems and solar lighting systems to provide a comprehensive energy-saving solution for a building. Often, tax subsidies and utility incentives are available to help reduce the installation cost of a combined system. It is often less expensive to install a combined system at one time than to install the different systems over a period of time.

When sized properly, a solar hot water heating system should be able to provide 100% of the hot water needs for a building during the summer. **See Figure 28-2.** In the wintertime, the short daylight hours cause the existing hot water heater to be used to supplement the solar hot water heating system.

Combination solar hot water and solar photovoltaic systems can provide hot water and electricity.

Residential Hot Water Usage and Conventional Tank Sizes

Use	Per Person Per Day*	Percentage of Total Daily Use	Gallons Used (Conventional)	Gallons Used (Water Saver)
Showers	11.6	16.8	7 – 10	2 – 4
Clothes washers	15.0	21.7	60	42
Dishwashers	1.0	1.4	15	7.5 – 10
Baths	1.2	1.7	36 – 50	30 – 40
Brushing teeth and shaving	7.5	10.8	30	4 – 8
Faucets	10.9	15.7	6 – 8	1 – 3
Other	3.6	5.2	18	10
Individuals	50 – 70	100.0	60 – 70	40 – 45

* in gallons

Use	Family Size	Electric (Size)*	Natural Gas (Size)*	Fuel Oil (Size)*
Vacation cottage - no dishwasher - no clothes washer	Up to 2 people	30	20	20
Small/medium home - 1 bathroom - no dishwasher - clothes washer	2 people	40	30	30
Medium home - 1.5 bathrooms - dishwasher - clothes washer	3 people	50	40	30
Medium/large - 2 bathrooms - dishwasher - heavy-duty clothes washer	4 people	65	40	40
Large home - 2 or more bathrooms - heavy-duty dishwasher - heavy-duty clothes washer	5 people	80	50	50
Large home (above) - whirlpool bath	6 people	120	75	75

* tank size in gallons

Figure 28-2. When sized properly, a solar hot water heating system should be able to provide 100% of the hot water needs for a building during the summer. In the wintertime, the short daylight hours cause the existing hot water heater to be used to supplement the solar system.

Solar hot water heating systems can be categorized as either active or passive. An active system uses a pump, while a passive system relies on a temperature differential and natural convection of the water for flow. Solar hot water heating systems typically include solar thermal collectors, piping, insulation, heat storage units, pumps, and controls. **See Figure 28-3.**

Several types of solar thermal collectors are used to produce heat. Flat-plate, solar air, evacuated-tube, and parabolic-trough collectors can be used in solar hot water heating systems.

The most popular types of solar thermal collectors for residential and light commercial use are flat-plate and evacuated-tube. **See Figure 28-4.** A *flat-plate collector* is a solar thermal collection device that consists of copper-pipe flow tubes attached to a dark-colored absorber plate in an insulated box underneath a hard glass or polymer cover. An *evacuated-tube collector* is a solar thermal collection device that consists of rows of parallel, transparent glass tubes from which air has been removed. Evacuated-tube collectors are very efficient heat exchangers. Evacuated-tube systems are used when higher temperatures or large amounts of water are required. This design works well when used for process heat and solar air conditioning systems.

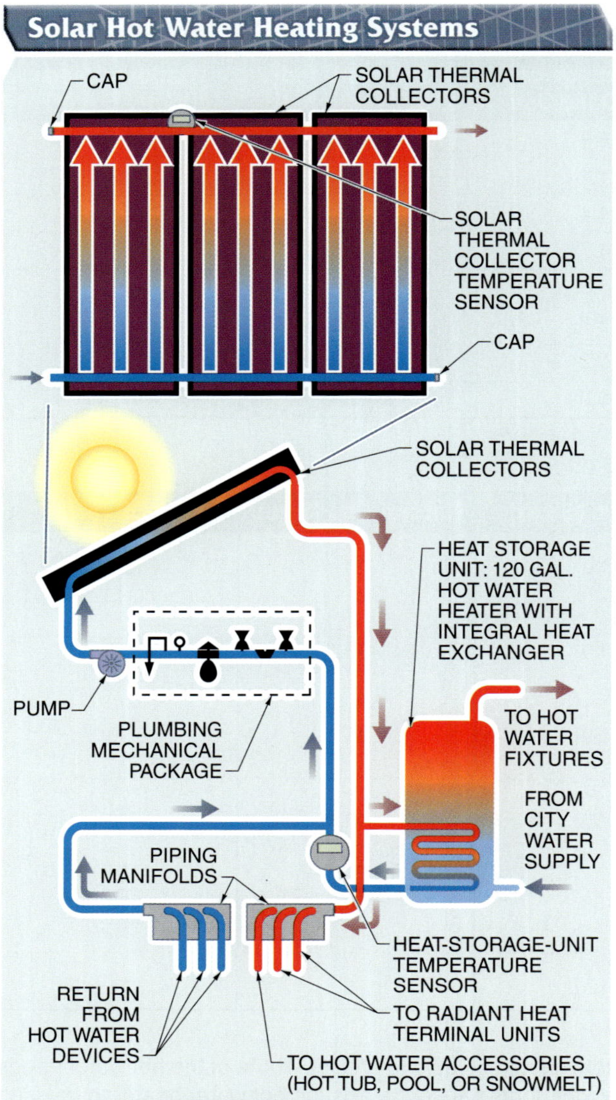

HEAT STORAGE OF MATERIALS

Material	Capacity*
Water	62.5
Cast iron	54.0
Concrete	31.7
Glass	27.7
Oak	26.8
Brick	24.8
Gypsum	20.3
Earth	20.0
Pine	18.1
Air	0.018

* in Btu/ft³/°F

Figure 28-3. Solar hot water heating systems include solar thermal collectors, piping, insulation, heat storage units, pumps, and controls.

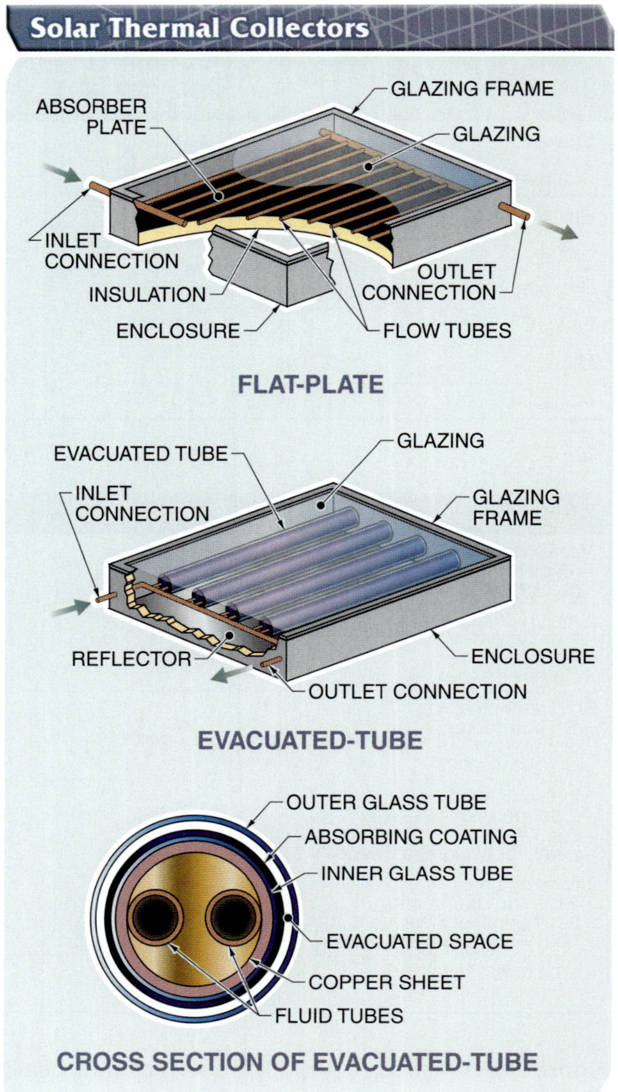

Figure 28-4. Flat-plate collectors and evacuated-tube collectors are solar thermal collectors, which collect heat. Evacuated-tube collectors are very efficient and used when higher temperatures and/or large amounts of water are required.

Solar Space Heating Systems

Solar space heating systems are similar to solar water heating systems. In general, the components are similar, but solar space heating systems use more solar collectors and larger storage units. In addition, the design and controls are more sophisticated. Solar space heating systems can use either water or another nontoxic liquid (that will not freeze) or air as a heat transfer medium from the solar collector. After collection, the heated liquid or air medium is then circulated throughout the building to provide space heating. **See Figure 28-5.**

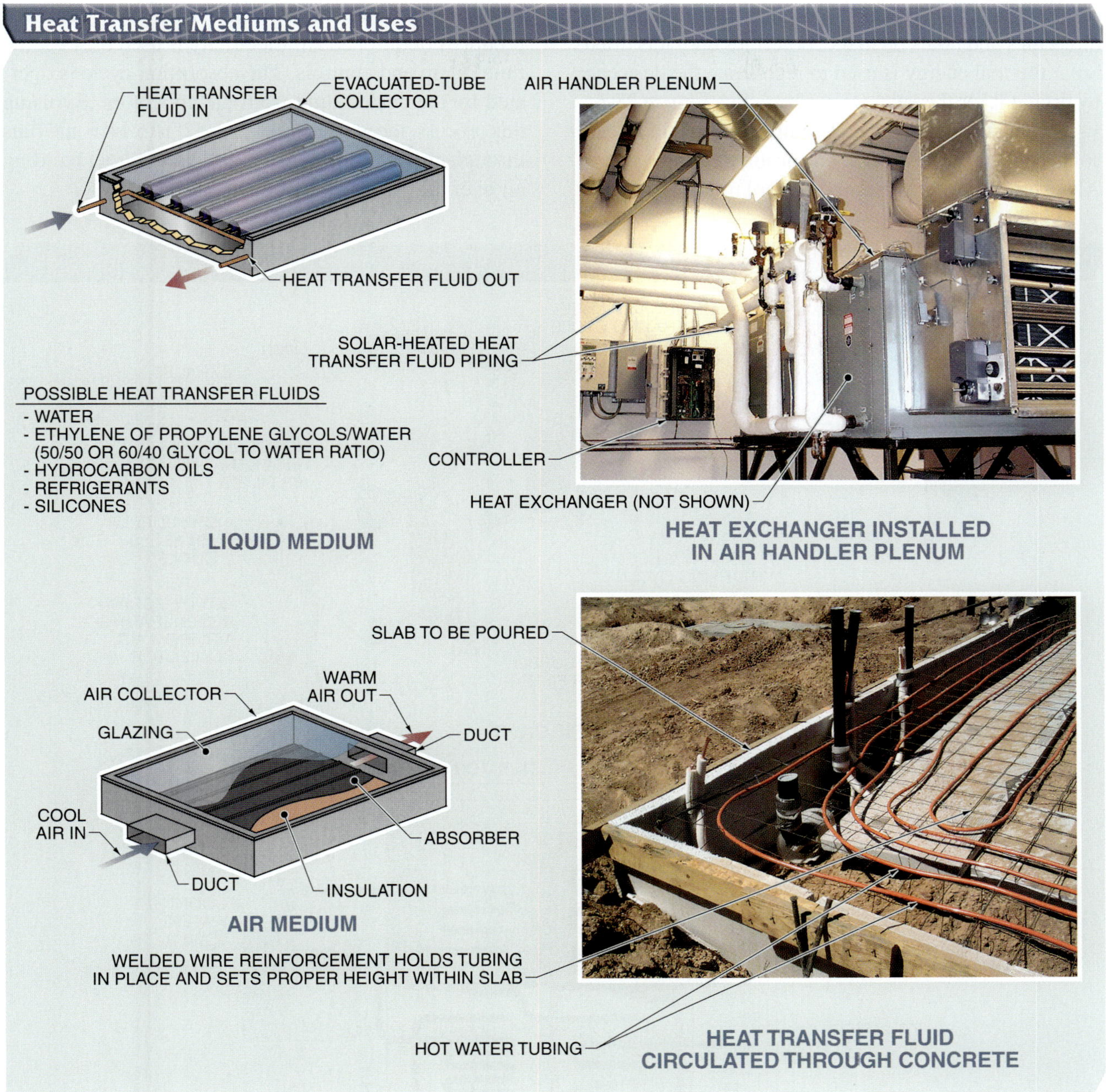

Figure 28-5. A solar space heating system can use either a nontoxic liquid or water or use air as a heat transfer medium from the solar collector to the building's heat exchanger.

One method used to circulate heat transfer fluid is to pump it through a heat exchanger installed in an air handler plenum. Also, the heated fluid may be circulated through tubing installed either within or under the subfloor or concrete slab of a house. Because of their high efficiency, evacuated-tube collectors are typically used for solar space heating systems.

Solar Cooling

In addition to providing space heating, solar energy can be used to make buildings cooler. The two types of solar cooling systems are desiccant and absorption chiller systems. A desiccant system for residential and light commercial applications uses air that passes through a desiccant wheel to remove moisture from the air to increase comfort. *Desiccant* is a solid substance that attracts water molecules and holds on to the molecules until it is dried. Solar thermal energy is used to regenerate the desiccant by drying it for the next cycle of use. **See Figure 28-6.**

The most common type of commercial and industrial solar cooling system is the absorption chiller system. Absorption chiller systems use solar water heating collectors and a thermal-chemical absorption process to produce conditioned air. Electricity is used only for the fluid pump and controls. The absorption cycle is operated by the heated fluid from the collectors. By using this process, the water is cooled to 42°F to 44°F and runs through copper piping. A fan is used to direct building air over the piping, which cools the air.

Solar Cooling Systems

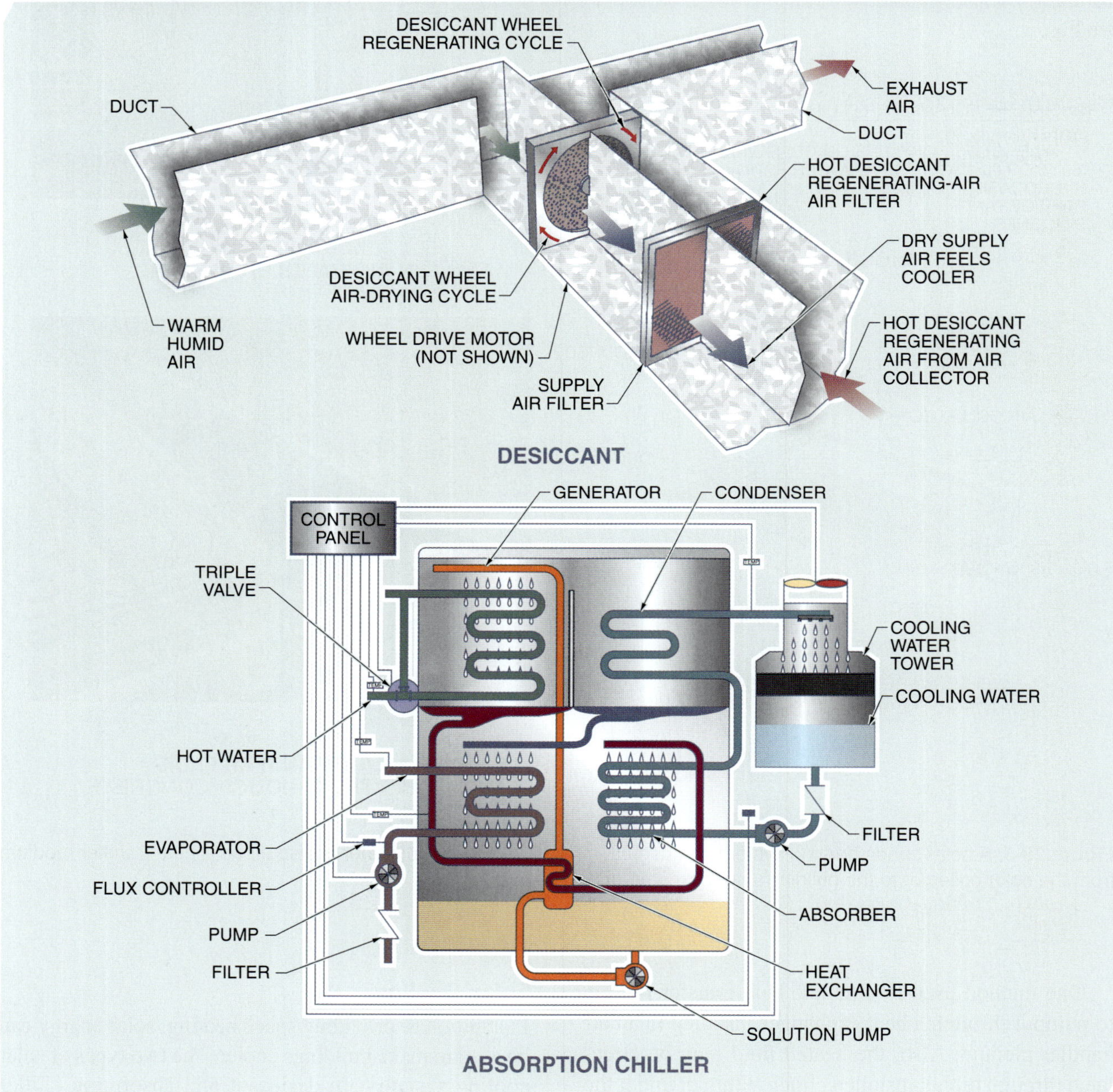

Figure 28-6. In some residential and light commercial applications, desiccant wheels are used to dry the air, and thermal solar panels are used to dry the desiccant. Absorption chiller systems are used with thermal solar panels to create cooling in commercial and industrial applications.

ELECTRIC HEAT

There are many types of fuels used and ways to heat a residence or light commercial space. Forced-air heat is an option. Boilers, either steam or hot water, are another option. These heating units typically use a fossil fuel such as natural gas or fuel oil. Another type of heat is generated by an air- or water-source heat pump. In addition to these types of heating units, there are other heating devices that must be added to the list such as electric baseboards (natural convection) and electric radiant ceiling panels. **See Figure 28-7.**

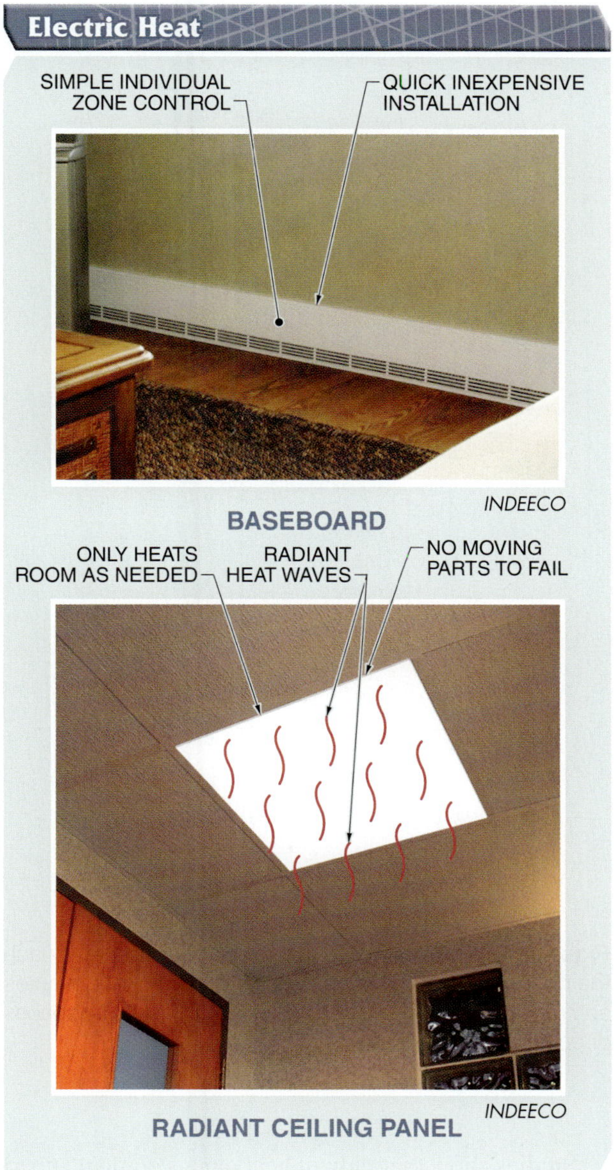

Figure 28-7. Electric baseboards (natural convection) are used for space heating but electric ceiling panel heat is typically used as back-up heating for a specific building space.

Advantages and Disadvantages of Electric Heat

While certainly not for every application in every climate, electric heating devices, such as electric baseboards and electric radiant ceiling panels, have a number of advantages and disadvantages that must be carefully weighed. The advantages of electric heat include the following:

- can be used anywhere electric power is available
- lower electric rates from some utilities for all electric heat homes or electric homes
- only heats rooms as needed
- simple on/off control
- blowers and air filters are not used
- no inefficient air leaks
- quick, inexpensive installation
- when one heating element fails, the whole system is not down
- no moving parts, which means radiant heat panels do not require maintenance or service
- individual zone control is simple, since each building space has a separate thermostat temperature control
- significant energy savings are obtained because only the occupied areas are heated, since each building space can be turned on and off as needed
- fast response times; unoccupied spaces can be brought to a comfortable temperature in minutes

The disadvantages of electric heat include the following:

- generally, electric heat costs more per Btu than natural gas or other fossil fuels
- electric rates may suddenly rise and are unpredictable, depending on location
- possibility of electrical shock in certain applications
- no built-in capability for air conditioning, must be added separately
- no built-in air filtration capability, must be added separately

Electric Baseboard Heat

Electric baseboard heat is popular in many locations and for many applications. These systems consist of one or more convectors mounted on or near the floor of a building space. An electric baseboard heater contains an element that heats rapidly when current passes through it. As the element warms, cold room air comes into contact with the element and is warmed. The warm air is then directed through a diffuser in the top of the heater and then passes into the room. **See Figure 28-8.**

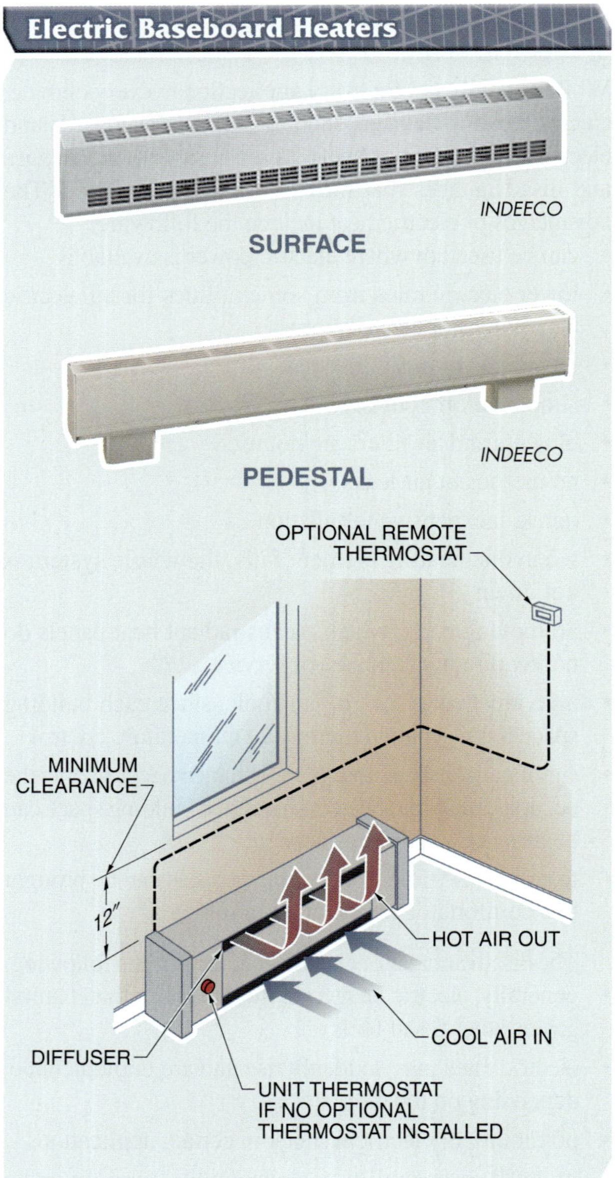

Figure 28-8. An electric baseboard heater contains an element that heats rapidly, warming cold room air as it comes into contact with the element, and then directing the air through a diffuser in the top of the heater to circulate through the space.

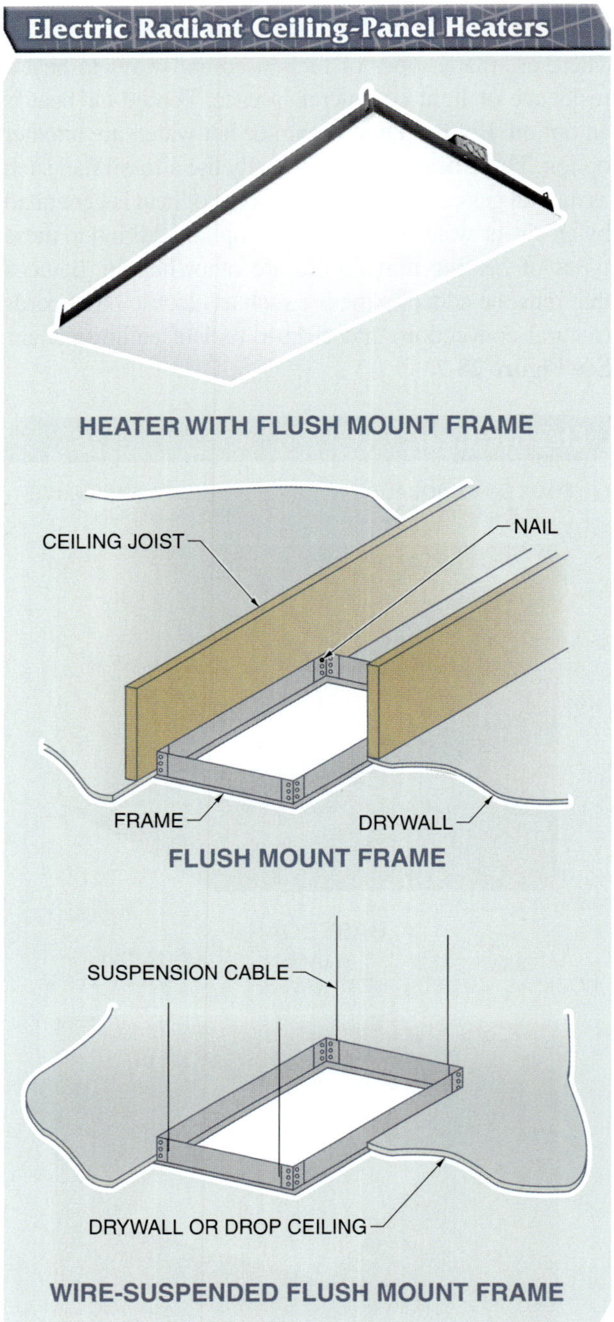

Figure 28-9. Electric radiant ceiling-panel heaters reach their normal operating temperatures in just a few minutes and radiate heat to objects and occupants in building spaces directly without heating the air.

Electric Radiant Ceiling-Panel Heat

Another type of electric heating element that is popular in some applications is the electric radiant ceiling-panel heater. These panels are popular because they are easy to install, warm up quickly, produce dust-free heat, and have a low cost relative to other retrofit heating systems. **See Figure 28-9.**

Studies show that radiant heat panels could be up to 50% less expensive to operate than electric baseboard heaters and considerably more efficient and effective than heat pumps. Electric radiant ceiling-panel heaters consist of ceiling panels with resistance heat elements embedded in them. The panels are about 1″ thick, and different ceiling panel sizes are available. They are textured to look like ceiling tiles and can be painted to match a room. Typically, the panels operate at 150°F to 170°F. They radiate heat to objects and occupants in the building space directly without heating the air. Radiant panels reach their normal operating temperatures in just a few minutes.

Radiant Heat Panel Applications

Radiant heat panels are excellent sources of heat for classrooms and light commercial applications, where minimal heat is needed. They are very popular in the southwest United States. They are commonly used in residential applications such as bathrooms, screened-in porches, and workshops.

Radiant Heat Panel Installation and Cost. An advantage of radiant heat panels is that they require low amperage. These low-amperage requirements can be met by using standard, existing residential 115 V or 230 V circuits. In most cases no new dedicated wiring is needed.

DUCTLESS MINI-SPLIT HVAC SYSTEMS

The prevalent type of residential air conditioning system installed in the United States today is the central air-conditioning unit. This unit typically consists of a large outdoor condensing unit, indoor evaporating unit with a circulating fan, ductwork, piping, and electrical system. Typically, there is one thermostat that controls the temperature in the space or part of the building.

Today, many of these systems are considered old and inefficient. Also, many buildings in the northeast and other parts of the United States have hot water heat with no air conditioning system at all. When air conditioning is desired, costly ductwork must be installed and technician labor and material costs and possible damage to the structure must be accepted. In addition, many older structures do not have either the interior or exterior room for a large outdoor unit and/or indoor unit. Some historic structures are not permitted by code to have a standard-type unit installed.

Fortunately, now there is an alternative available. For these applications and many others, ductless mini-split air conditioning units are very popular. **See Figure 28-10.** In other parts of the world, they hold a very large market share of all AC installations.

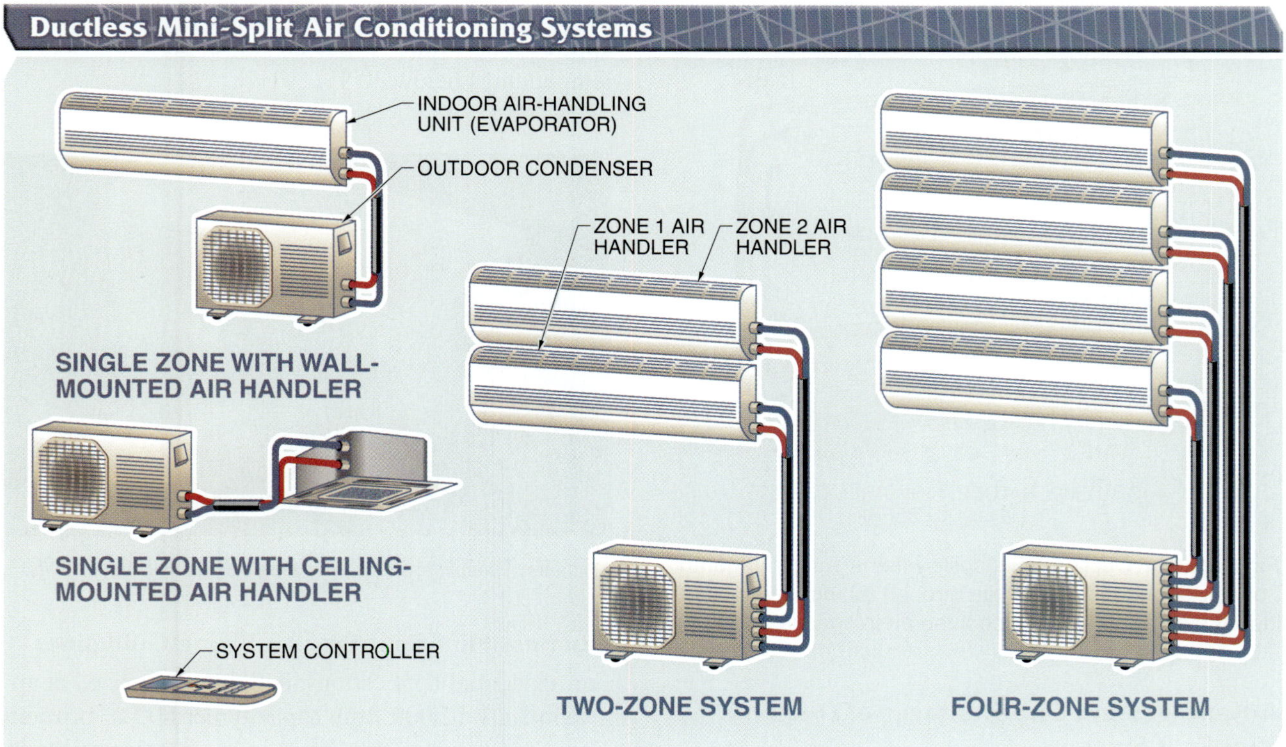

Figure 28-10. Mini-split air conditioning systems are typically used with older structures that do not have the interior or exterior room for a large outdoor and/or indoor unit or are required by local code to have a standard condenser unit installed.

A *ductless mini-split system* is an air conditioning system that consists of one or more indoor air handling units mounted onto the interior walls or ceilings of a building. The indoor unit is a housing with an evaporator coil inside. It also has a small outdoor condensing unit. The evaporator and condensing unit are connected together by copper tubing and an electrical wiring bundle. The bundle is routed through a 3″ hole. A handheld wireless remote or wall monitor controls the entire system.

Ductless Mini-Split System Operation

In a ductless mini-split system, refrigerant is pumped from the outdoor condenser coil and compressor through the tubing to wall-hung indoor evaporator units. The fan in the evaporator units draws room air across the evaporator coil, cooling the air in the space. The refrigerant is then sent back outdoors to the condenser to reject the heat absorbed from the space air. **See Figure 28-11.**

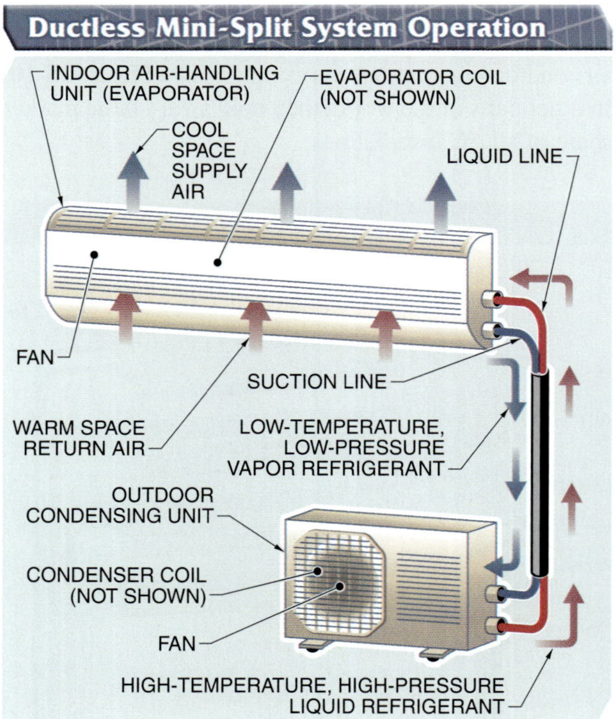

Figure 28-11. Ductless mini-split systems pump refrigerant from outdoor condenser coils through tubing to wall-hung indoor evaporator units to cool the air in spaces.

Advantages and Disadvantages of Ductless Mini-Split Systems

Many contractors and consumers are finding out that ductless systems include a number of important advantages. The advantages of ductless mini-split systems include the following:

- easy installation of copper tubing and wiring through one 3″ to 4″ opening in the wall or ceiling, which connects indoor and outdoor units without changing building appearance
- units quiet enough to be used in noise-sensitive areas such as study areas and meeting rooms
- simple TV-type remote control device used to control each zone
- attractive designs
- units are easily maintained, with periodic cleaning of filters
- various Btu sizes allow for the matching of systems to loads
- easily zoned for one, two, or any number of cooling zones, with individual thermostats in each zone
- efficiencies up to SEER-26 are possible, providing lower energy costs, with some models having ENERGY STAR® standard ratings

The many advantages of ductless mini-split systems far outweigh the major disadvantage. The major disadvantage of ductless mini-split systems is heating difficulties due to a lack of auxiliary or back-up electric heat. This absence of auxiliary or back-up heat limits applications to areas that do not experience outside temperatures below 30°F.

Ductless mini-split systems have small, efficient, outdoor units.

Ductless Mini-Split System Sizing Guidelines

For residential applications, bedrooms are sized at approximately 12,000 Btuh capacity per 600 ft^2 of room size. Extra components, such as low ambient air temperature operation kits, are typically not needed. **See Figure 28-12. See Appendix.**

Ductless Mini-Split System Sizing Guidelines

Building Space	9000 Btu	12,000 Btu	18,000 Btu	24,000 Btu
Space up to 300 ft² without situations B, C, D, or E, (space well-insulated, double-pane windows, and regular-size ceiling)	Yes	No	No	No
Space up to 300 ft² with situations B, C, D, or E, (space well-insulated, but has many windows)	Possible	Yes	No	No
Space up to 300 ft² with all situations (not insulated, and inefficient windows)	No	Yes	Possible	No
Space between 300 ft² and 500 ft² without situations B, C, D, or E, (space well-insulated, double-pane windows, and regular ceiling)	No	Yes	No	No
Space between 500 ft² and 700 ft² without situations B, C, D, or E, (space well-insulated, double-pane windows, and regular ceiling)	No	No	Yes	No
Space between 700 ft² and 1000 ft² without any situations B, C, D, or E, (space well-insulated, double-pane windows, and regular ceiling)	No	No	No	Yes
Space over 1000 ft²; more than one unit required	Single-Zone Unit	Two-Zone Unit	Three-Zone Unit	Four-Zone Unit

Situation A – one air handler per space

Situation B – space not well-insulated

Situation C – 2, 3, or more single-pane windows

Situation D – ceiling more than 10′ high

Situation E – any heat-generating source in space

Figure 28-12. Ductless mini-split air conditioning systems are sized according to the square footage of the space and the bad circumstances involved.

For light commercial applications, spaces are sized at approximately 12,000 Btuh capacity per 400 ft² of room size. Since there may be a large heat load due to the number of occupants, cooling may be required at a low ambient temperature (air temperature below 55°F), so a low ambient air temperature operation kit may be needed. **See Procedure for Sizing Ductless Mini-Split Air Conditioning Systems.**

Procedure

Sizing Ductless Mini-Split Air Conditioning Systems

1. Calculate the total square footage of the space to be air conditioned. Include the square footage of any other spaces that will typically be open to the space such as bathrooms and walk-in closets.

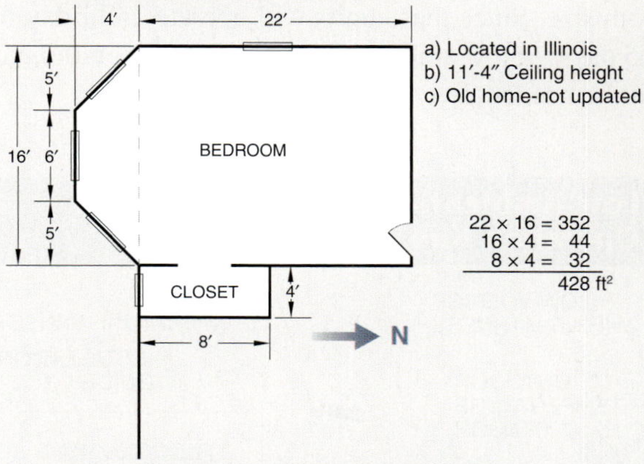

2. Determine the Btu required for the application.
   a. Multiply the square footage by 25 to find the Btu baseline.
      428 × 25 = 10,700 Btu baseline
   b. Increase the Btu capacity for each condition to find the Btu estimate.

Conditions
- climate (seasonal temperatures exceeding 90°F): **increase 30%**
- windows (south-facing or large): **increase 20%**
- ceilings (over 10′): **increase 20%**
- appliances (kitchen appliances and/or electronics): **increase 20%**
- occupants (more than two people in space): **increase 20%**
- insulation (poor): **increase 30%**

Condition Allowance

Climate:	Yes	30%
Windows:	Yes	20%
Ceilings:	Yes	20%
Appliances:	No	0%
Occupants:	No	0%
Insulation:	Yes	30%
Total Btu increase		**100%**

Increase Btu by 100%:
10,700 × 2 = 21,400 Btu estimate

3. If a zone system, add all Btu zone calculations together.
4. Select a mini-split air conditioning system that supplies at least the calculated Btu capacity. Always move up to the next higher Btu system rating.

System Sizes
9000 Btu: coverage area of 450 ft²
12,000 Btu: coverage area of 600 ft²
18,000 Btu: coverage area of 900 ft²
24,000 Btu: coverage area of 1200 ft²
Calculated 21,400 Btu requires 24,000 Btu system

Ductless Mini-Split System Applications

Ductless mini-split systems can be used in many applications. However, the application will affect the installation and operation of the unit, with some applications needing additional components, such as low ambient air temperature operation kits, or installation parameters.

In computer and data server facilities, it has been found that ductless mini-split systems are a great application for server rooms or closets. Since these applications generate a large amount of heat, they are sized for approximately 12,000 Btuh capacity per 250 ft^2 of room size. Since these units will operate 24 hr/day for 365 days/yr, low ambient air temperature operation kits will be needed. **See Figure 28-13.**

Low Ambient Air Temperature Operation Kits

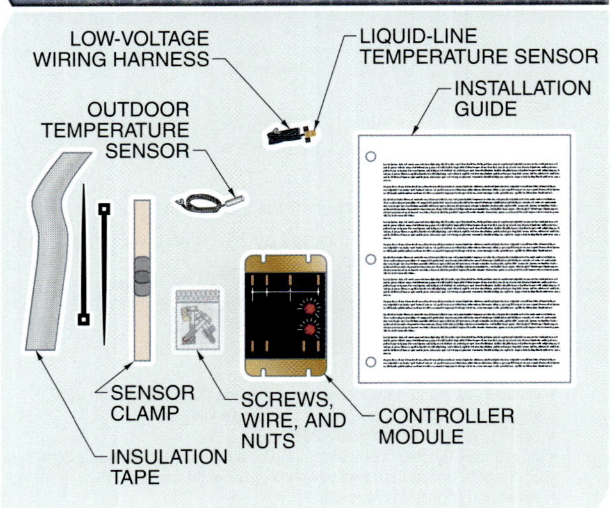

Figure 28-13. Low ambient air temperature operation kits allow air conditioning systems to produce conditioned air even when outdoor air temperature drops below 55°F.

In heavy commercial applications, multiple indoor units are connected to a single condensing unit. Each zone has a separate control. In most of these applications, since a variable amount of cooling is needed, the compressor is run by a variable-frequency drive (inverter). These large systems have three-phase (3ϕ) motors connected to the compressors. The inverter will speed up and slow down the compressor to provide only the correct amount of refrigerant flow. These systems are also known as variable-refrigerant-flow (VRF) systems. They have become very popular in heavy commercial applications such as college dormitories and churches.

DRAFT SYSTEMS

A *draft system* is a system in a furnace or boiler that regulates the flow of air into and out of the combustion chamber. For fuel to burn efficiently, the proper amount of oxygen must be provided. The proper amount of primary air and secondary air must be provided for complete fuel combustion and to discharge the gases of combustion. **See Figure 28-14.**

Boiler Draft

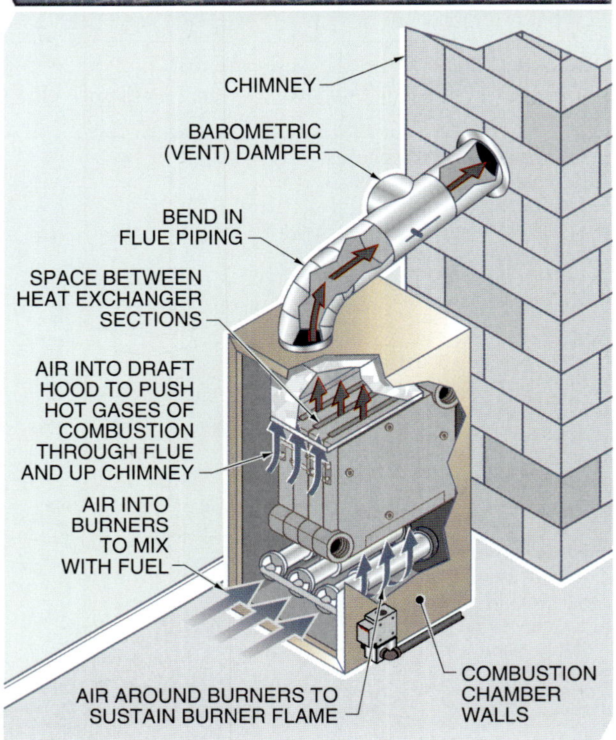

Figure 28-14. Draft systems in furnaces and boilers regulate the flow of air into and out of combustion chambers to achieve complete fuel combustion.

During boiler operation, a flow of air in the burners mixes with the fuel. A flow of air outside the burners sustain burner flame. Finally, a flow of air in the draft hood mixes with the hot gases of combustion. The gases are pushed through the flue and up the chimney to the atmosphere.

Each of these processes has resistance to flow from obstructions such as combustion chamber walls, tubes or spaces between heat exchanger sections, baffles, dampers, and bends in the flue piping. To overcome the resistance to flow, either natural draft or mechanical draft is employed. The type of draft used depends on the size and capacity of the boiler. The three types of mechanical draft include forced, induced, and balanced draft.

Draft System Operation Media Clip

Natural Draft

Natural draft is the draft produced by the natural action resulting from temperature and density differences between the cool atmospheric air entering a combustion chamber and the hot gases of combustion going up the chimney of a furnace or boiler. **See Figure 28-15.** Air for combustion is heated, expands, and becomes lighter. Colder, heavier air pushes into the heating unit to replace the hot gases of combustion going up the chimney. A difference in pressure is created as the cold air for combustion replaces the hot gases of combustion going up the chimney.

Natural Draft

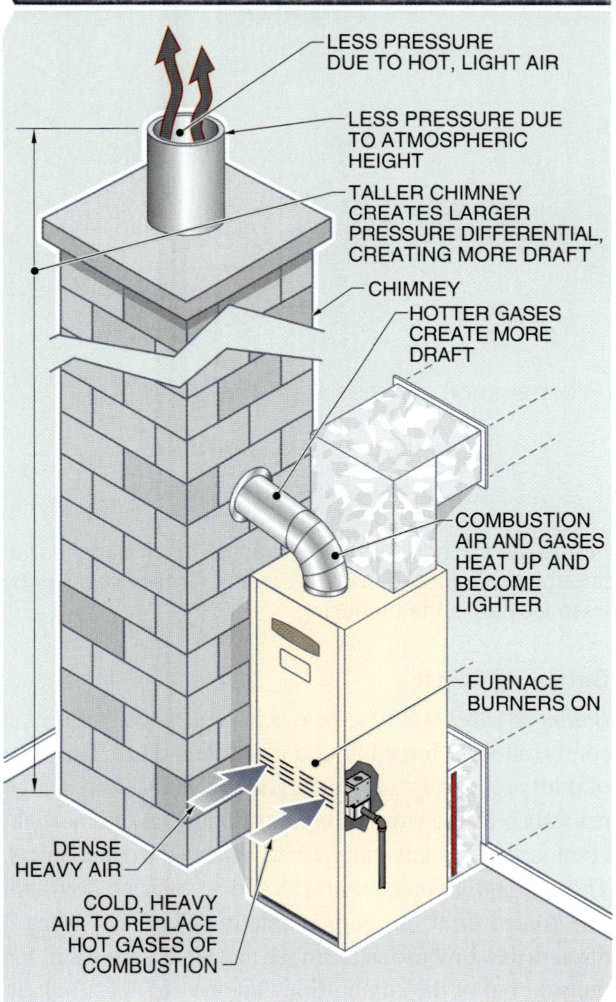

Figure 28-15. Natural draft is produced by air temperature and density differences between the cool atmospheric air entering the combustion chamber and the hot gases of combustion going up the chimney of a heating unit.

By controlling the flow of air and hot gases to the chimney, the amount of draft can be increased or decreased. The maximum amount of natural draft can be increased by increasing the chimney height. Controlling the temperature of the gases of combustion can also regulate natural draft. The more a gas is heated, the faster it rises. A column of heated gases of combustion weighs less than a column of cool air. Furnaces and boilers that employ natural draft are limited in size because of the limitations on chimney height.

Airflow proving switches are used to verify combustion fan operation.

Forced Draft

Forced draft is a type of mechanical draft created by air being pushed through a combustion chamber by a fan located on the outside of a furnace or boiler. Forced draft air is forced through burners to atomize the fuel in a heating unit. Therefore, the fire burns hotter because a greater quantity of air is available for combustion. A forced draft creates more pressure inside a combustion chamber than natural draft. However, if a crack appears in the wall of the combustion chamber, pressure is decreased and the poisonous gases of combustion could leak into the room, creating a hazard to occupants. **See Figure 28-16.**

TECH TIP

Carbon monoxide sensors are installed in buildings to sense the presence of carbon monoxide and alert occupants if the by-products of combustion are leaking from the combustion chamber.

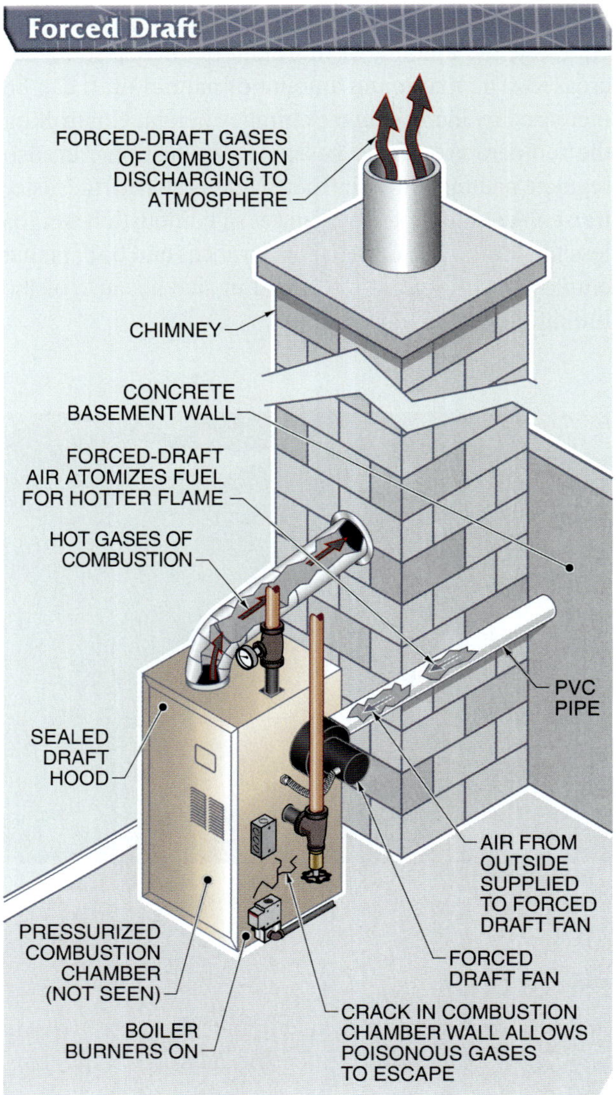

Figure 28-16. Forced draft is mechanical draft created by air being pushed through a combustion chamber by a fan located on the outside of the furnace or boiler.

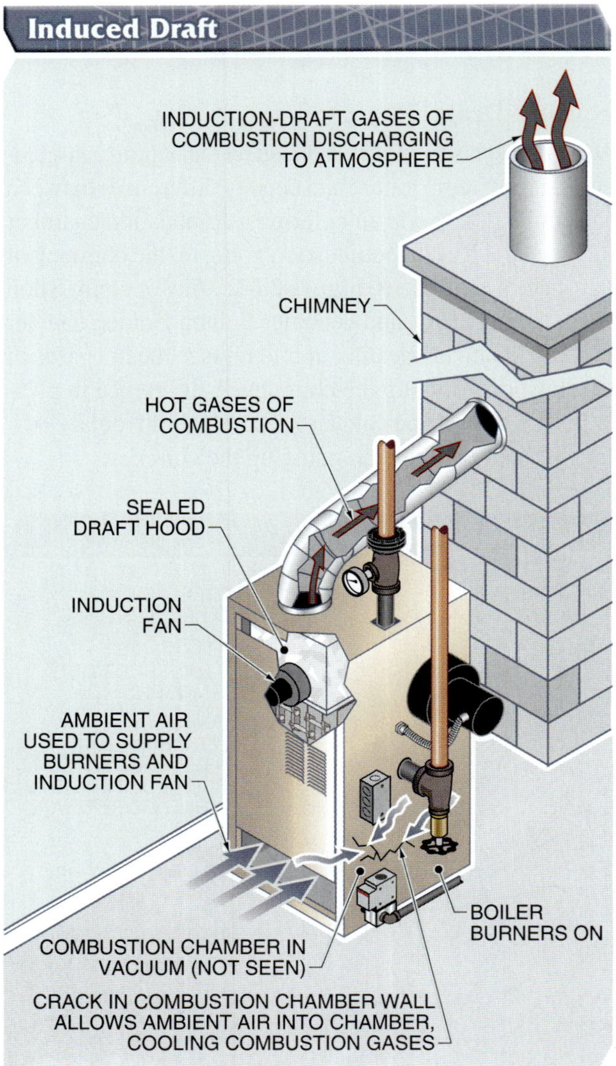

Figure 28-17. Induced draft is created by air being pulled through the combustion chamber of a furnace or boiler by a fan located in the breeching.

Induced Draft

Induced draft is a type of mechanical draft created by air being pulled through a furnace or boiler combustion chamber by a fan located in the breeching. The fan pulls air out of the combustion chamber and pushes it up the chimney. A vacuum is created in the combustion chamber, causing ambient air to take the place of the air that has been forced up the chimney. When there is a leak in the combustion chamber wall of a heating unit using induced draft, ambient air from the room will be allowed to enter the combustion chamber uncontrolled. This cools the gases of combustion and decreases boiler efficiency. **See Figure 28-17.**

Balanced Draft

Balanced draft is a type of mechanical draft that uses a combination of forced draft and induced draft. This type of draft is typically used on larger boilers that have a high resistance to the flow of combustion gases. Some light commercial designs incorporate balanced draft systems. There is both a forced draft fan and an induced draft fan. The forced draft fan pulls ambient air in and pushes it through the combustion chamber, and the induced draft fan pulls air out of the combustion chamber and pushes it up the chimney. The air pressure in the combustion chamber is slightly higher than the ambient air pressure. A leak in the combustion chamber wall would allow a slight amount of ambient air into the chamber. **See Figure 28-18.**

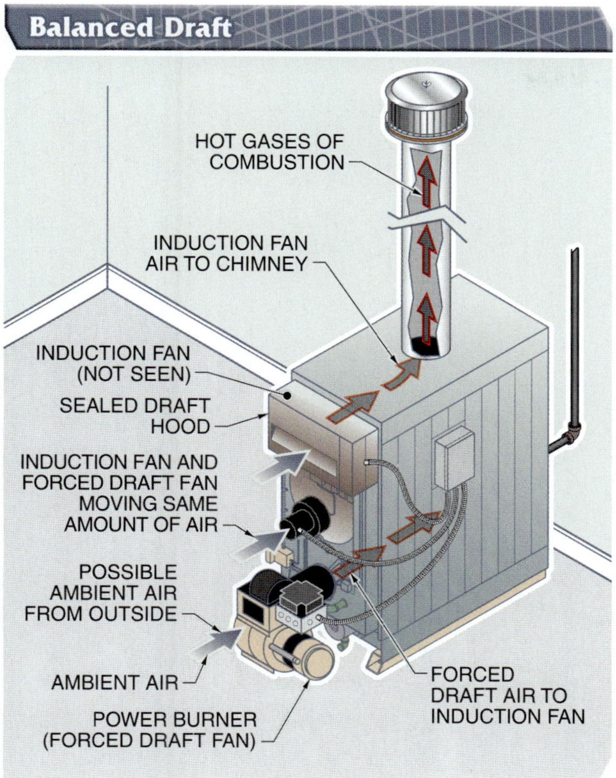

Figure 28-18. Balanced draft is a combination of forced draft and induced draft. It is typically used on commercial and industrial heating units.

Chimney and Vent Pipe Connections

All chimney and vent connections for gas furnaces and boilers must meet the requirements detailed in NFPA 54-2012 Part 7, *National Fuel Gas Code: Venting of Equipment.* Chimney sizing and all concerns related to vent installation must also meet the requirements. **See Figure 28-19.**

Heating Unit Chimney Sizes

Boiler Input*	Flue Size†		
	Height: 10 ft to 15 ft	Height: 15 ft to 25 ft	Height: 25 ft and Up
Up to 100,000	6 × 6	6 × 5	5 × 5
Up to 155,000	6 × 7	6 × 6	6 × 5
Up to 225,000	7 × 8	7 × 7	6 × 7
Up to 350,000	9 × 9	8 × 9	8 × 8

* in Btu per hour (Btuh)
† in inches (in.)

Figure 28-19. All chimney and vent connections for gas furnaces and boilers must meet the requirements detailed in NFPA 54-2012 Part 7, *National Fuel Gas Code: Venting of Equipment.*

Condensing Heating Unit Combustion Air and Vent Piping. Condensing heating units require fresh air for combustion as well as piping to vent the by-products of combustion outdoors. Various sizes are available for different types of equipment and locations. Manufacturer recommendations and installation procedures must always be followed. In many cases, installation documents are available from the manufacturer on the Internet. **See Figure 28-20.**

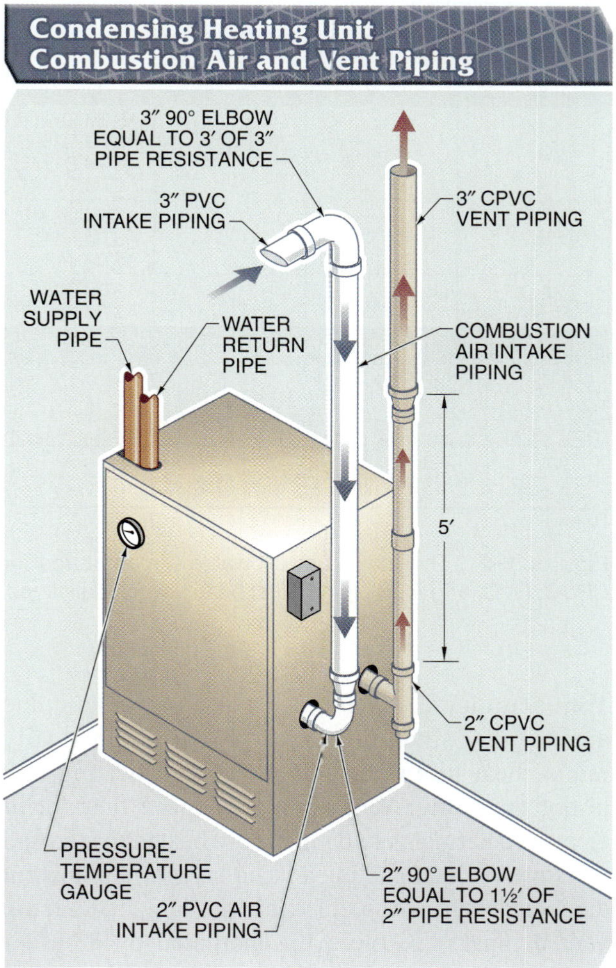

Figure 28-20. Condensing heating units require fresh air for combustion and piping to vent the by-products of combustion outdoors.

Combustion air piping and vent piping on condensing units are typically sized for 2″ CPVC pipe. Residential and light commercial pipe sizes can be 2″ or 3″ inside the cabinet of a heating unit. However, any changes to 3″ CPVC pipe must be made outside the cabinet. Since the hot gases of combustion are over 155°F, the first 5′ of the vent pipe must be CPVC pipe. After the first 5′ of CPVC piping, the remaining vent pipe can be PVC pipe. Any elbows or other fittings found within the first 5′ of the vent must also be of CPVC material. **See Figure 28-21.**

Condensing Unit Vent Pipe Specifications

Maximum Allowable Temperatures for Nonmetallic Vent Pipe			
Material	°F	°C	Standard
PVC	158	70	ASTM F891 ASTM D2665 ASTM D1785 ASTM D2241
CPVC	210	100	ASTM D2846 ASTM F441 ASTM 442
ABS	180	82	ASTM D2661 ASTM F628

Vent Piping Lengths				
Boiler Size	2″ Pipe Minimum Venting*	2″ Pipe Maximum Venting*	3″ Pipe Minimum Venting*	3″ Pipe Maximum Venting*
100	8	21	27	92
50 & 75	8	26	32	112

* in feet (ft)

Figure 28-21. The maximum allowable temperatures for CPVC, PVC, and ABS piping must be followed at all times.

Vent Dampers. A *vent damper* is an automatic flue damper that shuts off a flue pipe to dramatically reduce heat loss when the burner of a heating unit is not operating. Vent dampers can be thermal or electric. Thermal vent dampers are heat activated and are slow to open and close. The most common vent damper design is the electric vent damper. Electric vent dampers are typically energized by a burner ignition circuit. When the burner ignites, the vent damper opens. Vent dampers are always used with boilers that have draft hoods. Since most boilers still use draft hoods, vent dampers are used more often with boilers than furnaces, which do not require draft hoods. **See Figure 28-22. See Appendix.**

Vent and Barometric Damper Operation Media Clip

Vent Dampers

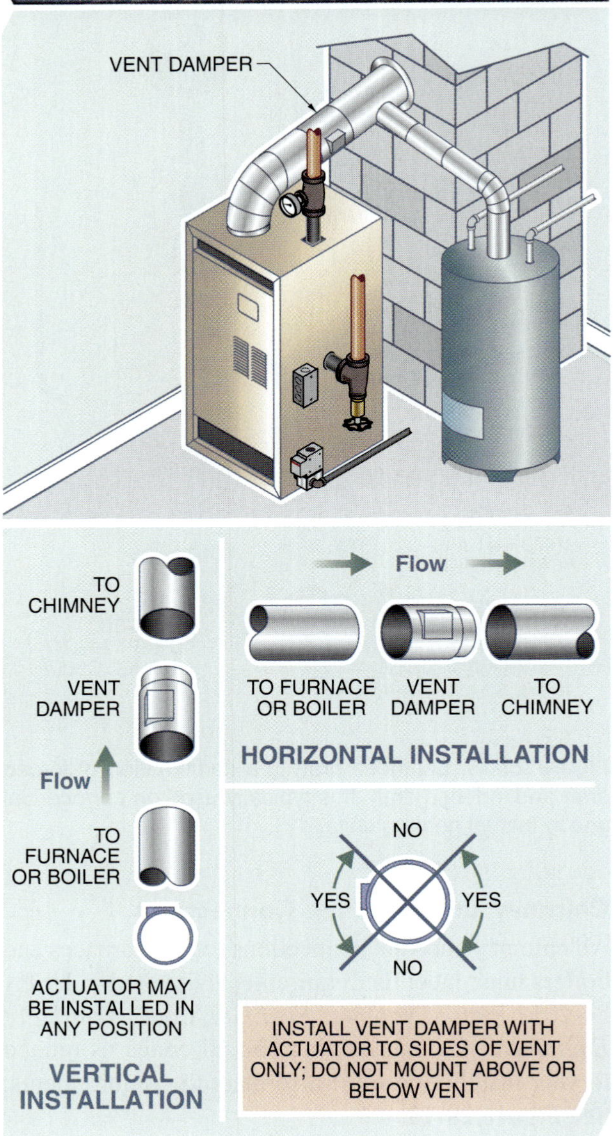

Figure 28-22. Vent dampers are automatic flue dampers that shut off flue pipes to dramatically reduce heat loss when a heating unit burner is not operating.

Barometric Dampers. A *barometric damper* is metal plate positioned in the opening of a flue that allows atmospheric pressure to balance the pressure at the bottom of a chimney to control the amount of airflow up through the combustion chamber and flue. Barometric dampers are typically used with oil-fired boilers. They may also be used with furnaces and boilers connected to very tall chimneys. Barometric damper operating pressure is measured with a draft gauge or manometer, and the damper is adjusted accordingly. **See Figure 28-23.**

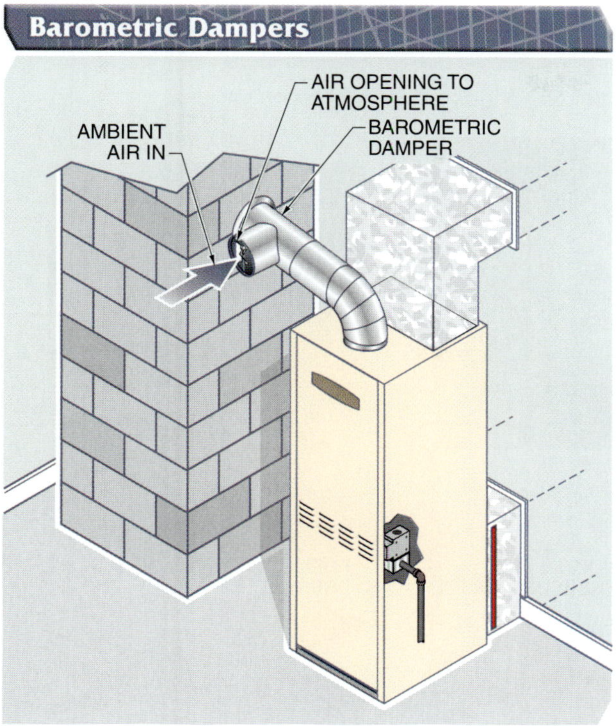

Figure 28-23. Barometric dampers allow air into flue systems to balance the pressure at the bottom of chimneys.

FUEL SYSTEMS

Fuel systems used with furnaces and boilers store, supply, and control the fuel supplied to the heating unit. In the combustion process, fuel mixed with air is burned to produce the heat necessary to operate the unit. The heat from the gases of combustion is transferred throughout the heat exchanger of the furnace or boiler for maximum efficiency. **See Figure 28-24.**

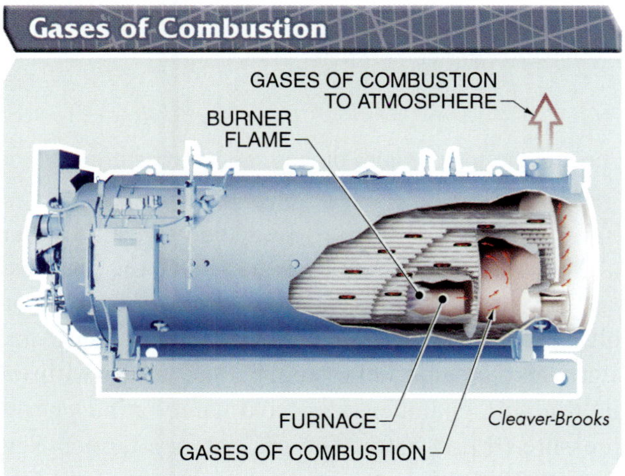

Figure 28-24. A fuel system prepares fuel for combustion to maximize heat transfer efficiency.

Fuels typically used in heating units include fuel oil, natural gas, propane, and coal. In some cases, electricity is used to produce heat in a heating system. The type of fuel used is determined by the design of the heating unit, the price and availability of the fuel, and its compliance with air pollution regulations. The type of fuel used determines the fuel system accessories. All fuels are combustible and dangerous if safety precautions are not followed.

Combination fuel oil and natural gas burners are popular in heating systems.

TECH TIP

While availability and price are determining factors when selecting a fuel for use in a heating installation, both may change over time. Prices fluctuate, and a fuel may become available due to improvements in infrastructure.

Fuel oil systems include all the devices required to safely and efficiently operate the fuel oil burner. Fuel oil accessories clean, control the temperature of, and regulate the pressure of fuel oil. **See Figure 28-25.** Fuel oil purity varies, and equipment to remove foreign matter is required to prevent buildup in the lines that restricts flow to the burner, causing inefficient combustion. Fuel oil accessories for a heavy fuel oil system require additional equipment for heating of the fuel oil for proper burner operation. Fuel oil accessories include storage tanks, pumps, strainers, and heaters. The main components of a gas fuel system include shutoff cocks, gas pressure regulators, main gas valves, butterfly valves, low and high gas pressure switches, and vent valves.

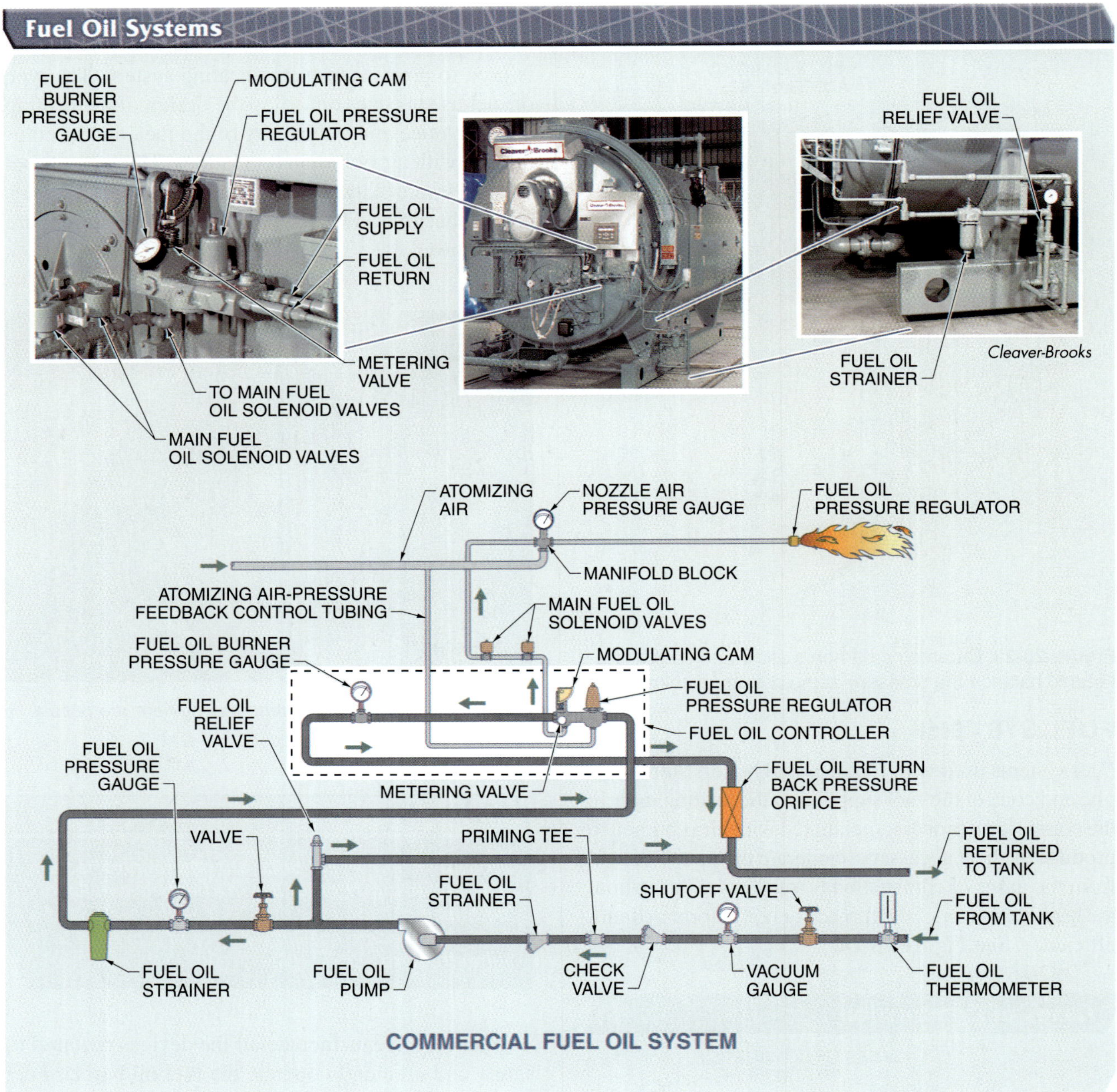

Figure 28-25. Fuel oil systems include all the devices to clean, control the temperature of, and regulate the pressure of fuel oil to safely and efficiently operate fuel oil burners.

TECH TIP

When available, natural gas has many advantages as a fuel. It has a very low carbon footprint, adapts well to automatic control, and burns very cleanly. Recently, large amounts of natural gas have been discovered, which should keep prices low for an extended time period.

When natural gas is used as a fuel for furnaces and boilers, either a low-pressure gas system or high-pressure gas system is used. A *low-pressure gas system* is a type of fuel system that mixes gas with air in a mixing chamber at approximately atmospheric pressure (0 psi) before the gas enters a burner. **See Figure 28-26. See Appendix.** A *high-pressure gas system* is a fuel system that mixes gas with air at pressures up to 5 psi. **See Figure 28-27.**

Low-Pressure Gas Systems

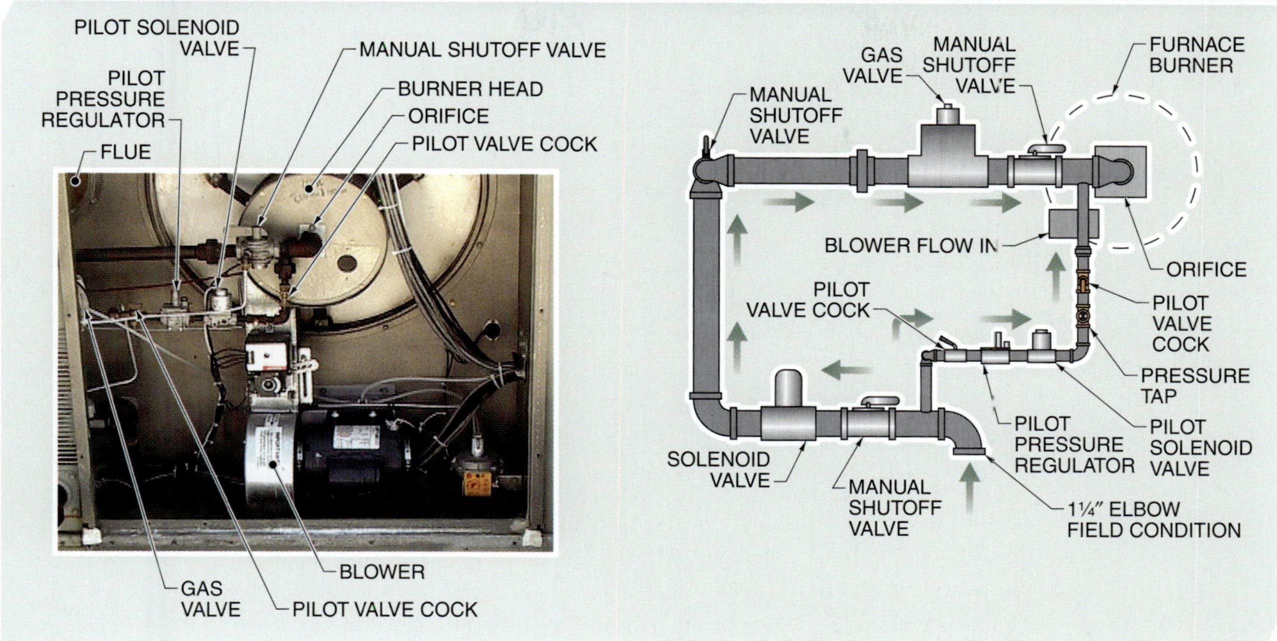

Figure 28-26. In a low-pressure gas system, the rate of combustion is controlled by the amount of air supplied to the orifice.

High-Pressure Gas System

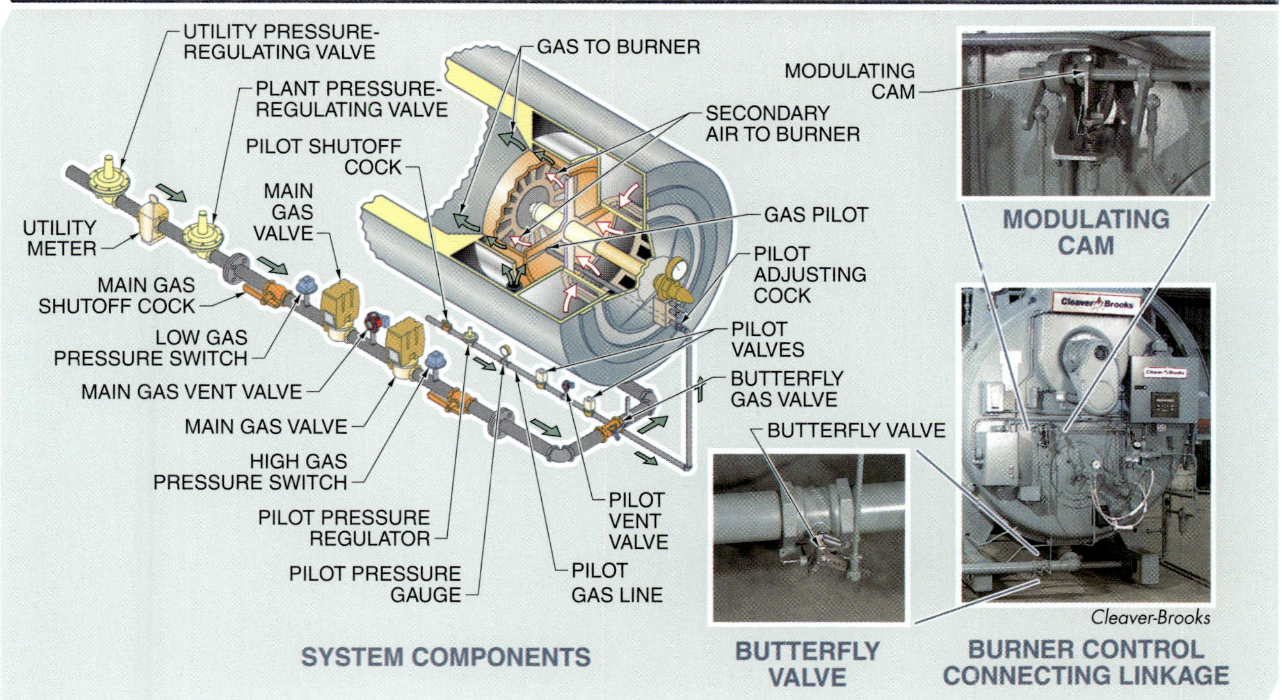

Figure 28-27. In a high-pressure gas system, the rate of combustion is controlled by the gas pressure regulator, safety shutoff valves, and firing rate valve.

Refrigeration Principles

UNIT 29

Refrigeration is the process of moving heat from where it is not wanted to a place where it can be properly rejected. Refrigeration is based on the second law of thermodynamics, which states that matter gains or loses heat when it changes state. Mechanical compression refrigeration systems use mechanical equipment to produce the refrigeration effect. All mechanical refrigeration systems use condensers to reject heat and evaporators to absorb heat.

OBJECTIVES

- List the main components of a mechanical compression refrigeration system that perform the heat-moving process.
- Describe the basic function of each refrigeration system component.

Index to Procedures

REFRIGERATION CYCLE

Refrigeration systems are heat-moving machines. According to the second law of thermodynamics, heat energy always moves from a substance at a higher temperature to a substance at a lower temperature. The refrigeration process is used to cool products or provide human comfort.

A refrigeration system consists of components that are controlled by devices that regulate temperature and pressure in order to facilitate the heat-moving process. The main components of a refrigeration system that perform the heat-moving process are the evaporator and the condenser. **See Figure 29-1.**

Evaporators

An evaporator is the part of a refrigeration system responsible for the absorption of heat from a product or building spaces into the fluid (refrigerant) of the system. Heat absorption occurs at a low temperature and pressure and is known as the low-pressure side of a refrigeration system.

Condensers

A condenser is the part of a refrigeration system that rejects heat. Heat rejection occurs at a high temperature and pressure and is known as the high-pressure side of a refrigeration system. The rejected heat energy can be captured and used to heat other processes or is rejected to the atmosphere.

TECH TIP

Air conditioning and refrigeration systems have differences that require different knowledge and experience when troubleshooting and servicing.

Refrigeration System Heat Movement

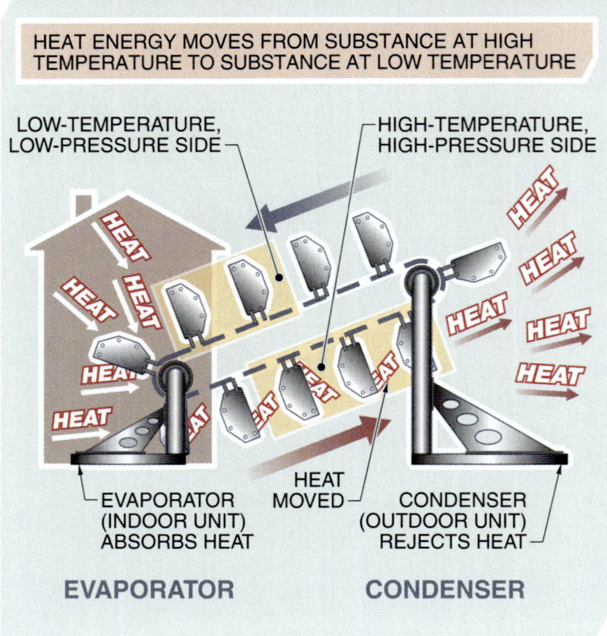

Figure 29-1. Heat is absorbed by the low-pressure side evaporator and moved into the refrigerant. The refrigerant transports the heat to the condenser. The high-pressure side condenser rejects the heat from the refrigerant to the atmosphere.

Mechanical Compression Refrigeration Systems

Before the twentieth century, the only way to provide refrigeration was to use a natural process. Natural refrigeration processes include placing a substance in contact with cold rain, storing ice blocks from a frozen lake or pond for later use, or using cool outside air circulation to provide human comfort. Natural processes are limited to those climates and locations where the proper conditions are present.

Today, mechanical compression refrigeration systems make the refrigeration process easier. Mechanical compression refrigeration system components include compressors, refrigerants, condensers, expansion devices (possibly valves), evaporators, piping, and accessories to produce the refrigeration effect. **See Figure 29-2.**

Mechanical Compression Refrigeration System Operation Media Clip

Mechanical Compression Refrigeration Systems

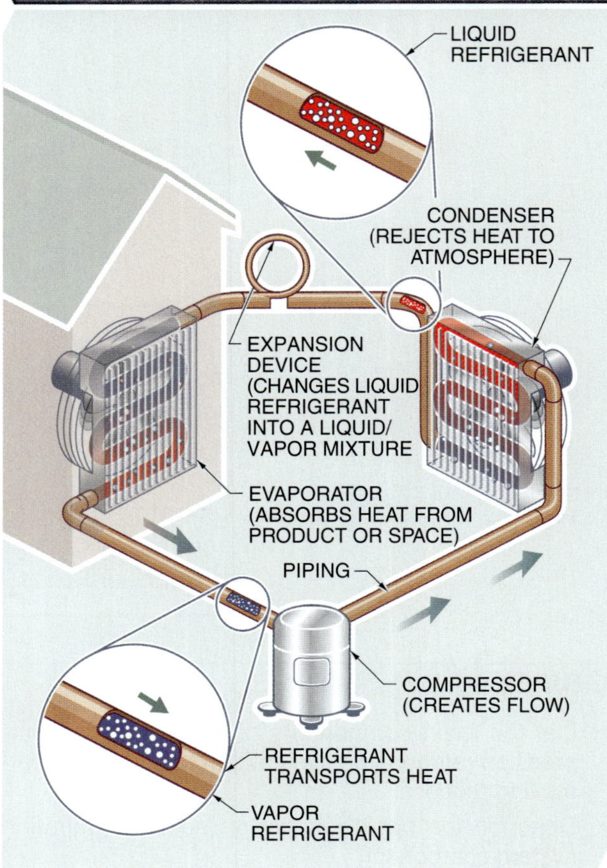

Figure 29-2. Mechanical compression refrigeration system components include compressors, refrigerants, condensers, expansion devices, evaporators, piping, and accessories.

TECH TIP

Compressor design and reliability have improved dramatically due to advanced manufacturing techniques and computer-aided design.

Compressors. A *compressor* is a refrigeration system component that creates refrigerant flow. A compressor generates a pressure difference, and the system resistance determines the flow rate. The greater the resistance, the greater the pressure. The increase in fluid pressure and temperature found downstream from the compressor allows heat to be rejected to a cooler medium, such as air or water. Mechanical compressors have different designs such as reciprocating, scroll, screw, and centrifugal. **See Figure 29-3.**

Compressor Designs

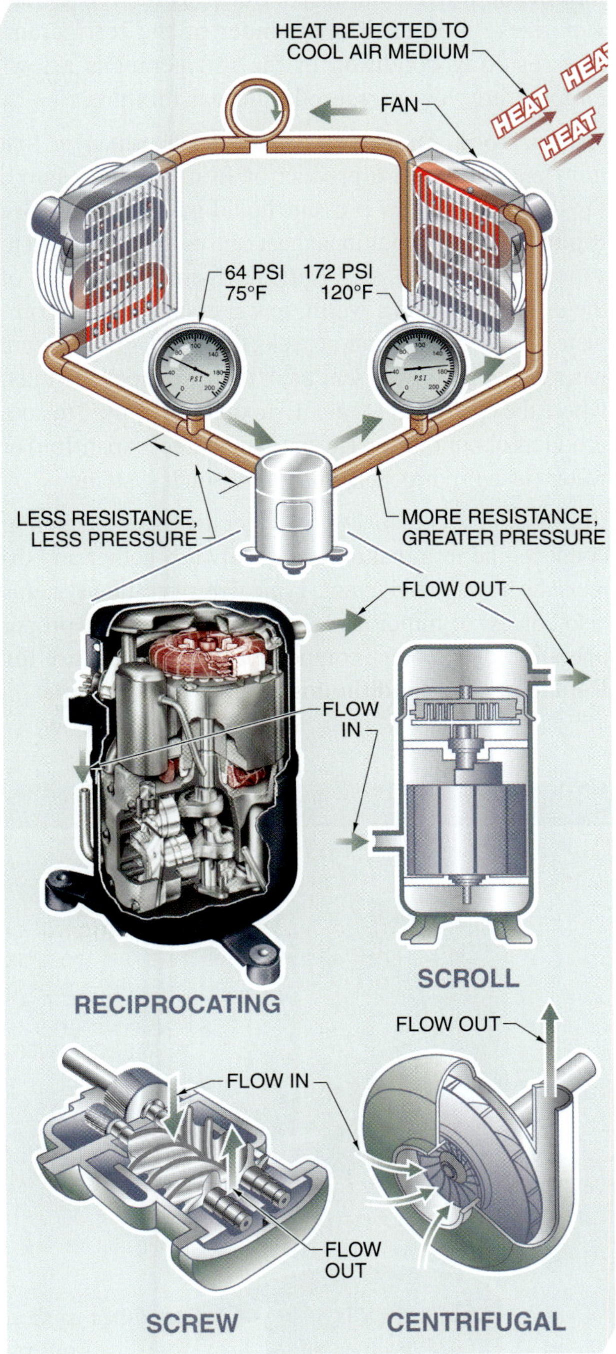

Figure 29-3. Compressors create refrigerant flow. Resistance to the flow creates pressure in the system.

Refrigerants. A *refrigerant* is a substance that moves heat by changing state from a liquid to a gas when absorbing heat and from a gas to a liquid when rejecting heat. **See Figure 29-4.** Refrigerants typically have a low boiling point, or vaporization point, that allows them to receive heat energy from the substance that needs to be cooled, such as room air when in use in an air conditioning system.

Refrigerants

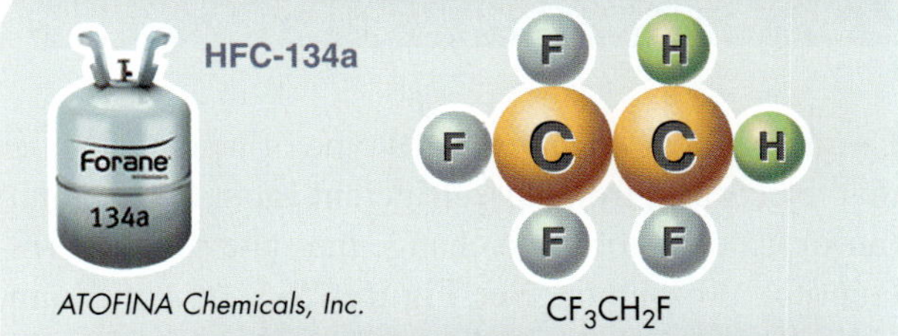

Properties	R-134a
Molecular weight	102.0
Normal boiling point (14.7 PS A)	-14.9
Critical temperature	214.1
ASHRAE safety group classification	A1
Ozone depletion potential (ODP)	0.0
Global warming potential (GWP)	1300

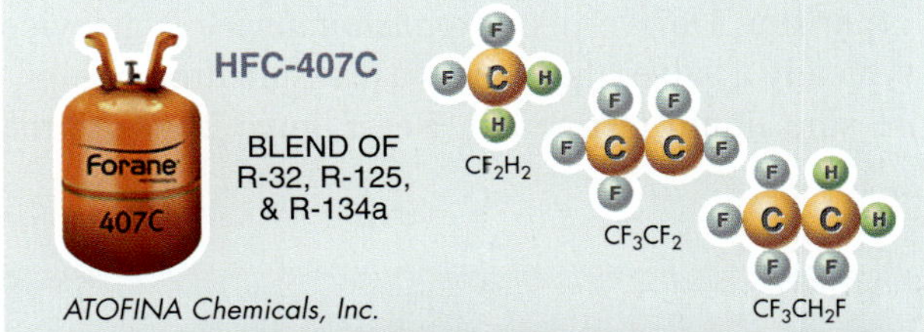

Properties	R-407C
Average molecular weight	86.2
Normal boiling point (14.7 PSIA)	-46.1
Critical temperature	187.2
ASHRAE safety group classification	A1
Ozone depletion potential (ODP)	0.0
Global warming potential (GWP)	1525

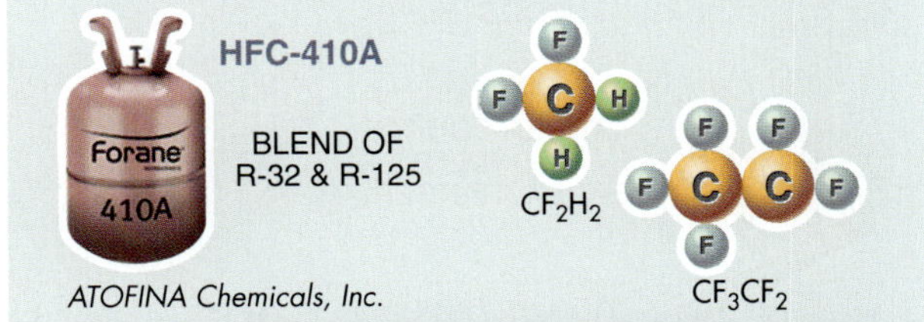

Properties	R-410A
Average molecular weight	72.6
Normal boiling point (14.7 PSIA)	-61.6
Critical temperature	162.0
ASHRAE safety group classification	A1
Ozone depletion potential (ODP)	0.0
Global warming potential (GWP)	1725

Figure 29-4. Mechanical compression refrigeration systems typically use refrigerants with low boiling points that easily change state from a liquid to a gas.

TECH TIP

Recent advances in condenser design have allowed condensers to become more efficient. A recent innovation allows more condenser surface area in a smaller area. The additional surface area permits higher efficiency.

Condensers. A *condenser* is a type of heat exchanger that rejects heat from hot refrigerant vapor to a medium (air or water) in order to change the state of the refrigerant to 100% liquid. **See Figure 29-5.** As a medium, such as air or water, circulates over the heat exchanger, heat is removed from the refrigerant. When the heat is removed, the refrigerant changes its state, or condenses fully, into a 100% liquid. The rejected heat is the heat that was originally absorbed from the building air or substance by the evaporator. Condensers are designed as air-source or water-source units.

Expansion Devices. An *expansion device* is a device (typically a valve) that causes a pressure drop of high-pressure liquid refrigerant. **See Figure 29-6.** As the pressure of the refrigerant drops, the boiling temperature of the refrigerant drops. Some of the refrigerant vaporizes, cooling the remainder of the refrigerant. The resulting condition of the refrigerant is a low-temperature, low-pressure liquid-gas mixture.

Evaporators. An *evaporator* is a heat exchanger that transfers heat from a product or from room air into a low-temperature, low-pressure liquid-gas refrigerant. **See Figure 29-7.** The additional heat causes the refrigerant to vaporize to a 100% gas. As the refrigerant moves out of the evaporator, it is a warm, low-pressure gas. To complete the basic refrigeration cycle, the warm, low-pressure gas refrigerant is then sent back to the compressor inlet, where the cycle starts again. Evaporators are used to cool products or building space air (using direct expansion) or water (using indirect expansion) in some systems.

Piping and Accessories. Refrigeration system piping connects the terminal units and valves together and directs the flow of refrigerant. Typical refrigeration systems also consist of minor accessories used to ensure proper operation of the other components. **See Procedure for Replacing Air Conditioning System Components.**

Condensers

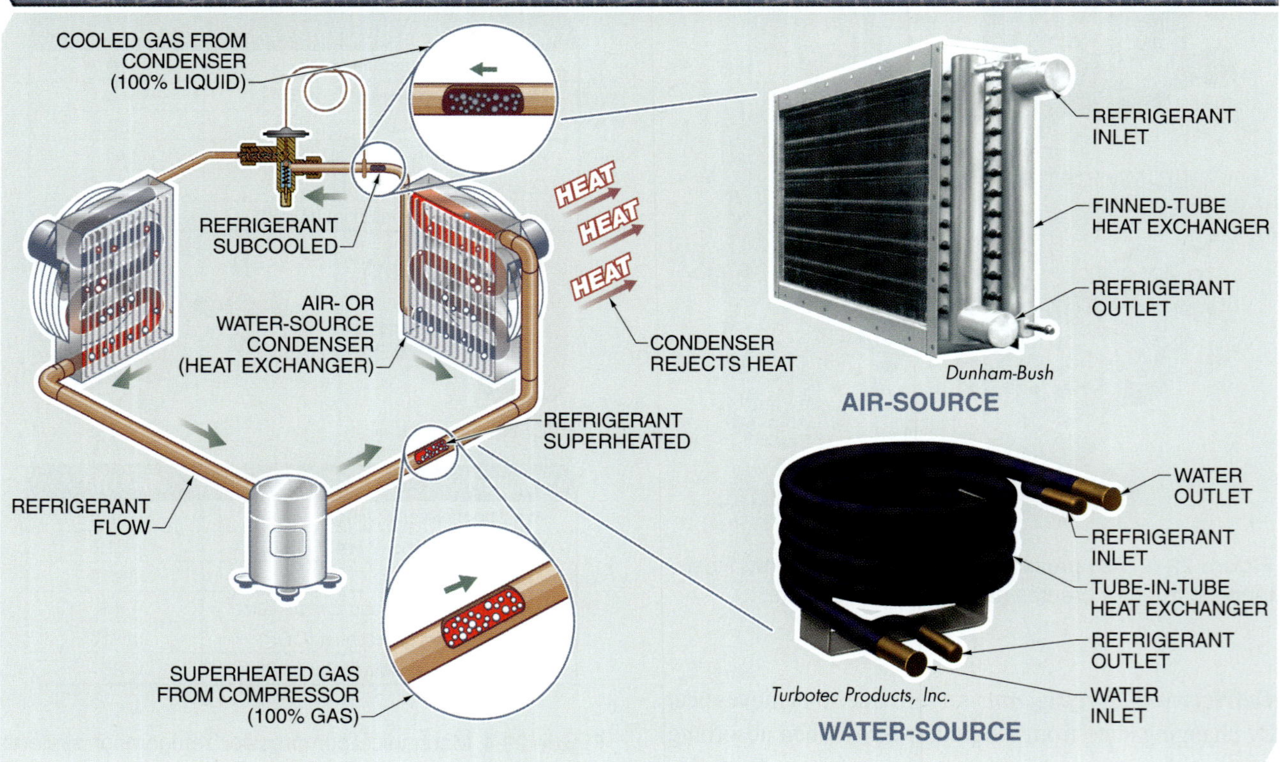

Figure 29-5. Condensers reject the heat in refrigerant, causing the refrigerant to become a 100% liquid.

Expansion Devices

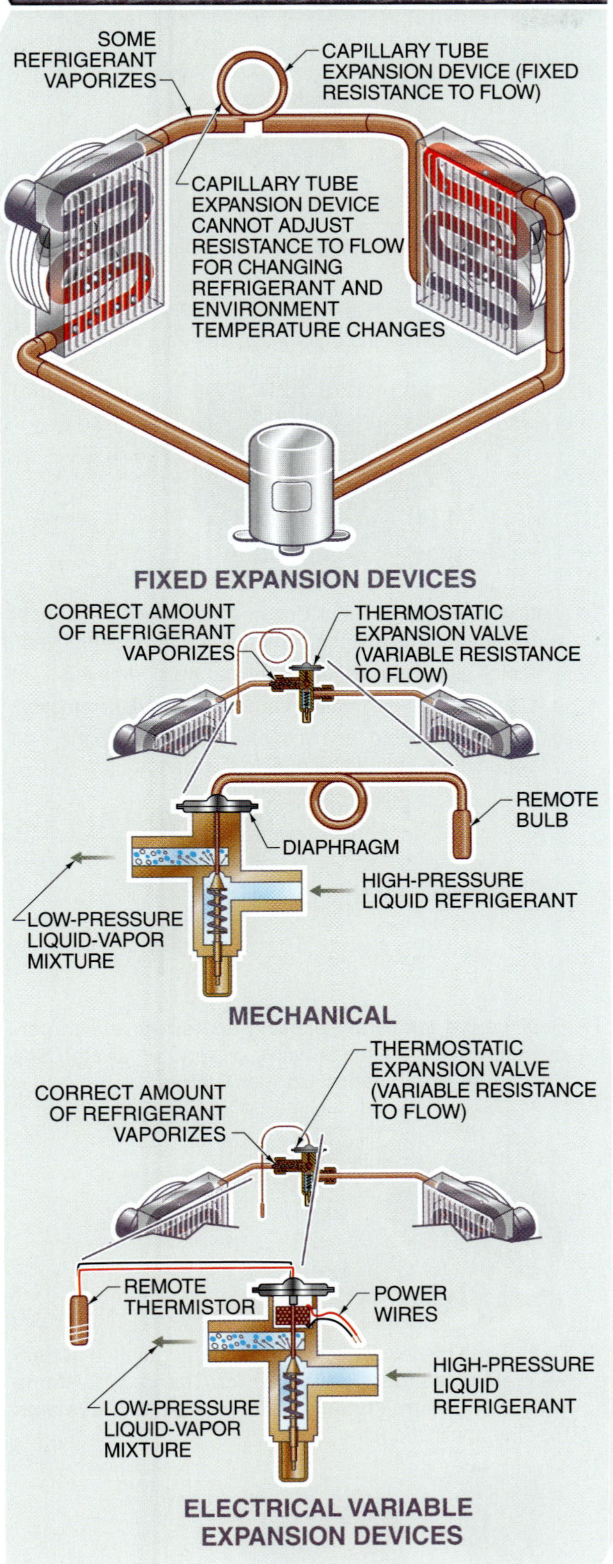

Figure 29-6. Fixed- or variable-resistance expansion devices change liquid refrigerant into an 80% liquid refrigerant.

Evaporators

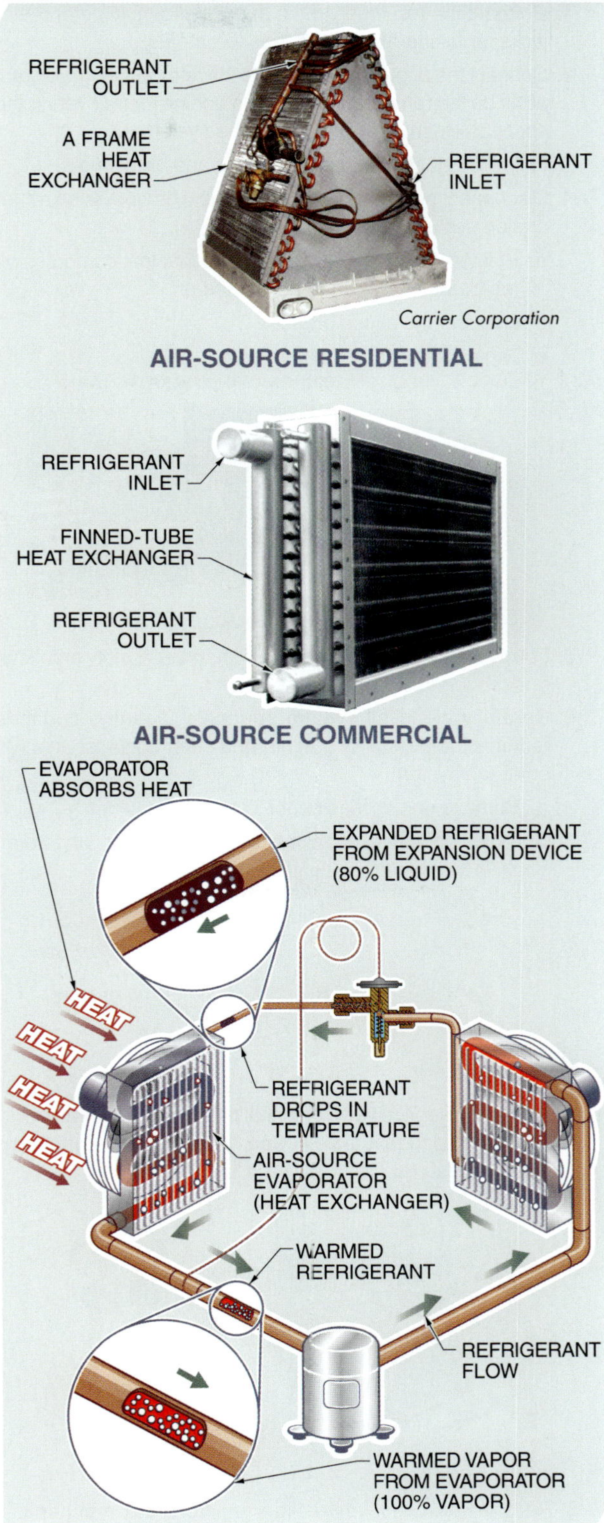

Figure 29-7. Evaporators are heat exchangers that transfer heat from a product or from room air into refrigerant that vaporizes to a 100% gas.

Procedure

Replacing Air Conditioning System Components

1. Connect the test leads to the appropriate multimeter meter jacks for taking AC voltage measurements.
2. Connect the appropriate hoses to the appropriate refrigerant ports on the refrigerant recovery machine for recovering the specific refrigerant contained in the system.
3. Follow all electrical safety practices and procedures.
4. Disconnect power from the mechanical compression refrigeration system.
   a. Pull the circuit breaker for the mechanical compression refrigeration system to be worked on.
   b. Lock out the circuit breaker.
   c. Tag out the circuit breaker.

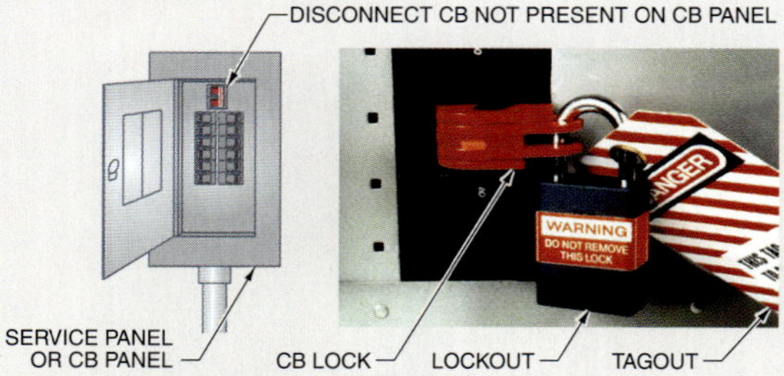

5. Follow all refrigerant recovery and handling safety practices and procedures.
6. Carefully clean all system surfaces. Cleaning surfaces reduces the possibility that dirt and other contaminants will enter the system.
   a. Remove unit access panel, covers, or cabinet.
   b. Clean the system components with a rag, soap, and water.

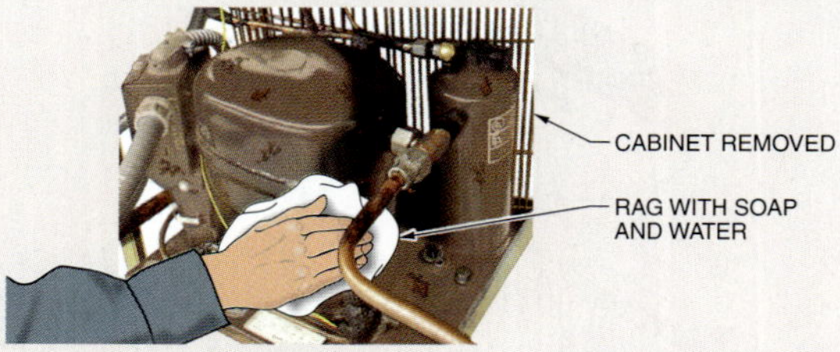

7. Uncap the service valve or install piercing access valve and connect to gauge manifold. **See Procedure for Installing Piercing Valves.**

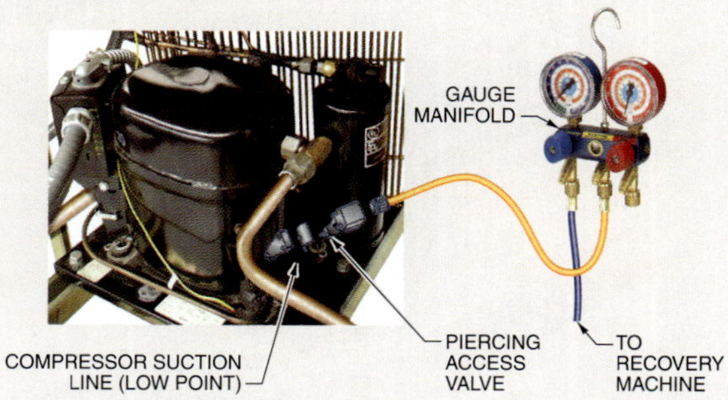

8. Remove and recover refrigerant from the mechanical compression refrigeration system. Follow all regulations set forth by Section 608 of the Clean Air Act and the Environmental Protection Agency (EPA). **See Procedure for Removing and Recovering Refrigerant—Type I Systems** or **Type II Systems** (per the application).

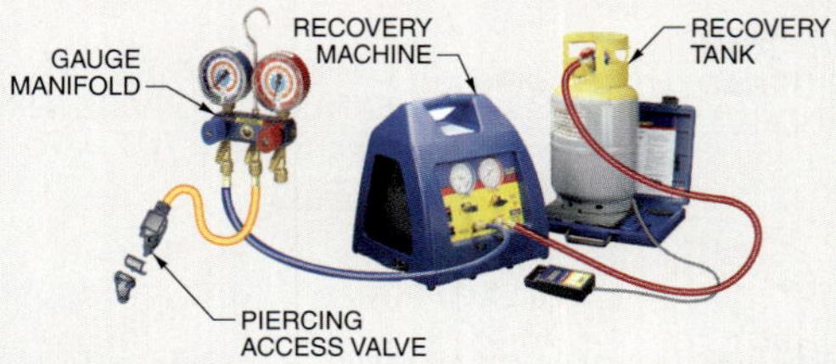

9. Unfasten or cut the tubing before and after the component to be replaced.
   a. Disconnect all wiring and/or wiring harnesses as needed.
   b. Remove all holding bolts, nuts, and pins as needed.

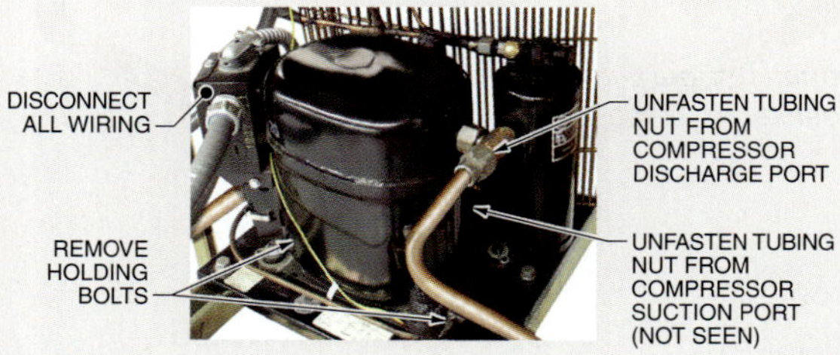

10. Install the new component. Ensure that the new component is designed to replace the component that was removed.
    a. Install all bolts, nuts, and pins that were removed.
    b. Connect all wires and wire harnesses that were removed.
    c. Install fittings into the suction and discharge ports, and connect the tubing and nuts to the fittings.

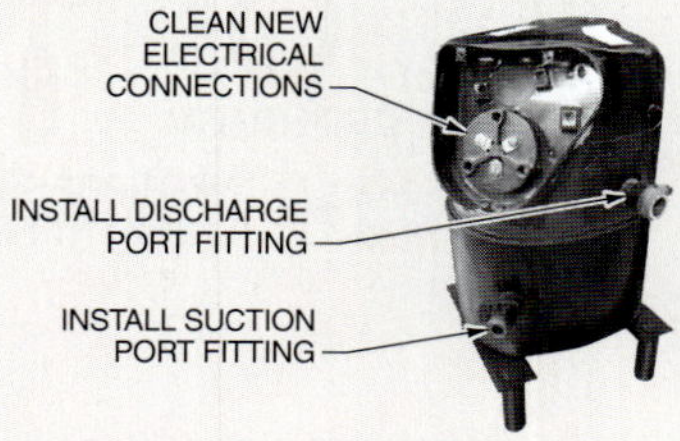

11. Replace the temporary piercing access valve with permanently soldered saddle valve or service valve(s). **See Procedure for Sweating Copper Tubing.**

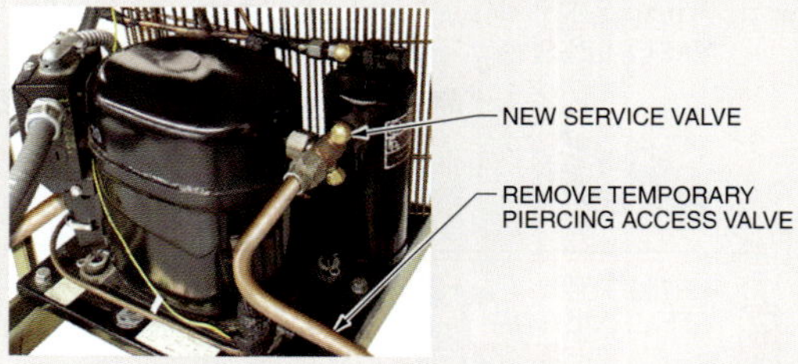

12. Charge the mechanical compression refrigeration system with refrigerant. **See Procedure for Charging Systems with Refrigerant—Type I Systems** or **Type II Systems** (per the application).

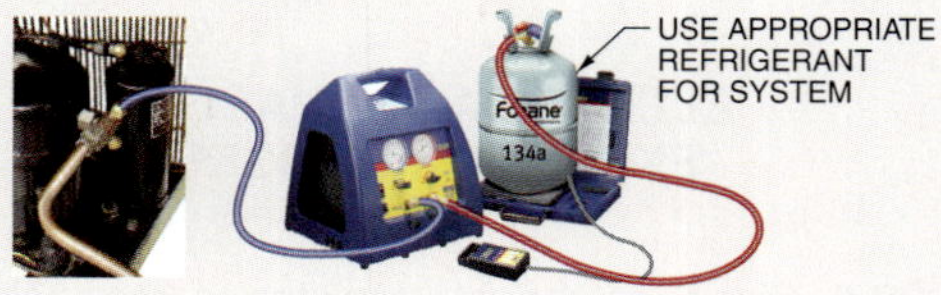

13. Connect power to the mechanical compression refrigeration system.

Product refrigeration systems often use a compressor and condenser mounted in a rack.

Heat Transfer

Several devices in a refrigeration system are capable of transferring heat. Most of the heat is transferred by the evaporator and condenser. **See Figure 29-8.**

Heat can enter a system by exposing the piping to hot ambient air temperatures and to the sun on a hot summer day. Heat is also added to the refrigerant by the compressor and flow through tight bends. Specialized heat exchangers may be added to a refrigeration system to recover unused heat or waste heat that is typically rejected to the atmosphere. The otherwise unused heat is utilized in a separate process such as heating water for domestic use.

Heat Transfer

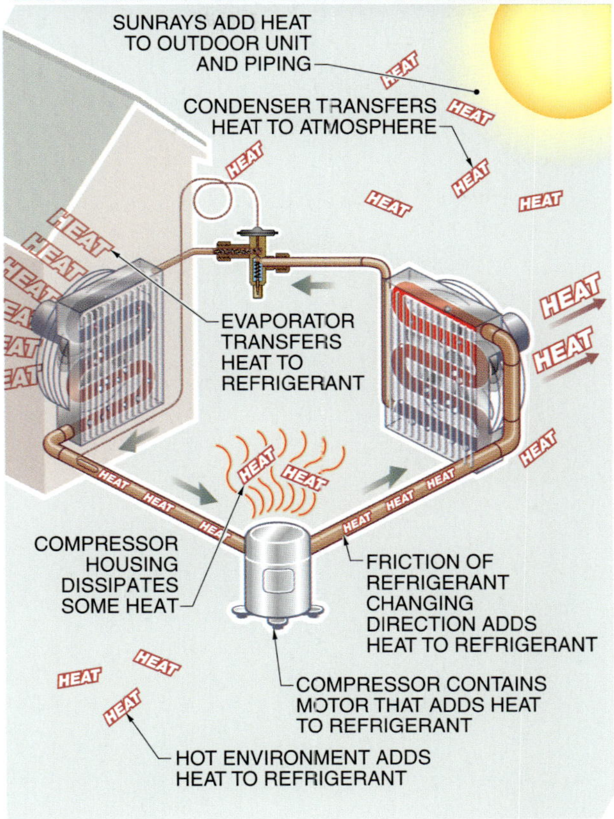

Figure 29-8. Hot environments, electric motors of compressors, flow changing direction, and sunrays all add heat into a mechanical compression refrigeration system.

UNIT 30

Low-Pressure-Side Evaporators

The low-pressure side of a refrigeration system cools the medium (air or water) and heats the refrigerant for later rejection. The low-pressure side is composed of the expansion device and evaporator and continues on to the suction side of the compressor. The evaporator transfers the heat of the medium (typically air) to the refrigerant, creating the beginning of the refrigeration effect.

OBJECTIVES

- Identify the low-pressure-side components of a refrigeration system.
- Describe the functions of an evaporator.
- List the different temperature range categories of evaporators.
- Identify the various types of evaporator construction.
- Describe methods of evaporator moisture removal and defrosting.
- Describe factors that affect evaporator performance.
- Determine the cooling capacity of an evaporator.

LOW-PRESSURE SIDE

The low-pressure side of a mechanical refrigeration system consists of an expansion device, evaporator, suction line, and number of accessories. **See Figure 30-1.** The expansion device regulates refrigerant flow into the evaporator. The cool, high-pressure liquid refrigerant passes through the expansion device. As the liquid refrigerant passes through the expansion device, a pressure drop occurs. This pressure drop allows some of the liquid refrigerant to vaporize. As the refrigerant vaporizes, the boiling refrigerant absorbs heat from the liquid refrigerant, allowing the remainder of the liquid refrigerant to become colder. The 20% vapor and 80% liquid refrigerant then enters the evaporator where it picks up heat from the medium (air or water) circulated across it. The refrigerant absorbs heat from the medium and becomes a 100% gas.

The result is that the air or water medium is cooled. The refrigerant may continue to pick up more heat from the evaporator if sized to do so, causing the refrigerant to become superheated. The compressor suction line allows the low-pressure area of the compressor's suction port to create a low-pressure area at the discharge of the evaporator. The compressor suction line then brings the gas refrigerant to the compressor.

TECH TIP

Refrigeration system performance can be measured at the evaporator and condenser. When troubleshooting a system, both pressures must be measured because any problem that affects one side of the system will affect the other. This should be kept in mind to avoid arriving at the wrong conclusion when troubleshooting a system.

Refrigeration System Low-Pressure Side

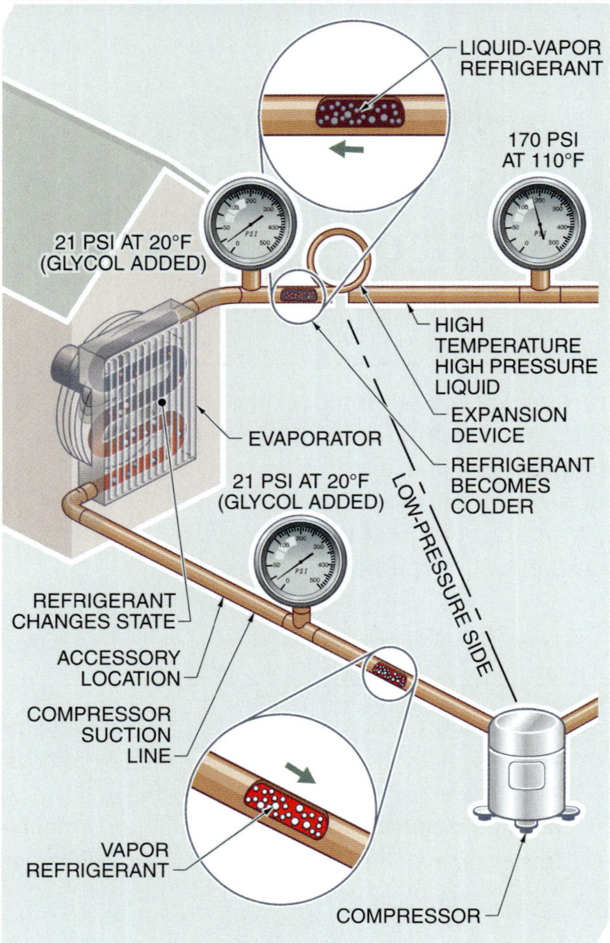

Figure 30-1. The low-pressure side of a mechanical refrigeration system contains refrigerant in a gaseous state and includes devices such as an expansion device, evaporator, suction line, and a number of accessories.

Evaporators

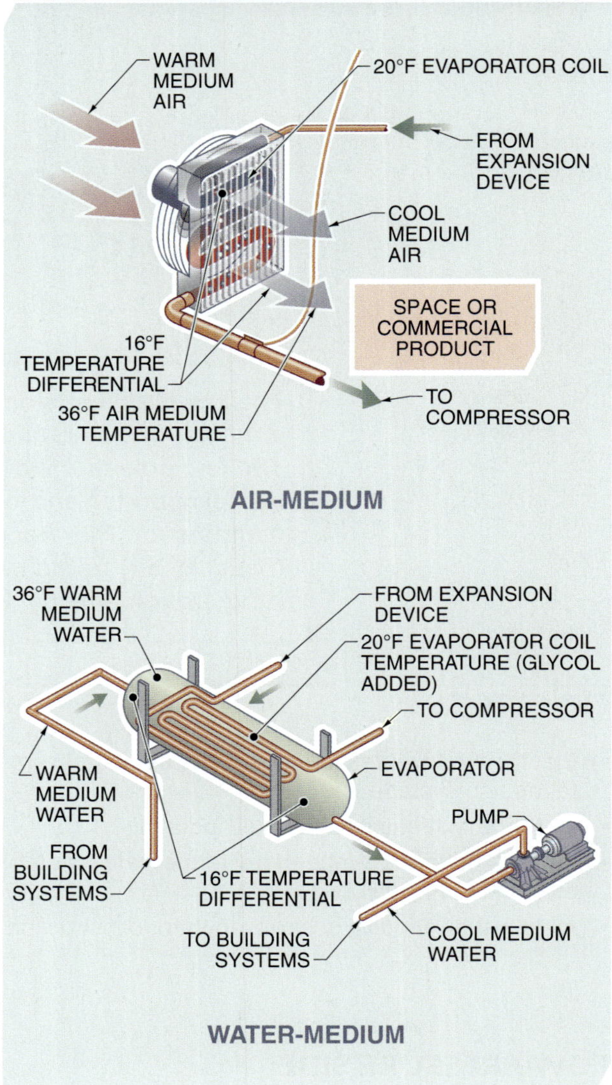

Figure 30-2. To cool a substance such as air or water, the evaporator must be at a lower temperature than the substance.

EVAPORATORS

The evaporator transfers heat from the medium being cooled (air or water) into the refrigerant. To accomplish this, the evaporator must be at a lower temperature than the air or water being cooled. **See Figure 30-2.** For example, when water is being cooled to 45°F, the evaporator must be at a lower temperature, such as 35°F. Keeping a product at 36°F means that the evaporator coil temperature should be approximately 20°F, providing a temperature differential of 16°F.

Evaporator Operation Media Clip

When air is the medium being cooled, the heat removed is in the form of both sensible and latent heat. The sensible heat removed is measured as a temperature change. The latent heat removed is measured as moisture removed from the medium air as part of the dehumidification process. The cool air circulates over the product, cooling the product. Different design parameters are used in evaporator design and sizing. These design parameters consist of physical size, heat transfer area, piping size, and area and spacing of the fins when a tube-and-fin evaporator is present.

Evaporator Temperature Range Categories

A properly operating evaporator contains boiling liquid refrigerant. The temperature of this boiling liquid refrigerant determines the temperature of the evaporator coil. The saturated relationship between temperature and pressure can be found on a pressure/temperature (P/T) chart. Temperature ranges are used by industry to categorize evaporator applications in refrigeration systems. **See Figure 30-3.** Common evaporator temperature range categories include high temperatures, medium temperatures, low temperatures, and ultra-low temperatures.

High-Temperature Applications. The most common high-temperature application is air conditioning, which strives to maintain a temperature of about 75°F. Product storage for perishables, such as flowers, is also considered a high-temperature application, with coil temperatures of 40°F to 50°F.

Evaporator coils can be located close to the ceiling in air conditioning applications.

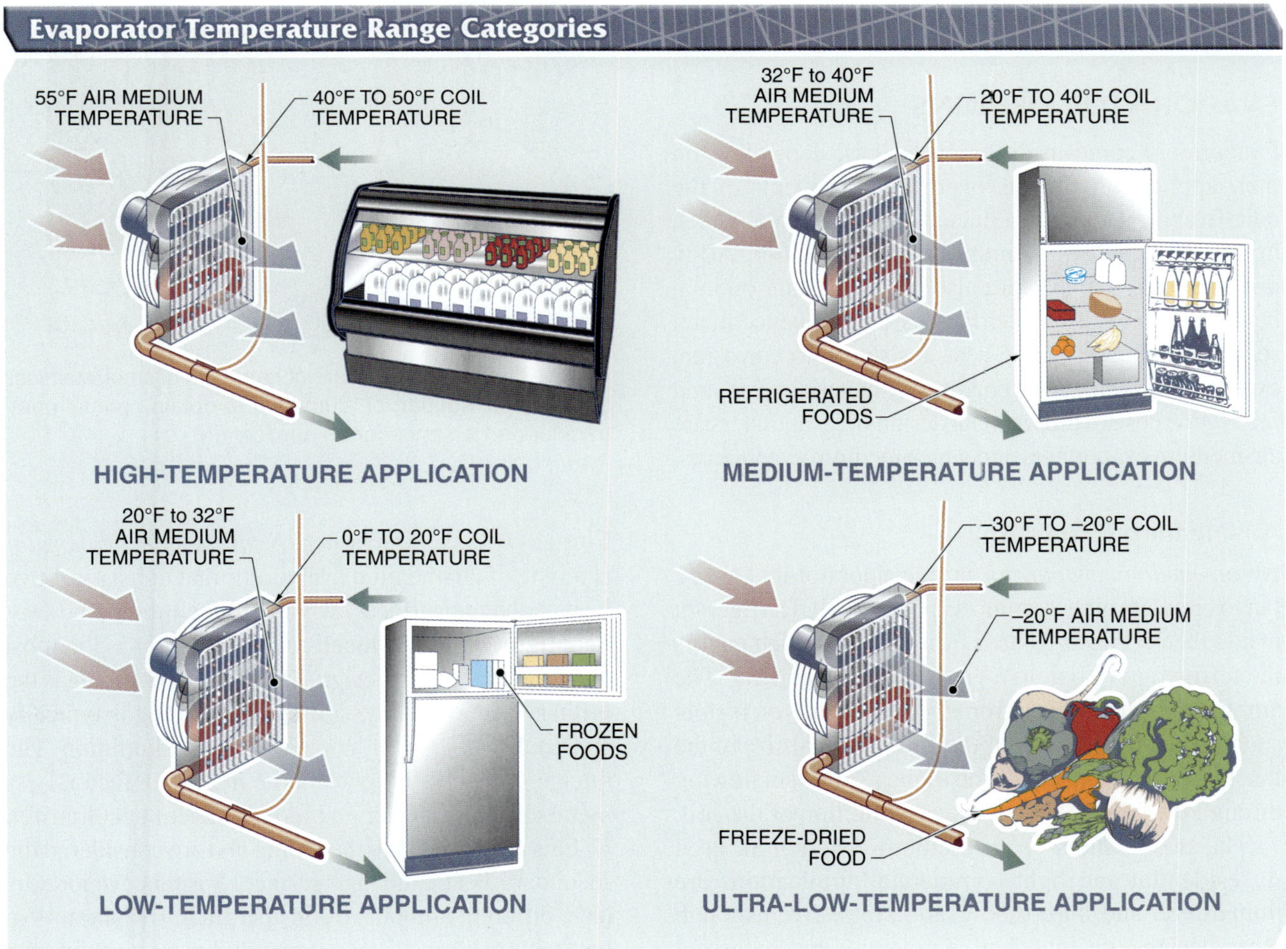

Figure 30-3. The most common evaporators are medium-temperature evaporators, which are used by product coolers to keep food fresh.

Medium-Temperature Applications. The most common medium-temperature application is domestic refrigeration. Medium-temperature applications are applications with coil temperatures ranging from 20°F to 40°F. The majority of medium-temperature applications are between 32°F and 40°F because this temperature range is slightly above the freezing temperature of water.

Low-Temperature Applications. The best example of a low-temperature application is frozen food. The corresponding temperature range of a low-temperature application is –20°F to 32°F. Coil temperatures are typically 0°F to 20°F.

Ultra-Low-Temperature Applications. Prime examples of ultra-low-temperature applications include freeze-drying, powder manufacturing, and cryogenic liquid gas storage. Any application with a temperature below –20°F is considered to be an ultra-low-temperature application. The vast majority of these applications are manufacturing-oriented or scientific. The refrigeration equipment for these applications is typically large, expensive, and complex.

EVAPORATOR DESIGNS

Evaporators come in different designs, depending on their application and the overall requirements of the unit. Evaporators are constructed of steel, stainless steel, brass, copper, or aluminum. **See Figure 30-4.** Alloy mixtures of the metals are also used to obtain optimal heat-transfer performance. Factors that influence metal selection include specific heat, resistance to corrosion, expected environmental conditions, material costs, and fabrication costs. The two main designs of evaporators are air-medium evaporators and water-medium evaporators.

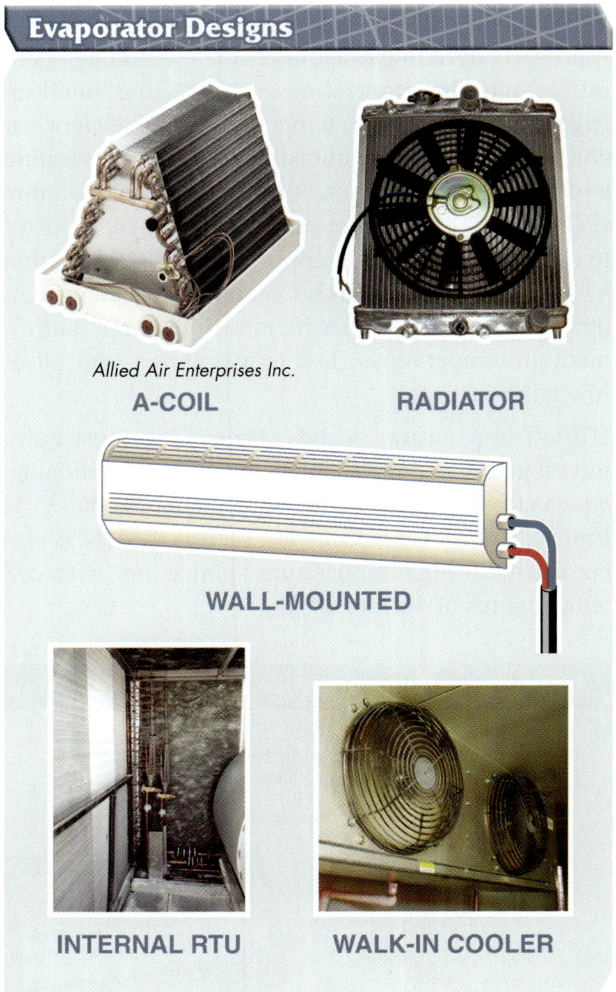

Figure 30-4. Evaporators are constructed of steel, stainless steel, brass, copper, or aluminum to obtain optimal heat-transfer performance for an application.

Air-Medium Evaporators

An *air-medium evaporator* is an evaporator that uses a fan to circulate air across an evaporator coil. Refrigerant inside the coil absorbs heat from the air, which causes the refrigerant to boil to a gas and then superheat. The amount of heat removed from the air depends on factors such as the physical size of the coil, physical size of the fan, the amount and condition of the refrigerant flowing through the coil, and the physical condition of the coil.

The two designs of air-medium evaporators used in residential and light commercial applications are finned-tube and bare-tube evaporators. Even though other designs are found in commercial and industrial applications, such as stamped plate, these types are not used in residential and light commercial buildings.

Finned-Tube Evaporators. A *finned-tube evaporator* is a type of air-medium evaporator that consists of two heat exchange surfaces, which are the tubing and fins. **See Figure 30-5.** A finned-tube evaporator is the most common evaporator design. The evaporator tubing is the primary heat exchange surface. The tubing is typically constructed of copper, copper alloy, or aluminum. The thickness of the tubing wall is an important factor. Less metal allows for faster and more efficient heat transfer.

Fins are attached to the tubing and are considered the secondary heat exchange surface. Various evaporators have different amounts of fins per inch. Evaporators in low-temperature systems typically have fewer fins per inch, meaning the fins are farther apart. This is done intentionally to reduce or slow down the formation of ice on the evaporator. Design engineers attempt to create the maximum amount of heat transfer in the smallest space to achieve the greatest efficiency. Even though maximum efficiency is desired, the level of efficiency that can be achieved is influenced by certain factors such as the weight, size, ruggedness, and manufacturing cost of the evaporator.

Finned-Tube Evaporators

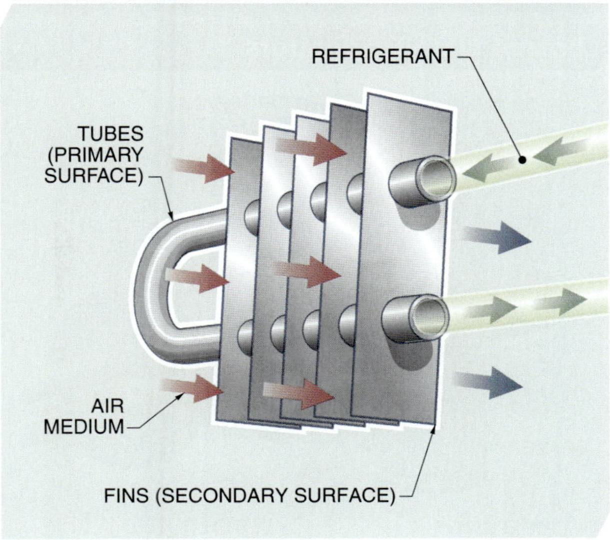

Figure 30-5. Finned-tube evaporators are air-medium-type evaporators that consist of a primary heat exchange surface (tubing) and a secondary heat exchange surface (fins).

Bare-Tube Evaporators. A *bare-tube evaporator* is a type of air-medium evaporator that consists of tubing only. **See Figure 30-6.** Large-capacity, low-temperature evaporators sometimes use bare-tube evaporators. Ice can form on bare-tube evaporators, which causes problems. Ice reduces airflow, blocks heat exchange, and must be removed through defrosting. In order to reduce the problem of ice formation, fins can be added on bare-tube evaporators where frost typically forms first. The main applications for bare-tube evaporators are cold storage, food processing, and scientific applications.

Water-Medium Evaporators

A *water-medium evaporator,* also known as a cooler, is an evaporator that uses a pump to circulate water across its coil to give up heat to the refrigerant. The cooled water is then pumped through the building to terminal units to achieve the cooling effect. Instead of cooling air with refrigerant, water is cooled and then pumped through the building or process to provide cooling by a secondary heat exchanger (terminal unit). The designs of water-medium evaporators used in residential and light commercial applications include tube-in-tube and brazed plate evaporators.

Bare-Tube Evaporators

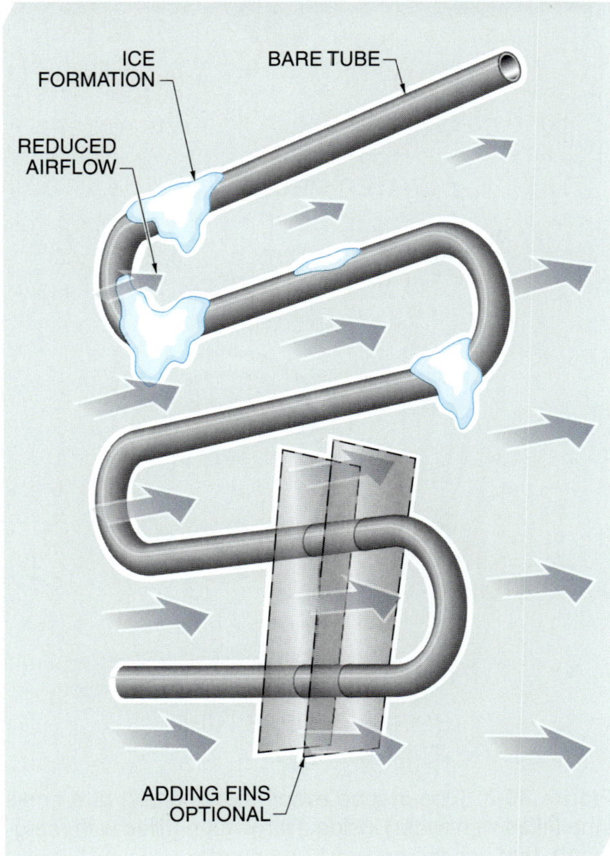

Figure 30-6. Large-capacity, low-temperature evaporators are typically air-medium, bare-tube evaporators that consist only of tubing.

Tube-in-Tube Evaporators. A *tube-in-tube evaporator* is a type of water-medium heat exchanger that consists of a small tube inside a large tube. Refrigerant is inside one tube and water is in the other tube. **See Figure 30-7.** The tubes are typically formed into spiral shapes to minimize their size. Tube-in-tube evaporators use the counterflow principle. Counterflow occurs when fluids flow in opposite directions.

Tube-in-Tube Evaporators

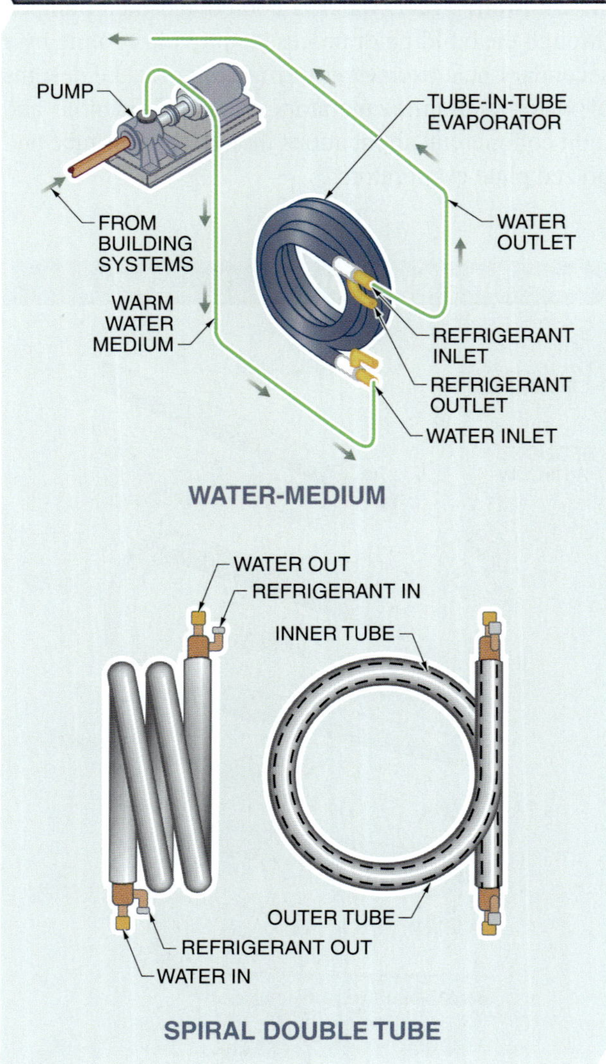

Figure 30-7. Tube-in-tube evaporators consist of a small tube (filled with water) inside a large tube (filled with refrigerant), that use the counterflow principle.

Tube-in-tube evaporators are used in water-source heat pumps. A major problem with this type of evaporator is that contamination can occur when the inner tube develops a leak into the outer tube. Because repairs are difficult, if not impossible, total system replacement and cleanup are typically required when tube-in-tube evaporators fail.

Brazed Plate Evaporators. A *brazed plate evaporator* is a heat exchanger that uses enhanced flat plates that are brazed together to prevent leakage. Brazed plate evaporators range in capacity from 1 ton to 100 tons. Brazed plate evaporators are most common in commercial and industrial applications but can be found in special large residential and light commercial applications. Advantages of brazed plate evaporators include low cost, ease of installation, small size, and low weight. A brazed plate evaporator offers a large contact area for efficient heat transfer. A brazed plate evaporator is typically 25% of the size of an equivalent-capacity shell-and-tube evaporator. **See Figure 30-8.**

Brazed Plate Evaporators

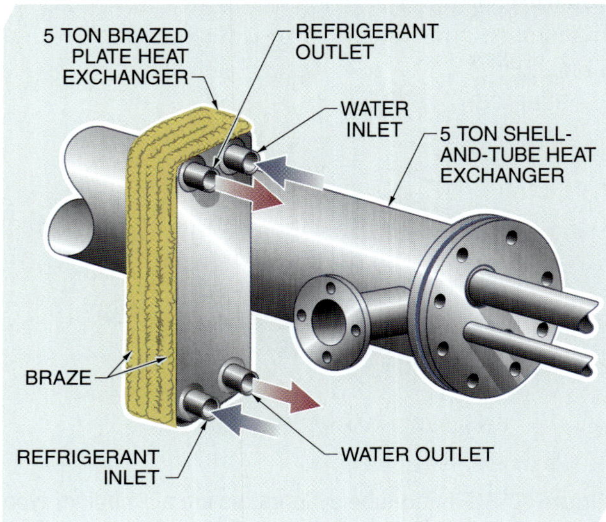

Figure 30-8. Brazed plate evaporators are flat plates that are brazed together to prevent leakage and are found in all applications because of their low cost, ease of installation, small size, and low weight.

Brazed plate evaporators are receiving wide acceptance in the refrigeration, heating, and cooling markets. The relative low cost and availability of brazed plate evaporators make them suitable for light commercial applications. Equipment can be designed smaller because brazed plate evaporators take up less space than comparable shell-and-tube evaporators. The compact design typically means lower final-construction costs. Brazed plate evaporator systems are also lighter than shell-and-tube evaporator systems. This lighter weight can mean smaller mounting systems, easier installation, and reduced shipping costs.

Also, less refrigerant is needed in a brazed plate evaporator system than in a shell-and-tube evaporator system. The piping of a brazed plate evaporator system holds less refrigerant, so fewer pounds of refrigerant are needed to charge the system. Older shell-and-tube systems can be upgraded to the new plate heat exchanger technology, eliminating the need to reinstall the larger shell-and-tube systems. In most cases, the original shell-and-tube condenser can be left in place, saving time and the cost of removal.

Brazed plate evaporators are easy to specify for most applications. Care must be taken concerning position when installing a brazed plate evaporator. Brazed plate evaporators may be position-sensitive in some applications, such as having a vertical orientation when used in a refrigeration application. Additionally, fluids must circulate in a counterflow pattern for optimal heat removal.

Flooded and Dry-Type Evaporators

Evaporators are also categorized as flooded or dry-type. A *flooded evaporator* is a heat exchanger in which a float or other device ensures that the entire evaporator is full of liquid refrigerant. **See Figure 30-9.** The major problem with flooded evaporators is that liquid refrigerant may easily get into the compressor and cause severe damage. True flooded evaporators are not found in residential and light commercial HVAC systems.

A *dry-type evaporator,* also known as a direct-expansion evaporator, is a heat exchanger in which refrigerant boils from a partial liquid/gas mixture to 100% gas before the refrigerant leaves the evaporator. This ensures that no damage is caused by liquid refrigerant entering the compressor (slugging of the compressor).

Poor evaporator maintenance can cause low efficiency and improper unit operation.

TECH TIP

Brazed plate evaporators have become popular in some applications. However, their advantages are often offset by their disadvantages. In some applications, such as package chillers, it may be easier to replace the evaporator with a backup unit than attempt to repair it.

Flooded and Dry-Type Evaporators

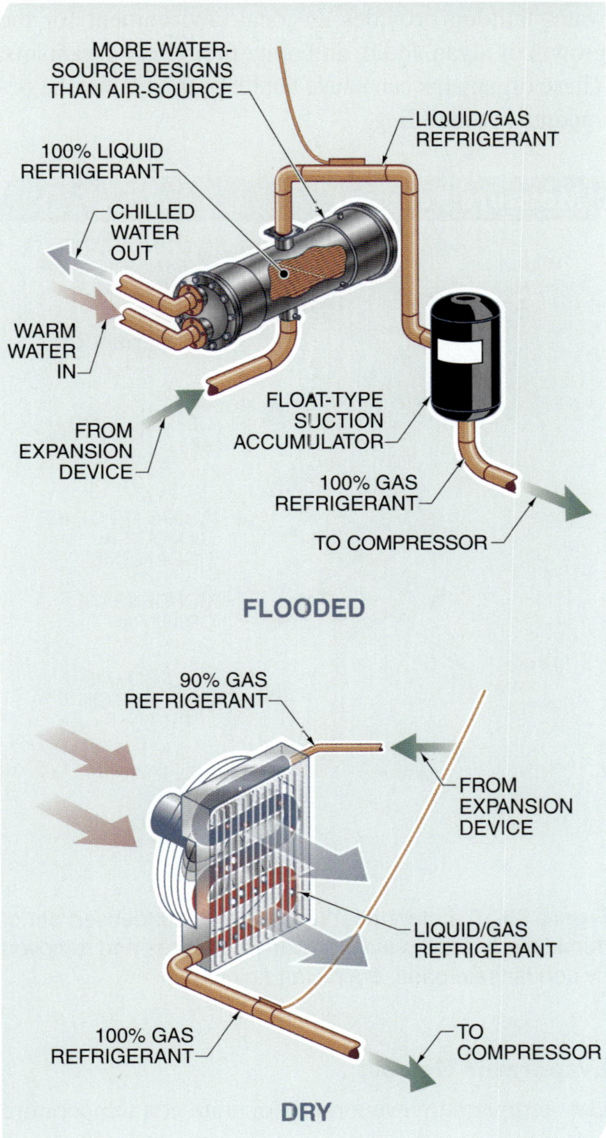

Figure 30-9. Flooded evaporators have a float or other device to ensure that the entire evaporator is full of liquid refrigerant. Dry-type evaporators boil refrigerant to a 90% gas before it leaves the evaporator.

Moisture Removal

As an evaporator works during cooling, the air medium flows across the heat exchanger. Latent heat is then removed in the form of moisture. As a result, dehumidification is as important as sensible cooling in air conditioning systems. **See Figure 30-10.** As the moisture condenses out of the air medium as a liquid, it must be removed to allow for proper system operation. Condensate pans, drains, and removal piping systems are typically provided for this purpose. The condensate drainage system must be properly maintained to prevent water buildup and overflow into the system. Excessive water buildup provides an ideal environment for the growth of algae, mold, and other biological organisms. These organisms can cause health problems for the occupants of a building.

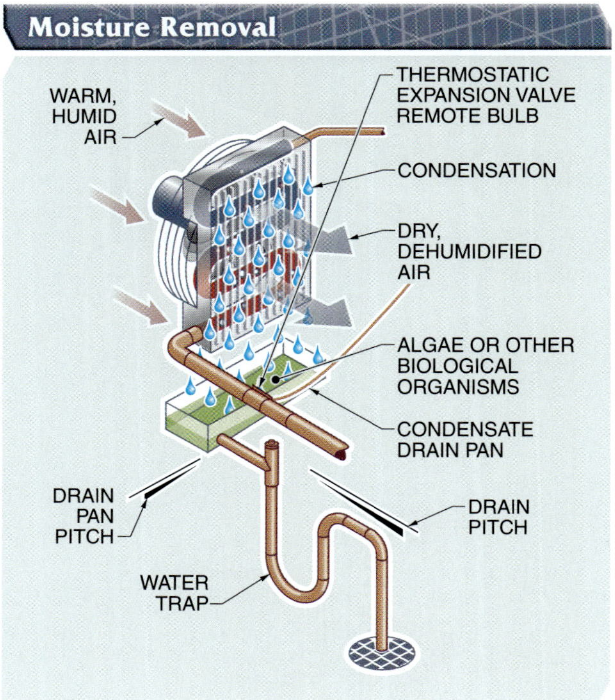

Figure 30-10. Latent heat (moisture) is condensed out of humid air during the air conditioning process and removed by condensate pans, drains, and piping.

Evaporator Defrost

Low-temperature evaporators operate at a temperature below freezing. Any moisture present on the evaporator coil will freeze and block the flow of air across the coil. Low-temperature systems must provide methods to sense and initiate a defrost cycle. **See Figure 30-11.** In the defrost cycle, heat is applied to the evaporator, melting the ice. The melted ice water must then be drained away. Defrost cycles are typically initiated by using time, temperature, or airflow, or a combination of these, as described in the following points:

- The simplest way to initiate a defrost cycle is to use a timer to set a known interval as determined by normal system operation.
- A thermostat is sometimes used to initiate a defrost cycle. As an evaporator is blocked by ice, heat transfer is lost and the evaporator becomes colder.
- As the evaporator coil is blocked, airflow is reduced. Therefore, an airflow switch can be used to sense the increased evaporator pressure drop and initiate the defrost cycle.
- Time, temperature, and airflow may be combined in a scheme that allows the system to defrost when a particular condition occurs.

Evaporator Defrosting Media Clip

Several different methods may be used to melt ice, including the following:

- Turning off the refrigeration unit allows ambient conditions to melt the ice. This process is not used often because the time that the unit is OFF is frequently too long and may affect the product. **See Figure 30-12.**
- Electric heating elements may be energized to defrost the coil. The indoor fan is OFF to avoid dumping heat into the building space during the defrost cycle.
- Hot gas from the compressor discharge may be diverted around the condenser through a bypass to the evaporator using a solenoid valve. When the hot gas melts the ice, the solenoid valve is closed to stop the hot gas flow, and normal system operation is restored. The indoor unit fan is switched off to avoid sending the warm air into building spaces.

EVAPORATOR PERFORMANCE

A technician must know the condition of the refrigerant inside the evaporator and in the suction line leading back to the compressor. Knowing the condition of the refrigerant in these places ensures proper diagnosis and troubleshooting of problems as they occur. The most common tools used are manifold gauges and thermometers. Low-pressure side pressure is measured and the resulting pressure converted to temperature on a T/P chart. A thermometer is used to measure evaporator inlet and outlet temperatures.

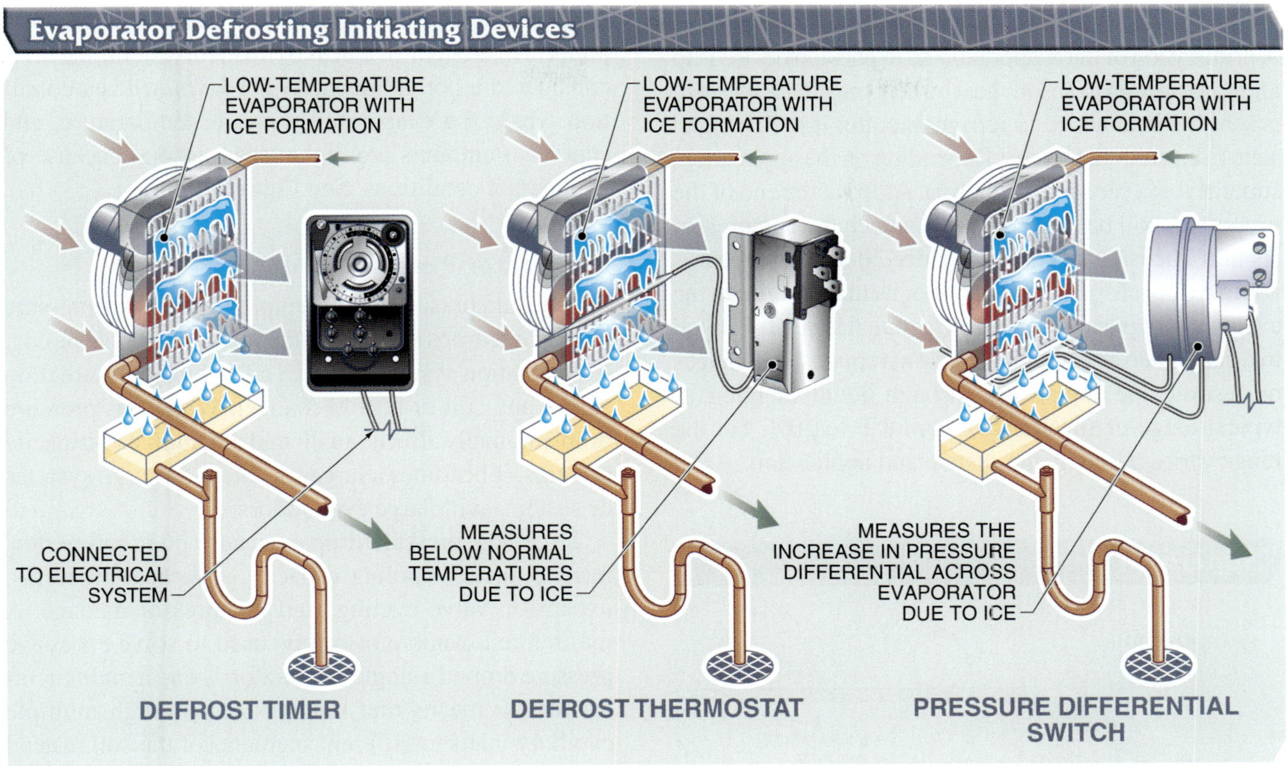

Figure 30-11. Defrost cycles are typically initiated by using timers, temperature sensors, pressure differential switches, or a combination of these devices.

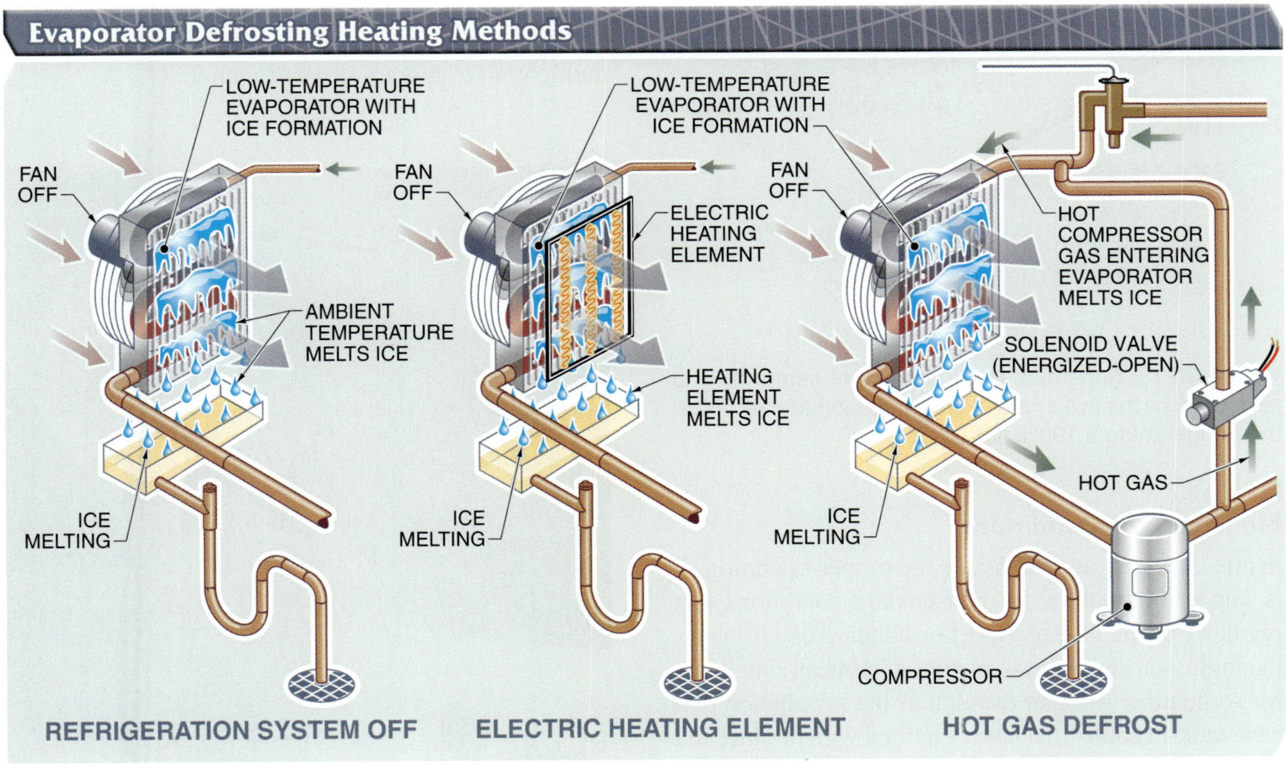

Figure 30-12. The three main methods of evaporator defrosting are shutting the refrigeration system off, using an electric heating element, or using a hot gas defrost.

Superheat

A primary job of the evaporator and expansion device is to allow refrigerant to absorb heat by heat transfer. *Superheat* is sensible heat added to refrigerant after it has changed state from a liquid to a gas. Depending on the application, superheat is a secondary function, added at the end of the evaporator coil before the point at which the compressor suction line is attached. **See Figure 30-13.** Superheat is added to the refrigerant at the end of its travels through the evaporator to ensure that all refrigerant has been changed to a gas. When any liquid refrigerant remains, it damages or destroys the compressor through liquid slugging. A typical range of superheat is from 8°F to 20°F, but the range varies based on the system and application.

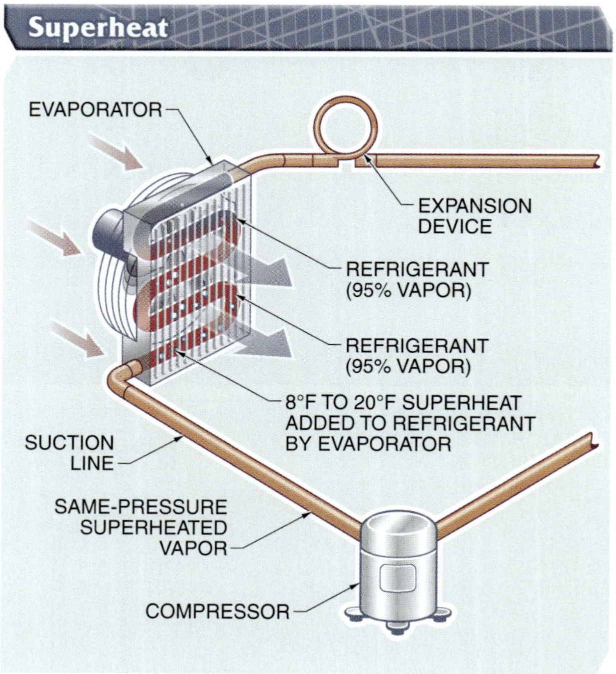

Figure 30-13. Superheat is the amount of heat added to refrigerant by the evaporator after the evaporator has boiled the refrigerant to a 100% gas.

Nonstandard Conditions

While checking an evaporator for proper operation, it is important to allow for nonstandard conditions that regularly occur. One of these nonstandard conditions is the initial startup of a system after it has been shut down for some time. Another example is the installation of a new walk-in cooler that has been filled with product and then started. The operation of the walk-in cooler is not the same as a system that has already pulled down the temperature to the design level. It can take hours for a pulldown to occur in a heavily loaded system, which is considered a hot pulldown. A *hot pulldown* is a condition where the evaporator pressure, temperature, and superheat amounts are higher than normal because of application conditions. **See Figure 30-14.**

Evaporator Pressure Drop

As any fluid passes through a piping system, the pressure of the fluid drops because of the resistance of the piping. Refrigeration systems have an additional pressure drop across any coil or heat exchanger. While this pressure drop minimally affects small- and medium-size-capacity systems, it becomes a larger problem for large systems or systems with large evaporators.

Excessive pressure drop can cause poor coil performance, uneven cooling capacity, superheat problems, expansion valve cycling, and compressor damage. A multicircuit evaporator can be used to solve excessive pressure drop in a single evaporator. Using a multicircuit evaporator means that liquid is fed through multiple capillary inlets to different segments of the coil instead of a single connection point. **See Figure 30-15.** When liquid is fed evenly into the different segments, there is only a small pressure drop by each particular segment. Problems with poor air distribution across the coil, a crimped capillary, or a dirty coil segment can cause a multicircuit evaporator to malfunction.

Refrigerant in the evaporator absorbs heat from the air.

Figure 30-14. Hot pulldown occurs when evaporator pressure, temperature, and superheat amounts are higher than normal because of extended system shutdown or when extremely hot air or product is cooled by the system.

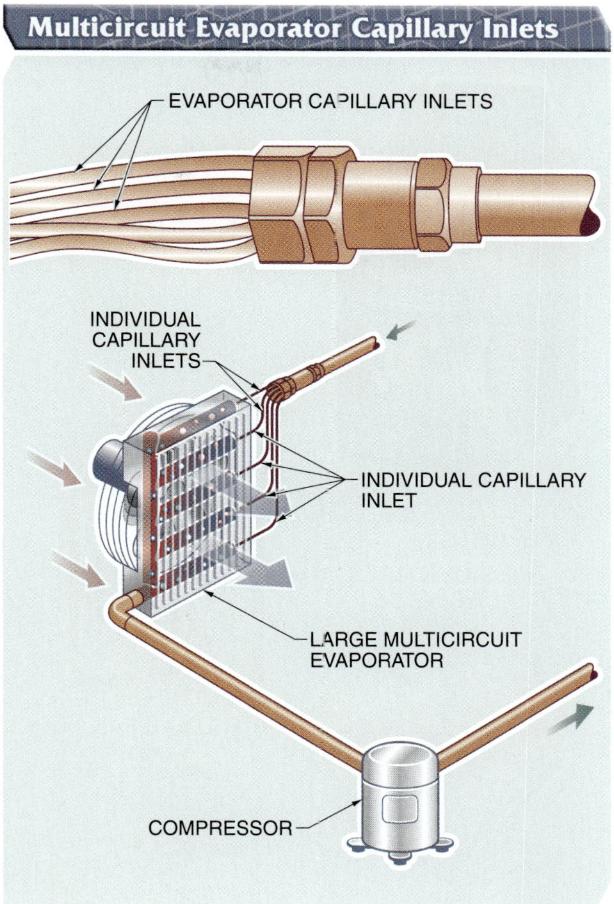

Figure 30-15. Evaporator capillary inlets allow liquid refrigerant to be fed evenly into the various segments of an evaporator, with only small engineered pressure drops being created by each capillary.

Blowers and Cabinets

In order to cool air, the air must be circulated across the evaporator with a fan. The fan is sized to provide adequate airflow across the coil to obtain the correct refrigeration effect at design conditions.

Most air conditioning systems use a centrifugal fan to circulate return room air across an evaporator. **See Figure 30-16.** Since many systems also use fans to provide heat to the space, two-speed fans are often used. The higher speed of a two-speed fan is used for cooling and circulates the air needed to cool rooms. Small package units and refrigeration applications use propeller-type axial fans that circulate the proper air amounts and need no ductwork. In either case, proper blower operation is critical for obtaining maximum evaporator performance.

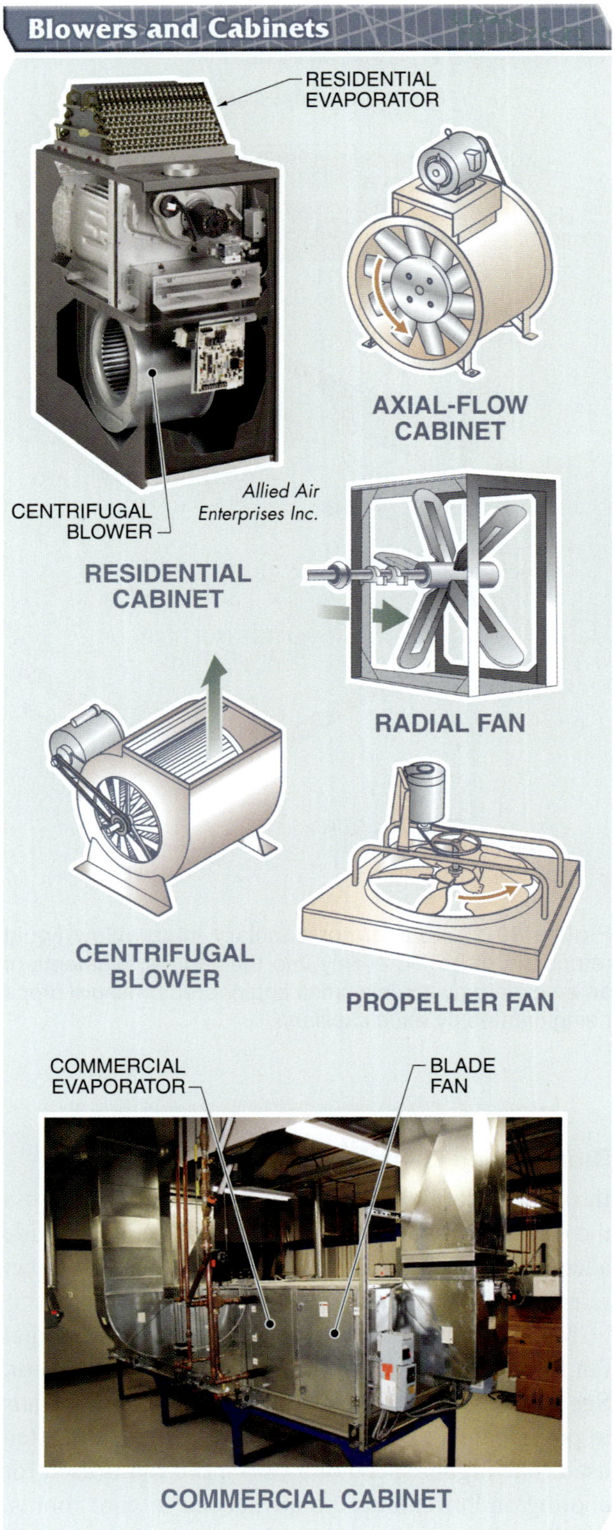

Figure 30-16. Most air conditioning systems use a centrifugal fan to circulate air across an evaporator, while small package units and refrigeration applications use propeller-type axial fans to circulate air. Large systems typically use blade fans.

In air conditioning systems and some other systems, the blower and fan controls are contained in a cabinet, which typically is a sheet metal panel or enclosure. Cabinets are coated to prevent rust, soundproofed to reduce objectionable noise, and designed and located to permit easy access and cleaning. Also, when they are exposed to public view, cabinets are painted a neutral color to blend in with their surroundings. Some cabinets provide hinges and locking mechanisms that allow the access door to remain locked in an open position. Adequate clearance for cleaning, maintenance, and replacement of unit parts must be factored into the placement and mounting of the cabinet.

TECH TIP

Over time, basic maintenance and operational issues can affect the operation and efficiency of an evaporator. This may reduce the evaporator output and cause the unit to run longer for the same load.

Cooling Capacity

Cooling capacity is the heat transfer capacity of an evaporator measured in British thermal units per hour. This value can then be converted into tons by using the conversion ratio of 12,000 Btu/hr/ton of refrigeration. The cooling capacity is a function of the type, flow rate, and effect of the refrigerant. The type of refrigerant is typically known, or easily discovered, but the flow rate and refrigeration effect must be found. *Refrigerant flow rate* is a measurement of the amount of refrigerant flowing through a system or component in one minute. The refrigerant flow rate may be measured in pounds per minute, resulting in the mass flow rate, or it may be measured in cubic feet per minute, resulting in the volumetric flow rate.

The refrigeration effect is the difference in enthalpy of the refrigerant between when it enters the evaporator and when it leaves. The values for calculating the refrigeration effect are found by consulting a pressure-enthalpy diagram or refrigerant property tables.

To determine the refrigeration effect of a refrigeration system, use the following formula:

$$RE = H_V - H_L$$

where

RE = refrigeration effect (in Btu/lb of refrigerant)

H_V = refrigerant vapor enthalpy (in Btu/lb of refrigerant)

H_L = refrigerant liquid enthalpy (in Btu/lb of refrigerant)

To determine the cooling capacity of an evaporator, use the following formula:

$$CC = RE \times w \times 60$$

where

CC = cooling capacity of evaporator (in Btu/hr)
RE = refrigeration effect (in Btu/lb of refrigerant)
w = mass flow rate (in lb/min)
60 = constant (to convert to hr)

Example: Determining the Cooling Capacity of Evaporators

A split-system air conditioning unit that uses R-22 has a low-pressure-side evaporator pressure of 68.55 psi. **See Figure 30-17.** The mass flow rate through the evaporator is 24 lb/min. To find the cooling capacity of the evaporator in British thermal units per hour and in tons, use the following procedure:

1. Use a refrigerant property table to find the values needed to calculate the refrigeration effect, which is the difference in enthalpy between the vapor and liquid refrigerant. **See Appendix.** The corresponding temperature of R-22 at 68.55 psi is 40°F. The enthalpy of the vapor is 107.94 Btu/lb and the enthalpy of the liquid is 21.474 Btu/lb. Refrigeration effect is found by applying the following formula:

$$RE = H_V - H_L$$
$$RE = 107.94 - 21.474$$
$$RE = \mathbf{86.466\ Btu/lb}$$

TECH TIP

The greater the temperature difference between the return air and the supply air, the more effective the heat loading of the evaporator coil. Also, the moisture in the hot or warm air will condense more with a higher temperature difference. The supply and return air grills should never be at the floor level because the cold air at the floor level will result in a loss of efficiency.

Finding Evaporator Cooling Capacity

R-22—PROPERTIES OF SATURATED LIQUID AND SATURATED VAPOR								
Temp °F	Pressure		Volume	Density	Enthalpy		Entropy	
	psia	psig	Vapor	Liquid	Liquid	Vapor	Liquid	Vapor
40	83.246	68.550	0.65571	79.049	21.474	107.94	0.046464	0.21951

Carrier Corporation

FIND THE EVAPORATOR REFRIGERATION EFFECT

$RE = H_V - H_L$
$RE = 107.94 - 21.474$
$RE = \mathbf{86.466\ BTU/LB}$

FIND THE EVAPORATOR COOLING CAPACITY

$CC = RE \times w \times 60$
$CC = 86.466 \times 24 \times 60$
$CC = \mathbf{124{,}511.04\ BTU/HR}$

$$\text{EVAPORATOR TONNAGE} = \frac{CC}{12{,}000}$$
$$\text{EVAPORATOR TONNAGE} = \frac{124{,}511.04}{12{,}000}$$
$$\text{EVAPORATOR TONNAGE} = \mathbf{10.376\ TONS}$$

Figure 30-17. Cooling capacity is the heat transfer capacity of an evaporator measured in British thermal units per hour.

2. The cooling capacity in British thermal units per hour is calculated as follows:

$$CC = RE \times w \times 60$$

$$CC = 86.466 \times 24 \times 60$$

$$CC = \mathbf{124{,}511\ Btu/hr}$$

3. The cooling capacity in tons is calculated as follows:

$$Evaporator\ tonnage = \frac{CC}{12{,}000}$$

$$Evaporator\ tonnage = \frac{124{,}511}{12{,}000}$$

$$Evaporator\ tonnage = \mathbf{10.376\ tons}$$

Digital Resources

ATPeResources.com/QuickLinks • Access Code: 482610

UNIT 31

Low-Pressure-Side Expansion Devices and Accessories

The low-pressure side of a refrigeration system consists of an expansion device, evaporator, and a number of accessories. The proper operation of the expansion device is critical to maintaining the correct conditions at the evaporator as well as providing the proper amount of cooling capacity. Accessories are used as needed to ensure proper low-pressure-side and system operation.

OBJECTIVES

- Describe the purpose and function of expansion devices.
- List the types of expansion devices.
- Identify the principles of the operation, setting, and selection of thermostatic expansion valves (TXVs).
- Describe typical problems associated with TXVs.
- Describe the operation of electronic expansion valves (EEVs).
- Explain the purposes of low-pressure-side accessories.

EXPANSION DEVICES

In the evaporator, refrigeration is accomplished by the boiling of low-temperature, low-pressure liquid refrigerant. The liquid supplied to the evaporator section is at a higher temperature and pressure before the expansion device. The expansion device is located in the liquid line, close to the evaporator inlet. The purpose of the expansion device is to cause a sudden pressure drop in the liquid refrigerant. **See Figure 31-1.** The sudden pressure drop causes some of the liquid refrigerant to flash to a vapor (gas). Flashing of the refrigerant to a gas state dramatically reduces the temperature of the remaining liquid refrigerant. It drops to a temperature that corresponds to the lower system pressure, which can be found on a temperature/pressure (T/P) chart.

The expansion device is one of the most important devices in a refrigeration system. When the expansion device malfunctions, the entire refrigeration system will not function properly. The expansion device is often misdiagnosed as the root of problems that were instead caused by another part of the system. Expansion devices operate by controlling the volume, pressure, and/or temperature of the refrigerant entering the evaporator.

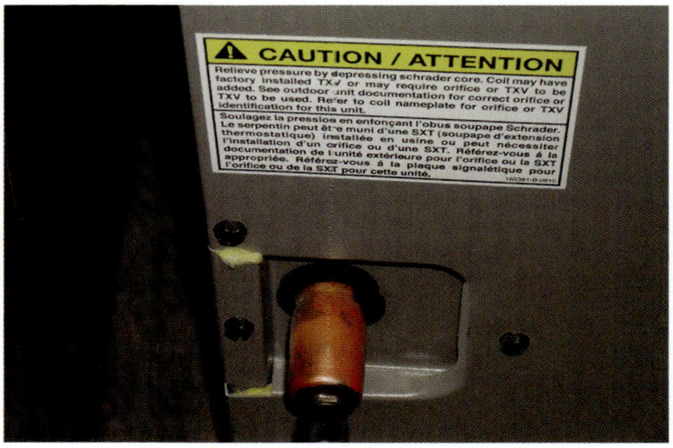

Expansion devices control the volume, pressure, and/or temperature of the refrigerant entering the evaporator.

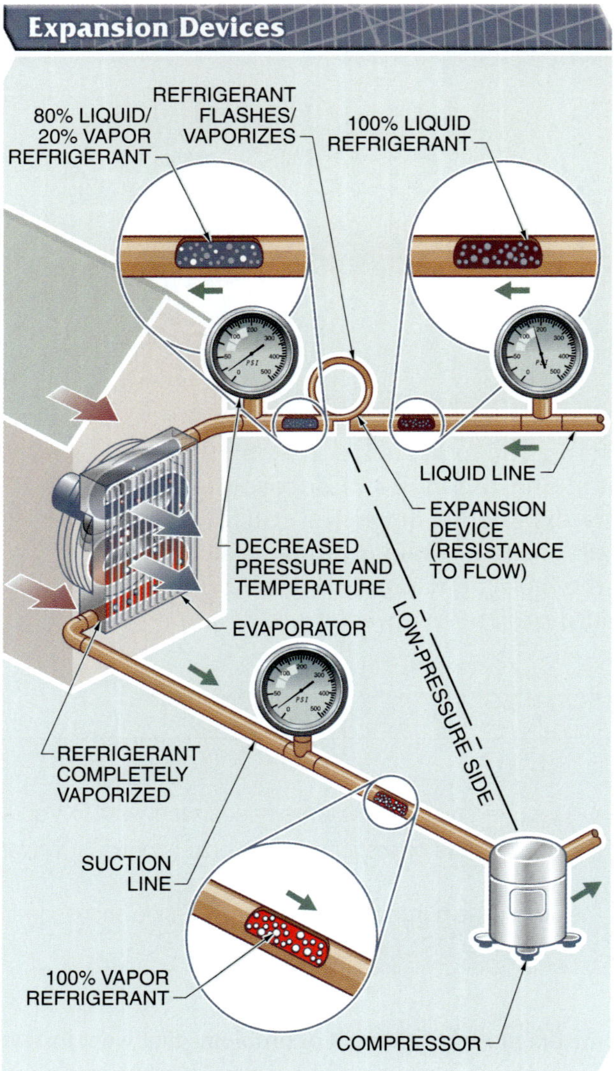

Figure 31-1. By causing a sudden pressure drop in the liquid refrigerant, expansion devices can control the volume, pressure, and/or temperature of the refrigerant entering the evaporator.

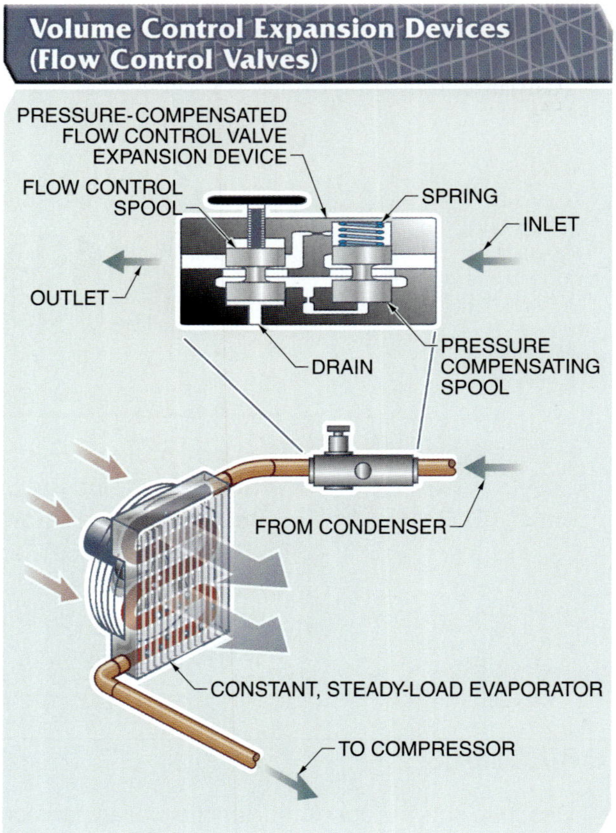

Figure 31-2. Volume control expansion devices are special flow control valves that maintain the flow of refrigerant into the evaporator at a constant rate (within parameters).

Expansion Device Operation Media Clip

Volume Control Expansion Devices

A *volume control expansion device* is an expansion device that controls the amount, or volume, of refrigerant that is able to pass through to the evaporator. **See Figure 31-2.** This pressure-compensated flow control valve causes a pressure drop of the refrigerant and, subsequently, the low-temperature, low-pressure liquid at the evaporator. This category of expansion device has a number of various designs and keeps the refrigerant volume flowing at a constant rate into the evaporator. Volume control expansion devices do not compensate or measure any effects or potential problems at the evaporator. This type of expansion device is typically used in systems that have a constant, steady load.

Capillary Tube Expansion Devices. The simplest and most common type of expansion device is a capillary (cap) tube. A *capillary tube expansion device* is a long length of copper tubing that produces a pressure decrease regardless of the direction of refrigerant flow through it. **See Figure 31-3.**

TECH TIP

Although a capillary tube is the simplest type of expansion device, it must be serviced properly. Capillary tube systems are critically charged, which means they contain only the correct amount of refrigerant.

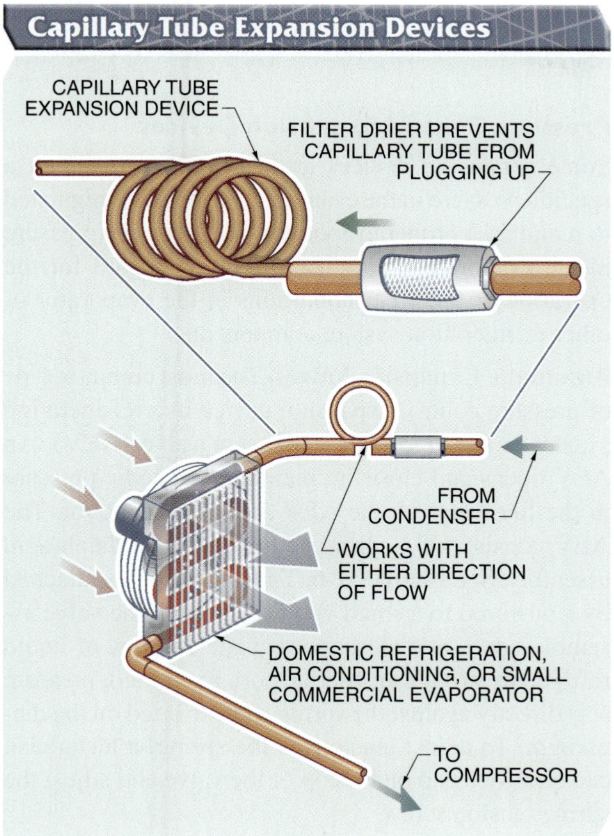

Figure 31-3. Capillary tube expansion devices are formed into a coil to reduce the amount of space needed for mounting and can be used effectively, regardless of the direction of flow through the device.

Refrigeration systems with capillary tube expansion devices are common in domestic refrigeration, air conditioning, and small commercial applications. Capillary tube expansion devices have also been used successfully in heat pump applications. Capillary tubes are easily manufactured and formed into a coil that reduces the amount of space needed for mounting. A capillary tube expansion device may be used effectively, regardless of the direction of flow through the device.

The design and operation of a capillary tube expansion device are extremely simple. However, the length and the exact bore diameters of capillary tubes are known only to refrigeration system manufacturers. Capillary tubes are easily plugged by contaminants, such as solder, flux, or copper chips, resulting from improper service techniques. They often have a filter drier installed ahead of them to prevent contaminants from plugging the tube. Capillary tubes are designed to allow only specific amounts of refrigerant to pass. When a system is overcharged, the system operation may be adversely affected. Small refrigeration systems contain only a few ounces of refrigerant, therefore the refrigerant must be weighed to avoid overcharging the system.

Fixed-Plate Orifice Expansion Devices. A *fixed-plate orifice expansion device* is an object that has a small fixed nozzle, or orifice, to create a predetermined resistance to flow (pressure drop). This orifice is installed in the liquid line ahead of the evaporator instead of a long tube with a gradual pressure reduction. **See Figure 31-4.** Fixed-plate orifice expansion devices are used on mass-produced central air conditioning systems.

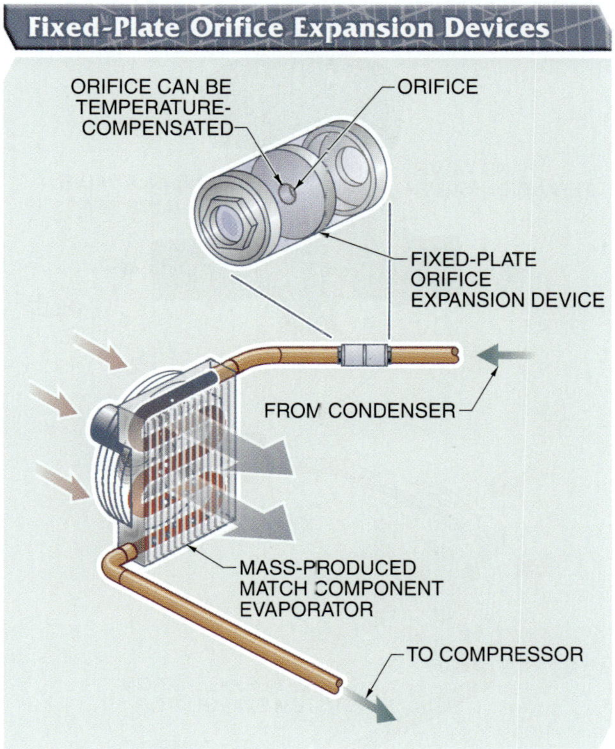

Figure 31-4. Fixed-plate orifice expansion devices work well because all components of the refrigeration system are designed by the manufacturer to match up and work together.

Fixed-plate orifice refrigeration systems work well because most of the components are designed by the manufacturer to match together out of the box. These matched components include the evaporator, compressor, condensing unit, air handler fan, expansion device, and a number of precharged refrigerant lines. Any change in any device selection will affect the operation of the refrigeration system.

Hand Valve Expansion Devices. Although not common, hand valves are used as expansion devices to control the flow of refrigerant into the evaporator. The most common types of hand valves in use are globe and needle designs. Both designs can be used in a system at the same time. **See Figure 31-5.** The globe valve is used to regulate large amounts of refrigerant flow. The needle valve is then used to make adjustments as needed for final control or fine-tuning. Both valves are operated by manual control. Hand-valve use is limited to systems that have full-time trained personnel available to make the proper adjustments under different load conditions as long as the equipment is operating.

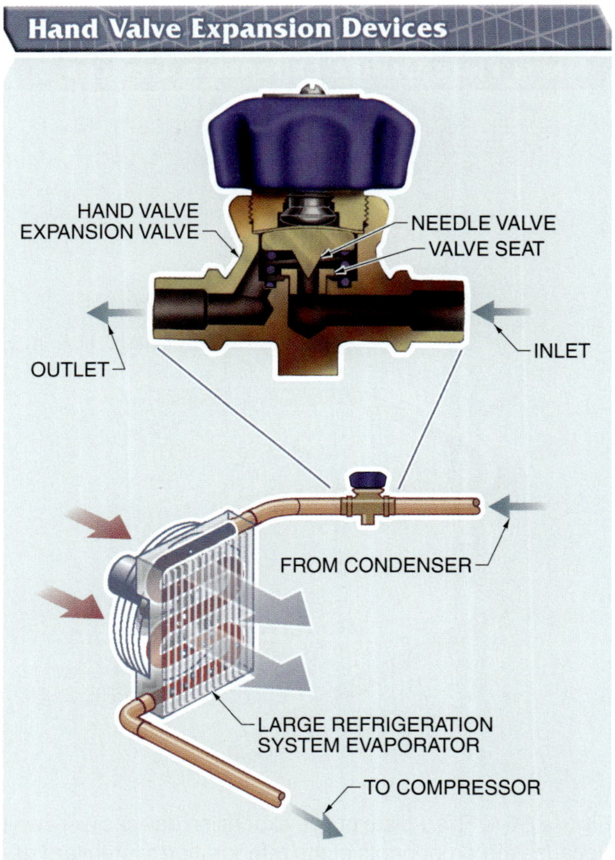

Figure 31-5. Hand valve expansion devices require that a technician manually adjust the globe valve (for large flow rate adjustments) and/or needle valves (for fine flow rate adjustments) as the system dictates.

Hand valve expansion devices are used in the large refrigeration systems of laboratories, manufacturing facilities, or process systems where trained staff is available to operate the equipment under the desired load conditions. Other volume control expansion devices that are used for large refrigeration systems include multiple orifice plate and float valve expansion devices.

Pressure Control Expansion Devices

Some expansion devices are designed to maintain a specific pressure in the evaporator. The flow is regulated to maintain a proper evaporator pressure. The pressure in the evaporator is regulated without regard for the operation of the other conditions in the evaporator or other refrigeration system components.

Automatic Expansion Valves. The most common type of pressure control expansion device in a refrigeration system is the automatic expansion valve (AEV). An AEV opens and closes to maintain a specific pressure in the line between the AEV and the evaporator. The AEV consists of a spring acting against a diaphragm assembly. **See Figure 31-6.** The diaphragm is attached by a push rod to a small valve assembly. The valve assembly opens and closes to regulate the flow of liquid refrigerant through an internal port. Evaporator pressure acts directly against the spring force exerted on the diaphragm. To adjust tension on the spring, a technician can remove a cap on the top of the valve and adjust the spring tension screw.

AEVs are most often used on small- to medium-size refrigeration equipment that operate under a steady load. Since the AEV responds only to evaporator pressure without reference to other conditions, it may not work well on equipment that is under a widely varying load or with changing operating conditions.

AEV Operation. When an AEV is operating in a perfectly balanced state, the spring tension on one side of the diaphragm is balanced by the evaporator pressure acting on the other side. **See Figure 31-7.** The push rod holds the valve at the proper position to allow just enough refrigerant to flow through the valve and orifice assembly and then into the evaporator to keep the system in balance.

When the cooling load on the evaporator increases, the pressure on the diaphragm from the system pressure increases. The pressure increase causes the valve to move toward the closed position, allowing a lower rate of refrigerant flow and causing the pressure in the evaporator to drop.

When the cooling load on the evaporator decreases, the pressure on the diaphragm from the system pressure also decreases. This pressure decrease causes the valve to move toward the open position, allowing a higher rate of refrigerant flow, which causes the evaporator pressure to increase.

Automatic Expansion Valves (AEVs)

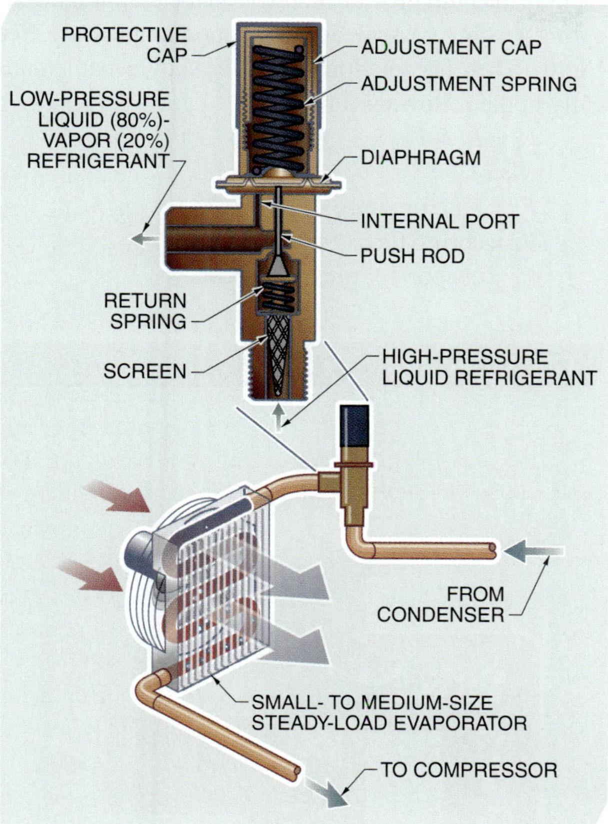

Figure 31-6. AEVs are controlled by the back pressure created by the refrigerant in the evaporator, which opposes the compression of the adjustment spring.

It is important to note that the designed operation of the AEV may cause the evaporator flow rate on system startup to be too low, causing a starved flow condition at the coil. The reverse is also true, with the refrigerant flow rate being too high and causing a flooded condition. These conditions illustrate that AEVs have a relatively limited application range.

Temperature Control Expansion Devices

A *temperature control expansion device* is an object that operates by sensing the temperature at a given point in the system (evaporator outlet) to control flow into the evaporator. The most commonly sensed temperature is the suction line as it leaves the evaporator on its way to the compressor. Expansion devices that sense temperature are sensitive to the load, because the temperature at that point is representative of the load in the system. The two expansion devices that fit in this category are the thermostatic expansion valve and the thermoelectric expansion valve.

TECH TIP

To troubleshoot an AEV system properly, the technician must have experience in working with the system. If not, the system problem may be misdiagnosed and the wrong part may be replaced.

AEV Operation

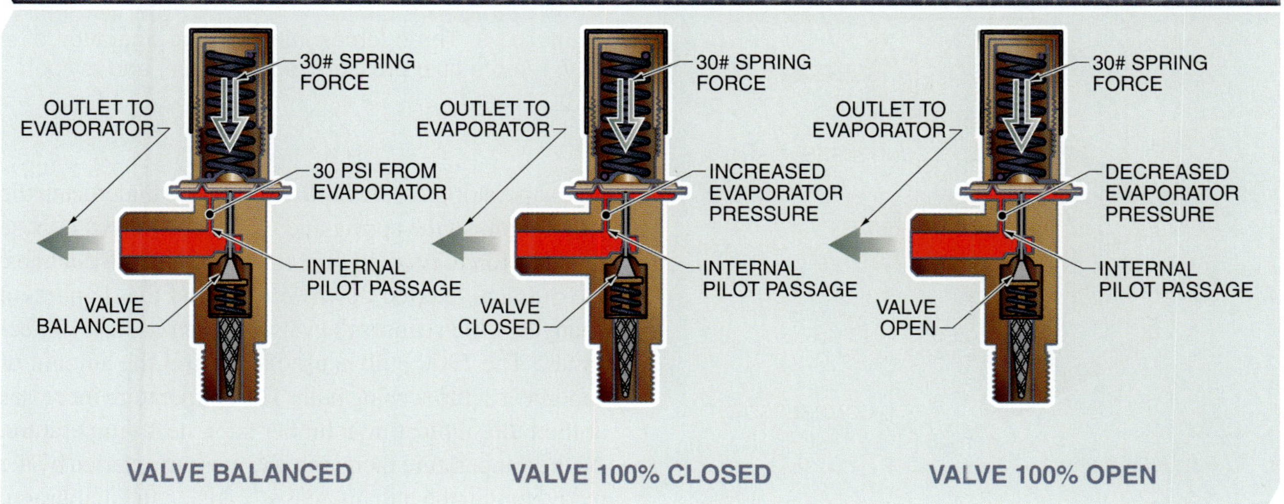

Figure 31-7. AEVs operate by controlling the refrigerant flow into the evaporator by reacting to an increasing or decreasing load (pressure) on the evaporator.

THERMOSTATIC EXPANSION VALVES

A *thermostatic expansion valve (TXV)* is an expansion device that uses pressure and temperature to control the amount of refrigerant entering the evaporator. TXVs are the most commonly used expansion devices found in refrigeration systems. **See Figure 31-8.** The temperature-sensing element consists of a refrigerant-charged bulb with a capillary tube attached. The capillary tube is attached to the power head, which is attached to the top of a diaphragm element on the TXV. The diaphragm is attached to a push rod and a valve assembly. Force created by the diaphragm is opposed by a spring that has a manual tension adjustment.

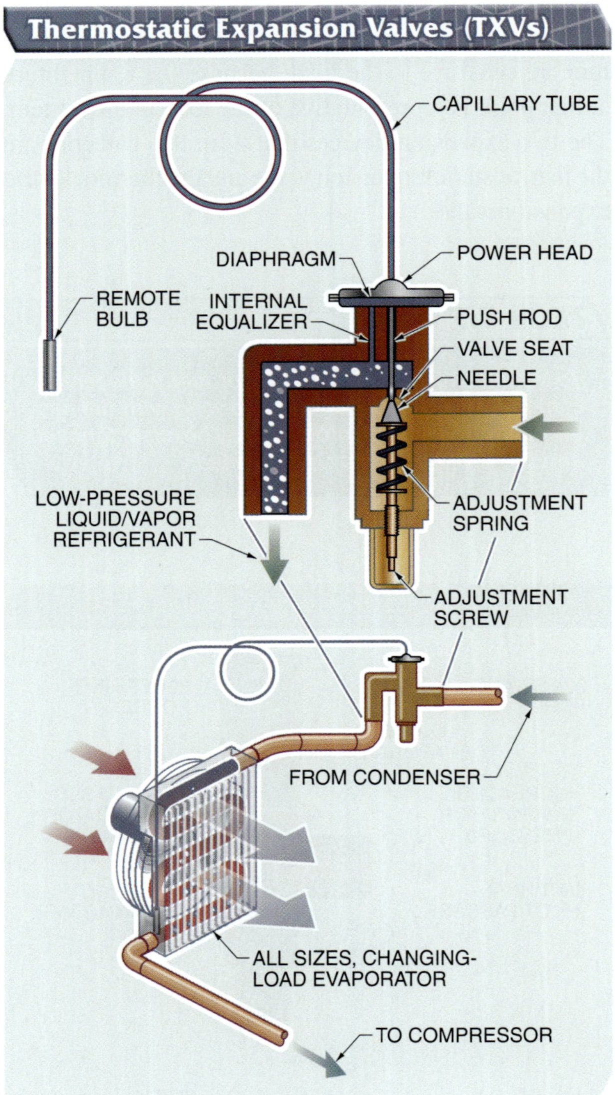

Figure 31-8. Thermostatic expansion valves can use the temperature sensed by a remote bulb to control the amount of superheat that a refrigerant absorbs in the evaporator.

The forces that influence the operation of a TXV include the bulb pressure exerted on the diaphragm, the spring tension pressure, and the evaporator pressure. **See Figure 31-9.** The relationship can be shown mathematically in the following formula:

$$P_1 = P_2 + P_3$$

where

P_1 = bulb pressure

P_2 = spring force

P_3 = evaporator pressure

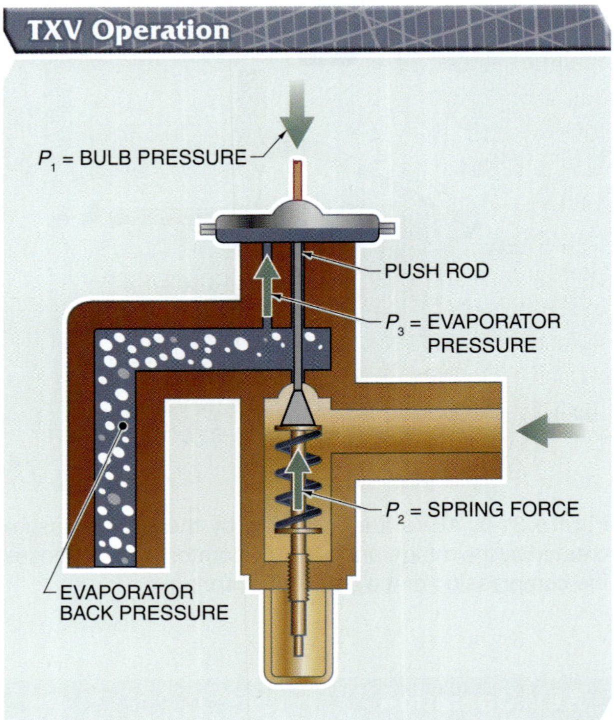

Figure 31-9. Three forces influence the operation of a TXV: remote bulb pressure, spring force, and evaporator pressure.

Evaporator pressure is also introduced underneath the diaphragm, so the spring and evaporator forces oppose the bulb force. As the temperature at the bulb changes, the pressure exerted downward against the diaphragm changes, which causes the valve to open or close as necessary. The TXV bulb is used to control the amount of superheat at the sensing bulb. The temperature increases at the bulb, indicating a higher superheat temperature. As the temperature increases, the pressure exerted by the refrigerant in the bulb increases, causing the diaphragm to overcome the force exerted upward by the spring, thus opening the valve.

More refrigerant is admitted into the evaporator, causing the superheat temperature to decrease. When the superheat temperature decreases below the TXV setpoint, the temperature at the bulb decreases, causing a smaller force to be exerted downward by the diaphragm. The spring then forces the valve upward, more toward the closed position, reducing refrigerant flow and causing the temperature to increase. The setpoint of the TXV is changed by adjusting the spring tension with a service wrench.

Setting TXVs

A split-system air conditioning unit that uses R-22 has a high-pressure-side pressure of 240 psi and a low-pressure-side pressure of 68.5 psi. The TXV is adjusted to maintain a desired 10°F of superheat. **See Figure 31-10.** Given these conditions, the pressure exerted below the diaphragm is the low-pressure-side pressure, or 68.5 psi. With a given superheat of 10°F, the bulb temperature is 50°F. The pressure in the remote bulb, which is the pressure above the diaphragm, is 84 psi. To find the required spring pressure setting for the TXV, use the following formula:

$$P_1 = P_2 + P_3$$

Modified to solve for P_2:

$$P_2 = P_1 - P_3$$

$$P_2 = 84 - 68.5$$

$$P_2 = \mathbf{15.5\ psi}$$

The spring pressure, 15.5 psi, becomes the spring tension setpoint. Once the TXV is adjusted to the proper setting, the TXV is a great expansion device for systems that have changing load conditions. The TXV compensates for the temperature changes that indicate varying loads on the evaporator.

TXV Selection

Many types of TXVs exist. A TXV is selected by a factory engineer based on a number of factors, including refrigerant type, tonnage of system, refrigerant mass flow rate, desired superheat, and expected evaporator pressure drop. When a TXV must be replaced, the same type of TXV must be used in its place. Typically, a manufacturer code is listed either on the TXV or on the information sheets that came with the unit. A replacement TXV cannot be chosen based on its appearance. The manufacturer code must be consulted. When the wrong TXV is used, the system does not perform properly.

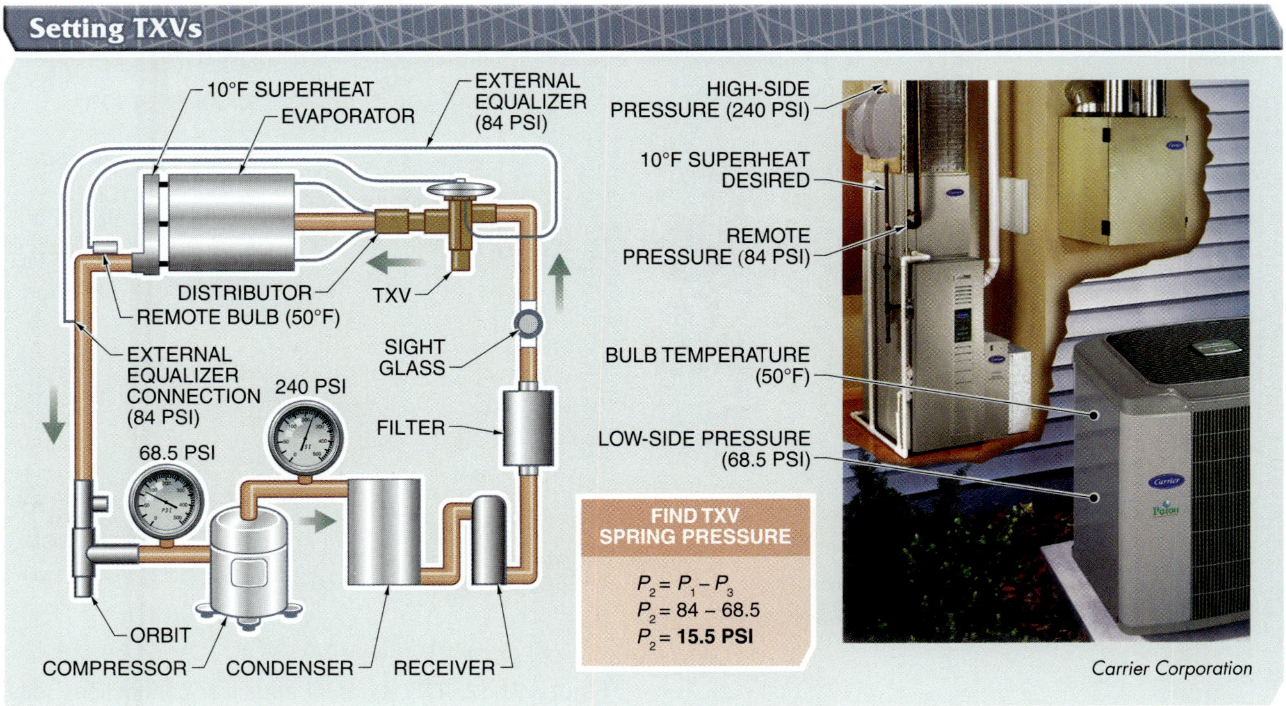

Figure 31-10. TXVs are adjusted to maintain a specific temperature (in °F) in the evaporator and superheat in the refrigerant.

Internally and Externally Equalized TXVs

On most TXVs, the evaporator pressure is ported from inside the TXV body underneath the diaphragm. This is called an internally equalized TXV. With some TXV applications, notably with the use of distributors, the evaporator and/or distributor has a larger pressure drop than normal. This larger pressure drop, which is required to be controlled, affects the TXV operation adversely because the conditions that are being controlled are typically those leaving the evaporator. In these cases, an externally equalized TXV is used. **See Figure 31-11.** An externally equalized TXV has equalization tubing connected to the output side of the evaporator. The evaporator is connected to the TXV opposite the external equalizer connection. This external equalization line compensates for inherent pressure drops in the evaporator.

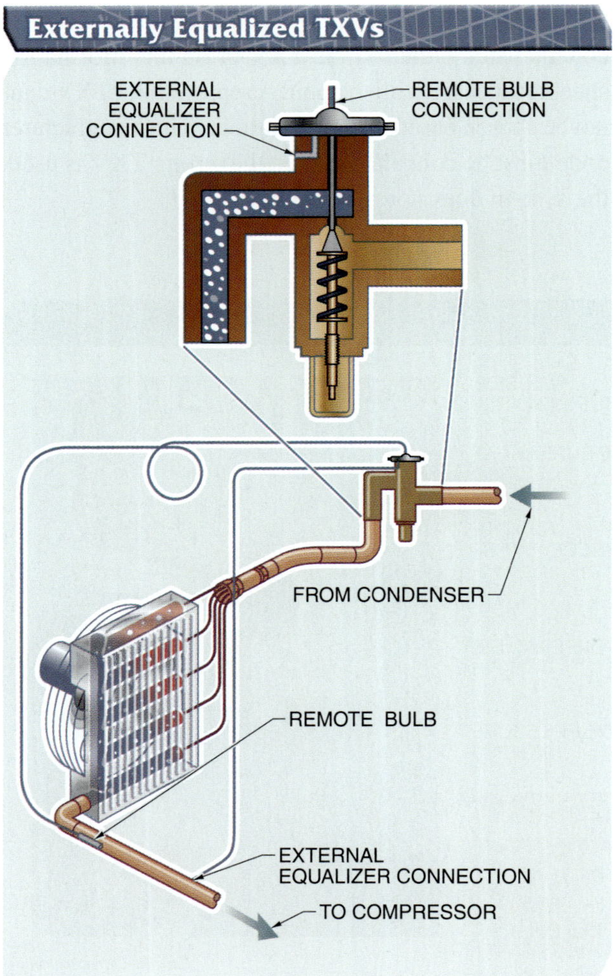

Figure 31-11. Externally equalized TXVs are used to compensate for excessive inlet pressure drops to evaporators.

TXV Equalizer Connections

A TXV should be mounted in a vertical position for proper operation. The TXV equalizer connection should be made on the suction line between the evaporator and the compressor. **See Figure 31-12.** The TXV connection should be attached to the suction line securely and insulated to adequately sense the line temperature. When a TXV is not attached securely, system vibrations may cause the capillary tube to weaken and break.

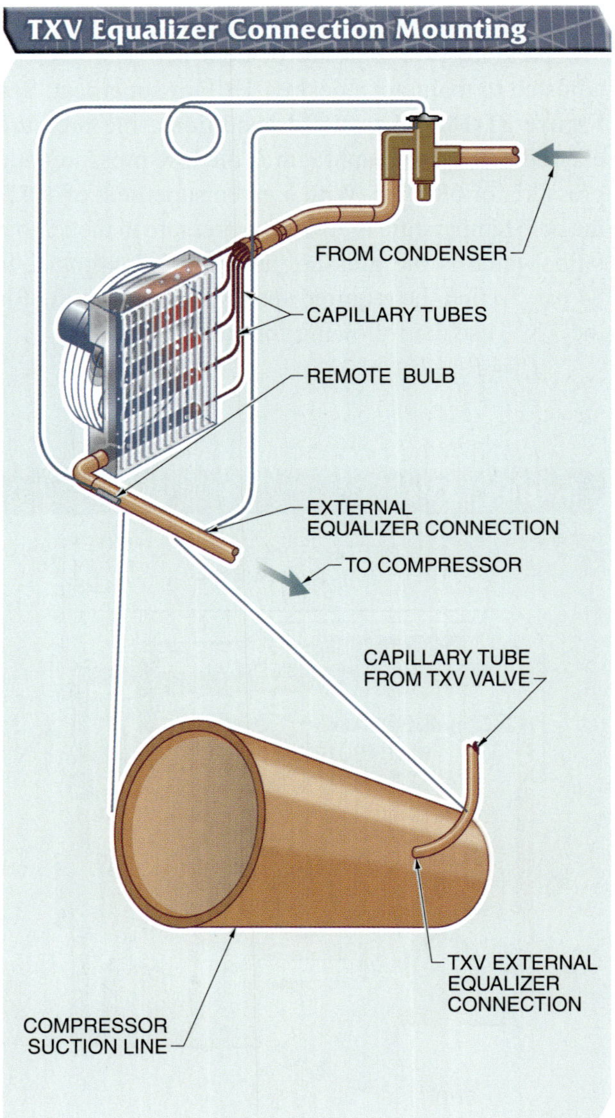

Figure 31-12. TXV external equalizer connections must be made on the suction line between the evaporator and the compressor.

TXV Troubleshooting and Repair

TXVs are often misdiagnosed as the source of a refrigeration system failure or the source of a problem in sections of the system that a TXV does not affect. Unnecessary adjustment of the TXV is a common mistake. The TXV is adjusted by removing the cap from the valve assembly and adjusting the tensioning screw. The TXV is adjusted to maintain the correct superheat. However, the TXV must not be used to compensate for refrigerant leaks, improperly operating evaporator fans, or other conditions. When the TXV cap is removed, the adjusting screw is exposed. Although the TXV is adjustable, some TXVs fail when adjusted multiple times.

TXV Hunting. *TXV hunting* is a condition where the thermostatic expansion valve opens and closes often, causing widely varying evaporator conditions. **See Figure 31-13.** The discharge air temperature from the evaporator swings wildly during TXV hunting. At one end of the cycle, the TXV could lose superheat altogether and slug the compressor. In some conditions, TXV hunting causes a frost line to move back and forth across the evaporator line as it alternately starves and then feeds the evaporator. The superheat reading may vary between 0°F and 20°F or more. One adjustment involves using the spring tension to open the valve, checking it, and readjusting at 15 min intervals. In extreme cases, refrigerant recovery, TXV replacement, system evacuation, and system recharging may be warranted.

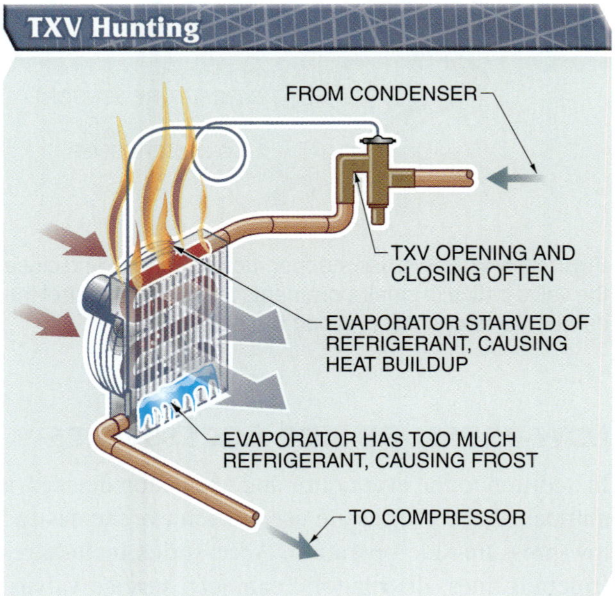

Figure 31-13. TXV hunting is when a TXV opens and closes quite often, causing widely varying discharge temperatures from the evaporator.

A TXV element should be attached securely and should not include kinks.

TXV Adjustment. In some systems, a TXV spring may be adjusted in an attempt to correct a problem that was not caused by the TXV, sometimes resulting in misadjustment of the TXV. When a TXV must be adjusted, one method for adjusting a TXV back to factory settings is to adjust it from one extreme to another and count the number of turns or threads in the process. Many TXVs are set at a center position from the factory. **See Figure 31-14.** Therefore, the count from one extreme to the other can be used to get the TXV close to its proper setting. The total number of turns divided in half is an approximate center setting. When a TXV is simply stuck in either the open or closed position, it must be replaced.

Load Problems. TXVs are also misdiagnosed when they do not maintain the correct conditions due to a constantly changing extreme load condition. Common problems include doors that are left open to a cooler or freezer, airflow problems due to a missing service panel, and instability when large amounts of product are taken in or out of location. These types of circumstances must be checked first before adjusting the TXV.

ELECTRONIC EXPANSION VALVES

An *electronic expansion valve (EEV)* is an expansion device that uses a stepper motor to open and close a valve orifice to control the amount of refrigerant entering an evaporator. EEVs have become more prevalent in various fields, including refrigeration control. EEVs come in different styles. Some EEVs use stepper motors or heat-actuated motors. **See Figure 31-15.**

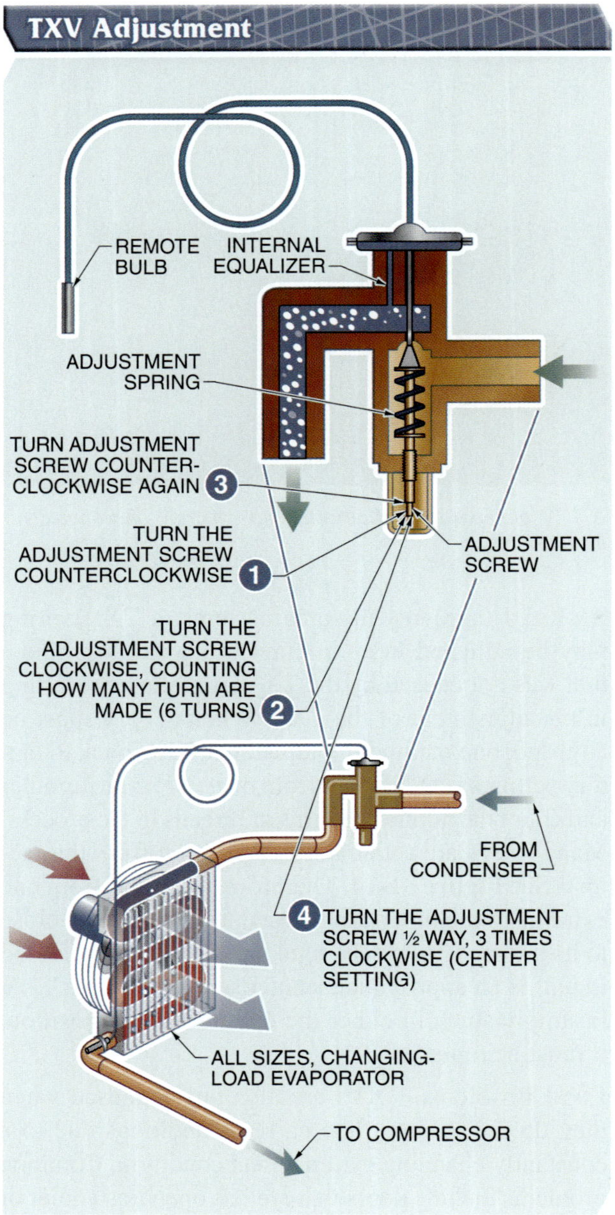

Figure 31-14. Most TXVs are set at a center position, which is determined by dividing in half the total number of turns required to open or close the TXV valve.

Stepper motors, used in advanced EEVs, have thousands of possible positions between fully open and fully closed. Stepper motor operation is similar to motors used in computer hard drives and variable-air-volume (VAV) boxes. Stepper motors provide precise positioning of the valve, which provides excellent control results. The stepper motor is operated by a controller, which uses a thermistor attached to the evaporator outlet to sense temperature. A *thermistor* is a solid-state sensor that changes resistance as temperature changes. Some controllers also have pressure-sensing and control capability.

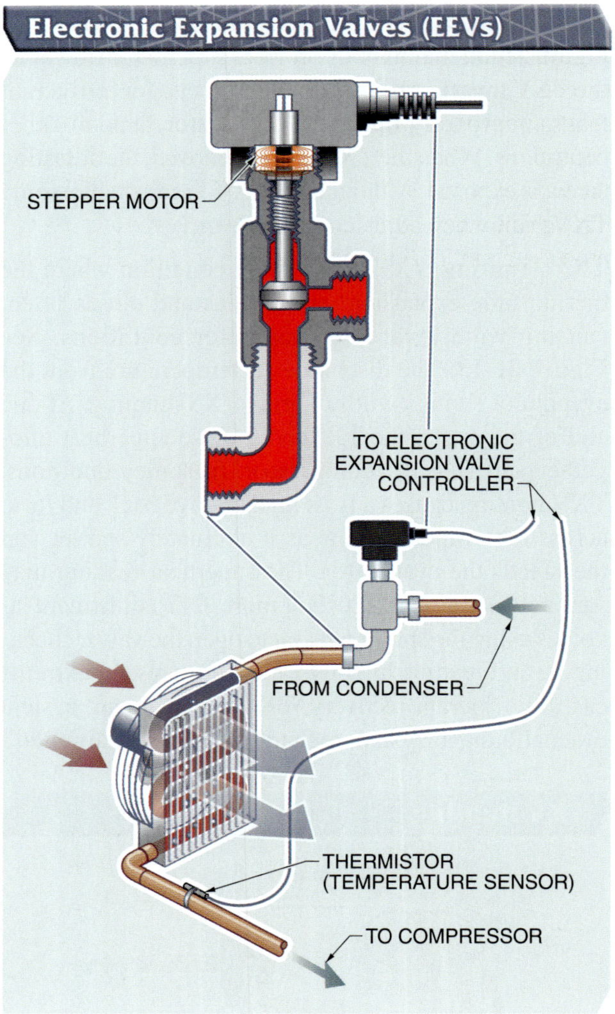

Figure 31-15. EEVs use stepper motors to open and close the valve orifice in small increments to precisely control the amount of refrigerant entering the evaporator.

LOW-PRESSURE-SIDE ACCESSORIES

In addition to the evaporator and expansion devices, a number of accessories are used to achieve the desired low-pressure-side operation. Accessories include refrigerant lines, distributors, cabinets, service valves, sight glasses, moisture indicators, suction accumulators, evaporator pressure regulators, and crankcase pressure regulators.

Refrigerant Lines

Refrigerant lines on the low side connect the expansion device to the evaporator, and the evaporator to the compressor. Proper sizing and installation of these lines are crucial for proper evaporator operation.

The connecting line is responsible for directing the low-pressure liquid-vapor refrigerant into the evaporator. **See Figure 31-16.** The tubing run must be as short as possible to reduce pressure drop created by the line. Sharp turns and twists must also be avoided. When evaporators are mounted vertically, it is common to introduce the refrigerant from the top in order to allow gravity to aid the refrigerant flow through the evaporator. The suction line is sized to allow the proper amount of flow out of the evaporator. The suction line is often sloped slightly toward the compressor, which aids in the oil returning back to the compressor.

Refrigerant Lines

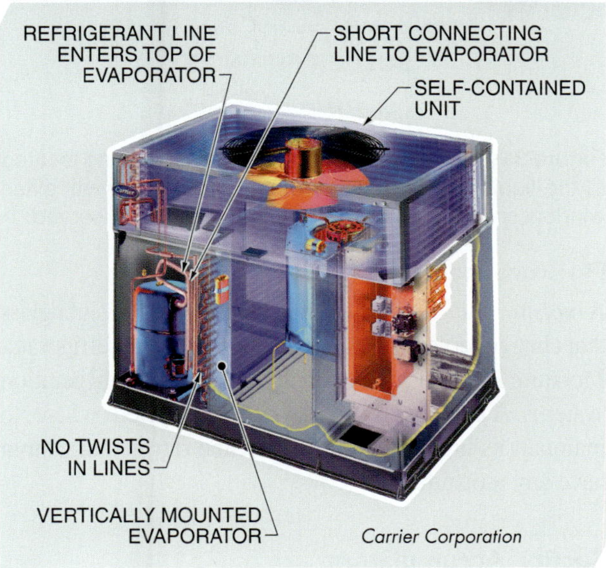

Figure 31-16. Refrigerant lines (piping) must be sized to pass the correct amount of refrigerant to and from an evaporator and compressor.

Distributors

Potential problems increase as the capacity and physical size of an evaporator increases. A *distributor* is a device used to evenly feed refrigerant from the expansion device into either multiple circuits on a single evaporator or separate evaporators. A large evaporator can suffer from an unusually high pressure drop and from hot and cold spots because of uneven refrigerant flow patterns in the coil. This may cause temperature variations as refrigerant leaves the coil, affecting comfort or product-cooling efficiency. **See Figure 31-17.** The distributor has a nozzle-shaped device attached to the expansion-device outlet. Small feeder lines are attached to each refrigerant port for feeding the refrigerant evenly into the evaporator.

Distributors

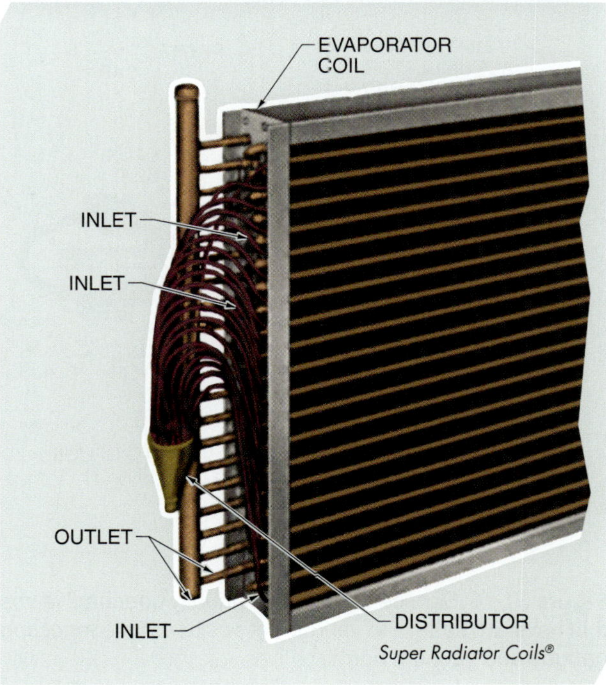

Figure 31-17. A distributor splits refrigerant flow entering an evaporator into several separate flow paths.

Cabinets

A *cabinet* is a sheet-metal enclosure that completely covers and provides support for the components of an HVAC device. Cabinets usually consist of an exterior shell of sheet metal, which may or may not be painted, depending on location and visibility. Insulation is sometimes added to reduce temperature drops and radiated noise levels. Fasteners are also added to allow secure operation and easy access.

Service Valves

Service valves are typically added to a refrigeration system at the suction side of a compressor or at the compressor discharge. A *service valve* is a type of valving system that allows technicians to attach manifold gauges to the system to record system pressures, charge the system, or recover refrigerant. Service valves come in different types, depending on unit type and cooling capacity. Schrader valves are often used on small-tonnage units. The cores of Schrader valves can be removed and replaced when leaks occur. On large units, the service valves are wrench-operated stem-valve types. **See Figure 31-18.**

Service Valves

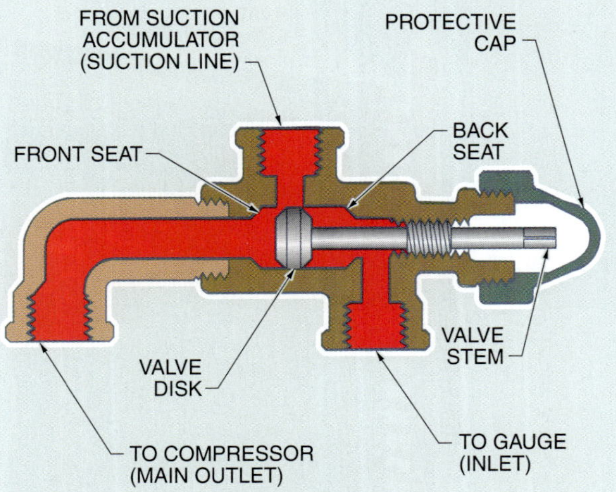

Figure 31-18. Service valves are manually operated valves that have two seats and various set positions for connecting gauges and vacuum pumps.

Covers over the valve stems prevent damage and help to limit access. Valve stems are designed to use a standard ACR ratchet wrench. The valve can be adjusted in one of three positions: front-seated, back-seated, or middle-position. When a service valve is front-seated, the compressor is isolated from the suction line. A front-seated valve can be used to isolate and service the compressor. Middle-position valves are used to allow the manifold gauges to have proper readings. A back-seated service valve is wide open to flow between the evaporator and compressor. Lightly oiling the valve stem and cap threads when using the valves reduces rust problems.

Sight Glasses

A *sight glass* is a special pipe fitting that contains a small glass window that provides a visual indication of a refrigerant's condition. The flow of refrigerant can be observed through the glass window. Sight glasses can be found in the liquid line before the metering device to indicate whether refrigerant vapor or liquid is entering the metering device. **See Figure 31-19.**

Sight Glasses

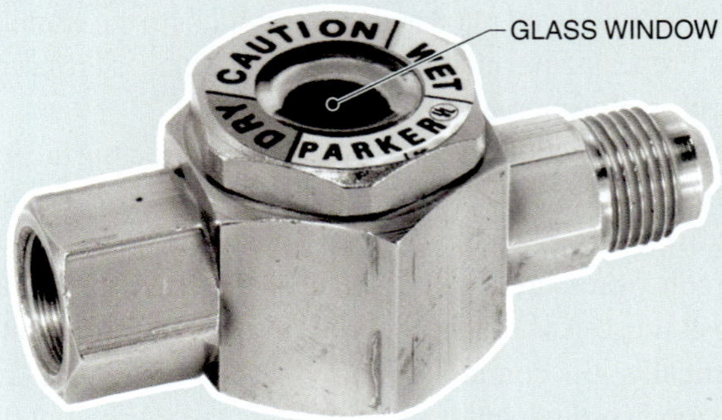

FLARE FITTINGS

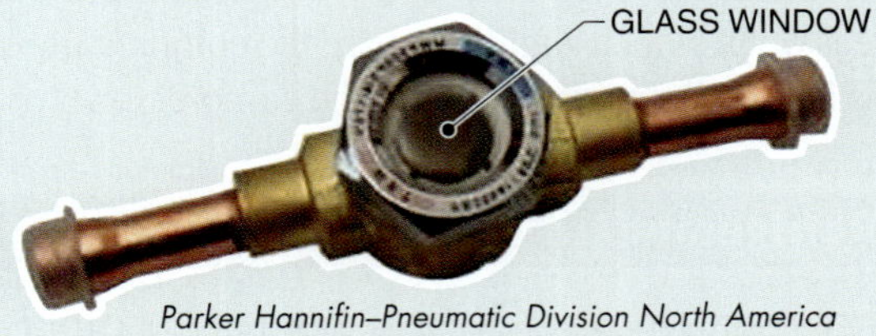

Parker Hannifin–Pneumatic Division North America

SOLDER FITTINGS

Figure 31-19. A sight glass is a special pipe fitting located in the liquid line of a system that contains a small glass window so that the flow of refrigerant can be observed.

Moisture Indicators

A *moisture indicator* is a chemical patch inside a sight glass that changes color when moisture is present in refrigerant. Moisture indicators are typically used in dry refrigeration systems. A dry refrigeration system is a system where, to maintain its proper characteristics, the refrigerant cannot have any moisture.

Suction Accumulators

Great care must be taken to prevent liquid refrigerant from entering the compressor. Many refrigeration systems operate under widely changing load conditions that may increase the possibility of liquid entering the compressor, also known as slugging the compressor. A *suction accumulator* is a device used to keep liquid refrigerant from entering a compressor. Suction accumulators are installed in the suction line after the evaporator and before the compressor. **See Figure 31-20.** A suction accumulator is a cylinder with an inlet tube and a U-shaped dip tube.

Liquid refrigerant stays in the bottom of the accumulator shell because it cannot change direction as easily as vapor, which is caused by the dip tube takeoff at the top of the accumulator. The liquid refrigerant remains in the bottom of the accumulator until it vaporizes and then is allowed to pass, whereas the vaporized refrigerant in the accumulator changes direction easily and passes on to the compressor. In order to promote lubricating oil return, a small orifice is installed in the bottom of the dip tube. The small orifice allows a small amount of liquid refrigerant, which mixes with oil, to be sucked up the outlet tube where the liquid refrigerant carrying oil flashes to a vapor before reaching the compressor.

Evaporator Pressure Regulators

It is often necessary to maintain a minimum evaporator pressure, regardless of load condition. An *evaporator pressure regulator (EPR),* also known as a holdback valve, is a valve that is designed to maintain a minimum evaporator pressure at all times. An EPR is located at the discharge opening of the evaporator. When the evaporator pressure drops too low, the EPR closes and keeps the pressure in the evaporator just above the minimum value. The EPR is often used to help prevent system freeze up. EPR designs include direct-acting, pilot-operated, and electronic.

Crankcase Pressure Regulators

A *crankcase pressure regulator (CPR)* is a device that maintains maximum pressure at the compressor inlet. While an EPR maintains a minimum pressure in the evaporator, the CPR maintains a specific maximum pressure at the suction port of the compressor. This protects the compressor from unusually high pressures that might be encountered during a system startup under a load or a hot pulldown, when large amounts of warm product are introduced into a cooler. The CPR often includes a gauge port that provides an attachment for manifold gauges.

Suction Accumulators

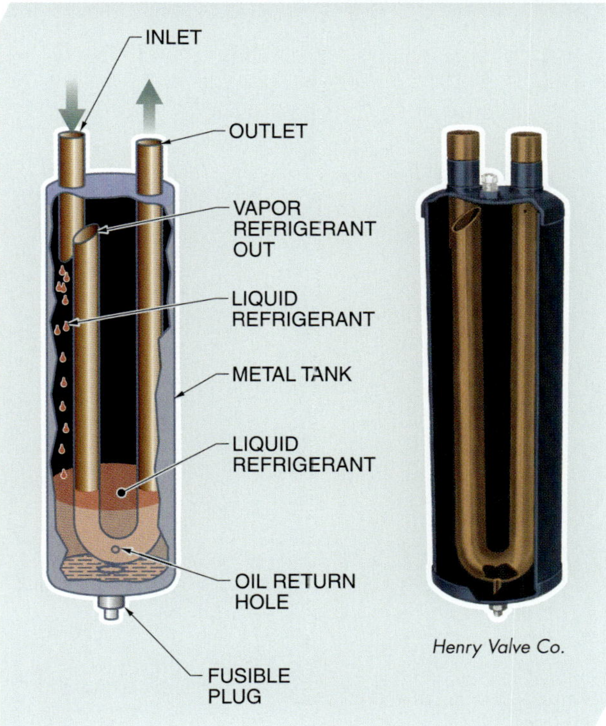

Figure 31-20. Suction accumulators store liquid refrigerant so that vapor refrigerant can enter but liquid refrigerant cannot enter the compressors.

UNIT

32 High-Pressure-Side Compressors

The high-pressure-side components of a refrigeration system are critical to its proper operation. Of the typical high-pressure-side components used, the compressor is considered the "heart and soul" of the refrigeration system. Today, there are many types of compressors and drives in use. To be effective, an HVAC technician must be familiar with the basic types and operation characteristics of the various compressors.

OBJECTIVES

- List the high-pressure components.
- Identify the purpose of a compressor.
- List the different types of compressor categories (housings).
- List the components of reciprocating compressors.
- Compare scroll, rotary, and large system compressors.
- Describe considerations for compressor motors in refrigerant atmospheres.
- Perform basic compressor calculations.

Index to Procedures

HIGH-PRESSURE SIDE

The high-pressure side of a refrigeration system consists of a compressor, a condenser, refrigerant lines, and a number of accessories. **See Figure 32-1.** The compressor receives cool, low-pressure refrigerant vapor. It raises the pressure and temperature of the refrigerant to a hot gas state. The condenser receives the hot, high-pressure gas refrigerant and rejects the heat to air or water. This heat rejection cools the refrigerant and allows it to condense back into a 100% liquid. The refrigerant lines carry the refrigerant between the compressor, condenser, and accessory components.

COMPRESSORS

The compressor is often considered to be the most important device in the refrigeration system. The purpose of the compressor is to create refrigerant flow and to allow pressures in the system. The pressures are due to the individual molecules of refrigerant vapor occupying a smaller volume than they do before the compressor. **See Figure 32-2.** For example, an R-22 split air-conditioning-system compressor may have a suction-side pressure of 68 psi and a discharge pressure of 267 psi. Each application and refrigerant has different pressure characteristics. Various methods are used among the different compressor designs to reduce the volume and increase the pressure of the refrigerant.

TECH TIP

Compressors are becoming more reliable and efficient as their design continues to advance. New manufacturing techniques and additions such as microprocessor controls help customers enjoy greater comfort.

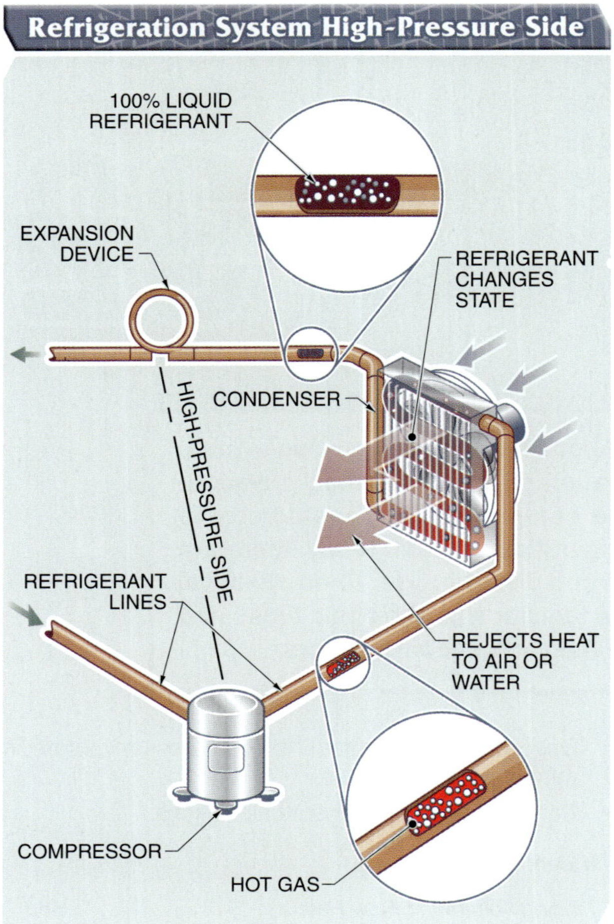

Figure 32-1. The high-pressure side of a refrigeration system includes a compressor, a condenser, refrigeration lines, and a number of accessories.

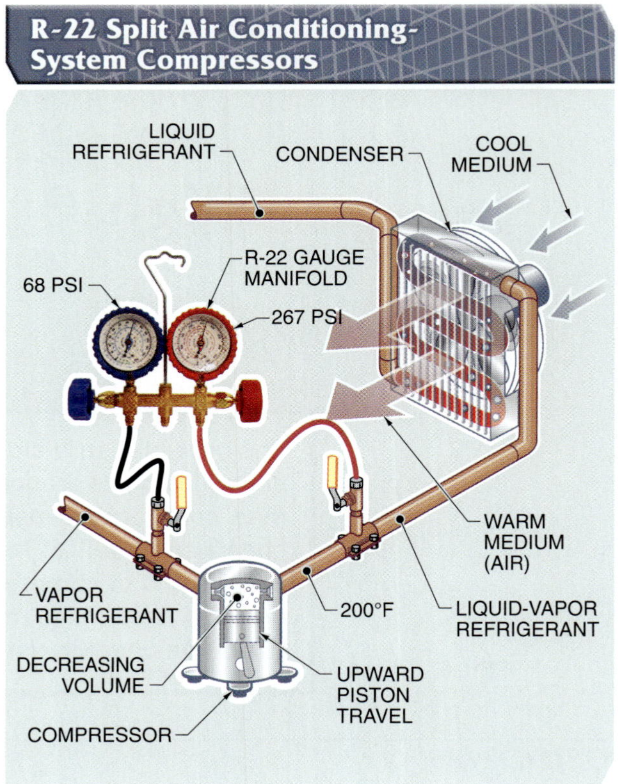

Figure 32-2. Compressors increase pressure in refrigeration systems (68 psi to 267 psi). Each application and each refrigerant has a different pressure characteristic.

A failed compressor is replaced with a new compressor after the system is flushed.

The compressor receives cool refrigerant vapor and increases the temperature and pressure of the refrigerant dramatically. The amount of heat added is increased by the conversion of energy due to compression. The discharge line of the compressor is typically the hottest part of a refrigeration system, with temperatures up to 200°F being common.

COMPRESSOR CATEGORIES

Compressors used in refrigeration systems can be classified by housing type, model, and application. Compressors can be categorized by their type of housing. **See Figure 32-3.** The term "housing type" refers to the type of metal shell containing the compressor such as fully welded hermetic (fully hermetic or hermetic), semihermetic (serviceable), and open-drive.

Compressor Housings Media Clip

Compressor Housings

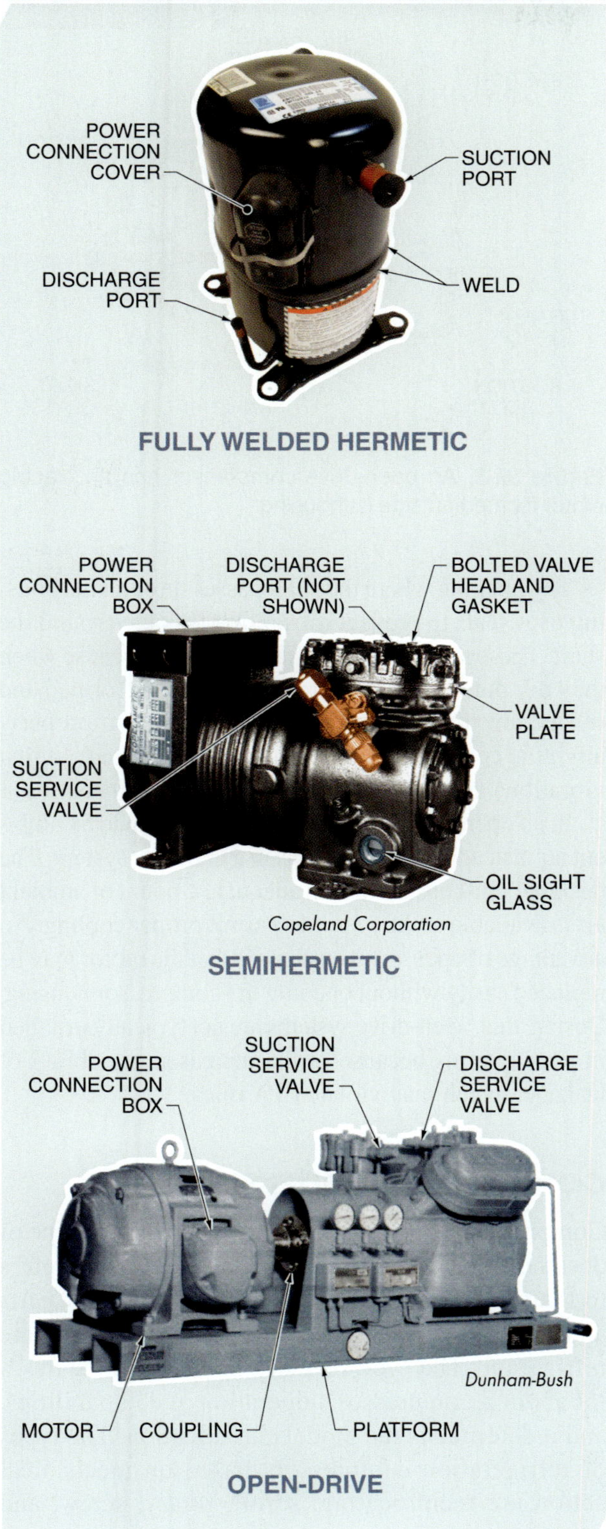

Figure 32-3. Compressors can be categorized by their type of housing such as fully welded hermetic, semihermetic, and open-drive.

Fully Welded Hermetic Compressors

One major difference between compressors is the construction of the compressor housings. Smaller-tonnage compressors often use fully welded hermetic housings (shells). The motor, lubricating oil, and compressor of a fully welded hermetic compressor are contained inside a fully sealed shell. **See Figure 32-4.** A fully welded hermetic compressor, also called a fully hermetic compressor or hermetic compressor, cannot be opened and serviced in the field. When a single component in the housing fails, total replacement is the only solution. The compressor motor inside the hermetic housing is exposed to and cooled by refrigerant. Oil for lubrication circulates through the system and back to the compressor crankcase.

Fully Welded Hermetic Compressors

Figure 32-4. A smaller-tonnage compressor typically has a hermetic housing that contains the electric motor and compressor inside a fully sealed shell that cannot be opened for servicing compressor components.

Semihermetic Compressors

Large-tonnage compressors provide access into the compressor housings for service. These compressors are called semihermetic or serviceable compressors. **See Figure 32-5.** Bolted heads, gasketed heads, and other access points permit the technician to replace or repair parts of the compressor without throwing away the entire assembly. While permitting repair, these compressor types are typically more expensive than similar-size fully hermetic compressors.

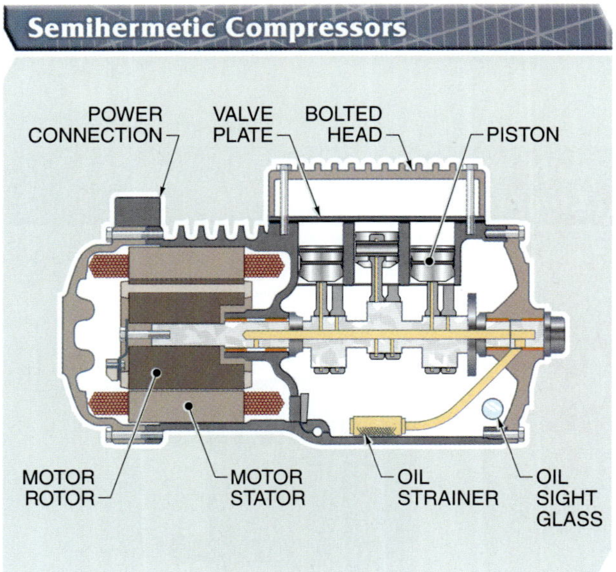

Figure 32-5. Semihermetic compressors are more expensive than similar-size fully hermetic compressors. However, semihermetic compressors have access points that permit technicians to replace or repair some parts instead of throwing away the entire assembly.

Open-Drive Compressors

In addition to the housing, a compressor can be distinguished by how the motor is connected to the pump. In hermetic and semihermetic motor designs, the motor is inside the housing with the compressor. Some smaller open-drive systems can be found in commercial applications. This type of compressor has the motor and pump as separate devices bolted to a base. The motor is mounted on a platform with the motor shaft connected to the crankshaft through the pump housing using a coupling. **See Figure 32-6.**

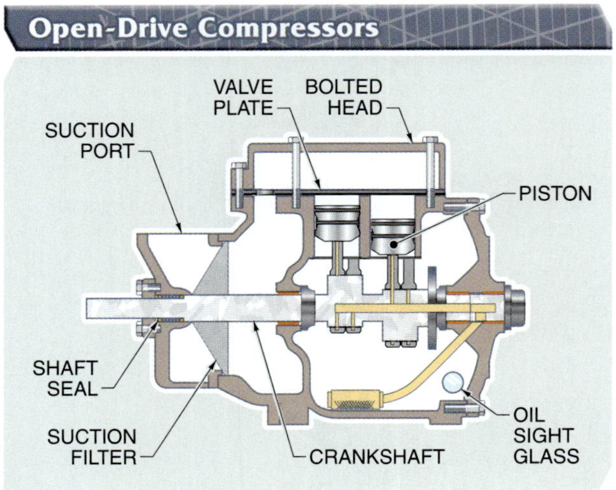

Figure 32-6. An open-drive compressor has an electric motor located outside its housing.

The opening where the shaft passes through the housing uses seals to prevent refrigerant leakage around the shaft. These shaft seals are often a concern because when they dry out, they crack, or are somehow deformed and refrigerant then leaks. Care must be used to properly align the compressor motor with the crankshaft so that vibrations do not occur and deform the shaft seals.

In an open-drive system, the motor is cooled by ambient air instead of refrigerant, as in a hermetic system. The designer must ensure that an adequate amount of ambient air is available at the motor to permit proper cooling. An advantage of open-drive systems is that the motor may be replaced easily without opening the compressor housing. Despite this, open-drive systems are not typically installed in new systems because of the increased possibility of leakage, which may violate EPA rules.

COMPRESSOR MODELS

Compressors can also be categorized by their type of designed mechanical action. Mechanical action refers to the movement of the individual compressor parts and pieces to achieve the compression action on the refrigerant. The type of mechanical action states a lot about a compressor's operation and capabilities, and it determines the model classification. The types of refrigeration compressors based on mechanical action are reciprocating, scroll, rotary, screw, and centrifugal compressors.

Compressor Types Media Clip

Reciprocating Compressors

One of the most commonly used compressors today is the reciprocating compressor. Reciprocating compressors compress refrigerant vapor by moving a piston up and down inside of a cylinder. **See Figure 32-7.** The design of a reciprocating compressor is quite similar to that of an automobile engine. While the automobile engine uses ignited fuel vapor that expands in the combustion process to create mechanical energy, a refrigeration compressor uses mechanical movement to compress refrigerant vapor.

The reciprocating compressor consists of a prime mover (typically an electric motor) that turns a crankshaft, which causes the up-and-down motion of the pistons. **See Figure 32-8.** The *piston stroke* is a measurement of the distance a piston moves up or down in a cylinder.

The reciprocating compressor cycle consists of two separate movements, or strokes, of the piston. **See Figure 32-9.** A *suction stroke* is the downward movement of a piston in a cylinder. The downward motion causes the volume of the cylinder to increase. This volume increase causes a rapid pressure decrease in the cylinder. As the pressure in the cylinder drops below that of the suction line, a suction valve (reed valve) opens and admits vapor refrigerant into the cylinder. A *compression stroke* is the upward movement of a piston in a cylinder. This upward movement reduces the volume of the cylinder, which increases the pressure. When the pressure is greater than that of the discharge line, a discharge valve opens and allows high-pressure vapor to flow from the cylinder to the discharge line.

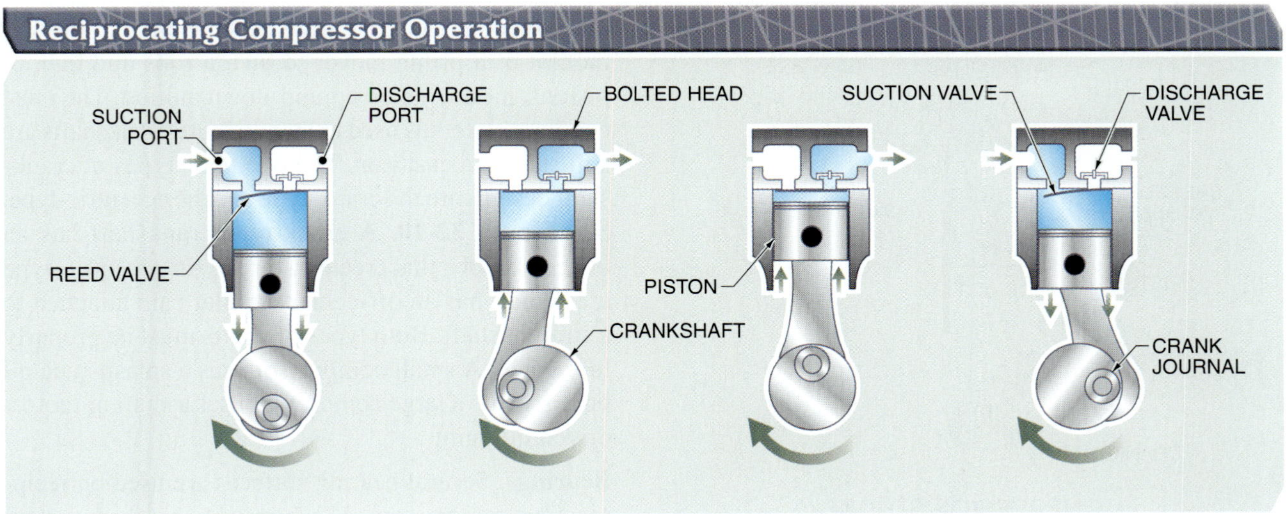

Figure 32-7. A reciprocating compressor uses a piston moving inside a cylinder to compress refrigerant.

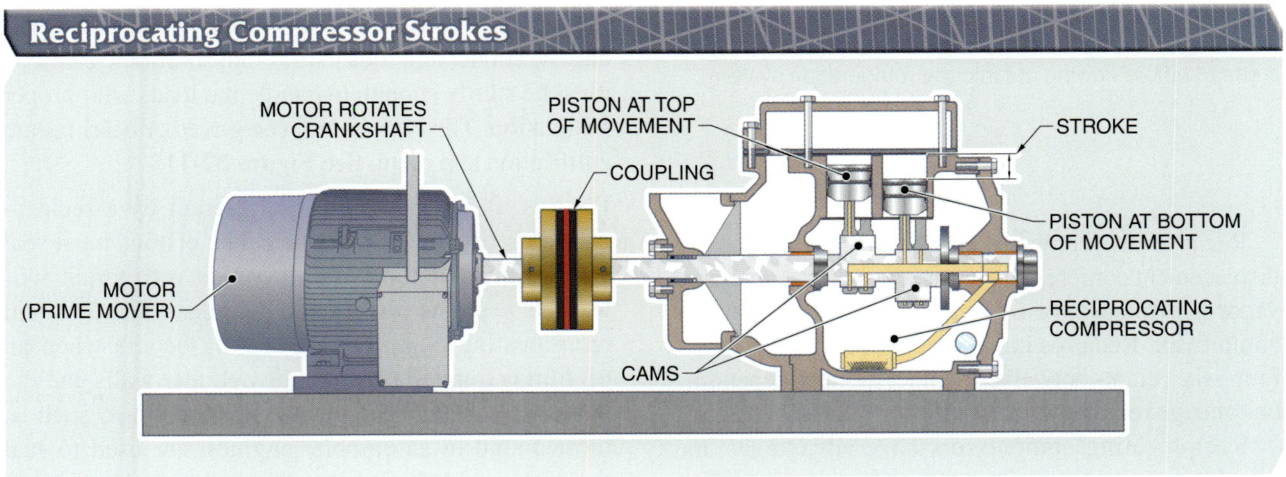

Figure 32-8. Reciprocating compressor pistons are driven up and down by crankshafts, which determine piston strokes.

Reciprocating Compressor—Suction and Compression Strokes

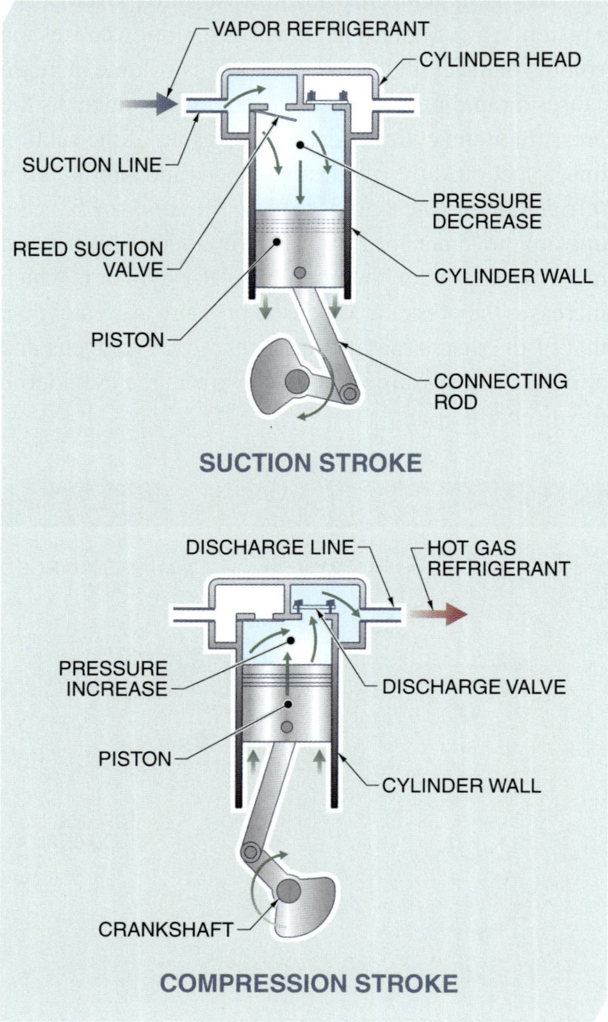

Figure 32-9. The suction stroke of a compressor allows refrigerant to enter the cylinder, and the compression (discharge) stroke allows refrigerant pressure to be increased before flowing out into the refrigeration system.

Reciprocating compressors are also known as positive displacement compressors because a specific amount of vapor is displaced, or moved, with each stroke of the compressor. Reciprocating compressors come in two-, four-, six-, eight- or twelve-cylinder models, depending on tonnage requirements.

Reciprocating compressors have advantages and disadvantages. One advantage of a reciprocating compressor is its ability to produce flow against high discharge pressures. The disadvantages of reciprocating compressors include the large number of moving parts and the relative difficulty of control under partial-load conditions. Reciprocating compressors also discharge the high-pressure vapor in pulsations, which must be dampened because pulsations produce undesirable vibration effects.

Reciprocating Compressor Components

The reciprocating compressor has several important components that must function properly in order to provide the correct refrigeration capacity. Reciprocating compressor components include crankshafts, bearings, connecting rods, pistons and piston rings, valves, mufflers, and pulsation chambers.

Crankshafts. A *crankshaft* is a reciprocating compressor component that transmits the circular turning motion of a prime mover to piston rods and then to pistons, producing an up-and-down motion. The most common materials used to manufacture crankshafts are soft steel and cast iron. The two main types of crankshaft designs are the crank-type and the eccentric-type. **See Figure 32-10.** A crank-type crankshaft has an off-center lobe that creates a throw. An eccentric-type crankshaft has an off-center circular cam attached to a straight shaft. Both types of shafts must be properly lubricated. A small compressor uses a splash-type oil pump, while a large compressor uses a shaft or motor-driven oil pump.

Bearings. Several bearing surfaces are used on reciprocating compressors. The bearings absorb the stress and the wear and tear of the components as they rotate at high speed. The shaft bearing surfaces, also called journals, typically include both ends of the crankshaft and the connecting rods to the crankshaft. The bearings must be sturdy enough to handle the loads with proper lubrication. The two general categories of bearings are antifriction and plain. **See Figure 32-11.**

Pistons. Pistons compress refrigerant by a reciprocating motion in the cylinders. The pistons must seal well to prevent the possibility that refrigerant will leak between the pistons and cylinder walls. In small reciprocating compressors, sealing occurs when an oil film is trapped between the cylinder walls and the pistons. In many compressors, piston rings, such as those found in automobile engines, are used to seal the pistons. Multiple sets of rings provide a higher amount of sealing.

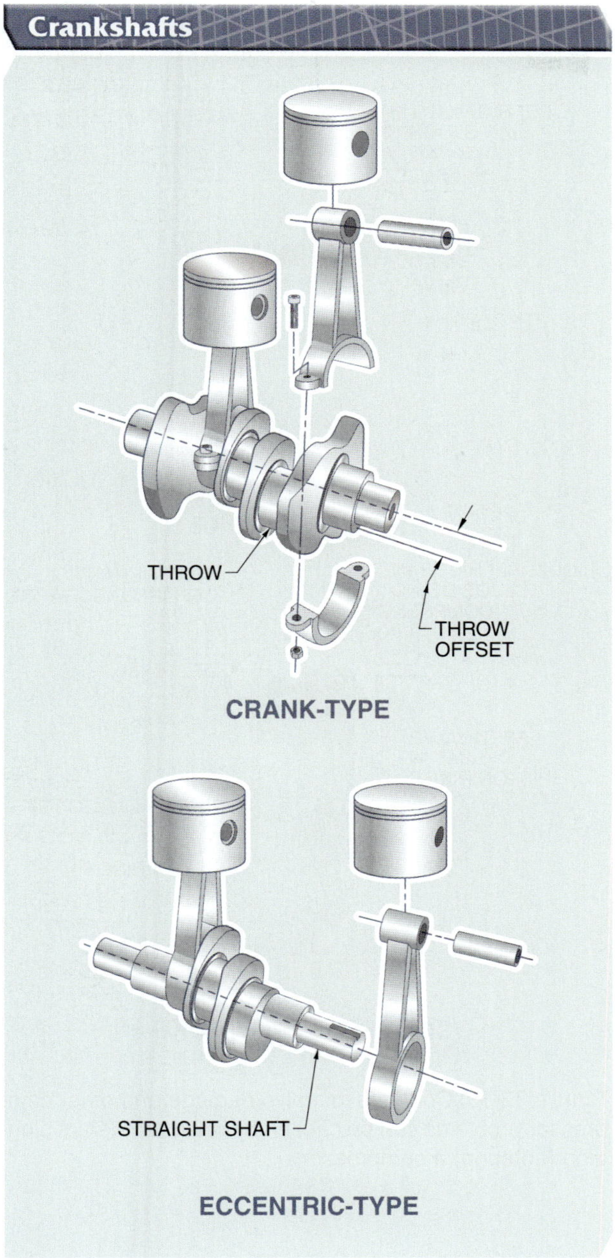

Figure 32-10. Compressors typically use either crank-type or eccentric-type crankshafts.

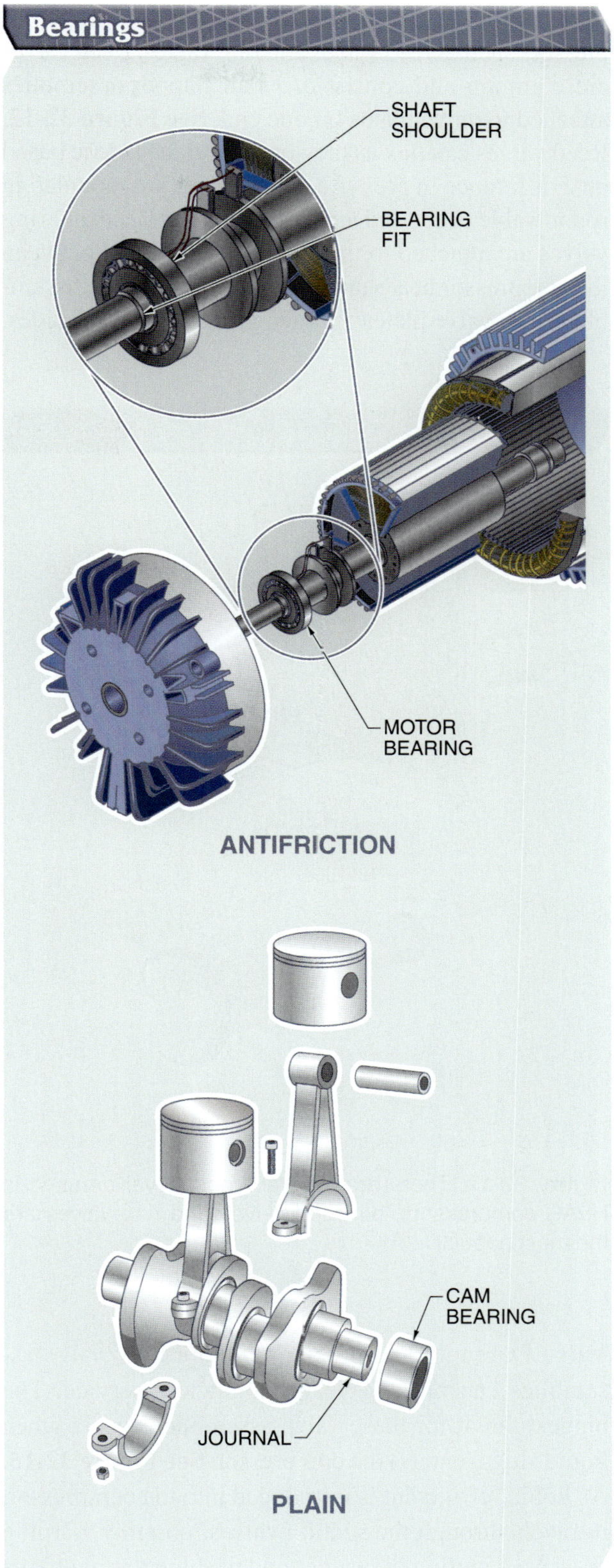

Figure 32-11. Bearings on compressor and motor shafts absorb the wear and tear of shaft rotation and can be classified as antifriction or plain.

Valve Types. Reciprocating compressors use valves to direct refrigerant flow into the compressor during the suction stroke and out of the compressor during the discharge stroke. Most valves for reciprocating compressors are constructed from a tough steel to resist wear and damage.

Several valve types are used in reciprocating compressors such as reed valves and ring valves. Reed valves are common and consist of small flapper assemblies attached to valve plates on one end. **See Figure 32-12.** Reed valves can flex as they open, and they close based on a difference in pressure. Ring valves are mounted in round valve cages and are spring loaded. Reed and ring valves are attached to the valve plates located between the compressor heads and the top of the cylinder assemblies. The valve plates are then gasketed on both sides.

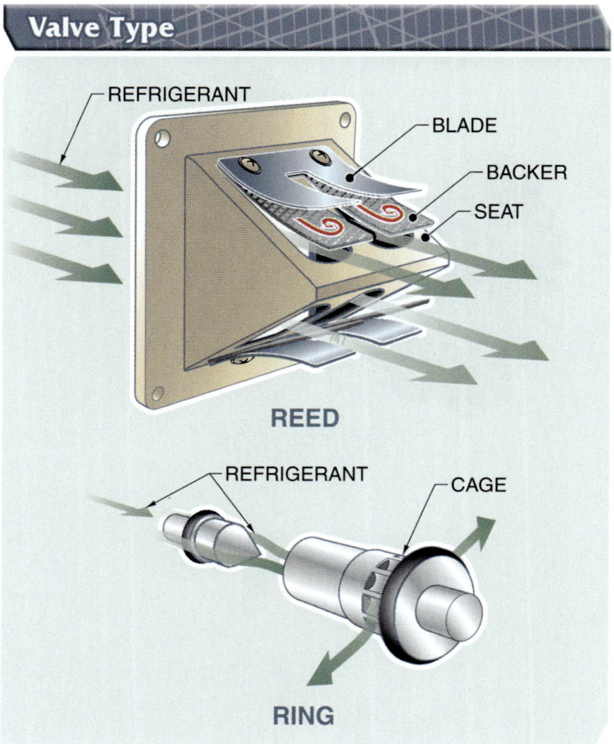

Figure 32-12. There are several types of valves used in HVAC compressors, but reed valves and ring valves are the most popular.

Valve Damage. Reciprocating compressor valve assemblies can provide years of flawless service. The biggest threat for these valve assemblies occurs when liquid slugs (enters) the compressor. **See Figure 32-13.** As liquid refrigerant is introduced into the compressor, it travels through the suction valves, possibly warping or even shattering the valves and pushing them into the cylinder. The broken pieces are then driven into the cylinder walls by the piston as it moves, deeply grooving the cylinder walls or gouging the piston.

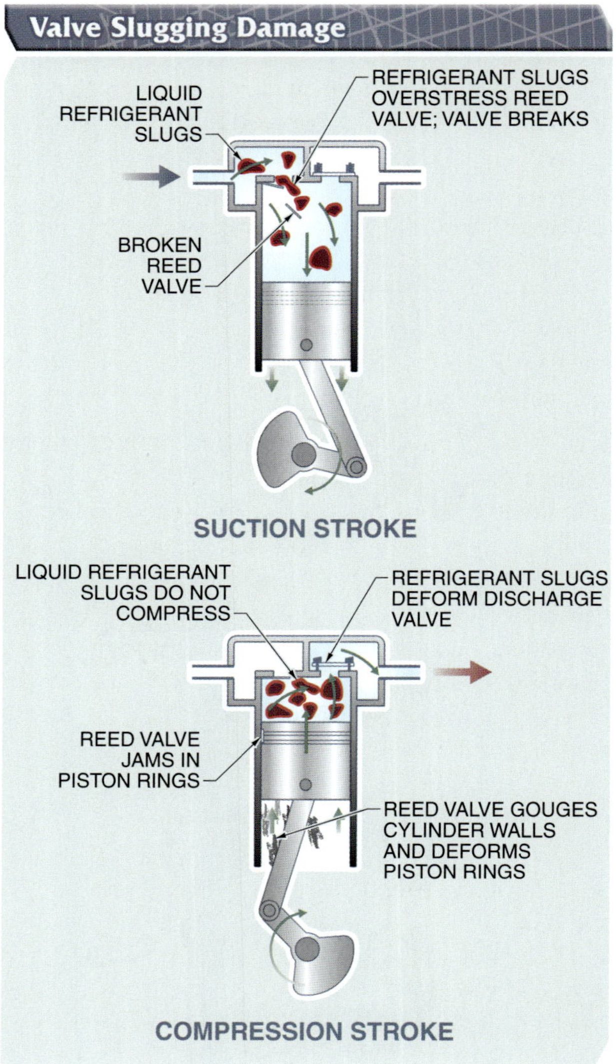

Figure 32-13. Compressor valve damage and other compressor problems can be caused by liquid refrigerant slugging (entering) a compressor.

Since liquid refrigerant slugs cannot be compressed, these slugs damage the valve assemblies when entering or discharging from the cylinder. Damage is seen in the form of warped valves, broken pieces lodged in the cylinder or system valves, and loss of compression because the valves can no longer operate properly. This slugging condition must be avoided. Therefore, superheat control and suction accumulators are used to dramatically reduce the possibility of liquid refrigerant leaking into, and therefore slugging, the compressor.

Compressor Slugging Media Clip

Mufflers and Pulsation Chambers. Reciprocating compressors discharge refrigerant in short pulsations. These pulsations can create undue vibrations and noise and possibly damage the system. A *muffler* is a device that reduces transmitted vibrations and noise. **See Figure 32-14.** A *pulsation chamber* is a short pipe in the discharge line of a compressor that is larger in diameter than the largest suction piping to the compressor. When used together, multiple pulsation chambers typically act like shock absorbers. Care must be taken to avoid creating an excessive pressure drop due to oversizing the pulsation chambers and/or muffler.

Volumetric Capacity. Reciprocating compressors are positive displacement compressors that move a specific volume of refrigerant. The amount of volume a compressor moves depends on the amount of volume available in each cylinder, the number of cylinders, and the speed (in rpm) of the compressor. The volumetric capacity of a single cylinder can be found by using the following formula:

$$V = \pi r^2 \times S$$

where

V = volumetric capacity of 1 cylinder (in in.^{3})

π = 3.14 (a constant)

r = radius of cylinder (in in.)

S = stroke length (in in.)

Example: Finding Cylinder Volume

A cylinder in a reciprocating compressor has a diameter of 6″ and a stroke of 2″. What is the volume of the cylinder? **See Figure 32-15.**

$$V = \pi r^2 \times S$$

$$V = 3.14 \times (3 \times 3) \times 2$$

$$V = \mathbf{56.52\ in.^3}$$

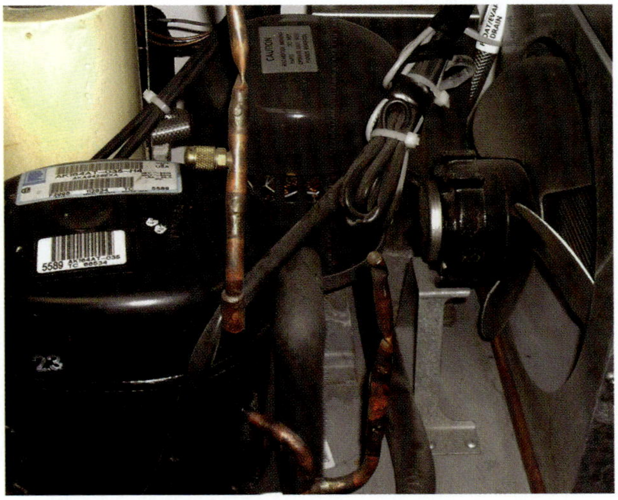

Compressors are often housed in tight locations, making them difficult to work on.

Mufflers and Pulsation Chambers

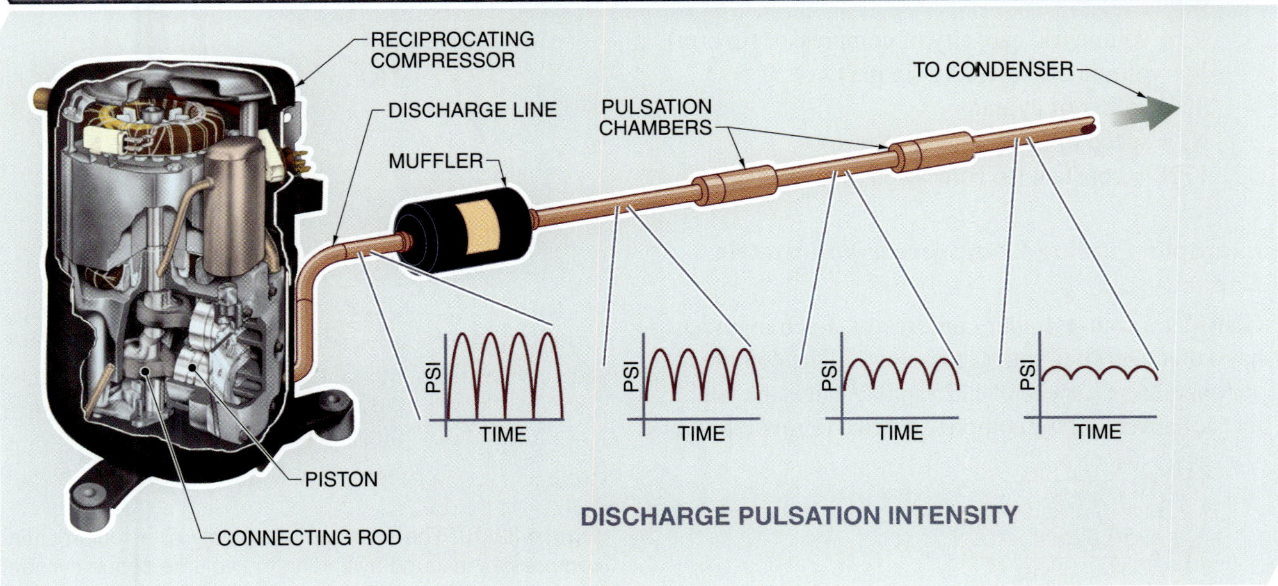

Figure 32-14. Compressor discharge pulsations and vibrations are greatly reduced by a muffler and multiple pulsation chambers being used in series.

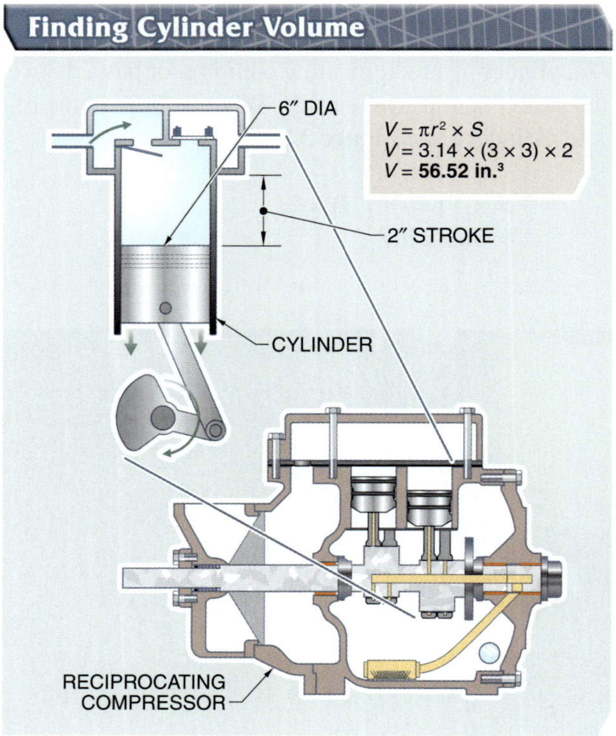

Figure 32-15. Cylinder volume must be known to find the volumetric capacity of a reciprocating compressor.

To determine the volumetric capacity of a compressor, apply the following formula:

$$V_C = \frac{V \times n \times N_C}{1728}$$

where

V_C = volumetric capacity of compressor (in cfm)

V = volume of 1 cylinder (in in.^{3})

n = number of cylinders

N_C = speed of compressor (in rpm)

1728 = constant (to convert cu in. to cu ft)

Example: Finding Compressor Volumetric Capacity

Consider a four-cylinder compressor. Each cylinder has a diameter of 6″ and a stroke of 2″. The compressor operates at a speed of 1725 rpm. What is the volumetric capacity of the compressor? **See Figure 32-16.**

$$V_C = \frac{V \times n \times N_C}{1728}$$

$$V_C = \frac{56.52 \times 4 \times 1725}{1728}$$

$$V_C = \mathbf{225.69\ cfm}$$

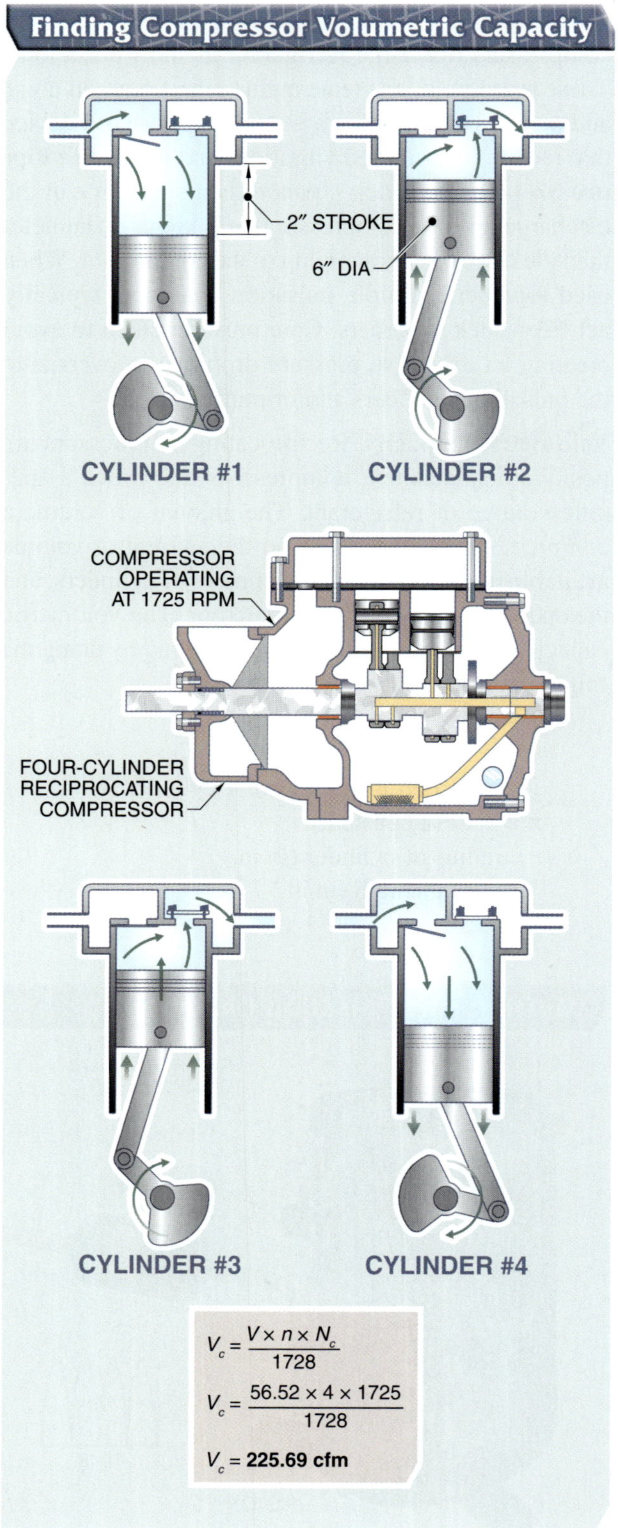

Figure 32-16. The volumetric capacity of a reciprocating compressor is found by finding the volume of one cylinder, multiplying this by the number of cylinders and the revolutions per minute of the compressor, and then dividing the product by 1728.

Scroll Compressors

Traditionally, scroll compressor designs have been introduced in small- and medium-size residential and commercial applications. The design for a scroll compressor differs significantly from that of a reciprocating compressor. The scroll compressor consists of a motor-driven shaft that turns a plate with a spiral of metal on it. **See Figure 32-17.** This rotating scroll orbits around another scroll plate that is stationary. Suction refrigerant is introduced at the outside of the two scrolls.

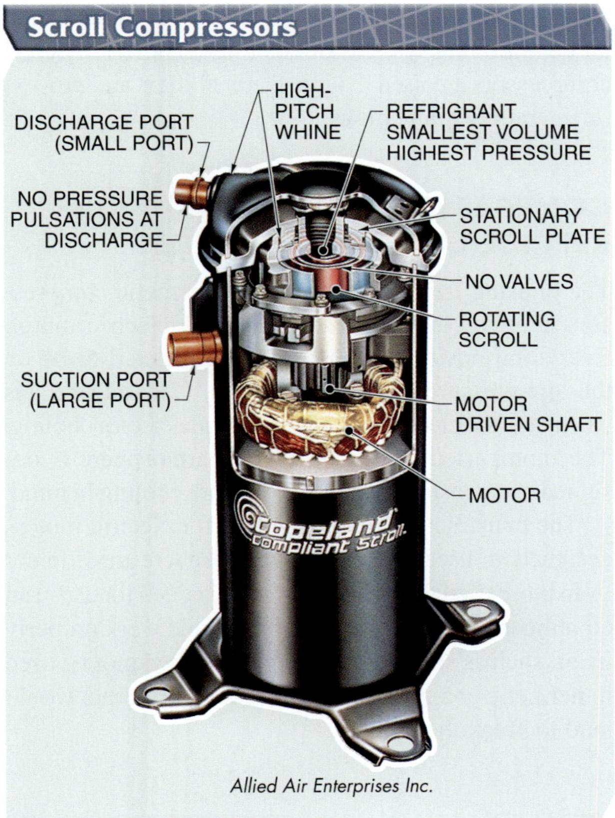

Figure 32-17. Scroll compressors consist of a stationary scroll and a rotating scroll, which discharge refrigerant from the center of the scrolls.

As the moving scroll orbits around the stationary scroll, a volume of gas is trapped between the two scrolls. The center of the scroll is where the gas occupies the smallest volume possible and at the highest pressure. The area at the center of the scroll is where the discharge is located. Different stages of compression are constantly occurring as one scroll orbits the other.

Scroll compressors have some advantages over reciprocating systems. For example, the refrigerant discharge does not pulsate, the noise factor is generally lower, there are fewer moving parts, and the scroll compressor is not as easily damaged by liquid refrigerant entering the compressor. A lower likelihood of damage exists because the rotating scroll is designed to have an allowance built in that permits the scroll to move slightly and allows liquid to move through the compressor without compressing the liquid or damaging the compressor. Still, it is not a good idea to introduce liquid refrigerant into any compressor, even a scroll compressor.

The sound of scroll compressors is different from that of reciprocating systems. The scroll compressor emits a high-pitch whine, but the reciprocating compressor emits a low-pitch pulsation sound. Also, scroll compressors have no suction or discharge valves that could break from being hit with liquid refrigerant. Many scroll compressors have a discharge check valve to prevent discharge gas from flowing back to the suction side when the motor is shut off.

Scroll compressors come in many sizes. Low-tonnage scroll compressors are used for central air or heat pump systems, and high-tonnage systems are used for water chiller systems. Scroll compressors are typically welded hermetic compressors and can be used in chilled water applications exceeding 100 tons, and some up to 300 tons.

Rotary Compressors

Rotary compressors are used in small domestic applications such as refrigerators, freezers, and window air conditioning units. The two main designs of rotary compressors are stationary-vane and movable-vane. **See Figure 32-18.**

A *stationary-vane compressor* is a type of rotary compressor that has a rotor to which the vanes are attached. The rotor is also mounted off-center within the steel cylinder. As the rotor rotates, it traps gas into a decreasing volume between the vanes to be discharged.

TECH TIP

Scroll compressors are being used in a greater number of HVAC and refrigeration applications. In large systems, it is not uncommon to have multiple scroll compressors staged to provide a greater tonnage than a single compressor. For example, three 10-ton scroll compressors may be staged in sequence to obtain a total system capacity of 30 tons.

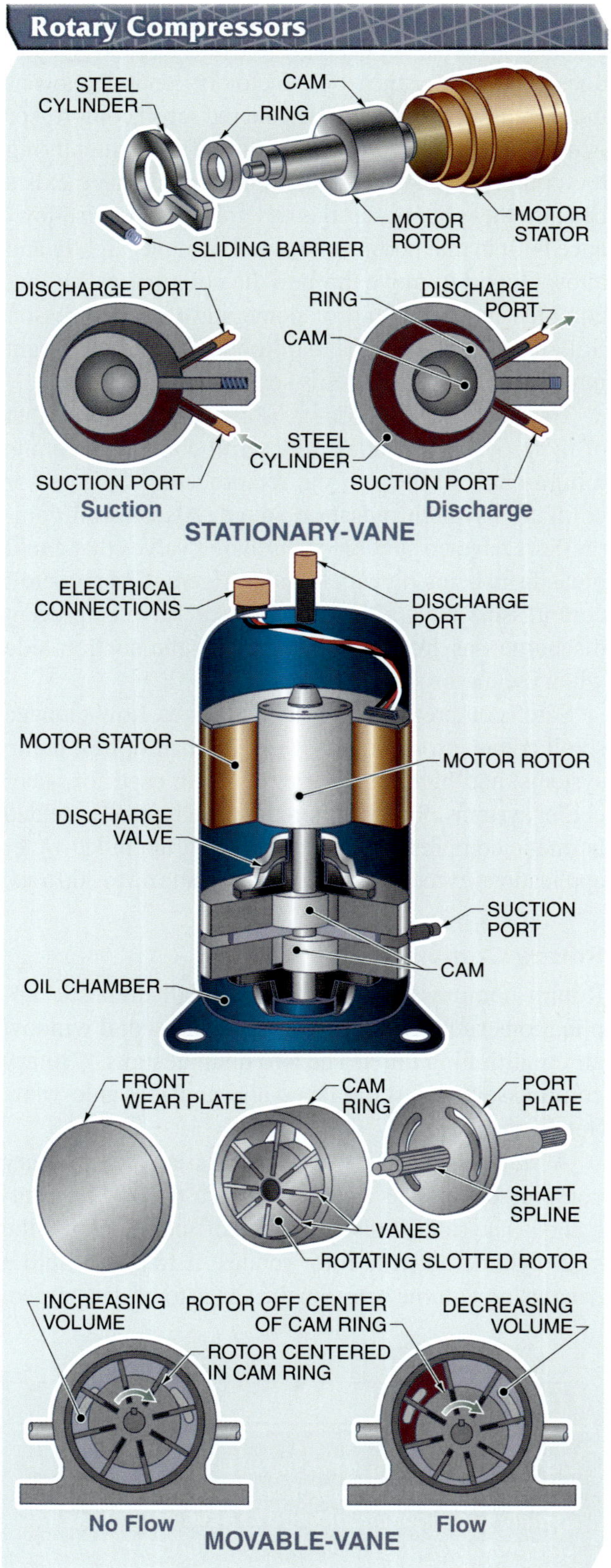

Figure 32-18. Rotary compressors use stationary or movable vanes inside cam rings to create the increasing and decreasing volumes of gas to create flow.

A *movable-vane compressor* is a type of rotary compressor that has a rotor, with sliding vanes, that is mounted off-center inside a cam ring. A motor turns the rotor inside the cam ring. Spring-loaded vanes connect the rotor to the inside of the cam ring. As the rotor rotates, the vanes trap suction gas into a decreasing volume as the gas is moved toward the discharge port. The vanes keep a tight fit against the cam ring to create a seal.

Large System Compressors

Centrifugal and screw compressors are used in large commercial and industrial refrigeration systems. **See Figure 32-19.** Centrifugal compressors are used in large air conditioning systems of 100 tons (t) or more. Screw compressors are used in large water chillers and process refrigeration applications.

COMPRESSOR MOTORS IN REFRIGERANT ATMOSPHERES

The housing for hermetic and semihermetic compressors contains the compressor and the electric motor. Everything inside the housing is considered to be on the low-pressure side of the system. Cool suction gas surrounds the electric motor and provides motor cooling. The motor used in this refrigerant atmosphere is designed and constructed with refrigerant cooling in mind.

The construction materials used for electric motors designed for use in a refrigerant atmosphere are different from those used for electric motors designed for use in an air atmosphere. The materials that would work properly in air, such as some rubber compounds, cannot be used in hermetic compressors because the refrigerant would tend to break them down.

The materials used in a hermetic compressor must be compatible because they are sealed inside a shell.

Large Compressors

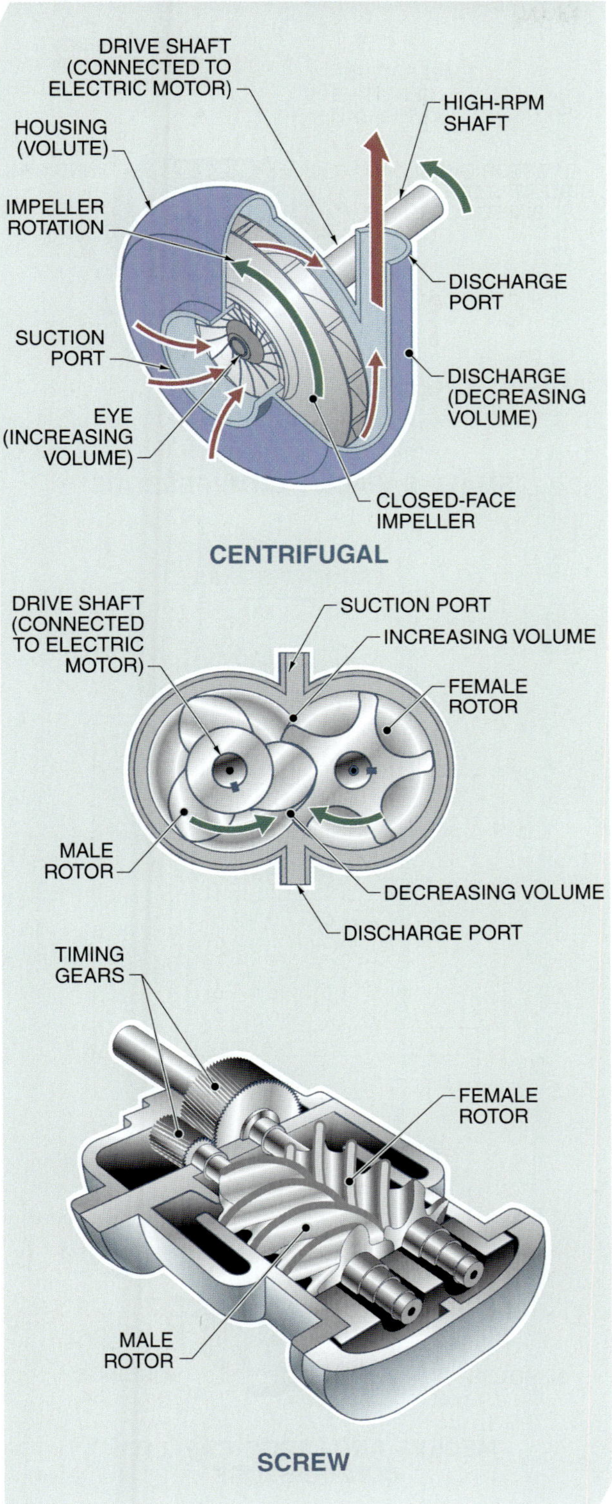

Figure 32-19. Centrifugal compressors change the velocity energy of refrigerant into pressure velocity. Screw compressors use male and female rotors to create large amounts of flow.

The proper design, installation, operation, and maintenance of fully hermetic and semihermetic compressors are crucial. Once installed, fully hermetic compressors cannot be serviced. Fully hermetic compressors are often assembled in factories in clean, dry areas to prevent contamination.

Motor Electrical Terminals

An electric motor used in a hermetic compressor must have electrical connections extending through the housing in order to provide power to the motor. These connections must insulate the compressor housing from the power connection, provide for secure electrical wiring connections, and allow for unit vibration and temperature extremes. Often an insulator with a stud bolt at the center is used. **See Figure 32-20.** The insulator ensures that the power conducted from the power wiring through the bolt or wiring terminals does not come into contact with the shell. Another method is to use a fiber insulator with O-ring connections. Care must be taken because high temperatures may damage or melt the insulating material and O-rings.

Motor Electrical Terminals

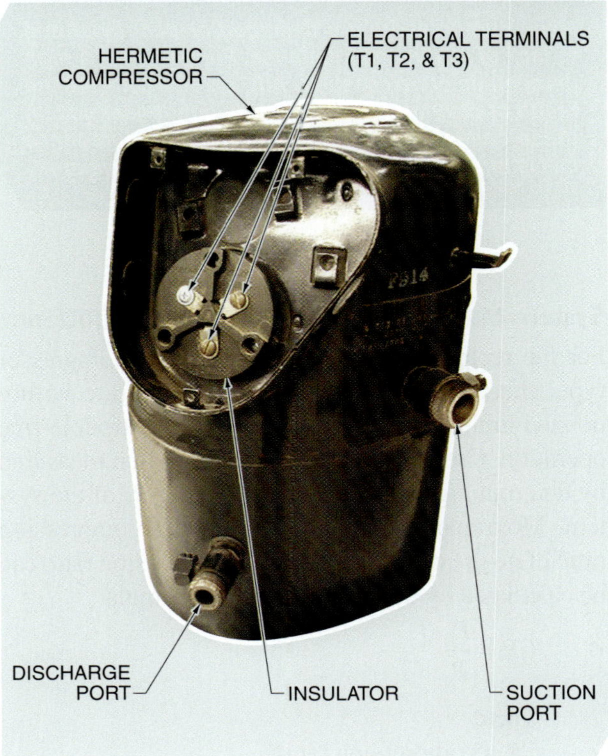

Figure 32-20. Hermetic compressors have wiring connections that penetrate the sealed compressor housings.

Internal Motor Overload Protection

Hermetic compressors have a motor overheating challenge, due to design. A method of protecting the motor against overheating must be provided. In general, two types of internal thermostat overload systems are used. One type of overload system is connected to the line voltage of the motor. This is used mainly in small-capacity compressors with single-phase (1ϕ) motors. A compact motor protection device (overload) is embedded in the motor windings and opens the power circuit of the motor at a specific temperature, stopping the motor. **See Figure 32-21.**

Another type of overload system used in medium- and large-capacity compressors with three-phase (3ϕ) motors is an external relay connected to an internal thermostat. In this overload system, the pilot relay is in the low-voltage control-circuit side of the coil circuit. The external relay can be replaced as needed while the thermostat in the motor on the line-voltage side cannot. To replace the internal winding thermostat of a hermetic compressor, the compressor must be replaced.

TECH TIP

Many compressors are damaged or destroyed due to lack of routine maintenance. For example, a high percentage of "excessively hot" service calls are the result of dirty air filters or a dirty condenser coil. In these cases, an inexpensive routine maintenance call is turned into a costly compressor replacement.

System Maintenance and Compressor Efficiency

For the reciprocating compressor, or any compressor type, the refrigeration system must operate within defined limits in order to provide years of trouble-free operation. Compressor performance is often measured by determining the compression ratio (*CR*) of the system. Most manufacturers recommend a compression ratio of no greater than 10:1. The compression ratio can be calculated by using the following formula:

$$CR = \frac{P_D}{P_S}$$

where

CR = compression ratio

P_D = discharge absolute pressure (in psia)

P_S = suction absolute pressure (in psia)

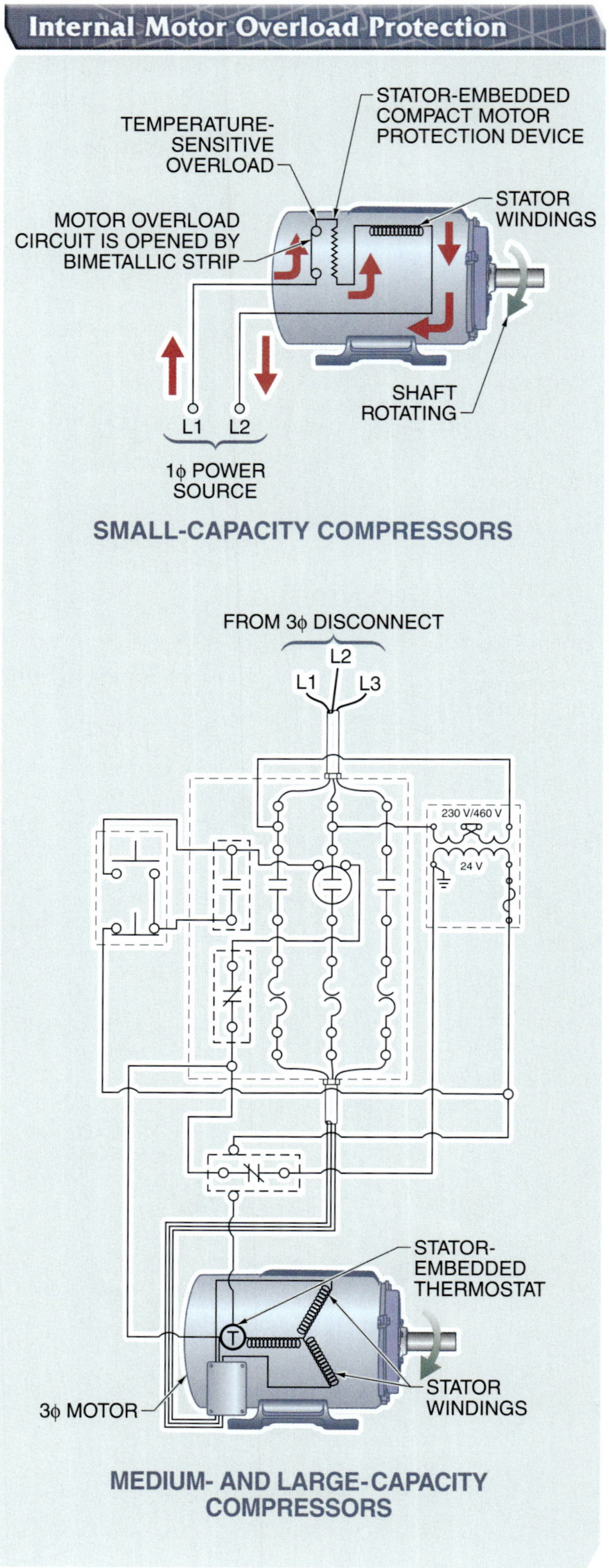

Figure 32-21. Compressor motors often have compact motor protection devices or thermostats embedded in the windings to protect the motors from overheating.

Example: Finding Compression Ratio

Using a suction pressure of 68 psi and discharge pressure of 267 psi, calculate the compression ratio of a reciprocating compressor. **See Figure 32-22.**

1. First, convert both pressures to absolute pressure values.

$$P_S = P_S + 14.7$$
$$P_S = 68 + 14.7$$
$$P_S = \mathbf{82.7\ psia}$$

$$P_D = P_D + 14.7$$
$$P_D = 267 + 14.7$$
$$P_D = \mathbf{281.7\ psia}$$

2. Solve for compression ratio.

$$CR = \frac{P_D}{P_S}$$
$$CR = \frac{281.7}{82.7}$$
$$CR = \mathbf{3.4:1}$$

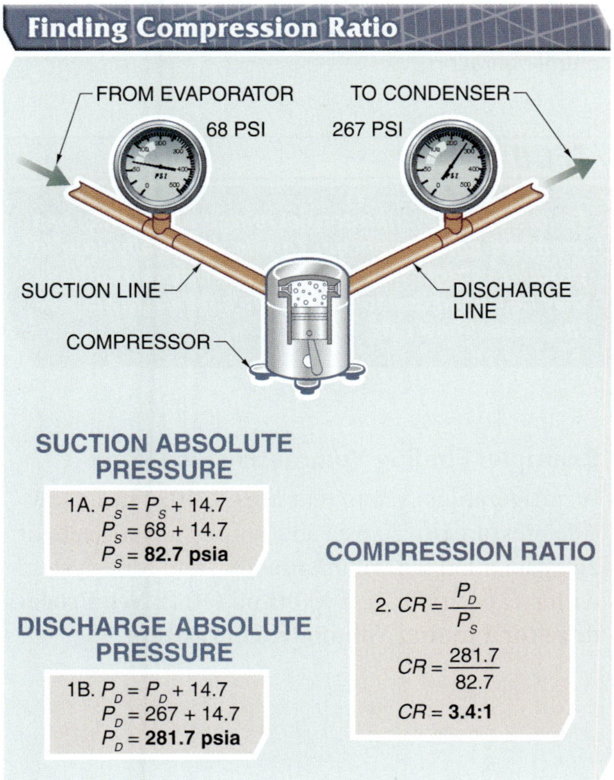

Figure 32-22. Most compressor manufacturers recommend a compressor ratio of no greater than 10:1, which will allow a compressor to provide years of trouble-free operation.

Any condition that increases the discharge pressure or decreases the suction pressure causes the compression ratio to increase. Common causes of this condition could include a dirty evaporator or condenser coil. As dirt accumulates, heat transfer is blocked. On a dirty evaporator coil, the dirt causes lower amounts of heat to be picked up by the refrigerant, lowering the suction pressure. The condenser also loses heat transfer capability, causing the discharge pressure to rise. Either or both of these conditions can cause the compression ratio to increase and over time can cause premature compressor failure.

COMPRESSOR PERFORMANCE

The performance of a compressor can be found by measuring the cooling capacity. The cooling capacity is determined by the volume of refrigerant that the compressor moves through the system.

Refrigerant Flow Rates

Refrigerant flow rate is a measurement of the amount of refrigerant flowing through a system or component in one minute. Refrigerant flow rate can be expressed either as weight (mass) or volume of refrigerant. The refrigerant flow rate is controlled by several factors including compressor design, ambient and load conditions, and capacity control operation, if present.

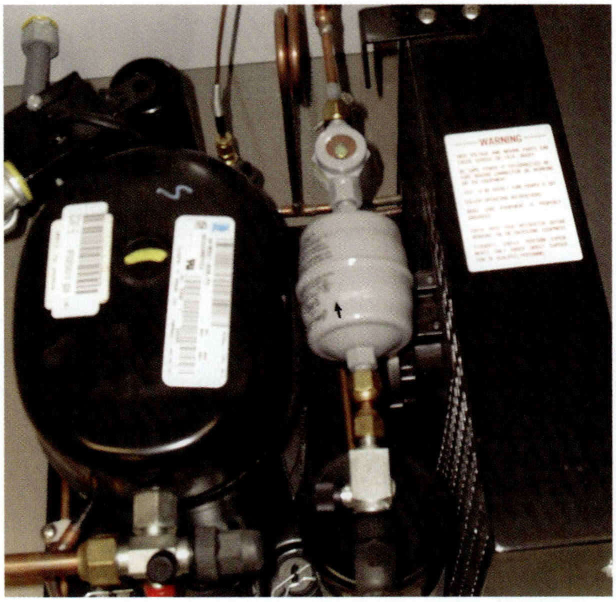

Refrigerant flow direction identifiers are especially important for installing components correctly and are a useful tool when troubleshooting to avoid confusion.

Mass Flow Rates

Mass flow rate is the rate of refrigerant flow in pounds per minute. This is the rate of refrigerant flow that must be circulated to provide the needed cooling capacity. The mass flow rate can be found by applying the following formula:

$$W = \frac{CC}{RE \times 60}$$

where

W = mass flow rate (in lb/min)

CC = cooling capacity (in Btuh)

RE = refrigeration effect (in Btuh)

60 = constant

Example: Calculating Mass Flow Rate

A refrigeration system that uses R-407c has a refrigeration effect of 68 Btuh. Find the mass flow rate for a system with a capacity of 120,000 Btuh (10 t). **See Figure 32-23.**

Apply the following formula to find the mass flow rate:

$$W = \frac{CC}{RE \times 60}$$

$$W = \frac{120{,}000}{68 \times 60}$$

$$W = \mathbf{29.41\ lb/min}$$

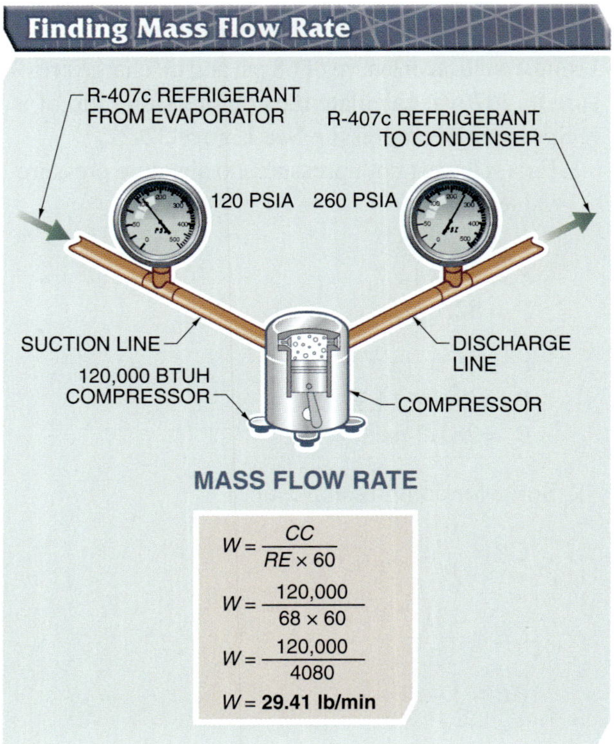

Figure 32-23. The mass flow rate of a refrigeration system is controlled by several factors such as compressor design, ambient and system load conditions, and capacity control operation.

TECH TIP

Failure to achieve the correct refrigeration effect has dramatic consequences. This is especially true in product refrigeration applications where products may be ruined due to the system's failure to achieve the correct flow rate.

Volumetric Flow Rates

Flow rate can be expressed in terms of volume. *Volumetric flow rate* is the rate of refrigerant flow expressed in cubic feet per minute. The volumetric flow rate can be found by applying the following formula:

$$Q = W \times V$$

where

Q = volumetric flow rate (in cfm)

W = mass flow rate (in lb/min)

V = specific volume (in ft^3/lb)

Example: Finding Volumetric Flow Rate

A refrigeration system that uses R-407c has a low-side pressure of 120 psia and a high-side pressure of 260 psia. Find the volumetric flow rate for a system with a capacity of 120,000 Btuh (10 t). **See Procedure for Finding Volumetric Flow Rate.**

Finding Volumetric Flow Rate

1. Calculate the refrigeration effect (*RE*) of the system.
   a. Use an R-407c thermodynamics properties table to determine the low-side enthalpy at 120 psia.

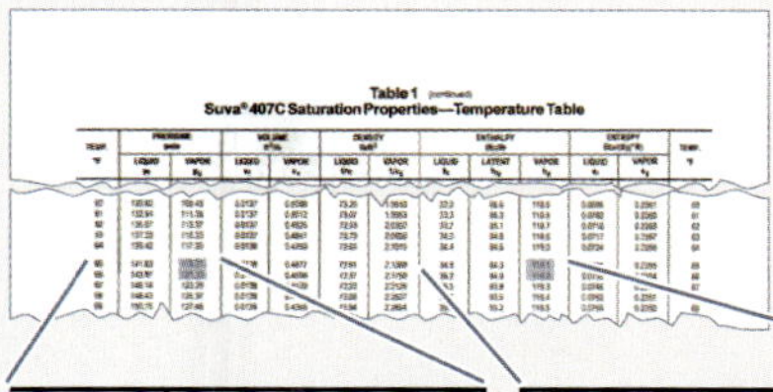

TEMP. °F	PRESSURE psia		ENTHALPY Btu/lb		
	LIQUID P_f	VAPOR P_g	LIQUID h_f	LATENT h_{fg}	VAPOR h_g
65	141.63	119.21	34.8	84.3	119.1
66	143.87	121.23	35.2	84.0	119.2

   b. Use an R-407c thermodynamics properties table to determine the high-side enthalpy at 260 psia.

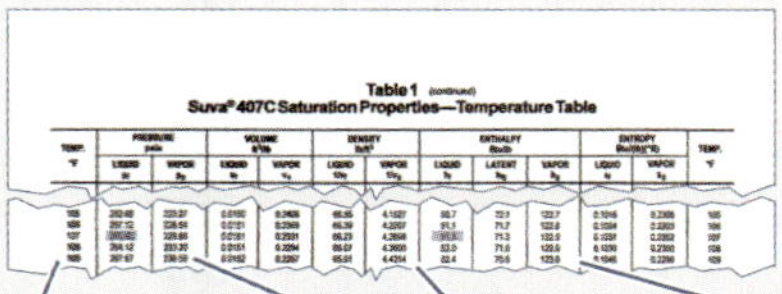

TEMP. °F	PRESSURE psia		ENTHALPY Btu/lb		
	LIQUID P_f	VAPOR P_g	LIQUID h_f	LATENT h_{fg}	VAPOR h_g
105	253.68	223.27	50.7	72.1	122.7
106	257.12	226.54	51.1	71.7	122.8
107	260.60	229.85	51.5	71.3	122.9
108	264.12	233.20	52.0	71.0	122.9
109	267.67	236.59	52.4	70.6	123.0

   c. Apply the following formula to find the refrigeration effect:

   $RE = E_L - E_H$

   $RE = 119.15 - 51.5$

   $RE =$ **67.65 Btuh**

2. Calculate the mass flow rate (*W*) by applying the following formula:

   $$W = \frac{CC}{RE \times 60}$$

   $$W = \frac{120{,}000}{67.65 \times 60}$$

   $W =$ **29.56 lb/min**

3. Use an R-407c thermodynamics properties table to determine the specific volume of R-407c as a vapor at 120 psia. **See Appendix.**

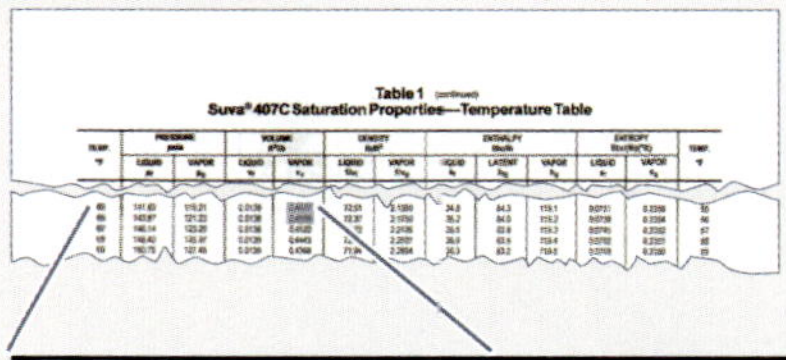

TEMP. °F	PRESSURE psia		VOLUME ft³/lb	
	LIQUID P_f	VAPOR P_g	LIQUID V_f	VAPOR V_c
65	141.63	119.21	0.0138	0.4677
66	143.87	121.23	0.0138	0.4598

4. Calculate the volumetric flow rate (*Q*) by applying the following formula:

   $Q = W \times V$

   $Q = 29.56 \times 0.4639$

   $Q =$ **13.71 ft³/min**

Procedure

UNIT 33

High-Pressure-Side Condensers

To thoroughly understand the operation of any refrigeration system, understanding both air- and water-medium condenser operations is critical. The condenser receives the hot high-pressure refrigerant from the compressor and rejects the heat to an air or water medium. The rejected heat leaves the refrigerant cooler, allowing the refrigerant to condense into a liquid. A malfunctioning condenser does not allow the expansion device or evaporator to function properly.

OBJECTIVES

- Explain the differences between the categories of condensers.
- Describe the components and operation of air-medium condensers.
- Describe the types and operation of water-medium condensers.
- Describe the maintenance that must be performed on water-medium condensers.
- Explain the function of wastewater systems.

Index to Procedures

HIGH-PRESSURE-SIDE CONDENSERS

Besides the compressor, another important high-pressure-side device is the condenser. A *condenser* is a type of heat exchanger that rejects heat from hot refrigerant vapor to a medium (air or water) in order to change the state of the refrigerant to 100% liquid. To accomplish this on a very hot summer day, some condensers are slightly oversized to guarantee that the refrigerant is a 100% liquid. **See Figure 33-1.**

The two main designs of condensers are air-medium condensers and water-medium condensers. **See Figure 33-2.** Air-medium condensers reject heat from the refrigerant to ambient air. Water-medium condensers reject heat to a water loop.

Air-cooled condensers contain many tightly packed fins for efficient heat transfer.

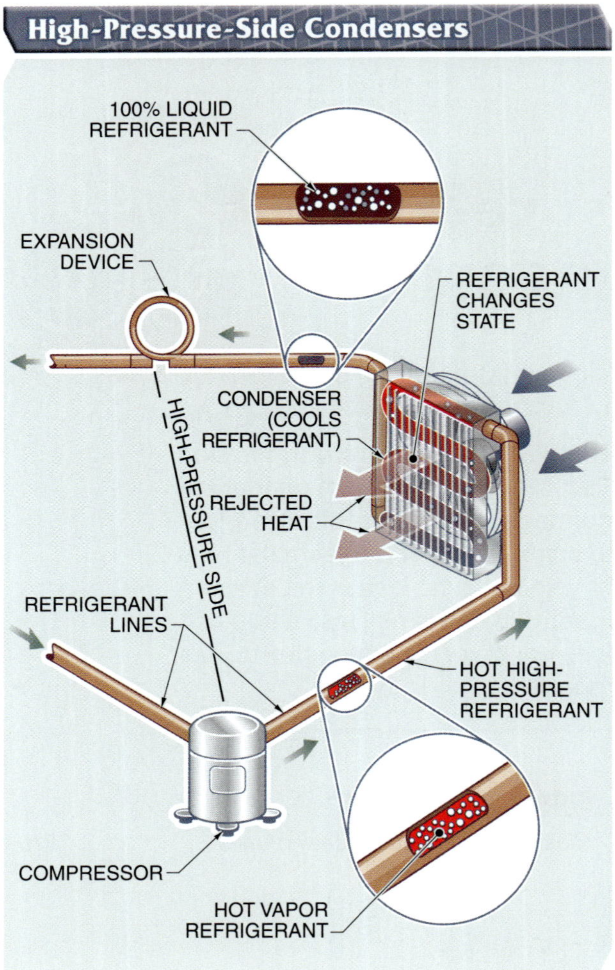

Figure 33-1. Condensers receive hot high-pressure refrigerant from the compressor and reject the heat to an air or water medium in order to condense the refrigerant to a liquid.

AIR-MEDIUM CONDENSERS

An *air-medium condenser* is a type of condenser that uses air as a heat-exchange medium. Air-medium condensers consist of a cabinet, fan, condenser coils, refrigerant lines, and accessories. It is common for the compressor to be mounted inside the condenser cabinet (outdoor unit) in order to reduce compressor noise, vibrations, and weight from the indoor unit.

Cabinets

Residential air-medium condenser cabinets are painted with a weather-resistant paint. The cabinets typically have removable access panels that are used for maintenance or service. The cabinet is constructed so that the fan circulates large amounts of ambient air across the condenser coils and then blows it out to the atmosphere. Several airflow patterns are used to run ambient air across the condenser coils. Residential air-conditioning condensing units typically bring air in from the side of the condenser and then blow it straight up and out of the unit. **See Figure 33-3.**

Large air-conditioning and refrigeration condensing units have cabinets mounted on legs. Mounting the condenser cabinet on legs allows air to enter the condenser from the bottom and then be expelled at the top. **See Figure 33-4.** The air flows in a straight line from the bottom to the top. Some product refrigeration condensers have a side horizontal discharge system that brings air in on one side of the unit and discharges it horizontally out the other side.

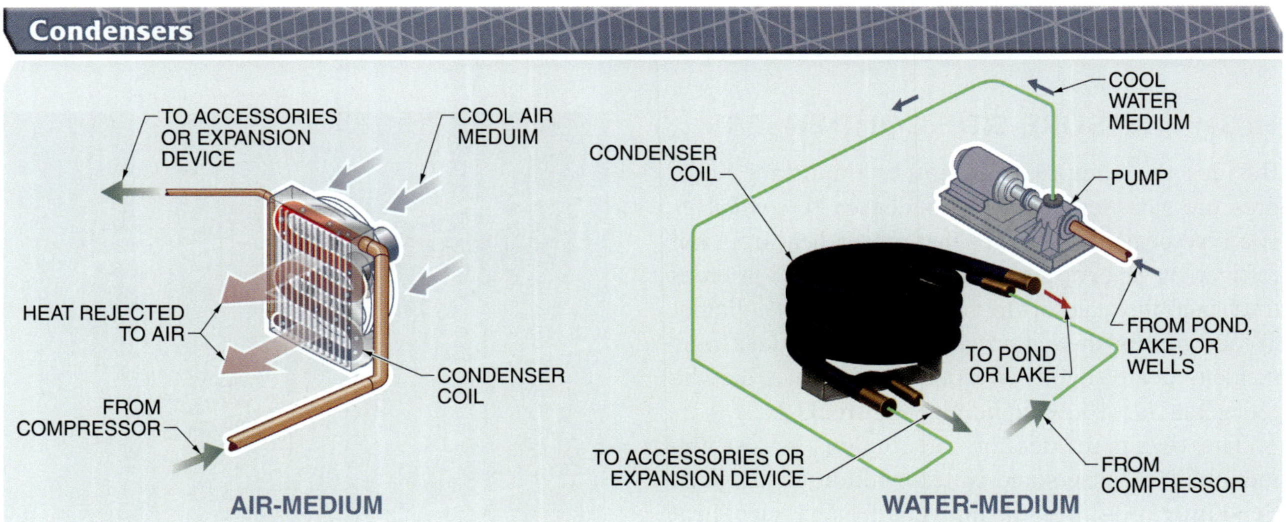

Figure 33-2. Refrigeration system condensers are categorized as air-medium or water-medium condensers.

Residential Air-Medium Condenser Cabinets

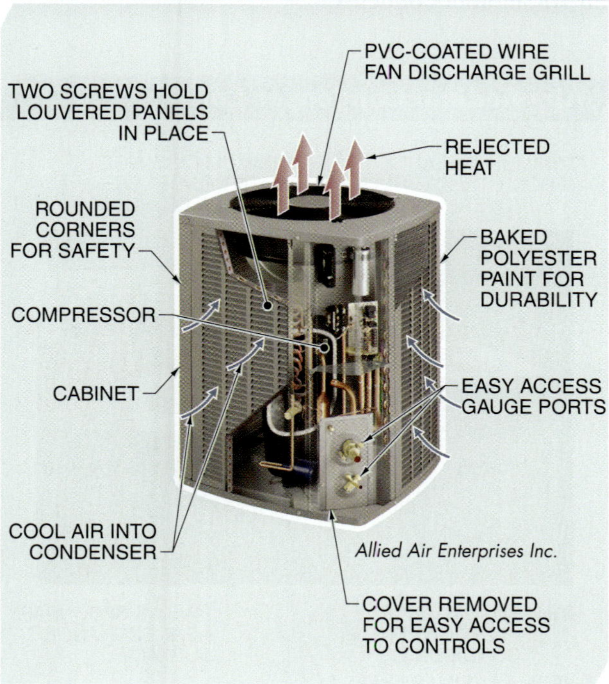

Figure 33-3. A residential condenser cabinet is constructed so that the fan circulates large amounts of ambient air across the condenser coils and blows the hot air up and out into the atmosphere.

Cabinets may also be mounted on a pad to keep them off the ground or off roofing material. Large air-conditioning or refrigeration air-medium condensers have vibration mounts that reduce transmitted vibrations to the building structure. Reducing vibrations is especially important when a unit is mounted on a roof.

Fans

The fans used in most condensing units are propeller-type axial fans. **See Figure 33-5.** This fan type is ideal for moving the large amounts of air necessary to provide maximum condensing capability. The condenser fan must move more air for high-pressure-side purposes than an evaporator fan does for low-pressure-side purposes. The vast majority of condensing units are kept outside so that the high noise generated by the condensing unit fan is not problematic. The fans and motors must be rated and constructed in a rugged fashion to withstand the rigors of outdoor use.

Large Condensers

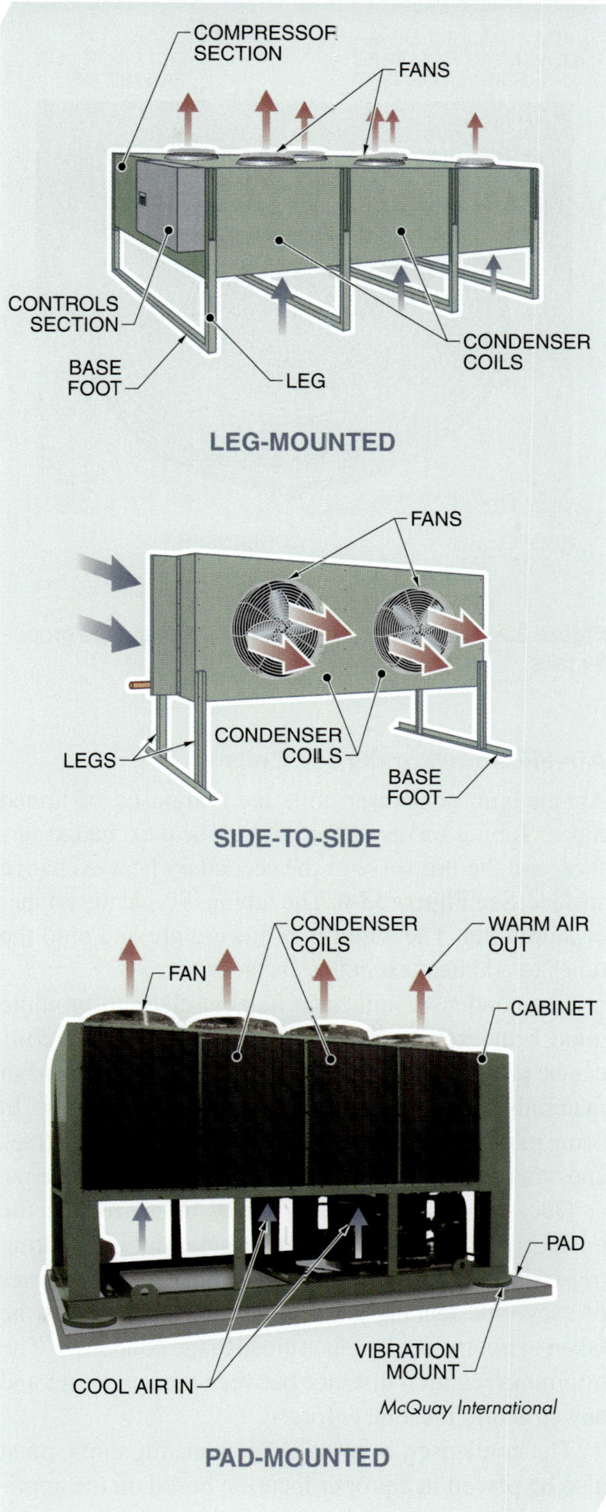

Figure 33-4. Commercial and industrial condenser units introduce air at the bottom of the condensers and discharge the hot air out the top of the units.

Figure 33-5. Most condensing units use axial-type propeller fans.

Air-Medium Condenser Coils

Air-medium condenser coils are composed of finned tubes. Tubing serves as the primary heat exchange surface, and the fins serve as the secondary heat exchange surface. **See Figure 33-6.** The tubing is typically copper or aluminum. The aluminum fins are pressed onto the tubes to add heat exchange surface area.

The condenser coil must be kept clean to promote good heat exchange characteristics. Condenser coils can be pressure washed to remove dirt and other foreign materials (precautions must be taken not to bend fins). In some parts of the country, seeds from cottonwood trees and other organic matter will cause additional problems.

Occasionally, landscapers may try to reduce the unsightly appearance of a condensing unit by planting trees, shrubs, or grass around the condenser. Sometimes privacy-type fencing is used, but precautions must be taken to not locate the fence close to the condenser. The minimum required distance between the condenser and any structure must be enforced.

The condenser, or multiple condensing units, must also be placed in a proper location based on the application. For example, when installing condensers under an overhang or obstruction, the result may be that hot air coming off the condenser gets "recirculated" and reduces proper airflow. When multiple condensers are used, care must be taken to ensure that the discharge air from one condenser does not become the supply air for another condenser.

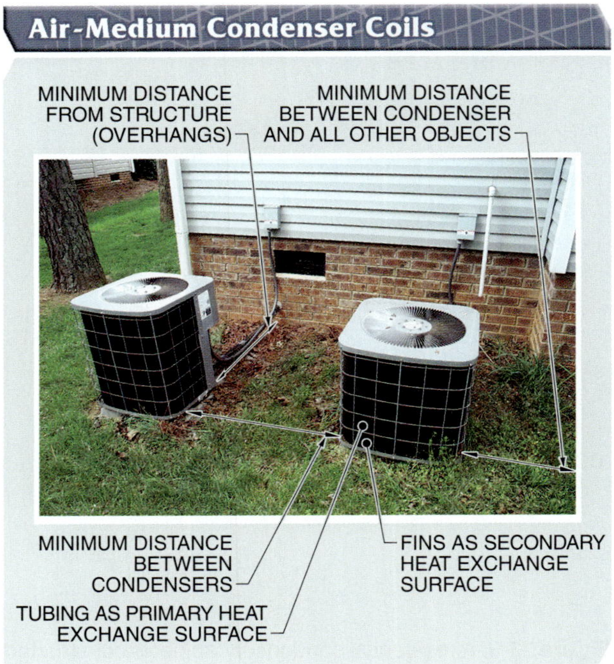

Figure 33-6. Air-medium condensers are of the finned-tube type. When multiple units are used, minimum distances must be observed between the units and other obstructions.

Superheat and Subcooling

The condenser is responsible for not only changing the state of refrigerant from a vapor to a liquid, but also eliminating the superheat in the refrigerant and allowing refrigerant subcooling. As the refrigerant is compressed, gains superheat. However, the heat of compression is added above and beyond the saturation temperature. The condenser must reject this superheat. **See Figure 33-7.** The first part of the condenser must remove the superheat from the refrigerant. After the superheat is removed from the refrigerant, the refrigerant is at saturation condition. Further removal of latent heat results in 100% of the refrigerant changing state from a liquid-vapor to a liquid by the time it reaches the end of the condenser coil.

Next, the refrigerant is subcooled. *Subcooling* is the process of removing heat from refrigerant after it has changed state from a vapor to a liquid. In most cases, subcooling the refrigerant is a normal and desirable function of the condenser. Subcooling refrigerant prevents the refrigerant from easily flashing back to a gas when the liquid encounters a small pressure drop, such as a resistance to flow in the piping. Subcooling also allows 100% of liquid refrigerant to be present at the expansion device. Typically, subcooled refrigerant is between 5°F and 15°F. After subcooling takes place and the refrigerant leaves the condenser, the refrigerant is a subcooled, high-pressure liquid.

Superheat and Subcooling

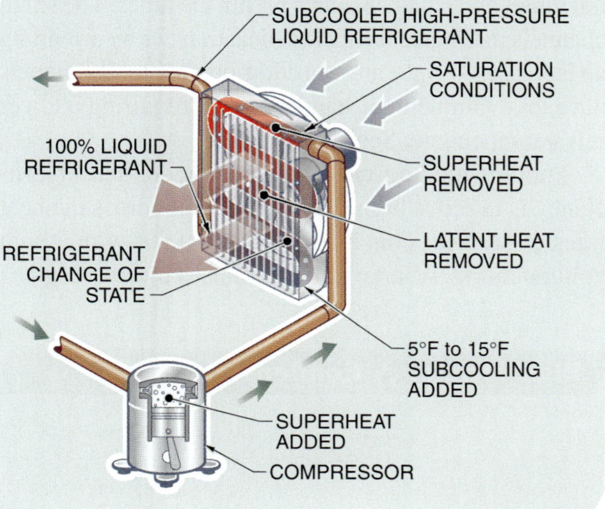

Figure 33-7. A condenser must be able to reject the superheat added by the compressor and subcool the refrigerant enough to keep it in a liquid state until it reaches the expansion device.

Condenser Operation Media Clip

Heat Transfer

The condenser rejects heat from the refrigerant to the condensing medium: air or water. The amount of rejected heat is the difference in enthalpy between the superheated vapor entering the condenser and the subcooled liquid leaving the condenser. The *heat rejection rate* is the rate at which heat is transferred from refrigerant in a condenser to a condensing medium. The heat rejection rate is a function of the mass flow rate of the refrigerant in the condenser and the heat of rejection. To calculate the heat rejection rate of the condenser, apply the following formula:

$$HR_R = HR \times W \times 60$$

where

HR_R = heat rejection rate (in Btuh)

HR = heat of rejection (in Btu/lb)

W = mass flow rate (in lb/min)

60 = conversion constant

Example: Calculating Condenser Heat Rejection Rate

A split-system air conditioning unit that uses R-22 refrigerant has a condenser with a heat rejection of 86.8 Btu/lb. **See Figure 33-8.** The mass flow rate through the system is 24 lb/min. Calculate the heat rejection rate of the system.

$$HR_R = HR \times W \times 60$$

$$HR_R = 86.8 \times 24 \times 60$$

$$HR_R = \mathbf{124{,}992\ Btuh}$$

Example: Calculating Condenser Heat Rejection Rate

Figure 33-8. The heat rejection rate of a condenser is the rate (in Btuh) at which heat is transferred from the refrigerant to the condensing medium (air or water).

Temperature Increases and Condenser Splits

As air flows across a condenser, it receives heat from the hot refrigerant, which causes an increase in the medium air temperature leaving the condenser. This temperature increase is a function of the design and operating characteristics of the condenser as well as the entire refrigeration system. **See Figure 33-9.** This difference in temperature is abbreviated as ΔT and is a known quantity for each system. For air-medium condensers, a ΔT of 20°F or more is common. *Condenser split* is the difference between the condensing medium temperature and the refrigerant temperature inside a condenser. The split is also known as the approach of the heat exchanger.

Condenser Splits

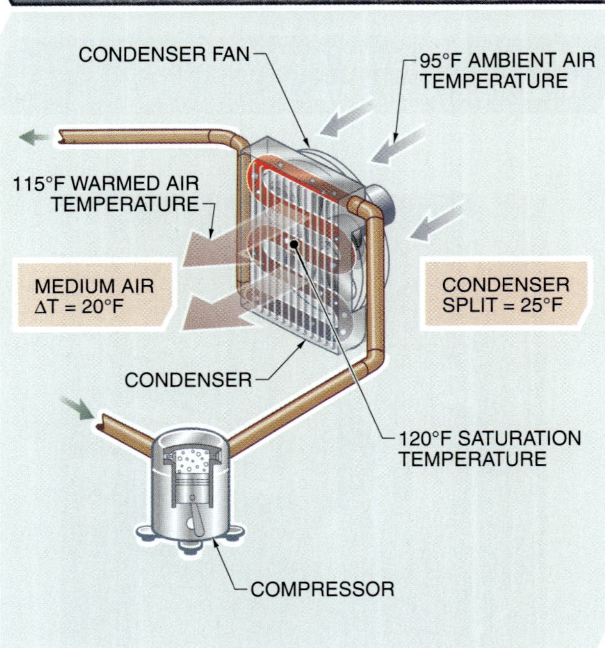

Figure 33-9. Condenser split is the difference between the condensing medium temperature and the refrigerant temperature inside the condenser.

A common standard is the ambient temperature plus 25°F, which means that the saturation temperature as derived from a T/P chart should correspond to a value of 25°F above the current ambient temperature. When the ambient temperature is 95°F, the saturation temperature should be around 120°F. Of course, condensers with other designs use slightly different figures.

High-Efficiency Condensers

As energy costs increase, the demand for energy-efficient refrigeration systems also increases. Multiple methods are available that allow energy savings when operating a refrigeration system. The most popular energy-savings method is to use a high-efficiency condenser. When condensing pressure can be dropped, the compressor does not have to work as hard, and energy is saved.

High-efficiency condensers consist of aluminum fins and multiple flat tubes containing small channels through which the refrigerant flows. The maximum amount of heat is transferred by the use of angled and louvered fins, flat tubes, and small channels within the tubes. The small channels maximize refrigerant heat transfer by providing an increased surface area. In addition, the small channels have less volume than standard coils, requiring a lower refrigerant charge. **See Figure 33-10.**

Another method used to increase condenser efficiency is to move higher volumes of air across the heat transfer surface. This involves using a fan with a high volume rating. A large fan in turn uses more energy.

High-Efficiency Condensers

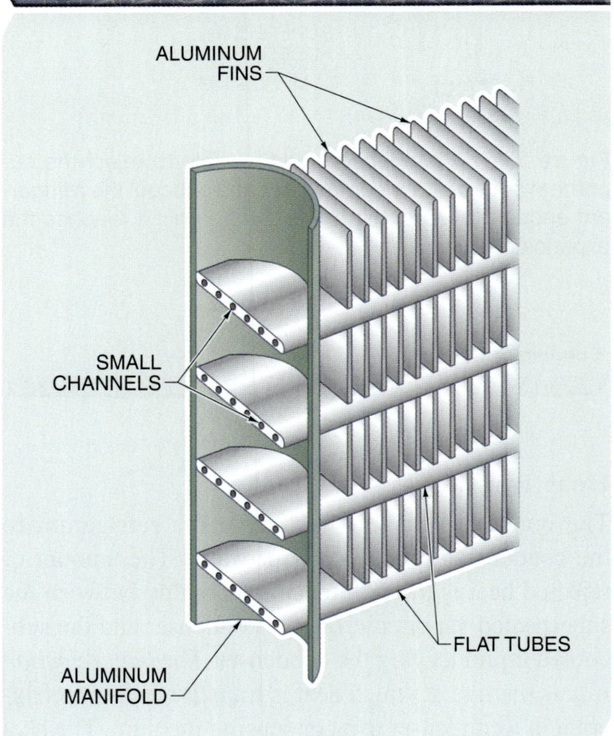

Figure 33-10. A high-efficiency condenser has aluminum fins, flat tubes, and small channels that provide a larger heat transfer surface with more tubing and more fins per inch than a standard condenser.

Low Ambient Temperature Condenser Conditions

A standard air conditioning system typically operates when the ambient air at the condenser is warm. Many other refrigeration systems, however, must operate even when the ambient temperature drops to cold temperatures. While it may seem desirable to operate a condenser at low ambient temperatures, it is not always desirable. **See Figure 33-11.** The low ambient temperature reduces the liquid refrigerant pressure at the expansion device. The reduced pressure fails to feed the proper amount of refrigerant to the evaporator. The result is a starved evaporator coil and improper evaporator conditions. A method of controlling the condenser head pressure must be implemented to prevent this occurrence.

TECH TIP

Manufacturers sometimes provide a low-ambient-temperature kit that enables a system to operate at lower-than-normal outside air temperatures by controlling the condenser fan speed or cycling operation.

Fan Cycling Devices

Many of the methods used to control condenser head pressure involve controlling the fan. Cycling the fan on and off to maintain a specific head pressure setting in the condenser is one method. This is accomplished by using a pressure-activated switch that is connected to the condenser by capillary tubing and wired to the control system to cycle the fan on and off at the desired setpoint. **See Figure 33-12.** This approach works well for control of single, small-horsepower fans.

Multiple multispeed fans are also used to control condenser head pressure. The speeds of the fans are designated as low speed and high speed. **See Figure 33-13.** A pressure-activated switch with multiple setpoints is also used. As the condenser pressure rises above the low-speed setpoint, the low-speed fan windings are energized. When the head pressure continues to rise, the high-speed fan windings are energized, and the refrigerant condensing pressure drops as more air moves across the condenser. In this type of refrigeration system, neither fan may be needed under low ambient conditions. A variation of this approach is the use of two separate fans, with one fan used under low ambient conditions and the second fan turned on as the refrigerant pressure increases in the condenser.

Low Ambient Temperature Condenser Conditions

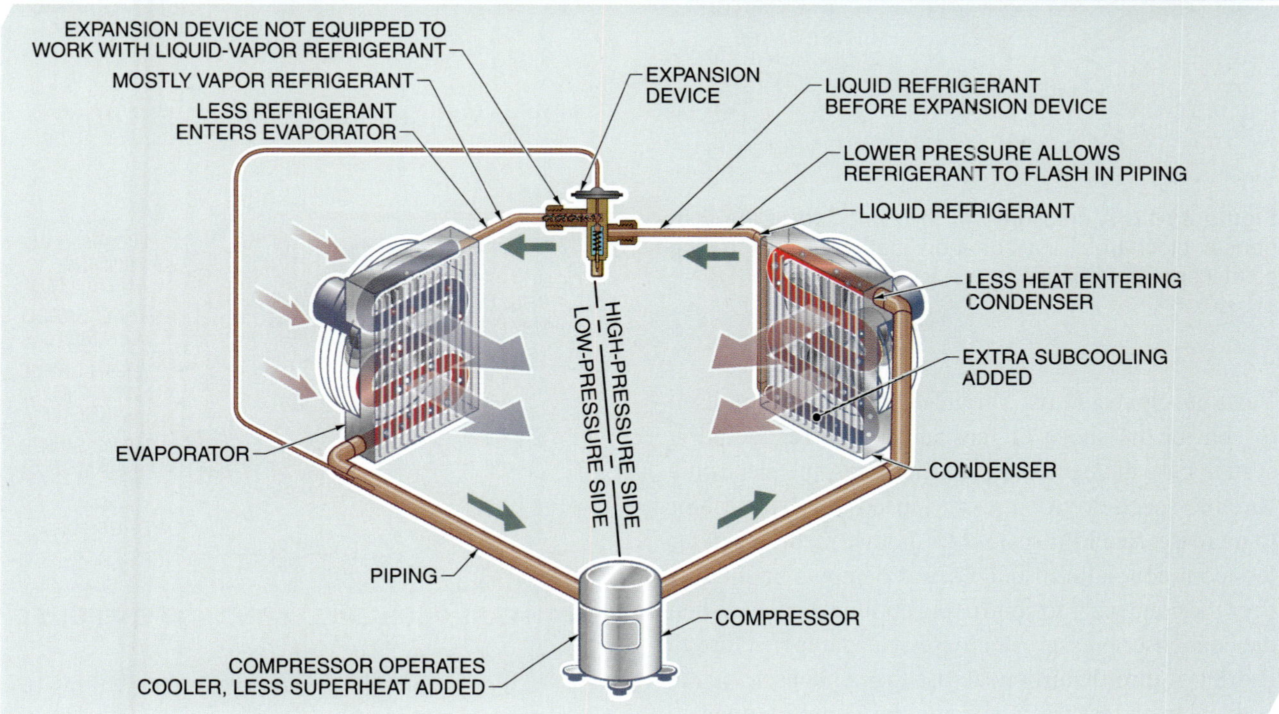

Figure 33-11. Low ambient medium temperatures at the condenser reduce the liquid refrigerant head pressure. This causes the expansion device to operate improperly, which starves the evaporator coil of refrigerant.

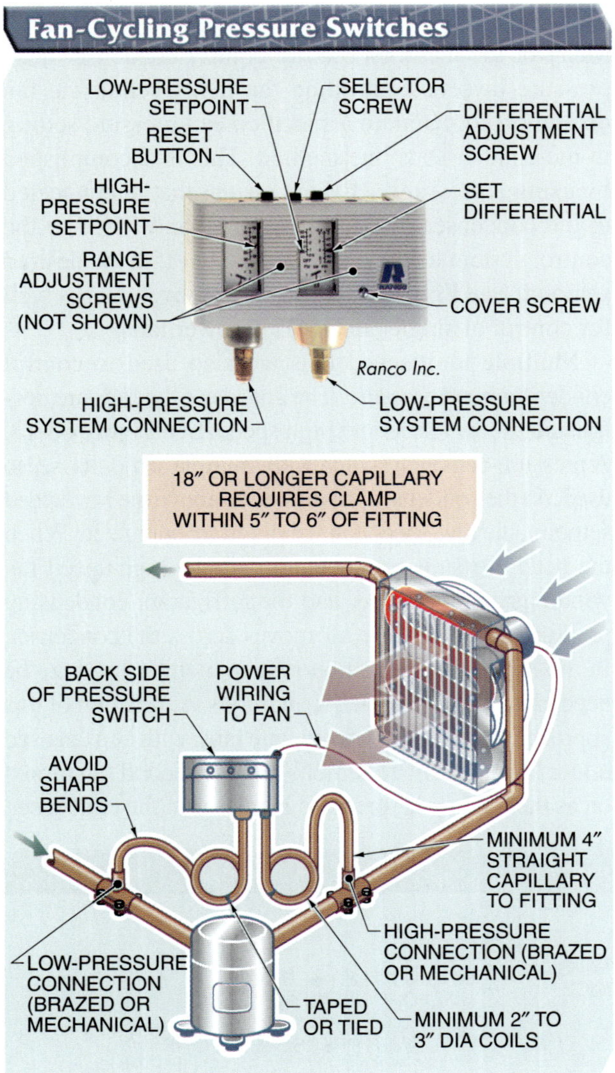

Figure 33-12. A pressure switch connected to the condenser discharge is a common method of controlling condenser fan operation.

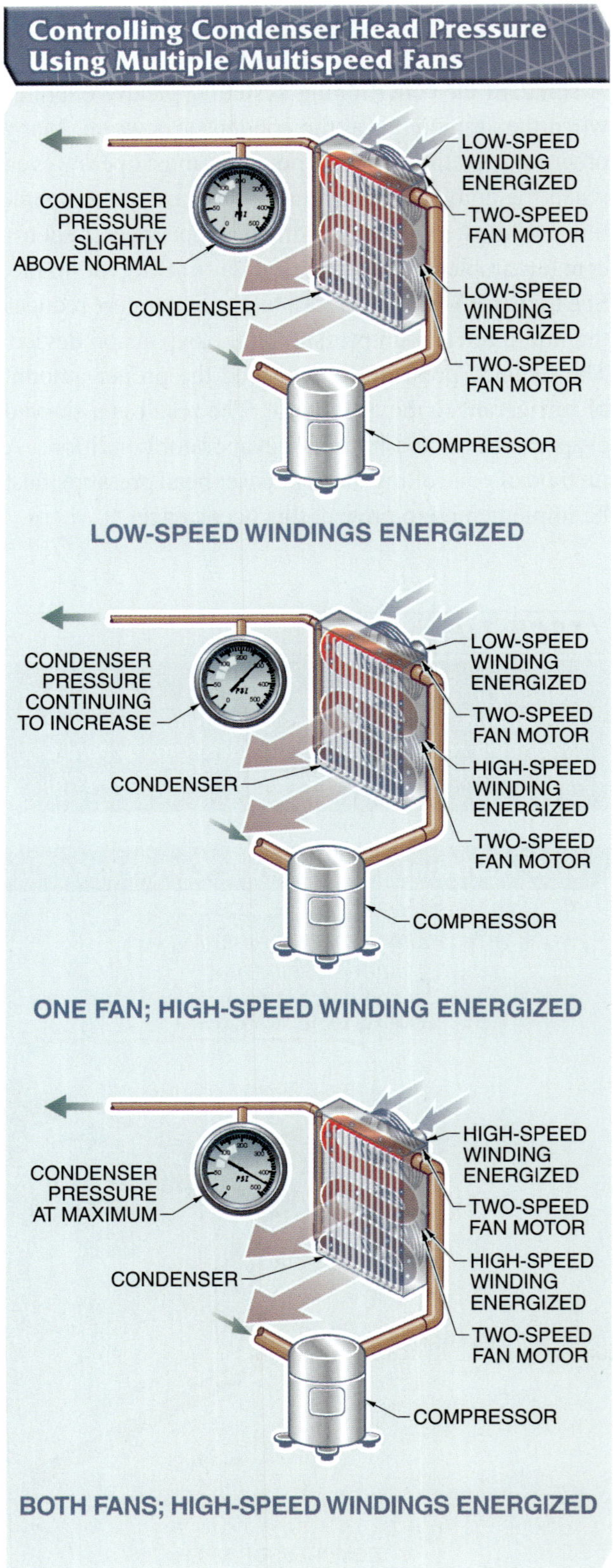

Figure 33-13. A multiple setpoint pressure-activated switch can be used to control two or more multispeed fans for condenser head pressure control.

Variable-Speed Fans. Variable-speed drives are used to control the speed of fans in modern refrigeration systems. As the head pressure increases, the electronic variable-speed drive causes the speed of the fan motor to increase. **See Figure 33-14.** Another method, but far less common, is to install a set of dampers or shutters over the condenser to control the cooling airflow. When the condenser pressure decreases, the dampers close to restrict or throttle airflow across the condenser. As the condensing pressure increases, the dampers open to admit more air. These systems typically use a pressure controller that is connected to the damper mechanism.

Controlling Condenser Head Pressure Using Variable-Speed Drives or Dampers

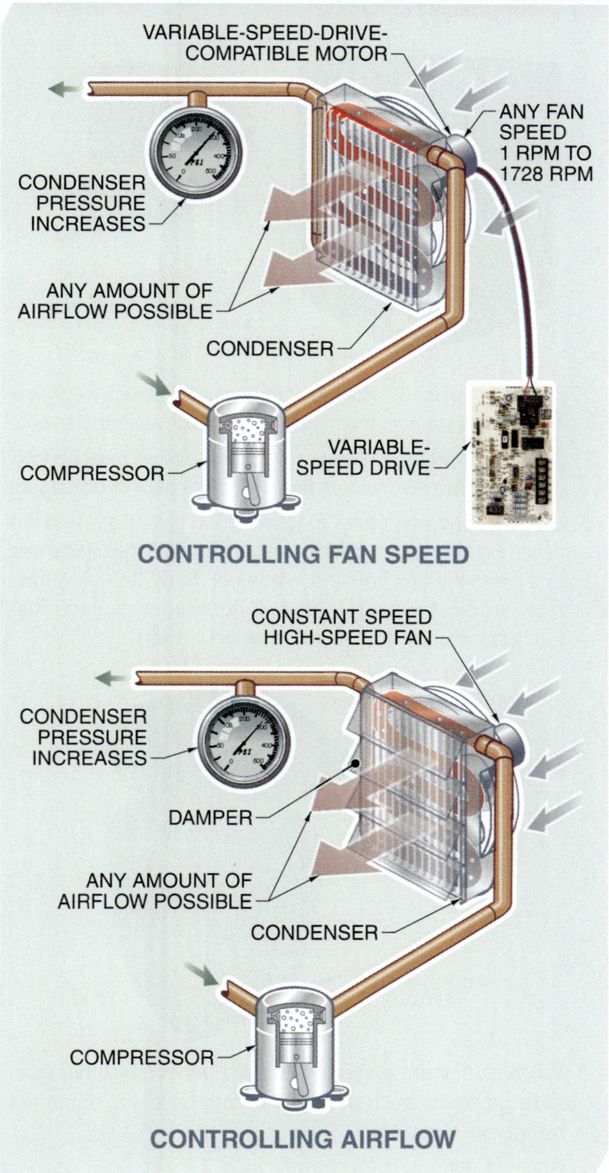

Figure 33-14. Variable-speed drives and condenser dampers are used to control the amount of airflow across condenser coils to control head pressure.

TECH TIP

Heat recovery systems are widely used in energy intensive systems such as product refrigeration systems. Supermarkets, for example, must refrigerate products and can use the condenser heat to raise the temperature of water needed elsewhere in the building.

Reclaiming Heat

The temperature of superheated refrigerant at the compressor is high. In most installations, this heat is considered a nuisance and is rejected to the outside atmosphere. This heat can be captured by a heat exchanger and reused in a facility. **See Figure 33-15.** The superheated refrigerant line can be passed through a water heat exchanger (condenser) and used to heat water for a residence or facility. This process reduces the heat energy needed for the normal water heating function, saving the residence or facility money. The discharge line can also be piped through ductwork, using the heat in the line to heat the air entering the facility air-handling system. This is yet another way to reduce energy expenditures. **See Procedure for Cleaning Outdoor Condensing Units.**

Reclaiming Heat

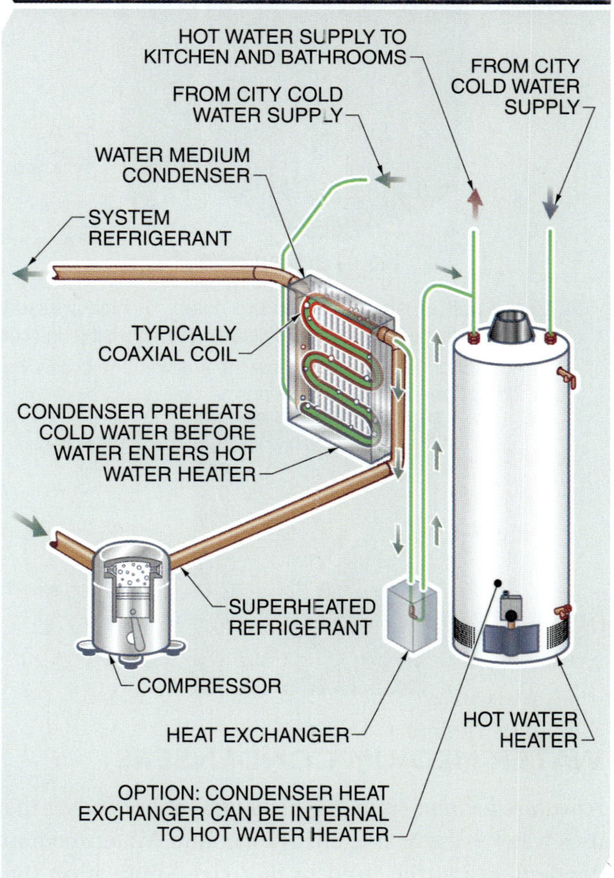

Figure 33-15. Superheat added to refrigerant by the compressor can be captured by a heat exchanger and reused in a residence or facility.

Cleaning Outdoor Condensing Units

1. Disconnect power from the mechanical compression refrigeration system.
   a. Pull the circuit breaker for the refrigeration system to be worked on.
   b. Lock out the circuit breaker.
   c. Tag out the circuit breaker.

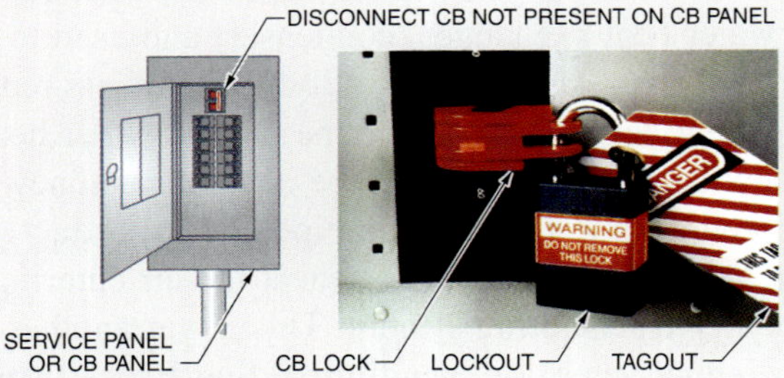

2. Follow all electrical and mechanical safety practices and procedures.
3. Remove condenser covers as needed.

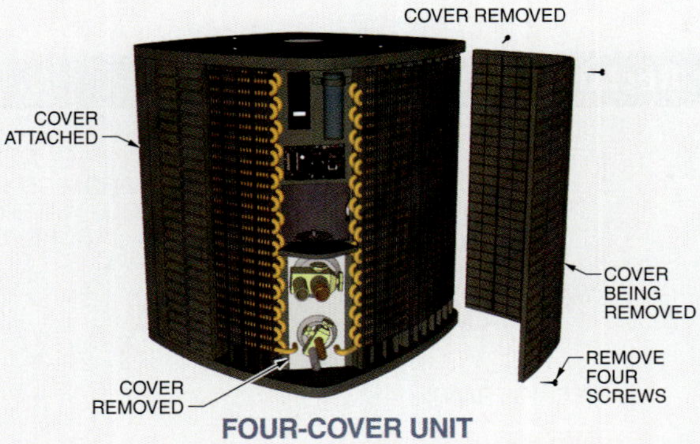

4. Use a soft brush, shop vacuum, foam, or high-pressure sprayer (not recommended) to clean the condensing unit coil.
   a. If a soft brush is used, follow the fins with the brush, but do not bend or distort the fins.

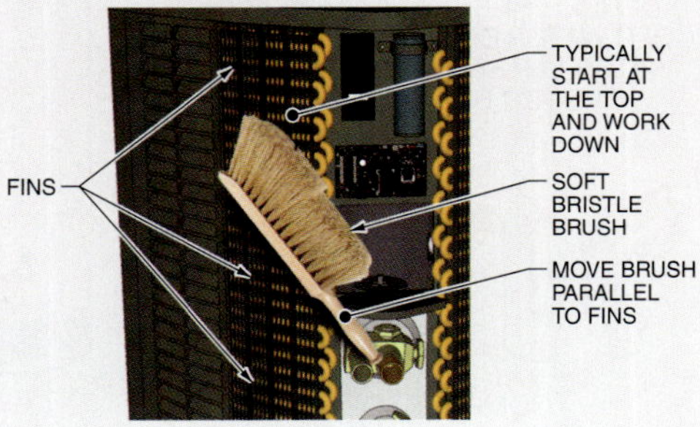

   b. If a shop vacuum is used, be careful not to impact any components or bend any fins.

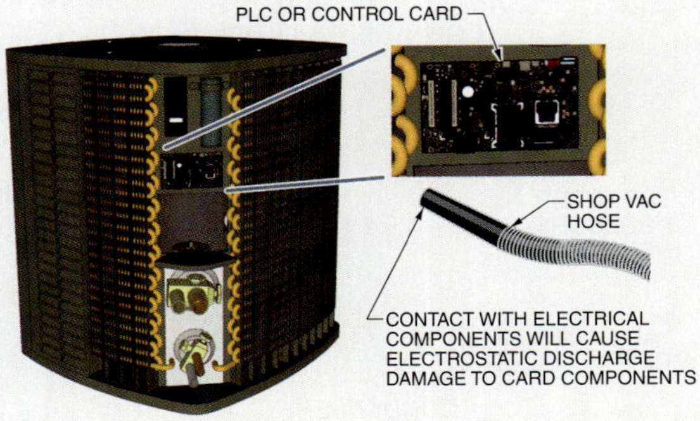

   c. If foam is used, follow all manufacturer instructions. Allow the unit to sit after foaming instructions have been followed.
   d. If a high-pressure sprayer is used, pressure must be lowered to the point where fins will not be bent or damaged.
5. Some technicians prefer to take the top grill or grill and fan assembly (depending on the manufacturer) off the condenser and clean the unit from the inside out. To do this, blow low-pressure compressed air (that will not bend or damage fins) through the coil from the inside to the outside.

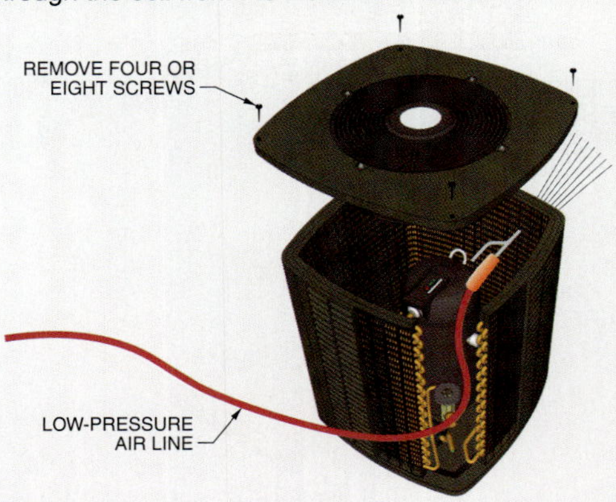

6. While cleaning the condenser coil, inspect for other possible problems such as loose wiring, oil leaks, damaged refrigerant lines, etc.
7. After cleaning, turn the power on.
   a. Remove the lockout/tagout.
   b. Reset the circuit breaker.

   Note: If power was shut off for more than 4 hr, restore power and wait a minimum of 24 hr before starting the unit.

WATER-MEDIUM CONDENSERS

A *water-medium condenser* is a type of condenser that uses water as the heat-exchange medium. Water-medium condensers can be used in any size application, but they are typically used in large applications of 15 t and larger. However, they are used in solar and geothermal applications of any size.

Water from an appropriate source is circulated through a heat exchanger and removes the heat from the hot high-pressure liquid refrigerant. While all water-medium condensers use water as the heat-exchange medium, there are many different designs and combinations of these types of condensers. Only a few are used in residential and light commercial applications.

Tube-in-Tube Condensers

A *tube-in-tube condenser* is a type of heat exchanger that consists of one small diameter metal tube concentrically placed inside a larger tube. **See Figure 33-16.** Copper is the material most commonly used. One of the fluids (typically refrigerant) is present in the inner tube, and the other fluid (typically water) is in the outer tube. Thus, the entire inner tube surface is able to exchange heat.

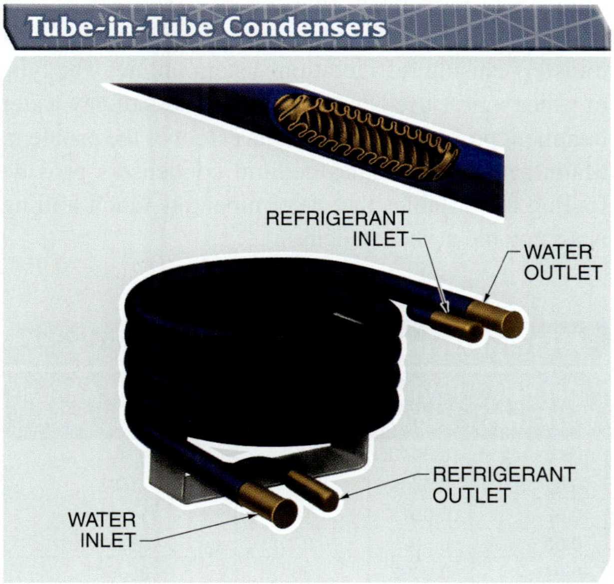

Figure 33-16. A tube-in-tube condenser is constructed of one metal tube of a small diameter placed concentrically inside a tube of a larger diameter. It is used in small-size refrigeration systems.

Tube-in-tube condensers use the counterflow principle for heat exchangers. In a tube-in-tube condenser, the fluids are introduced to the heat exchanger at opposite ends and flow in opposite directions. This helps ensure that the hottest refrigerant entering the condenser comes into contact with the coolest water. This increased difference in temperature, or ΔT, increases the condenser efficiency.

Tube-in-tube condensers are formed into a coil shape instead of one long tube. Coil shapes reduce overall size and allow a large condenser capacity to be manufactured into a small physical size. This reduces the size of the equipment and the installation footprint. Tube-in-tube condensers are often used on factory-packaged equipment.

The size and space used for tube-in-tube condensers is an advantage, but disadvantages are present as well. For instance, tube-in-tube condensers are not typically serviced or repaired. When a tube-in-tube condenser has internal problems, such as mineral deposits or a crack in the tubing, the whole condenser must be replaced. Hardly any diagnostic tests can be performed regarding the condition of the heat exchanger. The only warning that a tube-in-tube condenser has a problem is when the heat exchanger fails, causing cross contamination, and failure is always catastrophic.

Shell-and-Coil Condensers

Another type of water-medium condenser is a shell-and-coil condenser. **See Figure 33-17.** A *shell-and-coil condenser* is a type of heat exchanger that consists of a tubing coil placed inside a shell. Refrigerant passes through the coil, while water surrounds the coil. Instead of a set of straight tubes, the tubes are formed into a coil shape, which allows a large heat transfer surface in a small amount of space. Shell-and-coil condensers are used in small- and medium-size systems.

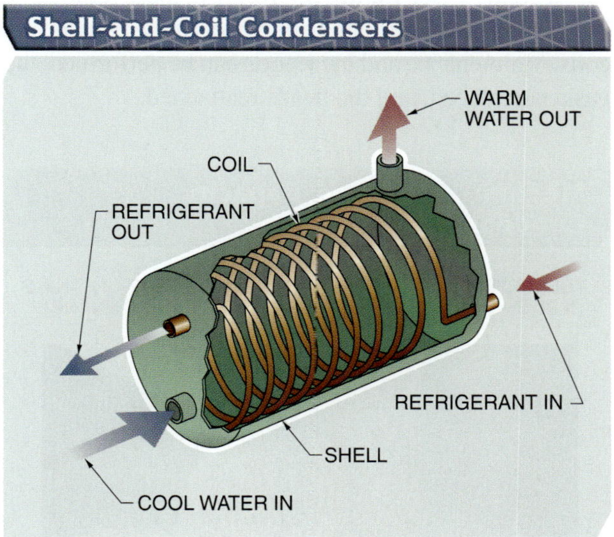

Figure 33-17. A shell-and-coil condenser is a heat exchanger that consists of a tubing coil placed inside a shell and is used in small- and medium-size refrigeration systems.

Shell-and-Tube Condensers

Another type of water-medium condenser is a tube-in-shell condenser. Shell-and-tube condensers are typically used in large water chillers and central plant equipment of 50 t or more. Shell-and-tube condensers consist of a number of tubes through which refrigerant is pumped. Water surrounds the tube bundle. Shell-and-tube condensers are available as one-, two-, or three-pass types. Each water pass adds greater efficiency to the condenser, but at the same time adds size, weight, and complexity. Each additional pass yields a smaller efficiency increase.

TECH TIP

Improper condenser maintenance can lead to degraded operation, higher-than-normal head pressures, increased compressor temperatures, and eventual failure.

Condenser and Heat Exchanger Maintenance

One of the problems with water-medium condensers in small systems is that they provide no opportunity for regular maintenance, cleaning, and inspection. Some allow external dirt and leaf maintenance but nothing else. Medium- and large-capacity condensers provide the ability to disassemble the heat exchanger so that maintenance can be performed. Most coil-in-shell and tube-in-shell condensers have heads bolted to the condenser shell. These heads have gaskets that prevent leakage and ensure a good fit. **See Figure 33-18.** Also, these removable heads provide access to the tube sheets, tube bundle, and coils. Maintenance and inspection can be performed, the gaskets replaced, and the heads reattached.

Medium- and Large-Capacity Condenser Maintenance

Figure 33-18. A water-medium condenser can be cleaned by removing the head cover(s) from the shell and exposing the tube sheet(s).

Cleaning Condensers Media Clip

Mineral Deposits. A major problem with any water-medium condenser is the potential buildup of mineral deposits. These minerals are in the water flowing through the condenser. When left untreated or unchecked, these minerals are deposited on the condenser tubing and reduce its efficiency.

Condensers of any size cannot tolerate deposits with a fouling factor greater than 0.0005 without the efficiency of the condenser being seriously reduced. **See Figure 33-19.** Eventually the tubes in the heat exchanger will become plugged and prevent water flow and heat transfer, causing refrigeration system failure. The type of water source used and the effectiveness of any water treatment program determine the extent of the problem. Manufacturers of water-medium condensers provide fouling factor tables that determine how much fouling is acceptable at the condenser.

Mineral Deposit Fouling Factors

Deposit Thickness*	Fouling Factor	Efficiency Loss†	Increased Annual Electrical Cost‡
0.00	0.0000	0	0
0.01	0.0008	9	8698
0.02	0.0017	18	17,936
0.03	0.0025	27	26,904
0.04	0.0033	36	35,872
0.05	0.0042	45	44,840

* in in.
† in %
‡ in $

Figure 33-19. Mineral deposits must be removed from a condenser in order for the condenser to operate efficiently. Fouling factor tables provide a baseline on how much inefficiency occurs due to fouling.

WASTEWATER SYSTEMS

A *condenser wastewater system* is a type of water-medium condenser water system in which a water source is introduced to the condenser, used once, and then removed or dumped into a drain. A municipal/city water connection is the most common water supply source. An advantage to this approach is that, in many municipalities, the water temperature supplied may range from 60°F to 70°F, which is a low temperature range for condenser inlet water. This temperature range provides for great condenser efficiency. Typically, little piping is needed because the municipal water connection is often close at hand. **See Figure 33-20.**

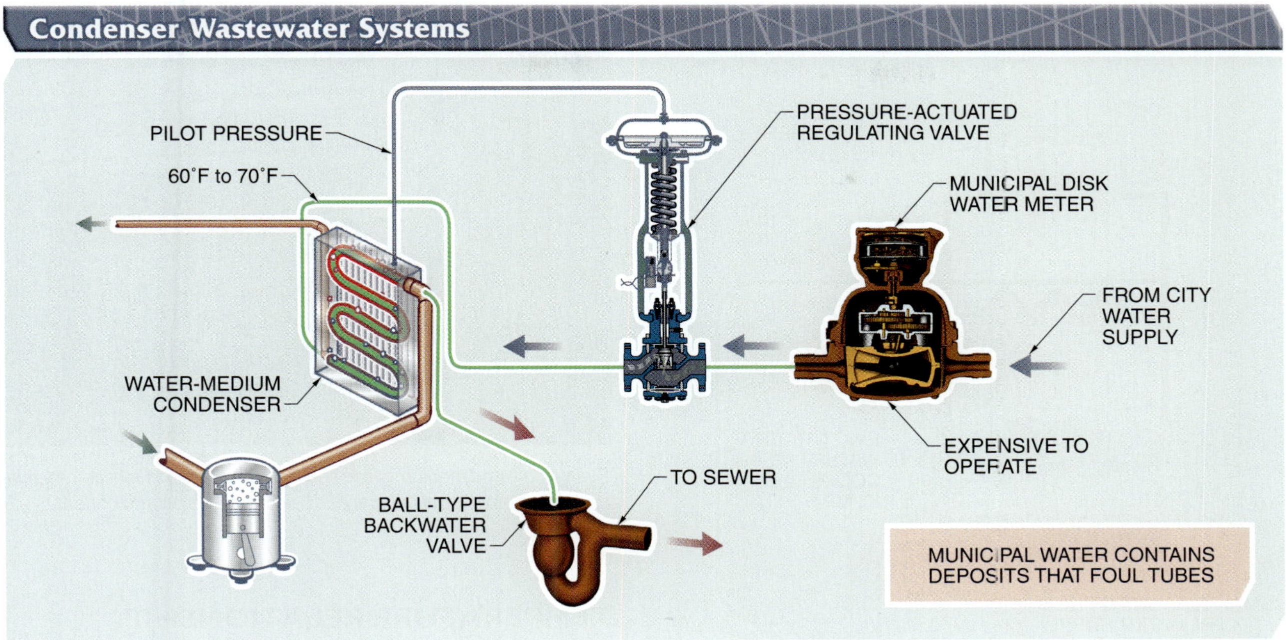

Figure 33-20. Wastewater systems are used on small- and some medium-size condensers because the water is used once then dumped into the sewer.

However, the disadvantages of a wastewater system far outweigh the advantages. The cost of using metered municipal water for heat rejection can be high in all but the smallest tonnage applications. Many municipal codes restrict or even ban the use of water for small heat exchange applications. Municipal water is also notorious for containing many types of deposits that foul condenser tubes. Without an effective water treatment program, the deposits quickly reduce condenser efficiency and plug the tubes.

The control of condenser conditions using a wastewater system is accomplished by using a pressure-actuated, water-regulating valve. As the condenser pressure increases above the setpoint, the condensing pressure causes the valve to open with a diaphragm mechanism. The valve then opens and admits cool water (or more cool water) into the condenser. Occasionally, another source of water is used, such as a well or pond. Wastewater systems are best suited for applications that only require sporadic cooling and where investment in the infrastructure has been minimal.

Recirculating Water Systems

In the vast majority of refrigeration and air conditioning systems, wastewater systems are undesirable and too expensive to operate. Typically, it is better to use a recirculating water system. A *recirculating water system* is a type of water-medium condenser water system in which water is sent into the condenser, moved to a location where the heat can be rejected, and then pumped back to the condenser.

A typical place for heat to be rejected is a cooling tower. A cooling tower, which is connected to the condenser water system, is located outside on the roof of a building or next to the building. **See Figure 33-21.** Cooling towers are the most common heat rejection device when using water. Cooling towers operate by having the warm water from the condenser come into contact with the ambient air. Some of the water evaporates, causing the remainder of the water to be cooled. Extra water is needed to replace the evaporated water, which is known as makeup water. Water treatment is as essential for cooling towers as it is for condensers.

TECH TIP

Cooling tower makeup water may use a separate water meter. This can lead to reduced costs because the makeup water evaporates, leading to reduced water sewer charges.

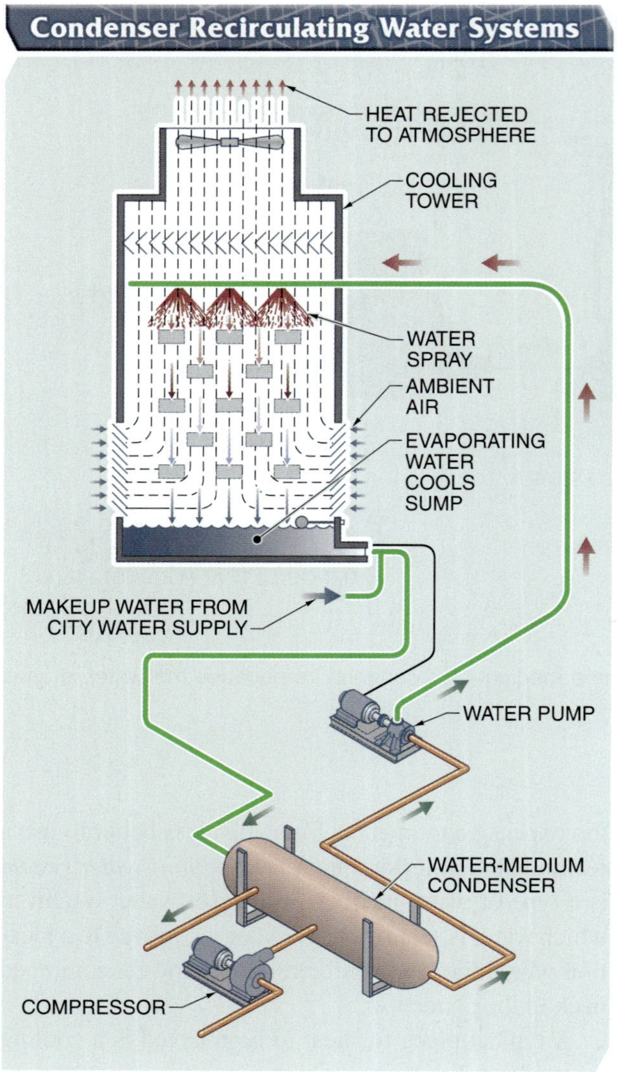

Figure 33-21. Recirculating water systems cool water by using fans to circulate ambient air through the water in the cooling tower and return the water to the condenser.

Even small- and medium-sized cooling towers used in light commercial and commercial applications can reject huge amounts of heat.

TEMPERATURE RELATIONSHIP

One of the advantages of a water-medium condenser is that the incoming water temperature has a known, stable value over time. The water contributes to a wide variation of temperatures in a water-medium condenser. Baseline operational values are established for the incoming and discharge water temperatures. The most common values used for water-medium condensers are 85°F water inlet temperature and 95°F discharge water temperature.

A relationship exists between the cooling tower water flow temperature and the heat transfer surface characteristics of the condenser. This relationship is such that the refrigerant enters as a superheated vapor and leaves as a subcooled liquid. While variations exist, technicians use these values for evaluation and troubleshooting purposes. In large systems, water and refrigerant inlet and outlet values are recorded on a log sheet.

UNIT 34

High-Pressure-Side Accessories

In addition to the condenser, a number of accessories are used to perform specific functions on the high-pressure side of a refrigeration system. The proper operation of these accessories is very important for the overall operation of the condenser, and thus the entire refrigeration system.

OBJECTIVES

- List the typical high-pressure-side accessories.
- Describe the types and sizing of refrigerant lines.
- Determine the proper size of the suction, discharge, and liquid lines using a pressure-drop chart.
- Determine the proper suction-line size that will return refrigerant-entrained oil to the compressor using a suction-line chart.
- Describe the function and operation of high-pressure-side accessories.

Index to Procedures

HIGH-PRESSURE-SIDE ACCESSORIES

The high-pressure side of a refrigeration system contains refrigerant lines and assorted accessories. **See Figure 34-1.** The refrigerant lines carry refrigerant between the compressor and the condenser and accessories, which include flow switches, status indicators, liquid receivers, king valves, filter-driers, sight glasses, oil separators, and solenoid valves. These accessories filter out impurities or perform other functions that allow the safe, efficient operation of the refrigeration system.

Refrigerant Lines

The high-pressure side of the system has two refrigerant lines. The two refrigerant lines are the hot-gas discharge line and the liquid line. **See Figure 34-2.** The refrigerant lines are fabricated using high-grade air conditioning and refrigeration (ACR) copper tubing. On residential systems, the lines are typically provided with the condensing units as a "line set" that includes the fittings, liquid and suction lines, and insulation. Care must be taken whenever working around refrigerant piping to not kink or damage the piping. Any rolled piping must be unrolled carefully.

Hot-Gas Discharge Lines. A *hot-gas discharge line* is the refrigerant line that connects a compressor to a condenser. In normal operation, this line is always the hottest part of the system. Due to the high pressure of the refrigerant, this line typically has a smaller diameter than the other lines of the system. An oil separator located on the compressor discharge line and return piping are used to maintain an adequate oil level in the compressor crankcase.

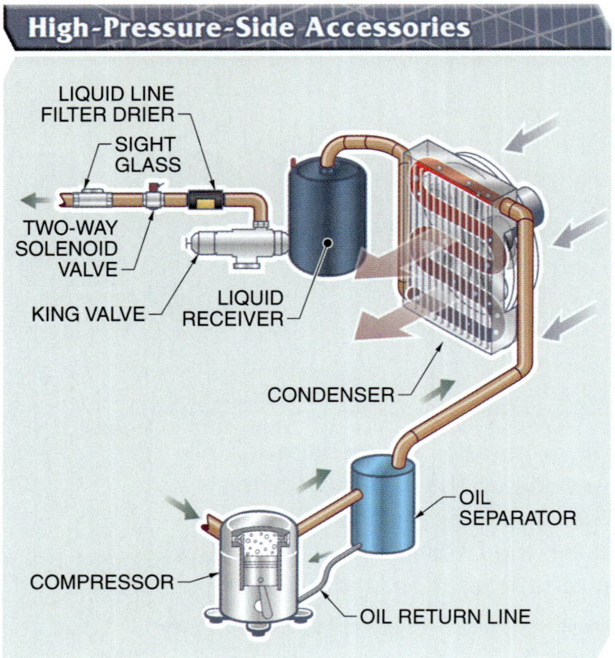

Figure 34-1. The high-pressure side of a refrigeration system typically includes lines (copper piping) that carry refrigerant between the compressor and the condenser and expansion device. It also includes assorted accessories such as the liquid receiver, filter-drier, sight glass, oil separator, solenoid valve, and king valve for larger volume systems.

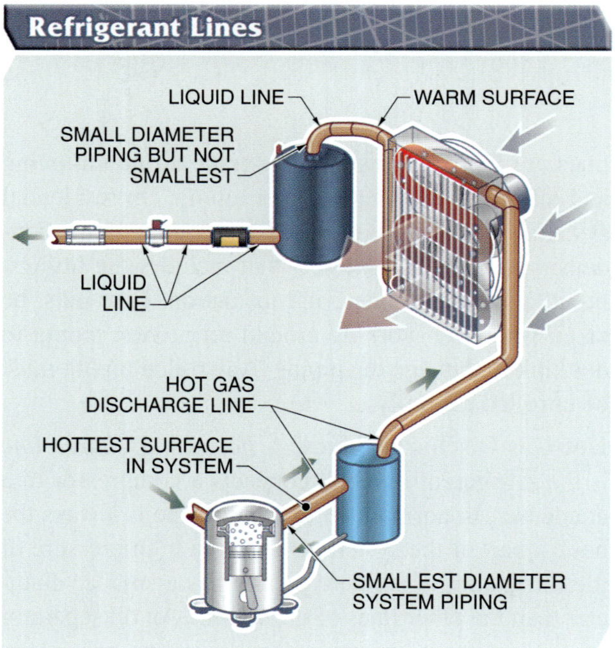

Figure 34-2. The two refrigerant lines used on the high-pressure side of a refrigeration system are the hot-gas discharge line and the liquid line.

Liquid Lines. A *liquid line* is the refrigerant line that connects a condenser to an expansion device. When the system has a receiver, the liquid line connects the condenser, liquid receiver, and expansion device. The refrigerant in the liquid line is a medium-temperature, high-pressure, subcooled liquid. In air conditioning systems, the line is warm but not hot. The liquid line is typically of a small diameter because the liquid refrigerant takes up less space.

Line Sizing

For proper refrigeration system operation, the refrigerant lines must be sized correctly. Proper sizing contributes to efficient system operation, lower installation costs, and lower operational costs. Equipment manufacturer recommendations must be followed when sizing lines. In addition, pressure-drop charts can be used to size refrigeration system lines.

A well-designed piping system will also ensure that the proper amount of lubricating oil remains in the compressor. Hermetic motor compressors use refrigerant to provide motor cooling. The refrigerant and oil come into contact inside the compressor. Some of the oil is carried out of the compressor discharge line along with the refrigerant. When proper steps are not taken, large amounts of oil are removed from the compressor crankcase, and eventually the compressor loses lubrication. Compressor damage is the result.

To prevent damage, horizontal suction lines may be pitched toward the compressor at a rate of ½″ for every 10′ in length to enhance oil return. **See Figure 34-3.** Vertical risers are especially important for rooftop-mounted condensers, in which the compressor and evaporator are on a much lower level.

Line Supports and Insulation. It is important to properly support refrigerant lines. This is especially important in long piping runs. The support may consist of straps that are attached to the building structure, wooden blocks that the lines are set upon, or more rigid supports as needed. Insulation is added as well to reduce heat transfer, allow higher efficiencies, and reduce any problems of sweating on the suction line. This is especially important where the lines pass through unconditioned spaces. The insulation is often provided with the copper tubing.

Pressure Drops

As refrigerant flows through piping or accessories (devices), resistance is encountered. A *pressure drop* is a loss of energy. Any resistance to flow causes a pressure drop in the refrigerant. Pressure drops increase as refrigerant lines become longer (more resistance) or as fittings, such as elbows, are added to the piping system. **See Figure 34-4.**

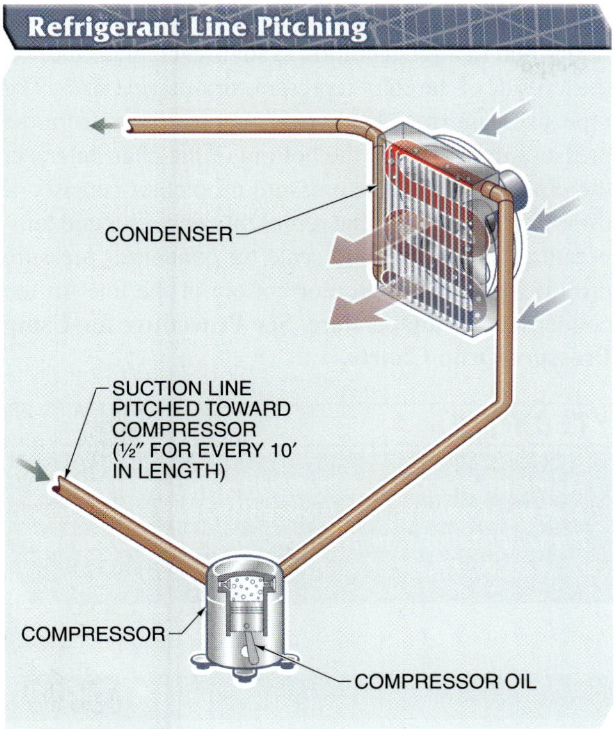

Figure 34-3. Horizontal suction lines may be pitched toward the compressor to enhance oil return.

Accessories mounted in the piping are resistances to flow and cause pressure drops. Regardless of the accessory being used, pressure drops must be minimized because they reduce system capacity. Pressure drops also cause the compressor to use more power to overcome the pressure drops (energy losses). Pressure drops are measured in either pounds per square inch (psi) or in degrees Fahrenheit (°F). The value in degrees Fahrenheit shows the effect of a specific pressure drop.

Every refrigerant has a specific pressure drop measurement when flowing through a resistance. Tables are used to relate the pressure drop information for each refrigerant. When evaluating pressure drops, special care must be taken with the liquid line. When the pressure drop in the liquid line is too high, at some point in the line the pressure will drop below the saturation condition at that refrigerant temperature. The result is that some of the refrigerant will prematurely flash to a gas. This flash is an undesirable condition. The expansion device is designed to meter a solid column of liquid refrigerant, not vapor or flash gas. This flash gas causes the expansion device to hunt (search for proper operating position) and operate improperly. System cooling capacity will also be reduced.

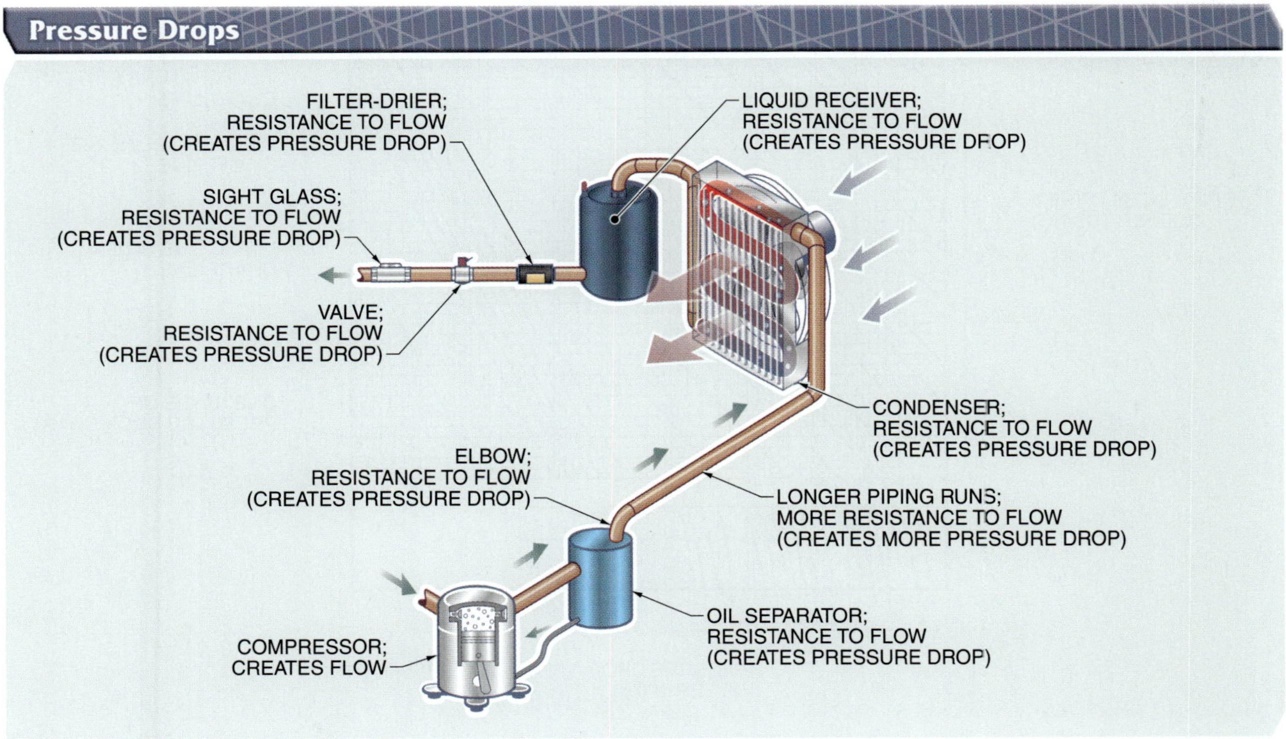

Figure 34-4. Pressure drop is a loss of energy and is caused by any resistance to flow in a system.

Pressure-Drop Charts. A refrigeration system operating at the correct capacity must have the correct amount of refrigerant flowing through the system. A *pressure-drop chart* is a chart used by a designer to properly size each line section of a refrigeration system to produce the required pressure drop. Using this method ensures the proper refrigeration system capacity. **See Figure 34-5. See Appendix.**

A pressure-drop chart has a scale in the top right corner that represents the cooling capacity of the system. The measurement is in tons of refrigeration. The cooling capacity needed in the system is read from this part of the chart. The diagonal lines in the lower right side of the chart are used to indicate the evaporator coil temperature, condenser coil temperature, and the chart usage lines. Compressor suction lines are sized using lines that represent the temperature of the evaporator coil. Compressor discharge lines are sized using lines that represent the temperature of the condenser coil.

The liquid-line section of the chart is used to size the liquid line in a refrigeration system. Diagonal lines on the left side of the chart represent various pipe sizes. The pipe sizes run from ⅜″ to 6⅛″. Vertical lines from the pressure-drop chart at the bottom of the chart intersect the pipe size lines. The pressure-drop chart consists of three horizontal lines that represent condenser coil temperatures. The indicating scale for condenser pressure drop is located at the top or bottom of the line for the condenser coil temperature. **See Procedure for Using Pressure-Drop Charts.**

TECH TIP

Poor refrigeration system design is the leading cause of many problems. This is especially true of nonstandard applications, such as condensing units mounted on a second floor or roof.

Pressure-Drop Charts

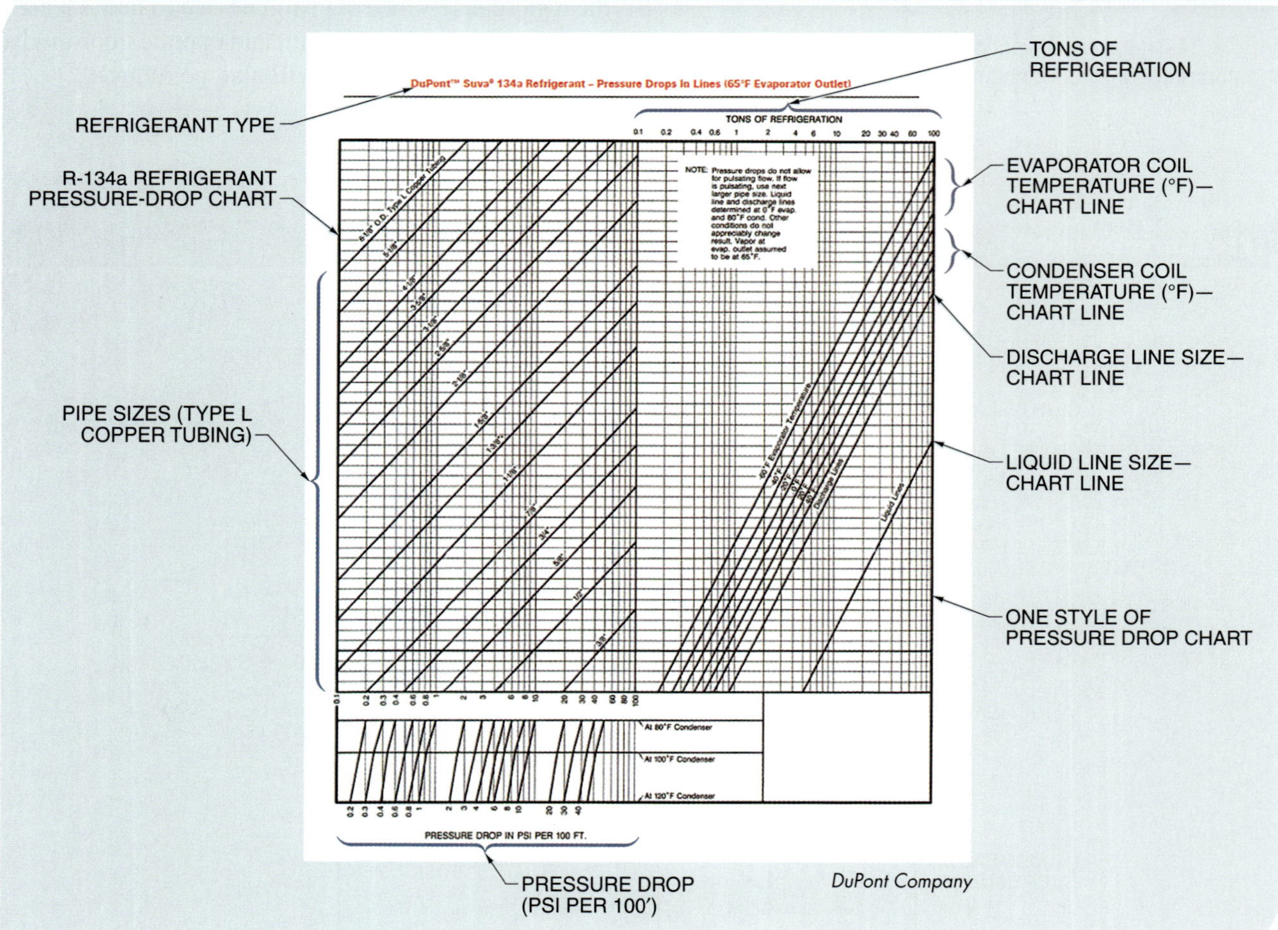

Figure 34-5. Pressure-drop charts are used to size the various piping in a refrigeration system and require the same procedure regardless of the manufacturer providing the chart.

Using Pressure-Drop Charts

Procedure

1. Find refrigerant used by using the correct pressure-drop chart.

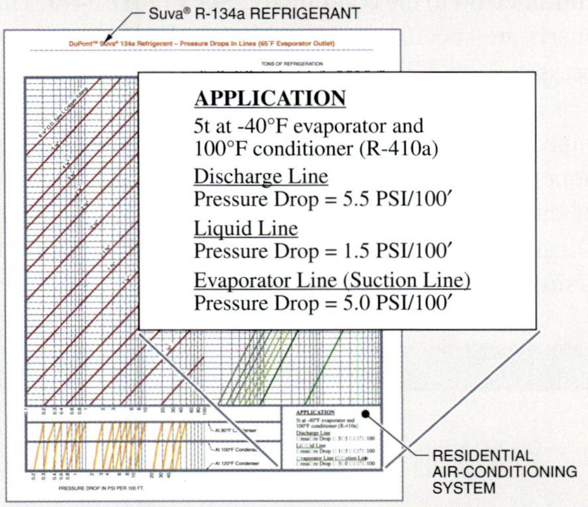

2. Find the system cooling capacity in the upper right corner on the chart.

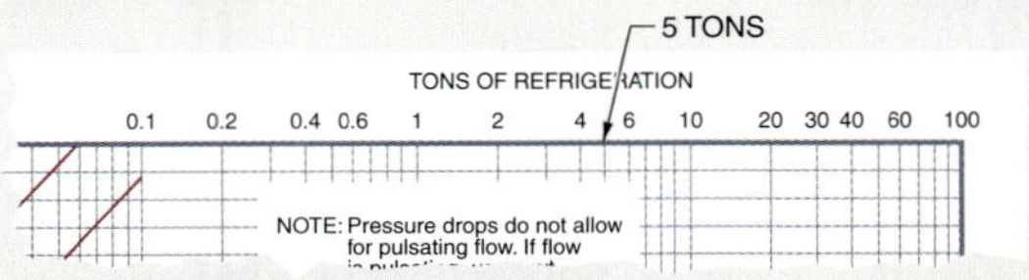

3. Draw a vertical line down from the 5 t point through the evaporator coil, condenser coil, and liquid-line temperature lines.

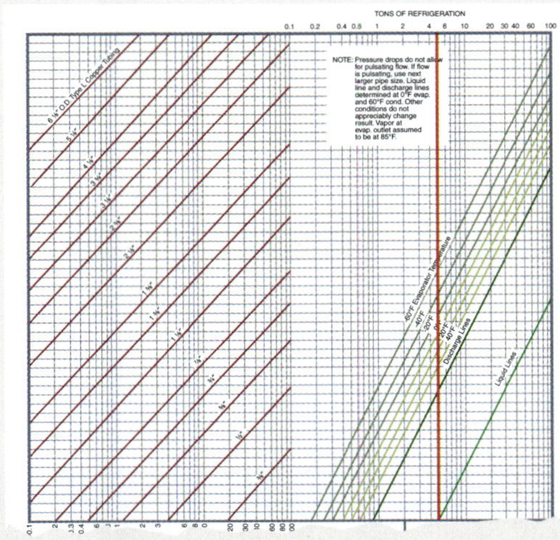

4. Draw a horizontal line from the correct evaporator temperature line, from the discharge line, and from the liquid line to the left side of the chart.

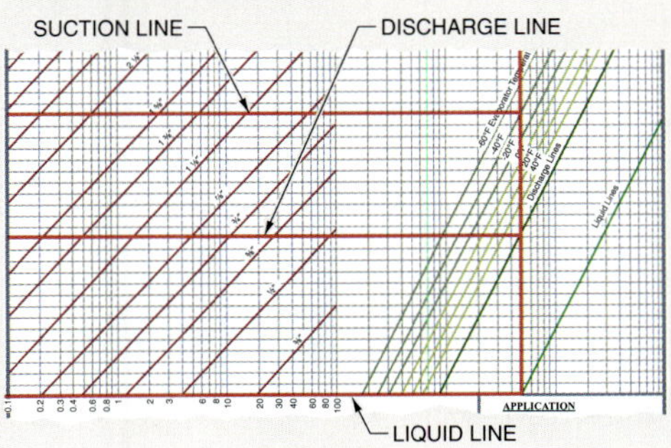

5. Draw a vertical line upward from the required pressure drop points for the evaporator pressure drop, condenser pressure drop, and the liquid-line pressure drop.

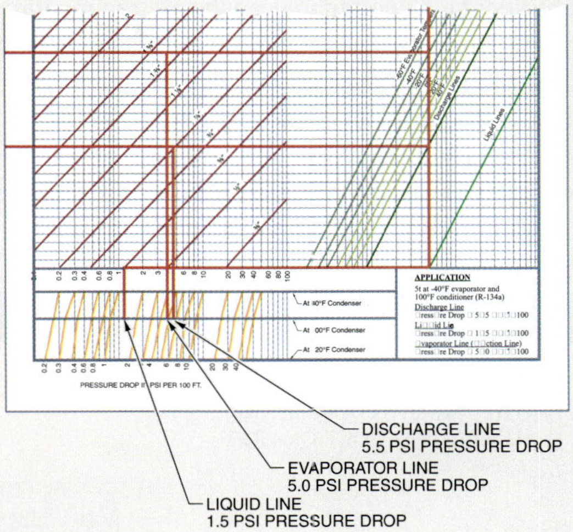

6. The intersection of the horizontal and vertical lines indicates the proper size of the evaporator line, discharge line, and liquid line.

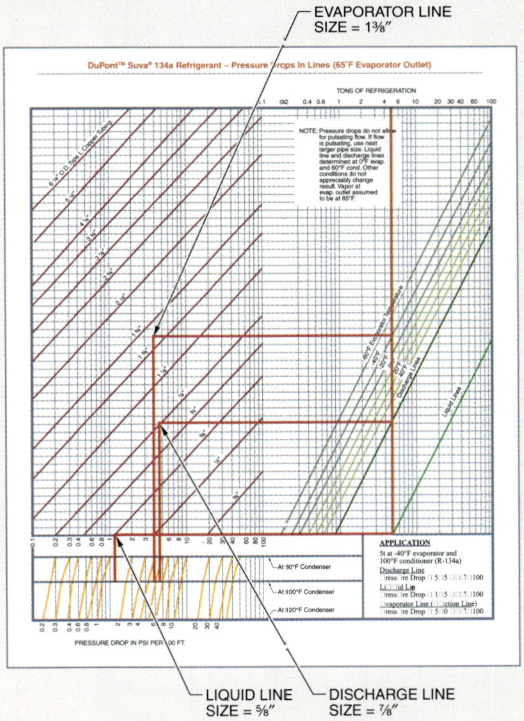

Velocity

Refrigerant velocity is the speed at which refrigerant flows through the piping of a refrigeration system and is expressed in feet per minute (fpm). Refrigerant velocity must be kept high enough to keep the oil moving through the system. Refrigerant velocities typically remain about 1200 fpm to keep the oil entrained in the refrigerant so that it does not drop out of the refrigerant somewhere in the system. The refrigerant velocity cannot be so high that excessive noise or vibration results, but at the same time, pressure drops must be kept to a minimum.

Maximum suction-line charts are used to indicate the proper suction line size that will return refrigerant entrained oil to the compressor. **See Figure 34-6.** These charts are specific to sizing the suction line to a compressor. Typically, it is recommended to keep suction-line loss at 3% or below. Horizontal lines indicate the capacity of the compressor in British thermal units per hour (Btuh). Vertical lines indicate the resistance to flow of the suction line in equivalent length of pipe in feet. Diagonal lines indicate the line size. **See Procedure for Using Suction-Line Charts.**

Suction-Line Charts

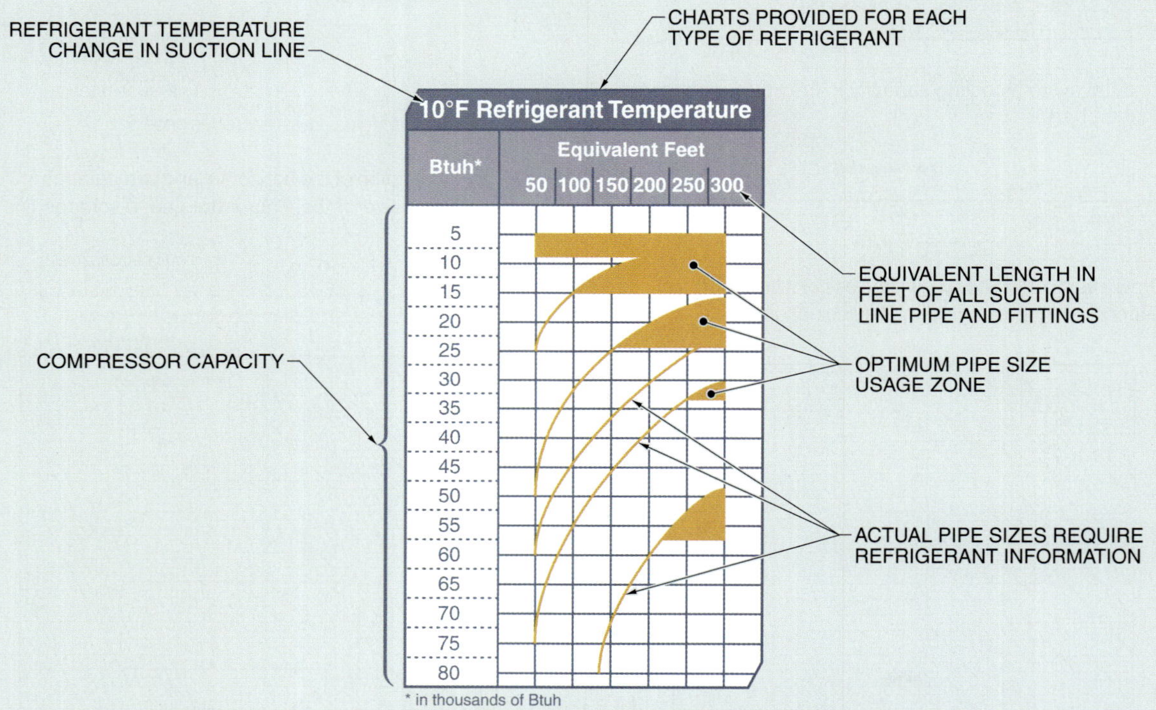

Equivalent Length of Pipe for Fittings and Valves

Line Size*	Globe Valve†	Angle Valve†	90° Elbow†	45° Elbow†	Tee, Sight Glass†	T-Branch†
½	9	5	0.9	0.4	0.6	2.0
⅝	12	6	1.0	0.5	0.8	2.5
⅞	15	8	1.5	0.7	1.0	3.5
1⅛	22	12	1.8	0.9	1.5	4.5
1⅜	35	17	2.8	1.4	2.0	7.0
2⅛	45	22	3.9	1.8	3.0	10.0
2⅝	51	26	4.6	2.2	3.5	12.0
3⅛	65	34	5.5	2.7	4.5	15.0
3⅝	80	40	6.5	3.0	5.0	17.0

* in in.
† in ft

Figure 34-6. Suction-line charts are quite exacting when sizing a refrigeration system suction line to a compressor.

Procedure

Using Suction-Line Charts

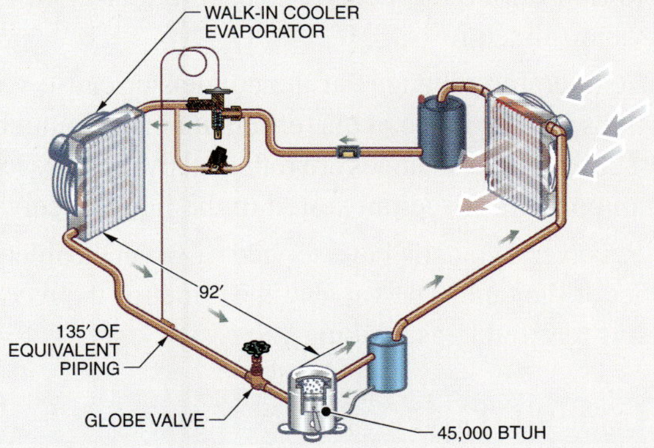

1. Find the compressor capacity on the left side of the chart and draw a horizontal line to the right.

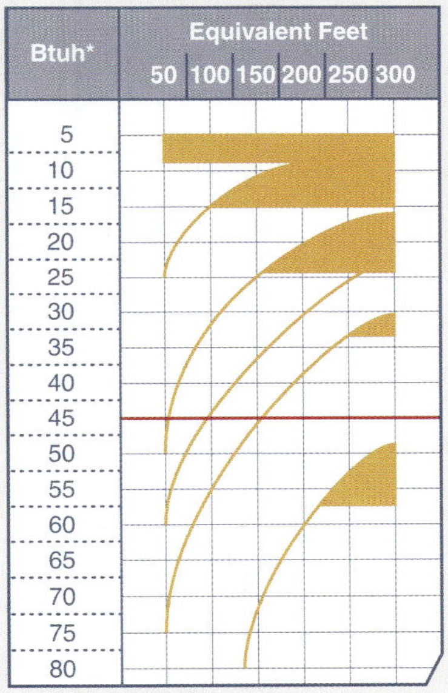

* in thousands of Btuh

2. Calculate the equivalent length in feet of the suction line.

92	LENGTH OF PIPE RUN
35	GLOBE VALVE EQUIVALENT
+ 8.4	(3) 90° ELBOW EQUIVALENT
135.4	EQUIVALENT FEET

3. Find the equivalent feet across the top of the chart and draw a vertical line down.

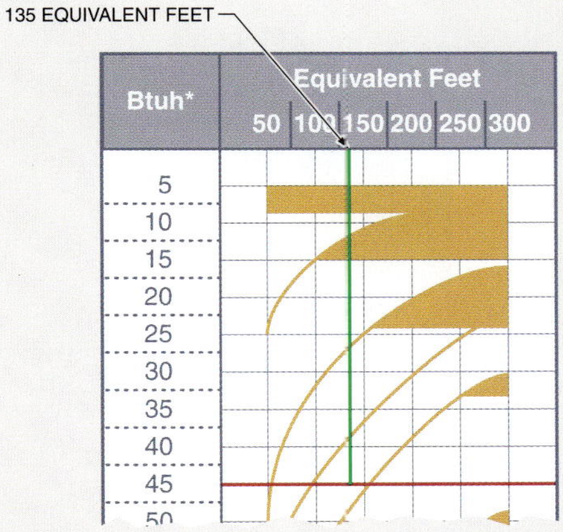

4. Find the intersection of the horizontal and vertical lines. The intersection between the two lines is the maximum-size piping to ensure oil return.

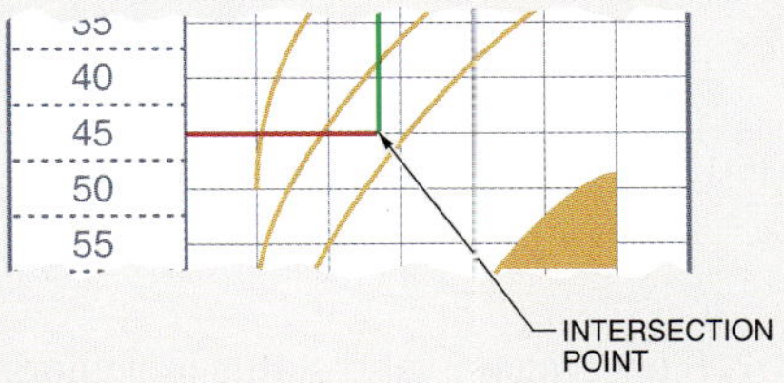

5. If the intersecting point is between two sizes, use the smaller of the two pipe sizes.

Flow Switches and Status Indicators

Flow switches are used on water-medium condensers to prove that water flow is available at the condenser. Flow switches and status indicators are used to prevent compressor operation in the event of a water pump failure. **See Figure 34-7.** They may also be used to energize a backup pump or send an alarm to appropriate staff members. Differential pressure switches or current sensing relays can also be used. All flow switches and status indicators have their advantages and disadvantages, but all of them must be properly located and calibrated, otherwise incorrect readings can occur.

Many condensing units use very short piping runs to minimize pressure and temperature piping problems.

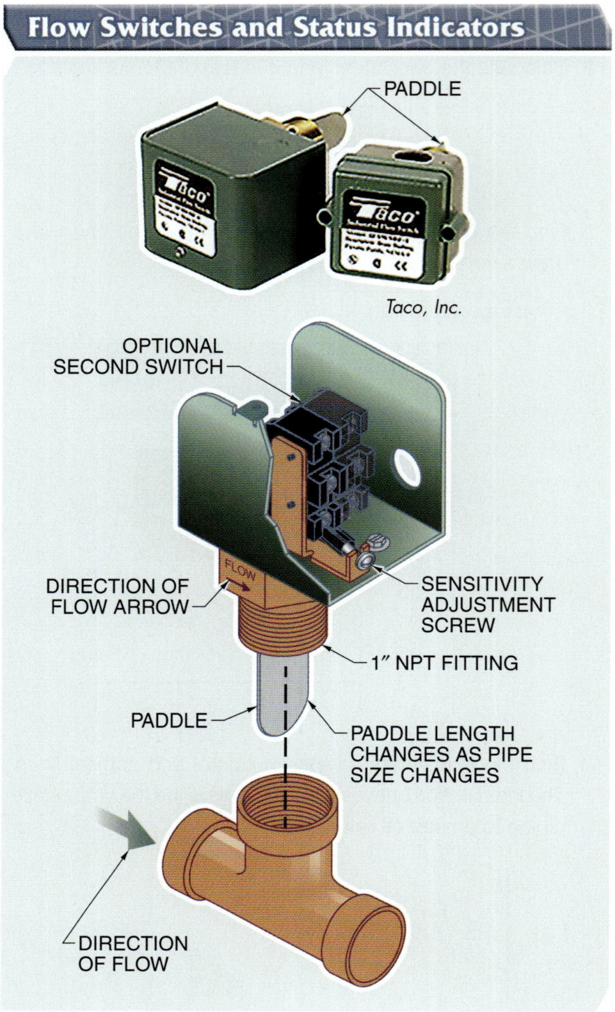

Figure 34-7. Flow switches and status indicators are used on water-medium condensers to prove that water flow is available at the condenser and to prevent compressor operation in the event of a water pump failure.

Liquid Receivers

A *liquid receiver* is a liquid-line storage tank for excess liquid refrigerant. It is located between the condenser and expansion device. **See Figure 34-8.** Liquid receivers are typically found in medium-size and large-size refrigeration systems and perform the following functions:

- When operating at light load, refrigerant not needed is stored in the liquid receiver tank.
- When servicing large systems, the refrigerant may be stored in the liquid receiver, which is then isolated from the system by valves.
- The liquid receiver helps ensure that a solid column of liquid refrigerant is present at the expansion device, regardless of the load on the system. Typically, the liquid to the expansion device is taken from the bottom of the liquid receiver to ensure that there is no vapor mixed in.
- Temperature changes at the condenser cause variations in the volume of the refrigerant in the liquid line. These volume changes are reflected by changes in the amount of refrigerant stored in the liquid receiver.

Receivers typically have service valves at the inlet and outlet of the tank. The valve at the outlet of the receiver tank is referred to as the king valve.

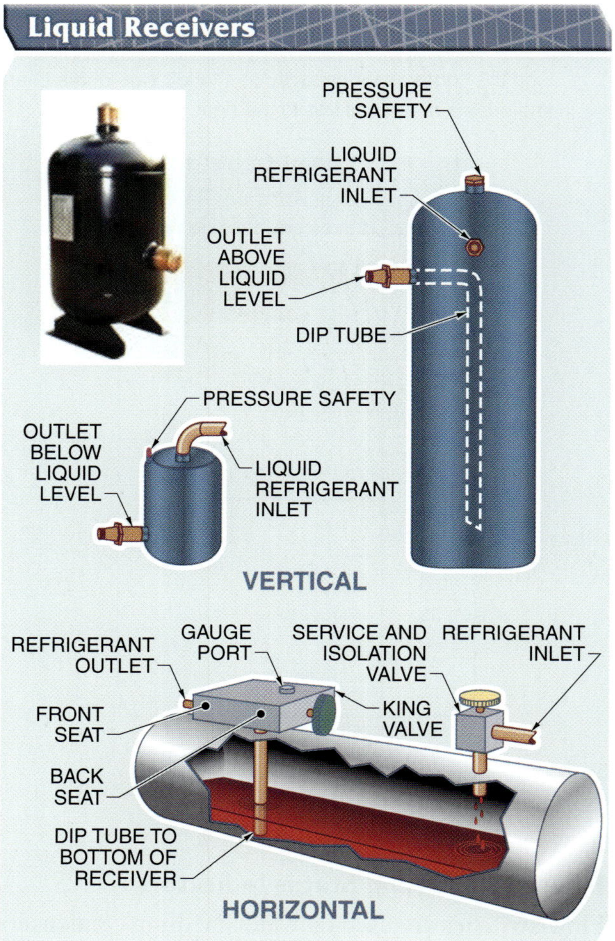

Figure 34-8. Liquid receivers store excess liquid refrigerant and are located between the condenser and expansion device to guarantee that the expansion device only receives liquid refrigerant.

King Valves

When the compressor is allowed to operate and the king valve (receiver valve) is front-seated (closed), the condenser and receiver will then fill with refrigerant. **See Figure 34-9.** The low side of the system may then be opened and serviced with little or no loss of refrigerant.

After service, the king valve is opened and the system is placed into operation again. Gauge connections can also be attached to the king valve to allow technicians to measure system pressure.

King Valves

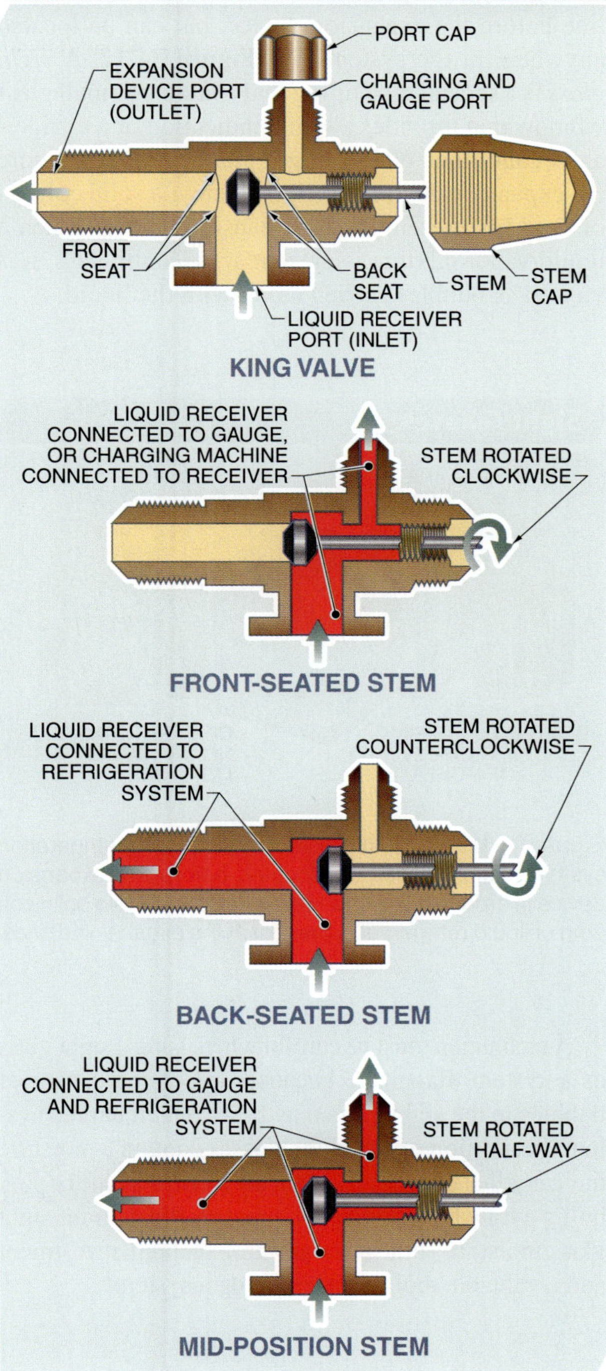

Figure 34-9. The output line of a liquid receiver typically has a king valve, which is used for troubleshooting and to isolate the liquid receiver from the rest of the system.

Liquid-Line Filter-Driers

Refrigerant filter-driers can be located anywhere in the liquid line. A *filter-drier* is a device that removes particulate matter, moisture, and acids from the refrigerant in a system. **See Figure 34-10.** Particulate matter consists of dirt, flux, solder globules, and metal filings. These particulates are often caused by component wear and improper service techniques. Moisture may be in the system due to inadequate drying capabilities of the system or improper service procedures that allowed moisture to get into the system. Acids are created by the reaction of moisture, oil, and refrigerant being together under high temperatures. When the acids are not removed, they will corrode metal parts of the system and damage or destroy the compressor motor windings. Motor windings are destroyed because they are exposed to the refrigerant containing the acids.

Filter-Driers

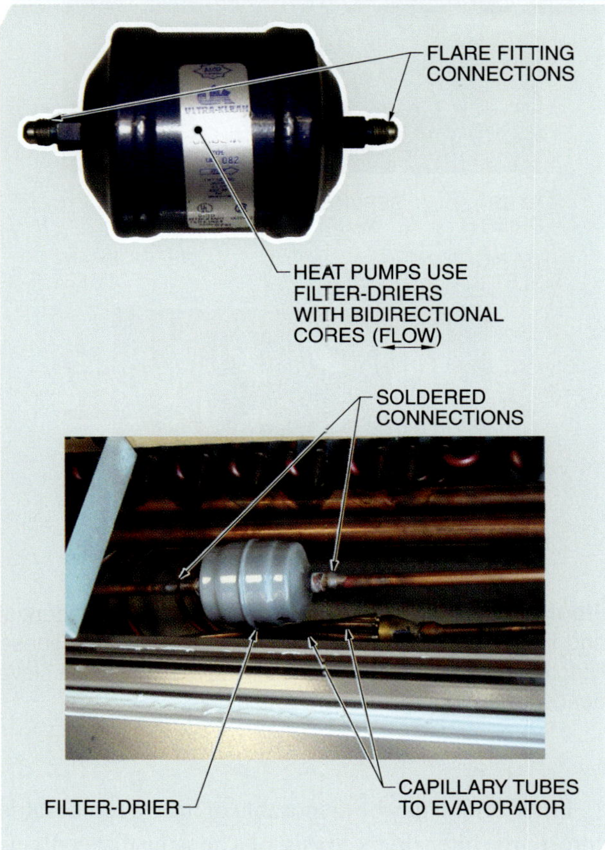

Figure 34-10. Filter-driers remove particulate matter, moisture, and acids, which can damage the compressor and system valves, from refrigerant.

Filter-driers contain metal screens that trap particulates, a desiccant that absorbs moisture, and chemicals that neutralize any acids present. **See Figure 34-11.** The desiccants in filter-driers are typically silica gel, activated alumina, or a molecular sieve. Filter-driers may be attached to the liquid line by flare fittings or solder connections. Flare fittings are used on smaller size lines up to ⅝″. Soldered connections are used on line sizes that are larger than ⅝″.

Figure 34-11. Small filter-driers used in most residential and light commercial applications are disposable. These filter driers include screens for cleaning and molecular sieve beads for collecting moisture and acids.

Filter-driers may be disposable or have a replaceable core. Large filter-driers consist of a housing that contains a replaceable core element that is changed when needed. The refrigeration system must be shut down and the filter-drier isolated when changing a core element. In systems containing a king valve, the king valve can be used to isolate the filter-drier. Small systems typically use throwaway filter-driers.

Sight Glasses

Refrigerant sight glasses are often located in the liquid line before the expansion device, but can be located anywhere in the system. **See Figure 34-12.** A *sight glass* is a special pipe fitting that contains a small glass window that provides a visual indication of a refrigerant's condition. When a sight glass is located before the expansion device, it will indicate if a solid column of liquid is present at the expansion device. When a liquid/vapor mixture is present at the sight glass, it is viewed as bubbles carried along with the liquid.

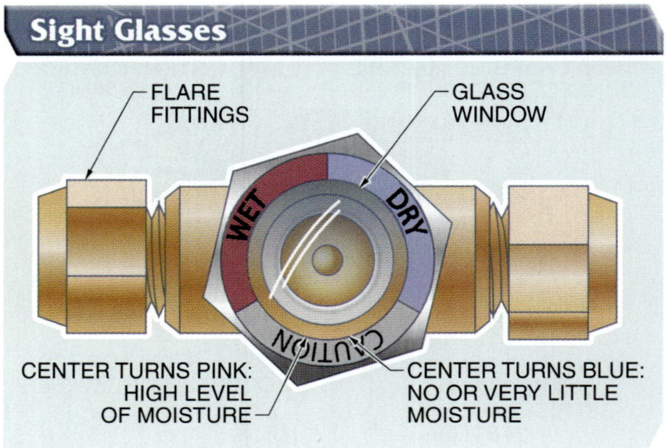

Figure 34-12. On the high-pressure side of a refrigeration system, sight glasses are located before the expansion device in order to provide a visual indication that a solid column of liquid refrigerant is present at the expansion device.

A technician must be careful when using a sight glass as a system diagnostic because one of the causes of bubbles in the sight glass may be an undercharged system. In addition, many sight glasses contain a moisture indicator that changes color (in the form of a blue or pink bull's-eye) when moisture is present. Refrigerant sight glass/moisture indicators are an excellent investment and a valuable tool when servicing a system.

Sight Glass/Moisture Indicator Operation
Media Clip

TECH TIP

Compressor oil loss is a major cause of compressor failure. Oil loss is often the result of poor refrigeration piping system design. Oil separators can help compensate for a poor system design.

Oil Separators

Keeping the proper oil level in the compressor crankcase is crucial to compressor life. Oil separators and the return system are used to maintain an adequate oil level in the compressor crankcase. Oil separators are located in the compressor discharge line. **See Figure 34-13.** An *oil separator* is a container that allows hot discharge refrigerant vapor to enter the container and then allows the refrigerant to discharge from the container to the condenser, while the oil is piped to the compressor crankcase. Any oil mixed with the vapor is unable to change direction as easily as the vapor and precipitates toward the bottom of the container, where it accumulates. A float mechanism opens and drains the oil back to the crankcase.

Oil separators are typically used on refrigeration systems that may experience poor oil return conditions. Oil separators are built into screw compressor applications, where any loss of lubrication will cause deterioration in compressor (ultimately system) performance and catastrophic compressor failure.

High-Pressure-Side Solenoid Valve Applications

Solenoid valves are used in many applications, including ON/OFF control and light-load control. In a simple ON/OFF refrigeration application, the liquid-line solenoid (LLS) valve is opened and closed to turn refrigerant flow ON and OFF. **See Figure 34-14.** When the cooling load has been satisfied, a thermostat closes the LLS. When the solenoid valve is closed, the refrigerant flow is shut off. Since liquid flow is discontinued, the pressure in the evaporator rapidly decreases. A low-pressure control (LPC) stops the compressor at a specific predetermined pressure. When cooling is required again, the process is reversed. The LLS opens, the evaporator pressure increases as liquid boils to a gas, and the LPC energizes the compressor.

Oil Separators

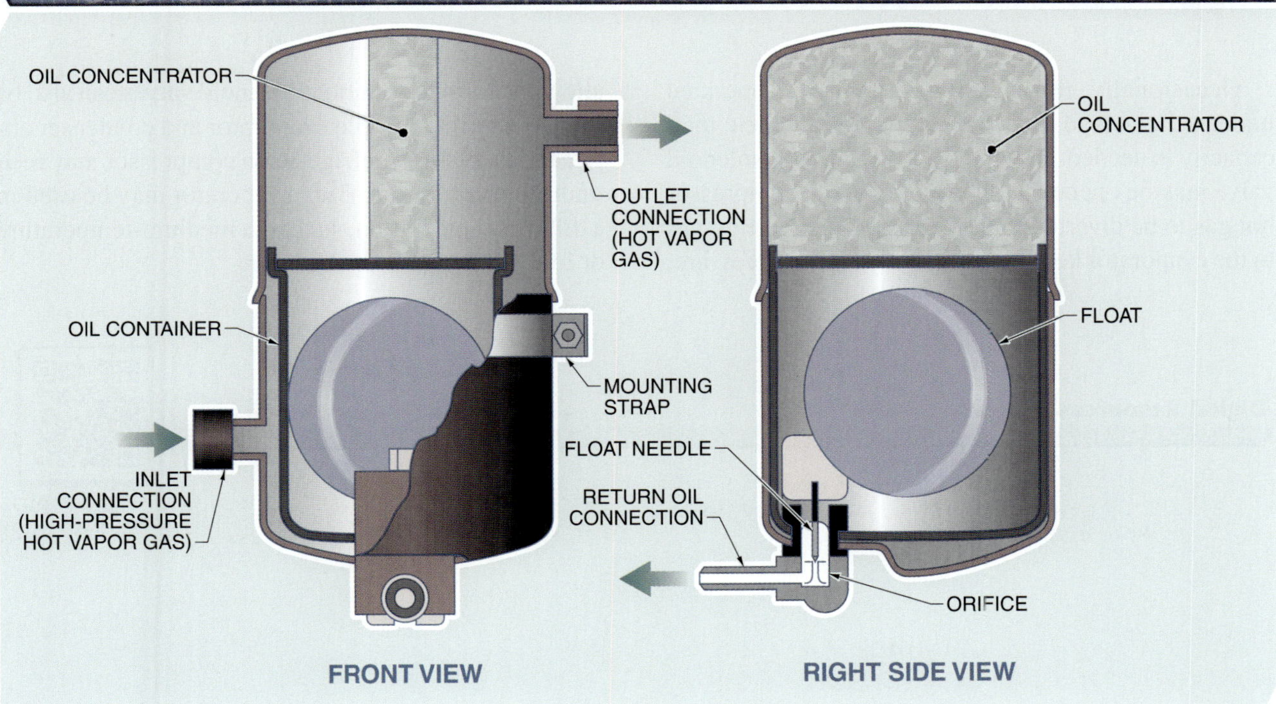

Figure 34-13. Oil separators located on the hot-gas discharge line remove oil that has mixed with the refrigerant and returns the oil back to the sump of the compressor.

Pump-Down Cycle Applications

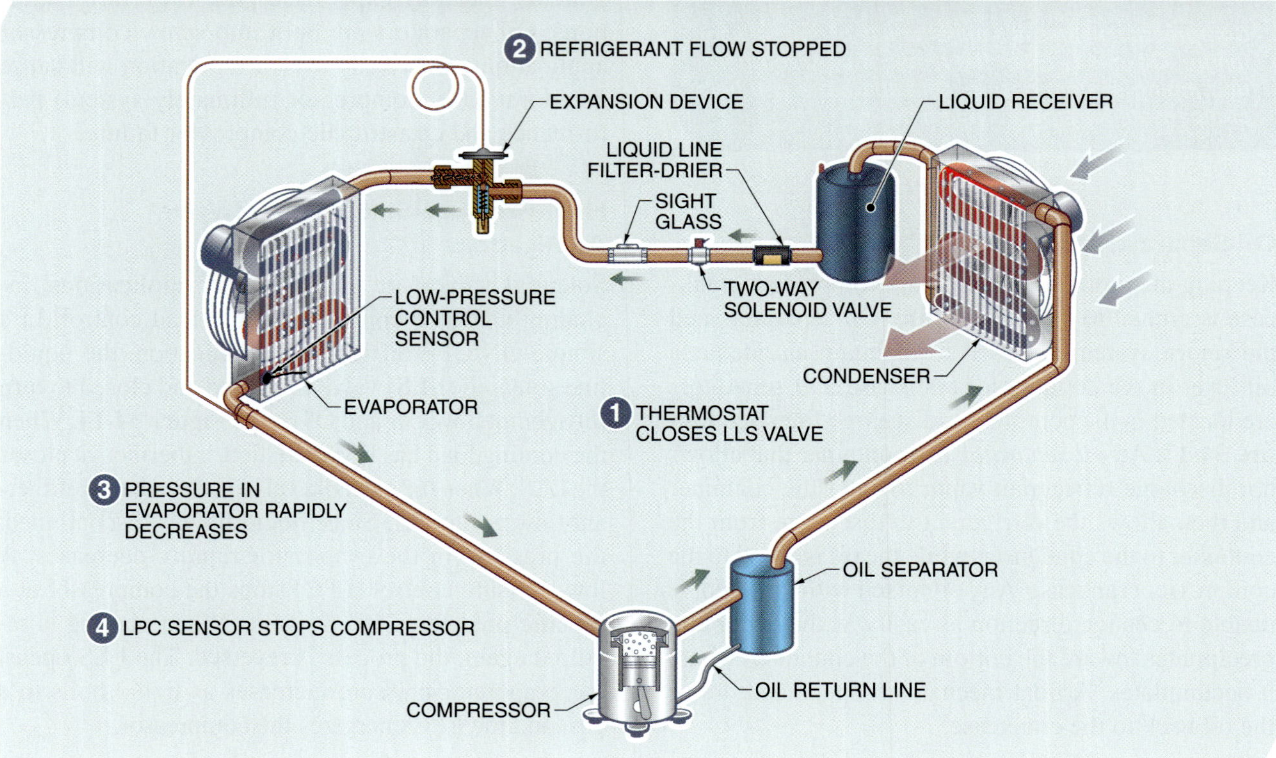

Figure 34-14. High-pressure-side solenoid valves located in the liquid-line determine whether the expansion device and evaporator receive refrigerant for system operation.

Occasionally, refrigeration systems are operated under a light load, and only a portion of their total capacity is needed. In light-load situations, a solenoid valve may be opened to allow some of the compressor hot gas to be diverted to the evaporator. Diverting gas to the evaporator keeps the gas at a minimum pressure, allowing normal operation. Solenoid valves can also be used to control multiple evaporator and condenser applications. A single refrigeration compressor may feed multiple evaporators. Each evaporator may be used in a different application, such as a medium-temperature or low-temperature application.

UNIT 35

Refrigeration Operating Controls

The installation and operation of refrigeration systems can be costly. The failure of a system can also be quite costly because of lost product from machinery not operating at full capacity, equipment downtime, and the cost of repairs to the system. For a refrigeration system to operate properly, it must be controlled properly. Even when a refrigeration system is sized and installed properly, an improper or poorly functioning control system will cause major problems and system failure. It is just as important to properly select and install the unit controls as any major mechanical part.

OBJECTIVES

- List the typical refrigeration operating controls.
- Identify the purpose and application of each operating control.
- Describe the operation of typical safety controls.

Index to Procedures

REFRIGERATION SYSTEM CONTROLS

The two categories of controls for refrigeration systems are operating controls and safety controls. Operating controls maintain the desired conditions in a refrigeration system at the condenser, expansion device, evaporator, and compressor. Safety controls must be provided to ensure that excessively low or high temperatures do not damage the refrigeration unit or endanger the safety of building occupants. **See Figure 35-1.** Some controls can be used as both operating and safety controls depending on the application.

OPERATING CONTROLS

An *operating control* is a device used to cycle on and off or regulate the various components of a refrigeration unit and/or air conditioning system to meet specific conditions for temperature, pressure, and humidity. Operating controls vary widely but typically consist of thermostats, low-pressure controls, airflow controls, and low ambient temperature controls.

Thermostats

One of the most important controls in a refrigeration system is the thermostat. The thermostat is often located at the inlet to the evaporator in a refrigerated cabinet. Therefore, the thermostat senses the cabinet air temperature as the air flows over it due to the operation of the evaporator fan. Care must be taken when locating a thermostat to ensure it is not easily damaged by anyone walking by in the space or by moving product inside the refrigerated cabinet. **See Figure 35-2.**

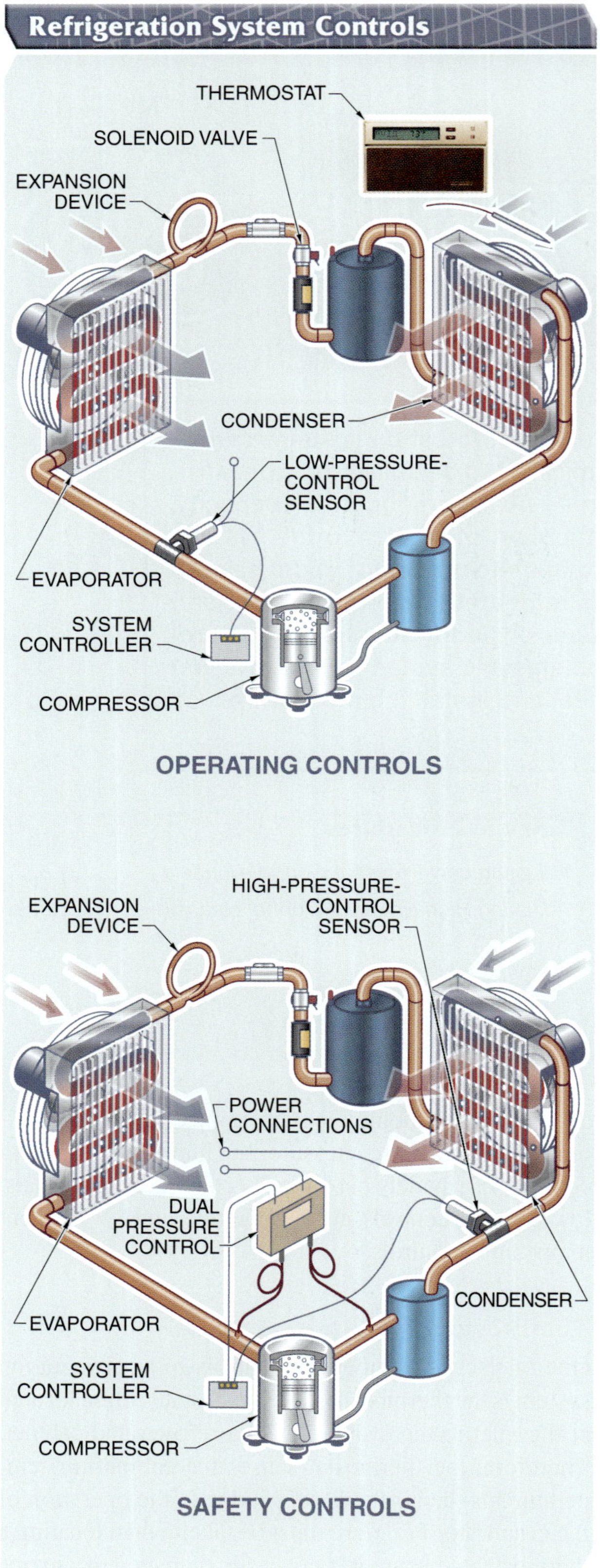

Figure 35-1. Refrigeration controls are classified as operating controls, safety controls, or both.

Figure 35-2. The most common refrigeration operating control is the thermostat.

A thermostat typically consists of a bulb or capillary element that actuates a set of contacts through a bellows mechanism. The temperature range of a thermostat must match that of the application. Depending on the type, thermostats may be either a fixed or adjustable differential. Typically, they are of a snap acting design. To check for proper thermostat operation, a thermometer is mounted at the door of a refrigerated cabinet so that the owner can easily check cabinet air temperature.

Low-Pressure Controls

Another very important refrigeration operating control is the low-pressure control (LPC). A *low-pressure control (LPC)* is a pressure-activated electromechanical switch used to shut off a compressor when system pressure becomes too low. LPCs are typically located near the condensing unit. An LPC typically has two adjustable setpoints: the cut-in and differential. A cut-in pressure setpoint turns on the compressor, while the differential setpoint causes the unit to cycle off after a certain pressure change has occurred. **See Figure 35-3.** This is called the cut-out pressure. Because the cut-out pressure is only adjusted by changing the cut-in or differential setpoint, typically a formula is used. The cut-out pressure is equal to the cut-in pressure minus the differential ($cut_{out} = cut_{in} - differential$). For proper operation, both setpoints must be properly adjusted.

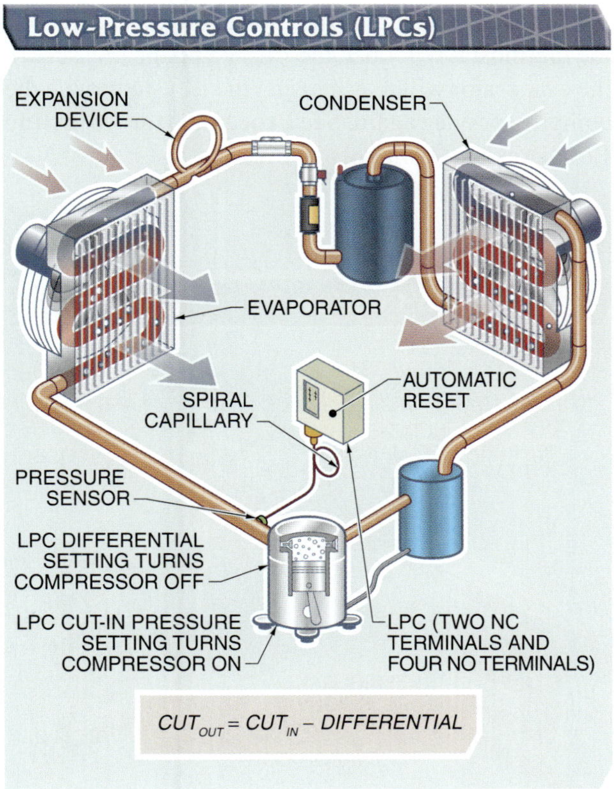

Figure 35-3. An LPC is a pressure-operated switch used to control the compressor.

Low-Pressure-Control (LPC) Operation Media Clip

Example: LPC Pressure Settings

A system uses R-134a refrigerant. The compressor should turn on at an evaporator coil temperature of 36°F and turn off at a temperature of 15°F. At 36°F, the evaporator pressure will be at 31 psi. The corresponding pressure at 15°F will be 15 psi. Therefore, the cut-in pressure setting is set at 31 psi, and the differential pressure setting is set at 16 psi (31 – 15 = 16). **See Figure 35-4.**

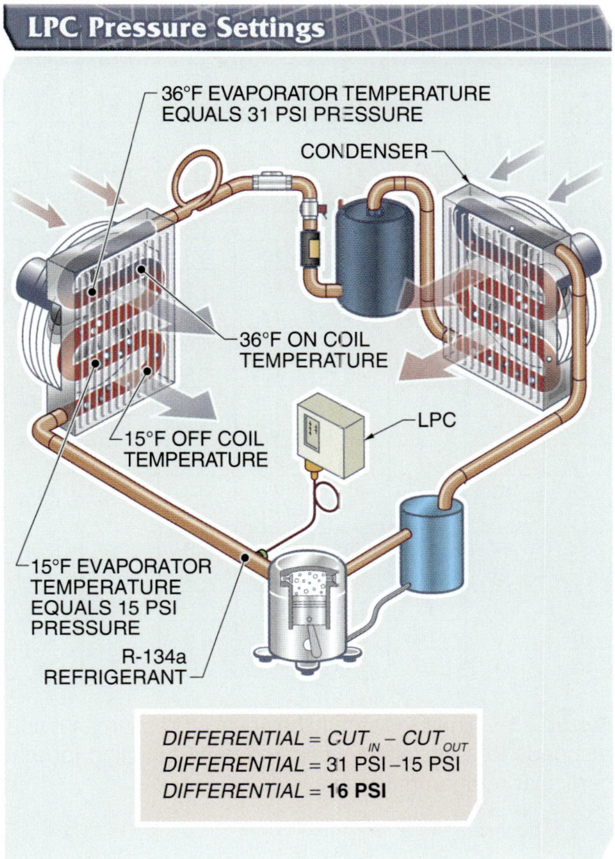

Figure 35-4. An LPC should turn on the compressor at an evaporator coil temperature of 36°F and turn off the compressor at a temperature of 15°F when R-134a refrigerant is being used in the system.

LPC Functions

Low-pressure control can act as an operating control and/or a safety control. When the system loses its charge, the LPC will open and stop operation of the compressor. In some systems this may prevent the low-pressure side of the system from going into a vacuum, stop system operation due to a failed compressor, operate other devices, and sound alarms. The LPC should be set at a pressure that corresponds to a typical loss of refrigerant condition for that particular refrigerant in the system. **See Figure 35-5. See Appendix.**

Refrigerant Pressure-Temperature Charts

PSI	R-12	R-22	R-124	R-134a	R-502	R-507	R-717
-5	-29	-48	3	-22	-57	-60	-34
-4	-28	-47	4	-21	-55	-58	-33
-3	-26	-45	6	-19	-54	-57	-32
-2	-25	-44	7	-18	-52	-55	-30
-1	-23	-43	9	-16	-51	-54	-29
0	-22	-41	10	-15	-50	-53	-28
1	-19	-39	13	-12	-47	-50	-26
2	-16	-37	16	-10	-45	-48	-23
3	-14	-34	18	-8	-42	-46	-21
4	-11	-32	21	-5	-40	-44	-19
5	-9	-30	23	-3	-38	-41	-17
11	4	-19	36	8	-27	-31	-7
12	5	-17	38	10	-25	-29	-5
13	7	-15	40	12	-24	-27	-4
14	9	-14	41	13	-22	-26	-2
15	11	-12	43	15	-20	-24	-1
16	12	-11	45	16	-19	-23	1
17	14	-9	46	18	-18	-21	2
18	15	-8	48	19	-16	-20	3
19	17	-7	49	21	-15	-19	4
20	18	-5	51	22	-13	-17	6
34	37	11	69	39	3	-1	20
35	38	12	71	40	4	0	21
36	39	13	72	41	5	1	22
37	40	14	73	42	6	2	23
38	41	15	74	43	7	3	24

Figure 35-5. The LPC should be set at a pressure that corresponds to a typical loss of refrigerant condition for that particular refrigerant in the system.

Low-pressure controls are often used along with defrost timer controls.

The LPC can also be used as a thermostat in a refrigerated cabinet for temperature control. Since a pressure-temperature relationship exists for any refrigerant, adjusting the LPC will, by definition, adjust the temperature of the refrigeration system.

Pump-Down Controls. When a thermostat controls a liquid-line solenoid valve (LLSV), the pressure in the evaporator will drop as the LLSV shuts off refrigerant flow to the evaporator. The refrigerant is then pumped into the liquid-line receiver by the compressor. **See Figure 35-6.** The LPC will then shut off the compressor when a low pressure is achieved at the evaporator. When the temperature in the cabinet or space increases, the thermostat will open the LLSV and the pressure in the evaporator will rise, causing the LPC to turn on the compressor once again. **See Procedure for Adjusting Low-Pressure Controls in Field.**

Pump-Down Controls

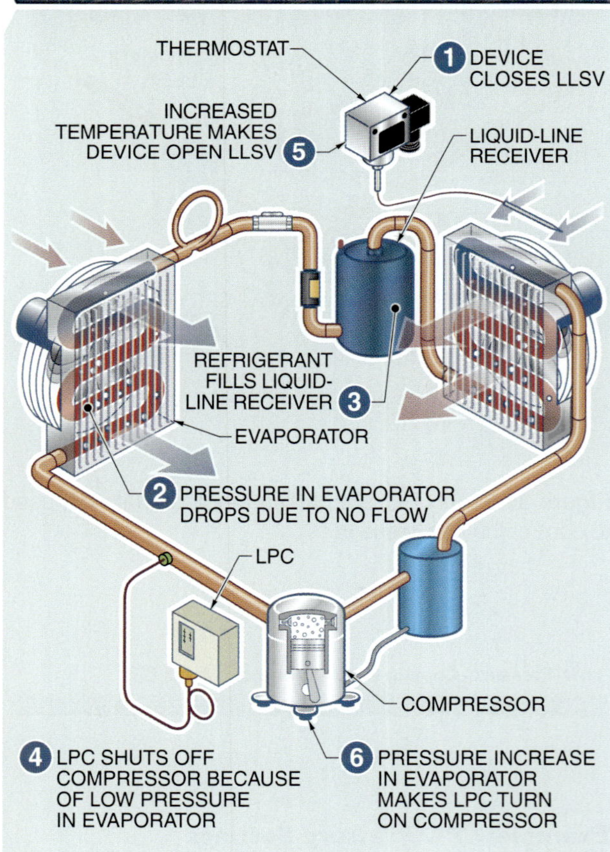

Figure 35-6. An LLSV typically has a manual override that is used during troubleshooting procedures.

Adjusting Low-Pressure Controls in Field

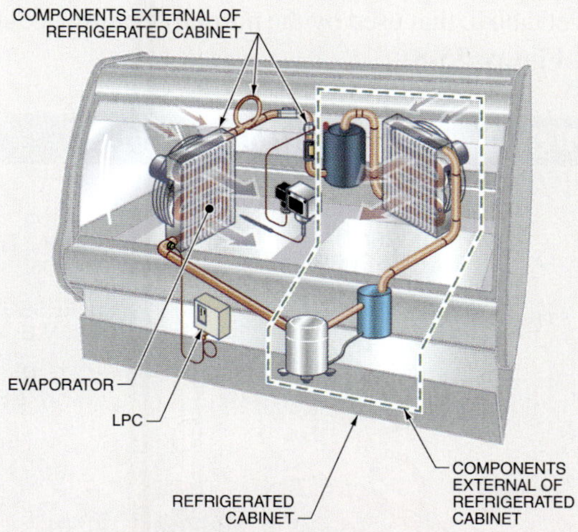

1. Set the cut-in value of the low-pressure control (LPC) to the value determined by the refrigerant and system.

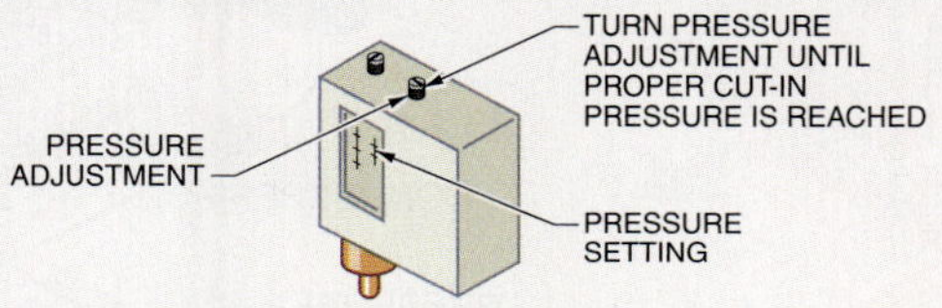

2. Adjust the differential to a very high value.

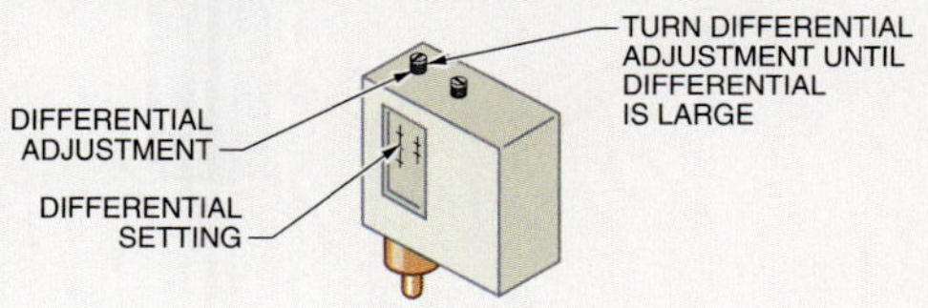

3. Start the system and let it run. Check the temperature of the refrigerated cabinet.

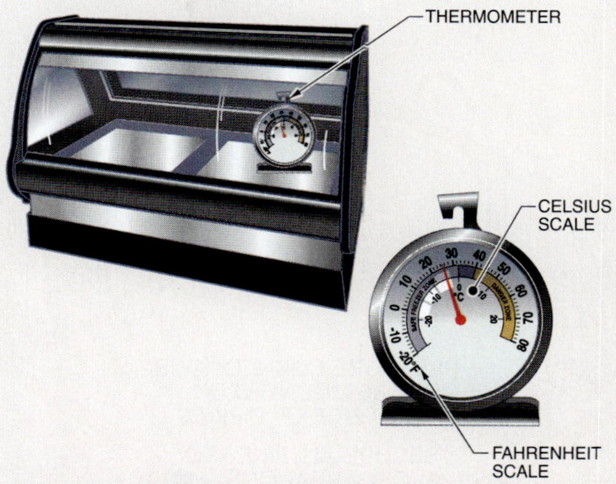

4. When the refrigerated cabinet is at its lowest required temperature, slowly adjust the differential smaller until the compressor turns off.

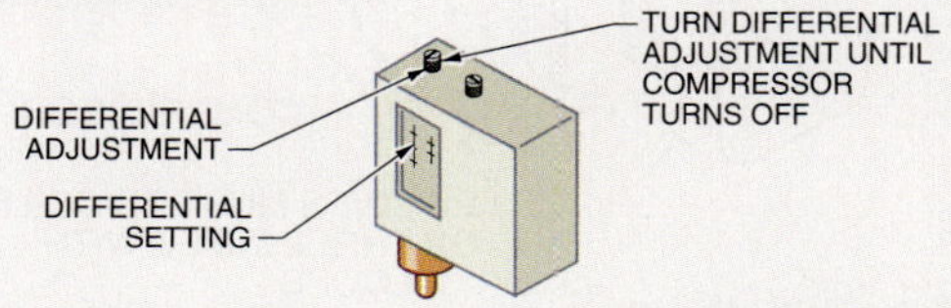

5. Check the operation of the refrigerated cabinet for a few cycles to ensure the system operates properly.

LPC Installation Tips

Tips for properly installing refrigeration system LPCs are as follows: **See Figure 35-7.**

- Secure all lines to avoid excessive vibration.
- Avoid any tight, straight runs of tubing. All LPC lines must be coiled two or three times with a 2″ or 3″ loop to help dampen vibration.
- Avoid any rubbing of the lines due to vibration.
- Mount pressure connections on the top or side of the suction line.
- System pressure controls use flexible piping.

Condenser Fan Speed Controls

Many condenser fans use simple ON/OFF control. For commercial systems, this may make head-pressure control more difficult due to variations in the ambient air temperature. Head pressure is a more energy-efficient refrigeration system that uses a speed control on the condenser fan motor. The speed control uses a variable speed drive that is operated by a refrigerant pressure controller on the high-pressure side of the system. As the head pressure increases above a specific setpoint, the condenser fan will start. As the pressure continues to increase, the fan speed will also be increased. Therefore, close proportional control of the head pressure can be achieved.

Condenser speed controls are traditionally designed for use with single-phase, permanent, split-capacitor motors that are approved by the motor manufacturer for variable-frequency drive applications. This type of speed control regulates the supply voltage to the fan motor in response to the refrigerant pressure. Variable-speed fans provide direct responses to changes in condenser pressure, regardless of the variations in air delivery curves. They also provide a great way to accommodate low ambient temperature operations when needed. ECM motors are up and coming.

LPC Installation Tips

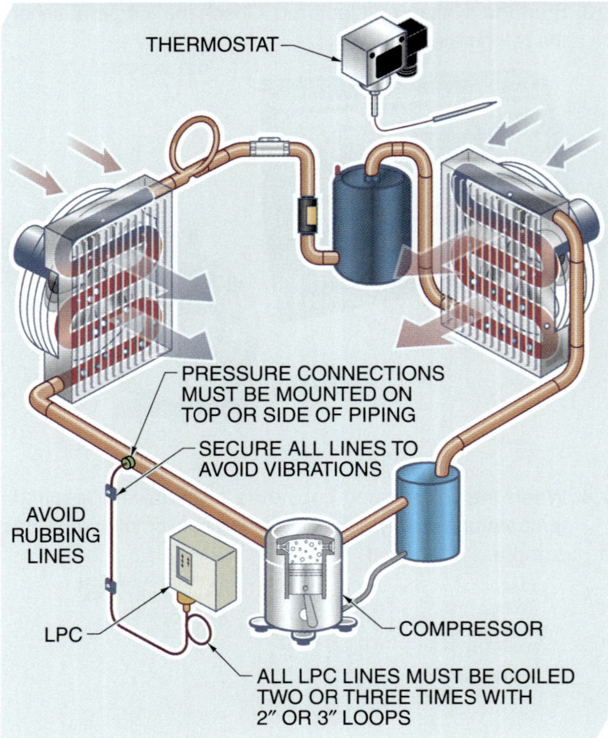

Figure 35-7. All LPC capillary lines must be coiled and protected against vibrations.

Typical applications include computer room air conditioning, commercial air conditioning, and commercial refrigeration. Some models of high-pressure control have two pressure connections and can control to the highest of either pressure. The setpoints for the fan speed controls are adjustable to match the application and the refrigerant being used.

SAFETY CONTROLS

Safety controls are used to protect refrigeration systems from abnormal conditions that may cause damage. Examples of safety controls include high-pressure controls, anticycling timers, and lube-oil pressure controls.

High-Pressure Controls

When the pressure in a system becomes too high, overheating and compressor burnout can result. A *high-pressure control (HPC)* is a pressure-activated electromechanical switch used to shut off a compressor when pressure in a system becomes too high. HPCs are connected to the high-pressure side of the system. High-pressure controls include a setpoint, differential, and either manual or automatic reset. The HPC setpoint is well above that used by the normal operating controls. **See Figure 35-8.**

High-Pressure Controls (HPCs)

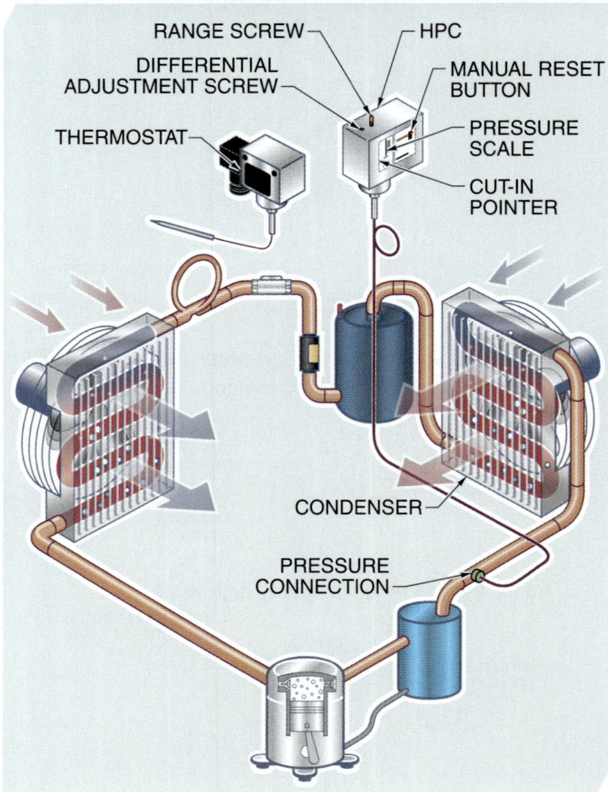

Figure 35-8. An HPC is a pressure-activated switch used to shut down a compressor when pressure in a system becomes too high.

High-Pressure-Control (HPC) Operation Media Clip

When a manual reset HPC is used, a person must reset the HPC by hand to start the refrigeration system. Manual reset controls are used when a fault prohibits further operation. A disadvantage of this type of HPC is that cooling does not occur while the HPC is tripped, which may cause the space or cabinet to heat up. An automatic reset HPC allows the compressor to operate after a specific pressure drop has occurred and the refrigeration unit has stopped. A possible advantage of this type is that the refrigeration system may cycle on the HPC, permitting at least some cooling to occur until the condition is fixed. HPCs are also used to cycle condenser fans. **See Procedure for Adjusting High-Pressure Controls in Field.**

Adjusting High-Pressure Controls in Field

1. Attach a gauge manifold to the high-pressure-side service port of the unit.

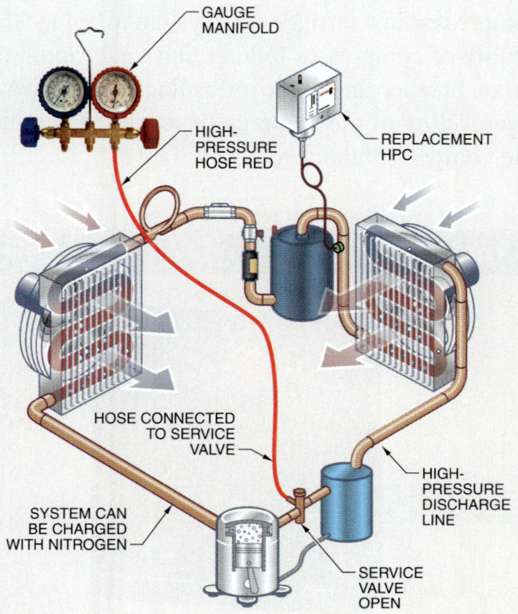

2. Use a piece of cardboard to restrict the airflow through the air-medium condenser or deactivate the fan. If it is a water-medium condenser, reduce the water flow. In both cases, the discharge line pressure will increase.

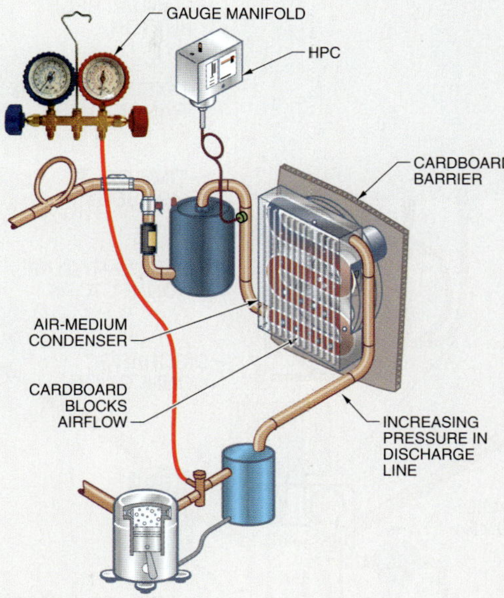

3. Check the pressure at which the high-pressure control (HPC) stops the unit. Adjust the HPC as needed and retest.

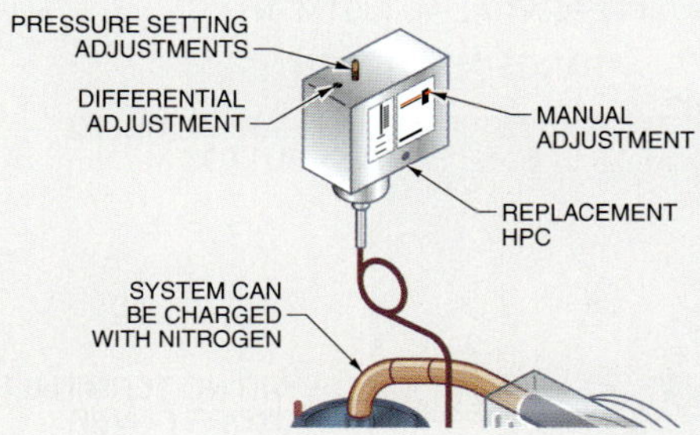

TECH TIP

Refrigeration system safety controls, such as high-pressure controls and lube oil pressure controls, should never be bypassed. Bypassing refrigeration safety controls can cause serious problems and lead to property damage, personnel injury, and death. A unit that has the system safety controls bypassed continues to operate in an unsafe condition and may fail catastrophically. In addition, high temperatures and high amperages can create a fire hazard.

Lube Oil Pressure Safety Controls

Air conditioning and refrigeration system compressors must have lubricated bearings to function properly. A *lube oil pressure safety control* is an electromechanical safety switch that is designed to open a compressor circuit when oil pressure falls below a specific setpoint. The oil pump is driven by the compressor shaft. The lube oil pressure safety control senses the crankcase and oil discharge pressure. The net oil pressure is the difference between the pressure at the pump discharge and the pressure in the compressor crankcase. For example, if the oil pump discharge reading is 100 psi and the crankcase pressure is 70 psi, the net oil pressure is 30 psi.

A lube oil pressure safety control must have a built-in time delay feature to allow the compressor to come up to speed and build oil pressure upon startup. Otherwise, the unit would trip immediately. Typical factory time delays of 30 sec, 45 sec, 60 sec, 90 sec, and 120 sec are available. **See Figure 35-9.**

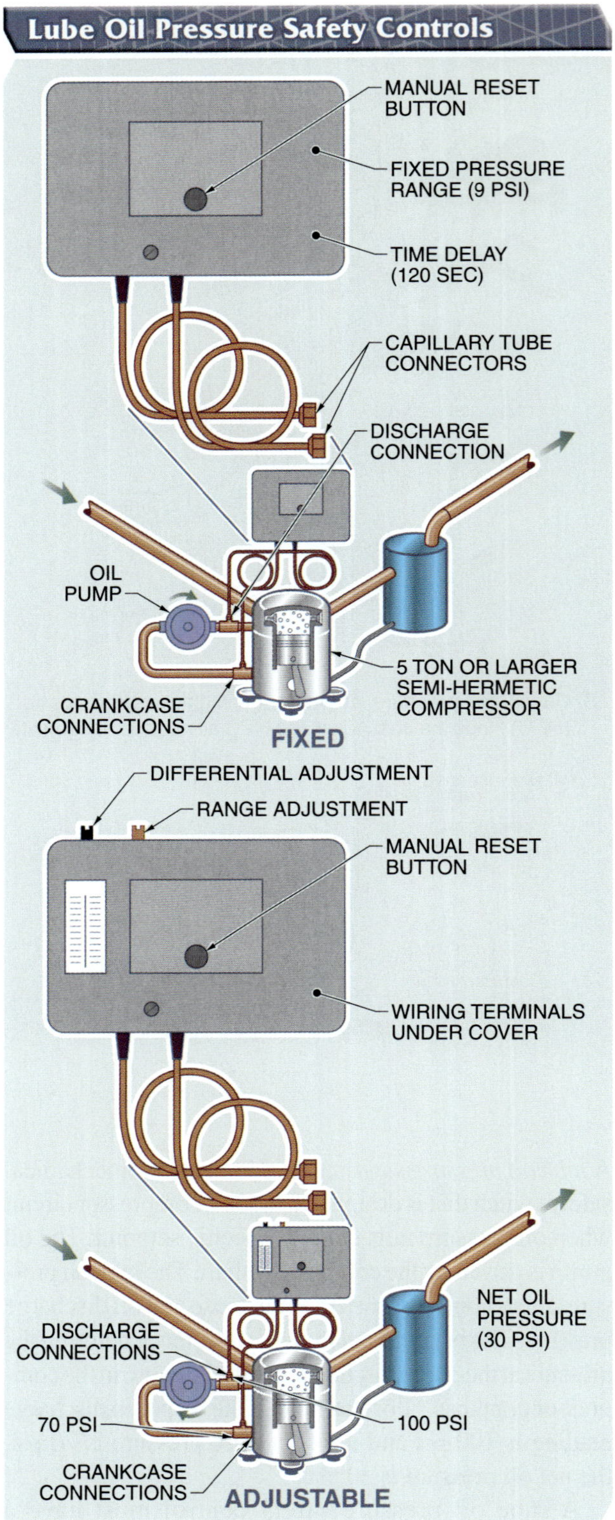

Figure 35-9. Lube oil pressure safety controls use the net oil pressure of the compressor to determine when the compressor should be shut down. All lube oil controls have a time delay of 30 sec, 45 sec, 60 sec, 90 sec, or 120 sec to avoid nuisance tripping.

Anticycle Timers

Another safety control used in a refrigeration system is an anticycle timer. An *anticycle timer,* also known as an antishort cycle timer, is a solid-state safety device that prevents premature cycling (overcycling) of the compressor in a refrigeration system to reduce the possibility of compressor failure, nuisance tripping of the circuit breaker, and fuse overloading. **See Figure 35-10.** Overcycling of the compressor results in heat buildup and compressor damage.

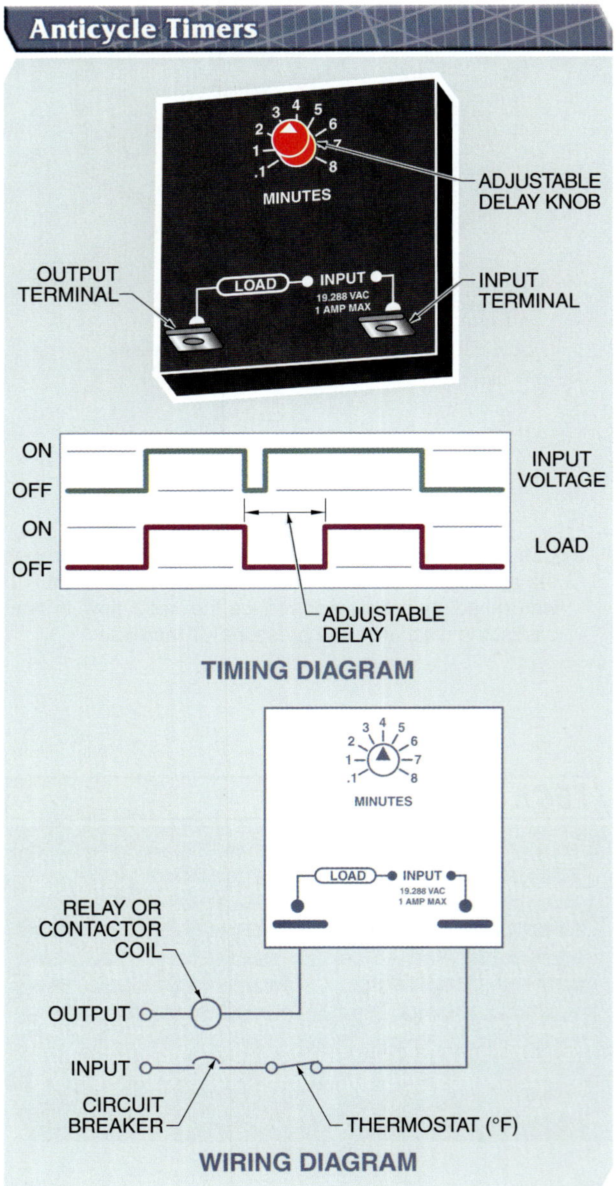

Figure 35-10. To avoid common compressor burnouts, anticycle timers are used to prevent compressors from starting more than five or six times an hour.

Often the timer is energized when the compressor stops and will not allow the compressor to start until a specific time period has elapsed. The timer can be a fixed time period or adjustable, from 15 sec to 8 min, time period design.

The anticycle timer must never be jumped out or bypassed as it protects the compressor motor. A common factory recommendation is that a compressor not be cycled more than six times an hour. Modern commercial refrigeration control systems have electronic boards that include anticycle timers as part of their programming.

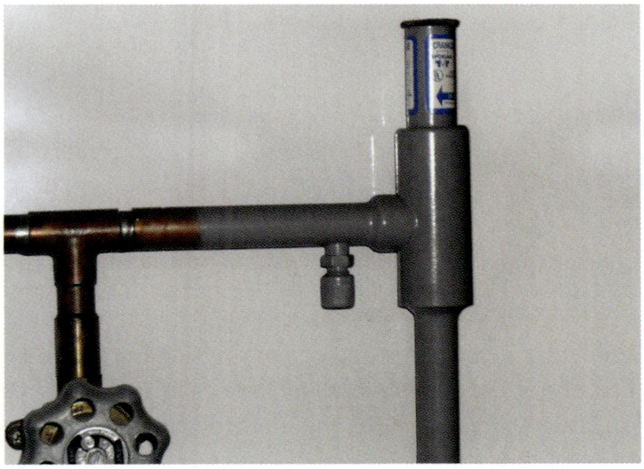

Pressure regulators keep the coil at a specific pressure in order to maintain the correct system conditions.

Water Regulating Valves

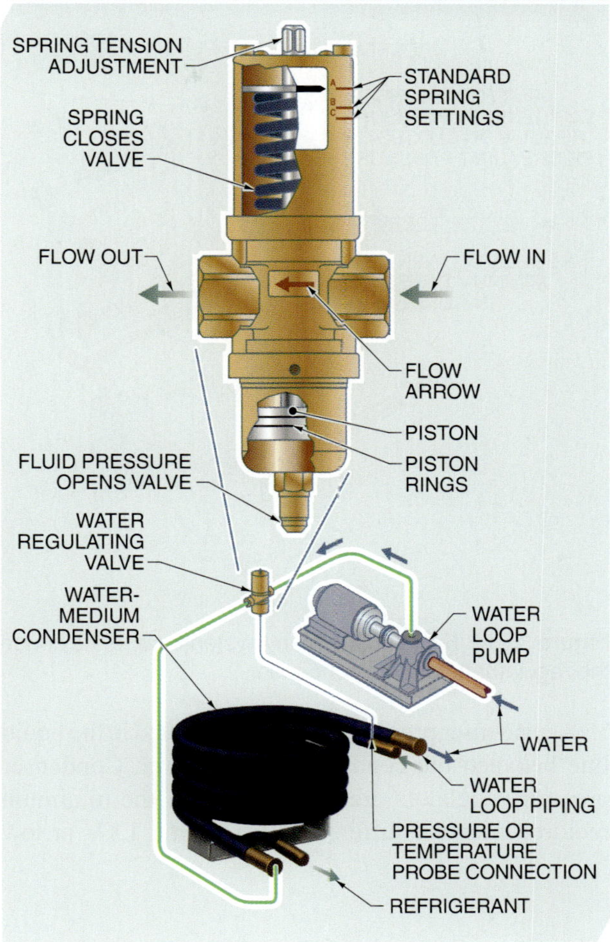

Figure 35-11. As the temperature and pressure of a water-medium condenser increase, the actuator of the water regulating valve opens and permits more water flow to cool the condenser.

Water Regulating Valves

Another operating control is the water regulating valve used on water-medium condenser systems. They are often used on medium-size package refrigeration systems. A water regulating valve consists of a valve body in the water loop piping to the condenser. The fluid power actuator on the valve body consists of a diaphragm or piston and an adjustable spring that is used to close the valve. **See Figure 35-11.**

Either a temperature or pressure connection from the condenser is connected to the valve actuator. As the temperature and pressure of the condenser increase, the valve actuator opens the valve and permits more water flow to cool the condenser. The setpoint can be adjusted with the manifold gauge attached to the system.

A water regulating valve may become fouled by sludge and sediment from the water loop piping. A strainer is often installed in the water loop piping to catch any contaminants.

Condensing Pressure Regulators

Some commercial air conditioning and refrigeration systems with air-medium condensers operate at low ambient temperatures. This presents a problem because the condenser pressure will drop and the column of liquid refrigerant at the thermostatic expansion valve (TXV) could be lost. Therefore, condenser pressure regulators are used to regulate the pressure. A *condenser pressure regulator* is a device used to maintain a sufficiently high condenser and receiver pressure when low ambient temperatures are present to guarantee that an expansion device receives liquid refrigerant. **See Figure 35-12.**

Condensing Pressure Regulators

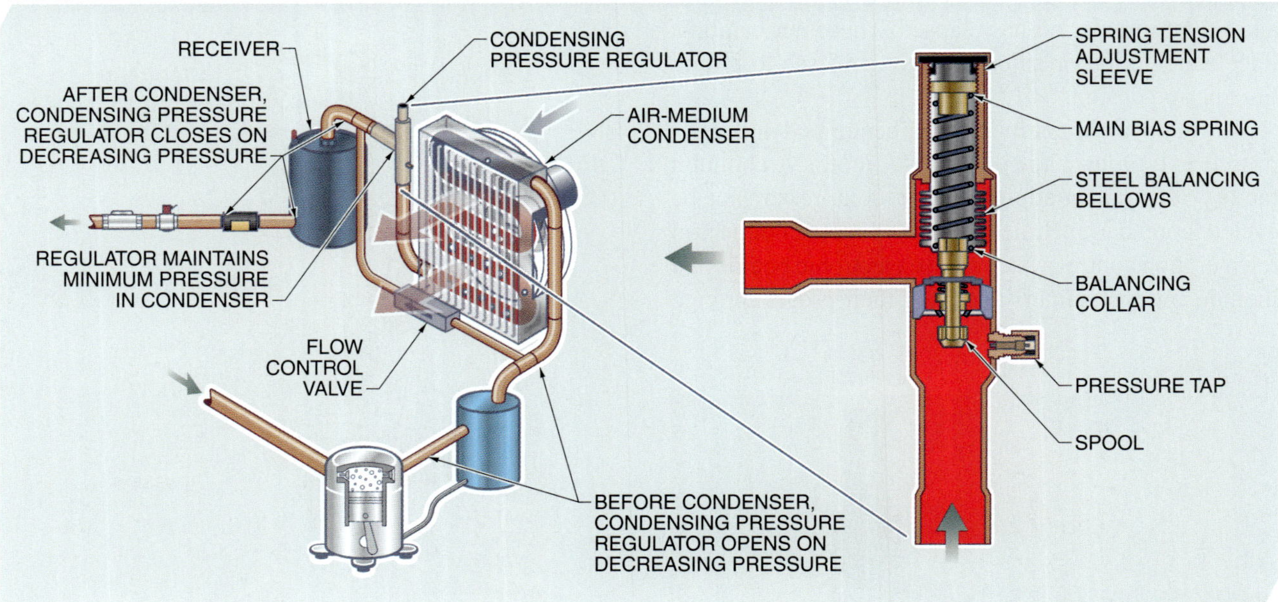

Figure 35-12. Even though many systems place the condensing pressure regulator after the condenser, it is possible to find the regulator before the condenser.

A condenser pressure regulator is located in the liquid line between the condenser and receiver. Condenser pressure regulators are set to maintain the minimum required pressure differential across the TXV at low ambient temperatures and/or varying load conditions. Therefore, the temperature in the space or in the refrigerated cabinet can be maintained regardless of the ambient temperature.

UNIT 36

Refrigerant Principles

An HVAC technician must understand the characteristics of refrigerants. Refrigerants are identified according to their molecular makeup. Refrigerants are classified by thermodynamic properties, chemical structure, flammability, and toxicity. Due to the harmful effects of refrigerants, the Environmental Protection Agency (EPA) regulates all refrigerants. Lubricating oils used in a refrigeration system must be compatible with the refrigerant and components of the system.

OBJECTIVES

- Explain the importance of having desirable characteristics in refrigerants.
- Describe refrigerant types and commonly used refrigerants.
- Identify features of refrigerant oils.

REFRIGERANTS

A *refrigerant* is a substance that moves heat by changing state from a liquid to a gas when absorbing heat and from a gas to a liquid when rejecting heat. Refrigerants absorb heat by changing state from a liquid to a vapor and reject heat by changing state from a vapor to a liquid. **See Figure 36-1.** Refrigerants typically have a low boiling point, which allows them to receive heat energy from a substance that needs to be cooled, such as the air in a building space or refrigerated cabinet.

Refrigerant Properties

The first refrigerant used was ice. Ice was used to keep food cold in the summer. To prevent ice from melting quickly, it was often stored underground and insulated with sawdust or other substances. Cold water was also used to cool products.

When mechanical refrigeration was developed in the early 1900s, other substances were needed to transfer heat in a refrigeration system. Today, there are dozens of refrigerant types with different chemical components and functions. In general, all refrigerants should meet the following criteria:

- low boiling point
- low freezing point
- nontoxic
- high critical point
- nonflammable
- noncorrosive
- chemically stable
- miscible with oil
- low compression ratio
- high latent heat
- low specific volume

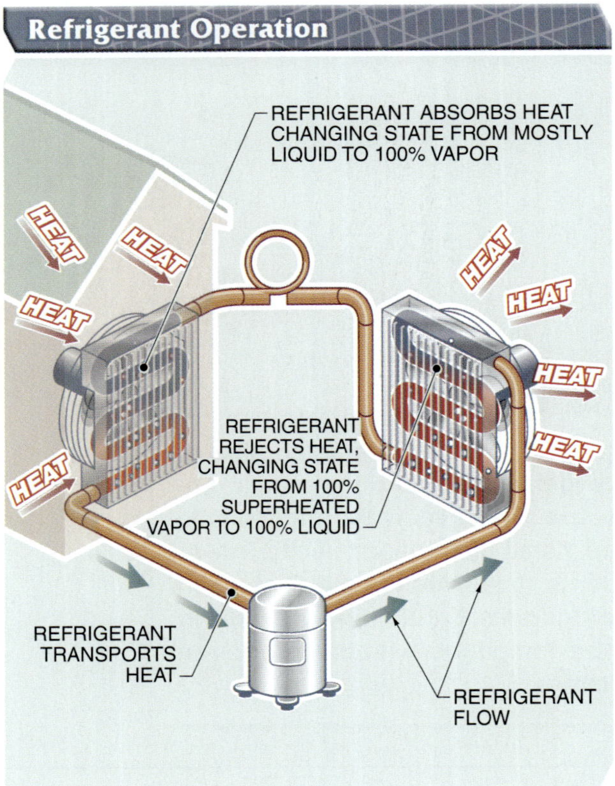

Figure 36-1. Refrigerants transport heat but must change state from a liquid to a vapor when absorbing heat and change state from a vapor to a liquid when rejecting heat.

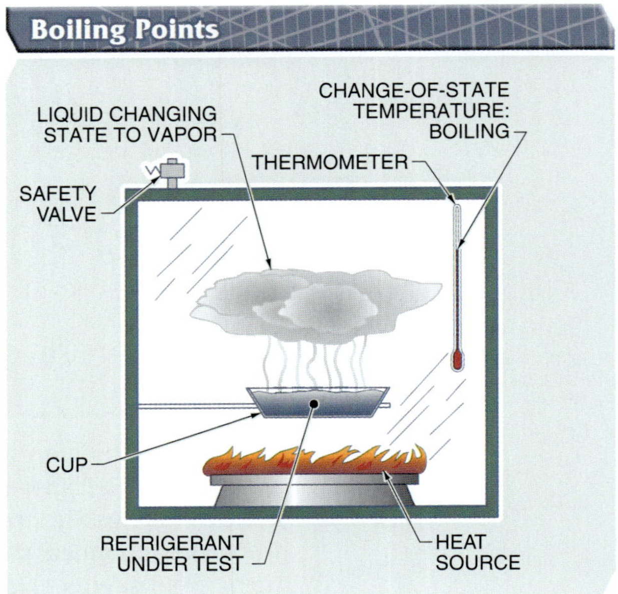

Figure 36-2. The refrigerant boiling point (temperature at which a refrigerant changes state from a liquid to a gas) should be 20°F or more below the temperature of the air or water medium to be cooled at the evaporator.

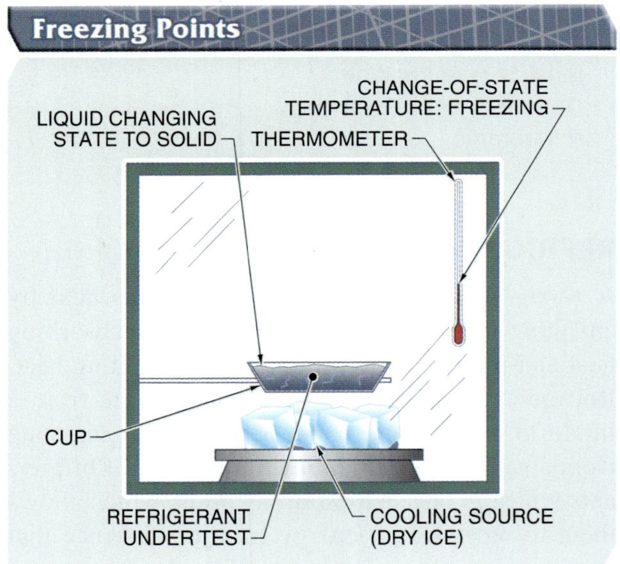

Figure 36-3. The refrigerant freezing point (temperature at which a refrigerant changes state from a liquid to a solid) should be low to ensure that a refrigerant cannot become a solid during system operation.

Low Boiling Point. A *boiling point* is the temperature at which a substance changes state from a liquid to a gas (vapor). The boiling temperature of a substance depends upon the pressure that the liquid exists in. **See Figure 36-2.** A refrigerant must have a boiling temperature that is lower than the air or medium it is being used to cool. In a product-freezing system, this temperature and pressure may be extremely low. In many cases, the boiling point of the refrigerant must be 20°F or more below the temperature of the air or water to be cooled at the evaporator. The design of most systems keeps the evaporator pressure above atmospheric pressure to eliminate the possibility of air leaking into the refrigeration system.

Low Freezing Point. A *freezing point* is the temperature at which a substance changes state from a liquid to a solid. **See Figure 36-3.** Refrigerants should have low freezing points. Refrigeration systems are designed so that the refrigerant is either a liquid or gas (vapor) but never a solid. A low freezing temperature helps to ensure that a refrigerant does not become a solid.

Nontoxic. *Toxicity* is a condition that occurs when a substance contains harmful components that may cause health problems. Refrigerants should be nontoxic, meaning that exposure to refrigerants should not adversely affect human health. To facilitate safety, all refrigerants, regardless of amount or type, must be handled with care. **See Figure 36-4.** When refrigerants are permitted to leak out and mix with air, they can become dangerous. Common refrigerant safety rules include the following:

- Store refrigerant cylinders in clean, dry areas with no direct sunlight. Do not allow refrigerant containers to be heated above 125°F.
- Comply with all codes regarding storage, cylinder types, quantities, and proper labeling.
- Make sure that refrigerant cylinders are properly secured at all times.
- Do not allow refrigerants to come into contact with skin, which would cause frostbite. Use proper gloves and face protection.
- Keep refrigerant vapors from accumulating in a confined or low area. Refrigerant vapors can displace oxygen and cause severe health problems, including asphyxiation and death, if the oxygen levels become too low.
- Do not use oxygen or compressed air to pressurize refrigeration systems. When combined with system oil, the mixture may explode.
- Oil spills must be cleaned up immediately.

To sense any drop in the oxygen levels due to a refrigerant leak, refrigeration mechanical rooms have oxygen-deprivation sensors that sense and monitor oxygen levels. Since refrigerants are heavier than air, the inlet ports from the sensors are often at floor level. Self-contained breathing apparatus (SCBA) equipment must be available for use in the case of a serious refrigerant leak.

Figure 36-4. Some refrigerants are toxic, and when permitted to leak out of containment and mix with the air, they become dangerous.

TECH TIP

Refrigerants must be stored inside fire-resistant tanks and kept dry and cool. Extremely hot temperatures cause a change in refrigerants that could lead to an explosion. Storage vessels, pumps, and pipes must be designed to be leakproof.

Refrigerants are categorized according to their flammability as well as toxicity. Per ANSI/ASHRAE Standard 34-2010, *Designation and Safety Classification of Refrigerants,* refrigerants are classified into matrix groups depending on flammability and toxicity. **See Figure 36-5.** In this matrix, the letters A and B indicate toxicity, with B being the higher level of toxicity. The numbers 1, 2, and 3 indicate flammability, with 3 being the highest level of flammability.

Refrigerant Safety Groups

	Lower Toxicity	Higher Toxicity
No flame propagation	A1	B1
Lower flammability	A2	B2
Higher flammability	A3	B3

Figure 36-5. Per ANSI/ASHRAE Standard 34-2010, *Designation and Safety Classification of Refrigerants,* refrigerants are classified into safety (matrix) groups depending on flammability and toxicity.

High Critical Point. In a substance, the *critical point* is the temperature and pressure above which the substance does not change state, regardless of how much heat is added or removed. A high critical point is needed in a refrigeration system to conserve energy. When a system operates close to the critical point, the compressor must perform more work. **See Figure 36-6.**

Critical Points

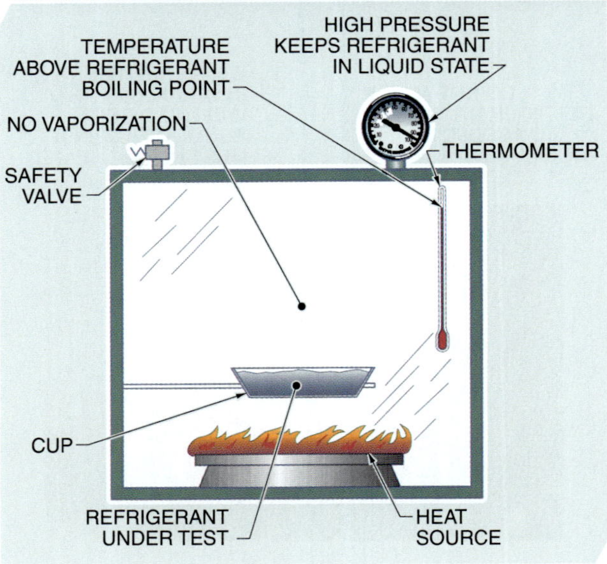

Figure 36-6. A critical point is the temperature and pressure above which a substance does not change state. A high critical point saves energy, but too high of a critical point makes the compressor work especially hard.

Nonflammable. *Flammability* is the ability of a substance to support ignition. A *nonflammable substance* is a substance that is not easily set on fire or is incapable of burning. Refrigerants should not be easily ignited. Ignition of a refrigerant should only be possible if a refrigerant comes into direct contact with an open flame, for example, in the soldering or brazing process.

Noncorrosive. *Corrosion* is a condition that occurs when oxygen in air reacts with and breaks down metals (electrochemical oxidation). Refrigerants should be noncorrosive so that they do not cause the deterioration of internal metallic parts. Corrosion may be caused not only by the presence of a refrigerant, but also in a reaction with other substances such as refrigerant oil.

Chemically Stable. *Stability* is a condition that occurs when a substance can withstand changing conditions without breaking down. Refrigerants in a system are expected to operate under varying temperature and pressure conditions for years. **See Figure 36-7.** When a refrigerant is not chemically stable, it may break down into its separate components. These chemical components can damage or even destroy the refrigeration system and will cause system inefficiency.

Chemically Stable Substances

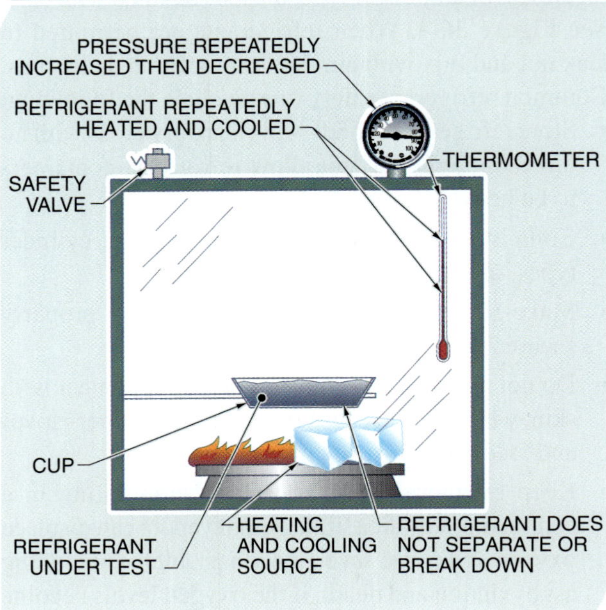

Figure 36-7. Substances that are chemically stable can withstand changing conditions without breaking down into separate chemicals. Separate chemical components can damage or destroy a refrigeration system.

Oil Miscibility. Refrigerants must also be miscible with the lubricating oil in a refrigeration system. *Miscibility* is the ability of a substance to mix with another substance. When a refrigerant is not miscible with oil, oil does not readily circulate through the refrigeration system. Circulation issues can cause problems such as a lack of compressor lubrication or the accumulation of oil in one part of the system. **See Figure 36-8.** Such problems can cause system damage and equipment inefficiency.

Low Compression Ratio. Refrigerant is introduced into a compressor at a low pressure and discharged at a high pressure. *Compression ratio* is the mathematical relationship between the pressure on the high-pressure side to the pressure on the low-pressure side of a refrigeration system. **See Figure 36-9.** This relationship is expressed as a ratio, such as 3.5:1.

As the compression ratio increases, the compressor must do more work. While a compressor is designed to operate at a specific compression ratio, system problems may cause the compression ratio to increase and cause eventual compressor damage or failure.

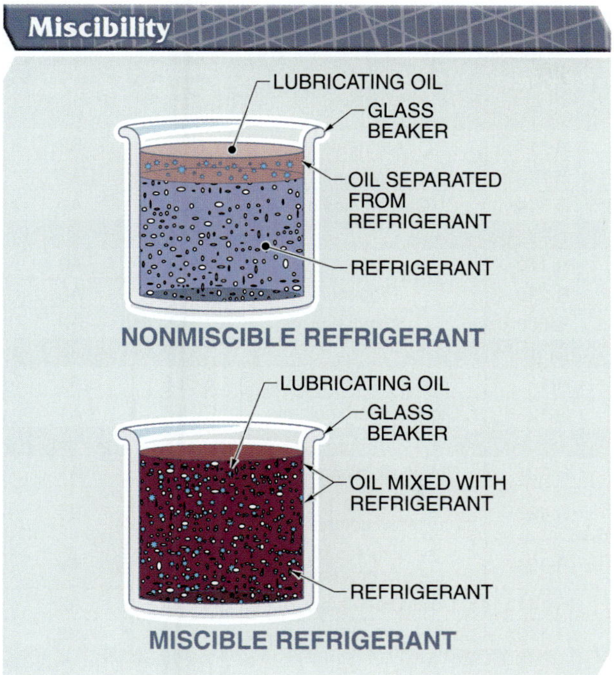

Figure 36-8. Miscibility is the ability of substances, such as refrigerants and oils, to mix together.

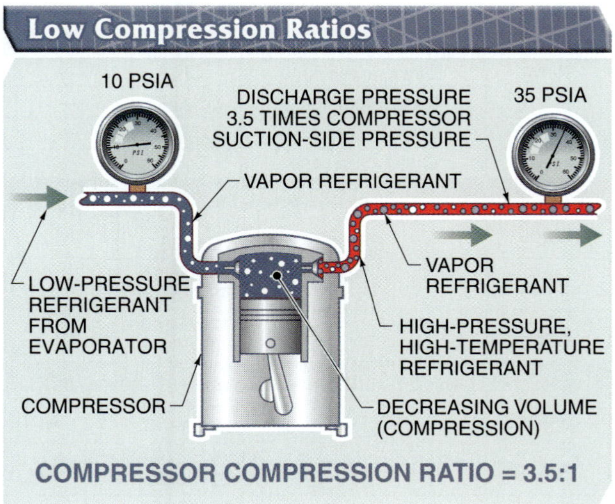

Figure 36-9. The higher the compression ratio with which a compressor must operate, the harder the compressor must work.

High Latent Heat. Latent heat is the heat applied to a substance (refrigerant) that causes it to change state. Refrigerants absorb heat in the evaporator by changing state from a liquid to a gas. Refrigerants reject heat in the condenser by changing state from a gas to a liquid. **See Figure 36-10.** These state changes require a change in latent heat. Efficient refrigerants can contain large amounts of latent heat. Refrigerants with high latent heat properties can perform the same amount of refrigeration effect with smaller amounts of refrigerant than refrigerants with low latent heat properties.

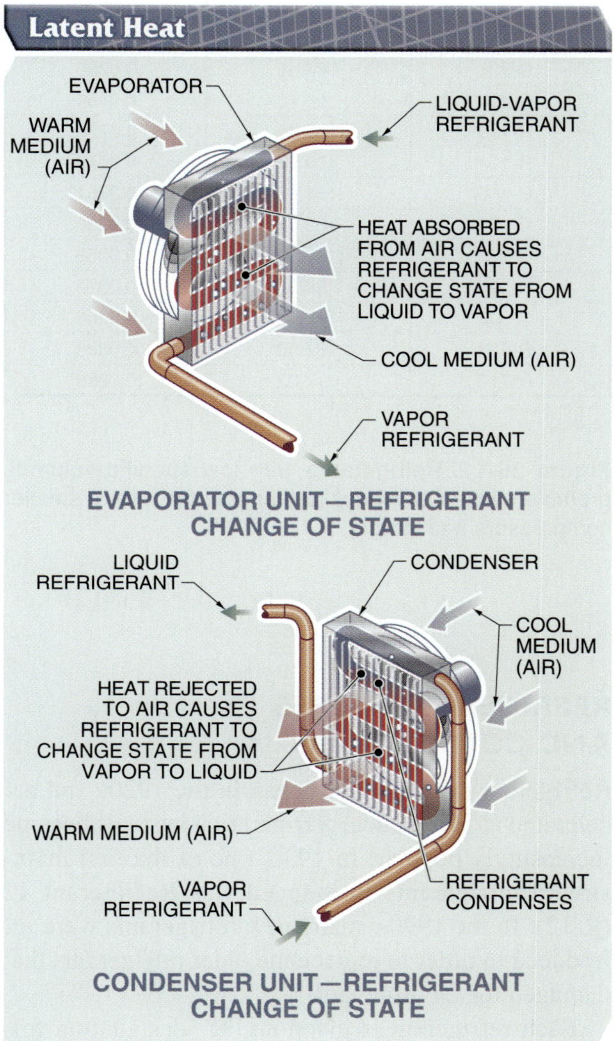

Figure 36-10. Latent heat is the heat applied to a refrigerant that causes the refrigerant not to change temperature, but to change state from a liquid to a gas or from a gas to a liquid.

Low Specific Volume. *Specific volume* is the volume that one pound of a gas substance occupies at a particular temperature and pressure condition; expressed in cubic feet per pound (ft^3/lb). When a refrigerant has a low specific volume, the compressor is required to circulate smaller volumes of the refrigerant to move the same weight as a refrigerant with a high specific volume. When using refrigerant with low specific volume, the refrigerant promotes system efficiency and enables the use of smaller compressors and motors. **See Figure 36-11.**

Specific Volume

Refrigerant	Molecular Mass*	Specific Volume†
R-11	137.37	0.0289
R-12	120.91	0.0287
R-13	104.46	0.0277
R-14	88.00	0.0256
R-22	86.46	0.0305
R-40	50.48	0.0454
R-113	187.39	0.0278
R-114	170.92	0.0275
R-115	154.47	0.0261
R-134a	102.03	0.0290
R-142b	100.50	0.0368
R-500	99.31	0.0323
R-502	111.63	0.0286
R-503	87.50	0.0326
R-717	17.02	0.0680

* in g/molecule
† in ft^3/lb

Figure 36-11. Refrigerants with low specific volumes promote system efficiency and enable the use of smaller compressors and motors.

REFRIGERANT TYPES AND COMPOSITION

Refrigerant development began in the 1920s and accelerated as mechanical refrigeration systems became increasingly popular. In 1930, one of the first mainstream refrigerants introduced was Refrigerant 12 (R-12). In the 1990s, many new refrigerants were introduced in order to replace the older refrigerants that damaged the environment.

Each refrigerant is given an "R" designation followed by a set of numbers. However, a manufacturer may give a name to its version of a particular refrigerant, such as ACME-12 or PUREBLEND-22. Despite any name differences, refrigerants are the same when the numbers following the name or the "R" are the same. Different refrigerants cannot be introduced into the same system at the same time, and no one refrigerant is perfect for use in all systems.

Most refrigerants consist of different combinations of hydrogen, carbon, chlorine, and fluorine. **See Figure 36-12.** R-717, also known as ammonia, is radically different from other refrigerants and is used in commercial deep freezers. R-717 has its own set of safety rules, handling certifications, and precautions.

Refrigerants

Refrigerant	Chemical Name	Molecular Formula	Safety Classification
Inorganic Compounds			
R-717	Ammonia	NH_3	B2
R-718	Water	H_2O	A1
R-744	Carbon dioxide	CO_2	A1
Organic Compounds			
R-170	Ethane	CH_3CH_3	A3
R-290	Propane	$CH_3CH_2CH_3$	A3
R-600A	Isobutane	$CH(CH_3)_2CH_3$	A3
Chlorofluorocarbons			
R-11	Trichlorofluoromethane	CCl_3F	A1
R-12	Dichlorofluoromethane	CCl_2F_2	A1
Hydrochlorofluorocarbons			
R-22	Chlorodifluoromethane	$CHClF_2$	A1
R-123	2, 2-dichloro-1, 1, 1-trifluoroethane	$CHCl_{12}CF_3$	B1
R-124	2-chloro-1, 1, 1, 2-tetrafluoroethane	$CFC_1F\text{-}CF_3$	A1
R-141b	1, 1-dichloro-1-fluoroethane	CH_3CCl_2F	A2
R-142b	1-chloro-1, 1-difluoroethane	CH_3CClF_2	A2
Hydrofluorocarbons			
R-23	Trifluoromethane	CHF_3	A1
R-32	Difluoromethane	CH_2F_2	A2
R-125	Pentafluoromethane	CHF_2CF_3	A1
R-134a	1, 1, 1, 2-tetrafluoroethane	CF_3CH_2F	A1
R-152a	1, 1-Difluoroethane	CF_3CHF_2	A2
Perfluorocarbons			
R-14	Tetrafluoromethane	CF_4	----
R-116	Hexafluoromethane	CF_3CF_3	----
R-218	Octafluoropropane	C_3F_8	----
R-318	Octafluorobutane	C_4F_8	----
Azeotropic Mixtures			
R-500	----	R12/R152a (73.8/26.2)	A1
R-502	----	R22/R115 (48.8/51.2)	A1
R-507	----	R125/R143a (50/50)	A1
Zeotropic Mixtures			
R-404A	----	R125/R143a/R134a (44/52/4)	A1
R-407C	----	R32/R125/R134a (23/25/52)	A1
R-410A	----	R32/R125 (50/50)	A1

Figure 36-12. Refrigerants consist of various combinations of hydrogen, carbon, chlorine, and fluorine. Despite name differences, refrigerants are the same when the numbers following the name or the letter "R" are the same.

Chlorofluorocarbon Refrigerants

A *chlorofluorocarbon (CFC) refrigerant* is a type of refrigerant that consists of chlorine, fluorine, and carbon. From the 1930s until the early 1990s, CFC refrigerants were the most common refrigerants in use. **See Figure 36-13.** Common examples of CFC refrigerants include R-11 and R-12. R-11 was used as a refrigerant in low-pressure water chillers. R-11 was also used in other industries as a solvent and cleaner of electrical parts.

Chlorofluorocarbon (CFC) Refrigerants

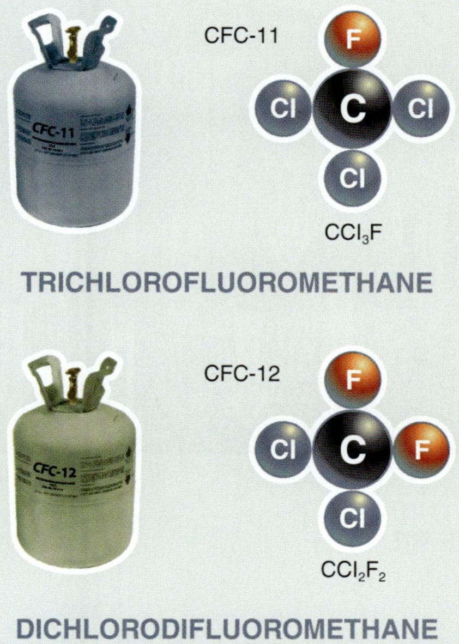

Refrigerant	Name
R-13	Chlorotrifluoromethane
R-113	Trichlorotrifluoroethane
R-500	Dichlorodifluoromethane/ Difluoroethane
R-502	Chlorodifluoromethane/ Chloropentafluoromethane
R-503	Chlorotrifluoromethane/ Trifluoromethane

Figure 36-13. CFC refrigerants consist of chlorine, fluorine, and carbon atoms, but by law they are no longer allowed to be manufactured.

TECH TIP

In the past, R-11 refrigerant was used in large quantities in large, low-pressure centrifugal chiller systems. Unfortunately, these systems had high leak rates, with a 30% refrigerant loss rate per year being common. These systems were targeted by the international community (Montreal Protocol) in the 1990s and have been replaced or retrofitted with improved purge systems.

R-12 was used for years in domestic and commercial refrigerators and freezers, as well as in automotive air conditioning. Both R-11 and R-12 are outstanding in performance, have low toxicity, and are chemically stable. However, the Montreal Protocol categorizes these refrigerants as Class I, meaning they have the highest ozone-depleting potential (ODP) value.

R-11 and R-12 refrigerants can be recovered into marked recovery cylinders and reused if added to a system in the same building from which the refrigerant was recovered.

The manufacturing of Class I refrigerants was stopped in the United States in 1995 because they cause great damage to the Earth's ozone layer when released into the atmosphere. **See Figure 36-14.** While CFC refrigerants are no longer manufactured, they are legal for continued use. However, any significant leaks to a system must be documented and the EPA notified.

Depleting Earth's Ozone Layer Media Clip

Hydrochlorofluorocarbon Refrigerants

A *hydrochlorofluorocarbon (HCFC) refrigerant* is a type of refrigerant that consists of hydrogen, chlorine, fluorine, and carbon. **See Figure 36-15.** R-22, a type of HCFC refrigerant, has been used throughout the commercial and residential air conditioning sector for many years. The vast majority of residential central air conditioning units used R-22 until recently. Other examples of HCFC refrigerants include R-123 and R-124. R-124 is not often used, but R-123 is used as a replacement refrigerant for R-11.

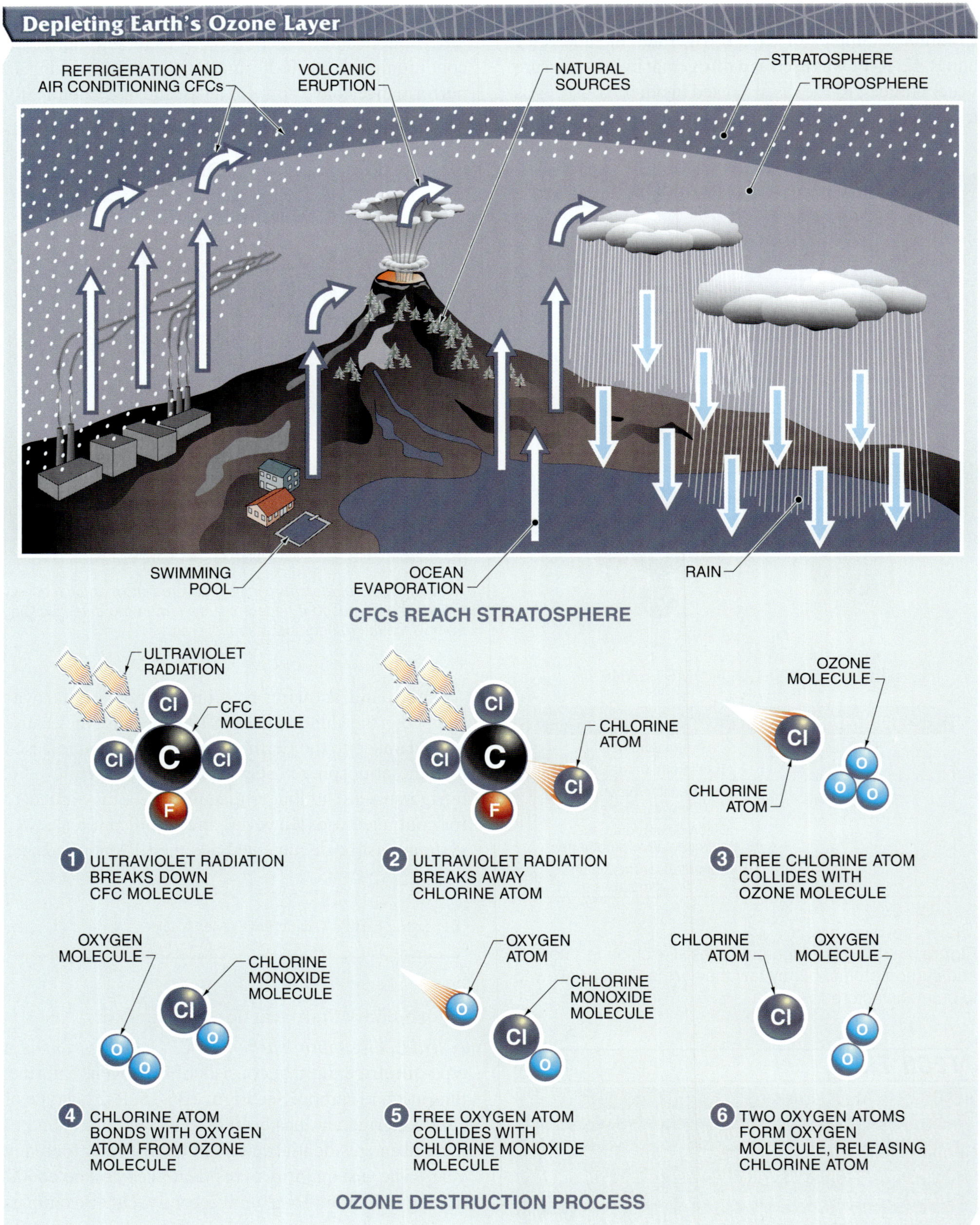

Figure 36-14. In the 1970s, CFC refrigerants were discovered to create great damage to the Earth's ozone layer when released into the atmosphere.

Hydrochlorofluorocarbon (HCFC) Refrigerants

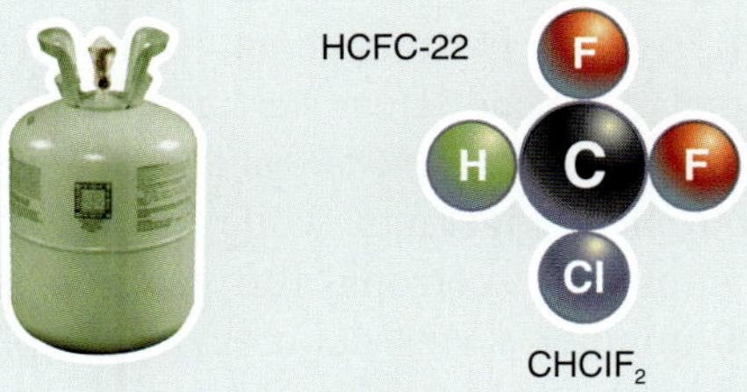

CHLORODIFLUOROMETHANE

Refrigerant	Name
R-123	Dichlorotrifluoroethane
R-124	Chlorotetrafluoroethane
R-401A	R-22 (53%)/R-152a (13%)/ R-124 (28%)
R-402A	R-22 (38%)/R-125 (60%)/ R-290 (2%)
R-403B	R-22 (56%)/R-218 (39%)/ R-290 (5%)
R-406A	R-22 (56%)/R-600A (4%)/ R-142b (41%)
R-408A	R-22 (47%)/R-143a (46%)/ R-125 (7%)
R-409A	R-22 (60%)/R-124 (25%)/ R-142b (15%)

Figure 36-15. HCFC refrigerants consist of hydrogen, chlorine, fluorine, and carbon atoms that are not as damaging to the ozone layer as CFC refrigerants.

Hydrofluorocarbon (HFC) Refrigerants

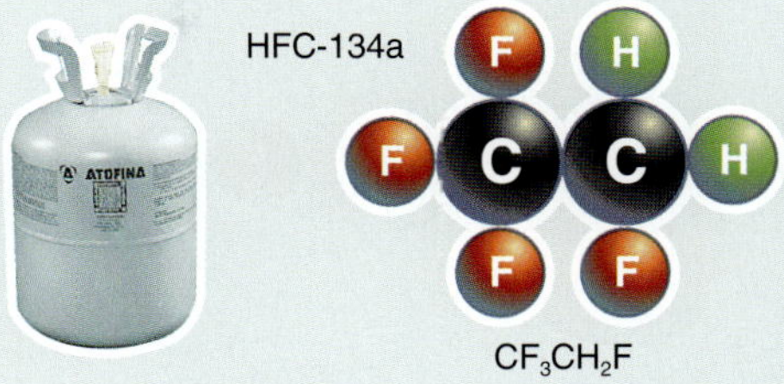

TETRAFLUOROETHANE

Refrigerant	Name
R-23	Trifluoromethane
R-404A	R-125 (44%)/R-143a (52%)/ R-134a (4%)
R-407A	R-32 (20%)/R-125 (40%)/ R-134a (40%)
R-410A	R-32 (50%)/R-125 (50%)
R-416A	R-134a (59%)/R-124 (39%)/ R-600 (2%)
R-507	R-125 (50%)/R-143a (50%)
R-508A	R-23 (39%)/R-116 (61%)

Figure 36-16. HFC refrigerants consist of hydrogen, fluorine, and carbon atoms that pose a small risk to the environment and are not categorized as Class I or Class II refrigerants.

The EPA has determined that HCFCs are Class II refrigerants (medium danger), meaning HCFCs are not as damaging to the ozone layer as Class I refrigerants. HCFC refrigerants are being phased out over time and will no longer be manufactured in the United States as of 2020.

Hydrofluorocarbon Refrigerants

A *hydrofluorocarbon (HFC) refrigerant* is a type of refrigerant that consists of hydrogen, fluorine, and carbon. **See Figure 36-16.** R-134a is the most widely known refrigerant in the HFC group. R-134a is used as a replacement refrigerant for the CFC refrigerant R-12. Other HFC refrigerants include R-23, R-32, and R-152a. According to the EPA, HFCs pose a small risk of damage to the environment and are not categorized as Class I or Class II refrigerants.

Perfluorocarbon Refrigerants

A *perfluorocarbon (PFC) refrigerant* is a type of refrigerant that consists of only fluorine and carbon. Examples of PFC refrigerants include R-14, R-116, R-218, and R-318. PFC refrigerants do not destroy the ozone layer but have an extremely high global warming potential. **See Figure 36-17.** PFC refrigerants are not typically used in commercial systems.

Blended Refrigerants

A blended refrigerant is created when two or more refrigerants are mixed together. A *binary blend* is a blend of two refrigerants. **See Figure 36-18.** A *ternary blend* is a blend of three refrigerants. Binary blends, such as R-500 and R-502, have been used in product freezer applications. R-500 is a blend of R-12 and R-152a. R-502 is a blend of R-22 and R-115. R-407c is a ternary blend consisting of R-32, R-125, and R-143a. R-407c is a refrigerant used as a replacement for R-22.

Perfluorocarbon (PFC) Refrigerants

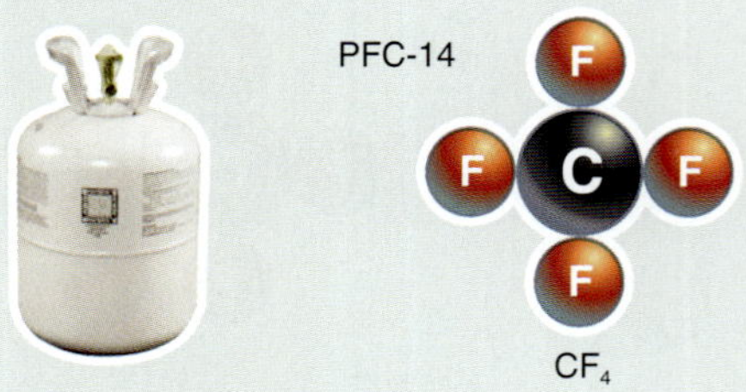

TETRAFLUOROMETHANE

Refrigerant	Name
R-116	Hexafluoroethane
R-218	Octafluoropropane
R-318	Octafluorobutane

Figure 36-17. PFC refrigerants consist of fluorine and carbon atoms that do not destroy the ozone layer but have the potential of contributing to global warming.

Blended Refrigerants

Refrigerant	Name
R-401A	Chlorodifluoromethane (HCFC-22) Difluoroethane (HFC-152a) Chlorotetrafluoroethane (HCFC-124)
R-402A	Chlorodifluoromethane (HCFC-22) Pentafluoroethane (HFC-125) Propane (R-290)
R-407C	Difluoromethane (HFC-32) Pentafluoroethane (HFC-125) Tetrafluoroethane (HFC-134a)
R-408A	Pentafluoroethane (HFC-125) Trifluoroethane (HFC-143a) Chlorodifluoromethane (HCFC-22)
R-500	Dichlorodifluoromethane (CFC-12) Difluoroethane (HFC-152a)
R-502	Chlorodifluoromethane (HCFC-22) Chloropentafluoroethane (CFC-115)

Figure 36-18. Blended refrigerants are refrigerants made from two or more refrigerants mixed together. A binary blend is a blend of two refrigerants. A ternary blend is a blend of three refrigerants.

R-410a is the popular replacement for R-22 in modern residential air conditioning systems. R-410a is a binary blend of R-32 and R-125. R-410a does not contribute to ozone depletion. It does however, have a much higher pressure than R-22. For instance, R-410a has a pressure of 416.9 psi at 120°F instead of 260 psi as does R-22. The components and gauges used in the system must be designed to handle this increased pressure. Safety training is recommended for all technicians who use R-410a.

Azeotropic and Zeotropic Refrigerants. An *azeotropic refrigerant* is a refrigerant that is formed when refrigerants are blended together and act like an entirely new and separate refrigerant. **See Figure 36-19.** A *zeotropic refrigerant* is a refrigerant that is formed when refrigerants are blended together but remain in their separate component states. R-500 and R-502 are examples of azeotropic refrigerants, while R-410a and R-407c are examples of zeotropic refrigerants. In general, the 500-number series denotes azeotropic blends, while the 400-number series denotes zeotropic blends.

Azeotropic and Zeotropic Refrigerants

Refrigerant	Molecular Formula	Composition by Weight-Percentage
Azeotropic		
CFC-500	CCl_2F_2/CH_3CHF_2	R-12 (73.8%)/ R-152 (26.2%)
CFC-502	$CHClF_2/CClF_2CF_3$	R-22 (48.8%)/ R-115 (51.2%)
CFC-503	$CHF_3/CClF_3$	R-23 (40.1%)/ R-13 (59.9%)
HFC-507A	CHF_2CF_3/CH_3CF_3	R-125 (50%)/ R-143a (50%)
PFC-508A	CHF_3/CF_3CF_3	R-23 (39%)/ R-116 (61%)
PFC-508B	CHF_3/CF_3CF_3	R-23 (46%)/ R-116 (54%)
PFC-509A	$CHCF_2/CF_3CFCF_3$	R-22 (44%)/ R-218 (56%)
Zeotropic		
HCFC-401A	$CHClF_2/CH_3CHF_2/$ $CHClFCF_3$	R-22 (53%)/ R-124 (34%)/ R-152a (13%)
HCFC-401B	$CHClF_2/CH_3CHF_2/$ $CHClFCF_3$	R-22 (61%)/ R-124 (28%)/ R-152a (11%)
HCFC-402A	$CHClF_2/CHF_2CF_3/$ C_3H_8	R-22 (38%)/ R-125 (60%)/ R-290 (2%)
HFC-407C	$CH_2F_2/CHF_2CF_3/$ CF_3CH_2F	R-32 (23%)/ R-125 (25%)/ R-134a (52%)
HFC-408A	$CHF_2CF_3CH_3CF_3/$ $CHClF_2$	R-22 (47%)/ R-125 (7%)/ R-143a (46%)

Figure 36-19. An azeotropic refrigerant is a blended refrigerant that forms an entirely new and separate refrigerant. A zeotropic refrigerant is a blended refrigerant in which the individual refrigerants remain in their separate component states.

Fractionation and System Glide. When zeotropic refrigerant leaks from a system, fractionation occurs. *Fractionation* is a problem that occurs when one refrigerant of a blend leaks out of a refrigeration system at a different rate than the other refrigerants of the blend due to the varying temperature-pressure characteristics of the refrigerants. As one refrigerant leaks out before another, the temperature-pressure characteristic of the remaining refrigerant changes the operation of the system. **See Figure 36-20.**

Fractionation and System Glide

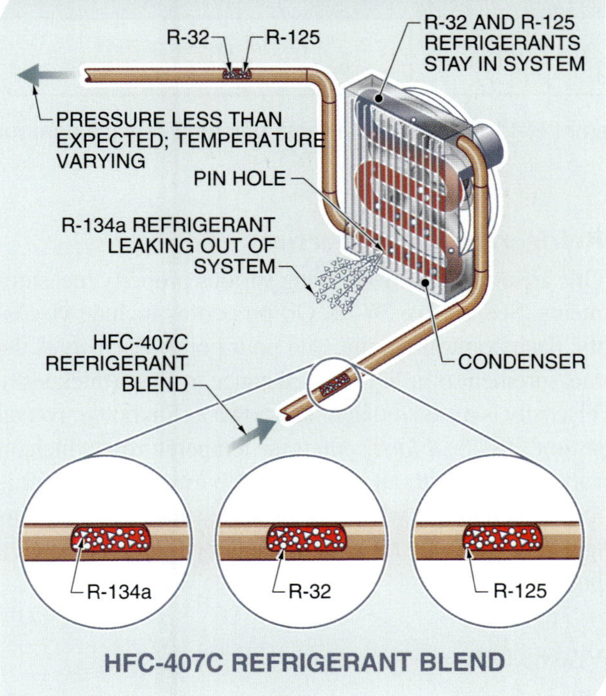

Figure 36-20. Fractionation is a problem that occurs when one refrigerant of a blend leaks out of a system at a different rate than other refrigerants of the blend due to their different temperature-pressure characteristics. System glide is a temperature-pressure change in the system due to fractionation.

System glide is a temperature-pressure change in a system due to fractionation. Dealing with system glide often requires adding the liquid of the missing refrigerant to the system to bring the blend characteristic back to where it should be. This only occurs in large volume refrigerant systems (heavy commercial and industrial). Only an experienced technician qualified to work on the system with the problem and with the refrigerant blend in use should attempt to add refrigerant to a system to repair a blend.

Fractionation and System Glide Media Clip

Substitute Refrigerants

Because CFC refrigerant production has ceased, it may be desirable to replace the CFC refrigerants in many systems, especially systems that use R-11 and R-12. The most common replacement refrigerant for R-11 is R-123. **See Figure 36-21.** The most common replacement refrigerant for R-12 is R-134a.

In large systems, the replacement is often in the off-season or while back-up equipment is available. In extreme cases, chillers and compressors may be rented while the refrigerant replacement occurs. Concerns and issues to look for during refrigerant replacement include a decrease in performance, oil that needs to be replaced, and any adverse effects on the system from the new refrigerant and oil.

Some systems may lose 10% or more of their performance capabilities after the old refrigerant is replaced. This should be evaluated on a case-by-case basis. Decreasing performance leads to longer equipment run times, higher energy use, and equipment wear and tear.

Each refrigerant type uses a specific oil type. A refrigerant change-out may necessitate an oil change-out as well. An oil change-out means that the old oil type has to be thoroughly removed before the new oil is added. Internal system parts, such as O-rings and metals, may need to be checked to ensure that they will not be damaged by the new replacement oil or refrigerant.

REFRIGERATION SYSTEM OILS

In addition to refrigerants, oil is present in a refrigeration system. Oil is needed to lubricate the moving parts of the compressor. When the required amount of oil is not present, the compressor may lose lubrication and eventually fail.

While some oil remains in the crankcase, oil also circulates with the refrigerant in the system. Both the refrigerant and oil must be compatible to ensure proper circulation. Proper piping and trap installation are needed to reduce the possibility of oil collecting in a specific part of the refrigeration system.

Various types of oils are used for the different categories of refrigerants. **See Figure 36-22.** For example, CFC refrigerants, which are no longer manufactured, use mineral oils. HCFC refrigerants use alkylbenzene oils. HFC refrigerants use polyalkylene glycol or polyol esters as lubricants. Polyalkylene glycol lubricants absorb moisture, which is undesirable in some systems. It is a violation of Section 608 of the Clean Air Act to change refrigerant oil in an open system at a pressure higher than 5 psi.

Substitute Refrigerants

Section 608 Refrigerant	Application	Substitute Refrigerant Product Manufacturers	Type	Comments
Air Conditioning				
R-11	Low-pressure centrifugal chillers	DuPont — Suva 123 Honeywell — Genetron 123 Atofina — Forane 123	HCFC pure fluid	None
Low and Medium Temperature (Interim)				
R-502	Ice machines and open drives	DuPont — Suva HP80 Honeywell — Genetron HP80	HCFC/HFC blend	Higher discharge temperature and average evaporator temperature above −20°F
Medium Temperature (Interim)				
R-12	High-capacity; evaporator temperature below 15°F	DuPont — Suva MP39 Honeywell — Genetron MP39	HCFC/HFC blend	No oil change above 20°F evaporator temperature unless long piping exist
Medium Temperature (Long Term)				
R-12	New equipment	DuPont — Suva 134a Honeywell — Genetron 134a Atofina — Forane 134a	HFC pure fluid	Performs well with 20°F evaporator temperature or higher

Figure 36-21. The most common replacement for CFC refrigerant R-11 is R-123. The most common replacement for refrigerant R-12 is R-134a.

Refrigerant Oils

Refrigerant	Oil*
R-11	Polyol ester
R-12	Alkylbenzene or mineral oil
R-13	Polyol ester, alkylbenzene, or mineral oil
R-22	Polyol ester, alkylbenzene, or mineral oil
R-23	Polyol ester
R-123	Alkylbenzene or mineral oil
R-124	Alkylbenzene
R-125	Polyol ester
R-134a	Polyol ester
R-401A	Polyol ester, alkylbenzene
R-401B	Polyol ester, alkylbenzene
R-401C	Polyol ester, alkylbenzene
R-402A	Polyol ester, alkylbenzene, or mineral oil
R-402B	Polyol ester, alkylbenzene, or mineral oil
R-403B	Polyol ester, alkylbenzene, or mineral oil
R-404A	Polyol ester
R-407A	Polyol ester
R-407B	Polyol ester
R-407C	Polyol ester
R-410A	Polyol ester
R-414A	Polyol ester
R-414B	Polyol ester, alkylbenzene, or mineral oil
R-416A	Polyol ester
R-500	Polyol ester, alkylbenzene, or mineral oil
R-502	Polyol ester, alkylbenzene, or mineral oil
R-503	Polyol ester, alkylbenzene, or mineral oil
R-507A	Polyol ester
R-508A	Polyol ester
R-508B	Polyol ester
R-717	Mineral oil

* Check with compressor manufacturer for recommended lubricant

Figure 36-22. Because lubricating oil for the compressor mixes with the system refrigerant, the oil must be compatible with the refrigerant.

Refrigerant Oil Properties

Oils are evaluated by applying various property measurements. **See Figure 36-23.** Oil properties include viscosity, flash point, fire point, and pour point. *Viscosity* is the measurement of a liquid's resistance to flow (thickness). Viscosity is measured in units known as Saybolt universal seconds (SUS). *Flash point* is the temperature at which oil vapors ignite without the oil itself burning. *Fire point* is the temperature at which a substance, such as oil, sustains ignition. *Pour point* is the lowest temperature at which a liquid can flow.

Waste Oils

Oil can be added and removed from a refrigeration system as needed. Oil is typically added using a manual or mechanical oil pump. Oil is removed by draining it through an oil-removal plug in the compressor crankcase or by removing all refrigerant from the system. In most cases, the used oil that is removed is not hazardous. **See Figure 36-24.** However, when the oil has been contaminated with acids or other materials, it may be considered hazardous. Before waste oil is disposed of or burned, the EPA must be consulted regarding the oil type, contents, waste site that should be used, and whether the oil is suitable for burning. All applicable local and state codes must also be followed.

TECH TIP

Care must be taken when replacing refrigerants as the old oil may not be compatible with the new refrigerant.

Refrigerant Oil Property Tests

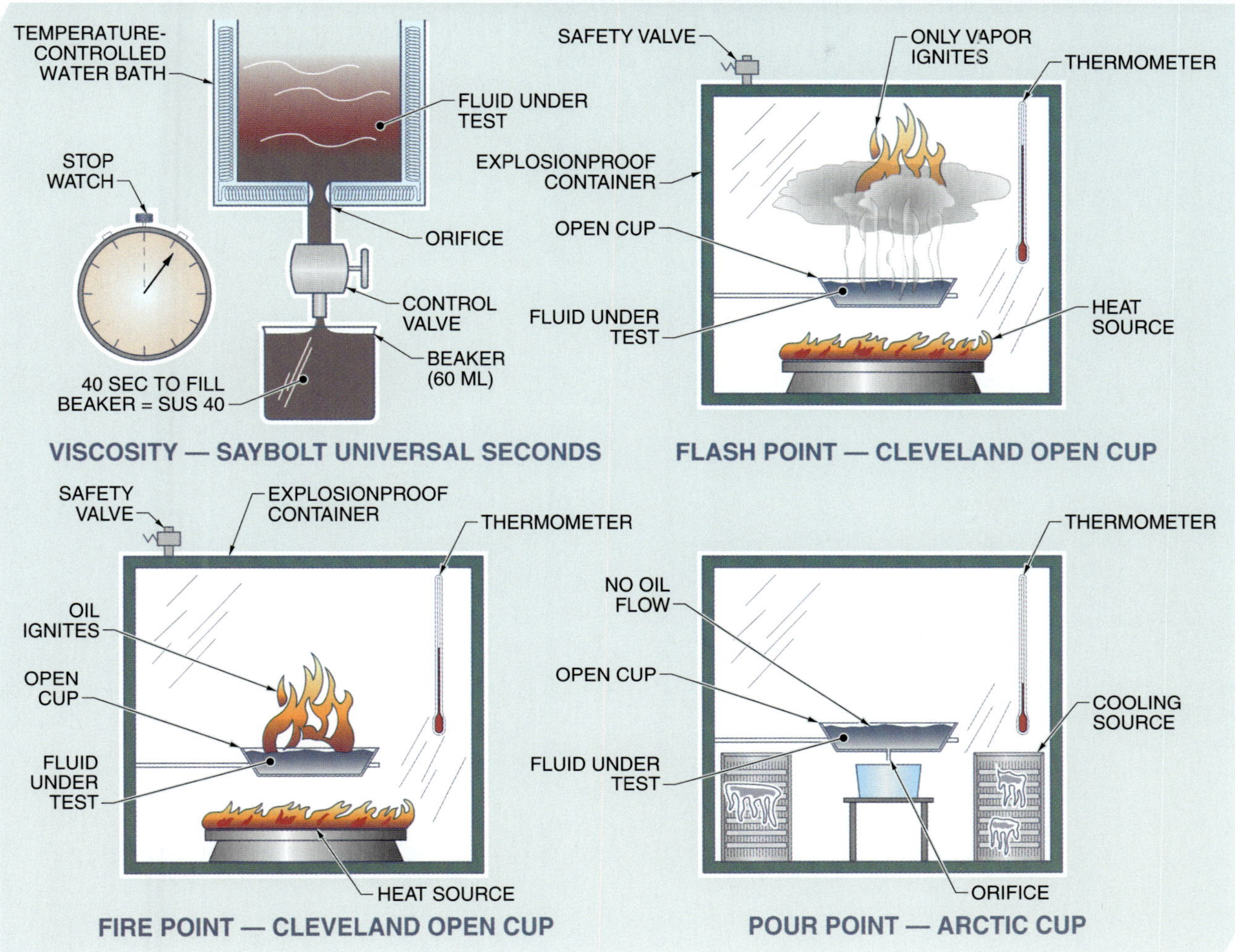

Figure 36-23. Refrigerant oil properties include viscosity, flash point, fire point, and pour point, all of which have standardized tests for measuring.

EPA Used Oil Identification

Used Oil*		Not Used Oil
• Synthetic oils—usually derived from coal, shale, or polymer-based starting material • Engine oils—typically includes gasoline and diesel engine crankcase oils and piston-engine oils for automobiles, trucks, boats, airplanes, locomotives, and heavy equipment • Transmission fluids • Refrigeration oils	• Compressor oils • Metalworking fluids and oils • Laminating oils • Industrial hydraulic fluids • Copper and aluminum wire drawing solutions • Electrical insulating oils • Industrial process oils • Oils used as buoyants	• Waste oils that are bottom clean-out wastes from virgin-fuel storage tanks, virgin-fuel oil-spill cleanups, or waste oils have not actually been used • Products such as antifreeze and kerosene • Vegetable and animal oils, even when used as a lubricant • Petroleum distillates used as solvents

* Oil that does not meet the EPA's definition of used oil can still pose a threat to the environment when disposed of and could be subject to the RCRA regulations for hazardous waste management.

Figure 36-24. Typically, oil removed from a refrigeration system is not considered hazardous. However, when the oil has been contaminated with acids or other substances, it may be classified as hazardous.

Digital Resources ATPeResources.com/QuickLinks • Access Code: 482610

UNIT 37

Management of Refrigerants and Refrigeration Systems

Each country that signed the Montreal Protocol has different legislation covering the use and handling of refrigerants. In the U.S., regulations issued by the Environmental Protection Agency (EPA) establish requirements under Sections 608 and 609 of the Clean Air Act for the recycling and disposal of Class I and Class II substances. In other countries, technicians must comply with local jurisdiction. The Clean Air Act of 1990 requires that all HVAC service technicians be certified with Type I, Type II, Type III, or Universal certifications to buy and handle refrigerants. The EPA requires that owners of refrigeration and air conditioning equipment with refrigerant charges over 50 lb repair system leaks.

OBJECTIVES

- List the refrigerant regulations.
- Identify proper refrigerant handling techniques and containers.
- Describe service technician refrigerant certification requirements.
- Describe basic refrigerant procedures such as detecting leaks and recovering, recycling, and reclaiming refrigerants.
- Identify factors that affect the evacuation of refrigeration and air conditioning equipment.

Index to Procedures

REFRIGERANT MANAGEMENT

Section 608, *National Recycling and Emissions Reduction Program,* of the Clean Air Act of 1990 requires management of refrigerants and refrigeration systems to maximize recovery and recycling of ozone-depleting substances during the servicing and disposal of air conditioning and refrigeration equipment. The Environmental Protection Agency (EPA) is responsible for establishing and enforcing Section 608, which was created for the protection of the environment. EPA 40 Code of Federal Regulations (CFR) 82 Subpart (f)—*Protection of Stratospheric Ozone,* is available for additional refrigerant management information. **See Figure 37-1.**

The EPA issued regulations establishing requirements under Section 608 of the Clean Air Act of 1990 for the recycling and disposal of Class I and Class II substances and reduction of emissions. Class I substances include chlorofluorocarbons (CFCs) and halons, and Class II substances include hydrochlorofluorocarbons (HCFCs). The five requirements are service practices, technician certification, equipment and reclaimer certification, leak repair of systems containing more than 50 lb of refrigerant charge, and safe disposal of equipment and appliances.

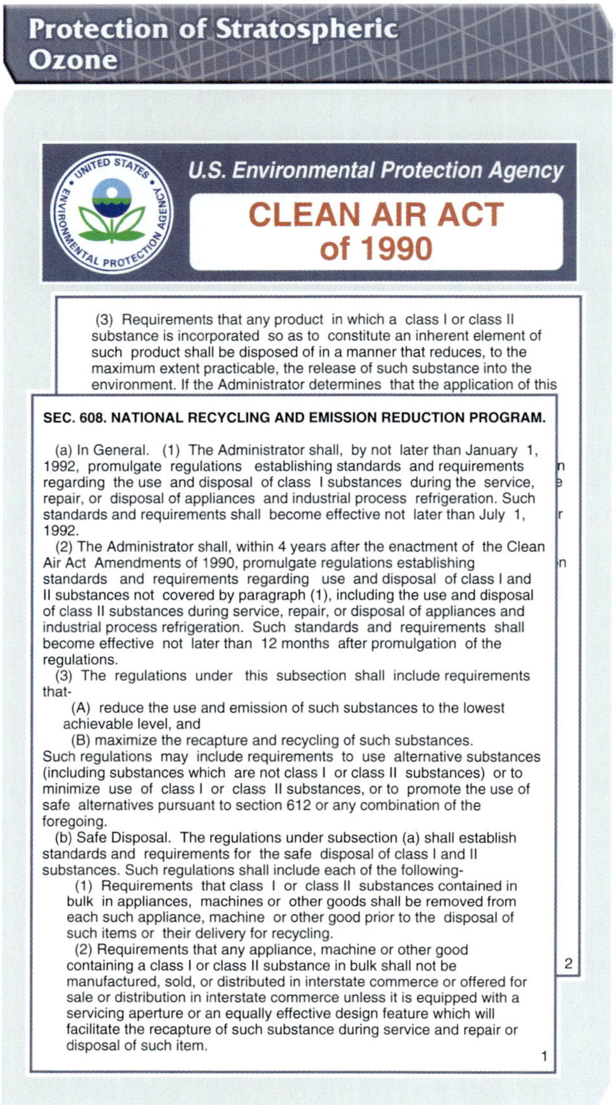

Figure 37-1. The EPA has issued regulations under Section 608 of the Clean Air Act for the recycling and disposal of Class I and Class II substances and the reduction of emissions.

A *service practice* is a procedure created by the EPA and used by technicians who service and dispose of refrigeration and air conditioning equipment for the reduction of refrigerant emissions. Technician certification is a credential created by the EPA and approved testing organizations that technicians must acquire before buying refrigerants and servicing refrigeration and air conditioning equipment. *Equipment and reclaimer certification* is a credential required by the EPA that contractors and/or technicians must acquire for their particular recovery and/or recycling equipment before use. **See Figure 37-2.**

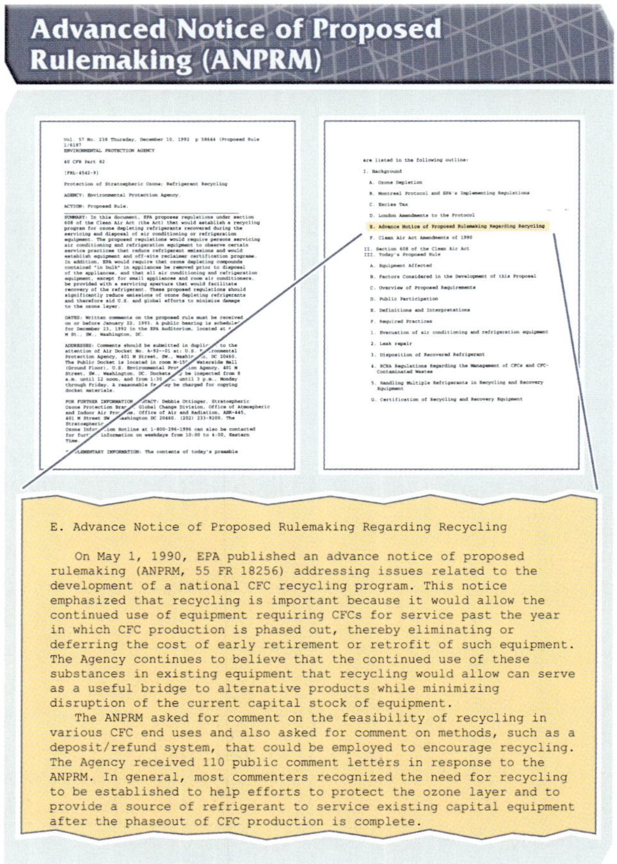

Figure 37-2. ANPRM 55 FR 18526 Notice emphasizes the importance of recycling CFC refrigerants and allowing the continued use of equipment requiring CFCs.

The EPA states that all substantial leaks in systems with more than 50 lb of refrigerant charge must be repaired as soon as possible, according to EPA Section 608 regulation *Leak Repair of Systems Containing More Than 50 lb of Refrigerant Charge*. The EPA requires the evacuation of all ozone-depleting refrigerants in appliances, machines, and other goods prior to disposal or dismantling according to EPA Section 608 regulation *Safe Disposal of Equipment and Appliances*.

Montreal Protocol

The United States and 22 other countries signed the Montreal Protocol on substances that deplete the ozone layer on September 16, 1987. This international environmental agreement established requirements to end the production and future use of CFC refrigerants. In 1992, an amendment was added to the Montreal Protocol that prohibited production of HCFC refrigerants and their future use.

The amended Montreal Protocol was agreed upon by the United States through Title VI of the Clean Air Act and enforced by the EPA. As of February 2011, 196 nations and the European Union have signed the Montreal Protocol. **See Figure 37-3.** Since then, it has had several revisions: in 1990 (London), 1991 (Nairobi), 1992 (Copenhagen), 1993 (Bangkok), 1995 (Vienna), 1997 (Montreal), and 1999 (Beijing).

TECH TIP

The Montreal Protocol of 1993 had a great impact on the air conditioning industry. Previous to these regulations, it was not uncommon for technicians to use R-12 or R-22 to blow sediment out of a reheat water coil or to dump the entire charge of refrigerant into the atmosphere when repairing a system. These common practices stopped when the new regulations were implemented.

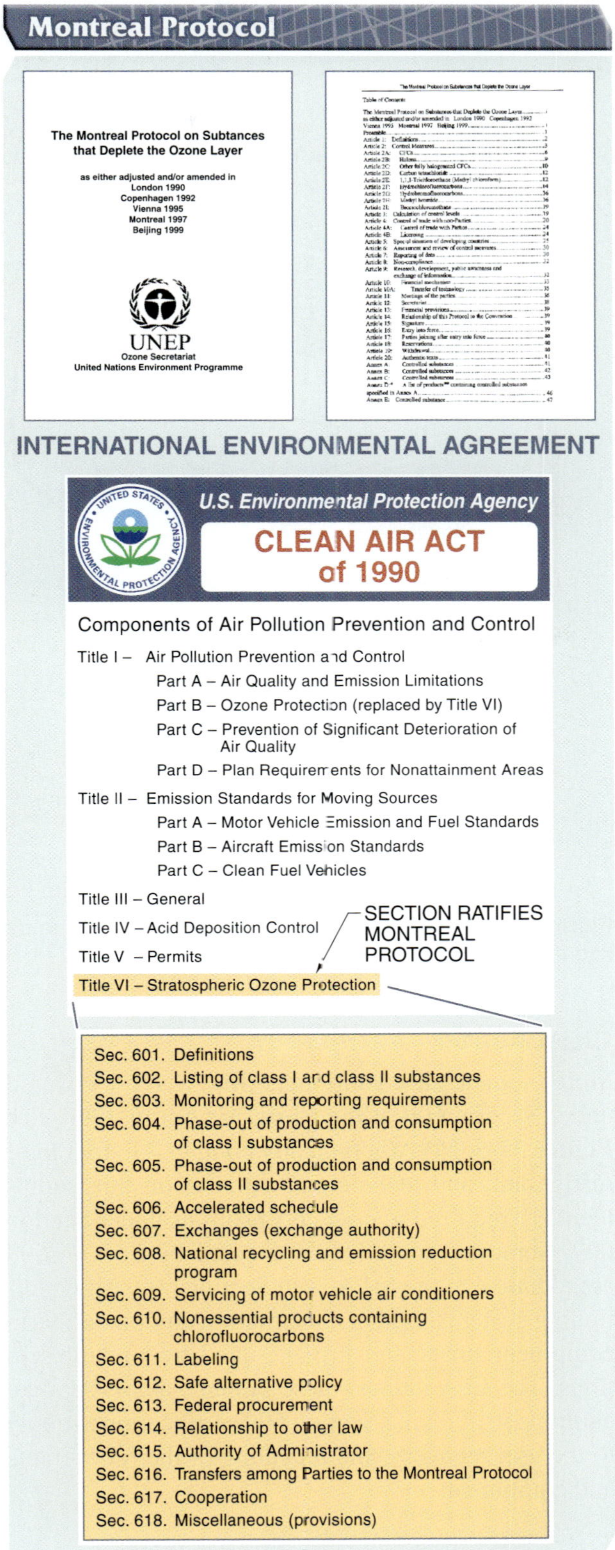

Figure 37-3. The Montreal Protocol is an international environmental agreement that establishes requirements for phasing out ozone-depleting CFCs and an amendment that phases out HCFCs worldwide.

Clean Air Act—Section 608

The final version of the Section VI regulations, including the 1990 amendments, was implemented by the Clean Air Act on May 14, 1993. **See Figure 37-4.** Under Section 608 of the Clean Air Act, the EPA established regulations that do the following:

- require service practices that maximize the recycling of ozone-depleting substances, such as CFCs and HCFCs, during the servicing and disposal of air conditioning and refrigeration equipment
- set certification requirements for recycling and recovery equipment, technicians, and reclaimers
- restrict the sale of refrigerants only to certified technicians
- require persons or technicians servicing or disposing of air conditioning and refrigeration equipment to prove to the EPA that recycling and recovery equipment being used is in compliance with EPA rules
- require the immediate repair of substantial leaks in air conditioning and refrigeration equipment with a charge greater than 50 lb
- establish safe disposal requirements to ensure the safe removal of refrigerants from goods, such as motor vehicle air conditioners, and prevent refrigerants from entering the waste stream with the refrigerant charge intact

Refrigerant Recovery—Type I System
Interactive Animation

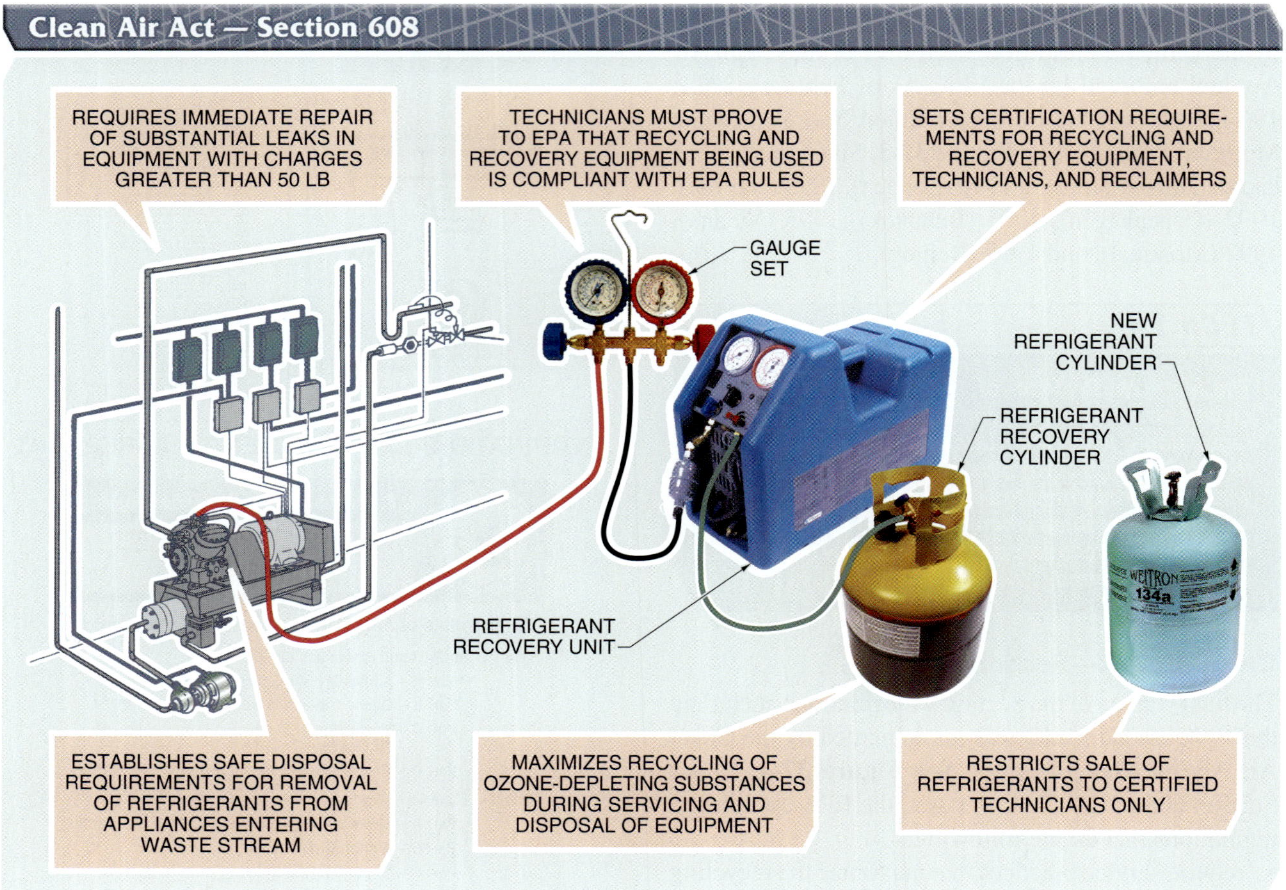

Figure 37-4. The EPA has established regulations under Section 608 of the Clean Air Act to regulate the handling of ozone-depleting substances.

Clean Air Act—Section 609

Motor vehicle air conditioners (MVACs) are covered under Section 609 of the Clean Air Act of 1990. **See Figure 37-5.** MVACs typically contain 2 lb to 3 lb of refrigerant, and technicians working on this equipment are required to have certifications under Section 609. Some air conditioners covered by Section 608 are identical to the MVACs in Section 609. Air conditioning systems used in construction equipment, farm equipment, boats, and airplanes are listed in both sections. The EPA defines these systems as MVAC-like equipment, allowing Section 609 certification to apply to the use, recovery, and recycling of the refrigerants and equipment.

Prohibition on Venting

As of July 1, 1992, Section 608 of the Clean Air Act prohibits anyone from knowingly venting CFC and HCFC ozone-depleting refrigerants into the atmosphere while maintaining, servicing, repairing, or disposing of air conditioning or refrigeration equipment. **See Figure 37-6.** The only four types of releases allowed under Section 608 are the following:

- "de minimus" (minimal) quantities of refrigerant released in the course of making good faith attempts to recover, recycle, or safely dispose of refrigerant
- refrigerant releases during the normal operation of air conditioning and refrigeration equipment
- releases of CFC and HCFC substances that are not used as refrigerants, such as mixtures of nitrogen and R-22 used for holding charges or for leak testing (*Note:* Technicians may not avoid recovering refrigerant by adding nitrogen to a charged system.)
- small releases of refrigerant that result from purging hoses or from connecting or disconnecting hoses (*Note:* As of November 15, 1993, all recovery and recycling equipment manufactured must include low-loss fittings.)

EPA Clean Air Act—Section 609

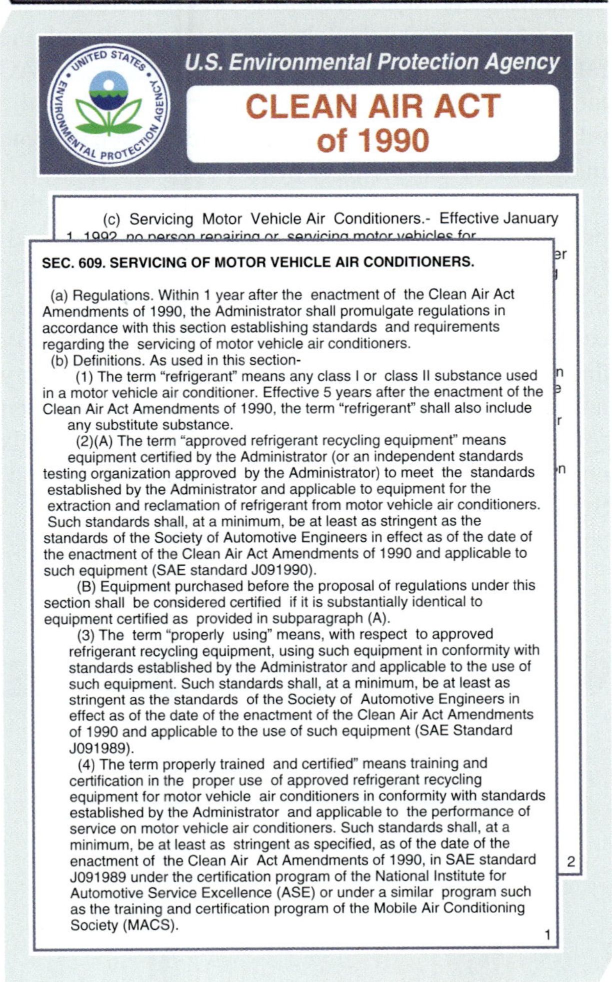
U.S. Environmental Protection Agency

CLEAN AIR ACT of 1990

(c) Servicing Motor Vehicle Air Conditioners.- Effective January

SEC. 609. SERVICING OF MOTOR VEHICLE AIR CONDITIONERS.

(a) Regulations. Within 1 year after the enactment of the Clean Air Act Amendments of 1990, the Administrator shall promulgate regulations in accordance with this section establishing standards and requirements regarding the servicing of motor vehicle air conditioners.

(b) Definitions. As used in this section-

(1) The term "refrigerant" means any class I or class II substance used in a motor vehicle air conditioner. Effective 5 years after the enactment of the Clean Air Act Amendments of 1990, the term "refrigerant" shall also include any substitute substance.

(2)(A) The term "approved refrigerant recycling equipment" means equipment certified by the Administrator (or an independent standards testing organization approved by the Administrator) to meet the standards established by the Administrator and applicable to equipment for the extraction and reclamation of refrigerant from motor vehicle air conditioners. Such standards shall, at a minimum, be at least as stringent as the standards of the Society of Automotive Engineers in effect as of the date of the enactment of the Clean Air Act Amendments of 1990 and applicable to such equipment (SAE standard J091990).

(B) Equipment purchased before the proposal of regulations under this section shall be considered certified if it is substantially identical to equipment certified as provided in subparagraph (A).

(3) The term "properly using" means, with respect to approved refrigerant recycling equipment, using such equipment in conformity with standards established by the Administrator and applicable to the use of such equipment. Such standards shall, at a minimum, be at least as stringent as the standards of the Society of Automotive Engineers in effect as of the date of the enactment of the Clean Air Act Amendments of 1990 and applicable to the use of such equipment (SAE Standard J091989).

(4) The term properly trained and certified" means training and certification in the proper use of approved refrigerant recycling equipment for motor vehicle air conditioners in conformity with standards established by the Administrator and applicable to the performance of service on motor vehicle air conditioners. Such standards shall, at a minimum, be at least as stringent as specified, as of the date of the enactment of the Clean Air Act Amendments of 1990, in SAE standard J091989 under the certification program of the National Institute for Automotive Service Excellence (ASE) or under a similar program such as the training and certification program of the Mobile Air Conditioning Society (MACS).

1

2

Figure 37-5. Motor vehicle air conditioners (MVACs) are covered under Section 609 of the Clean Air Act of 1990.

Permitted Refrigerant Releases (USA)

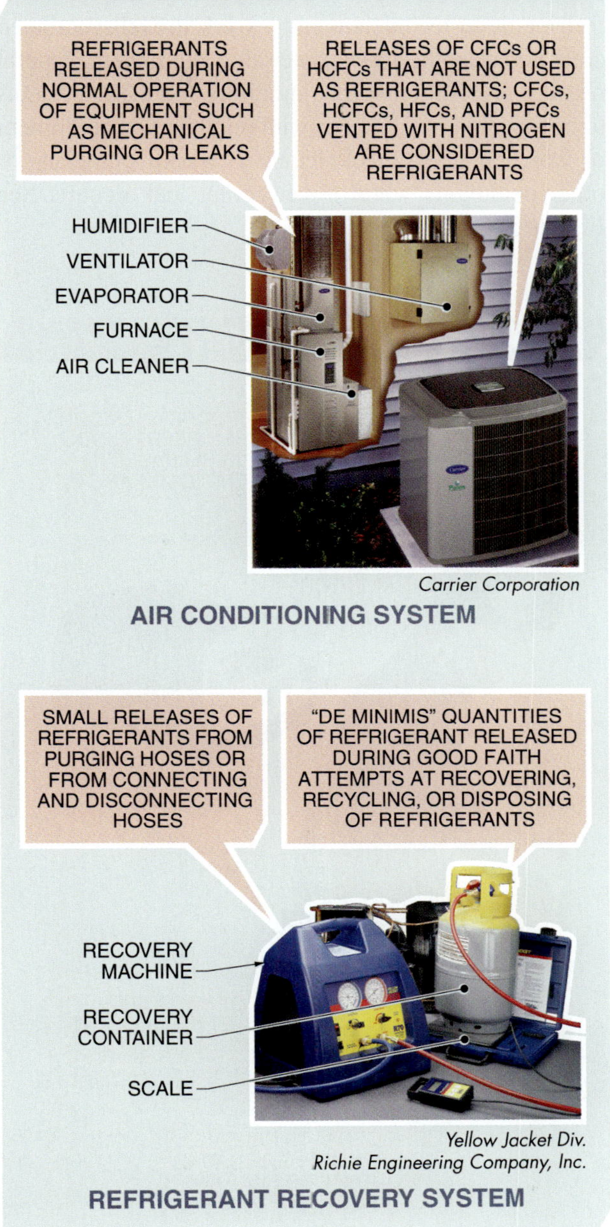

Figure 37-6. Four types of refrigerant releases are allowed under the Clean Air Act venting prohibition.

TECH TIP

Care must be taken and the correct PPE worn when working around refrigerants. It is possible that fittings may fail or coils may rupture, releasing refrigerant under pressure to the atmosphere. This can cause refrigerant burns and/or eye injuries.

HANDLING REFRIGERANTS

There are many hazards in the refrigeration and air conditioning industry. Some hazards are associated with the handling and storage of refrigerants. There have been improvements in the past few years because of new environmentally safe refrigerants, but Canada has a zero refrigerant release policy.

The newer refrigerants are classified by the EPA as nontoxic and nonflammable, but handling problems still exist because most refrigerants are stored in pressurized containers or they are under pressure in a system that requires service. When refrigerants are released to the atmosphere, they can cause frostbite to unprotected skin and freezing of the eyes. Suffocation can also occur if a large amount of refrigerant is released in an enclosed area. Refrigeration technicians must work in well-ventilated areas and wear the appropriate safety equipment. Federal regulations require oxygen monitors and alarm systems to be installed in all spaces with refrigeration equipment or worn on HVAC technicians to warn technicians or other occupants when a refrigerant leak occurs. **See Figure 37-7.**

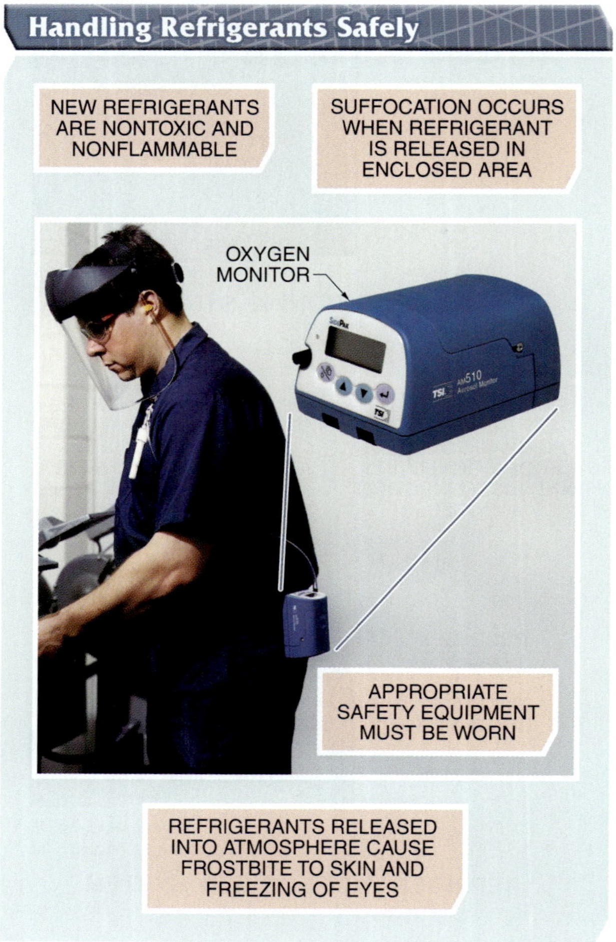

Figure 37-7. Federal regulations require oxygen monitors and alarm systems to be installed in all spaces with refrigeration equipment or worn on HVAC technicians themselves to warn technicians or other occupants of a refrigerant leak.

When removing refrigerant from or adding refrigerant to pressurized containers or systems, low-loss fittings are used to minimize the release of refrigerant to the atmosphere. The proper personal protective equipment (PPE) should always be worn when working with machinery and systems that contain refrigerants to avoid frostbite. When a short occurs in a compressor motor, for example, the lubricating oil mixed with the refrigerant is burnt and becomes very acidic. HVAC technicians can receive chemical burns from the acidic oil. Refrigerants and oil must never be allowed to come in contact with the skin or eyes.

Federal, state, and local regulations require that proper lockout and tagout procedures be followed. It must be ensured that all valves, switches, and any other energy sources are secured and the proper procedures are followed. The guidelines on the safety data sheets (SDSs) must be read before using any harmful or toxic materials (refrigerants) and then followed. In addition, all warning and caution labels on equipment, tools, and materials must be followed. **See Figure 37-8.**

Figure 37-8. Technicians must follow all warning and caution labels found on HVAC equipment, tools, and materials.

A system that contains refrigerant must not be welded, brazed, soldered, or have any heat applied to it. The refrigerant must first be recovered, and the system or container must be opened to the atmosphere. Applying heat to a component in a refrigeration system of which the refrigerant has not been recovered and the system opened can cause a refrigerant fire or the refrigerant to expand and rupture the weakest component in the system.

Handling Cylinders Containing Refrigerant

Handling the cylinders that refrigerants and other gases are transported in can be dangerous. Refrigerant and gas cylinders are heavy and difficult to handle. A gas cylinder must never be moved without a protective cap over the valve. Without the protective cap, the gas cylinder valve can break off or become damaged. When the valve is broken off a cylinder, the cylinder becomes an uncontrolled projectile that can cause serious injury or extensive damage to a building. Refrigerant cylinders have a protective collar. **See Figure 37-9.** The collar typically has built-in hand holes for easy lifting. A cylinder that is being transported or is in use must be secured in the upright position with a chain or rope.

Recovery Cylinders. When recovering refrigerant from a system, refrigerant recovery cylinders that are designed for this purpose must be used. Most refrigerant recovery cylinders are equipped with devices that limit the amount of refrigerant that can be pumped into the cylinder. When filling a recovery cylinder with liquid refrigerant, 80% of the cylinder volume (by weight) should not be exceeded. The refrigerant will expand as the refrigerant and the cylinder warm up. Space must be available inside the recovery cylinder to allow for the expansion. **See Figure 37-10.** Additional safety recommendations for handling refrigerant recovery cylinders are as follows:

- Verify that the recovery cylinder has a current hydrostatic test date (if older than 5 years, do not use).
- Verify that the pressure rating of the recovery cylinder is correct for the refrigerant being recovered.
- Do not mix different types of refrigerants in a refrigerant recovery cylinder. It is impossible to separate refrigerants once they have been mixed. The mix of refrigerants must be disposed of as hazardous waste.
- Do not fill a damaged recovery cylinder.
- After filling, verify that all valves are properly closed for transport and storage.
- Because refrigerant vapor is heavier than air, avoid low or confined areas without proper ventilation when filling recovery cylinders.

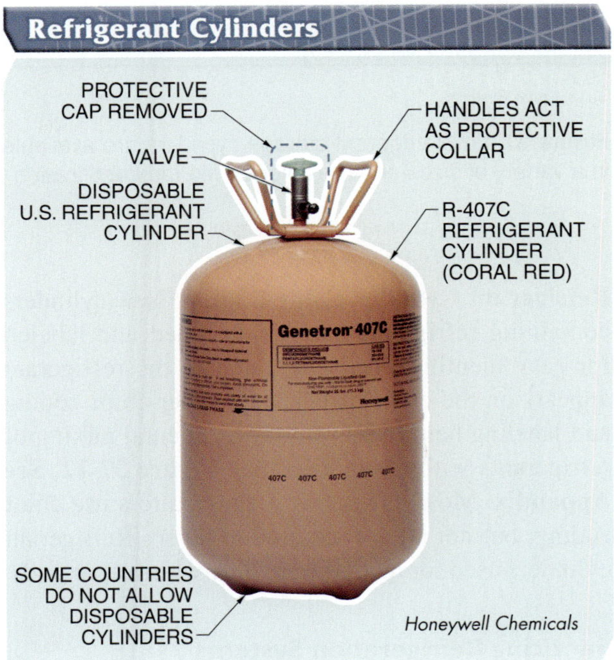

Figure 37-9. Colored refrigerant cylinders are typically disposable and must be evacuated prior to disposal in the appropriate waste stream.

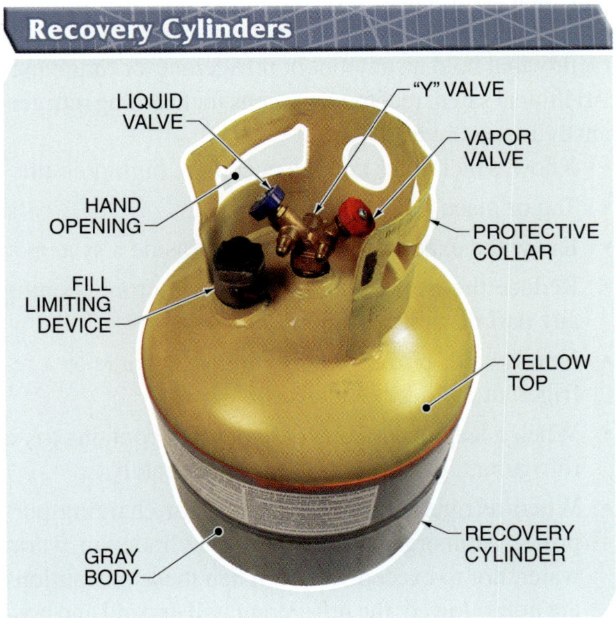

Figure 37-10. Most refrigerant recovery cylinders are equipped with devices that limit the amount of refrigerant that can be pumped into the cylinder (not to exceed 80% by weight).

Honeywell Chemicals

New refrigerant cylinders typically arrive in a box to protect the cylinder from dents and dings and wrapped in plastic to protect the box and cylinder from humidity and dirt.

New Refrigerant Cylinders. Refrigerant cylinders are available in a variety of sizes for refrigerant storage. Disposable cylinders must not be refilled. **See Figure 37-11.** Disposable cylinders are designed for the transport of new refrigerants only, and when they are totally empty, they must be rendered useless and disposed of properly. Returnable service cylinders are emptied of refrigerant and then returned to the supplier empty for refilling. Storage cylinders are typically large cylinders inside a facility that hold an amount of refrigerant for future use. Additional safety recommendations for handling refrigerant cylinders are as follows:

- Know which lines and tanks contain high pressure.
- Tag or place signs on hot piping.
- Know the exact refrigerant that is inside a system.
- Reduce the gauge pressure to 0 psi before opening any part of a refrigeration system.
- Never weld, braze, or apply an open flame to a refrigerant cylinder of any kind.
- When a leak occurs in a cylinder that contains toxic refrigerant, evacuate the area immediately.
- When refrigerant must be heated for charging purposes, submerge the refrigerant cylinder in warm water, not to exceed 125°F. When these precautions are not followed, the refrigerant will expand and possibly rupture the container. Devices are available that safely allow heating of a refrigerant cylinder only up to 125°F.

Refrigerant Cylinder Sizes

ASHRAE Number Type	Cylinder Size*	ASHRAE Number Type	Cylinder Size*
R-11 (CFC)	100	R-402A (HCFC)	27†
	200		110
R-12 (CFC)	30†	R-402B (HCFC)	15
	145	R-404A (HFC)	24†
	1000		100
R-13 (HFC)	9		1000
	23	R-407A (HFC)	24†
	80		110
R-22 (HCFC)	30	R-407B (HFC)	24†
	125		110
	1000	R-407C (HFC)	24†
R-23 (HFC)	20		100
	70	R-408A (HCFC)	24†
R-113 (CFC)	100		100
	200	R-409A (HCFC)	30†
R-114 (CFC)	30†		125
	150	R-410A (HFC)	25†
R-123 (HCFC)	100		125
	200	R-500 (CFC)	30†
R-124 (HCFC)	30†		125
	145	R-502 (CFC)	30†
R-134a (HFC)	30†		125
	30	R-503 (CFC)	5
	125		9
	1000		20
R-410A (HCFC)	30†		80
	125	R-507 (HFC)	25†
R-4021B (HCFC)	30†		100
	125		1000

* in lb
† disposable cylinder

Figure 37-11. Refrigerant storage cylinders are available in a variety of sizes and are color-coded for identification.

Refrigerant Cylinder Color Coding. New cylinders containing refrigerants are color-coded and labeled for easy identification. The name of the refrigerant appears on the cylinder label. Cylinder color coding and labeling help to prevent the accidental mixing of refrigerants within a system. **See Figure 37-12. See Appendix.** Most refrigerant manufacturers use color coding, but not all are required to do so. Refrigerant cylinders used for recovery are gray with yellow ends.

Servicing Refrigeration Systems

The proper methods and procedures must be used when servicing refrigeration systems to avoid personal injury and system damage. When pressurizing a system, an inert gas, such as nitrogen or carbon dioxide, must be used. It must be ensured that there is a regulator on the inert gas tank to limit the pressure in the system. Oxygen, acetylene, propane, or air should never be used. Pressure should not be raised above the specified low-side working pressure (typically 150 psi). **See Figure 37-13.**

Refrigerant Cylinder Color Coding

Refrigerant Number	Cylinder Color	Refrigerant Name	Application
CFC Refrigerants			
R-11	Orange	Trichloromonofluoromethane	Centrifugal chillers
R-12	White	Dichlorodifluoromethane	Reciprocating and rotary equipment
HCFC Refrigerants			
R-22	Light green	Monochlorodifluoromethane	Residential and commercial
R-123	Light gray	Dichlorotrifluoromethane	R-11 replacement for centrifugal chillers
HFC Refrigerants			
R-23	Light gray	Trifluoromethane	Low-temperature systems
R-134a	Light sky blue	Tetrafluoroethane	Automotive industry and refrigeration systems

Figure 37-12. Refrigerant cylinders (containers) are color-coded according to the type of refrigerant contained within.

Pressurizing Systems

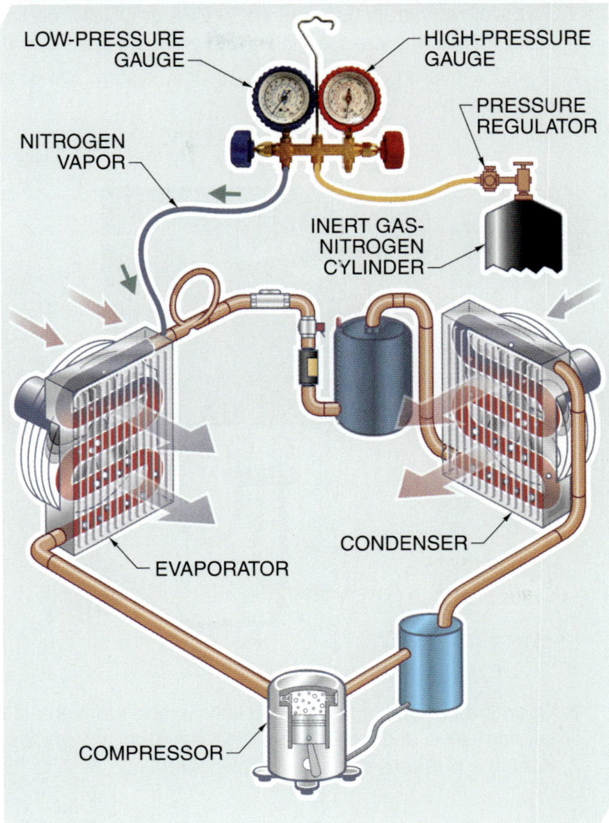

Figure 37-13. A system is pressurized using a pressure source such as nitrogen.

TECH TIP

Extra care must be taken when charging a system with refrigerant from a cylinder that has been in a hot location for an extended period of time. The cylinder temperature may be 120°F or more, which dramatically increases the refrigerant pressure. The cylinder should be allowed to cool before use.

When charging a system with refrigerant on the low side of the system, refrigerant must always be charged as a vapor. When liquid refrigerant is allowed to enter the compressor, the compressor may be damaged beyond repair and need to be replaced. To speed up the charging process, liquid may be charged on the high side. Some refrigerants may need to be charged as a liquid in order to retain the proper refrigerant blend. **See Procedure for Charging Systems with Refrigerant—Type II Systems (High-Pressure Side).**

Charging Systems with Refrigerant (High-Pressure Side)

1. Recover refrigerant as required by EPA or ODSR regulations. Repair any leaks identified and pressure test system.
2. Install a micron gauge directly on the system low-side service valve.

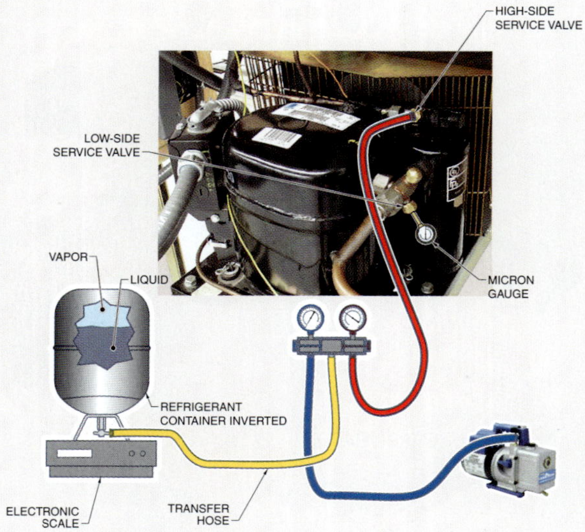

3. Open the red and blue manifold valves, and the high-side service valve and low-side service valve on the system. Keep the refrigerant container valve shut.

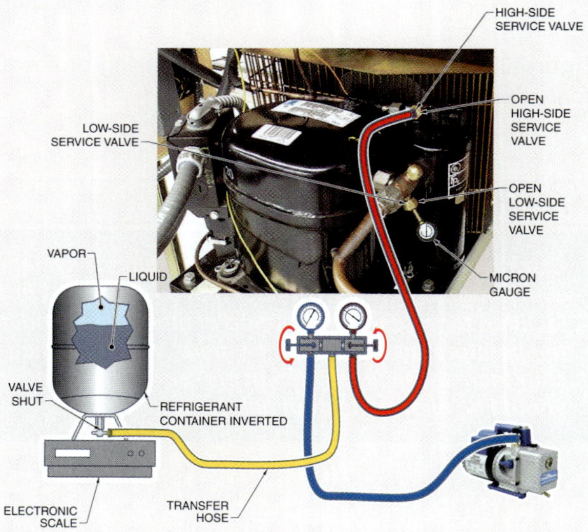

4. Evacuate the system to the proper level as required (500 microns). To check for proper evacuation level, close the blue service manifold valve (low pressure), loosen the blue hose connected to the vacuum pump before shutting off the pump to prevent the vacuum pump oil from being sucked out. Shut off the vacuum pump. Wait several minutes to see if the vacuum gauge shows that the vacuum level has not risen which means that there are no leaks and no moisture in system.
5. Shut off the low-side service valve isolating the micron gauge.
6. Shut off the high-side service valve closing off the red hose.

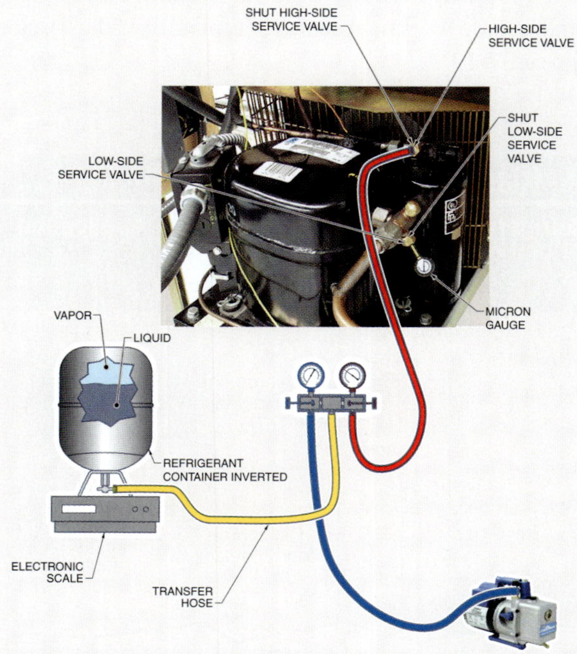

7. Note the scale reading and open the refrigerant cylinder valve allowing liquid refrigerant to flow into the yellow and red hoses up to the high side service valve.
8. Note the scale reading again. The difference is the weight of refrigerant that has entered the yellow and red hoses.

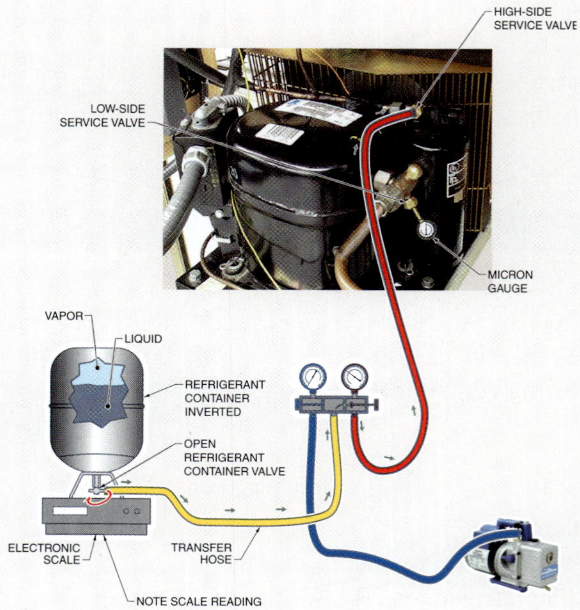

continued on next page

Procedure

Charging Systems with Refrigerant (High-Pressure Side) *(continued)*

9. Open the high-side service valve and watch the container scale while leaving the container valve and the red manifold valve open.

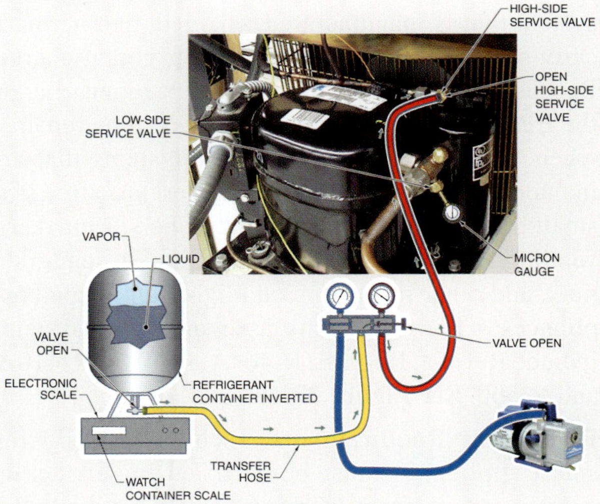

10. Shut off the high-side service valve when the proper weight of refrigerant has flowed as indicated by the scale. The system now has the exact required amount of refrigerant by weight.

11. Shut off the refrigerant container valve and the manifold valves, and disconnect the hoses using the proper PPE (gloves) to prevent frostbite.

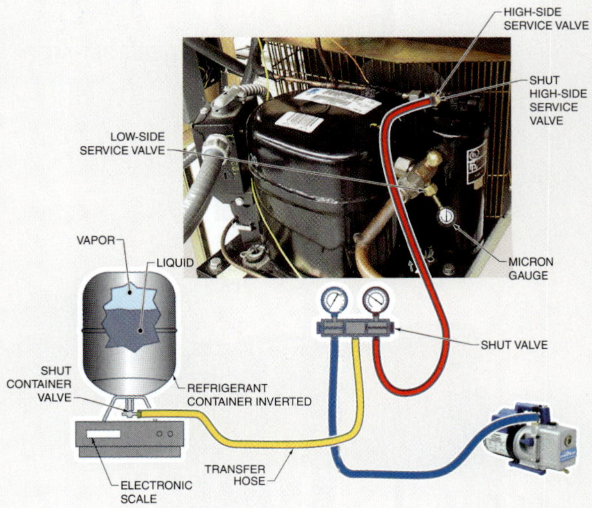

An open flame must never be used when refrigerant is present. When a flame must be used, the area must be ventilated with fresh outside air, and no one must be allowed to stand downwind from the refrigeration system. *Note:* Fluorocarbon-based refrigerants can break down and create poisonous phosgene gas.

When servicing a refrigeration system that has been opened, precautions must be taken to prevent moisture from entering the system. Moisture can cause considerable damage to a refrigeration system. **See Figure 37-14.**

Unused refrigerant oil must be kept in sealed containers when not in use. Refrigerant oil must not be used from containers that have been left open for a period of time. Open refrigerant oil containers have a strong attraction for moisture suspended in the air, allowing contamination of the oil.

Service Precautions

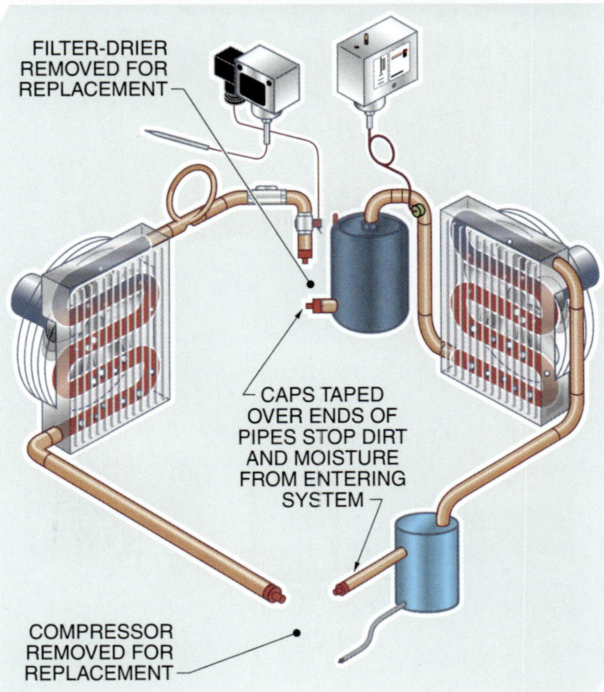

Figure 37-14. The moisture of the atmosphere on a very dry day (15% humidity) is about 1000 ppm. In a refrigeration system, moisture content of 100 ppm is considered high.

TECHNICIAN CERTIFICATION

In the United States, the Clean Air Act of 1990 requires that service technicians be certified to purchase and handle refrigerants. There are four certification types available for refrigeration service technicians: Type I, Type II, Type III, and Universal. **See Figure 37-15.**

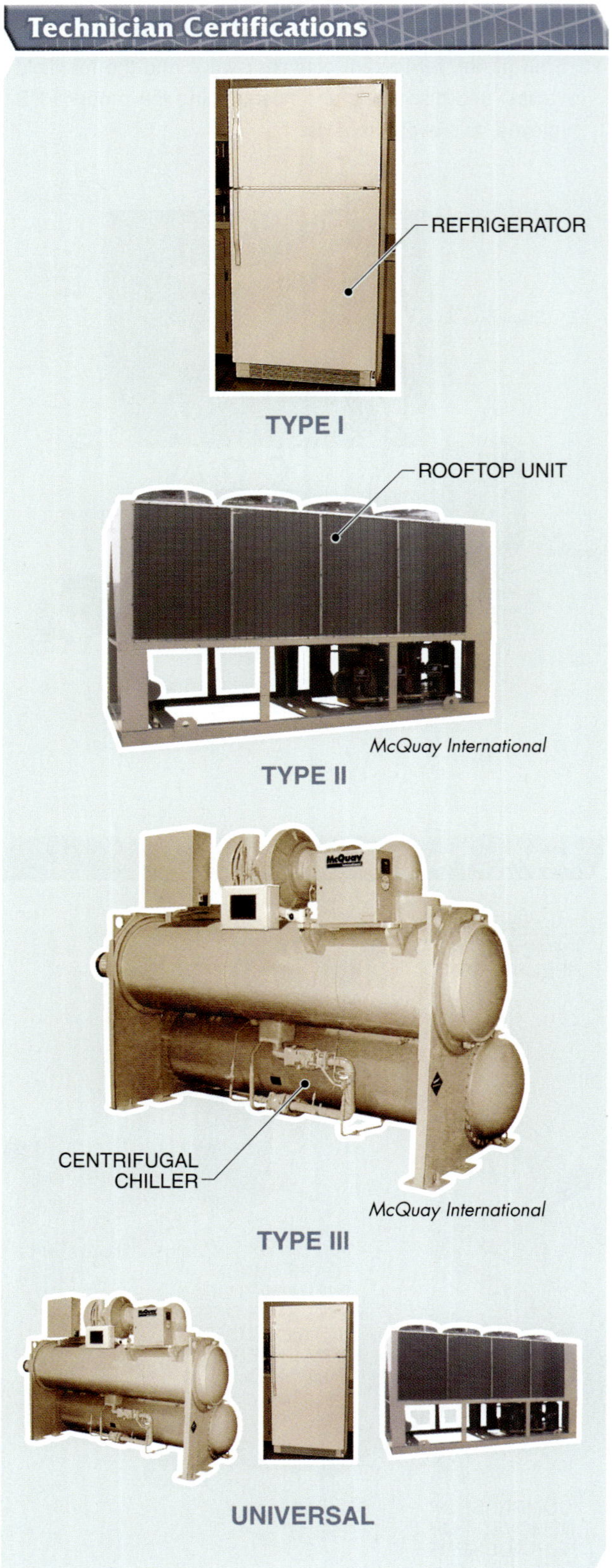

Figure 37-15. There are four certification types available for refrigeration service technicians: Type I, Type II, Type III, and Universal.

Type I Certification

Type I certification is required for the technician who services small appliances. According to the EPA, a small appliance is a refrigeration or air conditioning system that is manufactured, charged, and hermetically sealed with 5 lb or less of refrigerant charge. Because of the small amount of refrigerant charge, the EPA allows small appliance leaks to be repaired whenever possible. Refrigeration systems that are included in this category are refrigerators, freezers, small heat pumps, room air conditioners, icemakers, vending machines, water coolers, small dehumidifiers, and other residential and light commercial appliances. The refrigerants used in Type I systems include R-134a, R-402B, R-404A, R-407C, R-410A, and R-500. **See Figure 37-16.**

Regulatory Requirements. Technicians servicing small appliances must be certified in refrigerant recovery to perform work on sealed systems after November 14, 1994. If EPA regulations change after a technician becomes certified, it is the responsibility of the technician to comply with any changes to the law.

Technicians are required to pass EPA-approved tests (core and Type I) given by an EPA-approved certifying organization to become certified under the Section 608 program. Section 608 of the Clean Air Act requires that all persons who sell CFC and HCFC refrigerants retain invoices that indicate the name of the purchaser, date of sale, and quantity of refrigerant purchased. The records kept by sellers must be maintained for three years.

Enforcement. The EPA performs inspections and pursues potential cases against violators of Title VI of the Clean Air Act. Under the Clean Air Act, the EPA is authorized to assess fines of up to $32,000 per day per violation.

Type II Certification

According to the EPA, Type II certification is required for technicians who service high-pressure equipment. A *high-pressure appliance* is a refrigeration or air conditioning system that has a refrigerant charge over 5 lb, has a boiling point between –58°F and 50°F at atmospheric pressure, and is not categorized as low-pressure. Most refrigeration systems that are not Type I fall into this category. The refrigerants used in Type II systems include R-12, R-22, R-114, R-134a, R-500, and R-502. **See Figure 37-17.**

Type I (Small Appliance) Systems

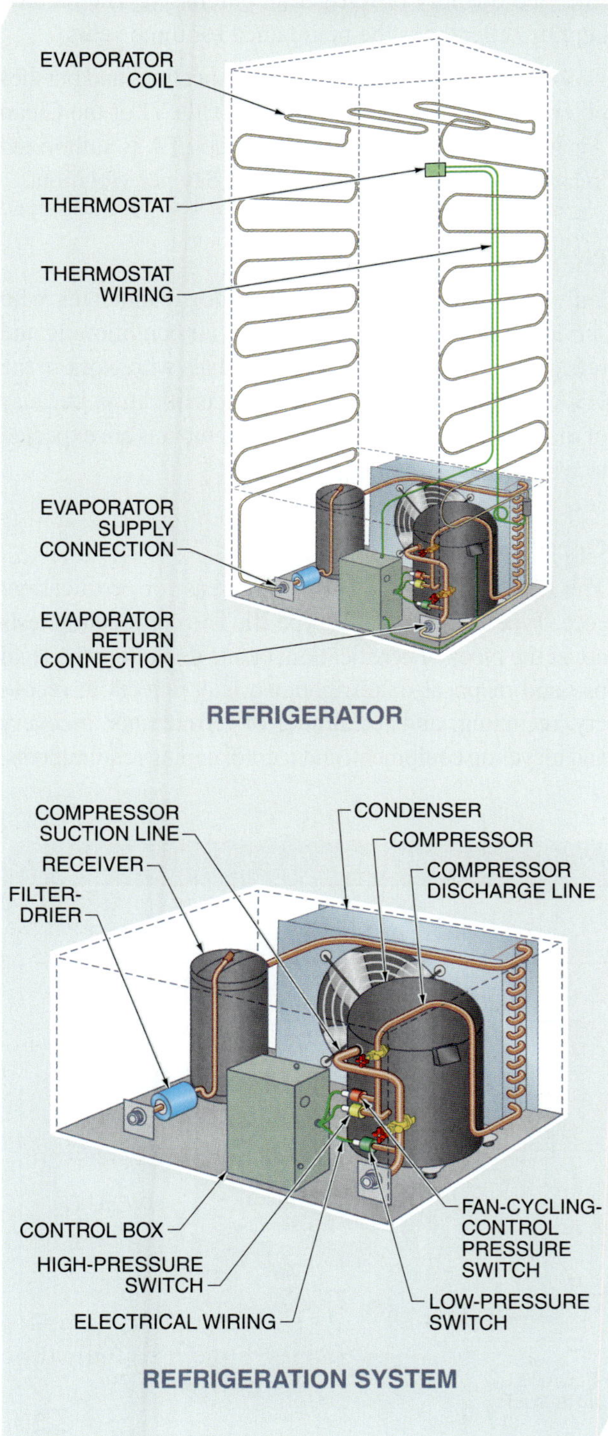

Figure 37-16. Technicians who pass the core and Type I tests for the EPA are certified to recover refrigerant during the maintenance, service, and repair of small appliances such as refrigerators, freezers, room air conditioners, small heat pumps, icemakers, vending machines, and water coolers.

Type II (High-Pressure) Systems

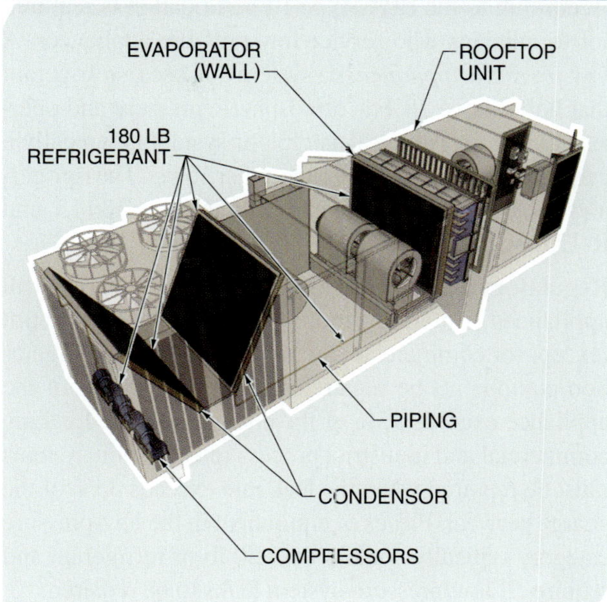

Figure 37-17. The nameplate on a Type II system provides the technician with information such as the type of refrigerant, type of compressor(s), maximum working pressure, and normal system charge.

Regulatory Requirements. The EPA requires that all appliances containing more than 50 lb of refrigerant (except for commercial and industrial process refrigeration equipment) be repaired when the leak rate of the appliance exceeds 15% of the charge per year. Leaking commercial and industrial process refrigeration systems must be repaired when the leak rate exceeds 35% of the charge per year.

Technicians are required to pass EPA-approved tests (core and Type II) given by an EPA-approved certifying organization to become certified under the Section 608 program. Section 608 of the Clean Air Act requires that all persons who sell CFC and HCFC refrigerants retain invoices that indicate the name of the purchaser, date of sale, and quantity of refrigerant purchased. The records kept by sellers must be maintained for three years.

Enforcement. The EPA performs inspections and pursues potential cases against violators of Title VI of the Clean Air Act. Under the Clean Air Act, the EPA is authorized to assess fines of up to $32,000 per day per violation.

Type III Certification

According to the EPA, Type III certification is required for technicians who service low-pressure appliances. A *low-pressure appliance* is a system that uses a refrigerant that boils above 50°F at atmospheric pressure and operates at low pressures. This type of system is typically a centrifugal chiller that operates under 15 psi. The refrigerants used in Type III systems include R-11, R-113, and R-123. **See Figure 37-18.**

Regulatory Requirements. The EPA requires that all appliances containing more than 50 lb of refrigerant (except for commercial and industrial process refrigeration equipment) be repaired when the leak rate of the appliance exceeds 15% of the charge per year. Leaking commercial and industrial process refrigeration systems must be repaired when the leak rate exceeds 35% of the charge per year. Pieces of equipment in the low-pressure category typically contain over 50 lb of refrigerant and require all low-pressure-system leaks to be repaired.

Technicians are required to pass EPA-approved tests (core and Type III) given by an EPA-approved certifying organization to become certified under the Section 608 program. Section 608 of the Clean Air Act requires that all persons who sell CFC and HCFC refrigerants retain invoices that indicate the name of the purchaser, date of sale, and quantity of refrigerant purchased. The records kept by sellers must be maintained for three years.

Enforcement. The EPA performs inspections and pursues potential cases against violators of Title VI of the Clean Air Act. Under the Clean Air Act, the EPA is authorized to assess fines of up to $32,000 per day per violation.

Universal Certification

Universal certification is required for technicians who service Type I, Type II, and Type III air conditioning and refrigeration systems. Most technicians who work in the HVAC industry acquire Universal certification because of the large variety of equipment technicians are expected to work on day-to-day.

Certification Testing

The EPA requires testing in four areas for certification: core, Type I, Type II, and Type III. The certification tests cover the types of certification; ozone depletion; the legal use and disposal of refrigerants; leak detection; recovery, recycling, and reclaiming of refrigerants; recovery and recycling equipment; and recordkeeping requirements.

Type III (Low-Pressure) Systems

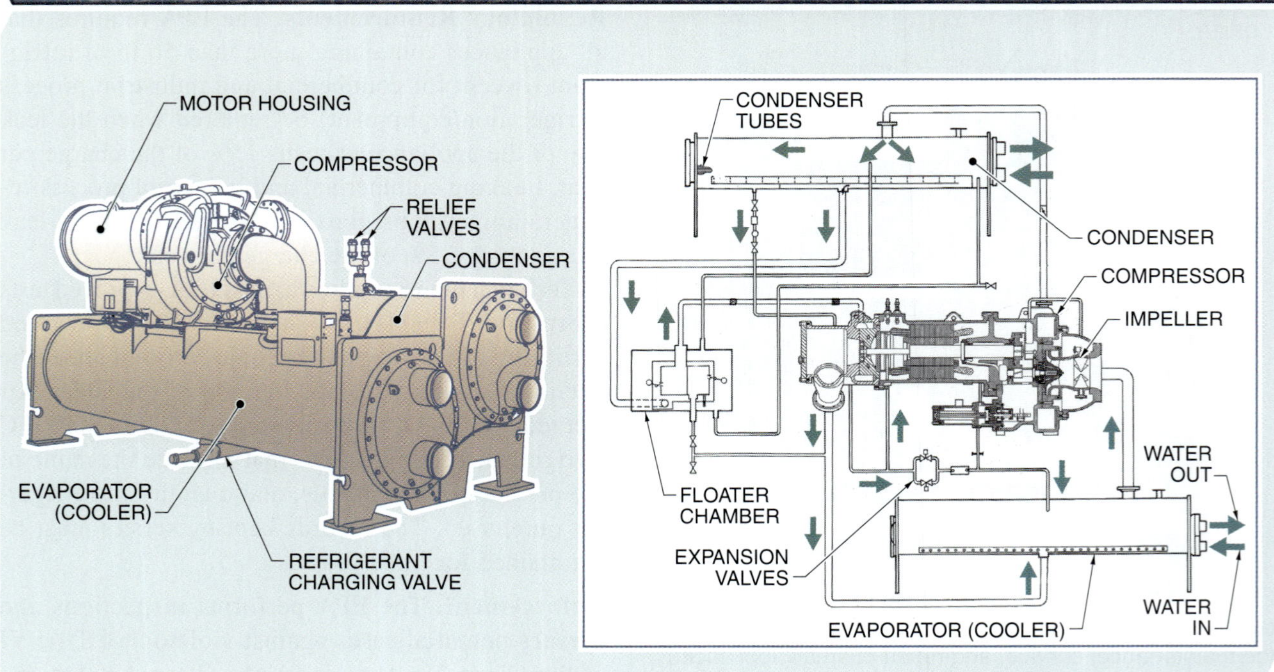

Figure 37-18. Low-pressure refrigeration systems (chillers) are the largest refrigeration and air conditioning systems and typically have the largest volume of refrigerant charge.

All technicians seeking certification are required to pass the core test with a score of 70% or better. The core test includes test questions on general information concerning ozone depletion, certification, recordkeeping, etc. All classes for technician certification cover core information. **See Figure 37-19.**

TECH TIP

The Type III certification test is often the most difficult to pass because technicians are rarely exposed to these systems. These systems are limited to large centrifugal chillers in large commercial buildings.

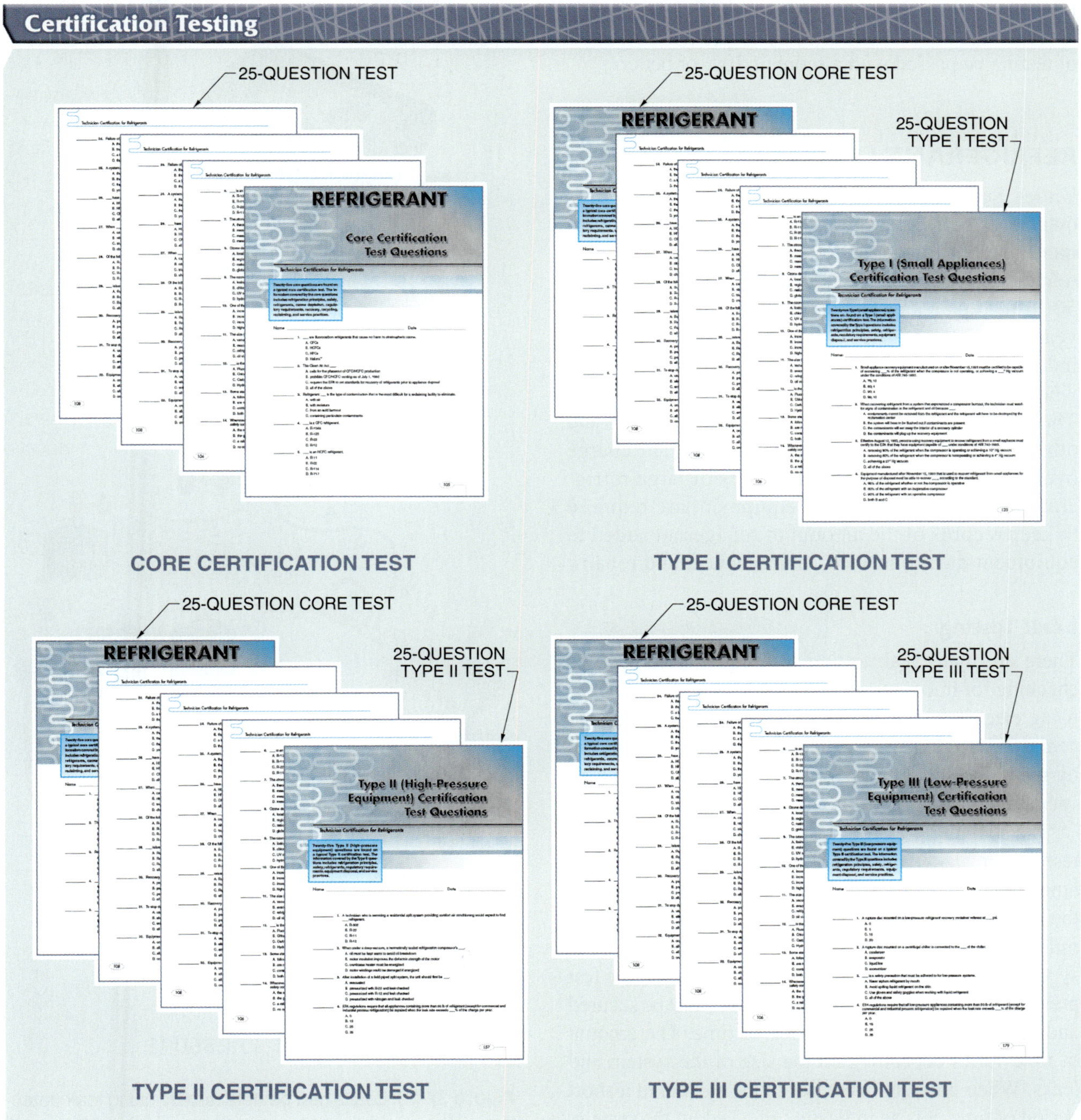

Figure 37-19. The EPA has established a technician certification program for technicians who perform maintenance, service, or repair, or who dispose of equipment that could be reasonably expected to release refrigerants into the atmosphere.

Technicians seeking Type I, Type II, or Type III certification must pass the core test and the appropriate Type I, Type II, or Type III test with a score of 70% or better. The Type I, Type II, and Type III tests include questions specific to that type of certification covering servicing, leak detection, refrigerant recovery, and refrigerant disposal. Some of this information may be covered in certification classes but most are acquired from on-the-job experience or from attending in-depth refrigeration classes. Universal certification requires that all four areas of testing be passed with a score of 70% or better.

REFRIGERANT LEAKS

Under Section 608 regulations, the EPA requires owners of refrigeration and air conditioning equipment (except for commercial and industrial process refrigeration equipment) with refrigerant charges over 50 lb to repair any leaks with a rate over 15% per year. The EPA requires that leaking commercial and industrial process refrigeration systems be repaired when the leak rate exceeds 35% of the charge per year. The leaks of all low-pressure equipment (chillers and other Type III equipment) with a refrigerant charge over 50 lb must be repaired. Owners of large refrigeration and air conditioning equipment are required to keep records of the amount of refrigerant added to equipment during service, maintenance, and repair.

Leak Testing

There are many leak testing techniques used such as checking for traces of oil on piping, using soap bubbles, using electronic leak detectors, or testing with fluorescent dyes. **See Figure 37-20.** Leaks can also be detected by using the standing pressure method or the standing vacuum method.

The standing pressure method is the best method for testing for leaks and is performed by adding a small amount of refrigerant to the system and then pressurizing the system using a pressure source such as nitrogen. Care must be taken not to exceed the working pressure of the system specified on the equipment label. When the test pressure is obtained, the pressure source must be secured and allowed to stand for a period of time. The amount of time varies depending on the size of the system and leaks. When pressure decreases in the system in a short period of time, it is an indication of a leak in the system. Typically an electronic leak detector is then used to find the location of the leak.

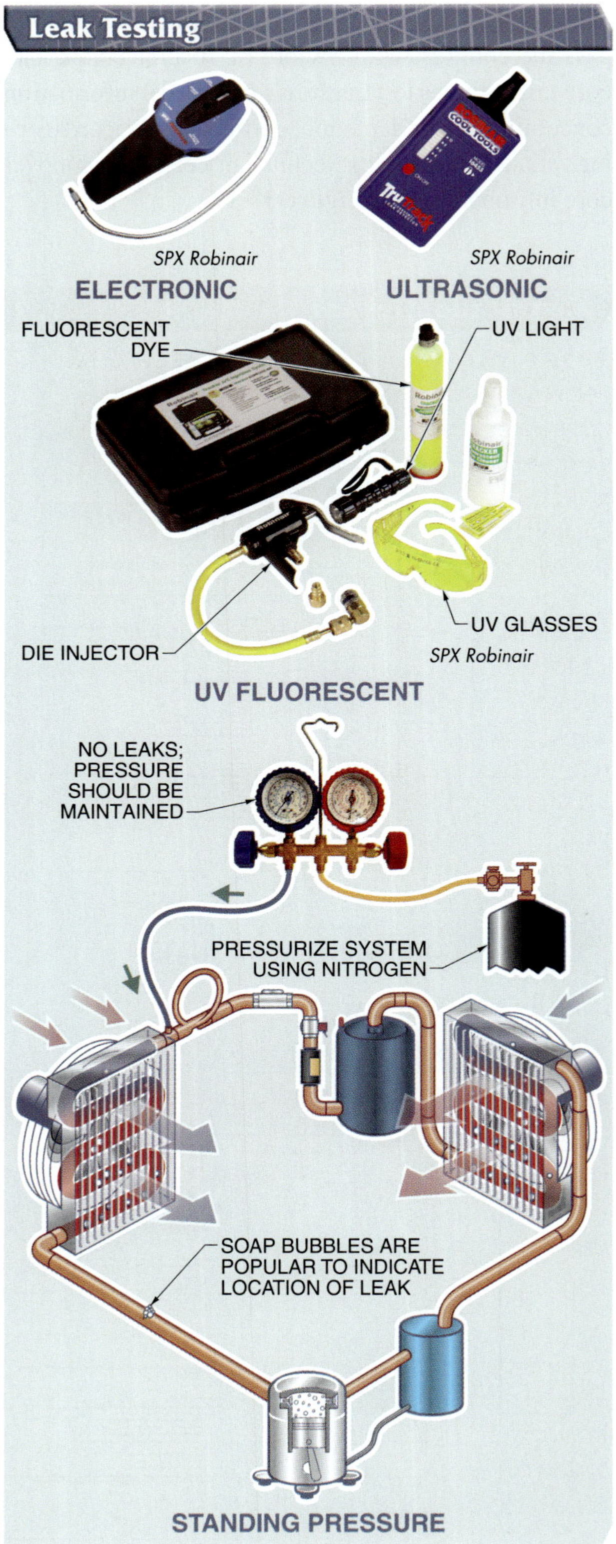

Figure 37-20. Leaks can be detected by using leak detectors or the standing pressure method, which adds a small amount of refrigerant into the system and pressurizes the system using a pressure source such as nitrogen.

It is not mandatory under the Clean Air Act to perform leak repairs on an appliance with a refrigerant charge less than 50 lb. However, technicians should repair all refrigerant leaks. The high cost of refrigerants, the potential for refrigerants to harm the environment, and customer satisfaction are all good reasons to repair refrigerant leaks.

RECOVERY, RECYCLING, AND RECLAIMING OF REFRIGERANTS

According to the Clean Air Act of 1990, technicians must not vent, release, or dispose of any substance used as a refrigerant in a manner that permits such a substance to enter the environment. Minute releases associated with good faith attempts to recapture and dispose of any refrigerant shall not be subject to the prohibition set forth in the preceding sentence.

A *minute release* is the unavoidable release of a very small amount of refrigerant during the service process while using low-loss fittings. For example, a minute release is the very small amount of refrigerant released when connecting or disconnecting service equipment to a refrigeration or air conditioning system while using low-loss fittings. **See Figure 37-21.** A *low-loss fitting* is a fitting used to prevent the release of refrigerant from a system to the atmosphere and to prevent air from entering the system.

In all other circumstances, the refrigerant must be captured, reused, and/or disposed of in a manner approved by the EPA. There are severe penalties and fines spelled out in the EPA regulations for contractors and technicians who do not abide by the regulations. The regulations also provide rewards to any person who furnishes information that leads to the arrest and conviction of personnel who intentionally vent refrigerants into the environment. To avoid such penalties and fines, contractors and technicians must properly recover, recycle, and reclaim refrigerants. **See Figure 37-22.**

Refrigerant Recovery

Refrigerant recovery is a process that allows refrigerant to be removed from a system in any condition and stored in an external container without necessarily testing or processing it in any way. Recovered refrigerant must not be reused in any equipment other than equipment having the same owner as the equipment from which the refrigerant was removed. For refrigerants to be used in any other systems, the refrigerant must be reclaimed to meet the Air Conditioning and Refrigeration Institute (ARI) Standard 700, *Specification for Fluorocarbon Refrigerants,* and be tested by a certified laboratory. **See Figure 37-23. See Procedure for Recovering Refrigerant—Type I Systems.**

Low-Loss Fittings and Valves

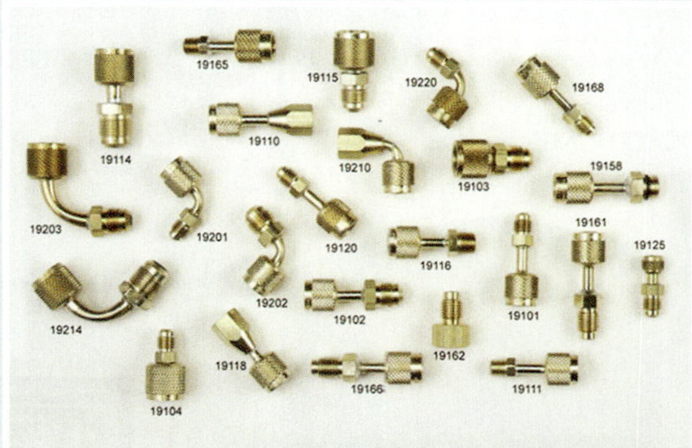

Yellow Jacket Div., Ritchie Engineering Company, Inc.

LOW-LOSS FITITNGS

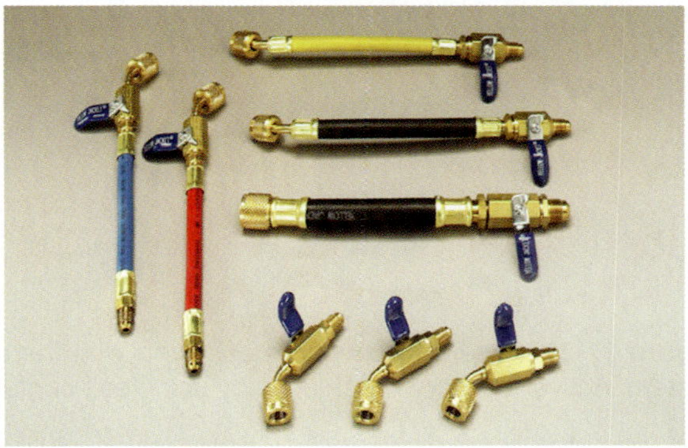

Yellow Jacket Div., Ritchie Engineering Company, Inc.

LOW-LOSS VALVES

Figure 37-21. Low-loss fittings are intended to prevent the release of refrigerant from a system to the atmosphere and to prevent air from entering the system.

Refrigerant Processes

Activity	Use
Recover	Method to use when refrigerant is in excellent condition
Recycle	Method to use when refrigerant is in excellent condition
Reclaim	Method to use when refrigerant is in poor condition
Destroy	Method to use when refrigerant is beyond reclaiming

Figure 37-22. Technicians can recover and recycle on site. Reclaiming and destroying a refrigerant can only be done by a refrigerant manufacturing facility.

Refrigerant Recovery

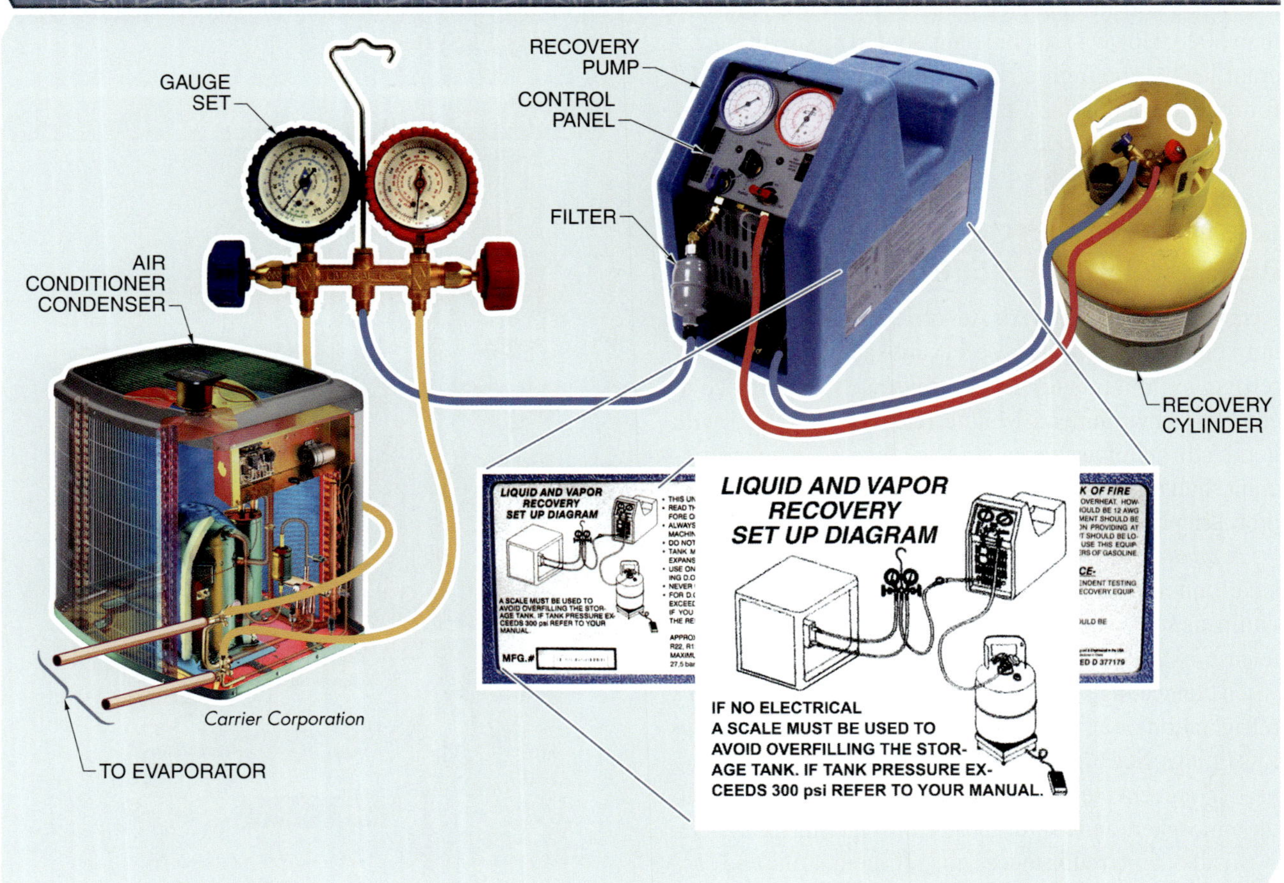

Figure 37-23. Passive recovery is achieved with the assistance of system components to remove refrigerant from the system. Active recovery is achieved by using self-contained recovery machines (most popular).

Procedure

Recovering Refrigerant—Type I Systems

1. Set up recovery equipment to recover refrigerant from the Type I system.
   a. Verify that the clean filter screen is installed behind the inlet fitting.
   b. Connect a hose from the outlet fitting of the recovery unit to the liquid port on the recovery cylinder.
   c. Connect a hose from the inlet fitting on the recovery unit to the output port on the manifold gauge set.
   d. Connect a hose from the liquid (low-pressure) side of the manifold gauge set to the liquid side of the system.
   e. Connect a hose from the vapor (high-pressure) side of the manifold gauge set to the vapor side of the system.
   f. Verify that the inlet and outlet valves on the recovery unit are closed.
   g. Place the recovery cylinder on a scale to avoid overfilling the cylinder.

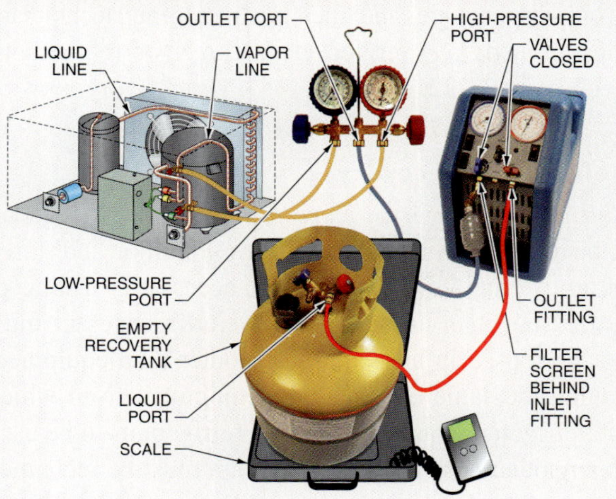

continued on next page

Recovering Refrigerant—Type I Systems *(continued)*

2. Plug the recovery unit into a 115 V power source.

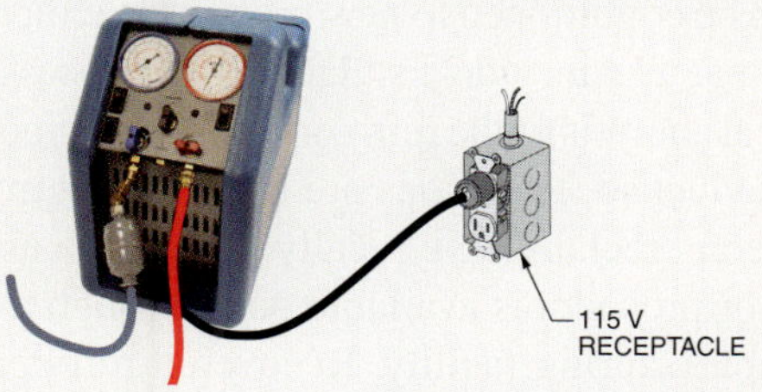

3. Slowly open the liquid valve of the recovery cylinder while watching the hoses and connections for leaks.

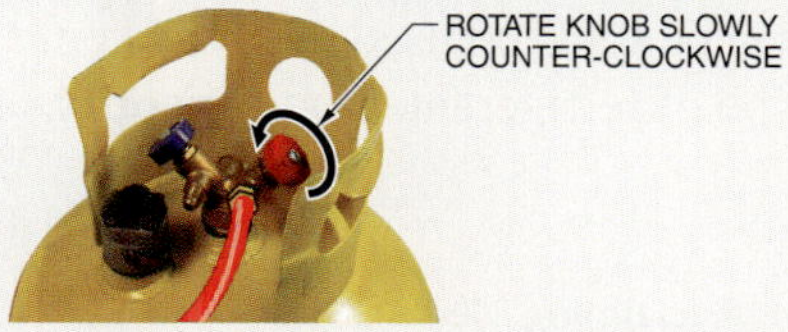

4. Set the recovery/purge valve on the recovery unit to RECOVER.

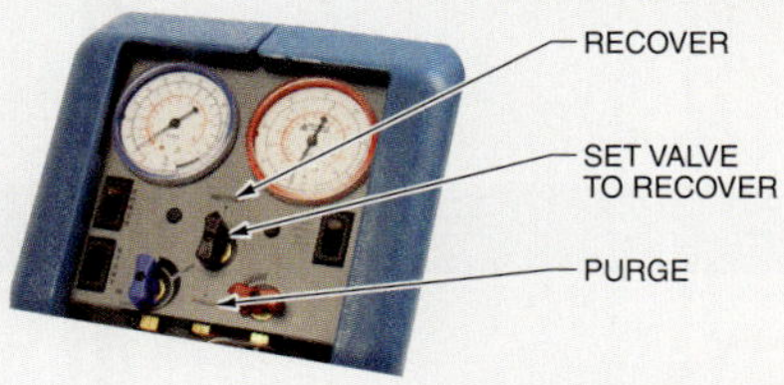

5. Open the liquid valve on the manifold gauge set. *Note:* Opening the liquid valve removes liquid refrigerant from the system first, greatly reducing recovery time.

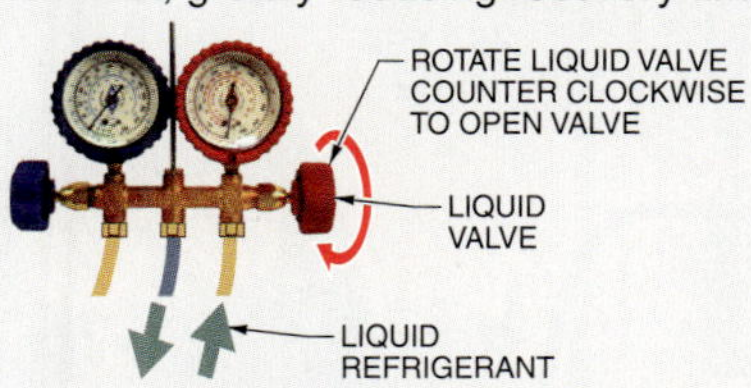

6. Open the outlet valve on the recovery unit.

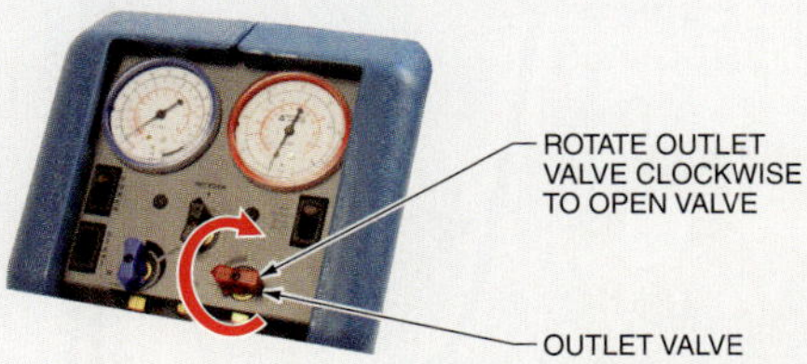

7. Toggle the power switch to the ON position.

8. Slowly open the inlet valve on the recovery unit. *Note:* If the recovery unit begins to "knock," slowly throttle back (close) the inlet valve until the noise stops.

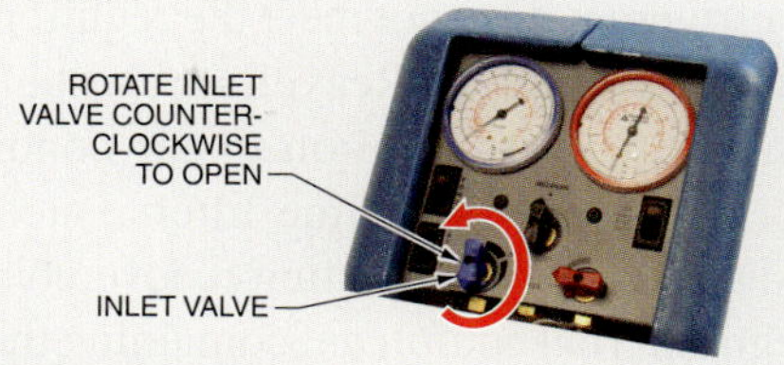

9. Toggle the start switch to RUN.

10. Once the liquid refrigerant has been removed from the system, open the vapor valve on the manifold gauge set to finish evacuating the system.

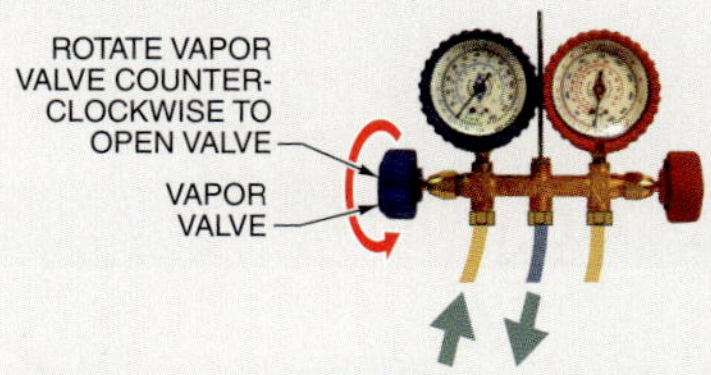

11. Run the recovery unit until the desired vacuum (typically 10 in. Hg vacuum for small Type I systems) is achieved.

12. Close the vapor and liquid valves on the manifold gauge set.

13. Rotate the inlet valve on the recovery unit to the closed position.

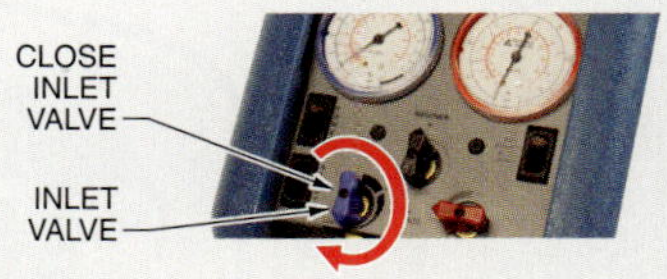

14. Toggle the start switch to the standby position and the power switch to the OFF position.

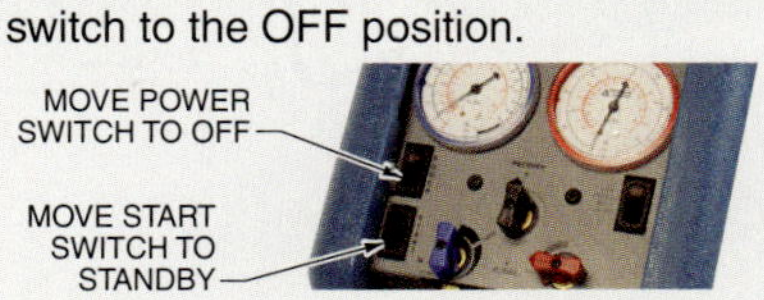

15. Always purge the recovery unit and hoses after a recovery procedure.

Refrigerant Recycling

Refrigerant recycling is a process that allows refrigerant to be cleaned by separation with single- or multiple-pass devices, such as replaceable-core filter-driers, which reduce moisture, acidity, and particulate matter. Refrigerant recycling typically applies to procedures implemented at the job site or at a local service shop. It is standard practice to recycle refrigerants that are suspected of being contaminated before recharging them into a system.

Since testing by a certified laboratory is required for compliance and recycled refrigerant is not tested, recycled refrigerant does not meet the requirements of ARI Standard 700-2006. Recycling can be accomplished inside the refrigeration or air conditioning system by installing liquid-line filters, suction-line filters, and driers or by circulating recovered refrigerants through external equipment containing the appropriate filters and driers. **See Figure 37-24.** Recycling machines use either the single-pass or multiple-pass method of cleaning refrigerant.

Refrigerant Reclaiming

Refrigerant reclaiming is a process that allows used refrigerant to be conditioned to new product specifications by means that may include distillation. The process requires chemical analysis of the refrigerant to determine whether new product specifications are met. **See Figure 37-25.** Refrigerant reclaiming typically implies the use of processes or procedures available only through a refrigerant manufacturing facility. Recovered refrigerants are typically sent to these facilities to be reclaimed. When a refrigerant is successfully reclaimed to ARI 700 standards, it can be sold and used in any system appropriate for that type of refrigerant. **See Appendix.**

Refrigerant Recycling Media Clip

Refrigerant Reclaiming Media Clip

Refrigerant Recycling

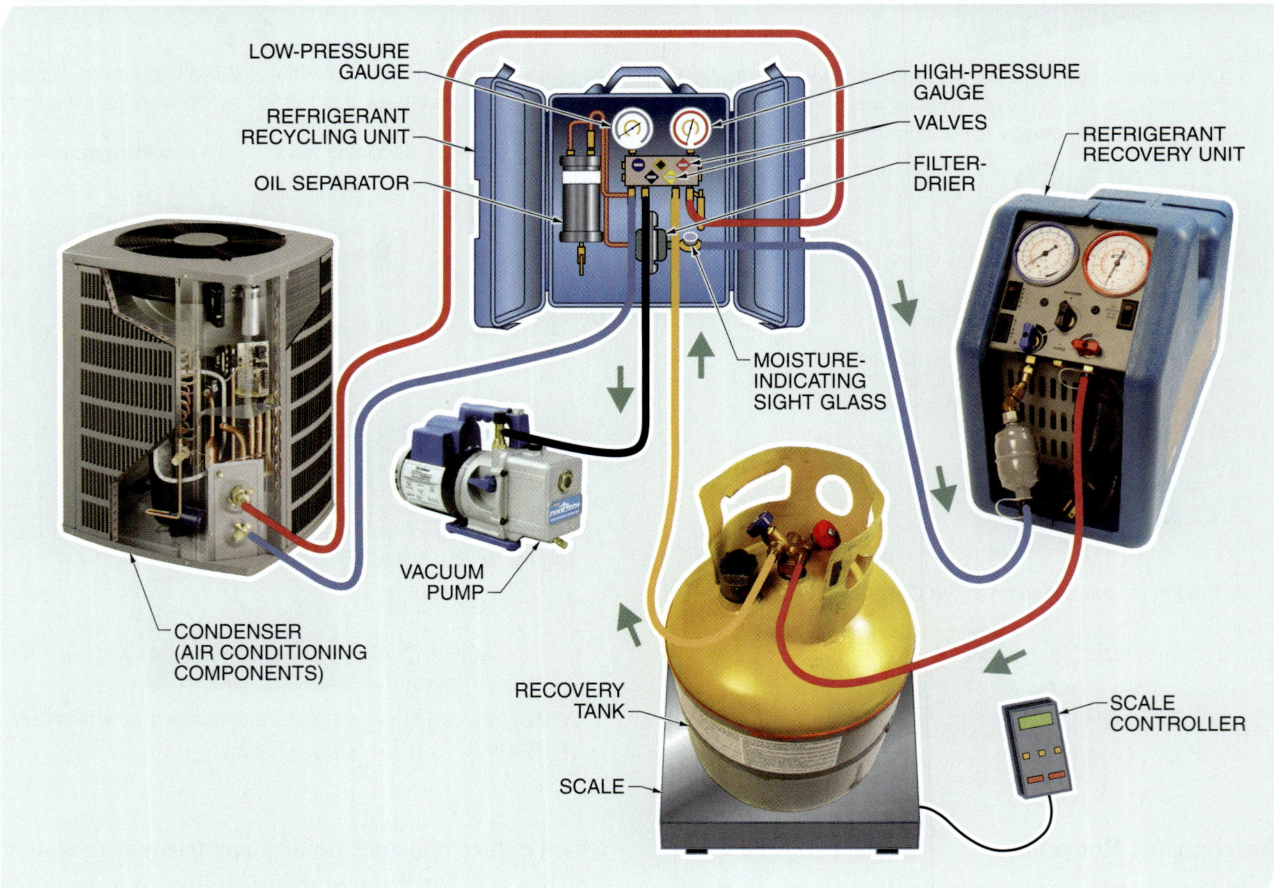

Figure 37-24. Refrigerant recycling is the removal of refrigerant from a system and cleaning of the refrigerant for reuse in the system.

Refrigerant Analysis

TEST RESEARCH INC.

P.O Box 56765 * 8225 S. River Road * Jefferson City, MO 65105
(555) 555-0000 * FAX (555) 555-0000 www.Testresearch.com

ARI 700 R-135a Analysis

Customer Name: AAA Mechanical   Customer PO: 98765
Reference ID: Batch 18   Receive Date: 3/7/12
Lab ID: XY200400600   Analysis Date: 3/8/12
Sample Number: BB01C0034-111   Analyzed By: MRB

Sample Data		ARI 700 Allowance Levels
Water	273 ppm	<10ppm
High Boiling Residue	0.3 vol%	<0.01 vol%
Chloride	Pass	<Pass
Particulate Solids	Pass	<Pass
Air	0.29 vol%	<1.5 vol%

Air analysis was performed on the vapor space of a liquid sample

Acidity	0.6 wppm as HCl	<1.0 wppm as HCl
Volatile Impurities	0.1618 wt%	<0.5 wt%
23	<0.0001 wt%	
32	0.0002 wt%	
1123	0.0019 wt%	
143a	0.0003 wt%	
125	<0.0001 wt%	
115	<0.0001 wt%	
134	<0.0001 wt%	
152a	<0.0001 wt%	
12	0.0196 wt%	
1122	0.0002 wt%	
124	0.0347 wt%	
31	<0.0001 wt%	
133a	<0.0001 wt%	
114	0.0035 wt%	
114a	<0.0001 wt%	
11	0.0682 wt%	
1112a	<0.0001 wt%	
1121C	<0.0001 wt%	
123	0.0260 wt%	
1121-T	<0.0001 wt%	
113	0.0073 wt%	

Notes
Noncondensable gases were analyzed from the vapor space of a liquid sample. Water is greater than 10 ppm. High boiling residue is greater than 0.01% by volume. Sample does not conform to ARI 700 requirements.

Figure 37-25. When it is reclaimed, refrigerant is reprocessed to meet new refrigerant standards and analyzed to ensure purity.

The EPA requires the evacuation of refrigeration and air conditioning equipment to established levels depending on the manufacture date of the evacuation equipment used. Evacuation levels for appliances are standard except for motor vehicles, motor-vehicle-like appliances, and small appliances.

The boiling point of refrigerants in high-pressure appliances is below 50°F at atmospheric pressure, making it much easier to evacuate refrigerants from these systems compared to low-pressure systems. These systems are easier to evacuate because a deep vacuum is not required. Low-pressure appliances have refrigerants with a boiling point above 50°F at atmospheric pressure, making it much harder to vaporize the refrigerant. Low-pressure appliances require a much deeper vacuum to vaporize the refrigerant and evacuate the system.

When using recovery and recycling equipment manufactured on or after November 15, 1993, technicians servicing small appliances are required by the EPA to recover 80% of the refrigerant when the system compressor is not operational and 90% of the refrigerant when the system compressor is operational. The required evacuation level of a system depends on the manufacture date of the evacuation equipment. **See Figure 37-26.**

Required Evacuation Levels

Type of Appliance	Manufacture Date	
	Before 11/15/93	On–After 11/15/93
HCFC-22 appliance* normally containing less than 200 lb of refrigerant	0.0†	0.0†
HCFC-22 appliance* normally containing 200 lb or more of refrigerant	4.0†	10.0†
Other high-pressure appliance* normally containing less than 200 lb of refrigerant (CFC-12, -114, -500, -502)	4.0†	10.0†
Other high-pressure appliance* normally containing 200 lb or more of refrigerant (CFC-12, -114, -500, -502)	4.0†	15.0†
Very-high-pressure appliance* (CFC-13, -503)	0.0†	0.0†
Low-pressure appliance* (CFC-11, HCFC-123)	25.0†	25.0‡

* or isolated component of appliance
† in in. Hg absolute
‡ in mm Hg absolute

Figure 37-26. To evacuate a system, 80% of the system refrigerant must be removed if the recycling and recovery equipment was manufactured after November 15, 1993.

TECH TIP

Evacuating systems to 0.0″ Hg is impossible. However, technicaians in all good faith must make every attempt to evacuate the system to that level.

Since evacuation levels may not be attainable because of leaks in a system or the recovery process contaminates the refrigerant due to leaks, technicians must isolate leaking components from nonleaking components. The nonleaking components must be evacuated to the required limits. Leaking components are evacuated to the lowest level possible. **See Procedure for Evacuating Type I Systems.**

Procedure

Evacuating Type I Systems

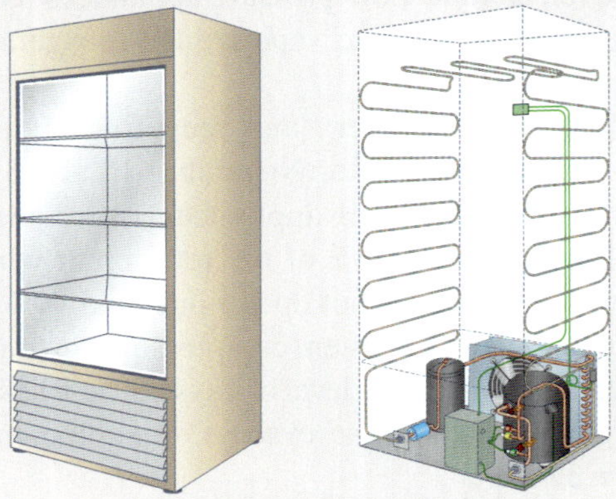

TYPE I SYSTEM (COOLER)

1. Shut off the Type I refrigeration system. Stop the compressor.

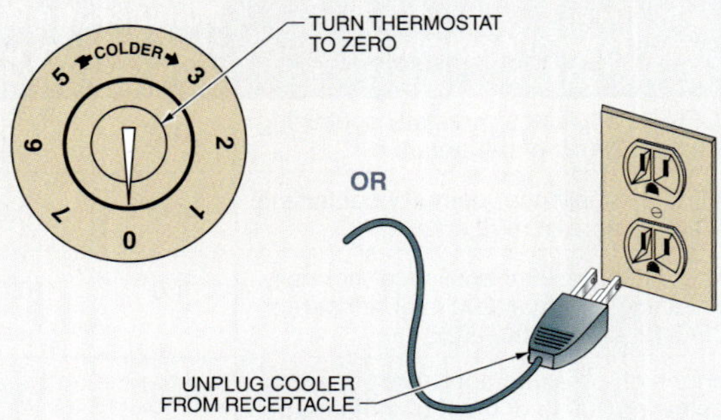

2. Install solderless service piercing valves and connect gauge set.
   a. Connect the high-pressure (red) gauge to the high-pressure liquid solderless piercing valve and the low-pressure (blue) gauge to the low-pressure vapor solderless piercing valve.
   b. Ensure that the valves on the gauge set are in the closed position.

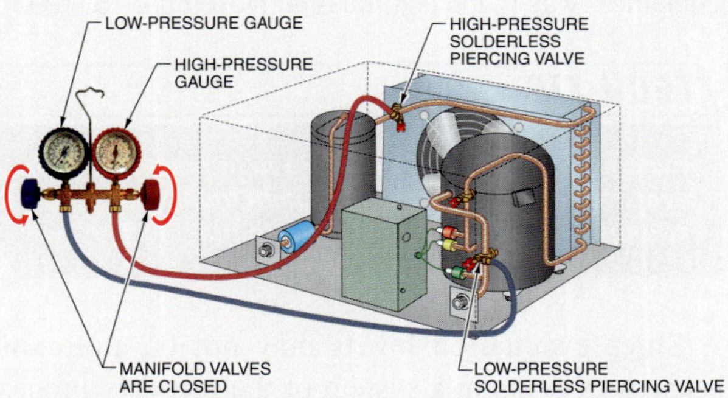

3. Connect a recovery machine to the center manifold (yellow) hose of the gauge set.
   a. The recovery machine should have a drier, so moisture can be removed from the evacuated refrigerant.
   b. A recovery cylinder is required to store evacuated refrigerant.

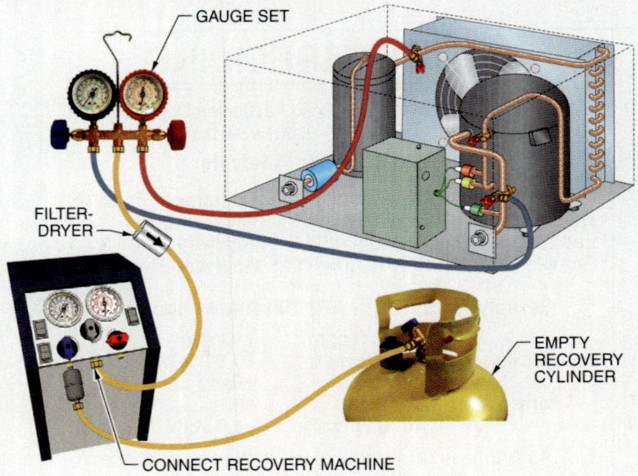

4. Open the low-pressure valve and the high-pressure valve on the gauge set and the valve on the recovery cylinder.

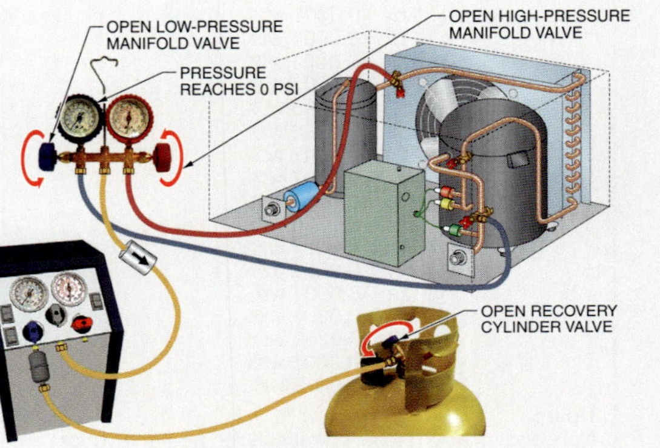

5. Operate the recovery machine to draw the refrigerant out of the refrigeration system.
6. Continue operating the recovery machine until the pressure in the system reaches 0 psi. Close the low-pressure and high-pressure valves on the manifold. Then, stop the recovery machine.

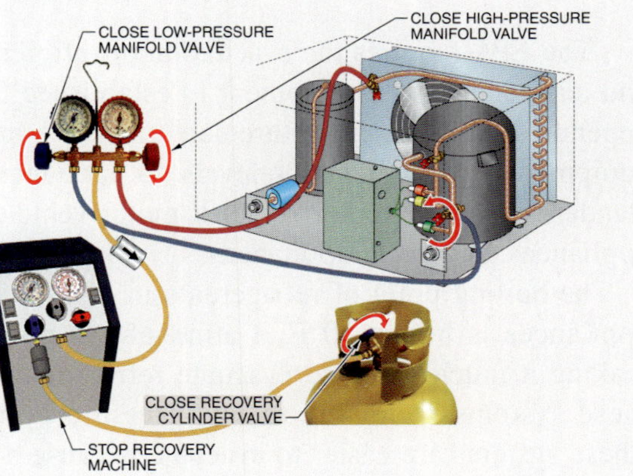

continued on next page

Evacuating Type I Systems *(continued)*

7. Observe the gauges for a pressure rise.
   - If pressure rises quickly (to atmospheric pressure), the hose connections could be loose. Tighten the connections, repeat steps 4 through 6, and observe.
   - If pressure continues to rise quickly, there is a leak in the system. Open the low-pressure valve and resume evacuating.
   - If pressure rises slowly, there is moisture in the system or refrigerant still in the system. Open the low-pressure valve and resume evacuating.
8. Hold system pressure at 0 psi for a few hours or half a day, depending on the size of the refrigeration system.
9. Close all valves and remove the gauge set from the system and recovery cylinder.
10. Remove temporary access p ercing valves and braze permanent access valves.
11. Perform a leak test once repairs have been made.
12. Evacuate the refrigeration system once again. Start with step 2 and continue to step 9.
13. Recharge the system using the evacuated refrigerant or new refrigerant.

Procedure

Digital Resources ATPeResources.com/QuickLinks • Access Code: 482610

UNIT

Pressure-Enthalpy Diagrams 38

Pressure-enthalpy diagrams define the relationships between the thermodynamic properties of a refrigerant as the refrigerant flows through a refrigeration system. The thermodynamic properties found on a pressure-enthalpy diagram are the pressure, enthalpy, constant temperature, constant quality, constant volume, and constant entropy of a refrigerant. The diagrams are graphic representations of the properties of a refrigerant as it flows through a refrigeration system. The graphic representation is helpful for visualizing and evaluating the performance of a refrigeration system.

OBJECTIVES

- Identify the purpose of a pressure-enthalpy diagram.
- Describe the properties that can be found on a pressure-enthalpy diagram.
- Plot an ideal refrigeration cycle on a pressure-enthalpy diagram.
- Plot an actual refrigeration cycle on a pressure-enthalpy diagram.
- Explain how to find the refrigeration effect, compression ratio, and coefficient of performance of a mechanical refrigeration system.

Index to Procedures

PRESSURE-ENTHALPY DIAGRAMS

A *pressure-enthalpy diagram,* also referred to as a Mollier diagram, is a graphic representation of the pressure/enthalpy relationship of a refrigerant. On a pressure-enthalpy diagram, pressure is plotted against enthalpy. The curves on a pressure-enthalpy diagram represent the absolute pressure (psia) and the enthalpy (Btu/lb) of a refrigerant at different properties. **See Figure 38-1.** All of the properties are plotted in relation to pressure and enthalpy.

When any two properties of a refrigerant are known, a pressure-enthalpy diagram can be used to find the other properties of the refrigerant. The properties that can be found on a pressure-enthalpy diagram are pressure, enthalpy, temperature, constant quality, constant volume, and constant entropy. Each refrigerant has a specific pressure-enthalpy diagram because the properties of each refrigerant are different. **See Figure 38-2.**

TECH TIP

Pressure-enthalpy diagrams are used to indicate design system operation. They can also be used to indicate abnormal system operation and to help troubleshoot individual components.

Pressure-Enthalpy Diagrams

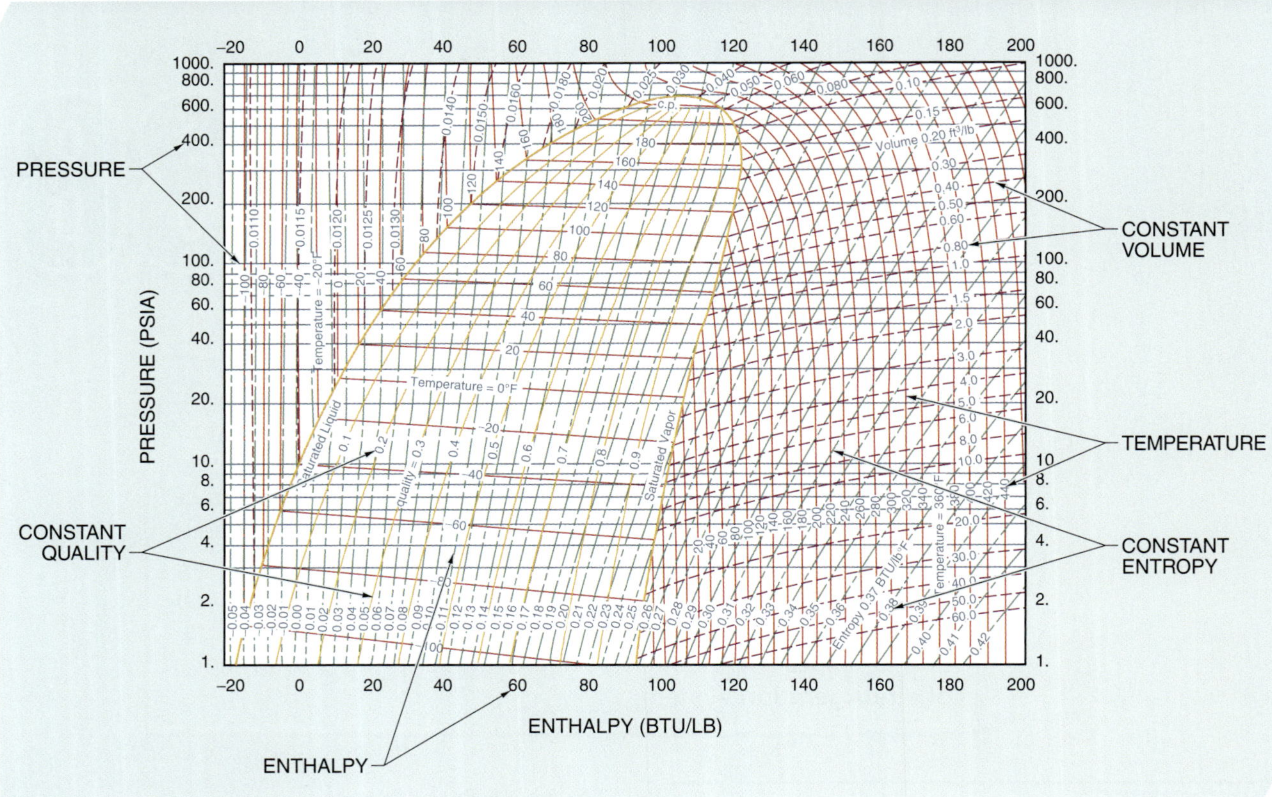

Figure 38-1. Pressure-enthalpy diagrams show the properties of a refrigerant as it flows through a refrigeration system.

By drawing a graph on a pressure-enthalpy diagram according to the pressure and enthalpy of a refrigerant, the other properties of the refrigerant at any point in the refrigeration system can be found. The graph on the diagram represents the properties of a refrigerant as it flows through a mechanical compression refrigeration system.

Pressure

System pressure is the pounds per square inch represented by the resistance to flow of refrigerant moving through a system. The system pressure of an operating refrigeration system depends on the type of refrigerant used and the temperatures required in the system. The pressure scales on a pressure-enthalpy diagram are located on the left and right sides of the diagram. On the diagram, system pressure is expressed in pounds per square inch absolute (psia). System pressure at different points is found on the horizontal lines that run across a pressure-enthalpy diagram.

Pressure gauges in psi that are permanently attached to a system allow the quick referencing of pressure-enthalpy diagrams during troubleshooting.

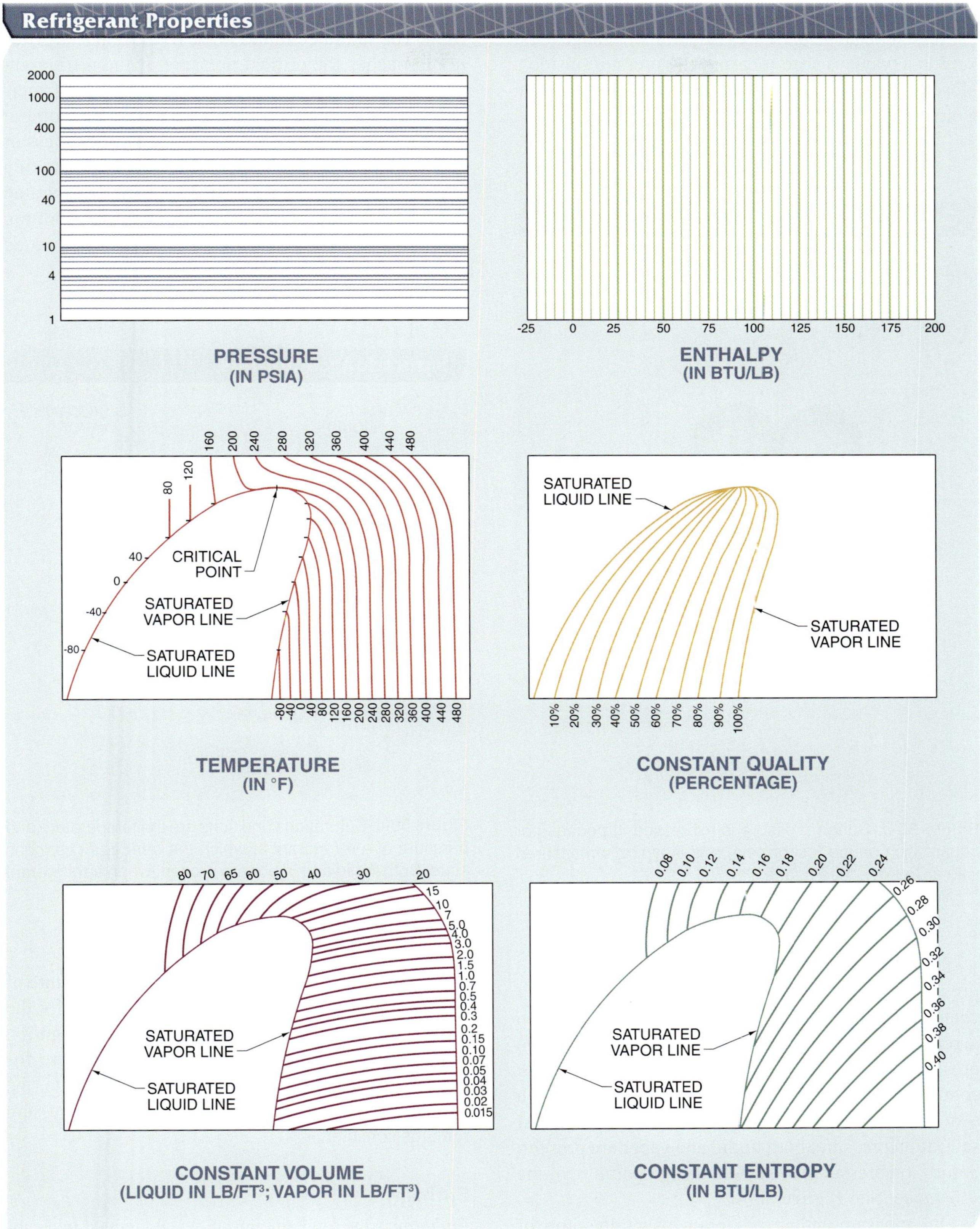

Figure 38-2. Each refrigerant has its own pressure-enthalpy diagram that consists of pressure, enthalpy, temperature, constant quality, constant volume, and constant entropy.

System pressure is measured with a gauge and is expressed in gauge pressure (psig or psi). Gauge pressure is always used to measure the pressure inside a closed system. Atmospheric pressure has no effect on a closed system. **See Figure 38-3.**

Gauge Pressure

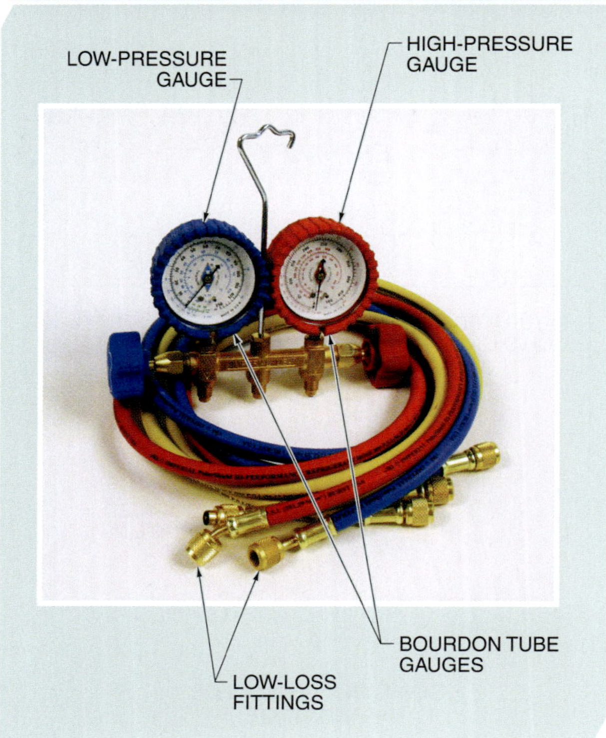

Figure 38-3. Gauge pressure is expressed in pounds per square inch gauge (psi or psig) and is typically measured with a Bourdon tube gauge.

Saturation conditions are the temperature and pressure of a refrigerant at which the refrigerant changes state. Both liquid and vapor are present when a refrigerant changes state, or reaches saturation pressure, which causes the refrigerant to vaporize or condense. *Saturation pressure* is the pressure at which a substance, such as a refrigerant, changes state. At saturation pressure, both liquid and vapor are present. Saturation pressure varies with the temperature of the substance. **See Figure 38-4.**

On a pressure-enthalpy diagram, the temperature of a saturated liquid refrigerant is found on the saturated liquid line, which is a curved line that begins near the lower left corner of the diagram and curves up to the right. The saturated liquid line curves right sharply as it approaches the critical pressure of the refrigerant and continues to the critical point for the refrigerant. The critical point of a refrigerant is the point at which refrigerant does not change state regardless of absorption or rejection of heat. The temperature of a saturated refrigerant vapor is found on the saturated vapor line, which begins at a point on the bottom of the diagram and curves left to meet the saturated liquid line at the critical point.

Saturation Conditions

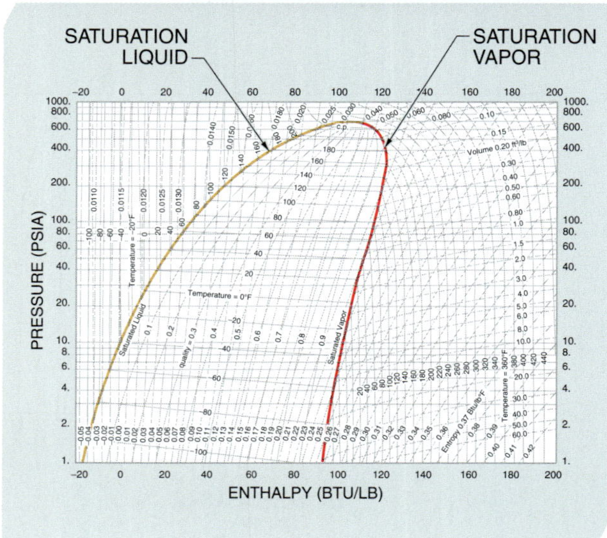

Figure 38-4. Saturation conditions are the temperature and pressure of a refrigerant at which the refrigerant changes state. Both liquid and vapor are present when a refrigerant changes state by vaporizing or condensing.

The saturated pressure and saturated temperature of a refrigerant are found on a table of properties for the saturated liquid and saturated vapor. **See Appendix.** A refrigerant has a specific saturation pressure for each temperature. As a refrigerant changes state, the temperature remains constant as long as the pressure remains constant.

Enthalpy

Enthalpy is the total amount of heat contained in a substance. Enthalpy includes sensible and latent heat. When calculating the difference in enthalpy between two temperatures or pressures, a base of 0 Btu at −40°F is commonly used. The enthalpy scale on a pressure-enthalpy diagram does not begin at 0 Btu because the enthalpy for refrigerants is considered to be 0 Btu/lb at −40°F. This is not technically correct, but because nearly all calculations are related to differences during a change of state, the results are satisfactory. Enthalpy is found on the vertical lines that run from the bottom to the top of a pressure-enthalpy diagram. Enthalpy values found on a pressure-enthalpy diagram are used to calculate the performance of a refrigeration system. **See Figure 38-5.**

Enthalpy

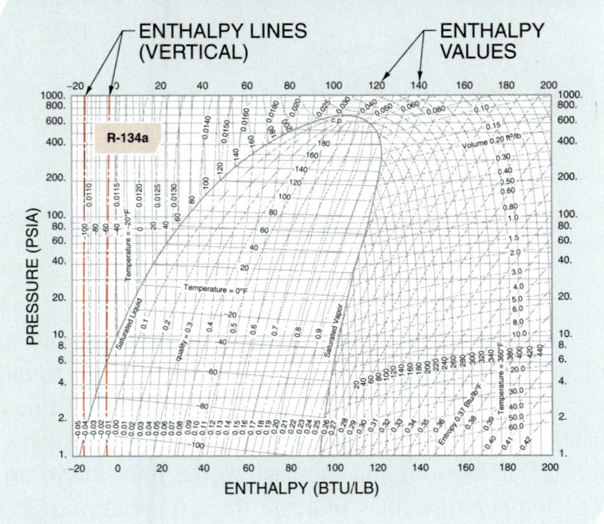

Figure 38-5. Enthalpy is the total heat (sensible and latent) contained in a substance. Pressure-enthalpy diagrams are used to calculate the performance of a refrigeration system.

Temperature

Temperature is the intensity of heat and is typically measured with a thermometer. A change in sensible heat causes a temperature change. *Saturation temperature* is the temperature at which a substance, such as a refrigerant, changes state. Saturation temperature varies with the pressure on the substance. At saturation temperature, both liquid and vapor are present.

The saturation temperature of a refrigerant is found on a table of properties for the particular refrigerant. In an evaporator, a refrigerant absorbs heat and vaporizes to a gas. While the refrigerant vaporizes, the liquid-vapor mixture is at the saturation temperature for that pressure in the evaporator. *Saturated liquid* is liquid at a certain temperature and pressure that will vaporize when pressure is decreased or temperature is increased. *Saturated vapor* is vapor at a certain temperature and pressure that will condense when the pressure is increased or temperature is decreased.

Superheat is sensible heat added to a refrigerant after it has changed state from a liquid to a gas. The amount of superheat is the amount of heat added above the saturation temperature. The temperature of a liquid remains constant during vaporization. For example, water absorbs heat as the water flows through a boiler. When the water has absorbed enough heat, it vaporizes and becomes steam. **See Figure 38-6.** When additional heat is added to the steam, the temperature of the steam will increase.

Heating Superheat

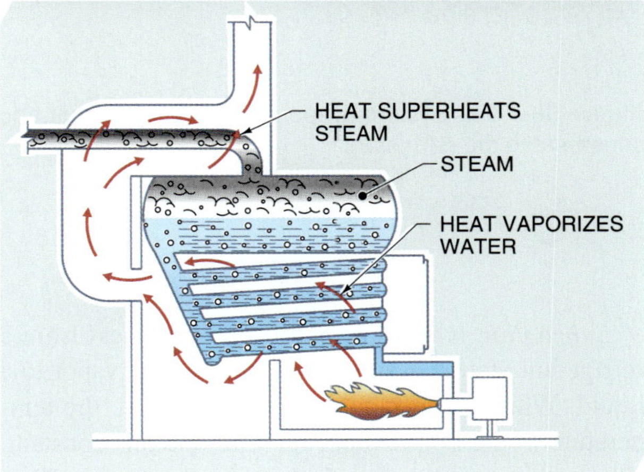

Figure 38-6. Superheat is sensible heat added to a substance after the substance has completely vaporized.

Also in refrigeration, superheat is typically added to the refrigerant by the evaporator. When the refrigerant leaves the evaporator, it flows to the compressor through the suction line. As the refrigerant flows through the compressor, heat of compression is also added. *Heat of compression* is the thermal energy equivalent to the mechanical energy expended by the motor that turns the compressor and the heat created by the compression of the refrigerant molecules. The heat of compression adds additional superheat to the refrigerant. **See Figure 38-7.** The amount of superheat is found by using the temperature difference between the temperature of the refrigerant and the saturation temperature at a given pressure.

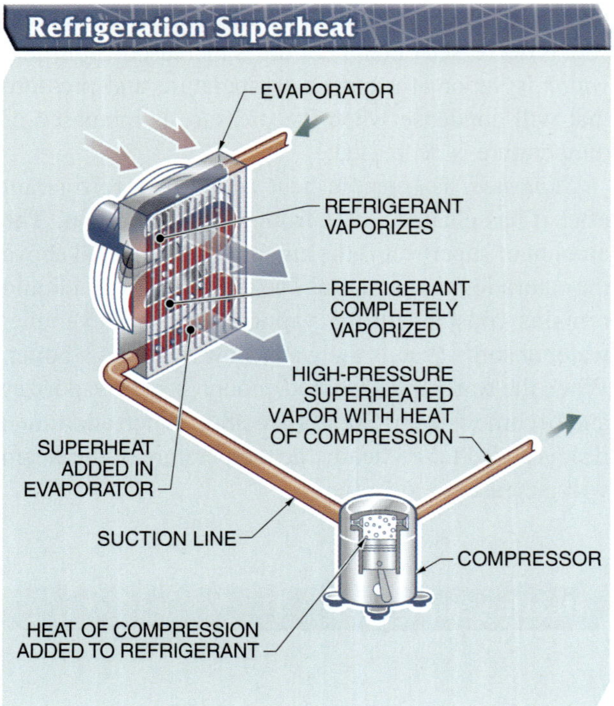

Figure 38-7. The system evaporator and compressor add superheat to the refrigerant.

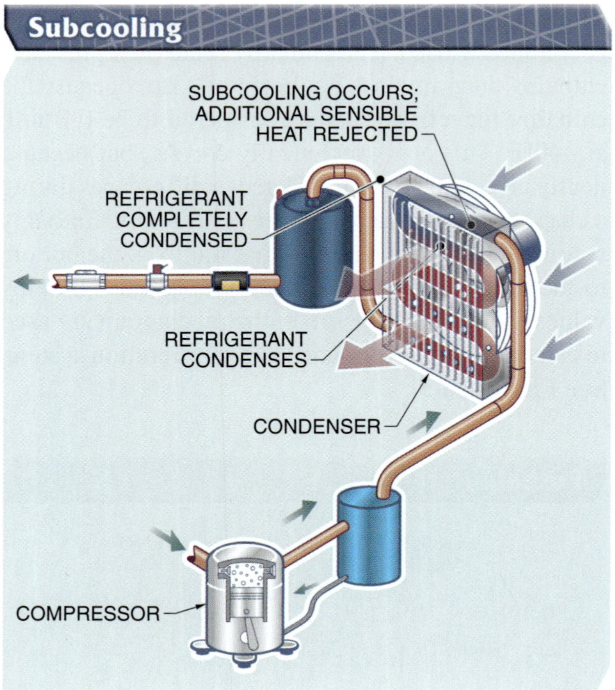

Figure 38-8. Subcooling typically occurs in the last one or two rows of the condenser coil.

Subcooling is the process of removing heat from a refrigerant after it has changed state from a vapor to a liquid. When a vapor condenses to a liquid, the temperature of the liquid-vapor mixture remains constant. When additional heat is rejected after a refrigerant has completely condensed, the temperature of the liquid refrigerant will decrease.

As vapor refrigerant flows through a condenser, the refrigerant is at the saturation temperature for the specific pressure. Heat is rejected by the refrigerant as it flows through the condenser. When the latent heat is rejected, the refrigerant changes state from a vapor to a liquid. After the refrigerant has completely condensed to a liquid, additional sensible heat is rejected in the last rows of the condenser coil. When the refrigerant leaves the condenser, the refrigerant is subcooled because the refrigerant rejects additional heat. **See Figure 38-8.**

For example, a refrigerant rejects heat as it flows through a condenser coil. The rejection of heat causes the refrigerant to condense back to the liquid state. The refrigerant condenses to a liquid in the first 90% of the condenser coil. The liquid refrigerant that flows through the last 10% of the condenser coil is subcooled as the liquid rejects heat.

Temperature scales that indicate the temperature of the saturated liquid and saturated vapor are found between the saturated liquid and saturated vapor lines. The temperature scales start with a negative temperature at the bottom and increase as the lines curve up. The temperature lines indicate the temperature of the refrigerant at any point on the lines. **See Figure 38-9.**

TECH TIP

Superheat and subcooling are shown on a pressure-enthalpy diagram. As the superheat and subcooling deviate from standard values, the deviation indicates that system efficiency and operation are being affected.

Constant Quality

Constant quality is a percentage that expresses the ratio of vapor to liquid as a refrigerant changes state. Constant quality is 0% at the saturated liquid line and 100% at the saturated vapor line. The area between the saturated liquid and saturated vapor lines is divided into ten segments. The two saturated liquid and saturated vapor lines curve up from the bottom of a pressure-enthalpy diagram and meet at the critical point for the refrigerant. **See Figure 38-10.**

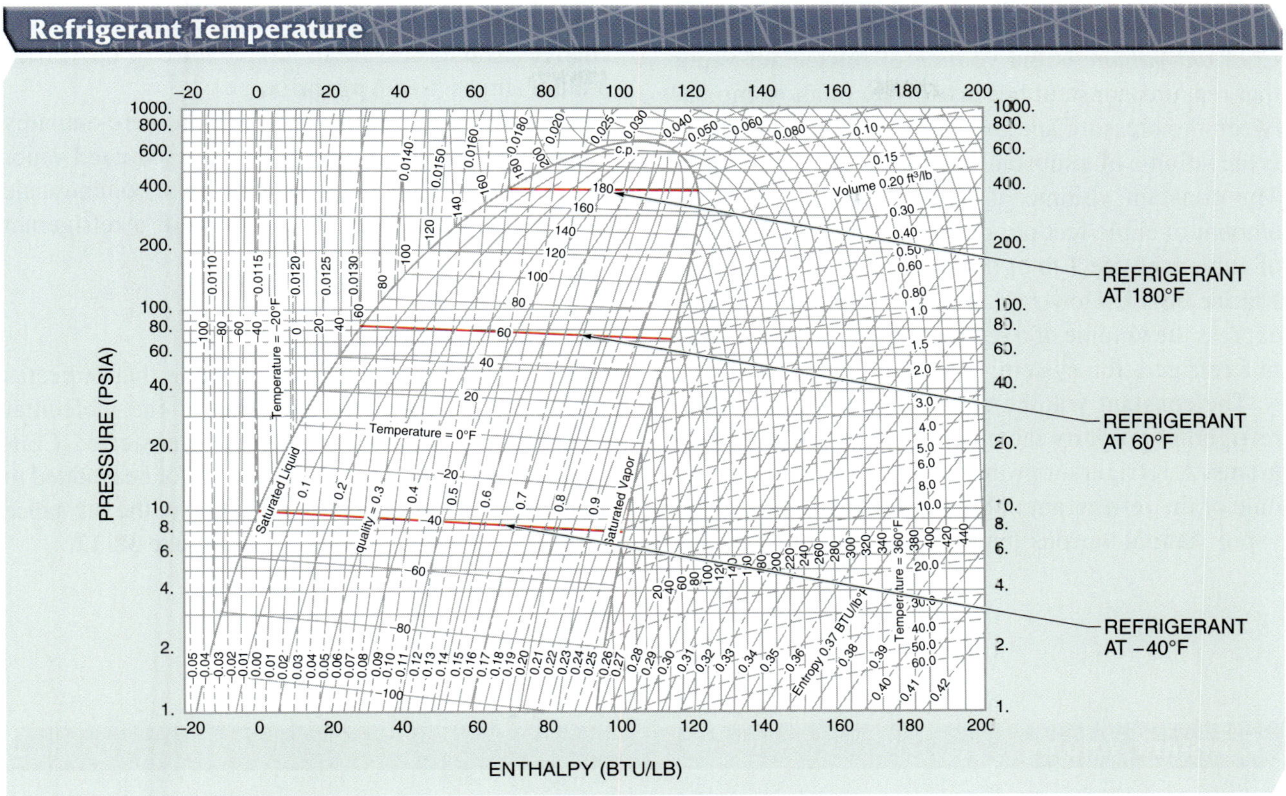

Figure 38-9. Temperature scales start with a negative temperature at the bottom and increase as the lines curve up to indicate the temperature of the saturated liquid and saturated vapor refrigerants.

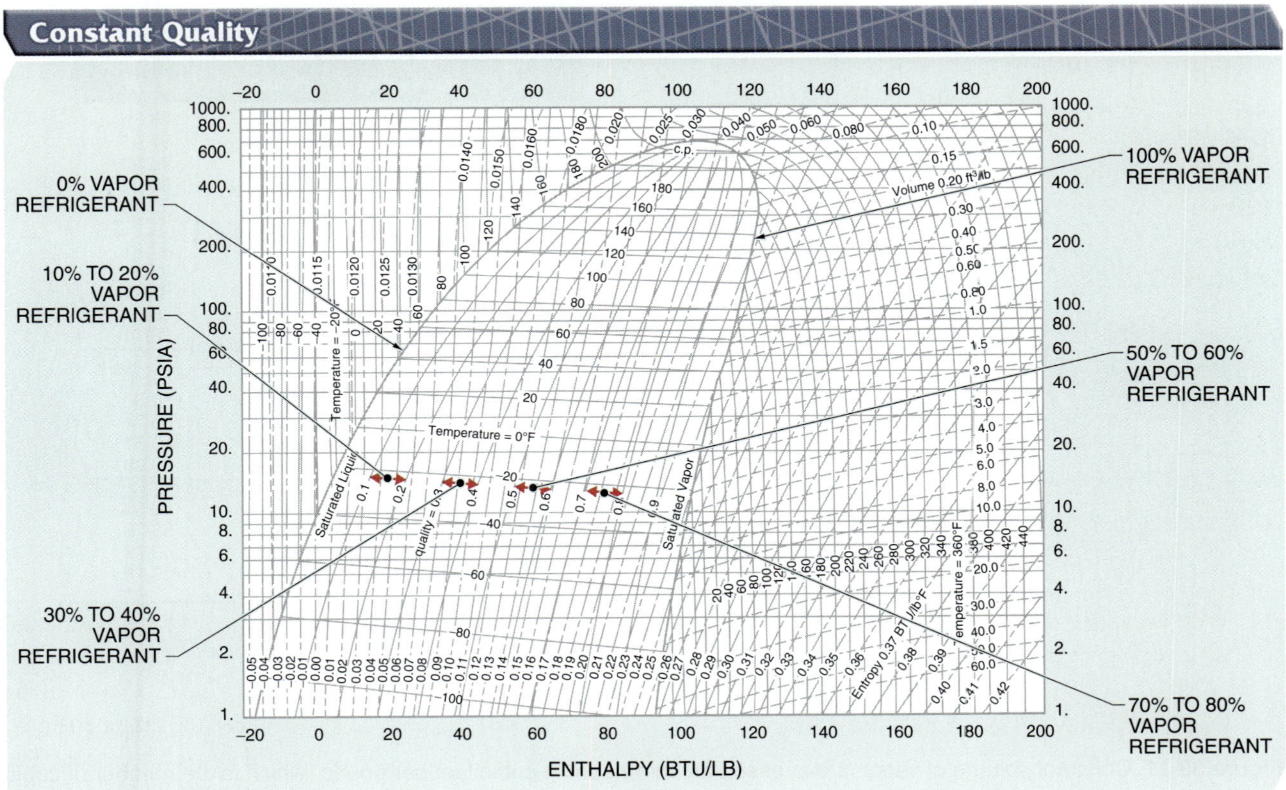

Figure 38-10. Constant quality, which is 0% at the saturated liquid line and 100% at the saturated vapor line, is a percentage that expresses the ratio of vapor to liquid as a refrigerant changes state.

Constant Volume

Constant volume is the volume of refrigerant vapor that remains constant because of the relationship between the pressure and enthalpy of the refrigerant. It is the volume of a substance per unit of the substance. The constant volume of vapor is expressed as an amount in cubic feet per pound, which is the number of cubic feet that 1 lb of the refrigerant occupies. **See Figure 38-11.** However, specific volume is used to express the volume of a refrigerant at a given pressure in a refrigeration system.

The constant volume of a refrigerant is found on refrigerant property tables or pressure-enthalpy diagrams. A refrigerant property table includes the volume of the refrigerant in both the liquid state and the vapor state at various temperatures. In an operating refrigeration system, the volume of the refrigerant changes as the pressure or temperature of the refrigerant changes. **See Appendix.**

Constant volume is found on a pressure-enthalpy diagram on the lines that begin on the saturated vapor line and run up to the right. The constant volume scale is found on these lines. The volume of a refrigerant decreases as pressure increases.

Constant Entropy

Constant entropy is a calculated value that indicates energy lost to the disorganization of the molecular structure of a substance when heat is transferred. Constant entropy is the ratio of the amount of heat added to a substance to the absolute temperature of the substance at the time the heat is added. **See Figure 38-12.**

Constant Volume

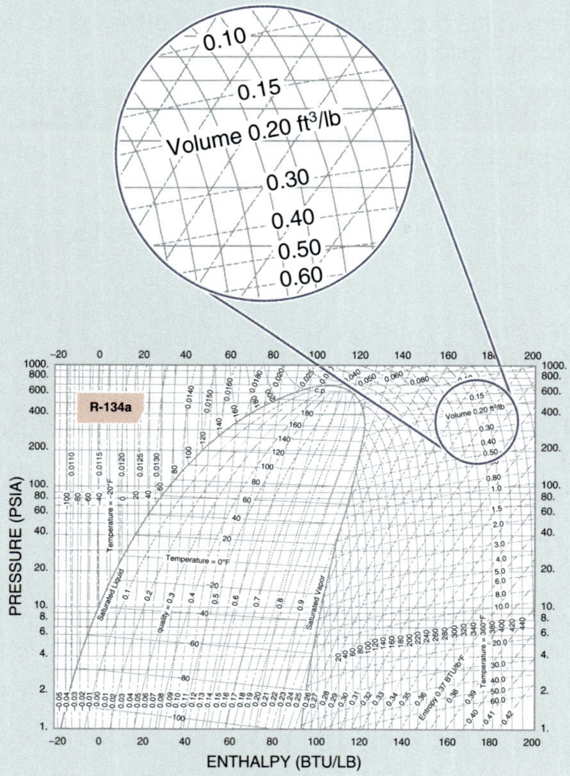

R-134a Properties of Saturated Liquid and Saturated Vapor

Temp. °F	Vapor Pressure*	In Cubic Feet Per Pound†	In Pounds Per Cubic Feet‡
−150	0.073	454.5455	98.86
−140	0.130	256.4103	97.91
−130	0.222	156.2500	96.96
−120	0.366	97.0874	96.01
−110	0.584	62.5000	95.06
−32	9.319	4.6729	87.51
−30	9.851	4.4366	87.31
−28	10.407	4.2141	87.11
−26	10.987	4.0032	86.91
−24	11.594	3.8066	86.71
28	39.166	1.2000	81.19
30	40.800	1.1538	80.96
32	42.486	1.1098	80.74
34	44.225	1.0676	80.51
36	46.018	1.0274	80.28
180	400.280	0.0964	56.83
190	449.384	0.0800	53.73
200	503.361	0.0645	49.73
210	563.037	0.0474	43.19

* in psia
† in ft³/lb
‡ in lb/ft³

Figure 38-11. Constant volume of vapor is expressed as an amount in cubic feet per pound, which is the number of cubic feet that 1 lb of the specific refrigerant occupies.

Constant Entropy

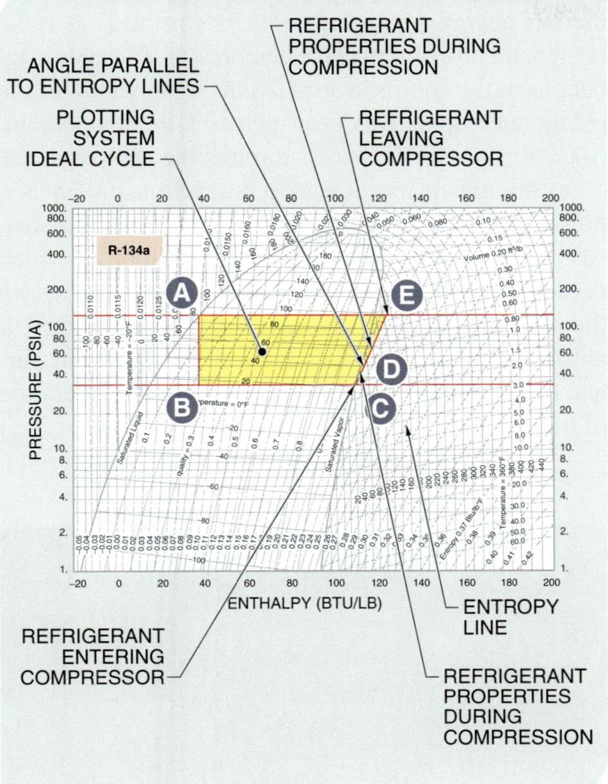

Figure 38-12. Constant entropy is included on a pressure-enthalpy diagram because it defines the properties of a refrigerant as it moves through an ideal compression cycle.

System Graphs

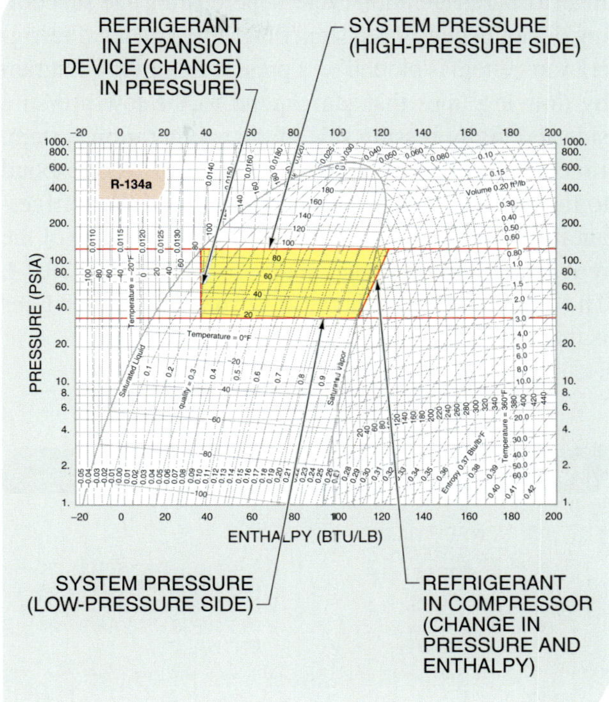

Figure 38-13. A system graph represents the performance of an operating refrigeration system using lines that correspond to the system pressure on the low-pressure side and high-pressure side of the refrigeration system, the change in pressure and enthalpy of the refrigerant in the compressor, and the change in pressure of the refrigerant in the expansion device.

Constant entropy lines begin on the saturated vapor line and run steeply up to the right. Constant entropy is only of value on a pressure-enthalpy diagram because it defines the properties of a refrigerant as it moves through an ideal compression cycle.

System Graphs

A *system graph* is a small four-sided graphic representation of the operation of a refrigeration system on a pressure-enthalpy diagram. System graphs are constructed by drawing lines that correspond to the system pressure on the low-pressure side and high-pressure side of the refrigeration system, the change in pressure and enthalpy of the refrigerant in the compressor, and the change in pressure of the refrigerant in the expansion device. All of the other properties of the refrigerant in a refrigeration system can be found on a system graph. **See Figure 38-13.**

TECH TIP

System graphs can be used by technicians to become familiar with new refrigeration systems and components. In many cases, technicians do not automatically know the proper system values. This is especially true as new refrigerants are introduced.

A system graph represents the performance of an operating refrigeration system. The properties of the refrigerant at different points are plotted on a pressure-enthalpy diagram and the resulting figure is the system graph, which is used to identify the other thermodynamic properties of a refrigerant as it flows through a refrigeration system. A system graph is a valuable tool for visualizing refrigeration system performance and obtaining data to calculate system performance.

IDEAL CYCLES

In an ideal refrigeration cycle, superheating and subcooling do not occur. The system diagram of an ideal refrigeration system is plotted on a pressure-enthalpy diagram by drawing lines that correspond to the low-pressure side and high-pressure side of the refrigeration system. Lines on the pressure-enthalpy diagram also correspond to the change in pressure and enthalpy of the refrigerant in the compressor and the change in pressure of the refrigerant in the expansion device. **See Figure 38-14.** The four-sided figure that results from drawing the lines is the ideal refrigeration cycle.

As a compressor compresses refrigerant, the mechanical energy produced by the compressor is converted to thermal energy. The refrigerant absorbs this heat as the refrigerant flows through the compressor. The additional heat from the compressor raises the temperature of the refrigerant vapor. This heat increases the temperature of the refrigerant as it flows through the compressor.

In the expansion device, a restriction produces a pressure decrease in the refrigeration system. The pressure decrease causes the refrigerant to vaporize, which causes the temperature of the refrigerant to change in the expansion device. As the refrigerant vaporizes, it absorbs sensible heat from the remaining liquid refrigerant. Removing this sensible heat causes the temperature of the refrigerant to decrease.

Ideal Cycles

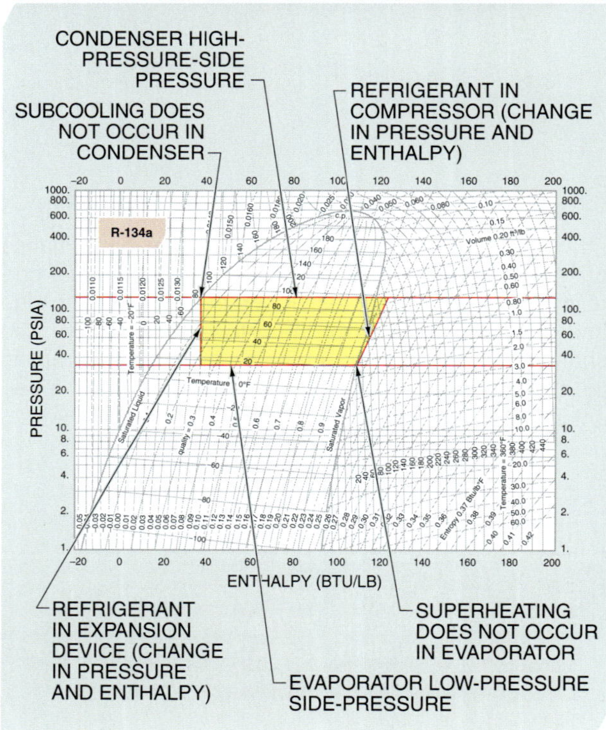

Figure 38-14. For an ideal cycle, the system graph of a refrigeration system is plotted on a pressure-enthalpy diagram without subcooling and superheat.

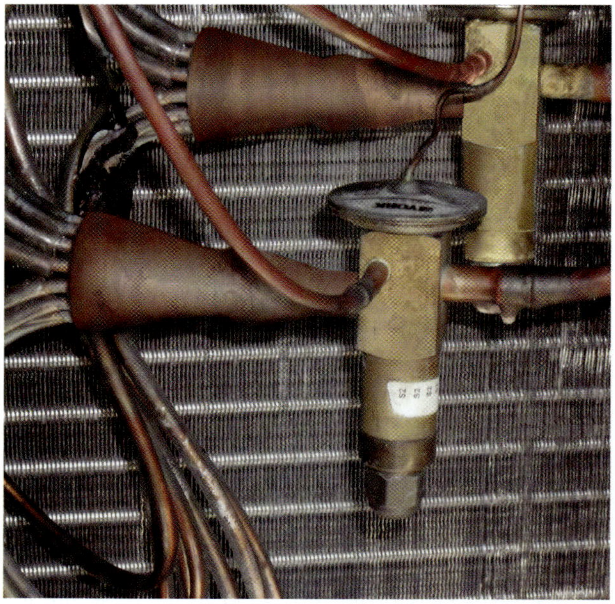

A pressure-enthalpy diagram can show the effect of the operation of thermostatic expansion valves.

TECH TIP

The effect of a TXV on system operation is shown on a pressure-enthalpy diagram. Any change or replacement of the TXV affects the system operation and is shown on the diagram.

Ideally, heat from outside a refrigeration system is not involved in the process of vaporizing the refrigerant as it flows through the expansion device, resulting in an adiabatic change. An *adiabatic change* is a change in the pressure and temperature of a substance in a closed system that occurs without heat transfer from outside the system. When system pressures on the low-pressure side and the high-pressure side of a refrigeration system and the type of refrigerant are known, the system diagram can be plotted. **See Procedure for Plotting Ideal Cycles on Pressure-Enthalpy Diagrams.**

Plotting Ideal Cycles on Pressure-Enthalpy Diagrams

A mechanical compression refrigeration system contains a refrigerant with characteristics as indicated in the chart below. The low-pressure side is 75 psi, and the high-pressure side is 185 psi. To plot the system graph of an ideal refrigeration cycle on a pressure-enthalpy diagram below, apply the following procedure:

1. First plot the pressure of the refrigeration system low-pressure side on a pressure-enthalpy diagram. The pressure must first be converted from gauge pressure to absolute pressure.

$$\begin{array}{r} 75.0 \text{ psi} \\ +\ 14.7 \\ \hline 89.7 \text{ psia} \end{array}$$

2. The line that represents the absolute pressure (psia) of the low-pressure side of the refrigeration system is located as a horizontal line. Draw this line from the left of the saturated liquid line to the right of the saturated vapor line. It will be the bottom line of the diagram.

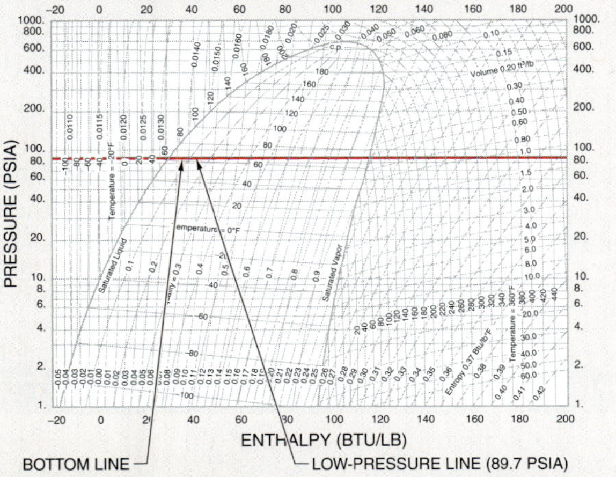

3. Plot the pressure of the high-pressure side on the pressure-enthalpy diagram after converting from gauge pressure to absolute pressure.

$$\begin{array}{r} 185.0 \text{ psi} \\ +\ 14.7 \\ \hline 199.7 \text{ psia} \end{array}$$

4. The line that represents the absolute pressure on the high-pressure side of the refrigeration system is a horizontal line. Draw the high-pressure line from the left of the saturated liquid line to the right of the saturated vapor line. It will be the top line of the graph.

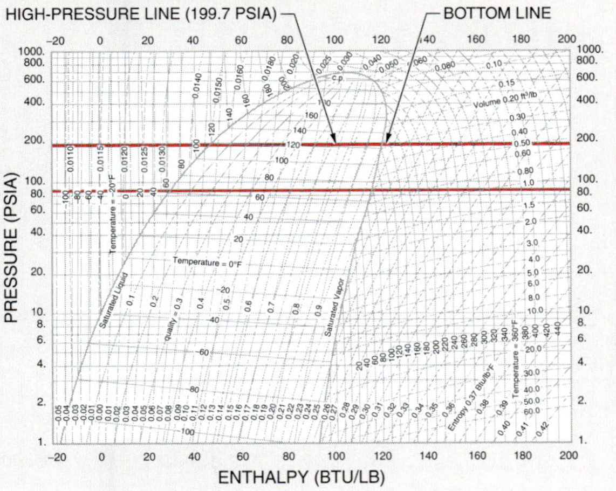

5. The temperature and pressure changes that occur in the compressor are also part of a system graph. A line begins at the point where the low-pressure line intersects the saturated vapor line and runs parallel with the closest constant entropy line until it intersects the high-pressure line. This line represents the refrigerant properties of temperature and pressure in the compressor. Points on the line represent the properties of the refrigerant as it flows through the compressor. Locate where the temperature line curves down across the high-pressure line to find the temperature of the refrigerant that leaves the compressor.

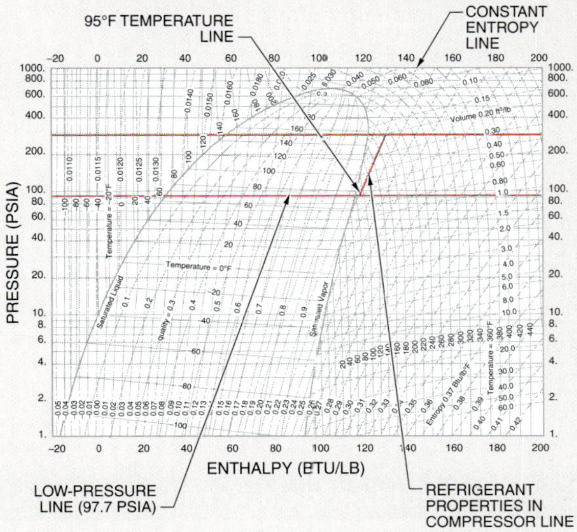

6. In the expansion device, a restriction produces a pressure decrease in the refrigeration system, which causes the refrigerant to vaporize. As the refrigerant vaporizes, it absorbs sensible heat from the remaining liquid refrigerant. The refrigerant properties in the expansion device are found on the line that begins at the intersection of the high-pressure line and the saturated liquid line. Draw the line vertically down to the low-pressure line.

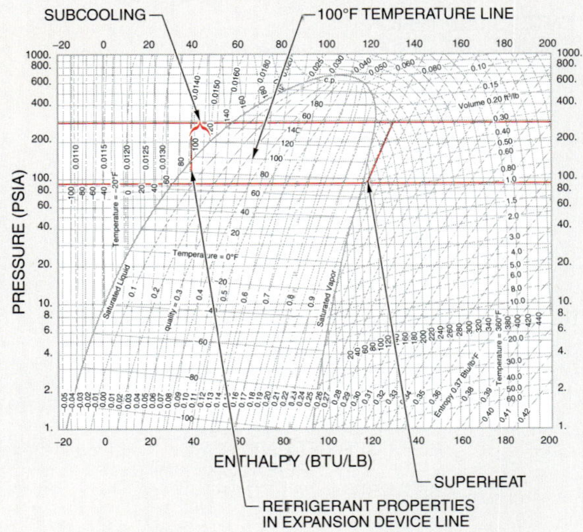

Example: Plotting Ideal Refrigeration System Cycles

A refrigeration system contains a refrigerant with characteristics as indicated in the chart below. Using the proper procedure, plot the ideal cycle for the refrigeration system on a pressure-enthalpy diagram. The pressure of the low-pressure side is 21 psi, and the pressure of the high-pressure side is 127 psi. **See Figure 38-15.**

1. Find the absolute pressure of the low-pressure side of the refrigeration system, and plot the value on a pressure-enthalpy diagram.

   $Pressure_{L\text{-}ABS}$ = **35.7 psia**

2. Find the absolute pressure of the high-pressure side of the refrigeration system, and plot the value on the pressure-enthalpy diagram.

   $Pressure_{H\text{-}ABS}$ = **141.7 psia**

3. Plot the flow through the compressor on the pressure-enthalpy diagram.
4. Plot the flow through the expansion device on the pressure-enthalpy diagram.

> ***TECH TIP***
>
> ***Improper refrigeration system heat transfer is an indicator of system problems and lack of comfort. Mismatched components in retrofit or repaired systems can be a major contributor to poor heat transfer and system problems. A pressure-enthalpy diagram can be used to indicate the poor operation of mismatched system components.***

Heat Transfer—Ideal Cycle

The amount of heat transfer in the various parts of a refrigeration system for an ideal cycle is found by using the enthalpy lines and the system graph drawn on a pressure-enthalpy diagram. **See Figure 38-16.** Heat transfer includes heat absorbed by the refrigerant in the evaporator, heat absorbed by the refrigerant in the compressor, and heat rejected by the refrigerant in the condenser. **See Procedure for Determining Amount of Heat Transfer from Refrigeration System Devices.**

Example: Plotting Ideal Refrigeration System Cycles

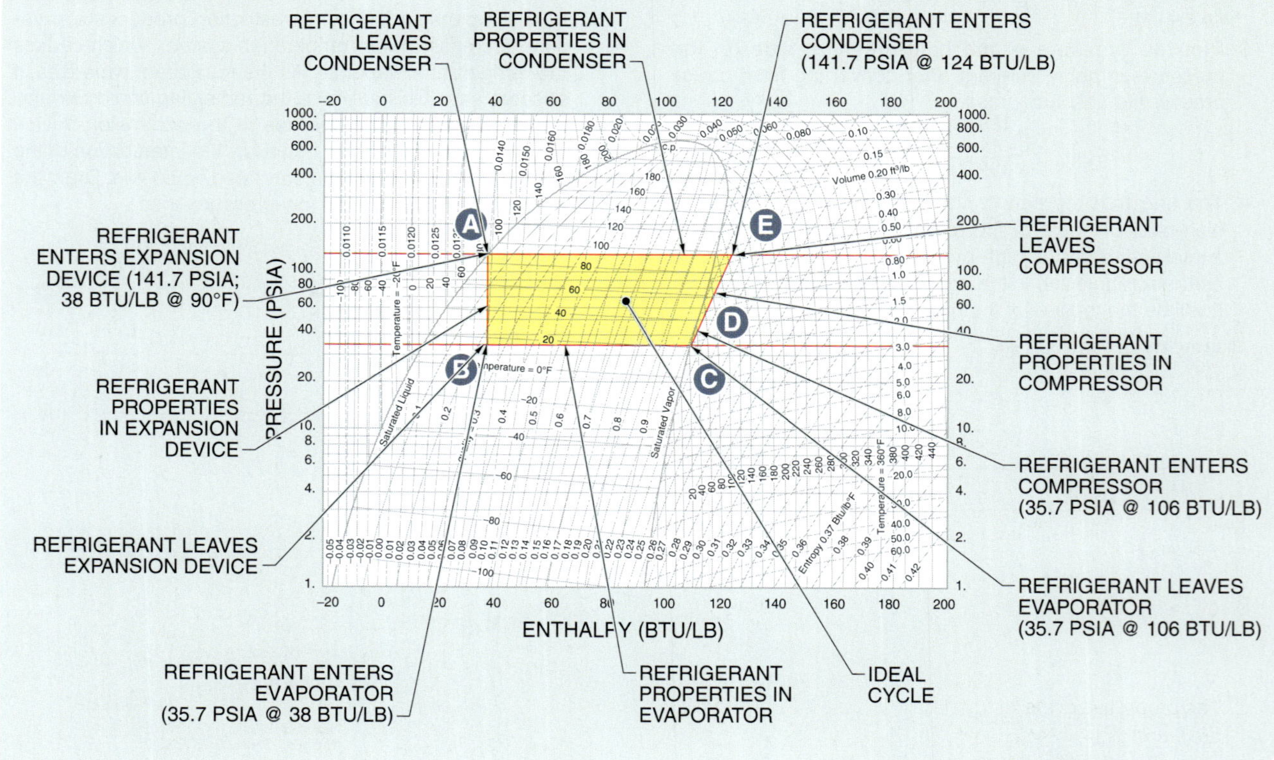

Figure 38-15. When gauge pressure on the low-pressure side and the high-pressure side of a refrigeration system are known, the system graph for the refrigerant can be plotted.

Heat Transfer — Ideal Cycle

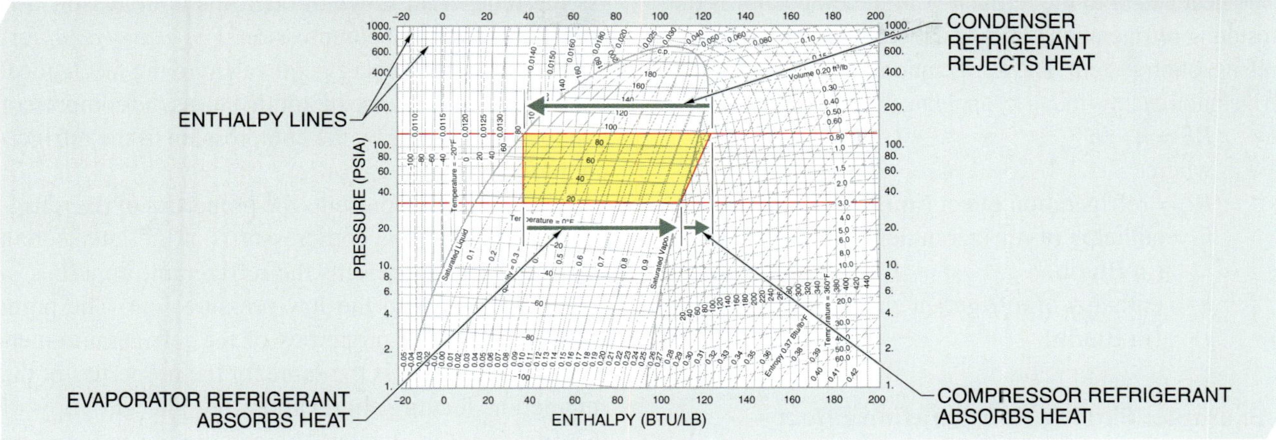

Figure 38-16. The amount of heat absorbed by the refrigerant in the evaporator, refrigerant in the compressor, and the amount of heat rejected by the refrigerant in the condenser can be found by using the enthalpy lines and the system graph drawn on a pressure-enthalpy diagram.

Determining Amount of Heat Transfer from Refrigeration System Devices

1. As refrigerant flows through an evaporator, it absorbs heat from the evaporating medium. The point that represents the properties of the refrigerant as it enters the evaporator is the intersection of the line that represents the properties of the refrigerant in the expansion device at the low-pressure line. Draw a line vertically down from this point to the enthalpy scale to find the enthalpy of refrigerant that enters the evaporator.

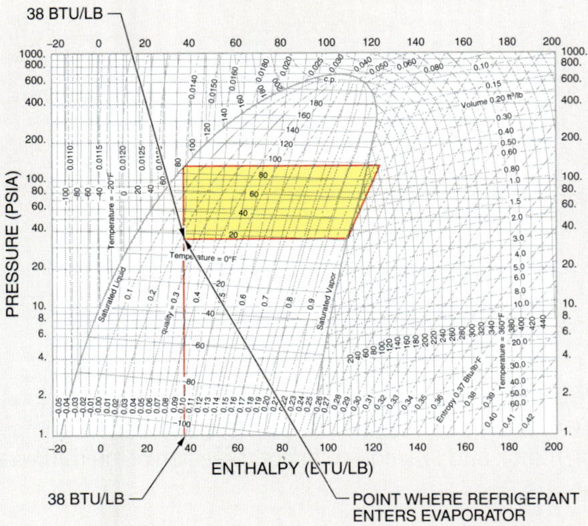

2. The properties of the refrigerant as it leaves the evaporator are found at the intersection of the low-pressure line and the saturated vapor line. To find the enthalpy of the refrigerant that leaves the evaporator, draw a line vertically from the intersection to the enthalpy scale.

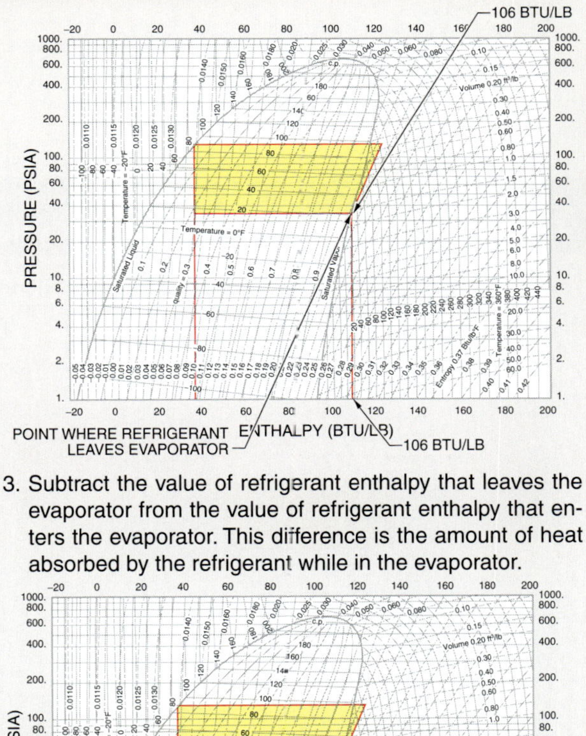

3. Subtract the value of refrigerant enthalpy that leaves the evaporator from the value of refrigerant enthalpy that enters the evaporator. This difference is the amount of heat absorbed by the refrigerant while in the evaporator.

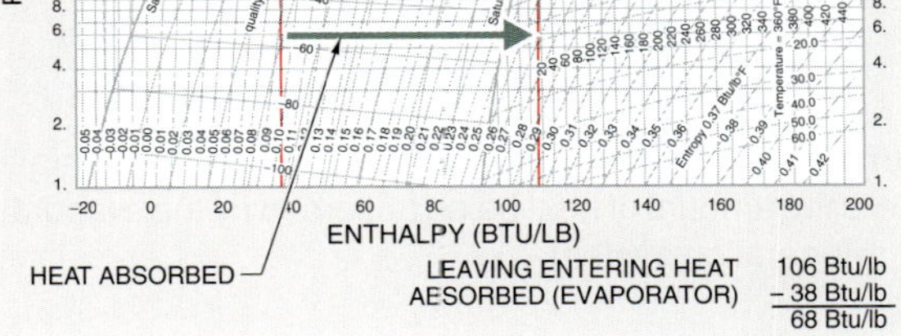

Procedure

Refrigeration Effect

The heat added to the refrigerant in the evaporator is the resulting refrigeration effect of a mechanical compression refrigeration system. The refrigeration effect of an ideal refrigeration cycle is found by applying the following formula:

$$RE = h_l - h_e$$

where

RE = refrigeration effect (in Btu/lb)

h_l = enthalpy of refrigerant leaving evaporator (in Btu/lb)

h_e = enthalpy of refrigerant entering evaporator (in Btu/lb)

Example: Finding Refrigeration Effect—Ideal Cycle

The enthalpy of the refrigerant that enters the evaporator in a refrigeration system is 38 Btu/lb. The enthalpy of the refrigerant as it leaves the evaporator is 106 Btu/lb. Find the ideal refrigeration effect of the system. **See Figure 38-17.**

$$RE = h_l - h_e$$

$$RE = 106 - 38$$

$$RE = \mathbf{68\ Btu/lb}$$

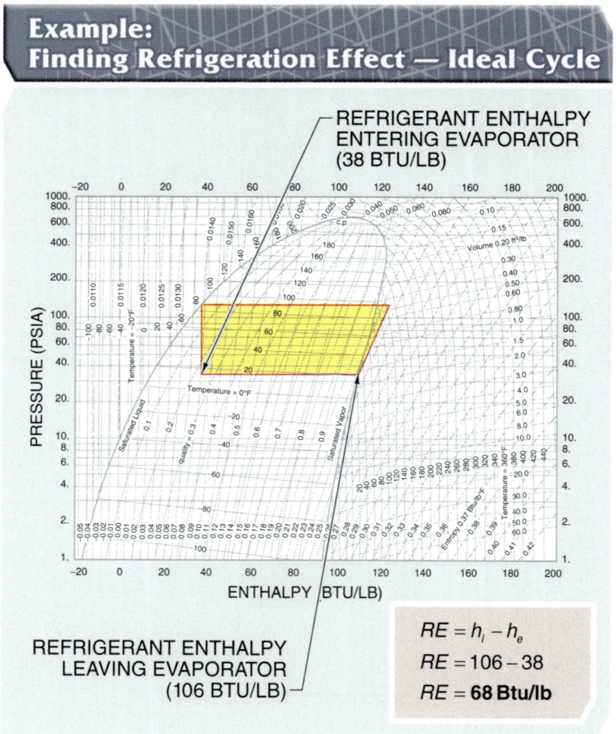

Figure 38-17. The refrigeration effect of an ideal refrigeration cycle is the amount of heat being absorbed by the evaporator of a refrigeration system.

Heat of Compression

As the refrigerant flows through the compressor, it is heated by the heat of compression. The *heat of compression* is the thermal energy equivalent to the mechanical energy expended by the motor that turns the compressor and the heat created by the compression of the refrigerant molecules.

The point that represents the properties of the refrigerant as it enters the compressor is at the intersection of the line that represents the refrigerant properties in the compressor and the low-pressure line. The point that represents the properties of the refrigerant entering the compressor is the same for the properties of the refrigerant leaving the evaporator. The enthalpy of the refrigerant as it enters the compressor is found on a line drawn vertically from this point to the enthalpy scale. **See Figure 38-18.**

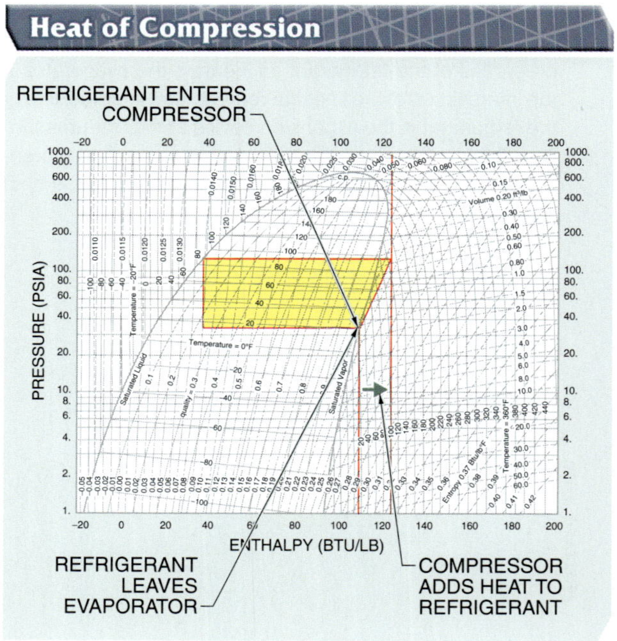

Figure 38-18. The heat of compression of an ideal refrigeration cycle is the heat added to the refrigerant by the compressor, motor, and the molecular compression of the refrigerant.

The point that represents the properties of the refrigerant as it leaves the compressor is at the intersection of the line that represents the refrigerant properties in the compressor and the high-pressure line. The enthalpy of the refrigerant as it leaves the compressor is found on a line drawn vertically from this point to the enthalpy scale. The ideal heat of compression is found by applying the following formula:

$$HC = h_l - h_e$$

where

HC = heat of compression (in Btu/lb)

h_l = enthalpy of refrigerant leaving compressor (in Btu/lb)

h_e = enthalpy of refrigerant entering compressor (in Btu/lb)

Example: Finding Heat of Compression—Ideal Cycle

The enthalpy of a refrigerant as it leaves the compressor in a refrigeration system is 124 Btu/lb. The enthalpy of the refrigerant as it enters the compressor is 106 Btu/lb. Find the ideal heat of compression. **See Figure 38-19.**

$$HC = h_l - h_e$$

$$HC = 124 - 106$$

$$HC = \mathbf{18\ Btu/lb}$$

Example: Finding Heat of Compression — Ideal Cycle

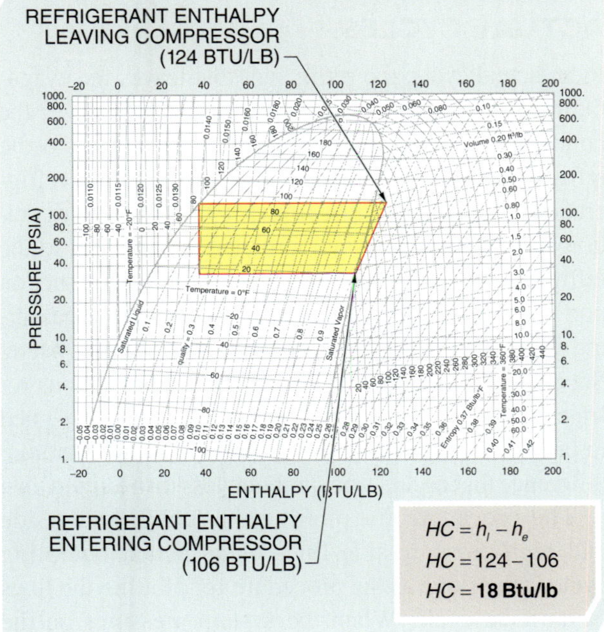

Figure 38-19. The heat of compression of an ideal refrigeration cycle is the difference between the enthalpy of the refrigerant entering the compressor and the enthalpy of the refrigerant leaving the compressor.

Heat of Rejection

The condenser in a refrigeration system must reject all of the heat absorbed in the evaporator and compressor. The *heat of rejection* is the amount of heat rejected by a condenser.

The point that represents the properties of the refrigerant as it enters the condenser is located at the intersection of the line that represents the refrigerant properties in the compressor and the high-pressure line. The point that represents the properties of the refrigerant entering the condenser is the same for the properties of the refrigerant leaving the compressor. The enthalpy of the refrigerant as it enters the condenser is found on a line drawn vertically from this point to the enthalpy scale. **See Figure 38-20.**

Heat of Rejection

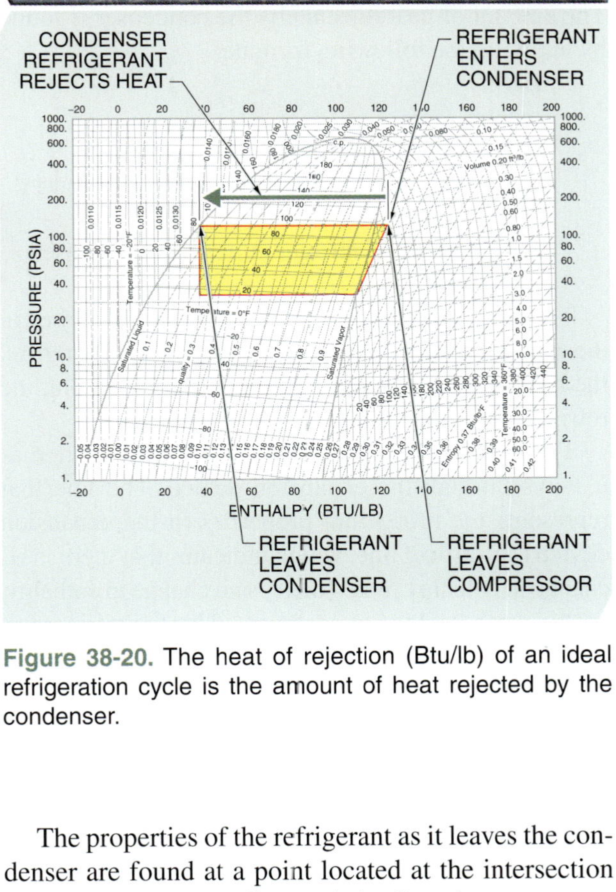

Figure 38-20. The heat of rejection (Btu/lb) of an ideal refrigeration cycle is the amount of heat rejected by the condenser.

The properties of the refrigerant as it leaves the condenser are found at a point located at the intersection of the high-pressure line and the line that represents refrigerant properties in the expansion device. The enthalpy of the refrigerant as it leaves the condenser is found on a line drawn vertically from this point to the enthalpy scale.

The condenser in a refrigeration system must reject all of the heat absorbed in the evaporator and compressor.

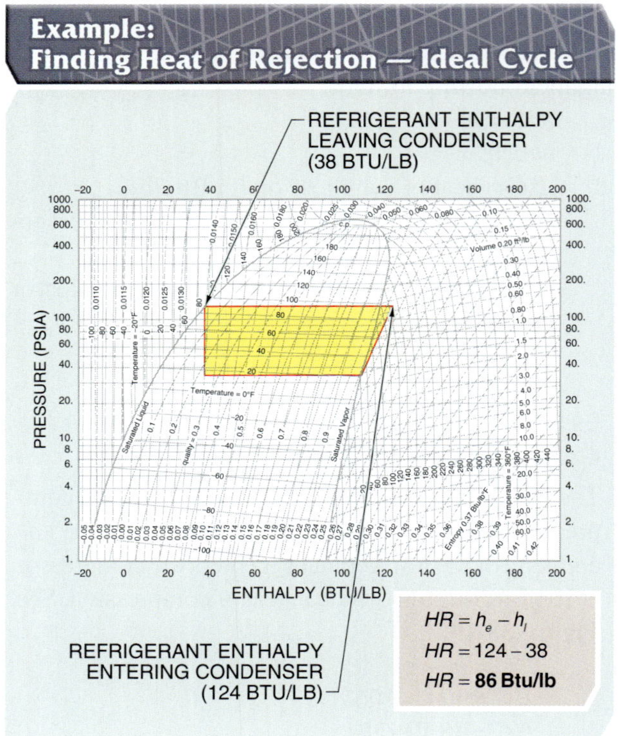

Figure 38-21. The heat of rejection of an ideal refrigeration cycle is found by subtracting the enthalpy of the refrigerant leaving the condenser from the enthalpy of the refrigerant entering the condenser.

The heat of rejection is found by subtracting the enthalpy of the refrigerant as it leaves the condenser from the enthalpy of the refrigerant as it enters the condenser. The amount of heat rejected by the condenser is found by applying the following formula:

$$HR = h_e - h_l$$

where

HR = heat of rejection (in Btu/lb)

h_e = enthalpy of refrigerant entering condenser (in Btu/lb)

h_l = enthalpy of refrigerant leaving condenser (in Btu/lb)

The heat of rejection must equal the heat absorbed in the evaporator plus the heat of compression. Therefore, the heat of rejection may also be found by adding the refrigeration effect to the heat of compression.

The enthalpy of the refrigerant does not change as it flows through the expansion device. The line that represents the refrigerant properties in the expansion device is a vertical line, which indicates that there is no change in enthalpy. When there is no change in enthalpy, heat is neither added to nor rejected by the refrigerant.

Example: Finding Heat of Rejection—Ideal Cycle

The enthalpy of a refrigerant as it enters the condenser is 124 Btu/lb. The enthalpy of a refrigerant as it leaves the condenser is 38 Btu/lb. Find the ideal amount of heat rejected by the condenser. **See Figure 38-21.**

$$HR = h_e - h_l$$

$$HR = 124 - 38$$

$$HR = \mathbf{86\ Btu/lb}$$

ACTUAL CYCLES

In an actual cycle, the refrigerant that leaves the evaporator is superheated, and the refrigerant that leaves the condenser is subcooled. The superheat added to the refrigerant in the evaporator varies among different refrigeration systems and ranges from 5°F to 30°F. A refrigeration system should always add enough superheat to the refrigerant to ensure that only vapor arrives at the suction port of the compressor. Superheat is found by measuring the operating temperature of a refrigeration system. Subcooling ranges from 5°F to 25°F among various refrigeration systems. Refrigeration systems are designed to provide enough subcooling to offset any temperature difference that occurs from pressure loss in the liquid line.

The procedure for plotting the low-pressure-side and high-pressure-side lines of an actual operating cycle is the same as the procedure for plotting the lines of an ideal cycle. When the system pressures on the low-pressure and high-pressure sides of a refrigeration system and the type of refrigerant are known, the system graph can be plotted. **See Procedure for Plotting Actual Cycles on Pressure-Enthalpy Diagrams.**

Plotting Actual Cycles on Pressure-Enthalpy Diagrams

A mechanical compression refrigeration system contains a refrigerant with the characteristics as indicated in the chart below. The pressure of the low-pressure side is 83 psi, and the pressure of the high-pressure side is 180 psi. To plot the system graph of the actual refrigeration cycle on the pressure-enthalpy diagram below, apply the following procedure:

1. First plot the actual low-pressure-side lines of a refrigeration system on a pressure-enthalpy diagram. The pressure of the low-pressure side must first be converted from gauge pressure to absolute pressure.

$$\begin{array}{r} 83.0 \text{ psi} \\ +\ 14.7 \\ \hline 97.7 \text{ psia} \end{array}$$

2. The line that represents the absolute pressure (psia) on the low side of the refrigeration system is a horizontal line. Draw a line from the left of the saturated liquid line to the right of the saturated vapor line. It will be the bottom line of the graph.

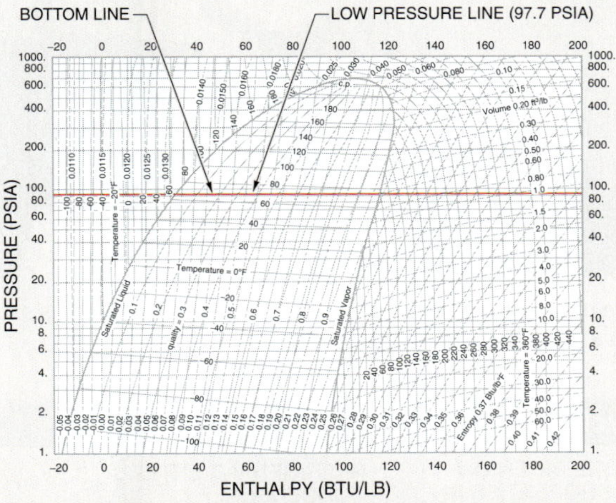

3. To plot the line that represents the actual high-pressure side of a refrigeration system on a pressure-enthalpy diagram, convert the pressure of the high-pressure side from gauge pressure to absolute pressure.

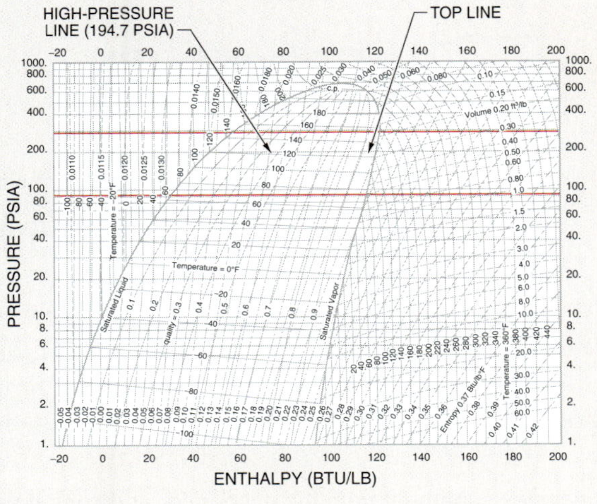

4. The temperature of the refrigerant leaving the evaporator is 95°F as it enters the compressor. Temperature is measured with a thermometer while the refrigeration system is operating. The point that represents the properties of the refrigerant as it enters the compressor is located at the intersection of the low-pressure line and the 95°F temperature line. Draw a line parallel with the nearest constant entropy line, which represents the refrigerant properties in the compressor. The line ends at the point where it intersects the high-pressure line.

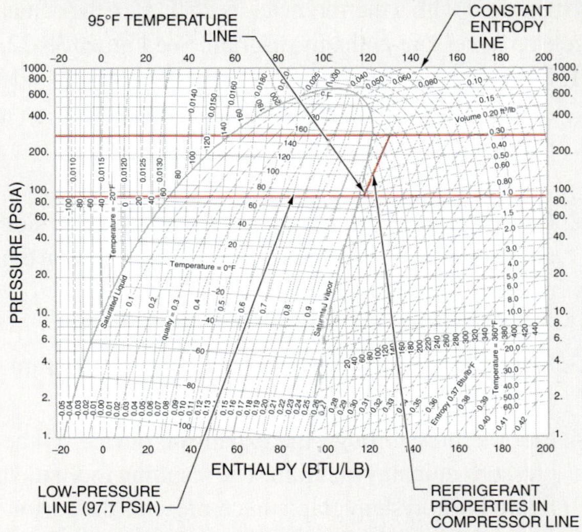

5. The restriction in the expansion device changes the pressure and temperature of the refrigerant. The temperature of the refrigerant is 98°F as it leaves the condenser. Temperature is measured with a thermometer while the refrigeration system is operating. To represent the refrigerant properties in the expansion device, draw a line vertically from 98°F on the saturated liquid line to the high-pressure and low-pressure lines. Find the properties of the refrigerant as it leaves the condenser at the intersection of the high-pressure line and saturated liquid line. Find the properties of the refrigerant as it enters the evaporator at the intersection of the expansion device line and low-pressure line.

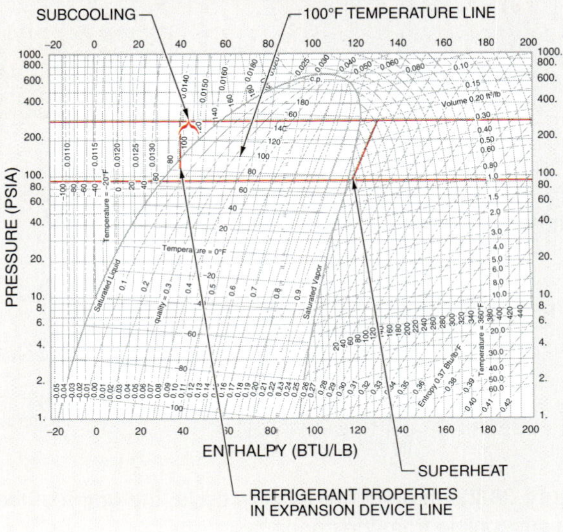

Example: Plotting Actual Refrigeration System Cycles

A mechanical compression refrigeration system contains a refrigerant with the characteristics as indicated in the example below. The pressure of the low-pressure side is 40 psi, and the pressure of the high-pressure side is 180 psi. A thermometer reads 105°F on the outlet of the evaporator. The temperature of the refrigerant leaving the condenser, as measured with a thermometer, is 75°F. Plot the actual cycle on a pressure-enthalpy diagram. **See Figure 38-22.**

1. Find the absolute pressure of the low-pressure side of the refrigeration system, and plot the value on a pressure-enthalpy diagram.

   $Pressure_{L\text{-}ABS} = \mathbf{54.7\ psia}$

2. Find the absolute pressure of the high-pressure side of the refrigeration system, and plot the value on the pressure-enthalpy diagram.

   $Pressure_{H\text{-}ABS} = \mathbf{194.7\ psia}$

3. Plot the flow through the compressor on the pressure-enthalpy diagram.
4. Plot the flow through the expansion device on the pressure-enthalpy diagram. The resulting four-sided figure is the system graph that represents the performance of the actual cycle of an operating refrigeration system.

Heat Transfer—Actual Cycle

The amount of heat transfer in the different parts of a refrigeration system for an actual cycle is found by using the system graph and the enthalpy lines on a pressure-enthalpy diagram. Heat transfer includes heat rejected by the evaporating medium in the evaporator, heat absorbed in the compressor, and heat rejected in the condenser. In order to find the amount of heat transfer for an actual cycle, the refrigeration effect, heat of compression, and heat of rejection for an actual cycle must be found.

A dirty condenser adversely affects unit operation, which is shown on a pressure-enthalpy diagram.

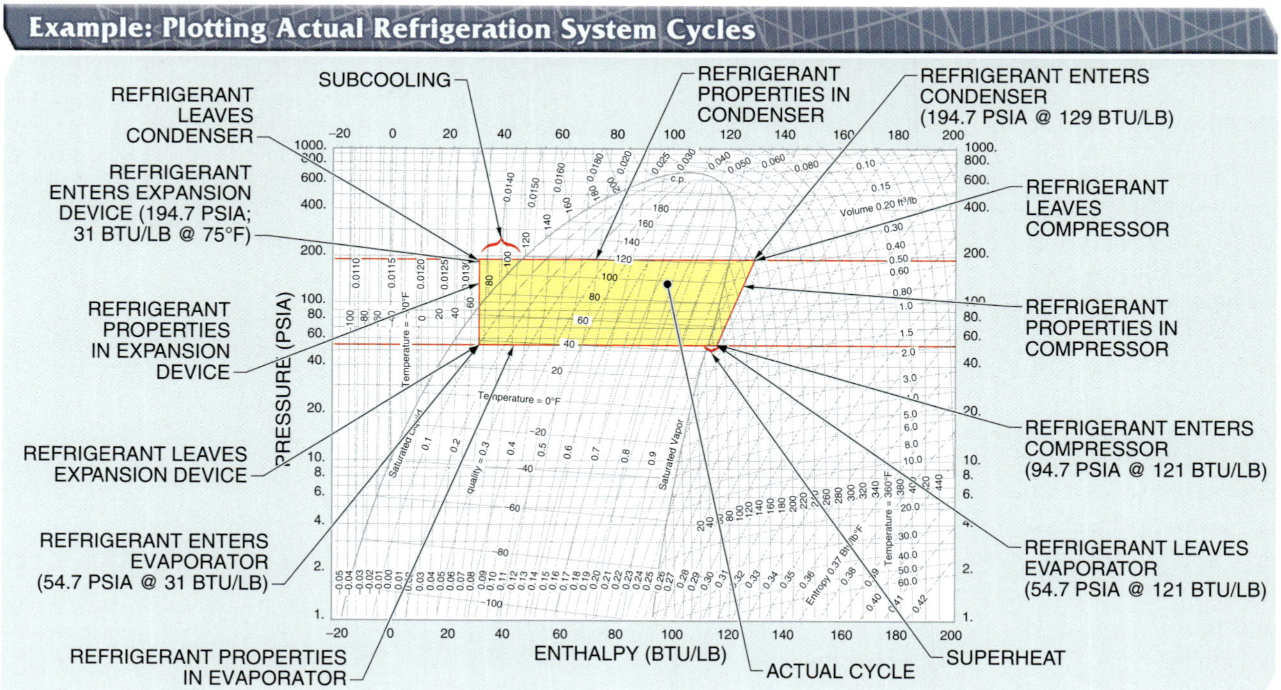

Figure 38-22. In an actual system cycle, the graph of the system includes superheat from the evaporator and compressor and subcooling from the condenser.

Refrigeration effect is the heat absorbed by the refrigerant as it flows through the evaporator. The point that represents the properties of the refrigerant as it enters the evaporator is located at the intersection of the line that represents the refrigerant properties in the expansion device and the low-pressure line. The enthalpy of the refrigerant as it enters the evaporator is found on a line drawn vertically from this point to the enthalpy scale. **See Figure 38-23.**

Heat Transfer — Actual Cycle

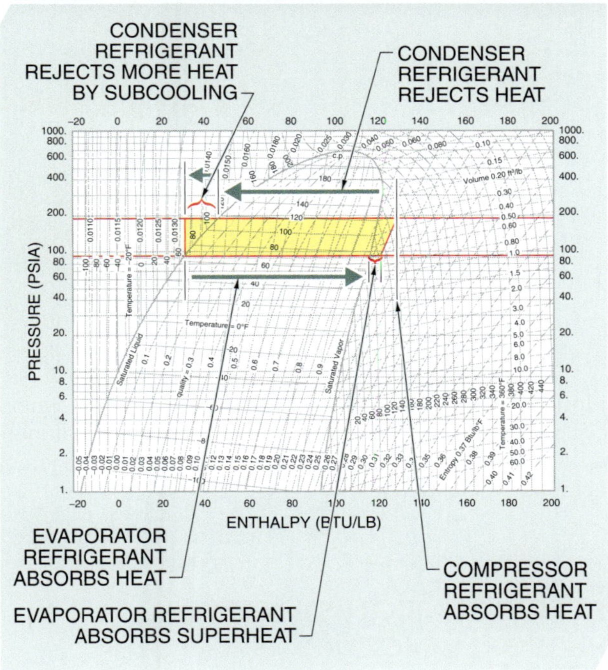

Figure 38-23. The heat transfer of an actual refrigeration cycle includes heat absorbed by the refrigerant in the evaporator, heat absorbed by the refrigerant in the compressor, and heat rejected by the refrigerant in the condenser.

The point that represents the properties of the refrigerant as it leaves the evaporator is located at the intersection of the low-pressure line and the line that represents the refrigerant properties in the compressor. The enthalpy of the refrigerant as it leaves the evaporator is found on a line drawn vertically from this point to the enthalpy scale. The difference between the enthalpy of the refrigerant as it enters the evaporator and the enthalpy of the refrigerant as it leaves the evaporator is the refrigeration effect.

Example: Finding Refrigeration Effect—Actual Cycle

The enthalpy of a refrigerant that leaves the evaporator is 121 Btu/lb. The enthalpy of the refrigerant that enters the evaporator is 31 Btu/lb. Find the actual refrigeration effect. **See Figure 38-24.**

$RE = h_l - h_e$

$RE = 121 - 31$

$RE =$ **90 Btu/lb**

Example: Finding Refrigeration Effect — Actual Cycle

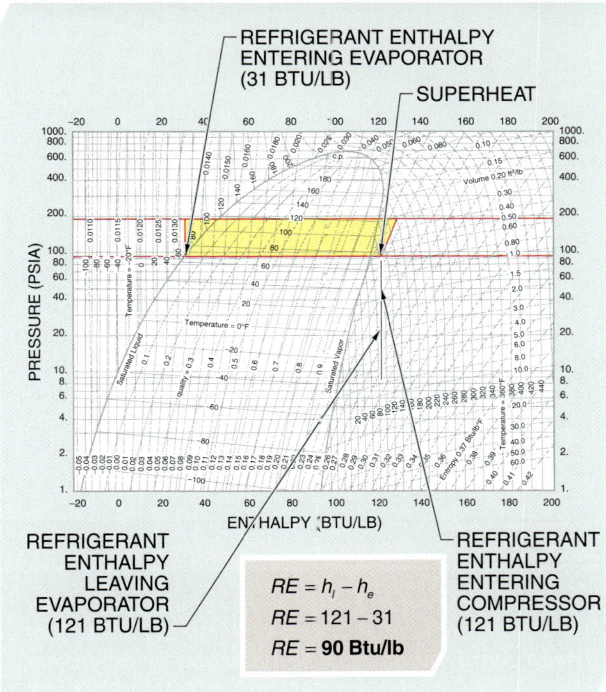

Figure 38-24. The refrigeration effect of an actual refrigeration cycle is the difference between the enthalpy of the refrigerant entering the evaporator and the enthalpy of the refrigerant leaving the evaporator.

Example: Finding Heat of Compression—Actual Cycle

The enthalpy of a refrigerant that leaves the compressor is 129 Btu/lb. The enthalpy of the refrigerant that enters the compressor is 121 Btu/lb. Find the actual heat of compression. **See Figure 38-25.**

$HC = h_l - h_e$

$HC = 129 - 121$

$HC =$ **8 Btu/lb**

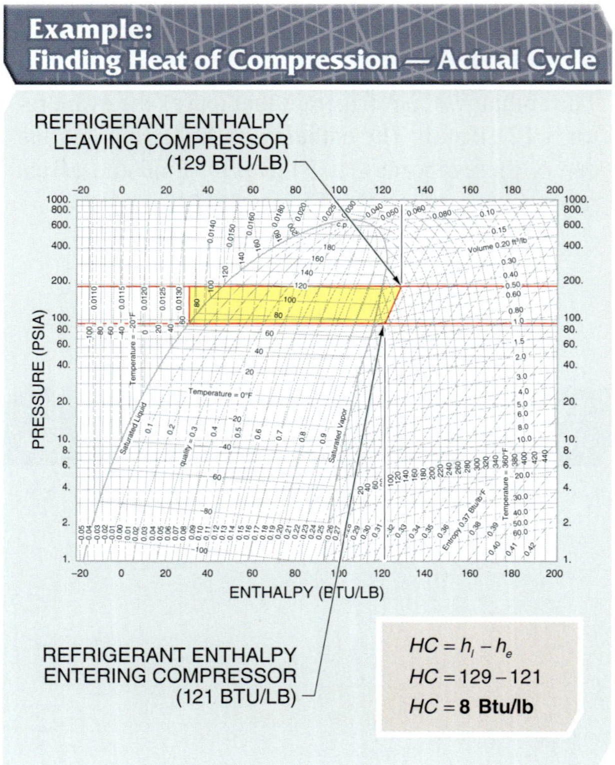

Figure 38-25. The heat of compression of an actual refrigeration cycle is the difference between the enthalpy of the refrigerant entering the compressor and the enthalpy of the refrigerant leaving the compressor.

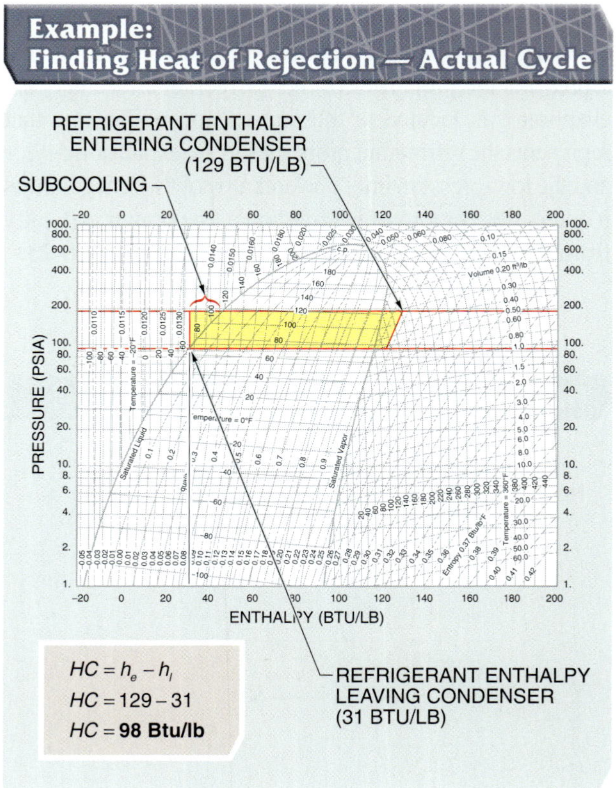

Figure 38-26. The heat of rejection of an actual refrigeration cycle is the difference between the enthalpy of the refrigerant leaving the condenser and the enthalpy of the refrigerant entering the condenser.

TECH TIP

In large commercial systems, the refrigeration effect is absolutely vital. In these systems, products that need refrigeration may be damaged or destroyed by the loss of the refrigeration effect. In addition, energy efficiency can be degraded to the point where thousands of dollars are wasted due to improper system operation.

Example: Finding Heat of Rejection—Actual Cycle

The enthalpy of the refrigerant that enters the condenser is 129 Btu/lb. The enthalpy of the refrigerant that leaves the condenser is 31 Btu/lb. Find the actual heat of rejection. **See Figure 38-26.**

$$HR = h_e - h_l$$
$$HR = 129 - 31$$
$$HR = \mathbf{98\ Btu/lb}$$

REFRIGERATION SYSTEM PERFORMANCE

The system graph of an operating refrigeration system drawn on a pressure-enthalpy diagram is used for collecting data about a refrigeration system. The data obtained from a system graph is used to calculate performance characteristics, which indicate the efficiency of the refrigeration system. The performance characteristics of an actual operating refrigeration system are the refrigeration effect, compression ratio, and coefficient of performance.

Refrigeration Effect

The refrigeration effect is the amount of heat absorbed by a refrigerant in the evaporator of a refrigeration system. The refrigeration effect is the amount of cooling that a refrigeration system can produce. The refrigeration effect for system performance is calculated by subtracting the enthalpy of the refrigerant that enters the evaporator from the enthalpy of the refrigerant that leaves the evaporator. The enthalpy at each point is found on the system graph drawn on a pressure-enthalpy diagram. **See Figure 38-27.**

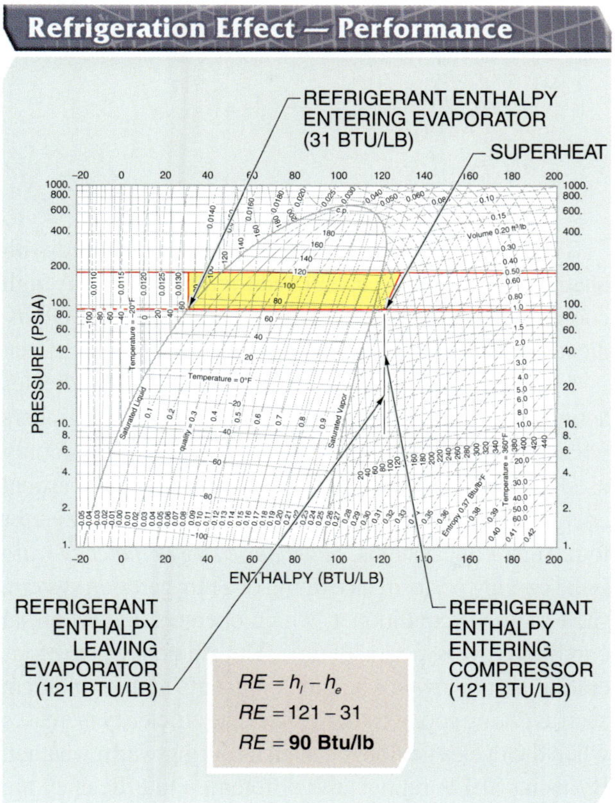

Figure 38-27. Refrigeration effect is the difference between the enthalpy of the refrigerant leaving the evaporator and the enthalpy of the refrigerant entering the evaporator.

Compression Ratio

A *compression ratio* is the mathematical relationship between the pressure on the high-pressure side to the pressure on the low-pressure side of a refrigeration system. The pressures must be absolute pressures. The compression ratio is equal to the pressure on the high-pressure side divided by the pressure on the low-pressure side. The compression ratio of an actual operating refrigeration system is found by applying the following formula:

$$C_r = \frac{P_H}{P_L}$$

C_r = compression ratio
P_H = absolute pressure on high-pressure side (in psia)
P_L = absolute pressure on low-pressure side (in psia)

Example: Finding Compression Ratio

A refrigeration system has a high-pressure side of 180 psi and a low-pressure side of 80 psi. Find the compression ratio. **See Figure 38-28.**

1. Convert the gauge pressure of the high-pressure side to absolute pressure.

   $P_H = 180 + 14.7$
   $P_H = \mathbf{194.7\ psia}$

2. Convert the gauge pressure of the low-pressure side to absolute pressure.

   $P_L = 80 + 14.7$
   $P_L = \mathbf{94.7\ psia}$

3. Find the compression ratio.

   $$C_r = \frac{P_H}{P_L}$$
   $$C_r = \frac{194.7}{94.7}$$
   $$C_r = \mathbf{2.06}$$

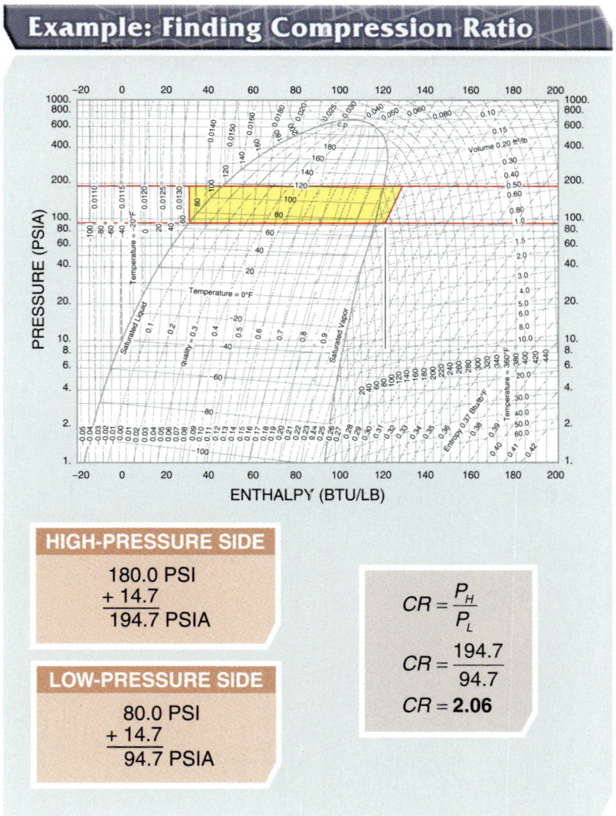

Figure 38-28. A compression ratio is used as a measurement of compressor performance, condition, and sizing and is the ratio of the absolute pressure of the high-pressure side to the low-pressure side.

A low compression ratio indicates efficient compressor operation. A high compression ratio indicates that the compressor uses more energy to raise the pressure of the refrigerant. Therefore, a refrigeration system with a high compression ratio uses more energy than a refrigeration system with a low compression ratio to produce the same refrigeration effect.

On a system graph for a refrigerant with a high compression ratio, the high-pressure and low-pressure lines move apart as the heat of compression increases. Energy used to operate the refrigeration system increases as the difference between the pressures increases, but the refrigeration effect does not increase proportionally.

Coefficient of Performance

The *coefficient of performance (COP)* of a mechanical compression refrigeration system is the theoretical efficiency of system operation. COP is the ratio of the refrigeration effect achieved in the refrigeration system to the energy used to achieve the refrigeration effect. COP is found by dividing the refrigeration effect by the kilowatt of energy per ton used by the system. COP is found by applying the following formula:

$$COP = \frac{3.516}{kW/ton}$$

where

COP = coefficient of performance

3.516 = constant (1 ton of refrigeration effect)

kW/ton = kilowatt per ton used by the system

Example: Finding Coefficient of Performance

A chiller uses 0.8 kW per ton of capacity. Find the COP of the refrigeration system.

$$COP = \frac{3.516}{kW/ton}$$

$$COP = \frac{3.516}{0.8}$$

$$COP = \mathbf{4.395}$$

This is cooling COP and should not be confused with heating COP, which is calculated for a heat pump. A high cooling COP indicates efficient operation. A low COP indicates that more energy is used to produce the same effect.

Cooling COP is a performance rating that indicates actual efficiency. COP is used primarily by engineers when designing a refrigeration system. It is not commonly used when referring to refrigeration equipment. Seasonal energy efficiency ratio (SEER) is used more often to refer to refrigeration systems. *Seasonal energy efficiency ratio* is the cooling performance rating of a refrigeration system, such as an air conditioner, which operates under normal conditions over a period of time. The efficiency of a refrigeration unit is lower when the unit starts and stops often, such as during cool weather cycles. The efficiency increases when the cycles get longer, such as during warm weather. By using SEER ratings, the differences in efficiency are averaged out over a longer period of operating time.

UNIT 39

Installing Residential Forced-Air Heating Systems

The ability to install any type of HVAC equipment is an important skill. There are a number of steps and procedures related to the installation process. For a forced-air heating system to perform properly and operate efficiently, these steps and procedures must be performed correctly.

OBJECTIVES

- Identify safe furnace installation practices.
- Explain how to determine the proper location to install a furnace.
- List the steps and procedures for installing a residential furnace.
- List tasks that must be performed during furnace maintenance.
- Explain how to convert furnace configurations and how to derate for high altitudes.

Index to Procedures

FORCED-AIR HEATING SYSTEMS

The most common type of residential forced-air heating system installed today is the furnace. This consists of a furnace air handler that typically contains a blower, filter rack, heat exchanger, electrical section, and possibly a separate air conditioning evaporator coil. **See Figure 39-1.** When a furnace is integrated with an air conditioning system, it is called a split system. This is because the air handler and heating sections are located inside the structure while the air conditioning condensing unit is located separately outside the structure.

Forced-Air Heating Systems

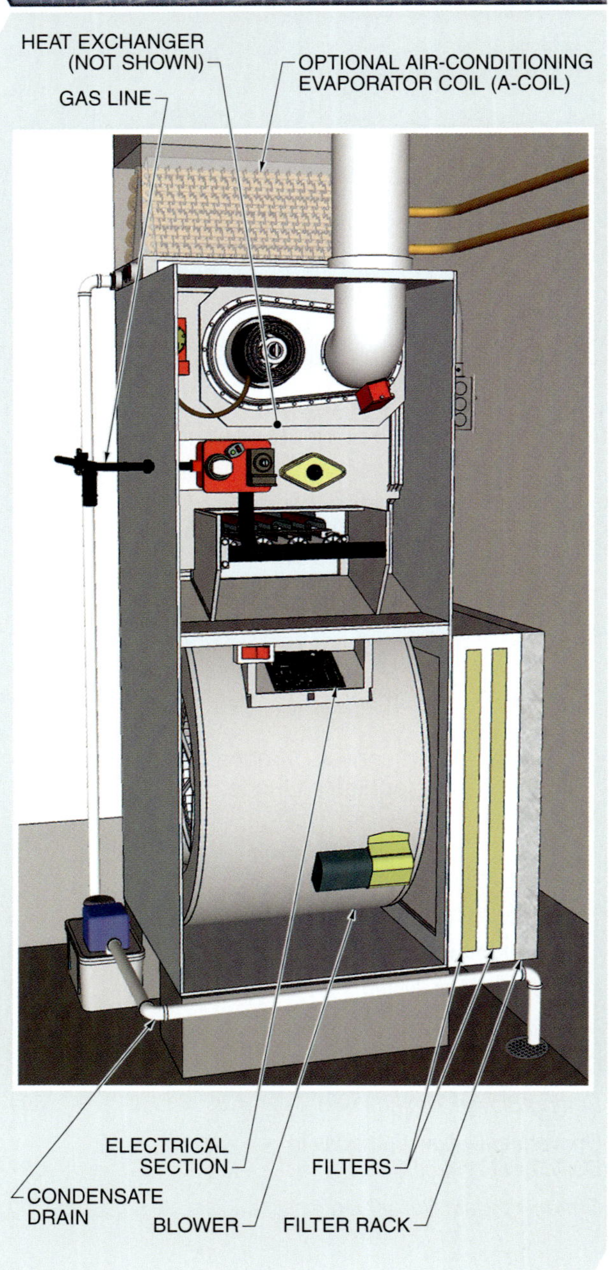

Figure 39-1. Catalogs and engineering specification sheets from manufacturers are the best sources of information for heating systems.

It is important that all HVAC technicians know how to properly install a furnace as well as a split-system air conditioning unit, as installing one of these systems may be the first thing a technician does on the job. **See Figure 39-2.** It is important to install a furnace properly, since a poor job will lead to customer dissatisfaction, multiple complaints, and service issues. While there are many furnace variations, the following general installation guidelines must be observed:

- All products must be installed according to manufacturer installation instructions and safety guidelines.
- All work must be performed according to all local, national, and trade codes.
- Technicians must wear the proper PPE when performing forced-air heating system installations.
- Technicians or other personnel must acquire all required permits for the project.
- All electrical work requires that qualified personnel perform high- and low-voltage electrical wiring.
- All sheet metal work must be performed safely (avoiding sharp edges and burrs).
- All forced-air heating system products must be properly installed and serviced to minimize serious injury or death.
- When a technician does not fully understand how to install a product after reading the manufacturer instructions, a supervisor must be sought for assistance.

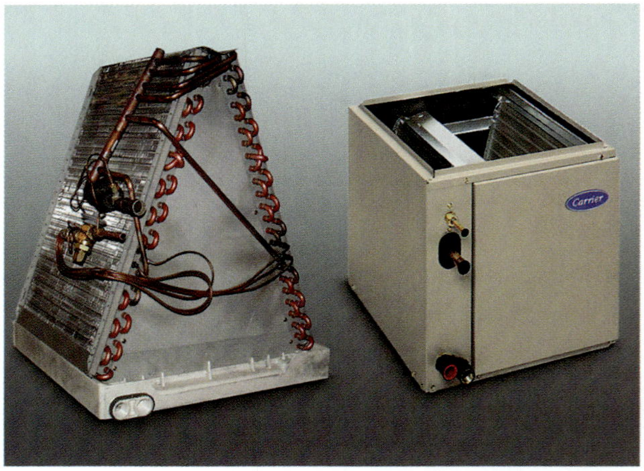

Carrier Corporation

A split-system evaporator coil can be installed as an A-coil in a condensate drain pan inside the ductwork above the furnace or as an evaporator coil unit above the furnace with ductwork attached.

Furnace and Air Handler Installation

To save costs on air conditioning installation, it is very common during furnace installation to install the evaporator for a split system even if the condensing unit is not installed at the same time. A split system consists of an indoor unit, which is typically a furnace, and an outdoor unit, which is the condenser. The furnace contains the system blower, with the indoor air conditioning coil (evaporator) in the discharge of the furnace. The condensing unit contains the compressor and controls. Installing a furnace or air handler involves a number of tasks. **See Figure 39-3.** A furnace installation list, which includes maintenance tasks, includes the following:

- determining furnace locations and clearances
- providing combustion air, ventilation air, and venting
- installing gas piping
- installing electrical wiring
- installing duct systems and filters
- checking the furnace before startup
- adjusting the gas supply, manifold pressure, and gas input rate
- checking blower speed and duct system static pressure
- measuring air temperature rise
- checking controls
- maintaining the furnace
- special installation applications and considerations

TECH TIP

Many residential and light commercial HVAC technicians start their careers as system installers. This provides experience for future service work as they can easily recognize problems caused by improper system installation.

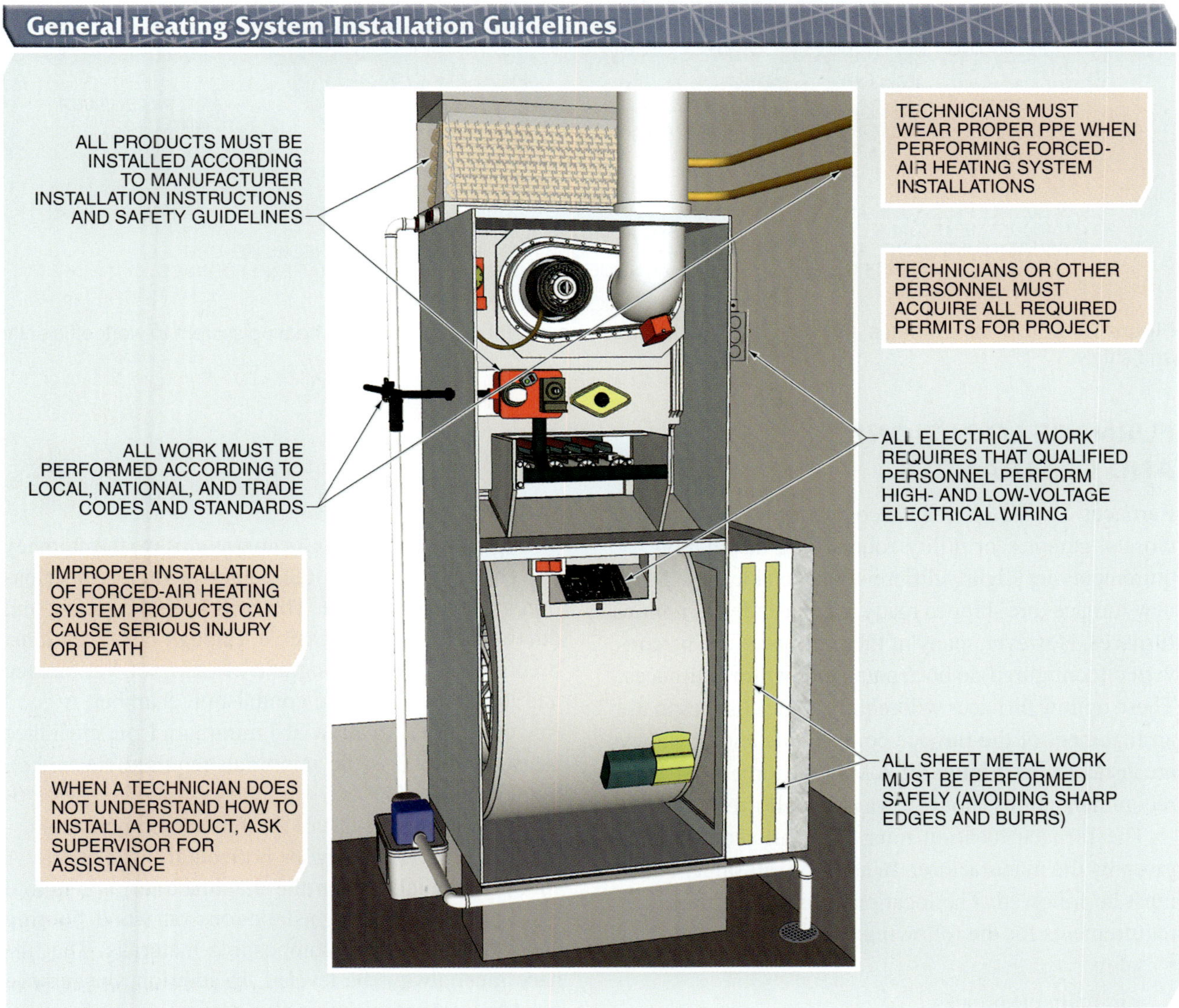

Figure 39-2. Following general installation guidelines help technicians properly and safely install forced-air heating systems.

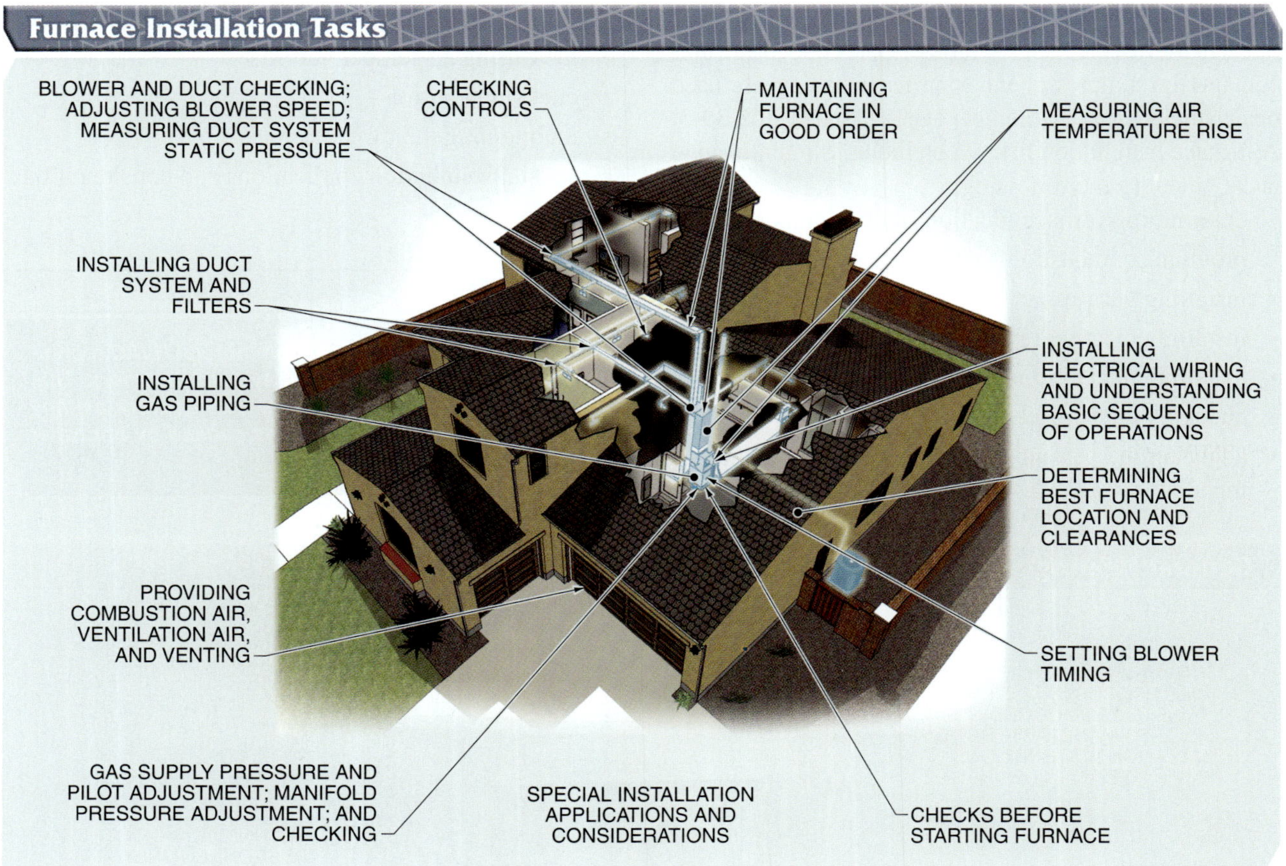

Figure 39-3. To install a furnace, many separate tasks must be performed in order for the heating system to work efficiently and safely.

FURNACE LOCATIONS AND CLEARANCES

Furnaces are installed in alcoves, attics, basements, closets, garages, or utility rooms. The installation requirements are slightly different for each location. Most new furnaces are shipped ready to be installed as upflow furnaces. However, many of these furnaces can be converted (configured) to horizontal or downflow furnaces. These upflow furnaces will require some changes to the arrangement of the furnace components and how they are installed. **See Figure 39-4.** Manufacturer installation recommendations should always be followed.

The furnace location must meet all requirements given by the manufacturer. In addition, all safety codes must be followed. The location chosen must meet the requirements for the following:

- safety
- minimum clearances
- combustion, ventilation air, and venting
- duct systems
- gas piping
- electrical wiring

Typically, a furnace is located as close to the chimney or vent and as near to the center of the air distribution system as possible. This will keep the piping and ductwork as short as possible. Enough clearance must be provided for servicing and cleaning of the furnace and to allow air into the combustion chamber. A good rule of thumb is to allow 36″ minimum front clearance for service (30″ is the minimum required clearance). **See Figure 39-5.**

Technicians can install the furnace on noncombustible flooring for upflow or horizontal installations. A subbase, available from the manufacturer, is required for downflow furnace installations on wood flooring and must be 1″ from combustible materials. The furnace must always be leveled. In addition, the furnace must be installed such that all electrical components are protected from water.

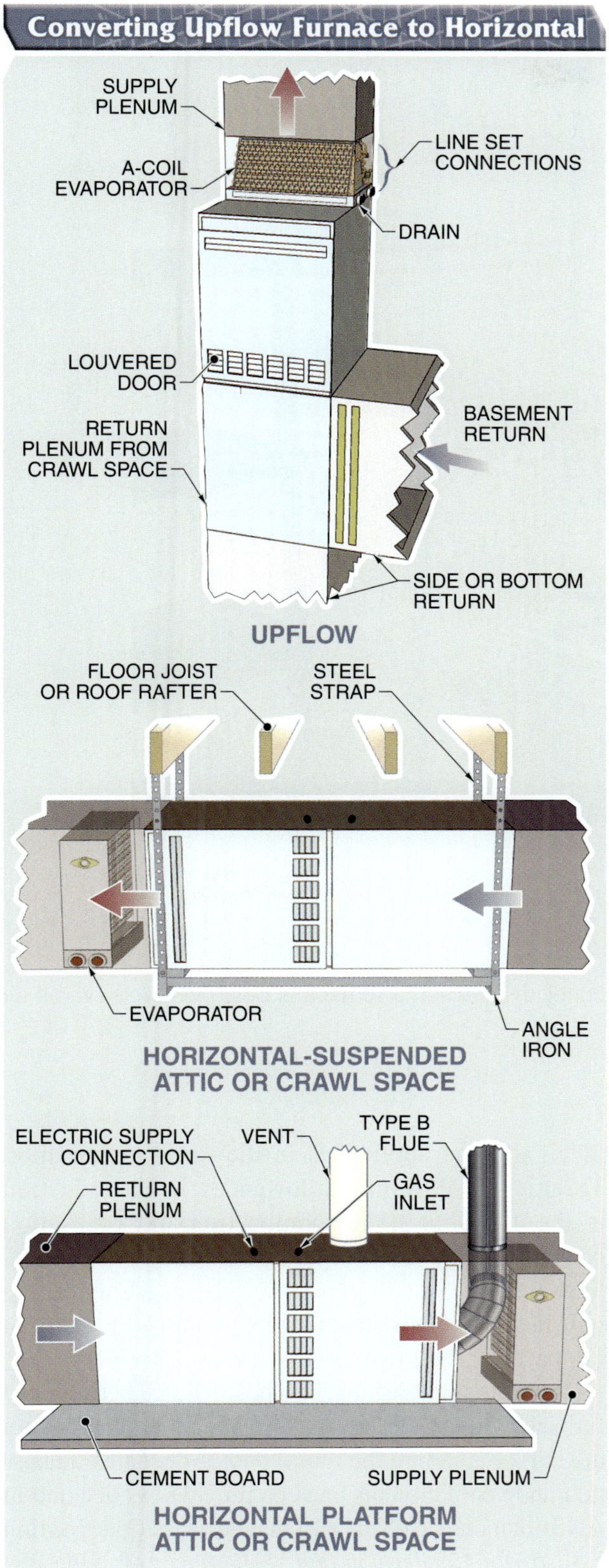

Figure 39-4. Upflow furnaces may require changes to the arrangement of the furnace components and how they are installed.

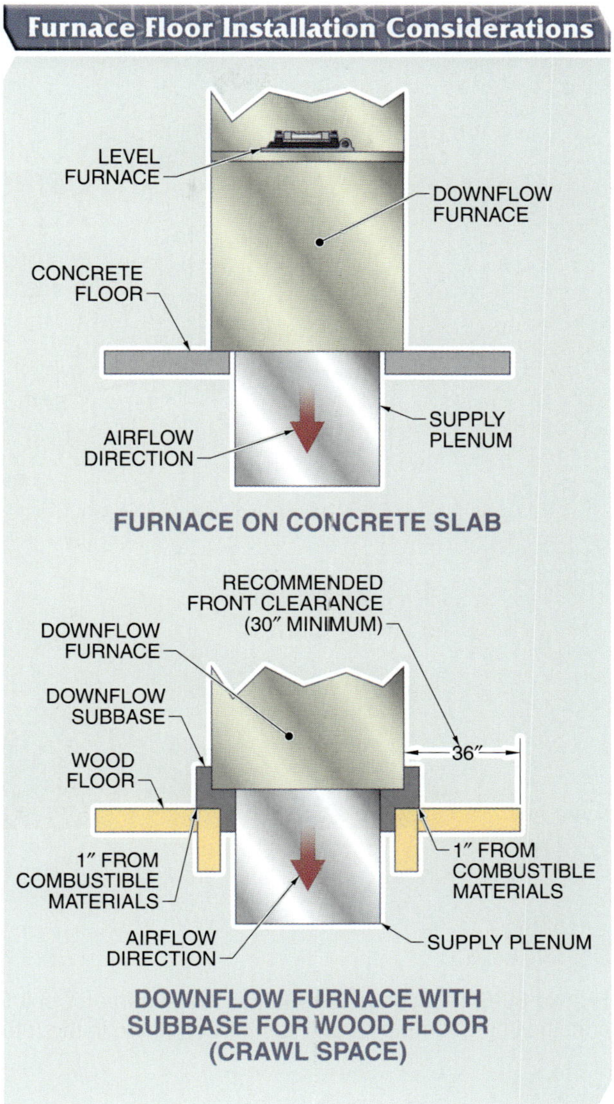

Figure 39-5. Furnaces must have at least 30″ of front clearance and at least 1″ of clearance from combustible materials including hardwood floors.

Most forced-air heating systems installed today are installed with direct expansion evaporator coils. When an evaporator coil is installed with the furnace, the furnace must be installed on the upstream side of the cooling unit to avoid condensation in the furnace heat exchanger. **See Figure 39-6.**

When a furnace is in a residential garage, it must be installed so that pilot, or igniter, and main burners are located no less than 18″ above the floor. Also, the furnace should be protected from physical damage by vehicles. This is typically accomplished by the installation of a concrete or metal post in front of the furnace.

Furnace Installation with Direct Expansion Evaporator Coils

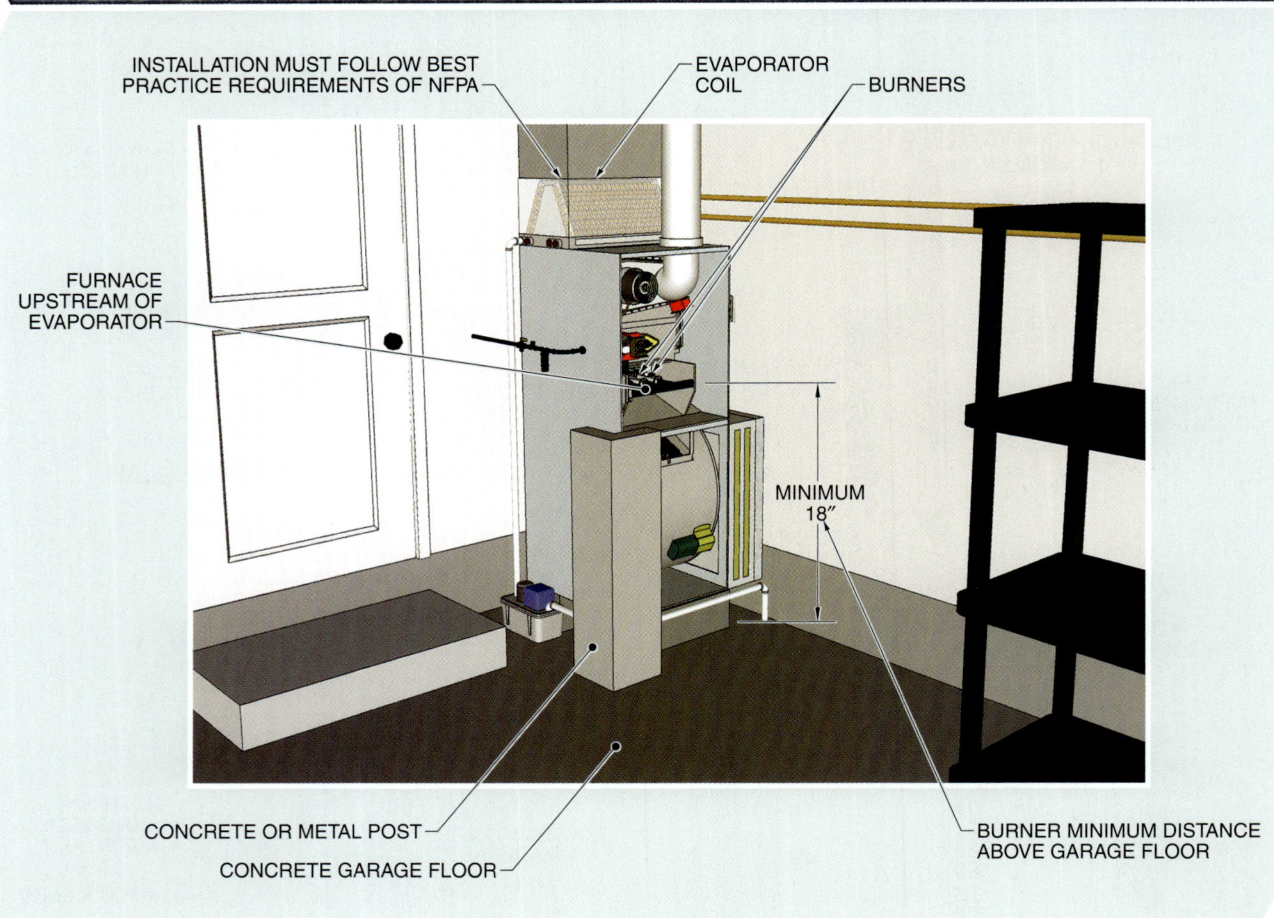

Figure 39-6. It is common to install the evaporator coil for air conditioning when a furnace is being installed even if air conditioning will not be installed until sometime in the future.

When the unit is in a public garage, airplane hangar, or other building that can have a hazardous atmosphere, the unit must be installed in accordance with the recommended good practice requirements of the National Fire Protection Association (NFPA). Local code authorities must be consulted for additional location requirements.

COMBUSTION AIR, VENTILATION AIR, AND VENTING

One of the most important aspects of any furnace installation is ensuring the availability of adequate combustion and ventilation air. In many cases, there is more than one gas appliance in a structure. The total amount of combustion and ventilation air provided within the room or building must equal the requirements of all the gas appliances in the room or building. Technicians must make allowances for the reduction in the available air for combustion and ventilation caused by exhaust fans, range hoods, and fireplaces. Sufficient air must be provided to ensure that there will never be negative pressure in the furnace room or space. **See Figure 39-7.**

In addition, there must be a positive seal between the furnace and the return air duct to avoid pulling air from the burner area into the return duct. Adequate combustion and ventilation air must be provided as detailed in ANSI Standard Z223.1, National Fuel Gas Code Section 5.3, *Air for Combustion and Ventilation,* and NFPA 54 Sections 7.2, 7.3, or 7.4, or in Canadian installation codes CAN/CGA B149.1 or B149.2. When in doubt, a supervisor or local inspector should be contacted.

Furnace Room Combustion and Ventilation Air

Sizing Louver or Grill Free Area

Heating Units in Confined Space	NFPA 54 All	NFPA 31 All	BOCA All
Minimum number openings required	2	2	2
Outdoor opening	1/4000*	1/4000*	1/4000*
Vertical ducts	1/4000*	1/4000*	1/4000*
Horizontal ducts	1/2000*	1/2000*	1/2000*
Ducts same as opening	Yes	Yes	-
Louver and grill free area	Opening	Opening	-

* in in²/Btuh

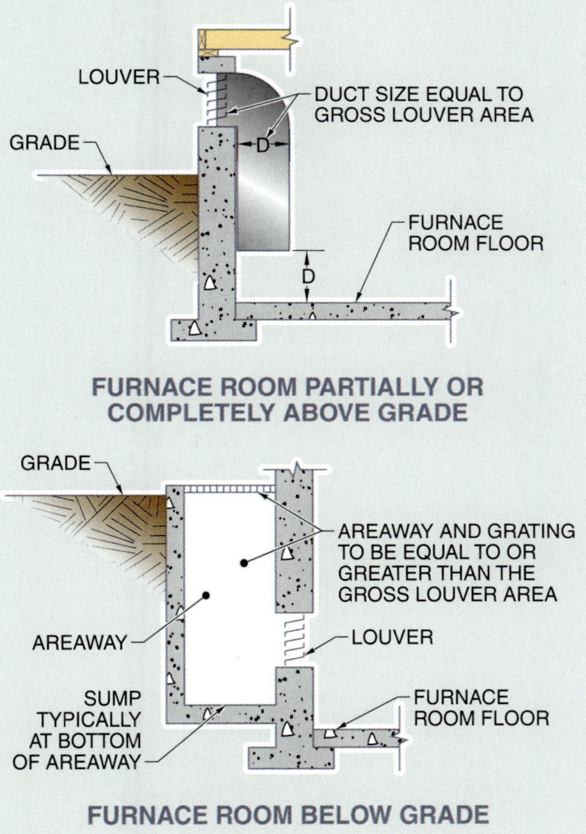

Figure 39-7. The total amount of combustion and ventilation air entering a furnace room must meet the requirements of all gas appliances in the room and be in accordance with National Fuel Gas Code Section 5.3, *Air for Combustion and Ventilation.*

Vents and Chimneys

When dealing with furnace venting, it is important to define the terms being used. The terms "vent" and "chimney" refer to open passageways that convey vent gases from the furnace or its vent connector to the outside. Older vents and chimneys typically run vertically or nearly vertical. When they serve only one gas appliance, they are known as dedicated vents or chimneys. When they serve multiple gas appliances, they are known as common vents or chimneys. The term "vent connector" refers to a pipe or duct that connects the furnace to a vent or chimney. Vent connectors typically run directly from the vent collar of the furnace to the vent or chimney. Vent connectors may have vertical and horizontal runs. **See Figure 39-8.**

Typical Vent and Chimney Location Considerations

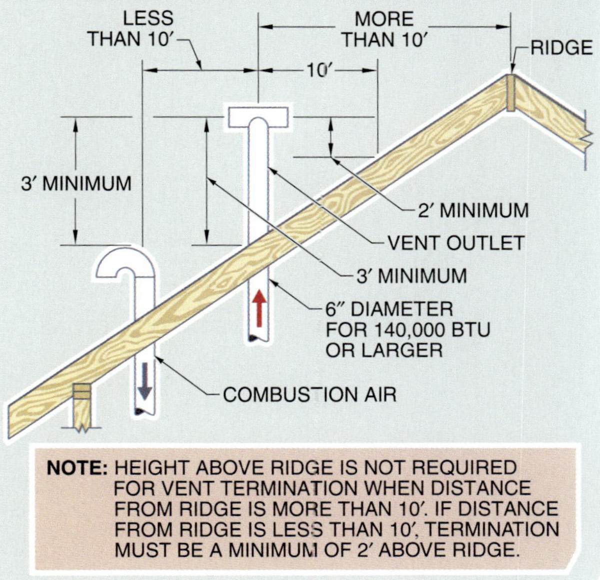

Figure 39-8. Many ordinances and building codes must be followed when installing vents and chimneys for forced-air heating systems.

A *venting system* is a continuous open passageway from a vent collar to the outside. Venting systems typically have at least one vent connector and a vent or chimney. Venting systems typically serve a single furnace or a single furnace and a water heater. Other multiple-appliance venting systems are less common.

Many high-efficiency furnaces today use a fan-assisted combustion system. This refers to an appliance equipped with an integral mechanical means to either draw or force products of combustion through the combustion chamber and/or heat exchanger. These types of furnaces use a draft inducer (induction fan) to draw combustion products through the heat exchanger. Category I furnaces (furnaces with efficiencies under 83%) with fan-assisted combustion systems must not be vented into single-wall metal vents.

General Design Considerations

In the past, low-efficiency furnaces were vented to an unlined masonry chimney. This was because the flue gases were too hot to condense inside the chimney. Condensation would cause the chimney to corrode and eventually fail. Today's higher efficiency furnaces recover as much heat as possible from the flue gases, lowering their dew point temperature. When vented directly into an unlined chimney, the gases condense and form droplets that corrode the chimney. **See Figure 39-9.**

Chimney Liner Considerations

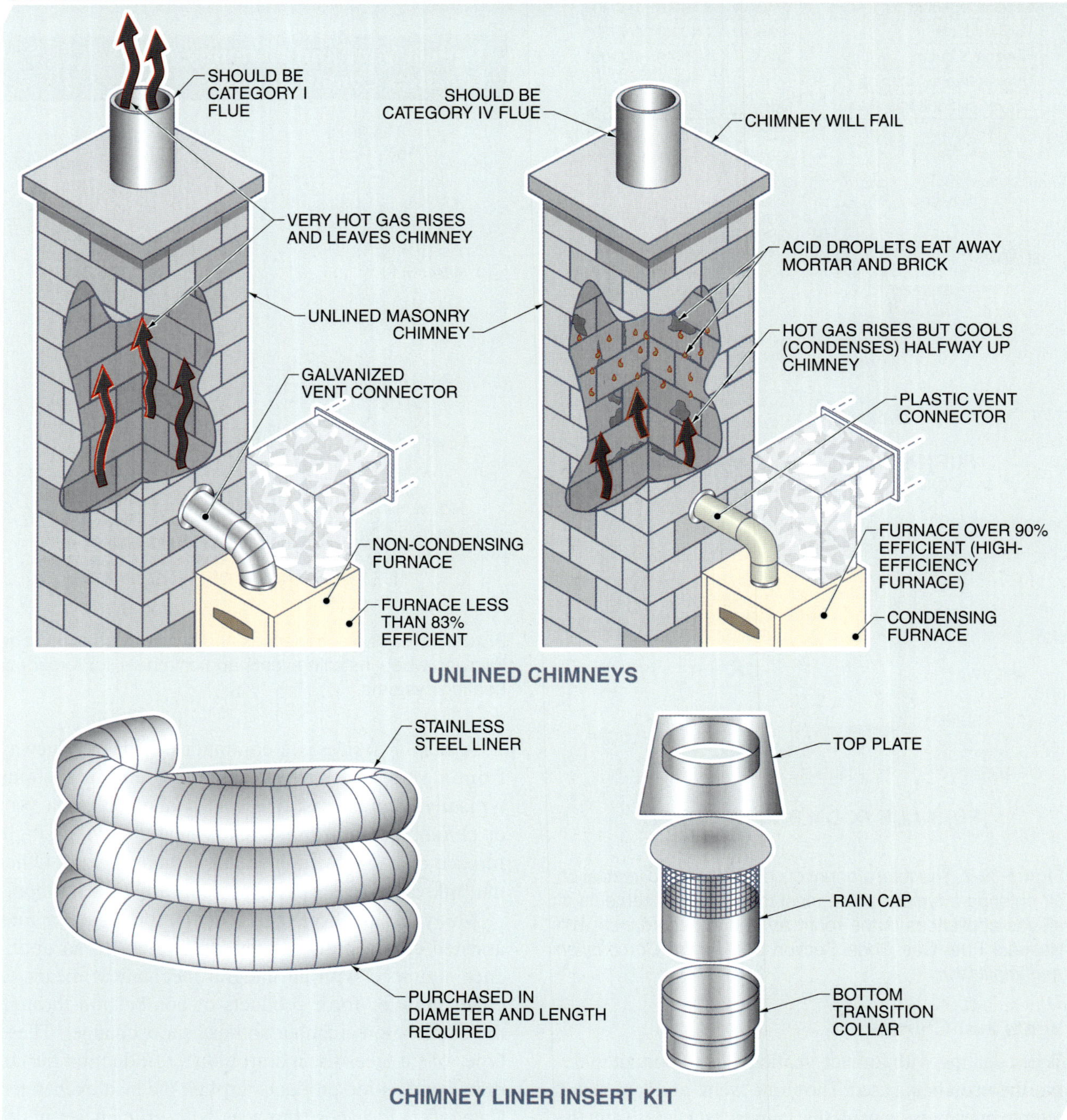

Figure 39-9. Two-, three-, and four-story masonry chimneys must be protected from the warm or cool flue gases found with modern furnaces.

Venting of combustion gases must be done according to code.

Technicians must connect today's higher-efficiency furnaces to lined chimneys or vents that comply with recognized standards. These chimneys are properly sized, constructed, and have the correct masonry lining. Chimney lining methods and materials must comply with local requirements. Liner kits are available that meet these standard requirements. A corrosion-resistant material should be used that meets nationally recognized standards for vent construction.

A non-condensing 80% efficient furnace must not be vented into an unlined masonry chimney or unlined concrete chimney. Also, it must not be a single appliance vented into a tile-lined chimney. The vent connectors must be made of listed single-wall or Type B double-wall components.

Type B double-wall vent connectors can be used in homes and commercial buildings at close clearance to combustible materials. They must be thick enough to withstand physical damage and be readily accessible for inspection, cleaning, and replacement. **See Figure 39-10.** Type B vent connectors must be used in attics, crawl spaces, or other cold areas. When vent connectors pass through walls or partitions of combustible material, thimbles are installed that meet local codes. Vent connectors must be kept as short as possible by locating the furnace as close as practical to the vent or chimney. Unnecessary turns or bends that create resistance to the flow of flue gases must be avoided. Adding an elbow adds resistance.

When replacing an existing furnace with a high efficiency model, the venting should be checked to see if it requires updating. It should not be assumed that an old vent system can be used with a new furnace. Also, it is a good idea to perform a carbon monoxide check in the building after installation to ensure proper combustion and venting.

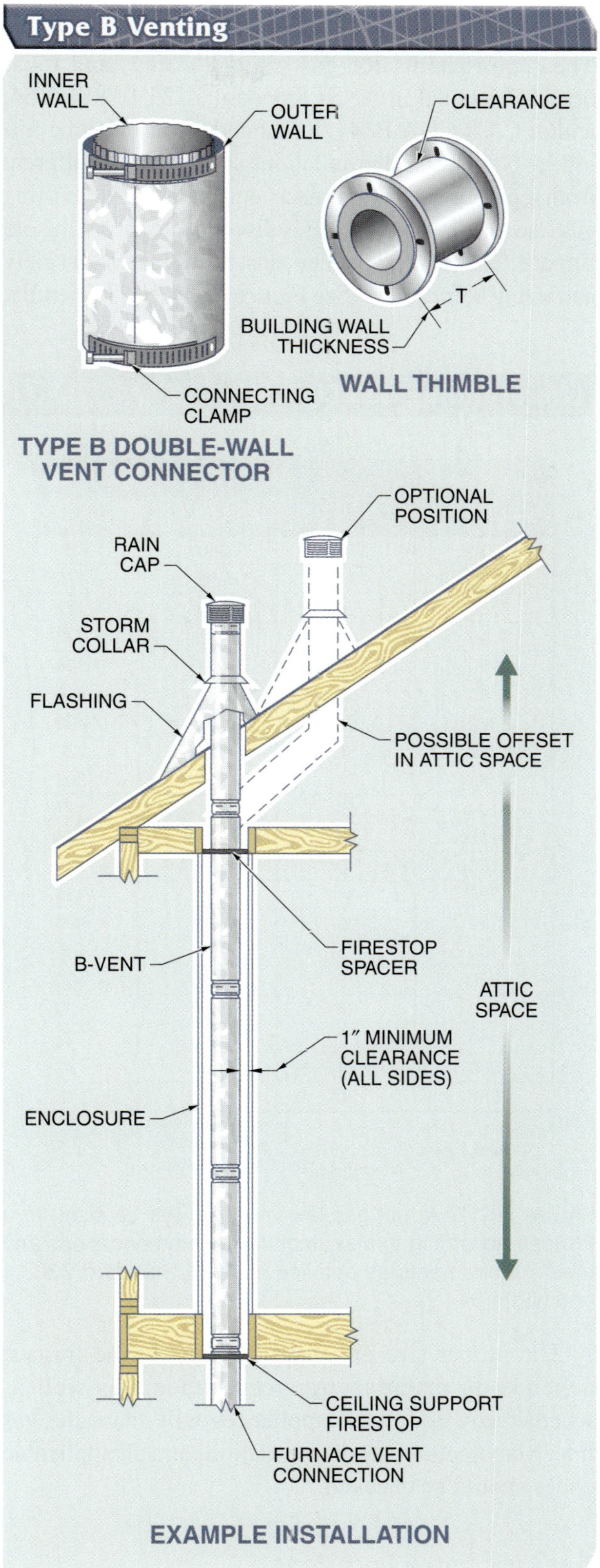

Figure 39-10. Type B double-wall vent connectors can be used in locations close to combustible materials.

GAS PIPING

The requirements for gas piping, sizing, and location are detailed in ANSI Standard Z223.1, NFPA 54, and/or CAN/CGA B149. Local codes must also be followed. New furnace installations are somewhat different from replacements. For furnace replacements, the existing gas line may be used after it is disconnected from the old furnace. The existing gas line must be evaluated for safety and sizing before reuse. **See Figure 39-11. See Appendix.**

Determining Gas Pipe Size

Pipe Length*	Pipe Size		
	½	¾	1
10	174†	363	684
20	119	249	470
30	96	200	377
40	82	171	323
50	73	152	286
60	66	135	259
70	61	127	239
80	56	118	222
90	53	111	206
100	50	104	197
125	44	93	174
150	40	84	158
200	34	72	136
250	30	64	120
300	28	58	109
350	25	53	100
400	24	49	93
450	22	46	87
500	21	44	82
550	20	42	78

* in ft
† ft^3 of natural gas/hr

Figure 39-11. A fuel gas line must always be sized to a furnace according to manufacturer recommendations and meet requirements as detailed in ANSI Standard Z223.1 and NFPA 54.

The proper size gas pipe that runs to the furnace depends on manufacturer requirements as well as whether any additional appliances will share the gas line. Manufacturer recommendations and all applicable codes should be checked.

Gas Line Entry Locations

Most modern furnaces have multiple possible gas entry locations. The one chosen may depend on furnace configuration, such as whether the furnace is upflow, downflow, or horizontal. Common pipe for residential furnace gas lines is ½″ NPT black iron pipe and ¾″ NPT for light commercial. It can be attached to the furnace gas control valve through a knockout hole provided in the casing. Other knockouts are provided for alternate piping. When needed, elbows and adapters may be used to attach the gas line to the control valve. **See Figure 39-12.**

Furnace Gas Lines

Figure 39-12. The most commonly used pipe for residential and light commercial furnace installations is ½″ NPT black iron pipe.

Also, consideration must be taken that accessories, such as air cleaners, may be installed in the future. Some large electronic air cleaners will interfere with the location of the incoming gas line. The air cleaner should be installed on the opposite side of the furnace from the gas entry, or the gas pipe should be routed over the top of the air cleaner through one of the alternate knockouts. **See Procedure for Installing Gas Lines.**

Installing Gas Lines

1. Install black iron gas pipe to the inlet side of the furnace gas control valve. If an inlet elbow is not used, use a holdback wrench on the square ends of the gas control valve. The holdback wrench will prevent twisting and damage to the gas control valve when the gas pipe is turned into the valve. If the inlet elbow is used, place a back-up wrench on the square ends of the inlet elbow rather than on the gas control valve.

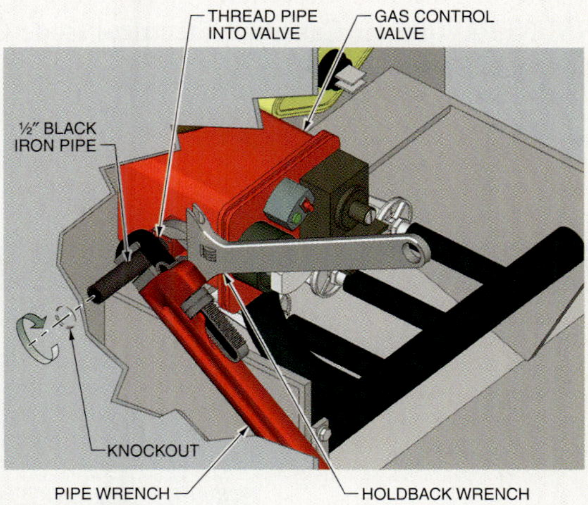

2. When using black iron pipe, install a ground joint union immediately upstream of the furnace to allow for easy servicing of the burner assembly and the gas control valve. The ground joint union must be listed by a nationally recognized testing laboratory. The union will allow for quick replacement and/or service of the gas control valve.

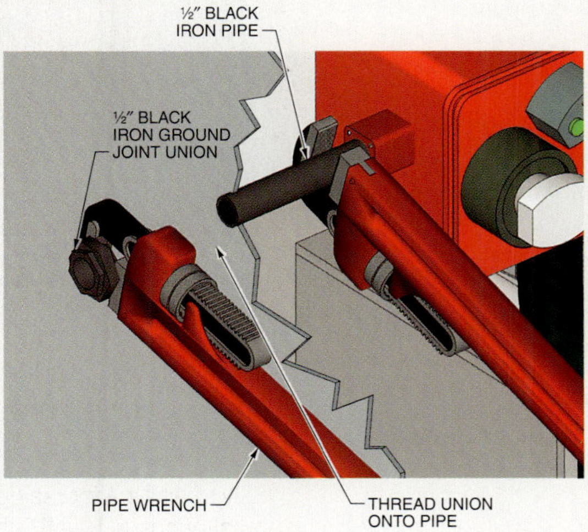

3. Upstream of the ground joint union, install a 90° elbow and drip leg (sediment trap). A convenient way to make a drip leg is to use a ½″ NPT to 1″ NPT tee. Then install a 1½″ long, 1″ NPT nipple in the tee with a 1″ NPT pipe cap to complete the drip leg. The drip leg will help prevent sediment from entering the gas valve.

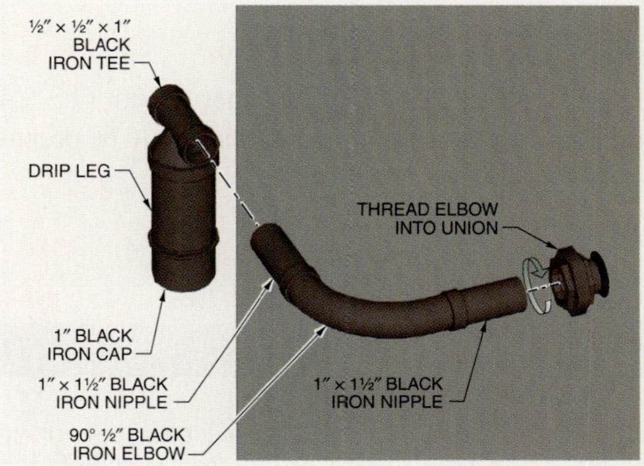

4. Install a gas shutoff valve (gas cock) in the gas supply line immediately upstream of the drip leg tee. The gas shutoff valve must be listed by a nationally recognized testing laboratory.

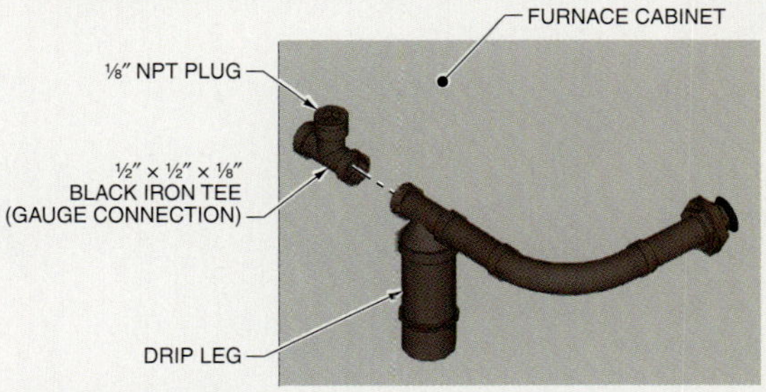

5. Install a ⅛″ NPT plugged tee in the gas supply line immediately downstream of the gas shutoff valve. The ⅛″ NPT plugged tee must be accessible for a test gauge connection. This will allow gas pressure testing without requiring disassembly.

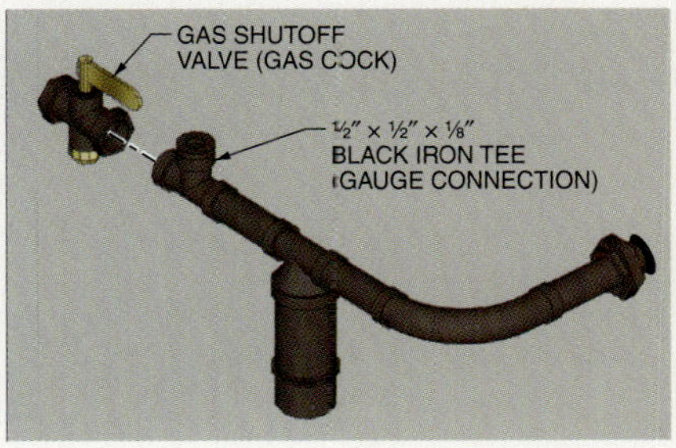

Procedure

Gas Line Leak Testing

After piping the gas line to the furnace, it is imperative to test for gas leaks. The furnace and its gas control valve must be isolated from the gas supply line during leak testing. The furnace and its gas control must be disconnected from the gas supply piping system at the ground joint union during any pressure testing of the piping at test pressures greater than 0.5 psi (14″ WC). **See Figure 39-13.**

During normal gas pressures, a commercial soap solution made to detect gas leaks must be used, and all gas piping connections must be checked. Bubbles indicate gas leakage. All leaks must be sealed before proceeding.

ELECTRICAL WIRING

The vast majority of furnaces operate on 115 VAC. Therefore, high-voltage connections should be prepared for 1ϕ. When a central air conditioning unit is to be installed, provisions for the installation are typically made. Central air conditioning units typically run on 208 V or 230 V of electricity.

The forced-air heating system installation may be either a replacement for an existing furnace or a new installation. When it is a replacement, electrical installation caution must be taken. When replacing original wire, the same type, color, or equivalent wire should be used. The wire ends should be renumbered before connecting to the new system, and proper operation should be verified after servicing. The furnace must be wired according to the wiring diagram provided by the manufacturer and meet local codes, the current National Electrical Code (NEC®), and ANSI/NFPA70E®, or Canadian Electrical Code CSA C22.1. **See Figure 39-14.**

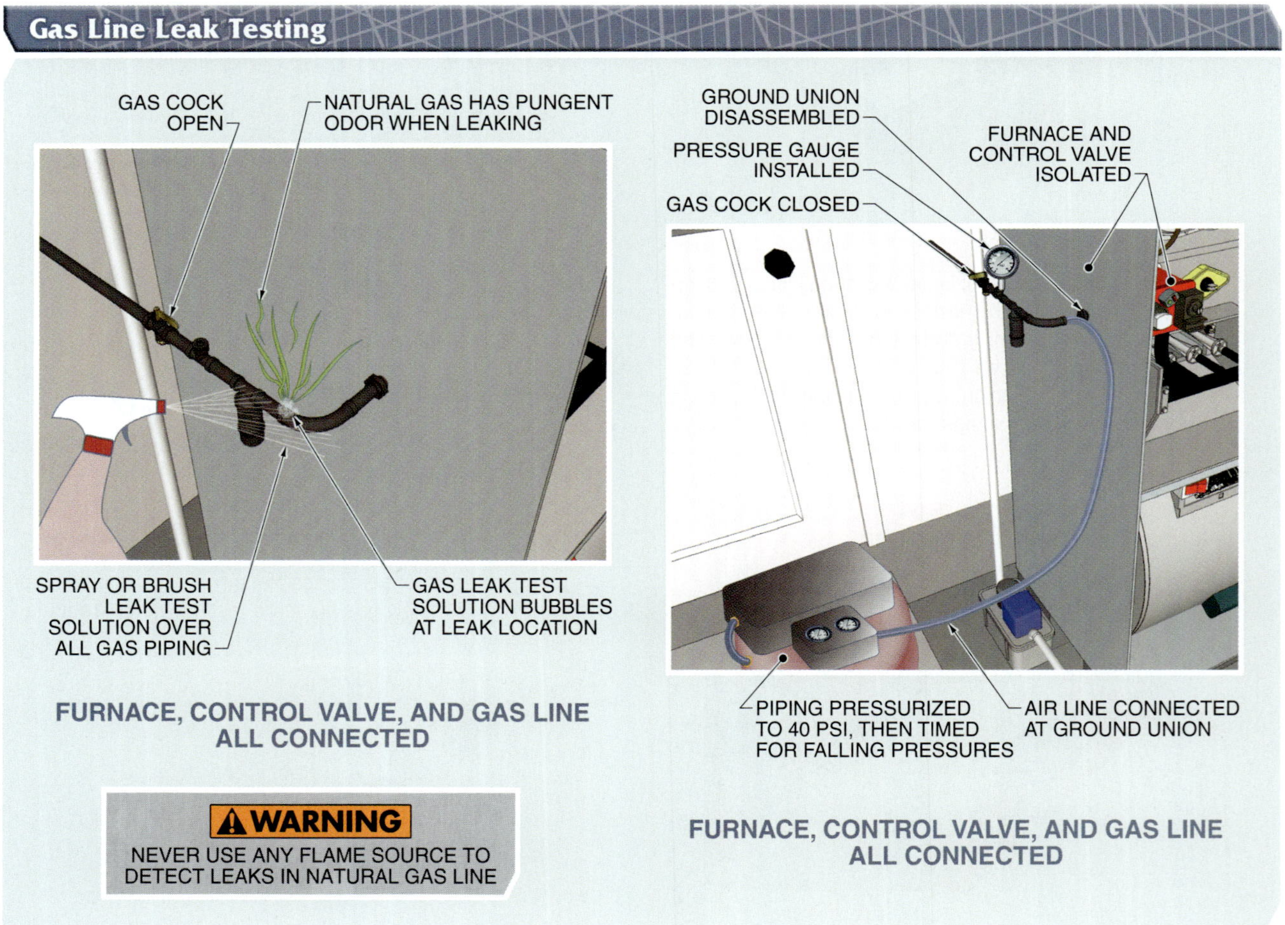

Figure 39-13. Commercial soap solutions are typically sprayed on gas piping to check for leaks. Bubbles indicate leakage.

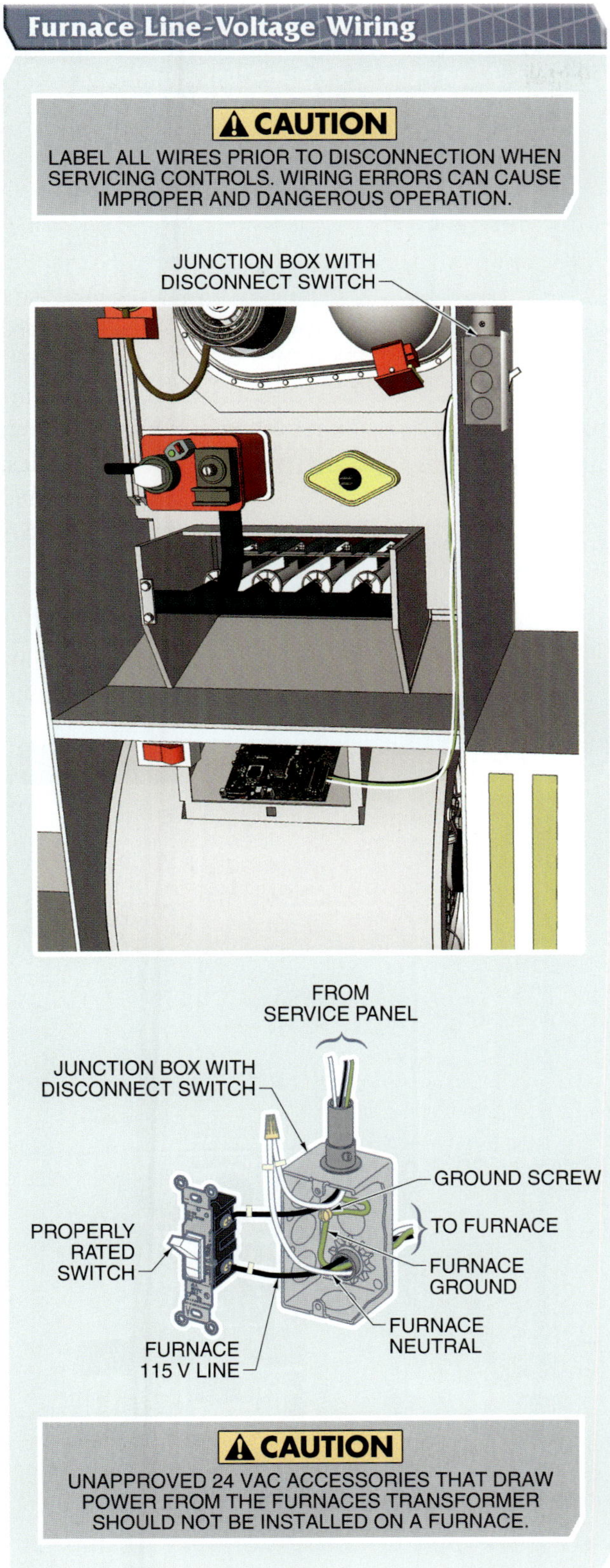

Figure 39-14. Furnaces and all other heating systems must have a separate means of disconnect within sight of the heating unit and be properly grounded.

Also, the furnace must be properly grounded by attaching a green wire to the grounding screw in the furnace junction box. All local codes or, in the absence of local codes, the current NEC®, ANSI/NFPA 70E®, and/or Canadian Electrical Code CSA C22.1 must be followed. One example is a humidifier added to the furnace. The addition of such accessories may cause improper furnace operation and overload the transformer. When accessories are added, separate transformers are required.

The furnace must be provided with its own separate electrical circuit and means of circuit protection. It must have an electrical disconnect switch located within easy reach of the furnace. Electrical wiring that meets local codes, the current NEC®, ANSI/NFPA 70E®, and/or Canadian Electrical Code CSA C22.1 must be used. Type T (63°C rise) wire or equivalent must also be used.

TECH TIP

It is absolutely vital to comply with all applicable codes and regulations during system installation. Inspectors that find problems with a system installation can demand that the work be done again. This can lead to excessive time spent on a job as well as profits being reduced or eliminated.

High-Voltage Wiring Procedures—115 V

A connection must be made from the main service panel in the home to the furnace in order for the furnace to operate. Anyone can prepare the wiring for hook-up, but only a qualified technician should complete the high-voltage wiring. Mistakes in the high-voltage wiring can cause electrical shock or damage to HVAC equipment or property. HVAC technicians who are electrically inexperienced must check with a supervisor or a licensed electrician must be used.

Technicians must select the correct thickness of wire and the corresponding breaker based on the current draw of the unit. Wires are rated by gauge. **See Figure 39-15. See Appendix.** The higher the gauge number, the thinner the wire. Most residential furnaces use 12 gauge wire. As always, the manufacturer specification sheets should be referred to for more information. In many cases, the distributor that provided the furnace will already have an installation kit available with the correct wire size and type for the area.

Selecting Wire Size

AWG	Copper 140° TYPES, TW, UF	Copper 167° TYPES RHW, THHW, THW, THWN, XHHW, USE, ZW	Copper 194° TYPES TBS, SA, SIS, FEP, FEPB, MI, RHH, RHW-2, THHN, THHW, THW-2, THWN-2, USE-2, XHH, XHHW, XHHW-2, ZW-2	Aluminum or Copper-Clad Aluminum 140° TYPES TW, UF	Aluminum or Copper-Clad Aluminum 167° TYPES RHW, THHW, THW, THWN, XHHW, USE	Aluminum or Copper-Clad Aluminum 194° TYPES TBS, SA, SIS, THHN, THHW, THW-2, THWN-2, RHH, RHW-2, USE-2, XHH, XHHW, XHHW-2, ZW-2
18	—	—	14	—	—	—
16	—	—	18	—	—	—
14*	15	20	25	—	—	—
12*	20	25	30	15	20	25
10*	30	35	40	25	30	35
8	40	50	55	35	40	45
				40	50	55

* conductor overcurrent protection limitations

NFPA-2011 NEC®

NOTE: ALLOWABLE AMPACITIES OF INSULATED CONDUCTORS RATED UP TO AND INCLUDING 2000 V, 140°F THROUGH 194°F, NOT MORE THAN THREE CURRENT-CARRYING CONDUCTORS IN RACEWAY, CABLE, OR EARTH, BASED ON AMBIENT TEMPERATURE OF 86°F.

Figure 39-15. Only a qualified technician should complete the high-voltage wiring to a furnace in accordance with manufacturer recommendations and the current NEC®.

Standard residential thermostat installations use 18-gauge, 6-wire (18/6) cable.

The first step in preparing high-voltage connections to a furnace is to install a local disconnect. Electrical codes require a local disconnect (safety switch) to be located near the furnace. Local electrical codes must be consulted for specifics. The local disconnect is a safety device that must be within eyesight of the forced-air heating unit per the NEC®. The disconnect requires that two separate wiring runs be created. **See Figure 39-16.** One wiring run travels from the service panel of the building to the disconnect, and the other run travels from the disconnect to the furnace. **See Procedure for Installing Wiring from Service Panels to Disconnects.**

Furnace Wiring Runs—115 V

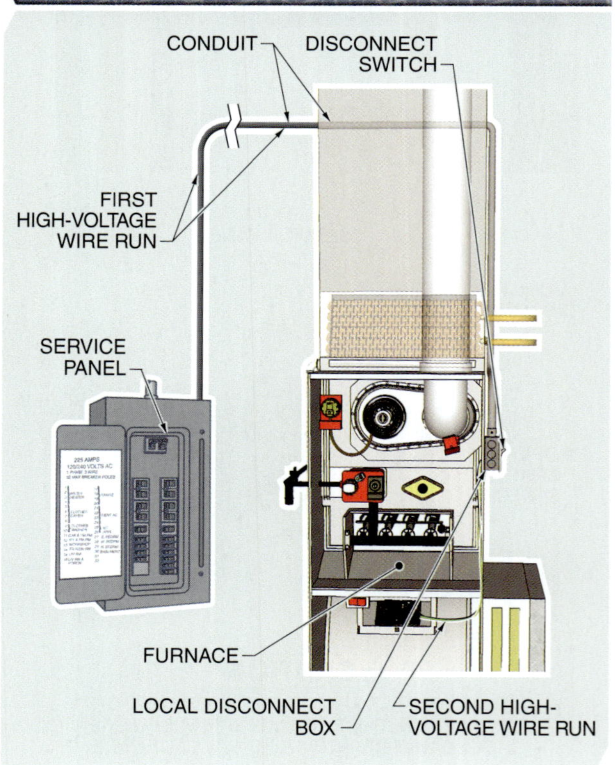

Figure 39-16. High-voltage wiring for a furnace has two parts. One wiring run is from the service panel to the local disconnect, and the other is from the local disconnect to the furnace (heating unit).

Installing Wiring from Service Panels to Disconnects

1. Remove the disconnect box cover to provide access to the inside. Small holes can be found in the corners to mount the box with screws.

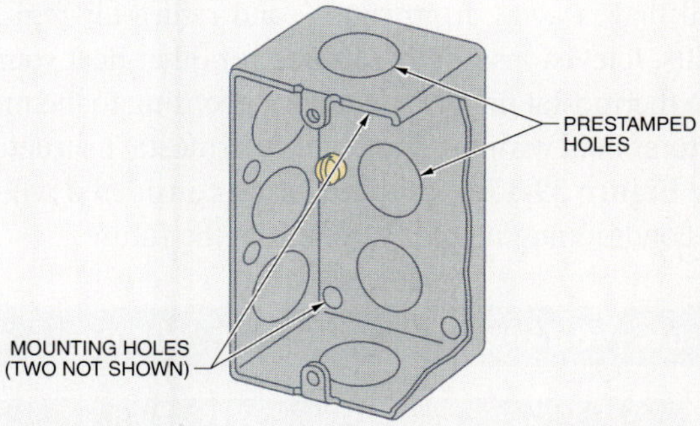

2. Punch out one prestamped hole on the appropriate side or top and one prestamped hole on the back of the disconnect box. Punch out the holes that are properly sized for the required wiring fittings.

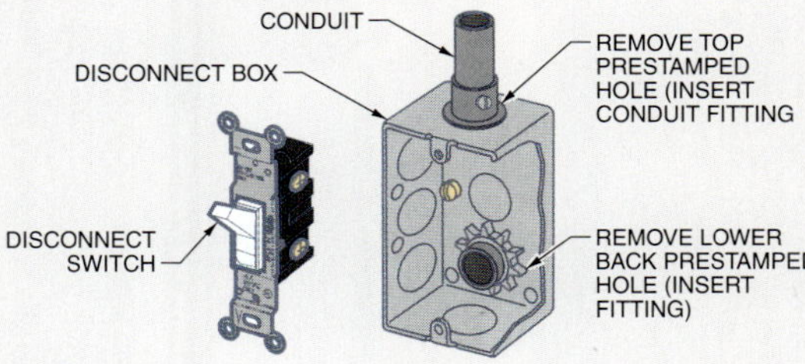

3. Mount the box to the appropriate surface with screws.
   a. Select a spot near the furnace to mount the disconnect box.
   b. Place the disconnect high enough to keep it away from any possible water conditions.
   c. See electrical codes for specifics on where to locate the disconnect box for the various types of installations.
   d. In many cases, the disconnect box will be attached directly to the side of the furnace in a spot that does not interfere with furnace operation.

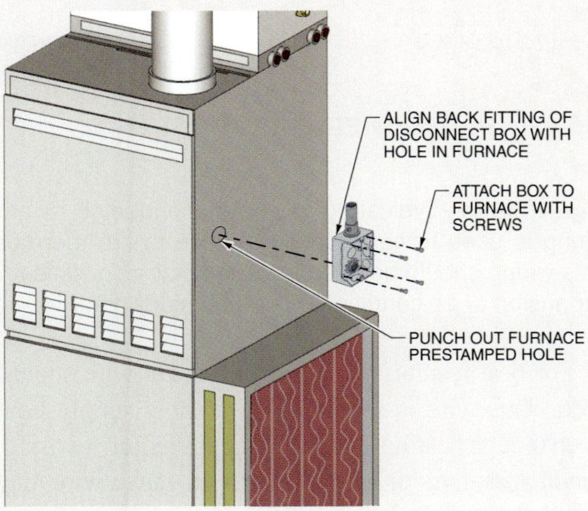

4. Slide wires through the fittings (conduit if required) in the service panel and tighten to ground connection, circuit breaker, and common busbar. The green ground wire should be connected to the grounding busbar terminal. The black hot wire should be connected to the circuit breaker terminal screw. The white common wire should be connected to the neutral busbar terminal.

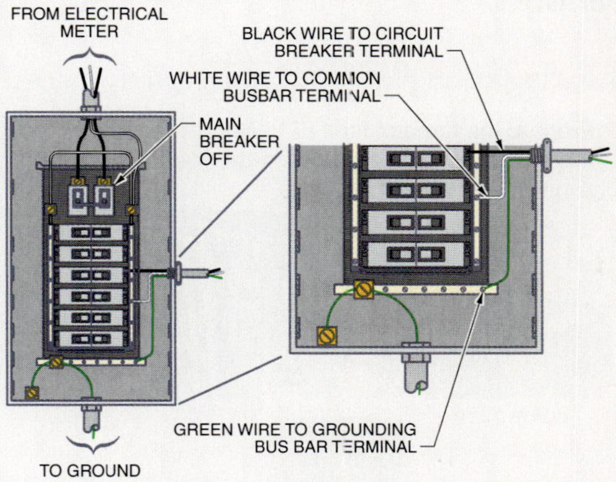

5. Run wire from the building service panel to the disconnect box. Use the correct wire gauge (based on the furnace current draw) and type. Wire such as NM cable may be used if building codes and electrical codes permit. Some local codes require the use of solid wire and conduit. Consult an electrician if needed.
6. Cut the two green wires to size, strip back a ½″ of insulation, connect to the screw in the center of the disconnect box, and tighten with a screwdriver. Ensure a good and solid ground connection.
7. Connect the power supply wires (black and white) to the disconnect switch and box. The black wire is the hot wire and should be connected to the disconnect switch. The white wire is the common wire and should be connected to the furnace common.

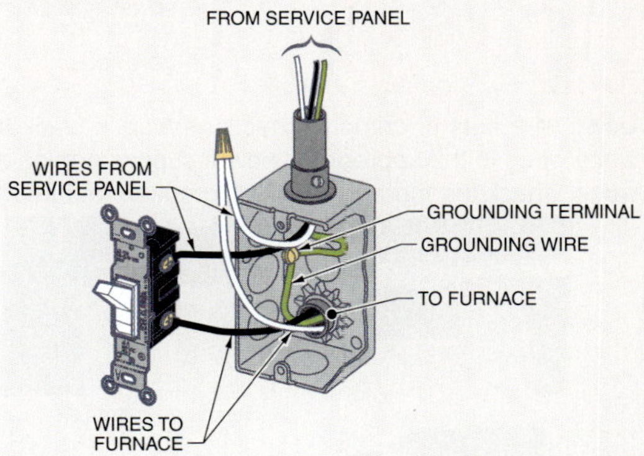

A furnace disconnect, according to the NEC®, must be mounted within eyesight of the furnace. Some local codes require that the disconnect be mounted within 3′ of the furnace. Most HVAC technicians mount the furnace disconnect directly to the furnace cabinet and run the wires to the furnace out the back of the disconnect box. **See Procedure for Installing Wiring from Disconnects to Furnaces.**

Procedure

Installing Wiring from Disconnects to Furnaces

1. Prepare the furnace for 115 V wiring. Most furnaces are already internally prepared to receive 115 VAC power supply wiring.

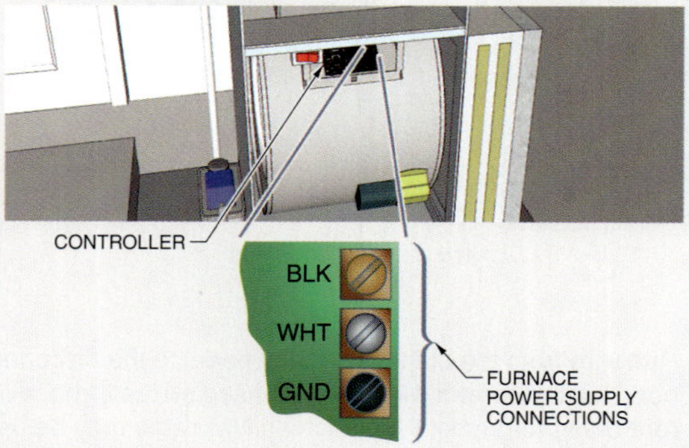

2. Run wires from the disconnect to the furnace. Typically, these wires are 12 gauge.

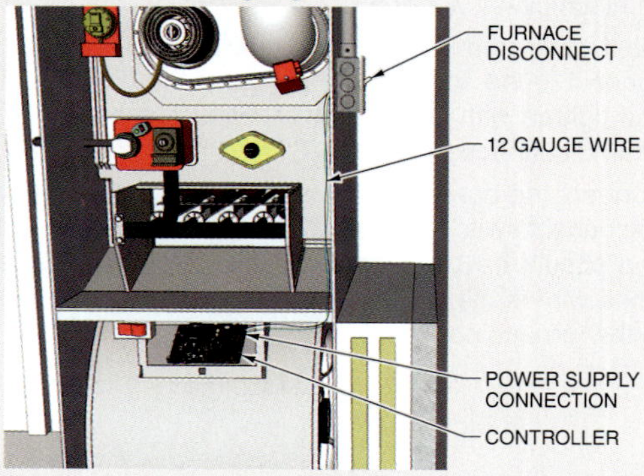

3. Using wire nuts or crimp connections, attach black and white wires to the appropriate power supply terminals or wires. Check the furnace wiring diagram as required.

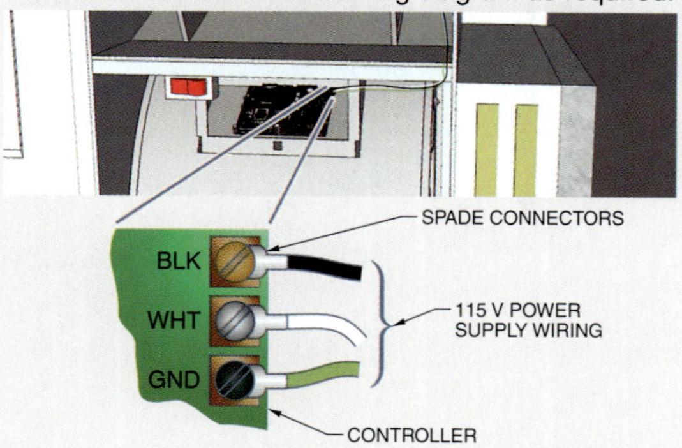

4. Leave the disconnect turned off until the unit is needed.

Low-Voltage Wiring Procedures—24 V

When installing a thermostat for a forced-air heating unit, 24 VAC wiring is to be used. A location should be selected for the room thermostat on a draft-free interior wall that is away from supply and return air registers, lights, televisions, direct sunlight, or other heat sources. The thermostat must be installed according to the manufacturer field wiring diagram and thermostat instructions. **See Figure 39-17.** It should also be considered whether air conditioning may be installed in the future.

Low-Voltage (24 V) Furnace Wiring

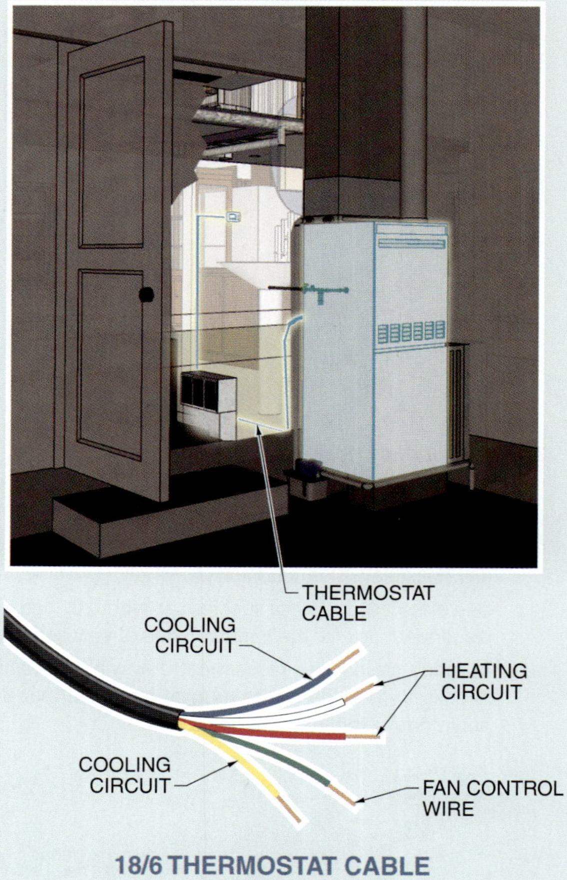

Figure 39-17. Typically, five- or six-conductor 18 gauge wiring is used to connect the furnace to the thermostat. This wiring should be used because of the possible future expansion of air conditioning and humidity controls.

For best results, 18 gauge thermostat wire should be used. This wire is sold in bulk at most supply houses. Even though heating-only applications require two or three conductors, five- or six-conductor 18 gauge wire may be installed. This wire is typically known as 18/6 wire or cable.

Most thermostats installed today are electronic and programmable. They are easily programmed for various temperature, time, and day run schedules that can help considerably with keeping heating and cooling costs down. Always follow the instructions provided by the thermostats manufacturer for installation and programming. **See Procedure for Running Low-Voltage Wiring to Furnaces.**

Procedure

Running Low-Voltage Wiring to Furnaces

1. Remove the access panels on the front of the furnace or air handler.

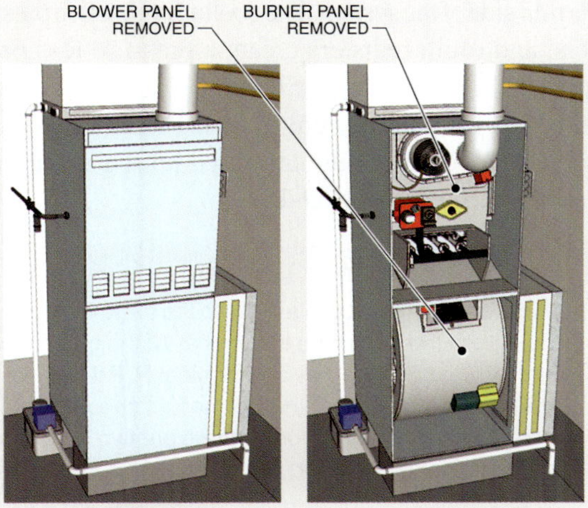

ACCESS PANELS IN PLACE

ACCESS PANELS REMOVED

2. Inside the furnace, run the 24 V thermostat cable and secure with plastic cable ties.

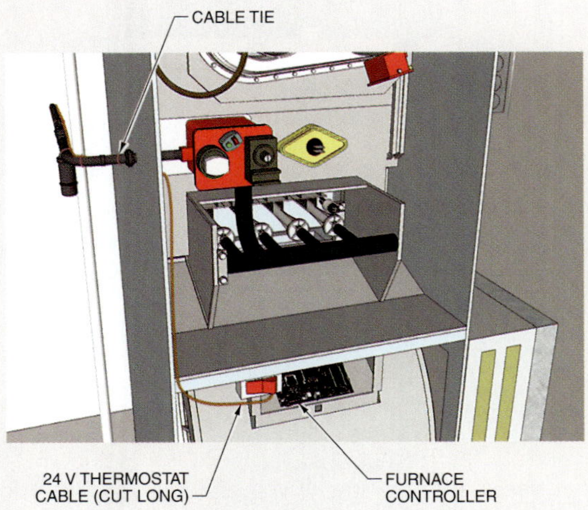

3. For a standard gas furnace, the control wiring area displays five or six terminals. Terminals are labeled either C, Y, Rc, Rh, R, and W, or C, Y, G, R, and W. The two wires used for the heating function are the R (red) and W (white) wires. If separate fan control is desired, connect the G (green) wire.

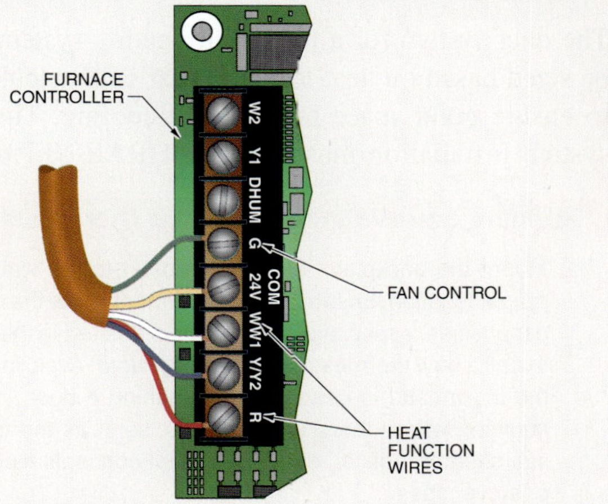

FURNACE LOW-VOLTAGE CONNECTIONS

4. Trim the 24 V wires so that they are long enough to reach the screw terminals, and strip the wire insulation back ½″.
5. Connect wires R, W (heating circuit), and G (fan control) to terminals R, W, and G of the furnace, and tighten with a screwdriver.

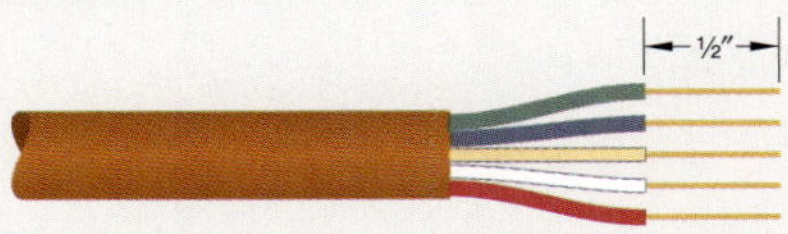

6. Replace the front access panels on the furnace or air handler.

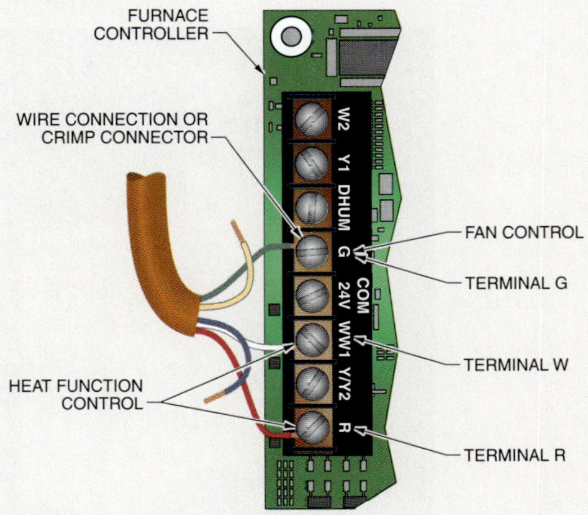

FURNACE LOW-VOLTAGE CONNECTIONS

The wiring from a furnace to a thermostat may have to carry current for 120 V systems. However, most residential and light-commercial forced-air heating systems use 24 V control systems. A 24 V thermostat control wire must not be bundled with other 115 V wiring. The higher voltage level can cause signal disturbances in the 24 V circuits. **See Procedure for Running Low-Voltage Wiring to Thermostats.**

DUCT SYSTEMS AND FILTERS

The duct system for a forced-air heating system must be sized based on heat loss and heat gain calculations to ensure good space heating and cooling. The duct system installation must meet ASHRAE/NFPA 90B, *Standard for the Installation of Warm Air Heating and Air-Conditioning Systems,* and local codes. **CAUTION:** Failure to follow these standards could reduce airflow or increase air leakage, resulting in reduced system performance or furnace damage.

Residential duct systems are typically designed so that the furnace will operate at a static pressure of 0.50″ WC or less. This total static pressure limitation includes the total pressure losses on both the supply air side and the return air side of the system. The cooling coil (when present), ducts, and room registers create a pressure loss on the supply side. The return grills, ducts, and filters create a pressure loss on the return side. Pressure losses are calculated based on a commonly used value of 400 cfm/ton of cooling. **See Figure 39-18.**

Procedure: Running Low-Voltage Wiring to Thermostats

1. Mount the backplate of the thermostat to the wall in the proper location. Ensure that the thermostat has the desired capabilities, especially if central air is desired in the future. Install a new thermostat if necessary. *Note:* Avoid installing the thermostat in dead air spots behind a door, in direct sunlight, where there are air drafts, such as the top of a staircase or register, or on a wall that conceals a chimney or stove pipes.

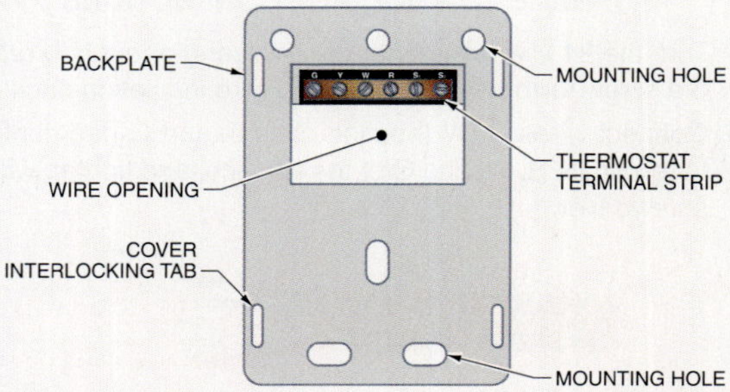

2. Run the thermostat cable from the furnace to the thermostat.

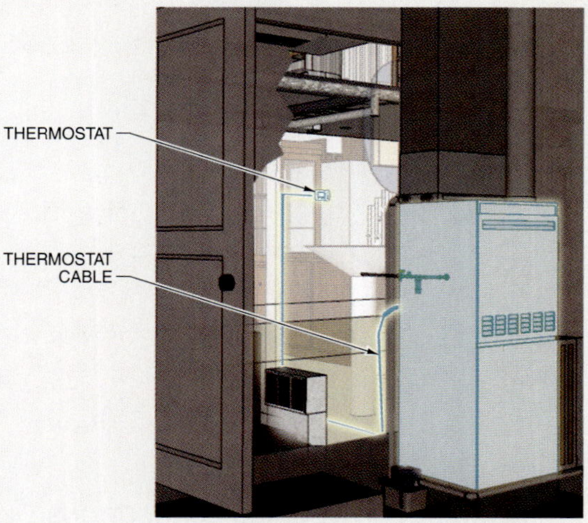

3. Attach the thermostat cable wires to the backplate terminal strip. Attach the white wire to terminal W, the green wire to terminal G, and the red wire to terminal R. If this is not the lettering configuration for the thermostat, check the manufacturer wiring diagram. *Note:* If there are two R terminals, be sure to install a jumper.

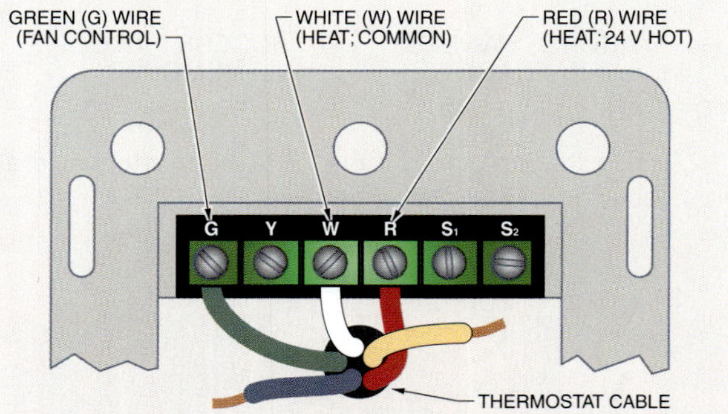

4. Replace the thermostat cover.
5. Check the thermostat operation after system startup per manufacturer instructions.

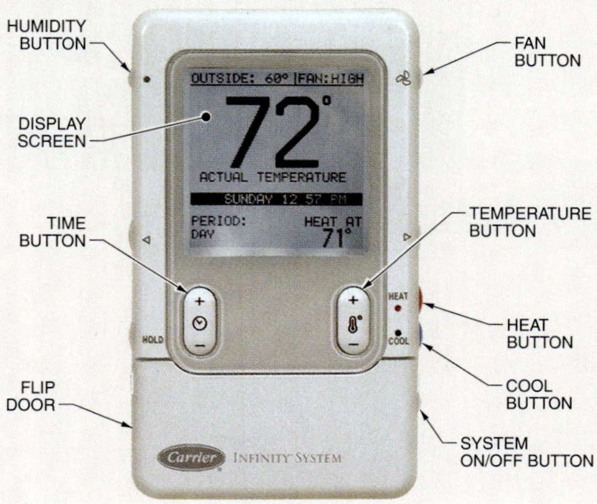

System Ducts and Filters

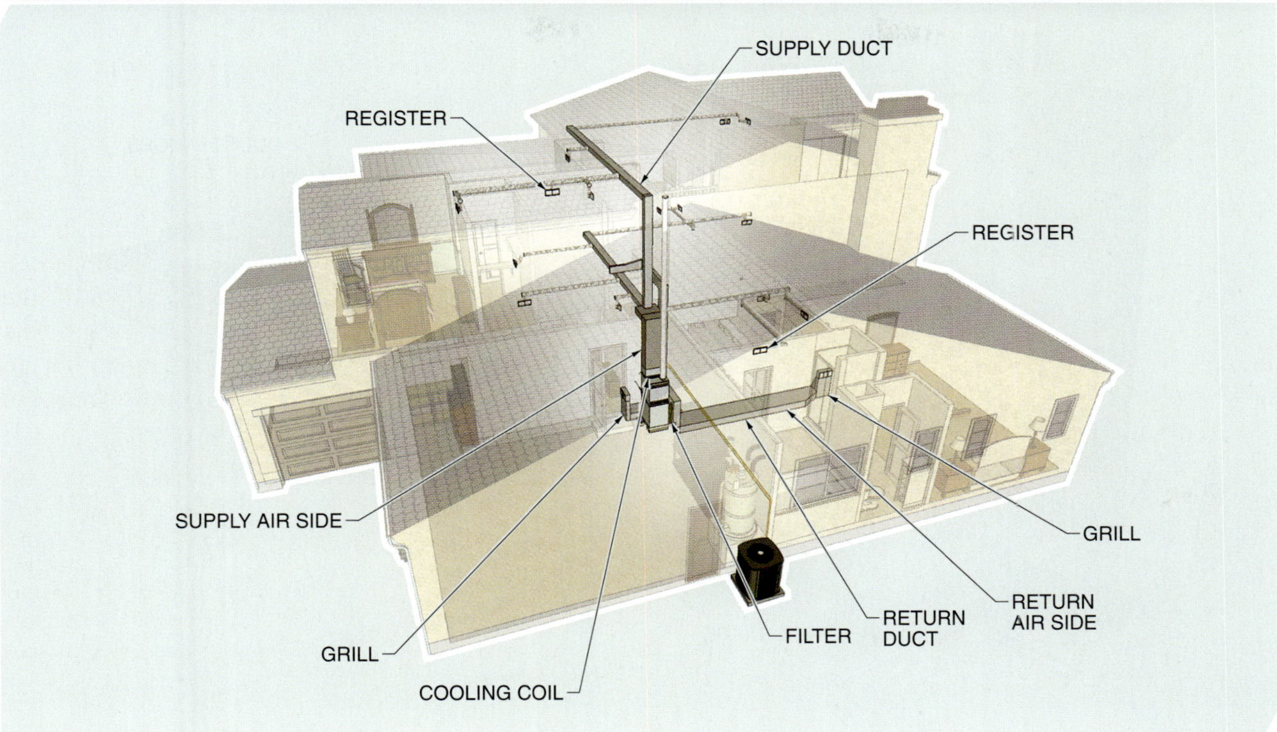

Figure 39-18. A residential duct system is designed so that the furnace will operate with the system at a static pressure of 0.50″ WC.

System filters are installed behind return grills in crawl space installations.

Supply Air Ductwork

Supply air duct plenum connections must be at least as large as the furnace supply opening. Supply duct flanges are provided at the furnace discharge. The supply air ductwork must be sealed to the furnace cabinet, walls, ceilings, or floors it passes through. When the furnace is equipped with a cooling coil, a removable access panel that is large enough to allow viewing of the cooling coil location should be provided in the supply duct. The access panel should be accessible when the furnace is installed and sealed to prevent air leaks. **See Figure 39-19.**

When a furnace is installed in the horizontal position with a cooling coil, an 18″ long transition between the furnace and cooling coil may be required to allow free airflow into the coil. The cooling coil installation instructions provided by the manufacturer must be consulted.

Return Air Ductwork

When a furnace is installed so that supply air ducts carry air to areas outside the space containing the furnace, return air must be handled by a duct sealed to the furnace cabinet and terminated outside the space containing the furnace. **See Figure 39-20.** For furnaces installed in an upflow position and the required air delivery is above a specific value (e.g., 1800 cfm), it is required that both sides of the furnace be used for return air or a combination of one side and the bottom. When bottom return air is not used, it is blanked off (closed off) using a bottom closure panel available from the manufacturer.

Supply Air Ductwork

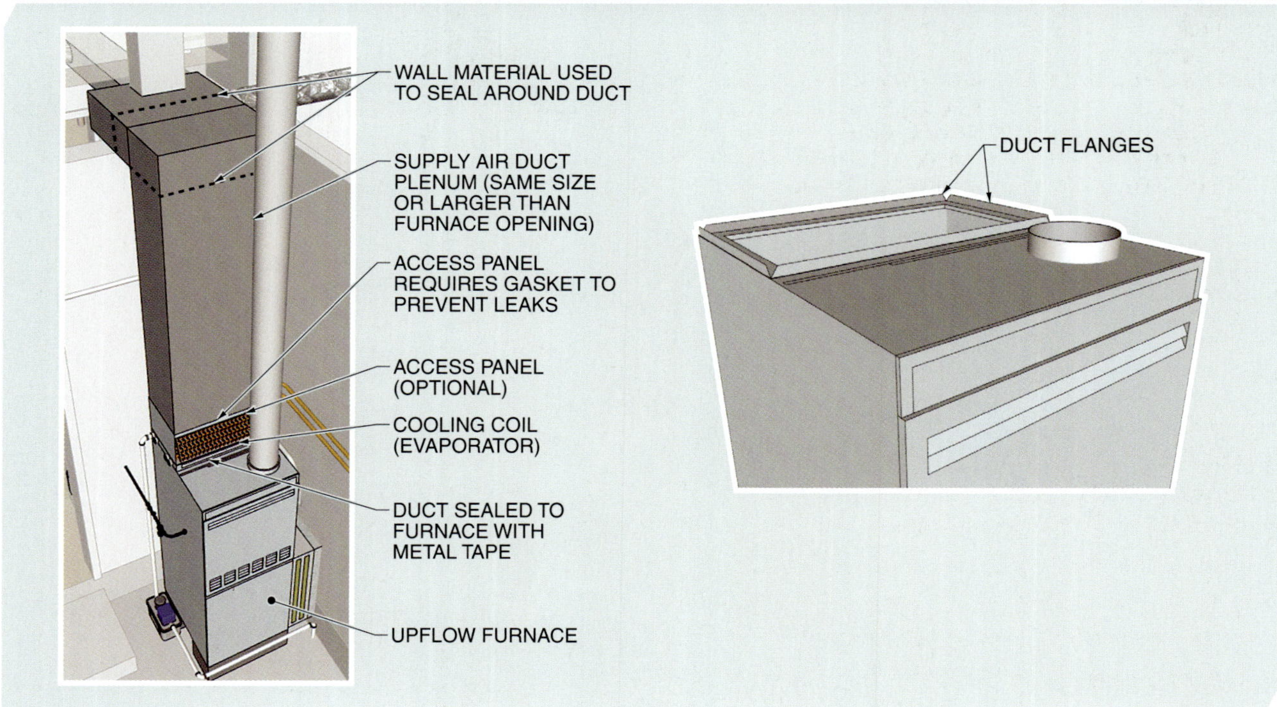

Figure 39-19. It is common to place an access panel in the location of a cooling coil even if the cooling coil does not exist. The time saved in the future when the coil will be installed is worth the effort.

Return Air Ductwork

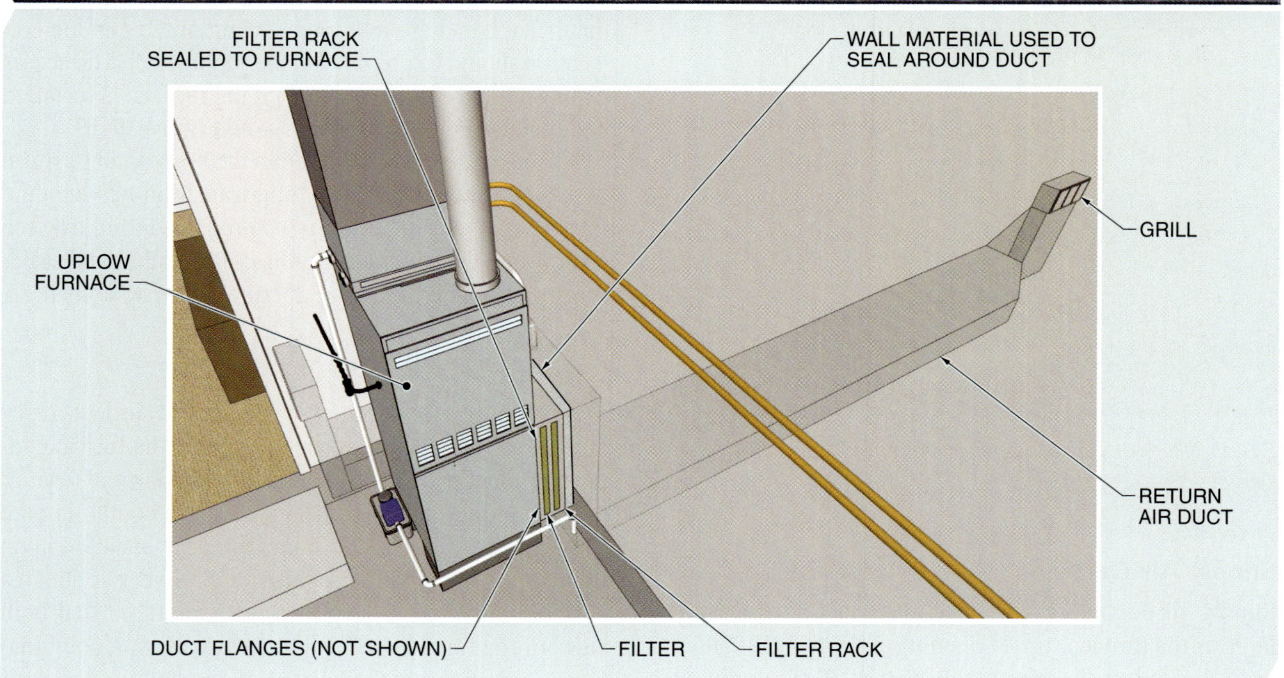

Figure 39-20. Most furnaces today allow return air in both sides of the furnaces simultaneously or a side and the bottom of the furnaces when required.

Duct Dampers

Dampers are required by code and are often installed in each branch run duct in order to balance airflow and adjust for even temperature throughout the heated space. **See Figure 39-21.** For proper furnace operation, the following must be ensured:

- Supply air registers and return air grills are open.
- Registers and grills are clear of rugs, carpets, drapes, and furniture.
- Size and shape of supply air plenum are correct.
- Number of supply air ducts is correct.

Cooling Coil Location

It is common for an air conditioning system to be installed at the same time as the furnace. Installing the cooling coil (evaporator) at the same time as the furnace saves much time and work later. When the furnace shares a common duct system with a separate cooling unit, the cooling coil should be installed downstream of the furnace. This prevents condensation from forming on the furnace heating element. Cooling coils should not be installed in the return air ductwork. **See Figure 39-22.**

Filters and Filter Racks

Filters are used to reduce the amount of particulate matter in the air. Filters are mounted in a filter rack to hold them in place and facilitate filter replacement as part of a routine maintenance program. The filter rack is often provided by the furnace manufacturer. If a filter rack is not provided, a rack must be field fabricated. The proper size and number of filters for the system should always be used per the manufacturer recommendations. **See Figure 39-23.**

Duct Dampers

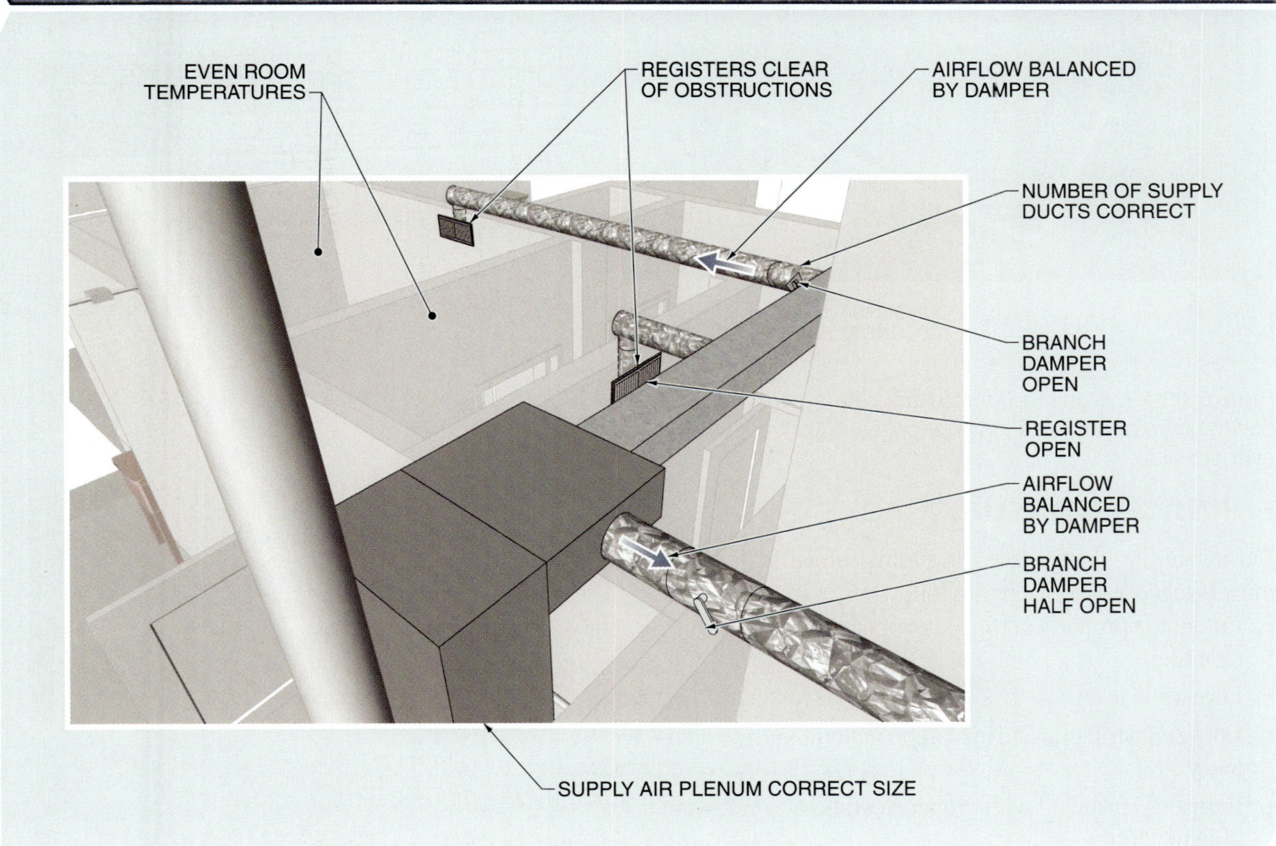

Figure 39-21. Dampers are installed in branch run ducts of supply air systems to balance airflow and temperatures throughout buildings.

Cooling Coil (Evaporator) Location

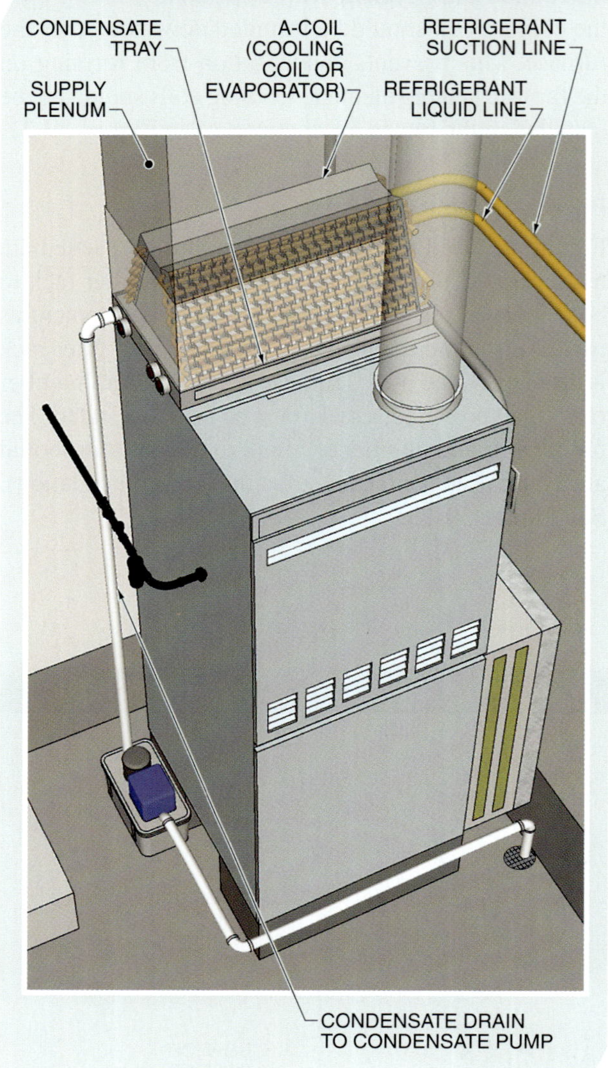

Figure 39-22. Cooling coils (evaporators) are located downstream of furnaces in the supply air ductwork whenever possible.

FURNACE STARTUP CHECKS

Before starting the furnace for the first time, the furnace must be checked to ensure the following:

- Furnace is properly equipped to operate with the available fuel.
- Furnace is level.
- Loose construction and insulation materials are cleared away.
- Furnace is installed with proper clearances to combustible materials.
- Furnace in a crawl space is sufficiently elevated above the ground.
- Sufficient combustion and ventilation air are available.
- Vent system meets current requirements per ANSI Standard Z223.1, NFPA 54, and/or CAN/CGA B149 as well as local codes.
- Vent connector is securely fastened to the vent collar.
- Gas pipe and controls have been completely checked for leaks.
- All electrical wiring follows the current NEC®, ANSI/NFPA 70E®, or Canadian Electrical Code CSA C22.1 as well as local codes.
- Furnace is grounded.
- Room thermostat is properly installed with its heat anticipator set correctly.
- Duct system is correctly sized and sealed.
- Proper air filter racks or filter supports have been installed.
- Air filter is in place and correctly sized.

Filters and Filter Racks

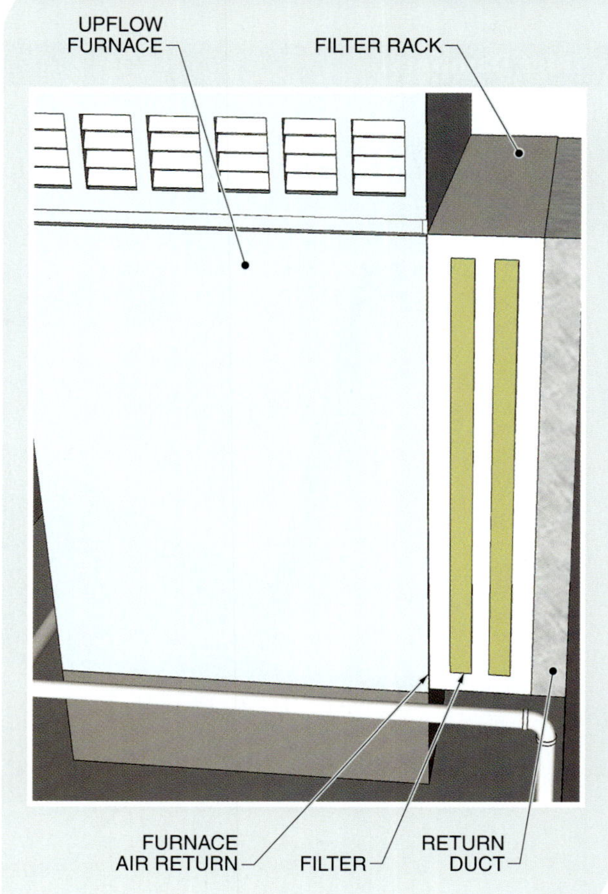

Figure 39-23. Filter racks are connected to the furnace duct flanges on the return-air side of the system. There are many different types of filter racks and filters.

GAS SUPPLY, MANIFOLD PRESSURE, AND GAS INPUT RATE

Natural gas is flammable and dangerous. Some of the most important tasks in the installation process are to check the gas supply system pressure, adjust the pilot, adjust the manifold pressure, and check the gas input rate. Any failure to perform these tasks can result in fire, explosion, damage, or death. It is important to use the correct tools when performing these tasks.

Gas Supply Pressure and Pilot Adjustment

It is very important before startup to check the gas supply pressure. A 0″ to 15″ water manometer with 0.1″ resolution is required along with a ⅛″ NPT manual shutoff valve to measure actual manifold gas pressure. Also, a plug used to seal the valve must be removed and a hose barb fitting (typically ⅛″) installed so that the manometer hose can be connected to the supply gas line. Today, many of these are electronic test instruments. **See Procedure for Testing Gas Pressure.**

Procedure

Testing Gas Pressure

1. Shut off the furnace gas shutoff valve (gas cock) in the gas supply line.

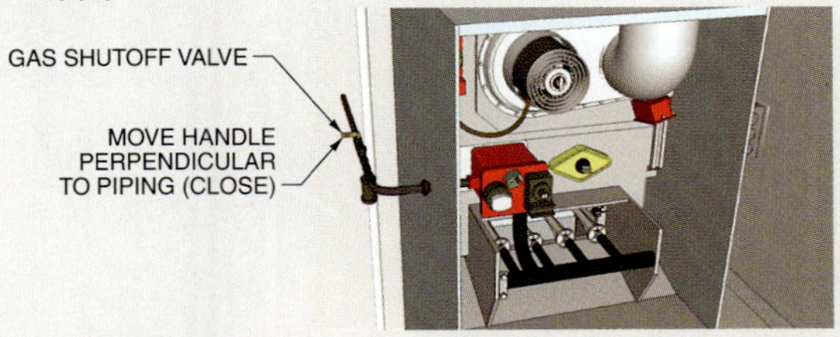

2. Remove the inlet pressure plug from the gas control tee.

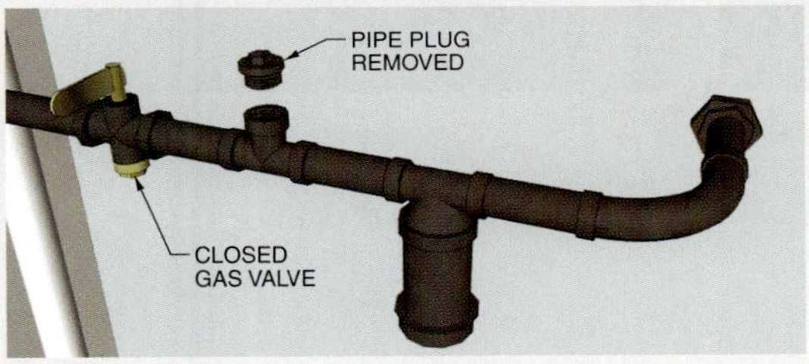

3. Install an ⅛″ NPT manual shutoff valve in the hole vacated by the pipe plug. Ensure the gas shutoff valve is in the OFF position.

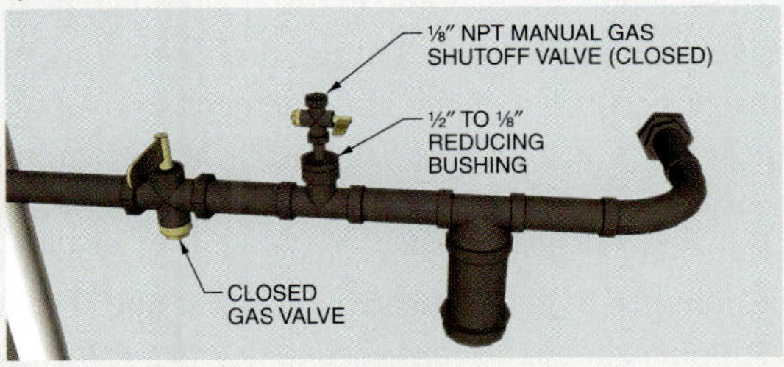

4. Attach a manometer to the ⅛″ NPT manual shutoff valve just installed.

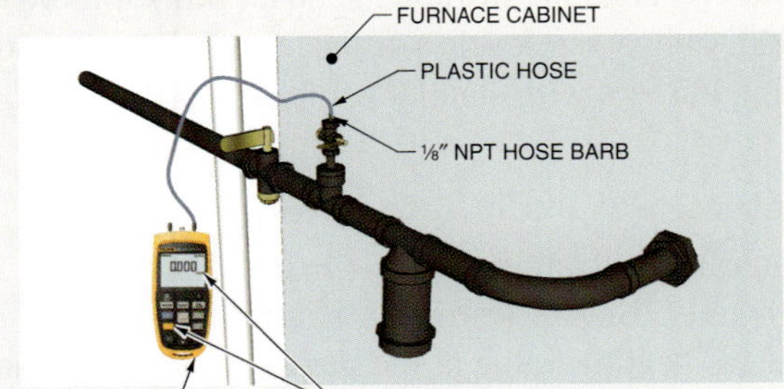

5. Slowly open the gas shutoff valve in the gas supply line.

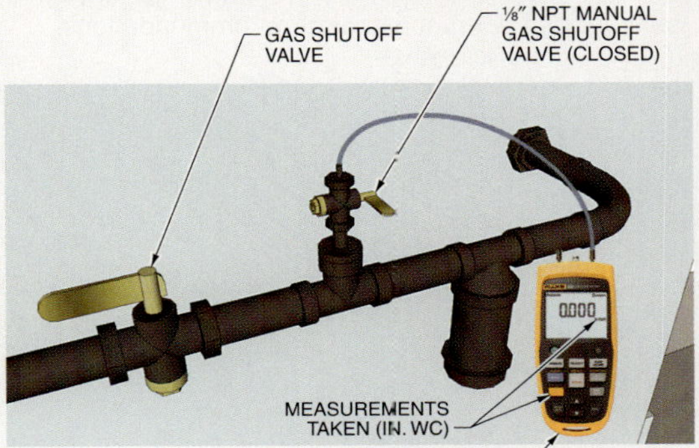

6. Slowly open the ⅛″ NPT manual shutoff valve leading to the manometer.

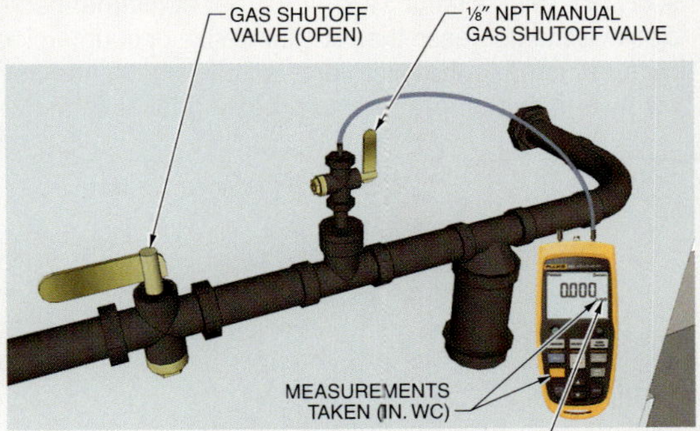

7. Turn on all gas appliances attached to the gas supply line.
8. With the furnace operating, read the gas supply pressure on the manometer.
   - Natural gas supply pressure must be between 5″ WC and 7″ WC.
   - Propane gas supply pressure must be between 11″ WC and 13″ WC.
9. If gas supply pressure is not within these limits, adjustment to the incoming gas supply will be required.
10. Shut off all gas appliances attached to the gas supply line.

Pilot Flame Adjustment

Before finishing a furnace installation, it is very important to check the pilot flame for proper operation. Prior to adjusting the pilot flame, it must be confirmed that gas supply pressure is correct. *Note:* The pilot flame is typically checked at the factory and should not require adjustment. However, pilot adjustment is possible when necessary. **See Procedure for Adjusting Pilot Flames.**

Procedure

Adjusting Pilot Flames

1. Start the furnace by following the operating instructions provided by the manufacturer. Sometimes the operating instructions are listed on the front door of the furnace.
2. The pilot flame should cover ½″ of the tip of the flame sensor or as per manufacturer recommendations.

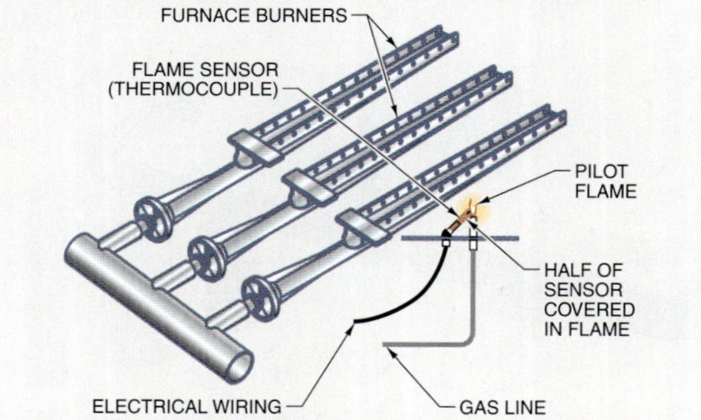

3. To adjust the pilot flame, the pilot adjustment cover screw on the gas control valve must be removed. Save the cover screw for reinstallation. Turn the inner adjustment screw clockwise to decrease the pilot flame size or counterclockwise to increase the pilot flame size. Reinstall the cover screw and tighten properly to prevent gas leakage.

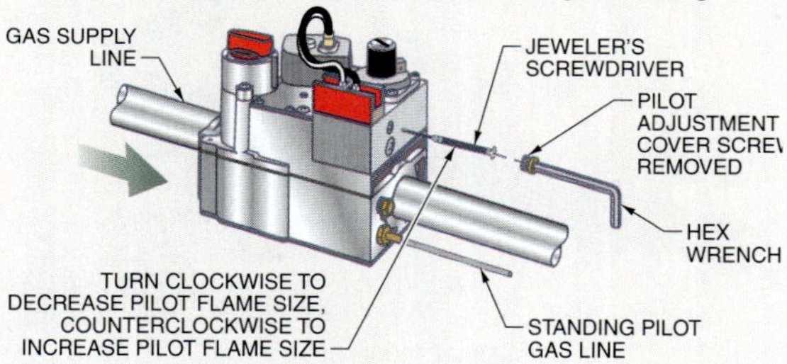

4. Shut off the furnace.
5. Shut off the furnace gas shutoff valve in the gas supply line. Remove the ⅛″ NPT shutoff valve from the gas-control-inlet pressure tap. Install the pressure tap plug and turn on the gas.

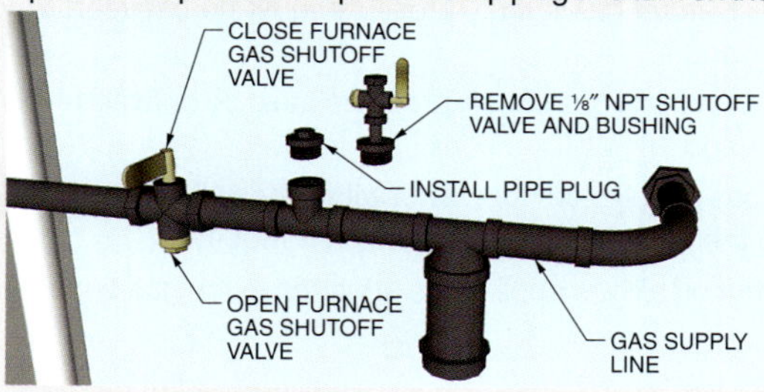

6. Check the pilot adjustment cover screw and the pressure tap plug for gas leaks. Use a commercial soap solution made for leak detection.

Manifold Pressure Adjustment

When the gas supply pressure is too high or low according to test results, the manifold pressure must be adjusted. Natural gas manifold pressure must be between 3.2″ WC and 3.8″ WC. Propane gas manifold pressure must be between 9.7″ WC and 10.3″ WC.

Note: For natural gas, the main burner orifices must be changed to obtain the proper gas flow rate when the gas flow rate cannot be properly set within the proper pressure ranges. Manufacturer instructions must be followed for the proper adjustment of burner orifices.

Carrier Corporation

A gas valve's internal pressure-reducing valve is used to ensure the correct gas pressure at the burners.

It is the responsibility of the installer to ensure that the gas input rate (in Btu/hr) of the furnace is properly adjusted. Underfiring could cause inadequate heat, excessive condensation, or ignition problems. Overfiring could cause sooting, flame impingement, or overheating of the heat exchanger. **See Figure 39-24.** Depending on the local gas heating value and elevation, the manifold pressure may need to be adjusted or the orifices may need to be changed to get the proper gas input rate. The local gas supplier should be consulted to determine the proper heating value (in Btu/ft³) of natural gas in the area. *Note:* When being installed at an altitude of more than 2000′ above sea level, a furnace must be derated by using burners with smaller orifices.

Natural Gas Input Rate

Natural gas heating values (in Btu/ft³) can vary significantly. Before starting a natural gas input check, the gas heating value at the particular location should be obtained from the local supplier. A stopwatch is needed to measure the actual gas input. **See Procedure for Measuring Actual Gas Input Rates.**

Gas Input Rate

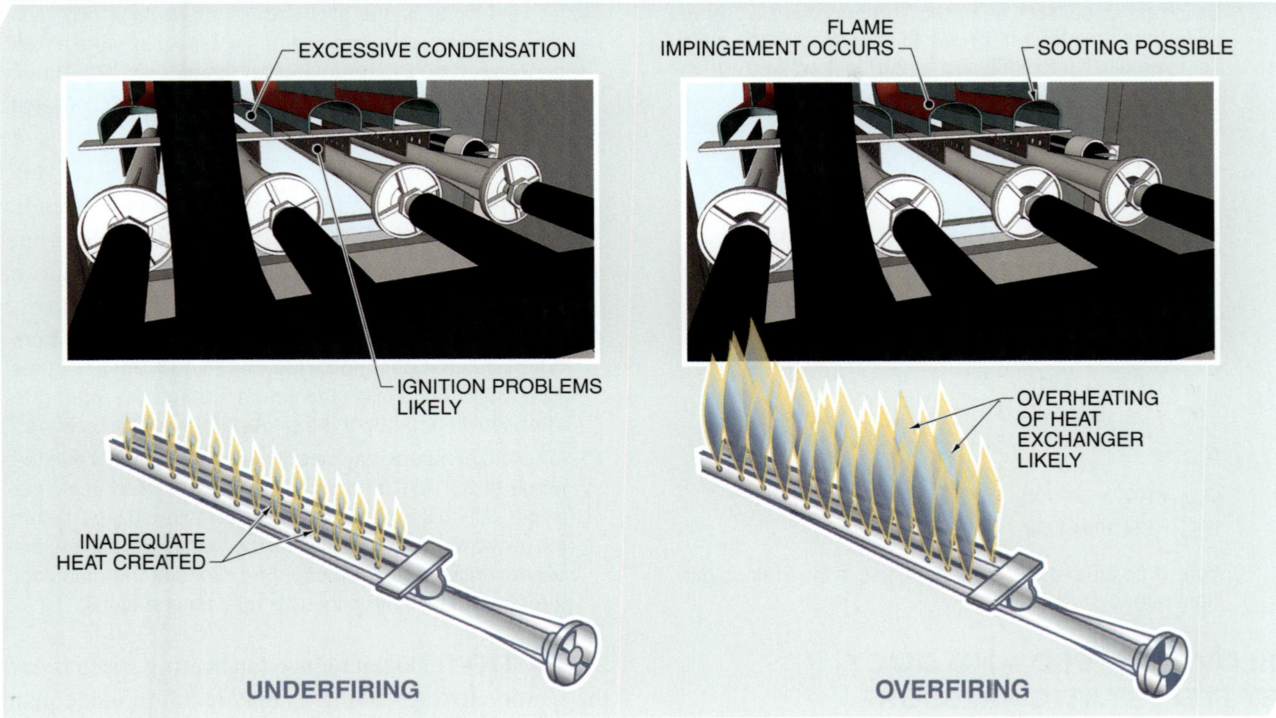

Figure 39-24. Technicians should be able to look at the burner flames of a furnace and determine whether the flames are the right color and whether the furnace is underfiring or overfiring.

Measuring Actual Gas Input Rates

1. The gas supply pressure must be between 5″ WC and 7″ WC for natural gas.

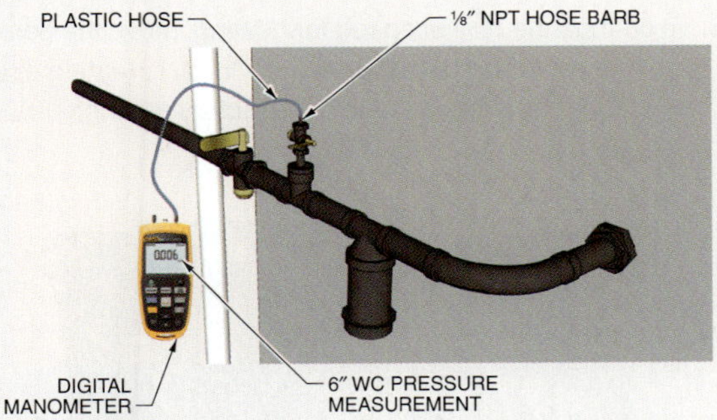

2. Turn off all other gas appliances. Leave on the pilot lights (flames).
3. Start the furnace following the manufacturer-provided operating instructions.
4. Let the furnace warm up for at least 6 min.
5. Locate the gas meter. Determine which dial represents the least cubic feet of gas and how many cubic feet per revolution it represents.

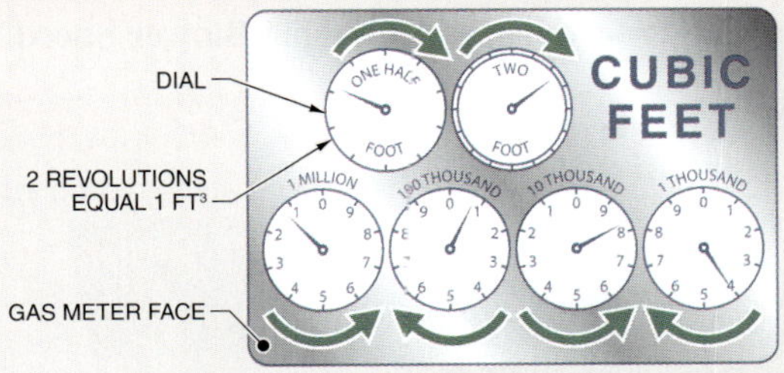

6. With the stopwatch, measure the time it takes to consume 2 ft³ of gas. If the dial is ½ ft³ per revolution, measure the time for four revolutions. If the dial is 1 ft³ per revolution, measure the time for two revolutions. If the dial is 2 ft³ per revolution, measure the time for one revolution.

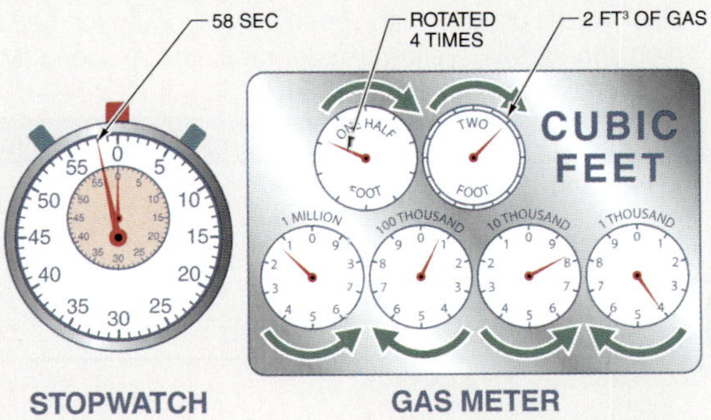

continued on next page

Procedure

Procedure

Measuring Actual Gas Input Rates *(continued)*

7. Divide the amount of time by two. This provides the average amount of time for 1 ft^3 of gas to flow through the meter. For example, if it took 58 sec for 2 ft^3 to flow, it would take 29 sec for 1 ft^3 to flow.
8. Calculate the gas input rate using the following formula:

$$G_{input} = \frac{g_{H\text{-}value} \times 3600}{t}$$

where

$G_{H\text{-}value}$ = 1 ft^3 heating value

3600 = constant (sec/hr)

t = time (in sec)

Therefore,

$$G_{input} = \frac{g_{H\text{-}value} \times 3600}{t}$$

$$G_{input} = \frac{1000 \times 3600}{29}$$

$$G_{input} = 1000 \times 124.138$$

$$G_{input} = \mathbf{124{,}138\ Btu/hr}$$

Note: If no other pilots are left on, this is the furnace gas input rate.

9. If a gas water heater pilot was left on, a dryer pilot, two range burner pilots, and one range oven pilot were all left on. Allow 1000 Btu/hr for a water heater pilot, 500 Btu/hr for a dryer pilot, 500 Btu/hr for each range burner pilot, and 500 Btu/hr for each range oven pilot.

Water heater pilot	1000 Btu/hr
Dryer pilot	500 Btu/hr
Two range burner pilots	1000 Btu/hr
Oven pilot	+ 500 Btu/hr
Total pilot usage	**3000 Btu/hr**

10. The total pilot usage of 3000 Btu/hr subtracted from the calculated gas input rate of 124,138 Btu/hr equals 121,138 Btu/hr. This is the actual furnace gas input rate after allowing for all of the pilots left on in the building.
11. Manifold pressure may be adjusted to 3.5″ WC, within the range of 3.2″ WC to 3.8″ WC to get the rated gas input within ±2%. If it is not possible to get the rated input within the allowable range because of the manifold pressure, the burner orifices must be changed. Consult the manufacturer operating instructions for recommendations.

BLOWER SPEED AND DUCT SYSTEM STATIC PRESSURE

The initial heating and cooling blower speeds required for an application should be determined at the system design stage. Airflow data may be provided by the manufacturer. Depending on the results of the tests performed on the system, technicians may need to change the speed of the blower motor. **See Procedure for Changing Blower Speed.**

CAUTION: Do not reduce fan heating speed below the factory setting. Doing so may result in inadequate air circulation and could cause excessive air temperature rise inside the furnace. This could cause a high temperature limit switch to cycle the burners on and off frequently, which will reduce furnace efficiency and shorten the life of the heat exchanger and blower motor.

Procedure

Changing Blower Speed

1. Turn off the electricity to the furnace at the disconnect.

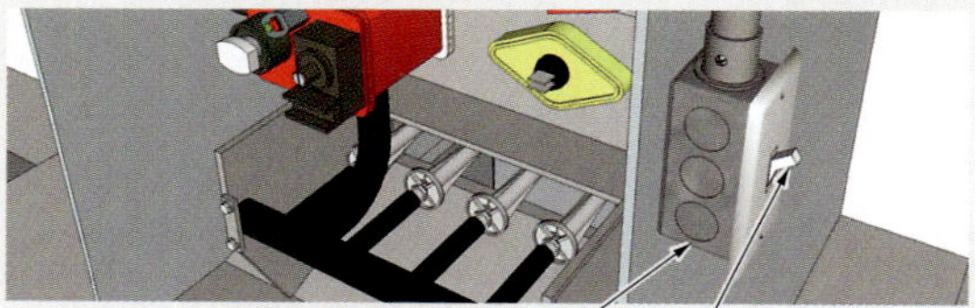

2. Although there are variations, many blower motors have the same four motor-speed and wiring designations. Normally, black wire is used for high speed (HI), blue wire for medium high speed (MH), yellow wire for medium low speed (ML), and red wire for low speed (LO).

Motor	Initial	RPMs
High speed	HI	1100
Medium high speed	MH	1000
Medium*	M	900
Medium low speed	ML	800
Low speed	LO	600

* not available from all manufacturers

3. To change the blower speed for heating, move the desired motor lead to the terminal marked "Heat." Heating speed should not be reduced below the factory setting.

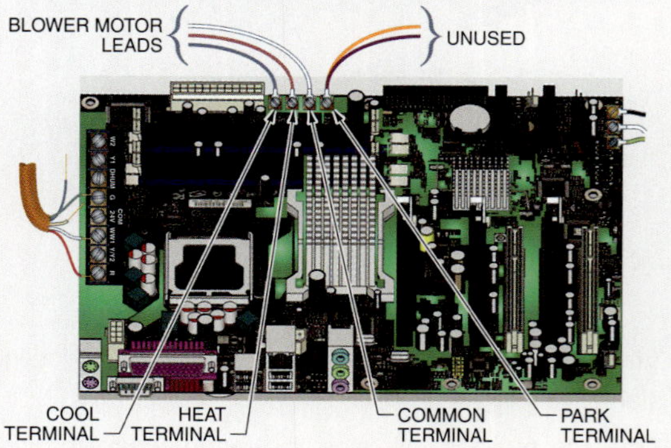

NOTE: NEVER REDUCE BLOWER SPEED BELOW FACTORY SETTINGS. LOWER SPEED EQUALS HIGHER TEMPERATURE RISE.

4. Connect the two unused motor leads to any terminals marked "Park," "Unused," or "Unused Motor Leads."
5. Turn on the electricity at the furnace disconnect.

Fan Static Pressure

System airflow can be determined from the product data sheet when duct system static pressure is known. Improper airflow in heating mode may result in poor heating performance and reduced heat exchanger life. Improper airflow in cooling mode may cause poor cooling performance or air conditioning coil freeze-up.

High duct-system static pressure is an indication of an overly restrictive duct system. Static pressure in excess of 0.5″ WC indicates a need for duct system redesign to ensure proper volume of airflow. Typically, a 0″ WC to 1″ WC gauge with 0.01″ resolution and two pressure measurement taps is used. **See Procedure for Measuring Duct System Fan Static Pressure.**

Measuring Duct System Fan Static Pressure

Procedure

1. Open the supply air registers and return air grills. Ensure that the registers and grills are free of obstruction from rugs, carpets, drapes, and/or furniture.

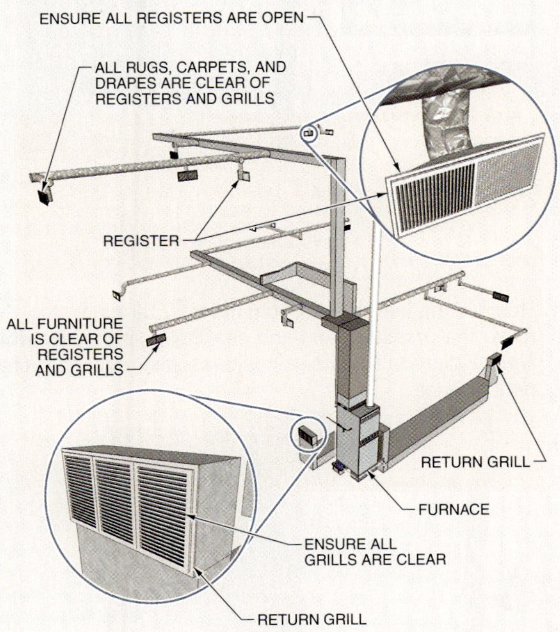

2. Set the balancing dampers in the supply duct system.

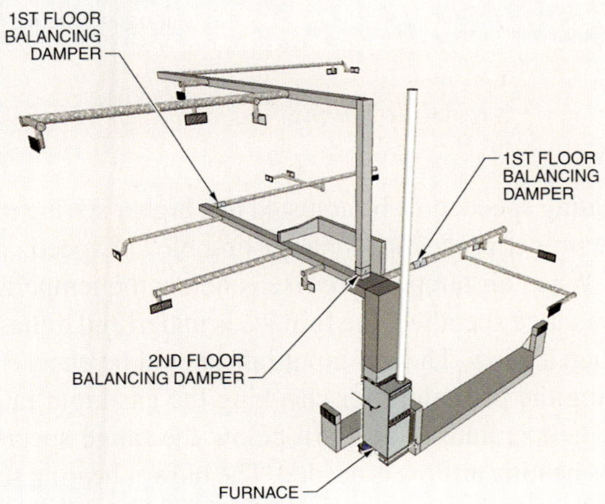

3. Check the ductwork for obstructions or leaks.

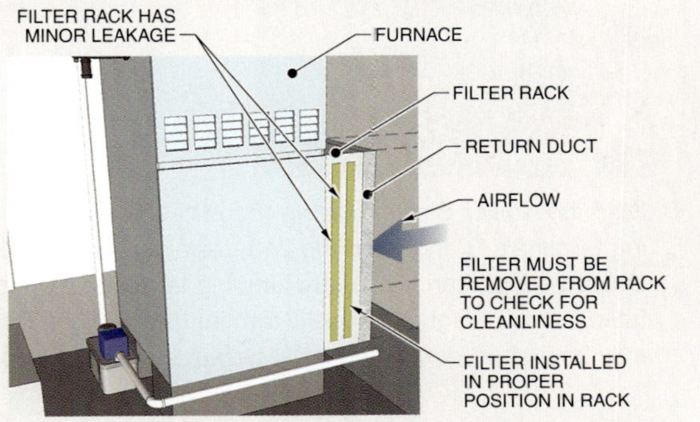

4. Ensure that the filters are clean and in the proper position.
5. Ensure that the blower speed taps are set for proper heating and cooling. Heating speed should be set to meet heating equipment requirements. See the manufacturer-provided data for heating airflow capacities.
6. Place the gauge or manometer near the furnace, and zero the scale to read 0.00″ WC.

7. Insert one static pressure tap into the supply air transition duct between the furnace and cooling coil or in the supply air plenum for heating only systems. Connect this pressure tap to the positive pressure side of the meter.

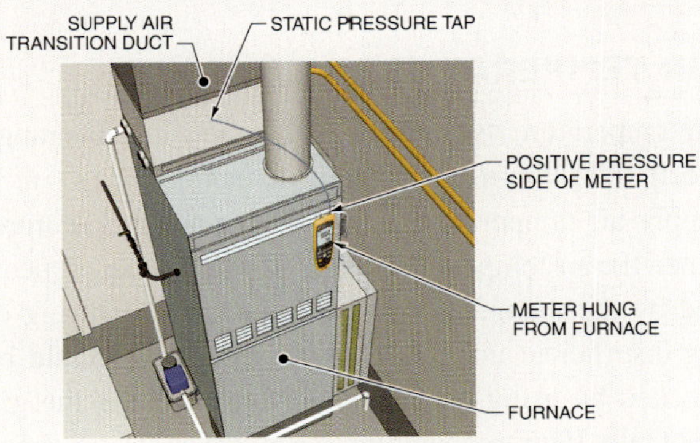

continued on next page

Measuring Duct System Static Pressure *(continued)*

8. Insert the other static pressure tap in the return air plenum. Connect this pressure tap to the negative pressure side of the meter.

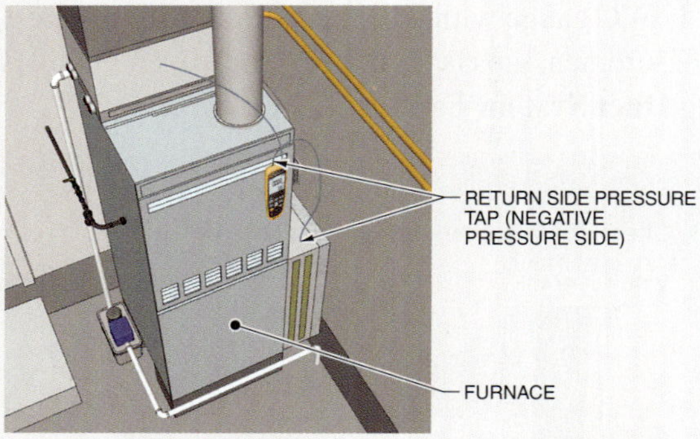

9. Start the blower. Blower cooling speed can be run by jumping terminals R and G on the 24 V terminal block. Blower cooling speed can be run by jumping terminals R and Y (starts compressor) on the 24 V terminal block. *Note:* There may be a short delay before the blower starts and stops.

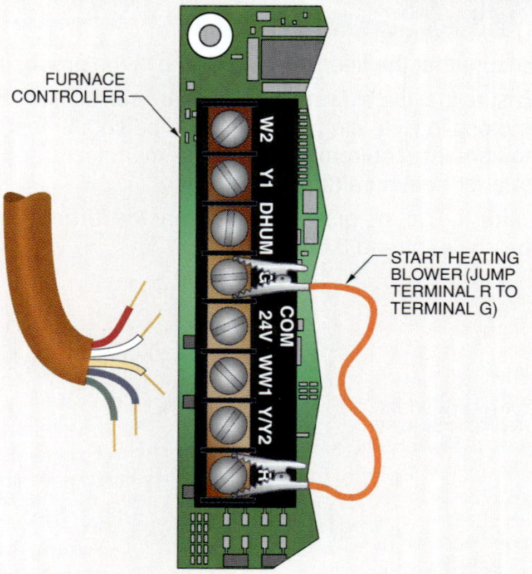

NOTE: THERE MAY BE A SHORT DELAY BEFORE THE BLOWER STARTS AND A DELAY AFTER THE BLOWER STOPS.

10. Read the duct system static pressure on the meter.

    Note: If the air filter location is upstream of the return air pressure tap, static pressure must be adjusted to exclude the filter pressure drop. Adjust static pressure by subtracting 0.08″ WC from the measured static pressure for the filter (duct system static pressure = measured static pressure −0.08″ WC).

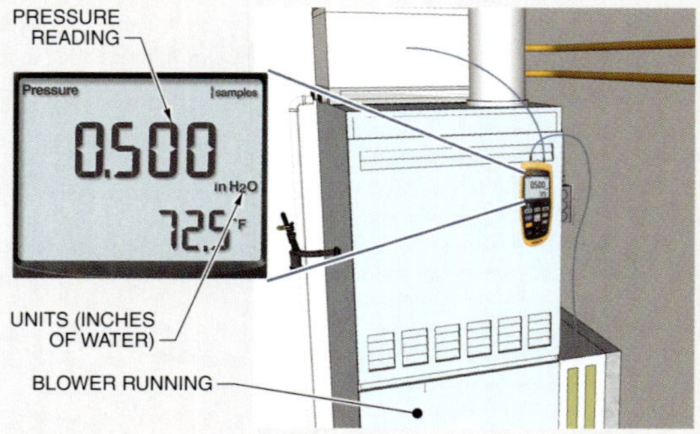

NOTE: IF AIR FILTER LOCATION IS UPSTREAM OF RETURN AIR PRESSURE TAP, STATIC PRESSURE MUST BE ADJUSTED TO EXCLUDE FILTER PRESSURE DROP. DO THIS BY SUBTRACTING 0.08″ W.C. FROM THE MEASURED STATIC PRESSURE.
DUCT SYSTEM STATIC PRESSURE = MEASURED − 0.08″ W.C. (FOR FILTER)

11. Remove the jumper wire from the 24 V terminal strip. Remove the pressure taps and seal holes in the ductwork. *Note:* Failure to seal holes could result in reduced system performance.

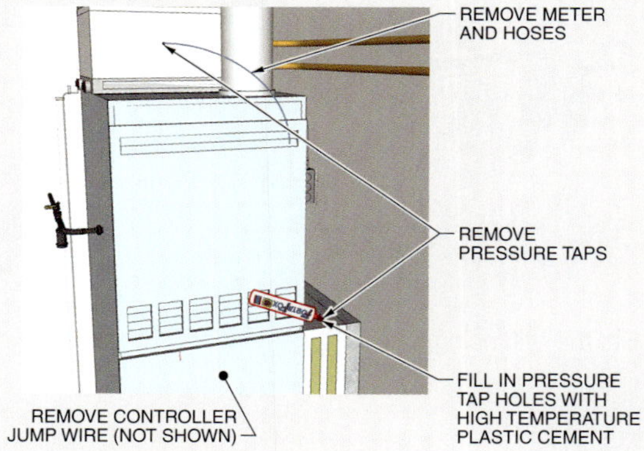

NOTE: FAILURE TO SEAL HOLES COULD RESULT IN REDUCED SYSTEM PERFORMANCE.

FURNACE LOW-VOLTAGE CONNECTIONS

AIR TEMPERATURE RISE

Air temperature rise must be within the allowable range specified by the manufacturer. *Air temperature rise* is supply air temperature minus return air temperature. When the air temperature rise is above the range specified by the manufacturer, the furnace is overfiring or has insufficient airflow. The gas input rate should be checked by using the procedure for checking the gas input rate. If air temperature rise is still above the range specified, more heating airflow is needed. The blower heating speed must be adjusted to a higher setting using the proper procedure for adjusting blower speed.

When air temperature rise is below the temperature rise range specified, the furnace is underfired or has too much airflow. The gas input rate should be checked by using the procedure for checking the gas input rate. If air temperature rise is still below the range specified, less heating airflow is needed. The blower heating speed must be adjusted to a lower setting using the proper procedure for adjusting blower speed. **See Procedure for Measuring Air Temperature Rise.**

Procedure

Measuring Air Temperature Rise

1. Open the supply air registers and return air grills. Ensure that the registers and grills are free of obstructions from rugs, carpets, drapes, and/or furniture.

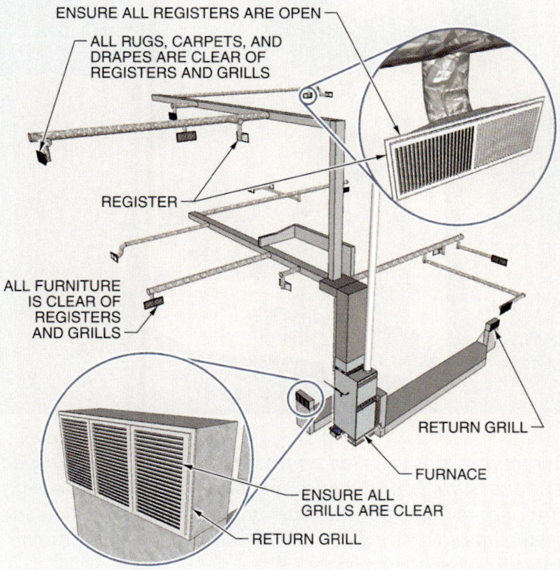

2. Set the balancing dampers in the supply duct system.

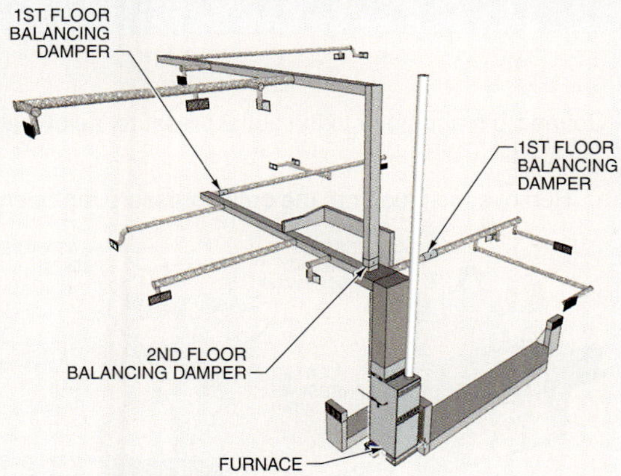

3. Check ductwork for obstructions and leaks.

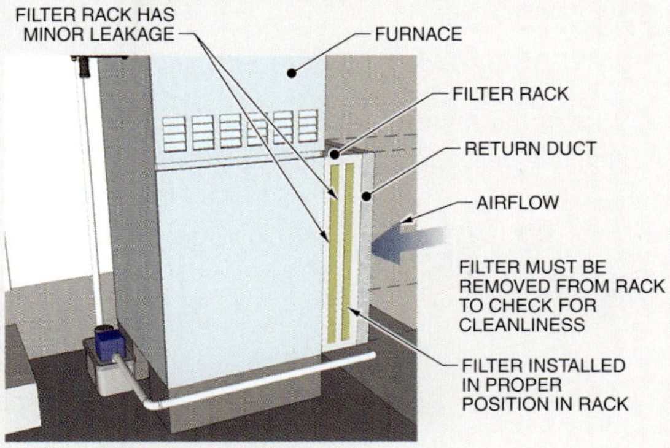

4. Ensure that the filters are clean and in the proper position.
5. Ensure that the blower speed taps are set properly for heating and cooling. Heating speed should be set to meet heating equipment requirements. See manufacturer-provided data for heating airflow capacities.
6. Place one thermometer in the supply air plenum approximately 2′ from the furnace. Locate the thermometer tip in the center of the plenum to ensure proper temperature measurement.

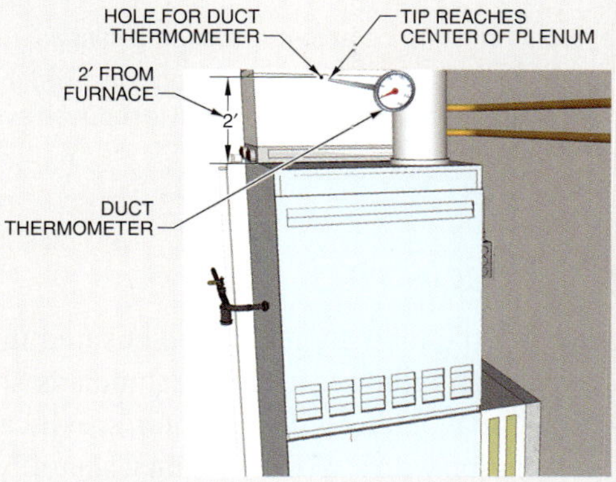

7. Place the second thermometer in the return air duct approximately 2′ from the furnace. Locate the thermometer tip in the center of the duct to ensure proper temperature measurement.

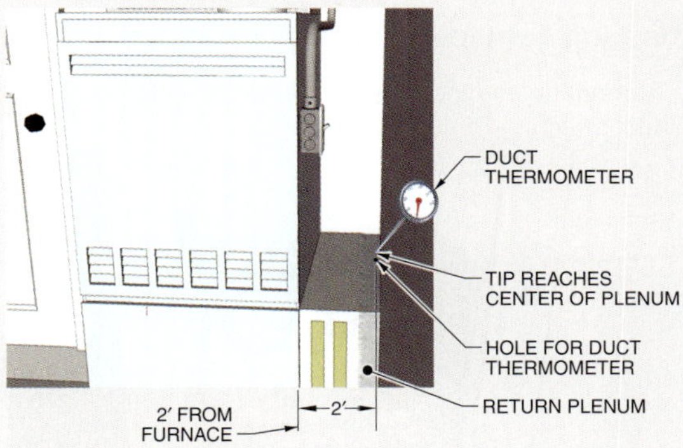

8. Set the room thermostat to the highest temperature setting. Operate the furnace for 6 min. Record the supply air and return air temperatures.

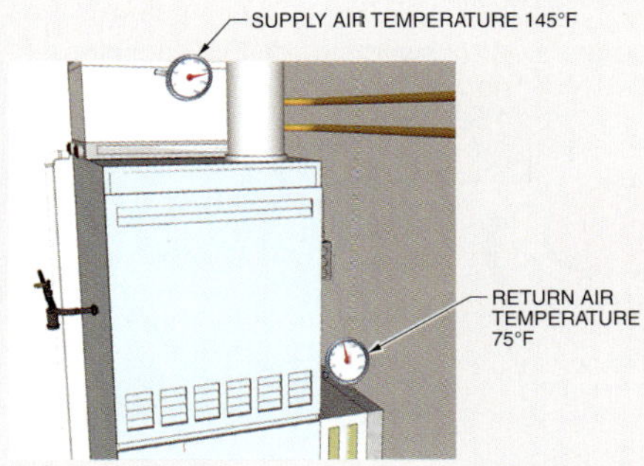

continued on next page

Procedure

Measuring Air Temperature Rise *(continued)*

9. Calculate the air temperature rise by subtracting the return air temperature from the supply air temperature. *Note:* If the air temperature rise is still outside the range specified by the manufacturer for temperature rise, check the duct system design with a qualified heating engineer. It may be necessary to resize the ductwork. Recheck the air temperature rise after resizing the duct systems.

SUPPLY AIR TEMPERATURE—145
– RETURN AIR TEMPERATURE— 75
TEMPERATURE RISE— 70°F TR

10. Set the room thermostat to the desired setting.
11. Remove the thermometers and seal any ductwork holes. *Note:* Failure to seal holes could result in reduced system performance.

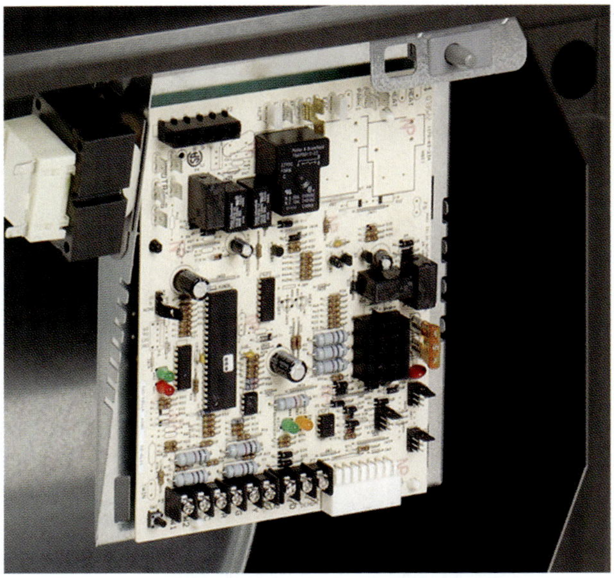

Allied Air Enterprises, Inc.

After a furnace is installed, the control board must be inspected for loose wire connections and bad or damaged electrical components. Also, it must be ensured that any time delay adjustments (dip switches) have been made before turning the furnace over to homeowners.

FURNACE CONTROLS

Before leaving the job site, it must be ensured that all controls are functioning properly. Technicians should use a 0″ to 15″ manometer with minimum specifications of a 0.1″ resolution with a ⅛″ NPT manual shutoff valve. **See Procedure for Checking Furnace Controls.**

Procedure

Checking Furnace Controls

1. Turn off the electricity using the furnace electrical disconnect switch.

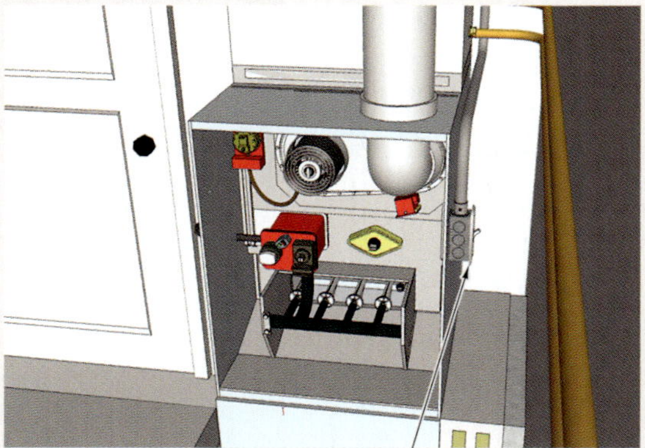

2. Turn the gas control switch to the OFF position.

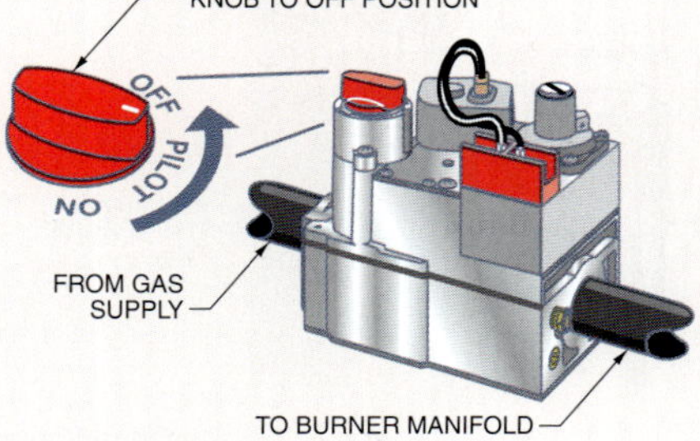

3. Connect a manometer to the outlet pressure tap of the gas valve.
   a. Remove the plug from the outlet pressure tap opening.

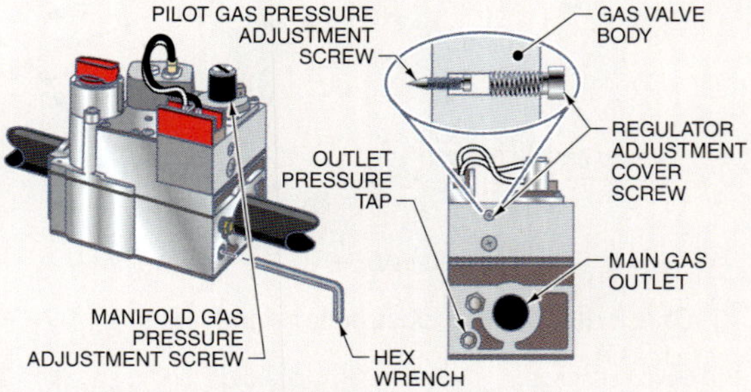

   b. Screw the manometer barb (hose-to-thread adapter) into the outlet pressure tap opening.

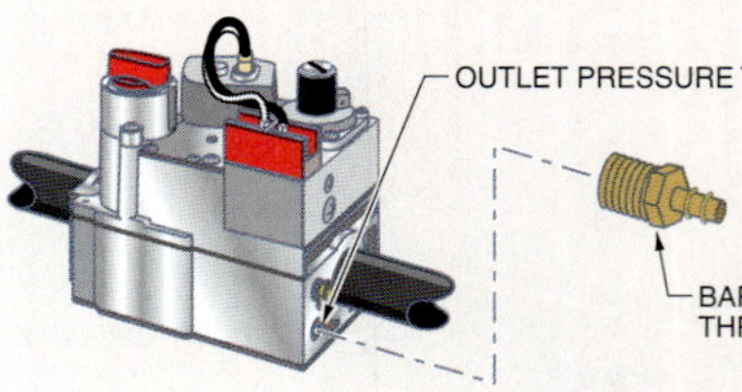

continued on next page

Checking Furnace Controls *(continued)*

c. Use an appropriate size hose to connect the barb in the outlet pressure tap to the input (+) port of the manometer. The reference (–) port is left unconnected (left open to ambient conditions).

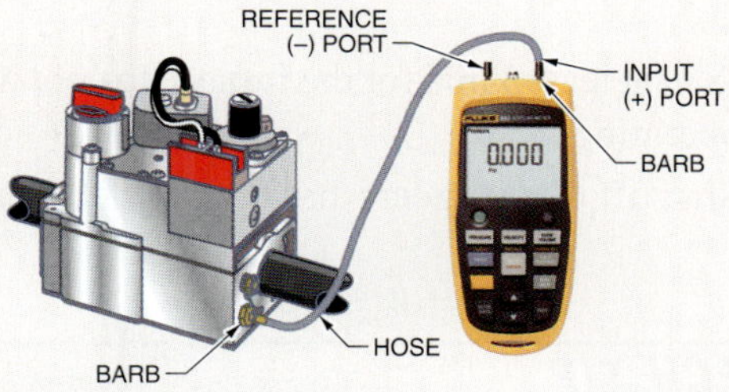

NATURAL-GAS-COMPATIBLE MANOMETER

4. Turn on the electricity at the electrical disconnect switch.
5. Set the space thermostat to its highest temperature.

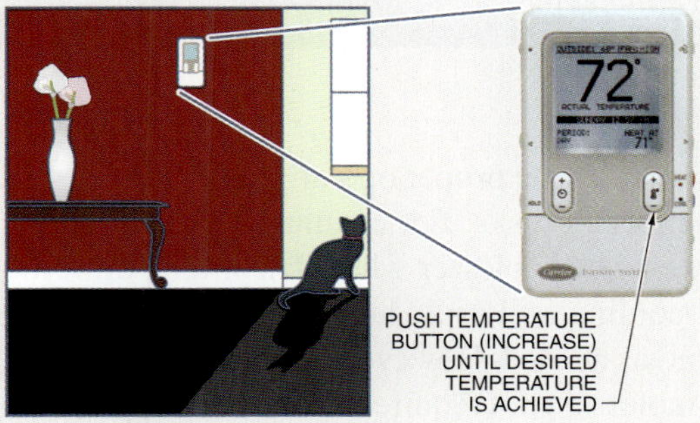

THERMOSTAT LOCATION

6. Check any diagnostic lights that the furnace may have.

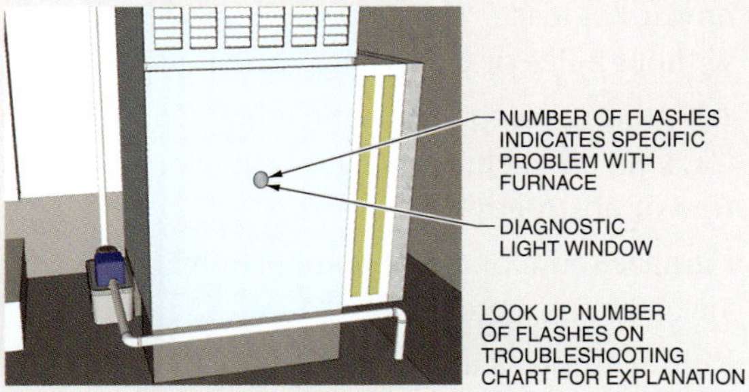

7. The draft induction fans should run and, if present, the pilot igniter should glow. However, the pilot will not ignite if the manifold gas pressure remains at zero.

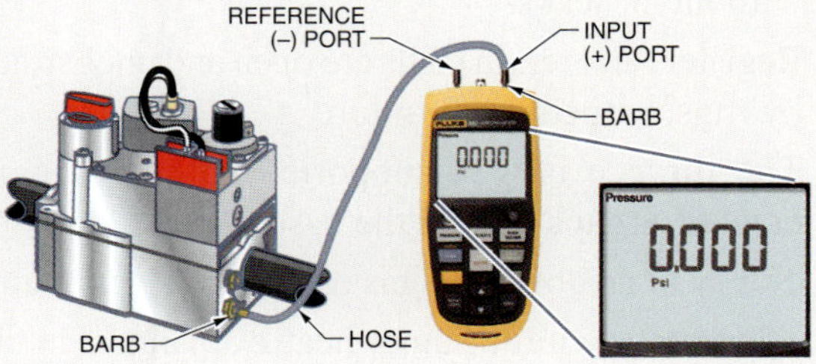

8. Shut system off at thermostat. Turn the gas control switch to the ON position.
9. Turn system on at thermostat. The draft induction fans should run, and the pilot igniter should glow or pilot flame should light and ignite the main burners. It may take several minutes to purge the gas line before the pilot flame lights. Wait 30 sec after main burner ignition for the main blower to start.
10. Cycle the furnace on and off at the thermostat. Watch at least three ignition cycles. The surface igniter or pilot flame should ignite the main burners smoothly.

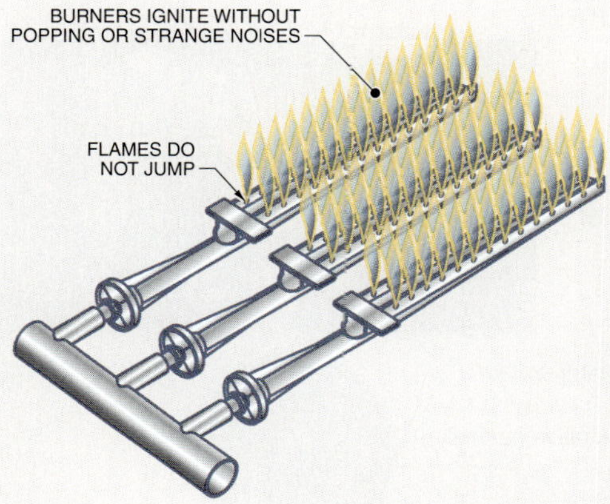

11. Burner flames should look the same with the circulation blower on or off.
12. Turn the gas control knob to the OFF position.
13. Remove the manometer from the gas control valve, and replace the outlet pressure tap plug. Turn the gas control knob to the ON position.

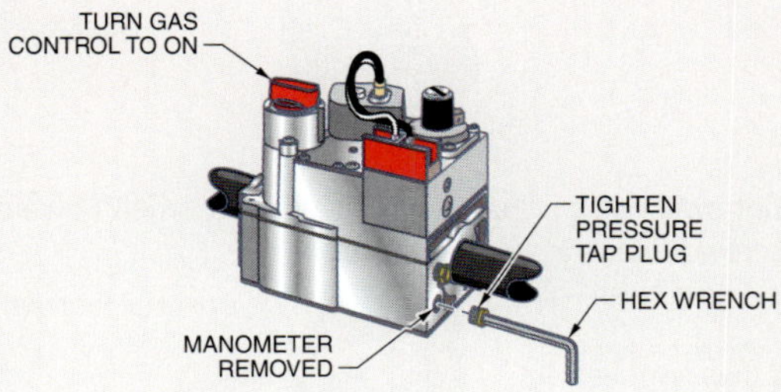

14. Use a commercial soap solution made to detect leaks, and check all gas piping connections. Bubbles indicate gas leakage. Seal all leaks before proceeding.
15. Set the room thermostat to the desired setting.

BLOWER TIMING

Heating blower on-delay is set by the factory in seconds from when the main valve opens. The on- and off-delay may not be adjustable. Many furnaces have fan delay timers. **See Procedure for Setting Blower Timing.**

Procedure

Setting Blower Timing

1. Find two small switches located near the 24 V terminal strip on the fan delay timer or controller.

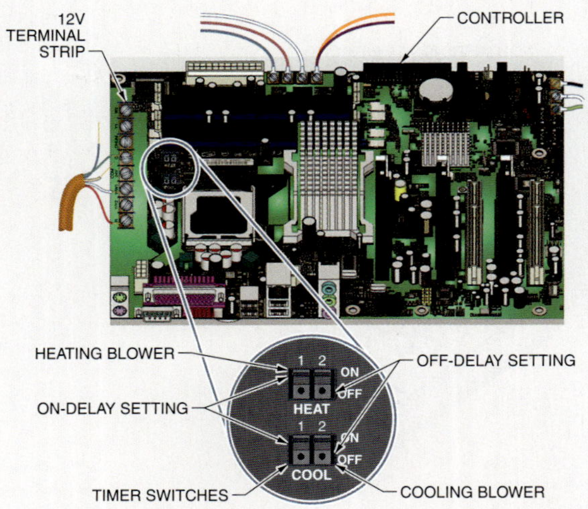

2. Determine which switch settings are for the heating blower off-delay.

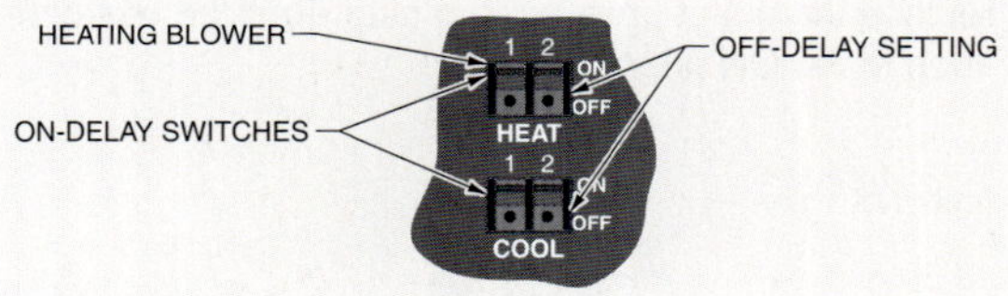

3. Set switches on the fan timer using a small (jeweler's) screwdriver.

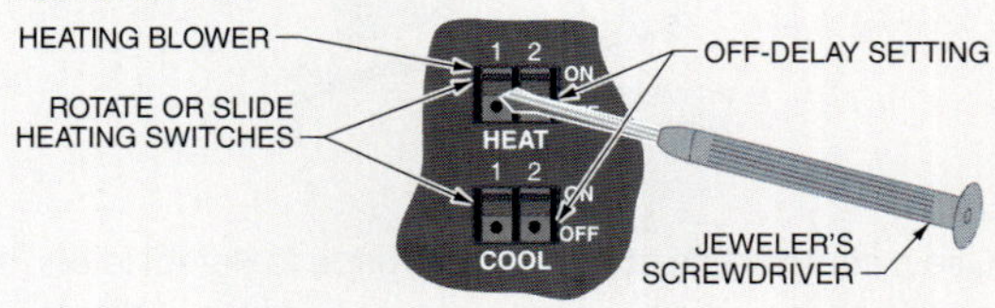

Note: Cooling blower timing is set in the same manner.

FURNACE MAINTENANCE

To ensure that the furnace is operating properly and safely, an operational check of the entire system must be performed. As part of maintenance, a qualified service person should inspect and clean the furnace once before each heating season and once during the heating season. **See Figure 39-25.** Before inspecting a furnace, apply the following procedure:

1. Turn the room thermostat to its lowest or off setting.
2. Shut off the furnace shutoff valve in the gas supply line.
3. Wait at least 5 min for the furnace to cool if recently operating.
4. Shut off (disconnect) the furnace electrical power.

TECH TIP

When starting a new furnace for the first time, a slight burnt odor may be detected. This is due to the coating applied to the heat exchanger at the factory. The building occupants should be notified so they do not overreact to the odor.

Ensuring the proper operation of a furnace involves a number of tasks. For example, the blower housing, blower motor, blower wheel, air filters, draft inducer, and cooling coil must be inspected, cleaned, and repaired as needed. However, most blower and induction fan motors do not require maintenance lubrication. **See Figure 39-26.** During a furnace maintenance inspection, the following must be ensured:

- The vent connector and vent pipe are in place, sloped upward at least ¼″ per foot, and physically sound, without holes or excessive corrosion.
- All venting components, including the vent connector, vent pipe, chimney, and vent cap, are clear and free of obstructions.
- Furnace room or closet combustion and ventilation air openings are clear.
- Return air ducts are physically sound, sealed to the furnace cabinet, and end outside the space containing the furnace.
- Supply air ducts and return air ducts are clear and without air leaks.
- Registers and return grills are open and unobstructed by rugs, carpets, or furniture.
- The furnace is well-supported without sagging, cracks, or gaps around the base.
- There are no obvious signs of furnace deterioration.
- The burners, pilot flame, heat exchanger, inducer pan, and draft inducer were all inspected.

- All electrical wiring, connections, and grounds were inspected.
- Gas pipes are checked for leaks using a commercial soap solution made to detect leaks.
- Main burners and the pilot burner are properly adjusted as shown in the information manual.
- Main burner flame is clear blue and centered within the heat exchanger openings.
- Flame holders, located inside each main burner, are free of lint and debris.
- Combustible, flammable, corrosive, and other such materials are 12″ from the furnace.
- There are no signs of high water levels in the furnace area.

SPECIAL INSTALLATION APPLICATIONS AND CONSIDERATIONS

Most furnaces shipped from a factory have an upflow configuration, with the vent collar extending out through the top plate at the supply end of the furnace. With an upflow configuration, the furnace can be installed as an upflow furnace or, in some cases, a horizontal furnace on its left or right side. With a downflow configuration, the induction fan assembly is repositioned. The vent collar is located inside the furnace vestibule near the manifold pipe. **See Figure 39-27.** The furnace can be installed as a downflow furnace or, in some cases, a horizontal furnace on its left or right side. **See Procedure for Converting Upflow Furnaces to Downflow Furnaces.**

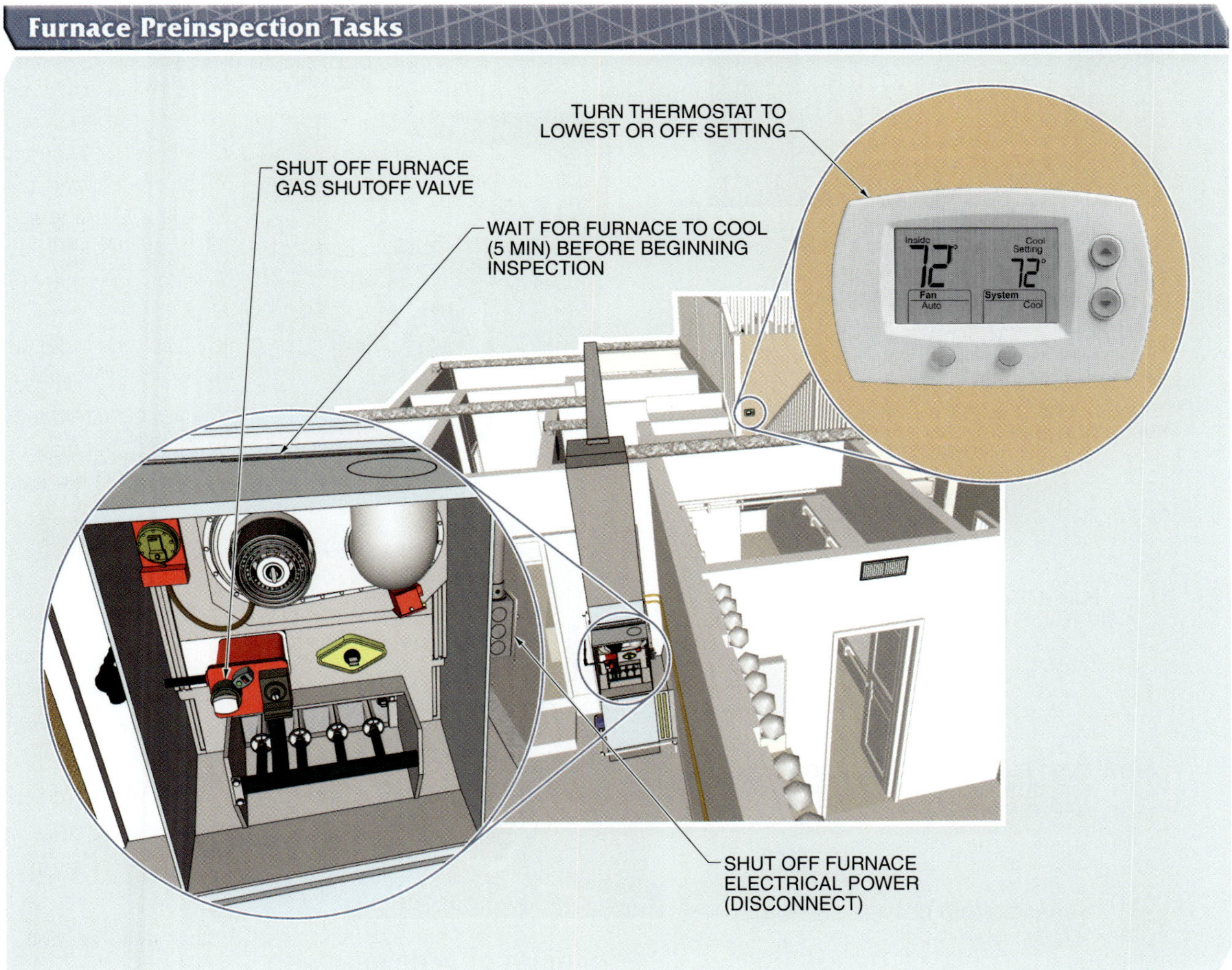

Figure 39-25. There are a number of tasks that must be completed before a furnace can be inspected.

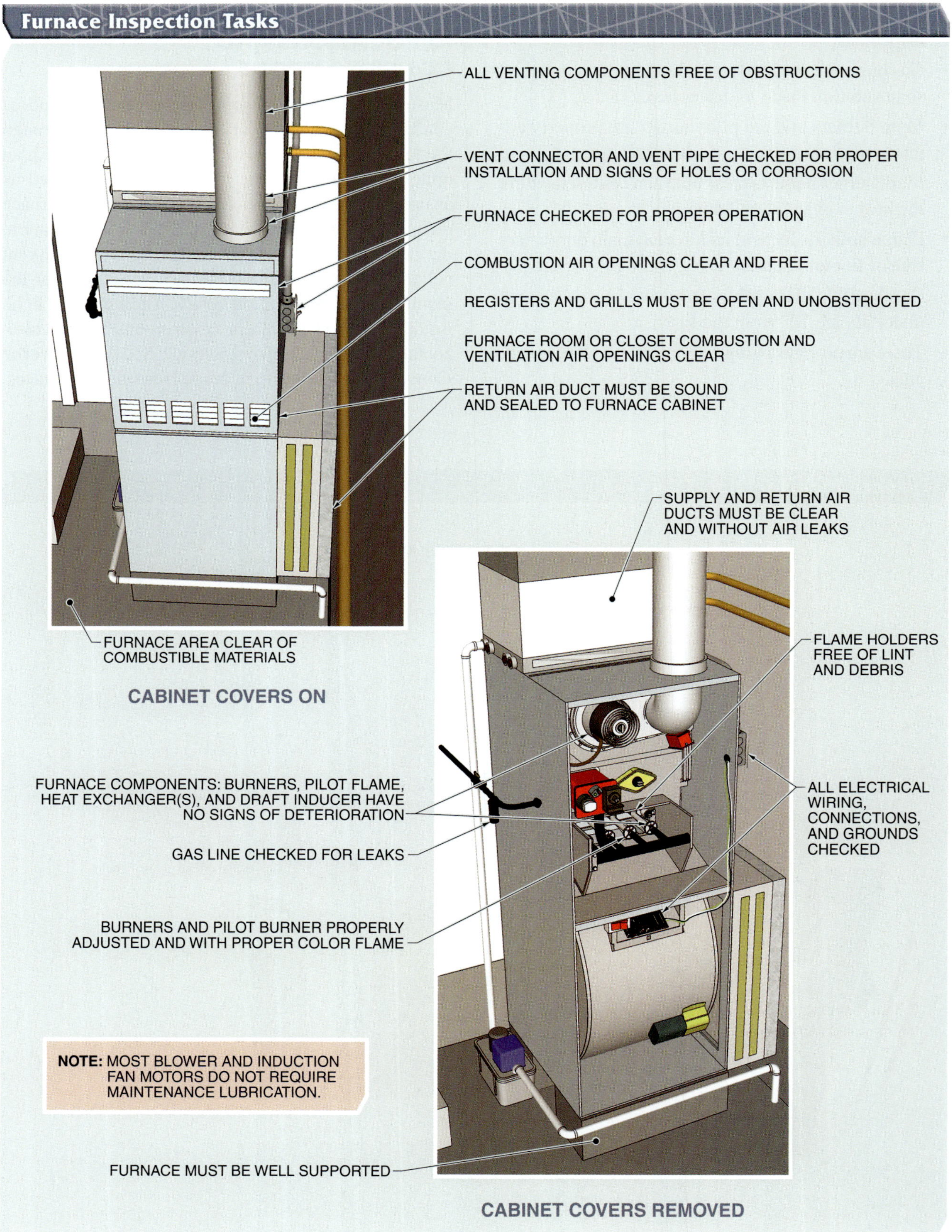

Figure 39-26. Using a checklist is quite helpful when inspecting a furnace.

Configuring Upflow Furnace to Downflow

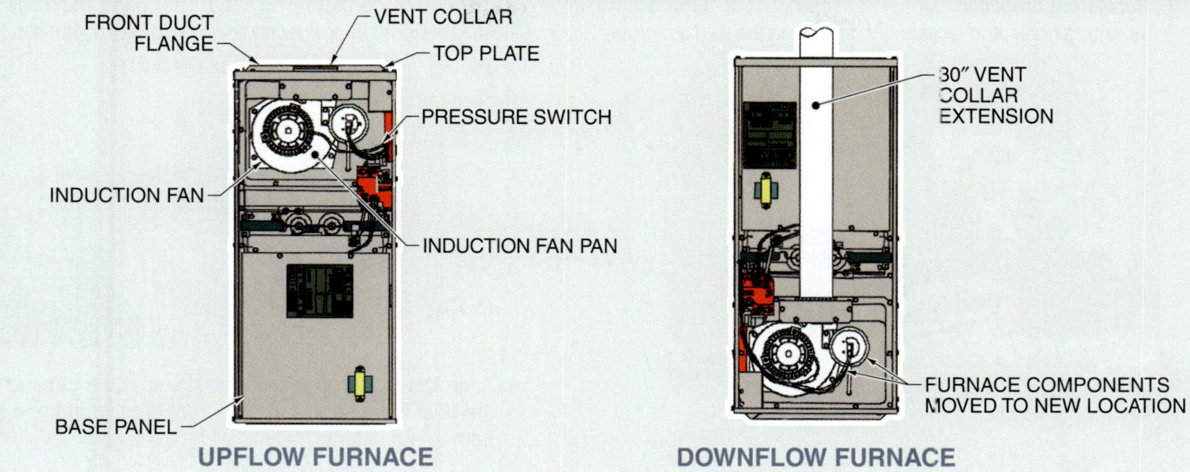

Figure 39-27. Variations exist in the methods used to change upflow furnaces to downflow furnaces. However, some general tasks are quite similar.

Converting Upflow Furnaces to Downflow Furnaces

1. Lay the furnace on its back and remove the front covers.

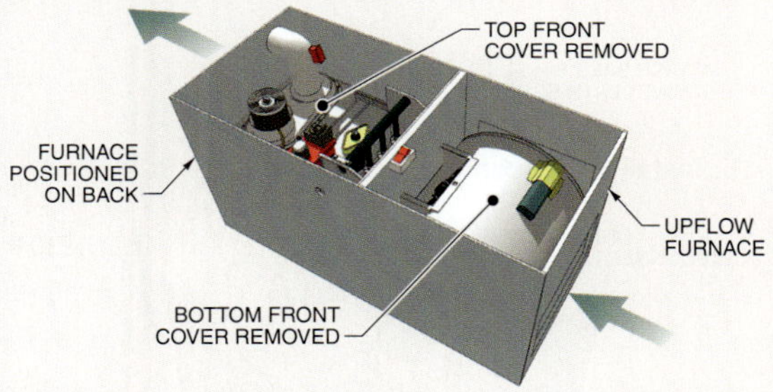

2. Decide on the direction of gas entry. This information is needed prior to assembly.

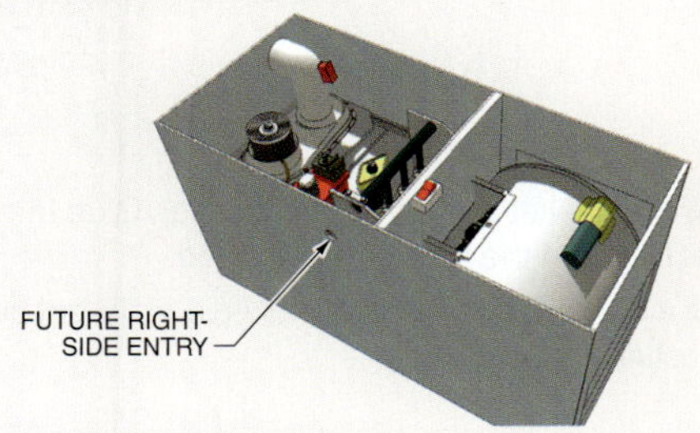

3. Remove the top plate and front duct flange. Save all screws removed during disassembly.

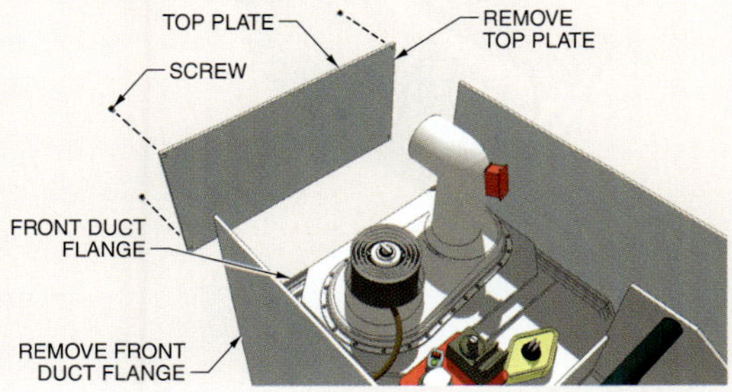

4. Disconnect the pressure switch wire(s) and the wires from the induction fan motor. Mark the wires for reconnection.

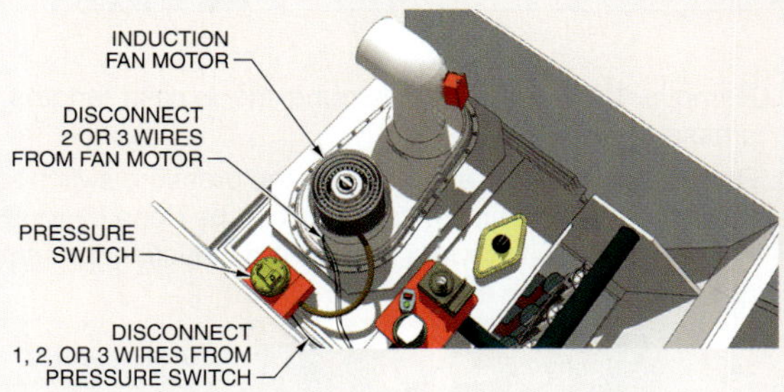

5. Remove the screws from the induction fan pan.

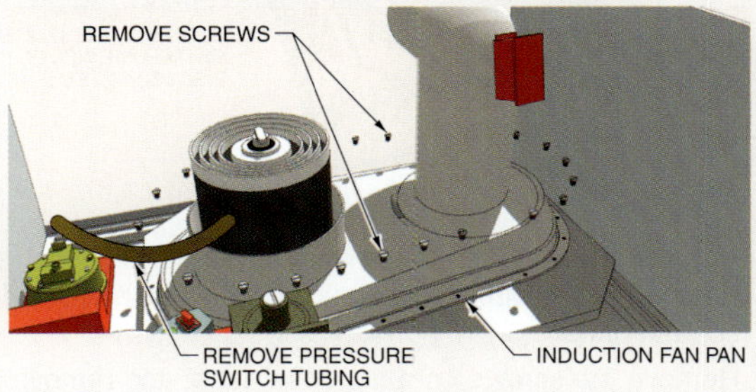

6. Lift the induction fan pan assembly about ¼″ and tilt left side up to clear the flanges. Remove the pan assembly, but keep in relative position.

continued on next page

Converting Upflow Furnaces to Downflow Furnaces *(continued)*

7. Rotate the induction fan pan assembly 180°. Line up the mounting holes and replace the screws in the pan assembly.

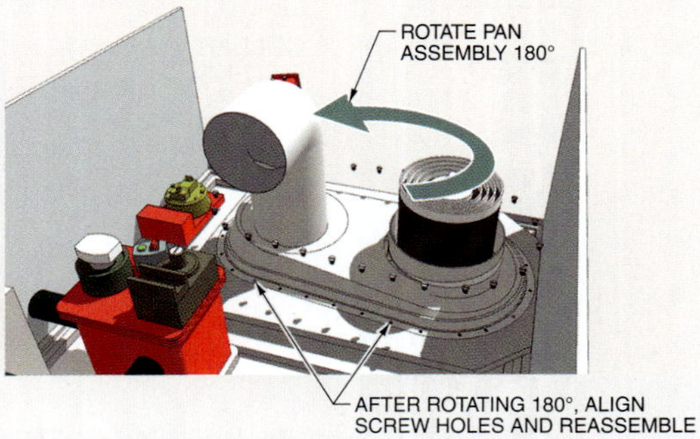

8. Rotate the furnace cabinet 180° on the floor.

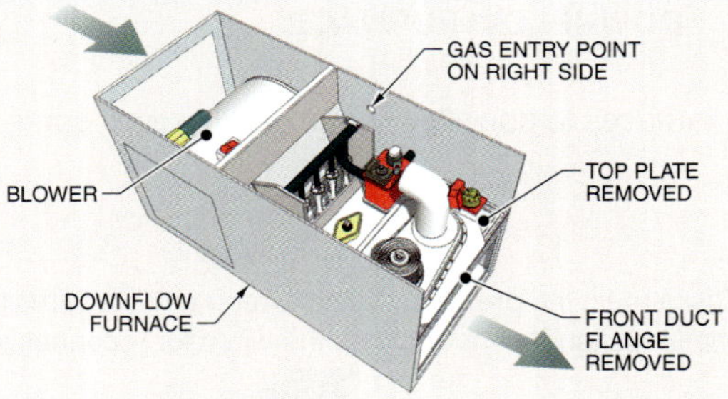

9. Remove the wire tie from around the excess lengths of pressure switch wires.
10. Reconnect the wires and tubing to the pressure switch and the wires to the induction fan. This can be done by routing the pressure switch wires around the induction fan motor and away from the hot flue box.

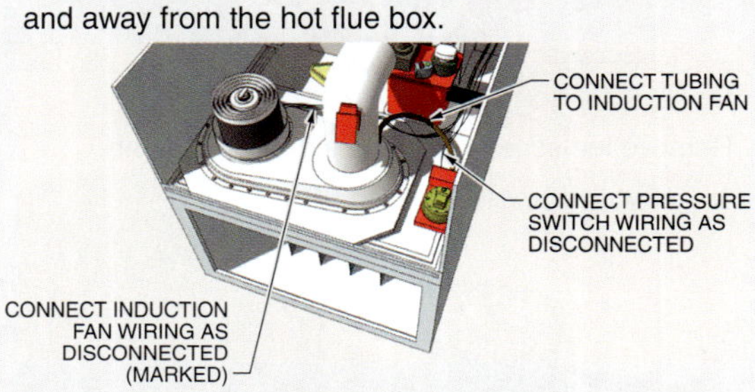

11. Replace the wire ties in the same area as removed, and ensure that the wires do not touch the hot flue box or induction fan pan.
12. Remove the screws from the base panel and remove the base panel.

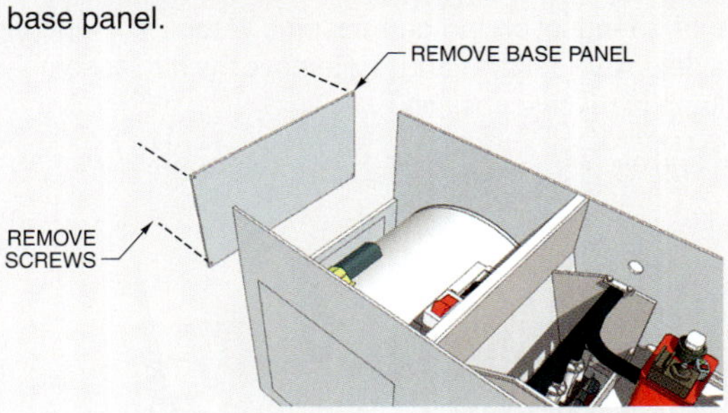

13. Install the top plate in place of the base panel. The vent collar is now aligned with the vent outlet in the top plate.

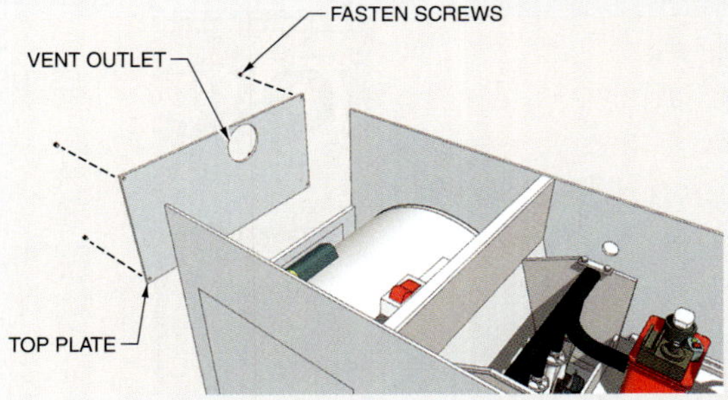

14. Align the duct flange with the holes in the cabinet. Place the duct base panel over the front duct flange and secure both to the cabinet.

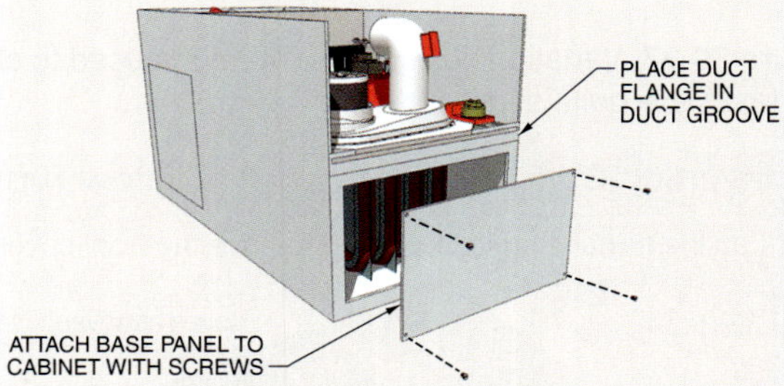

15. Install the disconnect box and switch.

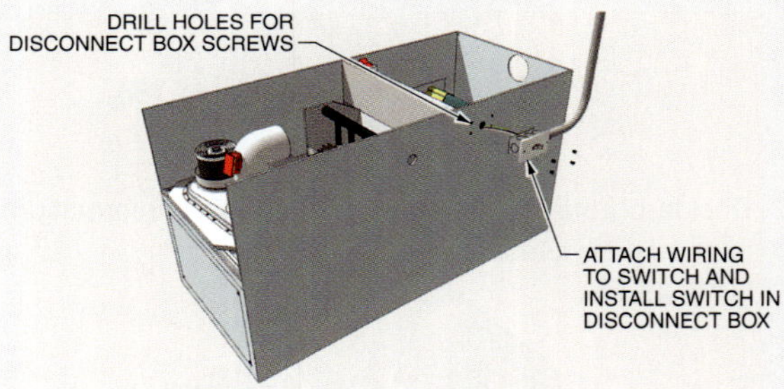

16. Install the gasket around the vent collar. Glue the gasket if loose. Replace the gasket if damaged.
17. Install a 30″ or longer section of single-wall vent pipe to the vent collar.

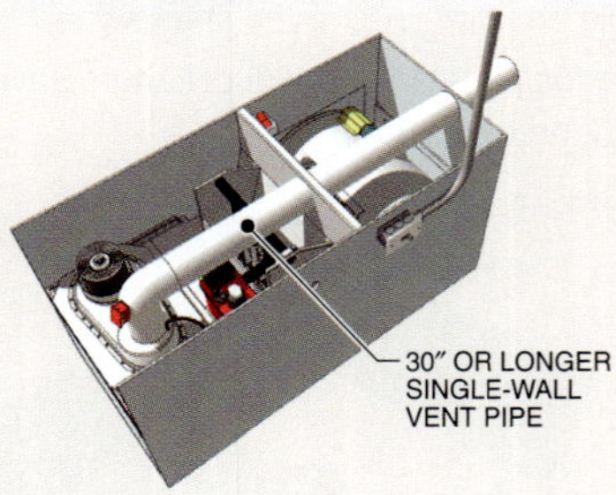

Derating for High Altitudes

As altitude increases, the density of the air is reduced. Therefore, for proper combustion, the quantity of gas burned in the furnace must also be reduced. This is called derating. Furnaces must be derated when installed at altitudes greater than 2000′ above sea level. A furnace is derated by adjusting the manifold pressure, which is accomplished by reducing the size of the main burner orifices according to manufacturer recommendations. **See Procedure for Changing Main Burner Orifices.** Typical values are for elevations above 2000′. The furnace must be derated 4% for every 1000′ above 2000′. The installer must ensure that the furnace input rate is adjusted properly. **See Figure 39-28.**

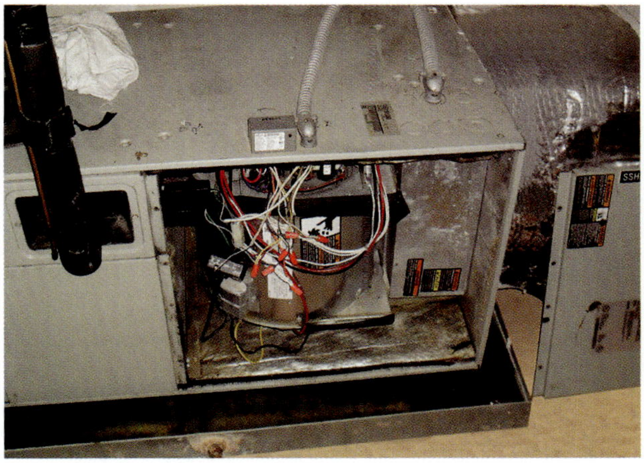

Many furnaces can be easily changed from horizontal to upflow or downflow configurations.

Furnace Derating for High Altitudes

Natural Gas								
Elevation	56,000 Btu Input*		70,000 Btu Input*		77,000 Btu Input*		90,000 Btu Input*	
	Orifice DIA	Drill Size	Orifice DIA	Drill Size	Orifice DIA	Drill Size	Orifice DIA	Drill Size
Sea level	0.136	29	0.154	23	0.161	20	0.180	15
2000	0.136	29	0.149	25	0.157	22	0.177	16
3000	0.128	30	0.149	25	0.157	22	0.173	17
4000	0.128	30	0.147	26	0.154	23	0.173	17
5000	0.128	30	0.144	27	0.152	24	0.169	18
6000	0.128	30	0.144	27	0.149	25	0.166	19
7000	0.120	31	0.140	28	0.147	26	0.161	20
8000	0.120	31	0.136	29	0.144	27	0.161	20
9000	0.120	31	0.136	29	0.140	28	0.157	22
10,000	0.116	32	0.128	30	0.136	29	0.152	24
Propane Gas								
Elevation	56,000 Btu Input*		70,000 Btu Input*		77,000 Btu Input*		90,000 Btu Input*	
	Orifice DIA	Drill Size	Orifice DIA	Drill Size	Orifice DIA	Drill Size	Orifice DIA	Drill Size
Sea level	0.082	45	0.093	42	0.098	40	0.106	36
2000	0.081	46	0.093	42	0.096	41	0.104	37
3000	0.078	47	0.089	43	0.093	42	0.101	38
4000	0.078	47	0.089	43	0.093	42	0.101	38
5000	0.078	47	0.089	43	0.093	42	0.099	39
6000	0.076	48	0.086	44	0.089	43	0.098	40
7000	0.076	48	0.086	44	0.089	43	0.096	41
8000	0.073	49	0.082	45	0.086	44	0.096	41
9000	0.073	49	0.081	46	0.086	44	0.093	42
10,000	0.070	50	0.078	47	0.082	45	0.089	43

* 4% input reduction per 1000′ elevation (per NFPA No. 54, ANSI Standard Z223.1, and National Fuel Gas Code)

Figure 39-28. Furnaces that are above 2000′ in elevation must be derated 4% for every 1000′ above sea level.

Procedure

Changing Main Burner Orifices

1. Shut off the gas supply and the electrical power to the furnace.

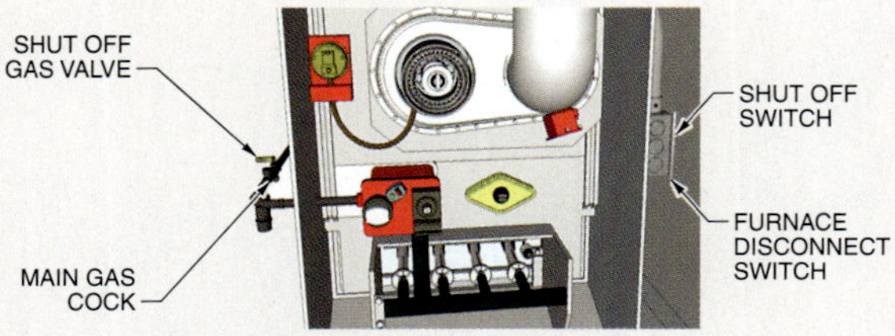

2. Disconnect the gas supply line from the gas valve.
   a. Disassemble the gas line union.
   b. Remove the pipe nipple from the gas valve.

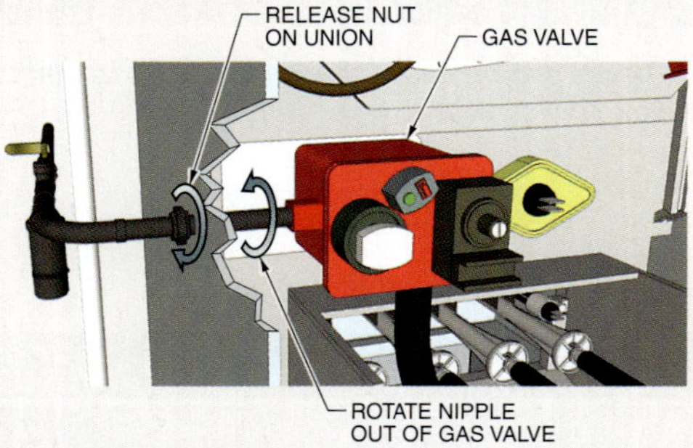

3. Remove the electrical wires from the gas valve.
4. Remove the screws attaching the burner mounting plate to the air box.

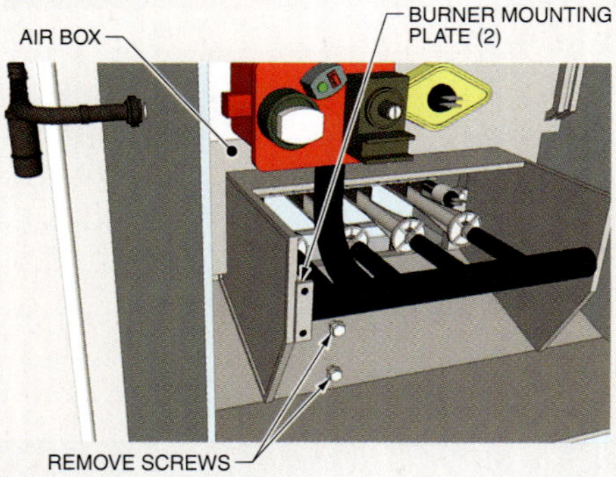

5. Slowly pull the burner assembly out of the furnace.

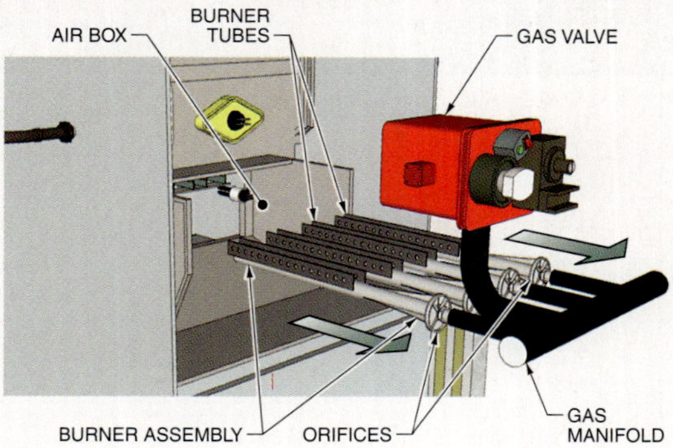

6. Remove the main gas orifice(s) from the spud(s).
   a. Remove the gas valve and manifold from the assembly.
   b. Rotate the orifice(s) counterclockwise to remove.

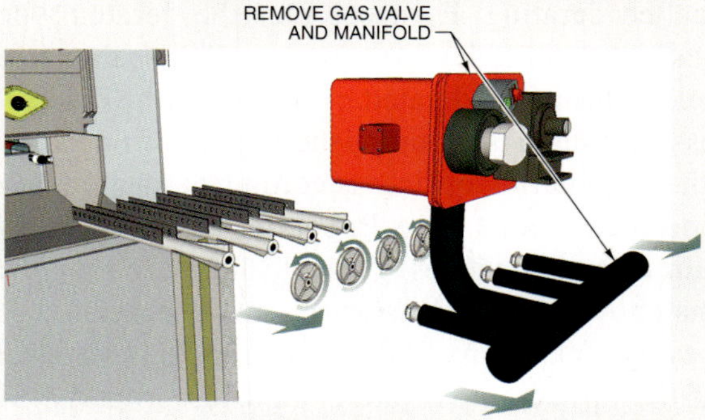

7. Screw the main gas orifice(s) into the burner spud(s).

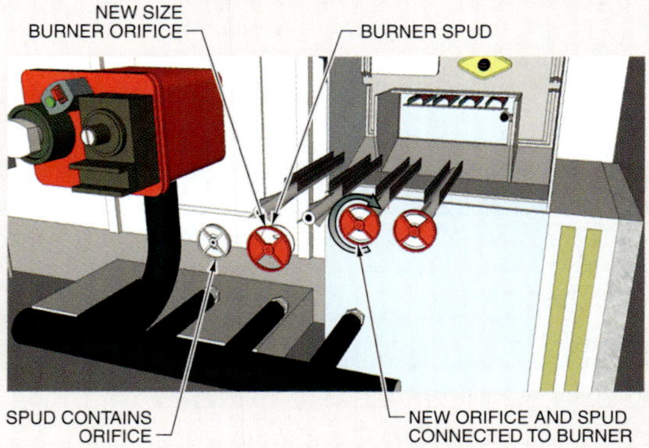

8. Install the modified burner assembly into the air box.

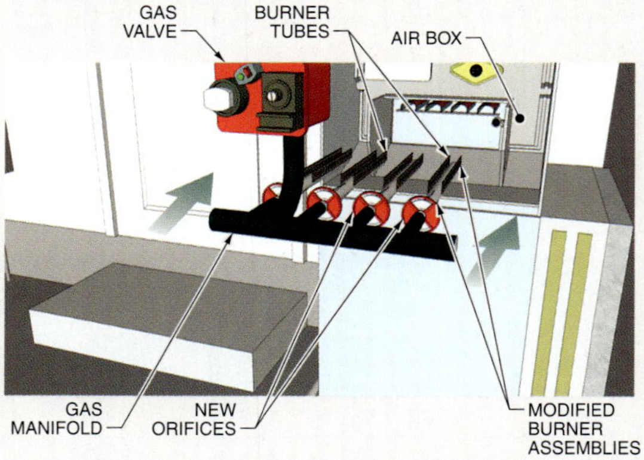

9. Connect the gas supply line and the electrical wires to the gas valve. Purge the gas supply line.
10. Turn on the gas supply and the electrical power to the furnace.
11. Recheck the manifold pressure.
12. Follow the lighting instructions as described on the furnace label.

UNIT

40

Installing Residential Forced-Air Cooling Systems

The installation of a residential furnace is often accompanied by the installation of a central air conditioning system. Together, they then become a split-system unit. It is preferable that provisions be made for central air conditioning during installation of the furnace. This makes things easier later. The R-410a condensing unit is the most commonly installed split-system air conditioning unit today.

OBJECTIVES

- Explain how to determine the proper location to install a condensing unit.
- List the typical installation procedures for residential forced-air cooling systems.
- List the typical procedures for flushing and leak testing a residential forced-air cooling system.
- Describe the evacuation, charging, and startup of residential forced-air cooling systems.
- Describe the operation of a typical forced-air cooling system.
- Identify the tasks required to perform normal maintenance on a forced-air cooling system.

Index to Procedures

INSTALLING SPLIT-SYSTEM AIR CONDITIONING

A split system consists of an indoor unit, which is typically a furnace, and an outdoor unit, which is the condensing unit. The furnace contains the system blower, heating section, and indoor air conditioning coil (evaporator) in the air discharge duct. **See Figure 40-1.** The condensing unit contains a fan, compressor, and outdoor air conditioning coil (condenser) as well as valves. The installation tasks may be categorized as follows:

- inspection of equipment
- electrical wiring
- refrigerant piping and installation
- flushing existing line sets and evaporator coils
- refrigerant metering devices
- manifold gauge sets
- leak testing
- evacuation, refrigerant charging, and startup
- operation
- maintenance

INSPECTION OF EQUIPMENT

Handlers of the condensing unit must be careful when handling the unit once it arrives on site. The condensing unit should be carefully removed from the truck using the proper lifting and moving procedures. It should then be carefully inspected for possible shipping damage. If damage is found, it must be reported to the carrier and distributor immediately. A damaged unit must not be installed since any claim may be denied. Also, the unit nameplate must be checked to ensure that it is the unit that was ordered. **See Figure 40-2.**

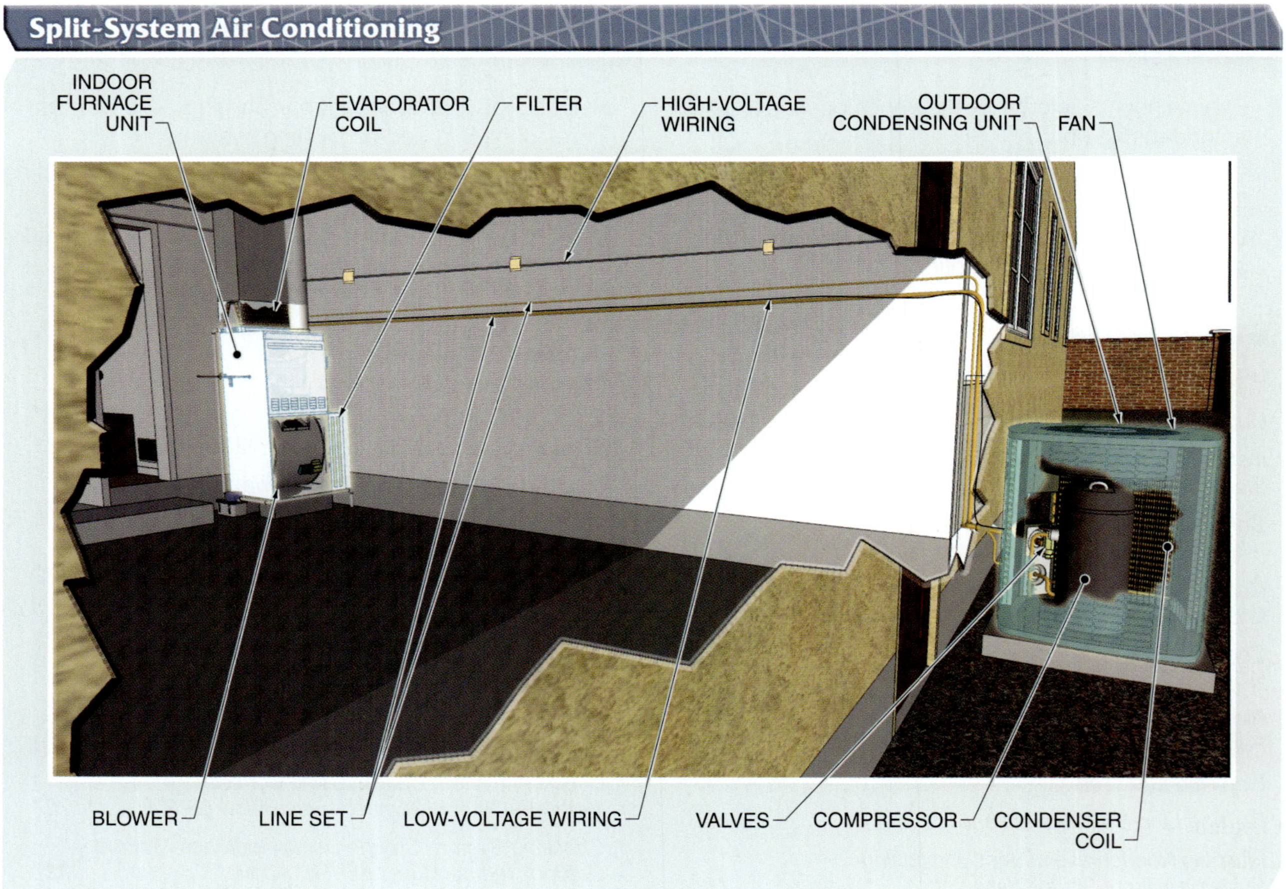

Figure 40-1. Split-system air conditioning systems include piping line sets and wiring to connect the indoor and outdoor units.

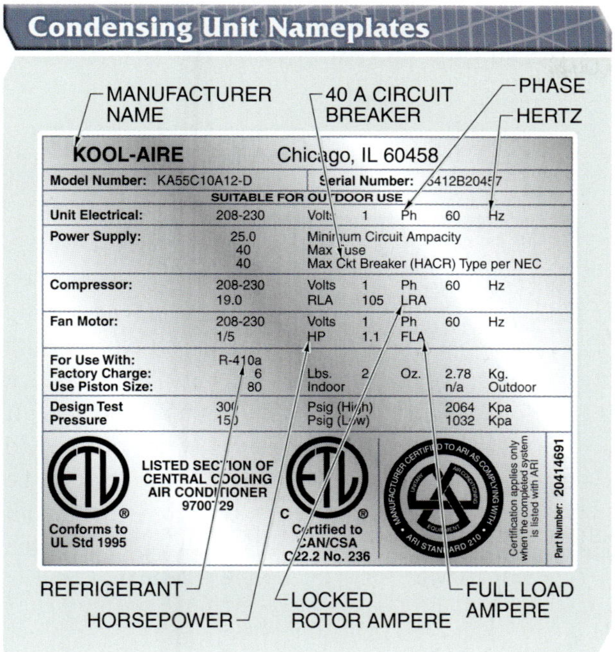

Figure 40-2. Using the nameplate of a condensing unit is the only reliable means of identifying the actual components of an outdoor unit.

When multiple condensing units are installed in an application, an adequate distance between them must be maintained for maintenance and airflow.

Location of Condensing Units

Condensing units are located outdoors and operate under a wide range of weather conditions. Therefore, several factors must be considered when positioning a condensing unit. The unit must be positioned to give adequate clearances for sufficient airflow and servicing. Manufacturer data must be referred to for installation clearances. The unit must be installed high enough above the ground or roof to allow adequate drainage of water and prevent snow buildup. **See Figure 40-3.** The condensing unit location may be already determined, but if not, the following guidelines must be applied when selecting a location:

- The ground is level and secure.
- The unit should be kept out of constant direct sunlight. It will run more efficiently in the shade.
- Rain typically does not hurt the unit, but it is best to keep it from needlessly getting wet. Therefore, the unit should not be positioned directly under a gutter or where excessive water can pour down on it.
- The unit should be as close as possible to an electrical circuit panel for an easier wiring run.
- The unit should not be placed near plants. The unit needs clearance around it for adequate airflow. In some cases, plants or bushes may have to be relocated.
- For aesthetic reasons, the unit should be placed in an area that is slightly hidden from direct view of the home.
- The unit must be 3′ away from a gas or electric meter. Local codes must be consulted.
- A 10″ to 12″ clearance exists between the condensing unit and the walls of the home.
- The unit must not be located where drifting snow will occur. The unit base should be elevated above the depth of average snows. *Note:* Elevation of the unit may be accomplished by constructing a frame using suitable materials. When a support frame is constructed, it must not block drain holes in the unit base. The support frame should not be constructed out of a material that will deteriorate.
- When installed in areas where low ambient temperatures exist, the unit must be located where winter prevailing winds do not blow directly into the outdoor coil.
- Some localities are adopting sound ordinances based on how noisy the unit is at a neighboring home instead of the original installation. The unit should be installed as far as possible from the property line. When possible, the unit should not be installed directly outside a bedroom window. Glass has a very high level of sound transmission.

Condensing Unit Location

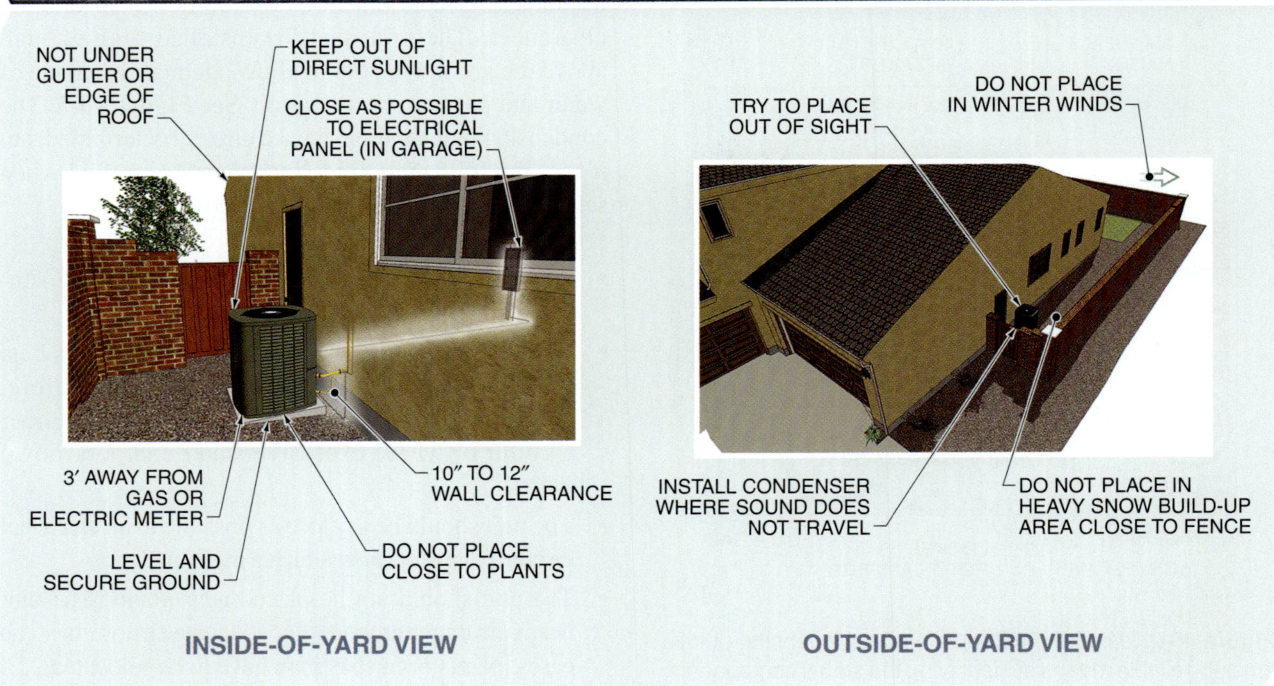

Figure 40-3. A condensing unit must be level, out of direct sunlight, close to an electrical panel, hidden, away from plants, at least 3′ away from electrical and gas meters, away from winter winds, and away from windows (especially neighbor windows).

TECH TIP

The curb appeal of a residence can be reduced by a poorly installed and/or located condensing unit. In addition, basic instructions should be provided to a homeowner to prolong equipment life, such as not bumping the condensing unit with yard equipment and not planting shrubbery in locations that block airflow to the unit.

Installing Condensing Units on Pads

A condensing unit should be installed on a slab high enough above grade level so that water from higher ground will not collect on or around the unit. **See Figure 40-4.** The pad should be level or slope away from the building approximately ¼″ per foot. Many contractors build up the area with dirt and gravel as necessary. This will prevent ice from building up under the unit. The roof mounting instructions supplied by the manufacturer should be consulted for barrier construction when the unit must face prevailing winter winds. **See Procedure for Placing Condensing Units on Pads.**

Installing Condensing Units on Slabs

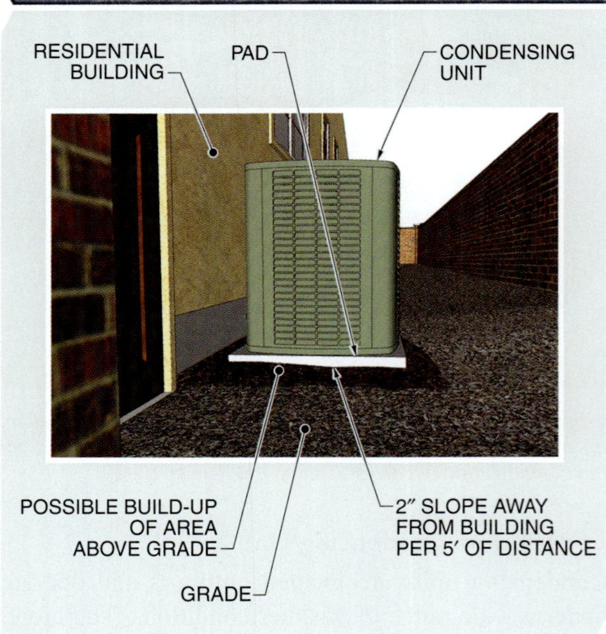

Figure 40-4. Condensing units should be placed on cement slabs that allow rainwater to drain away from the unit and structure.

Placing Condensing Units on Pads

1. Be careful to unload condensing unit safely and without damaging it or the property.
   a. Level the ground (possibly raise the ground) in area chosen for condensing unit pad. Use shovel and rake or equipment as needed to level the ground.

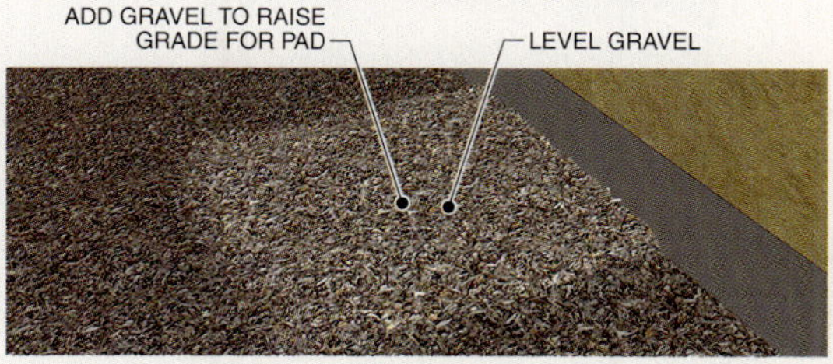

   b. Place pad in spot chosen for condensing unit, with 2° slope for drainage. Center condensing unit on pad.

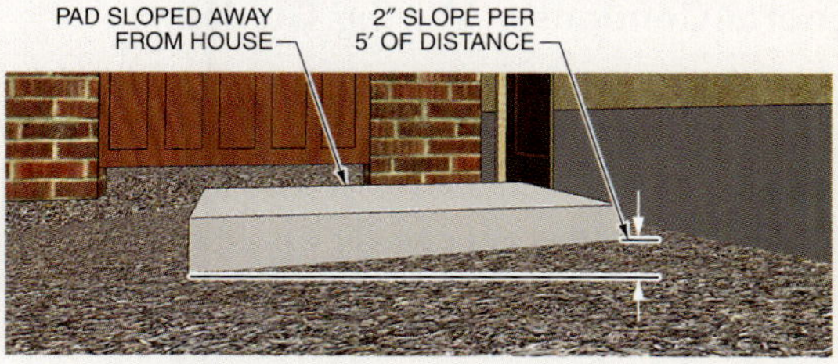

2. Check condensing unit with level for proper levelness (2° slope).

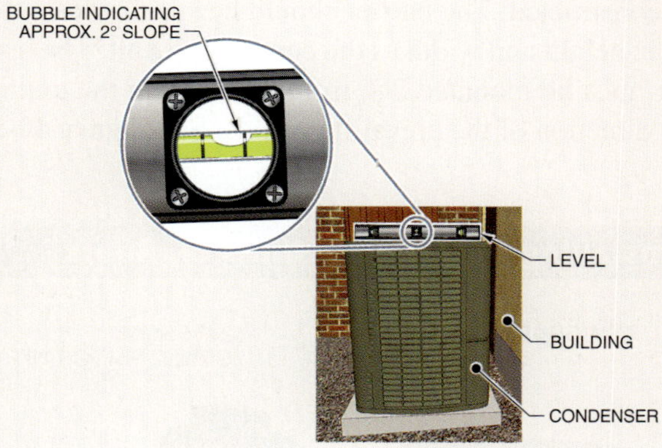

An object should be used both inside and outside of the home as a point of reference, such as a water spigot or electrical hole for phone wires, to ensure that the hole for wire or tubing is not drilled in the wrong spot. From the outside of the home, a ⅛″ or ¼″ drill is typically used to drill a pilot hole at a marked spot. A hole saw is then used to create a hole of the needed size. **See Procedure for Cutting Holes in Walls to Run Wires and Tubing.**

Cutting Holes in Walls to Run Wires and Tubing

1. Cut hole in wall at spot near condensing unit and near foundation of building.

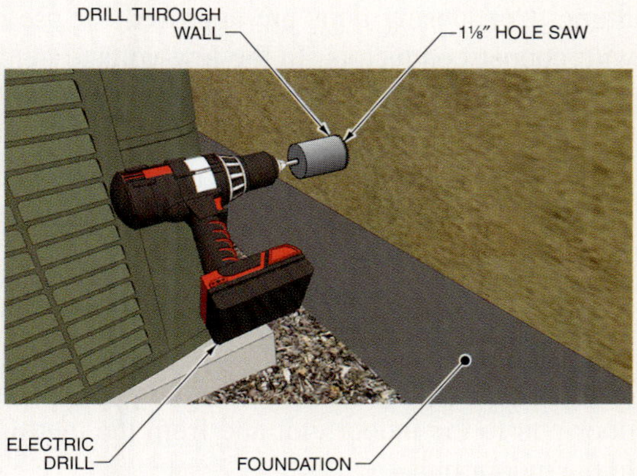

   a. Hole must allow wires to be run to main electrical panel.
   b. Hole must also allow refrigerant piping to be run to furnace or air handler evaporator coil.

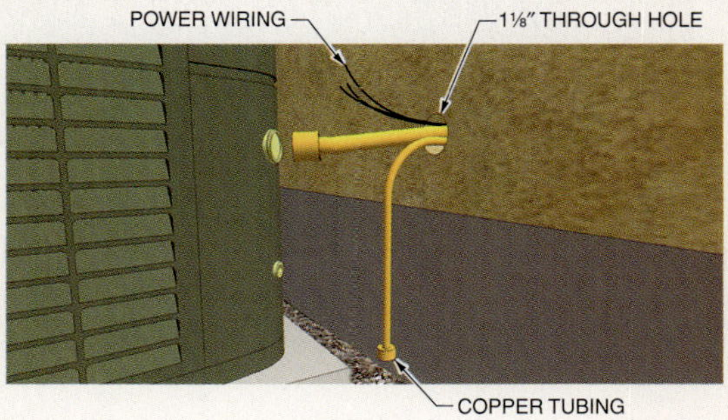

2. Seal around wiring and piping with caulk for weatherproofing.

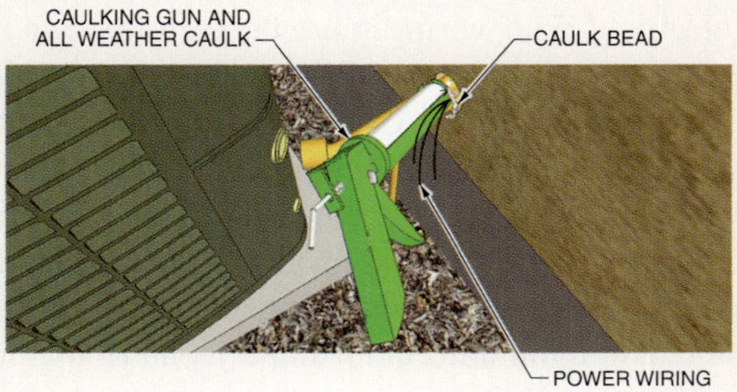

Installing Condensing Units on Roofs

In rare cases, it is necessary to install the condensing unit on the roof. In those cases, the unit should be installed at a minimum of 6″ above the roof surface to avoid ice buildup around the unit. The unit must be located above a load-bearing wall or area of the roof that can adequately support the unit. Local codes should be consulted for rooftop applications.

When the coil of the unit cannot be mounted away from prevailing winter winds, a wind barrier should be constructed. The barrier should be sized at least the same height and width as the condensing unit. The barrier must be mounted 24″ from the sides of the unit in the direction of the prevailing winds. **See Figure 40-5.**

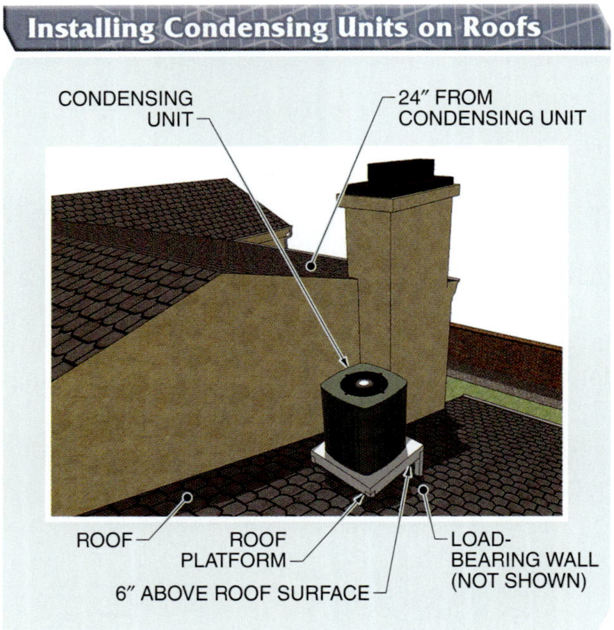

Figure 40-5. A condensing unit roof platform should be placed above a load-bearing wall and protected from prevailing winds as much as possible.

ELECTRICAL WIRING

All electrical field wiring must be performed in accordance with the National Electrical Code® (NEC®), the Canadian Electrical Code (CEC), and Canadian Standards Association (CSA) standards or local codes, where applicable. The furnace or blower installation instructions must be consulted for additional wiring application diagrams. The unit rating nameplate must be consulted for minimum circuit ampacity and maximum overcurrent protection size.

Roof-mounted condensing units are installed where they are not in the direct path of prevailing winds.

General Condenser Wiring Guidelines

Line voltage electrical service (typically 230 V, 1ϕ) is required for all condensing units. The power supply for the condenser is not required to be the same power supply as for the indoor unit. All control wiring (typically 24 V) for the condenser typically comes from the indoor unit control transformer. General guidelines for condenser wiring are as follows:

- Line voltage power must be connected to the condensing unit using a properly sized disconnect switch. Any excess high-voltage field wiring must be trimmed or secured away from the low-voltage field wiring. **See Figure 40-6.**
- The condensing unit must be grounded at the unit disconnect switch or to an earth ground. To facilitate conduit usage, a hole of the appropriate fitting size is in the bottom of the control box. Conduit should be connected to the control box using a proper conduit fitting. Condensing units are approved for use only with copper conductors. In the low-voltage junction box, 24 V Class II circuit connections are made. The manufacturer must be consulted for a high-voltage field wiring diagram. A complete unit wiring diagram is typically located on the cover of the unit control box.
- A room thermostat should be installed on an inside wall that is not subjected to drafts, direct sunshine, or other heat sources.
- Low-voltage wiring should be installed from the outdoor unit to the indoor unit and from the thermostat to the indoor unit.
- Excess 24 V control wire must not be bundled inside the control box. Control wire must run through the installed wire tie. The wire tie must be tightened to provide low-voltage strain relief and to maintain separation of field-installed low- and high-voltage circuits.

General Condenser Wiring Guidelines

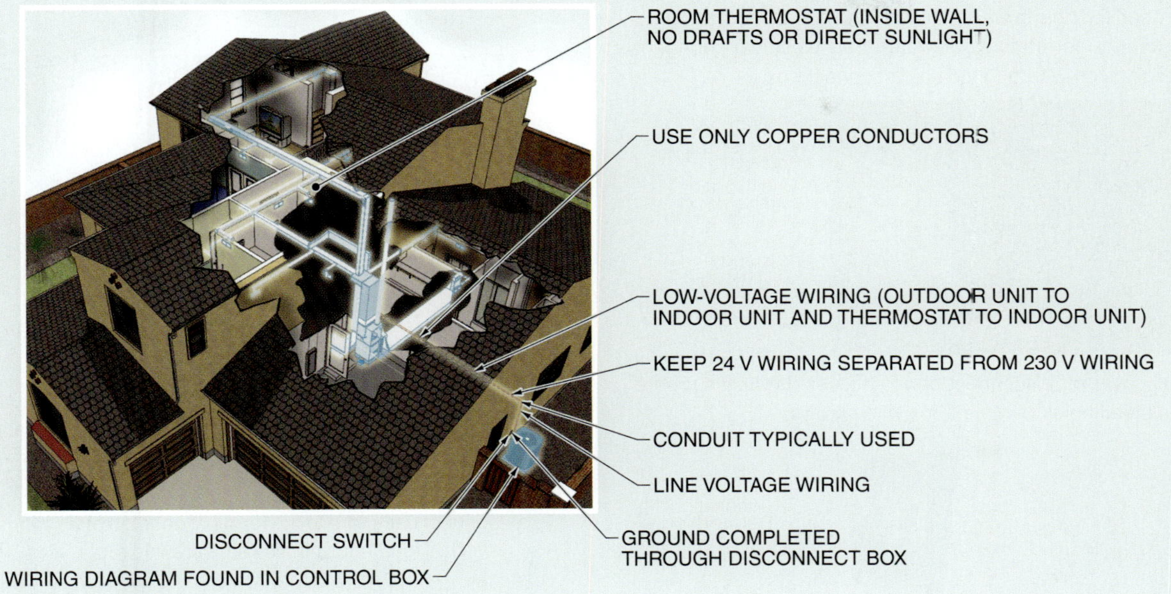

Figure 40-6. Condenser wiring must be connected to a properly sized disconnect switch, the unit must be properly grounded, a thermostat should be installed on an inside wall that is not subjected to drafts, and low-voltage wiring should not be bundled with high-voltage wiring.

Installing Condenser High-Voltage Wiring

A central air-conditioning unit typically operates with 230 V of electricity. A connection must be made between the main electrical circuit panel in the home and the condensing unit disconnect in order for the unit to function properly. Only a skilled contractor or electrician should complete the connection of the high-voltage wiring. **See Figure 40-7.**

Most air conditioning systems provide a standard kit of wire and connectors. When a kit is not available, the correct thickness of wire and the corresponding breaker based on the condensing unit ampacity must be selected. Most residential condensing units use 8 gauge or 10 gauge wire. The manufacturer specification sheets should be consulted for specific unit information.

The first step in preparing high-voltage connections is to install the disconnect box. The NEC® requires that a disconnect box be located near the condensing unit. Local electrical codes should be checked for specifics. The disconnect box has two electrical wiring bundles (whips) running in and out of it. One runs from the condensing unit to the disconnect box, and the other runs from the home service panel to the disconnect box. **See Procedure for Installing Disconnect Boxes and Electrical Whips.**

Installing Condenser High-Voltage Wiring

Figure 40-7. A central air-conditioning unit typically requires that a 230 V connection be made between the structure's main electrical panel and the condensing unit's disconnect switch. Only a skilled electrician should make the connection.

Procedure

Installing Disconnect Boxes and Electrical Whips

1. Select spot on wall to mount disconnect box. *Note:* Locate it near drilled hole in wall. Place it high enough off the ground to keep it out of snow or water. See electrical codes for specifics on where to locate disconnect box.

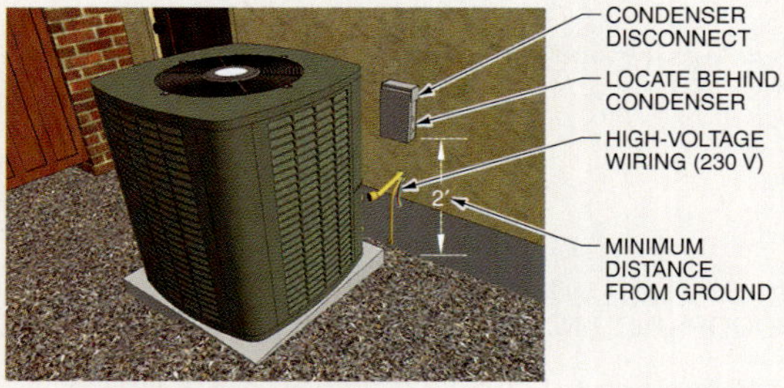

2. Pull cover off disconnect box to access the inside. There are small holes to mount box with screws.

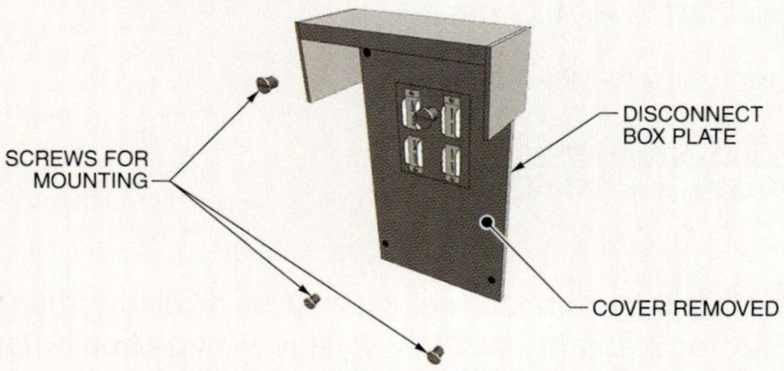

3. Remove on/off switch. (Punch out holes that are properly sized for electrical wire bundle. One bundle will be inserted on each side.)

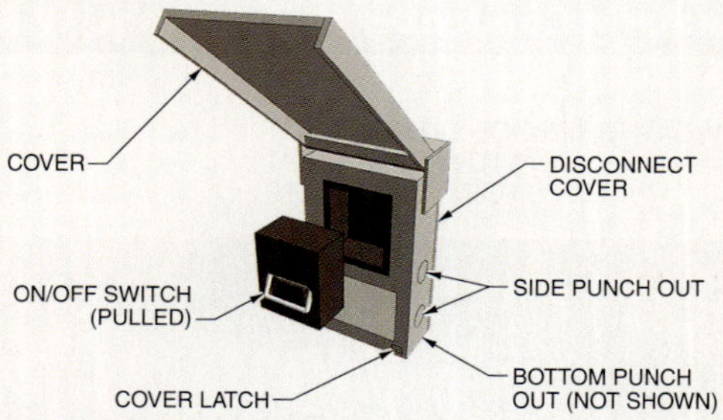

4. Mount disconnect box plate to wall.

5. Hook up ground wires to disconnect box plate.
   a. Green wire is for ground. Cut the two green wires to size, and strip back ½″ of insulation.
   b. Hook up both wires, with each one under either lug in center of disconnect box, and tighten with screwdriver. Ensure good connection.

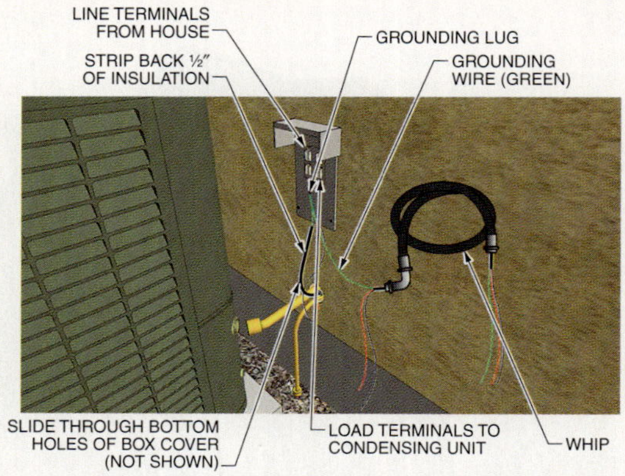

6. Hook up wires of supply wire bundle to supply switch terminals and wires of whip to condenser switch terminals.
   a. Cut red and black wires of bundle to size and strip off ½″ of insulation.
   b. Connect each wire to the terminal near top of disconnect switch. Tighten terminals with screwdriver.
   c. Cut red and black wires of whip to size and strip back ½″ of insulation.
   d. Connect each wire to the terminal near bottom of disconnect switch. Tighten terminals with screwdriver.

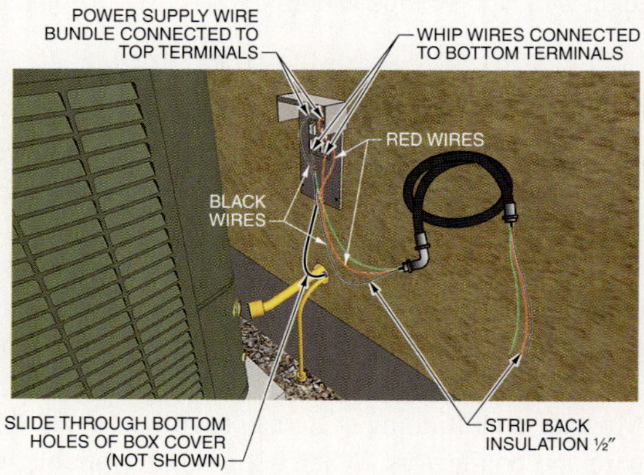

7. Attach box cover to disconnect box plate.

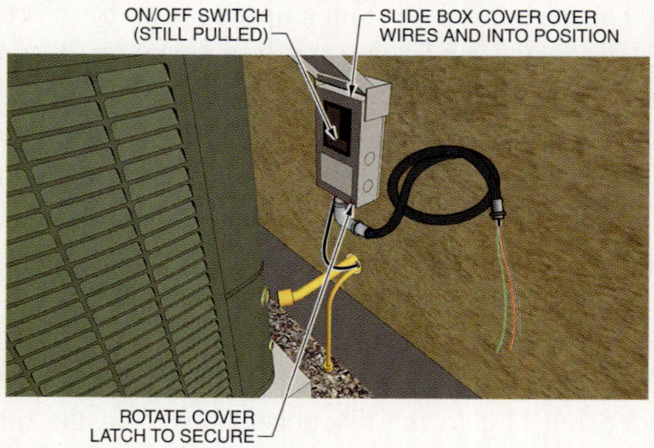

8. Seal hole with insulating material. Most contractors today use aerosol material.

continued on next page

Installing Disconnect Boxes and Electrical Whips *(continued)*

9. Hook up condenser whip wires to condenser ground and contactor.

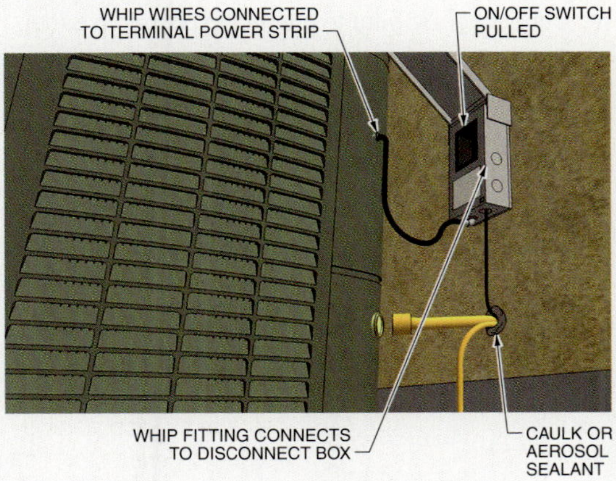

10. Replace on/off switch on disconnect box for air conditioner operation.
    a. Leave on/off switch pulled until system is ready to be started.

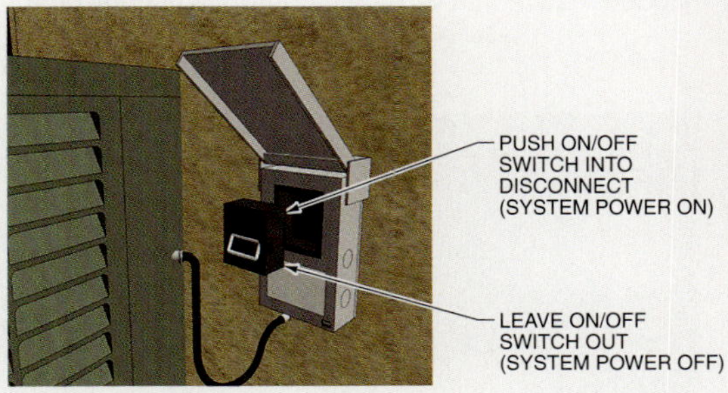

11. Close disconnect box door.

Procedure

Running and Connecting Wiring Bundles to Electrical Panels

Inside the home, the wiring bundle is typically mounted to the ceiling (sometimes following line set piping). One end is connected to the condenser, and the other end is run to the home's service panel. The end of the wiring bundle connected to the condenser is actually connected to the disconnect switch for the condenser. **See Procedure for Connecting Wiring Bundles (High-Voltage Wiring) to Service Panels.**

Connecting Wiring Bundles (High-Voltage Wiring) to Service Panels

1. Find spot on ceiling or out of the way to mount wire bundle. See local electrical and building codes for other possible placements.

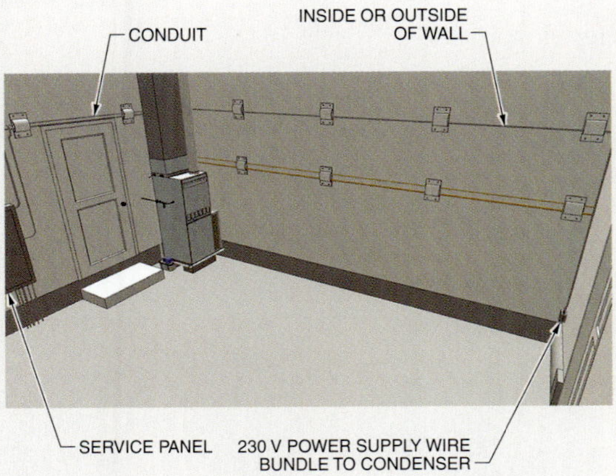

2. Use standard 4 × 4 junction box to run wire bundle into wall hole. Screw box into place.
3. Punch out appropriately sized holes on both sides or side and back of junction box.

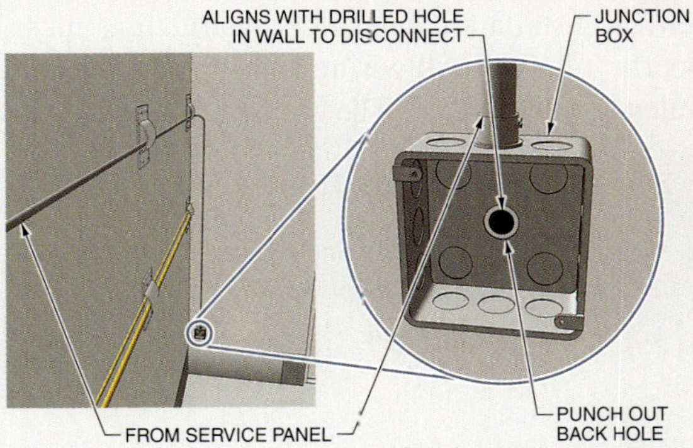

4. Run wire bundle into junction box mounted on inside wall of garage or building.

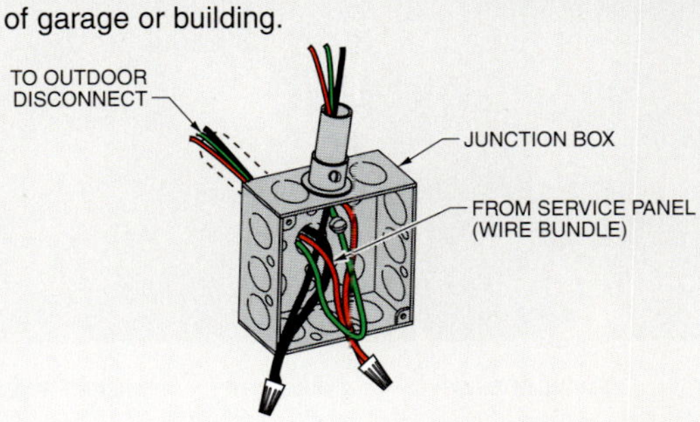

continued on next page

Procedure

Connecting Wiring Bundles (High-Voltage Wiring) to Service Panels *(continued)*

Procedure

5. Run conduit or flexible wire from junction box to home's electrical panel. Use correct wire gauge (based on condensing unit). Flexible wire, such as Romex, may be used if building and electrical codes permit. Some building and electrical codes require that wire be run through conduit pipe.

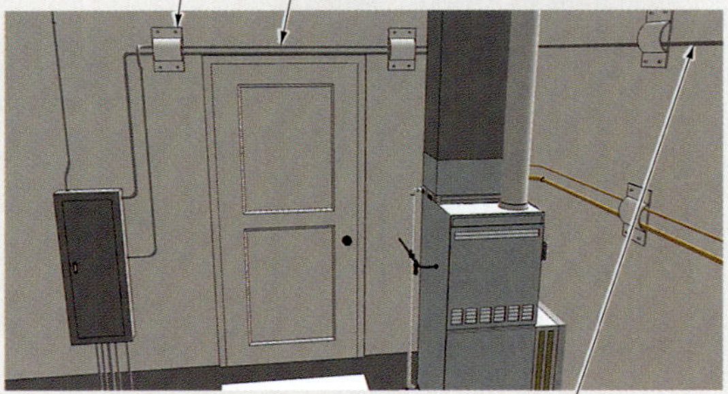

6. Hook up wires to service panel. *Note:* Only qualified technician or electrician should connect condenser circuit wires to service panel.
   a. Turn off main breaker and remove service panel cover.
   b. Cut back ½″ of insulation from each of the wires (red, black, and green).
   c. Connect red wire to circuit breaker (phase 1). A double-pole circuit breaker can be used.
   d. Connect black wire to circuit breaker (phase 2) or to second pole of double-pole circuit breaker.
   e. Connect green wire to grounding busbar in panel.
   f. Reattach service panel cover.

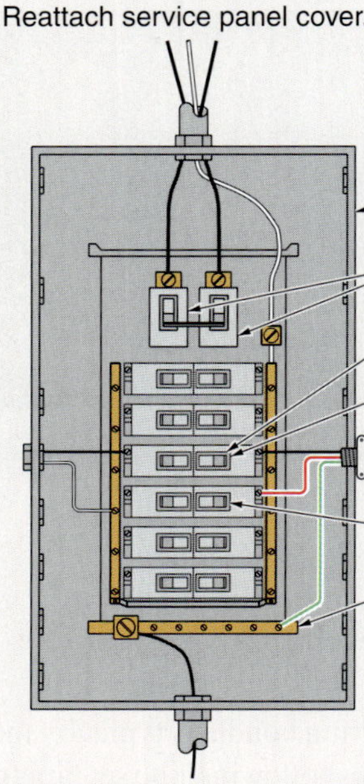

7. Turn on main service panel breakers.

The next major wiring step is to connect the wiring from the service disconnect to the condenser. All electrical installation and safety instructions must be properly followed. All wiring and supplies used must be rated for outdoor use. The wiring must be grounded. The correct size wiring must be used and must match the ampacity needed at the condensing unit. Only copper conductors should be used. **See Procedure for Connecting Wiring Whips (High-Voltage Wiring) to Condensing Units.**

Connecting Wiring Whips (High-Voltage Wiring) to Condensing Units

Procedure

1. While outside at condensing unit, remove screws from access panel on front of condensing unit.
   a. Remove access panel.

2. Hook up wire whip to access panel.
   a. Remove nut from electrical whip hanging from disconnect box.
   b. Run wiring whip up through high-voltage electrical hole that protrudes from corner of condensing unit. Slide nut over wires, and tighten down against access panel.

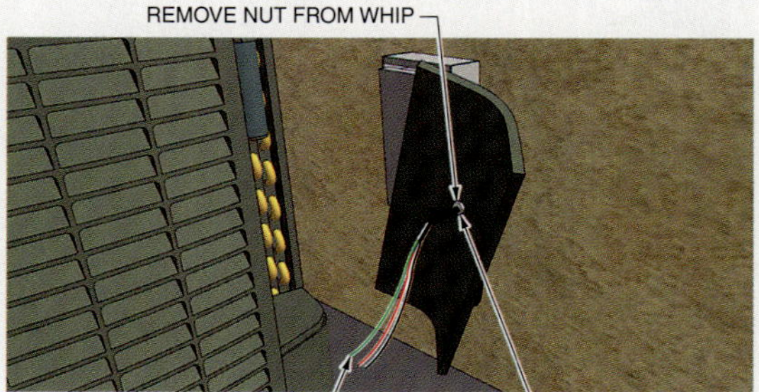

continued on next page

Connecting Wiring Whips (High-Voltage Wiring) to Condensing Units *(continued)*

3. Trim ground (green) wire and strip back insulation ½″.
4. Place ground wire under grounding lug and tighten down with screwdriver.

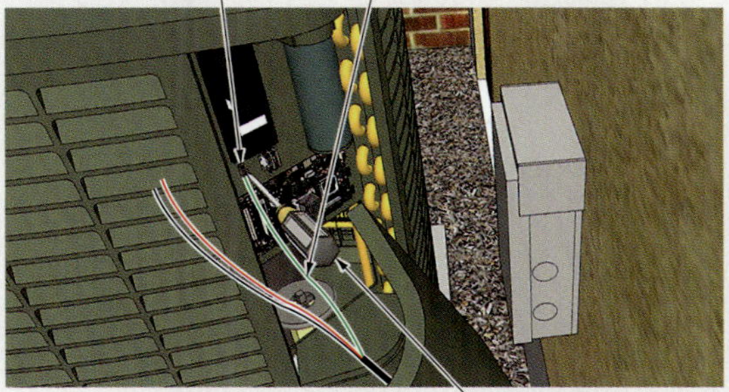

5. Connect red and black wires from wire whip to either side of contactor. Use spade connectors to make an easy connection.

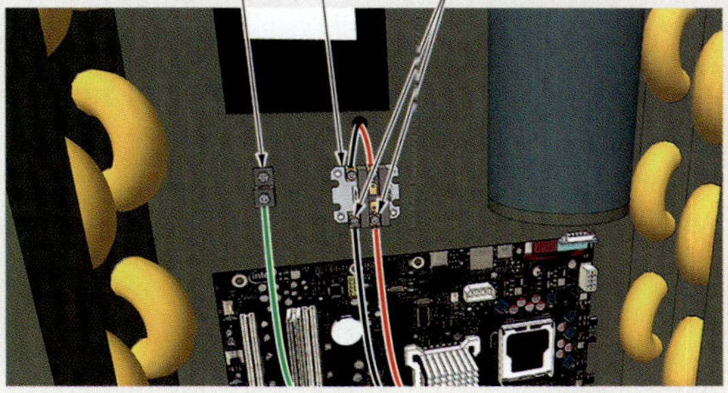

The next part of the wiring installation is to connect the low-voltage wiring from the indoor unit to the condensing unit. This allows the thermostat to energize the contactor and cycle the compressor on and off. Typically 18 AWG, two- or three-conductor jacketed wire is used. Since it typically has a plastic jacket, it does not need to be in the conduit. The low-voltage wiring is typically tied off to the line set piping or to the high-voltage conduit. Copper conductors must also be used. The wires from the disconnect box are terminated to the appropriate terminals of the contactor inside the condenser. **See Procedure for Connecting Low-Voltage (24 V) Wiring to Condensing Units.**

Connecting Low-Voltage (24 V) Wiring to Condensing Units

1. Run wires out hole on side of residence. When outside, pull wires through hole and run them toward condensing unit.

2. Run wires up through low-voltage electrical hole in access panel at front of condensing unit.
   a. Strip ½″ of insulation from end of both wires. Connect the two wires (blue) to 24 V terminal strip.

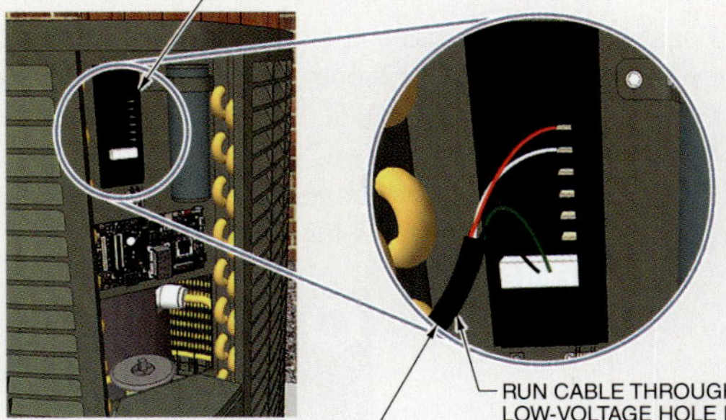

3. Replace access panel on condensing unit.

Another step in the electrical installation process is to install low-voltage wiring from the thermostat to the furnace or air handler. The wire used is always an 18 AWG, five- or six- conductor cable with a plastic jacket. It can be run anywhere inside the walls of the building or strung across the rafters in the attic. Care must be taken to not place the thermostat wire where it can be tripped over or become an obstruction. **See Procedure for Running Low-Voltage (24 V) Wiring to Furnaces or Air Handlers.**

Procedure

Running Low-Voltage (24 V) Wiring to Furnaces or Air Handlers

1. When inside residence, ensure that power to furnace or air handler is turned off.
2. Run 24 V wiring with line set piping and secure with plastic tie wraps.

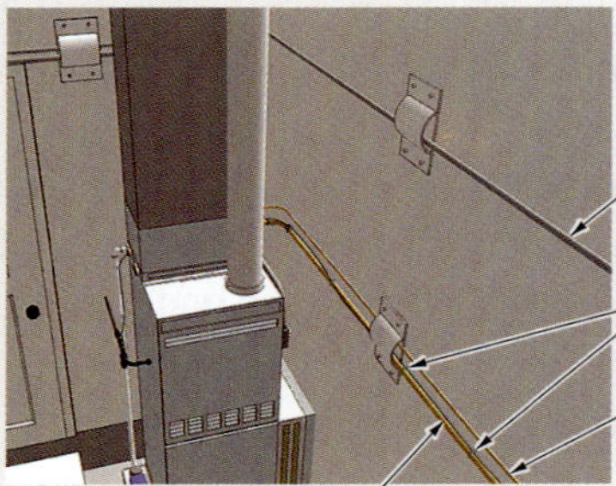

3. Remove lower cover on front of furnace or air handler.
4. For standard gas furnace, the control wiring area displays five or six terminals. Terminals are labeled either WW1, Y, G, R, and com or C, Y, G, R, WW1, and com. If furnace has already been installed, note that two wires from thermostat are already hooked up to R and WW1 (Wh) terminals. These wires are for heating function.

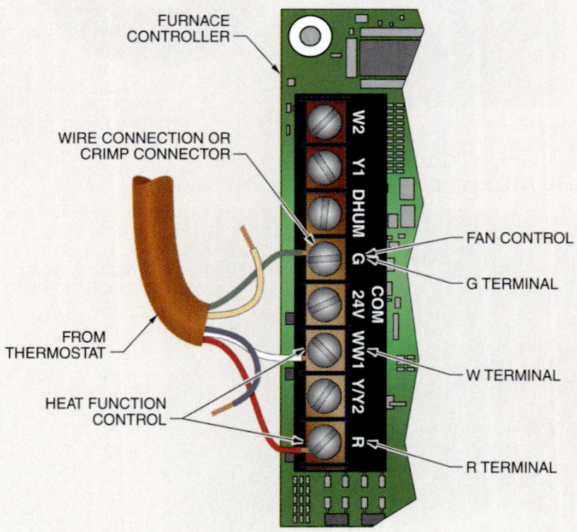

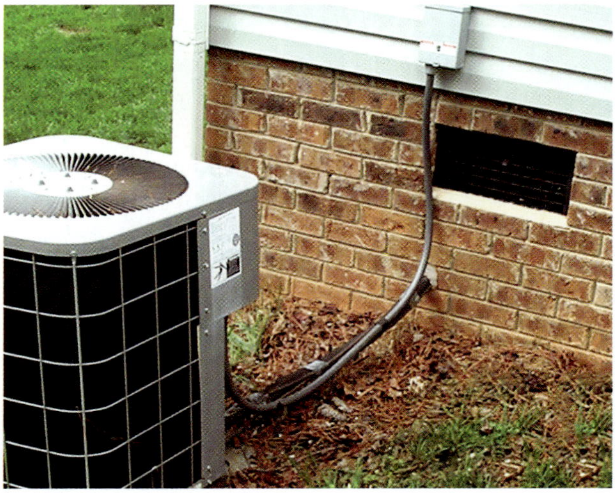

The line set insulated suction line and the uninsulated liquid line, control wiring, and power wiring are all bundled together for protection.

REFRIGERANT PIPING AND INSTALLATION

The line set consists of two refrigerant lines that connect the outdoor condensing unit to the indoor evaporator coil. To complete this part of the installation, the line set must be connected to the condensing unit and run inside along the ceiling (out-of-the-way path) toward the evaporator coil. *Note:* Technicians must be EPA-certified to complete the refrigerant installation.

TECH TIP

The proper connection of low-voltage wiring from the thermostat to the unit is required for correct equipment performance and ease when troubleshooting. Standard color coding should be used for all connections. Improperly installed wiring can make troubleshooting difficult and lead to increased service costs.

During the installation of an air conditioning system, it is important to properly isolate the refrigerant lines to prevent unnecessary vibration. **See Figure 40-8.** Line set contact with the structure (wall, ceiling, or floor) may cause objectionable noise when vibration is translated into sound. As a result, more energy or vibration can be expected. Close attention should be paid to line set isolation. **See Procedure for Installing Line Sets.**

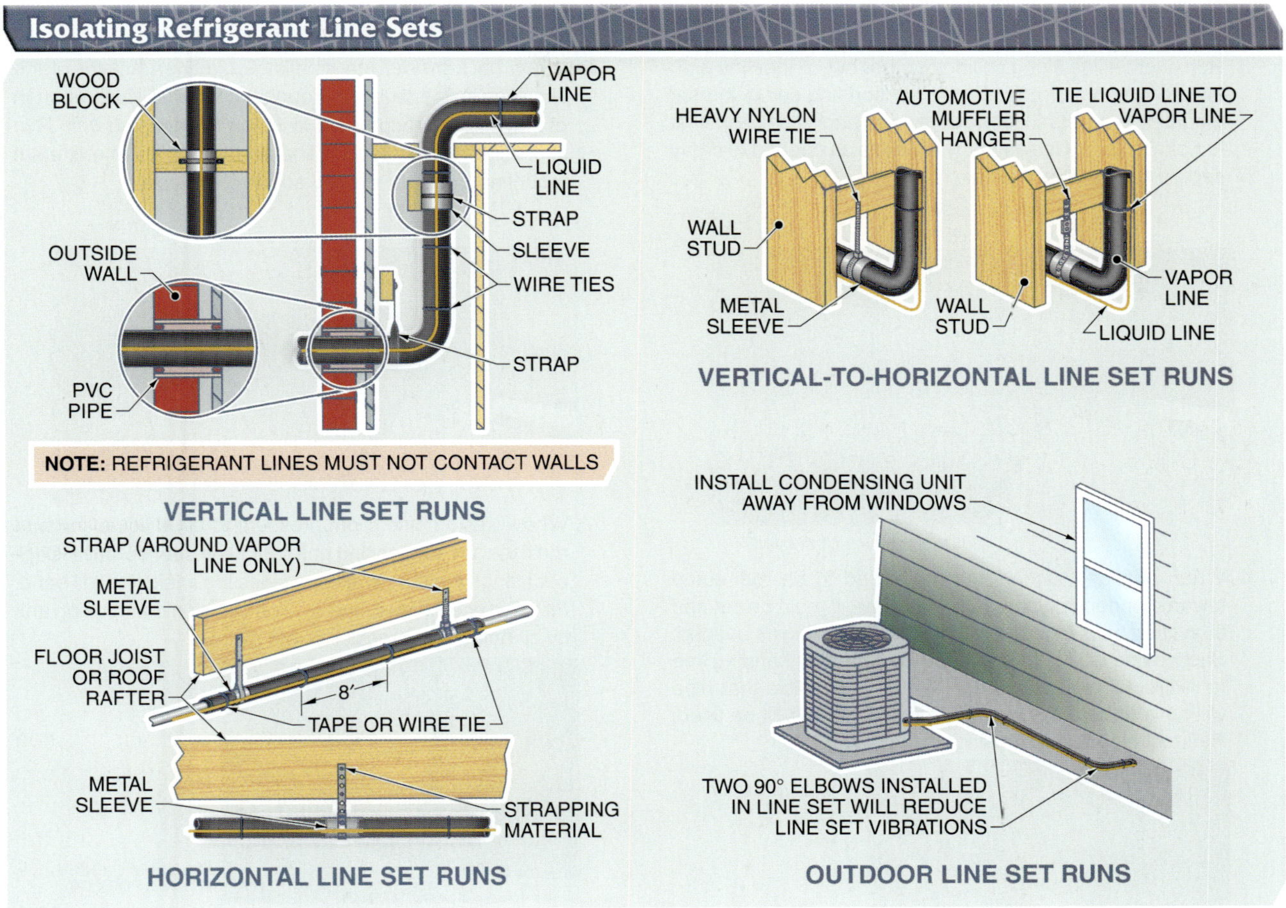

Figure 40-8. Proper line set installation and isolation guidelines should be followed to prevent objectionable noise from vibrations.

Installing Line Sets

Procedure

1. Roll out suction line (larger insulated line) holding one end down with foot.
   a. Peel back few feet of insulation from one end of line and cut off with utility knife. (Amount peeled back should be about same length as distance between hole in wall and condensing unit.)
   b. Set insulation aside.

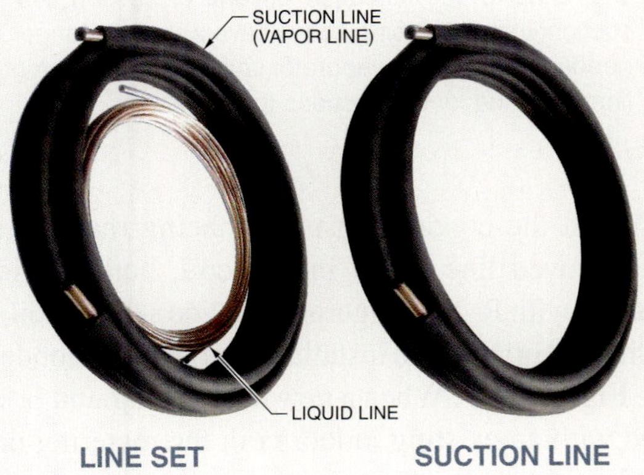

2. Push enough suction line through hole in wall to reach condensing unit. If too much is pushed through hole, it can be cut to length at later time.

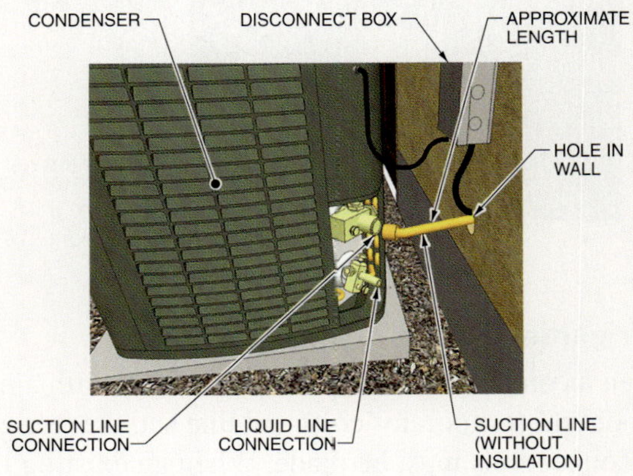

continued on next page

Installing Line Sets *(continued)*

3. Run suction line along ceiling or other out-of-the-way path toward evaporator coil. Secure suction line using line set mounting brackets. Use two screws on each side of bracket to hold in place. Use as many brackets as needed to firmly secure suction line in place.

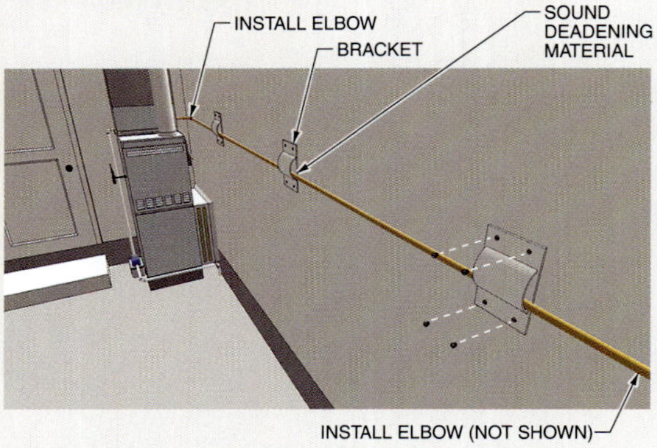

4. When outside, suction line may need to be redirected toward condensing unit. To redirect line, it must be cut and then fitted back together with 45° copper fitting as needed. Suction line is slightly malleable, but sharply bending line to redirect it may cause it to kink. It is possible that pipe will have to be cut twice and that two fittings will be used. Redirect pipe to condensing unit.

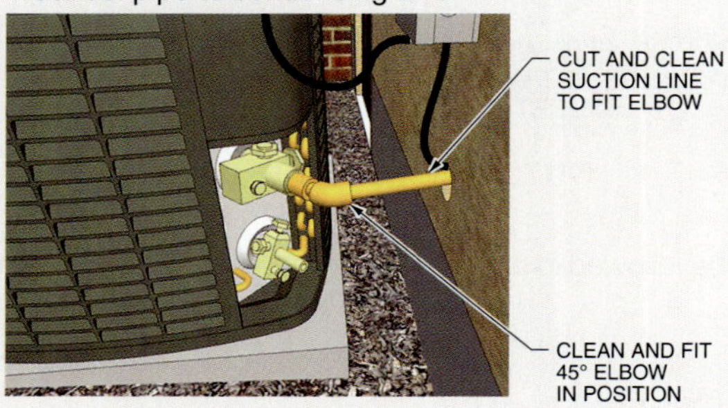

5. Ensure that suction line is pressed into 45° elbow installed on condensing-unit suction service valve and braze.

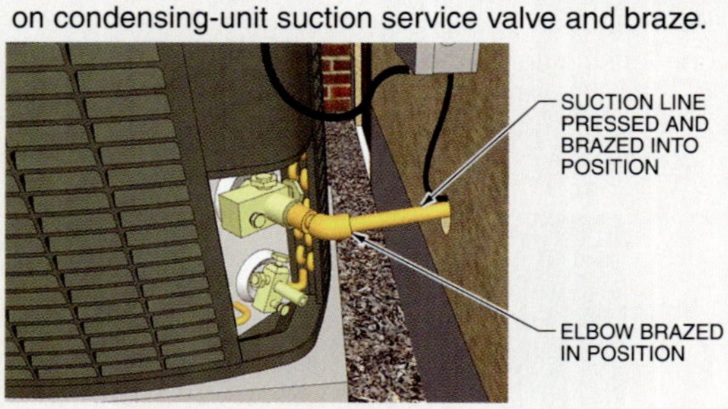

6. When back inside, run smaller liquid line (smaller of the two lines in line set) out through hole in wall. Push enough of liquid line through hole to reach condensing unit. Run liquid line next to suction line and through same line set mounting brackets already secured in place.

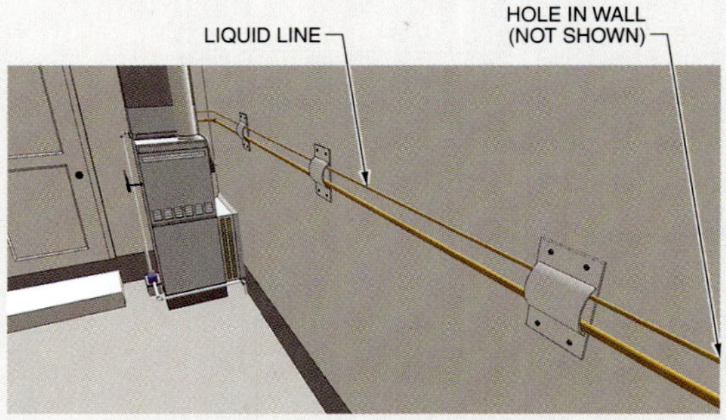

7. When outside, use slight pressure to bend liquid line and run it toward condensing unit liquid-line service valve. Copper fittings are not necessary since the liquid line will bend. Fit liquid line to liquid-line service valve on condensing unit by cleaning and pressing it into place.

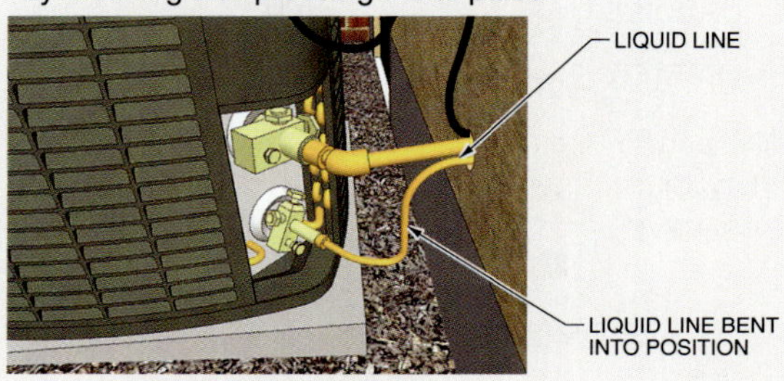

8. Braze liquid line into position.
9. When inside again, trim liquid line with tubing cutter to be several inches longer than necessary to reach evaporator coil.

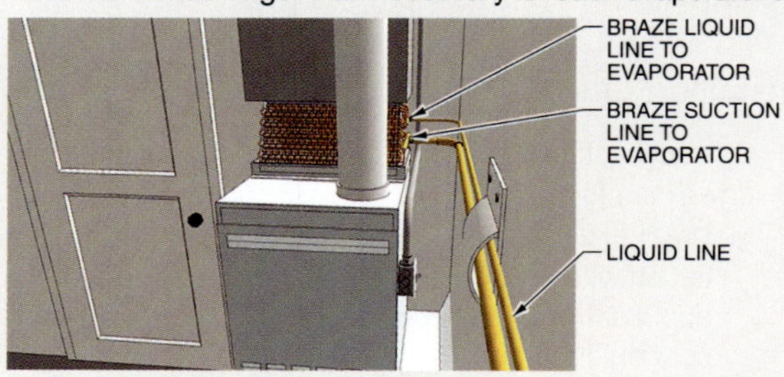

10. Use knife to cut insulation off end of suction line. Trim suction line with tubing cutter several inches longer than needed to attach to evaporator coil. Leave both refrigerant lines hanging for connections to be brazed.

Refrigerant Piping

When a condensing unit is being installed with a new indoor coil (evaporator coil) and line set, the refrigerant connections must be made. When an existing line set and/or evaporator coil will be used to connect the system together, the line set and indoor coil must also be flushed.

When the condensing unit is being matched with an approved line set or indoor coil, both previously charged with R-22 refrigerant, the line set and coil must be flushed prior to the installation of the new condenser. **See Figure 40-9.** When a new condensing unit is being used with an existing indoor coil, the metering device and liquid line must also be replaced.

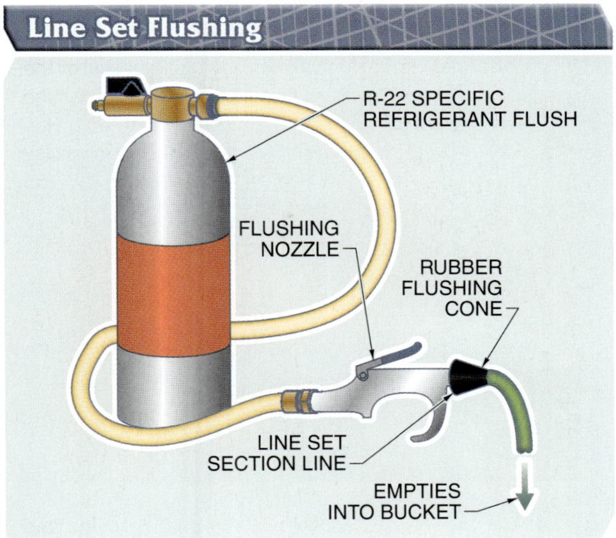

Figure 40-9. Each system and the refrigerant used require a specific flushing compound to be used because of the specific refrigerant in the system.

TECH TIP

When installing HVAC units, all installation must be done per manufacturer recommendations. Cutting corners, especially on multiple installation jobs such as a new subdivision, should be avoided because improper installation of new piping and equipment can lead to numerous service calls and system failures.

Field refrigerant piping consists of liquid and suction lines from the outdoor unit to the indoor coil. The proper line set diameters must be selected to ensure that the crankcase oil returns to the compressor. The vertical suction riser must be sized to maintain minimum velocity at minimum capacity. The recommended line length is 50′ or less. When more than a 50′ line set is required, the manufacturer should be contacted. **See Procedure for Brazing Line Set Connections.**

Brazing Line Set Connections

1. Cut ends of line set square (free from nicks or kinks), and debur ends. Tubing must remain round. Do not pinch ends of lines.

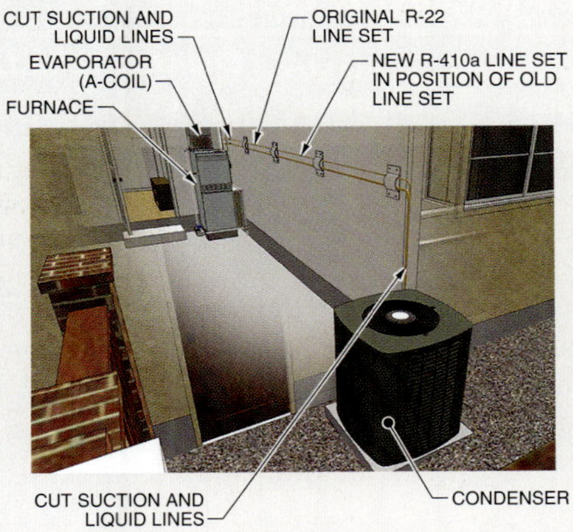

2. When making line set connections, use dry nitrogen to purge refrigerant piping. This will help prevent oxidation and introduction of moisture into system.
3. Use silver alloy brazing rods (5% or 6% silver alloy for copper-to-copper brazing or 45% silver alloy for copper-to-brass or copper-to-steel brazing), which are rated for use with R-410a refrigerant.

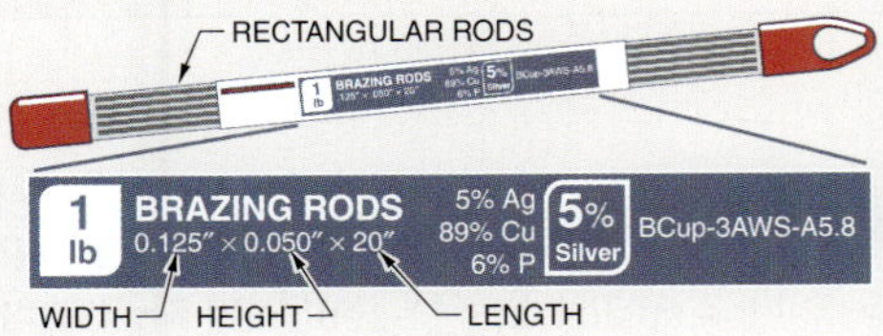

4. Remove core assemblies from any service or Schrader valves before brazing to protect them from damage due to extreme heat. Replace cores when brazing is complete.

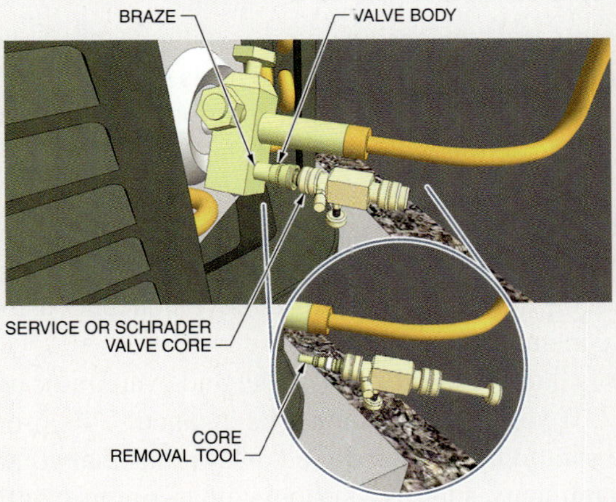

5. Remove washers from service valves, and shield them for protection during brazing. Wrap wet cloth around valve body and copper tube stub to protect it from heat damage.

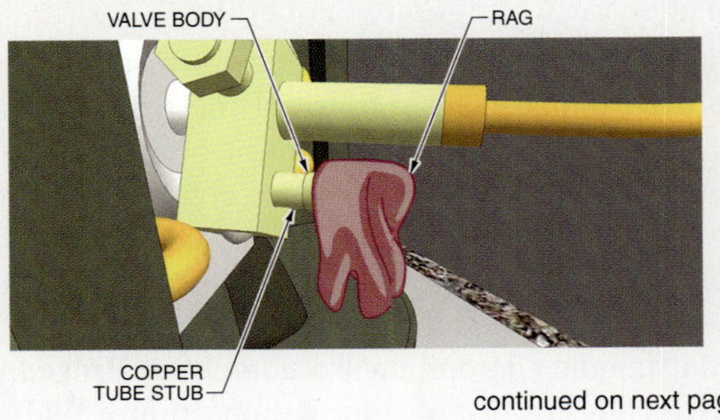

continued on next page

Brazing Line Set Connections *(continued)*

6. Braze line set to service valve. Quench joints with water or wet cloth to prevent heat damage to valve core and opening port. *Note:* Tube end must stay bottomed in fitting during final assembly to ensure proper seating, sealing, and rigidity.

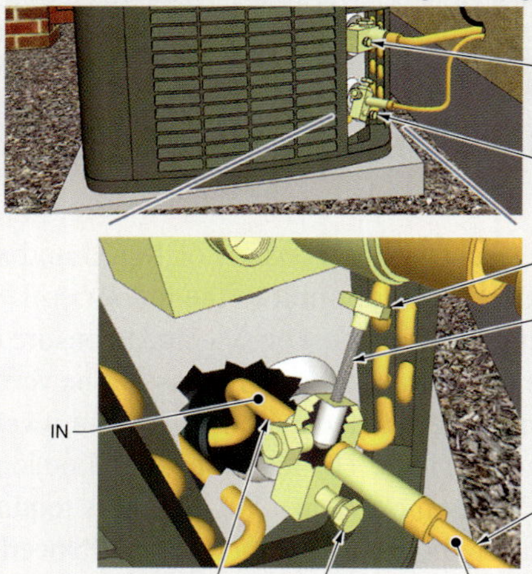

7. Install thermal expansion valve in liquid line at evaporator coil.

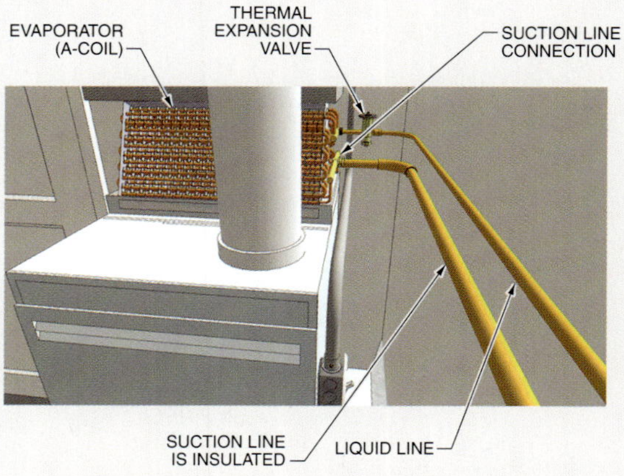

NOTE: THIS STEP IS SKIPPED IF SYSTEM IS TO BE FLUSHED AS ONE OF NEXT STEPS TO BE PERFORMED

CLEANING EXISTING LINE SETS AND EVAPORATOR COILS

Sometimes a condensing unit is installed with an existing line set and evaporator coil. This is not an ideal situation but it may occur. A better solution and much more preferable action would be to install a new evaporator and line set. **See Figure 40-10.** *Note:* System efficiency is affected by using an existing evaporator and line set with a new condensing unit. The line set and/or evaporator coil must be flushed to remove noncondensibles and contaminants.

It is likely that the old line set and evaporator used R-22. If so, the system must be cleaned. This procedure should not be performed on systems that contain contaminants, such as a system with a compressor that has burned out.

Refrigerant Changeout Procedure

When changing from one refrigerant to another, an approved replacement refrigerant must be used and the coil and refrigerant lines must be rated for the new refrigerant. When the existing condensing unit is not equipped with shutoff valves, or if the unit is not operational, the old refrigerant must be recovered from the system. The old refrigerant must always be recovered using the appropriate procedure. **See Procedure for Refrigerant Changeout without Shutoff Valves.**

Line Set Suction Line Cooling-Capacity Loss

Unit Nominal Size*	Vapor Line Diameter†	Equivalent Line Length					
		50	75	100	125	150	175
18,000	5/8	5‡	7	9	12	12	14
	3/4	1	3	4	5	5	7
24,000	5/8	6	9	13	16	19	22
	3/4	0	1	1	2	3	4
30,000	5/8	6	8	10	13	15	17
	3/4	2	3	4	5	6	7
36,000	3/4	7	10	14	17	21	NR
	7/8	2	4	6	8	10	11
42,000	3/4	7	10	13	17	20	23
	7/8	3	4	6	7	8	10
	1 1/8	0	0	1	1	2	2
48,000	3/4	10	14	18	22	NR	NR
	7/8	4	6	7	9	11	13
	1 1/8	0	0	1	1	2	2
60,000	7/8	7	9	11	14	16	19
	1 1/8	1	2	2	3	3	4

* in Btu/hr
† in in.
‡ loss in %

Figure 40-10. Most R-22 systems have a line set that is too small, which allows too much loss. Most system updates require that the suction line be replaced when R-22 refrigerant is replaced.

Refrigerant Changeout without Shutoff Valves

1. Disconnect all power to existing outdoor condensing unit.

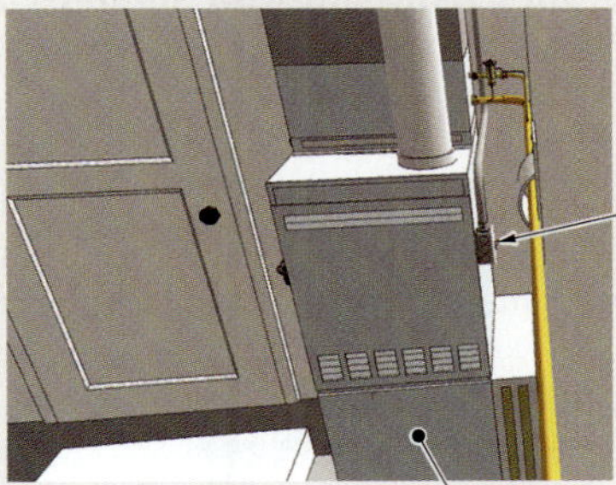

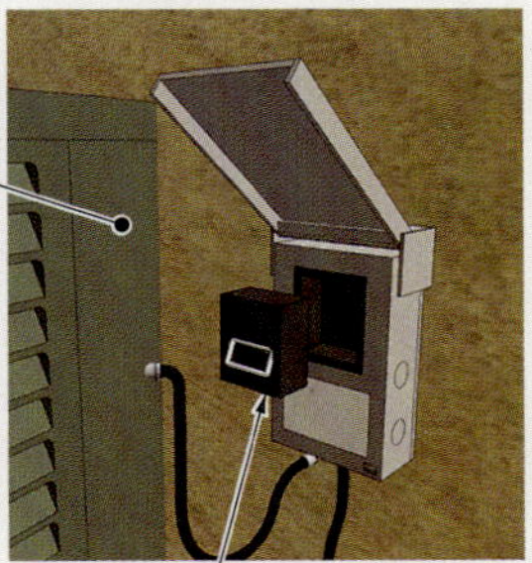

2. Connect existing air conditioning unit, and clean recovery cylinder to recovery machine according to manufacturer instructions.

3. Recover all old refrigerant from system. Use a scale to weigh recovery cylinder to prevent overfilling.
4. Refer to gauges after shutdown to confirm that entire system is completely void of refrigerant.

5. Prepare changeout equipment, line set, evaporator, and condenser for changeout. Flush piping with agent approved by the EPA and equipment manufacturer.
6. Purge piping with nitrogen to remove noncondensables.

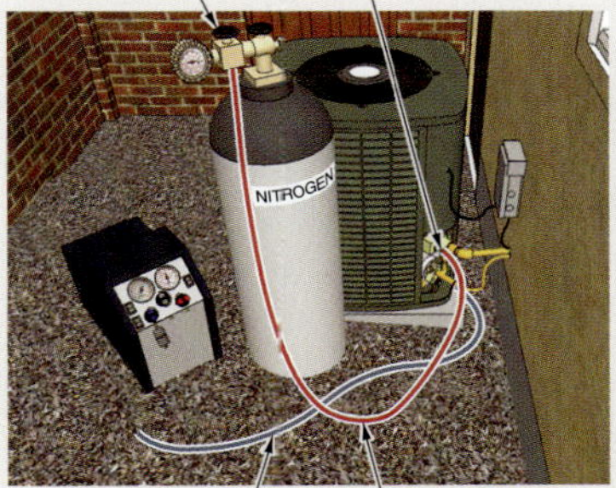

7. Connect yellow hose to vacuum pump, red hose to liquid line service valve, and blue hose to suction line service valve. Evacuate system as recommended by manufacturer.

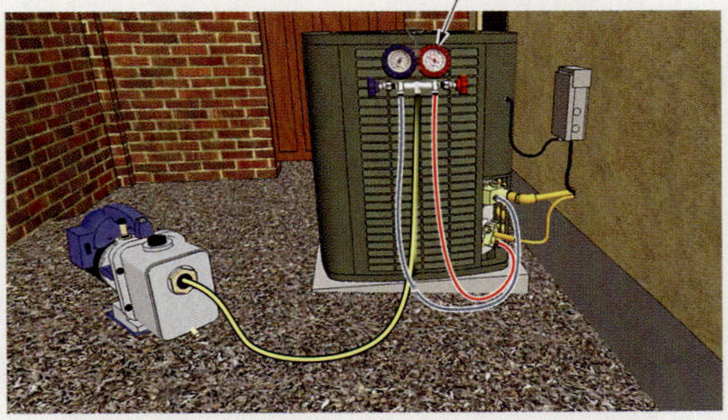

8. Close manifold gauge low- and high-side valves. Loosen yellow hose from vacuum pump (to avoid sucking the oil out from the pump) and shut off pump.
9. Connect new refrigerant cylinder to manifold gauge yellow hose, loosen the hose connection at gauge, and open cylinder valve to purge yellow hose. Tighten connections, open valves, and charge system per equipment manufacturer guidelines.

When an existing old outdoor condensing unit is equipped with manual shutoff valves, the old refrigerant must be recovered and the system must be evacuated. New refrigerant is then used to charge the system. The old TXV must not be replaced until the system is ready for charging. **See Procedure for Refrigerant Changeout with Shutoff Valves.**

Refrigerant Changeout with Shutoff Valves

1. Disconnect all power to existing outdoor condensing unit.

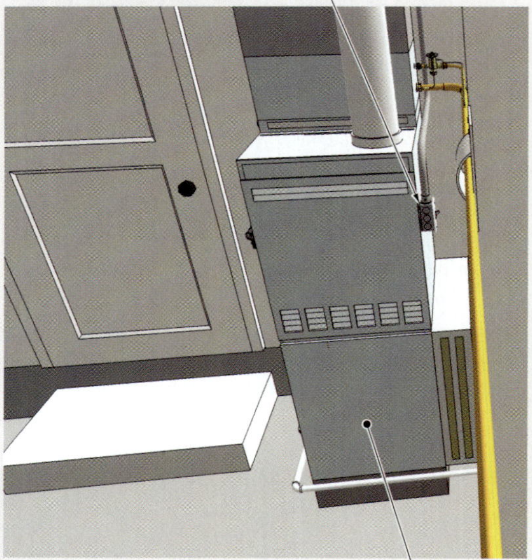

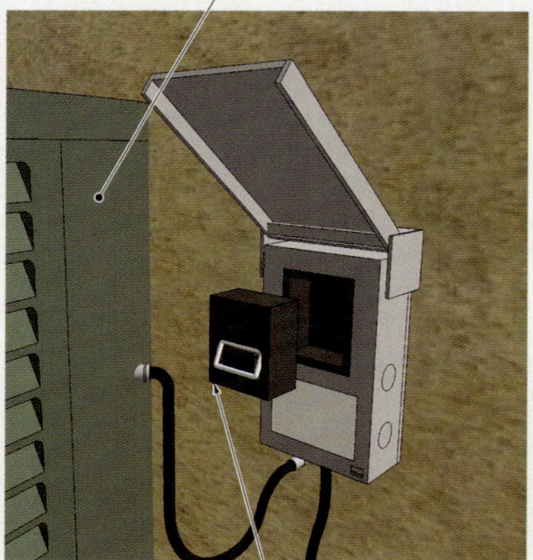

2. Connect existing air conditioning unit, and clean recovery cylinder to recovery machine according to manufacturer instructions.

3. Recover all old refrigerant from system. Use a scale to weigh recovery cylinder to prevent overfilling.
4. Refer to gauges after shutdown to confirm that entire system is completely void of refrigerant.

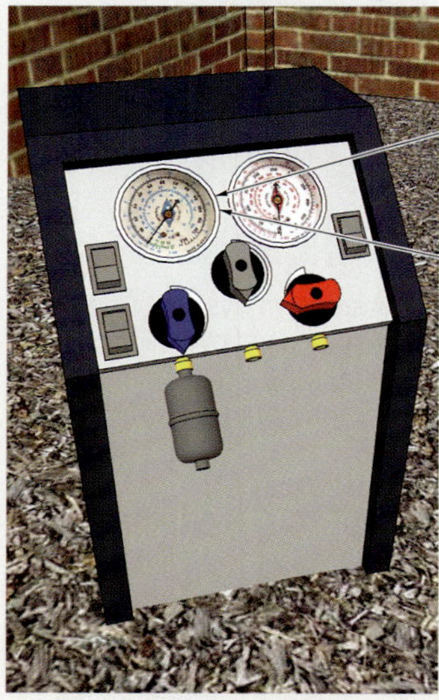

5. Prepare changeout equipment, line set, evaporator, and condenser for changeout. Flush piping with agent approved by the EPA and equipment manufacturer.

continued on next page

Refrigerant Changeout with Shutoff Valves *(continued)*

6. Purge piping with nitrogen to remove noncondensables.

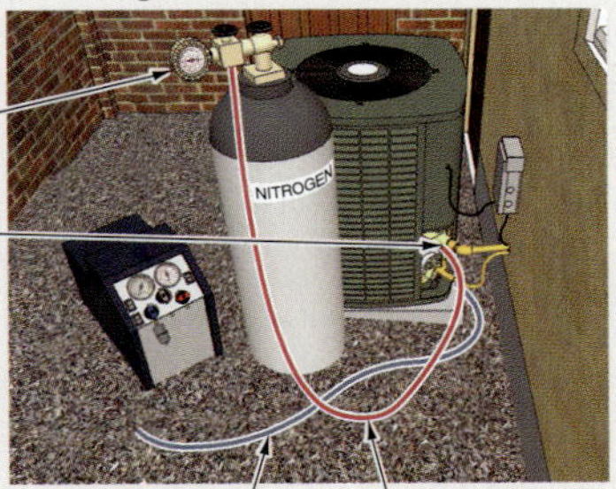

7. Connect yellow hose to vacuum pump, red hose to liquid line service valve, and blue hose to suction line service valve. Evacuate system as recommended by manufacturer.

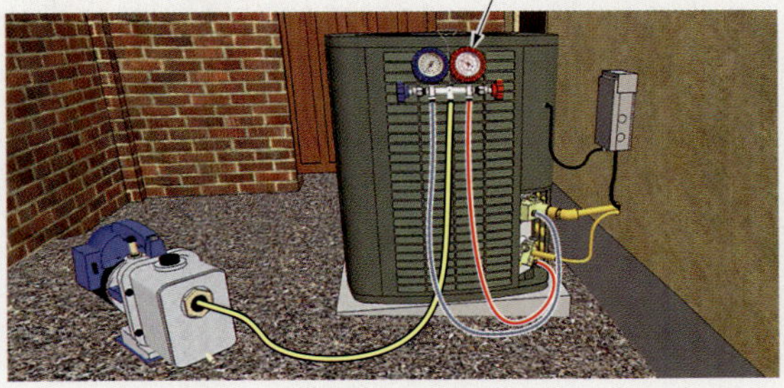

8. Close manifold gauge low- and high-side valves. Loosen yellow hose from vacuum pump (to avoid sucking the oil out from the pump) and shut off pump.
9. Connect new refrigerant cylinder to manifold gauge yellow hose, loosen the hose connection at gauge, and open cylinder valve to purge yellow hose. Tighten connections, open valves, and charge system per equipment manufacturer guidelines.

REFRIGERANT METERING DEVICES

With some exceptions, modern condensing units are designed for use with TXV systems. Typically, air conditioning systems charged with R-410a will not operate properly with an R-22 metering device. **See Procedure for Installing Air Conditioning System Metering Devices.**

Installing Air Conditioning System Metering Devices

1. Install refrigerant metering device (fixed orifice, capillary tube, or TXV) according to manufacturer recommendations.

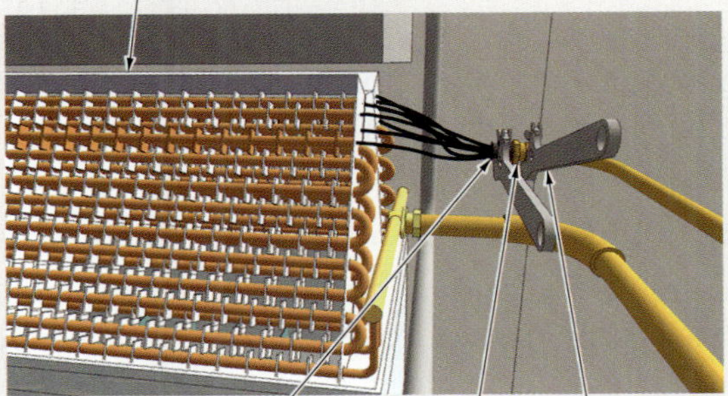

NOTE: DO NOT TWIST CAP TUBES WHEN LOOSENING SEAL NUT FROM ORIFICE HOUSING. ALWAYS USE BACK-UP WRENCH ON DISTRIBUTOR.

2. To install TXV, separate distributor assembly and remove piston orifice and used Teflon® seal. Insert nozzle end of TXV along with new Teflon seal into distributor and tighten to 20 ft-lb to 30 ft-lb. A back-up wrench should be used on all wrench flats. Overtightening and crushing the Teflon seal will cause leaks.

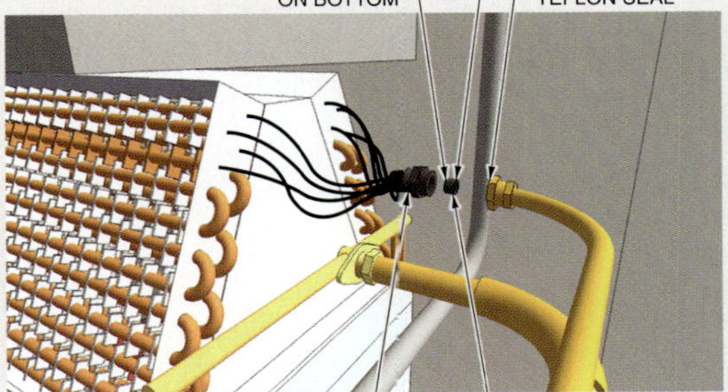

continued on next page

Procedure

Installing Air Conditioning System Metering Devices *(continued)*

3. Attach liquid line portion of distributor assembly along with new Teflon seal to inlet of expansion valve. Tighten to 20 ft-lb to 30 ft-lb. Use back-up wrench on all wrench flats. Overtightening will crush the Teflon seal and may cause a leak.

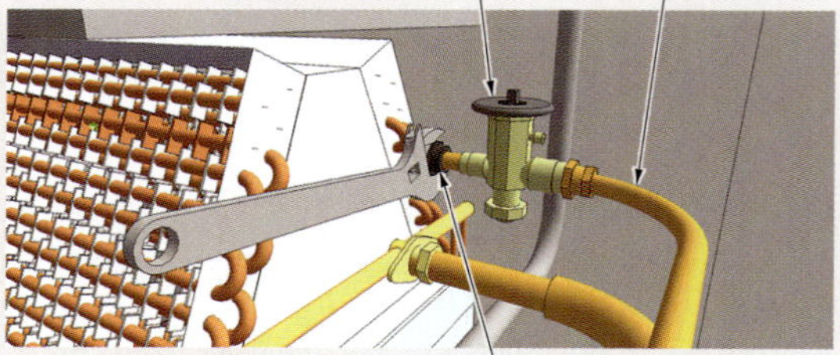

4. When an externally equalized TXV is being used, connect external equalizer line, if present, to equalizer port on suction line and tighten to 8 ft-lb.

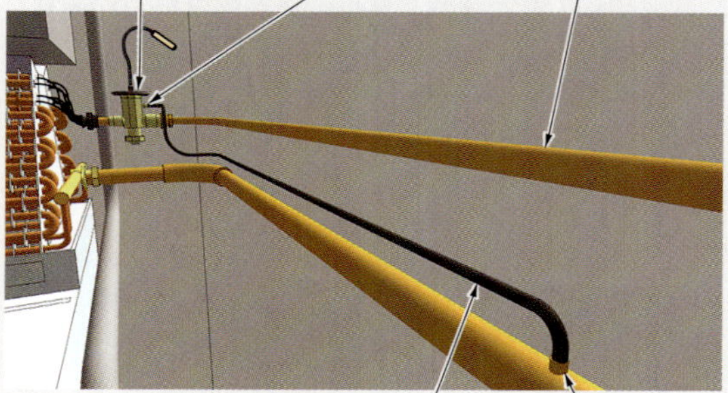

5. When an externally equalized TXV is being used, strap superheat sensing bulb to discharge pipe of evaporator (compressor suction line).

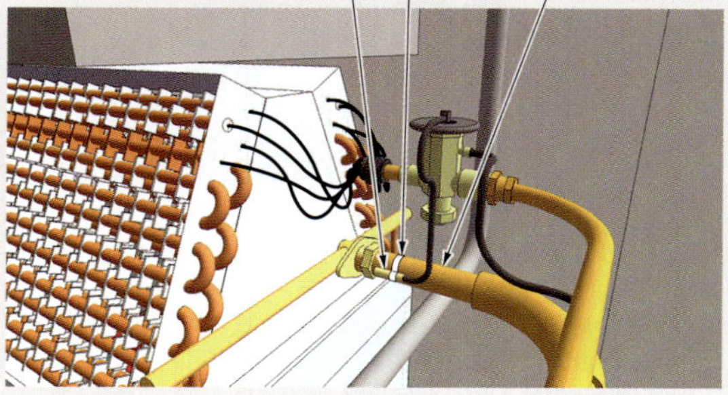

6. When an externally equalized TXV is being used on indoor coil that previously used a fixed orifice, be sure to remove existing fixed orifice.

NOTE: FAILURE TO REMOVE FIXED ORIFICE WHEN INSTALLING TXV TO INDOOR COIL WILL RESULT IN IMPROPER OPERATION AND DAMAGE TO SYSTEM.

R-410a systems require the use of different manifold gauges to handle the higher pressure.

MANIFOLD GAUGE SETS

Manifold gauge sets used with systems that are charged with R-410a refrigerant must be capable of handling the higher system operating pressures. The gauges must be rated for use with pressures from 1 psi to 800 psi on the high-pressure side and a 30″ Hg vacuum to 250 psi on the low-pressure side. Gauge hoses must be rated for use up to 800 psi with a 4000 psi burst rating.

Liquid-Line and Suction-Line Service Valves

The liquid-line and suction-line service valves and service ports are used for leak testing, system evacuation, charging, and system charge checks. Each valve is equipped with a service port, which has a factory-installed Schrader valve. A service port cap protects the Schrader valve from contamination and serves as the primary leak seal. **See Procedure for Accessing Service Port Schrader Valves.**

Accessing Service Port Schrader Valves

1. Remove service port cap with adjustable wrench.

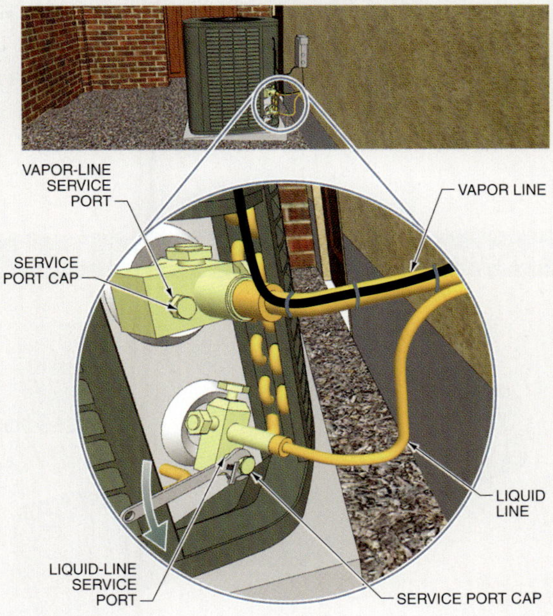

2. Connect gauge to service port.

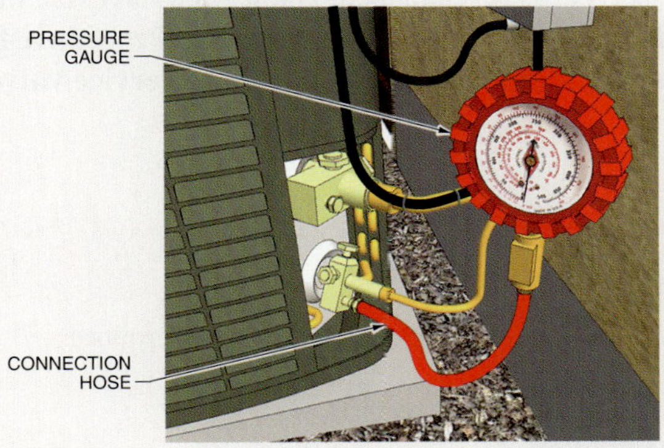

3. When testing is complete, replace service port cap. Tighten fingertight and then tighten an additional ⅛ turn.

Once the cap to the service port is removed, the service port can be accessed. The service valve is opened to allow refrigerant flow to the gauge set. This allows the manifold gauges to measure system pressure on the high- and low-pressure sides. **See Procedure for Opening Liquid-Line or Suction-Line Service Valves.**

Opening Liquid-Line or Suction-Line Service Valves

1. Gain access to service valve.

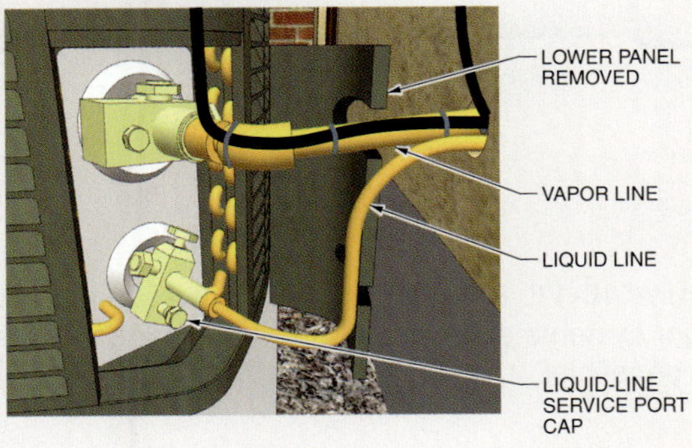

2. Remove service port cap with adjustable wrench.

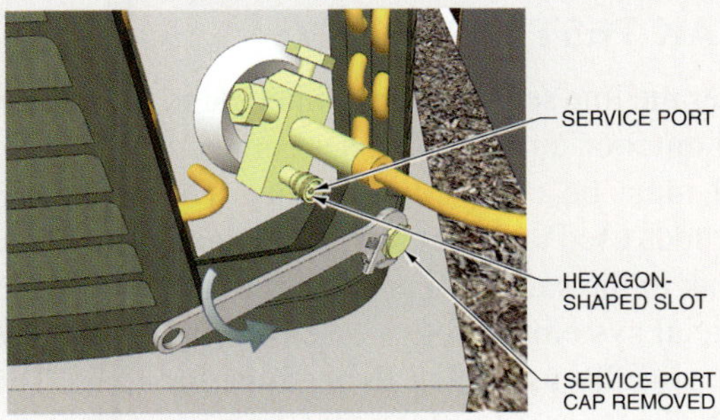

3. Use service wrench with hex-head extension to back stem out counterclockwise as far as it will go.

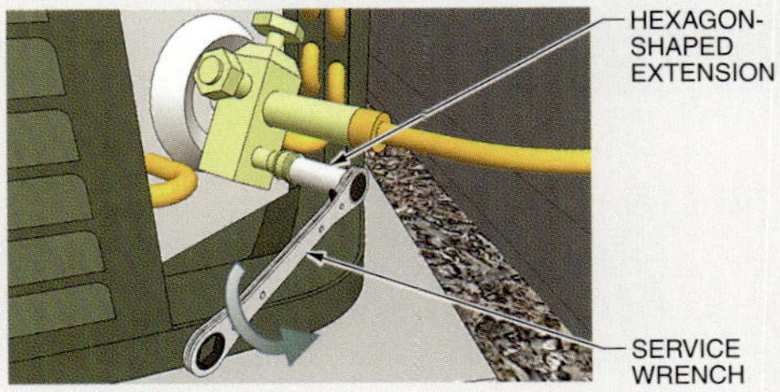

NOTE: USE 3⁄16″ HEX-HEAD EXTENSION FOR LIQUID-LINE SERVICE PORTS AND 5⁄16″ HEX-HEAD EXTENSION FOR SUCTION-LINE SERVICE PORTS

4. Replace service port cap. Tighten fingertight and then tighten an additional ⅛ turn.

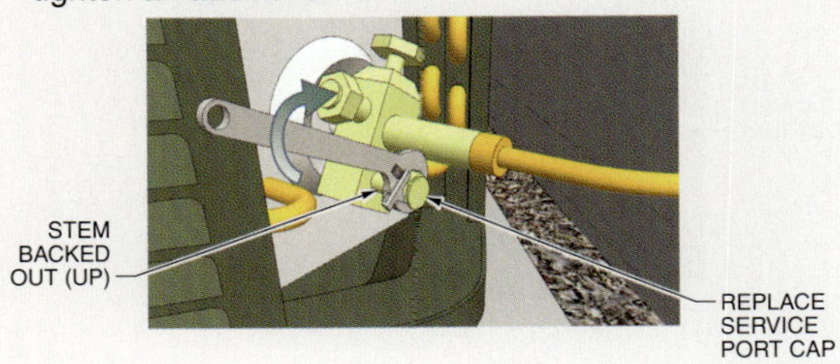

After system pressure is measured and the system is serviced, the service valves are backseated and the service port caps are reattached. The service valves are then ready for normal system operation. **See Procedure for Closing Liquid-Line or Suction-Line Service Valves.**

Closing Liquid-Line or Suction-Line Service Valves

1. Remove service port cap with adjustable wrench.

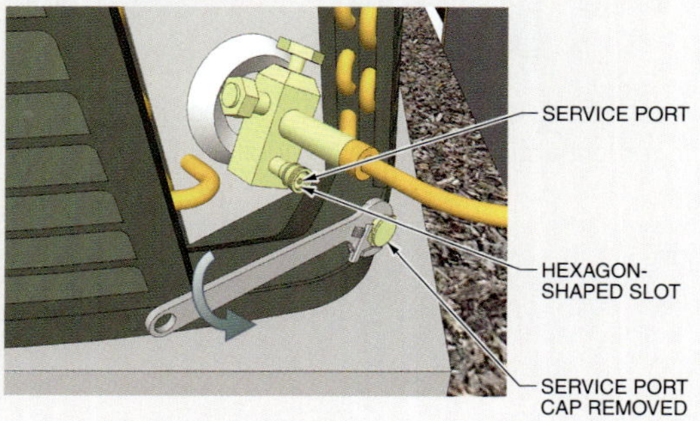

2. Use service wrench with hex-head extension to turn stem clockwise to seat valve. Tighten firmly.

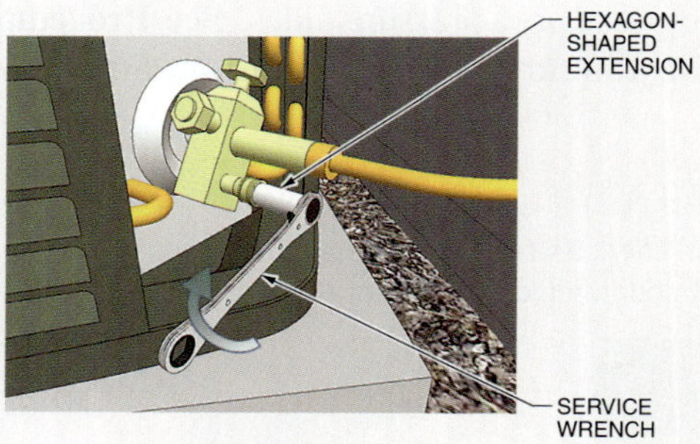

NOTE: USE 3/16″ HEX-HEAD EXTENSION FOR LIQUID-LINE SERVICE PORTS AND 5/16″ HEX-HEAD EXTENSION FOR SUCTION-LINE SERVICE PORTS

3. Replace service port cap. Tighten fingertight and then tighten an additional 1/6 turn.

Suction-Line Service Valves

A suction-line service valve functions the same way as other service valves. The difference is in the construction. The service valve is equipped with a service port that has a factory-installed Schrader valve. A service port cap protects the Schrader valve from contamination and serves as the primary seal. **See Figure 40-11.**

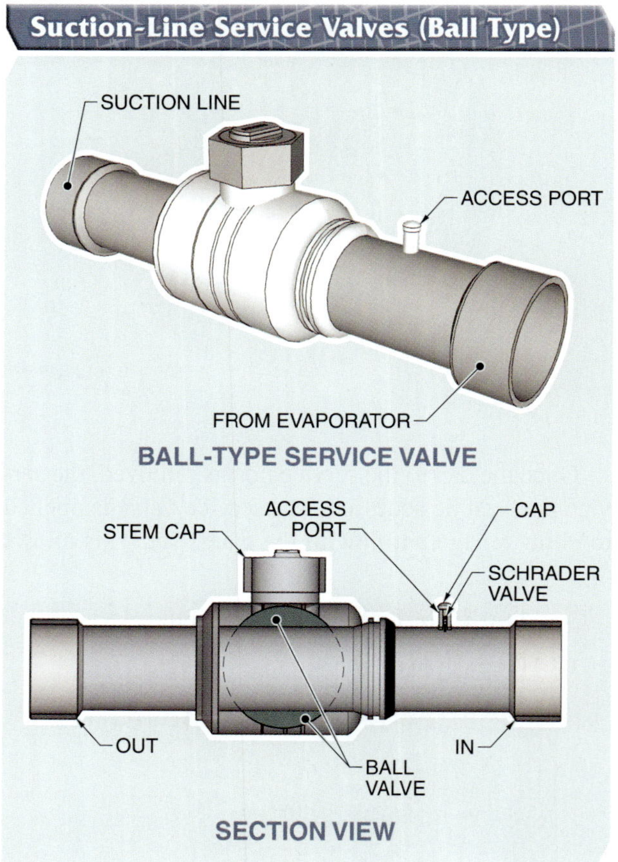

Figure 40-11. Ball-type service valves are used on larger systems and systems in which fast acting valves are beneficial.

LEAK TESTING

After the line set has been connected to the indoor coil and outdoor unit, the line set connections and indoor coil must be checked for leaks. There are various methods used to detect refrigerant leaks. For detecting a leak in the suction line of a residential or light commercial system, nitrogen or electronic leak detectors are typically used. **See Procedure for Leak Testing with Electronic Leak Detectors.**

Procedure

Leak Testing with Electronic Leak Detectors

1. Connect high-pressure hose of manifold gauge set to suction valve service port.

NOTE: TYPICALLY HIGH-PRESSURE HOSE IS CONNECTED TO LIQUID LINE PORT; HOWEVER, CONNECTING IT TO SUCTION PORT HELPS TO PROTECT MANIFOLD GAUGE SET FROM DAMAGE CAUSED BY HIGH PRESSURES

2. With both manifold valves closed, connect cylinder of R-410a refrigerant to charging machine.
   a. Connect charging machine to manifold gauge set.
   b. Open valve on R-410a cylinder (vapor only).

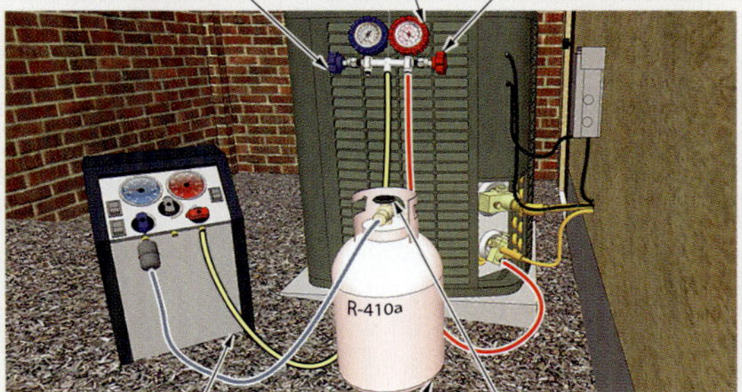

3. Open high-pressure side of manifold to allow R-410a into line set and evaporator coil. Weigh in trace amount of R-410a (maximum of 2 oz of refrigerant or 3 psi of pressure).
4. Close valve on R-410a cylinder and valve on high-pressure side of manifold gauge set.
5. Disconnect charging machine from manifold gauge set and remove inlet hose.
   a. Disconnect R-410a cylinder.

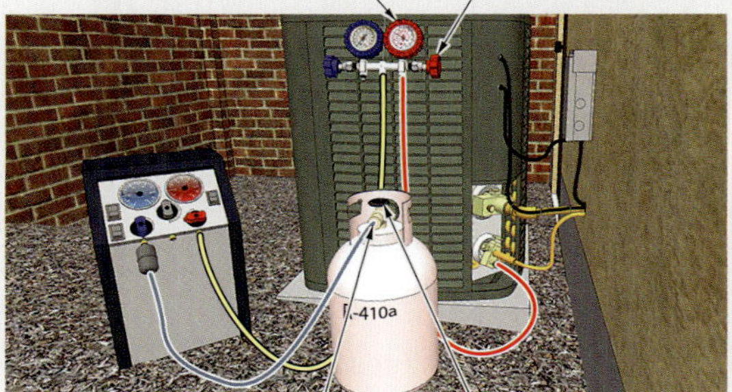

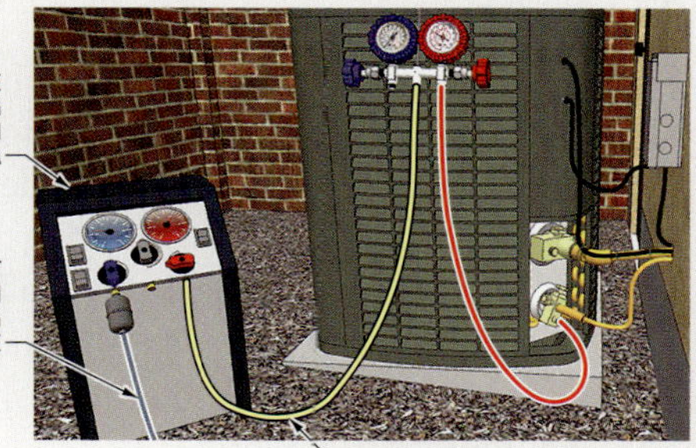

6. Connect cylinder of dry nitrogen with pressure regulating valve to center port of manifold gauge set.

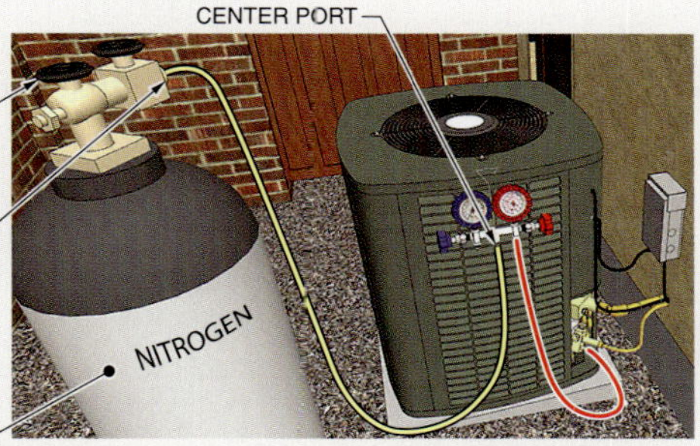

NOTE: WHEN USING HIGH-PRESSURE GAS SUCH AS NITROGEN FOR PRESSURE PURPOSES, BE SURE TO USE REGULATOR THAT CAN CONTROL PRESSURE DOWN TO 1 PSI OR 2 PSI

7. Open dry nitrogen cylinder valve. Adjust nitrogen pressure to 150 psi. Open valve on high-pressure side of manifold gauge set to pressurize line set and evaporator coil.

8. Check all joints for leaks.
   a. After short period of time, open refrigerant port to ensure that an adequate amount of refrigerant has been added for detection (refrigerant requirements will vary with piping lengths).
9. Purge dry nitrogen and R-410a mixture. Correct any leaks and recheck.

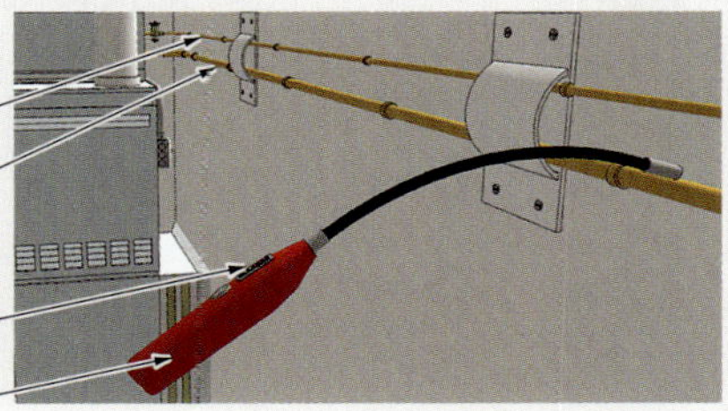

EVACUATION, CHARGING, AND STARTUP

Evacuating the system of noncondensables is critical for the proper operation of a condensing unit. A *noncondensable* is any gas that will not condense under the temperatures and pressures present during the normal operation of an air conditioning system. **See Figure 40-12.** Noncondensable gases get trapped in the upper tubes of the condenser by not being able to flow with the condensed liquid refrigerant.

When noncondensables and water vapor combine with refrigerant, substances are produced that corrode copper piping and compressor parts. To measure the required vacuum during system evacuation, a thermocouple or electronic vacuum gauge that is calibrated in microns (an instrument that reads down to at least 50μ) must be used.

Refrigerant Charging

Most residential air conditioning systems today are charged with R-410a refrigerant, which operates at much higher pressures than R-22. The liquid-line dryer provided with the unit must be approved for use with R-410a. It must not be replaced with one designed for use with R-22 refrigerant. Many new units are not approved for use with coils, which use capillary tubes as a refrigerant metering device.

R-410a refrigerant cylinders are rose colored. Refrigerant should be drawn from the cylinder in a liquid state and admitted to the system through the suction valve in a vapor state. Certain R-410a refrigerant cylinders are identified as being equipped with a dip tube. **See Figure 40-13.** The dip tube allows liquid refrigerant to be drawn from the bottom of the cylinder without inverting the cylinder. *Note:* Do not turn this type of cylinder upside down to draw refrigerant.

System Noncondensables

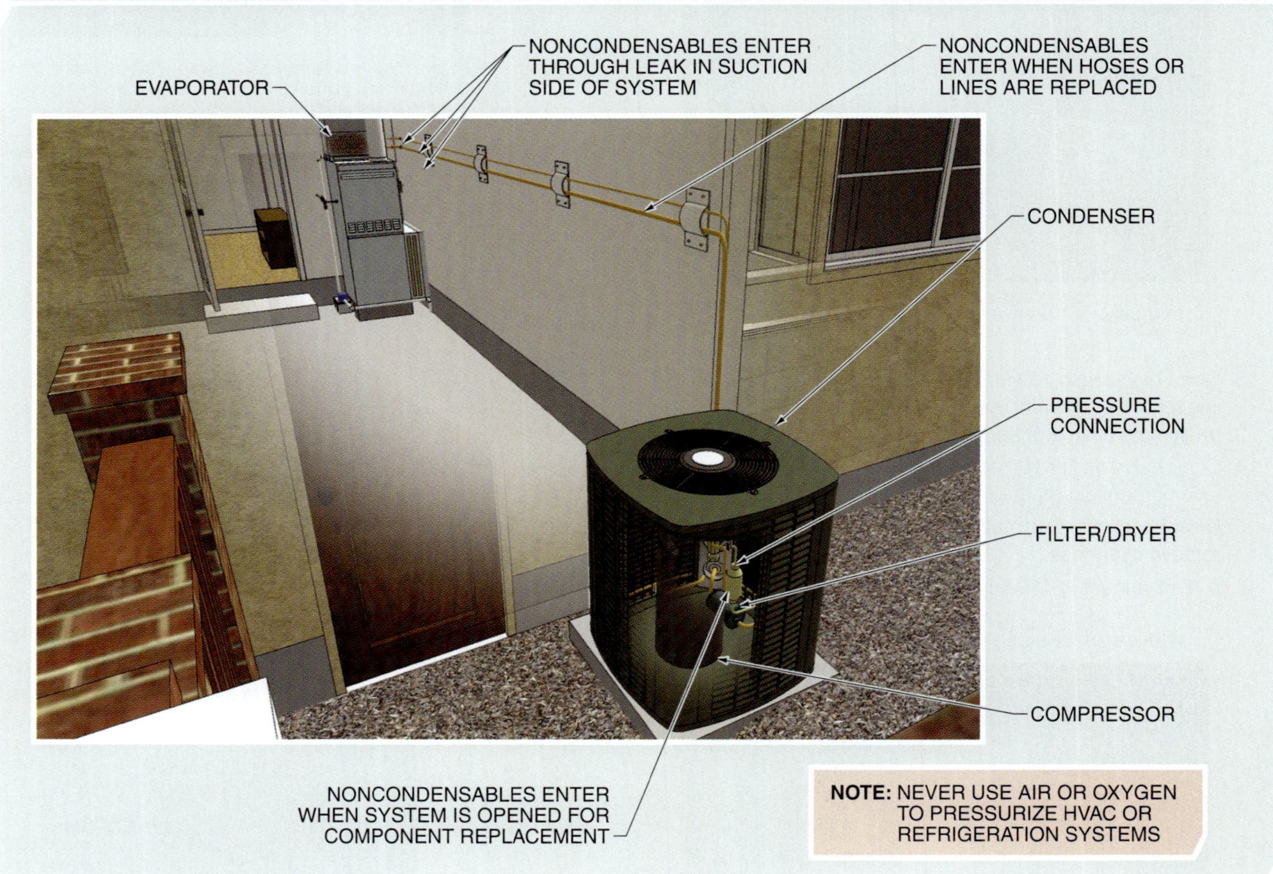

Figure 40-12. A noncondensable is any gas that will not condense under the temperatures and pressures present during the normal operation of an air conditioning system. When mixed with water vapor and combined with refrigerant, it produces substances that corrode copper piping and compressor parts.

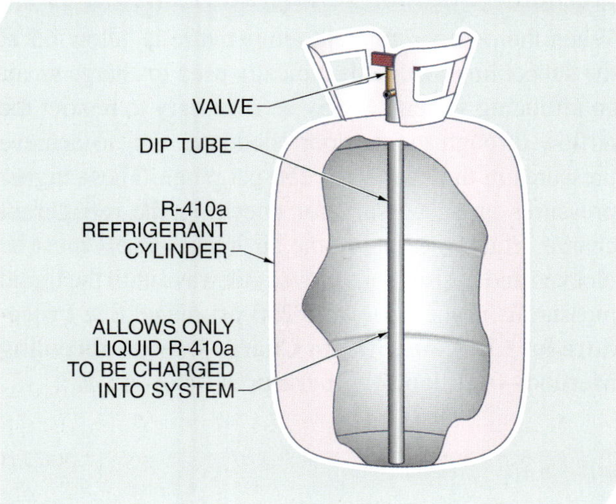

Figure 40-13. Some R-410a refrigerant cylinders contain a dip tube that allows liquid refrigerant to be taken from the cylinder without requiring the cylinder to be inverted.

TECH TIP

Most new condensing units installed today use R-410a refrigerant. The correct amount of R-410a is precharged into condensing units for the unit, evaporator coil, and a 25′ line set.

Units are factory charged with the amount of R-410a refrigerant indicated on the unit rating plate. This charge is based on a matching indoor coil and outdoor coil with 15′ of line set. A blank space is provided on the unit rating plate to list the actual field charge.

When void of refrigerant, an air conditioning system must be cleaned out. The air conditioning system is evacuated to clean out all of the moisture. Nitrogen is used to break the vacuum in the system. The system is evacuated again and then charged with R-410a refrigerant. **See Procedure for Removing Moisture from Systems.**

Removing Moisture from Systems

1. Use dry nitrogen to pressurize system and check for leaks. Do not add refrigerant unless leaks are repaired.

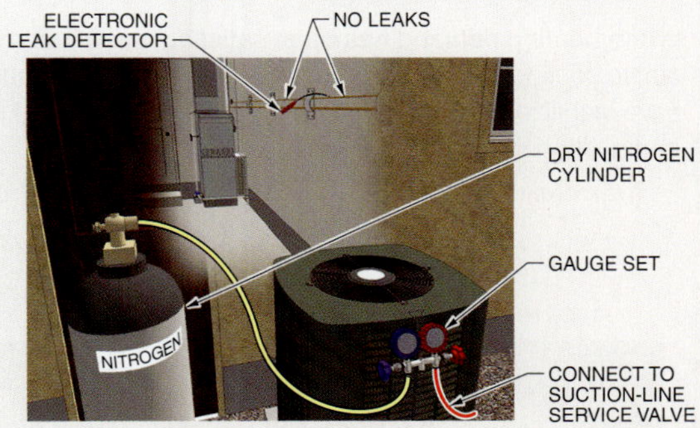

2. Evacuate system to remove as much moisture as possible.

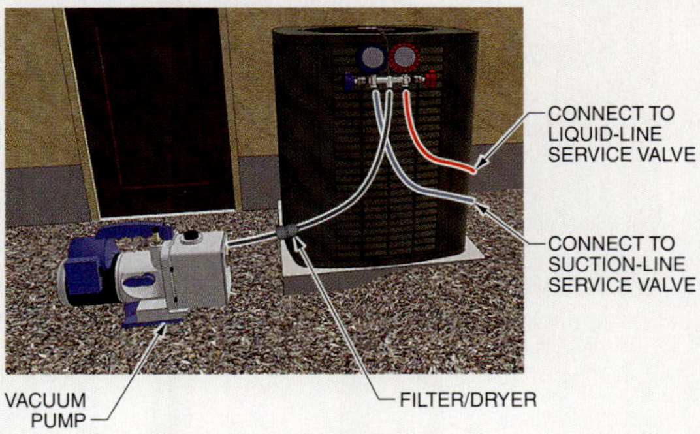

3. Use dry nitrogen to break the vacuum.
4. Evacuate system again.
5. Charge by weight appropriate amount of R-410a refrigerant (listed on unit nameplate) into system.

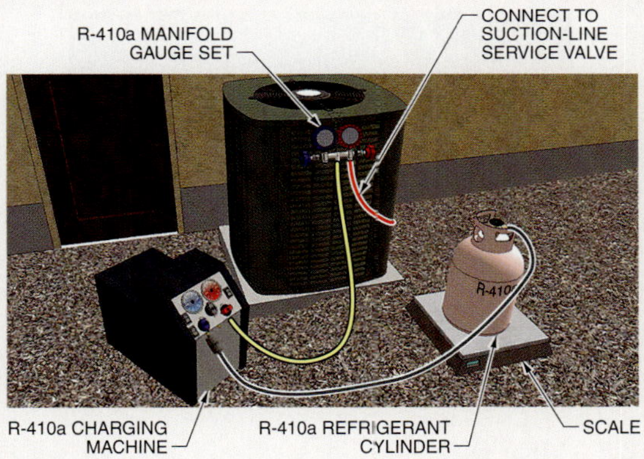

6. Monitor system to determine amount of moisture remaining in the oil. Use test kit to verify that moisture content is within dry color range of kit.

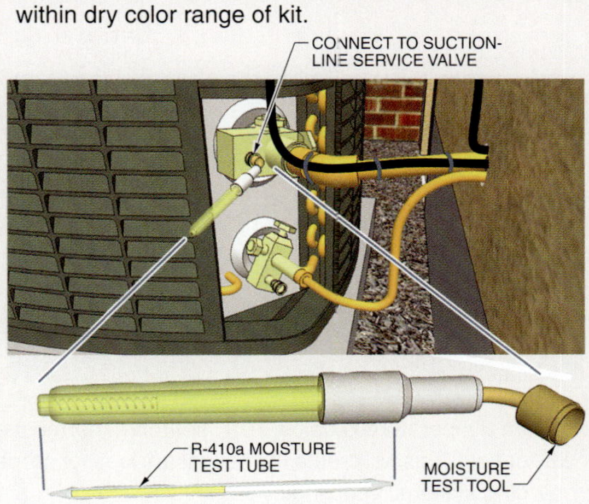

It may be necessary to replace the filter/dryer several times to achieve the required dryness level. When system dryness cannot be verified, the compressor has a greater chance of failing in the future.

The outdoor unit should be charged during warm weather. However, applications arise in which charging must occur in the colder months. The method of charging is determined by the unit's refrigerant metering device (fixed orifice, automatic, capillary tube, or thermostatic expansion valve) and the outdoor ambient temperature. **See Procedure for Measuring Line Set Temperatures and Outdoor Ambient Temperatures.**

Checking System Charge Using Subcooling Method—Outdoor Temperatures Below 65°F

When the outdoor ambient temperature is below 65°F, the subcooling method is typically used to charge an air conditioning system. It may be necessary to restrict the airflow through the outdoor condenser coil to achieve pressures in the 200 psi to 250 psi range. These higher pressures are necessary for checking the refrigerant charge. Equal sections of the air intake panels must be blocked and obstructions moved sideways until the liquid pressure is in the 200 psi to 250 psi range. **See Procedure for Checking System Charge Using Subcooling Method—Outdoor Temperatures Below 65°F.**

Procedure

Measuring Line Set Temperatures and Outdoor Ambient Temperatures

1. Ensure manifold gauge set valves are closed. Connect manifold gauge set to service valves access ports.
   a. Connect low-pressure gauge to suction-line service valve.
   b. Connect high-pressure gauge to liquid-line service valve.

2. Connect center-manifold gauge set hose to upright cylinder of R-410a.

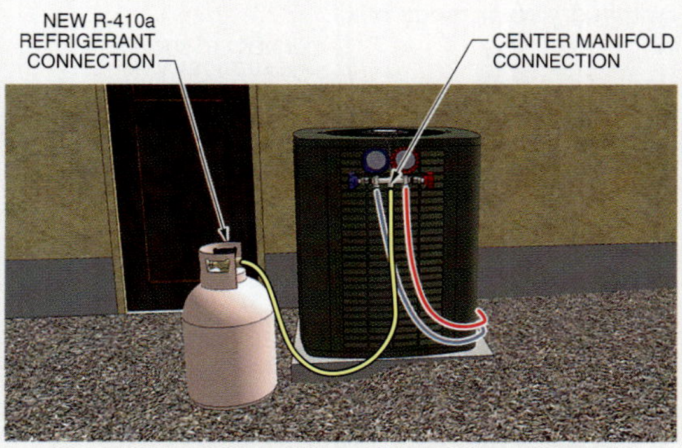

3. If room temperature is below 70°F, set room thermostat to call for heat. This will create the necessary load for properly charging system in cooling cycle.

4. Use digital thermometer to record outdoor ambient temperature.

5. When heating demand has been satisfied, switch thermostat to cooling mode with setpoint of 68°F. When pressures have stabilized, use digital thermometer to record liquid line and suction line temperatures.

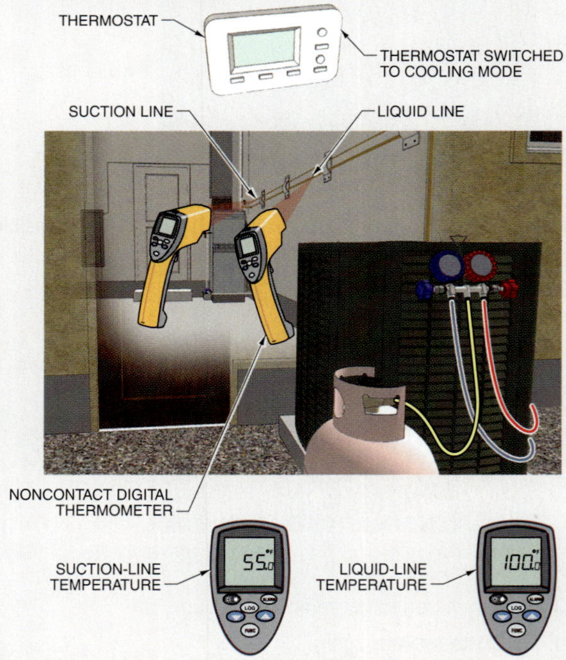

6. Outdoor temperature will determine appropriate charging method to use. Proceed with appropriate charging method.

Checking System Charge Using Subcooling Method—Outdoor Temperatures Below 65°F

1. With manifold gauge hose on liquid service port and air conditioning system operating stably, use digital thermometer to record liquid-line temperature.

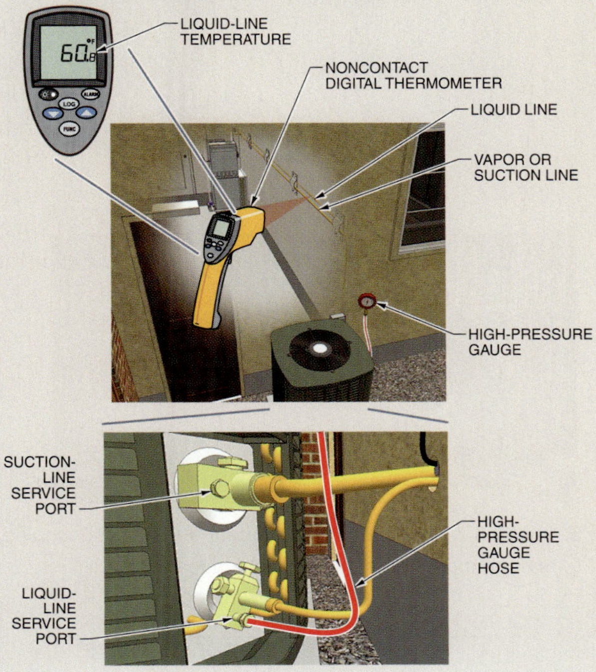

2. At same time, record liquid-line pressure reading (208.03 psi).
3. Use temperature/pressure chart for R-410a refrigerant to determine saturation temperature for liquid-line pressure reading.

Suva®410a Saturation Properties

Temp. °F	Pressure psia Liquid p_f	Volume ft³/lb Vapor v_g	Density lb/ft³ Liquid $1/v_f$	Enthalpy Btu/lb Liquid h_f
55	170.76	0.3505	69.99	34.0
56	173.52	0.3446	69.83	34.4
57	176.32	0.3388	69.67	34.8
58	179.15	0.3332	69.51	35.2
59	182.01	0.3276	69.35	35.6
60	184.91	0.3221	69.19	36.0
61	187.84	0.3168	69.03	36.3
62	190.81	0.3115	68.87	36.7
63	193.82	0.3064	68.70	37.1
64	196.86	0.3013	68.54	37.5
65	199.93	0.2963	68.37	37.9
66	203.04	0.2914	68.21	38.3
67	206.19	0.2866	68.04	38.7
68	209.38	0.2819	67.87	39.1

4. Subtract liquid-line temperature from saturation temperature to determine amount of subcooling.

SATURATION TEMPERATURE	67.6°F
LIQUID-LINE TEMPERATURE	– 60.8°F
SUBCOOLING	**6.8°F**

5. Compare subcooling value with those given by manufacturer. If subcooling is greater than shown, recover some refrigerant. If subcooling is less than shown, add some refrigerant.

Procedure

Checking System Charge Using Approach Method—Outdoor Temperatures 65°F or Above

The following procedure is intended as a general guide and is for use on expansion valve systems only. For best results, indoor temperature must be above 65°F, typically 70°F to 80°F. System pressures must always be monitored while charging. **See Procedure for Checking System Charge Using Approach Method—Outdoor Temperatures 65°F or Above.**

Checking System Charge Using Approach Method—Outdoor Temperatures 65°F or Above

1. Record outdoor ambient temperature using digital thermometer.

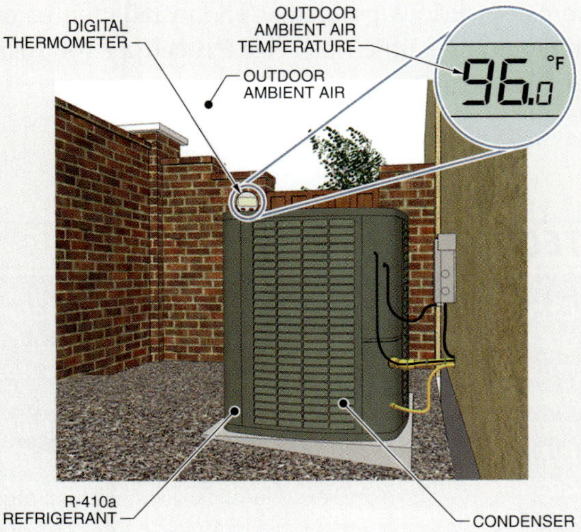

2. Attach high-pressure gauge set and operate unit for several minutes to allow system pressures to stabilize.

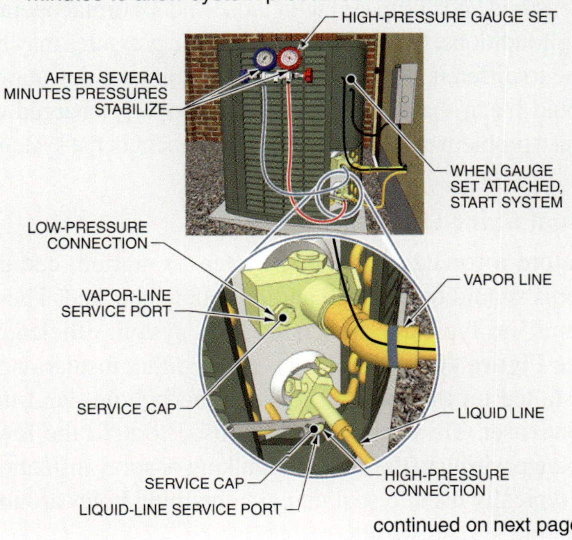

continued on next page

Procedure

Checking System Charge Using Approach Method—Outdoor Temperatures 65°F or Above *(continued)*

3. Compare stabilized pressures with those provided by manufacturer. Minor variations in these pressures may be expected due to differences in installations.

Suva®410a Saturation Properties

Temp. °F	Pressure psia	Volume ft³/lb	Density lb/ft³	Enthalpy Btu/lb
	Liquid p_f	Vapor v_g	Liquid $1/v_f$	Liquid h_f
90	289.29	0.1964	63.94	48.1
91	293.40	0.1932	63.74	48.5
92	297.55	0.1901	63.55	48.9
93	301.75	0.1870	63.36	49.4
94	305.99	0.1840	63.16	49.8
95	310.27	0.1810	62.96	50.2
96	314.60	0.1781	62.76	50.7
97	318.98	0.1752	62.56	51.1
98	323.40	0.1723	62.36	51.5
99	327.87	0.1695	62.16	52.0
123	449.68	0.1136	56.64	63.2
124	455.40	0.1116	56.37	63.7
125	461.18	0.1097	56.11	64.2
126	467.01	0.1078	55.83	64.8
127	472.89	0.1060	55.56	65.3
128	478.83	0.1041	55.28	65.8
129	484.83	0.1023	54.99	66.3
130	490.88	0.1005	54.70	66.9
131	496.98	0.0987	54.41	67.4
132	503.15	0.0970	54.11	68.0
133	509.37	0.0952	53.80	68.6
134	515.65	0.0935	[illegible]	69.1

NOTE: SIGNIFICANT DIFFERENCES COULD MEAN THAT THE SYSTEM IS NOT PROPERLY CHARGED OR THAT A PROBLEM EXISTS WITH SOME SYSTEM COMPONENT. PRESSURES HIGHER THAN THOSE LISTED INDICATE THAT THE SYSTEM IS UNDERCHARGED. VERIFY ADJUSTED CHARGE USING THE APPROACH METHOD.

4. Use noncontact digital thermometer to check temperature of liquid line.

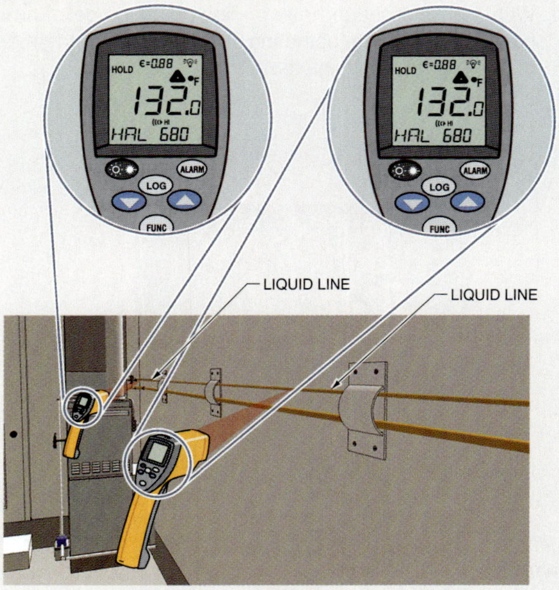

5. Subtract outdoor ambient temperature from subcooled temperature to determine approach temperature.

SUBCOOLED TEMPERATURE	132°F
OUTDOOR AMBIENT TEMPERATURE	− 96°F
APPROACH TEMPERATURE	**36°F**

6. Compare approach temperature with those provided by manufacturer. If values are different, add refrigerant to system to lower approach temperature or recover refrigerant from system to raise approach temperature.

Checking Charge Using Normal Operating Pressures

As part of system maintenance, system pressures should be measured and compared with those under normal operating conditions. Minor variations in these pressures may be due to differences in installations. Significant deviations could mean that the system is not properly charged or that a problem exists with some component in the system.

Final Start-Up Checks

Before a forced-air cooling system is started, certain items should be checked and/or double checked. These items are typically associated with system efficiency. **See Figure 40-14.** It must be ensured that insulation is installed on the line set between the building and the condenser. Tie wraps should be used to hold the low-voltage wiring to the line set. Caulking or spray insulation is typically used to seal up any remaining holes around the line set and wiring.

One task that must be completed before starting a forced-air cooling system is installing the condensate drain tubing. When the condensate drain tubing is not installed properly, the pan will overflow into the building and cause serious problems. The installation of drain tubing requires little time. **See Procedure for Installing Drain Tubing.**

TECH TIP

Excessive warranty costs have driven contractors out of business. For this reason, start-up procedures must be performed and documented on all new installations. It is the responsibility of the service manager to ensure that the start-up documentation is filled out legibly and stored in a safe location. The documentation should not be left in the service truck as it can be lost easily.

Final Start-Up Checks

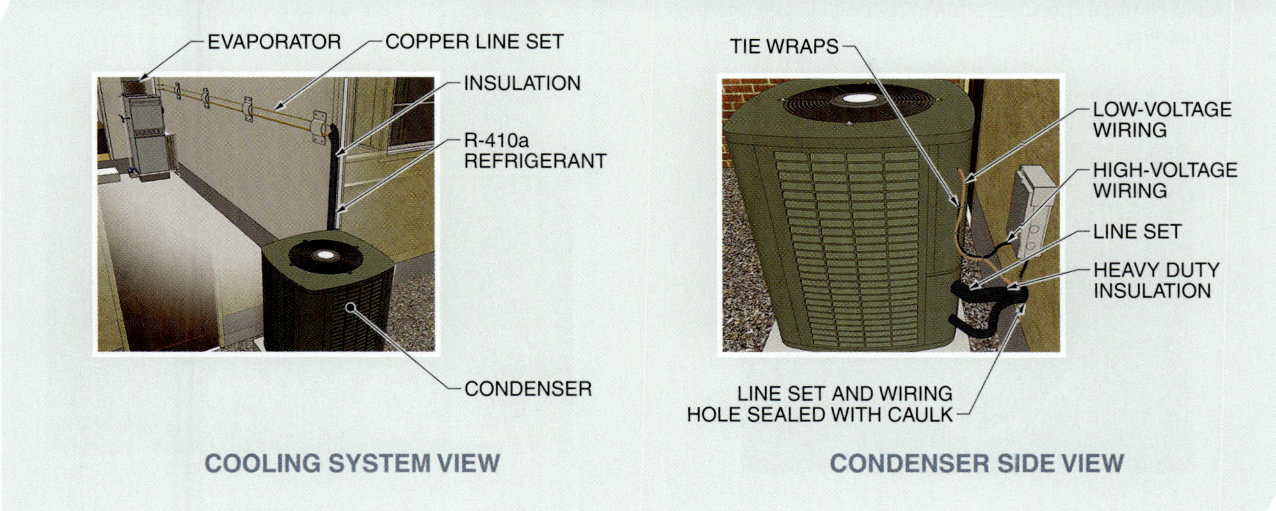

Figure 40-14. During final start-up checks it must be ensured that insulation is on the outside section of the line set, nylon tie wraps or tape is used to secure the low-voltage wiring to the line set, and caulk or the appropriate spray insulation is used to seal holes going through walls.

Installing Drain Tubing

1. Use plastic drain fitting to connect drain piping to evaporator coil condensation pan. This allows water that condenses from coil to be caught in condensate pan and then emptied into floor drain.
2. Screw connector into drain hole already available in condensation pan.
   a. Fit drain tubing up into connector. Ensure tubing fits snugly.

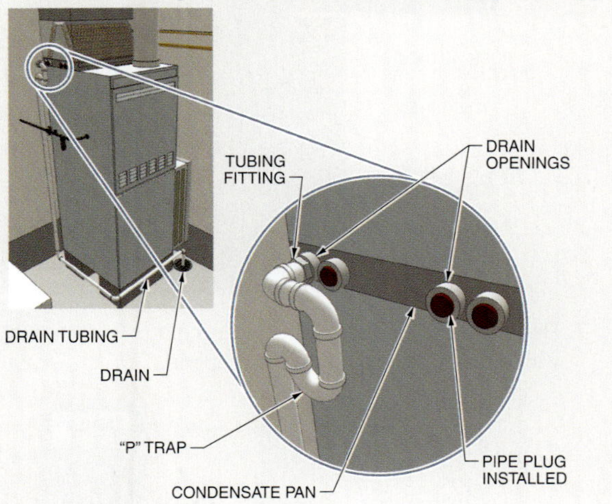

3. If floor drain is not readily available, a condensate pump can be used. When installed, the condensate pump allows water to be pumped up or down into an area that allows drainage.

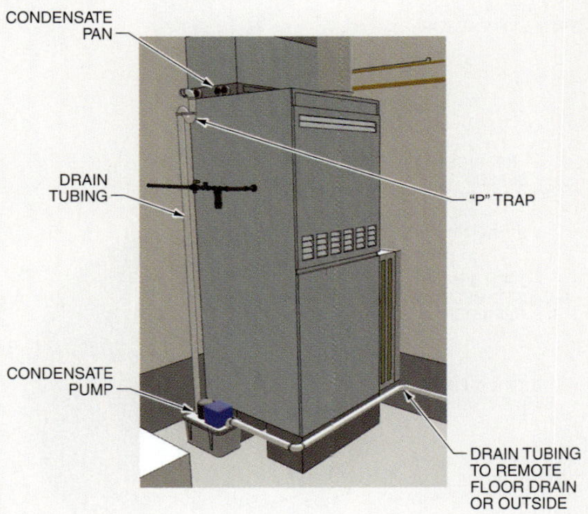

Procedure

Before the system can be turned over to the homeowner or customer and operated (when billing occurs), the system must be started and proven to operate properly. During a properly executed initial startup process, any installation and equipment problems are detected. Most manufacturers provide startup checklists for their equipment. The checklists must be kept for future reference. If equipment is not started properly, the warranty of the HVAC equipment may become void. **See Procedure for Starting Up Cooling Systems.**

Procedure

Starting Up Cooling Systems

1. Manually rotate condenser fan to check for frozen bearings or binding.

2. Inspect all factory and field-installed wiring for loose connections.
3. After evacuation is complete, open liquid-line and suction-line service valves to release refrigerant charge (contained in outdoor unit) into system (line set and evaporator).

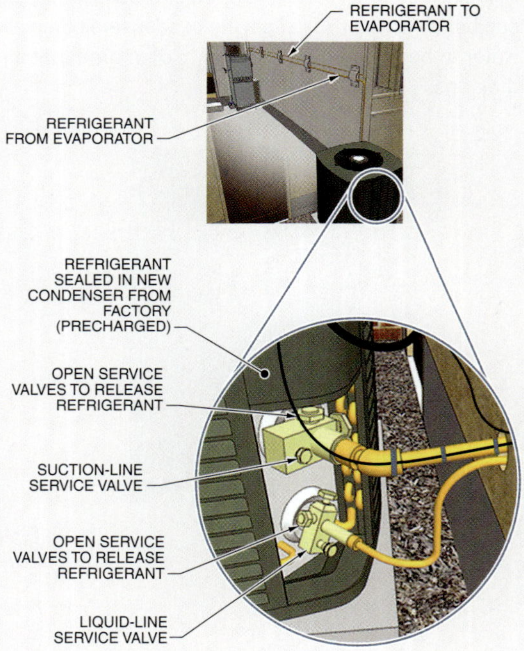

4. Replace stem caps and secure fingertight, then tighten an additional ⅙ of a turn.
5. Check voltage supply at disconnect switch. Voltage must be within range listed on unit nameplate. If not, do not start equipment until power company has been consulted and voltage condition is corrected.

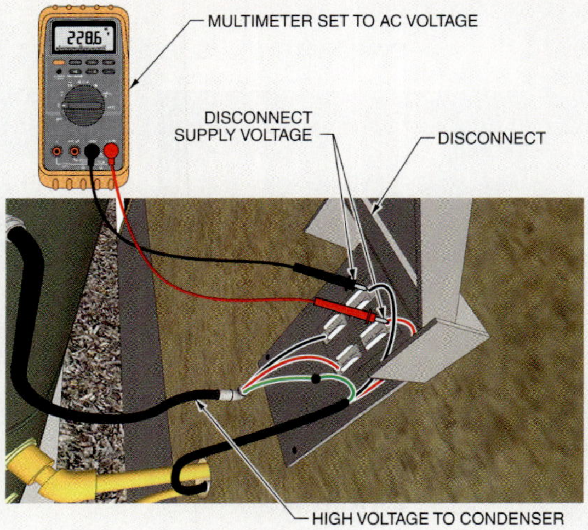

6. Set thermostat for cooling demand.
   a. Turn on power to indoor blower.
   b. Close door to outdoor unit disconnect switch to start unit.

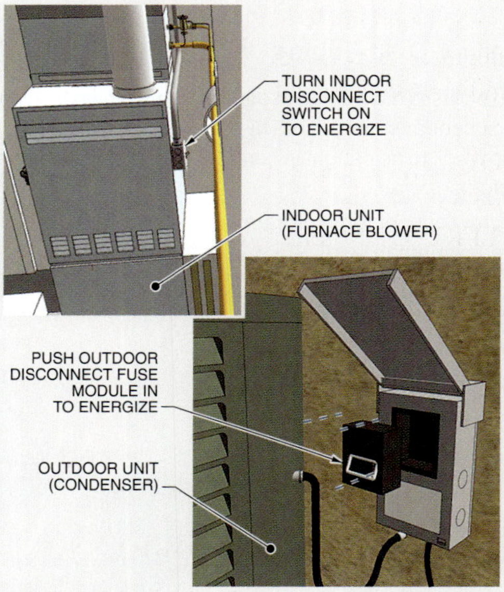

7. Recheck unit voltage with unit running. Power must be within range shown on unit nameplate.

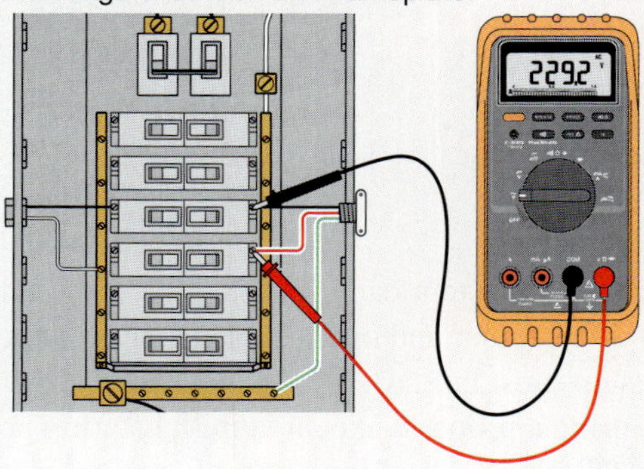

OPERATION

The outdoor unit and indoor blower cycle are on demand through the room thermostat. When the thermostat blower switch is moved to the ON position, the indoor blower operates continuously.

System Diagnostic Modules

Many modern air conditioning units contain a diagnostic module for troubleshooting air conditioning system failures. By monitoring and analyzing data from the compressor and thermostat demand, the module can accurately detect the cause of electrical and system related failure without any sensors. When a system problem occurs, a flashing light-emitting diode communicates the failure codes. The codes are indicated by a series of short and long flashes. **See Figure 40-15. See Appendix.**

Light-Emitting Diode Description

While each manufacturer uses a different diagnostic module, many have similar light-emitting diode (LED) fault indicators and functions. A few of these LED functions are as follows:

- "POWER" LED (green) indicates voltage is present at the power connection of the module.
- "ALERT" LED (yellow) communicates an abnormal system condition through a unique flash code. The pattern and number of consecutive flashes correlate to a particular abnormal condition. The pattern and fault are given in the manufacturer's literature.
- "TRIP" LED (red) typically indicates that the compressor protector is open or may indicate missing supply power to the compressor. This LED flashes to correspond with the troubleshooting chart.

24 VAC Power Wiring

The diagnostic module requires a constant nominal 24 VAC power supply. The wiring to the module's R and C terminals must be directly from the indoor unit or thermostat.

MAINTENANCE

Condensing unit maintenance and service must be performed by a qualified installer or service agency. At the beginning of each cooling season, apply the following preventive maintenance procedure:

1. Clean and inspect the condenser coil. The coil may be cleaned by using a cleaning agent and flushed with a water hose. Ensure that the power is OFF before using water to clean the coil.
2. The outdoor fan motor is prelubricated and sealed. No further lubrication is needed.
3. Visually inspect the connecting lines and coils for evidence of oil leaks.
4. Check the wiring for loose connections.
5. With the unit operating, check for correct voltage.
6. Check the amp-draw of the outdoor fan motor and compressor.

Note: When an owner complains of insufficient cooling, the unit should be gauged and the refrigerant charge should be checked.

Typical Light-Emitting Diode (LED) Cooling Failure Codes

Operation	Fault	Amber LED Flash Code	Possible Cause and Action
Standby — no call for unit operation	None	On solid, no flash	Normal operation — Three thermostat wires or 4-wire infinity control
Standby — no call for unit operation	None	Off	Normal operation — No call for cooling with 2-wire connection or indoor unit not powered
Cooling operation	None	1, pause	Normal operation
	System communications failure	16	Communication with user interface lost; check wiring to UI, indoor units, and outdoor units
	High-pressure switch open	31	High-pressure switch trip; check refrigerant charge, outdoor fan operation, and coils for airflow restrictions
	Low [illegible]	[illegible]	Low-pressure [illegible] refrigerant charge and [illegible]

Figure 40-15. The last pages of an operations manual typically include the failure code explanation document for the cooling system.

Indoor Coils

Indoor (evaporator) coils require periodic maintenance. Typically, indoor coil maintenance is not extensive because the coil is in a clean air stream and is protected from damage inside the supply plenum. Indoor coil periodic maintenance involves the following steps:

1. Clean the coil if necessary.
2. Check the connecting lines and coils for evidence of oil leaks.
3. Check the condensate pan line and clean if necessary.

TECH TIP

New aluminum condenser coils have smaller channels than older coils and require different cleaning materials.

Indoor Units

Maintenance of the indoor unit is typically performed at the same time as the indoor coil and condensing unit. Proper maintenance is required for the system to provide comfort and operate at peak efficiency. To maintain an indoor unit, apply the following procedure:

1. Clean or change the filters.
2. Adjust the blower speed for cooling. Measure the pressure drop over the coil to determine the correct blower CFM.
3. Check the belt of the belt drive blowers for wear and proper tension.
4. Check all wiring for loose connections.
5. With the unit operating, check for correct voltage.
6. Check the amp-draw of the blower motor.

Digital Resources

ATPeResources.com/QuickLinks • Access Code: 482610

UNIT 41

Servicing and Troubleshooting Residential Forced-Air HVAC Systems

Forced-air systems comprise the vast majority of residential HVAC systems in use today. Because there are so many forced-air systems in use, HVAC technicians must be familiar with a variety of systems. Depending on location and climate, technicians can spend most of their time working on forced-air systems one after the other. There are an enormous number of forced-air systems in existence, each with its own small differences. The majority of service calls are the result of the simplest problems, such as power being turned off accidently.

OBJECTIVES

- Identify and explain the appropriate safety guidelines when troubleshooting.
- Describe the courtesy guidelines for service calls.
- Discuss the basic questions to ask when troubleshooting a forced-air system.
- Identify the manuals that should be used when servicing and troubleshooting forced-air HVAC systems.
- Explain the typical furnace sequence of operation.
- List the most common troubleshooting problems for each type of residential forced-air HVAC system.

SERVICING FORCED-AIR SYSTEMS

As with all HVAC mechanical equipment, any time a technician is performing troubleshooting, proper PPE must be worn, a safe work environment must be maintained, and safety practices must be followed. **See Figure 41-1.** It only takes a few seconds of carelessness to cause an injury.

When troubleshooting an HVAC forced-air system, there are so many variations it is difficult for a technician to know every answer. Since there are such a wide variety of manufacturers and types of HVAC equipment, troubleshooting and other technical installation manuals from manufacturers must be consulted, and any safety references, hazards, and safety warnings must be carefully noted. **See Figure 41-2.** Many of these manuals can be found online. When a technician runs across a system type that they have never worked on before, extra caution is warranted.

Safety issues arise when troubleshooting an HVAC system. These issues involve the electrical system, fuel system, and mechanical system. The electrical system is of concern during troubleshooting because 115 V line voltage is present at various components in a furnace, along with 24 V control voltage. The fuel system is an issue because when mishandled, any natural gas or fossil fuel can explode or at least burn quickly. The mechanical system is of concern because of moving and rotating parts that start and stop. Technicians must be familiar with how to work with all of these systems safely when troubleshooting.

Troubleshooting and PPE

- PROPERLY CARE FOR ALL PPE ASSIGNED TO YOU
- KEEP ALL PPE CLEAN AND IN GOOD REPAIR
- INSPECT PPE BEFORE EACH USE
- IMMEDIATELY REPORT TO YOUR SUPERVISOR ANY PPE THAT BECOMES DEFECTIVE OR NO LONGER PROVIDES THE NECESSARY PROTECTION
- DO NOT MAKE ANY MODIFICATIONS OR ALTERATIONS TO PPE THAT WOULD VOID ITS WARRANTY OR COMPROMISE ITS FUNCTION, SUCH AS PAINTING HARDHATS OR COATS OR REMOVING SIDE SHIELDS FROM EYEWEAR

OSHA Office of Training and Education

Figure 41-1. Modifications, alterations, or changes, such as painting, should never be made to any PPE.

Manufacturer Installation Instructions Safety References

⚠ WARNING
Installation, adjustment, alteration, service or maintenance can cause property damage, personal injury or loss of life. Installation and service must be performed by a qualified installer or service agency.

⚠ IMPORTANT
The Clean Air Act of 1990 bans the intentional venting of refrigerant (CFC's and HCFC's) as of July 1, 1992. Approved methods of recovery, recycling or reclaiming must be followed. Fine and/or incarceration may be levied for noncompliance.

⚠ WARNING
This product and/or the indoor unit it is matched with may contain fiberglass wool.
Disturbing the insulation during installation, maintenance, or repair will expose you to fiberglass wool dust. Breathing this may cause lung cancer. (Fiberglass wool is known to the State of California to cause cancer.)
Fiberglass wool may also cause respiratory, skin, and eye irritation.
To reduce exposure to this substance or for further information, consult material safety data sheets.

Page 1

⚠ CAUTION
In order to avoid injury, take proper precaution when lifting heavy objects.

⚠ CAUTION
Sharp sheet metal edges can cause injury. When installing the unit, avoid accidental contact with sharp edges.

⚠ WARNING
Unit must be grounded in accordance with national and local codes. Electric shock hazard. Can cause injury or death.

Page 3

⚠ IMPORTANT
Failure to remove the fixed orifice when installing an expansion valve on the indoor coil will result in improper operation and damage to the system.

Page 7

⚠ WARNING
Danger of fire. Bleeding the refrigerant charge from only the high side may result in the low side shell and suction tubing being pressurized. Application of a brazing torch while pressurized may result in ignition of the refrigerant and oil mixture - check the high and low pressures before unbrazing.

Page 10

⚠ WARNING
Refrigerant can be harmful if it is inhaled. Refrigerant must be used and recovered responsibly.
Failure to follow this warning may result in personal injury or death.

⚠ WARNING
Danger of explosion: Can cause equipment damage, injury or death. Never use oxygen to pressurize a refrigeration or air conditioning system. Oxygen will explode on contact with oil and could cause personal injury.

⚠ WARNING
Danger of explosion: Can cause equipment damage, injury or death. When using a high pressure gas such as dry nitrogen to pressurize a refrigeration or air conditioning system, use a regulator that can control the pressure down to 1 or 2 psig (6.9 to 13.8 kPa).

Page 10

⚠ IMPORTANT
Use a thermocouple or thermistor electronic vacuum gauge that is calibrated in microns. Use an instrument that reads from 50 microns to at least 10,000 microns.

Page 12

⚠ CAUTION
Danger of equipment damage.
Avoid deep vacuum operation. Do not use compressors to evacuate a system.
Extremely low vacuums can cause internal arcing and compressor failure.
Damage caused by deep vacuum operation will void warranty.

⚠ IMPORTANT
If unit is equipped with crankcase heater, it should be energized 24 hours before unit start-up to prevent compressor damage as a result of slugging.

Page 12

⚠ IMPORTANT
The Clean Air Act of 1990 bans the intentional venting of refrigerant (CFC's and HCFC's) as of July 1,1992. Approved methods of recovery, recycling or reclaiming must be followed. Fine and/or incarceration may be levied for noncompliance.

Page 13

⚠ WARNING
Electric shock hazard. Can cause injury or death. Before attempting to perform any service or maintenance, turn the electrical power to unit OFF at disconnect switch(es). Unit may have multiple power supplies.

Page 16

16-PAGE INSTALLATION INSTRUCTION DOCUMENT

Figure 41-2. All safety references, hazards, and safety warnings must be understood before beginning installation, servicing, or troubleshooting residential and light commercial forced-air systems.

Courtesy Guidelines for Effective Forced-Air System Service Calls

Before looking at troubleshooting forced-air HVAC systems, basic common courtesy rules must be practiced. Good customer service creates a positive and professional relationship with the customer. Service technicians must make a good first impression and always act professionally. The choice a customer makes of one HVAC service contractor over another may be decided on how well the customer was treated the last time.

Because HVAC technicians who work on heating and cooling systems typically interact with the customer on a one-on-one basis, good customer service skills are essential. A successful customer service call includes some basic rules that, when followed, help keep the customer happy. **See Figure 41-3.**

- Call ahead: Make sure to call the customer while driving to their location to ensure they are home or access to the home has been arranged.
- Park where allowed: Do not block anyone's driveway. Park on the street whenever possible.
- Dress and groom neatly. Smile when you are with a customer.
- Do not smoke in the customer's home or on their property.
- Be honest with the customer. Make sure the customer is told everything that was discovered during the check.
- If you must leave to pick up parts or come back the next day, let the customer know when to expect you back. Do not leave panels, tools, or equipment at the customer's home overnight.
- Do not track dirt or bring HVAC system contaminants into the customer's home. Wear foot coverings when necessary.
- Do not use a customer's bathroom or facilities.
- Remember it costs five times more in marketing dollars to obtain a new customer than it takes to keep an existing one.
- Step outside to make personal calls.
- Maintain a positive attitude.

Dress and Groom Neatly. A service technician should always maintain a good personal appearance. Service technicians work in many different locations. Many HVAC contractors furnish uniforms for their employees, creating a professional image. The uniform must always be clean and neat. Photo identification badges are commonly issued to service technicians. A photo identification badge should always be worn when working on a service call.

Courtesy Guidelines for Effective Service Calls

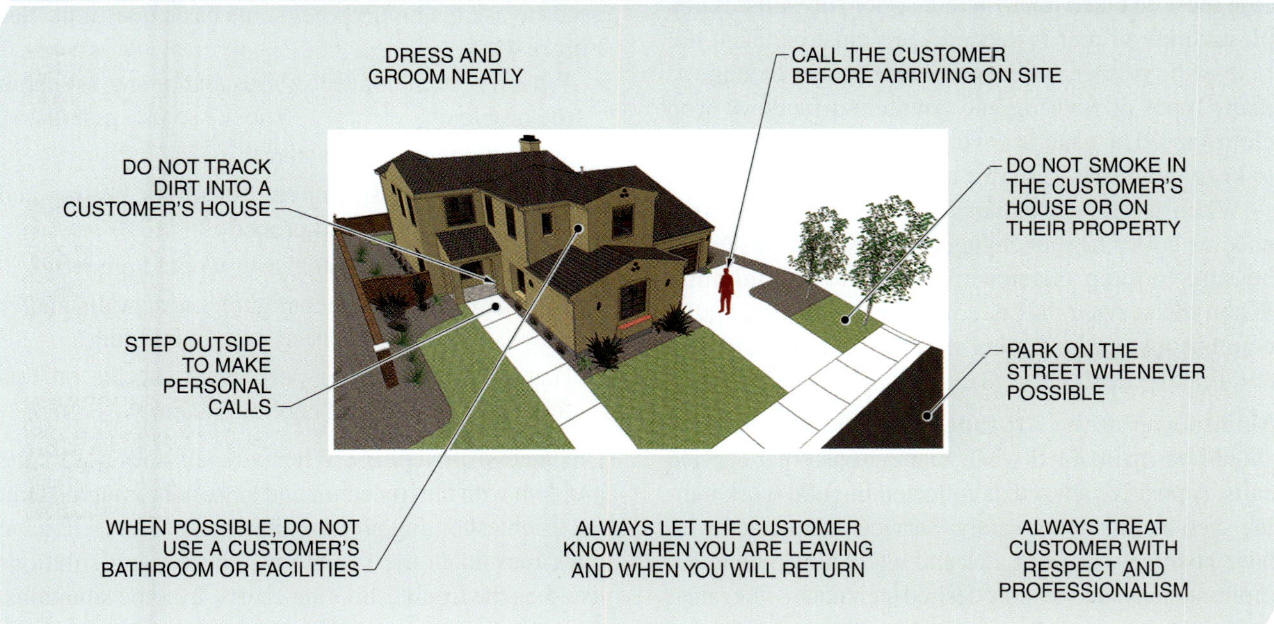

Figure 41-3. It is important to let the customer know they are valued by giving good customer service when making service calls.

Work with the Customer. The customer must be treated with respect and a professional approach. A service technician should always answer a customer completely and honestly. Technicians must never criticize customers for attempting to perform their own HVAC repairs.

When repairing an item, care must be taken not to damage the item any more than it is already damaged. If a service technician breaks an item in need of repair and it cannot be properly repaired, a problem arises. The technician or contractor can lose time and money if the item needs to be replaced.

The efforts of the service technician should be focused on equipment problems. The technician should take clear notes and identify for the customer the specific steps to be taken and the time frame required to repair individual system components and get the system back in service.

Pick Up Parts or Come Back the Next Day. When a technician must leave a customer's home before a call is complete, the technician should let the customer know the reason (a run for parts or the end of the workday) and the time of return. It does not matter if the absence is for 15 minutes, 15 hours (overnight), or for a day. The customer should always know the plan and how the plan is to be worked.

Do Not Track Dirt into Customer's Home. Service technicians must be careful when working anywhere in a customer's home. When carrying toolboxes, refrigerant cylinders, equipment parts, or filters into a building, care must be taken not to scratch doors or walls. When plastic pipe primer is needed, care must be taken not to drip the primer, which can cause severe damage to many types of flooring and counter surfaces. A drop cloth should be used to cover floors and other surfaces to keep them free from dirt and damage.

When a service call involves cleaning the furnace or boiler heat exchanger, the technician should be sure to keep cleaner spray use to a minimum. When the service call is complete, all floors, walls, countertops, and cabinets should be cleaned of dirt and fingerprints.

Maintain a Positive Attitude. A positive work attitude should be maintained when making customer service calls. A positive attitude is reflected in good work habits, such as practicing safety. Service technicians who have positive work attitudes and who know how to act professionally are the most desired technicians. Negative attitudes result in performance errors that cost time and money, as well as lost customers.

Interpersonal Skills

An *interpersonal skill* is a character trait that allows an individual to communicate effectively with others in a variety of situations. Forced-air heating and cooling equipment failure is costly and frustrating at times for the homeowner, and this frustration can be directed at the service technician.

When the customer is upset, the service technician must stay calm and listen carefully to the customer. The service technician must work with the information provided by the customer and any acquired troubleshooting information to find a solution. Clear communication of the facts helps eliminate misunderstandings and can reduce the troubleshooting time for the technician.

Advancement in HVAC Field Service

A successful forced-air heating and cooling service technician practices lifelong learning. When a skill has been mastered, higher levels of skill mastery are required for newly designed equipment. The ability to adapt as the HVAC industry changes is crucial for success. Reading trade publications, attending classes and seminars, and participating in professional organizations are activities that provide valuable information on current topics and trends. New skills are required to remain current with advancing heating and cooling technologies and to grow in the HVAC field.

Basic Troubleshooting Questions

Before troubleshooting an HVAC system, the technician needs to ask the homeowner some basic questions. **See Figure 41-4.**

- What has been noticed (noises and/or smells) about the problem?
- When did the problem start?
- When was the last time the forced-air heating and cooling system worked properly?
- Has the forced-air system ever worked properly?
- Have any changes (reconfiguration of walls and/or zones) been made to the system or building?
- Has anyone else ever performed service on this forced-air heating and cooling system?

Unsuccessful Repairs. When a repair does not fix the problem with the forced-air and air conditioning system, the troubleshooting process must begin. The technician must reevaluate the symptoms, causes, and solutions listed on the troubleshooting charts. In some situations, the entire process of gathering information about the system must begin again.

Basic Troubleshooting Questions

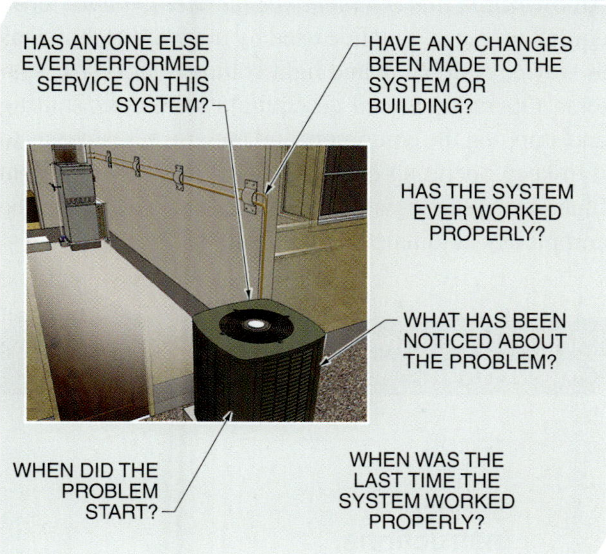

Figure 41-4. Before troubleshooting a problem in a residential system, the technician should ask the homeowner certain specific questions about the system.

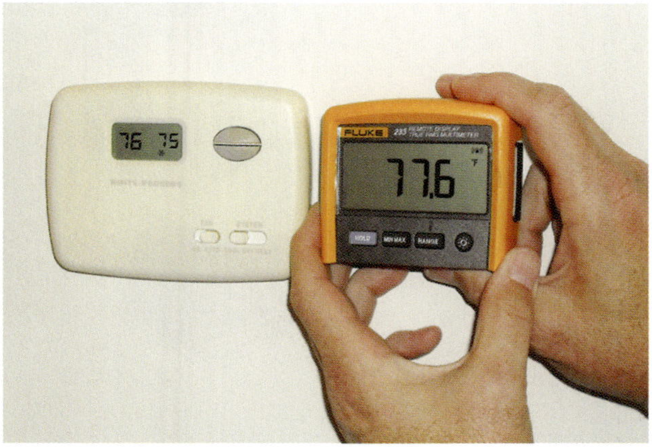

Some modern DMMs have a wireless display capability that can be used when checking control system accuracy.

Successful Repairs. When a repair is successful and the forced-air equipment operation has been restored, the troubleshooting process is recorded on a troubleshooting report or equipment maintenance log. Everything is recorded in a maintenance log including symptoms, causes, tests conducted, repairs made, and the condition of the forced-air equipment after the repair. **See Figure 41-5. See Appendix.** The spare parts used in the repair are recorded and replacement parts are ordered as needed. Suggestions on modifying procedures that could prevent similar problems are also recorded.

Troubleshooting Reports and Maintenance Logs

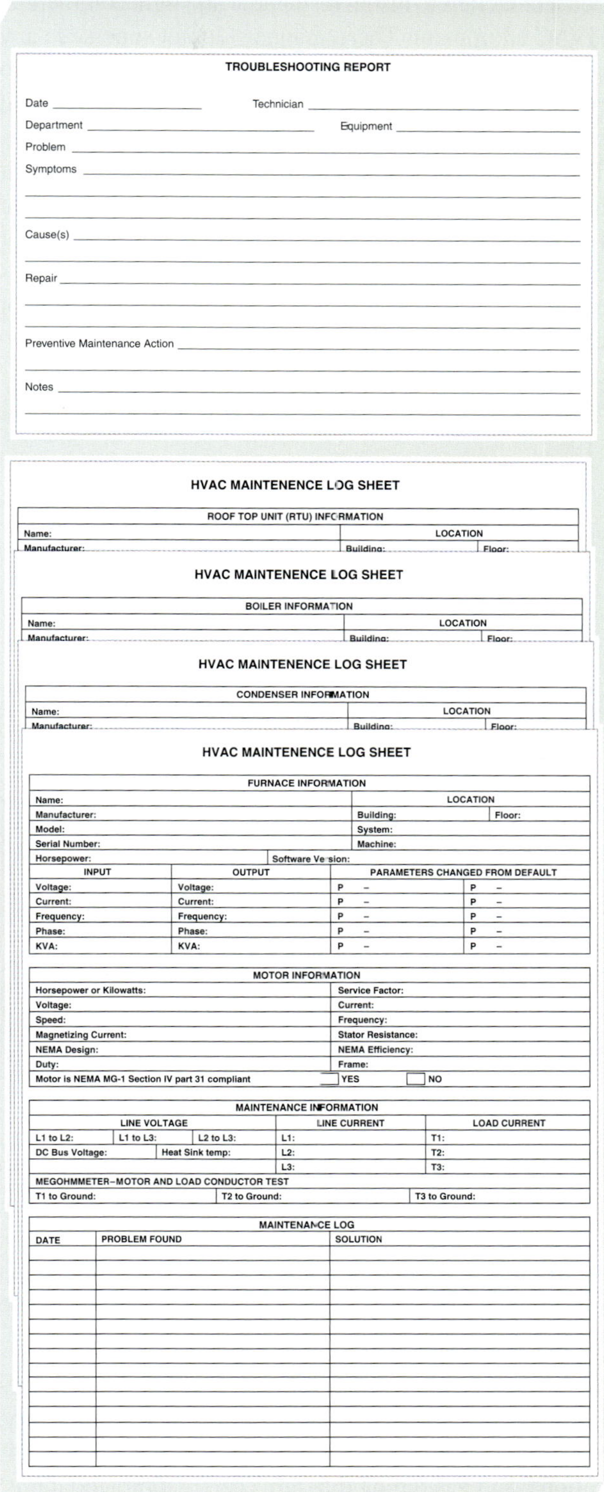

TROUBLESHOOTING REPORT

Date ______ Technician ______
Department ______ Equipment ______
Problem ______
Symptoms ______
Cause(s) ______
Repair ______
Preventive Maintenance Action ______
Notes ______

HVAC MAINTENENCE LOG SHEET

ROOF TOP UNIT (RTU) INFORMATION — Name: | LOCATION | Manufacturer: | Building: | Floor:

HVAC MAINTENENCE LOG SHEET

BOILER INFORMATION — Name: | LOCATION | Manufacturer: | Building: | Floor:

HVAC MAINTENENCE LOG SHEET

CONDENSER INFORMATION — Name: | LOCATION | Manufacturer: | Building: | Floor:

HVAC MAINTENENCE LOG SHEET

FURNACE INFORMATION			
Name:		LOCATION	
Manufacturer:		Building:	Floor:
Model:		System:	
Serial Number:		Machine:	
Horsepower:	Software Version:		
INPUT	OUTPUT	PARAMETERS CHANGED FROM DEFAULT	
Voltage:	Voltage:	P –	P –
Current:	Current:	P –	P –
Frequency:	Frequency:	P –	P –
Phase:	Phase:	P –	P –
KVA:	KVA:	P –	P –

MOTOR INFORMATION	
Horsepower or Kilowatts:	Service Factor:
Voltage:	Current:
Speed:	Frequency:
Magnetizing Current:	Stator Resistance:
NEMA Design:	NEMA Efficiency:
Duty:	Frame:
Motor is NEMA MG-1 Section IV part 31 compliant	☐ YES ☐ NO

MAINTENANCE INFORMATION				
LINE VOLTAGE			LINE CURRENT	LOAD CURRENT
L1 to L2:	L1 to L3:	L2 to L3:	L1:	T1:
DC Bus Voltage:	Heat Sink temp:		L2:	T2:
			L3:	T3:
MEGOHMMETER–MOTOR AND LOAD CONDUCTOR TEST				
T1 to Ground:	T2 to Ground:	T3 to Ground:		

MAINTENANCE LOG		
DATE	PROBLEM FOUND	SOLUTION

Figure 41-5. All troubleshooting and service information must be recorded in troubleshooting reports and/or maintenance logs for future reference.

Installation Manuals

An *installation manual* is a general guide used by qualified technicians to install residential and light commercial HVAC equipment. These manuals do not supersede any national or local codes. When installation manuals do not arrive with delivered equipment, the manuals should be obtained from the manufacturer as soon as possible. The installation of the furnace and forced-air equipment described in the manual must comply with all state and local codes as well as national codes such as the NEC® or CEC™. Failure to comply with manufacturer procedures can void warranties and reduce the energy efficiency of the equipment. **See Figure 41-6.**

Installation Manuals

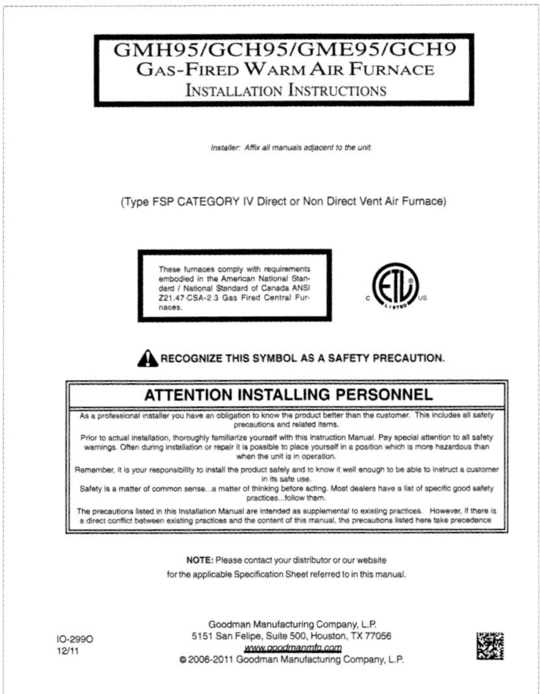
GMH95/GCH95/GME95/GCH9
GAS-FIRED WARM AIR FURNACE
INSTALLATION INSTRUCTIONS

Installer: Affix all manuals adjacent to the unit.

(Type FSP CATEGORY IV Direct or Non Direct Vent Air Furnace)

These furnaces comply with requirements embodied in the American National Standard / National Standard of Canada ANSI Z21.47·CSA-2.3 Gas Fired Central Furnaces.

RECOGNIZE THIS SYMBOL AS A SAFETY PRECAUTION.

ATTENTION INSTALLING PERSONNEL

As a professional installer you have an obligation to know the product better than the customer. This includes all safety precautions and related items.

Prior to actual installation, thoroughly familiarize yourself with this Instruction Manual. Pay special attention to all safety warnings. Often during installation or repair it is possible to place yourself in a position which is more hazardous than when the unit is in operation.

Remember, it is your responsibility to install the product safely and to know it well enough to be able to instruct a customer in its safe use.

Safety is a matter of common sense...a matter of thinking before acting. Most dealers have a list of specific good safety practices...follow them.

The precautions listed in this Installation Manual are intended as supplemental to existing practices. However, if there is a direct conflict between existing practices and the content of this manual, the precautions listed here take precedence

NOTE: Please contact your distributor or our website for the applicable Specification Sheet referred to in this manual.

IO-299O
12/11

Goodman Manufacturing Company, L.P.
5151 San Felipe, Suite 500, Houston, TX 77056
© 2006-2011 Goodman Manufacturing Company, L.P.

Goodman Global Group, Inc.

Figure 41-6. Installation manuals contain all information required to install forced-air equipment but do not supersede national and local codes.

Installation manuals for furnaces typically contain sections on safety, how to inspect the furnace upon delivery, where to locate the furnace, wiring diagrams and specifications, and gas piping diagrams and recommendations. They also include cleaning and/or flushing procedures, charging procedures and charts, procedures for setting up valves and adjusting sensors, and start-up procedures.

Operation and Maintenance Manuals

An *operation and maintenance (service) manual* is an equipment-specific guide used by qualified technicians to service residential and light commercial HVAC systems. Operating forced-air equipment involves starting and stopping the equipment and making adjustments to it to keep operating correctly. Many forced-air system functions are automated, but some functions cannot be completely automated. **See Figure 41-7.**

Operation and Maintenance Manuals

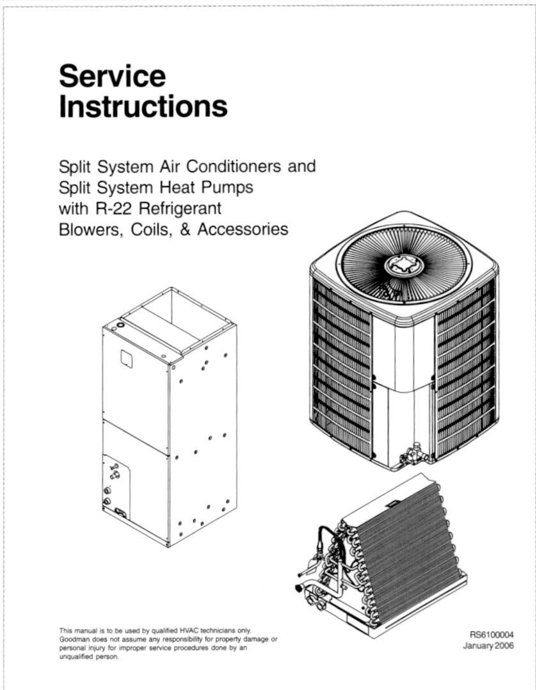
Service Instructions

Split System Air Conditioners and
Split System Heat Pumps
with R-22 Refrigerant
Blowers, Coils, & Accessories

This manual is to be used by qualified HVAC technicians only. Goodman does not assume any responsibility for property damage or personal injury for improper service procedures done by an unqualified person.

RS6100004
January 2006

Goodman Global Group, Inc.

Figure 41-7. Operation and maintenance manuals contain all the information needed to operate specific forced-air equipment in different applications, as well as the preventive maintenance information needed to keep the equipment running for years.

Operation and maintenance manuals for forced-air equipment typically contain safety information, sequence of operations, adjustment procedures for specific applications, troubleshooting matrixes, flash code charts, preventive maintenance procedures, performance checklists, wiring schematics, piping schematics, and warranty information.

Flash Code Charts. Furnaces typically have a self-diagnosing control module with an LED that flashes red to let technicians know when the furnace is having a problem. The fault indicator LED is behind the blower access door's observation window on the furnace. The number of flashes the control module displays is the error code for the problem the furnace is encountering. The technician counts the number of flashes and checks the error code against the flash code chart. Once the problem is identified, the troubleshooting section of the chart can be used to solve the problem. **See Figure 41-8. See Appendix.**

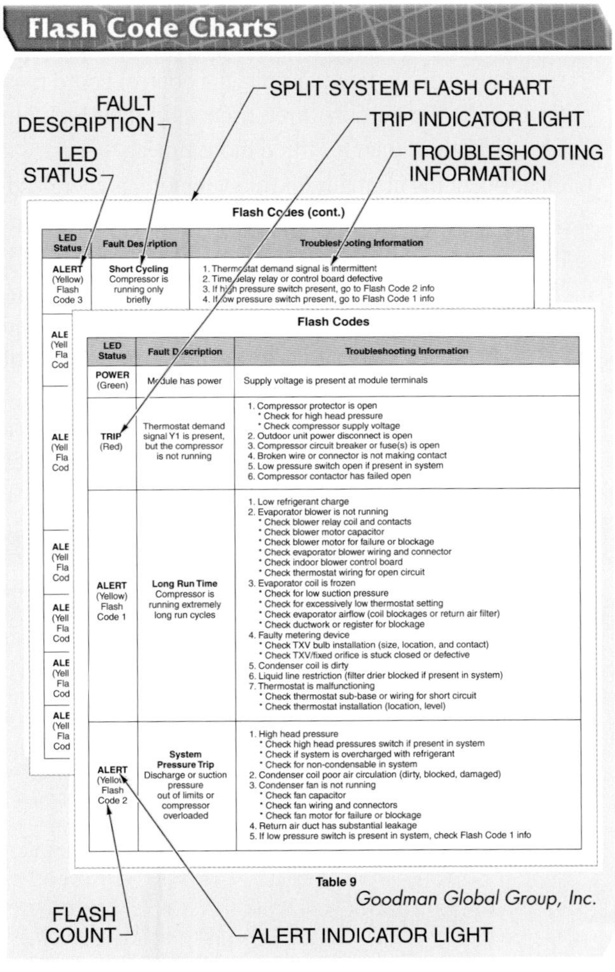

Figure 41-8. Flashing LEDs on furnaces or split systems convey fault information to technicians, who check the flash count against the flash code chart.

Troubleshooting Charts. Troubleshooting charts are only intended to be used on furnaces serviced by technicians who have successfully completed the manufacturer training. Troubleshooting charts must only be used in conjunction with the appropriate furnace installation or operation manual, which is also provided by the furnace manufacturer. These charts are not intended to be a complete list of all the possible failures a furnace may experience. There are many reasons why a furnace might not be working. However, many of the solutions can be found on furnace troubleshooting charts. **See Figure 41-9. See Appendix.**

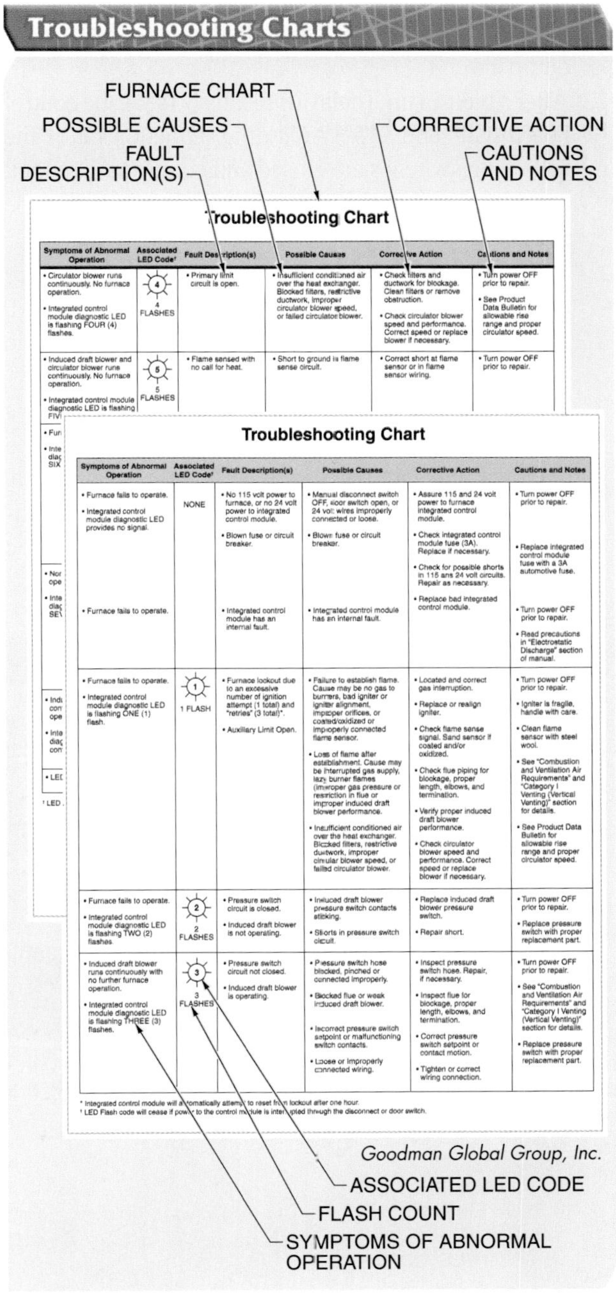

Figure 41-9. Troubleshooting charts list the most common faults that a furnace encounters and also list the standard corrective actions to take to eliminate the fault.

Typical Furnace Sequence of Operations

Furnaces are designed with multiple safety features. These safety features prevent the fire from escaping the firebox, and the buildup of fuel (natural gas, fuel oil, or propane) on the burners. The buildup of fuel can cause an explosion or allow flue gases to escape into the room. The sequence of operation is a valuable tool for diagnosing furnace problems. **See Figure 41-10.**

1. Upon a call for heat, the control turns on the appropriate damper actuators. Typically these consist of zone dampers.
2. After a built-in time delay (typically 30 sec), the control board will send 115 VAC to the induction fan if the following switches are closed and conditions are met:
   a. If a gas pressure safety switch is present, it must be closed.
   b. The pressure switch attached to the flue must be proved during induction fan operation. Once proved, the furnace will then provide 24 VAC to its enable relay for burner ignition.
   c. In general, any other limit switches such as a heat exchanger high-temperature limit switch must be closed before the board will energize the induction fan.
3. If present, the control board will energize the combustion air blower and then start a 10 sec to 30 sec prepurge cycle.
4. Once the prepurge cycle is complete and all limits are closed, the control board will attempt to start the burners by beginning a 5 sec to 10 sec trial for ignition. Starting the burner flame is accomplished by sending a voltage to the hot surface igniter and then opening the gas valve. Some furnaces have standing pilot lights.
5. If the control board does not detect a flame (flame detector sensor) by the end of the trial for ignition, the control closes the gas valve and performs a 10 sec to 30 sec postpurge, then another prepurge, and then it will try to light the burners again. Most furnaces will try to ignite the burners a set number of times, typically three to five times, and then go into lockout. The lockout may need to be manually reset or it may be subject to a timer. For example, the control board may try to light the burners three times and then lock out for a half hour, then try three more times.
6. Once operating normally, the blower motor is energized.
7. The burners will then operate until the setpoint temperature is reached or a safety limit opens.
8. When the system call for heat is satisfied, the control board will turn the burners off. The blower and induction fan will operate on a postpurge cycle of 10 sec to 30 sec.
9. The dampers will typically continue to operate for a specific time delay of 30 sec to 120 sec, then turn off.

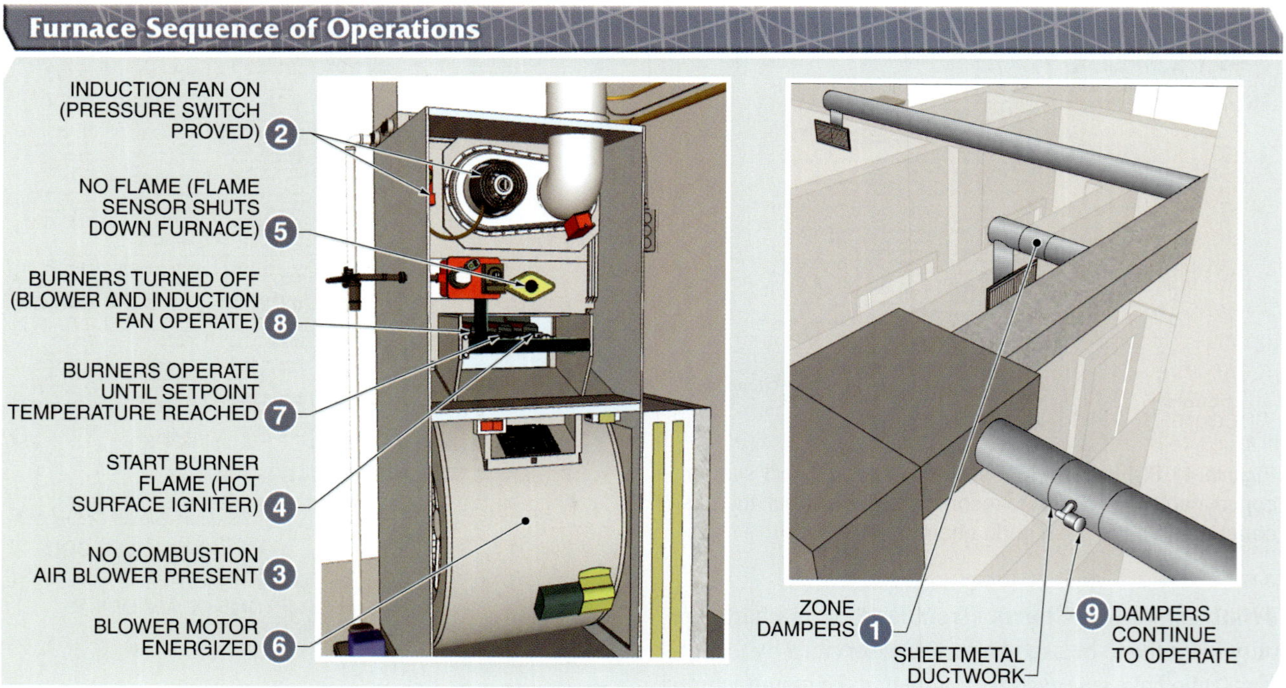

Figure 41-10. A sequence of operations indicates when safety features are playing a role in the operation of the furnace and contains valuable information for troubleshooting.

TROUBLESHOOTING FORCED-AIR HVAC SYSTEMS

The ability to troubleshoot effectively is a skill that combines technical expertise and logical and creative thought processes. Troubleshooting is a skill that a technician becomes better at through experience with equipment failures. Also, when the causes of HVAC equipment failures can be determined, modifying the PM work for equipment may prevent future failures.

The majority of forced-air heating systems in the US utilize natural gas. However, some forced-air heating systems use fuel oil and others use electric heat. Since each of these systems has its own component arrangements, it also has troubleshooting scenarios specifically designed for it.

Forced-air cooling systems and humidifiers have their own component arrangements, and they also have troubleshooting scenarios specifically designed for them. Filters have their own troubleshooting scenarios that deal with filters and racks.

TECH TIP

HVAC system troubleshooting is complicated by the wide range of systems in use. A technician is expected to understand the operation of new systems and systems that were installed 50 years ago. HVAC technicians must adopt a continuing education philosophy of learning from experienced technicians and from equipment manufacturer training classes.

Natural Gas Troubleshooting Scenario #1 — Bad Thermocouple

A technician receives a complaint from a residential customer that the customer's house is cold. The customer says that their furnace quit working earlier in the morning. The technician finds that the unit has a standing pilot with a thermocouple. The technician notices that the gas valve is turned on, but the pilot is not lit. The technician holds the button on the gas valve down to open the gas valve to the pilot position while manually lighting the pilot. The pilot flame is good blue flame that covers the thermocouple properly. After the technician holds the button in for a minute or so, the button is released and the pilot flame goes out. **See Figure 41-11.**

Service Work Performed: The thermocouple is replaced and the furnace operates properly.

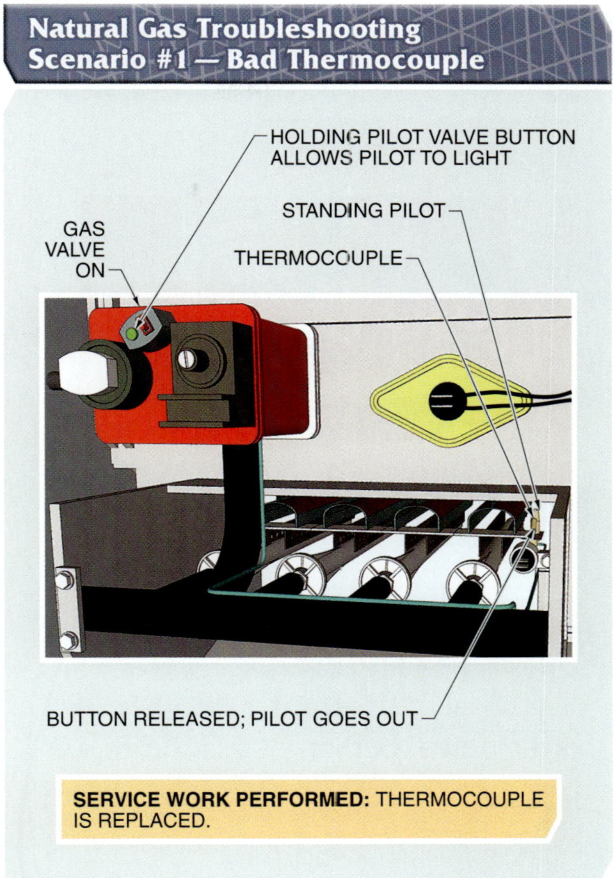

Figure 41-11. Thermocouples must be replaced with thermocouples of the same voltage and length.

Natural Gas Troubleshooting Scenario #2 — Bad Transformer

The customer calls saying their townhome is too cold. Upon arrival, the technician checks the unit and finds that there is no burner flame. The fan will not run even with the baseplate switch for the fan in the ON mode. The technician checks the control voltage and finds that there is no 24 VAC present. The line voltage is checked with a DMM and found to be fine. After taking the required steps to shut off power to the system, the secondary side of the step-down control transformer is checked with an ohmmeter and is found to be open. The reason for the failed transformer must be identified. **See Figure 41-12.**

Service Work Performed: A new transformer is installed and the control (secondary side) voltage is checked and found to be 24 VAC. *Note:* In some situations, a shorted gas valve can be the cause of an overheated and failed transformer.

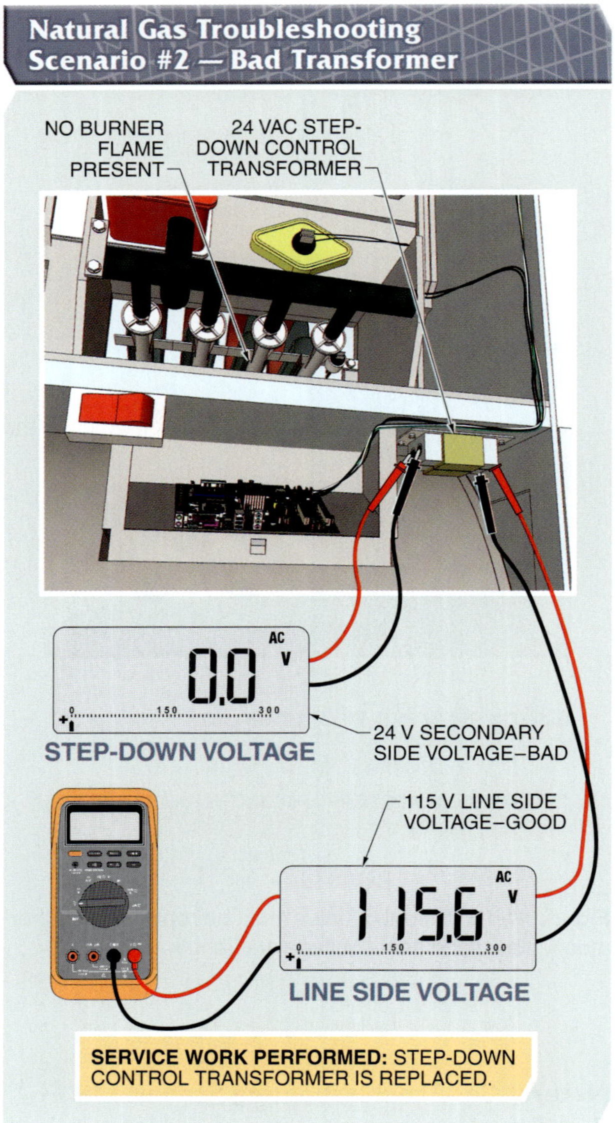

Figure 41-12. A transformer with the same KVA rating must be used to replace a bad step-down control transformer.

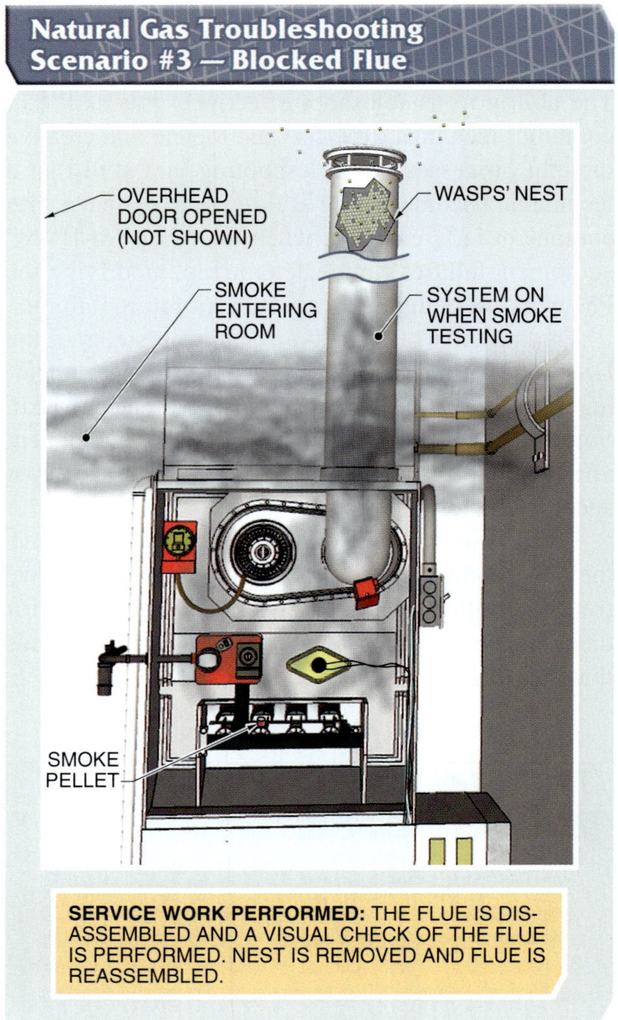

Figure 41-13. After cleaning out the flue, any openings or screw holes must be filled in with heat-resistant caulking or covered with metal heat tape.

Natural Gas Troubleshooting Scenario #3—Blocked Flue

A customer calls in complaining about fumes in the utility area where the furnace is located. The technician responds as quickly as possible because of the safety issue. The technician also calls and tells the homeowner to open a window or door and shut off the heat at the thermostat. Upon arrival, the technician starts the furnace and checks the unit. A quick check with smoke pellets indicates that the by-products of combustion are flowing out into the room instead of up the flue. The system is then shut down. **See Figure 41-13.**

Service Work Performed: The flue is disassembled and a visual check of the flue is performed. The inspection indicates that the flue is blocked due to a wasps' nest. The nest is removed, the flue reassembled, and the system restarted. The same smoke test as before is performed and indicates that the by-products of combustion are now moving up through the flue.

Natural Gas Troubleshooting Scenario #4—Pilot Will Not Stay Lit

A technician responds to a "too cold" call. Upon arrival, the technician notices that the pilot flame is out. The technician relights the pilot, but after the fan starts, the pilot flame is blown out. The system is shut down, and a visual check indicates that there is a hole in the heat exchanger. The hole is permitting air from the outside to enter the combustion chamber and blow out the pilot flame. **See Figure 41-14.**

Service Work Performed: The homeowner has the choice of having a new heat exchanger or a brand new furnace installed.

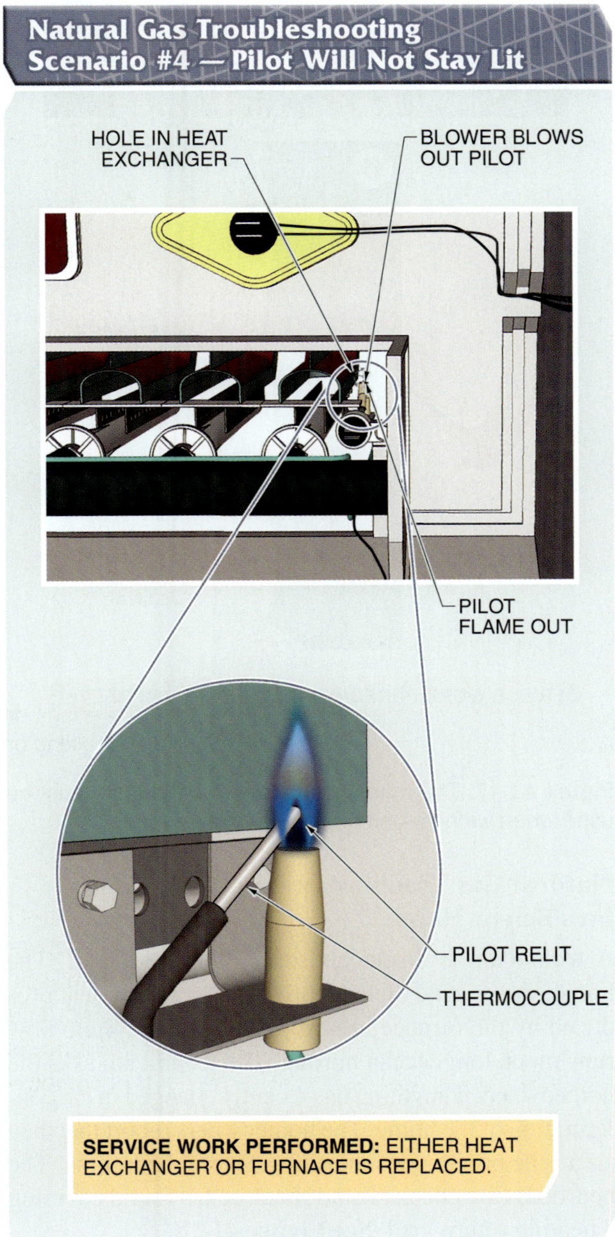

Figure 41-14. Typically, it is more economical and quicker to replace an entire furnace than to replace a bad heat exchanger in the furnace.

Natural Gas Troubleshooting Scenario #5— Blower Not Running

A homeowner contacts an HVAC contractor because of a nonfunctioning furnace. The technician arrives and finds that the furnace blower is not running. This is causing the gas valve to cycle on high limit because the heat is not being removed. **See Figure 41-15.** *Note:* There are a number of things that may cause a blower not to operate properly:

- The blower limit switch may not be set or working properly. This results in the blower coming on at the wrong temperatures or not coming on at all. The blower limit switch setpoint should be checked and the operation of the switch checked with a voltmeter. In either case, the switch must be adjusted or changed as needed.
- The blower relay is connected to the blower switch of the thermostat as well as the heating circuit. It is possible that the blower relay has failed or is malfunctioning. This is not allowing the blower to come on properly. The relay should be checked in operation and then powered down and checked with a DMM set for resistance.

Service Work Performed: The bad relay is replaced.

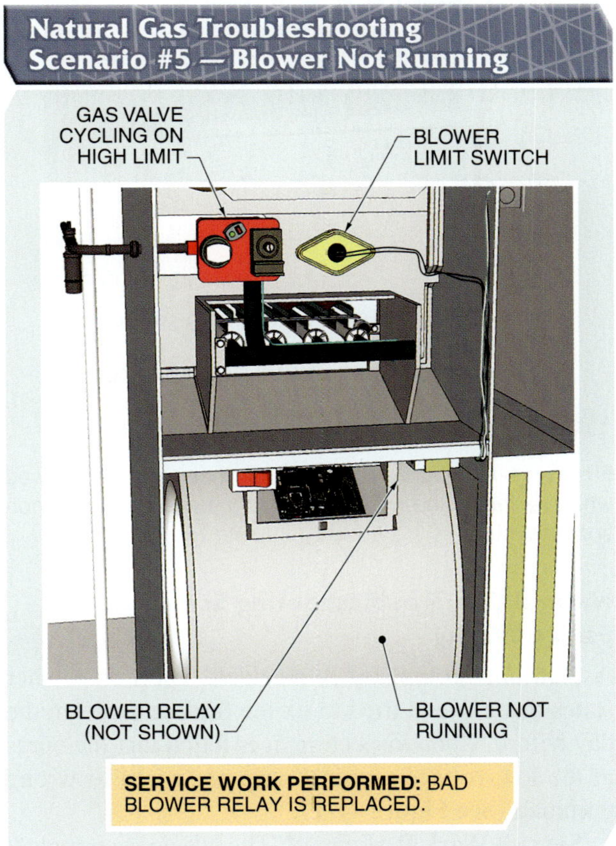

Figure 41-15. The use of a good dependable DMM is critical to the adjustment of a blower limit switch.

Natural Gas Troubleshooting Scenario #6—Defective Circuit Board

A cold complaint is called in by a homeowner with a newer high-efficiency furnace. The technician notices that the fault indicator light on the control board is flashing a fault code. The manufacturer literature is checked, and it is found that the flashing code indicates that the control circuit board has a nonrecoverable failure. The cover of the furnace is removed and the control board is inspected. **See Figure 41-16.**

Service Work Performed: The control circuit board is replaced and the system is checked for normal operation. It is found to be working correctly.

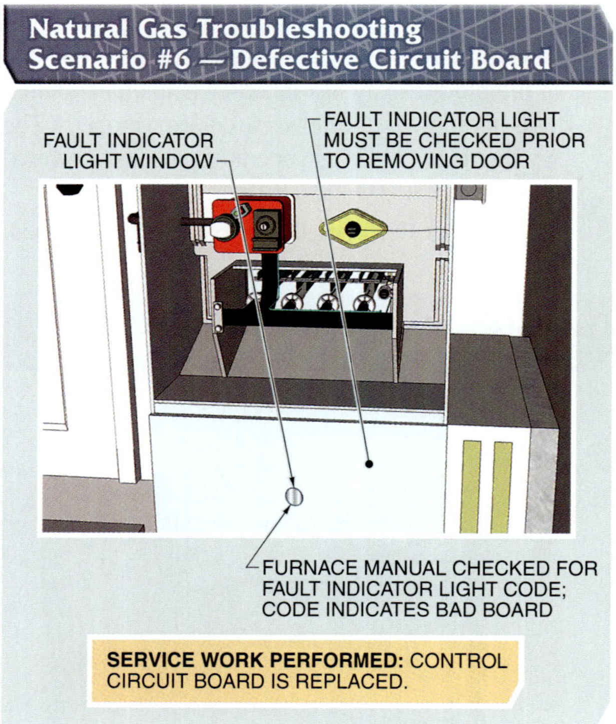

Figure 41-16. Static electricity guidelines must be followed when replacing the circuit board of a furnace or air conditioning system.

Natural Gas Troubleshooting Scenario #7—Faulty Wiring

A "too hot" complaint is received. The homeowner states that a friend tried to fix the furnace problem the day before. Upon inspection, it is found that the wires at the low-voltage terminal strip are wired to wrong terminals. **See Figure 41-17.**

Service Work Performed: The wiring is corrected, and the system is started up and checked. It is found to be working properly.

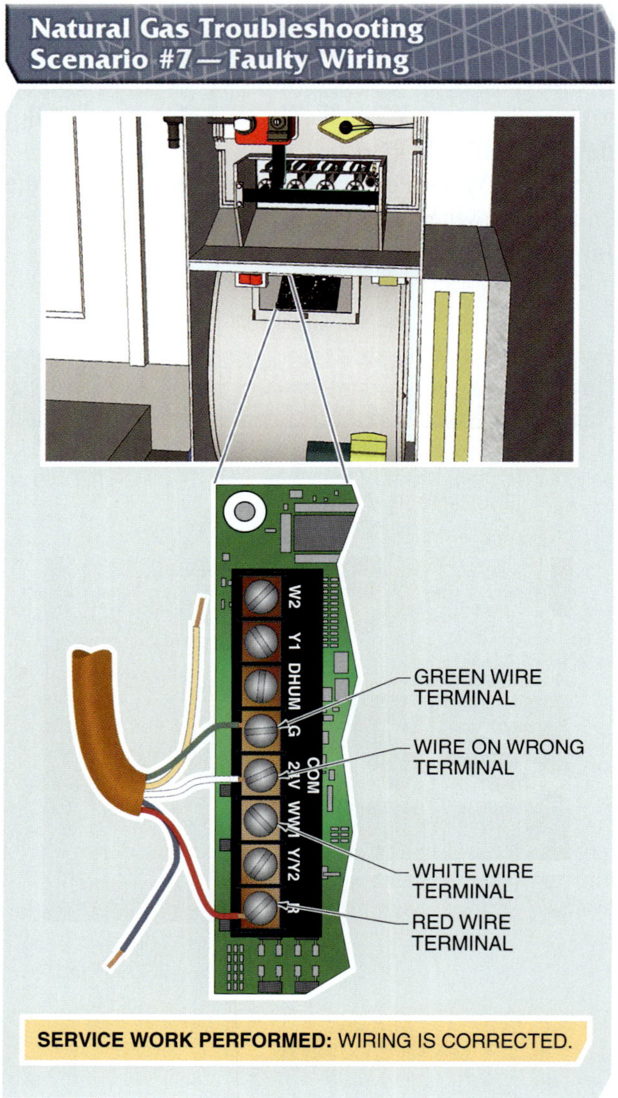

Figure 41-17. The manufacturer's wiring diagram must be used when wiring a unit during service work.

Natural Gas Troubleshooting Scenario #8—Insufficient Heat

A technician is dispatched on a "too cold" call. The homeowners state that recently the amount of heat produced by the furnace has dropped and the system now runs much longer than normal. The technician asks the homeowners if anything has recently changed in the configuration of the home. The homeowners report that they are in the process of renovating several bedrooms. The bedrooms are checked, and sheetrock is found covering a heating return grill. **See Figure 41-18.**

Service Work Performed: The sheetrock is removed, and adequate airflow is restored to the return grill. The system operates normally.

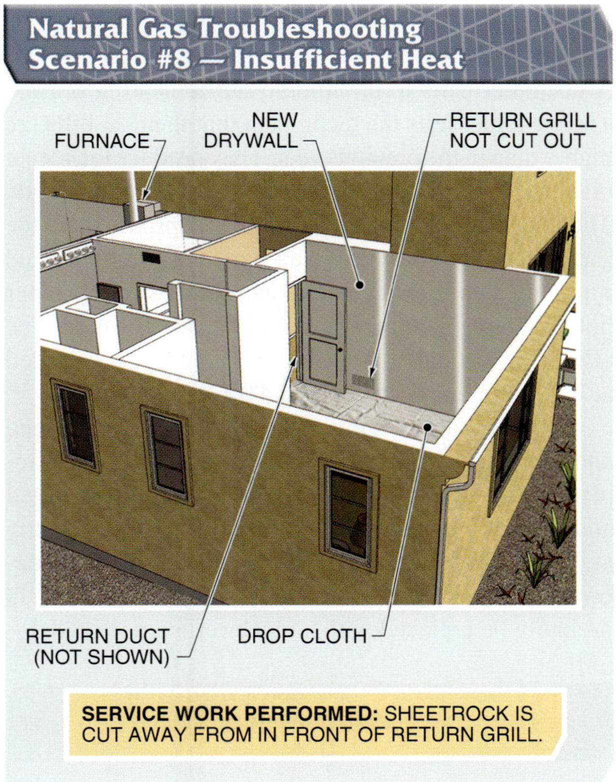

Figure 41-18. When performing service work or other work on an HVAC system, the technician should make sure that rags, wood, and other construction supplies do not interfere with system operation.

Natural Gas Troubleshooting Scenario #9—Cycles Too Much

A house sitter complains that the furnace has recently been switching on and off too much. It is discovered upon inspection that the anticipator on the thermostat was not adjusted properly and the unit is overcycling. **See Figure 41-19.**

Service Work Performed: The amperage of the heating circuit is measured with a clamp-on digital ammeter. This measured value is used to adjust the anticipator setting on the thermostat properly. The unit is placed back into operation and cycles correctly.

TECH TIP

Many HVAC system problems are caused by homeowners. Controls can be purchased and installed improperly by homeowners who have little technical knowledge, leading to improper system operation.

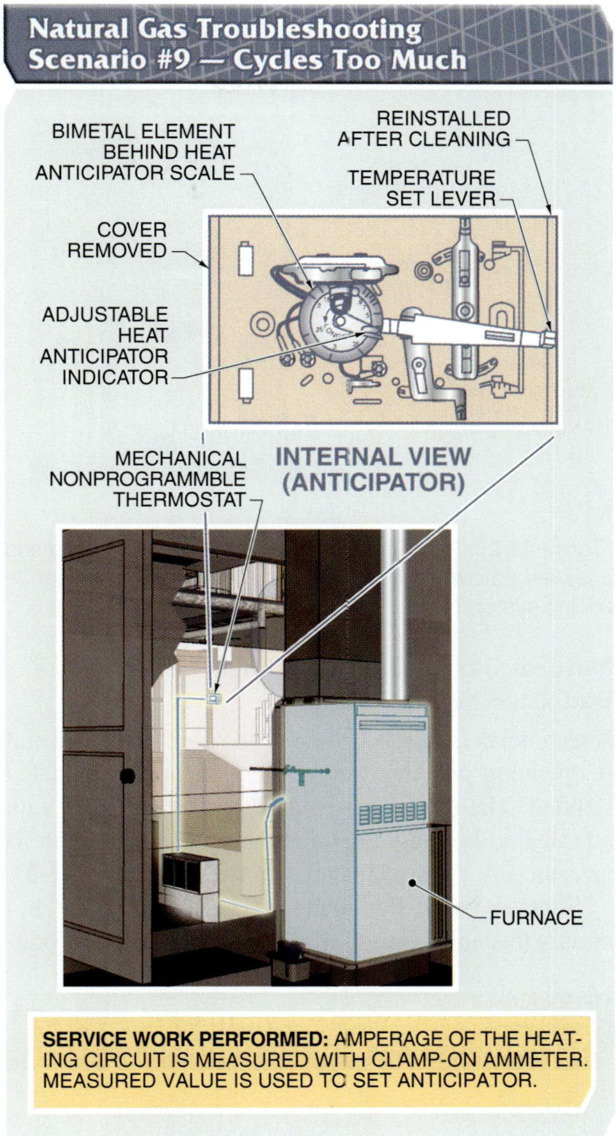

Figure 41-19. After the correct meter is used to acquire a current measurement, the anticipator of a thermostat can be set using the measurement.

Natural Gas Troubleshooting Scenario #10—Terminal Wire Loose

A customer calls in a complaint of a furnace operating erratically. The technician arrives and checks the unit. It is discovered that the furnace starts up and works properly for a short length of time and then abruptly stops. The system board is checked and is flashing a specific fault code. The code indicates that part of the control board has appeared to have failed. **See Figure 41-20.**

Service Work Performed: The loose terminal wire is tightened. The unit is placed back in service and operates properly.

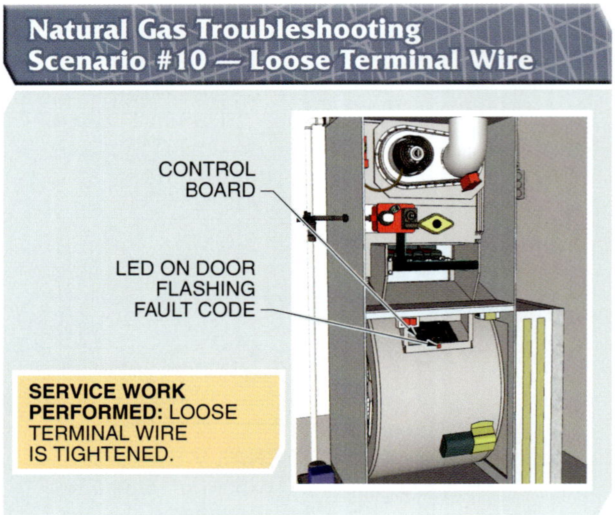

Figure 41-20. Fault indicator lights are useful in identifying a faulted part when troubleshooting a furnace or air conditioning system.

Natural Gas Troubleshooting Scenario #11 — Bad Thermostat

A technician is dispatched on a "too cold" call. The unit is operating properly, but at 2°F lower than the stated setpoint. The thermostat is checked for operation with a DMM at the same time that the space temperature is measured. A 2°F discrepancy is noted. **See Figure 41-21.**

Service Work Performed: The thermostat is not of an adjustable type, so it is replaced with a new one.

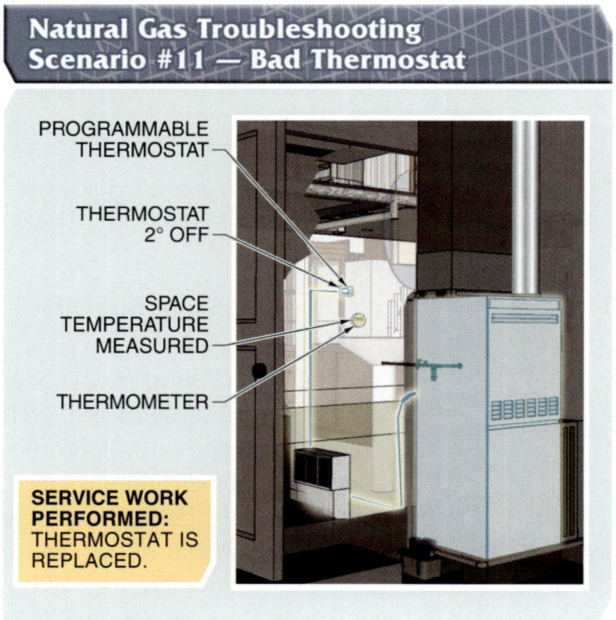

Figure 41-21. A dependable infrared thermometer is needed when diagnosing a problem with a thermostat in a house.

Oil-Fired Troubleshooting Scenario #1 — Low Efficiency

A customer calls in the middle of the heating season. The customer tells the technician that their oil bills are higher than in the previous year. They produce the recent bills and the bills from the same time the year before. The price of oil and the weather were approximately the same each year. The technician starts the oil furnace and runs it for awhile to get the flue and combustion chamber hot. A sample is drawn from the flue and checked. The efficiency is found to be 60%. This type furnace is rated at 70% efficiency. **See Figure 41-22.**

Service Work Performed: The furnace is cleaned and a somewhat defective ignition transformer is replaced, as well as the oil nozzle. The furnace is restarted and found to be right at 68% efficient. While not exactly at its rating, the efficiency has improved. The homeowner is informed that the unit is very old and should probably be replaced with a higher efficiency model.

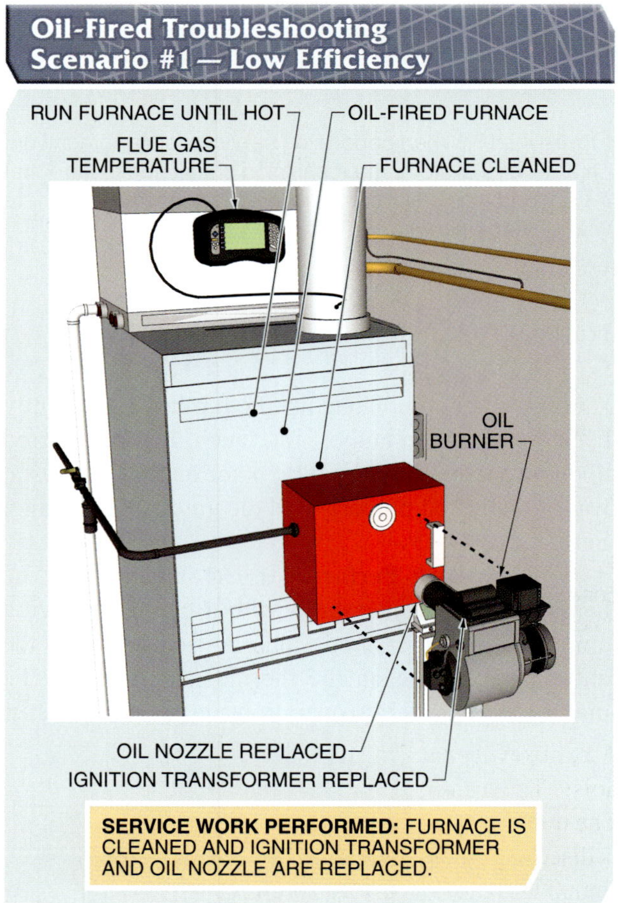

Figure 41-22. Cleaning the furnace and replacing the ignition transformer and oil nozzle will increase the efficiency of the furnace or air conditioning system.

Oil-Fired Troubleshooting Scenario #2—Low Capacity

It is very cold outside and a customer calls in with a complaint that the furnace is running too much. The customer tells the technician that until recently the furnace cycled on and off at regular intervals but now seems to be running almost continuously. The customer lets the technician know that the furnace has not been serviced in a long time. The technician locks out the furnace and performs a clean and inspect. During the clean and inspect, it is discovered that the oil nozzle is almost completely plugged. **See Figure 41-23.**

Service Work Performed: The oil nozzle is replaced and the furnace is placed back in service. It runs and cycles normally.

Figure 41-23. Cleaning or replacing dirty or plugged components will always make an oil-fired furnace operate more efficiently.

Oil-Fired Troubleshooting Scenario #3—Noise

A homeowner calls in and says that the furnace is very noisy. The technician arrives and finds that this is true. The unit is shut down and given a thorough cleaning. It is discovered that the ignition probes and oil nozzle are fouled and operating intermittently. **See Figure 41-24.**

Service Work Performed: The ignition probes and oil nozzle are replaced. Upon startup, an efficiency test and a spot test are performed. The furnace operates properly.

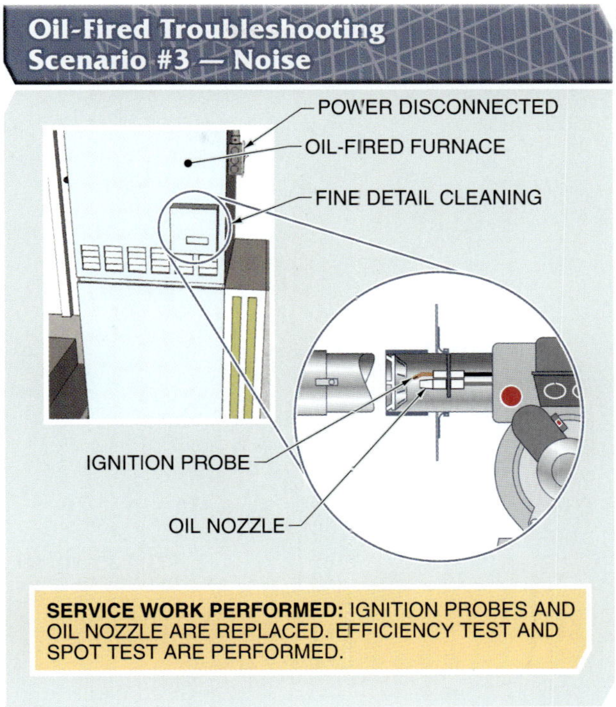

Figure 41-24. Many components of an oil-fired system must be inspected and cleaned to keep the system operating without problems.

Electric Heat Troubleshooting Scenario #1—Bad Sequencer

A customer with an electric furnace calls in a "no heat" complaint. The technician arrives and checks the furnace. The fan is running properly but there is no heat. The circuit is checked with a DMM, and the heating element sequencer circuit is found to be open when it should be closed. **See Figure 41-25.**

Service Work Performed: The sequencer is replaced and checked during normal operation. The furnace is working properly.

Electric Heat Troubleshooting Scenario #2—No Blower

A customer calls in with a "no heat" complaint. The technician checks the furnace and finds that the blower is not running. The blower motor is checked with a DMM and is found to have a defective capacitor. **See Figure 41-26.**

Service Work Performed: The defective capacitor is replaced. The unit is cleaned, and the air filter is replaced. The unit is restarted, and the heat elements cycle normally.

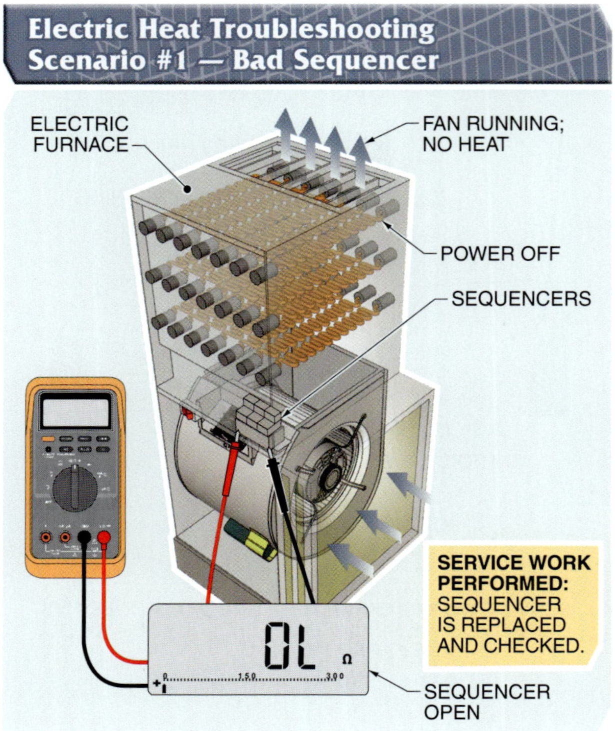

Figure 41-25. The voltage of the sequencer circuit is checked with the power ON, and then the resistance of the sequencer circuit is checked with the power OFF.

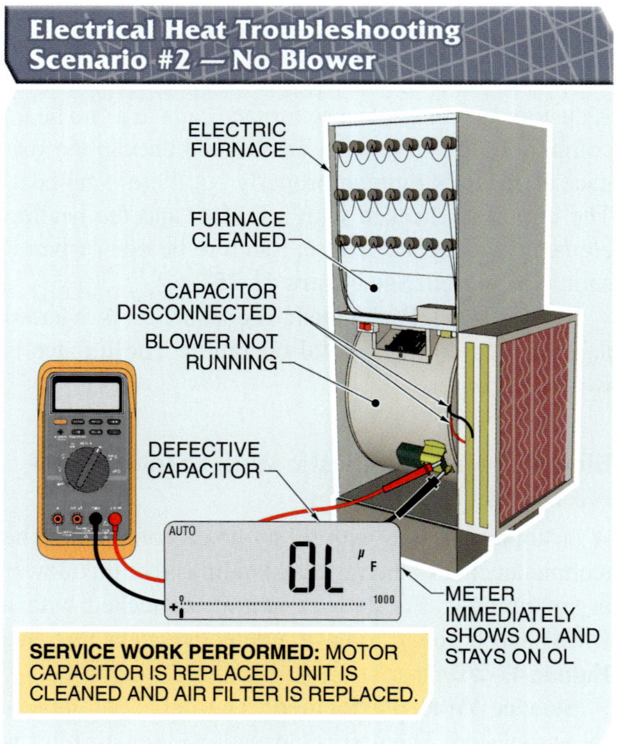

Figure 41-26. A DMM with a capacitor check setting is typically used to test the capacitor for a blower motor.

Electric Heat Troubleshooting Scenario #3—Low Capacity

A customer calls in with a "too cold" complaint. The technician arrives and checks the furnace to find that only one stage of heat is operating. The sequencer is checked and discovered to be damaged. Burn marks are seen on the sequencer and the wiring. **See Figure 41-27.**

Service Work Performed: The wiring is changed and the sequencer replaced. The unit is also thoroughly cleaned and the filter replaced.

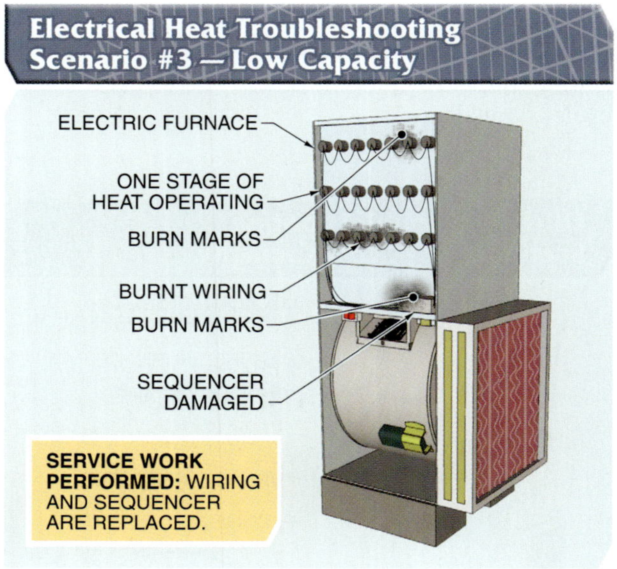

Figure 41-27. When electrical components and wiring look bad, a visual inspection is not enough. The circuits must be checked with a DMM to ensure that all bad components and wiring are found.

Cooling System Troubleshooting Scenario #1—Bad Transformer

A technician is dispatched on a "too hot" call. Upon checking with the customer, the technician discovers that until the day before, the system operated properly. The technician attempts to cycle the indoor blower on and off with the thermostat to no avail. The line voltage is checked at the condenser unit and found to be 230 VAC, which is nominal. The primary side of the control transformer is checked and found to be 120 VAC. The secondary side of the transformer is checked and found to be zero. The power is disconnected, and locked and tagged out. The secondary windings of the transformer are checked with an ohmmeter and are found to be open. **See Figure 41-28.**

Service Work Performed: The transformer is replaced and the system restarted. It is found to be operating properly.

Cooling System Troubleshooting Scenario #1 — Bad Transformer

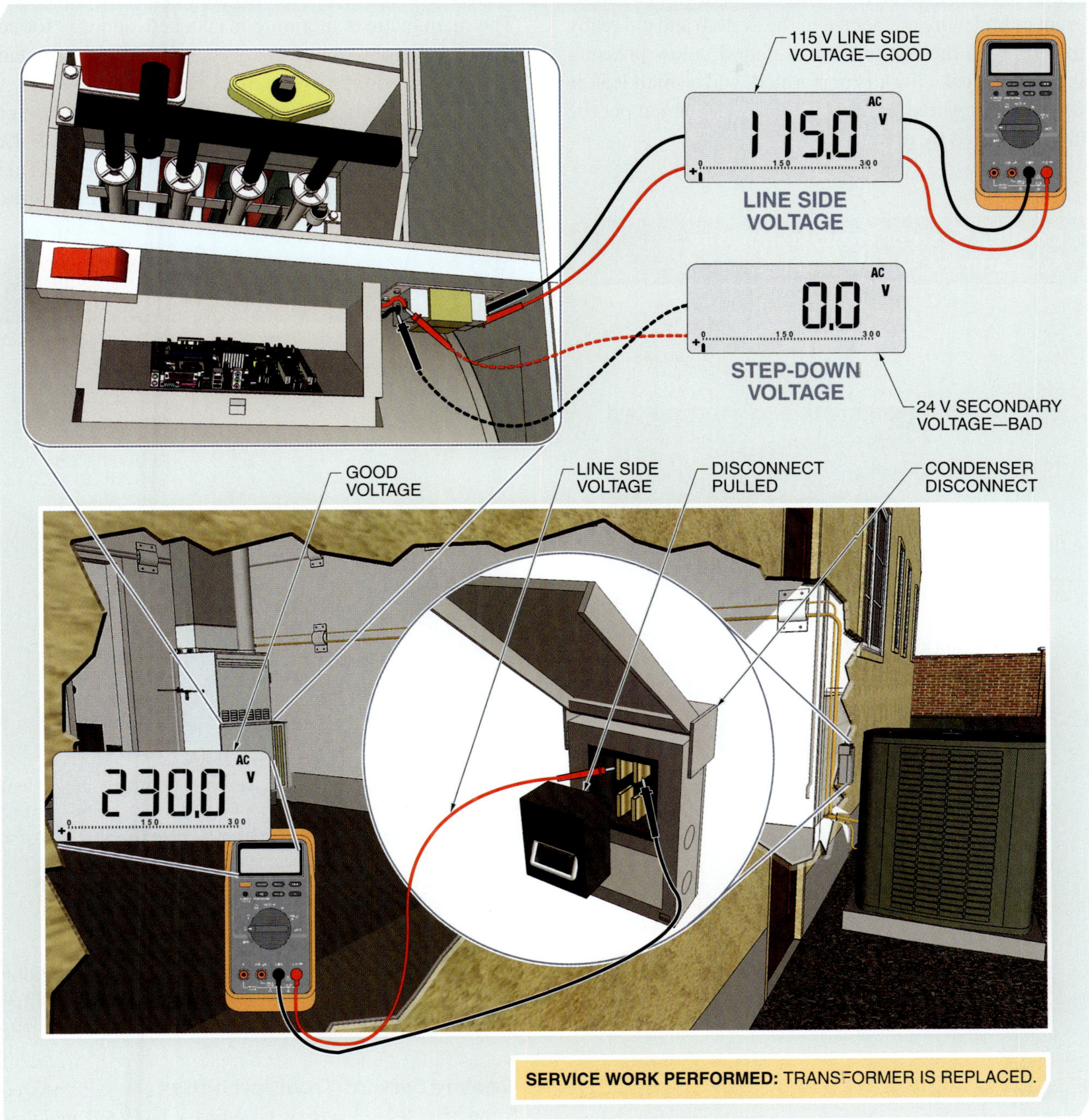

Figure 41-28. The line side of the transformer is initially checked using a DMM set on voltage at the transformer or anywhere line voltage is present, and the secondary side voltage is measured at the transformer. After power is removed, the transformer is checked using the meter set on resistance.

Cooling System Troubleshooting Scenario #2—Low Charge

A "too hot" complaint is called in. The technician arrives to find the indoor blower running with the compressor off. The compressor is very hot to the touch. The thermostat is set at 73°F, but the space temperature is 80°F. The discharge temperature is warm. The technician finds that the thermal overload circuit is open on the compressor starter. The compressor is allowed to cool, and the technician performs PM on the unit while this is occurring. **See Figure 41-29.**

Service Work Performed: The compressor motor starter overload circuit is reset. The gauge set and refrigerant cylinder are connected to the system and are ready to go. When the compressor is started, a low pressure is noted. A leak check is performed, and a small leak is found and repaired. The system is then charged per the chart supplied by the manufacturer.

Figure 41-29. The thermal overload circuit is reset by pressing the reset button, and once the system is operating, pressure measurements are taken that indicate more refrigerant is needed in the system.

Cooling System Troubleshooting Scenario #3—Improper Stage Wiring

The customer calls in a "too hot" complaint. It is the first warm day of the spring. Upon arrival, the customer lets the technician know that during the winter, the indoor blower was replaced by another service company. The blower is found to be running, but at the wrong speed. The manufacturer literature is checked, and it is found that the control wiring was not connected for the second stage of the cooling cycle. **See Figure 41-30.**

Service Work Performed: The control wiring for the second stage of the cooling cycle is connected, the system is restarted, and the blower is running at a higher speed. The air conditioning unit cools the space properly.

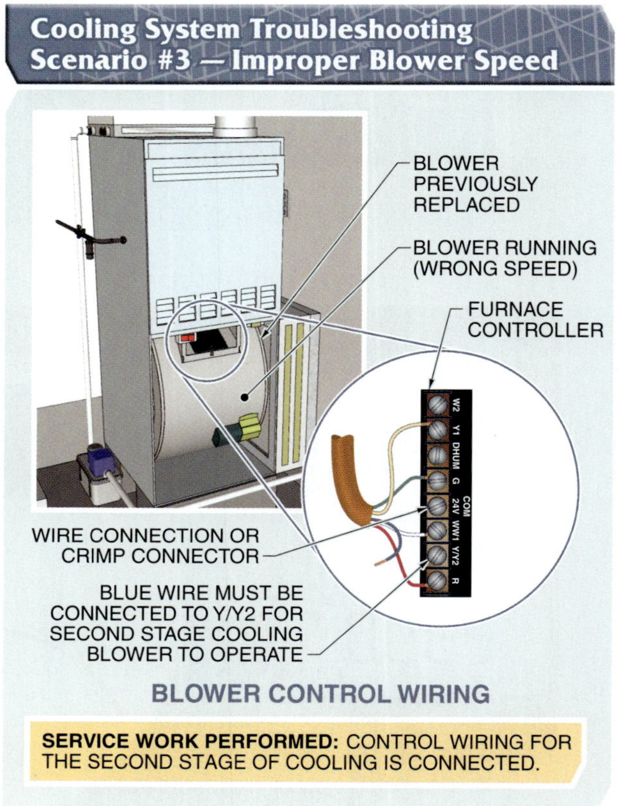

Figure 41-30. When wiring a cooling system together, the technician must use the manufacturer's wiring diagram to ensure the heating and cooling circuits are wired correctly.

Cooling System Troubleshooting Scenario #4—Loss of Charge

A customer calls in a "too hot" complaint on a very hot summer afternoon. The blower and compressor are both off. Pressures are checked, and both the high- and low-side pressures are at 0 psi. The system is inspected, and a hole in the line set is discovered close to the evaporator. It has occurred through vibration. **See Figure 41-31.**

Service Work Performed: The hole in the line set is repaired. The system is pressure tested, evacuated, and then charged with refrigerant. It operates normally.

Figure 41-31. When brazing or soldering the hole in the line set (depending on location) precautions must be taken so that the heat caused by brazing or soldering does not damage any valves or remote bulbs in the system.

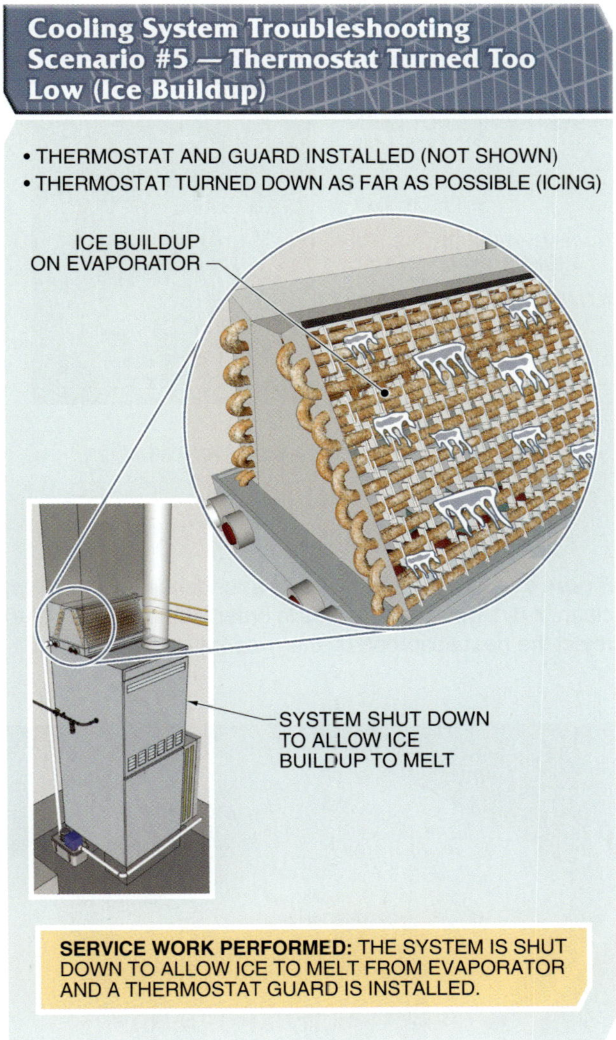

Figure 41-32. During technician inspection of the air conditioning system, ice is allowed to melt off of the evaporator.

Cooling System Troubleshooting Scenario #5—Thermostat Turned Too Low (Ice Buildup)

A "too hot" complaint is received on a summer morning. The technician arrives to find that the thermostat was turned down as far as possible by an occupant the previous afternoon. The air conditioning system has accumulated an ice buildup on the evaporator. **See Figure 41-32.**

Service Work Performed: The air conditioning system is shut down and the ice is allowed to melt off of the evaporator. The system is turned back on and the thermostat setting changed to normal. A lockable thermostat guard is used to prevent future tampering with the setting of the thermostat.

Cooling System Troubleshooting Scenario #6—Dirty Condenser

A customer calls in and complains that their air conditioning system is not cooling sufficiently. It is the first really warm day of spring. The technician notices that the high-side pressure is higher than normal. The customer states that the unit has not been serviced for some time. **See Figure 41-33.**

Service Work Performed: The condenser (an outdoor unit) is inspected and found to be very dirty. It is cleaned, and the unit is put back into operation. The head pressures are now lower, capacity is increased, and the system cools properly. As a result, the customer is signed to a maintenance contract.

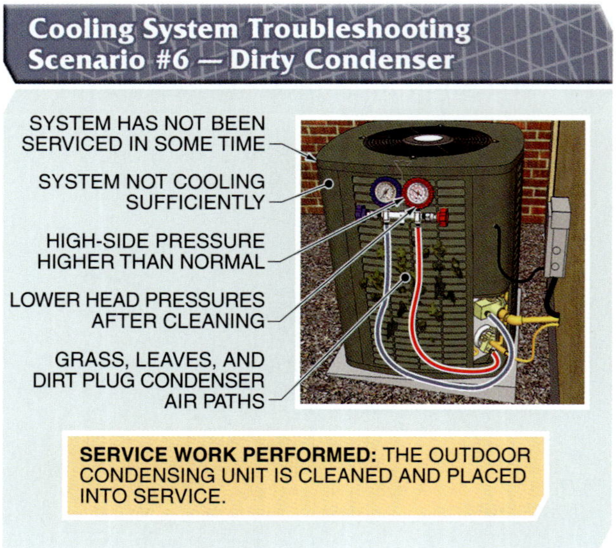

Figure 41-33. The air pathways of a condenser must be kept clean of dirt, grass, and leaves in order for the condenser to reject the heat absorbed by the evaporator.

Condenser air inlet and outlet temperatures can be checked by using a DMM with a temperature probe.

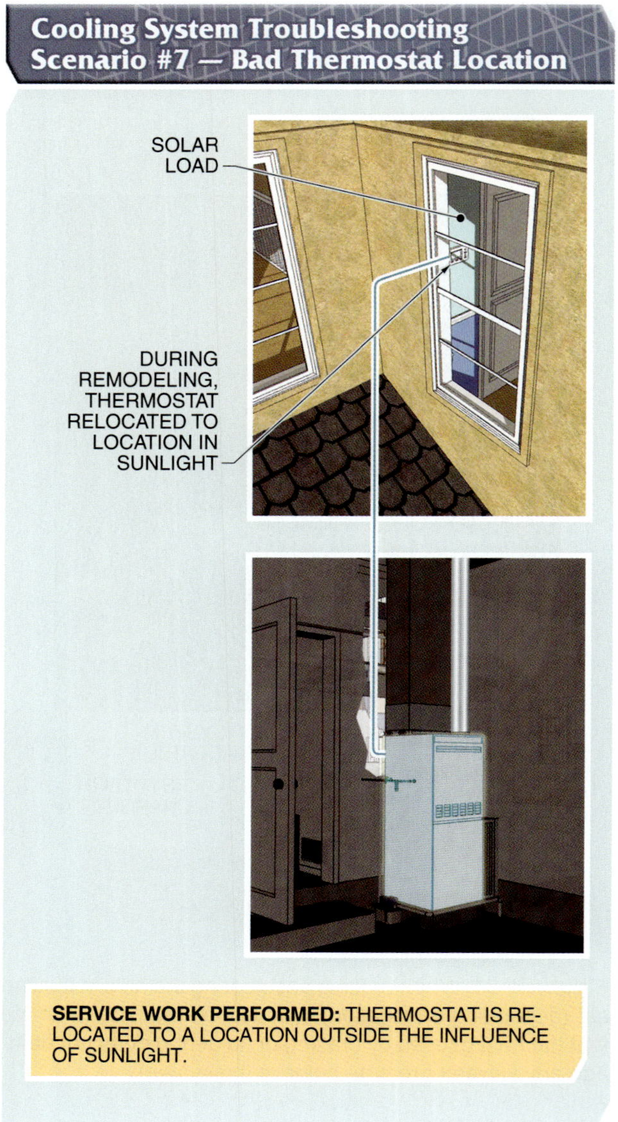

Figure 41-34. Thermostats must be located where average temperatures are found, but never in direct sunlight.

Cooling System Troubleshooting Scenario #7—Bad Thermostat Location

A long-term customer calls in a "too cold" complaint. When the technician arrives, the customer tells the technician that the system was operating properly until a recent remodeling project. It turns out that the thermostat was moved to a spot that is in direct sunlight. The solar load is causing the thermostat to call for cooling almost continuously in the afternoon. **See Figure 41-34.**

Service Work Performed: The thermostat is relocated to a better spot outside of the influence of the sunlight, allowing the system to operate properly.

Cooling System Troubleshooting Scenario #8—Defective TXV

A customer calls in and says that their air conditioning system is operating intermittently. The technician arrives and inspects the system. The suction line has frost on it, and the frost line is moving up and down the suction line. The manifold gauges indicate wildly fluctuating suction pressures. The TXV is the problem. **See Figure 41-35.**

Service Work Performed: The air conditioning system is pressure tested, evacuated, and the TXV is replaced. The system is evacuated again, and refrigerant is charged back into the system. The unit stabilizes and operates properly.

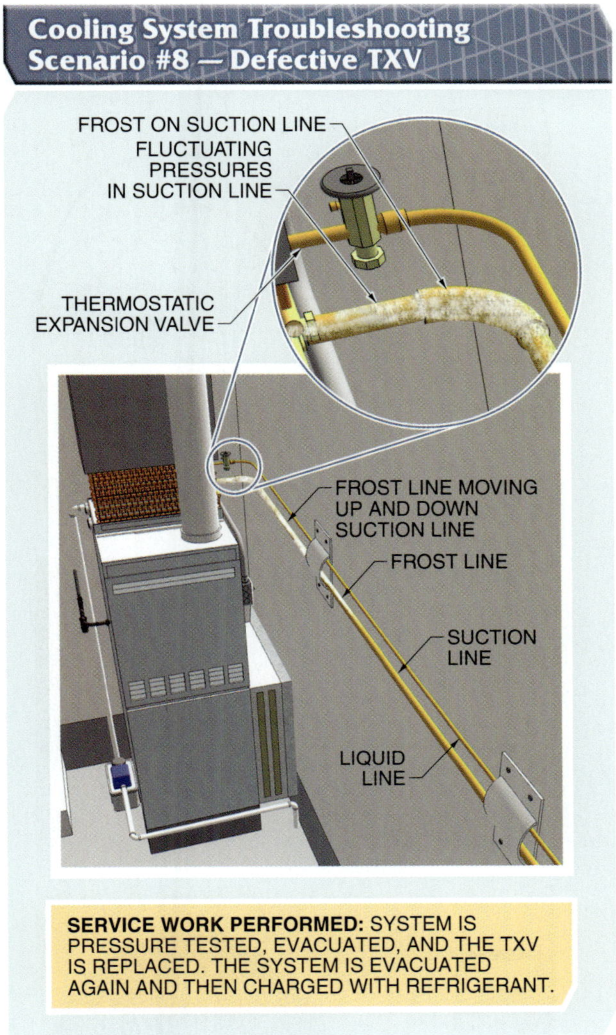

Figure 41-35. The system is evacuated to replace the TXV, and then the system is evacuated for hours to prepare the system for charging.

Cooling System Troubleshooting Scenario #9—Storm Caused Tripping

The office dispatches a technician on a "too hot" complaint. The air conditioning system is a new high-efficiency type with an electronic board to which the thermostat is wired. The board is checked and is found to be flashing a code that indicates a high-voltage situation. The night before, a line of severe thunderstorms came through the area. **See Figure 41-36.**

Service Work Performed: The thermostat is placed in the OFF mode and then power to the condensing unit and furnace are shut off for approximately one minute. The power is then turned on and the board reboots. The thermostat is placed in the cooling mode and the system cycles normally.

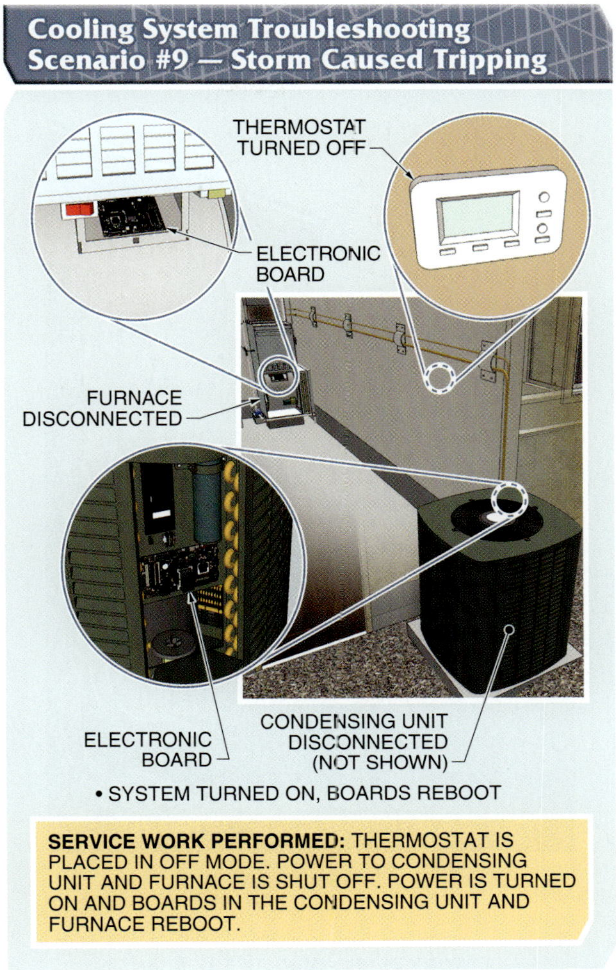

Figure 41-36. Digital electronics that trip can typically be reset by removing power for 30 sec and then turning power back on.

Cooling System Troubleshooting Scenario #10—Recirculated Condenser Air

A customer calls in a "too hot" complaint. The system was replaced by another HVAC company and is not cooling properly. The technician notes that the indoor blower is running and that the high-side pressure is higher than the ambient temperature would indicate. It is discovered that the condensing unit was placed under an overhang. The condenser air is being diverted and recirculated back onto the condenser. The only way to correct the problem is to relocate the condenser a few feet from its present location. **See Figure 41-37.**

Service Work Performed: The refrigerant is recovered and the condensing unit is moved, reconnected, pressure tested, evacuated, and refilled with refrigerant. The unit is placed back into operation. System pressures are checked and found to be normal.

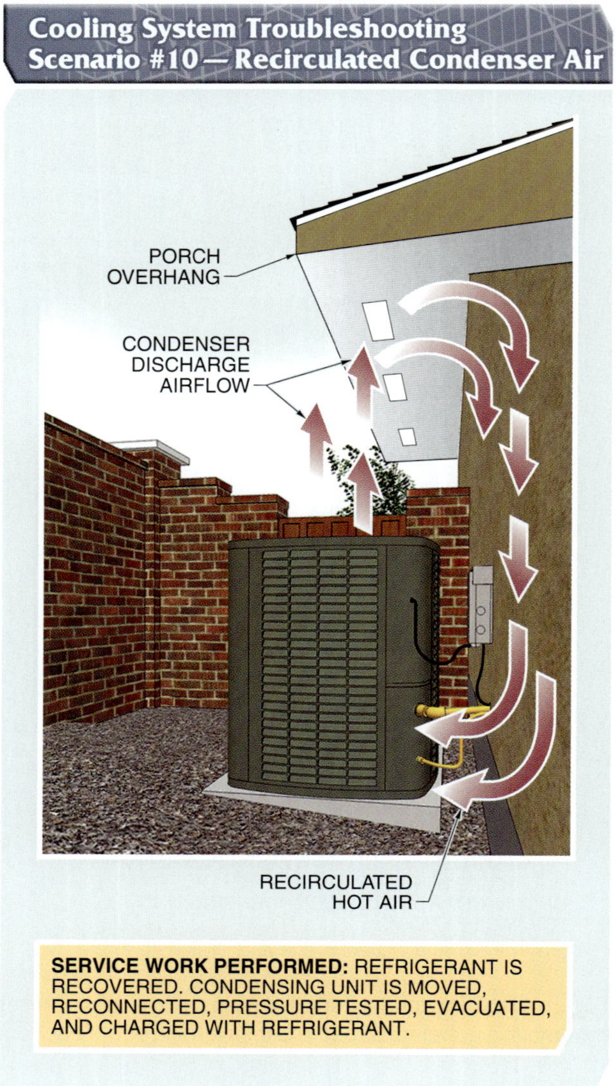

Figure 41-37. Airflow into and out of a condenser must not be obstructed. When the hot rejected air from the condenser is directed back to the condenser, it causes overheating.

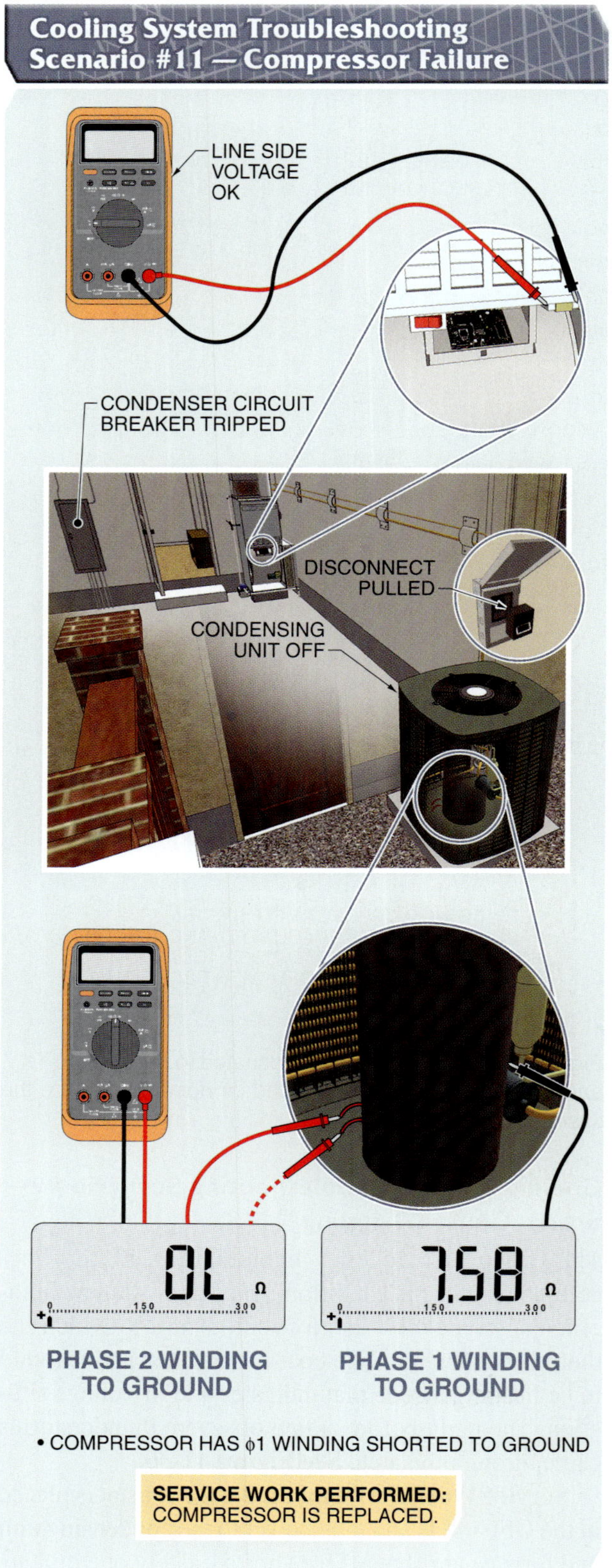

Figure 41-38. With power locked out and a DMM set on resistance, the leads of the compressor motor are found to be shorted to ground.

Cooling System Troubleshooting Scenario #11—Compressor Failure

A customer calls in a "too hot" complaint. The technician finds that the condensing unit is off and the breaker is tripped. The line-side voltage is checked and found to be okay. The condensing unit is locked out and the resistance at the compressor terminals is checked. The compressor is found to have a winding shorted to ground. **See Figure 41-38.**

Service Work Performed: The compressor is replaced. The unit is operating properly.

Cooling System Troubleshooting Scenario #12—Wrong Voltage

A customer calls in and says that their air conditioning is behaving abnormally. On arrival, the technician notices that the lights in the rear of the house appear to be dimming and then growing brighter on a regular basis. The voltage supply to the condenser is checked with a min/max-capable DMM and found to be alternately too high and too low. As soon as this is discovered, the disconnect at the condensing unit is pulled to prevent damage to the compressor. The utility company is contacted and utility workers are sent to check the incoming power. A loose neutral is discovered on the transformer at the pole. **See Figure 41-39.**

Service Work Performed: As soon as the loose neutral is fixed, the unit is placed back into service and found to operate properly.

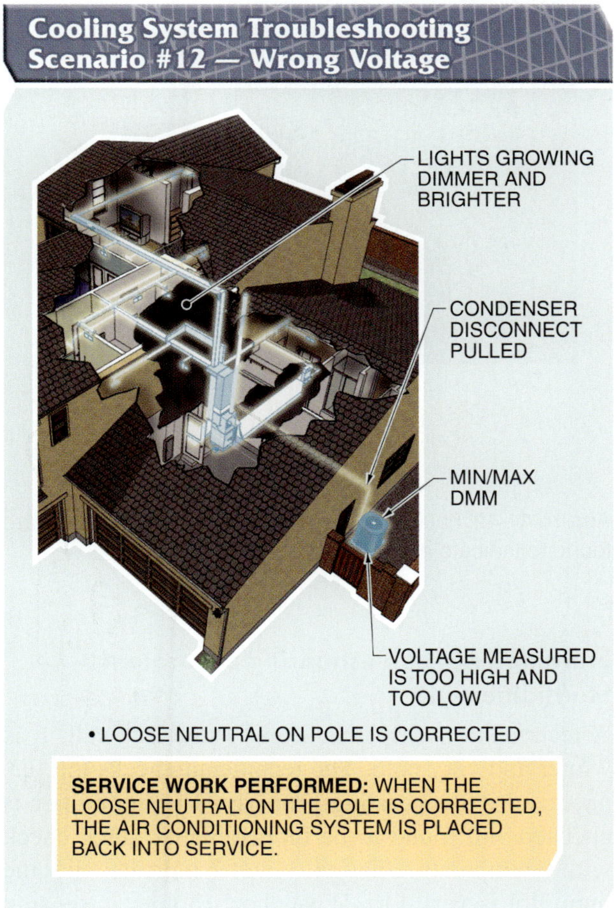

Figure 41-39. A min/max DMM finds voltage too high and too low, indicating a problem with the supply power, such as a loose neutral on the pole transformer.

Cooling System Troubleshooting Scenario #13—Defective Capacitor on Blower Motor

A customer calls in a "too hot" complaint. It is a hot summer day. The indoor temperature is 80°F while the thermostat is set at 72°F. No air is blowing out of the registers. The air-handling unit is in the garage. The blower motor is not running and the evaporator is iced over. The blower motor capacitor is disconnected and checked with a DMM that includes a capacitor check setting. The capacitor is found to be defective. **See Figure 41-40.**

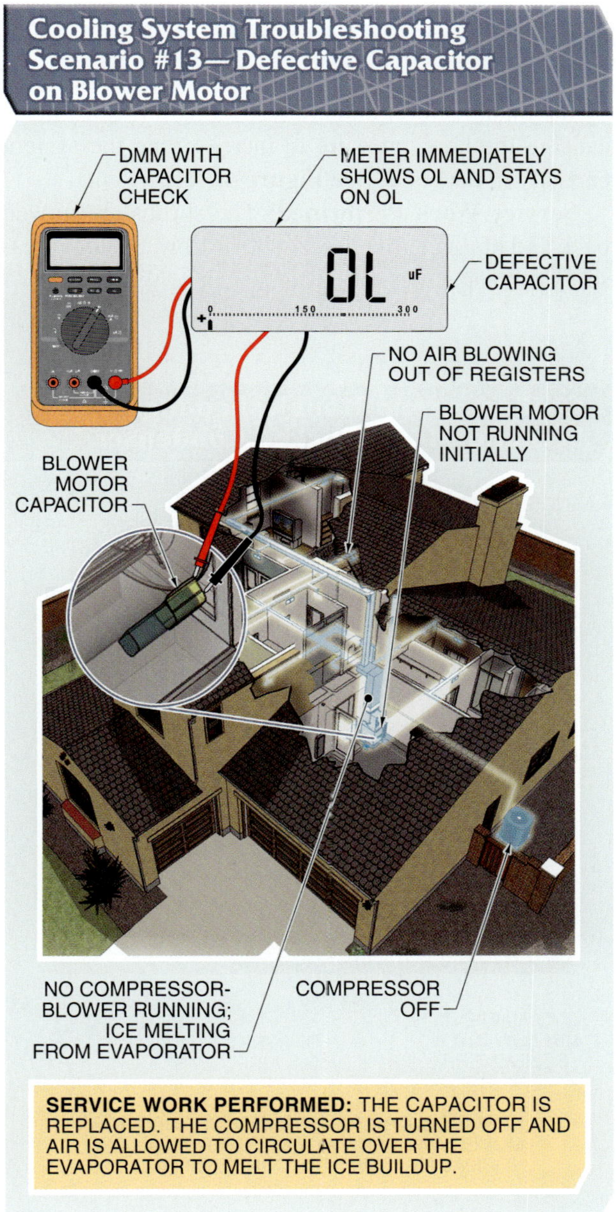

Figure 41-40. A DMM with a capacitor check setting is used to test the blower motor's capacitor, which is found to be bad.

Service Work Performed: The capacitor is replaced and the air-handling unit is placed back into service. The compressor is turned off, and the fan circulates air over the evaporator to melt the ice. After the ice is melted, the compressor is energized and the system runs normally.

Cooling System Troubleshooting Scenario #14—Blocked Condenser

The office dispatches a technician on a "too hot" complaint. The customer states that until recently, the unit worked fine but now suddenly does not. A visual inspection of the system shows that the customer has put in new shrubs to "hide" the ugly condensing unit. The shrubs are planted too close to the unit as determined by the manufacturer and are restricting airflow. This is causing the head pressure to increase and the system capacity to decrease. **See Figure 41-41.**

Service Work Performed: The shrubs are moved back and no longer interfere with airflow. System pressures and capacities return to normal.

Figure 41-41. When cooling air cannot enter a condenser freely, the condenser will continue to overheat over time.

Cooling System Troubleshooting Scenario #15—Bad Board

A technician is dispatched on a "too hot" call. Upon arrival, the technician finds that the system is not operating at all. The condensing unit is a newer model with an electronic board. The technician removes the cover to the unit and finds the board fault lights flashing. **See Figure 41-42.**

Service Work Performed: The manufacturer literature is checked and it is found that the flashing lights indicate a failed board. The board is replaced. The unit is restarted and operates properly.

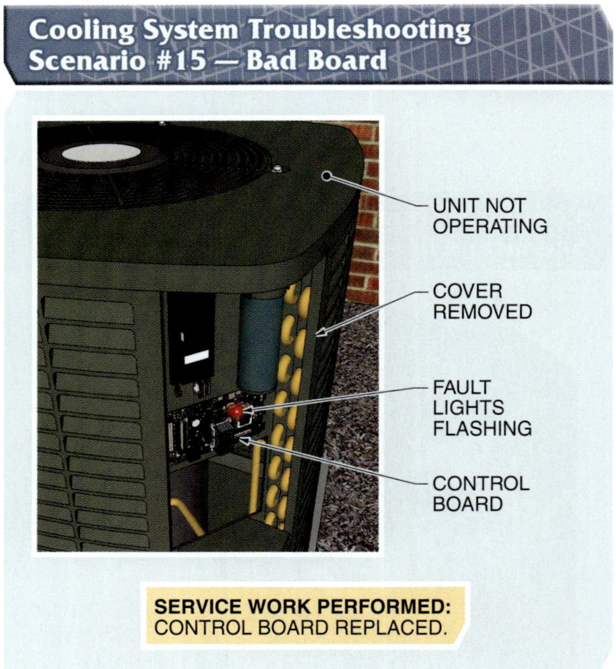

Figure 41-42. Fault indicator lights are investigated and are found to indicate a failed condenser board.

Humidifier Troubleshooting Scenario #1—Humidifier Off

A technician is dispatched on a "too dry" call. It is winter and very cold. The technician checks the furnace and it is operating properly. The humidifier is checked and found to be inoperable. A quick check with a temperature/humidity meter indicates that the humidity is very low. What the customer is sensing is not a temperature issue but a humidity issue. **See Figure 41-43.**

Service Work Performed: The humidifier is replaced.

Figure 41-43. Using a temperature/humidity meter, it is found that the humidity in the house is extremely low because the humidifier is not working, creating an extremely cold feeling.

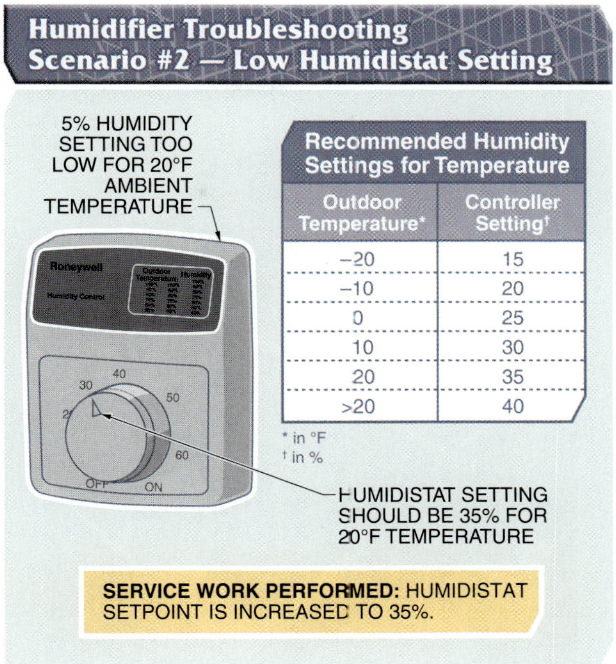

Recommended Humidity Settings for Temperature

Outdoor Temperature*	Controller Setting†
−20	15
−10	20
0	25
10	30
20	35
>20	40

* in °F
† in %

Figure 41-44. To feel comfortable, humidistat settings must change as the ambient temperature changes outside a residence.

TECH TIP

All HVAC technicians should be equipped with quality digital test instruments. This is especially true when diagnosing humidity problems. Without an accurate humidity test instrument, system performance cannot be accurately measured. Quality test instruments can provide guaranteed accuracy and enable the readings to be sent to a laptop if necessary.

Humidifier Troubleshooting Scenario #2—Low Humidistat Setting

During a cold winter day of 20°F, a technician is dispatched on a low humidity call. The customer indicates to the technician that they have no idea on how to set the humidistat. The technician notices that the humidistat is set at the lowest possible setting of 5%. **See Figure 41-44.**

Service Work Performed: The humidistat setpoint is increased to 35%, the humidifier comes on, and the humidity level improves. The owner is instructed how to set the humidistat.

Humidifier Troubleshooting Scenario #3—Moisture on Windows

On a cold winter day, a homeowner calls an HVAC contractor in a panic and requests that a technician come out immediately. When the technician arrives, they find that the homeowner has increased the humidistat setpoint to the maximum setting. The humidity level is so high that frost has formed on the inside of the windows. **See Figure 41-45.**

Service Work Performed: The humidistat setpoint is lowered immediately and the thermostat setpoint is increased. The frost starts to melt. The homeowner is instructed to let it melt on its own and then adjust the humidistat and thermostat to the proper levels.

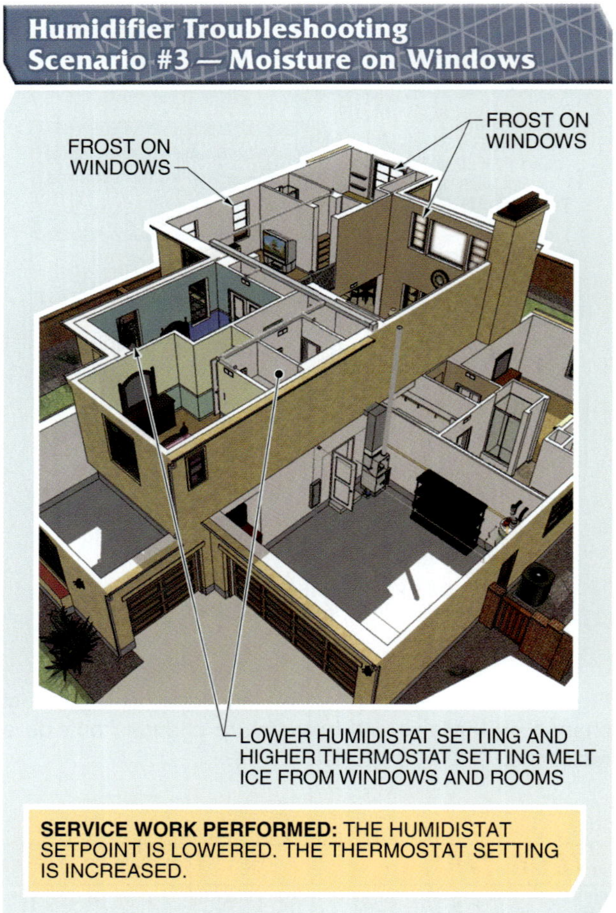

Figure 41-45. To feel warmer, a homeowner increased a humidistat setting, causing too much humidity to build up in the house. Lowering the setting on the humidistat will lower the humidity in the house.

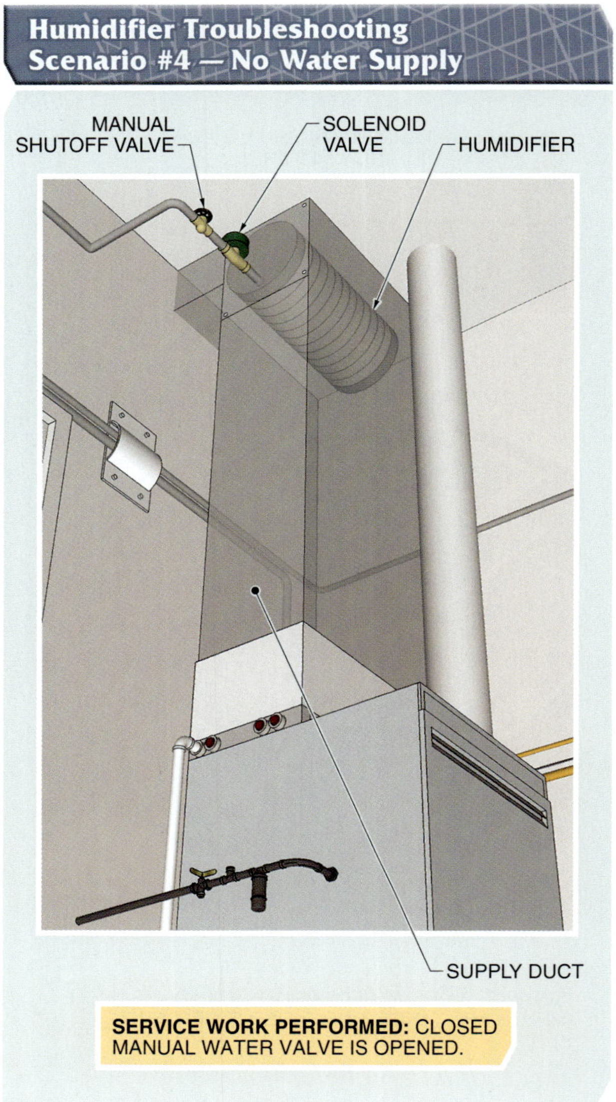

Figure 41-46. The control valves for a humidifier should be checked any time a humidifier does not have water.

Humidifier Troubleshooting Scenario #4—No Water Supply

At the beginning of the heating season, a customer calls in with a "low humidity" complaint. The technician arrives and finds that the humidistat is set correctly, but the humidifier is not working. The solenoid valve is open, but no water is present in the humidifier. It is discovered that, during the summer, some plumbing work was done, and the manual water valve to the humidifier was shut off. **See Figure 41-46.**

Service Work Performed: The water valve is opened and water begins to flow into the humidifier through the open solenoid valve. The technician allows the humidifier to operate for some time and the humidity begins to increase. The technician adjusts the humidistat to the proper level.

Filter Troubleshooting Scenario #1—Wrong Filters

It is very hot outside and the customer calls in a "too hot" complaint. The customer tells the technician that the filters in the system were recently changed to a different type. The listed pressure drop of the new filters is twice the recommended maximum. This is causing low airflow in the system and loss of cooling capacity. **See Figure 41-47.**

Service Work Performed: The filters are replaced with filters that have the appropriate pressure drop rating. Cool airflow returns to normal.

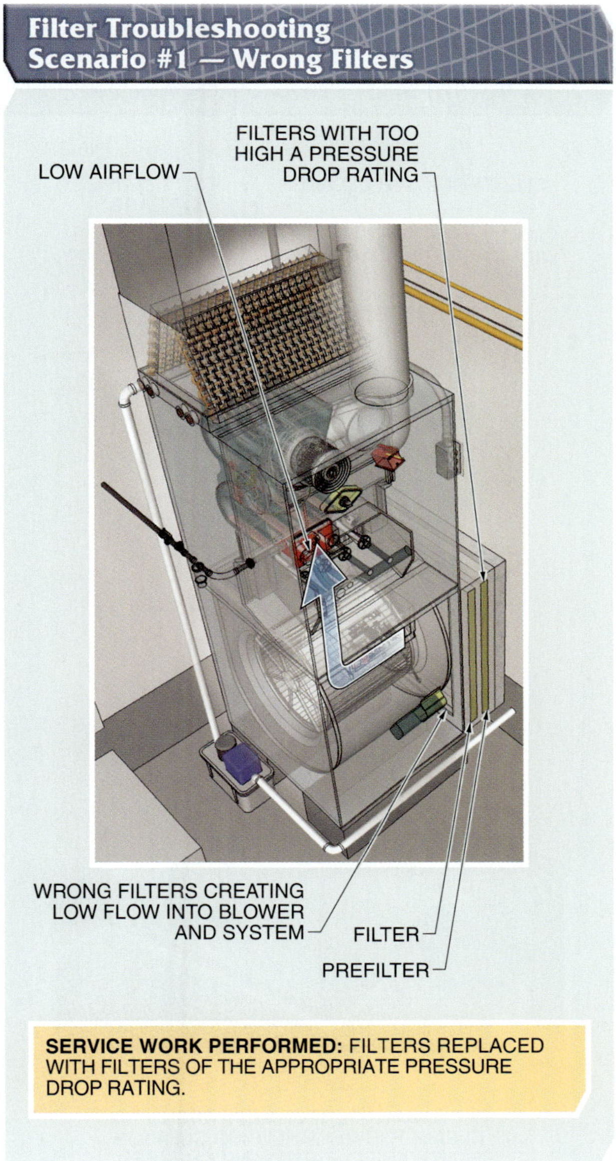

Figure 41-47. Furnaces and/or air handlers must use the correct resistance filters on the suction side in order to guarantee the proper airflow through the unit.

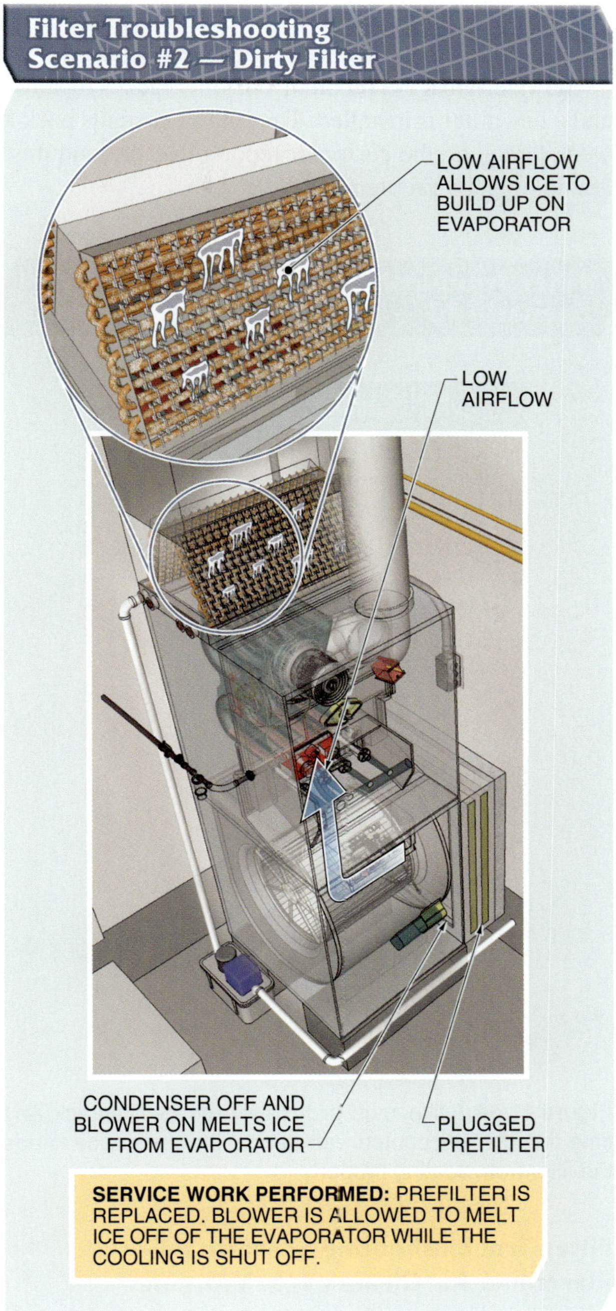

Figure 41-48. Because of the low airflow through the dirty filter, the blower cannot push enough warm air across the evaporator to keep ice from building up.

Filter Troubleshooting Scenario #2—Dirty Filter

During the summer, a customer calls in a "too hot" complaint. The technician arrives and finds that the evaporator is iced up. The technician also discovers that the prefilter is totally plugged. **See Figure 41-48.**

Service Work Performed: The prefilter is replaced, and the blower is allowed to melt the ice on the evaporator while the cooling is shut off.

Filter Troubleshooting Scenario #3—Air Bypassing Filter

A customer complains that dirt and dust are accumulating on items in their home. The technician checks and finds that the filter has been installed incorrectly and is only covering 50% of the duct opening. The filter in the system was installed incorrectly because the rack has a burr that catches and tears the filter. **See Figure 41-49.**

Service Work Performed: The filter rack is repaired and a new filter is installed. The technician calls back a week later, and the customer reports that dirt and dust are no longer a problem.

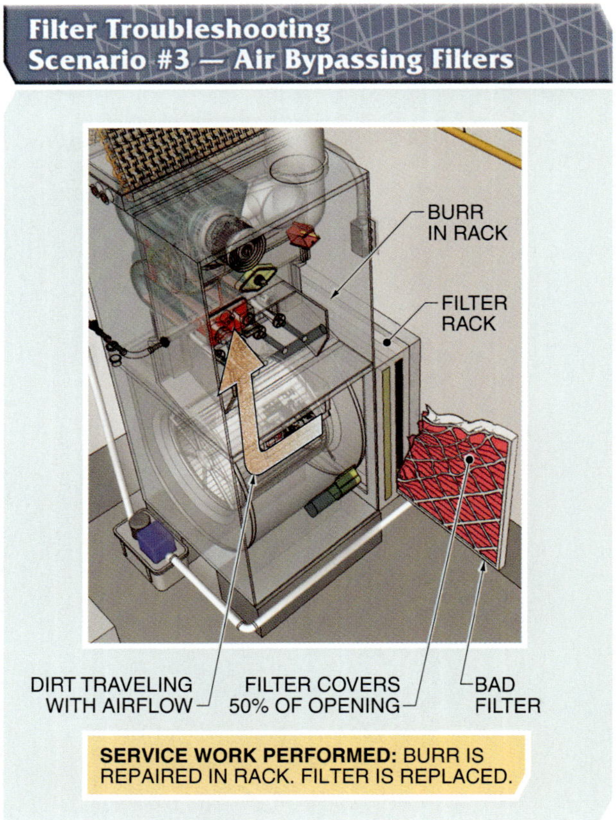

Figure 41-49. A filter must slide freely into the filter rack. Any time the filter encounters resistance, the resistance keeps the filter from seating properly.

Filter Troubleshooting Scenario #4—Electronic Air Cleaner Not Working

A customer with an electronic filter complains of excessive dust and dirt in their home. The technician arrives, and the customer explains that the electronic air cleaner has not had any problems and thus has not been serviced in years. The technician checks the electronic air cleaner and finds that the wiring to the plates has shorted. **See Figure 41-50.**

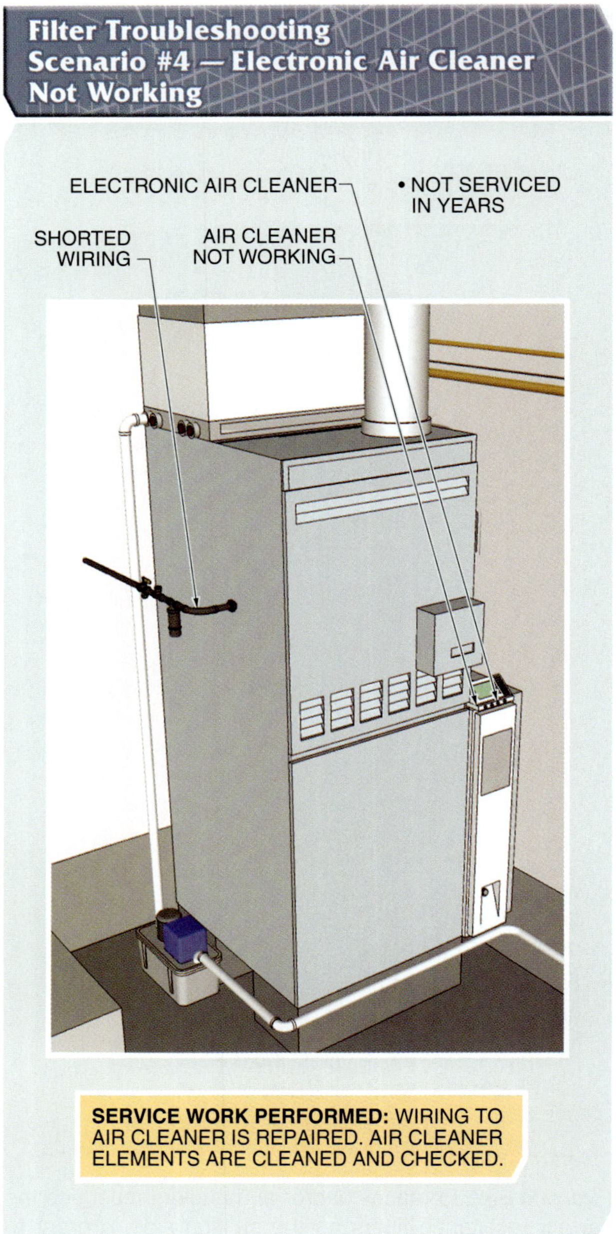

Figure 41-50. Shorted wires to the air cleaner are found, verified with a DMM, and replaced. All other wires are cleaned to avoid shorting.

Service Work Performed: The wiring is repaired and the air cleaner elements are checked and cleaned thoroughly. The electronic air cleaner is placed back in service and operates properly.

UNIT

42

Installing Residential Hydronic Systems

In many areas of the country, hydronic boilers are a viable alternative to residential forced-air furnaces. It is important that every HVAC technician know how to properly install the various types of hydronic systems. Many problems arise due to improperly installed equipment. Some manufacturers estimate that one-third or more of all the equipment installed has some type of installation error. These errors result in customer dissatisfaction, numerous callbacks, warranty claims, and energy waste. Many of these errors can be avoided by properly installing hydronic systems.

OBJECTIVES

- Identify safety requirements for installing a residential boiler and hydronic system.
- Explain how to prepare and properly determine the installation location of a boiler.
- List the steps and procedures for proper installation of a hydronic system.
- List the startup and preventive maintenance procedures used after installing a hydronic system.

Index to Procedures

HYDRONIC BOILER INSTALLATIONS

In general, there are two different types of installations used for residential hydronic (hot water) boilers. In the first, the installation may be a new one, possibly while the house is still under construction. This means that there are no previously installed devices or system components to remove or replace. Also, accessibility may be easier because the building is still under construction. **See Figure 42-1.** The other type of installation is a retrofit, where an old boiler is being replaced by a new one. In the case of a retrofit, inspections of existing components, piping, and any new installations will be necessary.

Hydronic Boiler Installations

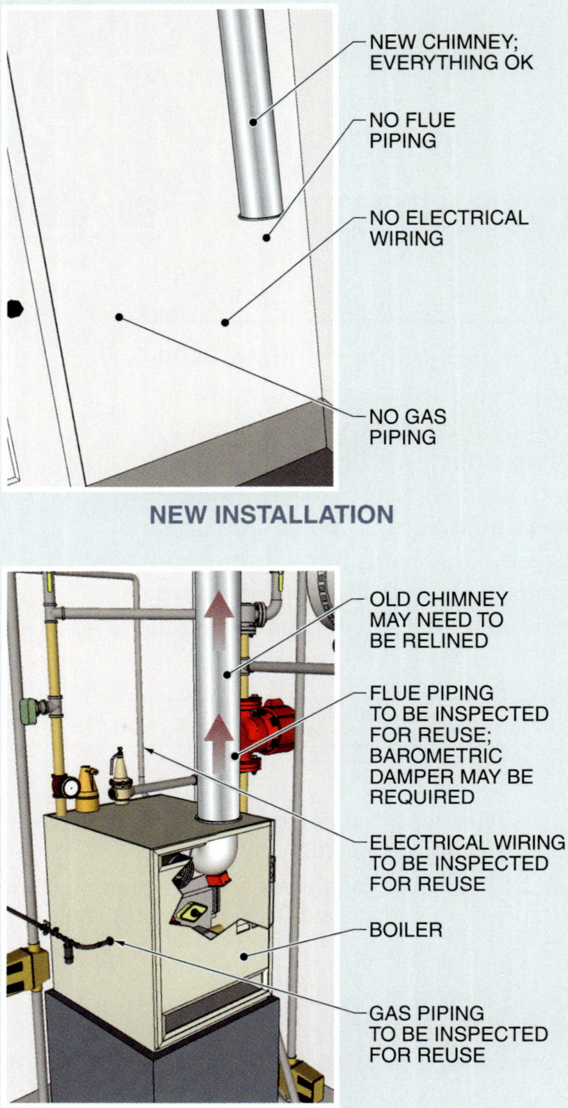

Figure 42-1. Retrofitting an older boiler requires close inspection of water and fuel piping, high- and low-voltage wiring, flue, and possibly the combustion air duct to ensure they meet present codes.

Warranties

It is critical for an HVAC contracting business to minimize warranty problems. When a boiler is installed improperly, it is possible and even likely that the manufacturer warranty will be voided. The warranty does not apply to boilers improperly installed or improperly operated. When a boiler fails and an additional boiler must be purchased from the manufacturer, it causes a huge financial impact on the HVAC contractor. Therefore the manufacturer installation instructions must be closely followed. **See Figure 42-2.** Improper installation or system design, rather than faulty equipment, is the cause of many operating problems.

Warranty Information

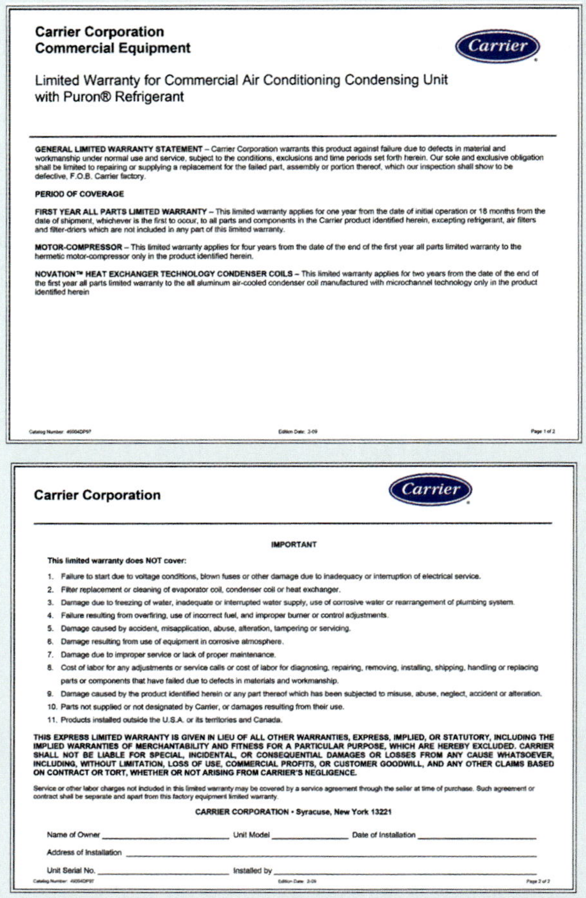

Carrier Corporation
Commercial Equipment

Limited Warranty for Commercial Air Conditioning Condensing Unit with Puron® Refrigerant

GENERAL LIMITED WARRANTY STATEMENT – Carrier Corporation warrants this product against failure due to defects in material and workmanship under normal use and service, subject to the conditions, exclusions and time periods set forth herein. Our sole and exclusive obligation shall be limited to repairing or supplying a replacement for the failed part, assembly or portion thereof, which our inspection shall show to be defective, F.O.B. Carrier factory.

PERIOD OF COVERAGE

FIRST YEAR ALL PARTS LIMITED WARRANTY – This limited warranty applies for one year from the date of initial operation or 18 months from the date of shipment, whichever is the first to occur, to all parts and components in the Carrier product identified herein, excepting refrigerant, air filters and filter-driers which are not included in any part of this limited warranty.

MOTOR-COMPRESSOR – This limited warranty applies for four years from the date of the end of the first year all parts limited warranty to the hermetic motor-compressor only in the product identified herein.

NOVATION™ HEAT EXCHANGER TECHNOLOGY CONDENSER COILS – This limited warranty applies for two years from the date of the end of the first year all parts limited warranty to the all aluminum air-cooled condenser coil manufactured with microchannel technology only in the product identified herein

Page 1 of 2

Carrier Corporation

IMPORTANT

This limited warranty does NOT cover:

1. Failure to start due to voltage conditions, blown fuses or other damage due to inadequacy or interruption of electrical service.
2. Filter replacement or cleaning of evaporator coil, condenser coil or heat exchanger.
3. Damage due to freezing of water, inadequate or interrupted water supply, use of corrosive water or rearrangement of plumbing system.
4. Failure resulting from overfiring, use of incorrect fuel, and improper burner or control adjustments.
5. Damage caused by accident, misapplication, abuse, alteration, tampering or servicing.
6. Damage resulting from use of equipment in corrosive atmosphere.
7. Damage due to improper service or lack of proper maintenance.
8. Cost of labor for any adjustments or service calls or cost of labor for diagnosing, repairing, removing, installing, shipping, handling or replacing parts or components that have failed due to defects in materials and workmanship.
9. Damage caused by the product identified herein or any part thereof which has been subjected to misuse, abuse, neglect, accident or alteration.
10. Parts not supplied or not designated by Carrier, or damages resulting from their use.
11. Products installed outside the U.S.A. or its territories and Canada.

THIS EXPRESS LIMITED WARRANTY IS GIVEN IN LIEU OF ALL OTHER WARRANTIES, EXPRESS, IMPLIED, OR STATUTORY, INCLUDING THE IMPLIED WARRANTIES OF MERCHANTABILITY AND FITNESS FOR A PARTICULAR PURPOSE, WHICH ARE HEREBY EXCLUDED. CARRIER SHALL NOT BE LIABLE FOR SPECIAL, INCIDENTAL, OR CONSEQUENTIAL DAMAGES OR LOSSES FROM ANY CAUSE WHATSOEVER, INCLUDING, WITHOUT LIMITATION, LOSS OF USE, COMMERCIAL PROFITS, OR CUSTOMER GOODWILL, AND ANY OTHER CLAIMS BASED ON CONTRACT OR TORT, WHETHER OR NOT ARISING FROM CARRIER'S NEGLIGENCE.

Service or other labor charges not included in this limited warranty may be covered by a service agreement through the seller at time of purchase. Such agreement or contract shall be separate and apart from this factory equipment limited warranty.

CARRIER CORPORATION • Syracuse, New York 13221

Name of Owner ________ Unit Model ________ Date of Installation ________

Address of Installation ________

Unit Serial No. ________ Installed by ________

Page 2 of 2

Carrier Corporation

Figure 42-2. It is always a good idea to check the water a boiler will be using before installation because many properties of water are harmful to boiler life and may void the manufacturer warranty.

A few common conditions that may also void the warranty of a boiler are excessive water hardness, causing lime buildup, and excessive pitting and erosion on the inside of the tubes caused by too great a water velocity through the boiler.

When receiving boilers and boiler equipment, an inspection should be performed to check for signs of shipping damage. Particular attention should be paid to the parts accompanying the boiler, which may show signs of being hit or otherwise mishandled. The total number of pieces shown on the packing slip should be checked against the pieces actually received. If there is damage or a shortage, the carrier should be notified immediately.

Boiler Installation Safety

Boiler installation and service must be performed by a qualified service installer or service technician from a reputable company. The natural gas or propane connections to a boiler should be made by a licensed technician and/or certified gas supplier. Electrical connections may require that a qualified technician or certified electrician perform all electrical work. When an installation is not performed properly, a fire or explosion may result, causing property damage, personal injury, or loss of life. The boiler must not be installed in any location where gasoline or flammable vapors are likely to be present. **See Figure 42-3.**

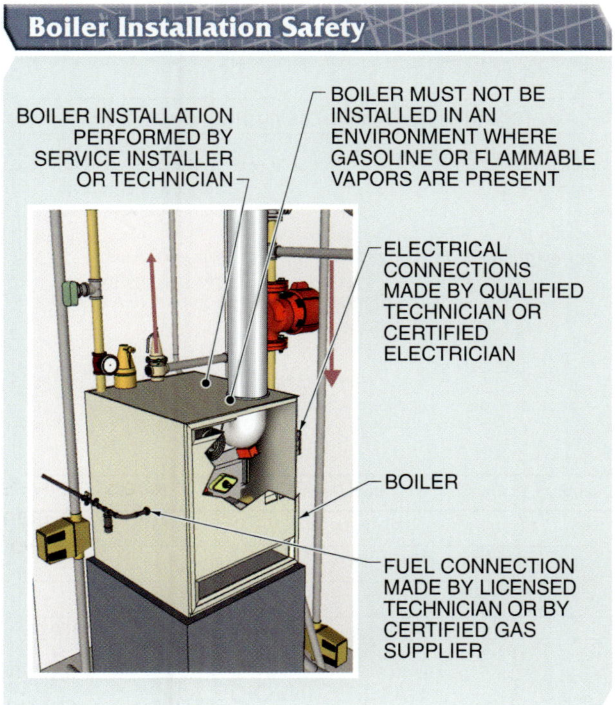

Figure 42-3. When the electrical, fuel, and mechanical systems of a boiler are not installed to code, explosion and/or fires can result.

Ratings

Boiler ratings should be checked to ensure that the boiler is the proper size for the application. Boiler ratings use 1 MBH as a designation to represent 1000 Btu/hr. Boiler ratings typically consist of the following:

- model number
- input MBH
- heating capacity MBH
- AFUE percentage

Codes

All local building codes, fire regulations, and other applicable safety codes should be consulted when installing a boiler. The boiler and accessories must be installed in accordance with the installation regulations in force in the area where the installation is made.

These regulations must be carefully followed in all cases. Authorities having jurisdiction must be consulted before installations are made. In the absence of such requirements, the installation must conform to the latest edition of recognized codes, such as the National Fuel Gas Code, ANSI standards, and/or the CAN/CGA-B149, Installation Code for Gas Burning Boilers and Equipment, as well as the ASME and other code authorities. **See Figure 42-4.**

Figure 42-4. Due to the complexity of boiler systems, various organizations have written codes for the installation of boilers.

The local gas utility company must authorize and inspect all gas and flue connections. Should overheating occur or the gas supply fail to shut off, the electrical supply to the circulating pump should not be turned off or disconnected. Instead, the gas supply at a location external to the boiler should be shut off. To minimize the possibility of serious personal injury, fire, or damage to the boiler, these safety rules should never be violated.

Boiler Preparation

After arrival at the job site, the boiler must be prepared for installation. The boiler must first be removed from the shipping and packaging material. If not immediately installed, the boiler should be stored in a protected space. The shipping invoice should be checked against the order to ensure that the correct boiler has been shipped. **See Figure 42-5.** If there is any shipping damage, a report should be filed immediately.

When bolts are used in the packaging, they must all be removed. If gas and water piping and/or electrical connectors have plugs installed, the plugs should be left in place until removal is necessary. Also, if the boiler needs to switch from natural gas to propane, the work should be accomplished prior to installing the unit. While it may be possible later to perform the conversion, it is often more convenient to do it earlier because of ease of access. Once the boiler is mounted in the desired location, the conversion may be more difficult due to tighter clearances.

Tight clearances in packaged boiler installations may make later service difficult.

Boiler Preparation

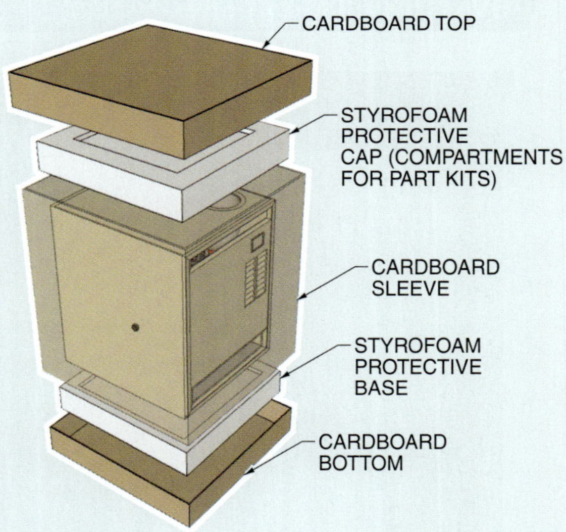

SHIPPING AND PACKAGING MATERIALS

RTG Boiler Manufacturing Inc. Invoice

Invoice #: 000001
Date: 3/4

Installing Contractor: ABC HVAC Contracting
789 Perfect St.
Somewhere, AB 88888

Sold To: William P. Sample
987 Low Energy Blvd.
Somewhere, AB 99999

Ship To: William P. Sample
987 Low Energy Blvd.
Somewhere, AB 99999

Project Name: Wonderful House Inc.
(If different than above)

Quantity	Model #	Item Description	Serial #	Cost
1	TR120RHW	True Natural Gas Hot Water Boiler	1010101	1200.00
1	TRPED24	TR Boiler Pedestal 24″ High	2121212	150.00
			Total Due:	**1350.00**

SAMPLE INVOICE

Figure 42-5. Boilers should be checked when picked up or delivered. Verifying that the correct boiler has been ordered requires checking it against the invoice.

TECH TIP

In some extreme cases, it may not be possible to remove an old boiler. Instead, the boiler may need to be retired in place (RIP). A boiler may be RIP due to excessive removal costs.

Boiler Location

For many boiler installations, the designer or architect has already determined where the boiler will be installed. Nevertheless, it is important for the installer to know the basics of what makes a good location for installing a boiler. When the boiler being installed is a retrofit, the new boiler will be located in the position the old boiler has vacated. As each application is different, it is a good idea for technicians to be familiar with the various boiler installation possibilities and requirements. This is especially true with the new small package boilers being sold today. **See Figure 42-6.**

There are different procedures used for each type of boiler installation. However, the following preliminary considerations are applicable to all types of boilers:

- system water piping availability
- gas piping requirements and availability
- venting requirements and connections
- electrical power connections
- leakage potential (the boiler should not be located where water leakage would cause structural damage, such as an attic where a leak would cause heavy damage through the ceiling)
- area free of combustible materials such as gasoline, rags, chemicals, and other items that could cause a fire

Other installation considerations include clearances, combustion air openings to space, and mounting locations. For safety and aesthetic purposes, specific guidelines must be followed when locating a new boiler for installation. **See Procedure for Determining Location to Install Boiler.**

Boiler Locations

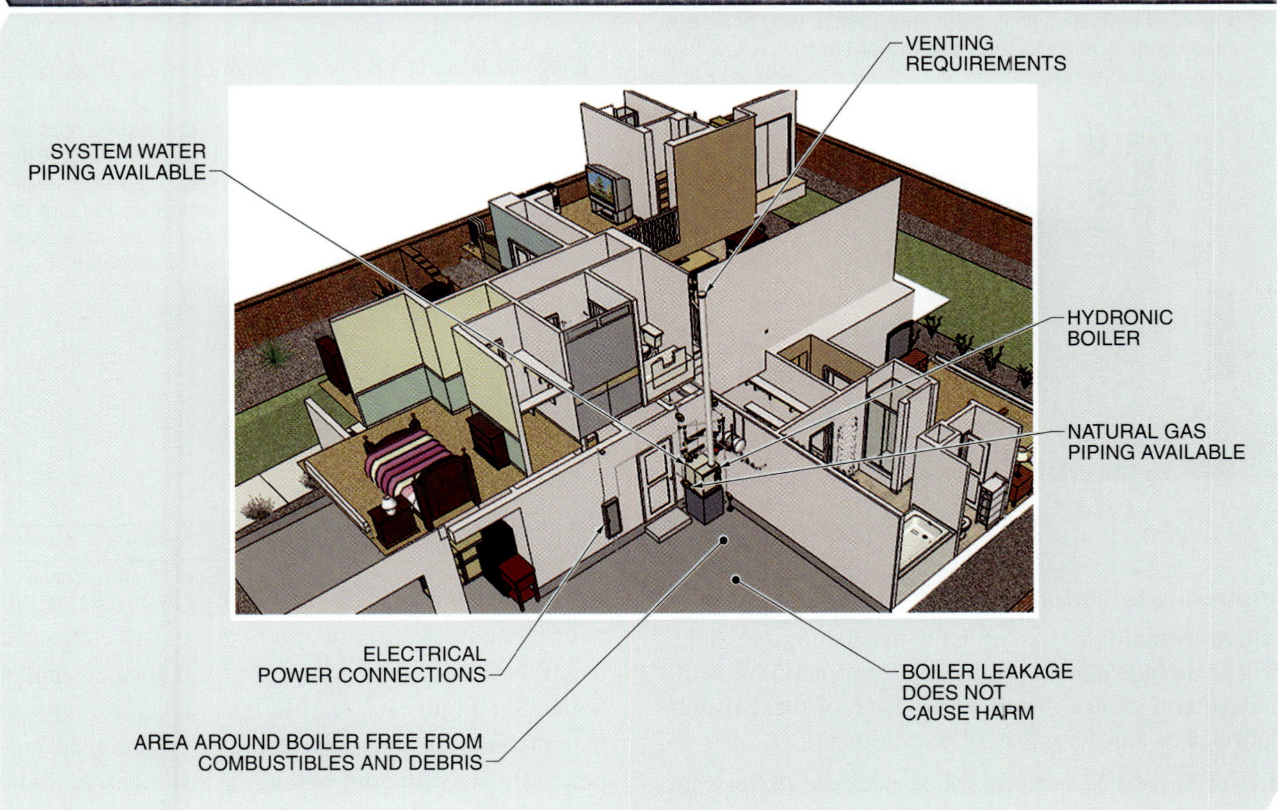

Figure 42-6. Boiler systems can be harmful to buildings and people when leaks occur or combustible gases are present.

Determining Location to Install Boiler

1. Locate the boiler so that, if a water connection or the boiler should leak, water damage will not occur. When such locations cannot be found, a suitable drain pan should be installed under the boiler. *Note:* Locating a boiler in the attic is typically a bad idea because a leak would cause heavy damage through the ceiling.

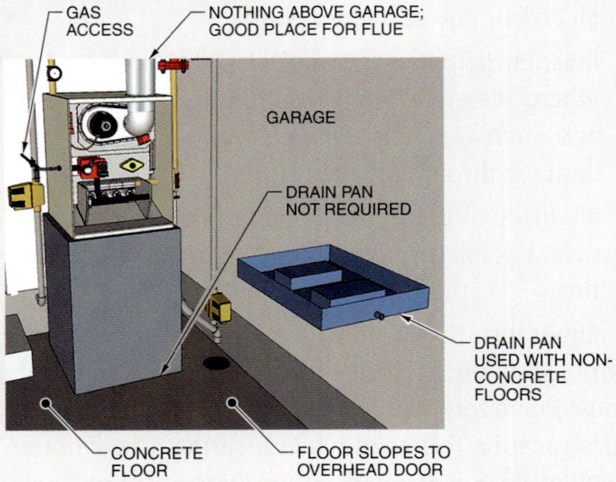

2. The boiler must be installed so that the ignition system components are protected from water during boiler operation and service.
3. Boilers located in a residential garage or in adjacent spaces that open to a garage and are not part of the living space must be installed so that all burners and burner ignition devices have a minimum clearance above the floor. Also, the boiler must be located or protected in such a way that it is not subject to physical damage by a moving vehicle.

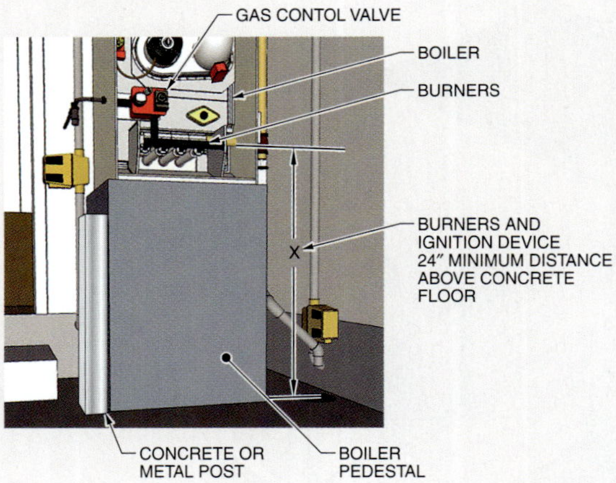

4. Do not install a boiler in any location where gasoline or flammable vapors are likely to be present.
5. The boiler must be installed on a level floor. Some boilers allow combustible wood flooring to be used without additional bases or special flooring buildup. Required clearances from combustible surfaces must be maintained.

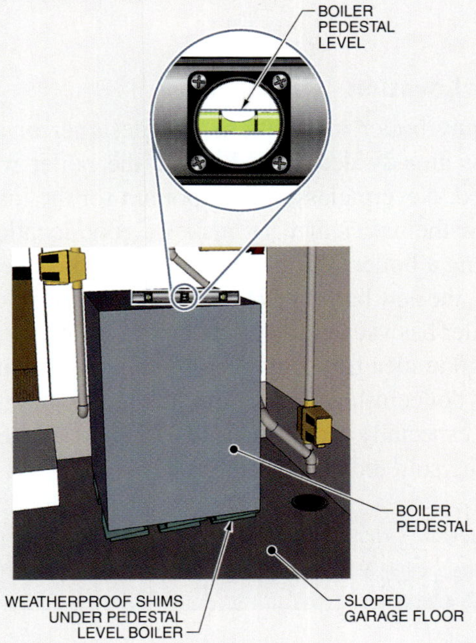

6. Do not install the boiler on carpet or other combustible material other than wood flooring.
7. Outdoor boiler models require the use of the outdoor kit supplied by the manufacturer. Outdoor models must not be installed directly on the ground. An outdoor boiler must be installed on concrete, brick, block, or wood flooring. Outdoor models also have additional special requirements. Always follow all manufacturer installation recommendations.

Mounting Locations

There are multiple types of boiler installations possible. These include existing utility/boiler room floor, wall, closet, and garage installations. Each of these installations has a different set of requirements.

Floor Mount Locations. Most package boilers are mounted on a concrete or wood floor. In the case of replacing an existing boiler, the old boiler is removed and the new boiler is installed in its place. The installation base used with the old boiler must be checked to ensure it is safe and compatible with the new model boiler. **See Figure 42-7.** Also, it should be recognized that installation codes have probably changed significantly since the old boiler was installed. In some instances, the old boiler may have been installed 40 to 50 years previously.

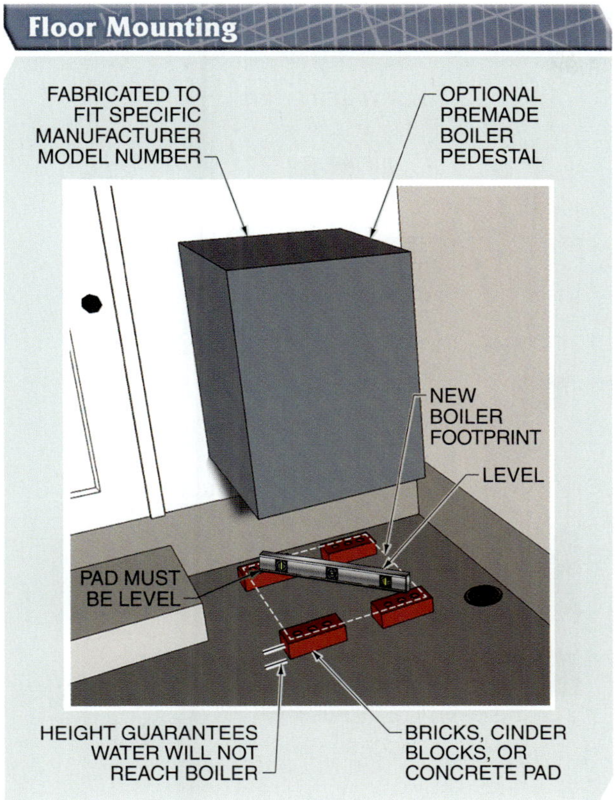

Figure 42-7. The installation base used with an old boiler must be checked to ensure it is safe and compatible with the new model boiler and present codes.

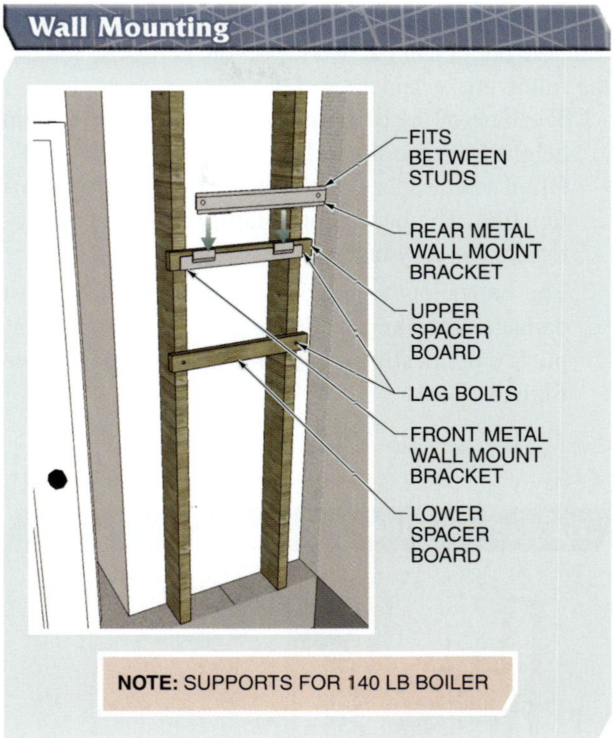

Figure 42-8. Most boilers that can be hung from a wall weigh at least 140 lb. Some models can weigh as much as 260 lb.

Wall Mounting Locations. When comparing old boiler installations with new boiler installations, it is evident that boilers today are much smaller in size. This smaller size allows for installations that were not possible with older boiler types. An example of this is the mounting of a package boiler directly onto a wall. The wall itself is typically constructed of brick, block, cement, or wooden studs on 16″ centers. No matter which type wall is used, the wall must be made to support the weight of the boiler and accessories. Many installations require that metal brackets be attached to the wall in order to attach the boiler. **See Figure 42-8.**

Closet Locations. Many new package boilers can be installed in a closet or in an alcove-type space. These can be very desirable as they reduce the space needed for the boiler and systems, which in turn increases the living space available. In the case of a closet installation, there must be an adequate volume of combustion air available for the boiler as well as minimum clearances from any adjacent combustible materials. Also, any leakage possibility must be considered and allowed for in the installation. **See Figure 42-9.**

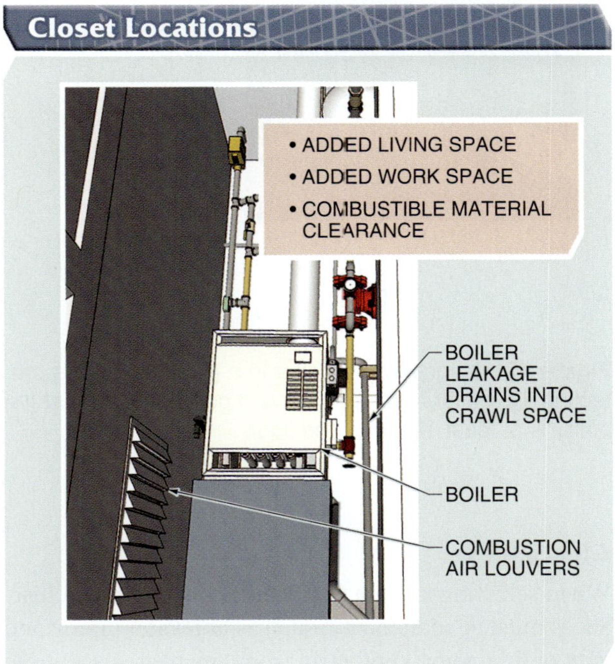

Figure 42-9. Closet installations must provide adequate combustion air for the boiler as well as minimum clearances from any adjacent combustible materials.

Residential Garage Locations. In new house construction, the boiler can be installed in the garage as long as the following considerations are met:

1. The base of the boiler must be a specific minimum height from the floor. This is typically 12″ to 24″. Often a pedestal or platform is used to perform this function. The pedestal or platform may be prefabricated for easy installation.
2. The boiler has to be located or protected against damage from any vehicles. This may be accomplished by locating a post in front of the boiler. **See Figure 42-10.**

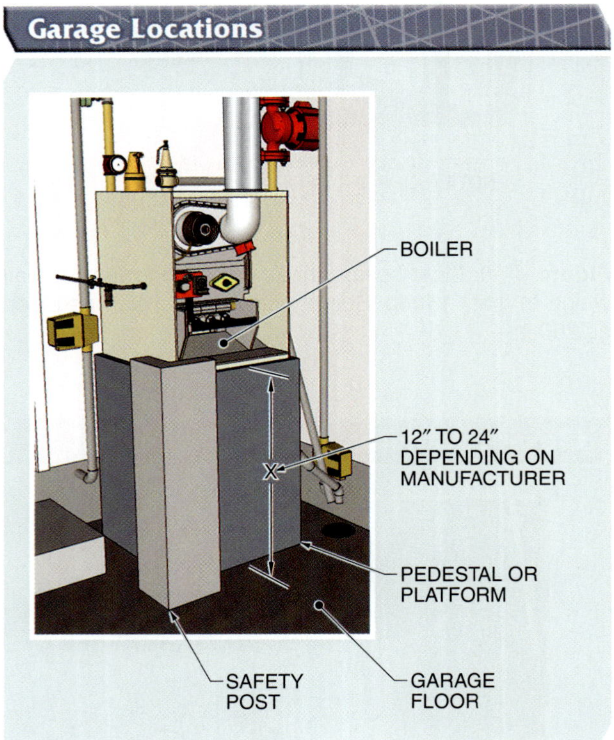

Figure 42-10. Installing a boiler in a garage involves protecting the boiler from vehicles and guaranteeing that the boiler is at least 12″ off of the floor.

Clearances

When selecting an area for installation, common clearances must be maintained, such as hot water piping and vent piping being kept away from any heat-susceptible or combustible materials. Always adhere to the recommended clearances found in the installation instructions provided by the manufacturer. **See Figure 42-11.**

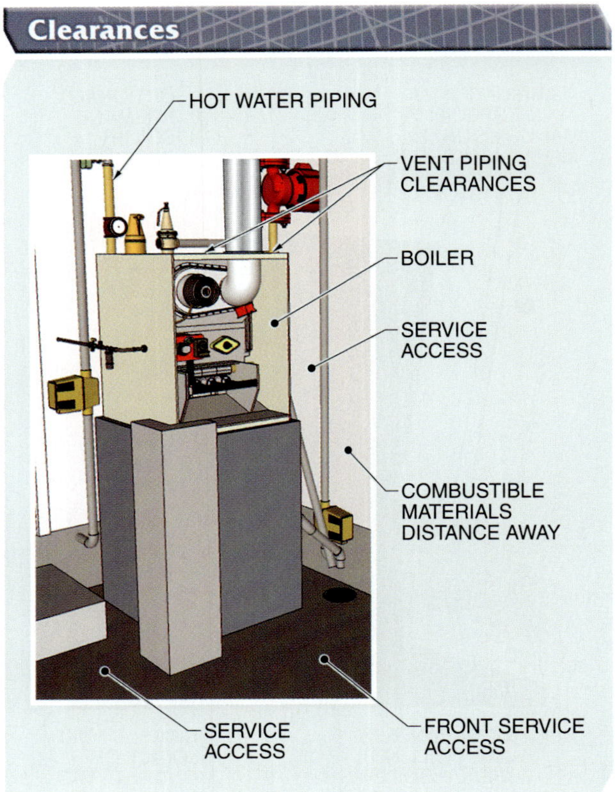

Figure 42-11. Clearances around vent piping and service access to all sides of the boiler must be maintained at all times.

Combustion Air

Air is needed for combustion purposes. The boiler manufacturer will specify air requirements for a boiler. When a boiler is located in the same area as other gas-fired appliances, the requirements for combustion air in the space will change. Typically, the amount of combustion air required for a boiler is measured in square inches per 1 MBH of boiler input. **See Figure 42-12.**

One of the most critical parts of a boiler installation is the provision of adequate combustion air and the venting of the by-products of combustion. The combustion air supply must be completely free of any chemical fumes, which are typically corrosive to the boiler. These include fluorocarbons and other halogenated compounds. Examples of these are refrigerants or solvents such as Freon, trichloroethylene, perchloroethylene, or chlorine. When burned, these chemicals form acids that quickly attack the boiler heat exchanger and burner tube surfaces. Chemical fumes reaching a boiler typically cause early boiler failure.

Combustion Air

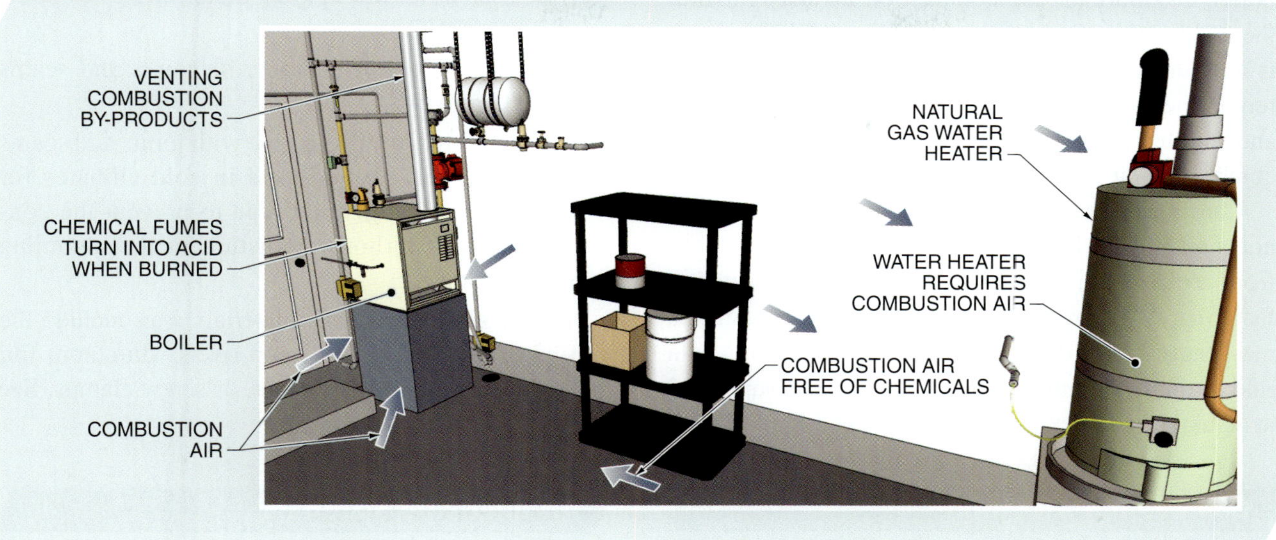

Figure 42-12. The combustion air supplied to a boiler must be free of any chemicals that, when burned, form acids that quickly attack the boiler heat exchanger and burner tube surfaces.

Combustion Air Inlet

Older boiler systems for residential applications typically used room air to supply air for combustion. The room had to have sufficient outside air openings for proper combustion airflow, so louvers in doors were used to provide this airflow. Failure to supply sufficient outside air caused incorrect combustion, unit inefficiency, property damage, and sometimes death. **See Figure 42-13.**

Combustion Air Openings

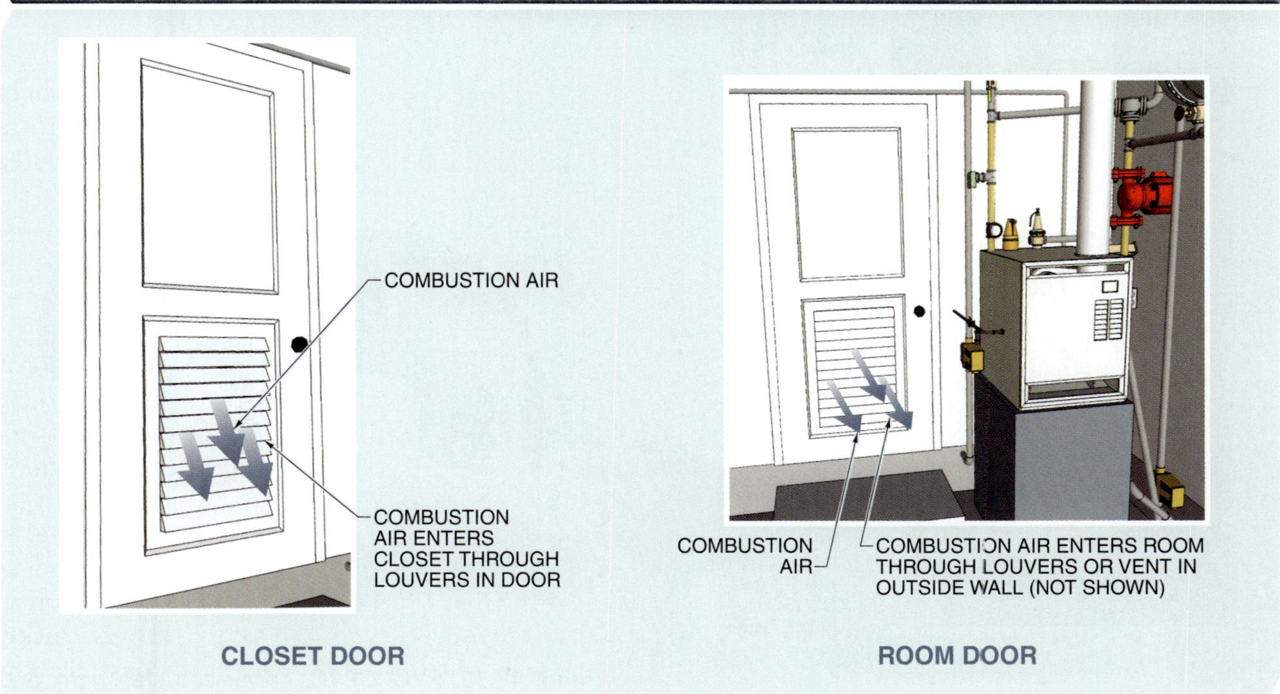

Figure 42-13. Door louvers are often used to provide airflow to a boiler for correct, efficient combustion.

Today, the vast majority of new high-efficiency boiler installations use combustion air brought into the combustion chamber via a PVC or CPVC pipe. It is absolutely vital to follow any and all manufacturer requirements concerning pipe. The location, pipe size, and length are all important. It is common that the inlet combustion air and vent are relatively close together.

Care must be taken to ensure that the combustion air is not contaminated by any sort of chemicals such as those from dry cleaning shops, auto body painting, or manufacturing. When mixed with the inlet air, these chemicals may cause boiler damage and unsafe operation. As a rule of thumb, combustion air piping can be supported at regular intervals at least every 10′. **See Figure 42-14.**

Air Inlet Pipe Materials. The following materials may be used as air inlet piping material:

- PVC or CPVC pipe
- galvanized steel vent pipe with joints and seams sealed properly
- type B double-wall vent pipe with joints and seams sealed properly (often used in cold climates for the combustion-air inlet pipe to prevent the condensation of airborne moisture in the incoming combustion air)

Proper sealing of these materials may include the use of sheet metal screws, pop rivets, aluminum foil tape, duct tape, silicone sealant, or screw clamps. **See Figure 42-15.**

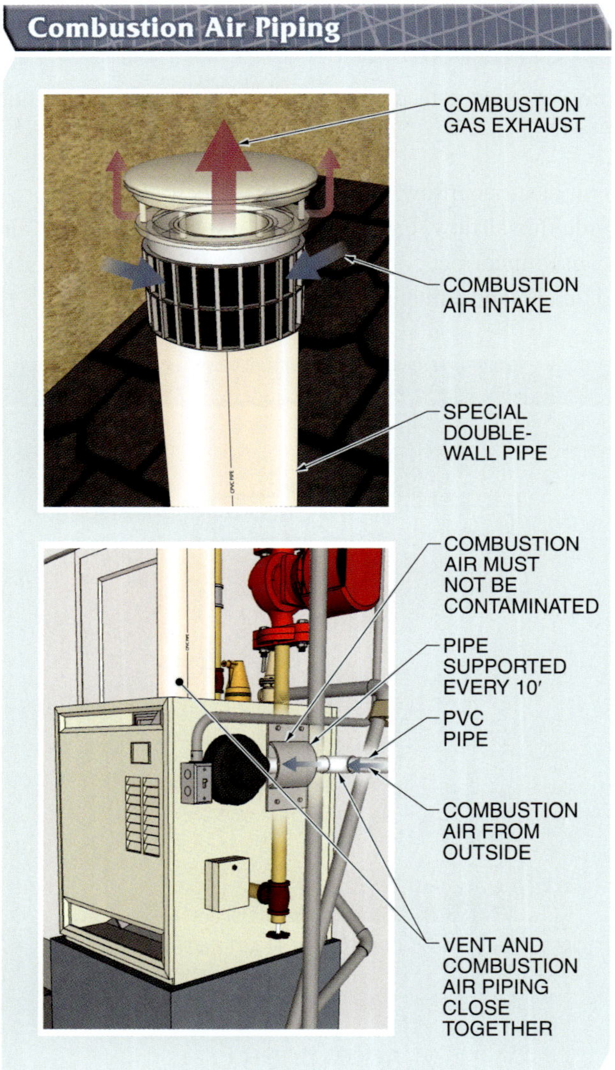

Figure 42-14. As a rule of thumb, combustion air piping should be supported at regular intervals at least every 10′.

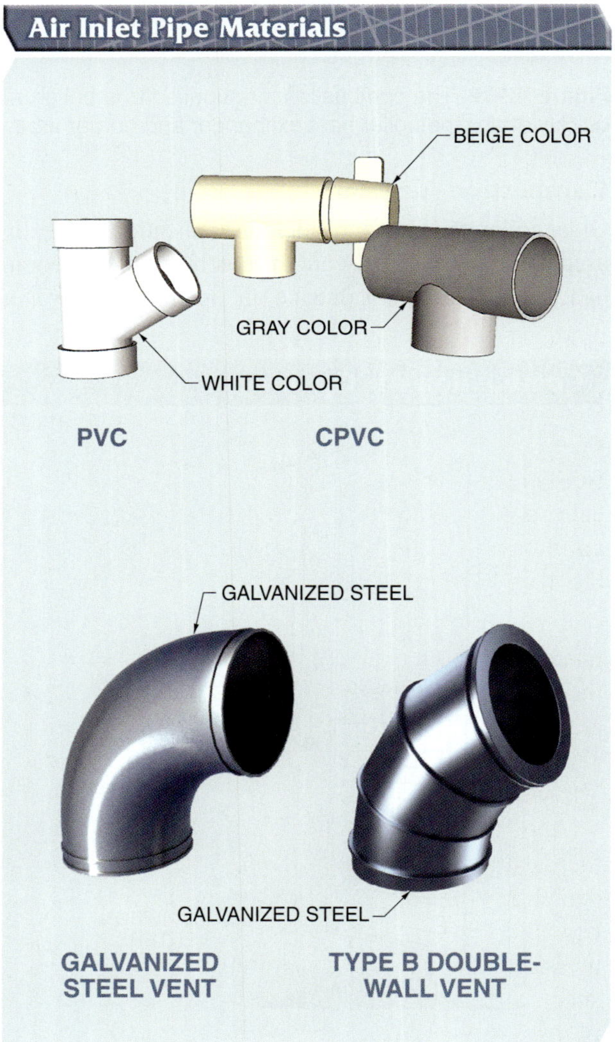

Figure 42-15. PVC, CPVC, galvanized steel vent, and type B double-wall vent are all used for air inlet piping to a hydronic boiler depending on the application.

Length of Air Inlet Pipes. The total equivalent length of the air inlet pipe must not exceed a maximum distance as documented by the manufacturer. Also, all local code requirements should be consulted for flue and combustion-air piping requirements.

> **TECH TIP**
>
> *The proper amount of combustion air and the proper venting of combustion gases are critical to the efficient operation of a boiler. All codes must be followed. Otherwise, a costly reinstallation may be demanded by the inspector. Also, an excessive amount of combustion air can produce a cold draft which may cause comfort problems.*

Combustion Air Pathways

The introduction of combustion air to a boiler can be accomplished in a number of different ways. A boiler can receive combustion air directly from outside, through ducts, or from an interior space. **See Figure 42-16.**

Combustion Air Directly from Outside. When air is taken directly from outside a building with no duct, there must be enough free area in the pathway to provide the amount of air recommended by the manufacturer. The boilers produced by a manufacturer will have different combustion air requirements.

Combustion Air through Ducts. When air is taken from outside using a duct to deliver it to the mechanical room or space, each of the two openings must be sized based on a minimum free area as determined by the manufacturer.

Combustion Air from an Interior Space. When air is taken from another interior space of a structure, each of the two openings specified above should have a net free area per manufacturer requirements. A rule of thumb is 1 sq in. for every 1 MBH of input, with a minimum of 100 sq in.

Note: Care must be taken when using a fan or equipment that exhausts air from the mechanical room, thus depleting the combustion air supply and/or causing a downdraft in the venting system. Spillage of flue products from the venting system into an occupied living space can cause a hazardous condition that must be immediately corrected. When a fan is used to supply combustion air to the boiler room, the installer must make sure that it does not cause drafts that could lead to operational problems with the boiler.

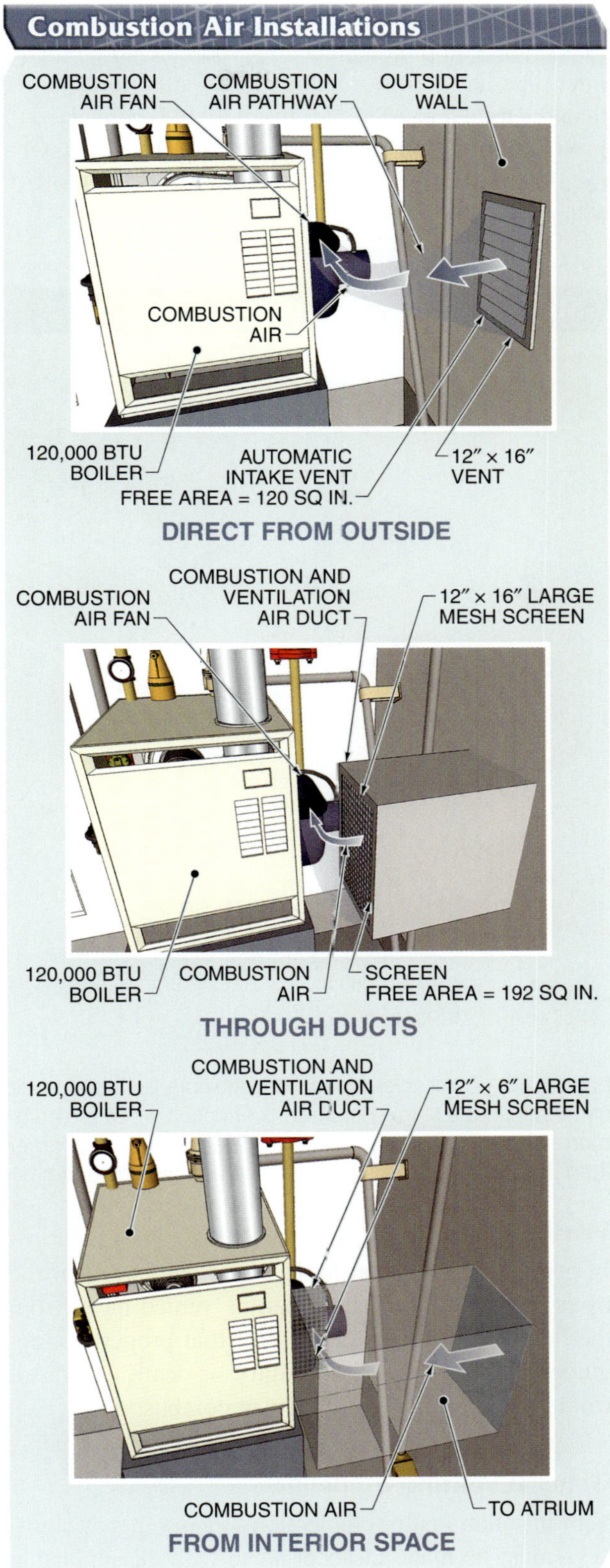

Figure 42-16. Combustion air not taken directly from outdoors must be analyzed for its effects on the boiler and other building systems.

Air inlet pipes from multiple boilers can be combined to a single common connection when the common air inlet pipe has a cross-sectional area equal to or larger than the total area of all the air inlet pipes connected to it. **See Figure 42-17.** Some type of air filtering may be required due to the volume of combustion air needed. Filtering will require the use of larger area intakes.

Multiple Boiler Air Inlets

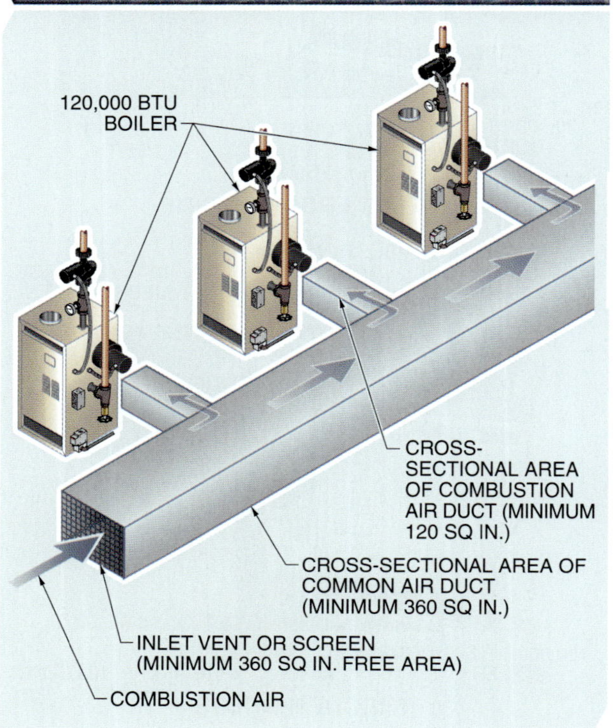

Figure 42-17. Air inlet pipes from multiple boilers can be combined into a single common connection when the common air inlet pipe has a cross-sectional area equal to or larger than the total area of all the air inlet pipes connected to it.

Venting

In general, there are a number of ways in which the by-products of combustion can be vented away from the home. It is absolutely critical that proper venting guidelines be followed, as injury or death can result from breathing the by-products of combustion.

General Venting Guidelines

Vent installations for connection to gas vents or chimneys must be made in accordance with all applicable codes. Examples are Part 7, "Venting of Equipment," of the latest edition of the National Fuel Gas Code, ANSI Z223.1, and applicable provisions of the local building codes. In Canada, examples include the latest edition of CAN/CGA B149, Installation Code for Gas Burning Boilers and Equipment, and applicable provisions of the local building codes.

Conventionally vented applications, where outside air is used, must have adequate combustion and ventilation air supplied to the mechanical room in accordance with the latest edition of the National Fuel Gas Code, ANSI Z223.1, the latest edition of CAN/CGA B149, Installation Code for Gas Burning Boilers and Equipment, or applicable provisions of the local building codes.

The vent connection is made directly to the top of the boiler. No additional draft diverter is required. The connection from the vent to the stack must be made as directly as possible. The negative draft in conventional vent installations is typically within the range of negative 0.02″ WC to 0.05″ WC to ensure proper operation. **See Figure 42-18.** All draft readings are made while the boiler is in stable operation (approximately 2 min to 5 min after firing). Boilers should be located as close as possible to the chimney or gas vent. Conventional venting systems use type B double-wall vent pipe.

General Venting Guidelines

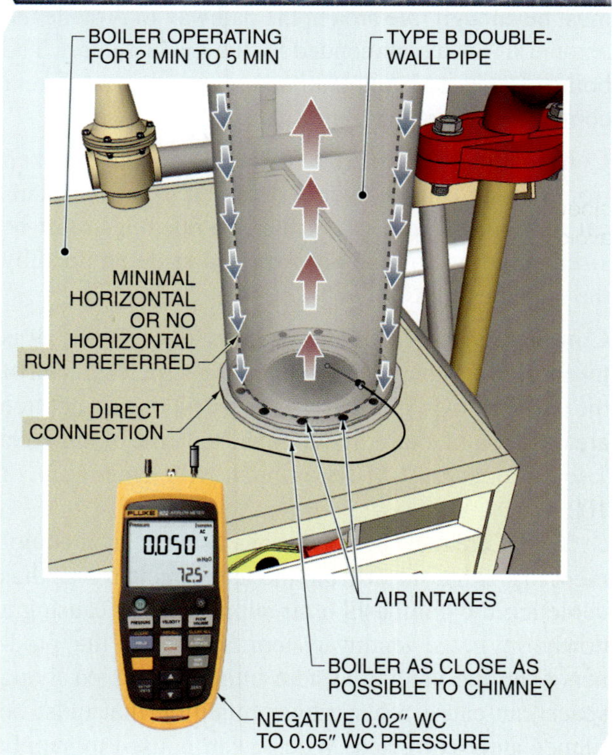

Figure 42-18. After installation, boiler venting must be tested to ensure that in conventional vent installations a negative draft between 0.02″ WC to 0.05″ WC is present.

Direct vent systems have specific vent kits and material requirements noted for each application. Any vent materials not provided or specified must be listed by a nationally recognized testing agency for use as vent material. Restrictions in the vent pipe, reduction of vent pipe size, 90° elbows, and long horizontal runs should be avoided. **See Figure 42-19.**

Avoiding Venting Difficulties

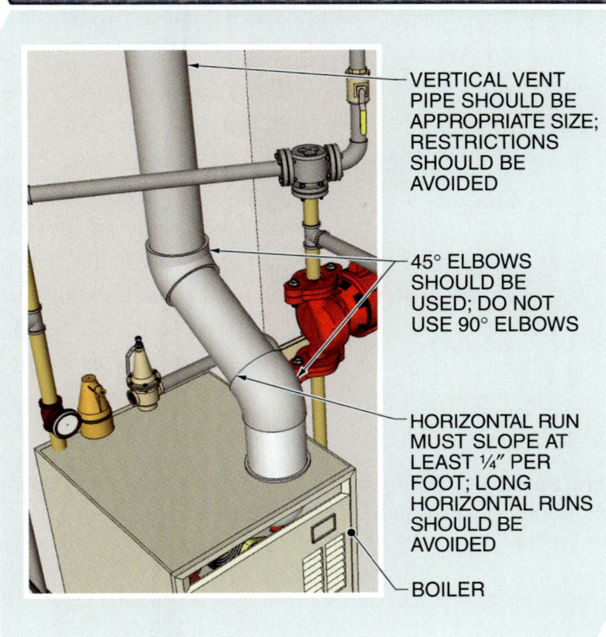

Figure 42-19. Restrictions, reductions in pipe size, 90° elbows, and horizontal runs without a slope should all be avoided in vent piping.

Some packaged boilers can be installed outdoors and thus can directly vent the by-products of combustion.

Horizontal portions of the venting system must be supported to prevent sagging. Horizontal runs must slope upwards not less than ¼″ per ft between the boiler and the vertical vent pipe. The weight of the venting system must not rest on the boiler. Adequate support of the venting system must be provided in compliance with applicable codes. All connections should be secured with rustproof sheet metal screws.

Vent Piping

One of the most critical items when installing a boiler is the vent piping. Improper installation of piping can result in carbon monoxide poisoning and death to building occupants.

Many new boiler installations are power vented. Power venting uses a combustion fan to push the by-products of combustion up the flue. Older and/or existing boilers often use natural draft venting, which uses a standard flue or chimney. In cases where boiler replacement is performed, the venting parameters generally change. **See Figure 42-20.**

Replacement Flues

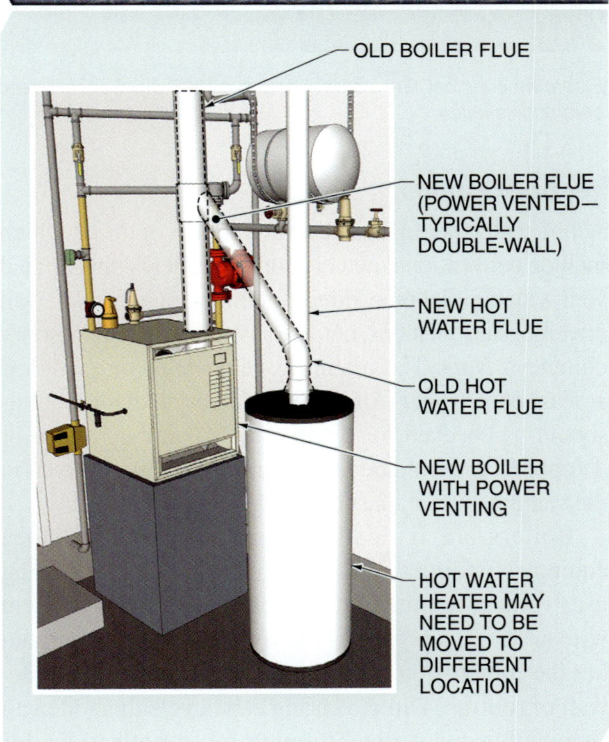

Figure 42-20. Flue parameters can never be taken for granted. Checks and testing must be performed so that the new boiler and all appliances operate properly.

When an old boiler is removed from an existing flue system, a larger issue arises. The issue is that other appliances might use the same flue or chimney as the existing boiler. In this case, when the existing boiler is removed from the flue system, the flue will no longer be the correct size for the existing application. The entire flue system may need to be changed and any remaining appliances, such as a hot water tank, checked for proper vent operation. Once this is done, the new boiler flue system can be installed.

When multiple boilers are used and must be piped, other venting issues arise. For this and other nonstandard applications, the manufacturer as well as local codes should be consulted. Also, kits that simplify both the venting and combustion air piping are available. These kits are highly recommended as they simplify the installation process.

Electromechanical flue components typically are bench tested before installation.

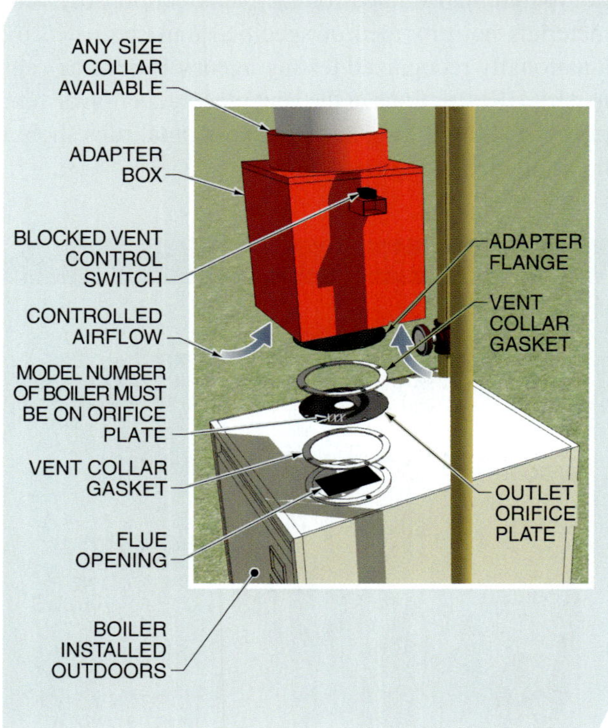

Figure 42-21. Some boiler applications place the boiler outdoors, which requires special flue kits to be used to guarantee that the gases of combustion go up the flue properly.

Venting Options

A hot water boiler has several venting options, which include conventional negative draft, vertical conventional vent with air inlet pipe, direct venting with either sidewall or vertical terminations, outdoor installations, and masonry chimneys. *Note:* The venting system should be examined at least once a year. All joints and vent pipe connections should be checked for tightness. Also, any corrosion or deterioration should be noted. Any problems observed in the venting system must be immediately corrected.

Boilers are typically shipped from the factory equipped for direct vertical venting. This type of venting uses vertical terminations. A vertical conventional vent with an air inlet pipe uses the vent for flue products and has the combustion air pipe connected to either the sidewall or rooftop. Direct venting can have either sidewall terminations or vertical through-roof terminations for flue products and combustion air. Outdoor installation requires the installation of a special kit provided by the manufacturer. **See Figure 42-21.**

When a high-efficiency gas-fired boiler is vented directly into a masonry chimney, the chimney must be properly sized and properly lined for the installation of such a boiler. Venting of a high-efficiency boiler into a cold or oversized masonry chimney can result in operational and safety problems.

Exterior masonry chimneys with one or more sides exposed to cold outdoor temperatures are more likely to have venting problems. The temperature of the flue products from a high-efficiency boiler may not be high enough to sufficiently heat the masonry structure of the chimney to generate proper draft. This will result in condensing of flue products, damage to the masonry flue/tile, insufficient draft, and possible spillage of flue products into an occupied living space. **See Figure 42-22.**

All chimney systems should be carefully inspected before installation. When there is any doubt about the sizing or condition of a masonry chimney and its lining, it must be relined with a properly sized and approved chimney liner system.

Masonry Chimneys and High-Efficiency Boilers

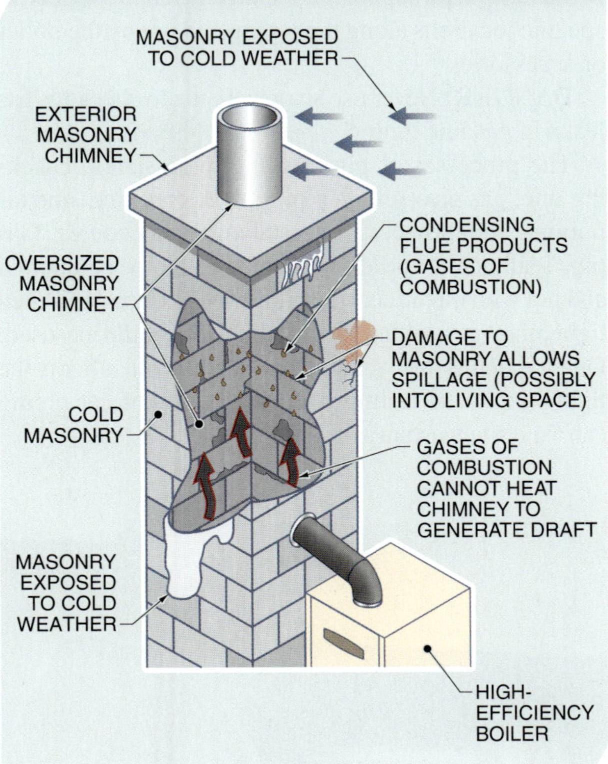

Figure 42-22. When installing a high-efficiency boiler, all chimney systems must be carefully inspected. If there is any doubt about the masonry chimney, it must be relined with a properly sized and approved chimney liner system.

A masonry chimney must be carefully inspected to determine its suitability for the venting of flue products from a high-efficiency boiler. A clay-tile-lined chimney must be structurally sound, straight, and free of misaligned tiles, gaps between liner sections, missing sections of liner, or any signs of condensate drainage at the breeching or cleanout. If there is any doubt about the condition of a masonry chimney, it must be relined. *Note:* An unlined masonry chimney must not be used to vent flue products from a high-efficiency boiler.

An unlined chimney must be relined with an approved chimney liner system when a new high-efficiency boiler is being attached to it. Metallic liner systems (type B double-wall flexible or rigid metallic liners) are recommended. Local code officials should be consulted to determine code requirements. **See Procedure for Venting Boiler to Outdoors.**

Venting Boiler to Outdoors

Procedure

1. Keep venting areas free of obstructions, combustibles, and flammable materials.

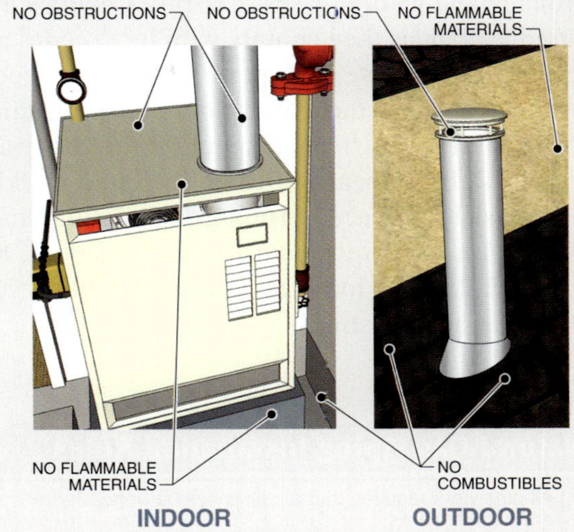

2. Maintain a minimum clearance from the rear of the outdoor vent cap to combustible surfaces.

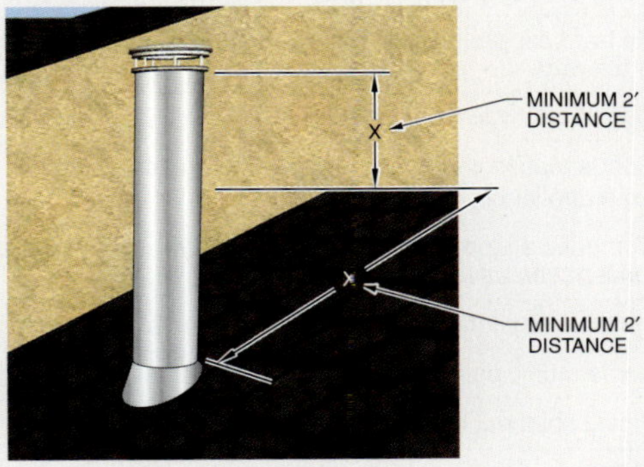

3. To avoid a blocked air inlet or blocked flue condition, keep the air inlet, flue outlet, and drain slot clear of snow, ice, leaves, and debris.
4. Do not locate boiler where high winds can deflect off of adjacent walls, buildings, or shrubbery and cause recirculation. Recirculation of flue products may cause operational problems, bad combustion, and/or damage to controls.

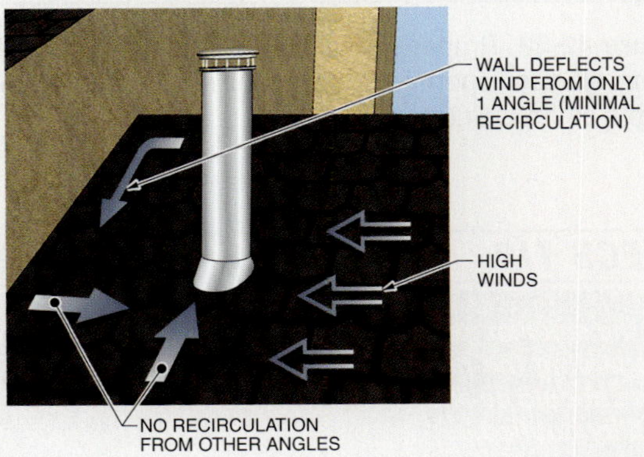

Gas Supply Piping

All gas connections must be made with pipe-joint compound that is resistant to the action of liquefied petroleum and natural gases. Existing gas lines may be used as long as they comply with local codes and ordinances. Because it is possible that the codes have changed since the existing boiler was installed, the gas piping must be inspected carefully and any pieces that appear not to be in prime working order must be replaced. Tubing installations must comply with approved standards and practices. **See Figure 42-23.** The local gas utility should be consulted with any questions.

Natural Gas Piping Installation Rules

1	All pipe work requires that a leak check be performed.
2	High-pressure pipes are required to have a pressure regulator at the location of attached appliance.
3	Three types of pipes can be used: flexible stainless steel, black iron, and copper (type k and type l).
4	No hard copper piping and in some locations, no black iron is allowed.
5	Gas shutoff valves must be located within 3′ of appliance.
6	Condensation catch drip leg must be present at any fixed appliance.
7	Compatible pipe joint compound must be used. Teflon tape must not be used.
8	Movable appliances require safety valves at appliances.
9	Some municipalities require safety valves at appliances.
10	Piping shall not be embedded in concrete floors.
11	All fittings and joints of piping larger than 2″ in diameter shall be welded.
12	Bending of metallic pipe in place of fittings is prohibited.
13	Residential piping systems shall be sized for a maximum pressure loss of 0.3″ WC.
14	Piping for commercial systems shall be air-tested at 40 psi for a period of 15 min with no loss of pressure.
15	Systems over 10 psi shall be tested at 100 psi.

Figure 42-23. Proper installation of gas piping is critical to the efficient operation of a boiler and the safety of all people in the building.

TECH TIP

If the smell of gas is present or there is any question regarding the proper installation of gas piping, the gas should be shut off immediately and the local gas utility notified.

The gas piping to a boiler includes a drip leg and piping support. A drip leg is a sediment trap that is created in the field. Pipe supports are hangers that are spaced at specific locations along the piping and not by the boiler or accessories.

DANGER: Never use an open flame to check for gas leaks in gas line piping.

The processes of pipe sealing, pipe sizing, checking inlet gas pressure on piping, leak checking, and air purging are all necessary installation procedures. Gas pipe sealing must be accomplished with pipe compound and not with thread seal tape (allowed in Canada). When tightening gas piping, two wrenches should be used. One wrench is the backup wrench while the other is the tightening wrench. Improper installation of gas piping can void the warranty of a boiler.

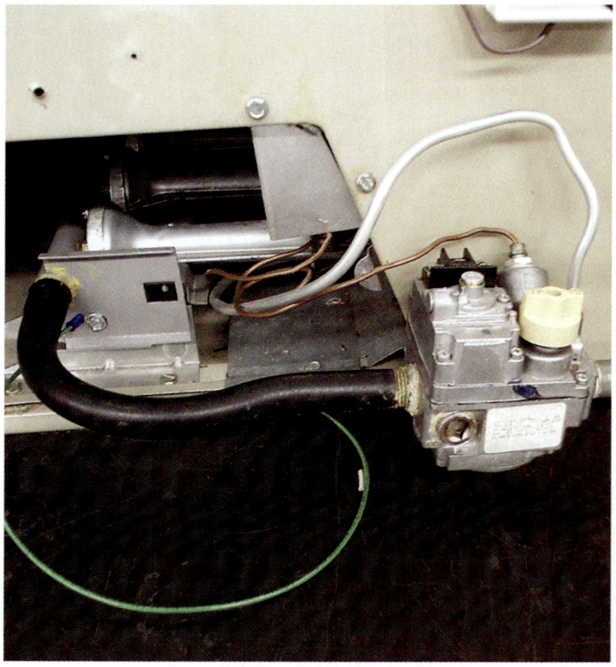

Packaged boilers may use combination gas valves for atmospheric burners.

The inlet gas piping must be sized for proper boiler capacity and, after the piping is installed, the inlet gas pressure must be checked. All gas piping, including the boiler gas control valve, must be checked for leaks. Air purging is a procedure that evacuates air from the gas line before placing the boiler in operation. **See Procedure for Purging Air from Boiler Gas Control Line.**

Purging Air from Boiler Gas Control Line

1. Open all windows and outside doors possible. Purging of gas must only be performed in a well-ventilated area or space by venting the contents of the gas line outside. *Note:* Natural gas is lighter than air and will rise. This process is illegal in Canada.
2. Identify and eliminate all sources of ignition (standing pilots, open flames, cigarettes, electric switches, appliances, and vehicles) in spaces where gas may be vented. *Note:* Do not purge into a confined space or where gas can accumulate.

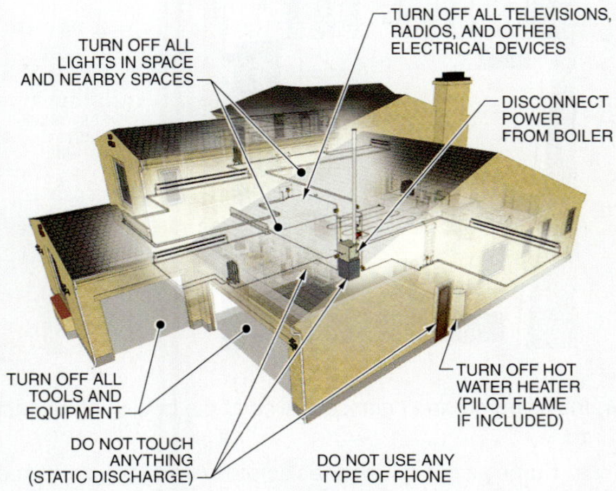

3. If the boiler does not have a pilot flame, partially loosen (to handtight) the gas line fitting into the gas control valve.

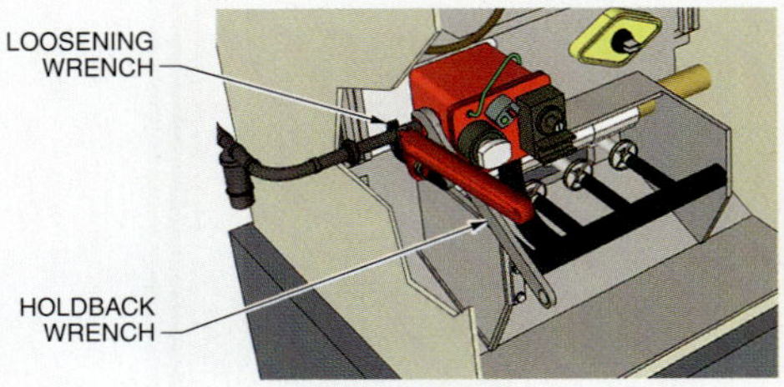

4. Open the gas shutoff valve.
5. If the boiler has a pilot flame, push down on gas control valve knob while it is set to pilot to bleed air slowly from the gas line until gas is released.

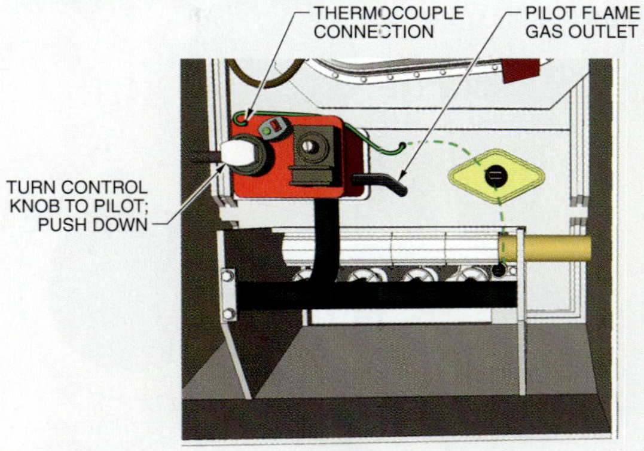

6. Loosen fitting a little further so air and gas are released from fitting. If possible, use gas detection meters during gas release operation to prevent gas from accumulating and creating a combustible condition. *Note:* Do not depend on sense of smell to detect the presence of natural gas.
7. Tighten fitting at gas control valve when gas is released.
   a. Release control knob when gas is released.
   b. Turn control knob to OFF.

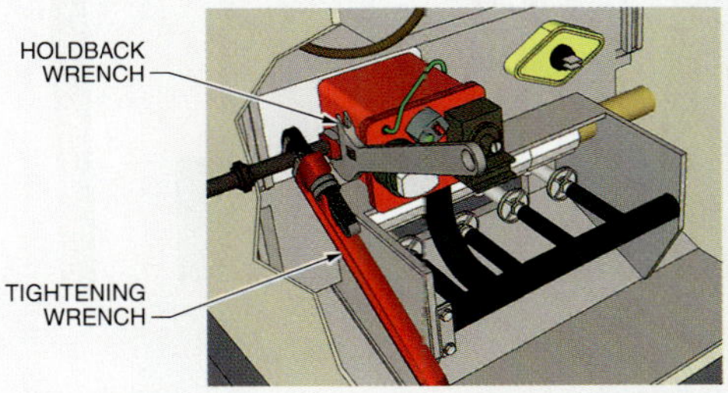

8. Close gas shutoff valve until boiler is placed into operation.

Procedure

Gas Pressure Measurement. For proper boiler operation, the gas pressure to the boiler must be within the proper limits. In the case of high gas pressure, a lockable pressure-reducing valve is installed ahead of the inlet to the gas control valve. This valve is installed a minimum distance from the gas control valve.

The gas pressure is checked at the tap on the gas control valve inlet with a manometer. Typically gas control valves are set at the factory and should not require adjustment in the field, but on some occasions, the valve may need to be adjusted.

Hydronic Piping

The most recommended type of water piping for a boiler is the primary-secondary piping arrangement. This piping system allows the boiler to operate at the proper temperature and pressure while avoiding temperature shutdowns and poor flow problems.

Water Inlet and Outlet Connections to Boiler. When the boiler installation is a retrofit, existing water connections may be used if they are found to be in acceptable condition after inspection. When a boiler is being installed for the first time and for ease of service, unions on the inlet and outlet connections of the boiler should be installed. The connection to the boiler marked "inlet" on the header must be used for return flow from the system. The connection on the header marked "outlet" must be connected to the supply side of the hydronic system. **See Procedure for Installing Hydronic Piping and Components to Boiler.**

Procedure

Installing Hydronic Piping and Components to Boiler

1. Inlet and outlet piping openings are clearly marked. Locate these openings and install the piping using the proper pipe sealing compound.

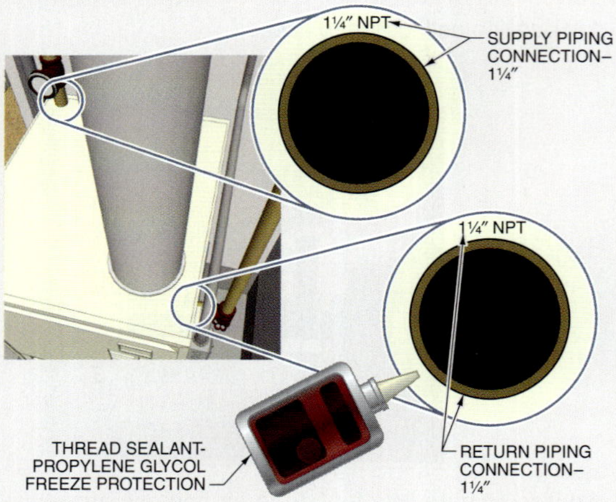

2. Install a low-water cutoff in the supply piping above the boiler. Electrode-probe-type cutoffs are typically used. Many jurisdictions require that a low-water cutoff be installed to ensure safe operation of the boiler in the event of a low-water condition.

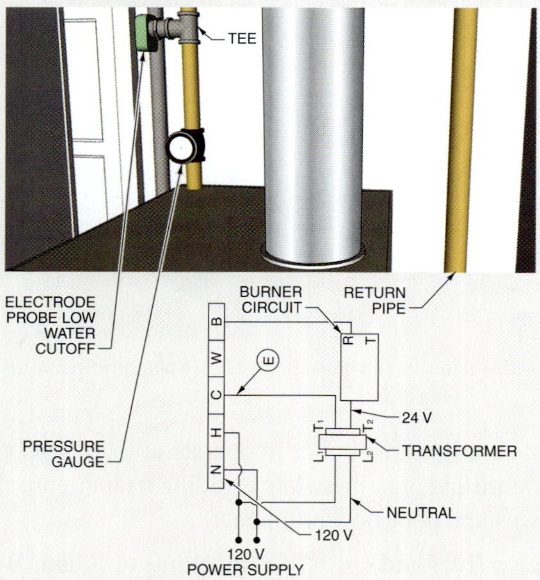

3. Add propylene glycol freeze protection to the hydronic system. Some systems require freeze protection in the event of system failure or shutdown because of cold ambient conditions. Adding propylene glycol consists of adding the proper antifreeze to the water. Only additives approved by the manufacturer can be added.

   CAUTION: Do not attempt to use automotive-type antifreeze in the system.

   a. Determine the appropriate number of gallons of propylene glycol (50-50 mixture) needed for the hydronic system.

   b. Turn boiler and pumps off. Open bottom drain, all valves, and air vents.

   c. Close bottom drain and remove air vent, cover, or pipe fitting from the top of the boiler and pour the correct amount of propylene glycol into the boiler.

   d. Replace any piping or devices that were removed from the boiler.

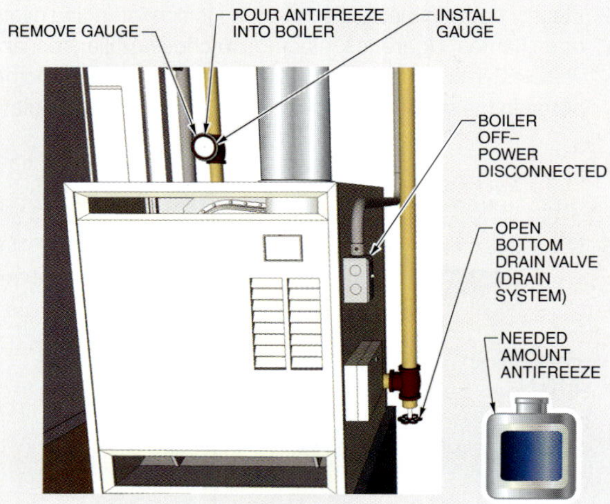

4. Install purge and balance valves at the beginning of each zone.

   a. Purge air out of the zones before boiler startup and after repairs.

   b. Use balance valves to ensure proper water flow to each zone.

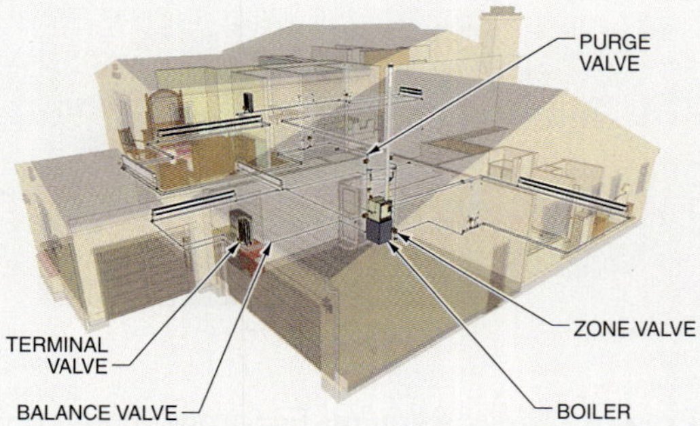

5. Install backflow preventer and pressure-reducing valve in water makeup line. Typically, the setting of the pressure-reducing valve is below 15 psi.

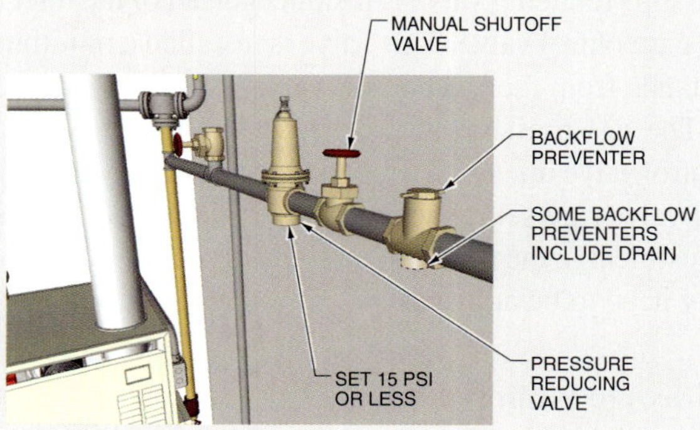

continued on next page

Installing Hydronic Piping and Components to Boiler *(continued)*

6. Install circulating pump before boiler. Various locations are possible.

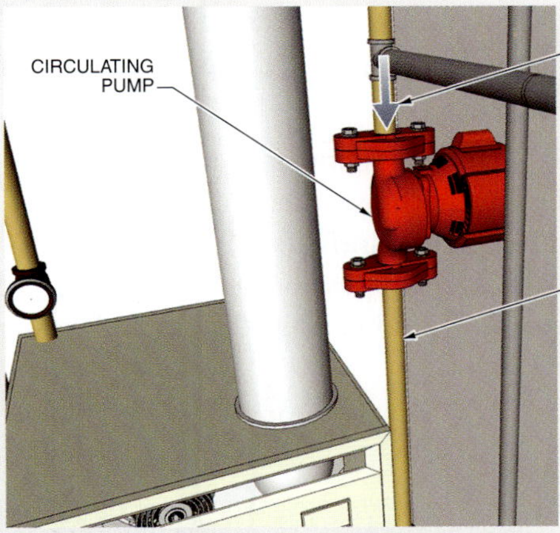

7. Size the expansion tank for the proper water volume of the hydronic system and install the tank on the suction side of the circulating pump. Manufacturer literature will provide guidelines.

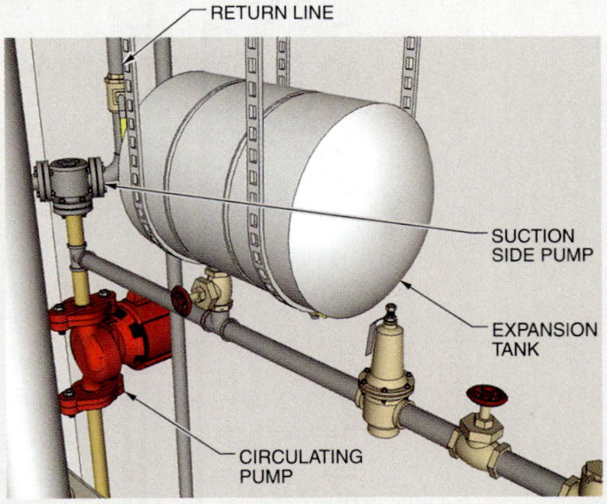

8. Install air eliminator device at the top of the boiler. For systems with air problems, there may be more than one eliminator. Eliminators should be installed at the highest place in the system.

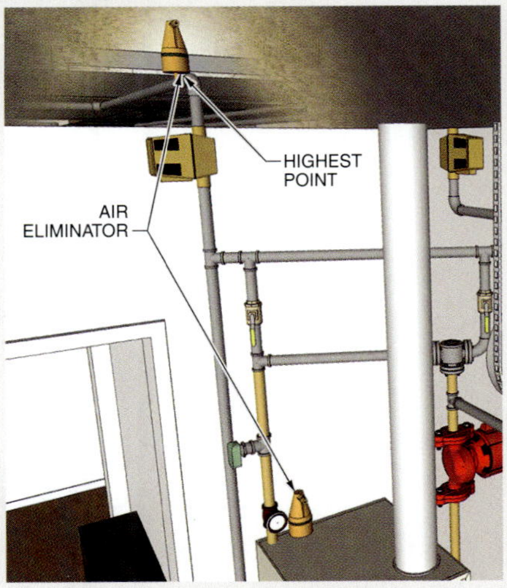

9. Install bottom drain valve. A drain valve is installed at the lowest point in the system to allow draining for cleaning and/or maintenance of the boiler.

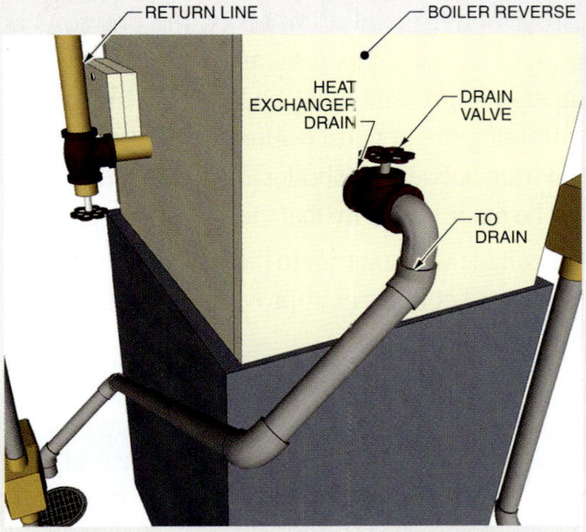

10. Install safety relief valve on top of boiler. A boiler relief valve is installed to prevent over-pressurization of the boiler.

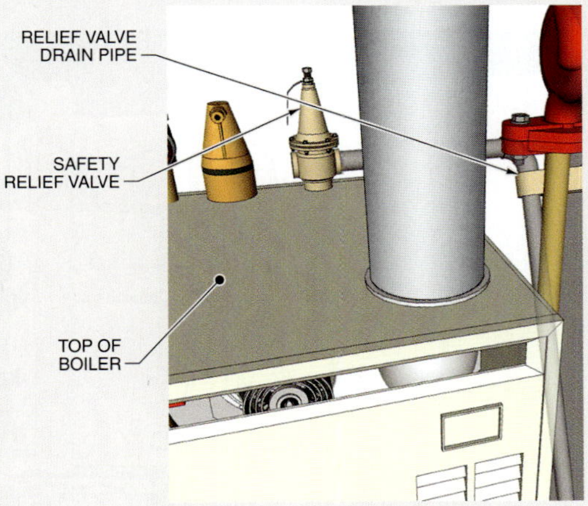

11. Install isolation valves before zone piping.

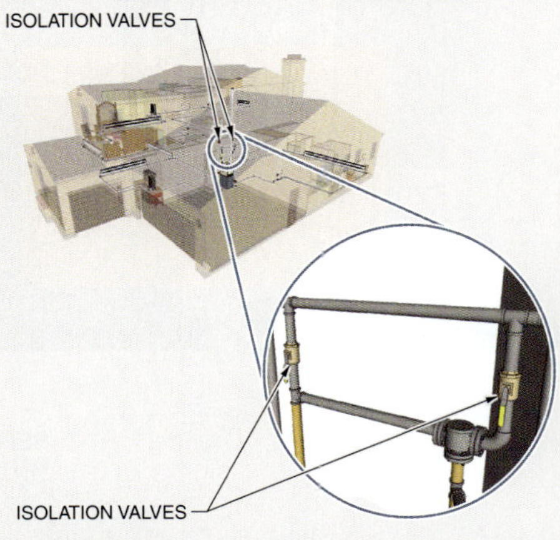

12. Install unions in locations that allow for easy service to the piping system at a later date.

Boiler Wiring

When making wiring connections to a boiler, all code requirements must be followed. A qualified electrician should make the electrical connections for a boiler. As with other boiler installation tasks, the previous boiler wiring may be reused as long as it is in good condition and meets all present local codes. **See Figure 42-24.** The following are general boiler wiring considerations:

- All wiring between the boiler and field-installed devices must be made with wire that meets all appropriate codes.
- Line voltage wire exterior to the boiler must be enclosed in approved conduit or approved metal-clad cable.
- The circulation pump must run continuously when the boiler is being fired.
- The system must be full of water before the boiler is energized, otherwise serious damage may result.
- The boiler must be provided with proper overload protection.
- The wall thermostat should be installed on an inside wall. The thermostat should not be installed in an area affected by drafts, sunlight, light fixtures, hot or cold water pipes, or near a fireplace. **See Procedure for Connecting Electrical Wiring to a Boiler.**

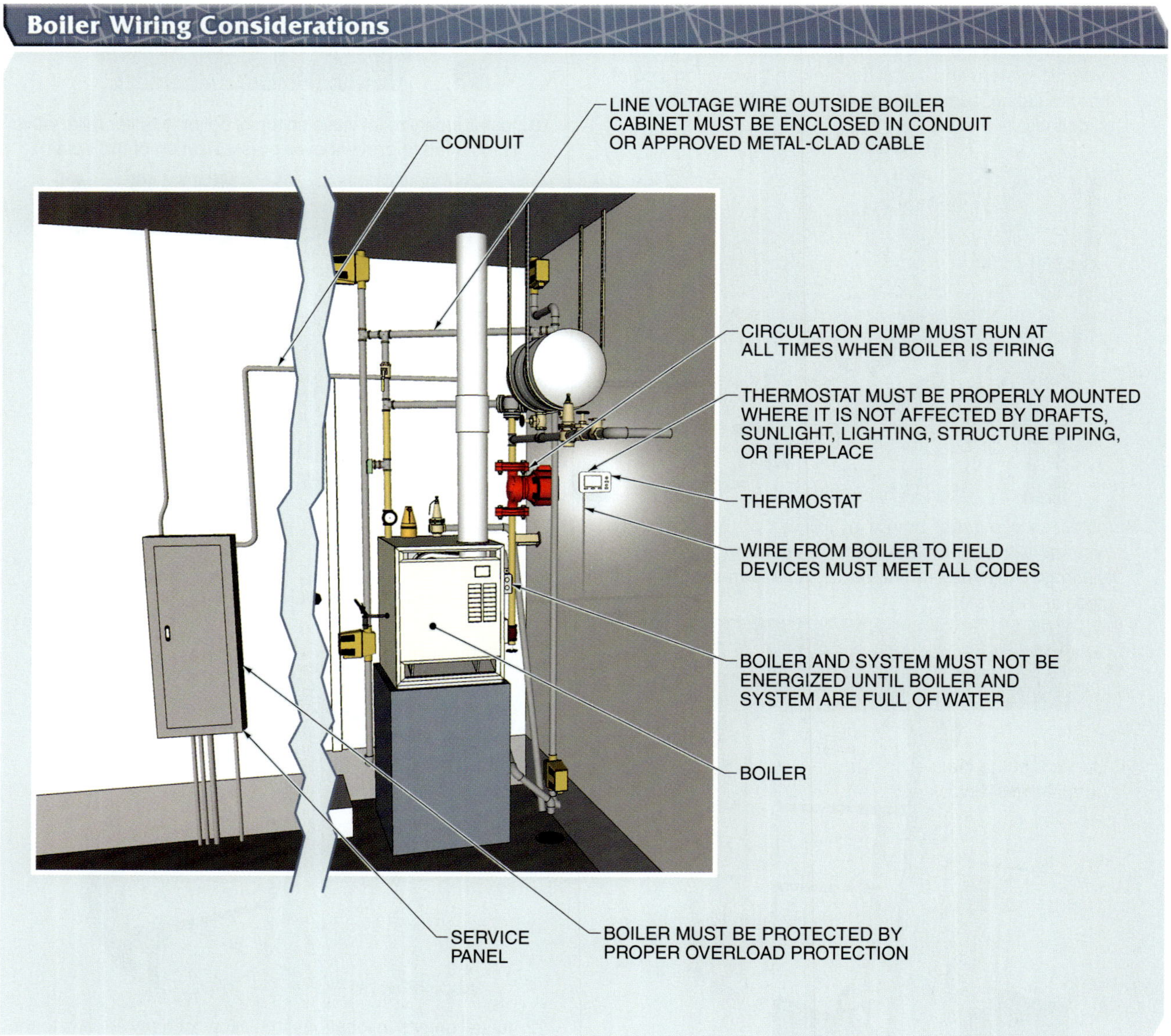

Figure 42-24. An electrician should follow all code requirements when making electrical connections to a boiler.

Connecting Electrical Wiring to a Boiler

1. Turn power off while wiring is performed and do not turn power on until boiler startup is performed.
2. Install disconnect to the boiler according to code.
   a. Mount disconnect box to boiler cabinet.
   b. Run conduit from service panel to disconnect box.
   c. Pull wire from service panel to disconnect box switch and connect wires to service panel and to disconnect switch.

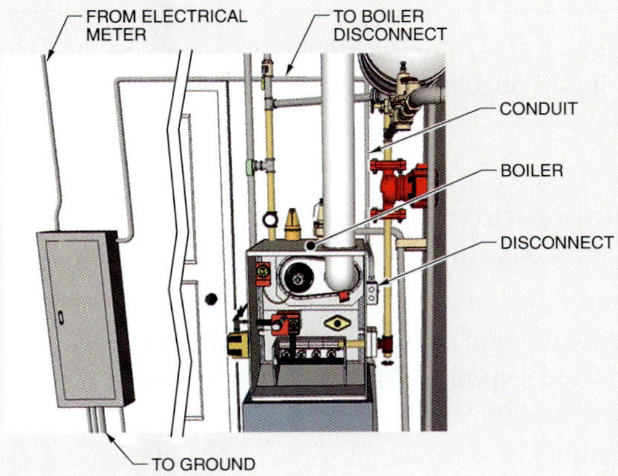

3. Boilers typically have two terminal strips: one for low voltage (24 V) and one for line voltage (115 V). Connect the wiring to the boiler via knockouts located on the sides or front of the boiler.

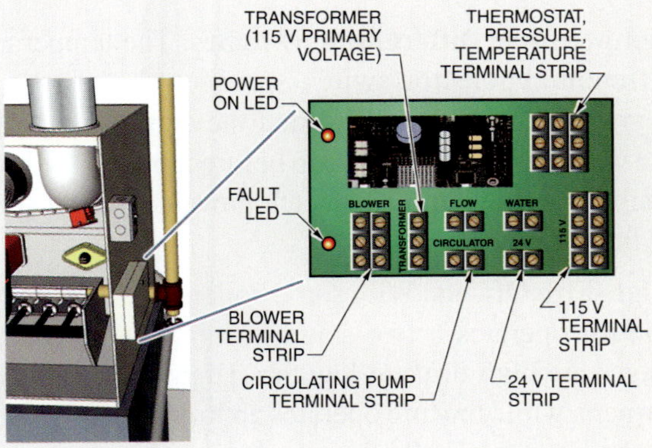

4. Check that the circulating pump is wired to the circulating pump terminal block or the line voltage terminal block.

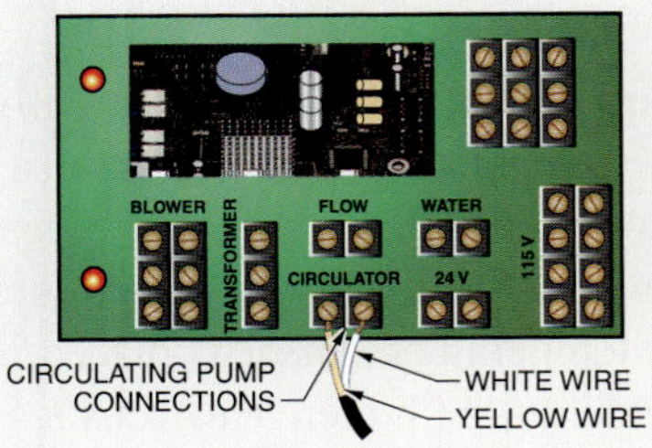

5. Verify that the entire electrical system is grounded properly.

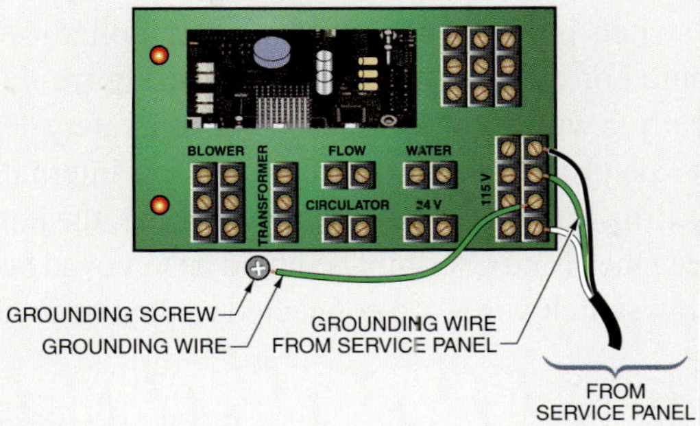

6. Establish the control system requirements. Typically the control units connected to the boiler include the following:
   a. System thermostat
   b. Outdoor temperature sensor
   c. Flow switch (used to prove water flow before boiler is permitted to function)
   d. Induction fan motor
   e. Boiler sequencer (It is not uncommon to have a main controller that will operate multiple boilers as needed to satisfy the load. Provision is also made to "lead/lag" boiler operation, such as operating one boiler as the first boiler to come on-line, then the other boilers. The sequencer will bring on a second boiler as needed if the first boiler fails or cannot meet load demand. It may also change the sequence as needed so that all the boilers have an equal amount of run time. As always, consult the manufacturer for specific control system requirements.)
   f. Auxiliary devices such as water temperature sensor and air pressure switch

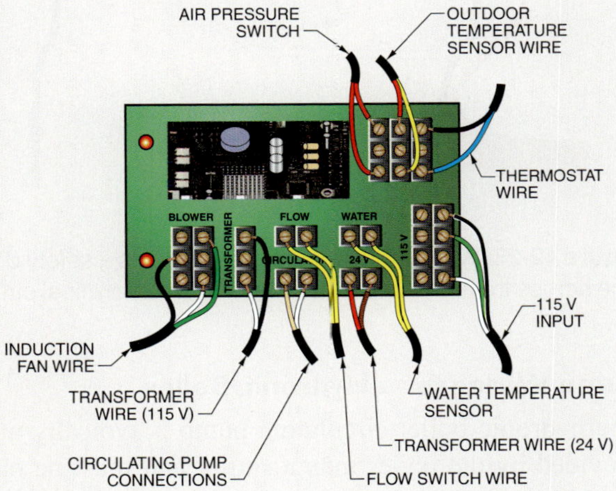

7. Check the wiring voltages and operation with a digital multimeter before system startup. *Note:* Checking voltages is performed with the gas supply valve to the boiler in the OFF position.

Water Flow Switch Wiring

A *water flow switch* is an electromechanical device that stops boiler operation when water flow is lost. A water flow switch is available as a factory supplied option on some hot water heating boilers. **See Figure 42-25.** Typically, a water flow switch is wired as a safety device across the proper terminals on the boiler's internal terminal strip. For some types of flow switches, the jumper between the H and C terminals should be removed before the flow switch wires are connected to the terminals.

Water Flow Switch Wiring

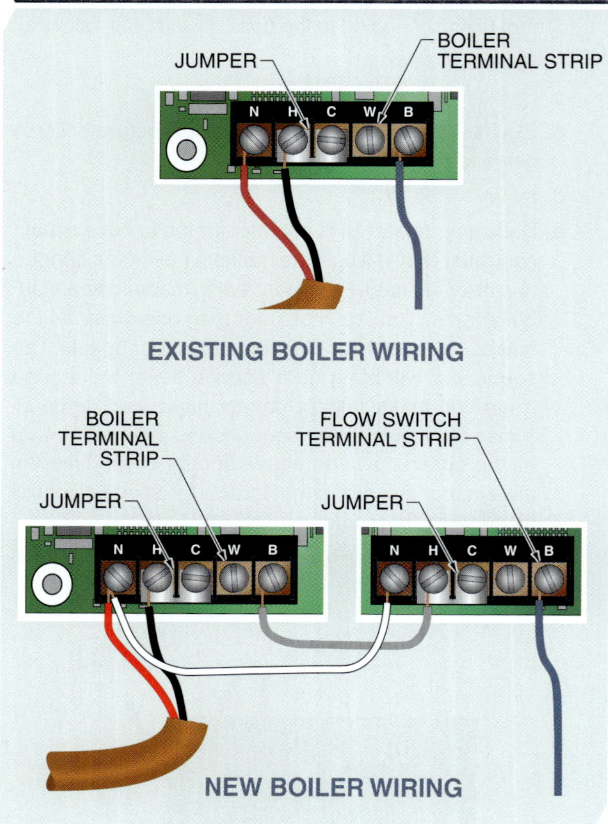

Figure 42-25. Water flow switches are wired as a safety device across the terminals on a boiler's internal terminal strip.

Pump Wiring for a Hydronic Boiler

The hydronic boiler circulating pump is typically not provided by the boiler manufacturer and must be obtained locally. Boiler manufacturer representatives are helpful in sizing the pump properly. The current draw for a field-installed 115 VAC pump must be added to the current draw of the boiler to determine the minimum wire size for the 115 V service. **See Figure 42-26.**

Typical Residential Circulating Pump Current Draw

Volts	Hz	Ph	Amps	RPM	HP
115	60	1	1.1	3250	1/20
115	60	1	1.17	3450	1/10
115	60	1	1.5	3450	1/8
115	60	1	2.0	3450	1/6
115	60	1	3.2	3450	1/3
115	60	1	5.0	3450	1/2

Figure 42-26. For residential applications, 1/8 HP and 1/6 HP are popular circulating pump sizes.

Temperature Control Settings

There are a number of methods that can be used to control the boiler water temperature. The first and most simple method is to control the boiler water temperature at a fixed setpoint. This temperature typically is between 190°F and 240°F. The temperature controller may have a number of settings. In some cases, these are preset at the factory.

Setpoints. The setpoint selector switch specifies the target water temperature in degrees Fahrenheit. After the water temperature reaches the setpoint, the temperature control shuts off the burners.

Temperature Differential Switches. The temperature differential selector switch specifies the number of degrees below the setpoint that the control will allow the water temperature to drop before it brings the boiler back on-line again. This switch prevents quick cycling of the boiler.

High-Fire Offsets. With some boilers, the temperature control operates a two-stage firing system. The two stages are high fire and low fire. High fire operates all burners, while low fire operates approximately one-half of the burners. The high-fire adjustment switch specifies the number of degrees below the setpoint at which the high-fire stage shuts down. At that point, the boiler will continue to operate at the low-fire stage until the setpoint is reached.

Water Temperature Sensors. Typically, hot water boilers have temperature sensors that must be installed. These sensors include inlet and outlet water temperature sensors as well as an outside air temperature sensor. The inlet water temperature sensor is placed into the thermal inlet well on the boiler. This sensor is sometimes installed by the factory in new boilers.

The outlet water temperature sensor must be installed by a technician in the water supply pipework leaving the boiler. **See Figure 42-27.** This sensor well is typically installed in a tee of the piping. The sensor must be inserted all the way into the well, leaving no air pocket between the front surface of the sensor and the back of the well. Air pockets are thermally nonconductive and will cause sensors to read inaccurately. The optional outdoor-air temperature sensor must be mounted under a sun shield and away from any heat sources.

Water Temperature Sensor Installation

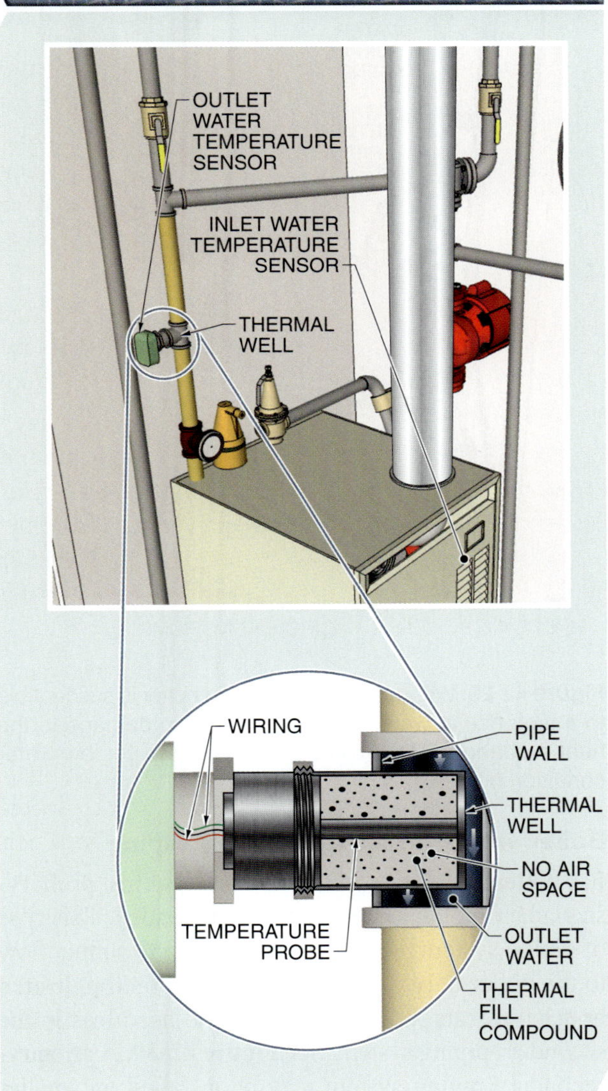

Figure 42-27. If water temperature sensors are not fitted into the thermal well properly, air gaps will cause the sensors to give false readings.

Outdoor-Air Reset Option. Another control option is to reset or change the hot water setpoint as the outside air temperature changes. This is called outdoor-air reset control. For boilers ordered with the outdoor-air reset option, there are additional controls that must be set. These include the adjustment selector switches for setting the differential, maximum temperature, and reset ratio. There is also a switch to turn the outdoor-air reset controller feature on or off. When the power switch is OFF, the thermostat specifies the maximum air temperature at which the boiler will operate. **See Figure 42-28.**

Outdoor-Air Reset Option

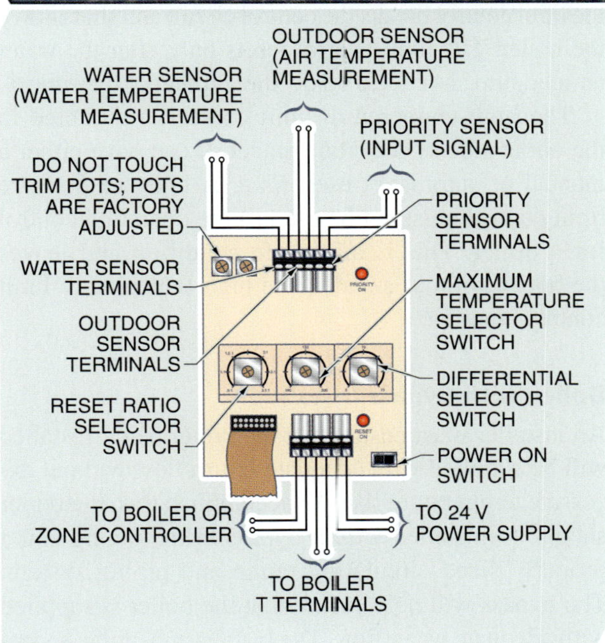

Figure 42-28. Outdoor-air resets are popular control devices used for improving the efficiency of a hydronic boiler and system.

When the differential trim pot is set to midscale (10), the boiler will turn off when the water is 10°F above the control point and turn back on when the water is 10°F below the control point temperature.

The maximum temperature switch sets the maximum water temperature that the outdoor-air reset controller specifies. When the outside air temperature is above the specified setting, the unit will not function in the reset mode but will continue to run at the selected setpoint temperature of the thermostat. The boiler's high-limit control must be set higher than the maximum temperature switch.

The reset ratio switch allows control over the ratio to be used during outdoor air reset. The ratios may be as low as 0.5:1 or as high as 1.5:1. Selecting the 0.5:1 ratio will increase the setpoint 0.5°F for every 1°F drop in outdoor air temperature up to the maximum setpoint temperature. Selecting the 1.5:1 ratio will increase the setpoint 1.5°F for every 1°F drop in outdoor air temperature up to the maximum setpoint temperature.

High-Temperature Limit Control

Hot water boilers are provided with a high-temperature limit control. Hydronic boilers typically have a high-temperature limit control that has a fixed setting of 240°F. When water temperature exceeds this setpoint, the limit control breaks the control circuit and shuts down the boiler. The limit control resets only after the water temperature has cooled below the setpoint of the control.

The high-temperature limit control is mounted in the outlet side of the front header. It can have either a manual or automatic reset. *Note:* A high-temperature limit control must not be used as the operating control for a boiler. This is an unsafe condition and leaves the boiler without an adequate high-temperature limit control function.

Boiler with Bypass Installed

An installer must ensure that the boiler being installed will be supplied with adequate return flow without excessive temperature rise. To accomplish this, the boiler should be installed with a bypass in the piping and a properly sized circulating pump and piping system. The bypass will help ensure that the boiler is supplied with adequate water flow. The installation of the system bypass will allow for increased flow through the boiler. **See Figure 42-29.**

If there is any doubt that the pump may not provide adequate flow, bypass piping must be provided at the time the boiler is installed. The bypass must be fully sized with a balancing valve to allow for proper adjustment. The balancing valve allows the flow through the bypass to be adjusted based on individual system requirements.

A bypass is recommended on all multiple-zone systems in addition to a properly sized circulating pump and system piping. When boiler temperature rise exceeds 40°F, it is an indication that the boiler may not be receiving adequate flow, and adjustment of the bypass is probably required. When a high temperature rise is experienced, the pump and system should be checked for proper operation.

Boiler Bypass Piping

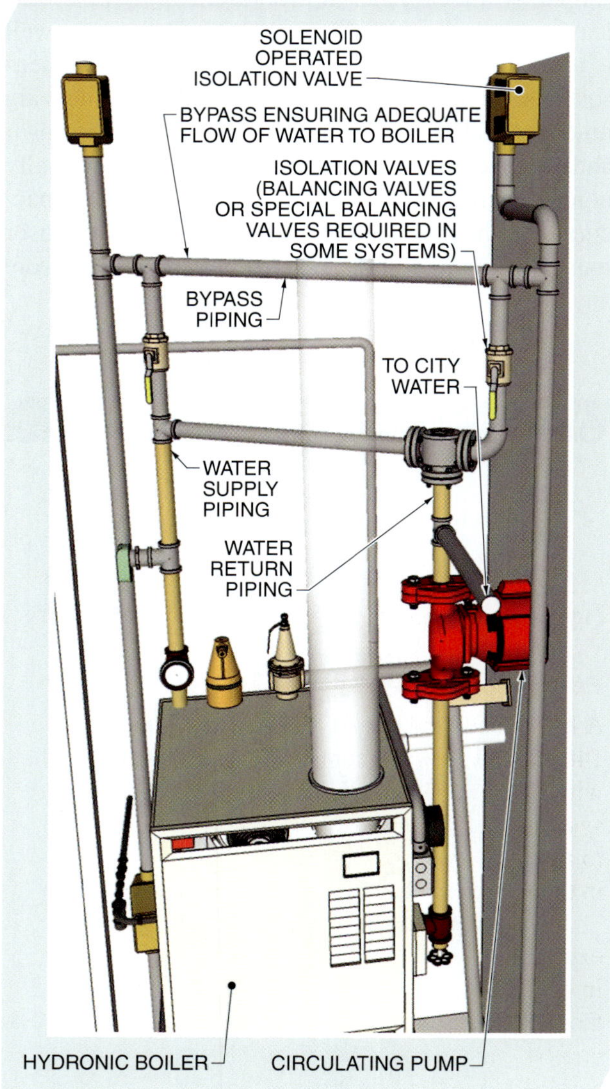

Figure 42-29. Without bypass piping, a boiler is susceptible to a high rise in water temperature and to damage to the heat exchanger from overheating when a minor low-water condition occurs.

Boiler with Primary-Secondary Piping

For boilers used with multiple-zone systems, primary-secondary piping systems are recommended. This type of piping system uses a dedicated pump to supply flow to the boiler only. The pump is sized based on desired boiler flow rate, boiler head loss, and head loss in the secondary piping system. **See Figure 42-30.** A primary-secondary piping system may be installed on smaller boilers for better uniform heating and to guarantee that the boiler has adequate flow. Typically, a properly sized system pump with a bypass will provide the flow needed.

Primary-Secondary Piping Systems

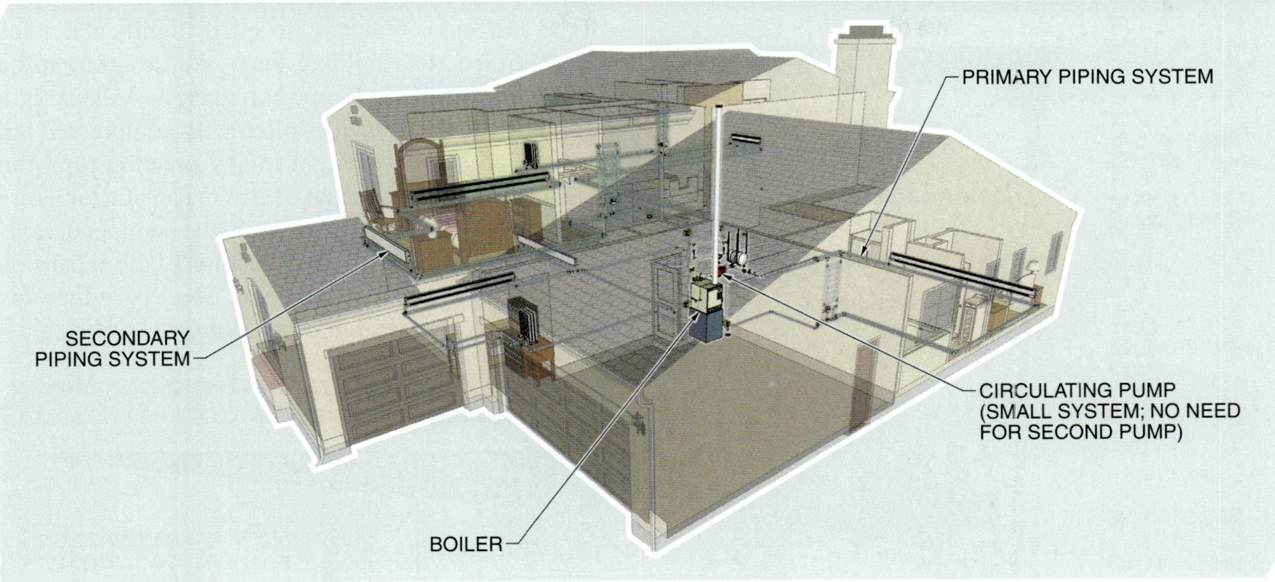

Figure 42-30. Primary-secondary piping systems provide uniform heating to buildings with multiple spaces.

Boiler with Low-Temperature Bypass Piping

A boiler operated with an inlet temperature lower than 140°F (60°C) must have a bypass to prevent problems with condensation. A bypass must be piped into the system at the time of installation. This piping is similar to the primary/secondary piping system with a bypass in the secondary boiler piping. **See Figure 42-31.**

Inlet water temperatures below 140°F (60°C) can excessively cool the products of combustion, resulting in condensation on the heat exchanger and in the flue. Condensation can cause operational problems, bad combustion, sooting, flue gas spillage, and reduced service life of the vent system and related components. The bypass allows part of the boiler discharge water to be mixed with the cooler boiler return water to increase the boiler inlet temperature above 140°F (60°C). This should prevent the products of combustion from condensing in most installations.

The bypass should be fully sized with a low-temperature valve (LTV) to allow for proper adjustment. A shutoff valve must also be provided on the boiler discharge line after the bypass. Closing the discharge shutoff valve forces water into and through the bypass. The bypass valve must be in the full-open position and the discharge shutoff valve must be half open when starting the boiler. A small amount of the higher temperature discharge water is mixed with the returning system water to maintain the temperature of the returning water above the minimum. **See Figure 42-32.**

Boiler with Low-Temperature Bypass Piping

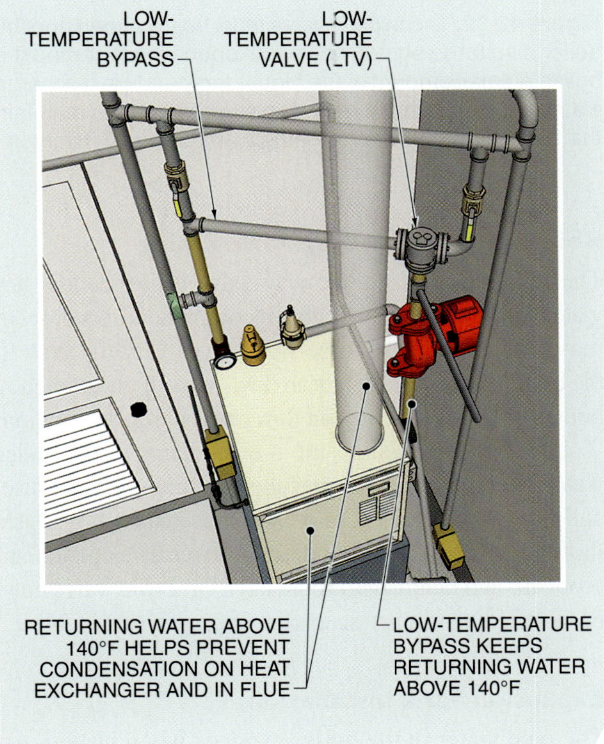

Figure 42-31. A low-temperature bypass prevents a condensing boiler and any boiler from having its inlet water temperature drop below 140°F.

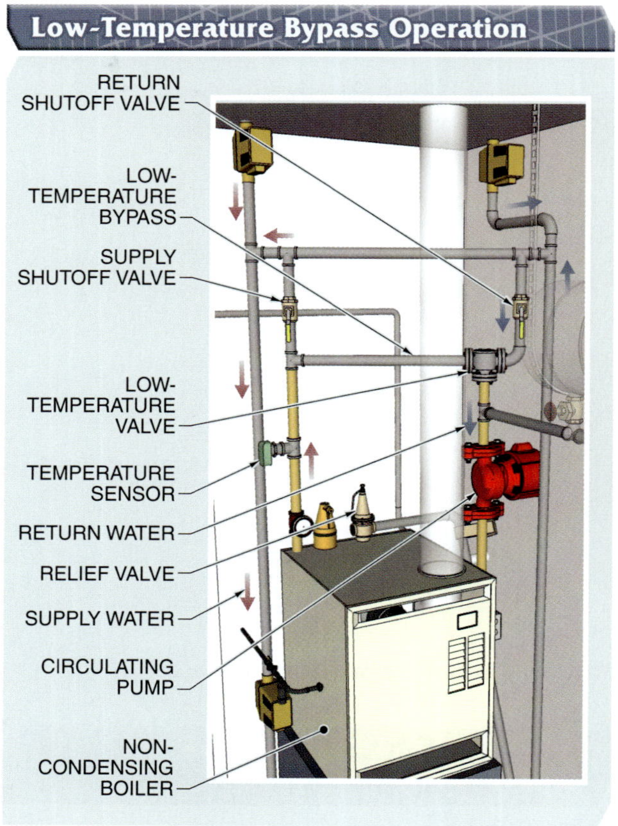

Figure 42-32. The bypass valve must be fully open and the boiler discharge shutoff valve half open when starting the boiler. A small amount of the higher temperature discharge water is mixed with the returning system water to maintain the temperature of the returning water above a minimum.

Boiler with Three-Way Valve Piping

The installation of a three-way valve on residential hot water boilers is not typically recommended because in most piping methods, three-way valves vary the flow to the boiler. Hydronic boilers are low-mass, high-efficiency boilers that require a constant flow rate for proper operation.

Low flow rates can result in overheating of the boiler water, which can cause burner short-cycling, system noise, and in extreme cases, a knocking noise caused by water flashing to steam. These conditions can result in operational problems and failure of the boiler. If a three-way valve must be installed, a primary-secondary piping system is used.

Expansion Tank Installation

The cold water in the boiler expands when heated. An expansion tank is used to allow for this expansion. The expansion tank must be installed near the boiler and on the suction side of the pump.

There are two types of expansion tanks. **See Figure 42-33.** A steel holding tank traps air in the top half of the unit and is typically installed slightly above the boiler. Some steel holding tanks are designed to be installed vertically and some horizontally. A diaphragm tank, which uses a bladder filled with compressed air, is installed at boiler height in any position (upright, upside down, or horizontal). Most expansion tanks are diaphragm tanks. The manufacturer recommendations regarding the location and installation of the expansion tank should always be followed. **See Procedure for Draining Water from a Steel Holding Tank.**

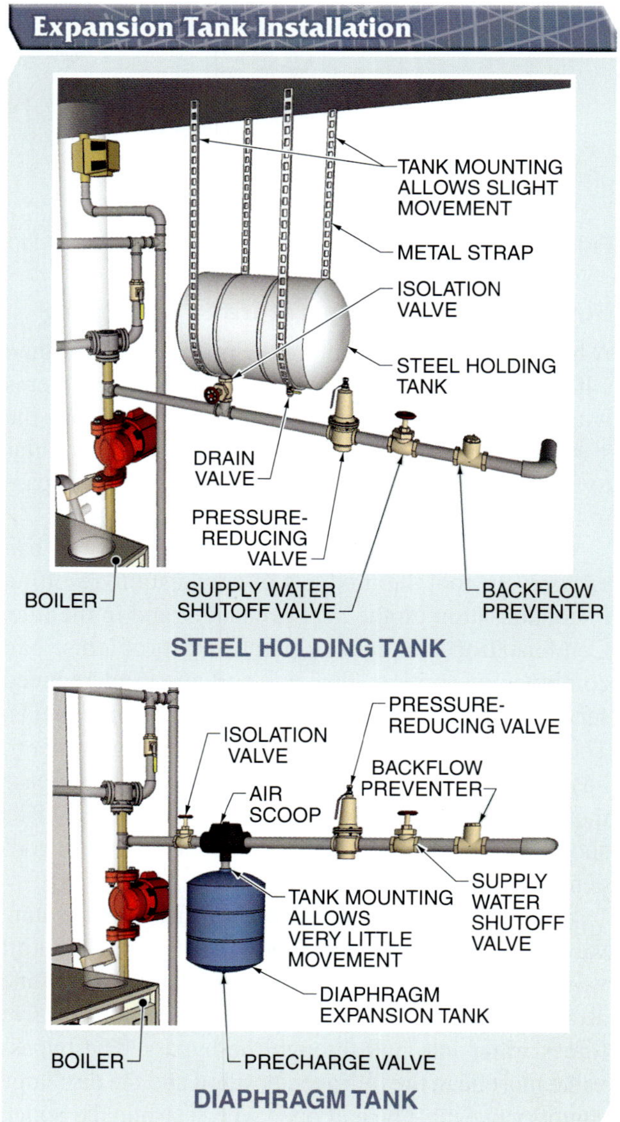

Figure 42-33. A steel holding tank is typically installed slightly above the boiler (vertically or horizontally), and a diaphragm tank is installed at boiler height in any position (upright, upside down, or horizontal).

Draining Water from a Steel Holding Tank

1. Disconnect power from the boiler.

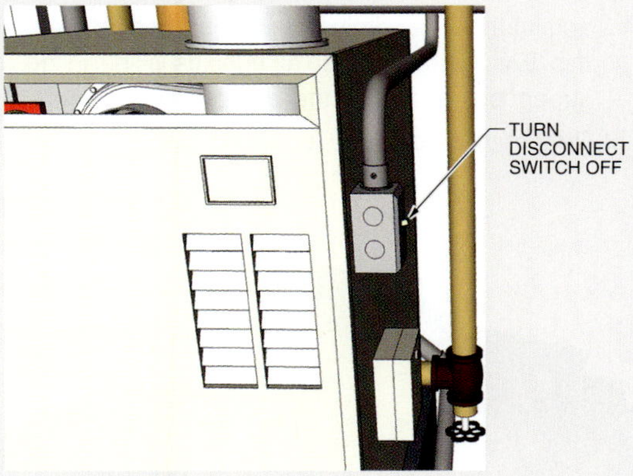

2. Attach a hose to the drain valve of the steel holding tank.

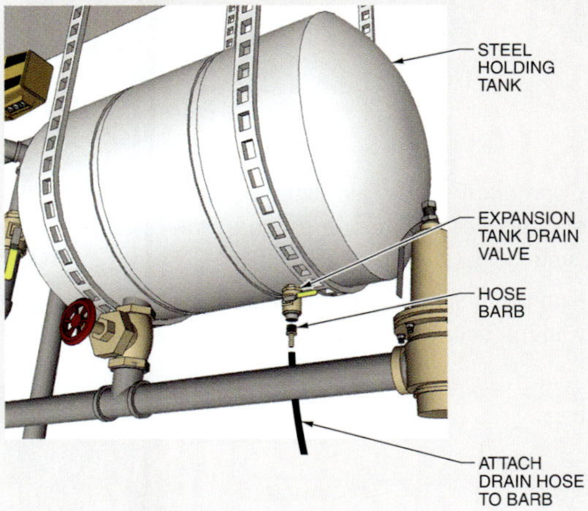

3. Close the valve that connects the expansion tank to the boiler and the water supply shutoff valve.

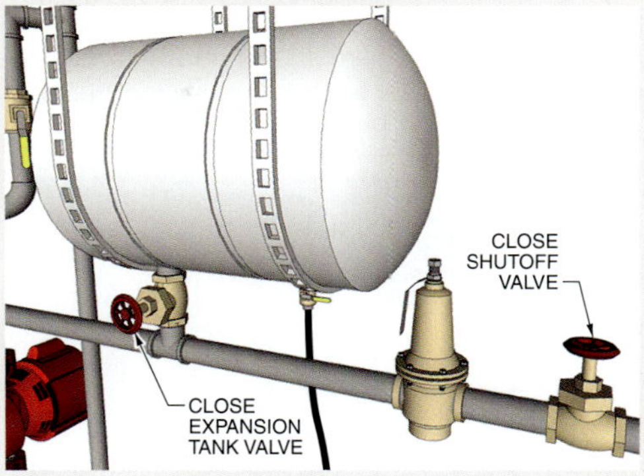

4. Open the drain valve on the expansion tank.

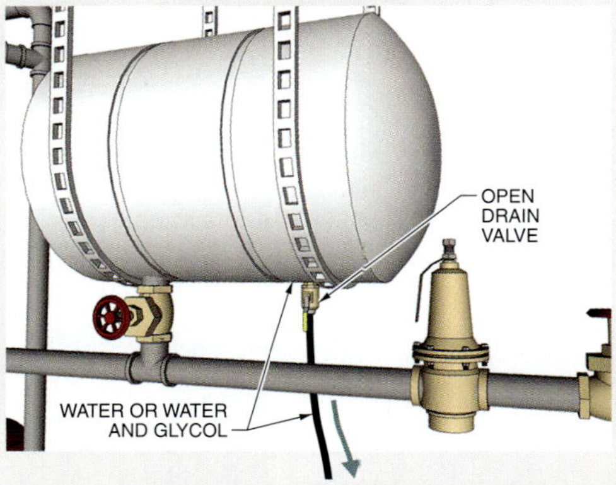

5. Drain the tank until it is empty. Then close the drain valve and open the valve to the boiler system.
6. The expansion tank should fill to the halfway mark or less and trap air in the top of the tank. If the tank fills again after a few days, there is an air leak in the tank.

Expansion tanks allow for a controlled increase in water volume as water is heated.

Filling a Boiler with Water

There are two types of hot water boiler systems in use: open systems and closed systems. Open systems are the more basic and older design. In an open system, the expansion tank is at the highest point in the system, typically the attic. The expansion tank is vented to the atmosphere, allowing water to escape through the vent pipe to a safe place.

A closed system is the type of system installed with today's boilers. A closed system has the expansion tank located close to the boiler, typically at the lowest point in the system. The expansion tank is closed and airtight. An altitude gauge is used to indicate when the boiler is filled with water. **See Procedure for Filling a Closed-System Boiler with Water.**

Filling a Closed-System Boiler with Water

Note: All water piping must be properly supported.

1. Purge all air from the system. All air must be purged for proper operation. An air scoop or air vent must be located close to the boiler outlet and there should be a minimum distance between the cold water feed and the system purge valve.

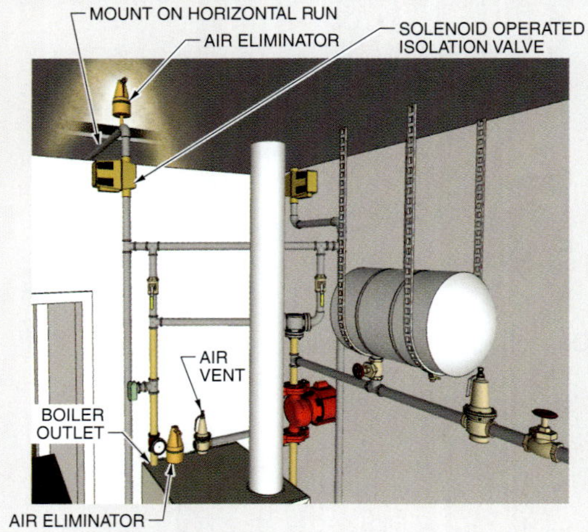

2. Close all drain cocks, valves, and air vents.
3. Open the makeup water valve and slowly fill the system.

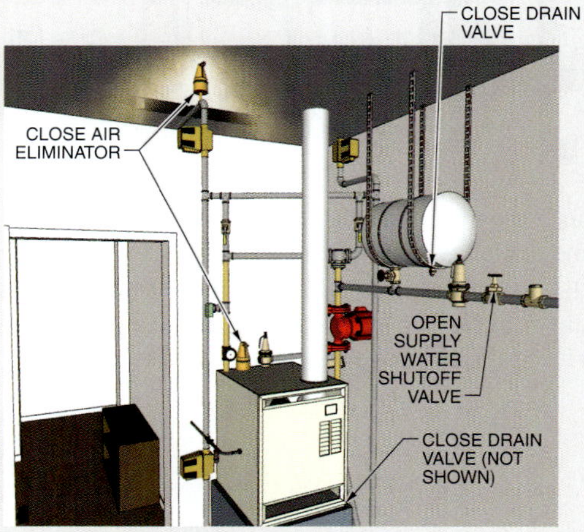

4. If a makeup water pump is employed, adjust the pressure to provide a minimum pressure of approximately 12 psi at the highest point in the system. If a pressure regulator is also installed in the line, it should be adjusted to the same pressure.

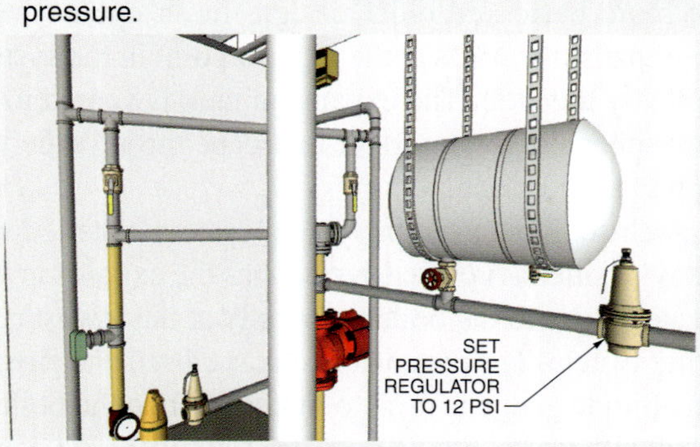

5. Purge air from system as follows:
   a. Open one branch air vent or air eliminator valve at a time and let the water drain from that branch for at least five min. Ensure there are no air bubbles visible in the water stream before closing the vent or valve.
   b. Repeat this procedure for each branch.

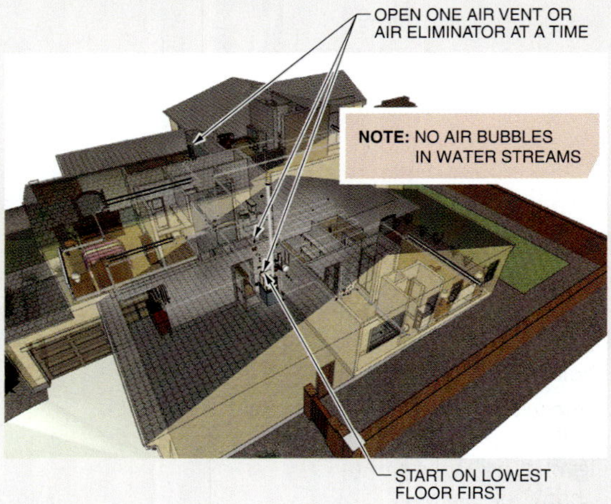

6. Open all air vents, air eliminators, and zone valves after all branches have been purged. Make sure there are no system leaks.
7. Run the system circulating pump for a minimum of 30 min with the boiler turned off.

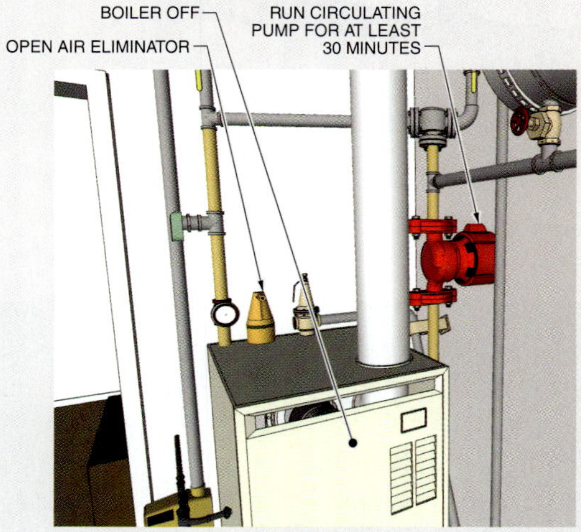

8. Open any strainers in the system and check for debris.
9. Recheck all air vents by opening one at a time again.
10. Inspect the liquid level in the steel holding tank. The system must be full and under normal operating pressure to ensure proper water level in the tank. If a diaphragm tank is used, ensure that the tank is properly charged and not waterlogged.

continued on next page

Filling a Closed-System Boiler with Water *(continued)*

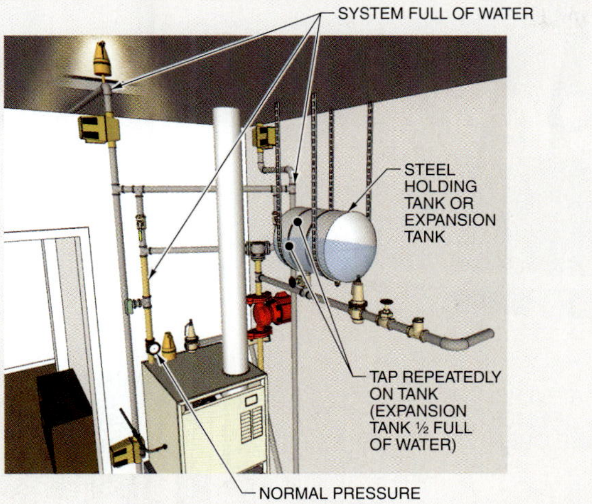

11. Start the boiler using the instructions in the installation and service manual. Operate the system, including the pump, boiler, and radiation units, for 1 hr.
12. Recheck the water level in the expansion tank. If it exceeds half the tank volume, open the tank to reduce the water level. If a diaphragm tank is used, recheck the pressure charge on the tank.
13. Shut down the entire system and vent all radiation units and high points in the system.

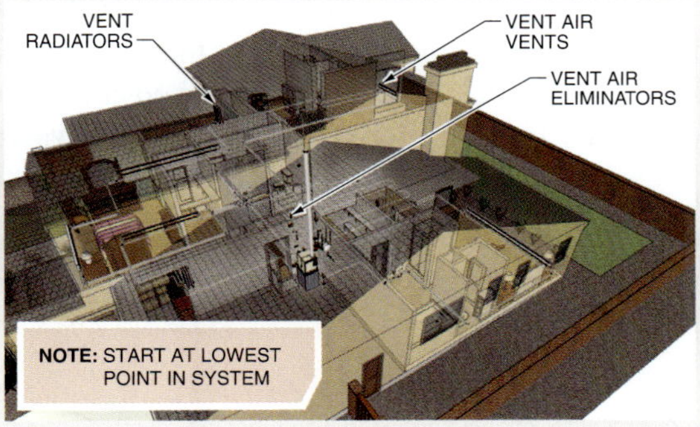

14. Close the water makeup valve and check the strainer and pressure-reducing valve for sediment or debris. Reopen the water makeup valve.

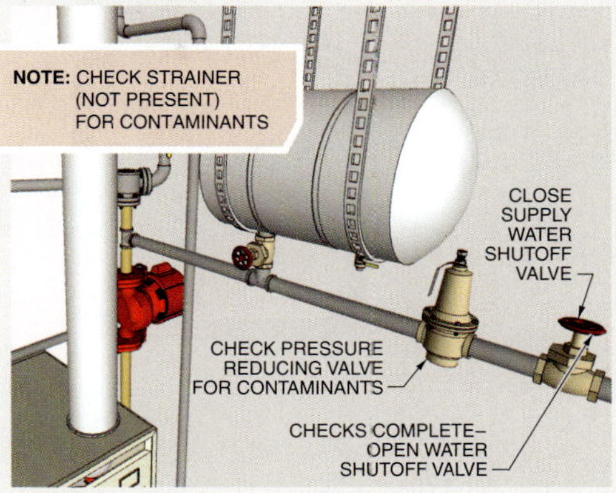

15. Verify system pressure with the boiler pressure gauge before beginning regular operation.

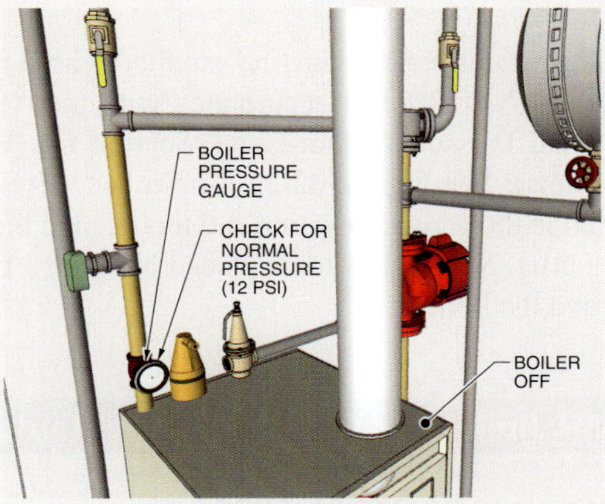

16. Within three days of startup, recheck and bleed all air vents and the expansion tank again.

Boiler Pressures. Typically, the temperature-pressure gauge on a boiler is located just above the burners or on the side of the manifold. For most residential and small commercial systems, this pressure will be 12 psi to 15 psi. The gauge should be read when the boiler is cold and the circulating pump is off. This will indicate the static water pressure of the boiler set by the water feed regulator (pressure-reducing valve in residential systems).

When pressure is too high, the regulator may be set too high or water may be leaking through the regulator. **See Figure 42-34.** When the feed regulator leaks even a small amount of water internally because of scale or sediment trapped in the seat, the pressure in the system will creep up. After a little time, pressure will increase to the relief valve setpoint, which causes the relief valve to open and blow off the water, decreasing the pressure.

Pressure-Relief Valves. Every hydronic boiler has at least one pressure-relief valve. When pressure climbs too high in the boiler, the relief valve will open, blow off the excessive pressure, and close again when the pressure falls below the setpoint.

There is typically a tag on the valve to identify at what pressure it will open. Most residential boiler relief valves are set at 30 psi. When the boiler is operating close to this setpoint, the pressure relief valve will leak slowly. The cause of the high boiler pressure must be determined. At least once a year the pressure-relief valve must be manually opened to blow off water and make sure the valve is not plugged up with sediment and scale.

Leaking Pressure-Reducing Valves

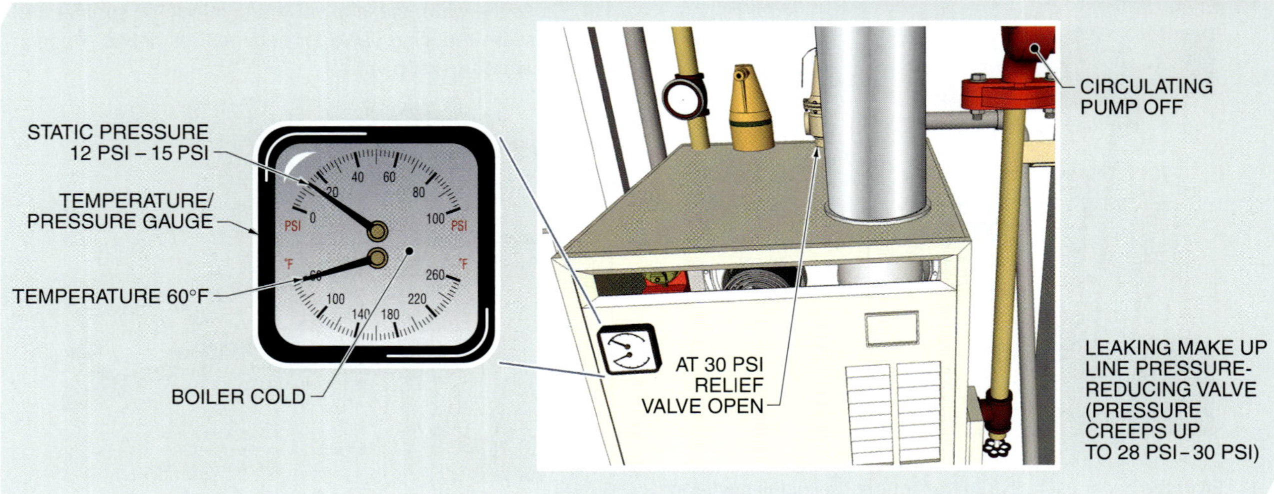

Figure 42-34. When pressure is too high, water may be leaking through the regulator, which causes pressure in the system to slowly rise.

All hot water boilers must be supplied with a relief valve or valves sized in accordance with the ASME Boiler and Pressure Vessel Code, Section IV (Heating Boilers). **See Figure 42-35.** The relief valves are installed in the vertical position and mounted in the hot water outlet. No valve is to be placed between a relief valve and the boiler.

To prevent water damage, the discharge from the relief valve must be piped to a suitable floor drain for disposal when relief occurs. No reducing couplings or other restrictions can be installed in the discharge line. The discharge line must allow complete drainage of the valve and line. Relief valves should be manually operated at least once a year. **See Procedure for Blowing Off Relief Valve.**

ASME Boiler and Pressure Vessel Code, Section IV (Heating Boilers Excerpts)

PRESSURE RELIEF DEVICES

General

(a) All pressure vessels within the Scope of this division, irrespective of size or pressure, shall be provided with pressure relief devices in accordance with these requirements.

(1) It is the responsibility of the user to ensure that the required pressure relief devices are properly installed prior to initial operation.

(2) It is the responsibility of the user or his/her designated agent to size and select the pressure relief device(s) based on its intended service.

Intended service considerations shall include, but not necessarily be limited to, the following:

(a) normal operating and upset conditions
(b) fluids
(c) fluid phases

(2) When a pressure vessel can be exposed to fire or other unexpected sources of heat, the pressure relief device(s) shall be capable of preventing the pressure from rising more than 21% above the maximum allowable working pressure.

(f) Vessels that are to operate completely filled with liquid shall be equipped with pressure relief devices designated for liquid service, unless otherwise protected against overpressure.

Installation

(a) Pressure relief devices intended for relief of liquids shall be connected below the liquid level.

(d) There shall be no intervening stop valves between the vessel and its pressure relief device or devices, or between the pressure relief device or devices and the point of discharge.

(e) The pressure relief devices on all vessels shall be installed so that their proper functioning will not be hindered by the nature of the vessel's contents.

(f) Discharge lines from pressure relief devices shall be designed to facilitate drainage or shall be fitted with drains to prevent liquid from lodging in the discharge side of the pressure relief device, and such lines shall lead to a safe place of discharge.

Figure 42-35. All hot water boilers must be supplied with a relief valve sized in accordance with ASME Boiler and Pressure Vessel Code, Section IV (Heating Boilers).

Blowing Off Relief Valve

1. Ensure boiler is OFF and cold.

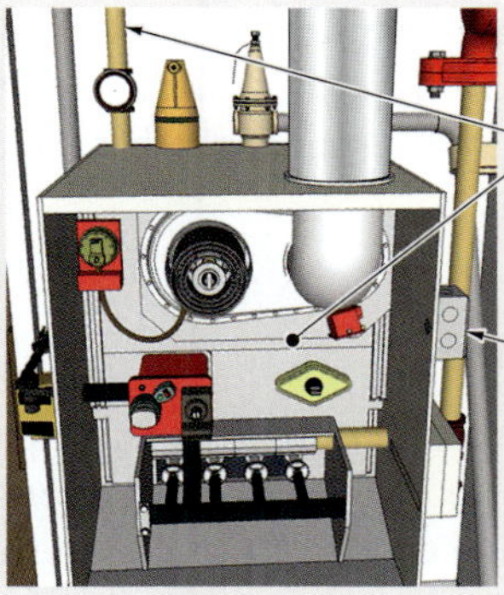

2. Lift the manual lever on the relief valve for a couple of seconds and then let the lever snap back into position. A gush of water will come out of the valve discharge pipe. The discharge flow from the valve should be fairly clear.

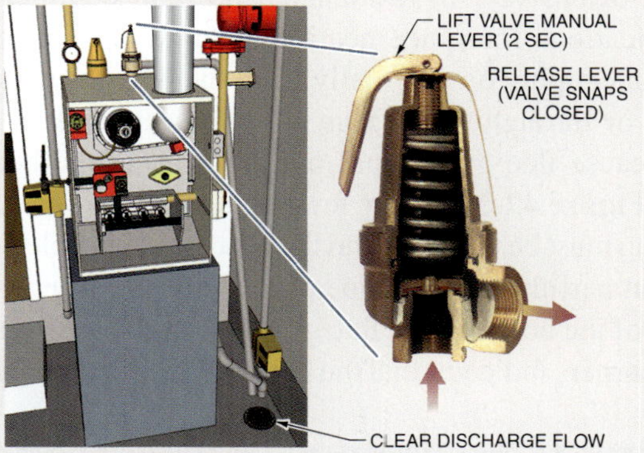

3. If the valve leaks a little afterwards:
   a. Lift the manual lever again for a second. Sediment can become trapped in the seat, causing a minor leak.
   b. Lift the manual lever as many times as required to clear the seat of sediment.
4. If the valve will not close at all:
   a. Close supply water shutoff valve.
   b. Remove and replace relief valve.
5. If no water comes out of the relief valve, the valve is plugged or there is no water in the boiler. *Note:* Do not use the boiler until the problem has been identified.

Circulating Pumps. Most hydronic system circulating pumps require lubrication for the motor and the pump. The motor typically requires a few drops of oil and the pump bearing may have oil cups. In this case, the oil cups must be filled to keep the oil wicks wet. The manufacturer instructions should be consulted for the correct pump lubrication and procedure.

There may be more than one pump in a system. Some of the smaller cartridge-style pumps do not require any maintenance. The volume of water circulated by the pump is very important to boiler operation. With too little water, the water will boil inside the heat exchanger and plug it with scale. This typically sounds like a crackling occurring inside the boiler when the burner is operating.

A single pressure gauge piped across the pump inlet and outlet is the best way to monitor pump performance. As the pressure differential across the pump increases, the flow rate decreases. As pressure differential across the pump decreases, the flow rate increases. **See Figure 42-36.** Another way to check water flow is by temperature. When the burner is operating, the temperature rise through the boiler can be checked. Most hydronic boilers are designed for a 10°F temperature differential (TD) with nominal water flow. When the TD is higher than 25°F, the water flow is too low, and the water can start to boil off in the heating coil.

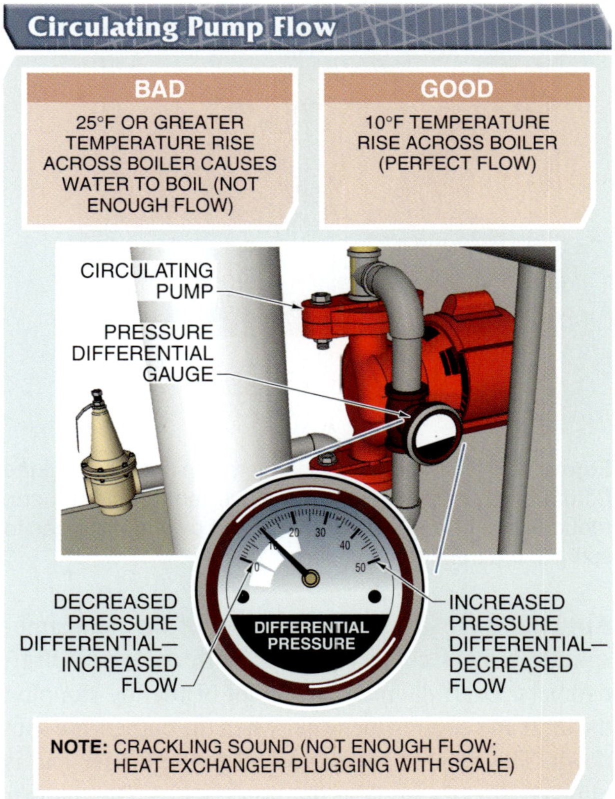

Figure 42-36. The higher the differential pressure across a pump, the lower the flow rate. The lower the differential pressure across a pump, the higher the flow rate.

Heat Exchanger Cleaning. Atmospheric boilers are used extensively in residential and light commercial applications for space heating. The heat exchangers in boilers plug up externally from the flue gas passing over or through them. This reduces heat transfer and can cause flue gas to spill out the front of the boiler. **See Figure 42-37.** Every few years the draft hood and casing must be removed and the heat exchanger cleaned. When performing a routine check, any staining on the face of the boiler or hot flue gas spilling out the front of the burners indicates that the heat exchanger is plugged.

Plugged Heat Exchangers

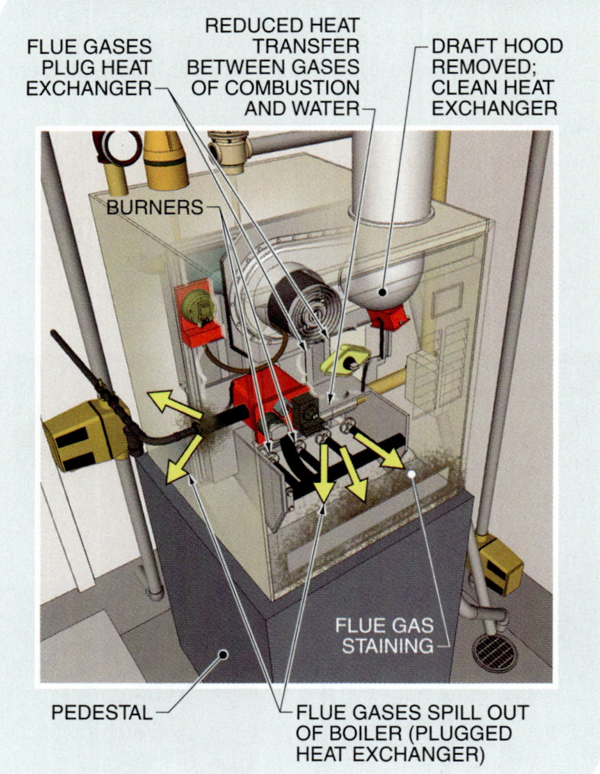

Figure 42-37. Boiler heat exchangers can become plugged from the flue gas that is passing over and through them, reducing their heat transfer ability and causing flue gas to spill out the front of the boiler.

Automatic Draft Dampers. Many of the new atmospheric boilers come equipped with automatic draft dampers. Draft dampers save a lot of fuel by stopping the draft and cooling the water when the burners are shut down. The damper is interlocked to the burner and is plugged directly into the ignition control. The damper should close when the burners are off and open before the burners ignite.

Over the summer months, the automatic draft damper can seize and not open. When the damper is stuck in the closed position, it will stop the burner from igniting or it may stick in the open position and not allow the burners to operate at all. Typically the damper actuator position can be used to determine whether the damper is closed or open.

BOILER STARTUP

Once the boiler is installed properly it may be started. Before startup, it is a good idea to inspect every aspect of the boiler installation and ensure that everything has been installed properly.

Water Quality

The water being supplied to hydronic boilers should be checked for hardness and chlorine content. Hydronic boilers are not to be used to heat pools, because pool water has a high chlorine content. In the event that freeze protection is used in the system, the freeze protection should be added to the system at manufacturer-recommended levels. After all of these requirements are met, the boiler may be filled with water. At this point, the vents and any drains should be closed. A minimum water pressure, typically above 12 psi, must be met as well. Once the boiler and system are filled, a leak check must be performed. **See Figure 42-38.**

Accepted Hydronic System Water Quality

HARDNESS EXPRESSIONS:

mg/L-milligrams per liter
ppm-parts per million
gpg-grains per gallon

1 ppm = 1 mg/L = 1gpg – 17.1 ppm

pH lower than 7.0 is acidic
pH higher than 7.0 is alkalinic (caustic)
pH = 7.0 is neutral

CONVERSION FORMULAS:

gpg × 17.1 = ppm
gpg × 17.1 = mg/L
ppm × 0.05833 = gpg
mg/L ÷ 17.1 = gpg

MAXIMUM HARDNESS:

Less than 20 ppm

Boiler pH should be between 10.5 and 12.5. Below 10.5 there is no protection from corrosion and hardness compounds. Above 12.5 there is possibility of caustic embrittlement.

ACCEPTED ALKALINITY:

200 ppm to 700 ppm

Figure 42-38. The water to be used in a boiler must be analyzed for hardness to determine if treatment is necessary.

Air Venting

Air trapped in the system prevents valves from operating correctly. All the air in the system must be vented out to allow proper system operation. This must be done thoroughly to ensure that all valves operate as specified.

When needed, a hose is connected to a purge valve or any valve in the system to direct water to a safe location. Water is directed through one zone at a time, with the other zone valves closed. In this way, each zone is purged individually. After that, the manual air vents are opened, starting with the lowest and ending with the highest. The air vents are typically left open until water comes out. The automatic air vents are then placed in operation and checked. **See Figure 42-39.**

Gas Leaks

Gas lines must be checked for gas leaks. The gas shutoff must be opened. Any gas smell must result in shutoff of the gas valves immediately. Any gas leaks must be repaired and rechecked with a solution or meter.

Exhaust Condensate System

When first installed, the exhaust condensate system must be filled with water manually. High-efficiency boilers produce condensate as a result of the exhaust gas temperature falling below the dewpoint. This exhaust condensate is removed to a drain. The drain line is sloped from the boiler flue to the drain. When the drain is above or at a distance from the boiler, a condensate pump must be installed. **See Figure 42-40.**

Exhaust Condensate Drain Lines

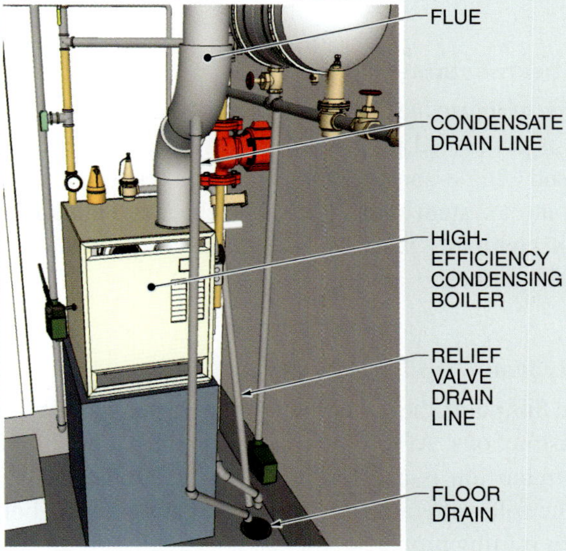

NOTE: CONDENSATE PUMP MAY BE REQUIRED

Figure 42-40. Exhaust condensate must be sent to a drain or acidic reactions in the flue system and heat exchanger will cause serious problems.

Venting Hydronic Systems

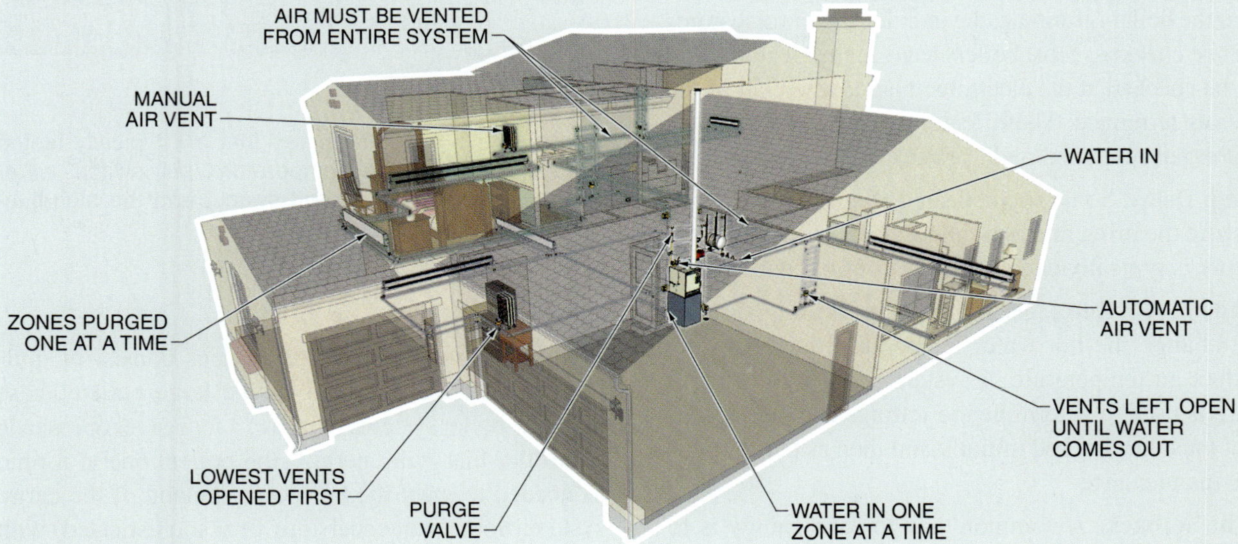

Figure 42-39. Air trapped in the system prevents valves from operating correctly, which means all the air in the system must be vented out to allow proper system operation.

The condensate is typically acidic with a pH in the range of approximately 5.0. Plastic piping must be used in this line. Also, the tee must be kept open in order to break any vacuum. Many modern boilers have a switch that will not allow the boiler to function when the condensate is not draining properly.

Thermostats

The wiring to the thermostat should be checked to ensure it is connected properly for the type of system being used (single-zone, multiple-zone, or economizer). The control system signal should be checked with a DMM for continuity when the thermostat or control system calls for heat.

Hydronic Boiler Startup

At this point, the boiler may be started per manufacturer instructions. After the boiler is running normally, the combustion is checked using an electronic analyzer. The following are some common causes of the boiler not starting properly:

- electrical supply problems such as the breakers turned off, faulty thermostat/wiring, or disconnect switch turned off
- electrical limit or safety open
- natural gas turned off or at wrong pressure

Operating and Control Features

After the boiler is running, the individual control features must be checked. Most boilers have control features that can be checked at the electronic board, where all of the controls terminate. These features include ramp delay, outdoor-air reset, and night setback.

Ramp Delay. Some sophisticated control systems will ramp up the firing rate as a call for heat is received. This feature is typically checked at the control board.

Outdoor-Air Reset. Many hot water heating systems will change the hot water temperature based on the outdoor air temperature. A system curve at the control board is used to determine the setting. The outdoor-air reset must be checked initially and then as part of periodic maintenance.

Night Setback. A common way to save energy is to reduce the water temperature in the unoccupied cycle. This feature can be checked at the control board as well and adjusted when necessary.

Testing Safety Features

The safety features of the boiler must be checked as thoroughly as possible. Safety features to be tested include the high-water-temperature limit switch, water flow switch, low-water cutoff switch, and the high-flue-temperature limit switch. These features can be checked at the control board, or they can be adjusted at the device to verify that the device trips the boiler off at the appropriate times. These safety tests must be performed while the technician is present. **See Figure 42-41.**

Testing Boiler Safety Features

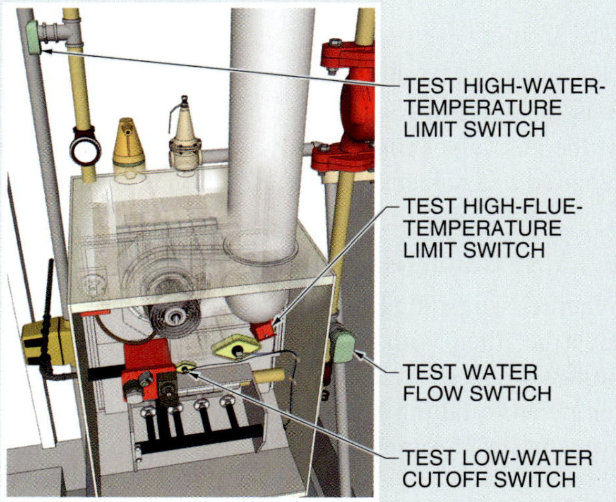

NOTE: NEVER LEAVE THE BOILER DURING A SAFETY TEST; ALWAYS BE READY TO TURN THE POWER AND/OR GAS OFF TO PREVENT AN UNSAFE CONDITION FROM OCCURRING

Figure 42-41. Safety features that are typically tested include the high-water-temperature limit switch, water flow switch, low-water cutoff switch, and the high-flue-temperature limit switch.

Sequencer Functions

Some light commercial installations consist of multiple small boilers instead of one large boiler. These multiple-boiler systems will have a sequencer or cascade controller that will energize the boilers one at a time as needed to meet the hot water demand of the entire system. Some sequencers are very sophisticated, with interstage differentials, minimum number of ON/OFF timers, and equalized run-time settings being some of the common features.

To check a sequencer, the boilers should be run through a number of cycles, and the timing of each stage and the conditions under which each stage energizes should be noted. The manufacturer literature must be consulted to ensure proper boiler operation. **See Figure 42-42.**

Sequencer Functions

Sequencer	Staging Control
Maximum number stages	8
Burner control outputs	Stage burners
System control	Water setpoint
	Hot water reset
Sensors	Outdoor temperature
	Heating system temperature
Remote communications	Enable/disable or Tsat
	4–20 mA setpoint input
	BACnet
Optional accessories	4–20 mA remote input
	Outdoor cutoff
	Reset ratio
	Offset
	Purge delay
	Min. system temperature
	Max system temperature
	Last stage hold
	Min. run time
	Adjustable stage rotation
	System pump
	DHW priority
	DHW pump relay output
	DHW temperature sensor input
	Night setback
	Night setback input
Staging software	Multistage
	Lo-lo/lo-high
	Oversize boiler

Figure 42-42. To check a sequencer, the boiler should be operated through a number of cycles, and careful notation should be made of when and under what conditions each stage is energized.

Freeze Protection

When freeze protection is used (inlet air damper), the control sequences may be checked at the control board and set up checked as necessary. The freeze protection package is typically pre-wired and directly interlocked to the boiler's sequencer through a series of relays and controllers. This makes freeze protection easy to install, because there is no separate control panel or wiring.

Error Recording

Many boilers can record faults at their control boards. Also, in many cases, previously recorded fault codes can be retrieved through a keypad or at the boiler control module. The manufacturer literature should be consulted for data retrieval information.

HYDRONIC BOILER MAINTENANCE

For proper operation and efficiency, preventive maintenance of a residential or light commercial boiler is necessary. Preventive maintenance must be done at least once a year. Manufacturer recommendations for proper maintenance procedures and operation of the boiler and hydronic system should always be consulted. The procedures described in manuals should only be performed by a qualified technician. Serious injury or death could result when tests and maintenance are improperly performed.

Most insurance companies require periodic boiler inspections to qualify for insurance in a commercial environment. Periodic inspections are typically scheduled when the boiler is down in the off season. This scheduled downtime is used to accomplish maintenance and replacement or repairs that cannot be done when the boiler is operating.

While periodic inspections pertain primarily to the water side and fire side sections of the boiler, they also provide the homeowner or technician with an excellent opportunity to service all boiler components and accessories, including piping, valves, pumps, gaskets, and any refractory materials. Many homeowners or facility managers will have an agreement in place that will ensure regular service and maintenance.

After a hydronic boiler has been down for any period of time, a complete inspection of the boiler must be performed on all controls, piping, wiring, and all fasteners such as nuts, bolts, and setscrews to ensure no damage has occurred. A preventive maintenance program prevents unnecessary downtime, costly repairs, and promotes safety.

Boiler manufacturers recommend that a boiler log for recording all activities associated with the boiler be maintained. A log is valuable in recognizing abnormalities and obtaining a long service life from equipment. It is common when performing preventive maintenance to keep residential and light commercial boilers in service for 40 to 50 years with only minor repairs needed.

Service and Maintenance Schedules

Each manufacturer will have a recommended maintenance schedule. This may be anywhere from once per year to monthly. Often the boiler maintenance is combined with the maintenance of other HVAC systems, such as chillers and cooling towers.

Almost all manufacturers recommend that the hydronic system receive maintenance at the beginning of the heating season, with another visit in the middle of the season being recommended as well. When performing this maintenance, the technician should address any problems reported by the owner. Often the owner is the source of the best information regarding the performance of the system.

Hydronic System Maintenance Tasks

A good place to start the service visit would be a good visual inspection of the boiler, both externally and internally. The area around the boiler must be checked for combustibles. Often leaks, corrosion, and other problems can be seen visually. An internal inspection can reveal combustion, water flow, and venting problems.

Checking for Leaks. A critical task in the boiler maintenance procedure is to check the gas and water piping for leaks. While either can be a safety hazard, gas leaks can be catastrophic. Soap bubbles are still recommended for checking for gas leaks. Checking around fittings such as tees and unions is especially important.

Cleaning the Heat Exchanger. Depending on the model, the heat exchanger should be cleaned thoroughly. This will improve system efficiency and help eliminate the possibility of gases escaping into the space, plus reduce any failures due to overheating.

Oiling the Circulating Pumps. When any of the circulator or system pumps require oiling, the oiling is performed as part of the maintenance. Most circulating pumps require oiling after a certain amount of hours of operation, but this typically is never more than twice a heating season.

Checking Air Openings. It is important to check the air openings for proper combustion-air flow rates. This would include the openings to the room where the boiler is located as well as the inlet to the boiler itself. The inlet is checked to ensure that there are no obstructions. Combustion inlet air piping should be checked for integrity and proper support.

Inspecting Flue Vent System. The flue vent system must be checked using the proper procedure. The flue system must be checked for corrosion, damage, support, and obstructions. Any problems must be repaired immediately as the flue system may leak by-products of combustion into the building space.

Inspecting Water System. The water system must be checked for leaks, pressure, and damage. All circulating pumps must be checked for proper operation as well.

Checking Expansion Tank. The expansion tank must be checked for leaks and proper water and air levels. Some tanks are equipped with a sight glass to facilitate checking levels. The wrong level may indicate that the inlet water regulating valve has failed or is set incorrectly. Leaks may cause this as well.

Checking Boiler Relief Valves. Boiler relief valves must be checked for proper operation. They may be exercised manually to clear any debris from the valve discharge. When the valve is released, it should close if not stuck open by debris. In this instance, the relief valve must be replaced.

Checking Boiler Wiring. The initial visual inspection should include inspecting the wiring. The technician should look for any signs of arcing or deterioration of insulation. Looseness of connections should also be checked. Any problems must be repaired.

Inspecting Ignition Systems. The ignition electrode and ground wiring must be inspected for corrosion and proper tightness. Any problems found require replacement or immediate repair.

UNIT 43

Servicing and Troubleshooting Residential Hydronic Systems

Hot water boilers are known to provide even levels of heat throughout a home and prevent some of the zone temperature problems associated with forced-air systems. Also, boilers do not cause many of the humidity problems associated with furnaces. HVAC technicians must know how these systems can fail and how to adequately troubleshoot and service these systems quickly and correctly.

OBJECTIVES

- Identify safety guidelines for troubleshooting.
- List the courtesy guidelines for service calls.
- Identify the typical questions to ask customers before performing service.
- Describe the features of installation, operation, and maintenance manuals.
- List the typical residential boiler sequence of operations.
- Describe the most common boiler troubleshooting problems.

SERVICING HYDRONIC SYSTEMS

As with all mechanical equipment, any time a technician is servicing or troubleshooting hydronic boilers and their associated systems, a safe work environment must be maintained and all safety practices must be followed. **See Figure 43-1.** It only takes a few seconds of carelessness to cause an injury.

When troubleshooting HVAC hydronic equipment, especially boilers, it is difficult for a technician to know all the answers. The manufacturer's troubleshooting and technical installation instructions must be consulted, with special attention given to any safety references, hazards, or safety warnings. **See Figure 43-2.**

TECH TIP

Local HVAC supply houses often host manufacturer training classes to demonstrate the latest system operation and service techniques to technicians.

Troubleshooting and PPE

- PROPERLY CARE FOR ALL PPE ASSIGNED TO YOU
- KEEP ALL PPE CLEAN AND IN GOOD REPAIR
- INSPECT PPE BEFORE EACH USE
- IMMEDIATELY REPORT TO YOUR SUPERVISOR ANY PPE THAT BECOMES DEFECTIVE OR NO LONGER PROVIDES THE NECESSARY PROTECTION
- DO NOT MAKE ANY MODIFICATIONS OR ALTERATIONS TO PPE THAT WOULD VOID ITS WARRANTY OR COMPROMISE ITS FUNCTION, SUCH AS PAINTING HARDHATS OR COATS OR REMOVING SIDE SHIELDS FROM EYEWEAR

Figure 43-1. Never make any modifications, alterations, or changes, such as sewing patches, to any PPE.

Manufacturer Installation Instructions Safety References

⚠ WARNING **If any clearance is less than in Figure 1, provide openings for combustion and ventilation air *located on the wall or door opposite the boiler FRONT (see Figure 2).***

These openings must be located as shown in Figure 2 to provide proper air flow around the boiler The free area of each opening (after deducting for louvers) must be at least **one square inch per 1,000 Btuh** of boiler input. If the building is of unusually tight construction (see page 10 for definition), the air openings must connect directly to outside or the building must have air openings to the outside as specified on page 10.
If clearances are equal to or greater than Figure 1, see pages 10 and 11 for location and sizing of combustion air openings. Failure to comply can result in severe personal injury, death or substantial property damage and reduced boiler life.

⚠ WARNING DO NOT mix components from different systems. The vent system could fail, causing leakage of flue products into the living space

⚠ WARNING **Never install the boiler in a space with clearances less than the minimum clearances shown in Figure 2**. Failure to comply can result in severe personal injury, death or substantial property damage and reduced boiler life

⚠ WARNING Do not install boiler on carpeting even if foundation is used. Fire can result, causing severe personal injury, death or substantial property damage.

Page 7

⚠ WARNING Failure to follow all instructions can result in flue gas spillage and carbon monoxide emissions, causing severe personal injury or death.

⚠ DANGER Inspect existing chimney before installing boiler. Failure to clean or replace perforated pipe or tile lining will cause severe personal injury or death.

⚠ DANGER **DO NOT COMMON VENT**—Do not install the CGs into a common vent with any other appliance. This will cause flue gas spillage or appliance malfunction, resulting in possible severe personal injury, death or substantial property damage.

Page 8

⚠ WARNING To prevent potential of severe personal injury or death, check for products or areas listed below before installing boiler. If any of these contaminants are found:
- remove contaminants permanently

— OR —
- isolate boiler and provide outside combustion air. See national, provincial or local codes for further information.

⚠ WARNING **CONSTRUCTION DUST HAZARD** — Airborne particulates, such as drywall dust or fiberglass dust, will cause blockage of the CGi burners, resulting in carbon monoxide production, a fire hazard, or building freeze damage. If the boiler is operated during construction, you must isolate the boiler air supply to provide clean air for combustion. Follow the instruction manual guidelines for piping intake air. **If you are unable to ensure uncontaminated air in the boiler air intake at all times, you must inspect the boiler at least once weekly.** When inspecting, clean the burners if necessary using the procedure given on page 39. Failure to follow these guidelines could result in severe personal injury, death or substantial property damage.

Page 9

NOTICE Air openings must be sized to handles all appliances and air movers (exhaust fans, etc.) using the air supply.
The sizing given in Figure 3 is based on the National Fuel Gas Code, ANSI Z223.1, allowing adequate air opening for gravity-vented gas appliances (Category I).

⚠ WARNING **Combustion air opening location and sizing requirements depend on the clearances around the boiler.** Check the boiler placement compared to Figure 1, page 7.
If all clearances are at least equal to Figure 1, page 7, apply the sizing and placement of openings given on pages 10 and 11.
If ANY clearance is less than Figure 1, page 7, you must provide air openings sized and located as shown in Figure 2, page 7. DO NOT apply the sizing and location information shown on page 10 or 11. The Figure 2 air openings are required even if the boiler is direct vented (outside air piped to the boiler air intake as described in this manual). These openings ensure adequate air circulation around the boiler for cooling, even if the boiler does not use inside air for combustion.

Page 10

NOTICE Do not drop boiler or bump jacket on floor or pallet. Damage to boiler can result.

⚠ DANGER Correctly sized manifold orifices must be used. Failure to do so will result in severe personal injury, death or substantial property damage.

⚠ WARNING Inlet air box top panel must be in position during boiler operation. Failure to do so could result in severe personal injury, death or substantial property damage.

⚠ WARNING Use only pipe dope compatible with propane gas, even if boiler is to be operated on natural gas. Failure to comply could result in severe personal injury, death or substantial property damage.

Page 12

⚠ WARNING **Do not leave boiler unattended.** A cold water fill could expand and cause excessive pressure, resulting in severe personal injury, death or substantial property damage.

⚠ WARNING **Leaks must be repaired at once.** Failure to do so can damage boiler, resulting in substantial property damage.

⚠ DANGER Do not use **petroleum-based** cleaning or sealing compounds in boiler system. Severe damage to boiler will occur, resulting in substantial property damage.

Page 13

⚠ WARNING To avoid water damage or scalding due to relief valve operation:
- Discharge line must be connected to relief valve outlet and **run to a safe place of disposal.** Terminate the discharge line to eliminate possibility of severe burns should the valve discharge.
- Discharge line must be as short as possible and be the **same size as the valve discharge connection** throughout its entire length.
- Discharge line must **pitch downward** from the valve and terminate at least 6" above the floor drain where any discharge will be clearly visible.
- The discharge line shall **terminate plain, not threaded,** with a material serviceable for temperatures of 375°F or greater.
- **Do not pipe the discharge to any place where freezing could occur.**
- **No shutoff valve** shall be installed between the relief valve and boiler, or in the discharge line. Do not plug or place any obstruction in the discharge line.
- **Failure to comply** with the above guidelines could result in failure of the relief valve to operate, resulting in possibility of severe personal injury, death or substantial property damage.
- **Test the operation of the valve** after filling and pressurizing system by lifting the lever. Make sure the valve discharges freely. If the valve fails to operate correctly, replace it with a new relief valve.

Page 14

⚠ CAUTION Undersized expansion tanks cause system water to be lost from relief valve and makeup water to be added through fill valve. Eventual section failure can result.

⚠ WARNING Use **Figure 4** or **Figure 5** only for single-zone systems designed for return water at least 130°F. For systems with low return water temperature possible, such as converted gravity systems and radiant heating systems, refer to the special piping suggestions of pages 18-23. Failure to prevent low return water temperature to the boiler could cause corrosion of the boiler sections or burners, resulting in severe personal injury, death or substantial property damage.

Page 15

⚠ WARNING Radiant heating system piping should include a means of regulating the boiler return water temperature and the system supply temperature (such as provided by an injection pumping control). Boiler return water temperature will be adequately controlled using the methods shown in this manual provided the system supply temperature is relatively constant.
DO NOT apply the methods in this manual if the system is equipped with an **outdoor reset control.** Instead, provide controls and piping which can regulate the boiler return water temperature at no less than 130°F regardless of system supply temperature. Contact your Weil-McLain representative for suggested piping and control methods. Failure to prevent cold return water temperature to the boiler could cause corrosion damage to the sections or burners, resulting in possible severe personal injury, death or substantial property damage.

⚠ WARNING Failure to prevent low return water temperature to the boiler could cause corrosion of the boiler sections or burners, resulting in severe personal injury, death or substantial property damage.

Page 16

⚠ WARNING For systems with possible low return-water temperature (such as converted gravity systems, radiant heating systems and heat pump systems), refer to the special piping suggestions of **Figures 8–12,** as applies. Failure to prevent sustained low return water temperature to the boiler could cause corrosion of the boiler sections, resulting in severe personal injury, death or substantial property damage.

Page 17

⚠ WARNING **Do not check for gas leaks with an open flame**— use bubble test. Failure to use bubble test or check for gas leaks can cause severe personal injury, death or substantial property damage.

Failure to apply pipe dope as detailed above can result in severe personal injury, death or substantial property damage.

Page 25

⚠ WARNING For your safety, **turn off electrical power supply at service entrance panel before making any electrical connections** to avoid possible electric shock hazard. Failure to do so can cause severe personal injury or death.

NOTICE **Wiring must be N.E.C. Class 1.**
If original wiring as supplied with boiler must be replaced, use only type **105°C** wire or equivalent. Boiler must be **electrically grounded** as required by National Electrical Code ANSI/NFPA 70 – latest edition.

NOTICE The CGs control module is polarity-sensitive. The hot and neutral wires must be connected to the correct leads. A flashing **POWER** light usually indicates reversed polarity of 120 VAC lead wires.

Page 26

Figure 43-2. All safety references, hazards, and safety warnings must be understood before beginning installation, servicing, or troubleshooting hydronic systems.

Most manufacturers post equipment installation and service manuals online. The technician should gather as much information as possible before troubleshooting. System information can include prints, manuals, wiring diagrams, and software. When a new type of boiler is being worked on for the first time, extra care must be taken.

Courtesy Guidelines for Effective Hydronic System Service Calls

Before looking at troubleshooting HVAC hydronic systems, basic common courtesy rules must be practiced. The following are basic guidelines for good customer service. **See Figure 43-3.**

- Call ahead: Make sure the customer is called before driving to their location to ensure they are home or access has been arranged.
- Park where allowed: Don't block their driveway. Park on the street whenever possible.
- Dress and groom neatly. Smile when you greet the customer.
- Do not smoke in the customer's home.
- Do not track dirt into the customer's home. Wear foot coverings when necessary.
- Do not use customer's bathroom or facilities.
- Be honest with the customer. Make sure the customer is told everything that is discovered when troubleshooting.
- Remember it costs five times more in marketing dollars to obtain a new customer than it takes to keep an existing one.
- When leaving to pick up parts or when coming back the next day, let the homeowner know when to expect you back. Do not leave piping, valves, or tools at the customer's home.
- Step outside to make personal calls.
- Maintain a positive attitude.

Call Ahead. Technicians must call customers prior to arriving at their home. The call may display on the customer's caller ID as unavailable, as private caller, or from a source the homeowner may not recognize because the technician is calling from a private cell phone. Caller ID in the house will not necessarily show the name of the HVAC company performing the work. Most mechanical companies supply their technicians with company-owned cell phones that identify the company name when used.

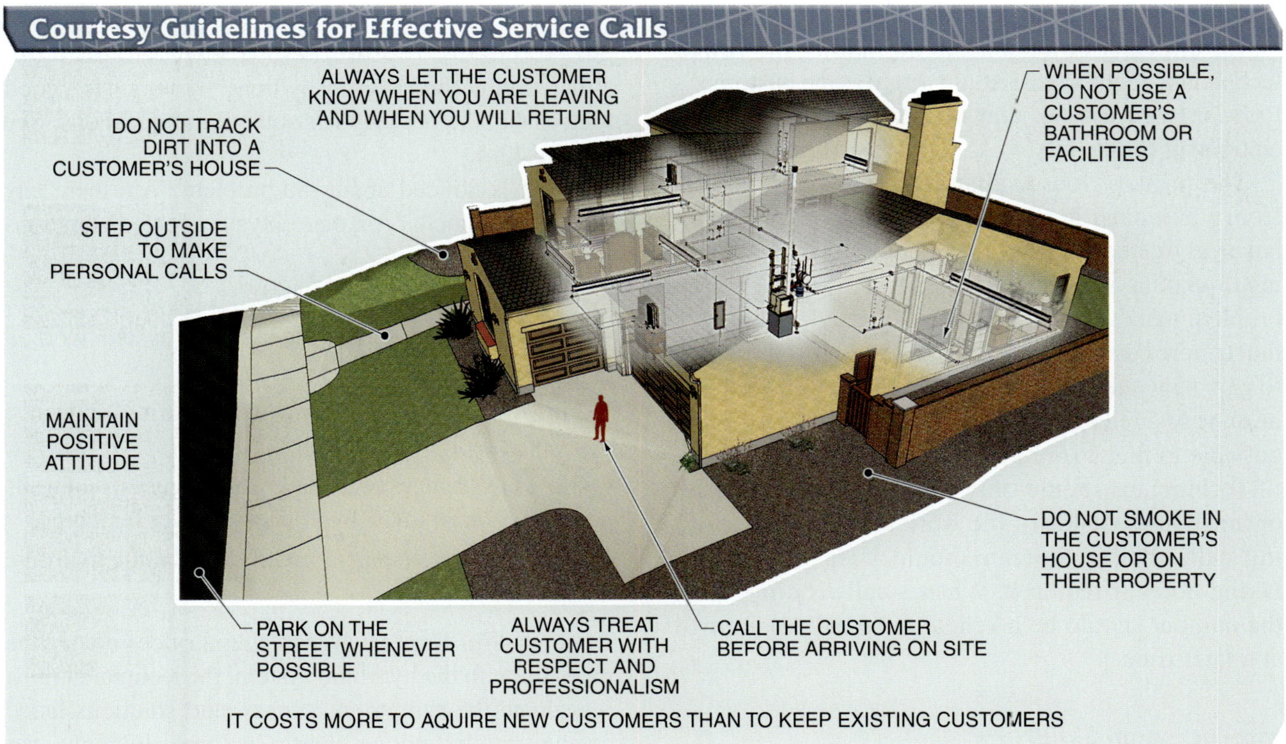

Figure 43-3. The customer should always know when the technician is coming for a service call. Also, the technician should be gracious and act in a professional manner when dealing with the customer.

When there is no answer on the call ahead, it will be assumed that no one is home. A busy signal is an exception to this policy. Technicians must still go to the appointment when a busy signal is received on a call ahead attempt. Customers must be home for the duration of the appointment. Someone over the age of 18 must be present for all field-service calls. The technician may ask for photo ID if the customer appears underage.

Keeping Existing Customers. When a mechanical or HVAC company does not give customers a good reason to stay, competitors will give them a reason to leave. It typically is less expensive to sell services to an existing customer than to seek a new single-transaction customer. Most surveys across the HVAC industry indicate that keeping an existing customer is five to seven times more profitable than trying to attract a new one. Even though expanding into new markets is a good thing, it should not become the main focus. Satisfied customers can tell four or five friends about the work performed, who in turn call the company for service.

Step Outside to Make Personal Calls. A typical day for an HVAC technician consists of several phone calls to and from customers. Technicians set up appointments and acquire materials based on these calls. The average technician will take two to four calls a day, depending on the difficulty of the job and the skill of the HVAC technician. Technicians should arrive at the customer's location with a truck that is carrying the necessary equipment for the job.

The primary reason that customers stop buying from a particular business is because someone at that business treated them poorly. Seeing a technician talking or making phone calls instead of working on their problem makes the customer feel cheated. Again, it is much more cost effective to retain loyal customers than to gain new ones. In order to create loyalty, customers must be assured that the technician is doing everything possible to find a solution to their HVAC problem.

Technicians should only make calls during break or lunch and outside of the work area. When receiving calls, the technician should keep them brief, taking notes if required. When a call requires time, the number should be taken, and the called returned at a later time.

Interpersonal Skills

Interpersonal skills are character traits that allow an individual to communicate effectively with others in a variety of situations. Hydronic equipment failure is costly and frustrating at times for the homeowner, and this frustration can be directed at the service technician. When the customer is upset, the service technician must stay calm and listen carefully to the customer.

The service technician must work with the information provided by the customer and any acquired troubleshooting information to find a solution. Clear communication of the facts helps eliminate misunderstandings and can reduce the troubleshooting time for the technician.

Advancement in the HVAC Field

A successful hydronic system service technician practices lifelong learning. When a skill has been mastered, higher levels of skill mastery are required for newly designed boiler equipment. The ability to adapt as the HVAC industry changes is crucial for success. Reading trade publications, attending classes and seminars, and participating in professional organizations are activities that provide valuable information on current topics and trends. New skills are required to remain current with advancing heating and cooling technologies and to grow in the HVAC field.

Basic Troubleshooting Questions

When troubleshooting any hydronic system, it is a good idea to ask the homeowner some basic questions. **See Figure 43-4.**

- What is noticed about the problem? Are there any noises or smells? Are there any unexpected vibrations occurring?
- When did the problem start?
- When was the last time the hydronic system worked properly?
- Has the boiler been following its normal sequence of operation?
- Have any changes been made (reconfiguration of walls and/or zones) to the hydronic system or building?
- Has anyone else performed service on the hydronic system?

Unsuccessful Repairs. When a repair does not fix the problem with the hydronic system, the technician must reevaluate the symptoms, causes, and solutions listed on the troubleshooting charts. In some situations, the entire process of gathering information about the system must begin again.

Basic Troubleshooting Questions

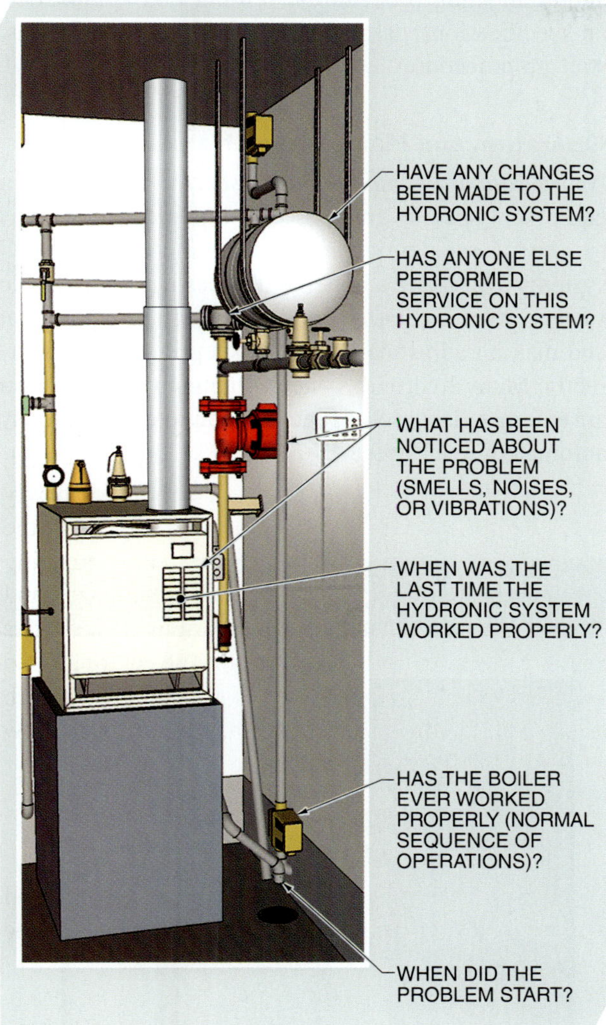

Figure 43-4. The technician should always begin the troubleshooting of a hydronic system by asking the owner specific questions about the system and how it has been operating.

Successful Repairs. When a repair is successful and the hydronic system operation has been restored, the troubleshooting process is recorded on a troubleshooting report or equipment maintenance log. Everything is recorded in a maintenance log including symptoms, causes, tests conducted, repairs made, and the condition of the hydronic system and boiler after the repair. **See Figure 43-5.** The spare parts used in the repair are recorded, and replacement parts are ordered as needed. Suggestions on modifying procedures that could prevent similar problems are also recorded.

Troubleshooting Reports and Maintenance Logs

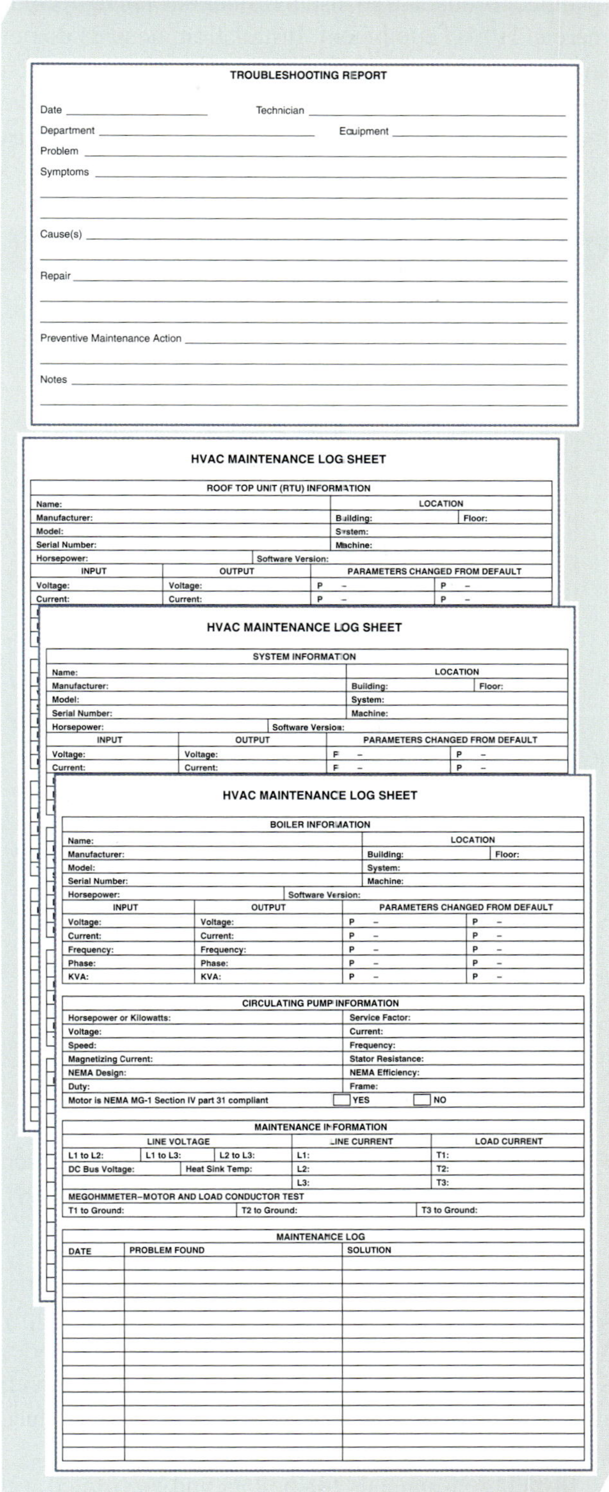

TROUBLESHOOTING REPORT

Date ______ Technician ______
Department ______ Equipment ______
Problem ______
Symptoms ______
Cause(s) ______
Repair ______
Preventive Maintenance Action ______
Notes ______

HVAC MAINTENANCE LOG SHEET

ROOF TOP UNIT (RTU) INFORMATION

Name:		LOCATION	
Manufacturer:		Building:	Floor:
Model:		System:	
Serial Number:		Machine:	
Horsepower:	Software Version:		
INPUT	OUTPUT	PARAMETERS CHANGED FROM DEFAULT	
Voltage:	Voltage:	P –	P –
Current:	Current:	P –	P –

HVAC MAINTENANCE LOG SHEET

SYSTEM INFORMATION

Name:		LOCATION	
Manufacturer:		Building:	Floor:
Model:		System:	
Serial Number:		Machine:	
Horsepower:	Software Version:		
INPUT	OUTPUT	PARAMETERS CHANGED FROM DEFAULT	
Voltage:	Voltage:	P –	P –
Current:	Current:	P –	P –

HVAC MAINTENANCE LOG SHEET

BOILER INFORMATION

Name:		LOCATION	
Manufacturer:		Building:	Floor:
Model:		System:	
Serial Number:		Machine:	
Horsepower:	Software Version:		
INPUT	OUTPUT	PARAMETERS CHANGED FROM DEFAULT	
Voltage:	Voltage:	P –	P –
Current:	Current:	P –	P –
Frequency:	Frequency:	P –	P –
Phase:	Phase:	P –	P –
KVA:	KVA:	P –	P –

CIRCULATING PUMP INFORMATION

Horsepower or Kilowatts:	Service Factor:
Voltage:	Current:
Speed:	Frequency:
Magnetizing Current:	Stator Resistance:
NEMA Design:	NEMA Efficiency:
Duty:	Frame:
Motor is NEMA MG-1 Section IV part 31 compliant	☐ YES ☐ NO

MAINTENANCE INFORMATION

LINE VOLTAGE			LINE CURRENT	LOAD CURRENT
L1 to L2:	L1 to L3:	L2 to L3:	L1:	T1:
DC Bus Voltage:	Heat Sink Temp:		L2:	T2:
			L3:	T3:
MEGOHMMETER–MOTOR AND LOAD CONDUCTOR TEST				
T1 to Ground:	T2 to Ground:		T3 to Ground:	

MAINTENANCE LOG

DATE	PROBLEM FOUND	SOLUTION

Figure 43-5. Hydronic systems have more components (sensors and valves) than forced-air systems, requiring that more information be kept for future use.

Installation Manuals

An *installation manual* is a general guide used by qualified technicians to install residential and light commercial HVAC equipment. Installation manuals do not supersede any national or local codes. When installation manuals do not arrive with the delivered equipment, the manuals should be obtained from the manufacturer as soon as possible. **See Figure 43-6.**

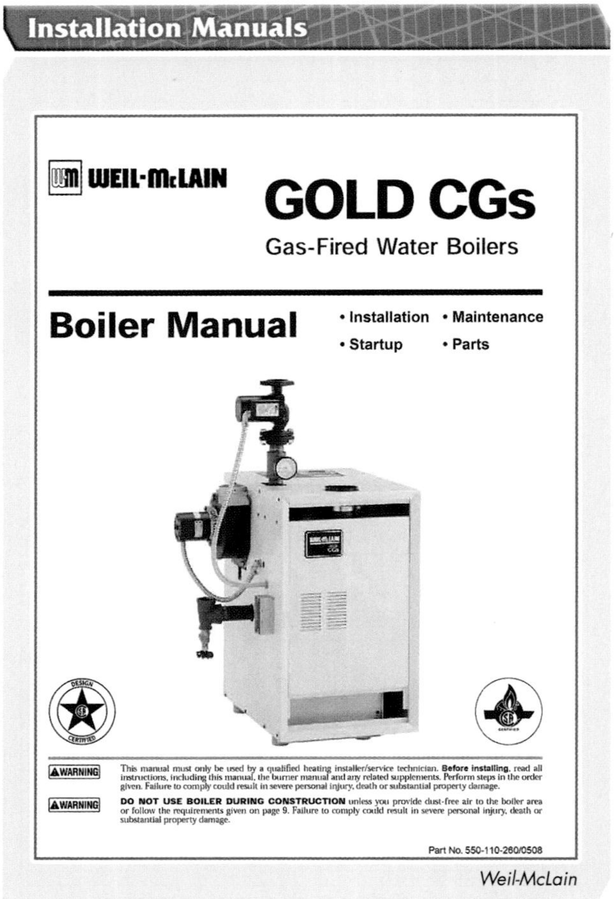

Figure 43-6. Installation manuals contain all information required to install hydronic systems but do not supersede national and local codes.

The installation of the boiler and associated hydronic system equipment described in the manual must comply with all state and local codes as well as national codes such as the NEC® or CEC™. Failure to comply with manufacturer procedures can void warranties and reduce the energy efficiency of the equipment.

Installation manuals for boilers and associated hydronic system equipment typically contain sections on safety, how to inspect the boiler upon delivery, where to locate the boiler, field wiring diagrams and specifications, water piping diagrams and recommendations, cleaning and/or flushing procedures, venting and combustion air specifications, gas piping installation specifications, start-up procedures, and check-out procedures.

Operation and Maintenance Manuals

An *operation and maintenance (service) manual* is an equipment-specific guide used by qualified technicians to service residential and light commercial HVAC systems. Operating a boiler and its associated hydronic system involves starting and stopping the equipment and making adjustments to it to keep it operating correctly. Many hydronic system functions for operation are automated, but some functions cannot be completely automated. **See Figure 43-7.**

Figure 43-7. Operation and maintenance manuals contain all the information needed to operate specific hydronic equipment in different applications, as well as the preventive maintenance information needed to keep the hydronic system running for years.

Operation and maintenance manuals for boilers and hydronic system equipment typically contain safety information, sequences of operations, burner adjustments and servicing information, preventive maintenance procedures, performance checklists, wiring schematics, flash code charts, troubleshooting guide, parts lists, and warranty information.

Flash Code Charts. Boilers typically have a self-diagnosing control module with an LED that flashes red to let technicians know when the boiler is having a problem. The fault indicator LED is behind the front access door observation window on the boiler. The number of flashes the control module displays is the error code for the problem the boiler is encountering. The technician counts the number of flashes and checks the error code against the flash code chart. Once the problem is identified, the troubleshooting section of the chart can be used to solve the problem. **See Figure 43-8. See Appendix.**

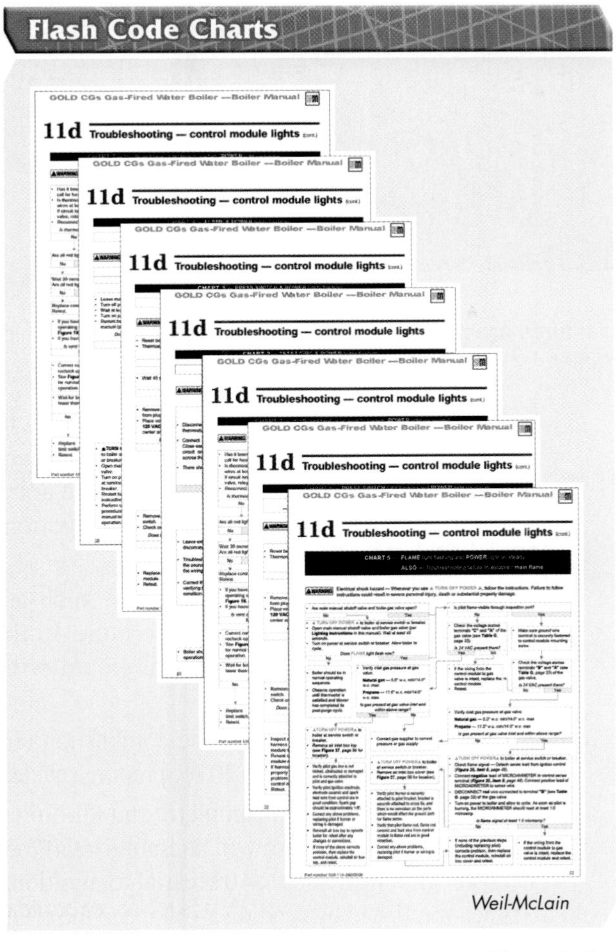

Figure 43-8. Flash code lights on the control board of a hydronic boiler convey fault information regarding specific boiler problems to the technician.

Troubleshooting Charts. Troubleshooting charts are only intended to be used on boilers serviced by technicians who have successfully completed the manufacturer training. Troubleshooting charts must only be used in conjunction with the appropriate boiler installation or operation manual, which is also provided by the boiler manufacturer. These charts are not intended to be a complete list of all the possible failures a boiler may experience. There are many reasons why a boiler might not be operating. Fortunately, many of the solutions are found on boiler troubleshooting charts. **See Figure 43-9. See Appendix.**

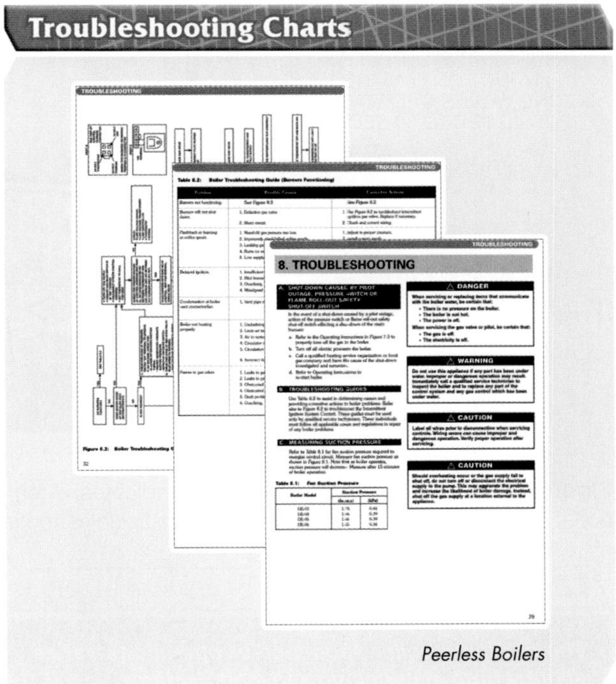

Figure 43-9. Troubleshooting charts list both the common faults that a hydronic boiler encounters and the standard corrective actions needed to eliminate these faults.

Typical Hydronic Boiler Sequence of Operations

Hot water boilers are designed with multiple safety features. These safety features prevent the fire from escaping the firebox, prevent the buildup of fuel (natural gas, fuel oil, or propane) on the burners, which can cause an explosion, and prevent flue gases from escaping into the room. Sequences of operations are valuable tools in diagnosing hydronic boiler problems. **See Figure 43-10.**

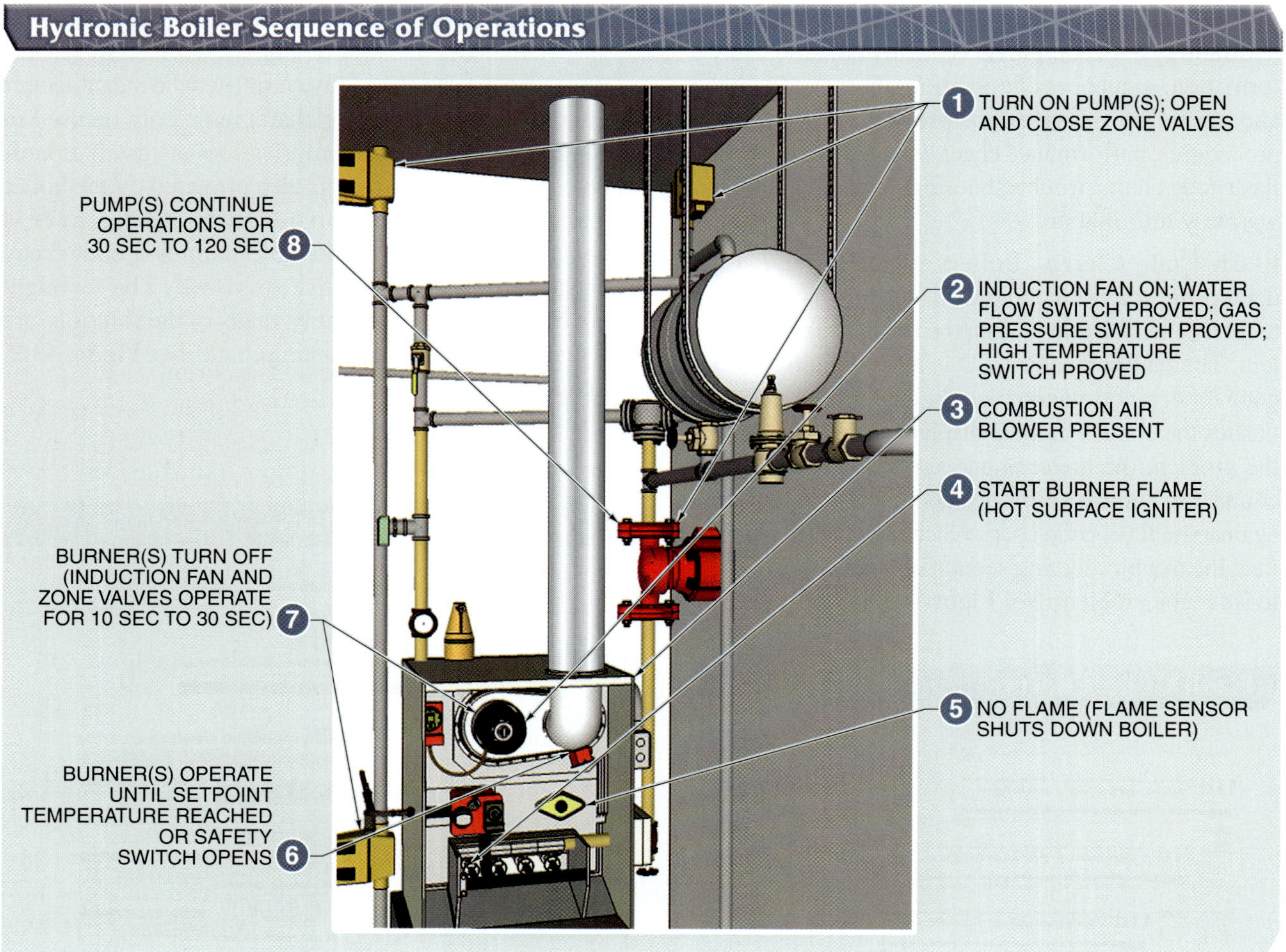

Figure 43-10. A sequence of operations indicates when safety features are playing a role in the operation of a hydronic boiler and contains valuable information for troubleshooting the boiler.

TECH TIP

The hydronic boiler sequence of operations varies slightly from manufacturer to manufacturer.

1. Upon a call for heat, the control turns on the appropriate pumps and opens and closes zone valves as needed. Typically, these consist of the boiler pump and/or zone pumps.
2. After a built-in time delay (typically 30 sec), the control board will send 115 VAC to the induction fan if the following conditions are met:
   a. The water flow switch or low-water cutoff must be closed.
   b. If a gas pressure safety switch is present, it must be closed.
   c. Some boilers have an auxiliary device, which must be proved before the induction fan can start. When this device is connected to the unit, the unit will provide 24 VAC to the induction fan enable relay.
   d. In general, any other limit switches, such as a manual reset high-water-temperature limit switch, must be closed before the board will energize the induction fan.
3. The control board will energize the combustion air blower and then start a 10 sec to 30 sec prepurge cycle.
4. Once the prepurge cycle is complete, and all limits are closed, the control board will attempt to start the burner by beginning a 5 sec to 10 sec trial for ignition. This is accomplished by sending a voltage to the pilot ignition device or hot surface igniter and opening the gas valve.

5. If the control board does not detect a flame by the end of the trial for ignition, the control performs a 10 sec to 30 sec postpurge, then another prepurge, and then it will try to ignite the burner again. Most boilers will try to ignite the burner a set number of times, typically three to five times and then lock out. The lockout may need to be manually reset or may be subject to a timer. For example, the control board may try to light the burner three times and then lock out for a half hour, then try three more times.
6. Once operating normally, the burner will operate until setpoint temperature is reached or a safety limit opens.
7. When the system call for heat is satisfied, the control board will turn the burner off. The induction fan will operate on a postpurge cycle of 10 sec to 30 sec.
8. The pumps and zone valves will typically continue to operate for a specific time delay of 30 sec to 120 sec, then turn off.

TROUBLESHOOTING HYDRONIC SYSTEMS

The vast majority of boiler and hydronic system problems can be broken down into two general categories: zone problems and system problems. Zone problems are those that occur in one room or zone. The overall system may be functioning properly, but one zone is too hot or too cold. Typically, zones include a heat exchanger, thermostat, zone valve, and possibly a circulating pump.

The majority of system problems are related to the hydronic boiler and the system itself. The system consists of the boiler, pump or pumps, controls, and accessories. Most hydronic system troubleshooting problems can be placed into four general groups:

- sizing and installation
- electrical and controls
- fuel, combustion, and venting
- water distribution and supply

Sizing and Installation

Often, mechanical systems are not sized or installed properly. This is true of residential hydronic systems. Sizing and installation problems may be diagnosed initially by asking the question, "Has the system ever worked properly?" A definitive answer may be difficult to come by. If at all possible, it is a good idea to look at any prints or documentation available. In general, a system that is dramatically oversized will cycle on and off too much, while one that is undersized will not be able to keep up under load at maximum design conditions.

Electrical and Controls

Another problem area is the electrical and control systems. Modern control systems consist of a microprocessor that energizes hydronic system components such as burners and pumps. Any control malfunction will cause the hydronic system to fail or operate improperly. The hydronic system must also have a reliable power source to operate properly. Any power interruptions will cause shutdowns and equipment failure.

Fuel and Combustion

Fuel can be another problem. When fuel is not provided at the correct pressure or temperature or is of the wrong quality or type, it will affect boiler operation. Another source of problems is in the combustion process. This may include problems with the airflow into the combustion chamber or problems with venting or draft.

Water Distribution and Supply

Water distribution and supply can be a problem area. Often, one or more zones are starved for water flow, which means they tend to be cold. In this case, the system may not have been balanced properly. The makeup water system valving can be a problem as well.

Zone Troubleshooting Scenario #1—Bad Thermostat

A "too cold" complaint is received from a homeowner about one room of a zoned residence. The other room temperatures are fine, and the hydronic boiler is working properly. The cold zone is checked and the thermostat is found to be defective, as the zone valve is closed but operates when given power. **See Figure 43-11.**

Service Work Performed: The thermostat is replaced and hot water flows into the zone.

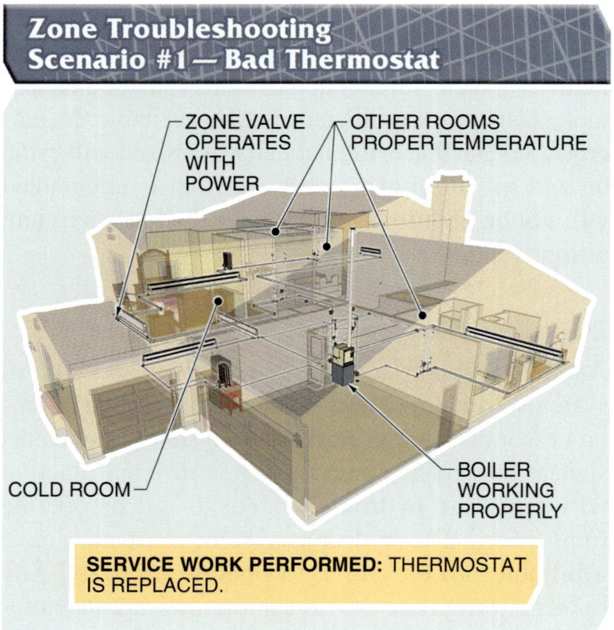

Figure 43-11. When a residence or light commercial building has multiple thermostats because of multiple zones, troubleshooting must lead to a thermostat or zone valve.

Zone Troubleshooting Scenario #2—Bad Zone Valve

A "too hot" complaint is received about a zone in a residence. All other zones of the hydronic system are working properly. It is found that no matter what the thermostat does, the zone valve remains open. When the valve is repaired, it closes immediately. **See Figure 43-12.**

Service Work Performed: The zone valve actuator (valve top) is replaced.

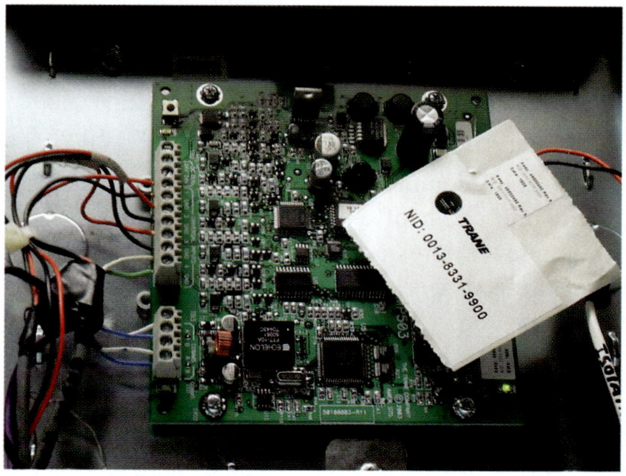

The electronic control board must be checked for proper operation.

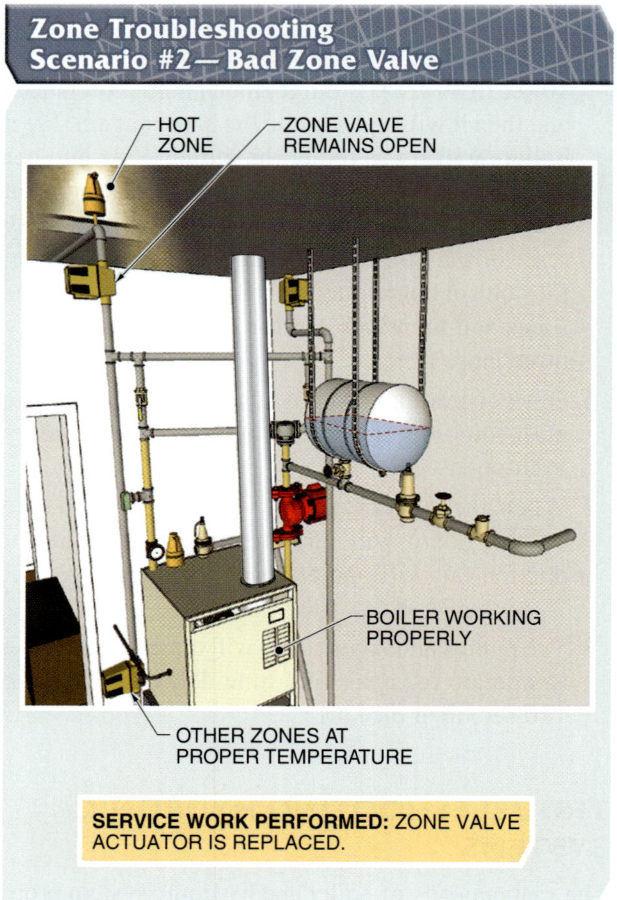

Figure 43-12. Many times the actuator of a valve is bad and not the valve itself. Replacing the actuator on a valve can solve many problems with zone and terminal unit valves.

Boiler Troubleshooting Scenario #1—Control Board Not Working

An HVAC technician receives a "too cold" complaint from a customer with a hydronic boiler system. When taking an initial look at the boiler, the main cover is removed, and the control board cover is removed. The board is off, so there is no power to indicate a fault. A DMM is used to ensure that 115 V is available on the primary side of the transformer and 24 V is available on the secondary side. A fuse on the board is checked and is found to be open. **See Figure 43-13.**

Service Work Performed: The reason for fuse failure is checked. The fuse is replaced and the unit starts and runs normally.

Note: Other causes of this problem could be an open breaker in the service panel, ON/OFF disconnect switch in OFF position, or possibly a loose power cable to the board.

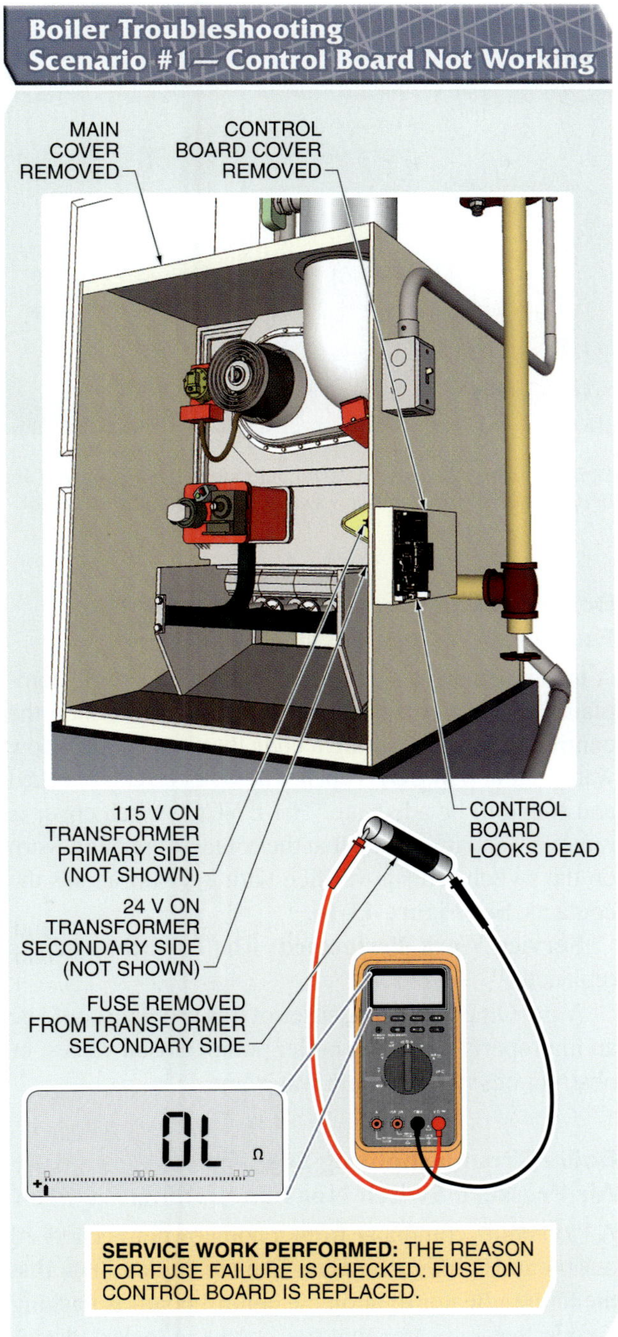

Figure 43-13. When the circuits of a system are not working, the power supplies should be checked. When power is present but there is no board operation, the fuses have probably failed.

Boiler Troubleshooting Scenario #2—Burners Not Operating

A homeowner calls in a "too cold" complaint. Upon arrival, the technician checks the system and finds that the power to the control board is on. The board indicates normal operation, no faults, and a satisfied system. The water temperature is checked and found to be 20°F lower than normal. The setpoint is raised at the board and the boiler cycles on. The technician checks the resistance of the water temperature sensor and the outdoor-air temperature sensor and finds that the outdoor-air temperature sensor is reading too high. This causes the setpoint to reset to a lower temperature. **See Figure 43-14.**

Service Work Performed: The outdoor-air temperature sensor is replaced and the system checked to ensure proper operation.

Note: A typical hot water boiler has a number of temperature sensors including inlet water, outlet water, system water, flue, and outdoor air. Temperature sensors are typically resistance-type devices. They change resistance as the temperature changes. Tables are provided by manufacturers that indicate the values for their sensors at various temperatures.

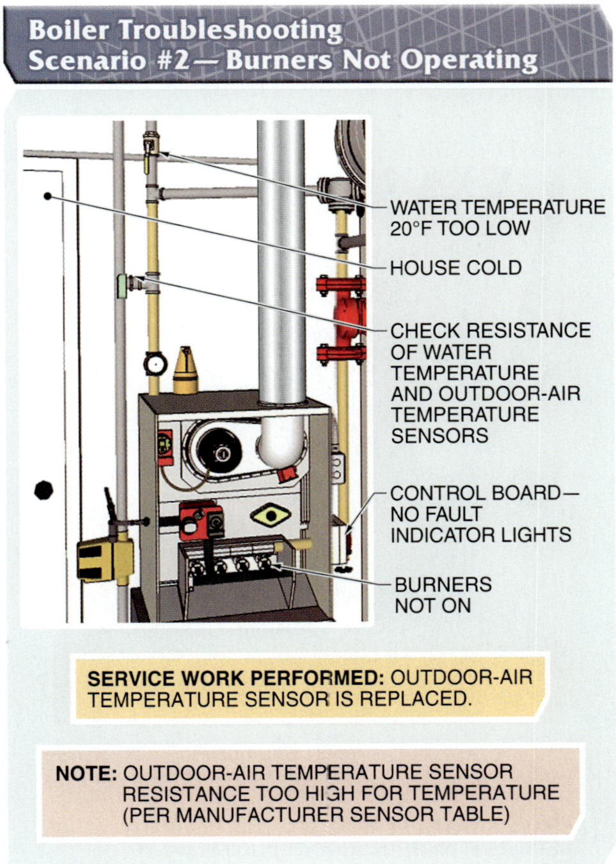

Figure 43-14. The outdoor-air temperature sensor must always be checked when a hydronic boiler is having problems with maintaining water temperature because of temperature reset.

Boiler Troubleshooting Scenario #3—Noisy Boiler

A homeowner calls in with a nonstandard complaint. While the temperature in the home seems fine, the boiler has become very noisy. The technician arrives and checks the natural gas pressure to find that it is 16″ WC, which is too high. The technician then checks the main gas supply and finds that the pressure to the home is too high. The gas company is called out to investigate, and it is discovered that the gas pressure-reducing valve has failed. **See Figure 43-15.**

Service Work Performed: The gas company repairs the valve and the pressure is corrected. The boiler operates properly.

Note: Other possible causes of a noisy boiler could be a gas/air mixture problem, a damaged or fouled burner, low water flow through the boiler, or excessive air in the water system.

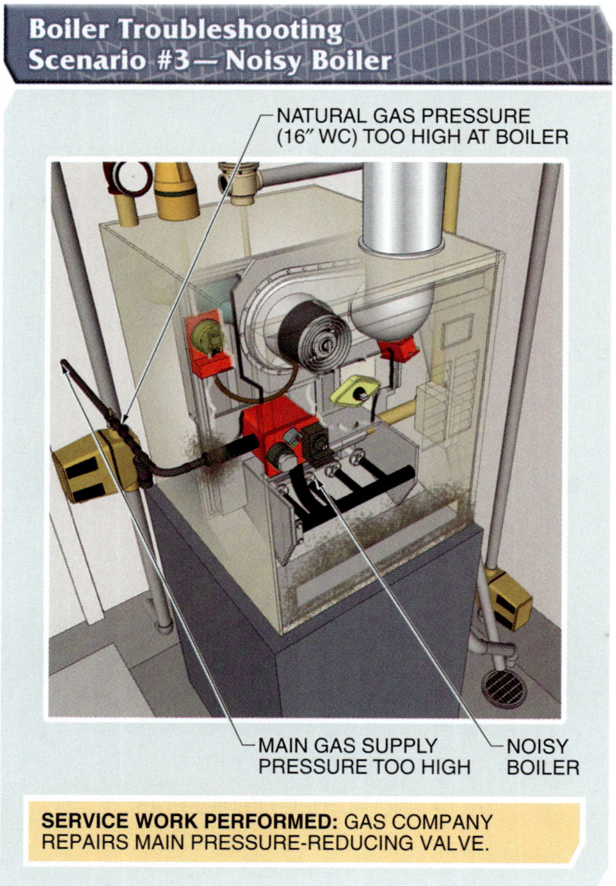

Figure 43-15. High gas pressure will cause the flames on a burner to jump and be erratic. The noise is the result of the flames jumping off the burner.

Parker Boiler Co.

Light commercial boilers installed outdoors have a complete set of difficulties associated with weather that indoor boilers do not.

Boiler Troubleshooting Scenario #4—Faulty Low Water Cutout (LWCO)

A technician arrives at a home with a "too cold" complaint. The boiler is not functioning. The LEDs on the control board are flashing a code that indicates the low water cutout (LWCO) is open. The water level is checked and found to be adequate. The LWCO switch cover is removed, and it is found that the contacts are quite worn on the switch. This is verified with a DMM across the contacts. **See Figure 43-16.**

Service Work Performed: The LWCO switch is replaced.

Note: Other possible causes of this problem could be an improperly operating boiler pump, closed valves, or obstructions in the system water line.

Boiler Troubleshooting Scenario #5—Air Pressure Switch Not Functioning

A "too cold" complaint from a homeowner is checked out by an HVAC technician. The technician finds that the fault indicator light on the control board is flashing a code that indicates that the burner is locked out. A check of the pressure switch terminals indicates that the airflow proving switch is open. The pressure switch is visually inspected, and it is found that the tubing is pinched. The owner recently cleaned around the boiler and inadvertently pinched the tubing. **See Figure 43-17.**

Service Work Performed: The tubing is freed and the airflow proving switch now operates properly.

Note: Other causes of this problem could be the switch being wired incorrectly, loose wiring, or a bad pressure switch.

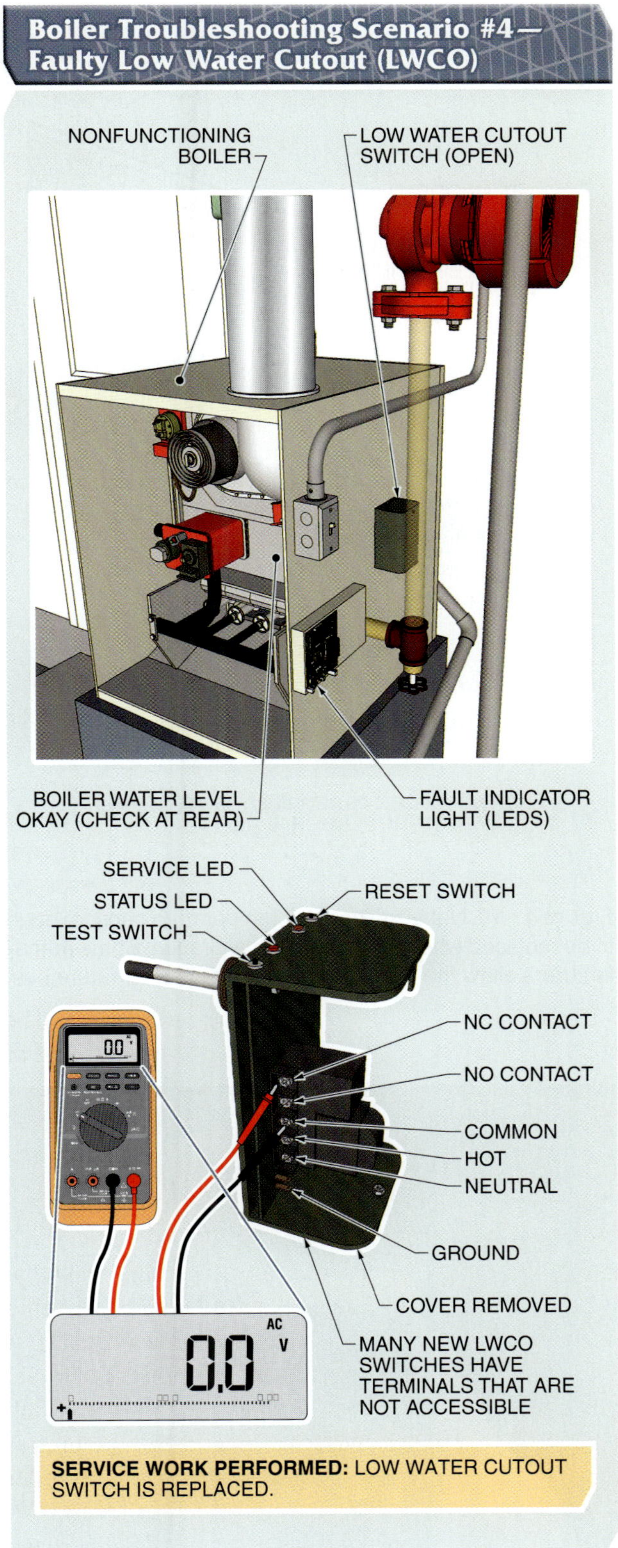

Figure 43-16. A technician must always verify that a boiler is in a low-water condition when the control board fault indicator light is flashing an LWCO fault. Because of the design of LWCO switches, these switches are known to fail.

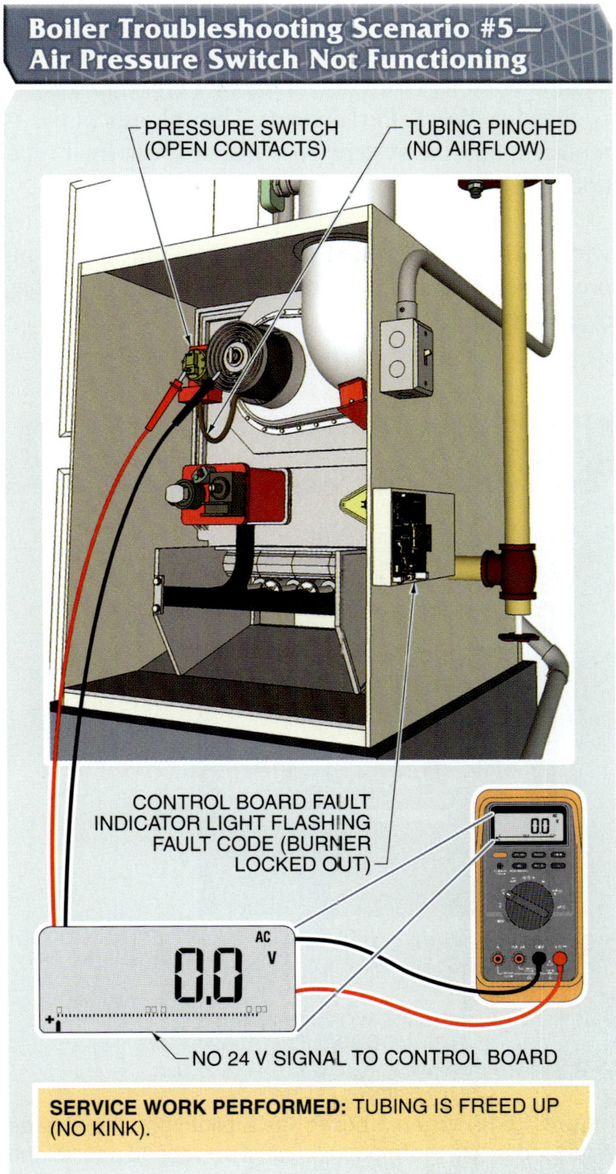

Figure 43-17. The air pressure switch, which verifies induction fan operation, is the sensor (switch) on the boiler that most commonly fails. Most technicians keep spare pressure switches on hand at all times.

Boiler Troubleshooting Scenario #6—No Burner Ignition

A customer calls in a "too cold" complaint. The technician arrives and finds that the boiler burners are off. The technician resets the control board and observes the boiler's attempt to go through its ignition sequence. The burners fail to light and are locked out by the control board. The technician inspects the pilot light thermocouple and finds it is faulty. Since the thermocouple is not operating, the gas valve has no power, and the pilot light and main burners cannot ignite. **See Figure 43-18.**

Service Work Performed: The thermocouple is replaced, and the system goes through a normal ignition sequence.

Note: Other causes of this problem could be igniter wiring, failure to ground the gas valve properly, improper gas pressure, lack of 24 VAC to the gas valve, or a faulty gas valve.

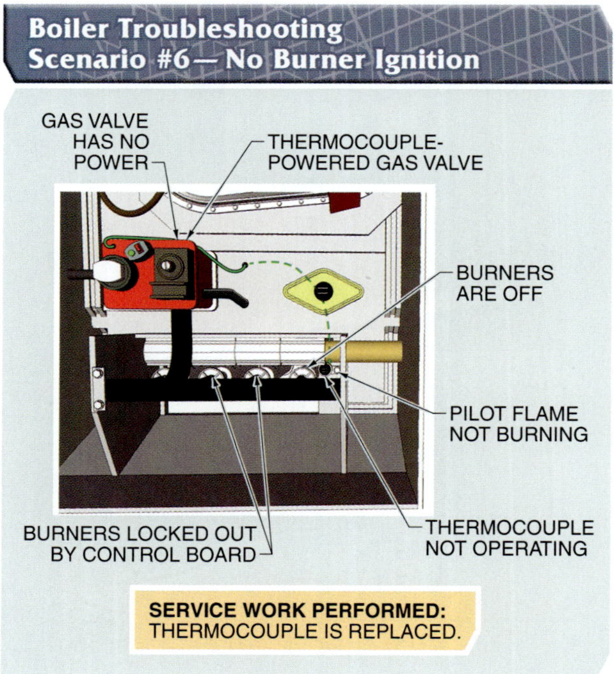

Figure 43-18. When a boiler has a pilot light, the voltage produced by the thermocouple must be verified against the voltage required by the gas valve.

Boiler Troubleshooting Scenario #7—Faulty Sensor

A "too hot" complaint is received from a customer. Upon arrival, the technician finds that the hydronic system water is 20°F higher than normal. This particular system utilizes an outdoor-air temperature reset for the hot water supply. The manufacturer's temperature/resistance chart is consulted, and the sensor is found to read 10°F too low. This is resetting the hot water temperature 20°F higher than normal. **See Figure 43-19.**

Service Work Performed: The outdoor-air temperature sensor is replaced.

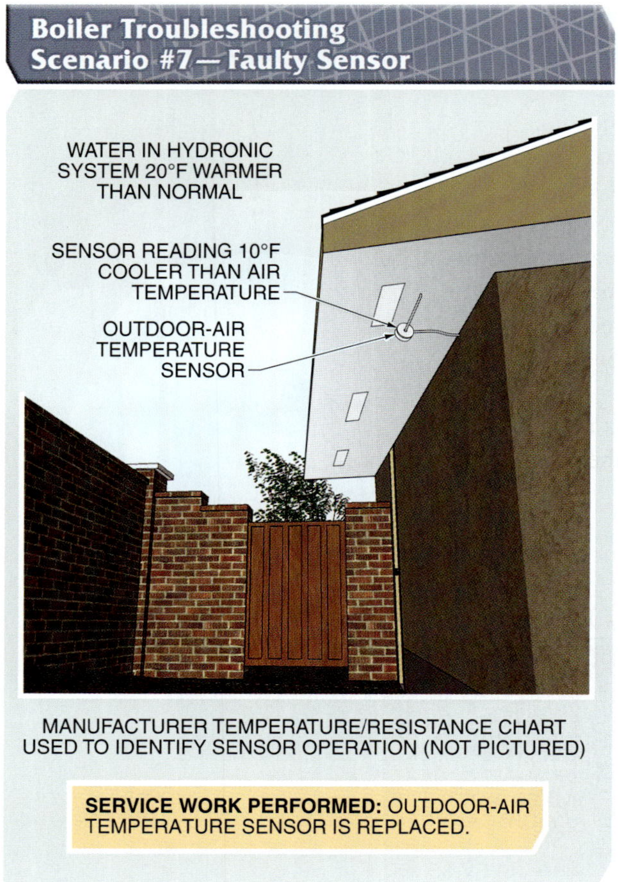

Figure 43-19. Not all outdoor-air temperature sensors need to be replaced when they go out of calibration. Some manufacturers allow the sensor to be recalibrated for future use.

Temperature sensors and limit controls are tested using a multimeter.

Boiler Troubleshooting Scenario #8—Low Water

A homeowner calls in a "too cold" complaint. The technician arrives and checks the control board. The fault code indicates that the boiler is locked out because of a low-water condition. The water level is checked for verification and found to be too low. The inlet makeup water pressure gauge is checked and found to indicate a low-pressure condition. The water regulator (pressure-reducing valve) is checked and is found to have failed. **See Figure 43-20.**

Service Work Performed: The inlet water regulator is replaced. Water pressure at the boiler increases, and the boiler operates properly.

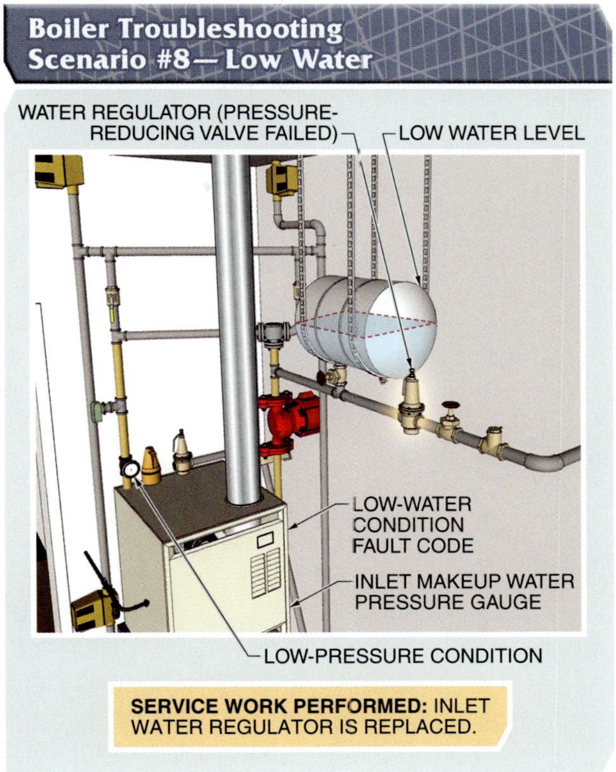

Figure 43-20. When supply water pressure is too high or too low, the water regulator (pressure-reducing valve) is typically the problem.

TECH TIP

All manufacturer installation instructions and safety references must be followed when servicing and troubleshooting hydronic boiler systems. Fuel, combustion gases, and electricity are all present and require different safety procedures. Improper service and troubleshooting techniques can damage a building and endanger its occupants.

UNIT 44

Installing Light Commercial HVAC Systems

The HVAC industry is large and is responsible for installing equipment in many environments, from small houses and apartments to large commercial facilities such as professional sports arenas, college campuses, and industrial plants. "Residential" is the industry term for HVAC equipment used in homes and residences, while larger buildings are known as "commercial." Commercial equipment can be broken down further into light commercial and heavy commercial, although in many cases, exact distinctions are difficult to make.

OBJECTIVES

- List the general installation considerations for all light commercial HVAC units.
- List the typical procedures for installing a rooftop single-zone unit.
- List the typical procedures for starting a rooftop single-zone unit.
- List the typical procedures for installing a rooftop VAV air-handling unit.
- List the typical procedures for starting a rooftop VAV air-handling unit.
- List the typical procedures for installing an air-cooled packaged chiller.
- List the typical procedures for starting an air-cooled packaged chiller.
- Explain the startup procedures for variable frequency drives.

Index to Procedures

INSTALLING LIGHT COMMERCIAL SYSTEMS

A general rule of thumb is that light commercial systems are roughly sized between 5 tons and 100 tons of cooling capacity. Three of the most commonly installed light commercial systems include rooftop single-zone units, rooftop VAV units, and in some cases, air-cooled packaged chillers. Since an HVAC technician may run across any of these systems in a light commercial building, it is important to understand how to work (install, service, and troubleshoot) these systems.

General Installation Procedures—All Units

While there are many different light commercial HVAC systems, installation procedures are similar, regardless of unit type. As always, the manufacturer guidelines as well as all applicable codes should be followed when installing, servicing, or troubleshooting any of these units.

Ground Level HVAC Unit Security

Before installing a light commercial system it is a good idea to check the site for access and clearances. Most light commercial units can be mounted on the roof or on a ground level pad outside the building. When a light commercial unit is mounted on the ground, adequate clearance as well as possible security of the unit must be considered. Many systems use a locked fence to protect the unit against theft and vandalism. **See Figure 44-1.** Typically, security is not as serious a consideration with rooftop units. **See Procedure for Checking and Preparing Site to Receive Delivery of HVAC Unit.**

HVAC Unit Security

Figure 44-1. In general, ground level HVAC equipment requires more security than rooftop HVAC systems.

Procedure

Checking and Preparing Site to Receive Delivery of HVAC Unit

1. Inspect ground conditions. Ensure they are suitable for supporting the weight of the crane and the HVAC unit to be lifted.
   a. Are net and gross loads known?
   b. Is ground firm and level?
   c. Is ground capable of withstanding the pressure of outrigger jacks? (Are outrigger matts or pads needed to stabilize any soft ground conditions?)

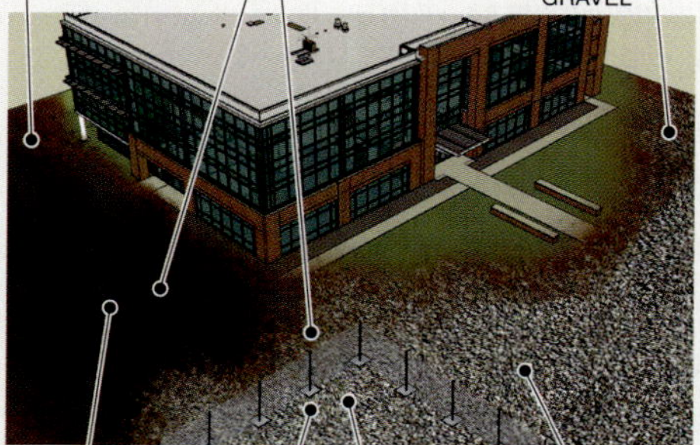

2. Plan the access and egress for flatbed truck carrying HVAC equipment and for crane.
   a. Can crane rotate 360° without coming into contact with an object?
   b. Is there space for outriggers to be fully extended?
3. Check for proximity to existing hazards.
4. Check for city restrictions on crane operations at the lift site.
5. Ensure that location is secure for storing HVAC equipment until lift to roof occurs.

Receiving Light Commercial Units

An important first step before actually installing a rooftop unit is properly taking delivery. Typically rooftop units come out on a flatbed truck. The order paperwork should be checked against the model number of the actual unit being delivered. It is not unheard of for the wrong unit to be delivered. Also, the unit should be checked for physical damage. Any insurance or delivery claims must be made before installation. Also, any loose parts, kits, or controls should be included in the delivery.

In some cases, it is possible that the unit was delivered well before the scheduled delivery. In this case, the unit should be stored using the manufacturer storage instructions to reduce any possible damage. Failure to follow the manufacturer instructions may cause the warranty to be voided. **See Figure 44-2. See Procedure for Receiving HVAC Equipment.**

TECH TIP

HVAC unit delivery is arranged before a unit is shipped to ensure that someone is available to sign for and receive the unit and that a location is available for storage of the unit.

Receiving Light Commercial HVAC Units

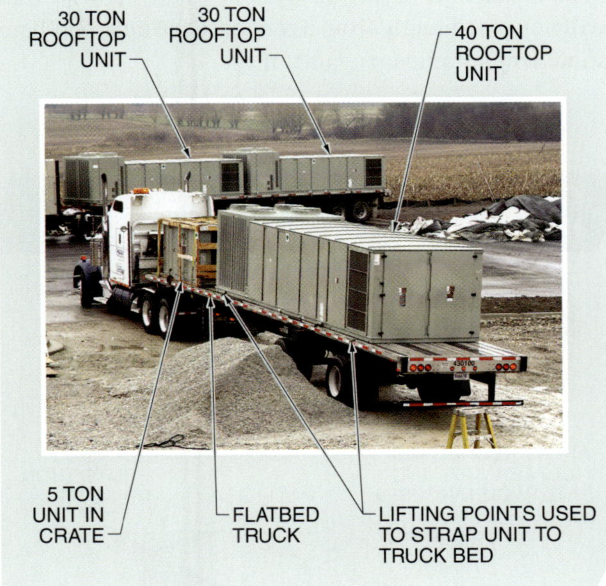

Figure 44-2. All HVAC equipment must be inspected thoroughly upon receipt to ensure that the equipment is in good shape. If the equipment is not in good shape, a claim must be filled out immediately.

Since most units are lifted onto the roof, clearance and safety distances must be checked beforehand. Many contractors will try to lift the unit after normal working hours to make it easier. Also, the ability of the ground to support the weight of the crane and unit is important. Any obstructions must also be checked. The site may have to be improved before it can be used for a lift.

Qualified riggers and cranes are required to place HVAC equipment on roofs.

Receiving HVAC Equipment

1. Always consult local codes and officials. Consult with building representatives regarding area to be used for equipment storage and future rigging.
2. Walk over storage area, carefully examining it for obstructions, interference from utilities, safety access issues, and whether soil type can handle heavy load.

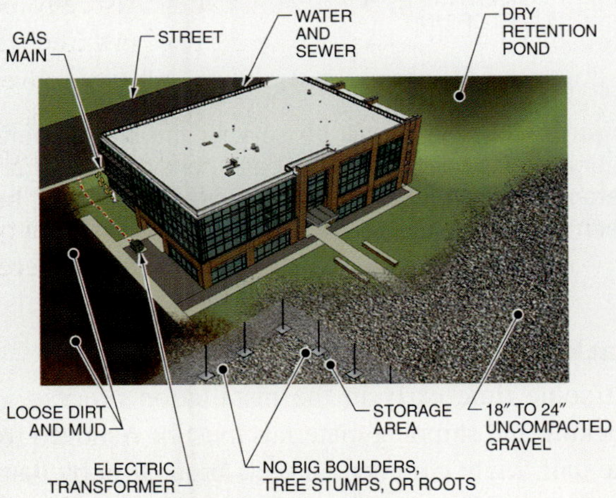

3. Inspect ground conditions and check weather.
   a. In some cases, soft soil may require a different area to be used for equipment storage or rigging.
   b. Weather conditions may be different from the conditions when the unit was delivered.
4. Coordinate with utility companies beforehand. In some instances, utilities may need to be relocated temporarily to allow access.

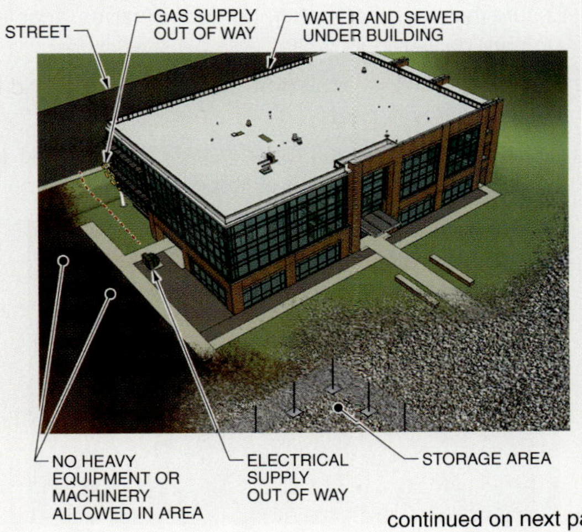

continued on next page

Procedure

Procedure

Receiving HVAC Equipment *(continued)*

5. Ensure the unit is delivered to the proper location at the proper time.

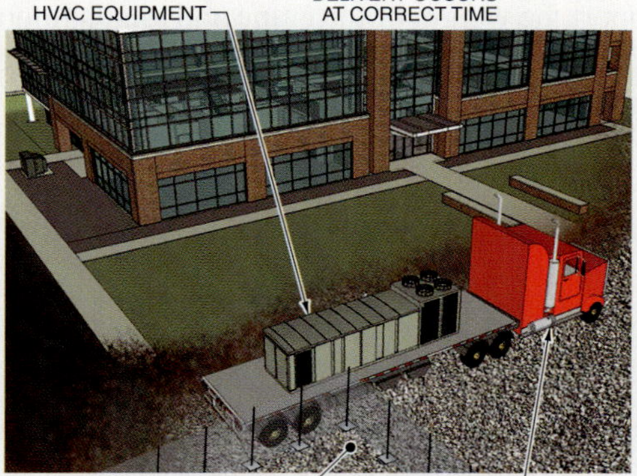

6. Ensure that a qualified person is present to receive the unit and sign the paperwork.
7. Carefully check and record the condition of the unit and check its model and serial number against bill of sale.
8. Remove the unit from the truck using appropriate material-handling equipment and factory-recommended attachment points only.

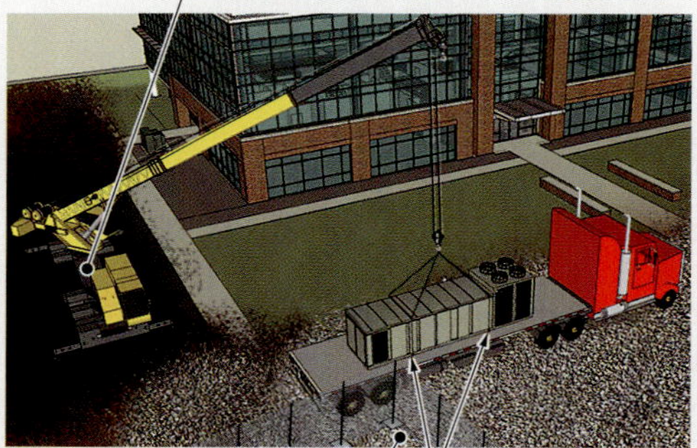

9. Ensure that any other materials needed (valves, adapters, etc.) are delivered as well.
10. Ensure that unit is located at a spot where it would not cause a safety hazard.
11. Ensure that the unit is properly sealed against the elements per factory recommendations.

Weather may also be a consideration, as rain or high winds may prevent a safe lift. **See Figure 44-3.** For units mounted on the ground outside the building, care must be taken to locate the system where snow or rain will not leak directly from a roof onto the unit, or where snow will not pile up against it.

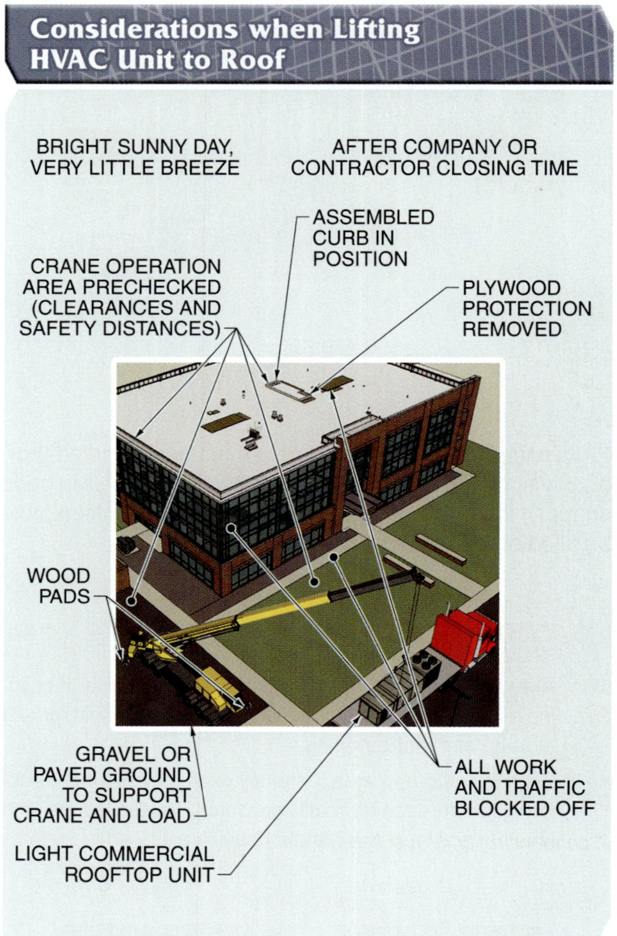

Figure 44-3. The lift will be safe for the crane operator after all lifting scenarios have been considered and all lift personnel and anyone not associated with the lift have been accounted for.

Packing and Shipping Materials

At some time early in the installation process, any packing and shipping materials must be removed from the unit. Light commercial units brought in by flatbed truck are typically packed with a minimum amount of materials. Piping is capped to prevent contamination. Doors may be taped when they are not lockable.

Before installation, any fan restraints that prevent damage to the fan during shipment need to be removed. Plugs must be left in until just before the actual connections are made. If the unit is to be installed at some point in the future, packing materials may be added to protect the unit from the weather.

Light Commercial Unit Pads

Units mounted on the ground must be installed on a pad. These light commercial unit pads are constructed at the job site. Location and drainage are considerations when installing the pad. Also, the pad must be level, and the soil must be able to support the weight of the pad and unit.

A condensing unit must be mounted on a pad that is level.

Roof Curbs

Rooftop units are placed on a curb. A *curb* is a metal frame with framed duct connections that allow a light commercial unit to be connected to the ductwork of a building. **See Figure 44-4.** Typically, a curb is prefabricated or preassembled at ground level and then lifted into position. Large unit curbs can sometimes be assembled in sections, with finished assembly occurring in position. The curb is attached via flashing and other weatherproofing materials to the roof structure of the building. The HVAC unit must fit snugly with screws and bolts to the curb. In some countries, local jurisdictions may require a roofer be part of the HVAC team. **See Procedure for Setting Curb.**

Setting Curb or Transition Fitting

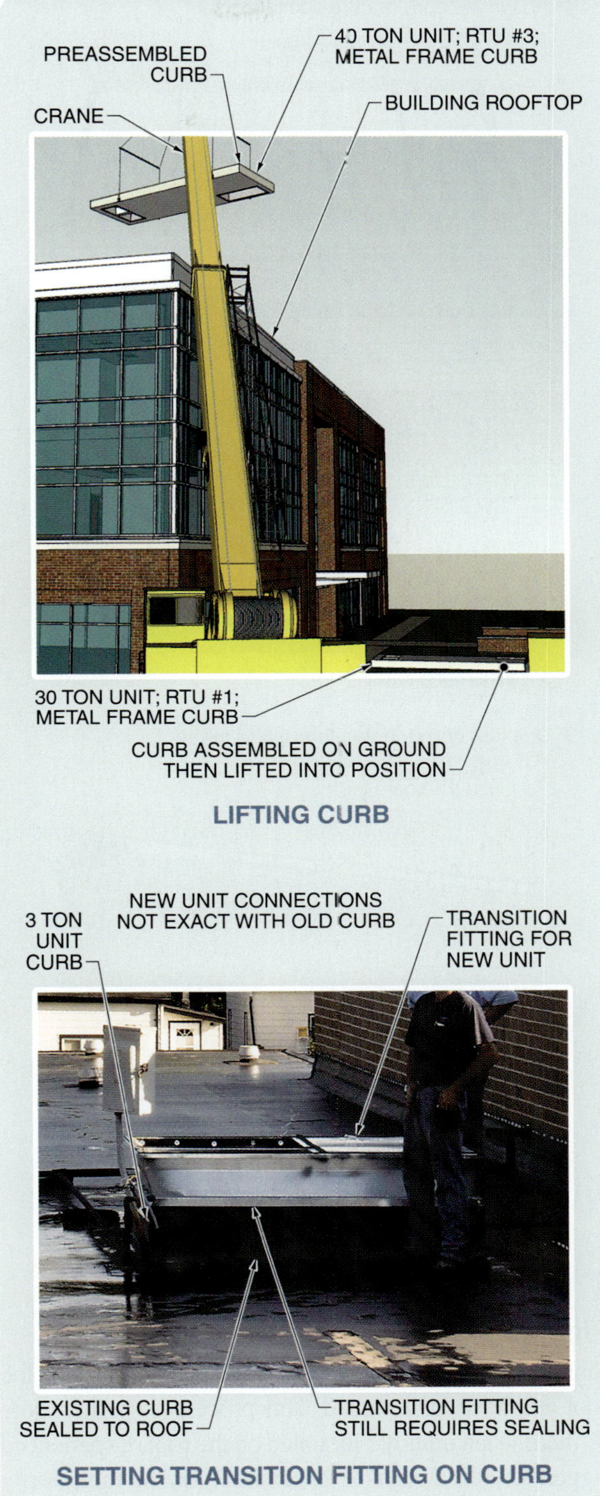

Figure 44-4. Curbs and transition fittings are typically assembled on the ground and lifted to the roof, where they must fit securely.

Procedure

Setting Curb

1. Assemble curb or curb sections.

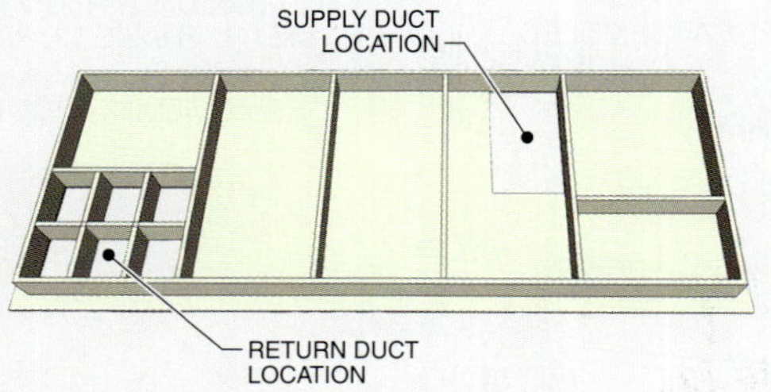

2. Identify curb position on ceiling.

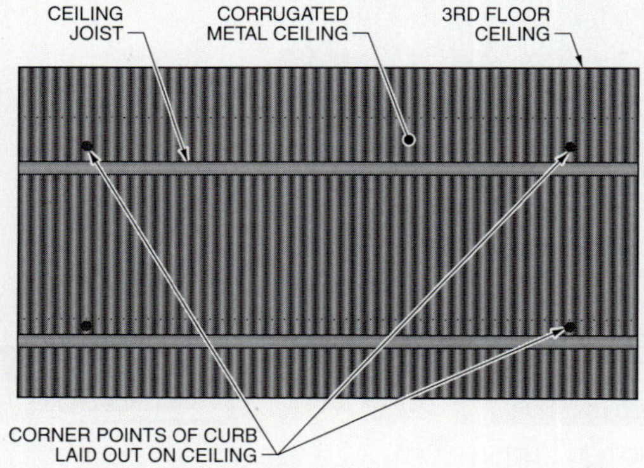

3. Drill four corner holes through to the roof.

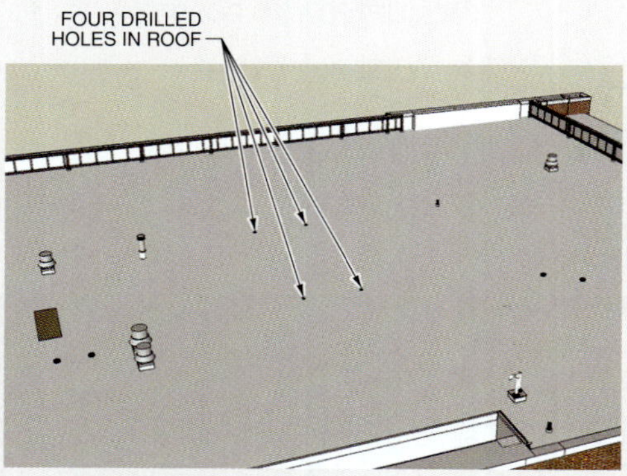

4. Cut out roof for curb installation.
   a. Cut out roofing material.
   b. Cut out insulation material.
   c. Cut out corrugated steel roof.

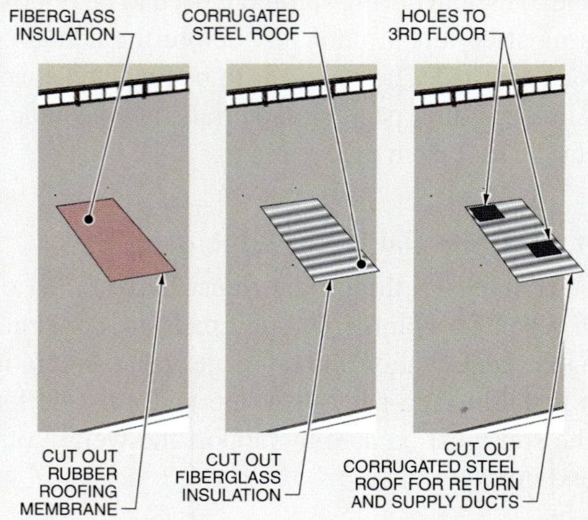

5. Place curb into position. (Curb may need sections assembled.)

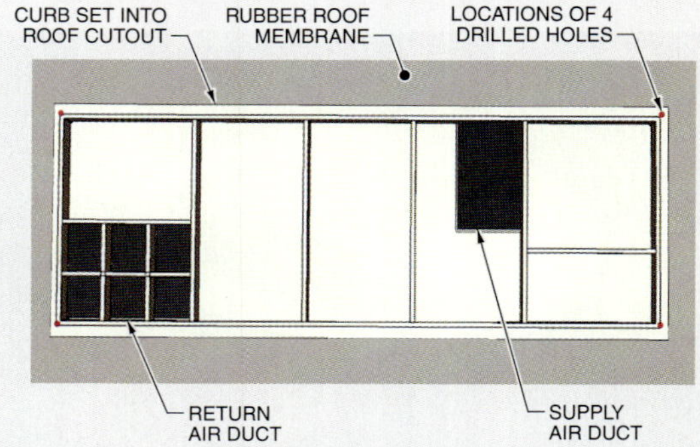

6. Level curb. Make appropriate taper shims to match slope of roof at curb position. Flash curb at roof. Follow roofing manufacturer's specifications for weather proofing the curb or contact a certified roofing contractor.

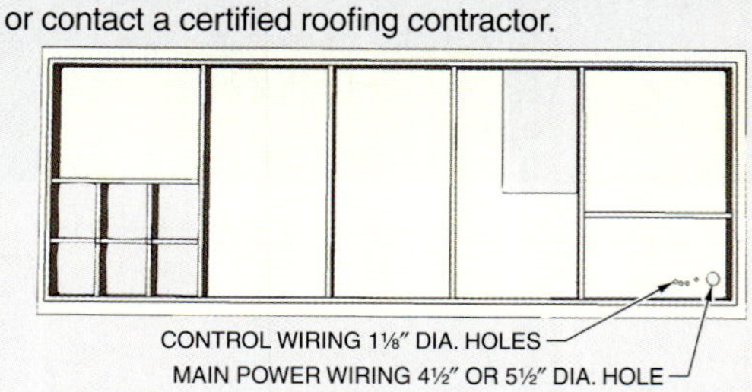

7. Cut round holes for utilities (2″, 4″, 6″, or whatever is required).
8. Cover curb with plywood to protect openings from weather.

Rigging

Single-zone or VAV units mounted on the ground or on the roof will have to be rigged. This process is slightly more difficult when units are mounted on the roof. Experienced riggers should always be used. Safety is the top priority when rigging. Many contractors will use someone with rigging and/or installation experience as a safety manager on the job site. The safety manager can shut down the lift at any time for any cause.

The packaged unit will come with lifting instructions, and specific points will be indicated for attaching the rigging equipment. These manufacturer recommendations should be followed at all times.

The unit is placed into the proper position on the roof curb. Often, before the new unit is lifted into position, the old unit is rigged, lifted, and moved to ground level by the same crane. **See Figure 44-5. See Procedure for Rigging HVAC Equipment for Lifting.**

Rigging for Lift

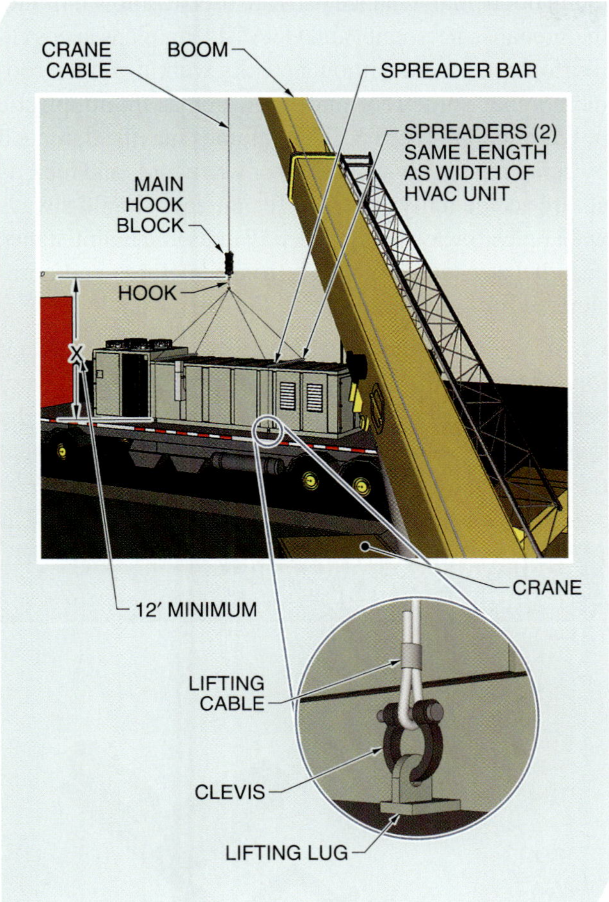

Figure 44-5. The manufacturer recommendations should always be consulted when lifting HVAC equipment.

Before installing any unit on a roof curb, the technician must ensure that the correct curb for the unit is being used, all necessary gaskets and hardware are in place, and the location allows clearances for proper operation. Also, the curb must be level and square. **See Procedure for Setting HVAC Unit on Roof Curb.**

TECH TIP

A technician should always follow the most up-to-date standards, including the latest appropriate safety standards, fall protection requirements, and roof barrier standards, when rigging and installing a unit. A few minutes saved by cutting a corner during a lift is not worth the potential liability of a preventable accident.

Rigging HVAC Equipment for Lifting

1. Unit should only be rigged by trained personnel.
2. Always follow all rigging and lifting safety requirements and codes.
3. Use only factory-approved attachment points.

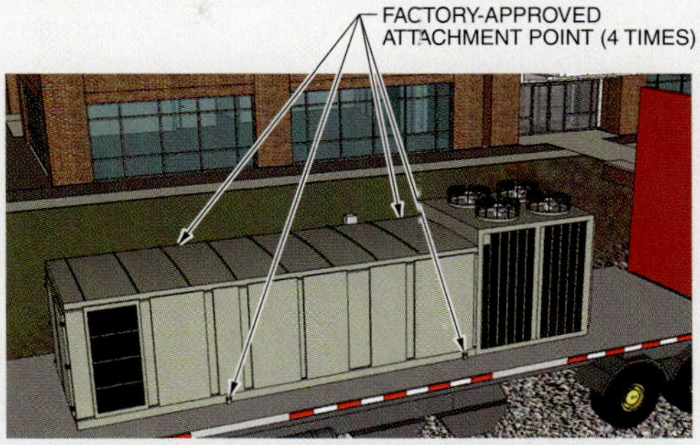

4. Rig the unit when building and other outside traffic is low. After-hours rigging is often preferable.

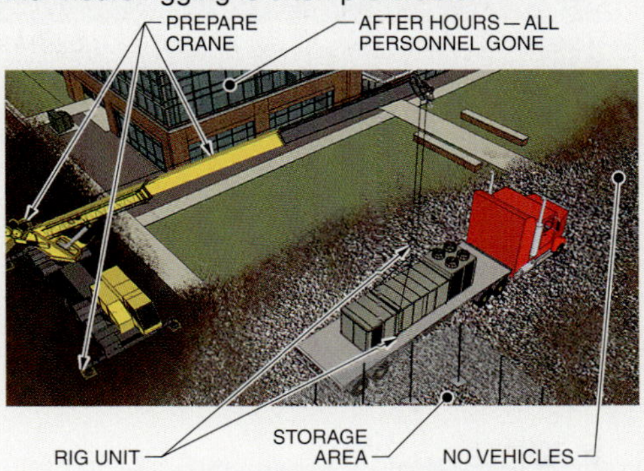

5. Choose a day with favorable weather conditions.
6. Remove any other components from the equipment to lighten the lift if needed or to allow rigging to be attached.
7. Lift unit slowly, using multiple spotters at different locations. Designate one person to signal the crane with either a radio or OSHA Federal Regulation 29 CFR Part 1926. All signal persons are third party certified by the National Commission for the Certification of Crane Operators (NCCCO).

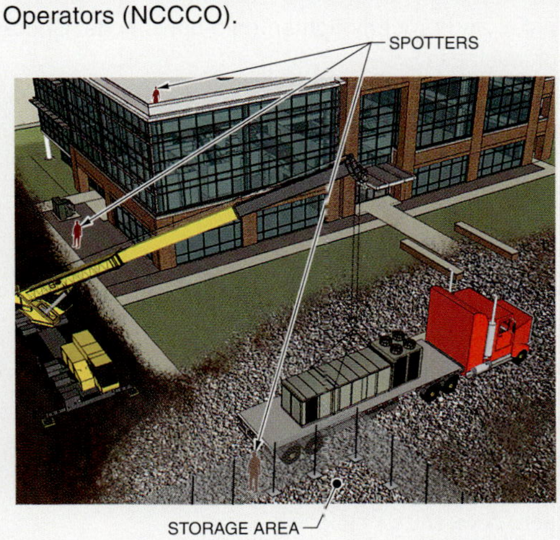

8. Place unit in storage area or over curb in position for attachment.

Procedure

Setting HVAC Unit on Roof Curb

1. Align the base rail of the unit's air handler section with the curb rail while lowering the unit onto the curb.
   a. Ensure that any dents or dings on the curb have been repaired.
   b. Ensure that the gasket on the curb is not damaged while positioning the unit.

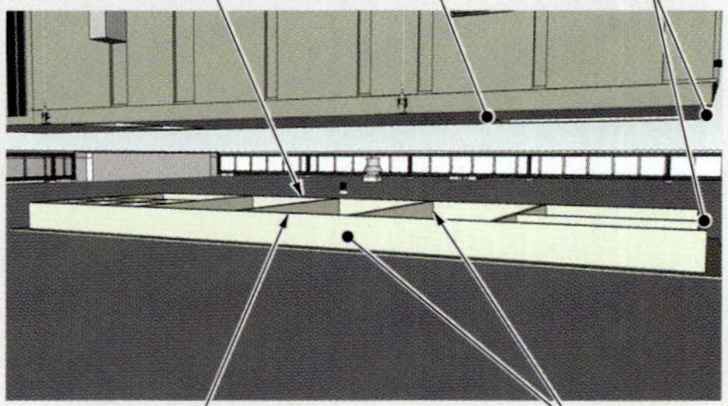

2. Lower the unit completely and ensure unit-to-curb seal is tight and without buckles or cracks.

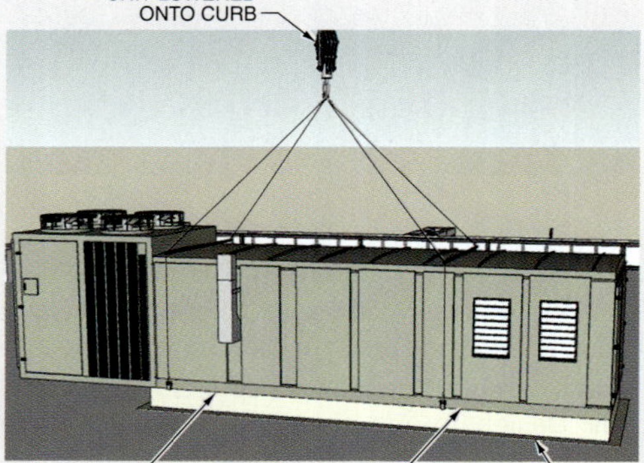

3. Install and properly tighten hold-down bolts. Check unit for levelness.

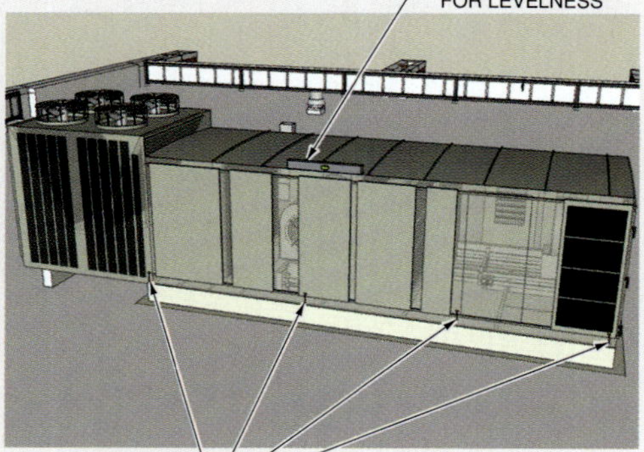

Roof Walkway Mats and Maintenance Pads

Quite often, maintenance pads are used to allow building maintenance personnel and HVAC technicians access to rooftop equipment without actually walking on the roof membrane. Some roof materials such as thermoplastic polyolefin (TPO) or PVC membranes may be damaged by being repeatedly walked on or when tools and materials are accidentally dropped. Technicians should always walk on walkway mats. Walkway mats and maintenance pads help ensure protection from punctures and abrasions to roof membranes. **See Figure 44-6.**

Clearances. Adequate clearances for maintenance and for connecting utilities must be maintained around and between all units. These distances are provided by the manufacturer and are found in the installation manual for the unit.

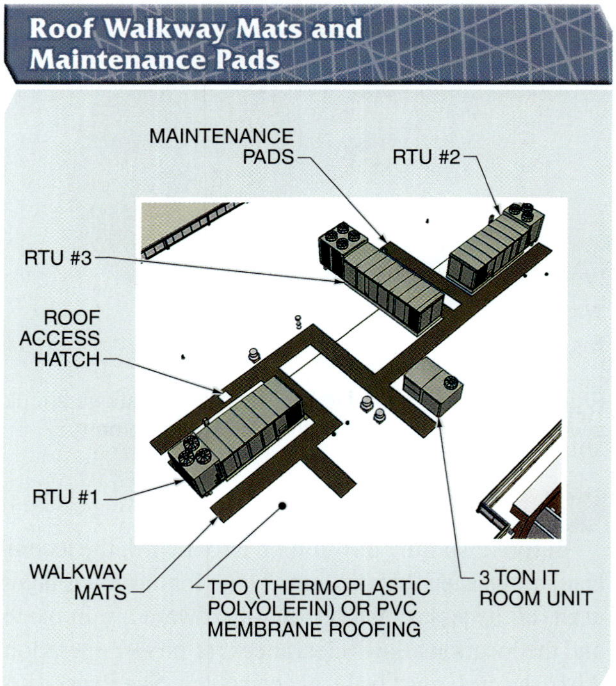

Figure 44-6. Walkway mats and maintenance pads help ensure the protection of roof membranes from punctures and abrasions.

Rooftop Unit Utilities—General

Installation of any HVAC unit requires electrical connections. All electrical connections must be made by a qualified electrician or technician. All codes as well as all manufacturer instructions must be followed. **See Procedure for Installing Electrical Wiring.**

Installing Electrical Wiring

1. Installation of all electrical wiring must be performed by a qualified electrician.
2. Electrical service voltage and amperage capabilities must be sufficient for the unit.
3. Overcurrent protection such as circuit breakers and overloads must be correctly sized for the unit.
4. The proper number and size of conductors must be run from the service panel to the unit's disconnect (internal or external), and the appropriate wires from the external disconnect (if available) must be run to the HVAC unit.

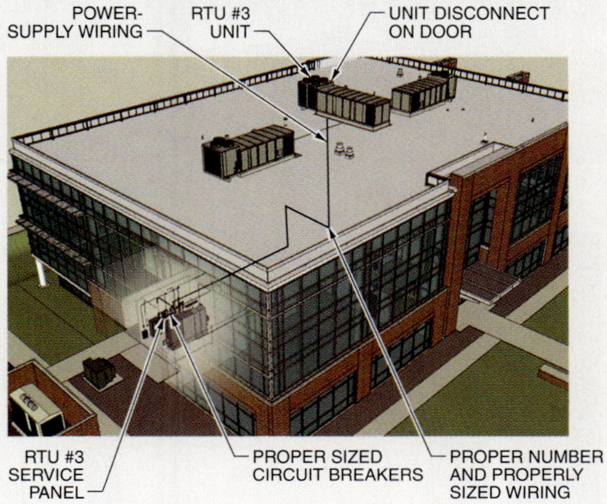

5. Check the voltage of each of the supply legs.

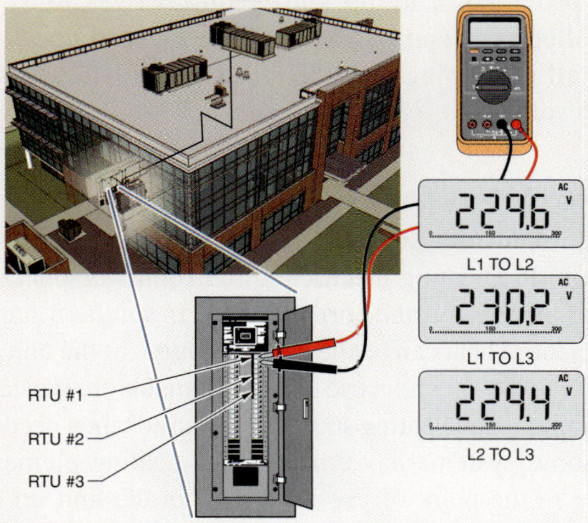

6. Lock out and tag out the unit until ready for startup.

Power Connections. Before installation, the power is typically run to a spot close to the unit. The supply power wire run is connected to a disconnect or junction box for each unit. The power to the units is locked out and tagged out at the source until installation is complete. Before termination, the source voltage and current available must be verified. **See Figure 44-7.**

> **TECH TIP**
>
> *Many units are damaged by incoming power problems. Always check the incoming power during unit startup and as part of a preventive maintenance (PM) program. Incoming power problems that are corrected early can prevent unit damage and costly repairs.*

Rooftop Unit Power Connections

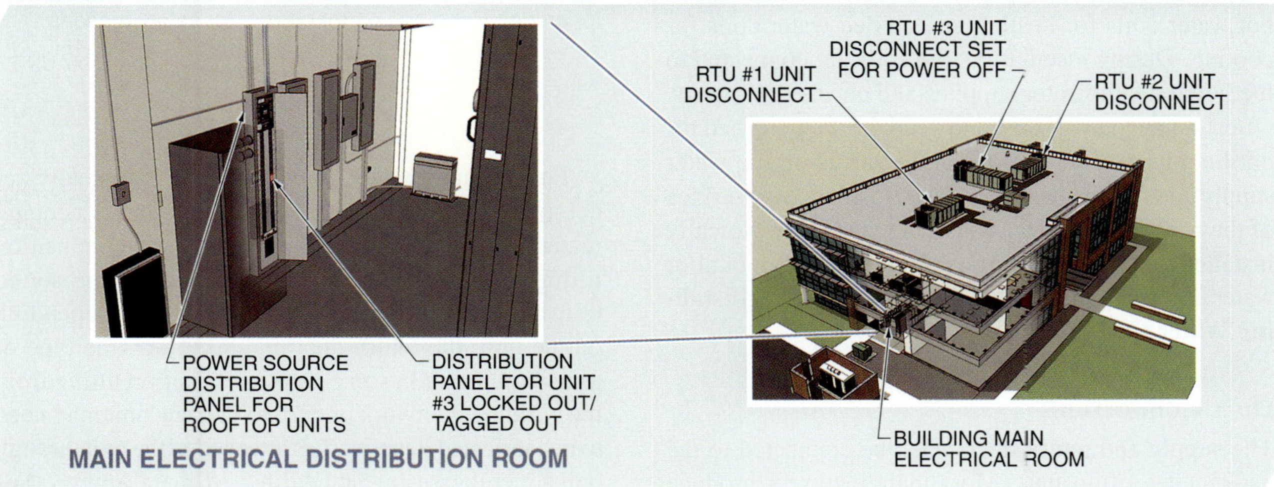

Figure 44-7. Power is typically run to a disconnect or junction box close to the HVAC unit. Before termination, the source voltage and current available must be verified.

Gas Piping. When the unit is natural gas fired, black iron piping must be used to connect the unit. When not provided, a union and shutoff valves must be installed. The supply gas pressure must be checked, and all gas piping must be checked for leaks with a soap solution.

Electric Heat Options

When a unit or system is equipped with an electric heat option, the heating electrical circuit must be checked for proper wiring and correct power. In southern states, the electric heat can be the heating source of the unit. In northern climates, electric heating elements are typically a supplemental heating source that is used when needed.

Some systems have the electric heating elements close to the point of use instead of in the unit on the roof. During the equipment start-up procedure, all safety components must be checked for proper installation and operation. These safety devices prevent electric element operation in case airflow is lost. **See Figure 44-8.**

TECH TIP

Safety components for electric heating coils must be tested during startup. If the safety components do not function properly, a fire due to overheating of the elements in the duct may occur.

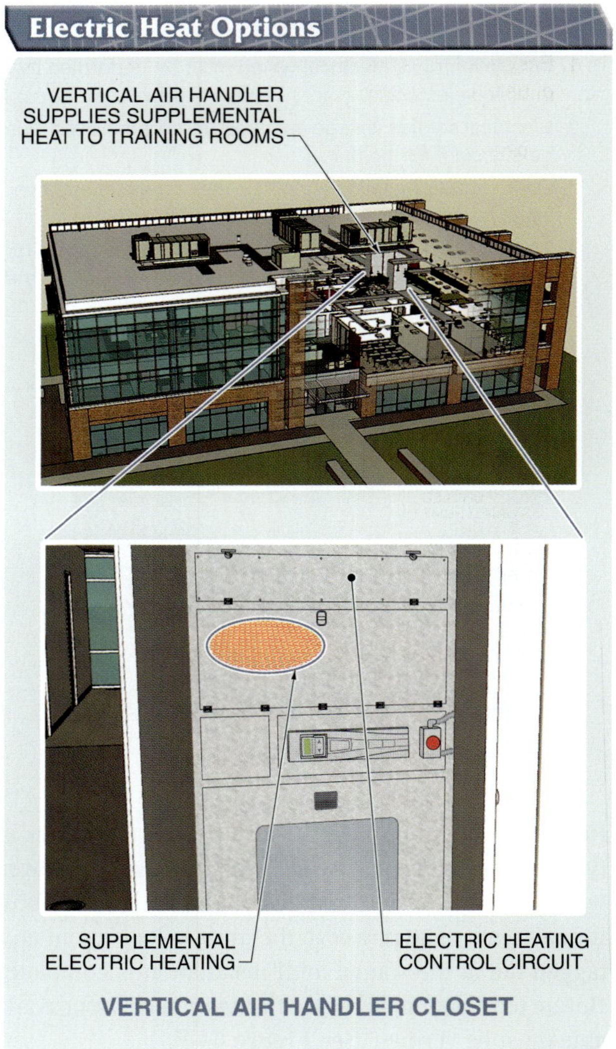

Figure 44-8. When a unit or system is equipped with an electric heat option, the heating electrical circuit must be checked for proper wiring and correct power for safety reasons.

Hot Water and Chilled Water Coils

One option for many types of rooftop units is the use of hot water coils for heating and chilled water coils for cooling. During installation, the coils are connected to the hot and cold water supplies and checked for leaks. Often, a hand valve is used to keep the water turned off until equipment startup. Isolation valves on the water supply and return lines can be used to allow easy service of the water coils. Water balancing valves are typically installed and adjusted to provide the proper amount of water flow through each coil. **See Procedure for Installing Water Coil Piping.**

Duct Connections

The supply and return ducts must be connected to the packaged rooftop unit. Most manufacturers produce units with several possible duct connections. The decision of which connections to use is made at the job site.

For the supply duct, the connection is typically accomplished with a metal duct with sound-deadening material inside. Duct board or a flexible duct transition fitting on the curb can be used to reduce the sound transmission from the fan into the building. Depending on the unit, the return duct may use the same type of transition piece. In some cases, a transition fitting from the building ductwork or curb to the new unit may need to be fabricated. Ductwork is attached to the unit through built-in collars using standard sheet metal screws. **See Figure 44-9. See Procedure for Making Duct Connections to Rooftop Unit.**

Procedure

Installing Water Coil Piping

1. Ensure all piping to rooftop unit is the correct size, is free of defects, and contains no contaminants.

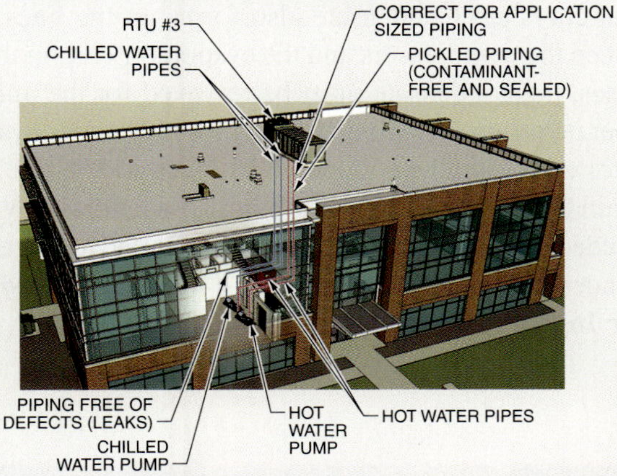

2. Use compatible piping materials throughout chilled and hot water coil installations.
3. Install valves and switches in piping as needed.

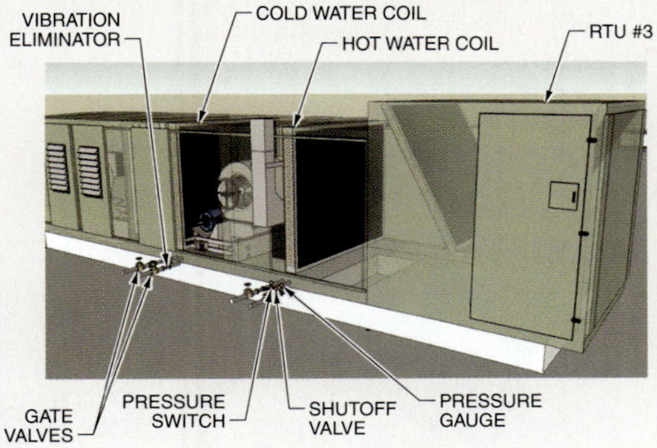

4. Install insulation on piping as needed.
5. If practicable, flush coils with water to remove any contaminants.

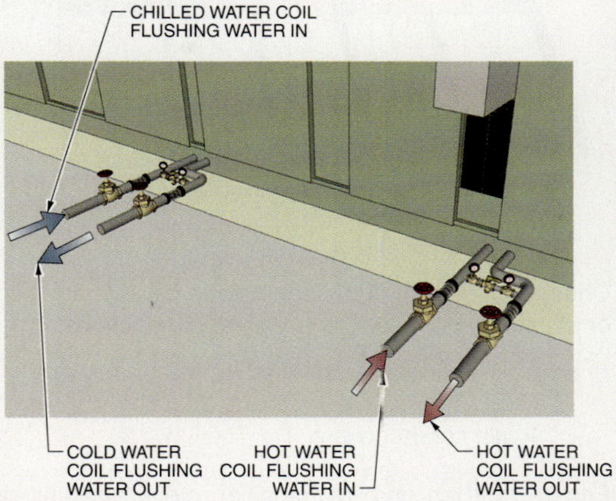

6. Ensure that coil connections are not used to support the weight of the piping.
7. Use adapters and piping materials as needed to connect hot and/or chilled water system piping.
8. Upon startup, check for leaks, proper flow, and pressure.

NOTE: In extremely cold-weather climates, it is uncommon to use water coil condensers in HVAC units instead of refrigerant coils because freezing may occur.

Duct Connections

Figure 44-9. Ductwork is attached to the unit through built-in collars using standard sheet metal screws to hold everything in place.

Procedure

Making Duct Connections to Rooftop Unit

1. Check curb or transition piece for any dirt, dents, or dings that will interfere with rooftop unit connections.

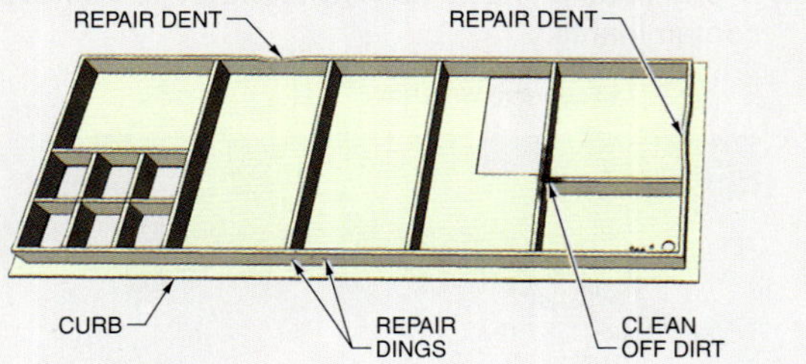

2. Check rooftop unit duct connections for any dirt, dents, or dings that will interfere with curb connections.

NOTE: THIS PROCEDURAL STEP TYPICALLY REQUIRES THAT SPECIAL SAFETY CONSIDERATIONS BE PLACED INTO EFFECT (MUST BE UNDER UNIT TO INSPECT UNIT'S DUCT CONNECTIONS).

3. Ensure that the curb or transition piece and unit duct connections are the correct size.

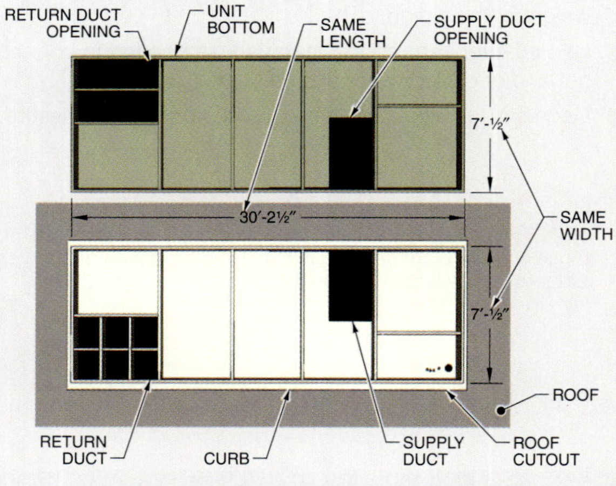

CURB DUCTS AND UNIT DUCT OPENINGS IN SAME LOCATION

4. Fabricate short pieces of supply and return duct, flanged and sized for the roof opening and long enough to extend from the top of the curb into the ceiling of the building. Install the supply and return duct after the roof curb is flashed in and before the unit is set by the crane.
5. Attach unit to curb or transition piece using screws.

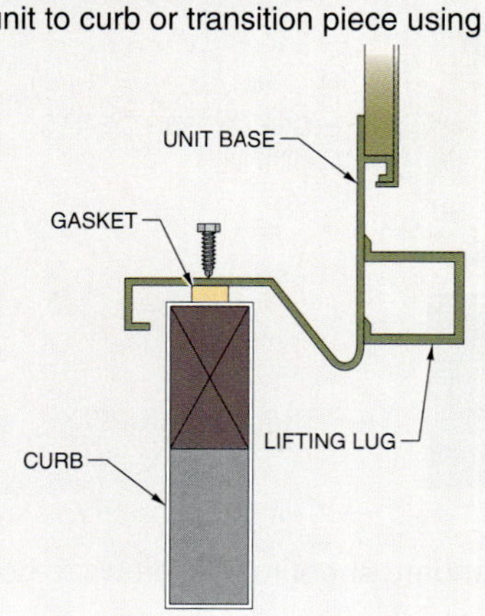

Condensate Drain Piping

Both heating and cooling units can form condensate. Condensation forms on the evaporator coil of any unit that cools air. Condensate also forms on the lines between the compressors and the evaporator coil. In these cases, the condensate must be removed for the unit to operate properly. A drain pan is provided under evaporator coils, and the condensate is piped to a drain. The drain line is pitched towards a proper roof drain. When needed, a condensate pump can be installed to force the condensate to a drain. **See Figure 44-10. See Procedure for Installing Condensate Piping.**

Condensate Drain Piping

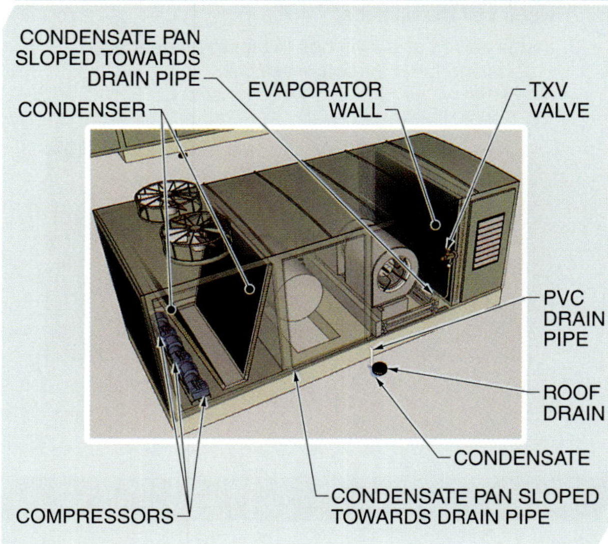

Figure 44-10. Drain pans are used under evaporator coils, and the condensate is piped on a pitch towards the drain.

Air Filters

Light commercial HVAC systems will circulate ventilation air or air that has been heated or cooled by the HVAC unit into the spaces of the building. Air filters are installed to remove particulate matter in the air and to promote a healthy indoor air environment. It is important during the installation process that the correct type, quantity, and size of air filters be installed. The unit should not be operated and/or started without these filters in place, otherwise damage and/or contamination of the evaporator coil can occur. **See Figure 44-11.**

Installing Condensate Piping

1. Locate condensate drain pan connections.

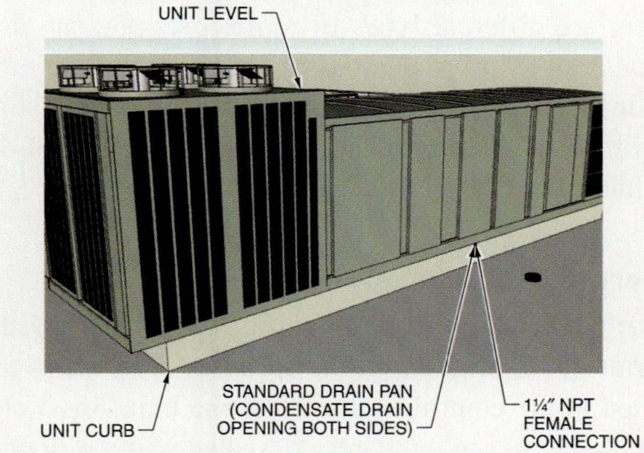

2. Using the proper diameter piping, run condensate line to the drain. Avoid bends as much as possible.

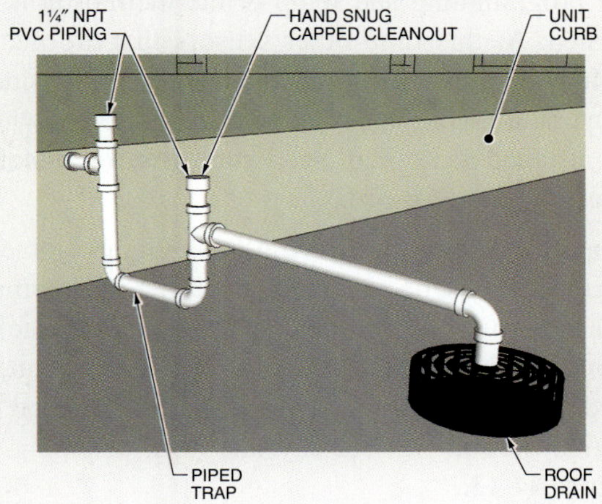

3. Ensure a proper trap location, pitch run towards the drain, and vent to atmosphere. Always follow manufacturer recommendations.

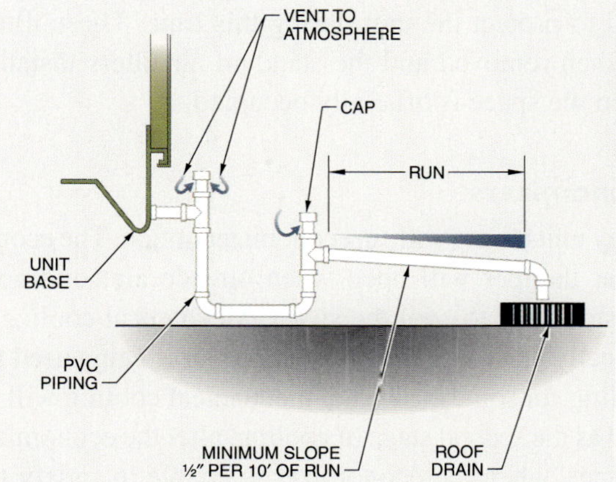

4. Check for free flow of condensate into drain.
   a. Energize unit cooling.
   b. Look for leaks.

Air Filters

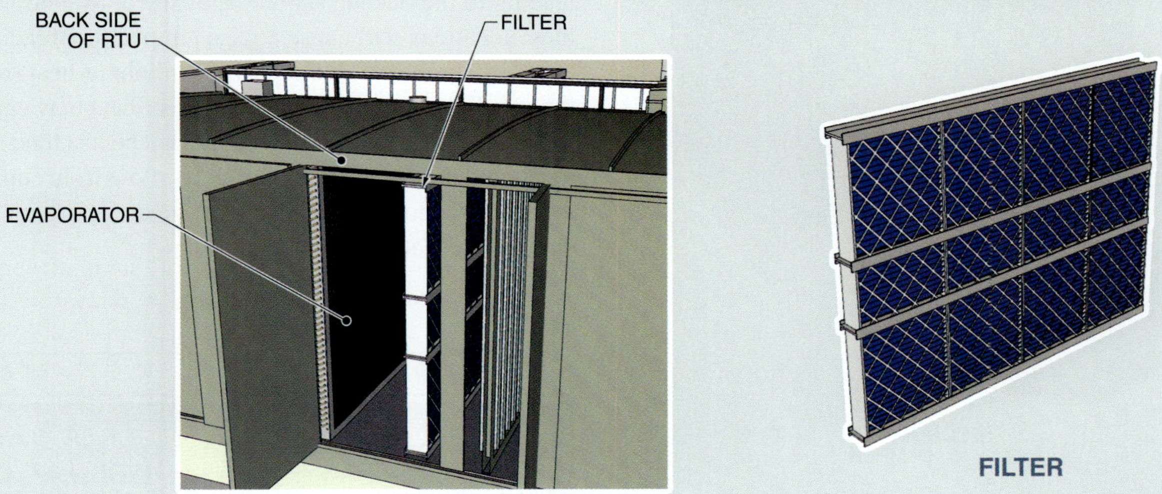

Figure 44-11. A light commercial HVAC unit should never be operated without its filters.

In new construction, there is a temptation to run the unit to provide cooling and/or heating to the building while other construction is taking place. This is done while many contaminants, such as dust and dirt, are in the air. Special inexpensive construction filters may be used to protect the unit during this time. These filters are then removed and the standard air filters installed when the space is officially occupied.

Economizers

Many units come with an economizer option. The economizer damper will open when outside air conditions are sufficient to cool the space. Mechanical cooling is typically disabled while the outside air is being used for cooling. In some cases, the mechanical cooling will be used as the second stage of cooling, after the economizer damper, when the economizer is unable to satisfy the cooling. Also, the controls of the economizer must be checked. When economizer controls are set incorrectly, the mechanical cooling may not run or air that is too hot or cold may be brought into the unit. This hot or cold air results in energy waste and improper temperature control. **See Figure 44-12.**

Economizers

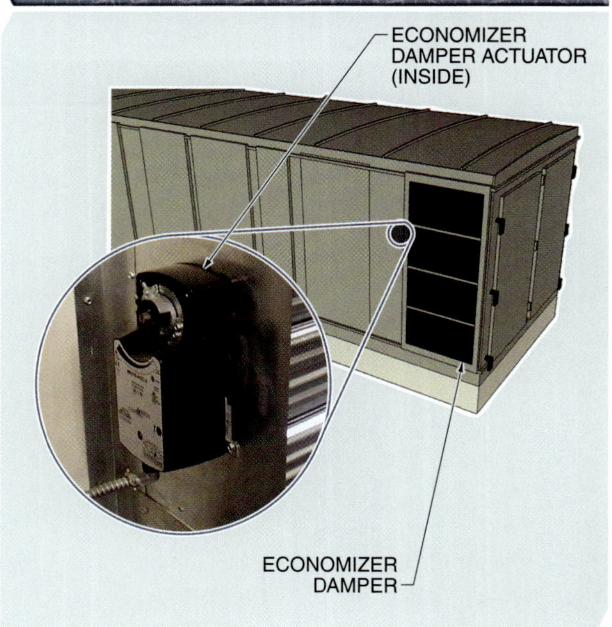

Figure 44-12. Mechanical cooling is typically disabled when outside air is used for cooling.

INSTALLING ROOFTOP SINGLE-ZONE UNITS

There are different types of rooftop single-zone unit installations. In some cases, the building will be new, while in others, the unit will be replacing an existing rooftop unit. Access to and geographic location of the building and type of roof will vary as well.

Thermostat and/or Other Sensors

After the packaged rooftop unit is set on the roof, the thermostat is installed in the space, and the wires are routed to the control enclosure of the unit. Any holes that are cut to run the wires are sealed with a silicone-based material. Depending on the type of thermostat, any programming and setup of the thermostat is performed. At this time other sensors that are needed, such as discharge-air temperature sensors, outside-air temperature sensors, or filter change indicator switches are installed and wired. **See Procedure for Installing Rooftop Unit Sensors.**

Proper Locations for Thermostats. A light commercial thermostat must be mounted properly. A light commercial thermostat should be mounted approximately 5′ above the finished floor (AFF on plans) in a frequently used room on an inside wall, out of sunlight, and on a section of wall without any ductwork or piping. **See Figure 44-13.**

Thermostat Locations to Avoid. There are certain locations to avoid when mounting a thermostat. Locations close to a window, on an outside wall, or next to a door leading to the outside should be avoided. Mounting in these locations will cause a false temperature reading. Also, locations exposed to direct sunlight or heat from the sun, lighting, or any other source of heat may cause a false reading. A thermostat should not be installed in a location that is in the line of direct airflow from supply registers or return-air grills. Also, areas with poor air circulation should be avoided.

TECH TIP

Always ensure that sensors are located properly in the building spaces and that the correct wires are attached to the sensors in each building space. Incorrectly connected unit wires and sensors can cause extensive service problems.

Installing Rooftop Unit Sensors

1. Determine proper location(s) of sensors (thermostat and others) using job layout drawings such as a ventilation plan.

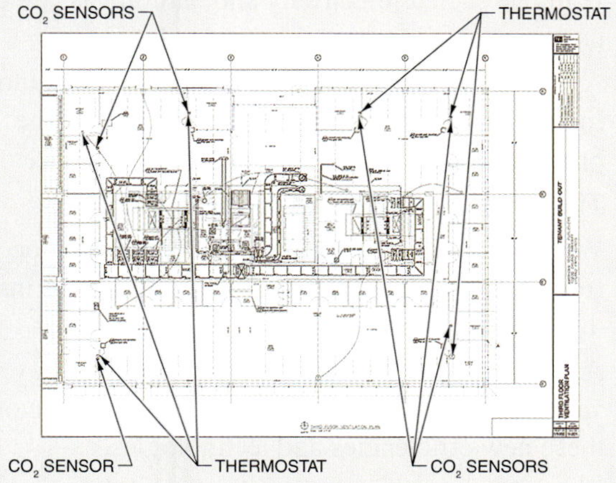

2. Check sensor part numbers against purchase order.

3. Securely attach sensors to duct or wall.

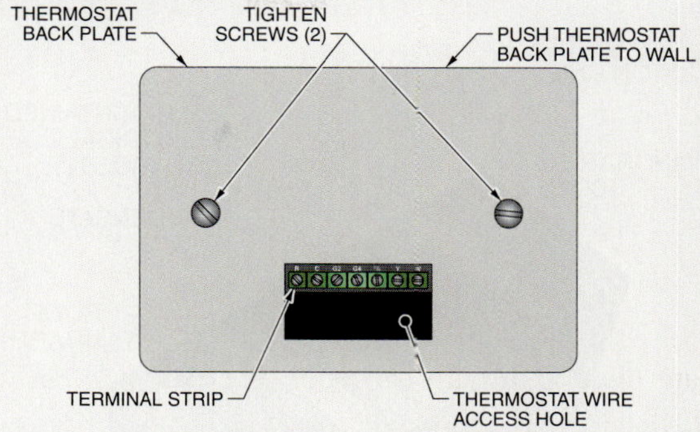

4. Run wire to sensors per code. Number wire as needed to aid in installation and service. Ensure wire is correct gauge and type (18/2 copper stranded wire is most common).
5. Attach wires to sensors. *Note:* Check sensor polarity.
6. Terminate wires at appropriate controller terminals.
7. Use DMM at controller to check for proper signal readings from sensors.

Thermostat Locations

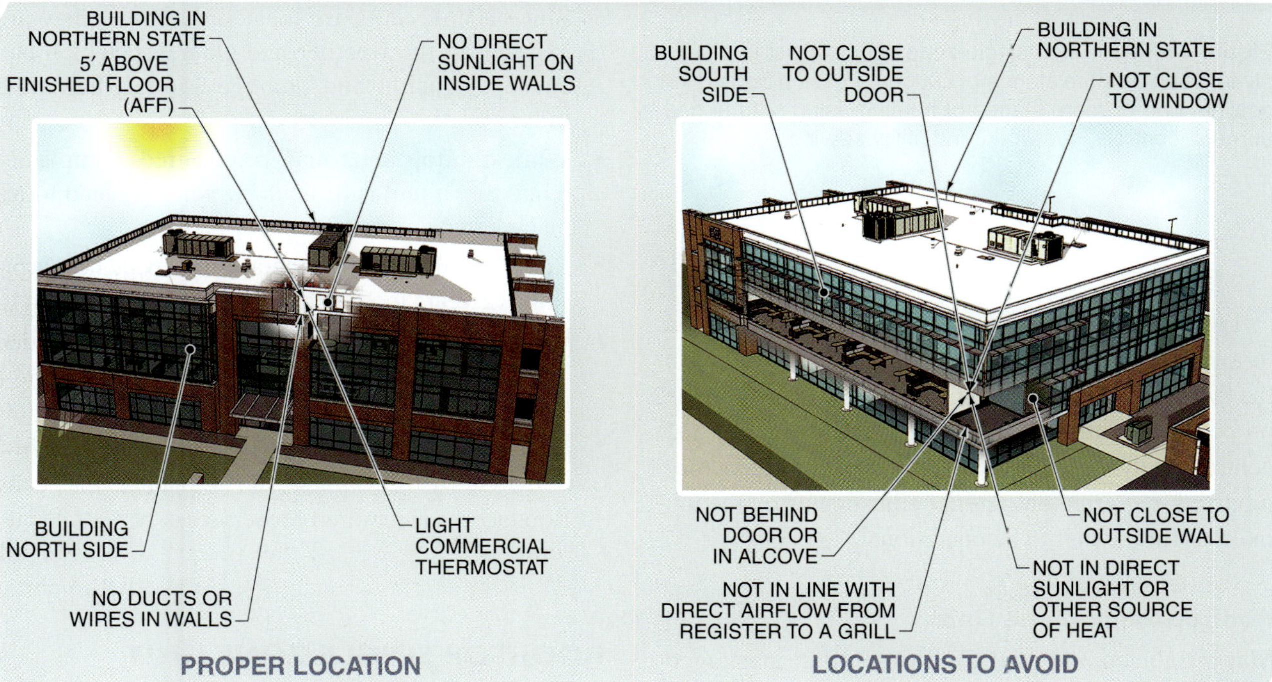

Figure 44-13. Thermostat location is of upmost importance to the proper operation of an HVAC system.

Light Commercial Rooftop Single-Zone Units

There are many types of rooftop single-zone units. In general, a typical rooftop single-zone unit consists of a supply fan, one or more stages of cooling, one or more stages of heating, controls, ductwork, and power connections. For cooling, most rooftop units use a standard DX compression cycle, with a few systems using chilled water coils. For the heating section, most utilize natural gas burners, with electric heating elements or hot water coils as possible options. **See Figure 44-14.**

Rooftop Single-Zone Units

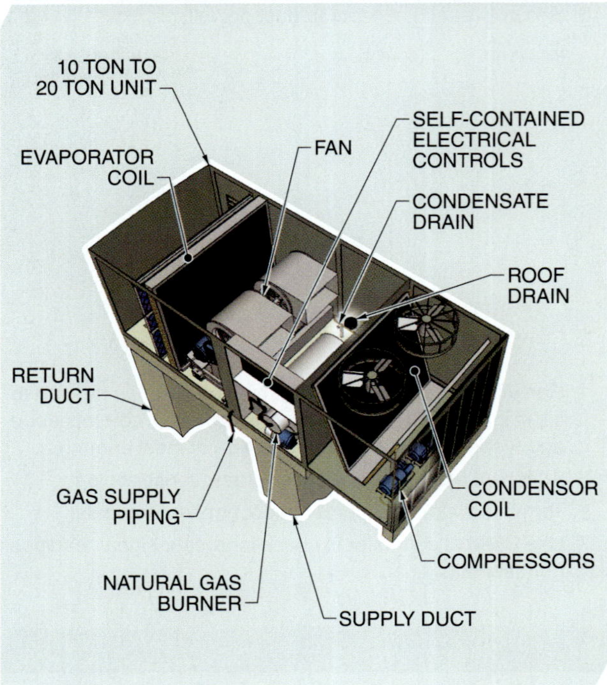

Figure 44-14. Rooftop single-zone units consist of a supply fan, one or more stages of DX compression cycling for cooling, one or more stages of heating using natural gas burners, controls, ductwork, and utility service.

For control, a light commercial thermostat is wired to a terminal strip. This type of thermostat will cycle the fan, heating, and cooling as needed to maintain the proper temperature in the building spaces. Today, the controls needed at the rooftop unit are typically factory mounted, requiring only minor adjustments on the job in order for the unit to be operational.

Rooftop Single-Zone Units

Many light commercial facilities utilize packaged rooftop single-zone units. Packaged rooftop single-zone units may range from 5 tons to approximately 75 tons, with 30 ton and 40 ton units being very popular. The facilities that use these include strip malls, big box retail stores, fast food restaurants, office buildings, and schools. Rooftop single-zone systems have a number of advantages over other system types, but also have disadvantages. **See Figure 44-15.**

Rooftop Single-Zone Unit Advantages. Using rooftop single-zone units offers many advantages:

- They do not require mechanical rooms, freeing up internal building space for other purposes.
- Utilities, such as electricity and natural gas, are outdoors and easily accessible to a technician.
- Any by-products of combustion are vented outdoors immediately, reducing any possibility of the by-products being introduced into occupied spaces.
- They often have a low initial-installed cost.
- When service or replacement is necessary, it occurs from outside the building, allowing functions inside the building to continue.
- As unit efficiencies improve and technology advances, the units can easily be replaced in favor of these new efficiencies and technologies.
- The failure of one unit will only affect the zone served by that unit and not the rest of the building.

Rooftop Single-Zone Unit Disadvantages. Rooftop single-zone units also come with many disadvantages:

- Since HVAC units are located outdoors, they are exposed to the weather and elements, which can cause corrosion and damage sooner than with indoor units.
- While rooftop units may be repaired, completely rebuilding a unit is difficult, which may lead to replacing units more often.
- Rooftop units have a general life expectancy, typically 12 years to 15 years, after which they need to be replaced. This means that units that were installed at the same time may need to be replaced at approximately the same time. This can be expensive.
- Replacement of a unit can only be done with a crane or helicopter from outside the building.
- Roof access is required for service.

ROOFTOP SINGLE-ZONE UNIT STARTUP

The installing contractor is often required to perform the startup procedure for a packaged rooftop single-zone unit. This is especially critical when a warranty claim arises later, as proper start-up documentation is required for a warranty claim. All manufacturers provide standard start-up documentation procedures and forms.

Single-Zone Unit Advantages and Disadvantages

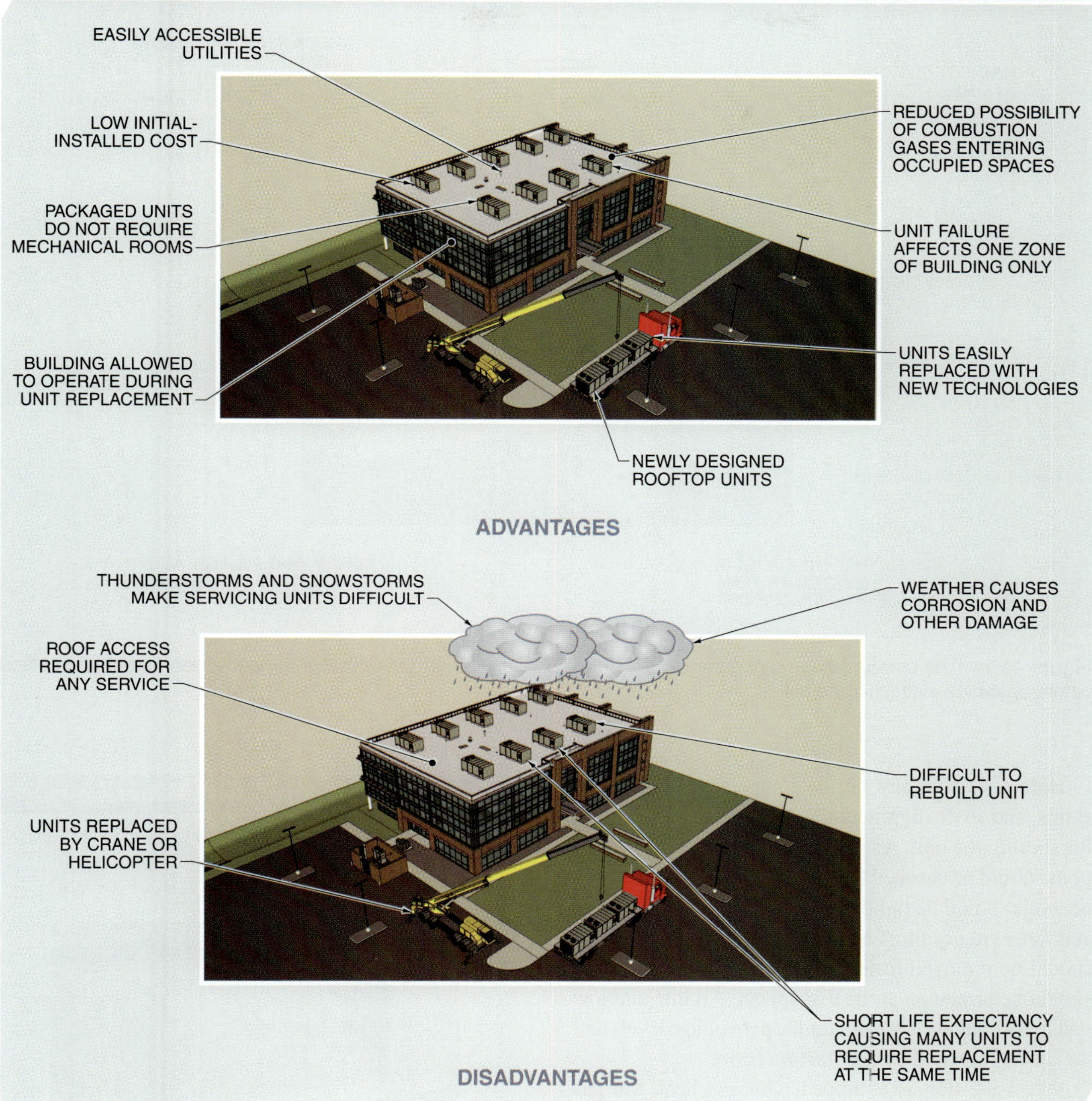

Figure 44-15. An advantage of rooftop single-zone units is that when service or replacement is necessary, it occurs from outside the building, allowing functions inside the building to continue. A disadvantage of these units is that when rooftop units require repair, completely rebuilding the unit is difficult, which may lead to replacing units more often.

Unit Startup

The rooftop single-zone unit must be installed properly using the manufacturer documentation and installation instructions, and all applicable codes must be followed. The rooftop unit must be commissioned to manufacturer requirements, with startup guidelines and procedures being followed and recorded. The correct size and number of filters must be installed before startup of the unit. In addition to the filters, the outdoor-air inlet screen must be in place before operating the unit. **See Figure 44-16.**

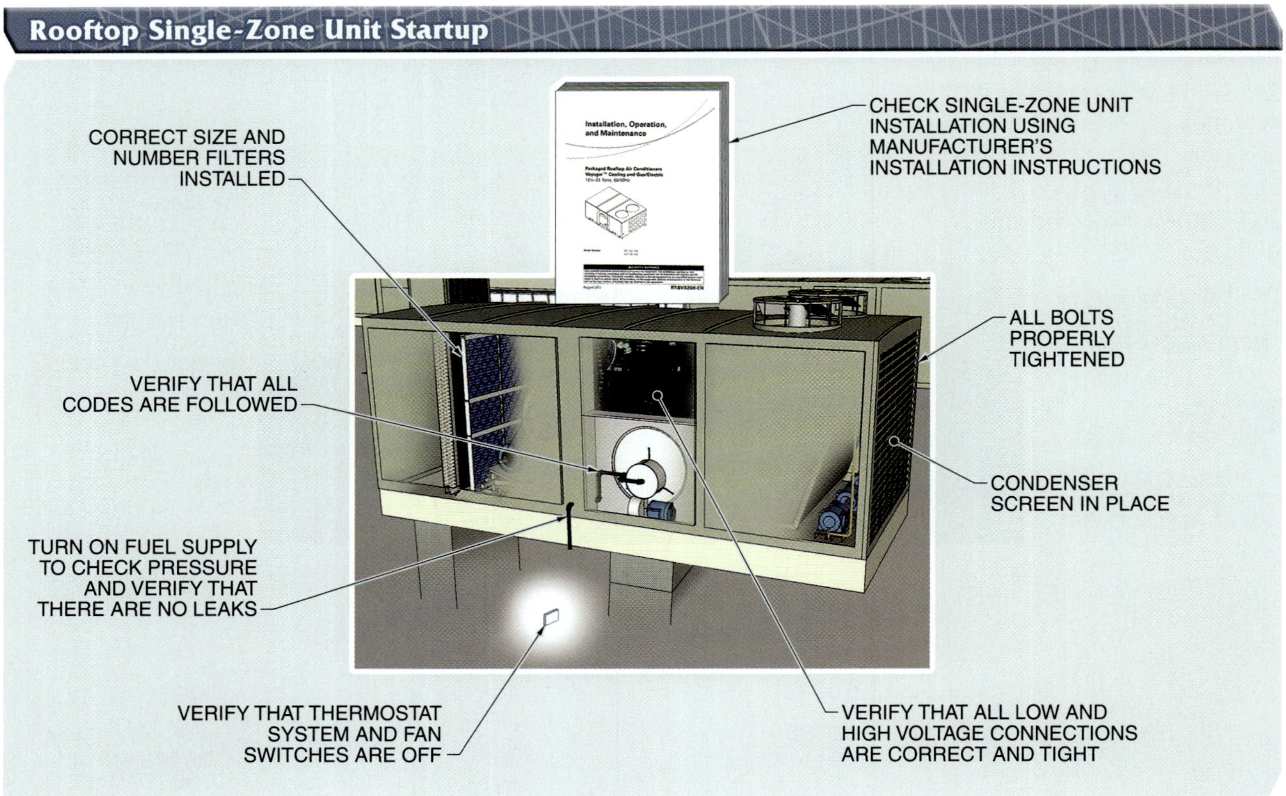

Figure 44-16. The correct size and number of filters must be installed and the outdoor-air inlet screen must be in place before operating a single-zone unit.

Before startup, any fan or compressor base, bolts, shims, and shaft alignment must be in proper condition for starting the unit. All low- and high-voltage connections should be checked for proper location, and all connections should be tight. The thermostat system switch and fan switches must be off. All lockouts and tagouts should be removed from the power supply, and power should be turned on at the disconnect. All line and low voltages should be checked for proper voltage levels and the levels recorded on the start-up form.

When the unit is equipped with gas heat, the gas supply should be turned on. The gas pressure should be checked, and the gas supply piping checked for leaks. All applicable codes must be followed. The unit must never be operated when any gas or gas smells are present.

When the hot and/or chilled water coil option is used, the shutoff valves for the hot and/or chilled water supplies to the unit's coil should be opened. The circulating pump or pumps should be working, and a source of hot and/or chilled water should be available. Piping should be checked for leaks. **See Figure 44-17.**

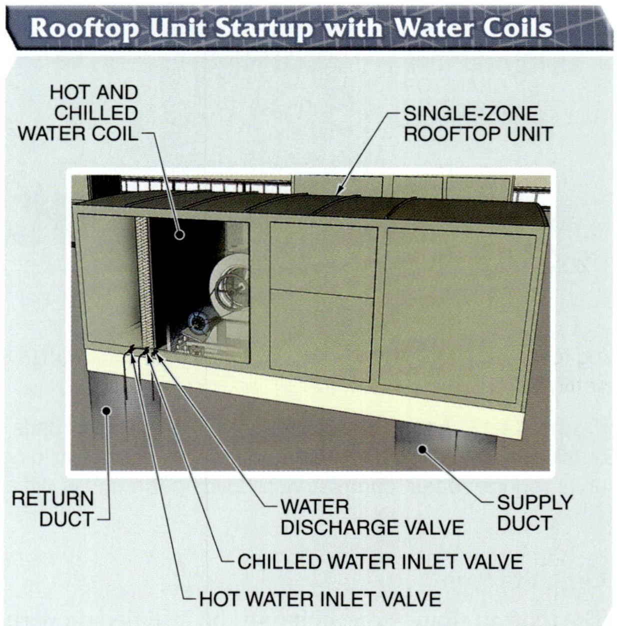

Figure 44-17. When starting up a unit with water coils, the shutoff valves for the hot and/or chilled water supplies to the unit's coil should be opened.

Economizer Testing. The economizer is tested by performing the steps in the manufacturer's service test guide. Typically, the minimum position setpoint for the economizer is set to the minimum ventilation percentage using the setpoint potentiometer. The economizer damper should move to its minimum position. The economizer should be jumped or set to full open. This will cause the economizer to stroke to the full-open position. **See Figure 44-18.**

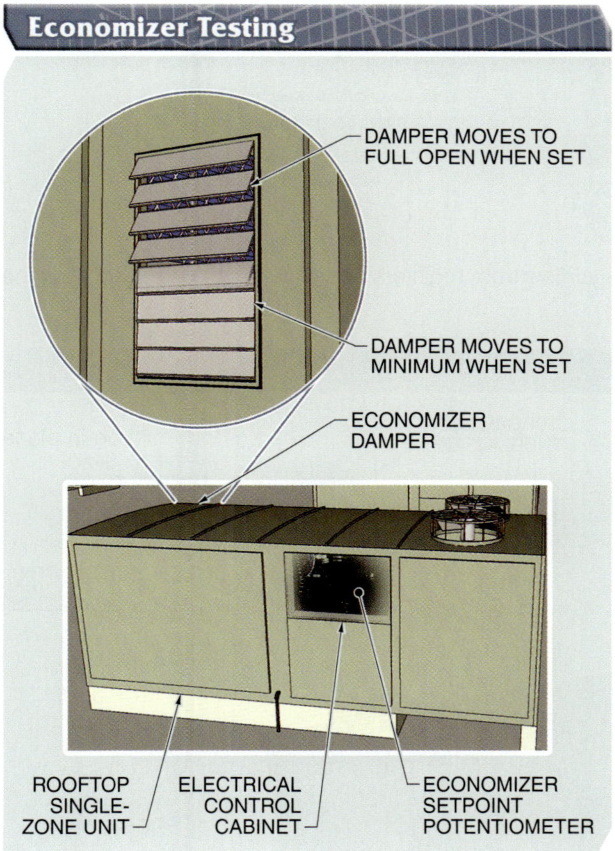

Figure 44-18. Verification of the economizer damper moving to its minimum and maximum positions by changing its settings is required when testing an economizer.

Compressor Testing. A set of service gauges are attached to the system. The rotation of each compressor should be verified, and it should also be verified that suction pressures go down while discharge pressures go up. The operating pressures in the system should be compared to the pressure curves in the manufacturer specifications.

The economizer damper actuator must be tested as part of the system start-up process.

Fan Testing. Depending on thermostat type, the unit should be cycled from the OFF mode to the ON mode. Then the fan should be cycled on. The fan should operate properly. A quick check of the airflow at the registers and diffusers should be made and compared against the design. **See Figure 44-19.** The supply fan should be checked to make sure no excessive vibrations are present. While some thermostats still have physical switches, touch screen models can be changed by simply moving a finger from one selection to another on the screen.

Test Cooling Cycle. The thermostat switch should be set to cooling mode. Also, the setpoint must be changed below the ambient temperature in the space. After a delay determined by the system, the supply fan will come on, and then, after another system delay, the first compressor will come on. Discharge air temperature, temperature differential across the cooling (evaporator) coil, supply fan amperage, and voltage must all be recorded. Low- and high-side pressures should be recorded as well. **See Figure 44-20.** When multiple stages of cooling are involved, a time delay between the stages is normal.

For chilled water systems, the valve for water flow will open and the chilled water will flow through the evaporator coil. Again, the temperature drops must be recorded. Also, a water flow balancing valve must be installed and adjusted for the correct water flow. The adjustment of the water flow should be performed by someone familiar with water-balancing principles and techniques.

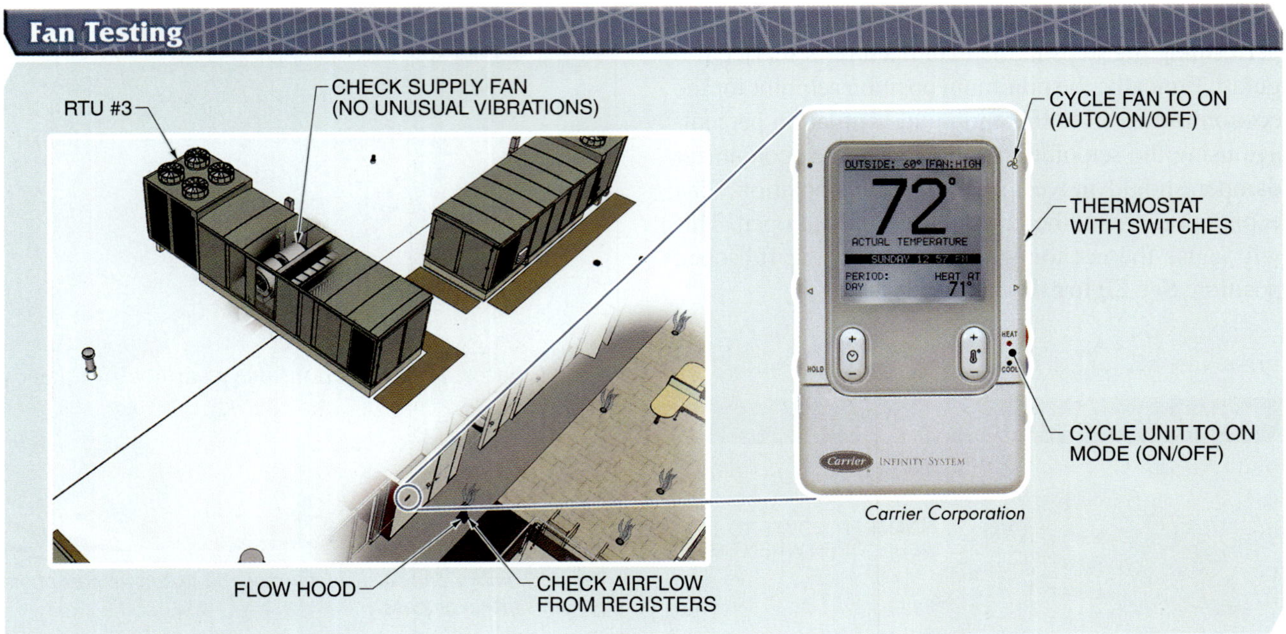

Figure 44-19. A common method of testing a fan is checking the airflow from registers and diffusers and verifying that the supply fan is not making any unusual vibrations.

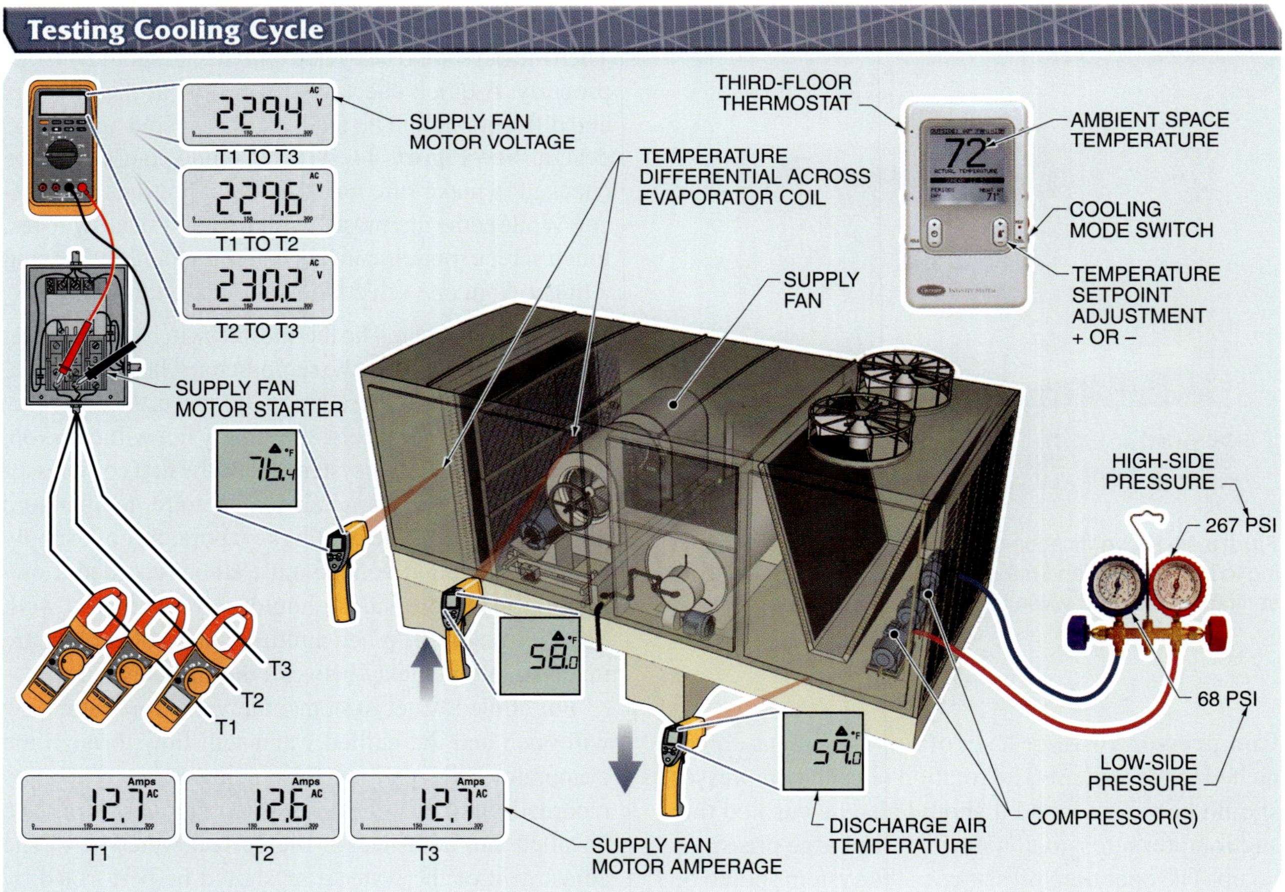

Figure 44-20. Discharge air temperature, temperature differential across the cooling (evaporator) coil, supply fan amperage, and voltage must all be recorded.

Test Heating Cycle. The thermostat switch is set or a touchscreen used to set the heating mode. Some delays may be incurred. It may be best to perform some other work while the delay runs out. At this point, the thermostat setpoint should be adjusted above the ambient room temperature. During the heating mode the supply fan will run. When the unit is gas fired, a normal ignition sequence will result. **See Figure 44-21.**

When electric heat is present, the element will energize as long as the safety devices are working properly. As with cooling, there may be multiple stages of heat, and a time delay between stages may exist. When the unit has hot water heat, the valve will open, and water will flow through the coil. As with a chilled water system, all valves must be open, the circulating pump must be running, and hot water must be available at the boiler or heat exchanger. At this time the balancing valve will be adjusted to obtain the correct flow in gpm through the coil. For startup of heating, temperature rise across the coil should be measured, along with amperage for any electric heat.

INSTALLING ROOFTOP VARIABLE-AIR-VOLUME UNITS

In many ways the installation of a variable-air-volume (VAV) rooftop unit is similar to the installation of a single-zone model. The biggest differences are the physical size, additional controls, and the larger capacity of VAV units for heating and cooling.

One of the major differences of a VAV unit is the complexity of the fan system. For a rooftop VAV unit, the fan is always driven by a three-phase induction motor. Today, almost all installations use a variable-frequency drive (VFD) to provide the fan capacity necessary to maintain the correct static pressure in the system ducts. Typically the VFD is factory installed and wired. During installation, the technician must ensure that the electrical connections are correct and should verify all voltages before startup. **See Figure 44-22.**

The wiring of a VAV unit is quite similar to a rooftop single-zone unit. The thermostat system switch and all fan switches must be off. The power to the VAV unit should be turned off temporarily to make sure that the unit control board powers up and that any indicator lights are normal.

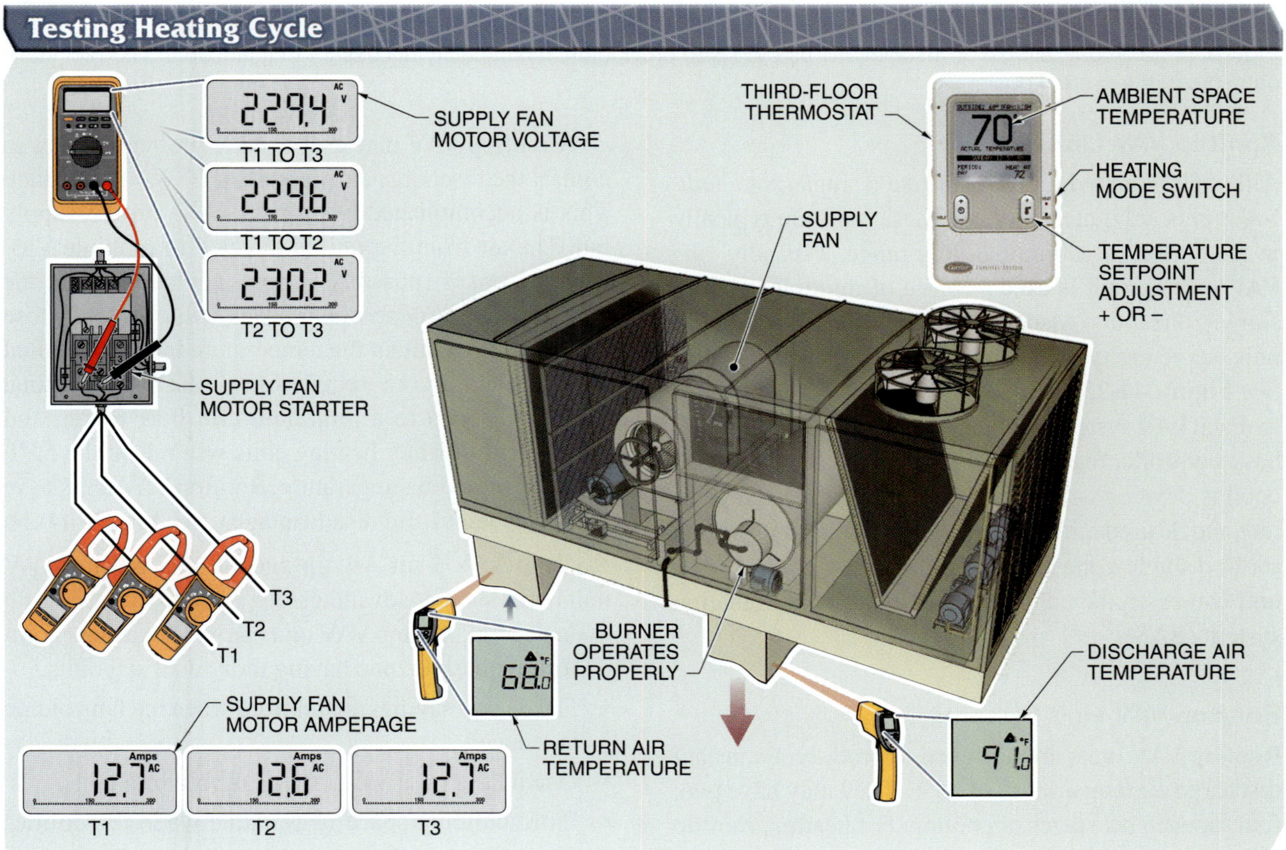

Figure 44-21. If the cooling cycle was previously energized, there may be a built in delay that will not allow the heating cycle to come on immediately.

VAV Unit Complexity

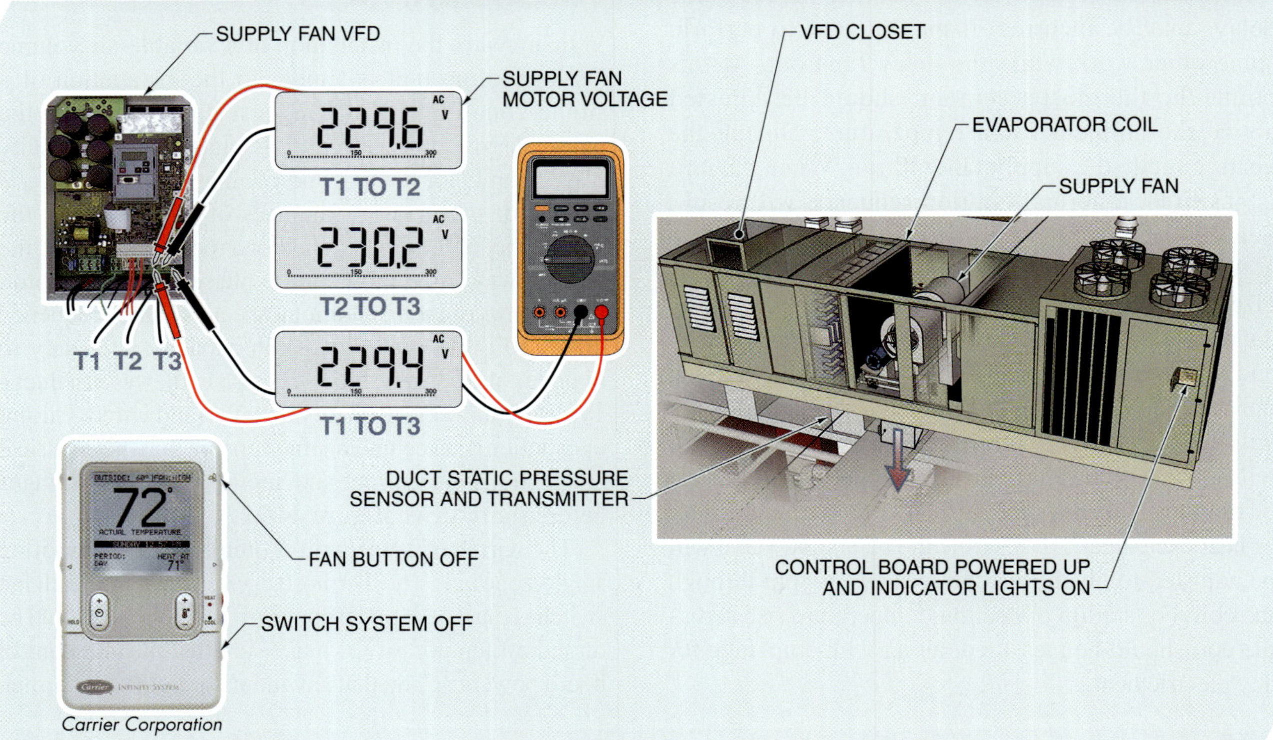

Figure 44-22. Today, almost all installations use VFDs to provide the fan capacity necessary to maintain the correct static pressure in the system ducts.

Rooftop VAV Units

Although mounted on the roof like a single-zone unit, VAV units are quite different. These units are typically in the 50 ton to 100 ton cooling range. Typically, one VAV rooftop unit is used in place of multiple, smaller single-zone units. Also, the buildings that VAV rooftop units serve are typically larger than single-zone types. **See Figure 44-23.**

Each VAV system terminal box or vertical air handler has a controller that opens the damper for the heating or cooling device as needed to maintain the zone temperature and desired air volume. Similar to the controls of a rooftop single-zone unit, the controls of a rooftop VAV unit can easily be connected to a building automation system (BAS).

Rooftop VAV Unit Systems

Rooftop VAV units are designed to produce a constant discharge air temperature of 55°F. They may have two, four, or even six stages of cooling. For heating, rooftop VAV units use multiple stages of gas, electric, or possibly a hot water coil.

A rooftop VAV unit uses a static pressure sensor to control the discharge air pressure at 1″ WC in the duct. This is accomplished by using a VFD on the supply fan. The air from the unit is directed to multiple VAV terminal boxes (possibly vertical air handlers) in the building. The VAV terminal boxes will open and close as needed to maintain the temperature in the individual zones. When a zone requires heat, the VAV terminal boxes will open to a minimum cfm flow rating, and electric or hot water heating coils will reheat the 55°F air to the required temperature. Rooftop VAV units have both advantages and disadvantages. **See Figure 44-24.**

Rooftop VAV Unit Advantages. Using rooftop VAV units offers many advantages:

- In general, rooftop VAV units provide excellent comfort, with each zone having individual setpoints.
- Fan energy savings are realized when the fan volume is reduced.
- One unit is in service instead of multiple units.
- Extra building space is available when the utilities and unit are on the roof.
- Internal building space is freed up.

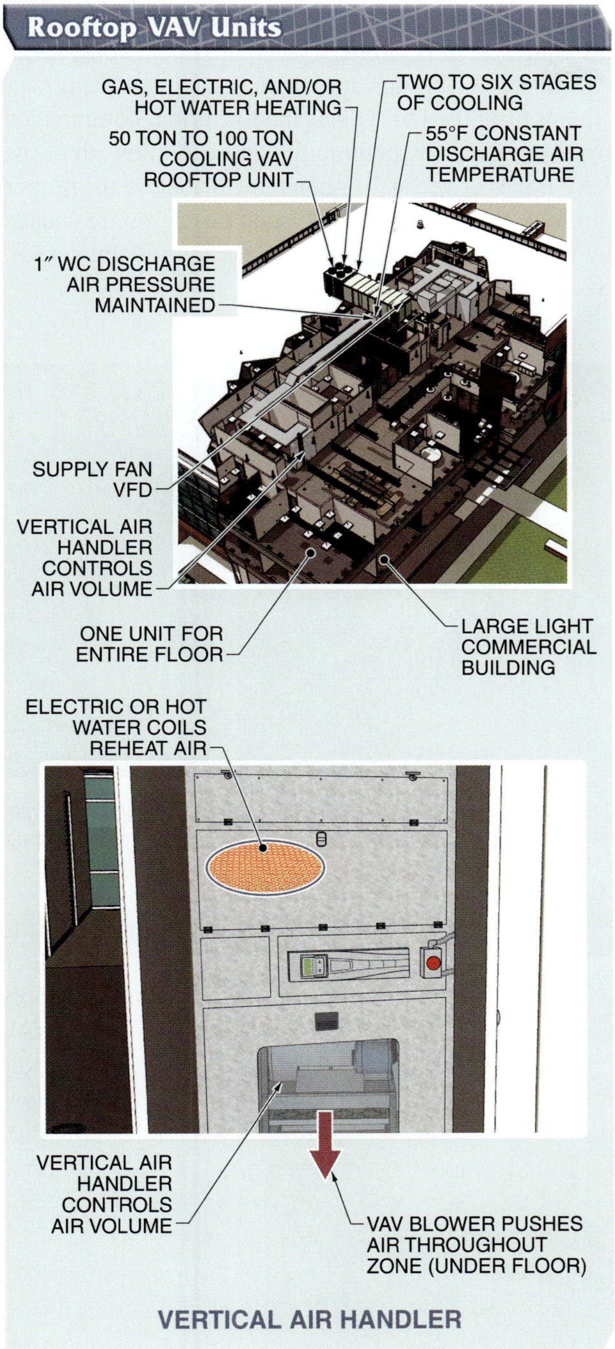

Figure 44-23. In most cases, one VAV rooftop unit is used in place of multiple, smaller single-zone units.

Rooftop VAV Unit Disadvantages. Rooftop VAV units also come with many disadvantages:

- When a unit fails, multiple zones are left without heating and cooling.
- Unit controls can be complex, requiring a high skill level to service and troubleshoot.
- Air balancing between zones is critical.
- VAV terminal boxes or vertical air handlers can be difficult to access.

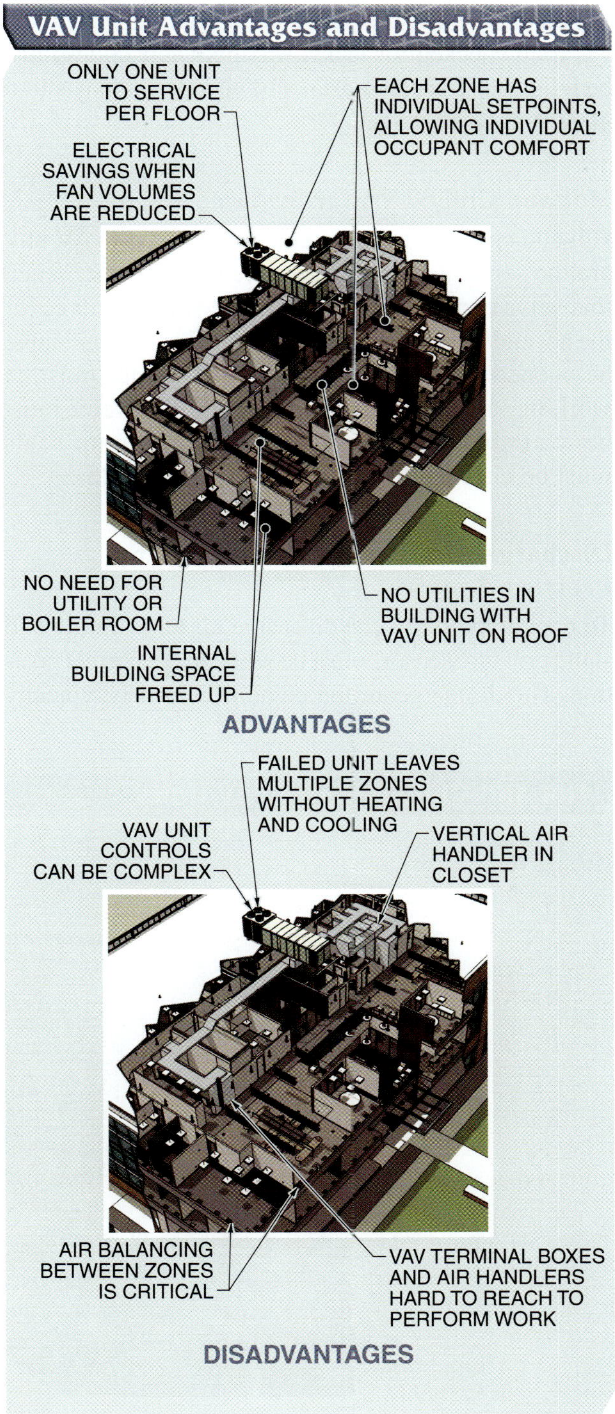

Figure 44-24. An advantage of a VAV unit is that there is one unit instead of multiple units to service. A disadvantage is that when a unit fails, multiple zones are left without heating and cooling.

Gas Piping

Gas piping for a rooftop VAV unit is virtually the same as for a single-zone rooftop unit. When a rooftop VAV unit is equipped with gas heat, the gas supply must be turned on. The gas pressure should be checked, and the gas train checked for leaks. All applicable codes must be followed. The unit must not be operated when natural gas aromas are present.

Hot and Chilled Water Systems

Hot and chilled water systems for a rooftop VAV unit are the same as for a single-zone rooftop unit. When this valve configuration is used, the shutoff valves for the hot and/or chilled water supply to the VAV unit must be opened. The circulating pump or pumps must be working, and a source of hot or chilled water must be available. The piping system and pump seals must be checked for leaks. **See Figure 44-25.**

Discharge Air Temperature and Static Pressure Sensors

To operate properly, the discharge air temperature and static pressure sensors must be wired in the correct position. The discharge air temperature sensor is typically wired about 6′ to 9′ downstream from the fan. The static pressure sensor is wired in the supply ductwork, about two-thirds of the distance of the longest duct, downstream from the fan. This is the location recommended by the engineering community and will allow all of the VAV boxes or vertical air handlers to achieve the proper air flow. *Note:* Many VAV system problems are caused by the improper location of the static pressure sensor. **See Figure 44-26.**

Inlet flow sensors at each VAV box are used to indicate airflow rate in cubic feet per minute (cfm).

Hot and Chilled Water VAV Units

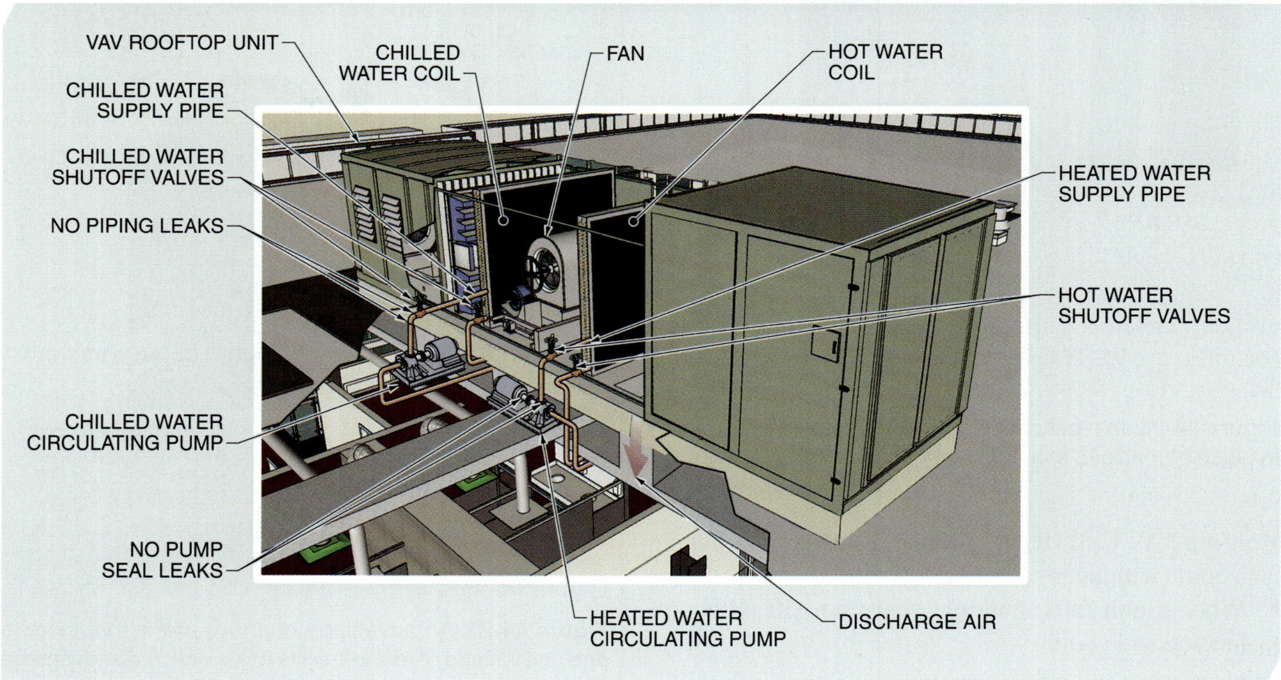

Figure 44-25. Hot and chilled water VAV unit coils and controls operate in the same manner as other water-filled systems.

Discharge Air Temperature and Static Pressure Sensors

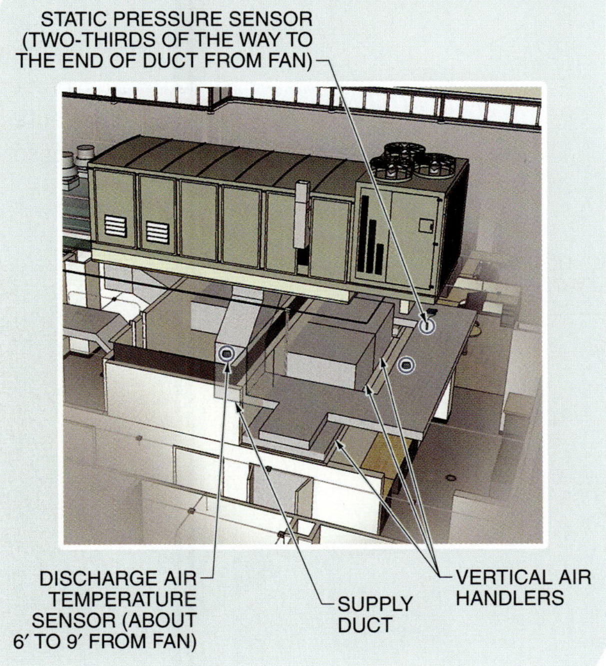

Figure 44-26. A discharge air temperature sensor is typically wired about 6′ to 9′ downstream from the fan, while a static pressure sensor is wired in the supply ductwork, about two-thirds of the distance of the longest duct, downstream from the fan.

Rooftop VAV Unit Controls

One of the large differences between a rooftop single-zone unit and a VAV unit are the controls. Today, VAV unit controls consist of a control board with a microprocessor. A laptop or display, along with software, is typically required to access the setpoints and parameters of a VAV system. Some control packages require field installation of sensors in the ductwork and/or in the occupied spaces. The manufacturer specifications and installation procedures should be followed carefully. For some installations, a controls specialist may be needed to complete this part of the installation. These units are often tied into the BAS of the facility using BACnet or LON protocols. **See Figure 44-27.**

VAV Boxes and Controls

In addition to the rooftop unit, individual VAV terminal units or vertical air handlers are installed as well. There are many possible VAV terminal box configurations. Some are cooling only, some have hot water reheat or electric reheat, and most are fan powered. They are attached to the supply duct, sometimes with a section of flex duct. As always, all manufacturer instructions must be followed. As with a rooftop VAV unit, a controls specialist may be needed to complete this part of the installation.

ROOFTOP VAV UNIT STARTUP

The installing contractor is often required to perform the start-up procedure for a packaged rooftop VAV unit. This is especially critical if a warranty claim arises later, as proper start-up documentation is required for a claim. All manufacturers provide standard start-up documentation procedures and forms.

VAV Unit Startup

The rooftop VAV unit must be installed properly using the manufacturer documentation and installation instructions, and all applicable codes must be followed. The correct size and number of filters must be installed before startup of the VAV unit. In addition to filters, the outdoor-air inlet screen must be in place before operating the unit.

Before startup, any fan or compressor base, bolts, shims, and shaft alignment must be in proper condition for starting the VAV unit. All low- and high-voltage connections should be checked for proper location, and all connections should be tight. The thermostat system switch and fan switches must be off. All lockouts and tagouts should be removed from the power supply, and power should be turned on at the disconnect. All line and low voltages should be checked for proper voltage levels and the levels recorded on the start-up form. All VFDs must be running the correct programs. **See Figure 44-28.**

When the VAV unit is equipped with gas heat, the gas supply is turned on. The gas pressure should be checked and the gas supply piping checked for leaks. All applicable codes must be followed. The VAV unit must not be operated when any gas or gas smells are present.

TECH TIP

The direction of rotation of 3ϕ motors depends on the correct phase wiring. The correct phase wiring for a 3ϕ fan motor can be checked by performing a bump test, in which the motor is momentarily energized to determine its direction of rotation.

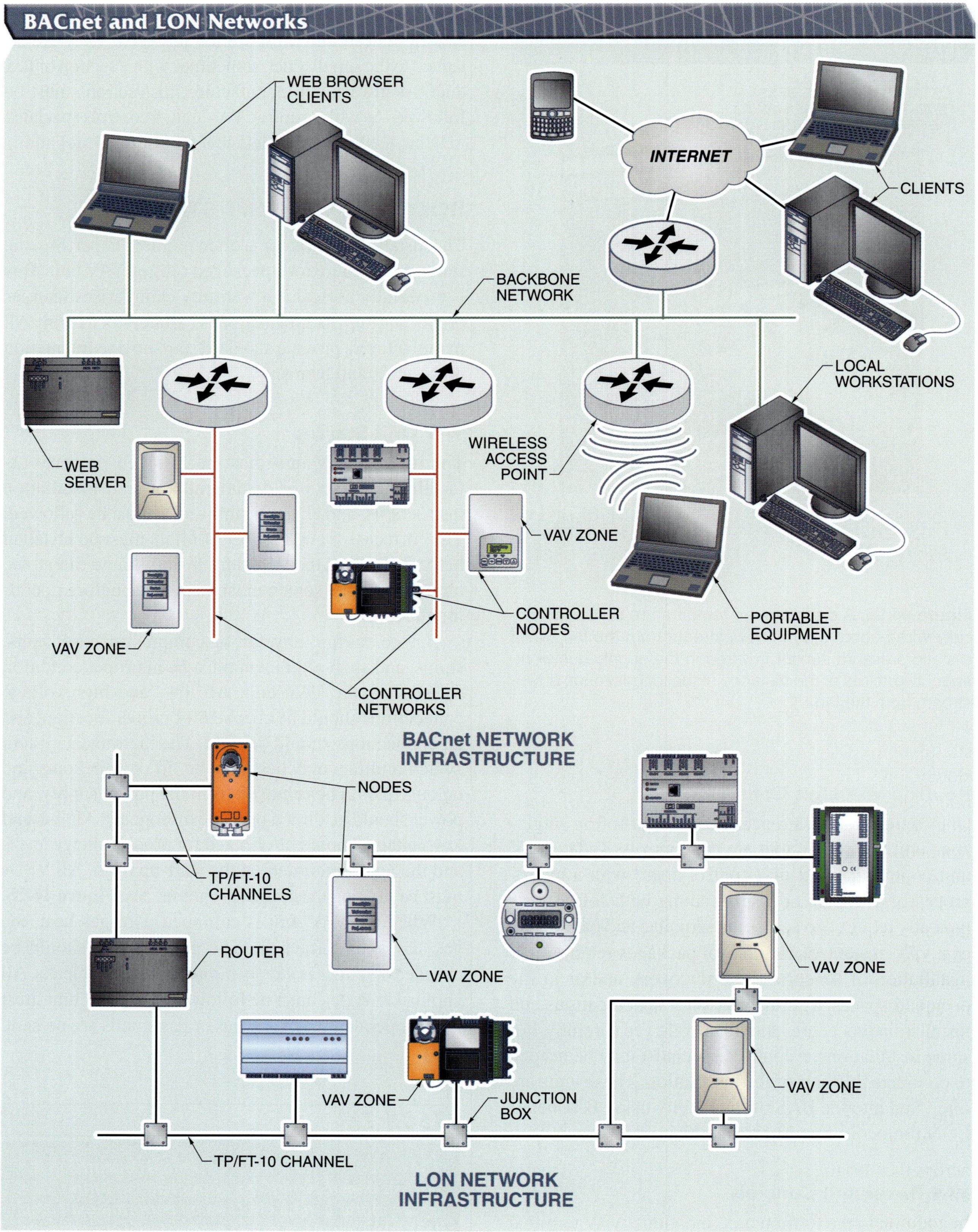

Figure 44-27. VAV unit systems typically require a lot of building information, so VAV units are tied into the BAS of the facility using BACnet or LON protocols.

Rooftop VAV Unit Startup

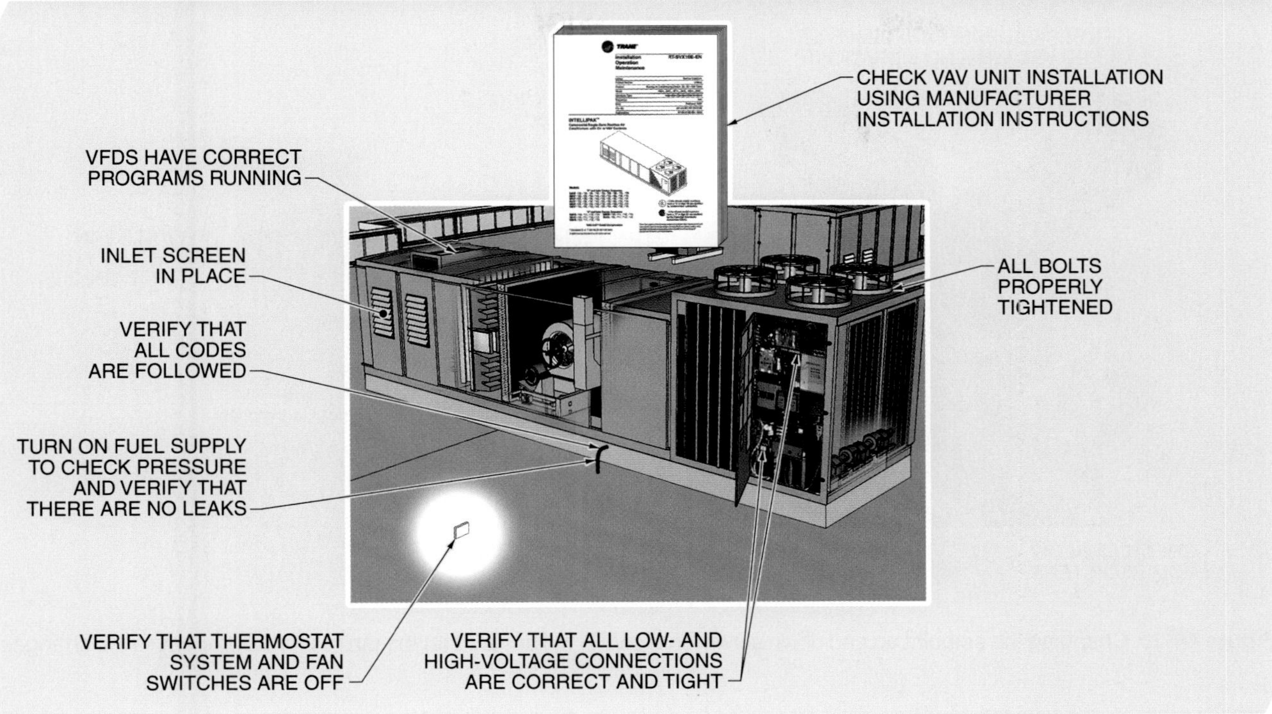

Figure 44-28. Besides the normal HVAC unit startup checks, VAV units must have the programs in the various VFDs checked.

VFD and Fan Testing. Using the control system, the startup of the VAV unit is initiated. The fan should start slowly, and the VFD display should be checked as it ramps up the fan. Typically, the duct pressure setpoint is 1″ of static pressure, so that pressure needs to be reached. A measuring instrument is needed to check the actual static pressure in the duct. The setpoint of the VFD should be changed up and down slowly to ensure the fan responds properly to the changes. A quick check should be made for proper airflow at outlet diffusers and registers. The fan should be checked to verify that no excessive vibrations are present. **See Figure 44-29.**

Testing Cooling Cycle. Using the control system, system cooling is initiated. One way to do this is to change the setpoint below the discharge temperature. After a system delay, the compressors should come on. Discharge temperature, temperature differential across the cooling coil, amperage, and voltage should all be recorded. Low- and high-side pressures should be recorded as well. When multiple stages of cooling are involved, a time delay between stages is normal. **See Figure 44-30.**

For chilled water systems, the valve for water flow will open and the chilled water will flow through the coil. A standard discharge setpoint is 55°F. The setpoint should be changed up or down by 2°F to verify that the system responds properly. Again, any temperature drops should be recorded.

Testing Heating Cycle. The control system is used to initiate system heating. One way to do this is to change the setpoint above the discharge temperature. *Note:* When cooling is initially energized, there may be a built-in delay that will not allow the heating to come on immediately.

It will be necessary to wait until the timer runs out. During the heating mode, the supply fan must be running for the heat to come on. At this point, a normal natural gas ignition sequence will result. When electric heat is present, the elements will energize as long as the safeties are working properly. Also, there may be multiple stages of heat, and a time delay between stages may exist. **See Figure 44-31.** All high- and low-voltages should be checked for proper voltage levels and recorded on the start-up form.

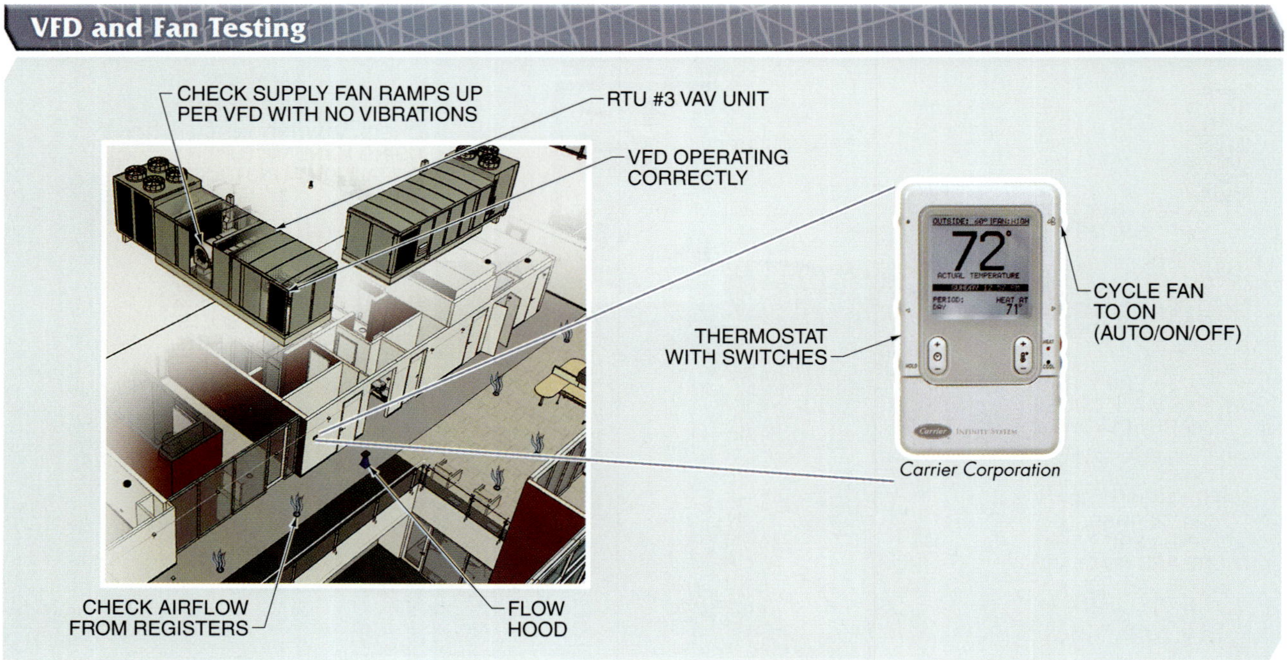

Figure 44-29. Changing the setpoint up and down slowly on the VFD will ensure that the fan responds properly to the changes.

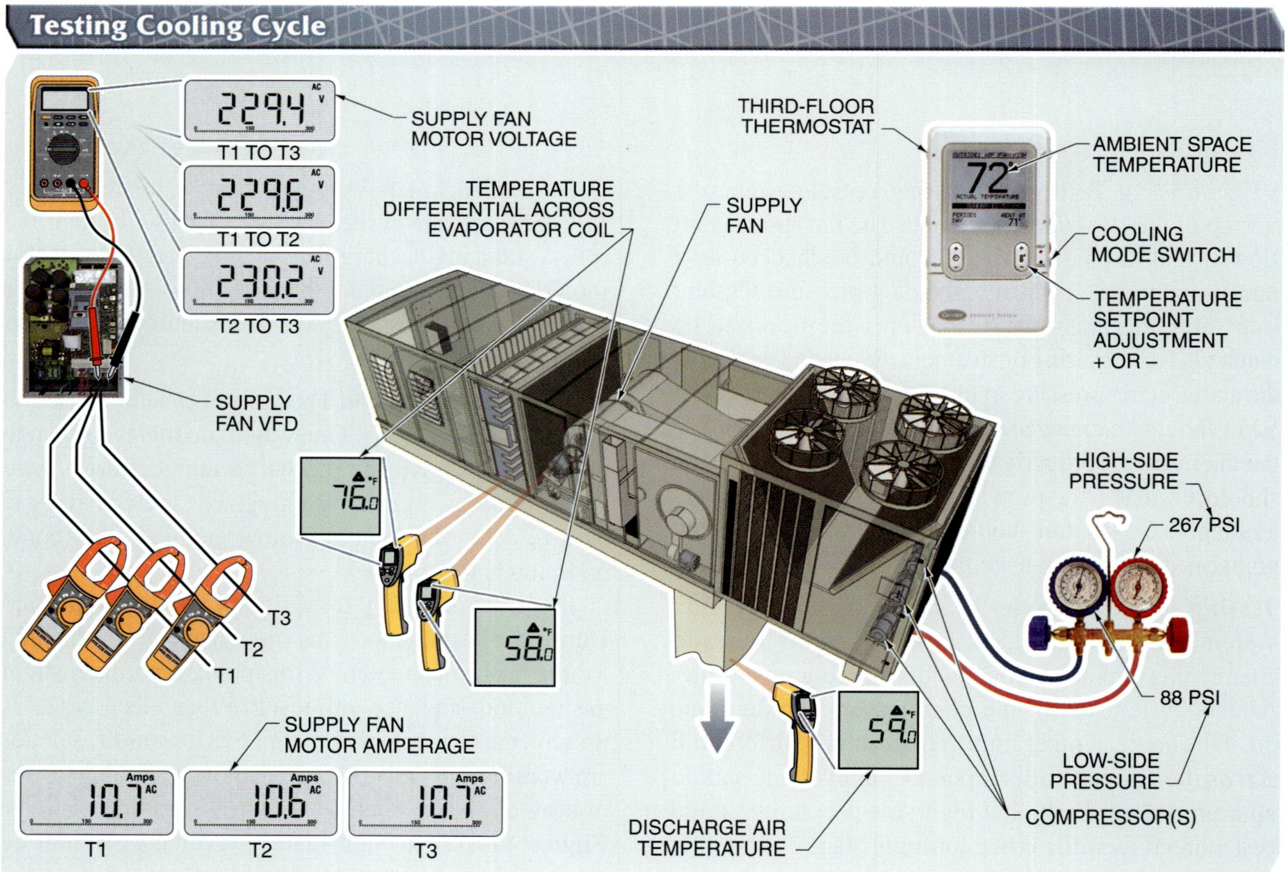

Figure 44-30. The cooling cycle of a VAV unit will be the same as the other units once the VFDs are checked for proper operation.

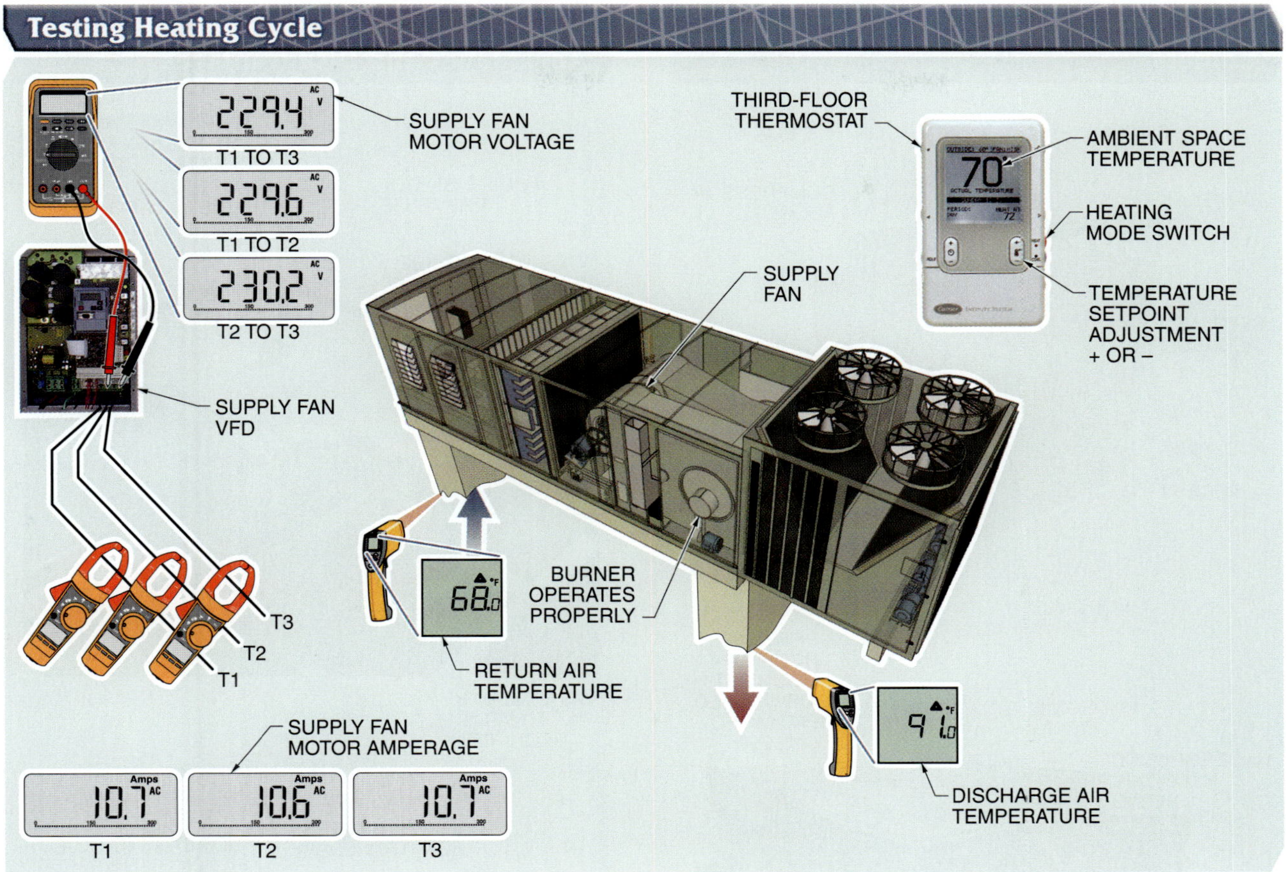

Figure 44-31. As with all HVAC units, during the heating mode, the supply fan must be running for the heat to come on.

When the unit has hot water heat, the valve will open, and hot water will flow through the heating coil. All valves should be open, the circulating pump running, and hot water available at the heat exchanger from the boiler. At this time, the balancing valve will be adjusted to obtain the correct flow in gpm through the coil. For startup of heating, temperature rise across the coil will be measured, along with amperage for electric heat. As with testing the cooling cycle, the setpoint should be changed up and down a few degrees to ensure proper response.

INSTALLING AIR-COOLED PACKAGED CHILLERS

An *air-cooled packaged chiller* is an all-in-one packaged unit shipped complete from the factory. This means that when it arrives at the building, the system has been evacuated, fully charged with refrigerant, and pressure tested. Air-cooled packaged chillers typically use R-410A refrigerant and are precharged with an initial oil charge as well. R-410A refrigerant is an efficient, nontoxic, environmentally correct blended refrigerant.

Light Commercial Air-Cooled Packaged Chillers

Most air-cooled packaged chillers are typically 8 tons to 125 tons in size with 40 ton to 90 ton units being very popular. Quiet scroll compressors can be found in these units, with a digital compressor as a common option. Because of their smaller footprint (floor space equipment takes up), these types of chillers can be located on the roof or at ground level. **See Figure 44-32.**

Air-cooled packaged chillers utilize an electronic expansion valve to increase efficiency. Controls provide capabilities such as chilled water temperature reset, demand limiting, compressor wear minimization, and protection. Temperature and pressure displays and diagnostic functions allow air-cooled packaged chillers to have a high reliability.

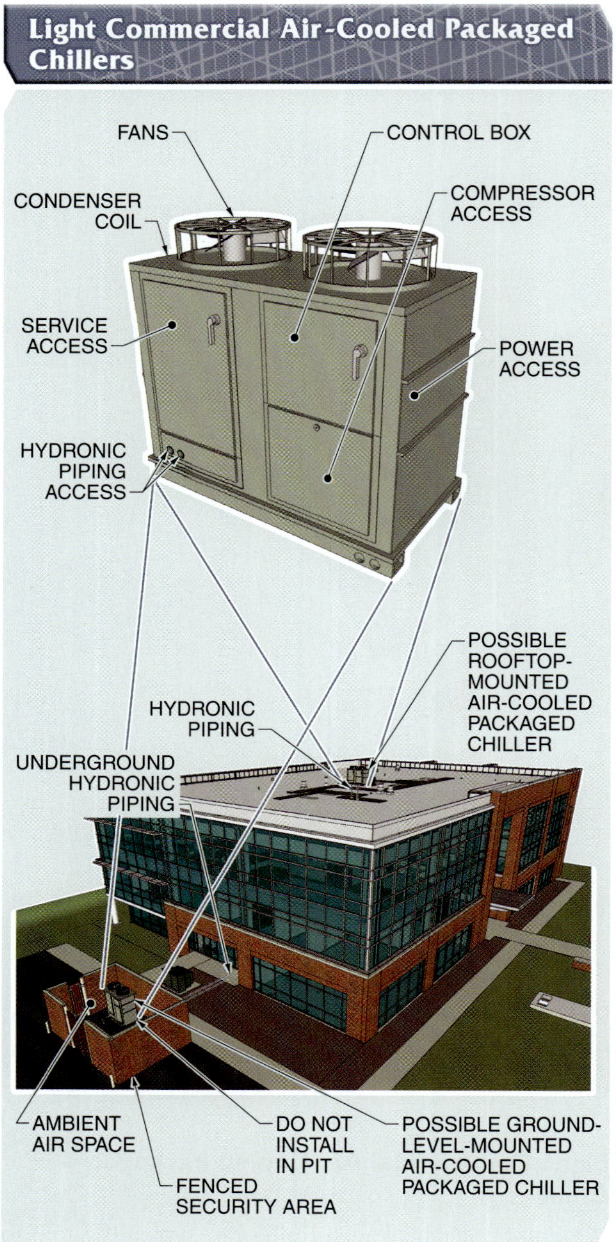

Figure 44-32. Air-cooled packaged chillers have a relatively small size footprint for the large tonnage of the unit.

Air-Cooled Packaged Chillers

The facilities that use air-cooled packaged chillers include large schools, small shopping malls, and large office buildings. Because air-cooled packaged chillers utilize open space outside, the building mechanical room only houses the air handling units. Walls can also inversely affect air-cooled packaged chillers during heat rejection. Air-cooled packaged chillers have both advantages and disadvantages. **See Figure 44-33.**

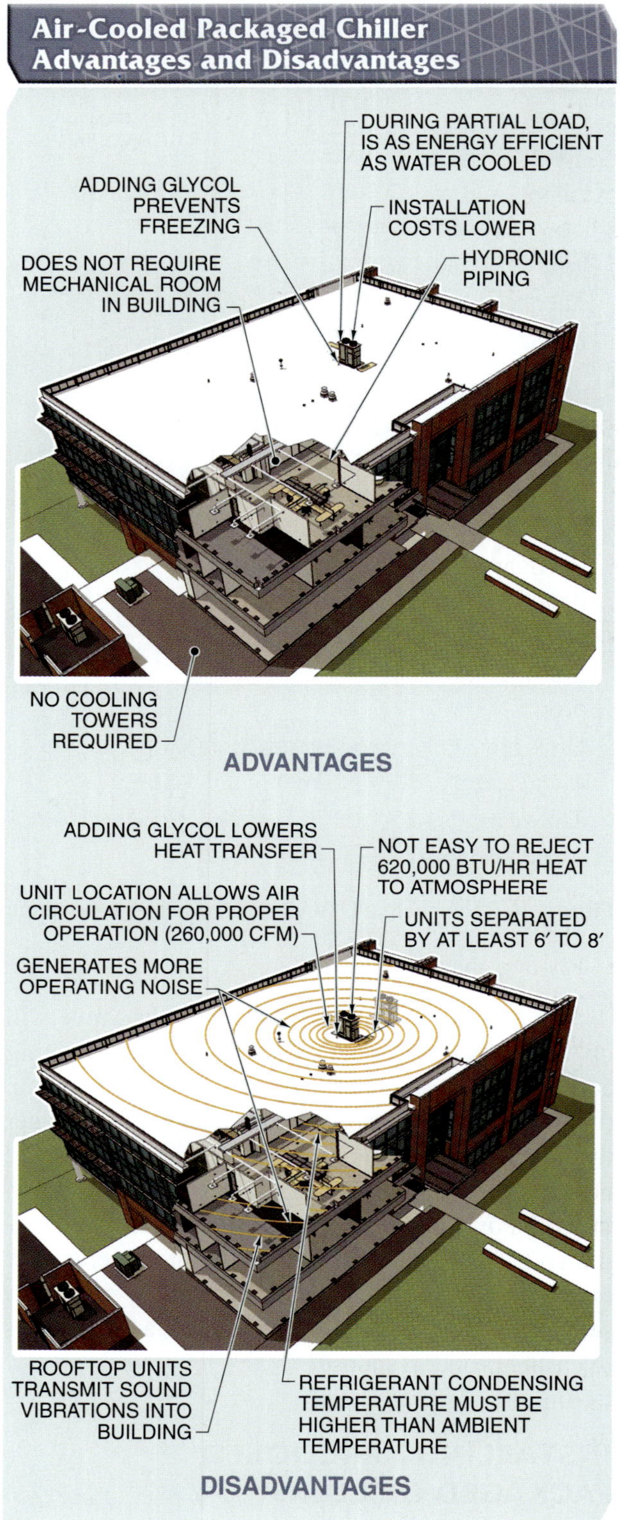

Figure 44-33. An advantage of air-cooled packaged chillers is that they do not require cooling towers or condenser water pumps. A disadvantage of air-cooled packaged chillers is that a 40 ton unit may have to reject upwards of 620,000 Btu/hr of heat to the atmosphere.

Air-Cooled Packaged Chiller Advantages. Using air-cooled packaged chillers offers many advantages:

- During part-load conditions, air-cooled chillers are as energy efficient as water-cooled chillers.
- They do not require cooling towers or condenser water pumps.
- They do not require a mechanical room for a chiller, freeing up building space for occupant use.
- The installation costs are typically lower.
- In winter, glycol can be added to prevent freezing.

Air-Cooled Packaged Chiller Disadvantages. Rooftop air-cooled packaged chillers also come with many disadvantages:

- A 40 ton unit may have to reject upwards of 620,000 Btu/hr of heat to the atmosphere.
- Refrigerant condensing temperature must be higher than the ambient temperature.
- The unit must be located so that enough air circulation (perhaps 260,000 cfm) is available to operate properly.
- Units have to be separated by 6′ to 8′ of clearance so as not to recirculate ambient air.
- They generate more operating noise than other units.
- Adding glycol lowers the heat transfer properties of water and increases each of the system pressure drops.
- Rooftop units transmit sound vibrations into buildings.
- When a rooftop unit is mounted above an air-handling unit (close-coupled), control stability problems occur because the fluid travels to the air handler and back in only a few seconds.

Wiring

The wiring of an air-cooled packaged chiller is very similar to the wiring of other HVAC units. The control system power must be off. All low- and high-voltage connections must be checked for proper locations, and all connections must be tight. Before performing startup, the lockouts and tagouts must be removed from the power source. Power should be turned on at the disconnect. All line and low voltages should be checked for proper voltage levels and the levels recorded on the start-up form.

The control system power switch is turned on, and the control board indicator lights are checked for normal operation. The crankcase heater needs to be energized for up to 24 hr before compressor startup. The manufacturer literature should be consulted for specific wiring information. See **Figure 44-34.**

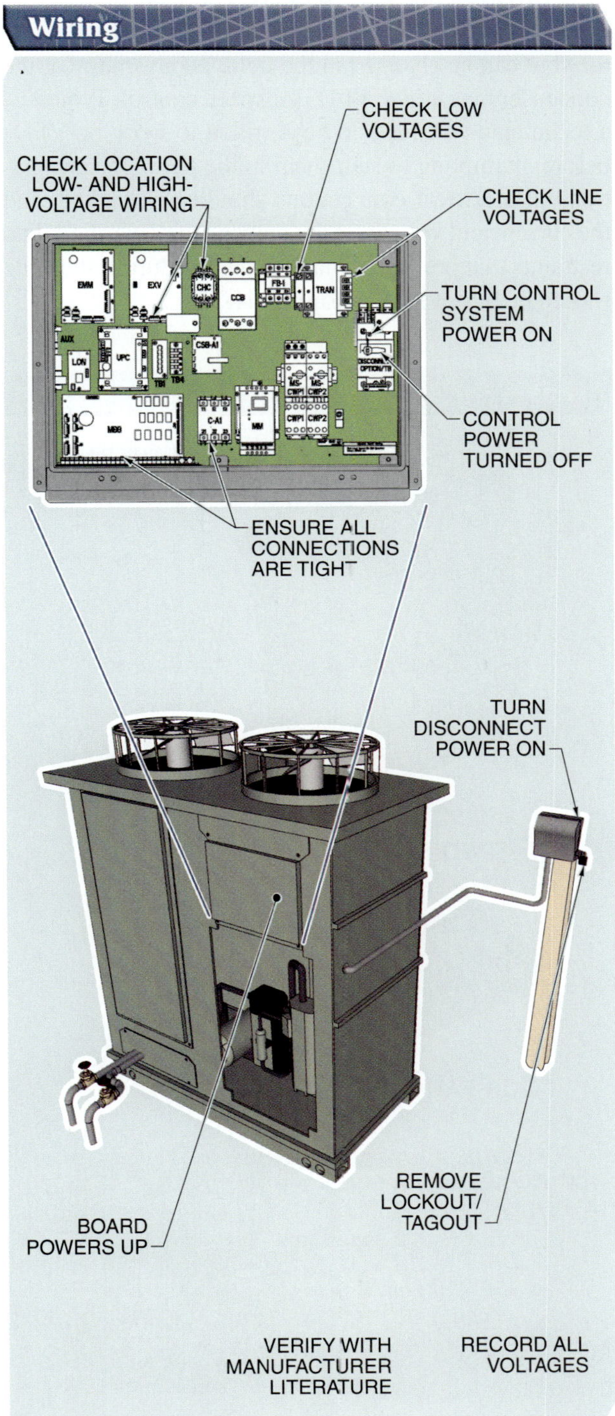

Figure 44-34. After normal HVAC unit wiring checks are made, the technician must verify that the control board has powered up and that the indicator lights are normal.

Fans

Air-cooled packaged chillers have a number of condenser fans that must be controlled to maintain the proper head pressure. Depending on unit size, there may be anywhere from two to eight condenser fans. The sequencing order of the fans is typically set by the control board at the factory but can be changed in the field. Today, many of the condenser fans utilize VFDs for speed control. Typically, a technician will have to be certified to work on VFDs before attempting to start them in the field. For installation, all electrical connections should be checked, and the current and voltage of the power source verified. Fan restraints that prevent damage to the fan during shipping should be removed. **See Figure 44-35.**

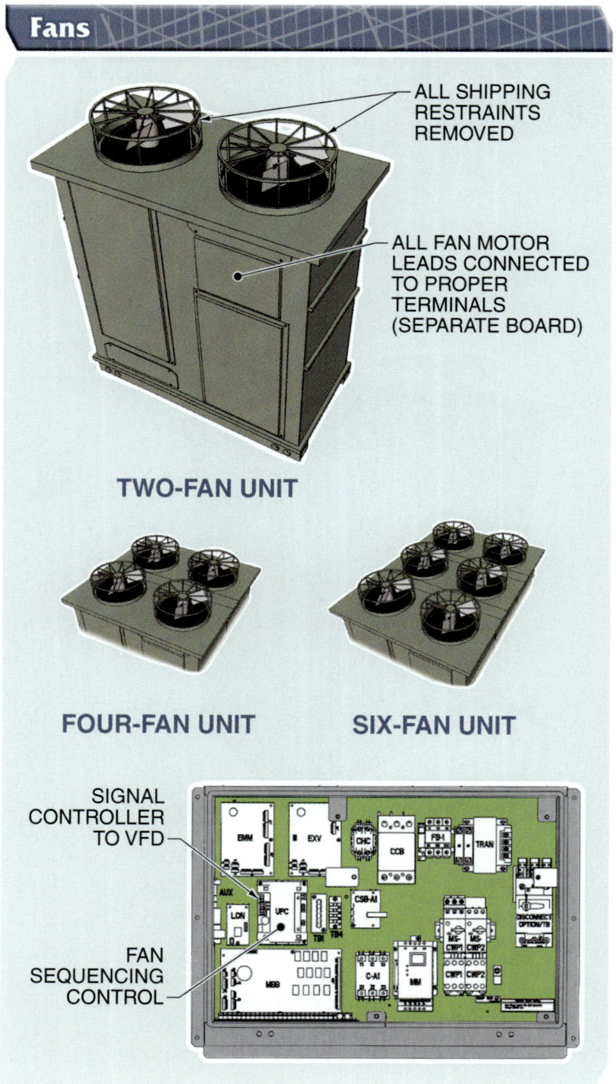

Figure 44-35. Air-cooled packaged chillers typically use a complex control system for the condenser fans.

Hydronic Pump and Piping Installation

After the air-cooled packaged chiller (without hydronic package) is mounted on the roof or the pad next to the building, the pumps are installed. Pumps are typically located in the mechanical room or next to the chiller (rarely in the chiller). The pumps were selected to match the system capacity requirements. Today, this is accomplished with software. **See Figure 44-36.**

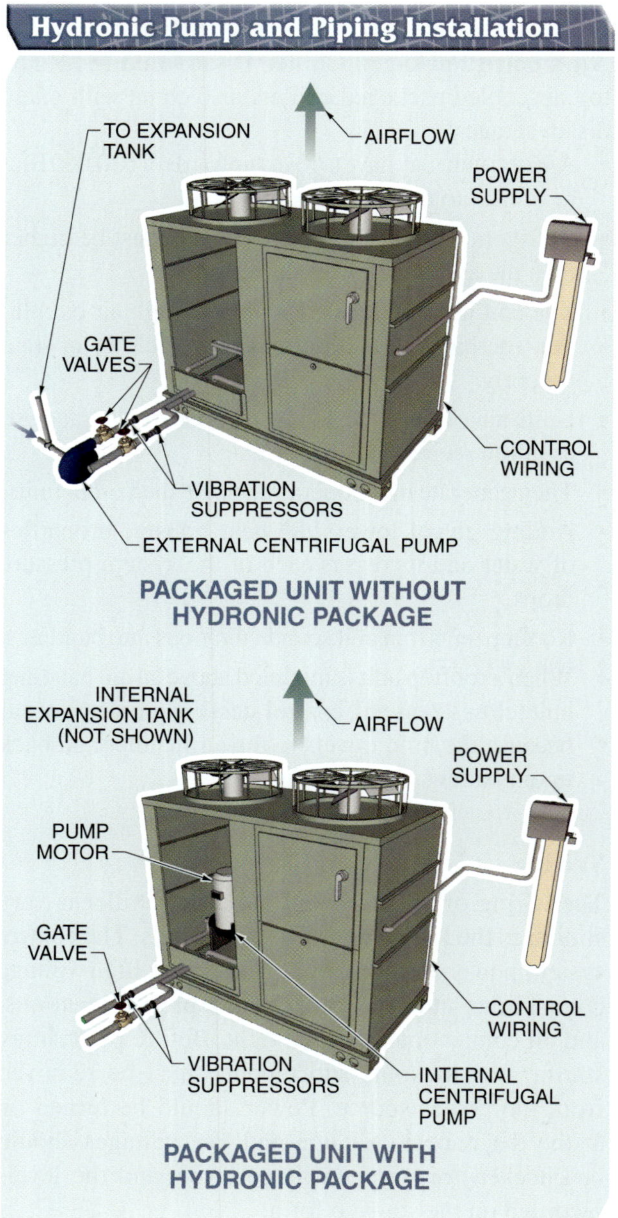

Figure 44-36. Air-cooled packaged chillers include a pump lead/lag software package that automatically switches the pumps on and off and switches pump operation in order to equalize run times.

While one pump may be sufficient for operation, two pumps are typically used in order to provide hot weather capacity and a backup pump in case of failure. In most cases, a pump lead/lag package is included to automatically switch the pumps on and off and to switch pump operation in order to equalize the run times.

The pumps are mounted using vibration isolators. The pumps are started and their pressures recorded before starting the packaged chiller. Typically a flow switch must prove flow before the chiller will start.

Control System

Today, virtually all packaged chillers use a microprocessor control system. Each board type will have a specific function set as determined by the manufacturer. While some boards will have a local display for the values, some may require a laptop computer with software to see the control functions. Some typical control system functions are as follows:

- unit on/off switch
- discharge chilled water setpoint
- return and discharge liquid temperature
- low discharge liquid temperature cutout setting
- low ambient temperature cutout setting
- outdoor air temperature
- suction pressure cutout setting
- each system suction and discharge pressure
- antirecycle timer status for each system
- compressor run status
- low cooling-load condition
- status of hot gas valves, evaporator heater, and fan operation
- number of compressors running
- liquid solenoid valve status
- load and unload timer status
- water pump status
- day, date, and time
- daily start/stop schedules and holidays
- compressor lead/lag control
- compressor starts and operating hours for each compressor

Options. Typical options for air-cooled packaged chiller installation may consist of sound blankets and ultra-quiet or whisper-quiet fans. This is especially true of unit compressors when they are located in an application or area where noise is of concern. Some installations will not utilize sound blankets initially, but after sound is found to be an issue, sound blankets will then be installed. Carefully choosing the location of the air-cooled packaged chiller can help reduce noise problems. **See Figure 44-37.**

Ultra-quiet fans typically reduce air noise because the fan blades are wide. The wider blades reduce generated sound (12 dBA to 18 dBA), vibrations, and make the fan lighter.

Air-cooled packaged chillers are often installed where their noise has minimal effect on the building occupants and surrounding environment.

AIR-COOLED PACKAGED CHILLER STARTUP

Using the manufacturer checklist will help ensure the proper startup of the packaged unit. The checklist provides a means of recording unit conditions, application requirements, and system information for later reference. Air-cooled packaged chillers typically have a prestart checklist that must be followed for proper startup to occur. Commissioning of the water side must occur according to ASHRAE standards prior to startup.

Test Pumps and Water Flow

The first major step in starting an air-cooled packaged chiller is to start the pumps and make sure that adequate water flow is occurring through the chiller. Any water treatment must be operating at this time. The water through the chiller must not be too hot or too cold as this may cause improper head pressures or operation when the chiller is started. The pumps should be checked for proper current draw, voltage, and system operational pressures. **See Figure 44-38.**

Sound Blankets and Quiet Fans

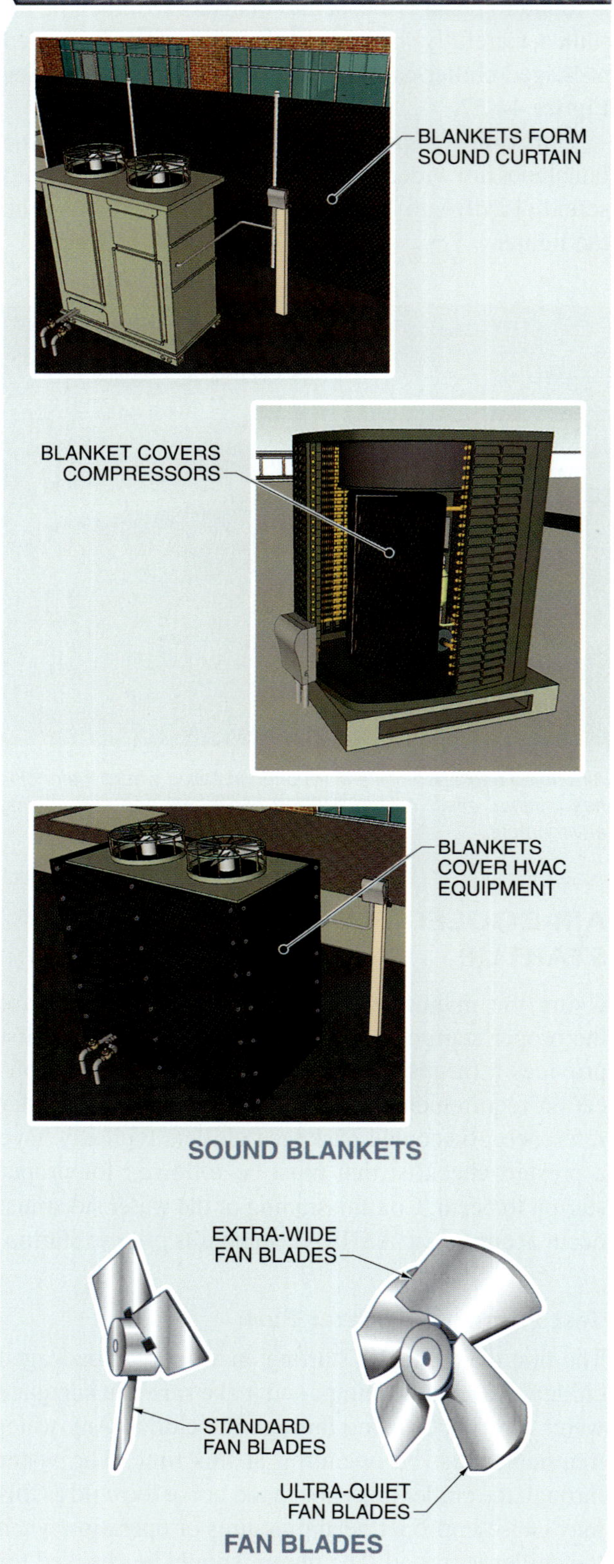

Figure 44-37. Sound blankets and quiet fans are typically added to air-cooled packaged chillers after sound from the chiller has been determined to be a problem.

Testing Pumps and Water Flow

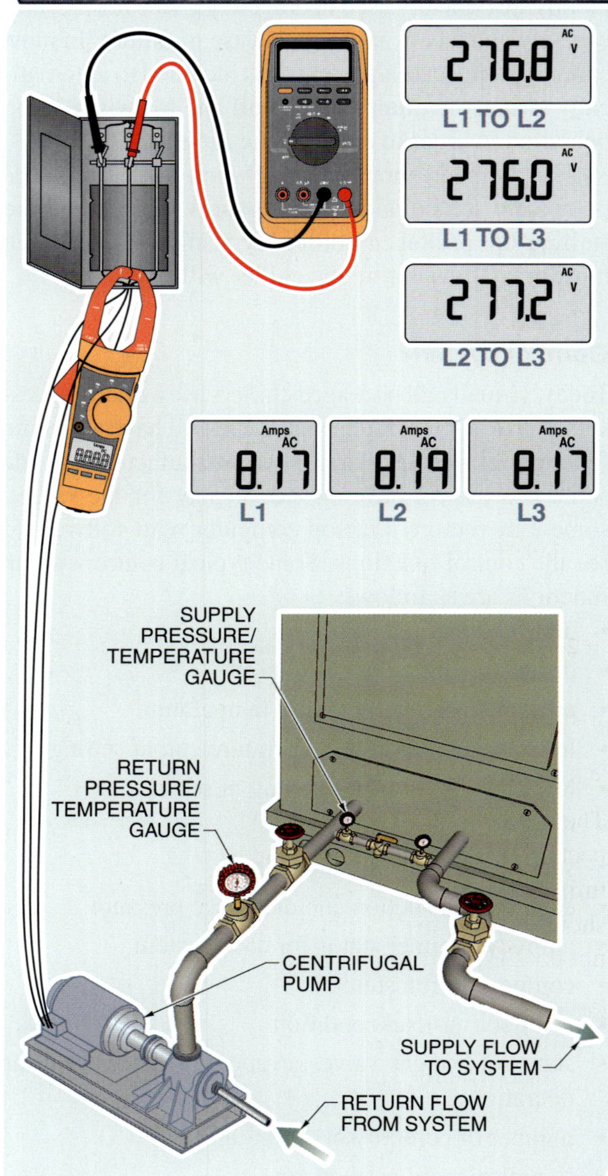

Figure 44-38. When testing the pumps and water flow, any water treatment systems must also be checked.

Bad Flow Switch. One item that can cause an air-cooled packaged chiller to go down is the water flow switch. The flow verification switch is wired as a safety circuit to the chiller control board. When the flow switch opens, the unit will shut down and the chilled water temperature will rise. On older units or units that use older style flow switches, switch malfunctions are known to cause chiller shutdown. **See Procedure for Testing Flow Switch Operation.**

Testing Flow Switch Operation

Procedure

1. Make sure pump (internal or external) is running, and check pressures and pump motor amperage to identify water flow rate.

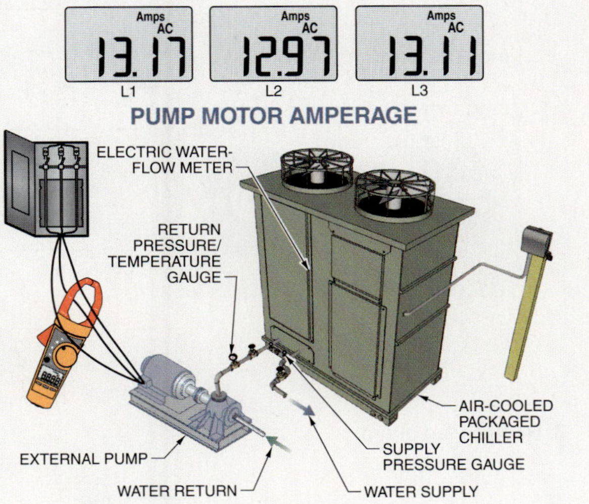

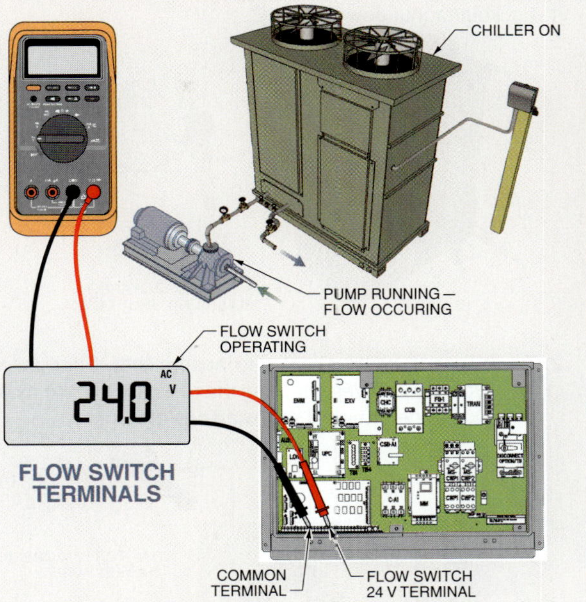

2. Use a DMM at the chiller control board to check whether flow switch is open or closed. Use manufacturer literature to locate appropriate terminals to use during test.

3. If water flow is present but switch indicates "open," replace the switch. *Note:* It is possible to use jumper wires temporarily to allow a chiller to run until repairs can be made.

Testing Condenser Fans

The technician should verify that all the electrical connections for the condenser fans have been made and then turn the power on. Where possible, the control system should be used to cycle the condenser fans. When this is not possible, fan operation can be tested during normal operation. When VFDs are present, they can be put into manual mode to allow for fan operation. The rotation direction, voltage, and amperage of each fan motor should be checked and recorded. **See Figure 44-39.** The running fans should be checked for unusual vibrations. **See Procedure for Replacing a Condenser Fan Motor.**

Testing Condenser Fans

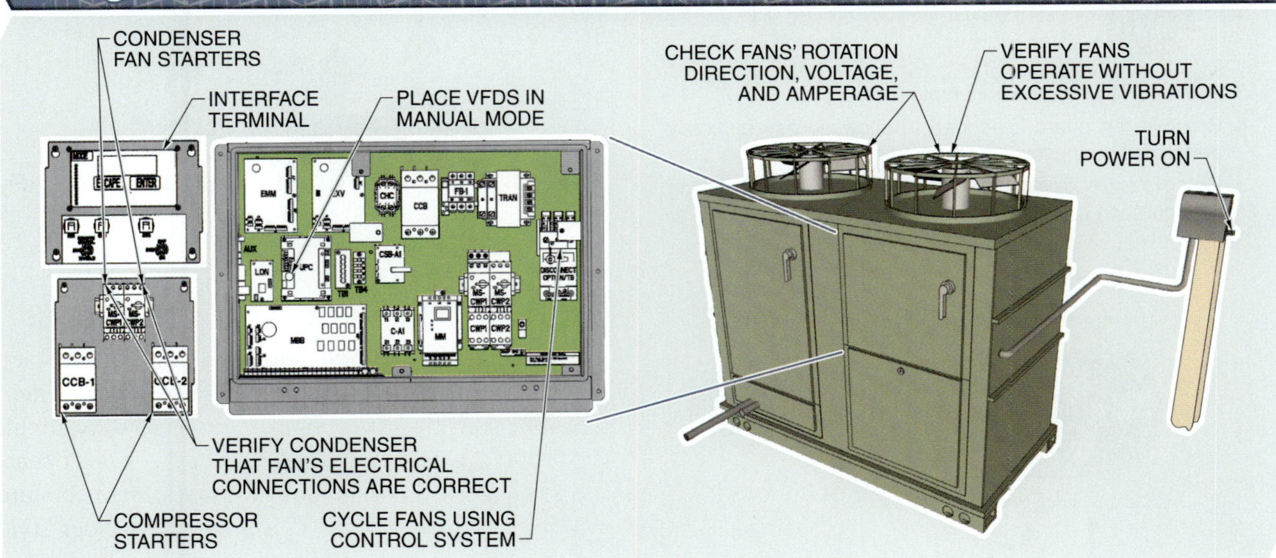

Figure 44-39. When testing condenser fans, any VFD associated with an air-cooled packaged chiller must be placed in manual operation to allow the fans to operate.

Procedure

Replacing a Condenser Fan Motor

1. Lockout/tagout power supply.

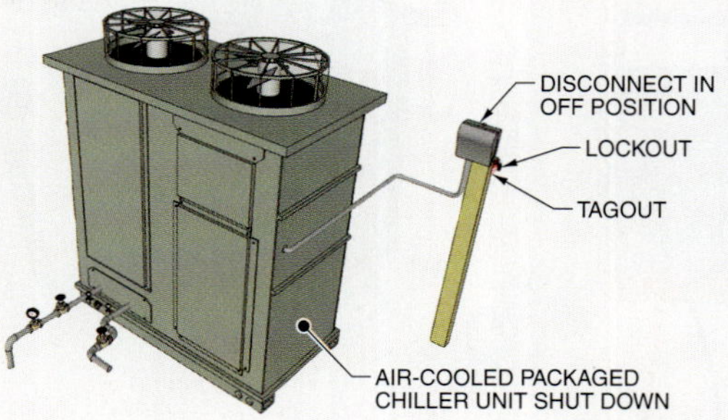

2. Troubleshoot existing motor to ensure that it has failed. This typically involves checking the resistance of the motor windings to determine if any windings are open.

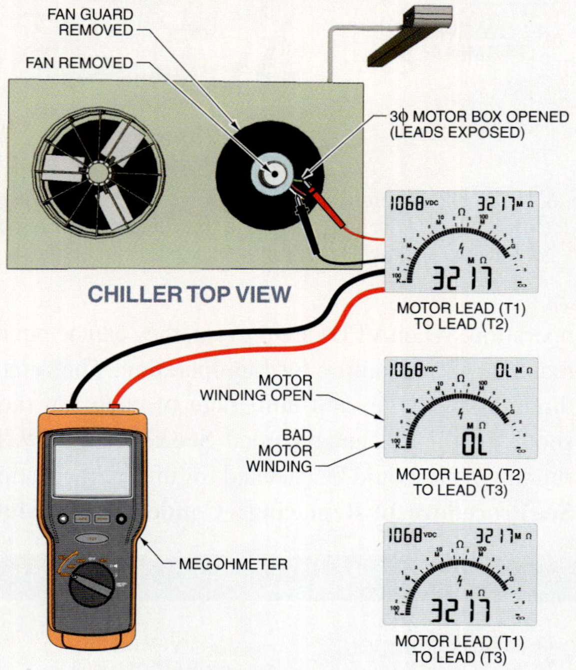

3. Order correct replacement motor.
4. Remove fan guard if needed to allow access (typically performed during insulation resistance test).
5. Remove existing motor from mounting bracket. Note wiring configuration using colored wires.

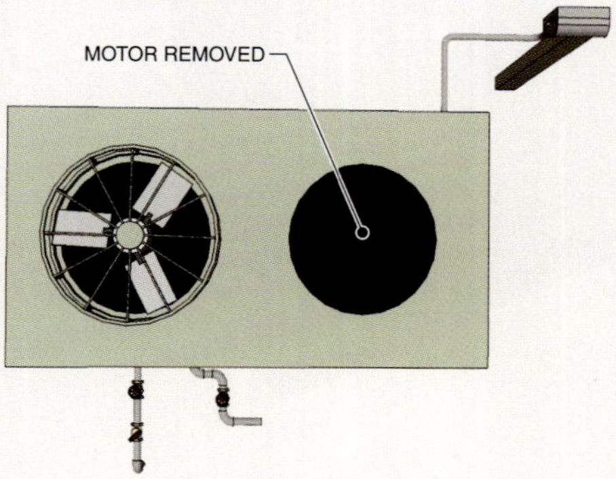

6. Install new motor.
   a. Attach mounting bracket.
   b. Wire motor.

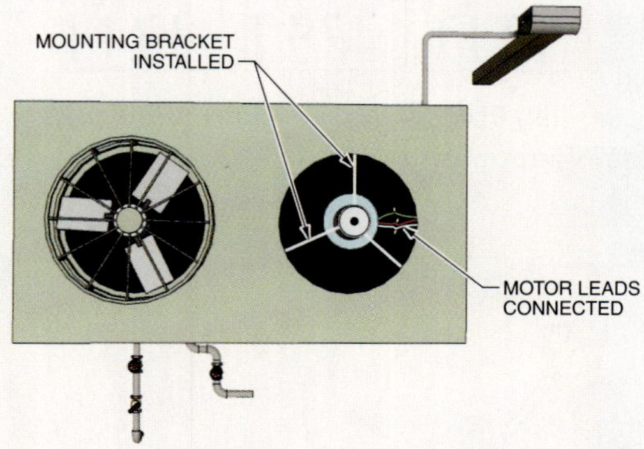

7. Replace fan guard.

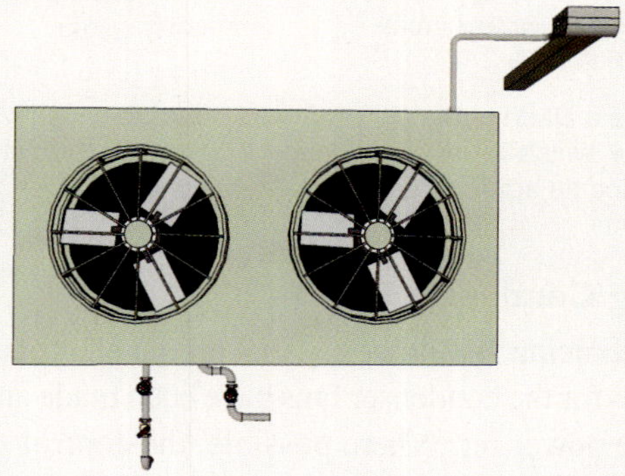

8. Remove lockout/tagout.
9. Start motor and check for actual operation. All 3ϕ motors must be checked for proper rotation.

Test Compressors and Compressor Staging

The compressors are now ready for startup. A good method for starting them is to use the control system and lower the discharge chilled water temperature setpoint a few degrees. The pumps (internal or external) must be running properly, and water flow in the system must be proved to enable compressor operation.

As the chilled water setpoint is lowered, the compressors should turn on in the proper sequence. Interstage timers must be checked for proper operation. Lead, lag, and any failure modes must be checked at this time. **See Figure 44-40.**

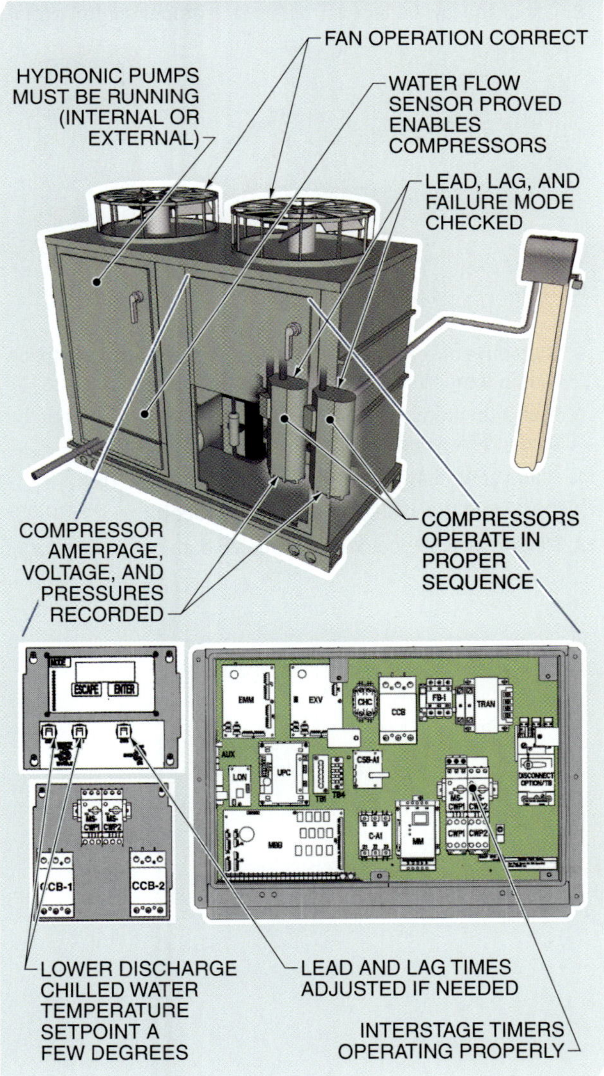

Figure 44-40. All compressor lead, lag, and failure modes must be checked and proven operational.

Amperages, voltages, and pressures of each compressor should be recorded on the start-up forms provided by the manufacturer. The unit is then placed in normal operation. Copies of the start-up forms are sent to the manufacturer, as they are needed for any warranty claims. Building staff may need job-site training in operation and maintenance.

Staging Programming and Parameters

Another common problem with a system that has VFDs on the motors is improper staging programming and parameters at the unit microprocessor board. These incorrect parameters may cause the unit compressors to cycle improperly. This may in turn lead to incorrect temperatures at the unit and consequently the space. The technician must have access to the unit microprocessor, the unit keypad or a laptop computer, and the unit microprocessor literature. **See Procedure for Changing Compressor Settings (Parameters).**

Changing Compressor Settings (Parameters)

Procedure

1. Use the keypad or laptop to access the microprocessor programming of the air-cooled packaged chiller.

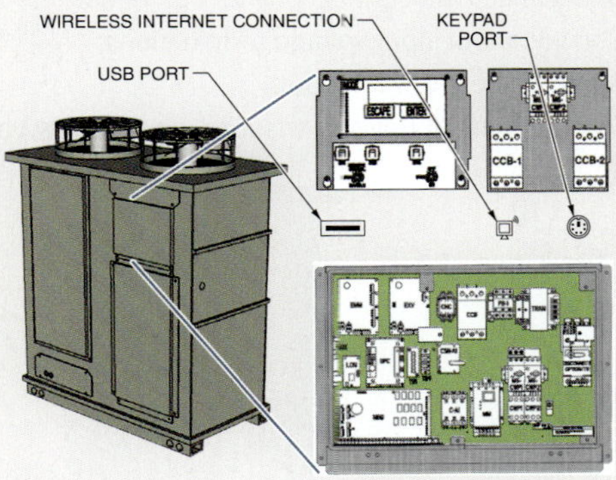

2. Locate the parameters dealing with the compressor staging. Examples would be the minimum on and off times and the staging temperature differential. Check these for correct application settings. If necessary, change to improve operation or reset to factory defaults.

MINUTES OF TIME—*CONFIGURATION→OPT2→DELY*

LEAD/LAG DETERMINATION—*CONFIGURATION→UNIT→ALTY=YES*
LAG—*CONFIGURATION→OPT2→LLCS*

LOAD SEQUENCE SELECT (STAGING)—*CONFIGURATION→OPT2→LOAD*

DEADBAND MULTIPLIER—*CONFIGURATION→SLCT→Z GN*

3. Record settings locally for use on future service visits.

VARIABLE FREQUENCY DRIVE STARTUP

With a VFD installed in the HVAC system, the speed of the pump motors, fan motors, and/or compressor motors can be controlled to meet the exact needs of the building's environment. VFDs are fully dedicated to the optimum operation of the HVAC applications. Drives offer energy savings, user friendliness, and built in functionality to meet the rigorous demands of today's HVAC applications. Motor control is based on a system that provides full motor power at rated speed without the need for derating. Today, HVAC drives have automatic motor adaptation features that ensure optimum energy conservation.

Controlling the flow of air and water in an HVAC system is the only effective way to meet the changing demands put on the system by the environment. Only VFDs can maximize the efficiency and energy savings in today's HVAC fans and pumping systems. **See Procedure for Starting a VFD.**

Procedure

Starting a VFD

1. Ensure the electrical supply is locked out and tagged out.

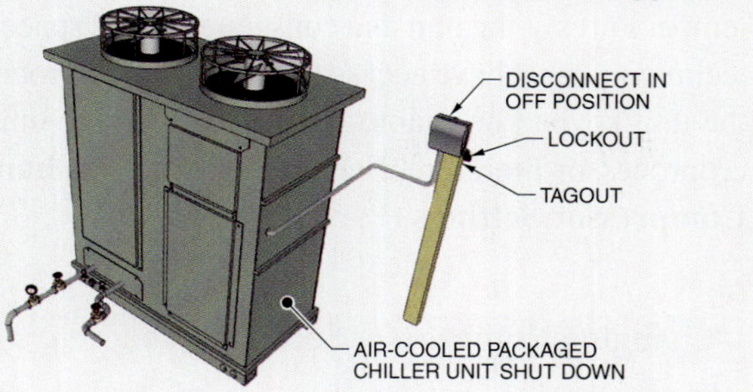

2. Check that all mechanical equipment is installed and ready to be operated.
3. Check electrical connections for tightness.
4. Check control input voltage terminations.
5. Remove the lockout/tagout.
6. Test voltages at drive to ensure proper supply voltage.

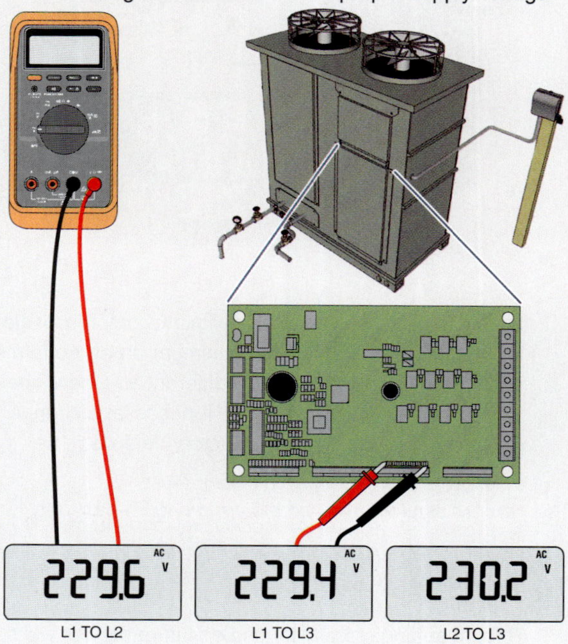

7. Enter all motor data in the drive by using the keyboard.
8. Check the low-voltage terminals for a proper control signal.

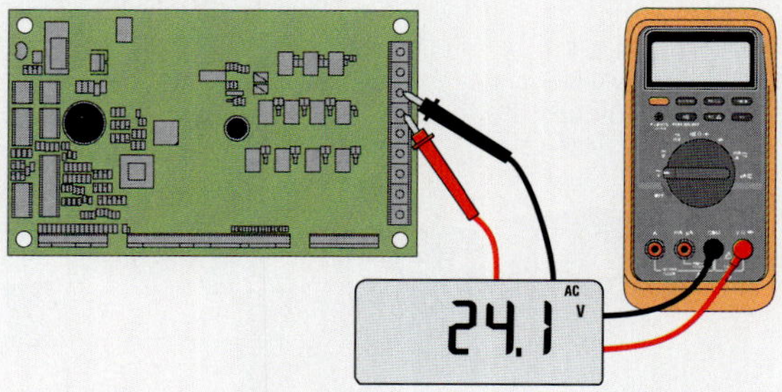

9. Start drive at very low speed to ensure proper motor rotation (bump test). Pump and fans must be ready to operate.
10. Ramp up motor to 100% to check full speed operation by using the keypad display.
11. Place control system in AUTO to operate the VFD.
12. Verify that the system is operating properly.
13. Record all electrical and operational data on start-up form.

The heart of any system that has a variable air volume is the delivery of the supply air. Typically, this is designed to be 55°F at 1″ WC static pressure, although individual systems may vary. The vast majority of light commercial systems installed in the last 10 years use variable VFDs. While specific VFD repair and in-depth electronics board procedures are typically performed by a manufacturer representative, a basic check can be performed easily by the technician. **See Procedure for Testing VFD Operation.**

TECH TIP

When a VFD is replaced, the new VFD parameters must be set to match the unit motor. This is done by using the keypad/display on the drive.

Testing VFD Operation

Procedure

1. Check control board for output to drive. Common examples are 0 VDC–10 VDC, 4 mA–20 mA, or 0 mA–20 mA DC.
2. Check drive output at drive display. For example, a 5 VDC signal to a drive should indicate a 30 Hz output to the motor.
3. If the drive output to the fan motor does not change or is incorrect, use the drive keypad to check basic setup parameters and faults. It is possible that the drive parameters were never entered properly.
4. A common simple procedure for resetting a drive is to shut off the power to the unit, allow the drive to power down (at least 20 sec), and then turn power back on. This works especially well after a thunderstorm or power problem that has caused the drive to malfunction.

UNIT 45

Servicing and Troubleshooting Light Commercial HVAC Systems

Light commercial HVAC systems are very common in many buildings and applications. These range from small convenience stores to larger retail and commercial buildings. A commercial HVAC technician will need to know how to troubleshoot and service different types of systems. While there are a large number of these types of systems available, in this chapter the focus is on rooftop single-zone, rooftop VAV, and air-cooled chiller package systems.

OBJECTIVES

- Identify safety issues and courtesy guidelines for light commercial service calls.
- List the typical troubleshooting questions for light commercial HVAC systems.
- Describe the manuals commonly used to service and troubleshoot HVAC systems.
- Explain the typical rooftop unit sequence of operation.
- Describe typical troubleshooting problems associated with light commercial HVAC equipment.

SERVICING LIGHT COMMERCIAL HVAC SYSTEMS

All appropriate safety PPE must be worn when working on light commercial HVAC systems. Since many of these units are on the roof, ladder access is needed, which means that the technician needs to be thoroughly familiar and follow all ladder safety rules. In some cases, fall protection may be necessary as well. **See Figure 45-1.**

Since there are such a wide variety of manufacturers and types of light commercial HVAC equipment, troubleshooting and other technical installation manuals from manufacturers must be read and any safety references, hazards, and safety warnings must be noted. **See Figure 45-2.** Many of these manuals can be found online if you are a registered technician with the manufacturer. When technicians run across a system type that they have never worked on before, extra caution is warranted.

Troubleshooting and PPE

Figure 45-1. Safety precautions and PPE requirements must be followed when troubleshooting HVAC systems.

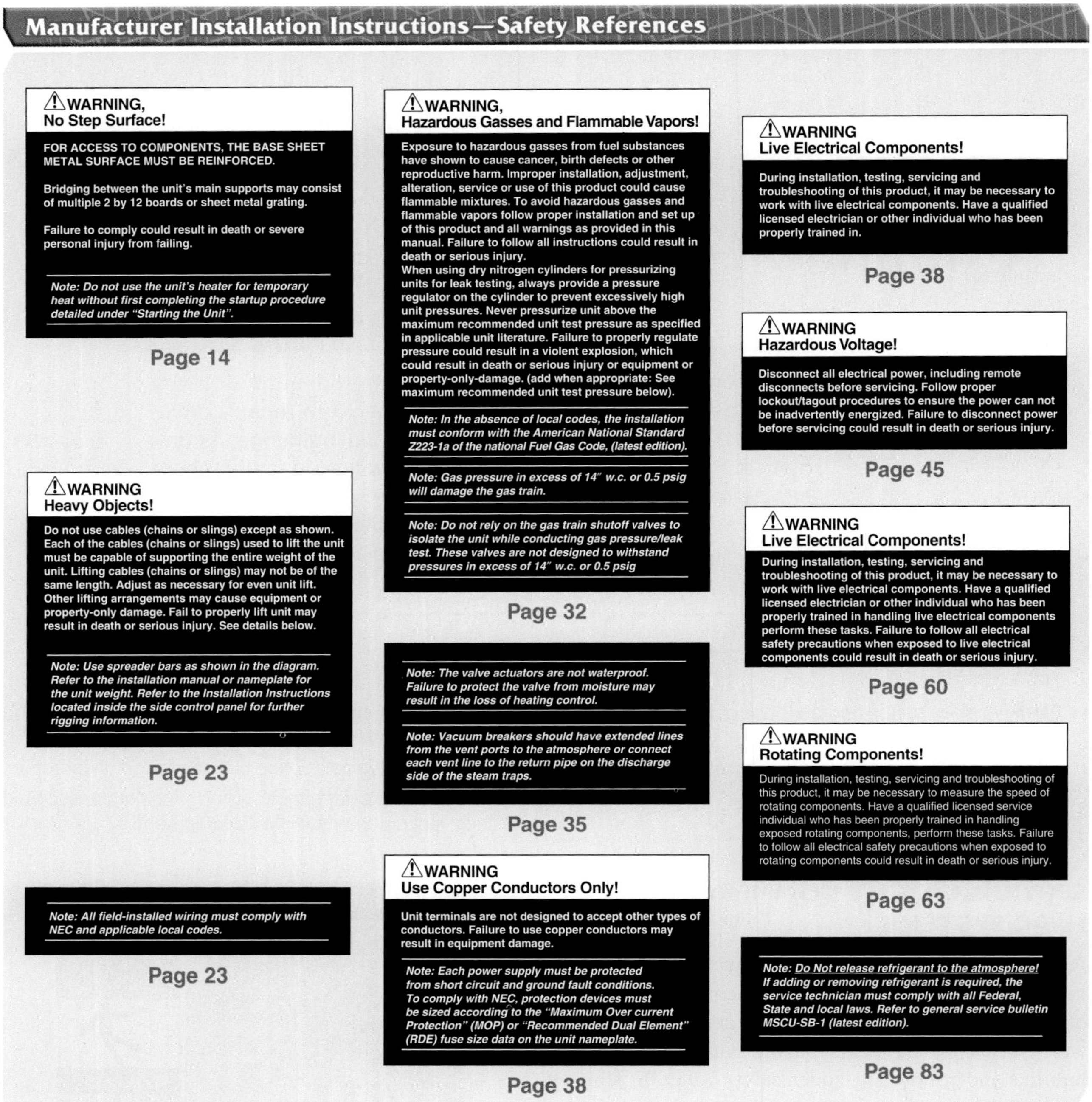

Manufacturer Installation Instructions—Safety References

⚠WARNING, No Step Surface!

FOR ACCESS TO COMPONENTS, THE BASE SHEET METAL SURFACE MUST BE REINFORCED.

Bridging between the unit's main supports may consist of multiple 2 by 12 boards or sheet metal grating.

Failure to comply could result in death or severe personal injury from failing.

Note: Do not use the unit's heater for temporary heat without first completing the startup procedure detailed under "Starting the Unit".

Page 14

⚠WARNING Heavy Objects!

Do not use cables (chains or slings) except as shown. Each of the cables (chains or slings) used to lift the unit must be capable of supporting the entire weight of the unit. Lifting cables (chains or slings) may not be of the same length. Adjust as necessary for even unit lift. Other lifting arrangements may cause equipment or property-only damage. Fail to properly lift unit may result in death or serious injury. See details below.

Note: Use spreader bars as shown in the diagram. Refer to the installation manual or nameplate for the unit weight. Refer to the Installation Instructions located inside the side control panel for further rigging information.

Page 23

Note: All field-installed wiring must comply with NEC and applicable local codes.

Page 23

⚠WARNING, Hazardous Gasses and Flammable Vapors!

Exposure to hazardous gasses from fuel substances have shown to cause cancer, birth defects or other reproductive harm. Improper installation, adjustment, alteration, service or use of this product could cause flammable mixtures. To avoid hazardous gasses and flammable vapors follow proper installation and set up of this product and all warnings as provided in this manual. Failure to follow all instructions could result in death or serious injury.
When using dry nitrogen cylinders for pressurizing units for leak testing, always provide a pressure regulator on the cylinder to prevent excessively high unit pressures. Never pressurize unit above the maximum recommended unit test pressure as specified in applicable unit literature. Failure to properly regulate pressure could result in a violent explosion, which could result in death or serious injury or equipment or property-only-damage. (add when appropriate: See maximum recommended unit test pressure below).

Note: In the absence of local codes, the installation must conform with the American National Standard Z223-1a of the national Fuel Gas Code, (latest edition).

Note: Gas pressure in excess of 14" w.c. or 0.5 psig will damage the gas train.

Note: Do not rely on the gas train shutoff valves to isolate the unit while conducting gas pressure/leak test. These valves are not designed to withstand pressures in excess of 14" w.c. or 0.5 psig

Page 32

Note: The valve actuators are not waterproof. Failure to protect the valve from moisture may result in the loss of heating control.

Note: Vacuum breakers should have extended lines from the vent ports to the atmosphere or connect each vent line to the return pipe on the discharge side of the steam traps.

Page 35

⚠WARNING Use Copper Conductors Only!

Unit terminals are not designed to accept other types of conductors. Failure to use copper conductors may result in equipment damage.

Note: Each power supply must be protected from short circuit and ground fault conditions. To comply with NEC, protection devices must be sized according to the "Maximum Over current Protection" (MOP) or "Recommended Dual Element" (RDE) fuse size data on the unit nameplate.

Page 38

⚠WARNING Live Electrical Components!

During installation, testing, servicing and troubleshooting of this product, it may be necessary to work with live electrical components. Have a qualified licensed electrician or other individual who has been properly trained in.

Page 38

⚠WARNING Hazardous Voltage!

Disconnect all electrical power, including remote disconnects before servicing. Follow proper lockout/tagout procedures to ensure the power can not be inadvertently energized. Failure to disconnect power before servicing could result in death or serious injury.

Page 45

⚠WARNING Live Electrical Components!

During installation, testing, servicing and troubleshooting of this product, it may be necessary to work with live electrical components. Have a qualified licensed electrician or other individual who has been properly trained in handling live electrical components perform these tasks. Failure to follow all electrical safety precautions when exposed to live electrical components could result in death or serious injury.

Page 60

⚠WARNING Rotating Components!

During installation, testing, servicing and troubleshooting of this product, it may be necessary to measure the speed of rotating components. Have a qualified licensed service individual who has been properly trained in handling exposed rotating components, perform these tasks. Failure to follow all electrical safety precautions when exposed to rotating components could result in death or serious injury.

Page 63

Note: Do Not release refrigerant to the atmosphere! If adding or removing refrigerant is required, the service technician must comply with all Federal, State and local laws. Refer to general service bulletin MSCU-SB-1 (latest edition).

Page 83

Figure 45-2. All safety references, notes, and safety warnings must be understood before beginning servicing or troubleshooting light commercial HVAC systems.

Safety issues arise when troubleshooting light commercial HVAC systems. These issues involve the electrical system, fuel system, and the mechanical systems. The electrical system is of concern during troubleshooting because 600 V, 460 V, or 230 V line voltages can be present at the various components of a rooftop unit along with 24 V control voltage. The fuel system is an issue because when even a small leak occurs in the large fuel piping or the system is mishandled, natural gas or fossil fuel can explode or at least burn quickly. The mechanical system is of concern because of large moving and rotating parts that are continuously starting and stopping. Technicians must be familiar with how to work safely with all of these systems when troubleshooting light commercial HVAC systems.

Courtesy Guidelines for Effective Light Commercial Service Calls

Before looking at light commercial system troubleshooting, some basic common courtesy rules must be practiced. The following are some basic guidelines for servicing HVAC customers. **See Figure 45-3.**

- Call ahead. Make sure the customer is called before driving to their location to ensure they are open and that access to the HVAC equipment has been arranged.
- Park where allowed. Do not block driveways or building entrances. Park in the loading dock area or contractor area.
- Dress and groom neatly. Smile when you greet the customer.
- Do not smoke in the customer's building.
- Do not track dirt or contaminants into the customer's building. Wear foot coverings when necessary.
- Be honest with the customer. Make sure that the customer is informed of everything that is discovered when troubleshooting.
- Remember it costs five times more in marketing dollars to obtain a new customer than it takes to keep an existing one.
- When you must leave to pick up parts or come back the next day, let the customer know when to expect you back. Do not leave panels or tools at the customer's place of business.
- Step outside to make personal calls.
- Replace the covers of all units checked, along with all of the proper fasteners.
- Maintain a positive attitude.

Park Where Allowed, Do Not Block Driveways or Building Entrances. Service technicians should avoid parking their service vehicles (trucks or vans) in the customer's driveway or near the building's entrance. Parked improperly, the service vehicle could make it difficult for personnel to enter or exit the building. Also, the service vehicle could be leaking engine oil, transmission fluid, or antifreeze, which can stain the customer's driveway or building entrance area.

Do Not Smoke in a Customer's Building. Tobacco smoke contains a deadly mixture of more than 7000 chemicals and compounds, hundreds of which are toxic, and at least 70 of which cause cancer. Exposure to smoke also decreases the benefits of chemotherapy and other cancer treatments. Also, tobacco odor is heavy and stays in an area for days or even weeks.

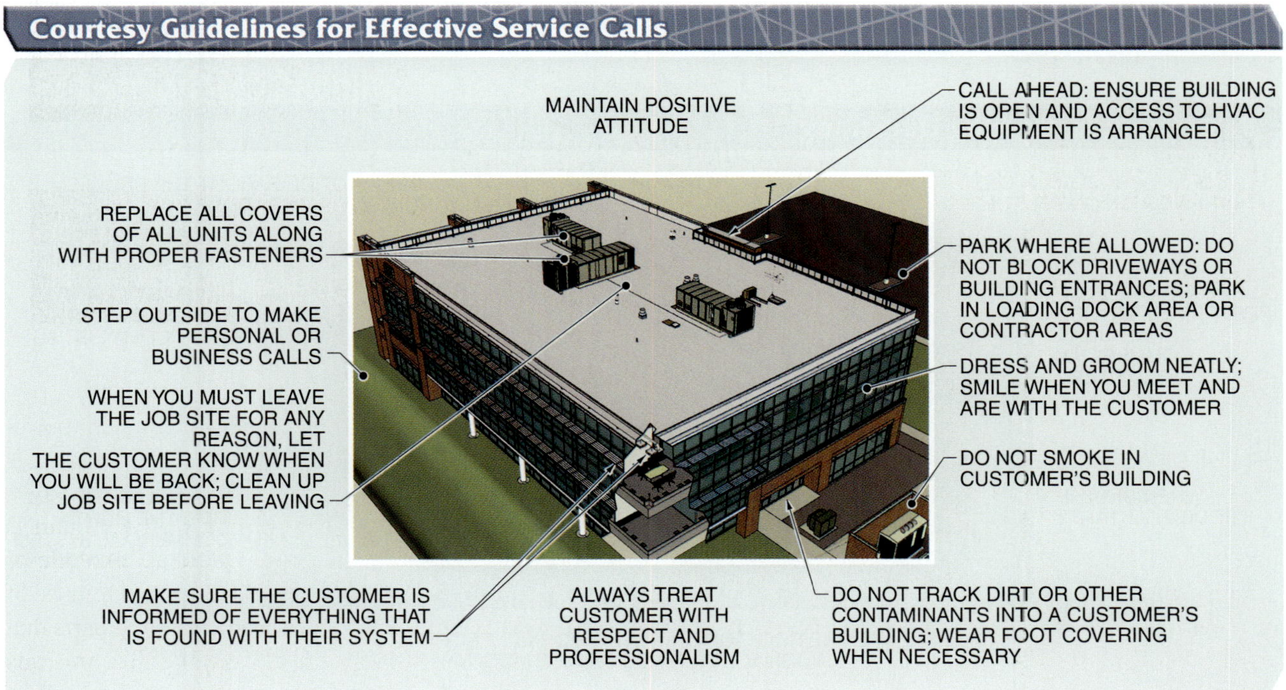

Figure 45-3. A technician's work must be thorough, such as placing all covers back on the HVAC unit with all fasteners that came off and verifying that all doors are closed and secured.

Replace Covers and Fasteners on All Units. The technician should always replace all covers on a unit with all the fasteners that came off the cover. Panels can vibrate, making noise when all fasteners are not replaced. Most light commercial units have doors instead of covers. The doors are generally locked or fastened closed. After being opened, all doors should be locked and fastened closed again.

Interpersonal Skills

Interpersonal skills are character traits that allow an individual to communicate effectively with others in a variety of situations. A service technician must use these skills when dealing with a customer. Light commercial equipment is costly and failures are frustrating at times for the business owner. This frustration can be directed at the service technician.

When the customer is upset, the service technician must stay calm and listen carefully to the customer. The customer should find the technician pleasant. The customer needs to know that the best work possible is being performed.

The service technician must work with the information provided by the customer and any acquired troubleshooting information to find a solution. Clear communication of the facts helps eliminate misunderstandings and can reduce the troubleshooting time for the technician.

Advancement in the HVAC Field

A successful light commercial system service technician practices lifelong learning. When a skill has been mastered, higher levels of skill mastery are required for newly designed light commercial equipment. The ability to adapt as the HVAC industry changes is crucial for success. Reading trade publications, attending classes and seminars, and participating in professional organizations are activities that provide valuable information on current topics and trends. New skills are required to remain current with advancing heating and cooling technologies and to grow in the HVAC field.

Basic Troubleshooting Questions

When troubleshooting any light commercial system, it is a good idea to ask the building owner, manager, or maintenance personnel some basic questions. **See Figure 45-4.**

- What has been noticed about the problem? Are there any noises or smells? Are there any unexpected vibrations occurring?
- When did the problem start?
- When was the last time the light commercial system worked properly?
- Have the systems (heating, cooling, and ventilation) been following normal sequences of operations?
- Have any changes been made (reconfiguration of walls and/or zones) to the light commercial system or building?
- Has anyone else performed service on this light commercial system?
- Is the problem in one area or on one floor?

Basic Troubleshooting Questions

Figure 45-4. The technician should begin the troubleshooting of a light commercial system by asking the owner specific questions about the system's operation when it stopped operating correctly.

TECH TIP

Most technicians carry a mobile phone. When confronted with an unknown problem, the phone can be used on location to call the service manager, tech support, or other HVAC technicians to rapidly and accurately diagnose and fix a problem.

Unsuccessful Repairs. When a repair does not fix the problem with the light commercial system, the technician must reevaluate the symptoms, causes, and solutions listed on the troubleshooting charts. In some situations, the entire process of gathering information about the system must begin again.

Successful Repairs. When a repair is successful and the light commercial system operation has been restored, the troubleshooting process is recorded on a troubleshooting report or equipment maintenance log. Everything is recorded in a maintenance log including symptoms, causes, tests conducted, repairs made, and the condition of the light commercial system after the repair. **See Figure 45-5.** The spare parts used in the repair are recorded, and replacement parts are ordered as needed. Suggestions on modifying procedures that could prevent similar problems are also recorded.

Installation Manuals

An *installation manual* is a general guide used by qualified technicians to install residential and light commercial HVAC equipment. Installation manuals do not supersede any national or local codes. When installation manuals do not arrive with the delivered equipment, the manuals should be obtained from the manufacturer as soon as possible. The installation of the light commercial (rooftop) units and any associated system equipment described in the manual must comply with all state and local codes as well as national codes such as the NEC® or CEC™. Failure to comply with manufacturer procedures can void warranties and reduce the energy efficiency of the equipment. **See Figure 45-6.**

Troubleshooting Reports and Maintenance Logs

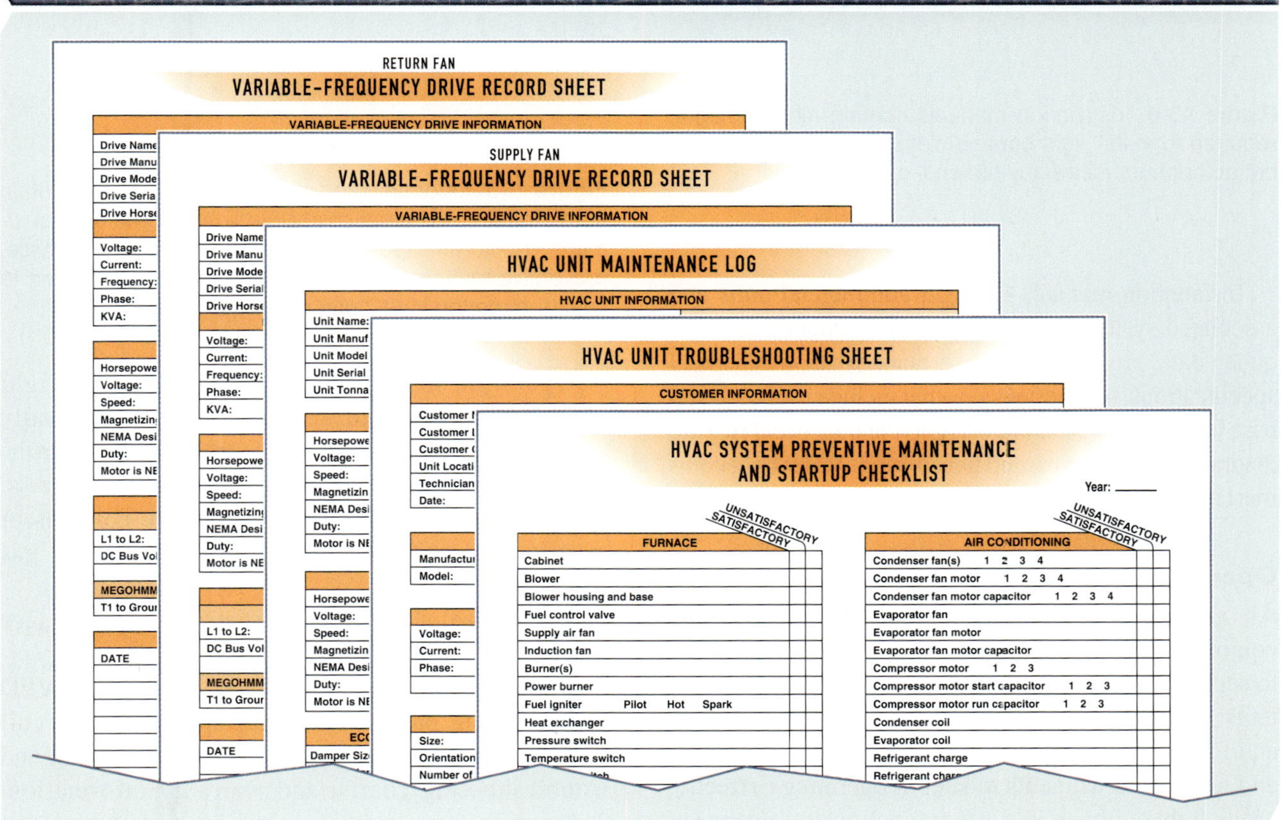

Figure 45-5. Light commercial systems have more components (sensors, valves, actuators, and controls) than forced-air systems, requiring that more information be kept for future use.

Installation Manuals

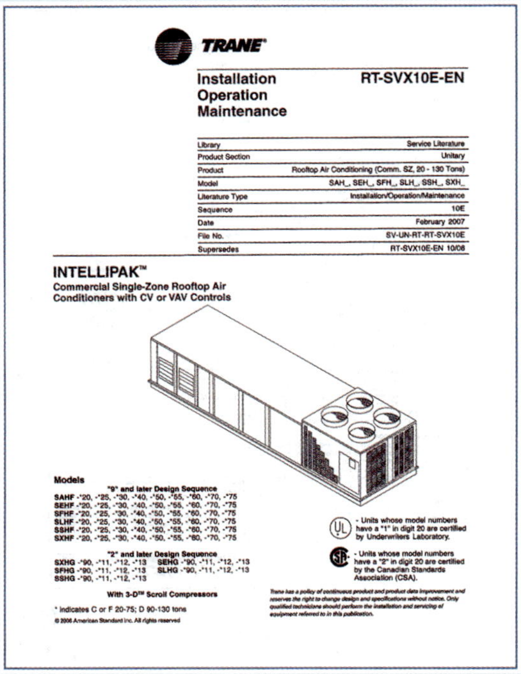

Trane

Figure 45-6. Installation manuals contain all information required to install light commercial units on roofs or pads but do not supersede any national or local codes.

Operation and Maintenance Manuals

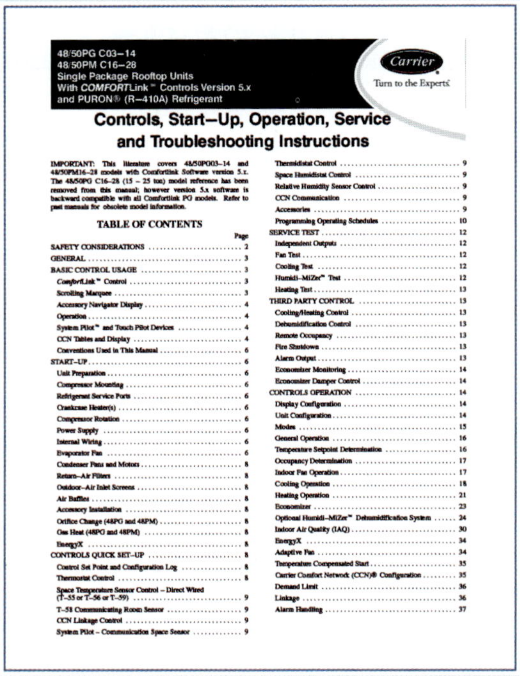

Carrier Corporation

Figure 45-7. Operation and maintenance manuals contain all information on operating specific light commercial equipment in different applications, plus have all the service, preventive maintenance, and maintenance information to keep the rooftop HVAC unit running for years.

Installation manuals for light commercial units and associated system equipment typically contain sections on tag data, product data, performance data, mechanical specifications, unit dimensions that include roof curb and base pan details, fan curve, weight (clearance and rigging diagrams), accessories, and field wiring. Many light commercial units also have installation options to consider.

Operation and Maintenance Manuals

An *operation and maintenance (service) manual* is an equipment-specific guide used by qualified technicians to service residential and light commercial HVAC systems. Operating a rooftop unit and its associated system equipment involves starting and stopping the equipment and making adjustments to keep it operating correctly. Many light commercial system functions for operation are automated, but some functions cannot be completely automated. **See Figure 45-7.**

Operation and maintenance manuals for light commercial units and system equipment typically contain information on the gas train, VAV controls, 24 V and analog input and output controls, fan controls, measuring system airflow, economizers, compressor startup, low-ambient-temperature dampers, and gas furnace startup.

The manual also includes service and maintenance procedures, fan belt adjustment procedures and specifications, compressor replacement procedures, VFD programming parameters, filter specifications, coil and unit maintenance procedures, system testing and troubleshooting charts, and warranty information. Today, the manuals for most major manufacturers of HVAC equipment can be downloaded from the web to a smartphone, tablet, or laptop.

Troubleshooting Manuals

System testing and troubleshooting sections typically include diagnostic menus. Diagnostic menus are used to view system failures within the unit. Typically there are two lists in the diagnostic menu of a light commercial rooftop unit: the active list and the diagnostic event log. At the end of the system testing and troubleshooting section, there is typically a system testing and troubleshooting chart. **See Figure 45-8.**

Troubleshooting Manuals

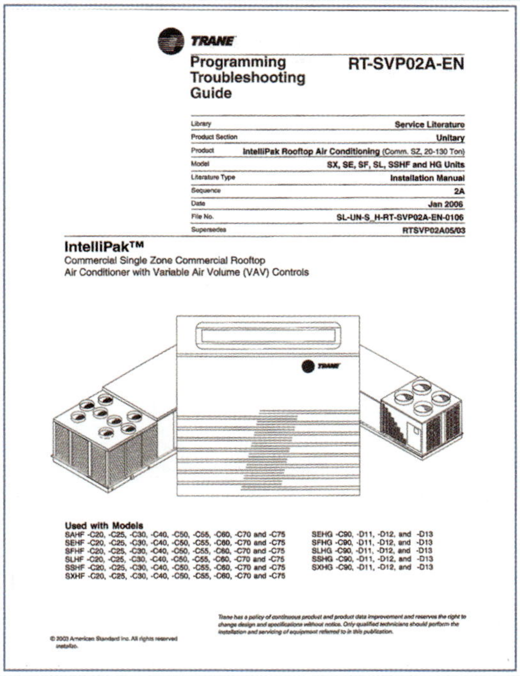

TRANE

Programming Troubleshooting Guide

RT-SVP02A-EN

Library	Service Literature
Product Section	Unitary
Product	IntelliPak Rooftop Air Conditioning (Comm. SZ, 20-130 Ton)
Model	SX, SE, SF, SL, SSHF and HG Units
Literature Type	Installation Manual
Sequence	2A
Date	Jan 2006
File No.	SL-UN-S_H-RT-SVP02A-EN-0106
Supersedes	RTSVP02A05/03

IntelliPak™

Commercial Single Zone Commercial Rooftop Air Conditioner with Variable Air Volume (VAV) Controls

Used with Models

SAHF -C20, -C25, -C30, -C40, -C50, -C55, -C60, -C70 and -C75
SEHF -C20, -C25, -C30, -C40, -C50, -C55, -C60, -C70 and -C75
SFHF -C20, -C25, -C30, -C40, -C50, -C55, -C60, -C70 and -C75
SLHF -C20, -C25, -C30, -C40, -C50, -C55, -C60, -C70 and -C75
SSHF -C20, -C25, -C30, -C40, -C50, -C55, -C60, -C70 and -C75
SXHF -C20, -C25, -C30, -C40, -C50, -C55, -C60, -C70 and -C75

SEHG -C90, -D11, -D12, and -D13
SFHG -C90, -D11, -D12, and -D13
SLHG -C90, -D11, -D12, and -D13
SSHG -C90, -D11, -D12, and -D13
SXHG -C90, -D11, -D12, and -D13

Trane has a policy of continuous product and product data improvement and reserves the right to change design and specifications without notice. Only qualified technicians should perform the installation and servicing of equipment referred to in this publication.

Trane

Figure 45-8. Troubleshooting manuals always cover the basics, but each manufacturer knows its equipment best and modifies its troubleshooting manuals accordingly.

Active List. An active list is a diagnostic tool used to view all active diagnostics (faults) and for clearing manually resettable diagnostics. Diagnostics cannot be automatically reset by the human interface because the condition that caused the diagnostic must be corrected for the diagnostic to be cleared from the active list.

Diagnostic Event Log. For most manufacturers, the diagnostic event log screens are found after paging through the active list or by pressing the diagnostic key. Diagnostic events are stacked in the log inversely, with the first screen being the most recent diagnostic reported. Any diagnostics in the event log that have not been viewed will be highlighted or will read "Not Viewed."

System Testing and Troubleshooting Chart. The chart contains a list of all the diagnostics (faults) that the unit's display will show in the active list and diagnostic event log. The chart explains what a specific diagnostic is, what caused the diagnostic to occur, what the unit's reaction typically is, and what is required to clear the diagnostic. **See Figure 45-9.**

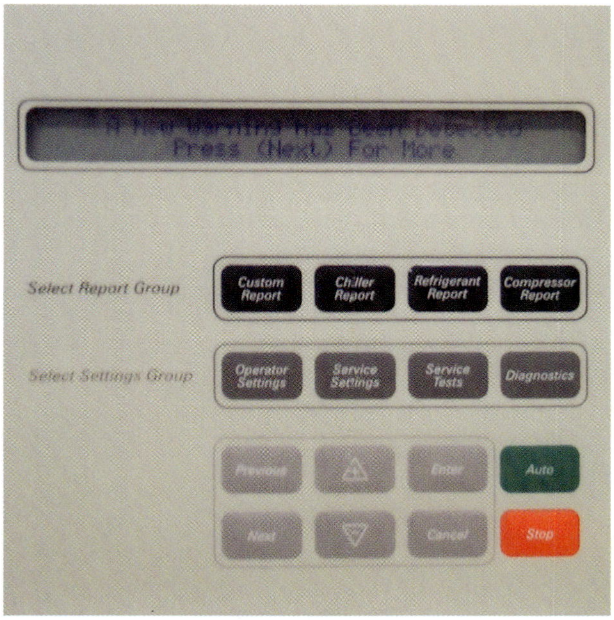

HVAC units often come from the factory equipped with control systems that can provide a record of system faults.

Rooftop Unit Sequence of Operations

Light commercial units are designed with multiple safety features. These safety features prevent the fire from escaping the firebox, prevent the buildup of fuel (natural gas, fuel oil, or propane) on the burners, which could cause an explosion, and prevent flue gases from escaping into the building's supply airflow. The sequence of operation is a valuable tool in diagnosing light commercial rooftop unit problems. **See Figure 45-10.**

System Testing and Troubleshooting Charts

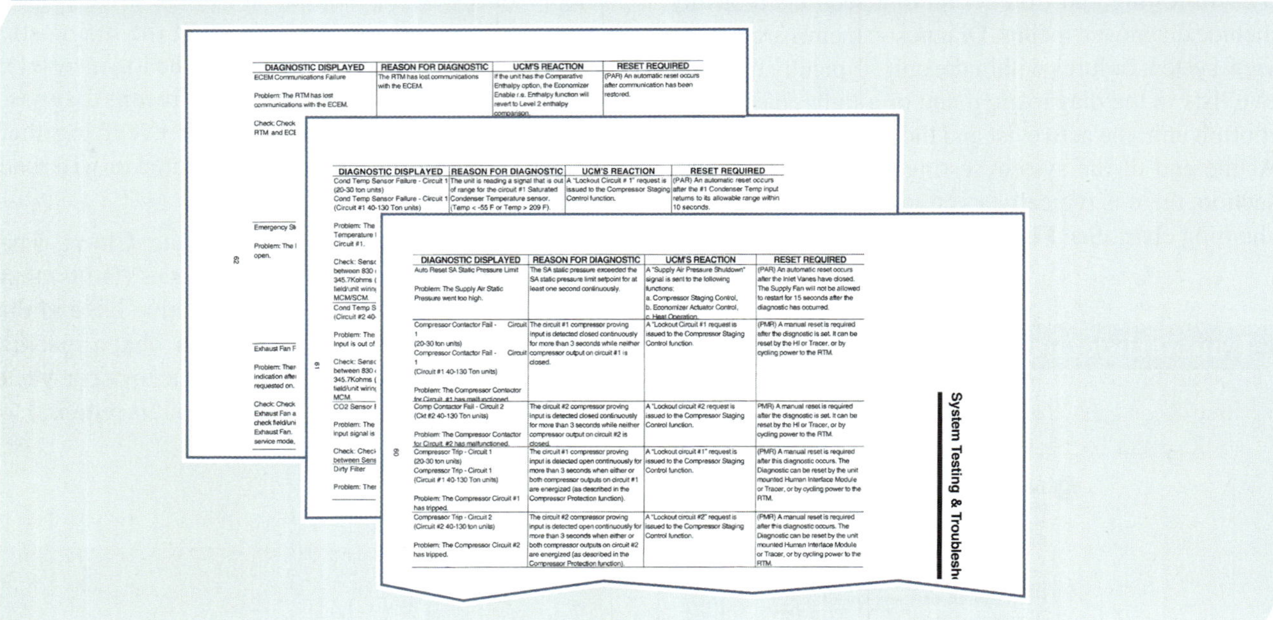

Figure 45-9. Troubleshooting charts list the most common faults that a light commercial unit encounters, how to test the fault for specific component failure, and the standard corrective actions needed to eliminate the fault.

Rooftop Unit Sequence of Operations

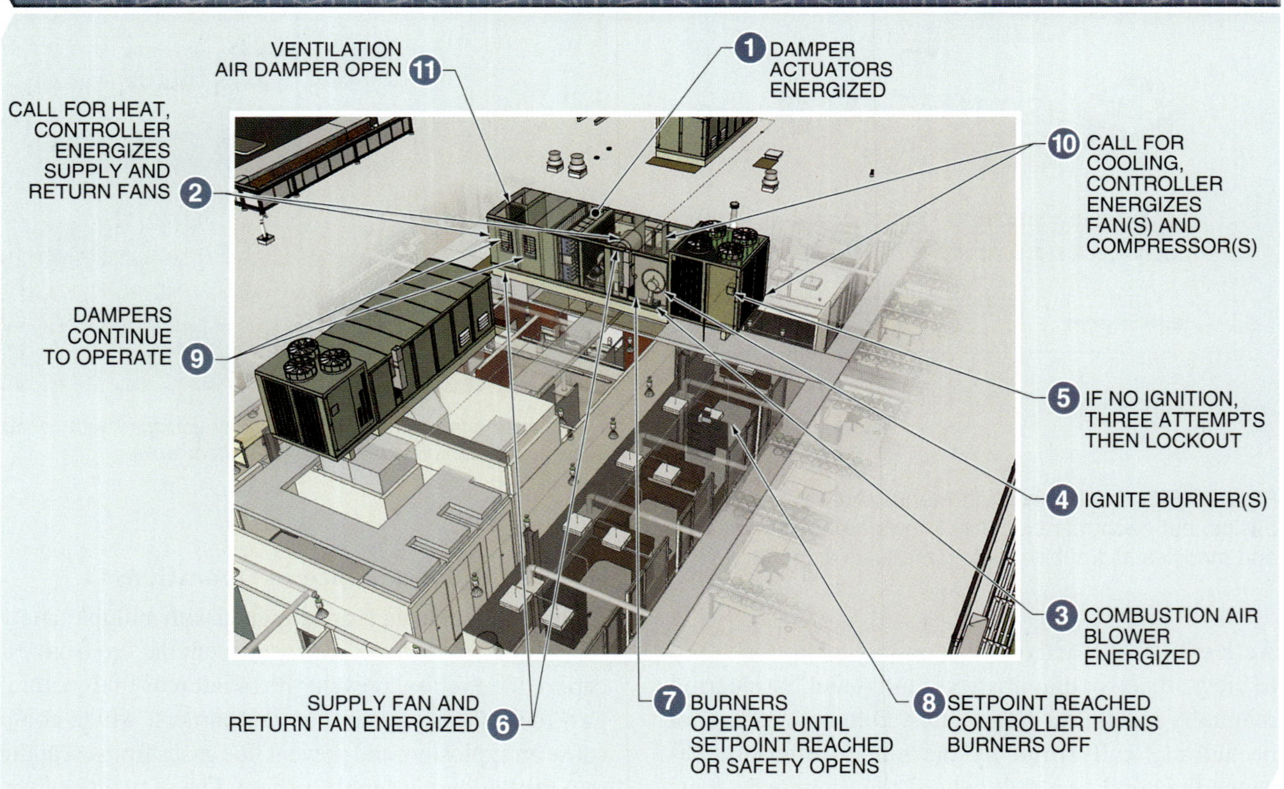

Figure 45-10. A sequence of operation indicates when safety features are playing a role in the operation of a light commercial unit and contains valuable information for troubleshooting the unit and system.

Rooftop units also control the cooling of building spaces and the introduction of the proper amount of ventilation air into various zones of the building according to the amount of people in those zones. Light commercial systems have many different possible sequences of operation. The following is a basic example of a sequence of operation for a light commercial unit (some sequences can be 120 to 150 steps).

1. Upon a call for heat or cooling, the light commercial unit controller turns on the appropriate damper actuators. These dampers can be ventilation dampers, exhaust dampers, economizer dampers, duct dampers, or zone dampers and allow the system to operate as requested by the thermostat and other building sensors such as CO_2 sensors.
2. When a call for heat is received and after a built-in time delay (typically 30 sec), the controller will send the proper VAC to the supply fan and return fan (if present) and to the furnace if the following conditions are met:
   a. The gas pressure safety switch must be closed.
   b. The pressure switch attached to the flue must be proved during fan operation. Once proved, the furnace will then provide 24 VAC to its enable relay for burner ignition.
   c. In general, any other limit switches, such as a heat exchanger high-temperature-limit switch, must be closed before the board will energize the burner.
3. If present, the controller will energize the combustion air blower and then start a 10 sec to 30 sec prepurge cycle.
4. Once the prepurge cycle is complete and all limits are closed, the controller will attempt to start the burners by beginning a 5 sec to 10 sec trial for ignition. Starting the burner flame is accomplished by sending a voltage to the hot surface igniter and then opening the gas valve. Some furnaces have standing pilot lights.
5. If the controller does not detect a flame by the end of the trial for ignition, the controller closes the gas valve and performs a 10 sec to 30 sec postpurge, then another prepurge, and then it will try to light the burners again. Most light commercial units will try to ignite the burners a set number of times, typically three times, and then go into lockout. The lockout may need to be manually reset.
6. Once operating normally, the supply fan and return fan are energized.
7. The burners will then operate until the setpoint temperature is reached or a safety limit opens.
8. When the system call for heat is satisfied, the controller will turn the burners off. The supply fan will operate on a postpurge cycle of 10 sec to 30 sec. The return fan is on a separate cycle according to system requirements (ventilation air can change parameters).
9. The dampers will typically continue to operate for a specific time delay of 30 sec to 120 sec, then turn off.
10. When a call for cooling is received and after a built-in time delay (typically 30 sec), the controller will send the proper VAC signal to the fan and to the compressors if the proper switches are closed and conditions are met.
11. When only ventilation air is called for, many different scenarios exist because of variables such as outside temperature, outside humidity levels, and whether the unit is checking for contaminants in the outside air.

TROUBLESHOOTING LIGHT COMMERCIAL HVAC SYSTEMS

Troubleshooting light commercial rooftop units requires technical expertise with the unit and system plus logical and creative thought processes. Manufacturer troubleshooting matrixes should be consulted and used due to the complexity of rooftop single-zone units and controls. Many problems found with light commercial systems are not with the rooftop units themselves, but with separate air-handling terminal units or vertical air handlers in the building. The sensors that the rooftop unit uses and the damper actuators that the unit controls can also be sources of system malfunctions.

Light Commercial Rooftop Single-Zone Units

A typical rooftop single-zone unit consists of a supply fan, one or more stages of cooling, one or more stages of heating, controls, ductwork, and utility service. For cooling, most rooftop units utilize a standard DX compression cycle with one or more stages, with a few systems using chilled water coils. In addition, a common retrofit add-on is an economizer damper and control package that will allow outdoor air to cool the space under the appropriate conditions. **See Figure 45-11.**

Light Commercial Rooftop Single-Zone Units

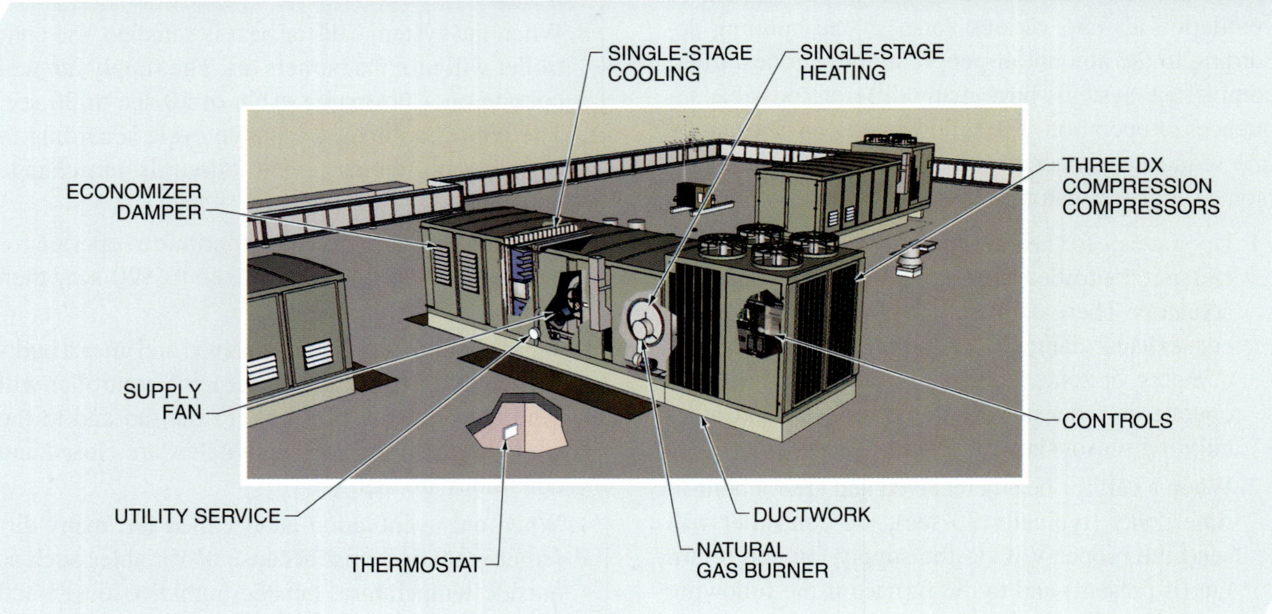

Figure 45-11. Light commercial single-zone HVAC units heat, cool, and provide filtered ventilation air. Many also include economizers and back-up electrical heat.

For the heating section, most single-zone units use natural gas burners, with electric heating elements or hot water coils as possible options. Again, multiple stages of heating are common. For control, a light commercial thermostat is typically wired to a terminal strip. It will cycle the fan, heating, and cooling as needed to maintain the proper temperature in the building space. Today there are an increasing number of multifunction electronic thermostats available with many different possible programming capabilities.

Rooftop Unit Troubleshooting Scenario #1—Power Turned Off

A customer calls in with a "too hot" complaint. The outdoor air temperature is very warm, and until the day of the call, the unit was operating properly. Upon arrival, the technician proceeds to the roof and finds that rooftop unit #3 (RTU #3), used for the 3rd floor of the building, is completely off. A quick power check finds that there is no power present. The power is traced back, and it is found that the breakers have been turned off in the main electrical room. Another contractor inadvertently killed the power while looking for the breaker for an outside lighting circuit. **See Figure 45-12.**

Service Work Performed: The power to RTU #3 is turned on at the circuit breaker, and the unit starts and runs normally.

Rooftop Unit Troubleshooting Scenario #2—Supply Fan Failure

On a very hot summer day, the customer calls in a "too hot" complaint. The temperature inside the building is 87°F. Upon arrival, the technician inspects the unit on the roof and finds that the supply fan is not running. A quick check shows that the breaker is tripped at the unit. The supply fan motor is checked and found to have a grounded winding. **See Figure 45-13.**

Service Work Performed: The supply fan motor is replaced and the unit operates normally.

TECH TIP

Heating and cooling systems in light commercial buildings are more sophisticated than ever. Many repairs require special diagnostic equipment. When a problem with refrigerant levels exists, the technician needs special equipment as well as a license to handle the refrigerant. When troubleshooting HVAC systems, the technician must remember to turn off the power to the system.

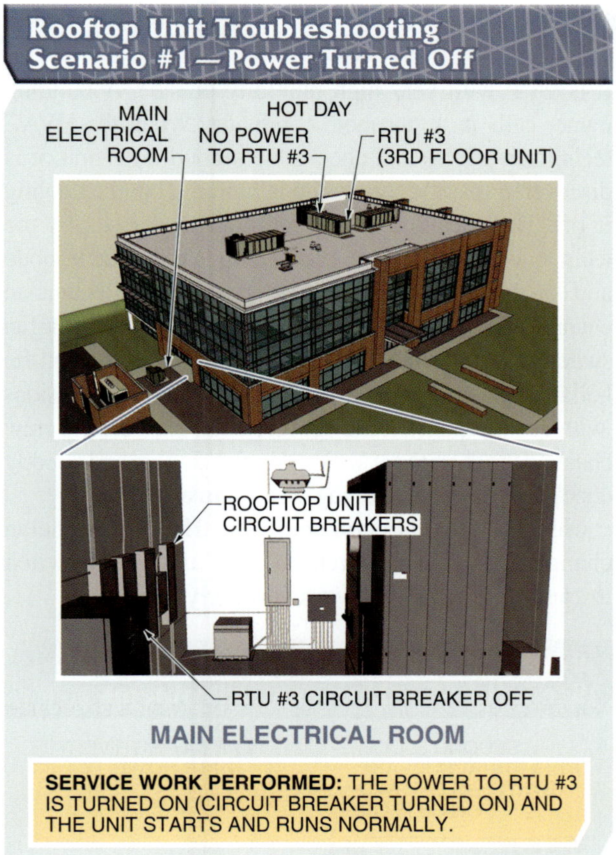

Figure 45-12. When troubleshooting a light commercial unit that is not running, the technician should begin by testing for incoming power.

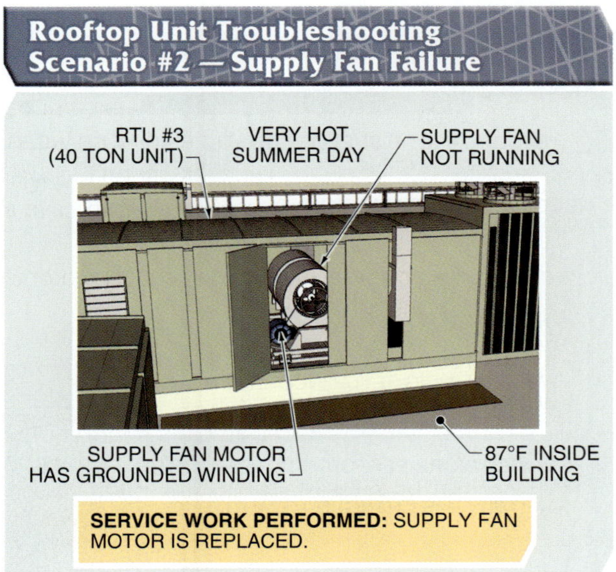

Figure 45-13. When testing any HVAC motor for a short, having a meter with a high-resistance test capability or having a megohmmeter is useful.

Rooftop Unit Troubleshooting Scenario #3—Economizer Controls Failed

A building with a number of rooftop units calls in a "too hot" complaint for floor #2. Upon arrival the technician notices that the #2 rooftop unit has been equipped with an economizer package and controls. In spite of the hot ambient conditions, the outdoor-air dampers are wide open on RTU #2. Too much hot humid air is entering the second floor of the facility. The technician checks the economizer controls and finds the controls are not adjusted properly. **See Figure 45-14.**

Service Work Performed: After adjusting the economizer controls, the damper closes to minimum position, and the temperature of the supply air to the second floor drops.

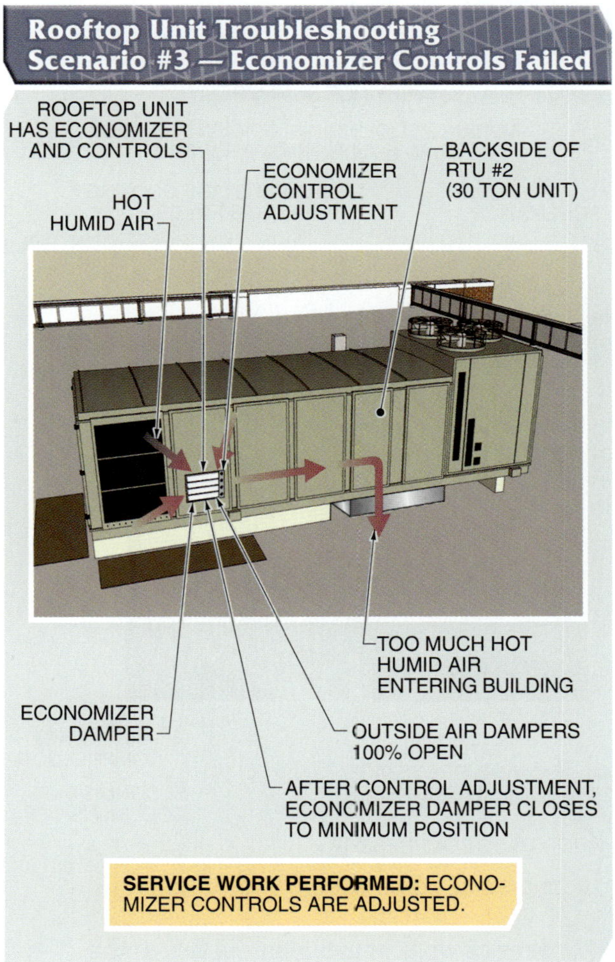

Figure 45-14. Economizer controls work with a damper that allows unconditioned air (warm or cool) into a building. With economizers, settings can change and outdoor-air temperature sensors can fail.

Rooftop Unit Troubleshooting Scenario #4—Staging Not Operating

On a very hot summer day, an office building manager calls in a "too hot" complaint. Upon arrival the technician finds that the RTU #3 supply fan and first stage of cooling are running. The second stage of cooling is off. The amount of working cooling is insufficient to cool the third-floor space properly. The technician checks and finds that the relay to energize the second-stage cooling compressor motor starter has failed. **See Figure 45-15.**

Service Work Performed: After replacing the second-stage cooling relay, the second-stage cooling compressor energizes and the temperature in the third-floor space begins to drop.

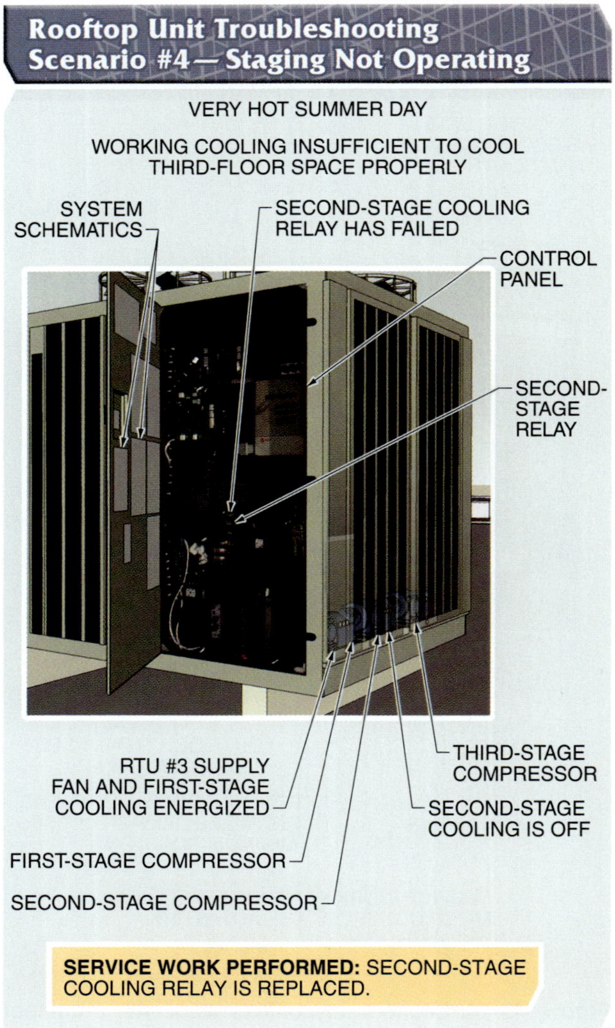

Figure 45-15. Any time staging is involved, relays must be checked for proper operation.

Rooftop Unit Troubleshooting Scenario #5—Thermostat Problems

It is 95°F with very high humidity outside. A building owner calls in a "too hot" complaint. When the HVAC technician arrives and checks the second-floor unit, it is discovered that the supply fan is running, but the cooling is off. The check of the second-floor thermostat by the technician shows that the unit switch is in the OFF mode and that the fan switch is in the ON position. The technician changes the subbase switch (unit switch) to "cool" and the fan switch to "auto". Immediately, the cooling energizes and the building space temperature begins to drop. From discussions with the occupants, it is discovered that the cleaning crew changed the thermostat setting the night before and nobody checked it the next morning. **See Figure 45-16.**

Service Work Performed: The HVAC technician changes the subbase switch to "cool" and the fan switch to "auto". The cooling works properly.

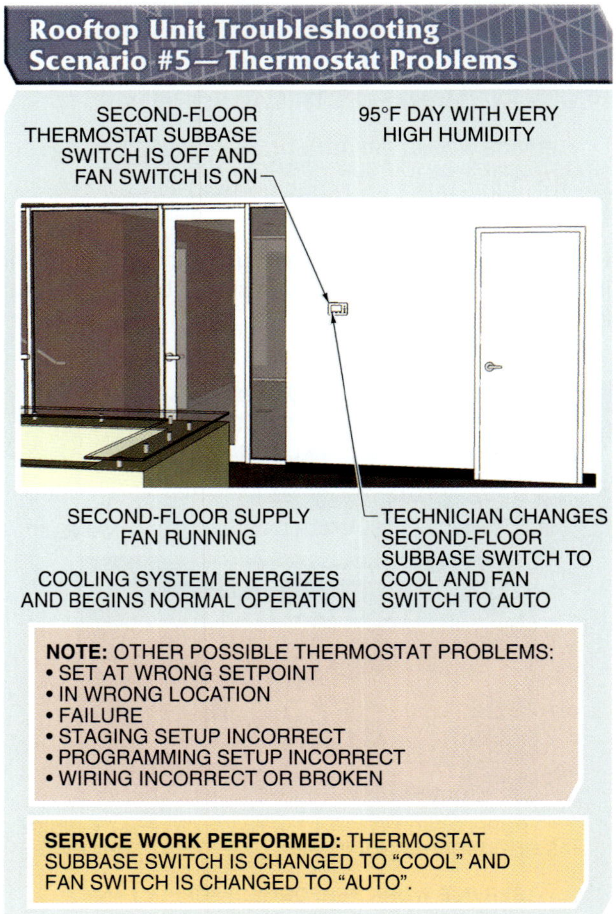

Figure 45-16. When an HVAC system is not operating but the fan is running, the technician should check the simplest possibility first, such as whether the thermostat is set for heating or cooling.

Rooftop Unit Troubleshooting Scenario #6—Dirty Coils

The weather has been unusually hot. A customer that owns a three-story building calls in a complaint of not enough cooling on the first floor. The technician arrives to find the supply fan in the rooftop unit for the first floor running. A quick glance of the unit's filter wall shows the filters are totally plugged.

The technician also inspects the evaporator and condenser coils and finds that they are caked with dirt. This has been causing the unit to shut down on high head pressure. A discussion with the customer reveals that, to their knowledge, the unit has not been serviced during the three years that they have occupied the first-floor space. The unit is shut down, locked and tagged out, and then both coils are thoroughly cleaned. The filters are replaced and the unit is restarted. The fan and compressor come on immediately, and the first-floor space temperature begins to drop. **See Figure 45-17.**

Service Work Performed: The condenser and evaporator coils are thoroughly cleaned. The filters are replaced, and the unit is restarted.

Rooftop Unit Troubleshooting Scenario #7—Bad Economizer Damper

It is the first cool day in the fall of the year. An office building customer calls in a "too cold" complaint. It seems that even though RTU #3 is in the heating mode, cold air is blowing out of the registers.

The technician checks the thermostat setting and sees that the temperature is close to the setpoint. A check of the rooftop unit reveals that the economizer outdoor-air damper is stuck wide open. The heat is cycling on and off to compensate for the cold air coming through the unit. When the heat is off, the cold outdoor air is being dumped directly onto the third-floor occupants. The economizer damper is found to be jammed open by sticking linkages. After repairs, the economizer damper is placed back in service and is working properly. **See Figure 45-18.**

Service Work Performed: The economizer damper is freed up and lubricated, the controls are checked, and the economizer is operating correctly.

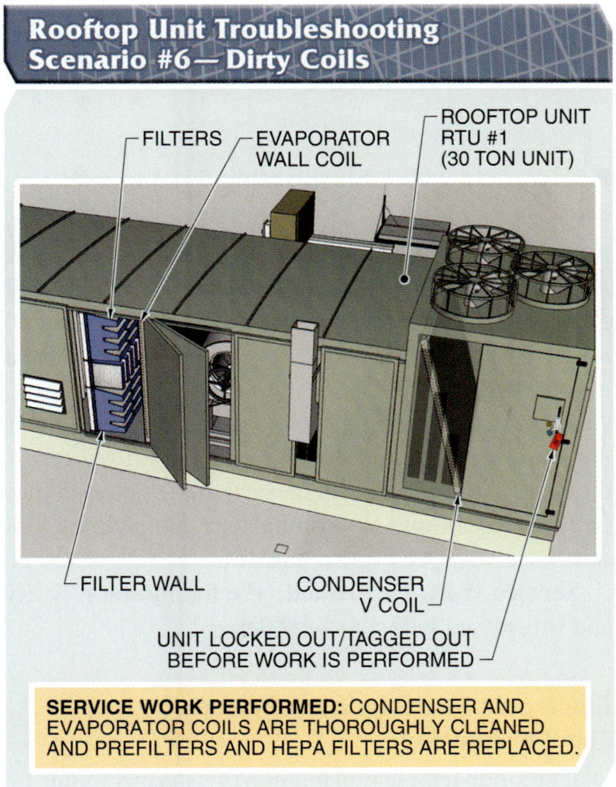

Figure 45-17. When a high head-pressure fault indication is found with a light commercial unit, the filters, condenser coil, and evaporator coil should be checked for excessive dirt.

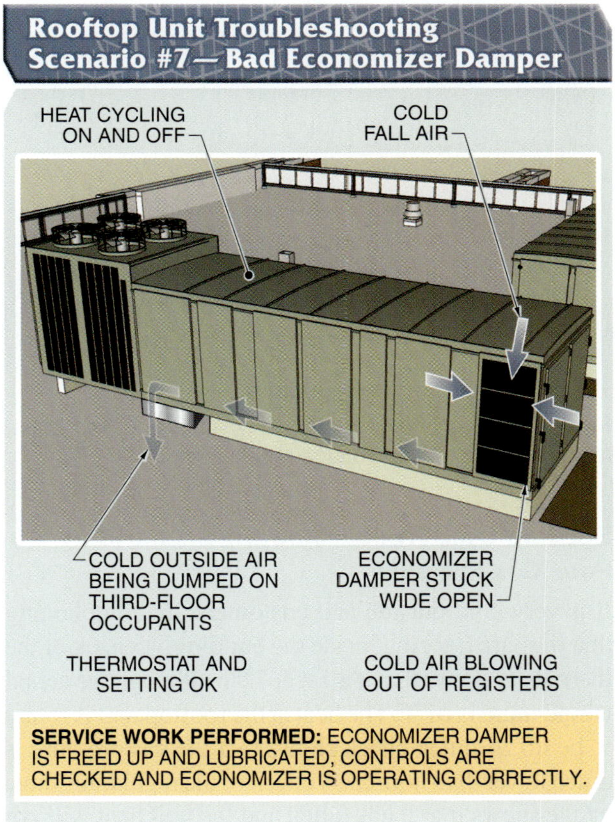

Figure 45-18. Intricate components in HVAC systems such as damper linkages must be kept clean and lubricated to allow for maximum operating efficiency.

Rooftop Unit Troubleshooting Scenario #8—Bad Gas Igniter

A technician is dispatched on a "too cold" complaint to a multistory building. Both the rooftop unit fan and heat are off. The thermostat is set at 70°F, but the actual temperature is 60°F. The thermostat is calling for heat. The technician checks the unit and finds that the heat will not come on. The igniter is checked and found to be defective. **See Figure 45-19.**

Service Work Performed: The igniter is replaced and the unit now works properly.

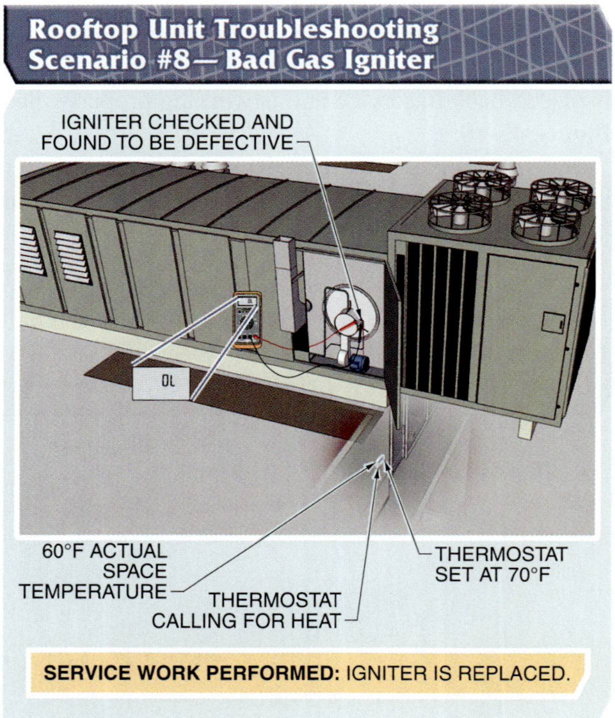

Figure 45-19. Igniter operation is typically checked with a multimeter set to take voltage measurements.

Rooftop Unit Troubleshooting Scenario #9—Poor Gas Pressure

It is very cold outside and customers are complaining that they are freezing inside the building. A check of the thermostat shows that it is set at 72°F although the actual temperature is 61°F. The heat at the rooftop unit is found to be off. A quick check shows that the gas pressure is much lower than normal. A check of the regulating valve gauge shows that it has failed and the unit heat will not come on. **See Figure 45-20.**

Service Work Performed: The gas pressure regulator is replaced and the unit runs normally.

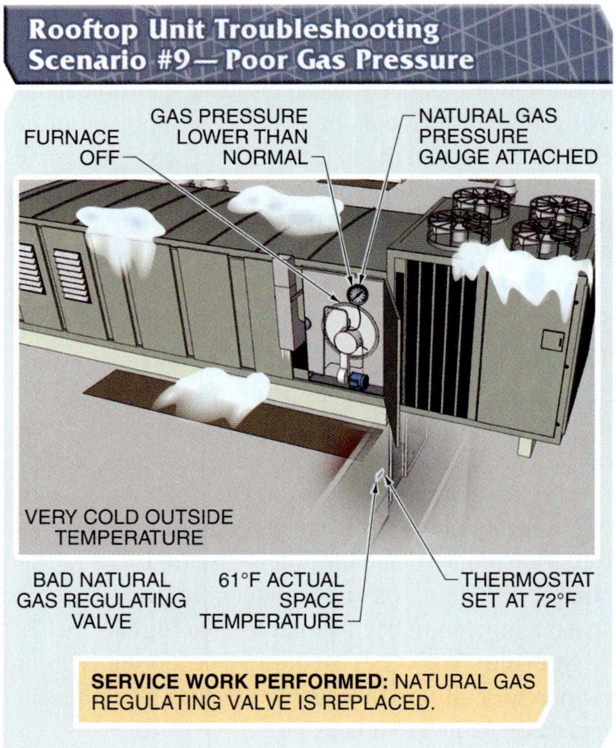

Figure 45-20. When checking the pressure of a natural gas line to a light commercial unit, the technician must ensure that proper ventilation is present in the cabinet or area of testing.

Rooftop Unit Troubleshooting Scenario #10—Bad Thermostat

A customer complains of the temperature being too cold. A check of the thermostat shows that it is set at 70°F although the space temperature is 62°F. The unit fan will run, but the heat will not cycle. The thermostat is forced into the heating mode, but the heat will not energize. A DMM is used to diagnose the problem, and it shows that the heating circuit is open. A jumper is used across the R and W terminals, and the heat immediately comes on. Once the R and W terminals are jumped, the unit runs normally. **See Figure 45-21.**

Service Work Performed: The thermostat is faulty and is replaced with a new model.

Electronic Spark Ignition System Troubleshooting

Older rooftop units use an ignition transformer to step up the voltage to an electrode. The higher voltage is used to create a spark to ground. The spark ignites the gas that is introduced through the pilot valve. The spark igniter is connected through a thick cable to ground. Once the flame is proven by a flame rod, the main gas valve is allowed to stay open. The spark electrode may become pitted and fail over time. In many cases the spark is visible, so ignition is easily verified. Typically, the electrode is simply replaced. The transformer may also fail, and this can be easily checked with a DMM. However, the technician must remember that the secondary voltages may be very high. Another possible problem is that the ground connection has become loose. Again, this can be easily checked with a DMM.

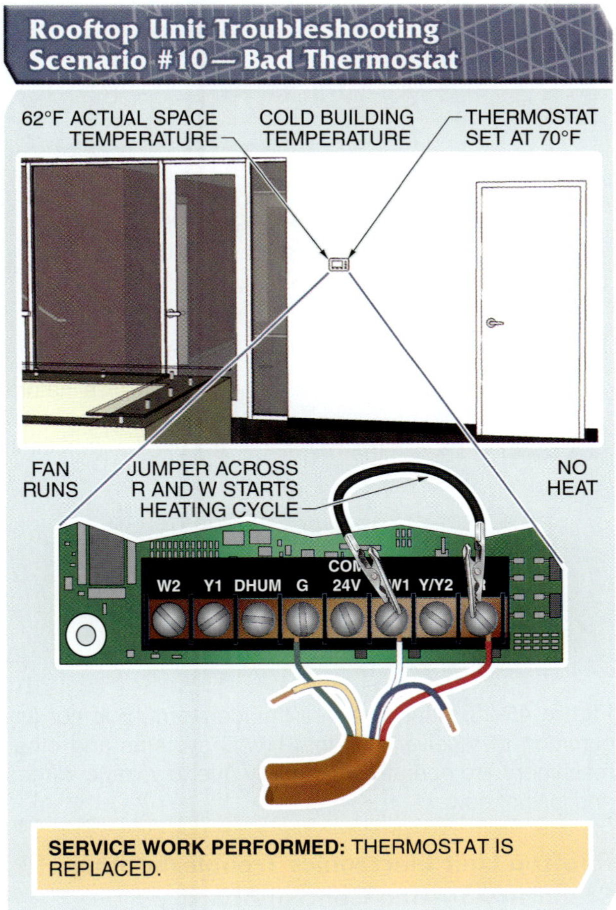

Figure 45-21. When proper precautions are taken, jumping out a component such as a thermostat may be the quickest way to determine if the component is faulty.

Hot Surface Ignition System Troubleshooting

The type of ignition system typically used today is a hot surface ignition system. These consist of a silicon carbide or silicon nitride element or filament. These elements will glow when an electrical current is passed through them. They are rugged and long-lasting when handled properly. However, after a number of years, they will crack and fail. Also, when hot surface igniters are handled with bare hands, oil from the fingers will cause them to fail prematurely. If doubt exists regarding the manufacturer of a particular igniter, some universal repair kits are available. Otherwise, the name or number of the part should be obtained and a local supply house contacted for a cross-reference.

Rooftop VAV Unit Operation

The operation of a rooftop VAV unit typically produces a constant discharge air temperature of 55°F. VAV units may have anywhere from two to four or even six stages of cooling or use chilled water. For heating, they may use multiple stages of gas or electric heat or have a hot water coil. A VAV rooftop unit will use a static pressure sensor to control the discharge air at 1″ WC in the duct. This is accomplished by using a variable-frequency drive (VFD) on the supply fan. The air from the unit is directed to multiple VAV terminal boxes or vertical air handlers in the building. When the zones need heat, the VAV boxes will open to a minimum cfm rating, or the vertical air handlers will rotate at a faster speed to whatever cfm is needed, and electric or hot water heating coils will reheat the air to whatever temperature is required.

Today, the controls needed at the rooftop unit are typically factory mounted, requiring only minor adjustments on the job in order to be functional. Each VAV terminal box has a controller that opens the damper and/or heating device as needed to maintain the zone temperature and desired air volume. The controls of both the rooftop units and the VAV boxes can be easily connected to a building automation system (BAS).

Rooftop Unit Electronics Troubleshooting Scenario #1—Bad Static Pressure

During the first hot spell of the summer, a light commercial building customer calls in a "too hot" complaint. The technician arrives and finds the rooftop unit running. The supply fan VFD and cooling stages are operating. The discharge air temperature is 55°F, which is normal. However, the static pressure is 0.75″ WC when it is supposed to be 1″ WC. The supply fan VFD is checked and is at 30 Hz output 0.75″ WC. The technician uses the VFD control board and display to change the setpoint to 1″ WC. Immediately the supply fan ramps up, and the static pressure approaches 1″ WC. **See Figure 45-22.**

Service Work Performed: The technician uses the control panel and display to change the supply fan VFD setpoint to 1″ WC.

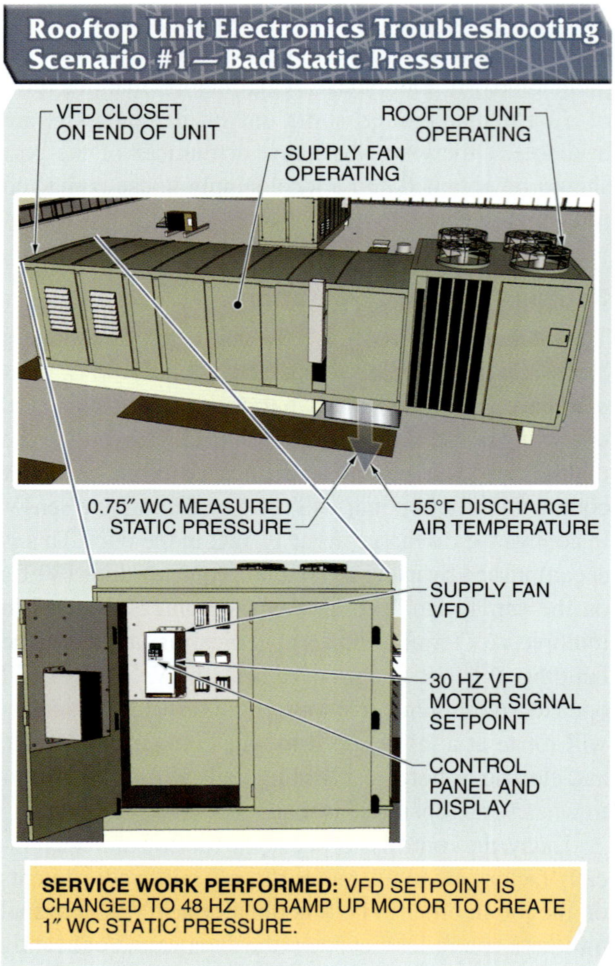

Figure 45-22. Many light commercial systems have VFDs that must be checked or tested for proper settings when seasons change.

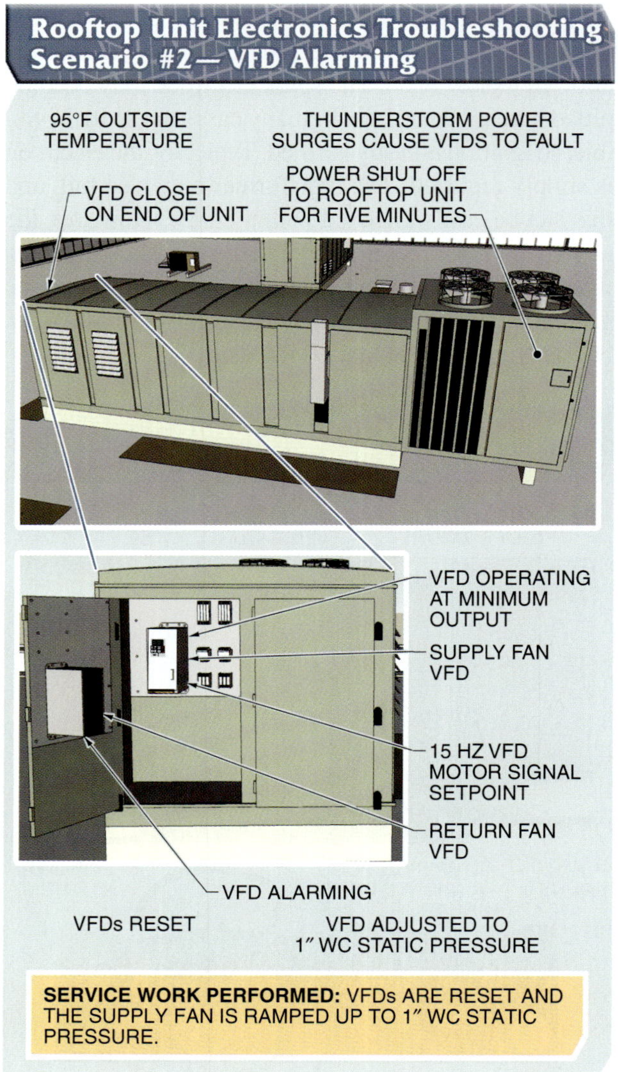

Figure 45-23. After a severe thunderstorm, it is not uncommon to receive calls that HVAC systems and other machinery are operating erratically due to voltage surges and loss of power.

Rooftop Unit Electronics Troubleshooting Scenario #2—VFD Alarming

On a 95°F day, a customer calls in a "too hot" complaint. It is the day following a thunderstorm. The technician arrives and finds that the supply fan VFD is only delivering 15 Hz to the fan. A check of the display on the supply and return fan drives indicates that the VFDs are operating at a minimum output, and that there are a number of faults and alarms present. The technician suspects that the thunderstorm activity has caused the unit's VFDs to fault. The technician shuts off the power to the rooftop unit. After five minutes the technician turns the power back on. The VFDs are reset, and the rooftop unit is now properly cooling the building spaces. **See Figure 45-23.**

Service Work Performed: The technician resets the rooftop unit's VFDs and ramps the supply fan up smoothly to a 1″ WC static pressure.

Rooftop Unit Electronics Troubleshooting Scenario #3—Bad Control Board

It is a very hot summer day when a customer reports that their building is too hot. The customer states that the rooftop unit ran fine the day before when the temperature was the same. The technician arrives and finds that the rooftop unit is off. The control board indicator lights are flashing a pattern that indicates board failure. The technician resets power to the unit and hopefully to the control board, but to no avail. The control board does not power up. **See Figure 45-24.**

Service Work Performed: The control board is replaced and the unit then cools properly.

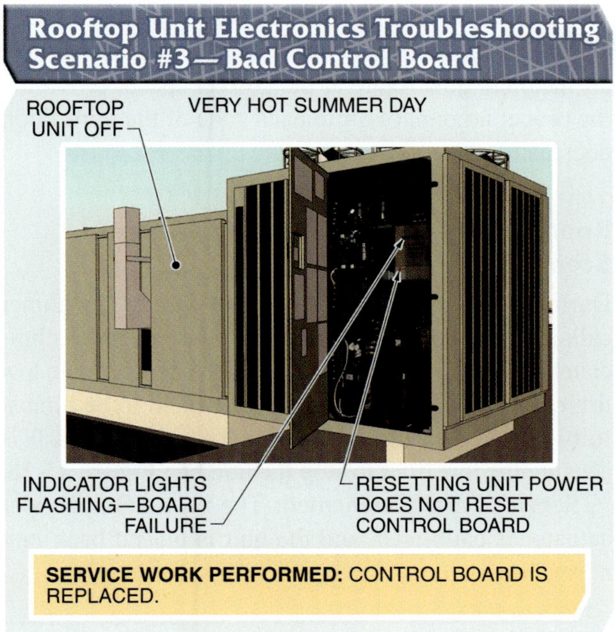

Figure 45-24. When an HVAC system is flashing a fault code, the control board is always the best place to start the troubleshooting procedure.

Rooftop Unit Electronics Troubleshooting Scenario #4—Wrong Compressor Settings

On the first hot day of the season, a technician arrives at a client's building to handle a "too hot" complaint. The supply fan is running and maintaining 1″ WC static pressure, but the discharge air temperature is fluctuating wildly between 50°F and 70°F. The technician notes that the cooling compressors are cycling incorrectly. The technician uses the controller display and keypad and finds that the cooling stage parameters are incorrect for the compressors. **See Figure 45-25.**

Service Work Performed: After resetting the cooling stage parameters to the factory defaults the cooling stages operate normally and maintain an approximately 55°F discharge temperature.

Rooftop Unit Electronics Troubleshooting Scenario #5—Wrong Economizer Settings

A technician is dispatched to a building with three rooftop HVAC units. The customer complains that the second floor of the building is too warm and humid. The technician inspects the second-floor RTU unit #2 and notes that the economizer damper is wide open. The controls are checked using the unit's display and are found to be set incorrectly. **See Figure 45-26.**

Service Work Performed: The changeover setpoint for the economizer is changed and the second-floor rooftop unit cools properly.

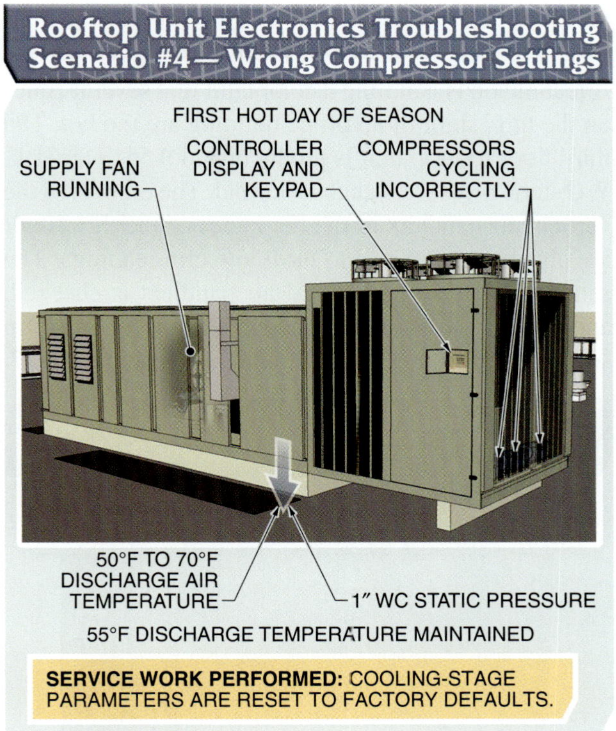

Figure 45-25. When an HVAC system has power and is partially operating, the controller display and software can be used to troubleshoot the system when the system parameters are known.

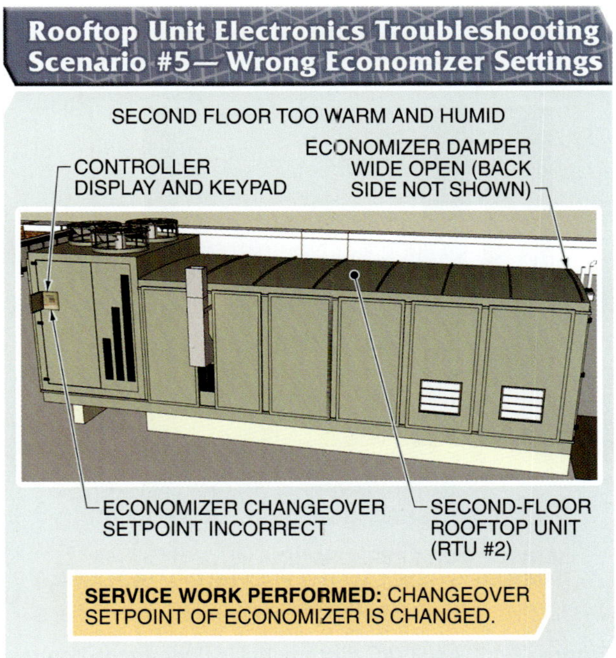

Figure 45-26. When a system is operating but not keeping the space properly warmed or cooled and the thermostat appears to be correct, the economizer settings or damper are typically the next item to check in the troubleshooting process.

Rooftop Unit Electronics Troubleshooting Scenario #6—Vertical Air Handler Problems

A technician is handling a complaint that several zones on the third floor of an office building are too hot. The third floor rooftop unit is producing cool 55°F air at 1″ WC static pressure, which is normal. The outputs of the vertical air handlers in the hot zones of the third floor are checked and found to have low cfm readings. The setup of each of the vertical air handlers is checked, and it is discovered that someone has lowered the cfm settings below the desired setpoints. **See Figure 45-27.**

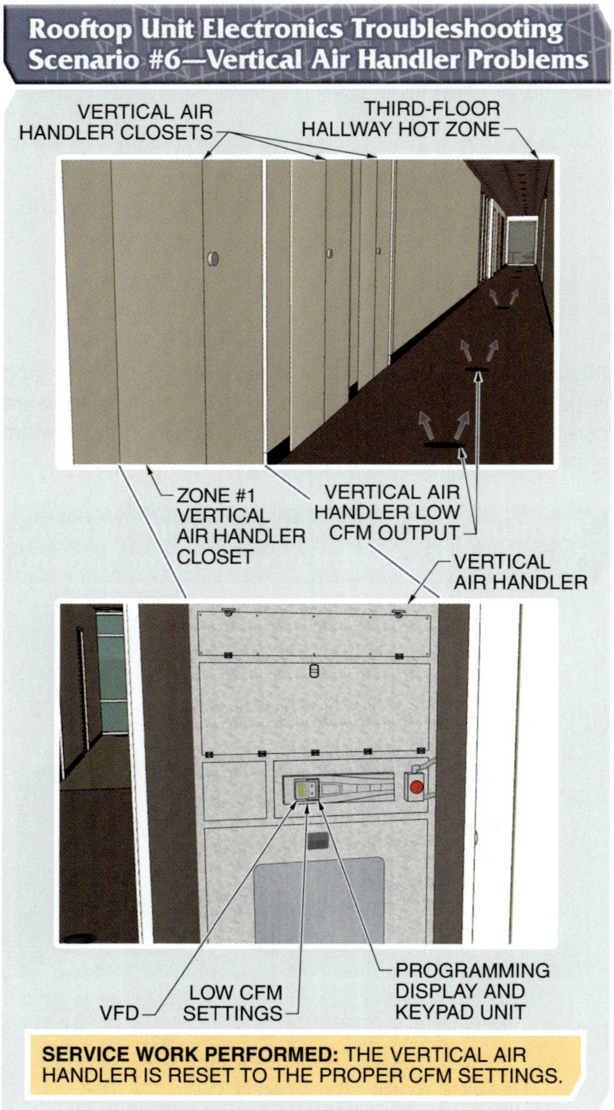

Figure 45-27. An HVAC system that includes zone controls and separate fan systems must always be checked and tested when one zone is not as comfortable as the other zones.

Service Work Performed: The cfm settings are changed back to the correct setpoints and the vertical air handlers now rotate at the correct speed, providing the proper amount of air to each zone of the third floor, decreasing the zone temperatures.

Rooftop Unit Electronics Troubleshooting Scenario #7—Low Limit

During a cold winter day, an office building customer calls in a "too cold" complaint. Upon arrival, the technician finds that the rooftop unit is off on a temperature low limit. The outdoor-air dampers are found to be almost fully open, admitting too much cold air in, which is causing the low limit to stop the fan. **See Figure 45-28.**

Service Work Performed: The ventilation damper actuator is calibrated, and the unit is placed back into normal operation.

TECH TIP

A low temperature limit should never be jumped in order to operate a unit. This may result in the piping for the water coil freezing and breaking.

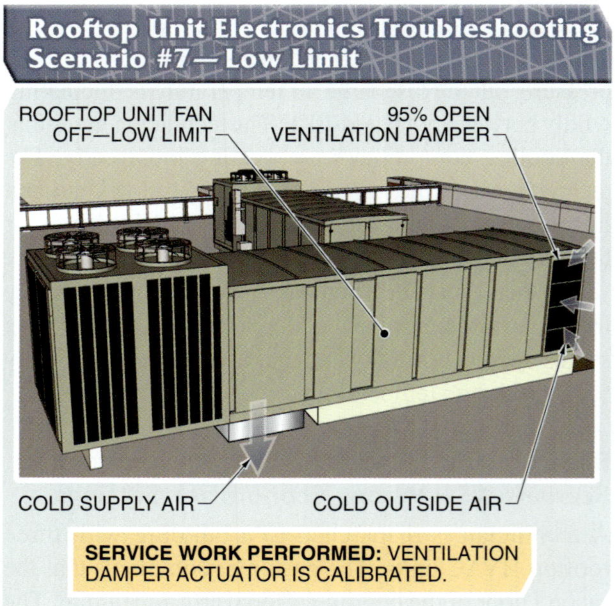

Figure 45-28. Analog sensors and actuator signaling systems must be checked periodically. It is not always the case that a component has failed, sometimes the analog sensors and actuators require recalibration to allow them to work at peak efficiency.

UNIT

46 Installing, Servicing, and Troubleshooting Air-Source Heat Pump Systems

Air-source heat pumps are one of the most commonly installed residential and light commercial HVAC systems today. This is especially true in the U.S. South and West, where air-source heat pumps are the predominant type of residential and light commercial HVAC system. As with any system, correct operation begins with proper installation. Also, a technician must be prepared to perform effective troubleshooting on these types of systems.

OBJECTIVES

- Identify the components of an air-source heat pump system.
- List the requirements for and configurations of air handlers.
- Describe the electrical and ductwork requirements for air-source heat pump systems.
- List the service checks that must be performed on an air-source heat pump system.
- Explain the use of manufacturer manuals when troubleshooting a heat pump system.
- Describe the sequences of operations during the cooling and heating modes for an air-source heat pump.
- Identify common air-source heat pump troubleshooting problems.

Index to Procedures

INSTALLING AIR-SOURCE HEAT PUMP SYSTEMS

There are a number of steps involved in the proper installation of an air-source heat pump. The two sections of a heat pump are the air handler (indoor unit) and the outdoor unit. The air handler consists of the cabinet, fan, and indoor coil. The outdoor unit consists of the cabinet, compressor, fan, outdoor coil, valves, and controls. **See Figure 46-1.** *Note:* Many air-source heat pump models are designed for indoor installation only.

Compliance to all codes and regulations during installation is required. Almost all new outdoor units use R-410A, which is an ozone-friendly HFC refrigerant. The outdoor unit must be installed with a matching indoor coil and line set. A filter drier approved for use with R-410A is typically provided by the manufacturer for the specific system.

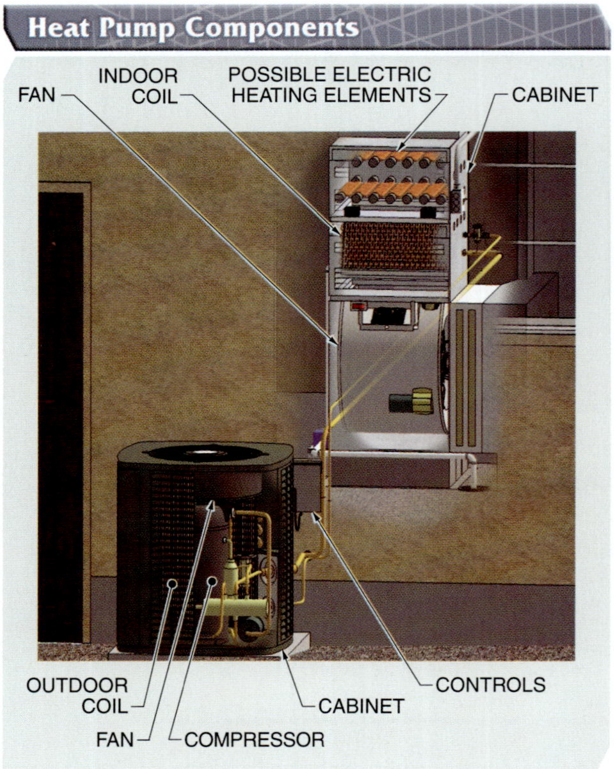

Figure 46-1. The two sections of a heat pump are the air handler (indoor unit), which consists of the cabinet, fan, and indoor coil, and the outdoor unit, which consists of the cabinet, compressor, fan, outdoor coil, valves, and controls.

Air Handler (Indoor Unit) Requirements

It is crucial in any air-source heat pump installation that the installation instructions be followed closely. However, the installation instructions are intended as a general guide only and do not supersede any national or local codes in any way. All local, state, or national codes pertaining to this type of equipment must be determined prior to installation. **See Figure 46-2.**

The installation of the air handler, field wiring, air ducts, and other components must conform to the requirements of the National Electrical Code® (NEC®) and ANSI/NFPA No. 70 (latest edition) in the United States and the Canadian Electrical Code (CEC™), CSA C22.2-12, in Canada.

Any local authorities having jurisdiction must be consulted before installation begins. In some cases, a permit may be required. In case of conflict, applicable regulations or requirements take precedence over the general instructions provided by the manufacturer. The conditioned air plenum, ducts, and air filters (not provided) must be installed in accordance with NFPA 90B, *Installation of Warm Air Heating and Air-Conditioning Systems* (latest edition).

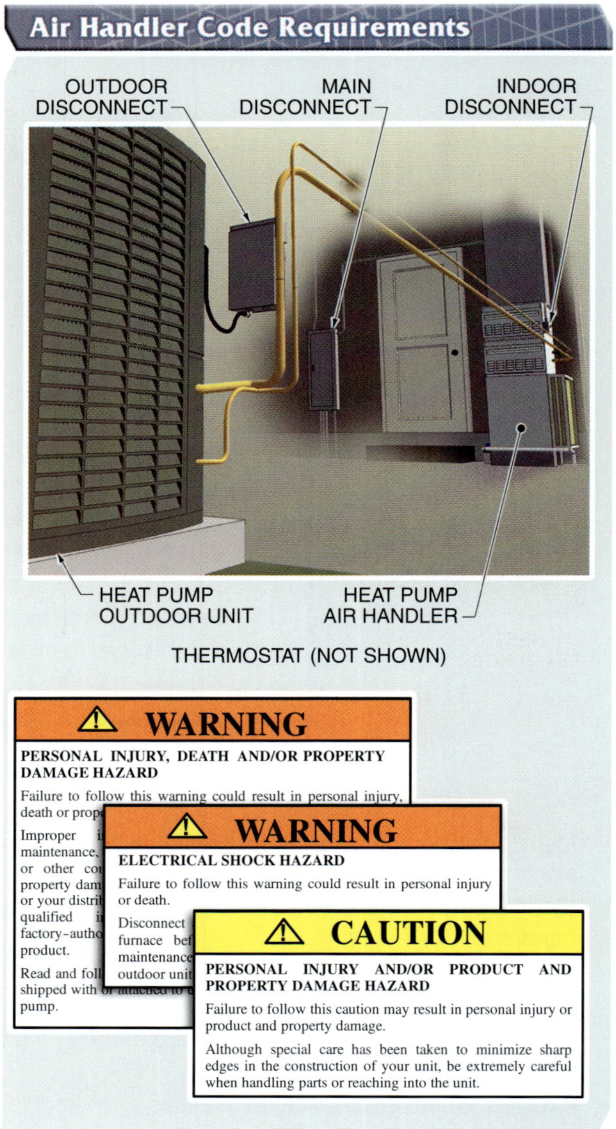

Figure 46-2. All local, state, national codes (ANSI/NFPA No. 70 and CSA C22.12), and any manufacturer instructions pertaining to heat pumps must be read before installation begins.

The air handler of the air-source heat pump is provided with flanges for the connection of the plenum and ducts. The filters must be listed as Class 2 furnace air filters. The manufacturer will ship the unit from the factory fully assembled. Some heat pump models are configured to allow different airflow patterns, with upflow and horizontal being the most common. Cabinet knockouts should not be removed until it has been determined which knockouts need to be removed for the installation. **See Figure 46-3.**

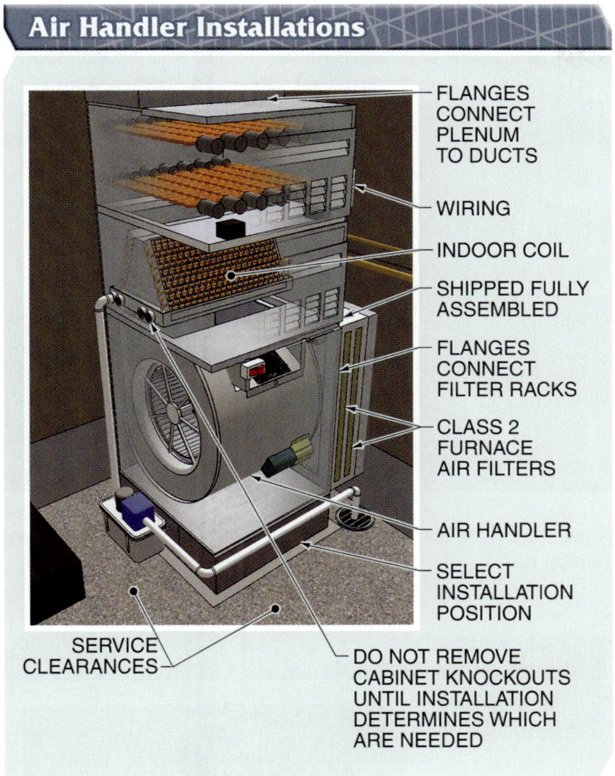

Figure 46-3. Heat pumps are shipped fully assembled, but modifications may be necessary to accommodate the specifications of the individual application.

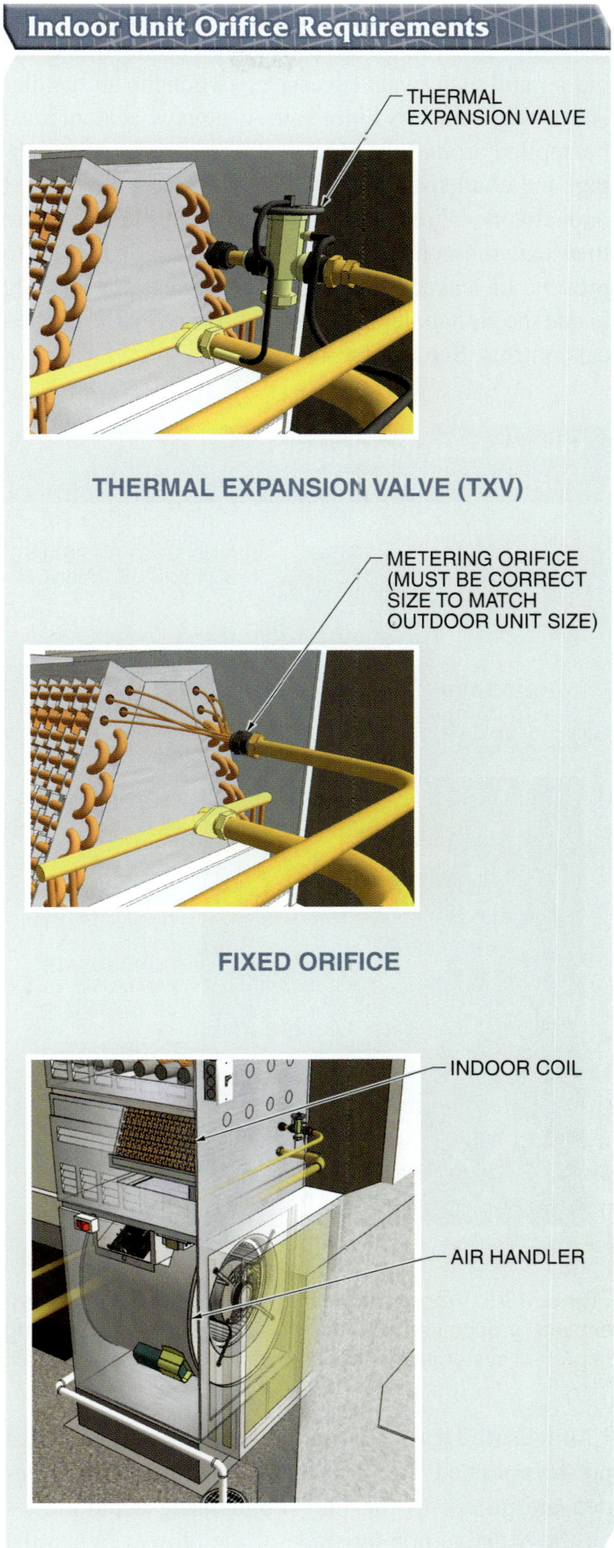

Figure 46-4. The correct orifice size may be included by the manufacturer, but it must always be sized to the rating of the outdoor unit or be replaced by a thermal expansion valve to obtain the system's rated performance.

The installation position that best suits the site conditions should be selected. Required clearances, space, and routing requirements for refrigerant lines, condensate disposal, filters, ductwork, wiring, and accessibility for service must also be considered. The rating plate on the air handler should be checked for specific information.

Indoor Unit Orifice Requirements. Manufacturers design air handlers to work with specific outdoor units. Typically, indoor units are manufactured with an interchangeable refrigerant-metering orifice to provide optimum refrigerant control and system performance with a variety of different capacities of outdoor units. The correct refrigerant-metering orifice size must be verified with the manufacturer.

The correct orifice size may be included or provided by the manufacturer with the outdoor unit. When the orifice is not provided, the correct size orifice must be purchased. In some cases, the rating of the outdoor unit may require that the air handler refrigerant-metering orifice be changed or replaced by a thermal expansion valve to obtain rated performance. **See Figure 46-4.**

Indoor Unit Piping and Wiring Requirements. When air-source heat pumps are used in the cooling mode, the indoor coil may sweat excessively when the air handler is working with very humid air. As always, sealant must be applied around the electrical wires, refrigerant tubing, and condensate lines where they enter the cabinet or ductwork. Wires should be sealed on the inside where they exit the conduit opening. Sealant is required to prevent air leakage into and condensate from forming inside the air handler and control box and on the electrical controls. **See Figure 46-5.**

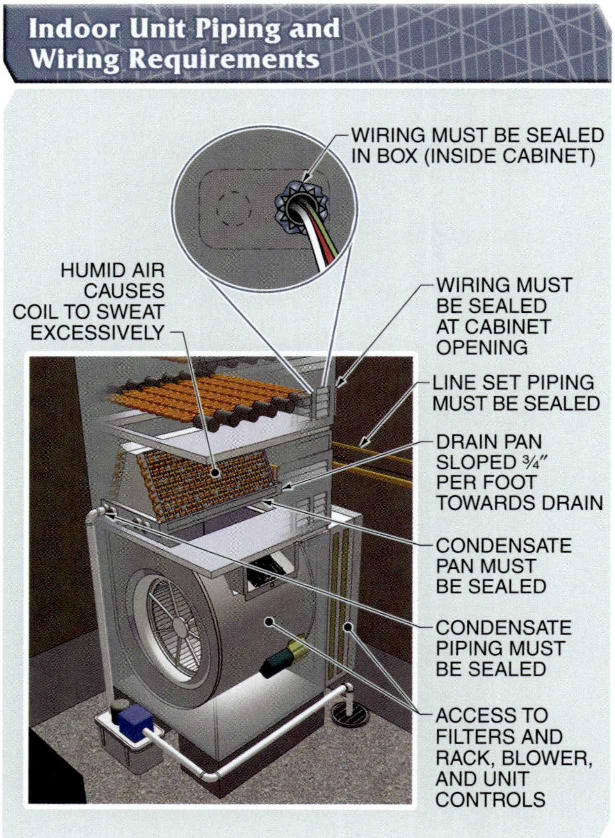

Figure 46-5. Whenever piping or wiring travels through a cabinet or duct, sealant must be used to fill in the opening to protect systems from condensate and unconditioned air.

Carrier Corporation

Air-source heat pumps require a condensate pan for the heat pump's coil.

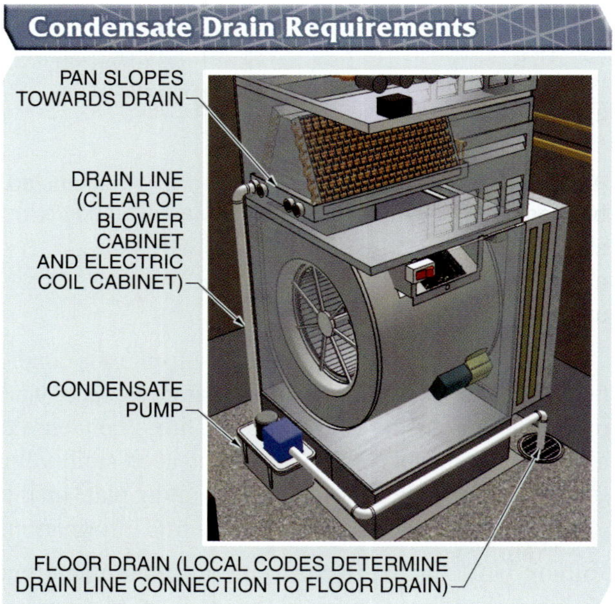

Figure 46-6. Besides following manufacturer clearance recommendations for condensate drain line installation, many cities and counties have ordinances covering the installation of condensate drain lines that the installer must observe.

Condensate Drain Requirements. Air handlers must not be operated without a working drain trap or condensate pump. In the unit, condensate accumulates on the negative pressure side of the blower. When the condensate line does not have a proper trap, air can be pulled through the condensate line, preventing drainage. The drain line is routed to an appropriate floor or roof drain. **See Figure 46-6.**

Drain lines must be installed in a manner that does not block service access to the front of the air handler. The manufacturer manuals should be consulted for further details. A minimum clearance is required for filter, coil, or blower removal and service access. *Note:* Local codes should be checked before the drain line is connected to an existing drainage system. **See Procedure for Testing the Condensate Drain Pan and Drain Line.**

Testing Condensate Drain Pan and Drain Line

1. Pour several quarts of water into the drain pan, enough to fill drain trap and line.

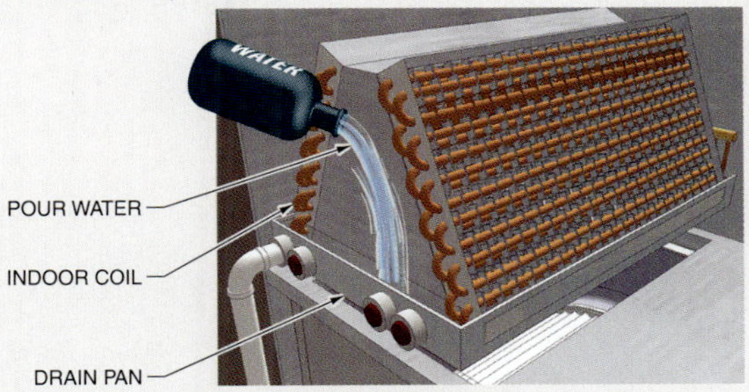

2. Ensure that the drain pan is draining completely. Ensure no leaks are found in the drain line fittings and that the water is draining from the end of the primary drain line.

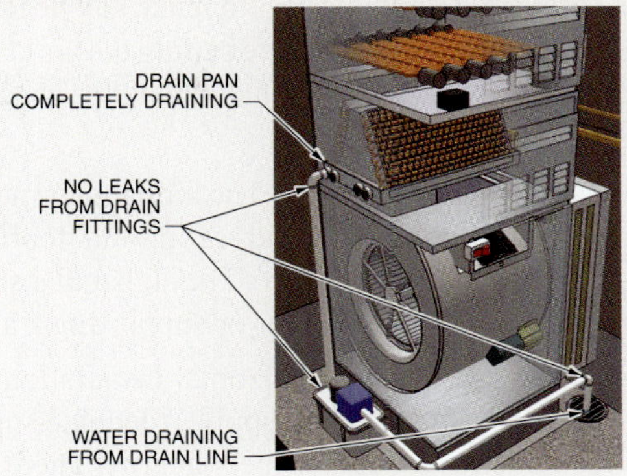

3. Correct any leaks found.

The air handler must be installed in such a way as to allow free access to the coil, filter rack, blower, and control compartments. The indoor coil must be installed with a minimum drop in the horizontal position towards the drain to ensure proper condensate drainage. Many manufacturers specify a ¾″-per-foot slope.

Note: Technicians installing or servicing heat pumps should know that the Clean Air Act of 1990 banned the venting of refrigerant (CFCs and HFCs) as of July 1, 1992. Approved methods of recovering and reclaiming refrigerant must be followed. Fines and/or incarceration for noncompliance may result.

TECH TIP

The amount of condensate produced varies based on the system size. In general, a gallon per hour of condensate is produced by a typical residential heat pump unit evaporator coil. All installations should include a properly wired auxiliary safety switch to shut down the air handler unit if the condensate pan has an overflow condition that could lead to property damage.

Closet Installation Requirements. The air handler can be installed in a closet with a false bottom as the return air plenum or with a return air plenum under the air handler. Louvers or return air grills are field supplied. Local codes may limit application of systems without a ducted return in single-story buildings. **See Figure 46-7.**

Closet Installation Requirements

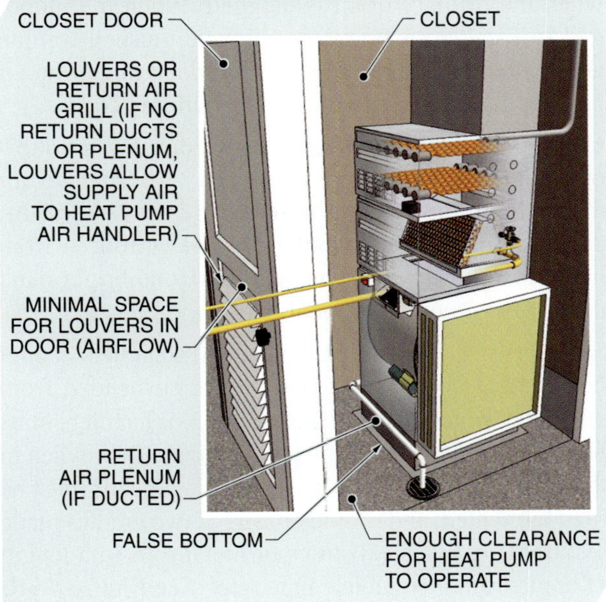

Figure 46-7. When an air handler is installed in a closet, the door to the closet typically requires modification or replacement to allow for louvers, and the closet must allow access to the unit for maintenance.

For a unit installed in a closet with a louvered return opening, there is a minimum amount of space for the louvers. The data from the manufacturer should be consulted for details. When a return air plenum is used, the return air grill must be immediately in front of the opening in the plenum to allow for the free flow of return air. When installed in a closet, the grill can be installed on the closet door and there must be adequate clearance around the air handler to allow for the free flow of return air.

Air Handler (Indoor Unit) Configurations

For ease of installation, it is best to make any necessary coil configuration changes before setting the air handler in place. The manufacturer manuals should be consulted for specific details.

Upflow Vertical Installations. The air handler must be supported on the bottom only and set on solid flooring or a field-supplied supporting frame. The air handler should be attached securely to the floor or supporting frame.

Horizontal Installations. Horizontal installations can be left-hand or right-hand air supply. Adequate support must be provided to ensure cabinet integrity. There must be adequate room to remove, service, and access panels when installing an air handler in the horizontal position.

Note: The indoor coil is provided with a secondary drain that should be tapped and piped to a location that will give the homeowner or technician a visual warning when the primary drain is clogged. When the indoor coil is installed in an attic or above a finished ceiling, an auxiliary drain pan must be provided under the air handler as specified by most local building codes.

Conversion from Vertical to Horizontal. An upflow only air handler may be converted to horizontal air discharge by installing a horizontal drain pan kit. An upflow/horizontal air handler may be converted from horizontal left-hand discharge to horizontal right-hand discharge without additional parts.

Suspended Cabinet Installation. Air handlers cannot be installed horizontally lying on or suspended from either the front or back of the cabinet. Air handlers must be positioned with one side parallel to the floor when in the horizontal position. The suspending means must be field fabricated, and should consist of two cradles made by attaching two steel straps or metal rods to a length of angle iron or suitable gauge steel. **See Figure 46-8.**

Electrical Requirements

The entire manufacturer installation manual should be read, as well as the instructions for each of the separate units, before starting the installation. All warnings, cautions, instructional labels, and tags should be observed. Failure to comply with the manufacturer's instructions could result in an unsafe condition and/or premature component failure. However, any instructions provided are intended as a general guide only and do not supersede any national or local codes in any way. The installation must comply with all provincial, state, and local codes, requirements of the UL, as well as the NEC® in the United States and the CEC™ in Canada.

Suspended Cabinet Installation

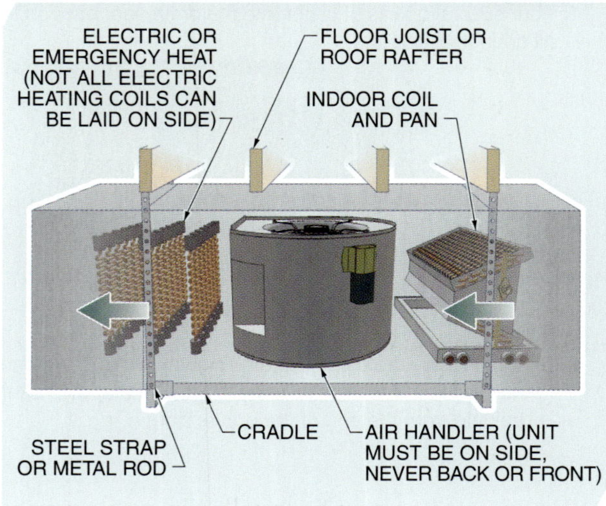

Figure 46-8. Air handlers installed in the horizontal position must be positioned with one side parallel to the floor.

Electrical wiring, disconnect means, and overcurrent protection are typically supplied by the installer. The air handler rating plate will state the maximum overcurrent protection required, minimum circuit ampacity, as well as the proper operating voltage. Heat pump power supplies must be sized and protected according to the specifications on the heat pump rating plate. **See Figure 46-9.** *Note:* Only copper conductors can be used.

Heat Pump Rating Plates

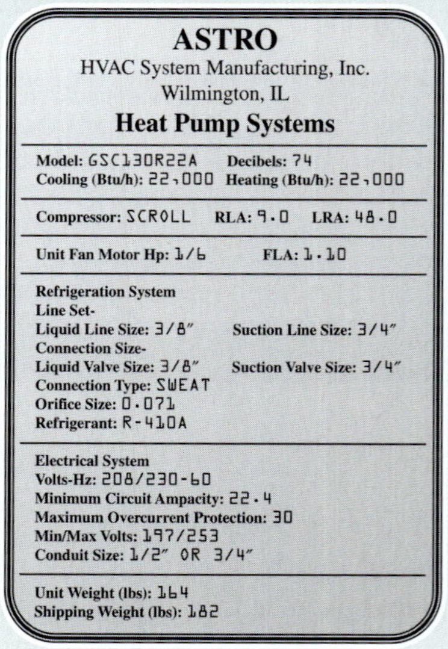
ASTRO
HVAC System Manufacturing, Inc.
Wilmington, IL
Heat Pump Systems

Model: GSC130R22A Decibels: 74
Cooling (Btu/h): 22,000 Heating (Btu/h): 22,000

Compressor: SCROLL RLA: 9.0 LRA: 48.0

Unit Fan Motor Hp: 1/6 FLA: 1.10

Refrigeration System
Line Set-
Liquid Line Size: 3/8″ Suction Line Size: 3/4″
Connection Size-
Liquid Valve Size: 3/8″ Suction Valve Size: 3/4″
Connection Type: SWEAT
Orifice Size: 0.071
Refrigerant: R-410A

Electrical System
Volts-Hz: 208/230-60
Minimum Circuit Ampacity: 22.4
Maximum Overcurrent Protection: 30
Min/Max Volts: 197/253
Conduit Size: 1/2″ OR 3/4″

Unit Weight (lbs): 164
Shipping Weight (lbs): 182

Figure 46-9. Rating plate information for heat pumps provides most of the information needed to choose components for the system.

Note: Heat pumps use electronically commutated motors (ECMs) or motors controlled by VSDs with DC filter capacitors that will cause a surge or inrush of current when power is applied. Power is to remain applied to the motor except during servicing. AC power must be removed before the input power to the motor is plugged or unplugged. The blower relays that interrupt the line voltage to the motor should not be installed. The manufacturer manuals should be consulted for specific details.

Making Electrical Connections. All power from the air handler must be disconnected and all covers removed to allow access to the air handler and electrical boxes. The ground wires must be in place and connected properly. **See Procedure for Wiring a Unit for Electric Heat.**

Fiber-insulated ducts are moisture resistant, help reduce heat loss or gain, and can be used in applications where equipment noise must be kept to a minimum.

Ductwork Requirements

The return and supply plenums, ducts, and air filters must be installed in accordance with NFPA 90B, *Installation of Warm Air Heating and Air-Conditioning Systems* (latest edition). The supply and return ductwork must be adequately sized to meet the system's air requirements and static pressure capabilities. Ductwork should be insulated with a minimum of 1″ thick insulation, with a vapor barrier in conditioned areas, or a 2″ thick minimum in unconditioned areas.

The supply plenum should be the same size as the flanged opening around the outlet of the air handler and should extend at least 3′ from the unit before turning or branching off into duct runs. This type of plenum forms an extension of the blower housing and minimizes air expansion losses from the blower. Residential air filters must be listed as Class 2 furnace air filters. **See Figure 46-10.**

Wiring a Unit for Electric Heat

Procedure

1. Determine the number of circuits needed to supply the heater with electrical power (1, 2, or 3 circuits). See the air handler accessory kit label for the number of circuits and ratings.
2. Disconnect all power supplies.
3. Knock out the correct number of knockouts (1, 2, or 3), and install UL listed wires and fittings.

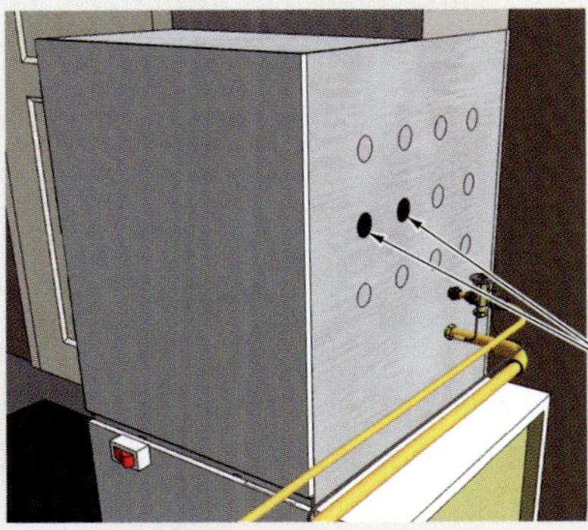

4. Connect appropriately sized wire to the circuit breaker terminals.
5. Connect green ground wire(s) to ground terminal(s) marked "GND". Connect to circuit breaker. *Note:* There may be 1, 2, or 3 ground terminals depending on the number of circuit breakers.
6. Install conduit-opening plugs in any unused openings.
7. If circuit breakers or pull disconnects are used, the front panel knockouts will need to be removed.
8. Reinstall the air handler blower access panels.

9. Reconnect the power.
10. Dispose of/recycle all remaining parts.

Air Filter Usage Classifications

Usage	Class	Performance	Particle Size
Primary	G1*	65%	>5.0 μm
	G2	65%–80%	>5.0 μm
	G3	80%–90%	>5.0 μm
	G4	90%	>5.0 μm
Secondary	M5†	40%–60%	>5.0 μm
	M6	60%–80%	>2.0 μm
	F7‡	90%–95%	>2.0 μm
	F8	90%–95%	>1.0 μm
	F9	95%	>1.0 μm
Semi HEPA	E10§	85%	>1.0 μm
	E11	95%	>0.5 μm
	E12	99.5%	>0.5 μm
HEPA	H13‖	99.95%	>0.3 μm
	H14	99.995%	>0.3 μm
ULPA	U15#	99.9995%	>0.2 μm
	U16	99.99995%	>0.2 μm
	U17	99.999995%	>0.1 μm

* general dust
† medium dust
‡ fine dust
§ suspended dust
‖ suspended medium dust
\# suspended fine dust

Figure 46-10. Air filters that catch small particles of contaminants, such as 0.1 μm to 5.0 μm in size, are classified as G4 through U17 filters. These filters create too much resistance to flow for normal residential furnaces.

Servicing Heat Pump Systems

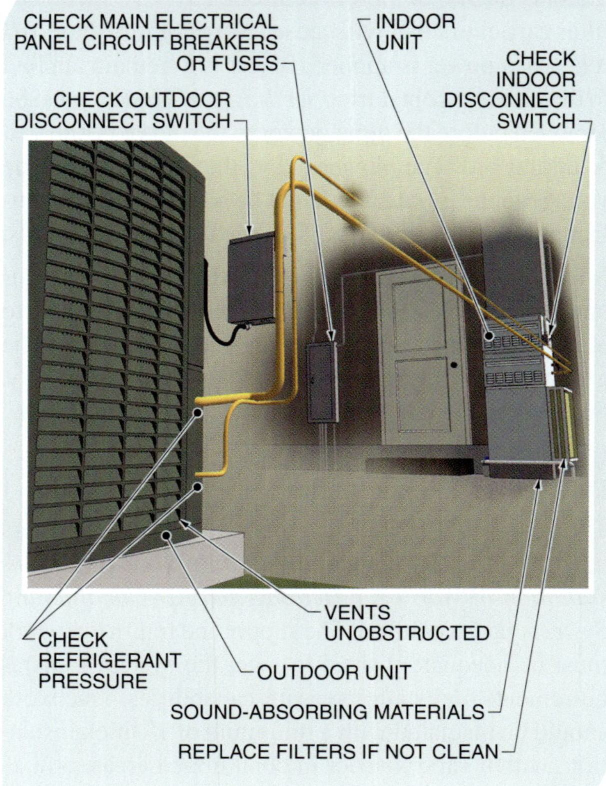

Figure 46-11. Certain service checks of the heat pump system must be made periodically to ensure continued operation.

SERVICING HEAT PUMP SYSTEMS

When servicing air-source heat pump systems, all fasteners and threaded piping must be appropriately tightened. A sound-absorbing material should be placed under the unit when it is in a location or position that would transmit sound or vibrations to other areas of the building.

For normal operation, certain service checks of the air-source heat pump system must be made to ensure continued operation. **See Figure 46-11.**

- Check the indoor and outdoor disconnect switches. Check main electrical panel circuit breakers or fuses.
- Check for sufficient airflow. Filters should be clean and vents should be unobstructed.
- Check thermostat settings. During cooling, the system/mode control must be set to COOL or AUTO. During heating, the system/mode control must be set to HEAT or AUTO.
- Inspect and clean blower wheel housing, indoor and outdoor coils, indoor drain pan, and drain piping.
- Check and clean condensate drain pan and line and possibly the auxiliary drain pan and line.

Basic Troubleshooting Questions

Before troubleshooting an air-source heat pump system, the technician needs to ask the building owner some basic questions:

- How long has the problem occurred?
- Does the problem occur during the heating or cooling mode?
- Does the reversing valve energize in the heating or cooling mode?
- Have there been any recent changes to the system, such as added internal load or a change in duct configuration?

Installation Manuals

An *installation manual* is a general guide used by qualified technicians to install residential and light commercial HVAC equipment. These manuals do not supersede any national or local codes. When installation manuals do not arrive with delivered equipment, the manuals should be obtained from the manufacturer as soon as possible.

The installation of the heat pump and air handler equipment described in the manual must comply with all state and local codes as well as national codes such as the NEC® or CEC™. Failure to comply with manufacturer procedures can void warranties and reduce the energy efficiency of the equipment. **See Figure 46-12.**

Installation Manuals

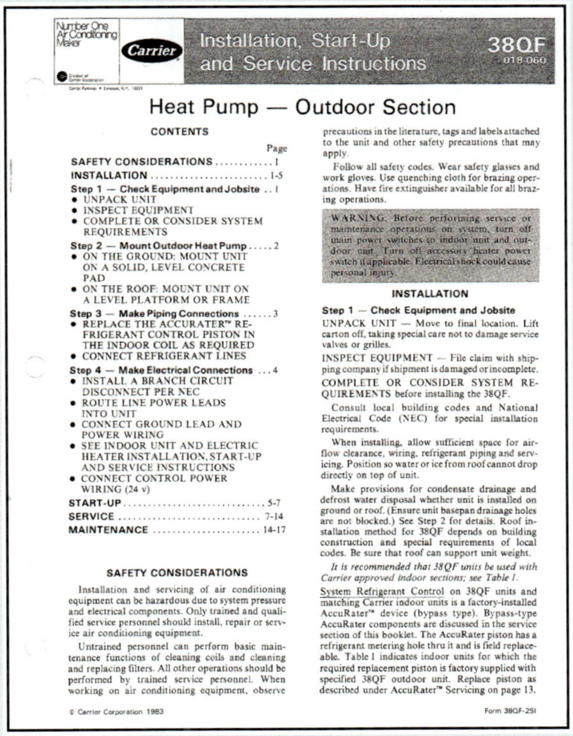

Carrier Corporation

Figure 46-12. Installation manuals provide specific manufacturer installation information for heat pump systems.

Owner Manuals

An *owner manual* is a booklet that typically offers the best way to use heat pump equipment for an application and contains specific preventive maintenance recommendations for the equipment. Operating heat pump equipment involves starting and stopping the equipment and making adjustments to it to keep it operating correctly. **See Figure 46-13.**

Owner Manuals

Carrier Corporation

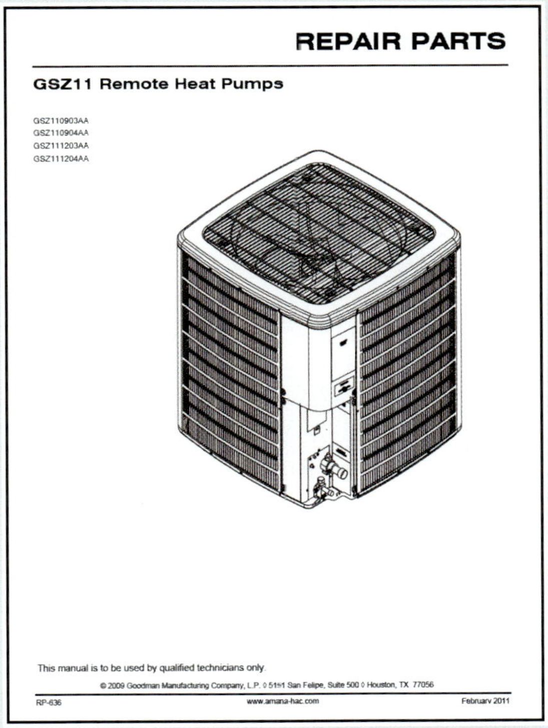

Goodman Global Group, Inc.

Figure 46-13. Owner manuals must be kept as references as long as the heat pump system is in operation.

Owner or repair manuals for heat pump equipment typically contain manufacturer specific heat pump facts, sequences of operations, adjustment procedures for specific applications, troubleshooting charts, flash code charts, preventive maintenance procedures, performance checklists, wiring schematics, piping schematics, and warranty information. The information included in owner or repair manuals can help the technician develop troubleshooting-type thinking, which is used when diagnosing system problems.

Troubleshooting Charts. Troubleshooting charts are only intended to be used on heat pumps serviced by technicians who have successfully completed manufacturer training. Troubleshooting charts must only be used in conjunction with the appropriate heat pump owner or repair manual, which is also provided by the heat pump manufacturer.

These charts are not intended to be complete lists of all the possible failures a heat pump may experience. There are many reasons why a heat pump might not be working correctly. However, many common solutions can be found on heat pump troubleshooting charts. **See Figure 46-14.**

Troubleshooting Charts

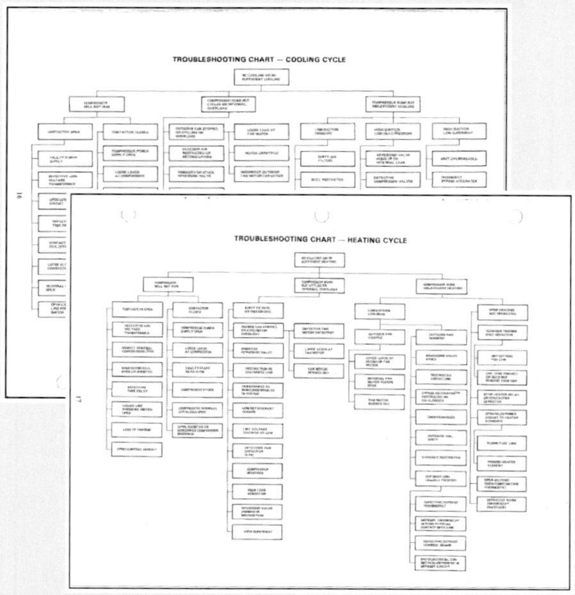

Carrier Corporation

Figure 46-14. Owner or repair manuals can be used with troubleshooting charts to help the technician understand the terminology and procedures used in the chart.

Typical Air-Source Heat Pump Sequence of Operations

For servicing and troubleshooting purposes, it is important that the technician understand the sequence of operation of the particular heat pump model being worked on. As there are important variations among manufacturers and models, this is a critical step in determining whether the heat pump is operating correctly or not.

Cooling Cycle. The cooling cycle on a heat pump functions as follows:

1. Setting the thermostat to the cooling mode automatically energizes the solenoid valve on the reversing valve. **See Figure 46-15.**
2. The compressor pumps high-pressure superheated refrigerant vapor through the system.
3. The superheated refrigerant vapor leaves the compressor and passes through the reversing valve.
4. The superheated refrigerant flows from the reversing valve through the outdoor vapor line to the finned outdoor coil. Air from the outdoor fan removes heat from the hot refrigerant vapor. When enough heat is removed, the vapor condenses into a high-pressure liquid. The liquid temperature is slightly warmer than the outside unit's ambient air temperature.
5. The warm high-pressure liquid leaves the outdoor coil and flows through the copper refrigerant liquid line into the building.
6. At the end of the liquid line, the refrigerant passes through a metering device, reducing its pressure and temperature and partially turning it into a vapor.
7. As the liquid/vapor refrigerant, at a reduced pressure, enters the indoor coil, it expands and absorbs heat from the indoor air passing over the finned surface of the indoor coil. Heat from the indoor air causes the low-pressure liquid/vapor refrigerant to evaporate while it cools the indoor air. The refrigerant is now a warm vapor.
8. The refrigerant vapor flows through the insulated suction line to the reversing valve. The reversing valve directs the vapor refrigerant into the accumulator.
9. The accumulator controls the liquid refrigerant by not allowing it to pass.
10. Refrigerant vapor flows from the accumulator, through the suction line, to the intake of the compressor.
11. Steps 1 through 10 are repeated.

Heat Pump Sequence of Operations — Cooling Cycle

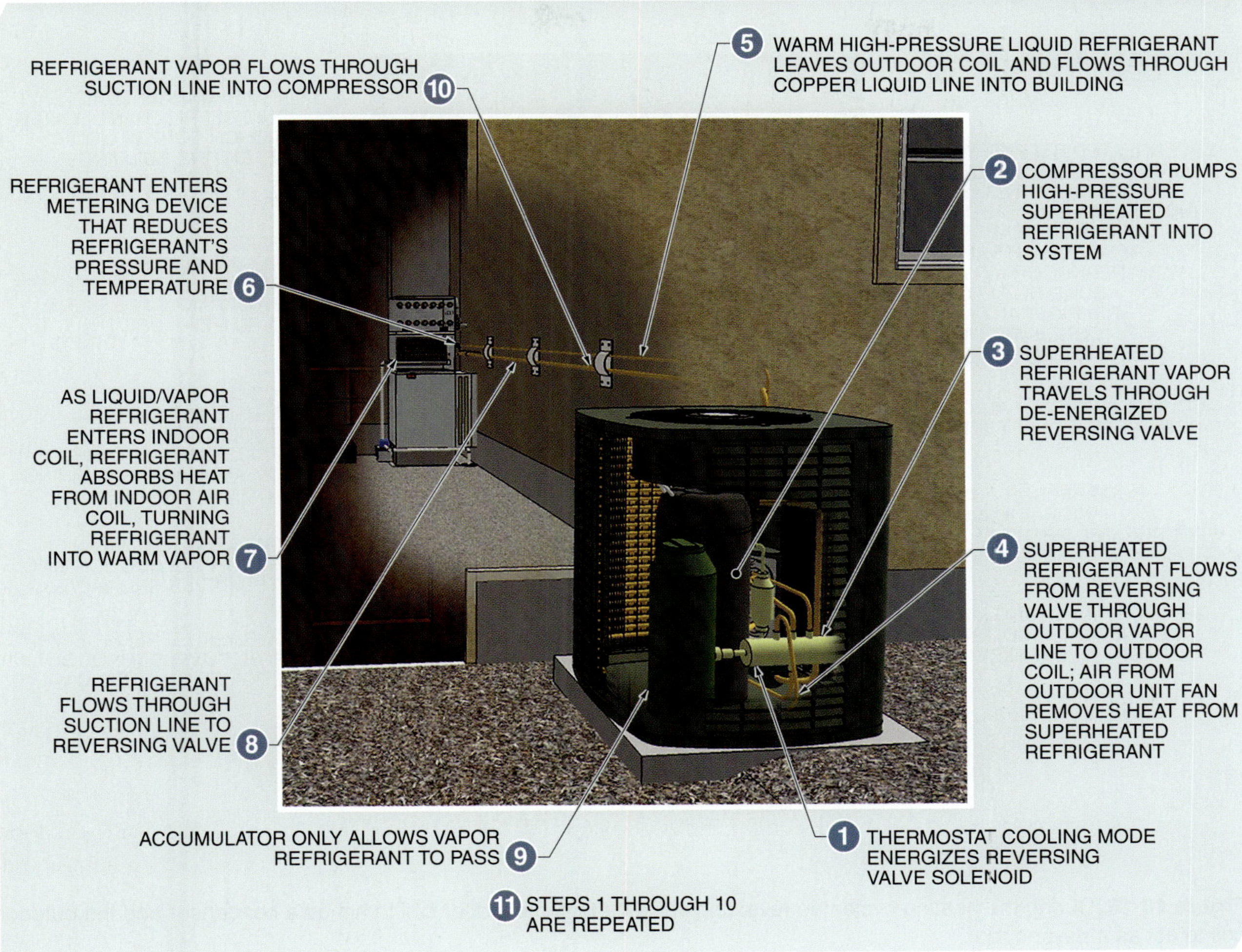

Figure 46-15. During the cooling cycle, a heat pump and air handler operate the same way a conventional split-system air conditioner operates.

Heating Cycle. The heating cycle on a heat pump functions as follows:

1. Setting the thermostat to the heating mode automatically de-energizes the solenoid valve to shift the reversing valve. **See Figure 46-16.**
2. The compressor pumps high-pressure superheated refrigerant vapor through the system.
3. The superheated refrigerant vapor leaves the compressor and passes through the reversing valve.
4. The superheated refrigerant flows from the reversing valve through the insulated indoor vapor line to the finned indoor coil. Air from the indoor blower removes heat from the refrigerant vapor, warming the indoor air and heating the building spaces. When enough heat is removed, the vapor condenses into a high-pressure liquid. The liquid refrigerant temperature is slightly warmer than the indoor air temperature.
5. The warm high-pressure liquid leaves the indoor coil and flows through the small copper refrigerant liquid line exiting the building.
6. At the end of the liquid line, the refrigerant passes through a metering device in the outdoor coil, reducing its pressure and temperature and partially turning into a vapor.
7. As the cool liquid/vapor refrigerant, at a reduced pressure, enters the outdoor coil, it expands and absorbs heat from the outdoor air passing over the finned surface. Heat from the outdoor air causes the low-pressure liquid to evaporate. The refrigerant is now a warm vapor.

Heat Pump Sequence of Operations — Heating Cycle

Figure 46-16. During the heating cycle, the reversing valve causes the indoor coil to act as a condenser and the outdoor coil to act as an evaporator.

8. The warm refrigerant vapor travels through the large outdoor vapor line to the reversing valve. The reversing valve directs the refrigerant into the accumulator.
9. The accumulator controls the liquid refrigerant by not allowing it to pass.
10. Refrigerant vapor flows from the accumulator, through the suction line, to the intake of the compressor.
11. Steps 1 through 10 are repeated.

Emergency Heat. The emergency heat on a heat pump air handler functions as follows:

1. Moving the thermostat selector switch to the emergency heat position breaks the cooling or heating electrical circuit to the compressor contactor and the indoor blower relay. **See Figure 46-17.**
2. Setting the thermostat to emergency heat causes the red emergency heat warning light to illuminate.
3. A thermostat emergency heat call energizes the electric heating element relay and sequencer.
4. The electric heating element relay contacts close, powering the electric heating elements and the indoor blower.
5. The emergency heat call ends, and the thermostat de-energizes the electric heating element relay and sequencer.
6. The electric heating element relay contacts open, de-energizing the electric elements and the indoor blower.
7. Moving the thermostat selector switch back to the heat position completes the circuit to the compressor contactor and indoor blower relay.
8. The red emergency heat warning light goes out.

Heat Pump Sequence of Operations — Emergency Heat

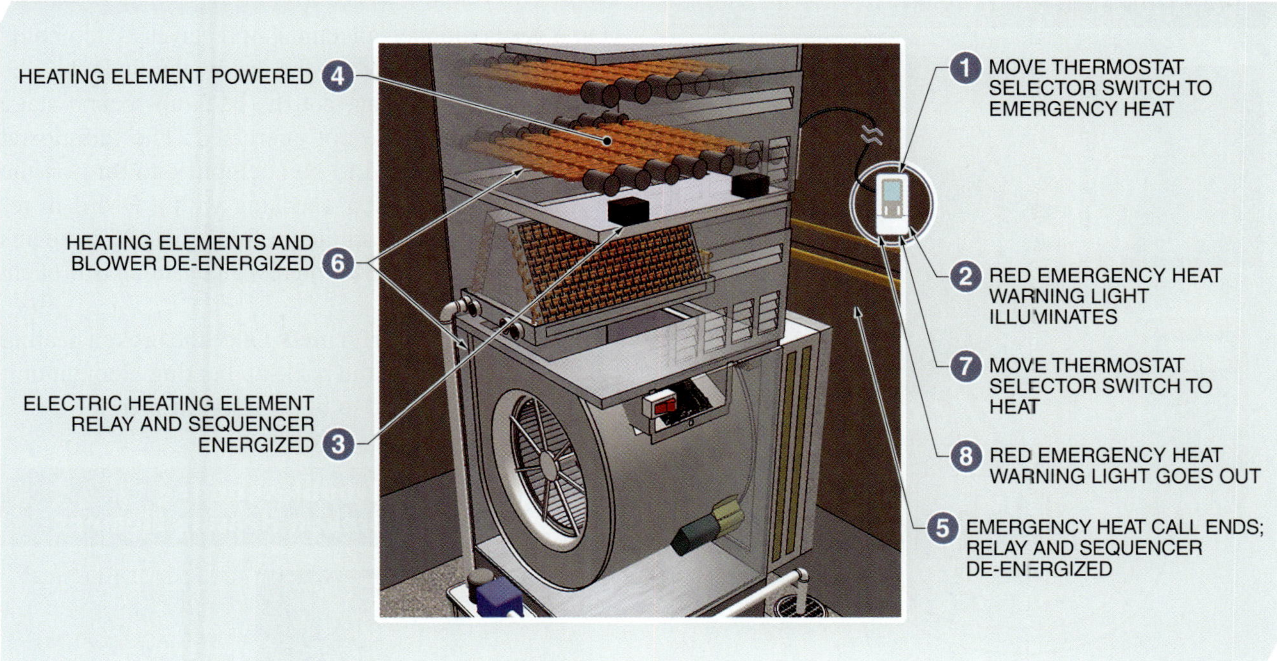

Figure 46-17. Emergency or backup heat for a heat pump is typically achieved using electrical heating elements.

TROUBLESHOOTING AIR-SOURCE HEAT PUMPS

To troubleshoot an air-source heat pump, the technician must know how to use diagnostic equipment such as a digital multimeter (DMM) and a gauge manifold. In addition, a knowledge of the basic sequences of operation is needed as well. The technician must be observant of local conditions and situations that may indicate the source of a problem.

As much as possible, technicians should use their senses. With any air-source heat pump system, 80% of the problems are simple and basic. An absolutely vital step is to ask the occupants some basic questions:

1. How long has the unit been malfunctioning?
2. When was the last time it worked correctly, or did it ever work correctly?
3. Was any service work or other work done to the unit or surrounding area?

Heating Mode Troubleshooting Scenario #1—Defective Transformer

A technician receives a "too cold" call from a residence. Upon arrival, the technician finds that neither the compressor, electric heaters, indoor fan, or outdoor fan is running. Incoming power is checked with a digital multimeter and is found to be 230 VAC. The control transformer is checked and has 0 VAC on the secondary side. It is also discovered that the control transformer is discolored and very hot to the touch. **See Figure 46-18.**

Service Work Performed: The technician replaces the transformer and then checks the unit to ensure it cycles properly.

Heating Mode Troubleshooting Scenario #2—Bad Thermostat

A technician receives a "too cold" complaint from a customer. Upon arrival the technician finds that the unit is operated by a very old thermostat that appears to be damaged. It is found that the compressor, indoor fan, and back-up heating elements are not operating. The technician removes the front plate of the thermostat and finds that the thermostat mechanism is not operating. **See Figure 46-19.**

Service Work Performed: The thermostat is replaced with a new model. The heating mode is checked and found to be working properly.

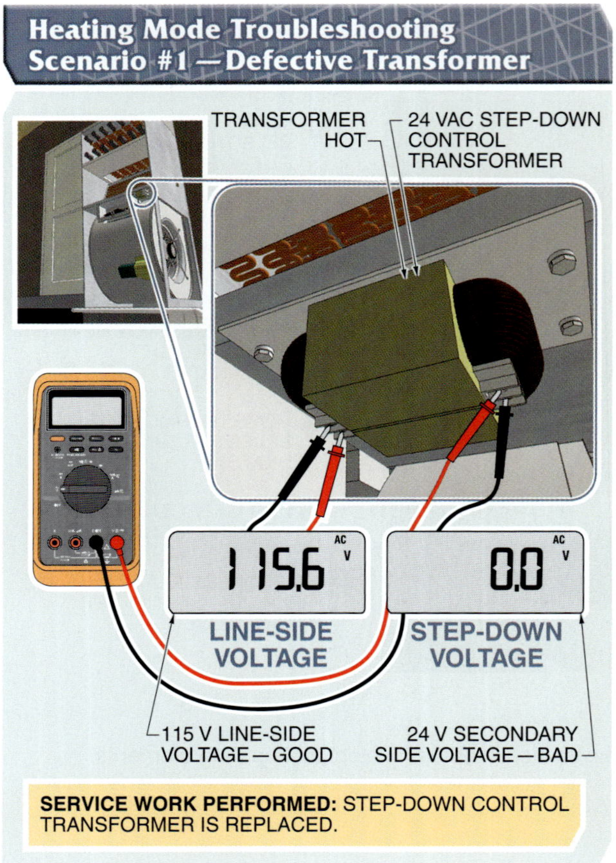

Figure 46-18. When nothing is running in a heat pump system, the best place to begin troubleshooting is with the power supply and control transformers.

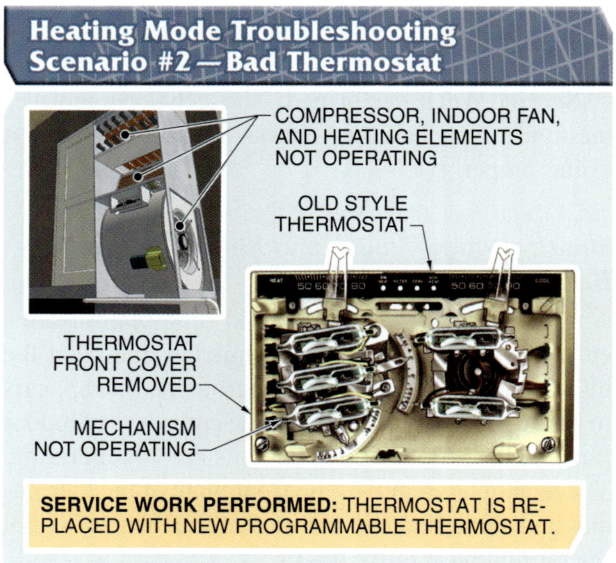

Figure 46-19. When power is evident, the thermostat should be checked to see if it is set for heat and is sending a signal calling for heat.

Heating Mode Troubleshooting Scenario #3—Broken Heating Element

On a very cold day, a technician receives a "too cold" complaint. After arriving, the technician finds that the compressor is running, but the back-up second stage electric elements are not energized. The thermostat is checked and found to be attempting to turn on the heating elements. After checking with a DMM in resistance mode, the technician finds that the elements are open. Further inspection reveals that one is broken. **See Figure 46-20.**

Service Work Performed: Once the broken heating element is replaced, the back-up heating element unit is checked and found to be working properly.

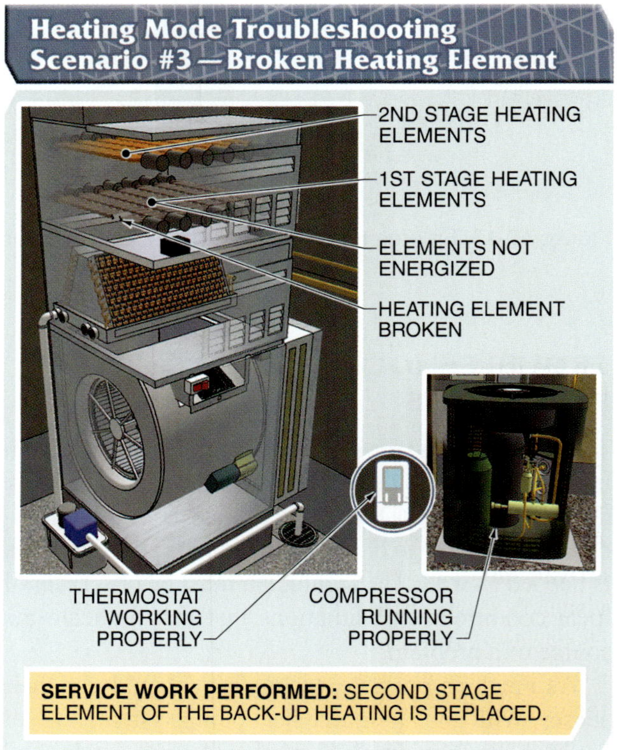

Figure 46-20. Broken electric heating elements can be found by manually looking for them, but using a multimeter set for resistance is a more reliable way of finding broken heating elements.

Heating Mode Troubleshooting Scenario #4—Faulty Start Relay

A customer complains that until recently, the heat pump was working but now their house is cold. The technician begins the troubleshooting process. The thermostat is calling for the compressor to run, but it is not running. The technician cycles the compressor off, then on again.

A momentary buzz is heard, but the compressor fails to start. The technician checks out the start relay for the compressor and finds it is not operating properly. **See Figure 46-21.**

Service Work Performed: The compressor start relay is replaced, and the compressor operates normally.

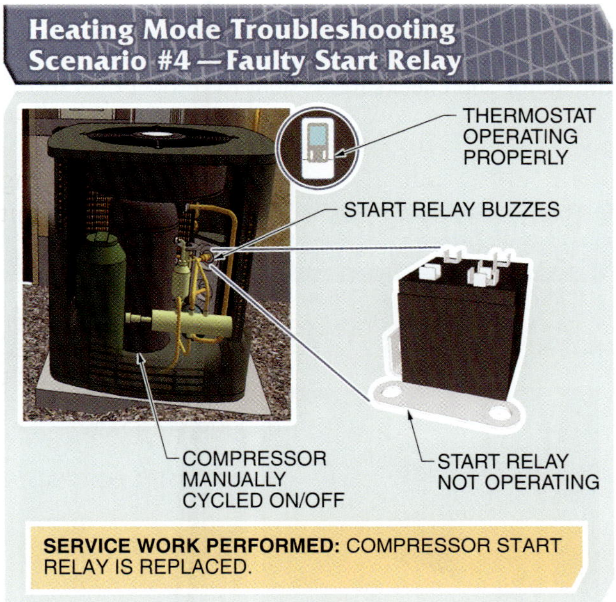

Figure 46-21. Relays, contactor coils, and starter coils are the most common electrical components to fail in an HVAC system.

Heating Mode Troubleshooting Scenario #5—Defrost Problems

A technician receives a "too cold" complaint. Upon arrival, the technician gives the outdoor unit a quick visual check and finds that it is iced over. The technician checks the defrost timer and finds that it has not initiated a defrost. The defrost timer is manually energized, and the ice begins to melt. **See Figure 46-22.**

Service Work Performed: The defrost timer is replaced and the unit is checked for normal operation.

Cooling Mode Troubleshooting Scenario #1—Dirty Air Filter

Homeowners complain that they are too hot. They have just moved into the house. A general system check shows that the compressor and indoor fan are running constantly. A visual check of the filters shows that they are completely plugged. **See Figure 46-23.**

Service Work Performed: The filters are replaced, and the homeowners are instructed to change them every 30 days.

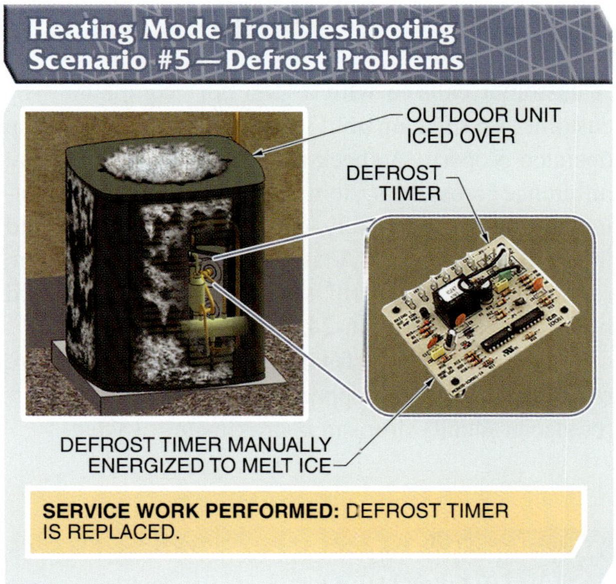

Figure 46-22. Mechanical and solid-state timers must always be checked when a heat pump system is not operating according to its sequence of operation.

Figure 46-23. Periodically changing air filters is the best preventive maintenance a homeowner can perform on their heat pump system.

Cooling Mode Troubleshooting Scenario #2—Duct Restriction

A customer calls in with a "too hot" complaint. The customer says that up until very recently, the heat pump operated properly. A check reveals that the temperature difference across the system is 25°F, with the fan operating properly. An unusually loud whistling noise is heard in the system ductwork. A visual check of the duct system reveals that the return grill is almost completely blocked by boxes. **See Figure 46-24.**

Service Work Performed: After the boxes are removed, the noise disappears and the temperature difference gradually drops to approximately 15°F.

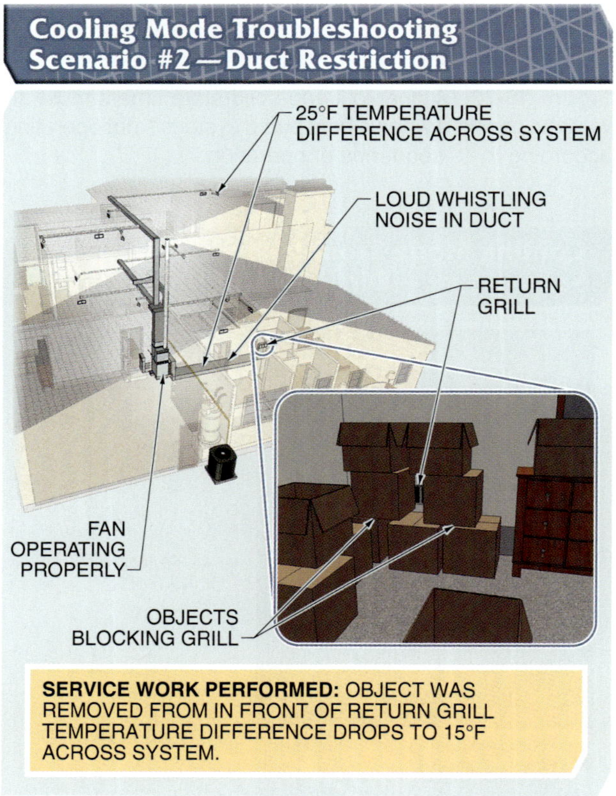

Figure 46-24. Any kind of noise coming from a system is a valuable piece of information when troubleshooting a heat pump system.

Cooling Mode Troubleshooting Scenario #3—Bad Fan Motor

A service technician arrives at a home on a very hot day. The compressor and outdoor unit fan are running, but the indoor fan motor is not. The technician uses a DMM to check the indoor fan motor and finds that a winding is open. **See Figure 46-25.**

Service Work Performed: The technician replaces the motor and checks the system for proper operation.

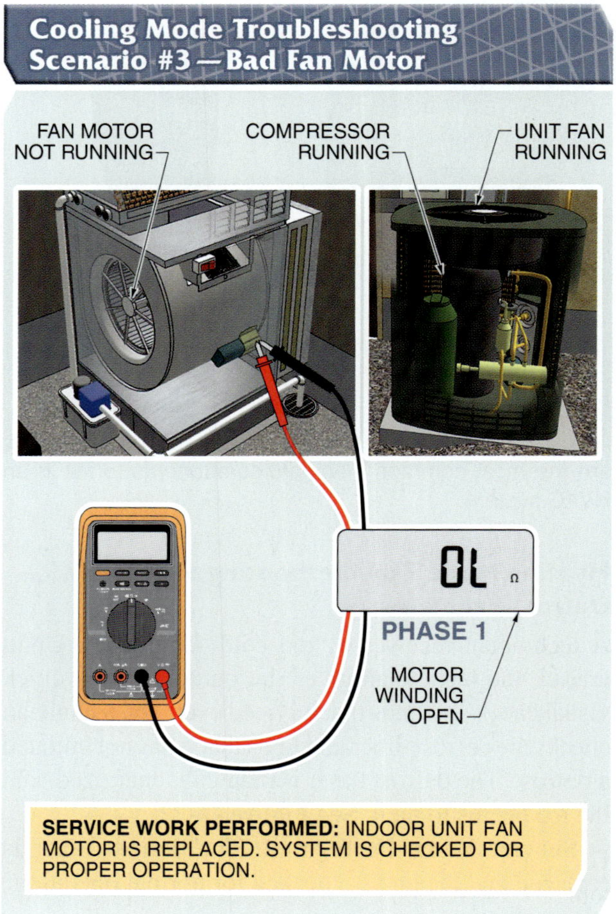

Figure 46-25. Using a multimeter set for resistance or a megohmmeter is the best method of troubleshooting the condition of an electric motor.

Installing, Servicing, and Troubleshooting Water-Source and Geothermal Heat Pump Systems

UNIT 47

Water-source and geothermal heat pumps are very popular in residential and light commercial HVAC applications. Because they use water instead of air as a heat transfer medium, they can be used in virtually any climate. Technicians must be familiar with water-source and geothermal heat pump systems in order to troubleshoot them properly. Geothermal heat pump installations are more popular than ever in the Midwest and Southwest of the United States.

OBJECTIVES

- Compare water-source and geothermal heat pumps.
- Identify common installation procedures and considerations for water-source and geothermal heat pumps.
- List the typical startup steps for water-source and geothermal heat pumps.
- Explain how to service water-source and geothermal heat pumps.
- Describe common troubleshooting problems for water-source and geothermal heat pumps.

Index to Procedures

WATER-SOURCE AND GEOTHERMAL HEAT PUMPS

There are many important differences between air-source heat pumps and water-source and geothermal heat pumps. These differences affect the installation, operation, service, and troubleshooting of these systems. Air-source heat pumps either reject or absorb heat from the ambient air outside a building. This is accomplished at the outdoor unit. The refrigerant from the outdoor unit is then piped to a refrigerant coil inside the indoor unit where air is either heated or cooled as needed. In an air-source heat pump, the compressor, valves, fan, and outdoor coil make up the outdoor unit.

Water-source and geothermal heat pumps do not have outdoor units. The two types of heat exchangers used in water-source and geothermal heat pumps are water-to-water and water-to-air. The water-to-water (typically larger tonnage) or water-to-air (small tonnage) heat exchangers are typically located inside the building. Ground water is pumped from a source to the heat pump, and the heat is rejected or absorbed from the water using a heat exchanger. **See Figure 47-1.**

The main difference between water-source heat pumps and geothermal heat pumps is the source of the water loop water. A water-source heat pump is typically connected to a water loop that has a cooling tower and a hot water boiler to cool and heat the water. A geothermal heat pump is typically connected to a geothermal ground loop that uses ground or pond water that is generally at a constant temperature.

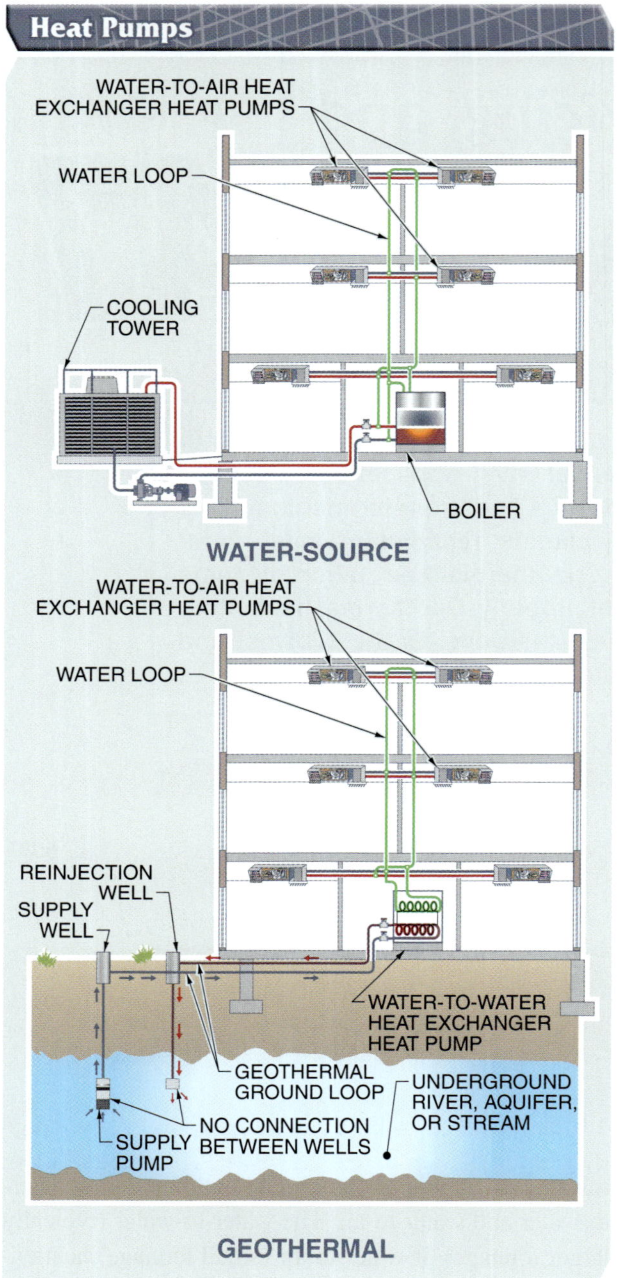

Figure 47-1. The two types of heat exchangers used in water-source and geothermal heat pumps are water-to-water and water-to-air. The differences between water-source heat pumps and geothermal heat pumps is the source of the water loop water.

Water-source and geothermal heat pump systems have water loop components that air-source heat pumps do not have. The water loop includes a pump, piping, and valves for control and possibly a cooling tower and boiler. The geothermal ground loop includes pumps, piping, valves, and ponds or wells. **See Figure 47-2.** *Note:* Water-source and geothermal heat pumps typically do not have electric back-up heating elements and the associated controls that air-source heat pumps have.

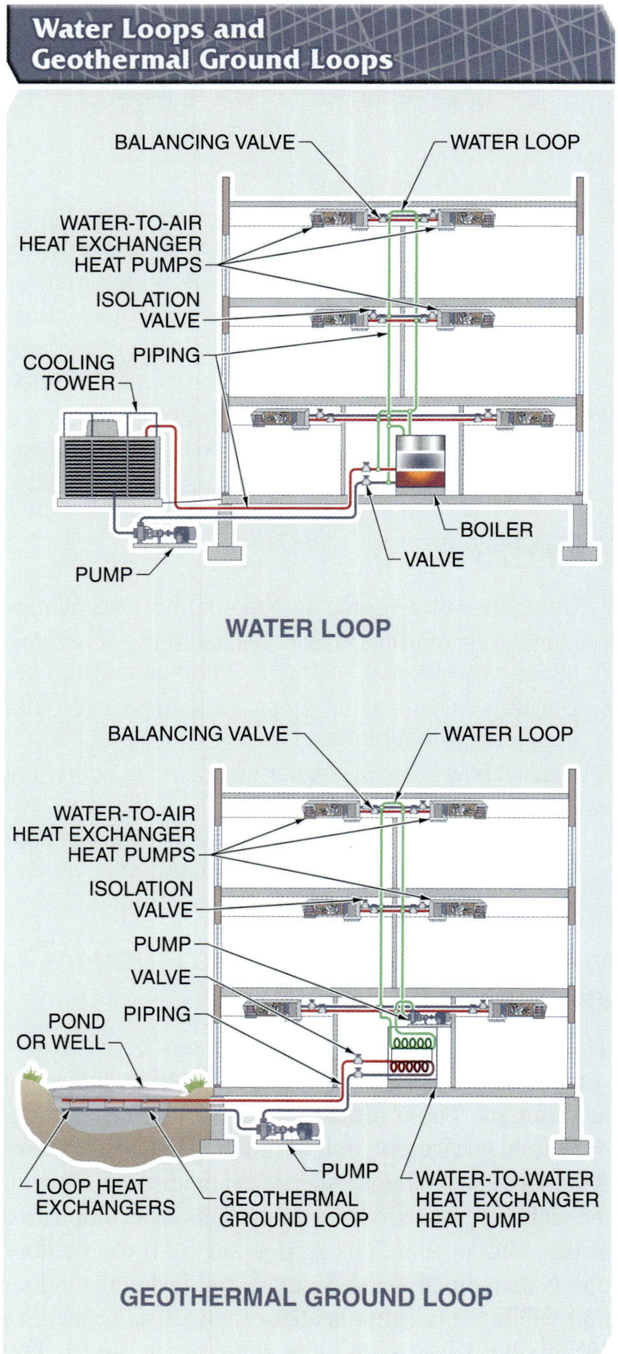

Figure 47-2. The water loop and the geothermal ground loop both require valves to isolate system components and to balance flow.

INSTALLING WATER-SOURCE AND GEOTHERMAL HEAT PUMP SYSTEMS

As with many packaged HVAC systems, water-source and geothermal heat pumps are shipped as a package from the supply house. There are a number of steps involved in the proper installation of water-source and geothermal heat pumps. Water-source and geothermal heat pumps consist of the cabinet, pump, water-to-water heat exchanger (may include second pump) or water-to-air heat exchanger (includes a fan), and controls. **See Figure 47-3.**

Water-Source and Geothermal Heat Pumps

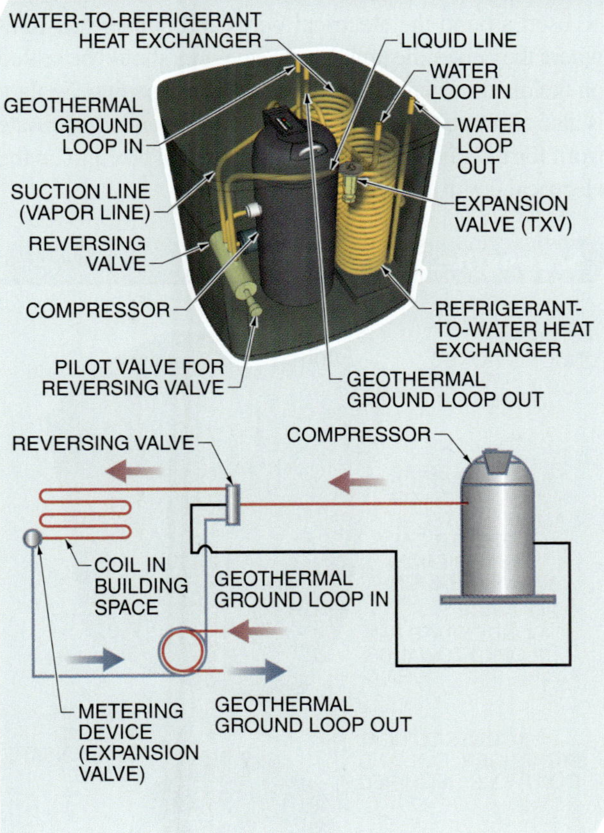

Figure 47-3. Water-source and geothermal heat pumps have many of the same components as the condenser of a split air conditioning system.

Compliance to all codes and regulations during installation is required. Water treatment of the indoor water loop and treatment of the outdoor geothermal ground loop are typically required to keep the system operating efficiently.

Heat Pump Requirements

The manufacturer's installation instructions must be followed closely when installing a water-source or geothermal heat pump. The installation instructions are intended as a general guide only and do not supersede any national or local codes in any way. All local, state, or national codes pertaining to water-source and geothermal heat pump systems must be determined prior to installation. **See Figure 47-4.** The manufacturer's installation instructions must be read completely before starting the installation.

Heat Pump Codes and Standards Examples

CODES AND STANDARDS

1. Heat pumps must comply with the latest edition of the *UL Standard for Safety for Heating and Cooling Equipment*, UL 1995 and CSA C22.2 No. 236.
2. With heat pumps intended for use in other than single-family dwellings, a clearly labeled emergency switch shall be provided as part of the installation. The switch shall be readily accessible to the occupants and shall be installed at least five feet away and within sight of the unit's cabinet.
3. Leave a minimum of three inches extra threaded rod below the double nuts or minimum three inches of clearance between the top of the unit and the ceiling when suspending the heat pump.
4. Many heat pumps contain HFC-(R-410A), an azeotropic mixture of R-32 and R-125. Do not vent HFC-(R-410A) to the atmosphere. The U.S. Clean Air Act requires the recovery of any residual refrigerant. Do not use R-22 service equipment or components on R-410A systems.

Figure 47-4. Most water-source and geothermal heat pump models are designed for indoor installation only.

With some exceptions, most water-source and geothermal heat pump models are designed for indoor installation only. The installation of the heat pump, field wiring, terminal unit warm air ducts, and other components must conform to the requirements of the National Electrical Code® (NEC®) and ANSI/NFPA No. 70 (latest edition) in the United States and the Canadian Electrical Code™ (CEC™), CSA C22.2-12, in Canada.

Any local authorities having jurisdiction should be consulted before installation is begun. In some cases, a permit may be needed. In case of conflict, applicable regulations or requirements take precedence over the general instructions provided by the manufacturer. Conditioned and return air plenums, ducts, and air filters must be installed in accordance with NFPA 90B, *Installation of Warm Air Heating and Air-Conditioning Systems* (latest edition).

Water-source or geothermal heat pump units are provided with pipe connections for the indoor water loop and the outdoor geothermal ground loop. Water-to-water heat exchangers will have two sets of pipe connections (geothermal ground loop and water loop), and water-to-air heat exchangers will have one set of pipe connections (water loop). **See Figure 47-5.**

Heat Pump Pipe Connections

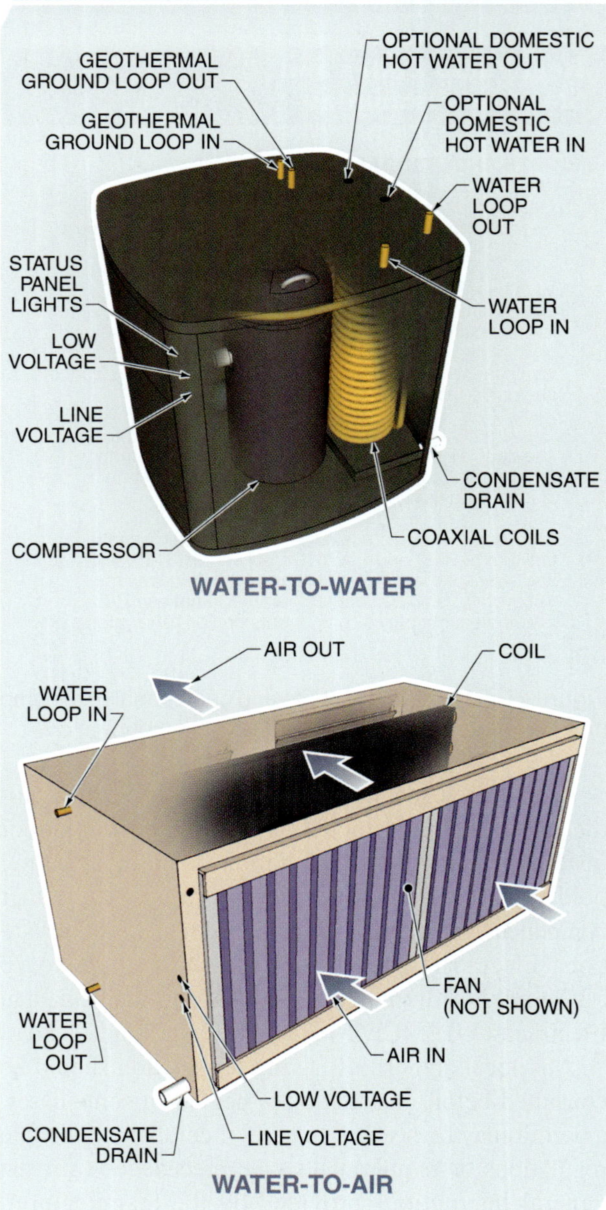

Figure 47-5. Water-to-water heat exchangers will have two sets of pipe connections (geothermal ground loop and water loop), and water-to-air heat exchangers will have one set of pipe connections (water loop).

The manufacturer ships the unit from the factory fully assembled. Cabinet knockouts must not be removed until it has been determined which knockouts need to be removed for installation. The installation position that best suits the site conditions must be selected. Required clearances, routing requirements for water piping, condensate disposal, control valves, wiring, and accessibility for service must all be considered. The rating plate on the heat pump should be checked for specific information.

Location Requirements

Water-source and geothermal heat pump units may sweat during cooling mode when the heat pump is operating in a very humid space. When water-source and geothermal heat pumps are installed in an unconditioned space, sealant must be used around the electrical wires and condensate lines where they enter the unit. Electrical wires should be sealed on the inside where they exit the conduit opening. Sealant is also required to prevent air leakage into and condensate from forming inside the heat pump control box and on the electrical controls. **See Figure 47-6.**

Sealing Heat Pump Cabinets

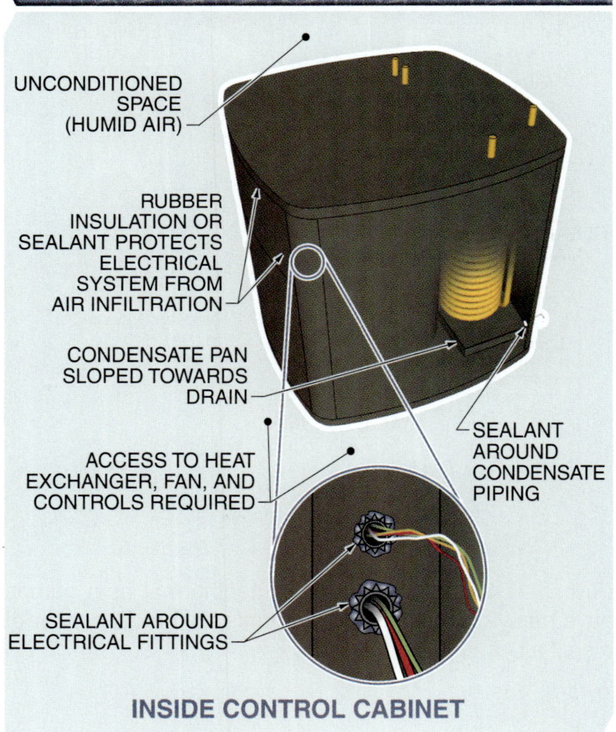

Figure 47-6. Heat pumps installed in an unconditioned space must have sealant around the electrical wires and condensate lines where they enter the unit. Electrical wires should be sealed on the inside of the unit where they exit the conduit opening to prevent air leakage into the control box.

Condensate Drain Requirements

Terminal units must not be operated without a working drain. The terminal unit drain line must be routed to an appropriate drain pipe. Drain lines must be installed so they do not block service access to the terminal unit. The heat pump's heat exchanger condensate drain pan must have a minimum slope (drop) in the horizontal position towards the drain to ensure proper drainage. A trap must be part of the drain to eliminate any possible siphon effect.

Many manufacturers specify a ¾″ per foot slope for the condensate pan. The manufacturer should be consulted for more details. **See Figure 47-7.** *Note:* All local codes should be checked before the drain line is connected to an existing drainage system.

Condensate Drain Requirements

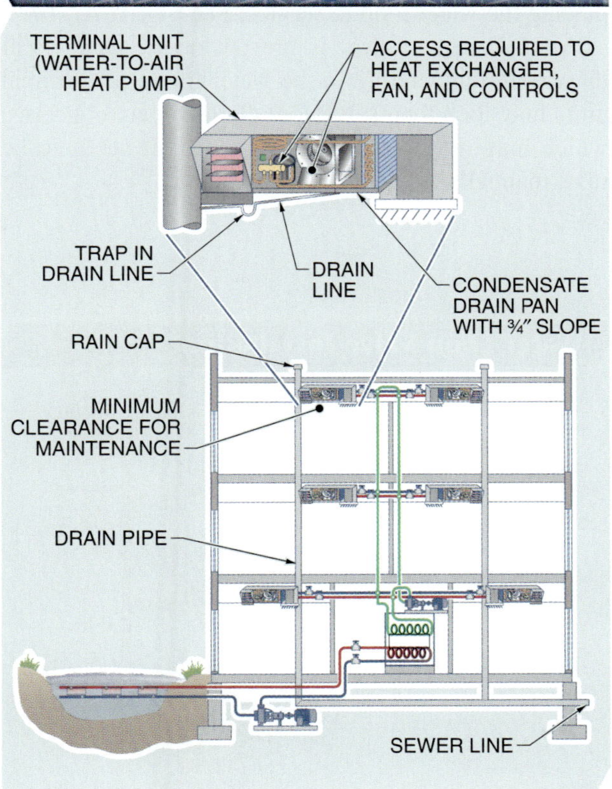

Figure 47-7. A heat pump's condensate drain pan must have a minimum slope (drop) in the horizontal position towards the drain to ensure proper drainage, such as a ¾″ per foot slope.

A minimum clearance is required for service access. The heat pump must also be installed in such a way as to allow free access to the heat exchanger, fan, and control compartments.

Electrical Requirements

The entire manufacturer installation manual should be read before starting the installation of a water-source heat pump. All warnings, cautions, instructional labels, and tags should be observed. Failure to comply with the manufacturer instructions could result in an unsafe condition and/or premature component failure.

Any instructions provided are intended as a general guide only and do not supersede any national or local codes in any way. The installation must comply with all provincial, state, and local codes as well as the National Electrical Code® (NEC®) in the United States and the Canadian Electrical Code™ (CEC™) in Canada.

Electrical wiring, disconnect means, and overcurrent protection are typically supplied by the installer. The rating plate on the unit will state the maximum overcurrent protection required, minimum circuit ampacity, and the proper operating voltage. Water-source and geothermal heat pump power supplies must be sized and protected according to the specifications supplied on the rating plate. **See Figure 47-8.** *Note:* Only copper conductors can be used.

Geothermal Heat Pump Rating Plates

Figure 47-8. The rating plate is used to size system electrical components and piping and provide low- and high-side pressures for the refrigerant used.

WATER LOOP INSTALLATION

It is critical that the water loop be installed properly. This installation can be split into two major categories depending on system type. These categories are boiler and cooling tower systems and geothermal systems. State and local codes must be followed and installation must conform to all applicable codes. It is the responsibility of the installing contractor to determine and comply with all applicable codes and regulations. **See Figure 47-9.**

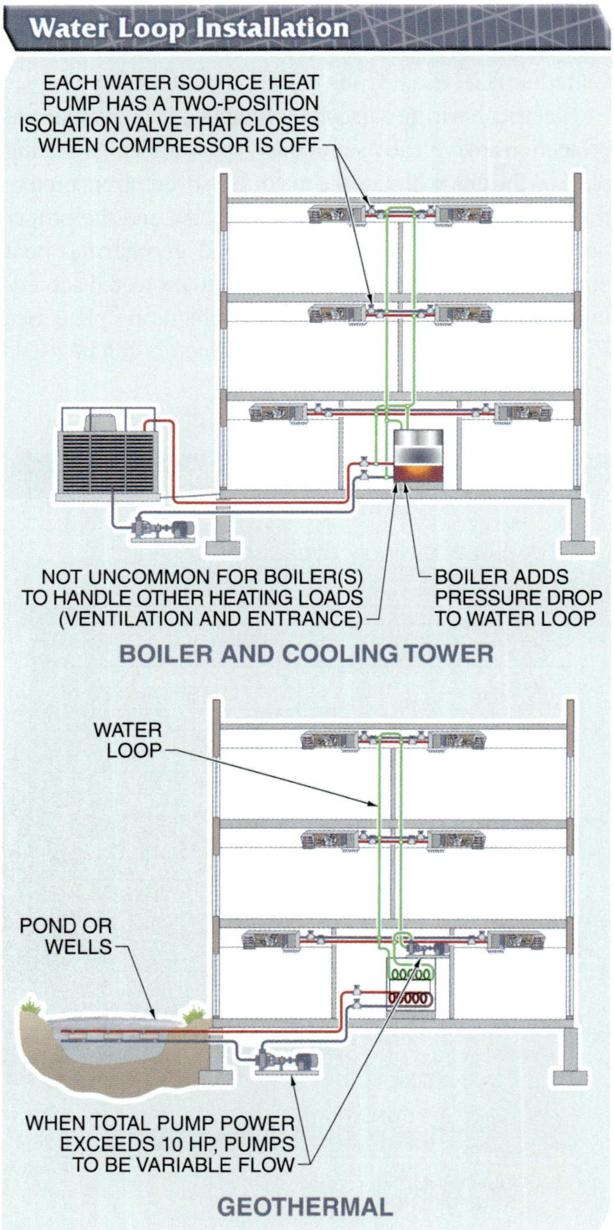

Figure 47-9. It is the responsibility of the installing contractor to determine and comply with all ordinances, state and local codes, and regulations.

The water loop for the boiler and cooling tower system should be installed according to manufacturer recommendations as these systems vary from manufacturer to manufacturer. After the water loop piping is installed, the system must be flushed and then filled with treated water.

Valve Installation

Shut-off valves are provided at the water-source or geothermal heat pump units that must be opened before operation begins. Balancing valves are used to adjust water flow to the various terminal units before placing the water loop in operation. The technician should keep in mind that supply and return stubs and shut-off valves are often installed in a manner that allows for future system expansion. The terminal unit pumps, condensate drains, and controls must be checked for proper operation before placing the water loop in service. **See Figure 47-10.**

Note: When no flow is present or the water is at the wrong temperature in the water loop, the terminal units may lock themselves out due to safety concerns, which may mean that each terminal unit needs to be reset manually.

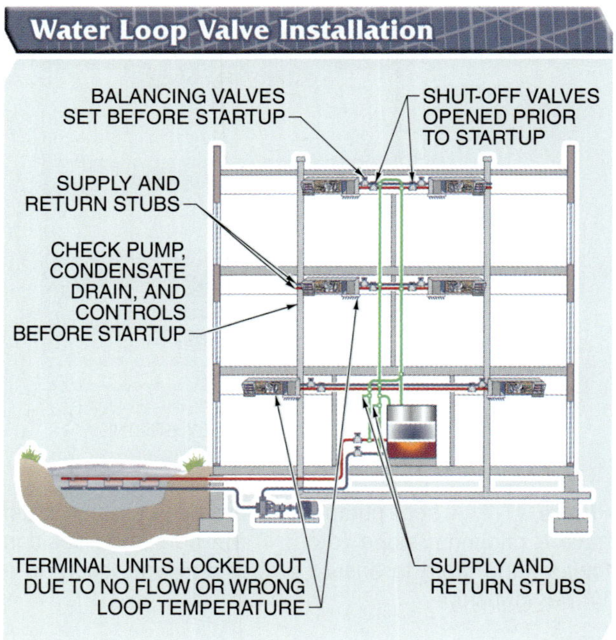

Figure 47-10. The valves in the water loop are used for isolating heat pumps, balancing flow, and shutting everything down when the system is locked out due to poor flow or incorrect temperature in the water loop.

Piping Connections

All piping connections to the water loop must be made using good plumbing practices and in accordance with all applicable local codes. Boiler and cooling tower systems are typically connected to supply and return piping in a two-pipe reverse-return configuration. A reverse-return system is inherently self-balancing and requires only trim balancing in the case of heat pumps with different flow and pressure drop characteristics that are connected to the water loop.

A simple way to check for proper water balance is to take a differential temperature measurement across the water connections when the terminal units are in the cooling mode. **See Figure 47-11.** A rule of thumb for proper water flow is a 10°F to 15°F temperature difference. The piping can be steel, copper, or PVC but must comply with local codes. Also, high-pressure flexible hoses can be used on the supply and return piping to the terminal units. These hoses also act as sound attenuators for an operating unit and for hydraulic pumping noise.

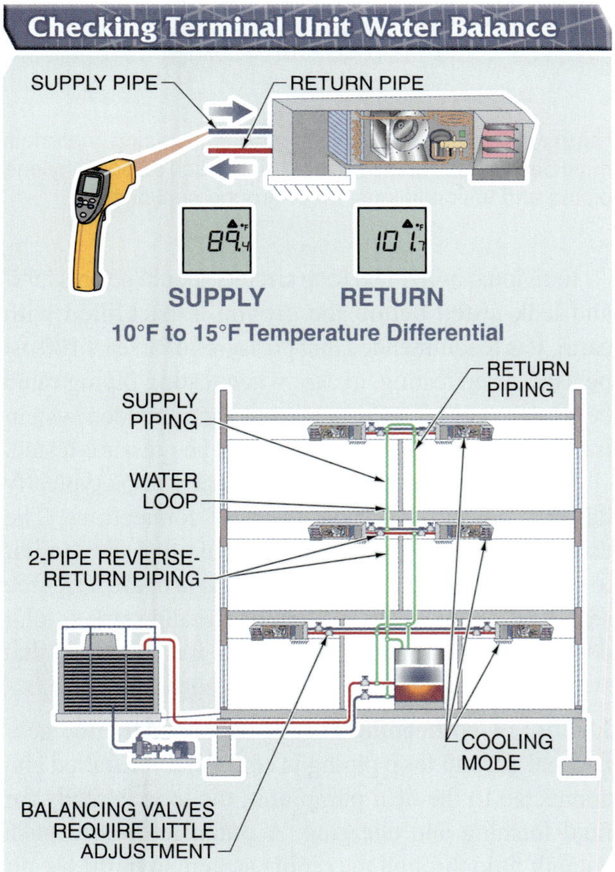

Figure 47-11. The simple way to check for proper water balance to a heat pump terminal unit is to take a differential temperature measurement across the water connections with the unit in cooling mode.

GEOTHERMAL GROUND LOOP INSTALLATION

Industry-accepted installation practices for closed-loop earth-coupled heat pump systems must be used. All existing underground utilities and piping must be located and marked to avoid damage and liability. **See Figure 47-12.** When possible, geothermal ground loops should be installed before construction of sidewalks, patios, and driveways has begun. All ground loop piping should be marked on the plot plan drawings as an aid to avoiding damage and any future problems.

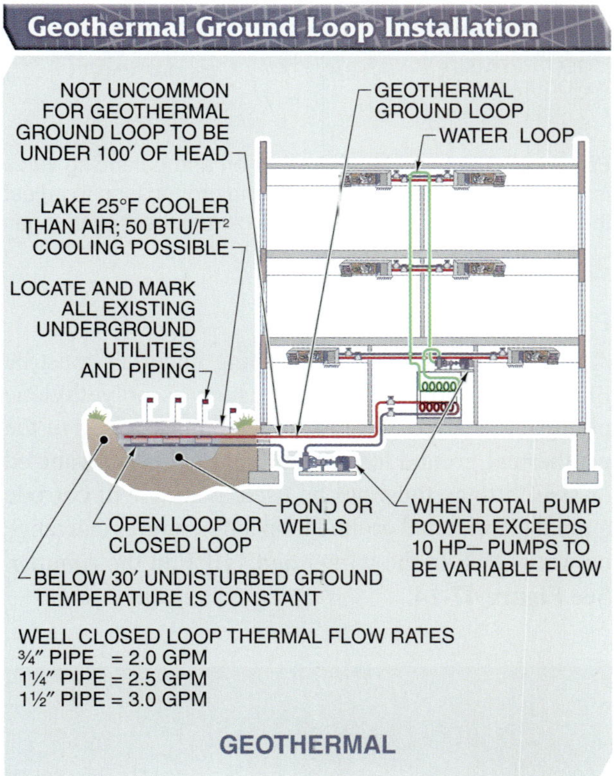

Figure 47-12. Along with marking the geothermal ground loop, all underground utilities and piping must also be located and marked to avoid damage and liability.

Valve Installation

The heat pump requires a shut-off valve on both the ground loop supply and return lines for easy servicing of the loop and heat pump. A combination shut-off and balancing valve is often used on the return side of the ground loop to adjust the system flow through the water-to-water heat exchanger. **See Figure 47-13.**

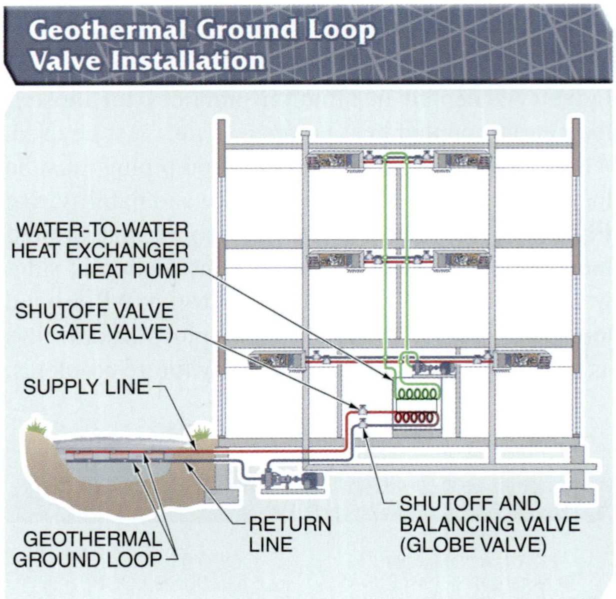

Figure 47-13. A combination shut-off and balancing valve is often used on the return side of the ground loop to adjust the system flow through the water-to-water heat exchanger.

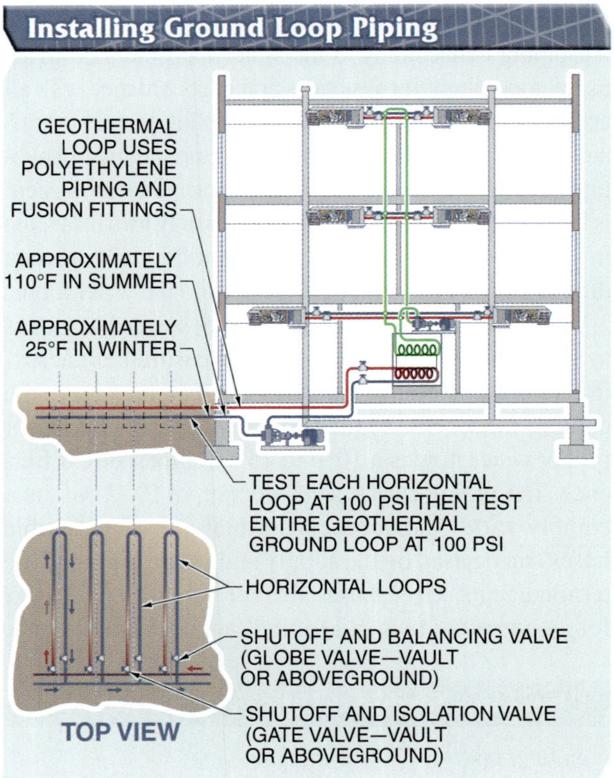

Figure 47-14. Geothermal ground loop piping materials must be limited to approved materials, such as polyethylene piping and fusion fittings, for in-ground sections.

Piping Installation

All geothermal ground loop piping materials must be limited to approved materials, such as polyethylene piping and fusion fittings, for in-ground sections of the geothermal ground loop. Materials such as galvanized or steel fittings must not be used as they will corrode quickly. Horizontal ground loop temperatures can range between 25°F in the winter and 110°F in the summer. **See Figure 47-14.**

Noritz America Corp.

Geothermal ground loops typically use copper or polyethylene piping and fittings. However, copper is susceptible to deterioration when exposed to some of the chemicals used in these ground loops.

Individual horizontal loop circuits should be pressure- and leak-tested before the ground is backfilled with earth. It is recommended that pressures of at least 100 psi be used when testing, except when testing piping rated below 100 psi. After the geothermal ground loop system is assembled, the entire system must be pressure-tested.

Water-source and geothermal heat pumps typically have swivel pipe fittings for water connections. The connections have rubber gasket seals that are similar to garden hose gaskets. These fittings provide leak-free seals, eliminating the need for thread sealing tape or joint compound. **See Figure 47-15.** *Note:* It is important that swivel pipe fittings not be overtightened.

Flushing Geothermal Ground Loops. After the geothermal ground loop piping is completely installed and connected to the heat pump unit, the loop is ready for final flushing and charging. A pump is used to push enough fluid through the piping system to purge the air and flush any dirt particles from the loop. Fittings can be loosened to allow air to vent. All air and debris must be removed from the ground loop piping before actual startup occurs. **See Figure 47-16.**

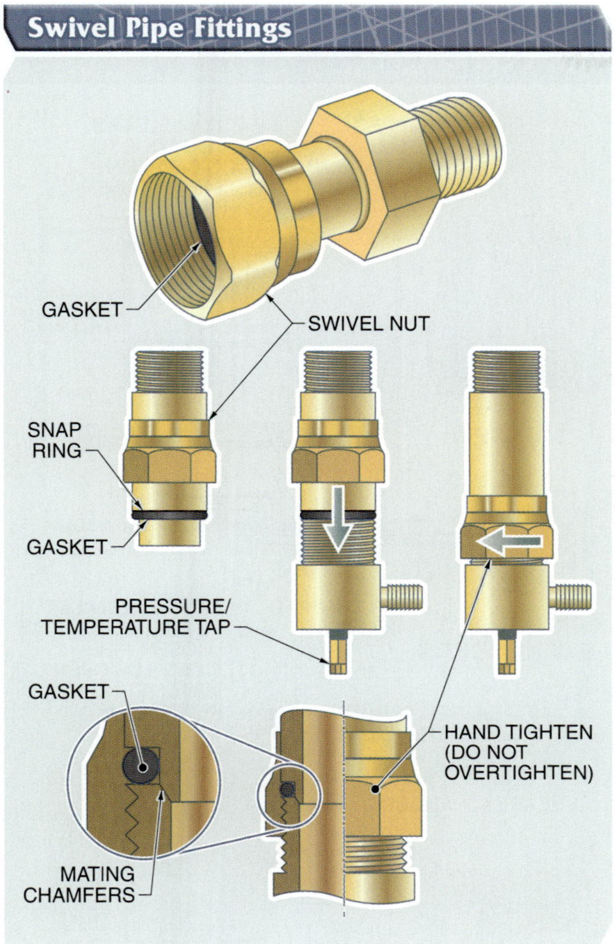

Figure 47-15. Swivel pipe fittings provide leak-free seals because they use rubber gasket seals that are similar to garden hose gaskets, eliminating the need for thread sealing tape or joint compound.

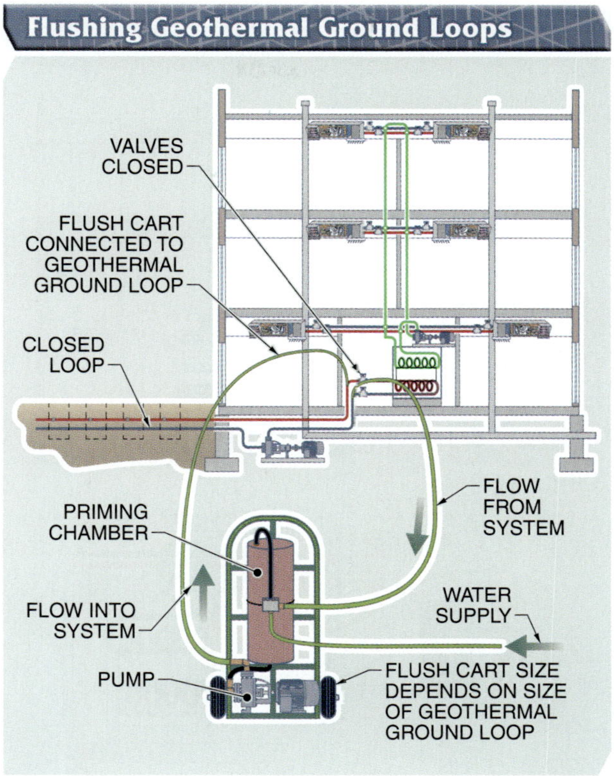

Figure 47-16. A pump is used to push enough fluid through the geothermal ground loop piping system to purge the air and flush any dirt particles from the loop.

Specialized technicians that perform geothermal loop flushing use custom-built flush carts that provide proper flow rates and velocities and flush the loop in both directions. Typically, when bubbles are no longer seen during loop flushing, the system flow is reversed to force any trapped air or sediment back through and out of the piping. It typically takes about two hours to perform a thorough loop flush during initial startup. Loop flushing continues until all air and sediment is removed from the geothermal heat exchanger.

Note: Before connecting the plastic ground heat exchanger loops to the header, each loop should be thoroughly flushed and purged in accordance with International Ground Source Heat Pump Association (IGSHPA) 21020 recommendations and filled with clean water. When a loop is not immediately joined to the header, it must be taped or capped.

An antifreeze solution is used in most areas to prevent the ground loop piping from freezing during the winter. The unit should be run in either heating or cooling for a number of minutes to condition the loop to a consistent temperature. The technician should ensure that there is adequate flow through the heat pump unit by checking the pressure drop across the heat exchanger and comparing the reading to the manufacturer specifications.

Open-Loop Geothermal Ground Loops. Open-loop geothermal ground loop systems will include shut-off valves for ease of servicing. Heat pump drains and/or other valves must be "tee'd" into the lines to allow acid flushing of the heat pump's heat exchanger. Shut-off valves must be positioned to allow acid flow through the tube-in-tube heat exchanger via the drains without allowing flow into the piping system. Pressure drop and temperature should be measured. **See Figure 47-17.**

Only copper or PVC SCH80 should be used as piping materials for the aboveground ground loop. Copper is typically used in closed-loop water systems and open-loop ground water systems where the soil is not high in mineral content or corrosiveness.

Acid Flushing a Heat Exchanger

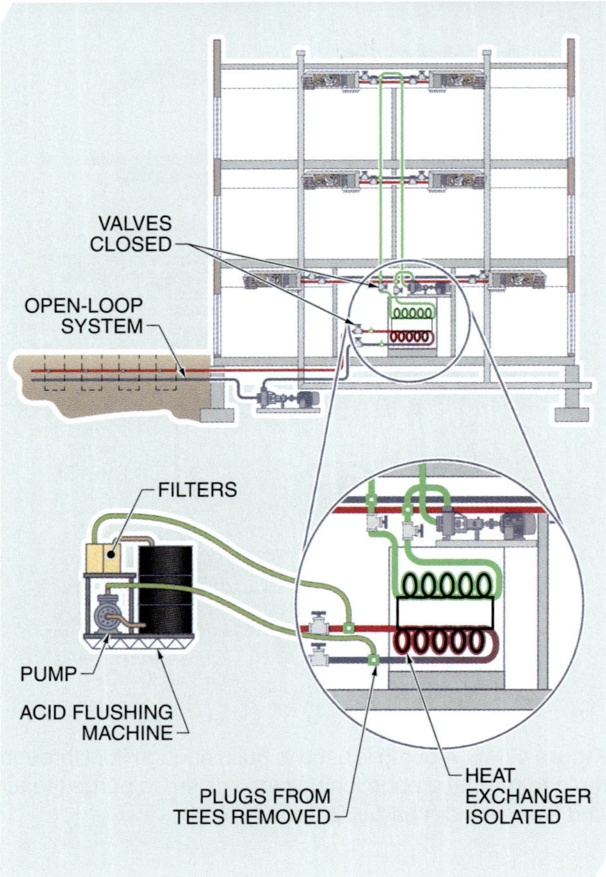

Figure 47-17. Shut-off valves must be positioned to allow acid flow through the tube-in-tube heat exchanger of the water-to-water heat pump via the drains without allowing flow into the piping system.

Air Trapped in Geothermal Heat Exchangers

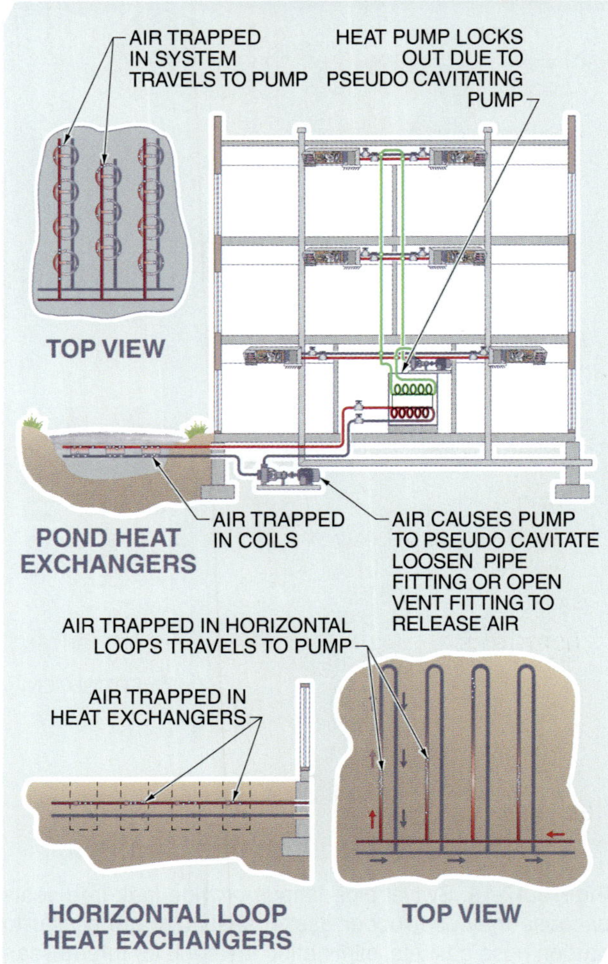

Figure 47-18. Air trapped in a closed-loop geothermal ground loop heat exchanger will cause one or more loops to become air-locked and can cause the circulating pump to pseudo cavitate.

When heavy internal scale formation is anticipated, a special cupronickel material heat exchanger is recommended. Heat exchanger coils will lose efficiency over time due to the buildup of mineral deposits. Acids and special pumping equipment are required to keep the heat exchanger clean. This work is typically performed by specialized technicians.

Closed-Loop Geothermal Ground Loops. If the water or water and antifreeze stop flowing, the geothermal heat pump will quickly lock out due to no fluid flow through the water-to-water heat exchanger. Air that is trapped in the closed-loop geothermal ground loop heat exchanger will cause one or more loops to become air-locked and can cause the circulating pump to pseudo cavitate. **See Figure 47-18.**

Heat Pump Thermostat and Controls Installation

Water-source and geothermal heat pumps are controlled by either a standard 24 VAC thermostat or a unit-mounted thermostat (console type) or are connected to a direct digital control (DDC) system.

Standard thermostat installation is accomplished in the same manner as with any packaged HVAC unit. The thermostat should be located on an interior wall in a large room, away from supply duct drafts. The thermostat should not be installed in areas subject to sunlight or drafts or on external walls. Sometimes the wire access hole behind the thermostat will need to be sealed to prevent erroneous temperature measurements. **See Figure 47-19.**

Thermostat Locations

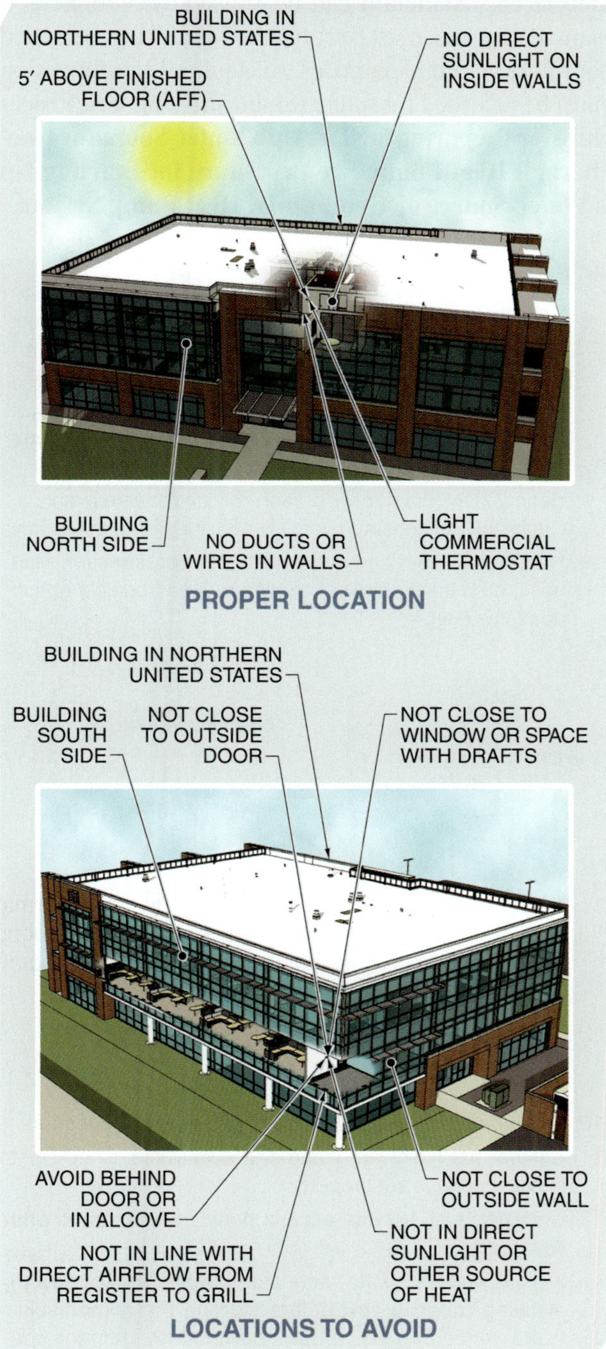

Figure 47-19. Thermostats should be located on an interior wall in a large room, away from supply duct drafts and sunlight, and the wire access hole behind the thermostat should be sealed to prevent erroneous temperature measurements.

The thermostat back plate should be positioned against the wall so that it appears level and so that the thermostat wires protrude through the middle of the back plate. Thermostat wire is typically 18 AWG. The low-voltage output from the low-voltage terminal strip is AC voltage to the wall thermostat. Most manufacturers make heat pump thermostats that will work with water-source and geothermal units, provided the units have the correct number of heating and cooling stages.

Most heat pumps will energize the reversing valve when set in the heating mode. Most circuit boards will have a fan interlock circuit to energize the fan whenever the compressor is on. Many thermostats use logic systems for these functions.

Also, many circuit boards today will have a number of safeties. Typically, they have lockout circuits to stop compressor operation when any one of the safety devices is triggered, such as the high-pressure switch, low-temperature sensor, condensate overflow sensor, or water flow switch. **See Figure 47-20.** When the condensate overflow sensor (float or water detection switch) detects a filled drain pan, the operation of the heat pump will be suspended during cooling mode.

Condensate Overflow Sensors

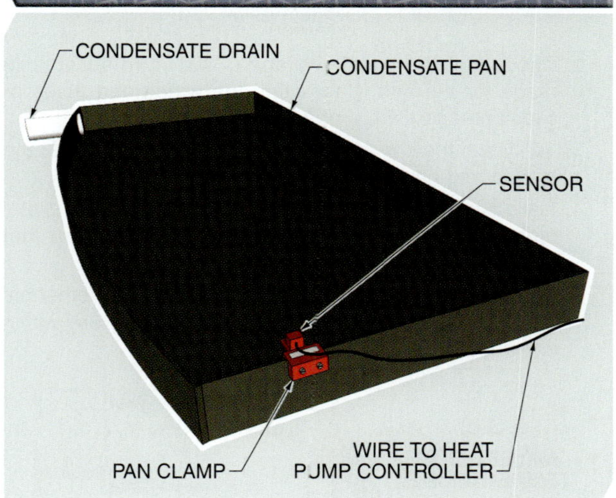

Figure 47-20. During cooling mode, when the condensate overflow sensor (float or water detection switch) detects a filled drain pan, the operation of a water-source heat pump will be suspended.

When any of the safety devices trip, the geothermal heat pump must be reset. This is accomplished by opening and closing the disconnect switch on the main power supply to the unit. Typically, a heat pump does not have to be reset on condensate overflow detection. Many thermostats will have an LED that alerts the occupants that a fault condition has occurred. Another LED is sometimes available on the thermostat to indicate a dirty filter condition.

DDC systems are often used in light commercial applications with multiple heat pumps. They are wired similarly to standard thermostats, using the terminal strip. The only additional wiring connection needed is the attachment of the heat pump digital controller to the DDC system network. This is typically accomplished using 18 AWG to 22 AWG twisted shielded wire. The heat pump controller must also be programmed so that it works properly with the DDC system.

STARTUP OF WATER-SOURCE OR GEOTHERMAL HEAT PUMPS

Before an installation can be considered complete, a start-up checklist must be completed and all data recorded for future warranty purposes. All equipment information must be recorded for future troubleshooting. **See Procedure for Preparing to Start Up a Water-Source or Geothermal Heat Pump. See Procedure for Starting Up a Water-Source or Geothermal Heat Pump System.**

Procedure

Preparing to Start Up a Water-Source or Geothermal Heat Pump

1. Ensure that all isolation valves are open and that all water control valves are in their proper positions for system startup.

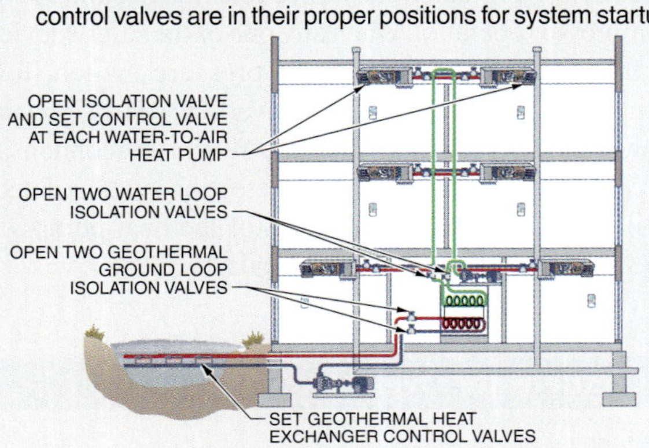

2. Check that the supply voltage is within an acceptable range for the unit. Check that the wiring and fuses or circuit breakers are properly sized.
3. Verify that the low-voltage wiring is complete and that the transformer has the correct secondary voltage.
4. Make sure that the ground loop water piping is complete, purged of air, and clean. If necessary, make sure that antifreeze has been added.
5. Check that the temperatures of the entering geothermal ground loop water and the water loop water are within operating limits.

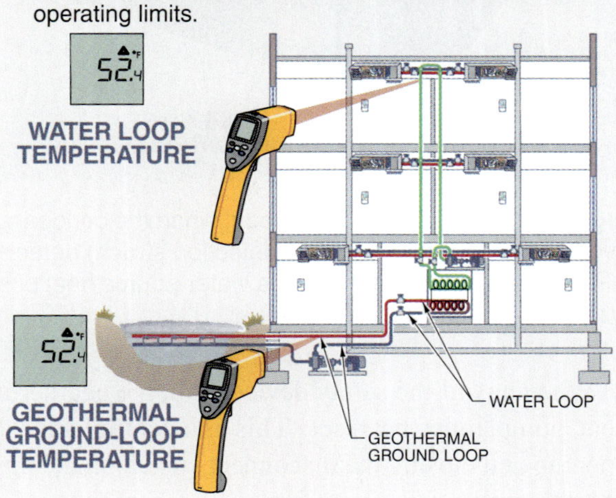

6. Verify that the low-water temperature control has the correct setpoint and is working properly.
7. Check that the fan of each heat pump unit can be rotated manually to verify free rotation and that each blower wheel is secured properly.
8. Check that the condensate line is piped properly, the trap is clear, and the drain line is properly pitched toward the drain. Check the condensate pump, if used, for proper operation.
9. Record both inlet and outlet water temperatures for each heat pump upon startup. This can reduce troubleshooting problems later.

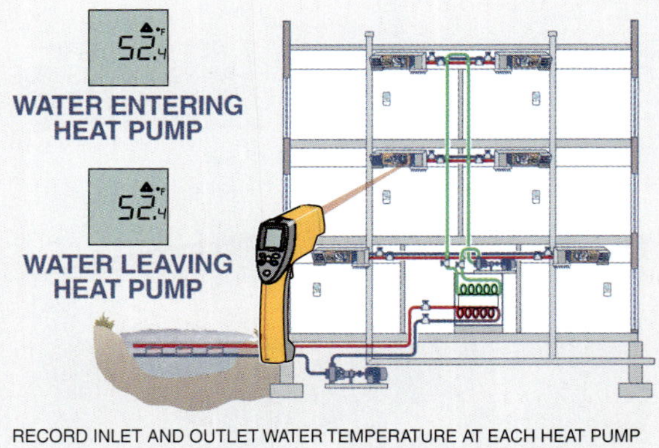

10. Clean air coils using a light detergent solution if needed.
11. Ensure that filters are clean and accessible. Check that blower speed is set properly.
12. Check that all service access panels are in place and secured.

Note: Make sure the water control valves are open and allowing water to flow before starting the compressor. If the water freezes, it may permanently damage the heat pump.

Starting Up a Water-Source or Geothermal Heat Pump System

1. Turn the thermostat fan to the ON position. The blower should start.
2. Adjust all valves to their full open position. Turn on the line power to all heat pump units.
3. Make sure that both the space air temperature and water loop temperature are within the minimum and maximum temperature ranges provided by the manufacturer.
4. Adjust the unit thermostat to a warm setting. Place the thermostat mode switch in the COOL position. Slowly increase the thermostat setting until the compressor activates.
5. Once the unit has been running for a few minutes, check for cool air delivery at the unit register.

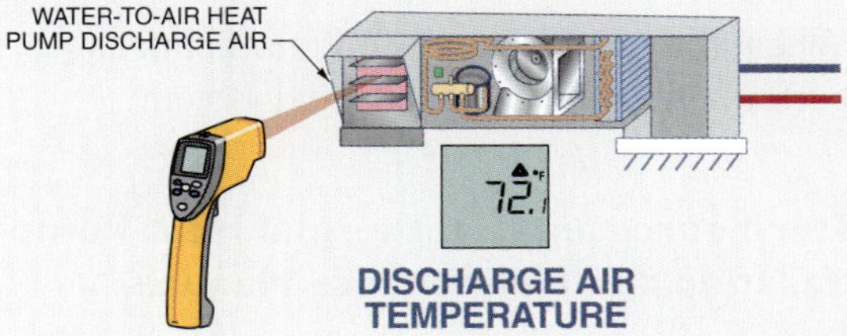

6. Check the air temperature drop across the air coil when the compressor is operating. For most water-source and geothermal heat pumps, the temperature drop should be between 15°F and 25°F. Turn the thermostat to OFF position. The reversing valve will click and hiss if operating properly.

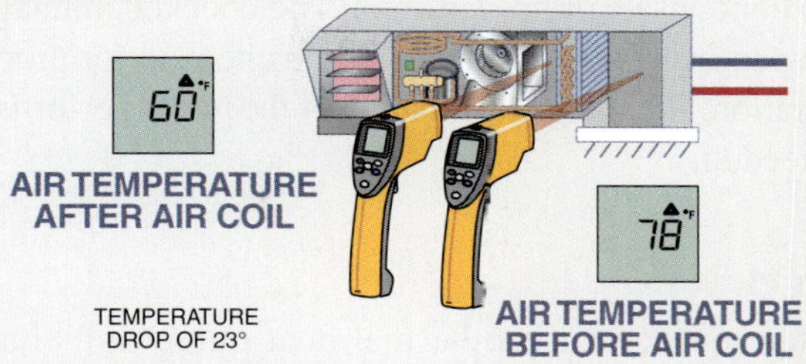

7. The compressor should be off a minimum of five minutes to allow the pressure to equalize before beginning the heating test.
8. Adjust the thermostat to the lowest setting. Place the thermostat mode switch in the HEAT position.
9. Slowly increase the setpoint temperature until the compressor starts.
10. Check for warm air delivery after the unit has been running for a few minutes.
11. Check the air temperature rise across the air coil when the compressor is operating. For most water-source and geothermal heat pumps, the air temperature rise should be between 20°F and 30°F.
12. Check for vibration, noise, and water leaks.
13. Check the elevation and cleanliness of the condensate lines and pan in the cooling mode.
14. Troubleshoot the unit if it operates improperly.
15. After testing, set the system to maintain the temperature desired for comfort.
16. Ensure that the warranty paperwork is properly filled out and the information recorded.

SERVICING WATER-SOURCE AND GEOTHERMAL HEAT PUMP SYSTEMS

Water-source and geothermal heat pumps require routine maintenance on system components such as filters, compressors, condensate drain pans, and fan motors. **See Figure 47-21.** Depending on the system, other components may also require routine maintenance.

Filters

Filters must be clean to obtain maximum performance from the unit. Filters should be inspected every month under normal operating conditions and be replaced as needed. Filters must not be used longer than 60 days. Heat pumps should never be operated without a filter.

Compressors

The amperage must be checked to ensure that current draw is no more than 10% greater than what is indicated on the data plate. A comparison of logged information with start-up and other annual information is useful as an indicator of compressor condition.

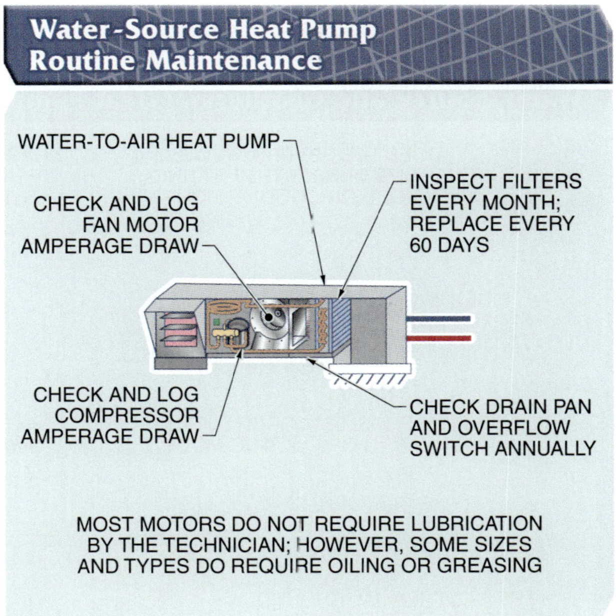

Figure 47-21. The manufacturer's owner or user manual should be consulted for the routine maintenance that must be performed on a water-source or geothermal heat pump.

Condensate Drain Pans

The condensate drain pan should be checked annually. The pan overflow switch should be checked for proper operation. Cleaning and flushing of the pan is performed as needed.

Fan Motors

Water-source and geothermal heat pump units have lubricated fan motors. They typically do not require additional lubrication by the technician. However, some larger sizes of motors have oil caps or grease fittings, so lubrication can be performed in accordance with the label on the motor or heat pump. The motor amperage should be checked to ensure current draw is not more than 10% greater than what is indicated on the data plate.

Basic Troubleshooting Questions

Before troubleshooting a water-source or geothermal heat pump, the technician must acquire as much information as possible about the equipment in question. **See Figure 47-22.** An absolutely vital step is to ask the equipment owner or technician basic water-source and geothermal heat pump troubleshooting questions such as the following:

- How long has the problem occurred?
- When was the last time the water-source or geothermal heat pump worked correctly, or did it ever work correctly?
- Does the heat pump problem occur during the heating or cooling mode?
- Is the problem with one area of the building or the entire building?
- Does the problem seem to be with the water loop?
- Does the problem seem to be with the geothermal ground loop?
- Does the reversing valve energize in the heating or cooling mode?
- Has any work been done to the system recently?
- Have there been any recent changes to the system, such as adding an internal load or a change in duct configuration?
- When was the last time routine maintenance was performed on the heat pump and system?

Water-Source and Geothermal Heat Pump Installation and Maintenance Manuals

Manufacturer manuals for water-source and geothermal heat pump equipment typically contain manufacturer specific heat pump and geothermal facts, sequences of operations, adjustment procedures for specific applications, general service guides, preventive maintenance procedures, wiring schematics, piping schematics, and warranty information. The information included in installation and maintenance manuals can help the technician develop troubleshooting-type thinking, which is used when diagnosing system problems.

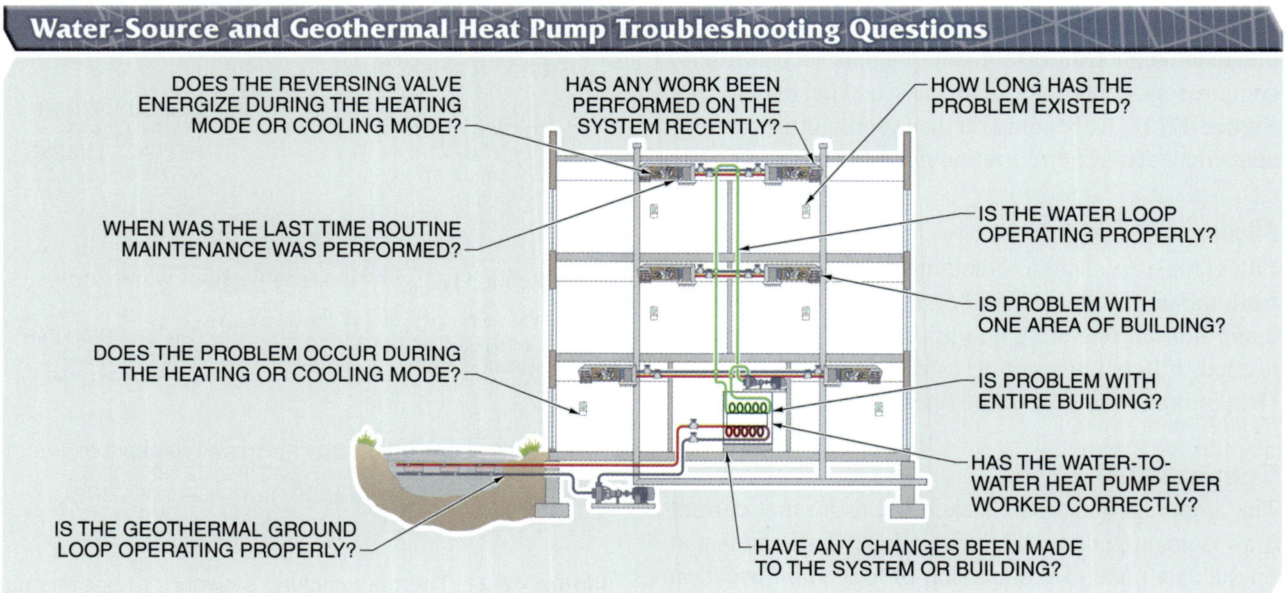

Figure 47-22. A technician should ask basic questions about the functioning of a water-source or geothermal heat pump before troubleshooting the system.

A *water-to-air installation and maintenance manual* is a general guide used by qualified technicians to install water-source heat pump systems. These manuals do not supersede any national or local codes. When installation manuals do not arrive with the equipment, the manuals should be obtained from the manufacturer as soon as possible. **See Figure 47-23.** The installation of the water-to-air heat pump (including the air handler) described in the manual must comply with all state and local codes as well as national codes such as the NEC® or CEC™. Failure to comply with manufacturer procedures can void warranties and reduce the energy efficiency of the equipment.

Water-to-Air Installation and Maintenance Manuals

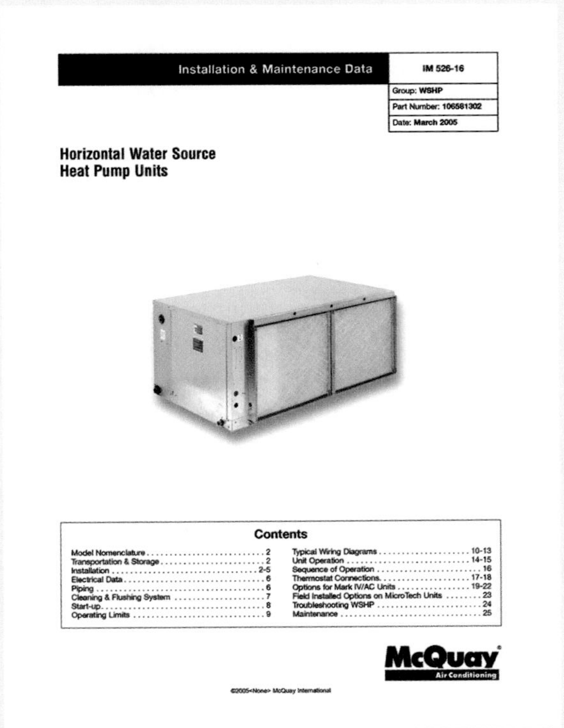

McQuay International

Figure 47-23. Water-to-air heat pumps have all of the valving and piping of a boiler plus the air-handling characteristics of a furnace.

A *water-to-water installation and maintenance manual* is a general guide used by qualified technicians to indicate the best way to use water-to-water heat pump equipment for an application and contains specific PM recommendations for the equipment. Operating geothermal heat pump equipment involves starting and stopping the equipment and making adjustments to it to keep it operating correctly. **See Figure 47-24.**

Water-to-Water Installation and Maintenance Manuals

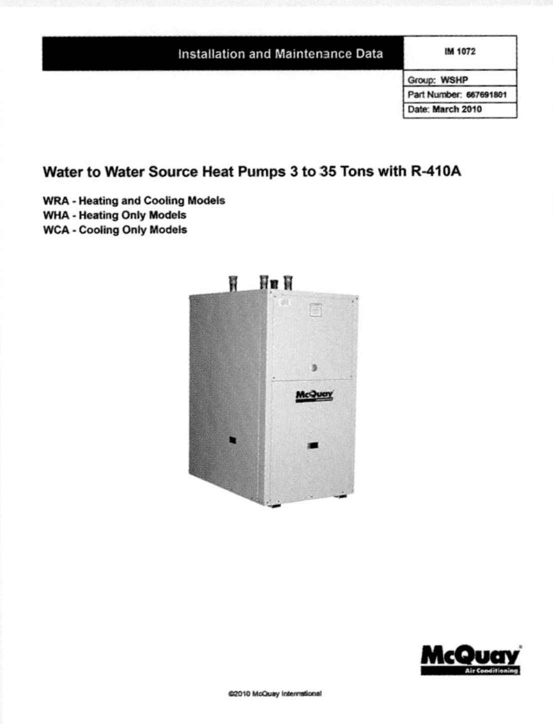

McQuay International

Figure 47-24. Operating geothermal heat pump equipment involves starting and stopping the equipment to make adjustments and performing PM to keep it operating correctly.

General Service Guides. General service guide charts are only intended to be used to troubleshoot water-source or geothermal heat pumps serviced by technicians who have successfully completed manufacturer training. Service guide charts must only be used in conjunction with the appropriate heat pump installation or maintenance manual, which is also provided by the heat pump manufacturer.

There are many reasons why a heat pump and system might not be working correctly. General service guide charts are not intended to be complete lists of all the possible failures a water-source or geothermal heat pump may experience. However, many common solutions can be found on these charts. **See Figure 47-25.**

Service Guide Charts

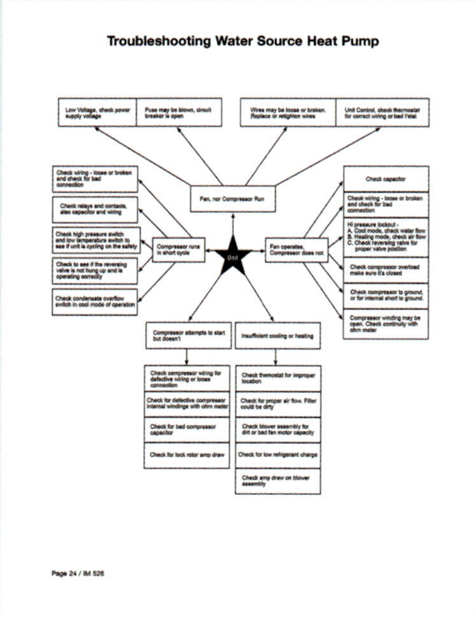

WATER-TO-AIR HEAT PUMP

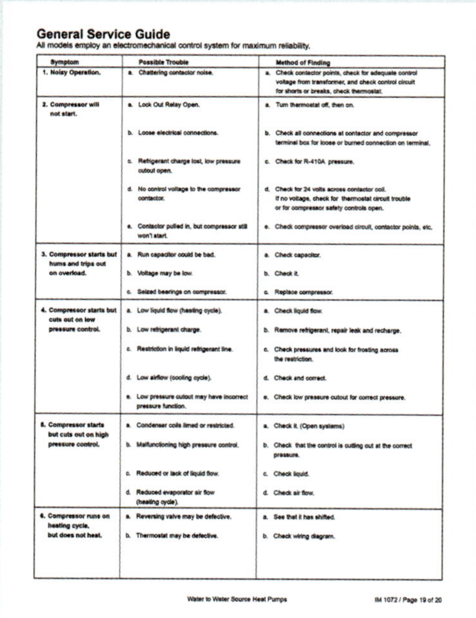

General Service Guide
All models employ an electromechanical control system for maximum reliability.

Symptom	Possible Trouble	Method of Finding
1. Noisy Operation.	a. Chattering contactor noise.	a. Check contactor points, check for adequate control voltage from transformer, and check control circuit for shorts or breaks, check thermostat.
2. Compressor will not start.	a. Lock Out Relay Open.	a. Turn thermostat off, then on.
	b. Loose electrical connections.	b. Check all connections at contactor and compressor terminal box for loose or burned connection on terminal.
	c. Refrigerant charge lost, low pressure cutout open.	c. Check for R-410A pressure.
	d. No control voltage to the compressor contactor.	d. Check for 24 volts across contactor coil. If no voltage, check for thermostat circuit trouble or for compressor safety controls open.
	e. Contactor pulled in, but compressor still won't start.	e. Check compressor overload circuit, contactor points, etc.
3. Compressor starts but hums and trips out on overload.	a. Run capacitor could be bad.	a. Check capacitor.
	b. Voltage may be low.	b. Check it.
	c. Seized bearings on compressor.	c. Replace compressor.
4. Compressor starts but cuts out on low pressure control.	a. Low liquid flow (heating cycle).	a. Check liquid flow.
	b. Low refrigerant charge.	b. Remove refrigerant, repair leak and recharge.
	c. Restriction in liquid refrigerant line.	c. Check pressures and look for frosting across the restriction.
	d. Low airflow (cooling cycle).	d. Check and correct.
	e. Low pressure cutout may have incorrect pressure function.	e. Check low pressure cutout for correct pressure.
5. Compressor starts but cuts out on high pressure control.	a. Condenser coils limed or restricted.	a. Check it. (Open systems)
	b. Malfunctioning high pressure control.	b. Check that the control is cutting out at the correct pressure.
	c. Reduced or lack of liquid flow.	c. Check liquid.
	d. Reduced evaporator air flow (heating cycle).	d. Check air flow.
6. Compressor runs on heating cycle, but does not heat.	a. Reversing valve may be defective.	a. See that it has shifted.
	b. Thermostat may be defective.	b. Check wiring diagram.

Water to Water Source Heat Pumps IM 1072 / Page 19 of 20

WATER-TO-WATER HEAT PUMP

Figure 47-25. Service guide charts are not intended to be complete lists of all the possible failures a water-source or geothermal heat pump may experience. However, many common solutions can be found on these charts.

Typical Water-Source and Geothermal Heat Pump Sequences of Operations

For servicing and troubleshooting purposes, it is important that the technician understand the sequence of operation of the particular heat pump model being worked on, regardless of whether it is an air-source, water-source, or geothermal type. As there are important variations among manufacturers and models, this is a critical step in determining whether the heat pump is operating correctly or not. Typical water-source or geothermal heat pump sequences of operations for cooling and heating are as follows:

Cooling Cycle. The cooling cycle of a water-source water-to-air heat pump functions as follows: **See Figure 47-26.**

1. The compressor pumps high-pressure, superheated refrigerant vapor into the system.
2. Hot refrigerant vapor leaves the compressor and passes through the reversing valve.
3. Refrigerant flows through the piping to the refrigerant-to-water heat exchanger (condenser). Water from the water loop removes heat from the refrigerant vapor. When enough heat is removed, the refrigerant vapor condenses into a high-pressure liquid. The liquid refrigerant temperature is slightly warmer than the ambient air temperature.
4. Warm high-pressure liquid refrigerant leaves the refrigerant-to-water heat exchanger and flows through the copper refrigerant liquid line.
5. At the end of the liquid line, the refrigerant passes through a metering device, reducing its pressure and temperature.
6. As the liquid, under reduced pressure, enters the heat pump refrigerant-to-air heat exchanger (evaporator), it expands and absorbs heat from the indoor air passing over the finned surface. Heat from the indoor air causes the low-pressure liquid to evaporate, cooling the indoor air. The refrigerant is now a warm vapor.
7. The refrigerant vapor travels through the insulated vapor line to the reversing valve. The reversing valve directs the refrigerant into the accumulator, if present.
8. The accumulator controls the liquid refrigerant by allowing only refrigerant vapor to pass.
9. The refrigerant vapor flows through the suction line to the compressor. The cycle then repeats.

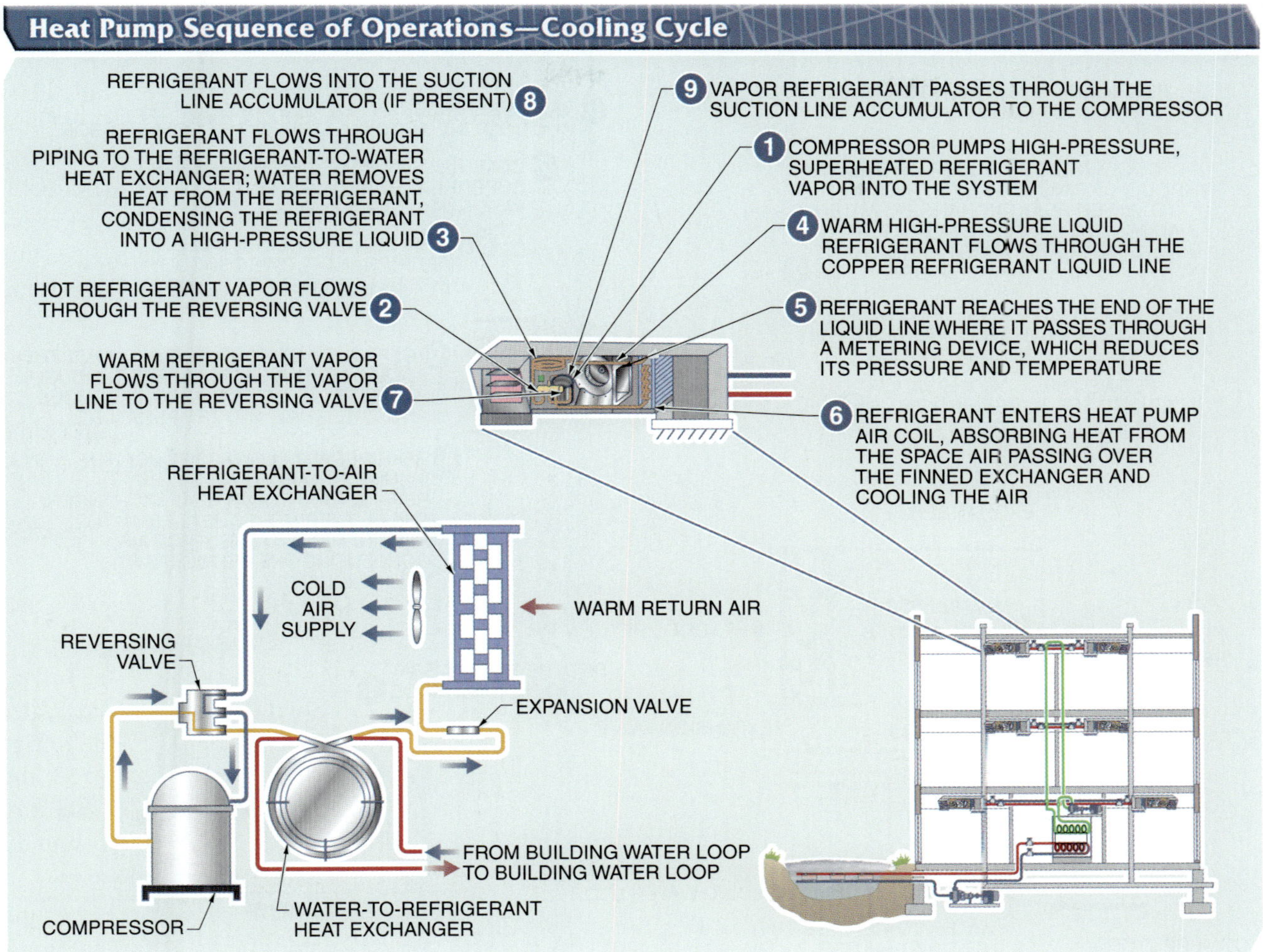

Figure 47-26. A water-source heat pump in cooling mode has no back-up or emergency cooling available.

Heating Cycle. The heating cycle of a water-source water-to-air heat pump functions as follows: **See Figure 47-27.**

1. Setting the thermostat to the heat mode automatically energizes the solenoid pilot valve on the reversing valve.
2. The compressor pumps high-pressure, superheated refrigerant vapor into the system.
3. Hot refrigerant vapor leaves the compressor and passes through the reversing valve.
4. Refrigerant flows through the insulated vapor line to the refrigerant-to-air heat exchanger. Air from the unit's blower removes heat from the refrigerant vapor, warming the indoor air. When enough heat is moved from the refrigerant into the space air, the refrigerant vapor condenses into a high-pressure liquid.
5. Warm high-pressure liquid refrigerant leaves the refrigerant-to-air air heat exchanger and flows through the copper refrigerant liquid line.
6. At the end of the liquid line, the refrigerant passes through a metering device, which reduces its pressure and temperature.
7. As the cool liquid refrigerant, under reduced pressure, enters the refrigerant-to-water heat exchanger, the refrigerant absorbs heat from the water loop. Heat from the water loop causes the low-pressure liquid to evaporate. The refrigerant is now a cold vapor.
8. The cold refrigerant vapor travels through the vapor line to the reversing valve. The reversing valve directs the refrigerant into the accumulator, if present.
9. The accumulator controls the liquid refrigerant by allowing only refrigerant vapor to pass.
10. Refrigerant vapor flows through the suction line to the intake of the compressor. The cycle then repeats.

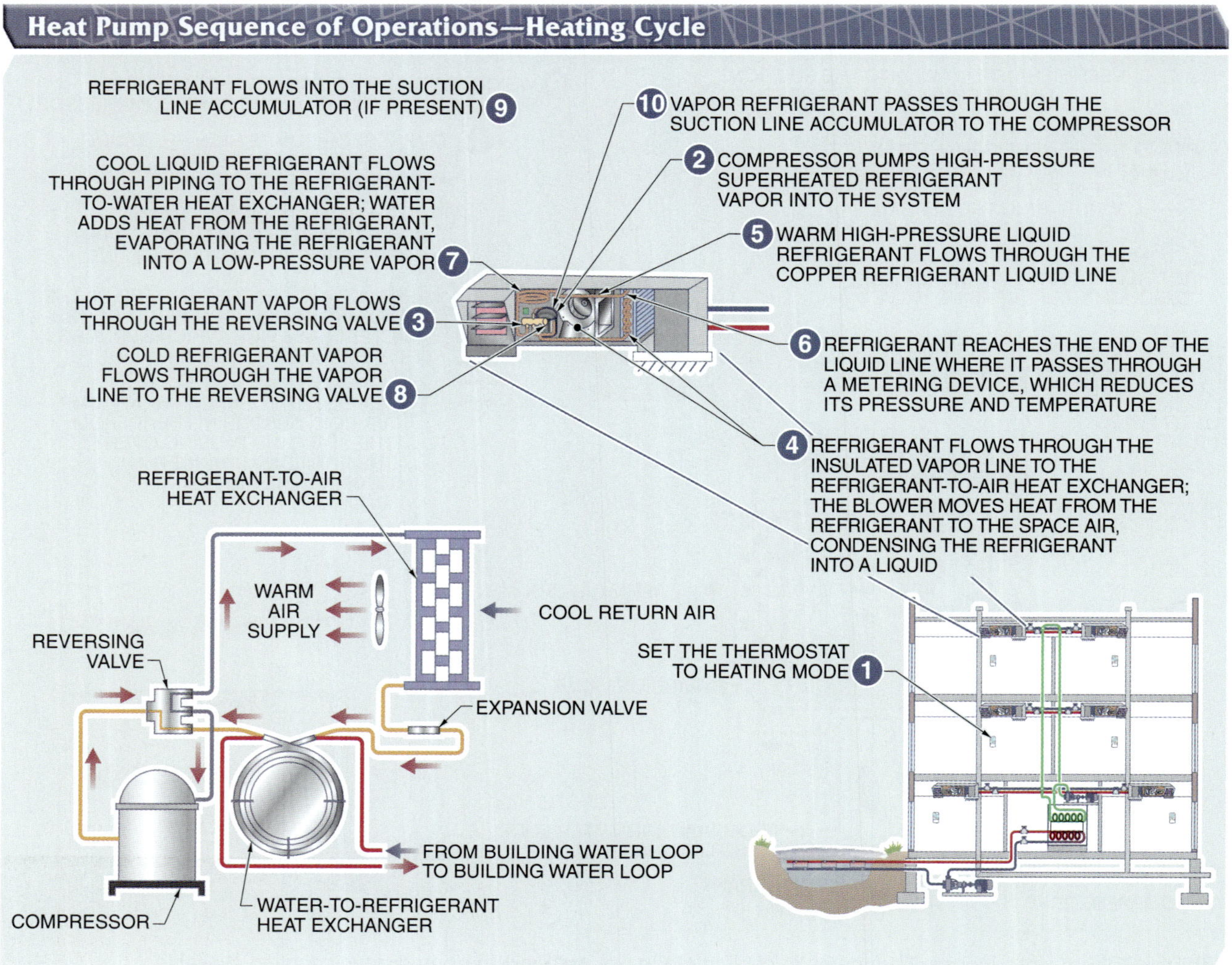

Figure 47-27. During a water-source heat pump's heating cycle, back-up or emergency heat may be used.

TROUBLESHOOTING WATER-SOURCE AND GEOTHERMAL HEAT PUMPS

To troubleshoot a water-source or geothermal heat pump, the technician must know how to use diagnostic equipment such as a DMM and a gauge manifold. As with most HVAC units, 80% of the problems are simple and basic. In addition, a knowledge of the basic sequence of operation is needed as well. The technician must be observant of local conditions and situations that may indicate the source of a problem.

Heating Mode Troubleshooting Scenario #1—Defective Transformer

A technician receives a call from a customer that an area of the office is too cold. Upon arrival, the technician finds that neither the compressor nor the fan is running on the water-to-air heat pump in the cold area of the office. Incoming power is checked with a DMM and is found to be 230 VAC. The control transformer is checked and has 0 VAC on the secondary side. The technician also notices that the control transformer is discolored and is very hot to the touch. **See Figure 47-28.**

Service Work Performed: The transformer is replaced and the water-to-air heat pump is checked to ensure it cycles properly.

TECH TIP

All manufacturers design systems for comparable discharge temperatures regardless of component selection.

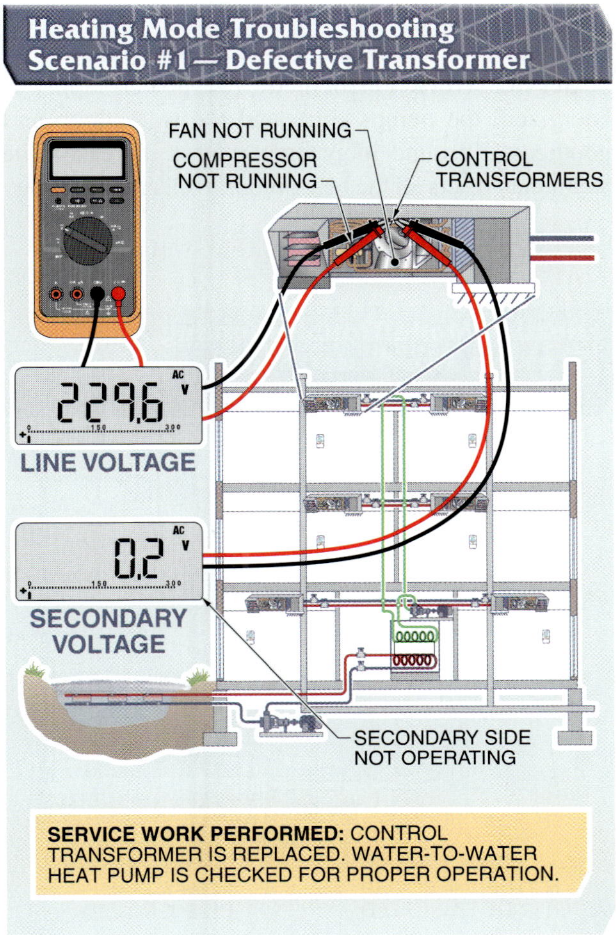

Figure 47-28. When nothing appears to be running in a water-source heat pump, the technician begins by checking the incoming power and then checks the secondary side of the transformer.

Heating Mode Troubleshooting Scenario #2—Wrong Thermostat

A maintenance technician receives a "too cold" call from the second floor of a building. Upon arrival, the technician finds out from the person in the corner office that the thermostat was replaced the morning before and that the heating has not worked correctly since then. The technician gets the model number from the thermostat and finds it is not the correct heat pump thermostat for the application. **See Figure 47-29.**

Service Work Performed: The thermostat is replaced with a model designed for water-source heat pumps and then the heating mode at the heat pump is checked and found to be working properly.

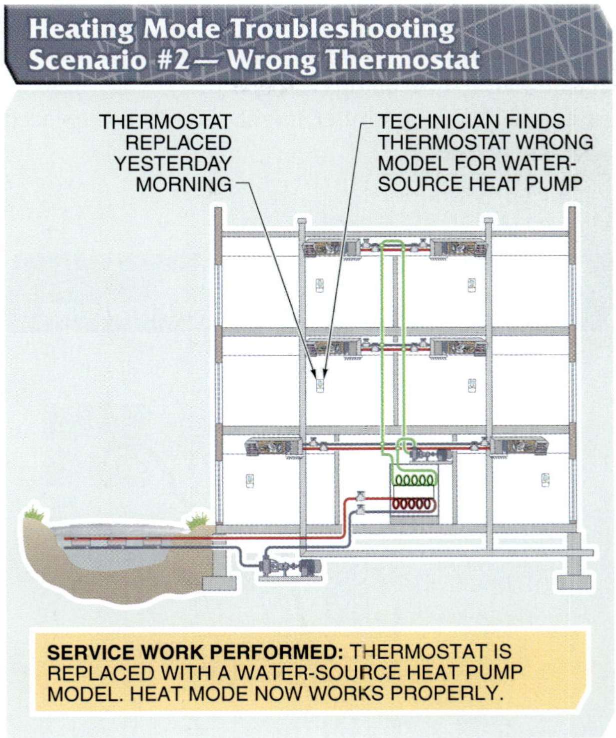

Figure 47-29. There are special thermostats that must be used for water-source heat pump systems that also provide increased system efficiency.

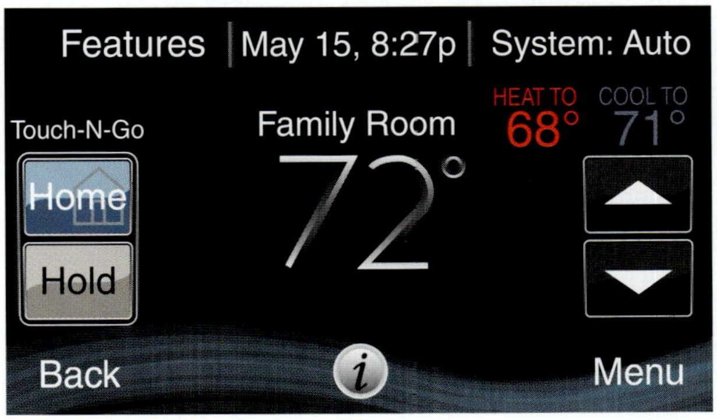

Carrier Corporation

Thermostats selected for a geothermal heat pump system today are touchscreen. Typically, the thermostat is capable of working with multiple zones.

Heating Mode Troubleshooting Scenario #3—DDC Controller Problems

A maintenance person at a school calls in a "too cold" complaint. The school has a geothermal heat pump system with 10 heat pumps. One of the zones is very cold, while the other nine are dropping in temperature. The technician uses a laptop to connect to the building's DDC controller and finds that the programming for the heat pumps has been corrupted. **See Figure 47-30.**

Service Work Performed: The original program, with modifications, is downloaded into the controller and now all 10 heat pumps work properly. *Note:* In some cases, the DDC controller might need to be replaced with a new one.

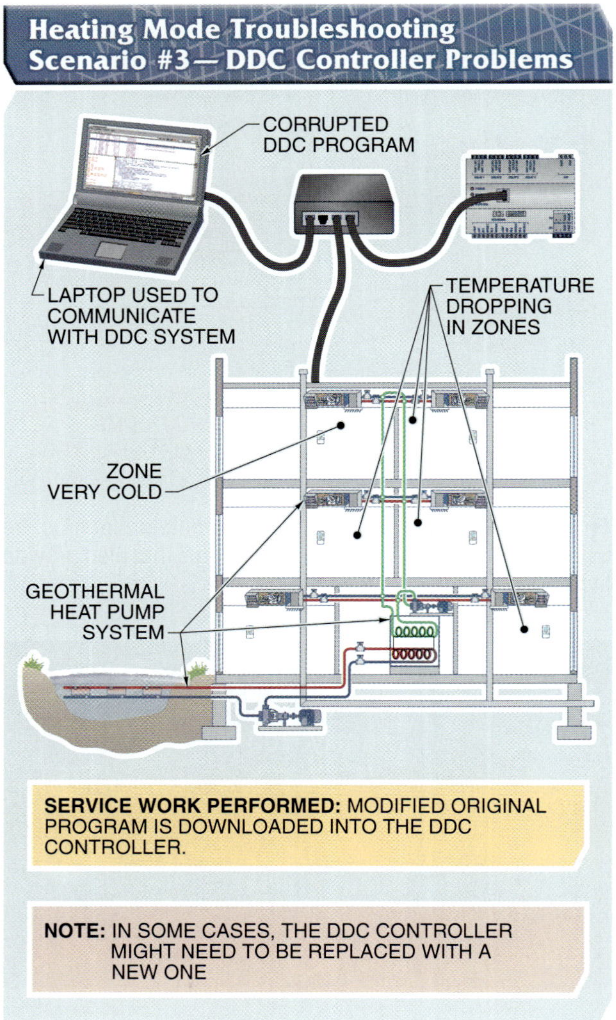

Figure 47-30. Laptop computers are typically required for troubleshooting DDC systems.

Heating Mode Troubleshooting Scenario #4—Water Loop Temperature Problems

A customer calls in and says the whole office building is cold. It is the first cool day in the fall. The technician arrives and finds that the water loop is cold and that the thermostats are locked out. A closer look reveals that the power supply to the water loop and geothermal ground loop pumps is off. It is determined that when PM was performed during the summer, the disconnects to the loop pumps were locked out. **See Figure 47-31.**

Service Work Performed: The power supply is energized, the pumps start, and the water loop and geothermal ground loop temperatures increase. The technician resets all the heat pumps to clear the lockout.

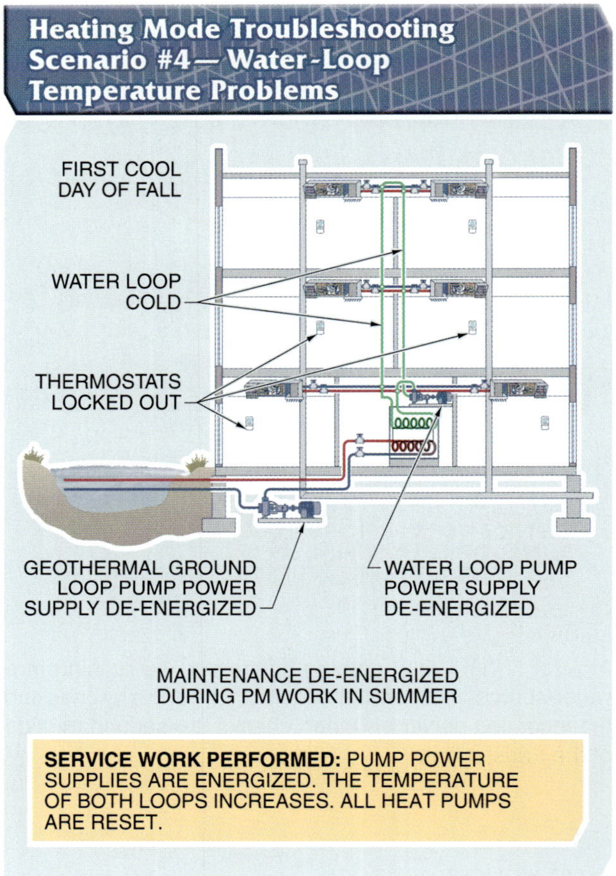

Figure 47-31. When the water loop has temperature problems, the heat pumps become locked out, and each heat pump must be reset in order to operate again.

Cooling Mode Troubleshooting Scenario #1—Dirty Air Filter

A customer calls complaining that some zones in an office building are too hot. The company has just recently occupied the space. A general system check shows that the compressors and fans on some water-to-air units are running constantly. A visual check of the unit filters shows that some are completely plugged. Inspection of the records indicates that the units were running during construction and that dust and dirt got into the filters. **See Figure 47-32.**

Service Work Performed: The filters are replaced and building maintenance is told to change them every 30 days.

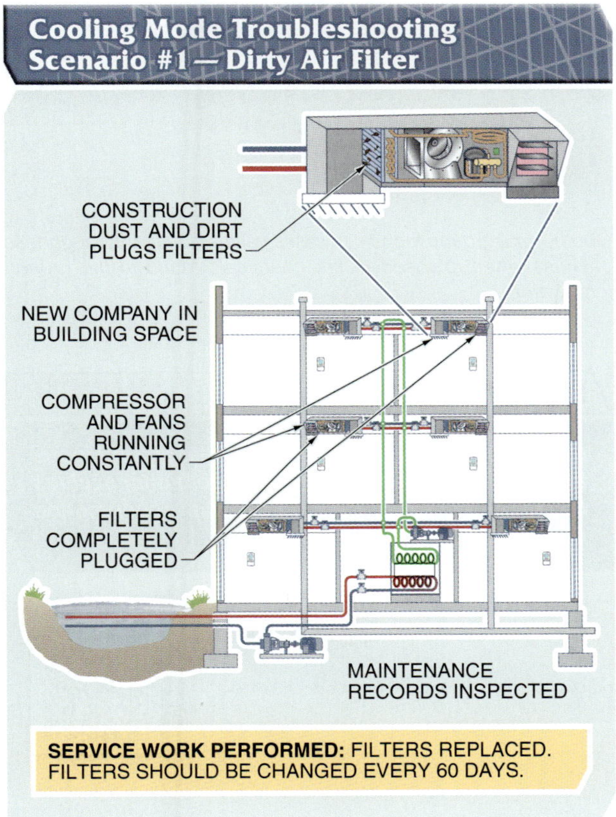

Figure 47-32. Dirty air filters cause a geothermal heat pump system to use more energy and can also cause the fan to fail.

Cooling Mode Troubleshooting Scenario #2—Water Loop Temperature Problems

A customer calls in and says that about half the building is hot and the problem is getting worse. The technician arrives and finds that the water loop is too hot. A visual inspection of the cooling tower shows that it is dirty and has not been cleaned in years. **See Figure 47-33.**

Service Work Performed: The heat pumps, cooling tower, and pumps are shut down, locked out, and tagged out. The cooling tower is thoroughly cleaned and put back in service. When the water temperature drops, the heat pumps are energized one at a time. The water-to-air heat pumps are checked and all are working properly.

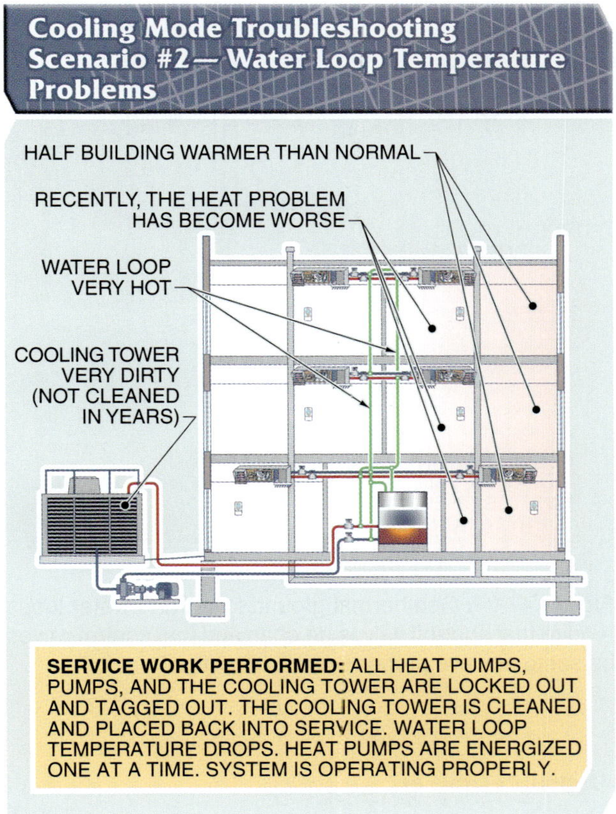

Figure 47-33. Dirty cooling towers can cause the water loop to overheat and the entire water-source heat pump systems to fail.

Cooling Mode Troubleshooting Scenario #3—Water Valves Shut Off

A customer calls in with a "too hot" complaint for one of the zones in the building. The technician checks the pumps, geothermal ground loop, and water loop. They all seem to be working properly. A visual check of the heat pump in the zone that is hot shows that the isolation valves were shut off when pipe work was performed in the ceiling. **See Figure 47-34.**

Service Work Performed: The technician opens the isolation valves to the heat pump and restarts the heat pump a few minutes later. It is checked and now operates properly.

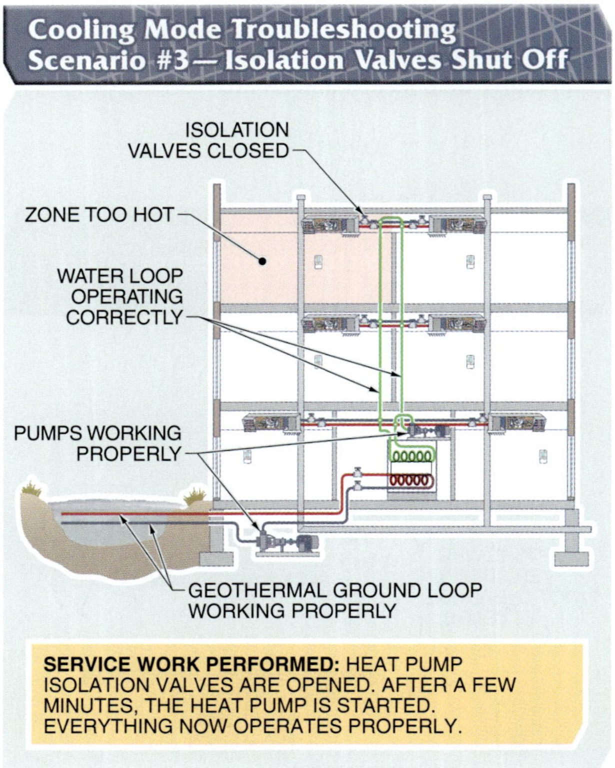

Figure 47-34. Geothermal ground loops and water loops require that shut-off valves be open and that control valves be set to achieve proper system operation.

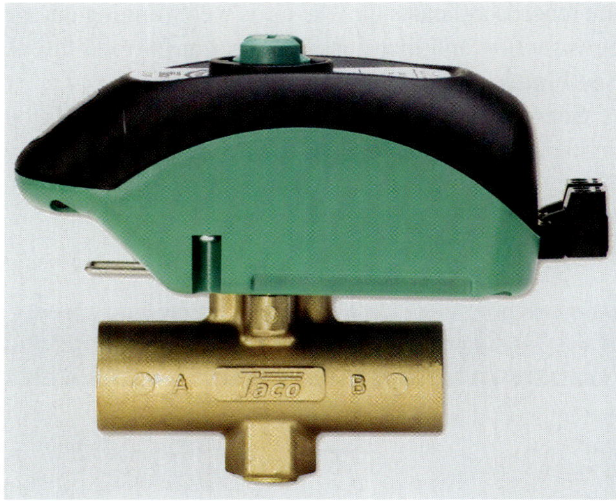

Taco, Inc.

Geothermal ground loops use brass valves due to the toughness of brass when exposed to the chemicals found in the ground loop water.

TECH TIP

Modulating valves used in geothermal ground loops allow for a more stable loop flow and higher savings because they lower the pump's watt consumption even at full operating speed.

UNIT 48

Installing, Servicing, and Troubleshooting Chiller Systems

Many heavy commercial buildings use chilled water systems for cooling because of the amount of heat produced by the building and/or processes. Building heating loads include equipment, electronics, manufacturing processes, as well as ambient conditions, solar load, and people. Chilled water systems typically produce 100 tons to more than 5000 tons of cooling.

HVAC technicians should be familiar with the basic operation of chillers and be able to troubleshoot simple problems. Advanced operation, maintenance, and troubleshooting of chillers requires specialized training and is typically performed by original equipment manufacturer (OEM) technicians.

OBJECTIVES

- List components typically used in chilled water systems.
- Compare common types of cooling towers.
- Identify typical chiller operating temperatures and pressures.
- Explain how to install chillers.
- Identify basic information an HVAC technician must gather to service or troubleshoot a chiller.
- Describe typical procedures for servicing chillers.
- Describe common chiller troubleshooting problems.

Index to Procedures

CHILLED WATER SYSTEM COMPONENTS

A *chilled water system* is a high-volume hydronic cooling system that typically includes chillers, pumps, piping, and accessories such as cooling coils and terminal devices. Cooling coils and terminal devices absorb heat, which is transferred to the chiller. Component number, size, and complexity depend on the size and design of the chilled water system. Chilled water system pumps, cooling towers, and auxiliary equipment are typically controlled by the building automation system (BAS).

Chillers

Chillers come in a variety of types and sizes. A *chiller* is a high-volume hydronic air conditioning device that cools circulating water for terminal device use. Chillers consist of components such as compressors, motors, evaporators, and condensers. **See Figure 48-1.**

A central chilled water system, depending on capacity, can consist of one chiller or more than ten chillers. Large facilities, such as college or university campuses, typically have multiple chilled water systems to serve all the buildings on campus. Some chilled water systems use absorption-type chillers instead of, or in addition to, compressor chillers. Common chiller compressor types include the following:

- centrifugal—chiller compressor used in installations of 150 tons or greater
- screw—chiller compressor used in newer buildings, especially with high-efficiency chillers using VFDs
- scroll—chiller compressor used in smaller installations of 150 tons or less
- reciprocating—chiller compressor used in older installations of 150 tons or less

Chillers

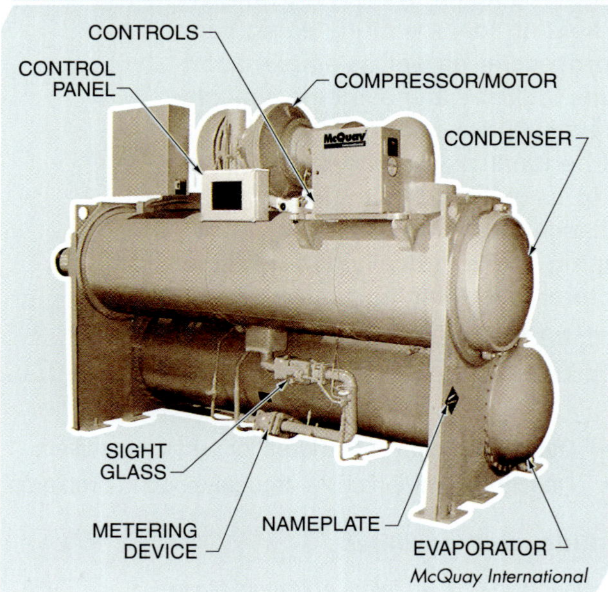

Figure 48-1. Chillers are high-volume hydronic air conditioning devices that include compressors, motors, evaporators, condensers, and controls.

Pumps

Pumps are used to circulate water from the chiller, through the chilled water system, and back to the chiller. Typically, centrifugal pumps are used to circulate water through the piping system and terminal devices. Some pump motors are controlled and driven by VFDs. VFD controls are used to maintain specific flow rates or differential pressures in a chilled water system. Larger chilled water systems typically have multiple pumps at various positions in the system, with some pumps used as backups in case of pump failure. Many pumps are labeled with arrows that indicate direction of water flow through the pump. **See Figure 48-2.**

Chilled Water System Pumps

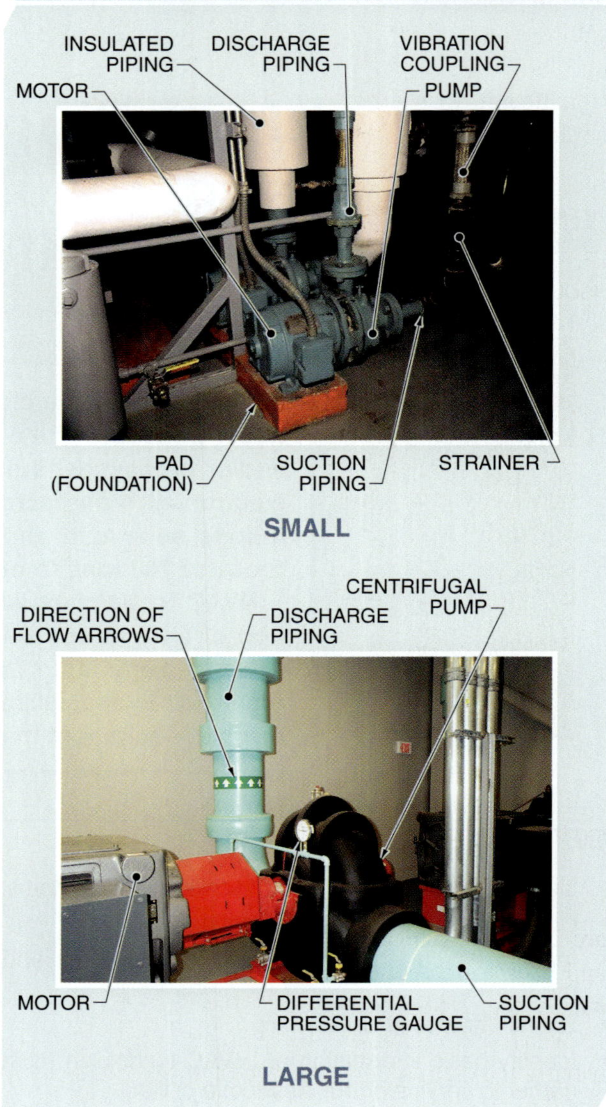

Figure 48-2. Pumps circulate chilled water to heat exchangers and terminal devices in order to remove heat from building spaces.

Piping and Valves

Chilled water systems use an extensive network of distribution piping and valves for flow control. Piping is used to circulate chilled water from the chillers, to the cooling coils or terminal devices of the building, and back to the chillers and to the cooling towers and back. Piping sizes are selected based on needed capacity, with large system piping typically 12″ in diameter or larger. Piping is typically insulated to reduce the amount of heat gain from the distribution system.

Valves are used to control the rate and direction of flow in piping and devices. Chilled water systems typically use motorized isolation valves to control water flow rate. **See Figure 48-3.** Isolation valves prevent water from flowing through cooling coils, terminal devices, or chillers when these components are not in use.

Chilled Water System Valves

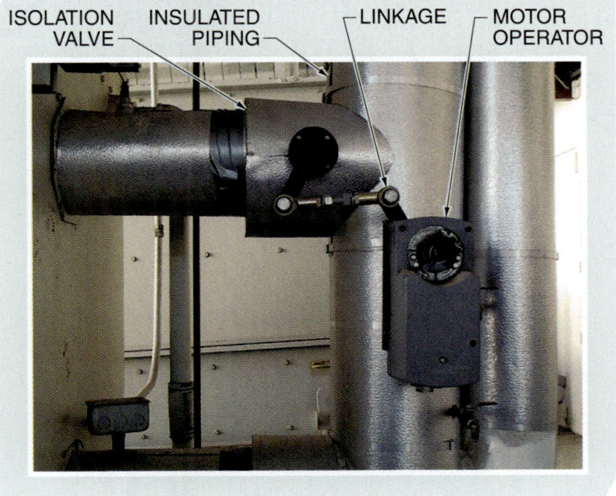

Figure 48-3. Valves control the flow of chilled water circulating through a building.

Primary/Secondary Piping Loops. Several methods are used to circulate chilled water to building spaces and back to the chiller. A typical distribution method used is the primary/secondary piping system.

A *primary piping loop* is a piping system that distributes the main supply of chilled water to the building and the main return of water to the chiller. The primary piping loop typically operates at a constant volume of water flow, which allows consistent chiller operation. A *secondary piping loop* is a piping system connected to the main supply and return piping loop used to circulate chilled water to building terminal units/cooling coils. Secondary piping loops use chilled water, as needed to maintain the temperature of building spaces.

Balancing Valves. Balancing valves are typically used to adjust flow rates into secondary piping loops. Balancing valves add resistance to flow to ensure proper amounts of flow to terminal devices. Water flow rates are adjusted by rotating the balancing valve knob towards the closed position to increase the resistance to flow. **See Figure 48-4.** A well-balanced system provides evenly distributed comfort during periods of high demand.

Balancing Valves

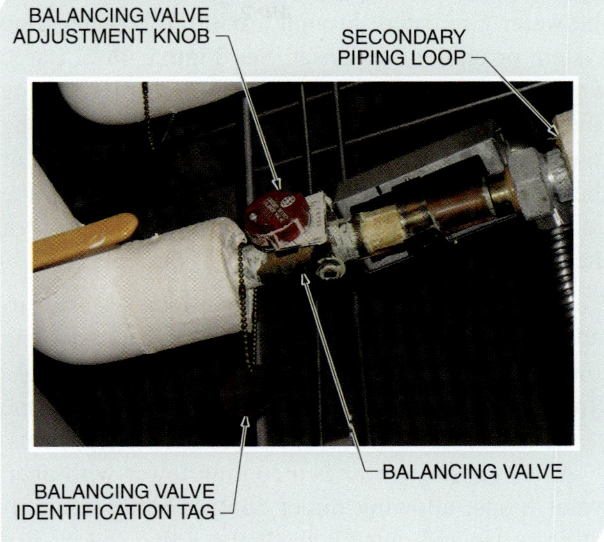

Figure 48-4. Balancing valves are used to adjust flow rates into the secondary piping loops to ensure proper amounts of flow to terminal devices.

Typically, terminal units and cooling coils have modulating valves used to regulate flow through their coils, which are used to maintain air temperature in building spaces. These modulating valves adjust the flow in the secondary loop by opening and closing. When needed, secondary loop pumps use a control system with differential pressure sensors in the piping system. VFDs are typically used to speed up or slow down the secondary pumps in order to maintain a specific differential pressure in the secondary loop.

There are many variations of piping systems, and each chilled water system is unique. When working with unfamiliar chilled water systems, a quick walk-through by the technician to locate piping, valves, pumps, and other components is critical.

WATER LOOP SYSTEM COMPONENTS

Smaller water loop systems typically use closed, air-cooled, circulating water loop systems to reject heat to the atmosphere. Larger water loop systems must discharge large amounts of heat and typically use cooling towers to reject heat into the air. The size and number of cooling towers depend on the amount of heat that needs to be rejected as well as local climate conditions.

Cooling Towers

A *cooling tower* is a device used to reject heat from the water circulated through a building's water loop system or chiller condenser. **See Figure 48-5.** Large water loop systems typically use multiple cooling towers and/or multiple chillers. One or more cooling towers may be needed to meet the cooling demand of the system. Maintenance and water treatment of the cooling water affects the operation and efficiency of the cooling towers.

Cooling Tower Water Loops. Water loop systems circulate water from the chiller condenser to the cooling tower and back to the chiller condenser. Water loops typically are open, or wet, circulating systems. In some cases, water loops are closed, or dry, circulating systems.

A *wet cooling tower* is a cooling tower with open water loops, allowing direct contact between circulating water and outside air. Wet cooling towers use natural evaporation to cool water. Open water loop systems provide greater heat transfer efficiencies with a compact design. Typically cooling towers use open water loops, although closed water loops are used in some locations.

A *dry cooling tower* is a cooling tower with closed water loops that uses piping with fins to exchange heat between circulating water and outside air. Closed water loop systems prevent circulating water from picking up airborne contaminants and losing circulating water to the air through evaporation.

Cooling Tower Drafts. Air circulates through the cooling tower when blown by fans located in the tower or by natural convection. A *forced-draft cooling tower* is a cooling tower in which fans are located in the airstream, typically at the top of the tower, in order to force the ambient air through. Typically, smaller cooling towers use fans to control the rate of airflow and maintain the system cooling capacity. Forced-draft cooling towers provide a compact design with great control of the cooling process. **See Figure 48-6.**

A *natural-draft cooling tower* is a cooling tower that uses natural draft to circulate the ambient airstream through the tower. *Natural draft* is a draft produced by the airflow resulting from temperature and density differences between the ambient air and the heated air. Air heated by the rejected heat in the tower is less dense and naturally rises, pulling cooler ambient air into the tower. Natural-draft cooling towers require less maintenance and have lower operating costs than forced-draft cooling towers.

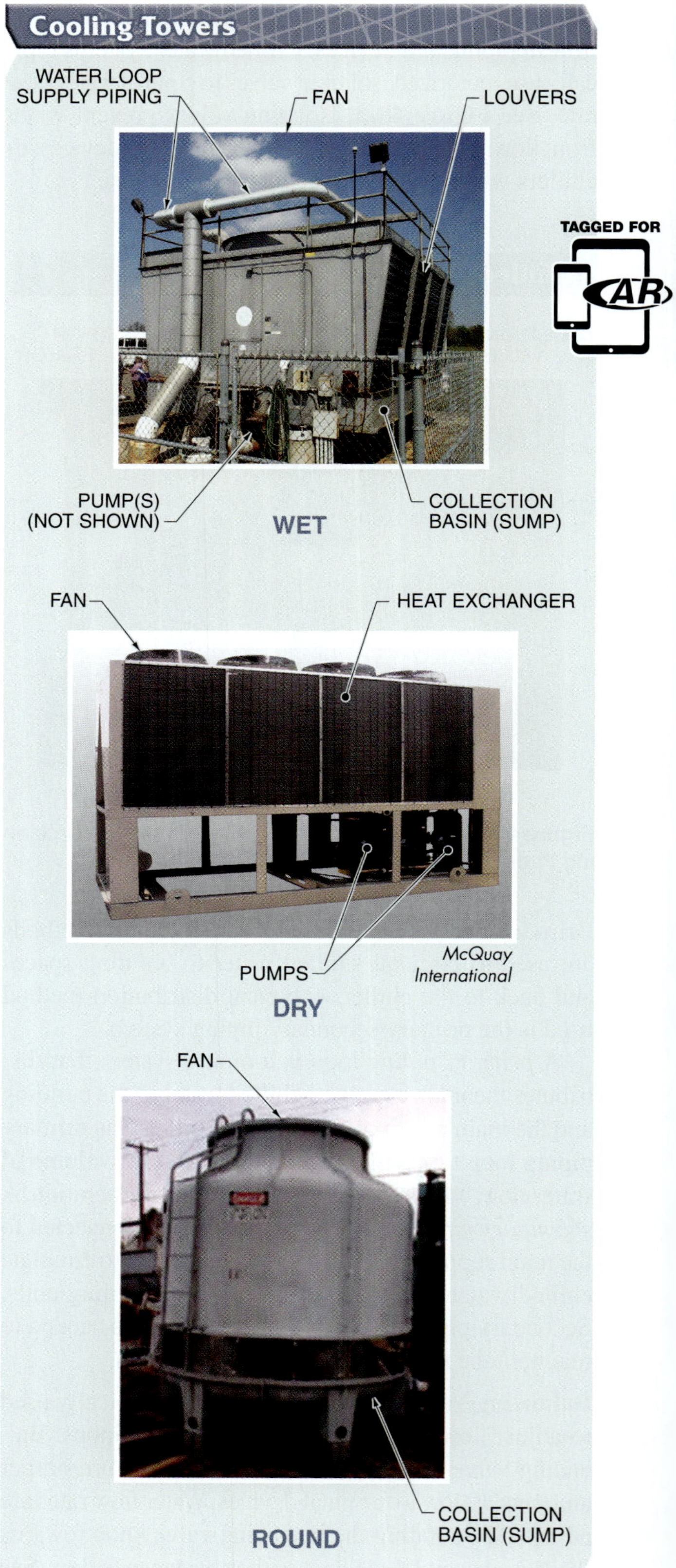

Figure 48-5. Cooling towers reject the heat absorbed by the chilled water system of a building to the atmosphere.

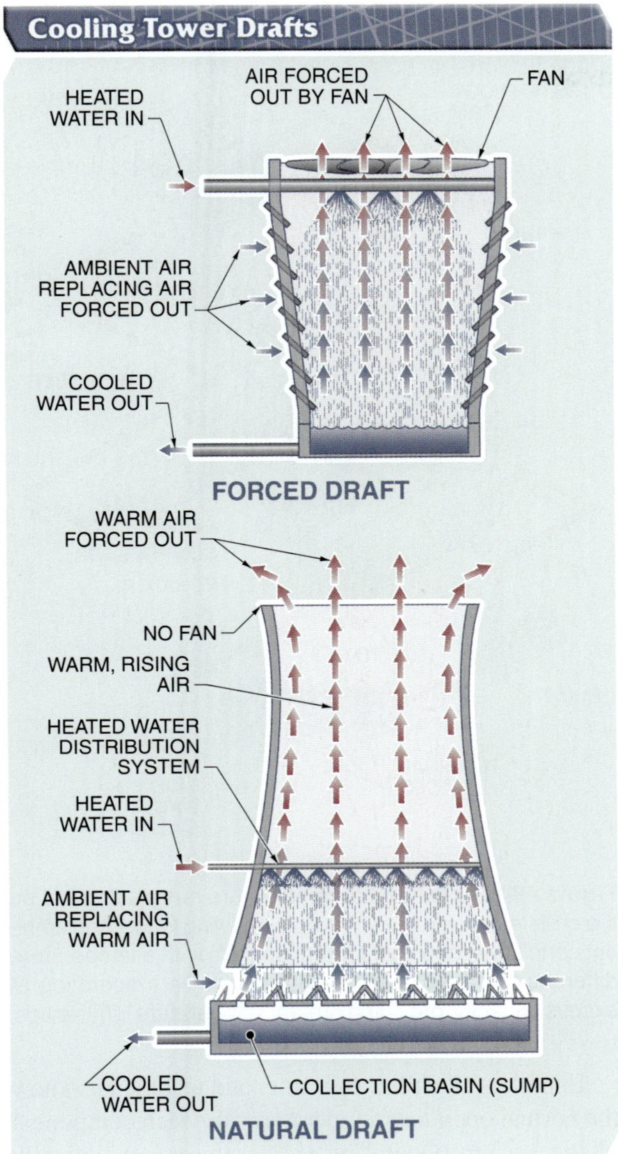

Figure 48-6. Air circulates through cooling towers by using forced-draft (fans located in the tower) or natural draft (airflow resulting from hot or warm air rising).

Cooling Tower Flow Designs. Cooling towers use standard flow designs such as crossflow, counterflow, and parallel flow. A *crossflow cooling tower* is a cooling tower that has a fill configuration through which the ambient air flows horizontally and the water flows downward. A *counterflow cooling tower* is a cooling tower in which the ambient air moves vertically upward through the fill, counter to the downward fall of water. A *parallel flow cooling tower* is a cooling tower in which the water and ambient air are introduced at the top of the tower and flow downward through the fill. **See Figure 48-7.**

Cooling tower fill is the heat transfer material within a cooling tower that distributes water, allowing it to spread out and increase its contact time with the air.

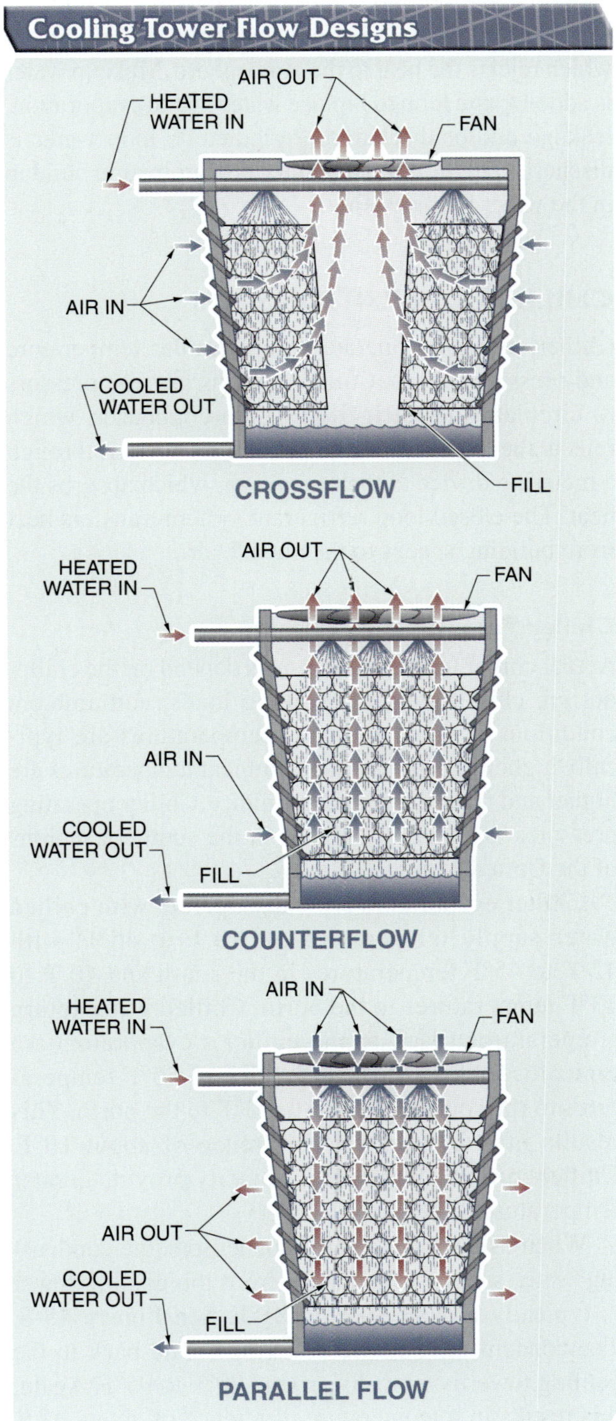

Figure 48-7. Cooling towers use standard flow designs such as crossflow, counterflow, and parallel flow to control the rate of heat rejected to the air.

Water Loop Systems

Pumps are used to circulate the water collected in cooling tower basins (sumps) through the building water loop. Pumps circulate water through the loop to the chiller condenser, which rejects heat to the outside water loop and discharges the water to the cooling tower, which rejects the heat to the atmosphere. Makeup water is added to the loop to replace water lost to evaporation, leakage and/or discharge. Periodically, loop water is discharged to the drain to reduce contamination buildup in the water loop system.

CHILLER OPERATIONS

Chillers typically operate within similar temperature and pressure ranges. Chiller systems use compressors to circulate the refrigerant to the condenser, which rejects the heat and discharges the refrigerant through a metering device to the evaporator, which absorbs the heat. The closed loop refrigerant system transfers heat from building spaces to the outside air.

Chiller Temperatures

Actual chiller temperature ranges depend on the chiller design, chiller capacity, cooling loads, and ambient conditions. Chiller operating temperatures are typically higher in regions where ambient temperatures are higher and that have higher humidity. Chiller operating pressures are typically higher in the southern regions of the United States.

Chiller evaporators typically operate with chilled water supply temperatures of 40°F to 45°F, with 42°F to 45°F temperatures in the south and 40°F to 43°F temperatures in the north. Chilled water return temperatures (back to the chiller's evaporator) are typically 50°F to 55°F, with 52°F to 55°F temperatures to the south and 50°F to 53°F to the north. This results in a temperature difference of about 10°F. Chillers operating at excess capacity provide greater temperature differences.

When using water-cooled condensers, the condensing water supply temperature from the cooling tower is typically around 80°F to 85°F. **See Figure 48-8.** The condensing water return temperature back to the cooling tower is typically around 90°F to 95°F. Again, this results in a temperature difference of about 10°F. The operating temperature depends on the number of cooling towers used, amount of heat rejected, and the ambient temperature.

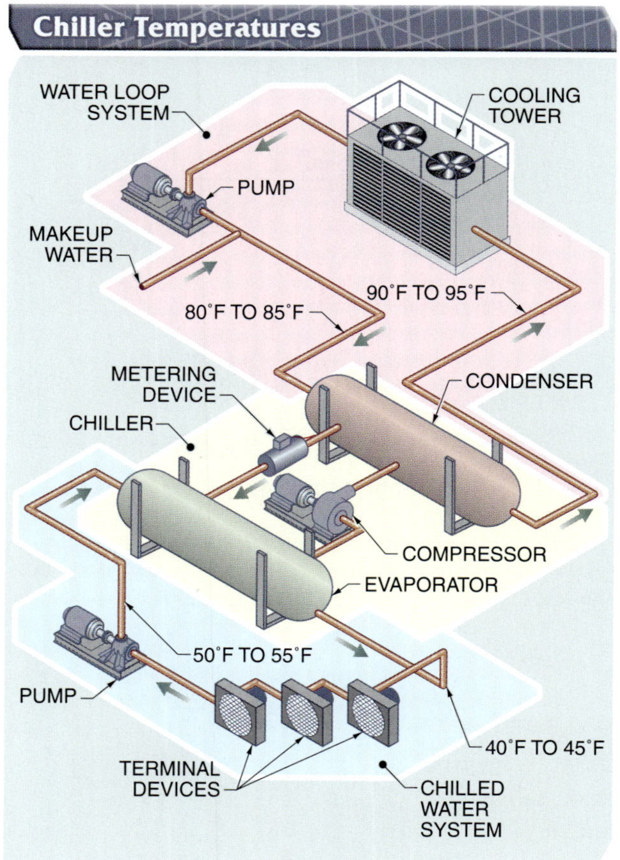

Figure 48-8. Actual chiller temperature ranges depend on the chiller design, capacity, building cooling loads, and ambient conditions, but most chillers operate with a temperature difference of about 10°F, with a few chillers operating at excess capacity, providing greater temperature differences.

Experienced chiller operators and technicians know the normal operating temperatures for each component of the system. Operating temperatures are typically recorded on log sheets, along with ambient air temperatures, and both are valuable information when troubleshooting chillers. Operating temperatures are often graphed to determine trends. **See Figure 48-9.**

Chiller Pressures

Operating pressures can be evaluated using a similar process. Many types of refrigerants are used in chilled water systems. High-pressure refrigerants, such as R-134 and R-22, are typically used in newer chilled water systems. High-pressure refrigerants allow the evaporator to operate above atmospheric pressure. Low-pressure refrigerants, such as R-11 and R-123, are typically used in older chilled water systems. Low-pressure refrigerants operate in a vacuum in the evaporator portion of the system.

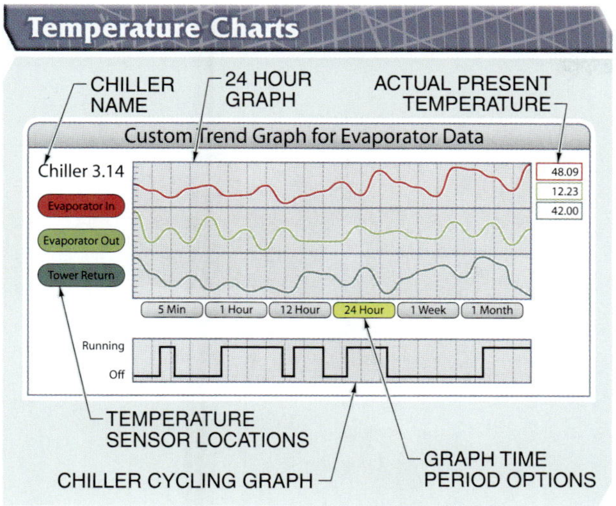

Figure 48-9. Operating temperatures are typically recorded on log sheets along with ambient air temperatures, and both are valuable information when troubleshooting chillers.

The actual pressures in each chilled water system can be found by consulting refrigerant pressure-temperature charts. **See Appendix.** Operating pressures are typically recorded on log sheets, and the information is used when troubleshooting chillers.

INSTALLING CHILLERS

Large chillers are typically very large and heavy. Some chillers can reach 50,000 lb in weight. Because of their size and weight, special preparations and safety procedures must be used when installing chillers. Components must be installed by experienced and trained personnel. **See Figure 48-10.**

In order to save space, some very large chillers have two compressors on one large heat exchanger bundle.

Figure 48-10. Some chillers are 12′ tall, 20′ in length, and weigh 50,000 lb.

Component delivery and rigging are complex and are typically scheduled to take place after normal working times, such as on weekends. Components typically come with lifting lugs or eyebolt connections for installation. **See Figure 48-11.** Some components are manufactured on frames for installation. The installation process is typically scheduled weeks in advance. Often, large access doors are included in the building mechanical spaces to allow future equipment installation, service, or replacement.

Lifting Lugs

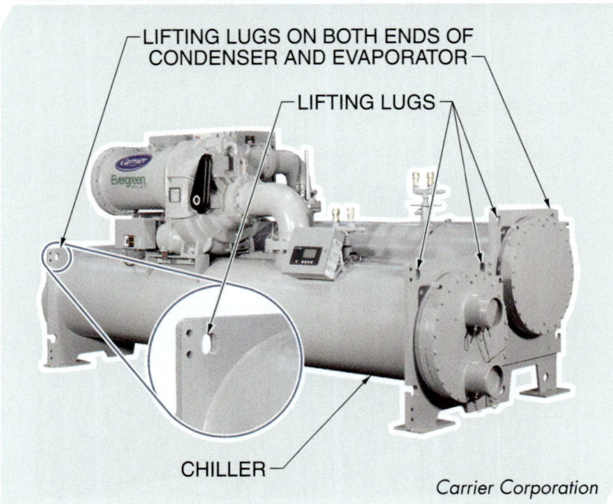

Figure 48-11. Chillers and components typically come with lifting lugs to accommodate rigging for installation and maintenance.

Microprocessor Control Panels (MCPs)

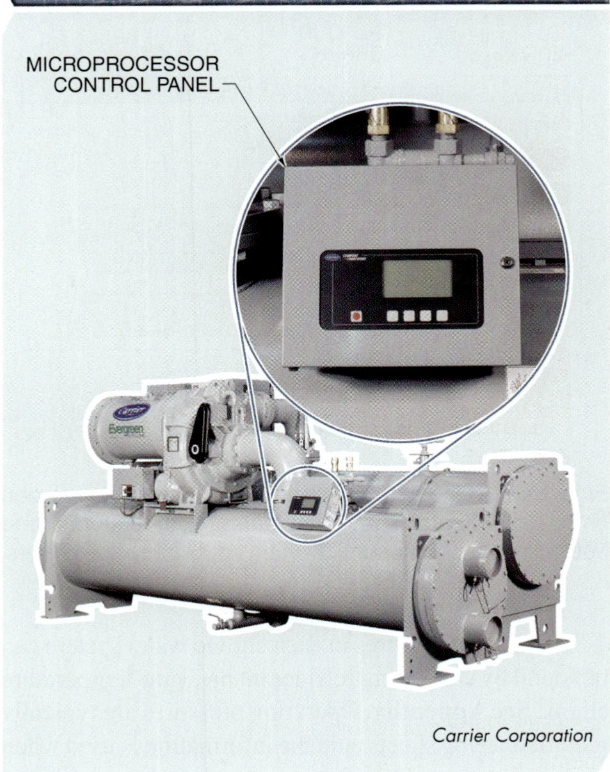

Figure 48-12. MCPs provide a high level of control.

Control of Chillers

Large chillers have very sophisticated control systems. New chillers are typically manufactured with a microprocessor control panel (MCP). **See Figure 48-12.** MCPs include the built-in logic needed to efficiently and safely operate the chiller. MCPs typically include a touch screen or keypad interface. New chillers typically include high-quality animated graphics on an LCD screen.

Chiller Sequencing. In larger chilled water systems, the BAS controls which chillers are in operation and in what order of operation, such as chillers 1, 2, 3, 4 or 2, 4, 1, 3. The chillers can be automatically selected based on capacity needed or by run hours. Operators can control the order of operation through the BAS. Operators can shut down a chiller for adjustment or maintenance.

Chiller BAS Controls. The BAS controls the primary and secondary pumps. Typically, pumps are installed in pairs that alternate run cycles, with one serving as a back-up pump. When one pump fails, a fault or drop in system pressure will turn on the back-up pump. Pumps will operate to maintain the required flow rate and pressure. Modern BASs issue alarms on chiller faults and provide problem identifications. Alarms can be issued as emails and sent to phones, tablets, or computers. Alarms can be set to identify a level of priority and any components or system parameters with large variations in operations.

The condenser pumps, bypass valves, and cooling tower fans are also controlled by the BAS. VFDs are used to modulate the flow created by the pumps and fans. Large chillers typically have multiple cooling tower units available, with the units that are on-line being selected by the BAS. **See Figure 48-13.**

Performance is typically recorded and graphed to indicate component and system level of operation. Trending analysis is used to schedule maintenance and plan for component replacement and system improvements.

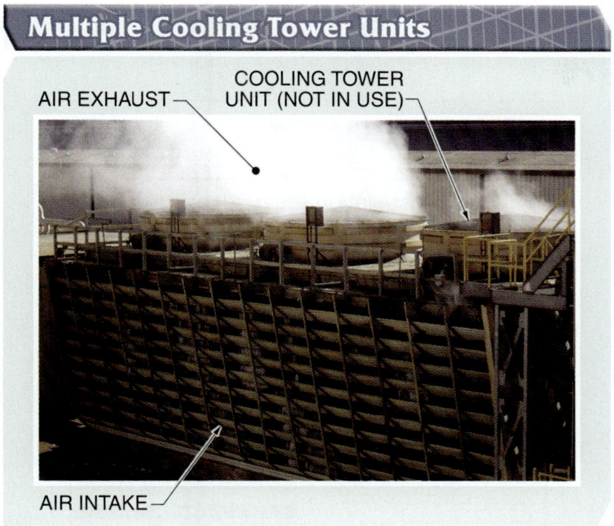

Figure 48-13. The BAS typically controls which cooling tower units are in operation at a specific point in time.

SERVICING AND TROUBLESHOOTING CHILLERS

As with all mechanical equipment, a safe work environment must be maintained any time a technician services or troubleshoots a chiller and its associated systems. All safety procedures must be followed at all times. An injury can result from a few seconds of carelessness. Typically, hazard warning stickers are provided in areas of hazardous conditions and on any materials used. **See Figure 48-14.**

Figure 48-14. Hazard warning labels and product labels provide information for the safe use of hazardous materials.

When troubleshooting HVAC equipment, especially chillers, it is difficult for HVAC technicians to know all the answers. The manufacturer's troubleshooting and technical installation instructions must be consulted, with special attention given to any safety references, hazards, or warnings. Most manufacturers provide equipment installation and service manuals online. The technician should gather as much information as possible before servicing or troubleshooting a system. System information can include prints, manuals, wiring diagrams, and software. Extra care must be taken when a new type of chiller is being worked on for the first time.

Basic Troubleshooting Questions

When troubleshooting chillers, the technician should ask the building owner some basic questions:

- How was this problem noticed? Are there any unexpected vibrations occurring?
- When did the problem start? Were there any other problems at the time?
- When was the last time the chiller worked properly?
- Has the chiller been following its normal sequence of operation?
- Have any changes been made to the chilled water system or building?
- Has anyone else performed service on the chiller or chilled water system?
- What provisions are in place to override the equipment or controls?
- Are written sequences of operation and system drawings available?

Unsuccessful Repairs. When a repair does not fix the problem with the chiller, the technician must reevaluate the symptoms, causes, and solutions listed on the troubleshooting charts. In some situations, the entire process of gathering information about the system must begin again. The repair efforts and results, even when the repair was unsuccessful, must be recorded.

Successful Repairs. When a repair is successful and the chiller operation has been restored, the troubleshooting process is recorded on a troubleshooting report or equipment maintenance log. Everything is recorded in a maintenance log including symptoms, causes, tests conducted, repairs made, and the condition of the chiller and chilled water system after the repair. **See Figure 48-15.** The spare parts used in the repair are recorded, and replacement parts are ordered as needed. Suggestions on modifying the equipment or procedures that could prevent similar problems are also recorded.

Chiller Maintenance Logs

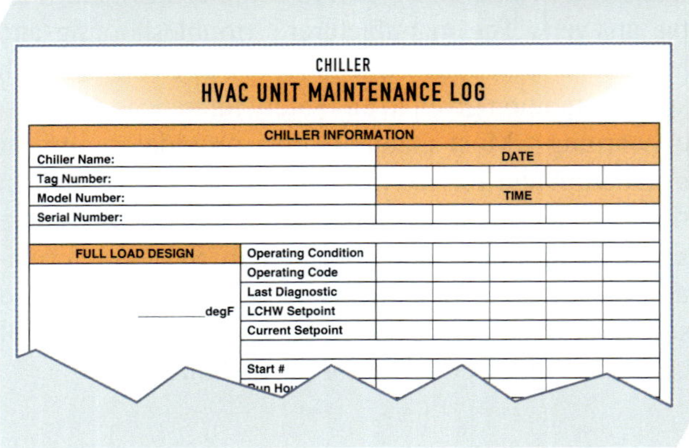

Figure 48-15. Maintenance logs are records of equipment conditions and can be used for future troubleshooting and trending.

TECH TIP

Upon customer acceptance, multiple copies of the installation and operation manuals are provided to the customer. These important documents should be read and understood by technicians and installers regardless of their familiarity with a system. Damage has occurred to HVAC systems as a result of technicians not consulting the available documentation.

Installation Manuals

Figure 48-16. Installation manuals provide information for the proper installation of the chiller and chilled water system components.

Installation Manuals

An *installation manual* is a general guide used by qualified technicians to install residential and light commercial HVAC equipment. Installation manuals do not supersede any national or local codes. When installation manuals do not arrive with the delivered equipment, the manuals should be obtained from the manufacturer as soon as possible. **See Figure 48-16.**

The installation of the chiller and chilled water system equipment described in the manual must comply with all state and local codes as well as national codes such as the NEC® or the CEC™. Failure to comply with manufacturer procedures can void warranties and reduce the energy efficiency of the equipment.

Installation manuals for chillers and chilled water system equipment typically contain sections on safety, how to inspect the equipment upon delivery, where to locate the equipment, field wiring diagrams and specifications, water piping diagrams and recommendations, cleaning and/or flushing procedures, start-up procedures, and check-out procedures.

Operation and Maintenance Manuals

An *operation and maintenance (service) manual* is an equipment-specific guide used by qualified technicians to service residential and light commercial HVAC systems. Operating a chiller and chilled water system involves starting and stopping the equipment and making adjustments to it to keep it operating correctly. Many chilled water system functions for operation are automated, but some functions cannot be completely automated. **See Figure 48-17.**

Operation and maintenance manuals for chillers and chilled water system equipment typically contain safety information, sequences of operations, adjustments and servicing information, preventive maintenance procedures, performance checklists, wiring schematics, flash code charts, troubleshooting guides, parts lists, and warranty information.

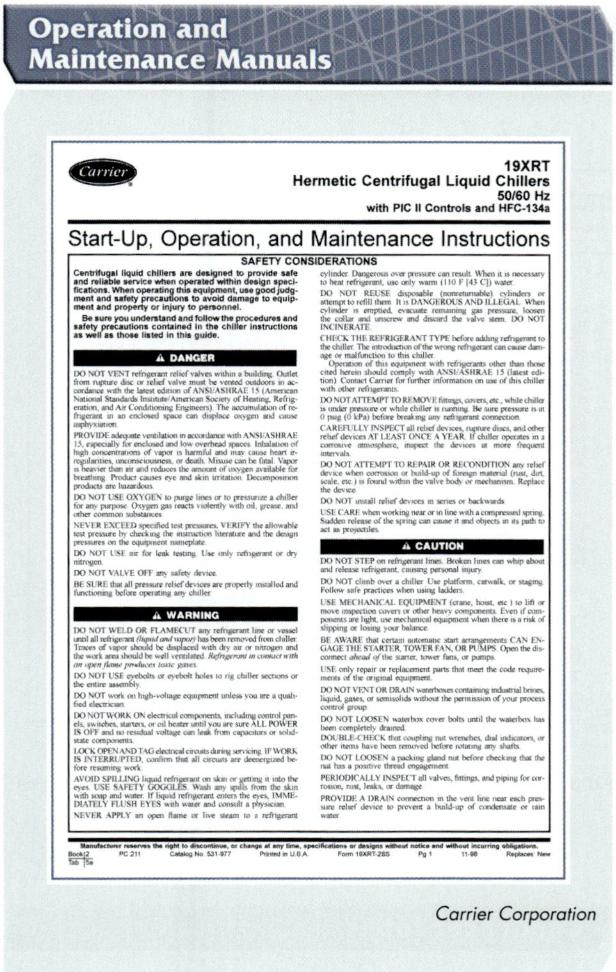

Figure 48-17. Service manuals provide information on the operation, maintenance, and troubleshooting of chiller equipment and system components.

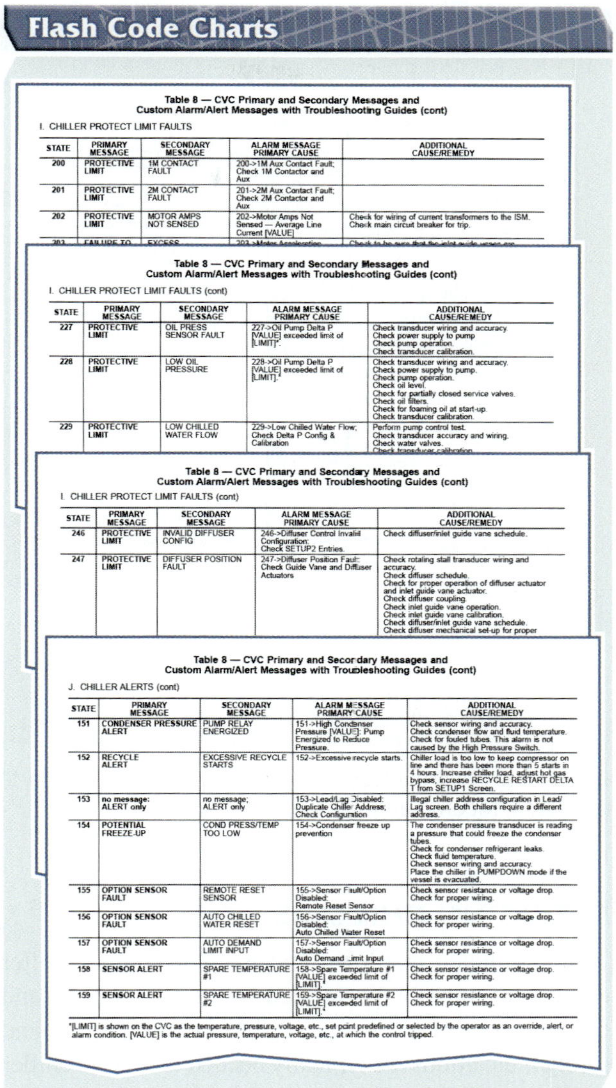

Figure 48-18. Flash code charts provide the meanings of LED codes, either by the number of flashes or by a numerical value display, to diagnose equipment problems.

Flash Code Charts. Chillers typically have a self-diagnosing control module with an LED that flashes red or a number-based code to let technicians know when the chiller has a malfunction. The fault indicator LED or number LED is behind the front access door observation window on the chiller. The number of flashes or code number that the control module displays is the error code for the problem the chiller is encountering. The technician counts the number of flashes or reads the code number and checks the error code against the flash code chart. Once the problem is identified, the troubleshooting section of the chart can be used to solve the problem. **See Figure 48-18.**

Troubleshooting Charts. Troubleshooting charts are only intended to be used on chillers serviced by technicians who have successfully completed the manufacturer training. Troubleshooting charts must only be used in conjunction with the appropriate chiller installation or operation manual, which is also provided by the chiller manufacturer. These charts are not intended to be a complete list of all the possible failures a chiller may experience. There are many reasons why a chiller might not be operating normally. Fortunately, many of the solutions are found on chiller troubleshooting charts. **See Figure 48-19.**

Troubleshooting Charts

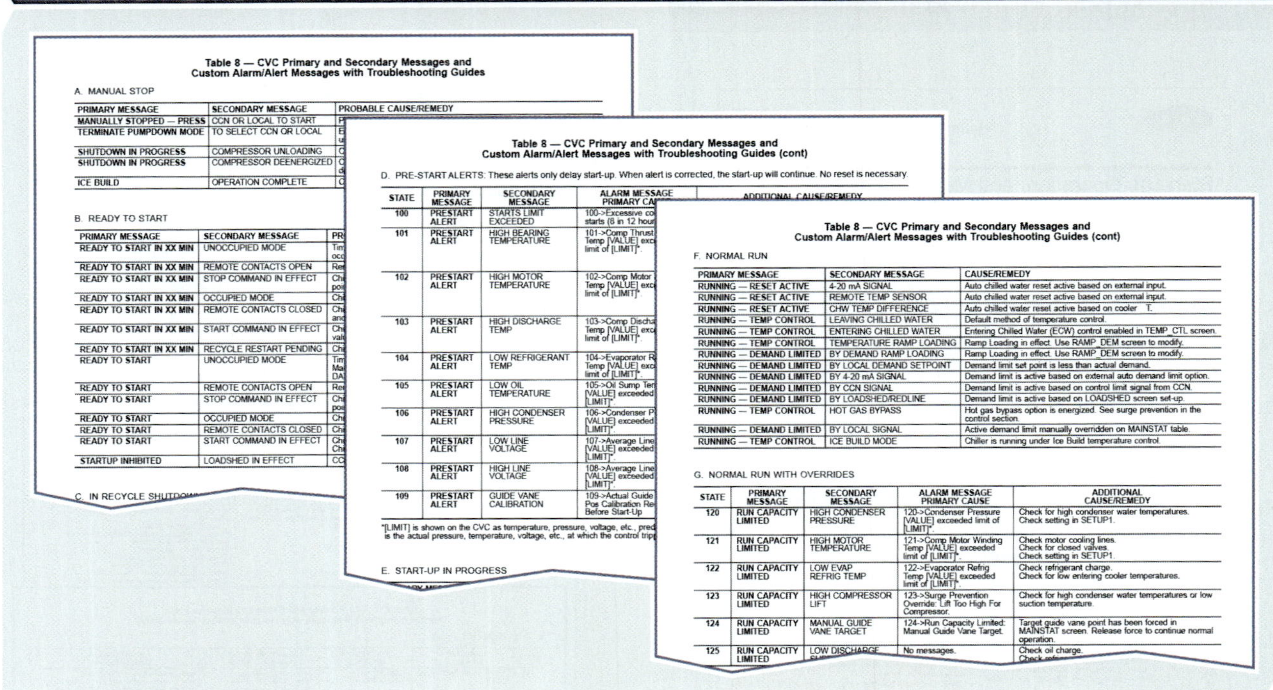

Table 8 — CVC Primary and Secondary Messages and Custom Alarm/Alert Messages with Troubleshooting Guides (cont)

F. NORMAL RUN

PRIMARY MESSAGE	SECONDARY MESSAGE	CAUSE/REMEDY
RUNNING — RESET ACTIVE	4-20 mA SIGNAL	Auto chilled water reset active based on external input.
RUNNING — RESET ACTIVE	REMOTE TEMP SENSOR	Auto chilled water reset active based on external input.
RUNNING — RESET ACTIVE	CHW TEMP DIFFERENCE	Auto chilled water reset active based on cooler T.
RUNNING — TEMP CONTROL	LEAVING CHILLED WATER	Default method of temperature control.
RUNNING — TEMP CONTROL	ENTERING CHILLED WATER	Entering Chilled Water (ECW) control enabled in TEMP_CTL screen.
RUNNING — TEMP CONTROL	TEMPERATURE RAMP LOADING	Ramp Loading in effect. Use RAMP_DEM screen to modify.
RUNNING — DEMAND LIMITED	BY DEMAND RAMP LOADING	Ramp Loading in effect. Use RAMP_DEM screen to modify.
RUNNING — DEMAND LIMITED	BY LOCAL DEMAND SETPOINT	Demand limit set point is less than actual demand.
RUNNING — DEMAND LIMITED	BY 4-20 mA SIGNAL	Demand limit is active based on external auto demand limit option.
RUNNING — DEMAND LIMITED	BY CCN SIGNAL	Demand limit is active based on control limit signal from CCN.
RUNNING — DEMAND LIMITED	BY LOADSHED/REDLINE	Demand limit is active based on LOADSHED screen set-up.
RUNNING — TEMP CONTROL	HOT GAS BYPASS	Hot gas bypass option is energized. See surge prevention in the control section.
RUNNING — DEMAND LIMITED	BY LOCAL SIGNAL	Active demand limit manually overridden on MAINSTAT table.
RUNNING — TEMP CONTROL	ICE BUILD MODE	Chiller is running under Ice Build temperature control.

G. NORMAL RUN WITH OVERRIDES

STATE	PRIMARY MESSAGE	SECONDARY MESSAGE	ALARM MESSAGE PRIMARY CAUSE	ADDITIONAL CAUSE/REMEDY
120	RUN CAPACITY LIMITED	HIGH CONDENSER PRESSURE	120->Condenser Pressure [VALUE] exceeded limit of [LIMIT]*.	Check for high condenser water temperatures. Check setting in SETUP1.
121	RUN CAPACITY LIMITED	HIGH MOTOR TEMPERATURE	121->Comp Motor Winding Temp [VALUE] exceeded limit of [LIMIT]*.	Check motor cooling lines. Check for closed valves. Check setting in SETUP1.
122	RUN CAPACITY LIMITED	LOW EVAP REFRIG TEMP	122->Evaporator Refrig Temp [VALUE] exceeded limit of [LIMIT]*.	Check refrigerant charge. Check for low entering cooler temperatures.
123	RUN CAPACITY LIMITED	HIGH COMPRESSOR LIFT	123->Surge Prevention Override: Lift Too High For Compressor.	Check for high condenser water temperatures or low suction temperature.
124	RUN CAPACITY LIMITED	MANUAL GUIDE VANE TARGET	124->Run Capacity Limited: Manual Guide Vane Target.	Target guide vane point has been forced in MAINSTAT screen. Release force to continue normal operation.

Figure 48-19. Troubleshooting charts provide service technicians and maintenance personnel with typical equipment problems, recommended checks, and repairs.

SERVICING CHILLERS

Any significant teardown and maintenance of chiller equipment must be performed by factory-trained and certified personnel. Specialized tools, equipment, and safety precautions are typically needed to service chiller equipment. Portable lifts or cranes are needed to lift large or heavy components.

Routine maintenance typically includes cleaning condenser tubes during annual shutdowns. Condenser tubes exposed to atmospheric contaminants from wet cooling towers typically need careful monitoring and extensive cleaning. Condenser tubes often become fouled with sediment and corrosion from circulating water. Sediment and corrosion on the condenser tubes reduces efficient heat transfer in chillers and can cause equipment failure.

To clean condenser tubes, the condenser cover is removed and the tubes are cleaned with a long brush. The condenser must be cleaned of residue and the cover reinstalled per manufacturer requirements. **See Figure 48-20.** Chiller maintenance also includes checking the motor and compressor. Maintaining the motor and compressor reduces unscheduled downtime.

Condenser Tubes

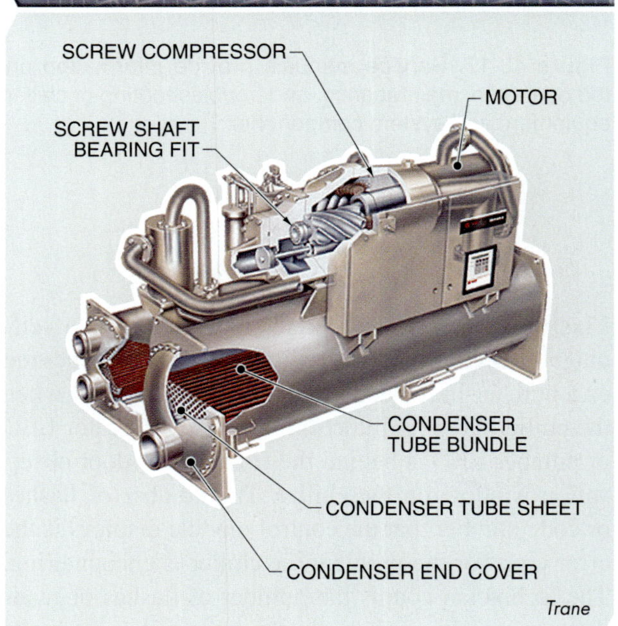

Figure 48-20. Condenser tubes must be checked and cleaned periodically to remove contaminants and increase chiller efficiency.

Water Treatment of Hydronic Systems

Chillers are paired with boilers to provide chilled and heated water for large hydronic systems. A characteristic of hydronic cooling and heating systems is that both use water to transfer heat. Any impurities in the water can accumulate on the inside of piping, valves, heat exchangers, boiler surfaces, and chiller components. These impurities can cause loss of system efficiency, reduced level of comfort, and equipment failure.

Water treatment in cooling and heating systems is critical for maintaining proper system operation. While both cooling and heating systems need water treatment, the problems that arise and their solutions are different for each system.

There are two approaches used to performing water treatment in heating or cooling systems. Larger facilities often train operating personnel to test the water and add the proper chemical treatments. Smaller facilities typically use an independent contractor to perform the tests and add the proper chemical treatments. Records are kept of the test results, the amount of chemicals added, and the date of treatment. **See Figure 48-21.**

Chemical Use Records

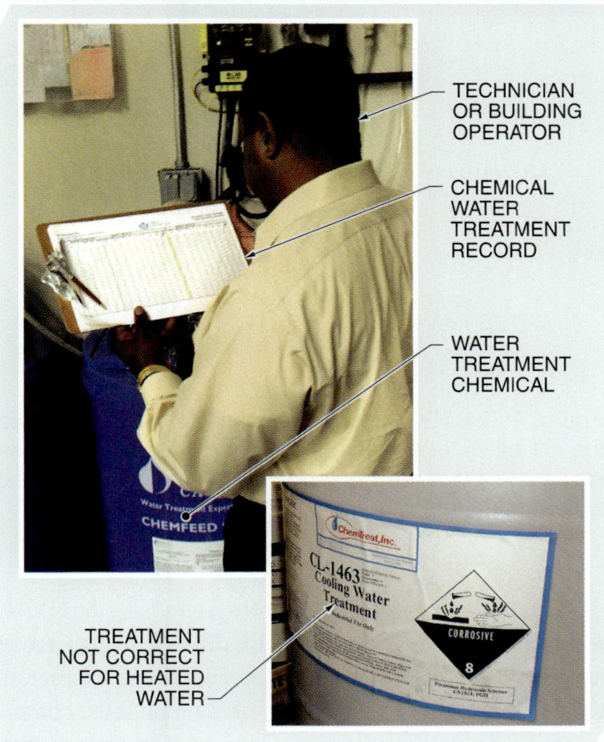

Figure 48-21. Records of water tests and chemicals used in chillers provide indicators of effective system operations.

Cooling Systems. Cooling systems often have two separate water piping loops: the chilled water loop and the circulating water loop (usually outdoors). The chilled water loop typically contains the 40°F to 45°F water that is pumped between the chiller evaporator and the building cooling coils/terminal devices. The chilled water loop is a closed system, meaning that, unless there is a leak or repairs are made, no water is lost or added to the loop during operation.

The chilled water loop is tested for contaminants on a periodic basis and chemicals are added through a chemical feed system. Chemicals are added to water loops to inhibit the corrosion of piping and chiller tubes and to remove mineral deposits from internal heat exchanger surfaces. **See Figure 48-22.**

Chemical Feed Systems

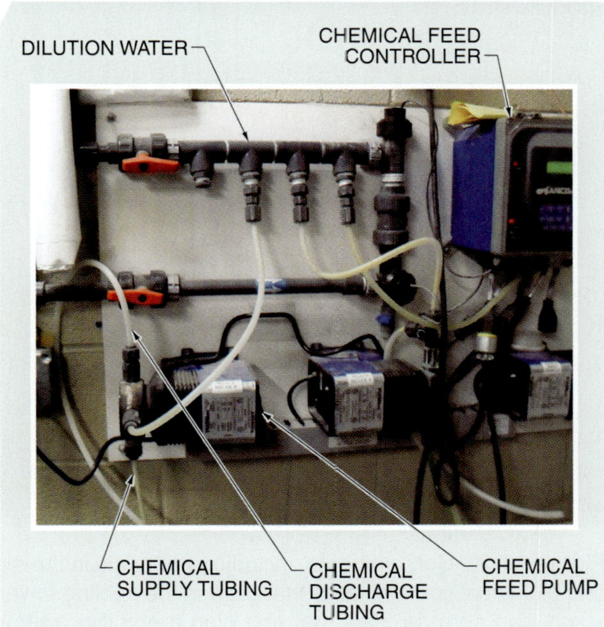

Figure 48-22. Chemicals are added to water loops to inhibit the corrosion of piping and chiller tubes and to remove mineral deposits from internal heat exchanger surfaces.

The circulating water loop pumps water to the cooling towers and back to the chiller. Typically, the water entering the condenser is 80°F to 85°F, and the water leaving is 90°F to 95°F. Water treatment of the circulating water loop is complex because the water is exposed to contaminants from outside air. Contaminants include the material in the air and any biological contaminants growing in the cooling tower sump. These contaminants can corrode, damage, and plug the chiller or heat exchanger tubes. **See Figure 48-23.** Biocides and other chemicals are added to the condenser water loop to control these contaminants. Also, UV lights can be used to kill biological contaminants.

Tube Corrosion

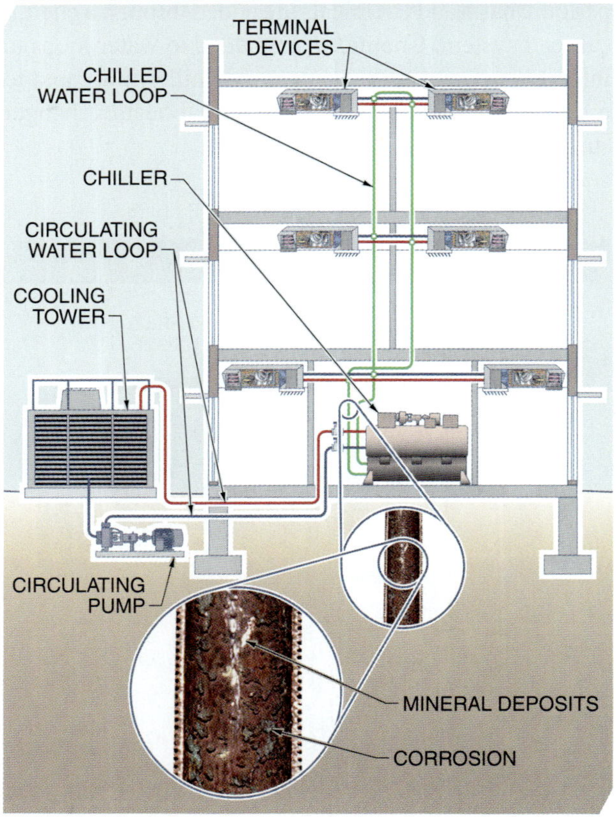

Figure 48-23. Contaminants, including materials in the air and biological contaminants that grow in the cooling tower sump, can corrode, damage, and plug the chiller and/or heat exchanger tubes.

Heating Systems. Water treatment in heating systems is needed because contaminants and minerals in the feedwater will precipitate out when heated. Oxygen, oil, and other contaminants can cause corrosion and coat the boiler heat exchanger surfaces. Contaminants decrease the efficiency of heating systems and cause damage to the components. Chemicals can be added to the water to remove the contaminants from the surfaces of the equipment.

Chiller Service Procedures

When chillers are inspected and serviced, the BAS must be checked for system operation status and errors. Typical chiller services performed include operating the VFD in manual mode and replacing the flow switch.

Placing a VFD in Manual Mode. VFDs are often used with chiller compressor motors to control capacity on both the compressor and condenser. HVAC technicians and operating personnel must be familiar with VFD basics and be able to operate a VFD using manual controls. **See Figure 48-24.** Chilled water is distributed by pumps and a piping system to the individual coils in the terminal units. Chiller systems of today typically use VFDs on pump motors and chiller compressor motors to improve system efficiency.

VFD Manual Operation

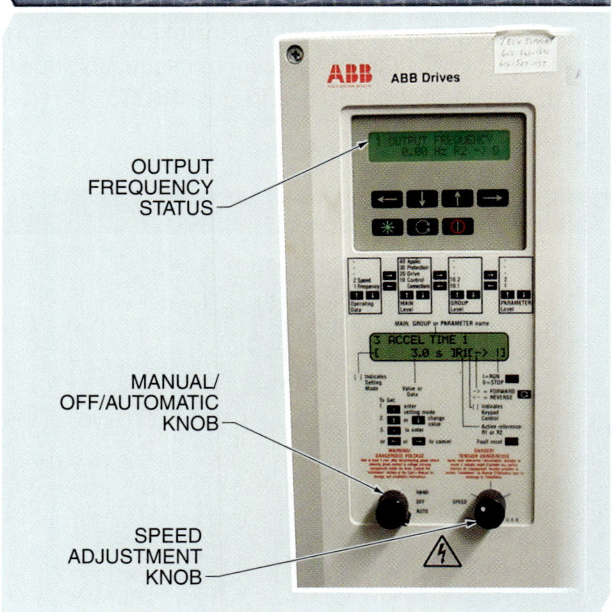

Figure 48-24. HVAC technicians must know how to place the chilled water system drives in manual mode and use keypads to speed up or slow down motors to keep the chilled water system running.

If the chiller's automatic control system fails, the HVAC technician must know how to put the drives in manual mode and use the keypad to speed up or slow down the motors to keep the chilled water system running until the control system is fixed. While keypads from different manufacturers vary, in general the method is similar. **See Procedure for Operating Chiller Motors in Manual Mode.**

Operating Chiller Motors in Manual Mode

1. Find and press the button on the touchscreen that changes the control system from AUTO to MANUAL.

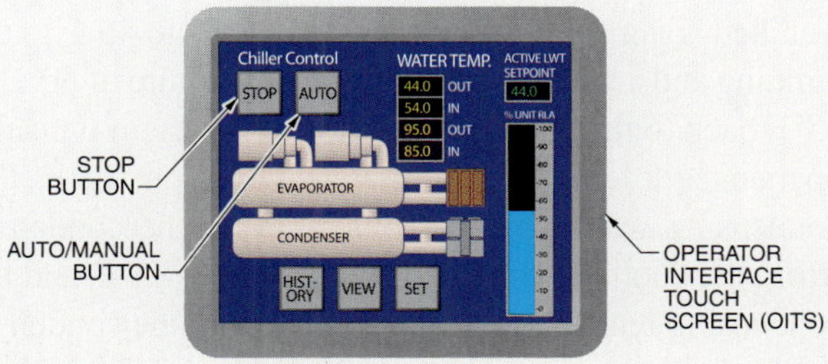

2. Use the home screen to press the SET button to get to the Motor VFD screen.

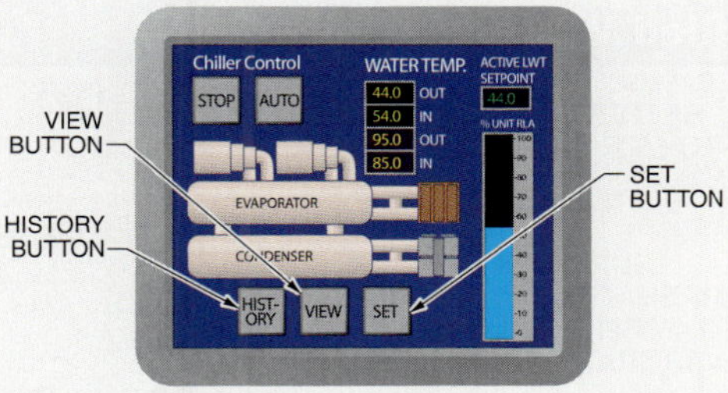

3. On the Motor Setpoint screen, push the 12 button.
   a. Push the change button (activates the arrow buttons).
   b. Use the up and down arrow buttons to change the compressor VFD (motor Hz) value to another value, such as 50% or 30 Hz.

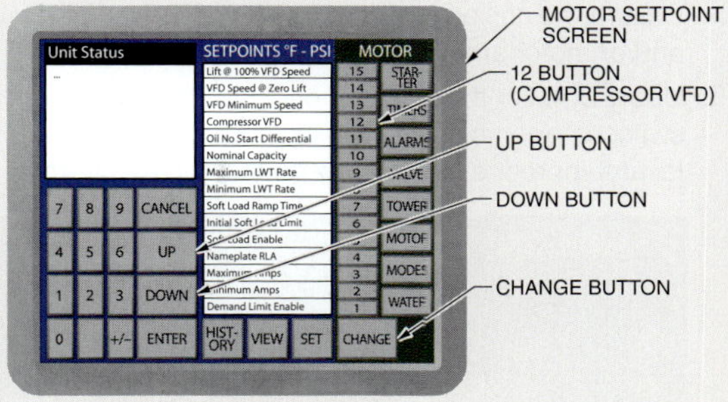

4. Observe system operation for temperature and/or pressure. Adjust the speed using the up or down buttons.
5. Determine the control problem and resolve it.
6. Use the home screen to place device back in AUTO mode.

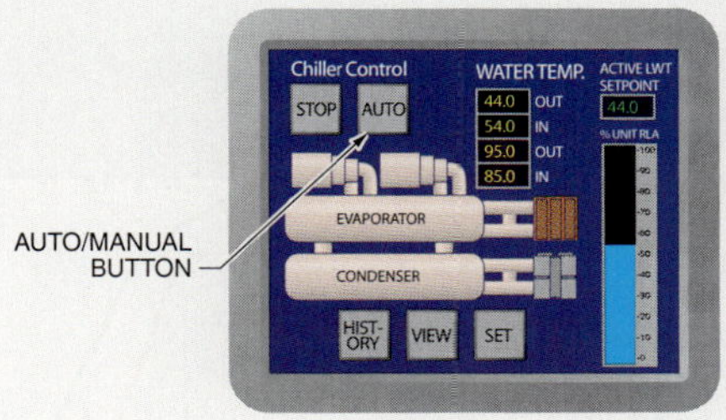

7. Check chiller operation.

TECH TIP

It is vital that procedures be in place to enable the manual operation of a chiller plant. Today, chiller plants are normally controlled automatically by building automation systems. Unfortunately, these systems can fail. In the event of failure, chilled water must still be supplied continually to crucial areas such as server rooms and hospital operating rooms.

Flow Switch Replacement. A typical problem that causes chillers to shut down is water flow switch failure. **See Figure 48-25.** The water flow switch is wired as a safety device to the chiller control board. When the flow switch opens (no flow), the chiller will shut down, and the chilled water temperature will rise. On older chillers, switch malfunction may cause chillers to shut down. A DMM is typically used to check the flow switch. **See Procedure for Checking Operation of Chiller Flow Switches.**

Water Flow Switches

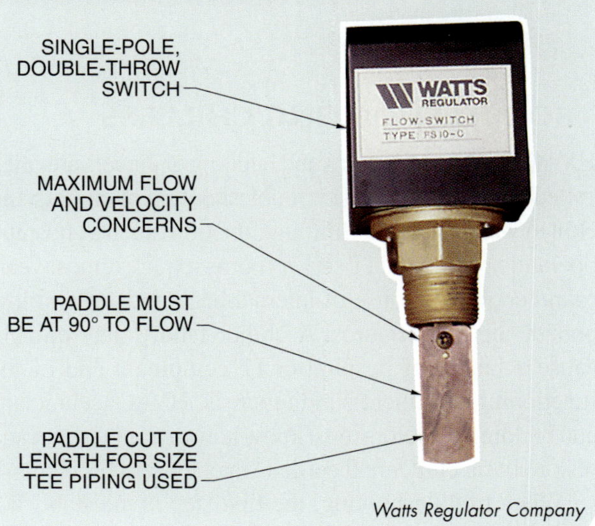

Watts Regulator Company

Figure 48-25. Chiller water flow switches are wired as safety devices to the chiller control board. When the flow switch opens (no flow) the chiller will shut down.

Procedure

Checking Operation of Chiller Flow Switches

1. Ensure the chiller pump is running and check pressures and/or motor amp draw to indicate water flow.
2. Use a DMM at the chiller control board to check whether switch is open or closed. Use manufacturer literature to locate appropriate terminals on the control board.

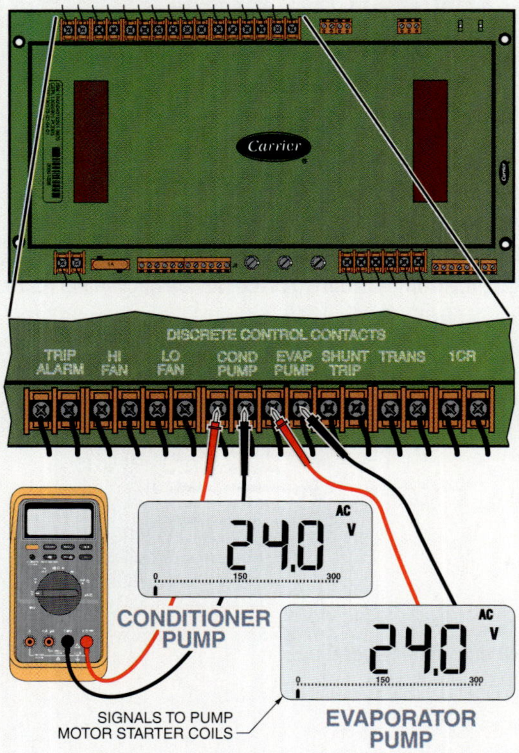

3. When water flow is present but switch indicates "open," the switch must be replaced. Note that a jumper wire can be used temporarily to allow the chiller to operate until repairs are made.

TROUBLESHOOTING CHILLERS

HVAC service technicians and building engineers are often called on to perform basic troubleshooting of chillers and chilled water systems. Building functions such as hospital operating rooms and IT server rooms require almost year-round cooling because of internal heat load, even at low outside-air temperatures. A chiller failure may quickly cause a failure of hospital or IT equipment and of the functioning of critical building areas. HVAC technicians and building operators must know how to troubleshoot and restore basic chiller and chilled water system problems.

When troubleshooting, the first step is to establish a basic knowledge of the particular chilled water system. The technician must be able to identify the system components, including the pumps, cooling towers, chillers, piping, and electrical and electronic controls.

When troubleshooting a chiller or chilled water system, it is best to start with the basics. Typical chilled water baseline temperature values should be kept in mind. **See Figure 48-26.** The chilled water evaporator operates with a chilled water supply temperature of 40°F to 45°F to the building and a chilled water return temperature of 50°F to 55°F back to the chiller evaporator. The system typically operates at a temperature difference of about 10°F. The condensing water supply temperature to the condenser (from the cooling tower) is around 80°F to 85°F, and the condensing return water temperature from the condenser (to the cooling tower) is around 90°F to 95°F. The system typically operates at a temperature difference of about 10°F.

Chilled Water Baseline Values

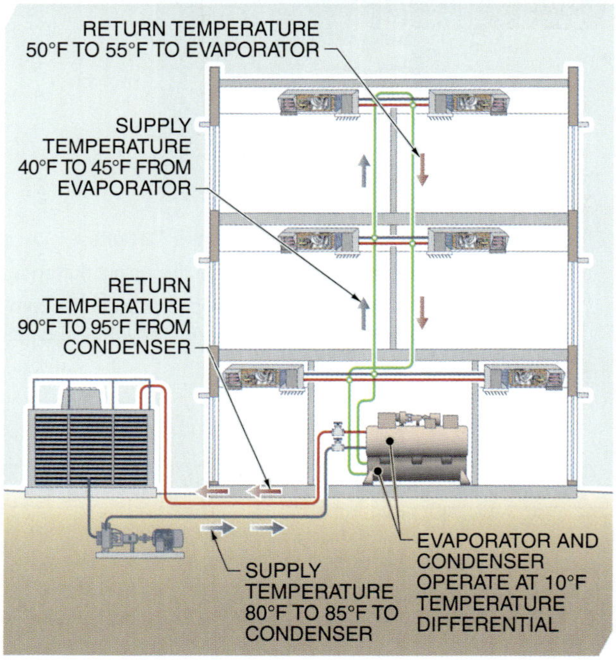

Figure 48-26. When troubleshooting a chiller or chilled water system, the chilled water baseline temperature values should be kept in mind.

Chiller Troubleshooting Scenario #1— Chiller Down

On a very hot summer day, a line of thunderstorms moved through the area. A customer calls in to the HVAC contracting office complaining about the heat. The dispatcher in turn calls the HVAC technician, who is sent to the facility. Upon arrival, the HVAC technician finds that the chilled water system is down. After a quick question to building maintenance personnel, the technician determines that there was a power surge and the chiller control system is not functioning. The HVAC technician checks the control transformer that supplies 24 VAC power to the chilled water control panel and determines that it was damaged by the power surge. **See Figure 48-27.**

Service Work Performed: The HVAC technician places the cooling tower pumps and fans in manual mode so the system can be returned to normal operation. The technician replaces the 24 VAC transformer, and the control system is rebooted. After the cooling tower pumps and fans are running, the chiller pump is started locally in manual mode to lower the chilled water loop temperature. After the control system starts, the cooling tower and chiller are placed in automatic mode.

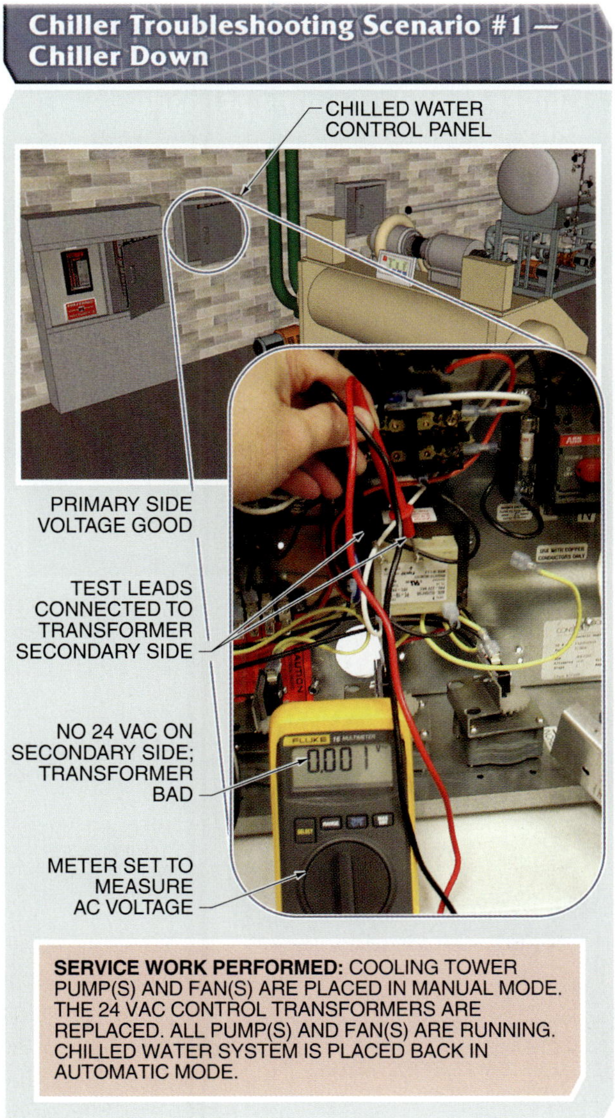

Figure 48-27. Chiller control systems, like all machinery control systems, are susceptible to power surges caused by thunderstorms.

Chiller Troubleshooting Scenario #2—Bad Water Flow Switch

On an extremely hot day, a customer calls and says it is too hot in the building. Upon arrival, the HVAC technician checks the chiller and finds that the compressor is off. A check of the control board finds that the flow switch is open, which caused the chiller to shut down. The pumps are checked and found to be off. Upon further checking, the HVAC technician determines that a maintenance technician inadvertently shut off power to the pumps. **See Figure 48-28.**

Service Work Performed: The HVAC technician restores power to the pumps. The pumps are restarted, and the chiller is energized. Immediately, the water in the chilled water system begins to cool down, providing cooling to the building.

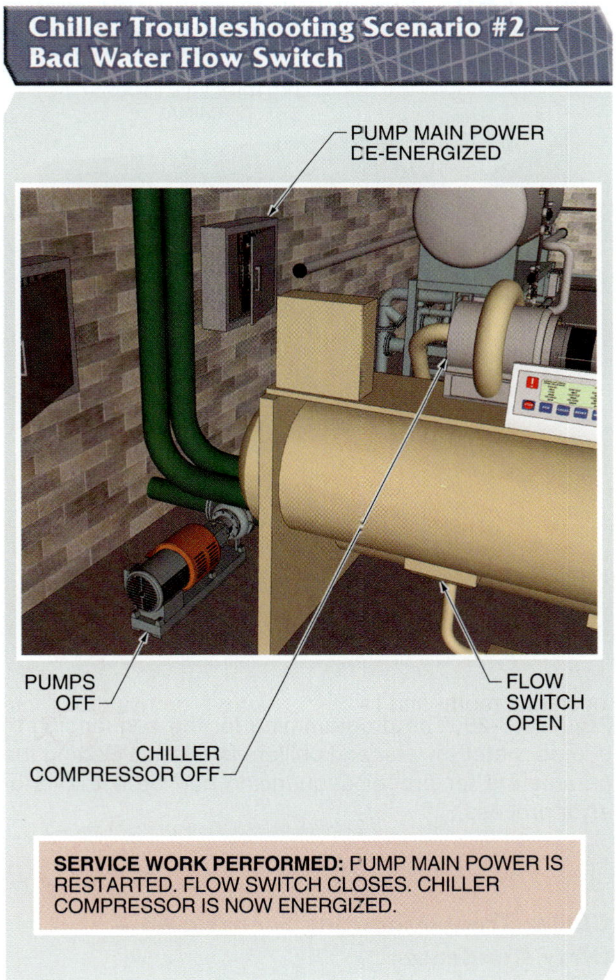

Figure 48-28. When troubleshooting, the simplest answer is typically the correct answer. A technician should always verify that power is on to the circuit or equipment in question.

Chiller Troubleshooting Scenario #3—Programming Out of Parameters

The dispatcher calls the HVAC technician with a service call that a building is not comfortable. The HVAC technician arrives at the building and checks the chilled water system BAS programming. The chilled water system programming is operating, so the technician checks the water temperatures. The technician determines that the system is not using enough chillers to meet the needed capacity. **See Figure 48-29.**

Service Work Performed: The HVAC technician overrides the next chiller in sequence to add more cooling capacity to the system. The chilled water system programming is checked and the sequence is adjusted to properly control the water temperature.

Figure 48-29. The programming for the sequencing of pumps, water towers, and chillers is difficult. Setting the parameters for proper sequencing can be a trial-and-error process.

Chiller Troubleshooting Scenario #4—Dirty Condenser

During the heat of the summer, a building chiller shuts down. The building owner calls in a "too hot" complaint. The HVAC technician arrives and checks the microprocessor display on the chiller for the fault. The display shows that the chiller is locked out on a high head pressure fault. A quick check of the inlet and outlet water pressures of the condenser indicates a high pressure differential. The condenser tubes are checked and found to be very dirty. A discussion with the building owner indicates that the condenser tubes have not been cleaned since the company moved in three years before. **See Figure 48-30.**

Service Work Performed: The HVAC technician shuts down the chiller and cleans the condenser tubes. After an inspection, the chiller is flushed and restarted. The chiller immediately begins to cool the water, and the building begins to cool down.

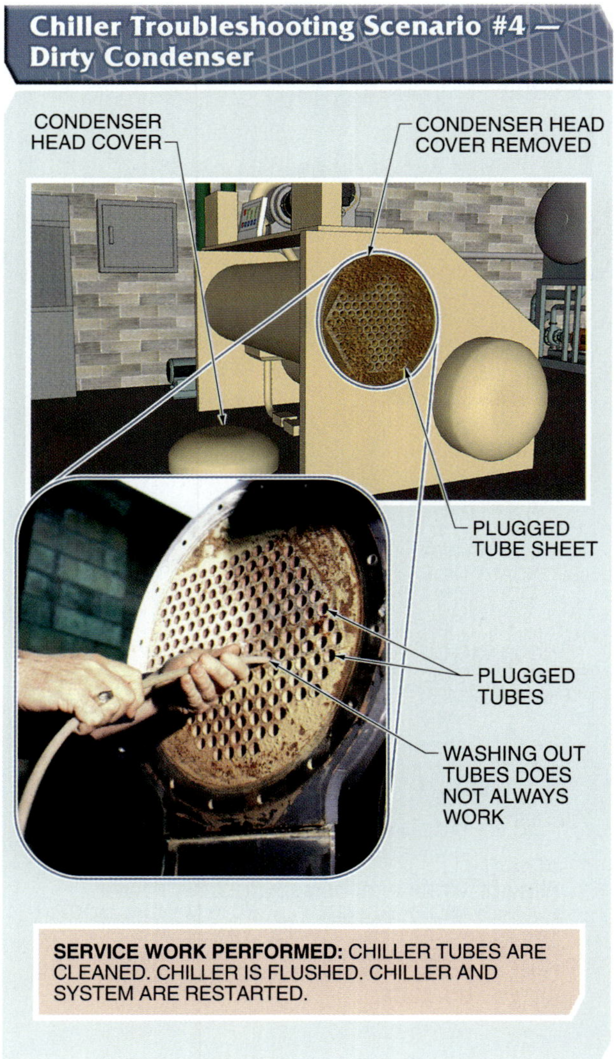

Figure 48-30. When a chiller is not as efficient as it once was, one common reason may be plugged tubes in the condenser.

Chiller Troubleshooting Scenario #5—Cooling Tower Fans Not Staging

A customer calls in and says that the chiller is operating erratically. At times it is too hot and at other times it operates properly. The HVAC technician observes the chiller operation. While the chiller compressor is running, the cooling tower fans are speeding up and slowing down erratically. The cooling tower fans are powered by a motor with a VFD. As the cooling tower fans change speed, the temperature of the water leaving the cooling tower fluctuates by 4°F. The HVAC technician checks the control system and discovers that the control system parameters for the cooling tower fans are set incorrectly. **See Figure 48-31.**

Service Work Performed: The HVAC technician checks the manufacturer specifications and adjusts the operating parameters of the cooling tower fans. The cooling tower fans operate properly, and the temperature of the water leaving the cooling tower holds steady.

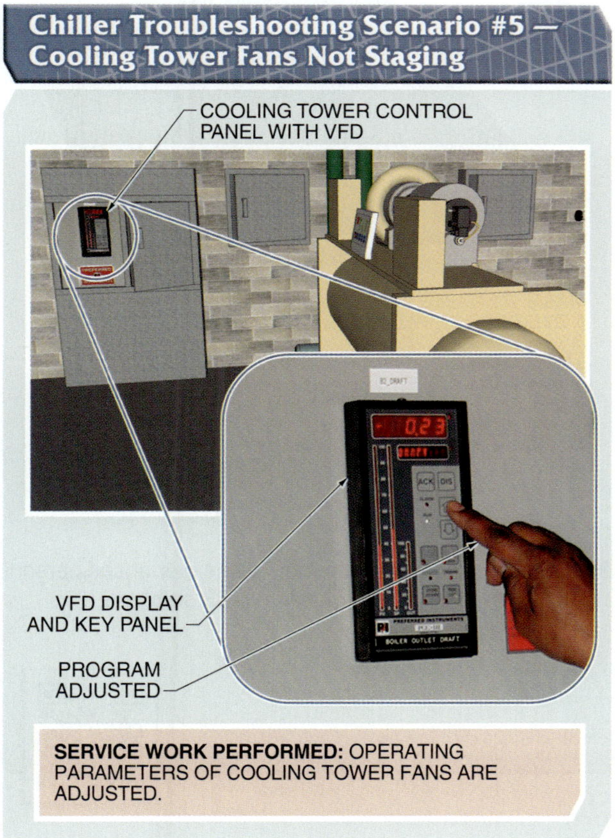

Figure 48-31. The VFD programming for motors that are controlled by VFDs, such as cooling tower fans, must be checked closely for proper settings.

Chiller Troubleshooting Scenario #6—Incorrect Temperature Setpoint

On a very hot day, a building is too hot for the occupants to work comfortably. The HVAC technician arrives and checks the chiller. It seems to be operating properly, but the temperature of the water leaving the chiller is 48°F instead of the normal 42°F. A check of the chiller control panel indicates that the temperature setpoint of the water leaving the chiller has been changed. A discussion with the building owner indicates that a maintenance technician attempted to adjust the setpoint but was unable to do so properly. **See Figure 48-32.**

Service Work Performed: The HVAC technician adjusts the chiller water temperature setpoint to manufacturer specifications. The building immediately begins to cool down.

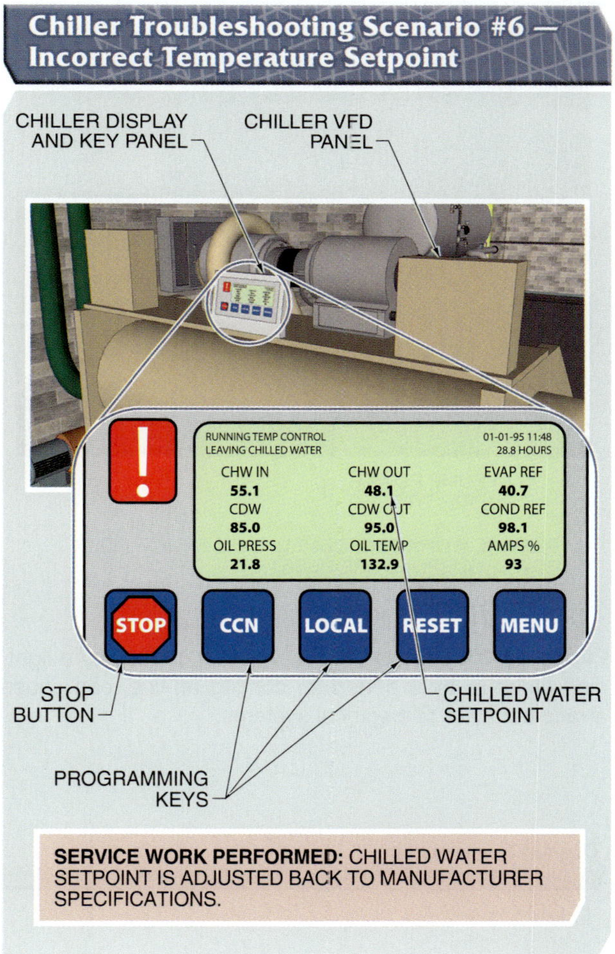

Figure 48-32. Inexperienced HVAC technicians and building personnel may find it difficult to set equipment and BAS programming properly.

Chiller Troubleshooting Scenario #7—Voltage Fault

A complaint is received from an office building that the chilled water system is down. The HVAC technician arrives quickly and begins to ask questions about the problem. It was determined that the chilled water system problem began immediately after a full load test is performed on the building's backup generator. A quick check of the chiller microprocessor panel display shows that the chiller stopped due to a voltage fault in the incoming power. **See Figure 48-33.**

Service Work Performed: The HVAC technician resets the chiller and chilled water pump power supply. The chiller and pump start normally, and the building begins to cool down properly.

Chiller Troubleshooting Scenario #8—Incorrect BAS Settings

A critical call comes in that a chilled water system is down. The HVAC technician arrives and finds that the entire chilled water system is shutdown. When asked, the building staff say the system shutdown occurred early that morning. The HVAC technician checks the BAS and finds that the chilled water system master enable switch was overridden and is in the OFF mode. A check with the building staff indicates that an inexperienced operator was checking the BAS at the time of the chilled water system shutdown. **See Figure 48-34.**

Service Work Performed: The HVAC technician removes the chilled water system master enable switch override. The chilled water system returns to normal operation.

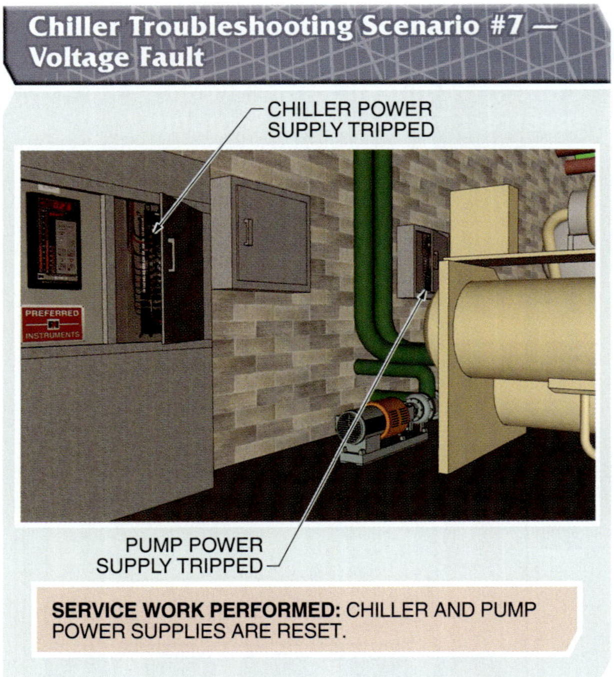

Figure 48-33. Voltage surges, whether caused by a lightning strike or by a generator coming on-line, can cause erratic behavior of electrical systems.

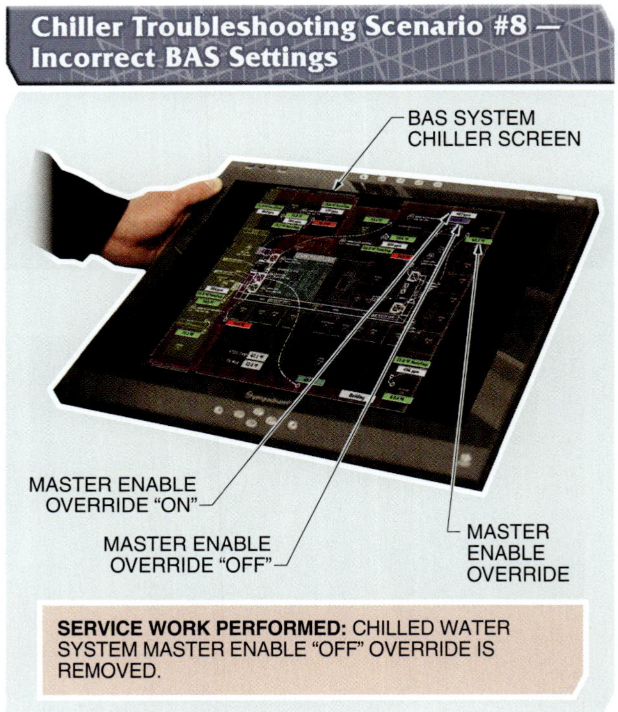

Figure 48-34. It is quite common for inexperienced personnel to enter incorrect settings into BAS software.

UNIT 49

Heating and Cooling Load Calculations

HVAC equipment is sized and selected based on the heating and cooling loads of a building. Correct amounts of heat must be added when heating or removed when cooling so that a comfortable building temperature is maintained. Many occupants may experience a lack of comfort due to incorrect heating or cooling load calculations. Heating and cooling loads are calculated using software applications or manual calculation methods.

OBJECTIVES

- Define indoor and outdoor design temperatures.
- Find summer and winter temperature differences.
- Explain how building structures affect heat loss and heat gain.
- Identify typical building variables.
- Describe the methods used for calculating, analyzing, and adjusting loads.

Index to Procedures

HEATING AND COOLING LOAD FACTORS

When the outdoor air is cold, heat must be added to a building to offset building space heat loss. When the outdoor air is hot, heat must be removed from a building to offset building space heat gain. Heating and cooling load calculations are used to determine the amount of heating or cooling needed to maintain a constant, comfortable temperature within a building.

Heating load is the amount of heat lost by a building. *Cooling load* is the amount of heat gained by a building. Factors that can affect heat loss and heat gain include the local design conditions (such as the difference between indoor temperature and outdoor temperature), building structure, building use, infiltration, and internal heat loads within the building design. Factors must be identified and calculated in order to determine the correct heating and cooling loads of a building.

Heating and cooling loads, design temperatures, and design factors are identified and shown on load calculation forms. Load calculation forms are developed by manufacturers and trade organizations, with local contractors adjusting the forms to fit local needs. Load calculation forms are used to collect design information and conditions and to calculate the needed capacity of the HVAC equipment and distribution systems.

Local Design Temperatures

Buildings will lose heat to the outdoor air in the winter and gain heat from the outdoor air in the summer. The primary heat loss and heat gain factors that must be considered in the calculations are the indoor and outdoor design temperatures.

Design temperature is the temperature of air at a predetermined set of conditions. *Indoor design temperature* is the temperature selected for the inside of a building. The typical indoor design temperature used for occupied buildings is 70°F for heating and 75°F for cooling.

Outdoor design temperature is the expected outdoor temperature that a heating or cooling load must balance. The outdoor design temperature is determined statistically by organizations like the American Society of Heating, Refrigeration, and Air-Conditioning Engineers (ASHRAE). Outdoor design temperatures are the coldest winter temperatures expected in a geographic area during 97.5% of the winter and the warmest summer temperatures expected in a geographic area during 95% of the summer.

An *outdoor design temperature table* is a table of data developed from records of temperatures taken in a geographic area over a period of years. **See Figure 49-1.** Outdoor design temperature tables are available from many sources. The tables include winter dry bulb temperature, summer dry bulb temperature and summer wet bulb temperature, daily temperature range, and latitude. **See Appendix.**

Outdoor Design Temperatures

State and City	Latitude*	Winter	Summer		
		DB†	DB†	DR‡	WB†
Alabama					
Alexander City	32	18	96	21	79
Auburn	32	18	96	21	79
Birmingham	33	17	96	21	78
Mobile	30	25	95	18	80
Montgomery	32	22	96	21	79
Talladega	33	18	97	21	79
Tuscaloosa	33	20	98	22	79
Alaska					
Anchorage	61	−23	71	15	60
Fairbanks	64	−51	82	24	64
Juneau	58	−4	74	15	61
Kodiak	57	10	69	10	60
Nome	64	−31	66	10	58

* in degrees
† in °F
‡ daily range in °F

Air Conditioning Contractors of America

Figure 49-1. Industry organizations provide tables for outdoor design temperatures based on temperature records over many years.

Winter dry bulb temperature is the coldest temperature expected to occur in a geographic area excluding the lowest temperatures that occur in 1% to 2½% of the total hours in the three coldest months of the year. Heating loads for winter conditions are calculated at the winter dry bulb temperature.

Summer dry bulb temperature is the warmest dry bulb temperature expected to occur in a geographic area excluding the highest temperature that occurs in 1% to 5% of the total hours in the three hottest months of the year. *Summer wet bulb temperature* is the wet bulb temperature that occurs concurrently with the summer dry bulb temperature. Wet bulb temperature is considered in cooling loads because of the effect of humidity on both the comfort level of occupants at higher temperatures and on cooling equipment capacity. Cooling loads are calculated for summer conditions at the summer dry bulb temperature.

Design temperature difference is the difference between the selected indoor design temperature and the expected outdoor design temperature for a particular season. Winter design temperature difference is found by subtracting the winter outdoor design temperature from the winter indoor design temperature. In some regions, the winter outdoor design temperature is negative. Winter design temperature difference is found by applying the following formula:

$$\Delta T_W = T_i - T_o$$

where

ΔT_W = winter design temperature difference (in °F)

T_i = indoor design temperature (in °F)

T_o = outdoor design temperature (in °F)

Summer design temperature difference is found by subtracting the summer indoor design temperature from the summer outdoor design temperature. Summer design temperature difference is found by applying the following formula:

$$\Delta T_S = T_o - T_i$$

where

ΔT_S = summer design temperature difference (in °F)

T_o = outdoor design temperature (in °F)

T_i = indoor design temperature (in °F)

TECH TIP

Some large contractors design HVAC systems for remote geographical areas. Outdoor design temperature charts must be consulted in order to size the system equipment properly.

Example: Finding Design Temperature Differences

Find the design temperature difference for a residential building located in Montgomery, Alabama. For indoor design temperatures, use 70°F for winter and 75°F for summer.

1. On a table of outdoor design temperatures, locate the winter dry bulb and summer dry bulb temperatures for Montgomery, Alabama. The winter outdoor design temperature is 22°F and the summer outdoor design temperature is 96°F.
2. Find the winter design temperature difference.

$$\Delta T_W = T_i - T_o$$
$$\Delta T_W = 70 - 22$$
$$\Delta T_W = \mathbf{48°F}$$

3. Find the summer design temperature difference.

$$\Delta T_S = T_o - T_i$$
$$\Delta T_S = 96 - 75$$
$$\Delta T_S = \mathbf{21°F}$$

Building Structures

Heat is lost or gained through building components that are exposed to outside air. A *building component* is an object that helps provide a structurally sound and environmentally safe space to protect occupants and contents from the outdoor environment. Heat loss or heat gain through each building component is calculated individually. Individual calculations can then be added together to find the heating load or cooling load for a building. Building components typically include the following:

- windows, glass doors, and skylights
- wood and metal doors
- walls, above and below grade
- ceilings and roofs
- floors, slabs, and basements

Exposed Surfaces. *Exposed surfaces* are building surfaces that are exposed to outdoor temperatures. Exposed surfaces include windows, doors, walls, and roofs. Depending on the structure type, heat is also transferred through floors, slabs, and basement walls. The greater the area exposed to the outdoor air, the greater the heat loss or heat gain.

To calculate heat loss and heat gain, the areas of all exposed surfaces must be found. Areas of the exposed surfaces of a building, such as walls and ceilings, are found using building dimensions. Overhangs that do not affect the transmission of heat are not included.

For sloping ceilings such as cathedral ceilings, the exposed area, not the area of the horizontal plane, is used for calculating heat loss or heat gain. For cooling loads, the wall areas should also be separated according to their geographic direction of exposure. The geographic direction of exposure is important because the solar gain on the surfaces of exterior walls, windows, and doors varies depending on the direction they are facing. **See Figure 49-2.**

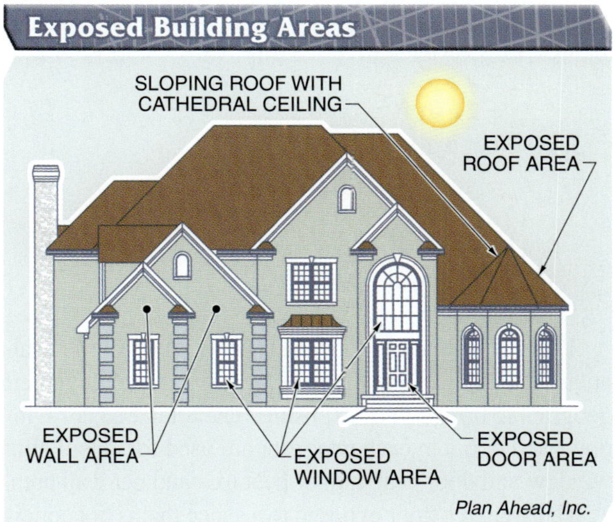

Figure 49-2. Areas exposed to outside air are used for calculating heat loss and heat gain.

Gross wall area is the total area of a wall including the areas of windows, doors, and other openings. Gross wall area is calculated separately for each exposed wall of a building. Gross wall area is found by applying the following formula:

$$A_g = w \times h$$

where

A_g = gross wall area (in sq ft)

w = width (in ft)

h = height (in ft)

Net wall area is the area of a wall after the areas of windows, doors, and other openings have been subtracted. Net wall area is used for calculating heating and cooling loads. Net wall area is found by applying the following formula:

$$A_n = A_g - A_o$$

where

A_n = net wall area (in sq ft)

A_g = gross wall area (in sq ft)

A_o = area of opening (in sq ft)

Example: Finding Net Wall Areas

The south wall of a building is 30′ wide and 8′ high. The wall has a door 3′ wide and 7′ high. Find the net wall area by subtracting the area of openings from the gross wall area.

1. Find the gross wall area.

$$A_g = w \times h$$
$$A_g = 30 \times 8$$
$$A_g = \mathbf{240\ sq\ ft}$$

2. Find the area of the opening.

$$A_o = w \times h$$
$$A_o = 3 \times 7$$
$$A_o = \mathbf{21\ sq\ ft}$$

3. Find net wall area.

$$A_n = A_g - A_o$$
$$A_n = 240 - 21$$
$$A_n = \mathbf{219\ sq\ ft}$$

The areas of each window and door of a building are calculated separately, and the results are categorized by their geographic direction of exposure. The total areas of windows and doors in each exposure are used for calculating window and door heat gain or heat loss and net wall area.

The area of a floor exposed to a space that is not conditioned is used to calculate heat loss or heat gain. However, in buildings with concrete slab floors on the ground, most of the heat loss from the floor occurs around the edge. Heat loss for a concrete slab floor is calculated as the distance around the perimeter of the exposed edge of the floor. Concrete slab floors have no heat gain.

Conduction. According to the second law of thermodynamics, when the temperature on two sides of a material is different, heat flows from the warmer side to the cooler side of the material. This applies to the outside surfaces of a building, such as walls, windows, doors, floors, and ceilings. When air outside of a building is cooler than the air inside, heat will flow from the inside to the outside. When air outside a building is warmer than the air inside, heat will flow from the outside to the inside.

A *conduction factor (U-factor)* is an expression of the amount of heat that flows through a building component due to a temperature difference between the inside and outside air. Conduction factors are expressed in Btu/hr per square feet of material per the temperature difference in °F through the material. **See Figure 49-3.** These factors are based on the sensible heat of air, the amount of air moved, and the difference between the indoor temperature and outdoor temperature.

Conduction Factor

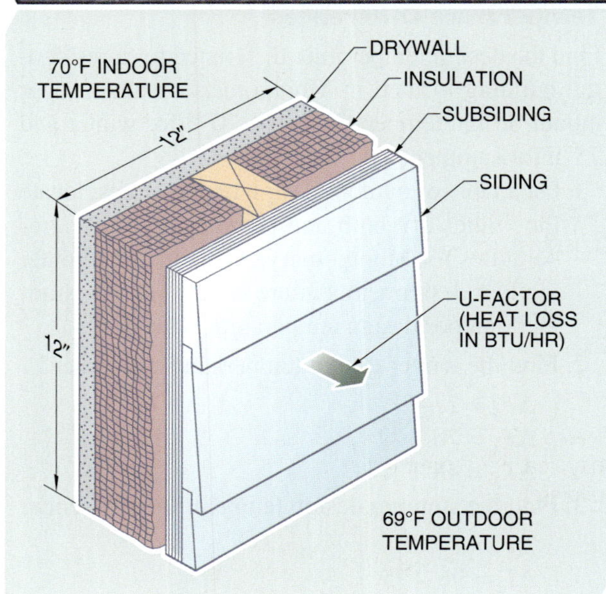

Figure 49-3. A conduction factor is an expression of the amount of heat that flows through a building component due to a temperature difference between the inside air and the outside air.

Heat transfer factors have been calculated for most building components and materials used in the construction of residential and light commercial buildings. A *heat transfer factor* is a conduction factor multiplied by a design temperature difference. Heat transfer factors are found in tables of conduction factors of various building materials. To simplify calculations, conduction factors and either outdoor design temperatures or design temperature differences are combined in a table of heat transfer factors. **See Appendix.**

Infiltration. *Infiltration* is the leaking of outdoor air into a building. *Infiltration air* is air that flows into a building when exterior doors or windows are open or when air leaks in through cracks around doors, windows, or other openings. The cold or hot air, depending on ambient conditions, adds to the heating or cooling load of a building. The amount of infiltration air entering a building is found using the building infiltration method or building air changes method.

The building infiltration method is calculated based on the impact that the air leaking into a building has on heating and cooling loads. The building infiltration method is typically used with modern residential buildings that are built to tight construction standards, minimizing infiltration. Factors for calculating heat loss and heat gain from infiltration air at different temperatures are found in tables of infiltration factors. **See Figure 49-4.** The building space heat loss or heat gain is multiplied by the infiltration factor to include the heat loss or heat gain from infiltration air.

Infiltration Factors*

Outdoor design temperature†	85	90	95	100	105	110
Infiltration factors‡	0.7	1.1	1.5	1.9	2.2	2.6

* indoor design temperature 75°F
† in °F
‡ in Btu/hr/sq ft

Air Conditioning Contractors of America

Figure 49-4. Tables of infiltration factors are often based on typical tight construction practices of modern residential buildings.

The building air changes method is based on the impact that air changes per hour has on heating and cooling loads. *Air changes per hour (ACH)* is a measure of the total hourly volume of infiltration air divided by the volume of building space. ACH is a value that represents the number of times per hour that air in a building space is completely replaced by outdoor air.

ANSI/ASHRAE Standard 62-1989, *Ventilation for Acceptable Indoor Air Quality,* specified a minimum ACH of 0.35 total or 15 cfm per occupant, whichever is greater, to provide adequate fresh air in buildings. While this typically is not measured, modern residential buildings built to tight construction standards have an ACH of approximately 0.50. Because of existing infiltration, HVAC systems in residential buildings typically do not provide ventilation air.

Specific building tightness requirements have been recently developed by the International Code Council® (ICC) and most recently published as the *2012 International Building Code®* (IBC). The 2012 IBC requirements include improved energy efficiency through limited infiltration. As a result, ANSI/ASHRAE Standard 62.1-2010 *Ventilation for Acceptable Indoor Air Quality* and ANSI/ASHRAE Standard 62.2-2010 *Ventilation and Acceptable Indoor Air Quality in Low-Rise Residential Buildings* were issued. ANSI/ASHRAE Standard 62.1-2010 sets ventilation standards for all buildings except low-rise residential buildings. ANSI/ASHRAE Standard 62.2-2010 requires low-rise residential buildings to have a minimum amount of outdoor air ventilation based on the building infiltration rate, use, and number of occupants.

As residential building envelopes are increasingly tightened per the IBC, building infiltration rates may become low enough that alternative building ventilation becomes necessary. Blower door tests can be used to measure the infiltration rate of a building. Infiltration rates of new residential buildings should be measured to confirm that the requirement (ACH of 0.35 or 15 cfm per occupant) is met. When the infiltration rate is below the required limit, ventilation must be provided.

To find the infiltration airflow rate (in cfm) using ACH, the volume of the building or building space (V_{bs}) must be found. The volume is multiplied by the ACH to find the infiltration airflow rate. Typical ACH values for selected buildings are found in air change factors tables. **See Figure 49-5.** The volume of a building space is found by applying the following formula:

$$V_{bs} = floor\ area \times h$$

where

V_{bs} = volume of building space (in cu ft)

floor area = area of occupied building space (in sq ft)

h = wall height of building space (in ft)

The infiltration airflow rate is found by applying the following formula:

$$Q_{\text{inf}} = V_{bs} \times \frac{ACH}{60}$$

where

Q_{inf} = infiltration airflow rate (in cfm)

V_{bs} = volume of building space (in cu ft)

ACH = air changes per hour

60 = constant (min/hr)

Air Change Factors

Building Type	Number of Air Changes per Hour — Construction Type*		
	Loose	Medium	Tight
Residential	1.5	1.0	0.5
Commercial	3.0	1.5	0.5
Industrial	5.0	3.0	1.0

* refers to type fit of doors and windows

Air Conditioning Contractors of America

Figure 49-5. Tables of air changes per hour are based on building type and construction type.

When using a table of infiltration factors based on the infiltration airflow rate, a factor of 1.08 and the winter design temperature difference are multiplied to find heat loss. The infiltration airflow rate is the cooling load due to infiltration air and can be found on a psychrometric chart.

The psychrometric chart gives a cooling load that includes the latent heat load for humidity. Latent heat load for humidity does not have to be considered in the heating load, but it must be considered in the cooling load. The heat loss from infiltration air is found by applying the following formula:

$$H_{inf} = Q_{inf} \times 1.08 \times \Delta T$$

where

H_{inf} = infiltration heat loss (in Btu/hr)

Q_{inf} = infiltration airflow rate (in cfm)

1.08 = constant

ΔT = temperature difference (in °F)

Example: Finding Heat Loss from Infiltration

A Chicago residence has a floor area of 970 sq ft and walls 8 ft high. The building is a medium-construction type. The outdoor air has a dry bulb temperature of –9°F. The dry bulb temperature of the indoor air is 70°F. Find the heat loss from infiltration air.

1. Find the volume of the building space.

$$V_{bs} = floor\ area \times h$$

$$V_{bs} = 970 \times 8$$

$$V_{bs} = \mathbf{7760\ cu\ ft}$$

2. Find the infiltration airflow rate.

$$Q_{\text{inf}} = V_{bs} \times \frac{ACH}{60}$$

$$Q_{\text{inf}} = 7760 \times \frac{1.00}{60}$$

$$Q_{\text{inf}} = 7760 \times 0.01666$$

$$Q_{\text{inf}} = \mathbf{129.33\ cfm}$$

3. Find the heat loss from infiltration air.

$$H_{inf} = Q_{inf} \times 1.08 \times \Delta T$$

$$H_{inf} = 129.33 \times 1.08 \times [70 - (-9)]$$

$$H_{inf} = 129.33 \times 1.08 \times 79$$

$$H_{inf} = 129.33 \times 85.32$$

$$H_{inf} = \mathbf{11{,}034\ Btu/hr}$$

Building Variables

Heat loss and heat gain are calculated using the building variables that are unique to a particular building. A *building variable* is the specific use, level of use, and operation of a building that increases or decreases the heating and cooling loads on the building. Heat loss or heat gain through each building variable is calculated individually. Building variables include the number of occupants, lights and appliances, solar gain, and ventilation rates.

Number of Occupants. In residential buildings, the estimated number of occupants is based on the number of bedrooms in the building. Occupant numbers are calculated based on two people for a main bedroom and one person for each additional bedroom. To include guests in the number of occupants, the typical number of guests for the building type is used.

In light commercial buildings, the estimate of the number of occupants is based on the area of floor space required per person and the activity level of each person. Tables of occupancy estimates for light commercial buildings show the area required per person for various activities. Care should be taken to only use space that a person can actually occupy. For example, approximately one-half of the floor space in light commercial buildings is typically used as support rooms and building facilities, such as lobbies, hallways, meeting rooms, and bathrooms.

TECH TIP

One of the more difficult problems encountered when performing heating and cooling load calculations is determining the heating and cooling loads for areas in which the number of occupants fluctuates widely, such as in churches, which may have hundreds of occupants one day but very few on other days.

The human body produces heat and moisture. Heat is given off by the skin. Moisture is expelled from the body through respiration and perspiration. The amount of heat given off varies with the level of activity. Vigorous activity produces more heat than passive activity. A table of occupant heat gain values is used to identify the amount of heat produced by people when engaged in various levels of activities. **See Figure 49-6.**

Heat gain from occupants is not typically calculated as heating loads in building load calculations because the greatest heating load typically occurs at night when the activity level is low. For residential cooling loads, the occupant heat gain value typically used is 400 Btu/hr/person. This value is multiplied by the number of occupants, and the resulting amount is included in cooling load calculations for the corresponding spaces typically occupied during the day. For light commercial loads, the number of occupants is very important in calculating the heating and cooling loads. Heat gain or loss for light commercial buildings is calculated using Manual N from the Air Conditioning Contractors of America (ACCA).

Occupant Heat Gain*

Degree of Activity	Typical Application	Total Heat†	Sensible Heat†	Latent Heat†
Seated, at rest	Theater, grade school classroom	330	225	105
Seated, very light work	Office, hotel, apartment, high school classroom	400	245	155
Moderately active office work	Office, hotel, apartment, college classroom	450	250	200
Standing, light work	Drug store, bank	500	250	250
Sedentary work	Restaurant	550	275	275
Light bench work	Factory	750	275	475
Moderate dancing	Dance hall	850	305	545
Walking, moderately heavy work	Factory	1000	375	625
Bowling, heavy work	Bowling alley, factory	1450	580	870

* based on 75°F room dry bulb temperature.
† in Btu/hr

Air Conditioning Contractors of America

Figure 49-6. Heat gain from occupants is found by multiplying the number of people by a value based on their degree of activity.

Lights and Appliances. Heat gain values for lights and appliances are only included for cooling loads. For residential buildings, the heat gain values for lights are not typically included since the use of lights during the day is minimal. For residential buildings, the appliance heat gain value typically used is 1200 Btu/hr for kitchen appliances.

For light commercial buildings, heat gain values for lighting must be factored into cooling load calculations. To calculate the heat gain from lighting, the total wattage of the lights is multiplied by 3.41 Btu/W to convert electrical energy to thermal (heat) energy. When fluorescent lights are used, the conversion factor is 4.25 Btu/W. Fluorescent lights have a larger factor because their ballasts radiate larger quantities of heat than other types of lights. A *ballast* is an electrical device used for starting and regulating fluorescent lamps, which generate heat.

For light commercial buildings, heat gain values used for appliances in cooling load calculations will depend on the typical use of the appliances during the day. Most appliances are used for a part of the day, so the heat gain can be reduced to the rate of use. Heat gain values for appliances are calculated by multiplying the wattage of an appliance by 3.41 Btu/W.

Calculating heat gain for appliances that use fuel instead of electricity is based on the rate of fuel used and the heat output for the fuel. Heat gain values are also found on appliance heat gain tables. **See Figure 49-7.** Light commercial buildings typically include a greater variety of appliances and equipment than residential buildings.

Solar Gain. *Solar gain* is heat gain caused by radiant energy from the sun striking opaque objects. Sunlight shines through windows and heats interior building spaces. The thermal energy is then absorbed by the building. For residential building cooling loads, the solar heat gain values are typically not included because the window areas are small, and blinds or curtains are often used during sunny days. For light commercial building cooling loads, the heat from solar energy is added to the cooling load of the building so the system can compensate for the additional heat.

The amount of window area must be included when calculating the cooling load for a building.

Equivalent temperature difference is the design temperature difference adjusted for solar gain. The equivalent temperature difference is used when calculating heat gain through walls or roofs. It is used only for calculating cooling loads and mostly for light commercial buildings. Using an equivalent temperature difference is one way to include the solar gain increase in the cooling load. Tables of equivalent temperature differences show the equivalent temperature difference for sunlit walls and roofs. **See Figure 49-8.**

Appliance Heat Gain Values

Type of Appliance	Electric				Gas			
	Without Hood			Hood	Without Hood			Hood
	Sensible*	Latent*	Total*	Sensible*	Sensible*	Latent*	Total*	Sensible*
Broiler-griddle 31″ × 20″ × 18″					11,700	6300	18,000	3600
Coffee brewer/warmer					1750	750	2500	500
per burner	770	230	1000	340				
per warmer	230	70	300	90				
Coffee urn								
3 gal.	2550	850	3400	1000	3500	1500	5000	1000
5 gal.	3850	1250	5100	1600	5250	2250	7500	1500
8 gal.	5200	1600	6800	2100	7000	3000	10,000	2000
Deep fat fryer								
15 lb fat	2800	6600	9400	3000	7500	7500	15,000	3000
21 lb fat	4100	9600	13,700	4300				
Dry food warmer per sq ft top	320	80	400	130	560	140	700	140
Griddle, frying per sq ft top	3000	1600	4600	1500	4900	2600	7500	1500
Hot plate					5300	3600	8900	2800
Short order stove per burner					3200	1800	5000	1000
Toaster								
Continuous								
360 slices per hour	1960	1740	3700	1200	3600	2400	6000	1200
720 slices per hour	2700	2400	5100	1600	600	4000	10,000	2000
Pop-up (4 slice)	2230	1970	4200	1300				
Waffle iron 18″ × 20″ × 13″	1680	1120	2800	900				
Hair dryer, blower type	2300	400	2700					
Lab burners								
Bunsen					1680	420	2100	
Meeker	60				3360	840	4200	
Neon sign per foot of tube			60					
Sterilizer	650	1200	1850					
Vending machines								
Hot drink			1200					
Cold drink			625					

*in Btu/hr

Air Conditioning Contractors of America

Figure 49-7. Heat gain from appliances is found by adding the values for each appliance in a building.

Equivalent Temperature Differences

Design temperature*	85	90	95		100		105	110
Daily temperature range	L or M	L or M	M	H	M	H	H	H
Walls and Doors								
Wood frame walls and doors	13.6	18.6	23.6	18.6	28.6	23.6	28.6	33.6
Solid masonry, block, or brick walls	6.3	11.3	16.3	11.3	21.3	16.3	21.3	26.3
Partitions	5.0	10.0	15.0	10.0	20.0	15.0	20.0	25.0
Ceilings and Roofs								
Dark exterior	34.0	39.0	44.0	39.0	49.0	44.0	49.0	54.0
Light exterior	26.0	31.0	36.0	31.0	41.0	36.0	41.0	46.0
Floors								
Over unconditioned, vented, or open space	5.0	10.0	15.0	10.0	20.0	15.0	20.0	25.0
Over conditioned space or on or below grade	0	0	0	0	0	0	0	0

* in °F

Air Conditioning Contractors of America

Figure 49-8. Equivalent temperature difference is the design temperature difference adjusted for solar gain.

Tables of heat transfer factors for windows and doors include the solar gain in Btu/hr/sq ft of glass surface for typical windows. The tables also provide data for modifying the factor values for different types of glass. **See Figure 49-9.** Solar gain on walls and ceilings is combined with conduction gain in some heat transfer factor tables. In other tables, solar gain is included with an equivalent temperature difference for the heating and cooling loads. When using a particular set of tables for calculating heat gain, the method for including solar gain through the building surfaces must be determined.

Ventilation. *Ventilation* is the process of reducing air pollution by introducing enough fresh outdoor air into a building to dilute indoor pollutants. *Ventilation air* is outdoor air that is brought into a building to improve indoor air quality. Ventilation air must be heated when the indoor air is warmer than the outdoor air and cooled when the indoor air is cooler than the outdoor air. Ventilation air is typically provided for light commercial buildings but not for residential buildings.

In light commercial buildings, the ventilation air airflow rate required for a building is based on the airflow rate required per person in the building. The airflow rate required per person is found using the applicable ASHRAE standards. The ASHRAE standards provide airflow rates required per building area or per person for various building uses and activity levels. To find the ventilation airflow rate required, the number occupants of the building is multiplied by the ventilation airflow rate required per person. To find the ventilation airflow rate required for a building, apply the following formula:

$$Q_{ven} = p \times Q_p$$

where

Q_{ven} = ventilation airflow rate required (in cfm)
p = number of people
Q_p = ventilation airflow rate required per person (in cfm)

Example: Finding Ventilation Airflow Rates

A light commercial building is occupied by 70 people. ASHRAE standards require 30 cfm of ventilation airflow rate per person in a light commercial building for the activities being performed. Find the ventilation airflow rate.

$$Q_{ven} = p \times Q_p$$
$$Q_{ven} = 70 \times 30$$
$$Q_{ven} = \mathbf{2100\ cfm}$$

To find the heat loss from ventilation air, the airflow rate is multiplied by both 1.08 and the difference between the indoor and outdoor temperatures. The heat loss from ventilation air is found by applying the following formula:

$$H_{ven} = Q_{ven} \times 1.08 \times \Delta T$$

where

H_{ven} = heat loss from ventilation air (in Btu/hr)
Q_{ven} = ventilation airflow rate required (in cfm)
1.08 = constant
ΔT = temperature difference (in °F)

Heat Transfer Factors for Windows and Doors

Exposure	Temperature Difference								
	Single Glass			Double Glass			Triple Glass		
	15°	20°	25°	15°	20°	25°	15°	20°	25°
Windows									
N	18	22	26	14	16	18	11	12	13
NE & NW	37	41	46	31	33	35	26	27	28
E & W	52	56	60	44	46	48	38	39	40
SE & SW	45	49	53	39	41	43	33	34	35
S	28	32	36	23	25	27	19	20	21
Doors									
Wood	8.6	10.9	13.2	8.6	10.9	13.2	8.6	10.9	13.2
Metal	3.5	4.5	5.4	3.5	4.5	5.4	3.5	4.5	5.4

Air Conditioning Contractors of America

Figure 49-9. Tables of heat transfer factors for windows and doors show the Btu/hr solar gain per square foot of surfaces for typical windows and doors.

Example: Finding Heat Loss from Ventilation Airflow Rates

A light commercial building is occupied by 35 people. ASHRAE standards require 30 cfm of ventilation airflow per person for the activities being performed. The difference between the indoor and outdoor temperatures is 40°F.

1. Find the ventilation airflow rate required.

$$Q_{ven} = p \times Q_p$$
$$Q_{ven} = 35 \times 30$$
$$Q_{ven} = \mathbf{1050\ cfm}$$

2. Find the heat loss from ventilation air.

$$H_{ven} = Q_{ven} \times 1.08 \times \Delta T$$
$$H_{ven} = 1050 \times 1.08 \times 40$$
$$H_{ven} = \mathbf{45{,}360\ Btu/hr}$$

The heat gain for a cooling load is calculated using the same procedure, except humidity in the building air must be considered. Latent heat gain from the removal of excess humidity must be added to the sensible heat gain during cooling. Typically the latent heat gain is about 30% of the sensible heat gain. However, this varies with regional conditions. The most accurate way to calculate heat gain from ventilation air is by using a psychrometric chart.

Distribution Ductwork and Piping

Heat lost or gained through the ductwork or piping that distributes air or water from the heating and/or cooling system is part of the total load for buildings. When any part of the HVAC ductwork or piping system runs through unconditioned space, heat is lost or gained. This heat loss and heat gain is added to the total heating load and cooling load of the building.

Heat Loss from Ductwork and Piping. Heat loss from ductwork or piping is usually more significant than heat gain because of the greater temperature difference between the air or water and the material around it. Ductwork and piping must be insulated to reduce heat loss. Adjustments for heat loss from ductwork or piping are made before heating equipment is selected. Adjusted heat loads for ductwork and piping are found by using a table of duct heat loss factors that shows the approximate loss through the ductwork surfaces. Duct heat loss factors show the increase in heat load as a percentage of the calculated building heat load. **See Figure 49-10.**

Duct Heat Loss Factors

Duct Location and Insulation Value	Duct Loss Factors Winter Design	
	Below 15°F	Above 15°F
Exposed to Outdoor Ambient Air — Attic, Garage, Exterior Wall, Open Crawl Space		
None	1.30	1.25
R-2	1.20	1.15
R-4	1.15	1.10
R-6	1.10	1.05
Enclosed in Unheated Space — Vented or Unvented Crawl Space or Basement		
None	1.20	1.15
R-2	1.15	1.10
R-4	1.10	1.05
R-6	1.05	1.00
Buried in or Under Concrete Slab — Edge Insulation		
None	1.25	1.20
R-3 to R-4	1.15	1.10
R-5 to R-7	1.10	1.05
R-8 to R-9	1.05	1.00

Air Conditioning Contractors of America

Figure 49-10. Tables of duct heat loss factors show the factors for the increase in heating load as a percentage of the calculated building heat load.

To use duct heat loss factor tables, the location and insulation of the ductwork and the winter outdoor design temperature are selected. The ductwork heat loss factor is multiplied by the building heat load. The result is an adjusted heat load that includes heat loss from ductwork that runs through unconditioned space. The adjusted heat load is found by applying the following formula:

$$AHL = BHL \times HL_{DW}$$

where

AHL = adjusted heat load (in Btu/hr)

BHL = building heat load (in Btu/hr)

HL_{DW} = ductwork heat loss factor

Example: Finding Adjusted Heat Loads

A building has a heat load of 57,194 Btu/hr. Ductwork runs the full length of the building through an unheated crawlspace. The ductwork is insulated with R-4 insulation. The outdoor design temperature is below 15°F. From a table of duct heat loss factors, the factor is 1.10. Find the adjusted heat load.

$$AHL = BHL \times HL_{DW}$$
$$AHL = 57{,}194 \times 1.10$$
$$AHL = \mathbf{62{,}913\ Btu/hr}$$

Heat Gain from Ductwork and Piping. Heat gain from ductwork or piping is usually less significant than heat loss. Ductwork and piping are typically insulated to reduce heat gain, especially when the same duct or piping system is used for heating. Adjustments for heat gain through ductwork or piping are made before cooling equipment is selected. Adjusted cooling loads for ductwork and piping heat gain are found by using a table of duct heat gain factors that shows the approximate heat gain through the ductwork surfaces. Duct heat gain factors show the increase in cooling load as a percentage of the calculated building cooling load. **See Figure 49-11.**

Duct Heat Gain Factors

Duct Location and Insulation Value	Duct Gain Factors
Exposed to Outdoor Ambient Air — Attic, Garage, Exterior Wall, Open Crawl Space	
None	1.30
R-2	1.20
R-4	1.15
R-6	1.10
Enclosed in Unheated Space — Vented or Unvented Crawl Space or Basement	
None	1.15
R-2	1.10
R-4	1.05
R-6	1.00
Buried in or under Concrete Slab — Edge Insulation	
None	1.10
R-3 to R-4	1.05
R-5 to R-7	1.00
R-8 to R-9	1.00

Air Conditioning Contractors of America

Figure 49-11. Tables of duct heat gain factors show the factors for the increase in cooling load as a percentage of the calculated building cooling load.

To use duct heat gain factor tables, the location and insulation of the ductwork are selected. The ductwork heat gain factor is multiplied by the building cooling load. The result is an adjusted cooling load that includes heat gain from the ductwork that runs through unconditioned space. The adjusted cooling load is found by applying the following formula:

$$ACL = BCL \times HG_{DW}$$

where

ACL = adjusted cooling load (in Btu/hr)

BCL = building cooling load (in Btu/hr)

HG_{DW} = ductwork heat gain multiplier

Example: Finding Adjusted Cooling Loads

A building has a cooling load of 25,022 Btu/hr. Ductwork runs the full length of the building through an open crawlspace. The ductwork is insulated with R-4 insulation. From a table of duct heat gain factors, the duct heat gain factor is 1.15. Find the adjusted cooling load.

$$ACL = BCL \times HG_{DW}$$

$$ACL = 25{,}022 \times 1.15$$

$$ACL = \mathbf{28{,}775\ Btu/hr}$$

LOAD CALCULATIONS

Heating and cooling load calculations indicate the amount of cooling or heating required to maintain a constant temperature and comfort level in a building. Heat loss and heat gain variables and factors are used to calculate heat losses and heat gains for the various components of a building. Component values are added to determine the total heating and cooling loads for a building.

Load calculations for residential buildings follow a standard format based on local conditions. Light commercial buildings have greater variations in construction materials and building uses. Only experienced HVAC technicians should calculate heating and cooling loads for light commercial buildings.

Calculation Methods

Calculation methods include using commercially available computer programs or manually using a prepared load calculation form. Regardless of the method used, the same information is needed. Heat loss or heat gain is calculated for each room or zone so the distribution piping or ductwork system can function properly for each room or zone. Small rooms or adjoining rooms without partitions are often combined.

Computer Programs. Most load calculations today are performed using computer programs that run prepared applications. Computer programs provide prompts for the needed information. Most computer programs come with basic information of regional conditions, sample distribution systems, and standardized components. Some manufacturers develop their own programs based on the capabilities of their equipment. Computer programs can easily provide detailed calculations for large or complex systems.

Manual Calculations. Load calculations can also be performed manually using prepared load calculation forms. Most manufacturers provide basic information to support contractor load calculations for their equipment and components. Some contractors develop their own spreadsheets or use the load calculation forms for sample projects. Contractors also use prepared load calculation forms to verify the calculations performed by computer programs.

A prepared load calculation form typically has space at the top for the job name, date of the calculations, and the name of the person who prepared the form. Design temperature information can be added in the space provided. A load calculation form also has columns with spaces for room names, room dimensions, building components, or other elements of the loads.

Other columns are used for information relating to individual rooms or zones in a building. Room headings identify each room column. Heat loss and gain factors are entered in the correct column for each building component. Columns have spaces for heating and cooling factors and areas of individual rooms or zones. **See Figure 49-12.**

Load Calculation Forms

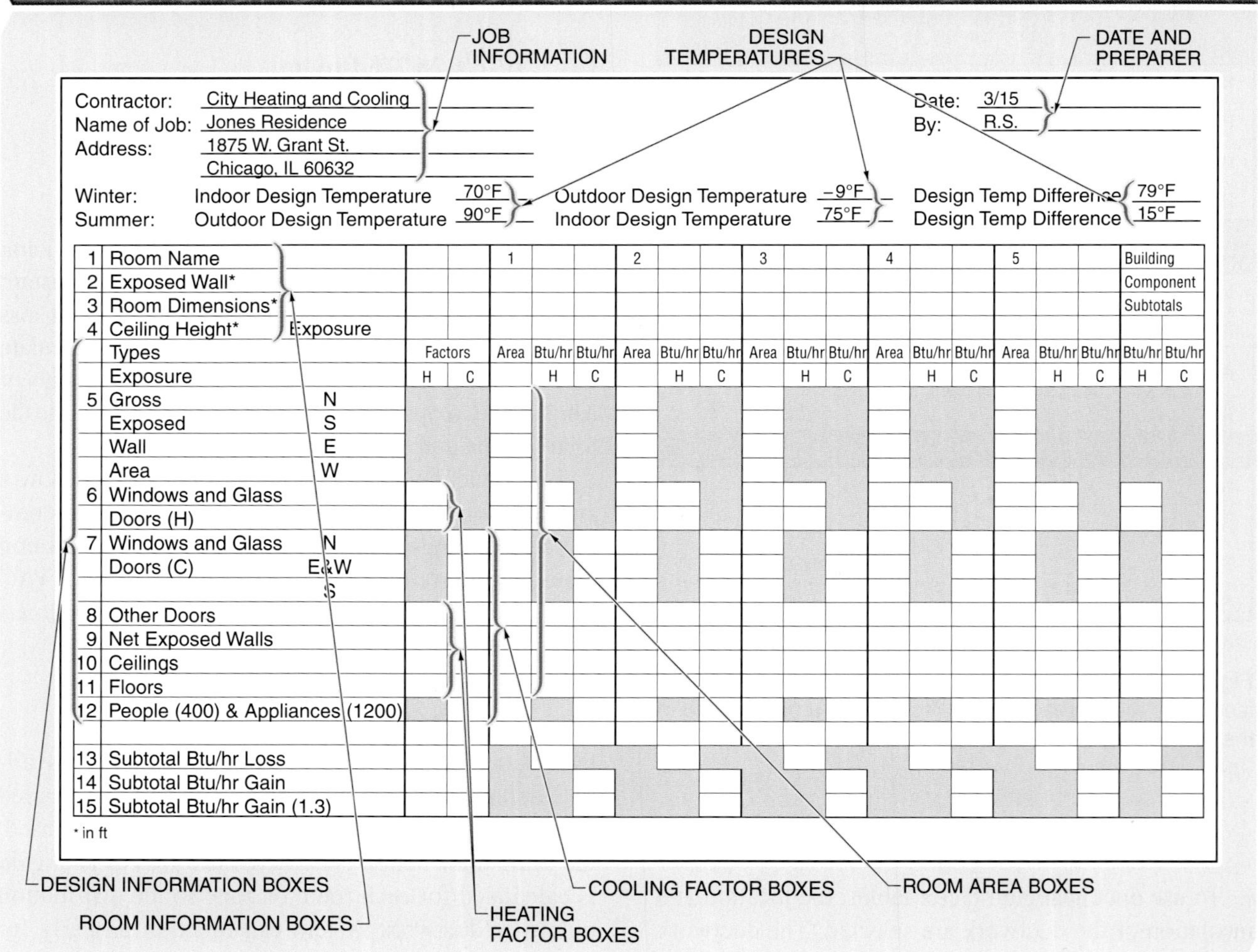

Figure 49-12. Prepared load calculation forms can be used for determining heating and cooling loads.

Building Data

Building data used for load calculations can be found on the plans and specifications for the building. New construction projects have building plans that show dimensions, construction materials, location, and arrangement of the spaces within the building. When plans are not available from the building contractor, HVAC technicians must measure the building and prepare a drawing with the needed information. **See Figure 49-13.**

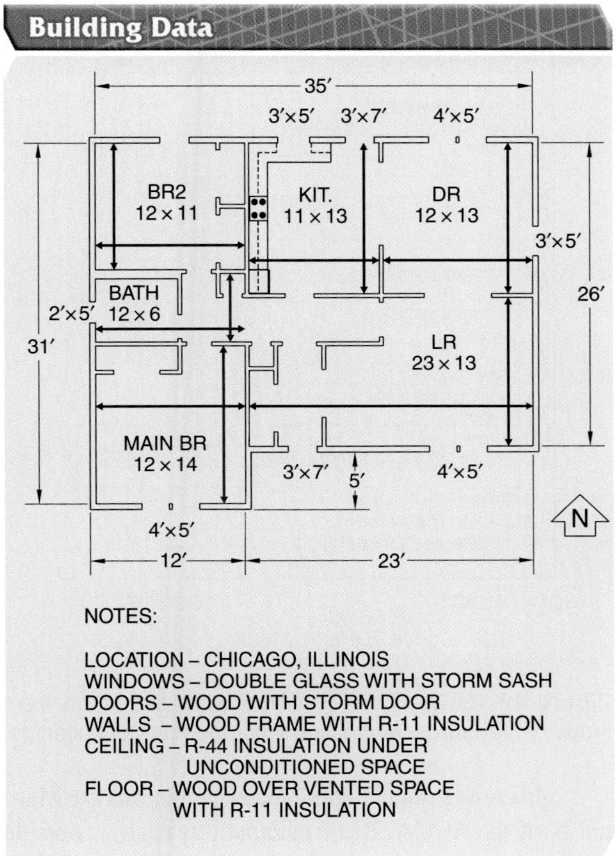

Figure 49-13. Information for load calculations is taken from the building plans and specifications or by inspecting and measuring the building.

Building Dimensions. Dimensions for walls, ceilings, and floors may be rounded to the nearest 6 in. Dimensions for windows and doors may be rounded to the nearest 3 in. Dimensions for rooms or zones should be taken from the outside of the exterior wall to the center of partitions between adjacent rooms or zones. Hallways, closets, stairways, and other spaces in the building not part of a room or zone are included in the adjacent rooms or zones. Sometimes, the dining room and kitchen of a residence are combined because they are not divided by a partition. Hallway and closet areas are combined with adjacent rooms.

Gross exposed wall area is the total area of a wall including windows, doors, and other openings. It is entered on the load calculation form along with the direction of exposure of the wall. Gross exposed wall dimensions are measured or taken from the outdoor dimensions of the building, which are found on plans of the building.

The areas of windows and glass doors are likewise measured or found on the plans of the building. For heating load calculations, the areas of windows and glass doors are entered on the load calculation form. For cooling load calculations, the areas of windows and glass doors are entered specifically according to direction of exposure. Areas of doors that are not glass are also entered. The net area of exposed walls is found by subtracting the areas of windows and doors from the gross exposed wall area for each room. The areas of the roof and floor are found by multiplying the dimensions of the rooms. **See Figure 49-14.**

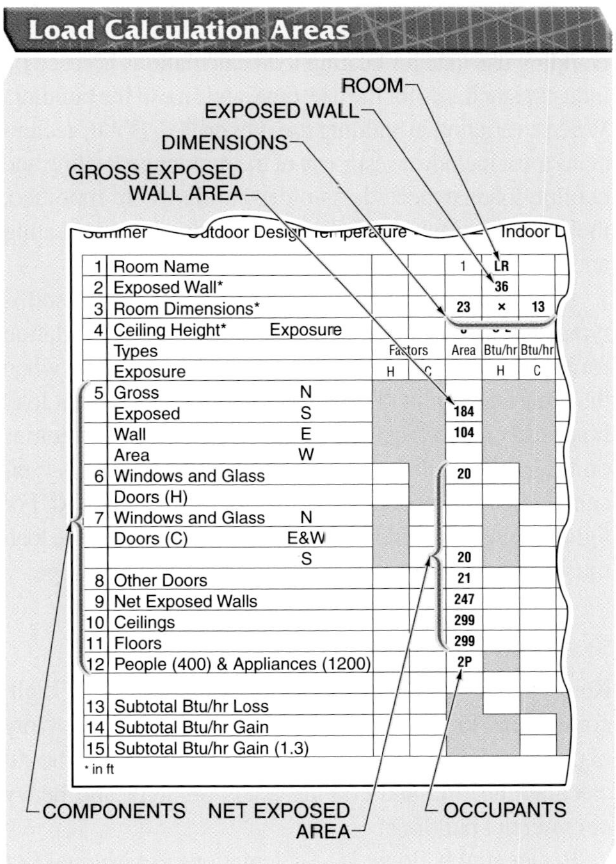

Figure 49-14. Information for room dimensions and exposed surfaces is taken from the building plans or calculated using building measurements.

Existing Buildings. When plans and specifications are not available for an existing building, the floor plans of the building are drawn, measurements taken, and notes included about its materials and construction methods. Building data for the load calculation form is then taken from these drawings and notes. The drawings are most commonly sketched on grid paper with measured dimensions and notes on the side.

When sizing replacement heating and cooling equipment, the existing equipment and distribution system must be evaluated. If the existing equipment adequately heated and cooled the building, then similarly sized replacement equipment can be used. If the building use changes or if the existing equipment is not adequate, the size needed for the replacement equipment must be calculated.

Many older buildings have been modified, so a blower door test should be used to determine the ACH. Typically blower door tests will identify locations of high infiltration. Locations of high infiltration should be sealed to the extent possible and the building retested.

Building Use Data

Building use data for heating load calculations is based on industry standards for the use, type, and size of the building. When a variation in building use is expected, HVAC technicians must include an estimate of the maximum heating and cooling loads expected. As older buildings are modified, their HVAC systems must be checked for adequate heating and cooling capacity and proper distribution.

For residential applications, 400 Btu/hr per person is typically entered into the spreadsheet or load calculation form for the rooms that are occupied during the day when the greatest amount of cooling is required. Occupant load is typically entered in the living room and kitchen columns on the load calculation form. Appliance heat gain is typically added as kitchen heat gain for the cooling load. For light commercial buildings, the lighting and appliance load must be determined based on estimated greatest usage.

Standard Load Calculations

Residential load calculations follow a basic format. Light commercial load calculations are far more complex. Only experienced HVAC and refrigerant technicians should calculate heating and cooling loads for light and heavy commercial buildings.

Residential building load calculations use Manual J of the ACCA. Basic calculations such as outside air exposure and heating and cooling loads are similar to other residential buildings within the region. Heating and cooling loads are based on factors such as local conditions, temperatures, building size and use, and type of construction.

Residential building structures are often made of common building materials. The number of occupants is typically low, and the time of use has less variation inside a residence. Heating and cooling load calculations use a standard format with basic information to provide system design requirements. **See Figure 49-15.**

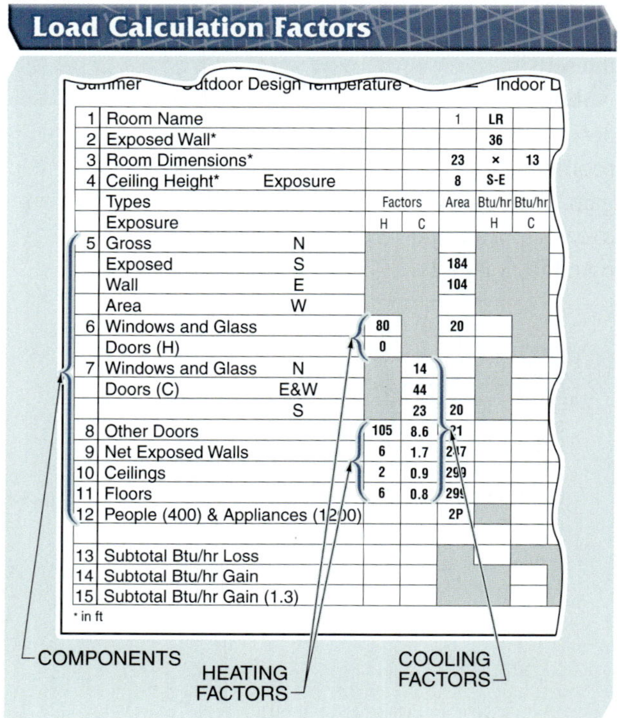

Figure 49-15. Load calculation factors are taken from industry standards for individual components of buildings.

Light commercial building load calculations use Manual N of the ACCA. Basic calculations such as outside air exposure are the same as with residential calculations, but equipment heating and cooling load calculations are more complex. Light commercial building structures typically use a greater variety of materials. The number of occupants is usually greater, and there are more variables involved with a commercial building. The types of equipment loads in a light commercial building are greater in number, size, and use.

Load Calculations. After the areas and factors have been determined and entered in the computer program or load calculation form, the calculations must be performed. Using the load calculation form, the areas in the area column are multiplied by the factors in the factors column. The heat losses must be kept separate from the heat gains. The results are the heating loads and cooling loads for each room or zone, broken down by building component. Each room or zone is calculated separately and added together to find the heating load and cooling load for the building. **See Figure 49-16.**

Load Subtotals. To find the subtotals for each room or zone, all of the heat losses or heat gains for the building components are added together. The sum in the heating load (heat loss) column is the heat loss for the room or zone. The sum in the cooling load (heat gain) column is the sensible heat gain for each room or zone.

For a typical residence, the latent heat gain is typically 30% of the sensible heat gain. The total heat gain for each room or zone is the sensible heat gain plus the latent heat gain. To simplify the mathematical process, 30% is added to each sensible load by multiplying the subtotal heat gain by a factor of 1.3.

Load Totals. Total heat loss or heat gain for a building is the sum of all the room or zone subtotals. The sum is shown on the far right bottom of the load calculation form. The total of all room or zone subtotals must equal the total of all building component subtotals.

Load Calculation Analysis

After the heating and cooling loads for a building have been calculated, the results must be checked for accuracy. This is important because the calculations are used to design the entire heating or cooling system of a building. If using a computer program, a check should be made to ensure the correct data has been entered. Results can also be verified by comparing the results of computer load calculations with estimated loads. The results should also be compared with load calculations of other nearby buildings that are the same type of construction, use, and size.

Load Calculation Results

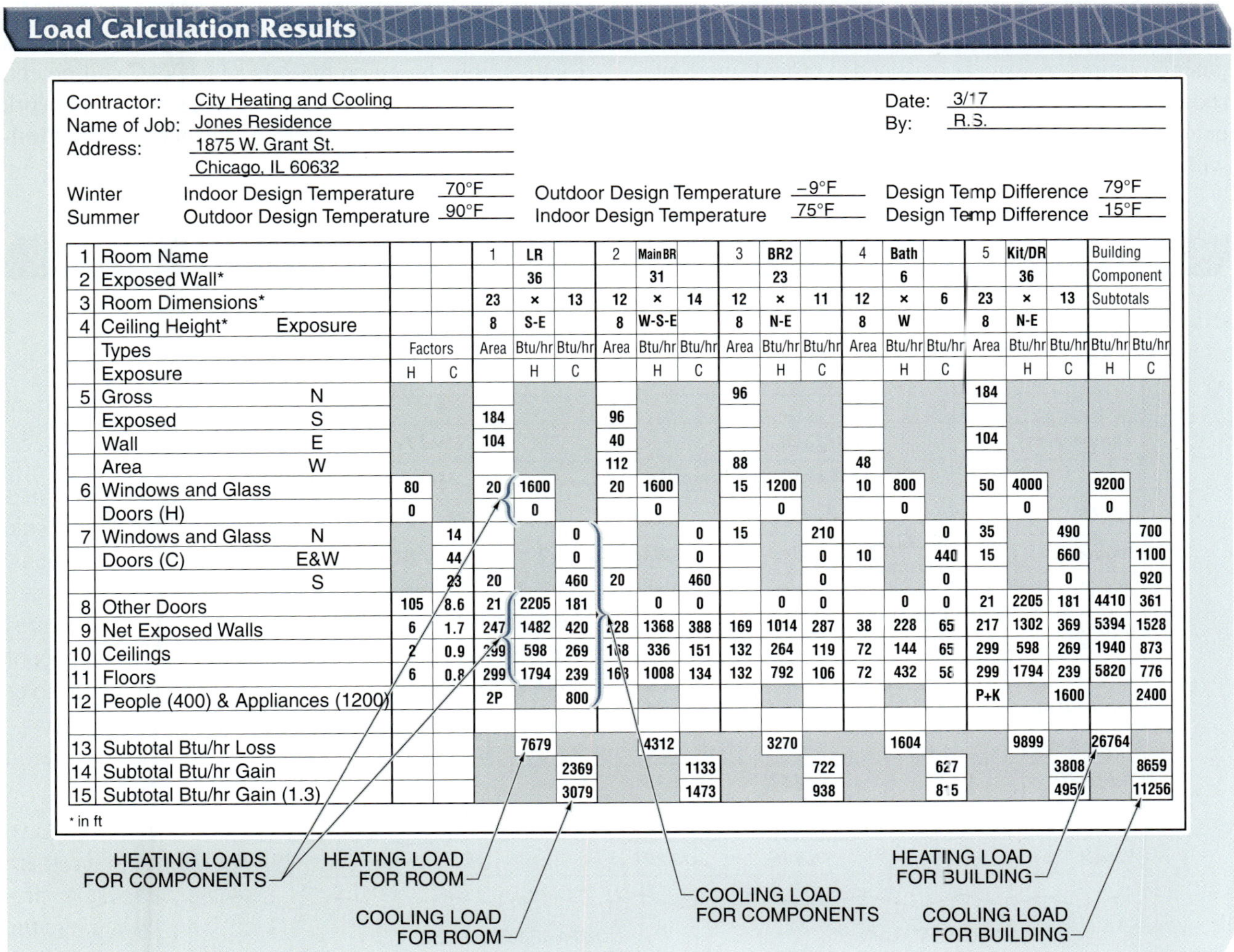

Contractor: City Heating and Cooling
Name of Job: Jones Residence
Address: 1875 W. Grant St.
Chicago, IL 60632

Date: 3/17
By: R.S.

Winter Indoor Design Temperature 70°F Outdoor Design Temperature −9°F Design Temp Difference 79°F
Summer Outdoor Design Temperature 90°F Indoor Design Temperature 75°F Design Temp Difference 15°F

1	Room Name			1	LR		2	Main BR		3	BR2		4	Bath		5	Kit/DR		Building	
2	Exposed Wall*				36			31			23			6			36		Component	
3	Room Dimensions*			23	×	13	12	×	14	12	×	11	12	×	6	23	×	13	Subtotals	
4	Ceiling Height* Exposure			8	S-E		8	W-S-E		8	N-E		8	W		8	N-E			
	Types	Factors		Area	Btu/hr	Btu/hr	Area	Btu/hr	Btu/hr	Area	Btu/hr	Btu/hr	Area	Btu/hr	Btu/hr	Area	Btu/hr	Btu/hr	Btu/hr	Btu/hr
	Exposure	H	C		H	C		H	C		H	C		H	C		H	C	H	C
5	Gross N									96						184				
	Exposed S			184			96													
	Wall E			104			40									104				
	Area W						112			88			48							
6	Windows and Glass	80		20	1600		20	1600		15	1200		10	800		50	4000		9200	
	Doors (H)	0			0			0			0			0			0		0	
7	Windows and Glass N		14			0			0	15		210			0	35		490		700
	Doors (C) E&W		44			0			0			0	10		440	15		660		1100
	S		23	20		460	20		460			0			0			0		920
8	Other Doors	105	8.6	21	2205	181		0	0		0	0		0	0	21	2205	181	4410	361
9	Net Exposed Walls	6	1.7	247	1482	420	228	1368	388	169	1014	287	38	228	65	217	1302	369	5394	1528
10	Ceilings	2	0.9	299	598	269	168	336	151	132	264	119	72	144	65	299	598	269	1940	873
11	Floors	6	0.8	299	1794	239	168	1008	134	132	792	106	72	432	58	299	1794	239	5820	776
12	People (400) & Appliances (1200)			2P		800										P+K		1600		2400
13	Subtotal Btu/hr Loss				7679			4312			3270			1604			9899		26764	
14	Subtotal Btu/hr Gain					2369			1133			722			627			3808		8659
15	Subtotal Btu/hr Gain (1.3)					3079			1473			938			815			4950		11256

* in ft

Figure 49-16. The total heat loss or heat gain for a building is the sum of all the room or zone subtotal heat losses or heat gains.

All data on the job information sheet must be verified and checked using computer software. Computer programs calculate component subtotals with the same method used on load calculation forms.

Checking Total Loads. An analysis of loads calculated using computer programs starts with checking the total building loads. The total loads can be compared with a standard for the type and location of the building or with manually calculated loads. Comparisons can be made by preparing a sheet showing the results from the computer program calculations and the manual calculations. **See Figure 49-17.**

Because different factors are used for each load calculation method, there is typically some difference between loads calculated by computer programs and those calculated manually. The totals of computer program load calculations should be within 10% of the totals taken from a reliable standard. When the totals are not close, the calculations must be examined to find the cause.

Checking Subtotal Loads. Commercially available computer programs are usually designed to provide printouts showing subtotals of heat losses and heat gains for rooms or zones in buildings. These subtotals must be compared with a reliable standard to find significant differences. A positive difference in one component can be offset by a negative difference in another. When there is a significant difference between the computer subtotals and the standard, the component factors and variables must be checked. The best way to compare subtotals is by listing them on a comparison sheet.

Checking Component Totals. When a computer printout shows component heat losses and heat gains, component totals can be checked to find significant differences. A quick check of the component loads may indicate a higher or lower heating or cooling load on one component in a room or zone. Any significant differences must be checked.

Final Load Adjustments

After the load calculations have been checked, the results must be adjusted using the actual capacity of available HVAC equipment. HVAC equipment is typically selected to provide the capacity required by the heating and cooling load calculations. Adjustments must also be made to the calculated heat loss and heat gain for each room or zone based on the selected HVAC equipment. Adjusted calculations are also used to size the ductwork and piping distribution system. **See Procedure for Finding Heating and Cooling Loads.**

Checking Calculated Results

Components		Heat Gain			Heat Loss	
		Manual	Computer		Manual	Computer
Walls	Includes solar gain and infiltration	1371	1128	Includes solar gain and infiltration	3652	4361
Windows		7155	6021		10,500	6191
Doors		158	756		3045	2923
Ceiling		1054	874		2108	3376
Floor		528	611		4216	4725
Infiltration			840			3240
People & Appl.		2400	1200 + 871			
Subtotals		12,666	12,301			
		x 1.3	x 1.3			
Totals		16,466	15,991		23,521	24,816

COMPARE HEAT GAIN

COMPARE HEAT LOSS

Figure 49-17. A worksheet can be used to compare computer program calculations with manually calculated results.

Procedure

Finding Heating and Cooling Loads

1. Review the construction plan of a residential building in Seattle, Washington. The plan contains the dimensions required for calculating the heating and cooling loads of the building. Notes on the plan describe construction features that affect the heating and cooling loads. The dining room and kitchen are combined because they are not divided by a partition. The baths and utility room are combined because they are both small.

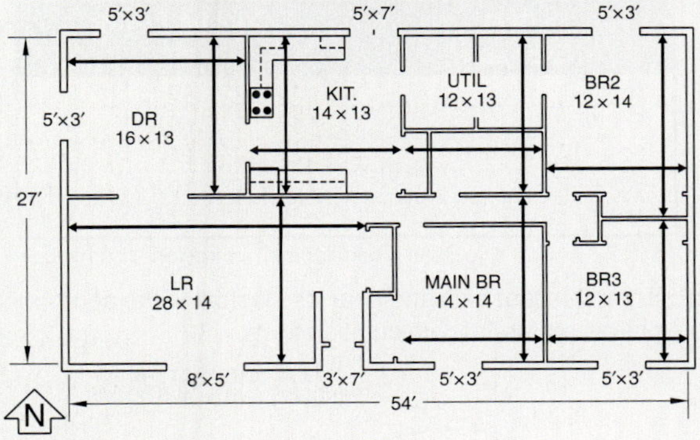

NOTES:
LOCATION – SEATTLE, WASHINGTON
WINDOWS – HORIZONTAL SLIDE, DOUBLE GLASS
DOORS – WOOD WITH STORM DOOR IN LIVING ROOM DOUBLE SLIDING GLASS DOOR IN KITCHEN
WALLS – WOOD FRAME WITH SHEATHING AND SIDING WITH R-11 INSULATION
CEILING – UNDER UNCONDITIONED SPACE WITH R-38 INSULATION
FLOOR – WOOD OVER VENTED SPACE WITH R-11 INSULATION

2. Enter the job data at the top of the load calculation form.

Contractor: Best Heating and Cooling
Name of Job: Patrick Residence
Address: 1541 N. Johnson St.
Seattle, WA 98031

3. Find the outdoor design temperatures. The outdoor design temperatures are taken from an outdoor design temperature table under the listing for Seattle, Washington. The winter design temperature is 17°F DB and the summer design temperature is 84°F DB.

Outdoor Design Temperatures

State	Latitude*	Winter DB†	Summer DB†	Summer DR‡	Summer WB†
Illinois					
Aurora	41	−6	93	20	79
Champaign	40	−3	95	21	78
Chicago	41	−9	90	15	79
Galesburg	40	−7	93	22	78
Joliet	41	−5	93	20	78
Kankakee	41	−4	93	21	78
Macomb	40	−5	95	22	79
Washington					
Aberdeen	46	25	80	16	65
Olympia	46	16	87	32	67
Seattle-Tacoma	47	17	84	22	66
Spokane	47	−6	93	28	65
Walla Walla	46	0	97	27	69

* in degrees
† in °F
‡ daily range in °F

Air Conditioning Contractors of America

OUTDOOR DESIGN TEMPERATURE FOR SEATTLE-TACOMA AREA

4. Enter the indoor design temperatures and the outdoor design temperatures at the top of the load calculation form. Find the design temperature differences and enter them at the top of the load calculation form.

Winter Indoor Design Temperature 70 °F
Summer Outdoor Design Temperature 84 °F

Outdoor Design Temperature 17 °F
Indoor Design Temperature 75 °F

Design Temp Difference 53 °F
Design Temp Difference 9 °F

5. Enter names of rooms, running feet of exposed wall, room dimensions, ceiling height, and the exposure of the individual rooms in the appropriate places on the form.
6. Find the gross exposed wall areas for each room and enter them in the appropriate places on the form.
7. Find the areas of windows and glass doors. Enter the areas on the form for the heating load. Enter the areas on the form according to exposure for the cooling load. Calculate the areas of the other doors and enter them in the appropriate places on the form.
8. Find net exposed wall areas for the individual rooms by subtracting the areas of the windows from the gross exposed wall areas and entering them in the appropriate places on the form. Find the areas of the ceilings and floors and enter them in the appropriate places on the form.

ROOM NAME
EXPOSED WALLS
ROOM DIMENSIONS

Summer Outdoor Design Temperature ... Indoor D

			Factors H	Factors C	1		
1	Room Name					LR	
2	Exposed Wall*					42	
3	Room Dimensions*				28	X	14
4	Ceiling Height*	Exposure			8	S-W	
	Types		Factors		Area	Btu/hr	Btu/hr
	Exposure		H	C		H	C
5	Gross	N					
	Exposed	S			224		
	Wall	E					
	Area	W			112		
6	Windows and Glass				40	0	
	Doors (H)					0	
7	Windows and Glass	N					0
	Doors (C)	E&W					0
		S			40		0
8	Other Doors				21	0	0
9	Net Exposed Walls				275	0	0
10	Ceilings				392	0	0
11	Floors				392	0	0
12	People (400) & Appliances (1200)						0
13	Subtotal Btu/hr Loss					0	
14	Subtotal Btu/hr Gain						0
15	Subtotal Btu/hr Gain (1/3)						0

* in ft

EXPOSED COMPONENTS

continued on next page

Procedure

Finding Heating and Cooling Loads *(continued)*

9. Find the component factors for the heating load from the table of heat gain factors.

Heat Gain Factors

Item	Design Temp Difference*						
	50	55	60	65	70	75	80
Windows							
Double-hung							
Single glass	75	80	90	95	105	110	120
Double glass	50	55	60	65	70	75	80
Fixed or picture							
Single glass	70	75	85	90	95	105	110
Double glass	45	45	50	55	60	65	70
Doors							
Sliding glass							
Single glass	125	140	150	165	175	190	200
Double glass	100	110	120	130	140	150	160
Other doors							
Weatherstrip and storm door	65	75	80	85	90	100	105
Weatherstrip or storm door	120	135	145	155	170	180	195
No weatherstrip	225	250	270	290	315	340	360
Walls and Partitions							
No insulation	13	14	15	16	18	19	20
R-11 batt insulation	4	4	4	5	5	5	6
Ceilings and Roofs							
No insulation	30	33	36	39	42	45	48
R-38 insulation	1	1	2	2	2	2	2
Floors							
No insulation	7	8	8	9	10	11	11
R-11 insulation	4	4	4	5	5	5	6

* in °F

Air Conditioning Contractors of America

10. Find the component factors for the cooling load from the table of heat loss factors.

Heat Loss Factors

Item	Cooling Factor		
	15	20	25
Walls			
No insulation	5.0	6.4	7.8
R-11 batt insulation	1.7	2.1	2.6
Ceilings and Roofs			
No insulation	17.0	19.2	21.4
R-38 insulation	1.0	1.1	1.2
Floors			
No insulation	3.9	5.8	7.7
R-11 insulation	0.8	1.3	1.7

Air Conditioning Contractors of America

11. Find the component factors for the cooling load for windows and glass doors from the table of heat loss factors for windows and glass doors.

Heat Loss Factors for Windows and Glass Doors

Item	Cooling Factor		
Exposure*	15	20	25
N	14	16	18
E & W	44	46	48
S	23	25	27
Wood	8.6	10.9	13.2
Metal	3.5	4.5	5.4

* in °F

Air Conditioning Contractors of America

12. Enter the heat gain and heat loss factors in the appropriate places on the load calculation form.

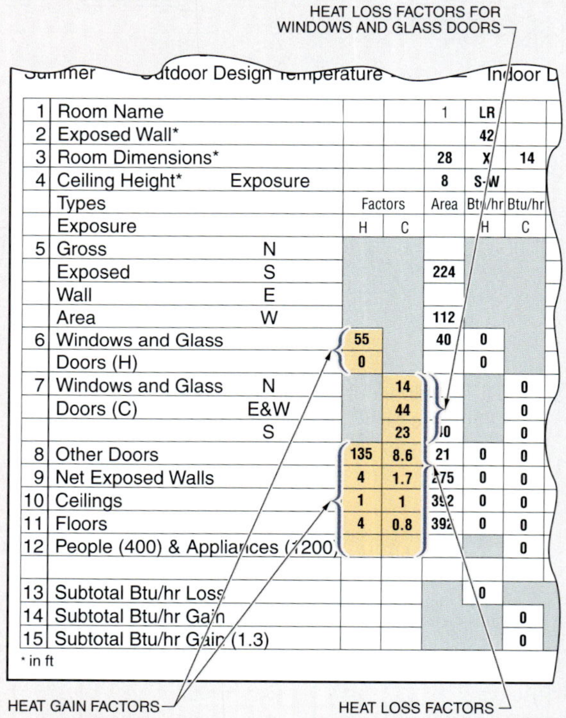

13. Determine the number of people occupying the building. In this case, there are four occupants. A heat gain factor of 400 Btu/hr/person is added for each of the two occupants in the dining room and the two occupants in the living room.
14. Determine the heat gain from appliances. A heat gain factor of 1200 Btu/hr is added to the kitchen.
15. Find the heating and cooling load for each room by multiplying the heating and cooling factors by the variables for each room.
16. Find building component subtotals by adding the heating and cooling loads for all components in the building.
17. Find the individual room heating and cooling loads subtotals by adding the component heat losses and heat gains for each room.

continued on next page

Procedure

Finding Heating and Cooling Loads *(continued)*

Contractor: Best Heating and Cooling
Name of Job: Patrick Residence
Address: 1541 N. Johnson St.
Seattle, WA 98031

Date: 3/17
By: R.S.

Winter	Indoor Design Temperature 70°F	Outdoor Design Temperature 17°F	Design Temp Difference 53°F
Summer	Outdoor Design Temperature 84°F	Indoor Design Temperature 75°F	Design Temp Difference 9°F

#	Item	Factors H	Factors C	1 Area	1 Btu/hr H	1 Btu/hr C	2 Area	2 Btu/hr H	2 Btu/hr C	3 Area	3 Btu/hr H	3 Btu/hr C	4 Area	4 Btu/hr H	4 Btu/hr C	5 Area	5 Btu/hr H	5 Btu/hr C	6 Area	6 Btu/hr H	6 Btu/hr C	Building Btu/hr H	Building Btu/hr C
1	Room Name			1	LR		2	Kit/DR		3	Util		4	BR2		5	BR3		6	Main BR		Building	
2	Exposed Wall*				42			43			12			26			25			14		Component	
3	Room Dimensions*			28	×	14	30	×	13	12	×	13	12	×	14	12	×	13	14	×	14	Subtotals	
4	Ceiling Height* Exposure			8	S-W		8	N-W		8	N		8	N-E		8	S-E		8	S			
	Types	Factors		Area	Btu/hr	Btu/hr	Area	Btu/hr	Btu/hr	Area	Btu/hr	Btu/hr	Area	Btu/hr	Btu/hr	Area	Btu/hr	Btu/hr	Area	Btu/hr	Btu/hr	Btu/hr	Btu/hr
	Exposure	H	C		H	C		H	C		H	C		H	C		H	C		H	C	H	C
5	Gross N						240			96			96										
	Exposed S			224												96			112				
	Wall E												112			104							
	Area W			112			104																
6	Windows and Glass	55		40	2200		65	3575			0		15	825		15	825		15	825		8250	
	Doors (H)	0			0			0			0			0			0			0		0	
7	Windows and Glass N		14			0	50		700			0	15		210			0			0		910
	Doors (C) E&W		44			0	15		660			0			0			0			0		660
	S		23	40		920			0			0			0	15		345	15		345		1610
8	Other Doors	135	8.6	21	2835	181		0	0		0	0		0	0		0	0		0	0	2835	181
9	Net Exposed Walls	4	1.7	275	1100	468	279	1116	474	96	384	163	193	772	328	185	740	315	97	388	165	4500	1913
10	Ceilings	1	1	392	392	392	390	390	390	156	156	156	168	168	168	156	156	156	196	196	196	1458	1458
11	Floors	4	0.8	392	1568	314	390	1560	312	156	624	125	168	672	134	156	624	125	196	784	157	5832	1166
12	People (400) & Appliances (1200)			2P		800	2P+K		2000														2800
13	Subtotal Btu/hr Loss				8[illegible]95			6641			1164			2437			2345			2193		22875	
14	Subtotal Btu/hr Gain					3074			4536			444			841			940			8[illegible]3		10698
15	Subtotal Btu/hr Gain (1.3)					3996			5897			577			1093			1222			1122		13907

* in ft

OCCUPANTS AND APPLIANCES
HEATING LOAD FOR ROOM
COOLING LOAD FOR ROOM
COOLING LOAD FOR ROOM WITH LATENT HEAT
TOTAL HEAT LOAD FOR BUILDING
TOTAL COOLING LOAD FOR BUILDING

18. Find the heating equipment adjustment ratio. The calculated heating required for the building is 22,875 Btu/hr. The nearest larger heating equipment has an output of 30,000 Btu/hr. The heating equipment adjustment ratio is 30,000/22,875, or 1.31. This value is multiplied by the subtotal heat loss for each room and used for designing the heating system. Compare the heating load calculated to the heating loads of similar projects.

19. Find the cooling equipment adjustment ratio. The calculated cooling required for the building is 13,907 Btu/hr. The nearest larger air conditioning equipment has an output of 18,000 Btu/hr. The cooling equipment adjustment ratio is 18,000/13,907, or 1.29. This value is multiplied by the subtotal heat gain for each room and used for designing the cooling system. Compare the cooling load calculated to the cooling loads of similar projects.

Heating Equipment Adjustments. Once the total heating load is calculated and checked, the heating equipment for the building can be selected. The heating equipment selection is based on the fuel available, distribution system used, and size needed to meet the heating load of the building. Typically, the selection of the fuel used and the distribution system is determined by the general contractor or local conditions.

The capacity of the installed heating equipment must be greater than the building heating load. Usually, the available heating equipment increases in capacity by 5000 Btu/hr or 10,000 Btu/hr increments. The heating equipment with the next greater capacity should be selected. Once selected, the heating equipment adjustment ratio must be found.

In some areas, residences must have two air conditioning units, one for the upstairs and one for the downstairs. In these cases, units are sized separately.

The heating equipment adjustment ratio is the rated capacity of the equipment divided by the calculated heating load. The heating equipment adjustment ratio is multiplied by the heating load for each room or zone and used to size the distribution system. The heating equipment adjustment ratio is found by applying the following formula:

$$HEAR = \frac{HEC}{HL_{BLDG}}$$

where
$HEAR$ = heating equipment adjustment ratio
HEC = heating equipment capacity (in Btu/hr)
HL_{BLDG} = building heating load (in Btu/hr)

Example: Finding Heating Equipment Adjustment Ratio

A building has a heating load of 57,194 Btu/hr. Selected heating equipment has a capacity of 60,000 Btu/hr. Find the heating equipment adjustment ratio.

$$HEAR = \frac{HEC}{HL_{BLDG}}$$

$$HEAR = \frac{60{,}000}{57{,}194}$$

$$HEAR = \mathbf{1.05}$$

Cooling Equipment Adjustments. Once the total cooling load is calculated and checked, the cooling equipment for the building can be selected. The cooling equipment selection is based on its compatibility with the heating equipment and is sized to meet the cooling load of the building.

The capacity of the installed cooling equipment must be greater than the building cooling load. Typically, the available cooling equipment increases in capacity by 3000 Btu/hr or 6000 Btu/hr increments. The cooling equipment with the next greater capacity should be selected. Once selected, the cooling equipment adjustment ratio must be found.

The cooling equipment adjustment ratio is the rated capacity of the equipment divided by the calculated cooling load. The cooling equipment adjustment ratio is multiplied by the cooling load for each room or zone and used to check the size of the distribution system. The cooling equipment adjustment ratio is found by applying the following formula:

$$CEAR = \frac{CEC}{CL_{BLDG}}$$

where
$CEAR$ = cooling equipment adjustment ratio
CEC = cooling equipment capacity (in Btu/hr)
CL_{BLDG} = building cooling load (in Btu/hr)

Example: Finding Cooling Equipment Adjustment Ratio

A building has a cooling load of 25,022 Btu/hr. Selected cooling equipment has a capacity of 30,000 Btu/hr. Find the cooling equipment adjustment ratio.

$$CEAR = \frac{CEC}{CL_{BLDG}}$$

$$CEAR = \frac{30{,}000}{25{,}022}$$

$$CEAR = \mathbf{1.20}$$

UNIT

50

HVAC Unit Selection

Residential and light commercial HVAC systems should provide comfort and operate efficiently. To accomplish this, HVAC systems, including the distribution system, must be sized and installed properly. The basic functions of forced-air heating and cooling, hydronic heating, and chilled water systems must be understood. An HVAC technician must thoroughly understand HVAC systems and design principles to install and troubleshoot these systems effectively.

OBJECTIVES

- List the major components of typical residential forced-air heating systems.
- Describe the location and function of typical air distribution components.
- List the major components of typical residential forced-air cooling systems.
- Perform basic airflow calculations for residential air conditioning systems.
- List the components of hydronic heating and cooling systems.

Index to Procedures

FORCED-AIR HEATING SYSTEMS

Forced-air heating systems consist of furnaces, ductwork, registers, grills, fans, and controls. Air is heated by the furnace and distributed through ductwork to registers in building spaces. **See Figure 50-1.** Grills collect return air that the fan circulates through the furnace.

Furnaces

The furnace is the central component in a forced-air heating system. The fuel or energy used for producing heat, direction of airflow, and output capacity vary among different furnaces. Design technicians must be familiar with standard furnace sizes that are available from furnace manufacturers. A *design technician* is a person who has the knowledge and skill to plan heating and/or air conditioning systems. Design technicians select the proper equipment to meet the specifications of heating and/or air conditioning systems.

Catalogs and engineering specification sheets from manufacturers are the best sources of information about heating and air conditioning equipment. Catalogs and engineering specification sheets contain descriptions of and specifications for furnaces. **See Figure 50-2.**

Forced-Air Heating Systems

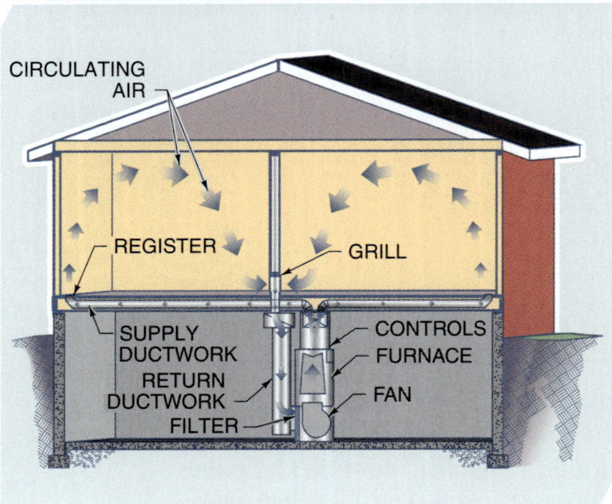

Figure 50-1. In a forced-air heating system, air is heated by the furnace and distributed through ductwork to registers in building spaces. Circulated air is collected by grills and returned through ductwork to the fan.

Fuel and Electricity. Furnace selection is based on the availability and cost of fuel and/or electricity. Furnaces that burn natural gas are often chosen because natural gas is available in most areas. Using natural gas is convenient because the fuel can be piped directly to the furnace. Natural gas is a clean-burning fuel and is reasonable in cost compared to other fuels.

When natural gas is not available, either liquefied petroleum (LP) gas or fuel oil is typically used. Furnaces that burn LP gas are similar to those that burn natural gas. Many furnaces that burn LP gas can be converted to burn natural gas when it becomes available. The cost of LP gas is typically higher than natural gas. LP gas is transported in tanks to rural areas where it is used for heating and standby commercial or industrial applications.

Furnaces that burn fuel oil are used for residential, commercial, and industrial heating. Fuel oil is also used in rural areas where there are no underground natural gas lines. Fuel oil is typically delivered by truck or railroad tank car and stored in tanks.

Electric furnaces are widely used because electricity is available in most areas. The combustion process is not required in electric furnaces, and the equipment is fairly simple. The heat produced by electricity does not contaminate the air at the point of use. Electricity is typically used for back-up heating in residences and light commercial applications.

Airflow Direction. Three categories of furnaces are based on the direction of airflow out of the furnace: upflow, downflow, and horizontal. Upflow furnaces are typically used in residential buildings with basements, where the supply ductwork is located above the furnace. Horizontal furnaces are typically used where the furnace and ductwork are located in an attic or crawlspace. Downflow furnaces are typically used where the ductwork is located below the furnace.

Furnace Manufacturer Specifications

Upflow Gas Furnaces							
Model Number	Dimensions* L	Dimensions* W	Dimensions* H	Input Ratings†	Output Ratings†	Shipping Weight‡	AFUE§
005-AX-10	28.5	14	46	40,000	30,000	100	75
050-AX-6	28.5	18	46	50,000	38,000	120	76
070-AX-5	28.5	20	46	75,000	58,000	120	77
58STX045	28.5	14.2	33.3	44,000	35,000	104	80
58DLX110	28.5	17.5	33.3	110,000	89,000	143	80
58MVP120	28.5	24	46	150,000	127,000	255	85
58MTA120	28.5	24.5	39.8	120,000	112,000	262	93
400-AX-0	27	24.5	39.8	120,000	112,000	270	96.6

* in in.
† in Btu/hr
‡ in lb
§ annual fuel utilization efficiency

Figure 50-2. Catalogs and engineering specification sheets from equipment manufacturers contain descriptions and specifications for furnaces.

Special Heating Units. Special-purpose heating units include unit, duct, space, and direct-fired heaters. Unit heaters and duct heaters are typically found in commercial and industrial applications where temperature control is not precise. Depending on the heater output, unit heaters and duct heaters may be used for residential and light commercial applications. Direct-fired heaters are used almost exclusively for industrial applications because the heaters require 100% outdoor air.

Heating Capacity. The heating capacity of a furnace is the heat output rating of the furnace at design conditions. A *design condition* is the condition of the air where heating or air conditioning equipment provides comfort in a building space. Catalogs and specification sheets for furnaces contain input and output ratings for furnaces. The input rating is the amount of heat produced per unit of fuel as the fuel burns. The output rating is the amount of heat a furnace produces in 1 hr. The heat exchanger in a furnace should transfer heat efficiently from the hot products of combustion to the air in the distribution system.

The input and output ratings of a furnace are different because of the heat lost when heat is allowed to flow up the flue to create a draft. When a furnace with a conventional burner and heat exchanger is used, 20% of the heat is lost to provide a draft up the flue. An older furnace relies on heat from the burner to provide a draft. A conventional heat exchanger extracts 80% of the heat from the products of combustion. The output rating of a furnace with a conventional heat exchanger is 80% of the input rating. **See Figure 50-3.**

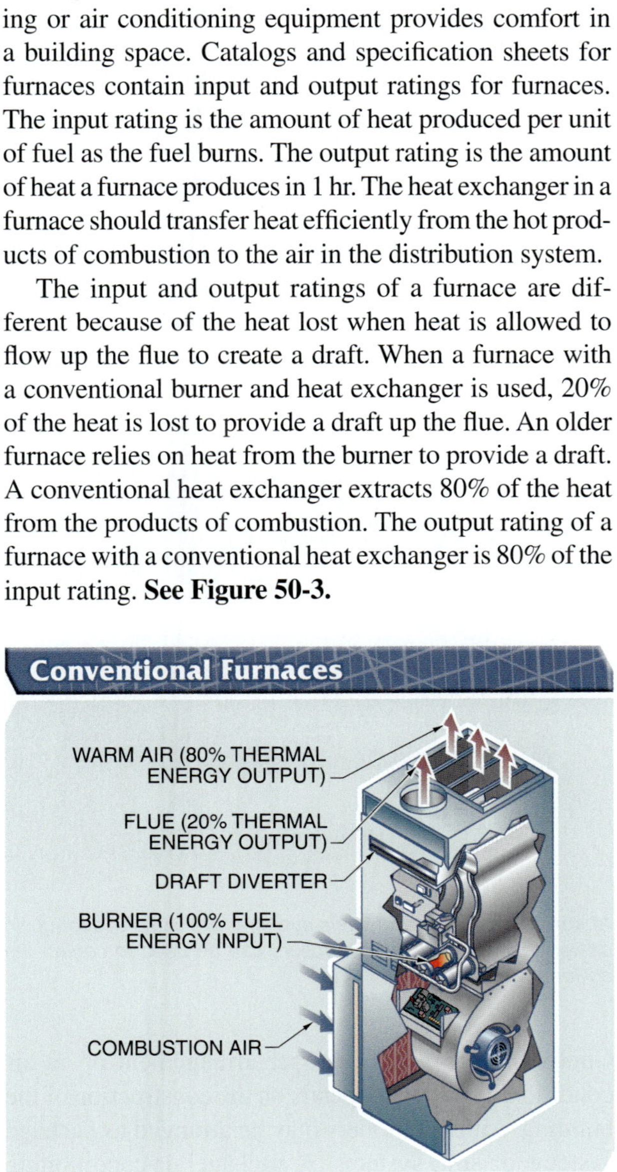

Figure 50-3. The input and output ratings of conventional furnaces differ because 20% of the heat is discharged up the flue.

High-efficiency furnaces with power burners and condensing heat exchangers have higher output ratings than conventional furnaces. The output rating of a high-efficiency furnace may be as high as 97% of the input rating. The output rating of a furnace must be equal to or higher than the adjusted heat loss of the building. There is no difference between the input and output ratings of electric furnaces because combustion is not involved and draft is not necessary. Some efficiency loss occurs in electric furnaces due to the energy lost in the resistance heating elements.

Distribution Systems

Distribution systems circulate conditioned air from HVAC units to building spaces and collect return air. Good airflow of conditioned air throughout building spaces provides occupant comfort. The distribution system consists of ductwork, registers, grills, fans, and a control system.

Ductwork. Considerations for selecting a ductwork system for a forced-air heating system include the furnace location, building spaces, and space available for the ductwork. In buildings that have no space for ductwork in the building frame, the ductwork may be placed under the floor. The ductwork in multistory buildings may be installed in a combination ceiling/floor configuration between floors. For buildings that cover large areas, ductwork may be located in the ceiling spaces or on roofs.

Registers and Grills. Registers are located based on the supply air pattern desired in each space and the location of the supply ductwork. When supply ductwork is under the floor, the floor or the low sidewall is the proper location for registers. When ductwork is overhead, the ceiling or high sidewall is the proper location for registers.

Grills must be located where a connection can be made to the return air ductwork. Grills are typically located on the opposite side of the building space from registers to improve air circulation. The air return efficiency from all building spaces must be considered.

Fans. Fans in forced-air heating systems are typically a part of the furnaces. In built-up heating systems, the fan must efficiently move the required amount of air. When the fan is located where noise should be low, reducing the noise level must be considered. Since noise also travels within the connecting ductwork, a few bends minimizes noise transmission.

Control Systems. The use of a building determines what kind of control system works best for the building. When the only requirement for a building space is to keep the temperature above the freezing point, the control system is simple. One thermostat located centrally in the building may be the only control required. When close control of the temperature in individual spaces in a building is required, such as within a few degrees, a control system that provides that level of sophistication will be required.

Heating Airflow

With forced-air heating systems, the furnace heats air and the fan circulates the heated air through the building. The airflow rate is the volumetric flow rate of air circulated by a fan. Airflow rate is related to the output rating of the furnace and the temperature increase of the air as it flows through the heat exchanger.

Air Temperatures. Forced-air heating equipment operates most efficiently with a certain temperature increase through the heat exchanger. Air temperature increase through a heat exchanger is the difference between the temperature of the air entering the heat exchanger and the temperature leaving the heat exchanger.

Combustion furnaces typically operate with a temperature increase of about 80°F to 100°F. Electric furnaces typically operate with a temperature increase of about 50°F or less. Heat pumps without auxiliary heat typically operate with a temperature increase of 20°F. Heat pumps with auxiliary heat typically operate with a temperature increase of 50°F or less.

Airflow Rates. After selecting a furnace that provides the amount of heat required to heat a building, the airflow rate must be found. Airflow rate is the amount of air in cubic feet per minute (cfm) that meets the heating requirements of a building. The airflow rate required for heating a building is a function of the furnace output rating and the air temperature increase through the heat exchanger.

FORCED-AIR COOLING SYSTEMS

In forced-air cooling systems, air is the evaporating medium. A forced-air cooling system contains an air conditioning system that cools air and a distribution system. Forced-air cooling distribution systems contain ductwork, registers, grills, fans, and controls that are similar to the equipment used in forced-air heating systems. Typically, the same distribution system is used by the forced-air heating system and air conditioning system.

Air Conditioners

The central component in forced-air cooling systems is the air conditioner unit. Air conditioner units produce a refrigeration effect, which causes heat transfer from air to a cooler substance. Typical air conditioner units produce a refrigeration effect by means of mechanical compression. These systems typically use electricity as the energy source.

Heat pumps are an alternative to air conditioning systems when the cooling demand is moderate. Heat pumps can be reversed to provide heating when the heating demand is moderate. Heat pumps exchange heat with the outdoor air or geothermal ground loop.

Some commercial chillers produce the refrigeration effect by means of absorption refrigeration. An absorption refrigeration system uses electricity to operate motors and controls, but it may use gas or oil to operate the generator. When some form of waste heat can be used to operate the system, an absorption refrigeration system may be better than a mechanical compression refrigeration system. Waste heat is steam or hot water that is discharged after a manufacturing process.

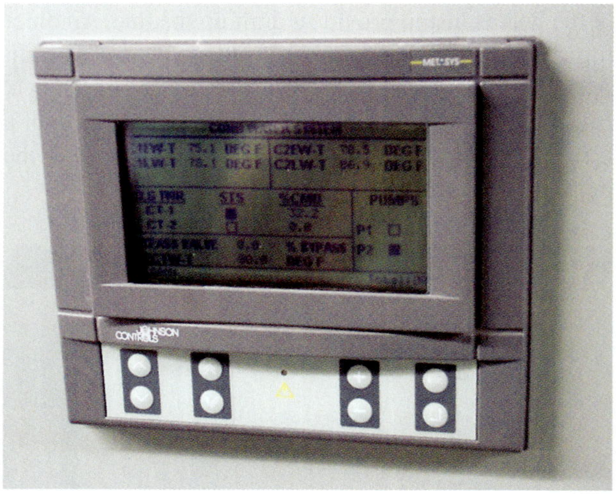

Absorption chillers and their systems are simple enough in design that small digital controllers can be used to control the chiller and system.

Cooling Systems. The proper arrangement of an air conditioning system depends on the construction of the building. Air conditioners may be arranged as package systems or split systems. A package air-conditioning system consists of a cabinet that holds all operating components. A split system consists of different cabinets that hold the operating components.

Package air-conditioning systems are used when ductwork to building spaces can be kept relatively short and individual control of building spaces is needed. A package system is often installed on the roof of a building space. For example, package air-conditioning units are used in large offices and retail stores that have large spaces with high internal heat gains. The supply ductwork is short in these applications. Also, multiple units cover the heat gain of the building spaces.

Split systems are used when condensers are placed outdoors and evaporators are located inside the buildings in ductwork. Many large air conditioning systems are delivered from factories as components, which are then assembled at the job site.

Cooling Capacity. The cooling capacity of an air conditioner is the refrigeration effect produced. The cooling capacity of an air conditioner must meet the cooling load of the building spaces. When selecting an air conditioner for an application, the cooling capacity and the airflow rate must be considered.

Rated Capacity. Nominal sizes for split air conditioning systems range from 1.5 tons to 30 tons incrementally. Larger refrigeration equipment sizes ranges up to 2000 tons in multiple ton increments. Equipment catalogs show the different types and sizes of air conditioning systems and include engineering specifications. **See Figure 50-4.**

Package air conditioners and components of built-up systems are rated at standard conditions per the Air Conditioning and Refrigeration Institute. The *Air Conditioning and Refrigeration Institute (ARI)* is an association of manufacturers of refrigeration and air conditioning equipment and allied products. An important function of the ARI is to establish product standards for air conditioning and refrigeration equipment. ARI standard conditions for different types of equipment are described in ARI bulletins. The conditions for standard equipment are described in Standard 210-81, *Unitary Air-Conditioning Equipment.*

When temperature swings above the desired temperature are acceptable, the next size smaller air conditioner unit may be selected. Smaller units are more efficient because of reduced system cycling, however comfort may be reduced during the highest temperatures. Temperature swing is a variation in the actual indoor temperature compared to the setpoint temperature. Since the highest temperatures and highest humidity occur for only a few hours a year, some temperature swing may be acceptable.

When temperature swings are acceptable, a table of temperature swing factors is used. The temperature swing factors are multiplied by the heat gain for a building. These factors are based on the outdoor design temperature and the acceptable temperature swing. When selecting an air conditioner using a temperature swing factor table, a smaller air conditioner may be used to offset the heat gain. **See Figure 50-5.**

Cooling Airflow

When designing an air conditioning system, the airflow rate across the evaporator must be high enough to transfer the correct amount of heat from the air to the refrigerant in the evaporator. The airflow rate through the building spaces must be high enough for occupant comfort.

Air Conditioning Manufacturer Specifications

Air Conditioning Systems								
Model Number	Dimensions* L	Dimensions* W	Dimensions* H	Nominal Tons	Nominal CFM	Operating Weight†	SEER
38TXA060	35	30	40	5	2000	337	13
38TKW060	35	35	37.8	5	2250	258	10.0
K250-Q	15	6	4	15	12,000	2200	9.2
K1000-Q	16	6	5	24	20,000	3976	8.7
K2400-P	16	6	5	30	24,000	4621	8.6
K3200-P	16	6	5	40	36,000	5307	8.2

* in ft
† in lb

Figure 50-4. Ratings for nominal sizes of air conditioning systems are listed in catalogs and on engineering specification sheets.

Temperature Swing Factors

Air-Cooled Units			
Outdoor Design Temperature*	Desired Indoor Temperature Swing*		
	6	4½	3
85–90	0.07	0.85	1.00
95	0.75	0.90	1.05
100	0.80	0.95	1.10
105	0.85	1.00	1.15
110	0.90	1.05	1.20

Water-Cooled Units															
Outdoor Design Temperature*	Leaving Water Temperature*														
	90	95	100	105	110	90	95	100	105	110	90	95	100	105	110
85–90	0.74	0.77	0.81	0.83	0.86	0.88	0.93	0.97	1.00	1.03	1.03	1.08	1.12	1.16	1.20
95	0.78	0.82	0.85	0.88	0.91	0.93	0.97	1.01	1.05	1.08	1.08	1.12	1.17	1.21	1.25
100	0.82	0.87	0.91	0.94	0.97	0.97	1.02	1.06	1.09	1.12	1.12	1.16	1.20	1.24	1.28
105	0.86	0.90	0.94	0.97	1.00	1.00	1.05	1.10	1.14	1.17	1.17	1.20	1.25	1.30	1.34
110	0.89	0.94	0.98	1.01	1.04	1.04	1.09	1.14	1.18	1.22	1.22	1.24	1.30	1.34	1.38

* in °F

Air Conditioning Contractors of America

Figure 50-5. When temperature swings are acceptable, a table showing temperature swing adjustments should be consulted before an air conditioner is selected.

Heat Removal. Because sensible heat and latent heat are involved in the cooling process, the actual performance of an air conditioner should be checked against a psychrometric chart. The performance is checked to ensure that the refrigeration effect produced can cool the building spaces. An air conditioner must remove heat and moisture from the air in order to decrease temperature. **See Procedure for Finding Air Conditioner Performance.**

Procedure

Finding Air Conditioner Performance

1. Air enters the evaporator of an air conditioner at a dry-bulb temperature of 75°F and 50% relative humidity. The air conditioner has a cooling capacity of 49,000 Btu/hr and airflow rate of 1400 cfm across the evaporator. Find the condition of the air as it leaves the evaporator and the performance of the air conditioner.
2. Use a psychrometric chart to find the point that represents the condition of air as it enters the coil. This point is located at the intersection of the dry-bulb temperature and relative humidity lines on the chart. When air entering the evaporator of an air conditioner has a dry-bulb temperature of 75°F and 50% relative humidity, the condition of the air will be located at the intersection of the 75°F dry-bulb temperature line and 50% relative humidity lines on the psychrometric chart.

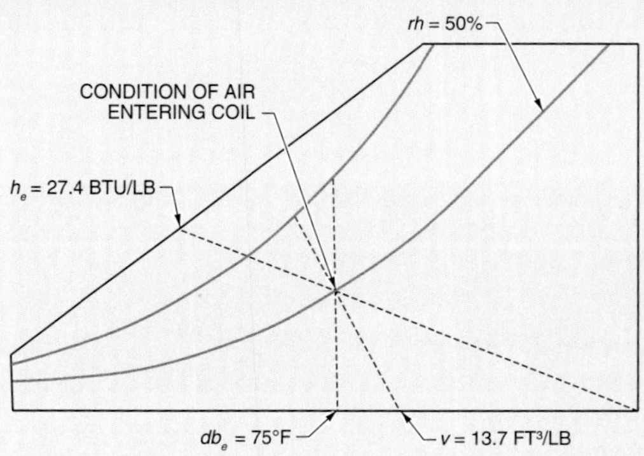

3. The specific volume of air at this point is found by drawing a line parallel with the specific volume lines on the chart and approximating the specific volume. The specific volume of air at this point is approximately 13.7 ft³/lb.
4. The enthalpy of air at this point is found by drawing a line parallel with the enthalpy lines that extend to the enthalpy scale on the left side of the chart. The enthalpy of air at this point is 27.4 Btu/lb.

continued on next page

Finding Air Conditioner Performance *(continued)*

5. The mass flow rate of air at this point is found by dividing the airflow rate by the specific volume of air and multiplying by 60. Multiplying by 60 converts mass flow rate from pounds per minute to pounds per hour. The mass flow rate of air is found by applying the following formula:

$$w = \frac{Q}{v} \times 60$$

where

w = mass flow rate (in lb/hr)

Q = airflow rate (in cfm)

v = specific volume of air (in ft³/lb)

60 = constant

The fan in the air conditioner circulates 1400 cfm of air, and the specific volume is 13.7 ft³/lb. Find the mass flow rate of the air.

$$w = \frac{Q}{v} \times 60$$

$$w = \frac{1400}{13.7} \times 60$$

$$w = 102 \times 60$$

$$w = \mathbf{6120\ lb/hr}$$

6. The refrigeration effect of the coil is found by dividing the cooling capacity of the air conditioner by the mass flow rate of the air. Refrigeration effect is found by applying the following formula:

$$RE = \frac{cc}{w}$$

where

RE = refrigeration effect (in Btu/lb)

cc = cooling capacity (in Btu/hr)

w = mass flow rate (in lb/hr)

The air conditioner has a cooling capacity of 49,000 Btu/hr, and the mass flow rate is 6120 lb/hr. Find the refrigeration effect.

$$RE = \frac{cc}{w}$$

$$RE = \frac{49{,}000}{6120}$$

$$RE = \mathbf{8.00\ Btu/lb}$$

7. The enthalpy of air as it leaves the evaporator is found by subtracting the refrigeration effect of the evaporator from the enthalpy of air as it enters the evaporator. The enthalpy of air as it leaves the evaporator is found by applying the following formula:

$$h_l = h_e - RE$$

where

h_l = enthalpy of refrigerant leaving evaporator (in Btu/lb)

h_e = enthalpy of refrigerant entering evaporator (in Btu/lb)

RE = refrigeration effect (in Btu/lb)

The refrigeration effect of the air conditioner is 8.0 Btu/lb. The refrigerant enters the evaporator with an enthalpy of 27.4 Btu/lb. Find the enthalpy of refrigerant as it leaves the evaporator.

$$h_l = h_e - RE$$

$$h_l = 27.4 - 8.0$$

$$h_l = \mathbf{19.4\ Btu/lb}$$

8. On the psychrometric chart, use the relative humidity and enthalpy of air as it leaves the evaporator to locate points that represent dry-bulb and wet-bulb temperatures of air leaving the evaporator. Air leaving the evaporator is at 50% relative humidity, and the enthalpy of air as it leaves the evaporator is 19.4 Btu/lb. The condition of air leaving the coil is located at the intersection of the 19.4 Btu/lb enthalpy line and 50% relative humidity line.

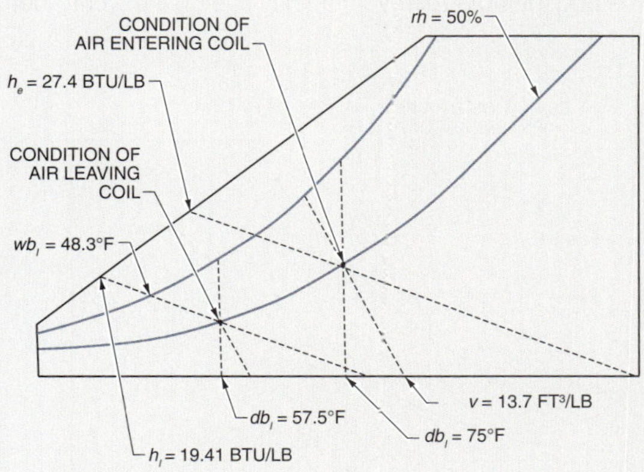

9. The dry-bulb temperature of air at this point is found by following the dry-bulb temperature lines to the dry-bulb temperature scale on the bottom of the chart. The dry-bulb temperature of air leaving the coil is 57.5°F.

10. The wet-bulb temperature of air at this point is found by following the wet-bulb temperature lines to the wet-bulb temperature scale on the curve of the chart. The wet-bulb temperature of air leaving the coil is 48.3°F.

11. Find the temperature decrease.

$$\Delta T = T_1 - T_2$$

$$\Delta T = 75 - 57.5$$

$$\Delta T = \mathbf{17.5°F}$$

This temperature decrease is less than 20°F, which is typically used for an air conditioning system. This illustrates the additional cooling required because of latent heat in the air.

Latent Heat Removal. Additional cooling is required to remove latent heat in the air. Latent heat is the result of moisture evaporating in the conditioned air and reducing its relative humidity. **See Procedure Finding Temperature Decrease of Air Conditioners.**

HYDRONIC HEATING SYSTEMS

Hydronic heating systems are heating systems in which water, steam, or other fluid is used as the heating medium. Water is used most often in hydronic heating systems because it has relatively high specific weight and specific heat.

Finding Temperature Decrease of Air Conditioners

Procedure

1. Air enters the evaporator of an air conditioner at a dry-bulb temperature of 80°F and 50% relative humidity. The air conditioner has a cooling capacity of 36,000 Btu/hr, and the airflow rate is 1200 cfm across the evaporator. Find the dry-bulb and wet-bulb temperatures of air leaving the coil when the relative humidity of air leaving the coil is 50%.
2. Locate the intersection of the 80°F dry-bulb temperature line and the 50% relative humidity line on a psychrometric chart.

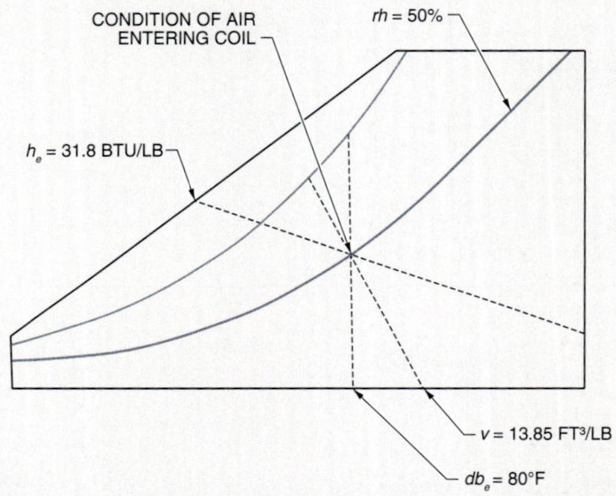

3. Find the specific volume for this point. Specific volume is 13.85 ft³/lb.
4. Find the enthalpy for this point. Enthalpy is 31.8 Btu/lb.
5. Find the mass flow rate.

$$w = \frac{Q}{v} \times 60$$

$$w = \frac{1200}{13.85} \times 60$$

$$w = 86.64 \times 60$$

$$w = \mathbf{5198\ lb/hr}$$

6. Find the refrigeration effect.

$$RE = \frac{cc}{w}$$

$$RE = \frac{36{,}000}{5198}$$

$$RE = \mathbf{6.93\ Btu/lb}$$

7. Find the enthalpy of air as it leaves the evaporator.

$$h_l = h_e - RE$$

$$h_l = 31.8 - 6.93$$

$$h_l = \mathbf{24.87\ Btu/lb}$$

8. Find the dry-bulb and wet-bulb temperatures of air leaving the evaporator at 50% relative humidity.

$db = \mathbf{68.5°F}$

$wb = \mathbf{57.5°F}$

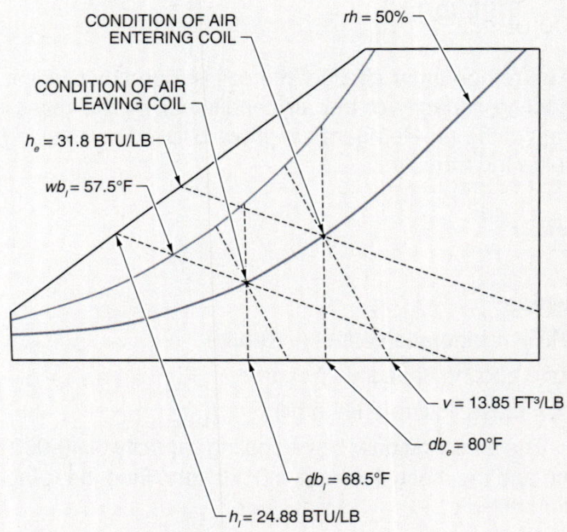

9. Find the temperature difference.

$$\Delta T = T_1 - T_2$$

$$\Delta T = 80 - 68.5$$

$$\Delta T = \mathbf{11.5°F}$$

The temperature decrease is 11.5°F. This is less than 20°F, which is usually used for air conditioning systems. This illustrates why additional cooling is required to remove latent heat.

Hydronic heating systems are used when the heating equipment is located far from the building spaces that require heat. An example is a multistory building in which the heating equipment is located on one floor, and hot water is distributed to terminal devices located on other floors. A hydronic heating system is also used when several buildings must be heated by a central heating plant. A hydronic heating system consists of a boiler, piping, terminal devices, circulating pumps, and controls. The boiler is the central component in a hydronic heating system. **See Figure 50-6.**

Boilers

A boiler is a pressure vessel that produces hot water in a hydronic heating system. Heat is produced in a boiler by burners or electric heating elements. When selecting a boiler, a design technician should consider the size of the boiler, the amount of heat required, and the availability and cost of fuels and electricity. Boilers are categorized according to boiler working pressures and temperatures, method of heat production, and heating capacity.

Pressures and Temperatures. Most boilers used for hydronic heating systems are low-pressure boilers. Low-pressure boilers operate at working pressures up to 15 psi for steam and 160 psi for water. Low-pressure steam boilers are typically used for heating warehouses, factories, schools, and apartments. Low-pressure hot water boilers are typically used for residential heating. Low-pressure hot water boilers operate with water temperatures up to 250°F.

High-pressure boilers operate at working pressures above 15 psi for steam and above 160 psi for water. High-pressure steam boilers are typically used for processing operations in industry and for generating electricity. High-pressure hot water boilers operate with water temperatures of 250°F or greater and are typically used for heating systems in which more than one building is heated. An example is a school campus where several buildings are heated by one boiler. In this application, the water is piped from the boiler to the various buildings.

Fuel and Electricity. Boilers burn fuel, use electricity, or use alternate sources of energy to produce heat. Combustion boilers burn fuel such as natural gas or fuel oil. Natural gas- and fuel-oil-fired boilers are typically used in residential and commercial buildings. The availability and convenience of using a fuel determines which fuel is used for an application. Electric boilers use heat produced by resistance heating elements. Electric boilers are also typically used in residential and commercial applications.

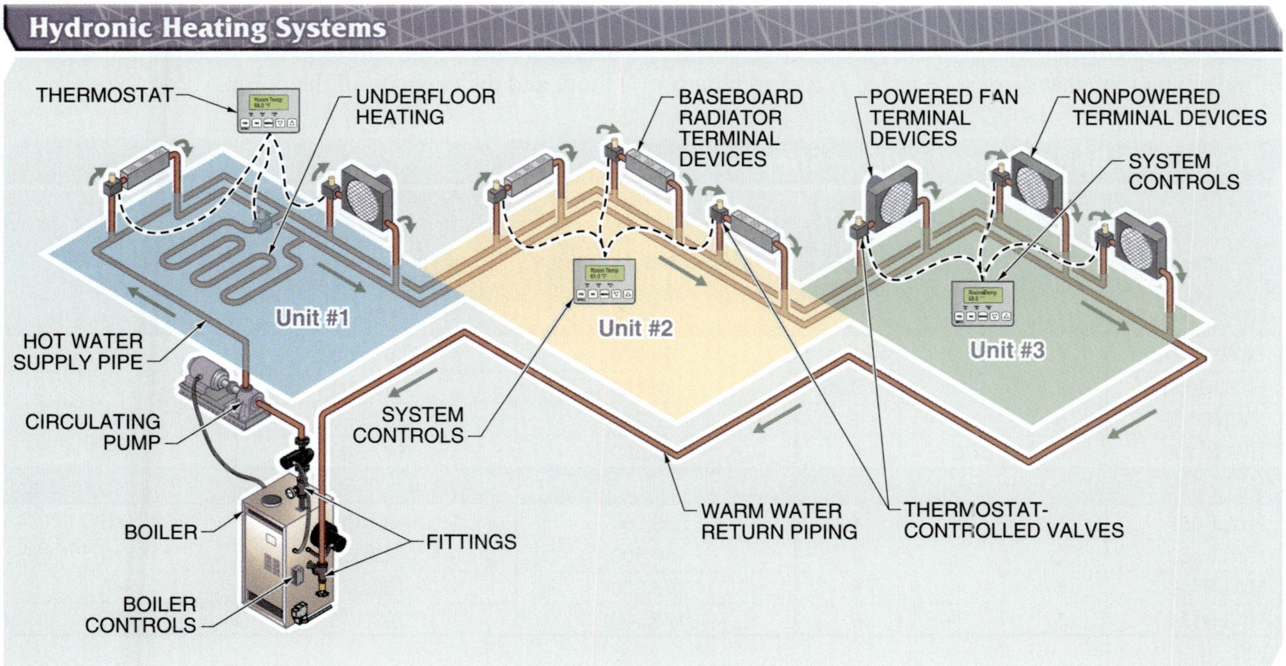

Figure 50-6. In hydronic heating systems, water is heated in a boiler and circulated through piping to terminal devices located in building spaces. Hydronic heating systems can provide heating to several floors or buildings.

Many hydronic heating systems can use solar energy, reclaim heat from exhaust air, or transfer heat from one part of a building to another. In hydronic solar heating systems, water is heated directly by the sun using collectors. Collectors, which are placed on the roof or other outside surface of a building, efficiently change radiant energy to thermal energy. The thermal energy is then transferred to the water. The heated water is stored in a tank and used for heating the building.

Heat reclaim systems collect heat from heated air or water that is exhausted from a building. For water, a hot water heat reclaim coil is placed in the exhaust piping. The waste heat is transferred to the water inside the coil. The heated water is then stored and used as a heat source.

Heating Capacity. Natural gas- and fuel-oil-fired boilers have heating capacities from about 100,000 Btu/hr for residential applications to millions of Btu/hr for commercial applications. A boiler that is selected for heating a building should have a heating capacity that is at least as large as the heating load on the building. Equipment catalogs and engineering specification sheets are used to select a boiler for an application. **See Figure 50-7.**

Piping Systems

A hydronic piping system distributes water from a boiler to terminal devices, where it is used to heat air. The pipes in a hydronic piping system are much smaller than the ductwork used to carry air in a forced-air heating system.

Piping systems are categorized by the arrangement of the piping loop that carries the water. A typical piping system has a supply line that carries water from the boiler to the terminal devices and a return line that carries water from the terminal devices to the boiler. Piping systems used for hydronic heating systems include one-pipe, two-pipe, three-pipe, and four-pipe systems. Each piping system consists of loops of pipe that include a supply and return line.

Terminal Devices

Terminal devices transfer heat from the hot water in a hydronic heating system to air in building spaces. Terminal devices for hydronic heating systems are combinations of water coils and other components. The water coils act as heat exchangers, and the other components control the flow of water and air through the terminal devices. The terminal devices used in hydronic heating systems include radiators, duct coils, convectors, blower coils, and unit ventilators.

Circulating Pumps

Circulating pumps move water in a hydronic heating system from the boiler through the supply piping, terminal devices, and return piping to the boiler. Selecting a circulating pump is a part of designing a hydronic heating system. The arrangement of zones in a building should be considered when selecting equipment. Each control zone, which is identified by a separate piping loop, typically has a circulating pump. A circulating pump is chosen for an application based on the location of the pump in the building, the location in the piping loop, and the capacity of the pump.

Hot Water Boiler Manufacturer Specifications

Hot Water Boilers						
Model Number	Dimensions* L	Dimensions* W	Dimensions* H	Heating Input†	Heating Capacity†	Shipping Weight‡
Standing Pilot — Natural Gas						
BW20037-SPK	6	3	4	60,000	51,000	220
HWNG050	2.5	2	3.25	38,500	30,000	227
HWNG100	6	3	5	127,000	107,950	300
HWNG150	6	3	6	149,000	127,500	375
Standing Pilot — LP Gas						
HWLP031	5	4	4	36,000	32,400	165
HWLP063	5	4	4	50,500	42,500	210
HWLP126	5	4	6	125,500	106,675	345
HWLP177	5	4	6	175,500	149,175	405

* in ft
† in Btu/hr
‡ in lb

Figure 50-7. Boiler catalogs and engineering specification sheets show the size, heating capacity, and weight of boilers.

Circulating pumps are often located near the boiler or chiller. The pump must deliver water at the flow rate required to overcome the resistance of the piping system. When the pump is too small, it will not move enough water. When the pump is too large, it will not deliver the water economically.

Control Systems

Control systems for hydronic heating systems consist of boiler controls and terminal device controls. Boiler controls regulate the temperature of the supply water by regulating the boiler burners. Terminal device controls regulate the valves or the fans in terminal devices, which control the temperature of the air in the building spaces.

The degree of control desired for an application determines the complexity of the control system. Control systems range from simple on/off manual controls to very sophisticated systems in which both water and airflow are controlled separately for individual zones in a building space. The degree of control for an application must be determined before a control system can be designed. Control systems vary in levels of sophistication and can be designed to achieve any level of control.

Heating Water Flow

In hydronic heating systems, water is heated by the burners in the boiler. Circulating pumps circulate hot water through the piping system to terminal devices. The heating capacity of the boiler must be at least as large as the heating load of the building. The water flow rate in a hydronic heating system must match the heating output capacity of the boiler for a given temperature drop through the system.

HYDRONIC COOLING SYSTEMS

A *hydronic cooling system* is a system in which water is chilled and circulated to terminal devices, where it cools air in building spaces. The piping system, terminal devices, circulating pump, and control system in a hydronic cooling system are similar to those in a hydronic heating system. Typically the same hydronic distribution system is used for heating and cooling. **See Figure 50-8.** A chiller is used that provides the needed cooling and is equivalent in flow capacity to the boiler in the hydronic heating system.

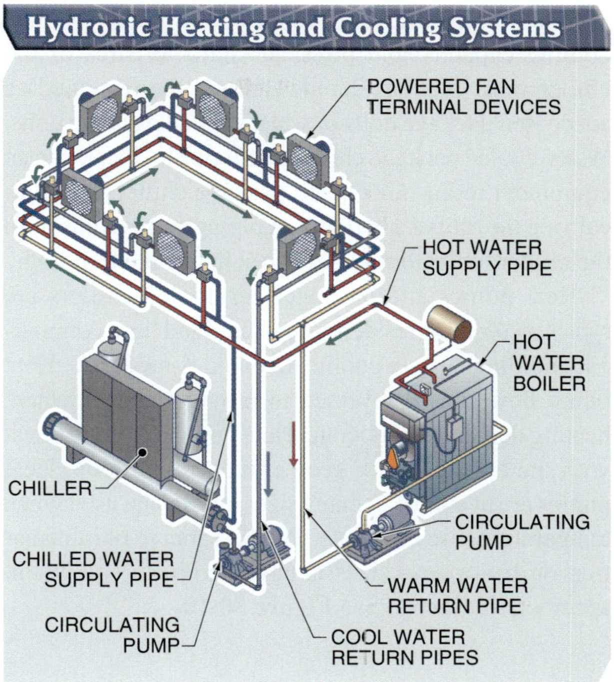

Figure 50-8. In hydronic heating and cooling systems, water is heated in a boiler and cooled in a chiller. Water is circulated through piping to terminal devices located in building spaces for heating and cooling purposes.

Chillers and Heat Pumps

The chiller is the component in a hydronic cooling system that cools water. Mechanical compression chillers are used in buildings of all sizes because they are compact and can be adapted to many conditions. **See Figure 50-9.**

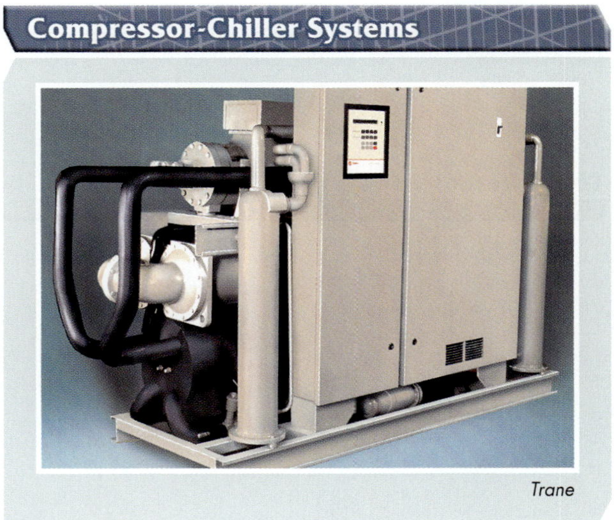

Figure 50-9. The components on factory-assembled compressor-chiller units are mounted for easy access during service and maintenance.

When selecting a chiller for an application, the cooling capacity and physical characteristics of the chiller must be considered. Chillers are designed as air-cooled package units or water-cooled package units. Water-cooled package chillers are typically located in an equipment room. Air-cooled package chillers are typically on the roof of a building. A piping system runs from the chiller to the terminal devices inside the building.

Heat pumps and packaged air-cooled chillers are typically used for residential systems and light commercial buildings where cooling demand is moderate. Heat pump flow can be reversed to provide heating where heating demand is moderate. Heat pumps exchange heat with the outside air or geothermal ground loop. Heat pumps are available in many sizes and can be used when temperatures are moderate. The best source of information on heat pump sizes is engineering or equipment specification sheets. **See Figure 50-10.**

Cooling Capacity. Chillers with a cooling capacity for almost any application are available. Equipment catalogs and engineering specification sheets should be used for selecting equipment for an application. The cooling capacity of a chiller is a function of the components of the chiller.

Cooling Water Flow

In hydronic air conditioning systems, the chiller cools water, and the circulating pump circulates the water through the piping system to the terminal devices in a building. The chiller must be able to provide enough cool water to offset the heat gain of the building at a given temperature increase through the piping system. The amount of water circulated through the piping system is related to the cooling load on the building and the temperature increase of the water as it flows through the terminal devices in the system.

Heat Pump Manufacturer Specifications

Model	Water Loop Heat Pump				Ground Water Heat Pump				Ground Loop Heat Pump			
	Cooling		Heating		Cooling		Heating		Cooling		Heating	
	Indoor 53.6°F Outdoor 86°F		Indoor 104°F Outdoor 68°F		Indoor 53.6°F Outdoor 59°F		Indoor 104°F Outdoor 50°F		Indoor 53.6°F Outdoor 77°F		Indoor 104°F Outdoor 32°F	
	Capacity Btu/hr	EER Btu/hr/W	Capacity Btu/hr	COP	Capacity Btu/hr	EER Btu/hr/W	Capacity Btu/hr	COP	Capacity Btu/hr	EER Btu/hr/W	Capacity Btu/hr	COP
TMW036	32,300	14.60	43,100	4.90	36,200	23.10	35,300	4.00	33,300	16.40	27,400	3.10
TMW060	52,800	14.30	72,700	4.70	56,600	20.30	60,300	4.00	55,600	16.20	48,500	3.10
TMW120	105,600	14.10	145,400	4.60	113,200	20.10	120,600	3.90	111,200	16.00	97,000	3.10

Figure 50-10. Catalogs and engineering specification sheets are used when selecting heat pump equipment.

UNIT

51 Forced-Air System Design

Most forced-air HVAC systems circulate air to heat, cool, humidify, or dehumidify building spaces. HVAC units also bring in outside air to provide ventilation and satisfy the indoor air quality requirements of building spaces. Light commercial HVAC systems can be single zone or variable-air-volume rooftop units. Circulated air offsets the heat gain or cooling load in each building space. Proper heating and cooling load calculations are necessary to design efficient forced-air systems.

OBJECTIVES

- Compare the types of supply plenum systems.
- Define the terms used to describe airflow and system pressure.
- Explain the use of equal friction charts when designing ductwork.
- Identify the components and information found on distribution system layout drawings.
- Perform typical forced-air system component sizing calculations.

Index to Procedures

FORCED-AIR DISTRIBUTION SYSTEMS

Forced-air distribution systems consist of supply and return ductwork, balancing dampers, fans, registers, and return grills. **See Figure 51-1.** Variable-air-volume (VAV) units also have terminal boxes in each zone to control the airflow. Forced-air distribution systems circulate conditioned air from HVAC units to building spaces through ductwork typically made of sheet metal, fiberglass, or fabric. Forced-air distribution systems for light commercial buildings typically include devices such as fire dampers and duct smoke detectors. Fire dampers prevent the spread of fire inside ductwork through dampers that close when a thermal element melts. The thermal element melts when high temperatures are reached inside the duct.

Properly designed forced-air distribution systems provide occupant comfort with high-performing HVAC distribution systems. Good airflow throughout building spaces increases energy efficiency and protects buildings from extreme temperature and humidity levels.

Supply Plenum Systems

A supply plenum system connects the HVAC unit supply opening to the ductwork used to distribute conditioned air. Plenum systems typically work best when heating and cooling units are centrally located. Residential plenum systems typically are located in basements, attics, or crawl spaces underneath a home. Light commercial building rooftop units are typically located near the center of the building with plenum systems located in ceiling spaces. Supply plenums are typically rectangular sheet metal enclosures, although other materials are used.

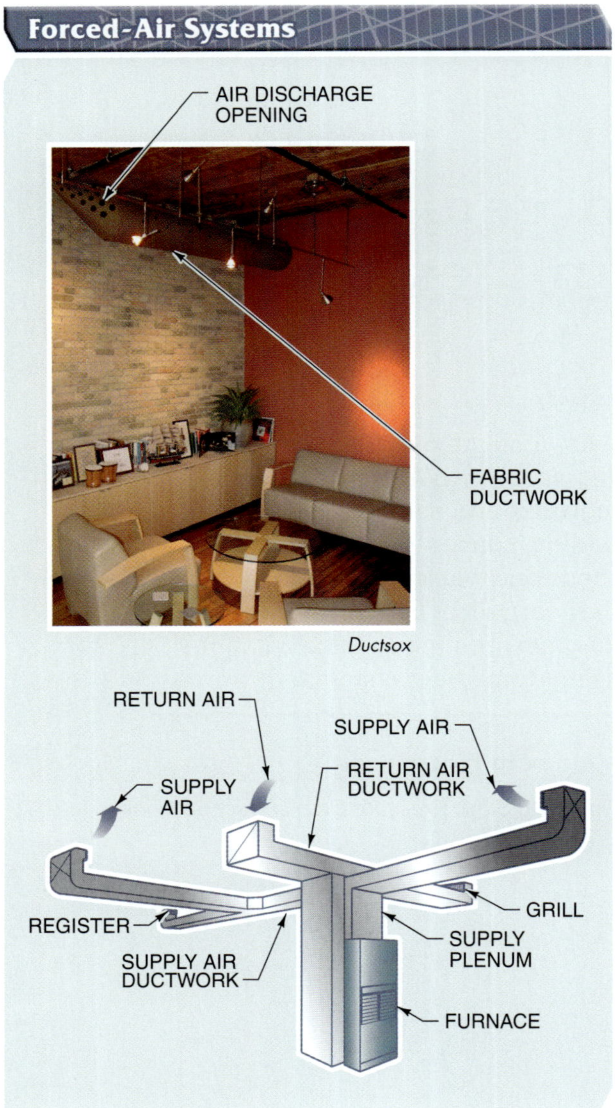

Figure 51-1. HVAC forced-air systems consist of supply and return ductwork, dampers, fans, registers, and grills.

Multiple factors determine the appropriate duct configuration that is to be used for a supply plenum system. Typical factors include cost, building type, structure accessibility, unit type, noise, and aesthetics.

Each type of supply plenum system has advantages and disadvantages. System selection is based on the building construction and type of building spaces. A supply plenum system includes a radial duct, extended plenum, reducing plenum, and perimeter loop. **See Figure 51-2.** Hybrid systems may include parts of each type of distribution system.

Radial Duct Systems. A radial duct system has a central supply plenum that feeds several individual ductwork branches with conditioned air. Radial ductwork systems are typically used where individual outlets are nearby and the heating and cooling unit is centrally located.

A box plenum is mounted at the discharge of the heating and cooling unit installed near the center of the building. Individual discharge ducts are connected to the plenum for distributing the conditioned air to registers at building spaces. Radial systems are easy to install and use inexpensive fittings.

Extended Plenum Systems. An extended plenum system has a central supply plenum of constant dimensions that extends from the central unit. It feeds several individual branch ducts with conditioned air. Typically, the extended plenum is limited to 24′ in length. The extended plenum is directed toward the discharge that is at the end of the building. Each branch for individual zones is connected to the extended plenum.

Reducing Plenum Systems. A reducing plenum system has a central supply plenum of dimensions that are reduced as it extends from the central unit. It feeds several individual ductwork branches with conditioned air. Reducing plenums are used when plenum lengths exceed 24′. Typically, plenum sizes are reduced when airflow rates are reduced by about half as airflow discharges into individual ductwork branches. Ductwork branches extend from trunk supply ducts to building space registers.

Typically, a reducing plenum system has one long plenum that extends the length of the building with the HVAC unit located on one end of the structure. Since air pressure drops through a long run of duct, the duct progressively gets smaller in order to keep the system pressure the same in each section. A reducing plenum requires that multiple sections of various-size duct be joined together, which results in a system that is more expensive to fabricate and harder to install.

Perimeter Loop Systems. A perimeter loop system has a central supply plenum connected to several feeder ductworks that feed a ductwork loop around the perimeter of the building with conditioned air. The ductwork loop discharges directly into building space registers. Perimeter systems are typically used in concrete slab-on-grade construction in colder climates. The building slab is warmed or cooled by the circulating air while the building space is heated or cooled by the discharged conditioned air.

Forced-Air Distribution Systems

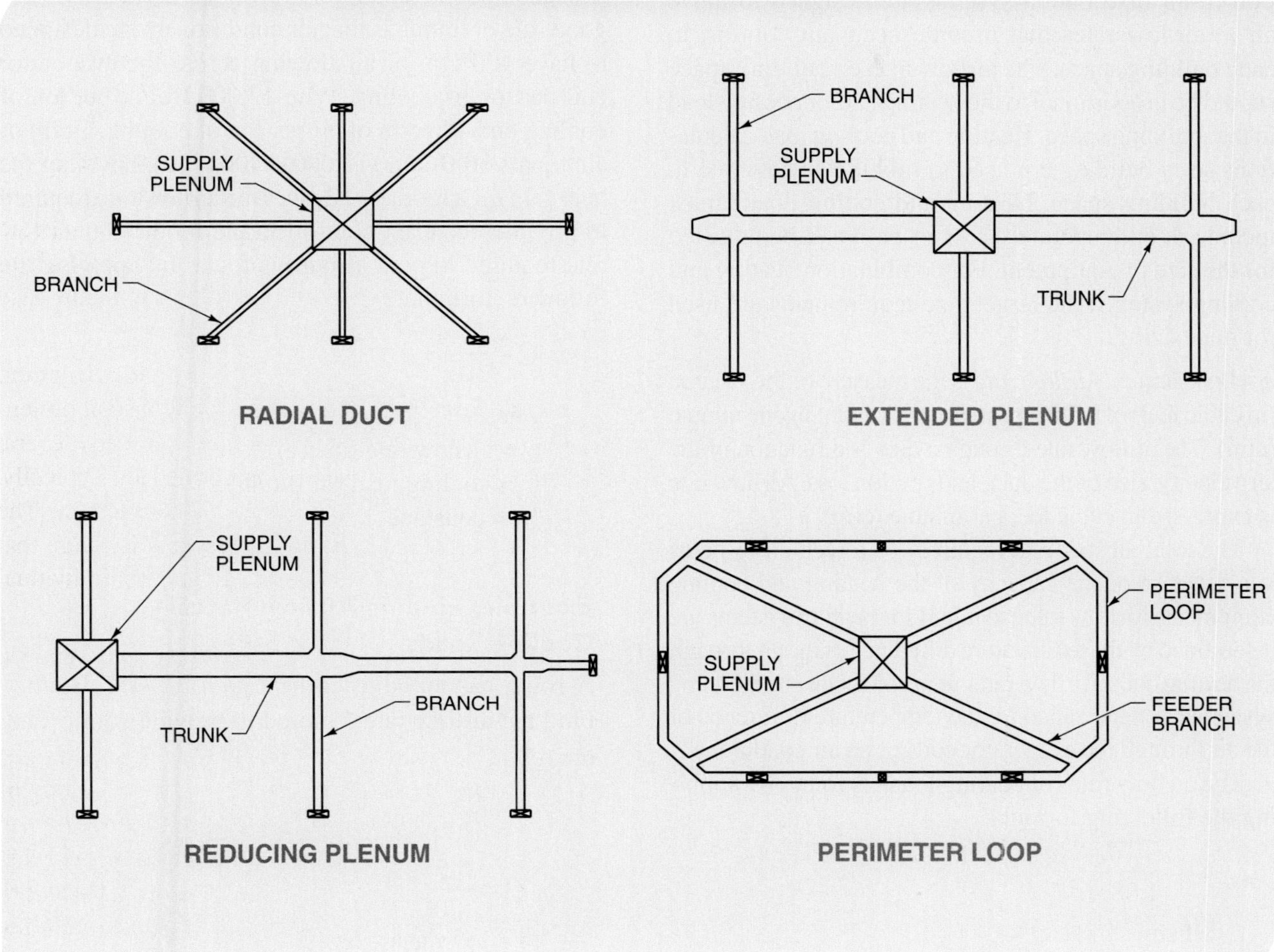

Figure 51-2. Distribution systems are categorized by the pattern of ductwork distribution layout.

An advantage of this system is that the pressure is constant almost everywhere in the system. Perimeter loop systems are more expensive to install. However, the entire loop and feeders have typically the same size duct.

Return Duct Systems and Open Returns

Some HVAC systems use ducts to collect return air, while other systems use open returns. In return duct systems, grills are connected to ducts that return air back to the units. This ensures that the proper amount of air goes back to the units from building spaces. Typically, there is one return duct per floor or area of the building. With open returns, ducts are minimized and the HVAC units are open to the surrounding areas. Return air is brought back to the units through slots in utility room doors and other openings. Typically, open returns tend to have excessive fan noise, so care must be taken when designing and placing the unit.

A variation of the open return is an open-ceiling return system, which is used in light commercial systems. Open-ceiling return systems use drop ceilings with return grills. The space above the drop ceiling is used as a common return ducted back to the HVAC unit. It is typically used on packaged rooftop air handlers and large VAV rooftop systems.

PRINCIPLES OF AIRFLOW

Air distribution systems are designed to control the movement of conditioned air to and from building spaces. Airflow and air pressure are controlled by fans, ductwork, fittings, accessories, and dampers.

Airflow

Forced-air distribution systems are designed to move air at airflow rates that provide occupant comfort to each building space. The airflow in each building space should be proportional to the heating and/or cooling load in that building space. Heating and cooling load calculations for a building are used to find the airflow rate in each building space. Heating and cooling loads must include duct temperature loss or gain and be adjusted for the size of equipment. For combination heating and cooling systems, the larger size requirements are used for duct sizing.

Airflow Rates. *Airflow rate* is the measure of the volume (in cubic feet) of air that passes a given point in one minute (cfm). The airflow rate through a duct is a function of the air velocity, size of the duct, and friction loss. Airflow rate is expressed in cubic feet per minute (cfm).

The total airflow rate required for forced-air systems is a function of the capacity of the heating and cooling equipment. Airflow rates required for heating systems are determined by the temperature difference of the air through the heating coils. Airflow rates required for air conditioning systems are determined by the temperature difference of the air through the evaporator coils of an air conditioner.

The airflow rate for heating loads is found by applying the following formula:

$$Q = \frac{L_h}{1.08 \times \Delta T}$$

where
Q = airflow rate (in cfm)
L_h = heating load (in Btu/hr)
1.08 = constant
ΔT = change in temperature (in °F)

Example: Finding Airflow Rates—Heating Loads

A room has an adjusted heating load of 13,300 Btu/hr. A temperature difference of 80°F is desired. Find the airflow rate required to provide comfort in the room.

$$Q = \frac{L_h}{1.08 \times \Delta T}$$

$$Q = \frac{13,300}{1.08 \times 80}$$

$$Q = \frac{13,300}{86.4}$$

Q = **153.9 cfm**

To find the airflow rate required to cool a room, the heat gain (cooling load) for the room is divided by 30. A good rule of thumb is that air conditioners are designed to have 400 cfm of air flowing across the evaporator coil per ton of cooling. With 12,000 Btu/hr per ton of cooling and 400 cfm of air per ton of cooling, 1 cfm of air equals 30 Btu/hr (12,000 ÷ 400 = 30). Dividing the heat gain for a room by 30 gives the airflow rate required to provide occupant comfort in the room. The airflow rate required to cool a room is found by applying the following formula:

$$Q = \frac{L_c}{30}$$

where
Q = airflow rate (in cfm)
L_c = cooling load (in Btu/hr)
30 = constant

Example: Finding Airflow Rates—Cooling Loads

A room has an adjusted heat gain of 7135 Btu/hr. Find the airflow rate required to provide comfort in the room.

$$Q = \frac{L_c}{30}$$

$$Q = \frac{7135}{30}$$

Q = **237.8 cfm**

For spaces occupied by people, air velocity must not be above 40 fpm. Air that strikes a person at a rate above 40 fpm feels uncomfortable because of the cooling effect of moving air. For a typical application, the airflow rate in a building space should be less than 1 cfm/ft² of floor space. This flow rate can be increased if the area is not occupied by people. The airflow rate required to heat a building space without causing drafts is a function of the airflow rate and area of the building. The proper airflow rate per unit area of building space is found by applying the following formula:

$$Q_A = \frac{Q}{A_{bs}}$$

where
Q_A = airflow rate per unit area (in cfm/ft²)
Q = airflow rate (in cfm)
A_{bs} = area of building space (in ft²)

Example: Finding Airflow Rates per Unit Area

A room has an area of 1200 ft². A furnace fan circulates 1000 cfm of air. Find the proper airflow rate for heating the room.

$$Q_A = \frac{Q}{A_{bs}}$$

$$Q_A = \frac{1000}{1200}$$

$$Q_A = \mathbf{0.83\ cfm/ft^2}$$

The airflow rate of this heating system is acceptable because it is less than 1 cfm/ft². Therefore, the heating system will not create a draft.

Air Velocity. *Air velocity* is the speed air moves from one point to another through ductwork, a register, or a grill and is measured in feet per minute (fpm). Maintaining suitable air velocity is required for proper air circulation in building spaces. Proper air circulation prevents temperature stratification and drafts. Temperature stratification is the variation of air temperature in a building space that occurs when warm air rises to the ceiling and cold air drops to the floor. When temperature stratification occurs, individuals become uncomfortably warm due to insufficient air circulation. Drafts typically occur when the air velocity is above 40 fpm. Individuals become uncomfortably cool due to the high rate of evaporation of skin perspiration.

The air velocity in a duct is a function of the airflow rate and duct size. **See Figure 51-3.** Air velocity is found by applying the following formula:

$$V = \frac{Q}{A}$$

where

V = air velocity (in fpm)

Q = airflow rate (in cfm)

A = area of duct (in ft²)

TECH TIP

Problems affecting building space comfort can occur when rooms and/or floor plans are modified. For example, a room may be divided into several smaller rooms or several small rooms may be opened into one large room. In most of these situations, the HVAC system is not redesigned to account for the changes. This usually results in loss of comfort.

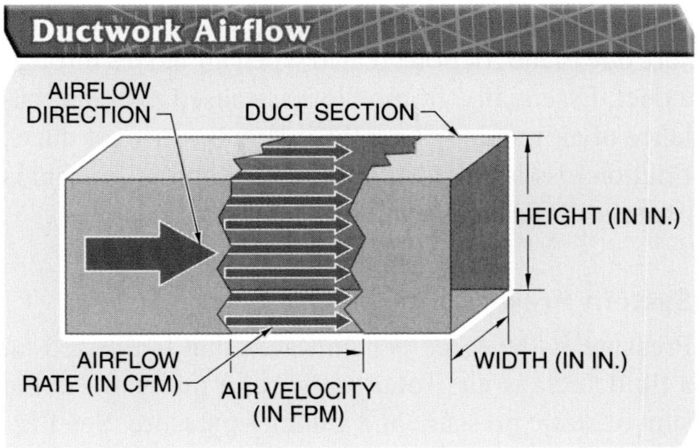

Figure 51-3. Airflow rate is the air velocity multiplied by the duct area.

Example: Finding Air Velocity

The fan in a forced-air distribution system moves 1200 cfm of air through a 24″ × 12″ duct. Find the air velocity.

1. Find the area of a cross-section of the duct in square feet. Multiply the width and height of the duct in inches and divide by 144. Dividing by 144 converts the area from square inches to square feet. When the dimensions of a duct are 24″ × 12″, find the area of a cross-section of the duct.

$$A = \frac{w \times h}{144}$$

$$A = \frac{24 \times 12}{144}$$

$$A = \frac{288}{144}$$

$$A = \mathbf{2.00\ ft^2}$$

2. Find the air velocity in a duct. Divide the airflow rate by the duct size to find the air velocity.

$$V = \frac{Q}{A}$$

$$V = \frac{1200}{2}$$

$$V = \mathbf{600\ fpm}$$

Duct Sizes. Ducts are sized to carry the required airflow rate to building spaces for occupant comfort. Duct size is the size of a duct, which is expressed in inches of diameter (in. DIA) for round ducts and in inches (in.) of width and height for rectangular ducts. Dimensions for rectangular ducts typically note the width first. For example, a 20″ × 8″ duct is a duct 20″ wide and 8″ high.

Friction Loss. *Friction loss* is the decrease in air pressure due to the friction created when air moves through a duct. Essentially, friction loss is caused by the turbulence of air moving along the internal surface of ducts. Friction loss is indicated by static pressure drop and is expressed in inches of water column (in. WC).

System Pressure

Pressure is the force per unit area that is exerted on a fluid such as air. Total pressure in ductwork is the sum of static pressure and velocity pressure. **See Figure 51-4.** Positive pressure exerted by the discharge side of the fan pushes the conditioned air through the supply plenum to the registers in the building spaces. Negative pressure exerted by the suction side of the fan pulls the return air from the building spaces through the grills and back to the fan.

Ductwork Air Pressure

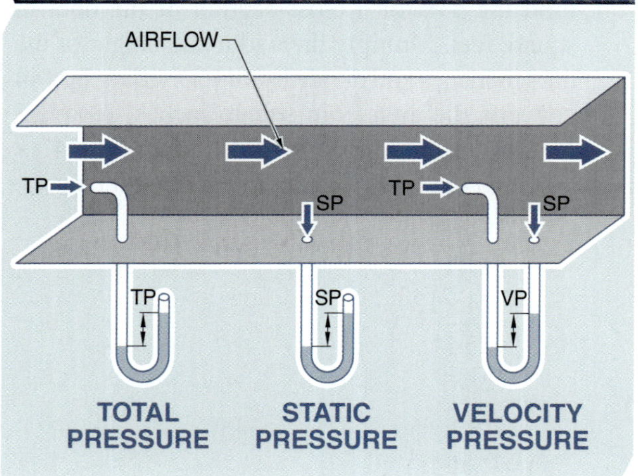

Figure 51-4. Total air pressure in ductwork is the sum of static air pressure and velocity air pressure.

Static Pressure. *Static pressure* is the force that nonmoving air exerts against surfaces such as ducts; measured in inches of water column (in. WC). Static pressure is the pressure that has a tendency to burst a duct. Ductwork is classified by the maximum static pressure difference between the air inside and outside of the duct.

Ductwork classes are based on static pressure at levels of ±0.5″ WC, 1.0″ WC, 2.0″ WC, 3.0″ WC, 4.0″ WC, 6.0″ WC, and 10.0″ WC. Residential buildings typically operate at ±0.5″ WC or ±1.0″ WC static pressure differentials. Commercial buildings typically operate at ±1.0″ WC to ±3.0″ WC static pressure differentials. Industrial buildings typically operate at higher static pressure differentials. Ductwork must be designed to withstand the operating pressure differential.

Static pressure is measured with a manometer. Manometers measure the pressures of vapors and gases. The types of manometers commonly used to measure air pressure in ducts are U-tube, inclined, and digital manometers. **See Figure 51-5.**

Manometers

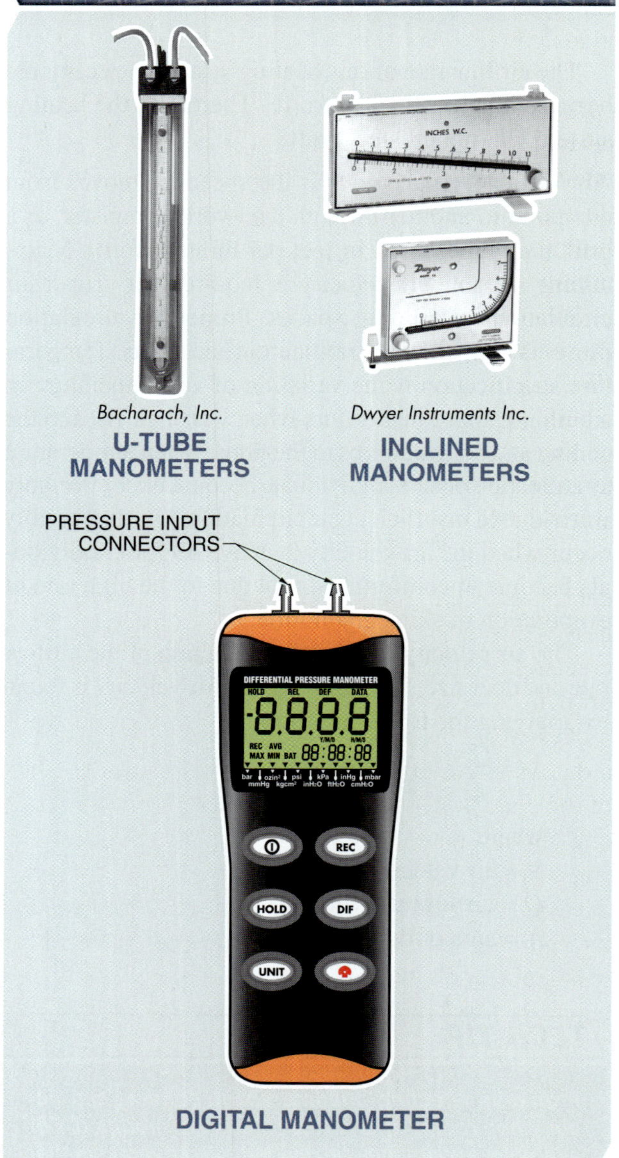

Figure 51-5. Tube manometers measure air pressure by the difference between the liquid levels in the two legs. Digital manometers are portable test instruments used to measure pressure in air systems.

U-tube manometers are U-shaped sections of glass or plastic tubing that are partially filled with water or mercury. The liquid in the two legs of U-tube manometers remains level due to gravity when the pressure on the two legs is equal. When more pressure is exerted on one leg, the liquid in that leg will be pushed down. As the liquid in one leg goes down, the liquid in the other leg rises. The difference in the liquid levels is a measurement of air pressure.

Inclined manometers are similar to U-tube manometers, except that one leg is sloped. The sloped leg exaggerates the changes in liquid level, which provides greater accuracy when measuring air pressure.

Digital manometers can measure air pressure accurately. Digital manometers are typically used in portable applications. Similar to U-tube and inclined manometers, digital manometers measure pressure differential and typically use pitot tubes. The memory function of typical digital manometers allows air pressure values to be stored for reference.

When using manometers, air pressure is measured in inches of water column (in. WC). Water column (WC) is the pressure exerted by a column of water and is typically expressed in inches. The equivalent of 1″ WC is 0.036 psi. This unit of measure is used to express small pressures above and below atmospheric pressure.

Static Pressure Drop. Static pressure drop is the decrease in air pressure caused by friction as air moves through the internal surfaces of ductwork and creates turbulence. **See Figure 51-6.** Static pressure drop in a duct reduces the pressure of the air as the air moves along the duct. To measure static pressure in a duct, a small hole is drilled or punched in the side of the duct, and a flexible tube is run from the hole to one leg of a manometer. The difference between the static pressure readings at two points in a duct is the friction loss that has occurred in the duct between the two points. Static pressure drop in the ductwork of residential and light commercial buildings is typically between 0.08″ WC per 100′ to 0.60″ WC per 100′.

Static pressure drop in a duct section is found by dividing the length of the duct section by 100. The result is multiplied by the design static pressure drop. The *design static pressure drop* is the pressure drop per unit length of duct for a given size of duct at a particular airflow rate. The length of the duct section is divided by 100 because the design static pressure drop for the distribution system is measured in lengths of 100′ of duct. The actual static pressure drop in a duct section is found by applying the following formula:

$$PD_d = \frac{L}{100} \times pd_d$$

where

PD_d = actual static pressure drop (in in. WC)
L = length of duct section (in ft)
100 = constant
pd_d = design static pressure drop (in in. WC)

Static Pressure Drop

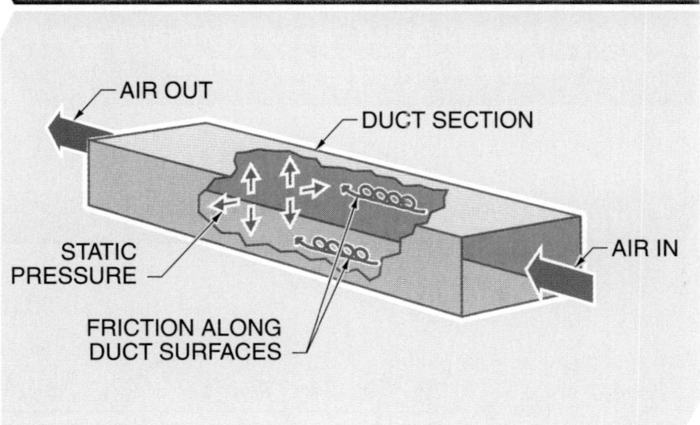

Figure 51-6. Friction loss caused by airflow along duct internal surfaces reduces the static air pressure within the duct.

Example: Finding Duct Static Pressure Drop

The airflow rate in a section of duct 140′ long is 1500 cfm. The duct section has a design static pressure drop of 0.08″ WC per 100′. When the dimensions of the duct are 20″ × 12″, find the static pressure drop of the duct section.

$$PD_d = \frac{L}{100} \times pd_d$$

$$PD_d = \frac{140}{100} \times 0.08$$

$$PD_d = 1.4 \times 0.08$$

$$PD_d = \mathbf{0.112'' \ WC}$$

Friction losses in ducts are directly related to the surface area exposed to airflow and the airflow rate. An *equal friction chart* is a chart that shows the relationship between the airflow rate (in cfm), static pressure drop (in in. WC/100′), duct size (in in. DIA), and air velocity (in fpm). **See Figure 51-7.** Equal friction charts are used to size ducts when two values are known. Equal friction charts show the sizes of round ducts. A conversion table must be used to find the size of a square or rectangular duct that provides the same static pressure drop. **See Appendix.**

TECH TIP

Flex duct is much easier to install than sheet metal duct. However, it has a much higher friction loss than other duct materials and can result in significantly higher energy costs and loss of comfort.

Example: Using Equal Friction Charts

The airflow rate in a section of duct is 1200 cfm and the static pressure drop is 0.10″ WC/100′. Find the size of the duct using an equal friction chart. **See Figure 51-8.**

1. Locate 1200 cfm on the scale at the bottom of the chart. Draw a vertical line from this airflow rate.
2. Locate 0.10″ WC/100′ on the scale on the left side of the chart. Draw a horizontal line across the chart from this static pressure drop.
3. From the point where the lines intersect, draw a line parallel with the diagonal duct diameter lines to the scale at the top of the chart. The size of a round duct is approximately 15″.

Equal Friction Charts

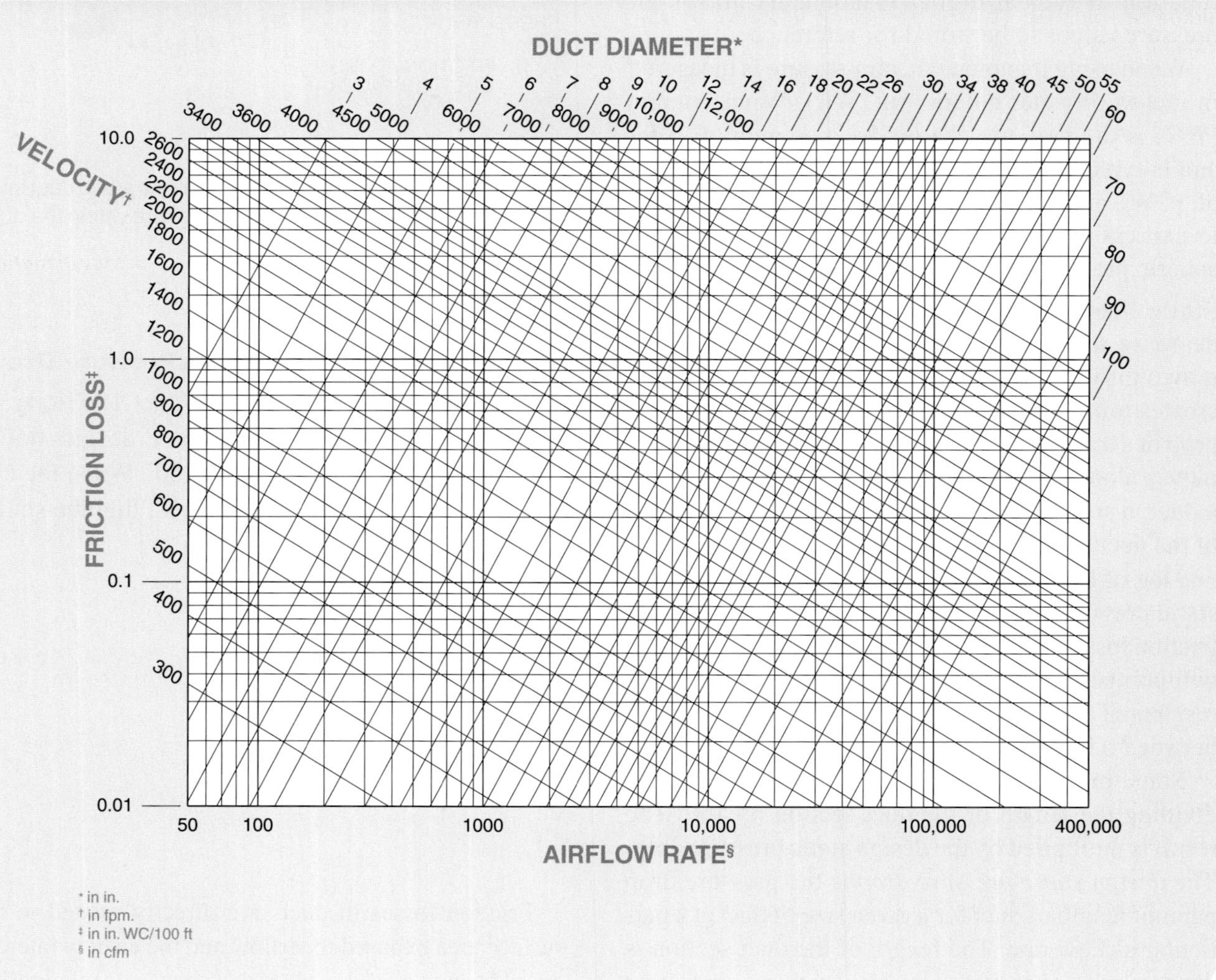

Figure 51-7. Equal friction charts show the relationship between duct size, flow rate, velocity, and static pressure drop.

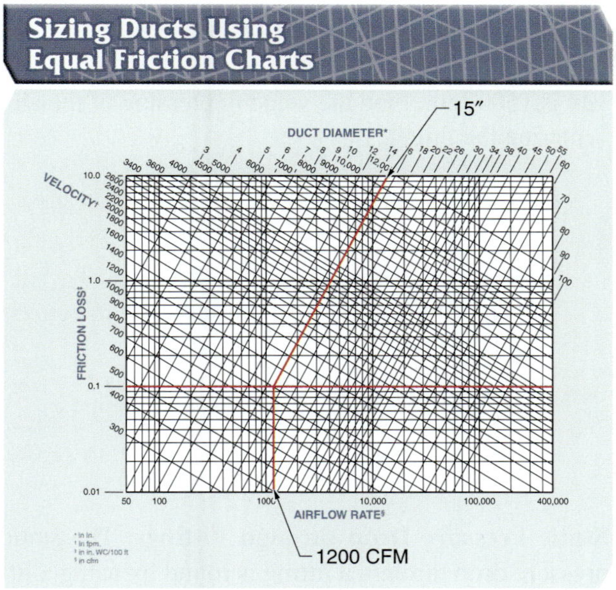

Figure 51-8. Duct size is found at the intersection of the airflow rate and the static pressure drop lines on an equal friction chart.

Ducts are typically sized using slide charts, commonly called ductulators, which were developed to show the relationship between duct size and airflow capacity. Digital calculators and computer programs can also be used for sizing ductwork. These devices show the same relationships as equal friction charts.

System Pressure Drop. *System pressure drop* is the total static pressure drop in a distribution system. The total static pressure drop in a forced-air distribution system is the sum of the static pressure drop through all duct sections. Duct sections include fittings, transitions, and accessories such as filters or coils, which add resistance to airflow in the distribution system.

Major duct sections must be considered when system pressure drop is calculated. A *major duct section* is an independent part of a forced-air distribution system through which all or part of the fan air supply flows. Many forced-air distribution systems have several major duct sections. An *individual duct section* is part of a distribution system that is between two fittings or transitions. Fittings and transitions are duct sections that change airflow direction or duct sizes. At the fittings, the airflow, direction, or velocity changes due to the configuration of the ductwork. **See Figure 51-9.**

Dynamic Pressure Drop. *Dynamic pressure drop* is the pressure drop in a duct fitting or transition caused by air turbulence as the air flows through the fitting or transition. Turbulence occurs whenever the airflow pattern is disturbed by a change in the airflow rate, direction of airflow, or size of the duct. Turbulence decreases the velocity pressure in the duct at the point where turbulence occurs.

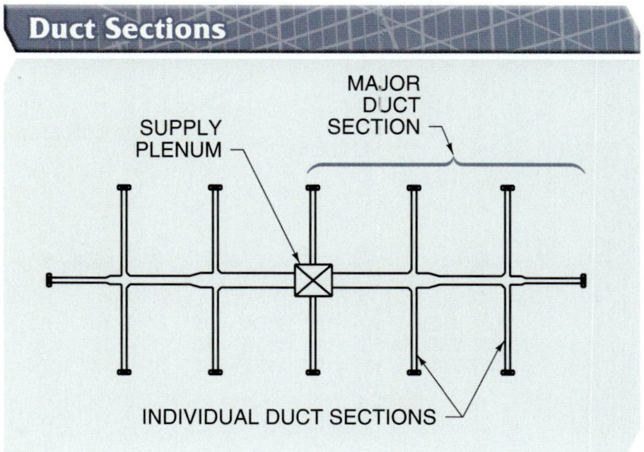

Figure 51-9. Duct sections are parts of a distribution system between fittings. At the fittings, the airflow, direction, or velocity changes due to the configuration of the ductwork.

Dynamic pressure drop through a fitting or transition is calculated using ductwork equivalent lengths or fitting loss coefficients. The *ductwork equivalent length method* is a method used to convert dynamic pressure drop to friction loss in feet of duct. The ductwork equivalent length method is used to determine the length of duct that gives the same static pressure drop as a particular metal fitting or transition. **See Figure 51-10.**

The velocity pressure of air in a duct is the air pressure that is measured parallel to the direction of airflow. Measuring velocity pressure in a duct without first measuring static pressure is difficult. The velocity pressure of air in a duct is typically found by measuring total pressure and static pressure and then subtracting the static pressure from the total pressure. Velocity pressure decreases as air velocity decreases.

As air enters a fitting, the velocity pressure of the air must be determined. The velocity pressure of the air entering a fitting is found by applying the following formula:

$$P_v = \left(\frac{V}{4005}\right)^2$$

where

P_v = velocity pressure (in in. WC)

V = air velocity (in fpm)

4005 = constant

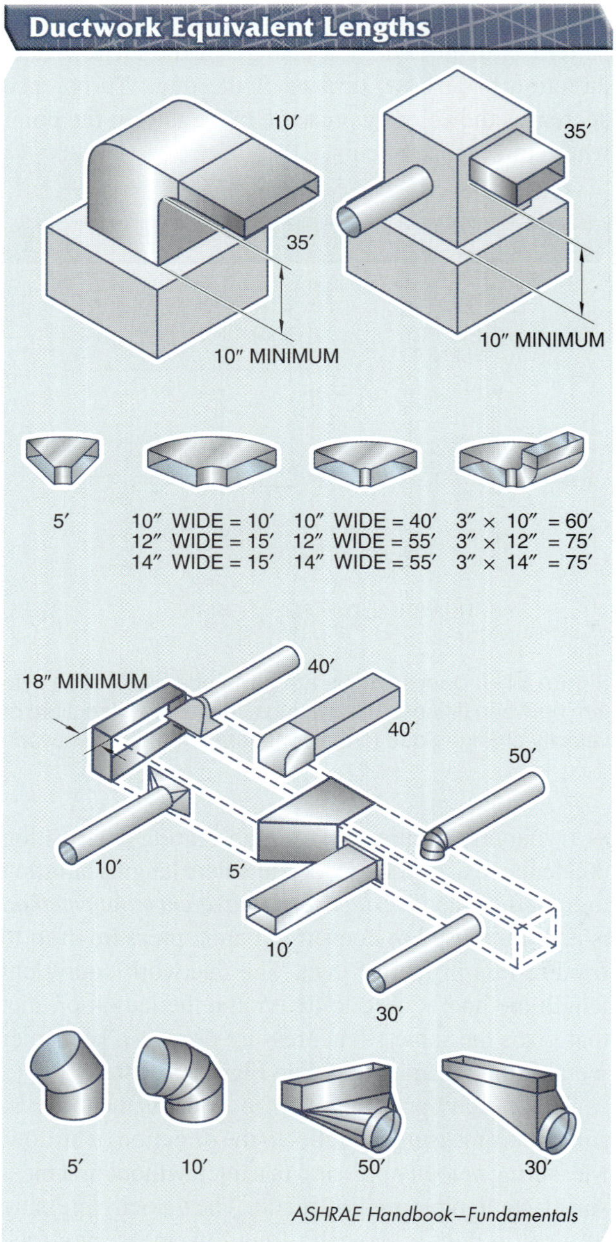

Figure 51-10. An equivalent length is the length of a duct section that gives the same static pressure drop as a particular fitting or transition piece.

TECH TIP

The minimum number of fittings should be used when laying out a duct system. Using an excessive number of duct fittings leads to an unsightly system and produces a higher-than-normal pressure drop. The higher-than-normal pressure drop can lead to reduced airflow and loss of comfort.

Example: Finding Velocity Pressure

A duct fitting handles 1200 cfm of air. The air velocity is 1500 fpm. Find the velocity pressure of the air entering the fitting.

$$P_v = \left(\frac{V}{4005}\right)^2$$

$$P_v = \left(\frac{1500}{4005}\right)^2$$

$$P_v = 0.3745^2$$

$$P_v = \mathbf{0.1403'' \ WC}$$

Static Pressure Drop through Fittings. The static pressure drop through a fitting is found by using a fitting loss coefficient. A fitting loss coefficient, or coefficient C, is a value that represents the ratio between the total static pressure loss through a fitting and the dynamic pressure at the fitting. Coefficient C is found using a fitting loss coefficient table. To find the static pressure drop through a fitting, coefficient C for the fitting is multiplied by the velocity pressure of the air entering the fitting. The static pressure drop through a fitting using coefficient C is found by applying the following formula:

$$P_d = P_v \times C$$

where

P_d = static pressure drop (in in. WC)

P_v = velocity pressure (in in. WC)

C = fitting loss coefficient

Example: Finding Static Pressure Drop through Fittings

An elbow with a 12″ radius has a velocity pressure of 0.1403″ WC. The fitting is an 18″ × 12″ smooth-radius elbow. Find the pressure drop through the fitting.

1. Find the coefficient for the smooth-radius elbow on a fitting loss coefficient table.
2. Multiply the velocity pressure by coefficient C. The coefficient C value of an 18″ × 12″ smooth-radius elbow is 0.19. **See Figure 51-11.**

$$P_d = P_v \times C$$

$$P_d = 0.1403 \times 0.19$$

$$P_d = \mathbf{0.0267'' \ WC}$$

Ductwork Coefficients

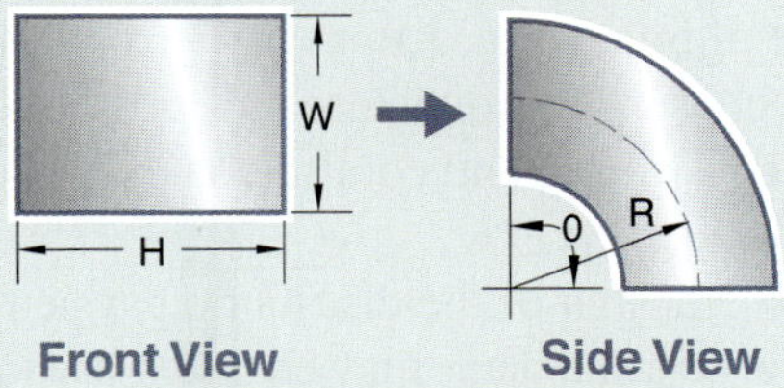

$C_o = K_\theta K_{Re} C_o'$

SMOOTH RADIUS

Coefficients for 90° Elbows (C_o')											
	H/W										
r/W	0.25	0.5	0.75	1	1.5	2	3	4	5	6	8
0.5	1.3	1.3	1.2	1.2	1.1	1.0	1.0	1.1	1.1	1.2	1.2
0.75	0.57	0.52	0.48	0.44	0.4	0.39	0.39	0.4	0.42	0.43	0.44
1.0	0.27	0.25	0.23	0.21	0.19	0.18	0.18	0.19	0.20	0.21	0.21
1.5	0.22	0.20	0.19	0.17	0.15	0.14	0.14	0.15	0.16	0.17	0.17
2.0	0.20	0.18	0.16	0.15	0.14	0.13	0.13	0.14	0.14	0.15	0.15

Angle Correction Factors											
θ	0	20	30	45	60	75	90	110	130	150	180
K_θ	0	0.31	0.45	0.60	0.78	0.90	1	1.13	1.20	1.28	1.40

Reynolds Number Correction Factors (K_{Re})									
	Re × 10^{-4}								
r/W	1	2	3	4	6	8	10	14	≥20
0.5	1.40	1.26	1.19	1.14	1.09	1.06	1.04	1	1
≥0.75	2	1.77	1.64	1.56	1.46	1.38	1.30	1.15	1

ASHRAE Handbook–Fundamentals

Figure 51-11. Fluid resistance coefficients provide the ratio of the total pressure loss to dynamic pressure loss at particular fittings or transitions.

DUCTWORK DESIGN

A forced-air heating system or air conditioning system is designed using a floor plan or sketch of the building. Each floor that has ductwork, registers, or grills requires a floor plan. To design the distribution system, the heating and cooling loads are calculated and the ductwork is sized.

Layout Drawings

A *layout drawing* is a drawing that shows the walls, partitions, windows, doors, fixtures, and other details of a building that affect the location of ductwork, registers, and grills. Layout drawings show the heating and cooling loads and required airflow rate for each room or zone. Because forced-air distribution systems are sized for a required airflow rate, the heating and cooling loads for a building must be converted to an airflow rate for building spaces. **See Figure 51-12.**

Layout Drawings

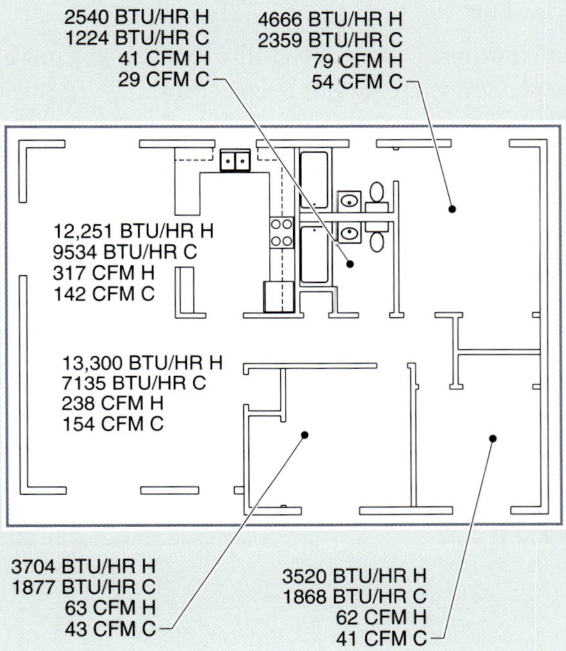

Figure 51-12. Layout drawings show the heating and cooling loads for each zone in a building.

Layout drawings include several overlays that show the placement of ductwork, registers, and grills on each building floor. Tentative locations of the registers and grills are shown on the overlay for each floor.

Register and Grill Locations. Registers are located based on the airflow pattern desired out of each register. Registers and grills are located so that all of the air supplied by the registers is returned to the system fans. Grills must be located where they are not covered by furniture or fixtures. Air must get back to the fans through the return ductwork. Therefore, grills must be located so that they are open at all times. Ceilings and walls are the best locations for grills because floor grills are easily covered with furniture or carpeting and collect dirt.

Grills should not be located near registers. When grills are located near registers, air from the registers will flow directly into the grills. This airflow upsets the distribution pattern of the registers. Grills should be located centrally in the building space. One or more grills may be used in a zone or building space to guarantee sufficient circulation. When some building spaces are isolated by partitions or doors, grills should be located in each building space. Grills may be connected to the same return ductwork. However, they must be located where they can return air to the system fans from all parts of the building.

Distribution Systems. The distribution system on a layout drawing is drawn on a separate overlay. The distribution system overlay shows the ductwork layout in relation to the registers and grills. Each register is connected to the supply ductwork, and each grill is connected to the return ductwork. The actual layout of a distribution system depends on the type of system chosen. On the drawings, ductwork must be located where it does not run through beams, columns, stairwells, elevator shafts, or where plumbing or electrical devices may be located. **See Figure 51-13.**

Distribution System Overlays

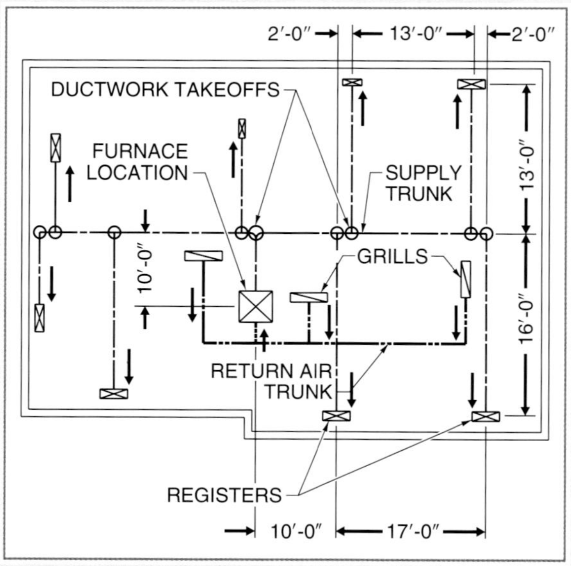

Figure 51-13. An air distribution system overlay shows the ductwork layout in relation to the registers and grills.

After layout drawings and overlays have been prepared, the distribution system is chosen based on the temperature control and sophistication level required for each building space. After the distribution system has been chosen, the registers and grills are chosen based on the heating and cooling loads of each building space.

When the design of the distribution system is complicated, large-scale plans may be necessary to identify the actual location of each duct section. Ductwork layout drawings should show all fittings and transitions in the distribution system. Fittings and transitions are indicated on a single-line drawing by an "X" marked on the duct drawing.

A *single-line drawing* is a drawing in which the walls and partitions are shown by a single line and does not indicate actual wall thickness. A *double-line drawing* is a drawing in which walls and partitions are shown using double lines and wall thickness is drawn to scale. Fittings and transitions are shown on double-line drawings in plan view as they actually appear.

Duct Section Identification. It is necessary to identify each duct section and keep track of the proper size ducts, dampers, and fittings of each section. A *duct section* is a section of ductwork between two fittings where the airflow rate changes. Duct sections are between takeoffs where air enters or leaves the duct. Duct sections are identified using letters or numbers. It is common to identify duct sections by marking them or marking the fittings at the end of each section.

When letters are used to identify fittings and duct sections, the duct sections between fittings are identified by the letters on the fittings. For example, all fittings are marked with a different letter. The first fitting, which is at the furnace or air conditioner, is identified as A. The fitting at the end of the first duct section is identified as B. The duct section between fittings A and B is identified as section A-B. The letters "A-B" are shown on the distribution system overlay for that duct. **See Figure 51-14.**

Duct Sizing

Ducts are sized to allow the required amount of airflow to each building space for proper heating or cooling. Airflow rates are controlled by the duct size, and ducts must be properly sized to carry the air at a given pressure loss.

Forced-air distribution systems often consist of major duct sections and individual duct sections. The major duct sections are connected to the supply plenum. The individual duct sections run off the major duct sections. When sizing duct and calculating pressure drop through a distribution system, each major duct section can be treated separately.

Distribution System Identification

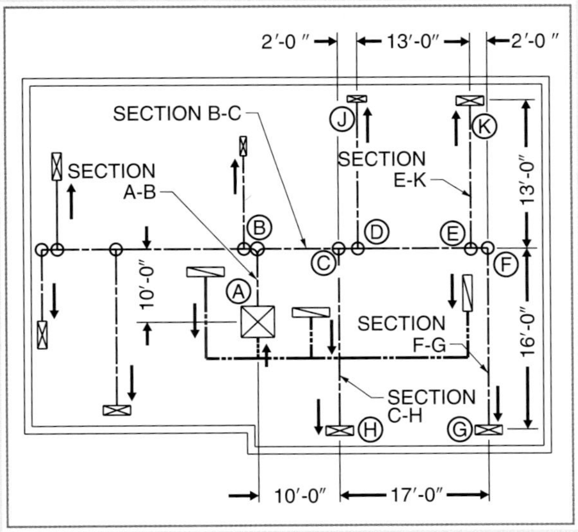

Figure 51-14. The duct sections between fittings are identified by the fitting letters.

The most common methods for sizing HVAC ductwork include the equal friction, velocity reduction, and static regain methods. Small distribution systems used in heating and air conditioning systems as well as small ventilation and exhaust systems are sized using either the equal friction method or velocity reduction method. Large sophisticated distribution systems are sized using the static regain method. In addition, the equal velocity method is typically used for sizing exhaust systems that carry particulate matter.

Supply plenums are typically sized to fit heating and cooling units. Trunk ducts are sized for the needed capacity. Branch ducts to individual registers are sized for the building space. Typically, branch ducts are round with a minimum diameter of 6″.

In the past, duct sizes were calculated manually or with duct sizing charts, graphs, or ductulators. Today, computer programs are typically used for sizing ducts. Small jobs are easily sized using conventional methods, but large jobs are more easily sized using computer duct-sizing programs. After all sizing is performed, the drawings and sizing information are given to the sheet metal fabricators and installers.

Equal Friction Method. The *equal friction method* is a method of sizing ductwork that sets static pressure approximately equal at each branch takeoff in a distribution system. The equal friction method involves using an equal friction chart to size the ductwork for a given static pressure drop.

To use an equal friction chart for sizing ductwork, the static pressure drop for the system is determined, and the chart is used to find the size of the duct. The static pressure drop can be selected based on the average pressure drop for the system. For larger systems, the static pressure drop is found by selecting an average air velocity and duct size for the first section of duct off the furnace or air conditioner. This data is used on equal friction charts to find the static pressure drop. This static pressure drop is used for sizing the other ducts in the distribution system.

Small distribution systems are typically designed for about 0.08″ WC to 0.10″ WC static pressure drop per 100′ of duct. In larger distribution systems, the static pressure drop per 100′ of duct may be greater.

Example: Duct Sizing—Equal Friction Method

The fan of a furnace used to heat a three bedroom residence moves 800 cfm of air. Size the ductwork using the equal friction method.

1. Prepare layout drawings and overlays that show the heating and cooling loads and airflow rates for each building space. **See Figure 51-15.**
2. Locate registers and grills on the first overlay.
3. Lay out the ductwork on the second overlay.
4. Enter the identifying letters for the duct sections in the first column of a columnar form.
5. Enter the airflow rate for each duct section in the second column. **See Figure 51-16.**
6. Select a design static pressure drop of 0.1″ WC for the forced-air distribution system.
7. Find the size for each duct section using an equal friction chart. Enter the values in the third column of the columnar form.
8. For the duct sections that should be rectangular, convert the round duct sizes to equivalent rectangular sizes. Enter the values in the fourth column of the columnar form.

Layout Drawings and Overlays

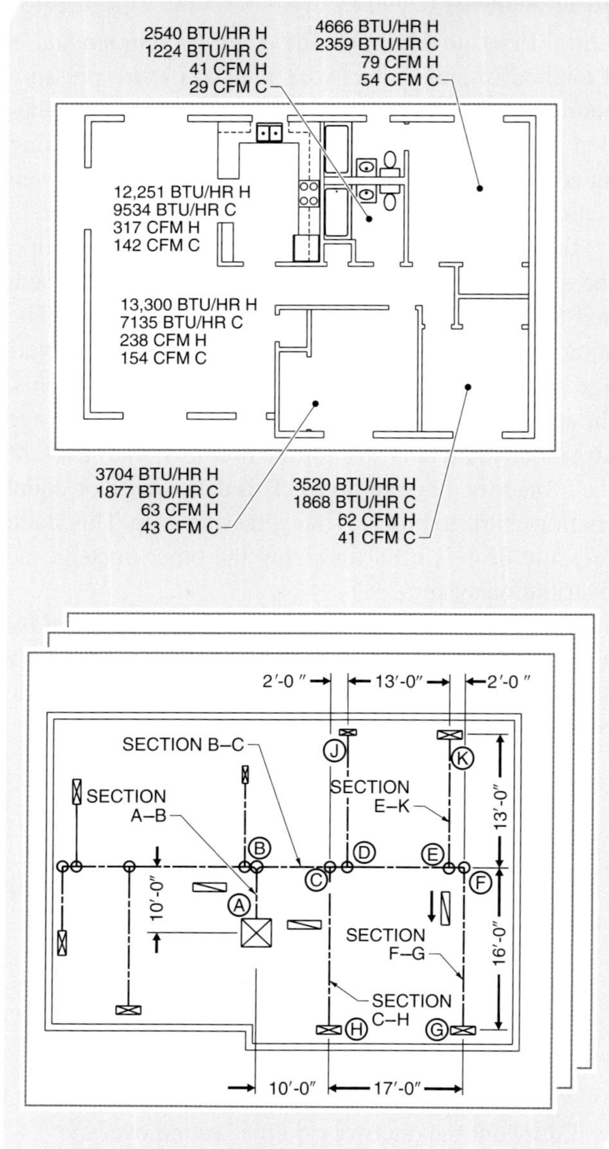

Figure 51-15. Layout drawings show the airflow rates for each building space. Overlays show the locations of the ductwork, registers, and grills.

Distribution System Forms

Project Name: Three Bedroom Residence
Location: Chicago, IL
HVAC System: Forced-Air Heating/Cooling System

Section	Airflow Rate*	Duct Size†	Actual Duct Size‡	Design Static Pressure Drop§
A-B	800	12.7	20 × 8	0.001
B-C	245	8.1	10 × 8	0.001
C-D	182	7.3	10 × 8	0.001
D-E	141	6.6	8 × 8	0.001
E-F	62	4.8	8 × 8	0.0004
F-G	62	4.8	6R	0.0004
C-H	63	4.9	6R	0.0004
D-J	41	4.1	6R	0.0003
E-K	79	5.3	6R	0.0006
TOTAL				

* in gpm
† in in. DIA
‡ in in.
§ in in. WC/100 ft
‖ in ft
in in. WC

Figure 51-16. Duct sections, airflow rates, and duct sizes are recorded in columns on distribution system forms or in spreadsheets.

Rectangular ducts are used in distribution systems where headroom is limited. The width of a rectangular duct typically is greater than the height. This allows the bottom of the duct to be higher from the floor of a building space. Trunk ducts are typically rectangular ducts. To function most efficiently, the aspect ratio of a rectangular duct should be less than three to four times the height. An *aspect ratio* is the ratio between the height and width of a rectangular duct.

When duct sizes are taken from an equal friction chart or ductulator, the actual sizes are recorded on the columnar form. However, for practical use, ducts of nominal size that are readily available and more economical to manufacture typically are used. When the actual size of a duct is not the same as the nominal size of an available duct, the next larger nominal size is used.

Velocity Reduction Method. The *velocity reduction method* is a method of sizing ducts that causes air velocity to be reduced at each branch takeoff. The air velocity in the first duct section nearest to the furnace and the size of the first duct section are selected based on the airflow needs of the system. The other duct sections are sized so that the air velocity is reduced proportionately in each section. Tables show recommended air velocities in different duct sections. When a distribution system is sized using the velocity reduction method, the design static pressure drop may be different in each duct section. **See Figure 51-17.**

Duct Air Velocity

Velocity	Use	Recommended Duct Air Velocities*					
		Residential Buildings		Commercial Buildings		Industrial Buildings	
		Recommended	Maximum	Recommended	Maximum	Recommended	Maximum
Total face area	Filters	250	300	300	350	350	350
	Coils	450	500	500	600	600	700
	Intakes	500	800	500	900	500	1200
	Air washers	650	650	650	650	650	650
Net free area	Ducts						
	Main	800	1000	1500	1600	1500	2200
	Branch	600	850	750	1300	900	1800
	Risers	500	800	650	1200	800	1300
	Outlets	1300	1700	1650	2200	2000	2800

* in fpm

Figure 51-17. Recommended duct air velocities are used when sizing ductwork using the velocity reduction method.

Static Regain Method. The *static regain method* is a method of sizing ducts in which each duct section is sized so that the static pressure increase at each takeoff offsets the friction loss in the preceding duct section. The static regain method is used for sizing any duct in long distribution systems with many takeoffs or with registers located on the duct itself.

The static regain method causes approximately the same static pressure at each branch takeoff and can result in large ducts with low air velocity at the ends of long runs. The static regain method is used for larger distribution systems and requires many calculations for each fitting and duct section.

Equal Velocity Method. The *equal velocity method* is a method of sizing ducts that causes each duct section to have the same air velocity. Ducts are sized with a ductulator and velocity scale. Sizing may be done mathematically using air velocity and a cross-sectional area of each duct. The equal velocity method is more suitable for sizing exhaust systems or material collection systems than it is for sizing distribution ducts of heating or air conditioning systems.

Fan Sizing

When designing forced-air distribution systems, fans must have the capacity to move air at the required airflow rates against the resistance in distribution systems. When sizing fans for distribution systems, the required airflow rates and total allowable static pressure drop for the systems are found on fan performance tables. The velocity of the fan wheel and the motor horsepower required to turn the fan to produce the required airflow rate are determined from the fan performance table.

Fan Airflow Rates. The fans in forced-air distribution systems must move air at the airflow rates required for the system and against airflow resistance. The airflow rate required is determined when heating and air conditioning equipment for the system is chosen.

Resistance to airflow in distribution systems is due to friction in the ducts and dynamic pressure losses through fittings and transitions in the ducts. Resistance to airflow occurs in both the supply and return ducts of distribution systems. Fans must be large enough to move the required airflow in the major section of supply and return ductwork having the greatest resistance to flow. The total resistance to airflow in a distribution system is calculated as static pressure drop in the system.

When fans circulate air through more than one major section of supply ductwork and one major section of return ductwork in a system, the fan does not have to overcome the sum of the static pressure drops in all of the sections. Fans must overcome the static pressure drops in the supply and return sections having the largest static pressure drops.

Fan Pressure Drops. When sizing fans for forced-air distribution systems, it is not necessary to calculate the pressure drop through all the sections of a distribution system having more than one major duct section running to or from the fan. It is necessary to find the supply duct section and return duct section with the highest static pressure drop. The two static pressure drop values are added together to find the total static pressure drop against which the fan has to work.

To calculate the total static pressure drop through a major duct section that has been designed using the equal friction method, the actual length of the duct section that has the greatest resistance to airflow is found. The length is taken from the layout drawing. The equivalent length of each fitting or transition in the section is found from ductwork equivalent lengths. The equivalent length of each fitting is added to the actual length of duct in the section. The result is the total equivalent length of that section of the distribution system.

The total equivalent length of the duct section is divided by 100′ and multiplied by the design static pressure drop that was used to size the duct. The result is the static pressure loss for that duct section.

To find the total static pressure drop through one section of a forced-air distribution system, the columnar form is first used to size the duct. A column is added for the actual length of each duct section and fitting equivalent lengths. **See Figure 51-18.** The actual duct length of each section from the overlays is entered into the column.

In the next column, the equivalent lengths of the fittings in each duct section are entered. Next, the lengths of the duct sections are added. The result is the total actual duct length in the section. The equivalent lengths of the fittings are added to find the total equivalent length of fittings. The total actual duct length and total equivalent fitting length are added to get the subtotal equivalent length of the major duct section.

The total equivalent length is divided by 100 and multiplied by the design static pressure drop to get the static pressure drop through the duct section. The sections are added together to calculate the greatest static pressure drop. This step is repeated for the section of return ductwork that appears to have the greatest pressure drop. The sum of the pressure drop for the supply ductwork section and the return ductwork section is the total pressure drop for the ductwork in the system.

The static pressure drop for filters, coils, dampers, and other devices that add resistance to airflow in the distribution system should be added to the duct pressure drop. The pressure drop through an accessory is found on the specifications sheets. The total static pressure drop for a major duct section is the duct pressure drop plus the pressure drop through accessories in that section.

Distribution System Duct Lengths

Project Name: Three Bedroom Residence
Location: Chicago, IL
HVAC System: Forced-Air Heating/Cooling System

Section	Airflow Rate*	Duct Size†	Actual Duct Size‡	Design Static Pressure Drop§	Section Length‖	Fittings	Fitting Equivalent Length‖	Total Equivalent Length‖	Subtotal Pressure Drop#
A-B	800	12.7	20 × 8	0.001	10	PT	10	20	0.020
B-C	245	8.1	10 × 8	0.001	10	Splitter	40	50	0.050
C-D	182	7.3	10 × 8	0.001	2		0	2	0.002
D-E	141	6.6	8 × 8	0.001	13	Reducer	5	18	0.018
E-F	62	4.8	8 × 8	0.0004	2		0	2	0.001
F-G	62	4.8	6R	0.0004	16	BT, Br, Boot	60	76	0.031
C-H	63	4.9	6R	0.0004	16	BT, Br, Boot	60	76	0.031
D-J	41	4.1	6R	0.0003	13	BT, Br, Boot	60	73	0.022
E-K	79	5.3	6R	0.0006	13	BT, Br, Boot	60	73	0.044
TOTAL					95		295	390	0.166

* in gpm
† in in. DIA
‡ in in.
§ in in. WC/100 ft
‖ in ft
\# in in. WC

PT = Plenum takeoff
BT = Branch takeoff
Br = Branch

Figure 51-18. The sum of the lengths of duct sections and the fitting equivalent lengths is the total duct length, which is used to determine the total static pressure drop.

When a distribution system has been designed using a method other than the equal friction method, the static pressure drops in each duct section between fittings may be different. To find the total static pressure drop in a major duct section, the static pressure drop in each individual duct section must be found first. The individual section pressure drops are then added together.

The pressure drops through fittings and transitions are calculated by using the equivalent length method or coefficient *C*. The result is added to the sum of the section pressure drops. The pressure drop through accessories is added to the duct section and the fitting pressure drop to find the total major section pressure drop. **See Procedure for Finding Static Pressure Drop in Duct Sections.**

The result is the static pressure drop in that section of the distribution system. The pressure drop of accessories in the section, which add resistance to airflow, must be added to the duct section pressure drop to determine the total static pressure drop in the section.

Register and Grill Sizing

Registers and grills are selected and sized to allow the proper amount of airflow into each building space to offset heat loss or gain. Registers and grills are located to efficiently distribute the air into building spaces.

Register Sizing. Sizes of registers are determined by airflow rates required from the register and the required air velocity in building spaces. Airflow rates from registers are a function of the size of the free area of the register. *Register free area* is the face area of a register minus the area blocked by the frame or vanes. The air velocity at a register is a function of the register free area and the airflow rate. The air velocity at the register determines the distance the air is thrown from the register.

TECH TIP

Registers may be considered unsightly in some buildings. For this reason, contractors may install registers that are too small, which degrade system performance, increase energy use, and reduce comfort.

Finding Static Pressure Drop in Duct Sections

1. Find the static pressure drop through an individual duct section with duct that is 16′ long and has a design flow rate of 85 cfm. Using an equal friction chart, find the 85 cfm flow rate along the bottom of the chart and draw a line up.
2. Set the friction loss at 0.10″ WC/100′ as the design static pressure drop. Find the 0.10″ WC/100′ along the left side of the chart and draw a line across.

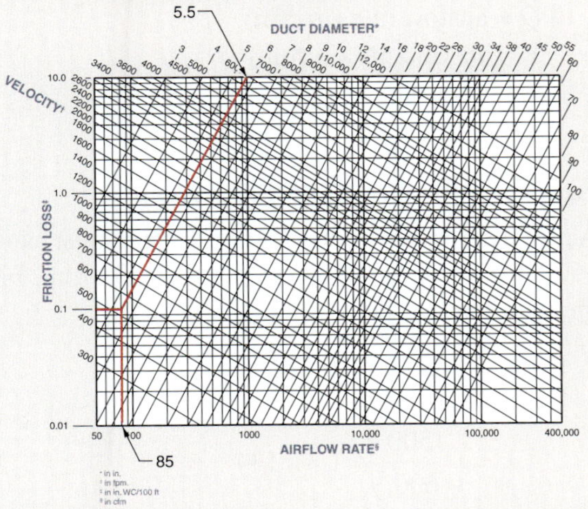

3. The intersection of the two lines is the design point of the duct. Draw a line upward to the right to find the duct size. In this case, the duct size is 5.5″. Draw a line upward to the left to find the air velocity. In this case, the air velocity is 540 fpm.
4. In this case, the duct size is less than the minimum size. Use a minimum size 6″ duct. Note that the 6″ duct results in a static pressure drop of about 0.07″ WC/100′ and an air velocity of 450 fpm. This is acceptable.
5. Calculate the actual length of the duct section from the duct layout drawing. This is the length of the duct measured from the centerline of the fitting at the beginning of the duct section to the centerline of the fitting at the end of the duct section. In this case, the duct is 16′ from the trunk to the register.
6. Determine the equivalent length of fittings in the section using an equivalent length chart.

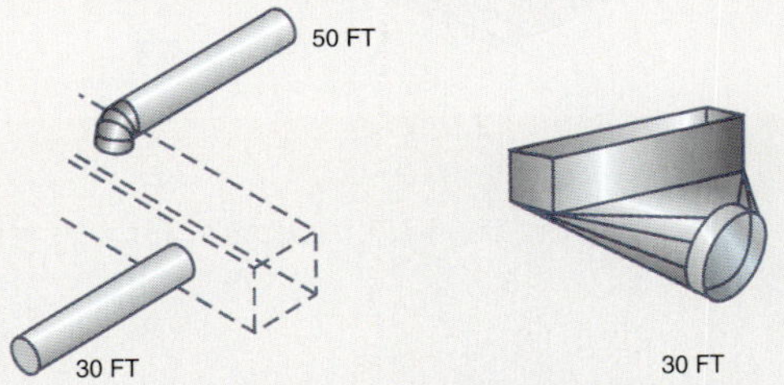

7. Add the equivalent lengths of the fittings to the actual lengths of the section. In this case, there is a riser connection to the trunk (50′) and a register boot (30′). The sum is the total equivalent length of the section, or 96′.
8. Divide the total equivalent length of the section by 100 and multiply by the design static pressure drop for the system. In this case, (96/100) × 0.10 = 0.096″ or 0.10″ WC.

Registers are designed to deliver air in various patterns. Air patterns are defined by throw and spread. Throw is the distance that air travels directly away from a register. Throw is measured to the point where the air velocity drops below 50 fpm. Spread is the distance measured across the envelope of air that spreads out from a register. Spread is the distance on each side of the air envelope where the air velocity drops below 50 fpm. The pattern of airflow out of registers is controlled by the vanes on the panel. **See Figure 51-19.**

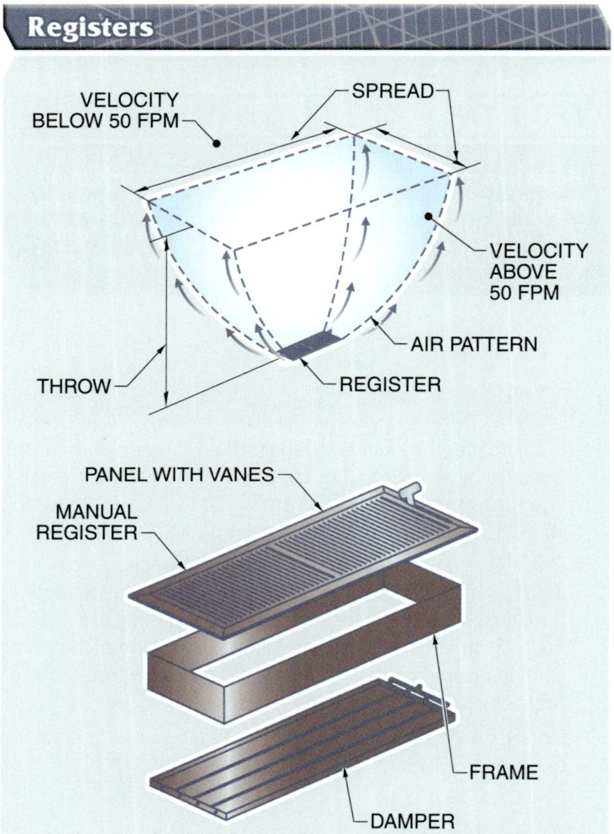

Figure 51-19. Registers distribute supply air into building spaces.

To size a register, the general type of register is chosen. The airflow rate required from the register, throw and spread characteristics, and the air velocity required to achieve those characteristics are determined. The register that produces these characteristics is found in a catalog. Various types and sizes of supply registers may be used in each building space. Registers should be chosen for air pattern and sized to provide complete coverage of the room with conditioned air. Registers are sized so that the face velocity is not high enough to produce noise. Catalogs are typically marked to indicate the air delivery rate at which noise is produced.

Grill Sizing. Grills are sized according to the velocity of the air at the grill face. Each grill must return air from building spaces at a low face velocity to prevent noise. **See Figure 51-20.** Face velocities of 500 fpm to 750 fpm are typically allowed when sizing grills.

To size a grill, the airflow rate through a grill is divided by the desired face velocity (500 fpm to 750 fpm) and multiplied by a factor of 1.3. The factor of 1.3 allows for 30% of the grill face area to be taken up by the frame and vanes. The result is the free area of the face of the grill in feet. Grills are sized in inches. To find the grill size in inches, the grill free area in square feet must be converted to square inches. The constant 144 is used to convert square foot to square inches. Grill free area is found by applying the following formula:

$$A_g = \left(\frac{Q}{V}\right) \times 1.3 \times 144$$

where

A_g = grill free area (in in^2)

Q = airflow rate (in cfm)

V = air velocity (in fpm)

1.3 = constant

144 = constant

Example: Finding Grill Free Area

A grill in a forced-air distribution system must handle 1500 cfm of air at a face velocity of 600 fpm. Find the required free area.

$$A_g = \left(\frac{Q}{V}\right) \times 1.3 \times 144$$

$$A_g = \left(\frac{1500}{600}\right) \times 1.3 \times 144$$

$$A_g = 2.5 \times 1.3 \times 144$$

$$A_g = 3.25 \times 144$$

$$A_g = \mathbf{468\ in^2}$$

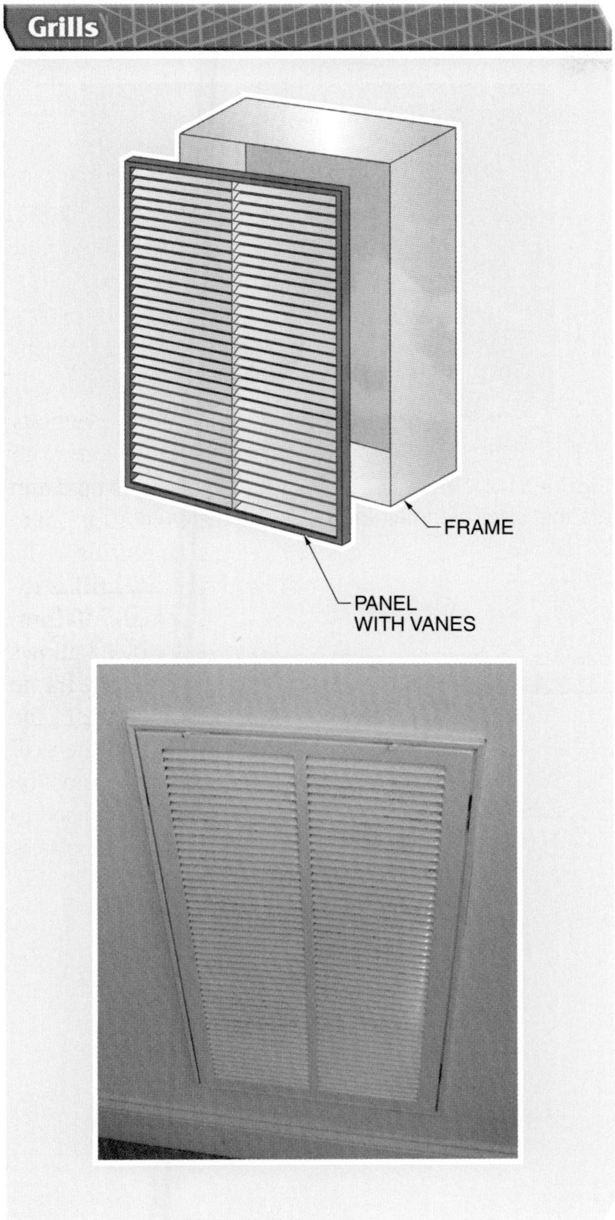

Figure 51-20. Grill sizing is based on air velocity.

When a square grill is required, the size of the grill is found by taking the square root of the grill free area. The size of a square grill is found by applying the following formula:

$$G_S = \sqrt{A_g}$$

where

G_S = size of square grill (in in.)

A_g = grill free area (in in^2)

Example: Finding Square Grill Size

A grill has a free area of 468 in^2. A square grill is desired. Find the size of the grill.

$$G_S = \sqrt{A_g}$$

$$G_S = \sqrt{468}$$

$$G_S = \mathbf{21.6''}$$

Note: The nominal grill size is 22″ × 22″.

When a rectangular grill is required, the area in square inches should be divided by the dimension of one side of the grill, and the result is the dimension of the other side of the grill. The size of a rectangular grill is found by applying the following formula:

$$w = \frac{A_g}{h}$$

where

w = width (in in.)

A_g = grill free area (in in^2)

h = height (in in.)

Example: Finding Rectangular Grill Size

A grill in a distribution system must handle 1000 cfm of air at a face velocity of 500 fpm. The height must be 14″. Find the width of the grill.

1. Find the free area of the grill.

$$A_g = \left(\frac{Q}{V}\right) \times 1.3 \times 144$$

$$A_g = \left(\frac{1000}{500}\right) \times 1.3 \times 144$$

$$A_g = 2 \times 1.3 \times 144$$

$$A_g = 2.6 \times 144$$

$$A_g = \mathbf{374\ in^2}$$

2. Find the width of the grill.

$$w = \frac{A_G}{H}$$

$$w = \frac{374}{14}$$

$$w = \mathbf{27''}$$

After the duct, register, and grill sizing for a building is complete, the sizes are shown on the distribution system plans. The duct size in each section and the sizes of all fittings and transitions are shown on the plan.

VAV Terminal Sizing

Light commercial HVAC systems that use VAV systems can control duct static pressures at around 1″ WC of static pressure. **See Figure 51-21.** The discharge air temperature stays constant, usually around 55°F. VAV terminal boxes are primarily sized for cooling loads of the building space. VAV terminal boxes are sized for the airflow (in cfm) needed at maximum cooling loads. Heating coils are typically added to VAV boxes located on the outside, or perimeter, of buildings. Heating coils typically use hot water as a heating source. However, electric coils are increasing in use. A minimum airflow across the heating coil must be maintained in order to transfer enough heat for building spaces.

VAV Dampers

Figure 51-21. VAV terminal units include motorized dampers that control the airflow into building spaces.

Digital Resources ATPeResources.com/QuickLinks • Access Code: 482610

UNIT 52

Hydronic System Design

Hydronic systems use water as a heat transfer medium. For heating, heat is generated in boilers and transferred to the water circulating in closed pipe systems. For cooling, heat is transferred to chillers from the water in the circulating piping of hydronic air conditioning systems. Hydronic system components must be properly selected and sized for the efficient operation of heat transfer systems.

OBJECTIVES

- Define the terms used to describe water flow and water pressure.
- Explain the use of equivalent length of pipe charts when installing hydronic distribution systems.
- Identify the temperature differences in common hydronic system designs.
- Explain how pipe sections are identified on building plans.
- Perform typical hydronic system component sizing calculations.
- Describe the typical components of hydronic control systems.

Index to Procedures

WATER DISTRIBUTION SYSTEMS

Water distribution for hydronic systems is circulated through piping systems. Piping systems distribute water from the boiler or chiller to the terminal devices and back. Piping systems typically consist of iron or steel pipe or copper tubing. Plastic tubing is also used when embedded in concrete flooring. Pumps, fittings, and valves control the water flow. The piping system is typically installed in the frame of a building. The piping system must be designed to distribute the required amount of water to each building space or zone to offset the heating or cooling loads. **See Figure 52-1.**

Principles of Water Flow

The rate of water flow through a pipe or an orifice is a function of the pressure exerted on the water and the size and material of the pipe. Water flow is expressed in gallons per minute (gpm). The water flow velocity depends on the rate of water flow and the cross-sectional area of the pipe. Water flow velocity is expressed in feet per second (fps). For a given size of pipe, greater water pressure increases water flow velocity. This is important to note in individual devices and sections of pipe. Water pressure is typically expressed in pounds per square inch (psi) or feet of head.

Hydronic Systems

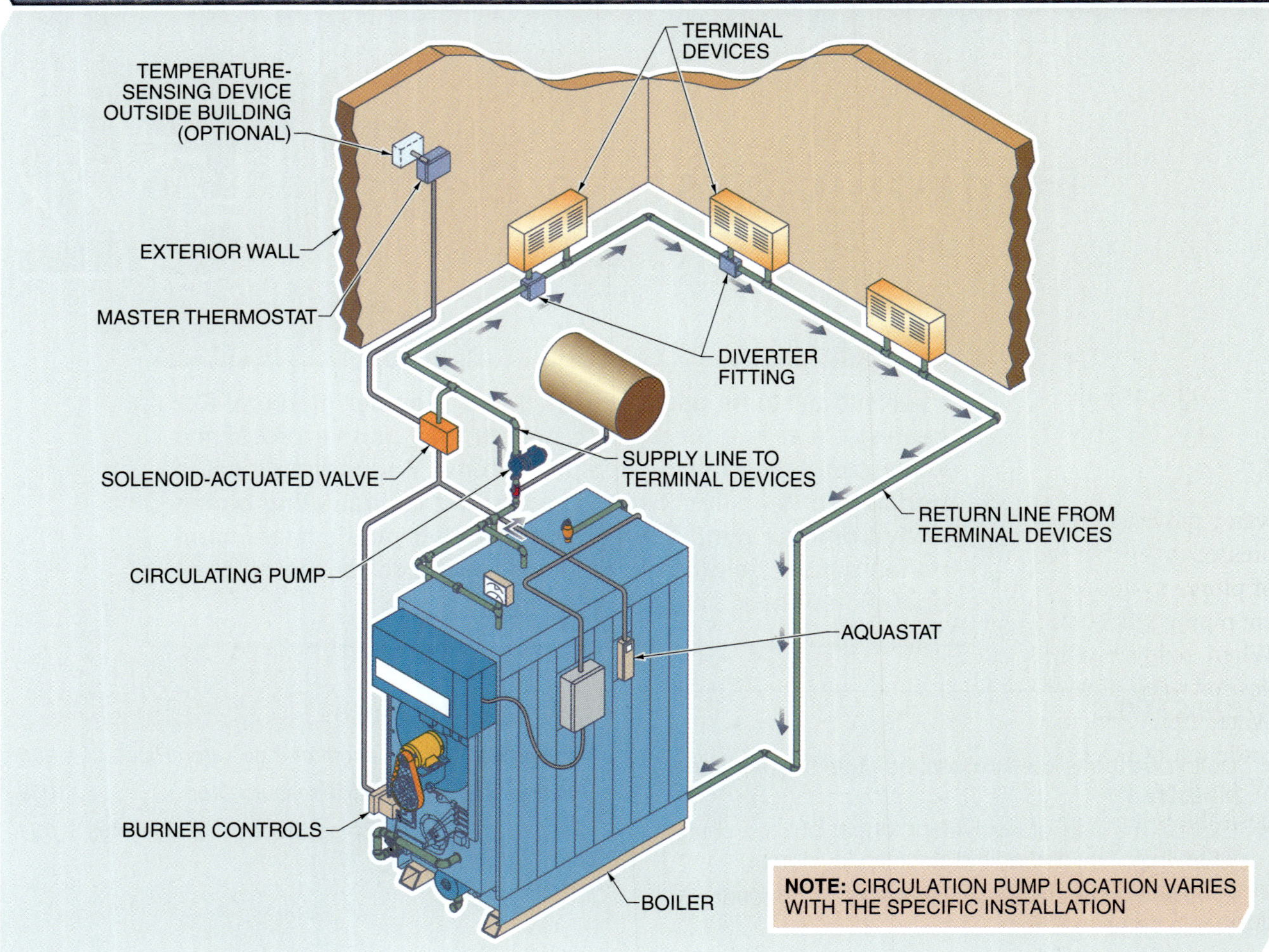

Figure 52-1. To heat building spaces, hydronic systems produce and regulate the circulation of hot water from boilers to terminal devices. Hydronic systems can be designed for heating and/or cooling.

Pipe or tube size is indicated by the nominal diameter of the pipe or tube and is typically expressed in inches (in.). *Nominal diameter* is the approximate distance from outside edge to outside edge of pipes or inside edge to inside edge of tubes. Manufacturer charts provide the actual dimensions of pipes and tubes.

Water Flow Rates. *Water flow rate* is the volumetric flow rate of water as it moves through a given pipe section. Depending on the type and configuration of the piping system, the flow rate may vary in different pipe sections within a piping distribution system.

When a piping system is sized for heating and cooling, the larger required flow rate is used for sizing the components for the system. Heating and cooling loads for buildings are calculated based on the local climate. Either the heating or the cooling load may require the larger flow rate.

Commercial water suppliers measure water flow rates with meters. For design purposes, water flow rates are found mathematically. The water flow rates needed in hydronic systems are based on the temperature difference of the water at two points in the system and the amount of heat transferred during the change of temperature. The water flow rate is found by applying the following formula:

$$Q = \frac{H}{500 \times \Delta T}$$

where

Q = water flow rate (in gpm)

H = heat transferred (in Btu/hr)

500 = constant

ΔT = temperature difference (in °F)

Example: Finding Water Flow Rate—Heating

A heating system has a total adjusted heat loss of 350,000 Btu/hr. A temperature difference of 20°F is typically used. Find the water flow rate.

$$Q = \frac{H}{500 \times \Delta T}$$

$$Q = \frac{350,000}{500 \times 20}$$

$$Q = \frac{350,000}{10,000}$$

$$Q = \mathbf{35\ gpm}$$

Water Flow Velocity. *Water flow velocity* is the speed water moves through a piping section, typically expressed in feet per second (fps). Depending on the type of piping system, the flow velocity may vary in different piping sections within a piping distribution system. When sizing piping based on design flow rates, the desired water flow velocity is between 2.5 fps and 8 fps. Water flow velocities below 2 fps allow sediment to settle out of the water and build up within the system. Water flow velocities above 10 fps typically create undesirable sounds and may damage the piping.

Typically, ½″ piping is the smallest size allowed to avoid plugging. When systems have little flow, the desired flow velocity cannot be maintained. Typically, small piping is sized based on the head loss range.

Water flow velocity in hydronic systems is based on the water flow rate between two points in the system and the size of piping. Water flow velocity is found by applying the following formula:

$$V = 0.4085 \times \frac{Q}{D^2}$$

where

V = water flow velocity (in fps)

0.4085 = constant

Q = water flow rate (in gpm)

D = diameter of piping (in in.)

TECH TIP

Chemicals added to the water in hydronic systems change the characteristics of the water and attention must be given to any resulting changes in system operation. An example is the glycol used in some chiller systems.

Example: Finding Water Flow Velocity

A heating system with a 2″ steel pipe has a water flow rate of 35 gpm. Find the water flow velocity in the pipe.

$$V = 0.4085 \times \frac{Q}{D^2}$$

$$V = 0.4085 \times \frac{35}{2^2}$$

$$V = 0.4085 \times \frac{35}{4}$$

$$V = 0.4085 \times 8.75$$

$$V = \mathbf{3.57\ fps}$$

Water Pressure

Water pressure is force exerted by water per unit of area. Water pressure is typically expressed in pounds per square inch (psi) or feet of head. Feet of head is the height of a column of water that would be supported by a given pressure. For example, one foot of head is the pressure equal to the weight of a column of water one foot high. **See Figure 52-2.** Water pressure in feet of head is typically measured using manometers or pressure gauges.

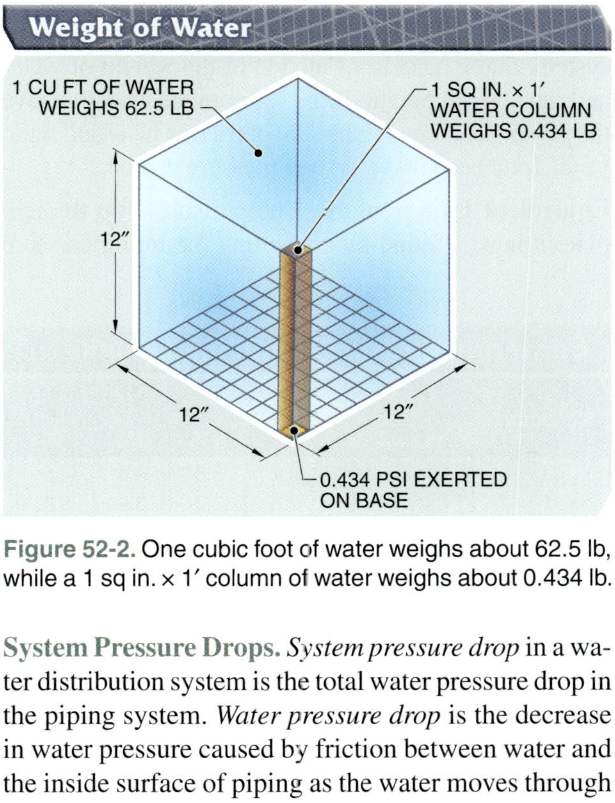

Figure 52-2. One cubic foot of water weighs about 62.5 lb, while a 1 sq in. x 1′ column of water weighs about 0.434 lb.

System Pressure Drops. *System pressure drop* in a water distribution system is the total water pressure drop in the piping system. *Water pressure drop* is the decrease in water pressure caused by friction between water and the inside surface of piping as the water moves through the piping. System pressure drop is the difference in water pressure between the points where water enters the piping system and where water leaves the piping system. System pressure drop is the resistance against which pumps move the water. System pressure drop is typically expressed in feet of head loss or psi.

For water to flow through a pipe, pressure must be exerted on the water at the inlet end of the pipe. The pressure must be great enough to raise the water upward in vertical columns of pipe to the highest point in the system, or the static head of the system. This pressure must also overcome the system pressure drop. The sum of the friction head and static head forces is the total head of the system.

Friction head is the effect of friction in a pipe. Friction occurs between the water moving through the pipe and the interior surfaces of the pipe. In a straight pipe section, the friction is a function of the size of the pipe, amount of water flow, and the roughness of the inner surface of the pipe. The amount of friction is proportional to the length of pipe and increases through fittings. Friction head can be calculated mathematically or found on tables that give pipe sizes relative to water flow rate and pressure drop.

Static head is the weight of water in a vertical column above a datum line. A *datum line* is the elevation at which the water pressure in a system is the same as atmospheric pressure. The pressure exerted by water in a vertical column, which is represented by the weight of the water above the datum line, is the static head of the system. Static head is a function of the weight of water and the vertical height of the water in the system above the pump. *Total head* is the sum of friction head and static head. Total head is the system pressure drop.

Equivalent Pipe Lengths. The pressure drop through pipe fittings is found by converting the fitting pressure drop to the equivalent length of pipe that would produce the same pressure drop. A table of elbow equivalents is typically used. **See Figure 52-3.**

The equivalent length of pipe is based on the pipe size and velocity of the water. The elbow equivalent value found on the table is multiplied by the equivalent length of pipe for standard 90° elbows. **See Figure 52-4.** The equivalent length of pipe is found by applying the following formula:

$$L = E \times N$$

where

L = equivalent length of pipe for fittings (in ft)

E = equivalent length of pipe for standard 90° elbows (in ft)

N = equivalent number of standard 90° elbows

Converting Fittings to Equivalent Elbows

Fitting	Iron Pipe	Copper Tubing
Elbow, 90°	1.0	1.0
Elbow, 40°	0.7	0.7
Elbow, 90° long turn	0.5	0.5
Elbow, welded, 90°	0.5	0.5
Reduced coupling	0.4	0.4
Open retun bend	1.0	1.0
Angle radiator valve	2.0	3.0
Radiator or convector	3.0	4.0
Boiler or heater	3.0	4.0
Open gate valve	0.5	0.7
Open globe valve	12.0	17.0

ASHRAE Handbook–Fundamentals

Figure 52-4. Tables can be used to convert elbow and fitting pressure drop equivalents to the number of 90° elbows with the same pressure drop.

Converting to Equivalent Pipe Lengths

Velocity†	Pipe Size														
	½	¾	1	1¼	1½	2	2½	3	3½	4	5	6	8	10	12
1	1.2*	1.7*	2.2*	3.0*	3.5*	4.5*	5.4*	6.7*	7.7*	8.6*	10.5*	12.2*	15.4*	18.7*	22.2*
2	1.4	1.9	2.5	3.3	3.9	5.1	6.0	7.5	8.6	9.5	11.7	13.7	17.3	20.8	24.8
3	1.5	2.0	2.7	3.6	4.2	5.4	6.4	8.0	9.2	10.2	12.5	14.6	18.4	22.3	26.5
4	1.5	2.1	2.8	3.7	4.4	5.6	6.7	8.3	9.6	10.6	13.1	15.2	19.2	23.2	27.6
5	1.6	2.2	2.9	3.9	4.5	5.9	7.0	8.7	10.0	11.1	13.6	15.8	19.8	24.2	28.8
6	1.7	2.3	3.0	4.0	4.7	6.0	7.2	8.9	10.3	11.4	14.0	16.3	20.5	24.9	29.6
7	1.7	2.3	3.0	4.1	4.8	6.2	7.4	9.1	10.5	11.7	14.3	16.7	21.0	25.5	30.3
8	1.7	2.4	3.1	4.2	4.9	6.3	7.5	9.3	10.8	11.9	14.6	17.1	21.5	26.1	31.0
9	1.8	2.4	3.2	4.3	5.0	6.4	7.7	9.5	11.0	12.2	14.9	17.4	21.9	26.6	31.6
10	1.8	2.5	3.2	4.3	5.1	6.5	7.8	9.7	11.2	12.4	15.2	17.7	22.2	27.0	32.0

* in ft
† in fps

ASHRAE Handbook–Fundamentals

Figure 52-3. Tables can be used to convert the pressure drop of 90° elbows to equivalent pressure drop per length of straight pipe.

For example, each standard 90° elbow is equivalent to 8.9′ of iron pipe. Each open globe valve is equivalent to 12.0 standard 90° elbows. The equivalent length of pipe for an open globe valve is calculated as follows:

$$L = E \times N$$
$$L = 8.9 \times 12.0$$
$$L = \mathbf{106.8'}$$

The equivalent length of an entire piping system can be found by converting all fittings to elbow equivalents. The elbow equivalents are converted to equivalent lengths of pipe. The equivalent length of a piping system is the sum of the equivalent lengths of pipe and the straight lengths of pipe. **See Procedure for Finding Equivalent Length of Pipe—Iron Pipe.**

The piping to hydronic units must be sized correctly to provide the proper amount of heating and cooling.

Finding Equivalent Length of Pipe—Iron Pipe

Procedure

1. Find the number of each type of fitting in a piping system with the same pressure drop as a system with a water flow rate of 6 fps in 3″ iron pipe.

   standard 90° elbows = 10 fittings

   globe valves = 1 fittings

   gate valves = 2 fittings

   straight pipe length = 400′

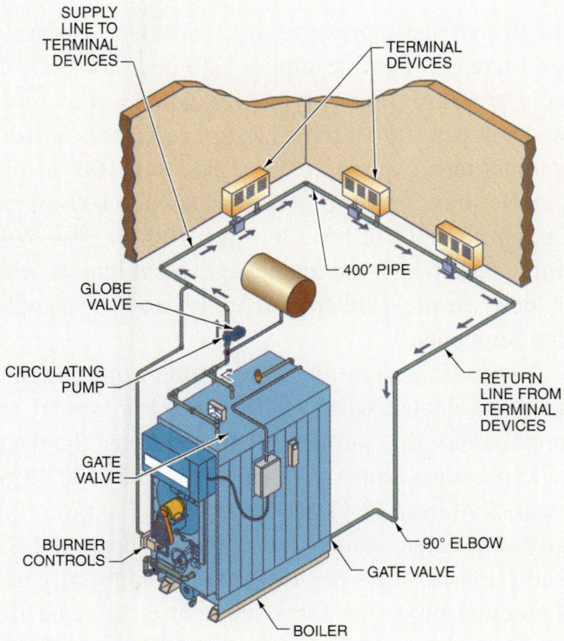

2. Find the equivalent length of pipe for a standard 90° elbow on a table of equivalent lengths. According to the table, each 90° elbow equals 8.9′ of pipe. Multiply the result by the number of equivalent elbows, which is also found on the table.

   $L_{elbow} = E \times N$

   $L_{elbow} = 8.9 \times 1.0$

   $L_{elbow} = \mathbf{8.9'}$

CONVERTING LENGTH OF PIPE FOR 90° ELBOWS

Velocity†	Pipe Size							
	½	¾	1	1¼	1½	2	2½	3
1	1.2*	1.7*	2.2*	3.0*	3.5*	4.5*	5.4*	6.7*
2	1.4	1.9	2.5	3.3	3.9	5.1	6.0	7.5
3	1.5	2.0	2.7	3.6	4.2	5.4	6.4	8.0
4	1.5	2.1	2.8	3.7	4.4	5.6	6.7	8.3
5	1.6	2.2	2.9	3.9	4.5	5.9	7.0	8.7
6	1.7	2.3	3.0	4.0	4.7	6.0	7.2	8.9
7	1.7	2.3	3.0	4.1	4.8	6.2	7.4	9.1
8	1.7	2.4	3.1	4.2	4.9	6.3	7.5	9.3
9	1.8	2.4	3.2	4.3	5.0	6.4	7.7	9.5
10	1.8	2.5	3.2	4.3	5.1	6.5	7.8	9.7

Velocity†	Pipe Size						
	3½	4	5	6	8	10	12
1	7.7	8.6	10.5	12.2	15.4	18.7	22.2
2	8.6	9.5	11.7	13.7	17.3	20.8	24.8
3	9.2	10.2	12.5	14.6	18.4	22.3	26.5
4	9.6	10.6	13.1	15.2	19.2	23.2	27.6
5	10.0	11.1	13.6	15.8	19.8	24.2	28.8
6	10.3	11.4	14.0	16.3	20.5	24.9	29.6
7	10.5	11.7	14.3	16.7	21.0	25.5	30.3
8	10.8	11.9	14.6	17.1	21.5	26.1	31.0
9	11.0	12.2	14.9	17.4	21.9	26.6	31.6
10	11.2	12.4	15.2	17.7	22.2	27.0	32.0

* in ft
† in fps

ASHRAE Handbook—Fundamentals

continued on next page

Procedure

Finding Equivalent Length of Pipe—Iron Pipe *(continued)*

3. Find the equivalent length of pipe for the fully open globe valve by multiplying the equivalent length of pipe for a 90° elbow by the number of equivalent elbows. Each fully open globe valve equals 12.0 standard 90° elbows.

$L_{globe} = E \times N$

$L_{globe} = 8.9 \times 12.0$

$L_{globe} = \mathbf{106.8'}$

CONVERTING LENGTH OF PIPE FOR 90° ELBOWS		
Fitting	**Iron Pipe**	**Copper Tubing**
Elbow, 90°	1.0	1.0
Elbow, 40°	0.7	0.7
Elbow, 90° long turn	0.5	0.5
Elbow, 90°, welded	0.5	0.5
Reduced coupling	0.4	0.4
Open retun bend	1.0	1.0
Angle radiator valve	2.0	3.0
Radiator or convector	3.0	4.0
Boiler or heater	3.0	4.0
Open gate valve	0.5	0.7
Open globe valve	12.0	17.0

ASHRAE Handbook–Fundamentals

4. Find the equivalent length of pipe for an open gate valve by multiplying the equivalent length of pipe for a 90° elbow by the number of equivalent elbows. Each open gate valve equals 0.5 standard 90° elbows.

$L_{gate} = E \times N$

$L_{gate} = 8.9 \times 0.5$

$L_{gate} = \mathbf{4.45'}$

CONVERTING LENGTH OF PIPE FOR 90° ELBOWS		
Fitting	**Iron Pipe**	**Copper Tubing**
Elbow, 90°	1.0	1.0
Elbow, 40°	0.7	0.7
Elbow, 90° long turn	0.5	0.5
Elbow, 90°, welded	0.5	0.5
Reduced coupling	0.4	0.4
Open retun bend	1.0	1.0
Angle radiator valve	2.0	3.0
Radiator or convector	3.0	4.0
Boiler or heater	3.0	4.0
Open gate valve	0.5	0.7
Open globe valve	12.0	17.0

ASHRAE Handbook–Fundamentals

5. Find the total equivalent length of the piping system by adding the equivalent lengths of each fitting type and the total length of straight pipe. To do this, multiply the number of each fitting type by its equivalent length. Then add the values and the length of straight pipe. In this example, there are 10 standard 90° elbows, 1 globe valve, 2 gate valves, and 400′ of straight pipe.

$L_{total} = (10 \times L_{elbow}) + (1 \times L_{globe}) + (2 \times L_{gate}) + L_{pipe}$

$L_{total} = (10 \times 8.9) + (1 \times 106.8) + (2 \times 4.45) + 400$

$L_{total} = 89.0 + 106.8 + 8.9 + 400$

$L_{total} = \mathbf{604.7'}$

Example: Finding Equivalent Length of Pipe—Copper Tubing

The copper tubing in a hydronic piping system contains an open gate valve. The copper tubing is 1″ and the velocity of the water in the valve is 2 fps. Find the equivalent length of pipe that will give the same pressure drop.

1. Find the equivalent length of pipe for a standard 90° elbow. On the table for equivalent lengths, the equivalent length of pipe is 2.5′.
2. Find the equivalent number of standard 90° elbows for a gate valve. On the table for elbow equivalents, a gate valve is equal to 0.7 standard elbows.
3. Calculate the equivalent length of pipe.

$L = E \times N$

$L = 2.5 \times 0.7$

$L = \mathbf{1.75'}$

Design Static Pressure Drops. *Hydronic design static pressure drop* is the pressure drop per unit length of piping for a given size of piping at a given flow rate. The design static pressure drop increases as the flow rate increases and the piping size decreases. Design static pressure drop is typically expressed as feet of head loss per 100′ of pipe. Design static pressure drop is sometimes expressed as psi loss per 100′ of pipe. A static pressure drop of 1 psi loss per 100′ of pipe is equal to 2.30′ of head loss per 100′ of pipe. *Note:* Piping length is based on the equivalent length, which includes the pressure drop in fittings and valves within pipe segments.

A specific design flow rate within a pipe segment must be selected when sizing pipe. For typical residential and light commercial applications, the design static pressure drop varies from 1′ of head (0.43 psi) loss to 5′ of head (2.17 psi) loss per 100′ of pipe. Typically, a selected design static pressure drop of 2.5′ of head (1.09 psi) loss per 100′ of pipe is initially used. The actual pipe size for a segment is selected from commercially available pipe sizes. The actual static pressure drop within a segment is based on the selected pipe size. **See Figure 52-5.**

Converting Flow of Water in Tubing Sizes*

Friction Loss‡	Actual Tubing Size†								Friction Loss‡
	3/8	1/2	5/8	3/4	1	1 1/4	1 1/2	2	
0.83	—	0.53	0.96	1.44	3.1	5.3	8.5	18.2	0.83
1.04	—	0.59	1.05	1.63	3.5	6.0	9.6	20.5	1.04
1.25	—	0.65	1.15	1.79	3.8	6.6	10.6	22.6	1.25
1.46	—	0.71	1.26	1.95	4.2	7.3	11.5	24.6	1.46
1.67	0.41	0.76	1.35	2.10	4.5	7.8	12.4	26.5	1.67
1.88	0.44	0.81	1.44	2.24	4.7	8.3	13.2	28.3	1.88
2.08	0.46	0.86	1.53	2.36	5.0	8.7	14.0	30.0	2.08
2.29	0.48	0.91	1.61	2.49	5.3	9.2	14.7	31.5	2.29
2.50	0.51	0.95	1.68	2.61	5.6	9.6	15.3	33.0	2.50
2.71	0.53	0.99	1.75	2.70	5.8	10.0	16.0	34.5	2.71
2.92	0.56	1.03	1.82	2.82	6.1	10.5	16.7	35.8	2.92
3.13	0.58	1.07	1.90	2.93	6.3	10.8	17.3	37.0	3.13
3.33	0.59	1.11	1.96	3.05	6.5	11.1	18.0	38.3	3.33
3.54	0.61	1.15	2.03	3.15	6.7	11.5	18.5	39.6	3.54
3.75	0.63	1.18	2.10	3.24	6.9	11.9	19.1	40.9	3.75

* in gpm
† in in.
‡ per 100 ft

ASHRAE Handbook—Fundamentals

Figure 52-5. A tubing sizing chart shows the size of tubing used for a given water flow rate at design pressure drops.

The segment static pressure drop of piping is found by multiplying the equivalent length of the segment by the design static pressure drop. To find the segment static pressure drop, apply the following formula:

$$PD_p = \frac{L}{100} \times pd_p$$

where

PD_p = piping segment static pressure drop (in ft of head loss)

L = equivalent length of pipe for fittings (in ft)

100 = constant

pd_p = piping design static pressure drop (in ft of head loss/100′)

For example, a 250′ tubing segment has a design static pressure drop of 2.71′ of head loss per 100′. Find the static pressure drop for the tubing segment.

$$PD_p = \frac{L}{100} \times pd_p$$

$$PD_p = \frac{250}{100} \times 2.71$$

$$PD_p = 2.5 \times 2.71$$

$$PD_p = \mathbf{6.78' \text{ head loss}}$$

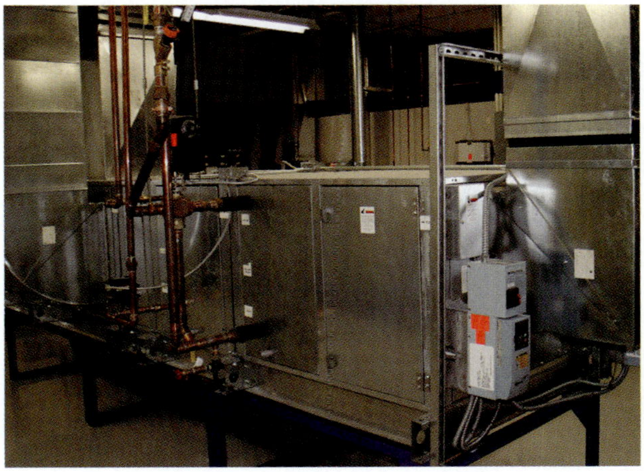

Hydronic systems deliver water to terminal units and air handlers to provide heating and cooling to building spaces.

When a piping system includes multiple segments, the pressure drop for the initial segment is found first. The process continues through the remaining segments of the entire system. The total pressure drop for all segments is the pressure drop for the system. **See Procedure for Finding Total Pipe Static Pressure Drop.**

Procedure

Finding Total Pipe Static Pressure Drop

1. A hydronic boiler system produces heated water at 140,000 Btu/hr. The water flows through a tube and discharges into a holding tank with a boiler pump for reuse. When the system has a temperature difference of 20°F, find the water flow rate.

$$Q = \frac{H}{500 \times \Delta T}$$

$$Q = \frac{140{,}000}{500 \times 20}$$

$$Q = \frac{140{,}000}{10{,}000}$$

$$Q = \mathbf{14\ gpm}$$

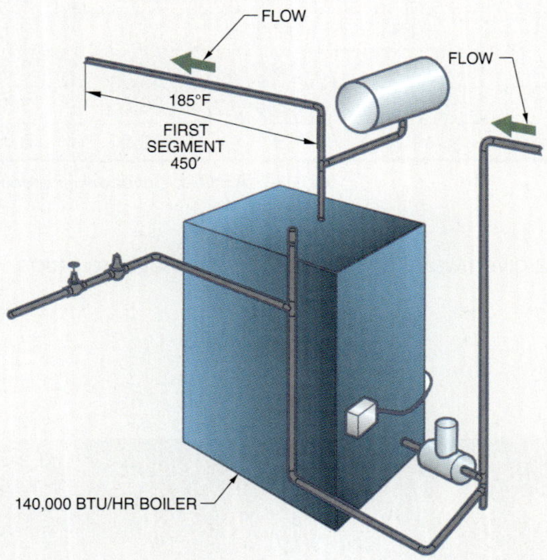

2. Find the estimated average pressure drop in the tube and select a tube size for this segment. Using the general guidelines of estimated pressure drop for a tube with 2.5′ of head loss per 100′ of tube, the minimum tube size according to the chart is 1½″. The 1½″ tube has the needed flow rate with a friction loss of 2.08′ of pressure drop per 100′ of tubing.

CONVERTING TUBING SIZES*

⅜	½	⅝	¾	1	1¼	1½	2	Friction Loss†
—	0.53	0.96	1.44	3.1	5.3	8.5	18.2	0.83
—	0.59	1.05	1.63	3.5	6.0	9.6	20.5	1.04
—	0.65	1.15	1.79	3.8	6.6	10.6	22.6	1.25
—	0.71	1.26	1.95	4.2	7.3	11.5	24.6	1.46
0.41	0.76	1.35	2.10	4.5	7.8	12.4	26.5	1.67
0.44	0.81	1.44	2.24	4.7	8.3	13.2	28.3	1.88
0.46	0.86	1.53	2.36	5.0	8.7	14.0	30.0	2.08
0.48	0.91	1.61	2.49	5.3	9.2	14.7	31.5	2.29
0.51	0.95	1.68	2.61	5.6	9.6	15.3	33.0	2.50
0.53	0.99	1.75	2.70	5.8	10.0	16.0	34.5	2.71
0.56	1.03	1.82	2.82	6.1	10.5	16.7	35.8	2.92
0.58	1.07	1.90	2.93	6.3	10.8	17.3	37.0	3.13
0.59	1.11	1.96	3.05	6.5	11.1	18.0	38.3	3.33
0.61	1.15	2.03	3.15	6.7	11.5	18.5	39.6	3.54
0.63	1.18	2.10	3.24	6.9	11.9	19.1	40.9	3.75

* in gpm
† in ft/100

ASHRAE Handbook–Fundamentals

3. Find the velocity of the water in the tube and confirm that it is acceptable.

$$V = 0.4085 \times \frac{Q}{D^2}$$

$$V = 0.4085 \times \frac{14}{1.5^2}$$

$$V = 0.4085 \times \frac{14}{2.25}$$

$$V = 0.4085 \times 6.222$$

$$V = \mathbf{2.54\ fps}$$

Design velocity is between 2.5 fps and 8 fps, so it is acceptable.

4. Find the total static pressure drop for the tubing in the segment. In this case, the segment has an equivalent length of 450′. The total pressure drop is the design static pressure drop per 100′ multiplied by the total equivalent length of the entire segment.

$$PD_p = \frac{L}{100} \times pd_p$$

$$PD_p = \frac{450}{100} \times 2.08$$

$$PD_p = 4.50 \times 2.08$$

$$PD_p = \mathbf{9.36'\ head\ loss}$$

Temperature Difference

Temperature difference is the difference between the initial and final temperature of a material through which heat was transferred. When designing piping systems, the temperature difference is the difference in temperature of the water at the beginning and end of a heating or cooling process. The temperature difference through each piping section and terminal device is important because temperature differences represent heat loss or gain. In some piping systems, the temperature differences vary through each piping section and terminal device.

The total system temperature difference for heating applications is the difference between the temperature of the water entering the piping system at the boiler and the water returning to the boiler. For cooling applications, the total system temperature difference is the difference between the temperature of the water entering the piping system at the chiller and the water returning to the chiller.

Terminal Devices. For typical residential or light commercial applications, a temperature difference of about 20°F may be used for both heating and cooling applications. When terminal devices are used in residential or light commercial buildings, the temperature of water leaving a boiler typically is 120°F to 180°F and the water returning to the boiler is 100°F to 140°F. **See Figure 52-6.** Lower system temperatures improve system safety but can require larger terminal devices for proper heat exchange. The temperature of water typically leaving a chiller is around 45°F and the water returning to the chiller is around 55°F. Terminal heating and cooling units exchange heat with building spaces.

Terminal Devices

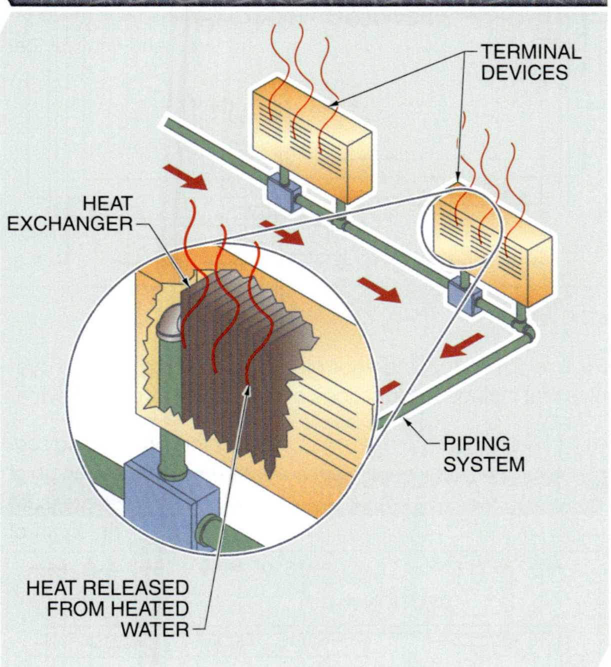

Figure 52-6. Terminal devices exchange heat with building spaces.

When terminal devices are used in heavy commercial buildings, the temperature of water leaving a boiler typically is 160°F to 200°F and the water returning to the boiler is 140°F to 180°F. Heavy commercial systems are typically designed for the specific application, which can include other processes using the same system for heating.

Radiant Flooring. When radiant floor heating is used in residential or light commercial buildings, the temperature of water leaving a boiler typically is 90°F to 120°F and the water returning to the boiler is 80°F to 110°F. Radiant floor heating applications typically have a temperature difference of about 10°F and are used for heating only. Designs for radiant floor heating systems vary by application and region. Equipment manufacturers and suppliers typically provide specific recommendations for the sizing and spacing of these systems.

SYSTEM DESIGN

Building floor plans, including information about spaces with terminal devices, are required when designing hydronic systems. The heating and cooling loads must be calculated for each of the building spaces. HVAC systems must be able to supply the total amount of heat and cooling energy for all heating and cooling loads. When the plans are made and the loads calculated, the pipe system sizes are selected. The remaining components of the system, which include pumps, fittings, valves, and controls, are selected. Each building space is reviewed to select the terminal device of appropriate size.

Piping System Design

A floor plan or sketch of the building is required when designing the piping system for a hydronic system. Each floor that has piping or terminal devices requires a floor plan. The heating and cooling loads must be calculated for all building spaces. Pipe locations, sizes, and fittings are chosen after the plans are complete and the loads are calculated.

Layout Drawings. Layout drawings are drawings of the floor plan of a building that show walls, partitions, windows, doors, fixtures, and other details that affect the placement of piping and terminal devices. Layout drawings also indicate the heating and cooling loads for each room or zone and the required water flow rate. Piping systems are sized for water flow rates in gallons per minute. Heating and cooling loads for each building must be converted to the water flow rate for each section. **See Figure 52-7.**

Layout drawings contain several overlays, which show the placement of the terminal devices and piping system. The overlays show the floors in the building that contain terminal devices and the places where pipes are run. Locations for terminal devices are on the overlay for each floor of the building. With this arrangement, the piping layout can be seen in relation to the terminal devices. On the piping layout, each terminal device is connected to the supply pipe and the return pipe. The actual layout of the piping system depends on the type of piping system chosen.

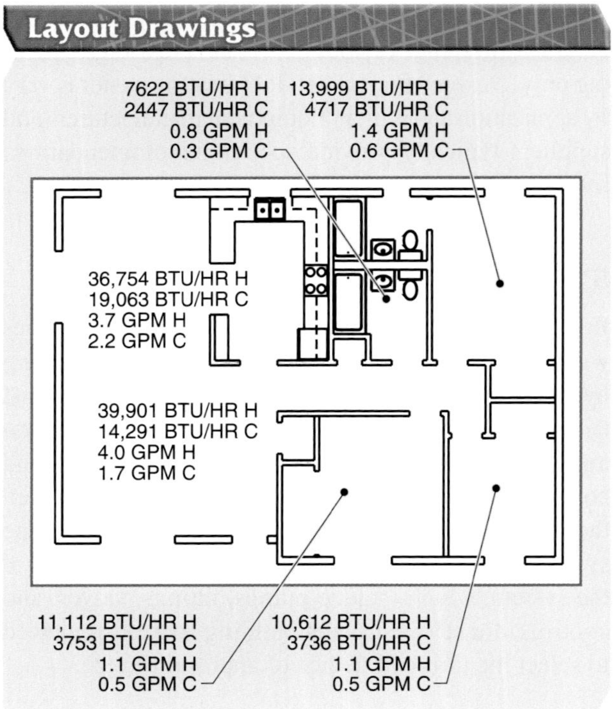

Figure 52-7. Layout drawings are drawings of the floor plan of a building. They indicate the heating and cooling loads for each room or zone.

On the drawings and in a building, piping must not run through beams, columns, stairwells, elevator shafts, or areas where plumbing or electrical devices may be located. When the layout drawings and overlays are ready, the piping system is chosen based on the temperature control level required in each building space. Once the piping system has been chosen, the terminal devices for each building space are chosen based on the heating and cooling requirements for each building space. **See Figure 52-8.**

Pipe Section Identification. It is necessary to identify each pipe section and keep track of the proper size pipes, valves, and fittings of each section. A *pipe section* is a length of pipe that runs from one fitting to another fitting. The length of a pipe section is measured from the outlet of one fitting to the outlet of the previous fitting. Pipe sections are identified by letters or numbers, which are marked on the pipe sections themselves or the fittings at the end of each pipe section. **See Figure 52-9.**

When letters are used to identify pipe sections, the pipe section between fittings is identified by the letters marked on the fittings. For example, all fittings are marked with a different letter. The first fitting in a system, which is on the boiler or chiller, is identified as A. The fitting at the other end of this first pipe section, which is at the beginning of the second pipe section, is identified as B. The pipe section between fittings A and B is identified as section A-B. The letters that identify the sections and fittings are shown on the piping overlay.

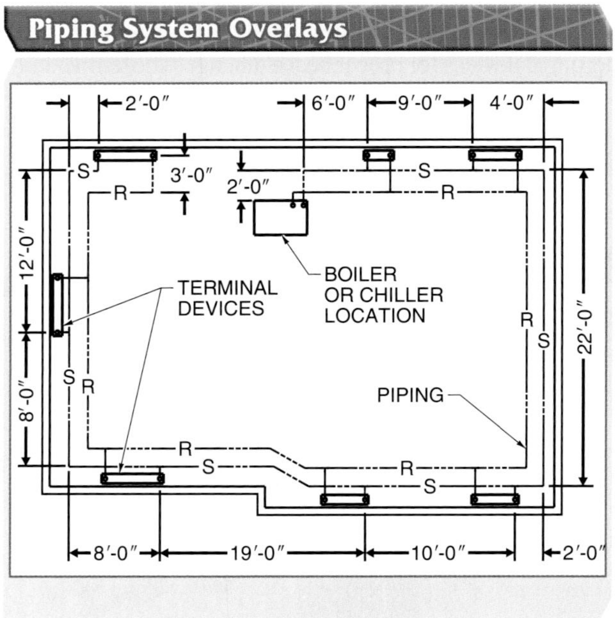

Figure 52-8. Building floor plans have piping system overlays that include the locations of piping and terminal devices.

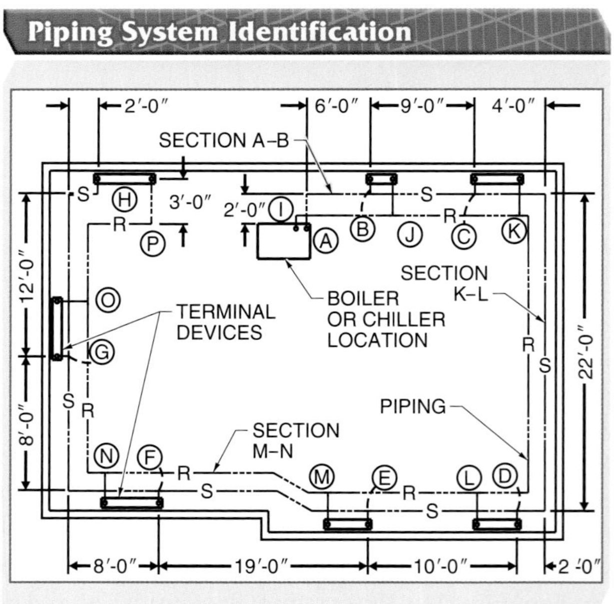

Figure 52-9. Letters or numbers are used to identify fittings at the end of each pipe section. Pipe sections are identified by the letters or numbers on the fittings.

Pipe Sizing

Piping is sized based on the piping material used. The roughness of the piping inside surface, which affects friction loss, varies depending on the piping material. Since friction loss must be considered, the type of piping used is a factor in determining the proper pipe size. The piping used in hydronic systems is typically copper tubing as well as Schedule 40 wrought iron or galvanized steel pipe. Copper tubing is typically used for residential applications, and iron or steel pipe is typically used for commercial applications.

Pipe sizing also is affected by additives in the system water and the operating temperature. Additives are typically added to the water of hydronic systems to protect the systems from corrosion. The additives can change the flow resistance of the fluid. Typically, the amount of additives is small and the change to fluid flow is minor. However, this must be checked prior to use.

When water temperature rises, the flow resistance of water reduces, and the pressure drop in pipe segments decreases. This decrease in pressure drop in hydronic systems could be as much as 30%, which could result in oversizing of the piping and circulating pumps. Hydronic systems operating at higher temperatures, especially large hydronic systems, must be designed for their actual operating conditions to avoid oversizing the piping and pump.

Typical residential hydronic systems with terminal devices are limited by the standard sizes of piping and equipment. Some oversizing is expected, which compensates for any system installation and operating problems. Typical residential hydronic systems with radiant heating have multiple long distribution tubing in standard sizes and lengths. Circulating pumps for radiant floor heating must be properly sized for the size and length of tubing.

Pipe Sizing Charts. Proper pipe sizes for distribution systems are found by selecting sizes that allow the necessary capacity for the water flow rate and that are in the acceptable range of the system pressure drop. Pipe sizing charts show the relationship between the pipe size, flow rate, pressure drop, and flow velocity of various types of pipes. **See Figure 52-10.** The pipe sizing chart for the type of piping material used must be selected. Relationships shown on pipe sizing charts are based on the normal water temperature of 60°F for specific piping roughness. Pipe sizing charts are useful for small systems at moderate operating temperatures. Hotter or larger piping systems must be designed for the actual operating conditions.

To find the required water flow rate on a pipe sizing chart, the design flow rate located on the bottom or side of the chart must be found. Then, a line is drawn on the chart at the design flow rate. Pressure drop is found on the bottom or side of the chart. Pressure drop on the chart is expressed as feet of head loss or psi loss per 100′ of pipe. The typical range of pressure drop in hydronic systems is between 1′ of head (0.43 psi) loss to 5′ of head (2.17 psi) loss per 100′ of pipe. Initially a line is drawn on the chart at 2.50′ of head (1.09 psi) loss per 100′ of pipe.

Standard pipe sizes are shown on the chart as slanted lines. The flow rate line should intersect the 2.50′ of head (1.09 psi) loss per 100′ of pipe pressure drop line near a standard pipe size line. The nearest standard pipe size line from the point of intersection should be selected. When the point is midway between standard pipe lines, the larger pipe size should be selected. Increasing the pipe size reduces the pressure drop, flow velocity, and pumping costs but increases the cost of the piping system.

Selecting pipe sizes near 2.5′ of head (1.09 psi) loss per 100′ of pipe is a good balance. The design flow conditions for the segment of pipe are at the point where the flow rate line intersects a standard pipe size line. Once the pipe size is determined, a line is drawn at the design flow conditions point to find the pipe segment pressure drop. It must be ensured that the pressure drop in the pipe segment is between 1′ of head (0.43 psi) loss and 5′ of head (2.17 psi) loss per 100′ of pipe. Flow velocities are shown on the chart as slanted lines. It must be ensured that the flow velocity in the pipe segment is between 2.5 fps and 8 fps.

The selected pipe size is the size of the initial pipe segment for the system. This process should be repeated to determine the pipe sizes for the remaining segments of pipe until the end of the piping system is reached.

Piping Loops. The system pressure drop is based on the design conditions of the total equivalent length of piping loops within the system. Piping loops are the series of pipe segments through which water flows from the heating/cooling unit discharge to the unit return. Pipe sizes and design flow velocities are used to determine the equivalent lengths of the pipe segments. All equivalent lengths of pipe segments in a piping loop are added to determine the equivalent length of the piping loop.

Piping loops typically include balancing valves that are used to increase the pressure drop of the loop, thereby reducing flow through the loop. This adjustment in the flow is used to balance the loop flow within the system and evenly distribute building space heating and cooling.

Pipe Sizing Charts

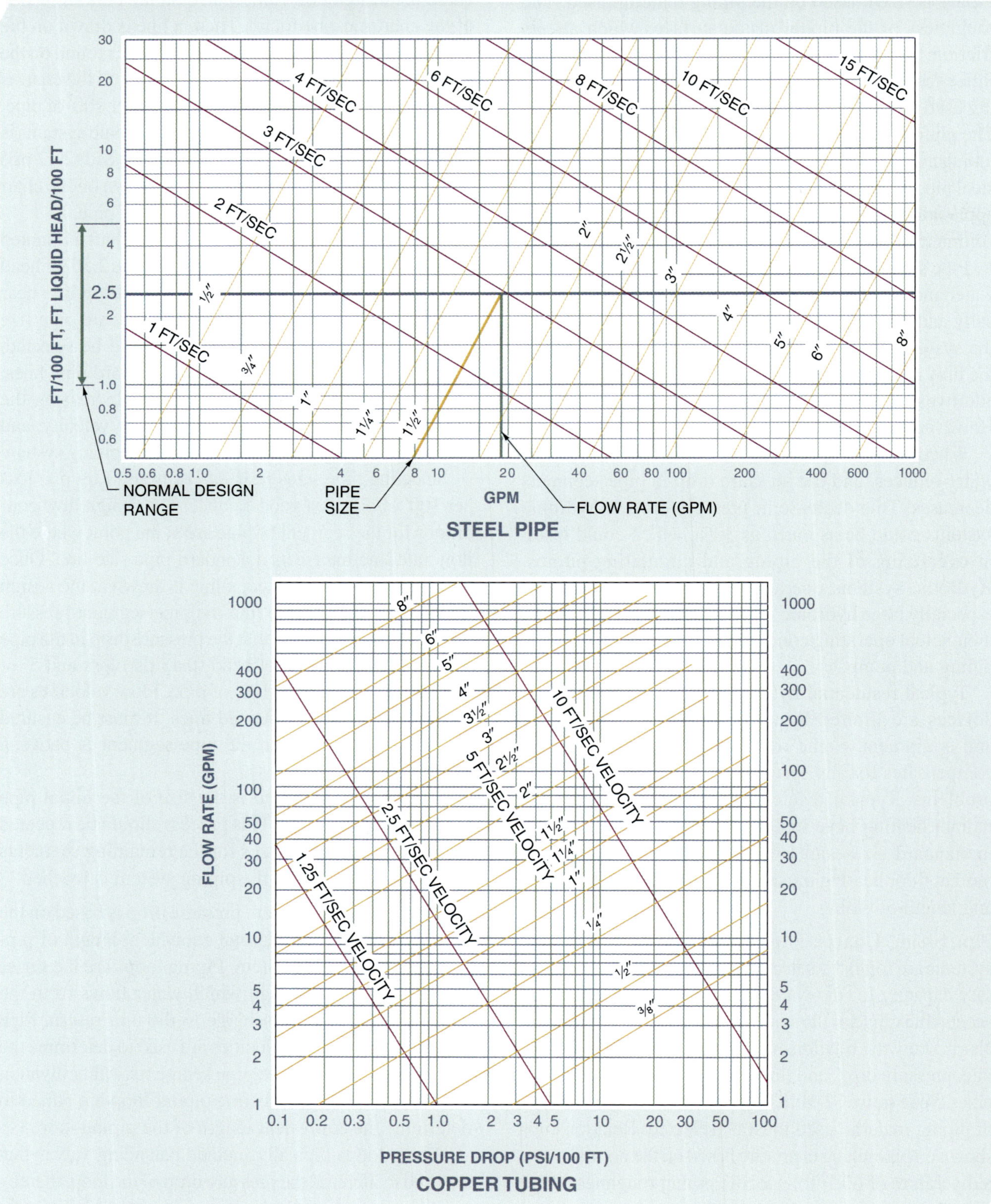

ASHRAE Handbook—Fundamentals

Figure 52-10. Pipe sizing charts show the relationship between pipe size, flow rate, pressure drop, and flow velocity.

Hydronic Piping Systems. The procedure for sizing a piping system varies depending on the type of piping system used. In one-pipe series systems, the pipe size remains the same throughout the system because the flow rate remains the same. In one-pipe primary-secondary systems, the supply and return pipe sizes change between each terminal device because the flow rate changes. **See Figure 52-11.**

In two-pipe direct-return systems, the supply pipe and return pipe are typically the same size in each pipe section between terminal devices. The return pipe gets progressively smaller as terminal devices are added. In two-pipe reverse-return systems, the supply pipe gets progressively smaller as it goes out from the boiler or chiller. Also, the return pipe gets progressively larger as it comes back to the boiler or chiller from the terminal device at the end of the piping loop. **See Figure 52-12.**

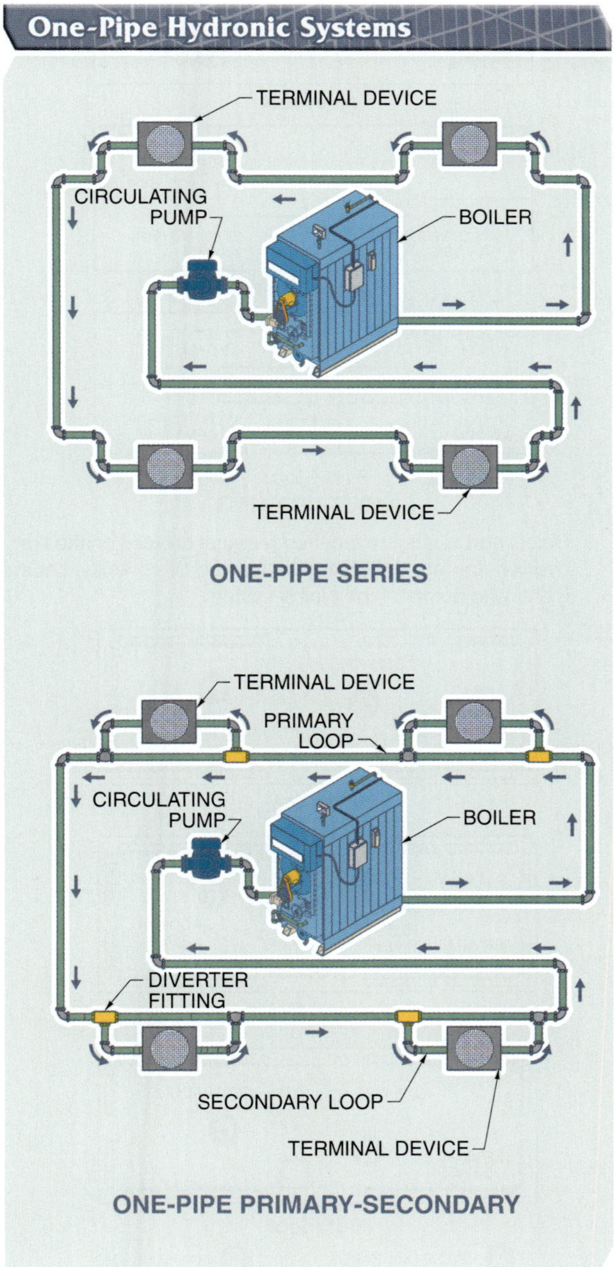

Figure 52-11. One-pipe hydronic systems are typically used in small-size residential and commercial buildings with a limited number of terminal units.

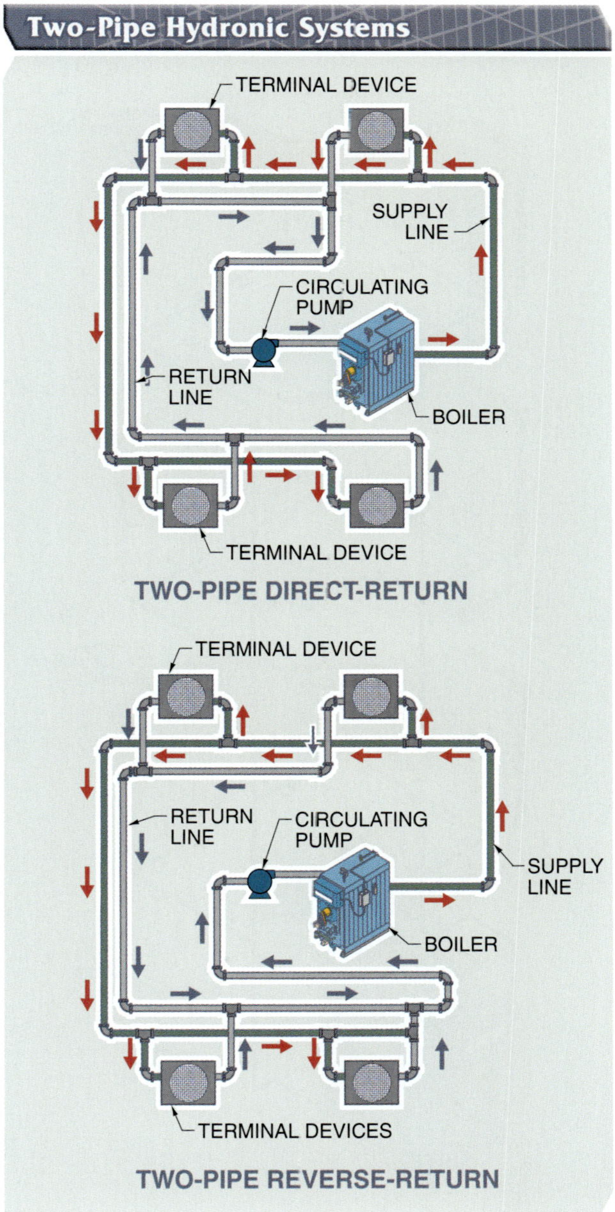

Figure 52-12. Two-pipe hydronic systems are typically used in medium-size residential and commercial buildings. One pipe supplies water and another pipe returns the water to the central unit.

Large Hydronic Systems. Large hydronic systems typically have a main supply pipe that branches to serve different areas of a building. The main supply pipe is designed to distribute heated water throughout the building based on its operating conditions. Each branch piping is sized using the same process for sizing small hydronic systems. Typically, each loop uses a pump to circulate the water to terminal devices. **See Procedure for Designing Hydronic Piping Systems for Buildings.**

Procedure

Designing Hydronic Piping Systems for Buildings

1. Review building plans to visualize the heating and cooling system. For simplicity, this example is limited to a one-pipe heating system. The same method is used for additional pipe systems, including cooling systems.

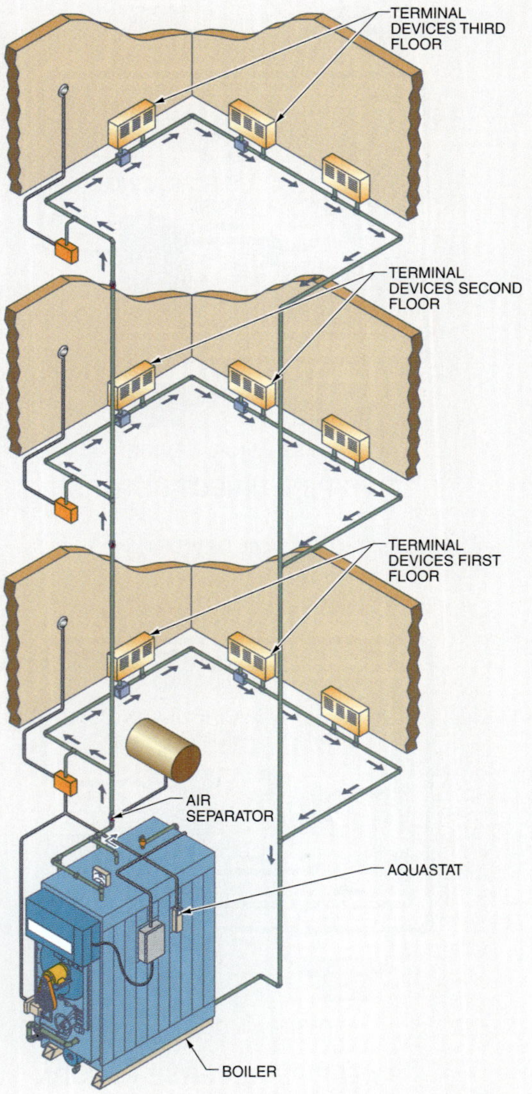

2. Prepare layout drawings and overlays, which show the heating and cooling loads and water flow rate for each building space.

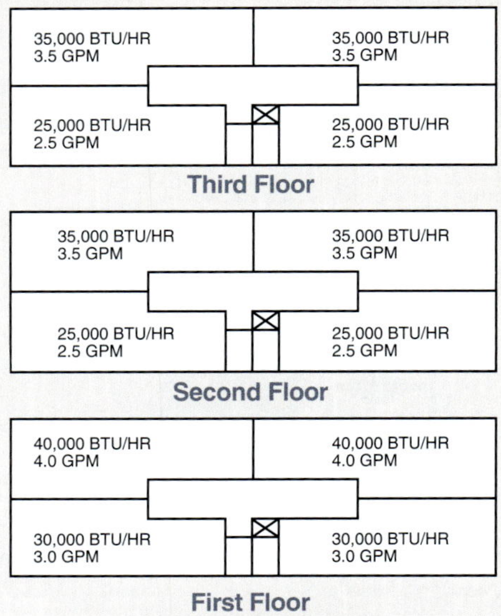

LAYOUT DRAWING

3. Select and locate the required terminal devices on the first overlay. Lay out the piping system on the overlay. Label identifying points of the piping system.

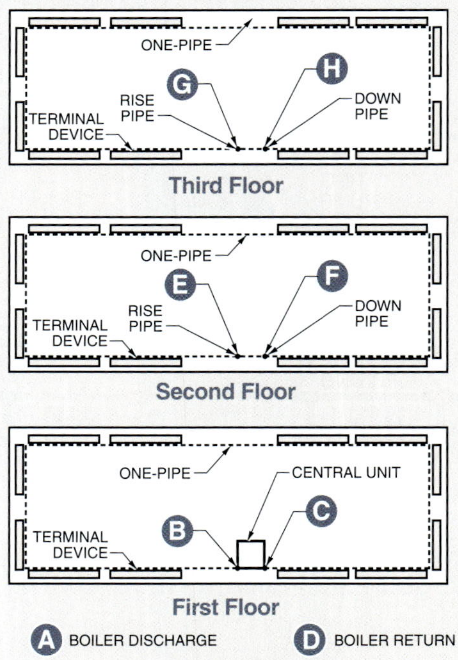

PIPE SYSTEM OVERLAY AND IDENTIFICATION

continued on next page

Procedure

Designing Hydronic Piping Systems for Buildings *(continued)*

4. Prepare a piping design form for the project. On the overlay, find the names of the piping segments. Enter the names of the piping segments in the segment column of the piping design form. Identify the piping loop when more than one loop is used. List the piping segments in order of the direction of flow and in order of the loops. Enter the appropriate piping loop next to the segment name. In this example, the first segment is A-B, which is in piping loops 1, 2, and 3.
5. Enter the water flow rate for each pipe segment in the flow rate column.

Project Name: Office Building
Location: Local Building
HVAC System: One-pipe Heating System

Section (Loop No)	Flow Rate*	Pipe Size†	Flow Velocity‡
A-B (1,2,3)	38.0		
B-C (1)	14.0		
C-D (1,2,3)	38.0		
B-E (2,3)	24.0		
E-F (2,3)	12.0		
F-C (2,3)	24.0		
E-G (3)	12.0		
E-H (3)	12.0		
H-F (3)	12.0		
Loop 1			
Loop 2			
Loop 3			

*in gpm
† in in.
‡ in fps
§ in ft head loss/100ft
|| in ft

Note: V=gate valve; GL=globe valve; E=90° elb

LOOP
SEGMENT

6. Using the pipe sizing chart, identify the flow rate for the first pipe. Select the pipe size that provides a design static pressure drop of approximately 1.0′ to 5.0′ of head loss per 100′ for that flow rate. Enter the pipe size in the pipe size column. Continue this process until all pipe sizes are selected.

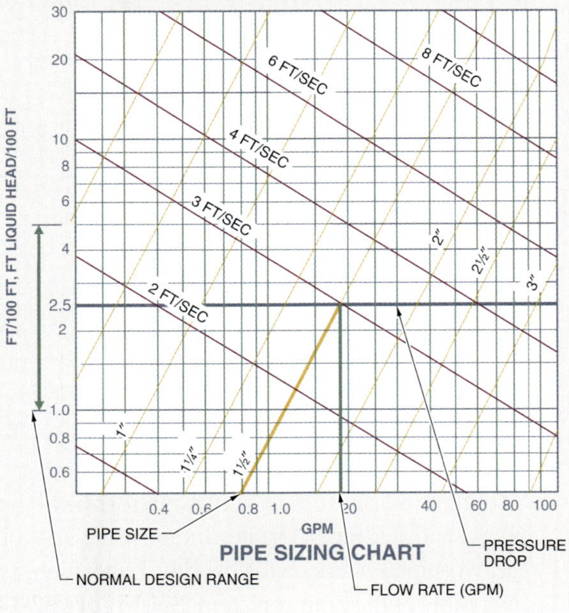

PIPE SIZING CHART

7. Using the pipe sizing chart, identify the flow velocity and design static pressure drop (design head loss) for the flow rate and selected pipe size. Enter the flow velocity and static pressure drop in the flow velocity and unit head loss columns. Continue this process until all pipe rows are complete.
8. Determine the length of piping in each pipe segment, and enter the results in the segment length column. In this example, the first pipe segment is 10′.

Project Name: Office Building
Location: Local Building
HVAC System: One-pipe Heating System

Segment (Loop No)	Flow Rate*	Pipe Size†	Flow Velocity‡	Unit Head Loss§	Segment Length||
A-B (1,2,3)	38.0	2	3.3	2.7	10
B-C (1)	14.0	1¼	3.2	3.3	180
C-D (1,2,3)	38.0	2	3.6	2.7	10
B-E (2,3)	24.0	1½	4.0	4.3	10
E-F (2,3)	12.0	1¼	2.7	2.6	180
F-C (2,3)	24.0	1½	4.0	4.3	10
E-G (3)	12.0	1¼	2.7	2.6	10
E-H (3)	12.0	1¼	2.7	2.6	180
H-F (3)	12.0	1¼	2.7	2.6	10
Loop 1					
Loop 2					
Loop 3					

*in gpm
† in in.
‡ in fps
§ in ft head loss/100ft
|| in ft

Note: V=gate valve; GL=globe valve

9. Determine the number and type of fittings in each pipe segment. In this example, the first segment has one gate valve and three 90° elbows.
10. Find the fitting equivalent length for each segment based on the pipe size and flow velocity. In this example, the first segment fittings have an equivalent length of 19.6′ (equivalent length 5.6 × elbow equivalent 3.5).
11. Find the segment equivalent length by adding the segment length to the fitting equivalent length.

EQUIVALENT LENGTH OF PIPE

Velocity*	Pipe Size					
	1	1¼	1½	2	2½	3
1	2.2	3.0	3.5	4.5	5.4	6.7
2	2.5	3.3	3.9	5.1	5.1	7.5
3	2.7	3.6	4.2	5.4	5.4	8.0
4	2.8	3.7	4.4	5.6	5.6	8.3
5	2.9	3.9	4.5	5.9	5.9	8.7
6	3.0	4.0	4.7	6.0	6.0	8.9
7	3.0	4.1	4.8	6.2	6.2	9.1
8	3.1	4.2	4.9	6.3	6.3	9.3
9	3.2	4.3	5.0	6.4	6.4	9.5
10	3.2	4.3	5.1	6.5	6.5	9.7

* in fps

FIRST SEGMENT

continued on next page

Designing Hydronic Piping Systems for Buildings *(continued)*

ELBOW EQUIVALENTS

Fitting	Iron Pipe	Copper Tubing
Elbow, 90°	1.0	1.0
Elbow, 40°	0.7	0.7
Elbow, 90° long turn	0.5	0.5
Elbow, welded, 90°	0.5	0.5
Reduced coupling	0.4	0.4
Open retun bend	1.0	1.0
Angle radiator valve	2.0	3.0
Radiator or convector	3.0	4.0
Boiler or heater	3.0	4.0
Open gate valve	0.5	0.7
Open globe valve	12.0	17.0

ASHRAE Handbook–Fundamentals

12. Find the total segment head loss for each pipe segment. Multiply the unit segment head loss by the segment equivalent length. In this example, the first segment has a total of 0.80′ of head loss.
13. Enter the above findings in the fittings column, the fittings equivalent length column, the segment equivalent length column, and the head loss column.

Project Nam
Location: Lo
HVAC Syste

Segment (Loop No)	Fittings	Fittings Eq Len[ǁ]	Segment Eq Len[ǁ]	Head Loss[ǁ]
A-B (1,2,3)	1V, 3E	19.6	29.6	0.80
B-C (1)	1GV, 5E	95.2	275.2	9.08
C-D (1,2,3)	1V, 2E	14.0	24.0	0.65
B-E (2,3)	1V	2.3	12.3	0.53
E-F (2,3)	1GV, 5E	61.2	241.2	6.27
F-C (2,3)	1V	2.3	12.3	0.53
E-G (3)	1V	1.8	11.8	0.31
E-H (3)	1GV, 5E	61.2	241.2	6.27
H-F (3)	1V	1.8	11.8	0.31
Loop 1				10.53
Loop 2				8.78
Loop 3				9.46

*in gpm
† in in.
‡ in fps
§ in ft head
ǁ in ft

E=90° elbow; Eq Len=Equivalent Length

14. Determine the piping loop head loss. This example has three piping loops, one for each floor. Add the segment head loss for each piping loop member and note them at the bottom of the head loss column.
    a. In this example, the first segment (A-B) is part of piping loops 1, 2, and 3, as noted in the first column. The head loss for the first segment is included in the head loss for piping loops 1, 2, and 3. The second segment (B-C) is part of piping loop 1.
    b. The head loss for the second segment is included in the head loss for loop 1.
    c. The third segment (C-D) is part of piping loops 1, 2, and 3. The head loss for the third segment is included in the head loss for piping loops 1, 2, and 3. In this example, the head loss for piping loop 1 is the sum of the head losses for segments A-B, B-C, and C-D.
15. Add the head losses for segments A-B, B-C, and C-D and enter the sum at the Loop 1 row and head loss column. Repeat steps 14 and 15 to find the head loss for piping loop 2 and piping loop 3.

Project Name: Office Building
Location: Local Building
HVAC System: One-pipe Heating System

Segment (Loop No)	Flow Rate*	Pipe Size†	Flow Velocity‡	Unit Head Loss§	Segment Length[ǁ]	Fittings	Fittings Eq Len[ǁ]	Segment Eq Len[ǁ]	Head Loss[ǁ]
A-B (1,2,3)	38.0	2	3.6	2.7	10	1V, 3E	19.6	29.6	0.80
B-C (1)	14.0	1¼	3.2	3.3	180	1GV, 5E	95.2	275.2	9.08
C-D (1,2,3)	38.0	2	3.6	2.7	10	1V, 2E	14.0	24.0	0.65
B-E (2,3)	24.0	1½	4.0	4.3	10	1V	2.3	12.3	0.53
E-F (2,3)	12.0	1¼	2.7	2.6	180	1GV, 5E	61.2	241.2	6.27
F-C (2,3)	24.0	1½	4.0	4.3	10	1V	2.3	12.3	0.53
E-G (3)	12.0	1¼	2.7	2.6	10	1V	1.8	11.8	0.31
E-H (3)	12.0	1¼	2.7	2.6	180	1GV, 5E	61.2	241.2	6.27
H-F (3)	12.0	1¼	2.7	2.6	10	1V	1.8	11.8	0.31
Loop 1									10.53
Loop 2									8.78
Loop 3									9.46

*in gpm
† in in.
‡ in fps
§ in ft head loss/100ft
ǁ in ft

Note: V=gate valve; GL=globe valve; E=90° elbow; Eq Len=Equivalent Length

Example: Sizing Piping Systems

Find the pipe sizes for a hydronic system in a three bedroom single-story house. The system is a two-pipe reverse-return system made of type M copper tubing. The boiler/chiller unit provides a 20°F temperature difference for the heating flow rate and a 17°F temperature difference for the cooling flow rate.

1. Prepare layout drawings and overlays. Determine the heating and cooling loads and water flow rate for each building space and note on the overlay. **See Figure 52-13.** *Note:* The flow rate for heating is greater than the flow rate for cooling. Therefore, the heating load is used to size the system.
2. Select and locate the required terminal devices, and add appropriate dimensions. Lay out the two-pipe system on the overlay. Label identifying points of the piping system. **See Figure 52-14.**

Overlay with Heating and Cooling Loads

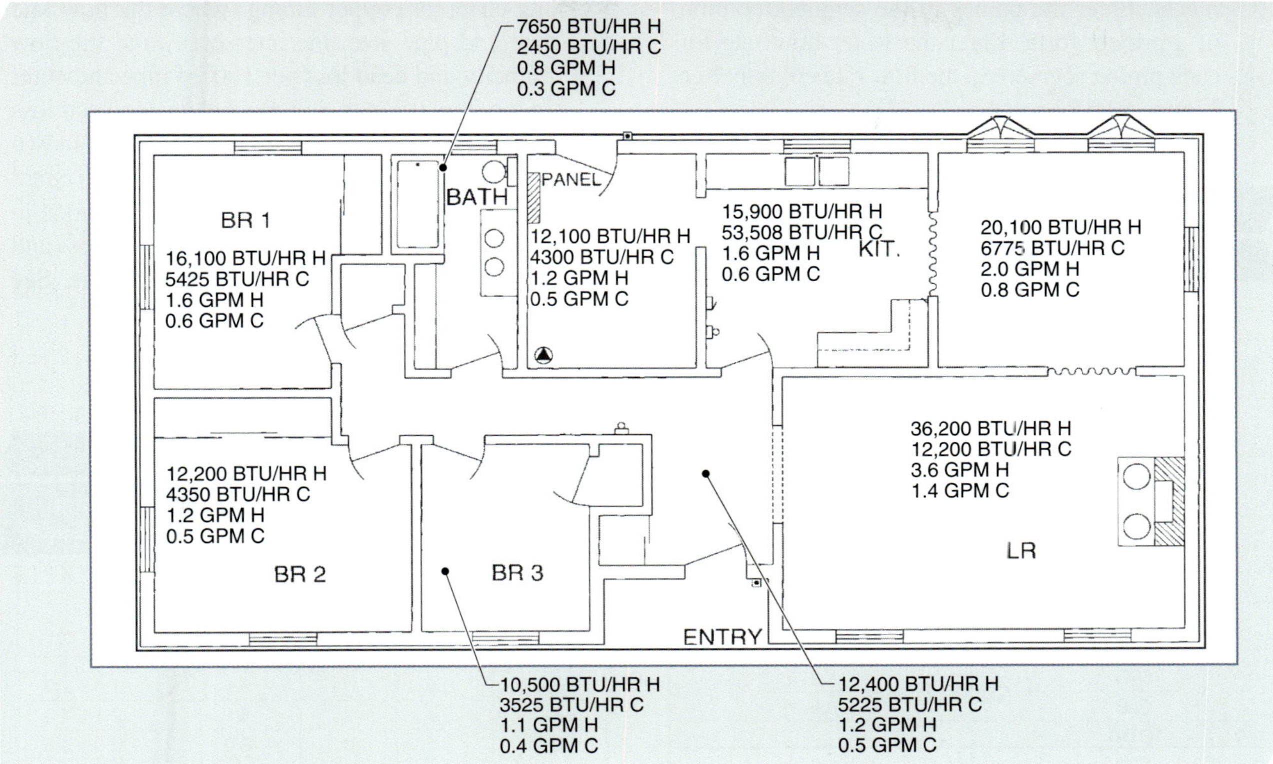

Figure 52-13. Layout drawings and overlays indicate heating and cooling loads as well as water flow rates.

Overlay with Systems Identification

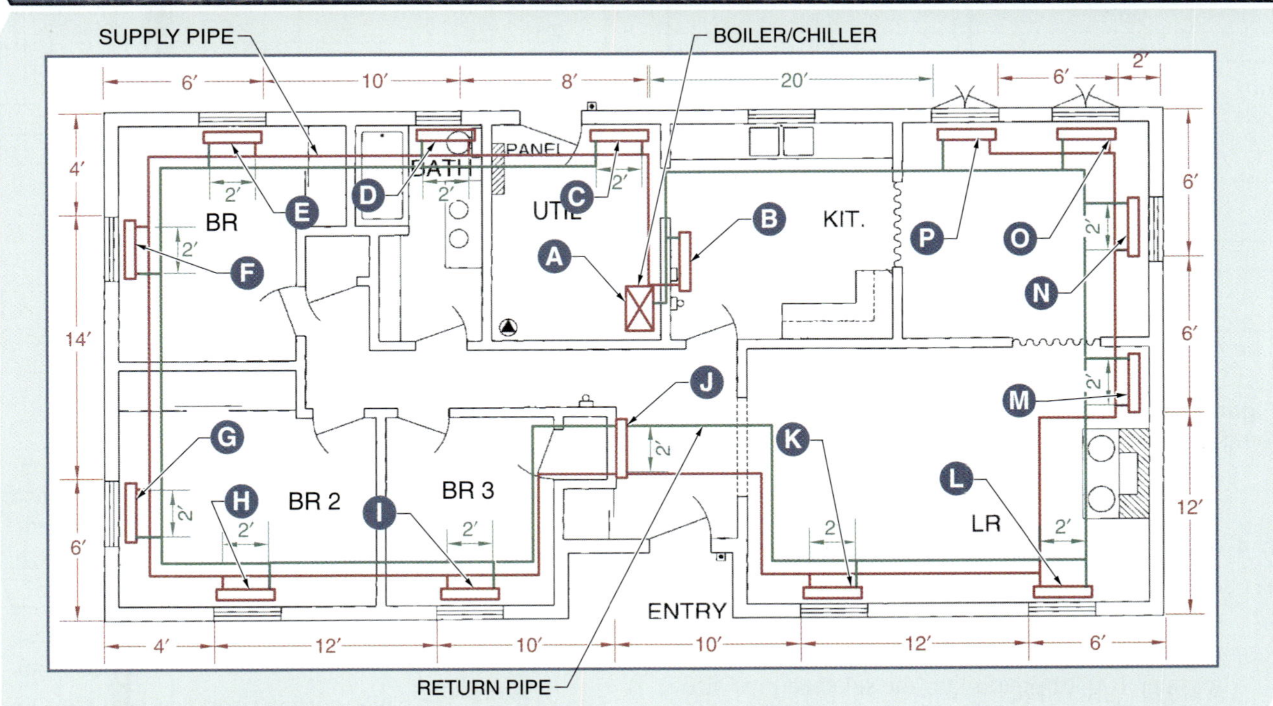

Figure 52-14. Layout drawings and overlays show the terminal device layouts and piping system layout.

3. On the overlay, find the names of the piping segments. Enter the names in the segment column of a project form. Enter the water flow rate for each piping segment in the flow rate column. **See Figure 52-15.**

Form with Piping Sections

Segment	Flow Rate*
A-B	14.3
B-C	12.7
C-D	11.5
D-E	10.7
E-F	9.9
F-G	9.1
G-H	8.5
H-I	7.9
I-J	6.8
J-K	5.6
K-L	4.3
L-M	3.0
M-N	2.0
N-O	1.2
O-P	0.6
TOTAL	
C-D	1.2
D-E	2.0
E-F	2.8
F-G	3.6
G-H	4.2
H-I	4.8
I-J	5.9
J-K	7.1
K-L	8.4
L-M	9.7
M-N	10.7
N-O	11.5
O-P	12.1
P-B	12.7
B-A	14.3
TOTAL	

* in gpm

Figure 52-15. The piping sections and flow rates are entered on a form.

4. On the pipe sizing chart for copper tubing, find the design flow rate for each pipe segment and select the appropriate piping size. Ensure that the design static pressure drop is between 0.4 psi loss and 2.5 psi loss per 100′ of pipe. Enter the selected pipe size for each pipe segment in the pipe size column.

5. Locate the segment design point on the pipe sizing chart for copper tubing (where the flow rate line and pipe size line intersect). Find the flow velocity and head loss per 100′ of pipe and enter in the flow velocity column and design head loss column. When the static pressure drop is shown in psi, convert to feet of head loss. (1 psi is equal to 2.30′ of head loss.) Because of limitations in the available piping sizes, the actual head loss and velocity may vary from the desired range. **See Figure 52-16.**

Form with Piping Flows

Segment	Flow Rate*	Pipe Size†	Flow Velocity‡	Design Head Loss§
A-B	14.3	1½	2.6	1.1 = 2.5
B-C	12.7	1½	2.4	0.9 = 2.1
C-D	11.5	1¼	2.9	1.4 = 3.2
D-E	10.7	1¼	2.5	1.1 = 2.5
E-F	9.9	1¼	2.4	1.0 = 2.3
F-G	9.1	1	3.1	2.1 = 4.8
G-H	8.5	1	3.0	2.0 = 4.6
H-I	7.9	1	2.7	1.7 = 3.9
I-J	6.8	1	2.4	1.3 = 3.0
J-K	5.6	1	1.9	1.0 = 2.3
K-L	4.3	¾	2.6	2.1 = 4.8
L-M	3.0	¾	1.9	1.2 = 2.8
M-N	2.0	½	2.1	2.5 = 5.8
N-O	1.2	½	1.5	1.2 = 2.8
O-P	0.6	½	1.0	0.7 = 1.6
TOTAL				
C-D	1.2	½	1.4	1.2 = 2.8
D-E	2.0	½	2.1	2.5 = 5.8
E-F	2.8	¾	1.8	1.0 = 2.3
F-G	3.6	¾	2.1	1.6 = 3.7
G-H	4.2	¾	2.6	2.1 = 4.8
H-I	4.8	¾	2.9	2.4 = 5.5
I-J	5.9	1	2.2	1.1 = 2.5
J-K	7.1	1	2.5	1.6 = 3.7
K-L	8.4	1	2.9	1.9 = 4.4
L-M	9.7	1	3.5	2.3 = 5.3
M-N	10.7	1¼	2.5	1.1 = 2.5
N-O	11.5	1¼	2.9	1.4 = 3.2
O-P	12.1	1¼	2.8	1.4 = 3.2
P-B	12.7	1½	2.4	0.9 = 2.1
B-A	14.3	1½	2.6	1.1 = 2.5
TOTAL				

* gpm
† in in.
‡ in fps
§ in pressure drop/100′ = ft head loss/100′

Figure 52-16. The piping section sizes, flow velocities, and unit head losses are entered on a form.

6. On the overlay, find the lengths of the piping segments. Enter the lengths in the segment length column of the form.
7. Enter the number and types of fittings for each pipe segment in the fittings column. Using the table for equivalent lengths and elbow equivalents, find the equivalent lengths of all fittings for each segment, and enter in the fittings equivalent length column. Find the total equivalent length of the segments by adding the segment length to the fitting equivalent lengths, and enter it in the segment equivalent length column. **See Figure 52-17.**
8. Find the head loss within each piping segment by dividing the segment equivalent length by 100 and multiplying by the design pressure drop in feet of head loss per 100′. Enter the segment head loss in the head loss column.
9. Find the system pressure drop by adding the section head losses for the supply piping and return piping. Enter the system head loss in the head loss column at the total row. In this example, water flows partially through the supply piping and partially through the return piping. Therefore, the actual piping pressure drop is the average pressure drop of the supply piping and return piping.

Form with Piping Fittings and Head Loss

Segment	Flow Rate*	Pipe Size†	Flow Velocity‡	Design Head Loss	Segment Length#	Fittings	Fittings Equ Len#	Segment Equ Len#	Head Loss
A-B	14.3	1½	2.6	1.1§ = 2.5‖	2	2E, 1V	2.7 × 4.2	13.3	0.33
B-C	12.7	1½	2.4	0.9 = 2.1	8			8.0	0.17
C-D	11.5	1¼	2.9	1.4 = 3.2	8	1R	3.0 × 3.6	18.8	0.60
D-E	10.7	1¼	2.5	1.1 = 2.5	10			10.0	0.25
E-F	9.9	1¼	2.4	1.0 = 2.3	10	1E	1.0 × 3.6	13.6	0.31
F-G	9.1	1	3.1	2.1 = 4.8	14	1R	3.0 × 2.7	22.1	1.06
G-H	8.5	1	3.0	2.0 = 4.6	10	1E	1.0 × 2.7	12.7	0.58
H-I	7.9	1	2.7	1.7 = 3.9	12			12.0	0.47
I-J	6.8	1	2.4	1.3 = 3.0	18	2E	2.0 × 2.7	23.4	0.70
J-K	5.6	1	1.9	1.0 = 2.3	18	1R, 2E	5.0 × 2.1	28.5	0.66
K-L	4.3	¾	2.6	2.1 = 4.8	12			12.0	0.58
L-M	3.0	¾	1.9	1.2 = 2.8	18	3E	3.0 × 1.9	23.7	0.66
M-N	2.0	½	2.1	2.5 = 5.8	6	1R	3.0 × 1.8	11.4	0.66
N-O	1.2	½	1.5	1.2 = 2.8	8	1E	1.0 × 1.4	9.4	0.26
O-P	0.6	½	1.0	0.7 = 1.6	6			6.0	0.10
TOTAL					160			224.9	7.39
C-D	1.2	½	1.4	1.2 = 2.8	8			8.0	0.22
D-E	2.0	½	2.1	2.5 = 5.8	10			10.0	0.58
E-F	2.8	¾	1.8	1.0 = 2.3	10	1E	1.0 × 1.5	11.5	0.26
F-G	3.6	¾	2.1	1.6 = 3.7	14			14.0	0.52
G-H	4.2	¾	2.6	2.1 = 4.8	10	1E	1.0 × 2.0	12.0	0.58
H-I	4.8	¾	2.9	2.4 = 5.5	12			12.0	0.66
I-J	5.9	1	2.2	1.1 = 2.5	18	2E	2.0 × 2.1	22.2	0.56
J-K	7.1	1	2.5	1.6 = 3.7	22	2E	2.0 × 2.7	27.4	1.01
K-L	8.4	1	2.9	1.9 = 4.4	12			12.0	0.53
L-M	9.7	1	3.5	2.3 = 5.3	18	3E	3.0 × 2.8	26.4	1.40
M-N	10.7	1¼	2.5	1.1 = 2.5	6			6.0	0.15
N-O	11.5	1¼	2.9	1.4 = 3.2	8	1E	1.0 × 3.6	11.6	0.37
O-P	12.1	1¼	2.8	1.4 = 3.2	6			6.0	0.19
P-B	12.7	1½	2.4	0.9 = 2.1	26	1E	1.0 × 4.2	30.2	0.64
B-A	14.3	1½	2.6	1.1 = 2.5	4	1E,1V	1.7 × 4.2	11.1	0.28
TOTAL					184			220.4	7.93

* gpm
† in in.
‡ in fps
§ in psi pressure drop
‖ ft. head loss/100′
\# in ft.

Note: V = valve gate; GL = globe valve; E = 90° elbow; e = 45° elbow; R = reducer; Eq Len = equivalent length

Figure 52-17. The fittings for each piping section are entered on a form. The equivalent lengths are found and the head loss is calculated. The sum of the section head losses is the system head loss.

Fittings and Valves

Pipe fittings and valves control the flow of water through the distribution system. Shutoff valves are located on each side of all major components of a system such as the boiler and terminal devices. Balancing valves are located on each branch line off the main supply line and on lines going directly to terminal devices. Balancing valves are used for balancing water flow through terminal devices to provide the proper heating or cooling effect. Data relating to valves and fittings, especially diverting and balancing valves, should be obtained from the manufacturer.

Some piping systems have modulating flow valves, which are located on the supply or return connections to terminal devices. Data and information for using and controlling these valves should be obtained from the manufacturer.

Terminal Devices

The type of terminal device used in hydronic heating or air conditioning systems depends on the degree of temperature control required in the building. Radiators or convectors are used as terminal devices with one-pipe series piping systems. In one-pipe series piping systems, all system sections, including each terminal device, have the same water flow rate. Each terminal device selected must be sized for that specific flow rate.

In one-pipe primary-secondary systems, the terminal devices are located on secondary loops that are in parallel with the main system loop. In this type of system, each terminal device can be of a different size. The piping to each terminal device must be sized for the flow rate required for the load handled by that terminal device only. The main piping loop is sized for the total load on the main loop. This load typically includes several terminal devices. In two-pipe systems, each terminal device may have a different flow rate.

Selecting Terminal Devices. The heating and cooling capacities of terminal devices are shown on specifications sheets. To choose the correct size, the heating and cooling load on each terminal device must be known. A terminal device specifications sheet that describes the terminal device is also required. The model number of each terminal device is recorded on the overlay that shows the terminal devices. **See Figure 52-18.**

Circulating Pump Sizing

Some distribution systems have one circulating pump on each piping loop in the system. Small piping systems typically have only one pump for the whole system. Larger systems may have several pumps. Circulating pumps are selected and sized to circulate the required amount of water against the total pressure drop of the piping system.

When one circulating pump is used with multiple piping loops, the piping loop with the greatest total pressure drop is used to size the pump. Typically, balancing valves are partially closed to increase the pressure drop of piping loops with lower pressure drops to balance the flow between loops.

Terminal Device Selection

Model	Flow Rate		Hydronic Air Conditioning			Hydronic Heating		
			Pressure Drop‡	45°F Water 80°F db, 67°F wb Air		Pressure Drop‡	140°F Water	
	Air*	Water†		Total§	Sensible§		50°F Air Total§	70°F Air Total§
Q-100	200	2	1.00	6500	5500	1.4	13,500	10,000
Q-160	250	2.5	1.60	8200	5900	1.6	15,500	12,000
Q-200	275	3	2.00	9500	6400	1.9	18,000	14,000
H-250	300	4.5	2.70	10,000	7600	2.7	21,000	16,000
H-750	400	4	2.80	13,500	10,200	3.9	27,000	21,000
H-1000	600	5	3.90	21,000	16,500	7.1	43,000	34,000

* heat in Btu/hr
† in ft of water
‡ in cfm
§ in gpm

Figure 52-18. Terminal devices are selected to provide the air and water flow rates needed for each building space. The model and size are chosen for each room or building space from specifications sheets, which are printed by manufacturers.

Total Pressure Drop. Pressure drops through terminal devices, boilers, chillers, and any other devices that would add resistance to water flow must be added to the piping pressure drop to find the total pressure drop of the system. **See Figure 52-19.** The total pressure drop in a piping system is the sum of all system pressure drops in the supply side and return side of a piping loop or system. Total pressure drop is found by applying the following formula:

$$PD_t = PD_p + PD_{td} + PD_{hu}$$

where

PD_t = total pressure drop (in ft of head)

PD_p = piping pressure drop, including fittings (in ft of head)

PD_{td} = terminal device pressure drop (in ft of head)

PD_{hu} = heating unit pressure drop (in ft of head)

System Pressure Drop

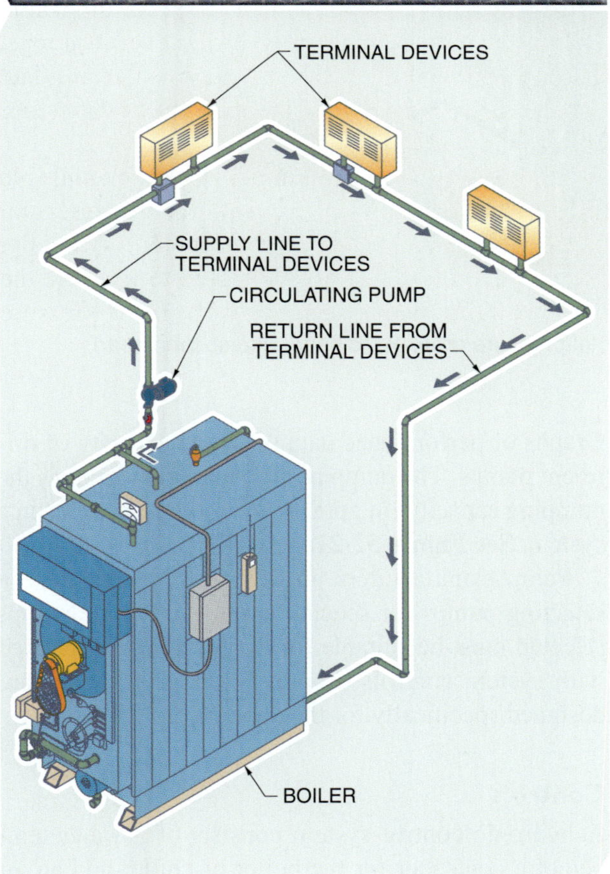

Figure 52-19. Pressure drops through terminal devices, boilers, chillers, and any other devices are added to the piping pressure drop to find the system pressure drop.

Example: Finding Total Pressure Drop—Piping

The pressure drop through a piping system is 3.52′ of head. The water must flow through six terminal devices with a pressure drop of 0.65′ of head per device. The water also flows through a boiler with a pressure drop of 1.25′ of head. Find the total pressure drop through the system.

$$PD_t = PD_p + PD_{td} + PD_{hu}$$

$$PD_t = 3.52 + (6 \times 0.65) + 1.25$$

$$PD_t = 3.52 + 3.9 + 1.25$$

$$PD_t = \mathbf{8.67' \text{ of head}}$$

Hot water circulating pumps are sized to move the correct volume of water at the correct temperature.

Total Dynamic Head. The total dynamic head of a hydronic system pump is the sum of the total pressure drop in the system, the static head of the system, and the suction head. **See Figure 52-20.** The static head is the sum of the lengths of the vertical risers in the loop or system. The suction head is the vertical lift from the lowest point of the system to the pump. Typically, the pump is located near the bottom of the system, and the suction head is small. The total dynamic head of a pump is found by applying the following formula:

$$TDH = PD_t + SH + sh$$

where

TDH = total dynamic head (in ft of head)

PD_t = total pressure drop (in ft of head)

SH = static head (in ft of head)

sh = suction head (in ft of head)

Total Dynamic Head of Pumps

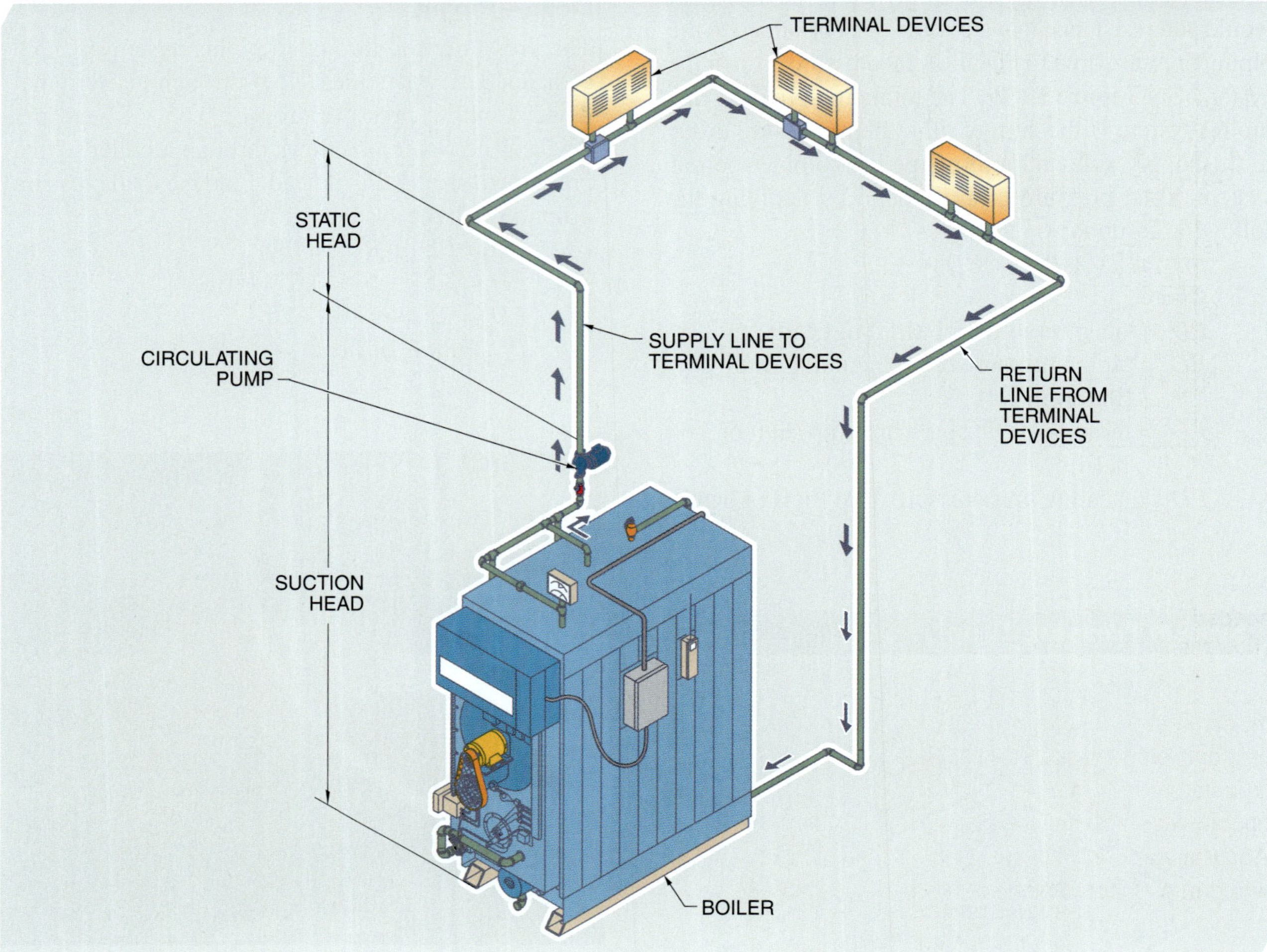

Figure 52-20. The total dynamic head of a pump includes the total pressure drop, static head, and suction head.

Example: Finding Total Dynamic Head

The total system pressure drop through a piping system is 8.67′ of head. The water must flow up vertically 12.5′ to terminal devices on the third floor. The water also flows through a boiler located 4.5′ below the pump. Find the total dynamic head of the pump.

$TDH = PD_t + SH + sh$

$TDH = 8.67 + 12.5 + 4.5$

$TDH = \mathbf{25.67' \text{ of head}}$

Selecting Circulating Pumps. The required size of the circulating pump for each piping loop or system is found by using the total flow rate and the total head of the pump. The circulating pump is selected based on the flow requirement of each piping loop or system. Graphs of performance data show the capacity of different pumps. The pump performance curve shows the pumping capacity (in gpm) as the total head varies in a system. **See Figure 52-21.**

Pump manufacturers provide recommendations in selecting pumps for specific applications. The pumps selected must be suitable for the service and matched with system controls. Pumps for large facilities are designed specifically for the application.

Controls

A hydronic control system consists of a water temperature controller for the boiler or chiller and an air temperature controller for the building spaces or zones being conditioned. The controllers ensure that the setpoint temperatures are maintained.

Circulating Pump Selection

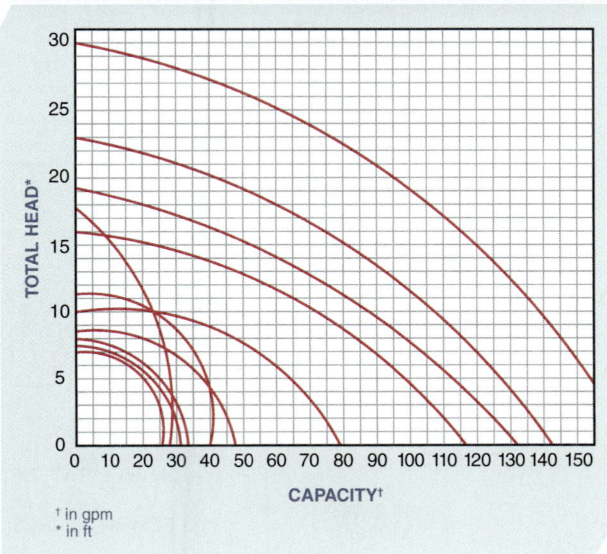

Figure 52-21. Pumps are selected to provide the flow rate needed for the total system head. Performance data tables show the capacity of circulating pumps.

Water Temperature Controls. Proper water temperature in a hydronic system is maintained by an aquastat. An aquastat senses water temperature in a boiler or chiller and operates equipment to maintain the setpoint temperature. Aquastats are thermostatically controlled switches that are wired into electrical control circuits. **See Figure 52-22.**

Boiler Temperature Controls

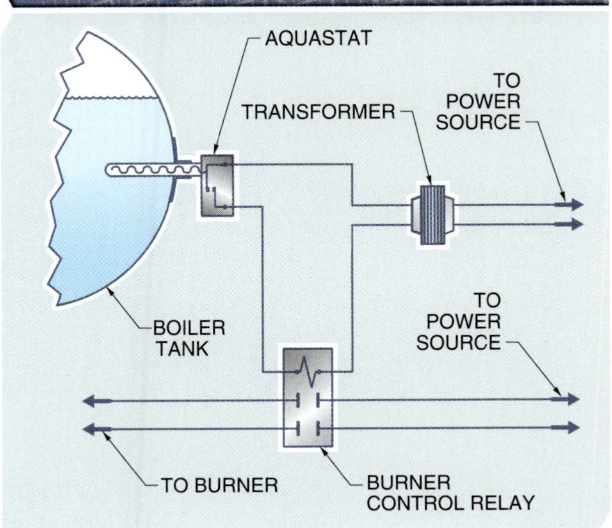

Figure 52-22. Aquastats are thermostatically controlled electric switches that regulate burners to maintain boiler temperature.

The control system for any application provides the necessary control for the application. Some control systems are more sensitive than others. An outdoor reset control is a sensitive controller that automatically resets the boiler water temperature in response to changes in the outdoor temperature. Sensitive controls help hydronic systems offset variations in loads caused by factors that are difficult to include in load calculations. Variable solar loads and rapid changes in outdoor temperature can be accounted for by using outdoor reset controls.

Air Temperature Controls. Air temperature controls may be as simple as water flow valves, which control the flow of water to radiators. This type of simple control ensures constant air temperature when the water flow rate is constant. This relationship is possible because the temperature of the air leaving a terminal device is a function of the temperature of the water inside the terminal device.

Some components of hydronic air-temperature control systems are installed in control panels within units. The control typically includes a thermostat, blower speed control, and a ventilation damper control. Other components are mounted on the walls of building spaces.

With a sophisticated hydronic control system, there is a modulating valve located on the supply or return side of each terminal device. Modulating valves are controlled by thermostats that monitor room temperatures. When room temperatures vary, the valves open or close to allow more or less water to flow through the terminal devices. When the system is a heating and cooling system, the valves and thermostat used must be capable of working in both modes. **See Figure 52-23.**

Water Flow Controls

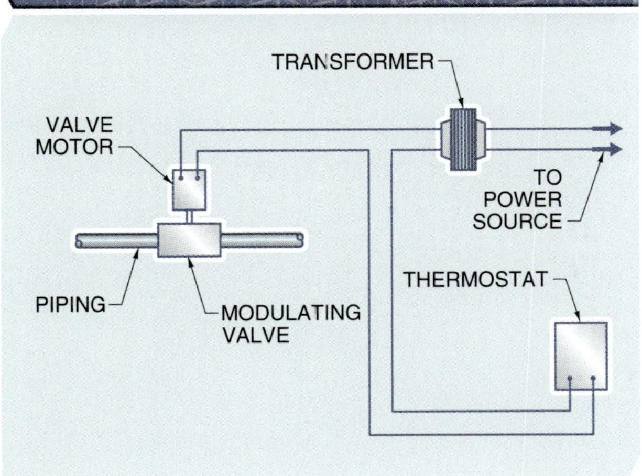

Figure 52-23. A modulating water valve opens and closes to allow more or less water to flow through a terminal device.

TECH TIP

The advanced digital control systems used in today's residential and light commercial hydronic systems provide excellent temperature control accuracy, temperature trending, remote alarm capabilities, and energy-saving features such as night setback setpoint control. In addition, these control systems often allow remote access from a computer or mobile device.

In systems using terminal devices with blowers, the control systems may control the operation of blowers instead of water flow. Control systems that control water flow and airflow provide excellent control of air temperature. In this type of system, a room thermostat controls a modulating control valve on the water line, and a manual or automatic switch controls the blower that circulates the air through ductwork. **See Figure 52-24.**

Terminal Controls

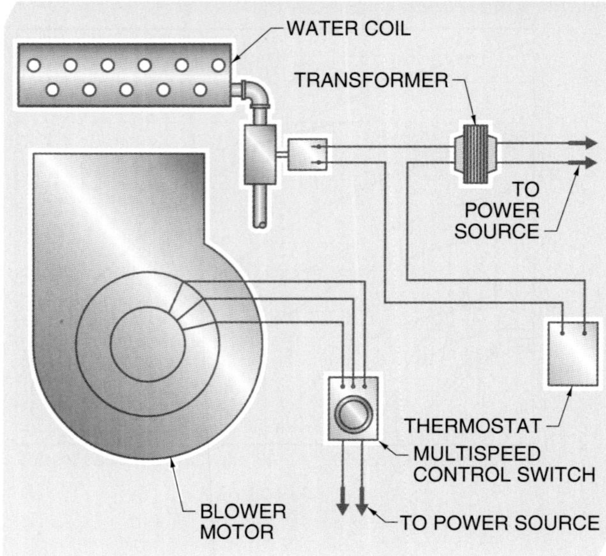

Figure 52-24. Terminal device controls may control blower operation or water flow to maintain building space temperature.

UNIT

53 Residential and Light Commercial System Testing and Balancing

A few of the most important tasks when installing or troubleshooting HVAC systems are to test, adjust, and balance the systems. Well-performed system testing, adjusting, and balancing can produce an effective HVAC system that provides a comfortable environment for building occupants. A well-balanced HVAC system provides the correct distribution of airflow or water flow to building spaces for even heating, cooling, and ventilation.

OBJECTIVES

- Explain the general testing, adjusting, and balancing processes.
- Describe the testing, adjusting, and balancing processes for forced-air HVAC systems.
- Describe the testing, adjusting, and balancing processes for hydronic HVAC systems.

Index to Procedures

TESTING, ADJUSTING, AND BALANCING

HVAC systems must be tested, adjusted, and balanced (TAB) to operate effectively. *Testing* is a procedure used to critically examine, observe, or evaluate the condition or operation of HVAC equipment. Testing an HVAC system is repeated during the adjusting and balancing processes until the system operates efficiently. Test results are documented for comparison to operating conditions during normal operation. *Adjusting* is the process of bringing parts or HVAC systems to more effective relative positions or satisfactory states. A well-adjusted HVAC system includes components that are set to perform properly. *Balancing* is the process of bringing HVAC systems into a state of stability produced by proportional distributions or functions.

TAB Process

TAB procedures should be performed by a technician who has experience with the specific type of HVAC equipment or system. Proper HVAC system TAB can prevent or alleviate many operational problems. New HVAC systems must have TAB as part of the commissioning process. Existing systems should have TAB performed after equipment retrofits or upgrades are made.

The HVAC system TAB process is typically accomplished quickly. Many HVAC technicians fail to properly TAB systems. Unfortunately, the failure to adequately TAB systems causes many operational problems. Typical deficiencies include lack of occupant comfort, poor HVAC system performance, and low system efficiency. In addition, the HVAC system service life is typically shortened.

Equipment Startup

On newly installed HVAC systems, the first step toward proper TAB is for the technicians or manufacturer representatives to properly startup the equipment. Startup typically includes providing basic system information such as fan operation, correct current draw for fans and pump motors, and correct refrigerant amounts and temperatures. Procedures and checklists are typically provided by the manufacturer. The HVAC system must go through the TAB process to ensure that air will circulate completely through occupied areas of the building and to flush out contaminants from the space. **See Figure 53-1.**

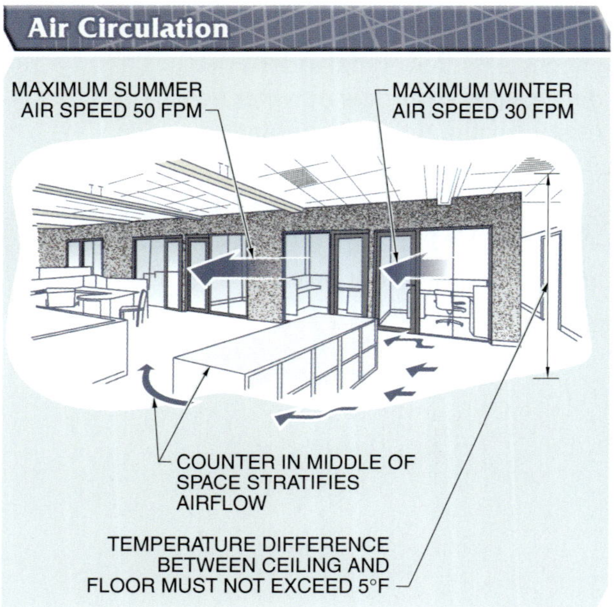

Figure 53-1. Proper system testing, adjusting, and balancing (TAB) procedures ensure that adequate air distribution is maintained in all building spaces.

Equipment warranties may be voided when the proper startup procedures are not followed. The warranty period, typically one year, starts on the date of equipment startup. Completed startup paperwork must be preserved for reference. The TAB process for HVAC units begins with proper equipment startup.

Properly performing TAB procedures on forced-air or hydronic HVAC distribution systems provides the required amount of heating or cooling to each part of a building. It also ensures the circulation of the proper amount of design air or water to building spaces.

FORCED-AIR DISTRIBUTION SYSTEM TAB

Forced-air distribution systems are balanced so that they can provide the amount of air that satisfies the heating, cooling, and ventilation requirements of building spaces. In forced-air distribution systems, fans are adjusted in order to provide the total airflow rate required. Balancing dampers in the main trunks and branches of ductwork are adjusted in order to deliver the proper amount of air to each part of the system. Register dampers are adjusted in order to deliver the proper amount of air from each register to building spaces.

Installing distribution systems exactly as designed is almost impossible. Changes in the building design, locations of other equipment in the building, or locations of plumbing or electrical wiring may require adjustments to the ductwork design. Nominal sizes of registers and grills are often oversized for operating conditions. These variables affect the operation of distribution systems. To offset the difference between the actual operation of the system and the designed operation, the system must be balanced after all equipment, ductwork, accessories, and controls are installed and operating.

Two types of forced-air distribution systems used in heating and air conditioning applications are constant-air-volume and variable-air-volume systems. A *constant-air-volume (CAV) system* is an air distribution system in which the airflow rate in each building space remains constant but the supply-air temperature is varied to maintain a setpoint. The air temperature is controlled by cycling the heating or air conditioning equipment on and off. **See Figure 53-2.** CAV systems are common in small residential applications.

TECH TIP

Technicians specializing in system testing, adjusting, and balancing (TAB) can perform this service only after training and certification.

A *variable-air-volume (VAV) system* is an air distribution system that provides air at a constant temperature but varies the amount of supplied air in order to maintain the desired setpoints within the building space. **See Figure 53-3.** VAV systems are common in light commercial applications. Airflow rates to each building space are varied in a VAV system by using dampers. Dampers are located in the branch ductwork or near the variable-frequency drives connected to fan motors.

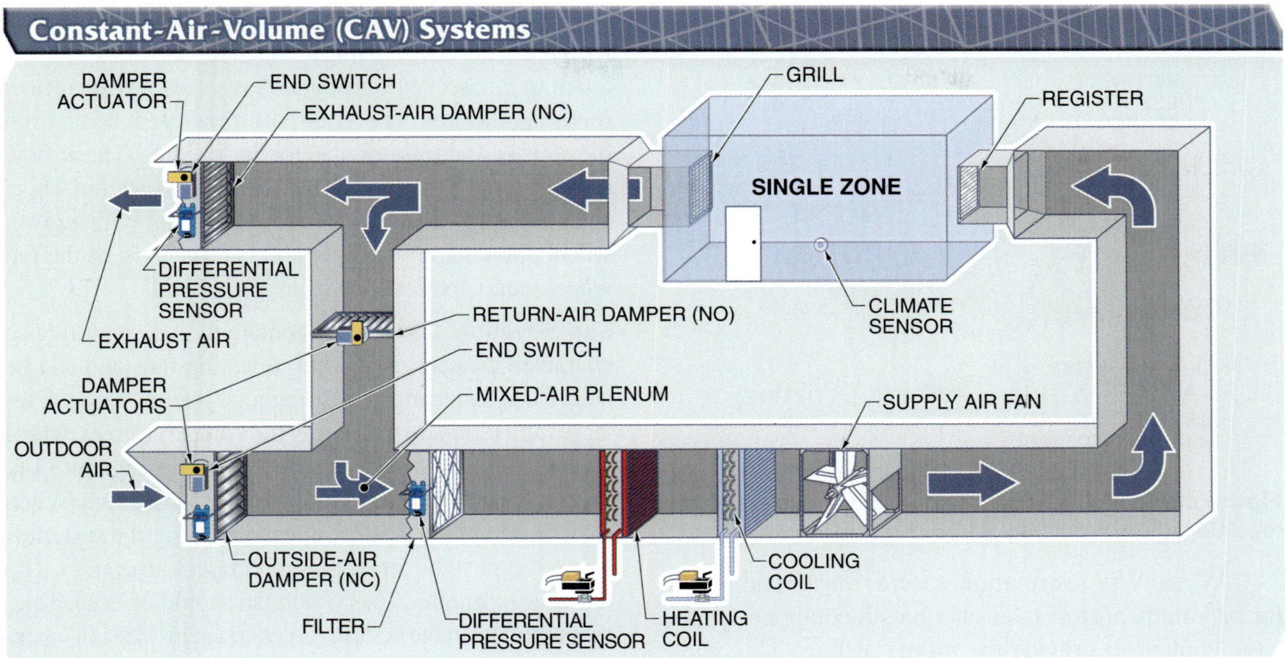

Figure 53-2. Most CAV systems serve a single zone or small buildings.

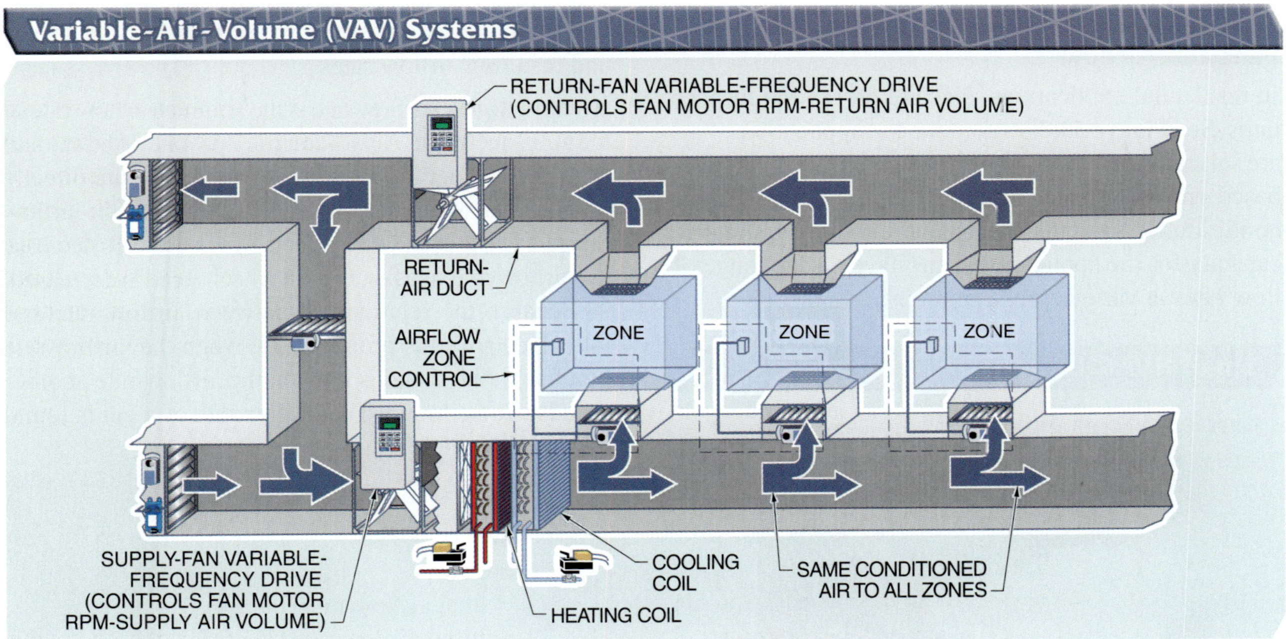

Figure 53-3. Most VAV systems serve multiple zones in larger buildings using variable-frequency drives to regulate static pressure in the supply duct.

CAV and VAV distribution systems are balanced using many of the same procedures. Fans are adjusted to provide the total airflow required for the building or building spaces. Balancing dampers in the distribution systems must be adjusted to allow the proper airflow to each zone in the building. A *balancing damper* is an airflow damper located in the branch takeoff from the trunk duct in a forced-air distribution system. The dampers in the registers or the variable-volume dampers at the registers must also be adjusted for proper airflow. A *variable-volume damper* is a damper that controls the amount of airflow at a terminal device in a VAV system. **See Figure 53-4.**

VAV Terminal Boxes

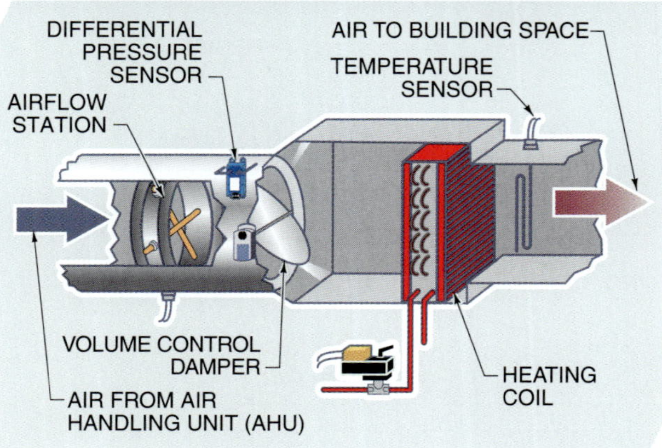

Figure 53-4. VAV terminal boxes include dampers to regulate airflow.

CAV and VAV distribution systems must be adjusted for maximum airflow to ensure that all components are large enough to provide the proper airflow. This may simulate the airflow rates required during the highest heating and cooling demand days of the year.

Fans and Airflow

In residential applications, fans are part of the furnace units. **See Figure 53-5.** In commercial applications, fans are selected and placed in the rooftop unit or air handler based on performance criteria. Fans for heating or air conditioning systems must have enough volumetric capacity for the application. Fans have a wide range of flow rates at various pressure drops.

Small HVAC Units

Figure 53-5. Small package heating and air conditioning units include fans.

Air distribution systems are balanced by setting the required airflow rates using dampers. The airflow rate of a fan must be checked and adjusted to ensure proper airflow for the application. The required airflow rate is taken from the register or ductwork plan for the building. The airflow rate of a fan is a function of fan wheel diameter, fan wheel speed, and system pressure drop. For a fan with a given wheel diameter, the airflow rate is a function of the fan wheel speed and resistance against the fan.

Fan Selection. Data sheets containing fan performance characteristics are used when selecting fans and can be used when balancing air distribution systems. Fans are selected and tested by measuring the voltage, current, wheel speed (in rpm), and output (in cfm). **See Figure 53-6.** Data is presented on a table or as curves on a graph. When selecting a fan, the airflow rate required for the system is plotted against the total pressure drop in the system. The fan table or graph shows which fan should be used. Total pressure drop in the system is plotted against the fan wheel speed to find the airflow rate from the fan.

Fan speed control is often performed at the control board of the unit. **See Figure 53-7.** Multiple winding taps are provided in the fan motor to adjust fan speeds and resulting airflow rates.

Airflow Rates. Airflow rate is the volumetric flow rate of the air in ductwork. Air volumetric flow rate is the amount of air moved per unit of time. Airflow rates are directly proportional to fan wheel speeds. To increase the airflow rate of a fan, the fan wheel speed is increased. To decrease the airflow rate of a fan, the fan wheel speed is decreased.

Because the relationship between airflow rate and fan wheel speed is proportional, when the airflow rate of a fan at one speed is known, the airflow rate at other speeds can be found. The airflow rate of a fan is found by applying the following formula:

$$Q_F = Q_I \times \frac{N_F}{N_I}$$

where

Q_F = final airflow rate (in cfm)
Q_I = initial airflow rate (in cfm)
N_F = final wheel speed (in rpm)
N_I = initial wheel speed (in rpm)

TECH TIP

In many VAV systems, the terminal box controllers have a flow inlet sensor that measures differential pressure. The controller software scales the differential pressure reading to cfm (cubic feet per minute).

Fan Selection

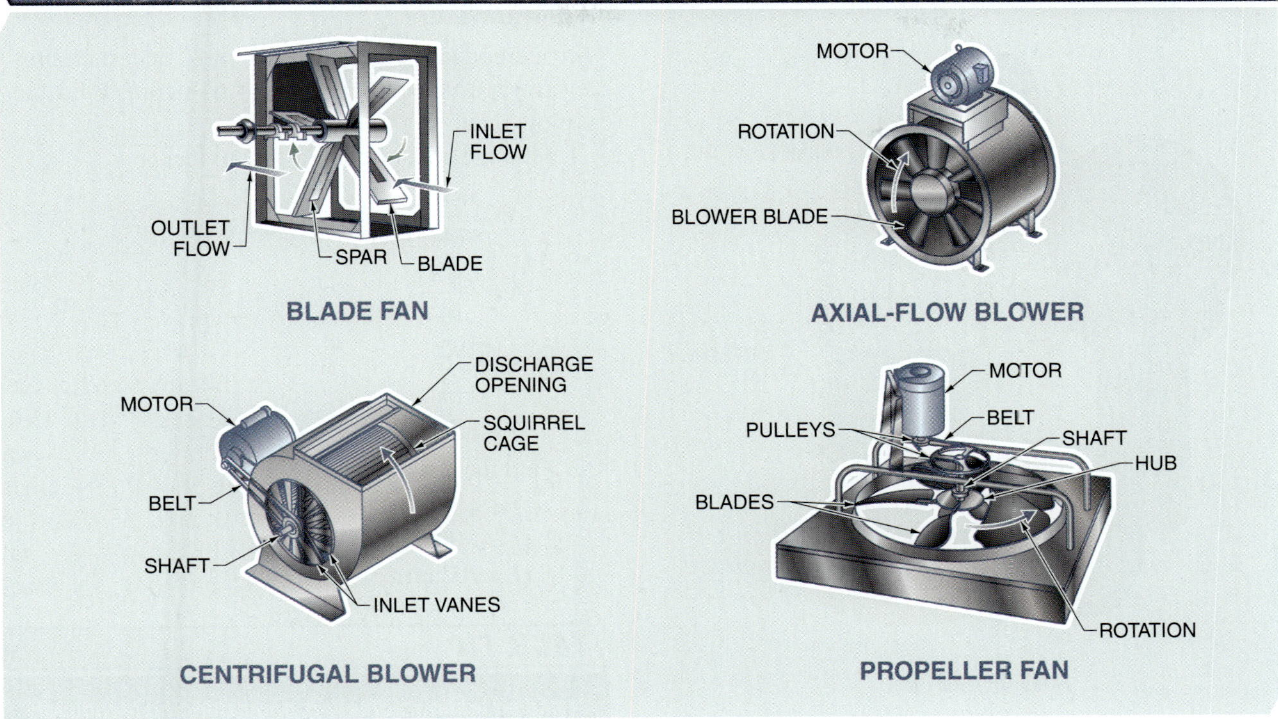

Figure 53-6. Fan type and capacity are selected based on the system and required airflow.

Control Boards

Figure 53-7. The fan speed of a package unit can be adjusted at the control board.

Example: Finding Airflow Rates—Fans

A fan produces 1800 cfm at a speed of 978 rpm. Find the airflow rate of the fan at a speed of 1200 rpm.

$$Q_F = Q_I \times \frac{N_F}{N_I}$$

$$Q_F = 1800 \times \frac{1200}{978}$$

$$Q_F = 1800 \times 1.227$$

$$Q_F = \mathbf{2209\ cfm}$$

Measuring Airflow Rates. Fan airflow rates are measured with anemometers or with pitot tubes and manometers. When the pressure difference of the air through the distribution system is known, the fan airflow rate can be calculated. Pressure drops, fan wheel speeds, and fan tables or graphs can also be used to calculate fan airflow rates.

An anemometer measures average air velocity and average airflow rate. An anemometer consists of a fan probe, which is inserted into the air stream inside a duct. The average velocity or average airflow rate is shown on a dial or scale. **See Figure 53-8.** When the velocity is read, the velocity and the duct size are used to find the airflow rate. **See Procedure for Finding Average Airflow Rates Using Anemometers.**

Measuring Airflow

Figure 53-8. Anemometers are used to measure airflow.

Example: Measuring Airflow Rates—Anemometers

An anemometer placed in a 24″ × 8″ duct measures an average velocity reading of 678 fpm. Find the airflow rate.

1. Find the area of duct in square feet.

$$A = \frac{w \times h}{144}$$
$$A = \frac{24 \times 8}{144}$$
$$A = \frac{192}{144}$$
$$A = 1.33 \text{ ft}^2$$

2. Find the airflow rate.

$$Q = A \times V$$
$$Q = 1.33 \times 678$$
$$Q = \mathbf{902\ cfm}$$

TECH TIP

A skilled testing, adjusting, and balancing technician must be able to properly use many types of electronic test instruments.

Procedure: Finding Average Airflow Rates Using Anemometers

1. When duct dimensions are 24″ × 10″, find the area of a duct cross-section in square feet. To do this, multiply duct width and height dimensions in inches, and divide by 144. Dividing by 144 converts the area from square inches to square feet.

$$A = \frac{w \times h}{144}$$
$$A = \frac{24 \times 10}{144}$$
$$A = \frac{240}{144}$$
$$A = \mathbf{1.67\ ft^2}$$

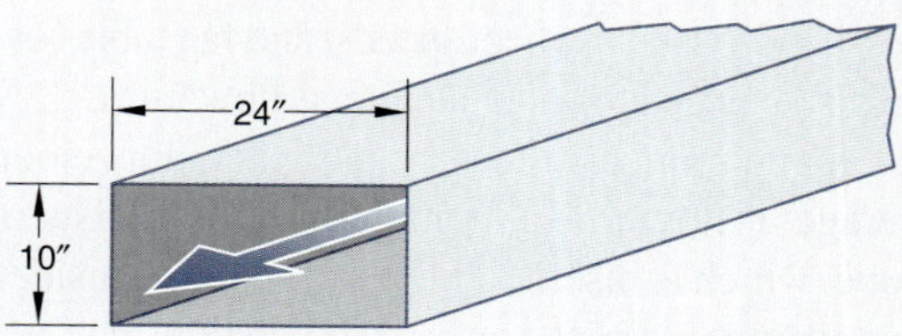

2. When an anemometer reads 53.7 fpm, find the duct average airflow. Multiply duct area in square feet by air velocity in feet per minute (fpm) to find average airflow rate in cubic feet per minute (cfm).

$$Q = A \times V$$
$$Q = 1.67 \times 53.7$$
$$Q = \mathbf{89.68\ cfm}$$

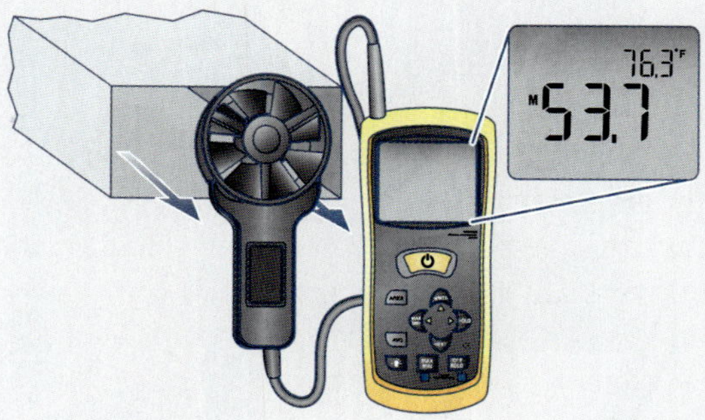

Measuring Airflow Pressure. Fan airflow pressures can be measured with pitot tubes and manometers. A *pitot tube* is a device that senses static pressure and total pressure in ductwork. A pitot tube consists of one tube inside another. The two tubes are arranged with a 90° bend near one end. On the bent end of the tube, the inside tube is open on the end, and the outside tube has a ring of small holes around it. Duct static pressure is sensed by the outside tube, while total pressure is sensed by the inside tube. **See Figure 53-9.**

Measuring Air Pressure

Figure 53-9. Pitot tubes are inserted into the duct to measure air pressure, which can be used to find airflow.

By connecting a pitot tube to a manometer, the two pressure readings oppose each other. This action indicates the velocity pressure of the fan. Velocity pressure is the air pressure caused by the flow of air. Manometers cannot read velocity pressure directly, but they can read total pressure and static pressure. Velocity pressure is found by subtracting static pressure from total pressure. **See Figure 53-10.** Velocity pressure in ductwork can be found by applying the following formula:

$$P_v = P_t - P_s$$

where

P_v = velocity pressure (in in. WC)
P_t = total pressure (in in. WC)
P_s = static pressure (in in. WC)

Velocity Air Pressure

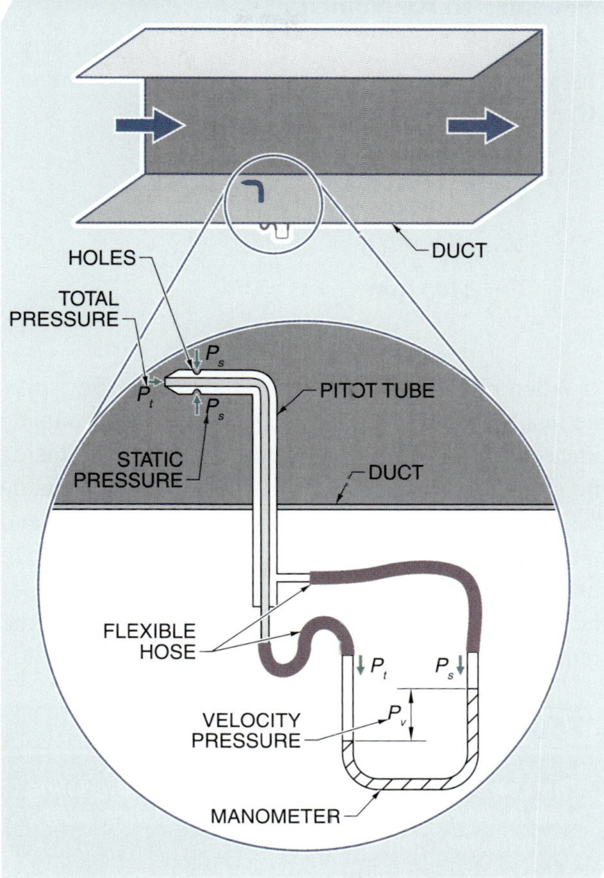

Figure 53-10. Pitot tubes measure total air pressure and static air pressure. The difference is the velocity air pressure.

Example: Finding Velocity Pressure

Pressure readings were taken using a pitot tube and manometer in a 20″ × 8″ duct. The total pressure reading is 0.68″ WC, and the static pressure reading is 0.32″ WC. Find the velocity pressure.

$$P_v = P_t - P_s$$
$$P_v = 0.68 - 0.32$$
$$P_v = \mathbf{0.36'' \ WC}$$

When the velocity pressure is known, the velocity of the airflow can be found. Air velocity can be found by applying the following formula:

$$V = 4005 \times \sqrt{P_v}$$

where

V = air velocity (in fpm)
4005 = constant
P_v = velocity pressure (in in. WC)

Example: Converting Velocity Pressure to Air Velocity

A velocity pressure reading of 0.36″ WC was taken using a pitot tube and manometer in a 20″ × 8″ duct. Convert velocity pressure to air velocity.

$$V = 4005 \times \sqrt{P_v}$$
$$V = 4005 \times \sqrt{0.36}$$
$$V = 4005 \times 0.6$$
$$V = \mathbf{2403\ fpm}$$

When pitot tubes are used to measure airflow rate, the inside tube senses total pressure while the outside tube senses static pressure. Both tubes run from inside the duct to an inclined manometer, which is outside the duct. The inside tube is connected to the high end of the manometer, and the outside tube is connected to the low end of the manometer. The static pressure of the outside tube is automatically subtracted from the total pressure of the inside tube.

TECH TIP

The correct airflow rate across heating or cooling coils is critical for proper system performance, safety, and occupant comfort.

Finding Airflow Rates. Airflow rates can also be found using the heating capacity and the temperature difference of the air as it flows through a furnace with or without an evaporator. Temperature readings are used to find the temperature difference of the air.

With the temperature difference method, the airflow rate from a heating system fan is found by using the furnace output rating and the temperature increase of the air through the furnace. Equipment catalogs and data sheets contain output ratings for particular furnaces at standard conditions. Temperature increase is found by measuring the inlet and outlet air temperatures of the furnace. The airflow rate for a heating system is found by applying the following formula:

$$Q = \frac{OR}{1.08 \times \Delta T}$$

where

Q = airflow rate (in cfm)
OR = output rating (in Btu/hr)
1.08 = constant
ΔT = change in temperature (in °F)

Example: Finding Airflow Rates—Furnaces

A furnace has an output rating of 100,000 Btu/hr. The air temperature increase through the furnace is 60°F. Find the airflow rate for the furnace.

$$Q = \frac{OR}{1.08 \times \Delta T}$$
$$Q = \frac{100{,}000}{1.08 \times 60}$$
$$Q = \frac{100{,}000}{64.8}$$
$$Q = \mathbf{1543.2\ cfm}$$

Airflow rates through the evaporator in a furnace can be found if the enthalpy of the air entering and leaving the evaporator and the cooling capacity of the evaporator are known. Enthalpy values are found on a psychrometric chart, and the cooling capacity of the evaporator is found in an equipment catalog or on a data sheet. The cooling capacity of the evaporator at existing conditions should be used. Airflow rates for evaporators, or cooling systems, are found by applying the following formula:

$$Q = \frac{OR}{4.5 \times (h_e - h_l)}$$

where

Q = airflow rate (in cfm)
OR = output rating (in Btu/hr)
4.5 = constant
h_e = enthalpy of air entering evaporator (in Btu/lb)
h_l = enthalpy of air leaving evaporator (in Btu/lb)

Example: Finding Airflow Rates—Evaporators

An air conditioner has an output rating of 36,000 Btu/hr. The enthalpy of air entering the evaporator is 30.9 Btu/lb. The enthalpy of air leaving the evaporator is 24.5 Btu/lb. Find the airflow rate through the evaporator.

$$Q = \frac{OR}{4.5 \times (h_e - h_l)}$$
$$Q = \frac{36{,}000}{4.5 \times (30.9 - 24.5)}$$
$$Q = \frac{36{,}000}{4.5 \times 6.4}$$
$$Q = \frac{36{,}000}{28.8}$$
$$Q = \mathbf{1250\ cfm}$$

Measuring Pressure Drops. Airflow rate can also be measured using the pressure drop in the system, the fan wheel speed, and a fan performance chart. A manometer is used to measure system pressure drop, and a tachometer is used to measure fan wheel speed. A *tachometer* is an instrument used to measure the speed of rotation of a shaft or wheel.

The total static pressure drop in a system is found by taking measurements with a manometer at the supply side (outlet) and return side (inlet) of the fan. **See Figure 53-11.** The return side pressure of the fan is negative, and the supply side pressure is positive.

Measuring Pressure Drops

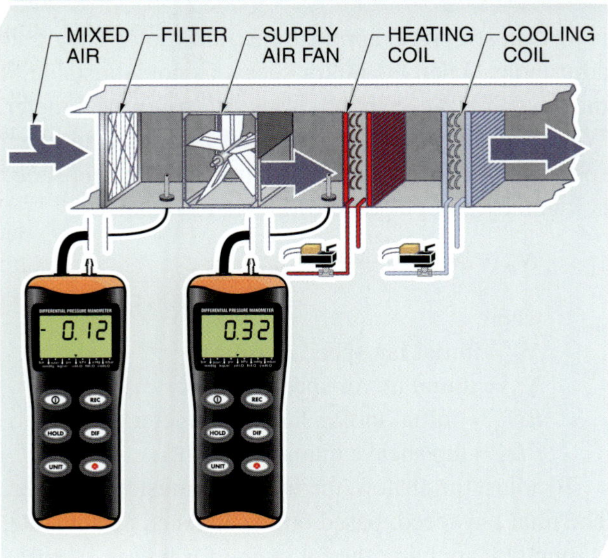

Figure 53-11. Manometers are used to measure duct air pressure in the supply side and return side of the system. The total air pressure drop for the system is the sum of the return side reading (negative pressure) and supply side reading (positive pressure).

The total static pressure drop in the system is the difference between the supply side and return side readings. Since the return side pressure is negative, the values are added, instead of subtracted, by changing the negative sign of the return side pressure to positive. The result is the total static pressure drop between the outlet and inlet of the fan. To find total static pressure drop, apply the following formula:

$$P_T = P_S - (P_R)$$

where

P_T = total static pressure drop (in psi)

P_S = supply side static pressure (in psi)

P_R = return side static pressure (in psi)

Example: Finding Total Static Pressure Drop

The pressure reading on the return side (inlet) of the fan is –0.12 psi, and the pressure on the supply side (outlet) of the fan is 0.32 psi. Find the total static pressure drop.

$$P_T = P_S - (P_R)$$
$$P_T = 0.32 - (-0.12)$$
$$P_T = 0.32 + 0.12$$
$$P_T = \mathbf{0.44\ psi}$$

As the fan wheel turns, the speed of the wheel is measured with a strobe light tachometer. A *strobe light tachometer* is a tachometer that contains a strobe light that flashes a beam of light at specified time intervals. The time interval is selected and adjusted with a dial. A spot of white reflective tape is attached to the rim of a motor sheave or fan wheel sheave. As the strobe light flashes, the time interval of the flashes is adjusted until the spot appears to be stationary. At that speed, the flashes are synchronized with the wheel rotation. The wheel speed is read from the dial on the strobe light tachometer.

CAUTION: The measured rotation could be half the actual rotation, which is checked by adjusting the strobe to twice the speed. When the location of the tape on the rim appears to alternate, the original measured rotation speed is correct.

The total static pressure drop and the fan wheel speed are necessary for finding the airflow rate of the fan. When these two values are known, the fan airflow rate is found on a table or graph.

Adjusting Airflow Rates. When the fan airflow rate is lower than the required rate for an application, the fan speed must be adjusted. Airflow rate is directly proportional to fan wheel speed. To change the airflow rate of a fan, the fan wheel speed must be changed. Fan wheel speed can be changed by changing the fan drive or selecting a different winding tap on the motor.

A fan drive is the connection between the fan motor and fan wheel. The two basic types of fan drives are belt-driven and direct-driven. **See Figure 53-12.** For belt-driven fans, a motor sheave is mounted on a motor shaft and a fan sheave is mounted on a fan wheel shaft. A belt or set of belts connects the two sheaves.

To change the airflow rate of a belt-driven fan, the fan wheel speed is changed by changing the size, or diameter, of the motor sheave. Fan wheel speed is a function of motor speed and sheave diameter. When the fan sheave and motor sheave on a belt-driven system are the same diameter, the fan wheel speed will be the same as the motor speed. On most fan drives, the motor sheave is smaller than the fan sheave. Thus, fan wheel speed is indirectly proportional to the diameters of the two sheaves.

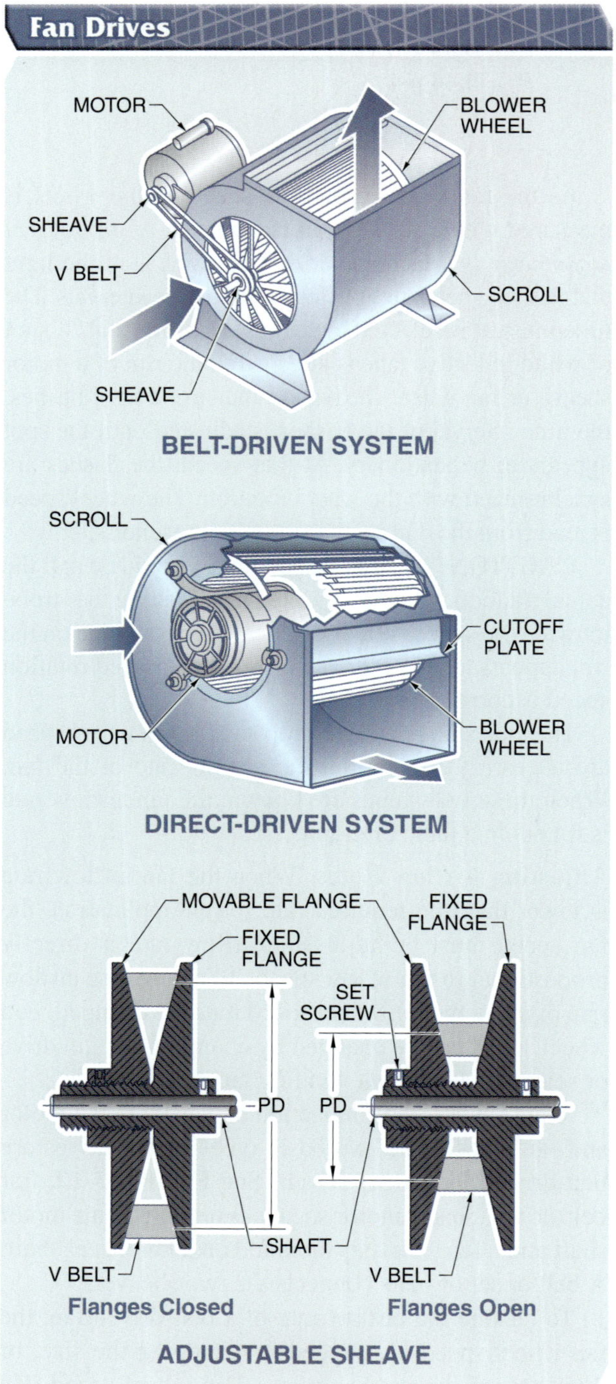

Figure 53-12. Split-system HVAC units use belt-driven or direct-driven fans.

Worn or cracked fan belts prevent a fan from producing the correct airflow volume and may fail at any time.

On most HVAC applications, the motor sheave is adjustable. When the motor sheave is not adjustable, it must be replaced with a sheave of the proper diameter. When adjusting the fan wheel speed, the initial fan wheel speed must be determined. The initial fan wheel speed is found by applying the following formula:

$$N_{FI} = N_{MI} \times \frac{PD_{MI}}{PD_F}$$

where
N_{FI} = initial fan speed (in rpm)
N_{MI} = initial motor speed (in rpm)
PD_{MI} = initial motor sheave diameter (in in.)
PD_F = fan sheave diameter (in in.)

To adjust the airflow, the fan speed must be adjusted. The final fan speed, based on the adjusted airflow rate, is found by applying the following formula:

$$N_{FF} = N_{FI} \times \frac{Q_F}{Q_I}$$

where
N_{FF} = final fan speed (in rpm)
N_{FI} = initial fan speed (in rpm)
Q_F = final airflow rate (in cfm)
Q_I = initial airflow rate (in cfm)

To adjust the fan speed, the motor sheave must be adjusted. The final motor sheave diameter, based on the final fan speed, is found by applying the following formula:

$$PD_{MF} = PD_{MI} \times \frac{N_{FF}}{N_{FI}}$$

where
PD_{MF} = final motor sheave diameter (in in.)
PD_{MI} = initial motor sheave diameter (in in.)
N_{FF} = final fan speed (in rpm)
N_{FI} = initial fan speed (in rpm)

Example: Finding Required Fan Speed

A fan produces 4500 cfm with a 6″ motor sheave and a 10″ fan sheave. The fan is driven by a 1725 rpm motor. An airflow rate of 5200 cfm is required from the fan. Find the required fan speed and the required motor sheave diameter.

1. Find the initial fan speed.

$$N_{FI} = N_{MI} \times \frac{PD_{MI}}{PD_F}$$

$$N_{FI} = 1725 \times \frac{6}{10}$$

$$N_{FI} = 1725 \times 0.6$$

$$N_{FI} = \mathbf{1035\ rpm}$$

2. Find the final fan speed to determine the required airflow rate.

$$N_{FF} = N_{FI} \times \frac{Q_F}{Q_I}$$

$$N_{FF} = 1035 \times \frac{5200}{4500}$$

$$N_{FF} = 1035 \times 1.156$$

$$N_{FF} = \mathbf{1196\ rpm}$$

3. Find the final motor sheave diameter to determine the required airflow rate.

$$PD_{MF} = PD_{MI} \times \frac{N_{FF}}{N_{FI}}$$

$$PD_{MF} = 6 \times \frac{1196}{1035}$$

$$PD_{MF} = 6 \times 1.156$$

$$PD_{MF} = \mathbf{6.933''}$$

With a direct-driven fan, the fan wheel is mounted directly on the motor shaft. The fan wheel rotates at the same speed as the motor. To change the fan wheel speed of a direct-driven fan, the motor speed must be changed. A variable-frequency drive (VFD) varies the speed of a motor and therefore, a direct-driven fan. Different motor speeds are attained by changing the motor voltage and frequency. **See Figure 53-13.**

The speed of a VFD-controlled motor is shown on a drive and motor manufacturer chart. Typically, the drive is adjusted to 60 Hz output, and the static air pressure in the system is checked. When static air pressure adjustments are needed, the VFD output is changed to produce the needed static air pressure. This adjustment is typically performed using an automatic control system.

Variable-Frequency Drives (VFDs)

LEESON Electric Corporation

Figure 53-13. VFDs are used in both forced-air and hydronic systems to control flow.

Ductwork and Airflow

The ductwork of a forced-air distribution system is sized to carry the required amount of airflow through each part of the system. The actual airflow in sections of new distribution systems differs from the design flow rate because of changes in ductwork design and the use of nominal size ducts. Most distribution systems have balancing dampers in branch takeoffs and at the supply registers. These balancing dampers are adjusted to discharge the design airflow to various building spaces.

Measuring Duct Airflow Rates. Airflow rates in ductwork can be measured using anemometers or pitot tubes and manometers. The airflow rate required in each trunk or branch of a distribution system is found in the data used for the system design. Balancing dampers in the ductwork are adjusted to modify the flow rates in the duct.

A balancing damper is typically manually operated and opened or closed with a handle that is connected to the damper shaft. The handle extends through one side of the duct. Required airflow rates through ductwork are found in the original design plans. The airflow is balanced by having the balancing damper wide open with the fan set for the proper airflow rate. The balancing damper is slowly closed until the proper flow rate is established.

Measuring Register Airflow Rates. Airflow rates are measured at each register using velometers (capture hoods). **See Figure 53-14.** A velometer measures the velocity of air flowing out of a register. A capture hood measures the amount (cfm) of air flowing out of a register. To find the airflow rate at a register, the air velocity is converted to airflow rate.

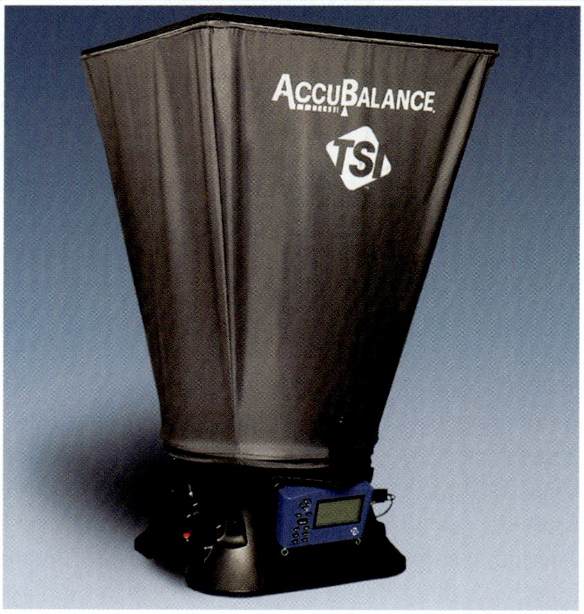

TSI Incorporated

Figure 53-14. Velometers (capture hoods) are used to measure airflow at registers.

To measure the air velocity at a large register, several velometer readings are taken across the surface of the register. The readings are then averaged. After the average velocity of the air leaving a register is found, the free area of the register is used to find the airflow rate. Alternatively, a capture hood can be used to find the airflow rate. Free area is the area of a register that allows air to pass through. The free area of a register is found in equipment catalogs and specifications sheets. To find the register airflow rate, apply the following formula:

$$Q_R = A_R \times V$$

where

Q_R = register airflow rate (in cfm)

A_R = register free area (in sq ft)

V = air velocity (in fpm)

Example: Finding Register Airflow Rates

A register with a nominal size of 16″ × 8″ has a free area of 0.684 sq ft. Air velocity at the register is 475 fpm. Find the register airflow rate.

$$Q_R = A_R \times V$$

$$Q_R = 0.684 \times 475$$

$$Q_R = \mathbf{325\ cfm}$$

Adjusting Register Airflow. The airflow at each register is regulated by the damper in the register. Airflow rate readings are taken, adjustments are made at the register damper, and readings are taken again to verify the results. When several registers are located in one duct section, all registers must be adjusted and the airflow rate at each register must be verified before proceeding to the next duct section.

Typically, all registers are fully opened at the start of the balancing process. The register closest to the supply is adjusted first, and adjustments continue to the farthest register. Changing the damper setting at one register typically causes the pressure drop to change in that duct section. The airflow rate also changes at other registers in that duct section. As the registers are readjusted, the system airflow should start to balance. Adjusting the register dampers for the proper airflow rate is a matter of trial and error.

TECH TIP

A register may be incorrectly sized if a room is reconfigured after system installation.

HYDRONIC SYSTEM TAB

For hydronic heating and cooling systems to work properly, water flow through the distribution systems must be tested, adjusted, and balanced. The circulation pump and motor are tested to ensure proper operation. The valves and control settings are adjusted to balance the water flow. Balanced water distribution systems allow the proper amount of water flow through each terminal device. Therefore, the required amount of heating or cooling is provided to building spaces. The pressure drop in each section of hydronic distribution system piping affects the water flow.

When hydronic distribution systems are balanced, the balancing valves in the piping must be adjusted. This ensures the proper water flow rate in each part of the distribution system. Water flow rate is the volumetric flow rate of water moving through pipe sections. Water flow rate is expressed in gallons per minute (gpm). Proper water flow rates are required in hydronic heating and cooling systems to provide the desired heating or cooling effect. Balancing, mixing, and diverting valves are used to balance hydronic distribution systems. These valves are typically located at strategic points in the system.

Hydronic distribution systems must be balanced because they typically cannot be installed as designed. Actual pressure drops differ from design pressure drops in distribution systems because of alterations made during the construction of the building. The components of hydronic distribution systems are typically larger than required because they are sized for maximum load conditions.

Water flow rates in hydronic distribution systems are balanced by checking the rates for one set of conditions and adjusting the settings of the balancing valves in the systems. Water flow is checked by measuring the temperature of the water at various points as it flows through sections of the piping system. Water flow rates are found by checking the pressure drop through sections of the system and using the pipe size and pressure drop.

Hydronic distribution systems are balanced by measuring the water flow rates of the circulating pump when all balancing valves in the system are fully open. Balancing valves are initially adjusted to achieve the maximum water flow rate from the pump. Balancing valves are adjusted in each branch in the piping system to ensure an adequate water flow rate in each branch. Water flow rates are adjusted at each terminal device to ensure the required water flow rate through each terminal device. The desired water flow rates are taken from the original design plans.

Hot water circulating pumps must be properly rated for pressure to overcome the total head of the system.

Circulation Pump Testing and Water Flow Rates

Circulating pumps must circulate water through the distribution system at required flow rates. Pumps are tested to ensure that the voltage and current at the pump motor are correct. Pump readings are checked against the pump curve to ensure that the pump operating point is correct. Electrical readings are checked to ensure they match the motor nameplate and system characteristics. **See Figure 53-15.** Pressures created by the pump must be equal to or greater than the pressure required to overcome the total head of the system. Hydronic distribution systems contain direct-drive pumps. Impellers of direct drive pumps turn at the same speed as the electric motor driving the pump.

Distribution systems that contain direct drive pumps are balanced by changing the resistance or head in the various sections of the distribution system. The pump has to work against the resistance in the system. Balancing valves in hydronic systems are adjusted to make the resistance in the distribution system sections the correct amount to provide the flow required in that section.

Measuring Pump Water Flow Rates. When measuring the water flow rates of distribution systems, all balancing valves in the system are opened to allow full flow. The pump is turned on and either the temperature difference or the pressure drop through the system is measured. Temperature difference is found by measuring the temperatures of the supply water and return water at the boiler or chiller.

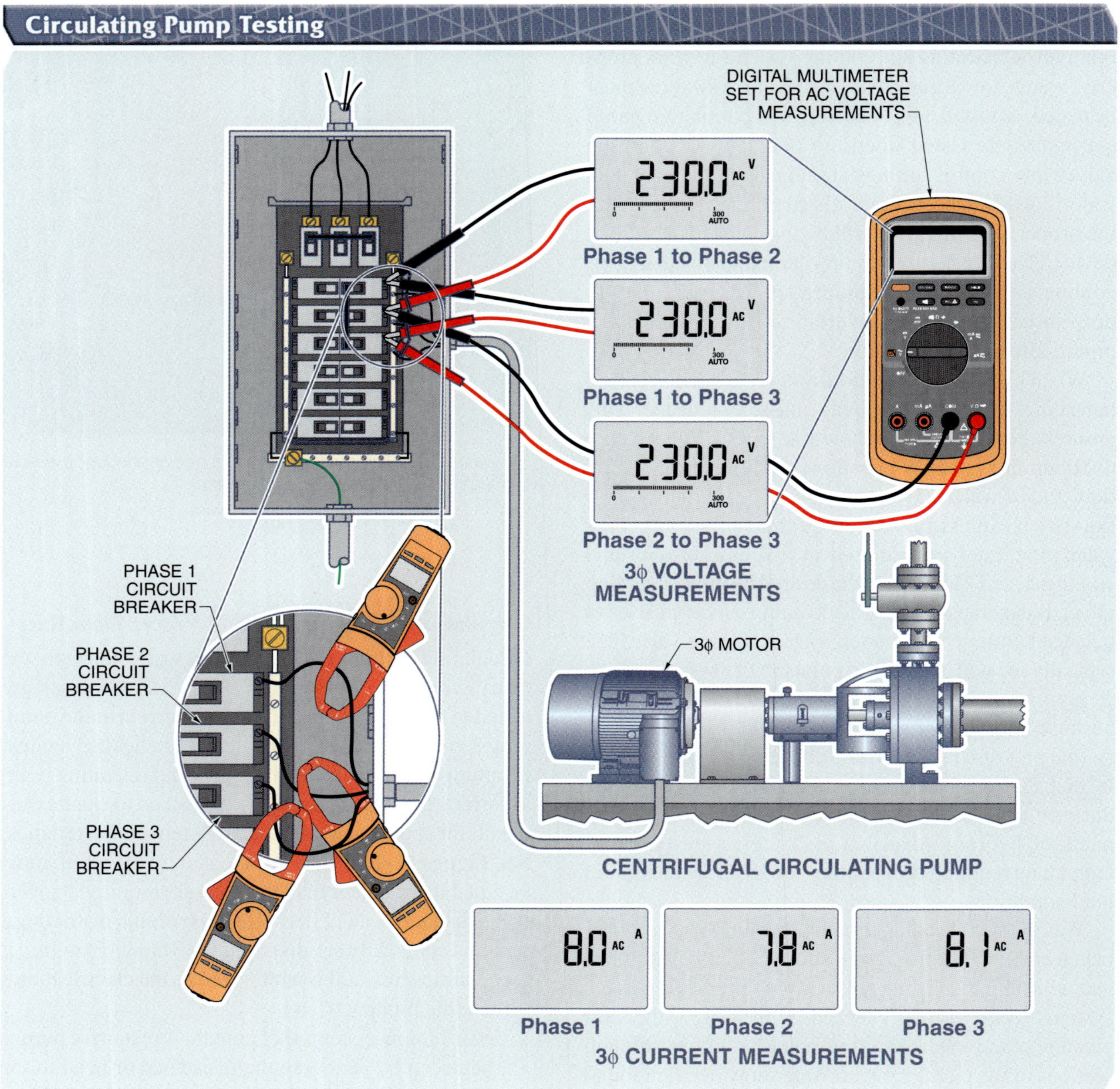

Figure 53-15. Electrical tests on the circulating pump are critical when performing a TAB on hydronic systems.

To find the water flow rate in the system by temperature difference, the temperature of the water entering and leaving the boiler or chiller is checked with thermometers located on the supply and return water pipes. The difference between the readings is the temperature decrease or increase in the system.

Finding Water Flow Rates. The water flow rate in hydronic systems is calculated by dividing the developed rate of heat transfer by the temperature difference. To find the flow rate for a given output rating and temperature difference apply the following formula:

$$Q = \frac{OR}{500 \times \Delta T}$$

where

Q = flow rate (in gpm)

OR = output rating (in Btu/hr)

500 = constant

ΔT = temperature difference (in °F)

Example: Finding Water Flow Rates—Temperature Differences

A boiler in a hydronic heating system has an output rating of 150,000 Btu/hr. The temperature difference is 26.4°F. Find the required water flow rate.

$$Q = \frac{OR}{500 \times \Delta T}$$

$$Q = \frac{150{,}000}{500 \times 26.4}$$

$$Q = \frac{150{,}000}{13{,}200}$$

$$Q = \mathbf{11.36\ gpm}$$

Finding Water Pressure Drops. Pressure drops are calculated by measuring the pressure at service ports on the supply and return pipes on each side of boilers or chillers. Service ports are typically located on balancing valves or other flow control components used in the system. **See Figure 53-16.** A service port is a valve port to which a pressure gauge or hose can be connected. The difference between the pressure at the beginning and end of the piping system is the pressure drop of the system.

Flow Balancing Valves

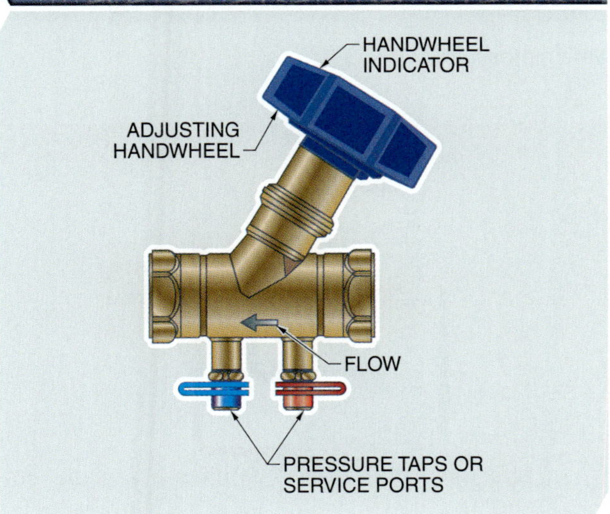

Figure 53-16. Flow balancing valves are used to adjust water flow in each branch of the piping system.

The water flow rate in a distribution system is found by measuring the pressure drop of the system. The pump flow rate for a given pump is based on the system pressure drop, which is found on a pump sizing chart for the system circulating pump. **See Figure 53-17.** A horizontal line is drawn across the chart to represent the pressure drop. The system operating condition is where the horizontal line meets the pump curve for the pump used. A vertical line is drawn to the bottom of the chart to the water flow rate. This point represents the water flow rate through the pump and the total water flow rate for the system.

Pump Curves

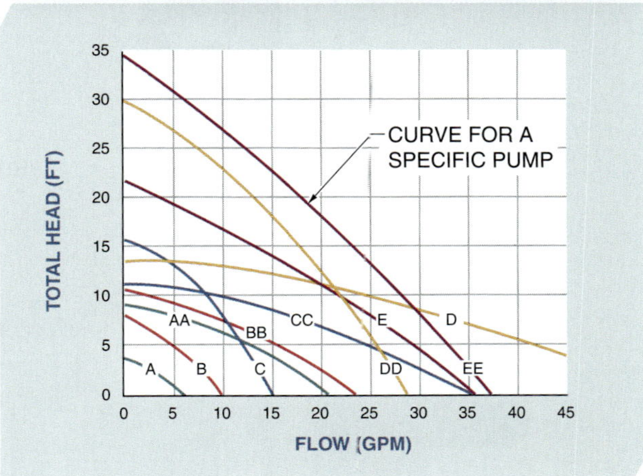

Figure 53-17. Pump curves show the relationship between the water flow rate the pump provides and the system pressure drop.

Adjusting Water Flow Rates. Water flow rates must be adjusted when the water flow rate in piping systems is greater than or less than the water flow rate required for comfort. The speed of circulating pumps can be adjusted to change the total water flow rate using VFDs. Water flow rates in the distribution systems without VFDs can be adjusted by changing the total system pressure drop. The total system pressure drop is changed by opening or closing the throttling valve. Branch line pressure drops are changed by opening or closing balancing valves in the branch line.

When the pump is the right size for the application, the water flow rate will be greater than the water flow rate needed for the application. To reduce the water flow rate to the required water flow rate, a throttling valve is partially closed. The throttling valve near the pump discharge is the primary balancing valve for the entire hydronic system. Digital hydronic manometers are used to check the pressure in hydronic systems. **See Figure 53-18.** When the hydronic pressure is read, the pressure drop through the throttling valve is used to find the water flow rate. **See Procedure for Finding Hydronic Pressure.**

Measuring Hydronic Pressure

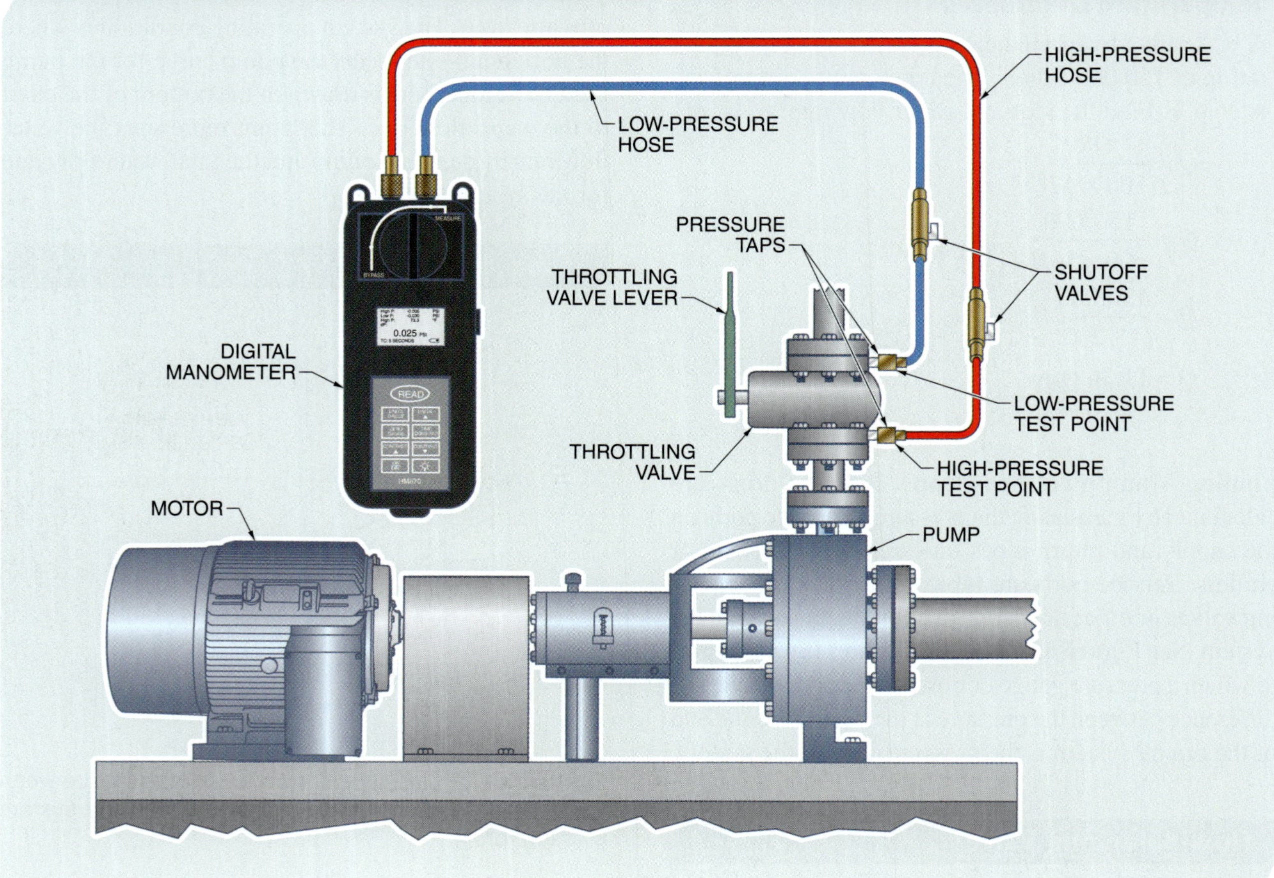

Figure 53-18. Digital manometers are used to check hydronic system pressure.

Finding Hydronic Pressure

Procedure

1. Turn off the motor/pump and secure.
2. Remove the caps from the test points. *Note:* Some water may leak from the test point during the hose connection.

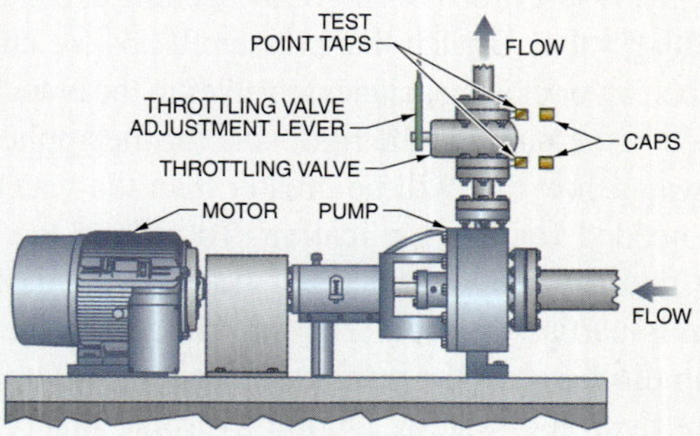

3. Connect the high-pressure hose to the valve high-pressure test point and the low-pressure hose to the valve low-pressure test point. Close hose shutoff valves.

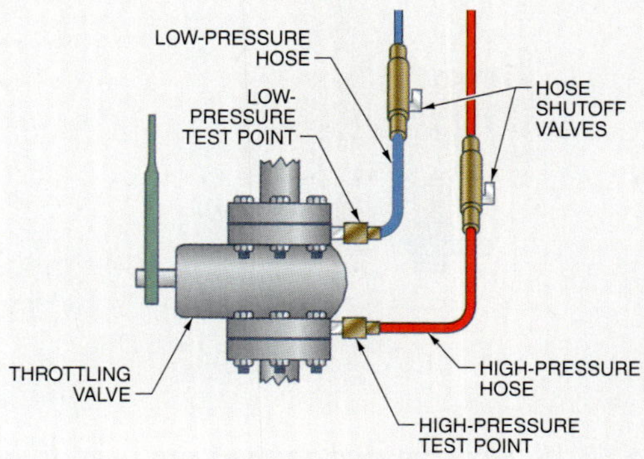

4. Turn on the motor/pump.
5. Open hose shutoff valves slowly to allow water to fill the hoses and bleed the air, then close the valves.
6. Turn on the hydronic manometer and zero the pressure setting.

continued on next page

Finding Hydronic Pressure *(continued)*

7. Connect the high-pressure hose to the manometer high-pressure test port and the low-pressure hose to the manometer low-pressure test port. Open hose shutoff valves.

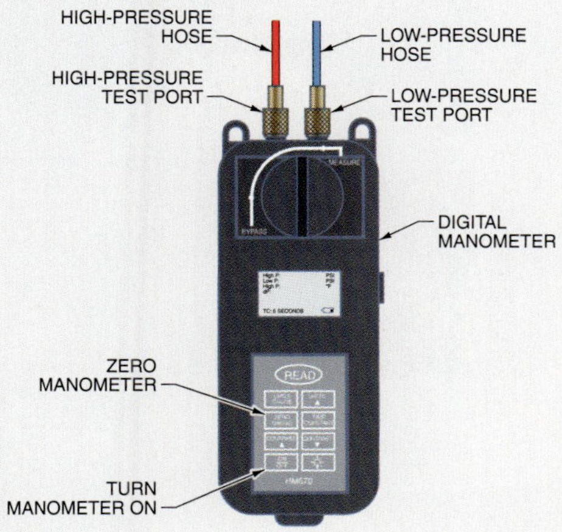

8. Read and record measured pressure readings.

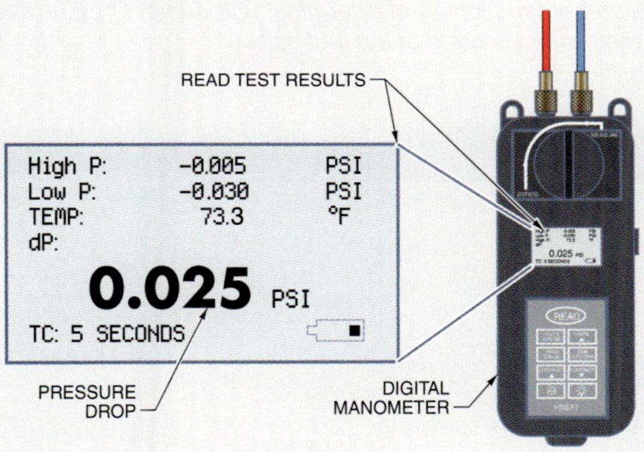

9. Adjust the throttling valve lever to adjust the water flow rate, if necessary. Read and record adjusted measured pressure readings, if necessary.

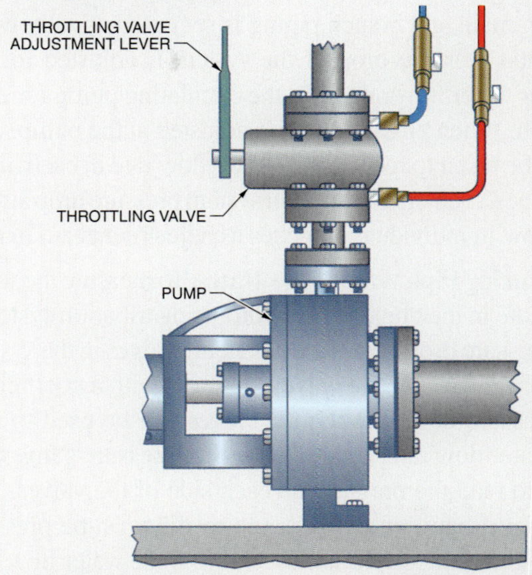

10. Turn off the hydronic manometer and close the hose shutoff valves.
11. Turn off the motor/pump and secure.
12. Disconnect the high-pressure hose from the valve high-pressure test point and the low-pressure hose from the valve low-pressure test point. Replace the caps to the test points. *Note:* Some water may leak from the test point during the hose disconnection process.
13. Open hose shutoff valves, and collect the test water from the hoses for safe disposal. Disconnect the high-pressure hose from the manometer high-pressure test port and the low-pressure hose from the manometer low-pressure test port.
14. Turn on the motor/pump to normal operation.

Balancing valves are sized by the flow rates and needed pressure drops. The balancing valves for any part of a piping system are selected for the pressure drop required at that point. The pressure drop of a valve is indicated by the flow coefficient (C_v) of the valve.

Flow coefficient is the water flow rate through a valve or any control mechanism in a piping system that causes a pressure drop of 1 psi through the valve or mechanism. To find the flow rate through a valve apply the following formula:

$$Q = C_v \times \sqrt{PD}$$

where
Q = water flow rate (in gpm)
C_v = flow coefficient
PD = pressure drop across valve (in psi)

Example: Finding Water Flow Rate Through Valves

A valve has a flow coefficient of 21. The pressure drop across the valve is 4.58 psi. Find the maximum water flow rate through the valve.

$$Q = C_v \times \sqrt{PD}$$
$$Q = 21 \times \sqrt{4.58}$$
$$Q = 21 \times 2.14$$
$$Q = \mathbf{44.9\ gpm}$$

To adjust the water flow rate through a balancing valve, the valve is opened or closed by turning the valve wheel or lever. When several branches lead off a main supply line, the balancing valves in each branch must be adjusted to provide the proper flow in that branch.

Water Flow in Mains or Branches

Water flow rates in main supply or return pipes and in each branch of hydronic distribution systems are regulated by adjusting balancing valves in the piping. The water flow rate in main and branch piping is set automatically when the total pressure drop of the system is adjusted for the proper water flow rate from the circulating pump for each branch. When pressure drop is adjusted at the pump, care must be taken to recheck the water flow rate in each of the branches. A change in the total system pressure drop affects the flow in individual branches as adjustments are made.

Measuring Pipe Water Flow Rates. To measure the water flow rate in the branches of hydronic distribution systems, the pressure drop across the balancing valves in the system must be checked. The valves are rated with a coefficient, so pressure drop through the valves can be used to find the water flow rate. Calibrated sets of pressure gauges are used to read the pressure on each side of the valves. The difference between the pressure readings is the pressure drop. Tables or graphs are used to find the water flow rate for a given pressure drop. **See Figure 53-19.**

The proper functioning of a terminal unit depends on having the correct water flow rate at the unit coil.

Finding Water Flow Rates

Flow Coefficient Values (C_v) at Various Handwheel Settings

Handwheel Setting	½″* DN† 15	¾″ DN 20	1″ DN 25	1¼″ DN 32	1½″ DN 40	2″ DN 50	2½″ DN 65	3″ DN 80	4″ DN 100	5″ DN 125	6″ DN 150
1.0	0.21	0.39	0.56	0.92	1.39	2.32	3.2	6.4	9.3	11.6	20.9
1.5	0.29	0.56	0.75	1.28	1.97	3.25	4.6	8.7	12.8	19.7	29.0
2.0	0.37	0.70	0.89	1.53	2.38	4.18	5.9	11.0	15.7	25.5	38.3
6.0	1.36	2.44	4.18	6.84	10.3	19.4	59.2	56.8	115	174	285
6.2	1.44	2.60	4.47	7.25	11.0	20.4	62.2	61.5	122	183	298
6.4	1.52	2.76	4.76	7.66	11.8	21.5	66.1	66.1	129	194	311
6.6	1.62	2.96	5.10	8.12	12.5	22.5	69.6	70.8	135	204	322
6.8	1.74	3.16	5.45	8.58	13.2	23.5	73.1	75.4	140	215	332

* North American nominal size in in.
† ISO nominal size in mm

1. DETERMINE BALANCING VALVE PRESSURE DROP IN FEET OF WATER COLUMN (6.37 PSI PRESSURE DROP = 14.7′ WC)
2. IDENTIFY VALVE SIZE (½″)
3. IDENTIFY PERCENT VALVE OPEN (HANDWHEEL SETTING: OPEN 6.8 TURNS)
4. DRAW HORIZONTAL LINE ON CHART FROM VALVE SIZE; OPEN SETTING TO C_V SCALE (1.7)
5. DRAW A LINE CONNECTING THE PRESSURE DROP OF VALVE THROUGH THE C_V SCALE AND EXTEND TO GPM SCALE (4.5)
6. READ AND RECORD GPM SCALE VALUE

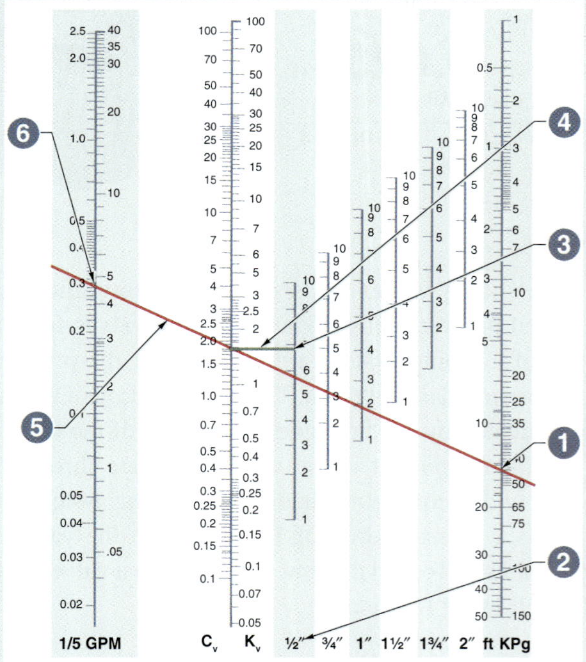

Figure 53-19. A chart can be used to calculate the water flow rate (in gpm) within a system based on pressure drop.

Adjusting Pipe Water Flow Rates. The balancing valve used for setting water flow rate by pressure drop is also used to adjust the pressure drop for the proper water flow rate in each branch of piping systems. Balancing valves have valve wheels or levers that are used for adjusting water flow rates.

Terminal Devices and Water Flow Rates

The required water flow rate in various terminal devices in hydronic systems depends on the type of system and the type of terminal device used. In one-pipe series systems, the water flow rate through each terminal device is the total water flow rate for the loop. In two-pipe systems, the water flow rate may be different through each terminal device. The water flow rate through each terminal device is based on the amount of heat required from the terminal device. The layout drawing of the system shows the water flow rate designed for each unit.

Measuring Water Flow Rates. The water flow rate through terminal devices is typically measured by the temperature difference method. The difference between the temperature of the water entering and leaving the system is a function of the heating or cooling capacity of the terminal devices. The heating capacity of a terminal device is used to find the water flow rate through the terminal device.

The water flow rate through a terminal device is found by checking the temperature drop created by the device. Inlet and outlet water temperatures are read with thermometers. Water flow rates are calculated from the known heating capacity of the terminal device and the temperature difference. **See Figure 53-20.**

Measuring Water Temperatures

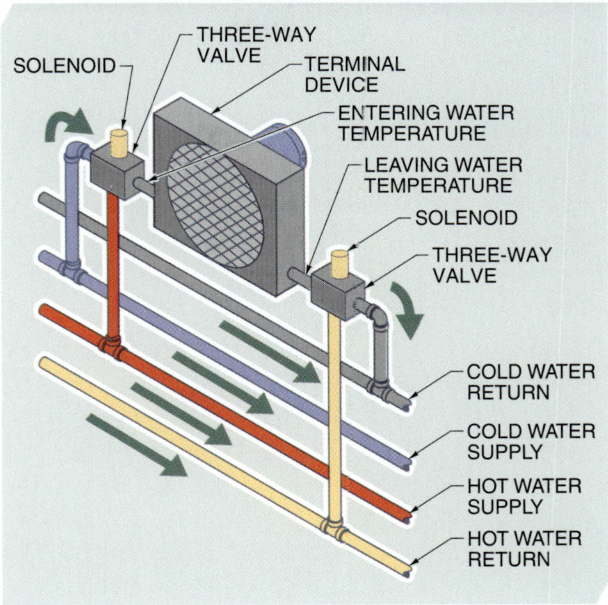

Figure 53-20. Terminal device water temperatures are measured as water enters and leaves the device.

Adjusting Water Flow Rates. The heating or cooling capacities of terminal devices are adjusted by adjusting the water flow rate through the device. The proper water flow rate for the application is selected for each terminal device. To adjust distribution systems properly, the water flow rate at each terminal device is set to the design water flow rate. Balancing valves used for balancing flow rates are located on the inlet or outlet piping to terminal devices. Water flow rates are found by pressure readings taken at pressure taps located at the valves. Valves are adjusted to maintain the proper water flow rates for that part of the system.

UNIT 54

Building Automation Principles

Today, many commercial and some residential buildings have building automation systems installed. HVAC technicians are expected to understand the basic operation of these systems. Technicians should understand the system architecture, controllers, types of input and output devices, and typical control strategies used.

OBJECTIVES

- Define the functions of a building automation system (BAS).
- Describe features of direct digital control (DDC) systems.
- List typical BAS controllers.
- Identify typical types of BAS input and output devices.
- Describe DDC control strategies and algorithms.

BUILDING AUTOMATION SYSTEMS

A *building automation system (BAS)* is a computer-based system that uses a distributed system of microprocessor-based controllers to automate any combination of building systems. A BAS may be used to control the energy-using devices in residential and small commercial buildings, but it is typically used in large commercial and industrial buildings. The BAS is the most common control system installed to control HVAC equipment, lighting, security, and other essential functions of buildings.

Sensors or controllers measure system conditions and communicate to monitoring computers in the building maintenance office. BAS controllers are connected by a wireless network or wired communication bus. A BAS can also indicate abnormal conditions, or alarms, and provide information regarding energy consumption and equipment maintenance.

Direct Digital Control Systems

A *direct digital control (DDC) system* is a control system in which electrical signals are used to measure and control system parameters from multiple microprocessor-based controllers. A central DDC system is a BAS system that uses a centrally located microprocessor-based controller to make operating decisions.

A *distributed DDC system* is a control system in which operating decisions are made by local microprocessor-based controllers with the decision-making ability distributed throughout the controlled system. Distributed DDC systems typically have a supervisory or master controller. **See Figure 54-1.** Distributed DDC systems are the BAS systems typically used today and replace central DDC systems.

TECH TIP

HVAC technicians who can work effectively on building automation systems are in high demand. Technicians with both mechanical and building automation system skills command high wages in many areas of the country.

Distributed Direct Digital Control (DDC) Systems

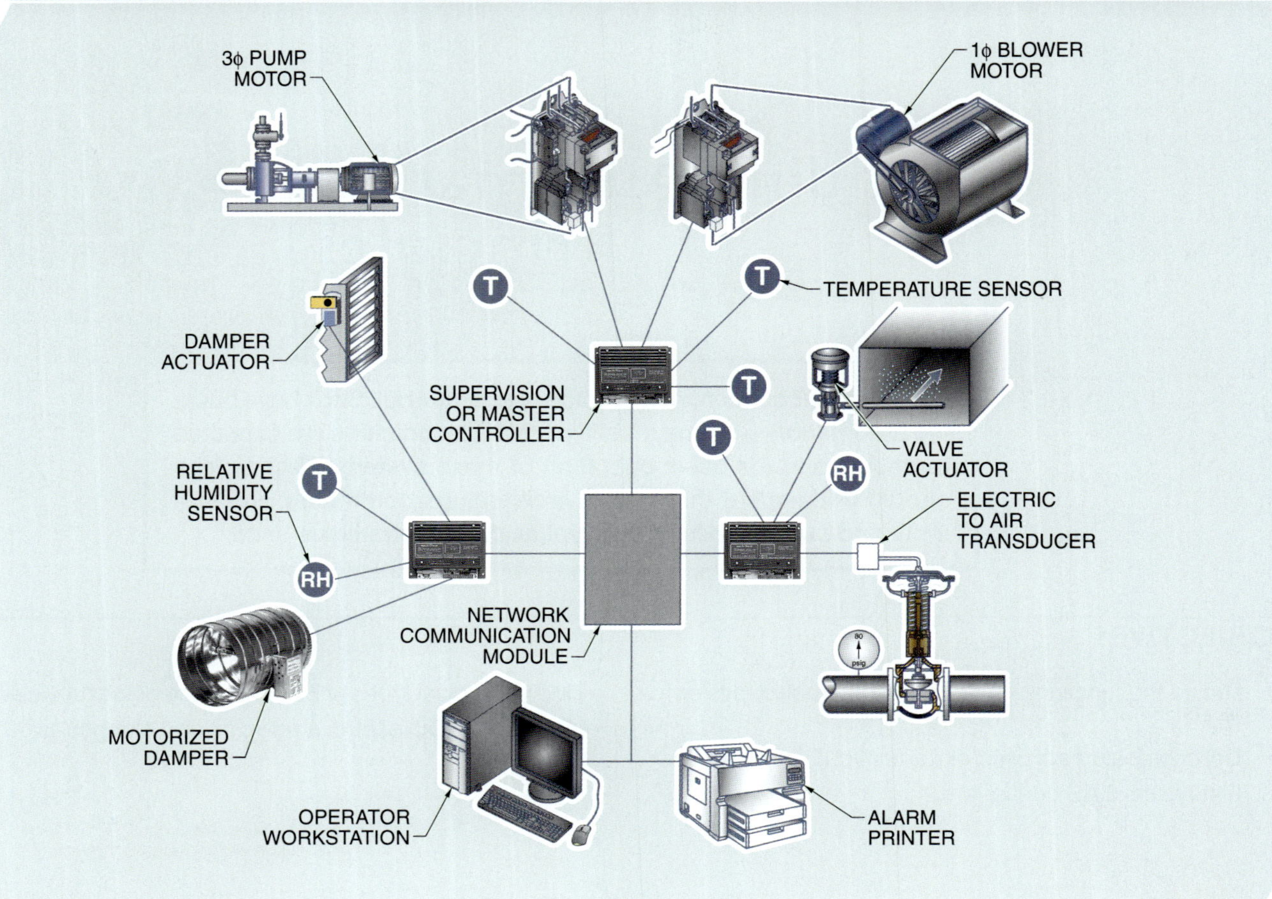

Figure 54-1. In a distributed DDC system, microprocessors at the local level make operational decisions for the local devices.

DDC System Architecture. The supervisory controller of a distributed DDC system provides communication support, a convenient connection to a personal computer (PC) or Internet network for monitoring purposes, and connections for global data. *Global data* is data that is needed by all controllers in a network. Global data includes outside air temperature and electrical demand.

Stand-alone control is the ability of a microprocessor-based controller to function on its own. Stand-alone control is less risky because a controller failure is local and has minimal effect on the entire system. Stand-alone control enables a system to function without the operation of a failed controller, which increases the reliability of the BAS.

Controllers in a network commonly share information using communication protocols. A controller that needs information from a failed controller in the network is unable to access that information. In that case, the controller continues to function at a reduced level or efficiency.

Stand-Alone Controllers Media Clip

DDC System Functions. Distributed DDC systems have the same control functions as central DDC systems, which include duty cycling, electric demand control, and time clock functions. Distributed DDC systems also can perform precise closed loop temperature control, humidity control, and pressure control operations.

Distributed DDC systems have improved reliability and increased capacity over central DDC systems. Because the controllers in the distributed DDC system can stand alone, distributed DDC systems are modular and easily expandable. A BAS can be started with a minimum number of controllers and expanded as required.

BAS CONTROLLERS

BASs such as distributed DDC systems require several controllers. When connected by a communications network, the controllers provide comprehensive control and monitoring of the HVAC equipment in a building. Controllers are often referred to as control modules, modules, and panels. BAS controllers include application-specific controllers, universal input-output controllers, and network communication modules.

Application-Specific Controllers

An *application-specific controller (ASC)* is a controller designed to control only one type of application. ASCs are typically used to control HVAC systems. **See Figure 54-2.** ASCs are more cost-effective than other types of controllers because most of the software programming is done by the manufacturer. Typical ASCs include unitary, air-handling unit, and variable-air-volume box controllers.

Application-Specific Controllers (ASCs)

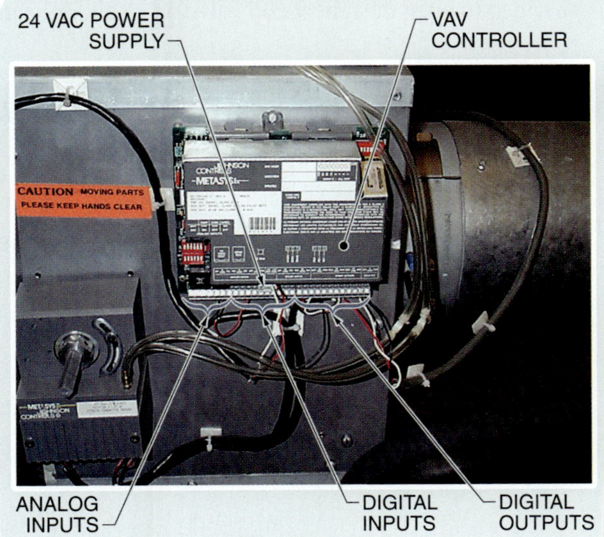

Figure 54-2. ASCs control one type of HVAC application, so they are usually easy to program.

Unitary Controllers. A *unitary controller* is a controller designed for basic zone control using a standard wall-mount temperature sensor. Unitary controllers are designed to control packaged HVAC equipment such as rooftop units, heat pumps, and fan coil units. Unitary controllers are typically compact to allow mounting at the packaged HVAC equipment. Weathertight and heated enclosures are often used, which allow operation of multiple stages of heating and cooling, economizer dampers, heat pump reversing valves, and supply fans. Some unitary controllers provide supply-air sensing capabilities, switches and/or sensors to determine airflow, and dirty-filter-condition switches (differential pressure switches).

Air-Handling Unit Controllers. An *air-handling unit (AHU) controller* is a controller that contains input and output signals required to operate large central-station AHUs. AHU controllers control humidity levels, static pressure, and indoor air quality. **See Figure 54-3.** Many central-station AHUs have complex control sequences. Dual-duct and multizone AHUs have needs that cannot be met by a unitary controller. AHU controllers are larger and more complex than unitary controllers. Also, AHU controller software is more complex than unitary controller software to allow the handling of the complex control sequences.

Air-Handling Unit (AHU) Controllers

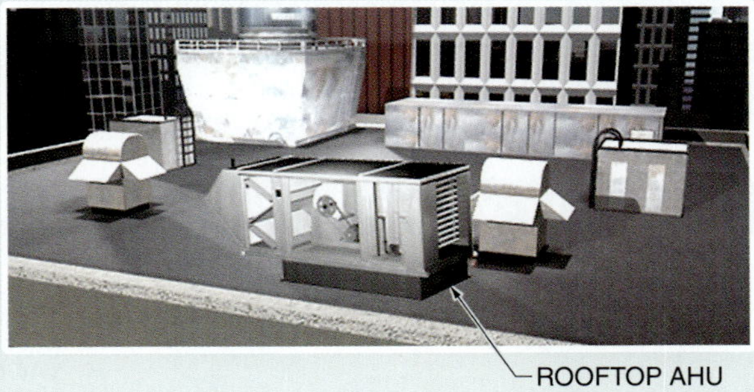

Figure 54-3. AHU major functions are controlled by air-handling controllers.

Air-Handling Unit Controllers Media Clip

A typical AHU controller application is the control of a variable-air-volume unit to maintain a minimum static pressure in the supply duct. **See Figure 54-4.** In this application, heating and cooling control is performed to maintain a constant discharge air temperature of 55°F. An output (static pressure sensor) from the AHU controller is provided to increase the volume of air provided by the supply fan. The air volume increase is produced by controlling the speed of the fan motor using a variable-frequency drive (VFD) or by opening dampers on the fan to admit a greater amount of airflow.

Variable-Air-Volume Box Controllers. A *variable-air-volume (VAV) box controller* is a controller that modulates the damper inside a VAV terminal box to maintain a specific building space temperature. VAV box controllers are similar in appearance to unitary controllers. The difference is in the manufacturer programming of the EPROM chip. **See Figure 54-5.**

VAV box controllers modulate the primary damper inside the VAV terminal box. Reheat valves can also be controlled. A flow input to the controller can be set up to indicate airflow (in cfm). The controller turns the fan inside the VAV terminal box on and off based on the input. The software of many VAV box controllers allows VAV terminal boxes to be checked for proper air balancing (proper cfm of air).

VAV Box Controllers Media Clip

VAV box controllers are also used in pressure-independent VAV terminal boxes with reheat applications. VAV box controllers are used to control building space temperatures and measure and control the amount of airflow. **See Figure 54-6.** In addition to controlling building space temperature, a flow sensor is installed at the inlet to the VAV terminal box. The flow sensor (differential pressure sensor) is wired to the VAV box controller to measure airflow. Information on airflow is used by the controller to position the VAV damper. An output connection from the controller is provided to open the reheat coil valve on a call for heat.

Variable-Air-Volume (VAV) AHU Applications

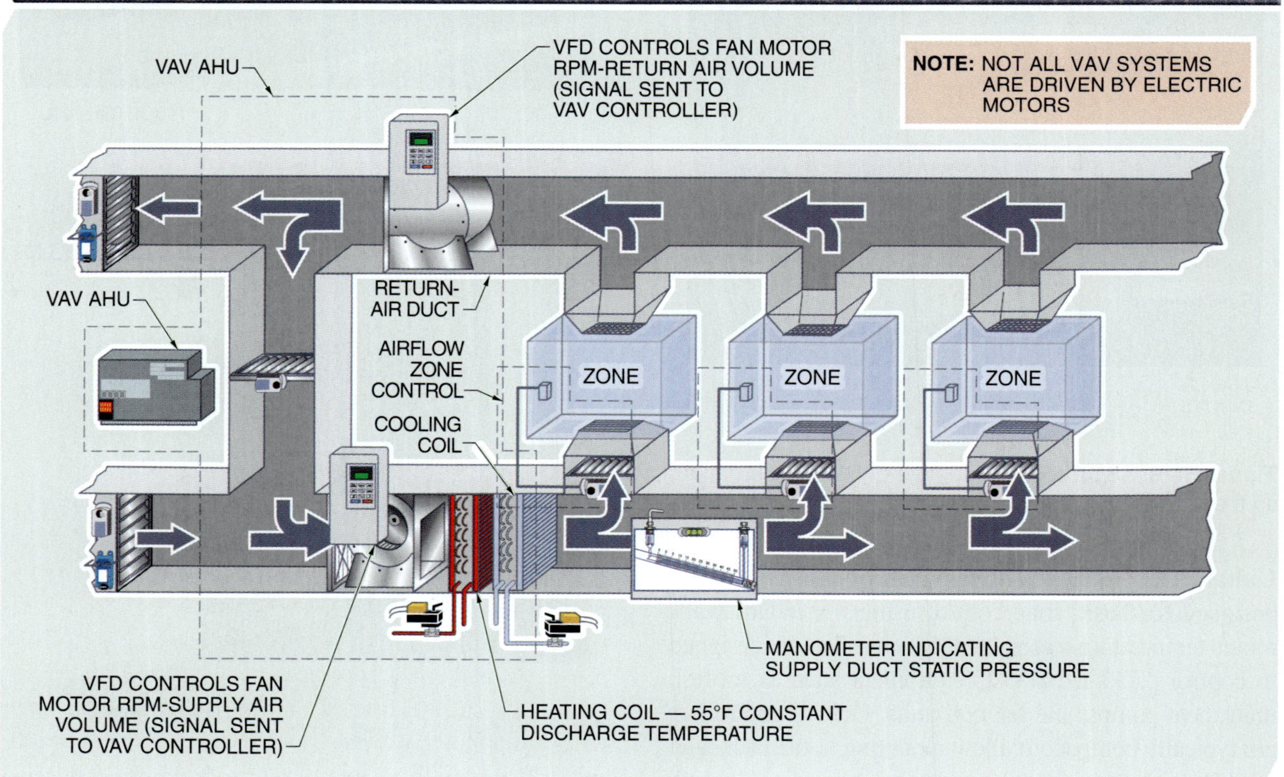

Figure 54-4. VAV AHUs adjust airflow while discharging air at 55°F.

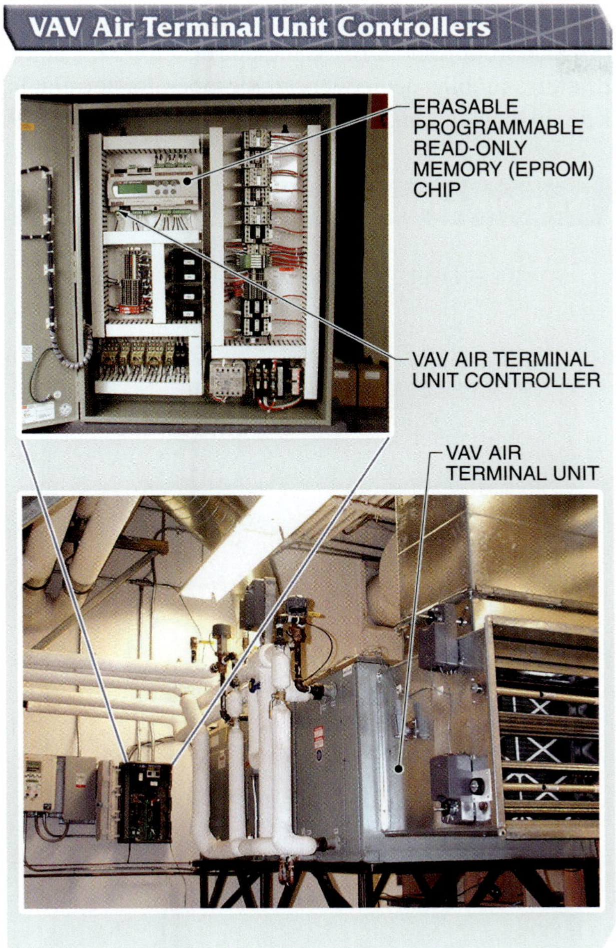

Figure 54-5. The VAV box controller program is in the EPROM memory chip.

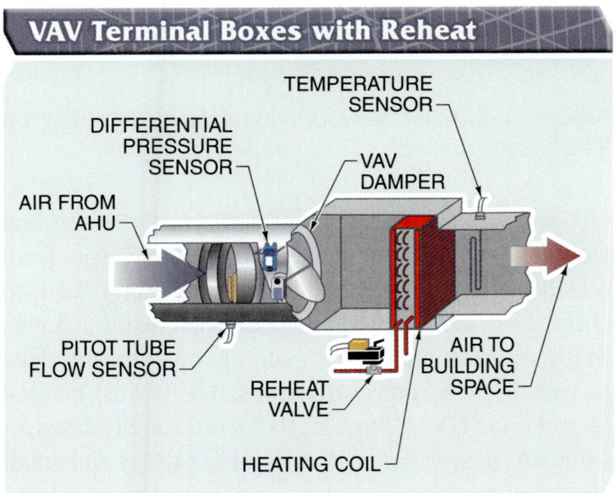

Figure 54-6. VAV controllers control air dampers and heating coils when reheat is needed.

Other ASCs. ASCs are available for computer room air conditioning (CRAC), reciprocating or centrifugal chillers, and indoor air-quality calculations. Most manufacturers install ASCs on equipment, typically referred to as an original equipment manufacturer (OEM) installation. Field installation and retrofit problems are minimized by purchasing HVAC equipment with ASCs installed by the manufacturer.

Universal Input-Output Controllers

A *universal input-output controller (UIOC)* is a general controller designed to control most HVAC equipment. Many UIOCs have the same two-piece construction as ASCs. UIOCs have a fixed number of inputs and outputs, with 8 and 16 being the most common. Most UIOCs use a 24 VAC power supply.

ASCs are manufacturer programmed and can control only the equipment for which they are designed. UIOCs are used to control most HVAC equipment. UIOC programming is blank when shipped from the manufacturer and must be custom programmed in the field. Usually the manufacturer provides software tools and interface cables. UIOC programming requires a high level of familiarity with the manufacturer's software.

Controller Programming Media Clip

TECH TIP

Universal input-output controllers from different manufacturers use different software packages and programming tools. These devices require familiarity with the software and can be difficult to program after the system is installed.

Network Communication Modules

A *network communication module (NCM)* is a controller that coordinates communication between controllers in a network and provides locations for operator interface. **See Figure 54-7.** Typically, BASs do not allow individual controllers to communicate directly with each other. In most applications, an NCM collects information from individual controllers and distributes it to other controllers in the network. One NCM is typically required in small- to medium-size buildings that have up to 200 controllers.

Controller Network Communication Modules (NCMs)

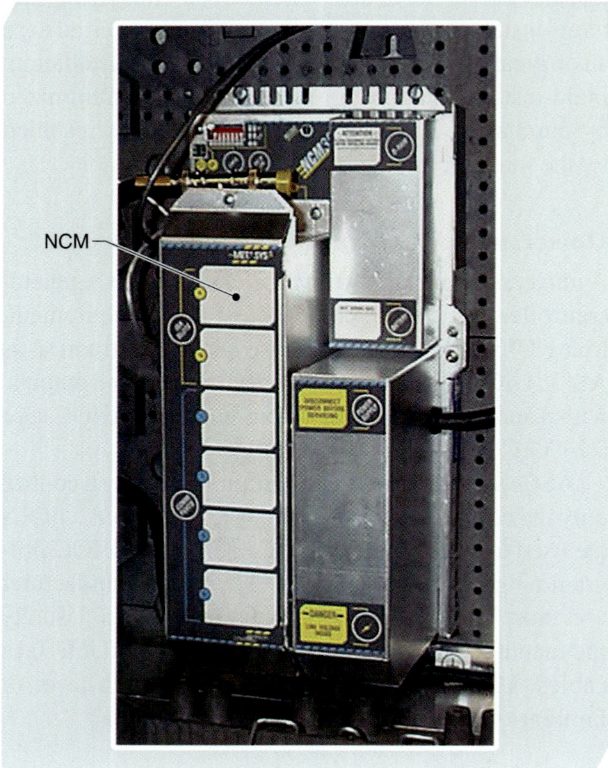

Figure 54-7. The primary function of NCMs is communication between controllers in the network.

Large installations may have several NCMs that are networked together. Typically, large NCMs, also known as web-based supervisory controllers, use web browsers instead of manufacturer specific software. See Figure 54-8.

There are many methods by which individuals can use and access information from a BAS such as the use of an operator interface. An *operator interface* is a device that allows individuals to access and respond to BAS information. Operator interface methods include the use of on-site and off-site devices and retained server data. Proper performance of the system depends on correctly using the appropriate operator interface device.

Operator Interfaces Media Clip

On-Site Operator Interface Methods. Maintenance or management personnel of medium and large buildings, such as hospitals, colleges, and office buildings, commonly access and troubleshoot BASs as part of their normal work responsibilities. On-site devices used to access and troubleshoot BASs include desktop PCs, alarm printers, mobile devices, portable operator terminals, keypad displays, and dumb terminals. **See Figure 54-9.**

Network Control Modules

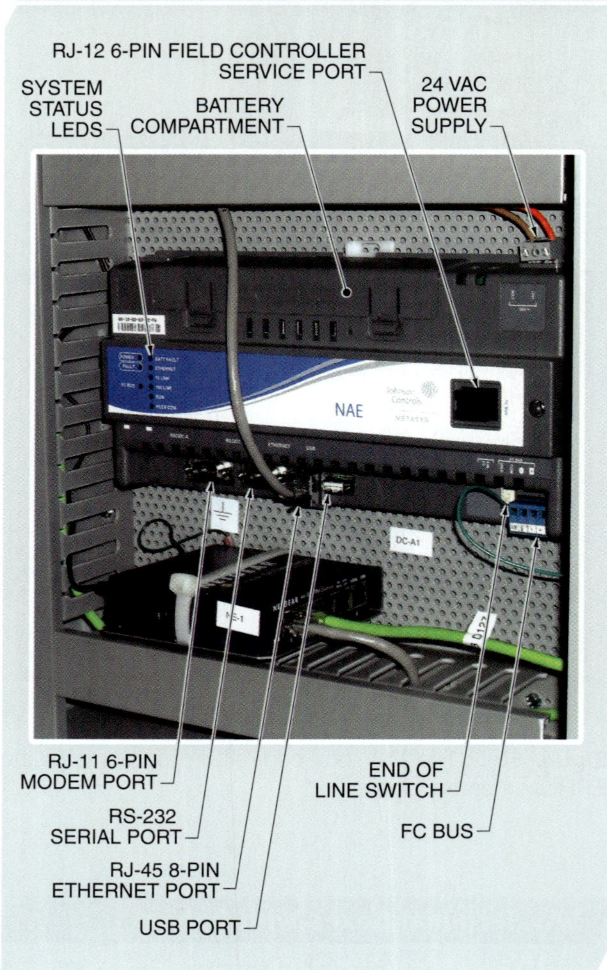

Figure 54-8. Most network control modules today are web-based.

Off-Site Operator Interface Methods. Operators and technicians may need to access HVAC systems from outside the building. For example, an operator needs to check the system status during a weekend while at home. Many methods are used for off-site operator interface to view system control functions. The typical method of accessing HVAC systems today is through a browser using an Internet connection. Text messages and email alarm functions are commonly sent to system operators. Browser-based tablets and smartphones are increasingly used to receive system updates. **See Figure 54-10.**

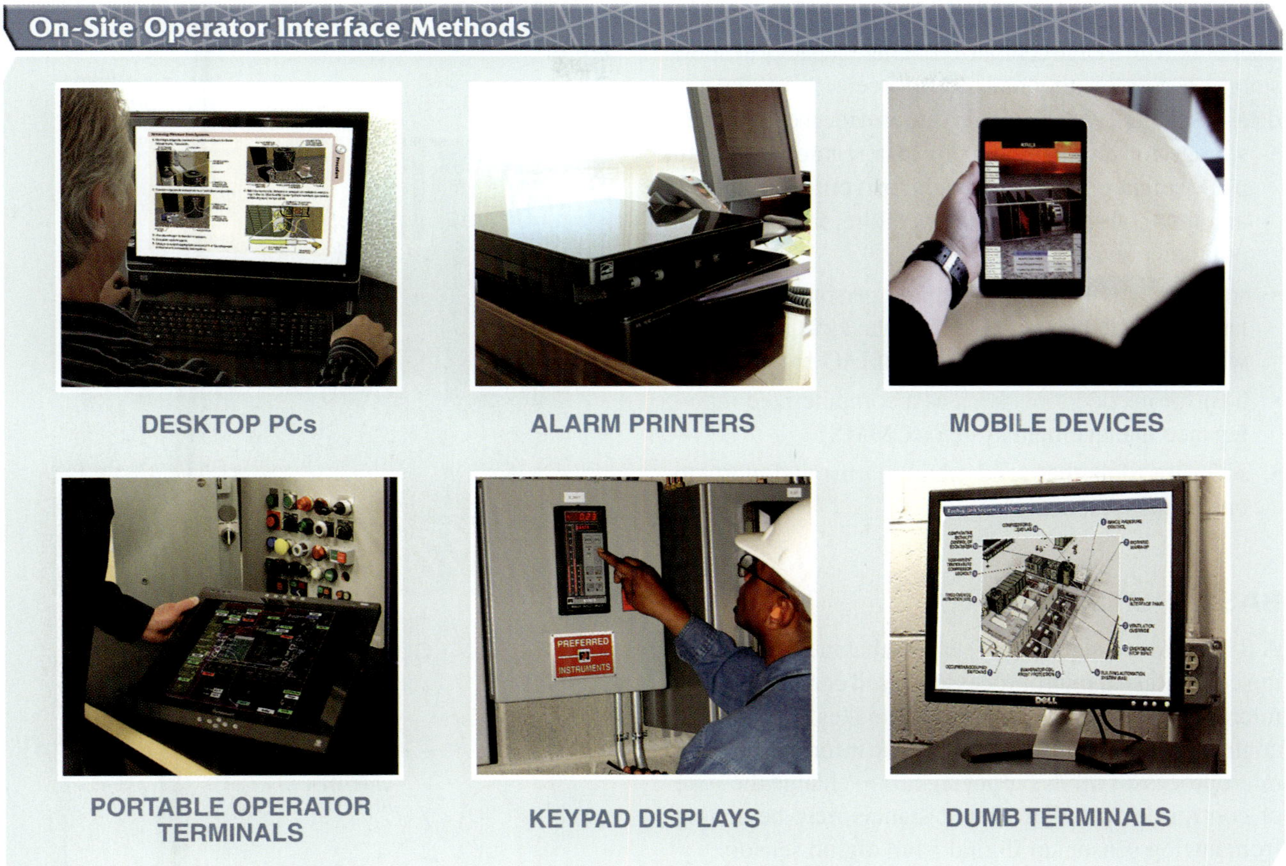

Figure 54-9. There are many different types of on-site operator interfaces available to access and troubleshoot BASs.

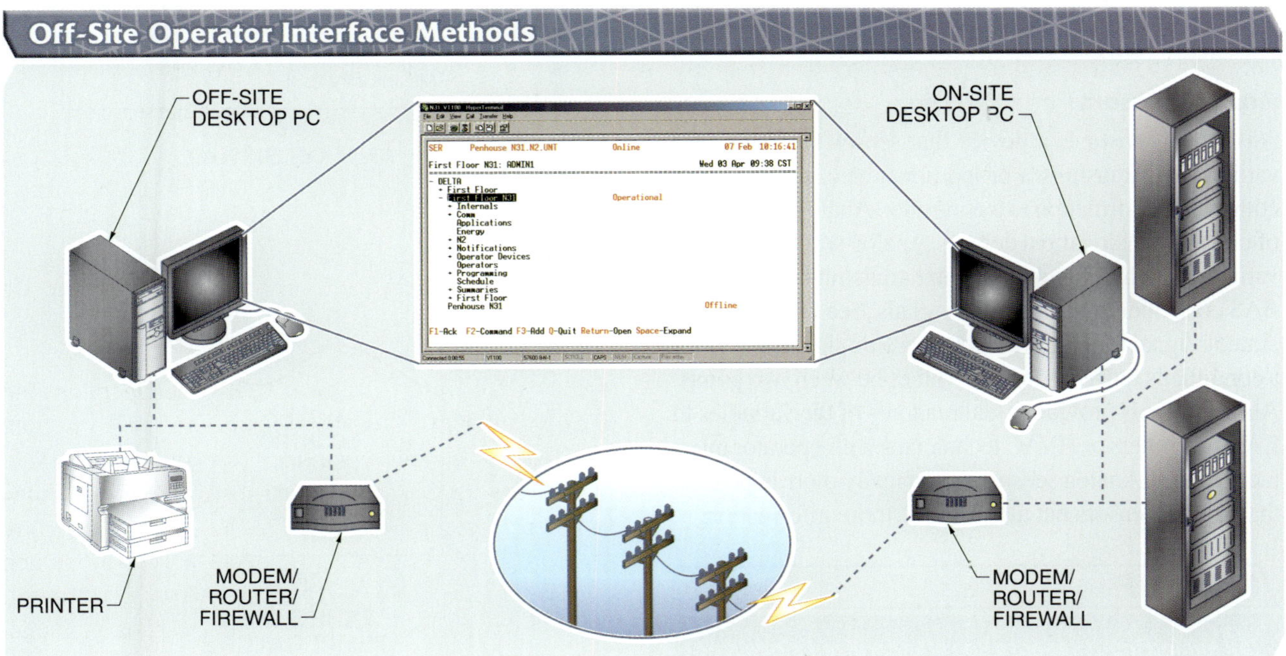

Figure 54-10. A BAS with a web browser software package allows technicians to communicate, troubleshoot, and change settings from anywhere.

Server Data. A server computer is a computer attached to a network with a large centralized storage capacity to allow storage, access, security, and backup for electronic files and data. When used with BASs, servers provide the following:

- storage for large amounts of data regarding system performance such as trends in ventilation flow amounts, fan operation, temperature, humidity, and pollutant measurements
- software tools that permit operating personnel and regulatory agencies access to stored data; especially useful when trying to determine causes of IAQ problems
- host for needed tools such as a computerized maintenance management system (CMMS)
- software tools that send email alarms and alerts to operating personnel

BAS INPUTS

A *BAS input* is a signal from sensors that indicates building conditions to microprocessor-based controllers. The microprocessor of BAS controllers makes decisions on controlling a system based on input information. The microprocessor sends output signals to change the state of control devices. BAS input signals may be analog from analog sensors or digital from digital sensors.

I/O Wiring Media Clip

Analog Sensors

An *analog sensor* is a device that senses the value of a variable and transmits a proportional electrical signal to convey the information to a controller. Analog sensors are often called analog input devices. Analog sensors measure variables, such as temperature, and transmit signals to the BAS controller as analog input signals. **See Figure 54-11.** Analog signals are transmitted electronic signals that have a continuous range of possible values between two points. Analog signals provide actual readouts of the variables to BAS controllers or HVAC technicians with operator interface devices. Analog sensors are typically more expensive than digital sensors but also provide more information.

TECH TIP

All analog input sensors use standardized signals of either resistance at 1 kΩ or 10 kΩ, current at 0 mA or 4 mA to 20 mA, or voltage at 0 VDC to 10 VDC.

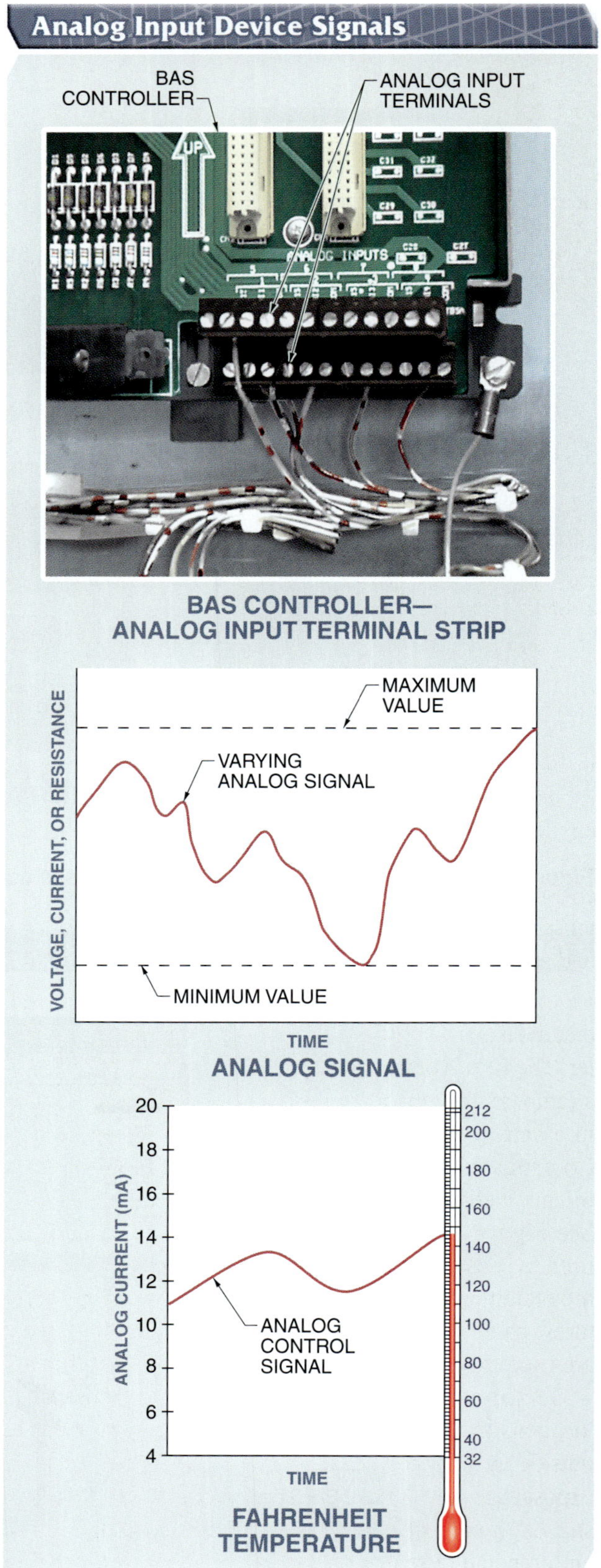

Figure 54-11. Analog input signals are proportional electrical signals of resistance, voltage, or current.

Temperature and humidity sensor accuracy can be checked with a calibrated handheld measuring device.

Analog sensors must be selected for particular tasks in buildings. Factors considered include the price and reliability of the sensor, proposed location, operating environment, dirt, and humidity. Once the sensor is selected, following manufacturer installation procedures ensures proper device operation. Analog sensors include temperature sensors, humidity sensors, pressure sensors, and specialized analog sensors.

Temperature Sensors. A temperature sensor is a device that measures temperature and sends a signal to a controller. The most commonly used analog sensor in a BAS is a temperature sensor. Temperature sensors are available in a variety of packages and mounting configurations. Common temperature sensors include wall-mount, duct-mount, well-mount temperature, and averaging sensors. **See Figure 54-12.** Wall-mount temperature sensors are used to sense air temperatures in building spaces. Duct-mount temperature sensors are used to sense air temperatures inside ductwork. Well-mount temperature sensors are used to sense water temperature in piping systems.

Temperature sensors must be placed in appropriate locations to sense the desired temperature. For example, outside air temperature sensors must be protected from exposure to direct sunlight. Manufacturers provide sunshields to prevent direct sunlight from affecting outside air temperature sensor operation. Averaging temperature sensors are used in large duct systems to obtain accurate temperature readings.

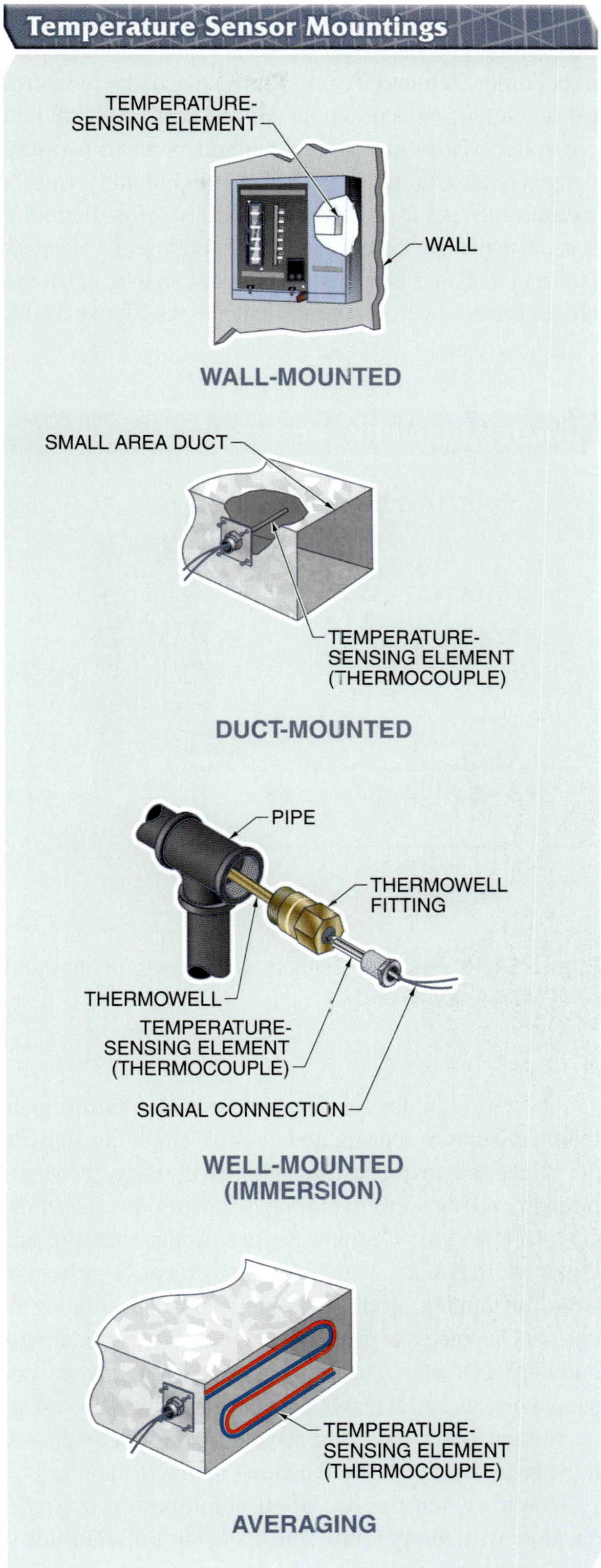

Figure 54-12. The temperature sensor used is determined by the application.

Humidity Sensors. Human comfort, product integrity, and corrosion prevention require control of the humidity level in buildings. A humidity sensor is a device that measures the amount of moisture in the air and sends a signal to a controller. The most common humidity sensors measure percent relative humidity (% RH). Other humidity sensors measure dewpoint or absolute humidity. Most humidity sensors use hygroscopic elements. A *hygroscopic element* is a material that changes its physical size or electrical characteristics as the humidity changes. **See Figure 54-13.**

Analog Humidity Sensors

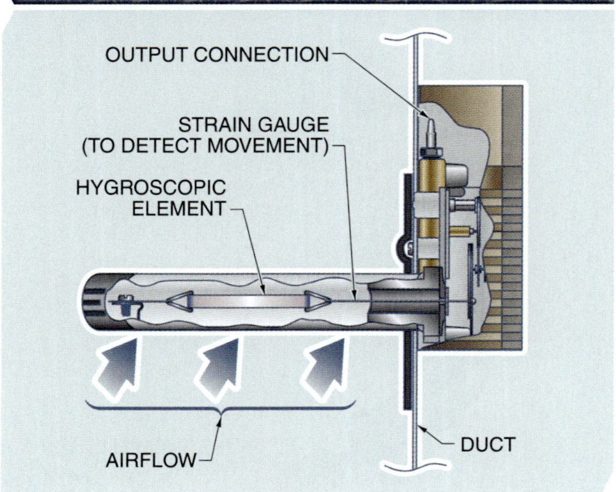

Figure 54-13. Humidity sensors are used to control and alert of humidity levels.

Accurate humidity sensing and control is more difficult than temperature sensing and control. Humidity sensors are susceptible to dirt accumulation on the element. Many humidity sensors drift over time, with common inaccuracies of 1% per year. Certain gases or solvents, when present in the air, may damage the sensing element or prevent it from functioning properly. Proper sensor functioning is essential in process or product storage environments where humidity accuracies must be maintained. Humidity sensors should be checked at least once per year for accuracy. Any instrument used to check the accuracy of humidity sensors must be regularly checked against a known standard.

Humidity sensors are often combined in a single package with temperature sensors to simplify humidity and temperature sensing. Humidity sensors use the same 4 mA to 20 mA and 0 VDC to 10 VDC standard signal values as temperature sensors.

Pressure Sensors. A pressure sensor is a device that measures the pressure exerted by a fluid, such as air or water, and sends a signal to a controller. Pressure sensors are used in HVAC applications to provide BAS controllers with an indication of the actual pressure inside a duct or pipe. Pressure sensors can also use piezoelectric elements (pressure-sensitive crystals). Pressure sensors use bellows elements in piping applications. **See Figure 54-14.** Standard pressure-sensor output signal values include 4 mA to 20 mA and 0 VDC to 10 VDC.

Pressure sensors, including the tubing and accessories, must be installed according to manufacturer specifications. Pressure sensors located or installed incorrectly sense pressure incorrectly. Pressure sensors must have an adequate range for the expected system pressures. For example, a sensor with a 0″ WC to 2″ WC range should not be used when the duct static pressure is higher upon startup. Some pressure sensors require an additional power supply and may require three or four wires instead of two. Many pressure sensors are polarity-sensitive.

Indoor Air Quality Analog Sensors. Specialized analog sensors are available that measure variables needed for effective indoor air quality (IAQ) control. For example, gas sensors are used to detect the amount of explosive or dangerous gas in the air. Refrigerant sensors are used to detect the amount of refrigerant in the air caused by leaks. Carbon dioxide sensors are used to measure the level of carbon dioxide due to human activity. Carbon monoxide sensors are used to measure the amount of toxic exhaust fumes inside parking garages. Analog sensors from devices such as VFDs are often used to indicate the motor speed (in rpm), current, and frequency (in hertz) of a BAS.

One IAQ application of a pressure sensor in a duct is an airflow measuring station. An airflow measuring station includes a number of sensors at regular intervals in a duct. The sensor location is the same as that used when performing a duct traverse. There are different types of airflow measuring stations available. For example, a pitot tube type has pressure outlets at the array that are connected to a differential pressure transmitter. The differential pressure transmitter is wired to the BAS.

After mathematical scaling is performed at the controller, direct readouts are available. Readouts from airflow measuring stations can be used for fan control, alarming, and long-term data storage. A second type of airflow measuring station uses electronic resistance temperature sensors in a duct. Current is passed through wires in the duct. Resistors are placed at intervals corresponding to the pitot tube traverse measuring points. The temperature at the sensors is related to the airflow across the array.

Analog Pressure Sensors

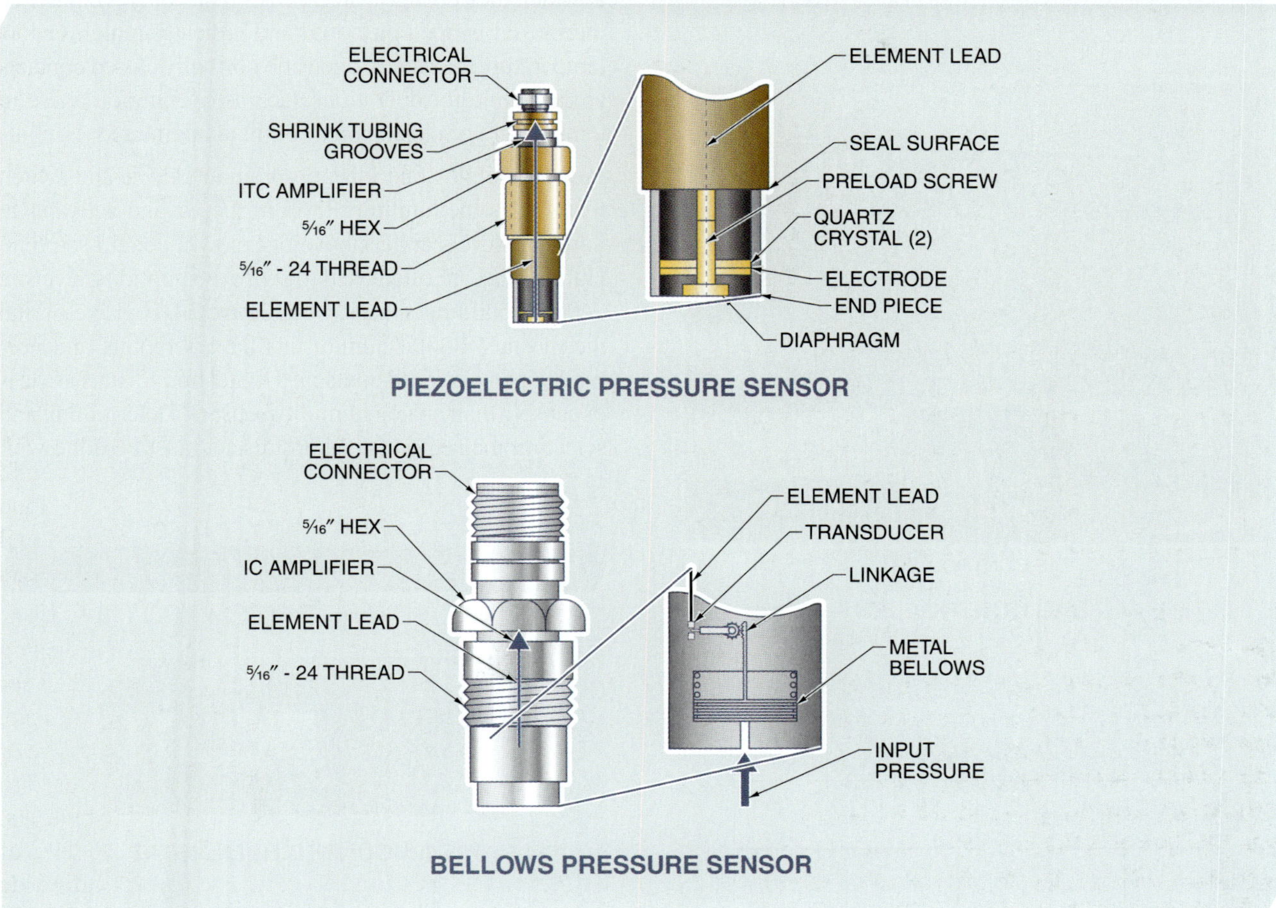

Figure 54-14. Pressure sensors are used to sense and control pressure in HVAC applications.

Digital Input Devices

A *digital input device* is a system component that acquires the ON or OFF state of a variable and transmits a high or low signal to convey the information to a controller. Digital input signals are also known as binary signals or discreet signals. Digital input signals are transmitted electronic signals that have only two possible values. **See Figure 54-15.**

Analog signals provide readouts of the actual value, while digital input signals indicate when a value is above or below a certain setpoint. For example, an analog sensor may provide a reading of 74.5°F. A digital input signal may indicate that a temperature is above or below 70°F.

Digital input devices are less expensive than analog sensors and are typically used as status points in a system. For example, digital input devices are used to indicate fan status (ON/OFF), pump status (ON/OFF), flow through filter status (sufficient/insufficient), or room status (occupied/unoccupied).

Alarm Panels Media Clip

The input device, whether analog or digital, selected for an application is often based on cost and functionality. For example, an analog sensor is used when the value of a variable, such as 72°F, is required. A digital input device is used when only status, such as ON/OFF, is required. Digital input devices are typically designated as having dry contact terminals. Dry contact terminal input devices require an external power supply. Dry contact terminal input devices also require wiring to the controller terminals to function. Digital input devices include thermostats, humidistats, differential pressure switches, flow switches, current-sensing relays, and other specialized digital devices.

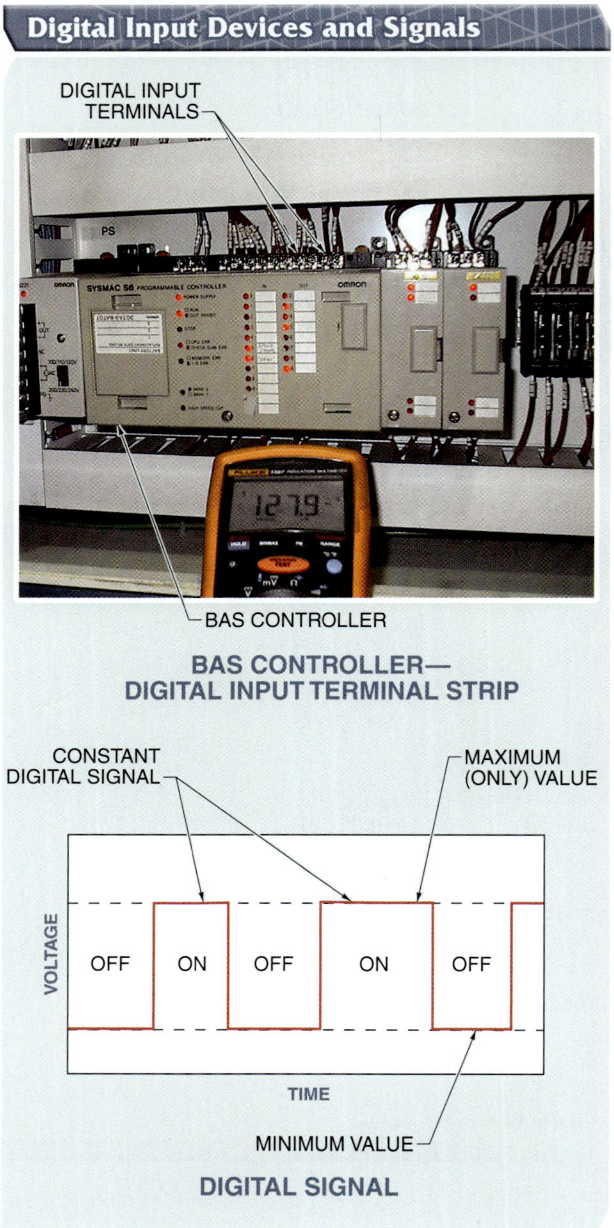

Figure 54-15. Digital input signals to a controller are either on or off.

Thermostats. A thermostat is a digital temperature input device and switch that activates at temperatures either above or below a certain setpoint. Thermostats are typically used as safety devices to control temperatures in a room. A limit thermostat is a thermostat that indicates whether a temperature has reached a certain value. Limit thermostats are typically used as safety devices to indicate improper temperature levels rather than as primary temperature controllers. Limit thermostats are inserted into ducts or connected to pipes.

Limit thermostats have a setpoint set at the temperature at which they change state. A BAS reads only whether the thermostat is open or closed and indicates a high or low temperature. Normally open and normally closed contacts are available for operations such as opening on a rise in temperature or closing on a rise in temperature.

Humidistats. A humidistat is an input device and switch that senses the humidity level in the air and activates at humidity levels either above or below a certain setpoint. Humidistats are often used to sense the humidity levels in ducts or building spaces. **See Figure 54-16.** Like digital thermostats, digital humidistats have setpoints and normally open or closed contacts. Digital humidistats are less expensive than analog humidity sensors. Determining and selecting the setpoints of humidistats must be done with care.

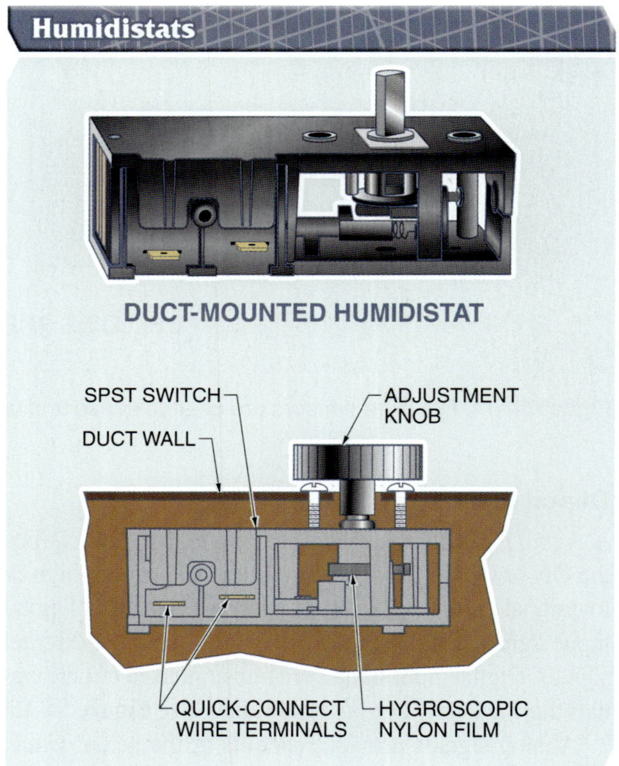

Figure 54-16. Humidistats are often used to indicate when a humidity setpoint has been reached in a high or low humidity alarm configuration.

Differential Pressure Switches. A differential pressure switch is a digital pressure input device and switch that activates at a differential pressure either above or below a certain value. Differential pressure switches often use bellows or diaphragm elements. The setpoint of a differential pressure switch must be adjusted correctly. The correct contacts (normally open or closed) also must be selected for a given application.

Differential pressure switches are often used to indicate a difference in pressure across a fan or pump or to indicate AHU filter condition. Differential pressure switches typically have a section of tubing located on the suction (inlet) side of a fan or pump and another section of tubing located on the discharge side. **See Figure 54-17.** Differential pressure switch tubing must be installed to avoid close bending, kinking, or long piping runs. Any air leaks must be located and corrected before differential pressure switches are put in service.

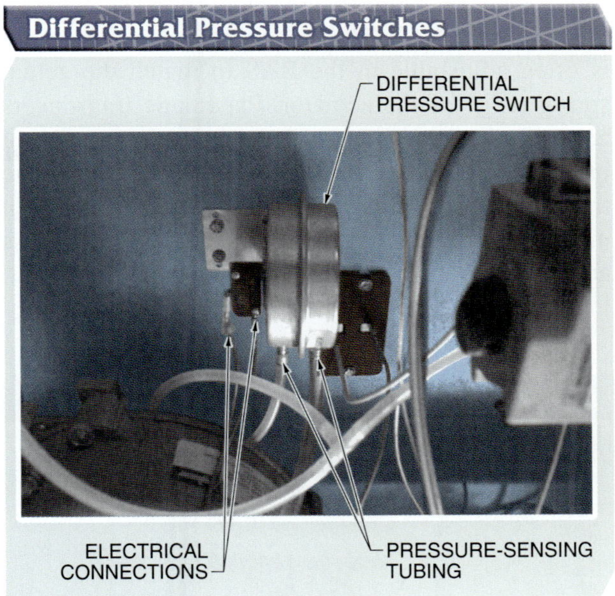

Figure 54-17. Differential pressure switches open or close circuits based on the difference between two pressures.

Differential pressure switches must not be located too close to fans or pumps where high pulsation and vibration occur. Also, differential pressure switches should not be installed too close to ductwork elbows, tees, or other obstructions that affect the pressure sensed. Manufacturer installation recommendations must always be followed when installing differential pressure switches.

Flow Switches. A flow switch is a switch with a paddle that moves from the force exerted by the air or water flowing within a duct or pipe. Flow switches can be used to indicate the flow status of a fan or pump. Differential pressure switches are also typically preferred to flow switches. Flow switches may be affected by fan or pump flow pulsations. The flow pulsation may lead to false indications being sent to the BAS controller. Alarms connected to a flow switch may provide false indications due to flow pulsation.

Current-Sensing Relays. A current-sensing relay is an input device that surrounds a wire and detects the electromagnetic field caused by electricity passing through the wire to determine current flow. Current-sensing relays are used to obtain the operational status of pumps, fans, compressors, and lighting loads. **See Figure 54-18.**

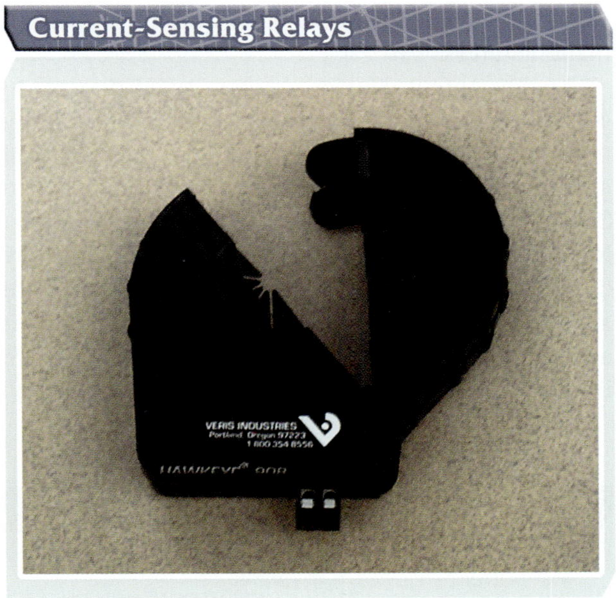

Figure 54-18. Current-sensing relays clamp around motor leads and provide on or off signals based on current flow.

Current-sensing relays are installed so that the current-carrying wires of a motor pass through the center of the relay. The relay trips when the motor starts, indicating to the BAS controller that current is flowing. The current flow indicates that the motor is doing work. A fan motor that has a broken or missing belt drive draws a low amount of current and does not trip the relay. The current setpoint must be set to avoid nuisance trips.

BAS OUTPUTS

A *BAS output* is a signal from a microprocessor-based controller to change the state of a controlled device. The microprocessor of BAS controllers makes decisions on controlling a system and transmits output signals to controlled devices to adjust the system. BAS output signals may be analog or digital.

Analog Output Devices

An *analog output device* is a device that produces and transmits electronic signals within a continuous range of values between two points. Analog output devices produce signals in the ranges of 0 mA to 20 mA, 0 VDC to 5 VDC, or 0 VDC to 10 VDC. Analog output devices provide proportional control of dampers, valve actuators, and variable-frequency drives. Analog output devices lack industry standardization. For example, a manufacturer may produce output devices with a specific voltage range that may be incompatible with some actuators or other devices from other manufacturers.

Analog output devices are connected directly to controlled devices or need their signals changed by transducers. VFDs are often connected directly to an analog output device and controlled with a 0 mA to 20 mA or 0 VDC to 10 VDC signal. Controller analog output signals cause VFDs to speed up or slow down connected motors. Analog output signals are also sent directly to motor-driven valves and damper actuators to make them open, close, or stop anywhere in between.

Transducers. A *transducer* is a device that changes one form of analog signal into another form. Most transducers are connected directly to controlled devices such as valve actuators, damper actuators, and VFDs. These controlled devices use common current and voltage ranges such as 4 mA to 20 mA and 0 VDC to 10 VDC. Transducers are available that receive these input currents and voltages and output a variable resistance such as 0 Ω to 135 Ω. Transducers allow the use of actuators that require a variable-resistance output. Other transducers are available that use analog input signals to activate multistage heating and cooling units.

Pneumatic transducers typically receive analog signals, such as 4 mA to 20 mA, from the BAS controllers and send air pressure signals of 0 psi to 20 psi. **See Figure 54-19.** The transducer output pressure is often adjustable in the BAS to match the actual spring range of the actuator. Pneumatic transducers are typically used in applications that have existing pneumatic valve and damper actuators. The pneumatic transducers enable pneumatic actuators to be used in modern BASs. This simplifies the installation process and reduces job costs.

Electric/Pneumatic Transducers

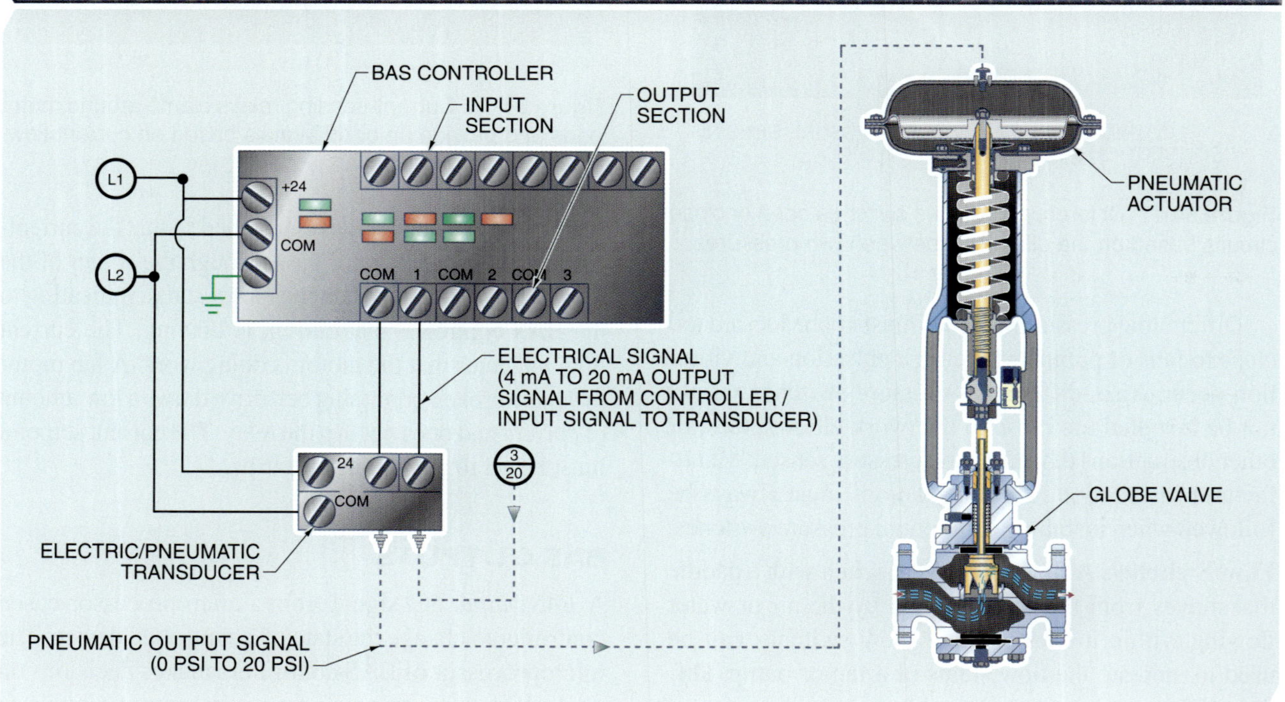

Figure 54-19. Electric/pneumatic transducers have 0 VDC to 10 VDC or 0 mA to 20 mA inputs and produce pneumatic output signals.

Digital Output Devices

A *digital output device* is a device that produces and transmits signals to a controlled device that accepts an on or off signal. BAS controllers evaluate the signals received and, depending on the software, send signals to change the state of controlled devices. Several output devices are available for various controlled devices.

Most digital output devices used in BASs are triacs, or triodes for alternating current. A *triac* is a solid-state switching device used to switch alternating current. **See Figure 54-20.** Triacs, which require an external power supply, are rated as Class II (low-voltage) devices. Digital output devices are connected directly to controlled devices or need relays, incremental outputs, or pulse width modulated outputs to change the signals.

Controller Power Supply Media Clip

ASCs are commonly used to control package rooftop units and heat pumps. Rooftop units and heat pumps that typically contain low-voltage control systems that are rated at 24 VAC use triac switches. ASCs can directly control only low-voltage control systems. The output signals of an ASC are wired directly to the terminal board of the rooftop unit or heat pump. In many cases, existing rooftop unit or heat pump transformers can be used to provide power for the ASC output signals. Care must be taken to avoid interference problems from spark ignition devices and to ensure that the transformer is of the proper size. HVAC technicians should use the standard color coding of the wires to make the proper connections.

Relays. The majority of controlled devices, such as motors, are line-voltage devices, which operate at line voltage of 115 VAC. BAS controller output devices are typically rated at low voltage. Therefore, a relay must be used to allow the low-voltage-rated triacs in BAS output devices to switch line-voltage devices. Relays are electrical switches that are actuated by separate electrical circuits. **See Figure 54-21.** Triac output signals are not powered internally, so an external power supply, such as a 24 VAC transformer, must be used. The relay can be supplied by the manufacturer or purchased separately.

BAS Controllers—Triac-Switched Controller Output Terminals

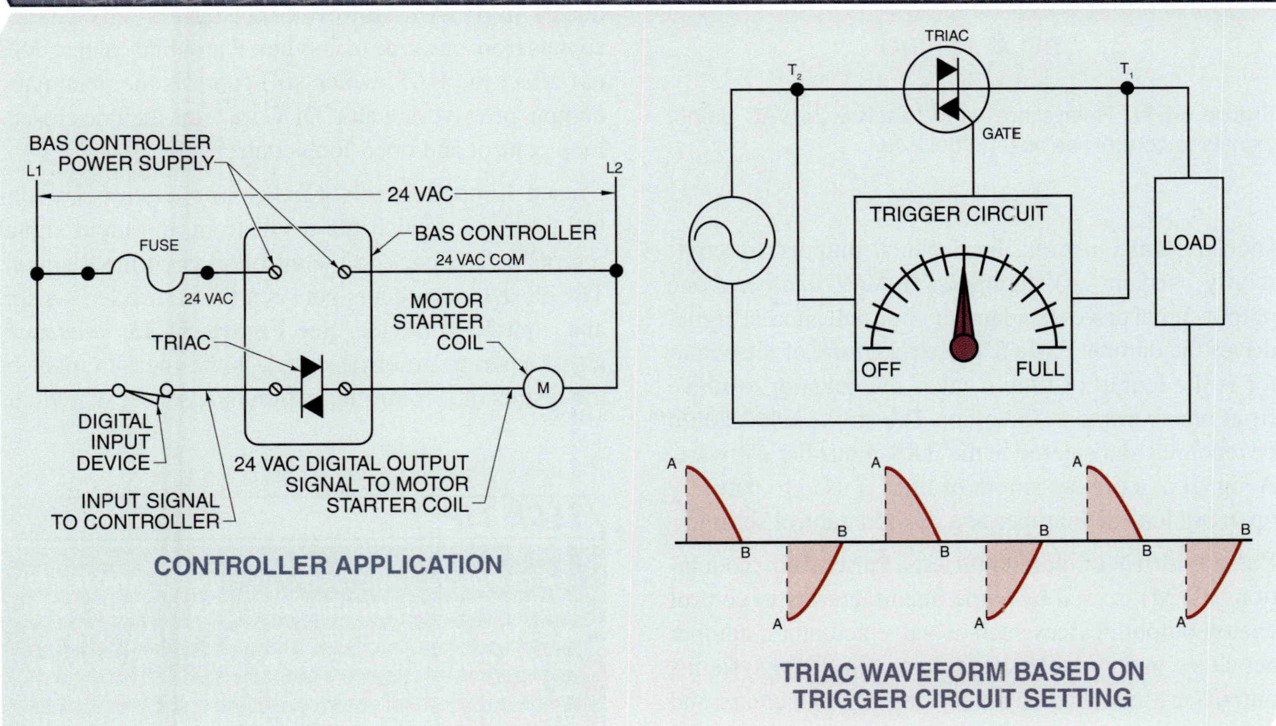

Figure 54-20. Almost all direct digital controllers today use triacs to switch digital output signals.

Relays

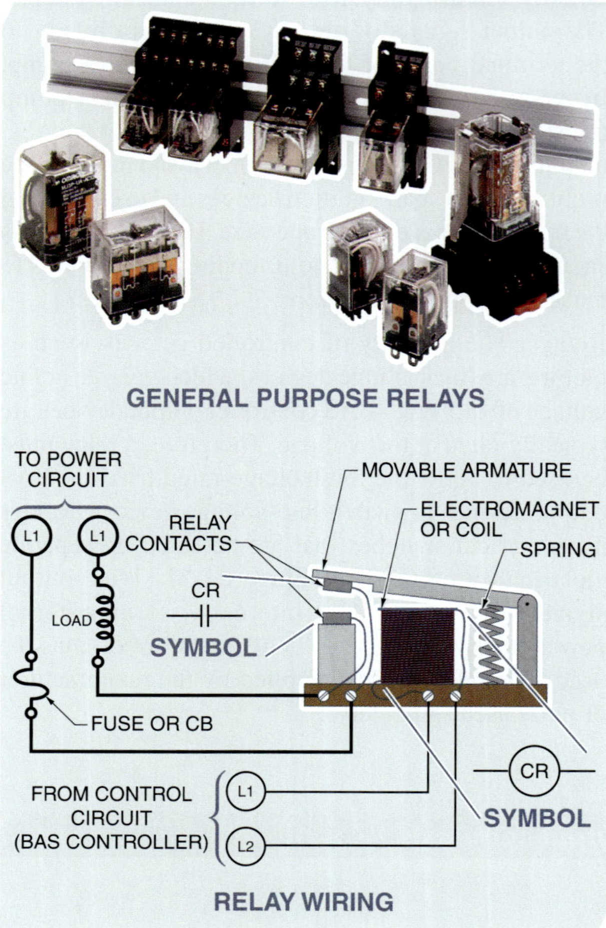

Figure 54-21. Relays are used to allow 24 VAC output signals to control line voltage devices.

Incremental Outputs. Incremental outputs are commonly used in VAV dampers. In VAV dampers, one output signal drives the damper open and another signal drives the damper closed. The stroke time of the actuator is the length of time it takes the actuator to move from one extreme to the other. The stroke time should be recorded and entered in the BAS controller software. A signal at a pre-set length of time is used to partially open or close the damper to a desired control setting.

Pulse Width Modulated Outputs. Pulse width modulation (PWM) is used by some manufacturers to control analog output devices such as valve actuators, damper actuators, and variable-speed drives. In PWM systems, outputs are switched on and off rapidly to increase and decrease the output signal to the device through a special transducer that decodes the pulses.

DDC STRATEGIES

A control strategy is a BAS method used to control the energy-using equipment in a building. A DDC strategy uses the BAS to perform the closed-loop temperature, humidity, or pressure control. A DDC control system is wired directly to controlled devices and can turn them on or off or start a motion. **See Figure 54-22.**

DDC System Controls

DDC systems contain sensors, controllers, and controlled devices. Sensors measure temperature, pressure, or humidity and provide the values they measure as input signals to a controller. Controllers compare the actual value to the programmed setpoint parameters. Controllers calculate the desired positions of the controlled devices and the required output signals to achieve the position. Controlled devices change the positions of HVAC hardware to control heating, ventilation, air conditioning, and lighting. Most BASs are DDC systems.

DDC strategies are not designed to be final safety devices. For example, low-temperature limit controls, boiler high-temperature limit controls, and flame safeguard controls are required with DDC systems. In addition, the installation of a DDC system does not reduce the need for regular preventive maintenance.

DDC systems are often identified by the use of transducers in a BAS. However, transducers only change signals from one type to another. The signal change does not affect the BAS, which still controls the pneumatic damper or valve actuator. DDC systems include closed-loop control and open-loop control.

Closed-Loop Controls. *Closed-loop control* is a control system in which the result from an output signal returns as feedback to a controller as an input signal. The closed loop occurs between the controller, sensor, and controlled device. **See Figure 54-23.** *Feedback,* also known as an error signal, is the measurement of the results of a controller action by a sensor or switch.

TECH TIP

Improvements in electronic technology have led to improvements in actuator motors. For example, many actuators today can operate on a variety of input signals that are determined by setting DIP switches on the actuator. In addition, the direction and length of stroke can be changed to provide precise control of HVAC systems.

DDC Systems

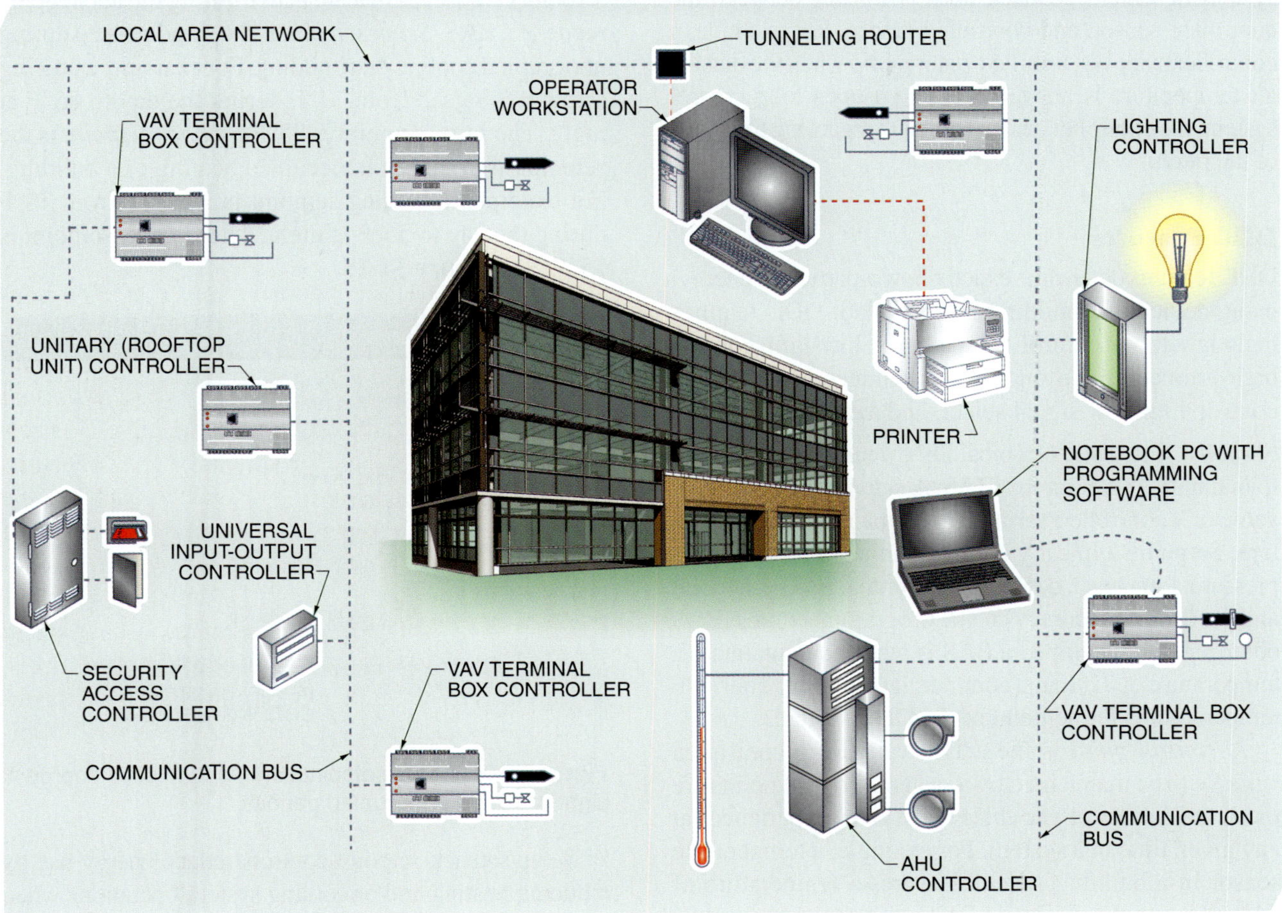

Figure 54-22. DDC systems are connected directly to devices and equipment in buildings that use energy such as HVAC and lighting equipment.

Closed-Loop Controls

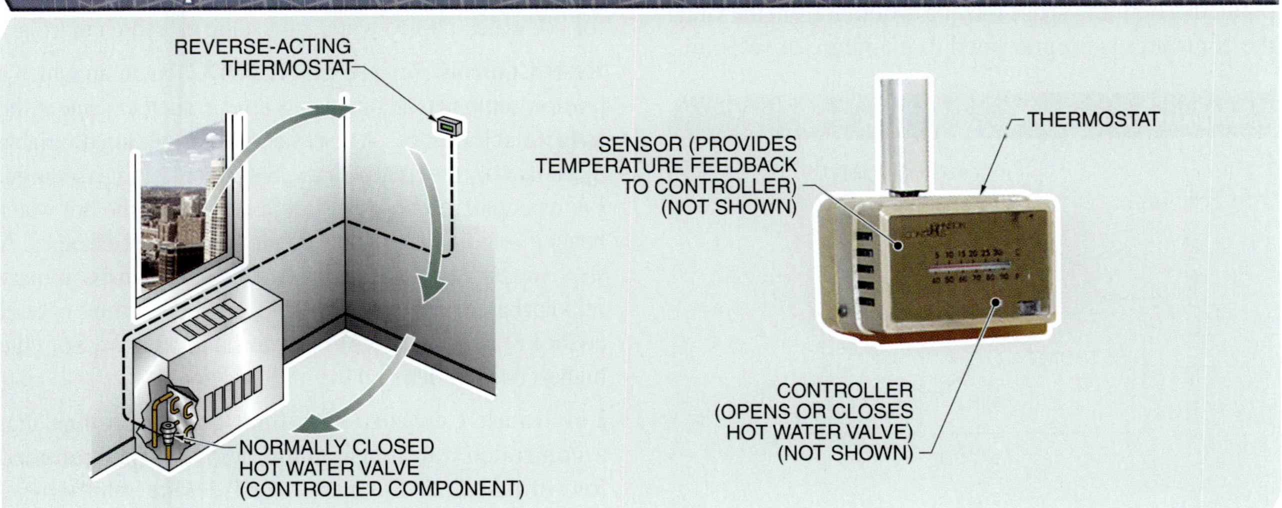

Figure 54-23. Closed-loop controls receive feedback about system conditions.

Open-Loop Controls. *Open-loop control* is a control system in which feedback does not occur between the controller, sensor, and controlled device. For example, a controller cycles on a chilled water pump when the outside air temperature is above 65°F. In an open-loop control system, the controller has no feedback regarding the status of the pump.

DDC Features

DDC features determine exactly how a particular energy-using device in a building is controlled. DDC features include setpoint control, reset control, low-limit control, high-limit control, economizer lockout control, lead/lag control, high/low signal select, and averaging control.

Setpoint Controls. A typical DDC feature is the ability to maintain a setpoint in a BAS. A *setpoint* is the desired value of a controlled variable to be maintained by a system. Setpoints typically include temperature, humidity, pressure, light level, dewpoint, and enthalpy. The setpoint and the desired accuracy can be programmed into a BAS controller. For example, a BAS is required to maintain a temperature of 72°F in a commercial building. The 72°F temperature is the setpoint of the DDC system.

A *control point* is the actual value of a controlled variable to be maintained by a system. Control points are the actual values of variables that a BAS experiences at any given time in a system. For example, a temperature sensor in a building space measures a temperature of 74°F. The control point is 74°F. At any given time, a control point may differ from a setpoint. The difference in the value of a variable between a control point and a setpoint is the offset. **See Figure 54-24.** Depending on the accuracy of the controller, the actual value, or control point, in a building space may be different from the value the controller is programmed to maintain, or setpoint.

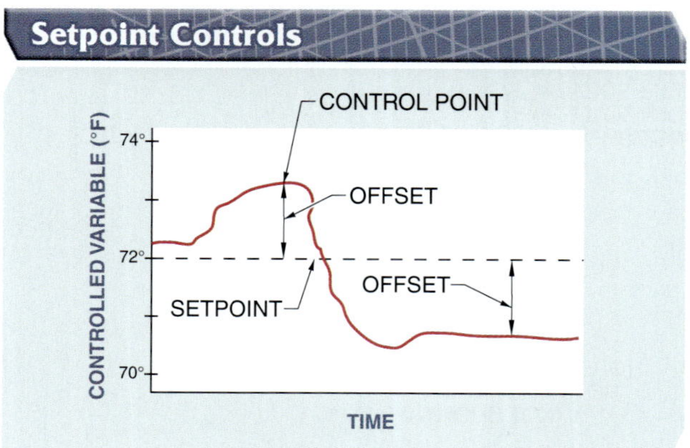

Figure 54-24. Offset is the difference between the actual temperature or variable and setpoint.

Setup and Setback Setpoint Controls. Setup/setback setpoints are values that are active during the unoccupied mode of a BAS. *Setup setpoint* is the setpoint used during unoccupied cooling of a building. For example, a cooling setpoint is raised from 74°F during the day to 85°F at night. The setup setpoint is 85°F. *Setback setpoint* is the setpoint used during unoccupied heating of a building. For example, a heating setpoint is lowered from 70°F during the day to 55°F at night. The setback setpoint is 55°F. **See Figure 54-25.**

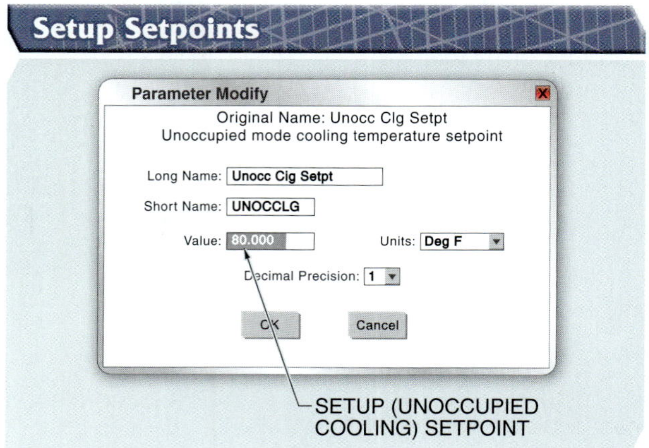

Figure 54-25. Setup setpoints are used to save energy at night or during unoccupied periods.

Setup/setback setpoint controls reduce energy use by reducing heating and/or cooling system operations when a building is unoccupied. Setup/setback setpoint control also protects buildings from abnormal temperatures by enabling the heating and cooling systems to operate when the building space temperature becomes excessively hot or cold. The setup/setback setpoint control feature is also used for hot water, chilled water, and static pressure controls.

Reset Controls. *Reset control* is a DDC feature in which a primary setpoint is automatically reset to another value as the reset variable changes. A reset variable is a measured variable that is used by the controller as a basis for resetting a setpoint. For example, reset controls are used to reset the hot water heating setpoint as the outside air temperature changes. A *setpoint schedule* is a description of the change in the primary setpoint based on the reset variable. Another example of reset control is the reset of VAV system static pressure from the highest cooling demand from several zones.

Low-Limit Controls. Low-limit control ensures that a controlled variable remains above a programmed low-limit value. The controlled device maintains a minimum value when the controlled variable does not remain above the low-limit value. Low-limit controls are typically used with mixed air damper controls. An excessively low outside air temperature causes a low mixed air temperature. **See Figure 54-26.** For example, when the mixed supply air temperature of buildings drops below 45°F, the controller overrides the normal control logic and forces the outside air dampers closed.

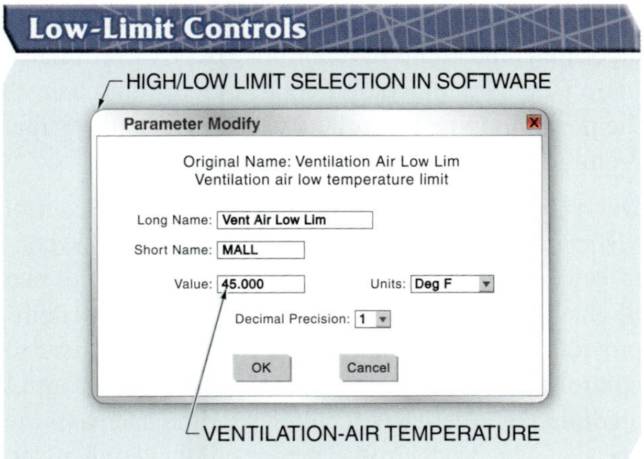

Figure 54-26. Low-limit setpoints are used to protect equipment against low temperatures.

High-Limit Controls. High-limit control provides capabilities similar to low-limit control. High-limit controls are typically used in applications that require a temperature, pressure, or humidity not to exceed a programmed value. For example, high-limit controls can be used to prevent the water temperature of a hot water heating system from exceeding 210°F.

Economizer Lockout Controls. On a direct expansion (DX) HVAC air conditioning package unit or DX AC system, an economizer is a device used to add cool outside air to the supply air of a building. Every mechanical system accepts minimum requirements of outdoor air into the building through the economizer. Outside air has the ability to cool a building when the outside air temperature is between 45°F and 65°F; however, humidity is a factor.

The economizer consists of a set of dampers in the return ductwork that permits outside air to mix with the return air prior to being treated by the system. The dampers are configured to allow 0% to 100% of either return air or outdoor air. If the ambient air is introduced in the same condition as the supply air from the system, the compressor can remain off and the outdoor air will cool the building. The mechanical cooling system is disabled when the economizer is used to provide cooling. *Economizer lockout control* is a DDC feature in which the economizer damper function is locked out or discontinued. Economizer lockout control methods include dry bulb and enthalpy economizer lockout control.

With the dry bulb economizer method, an outside air temperature sensor is used to control economizer operation. With the enthalpy economizer method, the enthalpy of the outside air and return air is measured and the lowest value is used for cooling. The economizer is used above 68°F and is locked out below 68°F.

Lead/Lag Controls. Lead/lag control is the control of the operation alternation between two or more similar pieces of equipment. A typical application of lead/lag control is the control of multiple hot water or chilled water pumps. For example, a primary (lead) pump is energized by a BAS when a pump operation is needed. A backup (lag) pump is energized when the primary pump fails to start. Often the lead/lag pump operation alternates on a time schedule to have equal run time on the two pumps. Refrigeration and air compressors are also common applications of lead/lag control.

High/Low Signal Select. *High/low signal select* is a DDC feature in which a building automation system selects among the highest or lowest values from multiple inputs. A typical application of high/low signal select is the control of building temperature using multiple zone temperature sensors at different locations in a building. The highest signal represents the warmest building space, and the lowest signal represents the coolest building space.

The signal selection results are communicated to the BAS reset control to determine the setpoint. The highest signal may be used to reset a cooling function to satisfy the warmest building space. The lowest signal may be used to reset a heating function to satisfy the coolest building space. HVAC technicians typically use high/low signal select instead of outside air temperature reset control.

Averaging Controls. *Averaging control* is a DDC feature that calculates an average value from all selected inputs. Averaging control is used in applications in which the highest and lowest values do not represent the actual conditions in a building. For example, a temperature sensor located in a foyer experiences outside air conditions whenever the door is opened.

This sensor typically senses the coldest building space temperature and controls the heating reset setpoint most of the time. An average from several temperature sensors located throughout the building is more representative of the actual building conditions. Correct placements of sensors are required for accurate results.

DDC Algorithms

An *algorithm* is a sequence of instructions for producing the desired result of an operation. Mathematical equations are used by BAS controllers to determine desired setpoints. DDC algorithms enable a BAS to achieve a high level of accuracy. Common algorithms used in DDC systems include proportional, integration, derivative, and adaptive control algorithms.

Proportional Control Algorithms. A *proportional control algorithm* is a control algorithm that positions the controlled device in direct response to the amount of offset in a BAS. Proportional control algorithms are the most commonly used algorithms in DDC systems. Proportional control algorithms are also known as P or P-only control. **See Figure 54-27.** The setpoint and the desired accuracy, or throttling range, are entered into the controller. The controller is programmed to have a specific output when the controller reaches the setpoint. This is typically done using the midstroke position of the controlled device. Proportional control algorithms do not provide precise control.

Integration Control Algorithms. An *integration control algorithm* is a control algorithm that positions the controlled device in response to the sum of the offset over a certain length of time. Integration control algorithms are commonly used in conjunction with proportional control algorithms. Proportional/integration control, or PI control, is the combination of proportional and integration control algorithms. When an offset remains after a specific length of time, the integration control algorithm repositions the controlled device to eliminate the offset. The integration control algorithm changes the position of the controlled device to meet the actual requirements of the load at a given time.

Derivative Control Algorithms. A *derivative control algorithm* is a control algorithm that positions the controlled device in direct response to the instantaneous rate of change of a variable. Derivative control algorithms provide real-time data to BAS controllers. The derivative control algorithm acts against the integration control algorithm. The derivative control algorithm increases the speed at which a controlled device eliminates an offset.

Proportional Control Algorithms

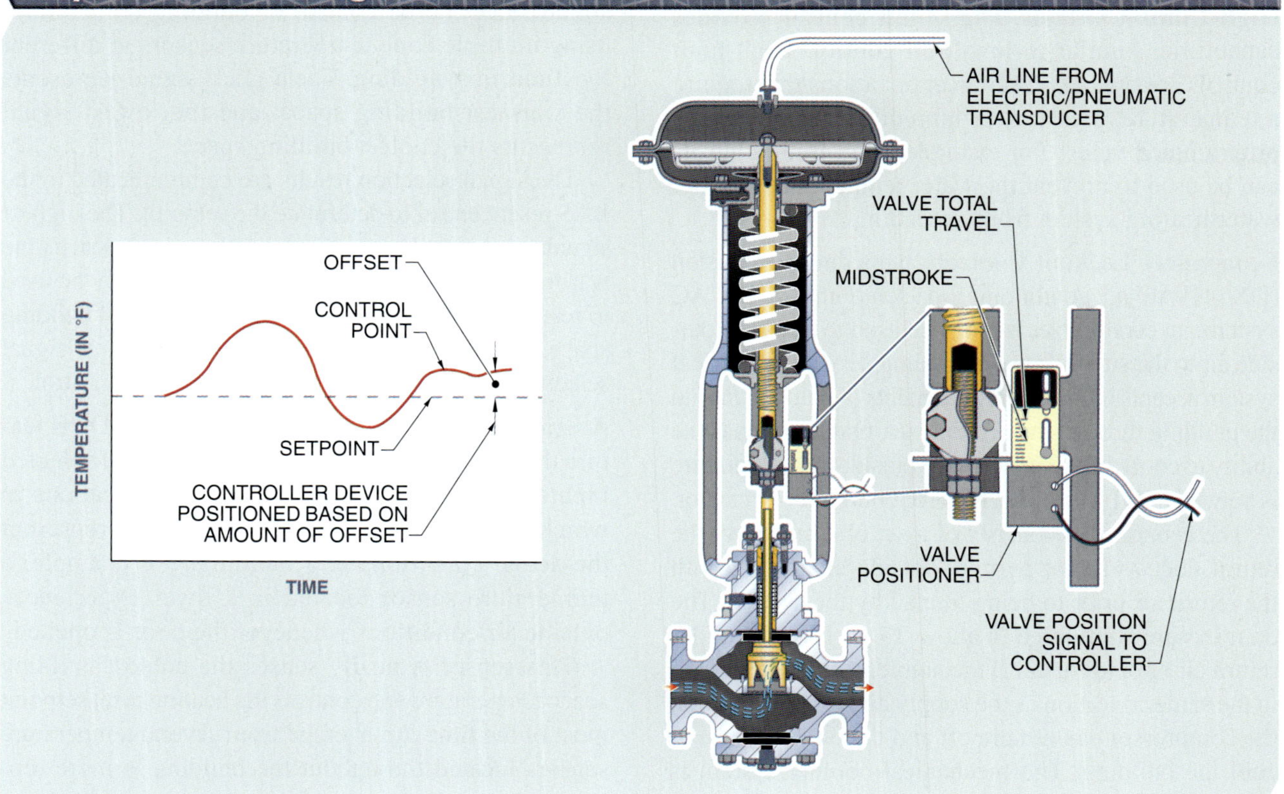

Figure 54-27. Proportional control algorithms can provide very accurate control by adjusting for the offset.

A proportional/integration/derivative controller, or PID controller, is a DDC system that uses proportional, integration, and derivative algorithms. In the HVAC industry, only extremely sensitive control applications require PID control. PI control is normally sufficient to achieve a setpoint.

Adaptive Control Algorithms. The most sophisticated control algorithms used today are adaptive control algorithms. An *adaptive control algorithm* is a control algorithm that positions a controlled device by automatically adjusting its response time based on environmental conditions. **See Figure 54-28.** Adaptive controllers result in increased accuracy, stability, and simplicity. Adaptive control algorithms are a self-tuning form of PID control.

Older BASs require HVAC technicians to accurately tune PI and PID controls by adjusting the response times in the BAS software to ensure accuracy and stability. Adaptive control algorithms have the ability to readjust to load changes or incorrectly programmed response times. Adaptive control algorithms reduce or eliminate the amount of time required to tune a control loop. Also, the programmed response times do not have to be accurate because adaptive control algorithms have the ability to change response times to the proper value. Adaptive control algorithms are not provided by all BASs.

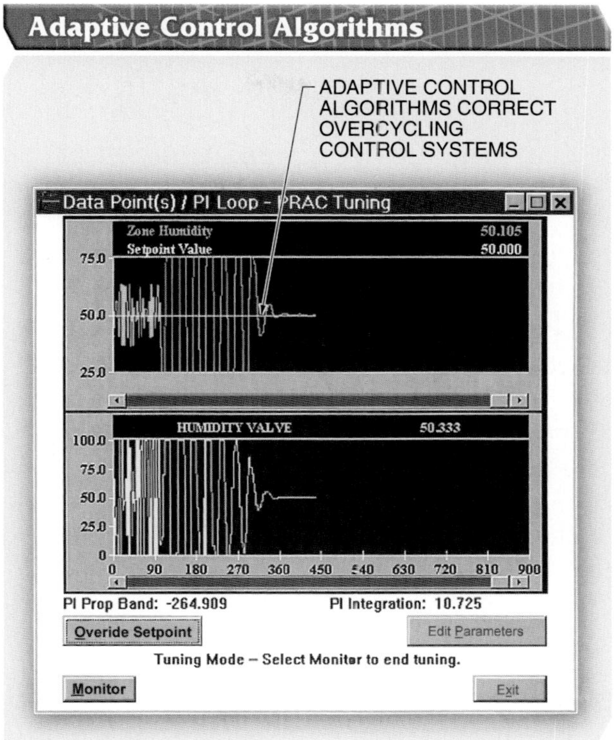

Figure 54-28. Adaptive controllers provide very high accuracy as they "learn" system characteristics to prevent overcycling of the control systems.

Digital Resources

ATPeResources.com/QuickLinks • Access Code: 432610

U N I T 55

Terminal Unit Heating and Cooling Control Methods

Large residential and commercial buildings use air-handling equipment to provide HVAC capability. Typically, air-handling equipment is controlled using building automation systems. HVAC technicians must be familiar with the basic operations of air-handling equipment in order to troubleshoot and fix problems that will resolve customer complaints.

OBJECTIVES

- Identify air-handling system characteristics.
- Describe features of air-handling system ventilation.
- List the different types of air-cleaning equipment.
- Compare common air-hancling distribution system designs.
- List the different types of terminal units and life safety functions.

AIR-HANDLING SYSTEM CHARACTERISTICS

HVAC systems must be geared to meet both specific and varying application requirements. Larger residential buildings typically contain common areas such as foyers, community rooms, and support facilities. Commercial buildings typically contain public areas such as offices, retail space, and cafeterias or restaurants. Air distribution systems for public areas of buildings typically operate for a minimum of 12 hr a day, 5 days a week, and up to 24 hr a day, 7 days a week. Air distribution systems for private residential areas of buildings typically cycle 24 hr a day, 7 days a week.

To reduce energy costs, some buildings use a main HVAC system at partial load and meet specific occupant needs through smaller supplemental systems. Despite the type of system used, HVAC systems condition air and distribute it to building spaces for occupant comfort. A thorough understanding of the equipment, functions, and controls of air-handling systems in buildings is necessary for HVAC technicians to be able to work on HVAC systems.

Air-Handling System Comfort Controls

HVAC systems must maintain building comfort. Reliability is an obvious characteristic required in all HVAC systems. HVAC equipment must operate at peak performance in order to provide occupant comfort at the lowest energy cost. Other desired characteristics include maintaining space temperature, relative humidity, and cleanliness. **See Figure 55-1.**

Comfortable Building Temperatures for Relative Humidity

Winter Temp*	Summer Temp*	Relative Humidity†
68.5–76.0	74.0–80.0	30
68.5–75.5	73.5–79.5	40
68.5–74.5	73.0–79.0	50
68.0–74.0	72.5–78.0	60

* in °F
† percentages

Figure 55-1. Comfortable building temperature ranges depend on the relative humidity and the season.

Air Temperature. Temperatures in occupied building spaces should be as close to the design setpoints as possible. When people feel uncomfortably hot or cold, they are more likely to complain. Some building leases specify required ranges for temperature and relative humidity. When HVAC systems cannot maintain comfortable temperatures within building spaces, occupants may not renew their building leases. Areas within building spaces may have varying demand for heating or cooling. Radiant heating may provide sufficient comfort for one occupant but not another. Cold exterior walls may cause convection currents that create drafts. Proper airflow distributes warm and cool air where required.

Air Relative Humidity. Relative humidity (RH) is as critical as temperature for occupant comfort. RH above 60% promotes microbial growth. RH below 30% causes respiratory problems and can lead to irritated mucous membranes in humans. All HVAC humidifying and dehumidifying equipment must operate correctly for occupant comfort. Specific spaces within buildings that have high or low RH must be conditioned with separate humidity control equipment that does not conflict with the RH in other parts of the facility.

Air Cleanliness. The inside and outside of all HVAC equipment must be kept clean and dry to minimize dirt and dust buildup. Dust buildup can become airborne and distributed by HVAC systems. In the presence of water, even small collections of dirt and dust provide breeding grounds for microbial contaminants. Drip pans, humidifiers, ducts, and coils must be checked periodically for microbial growth and cleaned.

Personnel using cleaning substances and equipment must follow all manufacturer instructions. Cleaning and other maintenance work must be scheduled when the building or specific space is not occupied. HVAC technicians must control the introduction of contaminants into the air and minimize exposure to building occupants. During large cleaning operations that result in fumes or objectionable odors, the space or building must be purged with fresh air to remove the contaminants.

Air-Handling System Designs

HVAC systems are designed to condition and deliver air to building spaces. HVAC systems use automatic controls that can be pneumatic, electric, or electronic. HVAC systems and sequences are similar within buildings regardless of the type or brand of control system. **See Figure 55-2.** Three general types of HVAC systems are all-air, all-water, and combined air-water. Most HVAC systems include outdoor-air, return-air, and exhaust-air dampers. HVAC systems are categorized as either 100% outdoor-air systems or mixed-air systems. Mixed-air systems mix outdoor air with return air and typically use economizers.

Heat Transfer Media Clip

100% Outdoor-Air Systems. A *100% outdoor-air system* is an HVAC system that does not recirculate any return air for building use. These systems only use air from outside of buildings for heating and cooling purposes. Using 100% outdoor air prevents any contaminants already in indoor air from being recirculated back into building spaces. Places where 100% outdoor-air systems are typically used include cafeterias, laboratories, and hospital operating rooms.

With few exceptions, places that use 100% outdoor-air systems do not have indoor air quality problems. The downside of 100% outdoor-air systems is the amount of energy needed to heat, cool, dehumidify, and filter all of the air, all of the time. Typically, 100% outdoor-air systems are required to have units with both preheat and precool coils to condition the incoming air during extreme conditions.

Mixed-Air Systems. A *mixed-air system* is an HVAC system that brings a controlled amount of return air into contact with outdoor air for building use. Mixed-air systems are used when the return air is relatively free of contaminants. The amount of outdoor air mixed with return air is determined by the control system. Outdoor-air dampers and return-air dampers are positioned to obtain the proper mix of airflow. A minimum open setting for outdoor-air dampers is used to ensure that a specified amount of outdoor air is introduced into the system for ventilation purposes. **See Figure 55-3.** Some mixed-air systems can also be used with economizers.

Economizer Systems. An *economizer system* is an HVAC system that uses outdoor air for cooling when the outdoor air temperatures are cool. Outdoor air is used instead of mechanical cooling mixed air from chilled water or direct expansion air conditioning coils. A direct expansion system is an HVAC system in which the refrigerant expands directly inside a coil in the main airstream to affect the cooling of the air.

HVAC Systems

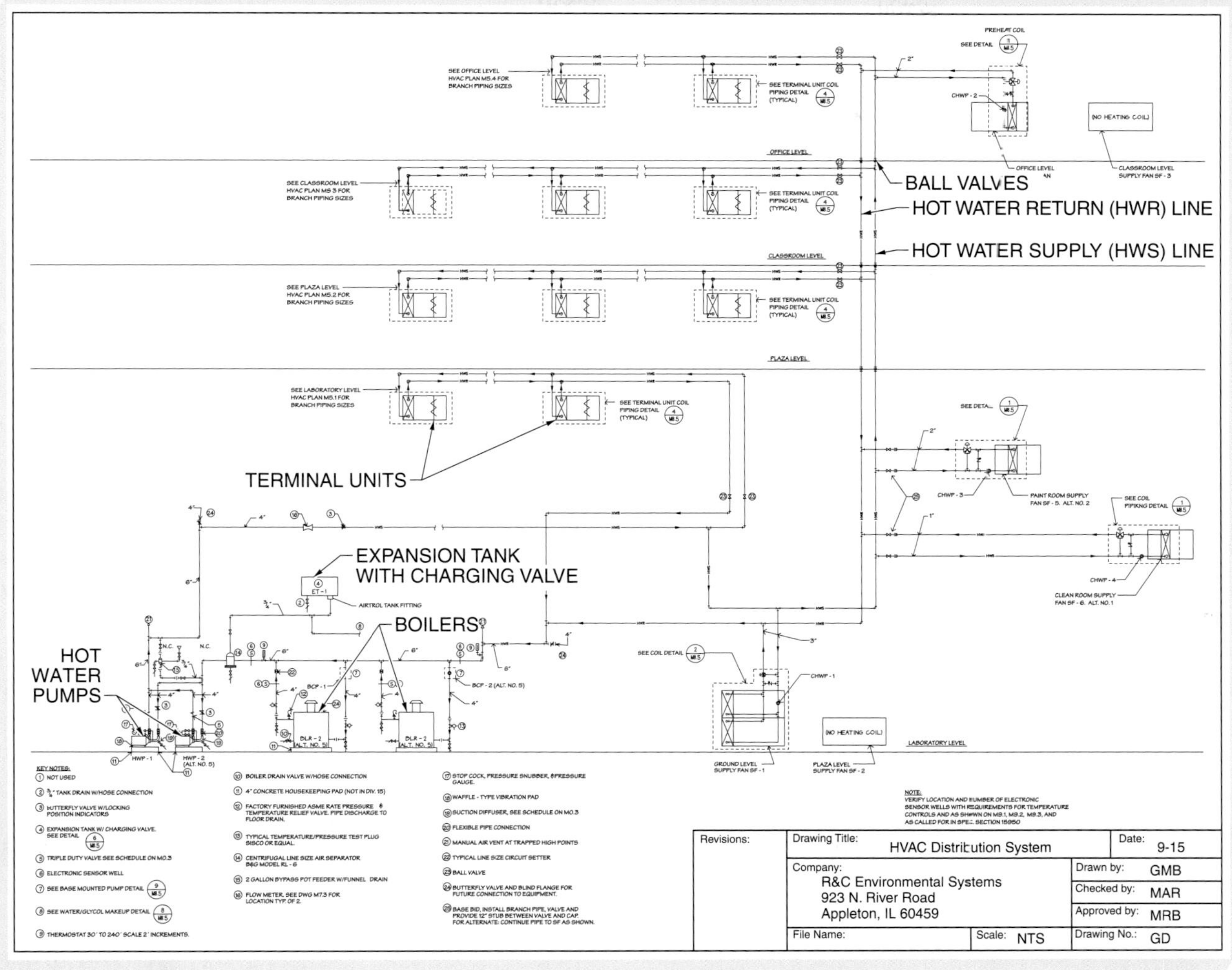

Figure 55-2. Conditioned air distribution systems are designed to operate over a wide range of conditions.

Mixed-Air Systems

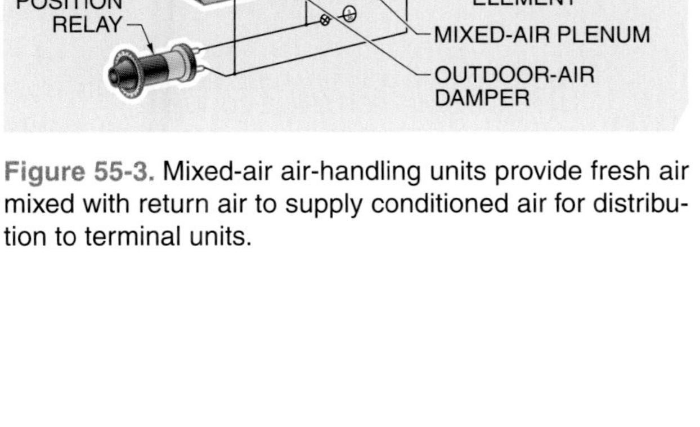

Figure 55-3. Mixed-air air-handling units provide fresh air mixed with return air to supply conditioned air for distribution to terminal units.

Buildings with high internal heat gain require cooling even when it is cold outside. An economizer system allows cool outdoor air to be used for cooling buildings with high internal heat gain. For economizer systems to operate effectively, special control schemes must be used that allow economizer operation only when outdoor air is within a proper temperature range. Building-economizer systems fall into two groups: dry bulb economizers and enthalpy economizers.

Dry Bulb Economizer Systems. A *dry bulb economizer system* is an economizer system that operates strictly in proportion to the outdoor-air temperature, with no reference to humidity values. When the outdoor-air temperature is below a specific value, the outdoor air can be used for economizer cooling.

See Figure 55-4. A typical setpoint value is 65°F, with the exact setting dependent on the prevailing climate in specific geographical regions. When the outdoor-air temperature rises and is too warm for economizer cooling, the economizer cycle is ended. Ending the economizer cycle forces the outdoor-air damper to a minimum setting, which allows enough outdoor airflow for ventilation purposes only.

Enthalpy Economizer Systems. An *enthalpy economizer system* is an economizer system that uses the temperature and humidity levels of the outdoor air to control HVAC operation. Enthalpy is the total heat content of a substance. For enthalpy control to work, both temperature and humidity sensors are mounted in the outdoor air. Because of problems associated with humidity sensors in the outside environment, humidity readings are typically taken in the mixed-air airstream using algorithms for the controller. **See Figure 55-5.** A controller uses the temperature and humidity values to calculate the current enthalpy of the outdoor air in British thermal units per pound (Btu/lb). When the enthalpy of the outdoor air is low enough, the economizer cycle of the HVAC system is used.

Enthalpy economizer systems allow for variations in temperature and humidity of the air. Enthalpy controlled systems allow outdoor air to be used under more conditions and an increased use of the economizer cycle, which promotes energy efficiency. When the outdoor-air enthalpy value is above the setpoint, the economizer cycle is ended and the outdoor-air damper is forced to a minimum setting for ventilation purposes only.

Air-Handling System Pressures

Building ventilation is typically operated at a slightly positive pressure so that contaminants do not leak into the building. Positive pressure is especially important on lower floors of taller buildings because of the stack effect. The stack effect occurs when hot air rises through the building elevator shafts, stairwells, and service columns. Outdoor air is drawn into lower floors, bringing in contaminants from the street level. The stack effect becomes larger as temperature differences between the indoor and outdoor environment increases. The stack effect can be minimized by keeping lower floors at a positive pressure, which limits the inflow of unfiltered, possibly contaminated air. **See Figure 55-6.**

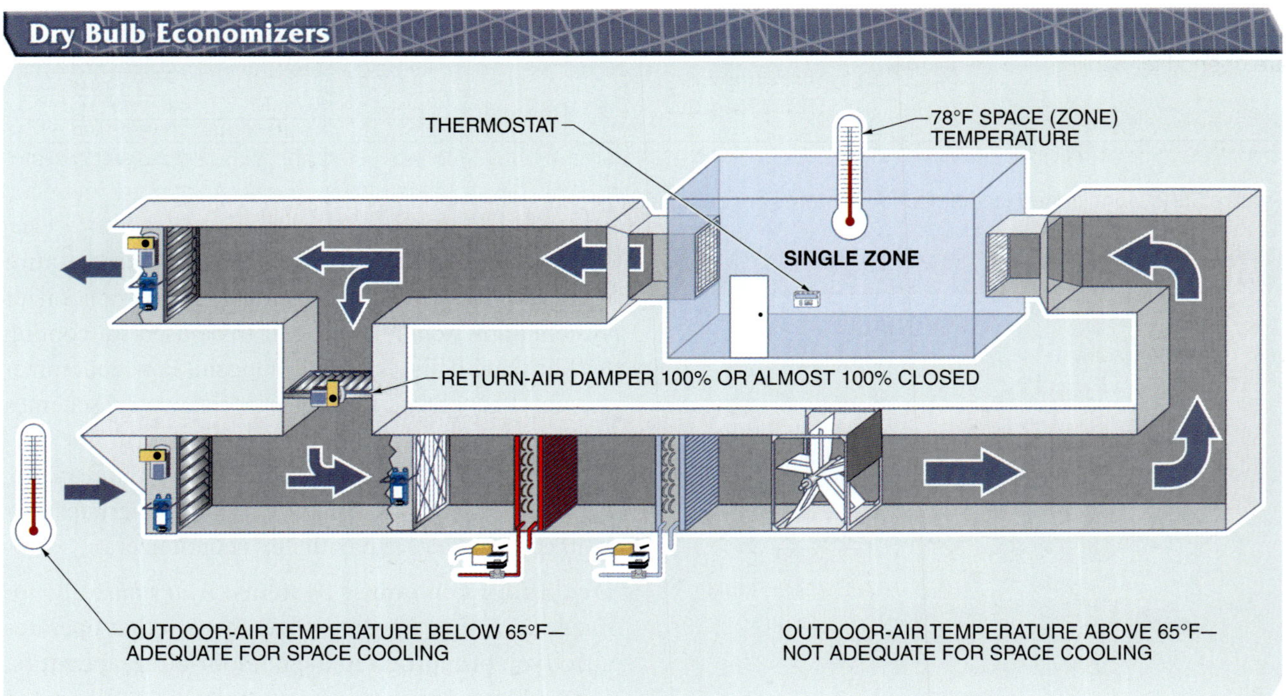

Figure 55-4. Dry bulb economizers measure outdoor-air temperatures to determine when the economizer dampers are opened for cooling.

Enthalpy Economizers

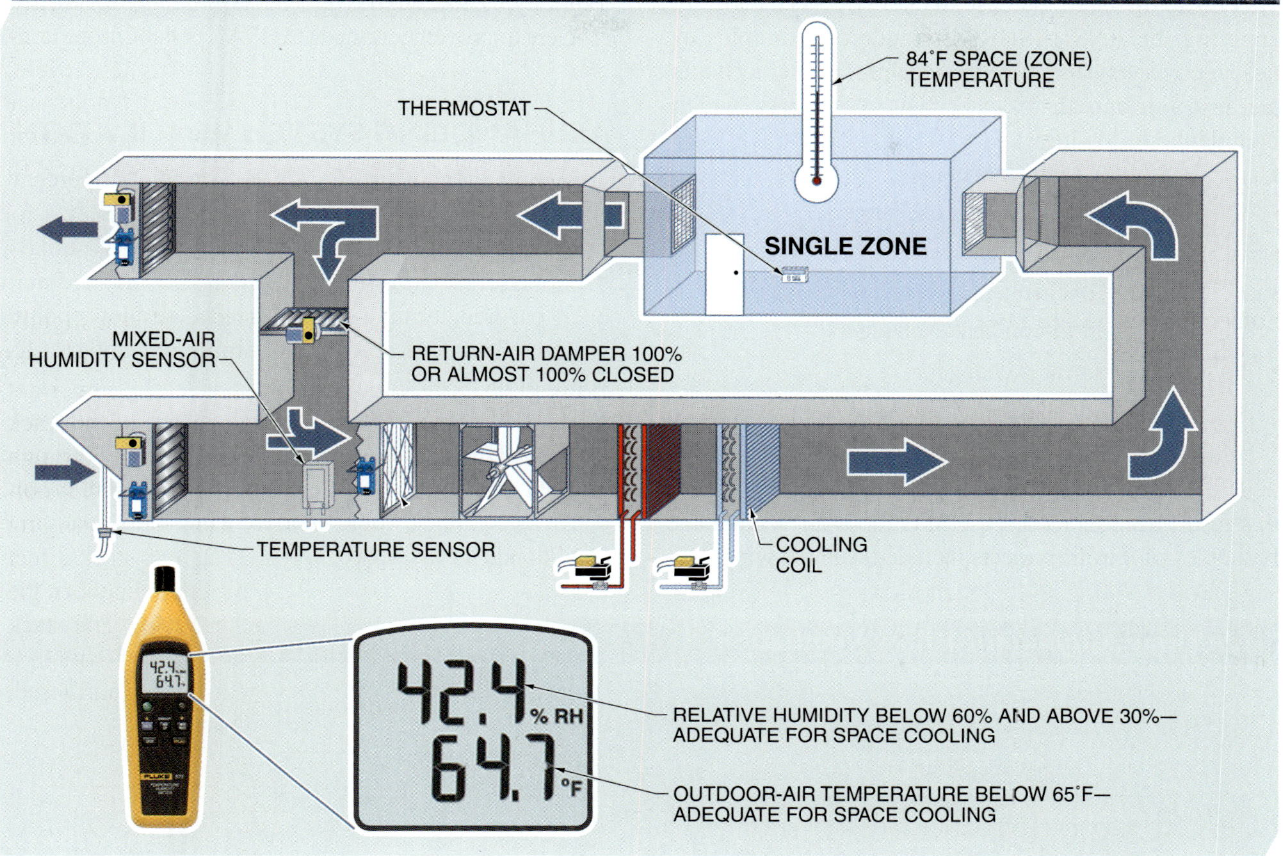

Figure 55-5. Enthalpy economizers measure both temperatures and humidity to determine when the economizer dampers are opened for cooling.

Stack Effect

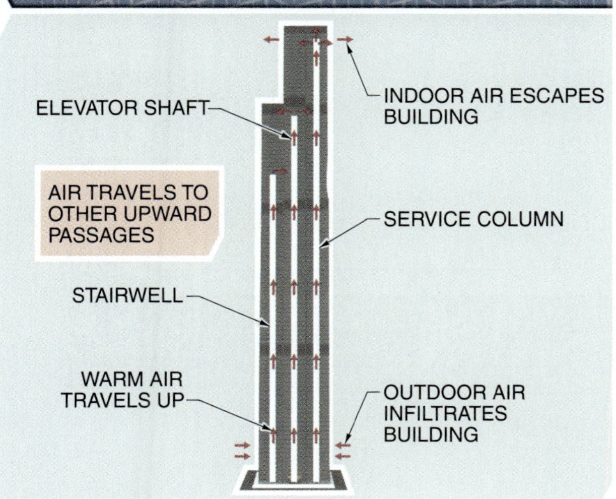

Figure 55-6. The stack effect occurs when warm building air rises up within stairwells, elevator shafts, and service columns. The stack effect lowers the air pressure near the bottom and pulls in air from the outdoors.

Positive pressure is created by supplying enough air to building spaces so that the building has a higher air pressure in relation to the atmospheric pressure. Pressure monitors measure the differential pressure between indoor space and outdoor or atmospheric pressure. Pressure monitors are used to maintain positive pressure in building spaces by controlling fan speed or volume control dampers.

Air-Handling System Heat Recovery

The American National Standards Institute (ANSI) and the American Society of Heating, Refrigerating and Air-Conditioning Engineers (ASHRAE) have set ventilation standards as a guide for maintaining proper indoor air quality. Per ANSI/ASHRAE Standard 62.1-2007, *Ventilation for Acceptable Indoor Air Quality,* the use of outdoor air should be increasingly used to provide a dilution effect to indoor airborne contaminants. Applying the fresh air standard and increasing fan static pressure required for high-efficiency filters increases energy costs.

Heat recovery equipment is used in air distribution systems to offset an increase in energy costs caused by applying the ANSI/ASHRAE standard. Examples of heat recovery systems integrated into HVAC systems are air-to-air and air-to-water heat exchangers, which include the following:

- rotary wheel heat exchangers
- fixed-plate counterflow heat exchangers
- crossflow plate heat exchangers
- heat pipe heat exchangers
- run-around coil-to-coil heat exchangers
- boiler flue-gas heat exchangers
- steam condensate heat exchangers
- wastewater recovery heat exchangers

Retrofit projects help offset any increase in energy costs that result from increased use in outdoor-air ventilation. Examples of retrofit projects include insulated windows, thermal window films, and improved building insulation. Typically, electric utilities offer rebate programs for energy-efficient improvements made to HVAC or lighting systems.

AIR-HANDLING SYSTEM VENTILATION

HVAC technicians must ensure that a correct percentage of outdoor air is brought into HVAC systems for occupied spaces. Variable-air-volume (VAV) boxes, economizers, and outdoor-air dampers must remain open to a predetermined minimum level to allow enough airflow of outdoor air to meet minimum standards for ventilation. **See Figure 55-7.**

The entire HVAC system should be compared to specifications. Diffusers and return-air grills must not become blocked or obstructed. Room partitions must not extend above a height that would interfere with the circulation of conditioned air in a space.

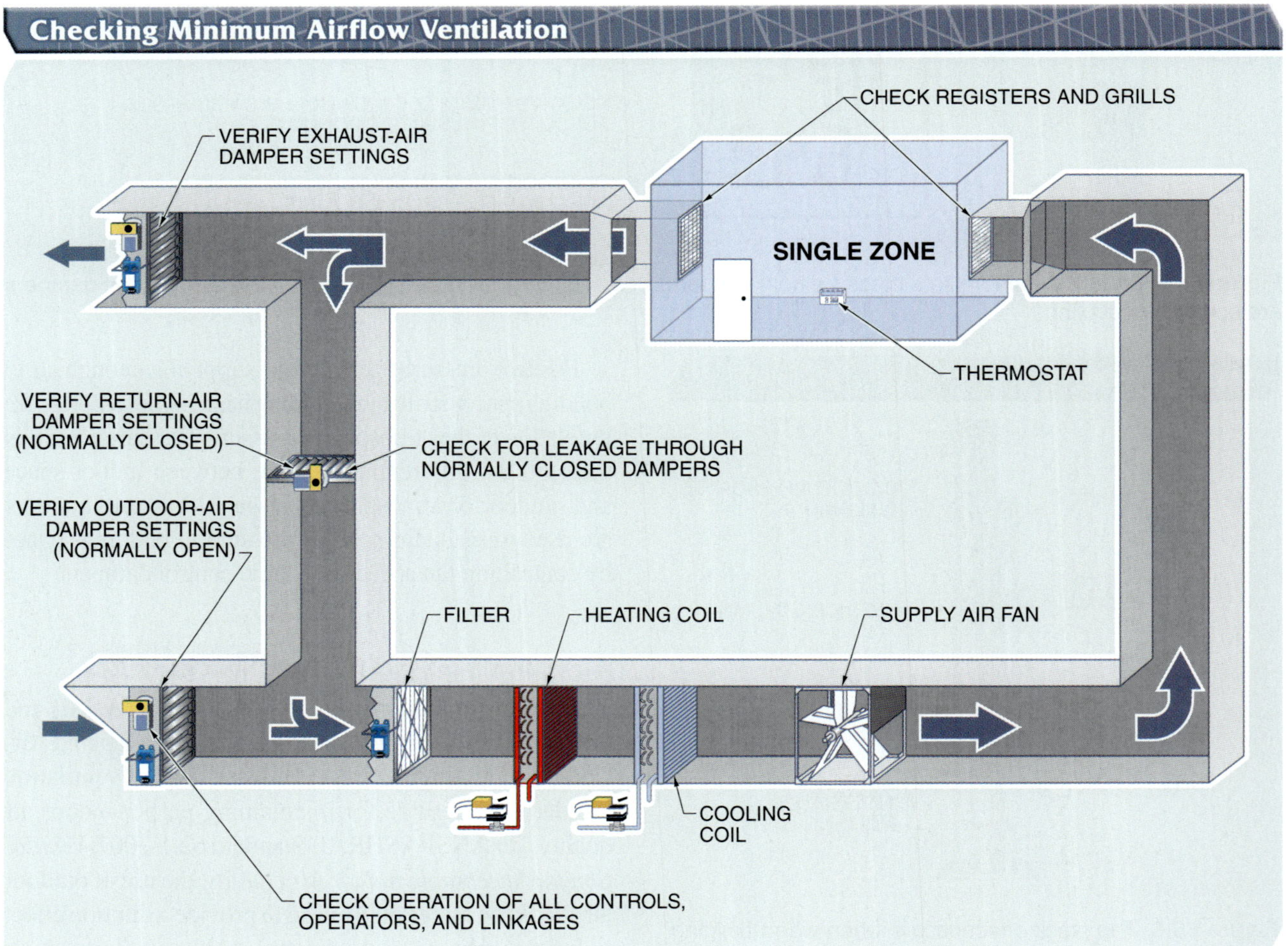

Figure 55-7. All HVAC system dampers must be periodically checked for proper operation to ensure good IAQ and efficient operation.

HVAC System Air Control Components

Air-handling units (AHUs) are part of HVAC systems that operate to distribute conditioned air throughout buildings. An AHU consists of a fan or blower, mixing plenum, filter rack, heating and/or cooling coil, and damper section. Ducts supply conditioned air to terminal units throughout buildings. **See Figure 55-8.** Large AHUs that operate as makeup-air units typically condition 100% of outdoor air for use throughout buildings. AHUs designed for outdoor installation are typically installed on roofs and are therefore known as rooftop units. A small AHU, or terminal unit, typically includes a filter, coil, and fan or blower.

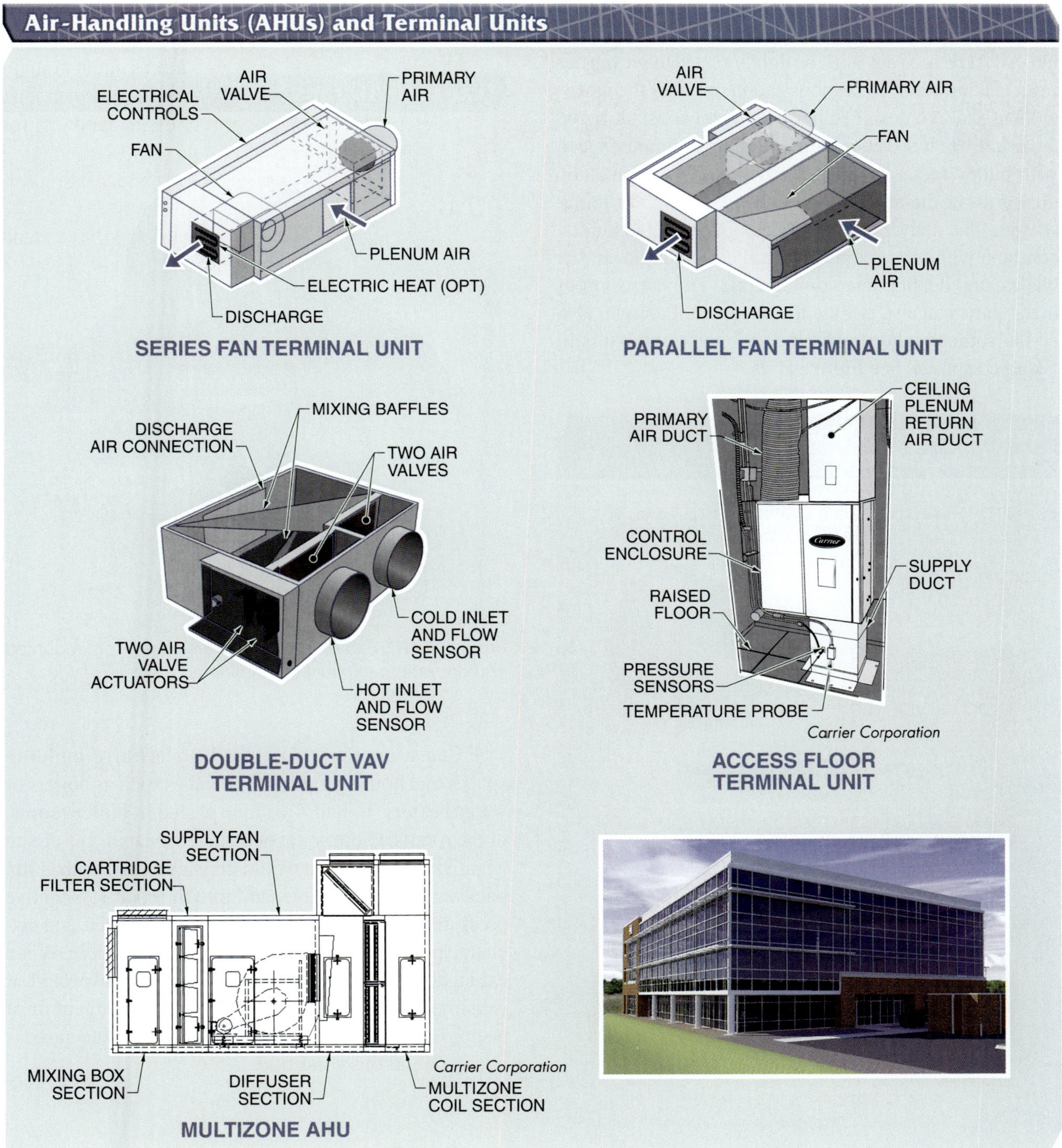

Figure 55-8. AHUs provide supply air for distribution to terminal units throughout buildings.

Fans. A *fan* is a mechanical device with rotating blades that moves air. Fans do not discriminate as to the quality of the air being distributed. When conditioned air is moving, fans provide the necessary energy to overcome the resistances for air to flow in HVAC systems. Resistance to airflow comes from terminal equipment, heating and cooling coils, louvers, grills, diffusers, and ductwork. Fans impose negative pressures in the duct on the suction side of the fan and positive pressures in the duct on the outlet side of the fan. All AHU sections must be tightly sealed to prevent air from leaking between sections and in or out of the unit.

Fans that are designed to operate at a constant speed are usually belt driven because the standard motor speed with pulley ratios is enough to overcome system static pressures, or the resistances to flow. During the cooling season, fans deliver cold saturated air. Saturated air can combine with airborne dirt and collect on dampers, fan blades, and heating and cooling coils. The corrosion of metal parts can also pollute the airstream. Fans can also be the source and/or transmitters of noise and odors to occupied spaces. **See Figure 55-9.**

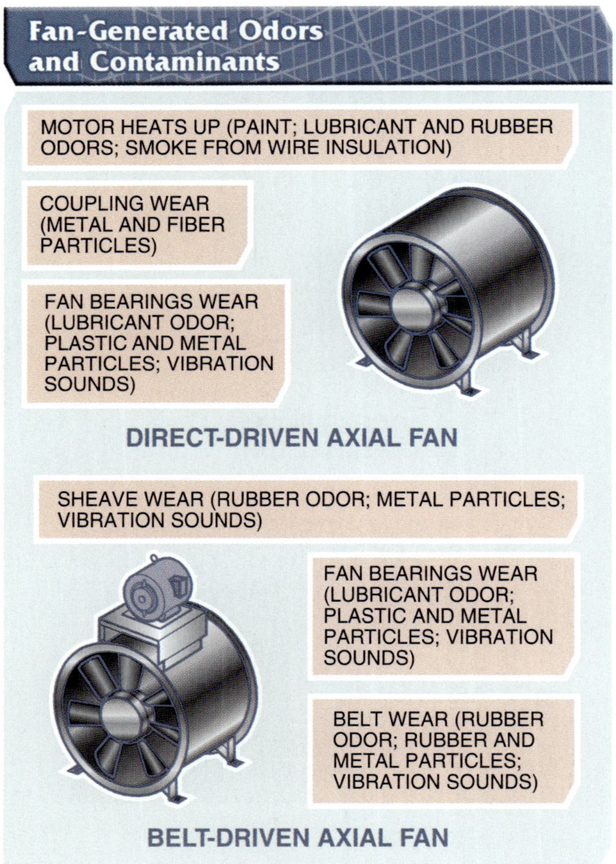

Figure 55-9. Fans generate odors and contaminants that are circulated in the supply air.

Most spaces in buildings are maintained at slightly positive pressures to prevent the infiltration of air from outside the building from entering occupied conditioned areas. Some spaces, such as rest rooms, are maintained under negative pressures to ensure that airborne contaminants are exhausted. Negative pressures are maintained using localized systems of exhaust-air fans and ducts. **See Figure 55-10.**

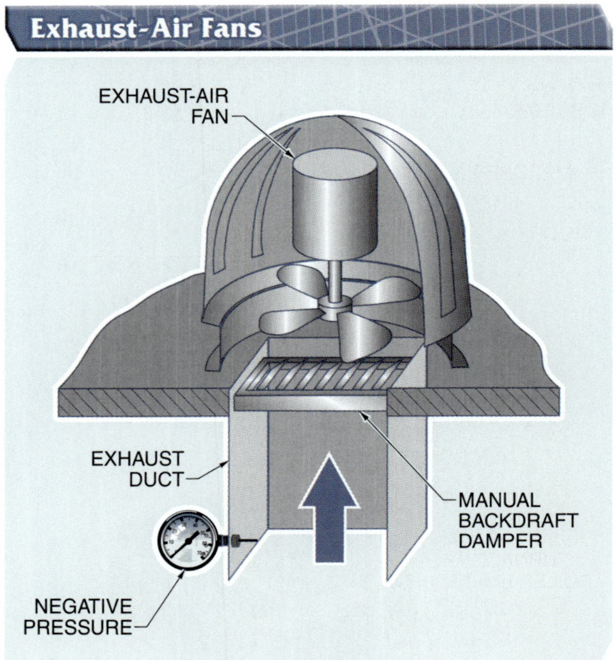

Figure 55-10. Exhaust-air fans are typically roof mounted and operate to exhaust contaminated air.

Exhaust-air fans must not be placed in building spaces that house gas-fired appliances such as boilers or water heaters. Exhaust-air fans placed in boiler rooms, or the rooms where water heaters are located, can create negative pressures that would draw gases of combustion backward, down the flue, and into the rooms. Negative pressures interfere with the proper gas flue draw and discharge poisonous carbon monoxide into the room. When exhaust-air fans or other exhaust fans are suspected of causing a reverse flow of flue gases, management must be consulted immediately. Reverse flow of flue gases is a very dangerous condition.

Fans Media Clip

Mixing Plenum and Dampers. The ANSI/ASHRAE standards provide for diluting indoor contaminants by introducing the proper amount of outdoor air into building spaces. It cannot be assumed that outdoor air is free of pollutants because outdoor air may contain vehicle exhaust fumes, boiler or building exhaust gases, acid-laden air, construction dust, trash compactor odors, and other contaminants and irritants. **See Figure 55-11.** Air-sample testing of the outdoor air may be performed to measure these contaminants.

Outdoor-Air Pollutants

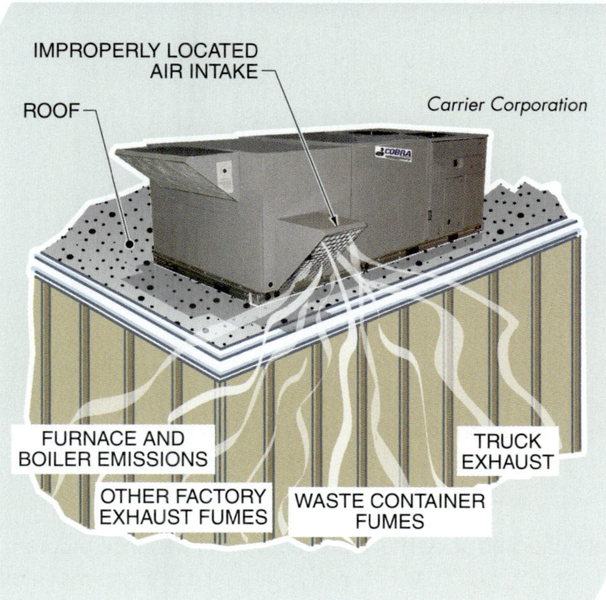

Figure 55-11. Outdoor air may contain many pollutants from exhaust gases and dust that can be pulled into AHUs.

According to ANSI/ASHRAE standards, when outdoor air is relatively free of contaminants, providing proper ventilation air is a function of the plenum, or mixing box, and its interrelated dampers. Because of the close proximity of the return-air damper to the outdoor-air damper, it is possible that the return air is short cycling into the outdoor-air damper, which defeats the purpose of the intended addition of fresh air. HVAC technicians typically use smoke tests to confirm the airflow.

Dampers Media Clip

A mixed air plenum has damper actuators that control the volume of outdoor, return, and exhaust airflow.

The function of outdoor-air, return-air, and exhaust-air dampers is an interrelated operation requiring periodic inspection, maintenance, and testing for proper operation. **See Figure 55-12.** For example, the following tasks should be performed during damper inspection or service:

- Remove the debris restricting airflow.
- Repair any damage to bird and insect screens.
- Clean the damper blades.
- Replace the deteriorated damper blade seals (gaskets).
- Clean and lubricate all moving parts.
- Observe the damper operation through a complete cycle.
- Check the normally open and normally closed dampers.
- Check for air leakage through the damper when the damper is closed.
- Verify the operation of all controls, operators, and linkages.
- Verify the operation of the gravity-relief damper, if installed.

The plenum, or mixing box, must properly blend the correct percentage of outdoor air with return air as the air moves to the next section of the AHU. It is not unusual for a poor mix to occur when parallel-type dampers are installed. Parallel blade dampers may be replaced with opposing blade dampers for better control of airflow. Instead of replacing the dampers, it may be sufficient to add a baffle screen to the mixing box section to increase the mixing of air. The proper percentage of blended outdoor air and return air can be verified by taking direct temperature readings of the mixed-air temperature on the downstream side of the filter section.

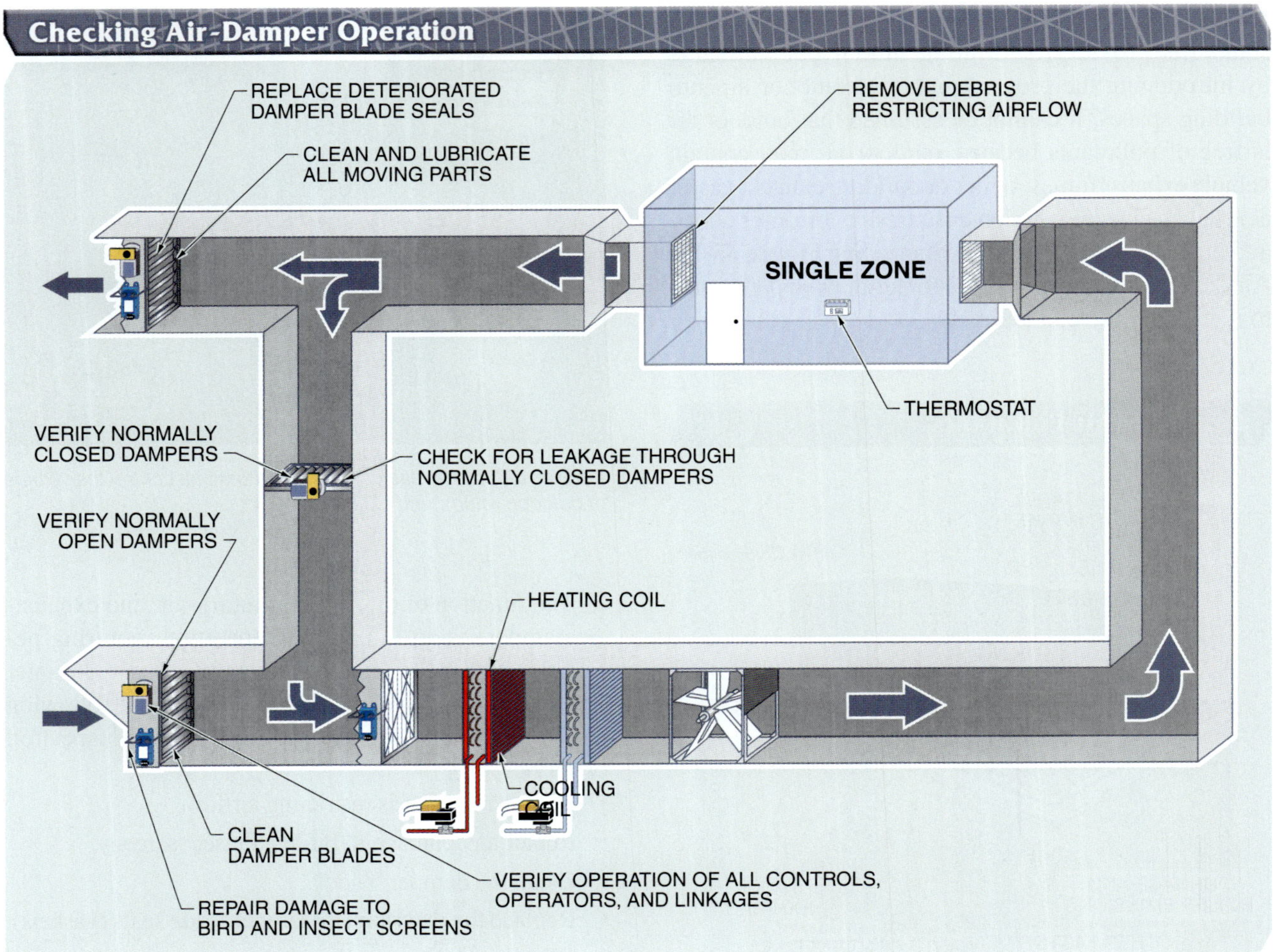

Figure 55-12. All dampers, heating coils, cooling coils, and fans are checked as part of a preventive maintenance program.

Summer Operation. During summer operation, the amount of outdoor air used is determined by the outdoor-air temperature and the air quality of the indoor air. The percentage of outdoor air for cooling during the summer can be determined by using the following formula:

$$A_S = \frac{(T_M - T_R)}{(T_O - T_R)} \times 100\%$$

where

A_S = summer outdoor air (in %)
T_M = average mixed-air temperature (in °F)
T_R = return-air temperature (in °F)
T_O = outdoor-air temperature (in °F)
100% = constant

TECH TIP

Temperatures in the continental United States typically are the highest during the month of August.

Example: Determining Percentage of Outdoor Air for Cooling

A building operating in the summer has a mixed-air temperature of 85°F, return-air temperature of 75°F, and outdoor-air temperature of 100°F. **See Figure 55-13.** Calculate the percentage of summer outdoor air being distributed to building spaces.

$$A_S = \frac{(T_M - T_R)}{(T_O - T_R)} \times 100\%$$

$$A_S = \frac{(85 - 75)}{(100 - 75)} \times 100\%$$

$$A_S = \left(\frac{10}{25}\right) \times 100\%$$

$$A_S = \mathbf{40\%}$$

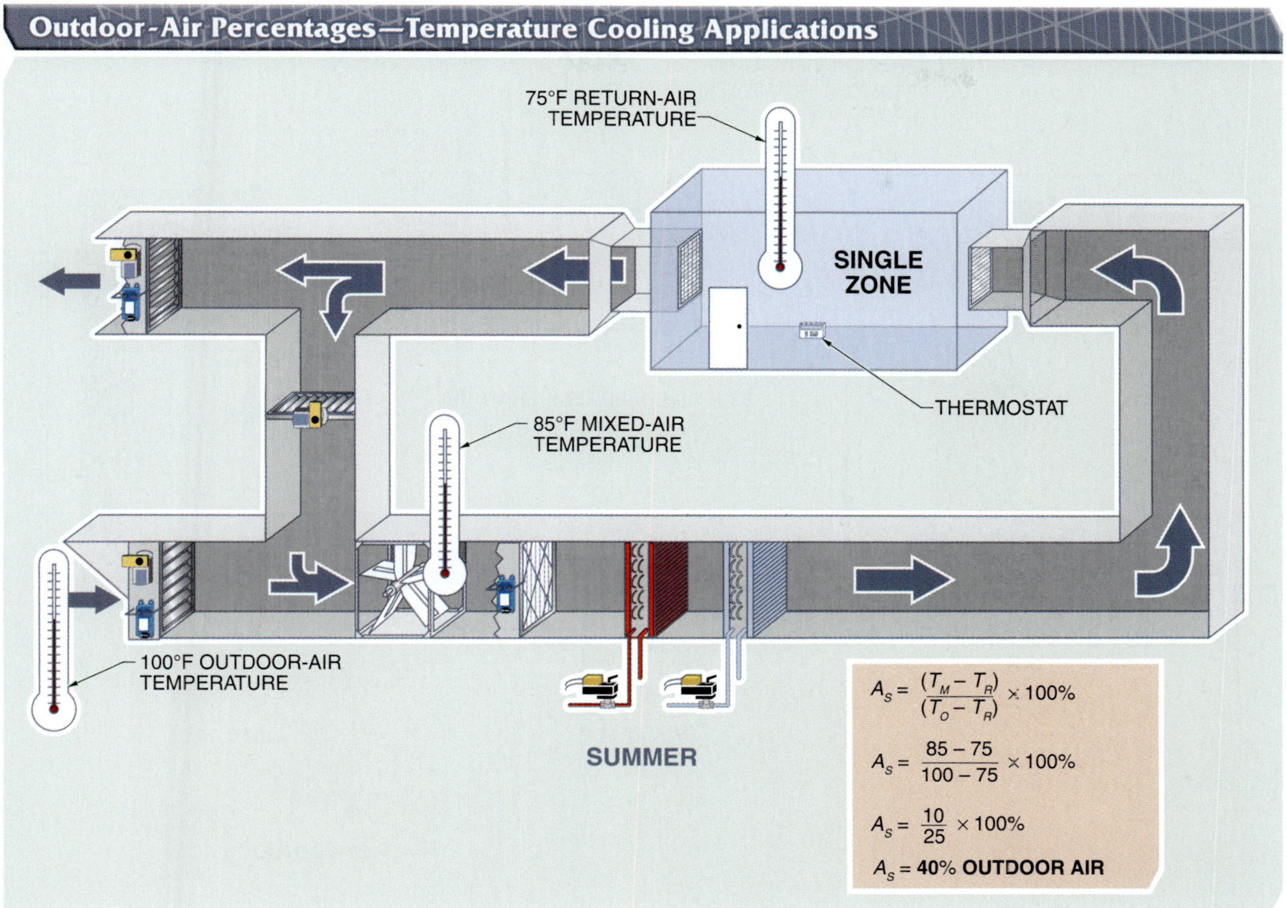

Figure 55-13. The percentage of outdoor air used for cooling depends on the temperature of the outdoor air, return air, and average mixed air.

Winter Operation. During winter operation, the amount of outdoor air used is determined by the outdoor-air temperature and the cooling needs of the building. The percentage of outdoor air for heating during the winter can be determined by using the following formula:

$$A_W = \frac{(T_R - T_M)}{(T_R - T_O)} \times 100\%$$

where
A_W = winter outdoor air (in %)
T_R = return-air temperature (in °F)
T_M = average mixed-air temperature (in °F)
T_O = outdoor-air temperature (in °F)
100% = constant

TECH TIP

Temperatures in the continental United States typically are the lowest during the month of January.

Example: Determining Percentage of Outdoor Air for Heating

A building operating in the winter has a mixed-air temperature of 65°F, return-air temperature of 75°F, and outdoor-air temperature of 30°F. **See Figure 55-14.** Calculate the percentage of winter outdoor air being distributed to building spaces.

$$A_W = \frac{(T_R - T_M)}{(T_R - T_O)} \times 100\%$$

$$A_W = \frac{(75-65)}{(75-30)} \times 100\%$$

$$A_W = \left(\frac{10}{45}\right) \times 100\%$$

$$A_W = \mathbf{22\%}$$

Outdoor-Air Percentages—Temperature Heating Applications

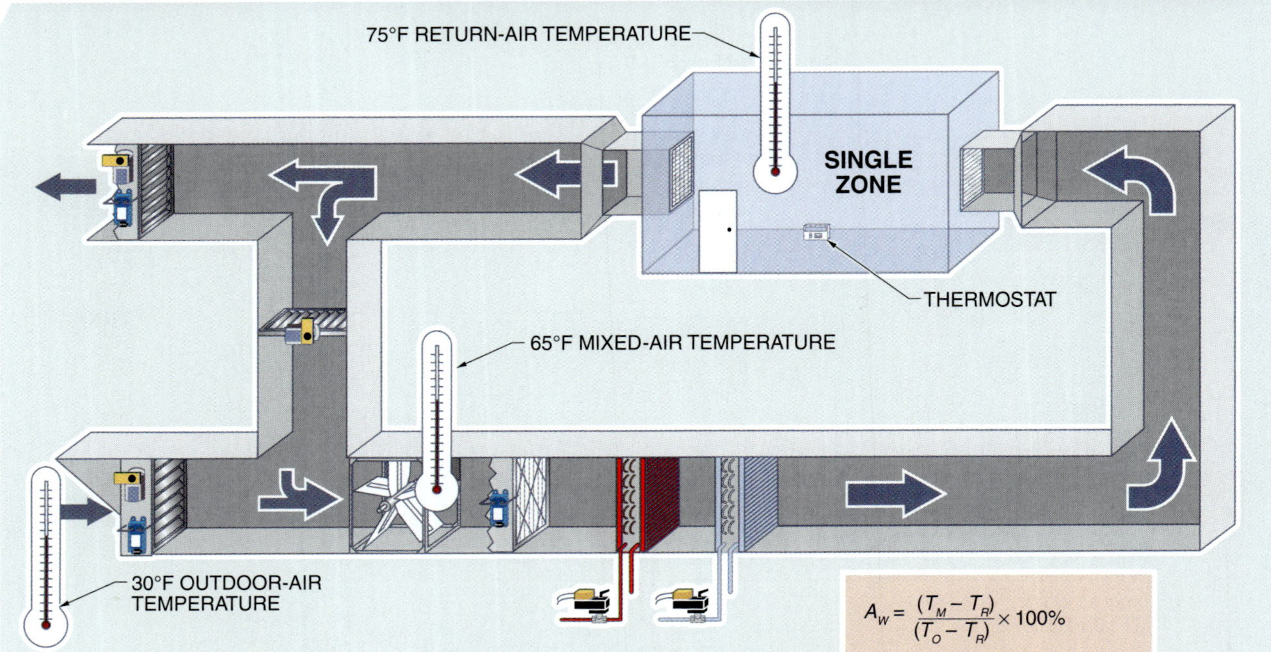

$$A_W = \frac{(T_M - T_R)}{(T_O - T_R)} \times 100\%$$

$$A_W = \frac{75 - 65}{75 - 30} \times 100\%$$

$$A_W = \frac{10}{45} \times 100\%$$

A_W = 22% **OUTDOOR AIR**

Figure 55-14. The percentage of outdoor air used for heating depends on the temperature of the outdoor air, return air, and average mixed air.

Carbon Dioxide Control. In many cases, it is desirable to determine the percentage of outdoor air from the percentage of carbon dioxide (CO_2) present. This can be determined by using the following formula:

$$OA = \frac{(C_R - C_S)}{(C_R - C_O)} \times 100\%$$

where

OA = outdoor air (in %)

C_R = amount of CO_2 in return air (in ppm)

C_S = amount of CO_2 in supply air or mixed air (in ppm)

C_O = amount of CO_2 in outdoor air (in ppm)

TECH TIP

Either temperature or CO_2 levels can be used to calculate air mixture percentages. Facilities with known indoor air quality issues typically use CO_2 level measurements to help ensure proper airflow.

Example: Determining Percentage of Outdoor Air for CO_2 Control

A building operating in the summer has a mixed-air amount of CO_2 at 845 ppm, return-air amount of CO_2 at 1100 ppm, and outdoor-air amount of CO_2 at 530 ppm. **See Figure 55-15.** Calculate the percentage of outdoor air being distributed to building spaces.

$$OA = \frac{(C_R - C_S)}{(C_R - C_O)} \times 100\%$$

$$OA = \frac{(1100 - 845)}{(1100 - 530)} \times 100\%$$

$$OA = \left(\frac{255}{570}\right) \times 100\%$$

$$OA = 0.447 \times 100\%$$

OA = **44.7%**

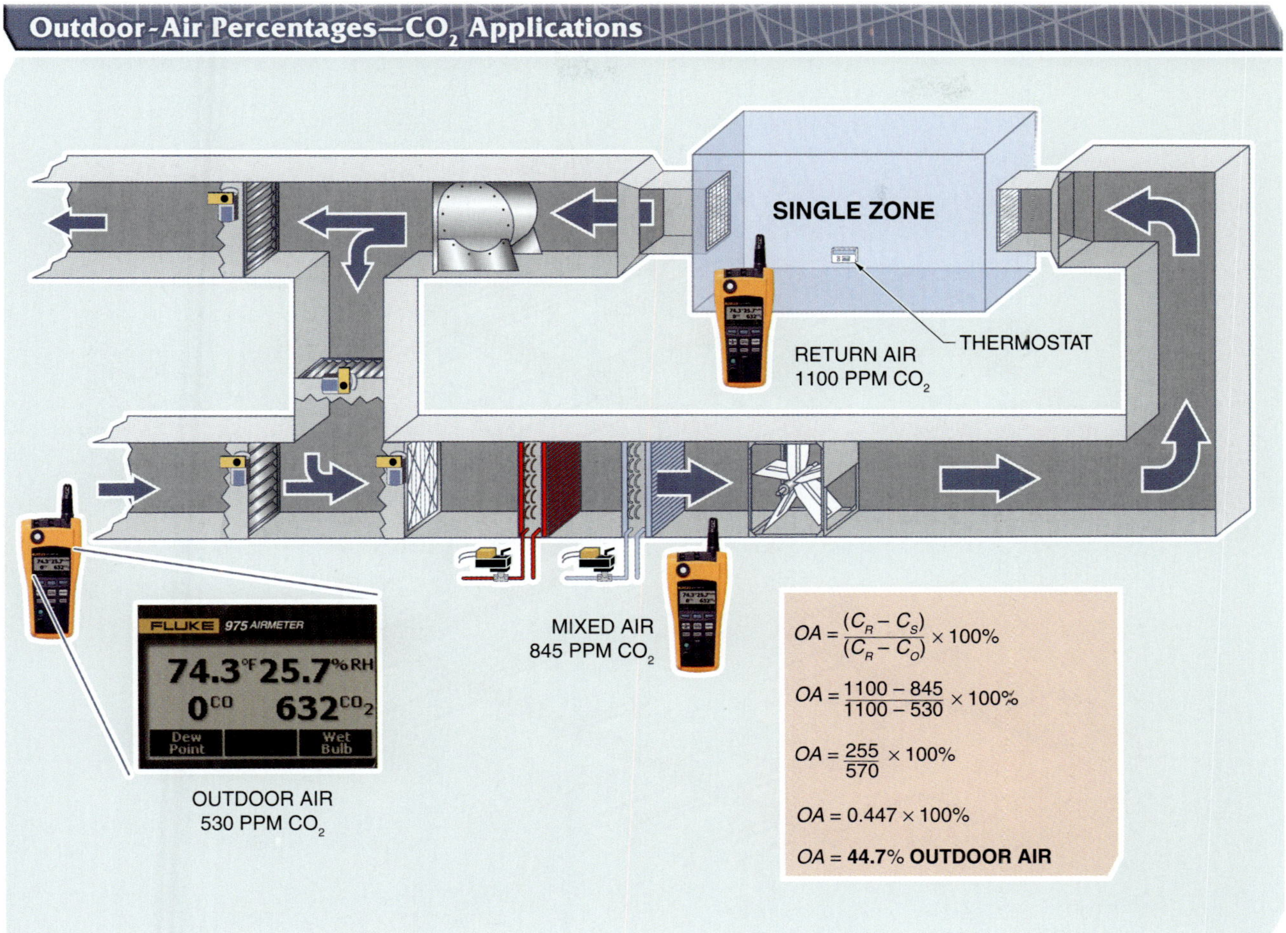

Figure 55-15. The percentage of outdoor air used for IAQ depends on the CO_2 levels of the outdoor air, return air, and average mixed air.

HVAC System Ductwork

AHUs must be connected to good distribution systems of ducts in order to distribute conditioned air properly. Space limitations can compromise the design of air distribution ducts. Control devices such as main and branch volume dampers or blowers controlled by variable-frequency drives distribute air according to the design engineer's load calculations. **See Figure 55-16.** The proper sizing, location, and aerodynamic design of supply diffusers help distribute air evenly throughout building spaces. Properly locating return-air grills ensures a sweeping effect of distributed air and negates the short cycling of conditioned supply air.

Return-Air Ductwork Media Clip

Damaged or separated duct systems can leak conditioned air to unwanted areas such as ceiling plenums, wall cavities, and crawl spaces. Leaks in ducts decrease the effectiveness of HVAC distribution systems and deprive occupants of the amount and quality of conditioned air intended for their comfort. Air-balancing reports provide an overview of intended supply air and resulting distributed air. Differences between the intended supply air and resulting distributed air is the air lost in the system to unintended areas. In some buildings, finding HVAC distribution system leaks can be difficult.

Comfort has a direct effect on how occupants perceive the quality of air being supplied to their spaces. Temperature and humidity are key factors to comfort. The velocity of air moving through an occupied space is also important for comfort. ASHRAE recommends air movement in a room to be around 50 fpm. Velocities that are too high can cause drafts, and velocities that are too low can be perceived as a comfort problem.

HVAC System Ductwork

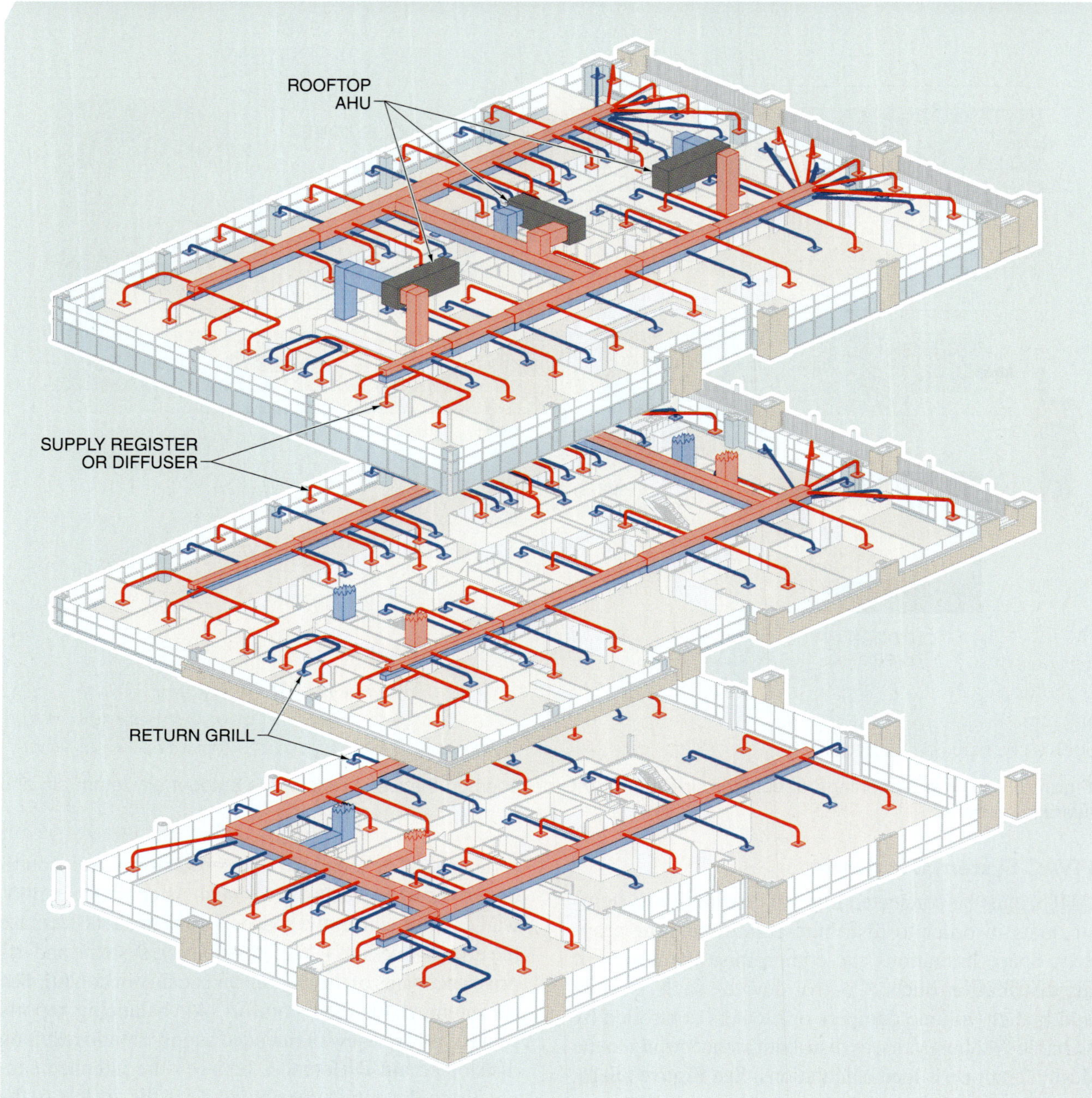

Figure 55-16. HVAC ducts distribute conditioned air and collect return air throughout the building using dampers to control the airflow.

Work performed on a duct system can also generate indoor air quality (IAQ) problems. HVAC technicians need to limit the jostling or movement of ducts as much as possible. Work that disturbs system ducts can introduce or loosen and convey dirt particles to occupied spaces with the airflow. Extreme care must be used when performing maintenance or inspections within the duct system. Replacing portions or all of an AHU, which involves disconnection and reconnection of ductwork, can easily release contaminants into the HVAC system. **See Figure 55-17.**

HVAC Duct Dirt and Debris

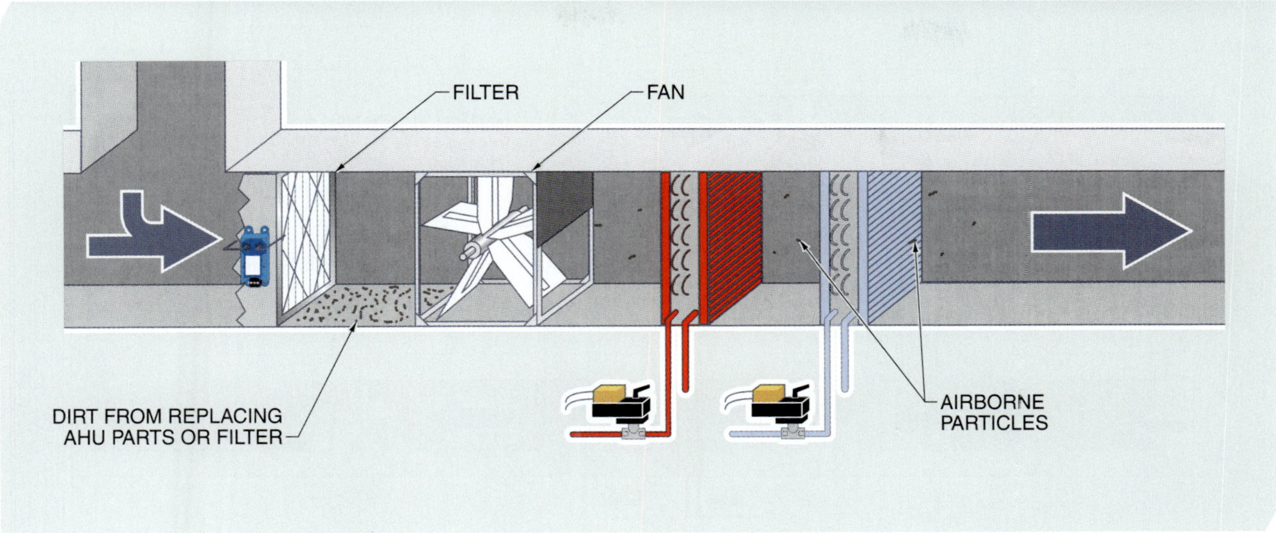

Figure 55-17. HVAC ducts distribute airborne particles throughout the building from dirt and debris left while servicing the HVAC system.

When ducts are exposed, the area should be cleaned and the inside of the duct vacuumed to the greatest extent possible. Typically, temporary filters are installed at the air diffusers while running the system to catch any remaining loose dirt and particles. During this temporary filtering process, airflow to affected spaces will be reduced. Temporary filtering should be used when scheduling work and/or cleaning ducts.

Heating/Cooling Coils Media Clip

Temporary filters should be used to remove loose dirt and particles from the duct airflow during scheduled work.

Cooling Coils. The cooling coils of an HVAC system allow chilled water or air to be used for cooling air. Cooling coils typically consist of finned coils, copper tubes, and aluminum fins. **See Figure 55-18.** Room air is cooled using chilled water or a refrigerant. While cooling coils cool the air, they are also removing moisture from the air as condensate. The cooled air is at a very high RH when leaving the cooling coil.

HVAC technicians recognize the condensation process that cooling coils produce, but they may not realize that the entire duct distribution system contains high RH air. The cold humid air traveling through the duct does not lower its humidity level until leaving the air diffuser to a room. In extreme cases, air passing over the drain pan can lift moisture out of the pan and spray it over the surfaces of the AHU. **See Figure 55-19.**

Cooling coil sections and the distribution ducts can have significant effects on IAQ. Dirt that attaches to wet cooling coils and/or stagnant water in the drain pan of coil sections can lead to bacteria growth, mold formation, and mildew. Mold and mildew can become airborne and thrive in the cool and damp surroundings of the duct system. The airborne mold and mildew eventually enter the conditioned spaces and become absorbed in the lungs of the occupants. Wet insulation in the cooling coil section does not dry out during the cooling season and may breed bacteria.

Cooling Coils

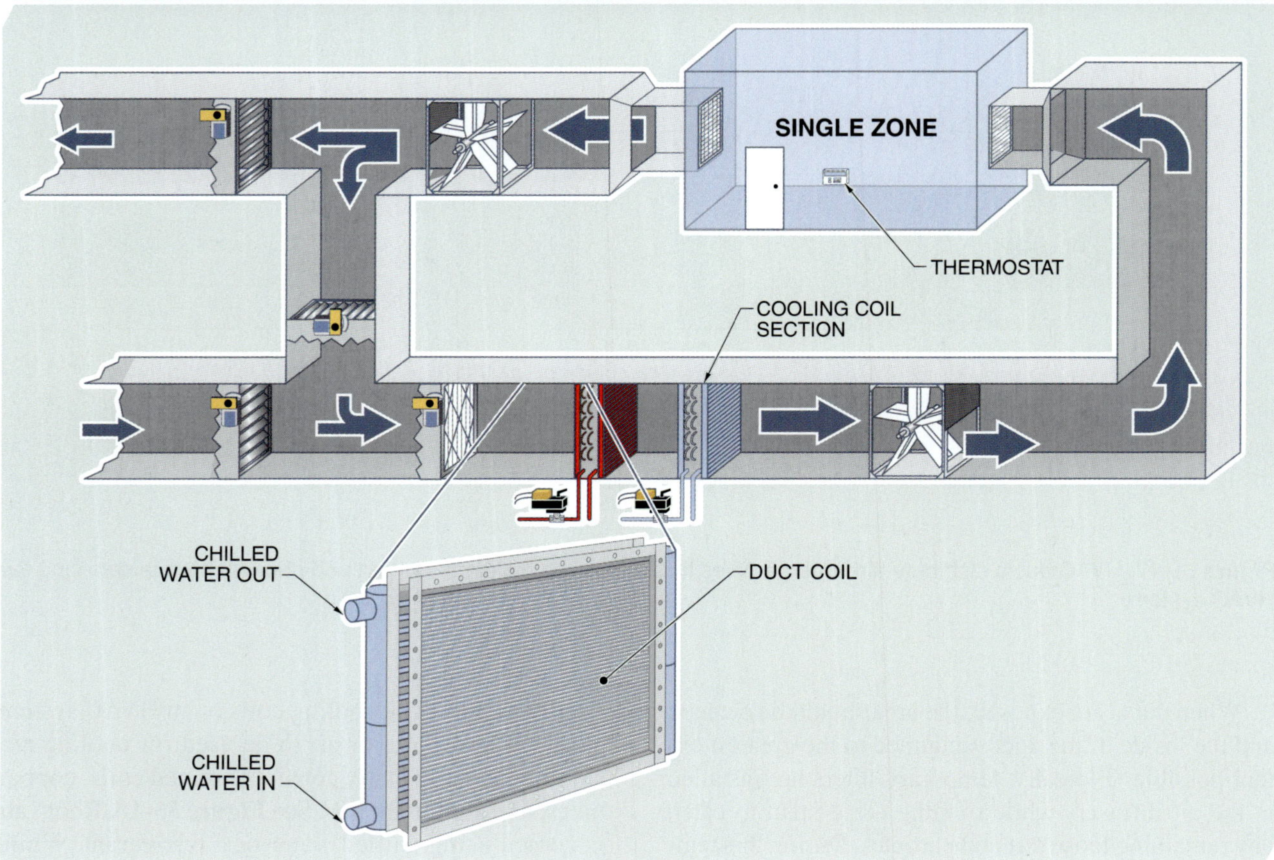

Figure 55-18. Duct cooling coils cool and remove moisture from the supply air.

Cooling Coil Drainage

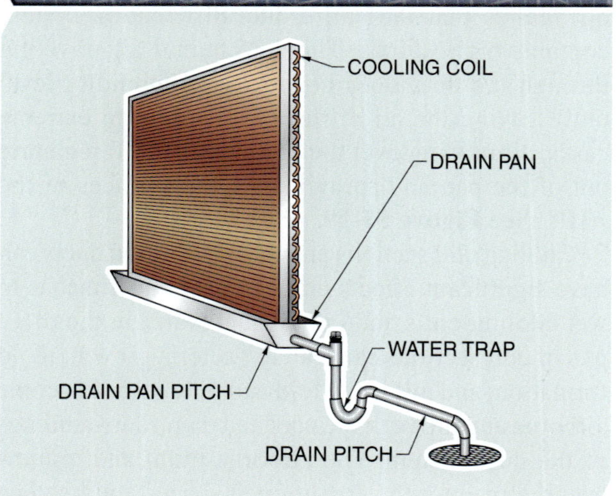

Figure 55-19. Drain pans collect moisture from the cooling coils and pass the draining water through a water trap to the drain.

Periodic inspection and cleaning of the interior parts of cooling coil sections, including the main drain pan, intermediate coil drain pans, water traps, and cooling coils, are essential. **See Figure 55-20.** Drain pans must be pitched properly for positive drainage. Even when drain pans are properly positioned, the drainage water trap must prevent air from entering a coil section that is under vacuum and from escaping a coil section that is under pressure.

Cooling Coil/Dehumidifier Sections. Spray coil dehumidifiers are rarely designed for new buildings, but many exist in older buildings. Systems that have standing water in the sumps are havens for bacteria growth. The evaporative cooling effect with recirculated sprayed water assists in the cooling of air. However, bacteria in water end up being directly injected into the airstream. Eliminating the spray on the chilled water coil may mean sacrificing cooling capacity. Care must be taken when monitoring the condition of the water in these sumps.

Cooling Coil Inspection

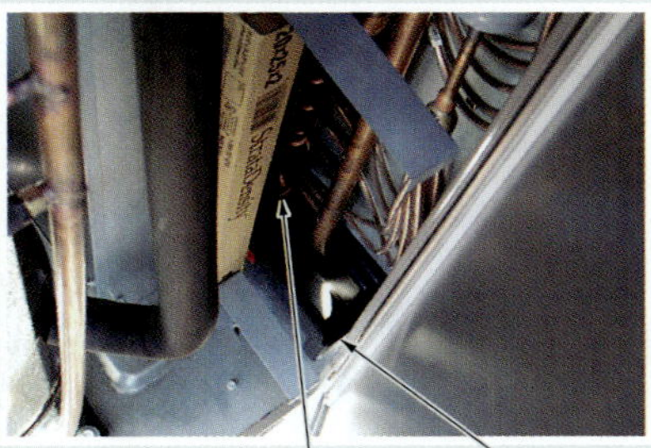

Figure 55-20. Drain pans and drainage piping must be inspected periodically to ensure proper drainage.

Raising chilled water temperatures is an energy-conservation measure often used to reduce costs. As chilled water temperatures increase, the amount of moisture in supply air also increases. The RH in occupied spaces must not be increased to the point where mold, mildew, and bacteria can thrive.

To provide enough conditioned air to occupants, fan speeds are sometimes increased, which causes the carryover of water particles into the duct system. According to ANSI/ASHRAE guidelines, an AHU mounted horizontally should not have a coil-face airflow velocity of more than 500 fpm. A vertically mounted unit can have an airflow velocity up to 600 fpm without undue carryover of moisture into the duct system.

Coil fin spacing can range from 2 fins per inch to 14 fins per inch. Closer fin spacing or higher airflow velocity across the coil results in greater carryover of water into the duct system. Since most cooling coils produce water condensate and distribute moist air into the duct system, special attention must be paid to the cooling coil section of an air distribution system.

Heating Coils. The heating coils of an HVAC system allow heated water or air to be used for heating air. Heating coils lower the RH of air as the coils increase air temperature (sensible heat). When a conditioned space has enough latent heat load, there is no need to add humidity to the air. A conditioned space should have an RH between 30% to 60% to minimize static electricity as well as bacterial, viral, and fungal growth. Humidity levels below 30% tend to extract moisture from occupants, especially from the lungs and mucous membranes, which can result in respiratory infections, dry skin, and general discomfort.

Heating Coil/Humidifier Sections. Most air-handling systems use some form of humidity control to add moisture to building air during the winter heating periods. Typically, water is added into the airstream at the discharge side of the heating coils. **See Figure 55-21.** Boiler steam must not be used to humidify the HVAC airstream under any circumstance. Steam from boilers typically contains boiler water treatment chemicals.

Controllers measure the humidity in rooms and ducts using sensors and adjust the amount of moisture introduced into the airstream. The humidity-control loop must be set up properly to ensure that the system does not add too much or too little moisture to the supply air. The airstream must absorb 100% of the moisture so that no condensation accumulates in the duct system.

Water pan humidifiers pose the same problem as standing water reservoirs by providing an ideal growth environment for bacteria, viruses, and fungi. For example, warm or hot standing water provides an ideal environment for Legionella bacteria to grow.

AIR-CLEANING METHODS

Three basic strategies used to reduce pollutant concentrations in indoor air are source control, ventilation, and air cleaning. *Source control* is a method of reducing air pollution by reducing the origin of pollutants. Reducing pollutant origins may involve the use of exhaust systems at specific points where pollutants are present. The source control air-cleaning method requires the correct handling and storage of all chemicals and other pollutants found in buildings.

Ventilation is the process of reducing air pollution by introducing enough fresh outdoor air into a building to dilute indoor pollutants. *Air cleaning* is a method of reducing air pollution by removing pollutants using electronic devices or filters. A preventive maintenance (PM) program must be used to periodically inspect and clean the air cleaning devices. Air cleaning methods typically include electronic air cleaning, ultraviolet air cleaning, and the use of carbon and mechanical filters.

Air Cleaners

Air cleaners use several methods to clean the air. Electronic air cleaners impart a charge to particles in the air that are attracted to plates with the opposite charge. Ultraviolet systems use ultraviolet light-generating equipment to kill biological contaminants. Carbon filters use activated carbon-filtering mediums to remove most odors, smoke, and smog from the air.

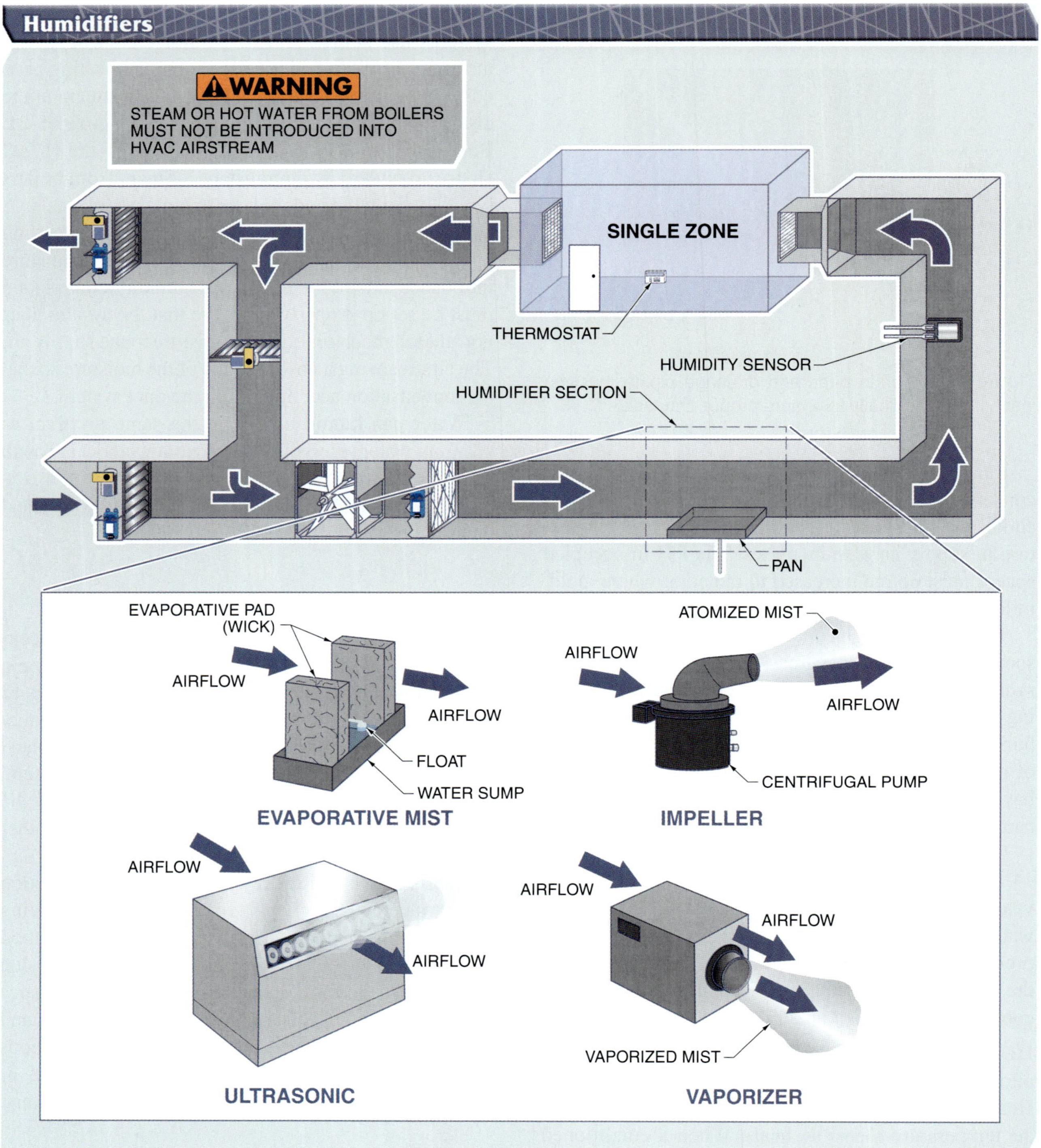

Figure 55-21. Humidifiers add moisture to the conditioned airstream after passing through the heating coils.

Electronic Air Cleaners. An *electronic air cleaner* is an air filtering system that passes an electrical charge through a wire grid inside the filter to charge particles and then attracts those particles to plates charged with the opposite charge, where they are trapped. Airborne particles receive a positive charge while passing through an ionizing section. Then, as the positively charged particles pass through negatively charged collecting areas, the dirt particles are attracted to the collector plates. **See Figure 55-22.** Depending on the system, particles clinging to the collector plates are washed off with water or blown off with compressed air.

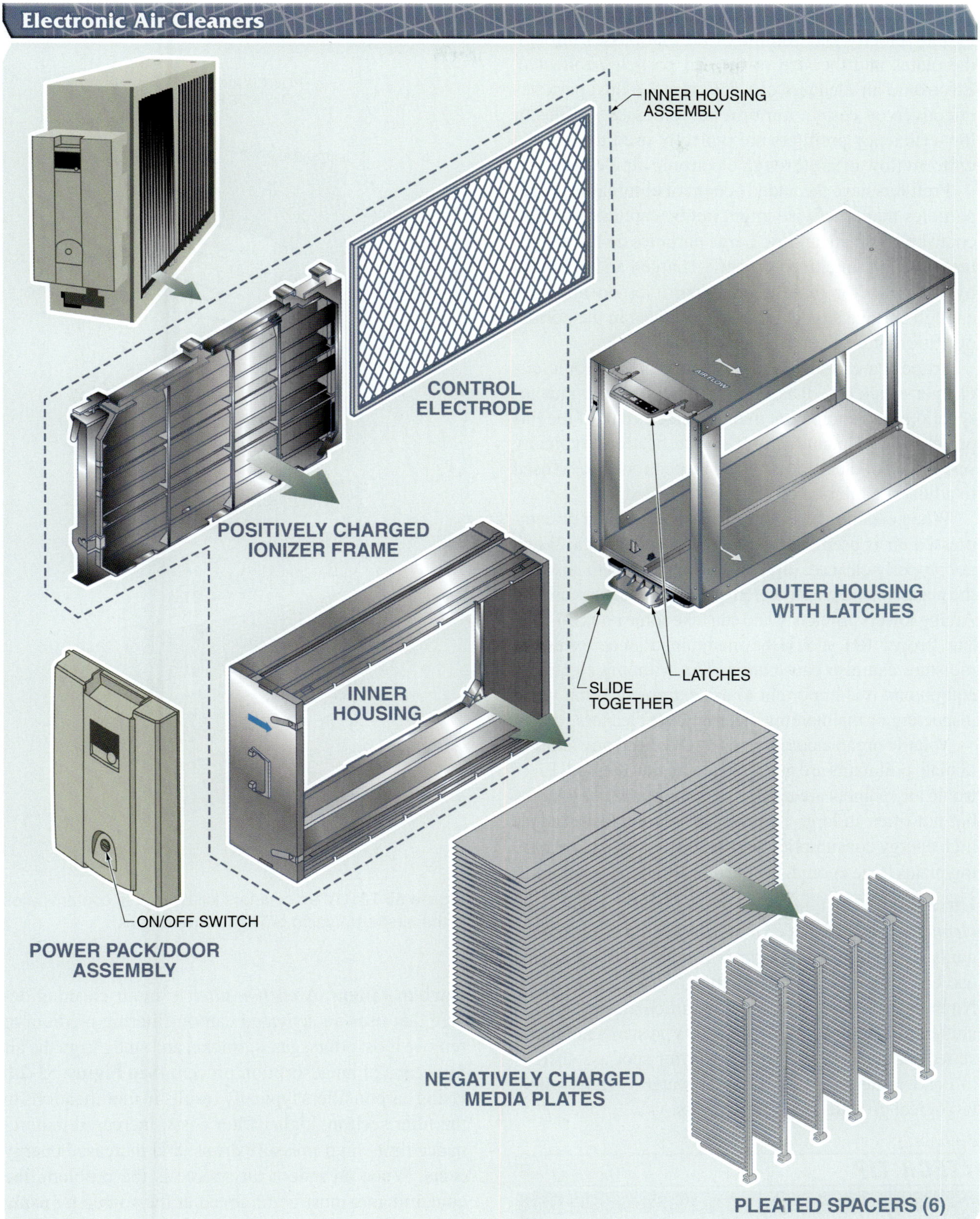

Figure 55-22. To remove particles from the airstream, electronic air cleaners use positively charged frames to charge dirt particles that are attracted to negatively charged plates.

The efficiency of electronic air cleaners is determined by the velocity and volume of air that passes through the plates and the size of the dirt particles collected. Electronic air cleaners can only remove dust particles effectively at certain uniform air velocities. Baffles or low-efficiency prefilters are typically used to ensure a uniform flow of air through electronic air cleaners.

Prefilters have the added benefit of eliminating larger particles that otherwise might not be captured and that contribute to plate arcing. Large particles do not always have enough time to be properly charged and collected by the collector plates. Small particles do not have enough surface area to cling to the plates in the normal operation of an electronic air cleaner.

Proper care is also a major factor in electronic air-cleaner efficiency. It is important to schedule unit inspections and equipment maintenance as part of the PM program. The manufacturer's specifications typically serve as a good baseline. Visual inspections can be used to adjust the interval of service as needed.

When cleaning electronic air cleaners, water or compressed air is used to clean the whole plate area. When not properly cleaned, dirt buildup creates a path between the negative and positive elements, which causes arcing. Arcing lowers efficiency and can take some units off-line. The proper RH should be maintained since excessive moisture can also cause arcing. As with any electronic equipment, it is important to use extreme caution when inspecting or maintaining electronic air cleaners.

Volatile organic compounds (VOCs) and any nonparticulate pollutants are not removed by this method. Electronic air cleaners are used widely in residential systems but not often in large commercial systems due to their high energy consumption and the difficulty of constructing plates large enough for the much higher air volumes.

Ultraviolet Air Cleaners. An *ultraviolet (UV) air cleaner* is an air cleaner that kills biological contaminants using a specific light wavelength. UV systems use UV light-generating equipment. **See Figure 55-23.** Nonbiological pollutants, such as particulates and VOCs, are not affected by UV systems. UV systems are better suited for small HVAC systems or for spot use such as in condensate pans or condenser water systems where biological growth commonly occurs.

TECH TIP

Air filter performance is only as good as the ongoing maintenance performed. Air filters must be checked and changed regularly as part of a PM program.

Ultraviolet (UV) Air Cleaners

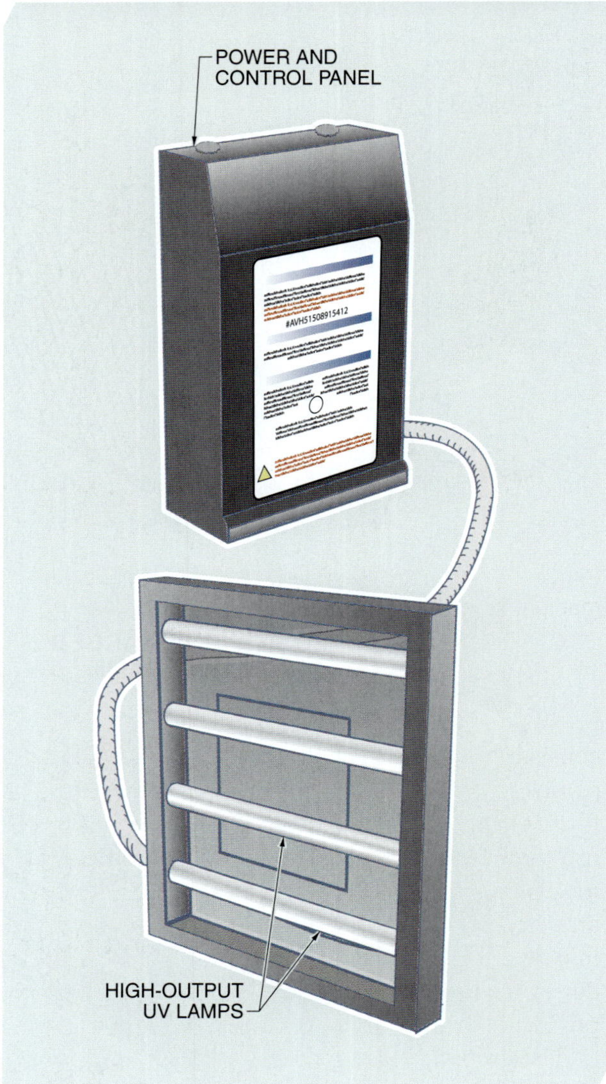

Figure 55-23. UV air cleaners kill biological contaminates in the airstream using UV light wavelengths.

Carbon Filters. A *carbon filter* is an air cleaning device that uses an activated carbon-filtering medium to remove most odors, gases, smoke, and smog from the air by means of an adsorption process. **See Figure 55-24.** Using carbon filters typically results in modifications to the filter section, higher filter costs, increased maintenance time, high-pressure drops, and increased energy costs. When an indoor air source is the problem, the contaminants must be removed at the source by using a localized exhaust fan. Using exhaust fans to remove contaminants solves the problem without having to redesign the air-handling system.

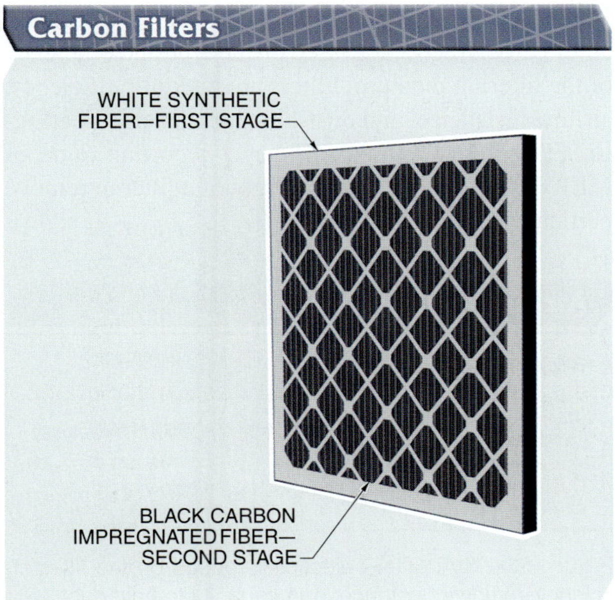

Figure 55-24. Carbon filters use activated carbon as a filter medium to remove most odors from the airstream.

Mechanical Filters. A *filter* is an air cleaning device consisting of fine-fiber fabric that is used to remove particles from the airstream. A mechanical filter contains fibers that trap and attach particulate matter to the filter medium. The filter medium traps particulate matter in multiple layers using several methods. Fibers may be arranged in layers, or interwoven, using alternating thicknesses of fiber. Sometimes an adhesive is applied during the manufacturing of filters or applied as a spray product at the time of installation. **See Figure 55-25.**

The proper care of mechanical filters prevents potential IAQ problems. Guidelines in caring for filters include the following:

- Ensure that filters meet the specifications of the air-handler manufacturer for size and thickness. Filters that are too thick impair the function of mechanical equipment. Filters too thin for the application will not stop enough particles to effectively clean the airstream.
- Inspect and maintain/replace filters during the PM program according to the manufacturer specifications.
- Keep filters dry by checking and adjusting water contamination and RH settings.
- Install filters snug to the frame with no leaks around the edges. Air-handler seams must be properly sealed. Most filters have arrows indicating how the filter must be installed in relation to the direction of airflow.
- Wear gloves and a respirator when handling dirty filters to protect against particles released. Handle filters with care to avoid disturbing the dirt when moving the filters.
- Dispose of filters as prescribed by maintenance procedures or local regulations.

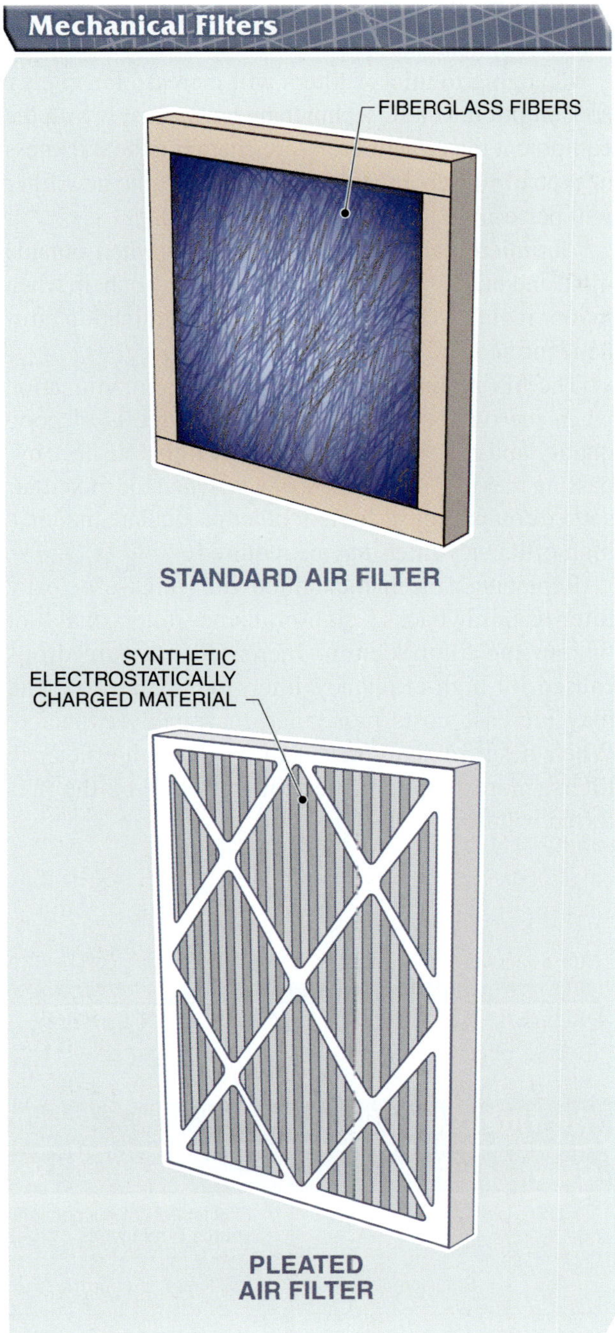

Figure 55-25. Mechanical filters use fine-fiber fabric to remove particles from the airstream.

Various grades of filters are available. Equipment manufacturer specifications must be checked for the recommended size, thickness, and flammability classification. Filters are classified as Class 1, which does not burn with negligible amounts of smoke; Class 2, which burns moderately with some smoke; and Class 3, which burns. **See Figure 55-26.** Class 2 filters are typically used unless HVAC systems require a low flammability classification. Class 3 filters are used in very limited applications.

Changing to thicker filters will increase fan energy consumption. HVAC technicians must consult with the equipment manufacturer before changing the thickness or type of filter being used to ensure that the new filter will perform in the equipment as needed.

Antimicrobial filters are appropriate when outside mold and mildew are problems. Filters do not help when indoor mold and mildew exist. The contamination must be identified and eliminated.

The filtration system does not require modification when the outdoor-air or indoor-air source is of good quality and the filters are working without airflow bypassing the filter medium. However, when the mixed-air source contains dust, soot, or other particulate matter, a high-efficiency filter may be required.

Replacing a mechanical filter with a high-efficiency filter has drawbacks. High-efficiency filters may not fit into the filter section. Increased pressure drops caused by high-efficiency filters reduce airflow and may increase costs more than reasonably expected. When these problems prevent the use of high-efficiency filters, more frequent filter changes may be the preferred solution.

Filters Media Clip

Filter Selection and Ratings. The quality of filtered air directly affects IAQ based on the type and efficiency of the filtering medium. Filters are rated for efficiency, airflow resistance, and dust-holding capacity. **See Figure 55-27.** The minimum efficiency reporting value, or MERV rating, indicates the ability of a filter to remove particulate matter from air.

Air Filter Selection

Efficiency	Size and amount of particles removed from airflow
Airflow resistance	Static-pressure drop at specified air velocity
Dust-holding capacity	Amount of dust filter can hold while operating at specified airflow resistance

Figure 55-27. Air filter selection is based on the filter efficiency, airflow resistance, and capacity to hold dust.

Filter airflow resistance is the pressure drop across a filter at a given velocity. An airflow velocity of 500 fpm is typically used, although some manufacturers use slower velocities. The *dust-holding capacity* is the measure of a filter's ability to hold dust without significantly reducing its efficiency. The type and design of filters determine their efficiency at removing particles of a given size and the amount of energy required to pull or push air through the filters. Filter media types include dry, viscous (sticky), cleanable, activated carbon, high-efficiency, and electronic.

Filters must be selected for their ability to protect both the HVAC system components and general air quality. In many buildings, the best choice is a medium-efficiency, pleated filter with a prefilter. Medium-efficiency, pleated filters have higher levels of removal efficiency than low-efficiency filters. Medium-efficiency filters last for longer periods of time without clogging than high-efficiency filters.

Air Filter Classifications

Class	Fibers	Note	Code
1	Fire-resistant	May ignite, but will not continue to burn once ignition source is removed	ASME AG-1, Section FC 2
2	Semicombustible	Will ignite and burn; less expensive	UL 586.20
3	Combustible	Readily combustilbe; limited use	UL 900, Class 1.21

Figure 55-26. Air filters are classified by their combustibility.

Filter Media Ratings. *Dust spot efficiency* is a filter rating that indicates the ability of a filter to remove from the airflow large particles that tend to soil building interiors. Dust-holding capacity is a measure of the total amount of dust that filters are able to hold during dust-loading tests. In the past, filters were rated by the ASHRAE dust spot rating.

Filter Replacement. Filters are scheduled to be replaced either when the filter has reached a certain pressure drop or after a given time interval such as every 90 days. To determine whether a filter must be replaced, an electronic manometer is used to measure the increase in pressure drop across the filter and/or a visual inspection is performed to assess the amount of dirt clogging the filter. **See Figure 55-28.**

Periodic visual inspections establish a baseline PM program with filters changed at preset intervals. Intervals can be set up by month, quarter, or total hours of operation. The interval serves as a method to approximate how often filters should be changed. The intervals can be lengthened or shortened as required to conform to varying conditions of operation and outdoor-air quality.

For example, a PM program may call for a filter to be changed when a manometer indicates that the pressure drop across the filter reads 0.9″ WC (water column). Other pressure drop values are used based on the filter design, airflow resistance, and dust-holding capacity of the filter. This may make economic sense, but other factors must be considered. In some areas, excessive pollen accumulation in the spring season will require that the filter be replaced more frequently.

Other conditions may require a filter to be changed outside of the regular time interval. For instance, water from a nearby humidifier might accidentally dampen the filter. Adjustments to the filter time interval will help prevent IAQ problems such as excessive particles in the air, reduced airflow to occupants, and microbial growth from molds or mildew.

AIR-HANDLING DISTRIBUTION SYSTEMS

Several types of air-handling distribution systems meet the environmental requirements of buildings. Common AHUs are centrally located and serve areas or zones of buildings. Common air-handling systems include constant-air-volume and variable-air-volume systems.

Constant-Air-Volume Systems

A *constant-air-volume (CAV) system* is an air distribution system in which the airflow rate in each building space remains constant but the supply-air temperature is varied to maintain a setpoint. CAV systems operate at their rated airflow capacity at all times. Common CAV systems include single-zone, multizone, dual-duct, and reheat systems.

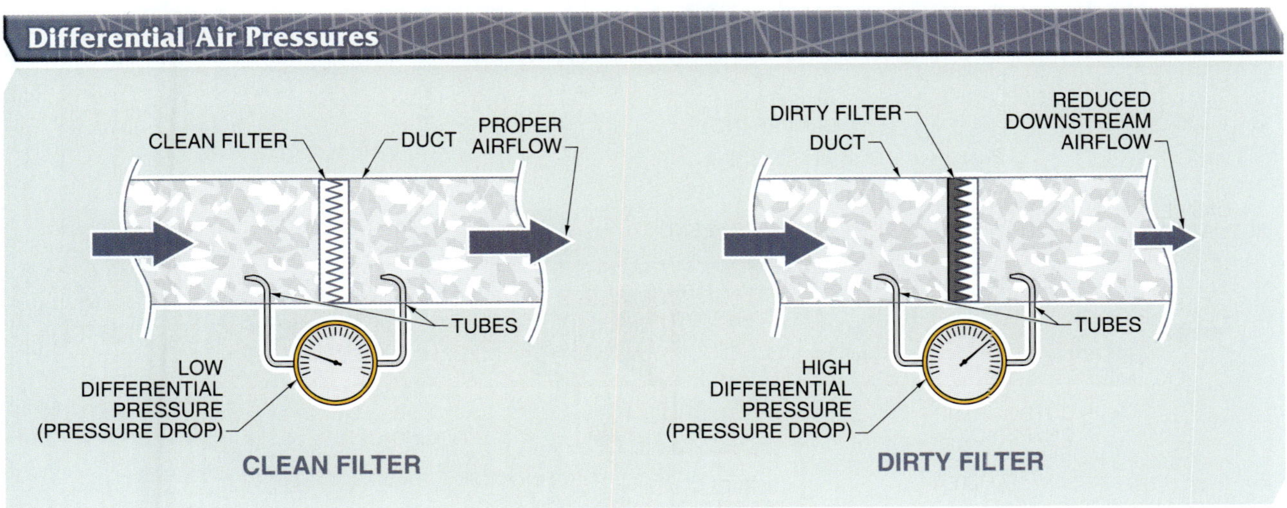

Figure 55-28. Differential air pressure between the upstream and downstream sides of a filter is a measure of particulate clogging and is used to indicate the need to service the filter.

Single-Zone Systems. A *single-zone system* is an air distribution system that serves a single temperature control zone and is the simplest form of air-handling system. **See Figure 55-29.** Single-zone systems can be shut down without affecting the operation of adjacent areas. Economizer control arrangements can be used with unit ventilators and central station systems. Single-zone systems supply a constant volume of air to the conditioned space. Supply air temperatures can be held constant and the fan cycled to control space temperature, or the fan can be run continuously and the air temperature reset to maintain conditions. There are a variety of single-zone systems, including room fan coil units, unit ventilators (with a special sequence control), induction units, and central station fan systems.

Multizone Systems. A *multizone system* is an air distribution system that serves a relatively small number of zones from a single, central AHU. **See Figure 55-30.** Multizone systems deliver a constant volume of air. In response to zone thermostats, the requirements of the various zones are met by mixing cold and warm air through zone dampers at the central air handler. Cold deck and hot deck temperatures may be varied in response to zone loads. Mixed conditioned air is distributed throughout the building by a system of single-zone ducts. All of the various economizer control arrangements can be used with multizone systems.

Fans of multizone systems run continuously when building spaces are occupied. Averaging thermostats in the hot deck modulate the heating valve to maintain the setpoint temperature, which may be reset according to the outdoor temperature. Averaging thermostats in the cold deck modulate the cooling coil valve to maintain the setpoint temperature. Space thermostats modulate the mixing dampers at the central fan to supply a mix of hot and cold air that satisfies the space load.

Dual-Duct Systems. A *dual-duct system* is an air distribution system in which all the air is conditioned in a central fan system and distributed to conditioned spaces through two parallel main ducts. **See Figure 55-31.** One duct carries cold air and the other carries warm air, thus providing air sources for both heating and cooling at all times. Dual-duct systems are used in buildings that also have multiple zones and can serve a wide variety of loads. However, they are considered to be energy inefficient and are very expensive to install.

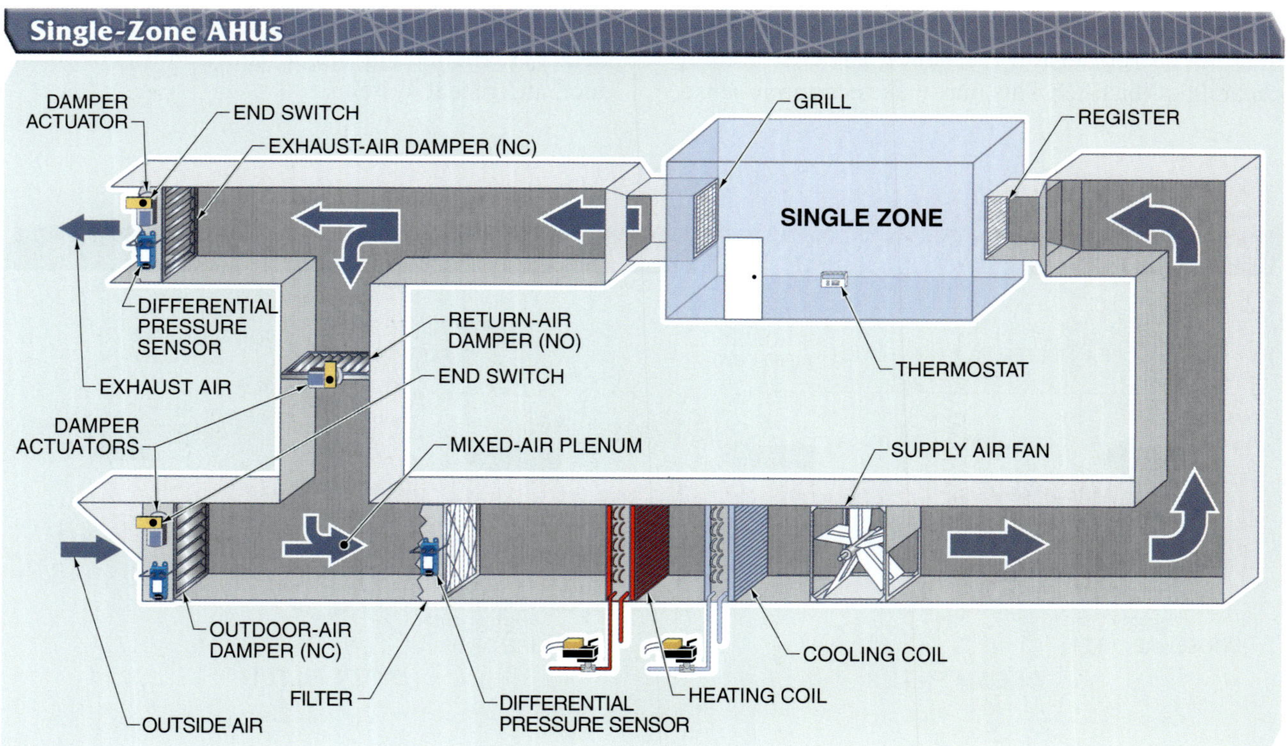

Figure 55-29. Single-zone AHUs provide conditioned air to one zone or area.

Multizone AHUs

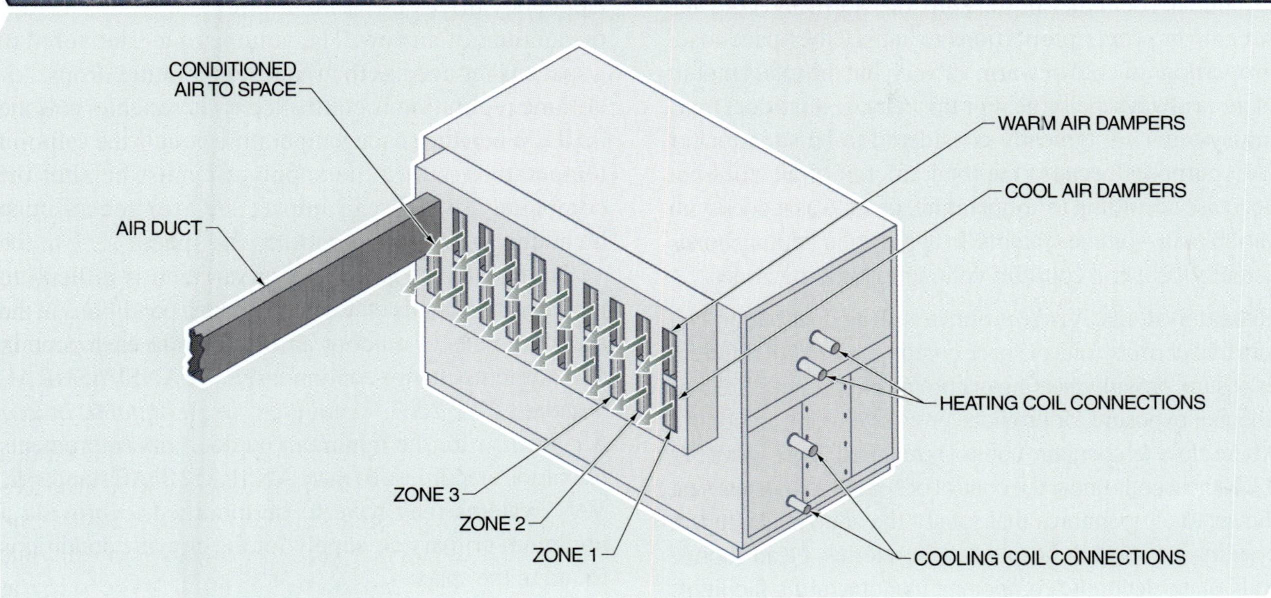

Figure 55-30. Multizone AHUs distribute centrally mixed conditioned air to several zones within a building using dampers to control the airflow.

Dual-Duct AHUs

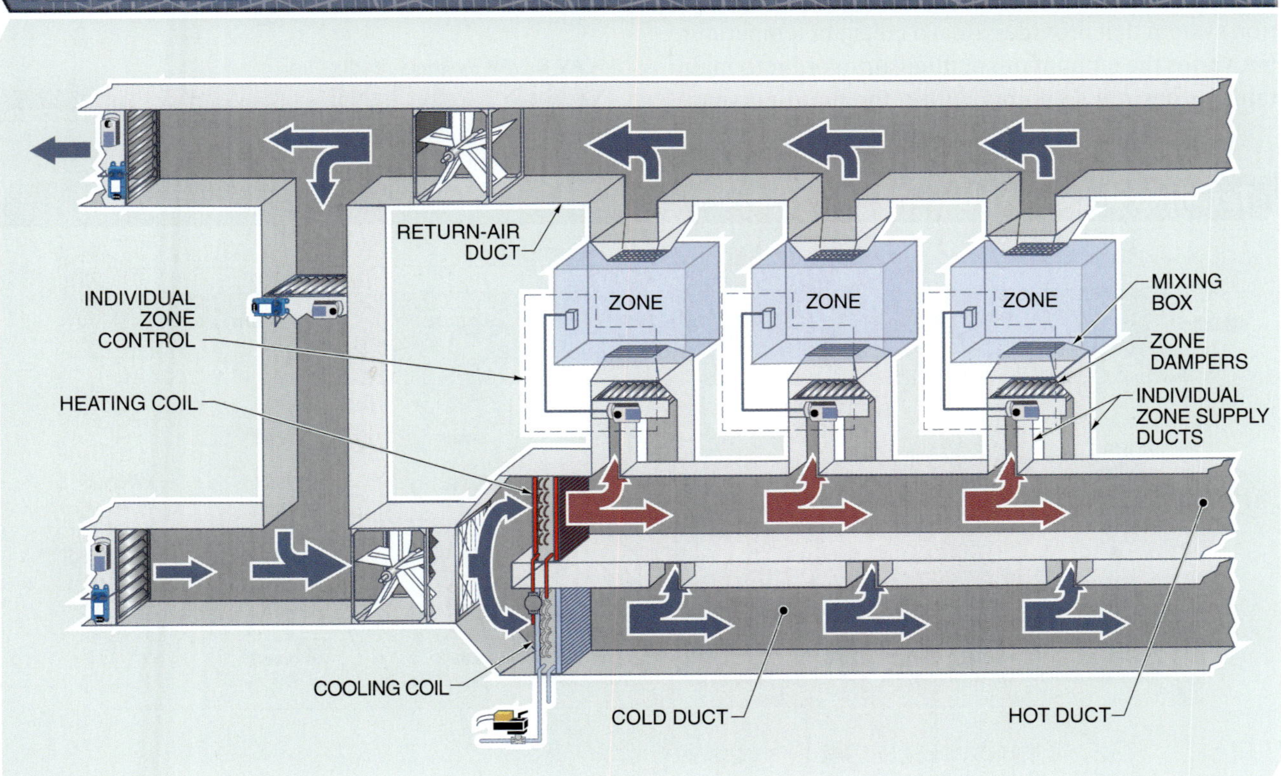

Figure 55-31. Dual-duct systems use separate hot and cold ducts to distribute conditioned air to mixing boxes at each zone. Zone dampers control the mixture of hot and cold conditioned air to provide the specific comfort needs of that zone.

In each conditioned space or zone, mixing valves that respond to room thermostats mix the warm and cold airstreams in proper proportions to satisfy the space load. Proportions of cold or warm air vary, but the total amount of air is always the same. For this reason, dual-duct mixing systems are typically considered to be superior for IAQ purposes because the total amount of air does not decrease according to temperature demand as it does with variable-air-volume systems. In application, mixing boxes actually deliver a constant volume to building zones.

Reheat Systems. A *reheat system* is an air distribution system that permits zone or space control for areas of different exposure, provides heating or cooling of perimeter areas of unequal exposure, or provides process or comfort control where close temperature control is required. **See Figure 55-32.** Reheat coils under the control of space thermostats heat the air to temperatures that satisfy the zone load. Reheat systems are used in hospitals, laboratories, clean rooms, solid-state electronic component manufacturing facilities, and other such spaces. All of the various economizer control arrangements can be used with reheat systems.

Variable-Air-Volume Systems

A *variable-air-volume (VAV) system* is an air distribution system that provides air at a constant temperature but varies the amount of supplied air in order to maintain the desired setpoints within the building space. **See Figure 55-33.** VAV systems typically use variable-frequency drives or variable-pitch fan blades to vary the amount of airflow. The volume of air delivered to a space is reduced as the space temperature drops. Air volume regulation is controlled by a variable-volume AHU. When the space temperature equals the setpoint temperature, either the supply air must be shut off completely, which can impact IAQ, or reheat must be added to avoid overcooling the space.

The volume of air per space variation is difficult to predict and depends on varying climatic conditions in the area. Amounts of outdoor air delivered to each zone in VAV systems change continuously. See ANSI/ASHRAE Standard 62.1-2007, *Ventilation for Acceptable Indoor Air Quality,* for the minimum outdoor-air requirements for various facilities. To meet ANSI/ASHRAE standards, VAV systems may have to be modified to provide a minimum primary air supply for the climatic conditions found in the space.

VAV systems are relatively energy efficient when compared with reheat, multizone, or dual-duct systems. All of the various economizer control arrangements are used with VAV systems. Several variations of VAV systems exist, with each variation having different impacts on IAQ.

CAV & VAV Systems Media Clip

Terminal Reheat AHUs

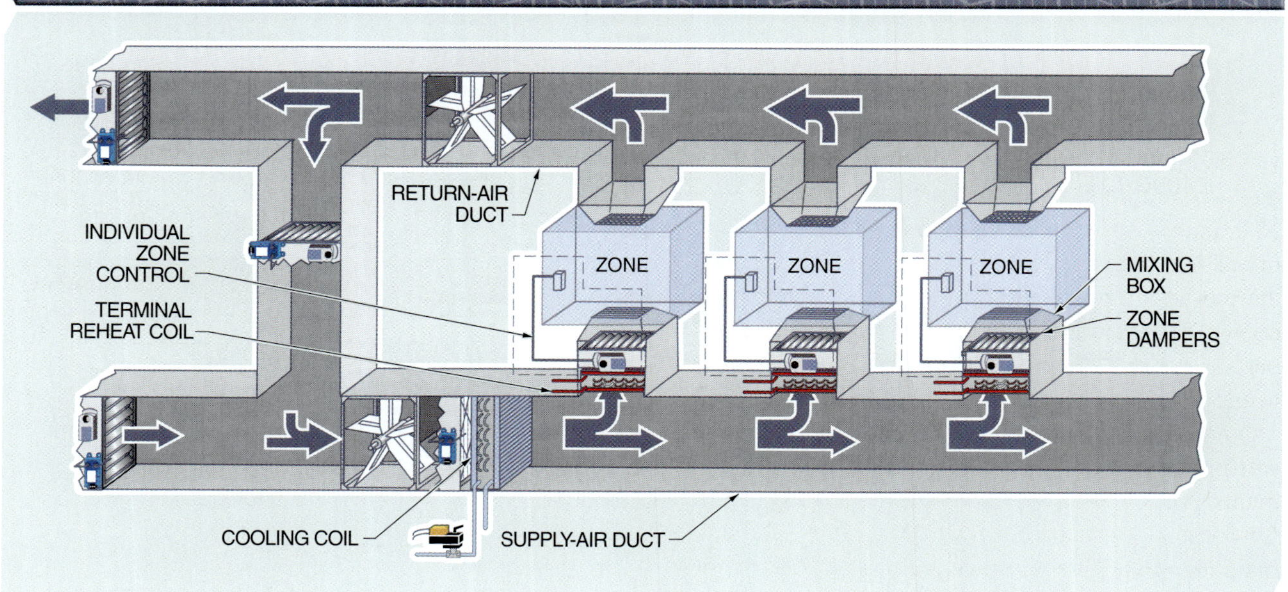

Figure 55-32. Independent heating coils in the duct of each zone are controlled by a zone thermostat to reheat conditioned air.

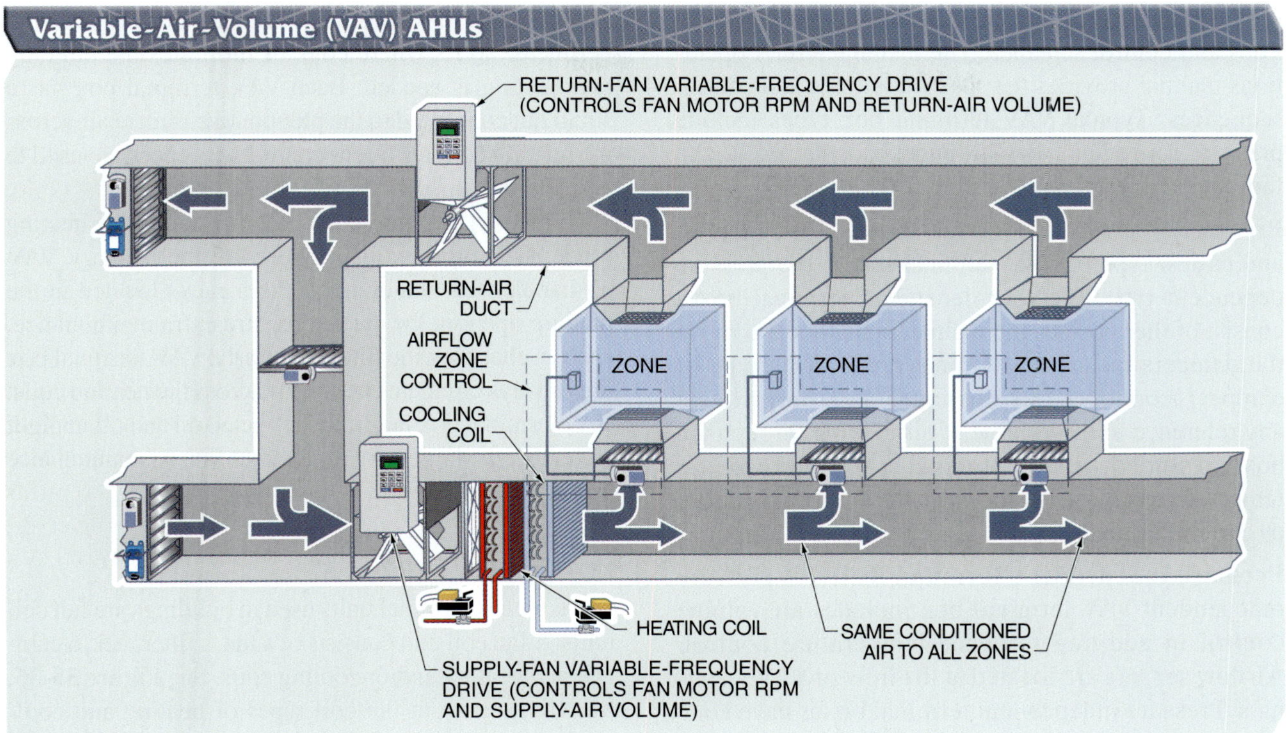

Figure 55-33. Fans controlled by a variable-frequency drive are typically used to vary air volume in ducts to maintain the desired static pressure.

TERMINAL UNITS

AHUs provide properly conditioned air to building spaces. Each room has a terminal unit that controls the environment in that particular space. The most common terminal units are VAV terminal boxes, fan coil units, and unit ventilators.

VAV Terminal Boxes

VAV terminal boxes provide temperature control and ventilation air to building zones. For typical air-handler units, conditioned air enters the VAV terminal box at 55°F and 1″ WC static pressure. Dampers in VAV terminal boxes are driven by actuators controlled by zone thermostats. As zone temperatures rise, the dampers will open to allow more cool air into the space. As zone temperatures drop, zone thermostats will close the dampers to allow the space temperature to rise. **See Figure 55-34.**

VAV terminal boxes always have a minimum damper setting determined by building occupancy requirements. Airflow is throttled from 100% to the minimum flow required to satisfy ventilation requirements and to maintain IAQ. This arrangement can cause overcooling in some or all zones when space temperatures are satisfied but the minimum amount of air continues to enter the space. When the minimum amount of air for the space is more than that required for cooling, the reheat of that air accommodates the desired space temperature. Air reheating has the advantage of satisfying ventilation requirements and contributing to acceptable IAQ standards.

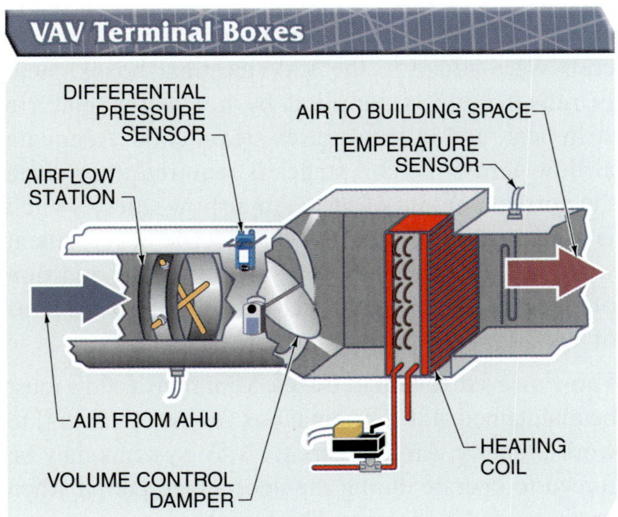

Figure 55-34. VAV terminal boxes adjust airflow to control air temperatures within building spaces.

VAV terminal boxes come in many different variations and control schemes. The manufacturer's operations manual provides the specific settings and control sequences. Typical VAV terminal box types include pressure dependent, pressure independent, reheat, and fan-powered terminal boxes.

Pressure Dependent Terminal Boxes. The simplest and easiest type of VAV terminal box is the pressure dependent type. Pressure dependent terminal boxes consist of thermostats and damper actuators that move the dampers inside the terminal boxes. This type of damper is controlled by space temperature only without any reference to the volume of air moving through the box. Pressure dependent terminal boxes work well for temperature control, but they lack the ability to measure or control airflow.

Pressure Independent Terminal Boxes. A pressure independent VAV terminal box includes air volume control in addition to space temperature control. Airflow sensors are located at the inlet of VAV terminals. Pressure independent terminal boxes have zone temperatures that change, or reset, the volume of air flowing through VAV terminal boxes. Monitoring and control of airflow is accomplished, which is important with ventilation and IAQ demands from the building spaces. Typical VAV terminal units used today are pressure independent types.

Reheat Terminal Boxes. Early types of VAV terminal boxes only provided cooling to building spaces. An internal building heat load was expected to provide the heat needed or an external heat source would be used such as baseboard perimeter heating. In many cases, this system did not work very well, so reheat coils were added to the VAV terminal boxes. Heat in reheat coils is supplied by hot water, electric strip heat, and in some cases steam coils. Adequate airflow across the coil stages is required to produce the correct amount of heat. An airflow safety switch is used to shut off power to the electric heat element when airflow drops below a safe level. The addition of reheat coils adds complexity to the control systems of VAV terminal boxes.

Fan-Powered Terminal Boxes. Minimum airflow must be maintained at the terminal box for the reheat coil to work properly. Air handlers for VAV systems may be forced to operate during the unoccupied period when heat is needed in one zone. This leads to excessive energy consumption and system use. Fan-powered VAV terminal boxes are used to reduce excessive VAV system use. Fan-powered VAV terminal boxes are typically used in perimeter zones where cold air infiltrates the building and heating is needed. Each VAV terminal box has a small fan connected to the plenum to circulate air across the heating coil. Different control sequences are used to regulate the airflow.

Typically, the small fans only operate in the heating mode. Cooling is still provided by the primary VAV air handler fan and damper. Filter racks located in the plenum opening for the fan require extra maintenance, such as changing the filters regularly. VAV terminal box fans always circulate enough air across the heating coils. VAV terminal box fans may be cycled on and off at night instead of the large VAV air handler fan, so maintenance problems are reduced.

Fan Coil Units

The simplest terminal units used in buildings are fan coil units. A fan coil unit consists of a fan, a filter rack, a cabinet, and heating and/or cooling coils. **See Figure 55-35.** Variations exist in fan coil types of heating and cooling, with the typical system using hot water and chilled water. The fan runs, circulating air across coils, which heats or cools the space. Thermostats control the heating or cooling valves as needed to provide comfort to the space. Filters help to clean the circulated air. A major disadvantage to fan coil units is that the fan only recirculates room air. Ventilation air must be provided to the space using separate fan systems.

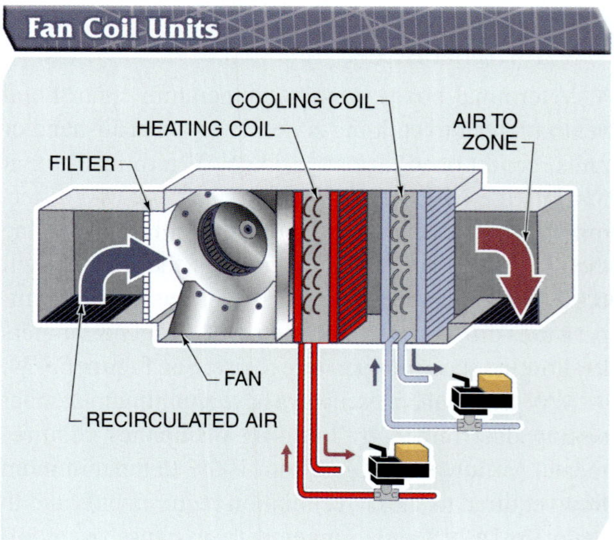

Figure 55-35. Fan coil units are self-contained systems that simply filter and heat/cool return air.

Unit Ventilators

Unit ventilators are very similar to fan coil units. Heating and/or cooling coils are located in a cabinet that contains a fan and filter element. The major difference is that the outdoor-air damper and actuator are included in the unit ventilators. Unit ventilators operate when space thermostats call for ventilation air. When the outdoor conditions are correct, cooling can also be accomplished by opening outdoor-air dampers. **See Figure 55-36.**

Unit Ventilators

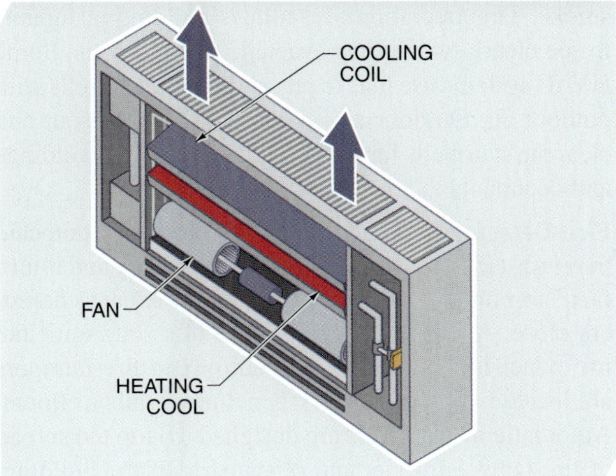

Figure 55-36. Unit ventilators are similar to fan coil units but include an economizing feature. These units can add cool outdoor air when zones require cooling.

While unit ventilators can be used in any type of building, they are typically used in schools. School buildings are not used in the summer for the most part. Spring and fall conditions often allow the use of outdoor air for cooling without the need for mechanical cooling systems. This application is heavily dependent on the local climate. In northern climates, low temperature limit switches are installed that force the dampers closed to outdoor air and force the heating to 100% when the low temperature limit is reached. Low limit conditions are due to the outdoor-air damper being stuck open and allowing too much cold air into unit ventilators.

Unit ventilators are notoriously difficult to work on because the components are packed into a very small space. Unit ventilators come in many different variations and control schemes, so the manufacturer manual should be checked for specifics.

LIFE SAFETY FUNCTIONS

The main function of AHUs is to provide temperature, humidity, pressure, and air-quality control. In addition, AHUs are typically connected to the life safety functions of buildings. These life safety functions typically include fire and smoke detection systems. **See Figure 55-37.**

Duct Smoke Detectors

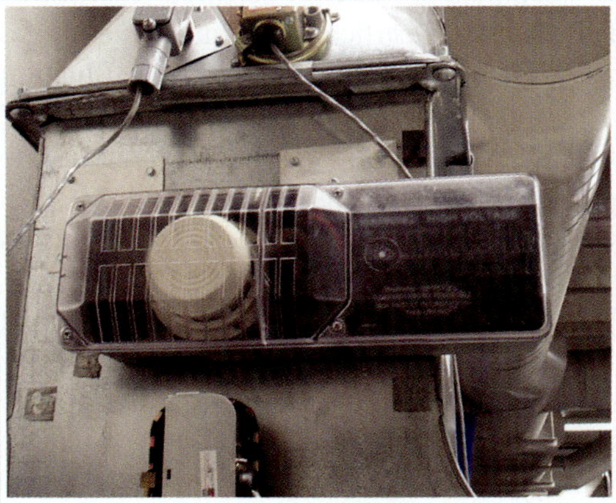

Figure 55-37. Smoke detectors are typically mounted outside the ductwork with sampling tubes or long sensors used to continuously monitor the air inside the duct for smoke particles.

Life safety systems are critical to protecting the lives of building occupants and the building itself. There are many codes and certifications involved with these systems. The primary code organization responsible for life safety functions is the National Fire Protection Association (NFPA). The NFPA creates and maintains standards for fire protection and suppression equipment, training, and related activities. Often, federal, state, and local governments incorporate NFPA standards into their own laws and codes. NFPA code requirements advance and change over time. NFPA standards incorporate the National Electrical Code® (NEC®). The NEC® was created to maintain standards for the safe installation of electrical wiring and equipment.

Equipment, procedures, testing, and training must meet all NFPA and other code and jurisdictional requirements. Failure to meet code requirements may result in injury and death of building occupants. Criminal and civil penalties may result from not adhering to these codes and standards.

Underwriters Laboratories (UL®) 864 is another standard for life safety systems. UL 864 determines the construction and use of products that provide all monitoring, control, and indicating functions of life safety systems. Documentation showing that components meet this standard must be provided during the installation.

Training and Certification

HVAC technicians must be certified to work on life safety systems. Certification requires passing a test from an organization such as the National Institute for Certification in Engineering Technologies (NICET). Verified job experience is typically a requirement to take the test. HVAC technicians usually possess state and/or local licenses as well. Unqualified technicians are not permitted to work on life safety systems.

Typically, HVAC technicians work with certified life safety technicians to check the systems. Life safety technicians troubleshoot and operate the life safety systems. HVAC technicians, in turn, ensure that the HVAC equipment responds properly as commanded by the life safety system.

Life Safety Sequence Commissioning

Life safety sequences and devices are typically commissioned by life safety technicians under the observation of local government representatives. Typically, local fire marshals or their representatives direct the technicians to simulate life safety conditions in order to observe the response of the HVAC equipment. Issuing of the building occupancy permit typically depends on passing these tests.

Life Safety and HVAC Controls. Typically, life safety controls and HVAC controls are installed and serviced by different companies. Life safety panels are typically connected to HVAC systems through relay-type inputs. Life safety conditions typically monitored are smoke and fire.

Smoke Detectors. Smoke is sensed in AHUs through smoke detectors located in the return-air ducts. Typically, ionization and photoelectric smoke detectors are used. The life safety panel takes over operation of the HVAC fans and dampers. When smoke is detected, the exhaust fans operate and the exhaust dampers open. This operation removes smoke from buildings and allows firefighters to enter and easily locate the source of the smoke. This operation also allows building occupants to see clearly while they evacuate the buildings. Some HVAC systems use fans to pressurize the stairwells with outdoor air. Outdoor air will force the smoke out and clear the stairwells for first responders to enter buildings and occupants to leave.

Fire Detectors. When high temperatures are detected in return-air ducts, AHUs shut down. Supply and return fans stop and the outdoor, return, and exhaust dampers close. AHU settings remain in this state until the fire panel is reset. Typically, automatic fire dampers are located where ductwork penetrates walls or floors. Automatic fire dampers are designed to stop the spread of fire between floors and other parts of the building. They are typically actuated by fusible links or other automatic mechanisms.

Typically, the fire damper has failed and closed when airflow is lost suddenly during normal system operation. Access doors are often provided in the duct at locations of the fire dampers to allow visual inspection of the damper position.

UNIT 56

Installing Building Automation Systems

The installation of HVAC controllers used with modern building automation systems includes mounting, wiring, and simple diagnostic techniques. HVAC technicians need to perform basic mounting and wiring procedures for devices and components in these systems. HVAC service technicians must understand the procedures and techniques used to maintain and troubleshoot HVAC equipment. Manufacturers have specific installation and testing recommendations for their products.

OBJECTIVES

- List the typical mounting requirements for controllers.
- Properly identify the most common type of power source.
- Explain how to size and wire power sources and back-up batteries.
- Identify wiring requirements for RS-232 and RS-485 systems.
- Describe types, features, and configurations of network wiring.
- Properly wire analog and digital inputs and outputs.
- Properly wire peripherals.
- List typical diagnostic procedures for checking wiring.

Index to Procedures

CONTROLLER MOUNTING

To successfully install a building automation system (BAS), all controllers must be properly mounted. The manufacturer recommendations must be followed for the installation of controllers and other support devices. Many BAS installations have ongoing problems after startup because the controllers were not secured properly.

Mounting Methods

BAS controllers are typically mounted inside available control panels, often in old pneumatic temperature control panels. Controllers must be mounted securely to firm surfaces with proper hardware. Many manufacturers design controllers and accessories to be mounted securely to DIN rails. **See Figure 56-1.** Thin DIN rails are typically used in control panels and can be mounted to most surfaces with anchors. For controllers mounted directly to an interior wall, the appropriate mounting hardware for that particular wall must be used. Some manufacturers require that vibration deadening mounts be used.

TECH TIP

When installing a BAS, the devices and wires should be labeled to aid in troubleshooting after installation.

Controllers Mounted on DIN Rails

Figure 56-1. Controllers and accessories are typically mounted on DIN rails.

Operator Interfaces

Figure 56-2. Operator interfaces are mounted in locations accessible to maintenance personnel.

Mounting Requirements

Many BAS controllers have operator interface requirements, which operators and technicians are required to monitor. **See Figure 56-2.** Typically, operator interface controllers are required to be at a specific distance above the finished floor such as 60″. Other controllers not requiring an operator interface are typically located in convenient locations. In some cases, it is desirable to locate controllers where they cannot be seen easily to prevent tampering.

Controller locations are typically chosen to reduce the overall length of wire runs. Experienced technicians keep the wiring simple to reduce installation costs. Manufacturers provide specific recommendations regarding minimum distances from high-voltage switching devices or other hazards to prevent electromagnetic interference to controllers.

Controllers must be installed with enough clearance on all sides. Installing controllers and other devices into panels that are too small dramatically increases the time needed for future servicing. It should not be required that equipment be removed to access space for servicing controllers. Control panels that are too small should be replaced with larger control panels or relocated to avoid future problems.

Requirements for mounting controllers vary widely because of the large variety of BAS controller applications. Controllers are typically installed on secure surfaces at convenient locations near the devices they control. The primary consideration for mounting controllers is whether the controllers are to be located in the interior or exterior of building spaces.

Interior-Mounted Controllers. Interior-mounted controllers must be located in accessible areas away from water lines that can leak or areas that are exposed to outdoor air. Typically, controllers are installed on walls with spacers to provide air gaps. Likely damage to controllers is reduced by air gaps when water leaks from nearby pipes. Typically, interior-mounted controllers are located where they cannot be accessed easily by occupants such as in mechanical spaces or utility closets.

Exterior-Mounted Controllers. Suitable waterproof enclosures must be provided when controllers are mounted outdoors such as in rooftop units. Wiring for exterior-mounted controllers must be run in weatherproof flexible conduit. Controllers must be protected from direct exposure and rated for exterior use.

POWER SOURCES AND WIRING

The power source typically used for most modern BAS controllers is 115 VAC stepped down with transformers to 24 VAC. **See Figure 56-3.** Proper wiring of the power source is critical to the proper operation of system controllers. The minimum wire requirement for a typical 24 VAC power source is double stranded No. 18 AWG copper wire with ground. BAS manufacturers typically use plug-in plastic connectors, screw-connected terminations, or spade terminations for power sources.

Power Sources

Figure 56-3. A building automation system (BAS) transformer typically steps down 115 VAC to 24 VAC for controllers.

Power Sources

Some manufacturers require that a specific power source be used. When determining the power source size for controllers, all power needs of the various devices should be added together. Manufacturer manuals should be consulted for power requirements. **See Figure 56-4.** The largest single transformer used in most BAS applications is 100 VAC. However, large transformers are used to power many controllers and/or devices. For example, it is common to use one power source for multiple VAV terminal box controllers. HVAC technicians must verify, according to manufacturer installation instructions, that multiple controllers can be powered from the same transformer.

Controller Power-Source Requirements

Actuator	Type	Requirements for 24 VAC Source*
IDA2020	Incremental	3
IL-7150	Incremental	2.7
IL-7200	Incremental	5.5
IL-8020	Incremental	4
IL-8050	Incremental	6
K Series Electric Zone Valve	ON/OFF	7
A9100-A, M9200-C	Floating	6.5
A9100-G, M9200-F	Proportional (voltage of current)	6.5
IL-7152	Voltage (0 VDC to 10 VDC)	4.7
IL-7202	Voltage (0 VDC to 10 VDC)	7.5
IL-8022	Voltage (0 VDC to 10 VDC)	4
IL-8052	Voltage (0 VDC to 10 VDC)	6

* in VA

Figure 56-4. Manufacturers provide charts for the power source requirements of control devices.

Power Source Isolation and Polarity. Depending on the manufacturer, the power sources used for controllers may need to be isolated. Power source isolation from ground is necessary to prevent ground loops, which can cause individual controllers to operate improperly. A *ground loop* is an unwanted circuit that has more than one point connected to ground with a voltage potential difference between the ground points high enough to produce a circulating current in the ground system. Older controllers are especially susceptible to ground loop problems.

Ground loops are caused by multiple grounding locations with different potentials, typically due to improperly installed equipment. While these differences may be less than 1 V, typical BAS components operate at less than 5 VDC. A small difference in potential between different locations can be significant.

Correct transformer polarity is essential because controller power sources, sensors, and relays operate properly only when connected to the correct polarity. The two wires from the step-down transformers are typically labeled 24 VAC and common. The wire with a slightly higher potential to ground is the 24 VAC wire. The polarity of the power source may be checked to confirm the 24 VAC wire and the common wire. **See Figure 56-5.**

Package Unit Power Sources. Many package units, such as rooftop heat pumps, have the power source built in. Instead of purchasing a separate power source, the existing power source can be used to power controllers. HVAC technicians must confirm that the transformer has the required capacity and is properly isolated. For example, transformers used to spark burner ignition in natural gas systems can cause serious electromagnetic interference. When HVAC technicians have any doubt about using an existing transformer, another transformer should be added and used.

Sharing One Power Source among Multiple Controllers. Many BASs have one large transformer that has the rated capacity for multiple controllers. Care must be taken to ensure that the controllers are isolated from each other to prevent communication problems.

Back-Up Batteries

A BAS controller typically has an onboard back-up battery for controller memory and clock functions. **See Figure 56-6.** Batteries often self-charge while normal 24 VAC power is applied. A controller is typically shipped from the manufacturer with the back-up battery disconnected to prevent power drain during shipment. Back-up batteries must be enabled by removing isolating barriers or inserting jumpers as part of the start-up procedure.

Figure 56-6. Batteries in controllers provide back-up power for the retention of memory and clock functions.

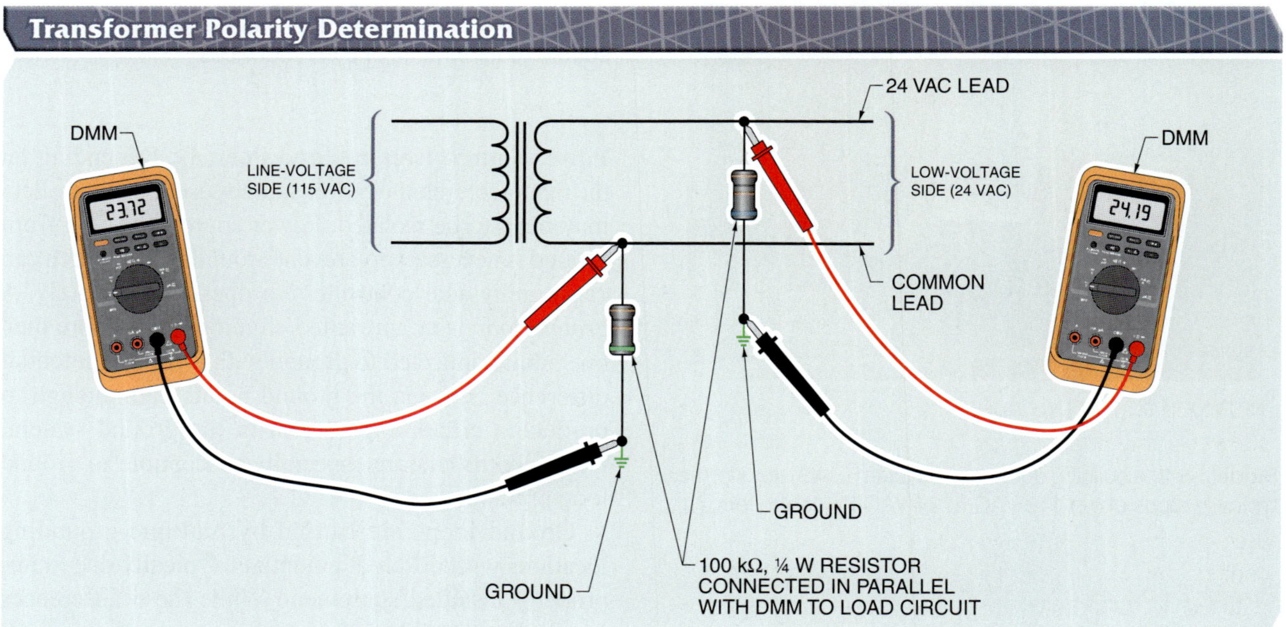

Figure 56-5. A DMM is used to determine the polarity of a controller power source by measuring the 24 VAC wire and the common wire.

COMMUNICATION WIRING

Communication trunk wiring enables controllers to talk to each other. This wiring must be installed properly in order to prevent communication problems and failure. BASs typically use communication standards created by the Electronics Industry Association. Data is transferred differentially on two wires twisted together for high noise immunity. Data communications operate at a maximum baud rate of data bits per second (BPS). A *baud rate* is a measure of the maximum data transmission rate measured in 0s or 1s per second.

Data Communications Standards

Data communications standards were developed by the Electronics Industry Association (EIA) to provide compatibility between devices from different manufacturers. Today the data communications standards are administered by the Telecommunications Industry Association (TIA). The standards are noted with the prefix "RS" for recommended standards but are sometimes listed with the prefix "EIA" or "TIA". RS standards create a uniform system of data communication. **See Figure 56-7.** Some variations exist between manufacturers that may cause difficulty during installation.

EIA-Recommended Standards and Specifications

Standards and Specifications	RS-232	RS-485
Mode of communication operation	Single-ended	Differential
Total number of transmitters and receivers (one transmitter active at a time)	1 Transmitter 1 Receiver	32 Transmitters 32 Receivers
Maximum cable length	50 ft	4000 ft
Maximum data rate (40 ft to 4000 ft for RS-485)*	20 kb/s	96 mb/s to 100 kb/s

* Data transmission rates for RS-485 are inversely proportional to the cable length

Figure 56-7. Data communications standards provide a uniform system of communication between controllers and devices.

RS-232 is a single-ended data transmission standard that allows communication from one transmitter to one receiver at data rates up to 20 kb/s and distances up to 50 ft. The RS-232 standard was introduced in 1962 and is still widely used in industry for small systems and local communication connections. RS-232 signals are changes in voltage levels with respect to the system common ground. With respect to ground, the active state is a positive signal and the idle state is a negative signal. RS-232 uses specific communication protocols that are primarily used with modems.

RS-485 is a differential data transmission standard that allows communication from 32 transmitters to 32 receivers at data rates up to 9600 kb/s and distances up to 4000 ft. RS-485 is a true multipoint communications network using a 2-wire or 3-wire connection. The actual transmitted data rate depends on the length of the connecting wire. Using automatic repeaters and high-impedance drivers/receivers, the limitation of receivers can be extended to hundreds of nodes on a network. RS-485 drivers remain in receive mode until data can be transmitted, so data congestion or bus fault conditions are avoided. The RS-485 standard is widely used in industry for larger networks.

Wire Types

Data communications protocol requires either a 2-wire or 3-wire twisted and shielded connection. **See Figure 56-8.** The minimum wire size is usually No. 22 AWG. However, wire this thin is easily kinked and damaged. Typically, thicker No. 20 AWG or No. 18 AWG wire is used.

Controller Wiring

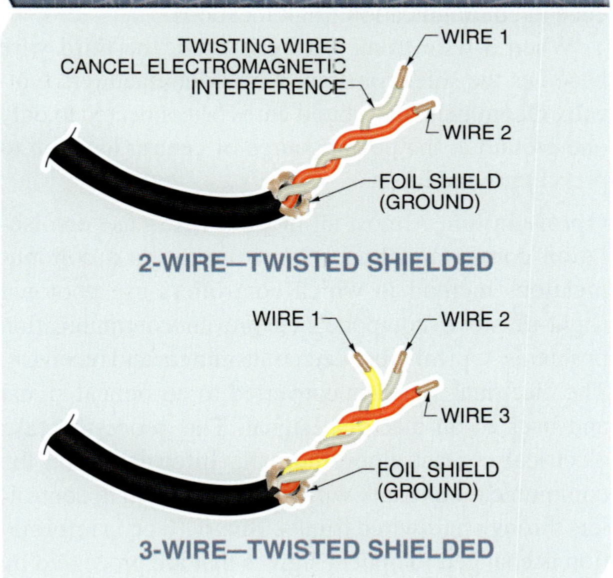

Figure 56-8. Twisted shielded cables are required for RS-485 communications.

The wires are twisted in opposite directions inside the cable to cancel out the electromagnetic interference on both wires. Color-coded wire insulation is often used to ensure that the proper wires are consistently connected to the same terminals at controllers. When used, RS-485 is polarity sensitive. The connection terminals at each controller are typically labeled RS-485+, RS-485–, and common. **See Figure 56-9.**

Typical Thermostat Wire Colors

Terminal	Color	Connection Description
C	White/Varies	Common
R	Red	24 VAC power source
Rc	Red	Cooling 24 V 2nd transformer
Rh	Red	Heating 24 V primary transformer
G	Green	Fan
Y1	Yellow	Cooling
Y2	Varies	Cooling 2nd stage
W1/B1	White/Blue	Heating
W2/B2	Varies	Heating 2nd stage

Figure 56-9. Typical wire colors used for thermostats are not standard. When replacing thermostats, wires should be labeled prior to removal. When installing new thermostats, manufacturer instructions must be followed.

Shielding and Grounding. Shielding is used to reduce the effect of electromagnetic interference on communication signals. Electromagnetic interference is caused by the operation of large motors, compressors, and some lighting loads. Runs close to these types of equipment should be avoided to eliminate electromagnetic interference in communication trunk locations.

When 3-wire trunk wiring is used, the third wire becomes the shield or drain wire. Manufacturers typically recommend that shield wires be connected to only one ground at the power source or central location to prevent ground loops.

Optoisolation. Almost all modern BASs use optoisolation communications. *Optoisolation* is a communications method in which controllers use photonic (light-sensitive) components to prevent communication problems, typically between transmitters and receivers. The electrical signal is converted to an optical signal and back to an electrical signal. This process breaks electrical contact along the path. Information on the communications trunk wiring is transmitted to controllers through hardwired trunks. Binary (0 or 1) information is changed to optical signals that are processed by controllers. Controllers are not connected physically to communication trunks. This results in electrical isolation of the controllers from the communication trunks.

Network Configurations. Many configurations are used to connect controllers to networks. Network types are distinguished by their shapes or connection configuration. Control networks are the physical connections that allow BAS controllers to communicate. The controller network configuration used depends on the building layout, devices connected, and manufacturer recommendations. **See Figure 56-10.**

Control Network Configurations

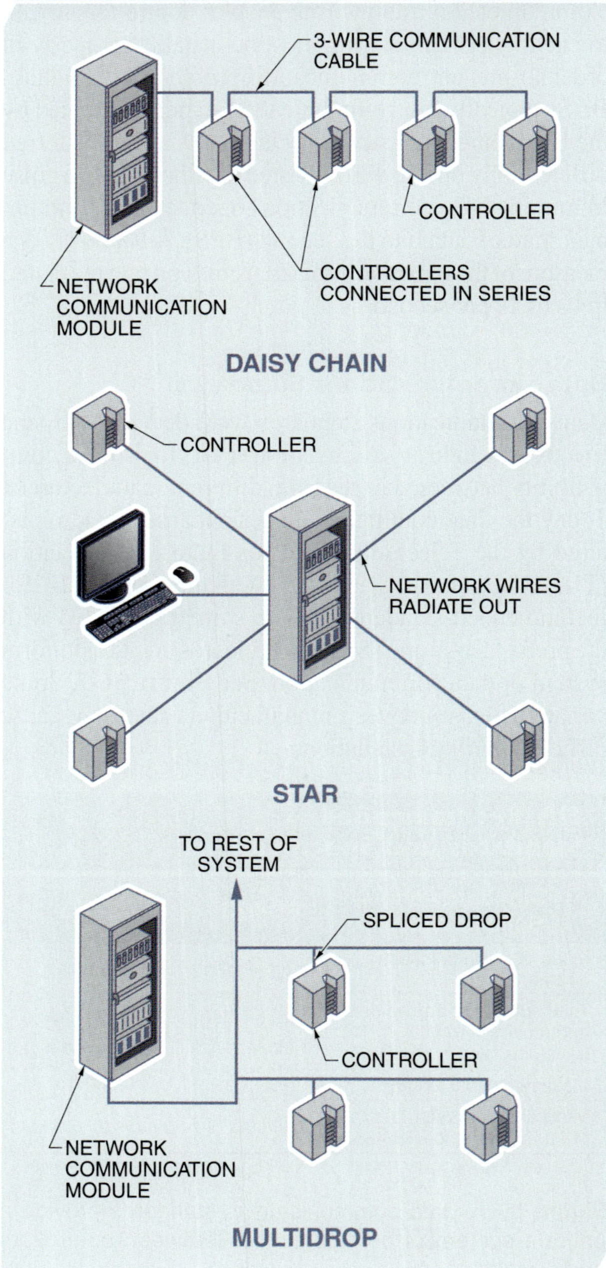

Figure 56-10. Control network configurations depend on the building layout, devices connected, and manufacturer recommendations.

A *daisy chain network* is a network of controllers linked in a series in which each controller is connected to the next controller. The network is one continuous chain of controllers. Daisy chain networks are typically used in small buildings to avoid wiring splices that can fail and cause communications failure. All wiring terminations are at easily accessible controllers.

A *star network* is a network of communication trunk wires extending to the remote controllers located in a building. Star networks are not used often because of the expensive long wire runs. It is suitable mainly in buildings that use a central point for the network communications module.

Many BASs in large buildings have multidrop networks. A *multidrop network* is a network of communications modules connected to a main trunk wire that runs throughout a building. Spliced drop connections are made to individual controllers at accessible locations. Secondary trunk lines, which have drop connections to individual controllers, may lead from the main trunk line. Multidrop networks come in a myriad of shapes. All splices must be made correctly to avoid failed communications.

Network Addresses. Controllers must have unique network addresses to properly communicate within a network. A *network address* is a unique number given to each controller on an RS-485 communication trunk line. RS-485 networks are limited to 255 possible addresses. HVAC technicians must be able to set controllers to the correct address as part of completing network installations.

Each controller has a set of dual-in-line-package (DIP) switches. A *dual-in-line-package (DIP) switch* is a manual on/off electrical switch that creates an address system for controllers. DIP switches are used on electronic devices as an alternative to terminal strips and jumpers. **See Figure 56-11.** Each switch represents a value in the binary number system such as 1, 2, 4, 8, 16, 32, 64, and 128. When that particular switch is in the ON position the corresponding value is added to the total. For example, when the switches that represent 1, 8, and 64 are turned on, the controller address is 73.

Figure 56-11. Controller network addresses are determined by the DIP switch settings.

Input Wiring

Devices such as temperature sensors and humidity sensors provide input signals to controllers. Input devices must be properly wired to a BAS controller to enable the controller to process the information. Insulation covering wires should be numbered or labeled for easy identification. Typically, stranded No. 20 wires are used. However, No. 18 AWG wires are preferred. When wires are thin and the wire runs are hundreds of feet long, the accuracy of the readings can suffer due to increased resistance. This is especially true for thermistor-type temperature sensors.

Analog Input Wiring. Pairs of wires connect analog input devices to appropriate terminals on termination boards. When analog inputs are polarity sensitive, the polarity must be checked and labeled. Terminals are typically labeled AI# and common. The letters "AI" stand for analog input, and the symbol "#" represents the analog input connection number on the termination board. **See Figure 56-12.** Some analog inputs require separate power sources that many BASs provide at the terminals of the controllers. Additional pairs of wires are required to provide power to the controllers.

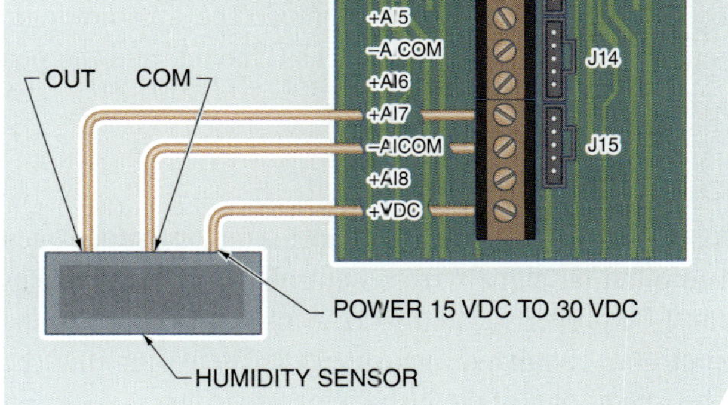

Figure 56-12. Wiring from an analog input device is connected to a controller terminal strip.

Digital Input Wiring. Pairs of wires connect digital switches to appropriate terminals on termination boards. **See Figure 56-13.** Typically, a minimum of No. 20 AWG wire is needed. Typical digital inputs are configured as dry contacts. A *dry contact* is a contact that requires external power for transmission. Controller terminals are typically labeled DI# and COM. The letters "DI" stand for digital input, and the symbol "#" represents the connection number on the termination board. COM completes the circuit for that input. When power is externally supplied, an isolation relay is required to prevent interference.

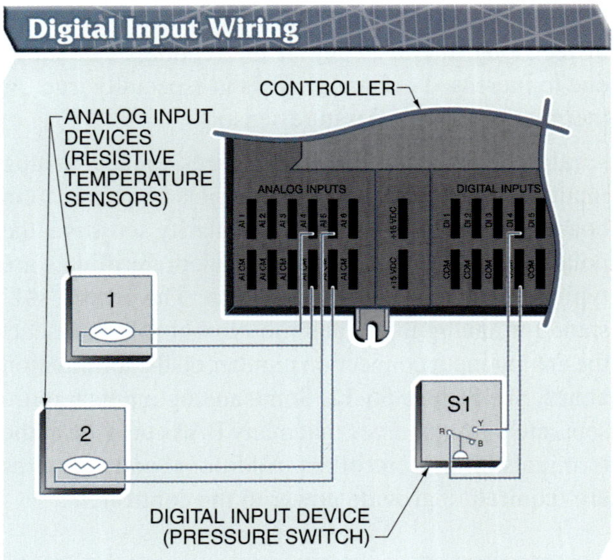

Figure 56-13. Wiring from a digital input device is connected to a controller digital input strip.

It must be ensured that the correct terminals are used at the switch. Controller software logic must agree with the switch contacts selected. For example, airflow differential pressure switches are configured to be normally open (NO) when the fan is running. The software logic must recognize a closed contact as "fan on" and an open contact as "fan off."

Output Wiring

Devices such as valves and damper actuators are activated using output signals from controllers. Output devices must be properly wired to BAS controllers so that the controller commands are executed. The wiring must be checked as part of the installation procedure.

Analog Output Wiring. Analog output wiring connects controllers to actuators, typically using two No. 20 AWG wires. Many analog outputs are polarity sensitive and must be wired accordingly. Typical terminals used are labeled AO# and COM. The letters "AO" stand for analog output, and the symbol "#" represents the connection number on the termination board.

Digital Output Wiring. The methods used to wire digital output devices depend on the application. Typically, a minimum of No. 20 AWG wire is recommended. Contacts typically provided by manufacturers are dry contact closures. External power sources must be provided because the contacts are not powered. These contacts typically are rated Class II (low voltage), and power is supplied by a 24 VAC transformer. Low-voltage connections can be wired directly to the unit terminal board of package HVAC units.

Other digital output terminals are wired as incremental contacts. One digital output drives the actuator motor open and the other closed. Digital outputs are typically wired to interface relays, which select the correct contacts for the application. Terminals are typically labeled DO# and COM. The letters "DO" stand for digital output, and the symbol "#" represents the connection number on the termination board.

Peripherals Wiring

HVAC technicians must properly wire controller inputs and outputs to peripheral devices. A *peripheral device* is a device that is connected to a personal computer or BAS controller to perform a specific function. Commercially purchased cables are used to connect most peripheral devices. The correct cable must be selected for each peripheral device. Many cables look similar and have the same connections but cannot be wired to the same connections internally.

Modem Wiring and Connections. Modems are commonly connected to network communication modules as serial devices. **See Figure 56-14.** A *serial device* is a device that transmits one bit of information (0 or 1) at a time. Serial devices are typically connected using 9-pin male connectors. Modems are also connected to phone lines using standard modular RJ11 cable.

Printer Wiring and Connections. Printers are typically connected to a network communications module as parallel devices. **See Figure 56-15.** A *parallel device* is a device that transmits multiple bits of information (0s or 1s) at the same time. Printers are typically connected using 25-pin female connectors. Standard printer cables are readily commercially available.

Dumb Terminal Wiring and Connections. Some older BASs use personal computers as dumb terminals. A *dumb terminal* is a display monitor and keyboard with no processing capabilities. Dumb terminals are typically connected with serial cables to modems connected to network communications modules using 9-pin RS-232 cables.

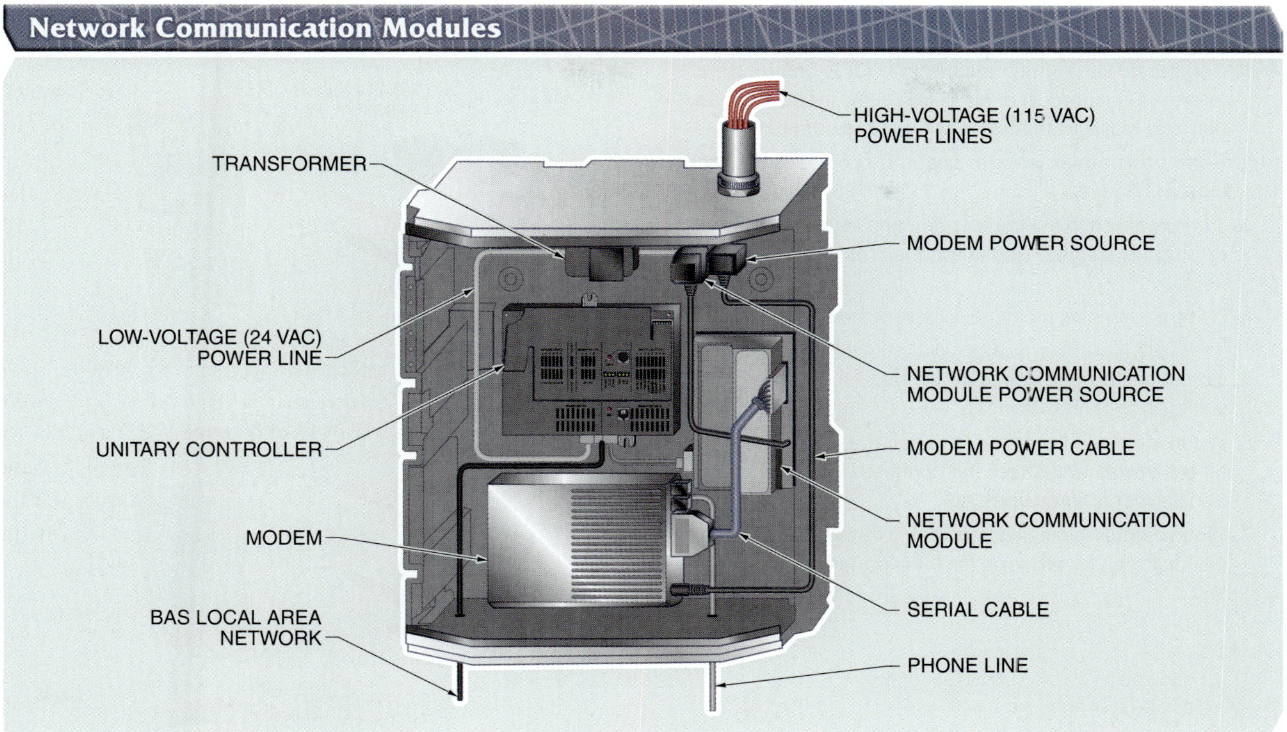

Figure 56-14. A network communication module typically includes a modem connected as a serial device.

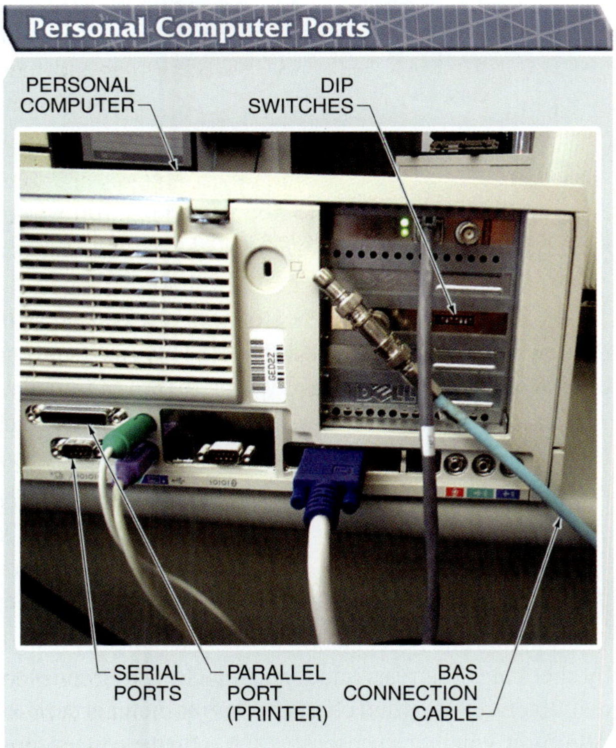

Figure 56-15. Parallel devices such as printers are typically connected to personal computers or network communications modules using parallel ports.

TECH TIP

Old BASs often use serial ports and dial-up modems. Adapters are needed to enable computers with USB ports to communicate with these old systems.

COMMON TESTING AND DIAGNOSTIC PROCEDURES

Common testing procedures are used to perform simple checks and diagnostics of system wiring upon completing the installation. HVAC technicians must be familiar with basic testing procedures. Good-quality digital multimeters (DMMs) provide accurate readings. Older analog meters lack the needed accuracy and should not be used.

Testing Transformer Isolation

A test typically performed involves checking whether the transformer power source is properly isolated from ground. When the power source is not properly isolated, controller communications may be disrupted. The transformer secondary should be tested to ground using a digital ohmmeter. Resistance between the secondary and ground should be above the manufacturer-specified levels. **See Procedure for Testing 24 V Transformers.**

Procedure

Testing 24 V Transformers

1. Confirm power to the transformer is OFF. Ensure transformer is secured and wired correctly. Inspect for any damage to the transformer, wiring, and connections.
2. Read transformer label to confirm the rated input and output voltages.
3. Turn power on to the transformer, and listen for humming or buzzing sounds. Replace transformer if there is any noise.
4. Using a voltmeter, measure voltage at the input and output terminals.
5. Compare the readings with the rated input and output voltages of the transformer.
6. When input readings do not match the rated input, turn off the power, and check the connections to confirm the transformer is wired correctly.
7. When output readings do not match the rated output, turn off the power, replace existing transformer, and repeat the test.

Testing RS-485 Communications

The voltage levels of RS-485 connections are typically tested. Communication cables typically use three wires: RS-485+, RS-485–, and common. RS-485 voltage levels are standardized with RS-485+ and RS-485– wires to common readings of about 1.5 VDC or 2.5 VDC. When RS-485+ and RS-485– are checked, the voltage should read about 3 VDC to 5 VDC. **See Figure 56-16.** When these voltage levels are not measured, the communication trunk wires and controllers should be checked for problems. **See Procedure for Testing RS-485 Communications.**

Procedure

Testing RS-485 Communications

1. Confirm power to the device is ON. Ensure device is secured and wired correctly. Inspect for any damage to the device, wiring, and connections.
2. Measure the voltage at each data line and record.
3. Check the polarity of each data line.
4. Check both sides of each twisted pair for isolation. Check each wire for isolation to ground.
5. Initiate bidirectional communications using known signals and observe results.
6. Disconnect the normal power source and repeat tests under battery power. Reconnect the normal power source after concluding tests.

Testing RS-485 Communications

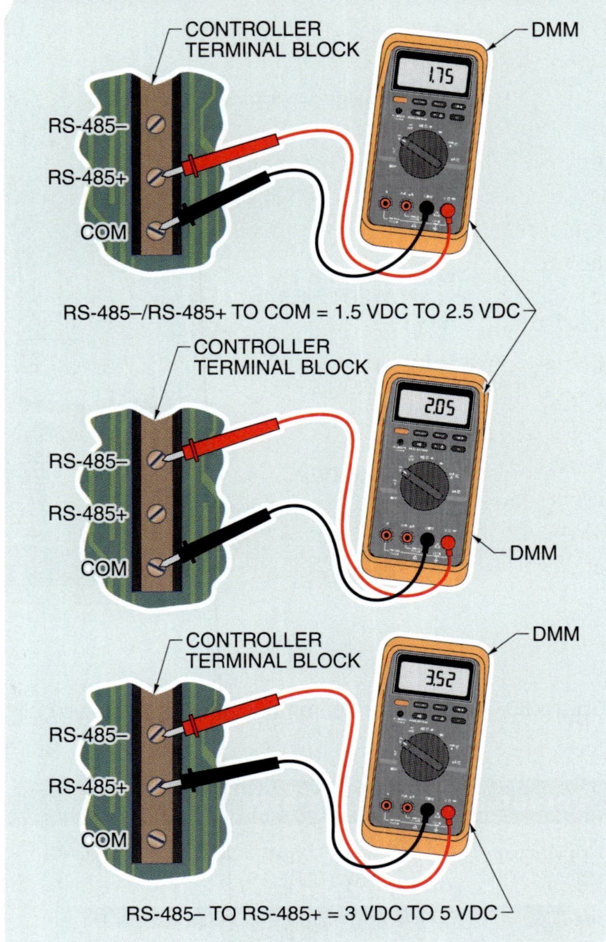

Figure 56-16. Voltage for RS-485 communication wiring is based on predetermined levels within a control system.

The communication trunk isolation test is similar to the power source isolation test. The RS-485 communications wire is tested to ground with a digital ohmmeter. The resistance must be below the manufacturer-specified levels. When the communications trunk line is not properly isolated, one of the inputs, outputs, or power source is improperly grounded.

Testing Analog Input Resistance Values

Thermistors and other analog inputs are typically tested by checking their resistances and matching the values to manufacturer-supplied charts. A thermometer is used to determine whether a sensor reading is on the proper spot of the chart. Thermistor-type sensors are either positive temperature coefficient (PTC) or negative temperature coefficient (NTC) sensors. The resistance value of a PTC sensor increases as temperatures increase and decreases as temperatures decrease. The resistance value of an NTC sensor decreases as temperatures increase and increases as temperatures decrease.

The coefficient of a thermistor-type sensor is the amount of resistance change in ohms per unit change in temperature (Ω/°F). For example, a thermistor-type sensor may have a coefficient value of 2.2 Ω/°F. This sensor resistance value will change by 2.2 Ω every time the temperature changes by 1°F. The base or nominal resistance value used for a typical sensor is 1000 Ω at 70°F. When a PTC sensor has a coefficient of 2.2 Ω/°F, the resistance value will increase by 2.2 Ω for every 1°F temperature change at the sensor.

HVAC technicians can easily determine whether sensors are reading properly by using an accurate DMM and thermometer. Since the 2.2 Ω/°F coefficient is typically used, HVAC technicians should understand how the wire resistance in long wire runs can affect the reading of sensors.

Testing Digital Input and Output Wiring

When testing the electrical continuity of wires, digital inputs and outputs can be checked easily. Wires can be disconnected from controllers and digital devices, and the resistances can be read on a good-quality meter. Normally open and normally closed contacts must be set up properly. When 24 VAC is present at a digital output, the voltage can be easily measured with a digital multimeter. **See Procedure for Testing Wall Thermostats.**

Testing Jumpers

Jumpers are often used on controllers for selection purposes. A *jumper* is a conductor used to connect pins on a controller or device. Jumpers are used on BAS controllers to select the type of analog input or output signal received by the controller. It is relatively common for analog input types such as thermistor, current, and voltage to be selected on the controller using jumpers. Jumpers are also used for the proper selection of analog outputs and to enable back-up batteries. A jumper selection label is normally attached to the back of a controller to aid in jumper selection. **See Figure 56-17.** HVAC technicians should check manufacturer instructions to make proper connections using jumpers.

Controller Jumper Usage

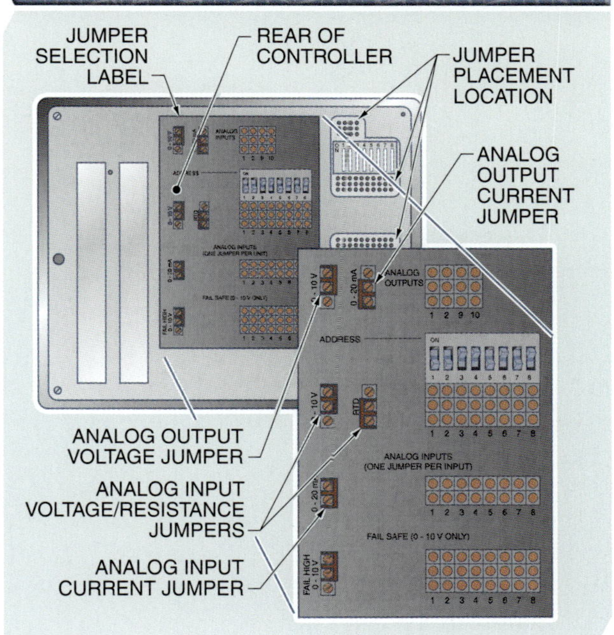

Figure 56-17. Jumpers are used to select the type of analog input or output signal controllers receive.

Procedure

Testing Wall Thermostats

1. Use a good-quality digital indoor thermometer to measure wall thermostats. Thermostats typically have small variations in temperature readings.
2. Close all external doors and windows, open inside doors, and place the digital thermometer next to the wall thermostat. Record both temperature readings. The values should be very close.
3. Turn thermostat to the heating setting, and adjust temperature setting up by 2°F. The furnace should respond. Confirm by sensing the heat airflow.
4. When the furnace shuts off, record both temperature readings. Again, the values should be very close.
5. Turn thermostat to the cooling setting, and adjust temperature setting down by 2°F. The air conditioner should respond. Confirm by sensing the cool airflow.
6. When the air conditioner shuts off, record both temperature readings. Again, the values should be very close.
7. When the temperature variations are significant, adjust the temperature setpoint within the thermostat and retest, or replace the thermostat.

UNIT 57

Servicing and Troubleshooting Building Automation Systems

Typically, only the most experienced HVAC technicians troubleshoot building automation systems. Technicians that can effectively troubleshoot these control systems are in very high demand. Troubleshooting building automation systems effectively combines technical expertise, logic, and the creative thought process. HVAC technicians must correct common problems in system operations and distinguish between mechanical system problems and control system problems.

OBJECTIVES

- List common BAS troubleshooting procedures and information needed.
- List the typical causes of and solutions for common system problems.

BUILDING AUTOMATION SYSTEM TROUBLESHOOTING

All control systems, including building automation systems (BASs), will have operational problems. Operating personnel typically do not perform in-depth system troubleshooting. Instead, HVAC service technicians are required to identify and correct problems. Most common HVAC and BAS operating problems can be corrected by experienced HVAC service technicians. However, some complex issues can only be diagnosed and corrected by consulting with manufacturer engineering personnel.

Standard Troubleshooting Procedures

Normally, BASs require very little maintenance. The majority of daily service work involves identifying and solving common BAS problems. Most problems that occur are mechanical problems unrelated to the automation system. For example, broken actuator linkages, loose connections, leaky ducts, and blocked airflow are common HVAC system problems. Most of these problems are easy to troubleshoot by on-site maintenance personnel.

Troubleshooting is the systematic elimination of the various parts of a system or process to locate a malfunction. Although each BAS manufacturer has different troubleshooting procedures, many procedures apply to most typical problems. Manufacturers should always be consulted regarding specific procedures for troubleshooting their equipment.

Most HVAC technicians use BAS computer interfaces to check the condition of HVAC equipment. A wealth of information about sensor measurements and control actions occurring throughout the building can be accessed through the BAS computer interface. **See Figure 57-1.**

Building Automation System (BAS) Software Indicators

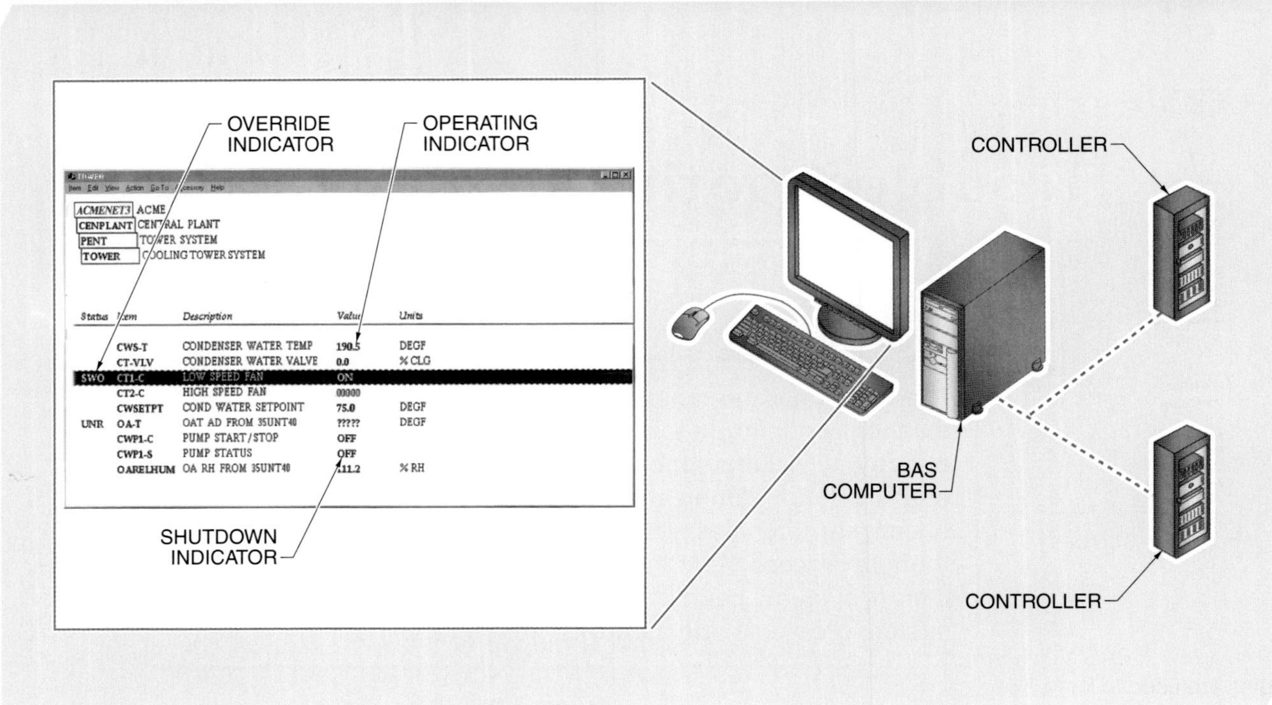

Figure 57-1. BAS software provides the status of an HVAC system, including any system overrides.

Troubleshooting Information

The primary problem regarding BAS troubleshooting is the complexity of the design and programming of the BAS. In order to properly troubleshoot a BAS, HVAC technicians typically must have the following:

- building HVAC prints and diagrams
- manufacturer cut sheets for each component
- knowledge of the building layout, device locations, and types of equipment
- in-depth knowledge of mechanical system operation
- a written sequence of operation for each HVAC system controlled by the BAS

There are many possible BAS troubleshooting scenarios. Scenarios vary based on the equipment and systems used to provide occupant comfort in buildings. HVAC technicians should be familiar with the most typical scenarios.

BAS Troubleshooting Scenario 1— HVAC Unit Does Not Operate

A technician is dispatched to an office complex with a BAS. The customer complains that one of the HVAC units should be running, but it is not. The technician checks the BAS for the status of the HVAC system.

Possible Cause 1—Overridden Unit in Control Software. When an HVAC unit fails to operate it is typically because a technician has overridden the unit operation to the OFF mode in the BAS control software. When performing routine maintenance on units, system overrides are typically performed using proper lockout/tagout procedures. Upon completion, the technician may have removed the lockout/tagout devices but not the computer override.

When an override in the BAS control software is suspected as the cause of the HVAC unit being OFF, the software must be checked. Most manufacturers provide an indicator, such as an O in the control software, that an override is in effect. HVAC technicians must confirm the reason for the override and that the HVAC unit can be returned to service. The override must be removed in order for the unit to resume functioning. **See Figure 57-2.** Follow-up communication should take place with the technician who initiated the override regarding the removal of future overrides at the completion of the service work.

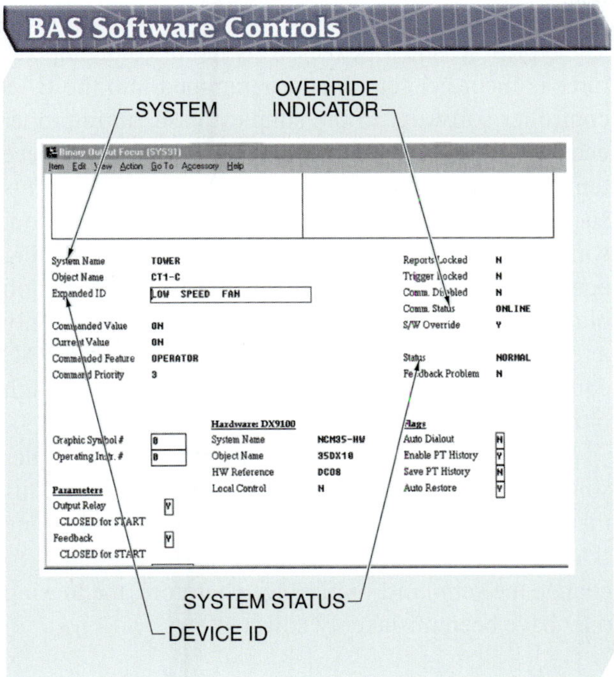

Figure 57-2. BAS software indicates which control strategies are used to keep a device on or off.

Possible Cause 2—Undetermined Software Control Strategy. Sometimes an undetermined software control strategy prevents an HVAC unit from turning on. Demand or duty cycling control strategies can cause the unit to remain off. For example, an improperly setup time schedule could keep the unit from operating.

When the technician suspects that an undetermined software control strategy is preventing the unit from operating, the technician must check the BAS control computer. Manufacturers typically provide indicators for the loads on each controller. The indicators show the control strategies that are keeping the unit off due to certain conditions. When controller indicators and settings of the appropriate control strategy are checked, the settings can be adjusted as required.

Solution: In this case, preventive maintenance was performed on the unit the previous day. The HVAC technician must use the computer interface to confirm that the "off" override placed on the unit was not removed. After it is ensured that the equipment is safe to operate, the override can be removed.

Note: When an HVAC service technician performs preventive maintenance on HVAC equipment, the technician must always check the equipment status on the BAS interface computer. The service technician may notice that one of the HVAC units is OFF, while the unit should be ON. After checking for planned maintenance and whether the equipment was placed out of service, the technician must inspect the equipment. Caution must be used when inspecting equipment because the HVAC unit may restart without warning.

Service Work Performed: The HVAC technician reviews the system BAS software for overrides and notes that the equipment override is in place. The technician inspects the equipment, concludes it is operable, and removes the override. The unit is checked for proper operation.

A unit that should be running, but is not, may be using a control strategy that is locking the unit out.

BAS Troubleshooting Scenario 2—Excessive Alarms

A customer calls in and states that the HVAC system has too many alarms. A technician is dispatched to the building and logs into the BAS.

Possible Cause—Incorrect Alarm Parameters. The technician checks the BAS alarm log. The technician notes that many temperature alarm parameters are improperly set, which results in too many alarms. Parameters set too close together create needless alarms, while parameters set too far apart create lower levels of confidence in equipment operations.

Solution: When excessively close alarm parameters are suspected, the BAS alarm folder should be opened for the device. The alarm setpoint and differential parameters must be adjusted as required. **See Figure 57-3.** The system parameters must match the actual operation of the mechanical equipment and system.

Service Work Performed: The BAS alarm parameters were reviewed and adjusted to reasonable levels to reduce the number of alarms. The unit was checked for proper operation.

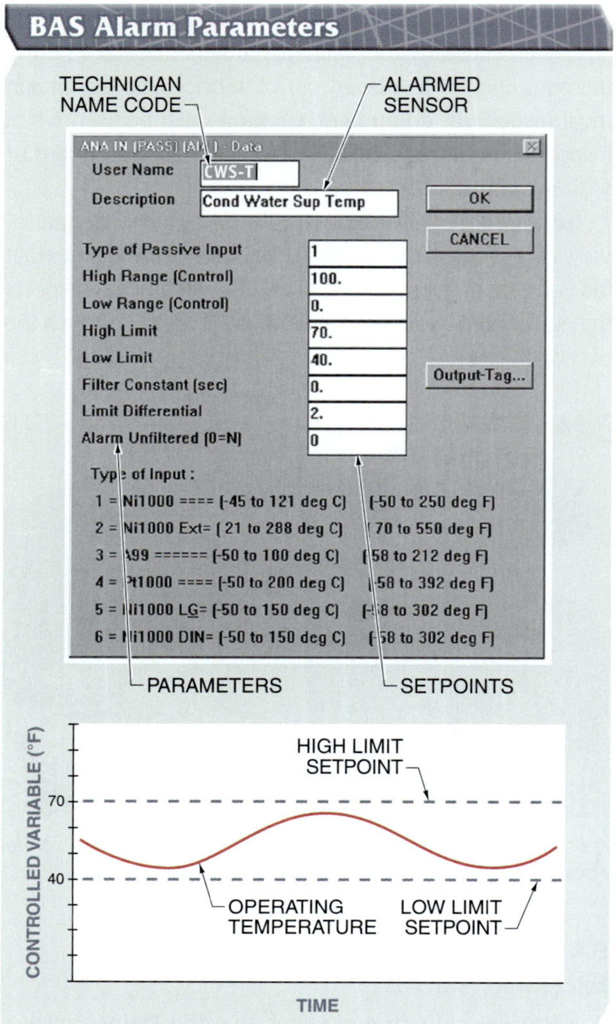

Figure 57-3. BAS alarm parameters include alarm setpoints and differentials. Improperly set parameters can lead to excessive alarms.

BAS Troubleshooting Scenario 3—Excessive Building Space Temperatures

A technician is dispatched to an apartment building. The apartment building manager complains that the building is too hot in some spaces and too cold in others. The technician checks the system and finds that someone has changed the temperature setpoints.

Troubleshooting calls often include occupant complaints about building spaces being too hot or cold. The actual building space temperature must be measured and compared to the BAS setpoints. There are many causes for temperatures to be out of the normal range, but common causes include incorrect temperature and pressure setpoints and malfunctioning mechanical equipment.

Possible Cause 1—Incorrect Temperature Setpoints. The typical reason for excessively hot or cold temperatures is incorrect setpoints programmed into the BAS controller software. In the simplest case, an improper occupied cooling setpoint, such as 80°F, causes the space temperature to be too hot. While some basic setpoints such as heating and cooling are easily understood, some are complex. For example, heating and cooling equipment is typically locked out or disabled based on outdoor air temperatures. When set improperly by only a few degrees, the lockout setpoint may prevent HVAC equipment from operating and cause excessively high or low space temperatures.

When incorrect setpoints are suspected, the controller software must be checked for any temperature setpoints that are excessively high or low. **See Figure 57-4.** Many BAS controllers provide methods for occupants to change the setpoints, or biasing. Therefore, the biasing may have been mistakenly adjusted.

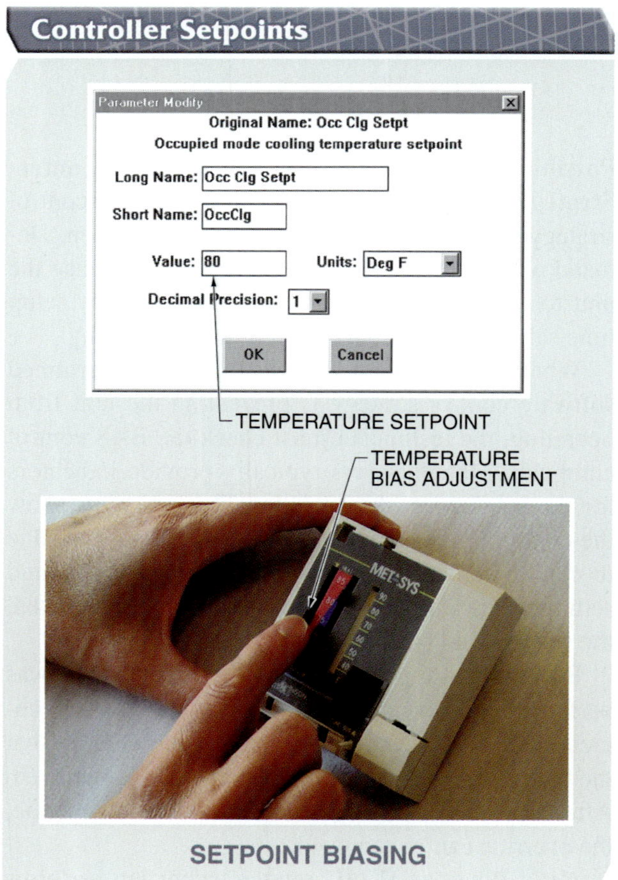

Figure 57-4. Controller setpoints must be set properly for occupant comfort. Setpoints that are too high or low lead to unwanted adjustments by the occupants.

Possible Cause 2—Incorrect Pressure Setpoints. Another example of a subtle setpoint problem is static pressure of a variable-air-volume (VAV) system that is set too low. The standard static pressure setpoint for VAV systems is 1″ WC. When the static pressure setpoint is set too low, the airflow volume in ducts is not enough to supply conditioned air to the VAV terminal boxes in all building spaces. The space temperature may become too hot, even though the VAV terminal box, controller, and software are operating properly.

Solution: The BAS operating conditions need to be checked. The actual system temperature and/or pressure must be measured and compared to the system design. The temperature and/or pressure setpoints should be adjusted to proper levels.

Service Work Performed: Temperature and/or pressure setpoints were measured and adjusted back to normal values. The unit was checked for proper operation.

BAS Troubleshooting Scenario 4—Malfunctioning Mechanical Equipment

A building manager calls the office with a complaint. An area in the building is too hot for the occupants. The technician arrives and checks the customer BAS computer. The BAS indicates the unit is ON, but an inspection of the unit reveals that the unit is not running.

Possible Cause—Nonoperating Mechanical Equipment. A typical reason for excessive hot or cold temperatures is malfunctioning mechanical equipment that is used to provide heating or cooling to building spaces. Sometimes mechanical contractors and BAS controls contractors blame the other for problems with occupant comfort. Even though they are not BAS problems, problems with HVAC equipment still lead to BAS complaints.

In most cases, occupants do not understand the difference between problems with control systems and mechanical equipment. Therefore, HVAC technicians must solve HVAC problems regardless of whether they are caused by malfunctioning building automation or mechanical systems. HVAC technicians must be familiar with the tools and procedures used to troubleshoot BASs and diagnose mechanical system problems.

Solution: When malfunctioning mechanical equipment is suspected, HVAC equipment is checked for heating or cooling devices that are ON or valves that are 100% open. Completely open valves may indicate that the HVAC equipment cannot satisfy the heating or cooling demands for the space. When the HVAC unit is a VAV system, the discharge air temperature at the air handling unit is checked to ensure that it is 55°F. The airflow is checked at the VAV terminal box to ensure that it is the design volume. Systems operating incorrectly result in temperatures excessively high or low in building spaces.

Service Work Performed: The technician shut down the HVAC unit and inspected the mechanical components. During inspection, broken fan belts were identified and replaced. The unit was turned on and resumes operation. The building spaces were measured for proper temperatures. The unit was checked for proper operation.

BAS Troubleshooting Scenario 5—Inaccurate or No Sensor Readings

A technician is dispatched to a job site. The technician is informed that a temperature sensor is not reading a value, and question marks are displayed on the screen.

BASs use temperature, pressure, and humidity sensors for HVAC control and to signal when improper conditions occur in buildings. Specialty sensors such as carbon dioxide and oxygen sensors are commonly used to check air quality. The proper selection, installation, and operation of these sensors are necessary for proper system control. Problems with sensor readings cause problems with system controls. **See Figure 57-5.** Sensor problems typically fall into two categories: erratic sensor readings and absent readings.

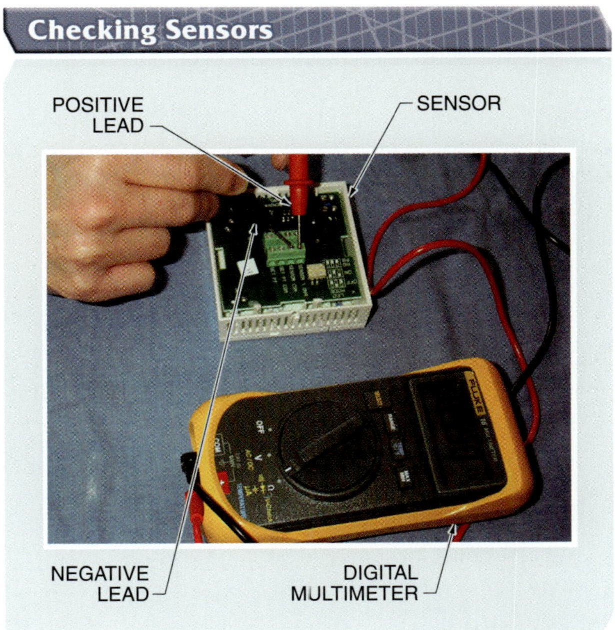

Figure 57-5. Sensors are checked using digital multimeters (DMMs) when readings are excessively high/low or are not present.

Possible Cause 1—Physical Damage to Sensors. Sensors that once measured temperature and pressure values properly are now measuring the same values improperly. This is typically the result of physical damage. **See Figure 57-6.** Damage can occur during collisions, when sensors are painted, or when sensors are cleaned with excessive force or improper chemicals.

Damaged Sensors

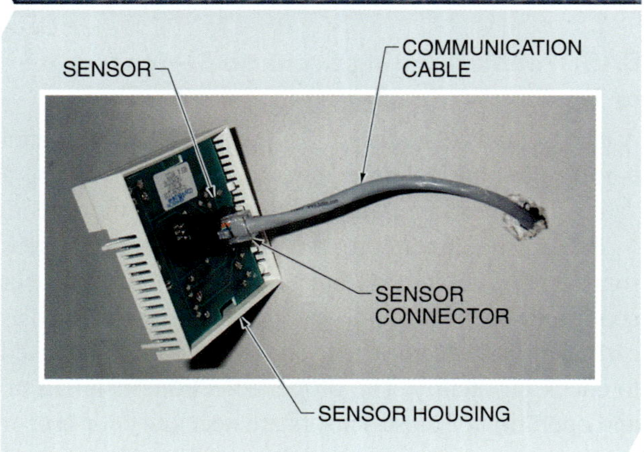

Figure 57-6. A damaged sensor can produce erratic sensor readings that lead to occupant discomfort.

Possible Cause 2—Incorrect Sensor Calibration. Sensors that measure variables in certain locations are exposed to a variety of environmental conditions. Extended time in harsh environments will cause the measuring ability of the sensor to drift. Sensors must be periodically checked for drift and recalibrated when necessary.

Possible Cause 3—Wiring Problems. Wiring and wire connections may be damaged and cause sensors to operate improperly. Typical wiring problems include loose and improperly terminated wires.

Solution: HVAC technicians should inspect the sensors and connections. Possible causes for sensor problems include physically damaged sensors, incorrect sensor calibration, damaged wires, and loose connections. Damaged sensors must be replaced. When necessary, sensors should be relocated to prevent damage. When local conditions contribute to sensor damage, more suitable replacement sensors should be considered.

Service Work Performed: The technician inspected the sensor location and found that electrical work was recently completed in the area. Sensor communication wires were inadvertently disconnected from the sensor. The technician reconnected the sensor wiring and checked the BAS to confirm that the sensor was reading correctly. The unit was checked for proper operation.

UNIT 58

Building Commissioning

Building commissioning is a process used to ensure that building systems function as designed, fulfill the written sequences of operation, and work with other building systems such as building automation systems. Buildings that are commissioned are more energy efficient and have lower energy usage and operating costs.

OBJECTIVES

- List the different types of building commissioning.
- Identify the basic features of commissioning checklists.
- Describe the commissioning procedures used with different equipment.
- Explain the basics of an operations and maintenance (O&M) plan and program.
- Compare the four levels of maintenance programs.
- Describe O&M training programs.

COMMISSIONING

When functioning properly, building HVAC systems provide occupants with comfort at low energy costs and a minimum level of maintenance. Unfortunately, many residential and light commercial building systems fail to meet these standards. Many buildings have HVAC and other systems that fail to work properly and control systems that were not checked to ensure correct operation. In some cases, wires were not terminated, fan motors run backward, and control systems were not programmed correctly. **See Figure 58-1.** It is difficult to achieve comfort in a building with such problems. These problems result in occupant complaints and loss of revenue due to cancelled leases.

Building commissioning is a quality-oriented management practice that delivers improved building systems. *Commissioning* is a process used to verify and document the performance of building systems and ensure that building systems achieve stated objectives for installation, testing, and operation. The building commissioning process is applied to many types of residential and light commercial buildings.

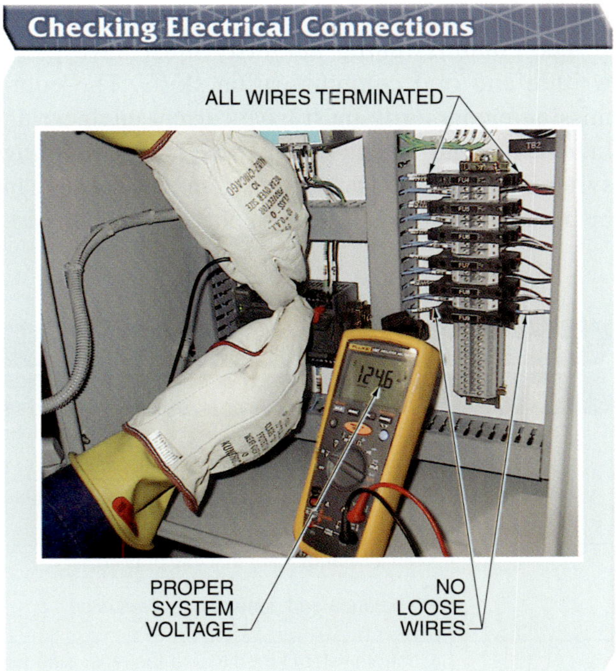

Figure 58-1. Electrical connections to equipment must be checked to ensure proper HVAC system operation.

Typically, the commission process follows a general progression. **See Figure 58-2.** Commissioning begins with checking, cleaning, and aligning the belts, fuses, starters, refrigerant charge, and control panels. Commissioning continues with calibrating the air handlers and adjusting air balance and zone controls. Commissioning concludes with fine adjustments of the control systems to operate within the setpoints with minimal energy use.

TECH TIP

Building automation systems are typically used to verify equipment operation as part of the commissioning process.

Commissioning Authorities

Building commissioning is performed by a commissioning authority. A *commissioning authority,* also known as a commissioning agent, is a person or organization that is responsible for planning, coordinating, and implementing the commissioning process. In many cases, experienced commissioning authorities have received specialized training on specifically installed equipment and are objective and independent agents. Some government or insurance agencies require that the building commissioning process be witnessed by their commissioning authority prior to issuing an operating permit or insurance policy.

The commissioning authority must have good written and oral communication skills. The commissioning authority must have current engineering knowledge and recent field experience in building systems and design. The commissioning authority can be brought in during the design phase, when needed. Typically, building commissioning is not needed until equipment installation and testing occurs. The commissioning authority should be involved in preparing the facility operations and maintenance plan. Typical responsibilities of the commissioning authority include the following:

- Create, with clear statements, the design intent for each building system.
- Write the commissioning specifications and incorporate them into the construction specifications.
- Test all equipment and systems that are commissioned, using standard checklist procedures.
- Review contractor-supplied operations and maintenance documents.
- Develop operations plans that ensure efficient equipment operations.
- Develop maintenance training plans that ensure effective equipment maintenance.
- Write final commissioning reports that include all documentation and recommendations to the building owners.

Building Types

The building commissioning process is adjusted to the installed equipment and systems, and it is based on the specific uses, type, and size of the building. Most small residential buildings have simple systems that are commissioned by the equipment installer. Larger residential and light commercial buildings have relatively simple systems that can be commissioned after contractor installation. Heavy commercial and industrial buildings typically need a building-specific commissioning plan.

Commissioning Performance Levels

Level	Equipment Condition	Effect on System Performance
1	Mechanical equipment (belts and shafts) cleaned and aligned Electrical devices (overloads, fuses, starters) checked Refrigeration (charge and superheat) adjusted Main refrigeration plant controls adjusted	Minimized mechanical failures
2	Main air-handler controls calibrated Air balance and zone controls adjusted	Proper temperature and airflow to zones Zones controlling thermostat settings Even temperature throughout building
3	Control system tuned to use the least amount of energy without losing control setpoints	Building energy usage is close to actual building load or energy losses

Figure 58-2. The commissioning of HVAC systems to increasing levels of efficiency follows a general progression.

Commercial buildings may be categorized by their use and size in square feet. For example, buildings may be classified as smaller than 40,000 ft^2, between 40,000 ft^2 and 100,000 ft^2, and larger than 100,000 ft^2. Building equipment determines the actual commissioning process, with light commercial buildings often using packaged units such as rooftop HVAC equipment. **See Figure 58-3.** Packaged equipment typically uses a simplified and shortened commissioning process. Specific building types, such as schools, have specific commissioning process requirements.

Commissioning Types

Commissioning is typically included with the initial operation of new building equipment. The commissioning process is also used to improve the performance of existing building systems. Commissioning is typically included as part of new building construction, a major system retrofit, or equipment replacement within a building.

New-Building Commissioning. The commissioning process of new buildings provides documentation to the owners that the systems operate as expected. *New-building commissioning* is a commissioning process applied to new systems and buildings during construction. **See Figure 58-4.** The installation, testing, startup, and operation of equipment and HVAC systems are required for new-building commissioning. Commissioning decreases the number of change orders during construction and lowers equipment operation and maintenance costs. When complicated equipment is involved, commissioning often pays for itself before the construction is completed.

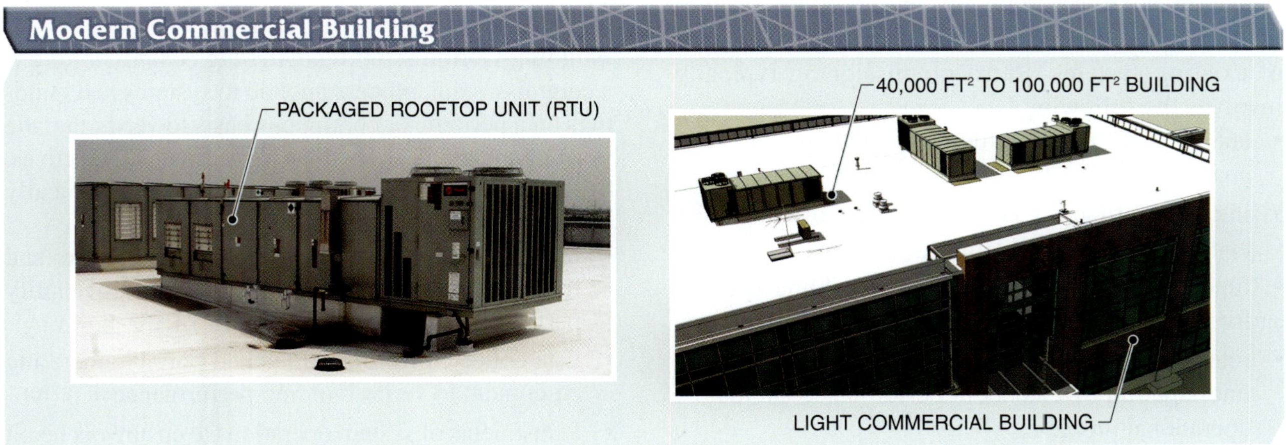

Figure 58-3. Modern commercial buildings have packaged HVAC units on the roof.

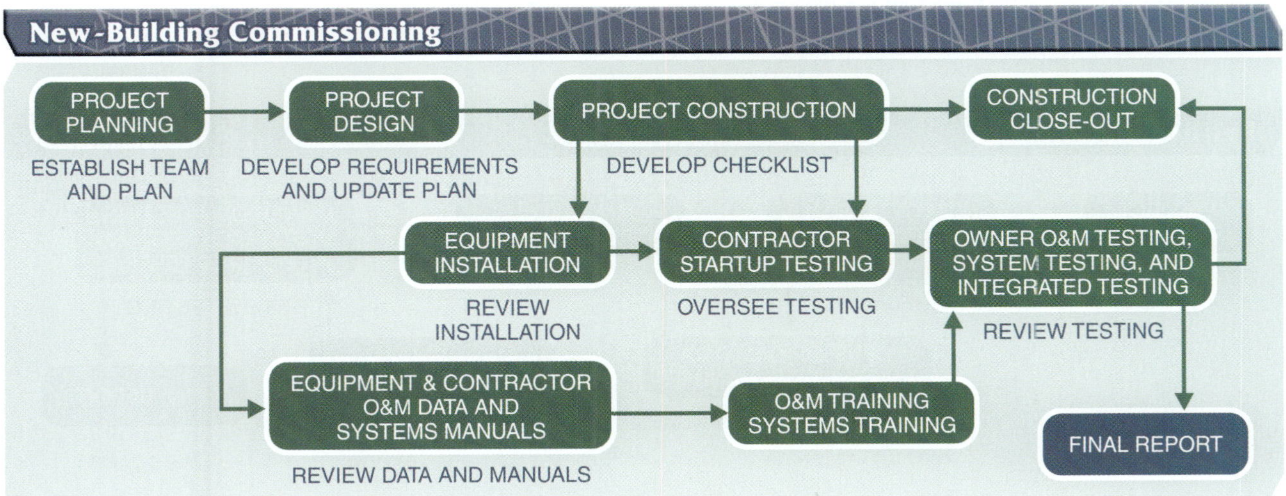

Figure 58-4. Building commissioning is part of the construction process.

When performed properly, new-building commissioning begins at the construction planning phase and continues through the entire building life cycle phase. Commissioning typically includes the construction, systems startup, building acceptance, operations and maintenance training, and the equipment warranty phase.

Retrocommissioning. *Retrocommissioning,* also known as existing building commissioning, is a commissioning process applied to existing systems or buildings that were not previously commissioned. **See Figure 58-5.** Retrocommissioning is a documented process that returns the operations of existing systems and buildings to the original intended design. Commissioning helps systems perform better and reduces the cost of building operations and maintenance. The retrocommissioning process typically covers mechanical equipment, lighting, and related controls. Instead of replacing existing equipment, the retrocommissioning process is focused on increasing the performance of existing systems. Retrocommissioning typically involves the following:

- energy audit of the building, including a detailed analysis of previous utility bills
- interview of facility personnel to identify operation and maintenance problems
- functional testing, monitoring of building systems for performance, and analysis of the collected data
- retesting and monitoring of building systems to fine-tune improvements in performance and adjust operational problems
- preparation of a final report that includes the commissioning plan, a schedule of the effort, and any additional problems identified in the process

Recommissioning. Buildings may need to be commissioned again when problems are encountered during operation of the facilities. *Recommissioning* is a commissioning process applied to systems or buildings that were previously commissioned and typically involves changes in their use or operation. **See Figure 58-6.**

Often problems develop with equipment manufactured by multiple vendors within a system and when insufficient maintenance is performed. In many commercial buildings, the operational staff receives constant complaints and demands from the occupants regarding comfort issues such as building temperature. As the building staff attempts to solve immediate complaints, system changes are implemented that often cause unintended equipment operation and sequencing problems. During recommissioning the systems can be readjusted and the operations of the building corrected by identifying the root cause of the original complaints. Some building systems are inspected, retrofitted, and recommissioned every five years.

Ongoing Commissioning. *Ongoing commissioning* is a commissioning process applied to systems and buildings on a periodic or continuous basis to verify that the systems and building continue to meet stated objectives. **See Figure 58-7.** Ongoing commissioning typically includes the following:

- review questionnaires and surveys of operations and maintenance staff and building occupants to identify performance concerns
- data collected from installed and portable metering equipment to verify building performance
- adjustments of system operations to ensure occupant comfort
- ongoing monitoring of the actual amount of energy used by a building

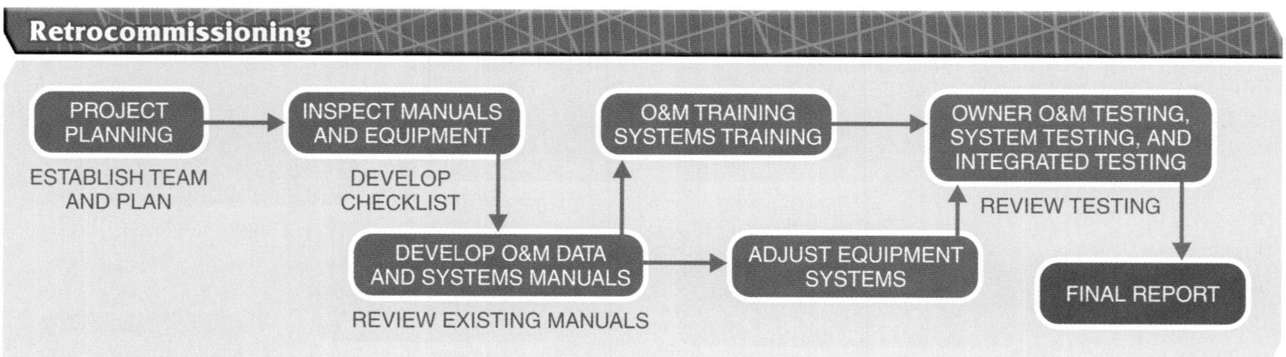

Figure 58-5. Buildings can be commissioned after completion of the construction process to improve the performance of building systems consistent with the original design.

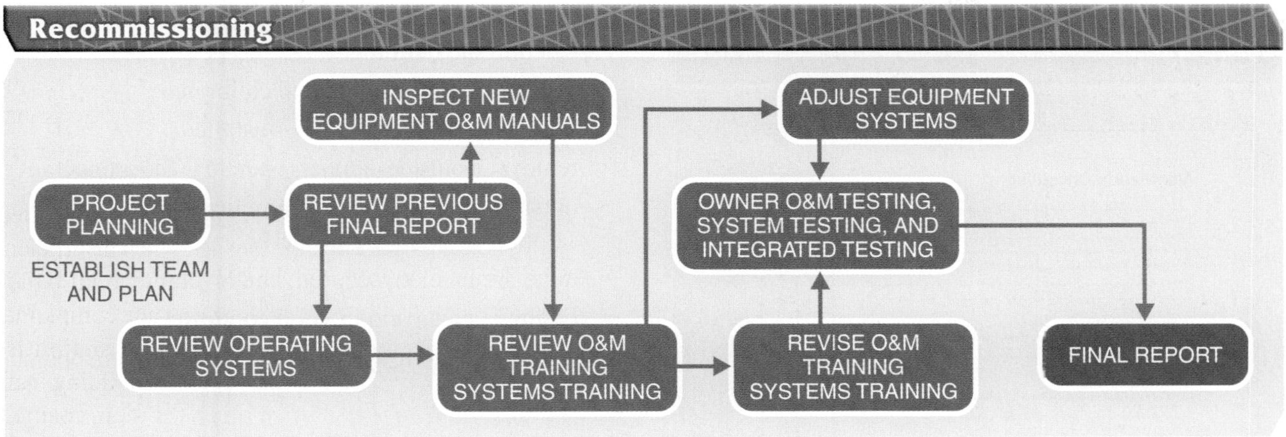

Figure 58-6. Buildings can be commissioned after building renovations to correct building system performance problems.

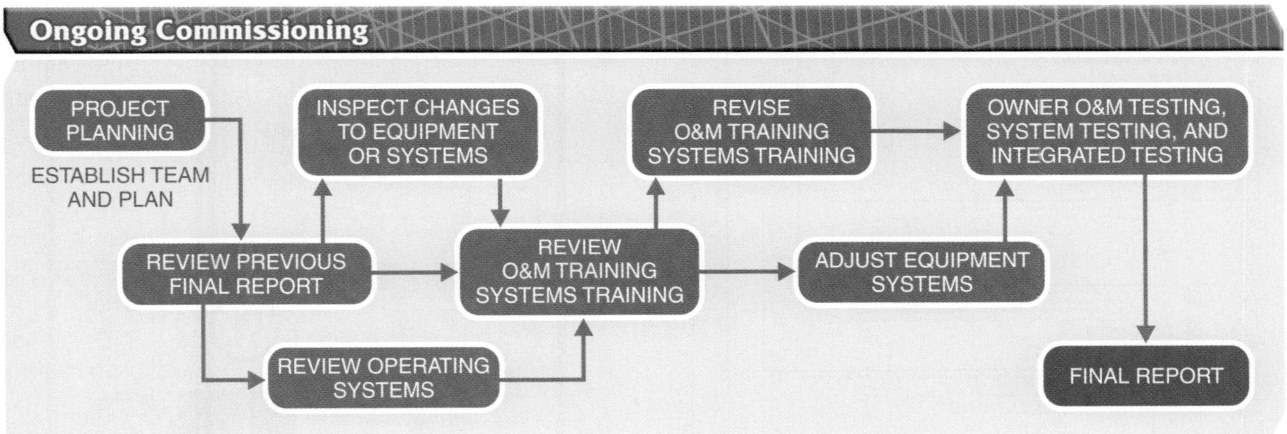

Figure 58-7. Buildings can be commissioned periodically or continuously to improve the performance of the building systems.

TYPICAL COMMISSIONING PROCESS

The building commissioning process includes major systems such as HVAC equipment and controls, hot water systems, and lighting systems. Other systems may be included such as life safety, fire suppression, and security systems. Governing authorities may require specific commissioning processes to be performed on some systems such as the ASME Code for boilers.

All specific commissioning processes are included in the building commissioning process. Full building commissioning covers all building systems. When done properly, building commissioning results in energy efficiency, better indoor air quality, and longer service life of mechanical systems.

Commissioning Checklists

A *commissioning checklist* is a composite list used to verify that all phases of the commissioning process meet requirements, including installation, testing, and training. Commissioning checklists typically include testing procedure requirements. **See Figure 58-8.** A *test procedure* is a description of the test methods, testing equipment, and expectations for tests conducted on components, equipment, systems, and buildings. Typically, certification documentation is required with equipment warranties to verify that the equipment installation and initial startup procedures were performed properly.

Equipment energy efficiency is decreased as a building automation system tries to compensate for problems due to equipment wear, improper installation, improper control settings, and excessive operation. Occupant health is at risk when temperature and humidity fluctuate due to ventilation system operation. Improperly operating HVAC building systems cause conflicts between owners and the contractors responsible for the installation of the HVAC systems. Building commissioning resolves many of these conflicts.

Building Commissioning Checklists

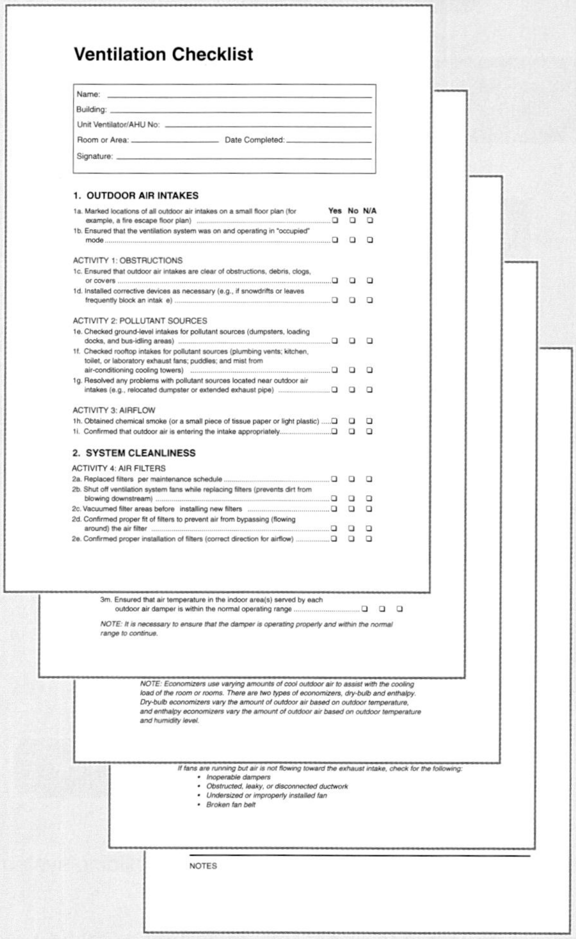

Ventilation Checklist

Name: ____
Building: ____
Unit Ventilator/AHU No: ____
Room or Area: ____ Date Completed: ____
Signature: ____

1. OUTDOOR AIR INTAKES

	Yes	No	N/A
1a. Marked locations of all outdoor air intakes on a small floor plan (for example, a fire escape floor plan)	❑	❑	❑
1b. Ensured that the ventilation system was on and operating in "occupied" mode	❑	❑	❑
ACTIVITY 1: OBSTRUCTIONS			
1c. Ensured that outdoor air intakes are clear of obstructions, debris, clogs, or covers	❑	❑	❑
1d. Installed corrective devices as necessary (e.g., if snowdrifts or leaves frequently block an intak e)	❑	❑	❑
ACTIVITY 2: POLLUTANT SOURCES			
1e. Checked ground-level intakes for pollutant sources (dumpsters, loading docks, and bus-idling areas)	❑	❑	❑
1f. Checked rooftop intakes for pollutant sources (plumbing vents; kitchen, toilet, or laboratory exhaust fans; puddles; and mist from air-conditioning cooling towers)	❑	❑	❑
1g. Resolved any problems with pollutant sources located near outdoor air intakes (e.g., relocated dumpster or extended exhaust pipe)	❑	❑	❑
ACTIVITY 3: AIRFLOW			
1h. Obtained chemical smoke (or a small piece of tissue paper or light plastic)	❑	❑	❑
1i. Confirmed that outdoor air is entering the intake appropriately	❑	❑	❑

2. SYSTEM CLEANLINESS

ACTIVITY 4: AIR FILTERS			
2a. Replaced filters per maintenance schedule	❑	❑	❑
2b. Shut off ventilation system fans while replacing filters (prevents dirt from blowing downstream)	❑	❑	❑
2c. Vacuumed filter areas before installing new filters	❑	❑	❑
2d. Confirmed proper fit of filters to prevent air from bypassing (flowing around) the air filter	❑	❑	❑
2e. Confirmed proper installation of filters (correct direction for airflow)	❑	❑	❑
3m. Ensured that air temperature in the indoor area(s) served by each outdoor air damper is within the normal operating range	❑	❑	❑

NOTE: It is necessary to ensure that the damper is operating properly and within the normal range to continue.

NOTE: Economizers use varying amounts of cool outdoor air to assist with the cooling load of the room or rooms. There are two types of economizers, dry-bulb and enthalpy. Dry-bulb economizers vary the amount of outdoor air based on outdoor temperature, and enthalpy economizers vary the amount of outdoor air based on outdoor temperature and humidity level.

If fans are running but air is not flowing toward the exhaust intake, check for the following:
- *Inoperable dampers*
- *Obstructed, leaky, or disconnected ductwork*
- *Undersized or improperly installed fan*
- *Broken fan belt*

NOTES

Figure 58-8. Checklists are developed to verify that all phases of the commissioning process have been completed.

HVAC Equipment Systems. HVAC equipment compliance stickers provide basic information, including energy consumption. The identification and/or designation of the unit, nameplate information, and unit's time schedule of use should be noted on the checklist. **See Figure 58-9.** HVAC equipment system checklists are typically used to ensure the following:

- electrical wiring to the unit, with disconnects, is properly installed and labeled
- thermostat is wired and properly functioning, including heating/cooling deadbands, setpoints, and time programming
- operating schedule override is in place
- thermostats are of the correct type and properly located in the building
- CO_2 sensor is installed and calibrated
- HVAC condensate drain is installed
- return and outside dampers operate in the proper range
- outside air economizer is functional and switchover setpoint is operating properly
- units are labeled, secured, and leveled
- casing is clean and in good condition
- unit fans, heating system, and cooling system operate properly

ASHRAE 90.1 Compliance

Figure 58-9. Manufacturers provide compliance stickers that include basic information about equipment.

ASHRAE 90.1 Media Clip

Domestic Hot Water Systems. Temperature tests should be conducted using digital infrared thermometers (heat guns) to safely obtain accurate temperature readings. **See Figure 58-10.** Domestic hot water system checklists are typically used to ensure the following:

- equipment is clean with all shipping attachments removed
- storage tanks were tested for leaks (visually) and response to varying loads and water temperature settings
- water piping was tested for leaks (visually) and response time for hot water to reach remote locations

- T&P relief valve was installed and the discharge pipe is routed to the nearest drain
- T&P relief valve was tested for proper operation (manual release), recovery time (control and burner operation observations), temperature overshoot, and drainage
- installed water heater meets specifications for efficiency
- gases of combustion properly vent from the gas water heater
- pipes for the cold and hot water were installed and sized in accordance with the design
- natural gas piping was installed and sized in accordance with the design
- water piping was flushed and pressure tested and is properly connected
- equipment is level and properly anchored
- maximum temperature of the aquastat is 120°F
- electricity, gas, and water utilities are properly connected
- controls are properly connected and programmed
- metal drain valve was installed instead of a plastic valve
- provisions were made for the combustion air of gas heaters such as the use of combustion air louvers
- there is open access for maintenance

Temperature Tests—Hot Water Systems

Figure 58-10. Hot water systems are tested to determine actual temperatures.

Indoor Lighting Systems. Indoor lighting systems are relatively easy to modify and can reduce electricity consumption in a building. Lamp electrical efficiency varies by the lamp type used. **See Figure 58-11.** Typical tasks on indoor lighting system checklists include verification of the following:

- all lighting loads and light fixture types, including lamps and ballasts, are in accordance with the design
- correct type and setting of controls
- occupancy sensor controls were installed for all lighting fixtures that have occupancy sensor control indicated in the construction documents (typically shown on lighting or reflective ceiling plans)
- correct sensor type (e.g., passive infrared, ultrasonic, dual mode) and proper location (e.g., ceiling or wall)
- proper sensor sensitivity, time delay, and functionality
- sensors have been properly installed and adjusted
- proper light levels, with a light meter

Lamp Electrical Efficiency

Lamp Type	Luminous Efficacy*
Conventional incandescent	12 to 18
Halogen	18 to 24
Fluorescent	55 to 100
Low-pressure sodium HID	190 to 200
Mercury vapor HID	50 to 60
Metal-halide HID	65 to 115
High-pressure sodium HID	100 to 150
Light-emitting diode	25 to 70

* in lm/W

Figure 58-11. Luminous efficacy is a measure of how efficiently a lamp produces light for each unit of electrical power.

Commissioning and Equipment Maintenance

Building commissioning involves development of the operations and maintenance (O&M) plan for the facility. An *operations and maintenance (O&M) plan* is a written course of action with specific tasks related to the scheduling of repairs to equipment, production systems, and facilities. **See Figure 58-12.**

Building commissioning also involves preparing system manuals and a training plan for the O&M staff. An *O&M systems manual* is a document used to describe the operations and maintenance information and the limitations of individual systems. An *O&M training plan* is a written course of action with specific tasks related to the training needs of operating and maintenance personnel and users of systems and buildings. The O&M training plan also includes scheduling and budget information.

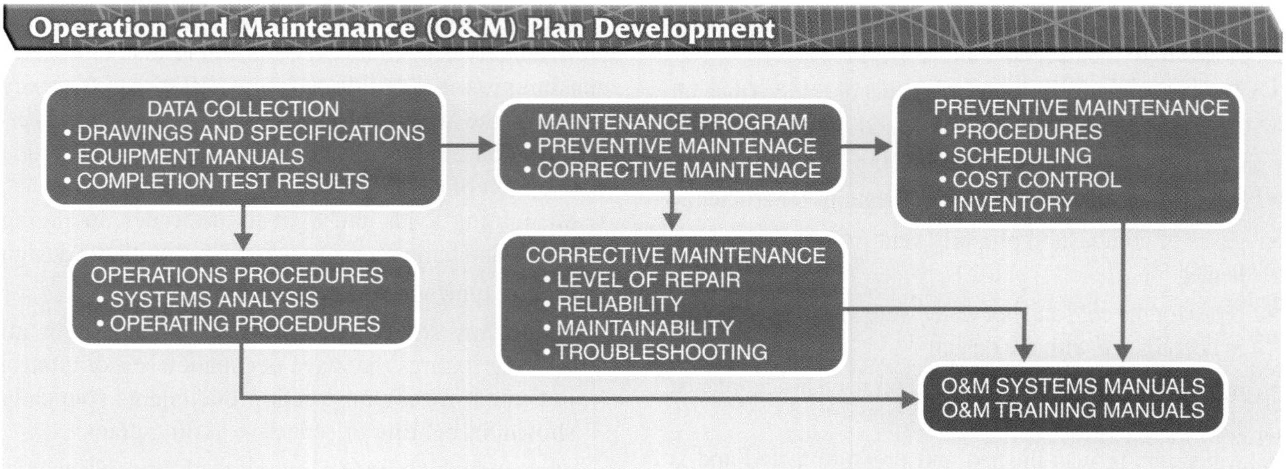

Figure 58-12. O&M manuals and training materials are developed from building drawings, equipment manuals, and completion test reports.

At the conclusion of the commissioning process, the commissioning authority creates the commissioning report for the owner. A *commissioning report* is a record of the activities and results of the commissioning process. The commissioning report provides a basis for the ongoing operations and maintenance of the equipment, systems, and building.

O&M Plans. O&M plans typically involve verifying proper device operation and recording readings from the devices. Owners may develop O&M plans for any equipment and processes critical to the operation of their facilities. However, government authorities require O&M plans for certain equipment and processes as part of a permit process. **See Figure 58-13.** Specifically, O&M plans and environmental permits are required for environmental controls, equipment, and operations. Typically, government authorities provide guidelines and forms for the required O&M plans.

An O&M plan includes general information, an operation plan, a maintenance plan, and additional information for the control device or system. General information included in the O&M plan is information about the facility, process being controlled, discharges being controlled such as pollutants, and the control device. Operation plans include the operating parameters with frequency and instrumentation readings. Maintenance plans include maintenance procedures with the performance frequencies, the permit conditions, training requirements, and supplemental information about the facility. Operation log sheets and maintenance checklists must be included in O&M plans.

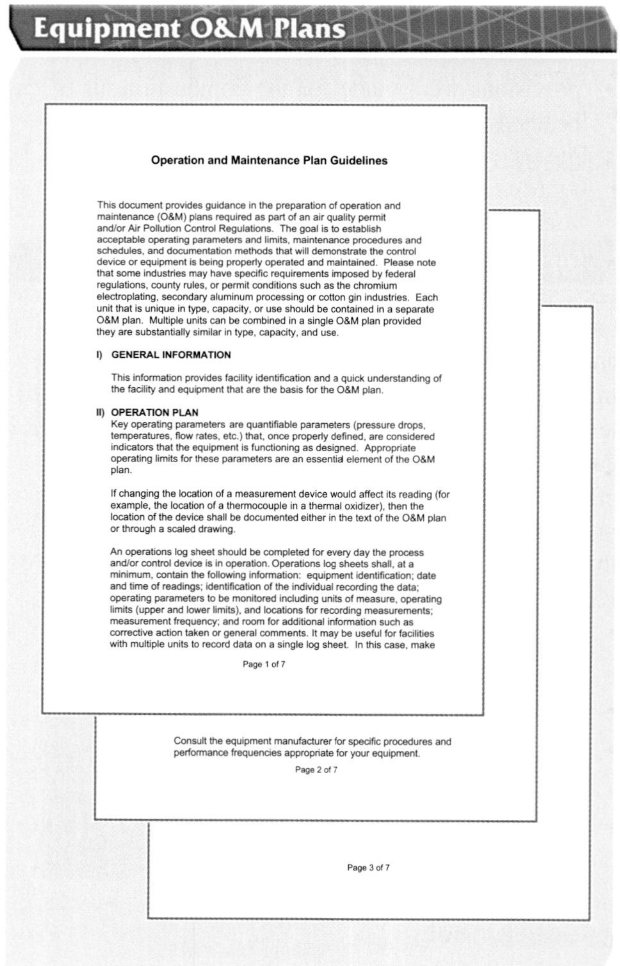
Equipment O&M Plans

Operation and Maintenance Plan Guidelines

This document provides guidance in the preparation of operation and maintenance (O&M) plans required as part of an air quality permit and/or Air Pollution Control Regulations. The goal is to establish acceptable operating parameters and limits, maintenance procedures and schedules, and documentation methods that will demonstrate the control device or equipment is being properly operated and maintained. Please note that some industries may have specific requirements imposed by federal regulations, county rules, or permit conditions such as the chromium electroplating, secondary aluminum processing or cotton gin industries. Each unit that is unique in type, capacity, or use should be contained in a separate O&M plan. Multiple units can be combined in a single O&M plan provided they are substantially similar in type, capacity, and use.

I) GENERAL INFORMATION

This information provides facility identification and a quick understanding of the facility and equipment that are the basis for the O&M plan.

II) OPERATION PLAN

Key operating parameters are quantifiable parameters (pressure drops, temperatures, flow rates, etc.) that, once properly defined, are considered indicators that the equipment is functioning as designed. Appropriate operating limits for these parameters are an essential element of the O&M plan.

If changing the location of a measurement device would affect its reading (for example, the location of a thermocouple in a thermal oxidizer), then the location of the device shall be documented either in the text of the O&M plan or through a scaled drawing.

An operations log sheet should be completed for every day the process and/or control device is in operation. Operations log sheets shall, at a minimum, contain the following information: equipment identification; date and time of readings; identification of the individual recording the data; operating parameters to be monitored including units of measure, operating limits (upper and lower limits), and locations for recording measurements; measurement frequency; and room for additional information such as corrective action taken or general comments. It may be useful for facilities with multiple units to record data on a single log sheet. In this case, make

Page 1 of 7

Consult the equipment manufacturer for specific procedures and performance frequencies appropriate for your equipment.

Page 2 of 7

Page 3 of 7

Figure 58-13. O&M plans are prepared for specific equipment requiring permits.

The O&M plan is documentation of the tasks necessary to properly operate and maintain an approved control device or system. The O&M plan also can be used to detect operation problems. The early detection of operation problems decreases repair and replacement costs, prevents malfunctions, and decreases downtime.

O&M Programs. An *O&M program* is a program that integrates specific O&M plans and general O&M procedures to ensure that control devices and equipment are properly operated and maintained. O&M programs require a significant investment of time and money. An effective O&M program can reduce annual utility costs without increasing the expenses of control devices and equipment. O&M programs identify improvements performed by in-house staff at low cost and have short payback periods. Some benefits of an effective O&M program include the following:

- identification of equipment problems for repair before building occupants notice them and complain
- reduction of the time spent by facility staff on callbacks
- increase in the useful life of equipment
- minimal equipment downtime due to repair and replacement
- better compliance with environmental, health, and safety regulations

Equipment diagnostic and metering tools are important tools used in the O&M program. Diagnostic tools measure the condition of the equipment, systems, and building. **See Figure 58-14.** For example, digital multimeters and other instruments are used to measure temperature, relative humidity, lighting levels, electrical current, liquid flow rate, pressure, and the combustion efficiency of furnaces and boilers. A facility may use qualified in-house staff or outside contractors to perform the diagnostics. Typical predictive diagnostic strategies include the following types of analyses:

- thermal
- oil
- ultrasound
- vibration
- flue gas
- mechanical

The O&M program typically includes the installation of meters on equipment and building systems. Meters measure the energy consumption of the entire facility, whereas submeters measure the isolated energy use of specific parts of mechanical systems or buildings. For example, submeters measure factors such as peak demand, time-of-use, and power outages. The energy use of specific parts is measured and recorded independently. This establishes separate energy costs between specific parts and can be used to diagnose problems.

Mechanical Equipment Diagnostics

The Gates Rubber Company

Figure 58-14. A pulley gauge is used to check a pulley for wear. A pulley with 1/16″ wear on either side of the gauge should be replaced.

Maintenance Levels. The commissioning process involves implementing an O&M program. The level of planned equipment maintenance determines the availability of the equipment. **See Figure 58-15.** There are four levels of maintenance programs, which include the following:

- Level I, reactive maintenance—This level of maintenance involves the repair or replacement of worn or failed equipment. Reactive maintenance is the most expensive type of maintenance.
- Level II, preventive maintenance—This level of maintenance involves scheduled routine maintenance to reduce the occurrence of failures and extend equipment life.
- Level III, predictive maintenance—This level of maintenance involves testing the current condition of equipment to determine what, if any, maintenance should be performed. The predictive approach provides an estimation of the equipment's remaining service life, reduces equipment downtime, and is less costly than the preventive approach.

- Level IV, reliability-centered maintenance—This level of maintenance is similar to predictive maintenance, but it takes into consideration that it is often more cost-effective to repair and replace specific inexpensive or unimportant parts only as needed. This approach is costly to implement but is the most economical over an extended period.

O&M Training Programs. Building commissioning efforts typically include the development and implementation of an O&M training program. An *O&M training program* is a program that identifies specific training objectives, the knowledge needed for staff training, and the methods for delivering the information. **See Figure 58-16.** Since sophisticated controls and mechanical equipment are involved, the O&M staff needs to be properly trained to effectively maintain the facility.

The commissioning authority reviews training programs created by equipment suppliers, system suppliers, and installation contractors. The curriculum, objectives, and course materials are reviewed by the commissioning authority to ensure their effectiveness. For specialized systems, training by a manufacturer-certified trainer is typically required. The O&M training for the individual building system components is integrated into a comprehensive training program for the O&M staff.

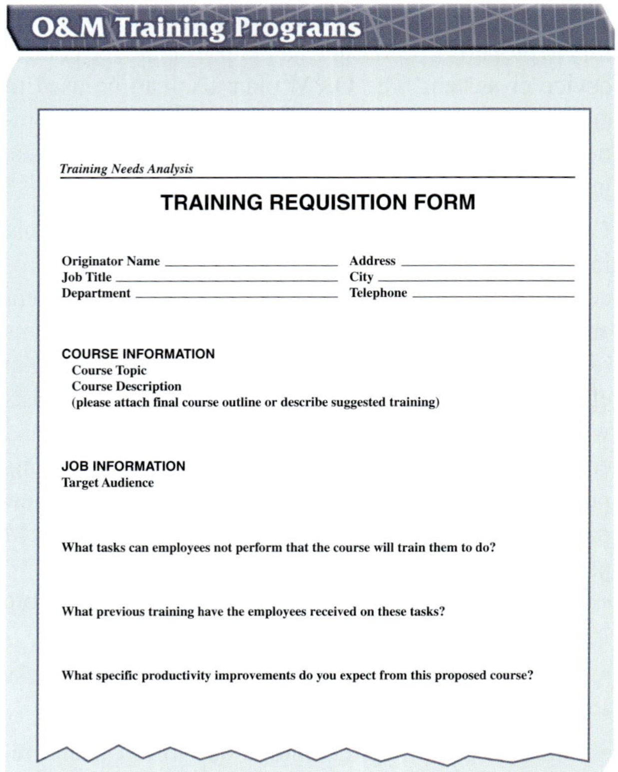
O&M Training Programs

Training Needs Analysis

TRAINING REQUISITION FORM

Originator Name ______ Address ______
Job Title ______ City ______
Department ______ Telephone ______

COURSE INFORMATION
Course Topic
Course Description
(please attach final course outline or describe suggested training)

JOB INFORMATION
Target Audience

What tasks can employees not perform that the course will train them to do?

What previous training have the employees received on these tasks?

What specific productivity improvements do you expect from this proposed course?

Figure 58-16. A training-needs analysis identifies the skills needed by building automation system personnel.

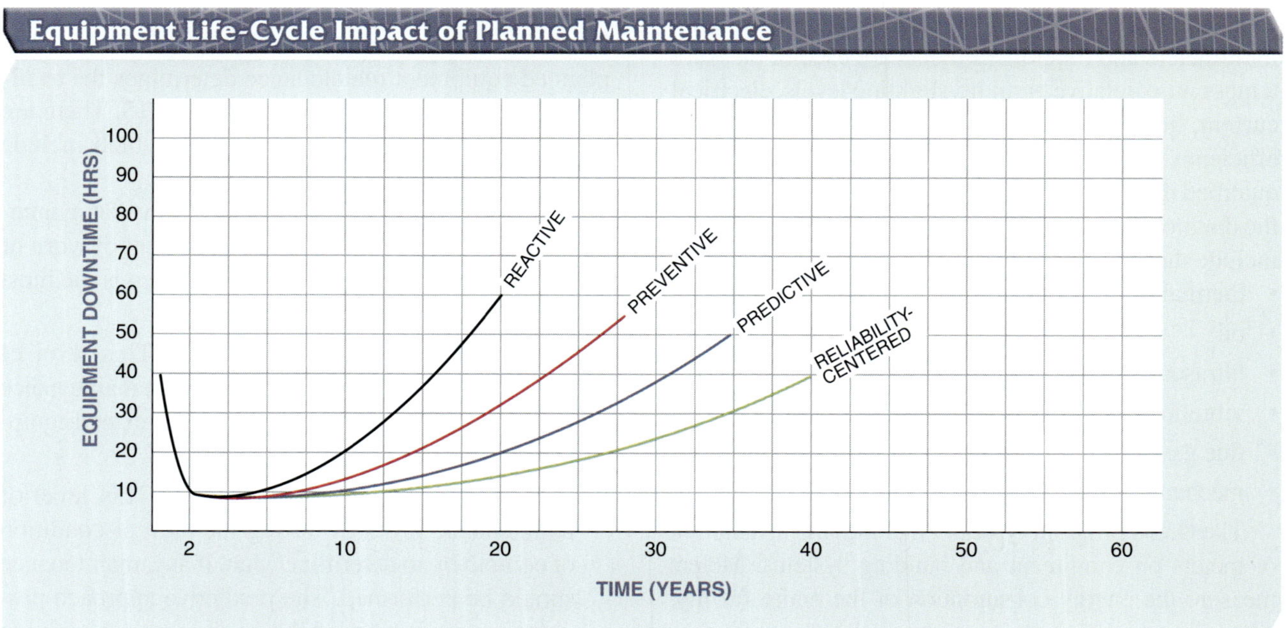

Figure 58-15. Maintenance programs are created to increase the availability of equipment.

Other Building Commissioning Goals. Other goals beyond ensuring mechanical system functionality are often included in building commissioning efforts. **See Figure 58-17.** Examples of other goals include the following:

- environmental requirements and goals related to industry
- indoor air quality standards and goals, which are typically set by codes or the owner
- Leadership in Energy Efficiency and Design (LEED) goals, which are typically included to reach specific targets such as the silver level; to meet a LEED goal, points are typically awarded by implementing a comprehensive commissioning process
- energy and water use targets, which include specific levels of the rate of use of energy and water and the total annual amount of energy and water used

Industry Regulations

Characteristic	Codes	Standards	Guidelines
Scope	Usually broad, addressing many trades	Usually narrow, addressing a specific group or requirements	Varies
Developer	Organization or government entity	Trade or industry organization	Trade or industry organization
Enforceability	Yes, when developed by federal government; otherwise, only when adopted by an authority having jurisdiction	Only if specified within a code that is mandatory for the jurisdiction; otherwise, voluntary	Never

Figure 58-17. Industry regulations include codes, standards, and guidelines, which vary in scope, development, and enforcement.

Digital Resources

ATPeResources.com/QuickLinks • Access Code: 482610

UNIT 59

Building Energy Audits

Building managers are under pressure to reduce costs related to personnel, maintenance, and energy. Energy cost reductions are given high priority. Energy audits are necessary to determine where and how energy is used in buildings. Once the devices and equipment using energy are identified, opportunities for possible reduction in energy use are evaluated. The success of energy cost savings programs depends on an accurate and comprehensive audit of equipment and building energy usage.

OBJECTIVES

- Compare simple, general, and detailed energy audits.
- List factors used to analyze electric and natural gas utility bills and rates.
- Perform common utility energy cost calculations.
- List information gathered during energy audits.
- Describe the auditing of HVAC and non-HVAC equipment.
- Identify features of energy audit reports and cost savings proposals.

AUDIT INFORMATION SOURCES

There are three levels of energy audits: simple, general, and detailed. Residential and light commercial building owners can perform simple energy audits using checklists that are available from utility companies, state agencies, and federal agencies. Some utility companies provide audit services to assist residential and light commercial building owners. Simple energy audits involve an overview of major energy consuming devices and equipment within a building and provide a yardstick comparison. Simple energy audits are easy to complete but typically uncover only the most obvious excessive uses of energy. **See Figure 59-1.**

General energy audits require a qualified energy auditor to review utility bills and measure the energy used in devices and equipment. **See Figure 59-2.** General energy audits are typically done prior to renovating or retrofitting residential or commercial buildings. General energy audits require a moderate effort in data collection and analysis to evaluate energy use reduction options. General energy audits identify and prioritize energy efficiency measures in the energy audit report.

Electronic demand meters are capable of recording many different power supply parameters and allow remote access to the data.

Home Energy Yardstick Forms

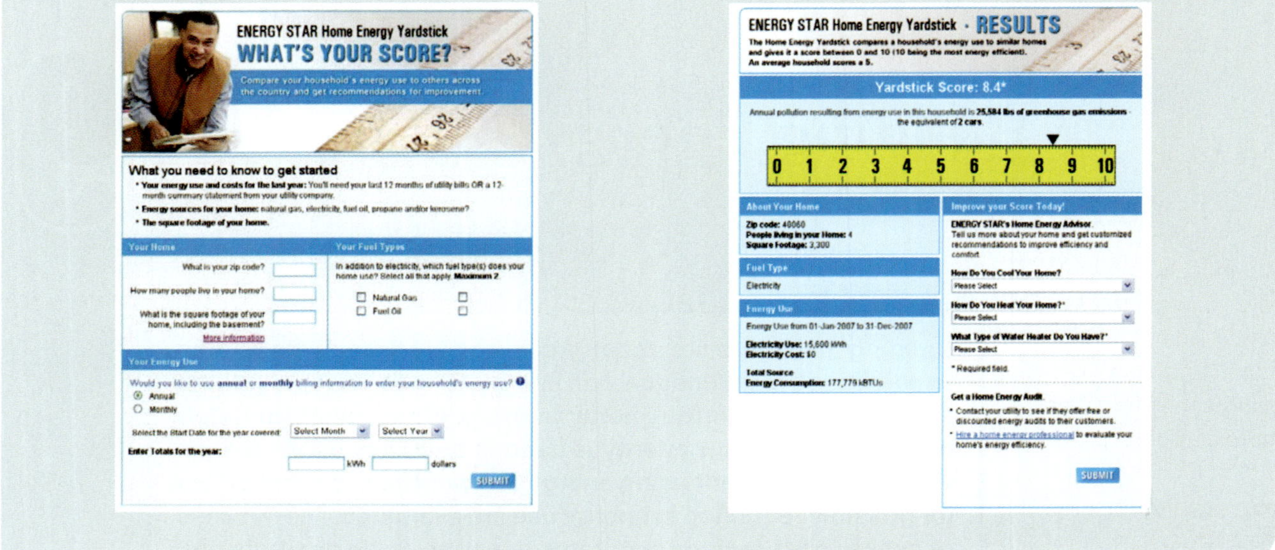

Figure 59-1. Documents such as the Home Energy Yardstick Form, developed by the Department of Energy, are incorporated into simple energy audits to compare the energy use of a home to a national standard.

Energy Use

ELECTRIC USE AND COST

Year: ______

Building:	Account No.:	Building Size (sq ft):
Address:	Meter No.:	Max kW Demand W/O Charge:
Previous Meter Reading:	Utility:	Min Power Factor W/O Charge:

Meter Reading		kWh used	kWh sq ft	Monthly EUI (1000 BTU/sq ft)	Energy Costs	Fixed Service Costs	Demand Costs	Power Factor Costs	Total Costs
Date	Valve								

Figure 59-2. During general energy audits, monthly utility readings are collected to track and evaluate actual energy use during a specific period of time.

Detailed energy audits are typically performed on heavy commercial or industrial buildings by specially trained expert auditors who work for an engineering or consulting firm. Detailed energy audits involve comprehensive measurements and an extensive analysis tailored to the specific project. Detailed audit recommendations are used to determine investments in proposed efficiency measures.

Utility Users

An understanding of energy rates for users is necessary when conducting a general energy audit. Rates for residential and light commercial buildings are typically set rates, regardless of the amount of energy used. Rates for heavy commercial and industrial buildings are usually more complex than residential rates.

Residential and Light Commercial Electricity Users. Residential and light commercial electricity rates are typically straight forward. The user is billed for the electricity consumed in kilowatt-hours (kWh) for that month. Some electric utility companies provide adjusted level billing in which the annual electricity costs are leveled on each monthly electric bill. Many electric utility companies provide a graph or chart of the annual use profile and average daily usage as part of the bill. Electric utility bills typically include account and meter information, the amount of electricity used, supply and delivery charges, utility tax, and other charges. **See Figure 59-3.** Some electric utility companies include rate adjustments, such as those for the summer/winter, time of day, or fuel recovery costs. Rate adjustments are reflected in the monthly electric bill.

Electric Utility Bills

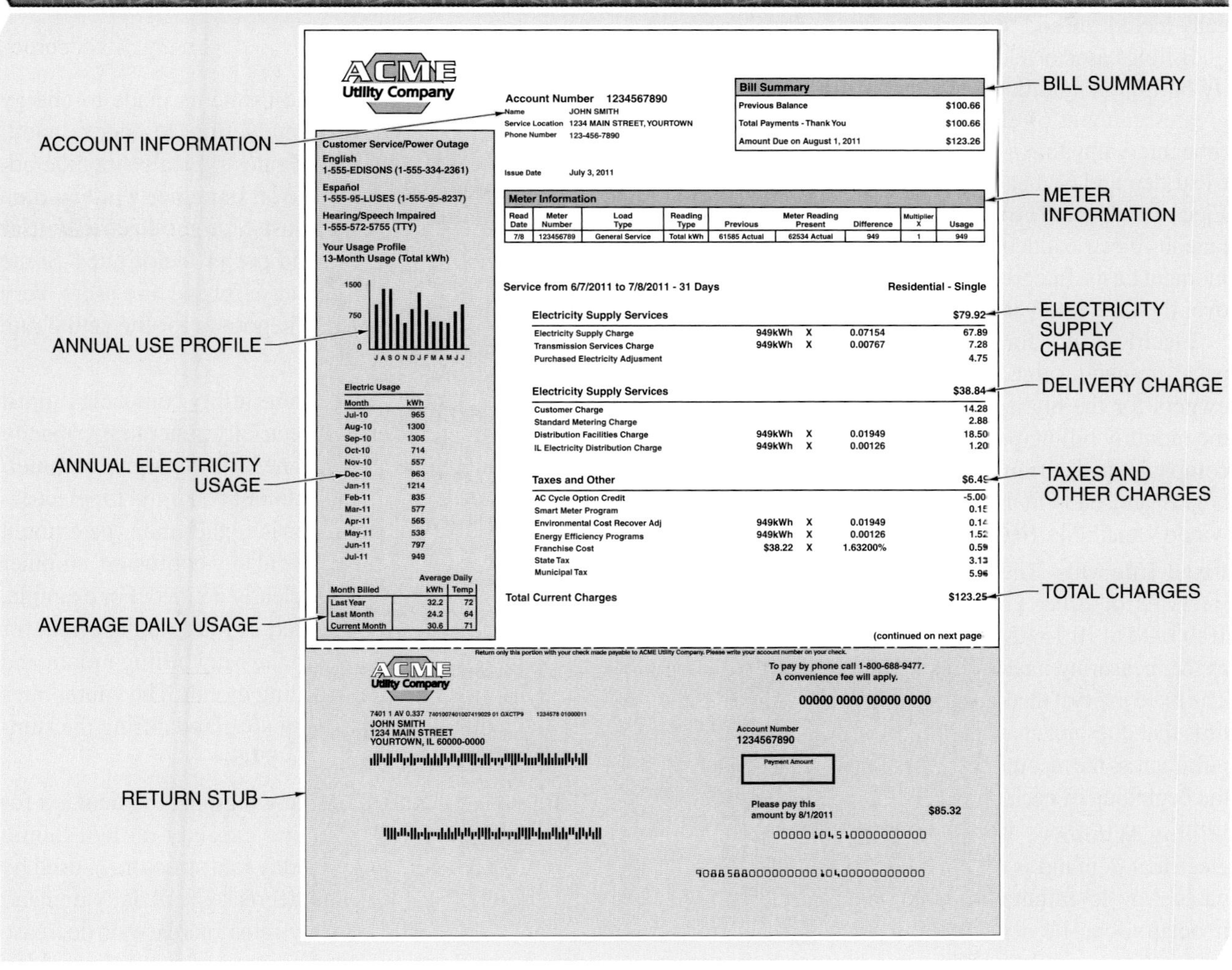

Figure 59-3. The total monthly electric bill for a building is based on the electricity consumed, delivery charges, and taxes.

Some electric utility companies provide an all-electric rate in locations where natural gas or propane is not used or available. All-electric rates are typically lower per kilowatt-hour than standard rates.

Heavy Commercial and Industrial Electricity Users. Heavy commercial and industrial electric utility bills are broken down into electricity consumed, electrical demand, power factor, and fuel recovery costs. Building and facility owners who understand the factors impacting rates for electricity use can adjust electricity usage to times where costs are lower.

Electrical Consumption and Demand

Electrical consumption is the total amount of electricity used by a building or facility during a billing period. Electrical consumption is measured in kilowatt-hours (kWh). Electricity usage in buildings commonly varies for the day and month. Electric bills include the total electricity consumed. Commercial and industrial electric bills typically include an additional charge for electrical demand.

Billing based on electrical demand is a type of electricity billing method in which customers are billed for the greatest amount of electricity used during a 15-minute time interval within a billing period. In some areas, electrical demand is measured in 30-minute time intervals. Electrical demand is measured in kilowatts (kW). Instantaneous electrical demand is calculated on a moment by moment basis. Integrated electrical demand is an average over the time interval.

Electric utility companies constantly meter the electrical demand value and charge building and facility owners for the highest measured demand value within the monthly billing period. Two methods used by utility companies to determine when the 15-minute time interval begins and ends include the fixed interval and sliding window methods. **See Figure 59-4.**

Fixed Intervals. The fixed interval method uses fixed intervals of time. For example, fixed intervals from 10:00 AM to 10:15 AM or 10:15 AM to 10:30 AM are used as the measuring interval to determine electrical demand. The fixed interval method uses a definite start time and stop time that does not change. Some utility companies send a pulse across the incoming power service line, which signals the beginning of each new fixed time period.

Sliding Windows. With the sliding window method, electrical demand is computed on a minute-by-minute basis. Any 15-minute time interval, regardless of when it occurs, can be a new peak demand period for the month. For example, a sliding window can be used from 10:01 AM to 10:16 AM or 10:02 AM to 10:17 AM.

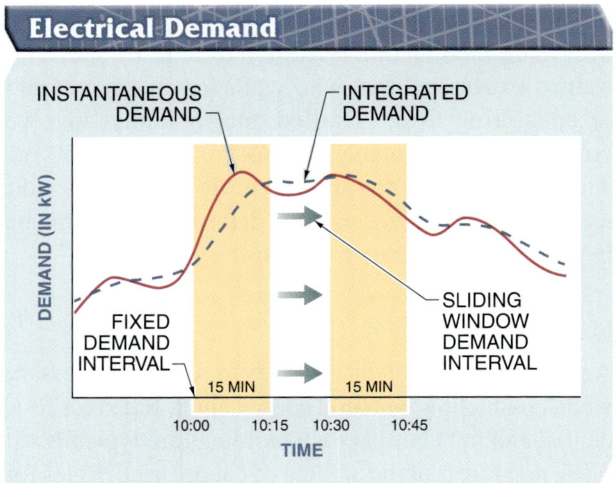

Figure 59-4. The total monthly electric bill for a commercial building is based on the electricity consumed, electrical demand, power factor, fuel recovery costs, and taxes. The electrical demand can be based on a fixed interval or sliding window interval.

Electric Utility Rates

An understanding of the adjustments made to energy rates is necessary when conducting an energy audit. Commercial and industrial rates typically include adjustments based on the season (summer/winter), time of day, and fuel recovery costs. Larger electricity user rates may include tiered and power factor rates. Some utilities include a ratchet clause for specific users. Very large electricity users may negotiate for the actual rate of the electricity used.

Summer/Winter Rates. Some utility companies adjust for the different levels of electricity generating capacity to meet fluctuating seasonal needs by charging summer/winter rates. Typically, the amount charged for electricity used during the summer is higher than the amount charged in the winter. To avoid any confusion, summer and winter periods must be clearly defined. For example, summer rates may be defined as rates taking effect for five months beginning with the May billing month and ending with the October billing month. The winter rates apply in all other months not specified during the summer rate period. **See Figure 59-5.**

Time-of-Day Rates. In many regions, the demand for electricity reaches generating capacity on hot, humid summer days. The time-of-day rate structure is used by some utility companies to decrease peak daily demand for electricity. Time-of-day rates increase or decrease the cost of electricity used depending on the time of day in which the electricity is used.

Summer/Winter Rates

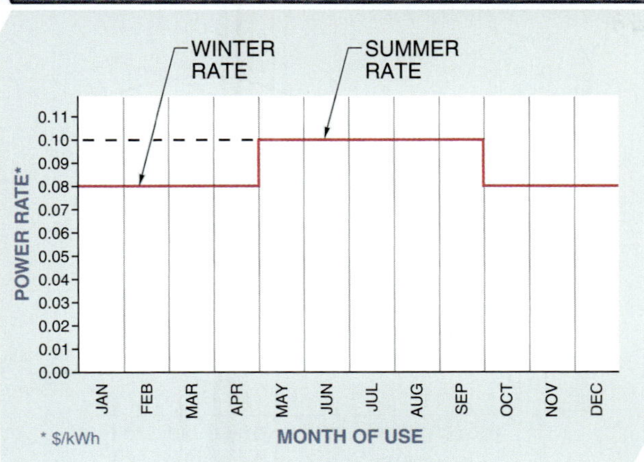

Figure 59-5. Summer and winter electricity rates are used to recover the costs of producing additional power during the higher summer usage period.

For a typical time-of-day rate structure, the peak (day) rate may be defined as 6 AM to 6 PM. The electricity used within this period is charged at a higher rate. From 6 PM to 12 PM the rate for electricity used decreases from the peak rate. During the period from 12 PM to 6 AM the rate is lowest (night) because of lower electricity usage.

Actual rates and effective time-of-day periods are set by regional limits and conditions. In some regions, the peak rates are so expensive that many customers use alternate equipment such as gas- or steam-fired chillers and thermal storage systems. **See Figure 59-6.**

Fuel Recovery Rates. Fuel recovery rates are often added to utility bills, along with state or local taxes. Fuel recovery rates are charged by utility companies as reimbursement for the fuel used to generate electricity. Fuel recovery charges are typically adjusted monthly. The fuel recovery rate is often calculated by multiplying the fuel recovery charge by the energy consumed. For example, if the cost of purchased fuel to generate electricity increased by 1.2% above the base rate, then the cost of each kWh of electricity would increase by 1.2%.

TECH TIP

When analyzing utility expenditures, enlist the help of utility company representatives. Utility representatives can often help diagnose bills, offer tips, and provide statistical information. Time should be taken to utilize these resources when undertaking a utility analysis.

Tiered Rates. A tier for electricity billing is a classification of customers based on the service type and the amount of electricity used. Tier rates are used to adjust the rate paid for electricity. Lower rates are paid by the customers who use more electricity. For example, a light commercial customer may pay $0.10/kWh, whereas a large industrial customer may only pay $0.05/kWh. A variation of tiered rates is graduated rates, which are based on the amount of electricity used. For example, the first 50 kW of electricity costs $10/kW, and from 50 kW to 100 kW electricity costs $8/kW. **See Figure 59-7.**

Time-of-Day Load Shifting

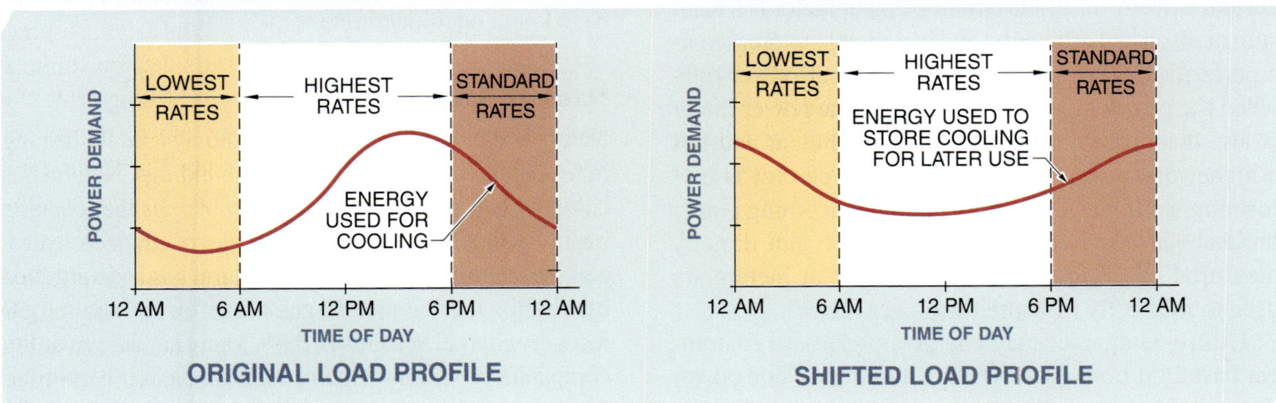

Figure 59-6. Electricity rates may vary for different times of the day. Shifting large loads from high-rate periods to low-rate periods can result in significant savings, even if the overall energy consumed remains the same.

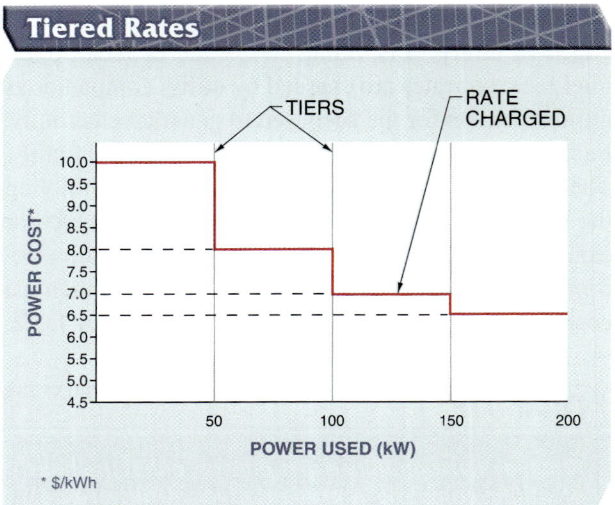

Figure 59-7. With tiered rates, the cost of power for large customers is discounted.

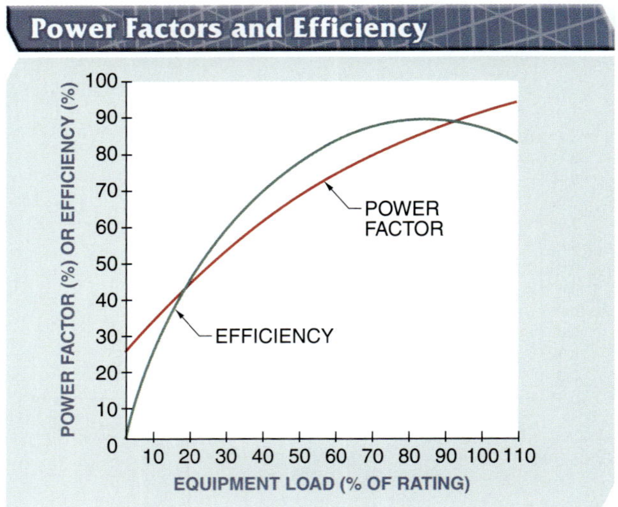

Figure 59-8. Power factors and equipment efficiency fall when an electric motor is loaded at less than its load rating.

Tiered rates are typically established based on customer and service type. Typically, most electric utility companies classify customers as residential, light commercial, heavy commercial, or industrial customers. Residential rates are set through a governmental agency review process. Light commercial customers pay the base rate per kilowatt-hour. Heavy commercial customers pay less than light commercial customers. Industrial customers that use very large amounts of electricity pay the smallest amount. For example, automotive plants may be classified as industrial based on their usage of 1,000,000 kWh/month. Each kilowatt-hour is less expensive than the amount charged to a customer in a different tier because of their classification as an industrial customer.

Power Factor. *Power factor (PF)* is the ratio of true power to apparent power in an alternating current (AC) circuit and describes the displacement of voltage and current waveforms in AC circuits. Power factor is a measure of electrical efficiency and is typically expressed as a percentage. True power equals apparent power only when the power factor is 100%. When the power factor is less than 100%, the circuit is less efficient and has a higher operating cost because not all current is performing work. **See Figure 59-8.** Some incoming power produces an electromagnetic field that is not directly measured or charged by a utility. Special meters are used to indirectly measure the power factor.

Utility companies may charge a penalty to customers based on poor power factor numbers. Poor power factor numbers typically occur in large energy-intensive facilities that use numerous single-phase AC motors and fluorescent lamps.

Ratchet Clauses. A *ratchet clause* is a penalty charged for the electricity demanded during an entire 12-month period based on the highest amount measured during a single month. Normally the demand charge on an electric bill reflects the interval for a single billing period. Some utility companies use a ratchet clause to set the demand charge for up to a year using the highest amount used during a month. When a building or facility sets a new demand usage peak, this amount may be the amount paid each month for the next year.

Negotiated Rates. Large industrial customers typically negotiate a rate directly with the utility company instead of paying a listed rate. Negotiated rates are often a fraction of the standard industrial customer rate. Negotiated rates are often tied to political and social factors such as economic development zones and tax negotiations. Decisions to relocate a firm to a particular area may be based on a negotiated rate.

Natural Gas Utility Rates

Natural gas is an energy source commonly used for heating in residential and light commercial buildings. Natural gas is a colorless and odorless combustible gas that consists mainly of methane with small quantities of ethylene, hydrogen, and other gases. Billing for natural gas is determined by multiplying the natural gas usage by the natural gas recovery charge. **See Figure 59-9.** Many natural gas utility companies provide previous weather and gas use information as part of the bill. Natural gas utility bills typically include account and meter information, natural gas used, supply and delivery charges, utility tax, and other charges.

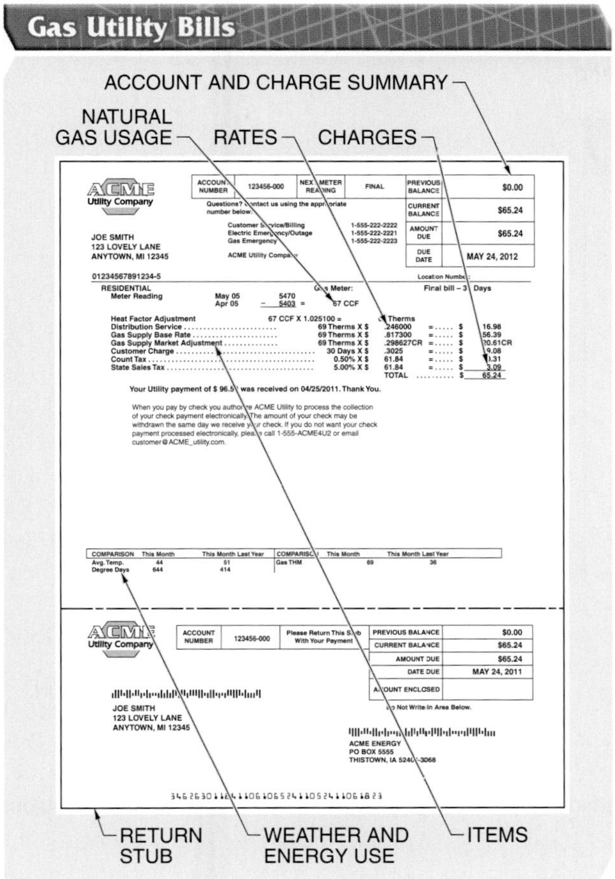

Figure 59-9. The total monthly natural gas bill is based on the amount of gas consumed, delivery charges, and taxes.

Units of measurement for natural gas are based on the cubic foot. Natural gas may be measured in units of a hundred cubic feet (CCF), thousand cubic feet (MCF), or million cubic feet (MMCF). Natural gas may also be measured in therms. A therm is a unit used to measure the heat content of natural gas and is equal to 100,000 Btu. A British thermal unit (Btu) is the quantity of heat required to raise the temperature of 1 lb of water by 1°F.

Natural gas billing may also be based on the type of customer and the tier in which the customer is classified. Large customers typically negotiate long-term rates directly with utility companies for natural gas similar to electric utilities.

UTILITY ENERGY COSTS

General energy audits include a review of utility bills to identify the energy used and its associated costs. By reviewing the utility cost, the amount of energy actually used by a building and the potential cost savings by reducing energy use can be calculated.

Statistical Review of Utility Bills

Utility bills disclose useful information on energy use when properly analyzed. Energy use costs are calculated when considering mechanical system retrofits and expenditures on new HVAC control systems. The actual energy use of existing equipment, such as furnaces, boilers, and air conditioning systems, can be used to determine the estimated energy use for new equipment systems.

For example, a new air conditioning unit with a seasonal energy efficiency ratio (SEER) of 16 will replace an existing unit with an SEER of 10. If the existing air conditioning unit consumed 20,800 kWh/yr, what is the reduction in electricity use?

$$\textit{reduced electricity use} = \textit{electricity use}\left(1 - \frac{SEER_{original}}{SEER_{new}}\right)$$

$$\textit{reduced electricity use} = 20{,}800\left(1 - \frac{10}{16}\right)$$

$$\textit{reduced electricity use} = 20{,}800(1 - 0.625)$$

$$\textit{reduced electricity use} = 20{,}800(0.375)$$

$$\textit{reduced electricity use} = \mathbf{7800\ kWh/yr}$$

If the electricity costs \$0.10/kWh, the amount saved is calculated as follows:

$$\textit{savings} = \textit{reduced electricity use} \times \textit{cost/kWh}$$

$$\textit{savings} = 7800\text{ kWh/yr} \times \$0.10\text{/kWh}$$

$$\textit{savings} = \mathbf{\$780/yr}$$

The cost of installing the new air conditioning unit is divided by the reduced annual energy cost to determine the number of years needed to pay for retrofitting the air conditioning unit. If the installed cost of the new unit is \$3900, the payback period is calculated as follows:

$$\textit{payback period} = \frac{\textit{cost of retrofit}}{\textit{savings in electricity costs}}$$

$$\textit{payback period} = \frac{\$3900}{\$780\text{/yr}}$$

$$\textit{payback period} = \mathbf{5.0\ yr}$$

The same basic calculation is used for other energy saving retrofits and measures. Typically, retrofits with shorter payback periods are done sooner than those with longer payback periods.

Utility bills are used to compare energy-efficient buildings to typical buildings. Regional comparisons are used to compare the cost of operation in one area to the cost of operation in another area. Statistical averages are used after an energy auditor analyzes the utility bills. **See Figure 59-10.**

Total Building Energy Use

Month	Electricity (kWh)	E Cost	Natural Gas (therm)	NG Cost
January	956	$ 119.50	67	$ 130.65
February	927	$ 115.88	58	$ 113.10
March	613	$ 76.63	35	$ 68.25
April	536	$ 67.00	22	$ 42.90
May	726	$ 105.27	17	$ 33.15
June	936	$ 135.72	13	$ 25.35
July	1257	$ 182.27	12	$ 23.40
August	1173	$ 170.09	13	$ 25.35
September	987	$ 143.12	20	$ 39.00
October	634	$ 79.25	29	$ 56.55
November	814	$ 101.75	38	$ 74.10
December	976	$ 122.00	53	$ 103.35
TOTAL	10,535	$1,418.46	377	$ 735.15

	ELECTRICITY	NATURAL GAS
CONVERSION FACTOR	3414 Btu/kWh	100,000 Btu/therm
EQUIVALENT ENERGY	35,945,420 Btu	37,700,000 Btu
TOTAL ENERGY	73,645,420 Btu	
COST	$ 2,153.61	
BUILDING AREA	1850 sq ft	
ENERGY USED / AREA	39,808 Btu/sq ft	
ENERGY USED / AREA	$ 1.16/sq ft	

ENERGY COST INDEX (ECI) — $ 1.16/sq ft

ENERGY UTILIZATION INDEX (EUI) — 39,808 Btu/sq ft

Figure 59-10. Energy auditors collect total energy usage for comparison to the average usage of similar users within the region.

Energy Utilization Index. The *energy utilization index (EUI)* is a ratio of the energy used in a building annually divided by the floor area of the building. The EUI is useful when comparing buildings within a region that have similar systems and uses. The EUI of a building can be compared during changing climate conditions and over time. The EUI of commercial buildings can be measured against the level of use of the building.

Energy Cost Index. The *energy cost index (ECI)* is a ratio of the dollars spent annually on energy used in a building divided by the floor area of the building. The ECI is useful when comparing buildings within a facility or organization that have similar uses. For instance, a school system may have an ECI of $4/sq ft. A neighboring school system may have an ECI of $5/sq ft. Proposed retrofit projects should include a review to determine why ECI discrepancies exist.

Utility Cost Calculations

Utility costs to a facility are typically billed monthly. In general, utility costs are calculated by multiplying the amount of energy consumed by the cost charged per unit. Other factors may be included in the utility billing, depending on the service type and the amount of energy used. Utility billing for electricity is typically more complicated than for natural gas.

Electricity Cost Calculations. The monthly electric bill total for a residential building is typically the cost of electricity consumed. The monthly electric bill total for a commercial building is typically the sum of the costs of the electricity consumed, electrical demand, power factor, and fuel recovery rate. Total usage is determined by multiplying the electricity consumed (in kWh) by the electricity rate, which sometimes is adjusted for the time of day, season, and/or usage rate.

A component in the usage charge is the total fuel recovery charge. The fuel recovery charge is the amount the utility company is permitted to charge, which reflects the constantly changing cost of energy. The total fuel recovery charge is determined by multiplying the total electricity usage by the fuel recovery charge (in ¢/kWh). Electricity demand rates sometimes are adjusted based on the time of year and/or the amount of electricity used. **See Figure 59-11.**

Large Industrial User Rates

ACME Electric Utility

Large Commercial Rate: This rate applies to customers having a demand of 100 kW or greater during the previous 12 months.

Usage:	Summer	Winter
First 40,000 kWh	9 cents/kWh	8 cents/kWh
Next 60,000 kWh	7 cents/kWh	6 cents/kWh
All Excess	5 cents/kWh	4 cents/kWh

Power Factor/Fuel Recovery Charge 0.12 cents/kWh

Demand:	Summer	Winter
First 50 kW	$9.50/kW	$8.50/kW
All Excess	$7.50/kW	$6.50/kW

Summer rates begin in May and end in October.

Figure 59-11. Larger electric users with high demand pay additional charges for the demand.

Example: Finding Cost of Electricity

Electric utility bills were obtained for a large facility. The electric usage charge is 8¢/kWh for the first 40,000 kWh, 6¢/kWh for the next 60,000 kWh, and 4¢/kWh for additional usage. A 0.12¢/kWh fuel recovery charge is included for the month. The electricity demand charge is $8.50/kW for the first 50 kW and $6.50/kW for additional demand. The bill for November included usage of 116,000 kWh and demand of 395 kW. Find the total cost of electricity for the facility.

1. Calculate the usage charge.

   First 40,000 kWh: 40,000 kWh × 8¢/kWh = $3200

   Next 60,000 kWh: 60,000 kWh × 6¢/kWh = $3600

   All excess kWh: 16,000 kWh × 4¢/kWh = $640

   usage charge = $7440

   fuel recovery = 116,000 kWh × 0.12¢/kWh = $13,920

   total usage charge = **$21,360**

2. Calculate the demand charge.

   First 50 kW: 50 kW × $8.50/kW = $425

   All excess: 345 kW × $6.50/kW = $2242.50

   total demand charge = **$2667.50**

3. Calculate the total electricity charge.

   total electricity cost = *total usage charge* + *total demand charge*

   total electricity cost = $21,360 + $2667.50

   total electricity cost = **$24,027.50**

Natural Gas Cost Calculations. The natural gas cost for residential and light commercial buildings is determined by multiplying the natural gas usage by the natural gas recovery charge. The natural gas recovery charge is the amount the gas utility charges per unit of natural gas. This cost is typically adjusted by applying different rates for natural gas use during the summer and winter.

Example: Calculating Cost of Natural Gas

Natural gas utility bills were obtained for a large facility. A $2.77/MCF fuel recovery charge is included for the month. The bill for February included usage of 20,150 MCF. Calculate the total cost of natural gas for the facility.

total natural gas cost = *natural gas usage* × *natural gas recovery charge*

total natural gas cost = 20,150 MCF × $2.77/MCF

total natural gas cost = **$55,815.50**

AUDIT PROCESS

Energy audits are performed by homeowners and business owners, in-house staff, and energy consulting firms. General energy audits require comprehensive building energy-use surveys to evaluate equipment and devices for retrofits. An *energy audit* is a comprehensive review of the energy use in a building. Energy audits start with identifying energy-transmission and energy-consuming equipment and devices. These audits are used to identify the characteristics of loads and determine appropriate energy use reduction strategies for the loads.

Energy audits are used to evaluate retrofitting alternatives and are a way of estimating savings in energy use costs. Retrofit evaluations are adversely affected when an energy use survey of a building is incomplete or flawed. Incomplete and inaccurate energy use surveys lead to uncertain financial evaluations for retrofit projects. Technicians performing energy use surveys must understand what is included and the effective methods of collecting the information. Energy audits begin with collecting information at the source of energy and continue to each point of use.

Energy Information

Energy audits begin with identifying all energy providers as well as listing all utility meters and their location, model numbers, and serial numbers. A review of the energy bills for the past three years will determine prior energy usage. Prior energy usage establishes a basis for the retrofit evaluations and estimations of future energy uses.

General Building Information. General building information is used to create a basis for comparison with other buildings and as a reference for retrofit projects. **See Figure 59-12.** General building information that is documented in an energy use survey includes the following:

- address
- construction type, general materials, and date
- dates and types of any renovations
- size in square feet that is heated and/or cooled

Building Characteristics. An energy audit includes a simple site plan showing the building size, shape, orientation, and shading. Shade is an important consideration for solar heat load. The building condition and construction characteristics, such as construction materials, amount and type of insulation, and number and type of windows and doors, should also be indicated.

Building Occupancy Patterns. Energy audits must list the occupancy patterns of buildings. The use of the building spaces and times of day of the use, including weekends or after-hours occupancy, should be documented. **See Figure 59-13.** Building occupancy patterns are used to implement strategies for time scheduling and energy use reductions.

HVAC Controls and Systems. The existing HVAC controls and equipment systems for the building should be documented. The type of controls, operational strategies used, previous retrofits, and general condition of the system should be included.

Building Information

BUILDING INFORMATION

Year: ______

GENERAL BUILDING INFORMATION	
Name:	Age:
Address:	Heating unit size and type:
Size:	Cooling unit size and type:

BUILDING CHARACTERISTICS					
Gross floor area:	sq ft	Ceiling height:	ft	Gross volume:	cu ft
Exterior door areas:	sq ft	Window areas:	sq ft	Other openings:	sq ft
Exterior wall areas:	sq ft	Type:		Insulation:	
Roof areas:	sq ft	Type:		Insulation:	
General condition:		Type:		Insulation:	

SKETCH OF BUILDING AND SYSTEM

Locate meters and HVAC units:

Figure 59-12. Energy auditors collect building information to develop energy use comparisons and evaluate alternatives for energy use reductions.

Building Occupancy Patterns

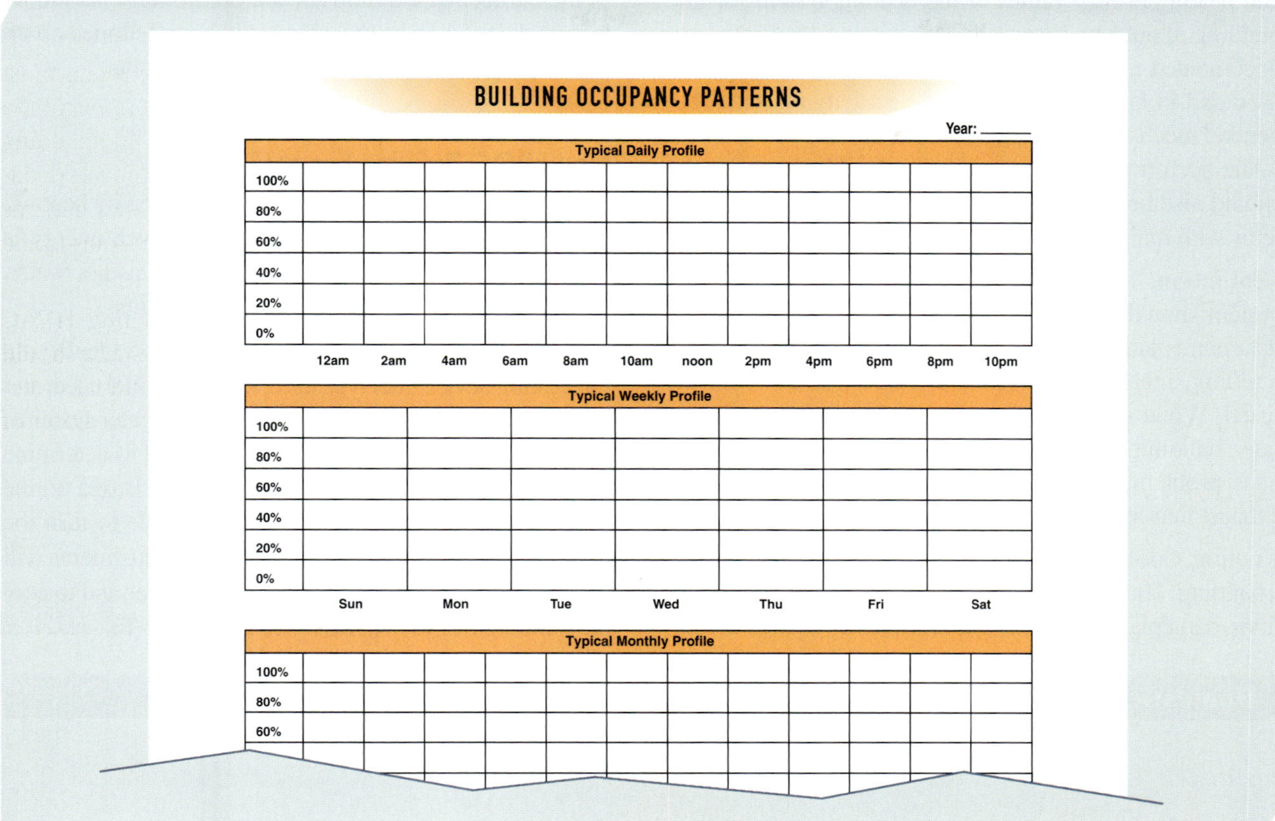

Figure 59-13. Building occupancy patterns are used to evaluate alternatives for energy use reductions.

HVAC Equipment

Packaged HVAC equipment, such as rooftop units, should be identified with the nomenclature used by the building maintenance personnel. **See Figure 59-14.** The name, type, and location of all HVAC equipment should be documented. Nameplates of HVAC units provide the voltage, current, phase, horsepower (HP), and air conditioning tonnage. The types of controls currently used for each load must be listed.

The maintenance technician should determine the importance of each load and the approximate times that each load must be ON. This information is used to develop time-based scheduling, optimization factors, electric demand control, and/or duty cycling scheduling strategies.

TECH TIP

Energy calculations are often made using British thermal units (Btu). The calculation results can then be converted into the energy units used by equipment.

Nameplates

Figure 59-14. Equipment nameplates provide power consumption information for equipment.

Heating. The number of heating devices, unit types, fuels, and heating energy values of the heating system for the building should be recorded. The heating load must be documented in British thermal units when natural gas is used and in kilowatt-hours when electricity is used. The general mechanical condition of the units should be noted while performing the energy audit. Building occupants should also be questioned about heating levels and problems with maintaining consistent heat in the building.

Ventilation. The components of the building ventilation system should be recorded. The number, size, type, and data nameplate information from the exhaust fans in the building, including the kitchen exhaust fan, should be listed. While determining the conditions of the exhaust fans, building occupants should be questioned about exhaust problems and building pressure issues. Occasionally, exhaust fans stop functioning due to lack of maintenance.

Cooling. Cooling systems typically use the most energy in a building. The number of cooling devices, unit types, locations, nameplate information, and efficiency ratings should be recorded. Information about the system equipment, such as the number of fan blowers and pumps, also should be recorded. Building occupants should be questioned about any cooling problems and prior system retrofits.

Non-HVAC Building Energy Use

A large amount of energy in a building is used by heating, ventilation, and cooling equipment. However, energy is also used for non-HVAC applications such as hot water, lighting, and plug loads for building occupants.

Domestic Hot Water. Energy audits involve identifying the heating device, fuel type used, amount of fuel used, and amount of hot water used by domestic hot water systems. The discharge water temperature is checked to determine whether it is too high for the building type. Based on the hot water usage patterns, it may be possible to turn the system temperature down or off at night. Adjustments will reduce the energy consumed to heat the water and to cool the building due to heat loss. **See Figure 59-15.**

Domestic Hot Water Use

DOMESTIC HOT WATER INFORMATION

Year: ______

HOT WATER SOURCE	
Location:	Age:
Energy source:	Major uses:
Size:	Other uses:

USE CHARACTERISTICS		
Daily usage: gal./day	Temperature at heater: °F	Temperature at farthest use: °F
No. of water use units:	Unit size: gpm	Farthest use distance: ft
Tanks wrapped:	Type:	Insulation:
Pipes wrapped:	Type:	Insulation:
General condition:	Type:	Insulation:

SKETCH OF BUILDING AND SYSTEM

Locate source and points of use:

Figure 59-15. Energy audits measure domestic hot water energy used compared to the total amount of water used. Domestic hot water requires energy to heat the water and energy to cool the building from heat lost from the hot water.

Lighting. Energy audits involve identifying the area served, the type of lighting, and the wattage, voltage, and current of each lighting service load. The general condition of the lighting loads and the type of controls used for each lighting load should be noted. Often replacing fixtures and adding light controls, such as occupancy sensors, will reduce lighting use and reduce energy costs. **See Figure 59-16.**

Plug Loads. Energy audits involve identifying the electric plug loads in the building or facility. Energy audits must include the number, type, and electrical value of each major plug load. Plug loads range from copier and printer machines to multiple electric heaters at the desks of occupants. Multiple heaters at occupant workstations often indicate problems with the HVAC or control system. **See Figure 59-17.**

Generators. Energy audits involve locating building generators and documenting their condition. Building generators are used when the electric supply from the utility company goes down. Building generators are often tested on a regular basis. Some users generate their own electricity rather than use electric power from an electric utility company.

Refrigeration. Energy audits must include an analysis of building product refrigeration systems. Some buildings use a considerable amount of energy to refrigerate products. Facility applications range from cafeterias with multiple freezers and coolers to large cold-storage facilities.

Lighting Information

LIGHTING INFORMATION

Year: ______

GENERAL ELECTRICITY SOURCE	
Location:	Age:
Controls:	Major uses:
Size:	Other uses:

LIGHTING CHARACTERISTICS	
Gross floor area: sq ft	Number of rooms:
General condition:	Number of fixtures:

Room	No. Fixtures	Fixture Types	Fixture Wattage	Condition

SKETCH OF BUILDING AND SYSTEM

Locate breakers and fixtures:

Figure 59-16. Energy audits can be used to evaluate the lighting used in a building. A significant reduction in energy use is possible by replacing standard light bulbs with low-energy-use fixtures.

Plug Load Information

PLUG LOAD INFORMATION

Year: ______

GENERAL ELECTRICITY SOURCE	
Location:	Age:
Controls:	Major uses:
Size:	Other uses:

PLUG LOAD CHARACTERISTICS	
Gross floor area: sq ft	Number of rooms:
General condition:	Number of plug loads:

Room	No. Plug Loads	Plug Load Types	Plug Wattage	Uses

SKETCH OF BUILDING AND SYSTEM

Locate breakers and fixtures:

Figure 59-17. Energy audits can be used to evaluate all of the plug loads in a building to determine the power usage and develop alternatives to decrease that usage.

Equipment Operation and Maintenance

Energy audits include information about energy using equipment. Energy auditors review the maintenance programs for the energy using equipment to determine the condition and remaining service life of the equipment. **See Figure 59-18.** The technician performing the building energy audit can obtain valuable information by asking occupants basic questions, listening carefully to the responses, and engaging in careful observation. Basic questions can often lead to useful information. Questions typically asked during a building energy audit include the following:

- What are the major heating/cooling problems?
- What parts of the building become excessively hot or cold?
- What are the likes and dislikes of the present HVAC control system?

Equipment Condition. Energy auditors document the condition of the current HVAC equipment. Items such as missing doors, loose access panels, and empty refrigerant cylinders next to equipment may indicate poorly functioning and poorly maintained equipment. Areas identified as continually too hot or too cold may have mechanical or control problems that require correction. Energy auditors must observe and ask whether the existing controllers, such as thermostats, are continually readjusted.

Equipment Operation. The continual adjustment of controllers results in poor temperature control and excessive energy use. Controller adjustment is corrected by using a building automation system with no access to change the setpoints. Maintenance technicians know whether occupants adjust the thermostat setpoint when it is time to leave for the day. With some systems, maintaining the occupied setpoints during unoccupied times (at night) needlessly uses energy. This can be corrected by installing a building automation system to automatically adjust the setpoints. Any windows open to the outside air is an indication of poor temperature control and leads to increased energy use.

Preventive Maintenance Programs

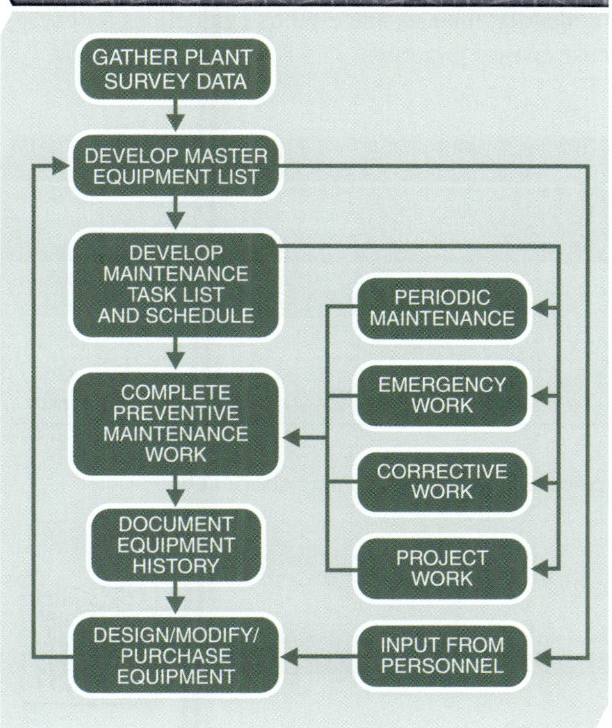

Figure 59-18. A preventive maintenance program can be continually improved through the knowledge and experience of maintenance personnel.

Occupant Comfort Levels. Energy auditors observe the dress of individuals in a building as an indicator of building comfort levels. Employees who are overly dressed indicate that the building space is too cold. The use of desk fans in the summer signifies that the building space is too hot.

When buildings are remodeled, mechanical systems are sometimes divided and reallocated, sometimes in an inconsistent manner. Poorly reallocated HVAC systems make it difficult to provide temperature control to various areas. Poorly remodeled buildings may prevent the mechanical system from maintaining proper temperatures regardless of system operation. Poor air distribution areas must be identified prior to the installation of a building automation system.

AUDIT REPORTS

An energy audit report is complete documentation of the energy audit and data analysis. Energy audit reports typically include comparisons of building energy use to industry averages and standards. Energy use comparisons can serve as guides for implementing projects designed to reduce energy use.

For example, the home energy use of different homes is compared using the home energy rating system (HERS) index. **See Figure 59-19.** The HERS index for a rated home is the ratio of its net energy use to the net energy use of a reference home. The reference home is a new home built to the minimum 2004 International Energy Conservation Code (IECC) and 2006 SEER standards set forth by the Department of Energy (DOE) for the location of the rated home.

Home Energy Rating System (HERS) Index

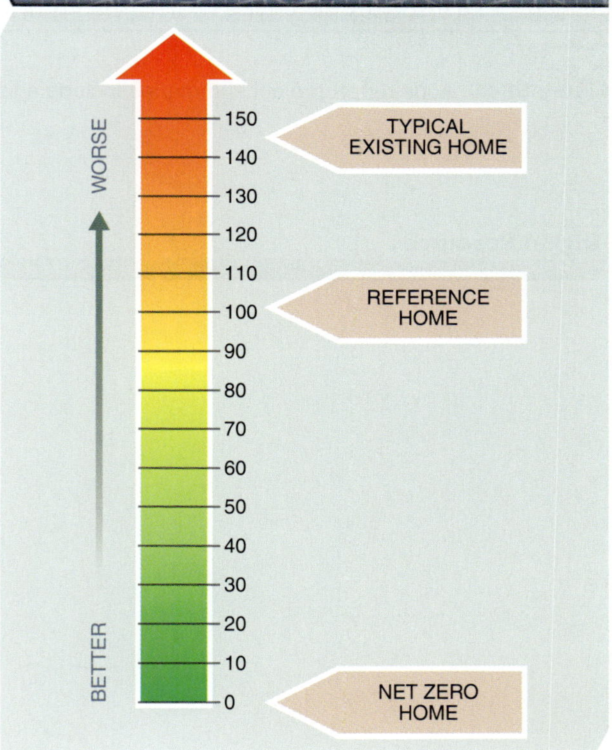

Figure 59-19. The HERS index is a method used to compare the energy used in a home to a referenced standard home.

Home Energy Rating System Media Clip

Energy Cost Savings Proposals

The energy audit report summarizes all of the opportunities to reduce energy costs. The report typically includes a prioritized action plan of retrofit projects with the expected savings for each project. A summary of identified energy savings projects are provided with appropriate graphs, charts, and supporting calculations. **See Figure 59-20.**

The completed energy audit is used to develop specific proposals for identified energy cost saving projects in the facility. The implementation costs and energy cost savings in dollars of each project are considered, and projects are ranked by the payback period. Projects with a payback period of less than three years are often implemented, and projects with a payback period of more than five years are typically deferred until other reasons warrant the retrofit. If the energy audit is performed by an energy savings performance contractor (ESPC), the savings may be guaranteed. ESPCs typically help to identify, finance, and establish guidelines for energy cost savings programs.

Audit Report Project Summaries

Priority	Project Description	Project Installed Cost	Projected Monthly Savings	Payback Period*	One-Year ROI
1	Replace 10 standard light bulbs with CFL bulbs	$ 15.00	$ 1.50	10.0	120%
1	Replace thermostat with programmable unit	$ 200.00	$ 15.00	13.3	90%
2	Retrofit 4 light fixtures with fluorescent lamps	$ 150.00	$ 6.00	25.0	48%
2	Retrofit CAV fan with VAV fan controls	$ 65.00	$ 18.00	36.1	33%
3	Retrofit existing SEER 10 AC with SEER 16 AC	$3900.00	$ 65.00	60.0	20%

* in months

Figure 59-20. Audit report project summaries include a list of projects with the expected payback periods.

Digital Resources

ATPeResources.com/QuickLinks • Access Code: 482610

UNIT

HVAC and Refrigeration System Efficiency and Retrofitting

60

The energy efficiency of HVAC equipment is vital in lowering building energy costs and a critical factor in the U.S. government energy policy. The selection of HVAC equipment and retrofits reflect the anticipated increase in the cost of energy. Retrofitting HVAC equipment usually includes removing old equipment and replacing it with more efficient new equipment. Often the extent of the retrofit and connections to the existing system must be carefully planned. Retrofitting HVAC equipment also includes adding components to an existing system to enhance its performance.

OBJECTIVES

- Describe the following acronyms: EER, SEER, COP, AFUE, and HSPF.
- Describe the Energy Star® program.
- List the different Leadership in Energy and Environmental Design® (LEED®) strategies and certification levels.
- List common HVAC system retrofits.
- Describe common methods of retrofitting air-handling units and building automation systems.

Index to Procedures

HVAC EQUIPMENT

HVAC systems are used to condition air by maintaining the proper temperature, humidity, and air quality. Properly conditioned air improves the comfort and health of building occupants. Equipment energy regulations have been updated over the years to raise the standard of efficient energy use. The American Society of Heating, Refrigerating, and Air-Conditioning Engineers (ASHRAE) developed several standards for indoor air. Energy savings measurements vary by equipment type.

HVAC systems are designed to operate at optimum energy efficiency while maintaining desired environmental conditions. Equipment efficiency ratios and standard measurements developed by the Air-Conditioning, Heating, and Refrigeration Institute (AHRI) include the following:

- energy efficiency ratio
- seasonal energy efficiency ratio
- coefficient of performance
- annual fuel utilization efficiency
- heating seasonal performance factor

Refrigeration Equipment

Refrigeration equipment is used in closed systems to control the pressure and temperature of refrigerant in order to regulate its absorption and rejection of heat. Refrigeration equipment energy-efficiency measurements include energy efficiency ratios, seasonal energy efficiency ratios, and coefficients of performance. **See Figure 60-1.**

Cooling Efficiency Ratings

Rating	Definition	Units
Seasonal energy efficiency ratio (SEER)	$\frac{\text{Total cooling output (Btu)}}{\text{Total electrical energy input (Wh)}}$	Btu/Wh
Energy efficiency ratio (EER)	$\frac{\text{Cooling output (Btu/h)}}{\text{Electrical power input (W)}}$	Btu/Wh
Coefficient of performance (COP)	$\frac{\text{Cooling output (W)}}{\text{Electrical power input (W)}}$	none

Figure 60-1. The cooling output is compared to the energy input of air conditioners and heat pumps to determine the efficiency ratings.

Energy Efficiency Ratios. An energy efficiency ratio (EER) is a ratio of the output cooling (in Btu/hr) of a cooling system to the input electrical power (in Wh) at a specific operating condition. The EER uses a value of 95°F outside temperature and a return air temperature of 80°F with 50% relative humidity. Since conditions are fixed, the EER does not reflect partial load conditions at lower temperatures.

Seasonal Energy Efficiency Ratios. A seasonal energy efficiency ratio (SEER) is a ratio of the annual output cooling (in Btu) of a cooling system to the annual input electrical power (in Wh) over the temperature and humidity for a typical year in a given location. Today, air conditioning system efficiency is typically rated using the SEER. SEER represents the overall performance of refrigeration equipment in a typical year in a given location. SEER uses the same indoor temperature but over a range of outside temperatures from 65°F to 104°F. Higher SEER numbers indicate that the unit is more energy efficient.

Carrier Corporation

Considerable energy savings are possible when replacing an older, low SEER HVAC unit with a new, high SEER unit.

Example: Finding Operating Costs

A 7000 Btu/hr air conditioning unit with a SEER of 14 runs a total of 1000 hr during an annual cooling season. Find the hourly operating cost.

1. Find the annual cooling output.

   annual cooling = *cooling rate* × *annual season usage*

   annual cooling = 7000 Btu/hr × 1000 hr/yr

   annual cooling = **7,000,000 Btu/yr**

2. With a SEER of 14, find the annual electric used.

   annual electric used = *annual cooling* ÷ *SEER*

   annual electric used = 7,000,000 Btu/yr ÷ 14 Btu/Wh

   annual electric used = **500,000 Wh/yr**

3. Find the average power used.

   average power used = *cooling rate* ÷ *SEER*

   average power used = 7000 Btu/hr ÷ 14 Btu/Wh

   average power used = **500 W or 0.5 kW**

4. Using the electricity cost of 16¢/kWh, find the hourly operating cost.

   hourly operating cost = *average power used* × *power cost*

   hourly operating cost = 0.5 kW × 16¢/kWh

   hourly operating cost = **8¢/hr**

Energy Efficiency Table Interactive Animation

Coefficients of Performance. The EER and SEER are related to the coefficient of performance. The coefficient of performance (COP) is the ratio of heat energy removed from an indoor environment per unit of energy used to operate the unit; measured in watts. A SEER of 13 roughly equals an EER of 11 and a COP of 3.43. This means that 3.43 units of heat energy are removed from the indoor environment per unit of energy used to operate the unit.

SEER Standards. SEER ratings more closely approximate system efficiency across an entire year, while the EER is measured at one operating condition. While both ratings are available and used, the SEER is a better measurement overall. The U.S. Government has mandated that all residential air conditioners, except window units, sold in the United States have a SEER of at least 13. An Energy Star® air conditioning unit has a minimum SEER of 14. Higher SEER units have larger coils and may have multiple compressors. Typically, higher SEER units are more expensive to purchase. **See Figure 60-2.**

Efficiency-Standard Comparisons

Three-Phase Equipment Size*	Heating Category	ASHRAE 90.1-2010	Energy Star May 2010
Less than 65	Electric	13 SEER	14 SEER
	Gas	13 SEER	14 SEER
65 to 135	Electric	11.2 SEER	11.7 SEER
	Gas	11.0 SEER	11.5 SEER
135 to 240	Electric	11.0 SEER	11.7 SEER
	Gas	10.8 SEER	11.5 SEER
240 to 760	Electric	10.0 SEER	†
	Gas	9.8 SEER	†
Greater than 760	Electric	9.7 SEER	†
	Gas	9.5 SEER	†

* in thousands of Btu/h
† future specifications

Figure 60-2. The SEER efficiency standard is based on the capacity of the equipment.

Older units may be rated at SEER 9 or below. Replacing an older unit with a new unit that has a SEER of 13 will result in hundreds of dollars a year in energy cost reductions. Actual costs will vary depending on the local climate and cost of electricity. Older units may deteriorate rapidly or fail without warning, which increases maintenance costs. All units must be adequately maintained to prevent drops in efficiency.

In some instances, it may be cost effective to keep a lower efficiency unit for the time being. Replacing well-maintained older units that are used infrequently does not result in enough energy cost savings to pay for the replacement unit. Some unit replacements require costly system modifications that are best done with major building renovation.

Today, units are available with a variety of SEER ratings. Higher efficiency units with SEER ratings above 20 are available, but they are expensive. Residential air conditioning units with SEER ratings up to 26 are available today. Advanced technologies like inverters, rotary compressors, DC brushless motors, variable-speed drives, and electronic control systems are used to create higher SEER-rated units. Careful analysis of the length of cooling season, climate, and electricity cost should be performed to justify the initial cost of ultra-high-efficiency models.

Heating Equipment

Heating equipment is used to raise the temperature in buildings. Heating equipment energy efficiency standards and measurements include the annual fuel utilization efficiency and heating seasonal performance factor.

Annual Fuel Utilization Efficiency. The annual fuel utilization efficiency (AFUE) is a measurement of the thermal efficiency of fossil fuel combustion heating equipment such as boilers, furnaces, and water heaters. AFUE determination methods are defined in ASHRAE standard 103. **See Figure 60-3.** Fuels include natural gas, fuel oil, propane, and liquefied petroleum gas. The AFUE is different from a standard thermal efficiency measurement in that it represents the equipment efficiency during the entire heating season at a given location. The AFUE for heating equipment is similar to the SEER for cooling equipment. Heating devices with higher efficiency have a higher AFUE.

HVAC Service Order/Invoice Forms and Documents

Furnace Annual Fuel Utilization Efficiency (AFUE)

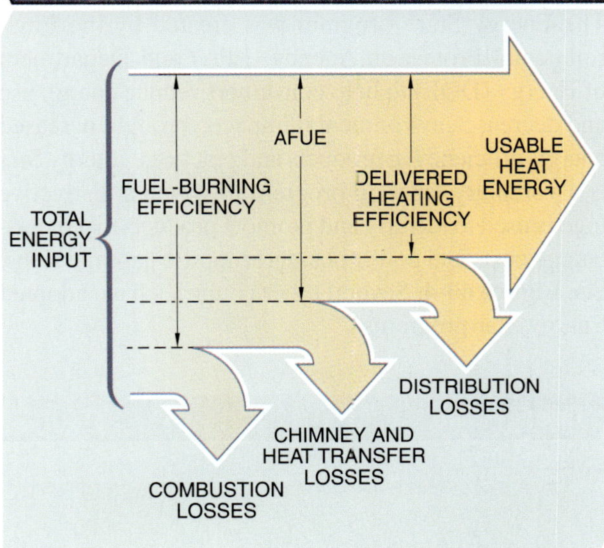

Figure 60-3. The AFUE of a furnace is the portion of the total energy input that results in hot supply air.

Heating Seasonal Performance Factor. The *heating seasonal performance factor (HSPF)* is a ratio of heat output (in Btu) to the electricity used (in Wh) in the efficiency measurement of air-source heat pumps. The HSPF is measured in units of Btu/Wh. The HSPF standard is defined in AHRI standard 210/240, *Performance Rating of Unitary Air-Conditioning and Air-Source Heat Pump Equipment.* Like the SEER and AFUE, HSPF is measured over the entire heating season. More efficient units have higher HSPF numbers. Heat pumps with an HSPF of at least 8.5 are eligible for a U.S. energy tax credit.

The HSPF can be converted to a COP across the entire heating season by converting both the Btu heat output and the electrical input to the same energy unit, which is Joules. The COP is primarily used by design engineers. The HSPF is the measurement used to evaluate efficiency. The COP provided by the manufacturer is an instantaneous measurement of one fixed condition, while the HSPF is based on an entire heating season at a given location.

ENERGY EFFICIENCY STANDARDS

Today, there are a number of construction and building standards used that encourage the evaluation and implementation of energy efficiency strategies. Two of the most important standards include Energy Star® and Leadership in Energy and Environmental Design® (LEED®).

Energy Star®

The Energy Star® program was created by the Environmental Protection Agency (EPA) and Department of Energy (DOE) to help consumers reduce energy use and decrease environmental impacts through increased energy-efficiency in products and practices. Energy Star is a voluntary labeling program in which a distinctive logo is used to identify and promote products that reduce energy costs and power plant greenhouse gas emissions. **See Figure 60-4.** Several other countries have adopted Energy Star programs.

Energy Star®

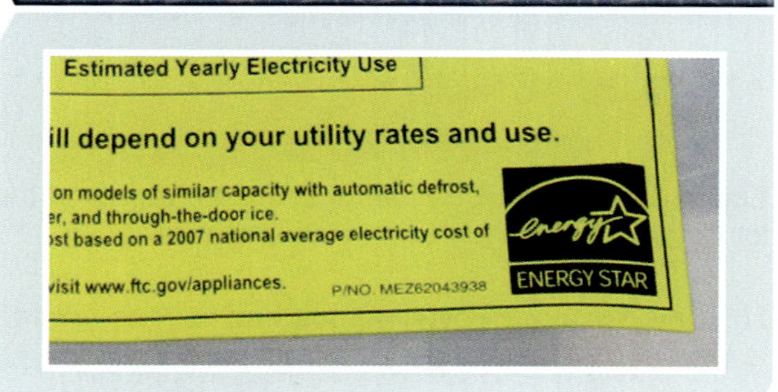

Figure 60-4. The Energy Star logo is placed on products that comply with Energy Star standards.

Products that meet the Energy Star standard typically reduce energy use by 20% to 30% when compared to non-Energy Star products. Energy Star standards apply to more than 60 product categories from residential-consumer electronics to buildings.

According to public estimates developed by the EPA, energy costs have been reduced by billions of dollars each year. The Energy Star program has increased the use of devices such as LED traffic lights and compact fluorescent lighting (CFL). The program also includes a component that encourages using renewable power sources. Energy Star specifications, which are different within each product category, are determined by either the EPA or DOE.

Products. Some of the products included in Energy Star programs are computers, servers, appliances, home electronics, imaging equipment/printers, lighting, and home office equipment. Products receive specific labels that provide consumers with performance ratings for product comparison. For example, the National Fenestration Rating Council (NFRC) certifies the energy performance values of windows, doors, and skylights. **See Figure 60-5.**

Product Performance Labels

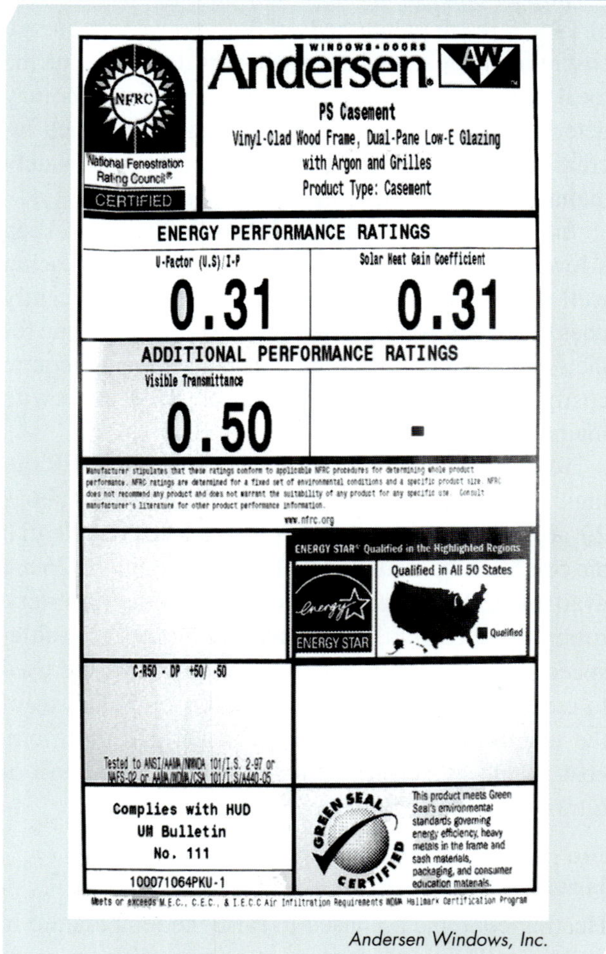

Andersen Windows, Inc.

Figure 60-5. The insolating value of window glazing is included on the NFRC label.

Buildings. Buildings can receive energy-use scores. Homes that meet strict guidelines for energy efficiency by using at least 15% less energy than standard homes can qualify for Energy Star certification. The DOE developed a home energy scoring tool that allows qualified assessors to generate credible energy assessments of homes at reasonable costs. **See Figure 60-6.** A home energy assessment helps consumers compare the energy use of different homes and provides recommendations on possible cost reduction upgrades. Examples of Energy Star buildings include residential buildings, commercial and industrial buildings, data centers, banks, government hospitals/health care centers, hotels, schools, offices, college buildings and dormitories, retail stores, supermarkets, and warehouses.

DOE Home Energy Scores

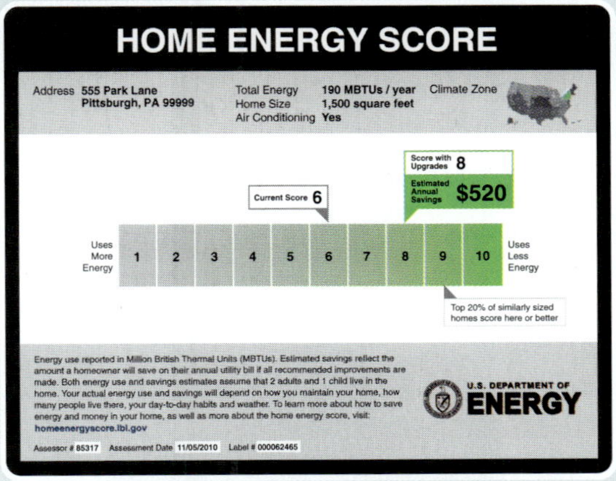

U.S. Department of Energy

Figure 60-6. The Home Energy Score is a rating system for the energy use of a home. Rating systems were developed to provide a method of comparing homes and a basis for possible improvements.

Home Energy Score Media Clip

Heating and Cooling Systems. Heating and cooling systems can receive Energy Star certification. Energy Star systems include heat pumps, air conditioners, furnaces, and boilers. Many consumers purchase room air conditioning units, with Energy Star labeling, that are at least 10% more energy efficient when compared to units that only meet the minimum energy standards.

Energy Star Issues. While the Energy Star program overall has been a success, there have been a number of controversies and inconsistencies. Controversies include inaccurate and unsubstantiated claims of the reduction in greenhouse gas produced and energy consumption rate. Inconsistencies include conditions under which energy consumption rate tests were performed and the self-testing allowed by manufacturers for Energy Star certification. The Energy Star program is evaluating changes to address these issues.

Leadership in Energy and Environmental Design®

Leadership in Energy and Environmental Design® (LEED®) is a green building certification system that awards points based on specific energy and environmental construction types and activities. The LEED standard was developed and is maintained by the U.S. Green Building Council® (USGBC®). LEED is one of the most important energy and environmental standards used in building construction today.

Building projects constructed to LEED standards include over 7000 buildings and over 1.5 billion sq ft of building space to date. **See Figure 60-7.** An online directory of U.S. LEED-certified projects is maintained by the USGBC. LEED has recently developed a standard for existing buildings known as LEED-EB. LEED standards are used in the United States, Canada, and several other countries.

LEED® Recognition

Figure 60-7. Buildings with high levels of energy efficiency may promote this achievement through standardized awards offered by independent organizations.

LEED standards mandate independent, third-party verification that the construction strategies have taken place. Each strategy has measurable targets and goals that are regularly evaluated. Examples of LEED construction strategies include the following:

- energy-use reduction
- water-use reduction
- CO_2 emissions reduction
- improved indoor environmental quality
- evaluation of environmental impact

The USGBC offers a series of exams for individuals to achieve accreditation in the LEED ratings system. Depending on the test they pass, individuals receive LEED Accredited Professional (LEED AP) or LEED Green Associate (LEED Green Assoc.) accreditation.

LEED Rating System. There are four LEED certification levels available for commercial buildings. In LEED 2009, also known as LEED version 3, the certification levels for commercial buildings are the following:

- certified—40 points to 49 points
- silver—50 points to 59 points
- gold—60 points to 79 points
- platinum—80 points and above

Some concentration areas that are covered in the LEED standard include the following:

- green building design and construction
- green interior design and construction
- green building operations and maintenance
- green neighborhood development
- green home design and construction

LEED certification levels are determined by the specific number of points achieved. LEED projects are awarded points in multiple categories such as sustainable sites, water efficiency, energy and atmosphere, materials and resources, and indoor environmental quality. **See Figure 60-8.**

LEED® Measurement Areas

- SUSTAINABLE SITES
- WATER EFFICIENCY
- ENERGY & ATMOSPHERE
- MATERIALS & RESOURCES
- INDOOR ENVIRONMENTAL QUALITY
- LOCATION & LINKAGES
- AWARENESS & EDUCATION
- INNOVATION IN DESIGN
- REGIONAL PRIORITY

Figure 60-8. The LEED rating system is used to reward the efficient use of energy in multiple categories.

LEED Certification Process. LEED certification for buildings can be obtained by submitting an application that includes documentation that requirements of the rating system have been met in specific areas and includes an administrative fee. Certification is granted by the Green Building Certification Institute (GBCI). The GBCI acts as an independent third party that ensures projects comply with LEED requirements.

Most of the certification process is completed online, including the application and review. The USGBC has web-based tools that allow easy collection of the required documentation. Online tools also permit fast communication between the facility representatives and the USGBC. There are many advantages and disadvantages to LEED certification. **See Figure 60-9.**

LEED®-Certified Buildings

ADVANTAGES

- ENERGY EFFICIENT
- LOWER OPERATING COSTS
- LOWER LIFE-CYCLE COSTS
- SAVES RESOURCES
- HEALTHIER WORK/LIVING ENVIRONMENT
- BETTER AIR AND WATER QUALITY
- INCREASED COMFORT AND CONTROL
- HIGHER VALUE, HIGHER RENT, AND HIGHER OCCUPANCY

DISADVANTAGES

- HIGHER DESIGN AND CONSTRUCTION COSTS
- LEED® PRINCIPLES NOT WELL-UNDERSTOOD BY PARTICIPANTS
- REQUIRES RESEARCH ON PRODUCTS
- ADDED FRICTION BETWEEN OWNER, DESIGNER, AND BUILDER
- SHORTAGE OF DESIRED MATERIALS
- COST TO OBTAIN CERTIFICATION
- NOT COMMON IN ALL REGIONS
- LEED® SYSTEM NOT FLEXIBLE
- COST OF COMPLIANCE INCREASING RAPIDLY

Figure 60-9. High-performance buildings receive LEED certification.

LEED Advantages. LEED-certified buildings have many advantages such as the following:

- savings due to energy efficiency; savings of 25% to 30% are commonly reported
- lower operational costs offset the higher initial costs of LEED construction
- lower total life-cycle cost of ownership when all aspects of building costs are included; investment typically pays for itself multiple times
- certification can be used as a real estate selling point
- resources are saved when compared to buildings built to minimum code requirements
- healthier work and living environments, which contribute to higher employee productivity, health, and comfort
- better air and water quality
- reduction in the amount of solid waste generated
- increased comfort due to better temperature control and higher quality lighting
- higher rental and occupancy rates

LEED Disadvantages. LEED certification and efforts to achieve it have many disadvantages such as the following:

- increased initial design and construction costs
- principles are often not well-understood by the design professionals undertaking the project
- construction may require time to be spent on research
- aspects of the process can lead to friction between the client, contractors, and design team
- possible shortage of required specific materials
- added costs related to correspondence, LEED consultants, and the hiring of contractors such as a commissioning authority
- slow acceptance in smaller areas compared to large metropolitan areas
- rating system does not provide enough flexibility to accommodate a wide variety of local climate conditions such as climates in the northeastern and southwestern regions of the United States
- requires funds that could be used in other aspects of the building construction process
- costs of compliance and certification have increased rapidly

Incentive Programs. There are many types of incentives implemented by federal, state, and local government agencies. These incentives include tax credits and breaks, grants, low-interest loans, reduced fees, and priority issued permits. **See Figure 60-10.**

Energy Efficiency Incentives

Type	Form	Description
Rebate	Single payment	Offsets portion of equipment purchase price
Grant	Single payment	Offsets portion of project costs, regardless of mix of material, equipment, and labor cost required for project
Loan	Low interest rate	Reduces cost of borrowing money for investing in efficiency projects
Tax incentive	Reduced tax liability	Reduces amount of taxes owed

Figure 60-10. Incentive programs are available to help building owners finance energy retrofit projects.

RETROFITTING RESIDENTIAL AND LIGHT COMMERCIAL HVAC SYSTEMS

When old HVAC equipment is removed, it may be possible to reuse some or all of the controls, gas piping, electrical devices, and venting for the new HVAC equipment. Care must be taken to ensure that all reused devices and system components comply with the current applicable codes and manufacturer recommendations. Failure to comply could lead to code violations and a revocation of the manufacturer's warranty.

Many contractors take the approach that, when there is doubt, an old device should be replaced with a new one. Contractors want to avoid starting up new HVAC equipment that fails due to an old, reused controller that failed. Any possible savings are negated by the cost of emergency service.

Older residential and light commercial HVAC systems are typically considered for upgrade by retrofitting the equipment. There are several mechanical systems that can be added to the heating or cooling system after initial installation that will improve operations. Retrofitting systems, or later adding to the HVAC systems, may improve the comfort, efficiency, or indoor air quality of buildings. Technicians need to understand the process of retrofitting an HVAC system, HVAC operation, and HVAC service and maintenance. Equipment can be retrofitted to improve building relative humidity, indoor air quality, and system controls.

Humidity

Standard residential and light commercial forced-air systems only provide passive control of the humidity in a space. When air is cooled during the cooling season, the indoor humidity level is lowered. When air is heated during the heating season, the indoor humidity level is lowered. During heating season, low-humidity cold air is drawn into the building and reduces the humidity inside. One of the major problems with forced-air heating systems is low humidity.

Low humidity can lead to many issues including health problems such as sore throats, cracked and itchy skin, and allergy problems. Colds and other infections can occur more readily when the nose and throat are dried out. Also, building fixtures and other objects, such as furniture and wood flooring, may be damaged by low humidity levels. For positive humidity control, humidifiers and/or dehumidifiers are required. **See Figure 60-11.**

Humidifiers. Humidifiers are used to increase the moisture content of indoor air. Humidifiers are typically integrated into the forced-air system of a building. Humidistats are automatic regulating devices that maintain the relative humidity in an area.

Residential and light commercial buildings up to 5000 sq ft typically use bypass-type humidifiers. A bypass-type humidifier uses a duct that connects the return air duct to the humidifier located on the discharge side of the unit. Water supply lines are connected to the humidifier. Some humidifiers have panels that allow water to drip across the airflow. Moisture is evaporated by hot dry air at the furnace discharge to increase the humidity of the air. Excess water is directed to a drain at the bottom of the panel. Humidistat controls are mounted in the conditioned space to control the humidity levels.

Residential and commercial buildings over 5000 sq ft may use steam humidifiers. Steam humidifiers use electrodes to heat water. The heated water creates steam inside a sealed container, and the steam is then introduced into the ductwork through a tube. Humidistat controls cycle the unit on and off as needed. Safety controls and fan interlocks are used to prevent humidifier operation when the fan is OFF or when the discharge humidity is too high.

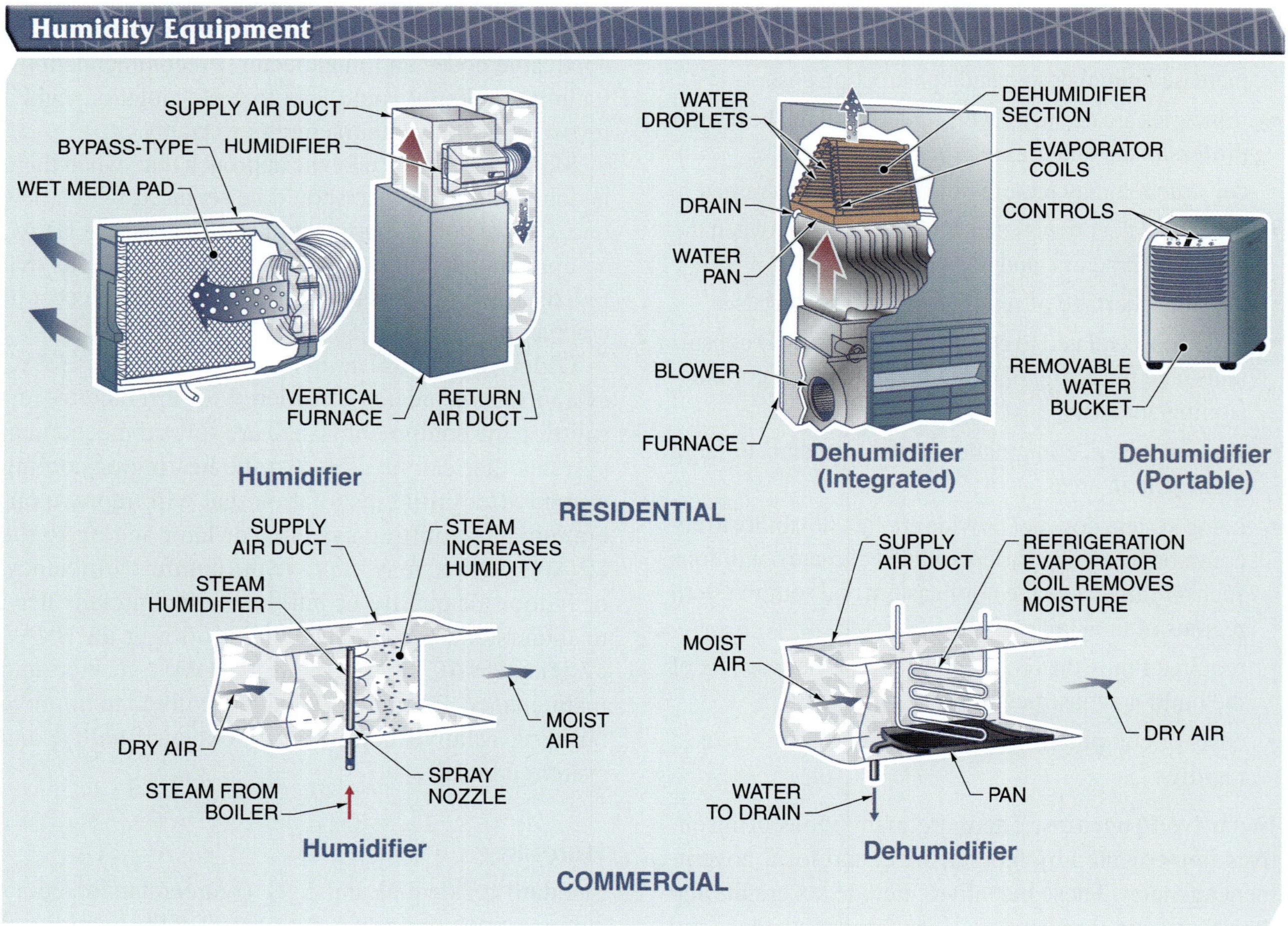

Figure 60-11. Humidifiers and dehumidifiers control the level of humidity in buildings.

Today, many humidistats are accurate digital devices with LCD readouts. The desired humidity setpoint in buildings is related to the current outside air temperature. As outside air temperatures decrease, the humidity setpoint in the building or space should decrease as well. Lower indoor humidity levels prevent condensation on windows, which is a sign of excessive humidity. Indoor humidity levels typically are maintained at 30% to 50% relative humidity. Maintaining controlled indoor air humidity levels during the heating season permits lower temperature setpoints while maintaining a good range of comfort.

Dehumidifiers. Dehumidifiers are used to decrease the moisture content of the air. Typically, dehumidifiers are portable units but can be integrated into the forced-air system of buildings. High daily relative humidity is typical in parts of the South and Southeast (United States) during summertime. High humidity in buildings causes mold growth and the deterioration of metal, wood, and carpeting. It also will lead to the discomfort of building occupants. In most cases, the cooling system will reduce the humidity levels in the space through latent heat removed by the evaporator as the air conditioner cycles normally to satisfy the temperature setpoint of the building. System effectiveness depends on proper equipment sizing and well-distributed airflow.

In some buildings, latent heat removed by the evaporator is insufficient due to extremely high ambient humidity levels, wrong mechanical system sizing, or inconsistent airflow distribution. Dehumidification is then provided using a portable or central dehumidifier. A dehumidifier consists of a small refrigeration system with a drain that removes moisture as it condenses on a coil. Portable dehumidifiers can be moved to a space with a humidity problem. Central dehumidifiers are centrally controlled and lower humidity levels in the entire building.

Indoor Air Quality

Efforts to achieve or maintain indoor air quality (IAQ) include the removal of pollutants from the air and pollution prevention. Pollutants include dust, dirt, animal dander, formaldehyde, radon, household chemicals, odors, cigarette smoke, and other contaminants. Pollutants may cause allergies, headaches, nausea, and possibly cancer.

The EPA and industry agencies have performed health studies to develop guidelines for HVAC manufacturers and installers. Buildings built in recent years are sealed tighter to reduce energy loss. This also reduces the natural air exchange between the outside and inside. Methods that improve IAQ in houses and light commercial facilities include the use of air filters, electronic air cleaners, energy recovery ventilator (ERV) systems, and ultraviolet (UV) light treatment systems.

Air Filters. Typical IAQ devices used today are air filters. An *air filter* is a device that contains a porous substance through which air can pass and is used to remove particles from the air. **See Figure 60-12.** The particle size is measured in microns. A micron (μ) is 39 millionths of an inch or 1/1000 of a millimeter. A strand of hair typically has a diameter of 70μ to 100μ. Filters are measured by the percentage of particulates they remove.

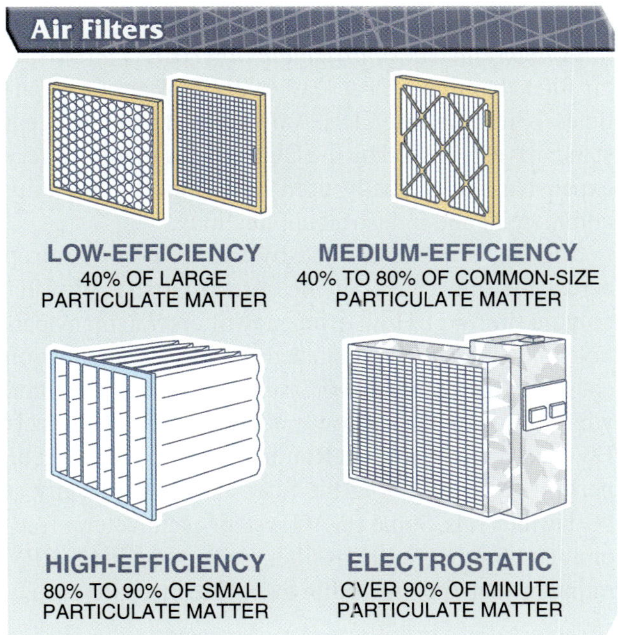

Figure 60-12. Air filters remove particulate matter from the air. The filter selected depends on the air quality requirements.

The minimum efficiency reporting value (MERV) is an ASHRAE rating system used to measure the particulate removal capacity of air filters. MERV ratings range from 1 to 16. High-efficiency filters have high MERV ratings. Higher MERV rated filters allow fewer dust particles and other airborne contaminants to pass through the filter. Some of the common particles related to MERV ratings are pollen, spores, pet dander, smog, dust, and bacteria. Testing agencies determine the MERV rating of a filter by adding particles of known sizes into a test system and measuring the filtering efficiency. **See Figure 60-13.**

MERV Ratings

MERV Rating	ASHRAE Std 50.1 Dust Spot Efficiency	Particle Size	Typical Applications	Typical Air Filter Type
13 to 16	80% to 95%	0.3μ–1μ Bacteria, smoke, droplet nuclei (sneeze)	General surgery, hospital care, superior commercial buildings	Bag filter 12″ to 36″ thick, box filter 6″ to 12″ thick, exceptional filters
9 to 12	40% to 75%	1μ–3μ Humidifier dust, auto emissions	Superior residential, better commercial, hospital laboratories	Pleated filter 6″ to 12″ thick, box filter 6″ to 12″ thick, very good filters
5 to 8	20% to 35%	3μ–10μ Mold spores, hair spray, dusting aids	Better residential, commercial buildings, industrial work space	Pleated filter 1″ to 4″ thick, cartridge filter, better filters
1 to 4	< 20%	> 10μ Pollen, dust fibers	Residential minimal filtration	Fiberglass 1″ thick, foam rubber, cheap and disposable filters

Figure 60-13. MERV ratings indicate the efficiency of a filter in removing particulate matter from the air.

A *high-efficiency particulate air (HEPA) filter* is an air filter that removes at least 99.97% of airborne particles 0.3μ in diameter. HEPA filters must meet stringent standards established by the DOE. HEPA filters are very expensive and typically used in specific high-quality commercial and industrial applications.

Air filters are also rated by the air pressure drop across the filter, which is measured in inches of water column (in. WC). High-efficiency filters trap many particles and become contaminated quickly. Contamination causes an increase in the pressure drop across the filter, which reduces the airflow rate and adversely affects HVAC system operation. Routine maintenance must be performed by changing the filter element, typically at 90 day intervals. Some HVAC system manufacturers recommend for their systems filters with a specific MERV rating or that operate within a specific pressure drop.

Electrostatic Air Filters. An *electrostatic air filter* is an air filtering system in which particles pass through electrostatic air filter media where friction gives them an electrical charge and then they are attracted to another layer of media that acts like a magnet. These filters trap smaller particles at higher efficiencies. Electrostatic air filters typically cost more than lower efficiency filters.

Electronic Air Cleaners. An *electronic air cleaner* is an air filtering system that passes an electrical charge through a wire grid inside the filter to charge particles and then attracts those particles to plates charged with the opposite charge, where they are trapped. This filter system works effectively but has a higher installation cost, a greater maintenance cost, and requires electricity to work. Electronic air cleaners effectively remove pet dander and other indoor air pollutants from buildings.

Energy Recovery Ventilators. An *energy recovery ventilator (ERV)* is a device used to introduce fresh outside air into a building and exhaust stale indoor air while minimizing the energy lost during the process. An ERV consists of two fans, filters, ductwork, and a heat exchanger that is connected to the existing HVAC ductwork. One fan brings fresh air into a building, while the other exhausts stale air to the outside. The heat exchanger transfers heat between the airstreams. During the heating season, warm exhaust air preheats the inlet air. During the cooling season, cool indoor air cools and dehumidifies the incoming outdoor air. Maintenance tasks for the ERV include changing the filters, servicing the motor, and cleaning the heat exchanger.

Trane

Electronic air cleaners use cleanable reuseable filters to remove up to 99.97% of allergens.

Ultraviolet Light Treatments. An *ultraviolet (UV) light treatment* is a treatment that involves using ultraviolet light to kill the microorganisms and biological growth that exist in the presence of moisture. UV light treatments are used in commercial and some residential buildings.

UV light treatments can be very effective in reducing biological growth inside HVAC systems such as in condensate drain pans or on cooling coils. However, UV light treatments have high installation and maintenance costs and are only effective in the direct line of sight of the lamps. A number of methods are used in which UV lamps are directed at the locations most likely to accumulate the growth of organisms. The lamps need to be checked routinely for proper operation.

System Controls

Typical HVAC system retrofits include replacing system controls such as thermostats. Many manufacturers provide a wide variety of controllers and units. Advances in microprocessor controller technology in recent years have increased system functionality and reduced costs.

Thermostats. Typical control retrofits involve replacing old thermostats with new ones. A thermostat is a temperature sensor inside a temperature-controlled space that sends a signal to a controller system in order to maintain a set temperature. In many cases, the old thermostat is an electromechanical type with limited accuracy and function. These units lack time schedule program capabilities or separate setpoints to reduce energy use.

Thermostat Functions. New thermostats have many functions and features. **See Figure 60-14.** Typical features of modern thermostats include the following:

- changeover selection—Thermostats must have a method of changing between the heating and cooling functions as outdoor air temperatures and/or load conditions change. Manual changeover systems require a switch or setting to be changed by a person. Automatic changeover systems can change the unit between heating and cooling as needed.
- touch screens—Manufacturers today offer touch screen thermostats instead of the buttons and sliders of older models. Most homeowners and building staff are used to operating touch screen devices.
- wireless operation—Wireless thermostats eliminate wire connections to HVAC units. As with any device, the manufacturer's instructions should be followed. Wireless systems consist of two devices: the wireless thermostat and the wireless receiver connected to the HVAC unit. Devices are located to limit obstructions, such as walls, and the total distance between the units. Some wireless units include a signal strength indicator.
- extra sensors—Many new thermostat models have the capability of reading extra temperature sensors that can be mounted remotely. These sensors typically sense outside and/or return air temperatures.
- filter alerts—Many new thermostats include features that will alert homeowners when the filter needs to be changed. This may be based on a set time interval such as every three months.
- time scheduling—One of the main features of modern thermostats is the ability to change setpoints based on a schedule, which reduces energy usage. New thermostats have separate time schedules available for each day of the week.
- selectable fan operation—New thermostats have a fan on/auto switch or touch screen selector that allows the selection of whether the fan will only run on a call for heating/cooling or all the time. Some occupants prefer continuous instead of intermittent air circulation.
- wide range of applications—Typical modern thermostats are factory equipped to handle a wide range of applications such as multiple stages of heating/cooling.

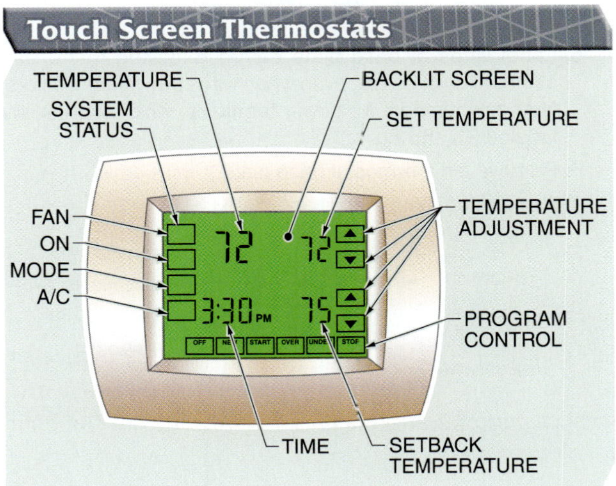

Figure 60-14. LCD thermostats provide many programmable functions and features.

Thermostat Retrofits. HVAC technicians must be familiar with the procedure for replacing an old thermostat with a new one. Technicians should read and follow the manufacturer instructions provided with the thermostats. **See Procedure for Retrofitting Thermostats.**

Procedure

Retrofitting Thermostats

1. Ensure new thermostat is correct model for replacing old one.
2. Turn off power to unit, and use DMM to ensure power is OFF.

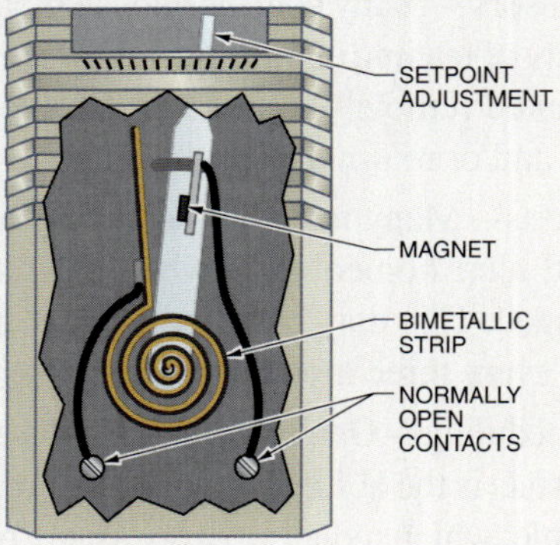

3. Remove old thermostat from base plate.
4. If old thermostat has mercury element, be careful not to break it. Mercury is a toxic substance and must be disposed of properly.

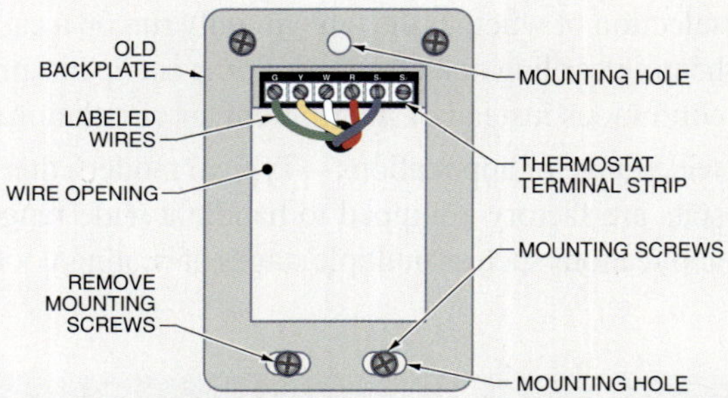

5. Label wires at base plate before removing them. Use standard color codes, even when wires are different colors. Note any jumpers for power terminals, which are usually labeled Rh and Rc.
6. Remove old thermostat base plate.
7. Anchor and mount new base plate.

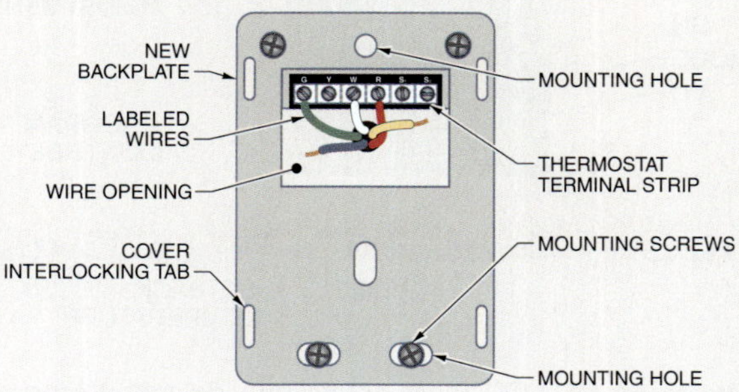

8. Using color codes and labeled wires in Step 5, attach wires to base plate. Connect any jumpers for power terminals, which are usually labeled Rh and Rc.
9. Attach any extra sensors needed to base plate.

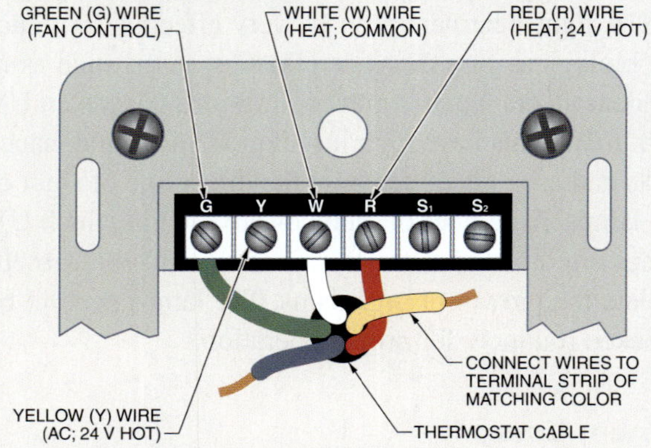

10. Many thermostats need batteries to back up setpoints and schedules in thermostat. Install new batteries in thermostat.
11. Install thermostat on base plate.
12. Turn on power.

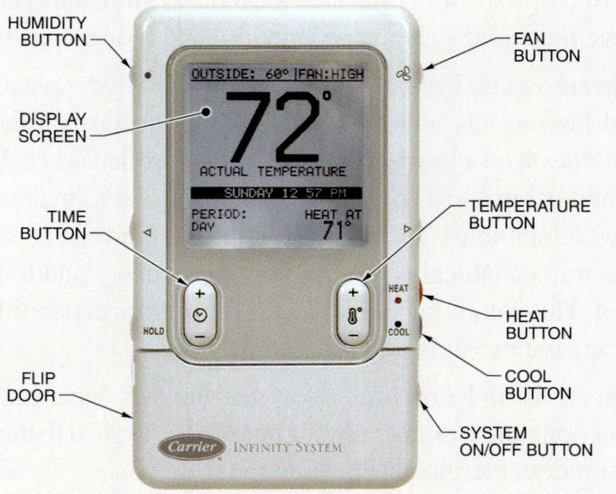

Carrier Corporation

13. Program thermostat using touch screen or buttons provided. Follow manufacturer instructions to program thermostat.
    a. Set date and time.
    b. Set time schedules and setpoints.
    c. Select system.
    d. Set heating/cooling changeover.
    e. Set filter alert.
    f. Set up any extra sensors.

Zoning

Another retrofit method for residential building systems is zoning, which involves installing simplified versions of the systems used in light commercial buildings. Zone-control systems are used in large houses when a single thermostat does not provide adequate temperature control responses. Separate controls in each individual room or area are often desired. Residential and light commercial zone-control systems have the same basic components: a central unit controller, zone thermostats, zone dampers, and a bypass system. **See Figure 60-15.**

VAV Air-Handling Units (AHUs)

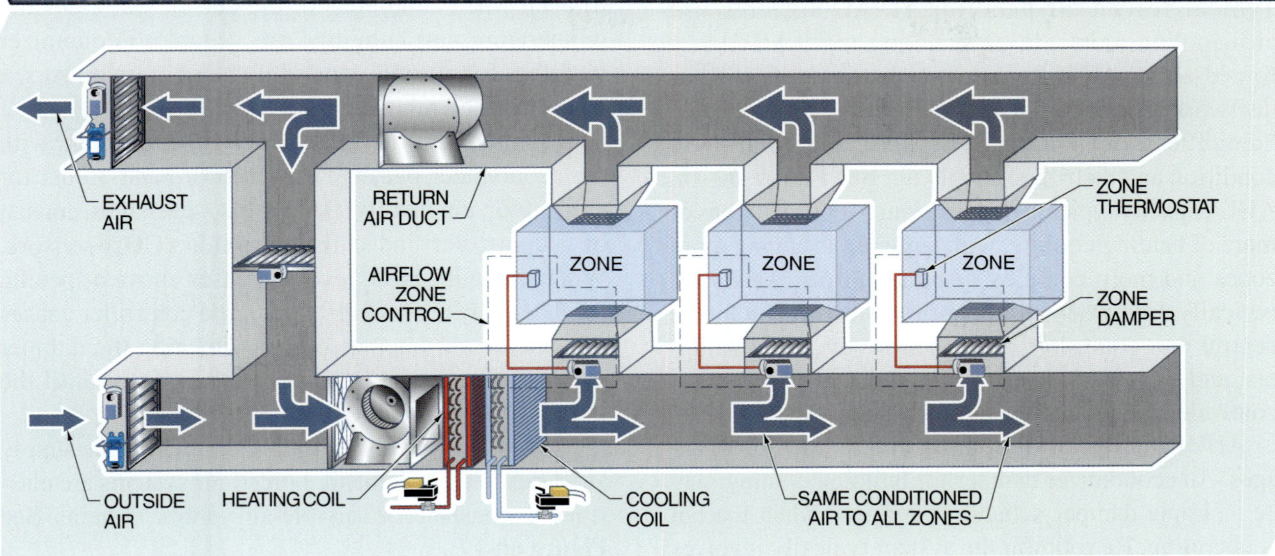

Figure 60-15. Air-handling units can be used to supply air for multiple zones. Each zone receives the same conditioned air.

Central Unit Controllers. Central unit controllers are connected to the HVAC system, which operates the fan, heating stage, and cooling stage. The controller also communicates with the zone thermostats in each space.

Zone Thermostats. Each room or area has a zone thermostat. Zone thermostats communicate with the central control unit of an HVAC system. Zone thermostats open and close zone dampers, as needed, to maintain the space temperature at its setpoint.

Zone Dampers. Zone thermostats are connected to zone damper actuators. Zone dampers are installed in the supply duct that delivers conditioned air to the zoned area. Dampers typically use 24 VAC motor-driven actuators and a controller output to open and close. Dampers have drive times ranging from 30 sec to a few minutes from fully open to fully closed.

Bypass Dampers and Controllers. Multiple zones may have their zone dampers mostly or fully closed at any given time. Multiple zone dampers that are closed result in the supply air having no discharge and an increase in static pressure within the duct. Air pressure is relieved using a bypass duct, damper, and controller. As the zone dampers close, the bypass controller opens the bypass damper and limits the increase in duct air pressure.

Zone Sequences of Operations. Zoned systems typically operate with a similar sequence as centrally controlled systems. Individual zone thermostats signal the central unit controller that heating or cooling is needed. The central unit controller evaluates the signals and then determines whether to cycle the HVAC unit into the heating or cooling mode.

When a zone signals heat and the HVAC unit is in a cooling cycle, that zone damper will close until the cooling zones are satisfied. The HVAC unit may then cycle into heating, and the cooling zones close while the heating zones open. During this time, bypass controllers will modulate bypass dampers to maintain a specific unit discharge air pressure.

TECH TIP

When retrofitting an older thermostat with a new one, the equipment should be tested in both the heating mode and cooling mode regardless of the season. This reduces callbacks when seasonal changeover occurs.

RETROFITTING AIR-HANDLING HVAC SYSTEMS

The retrofitting process for forced-air-handling systems is similar to that of forced-air systems for residential or light commercial buildings. Typically, forced-air-handling systems are larger, more expensive, and more complex than forced-air systems.

Air-Handling Units

Typical forced-air-handling HVAC systems use air-handling units. An air-handling unit (AHU) is a forced-air HVAC system device that consists of fans, ductwork, filters, dampers, heating coils, cooling coils, humidifiers, dehumidifiers, sensors, and controls to condition and distribute supply air. **See Figure 60-16.** AHU capacity typically ranges from 5 tons to 50 tons or more of heating/cooling, with larger units having more zones and more complex system controls. AHUs are typically used when buildings are zoned and include a central unit controller, zone thermostats, zone dampers, and bypass controllers and dampers. Central unit controller sequences are more complicated.

AHUs are typically retrofitted with economizer packages. An economizer damper for building cooling may be a simple damper actuator that opens when the fan is operating. Retrofitting the system typically involves adding a controls package that only allows the damper to open when conditions of the outside air, including temperature and humidity, are suitable. Economizer packages for retrofits consist of one or more sensors, a controller, and related wiring.

Common retrofits include adding controllers with the economizer package that measure and adjust for the indoor air quality (IAQ). IAQ packages consist of a controller and carbon dioxide (CO_2) sensors. When the indoor CO_2 level increases above a specific level, usually around 1100 ppm, the controller causes the outside air damper to open. Opening the damper increases the fresh air supply to the system until the indoor CO_2 level decreases.

Forced-air systems are classified by how the supply air varies to meet demand. Forced-air systems are classified as constant- or variable-air-volume systems. **See Figure 60-17.**

AHUs

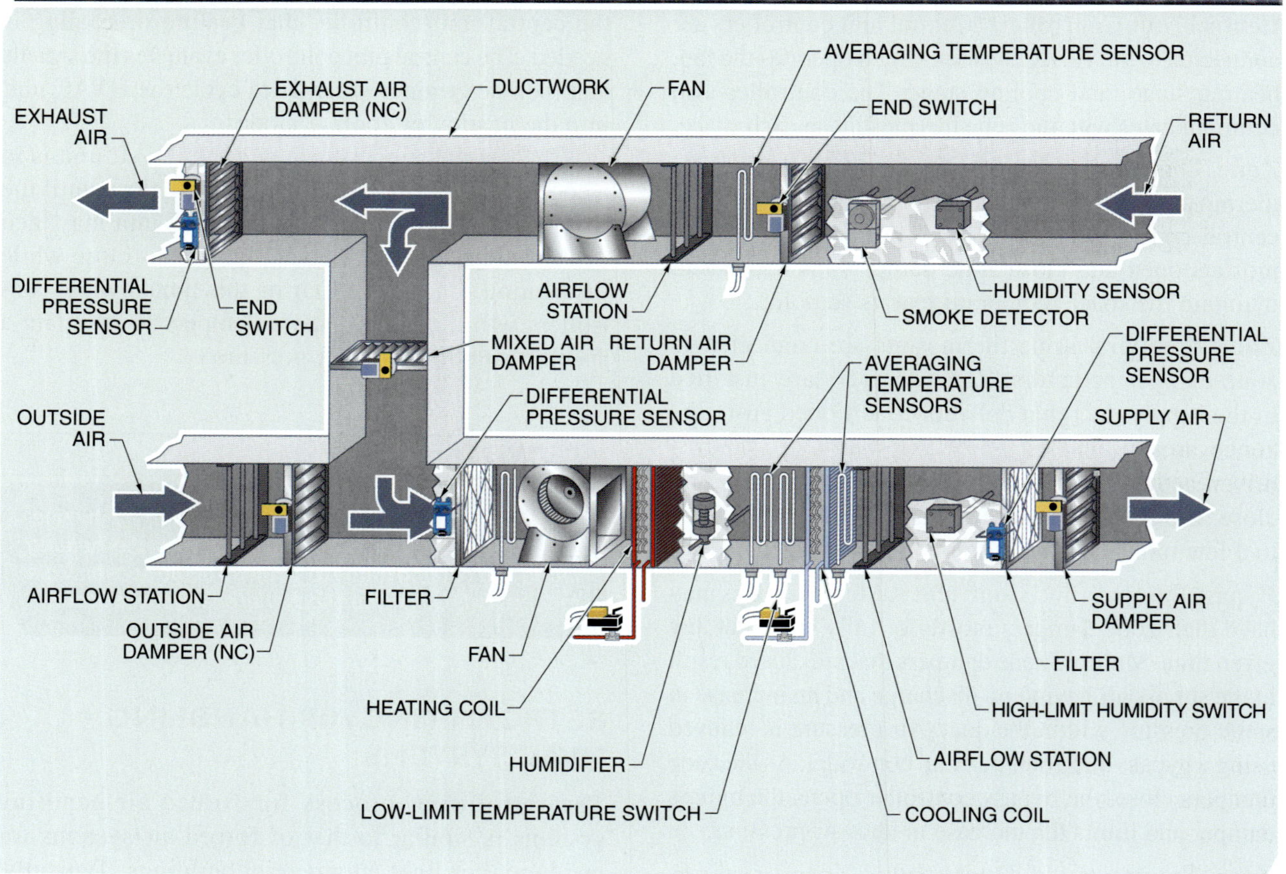

Figure 60-16. AHUs include all of the devices needed to condition air for a building. Ductwork distributes the supply air to the building and collects the return air.

Types of AHUs

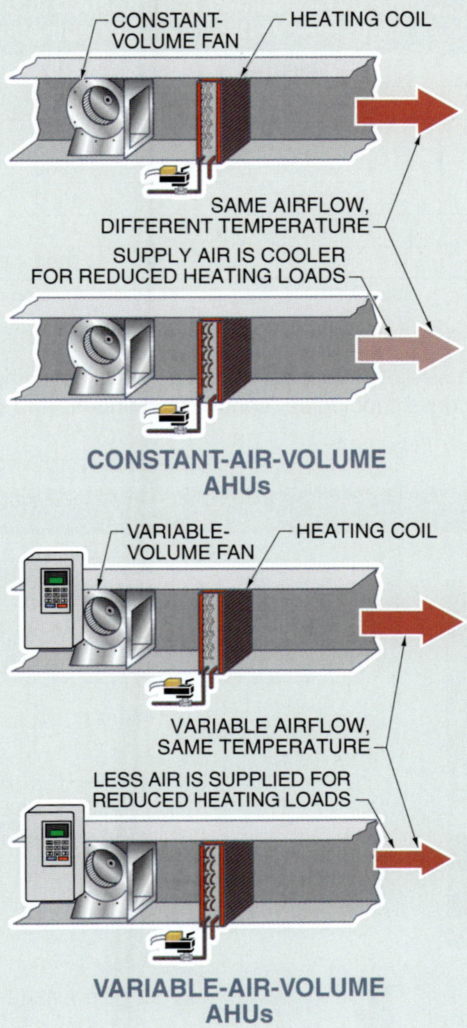

Figure 60-17. Constant-air-volume air-handling units vary the supply air temperature. Variable-air-volume air-handling units vary the supply airflow rate.

HVAC AHU Systems Media Clip

Constant Air Volume. A *constant-air-volume AHU* is an air-handling unit that provides a steady supply of air and varies the heating, cooling, or other conditioning functions, as necessary, to maintain the desired setpoints within a building zone. CAV systems are common in residential and light commercial buildings. They are simple and reliable but not energy efficient because the fan operates at maximum power 100% of the time. CAV capacity can be selected and manufactured to satisfy all ranges of heating/cooling. The temperature controls are simple, and the system cycles on or off as needed. Most CAV systems can be retrofitted to a variable-air-volume system using two speed fans and controls.

Variable Air Volume. A *variable-air-volume (VAV) system* is an air distribution system that provides air at a constant temperature but varies the amount of supplied air in order to maintain the desired setpoints within the building space. VAV capacities can be selected and manufactured to satisfy all ranges of heating/cooling. VAV systems consist of air handlers that provide air at a constant static pressure of 1″ WC and a temperature of 55°F. VAV terminals located in the zones open and close zone dampers to provide supply air. Since some exterior zones need heating, hot water or electric heating coils are located in the duct to add any needed heat.

Typically, a VAV system comes from the manufacturer with an economizer damper and controller. Typical VAV system retrofits include the installation of upgraded economizer control packages. Economizer controls consist of a temperature, humidity, or enthalpy (total heat content in Btu/lb) sensing package. The controllers only allow the economizer damper to open when the condition of the outside air is suitable. VAV systems typically can be used to measure return air conditions as well. However, this type of system is more complex and costly.

Typical VAV system retrofits include the installation of variable-frequency drives (VFDs). Many older VAV systems use inlet vanes to provide fan capacity control. VFDs can be added to allow improved static air-pressure control. Replacing an older VFD with a new VFD can result in a significant reduction in energy costs. **See Figure 60-18.**

BUILDING AUTOMATION SYSTEMS

High-level retrofit installations in buildings use a building automation system. A *building automation system (BAS)* is a computer-based system that uses a distributed system of microprocessor-based controllers to automate any combination of building systems. BASs integrate HVAC heating and cooling functions with other building system functions such as security, lighting, entertainment, and communications. **See Figure 60-19.**

BASs provide owners the capability of local or remote control over automated functions. The same concepts are used in residential and light commercial BASs during installation and operation.

Compare Old VFD to New VFD Energy Consumption

HVAC Airflow Volume*	Daily Operating Time†	Energy Consumed with 2003 VFD‡	Energy Consumed with New VFD‡	Difference in Energy Consumption‡
50	2	5000	4800	200
60	3	10,400	9800	600
70	6	28,100	26,800	1300
80	6	37,100	35,900	1200
90	4	33,600	32,600	1000
100	3	36,500	36,800	–300
Total	24	150,700	145,100	5600

* percent of maximum
† in hours
‡ in kWh/year

Source: Office of Energy Efficiency, Natural Resources of Canada

Figure 60-18. Power electronics technology has reduced VFD cost and size over the past decades. VFDs have improved performance through advances in semiconductor switching devices, drive topologies, control techniques, and control hardwares and softwares.

Building Automation Systems (BASs)

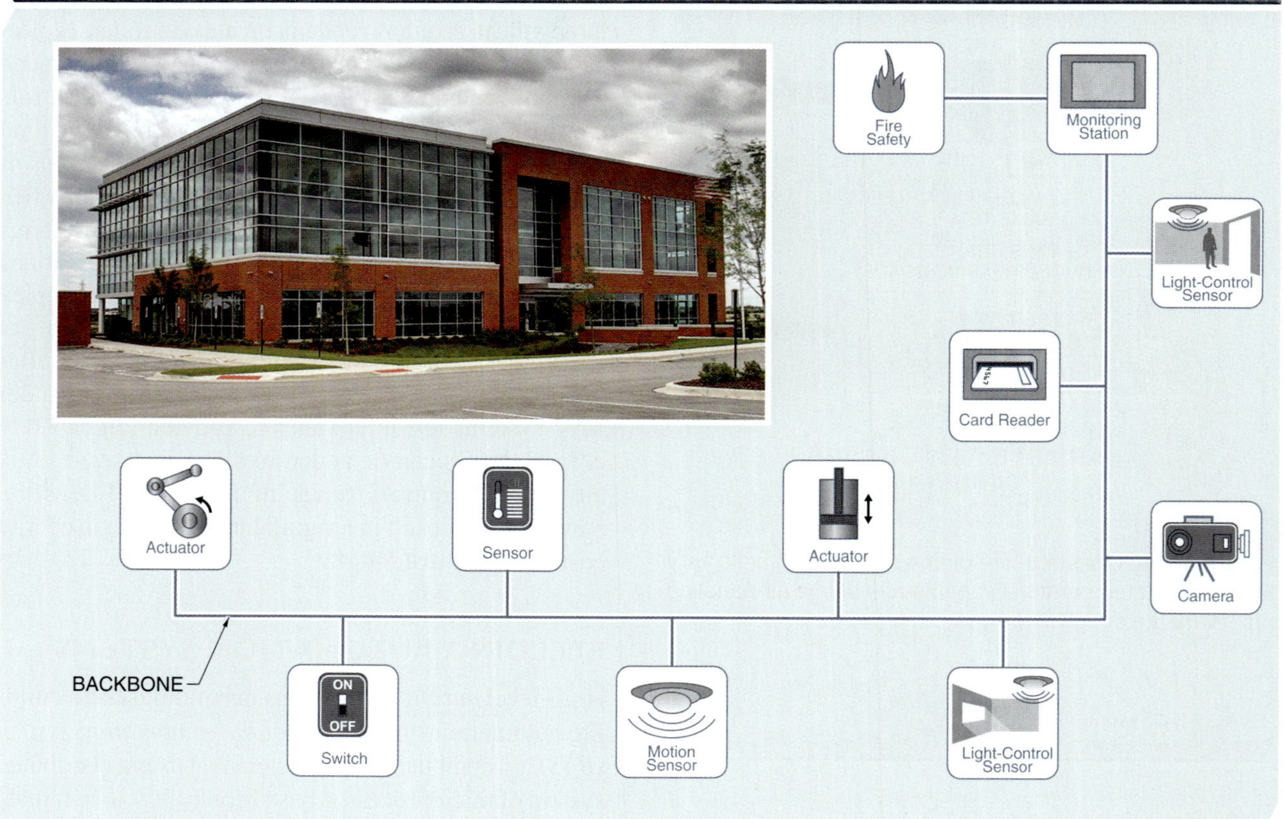

Figure 60-19. BASs typically use IT networks similar to those used for computer systems.

Typical AHU Retrofits

Most AHUs are controlled by temperature controllers that cycle system functions. Retrofitting a unit typically involves connecting the AHU to a BAS. The connection allows facility-wide network control of all AHUs in the facility and includes setting alarms, adjusting the central setpoint, remote access, and simple troubleshooting. Typical AHU retrofit tasks include replacing existing AHU thermostats with networkable thermostats.

Communication bus wires are run from AHU controllers to a web-based supervisory controller. An Internet browser is then used to connect AHU controllers and points to the network. Features, such as schedules and alarms, are then added to the points in the controller. BAS AHU controllers typically use an interoperable protocol, such as BACnet™ or LonWorks™, to facilitate this retrofit. **See Figure 60-20.**

Control sequences can be added or changed during a retrofit as well. Typically, AHUs are time scheduled to operate only when needed or to shut off individual units when the building electrical demand is too high.

CAV Retrofits. While many CAV systems operate with built-in control systems supplied by the manufacturer, the control systems are typically connected to the BAS. Microprocessor control boards that come with CAV systems are typically set up for integration and connection to the BAS.

VAV Retrofits. While many VAV systems operate with built-in control systems supplied by the manufacturer, control systems are typically connected to the BAS. The retrofit procedure for a VAV is similar to the procedure for a CAV. Many microprocessor control boards that come with VAV units are set up for integration and connection to the BAS.

Typically, the control sequences of VAV systems are changed. VAV systems respond to electrical demand, changes in the static pressure setpoint based on zone conditions, or time schedules to ensure that the systems run only when needed. Operating strategies are often controlled by using a BAS.

Typical Packaged Chiller Retrofits

Packaged chillers can be integrated into the BAS. This results in a significant reduction in energy use. The BAS allows variables from the chiller package to be viewed and trended on a computer system. Variables may include electrical consumption, internal compressor temperatures, and a log of electrical faults.

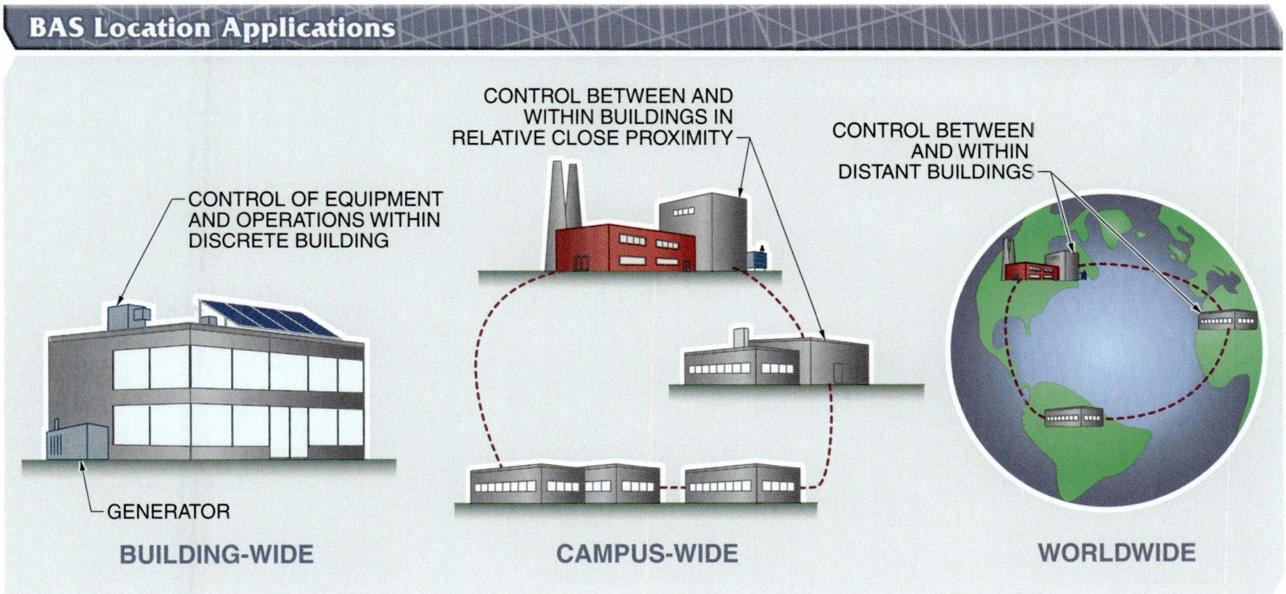

Figure 60-20. BASs can be extended beyond the building by using a suitable network infrastructure.

Appendix

Physical Properties of Vapors

Name	Normal Boiling Point*	Specific Heat†	Mass Density‡	Thermal Conductivity§
Alcohol, ethyl	173.3	0.362		0.0876
Alcohol, methyl	148.9	0.322		0.2088
Ammonia	– 28.0	0.525	0.0482	0.1536
Argon	–302.3	0.125	0.1114	0.1128
Acetylene	–118.5	0.377	0.0732	0.1296
Benzene	176.2	0.31	0.167	0.0492
Bromine	137.8	0.055	0.38	0.042
Butane	31.1	0.377	0.168	0.0948
Carbon dioxide	–109.3	0.20	0.123	0.1008
Carbon disulfide	115.2	0.1431		
Carbon monoxide	–312.7	0.25	0.078	0.1596
Carbon tetrachloride	169.8	0.206		0.45
Chlorine	–30.3	0.117	0.201	0.0552
Chloroform	143.1	0.126		0.0972
Ethyl chloride	54.2	0.426	0.1793	0.0605
Ethylene	–154.6	0.352	0.0783	0.1224
Ethyl ether	94.4	0.589		0.3276
Fluorine	–304.5	0.194	0.1022	0.1764
Helium	–452.1	1.241	0.0111	0.9876
Hydrogen	–423.0	3.40	0.00562	1.1664
Hydrogen chloride	–120.8	0.191	0.1024	0.0908
Hydrogen sulfide	–77.3	0.238	0.0961	0.0901
Heptane	209.2	0.476	0.21	0.1284
Hexane	154.0	0.449	0.21	0.1165
Isobutane	–10.9	0.376	0.154	0.0972
Methyl chloride	–11.6	0.184	0.1440	0.0648
Methane	–263.2	0.520	0.0448	0.2136
Napthalene	–360.4	0.313		
Neon	–412.6	0.246		0.3216
Nitric oxide	–241.6	0.238		
Nitrogen	–320.4	0.248		0.1656
Nitrous oxide	–127.3	0.203		0.1201
Oxygen	–297.4	0.218		0.1692
Phenol	358.5	0.34	0.16	0.1188
Propane	–43.73	0.3753	0.126	0.1044
Propylene	–53.86	0.349	0.120	0.0972
Sulfur dioxide	14.0	0.145	0.183	0.0588
Water vapor	212.0	0.489	0.0373	0 .1716

* in °F
† in Btu/lb × °F
‡ in lb/cu ft
§ in Btu/hr × ft × °F

Source: ASHRAE Handbook — Fundamentals

Physical Properties of Liquids

Name	Normal Boiling Point*	Specific Heat†	Mass Density‡	Thermal Conductivity§
Benzene	176.2	0.412	54.9	1.02
Calcium chloride brine		0.744	73.8	3.984
Carbon disulfide	115.3	0.240	78.9	1.116
Carbon tetrachloride	170.2	0.201	99.5	0.744
Chloroform	142.3	0.234	92.96	0.90
Ethyl acetate	170.8	0.468	52.3	1.212
Ethylene glycol	388.4		69.22	1.2
Formic acid	213.3	0.526	76.16	1.248
Glycerine (glycerol)	359.0		78.72	1.356
Heptane	209.2	0.532	42.7	0.8892
Hexane	154.0	0.538	41.1	0.864
Isobutyl alcohol	226.4	0.116	50.0	0.984
Kerosene	572.0	0.50	51.2	1.032
Linseed oil			58.0	
Methyl acetate	134.6	0.468	60.6	1.116
Nitric acid	186.8	0.42	94.45	1.92
Nitrobenzene	411.6	0.348	75.2	11.52
Octane	258.3	0.51	43.9	1.008
Petroleum		0.4	40.0	
Sodium chloride brine	220.8	0.745	71.8	4.044
Turpentine	303.0	0.42	53.9	0.876
Water	212.0	1.00	62.32	4.176
Zinc sulfate and water	0.90	0.90	69.2	4.044

* in °F at 14.7 psia
† in Btu/lb × °F at 68°F
‡ in lb/cu ft at 68°F
§ in Btu/hr × ft × °F

Source: ASHRAE Handbook — Fundamentals

Physical Properties of Solids

Name	Specific Heat*	Mass Density†	Thermal Conductivity‡
Aluminum	0.214	171	128
Ashes (wood)	0.20	40	0.041
Asphalt	0.22	132	0.43
Bakelite	0.35	81	9.7
Bronze	0.104	530	17
Cadmium	0.055	540	53.7
Cast Iron	0.12	450	27.6
Cement (Portland clinker)	0.16	120	0.017
Chalk	0.215	143	0.48
Charcoal (wood)	0.20	15	0.03
Chrome brick	0.17	200	0.67
Clay	0.22	63	
Coal	0.3	90	0.098
Coal tars	0.35	75	0.07
Coke (petroleum, powdered)	0.36	62	0.55
Concrete (stone)	0.156	144	0.54
Copper (electrolytic)	0.092	556	227
Cork (granulated)	0.485	5. 4	0.028
Cotton (fiber)	0.319	95	0.024
Diamond	0.147	151	27
Gold	0.0312	1208	172
Graphite (powder)	0.165		0.106
Gypsum	0.259	78	0.25
Hemp (fiber)	0.323	93	
Ice (32°F)	0.487	57.5	1.3
Lead	0.0309	707	20.1
Leather (sole)		62.4	0.092
Limestone	0.217	103	0.54
Magnesium	0.241	108	91
Marble	0.21	162	1.5
Nickel	0.105	555	34.4
Paper	0.32	58	0.075
Paraffin	0.69	56	0.14
Plaster		132	0.43
Platinum	0.032	1340	39.9
Porcelain	0.18	162	1.3
Sand	0.191	94.6	0.19
Sawdust		12	0.03
Silica	0.316	140	0.83
Silver	0.0560	654	245
Steel (mild)	0.12	489	26.2
Stone (quarried)	0.2	95	

* in Btu/lb × °F
† in lb/cu ft
‡ in Btu/hr × ft × °F

Approximate Gross Heating Value of Common Fuels

Fuel	Gas	Liquid	Solid
Natural gas	1030 Btu/cu ft	100,000 Btu/therm	
Propane	2500 Btu/cu ft	92,500 Btu/gal	
Methane	1000 Btu/cu ft		
Landfill gas	500 Btu/cu ft		
Butane	3200 Btu/cu ft	130,000 Btu/gal	
Methanol		57,000 Btu/gal	
Ethanol		76,000 Btu/gal	
Fuel oil			
Kerosene		135,000 Btu/gal	
#2		138,500 Btu/gal	
#4		145,000 Btu/gal	
#6		153,000 Btu/gal	
Waste oil		125,000 Btu/gal	
Biodiesel — waste vegetable oil		120,000 Btu/gal	
Gasoline		125,000 Btu/gal	
Wood			
Softwood			2-3000 lb/cord
Hardwood			4-5000 lb/cord
Sawdust - green			10-13 lb/cu ft
Sawdust - kiln dry			8-10 lb/cu ft
Chips - 45% moisture			10-30 lb/cu ft
Hogged			10-30 lb/cu ft
Bark			10-20 lb/cu ft
Wood pellets - 10% moisture			40-50 lb/cu ft
Coal			13,000 Btu/lb
Hard (anthracite)			12,000 Btu/lb
Soft (bituminous)			16,000 Btu/lb
Rubber - pelletized			18-20,000 Btu/lb
Plastic			
Corn			7800-8500 Btu/lb
Shelled			8000-8300 Btu/lb
Cobs		3412 Btu/kilowatt hour	
Electricity			

Fuel Gases — Heating Values

Gas	Heating Value*
Acetylene (ethyne) - C_2H_2	20,837
Benzene	17,418
Blast furnace gas	1178
Blue water gas	–
Butane - C_4H_{10}	19,976
Butylene (butene)	19,420
Carbon to CO_2	14,150
Carbon to CO	3960
Carbon monoxide - CO	4368
Carburetted water gas	10,566
Coal gas	–
Coke oven gas	15,266
Digester gas (sewage or biogas)	10,184
Ethane - C_2H_6	20,295
Ethyl alcohol saturated with water	–
Ethylene	20,525
Hexane	18,976
Hydrogen (H_2)	51,628
Hydrogen sulphide	–
Landfill gas	–
Methane - CH_4	21,433
Methyl alcohol saturated with water	–
Naphthalene	–
Natural gas (typical)	17,500 –22,000
Octane saturated with water	10,444
Pentane	19,322
Producer gas	–
Propane - C_3H_8	19,834
Propene (propylene) - C_3H_6	19,630
Propylene	19,683
Sasol	13,016
Sulphur	–
Toulene	17,301
Water gas (bituminous)	4469
Xylene	–

* in Btu/lb

Equivalents of psia

psia*	in. WC†	ft WC‡	in. Hg§
0.00	0.00	0.00	29.92
0.01	0.25	0.02	29.90
0.02	0.50	0.04	29.88
0.03	0.75	0.06	29.86
0.04	1	0.08	29.85
0.07	2	0.17	29.77
0.11	3	0.25	29.70
0.14	4	0.33	29.63
0.18	5	0.42	29.55
0.22	6	0.50	29.48
0.29	8	0.67	29.33
0.36	10	0.83	29.18
0.43	12	1.00	29.04
0.51	14	1.17	28.89
0.58	16	1.33	28.74
0.65	18	1.50	28.60
0.72	20	1.67	28.45
0.80	22	1.83	28.30
0.87	24	2.00	28.15
0.94	26	2.17	28.01
1	27.71	2.31	2.041
2	55.42	4.62	4.081
3	83.14	6.93	6.122
4	110.85	9.24	8.163
5	138.56	11.55	10.20
6	166.27	13.86	12.24
7	193.99	16.17	14.28
8	221.70	18.47	16.33
9	249.41	20.78	18.37
10	277.12	23.09	20.41
11	304.84	25.40	22.45
12	332.55	27.71	24.49
13	360.26	30.02	26.53
14	387.97	32.33	28.57
14.7	407.37	33.95	29.92
15	415.68	34.64	30.61
16	443.40	36.95	32.65
17	471.11	39.26	34.69
18	498.82	41.57	36.73
19	526.53	43.88	38.77
20	554.25	46.19	40.81
21	581.96	48.50	42.85
22	609.67	50.51	44.89
23	637.38	53.12	46.94
24	665.10	55.42	48.98
25	692.81	57.73	51.02

* pounds per square inch absolute
† inches of water column
‡ feet of water column
§ inches of Mercury

Properties of Commercial Propane, Butane, and Natural Gas

Properties	Propane	Butane	Natural Gas
Chemical formula	C_3H_8	C_4H_{10}	CH_4
Molecular weight	44.06	58.1	16
Freezing point, °F, at atmospheric pressure	-310	-260	-296
Boiling point, °F	-44	32	-260
Normal state, at 60°F, atmospheric pressure	Gas	Gas	Gas
Relative density of gas (air = 1)	1.52	2	0.55
Relative density of liquid (water = 1)	0.51	0.58	0.3
Mass per gal of liquid at 60°F, lb	5.1	5.8	59.8
Calorific value, Btu/ft^3 (vaporized)	2520	3260	1000
Btu/lb (vaporized)*	21,622	21,190	23,875
Btu/Imperial gal. (vaporized)	109,956	122,900	101,600
ft^3 of vapor from 1 lb of liquid at 60°F	8.5	6.5	23
ft^3 of vapor from 1 gal. of liquid at 60°F	44	38	59
Latent heat of vaporization at boiling point, Btu/lb	185	166	
Latent heat of vaporization at boiling point, Btu/gal.	944	968	—
Specific heat of vapor, Btu/lb/°F at 60°F	0.405	0.385	0.53
Specific heat of liquid, Btu/lb/°F at 60°F	0.59	0.55	—
Mass density of vapor, lb/ft^3	0.116	0.153	0.042
Mass density of liquid, lb/ft^3	31.7	36	—
ft^3 of air required to burn 1 ft^3 of gas	23.5	30	9.5
ft^3 of oxygen required to burn 1 ft^3 of gas	5	6.25	2
ft^3 of air required to burn 1 lb of gas	200	195	226
Approximate ignition temperature, °F	920	900	1300
Approximate maximum flame temperature, °F	3600	3600	3600
Limits of flammability, percentage of gas in air mixture:			
Percentage of gas at lower limit	2.4	1.9	4
Percentage of gas at upper limit	9.5	8.5	15
Approximate theoretical percentage of gas at maximum rate of flame propagation	5	4	0
Products of complete combustion:			
ft^3 of carbon dioxide (CO_2)/ft^3 of gas	3	3.9	1
ft^3 of water vapor/ft^3 of gas	4	5	2
ft^3 of nitrogen (N_2)/ft^3 of gas	18.5	23.6	8
lb of water (H_2O)/of gas	1.6	1.5	2.25
lb of nitrogen (N_2)/lb of gas	12	11.8	13.2
Ultimate carbon dioxide (CO_2) percentage by volume	13.9	14.1	11.7

Chemical Elements

Name	Symbol	Atomic Weight*	Atomic Number	Name	Symbol	Atomic Weight*	Atomic Number
Actinium	Ac	[227]	89	Neon	Ne	20.183	10
Aluminum	Al	26.9815	13	Neptunium	Np	[237]	93
Americium	Am	[243]	95	Nickel	Ni	58.71	28
Antimony	Sb	121.75	51	Noibium	Nb	92.906	41
Argon	Ar	39.948	18	Nitrogen	N	14.0067	7
Arsenic	As	74.9216	33	Nobelium	No	[255]	102
Astatine	At	[210]	85	Osmium	Os	190.2	76
Barium	Ba	137.34	56	Oxygen	O	15.9994	8
Berkelium	Bk	[247]	97	Palladium	Pd	106.4	46
Beryllium	Be	9.0122	4	Phosphorus	P	30.9738	15
Bismuth	Bi	208.980	83	Platinum	Pt	195.09	78
Boron	B	10.811	5	Plutonium	Pu	[244]	94
Bromine	Br	79.909	35	Polonium	Po	[210]	84
Cadmium	Cd	112.40	48	Potassium	K	39.102	19
Calcium	Ca	40.08	20	Praseodymium	Pr	140.907	59
Californium	Cf	[251]	98	Promethium	Pm	[145]	61
Carbon	C	12.01115	6	Protactinium	Pa	[231]	91
Cerium	Ce	140.12	58	Radium	Ra	[226]	88
Cesium	Cs	132.905	55	Radon	Rn	[222]	86
Chlorine	Cl	35.453	17	Rhenium	Re	186.2	75
Chromium	Cr	51.996	24	Rhodium	Rh	102.905	45
Cobalt	Co	58.9332	27	Rubidium	Rb	85.47	37
Copper	Cu	63.54	29	Ruthenium	Ru	101.07	44
Curium	Cm	[247]	96	Samarium	Sm	150.35	62
Dysprosium	Dy	162.50	66	Scandium	Sc	44.956	21
Einsteinium	Es	[254]	99	Selenium	Se	78.96	34
Erbium	Er	167.26	68	Silicon	Si	28.086	14
Europium	Eu	151.96	63	Silver	Ag	107.870	47
Fermium	Fm	[257]	100	Sodium	Na	22.9898	11
Fluorine	F	18.9984	9	Strontium	Sr	87.62	38
Francium	Fr	[223]	87	Sulfur	S	32.064	16
Gadolinium	Gd	157.25	64	Tantalum	Ta	180.948	73
Gallium	Ga	69.72	31	Technetium	Tc	[97]	43
Germanium	Ge	72.59	32	Tellurium	Te	127.60	52
Gold	Au	196.967	79	Terbium	Tb	158.924	65
Hafnium	Hf	178.49	72	Thallium	Tl	204.37	81
Helium	He	4.0026	2	Thorium	Th	232.038	90
Holmium	Ho	164.930	67	Thulium	Tm	168.934	69
Hydrogen	H	1.00797	1	Tin	Sn	118.69	50
Indium	In	114.82	49	Titanium	Ti	47.90	22
Iodine	I	126.9044	53	Tungsten	W	183.85	74
Iridium	Ir	192.2	77	Unnilennium	Une	[266]	109
Iron	Fe	55.847	26	Unnilhexium	Unh	[263]	106
Krypton	Kr	83.80	36	Unniloctium	Uno	[265]	108
Lanthanum	La	138.91	57	Unnilpentium	Unp	[262]	105
Lawrencium	Lr	[256]	103	Unnilquadium	Unq	[261]	104
Lead	Pb	207.19	82	Unnilseptium	Uns	[262]	107
Lithium	Li	6.939	3	Uranium	U	238.03	92
Lutetium	Lu	174.97	71	Vanadium	V	50.942	23
Magnesium	Mg	24.312	12	Xenon	Xe	131.30	54
Manganese	Mn	54.9380	25	Ytterbium	Yb	173.04	70
Mendelevium	Md	[258]	101	Yttrium	Y	88.905	39
Mercury	Hg	200.59	80	Zinc	Zn	65.37	30
Molybdenum	Mo	95.94	42	Zirconium	Zr	91.22	40
Neodymium	Nd	144.24	60				

Properties of Saturated Water and Steam

Saturated Conditions		Enthalpy (Btu/lb)			Specific Volume (Cu ft/lb)	
Pressure (psig)	Temp (°F)	Water (H_f)	Vaporization (H_{fg})	Vapor (H_g)	Water (V_f)	Vapor (V_g)
0	212.0	180.3	970.0	1150.3	0.01672	26.7990
5	227.2	195.6	960.5	1156.1	0.01683	20.8300
10	239.4	208.0	952.6	1160.6	0.01692	16.4910
15	249.8	218.5	945.7	1164.2	0.01700	13.8780
20	258.8	227.6	939.7	1167.3	0.01708	11.9660
30	274.0	243.1	929.1	1172.2	0.01721	9.4597
40	286.7	256.1	919.9	1176.0	0.01733	7.8261
50	297.7	267.3	911.9	1179.2	0.01743	6.6830
60	307.3	277.3	904.6	1181.9	0.01753	5.8369
70	316.0	286.2	897.9	1184.1	0.01761	5.1845
80	323.9	294.4	891.7	1186.1	0.01770	4.6656
90	331.2	301.9	885.9	1187.8	0.01778	4.2326
100	337.9	308.9	880.5	1189.4	0.01785	3.8911
110	344.2	315.5	875.3	1190.8	0.01792	3.5941
120	350.0	321.7	870.3	1192.0	0.01799	3.3397
130	355.6	327.5	865.7	1193.2	0.01806	3.1193
140	360.9	333.1	861.1	1194.2	0.01812	2.9266
150	365.9	338.4	856.7	1195.1	0.01816	2.7563
160	370.6	343.4	852.6	1196.0	0.01824	2.6050
170	375.2	348.3	848.5	1196.8	0.01830	2.4694
180	379.6	353.0	844.5	1197.5	0.01835	2.3473
190	383.8	357.5	841.5	1198.9	0.08141	2.2368
200	387.8	361.8	837.6	1199.4	0.01847	2.1361
210	391.7	366.0	834.0	1200.0	0.01852	2.0442
220	395.4	370.1	830.5	1200.5	0.01857	1.9598
230	399.1	374.0	826.9	1200.9	0.01863	1.8820
240	402.6	377.8	823.5	1201.3	0.01868	1.8102
250	406.0	381.5	820.2	1201.7	0.01873	1.7436
260	409.4	385.2	816.9	1202.1	0.01878	1.6816
270	412.6	388.7	813.7	1202.4	0.01882	1.6239
280	415.7	392.1	810.6	1202.7	0.01887	1.5700
290	418.8	395.5	807.5	1203.0	0.01892	1.5194
300	421.8	398.8	804.4	1203.2	0.01896	1.4720
310	424.7	402.0	801.4	1203.4	0.01901	1.4274
320	427.5	405.2	798.5	1203.6	0.01905	1.3854
330	430.3	408.2	795.6	1203.8	0.01910	1.3457
340	433.0	411.3	792.7	1204.0	0.01914	1.3084
350	435.7	414.2	789.9	1204.1	0.01919	1.2729
360	438.3	417.1	787.1	1204.2	0.01923	1.2392
370	440.8	420.0	784.3	1204.3	1.01927	1.2071
380	443.3	422.8	781.6	1204.4	0.01931	1.1767
390	445.7	425.5	778.7	1204.2	0.01936	1.1477
400	448.2	427.9	776.3	1204.2	0.01940	1.1200
410	450.5	430.6	773.7	1204.3	0.01944	1.0936
420	452.8	433.2	771.1	1204.3	0.01948	1.0684
430	455.1	435.8	768.5	1204.3	0.01985	1.0443
440	457.3	438.3	766.0	1204.3	0.01956	1.0212
450	459.5	440.8	763.5	1204.3	0.01961	0.9990
500	470.0	452.8	751.3	1204.1	0.01970	0.9278
600	488.8	474.6	728.3	1202.9	0.02010	0.7698
700	503.2	491.9	708.5	1200.4	0.02050	0.6554
800	517.6	509.3	688.6	1197.9	0.02090	0.5687
900	532.0	526.6	668.8	1195.4	0.02120	0.5006
1000	544.6	542.4	649.4	1191.8	0.02160	0.4456

Refrigerant Cylinder Color Coding

Refrigerant Number	Cylinder Color	Refrigerant Name	Application
CFC Refrigerants			
R-11	Orange	Trichloromono-fluoromethane	Centrifugal Chillers
R-12	White	Dichlorodi-fluoromethane	Reciprocating and rotary equipment
R-13	Light Blue	Monochlorotri-fluoromethane	Used in low stage of cascade systems
R-13B1	Coral	Bromotri-fluoromethane	Low- to medium-temperature applications
R-113	Purple	Trichlorotri-fluoroethane	Low-capacity centrifugal chillers
R-114	Dark Blue	Dichloromono-fluoroethane	High-capacity chillers
R-500	Yellow	Refrig. 152A/12	Industrial and commercial reciprocating compressors
R-502	Lt. Purple	Refrig. 22/115	Reciprocating and rotary equipment
R-503	Aquamarine	Refrig. 23/13	Low-temperature commercial refrigeration
HCFC Refrigerants			
R-22	Lt. Green	Monochlorodi-fluoromethane	Residential, commercial, and industrial
R-123	Lt. Gray	Dichlorotri-fluoroethane	R-11 replacement for centrifugal chillers

Refrigerant Number	Cylinder Color	Refrigerant Name	Application
HCFC Refrigerants			
R-124	Deep Green	Chlorotetra-fluoroethane	Medium-pressure chillers
R-401A	Coral Red	R-22 + R-152a + R-124	Medium-temperature systems
R-401B	Mustard Yellow	R-22 + R-152a + R-124	Transport refrigeration, domestic refrigerators
R-402A	Lt. Brown	R-22 + R-125 + R-290	Ice machine, vending, supermarket, and food service
R-402B	Green Brown	R-22 + R-125 + R-290	Supermarket, food service, and transport
HFC Refrigerants			
R-23	Light Gray	Trifluoromethane	Low-temperature systems
R-134a	Light Sky Blue	Tetrafluoroethane	Automotive industry and refrigeration systems
R-404A	Orange	R-125 + R-143a + R-134a	Medium and low temperature
R-407C	Chocolate Brn.	R-32 + R-125 + R-134a	R-22 replacement
R-410A	Rose	R-32 + R-125	Replacement for residential air conditioning
R-507	Lt. Brown	Refrig. 125/143a	Replacement for low-temp commercial refrigeration

Manufacturer Refrigerant Usage

Refrigerant	Carrier	ICOR	Trane	York
R-11	X		X	X
R-12	X	X	X	X
R-13				
R-22	X	X	X	X
R-23				
R-113				
R-114				X
R-123				X
R-134a	X	X	X	X
R-401A	X	X	X	
R-401B	X	X	X	
R-401C	X	X	X	
R-402A	X		X	
R-402B	X	X	X	
R-404A	X		X	
R-406A	X	X	X	
R-407A	X		X	
R-407B	X		X	
R-407C	X	X	X	
R-407D	X	X	X	
R-408A	X	X	X	
R-409A	X	X	X	
R-410A	X		X	
R-411A	X	X	X	
R-411B	X	X	X	
R-500	X	X	X	X
R-502	X	X	X	X
R-503				
R-508A				

Refrigerant Pressures

	Temperature*						
PSI	R-12	R-22	R-124	R-134a	R-502	R-507	R-717
-5	-29	-48	3	-22	-57	-60	-34
-4	-28	-47	4	-21	-55	-58	-33
-3	-26	-45	6	-19	-54	-57	-32
-2	-25	-44	7	-18	-52	-55	-30
-1	-23	-43	9	-16	-51	-54	-29
0	-22	-41	10	-15	-50	-53	-28
1	-19	-39	13	-12	-47	-50	-26
2	-16	-37	16	-10	-45	-48	-23
3	-14	-34	18	-8	-42	-46	-21
4	-11	-32	21	-5	-40	-44	-19
5	-9	-30	23	-3	-38	-41	-17
6	-7	-28	26	-1	-36	-39	-15
7	-4	-26	28	1	-34	-38	-13
8	-2	-24	30	3	-32	-36	-12
9	0	-22	32	5	-30	-34	-10
10	2	-20	34	7	-29	-32	-8
11	4	-19	36	8	-27	-31	-7
12	5	-17	38	10	-25	-29	-5
13	7	-15	40	12	-24	-27	-4
14	9	-14	41	13	-22	-26	-2
15	11	-12	43	15	-20	-24	-1
16	12	-11	45	16	-19	-23	1
17	14	-9	46	18	-18	-21	2
18	15	-8	48	19	-16	-20	3
19	17	-7	49	21	-15	-19	4
20	18	-5	51	22	-13	-17	6
21	20	-4	52	24	-12	-16	7
22	21	-3	54	25	-11	-15	8
23	23	-1	55	26	-9	-14	9
24	24	0	57	27	-8	-12	11
25	25	1	58	29	-7	-11	12
26	27	2	59	30	-6	-10	13
27	28	4	61	31	-5	-9	14
28	29	5	62	32	-3	-8	15
29	31	6	63	33	-2	-7	16
30	32	7	65	35	-1	-6	17
31	33	8	66	36	0	-4	18
32	34	9	67	37	1	-3	19
33	35	10	68	38	2	-2	19
34	37	11	69	39	3	-1	20
35	38	12	71	40	4	0	21
36	39	13	72	41	5	1	22
37	40	14	73	42	6	2	23
38	41	15	74	43	7	3	24
39	42	16	75	44	8	4	25
40	43	17	76	45	9	4	26
42	45	19	78	47	11	6	28
44	47	21	80	49	13	8	29
46	49	23	82	51	15	10	31
48	51	24	84	52	16	11	32
50	53	26	86	54	18	13	34
52	55	28	88	56	20	15	35
54	57	29	90	57	21	16	37
56	58	31	91	59	23	18	38
58	60	32	93	60	24	19	40
60	62	34	95	62	26	21	41
62	64	35	97	64	27	22	42
64	65	37	98	65	29	24	44
66	67	38	100	66	30	25	45
68	68	40	101	68	32	26	46
70	70	41	103	69	33	28	47
72	71	42	104	71	34	29	49
74	73	44	106	72	36	30	50
76	74	45	107	73	37	32	51
78	76	46	109	75	38	33	52
80	77	48	110	76	40	34	53
85	81	51	114	79	43	37	56
90	84	54	117	82	46	40	58
95	87	56	120	85	49	43	61
100	90	59	123	88	51	45	63
105	93	62	126	90	54	48	66
110	96	64	129	93	57	51	68
115	99	67	132	96	59	53	70
120	102	69	135	98	62	55	73
125	104	72	138	100	64	58	75
130	107	74	140	103	67	60	77
135	109	76	143	105	69	62	79
140	112	78	145	107	71	64	81
145	114	81	148	109	73	66	82
150	117	83	150	112	75	68	84
155	119	85	152	114	77	70	86
160	121	87	154	116	80	72	88
165	123	89	157	118	82	74	90
170	126	91	159	120	83	76	91
175	128	92	161	122	85	78	93
180	130	94	163	123	87	80	95
185	132	96	165	125	89	82	96
190	134	98	167	127	91	83	98
195	136	100	169	129	93	85	99

* in F°

Refrigerant Properties

Refrigerant No.	Refrigerant Name	Chemical Formula	Molecular Mass	Normal Boiling Point*	Freezing Point*	Critical Temperature*	Critical Pressure†	Critical Volume§
704	Helium	He	4.0026	–452.1	None	–450.3	33.21	0.2311
702	Hydrogen (normal)	H_2	2.0159	–423.0	–434.5	–399.9	190.8	0.5320
702	hydrogen (para)	H_2	2.0159	–423.2	–434.8	–400.3	187.5	0.5097
720	neon	Ne	20.183	–410.9	–415.5	–379.7	493.1	0.03315
728	Nitrogen	N_2	28.013	–320.4	–346.0	–232.4	492.9	0.05092
729	Air	-	28.97	–317.8	-	–221.3	547.4	0.04883
740	Argon	A	39.948	–302.6	–308.7	–188.1	710.4	0.02990
732	Oxygen	O_2	31.9988	–297.3	–361.8	–181.1	736.9	0.0375
50	Methane	CH_4	16.04	–258.7	–296	–116.5	673.1	0.099
14	Tetrafluoromethane	CF_4	88.01	–198.3	–299	–50.2	543	0.0256
1150	Ethylene	C_2H_4	28.05	–154.7	–272	48.8	742.2	0.070
170	Ethane	C_2H_6	30.07	–127.85	–297	90.0	709.8	0.0830
744	Nitrous oxide	N_2O	44.02	–129.1	–152	97.7	1048	0.0355
23	Trifluoromethane	CHF_3	70.02	–115.7	–247	78.1	701.4	0.0311
13	Chlorotrifluoromethane	$CClF_3$	104.47	–114.6	–294	83.9	561	0.0277
744	Carbon dioxide	CO_2	44.01	–109.2	–69.9	87.9	1070.0	0.0342
13	Bromotrifluoromethane	$CBrF_3$	148.93	–71.95	–270	152.6	575	0.0215
1270	Propylene	C_3H_6	42.09	–53.86	–301	197.2	670.3	0.0720
290	Propane	C_3H_8	44.10	–43.73	–305.8	206.3	617.4	0.0728
22	Chlorodifluoromethane	$CHClF_2$	86.48	–41.36	–256	204.8	721.9	0.0305
115	Chloropentafluoroethane	$CClF_2CF_3$	154.48	–38.4	–159	175.9	457.6	0.0261
717	Ammonia	NH_3	17.03	–28.0	–107.9	271.4	1657	0.068
12	Dichlorodifluoromethane	CCl_2F_2	120.93	–21.62	–252	233.6	596.9	0.0287
134	Tetrafluoroethane	CF_3CH_2F	102.03	–15.08	–141.9	214.0	589.8	0.029
152	Difluoroethane	CH_3CHF_2	66.05	–13.0	–178.6	236.3	652	0.0439
40	Methyl chloride	CH_3Cl	50.49	–11.6	–144	289.6	968.7	0.0454
600	Isobutane	C_4H_{10}	58.13	10.89	–255.5	275.0	529.1	0.0725
764	Sulfur dioxide	SO_2	64.07	14.0	–103.9	315.5	1143	0.0306
142	Chlorodifluoroethane	CH_3CClF_2	100.5	14.4	–204	278.8	598	0.0368
630	Methyl amine	CH_3NH_2	31.06	19.9	–134.5	314.4	1082	
318	Octafluorocyclobutane	C_4F_8	200.04	21.5	–42.5	239.6	403.6	0.0258
600	Butane	C_4H_{10}	58.13	31.1	–217.3	305.6	550.7	0.0702
114	Dichlorotetrafluoroethane	$CClF_2CClF_2$	170.94	38.8	–137	294.3	473	0.0275
21	Dichlorofluoromethane	$CHCl_2F$	102.92	47.8	–211	353.3	750	0.0307
160	Ethyl chloride	C_2H_5Cl	64.52	54.32	–216.9	369.0	764.4	0.0485
631	Ethyl amine	$C_2H_5NH_2$	45.08	61.88	–113	361.4	815.6	
11	Trichlorofluoromethane	CCl_3F	137.38	74.87	–168	388.4	639.5	0.0289
611	Methyl formate	$C_2H_4O_2$	60.05	89.22	–146	417.2	870	0.0459
610	Ethyl ether	$C_4H_{10}O$	74.12	94.3	–177.3	381.2	523	0.0607
216	Dichlorohexafluoropropan	$C_3Cl_2F_6$	220.93	96.24	–193.7	356.0	399.5	0.0279
30	Methylene chloride	CH_2Cl_2	84.93	104.4	–142	458.6	882	
113	Trichlorotrifluoroethane	CCl_2FCClF_2	187.39	117.63	–31	417.4	498.9	0.0278
1130	Dichloroethylene	CHCl=CHCl	96.95	118	–58	470	795	
1120	Trichloroethylene	$CHCl=CCl_2$	131.39	189.0	–99	520	728	
718	Water	H_20	18.02	212	32	705.6	3208	0.0501

* in °F
† in psia
§ in cu ft/lb

ASHRAE Handbook — Fundamentals

R-12—Properties of Saturated Liquid and Saturated Vapor

Temp. °F	Pressure psia	Pressure psig	Volume† Vapor	Density‡ Liquid	Enthalpy§ Liquid	Enthalpy§ Vapor	Entropy‖ Liquid	Entropy‖ Vapor
−140	0.25567	29.401*	110.72	104.03	−20.398	62.190	−0.055433	0.20292
−130	0.41131	29.084*	70.904	103.12	−18.380	63.258	−0.049218	0.19842
−120	0.64047	28.617*	46.858	102.21	−16.362	64.338	−0.043187	0.19440
−110	0.96829	27.950*	31.857	101.29	−14.341	65.430	−0.037326	0.19081
−100	1.4252	27.019*	22.220	100.36	−12.317	66.530	−0.031620	0.18760
−90	2.0473	25.753*	15.861	99.429	−10.287	67.638	−0.026058	0.18474
−85	2.4332	24.967*	13.509	98.958	−9.2699	68.194	−0.023326	0.18343
−80	2.8765	24.065*	11.563	98.486	−8.2506	68.751	−0.020626	0.18219
−75	3.3834	23.033*	9.9442	98.012	−7.2292	69.310	−0.017956	0.18102
−70	3.9604	21.858*	8.5912	97.535	−6.2054	69.869	−0.015314	0.17991
−65	4.6144	20.526*	7.4545	97.056	−5.1789	70.429	−0.012700	0.17887
−60	5.3526	19.023*	6.4950	96.575	−4.1497	70.990	−0.010112	0.17789
−55	6.1826	17.333*	5.6813	96.091	−3.1174	71.550	−0.007549	0.17697
−50	7.1124	15.440*	4.9884	95.605	−2.0818	72.111	−0.005010	0.17609
−45	8.1502	13.327*	4.3957	95.116	−1.0427	72.672	−0.002494	0.17527
−40	9.3045	10.977*	3.8868	94.624	0.0	73.232	0.0	0.17450
−38	9.8008	9.9666*	3.7035	94.426	0.4182	73.456	0.000992	0.17420
−36	10.318	8.9139*	3.5307	94.228	0.8370	73.679	0.001980	0.17391
−34	10.856	7.8180*	3.3677	94.029	1.2564	73.903	0.002965	0.17363
−32	11.416	6.6777*	3.2138	93.830	1.6766	74.127	0.003948	0.17335
−30	11.999	5.4916*	3.0684	93.631	2.0974	74.350	0.004927	0.17309
−28	12.604	4.2586*	2.9310	93.431	2.5189	74.574	0.005903	0.17282
−26	13.234	2.9773*	2.8011	93.230	2.9411	74.797	0.006876	0.17257
−24	13.887	1.6466*	2.6782	93.029	3.3641	75.020	0.007845	0.17232
−22	14.566	0.2651*	2.5618	92.827	3.7877	75.242	0.008813	0.17207
−21.62	14.696	0.0	2.5407	92.789	3.8673	75.284	0.008994	0.17203
−20	15.270	0.5739	2.4516	92.625	4.2121	75.465	0.009777	0.17184
−18	16.000	1.3043	2.3471	92.422	4.6372	75.687	0.010738	0.17160
−16	16.758	2.0615	2.2481	92.219	5.0631	75.909	0.011697	0.17138
−14	17.542	2.8463	2.1541	92.015	5.4898	76.131	0.012653	0.17116
−12	18.355	3.6592	2.0650	91.810	5.9173	76.353	0.013606	0.17094
−10	19.197	4.5011	1.9803	91.605	6.3456	76.574	0.014557	0.17073
−8	20.069	5.3725	1.8999	91.399	6.7747	76.795	0.015505	0.17053
−6	20.970	6.2742	1.8234	91.192	7.2046	77.015	0.016451	0.17033
−4	21.903	7.2067	1.7507	90.985	7.6354	77.236	0.017394	0.17014
−2	22.867	8.1710	1.6815	90.777	8.0670	77.456	0.018335	0.16995
0	23.863	9.1675	1.6157	90.569	8.4995	77.675	0.019274	0.16976
2	24.893	10.197	1.5530	90.360	8.9329	77.895	0.020210	0.16958
4	25.956	11.260	1.4933	90.150	9.3671	78.114	0.021144	0.16941
6	27.054	12.358	1.4364	89.939	9.8023	78.332	0.022075	0.16924
8	28.187	13.491	1.3821	89.728	10.238	78.550	0.023005	0.16907
10	29.356	14.660	1.3303	89.516	10.675	78.768	0.023932	0.16891
12	30.561	15.865	1.2809	89.303	11.113	78.985	0.024857	0.16875
14	31.804	17.108	1.2337	89.089	11.552	79.202	0.025780	0.16860
16	33.085	18.389	1.1887	88.875	11.992	79.418	0.026701	0.16845
18	34.405	19.709	1.1456	88.659	12.433	79.633	0.027620	0.16830
20	35.765	21.069	1.1045	88.443	12.874	79.849	0.028537	0.16816
25	39.341	24.645	1.0093	87.899	13.983	80.384	0.030821	0.16782
30	43.182	28.486	0.92401	87.349	15.098	80.916	0.033093	0.16751

continued

* in in. Hg. vacuum
† in cu ft/lb
‡ in lb/cu ft
§ in Btu/lb
‖ in Btu/lb × °R
Critical point

continued

R-12—Properties of Saturated Liquid and Saturated Vapor

Temp. °F	Pressure psia	Pressure psig	Volume† Vapor	Density‡ Liquid	Enthalpy§ Liquid	Enthalpy§ Vapor	Entropy‖ Liquid	Entropy‖ Vapor
35	47.300	32.604	0.84738	86.793	16.220	81.444	0.035354	0.16721
40	51.705	37.009	0.77838	86.231	17.348	81.968	0.037604	0.16693
50	61.432	46.736	0.65984	85.087	19.625	83.000	0.042075	0.16642
60	72.462	57.766	0.56254	83.912	21.931	84.012	0.046508	0.16597
70	84.900	70.204	0.48204	82.704	24.266	84.998	0.050906	0.16557
80	98.850	84.154	0.41495	81.458	26.633	85.955	0.055273	0.16520
90	114.42	99.725	0.35864	80.168	29.032	86.880	0.059613	0.16485
100	131.72	117.03	0.31106	78.830	31.466	87.766	0.063928	0.16452
110	150.87	136.18	0.27060	77.436	33.937	88.609	0.068224	0.16420
120	171.99	157.29	0.23597	75.978	36.448	89.400	0.072504	0.16385
130	195.20	180.50	0.20615	74.446	39.003	90.133	0.076777	0.16349
140	220.63	205.93	0.18032	72.827	41.608	90.795	0.081050	0.16307
150	248.43	233.73	0.15780	71.106	44.271	91.375	0.085335	0.16260
160	278.74	264.05	0.13804	69.264	47.002	91.856	0.089649	0.16203
170	311.74	297.04	0.12057	67.275	49.818	92.214	0.094013	0.16134
180	347.61	332.91	0.10499	65.103	52.741	92.418	0.098461	0.16049
190	386.56	371.86	0.090963	62.697	55.808	92.419	0.10304	0.15940
200	428.84	414.14	0.078132	59.979	59.075	92.139	0.10784	0.15796
210	474.75	460.06	0.066108	56.806	62.647	91.430	0.11299	0.15597
220	524.70	510.01	0.054263	52.851	66.748	89.942	0.11882	0.15295
230	579.33	564.63	0.040319	46.607	72.294	86.110	0.12662	0.14665
233.2#	598.3	583.6	0.02871	34.83	79.40	79.40	0.1368	0.1368

* in in. Hg. vacuum
† in cu ft/lb
‡ in lb/cu ft
§ in Btu/lb
‖ in Btu/lb × °R
Critical point

continued

continued

R-12—Properties of Saturated Liquid and Saturated Vapor

Temp. °F	Viscosity* Sat. Liquid	Viscosity* Sat. Vapor	Viscosity* Gas§	Thermal Conductivity† Sat. Liquid	Thermal Conductivity† Sat. Vapor	Thermal Conductivity† Gas‖	Specific Heat‡ Sat. Liquid	Specific Heat‡ Sat. Vapor	Specific Heat‡ Gas#	Temp. °F
-140	2.47			0.0655			0.199		0.1085	-140
-120	1.97			0.0631			0.202		0.1123	-120
-100	1.612			0.0608			0.204		0.1160	-100
-80	1.347			0.0585			0.207		0.1196	-80
-60	1.146			0.0561			0.209	0.126	0.1230	-60
-40	0.990			0.0538			0.212	0.133	0.1264	-40
-20	0.866	0.0249	2.49	0.0514	0.0040	4.00	0.214	0.139	0.1296	-20
0	0.767	0.0265	2.61	0.0490	0.0043	4.31	0.217	0.145	0.1327	0
20	0.687	0.0279	2.72	0.0467	0.0046	4.63	0.220	0.150	0.1356	20
40	0.620	0.0291	2.83	0.0443	0.0050	4.95	0.224	0.157	0.1385	40
60	0.564	0.0301	2.94	0.0420	0.0053	5.28	0.229	0.164	0.1413	60
80	0.517	0.0311	3.05	0.0397	0.0056	5.61	0.234	0.174	0.1439	80
100	0.477	0.0324	3.15	0.0373	0.0060	5.94	0.240	0.185	0.1465	100
120	0.441	0.0399	3.26	0.0350	0.0064	6.27	0.251	0.199	0.1490	120
140	0.409	0.0359	3.36	0.0326	0.0068	6.60	0.266	0.216	0.1513	140
160	0.370	0.0384	3.47	0.0302	0.0072	6.94	0.288	0.235	0.1536	160
180	0.329	0.0417	3.57	0.0276	0.0076	7.28	0.317	0.290	0.1558	180
200	0.273	0.0458	3.67	0.0246	0.0083	7.63	0.356	0.362	0.1579	200
220	0.200	0.051	3.77	0.0204	0.0093	7.98	0.406		0.1599	220
230	0.149	0.060	3.82	0.0161	0.0107	8.16			0.1609	230
234	0.075	0.075	3.84	0.0130	0.0130	8.23			0.1612	234**
240			3.87			8.34			0.1618	240
260			3.96			8.71			0.1637	260
280			4.06			9.08			0.1654	280
300			4.15			9.45			0.1671	300
320			4.25			9.82			0.1687	320
340			4.34			10.1			0.1703	340
360			4.43			10.5			0.1718	360
380			4.53			10.8			0.1732	380
400			4.62			11.2			0.1746	400

* in lb/ft × hr
† in Btu/hr × ft × °F
‡ in Btu/lb × °F
§ at pressure of 1×10^{-2} atm
‖ at pressure of 1×10^{-3} atm
at pressure of 0 atm
** critical temperature

ASHRAE Handbook — Fundamentals

R-12 Pressure — Enthalpy Diagram

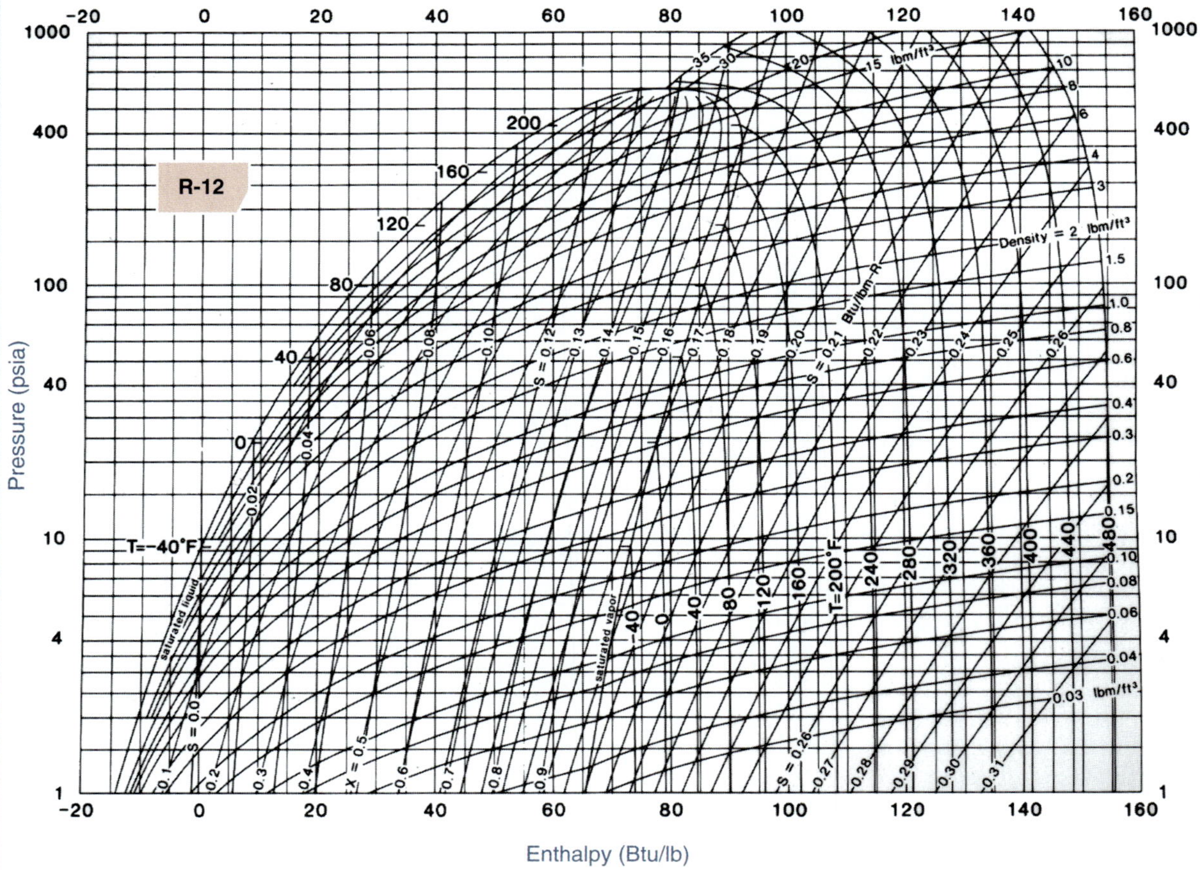

ASHRAE Handbook – Fundamentals

R-22—Properties of Saturated Liquid and Saturated Vapor

Temp °F	Pressure psia	Pressure psig	Volume† Vapor	Density‡ Liquid	Enthalpy§ Liquid	Enthalpy§ Vapor	Entropy‖ Liquid	Entropy‖ Vapor
–130	0.68858	28.519*	59.170	96.313	–24.388	89.888	–0.065456	0.28118
–120	1.0725	27.738*	39.078	95.416	–21.538	91.049	–0.056942	0.27452
–110	1.6199	26.623*	26.578	94.509	–18.738	92.211	–0.048818	0.26848
–100	2.3802	25.075*	18.558	93.590	–15.980	93.371	–0.041046	0.26298
–90	3.4111	22.976*	13.268	92.660	–13.259	94.527	–0.033590	0.25798
–80	4.7793	20.191*	9.6902	91.717	–10.570	95.676	–0.026418	0.25342
–75	5.6131	18.493*	8.3419	91.241	–9.2346	96.247	–0.022929	0.25128
–70	6.5603	16.564*	7.2139	90.761	–7.9050	96.815	–0.019500	0.24924
–65	7.6317	14.383*	6.2655	90.278	–6.5802	97.380	–0.016128	0.24728
–60	8.8386	11.926*	5.4641	89.791	–5.2593	97.942	–0.012808	0.24541
–50	11.707	6.0851*	4.2039	88.807	–2.6263	99.055	–0.006316	0.24189
–48	12.361	4.7548*	3.9962	88.608	–2.1007	99.275	–0.005039	0.24122
–46	13.042	3.3666*	3.8007	88.408	–1.5753	99.495	–0.003770	0.24056
–44	13.754	1.9186*	3.6168	88.208	–1.0501	99.714	–0.002507	0.23991
–42	14.495	0.4090*	3.4437	88.007	–0.5250	99.932	–0.001250	0.23927
–41.47	14.696	0.0	3.3997	87.954	–0.3865	99.990	–0.000920	0.23910
–40	15.268	0.5717	3.2805	87.806	0.0	100.15	0.0	0.23864
–38	16.072	1.3763	3.1267	87.604	0.5250	100.37	0.001244	0.23802
–36	16.910	2.2138	2.9816	87.401	1.0500	100.58	0.002482	0.23741
–34	17.781	3.0852	2.8446	87.197	1.5751	100.80	0.003714	0.23681
–32	18.687	3.9914	2.7152	86.993	2.1003	101.01	0.004940	0.23622
–30	19.629	4.9333	2.5930	86.788	2.6257	101.22	0.006161	0.23564
–28	20.608	5.9119	2.4774	86.582	3.1512	101.44	0.007377	0.23506
–26	21.624	6.9283	2.3680	86.375	3.6771	101.65	0.008587	0.23450
–24	22.679	7.9832	2.2645	86.168	4.2032	101.86	0.009792	0.23394
–22	23.774	9.0778	2.1664	85.960	4.7297	102.07	0.010993	0.23340
–20	24.909	10.213	2.0735	85.751	5.2566	102.28	0.012188	0.23285
–18	26.086	11.390	1.9854	85.542	5.7840	102.48	0.013379	0.23232
–16	27.306	12.610	1.9018	85.331	6.3119	102.69	0.014566	0.23180
–14	28.569	13.873	1.8225	85.120	6.8403	102.90	0.015748	0.23128
–12	29.877	15.181	1.7472	84.908	7.3693	103.10	0.016926	0.23077
–10	31.231	16.535	1.6757	84.695	7.8989	103.30	0.018100	0.23027
–8	32.632	17.936	1.6077	84.481	8.4292	103.51	0.019270	0.22977
–6	34.081	19.385	1.5430	84.266	8.9603	103.71	0.020436	0.22928
–4	35.579	20.883	1.4815	84.051	9.4921	103.91	0.021598	0.22880
–2	37.127	22.431	1.4230	83.834	10.025	104.10	0.022757	0.22832
0	38.726	24.030	1.3672	83.617	10.558	104.30	0.023912	0.22785
2	40.378	25.682	1.3141	83.399	11.093	104.50	0.025064	0.22738
4	42.083	27.387	1.2635	83.179	11.628	104.69	0.026213	0.22693
6	43.843	29.147	1.2152	82.959	12.164	104.89	0.027359	0.22647
8	45.658	30.962	1.1692	82.738	12.702	105.08	0.028502	0.22602
10	47.530	32.834	1.1253	82.516	13.240	105.27	0.029642	0.22558
12	49.461	34.765	1.0833	82.292	13.779	105.46	0.030779	0.22515
14	51.450	36.754	1.0433	82.068	14.320	105.64	0.031913	0.22471
16	53.501	38.805	1.0050	81.843	14.862	105.83	0.033045	0.22429
18	55.612	40.916	0.96841	81.616	15.405	106.02	0.034175	0.22387
20	57.786	43.090	0.93343	81.389	15.950	106.20	0.035302	0.22345
25	63.505	48.809	0.85246	80.815	17.317	106.65	0.038110	0.22243
30	69.641	54.945	0.77984	80.234	18.693	107.09	0.040905	0.22143

continued

* in in. Hg. vacuum
† in cu ft/lb
‡ in lb/cu ft
§ in Btu/lb
‖ in Btu/lb × °R
\# Critical point

continued

R-22—Properties of Saturated Liquid and Saturated Vapor

Temp. °F	Pressure psia	Pressure psig	Volume† Vapor	Density‡ Liquid	Enthalpy§ Liquid	Enthalpy§ Vapor	Entropy‖ Liquid	Entropy‖ Vapor
35	76.215	61.519	0.71454	79.645	20.078	107.52	0.043689	0.22046
40	83.246	68.550	0.65571	79.049	21.474	107.94	0.046464	0.21951
45	90.754	76.058	0.60258	78.443	22.880	108.35	0.049229	0.21858
50	98.758	84.062	0.55451	77.829	24.298	108.74	0.051987	0.21767
55	107.28	92.583	0.51093	77.206	25.728	109.12	0.054739	0.21677
60	116.34	101.64	0.47134	76.572	27.170	109.49	0.057486	0.21589
65	125.95	111.26	0.43531	75.928	28.626	109.84	0.060228	0.21502
70	136.15	121.45	0.40245	75.273	30.095	110.18	0.062968	0.21416
80	158.36	143.66	0.34497	73.926	33.077	110.80	0.068441	0.21246
90	183.14	168.44	0.29668	72.525	36.121	111.35	0.073911	0.21077
100	210.67	195.97	0.25582	71.061	39.233	111.81	0.079400	0.20907
110	241.13	226.44	0.22102	69.524	42.422	112.17	0.084906	0.20734
120	274.73	260.03	0.19118	67.901	45.694	112.42	0.090444	0.20554
130	311.66	296.96	0.16542	66.174	49.064	112.52	0.096033	0.20365
140	352.14	337.45	0.14300	64.319	52.550	112.47	0.10170	0.20161
150	396.42	381.72	0.12334	62.301	56.177	112.20	0.10749	0.19938
160	444.75	430.06	0.10590	60.068	59.989	111.67	0.11345	0.19684
170	497.46	482.76	0.090228	57.532	64.055	110.76	0.11970	0.19386
180	554.89	540.19	0.075819	54.533	68.504	109.30	0.12640	0.19018
190	617.52	602.82	0.061991	50.703	73.617	106.88	0.13399	0.18518
200	686.02	671.32	0.046923	44.671	80.406	101.99	0.14394	0.17666
205.07#	723.4	708.7	0.03123	32.03	91.58	91.58	0.1605	0.1605

* in in. Hg. vacuum
† in cu ft/lb
‡ in lb/cu ft
§ in Btu/lb
‖ in Btu/lb × °R
\# critical point

continued

continued

R-22—Properties of Saturated Liquid and Saturated Vapor

Temp. °F	Viscosity* Sat. Liquid	Viscosity* Sat. Vapor	Viscosity* Gas§	Thermal Conductivity† Sat. Liquid	Thermal Conductivity† Sat. Vapor	Thermal Conductivity† Gas‖	Specific Heat‡ Sat. Liquid	Specific Heat‡ Sat. Vapor	Specific Heat‡ Gas#	Temp. °F
–100	1.167			0.0789			0.255		0.1260	–100
–80	1.014			0.0757			0.256		0.1292	–80
–60	0.894			0.0725			0.259	0.139	0.1324	–60
–40	0.798	0.0245	2.45	0.0693	0.0040	4.04	0.262	0.146	0.1356	–40
–20	0.719	0.0257	2.57	0.0661	0.0044	4.43	0.266	0.152	0.1388	–20
0	0.654	0.0269	2.68	0.0630	0.0048	4.81	0.271	0.158	0.1420	0
20	0.599	0.0282	2.80	0.0598	0.0052	5.20	0.276	0.165	0.1452	20
40	0.553	0.0295	2.91	0.0566	0.0056	5.58	0.283	0.175	0.1484	40
60	0.513	0.0309	3.03	0.0534	0.0060	5.97	0.291	0.187	0.1515	60
80	0.480	0.0325	3.14	0.0502	0.0064	6.35	0.300	0.204	0.1546	80
100	0.449	0.0343	3.25	0.0471	0.0068	6.74	0.313	0.226	0.1577	100
120	0.427	0.0362	3.37	0.0439	0.0072	7.12	0.332	0.253	0.1608	120
140	0.392	0.0383	3.48	0.0407	0.0077	7.51	0.357	0.288	0.1638	140
160	0.344	0.0411	3.59	0.0371	0.0084	7.90	0.390	0.332	0.1668	160
180	0.285	0.045	3.70	0.0318	0.0105	8.28	0.433		0.1697	180
190	0.244	0.049	3.75	0.0288	0.0119	8.48			0.1712	190
200	0.182	0.058	3.81	0.0238	0.0140	8.67			0.1726	200
205**	0.074	0.074	3.83	0.0177	0.0177	8.76			0.1733	205**
220			3.92			9.05			0.1754	220
240			4.02			9.44			0.1782	240
300			4.34			10.59			0.1863	300
400			4.86			12.5			0.1983	400
440			5.06			13.3			0.2026	440

* in lb/ft × hr
† in Btu/hr × ft × °F
‡ in Btu/lb × °F
§ at pressure of 1 × 10^{-2} atm
‖ at pressure of 1 × 10^{-3} atm
at pressure of 0 atm
**critical temperature

ASHRAE Handbook — Fundamentals

R-22 Pressure—Enthalpy Diagram

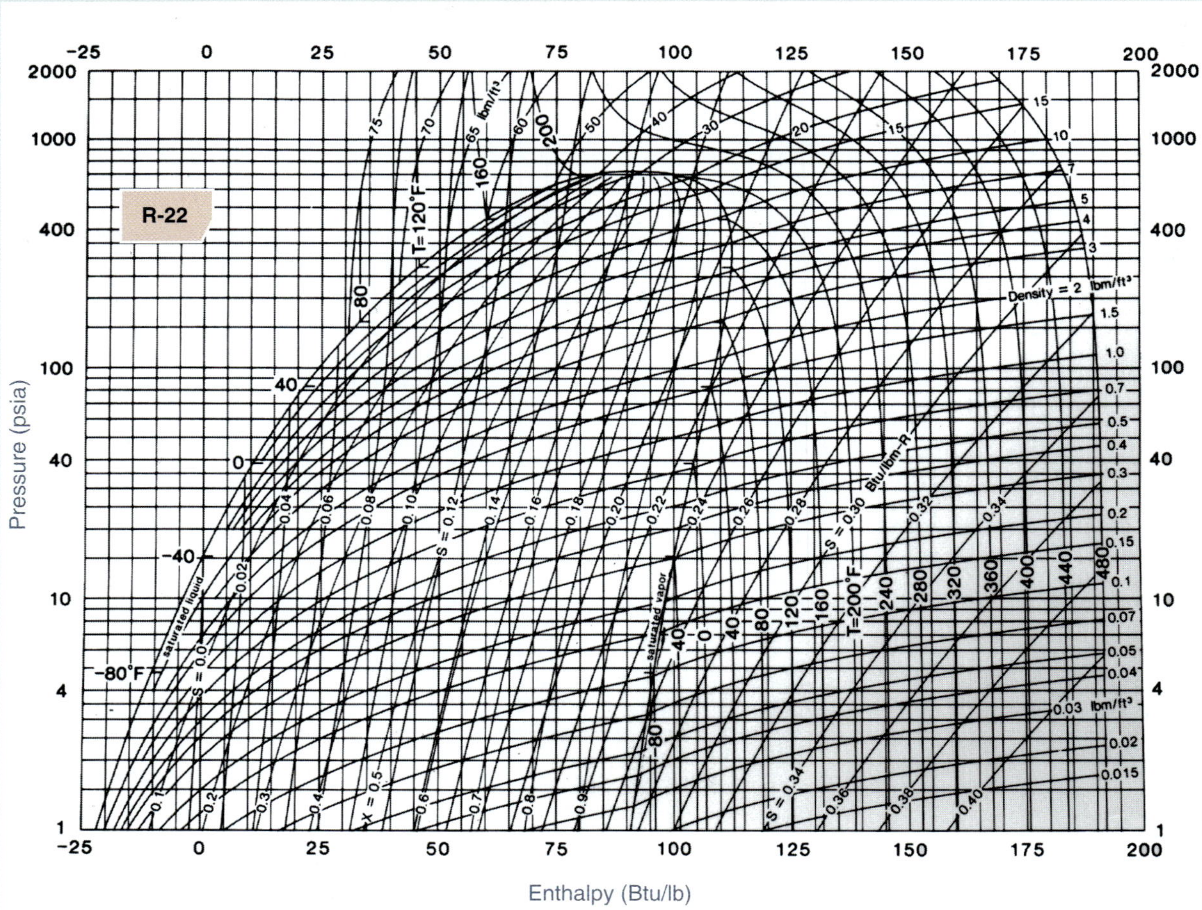

ASHRAE Handbook – Fundamentals

R-502—Properties of Saturated Liquid and Saturated Vapor

Temp. °F	Pressure psia	Pressure psig	Volume† Vapor	Density‡ Liquid	Enthalpy§ Liquid	Enthalpy§ Vapor	Entropy‖ Liquid	Entropy‖ Vapor
–75	7.2807	15.098*	4.9585	95.235	–7.6022	68.694	–0.018844	0.18020
–70	8.4343	12.749*	4.3241	94.699	–6.5643	69.570	–0.016169	0.17921
–65	9.7307	10.109*	3.7849	94.159	–5.5105	70.174	–0.013488	0.17828
–60	11.182	7.1539*	3.3248	93.615	–4.4406	70.777	–0.010802	0.17740
–55	12.802	3.8567*	2.9306	93.067	–3.3546	71.377	–0.008110	0.17656
–50	14.602	0.1906*	2.5915	92.514	–2.2525	91.975	–0.005412	0.17578
–49.75	14.696	0.0	2.5761	92.487	–2.1978	72.004	–0.005279	0.17574
–48	15.376	0.6804	2.4693	92.292	–1.8071	72.213	–0.004331	0.17547
–46	16.183	1.4865	2.3541	92.068	–1.3592	72.451	–0.003249	0.17518
–44	17.022	2.3256	2.2452	91.844	–0.9087	72.689	–0.002167	0.17489
–42	17.895	3.1986	2.1425	91.620	–0.4556	72.925	–0.001084	0.17461
–40	18.802	4.1064	2.0453	91.394	0.0	73.162	0.0	0.17433
–38	19.746	5.0500	1.9535	91.167	0.4582	73.398	0.001085	0.17406
–36	20.726	6.0303	1.8666	90.940	0.9189	73.633	0.002170	0.17380
–34	21.744	7.0482	1.7844	90.712	1.3822	73.867	0.003256	0.17354
–32	22.801	8.1048	1.7066	90.483	1.8481	74.101	0.004343	0.17329
–30	23.897	9.2010	1.6328	90.253	2.3165	74.335	0.005430	0.17304
–28	25.034	10.338	1.5629	90.022	2.7874	74.567	0.006518	0.17280
–26	26.212	11.516	1.4966	89.790	3.2608	74.799	0.007607	0.17257
–24	27.433	12.737	1.4336	89.557	3.7367	75.031	0.008696	0.17234
–22	28.697	14.001	1.3739	89.323	4.2152	75.261	0.009786	0.17211
–20	30.006	15.310	1.3172	89.088	4.6961	75.491	0.010876	0.17189
–18	31.361	16.665	1.2633	88.853	5.1795	75.720	0.011966	0.17168
–16	32.762	18.066	1.2120	88.616	5.6654	75.948	0.013057	0.17147
–14	34.211	19.515	1.1633	88.378	6.1537	76.175	0.014149	0.17126
–12	35.709	21.013	1.1169	88.139	6.6445	76.402	0.015240	0.17106
–10	37.256	22.560	1.0727	87.889	7.1377	76.627	0.016332	0.17087
–8	38.854	24.158	1.0307	87.658	7.6333	76.852	0.017425	0.17067
–6	40.504	25.808	0.99066	87.416	8.1313	77.075	0.018517	0.17049
–4	42.207	27.511	0.95248	87.172	8.6318	77.298	0.019610	0.17030
–2	43.964	29.268	0.91608	86.928	9.1346	77.520	0.020703	0.17012
0	45.776	31.080	0.88135	86.682	9.6397	77.741	0.021796	0.16995
2	47.644	32.948	0.84821	86.435	10.147	77.960	0.022889	0.16978
4	49.569	34.873	0.81657	86.187	10.657	78.197	0.023982	0.16961
6	51.552	36.856	0.87636	85.937	11.169	78.397	0.025075	0.16944
8	53.594	38.898	0.75749	85.686	11.684	78.613	0.026168	0.16928

* in in. Hg vacuum
† in cu ft/lb
‡ in lb/cu ft
§ in Btu/lb
‖ in Btu/lb × °R

continued

continued

R–502 — Properties of Saturated Liquid and Saturated Vapor

Temp. °F	Pressure psia	Pressure psig	Volume† Vapor	Density‡ Liquid	Enthalpy§ Liquid	Enthalpy§ Vapor	Entropy‖ Liquid	Entropy‖ Vapor
10	55.697	41.001	0.72989	85.434	12.200	78.828	0.027261	0.16912
15	61.226	46.530	0.66605	84.798	13.502	79.362	0.029992	0.16874
20	67.155	52.459	0.60884	84.153	14.818	79.887	0.032722	0.16838
25	73.503	58.807	0.55746	83.498	16.147	80.405	0.035450	0.16803
30	80.287	65.591	0.51121	82.833	17.490	80.913	0.038176	0.16770
35	87.523	72.827	0.46948	82.157	18.846	81.413	0.040897	0.16738
40	95.229	80.533	0.43175	81.470	20.216	81.903	0.043617	0.13707
45	103.42	88.726	0.39758	80.771	21.597	82.383	0.046331	0.16678
50	112.12	97.425	0.36656	80.059	22.991	82.852	0.049040	0.16649
55	121.34	106.65	0.33834	79.333	24.397	83.310	0.051743	0.46621
60	131.10	116.41	0.31264	78.592	25.814	83.755	0.054440	0.16594
65	141.42	126.73	0.28916	77.835	27.244	84.187	0.057131	0.16566
70	152.32	137.63	0.26769	77.062	28.685	84.606	0.059816	0.16539
75	163.81	149.12	0.24802	76.269	30.138	85.009	0.062494	0.16512
80	175.92	161.23	0.22994	75.457	31.602	85.397	0.065165	0.16484
85	188.66	173.97	0.21333	74.623	33.078	85.767	0.067829	0.16456
90	202.06	187.36	0.19802	73.765	34.566	86.118	0.070487	0.16427
95	216.13	201.43	0.18388	72.881	36.066	86.449	0.073139	0.16397
100	230.89	216.20	0.17079	71.968	37.578	86.758	0.075786	0.16366
105	246.38	231.68	0.15866	71.023	39.104	87.402	0.078429	0.16332
110	262.61	247.92	0.04740	70.042	40.644	87.298	0.081070	0.16297
115	279.61	264.92	0.13691	69.022	42.201	87.524	0.083711	0.16258
120	297.41	282.72	0.12711	67.956	43.774	87.716	0.086354	0.16216
125	316.05	301.35	0.11795	66.839	45.369	87.869	0.089005	0.16170
130	335.54	320.85	0.10935	65.661	46.987	87.977	0.091667	0.16118
135	355.94	341.25	0.10125	64.414	48.634	88.032	0.094350	0.16060
140	377.30	362.60	0.093586	63.083	50.316	88.024	0.097063	0.15994
145	399.65	384.95	0.086303	61.650	52.045	87.941	0.099822	0.15919
150	423.06	408.37	0.079335	60.092	53.834	87.763	0.10265	0.15830
155	447.61	432.92	0.072610	58.371	55.705	87.463	0.10558	0.15724
160	473.39	458.69	0.066037	56.429	57.698	86.997	0.10867	0.15595
165	500.50	485.81	0.059485	54.169	59.878	86.293	0.11202	0.15430
170	529.11	514.41	0.052715	51.391	62.386	85.198	0.11584	0.15207
175	559.42	544.72	0.045119	47.551	65.610	83.303	0.12073	0.14861
#179.9	591.0	576.3	0.02857	35.00	74.81	74.81	0.1346	0.1346

* in in. Hg vacuum
† in cu ft/lb
‡ in lb/cu ft
§ in Btu/lb
‖ in Btu/lb × °R
\# critical point

continued

continued

R-502 — Properties of Saturated Liquid and Saturated Vapor

Temp. °F	Viscosity†			Thermal Conductivity§			Specific Heat#			Temp. °F
	Sat. Liquid	Sat. Vapor	Gas‡	Sat. Liquid	Sat. Vapor	Gas‖	Sat. Liquid	Sat. Vapor	Gas**	
–100	1.39			0.0595			0.244			–100
–80	1.16			0.0570			0.248			–80
–60	1.00	0.0228		0.0545			0.253	0.138		–60
–40	0.86	0.0244	2.42	0.0519	0.0046	4.58	0.259	0.149	0.148	–40
–20	0.76	0.0258	2.54	0.0494	0.0050	4.95	0.264	0.155	0.151	–20
0	0.67	0.0270	2.66	0.0469	0.0053	5.31	0.271	0.160	0.154	0
20	0.60	0.0283	2.77	0.0444	0.0057	5.67	0.277	0.164	0.157	20
40	0.54	0.0295	2.89	0.0419	0.0060	6.03	0.285	0.171	0.160	40
60	0.487	0.0310	3.01	0.0394	0.0064	6.39	0.292	0.180	0.164	60
80	0.433	0.0327	3.12	0.0369	0.0068	6.76	0.300	0.195	0.167	80
100	0.380	0.0348	3.23	0.0344	0.0071	7.14	0.308	0.218	0.170	100
120	0.329	0.0373	3.34	0.0314	0.0075	7.52	0.316	0.249	0.173	120
140	0.284	0.039	3.45	0.0281	0.0083	7.91	0.326	0.310	0.176	140
160	0.243	0.045	3.56	0.0237	0.0090	8.31	0.335		0.178	160
170	0.207	0.053	3.62	0.021	0.0103	8.52	0.345		0.179	170
180*	0.074	0.074	3.67	0.014	0.014	8.73			0.181	180††
190			3.72			8.94			0.182	190
200			3.78			9.16			0.183	200
220			3.88			9.60			0.186	220
240			3.99			10.07			0.188	240
260			4.10			10.5			0.190	260
280			4.20			11.0			0.192	280
300			4.29			11.6			0.193	300
320						12.1				320
340						12.7				340
360						13.3				360
380						13.9				380
440						16.0				440
460						16.7				460
500						18.4				500

* critical temperature
† in Btu/hr × ft × °F
‡ at pressure of 1×10^{-2} atm
§ in Btu/hr × ft × °F
‖ at pressure of 1×10^{-3} atm
in Btu/hr × ft × °F
** at pressure of 1 atm
†† critical temperature

ASHRAE Handbook — Fundamentals

R-502 Pressure—Enthalpy Diagram

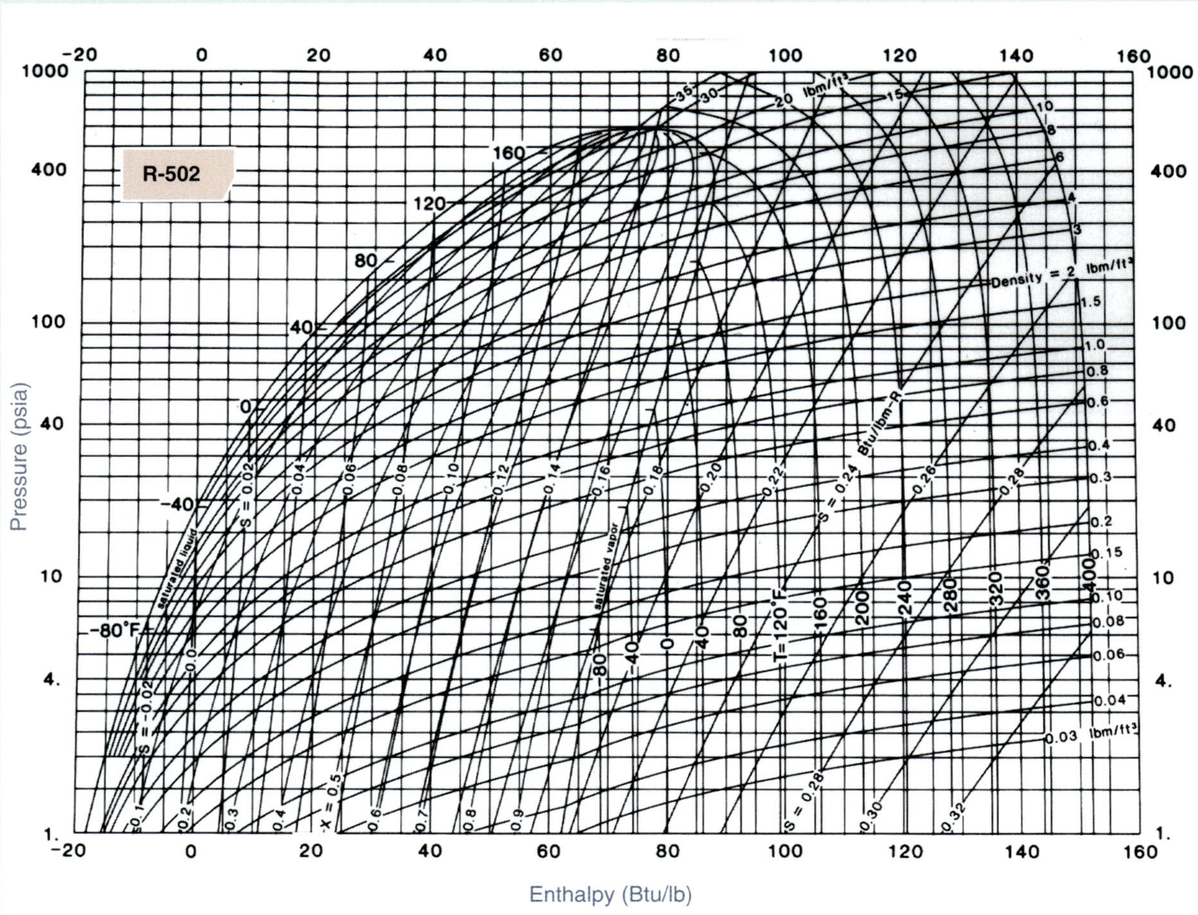

ASHRAE Handbook – Fundamentals

R–134a—Properties of Saturated Liquid and Saturated Vapor

Temp. °F	Vapor Pressure psia	Volume* Vapor	Density† Liquid	Enthalpy‡ Liquid	Enthalpy‡ Vapor	Entropy§ Liquid	Entropy§ Vapor
–150	0.073	454.5455	98.86	–31.2	80.9	–0.0859	0.2761
–140	0.130	256.4103	97.91	–28.4	82.3	–0.0771	0.2693
–130	0.222	156.2500	96.96	–25.6	83.8	–0.0686	0.2633
–120	0.366	97.0874	96.01	–22.9	85.2	–0.0604	0.2579
–110	0.584	62.5000	95.06	–20.1	86.7	–0.0523	0.2531
–100	0.903	41.6667	94.11	–17.3	88.2	–0.0444	0.2488
–90	1.358	28.4091	93.15	–14.5	89.6	–0.0367	0.2450
–85	1.649	23.6407	92.68	–13.1	90.4	–0.0329	0.2433
–80	1.991	19.8413	92.20	–11.6	91.1	–0.0291	0.2416
–75	2.389	16.7224	91.72	–10.2	91.9	–0.0254	0.2401
–70	2.850	14.1844	91.24	–8.8	92.7	–0.0217	0.2386
–65	3.384	12.0773	90.76	–7.3	93.4	–0.0180	0.2373
–60	3.996	10.3306	90.27	–5.9	94.2	–0.0143	0.2360
–55	4.696	8.8810	89.78	–4.4	94.9	–0.0107	0.2348
–50	5.493	7.6687	89.30	–3.0	95.7	–0.0071	0.2336
–45	6.397	6.6489	88.81	–1.5	96.4	–0.0035	0.2326
–40	7.417	5.7904	88.31	0.0	97.2	0.0000	0.2316
–38	7.860	5.4825	88.11	0.6	97.5	0.0014	0.2312
–36	8.325	5.1948	87.91	1.2	97.8	0.0028	0.2308
–34	8.811	4.9261	87.71	1.8	98.1	0.0042	0.2304
–32	9.319	4.6729	87.51	2.4	98.4	0.0056	0.2301
–30	9.851	4.4366	87.31	3.0	98.7	0.0070	0.2297
–28	10.407	4.2141	87.11	3.6	99.0	0.0084	0.2294
–26	10.987	4.0032	86.91	4.2	99.3	0.0098	0.2291
–24	11.594	3.8066	86.71	4.8	99.6	0.0112	0.2288
–22	12.226	3.6206	86.51	5.4	99.9	0.0126	0.2284
–20	12.885	3.4471	86.30	6.0	100.2	0.0139	0.2281
–18	13.572	3.2819	86.10	6.6	100.5	0.0153	0.2278
–16	14.289	3.1270	85.89	7.2	100.8	0.0167	0.2276
–14	15.035	2.9806	85.69	7.8	101.1	0.0180	0.2273
–12	15.812	2.8417	85.48	8.4	101.4	0.0194	0.2270
–10	16.620	2.7115	85.28	9.0	101.7	0.0208	0.2267
–8	17.461	2.5880	85.07	9.7	102.0	0.0221	0.2265
–6	18.334	2.4716	84.86	10.3	102.3	0.0235	0.2262
–4	19.242	2.3613	84.65	10.9	102.5	0.0248	0.2260
–2	20.184	0.4337	84.44	11.5	102.8	0.0262	0.2258
0	21.163	0.3999	84.23	12.1	103.1	0.0275	0.2255
2	22.178	0.3408	84.02	12.7	103.4	0.0288	0.2253
4	23.231	0.2912	83.81	13.4	103.7	0.0302	0.2251
6	24.322	0.2494	83.60	14.0	104.0	0.0315	0.2249

* in cu ft/lb
† in lb/cu ft
‡ in Btu/lb
§ in Btu/lb × °R

continued

continued

R-134a—Properties of Saturated Liquid and Saturated Vapor

Temp. °F	Vapor Pressure psia	Volume* Vapor	Density† Liquid	Enthalpy‡		Entropy§	
				Liquid	Vapor	Liquid	Vapor
8	25.454	2.2568	83.38	14.6	104.3	0.0328	0.2247
10	26.625	2.1580	83.17	15.2	104.6	0.0342	0.2244
12	27.839	2.0644	82.95	15.9	104.9	0.0355	0.2243
14	29.095	1.9755	82.73	16.5	105.2	0.0368	0.2241
16	30.395	1.8911	82.52	17.1	105.5	0.0381	0.2239
18	31.739	1.8113	82.30	17.7	105.7	0.0395	0.2237
20	33.129	1.7355	82.08	18.4	106.0	0.0408	0.2235
22	34.566	1.6633	81.86	19.0	106.3	0.0421	0.2233
24	36.050	1.5949	81.64	19.6	106.6	0.0434	0.2232
26	37.583	1.5298	81.41	20.3	106.9	0.0447	0.2230
28	39.166	1.4678	81.19	20.9	107.2	0.0460	0.2229
30	40.800	1.4090	80.96	21.6	107.4	0.0473	0.2227
32	42.486	1.3528	80.74	22.2	107.7	0.0486	0.2226
34	44.225	1.2995	80.51	22.8	108.0	0.0499	0.2224
36	46.018	1.2486	80.28	23.5	108.3	0.0512	0.2223
38	47.866	1.2000	80.05	24.1	108.6	0.0525	0.2221
40	49.771	1.1538	79.82	24.8	108.8	0.0538	0.2220
45	54.787	1.1098	79.24	26.4	109.5	0.0570	0.2217
50	60.180	1.0676	78.64	28.0	110.2	0.0602	0.2214
55	65.963	1.0274	78.04	29.7	110.9	0.0634	0.2211
60	72.167	0.9890	77.43	31.4	111.5	0.0666	0.2208
65	78.803	0.9523	76.81	33.0	112.2	0.0698	0.2206
70	85.890	0.8675	76.18	34.7	112.8	0.0729	0.2203
75	93.447	0.7916	75.54	36.4	113.4	0.0761	0.2201
80	101.494	0.7234	74.89	38.1	114.0	0.0792	0.2199
85	110.050	0.6622	74.22	39.9	114.6	0.0824	0.2196
90	119.38	0.6069	73.54	41.6	115.2	0.0855	0.2194
100	138.996	0.5570	72.13	45.1	116.3	0.0918	0.2190
110	161.227	0.5119	70.66	48.7	117.4	0.0981	0.2185
120	186.023	0.4709	69.10	52.4	118.3	0.1043	0.2181
130	213.572	0.2139	67.46	56.2	119.2	0.1106	0.2175
140	244.068	0.1835	65.70	60.0	119.9	0.1170	0.2168
150	277.721	0.1573	63.81	64.0	120.5	0.1234	0.2160
160	314.800	0.1344	61.74	68.1	120.8	0.1299	0.2150
170	355.547	0.1143	59.45	72.4	120.9	0.1366	0.2137
180	400.280	0.0964	56.83	76.9	120.7	0.1435	0.2119
190	449.384	0.0800	53.73	81.8	119.8	0.1508	0.2093
200	503.361	0.0645	49.73	87.3	118.0	0.1589	0.2054
210	563.037	0.0474	43.19	94.4	113.4	0.1693	0.1976

* in cu ft/lb
† in lb/cu ft
‡ in Btu/lb
§ in Btu/lb × °R

R-134a Pressure—Enthalpy Diagram

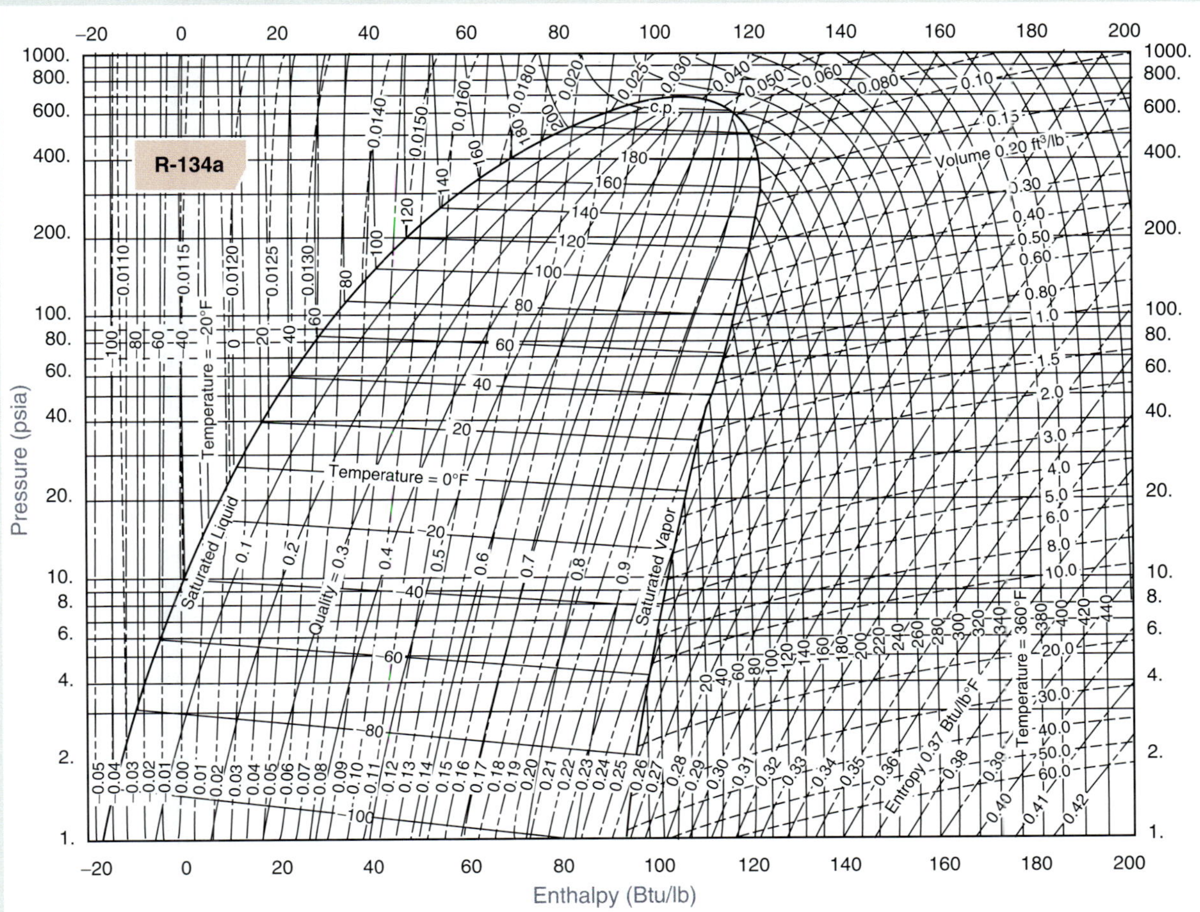

ASHRAE Handbook – Fundamentals

R–407c—Properties of Saturated Liquid and Saturated Vapor

Temp. °F	Vapor Pressure psia	Volume† Vapor	Density‡ Liquid	Enthalpy§		Entropy‖	
				Liquid	Vapor	Liquid	Vapor
–150	0.15*	250.0000	99.10	–31.0	87.8	–0.0852	0.3073
–140	0.26*	151.5152	97.94	–28.3	89.2	–0.0767	0.2993
–130	0.44*	93.4579	96.78	–25.6	90.7	–0.0684	0.2921
–120	0.71*	59.5238	95.62	–22.9	92.2	–0.0603	0.2985
–110	1.10*	39.3701	94.45	–20.1	93.7	–0.0524	0.2799
–100	1.66*	26.7380	93.27	–17.4	95.2	–0.0445	0.2747
–90	2.45*	18.6220	92.09	–14.6	96.8	–0.0368	0.2701
–85	2.95*	15.6740	91.50	–13.1	97.5	–0.0331	0.2679
–80	3.53*	13.2802	90.90	–11.7	98.3	–0.0293	0.2659
–75	4.19*	11.2994	90.31	–10.3	99.1	–0.0255	0.2639
–70	4.96*	9.6525	89.71	–8.9	99.9	–0.0218	0.2621
–65	5.84*	8.2919	89.11	–7.4	100.6	–0.0181	0.2604
–60	6.84*	7.1531	88.51	–5.9	101.4	–0.0145	0.2587
–55	7.98*	6.1996	87.90	–4.5	102.2	–0.0108	0.2572
–50	9.26*	5.3937	87.30	–3.0	103.0	–0.0072	0.2557
–45	10.71*	4.7103	86.69	–1.5	103.7	–0.0036	0.2543
–40	12.33*	4.1288	86.08	0.0	104.5	0.0000	0.2530
–38	13.03*	3.9200	85.84	0.6	104.8	0.0014	0.2525
–36	13.76*	3.7244	85.59	1.2	105.1	0.0029	0.2520
–34	14.53	3.5411	85.35	1.8	105.4	0.0043	0.2515
–32	15.32	3.3681	85.10	2.6	105.7	0.0060	0.2510
–30	16.16	3.2051	84.86	3.2	106.0	0.0074	0.2505
–28	17.02	3.0516	84.61	3.8	106.3	0.0088	0.2501
–26	17.93	2.9070	84.37	4.4	106.6	0.0102	0.2496
–24	18.87	2.7701	84.12	5.0	106.9	0.0116	0.2492
–22	19.85	2.6413	83.87	5.6	107.2	0.0130	0.2487
–20	20.87	2.5195	83.62	6.2	107.6	0.0144	0.2483
–18	21.94	2.4050	83.37	6.8	107.9	0.0158	0.2479
–16	23.04	2.2962	83.13	7.3	108.2	0.0168	0.2475
–14	24.19	2.1935	82.88	7.9	108.4	0.0183	0.2471
–12	25.38	2.0960	82.63	8.6	108.7	0.0197	0.2467
–10	26.62	2.0036	82.38	9.2	109.0	0.0211	0.2463
–8	27.91	1.9164	82.13	9.8	109.3	0.0225	0.2459
–6	29.24	1.8335	81.87	10.5	109.6	0.0239	0.2456
–4	30.63	1.7550	81.62	11.1	109.9	0.0253	0.2452
–2	32.06	1.6804	81.37	11.7	110.2	0.0266	0.2448
0	33.55	1.6098	81.12	12.4	110.5	0.0280	0.2445
2	35.09	1.5425	80.86	13.0	110.8	0.0294	0.2441

* in in. Hg vacuum
† in cu ft/lb
‡ in lb/cu ft
§ in Btu/lb
‖ in Btu/lb × °R

continued

continued

R-407c—Properties of Saturated Liquid and Saturated Vapor

Temp. °F	Vapor Pressure psia	Volume*† Vapor	Density† Liquid	Enthalpy‡		Entropy§	
				Liquid	Vapor	Liquid	Vapor
4	36.69	1.4786	80.61	13.7	111.1	0.0308	0.2438
6	38.34	1.4180	80.36	14.3	111.4	0.0322	0.2435
8	40.05	1.3604	80.10	15.0	111.7	0.0336	0.2432
10	41.82	1.3053	79.85	15.6	112.0	0.0350	0.2428
12	43.64	1.2531	79.59	16.2	112.2	0.0363	0.2425
14	45.53	1.2032	79.33	16.9	112.5	0.0376	0.2422
16	47.49	1.1558	79.07	17.6	112.8	0.0390	0.2419
18	49.51	1.1105	78.82	18.2	113.1	0.0404	0.2416
20	51.59	1.0674	78.56	18.9	113.4	0.0418	0.2413
22	53.74	1.0261	78.30	19.6	113.6	0.0432	0.2410
24	55.96	0.9868	78.04	20.2	113.9	0.0446	0.2407
26	58.25	0.9493	77.78	20.9	114.2	0.0460	0.2404
28	60.61	0.9134	77.51	21.6	114.5	0.0474	0.2402
30	63.05	0.8792	77.25	22.3	114.7	0.0487	0.2399
32	65.56	0.8465	76.99	23.0	115.0	0.0501	0.2396
34	68.15	0.8151	76.72	23.7	115.3	0.0515	0.2393
36	70.81	0.7852	76.46	24.4	115.5	0.0529	0.2391
38	73.55	0.7565	76.19	25.1	115.8	0.0543	0.2388
40	76.38	0.7291	75.93	25.8	116.1	0.0557	0.2386
45	83.81	0.6656	75.26	27.5	116.7	0.0591	0.2379
50	91.78	0.6082	74.58	29.3	117.3	0.0626	0.2373
55	100.31	0.5566	73.89	31.1	117.9	0.0661	0.2367
60	109.45	0.5099	73.20	32.9	118.5	0.0696	0.2361
65	119.21	0.4677	72.51	34.8	119.1	0.0731	0.2355
70	129.61	0.4294	71.80	36.7	119.6	0.0766	0.2349
75	140.70	0.3946	71.08	38.6	120.1	0.0801	0.2343
80	152.50	0.3629	70.36	40.5	120.7	0.0836	0.2337
85	165.04	0.3340	69.62	42.5	121.1	0.0872	0.2331
90	178.36	0.3076	68.88	44.5	121.6	0.0907	0.2324
100	207.44	0.2612	67.34	48.6	122.4	0.0980	0.2312
110	240.01	0.2220	65.74	52.8	123.0	0.1053	0.2298
120	276.37	0.1887	64.06	57.3	123.5	0.1129	0.2282
130	316.84	0.1602	62.28	62.0	123.8	0.1207	0.2264
140	361.78	0.1355	60.36	66.9	123.8	0.1287	0.2244
150	411.58	0.1139	58.23	72.3	123.5	0.1373	0.2219
160	466.74	0.0947	55.79	78.1	122.6	0.1464	0.2188
170	527.92	0.0773	52.78	84.8	121.0	0.1567	0.2146

* in cu ft/lb
† in lb/cu ft
‡ in Btu/lb
§ in Btu/lb × °R

R-407c Pressure—Enthalpy Diagram

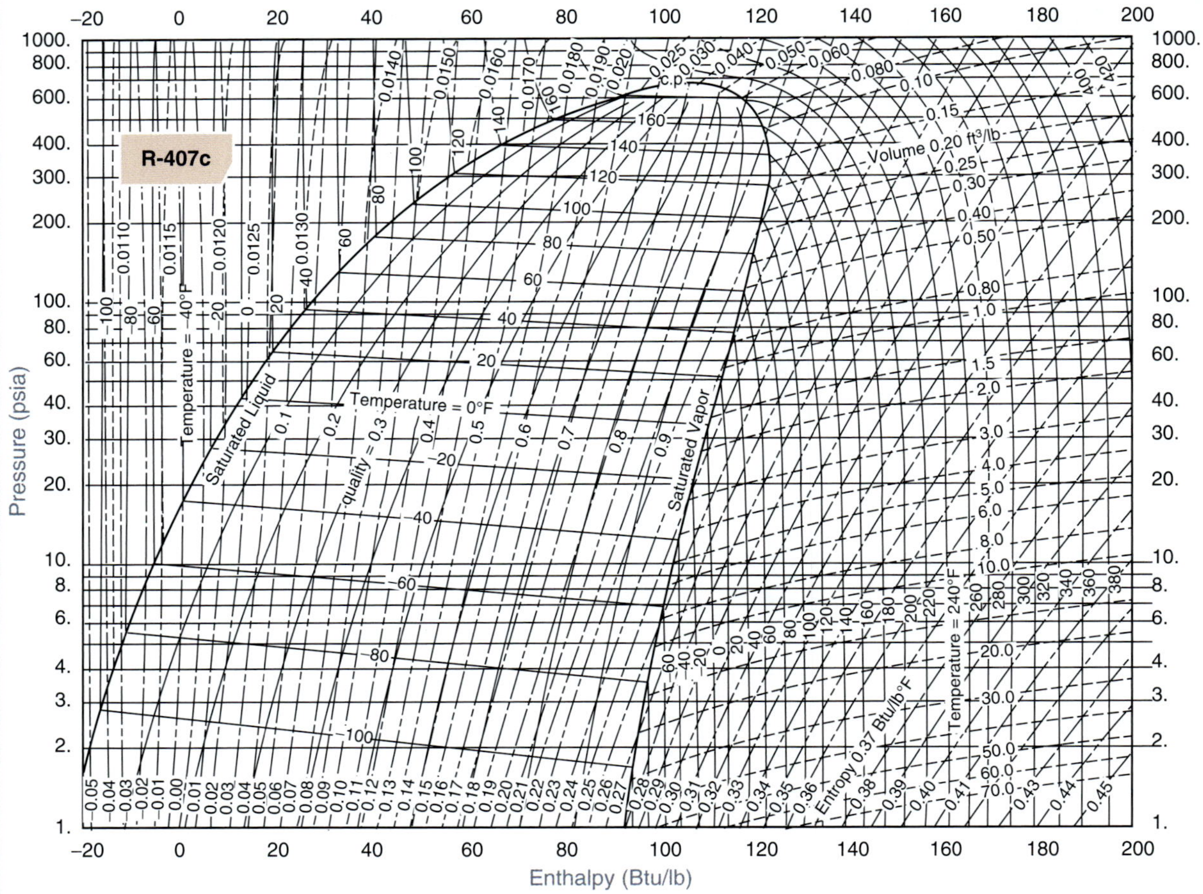

ASHRAE Handbook – Fundamentals

Table 1 – Suva® 410A Saturation Properties – Temperature Table

Temp. °F	Pressure (psia) Liquid (P_f)	Volume (ft³/lb) Vapor (v_g)	Density (lb/ft³) Liquid ($1/v_f$)	Enthalpy (Btu/lb) Liquid (h_f)	Enthalpy (Btu/lb) Vapor (h_g)	Entropy (Btu/(lb)(°R) Liquid (s_f)	Entropy (Btu/(lb)(°R) Vapor (s_g)
–150	0.49	94.0107	94.39	–34.1	100.2	–0.0946	0.3390
–149	0.52	89.5452	94.30	–33.8	100.3	–0.0937	0.3380
–148	0.55	85.3254	94.20	–33.5	10.5	–0.0928	0.3371
–147	0.58	81.3361	94.10	–33.2	100.6	–0.0919	0.3361
–146	0.60	77.5632	94.01	–32.9	100.7	–0.0910	0.3352
–145	0.64	73.9935	93.91	–32.6	100.9	–0.0900	0.3343
–144	0.67	70.6147	93.81	–32.4	101.0	–0.0891	0.3334
–143	0.70	67.4151	93.72	–32.1	101.2	–0.0882	0.3325
–142	0.74	64.3842	93.62	–31.8	101.3	–0.0873	0.3316
–141	0.77	61.5119	93.52	–31.5	101.4	–0.0864	0.3307
–140	0.81	58.7888	93.42	–31.2	101.6	–0.0855	0.3298
–139	0.85	56.2061	93.33	–30.9	101.7	–0.0846	0.3289
–138	0.89	53.7558	93.23	–30.6	101.8	–0.0838	0.3281
–137	0.93	51.4301	93.13	–30.3	102.0	–0.0829	0.3272
–136	0.98	49.2219	93.03	–30.0	102.1	–0.0820	0.3264
–135	1.02	47.1244	92.93	–29.7	102.3	–0.0811	0.3255
–134	1.07	45.1414	92.84	–29.5	102.4	–0.0802	0.3247
–133	1.12	43.2370	92.74	–29.2	102.5	–0.0793	0.3239
–132	1.17	41.4356	92.64	–28.9	102.7	–0.0784	0.3231
–131	1.23	39.7222	92.54	–28.6	102.8	–0.0775	0.3223
–130	1.28	38.0917	92.44	–28.3	102.9	–0.0766	0.3215
–129	1.34	36.5397	92.34	–28.0	103.1	–0.0757	0.3207
–128	1.40	35.0619	92.24	–27.7	103.2	–0.0748	0.3199
–127	1.46	33.6542	92.14	–27.4	103.4	–0.0739	0.3191
–126	1.52	32.3129	92.04	–27.1	103.5	–0.0730	0.3184
–125	1.59	31.0344	91.94	–26.8	103.6	–0.0721	0.3176
–124	1.66	29.8154	91.84	–26.5	103.8	–0.0713	0.3169
–123	1.73	28.6527	91.74	–26.2	103.9	–0.0704	0.3161
–122	1.81	27.5434	91.64	–25.9	104.0	–0.0695	0.3154
–121	1.88	26.4846	91.54	–25.6	104.2	–0.0686	0.3147
–120	1.96	25.4738	91.44	–25.3	104.3	–0.0677	0.3139
–119	2.04	24.5085	91.34	–25.0	104.5	–0.0668	0.3132
–118	2.13	23.5863	91.24	–24.7	104.6	–0.0660	0.3125
–117	2.22	22.7051	91.14	–24.4	104.7	–0.0651	0.3118
–116	2.31	21.8628	91.04	–24.1	104.9	–0.0642	0.3111
–115	2.40	21.0573	90.93	–23.8	105.0	–0.0633	0.3104
–114	2.50	20.2870	90.83	–23.5	105.1	–0.0624	0.3097
–113	2.60	19.5499	90.73	–23.2	105.3	–0.0616	0.3091
–112	2.70	18.8446	90.63	–22.9	105.4	–0.0607	0.3084
–111	2.81	18.1694	90.53	–22.6	105.5	–0.0598	0.3077
–110	2.92	17.5228	90.42	–22.3	105.7	–0.0589	0.3071
–109	3.03	16.9035	90.32	–22.0	105.8	–0.0580	0.3064
–108	3.15	16.3101	90.22	–21.7	105.9	–0.0572	0.3058
–107	3.27	15.7415	90.11	–21.4	106.1	–0.0563	0.3051
–106	3.40	15.1963	90.01	–21.1	106.2	–0.0554	0.3045
–105	3.53	14.6736	89.91	–20.8	106.3	–0.0546	0.3039
–104	3.66	14.1722	89.81	–20.5	106.5	–0.0537	0.3033
–103	3.80	13.6912	89.70	–20.2	106.6	–0.0528	0.3027
–102	3.94	13.2296	89.60	–19.9	106.7	–0.0520	0.3020
–101	4.08	12.7865	89.49	–19.6	106.9	–0.0511	0.3014
–100	4.23	12.3611	89.39	–19.3	107.0	–0.0502	0.3008
–99	4.39	11.9525	89.29	–19.0	107.1	–0.0494	0.3002
–98	4.54	11.5599	89.18	–18.6	107.3	–0.0485	0.2997
–97	4.71	11.1828	89.08	–18.3	107.4	–0.0476	0.2991

continued

continued

Table 1 – Suva® 410A Saturation Properties – Temperature Table

Temp. °F	Pressure (psia) Liquid (P_f)	Volume (ft³/lb) Vapor (v_g)	Density (lb/ft³) Liquid ($1/v_f$)	Enthalpy (Btu/lb) Liquid (h_f)	Enthalpy (Btu/lb) Vapor (h_g)	Entropy (Btu/(lb)(°R) Liquid (s_f)	Entropy (Btu/(lb)(°R) Vapor (s_g)
–96	4.88	10.8203	88.97	–18.0	107.5	–0.0468	0.2985
–95	5.05	10.4718	88.87	–17.7	107.7	–0.0459	0.2979
–94	5.23	10.1367	88.76	–17.4	107.8	–0.0450	0.2974
–93	5.41	9.8144	88.66	–17.1	107.9	–0.0442	0.2968
–92	5.60	9.5043	88.55	–16.8	108.1	–0.0433	0.2962
–91	5.79	9.2059	88.44	–16.5	108.2	–0.0425	0.2957
–90	5.99	8.9187	88.34	–16.2	108.3	–0.0416	0.2951
–89	6.19	8.6422	88.23	–15.9	108.4	–0.0408	0.2946
–88	6.40	8.3760	88.12	–15.5	108.6	–0.0399	0.2940
–87	6.62	8.1196	88.02	–15.2	108.7	–0.0390	0.2935
–86	6.84	7.8725	87.91	–14.9	108.8	–0.0382	0.2930
–85	7.07	7.6345	87.80	–14.6	109.0	–0.0373	0.2925
–84	7.30	7.4050	87.70	–14.3	109.1	–0.0365	0.2919
–83	7.54	7.1839	87.59	–14.0	109.2	–0.0356	0.2914
–82	7.79	6.9706	87.48	–13.7	109.4	–0.0348	0.2909
–81	8.04	6.7649	87.37	–13.3	109.5	–0.0339	0.2904
–80	8.30	6.5665	87.27	–13.0	109.6	–0.0331	0.2899
–79	8.56	6.3750	87.16	–12.7	109.7	–0.0323	0.2894
–78	8.84	6.1903	87.05	–12.4	109.9	–0.0314	0.2889
–77	9.12	6.0120	86.94	–12.1	110.0	–0.0306	0.2884
–76	9.40	5.8398	86.83	–11.8	110.1	–0.0297	0.2879
–75	9.70	5.6736	86.72	–11.4	110.2	–0.0289	0.2874
–74	10.00	5.5130	86.61	–11.1	110.4	–0.0280	0.2869
–73	10.30	5.3579	86.50	–10.8	110.5	–0.0272	0.2865
–72	10.62	5.2080	86.39	–10.5	110.6	–0.0264	0.2860
–71	10.94	5.0632	86.28	–10.2	110.7	–0.0255	0.2855
–70	11.28	4.9232	86.17	–9.8	110.9	–0.0247	0.2851
–69	11.61	4.7878	86.06	–9.5	111.0	–0.0238	0.2846
–68	11.96	4.6570	85.95	–9.2	111.1	–0.0230	0.2841
–67	12.32	4.5304	85.84	–8.9	111.2	–0.0222	0.2837
–66	12.68	4.4080	85.73	–8.5	111.4	–0.0213	0.2832
–65	13.05	4.2895	85.62	–8.2	111.5	–0.0205	0.2828
–64	13.43	4.1749	85.51	–7.9	111.6	–0.0197	0.2823
–63	13.82	4.0639	85.40	–7.6	111.7	–0.0189	0.2819
–62	14.22	3.9565	85.29	–7.3	111.8	–0.0180	0.2815
–61	14.63	3.8526	85.17	–6.9	112.0	–0.0172	0.2810
–60	15.05	3.7519	85.06	–6.6	112.1	–0.0164	0.2806
–59	15.47	3.6543	84.95	–6.3	11.2	–0.0155	0.2802
–58	15.91	3.5599	84.84	–6.0	112.3	–0.0147	0.2797
–57	16.35	3.4683	84.72	–5.6	112.4	–0.0139	0.2793
–56	16.81	3.3796	84.61	–5.3	112.6	–0.0131	0.2789
–55	17.28	3.2937	84.50	–5.0	112.7	–0.0122	0.2785
–54	17.75	3.2103	84.38	–4.6	112.8	–0.0114	0.2781
–53	18.24	3.1295	84.27	–4.3	112.9	–0.0106	0.2777
–52	18.73	3.0512	84.15	–4.0	113.0	–0.0098	0.2773
–51	19.24	2.9752	84.04	–3.7	113.2	–0.0090	0.2768
–50	19.76	2.9015	83.92	–3.3	113.3	–0.0082	0.2764
–49	20.29	2.8300	83.81	–3.0	113.4	–0.0073	0.2761
–48	20.82	2.7606	83.69	–2.7	113.5	–0.0065	0.2757
–47	21.38	2.6933	83.58	–2.3	113.6	–0.0057	0.2753
–46	21.94	2.6279	83.46	–2.0	113.7	–0.0049	0.2749
–45	22.51	2.5645	83.35	–1.7	113.8	–0.0041	0.2745
–44	23.10	2.5029	83.23	–1.3	114.0	–0.0033	0.2741
–43	23.69	2.4430	83.11	–1.0	114.1	–0.0025	0.2737

continued

continued

Table 1 – Suva® 410A Saturation Properties – Temperature Table

Temp. °F	Pressure (psia) Liquid (P_f)	Volume (ft³/lb) Vapor (v_g)	Density (lb/ft³) Liquid ($1/v_f$)	Enthalpy (Btu/lb) Liquid (h_f)	Enthalpy (Btu/lb) Vapor (h_g)	Entropy (Btu/(lb)(°R) Liquid (s_f)	Entropy (Btu/(lb)(°R) Vapor (s_g)
–42	24.30	2.3849	83.00	–0.7	114.2	–0.0017	0.2733
–41	24.92	2.3285	82.88	–0.3	114.3	–0.0008	0.2730
–40	25.56	2.2737	82.76	0.0	114.4	0.0000	0.2726
–39	26.20	2.2204	82.65	0.3	114.5	0.0008	0.2722
–38	26.86	2.1687	82.53	0.7	114.6	0.0016	0.2718
–37	27.53	2.1183	82.41	1.0	114.7	0.0024	0.2715
–36	28.22	2.0694	82.29	1.3	114.9	0.0032	0.2711
–35	28.92	2.0219	82.17	1.7	115.0	0.0040	0.2708
–34	29.63	1.9757	82.05	2.0	115.1	0.0048	0.2704
–33	30.35	1.9307	81.93	2.3	115.2	0.0056	0.2700
–32	31.09	1.8870	81.81	2.7	115.3	0.0064	0.2697
–31	31.85	1.8444	81.70	3.0	115.4	0.0072	0.2693
–30	32.61	1.8030	81.58	3.4	115.5	0.0080	0.2690
–29	33.39	1.7628	81.45	3.7	115.6	0.0088	0.2686
–28	34.19	1.7236	81.33	4.0	115.7	0.0096	0.2683
–27	35.00	1.6854	81.21	4.4	115.8	0.0104	0.2679
–26	35.83	1.6483	81.09	4.7	115.9	0.0112	0.2676
–25	36.67	1.6122	80.97	5.1	116.0	0.0120	0.2673
–24	37.52	1.5770	80.85	5.4	116.1	0.0127	0.2669
–23	38.39	1.5427	80.73	5.7	116.2	0.0135	0.2666
–22	39.28	1.5093	80.60	6.1	16.3	0.0143	0.2663
–21	40.18	1.4768	80.48	6.4	116.5	0.0151	0.2659
–20	41.10	1.4452	80.36	6.8	116.6	0.0159	0.2656
–19	42.03	1.4143	80.23	7.1	116.7	0.0167	0.2653
–18	42.98	1.3843	80.11	7.5	116.8	0.0175	0.2649
–17	43.95	1.3550	79.99	7.8	116.9	0.0183	0.2646
–16	44.93	1.3264	79.86	8.2	117.0	0.0191	0.2643
–15	45.94	1.2986	79.74	8.5	117.1	0.0198	0.2640
–14	46.95	1.2715	79.61	8.8	117.2	0.0206	0.2637
–13	47.99	1.2450	79.49	9.2	117.3	0.0214	0.2633
–12	49.04	1.2192	79.36	9.5	117.4	0.0222	0.2630
–11	50.11	1.1941	79.24	9.9	117.5	0.0230	0.2627
–10	51.20	1.1695	79.11	10.2	117.6	0.0237	0.2624
–9	52.31	1.1456	78.98	10.6	117.6	0.0245	0.2621
–8	53.43	1.1222	78.86	10.9	117.7	0.0253	0.2618
–7	54.57	1.0995	78.73	11.3	117.8	0.0261	0.2615
–6	55.74	1.0772	78.60	11.6	117.9	0.0269	0.2612
–5	56.92	1.0555	78.47	12.0	118.0	0.0276	0.2609
–4	58.12	1.0344	78.35	12.3	118.1	0.0284	0.2606
–3	59.34	1.0137	78.22	12.7	118.2	0.0292	0.2603
–2	60.58	0.9935	78.09	13.0	118.3	0.0299	0.2600
–1	61.83	0.9738	77.96	13.4	118.4	0.0307	0.2597
0	63.11	0.9546	77.83	13.7	11.5	0.0315	0.2594
1	64.41	0.9638	77.70	14.1	118.6	0.0323	0.2591
2	65.73	0.9175	77.57	14.5	118.7	0.0330	0.2588
3	67.07	0.8996	77.44	14.8	118.8	0.0338	0.2585
4	68.43	0.8821	77.31	15.2	118.9	0.0346	0.2582
5	69.81	0.8650	77.17	15.5	118.9	0.0353	0.2579
6	71.22	0.8483	77.04	15.9	119.0	0.0361	0.2576
7	72.64	0.8320	76.91	16.2	119.1	0.0369	0.2573
8	74.09	0.8160	76.78	16.6	119.2	0.0376	0.2570
9	75.56	0.8004	76.64	16.9	119.3	0.0384	0.2568
10	77.05	0.7852	76.51	17.3	119.4	0.0392	0.2565
11	78.56	0.7703	76.37	17.7	119.5	0.0399	0.2562

continued

continued

Table 1 – Suva® 410A Saturation Properties – Temperature Table

Temp. °F	Pressure (psia) Liquid (P_f)	Volume (ft³/lb) Vapor (v_g)	Density (lb/ft³) Liquid ($1/v_f$)	Enthalpy (Btu/lb) Liquid (h_f)	Enthalpy (Btu/lb) Vapor (h_g)	Entropy (Btu/(lb)(°R) Liquid (s_f)	Entropy (Btu/(lb)(°R) Vapor (s_g)
12	80.09	0.7557	76.24	18.0	119.5	0.0407	0.2559
13	81.65	0.7415	76.11	18.4	119.6	0.0414	0.2556
14	83.23	0.7276	75.97	18.7	119.7	0.0422	0.2554
15	84.84	0.7140	75.83	19.1	119.8	0.0430	0.2551
16	86.46	0.7007	75.70	19.5	119.9	0.0437	0.2548
17	88.11	0.6876	75.56	19.8	119.9	0.0445	0.2545
18	89.79	0.6749	75.42	20.2	120.0	0.0452	0.2543
19	91.49	0.6625	75.29	20.5	120.1	0.0460	0.2540
20	93.21	0.6503	75.15	20.9	120.2	0.0467	0.2537
21	94.96	0.6383	75.01	21.3	120.3	0.0475	0.2534
22	96.73	0.6267	74.87	21.6	120.3	0.0483	0.2532
23	98.53	0.6152	74.73	22.0	120.4	0.0490	0.2529
24	100.36	0.6041	74.59	22.4	120.5	0.0498	0.2526
25	102.20	0.5931	74.45	22.7	120.6	0.0505	0.2524
26	104.08	0.5824	74.31	23.1	120.6	0.0513	0.2521
27	105.98	0.5719	74.17	23.5	120.7	0.0520	0.2518
28	107.91	0.5616	74.03	23.8	120.8	0.0528	0.2516
29	109.86	0.5516	73.89	24.2	120.9	0.0535	0.2513
30	111.84	0.5417	73.74	24.6	120.9	0.0543	0.2510
31	113.85	0.5321	73.60	24.9	121.0	0.0550	0.2508
32	115.88	0.5227	73.46	25.3	121.1	0.0558	0.2505
33	117.94	0.5134	73.31	25.7	121.1	0.0565	0.2502
34	120.03	0.5043	73.17	26.1	121.2	0.0573	0.2500
35	122.15	0.4955	73.02	26.4	121.3	0.0580	0.2497
36	124.30	0.4868	72.88	26.8	121.3	0.0588	0.2495
37	126.47	0.4782	72.73	27.2	121.4	0.0595	0.2492
38	128.67	0.4699	72.58	27.5	121.5	0.0602	0.2489
39	130.91	0.4617	72.43	27.9	121.5	0.0610	0.2487
40	133.17	0.4537	72.29	28.3	121.6	0.0617	0.2484
41	135.46	0.4458	72.14	28.7	121.6	0.0625	0.2482
42	137.78	0.4381	71.99	29.1	121.7	0.0632	0.2479
43	140.13	0.4305	71.84	29.4	121.8	0.0640	0.2477
44	142.51	0.4231	71.69	29.8	121.8	0.0647	0.2474
45	144.92	0.4159	71.54	30.2	121.9	0.0655	0.2471
46	147.36	0.4087	71.39	30.6	121.9	0.0662	0.2469
47	149.83	0.4018	71.23	30.9	122.0	0.0669	0.2466
48	152.34	0.3949	71.08	31.3	122.0	0.0677	0.2464
49	154.87	0.3882	70.93	31.7	122.1	0.0684	0.2461
50	157.44	0.3816	70.77	32.1	122.2	0.0692	0.2459
51	160.04	0.3751	70.62	32.5	122.2	0.0699	0.2456
52	162.67	0.3688	70.46	32.9	122.3	0.0706	0.2454
53	165.33	0.3626	70.31	33.2	122.3	0.0714	0.2451
54	168.03	0.3565	70.15	33.6	122.3	0.0721	0.2448
55	170.76	0.3505	69.99	34.0	122.4	0.0729	0.2446
56	173.52	0.3446	69.83	34.4	122.4	0.0736	0.2443
57	176.32	0.3388	69.67	34.8	122.5	0.0743	0.2441
58	179.15	0.3332	69.51	35.2	122.5	0.0751	0.2438
59	182.01	0.3276	69.35	35.6	122.6	0.0758	0.2436
60	184.91	0.3221	69.19	36.0	122.6	0.0766	0.2433
61	187.84	0.3168	69.03	36.3	122.7	0.0773	0.2431
62	190.81	0.3115	68.87	36.7	122.7	0.0780	0.2428
63	193.82	0.3064	68.70	37.1	122.7	0.0788	0.2426
64	196.86	0.3013	68.54	37.5	122.8	0.0795	0.2423
65	199.93	0.2963	68.37	37.9	122.8	0.0802	0.2420

continued

continued

Table 1 – Suva® 410A Saturation Properties – Temperature Table

Temp. °F	Pressure (psia) Liquid (P_f)	Volume (ft³/lb) Vapor (v_g)	Density (lb/ft³) Liquid ($1/v_f$)	Enthalpy (Btu/lb) Liquid (h_f)	Enthalpy (Btu/lb) Vapor (h_g)	Entropy (Btu/(lb)(°R)) Liquid (s_f)	Entropy (Btu/(lb)(°R)) Vapor (s_g)
66	203.04	0.2914	68.21	38.3	122.8	0.0810	0.2418
67	206.19	0.2866	68.04	38.7	122.9	0.0817	0.2415
68	209.38	0.2819	67.87	39.1	122.9	0.0825	0.2413
69	212.60	0.2773	67.71	39.5	122.9	0.0832	0.2410
70	215.86	0.2727	67.54	39.9	123.0	0.0839	0.2408
71	219.15	0.2682	67.37	40.3	123.0	0.0847	0.2405
72	222.49	0.2638	67.19	40.7	123.0	0.0854	0.2402
73	225.86	0.2595	67.02	41.1	123.0	0.0861	0.2400
74	229.27	0.2553	66.85	41.5	123.1	0.0869	0.2397
75	232.72	0.2511	66.68	41.9	123.1	0.0876	0.2395
76	236.20	0.2470	66.50	42.3	123.1	0.0884	0.2392
77	239.73	0.2430	66.32	42.7	123.1	0.0891	0.2389
78	243.30	0.2390	66.15	43.1	123.1	0.0898	0.2387
79	246.91	0.2351	65.97	43.5	123.2	0.0906	0.2384
80	250.55	0.2313	65.79	43.9	123.2	0.0913	0.2381
81	254.24	0.2276	65.61	44.3	123.2	0.0921	0.2379
82	257.97	0.2239	65.43	44.8	123.2	0.0928	0.2376
83	261.74	0.2202	65.25	45.2	123.2	0.0936	0.2373
84	265.55	0.2166	65.06	45.6	123.2	0.0943	0.2371
85	269.40	0.2131	64.88	46.0	123.2	0.0950	0.2368
86	273.29	0.2097	64.69	46.4	123.2	0.0958	0.2365
87	277.23	0.2063	64.51	46.8	123.2	0.0965	0.2363
88	281.21	0.2029	64.32	47.3	123.2	0.0973	0.2360
89	285.23	0.1996	64.13	47.7	123.2	0.0980	0.2357
90	289.29	0.1964	63.94	48.1	123.2	0.0988	0.2354
91	293.40	0.1932	63.74	48.5	123.2	0.0995	0.2352
92	297.55	0.1901	63.55	48.9	123.2	0.1003	0.2349
93	301.75	0.1870	63.36	49.4	123.2	0.1010	0.2346
94	305.99	0.1840	63.16	49.8	123.2	0.1018	0.2343
95	310.27	0.1810	62.96	50.2	123.2	0.1025	0.2340
96	314.60	0.1781	62.76	50.7	123.2	0.1033	0.2338
97	318.98	0.1752	62.56	51.1	123.1	0.1040	0.2335
98	323.40	0.1723	62.36	51.5	123.1	0.1048	0.2332
99	327.87	0.1695	62.16	52.0	123.1	0.1056	0.2329
100	332.38	0.1668	61.95	52.4	123.1	0.1063	0.2326
101	336.94	0.1640	61.74	52.8	123.1	0.1071	0.2323
102	341.54	0.1614	61.53	53.3	123.0	0.1079	0.2320
103	346.20	0.1587	61.32	53.7	123.0	0.1086	0.2317
104	350.90	0.1561	61.11	54.2	123.0	0.1094	0.2314
105	355.65	0.1536	60.90	54.6	122.9	0.1102	0.2311
106	360.45	0.1511	60.68	55.1	122.9	0.1109	0.2308
107	365.29	0.1486	60.46	55.5	122.9	0.1117	0.2305
108	370.19	0.1462	60.24	56.0	122.8	0.1125	0.2302
109	375.13	0.1438	60.02	56.5	122.8	0.1133	0.2299
110	380.12	0.1414	59.79	56.9	122.7	0.1141	0.2296
111	385.17	0.1391	59.57	57.4	122.7	0.1148	0.2293
112	390.26	0.1368	59.34	57.8	122.6	0.1156	0.2289
113	395.40	0.1345	59.10	58.3	122.6	0.1164	0.2286
114	400.60	0.1323	58.87	58.8	122.5	0.1172	0.2283
115	405.84	0.1301	58.63	59.3	122.5	0.1180	0.2280
116	411.14	0.1279	58.39	59.7	122.4	0.1188	0.2276
117	416.49	0.1258	58.15	60.2	122.3	0.1196	0.2273
118	421.89	0.1237	57.91	60.7	122.3	0.1204	0.2270
119	427.34	0.1216	57.66	61.2	122.2	0.1213	0.2266

continued

continued

Table 1 – Suva® 410A Saturation Properties – Temperature Table

Temp. °F	Pressure (psia) Liquid (P_f)	Volume (ft³/lb) Vapor (v_g)	Density (lb/ft³) Liquid ($1/v_f$)	Enthalpy (Btu/lb) Liquid (h_f)	Enthalpy (Btu/lb) Vapor (h_g)	Entropy (Btu/(lb)(°R) Liquid (s_f)	Entropy (Btu/(lb)(°R) Vapor (s_g)
120	432.85	0.1195	57.41	61.7	122.1	0.1221	0.2263
121	438.41	0.1175	57.16	62.2	122.0	0.1229	0.2259
122	444.02	0.1155	56.90	62.7	121.9	0.1237	0.2256
123	449.68	0.1136	56.64	63.2	121.9	0.1246	0.2252
124	455.40	0.1116	56.37	63.7	121.8	0.1254	0.2249
125	461.18	0.1097	56.11	64.2	121.7	0.1263	0.2245
126	467.01	0.1078	55.83	64.8	121.6	0.1271	0.2242
127	472.89	0.1060	55.56	65.3	121.5	0.1280	0.2238
128	478.83	0.1041	55.28	65.8	121.4	0.1289	0.2234
129	484.83	0.1023	54.99	66.3	121.3	0.1298	0.2230
130	490.88	0.1005	54.70	66.9	121.2	0.1306	0.2227
131	496.98	0.0987	54.41	67.4	121.0	0.1315	0.2223
132	503.15	0.0970	54.11	68.0	120.9	0.1324	0.2219
133	509.37	0.0952	53.80	68.6	120.8	0.1334	0.2215
134	515.65	0.0935	53.49	69.1	120.7	0.1343	0.2211
135	521.98	0.0918	53.18	69.7	120.5	0.1352	0.2207
136	528.37	0.0901	52.85	70.3	120.4	0.1362	0.2203
137	534.83	0.0885	52.52	70.9	120.3	0.1372	0.2198
138	541.34	0.0868	52.18	71.5	120.1	0.1381	0.2194
139	547.90	0.0852	51.84	72.1	119.9	0.1391	0.2190
140	554.53	0.0835	51.48	72.8	119.8	0.1401	0.2185
141	561.22	0.0819	51.12	73.4	119.6	0.1412	0.2181
142	567.97	0.0803	50.74	74.1	119.4	0.1422	0.2176
143	574.77	0.0787	50.36	74.8	119.3	0.1433	0.2171
144	581.64	0.0771	49.96	75.4	119.1	0.1444	0.2167
145	588.57	0.0755	49.55	76.2	118.9	0.1455	0.2162
146	595.55	0.0739	49.13	76.9	118.7	0.1467	0.2157
147	602.60	0.0723	48.69	77.6	118.4	0.1479	0.2151
148	609.71	0.0706	48.23	78.4	118.2	0.1491	0.2146
149	616.88	0.0690	47.76	79.2	117.9	0.1504	0.2140
150	624.12	0.0673	47.26	80.0	117.7	0.1517	0.2135
151	631.41	0.0656	46.73	80.9	117.4	0.1531	0.2129
152	638.77	0.0639	46.18	81.8	117.1	0.1545	0.2122
153	646.19	0.0621	45.59	82.8	116.8	0.1560	0.2116
154	653.67	0.0602	44.96	83.8	116.5	0.1576	0.2109
155	661.22	0.0583	44.28	84.9	116.1	0.1593	0.2102
156	668.82	0.0562	43.54	86.0	115.8	0.1612	0.2095
157	676.49	0.0541	42.71	87.3	115.5	0.1632	0.2089
158	684.23	0.0518	41.77	88.7	115.2	0.1654	0.2083
159	692.02	0.0496	40.65	90.3	115.1	0.1680	0.2080
160	699.88	0.0476	39.25	92.3	115.2	0.1711	0.2081

R-410A Pressure — Enthalpy Diagram

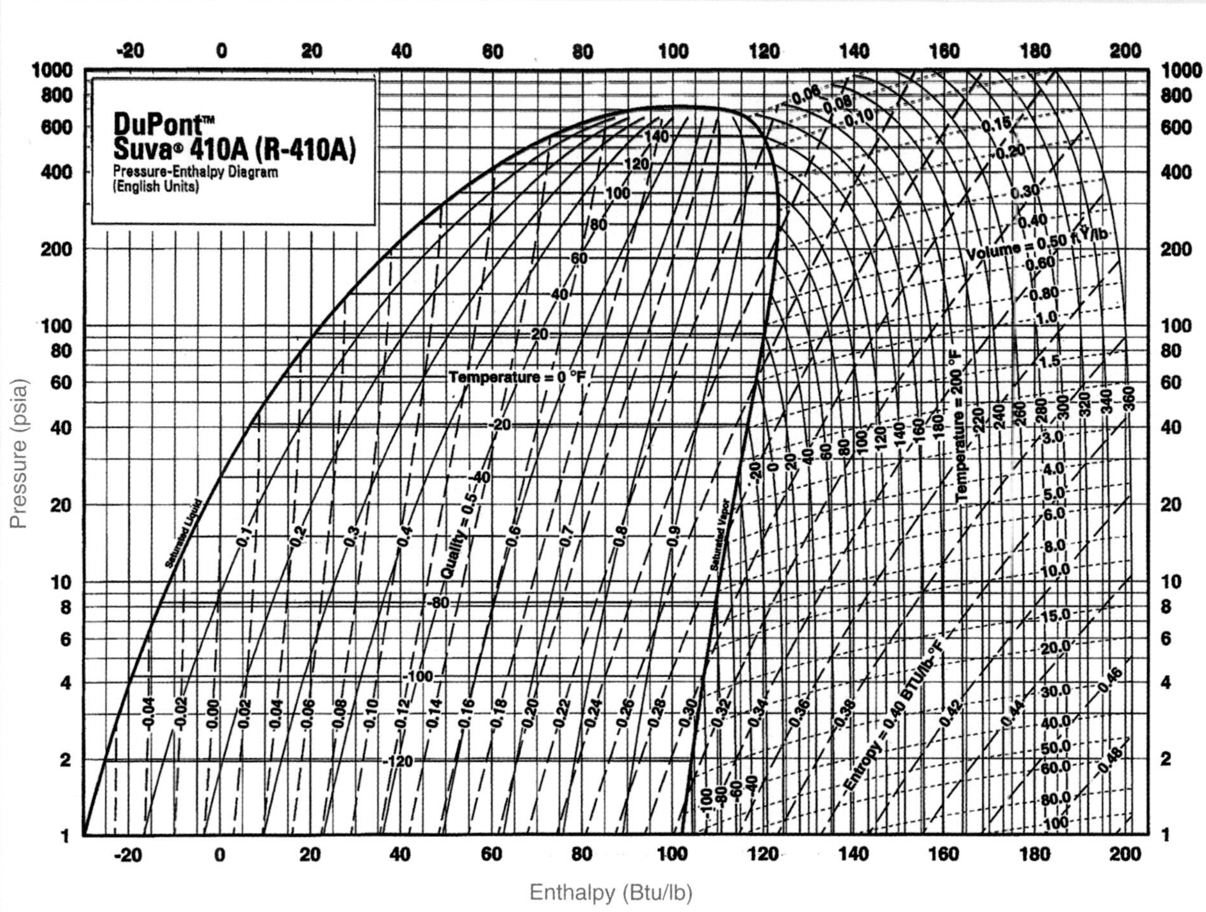

DuPont™ Suva® 134a Refrigerant – Pressure Drops in Lines {65˚F Evaporator Outlet}

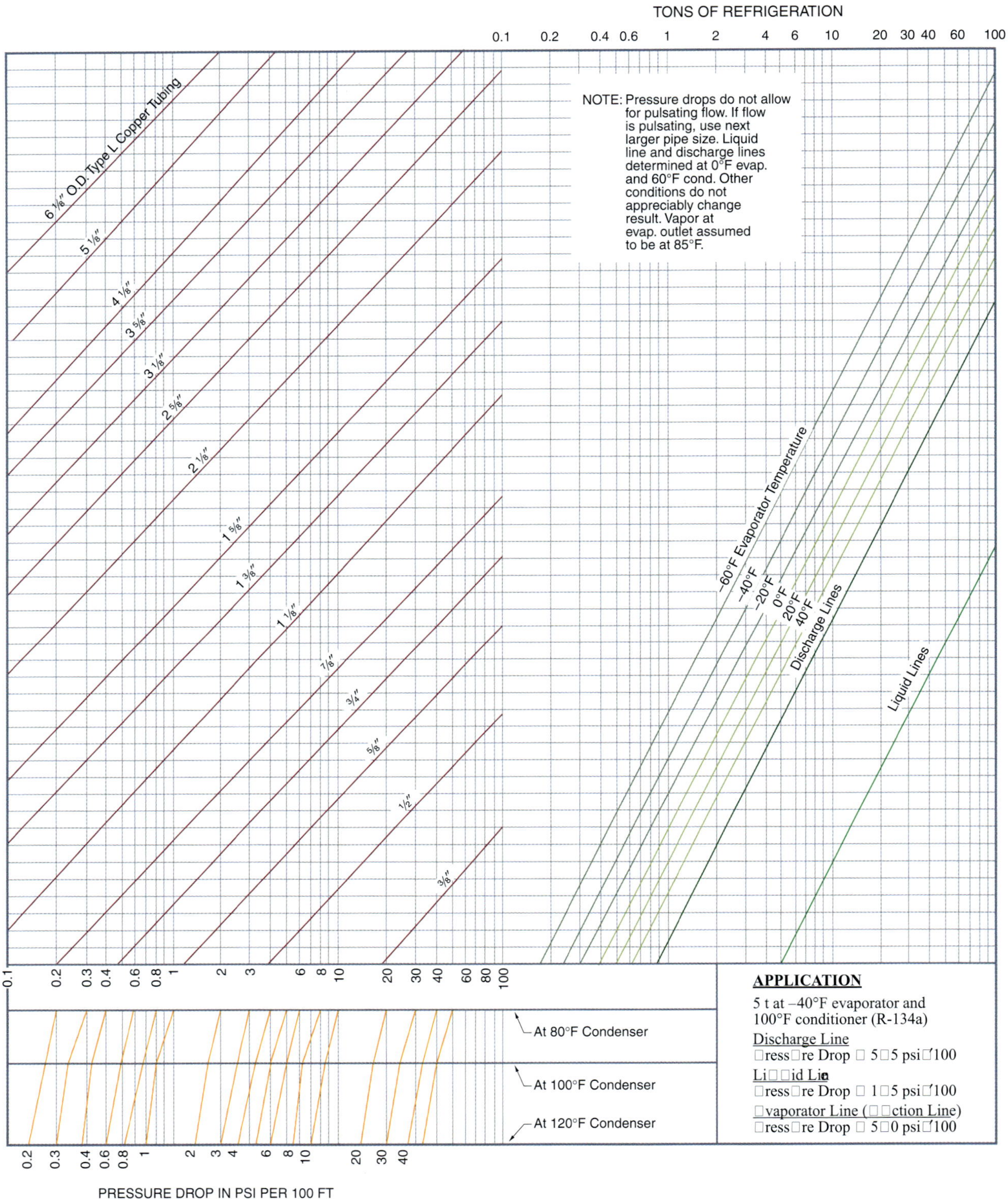

Suction Line Sizes

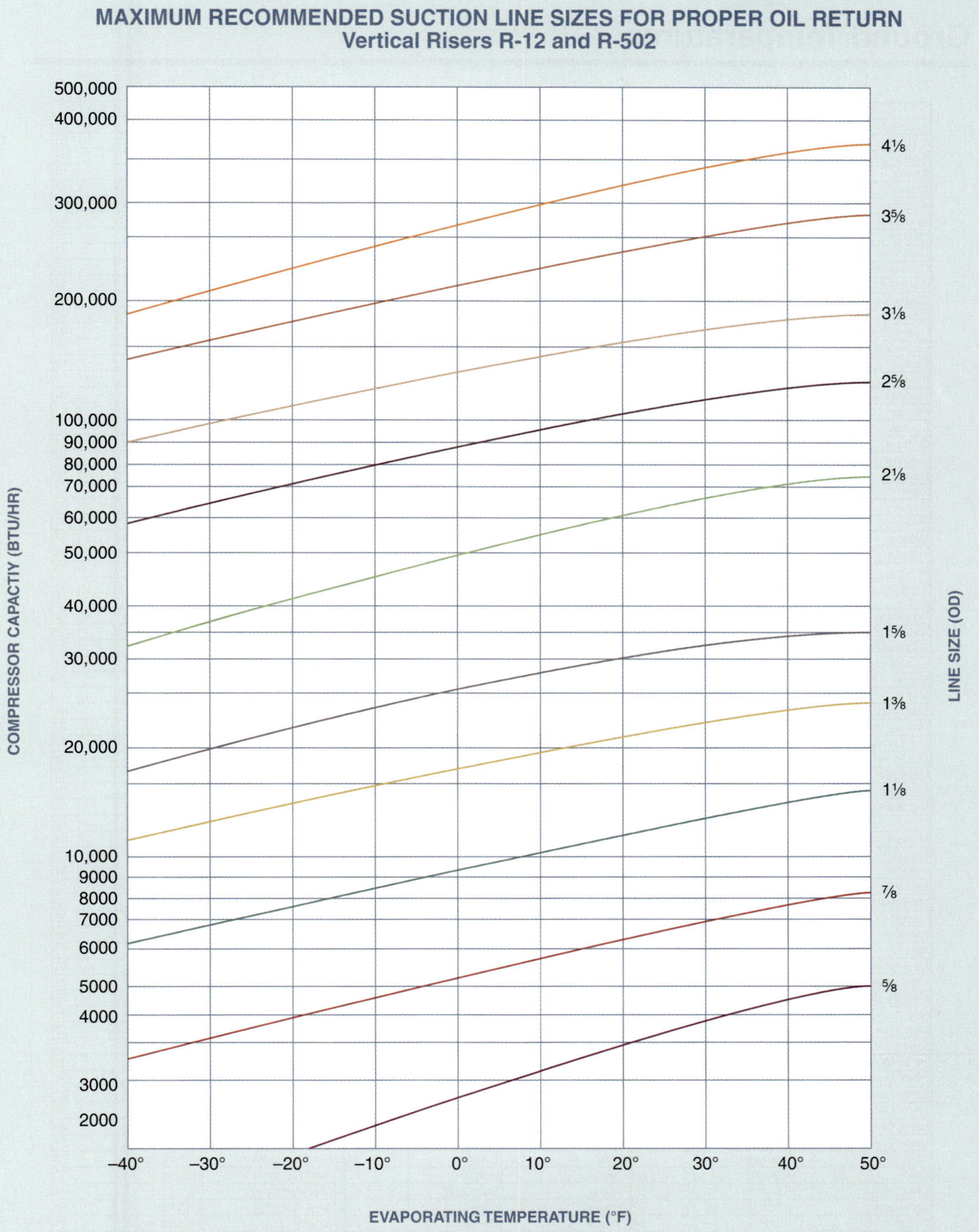

Ground Temperatures

State	City	GWT (F)	State/Country	City	GWT (F)	State/Country	City	GWT (F)
Alabama	Birmingham	65	South Carolina	Charleston	66	Germany	Frankfurt	52
Alabama	Mobile	70	South Carolina	Columbia	64	Germany	Hamburg	51
Alabama	Montgomery	67	South Carolina	Greenville	62	Germany	Munich	49
Alaska	Anchorage	40	South Dakota	Sioux Falls	51	Germany	Stuttgart	51
Alaska	Fairbanks	0	Tennessee	Knoxville	61	Greece	Athens	67
Arizona	Phoenix	73	Tennessee	Memphis	63	Hungary	Budapest	53
Arkansas	Little Rock	64	Tennessee	Nashville	60	Indonesia	Jakarta	83
California	Fresno	68	Texas	Austin	71	Ireland	Dublin	52
California	Los Angeles	64	Texas	Dallas	68	Israel	Jerusalem	63
California	Sacramento	67	Texas	Houston	71	Italy	Genova	60
California	San Diego	64	Texas	San Antonio	72	Italy	Milan	57
California	San Francisco	60	Utah	Salt Lake City	53	Italy	Naples	63
Colorado	Denver	52	Vermont	Burlington	46	Italy	Palermo	66
Connecticut	Hartford	51	Virginia	Norfolk	61	Italy	Rome	61
Delaware	Dover	57	Virginia	Richmond	60	Italy	Torino	58
Florida	Daytona Beach	70	Virginia	Roanoke	59	Italy	Trieste	58
Florida	Jacksonville	71	DC	Washington	57	Italy	Venice	58
Florida	Miami	78	Washington	Seattle	53	Japan	Nagoya	62
Florida	Tallahassee	69	Washington	Spokane	49	Japan	Osaka	63
Florida	Tampa	75	West Virginia	Charleston	58	Japan	Sapporo	51
Georgia	Atlanta	62	Wisconsin	La Crosse	48	Japan	Tokyo	64
Georgia	Savannah	67	Wisconsin	Milwaukee	47	Korea	Inch'on	56
Hawaii	Honolulu	79	Wyoming	Cheyenne	48	Korea	Pusan	59
Idaho	Boise	47	Alberta	Calgary	42	Korea	Seoul	57
Illinois	Chicago	51	Alberta	Edmonton	40	Kuwait	Kuwait City	80
Illinois	Springfield	56	British Columbia	Vancouver	53	Libyan Arab Jamahiriya	Tripoli	71
Indiana	Fort Wayne	53	Manitoba	Winnipeg	40	Malaysia	George Town	84
Indiana	Indianapolis	55	New Brunswick	Moncton	42	Malaysia	Kuala Lumpur	83
Iowa	Des Moines	53	Newfoundland	Saint John's	43	Mexico	Acapulco	78
Kansas	Wichita	59	Nova Scotia	Halifax	45	Mexico	Mexico City	65
Ohio	Cincinatti	57	Ontario	Ottawa	45	Mexico	Veracruz	75
Kentucky	Lexington	60	Ontario	Toronto	48	Morocco	Casablanca	66
Kentucky	Louisville	60	Prince Edward Island	Charlottetown	42	Netherlands	Amsterdam	52
Louisiana	New Orleans	70	Quebec	Montreal	46	New Zealand	Auckland	56
Louisiana	Shreveport	66	Saskatchewan	Regina	39	New Zealand	Christchurch	54
Maine	Caribou	46	Argentina	Buenos Aires	64	New Zealand	Wellington	57
Maine	Portland	48	Australia	Adelaide	64	Norway	Oslo	46
Maryland	Baltimore	57	Australia	Brisbane	72	Paraguay	Asuncion	64
Massachusetts	Boston	50	Australia	Canberra	67	Peru	Lima	70
Massachusetts	Worchester	50	Australia	Melbourne	59	Philippines	Manila	67
Michigan	Detroit	50	Australia	Perth	67	Poland	Krakow	48
Michigan	Flint	49	Australia	Sydney	67	Poland	Warsaw	49
Michigan	Grand Rapids	46	Austria	Salzburg	51	Portugal	Porto	64
Minnesota	Duluth	41	Austria	Vienna	50	Puerto Rico	San Juan	84
Minnesota	Minneapolis	47	Bahrain	Manama	80	Qatar	Doha	83
Mississippi	Jackson	67	Belgium	Brussels	52	Romania	Bucharest	54
Missouri	Kansas City	58	Bolivia	La Paz	50	Saudi Arabia	Riyadh	82
Missouri	St. Louis	58	Brazil	Belem	82	Singapore	Singapore	82
Montana	Billings	49	Brazil	Brasilia	80	Spain	Barcelona	62
Montana	Helena	47	Brazil	Recife	78	Spain	Madrid	60
Nebraska	Omaha	53	Brazil	Sao Paulo	78	Spain	Sevilla	67
Nevada	Las Vegas	69	Bulgaria	Sofia	55	Spain	Valencia	65
Nevada	Reno	50	Chile	Santiago	61	Sweden	Stockholm	47
New Hampshire	Concord	50	China	Beijing	58	Switzerland	Geneva	53
New Jersey	Trenton	55	China	Guangzhou	77	Syria	Damascus	71
New Mexico	Albuquerque	59	China	Harbin	53	Taiwan	Taipei	75
New York	Albany	50	China	Hong Kong	77	Thailand	Bangkok	85
New York	Buffalo	50	China	Shanghai	62	Tunisia	Tunis	69
New York	New York City	52	China	Shenyang	55	Turkey	Ankara	52
North Carolina	Asheville	59	Colombia	Bogota	58	Turkey	Istanbul	60
North Carolina	Charlotte	62	Cuba	Havana	77	Turkey	Izmir	64
North Carolina	Greensboro	60	Czech Republic	Prague	50	United Arab Emirates	Abu Dhabi	80
North Carolina	Raleigh	62	Denmark	Copenhagen	47	United Arab Emirates	Dubai	80
North Dakota	Bismarck	44	Ecuador	Quito	60	United Kingdom	Aberdeen	49
North Dakota	Fargo	42	Egypt	Aswan	78	United Kingdom	Belfast	51
Ohio	Cleveland	51	Egypt	Cairo	73	United Kingdom	Birmingham	52
Ohio	Columbus	55	Finland	Helsinki	47	United Kingdom	Edinburgh	50
Ohio	Dayton	50	France	Bordeaux	61	United Kingdom	Liverpool	54
Oklahoma	Oklahoma City	62	France	Lyon	57	United Kingdom	London	54
Oregon	Portland	54	France	Marseille	59	Uruguay	Montevideo	64
Pennsylvania	Harrisburg	52	France	Nantes	56	Venezuela	Caracas	78
Pennsylvania	Philadelphia	55	France	Paris	54	Vietnam	Hanoi	84
Pennsylvania	Pittsburgh	52	Germany	Berlin	52	Yugoslavia	Belgrade	55

Outdoor Design Temperatures

State and City	Lat.*	Winter	Summer		
		DB†	DB†	Daily Range†	WB†
ALABAMA					
Alexander City	32	18	96	21	79
Auburn	32	18	96	21	79
Birmingham	33	17	96	21	78
Huntsville	34	11	95	23	78
Mobile	30	25	95	18	80
Montgomery	32	22	96	21	79
Talladega	33	18	97	21	79
Tuscaloosa	33	20	98	22	79
ALASKA					
Anchorage	61	–23	71	15	60
Barrow	71	–45	57	12	54
Fairbanks	64	–51	82	24	64
Juneau	58	–4	74	15	61
Kodiak	57	10	69	10	60
Nome	64	–31	66	10	58
ARIZONA					
Flagstaff	35	–2	84	31	61
Phoenix	33	31	109	27	76
Tucson	32	28	104	26	72
Yuma	32	36	111	27	79
ARKANSAS					
Fort Smith	35	12	101	24	80
Hot Springs	34	17	101	22	80
Little Rock	34	15	99	22	80
Pine Bluff	34	16	100	22	81
CALIFORNIA					
Bakersfield	35	30	104	32	73
Burbank	34	37	95	25	71
Fresno	36	28	102	34	72
Laguna Beach	33	41	83	18	70
Long Beach	33	41	83	22	70
Los Angeles	33	41	83	15	70
Monterey	36	35	75	20	64
Napa	38	30	100	30	71
Oakland	37	34	85	19	66
Oceanside	33	41	83	13	70
Palm Springs	33	33	112	35	76
Pasadena	34	32	98	29	73
Sacramento	38	30	101	36	72
San Diego	32	42	83	12	71
San Fernando	34	37	95	38	71
San Francisco	37	38	74	14	64
San Jose	37	34	85	26	68
Santa Barbara	34	34	81	24	68
Santa Cruz	36	35	75	28	64
Santa Monica	34	41	83	16	70
Stockton	37	28	100	37	71
COLORADO					
Boulder	40	2	93	27	64
Colorado Spgs.	38	–3	91	30	63
Denver	39	–5	93	28	64
Pueblo	38	–7	97	31	67
CONNECTICUT					
Bridgeport	41	6	86	18	75
Hartford	41	3	91	22	77
New Haven	41	3	88	17	76
Waterbury	41	–4	88	21	75
DELAWARE					
Dover	39	11	92	18	79
Wilmington	39	10	92	20	77
DISTRICT OF COLUMBIA					
Washington	38	14	93	18	78
FLORIDA					
Cape Kennedy	28	35	90	15	80
Daytona Beach	29	32	92	15	80
Fort Lauderdale	26	42	92	15	80
Key West	24	55	90	9	80
Miami	25	44	91	15	79
Orlando	28	35	94	17	79
Pensacola	30	25	94	14	80
St. Petersburg	27	36	92	16	79
Sarasota	27	39	93	17	79
Tallahassee	30	27	94	19	79
Tampa	27	36	92	17	79
GEORGIA					
Athens	33	18	94	21	78
Atlanta	33	17	94	19	77
Augusta	33	20	97	19	80
Griffin	33	18	93	21	78
Macon	32	21	96	22	79
Savannah	32	24	96	20	80
HAWAII					
Honolulu	21	62	87	12	76
Kaneohe Bay	21	65	85	12	76
IDAHO					
Boise	43	3	96	31	68
Idaho Falls	43	–11	89	38	65
Lewiston	46	–1	96	32	67
Twin Falls	42	–3	99	34	64
ILLINOIS					
Aurora	41	–6	93	20	79
Bloomington	40	–6	92	21	78
Carbondale	37	2	95	21	80
Champaign	40	–3	95	21	78
Chicago	41	–9	90	15	79
Galesburg	40	–7	93	22	78
Joliet	41	–5	93	20	78
Kankakee	41	–4	93	21	78
Macomb	40	–5	95	22	79

* in degrees
† in °F

continued

continued

Outdoor Design Temperatures

State and City	Lat.*	Winter DB†	Summer DB†	Summer Daily Range†	Summer WB†
Peoria	40	–8	91	22	78
Rantoul	40	–4	94	21	78
Rockford	42	–9	91	24	77
Springfield	39	–3	94	21	79
Waukegan	42	–6	92	21	78
INDIANA					
Fort Wayne	41	–4	92	24	77
Hobart	41	–4	91	21	77
Indianapolis	39	–2	92	22	78
Kokomo	40	–4	91	22	77
Lafayette	40	–3	94	22	78
Muncie	40	–3	92	22	76
South Bend	41	–3	91	22	77
Terre Haute	39	–2	95	22	79
Valparaiso	41	–3	93	22	78
IOWA					
Ames	42	–11	93	23	78
Burlington	40	–7	94	22	78
Cedar Rapids	41	–10	91	23	78
Des Moines	41	–10	94	23	78
Dubuque	42	–12	90	22	77
Iowa City	41	–11	92	22	80
Keokuk	40	–5	95	22	79
Sioux City	42	–11	95	24	78
KANSAS					
Garden City	37	–1	99	28	74
Liberal	37	2	99	28	73
Russell	38	0	101	29	78
Topeka	39	0	99	24	79
Wichita	37	3	101	23	77
KENTUCKY					
Bowling Green	35	4	94	21	79
Lexington	38	3	93	22	77
Louisville	38	5	95	23	79
LOUISIANA					
Alexandria	31	23	95	20	80
Baton Rouge	30	25	95	19	80
Lafayette	30	26	95	18	81
Monroe	32	20	99	20	79
New Orleans	29	29	93	16	81
Shreveport	32	20	99	20	79
MAINE					
Augusta	44	–7	88	22	74
Bangor	44	–11	86	22	73
Caribou	46	–18	84	21	71
Portland	43	–6	87	22	74
MARYLAND					
Baltimore	39	10	94	21	78
Frederick	39	8	94	22	78
Salisbury	38	12	93	18	79

* in degrees
† in °F

Outdoor Design Temperatures

State and City	Lat.*	Winter DB†	Summer DB†	Summer Daily Range†	Summer WB†
MASSACHUSETTS					
Boston	42	6	91	16	75
Clinton	42	–2	90	17	75
Lawrence	42	–6	90	22	76
Lowell	42	–4	91	21	76
New Bedford	41	5	85	19	74
Pittsfield	42	–8	87	23	73
Worcester	42	0	87	18	73
MICHIGAN					
Battle Creek	42	1	92	23	76
Benton Harbor	42	1	91	20	75
Detroit	42	3	91	20	76
Flint	42	–4	90	25	76
Grand Rapids	42	1	91	24	75
Holland	42	2	88	22	75
Kalamazoo	42	1	92	23	76
Lansing	42	–3	90	24	75
Marquette	46	–12	84	18	72
Pontiac	42	0	90	21	76
Port Huron	42	0	90	21	76
Saginaw	43	0	91	23	76
Sault Ste. Marie	46	–12	84	23	72
MINNESOTA					
Alexandria	45	–22	91	24	76
Duluth	46	–21	85	22	72
International Falls	48	–29	85	26	71
Minneapolis	44	–16	92	22	77
Rochester	43	–17	90	24	77
MISSISSIPPI					
Biloxi	30	28	94	16	82
Clarksdale	34	14	96	21	80
Jackson	32	21	97	21	79
Laurel	31	24	96	21	81
Natchez	31	23	96	21	81
Vicksburg	32	22	97	21	81
MISSOURI					
Columbia	38	–1	97	22	78
Hannibal	39	–2	96	22	80
Jefferson City	38	2	98	23	78
Kansas City	39	2	99	20	78
Kirksville	40	–5	96	24	78
Moberly	39	–2	97	23	78
St. Joseph	39	–3	96	23	81
St. Louis	38	2	97	21	78
Springfield	37	3	96	23	78
MONTANA					
Billings	45	–15	94	31	67
Butte	45	–24	86	35	60
Great Falls	47	–21	91	28	64
Lewiston	47	–22	90	30	65

continued

continued

Outdoor Design Temperatures

State and City	Lat.*	Winter DB†	Summer DB†	Summer Daily Range†	Summer WB†
Missoula	46	–13	92	36	65
NEBRASKA					
Columbus	41	–6	98	25	77
Fremont	41	–6	98	22	78
Grand Island	40	–8	97	28	75
Lincoln	40	–5	99	24	78
Norfolk	41	–8	97	30	78
North Platte	41	–8	97	28	74
Omaha	41	–8	94	22	78
NEVADA					
Carson City	39	4	94	42	63
Las Vegas	36	25	108	30	71
Reno	39	6	96	45	64
NEW HAMPSHIRE					
Claremont	43	–9	89	24	74
Concord	43	–8	90	26	74
Manchester	42	–8	91	24	75
Portsmouth	43	–2	89	22	75
NEW JERSEY					
Atlantic City	39	10	92	18	78
Long Branch	40	10	93	18	78
Newark	40	10	94	20	77
New Brunswick	40	6	92	19	77
Trenton	40	11	91	19	78
NEW MEXICO					
Albuquerque	35	12	96	27	66
Carlsbad	32	13	103	28	72
Gallup	35	0	90	32	64
Los Alamos	35	5	89	32	62
Santa Fe	35	6	90	28	63
Silver City	32	5	95	30	66
NEW YORK					
Albany	42	–6	91	23	75
Batavia	43	1	90	22	75
Buffalo	42	2	88	21	74
Geneva	42	–3	90	22	75
Glens Falls	43	–11	88	23	74
Ithaca	42	–5	88	24	74
Kingston	41	–3	91	22	76
Lockport	43	4	89	21	76
New York City	40	11	92	17	76
Niagara Falls	43	4	89	20	76
Rochester	43	1	91	22	75
Syracuse	43	–3	90	20	75
Utica	43	–12	88	22	75
NORTH CAROLINA					
Charlotte	35	18	95	20	77
Durham	35	16	94	20	78
Greensboro	36	14	93	21	77
Jacksonville	34	20	92	18	80

* in degrees
† in °F

Outdoor Design Temperatures

State and City	Lat.*	Winter DB†	Summer DB†	Summer Daily Range†	Summer WB†
Wilmington	34	23	93	18	81
Winston–Salem	36	16	94	20	76
NORTH DAKOTA					
Bismark	46	–23	95	27	73
Fargo	46	–22	92	25	76
Grand Forks	47	–26	91	25	74
Williston	48	–25	91	25	72
OHIO					
Akron–Canton	40	1	89	21	75
Athens	39	0	95	22	78
Bowling Green	41	–2	92	23	76
Cambridge	40	1	93	23	78
Cincinnati	39	1	92	21	77
Cleveland	41	1	91	22	76
Columbus	40	0	92	24	77
Dayton	39	–1	91	20	76
Fremont	41	–3	90	24	76
Marion	40	0	93	23	77
Newark	40	–1	94	23	77
Portsmouth	38	5	95	22	78
Toledo	41	–3	90	25	76
Warren	41	0	89	23	74
OKLAHOMA					
Bartlesville	36	6	101	23	77
Chickasha	35	10	101	24	78
Lawton	34	12	101	24	78
McAlester	34	14	99	23	77
Norman	35	9	99	24	77
Oklahoma City	35	9	100	23	78
Seminole	35	11	99	23	77
Stillwater	36	8	100	24	77
Tulsa	36	8	101	22	79
Woodward	36	6	100	26	78
OREGON					
Albany	44	18	92	31	69
Astoria	46	25	75	16	65
Baker	44	–1	92	30	65
Eugene	44	17	92	31	69
Grants Pass	42	20	99	33	71
Klamath Falls	42	4	90	36	63
Medford	42	19	98	35	70
Portland	45	17	89	23	69
Salem	44	18	92	31	69
PENNSYLVANIA					
Allentown	40	4	92	22	76
Altoona	40	0	90	23	74
Butler	40	1	90	22	75
Erie	42	4	88	18	75
Harrisburg	40	7	94	21	77
New Castle	41	2	91	23	75

continued

continued

Outdoor Design Temperatures

State and City	Lat.*	Winter DB†	Summer DB†	Summer Daily Range†	Summer WB†
Philadelphia	39	10	93	21	77
Pittsburgh	40	1	89	22	74
Reading	40	9	92	19	76
West Chester	39	9	92	20	77
Williamsport	41	2	92	23	75
RHODE ISLAND					
Newport	41	5	88	16	76
Providence	41	5	89	19	75
SOUTH CAROLINA					
Charleston	32	25	94	13	81
Columbia	33	20	97	22	79
Florence	34	22	94	21	80
Sumter	33	22	95	21	79
SOUTH DAKOTA					
Aberdeen	45	−19	94	27	77
Brookings	44	−17	95	25	77
Huron	44	−18	96	28	77
Rapid City	44	−11	95	28	71
Sioux Falls	43	−15	94	24	76
TENNESSEE					
Athens	35	13	95	22	77
Chattanooga	35	13	96	22	78
Dyersburg	36	10	96	21	81
Knoxville	35	13	94	21	77
Memphis	35	13	98	21	80
Murfreesboro	34	9	97	22	78
Nashville	36	9	97	21	78
TEXAS					
Abilene	32	15	101	22	75
Alice	27	31	100	20	82
Amarillo	35	6	98	26	71
Austin	30	24	100	22	78
Beaumont	29	27	95	19	81
Big Spring	32	16	100	26	74
Brownsville	25	35	94	18	80
Corpus Christi	27	31	95	19	80
Dallas	32	18	102	20	78
El Paso	31	20	100	27	69
Fort Worth	32	17	101	22	78
Galveston	29	31	90	10	81
Houston	29	27	96	18	80
Huntsville	30	22	100	20	78
Laredo	27	32	102	23	78
Lubbock	33	10	98	26	73
McAllen	26	35	97	21	80
Midland	31	16	100	26	73
Pecos	31	16	100	27	73
San Antonio	29	25	99	19	77
Temple	31	22	100	22	78
Tyler	32	19	99	21	80

* in degrees
† in °F

Outdoor Design Temperatures

State and City	Lat.*	Winter DB†	Summer DB†	Summer Daily Range†	Summer WB†
Victoria	28	29	98	18	82
Waco	31	21	101	22	78
Wichita Falls	33	14	103	24	77
UTAH					
Cedar City	37	−2	93	32	65
Logan	41	−3	93	33	65
Provo	40	1	98	32	66
Salt Lake City	40	3	97	32	66
VERMONT					
Barre	44	−16	84	23	73
Burlington	44	−12	88	23	74
Rutland	43	−13	87	23	74
VIRGINIA					
Charlottesville	38	14	94	23	77
Fredericksburg	38	10	96	21	78
Harrisonburg	38	12	93	23	75
Lynchburg	37	12	93	21	77
Norfolk	36	20	93	18	79
Petersburg	37	14	95	20	79
Richmond	37	14	95	21	79
Roanoke	37	12	93	23	75
Winchester	39	6	93	21	77
WASHINGTON					
Aberdeen	46	25	80	16	65
Bellingham	48	10	81	19	68
Olympia	46	16	87	32	67
Seattle–Tacoma	47	17	84	22	66
Spokane	47	−6	93	28	65
Walla Walla	46	0	97	27	69
WEST VIRGINIA					
Charleston	38	7	92	20	76
Clarksburg	39	6	92	21	76
Parkersburg	39	7	93	21	77
Wheeling	40	1	89	21	74
WISCONSIN					
Beloit	42	−7	92	24	78
Fon Du Lac	43	−12	89	23	76
Green Bay	44	−13	88	23	76
La Crosse	43	−13	91	22	77
Madison	43	−11	91	22	77
Milwaukee	42	−8	90	21	76
Racine	42	−6	91	21	77
Sheboygan	43	−10	89	20	77
Wausau	44	−16	91	23	76
WYOMING					
Casper	42	−11	92	31	63
Cheyenne	41	−9	89	30	63
Laramie	41	−14	84	28	61
Rock Springs	41	−9	86	32	59
Sheridan	44	−14	94	32	66

continued

continued

Outdoor Design Temperatures

Province or Territory and City	Lat.*	Winter DB†	Summer DB†	Daily Range†	WB†
ALBERTA					
Calgary AP	51	–17	80	M	59
Cold Lake AP	54	–26	78	M	62
Coronation	52	–23	82	M	60
Edmonton AP	53	–23	78	M	62
Grand Prarie AP	55	–27	78	M	60
Jasper	52	–26	80	H	62
Lethbridge AP	49	–16	84	H	61
McMurray AP	56	–29	84	H	62
Medicine Hat AP	50	–19	87	H	62
Peace River AP	56	–27	78	M	60
Red Deer AP	52	–21	79	M	61
Rocky Mtn. House	52	–20	78	M	61
Vermillion AP	53	–25	80	M	62
Whitecourt	54	–24	77	M	60
BRITISH COLUMBIA					
Abbotsford AP	49	20	80	M	66
Cape St. James	52	29	62	L	58
Castlegar AP	49	9	88	H	63
Comox AP	49	25	76	M	62
Cranbrook AP	49	–8	85	H	60
Dawson Creek	55	–33	79	H	63
Fort Nelson AP	58	–30	78	M	60
Fort St. John AP	56	–25	76	M	59
Kamploops AP	50	–1	88	H	63
Namaimo	49	20	80	M	65
New Westminster	49	18	81	M	67
Penticton AP	49	10	87	H	64
Port Hardy AP	50	26	65	L	58
Prince George AP	53	–18	78	H	59
Prince Rupert AP	54	13	63	L	57
Quesnel AP	53	–14	81	H	60
Sandspit AP	53	25	65	L	59
Smithers AP	54	–12	77	M	59
Spring Island	50	31	66	L	59
Terrace AP	54	2	78	M	60
Tofino AP	49	29	68	L	60
Trail	49	0	89	H	65
Vancouver AP	49	24	74	M	64
Victoria CO	48	26	75	M	62
Williams Lake AP	52	–14	79	M	57
MANITOBA					
Brandon AP	49	–24	84	M	66
Churchill AP	58	–33	72	M	60
Dauphin AP	51	–23	84	M	66
Flin Flon	54	–37	81	M	66
Portage La Prairie AP	49	–21	85	M	67
The Pas AP	53	–28	79	M	64
Thompson AP	55	–34	79	M	62
Winnipeg AP	49	–23	84	M	67

* in degrees
† in °F

Outdoor Design Temperatures

Province or Territory and City	Lat.*	Winter DB†	Summer DB†	Daily Range†	WB†
NEW BRUNSWICK					
Campbellton CO	48	–14	82	M	67
Charlo AP	47	–10	79	M	62
Chatham AP	47	–7	83	M	67
Edmunston CO	47	–16	83	M	68
Fredericton AP	45	–7	83	M	68
Moncton AP	46	–5	80	M	67
Saint John AP	45	–4	75	M	64
NEWFOUNDLAND					
Battle Harbor	52	–10	60	L	55
Bonavista	48	7	71	L	63
Cartwright	53	–15	70	M	59
Corner Brook	48	0	73	M	63
Daniels Harbor	50	–3	66	L	62
Deer Lake AP	49	–7	77	M	64
Gander AP	48	0	76	M	63
Goose Bay AP	53	–20	77	M	61
Hopedale	55	–18	65	M	57
St. John's AP	47	7	73	M	64
Stephenville AP	48	3	71	L	64
Wabush Lake AP	52	–30	72	M	58
NORTHWEST TERRITORY					
Baker Lake	64	–39	65	M	55
Cambridge Bay AP	69	–35	57	L	51
Cape Pary AP	70	–33	54	L	50
Chesterfield	63	–34	60	M	52
Coral Harbour AP	64	–38	60	M	51
Fort Smith AP	60	–32	78	M	61
Frobisher AP	63	–36	57	L	48
Hall Beach AP	68	–38	52	L	47
Inuvik	68	–40	75	M	59
Norman Wells AP	65	–36	77	M	60
Resolute AP	74	–40	48	L	43
Yellowknife AP	62	–36	74	M	59
NOVA SCOTIA					
Amherst	45	–6	81	M	68
Greenwood AP	44	3	80	M	67
Halifax AP	44	2	78	M	66
Kentville	45	1	83	M	68
New Glasgow	45	–5	79	M	68
Sable Island	43	17	69	L	66
Shearwater AP	44	5	75	M	64
Sydney AP	46	3	78	M	67
Truro CO	45	–4	71	M	64
Yarmouth AP	43	10	71	L	64
ONTARIO					
Armstrong AP	50	–30	78	H	65
Atikokan	48	–27	80	M	66
Big Trout Lake	53	–30	75	M	62
Belleville	44	–7	84	M	72

continued

continued

Outdoor Design Temperature

Province or Territory and City	Lat.*	Winter DB†	Summer DB†	Summer Daily Range†	Summer WB†
Chatham	42	3	87	M	73
Cornwall	45	−9	87	M	72
Earlton AP	47	−21	81	M	67
Geraldton	49	−28	78	M	65
Gore Bay AP	45	−7	78	M	67
Hamilton	43	1	86	M	72
Kapuskasing AP	49	−25	80	M	65
Kenora AP	49	−22	81	M	65
Kingston	44	−7	84	M	72
Kitchener	43	−2	85	M	72
London AP	43	2	83	M	70
Mount Forest	43	−3	80	M	68
Muskoka AP	44	−11	80	M	68
North Bay AP	46	−13	78	M	66
Oshawa	43	−3	86	M	72
Ottawa AP	45	−8	83	M	69
Owen Sound	44	−2	83	M	70
Petersborough	44	−9	85	M	71
St. Catharines	43	3	85	M	72
Sarnia	42	3	86	M	72
Sault Ste. Marie AP	46	−8	79	M	67
Simcoe	42	3	83	M	70
Sioux Lookout AP	50	−25	80	M	65
Sudbury AP	46	−14	81	M	66
Thunder Bay AP	48	−18	80	M	65
Timmons AP	48	−23	81	M	65
Toronto AP	43	1	84	M	70
Trenton AP	44	−3	81	M	70
Wiarton AP	44	0	80	M	69
Windsor AP	42	6	86	M	71
PRINCE EDWARD ISLAND					
Charlottetown AP	46	−2	77	M	67
Summerside AP	46	−1	77	M	66
QUEBEC					
Bagotville AP	48	−19	80	M	65
Baie Corneau AP	49	−14	71	M	61
Chicoutimi	48	−22	83	M	68
Drummondville	45	−14	85	M	71
Granby	45	−14	85	M	71
Grindstone Island	47	3	70	L	65
Hull	45	−14	87	M	71
Kuujjuarapik AP	55	−30	70	M	59
Kuujuaq AP	58	−31	69	M	57
La Grande Riviere AP	53	−30	75	M	60
Lake Eon AP	51	−27	70	M	58
Megantic AP	45	−16	83	M	70
Mont Joli AP	48	−8	76	M	65
Montreal AP	45	−7	83	M	70
Montreal Mirabel AP	45	−11	81	M	69
Nitchequon	53	−31	69	M	58

* in degrees
† in °F

Outdoor Design Temperature

Province or Territory and City	Lat.*	Winter DB†	Summer DB†	Summer Daily Range†	Summer WB†
Quebec AP	46	−11	80	M	68
Rimouski	48	−12	79	M	66
Riviere Du Loup	47	−10	76	M	67
Roberval AP	48	−19	79	M	66
St. Jean	45	−11	86	M	72
Saint Jérôme	45	−13	86	M	71
St. Hubert AP	45	−7	83	M	69
Schefferville AP	54	−31	69	M	57
Sept-Iles AP	50	−15	69	M	59
Shawnigan	46	−14	84	M	70
Sherbrooke CO	45	−14	80	M	68
Ste. Agathe Des Monts	46	−15	78	M	66
Thetford Mines	46	−14	84	M	70
Trois Rivières	46	−13	85	M	70
Val d'Or AP	48	−22	80	M	65
Valleyfield	45	−10	86	M	72
SASKATCHEWAN					
Broadview	50	−25	83	H	64
Estevan AP	49	−21	86	H	65
Moose Jaw AP	50	−21	87	H	64
North Battleford AP	52	−26	82	M	63
Prince Albert AP	53	−29	81	M	64
Regina AP	50	−24	85	H	64
Saskatoon AP	52	−26	84	H	63
Swift Current AP	50	−21	84	M	62
Uranium City AP	59	−35	76	M	60
Wynyard	51	−25	81	M	63
Yorkton AP	51	−26	82	M	64
YUKON TERRITORY					
Burwash AP	61	−32	69	M	55
Whitehorse AP	60	−31	73	M	55

Air Conditioning Contractors of America

Heat Transfer Factors

Item	Design Temperature Difference*														
	30	35	40	45	50	55	60	65	70	75	80	85	90	95	100
Windows† — Wood or Metal Frame															
Double-hung, Horizontal-slide, Casement, or Awning															
Single glass	45	50	60	65	75	80	90	95	105	110	120	125	135	140	150
With double glass or insulating glass	30	35	40	45	50	55	60	65	70	75	80	80	85	90	95
With storm sash	25	30	35	40	45	50	55	60	60	65	70	75	80	85	90
Fixed or Picture															
Single glass	40	50	55	60	70	75	85	90	95	105	110	115	125	130	140
Double glass or with storm sash	25	30	35	40	45	45	50	55	60	65	70	75	75	80	85
Jalousie															
Single glass	225	265	300	340	375	415	450	490	525	565	600	640	675	715	750
With storm sash	65	75	90	100	110	120	135	145	155	165	175	190	200	210	220
Doors†															
Sliding Glass Doors															
Single glass	75	85	100	115	125	140	150	165	175	190	200	210	225	240	250
Double glass	60	70	80	90	100	110	120	130	140	150	160	170	180	190	200
Other Doors															
Weatherstripped and with storm door	40	45	55	60	65	75	80	85	90	100	105	110	120	125	130
Weatherstripped or with storm door	70	85	95	110	120	135	145	155	170	180	195	205	215	230	240
No weatherstripping or storm door	135	160	180	200	225	250	270	290	315	340	360	380	405	430	450
Walls and Partitions† — Wood Frame with Sheathing and Siding															
No insulation	8	9	10	11	13	14	15	16	18	19	20	21	23	24	25
R-5 polystyrene sheathing	3	4	4	5	6	6	7	7	8	8	9	9	10	10	11
R-7 batt insulation (2″ – 2¾″)	3	4	4	5	5	6	6	7	7	8	8	9	9	10	10
R-11 batt insulation (3″ – 3½″)	2	2	3	3	4	4	4	5	5	5	6	6	6	7	7
Partition Between Conditioned and Unconditioned Spaces															
Finished one side only, no insulation	17	19	22	25	28	30	33	36	39	41	44	47	49	52	55
Finished both sides, no insulation	9	11	12	14	16	17	19	20	22	23	25	26	28	29	31
Partition with 1″ polystyrene board R-5	4	4	5	5	6	7	7	8	8	9	10	10	11	11	12
R-7 insulation finished both sides	3	4	4	5	5	6	6	7	7	8	8	9	9	10	10
R-11 insulation finished both sides	2	3	3	4	4	4	5	5	6	6	6	7	7	8	8
Solid Masonry, Block, or Brick															
Plastered or plain	14	16	18	20	22	25	27	29	32	34	36	38	40	43	45
Furred, no insulation	9	10	12	13	14	16	17	19	20	22	23	25	26	28	29
Furred, with R-5 insulation	4	5	5	6	6	7	8	8	9	10	10	11	12	12	13
Basement or Crawl Space															
Above grade, no insulation	15	18	20	23	26	28	31	33	36	38	41	43	46	48	51
R-3.57 insulation (molded bead bd.)	5	6	7	8	9	10	11	12	13	14	14	15	16	17	18
R-5 insulation (ext. polystrene bd.)	4	5	6	6	7	8	9	9	10	11	12	12	13	14	14

* in °F
† in Btu/hr per sq ft (factors include heat loss for transmission and infiltration)
Note: R values on this chart refer to thermal resistance value.

continued

continued

Heat Transfer Factors

Item	Design Temperature Difference*														
	30	35	40	45	50	55	60	65	70	75	80	85	90	95	100
Basement or Crawl Space															
Wall of crawl space used as supply plenum, R-3.57 insulation	11	12	13	14	15	16	16	17	18	19	20	21	22	23	24
Wall of crawl space used as supply plenum, R-5 insulation	9	10	10	11	12	12	13	14	15	15	16	17	17	18	19
Below grade wall	2	2	2	3	3	3	4	4	4	5	5	5	5	6	6
Ceilings and Roofs† — Ceiling Under Unconditioned Space or Vented Roof															
No insulation	18	21	24	27	30	33	36	39	42	45	48	51	54	57	60
R-11 insulation (3″ – 3¼″)	2	3	3	4	4	4	5	5	6	6	6	7	7	8	8
R-19 insulation (5¼″ – 6½″)	2	2	2	2	2	3	3	3	4	4	4	4	4	5	5
R-22 insulation (6″ – 7″)	1	1	2	2	2	2	2	3	3	3	3	3	4	4	4
R-30 insulation	1	1	1	2	2	2	2	2	2	3	3	3	3	3	3
R-38 insulation	1	1	1	1	1	1	2	2	2	2	2	2	2	3	3
R-44 insulation	1	1	1	1	1	1	1	2	2	2	2	2	2	2	2
Roof on Exposed Beams or Rafters															
Roofing on 1½″ wood decking no ins.	10	12	14	15	17	19	20	22	24	26	27	29	31	32	34
Roofing on 1½″ wood decking 1″ insulation between roofing and decking	5	6	7	8	8	9	10	11	12	13	13	14	15	16	17
Roofing on 1½″ wood decking 1½″ insulation between roofing and decking	4	5	5	6	7	7	8	9	10	10	11	12	12	13	14
Roofing on 2″ wood plank	6	7	8	9	10	11	12	14	15	16	17	18	19	20	21
Roofing on 3″ wood plank	5	5	6	7	8	8	9	10	11	11	12	13	14	14	15
Roofing on 1½″ fiberboard decking	6	7	8	9	10	10	11	12	13	14	15	16	17	18	19
Roofing on 2″ fiberboard decking	4	5	6	7	8	8	9	10	11	11	12	13	14	14	15
Roofing on 3″ fiberboard decking	3	4	4	5	6	6	7	7	8	8	9	9	10	10	11
Roofing-Ceiling Combination															
No insulation	9	11	12	14	16	17	19	20	22	23	25	26	28	29	31
R-11 insulation	2	2	3	3	4	4	4	5	5	5	6	6	6	7	7
R-19 insulation	1	2	2	2	2	2	3	3	3	3	4	4	4	4	5
R-22 insulation (6″ – 7″)	1	1	2	2	2	2	2	3	3	3	3	3	4	4	4
Floors† — Floors Over Unconditioned Space															
Over unconditioned room	4	5	6	6	7	8	8	9	10	11	11	12	13	13	14
No insulation	8	10	11	13	14	15	17	18	20	21	22	24	25	27	28
R-7 insulation (2″ – 2¾″)	3	3	4	4	5	5	6	6	7	7	8	8	9	9	9
R-11 insulation (3″ – 3½″)	2	2	3	3	4	4	4	5	5	5	6	6	6	7	7
R-19 insulation (5¼″ – 6½″)	1	2	2	2	2	2	3	3	3	3	4	4	4	4	4
Floor of Room Over Heated Crawl Space															
Less than 18″ below grade	35	40	40	45	45	50	50	55	55	60	60	65	65	70	75
18″ or more below grade	15	20	20	25	25	30	30	35	35	40	40	45	45	50	50

* in °F
† in Btu/hr per sq ft (factors include heat loss for transmission and infiltration)

Air Conditioning Contractors of America

Cooling Heat Transfer Factors

Type of Construction	Cooling Factor*		
Walls	**15°**	**20°**	**25°**
Wood frame with sheeting, siding, and veneer or other finish			
No insulation, ½″ gypsum board	5.0	6.4	7.8
R-11 cavity insulation + ½″ gypsum board	1.7	2.1	2.6
R-13 cavity insulation + ½″ gypsum board	1.5	1.9	2.3
R-13 cavity insulation + ¾″ bead board (R-2.7)	1.3	1.7	2.0
R-19 cavity insulation + ½″ gypsum board	1.1	1.4	1.7
R-19 cavity insulation + ¾″ extruded poly	0.9	1.2	1.4
Masonry			
Above grade — No insulation	5.8	8.3	10.9
Above grade + R-5	1.6	2.3	3.1
Above grade + R-11	0.9	1.3	1.6
Below grade — No insulation	0.0	0.0	0.0
Below grade + R-5	0.0	0.0	0.0
Below grade + R-11	0.0	0.0	0.0
Ceilings			
No insulation	17.0	19.2	21.4
2″ – 2½″ insulation R-7	4.4	4.9	5.5
3″ – 3½″ insulation R-11	3.2	3.7	4.1
5¼″ – 6½″ insulation R-19	2.1	2.3	2.6
6″ – 7″ insulation R-22	1.9	2.1	2.4
10″ – 12″ insulation R-38	1.0	1.1	1.3
12″ – 13″ insulation R-44	0.9	1.0	1.1
Cathedral type (roof/ceiling combination)			
No insulation	11.2	12.6	14.1
R-11	2.8	3.2	3.5
R-19	1.9	2.2	2.4
R-22	1.8	2.0	2.2
Floors			
Over unconditioned space			
Over basement or enclosed crawl space (not vented)	0.0	0.0	0.0
Over vented space or garage	3.9	5.8	7.7
Over vented space or garage + R-11 insulation	0.8	1.3	1.7
Over vented space or garage + R-19 insulation	0.5	0.8	1.1
Basement concrete slab floor unheated			
No edge insulation	0.0	0.0	0.0
1″ edge insulation R-5	0.0	0.0	0.0
2″ edge insulation R-9	0.0	0.0	0.0
Basement concrete slab floor duct in slab			
No edge insulation	0.0	0.0	0.0
1″ edge insulation R-5	0.0	0.0	0.0
2″ edge insulation R-9	0.0	0.0	0.0

* in °F
Note: R values on this chart refer to thermal resistance value.

Air Conditioning Contractors of America

Round-to-Rectangular Duct Conversion Chart

Lgth. Adj.	4.0	4.5	5.0	5.5	6.0	6.5	7.0	7.5	8.0	9.0	10.0	11.0	12.0	13.0	14.0	15.0	16.0
	Length of One Side of Rectangular Duct*																
3.0	3.8	4.0	4.2	4.4	4.6	4.7	4.9	5.1	5.2	5.5	5.7	6.0	6.2	6.4	6.6	6.8	7.0
3.5	4.1	4.3	4.6	4.8	5.0	5.2	5.3	5.5	5.7	6.0	6.3	6.5	6.8	7.0	7.2	7.5	7.7
4.0	4.4	4.6	4.9	5.1	5.3	5.5	5.7	5.9	6.1	6.4	6.7	7.0	7.3	7.6	7.8	8.0	8.3
4.5	4.6	4.9	5.2	5.4	5.7	5.9	6.1	6.3	6.5	6.9	7.2	7.5	7.8	8.1	8.4	8.6	8.8
5.0	4.9	5.2	5.5	5.7	6.0	6.2	6.4	6.7	6.9	7.3	7.6	8.0	8.3	8.6	8.9	9.1	9.4
5.5	5.1	5.4	5.7	6.0	6.3	6.5	6.8	7.0	7.2	7.6	8.0	8.4	8.7	9.0	9.3	9.6	9.9

Lgth. Adj.	6	7	8	9	10	11	12	13	14	15	16	17	18	19	20	22	24	26	28	30	Lgth. Adj.
	Length of One Side of Rectangular Duct*																				
6	6.6																				6
7	7.1	7.7																			7
8	7.6	8.2	8.7																		8
9	8.0	8.7	9.3	9.8																	9
10	8.4	9.1	9.8	10.4	10.9																10
11	8.8	9.5	10.2	10.9	11.5	12.0															11
12	9.1	9.9	10.7	11.3	12.0	12.6	13.1														12
13	9.5	10.3	11.1	11.8	12.4	13.1	13.7	14.2													13
14	9.8	10.7	11.5	12.2	12.9	13.5	14.2	14.7	15.3												14
15	10.1	11.0	11.8	12.6	13.3	14.0	14.6	15.3	15.8	16.4											15
16	10.4	11.3	12.2	13.0	13.7	14.4	15.1	15.7	16.4	16.9	17.5										16
17	10.7	11.6	12.5	13.4	14.1	14.9	15.6	16.2	16.8	17.4	18.0	18.6									17
18	11.0	11.9	12.9	13.7	14.5	15.3	16.0	16.7	17.3	17.9	18.5	19.1	19.7								18
19	11.2	12.2	13.2	14.1	14.9	15.7	16.4	17.1	17.8	18.4	19.0	19.6	20.2	20.8							19
20	11.5	12.5	13.5	14.4	15.2	16.0	16.8	17.5	18.2	18.9	19.5	20.1	20.7	21.3	21.9						20
22	12.0	13.0	14.1	15.0	15.9	16.8	17.6	18.3	19.1	19.8	20.4	21.1	21.7	22.3	22.9	24.0					22
24	12.4	13.5	14.6	15.6	16.5	17.4	18.3	19.1	19.9	20.6	21.3	22.0	22.7	23.3	23.9	25.1	26.2				24
26	12.8	14.0	15.1	16.2	17.1	18.1	19.0	19.8	20.6	21.4	22.1	22.9	23.5	24.2	24.9	26.1	27.3	28.4			26
28	13.2	14.5	15.6	16.7	17.7	18.7	19.6	20.5	21.3	22.1	22.9	23.7	24.4	25.1	25.8	27.1	28.3	29.5	30.6		28
30	13.6	14.9	16.1	17.2	18.3	19.3	20.2	21.1	22.0	22.9	23.7	24.4	25.2	25.9	26.6	28.0	29.3	30.5	31.7	32.8	30
32	14.0	15.3	16.5	17.7	18.8	19.8	20.8	21.8	22.7	23.5	24.4	25.2	26.0	26.7	27.5	28.9	30.2	31.5	32.7	33.9	32
34	14.4	15.7	17.0	18.2	19.3	20.4	21.4	22.4	23.3	24.2	25.1	25.9	26.7	27.5	28.3	29.7	31.0	32.4	33.7	34.9	34
36	14.7	16.1	17.4	18.6	19.8	20.9	21.9	22.9	23.9	24.8	25.7	26.6	27.4	28.2	29.0	30.5	32.0	33.3	34.6	35.9	36
38	15.0	16.5	17.8	19.0	20.2	21.4	22.4	23.5	24.5	25.4	26.4	27.2	28.1	28.9	29.8	31.3	32.8	34.2	35.6	36.8	38
40	15.3	16.8	18.2	19.5	20.7	21.8	22.9	24.0	25.0	26.0	27.0	27.9	28.8	29.6	30.5	32.1	33.6	35.1	36.4	37.8	40
42	15.6	17.1	18.5	19.9	21.1	22.3	23.4	24.5	25.6	26.6	27.6	28.5	29.4	30.3	31.2	32.8	34.4	35.9	37.3	38.7	42
44	15.9	17.5	18.9	20.3	21.5	22.7	23.9	25.0	26.1	27.1	28.1	29.1	30.0	30.9	31.8	33.5	35.1	36.7	38.1	39.5	44
46	16.2	17.8	19.3	20.6	21.9	23.2	24.4	25.5	26.6	27.7	28.7	29.7	30.6	31.6	32.5	34.2	35.9	37.4	38.9	40.4	46
48	16.5	18.1	19.6	21.0	22.3	23.6	24.8	26.0	27.1	28.2	29.2	30.2	31.2	32.2	33.1	34.9	36.6	38.2	39.7	41.2	48
50	16.8	18.4	19.9	21.4	22.7	24.0	25.2	26.4	27.6	28.7	29.8	30.8	31.8	32.8	33.7	35.5	37.2	38.9	40.5	42.0	50
52	17.1	18.7	20.2	21.7	23.1	24.4	25.7	26.9	28.0	29.2	30.3	31.3	32.3	33.3	34.3	36.2	37.9	39.6	41.2	42.8	52
54	17.3	19.0	20.6	22.0	23.5	24.8	26.1	27.3	28.5	29.7	30.8	31.8	32.9	33.9	34.9	36.8	38.6	40.3	41.9	43.5	54
56	17.6	19.3	20.9	22.4	23.8	25.2	26.5	27.7	28.9	30.1	31.2	32.3	33.4	34.4	35.4	37.4	39.2	41.0	42.7	44.3	56
58	17.8	19.5	21.2	22.7	24.2	25.5	26.9	28.2	29.4	30.6	31.7	32.8	33.9	35.0	36.0	38.0	39.8	41.6	43.3	45.0	58
60	18.1	19.8	21.5	23.0	24.5	25.9	27.3	28.6	29.8	31.0	32.2	33.3	34.4	35.5	36.5	38.5	40.4	42.3	44.0	45.7	60

* in in.

ASHRAE Handbook – Fundamentals

Duct Heat Loss Multipliers

Duct Location and Insulation Value Exposed to Outdoor Ambient Air — Attic, Garage, Exterior Wall, Open Crawl Space	Duct Loss Multipliers	
	Winter Design Below 15°F	Winter Design Above 15°F
None	1.30	1.25
R-2	1.20	1.15
R-4	1.15	1.10
R-6	1.10	1.05
Enclosed in Unheated Space — Vented or Unvented Crawl Space or Basement		
None	1.20	1.15
R-2	1.15	1.10
R-4	1.10	1.05
R-6	1.05	1.00
Duct Buried in or under Concrete Slab — Edge Insulation		
None	1.25	1.20
R value = 3 to 4	1.15	1.10
R value = 5 to 7	1.10	1.05
R value = 7 to 9	1.05	1.00

Air Conditioning Contractors of America

Duct Heat Gain Multipliers

Duct Location and Insulation Value Exposed to Outdoor Ambient Air — Attic, Garage, Exterior Wall, Open Crawl Space	Duct Gain Multipliers
None	1.30
R-2	1.20
R-4	1.15
R-6	1.10
Enclosed in Unheated Space — Vented or Unvented Crawl Space or Basement	
None	1.15
R-2	1.10
R-4	1.05
R-6	1.00
Duct Buried in or under Concrete Slab — Edge Insulation	
None	1.20
R value = 3 to 4	1.10
R value = 5 to 7	1.05
R value = 7 to 9	1.00

Air Conditioning Contractors of America

Friction Loss Chart

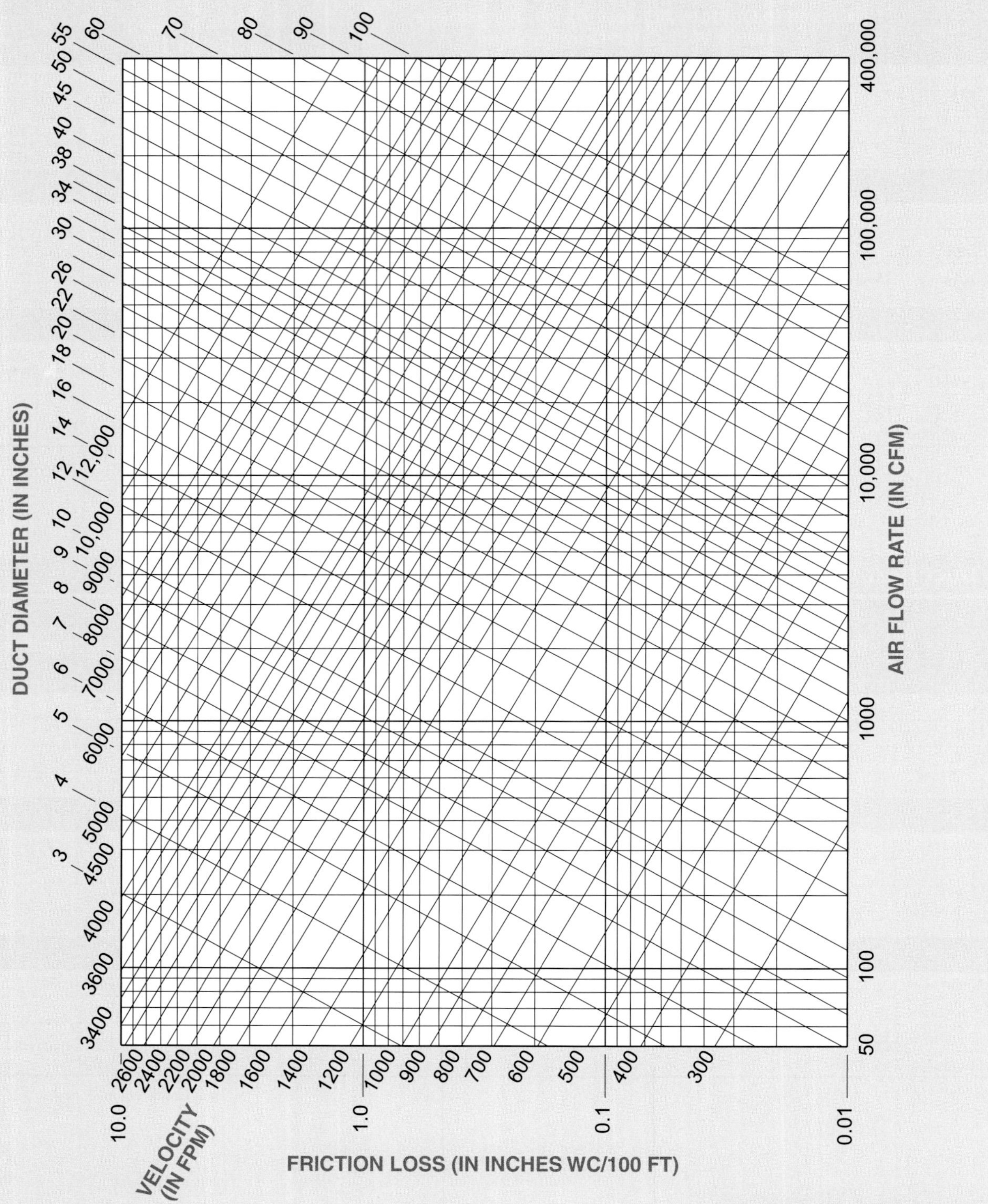

ASHRAE Handbook – Fundamentals

Ductless Mini-Split System Sizing Guidelines

Building Space	94,000 Btu	12,000 Btu	18,000 Btu	24,000 Btu
Space up to 300 ft^2 without any bad situations B, C, D, or E (space well-insulated, double-pane windows, and regular ceiling)	Yes	No	No	No
Space up to 300 ft^2 with either of the bad situations B, C, D, or E (space well-insulated, but has many windows)	Possible	Yes	No	No
Space up to 300 ft^2 with all bad situations (not insulated and bad windows)	No	Yes	Possible	No
Space between 300 ft^2 and 500 ft^2 without any bad situations B, C, D, or E (space well-insulated, double-pane windows, and regular ceiling)	No	Yes	No	No
Space between 300 ft^2 and 500 ft^2 with either of the bad situations B, C, D, or E (space well-insulated, double-pane windows, and regular ceiling)	No	Possible	Better Possible	No
Space between 300 ft^2 and 500 ft^2 with all bad situations (not insulated and bad windows)	No	No	Yes	Possible
Space between 500 ft^2 and 700 ft^2 without any bad situations B, C, D, or E (space well-insulated, double-pane windows, and regular ceiling)	No	No	Yes	No
Space between 500 ft^2 and 700 ft^2 with either of the bad situations B, C, D, or E (space well-insulated, but has many windows)	No	No	Yes	Possible
Space between 500 ft^2 and 700 ft^2 with all bad situations (not insulated and bad windows)	No	No	Possible	Yes
Space between 700 ft^2 and 1000 ft^2 without any bad situations B, C, D, or E (space well-insulated, double-pane windows, and regular ceiling)	No	No	No	Yes
Space between 700 ft^2 and 1000 ft^2 with either of the bad situations B, C, D, or E (space well-insulated, but has many windows)	No	No	No	Yes (Possible 30,000 Btu)
Space between 700 ft^2 and 1000 ft^2 with all bad situations (not insulated and bad windows)	No	No	Two 18,000 Btu required	No
Space over 1000 ft^2, more than one unit required	Single-Zone Unit	Two-Zone Unit	Three-Zone Unit	Four-Zone Unit

Good:
Situation A – one air handler per space

Bad:
Situation B – space not well insulated
Situation C – 2, 3, or more single pane windows
Situation D – ceiling more than 10 ft high
Situation E – any heat generating source in space

Selecting Wire Size

	Copper			Aluminum or Copper-Clad Aluminum		
	140°F	167°F	194°F	140°F	167°F	194°F
SIZE AWG	TYPE TW, UF	TYPES RHW, THHW, THW, THWN, XHHW, USE, ZW	TYPES TBS, SA, SIS, FEP, FEPB, MI, RHH, RHW-2, THHN, THHW, THW-2, THWN-2, USE-2, XHH, XHHW, XHHW-2, ZW-2	TYPES TW, UF	TYPES RHW, THHW, THW, THWN, XHHW, USE	TYPES TBS, SA, SIS, THHN, THHW, THW-2, THWN-2, RHH, RHW-2, USE-2, XHH, XHHW, XHHW-2, ZW-2
18	—	—	14	—	—	—
16	—	—	18	—	—	—
14*	15	20	25	—	—	—
12*	20	25	30	15	20	25
10*	30	35	40	25	30	35
8	40	50	55	35	40	45
6	55	65	75	40	50	55
4	70	85	95	55	65	75
3	85	100	115	65	75	85
2	95	115	130	75	90	100
1	110	130	145	85	100	115
1/0	125	150	170	100	120	135
2/0	145	175	195	115	135	150
3/0	165	200	225	130	155	175
4/0	195	230	260	150	180	205
250	215	255	290	170	205	230
300	240	285	320	195	230	260
350	260	310	350	210	250	280
400	280	335	380	225	270	305
500	320	380	430	260	310	350
600	350	420	475	285	340	385
700	385	460	520	315	375	425
750	400	475	535	320	385	435
800	410	490	555	330	395	445
900	435	520	585	355	425	480
1000	455	545	615	375	445	500
1250	495	590	665	405	485	545
1500	525	625	705	435	520	585
1750	545	650	735	455	545	615
2000	555	665	750	470	560	630

*conductor overcurrent protection limitations

Thermal Conductivity of Metals

Metal	Temperature t (°F)	Thermal Conductivity k (Btu/[hr °F ft])
Admiralty brass	68	64
Aluminum, pure	68	118
	200	124
	400	144
Aluminum bronze	68	44
Antimony	68	10.7
Beryllium	68	126
Beryllium copper	68	38
Bismuth	68	4.9
Cadmium	68	54
Carbon steel, max 0.5% C	68	31
Carbon steel, max 1.5% C	68	21
	752	19
	2192	17
Cartridge brass (UNS C26000)	68	69.4
Cast iron, gray	70	27–46
Chromium	68	52
Cobalt	68	40
Copper, pure	68	223
	572	213
	1112	204
Copper bronze (75% Cu, 25% Sn)	68	15
Copper brass (70% Cu, 30% Zi)	68	64
Cupronickel	68	17
Gold	68	182
Hastelloy B		6
Hastelloy C	70	5
Inconel	70–212	8.4
Incoloy	32–212	6.8
Iridium	68	85
Iron, nodular pearlitic	212	18
Iron, pure	68	42
	572	32
	1832	20

Thermal Conductivity of Metals

Metal	Temperature t (°F)	Thermal Conductivity k (Btu/[hr °F ft])
Iron, wrought	68	34
Lead	68	20
	572	17.2
Manganese bronze	68	61
Magnesium	68	91.9
Mercury	68	4.85
Molybdenum	68	81
Monel	32–212	15
Nickel	68	52
Nickel wrought	32–212	35–52
Niobium (columbium)	68	30
Osmium	68	35
Phosphor bronze (10% Sn, UNS C52400)	68	28.9
Platinum	68	42
Plutonium	68	4.6
Potassium	68	57.8
Red brass	68	92
Rhodium	68	86.7
Selenium	68	0.3
Silicon	68	48.3
Silver, pure	68	235
Sodium	68	77.5
Stainless steel	68	7–26
Tantalum	68	31
Thorium	68	24
Tin	32	36–39
Titanium	68	11–13
Tungsten	68	94–100
Uranium	68	14
Vanadium	68	35
Wrought carbon steel	32	34
Yellow brass	68	67
Zinc	–	67
Zirconium	–	145

System Components

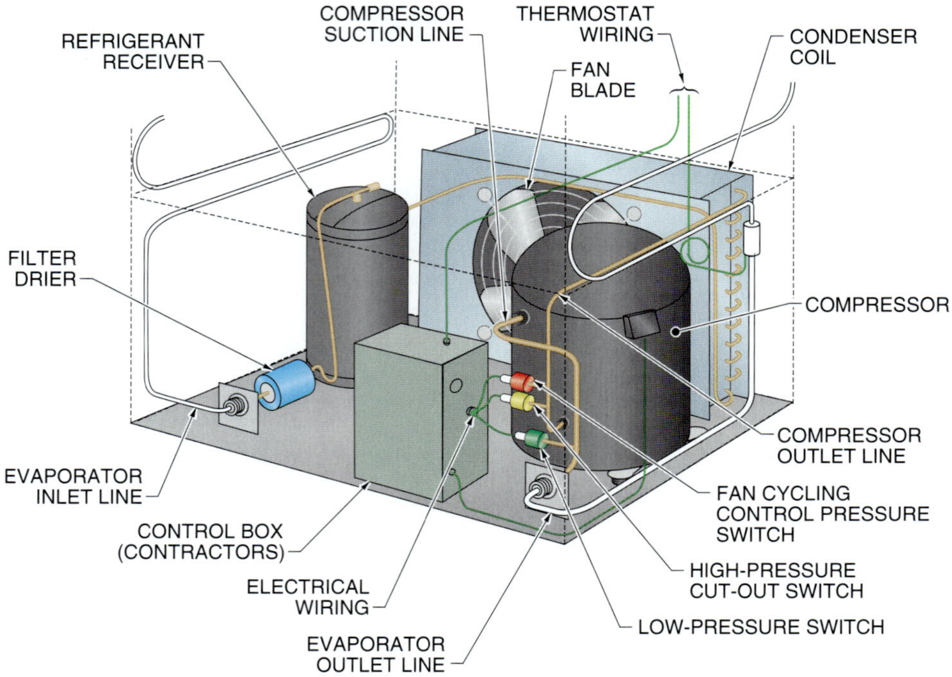

Thermostat Wiring Installation Guide—Conventional Systems

1H/1C SYSTEM

Rc	Power*
R	[R+Rc joined by jumper]
Y	Compressor contactor
C	24 VAC common‡
W	Heat relay
G	Fan relay

HEAT-ONLY SYSTEM

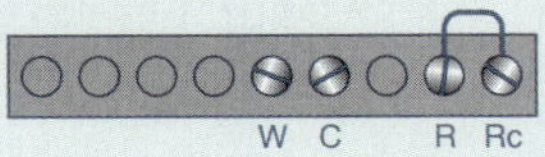

Rc	Power*
R	[R+Rc joined by jumper]
C	24 VAC common‡
W	Heat relay

HEAT-ONLY SYSTEM

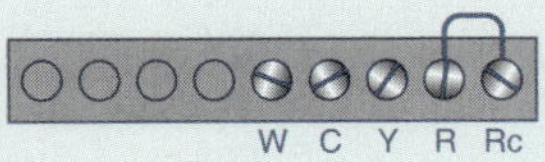

Rc	[R+Rc joined by jumper]
R	Series 20 valve terminal "R"*
Y	Series 20 valve terminal "W"
C	24 VAC common‡
W	Series 20 valve terminal "B"

HEAT-ONLY SYSTEM

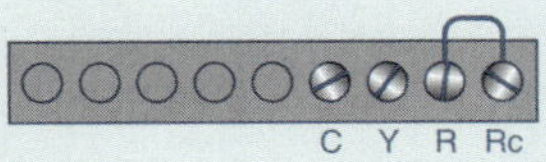

Rc	[R+Rc joined by jumper]
R	Power*
Y	Normally open zone valve
C	24 VAC common‡

1H/1C SYSTEM

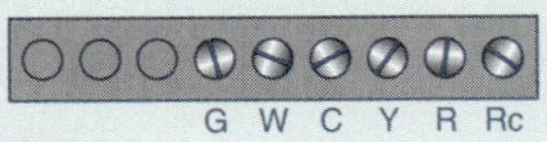

Rc	Power (cooling transformer)*†
R	Power (heating transformer)*†
Y	Compressor contactor
C	24 VAC common‡§
W	Heat relay
G	Fan relay

HEAT ONLY SYSTEM WITH FAN

Rc	Power*
R	[R+Rc joined by jumper]
C	24 VAC common‡
W	Heat relay
G	Fan relay

COOL-ONLY SYSTEM

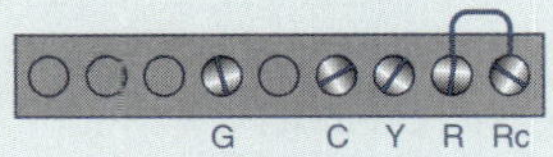

Rc	Power*
R	[R+Rc joined by jumper]
Y	Compressor contactor
C	24 VAC common‡
G	Fan relay

2H/2C SYSTEM

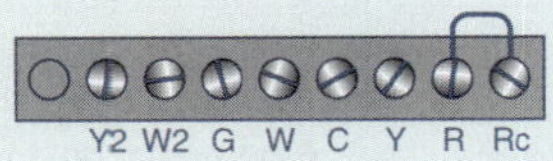

Rc	Power*
R	[R+Rc joined by jumper]
Y	Compressor contactor (stage 1)
C	24 VAC common‡
W	Heat relay (stage 1)
G	Fan relay
W2	Heat relay (stage 2)
Y2	Compressor contactor (stage 2)

2H/2C SYSTEM

Rc	Power (cooling transformer)*†
R	Power (heating transformer)*†
Y	Compressor contactor (stage 1)
C	24 VAC common‡§
W	Heat relay (stage 1)
G	Fan relay
W2	Heat relay (stage 2)
Y2	Compressor contactor (stage 2)

NOTE: USE 18- TO 22- GAUGE THERMOSTAT WIRE. SHIELDED CABLE NOT REQUIRED.

* power supply; provide disconnect means and overload protection as required
† remove jumper for two-transformer systems
‡ optional 24 VAC common connection
§ common connection must come from cooling transformer

Thermostat Wiring Installation Guide—Heat Pump Systems

1H/1C HEAT PUMP SYSTEM

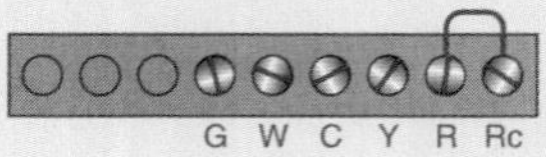

Rc	Power*
R	[R+Rc joined by jumper]
Y	Compressor contactor
C	24 VAC common†
O/B	Changeover valve
G	Fan relay

2H/1C HEAT PUMP SYSTEM

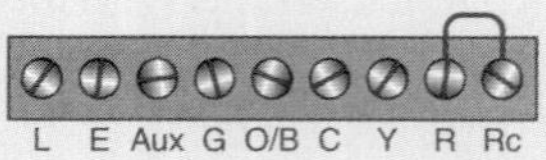

Rc	Power*
R	[R+Rc joined by jumper]
Y	Compressor contactor
C	24 VAC common†
O/B	Changeover valve
G	Fan relay
Aux	Auxiliary heat relay
E	Emergency heat relay
L	Sends output when set to Em. Heat

2H/1C HEAT PUMP SYSTEM

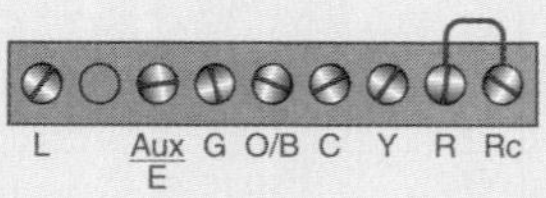

Rc	Power*
R	[R+Rc joined by jumper]
Y	Compressor contactor
C	24 VAC common†
O/B	Changeover valve
G	Fan relay
Aux/E	Auxiliary/Emergency heat relay
L	Sends output when set to em. heat

2H/2C HEAT PUMP SYSTEM

Rc	Power*
R	[R+Rc joined by jumper]
Y	Compressor contactor (stage 1)
C	24 VAC common†
O/B	Changeover valve
G	Fan relay
Y2	Compressor contactor (stage 2)
L	Sends output when set to em. heat

3H/2C HEAT PUMP SYSTEM

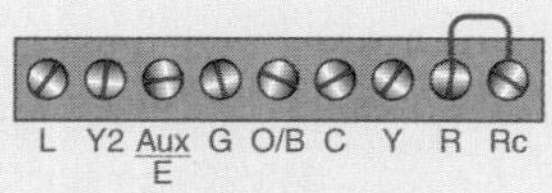

Rc	Power*
R	[R+Rc joined by jumper]
Y	Compressor contactor (stage 1)
C	24 VAC common†
O/B	Changeover Valve
G	Fan relay
Aux/E	Auxiliary/Emergency heat relay
Y2	Compressor contactor (stage 2)
L	Sends output when set to em. heat

* power supply; provide disconnect means and overload protection as required
† optional 24 VAC common connection

Determining Gas Pipe Size

Pipe Length*	Pipe Size							
	½	¾	1	1¼	1½	2	2½	3
10	174†	363	684	1404	2103	4050	6455	11412
20	119	249	470	955	1445	2784	4437	7843
30	96	200	377	775	1161	2235	3503	6299
40	82	171	323	663	993	1913	3049	5391
50	73	152	286	588	880	1696	2703	4778
60	66	135	259	532	795	1538	2449	4329
70	61	127	239	490	734	1413	2253	3983
80	56	118	222	456	688	1315	2096	3705
90	53	111	206	428	641	1234	1966	3476
100	50	104	197	404	605	1165	1857	3284
125	44	93	174	358	536	1033	1646	2910
150	40	84	158	324	486	936	1492	2537
200	34	72	136	278	416	801	1277	2257
250	30	64	120	246	369	710	1131	2000
300	28	58	109	223	334	643	1025	1812
350	25	53	100	205	307	592	943	1667
400	24	49	93	191	296	551	877	1551
450	22	46	87	179	268	517	823	1455
500	21	44	82	169	253	488	778	1375
550	20	42	78	161	241	463	739	1306

*in ft
† ft^3 natural gas/hr

HVAC Symbols

Equipment Symbols

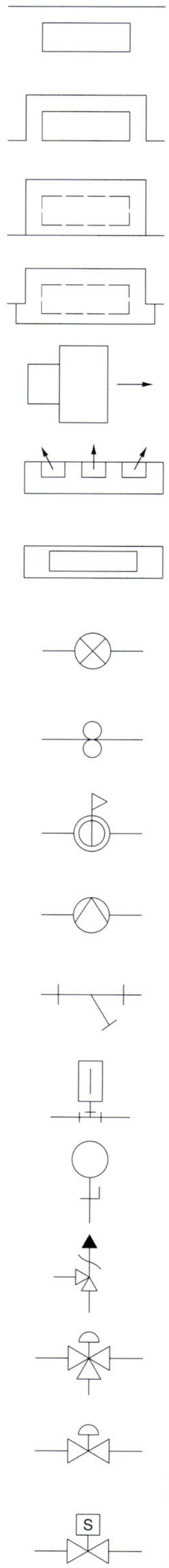

- Exposed radiator
- Recessed radiator
- Flush enclosed radiator
- Projecting enclosed radiator
- Unit heater (propeller) — plan
- Unit heater (centrifugal) — plan
- Unit ventilator — plan
- Steam
- Duplex strainer
- Pressure-reducing valve
- Air line valve
- Strainer
- Thermometer
- Pressure gauge and cock
- Relief valve
- Automatic 3-way valve
- Automatic 2-way valve
- Solenoid valve

Ductwork

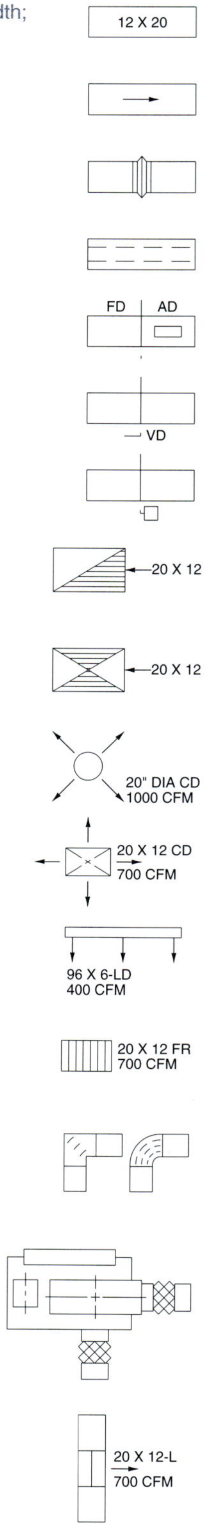

- Duct (1st figure, width; 2nd figure, depth)
- Direction of flow
- Flexible connection
- Ductwork with acoustical lining
- Fire damper with access door
- Manual volume damper
- Automatic volume damper
- Exhaust, return or outside air duct — section
- Supply duct — section
- Ceiling diffuser supply outlet
- Ceiling diffuser supply outlet
- Linear diffuser
- Floor register
- Turning vanes
- Fan and motor with belt guard
- Louver opening

Heating Piping

High-pressure steam	—— HPS ——
Medium-pressure steam	—— MPS ——
Low-pressure steam	—— LPS ——
High-pressure return	—— HPR ——
Medium-pressure return	—— MPR ——
Low-pressure return	—— LPR ——
Boiler blow off	—— BD ——
Condensate or vacuum pump discharge	—— VPD ——
Feedwater pump discharge	—— PPD ——
Makeup water	—— MU ——
Air relief line	—— V ——
Fuel oil suction	—— FOS ——
Fuel oil return	—— FOR ——
Fuel oil vent	—— FOV ——
Compressed air	—— A ——
Hot water heating supply	—— HW ——
Hot water heating return	—— HWR ——

Air Conditioning Piping

Refrigerant liquid	—— RL ——
Refrigerant discharge	—— RD ——
Refrigerant suction	—— RS ——
Condenser water supply	—— CWS ——
Condenser water return	—— CWR ——
Chilled water supply	—— CHWS ——
Chilled water return	—— CHWR ——
Makeup water	—— MU ——
Humidification line	—— H ——
Drain	—— D ——

Refrigeration Symbols

Name	Symbol	Name	Symbol	Name	Symbol
Gauge	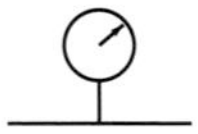	Pressure switch		Dryer	
Sight glass	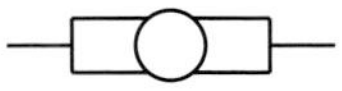	Hand expansion valve		Filter and strainer	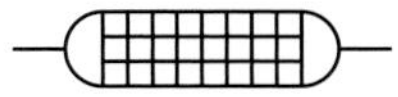
High side float valve	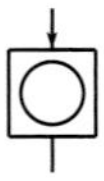	Automatic expansion valve		Combination strainer and dryer	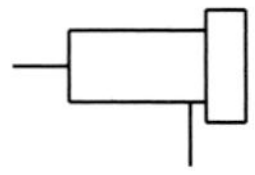
Low side float valve	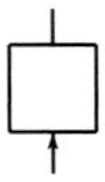	Thermostatic expansion valve	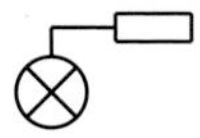	Evaporative condensor	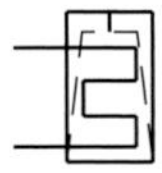
Immersion cooling unit		Constant pressure valve, suction	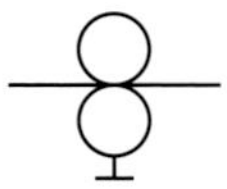	Heat exchanger	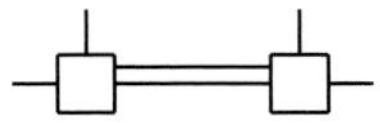
Cooling tower	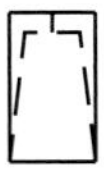	Thermal bulb	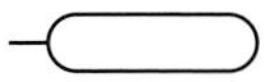	Air-cooled condensing unit	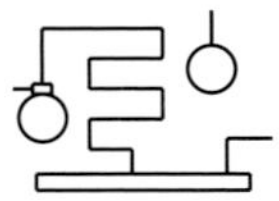
Natural convection, finned type evaporator	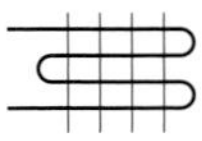	Scale trap	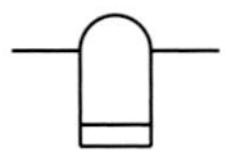	Water-cooled condensing unit	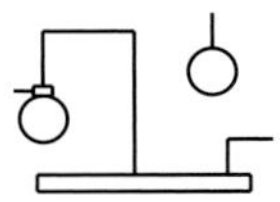
Forced convection evaporator	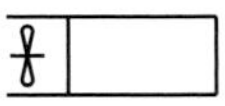	Self-contained thermostat			

Indoor Air Quality Complaint Form

This form can be filled out by the building occupant or by a member of the building staff.

Occupant Name: ______________________________ Date: ______________

Department/Location in Building: ______________________________ Phone: ______________

Completed by: ____________________ Title: ______________ Phone: ______________

This form should be used if your complaint may be related to indoor air quality. Indoor air quality problems include concerns with temperature control, ventilation, and air pollutants. Your observations can help to resolve the problem as quickly as possible. Please use the space below to describe the nature of the complaint and any potential causes.

May we contact you to discuss your complaint? What is the best time to reach you? ______________________

So that we can respond promptly, please return this form to: ______________________________
IAQ Manager or Contact Person

____
Room, Building, Mail Code

OFFICE USE ONLY

File Number: __________ Received By: ____________________ Date Received: ______________

Typical LED Cooling Failure Codes

Operation	Fault	Amber LED Flash Code	Possible Cause and Action
Standby — no call for unit operation	None	On solid, no flash	Normal operation — 3 thermostat wires or 4-wire Infinity Control
Standby — no call for unit operation	None	Off	Normal operation — No call for cooling with 2-wire connection or indoor unit not powered
Cooling operation	None	1, pause	Normal operation
	System communications failure	16	Communication with user interface lost; check wiring to UI; indoor and outdoor units
	High-pressure switch open	31	High-pressure switch trip; check refrigerant charge, outdoor fan operation and coils for airflow restrictions
	Low-pressure switch open	32	Low-pressure switch trip; check refrigerant charge and indoor airflow
	Control fault	45	Outdoor unit control board has failed; control board needs to be replaced
	Brown out (24 V)	46	The control voltage is less than 15.5 V for at least 4 seconds; compressor and fan operation not allowed until control voltage is minimum of 17.5 V; verify control voltage
	Outdoor air temp sensor fault	53	Outdoor air sensor not reading or out of range; ohm out sensor and check wiring
	Outdoor coil temp sensor fault	55	Coil sensor not reading or out or range; ohm out sensor and check wiring
	Thermistors out of range	56	Improper relationship between coil sensor and outdoor air sensor; ohm out sensors and check wiring
	Thermal cutout	72	Compressor voltage sensed after startup, then absent for 10 consecutive seconds while cooling demand exists; possible causes are internal compressor overload trip or loss of high voltage to compressor without loss of control voltage; control will continue fan operation and wait 15 minutes to attempt restart; fault will clear when restart is successful or low-voltage power is cycled
	Contactor shorted	73	Compressor voltage sensed when no demand for compressor operation exists; contactor may be stuck closed or there is wiring error
	No 230 V at compressor	74	Compressor voltage not sensed when compressor should be starting; disconnect may be open or contactor may be stuck open or there is wiring error
	Thermal lockout	82	Thermal cutout occurs in three consecutive cycles; unit operation locked out for 4 hours or until 24 V power recycled
	Low-pressure lockout	83	Low-pressure switch trip has occurred during five consecutive cycles; unit operation locked out for 4 hours or until 24 V power recycled
	High-pressure lockout	84	High-pressure switch trip has occurred during five consecutive cycles; unit operation locked out for 4 hours or until 24 V power recycled

TROUBLESHOOTING CHART

Symptoms of Abnormal Operation	Associated LED Code[2]	Fault Description(s)	Possible Causes	Corrective Action	Cautions and Notes
• Furnace fails to operate. • Integrated control module diagnostic **LED** provides **no signal**.	NONE	• No 115 volt power to furnace, or no 24 volt power to integrated control module. • Blown fuse or circuit breaker.	• Manual disconnect switch OFF, door switch open, or 24 volt wires improperly connected or loose. • Blown fuse or circuit breaker.	• Assure 115 and 24 volt power to furnace integrated control module. • Check integrated control module fuse (3A). Replace if necessary. • Check for possible shorts in 115 and 24 volt circuits. Repair as necessary.	• Turn power OFF prior to repair. • Replace integrated control module fuse with 3A automotive fuse.
• Furnace fails to operate.		• Integrated control module has an internal fault.	• Integrated control module has an internal fault.	• Replace bad integrated control module.	• Turn power OFF prior to repair. • Read precautions in "Electrostatic Discharge" section of manual.
• Furnace fails to operate. • Integrated control module diagnostic **LED** is flashing **ONE** (1) flash.	1 FLASH	• Furnace lockout due to an excessive number of ignition attempt (1 total) and "retries" (3 total)[1]. • Auxiliary Limit Open	• Failure to establish flame. Cause may be no gas to burners, bad igniter or igniter alignment, improper orifices, or coated/oxidized or improperly connected flame sensor. • Loss of flame after establishment. Cause may be interrupted gas supply, lazy burner flames (improper gas pressure or restriction in flue or improper induced draft blower performance. • Insufficient conditioned air over the heat exchanger. Blocked filters, restrictive ductwork, improper circulator blower speed, or failed circulator blower.	• Locate and correct gas interruption. • Replace or realign igniter. • Check flame sense signal. Sand sensor if coated and/or oxidized. • Check flue piping for blockage, proper length, elbows, and termination. • Verify proper induced draft blower performance. • Check circulator blower speed and performance. Correct speed or replace blower if necessary.	• Turn power OFF prior to repair. • Igniter is fragile, handle with care. • Clean flame sensor with steel wool. • See "Combustion and Ventilation Air Requirements" and "Category I Venting (Vertical Venting)" section for details. • See Product Data Bulletin for allowable rise range and proper circulator speed.
• Furnace fails to operate. • Integrated control module diagnostic **LED** is flashing **TWO** (2) flashes.	2 FLASHES	• Pressure switch circuit is closed. • Induced draft blower is **not** operating.	• Induced draft blower pressure switch contacts sticking. • Shorts in pressure switch circuit.	• Replace induced draft blower pressure switch. • Repair short.	• Turn power OFF prior to repair. • Replace pressure switch with proper replacement part.
• Induced draft blower runs continuously with no further furnace operation. • Integrated control module diagnostic **LED** is flashing **THREE** (3) flashes.	3 FLASHES	• Pressure switch circuit not closed. • Induced draft blower **is** operating.	• Pressure switch hose blocked, pinched or connected improperly. • Blocked flue or weak induced draft blower. • Incorrect pressure switch setpoint or malfunctioning switch contacts. • Loose or improperly connected wiring.	• Inspect pressure switch hose. Repair, if necessary, • Inspect flue for blockage, proper length, elbows, and termination. • Correct pressure switch setpoint or contact motion. • Tighten or correct wiring connection.	• Turn power OFF prior to repair. • See "Combustion and Ventilation Air Requirements" and "Category I Venting (Vertical Venting)" section for details. • Replace pressure switch with proper replacement part.

[1] Integrated control module will automatically attempt to reset from lockout after one hour.

[2] LED Flash code will cease if power to the control module is interrupted through the disconnect or door switch.

IO-242C 29 05/05

Goodman Global Group, Inc.

Troubleshooting Chart

Symptoms of Abnormal Operation	Associated LED Code[2]	Fault Description(s)	Possible Causes	Corrective Action	Cautions and Notes
• Circulator blower runs continuously. No furnace operation. • Integrated control module diagnostic **LED** is flashing **FOUR** (4) flashes.	4 FLASHES	• Primary limit circuit is open.	• Insufficient conditioned air over the heat exchanger. Blocked filters, restrictive ductwork, improper circulator blower speed, or failed circulator blower.	• Check filters and ductwork for blockage. Clean filters or remove obstruction. • Check circulator blower speed and performance. Correct speed or replace blower if necessary.	• Turn power OFF prior to repair. • See Product Data Bulletin for allowable rise range and proper circulator speed.
• Induced draft blower and circulator blower runs continuously. No furnace operation. • Integrated control module diagnostic **LED** is flashing **FIVE** (5) flashes.	5 FLASHES	• Flame sensed with no call for heat.	• Short to ground in flame sense circuit.	• Correct short at flame sensor or in flame sensor wiring	• Turn power OFF prior to repair.
• Furnace fails to operate. • Integrated control module diagnostic **LED** is flashing **SIX** (6) flashes.	6 FLASHES	• Rollout limit open.	• Flame rollout. • Misaligned burners, blocked flue and/or air inlet pipe, or failed induced draft blower. • Loose or improperly connected wiring.	• Check burners for proper alignment. • Check flue and air inlet piping for blockage, proper length, elbows, and termination. Correct as necessary. • Check induced draft blower for proper performance. Replace, if necessary. • Tighten or correct wiring connection.	• See "Vent/Flue Pipe" section for piping details. • Replace induced draft blower with proper replacement part.
• Normal furnace operation. • Integrated control module diagnostic **LED** is flashing **SEVEN** (7) flashes.	7 FLASHES	• Flame sense microamp signal is low.	• Flame sensor is coated/ oxidized. • Flame sensor incorrectly positioned in burner flame. • Lazy burner flame due to improper gas pressure or combustion air.	• Sand flame sensor is coated/oxidized. • Inspect for proper sensor alignment. • Check inlet air piping for blockage, proper length, elbows, and termination. • Compare current gas pressure to rating plate info. Adjust as needed.	• Turn power OFF prior to repair. • Clean flame sensor with steel wool. • See "Vent/Flue Pipe" section for piping details. • See rating plate for proper gas pressure.
• Induced draft blower runs continuously. No furnace operation. • Integrated control module diagnostic **LED** is flashing **continuously.**	C CONTINUOUS FLASHING	• Polarity of 115 volt power is reversed.	• Polarity of 115 volt AC power to furnace or integrated control module is reversed. • Poor unit ground.	• Review wiring diagram to correct polarity. • Verify proper ground. Correct if necessary. • Check and correct wiring.	• Turn power OFF prior to repair.
• LED is steady on	steady on	• Normal operation.			

[2] LED Flash code will cease if power to the control module is interrupted through the disconnect or door switch.

[3] Applies to **GMS8/GDS8/GHS8** Models only.

Flash Codes

LED Status	Fault Description	Troubleshooting Information
POWER (Green)	Module has power	Supply voltage is present at module terminals
TRIP (Red)	Thermostat demand signal Y1 is present, but the compressor is not running	1. Compressor protector is open * Check for high head pressure * Check compressor supply voltage 2. Outdoor unit power disconnect is open 3. Compressor circuit breaker or fuse(s) is open 4. Broken wire or connector is not making contact 5. Low pressure switch open if present in system 6. Compressor contactor has failed open
ALERT (Yellow) Flash Code 1	**Long Run Time** Compressor is running extremely long run cycles	1. Low refrigerant charge 2. Evaporator blower is not running * Check blower relay coil and contacts * Check blower motor capacitor * Check blower motor for failure or blockage * Check evaporator blower wiring and connector * Check indoor blower control board * Check thermostat wiring for open circuit 3. Evaporator coil is frozen * Check for low suction pressure * Check for excessively low thermostat setting * Check evaporator airflow (coil blockages or return air filter) * Check ductwork or registers for blockage 4. Faulty metering device * Check TXV bulb installation (size, location, and contact) * Check if TXV/fixed orifice is stuck closed or defective 5. Condenser coil is dirty 6. Liquid line restriction (filter drier blocked if present in system) 7. Thermostat is malfunctioning * Check thermostat sub-base or wiring for short circuit * Check thermostat installation (location, level)
ALERT (Yellow) Flash Code 2	**System Pressure Trip** Discharge or suction pressure out of limits or compressor overloaded	1. High head pressure * Check high pressure switch if present in system * Check if system is overcharged with refrigerant * Check for non-condensable in system 2. Condenser coil poor air circulation (dirty, blocked, damaged) 3. Condenser fan is not running * Check fan capacitor * Check fan wiring and connectors * Check fan motor for failure or blockage 4. Return air duct has substantial leakage 5. If low pressure switch present in system, check Flash Code 1 info

Table 9

48442L005

Page 17

Goodman Global Group, Inc.

Flash Codes (cont.)

LED Status	Fault Description	Troubleshooting Information
ALERT (Yellow) Flash Code 3	**Short Cycling** Compressor is running only briefly	1. Thermostat demand signal is intermittent 2. Time delay relay or control board defective 3. If high pressure switch present, go to Flash Code 2 info 4. If low pressure switch present, go to Flash Code 1 info
ALERT (Yellow) Flash Code 4	**Locked Rotor**	1. Run capacitor has failed 2. Low line voltage (contact utility if voltage at disconnect is low) * Check wiring connections 3. Excessive liquid refrigerant in compressor 4. Compressor bearings are seized *Measure compressor oil level
ALERT (Yellow) Flash Code 5	**Open Circuit**	1. Outdoor unit power disconnect is open 2. Compressor circuit breaker or fuse(s) is open 3. Compressor contactor has failed open * Check compressor contactor wiring and connectors * Check for compressor contactor failure (burned, pitted, or open) * Check wiring and connectors between supply and compressor * Check for low pilot voltage at compressor contactor coil 4. High pressure switch is open and requires manual reset 5. Open circuit in compressor supply wiring or connections 6. Unusually long compressor protector reset time due to extreme ambient temperature 7. Compressor windings are damaged * Check compressor motor winding resistance
ALERT (Yellow) Flash Code 6	**Open Start Circuit** Current only in run circuit	1. Run capacitor has failed 2. Open circuit in compressor start wiring or connections * Check wiring and connectors between supply and the compressor S terminal 3. Compressor start winding is damaged * Check compressor motor winding resistance
ALERT (Yellow) Flash Code 7	**Open Run Circuit** Current only in start circuit	1. Open circuit in compressor run wiring or connections * Check wiring and connectors between supply and the compressor R terminal 2. Compressor run winding is damaged * Check compressor motor winding resistance
ALERT (Yellow) Flash Code 8	**Welded Contactor** Compressor always runs	1. Compressor contactor has failed closed 2. Thermostat demand signal not connected to module
ALERT (Yellow) Flash Code 9	**Low Voltage** Control circuit less than 17VAC	1. Control circuit transformer is overloaded 2. Low line voltage (contact utility if voltage at disconnect is low) * Check wiring conditions

Table 10

Miswired Module Troubleshooting

Miswired Module Indication	Recommended Troubleshooting Action
Green LED is not on, module does not power up.	Determine if both R and C module terminals are connected. Verify voltage is present at module's R and C terminals. Review 24VAC Power Wiring section on page 16 for R and C wiring.
Green LED intermittent, module powers up only when compressor runs.	Determine if R and Y terminals are wired in reverse. Verify module's R and C terminals have a constant source. Review 24VAC Power Wiring section on page 16 for R and C wiring.
TRIP LED is on but system and compressor check OK.	Verify Y terminal is connected to 24VAC at contactor coil. Verify voltage at contactor coils falls below 0.5VAC when off.
TRIP LED and ALERT LED flashing together.	Verify R and C terminals are supplied with 19-28VAC.
ALERT Flash Code 3 (Compressor Short Cycling) displayed incorrectly.	Verify Y terminal is connected to 24VAC at contactor coil. Verify voltage at contactor coil falls below 0.5VAC when off.
ALERT Flash Code 5, 6, or 7 (Open Circuit, Open Start Circuit, or Open Run Circuit) displayed incorrectly.	Check that compressor run and start wires are through module's current sensing holes. Verify Y terminal is connected to 24VAC at contactor coil. Verify voltage at contactor coil falls below 0.5VAC when off.
ALERT Flash Code 6 (Open Start Circuit) displayed for Code 7 (Open Run Circuit or vice versa.	Check that compressor run and start wires are routed through correct module sensing holes.
ALERT Flash Code 8 (Welded Contactor) displayed incorrectly.	Determine if module's Y terminal is connected. Verify Y terminal is connected to 24VAC at contactor coil. Verify 24VAC is present across Y and C when thermostat demand signal is present. If not, R and C are reverse wired. Verify voltage at contactor coil falls below 0.5VAC when off. Review Thermostat Demand Wiring on page 16 for Y and C wiring.

Table 11

11d Troubleshooting — control module lights

CHART 1 — Troubleshooting POWER light status

— Usually indicates reversed 120 VAC polarity if POWER light flashes by itself —

⚠WARNING Electrical shock hazard — Wherever you see ▲TURN OFF POWER ▲, follow the instructions. Failure to follow instructions could result in severe personal injury, death or substantial property damage.

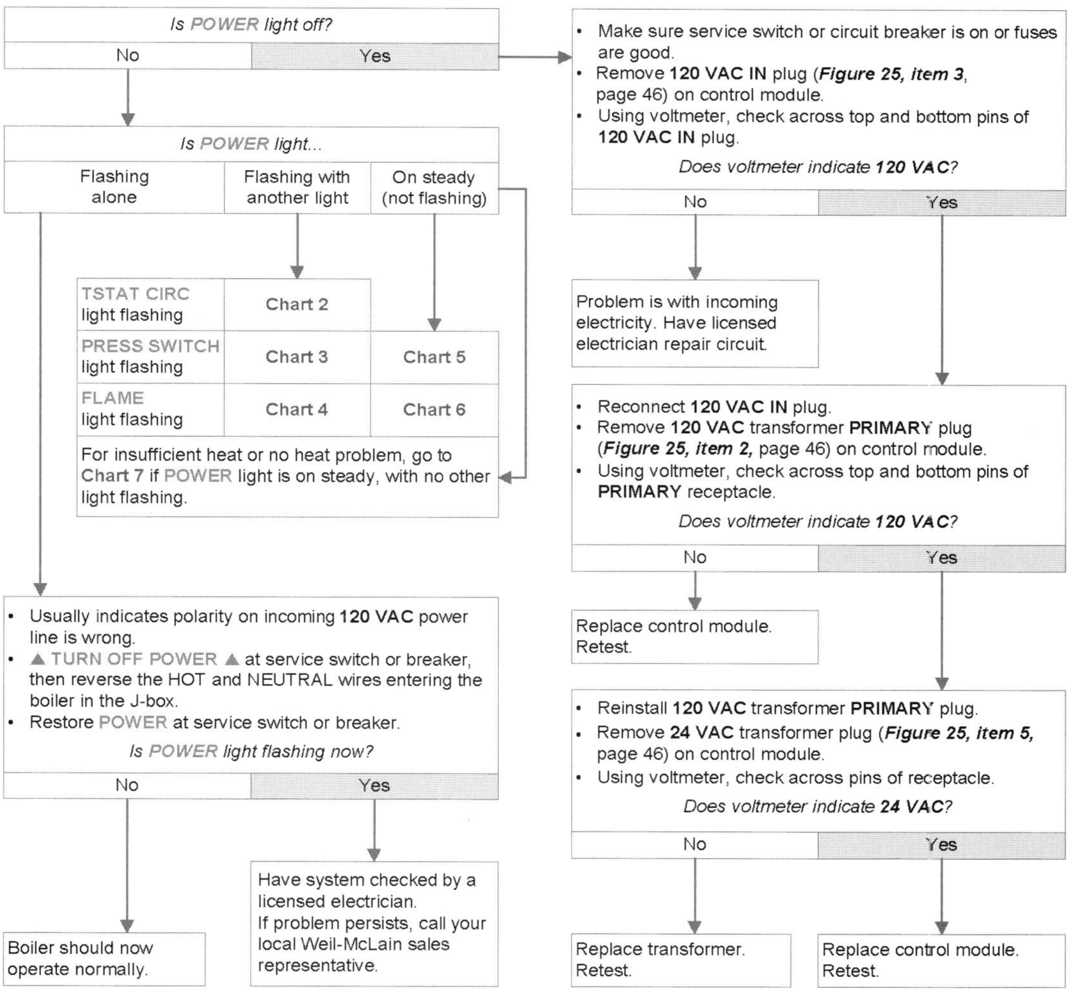

11d Troubleshooting — control module lights

CHART 2 — TSTAT CIRC & POWER lights flashing

— Usually indicates 48 VAC on thermostat circuit (stray voltage) —

⚠WARNING Electrical shock hazard — Wherever you see **▲TURN OFF POWER▲**, follow the instructions. Failure to follow instructions could result in severe personal injury, death or substantial property damage.

- Disconnect the two external wires connected to the boiler thermostat leads (two black low voltage leads in J-box).
- Connect a voltmeter across these two incoming wires. Close each thermostat, zone valve and relay in the external circuit one at a time and check the voltmeter reading across the wires.
- There should NEVER be a voltage reading.
- If a voltage does occur under any condition, check and correct the external wiring. (This is a common problem when using 3-wire zone valves.)
- Once the external thermostat circuit wiring is checked and corrected if necessary, reconnect the external thermostat circuit wires to the boiler thermostat wires and allow the boiler to cycle.

Did you find a voltage across the two external thermostat circuit wires?

Yes →

- Leave external boiler thermostat connection wires disconnected from boiler.
- Troubleshoot the external thermostat circuit until you find the source of the stray voltage. (Pay close attention to the wiring connections to 3-wire zone valves.)
- Correct the problem and repeat the voltmeter test above, verifying there is no longer a voltage reading under any condition in the external thermostat circuit.

→

- Boiler should now operate per the normal sequence of operation shown in **Figure 18**, page 31.

No →

- If no voltage is found under any condition of the external thermostat circuit, connect the two boiler thermostat connection leads together (or jumper the boiler aquastat T-T terminals).
- Turn off power to the boiler for 1 minute.
- Turn on power and allow boiler to cycle.

*Do the **TSTAT** and **POWER** lights still flash?*

No → (return to the steps under "Yes" above: troubleshoot the external thermostat circuit)

Yes →

- Replace control module.
- Retest.

11d Troubleshooting — control module lights (cont.)

CHART 3 — PRESS SWITCH & POWER lights flashing

— Usually indicates pressure switch stuck closed or failed to make within 5 minutes —

⚠WARNING Electrical shock hazard — Wherever you see ⚠TURN OFF POWER⚠, follow the instructions. Failure to follow instructions could result in severe personal injury, death or substantial property damage.

- Reset boiler control by turning off power at service switch or turning down thermostat for at least 45 seconds.
- Thermostat should call for heat and appropriate zone valves open.

Does inducer motor operate?

No →

- Wait 45 seconds.

Does inducer motor operate?

No →

- Remove **INDUCER** plug (***Figure 25, Item 6**, page 46*) from plug receptacle of control module.
- Place voltmeter leads across the **120 VAC N** and **120 VAC H** pins of the **INDUCER** receptacle (left column center and right column bottom pins).

*Does the voltmeter indicate **120 VAC**?*

Yes →

- Try reseating plug in module receptacle and restart.
- If inducer still does not work, replace inducer assembly.
- Retest.

No →

- Remove one of the wires connected to the pressure switch.
- Check continuity across the switch terminals.

Does the ohmmeter indicate the switch is closed?

No →

- Replace control module.
- Retest.

Yes →

- Check air pressure switch per manual Section 11b.
- Replace if necessary.
- Retest.

Yes (either inducer question) →

- Wait 5 minutes.

*Is **PRESS SWITCH** light flashing?*

No →

- Boiler should be in normal operating sequence.
- Observe operation until thermostat is satisfied and inducer has completed its post-purge cycle.

Yes →

Are vent or combustion air passages blocked?

No → Check air pressure switch per manual Section 11b. Replace if necessary. Retest.

Yes →

- Clear passages and recycle boiler.
- Recheck boiler operation.

GOLD CGs Gas-Fired Water Boiler —Boiler Manual

11d Troubleshooting — control module lights (cont.)

CHART 4 — **FLAME & POWER** lights flashing

— Usually indicates flame sensed when it shouldn't be there —

⚠WARNING Electrical shock hazard — Wherever you see **⚠ TURN OFF POWER ⚠**, follow the instructions. Failure to follow instructions could result in severe personal injury, death or substantial property damage.

Are manual main shutoff valve and gas valve open?

No →

- Leave main manual gas valve closed.
- Turn off power to boiler at service switch or breaker.
- Wait at least 45 seconds.
- Turn on power to boiler.
- Restart boiler, following **Operating instructions** in this manual (page 34, 35 or 36).

Do **FLAME** *and* **POWER** *lights still flash?*

- **No** →
  - **⚠TURN OFF POWER⚠** to boiler at service switch or breaker.
  - Open main manual gas valve.
  - Turn on power to boiler at service switch or breaker.
  - Restart boiler per lighting instructions.
  - Perform start-up procedures in boiler manual to verify proper operation.
- **Yes** →
  - Replace control module.
  - Retest.

Yes →

- Turn off power to the boiler at service switch or breaker.
- Wait at least 45 seconds.
- Turn on power to boiler.
- Restart boiler, following **Operating instructions** in this manual (page 34, 35 or 36).
- Look through the pilot inspection port (in front of air inlet box) to see if the pilot is burning during the prepurge period (first 30 seconds of operation).

Is pilot burning during prepurge period?

- **No** →
  - Allow boiler to continue cycling.

  Are ***FLAME*** *&* ***POWER*** *lights flashing?*

  - **No** →
    - Boiler may now be operating normally.
    - Perform start-up procedures in boiler manual to verify proper operation.
  - **Yes** →
    - Replace control module.
    - Retest.
- **Yes** →
  - Replace gas valve.
  - Retest boiler.

  Are ***FLAME*** *and* ***POWER*** *lights flashing?*

  - **No** →
    - Boiler should now operate normally.
    - Original flashing FLAME light caused by gas valve not operating properly.
    - See normal sequence of operation, **Figure 18**, page 31.
  - **Yes** →
    - Replace control module.
    - Retest.

Weil-McLain

11d Troubleshooting — control module lights (cont.)

CHART 5 — FLAME light flashing and POWER light on steady

ALSO — Troubleshooting failure to establish main flame

⚠WARNING Electrical shock hazard — Wherever you see ▲ TURN OFF POWER ▲, follow the instructions. Failure to follow instructions could result in severe personal injury, death or substantial property damage.

- Are main manual shutoff valve and boiler gas valve open?

No | Yes

- ▲ TURN OFF POWER ▲ to boiler at service switch or breaker.
- Open main manual shutoff valve and boiler gas valve (per **Lighting instructions** in this manual). Wait at least 45 seconds.
- Turn on power at service switch or breaker. Allow boiler to cycle.

Does FLAME light flash now?

No | Yes

- Boiler should be in normal operating sequence.
- Observe operation until thermostat is satisfied and blower has completed its post-purge cycle.

- Verify inlet gas pressure at gas valve:

Natural gas — 5.0" w.c. min/14.0" w.c. max

Propane — 11.0" w.c. min/14.0" w.c. max

Is gas present at gas valve inlet and within above range?

Yes | No

- ▲TURN OFF POWER▲ to boiler at service switch or breaker.
- Remove air inlet box top (see **Figure 27**, page 56 for location).

- Verify pilot gas line is not kinked, obstructed or damaged and is correctly attached to pilot and gas valve.
- Verify pilot ignition electrode, electrode ceramic and spark lead wire from control are in good condition. Spark gap should be approximately 1/8".
- Correct any above problems, replacing pilot if burner or wiring is damaged.
- Reinstall air box top to operate boiler for retest after any changes or corrections.
- If none of the above corrects problem, then replace the control module, reinstall air box top, and retest.

- Contact gas supplier to correct pressure or gas supply.

- Is pilot flame visible through inspection port?

No | Yes

- Check the voltage across terminals **"C"** and **"A"** of the gas valve (see **Table G**, page 33).

Is 24 VAC present there?

Yes | No

- If the wiring from the control module to gas valve is intact, replace the control module.
- Retest.

- Make sure ground wire terminal is securely fastened to control module mounting screw.

- Check the voltage across terminals **"B"** and **"A"** (see **Table G**, page 33) of the gas valve.

Is 24 VAC present there?

No | Yes

- Verify inlet gas pressure at gas valve:

Natural gas — 5.0" w.c. min/14.0" w.c. max

Propane — 11.0" w.c. min/14.0" w.c. max

Is gas present at gas valve inlet and within above range?

No | Yes

- ▲TURN OFF POWER▲ to boiler at service switch or breaker.
- Check flame signal — Detach sense lead from ignition control (***Figure 25, item 8**, page 46*).
- Connect **negative** lead of MICROAMMETER to control sense terminal (***Figure 25, item 8**, page 46*). Connect positive lead of MICROAMMETER to sense wire.
- DISCONNECT **red** wire connected to terminal **"B"** (see **Table G**, page 33) of the gas valve.
- Turn on power to boiler and allow to cycle. As soon as pilot is burning, the MICROAMMETER should read at least 1.0 microamp.

Is flame signal at least 1.0 microamp?

No | Yes

- ▲TURN OFF POWER▲ to boiler at service switch or breaker.
- Remove air inlet box cover (see **Figure 27**, page 56 for location).

- Verify pilot burner is securely attached to pilot bracket, bracket is securely attached to cross tie, and there is no corrosion on the parts which would affect the ground path for flame sense.
- Verify that pilot flame rod, flame rod ceramic and lead wire from control module to flame rod are in good condition.
- Correct any above problems, replacing pilot if burner or wiring is damaged.

- If none of the previous steps (including replacing pilot) corrects problem, then replace the control module, reinstall air box cover and retest.

- If the wiring from the control module to gas valve is intact, replace the control module and retest.

11d Troubleshooting — control module lights (cont.)

CHART 6 — **PRESS SWITCH** light flashing and **POWER** light on steady

— Usually indicates pressure switch opened during run cycle —
— May also be caused by wind gusts in excess of 31 mph for non-direct vent sidewall-vented boilers —

⚠WARNING Electrical shock hazard — Wherever you see ⚠ **TURN OFF POWER** ⚠, follow the instructions. Failure to follow instructions could result in severe personal injury, death or substantial property damage.

- Reset boiler control by turning off power at service switch or turning down thermostat for at least 45 seconds.
- Thermostat should call for heat and appropriate zone valves open.

Does blower motor operate?

No →

- Remove **INDUCER** plug (***Figure 25, Item 6***, *page 46*) from plug receptacle of control module.
- Place voltmeter leads across the **120 VAC N** and **120 VAC H** pins of the **INDUCER** receptacle (left column center and right column bottom pins).

*Does the voltmeter indicate **120 VAC**?*

- **Yes** →
  - Replace inducer assembly.
  - Retest.
- **No** →
  - Remove one of the wires connected to the pressure switch.
  - Check continuity across the switch terminals.

  Does the ohmmeter indicate the switch is closed?

  - **No** →
    - Inspect and check wire harness from control module to pressure switch.
    - Reseat connector in control module receptacle.
    - If harness is intact and properly seated, but problem persists, replace control module.
    - Retest.
  - **Yes** →
    - Check air pressure switch per manual Section 11.
    - Replace if necessary. Retest.

Yes →

- Wait 5 minutes.

*Is **PRES SWITCH** light flashing?*

- **No** →
  - Boiler should be in normal operating sequence.
  - Observe operation until thermostat is satisfied and inducer has completed its post-purge cycle.
- **Yes** →

  Are vent or combustion air passages blocked?

  - **No** →
    - Check air pressure switch per manual Section 11.
    - Replace if necessary. Retest.
  - **Yes** →
    - Clear passages and recycle boiler.
    - Recheck boiler operation.

Weil-McLain

11d Troubleshooting — control module lights (cont.)

CHART 7 — Insufficient heat or no heat to system (**POWER** light on steady)

⚠WARNING Electrical shock hazard — Wherever you see **⚠TURN OFF POWER⚠**, follow the instructions. Failure to follow instructions could result in severe personal injury, death or substantial property damage.

- Has it been at least 5 minutes since setting thermostat to call for heat? If not, wait 5 minutes. Is system heating?
- Is thermostat set to call for heat? Remove thermostat wires at boiler and check continuity across the two wires. If circuit isn't closed, check external thermostat (zone valve, relay, etc.) wiring. Correct problems and retry.
- Reconnect external thermostat wiring at boiler.

Is thermostat circuit closed (continuity across wires)?

No | Yes

Are all red lights off?

No | Yes

Wait 30 seconds.
Are all red lights off?

No | Yes

Replace control module.
Retest.

- Boiler is in standby.
- Set thermostat to call for heat and recheck operation.
- Verify **Sequence of operation**, **Figure 18**, page 31.

Is boiler System circulator operating?

No | Yes

- Remove CIRCULATOR harness plug from CIRCULATOR plug receptacle (***Figure 25, Item 7**, page 46*). Check with voltmeter across pins of control module receptacle.

Does voltmeter show 120 VAC across pins?

No | Yes

- Replace control module.
- Retest.

- Repair/replace circulator. Retest.
- Repair/replace wiring. Retest.

- Wait 5 minutes.

*Is **FLAME** light on?*

Yes | No

- If you have sufficient heat — boiler should be in normal operating sequence. (See normal sequence of operation, **Figure 18**, page 31.)
- If you have less than sufficient heat —

Is vent or combustion air piping free of blockage?

No | Yes

- Correct conditions and recheck operation.
- See **Figure 18**, page 31 for normal sequence of operation.

Contact your Weil-McLain sales representative.

*Is **LIMIT** light on?*

No | Yes

- Check limit switch contacts with continuity checker.

Are limit switch contacts closed?

No | Yes

- Replace control module.
- Retest.

Check rollout switch continuity — is switch closed?

No | Yes

- Wait for boiler water to cool to temperature 20 to 30 °F lower than temperature set on limit switch.

Are limit switch contacts closed?

No | Yes

- Replace limit switch.
- Retest.

- Boiler water temperature exceeded setting on limit switch with thermostat calling for heat. Boiler is OK. (See normal sequence of operation, **Figure 18**, p. 31.)
- Also check setting on limit switch (should be above 140°F.).

- Replace rollout switch.

⚠DANGER

If rollout thermal fuse element has opened, determine cause and correct condition. Failure to do so will cause severe personal injury, death or substantial property damage.

- Check any other limit controls wired into the limit circuit. If all are OK, then replace control module.
- Retest.

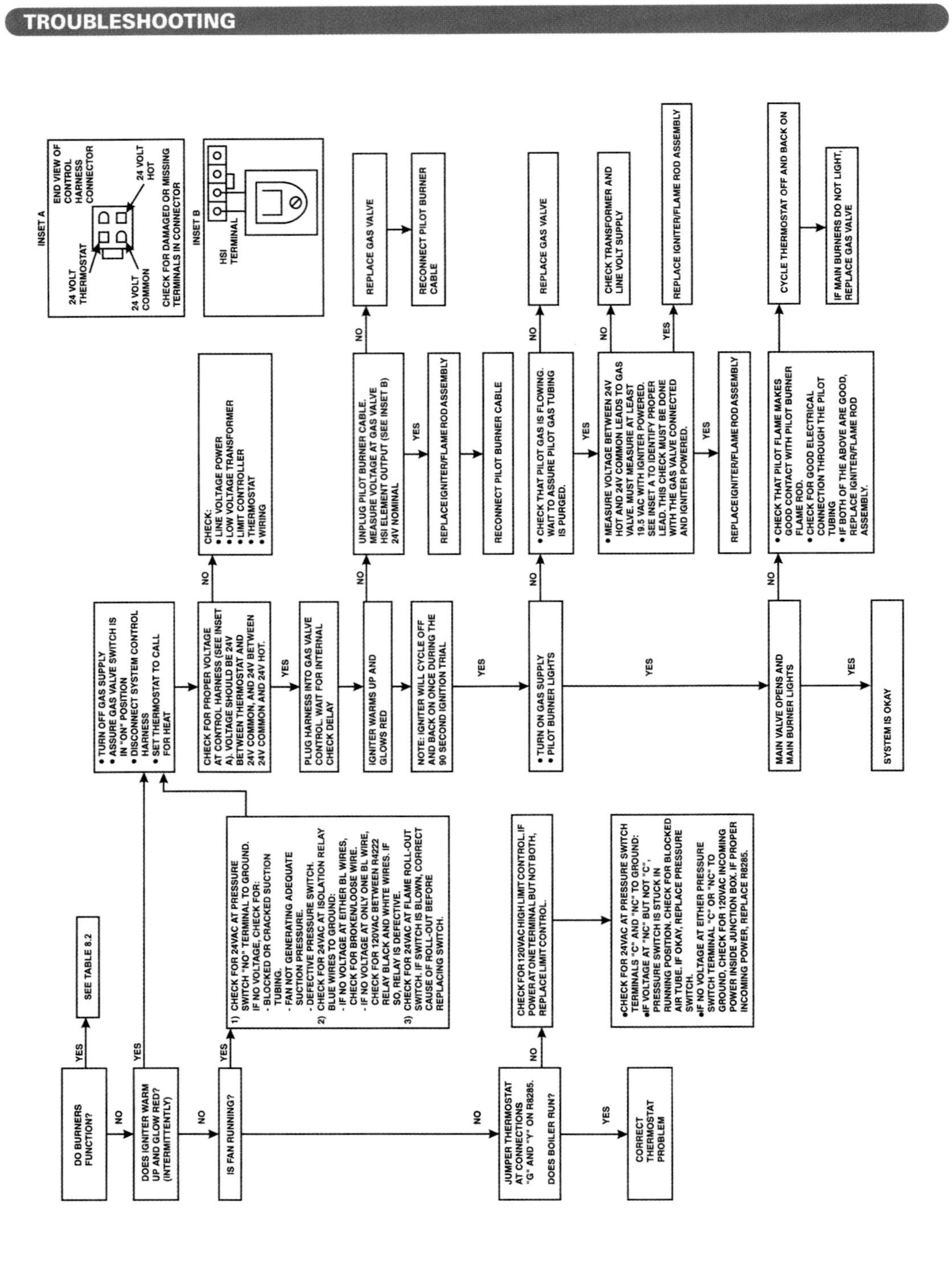

Figure 8.2: Boiler Troubleshooting Guide (Burners Not Functioning)

32

Peerless Boilers

TROUBLESHOOTING

Table 8.2: Boiler Troubleshooting Guide (Burners Functioning)

Problem	Possible Causes	Corrective Actions
Burners not functioning.	See Figure 8.2	See Figure 8.2
Burners will not shut down.	1. Defective gas valve. 2. Short circuit.	1. Use Figure 8.2 to troubleshoot intermittent ignition gas valve. Replace if necessary. 2. Check and correct wiring.
Flashback or burning at orifice spuds.	1. Manifold gas pressure too low. 2. Improperly sized/drilled orifice spuds. 3. Leaking gas valve. 4. Burrs on orifice. 5. Low supply gas pressure.	1. Adjust to proper pressure. 2. Install correct spuds. 3. Replace valve. 4. Remove burrs. 5. Contact gas supplier if natural gas. Adjust regulator if LP gas.
Delayed ignition.	1. Insufficient pilot flame. 2. Pilot burner/orifice clogged. 3. Overfiring. 4. Misaligned burners or pilot.	1. Increase pilot gas flow. 2. Clean pilot burner and orifice. 3. Reduce rate to input on rating label. 4. Realign burners or pilot.
Condensation at boiler vent connector/fan.	1. Vent pipe not sloped towards vent terminal.	1. Install condensate trap per vent manufacturer's instructions. 2. Slope vent pipe towards vent terminal.
Boiler not heating properly.	1. Underfiring. 2. Limit set too low. 3. Air in system. 4. Circulator malfunctioning. 5. Circulation system clogged. 6. Incorrect thermostat heat anticipator setting.	1. Increase rate to input on rating label. 2. Reset limit to higher setting. 3. Vent air from all points in system. 4. Check circulator, replace if necessary. 5. Shut down and cool boiler, drain and flush system. 6. Adjust heat anticipator to 0.2 Amp.
Fumes or gas odors	1. Leaks in gas piping or fittings. 2. Leaks in gas service line or meter. 3. Obstructed chimney. 4. Obstructed flueways or vent. 5. Draft problem in boiler room. 6. Overfiring.	1. Locate and repair or replace. 2. Shut down boiler and notify gas provider. 3. Check, repair and/or clean chimney. 4. Clean flueways or vent and remove obstructions. 5. Check air supply, ventilation and venting system. 6. Reduce rate to input on rating label.

8. TROUBLESHOOTING

A. SHUT-DOWN CAUSED BY PILOT OUTAGE, PRESSURE SWITCH OR FLAME ROLL-OUT SAFETY SHUT-OFF SWITCH

In the event of a shut-down caused by a pilot outage, action of the pressure switch or flame roll-out safety shut-off switch effecting a shut-down of the main burners:

a. Refer to the Operating Instructions in Figure 7.3 to properly turn off the gas to the boiler.

b. Turn off all electric power to the boiler.

c. Call a qualified heating service organization or local gas company and have the cause of the shut-down investigated and corrected.

d. Refer to Operating Instructions to re-start boiler.

B. TROUBLESHOOTING GUIDES

Use Table 8.2 to assist in determining causes and providing corrective actions to boiler problems. Refer also to Figure 8.2 to troubleshoot the Intermittent Ignition System Control. These guides must be used only by qualified service technicians. These individuals must follow all applicable codes and regulations in repair of any boiler problems.

C. MEASURING SUCTION PRESSURE

Refer to Table 8.1 for fan suction pressure required to energize control circuit. Measure fan suction pressure as shown in Figure 8.1. Note that as boiler operates, suction pressure will decrease. Measure after 15 minutes of boiler operation.

Table 8.1: Fan Suction Pressure

Boiler Model	Suction Pressure (In. w.c)	Suction Pressure (kPa)
DE-03	1.78	0.44
DE-04	1.56	0.39
DE-05	1.56	0.39
DE-06	1.35	0.34

⚠ DANGER

When servicing or replacing items that communicate with the boiler water, be certain that:

- **There is no pressure on the boiler.**
- **The boiler is not hot.**
- **The power is off.**

When servicing the gas valve or pilot, be certain that:

- **The gas is off.**
- **The electricity is off.**

⚠ WARNING

Do not use this appliance if any part has been under water. Improper or dangerous operation may result. Immediately call a qualified service technician to inspect the boiler and to replace any part of the control system and any gas control which has been under water.

⚠ CAUTION

Label all wires prior to disconnection when servicing controls. Wiring errors can cause improper and dangerous operation. Verify proper operation after servicing.

⚠ CAUTION

Should overheating occur or the gas supply fail to shut off, do not turn off or disconnect the electrical supply to the pump. This may aggravate the problem and increase the likelihood of boiler damage. Instead, shut off the gas supply at a location external to the appliance.

Carrier
HEATING & COOLING

38CKG Air Conditioning Unit 3 Phase

Visit www.carrier.com

Wiring Diagrams

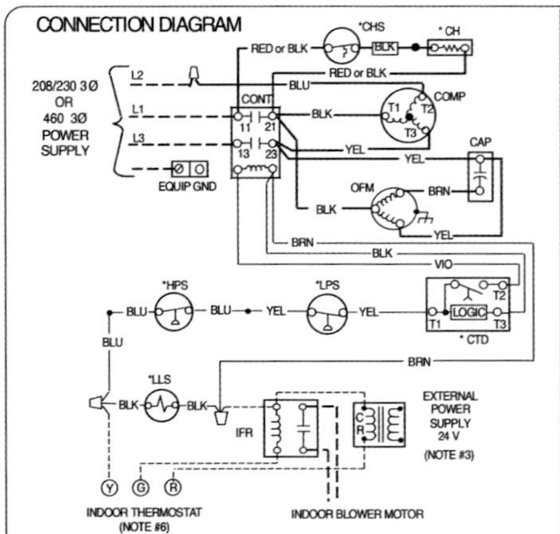

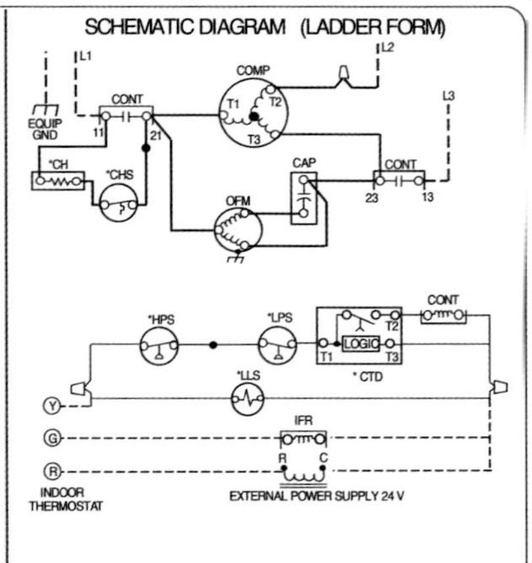

-LEGEND-

———	FACTORY POWER WIRING
———	FACTORY CONTROL WIRING
- - - - - -	FIELD CONTROL WIRING
— — —	FIELD POWER WIRING
○	COMPONENT CONNECTION
	FIELD SPLICE
—●—	JUNCTION
CONT	CONTACTOR
CAP	CAPACITOR (DUAL RUN)
*CH	CRANKCASE HEATER
*CHS	CRANKCASE HEATER SWITCH
COMP	COMPRESSOR
*CTD	COMPRESSOR TIME DELAY
*HPS	HIGH PRESSURE SWITCH
IFR	INDOOR FAN RELAY
*LLS	LIQ. LINE SOLENOID VALVE
*LPS	LOW PRESSURE SWITCH
OFM	OUTDOOR FAN MOTOR

* MAY BE FACTORY OR FIELD INSTALLED.

-NOTES-

1. Symbols are electrical representation only.
2. Compressor and fan motor furnished with inherent thermal protection.
3. To be wired in accordance with National Electric N.E.C. and local codes.
4. N.E.C. class 2, 24 V circuit, min. 40 VA required, 60 VA on units installed with LLS.
5. Use copper conductors only. Use conductors suitable for at least 75°C (167°F).
6. Connection for typical cooling only thermostat, for other arrangements see installation instructions.
7. If indoor section has a transformer with a grounded secondary, connect the grounded side to the BRN lead.
8. If any of the original wire, as supplied must be replaced, use the same or equivalent wire.
9. Check all electrical connections inside control box for tightness.
10. Do not attempt to operate unit until service valves have been opened.
11. Do not rapid cycle compressor. Compressor must be off 3 minutes to allow pressures to equalize between high and low side before starting.
12. It is imperative to connect 3Ø field power to unit with correct phasing. wrong phasing will cause reverse rotation of scroll compressor which will result in reduced current draw, elevated noise level and improper operation. If rotation is reversed, simply interchange any two of the three power connections on field side.

COOLING ONLY PROCEDURE

1. Operate unit a minimum of 10 minutes before checking charge.
2. Measure suction pressure by attaching a gage to suction valve service port.
3. Measure suction temperature by attaching an accurate thermistor type or electronic thermometer to the suction line at service valve.
4. Measure outdoor air dry-bulb temperature with a thermometer.
5. Measure indoor air (entering indoor coil) wet-bulb temperature with a sling psychrometer.
6. Refer to table I. Find outdoor temperature and evaporator entering air wet-bulb temperature. At this intersection note the superheat.
7. Refer to table II. Find superheat temperature located in step 6 and suction pressure. At this intersection note suction line temperature.
8. If unit has a higher suction line temperature than charted temperature, add refrigerant until charted temperature is reached.
9. If unit has a lower suction line temperature than charted temperature, remove and recover refrigerant until charted temperature is reached.
10. If outdoor air temperature or pressure at suction valve changes, charge to new suction line temperature indicated on chart.
11. This procedure is valid when indoor air flow is within ±21% of its rated cfm.

⚠ CAUTION

1. Compressor damage may occur if system is overcharged.
2. This unit is factory charged with R-22 in accordance with the amount shown on the rating plate. The charge is adequate for most systems using matched coils and tubing not over 15 feet long. The best performance will be achieved when the unit operates with a suction gas superheat at the compressor inlet of 5 °F at normal rating conditions of the air conditioning and refrigeration institute (ARI). This chart may be used to approximate the charge if ARI rating conditions cannot be obtained. ARI rating conditions are equivalent to DOE test "A" conditions. See product data literature for required indoor air flow rates and for use of line lengths over 15 Ft.
3. Relieve pressure and recover all refrigerant before system repair or final unit disposal. Use all service ports and open all flow-control devices, including solenoid valves.

TABLE I-SUPERHEAT CHARGING TABLE
(SUPERHEAT °F AT LOW-SIDE SERVICE PORT)

OUTDOOR TEMP °F	EVAPORATOR ENTERING AIR °F WB.													
	50	52	54	56	58	60	62	64	66	68	70	72	74	76
55	9	12	14	17	20	23	26	29	32	35	37	40	42	45
60	7	10	12	15	18	21	24	27	30	33	35	38	40	43
65	–	6	10	13	16	19	21	24	27	30	33	36	38	41
70	–	–	7	10	13	16	19	21	24	27	30	33	36	39
75	–	–	–	6	9	12	15	18	21	24	28	31	34	37
80	–	–	–	–	5	8	12	15	18	21	25	28	31	35
85	–	–	–	–	–	–	8	11	15	19	22	26	30	33
90	–	–	–	–	–	–	5	9	13	16	20	24	27	31
95	–	–	–	–	–	–	–	6	10	14	18	22	25	29
100	–	–	–	–	–	–	–	–	8	12	15	20	23	27
105	–	–	–	–	–	–	–	–	5	9	13	17	22	26
110	–	–	–	–	–	–	–	–	–	6	11	15	20	25
115	–	–	–	–	–	–	–	–	–	–	8	14	18	23

TABLE II-REQUIRED SUCTION TUBE TEMPERATURE °F
(MEASURED AT LOW-SIDE SERVICE PORT)

SUPERHEAT TEMP. °F	SUCTION PRESSURE AT SERVICE PORT PSIG								
	61.5	64.2	67.1	70.0	73.0	76.0	79.2	82.4	85.7
0	35	37	39	41	43	45	47	49	51
2	37	39	41	43	45	47	49	51	53
4	39	41	43	45	47	49	51	53	55
6	41	43	45	47	49	51	53	55	57
8	43	45	47	49	51	53	55	57	59
10	45	47	49	51	53	55	57	59	61
12	47	49	51	53	55	57	59	61	63
14	49	51	53	55	57	59	61	63	65
16	51	53	55	57	59	61	63	65	67
18	53	55	57	59	61	63	65	67	69
20	55	57	59	61	63	65	67	69	71
22	57	59	61	63	65	67	69	71	73
24	59	61	63	65	67	69	71	73	75
26	61	63	65	67	69	71	73	75	77
28	63	65	67	69	71	73	75	77	79
30	65	67	69	71	73	75	77	79	81
32	67	69	71	73	75	77	79	81	83
34	69	71	73	75	77	79	81	83	85
36	71	73	75	77	79	81	83	85	87
38	73	75	77	79	81	83	85	87	89
40	75	77	79	81	83	85	87	89	91

324475-101 REV. C

A03112

Fig. 1—38CKG 030-060

Book 1 4 | Tab 2a 3a | PC 101 | Printed in U.S.A. | Catalog No. 38CKG-1W | Pg 1 | 3-06 | Replaces: New

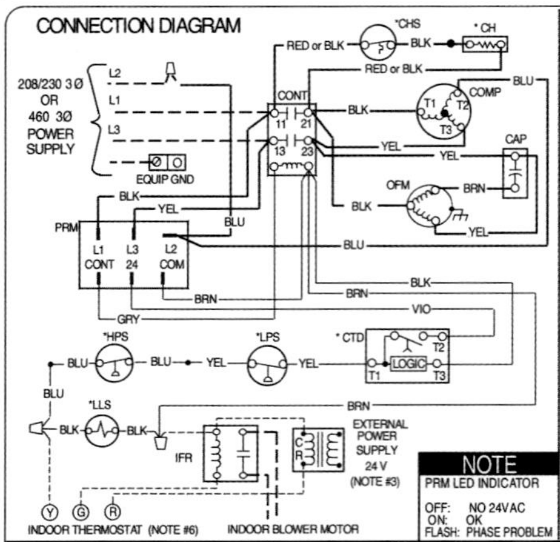

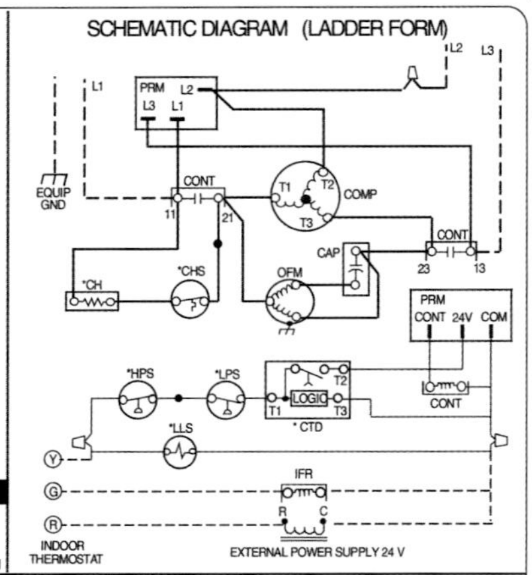

-LEGEND-

——— (heavy)	FACTORY POWER WIRING
——— (light)	FACTORY CONTROL WIRING
- - - - -	FIELD CONTROL WIRING
— — —	FIELD POWER WIRING
○	COMPONENT CONNECTION
(splice symbol)	FIELD SPLICE
—●—	JUNCTION
CONT	CONTACTOR
CAP	CAPACITOR (DUAL RUN)
*CH	CRANKCASE HEATER
*CHS	CRANKCASE HEATER SWITCH
COMP	COMPRESSOR
*CTD	COMPRESSOR TIME DELAY
*HPS	HIGH PRESSURE SWITCH
IFR	INDOOR FAN RELAY
*LLS	LIQ. LINE SOLENOID VALVE
*LPS	LOW PRESSURE SWITCH
OFM	OUTDOOR FAN MOTOR
PRM	PHASE ROTATION MONITOR

* MAY BE FACTORY OR FIELD INSTALLED.

-NOTES-

1. Symbols are electrical representation only.
2. Compressor and fan motor furnished with inherent thermal protection.
3. To be wired in accordance with National Electric N.E.C. and local codes.
4. N.E.C. class 2, 24 V circuit, min. 40 VA required, 60 VA on units installed with LLS.
5. Use copper conductors only. Use conductors suitable for at least 75°C (167°F).
6. Connection for typical cooling only thermostat, for other arrangements see installation instructions.
7. If indoor section has a transformer with a grounded secondary, connect the grounded side to the BRN lead.
8. If any of the original wire, as supplied must be replaced, use the same or equivalent wire.
9. Check all electrical connections inside control box for tightness.
10. Do not attempt to operate unit until service valves have been opened.
11. Do not rapid cycle compressor. Compressor must be off 3 minutes to allow pressures to equalize between high and low side before starting.
12. It is imperative to connect 3Ø field power to unit with correct phasing. The Phase Rotation Monitor will not allow the contactor to be energized if the phasing is not correct. If phasing is reversed, simply interchange any two of the three power connections on field side.

COOLING ONLY PROCEDURE

1. Operate unit a minimum of 10 minutes before checking charge.
2. Measure suction pressure by attaching a gage to suction valve service port.
3. Measure suction temperature by attaching an accurate thermistor type or electronic thermometer to the suction line at service valve.
4. Measure outdoor air dry-bulb temperature with a thermometer.
5. Measure indoor air (entering indoor coil) wet-bulb temperature with a sling psychrometer.
6. Refer to table I. Find outdoor temperature and evaporator entering air wet-bulb temperature. At this intersection note the superheat.
7. Refer to table II. Find superheat temperature located in step 6 and suction pressure. At this intersection note suction line temperature.
8. If unit has a higher suction line temperature than charted temperature, add refrigerant until charted temperature is reached.
9. If unit has a lower suction line temperature than charted temperature, remove and recover refrigerant until charted temperature is reached.
10. If outdoor air temperature or pressure at suction valve changes, charge to new suction line temperature indicated on chart.
11. This procedure is valid when indoor air flow is within ±21% of its rated cfm.

⚠ CAUTION

1. Compressor damage may occur if system is overcharged.
2. This unit is factory charged with R-22 in accordance with the amount shown on the rating plate. The charge is adequate for most systems using matched coils and tubing not over 15 feet long. The best performance will be achieved when the unit operates with a suction gas superheat at the compressor inlet of 5 °F at normal rating conditions of the air conditioning and refrigeration institute (ARI). This chart may be used to approximate the charge if ARI rating conditions cannot be obtained. ARI rating conditions are equivalent to DOE test "A" conditions. See product data literature for required indoor air flow rates and for use of line lengths over 15 Ft.
3. Relieve pressure and recover all refrigerant before system repair or final unit disposal. Use all service ports and open all flow-control devices, including solenoid valves.

TABLE I-SUPERHEAT CHARGING TABLE
(SUPERHEAT °F AT LOW-SIDE SERVICE PORT)

OUTDOOR TEMP °F	EVAPORATOR ENTERING AIR °F WB.													
	50	52	54	56	58	60	62	64	66	68	70	72	74	76
55	9	12	14	17	20	23	26	29	32	35	37	40	42	45
60	7	10	12	15	18	21	24	27	30	33	35	38	40	43
65	--	6	10	13	16	19	21	24	27	30	33	36	38	41
70	--	--	7	10	13	16	19	21	24	27	30	33	36	39
75	--	--	--	6	9	12	15	18	21	24	28	31	34	37
80	--	--	--	--	5	8	12	15	18	21	25	28	31	35
85	--	--	--	--	--	--	8	11	15	19	22	26	30	33
90	--	--	--	--	--	--	5	9	13	16	20	24	27	31
95	--	--	--	--	--	--	--	6	10	14	18	22	25	29
100	--	--	--	--	--	--	--	--	8	12	15	20	23	27
105	--	--	--	--	--	--	--	--	5	9	13	17	22	26
110	--	--	--	--	--	--	--	--	--	6	11	15	20	25
115	--	--	--	--	--	--	--	--	--	--	8	14	18	23

TABLE II-REQUIRED SUCTION TUBE TEMPERATURE °F
(MEASURED AT LOW-SIDE SERVICE PORT)

SUPERHEAT TEMP. °F	SUCTION PRESSURE AT SERVICE PORT PSIG.								
	61.5	64.2	67.1	70.0	73.0	76.0	79.2	82.4	85.7
0	35	37	39	41	43	45	47	49	51
2	37	39	41	43	45	47	49	51	53
4	39	41	43	45	47	49	51	53	55
6	41	43	45	47	49	51	53	55	57
8	43	45	47	49	51	53	55	57	59
10	45	47	49	51	53	55	57	59	61
12	47	49	51	53	55	57	59	61	63
14	49	51	53	55	57	59	61	63	65
16	51	53	55	57	59	61	63	65	67
18	53	55	57	59	61	63	65	67	69
20	55	57	59	61	63	65	67	69	71
22	57	59	61	63	65	67	69	71	73
24	59	61	63	65	67	69	71	73	75
26	61	63	65	67	69	71	73	75	77
28	63	65	67	69	71	73	75	77	79
30	65	67	69	71	73	75	77	79	81
32	67	69	71	73	75	77	79	81	83
34	69	71	73	75	77	79	81	83	85
36	71	73	75	77	79	81	83	85	87
38	73	75	77	79	81	83	85	87	89
40	75	77	79	81	83	85	87	89	91

326303-101 REV. A

A03101

Fig. 4—38CK(C, W) 036-060, 208/230 and 460v, 3 Phase, 60 Hertz, with Scroll Compressors

 38ckc4w

Book | 1 | 4 — Tab | 2a | 3a   PC 101   Catalog No. 533-80102   Printed in U.S.A.   Form 38CKC-4W   Pg 4   1-05   Replaces: 38CKC-3W

Carrier Corporation

58DL

the prepurge period, then shut off for 24 seconds then come back on at HEAT speed. When the W input signal disappears, the furnace control begins a normal inducer post-purge period and the blower switches to COOL speed after a 3 second delay. If the R-to-W-and-Y/Y2-and-G or R-to-W-and-Y1-and-Y/Y2-and-G signals disappear at the same time, the blower motor BLWM will remain on for the selected blower-OFF delay period. If the R-to-W-and-Y/Y2 or R-to-W-and-Y1-and-Y/Y2 signals disappear, leaving the G signal, the blower motor BLWM will remain on for the selected blower-OFF delay period then switch to continuous blower speed.

Wiring Diagrams

Refer to Fig. 59 for wiring diagram.

Troubleshooting

Refer to the service label (See Fig. 55). The Troubleshooting Guide (See Fig. 60) can be a useful tool in isolating furnace operation problems. Beginning with the word "Start," answer each question and follow the appropriate arrow to the next item. The Guide will help to identify the problem or failed component. After replacing any component, verify correct operation sequence.

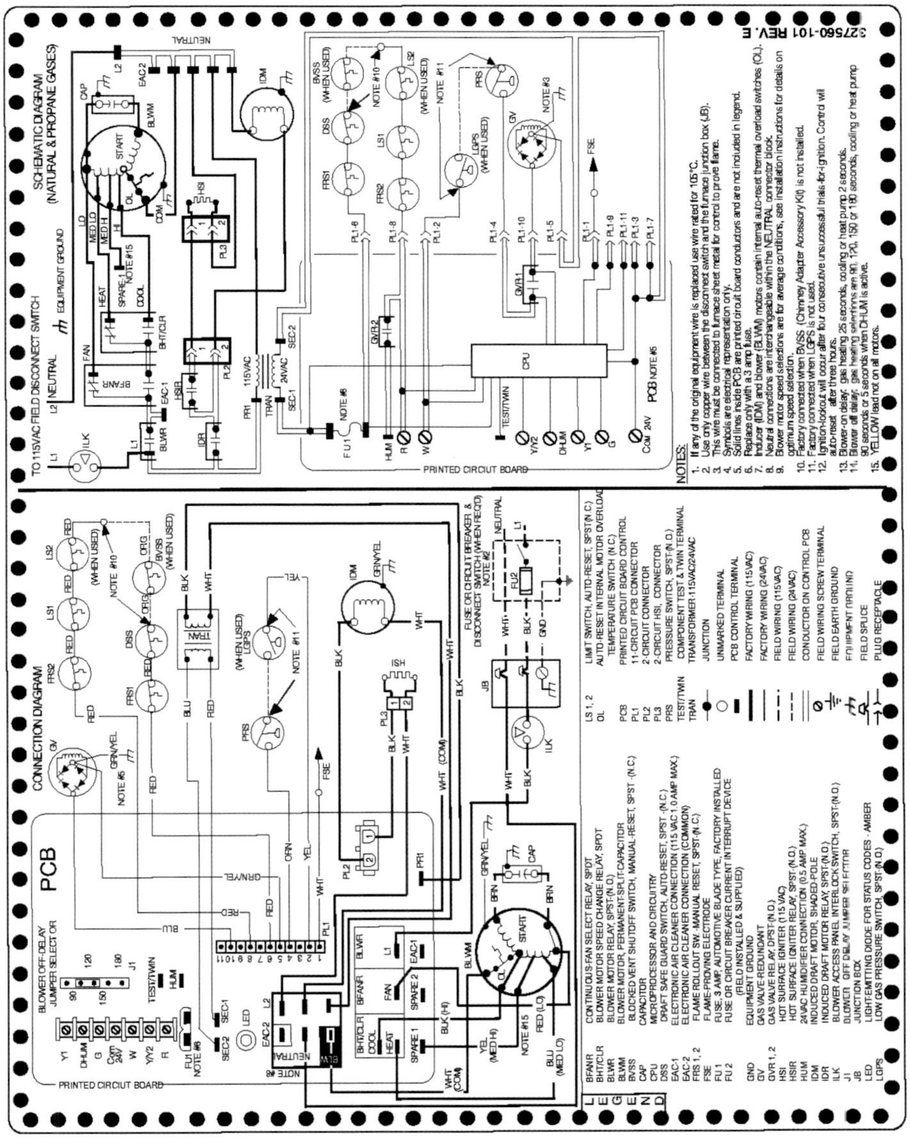

A04222

Fig. 59 - Wiring Diagram

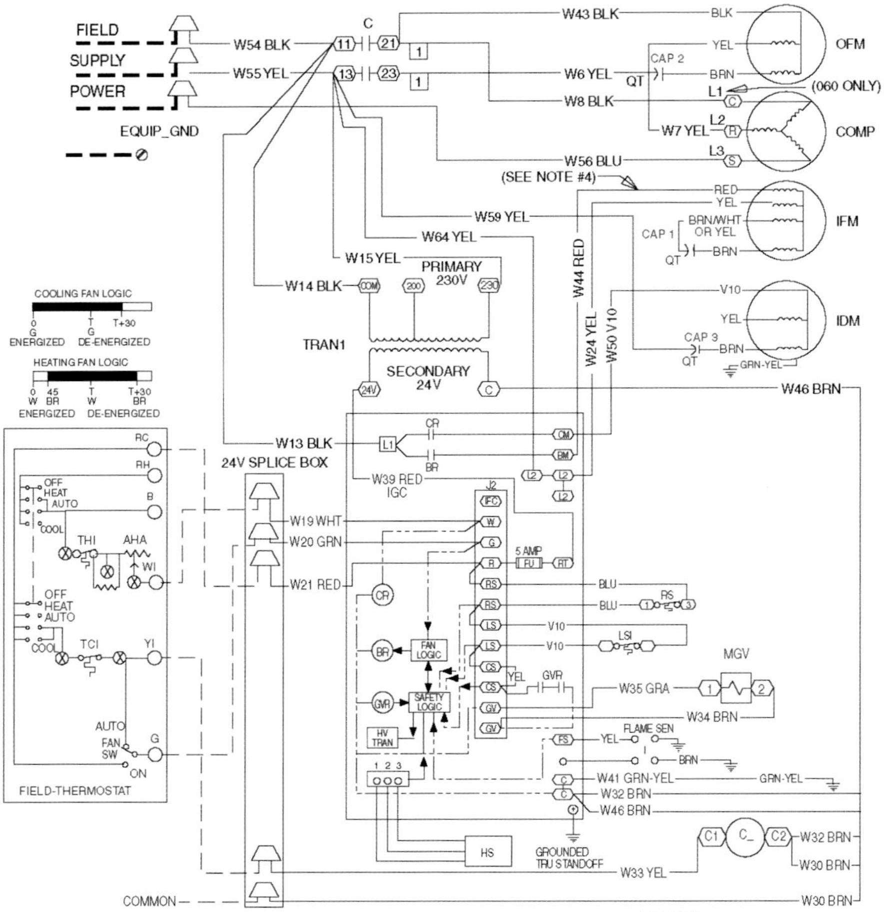

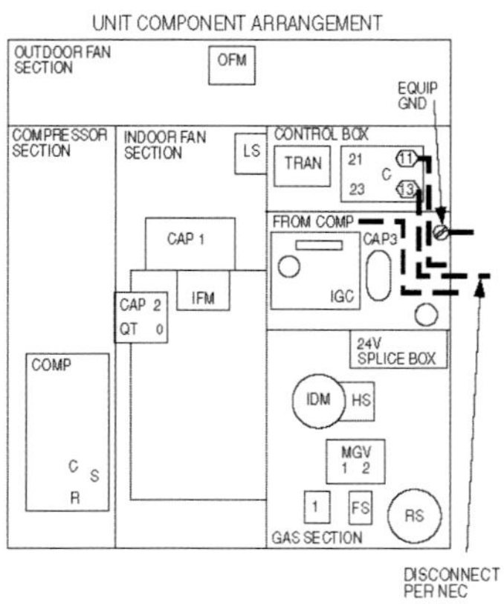

LEGEND

AHA – Adjustable Heat Anticipator
BR – Blower Relay
C – Contactor
CAP – Capacitor
COMP – Compressor Motor
CR – Combustion Relay
CS – Centrifugal Switch
EQUIP – Equipment
FS – Flame Sensor
FU – Fuse
GND – Ground
GV – Gas Valve
GVR – Gas Valve Relay
HS – Hall Effect Sensor
HV – High Voltage
TRAN – Transformer
I – Ignitor
IDM – Induced-Draft Motor
IFC – Indoor-Fan Contactor
IFM – Indoor-Fan Motor
IGC – Integrated Gas Unit Controller
L1 – Line
LS – Limit Switch
LS1 – Limit Switch
MGV – Main Gas Valve
NEC – National Electrical Code
OFM – Outdoor-Fan Motor
QT – Quadruple Terminal
RS – Rollout Switch
SEN – Sensor
SW – Switch
TRAN – Transformer

Field Splice
Terminal (Marked)
Terminal (Unmarked)
Splice
Splice (Marked)
Factory Wiring
Field Control Wiring
Field Power Wiring
Accessory or Optional Wiring
To Indicate Common Potential Only, Not to Represent Wiring

NOTES:
1. If any of the original wires furnished are replaced, they must be replaced with type 90 degree C wire or its equivalent.
2. See price pages for thermostat and subbases.
3. Use 75 degree C copper conductors for field installation.
4. For high speed IFM, disconnect RED wire from IGC terminal BM and connect BLK wire from IFM. For medium speed, disconnect RED wire from IGC terminal BM and connect BLU wire from IFM.

C00058

Fig. 15—208/230-3-60 Wiring Diagram, Units 48GS and 48GX

17

Carrier Corporation

High-Pressure Gas Systems

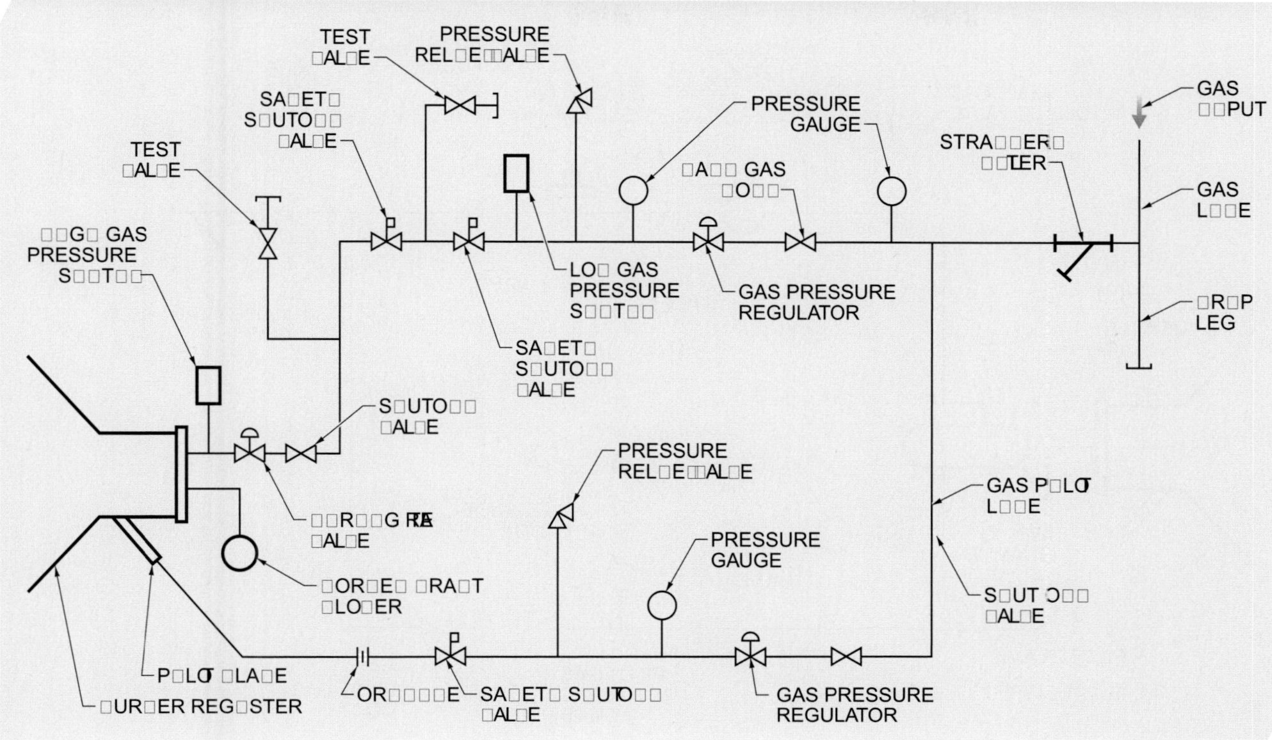

Low-Pressure Gas Systems

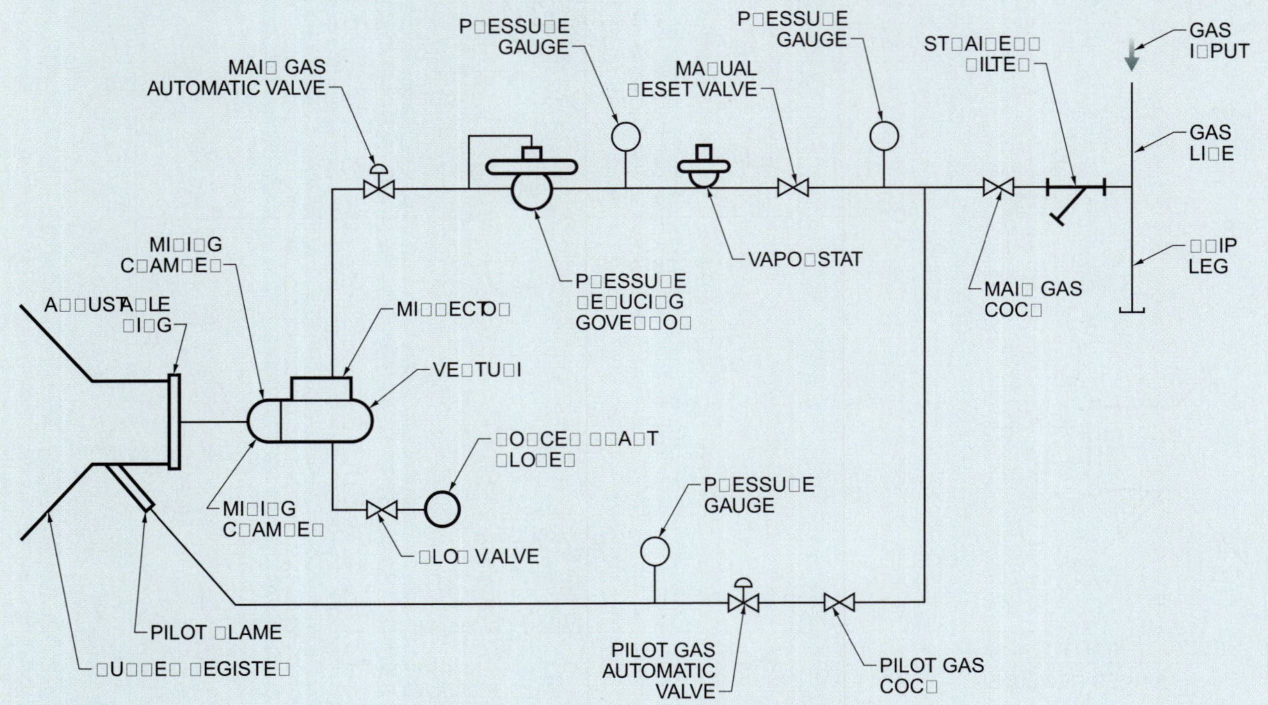

COMMERCIAL LOW-PRESSURE GAS SYSTEM

Direct-Fired Heater 950 H

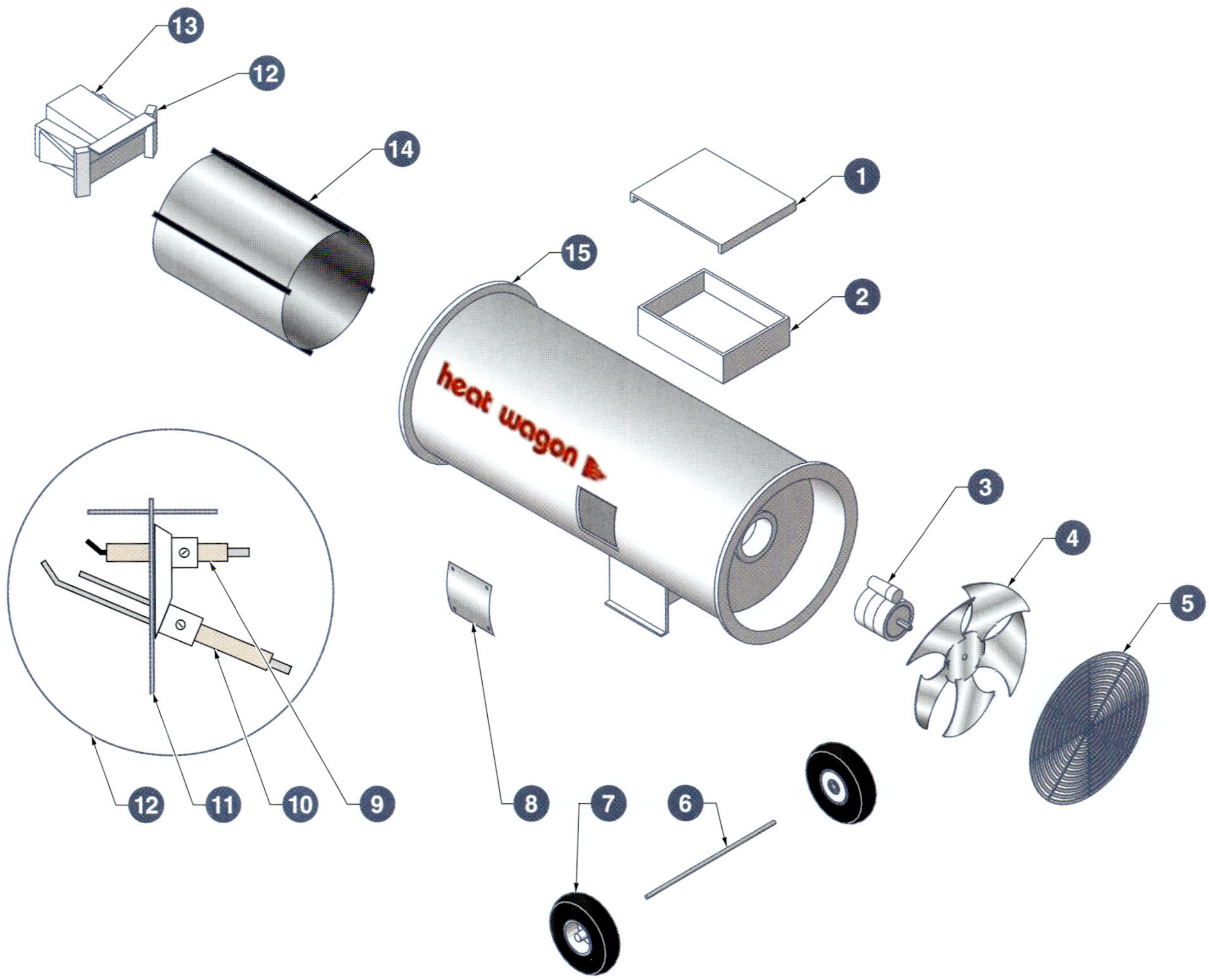

Item	Part No.	Description
1	HWP HC 1082	Control Box Lid
2	HWP HC 1083	Control Box
3	HWP HM1050	Motor
4	HWP HP1162	Fan Blade
5	HWP HG950A	22″ Fan Guard
6	HWP HW10656D	Axle (950H - Non-Stackable)
7	HWP 400017	Wheel (950H - Non-Stackable)
8	HWP HC1064	Burner Access Door

Item	Part No.	Description
9	HWP 134290	Igniter
10	HWP 136003	Flame Rod
11	N/A	Pilot Assy. Bracket
12	HWP 1190310	Pilot w/Spark Rod and Flame Rod
13	HWP HB1171B	Burner Complete
14	HWP HB1177C	Heat Shield
15	HWP HB950A	Heater Body Shell

Direct-Fired Heater DF 400

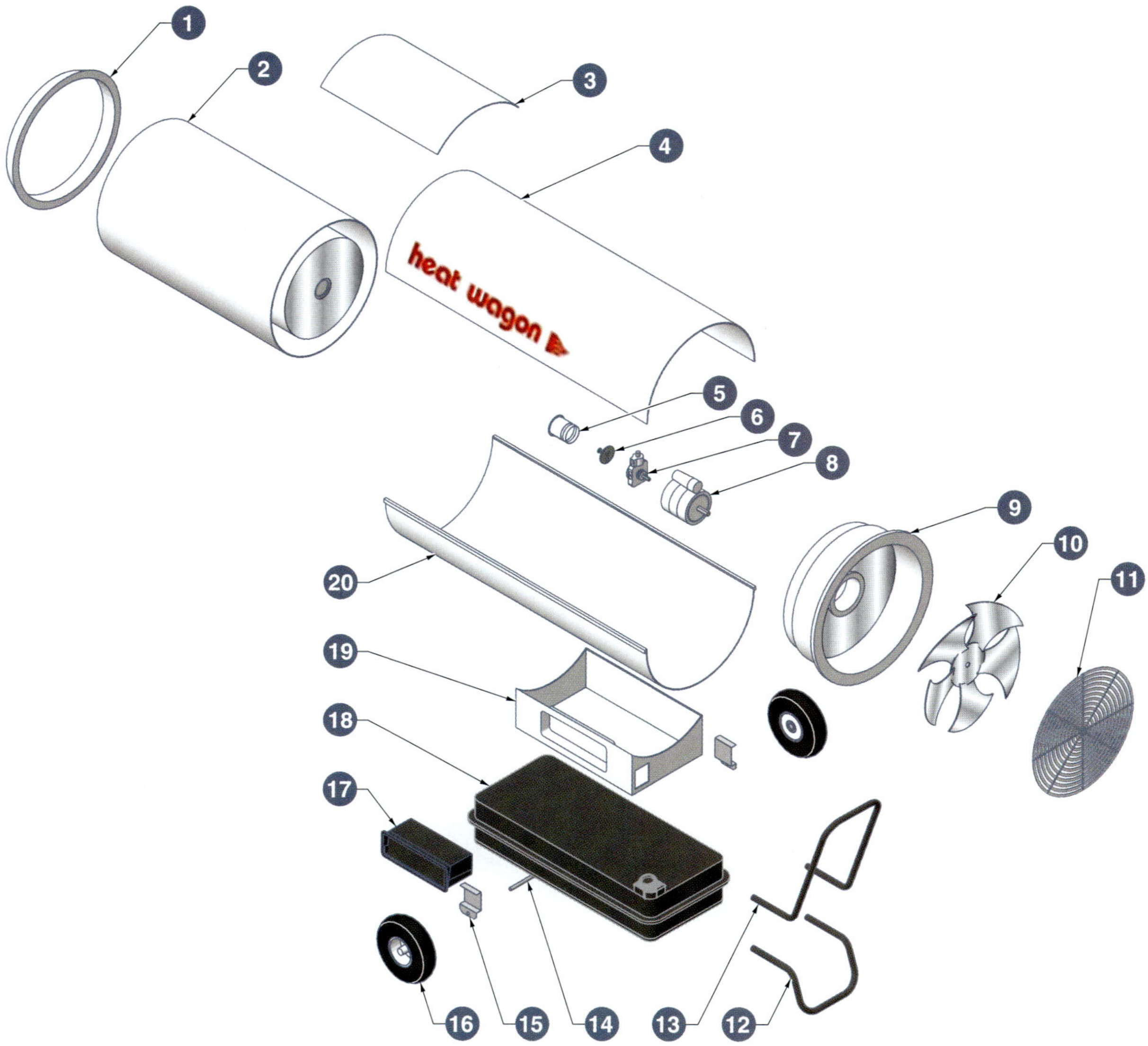

Item	Part No.	Description
1	G06139-9010	Outlet Cone
2	G06231	Combustion Chamber
3	G06143-9010	Protection
4	G06234-9010	Upper Body
5	E50102	Limit Thermostat
6	G01077	Turbo Disk
7	T20117	Solenoid Valve
8	E10679-110	Motor 750W with Condenser
9	G06239-9010	Motor Flange
10	T2015-B	Fan Ø 496 18°

Item	Part No.	Description
11	P30129	Inlet Grille
12	P20180-9005	Foot
13	P20176-9005	Handle
14	G06129-9005	Wheel Axle
15	G06107-9005	Wheels Axle Support
16	C10546	Wheel Ø 260 - Ø 25
17	P50127	Control Box Cover
18	P20146-9005	Fuel Tank
19	G06241-9010	Base
20	G06137-9010	Lower Body

Direct-Fired Heater VG 700

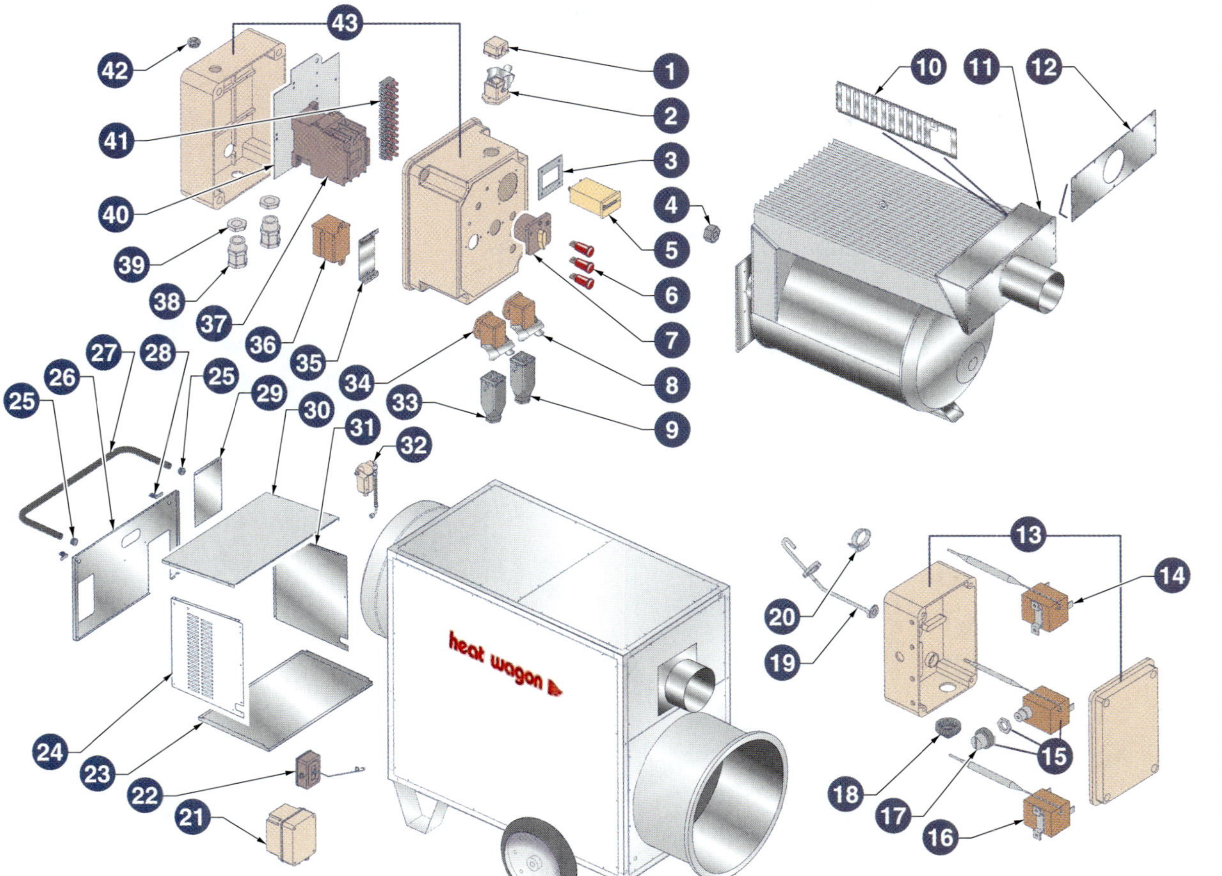

Heat Wagon

Item	Part No.	Description
1	E20665	Drain Plug
2	E20639	Thermostat Plug 4P + T
3	G04041	Hour Counter Support Plate
4	I25001	Female Plug 1″
5	DR205	Hour Counter
6	E11021	Red Pilot Lamp Ø12 V230
7	E10109	Control Knob
8	E20626	Thermostat Plug 3P + T
9	E20627	Plate Plug 3P + T
10	G01759, 1673	Baffle Plate
11	G01760, 1674	Chimney Cover
12	T10635	Chimney Seal 551 x 248 x 5
12	T10633	Chimney Seal 671 x 248 x 5
13	E20719-02	Electrical Components Box
14	E50748	Thermostat TY95 30/90 °C Campini
15	E50749	Thermostat TY95H 120 °C Campini
16	E50747	Thermostat TY95 0/60 °C Campini
17	E50750	Safety Thermostat Plastic Profile
18	C30343	Cable Protection Ø19
19	P30159	Bulbs Holder
20	C30712	Clip
21	G00221	Electric Control Box

Item	Part No.	Description
22	G00218	Electric Control Box
23	G04205-9010	Tank Casing Lower Panel
24	G04202-9010	Burner Casing DX Cover
25	C30328	Plug
26	G04190-9010	Casing Front Panel
27	G04050-9010	Handle
28	M20418	Wing Nut Lock
29	G04193-9010	Flap Door
30	G04196-9010	Burner Casing Top Cover
31	G04199-9010	Burner Casing SX Cover
32	02AC550	Kit Oil Pre-heaters Filter 1/4″
33	G00217	Plate Plug 3P + T
34	E20629	Thermostat Plug 3P + T
35	G04207	Relay Flange
36	E11120	Relay Finder 65.31 AC
37	E10419	Contactor Wimex KN16-10 V230
38	E20949	Cable Fastener PG 13.5
39	E20950	Ring Nut PG 13.5
40	G04042	Plate for Electrical Components
41	E20301	Terminal Board 12 el, mmg 10
42	C30343	Cable Protection Ø19
43	E20712	Electrical Components Box

Direct-Fired Heater VG 175

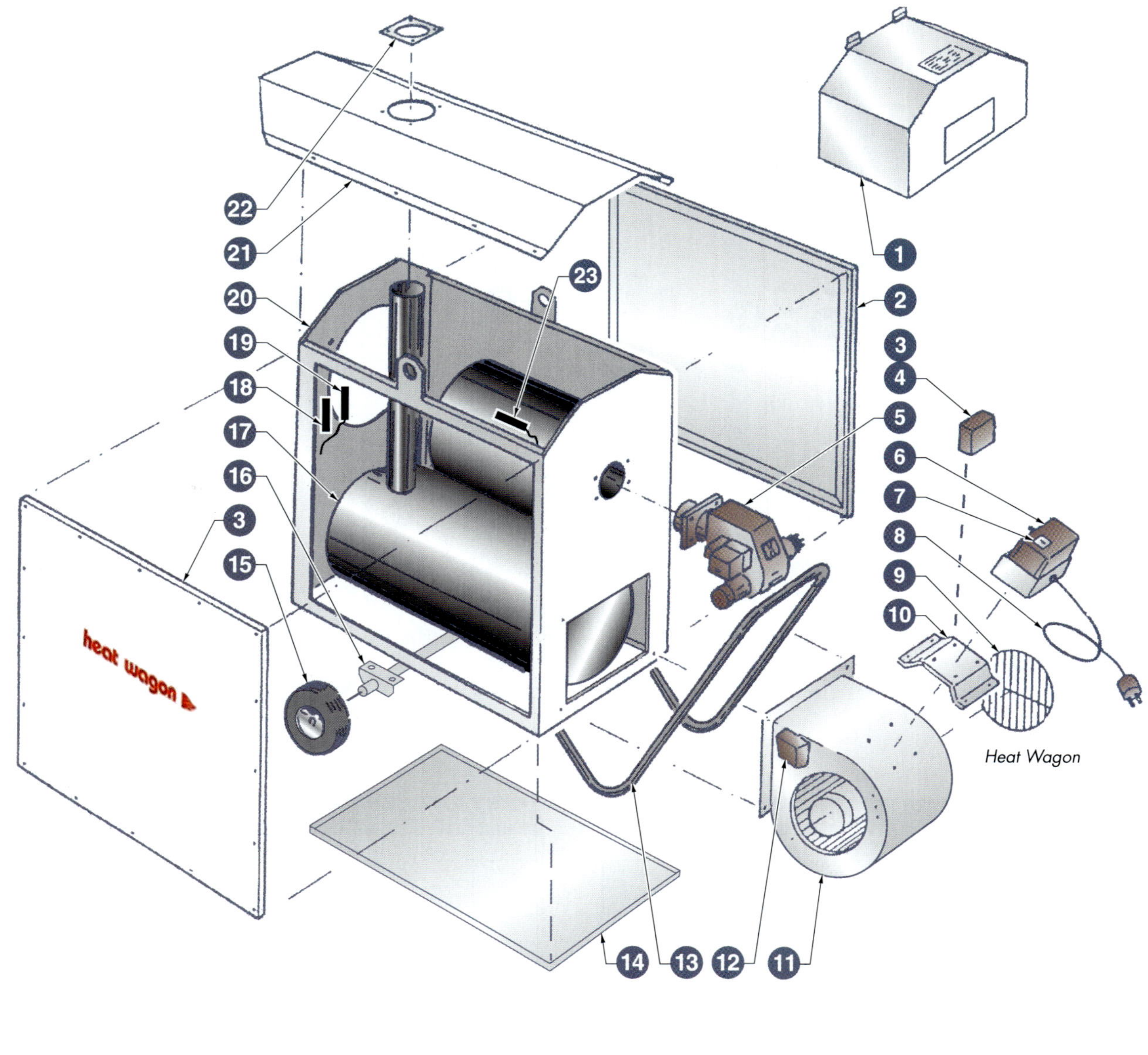

Item	Part No.	Description
1	HWP 2709016	Burner Cover
2	HWP 2709003	Side Panels (2)
3	HWP 20579	Heater Control Unit
4	HWP 40850	Thermostat Box
5	HWP 63956	Burner VG175
6	HWP 41000	Main Control Box
7	HWP 26400	Main Switch
8	HWP HC1020	Power Cord & Plug
9	HWP 2709011	Fan Guard, Right
10	HWP 1709023	Bracket, Main Control Box
11	HWP 2701244	Fan, Complete
12	HWP 40900	Fan Connection Box

Item	Part No.	Description
13	HWP 2709016	Rear Handle
14	HWP 2709005	Bottom Plate
15	HWP 2701065	Wheel
16	HWP 2709008	Wheel Shaft
17	HWP 2709009	Burner Chamber Assembly
18	HWP 21400	Overheat Limit Switch (Capillary)
19	HWP 20582	Burner Sensor (For Pos. #3)
20	HWP 2709001	Heater Frame
21	HWP 2709002	Cover Plate
22	HWP 120001	Flue Collar
23	HWP 20581	Fan Sensor (For Pos. #3)

Direct-Fired Heater S 405

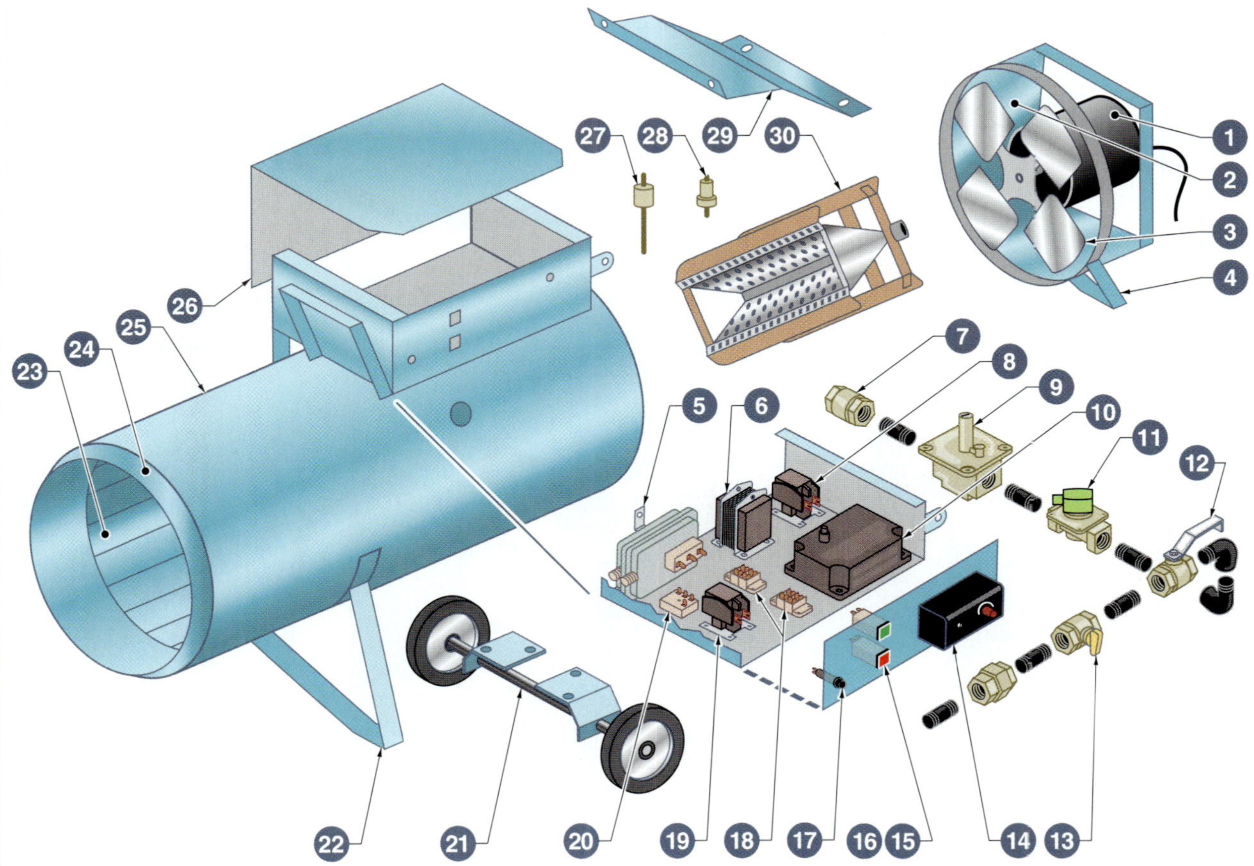

Item	Part No.	Description
1	SFP 2430	Motor 1/4 HP
2	SFP 3859	Motor Mount (with Screen & Rear Legs)
3	SFP 2420	Fan Blade
4	SFP S400-47	Rear Leg Assembly
5	SFP 5124	Air Switch
6	SFP 4510	Transformer 24 V
7	SFP 40SV01	Strainer (S400-80)
8	SFP 4519	Motor Relay 24 V
9	SFP 2524	Low-Pressure Regulator
10	SFP 5943	Direct Spark Ignition Control
11	SFP 4514	Solenoid Shut Off Valve 24 V
12	SFP 40SV04	Gas Selector Valve (S400-73)
13	SFP 2538	Main Shut Off Valve 3/4″
14	SFP 2453	Thermostat
15	SFP 3337R	Off Switch (Red)

Item	Part No.	Description
16	SFP 3337G	On Switch (Green)
17	SFP 4518	Indicator Light 24 V
18	SFP 5768	Terminal Block
19	SFP 4519	Control Relay 24 V
20	SFP 7465	24 VAC 5-Second Delay-On-Break Timer
21	ACC-SWK1	Wheel Kit (Optional)
22	SFP S400-57	Front Leg Assembly
23	SFP S400-2	Heat Shield
24	SFP S400-3	Nose Cone
25	SFP S400-53	Heater Body Includes 22, 23, and 24
26	SFP 4517	Control Box Lid
27	SL11B86	Flame Rod
28	SFP 2142	Spark Plug
29	SFP SC400-38	Valve Train Cover
30	SFP 40SB26	Burner

Direct-Fired Heater P4000

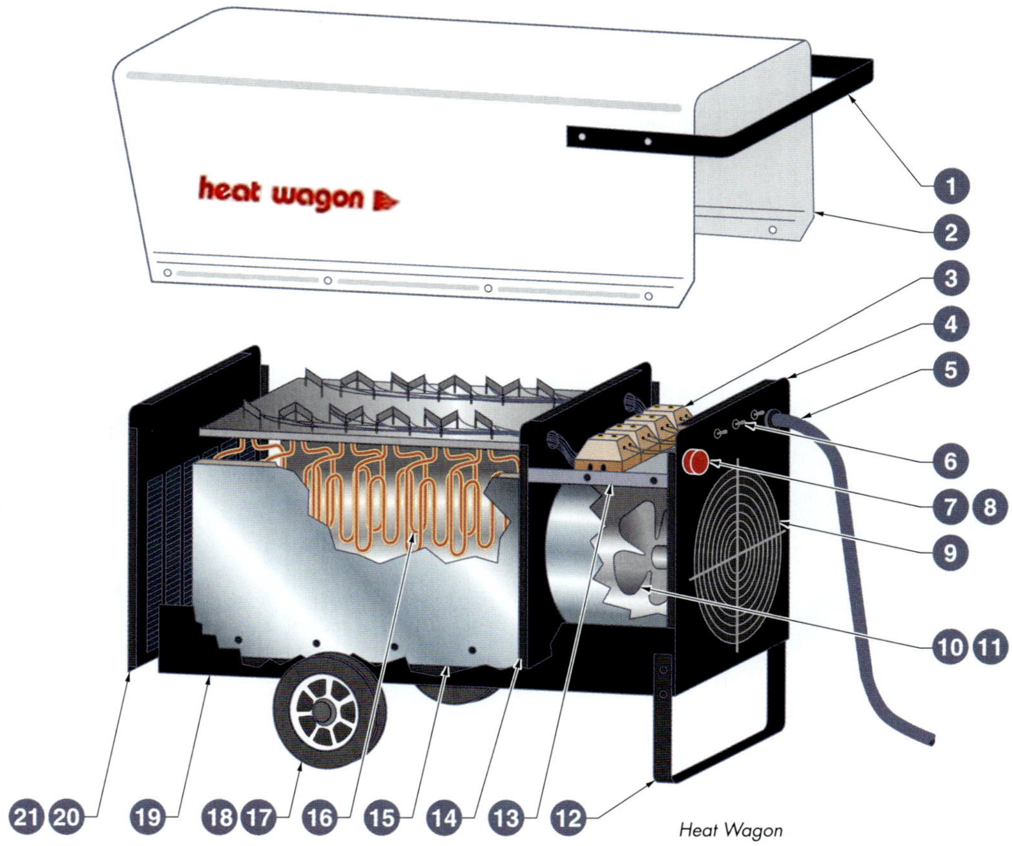

Heat Wagon

Item	Part No.	Description
1	HWP 400001	Handle
2	HWP 400002	Hood
3	HWP 400022	Contactors
4	HWP 400015	Rear Panel
5	HWP 400016	Strain Reliever
6	HWP 400021	Switches
7	HWP 400017	Ambient Thermostat
8	HWP 400018	Thermostat Knob
9	HWP 400019	Fan Guard Grill
10	HWP 400012	Fan Blade
11	HWP 400014	Fan Motor

Item	Part No.	Description
12	HWP 400011	Rear Leg
13	HWP 400010	Contactor Mounting Plate
14	HWP 400009	Bracket
15	HWP 400003	Heat Exchange Chamber
16	HWP 400026	Heating Element
17	HWP 400004	Wheel Assembly
18	HWP 400005	Wheel Shaft
19	HWP 400007	Bottom Frame
20	HWP 400006	Front Panel
21	HWP 400008	Guard Thermostat

Glossary

A

absolute pressure: Any pressure above a perfect vacuum (0 psia).

absolute zero: A theoretical condition in which a substance has no molecular movement.

acceptable indoor air quality: Air that either does not contain any known contaminants or contains contaminants at harmless concentrations and that does not cause dissatisfaction among a substantial majority of people.

accessory: A component not directly attached to the boiler but is needed for proper boiler operation.

accuracy: The degree to which a measurement conforms to a specific standard.

acute health effect: A health effect that occurs immediately after exposure to a substance and typically disappears soon after the exposure ends.

adaptive control algorithm: A control algorithm that positions a controlled device by automatically adjusting its response time based on environmental conditions.

adiabatic change: A change in the pressure and temperature of a substance in a closed system that occurs without heat transfer from outside the system.

adjusting: The process of bringing parts or HVAC systems to more effective relative positions or satisfactory states.

air changes per hour (ACH): A measure of the total hourly volume of infiltration air divided by the volume of building space.

air circulation: The movement of air through distribution systems and building spaces.

air cleaning: A method of reducing air pollution by removing pollutants using electronic devices or filters.

air cock: *See* steam vent valve *and* vent valve.

Air Conditioning and Refrigeration Institute (ARI): An association of manufacturers of refrigeration and air conditioning equipment and allied products.

air conditioning and refrigeration (ACR) tubing: Copper tubing used in HVAC and refrigeration applications.

air filter: A device that contains a porous substance through which air can pass and is used to remove particles from the air.

air filtration: The process of removing particles and contaminants, such as gases and odors, from air that circulates through air distribution systems.

airflow rate: The measure of the volume (in cubic feet) of air that passes a given point in one minute (cfm).

air-handling unit (AHU) controller: A controller that contains input and output signals required to operate large central-station AHUs.

air-medium condenser: A type of condenser that uses air as a heat-exchange medium.

air-medium evaporator: An evaporator that uses a fan to circulate air across an evaporator coil.

air short circuiting: A condition that results when air travels directly from a supply outlet (register or diffuser) to a return inlet (grill) without heating or cooling a building space or room.

air temperature rise: Supply air temperature minus return air temperature.

air tube: A gun-type fuel oil burner component that is used to change the angle of air introduced into the combustion chamber. Also known as a draft tube.

air velocity: The speed air moves from one point to another through ductwork, a register, or a grill; measured in feet per minute (fpm).

air vent: An automatic or manual device that vents air from a hydronic heating system without allowing water to escape or air to enter the system.

aldehyde: A toxic chemical compound that contains hydrogen, carbon, and oxygen.

algorithm: A sequence of instructions for producing the desired result of an operation.

allergen: A contaminant that can trigger an allergic reaction.

alternating current (AC): Current that reverses direction at regular intervals.

altitude pressure gauge: A needle gauge on a boiler that indicates when the system is properly filled.

ambient air: Unconditioned atmospheric air at normal atmospheric pressure.

American National Standards Institute (ANSI): A national standards organization that helps identify industrial and public need for national standards.

American Society of Heating, Refrigerating and Air-Conditioning Engineers (ASHRAE): An international organization established in 1894 whose mission is to advance the arts and sciences of heating, ventilation, air conditioning, and refrigeration to serve humanity and promote a sustainable world.

analog display: An electromechanical device that indicates a value by the position of a pointer on a scale.

analog multimeter: A meter that can measure two or more electrical properties and displays the measured properties along calibrated linear and nonlinear scales using a pointer.

analog output device: A device that produces and transmits electronic signals within a continuous range of values between two points.

analog sensor: A device that senses the value of a variable and transmits a proportional electrical signal to convey the information to a controller.

anemometer: A device that measures average air velocity or average airflow rate.

angular misalignment: A condition in which shafts are parallel in one plane but at different angles in another.

annual fuel utilization efficiency (AFUE) rating: A measure of the thermal efficiency of HVAC combustion equipment; expressed as a percent.

anticycle timer: A solid-state safety device that prevents premature cycling (overcycling) of the compressor in a refrigeration system to reduce the possibility of compressor failure, nuisance tripping of the circuit breaker, and fuse overloading. Also known as an antishort cycle timer.

antishort cycle timer: *See* anticycle timer.

apparatus dewpoint (ADP): The temperature in which a cooling coil is set to achieve proper cooling.

application-specific controller (ASC): A controller designed to control only one type of application.

aquastat: A device that controls the starting, stopping, and firing rate of a hot water boiler burner by measuring the temperature of the water in the boiler.

arc blast hood: An eye and face protection device that consists of a flame-resistant hood and face shield.

arc face shield: An eye and face protection device that protects against low-voltage arc hazards by covering the entire face with a tinted face shield.

area of influence: The area from the front of a register or diffuser to the point where air velocity drops below 50 fpm.

armature: The movable part of a solenoid.

asbestos: A collection of impure magnesium silicate mineral fibers that can pollute air and water and cause cancer or asbestosis when inhaled.

aspect ratio: The ratio between the height and width of a rectangular duct.

asphyxiate: A contaminant that prevents oxygen from entering the body, which causes drowsiness, headaches, and depressed respiration.

atmospheric burner: A type of burner that uses ambient air as combustion air.

atmospheric pressure: The force exerted by the weight of the atmosphere (gas molecules) on the earth's surface.

atom: The smallest unit of an element that holds the properties of the element.

atomization: The process of breaking a liquid into small droplets.

authority having jurisdiction (AHJ): An organization, office, or individual designated by a government agency with legal powers to administer, interpret, and enforce building codes.

automatic flow control valve: A check valve that opens and closes automatically.

automatic gas valve: A gas control valve used with direct spark ignition (DSI) systems or hot surface ignition systems.

automatic zone valve: A valve that is opened or closed automatically by an actuator such as a motor, solenoid coil, or pneumatic device.

averaging control: A DDC feature that calculates an average value from all selected inputs.

axial flow blower: A type of air moving device that consists of a blower wheel, which works like a turbine wheel.

azeotropic refrigerant: A refrigerant that is formed when refrigerants are blended together and act like an entirely new and separate refrigerant.

B

backpressure: The pressure produced by the ignition of an air-fuel mixture against the normal pressure of a gas flow.

balanced draft: A type of mechanical draft that uses a combination of forced draft and induced draft.

balance point temperature: The outdoor temperature at which the output of a heat pump equals the heat loss of a building.

balancing: **1.** The adjustment of the resistance to water flow through the piping loops in a system. **2.** The process of bringing HVAC systems into a state of stability produced by proportional distributions or functions.

balancing damper: An airflow damper located in the branch takeoff from the trunk duct in a forced-air distribution system.

ballast: An electrical device used for starting and regulating fluorescent lamps, which generate heat.

bare-tube evaporator: A type of air-medium evaporator that consists of tubing only.

bar graph: A graph composed of segments that function as an analog pointer.

barometric damper: A metal plate positioned in the opening of a flue that allows atmospheric pressure to balance the pressure at the bottom of a chimney to control the amount of airflow up through the combustion chamber and flue.

baseboard convector: A natural convection heater that is enclosed in a low cabinet that fits along a baseboard.

baseboard radiator: A radiator that has a section of finned horizontal tubes that are connected with headers at each end, which are concealed by an attractive housing.

BAS input: A signal from sensors that indicates building conditions to microprocessor-based controllers.

BAS output: A signal from a microprocessor-based controller to change the state of a controlled device.

baud rate: A measure of the maximum data transmission rate measured in 0s or 1s per second.

bearing: A mechanical device that allows free shaft movement by reducing the friction between moving surfaces.

belt drive system: A motor-to-wheel connection that has a motor mounted to a frame and the motor is connected to the blower wheel through a belt and pulley arrangement.

bimetallic element: An element consisting of a bimetallic strip that is usually wound into a spiral, helix, or coil to allow the element to be placed in a smaller space than the straight element requires.

bimetallic thermometer: A thermal expansion thermometer that uses a strip consisting of two alloys with different coefficients of thermal expansion that are fused together and formed into a single strip and a pointer or indicating mechanism calibrated for temperature reading.

binary blend: A blend of two refrigerants.

biological contaminant: A mass of particulate living organisms that can cause adverse health effects.

bipolar junction transistor (BJT): A transistor that controls the flow of current through the emitter (E) and collector (C) when the base (B) is properly biased.

blower: A mechanical device that consists of moving blades or vanes that force air to travel through ductwork.

blower control switch: A temperature-actuated switch that controls the blower motor of a furnace.

blower wheel: A sheet-metal cylinder with curved vanes along its perimeter.

boiler: A pressure vessel that transfers heat to water.

boiler combustion safety control: A device that controls and monitors the startup sequence, main flame, and shutdown sequence of a boiler burner.

boiler/tower system: A closed-loop hydronic system through which nonrefrigerated water circulates continuously through a building.

boiler vent: *See* vent valve.

boiling point: The temperature at which a substance changes state from a liquid to a gas (vapor).

bottom blowdown valve: A valve located on the blowdown line at the lowest point of the water side of a steam boiler that is used to release water and sediment from the bottom of the boiler.

bottom tap: A tap designed to cut threads at the bottom portion of a hole and should not be used for cutting threads in an unthreaded hole.

Bourdon tube pressure gauge: A mechanical pressure gauge with a curved tube made of thin metal that acts as a pressure-sensing element.

Boyle's law: A gas law stating that the volume of a gas decreases as pressure is applied.

brazed plate evaporator: A heat exchanger that uses enhanced flat plates that are brazed together to prevent leakage.

brazing: A copper tubing joining method in which nonferrous filler metal with a melting temperature of 840°F or higher is used.

bridge rectifier: A circuit containing four diodes that permits both halves of an input AC sine wave to pass.

British thermal unit (Btu): The quantity of heat required to raise the temperature of 1 lb of water 1°F.

building automation system (BAS): A computer-based system that uses a distributed system of microprocessor-based controllers to automate any combination of building systems.

building component: An object that helps provide a structurally sound and environmentally safe space to protect occupants and contents from the outdoor environment.

building-related illness (BRI): An identifiable illness, disease, or infirmity caused by exposure to contaminants in the air of a building.

building variable: The specific use, level of use, and operation of a building that increases or decreases the heating and cooling loads on the building.

burner: **1.** The heat-producing component of a combustion furnace. **2.** The heat-producing component of a combustion boiler.

burner face: The point at which a flame is established.

burner tube: An open or closed tube that has an opening on one end and burner ports located along the top.

burner vestibule: The area in a furnace where the burner and controls are located.

butane: A gas fuel composed of 95% to 99% butane gas and up to 5% butylene (C_4H_8).

bypass circuit: A refrigerant line that contains a check valve to allow refrigerant to flow in the opposite direction of an expansion device.

C

cabinet: A sheet-metal enclosure that completely covers and provides support for the components of an HVAC device.

cabinet convector: A convection heater that is enclosed in a cabinet.

cabinet heater: A forced convection heater that has a blower, hot water coils, filter, and controls in one cabinet.

calibration: The verification of graduations and incremental values of a precision measuring instrument for accuracy and adjustments.

calibration standard: A finely ground, precisely sized object that is used as a basis of dimensional comparison.

caliper: A hand tool with one fixed and one adjustable jaw used for making precision measurements.

capacitance probe: An anodized aluminum strip with a porous aluminum oxide layer used as a dielectric.

capacitor: An electrical device that consists of two thin metal plates with an insulating dielectric material separating them.

capacitor-run motor: A split-phase motor that has increased running torque due to a capacitor connected to the run windings.

capacitor start-and-run motor: A split-phase motor with two capacitors to allow greater torque during startup and running.

capacitor-start motor: A split-phase motor that has increased starting torque due to a capacitor connected to the start windings.

capillary tube expansion device: A long length of copper tubing that produces a pressure decrease regardless of the direction of refrigerant flow through it.

carbon dioxide (CO_2): An odorless, colorless gas that is typically present in indoor air.

carbon filter: An air cleaning device that uses an activated carbon-filtering medium to remove most odors, gases, smoke, and smog from the air by means of an adsorption process.

carbon monoxide (CO): An odorless, colorless, and toxic gas.

carcinogen: A contaminant that can cause cancer.

caution signal word: A word that indicates a potentially hazardous situation that, if not prevented, may result in minor or moderate injury or damage to equipment.

central forced-air heating system: A heating system that uses a centrally located heating unit to produce heat for a building.

centrifugal blower: A type of air moving device that consists of a scroll, blower wheel, shaft, and inlet vanes.

centrifugal force: The force that pushes a body outward when it is spinning around a center.

centrifugal pump: A pump that has a rotating impeller inside a cast iron or steel housing (volute).

Charles' law: A gas law stating that when the pressure on a gas is constant, the volume of the gas is directly proportional to absolute temperature; and when the volume of a gas is constant, the pressure on the gas is directly proportional to absolute temperature.

check valve: A valve that allows flow in only one direction.

chemical composition: The constitution of all elements that make up a compound.

chemical reaction: A chemical change of one or more substances that produces another substance.

chilled water system: A high-volume hydronic cooling system that typically includes chillers, pumps, piping, and accessories such as cooling coils and terminal devices.

chiller: A high-volume hydronic air conditioning device that cools circulating water for terminal device use.

chlorofluorocarbon (CFC) refrigerant: A type of refrigerant that consists of chlorine, fluorine, and carbon.

chronic health effect: A health effect that appears either years after exposure or only after long or repeated periods of exposure to a substance has occurred.

circuit breaker: An overcurrent protection device with a mechanism that automatically opens a circuit when an overload condition or short circuit occurs.

circulating pump: A type of pump that creates a flow of water through the boiler, system piping, and terminal devices of a hydronic heating system.

clamp-on ammeter: A meter that measures the current in a circuit by measuring the strength of the magnetic field around a single conductor.

clamshell heat exchanger: A heat exchanger that has multiple clam-shaped sections welded together.

class: The hazardous location category that refers to the generic hazardous material present.

closed-loop control: A control system in which the result from an output signal returns as feedback to a controller as an input signal.

closed loop system: A type of geothermal ground loop system that circulates water through piping in the ground to a water loop in order to exchange thermal energy.

close plenum duct system: A supply duct system consisting of a plenum box that is attached to the discharge end of a heating unit with individual small ducts connected from the plenum box to each zone.

code: A set of regulations or minimum requirements related to a particular trade or environment.

coefficient of performance (COP): The theoretical efficiency of operation of a mechanical compression refrigeration system.

cogeneration power plant: A power plant that generates both electricity and heat for use.

coil bypass factor: A number that represents the amount of air that bypasses the surface of a coil unchanged.

combination heating system: A type of heating system that uses hot water and steam to control building space temperature within a small range.

combination valve: A gas control valve that has a built-in fuel pressure regulator, which controls the pressure of fuel gas as it enters a burner.

combined gas law: A gas law stating that the volume of gas is directly proportional to absolute temperature and inversely proportional to pressure.

combustion: A chemical reaction in fuel that occurs when oxygen reacts with hydrogen (H) and carbon (C) at an ignition temperature.

combustion air: Air provided to a burner for the proper combustion of fuel.

combustion air blower: A blower used to provide positive pressure combustion air at a burner face.

combustion chamber: The area in a heating unit where combustion takes place.

combustion efficiency: A measure of how well chemical energy in a fuel is converted to thermal energy during combustion.

combustion safety control: A safety control that shuts down a burner when a malfunction occurs.

comfort: The condition that occurs when a person cannot sense a difference between themselves and the surrounding air.

commissioning: A process used to verify and document the performance of building systems and ensure that building systems achieve stated objectives for installation, testing, and operation.

commissioning agent: *See* commissioning authority.

commissioning authority: A person or organization that is responsible for planning, coordinating, and implementing the commissioning process. Also known as a commissioning agent.

commissioning checklist: A composite list used to verify that all phases of the commissioning process meet requirements, including installation, testing, and training.

commissioning report: A record of the activities and results of the commissioning process.

complete combustion: A type of combustion that occurs when enough oxygen is supplied by combustion air to combine with all elements in a fuel.

compound: A substance that can be broken down into elements.

compound gauge: A pressure gauge that indicates vacuum in inches of mercury (in. Hg) and pressure in pounds per square inch (psi).

compression ratio: The mathematical relationship between the pressure on the high-pressure side to the pressure on the low-pressure side of a refrigeration system.

compression stroke: The upward movement of a piston in a cylinder.

compressive force: The force that squeezes air together.

compressor: A refrigeration system component that creates refrigerant flow.

condensate pump: A pump that transports condensate from anywhere in a system back to the steam boiler or condensate return tank for reheating and reuse.

condensate return system: A system used to return condensate to a boiler after all useful heat has been removed.

condensate return tank: A steam system device that collects condensate returned from terminal units.

condensation: The formation of liquid (condensate) as moisture or other vapor cools below its dew point.

condenser: A type of heat exchanger that rejects heat from hot refrigerant vapor to a medium (air or water) in order to change the state of the refrigerant to 100% liquid.

condenser pressure regulator: A device used to maintain a sufficiently high condenser and receiver pressure when low ambient temperatures are present to guarantee that an expansion device receives liquid refrigerant.

condenser split: The difference between the condensing medium temperature and the refrigerant temperature inside a condenser.

condenser wastewater system: A type of water-medium condenser water system in which a water source is introduced to the condenser, used once, and then removed or dumped into a drain.

condensing boiler: A type of high-efficiency boiler that extracts heat from the condensed water vapor found in hot combustion gases before the gases enter a chimney.

condensing heat exchanger: A secondary heat exchanger that reduces the temperature of the flue gas below the dewpoint temperature of the heat exchanger.

conduction: A method of heat transfer caused by the direct physical contact of substances.

conduction factor (U-factor): An expression of the amount of heat that flows through a building component due to a temperature difference between the inside and outside air.

conductor: A material with low resistance that permits electrons to flow through it easily.

confined space: A space large enough that a technician can enter and perform assigned work but has limited or restricted means of entry and exit and is not designed for continuous occupancy.

confined space entry permit: A document issued by an employer to allow and control entry into a confined space.

constant-air-volume AHU: An air-handling unit that provides a steady supply of air and varies the heating, cooling, or other conditioning functions, as necessary, to maintain the desired setpoints within a building zone.

constant-air-volume (CAV) system: An air distribution system in which the airflow rate in each building space remains constant but the supply-air temperature is varied to maintain a setpoint.

constant entropy: A calculated value that indicates energy lost to the disorganization of the molecular structure of a substance when heat is transferred.

constant quality: A percentage that expresses the ratio of vapor to liquid as a refrigerant changes state.

constant volume: The volume of refrigerant vapor that remains constant because of the relationship between the pressure and enthalpy of the refrigerant.

constructive eviction: A legal issue that occurs when a tenant is unable to use a building space due to special circumstances and results in termination of the landlord-tenant relationship.

contactor: A large heavy-duty, switching device that uses low voltage to energize or de-energize the high-current, nonmotor load connected to it.

contact temperature probe: A device used for taking temperature measurements at a single point by direct contact with the area being measured.

continuity tester: An instrument that tests for a complete path for current to flow.

Continuity Test mode: A digital multimeter mode that detects a complete path for current to flow.

control point: The actual value of a controlled variable to be maintained by a system.

convection: A method of heat transfer in which heat is moved from one location to another through the circulation of a substance.

cooler: *See* water-medium evaporator.

cooling capacity: The heat transfer capacity of an evaporator measured in British thermal units per hour.

cooling load: The amount of heat gained by a building.

cooling tower: A device used to reject heat from the water circulated through a building's water loop system or chiller condenser.

cooling tower fill: The heat transfer material within a cooling tower that distributes water, allowing it to spread out and increase its contact time with the air.

corrosion: A condition that occurs when oxygen in air reacts with and breaks down metals (electrochemical oxidation).

counterflow cooling tower: A cooling tower in which the ambient air moves vertically upward through the fill, counter to the downward fall of water.

counterflow steam heating system: A type of one-pipe piping system that has steam and condensate flowing in the same pipe but in opposite directions.

crankcase pressure regulator (CPR): A device that maintains maximum pressure at the compressor inlet.

crankshaft: A reciprocating compressor component that transmits the circular turning motion of a prime mover to piston rods and then to pistons, producing an up-and-down motion.

critical point: The temperature and pressure above which a substance does not change state, regardless of how much heat is added or removed.

crossflow cooling tower: A cooling tower that has a fill configuration through which the ambient air flows horizontally and the water flows downward.

curb: A metal frame with framed duct connections that allow a light commercial unit to be connected to the ductwork of a building.

current: The amount of electrons flowing through a conductor.

D

daisy chain network: A network of controllers linked in a series in which each controller is connected to the next controller.

Dalton's law: A gas law stating that the total pressure of a gas mixture is the sum of the pressures of each gas in the mixture.

damper: A device that controls airflow.

danger signal word: A word that indicates an imminently hazardous situation that, if not prevented, will result in death or serious injury.

datum line: The elevation at which the water pressure in a system is the same as atmospheric pressure.

deadband: The temperature range between setpoints in which no mechanical heating or cooling occurs.

dead short: A short circuit that opens an overcurrent protection device as soon as the circuit is energized.

deburring tool: A hand tool used to remove burrs left inside the tubing by tubing cutters.

decibel (dB): A unit of measure used to express the relative intensity of sound.

defrost cycle: A mechanical procedure used to temporarily reverse the refrigerant flow in a heat pump to melt frost or ice on an outdoor unit coil.

dehumidifier: A device that removes moisture from air by causing the moisture to condense or be trapped.

density: The mass of a substance per unit of volume.

depth micrometer: A micrometer that measures component heights and slot depths.

derivative control algorithm: A control algorithm that positions the controlled device in direct response to the instantaneous rate of change of a variable.

desiccant: A solid substance that attracts water molecules and holds on to the molecules until it is dried.

design condition: The condition of the air where heating or air conditioning equipment provides comfort in a building space.

design static pressure drop: The pressure drop per unit length of duct for a given size of duct at a particular flow rate.

design technician: A person who has the knowledge and skill to plan heating and/or air conditioning systems.

design temperature: The temperature of air at a predetermined set of conditions.

design temperature difference: The difference between the selected indoor design temperature and the expected outdoor design temperature for a particular season.

dew point: The temperature below which moisture begins to condense.

dewpoint temperature (dp): The temperature at which water vapor in the air begins to condense.

dial caliper: A caliper with a dial indicating gauge.

die: A hand tool used to cut external threads on round rods or clean threads that have been disfigured by damage or improper operation.

digital caliper: A caliper that indicates measurements with a digital display.

digital display: An electronic device that displays readings on a meter as numerical values.

digital input device: A system component that acquires the ON or OFF state of a variable and transmits a high or low signal to convey the information to a controller.

digital logic probe: A special DC voltmeter that detects the presence or absence of a digital signal.

digital manometer: A portable electrical device that precisely measures positive, negative, or differential pressures in small increments.

digital micrometer: A micrometer with a digital electronic indicating gauge.

digital multimeter (DMM): A meter that can measure two or more electrical properties and displays the measured properties as numerical values.

digital output device: A device that produces and transmits signals to a controlled device that accepts an on or off signal.

digital valve: A two-position (ON/OFF) valve.

dilution air: Atmospheric air that mixes with, dilutes, and cools the products of combustion.

diode: A type of semiconductor device that allows current to flow in one direction only.

direct current (DC): Current that flows in one direction only.

direct digital control (DDC) system: A control system in which electrical signals are used to measure and control system parameters from multiple microprocessor-based controllers.

direct drive system: A motor-to-wheel connection that has a blower wheel mounted directly on a motor shaft.

direct-expansion evaporator: *See* dry-type evaporator.

direct-fired heater: A unit heater that does not have a heat exchanger.

discomfort: The condition that occurs when a person can sense a difference between themselves and the surrounding air.

disconnect: A switch that isolates electrical circuits from a voltage source to allow safe access into an electrical panel or equipment for maintenance or repair.

distributed DDC system: A control system in which operating decisions are made by local microprocessor-based controllers with the decision-making ability distributed throughout the controlled system.

distributor: A device used to evenly feed refrigerant from the expansion device into either multiple circuits on a single evaporator or separate evaporators.

diverter fitting: A tee that meters the water flow from a primary loop.

diverter valve (zone valve): A tee-shaped valve that meters the steam from a primary loop.

diverting valve: A three-way valve that has one inlet and two outlets to divert water flow to a boiler and/or chiller.

division: The hazardous location category that refers to the probability that a hazardous material is present.

doping: The addition of material to a base element to alter the crystal structure of the base element.

double-line drawing: A drawing in which walls and partitions are shown using double lines and wall thickness is drawn to scale.

downflow (counterflow) furnace: A furnace that heats air and distributes the heated air downward through ductwork.

draft: The movement of air across fire and through a heat exchanger.

draft diverter: A box made of sheet metal that runs the width of a clamshell heat exchanger to trap flue gases and then travels up a flue with dilution air.

draft inducer: A fan that overcomes the resistance to flow in the combustion chamber and provides positive pressure in the flue and vacuum at the burners, which allows the flame to travel through the heat exchanger.

draft induction fan: A fan designed to create the correct amount of flow through a combustion chamber and to handle the products of combustion.

draft system: A system in a furnace or boiler that regulates the flow of air into and out of the combustion chamber.

draft tube: *See* air tube.

drum heat exchanger: A round drum or tube that is located on a combustion chamber to make the products of combustion flow through it.

dry base boiler: A boiler that has no water alongside or below a firebox.

dry bulb economizer system: An economizer system that operates strictly in proportion to the outdoor-air temperature, with no reference to humidity values.

dry-bulb temperature: The temperature of air as taken with a standard thermometer.

dry contact: A contact that requires external power for transmission.

dry cooling tower: A cooling tower with closed water loops that uses piping with fins to exchange heat between circulating water and outside air.

dry-type evaporator: A heat exchanger in which refrigerant boils from a partial liquid/gas mixture to 100%

gas before the refrigerant leaves the evaporator. Also known as a direct-expansion evaporator.

dual-duct system: An air distribution system in which all the air is conditioned in a central fan system and distributed to conditioned spaces through two parallel main ducts.

dual-in-line-package (DIP) switch: A manual on/off electrical switch that creates an address system for controllers.

duct heater: A unit heater that is installed in a duct and supplied with air from a remote blower.

ductless mini-split system: An air conditioning system that consists of one or more indoor air handling units mounted onto the interior walls or ceilings of a building.

duct section: A section of ductwork between two fittings where the airflow rate changes.

ductwork equivalent length method: A method used to convert dynamic pressure drop to friction loss in feet of duct.

dumb terminal: A display monitor and keyboard with no processing capabilities.

dust-holding capacity: The measure of a filter's ability to hold dust without significantly reducing its efficiency.

dust spot efficiency: A filter rating that indicates the ability of a filter to remove from the airflow large particles that tend to soil building interiors.

dynamic pressure drop: The pressure drop in a duct fitting or transition caused by air turbulence as the air flows through the fitting or transition.

E

earmuff: An ear protection device worn over the ears.

earplug: An ear protection device made of moldable rubber, foam, or plastic that is inserted into the ear canal.

ECM (electronically commutated motor): A very high efficiency programmable brushless DC motor with a permanent magnet rotor and a built-in variable-frequency drive (VFD).

economizer lockout control: A DDC feature in which the economizer damper function is locked out or discontinued.

economizer system: An HVAC system that uses outdoor air for cooling when the outdoor air temperatures are cool.

efficiency rating: A number provided by a manufacturer that represents the percentage of heat that is delivered to the space versus what goes up the chimney.

elastic deformation element: A device consisting of flexible components such as pressure springs or pistons that flex, expand, or contract in proportion to the pressure applied within or against them.

electrical circuit: The interconnection of conductors and electrical elements through which current is designed to flow.

electrical consumption: The total amount of electricity used by a building or facility during a billing period.

electrical control system: A control system that uses electricity to operate.

electrical foot protection: Protection for feet in the form of insulated, rubber-soled shoes that is typically worn during electrical work to prevent electrical shock.

electrical hand protection: A set of hand coverings (gloves) that are worn to prevent injuries to hands caused by electrical shock.

electrical personal protective equipment (PPE): Gear worn by electricians and HVAC technicians to reduce the possibility of injury in the work area.

electrical safety glasses: Eye protection devices with plastic frames designed to minimize the shock hazard when working with electrical equipment.

electrical safety label: A label that indicates areas or tasks that can pose an electrical hazard to personnel and/or HVAC equipment.

electrical shock: The result of a body becoming part of an electrical circuit.

electrical trade association: An organization that represents the manufacturers of electrical equipment.

electrical warning symbol and signal word: A combination signal and symbol that indicates a high-voltage location and condition that could result in death or serious personal injury from electrical shock if proper precautions and procedures are not taken.

electric boiler: A type of boiler that uses heat produced by resistance heating elements to heat water.

electric furnace: A furnace that uses resistance heating elements to warm passing air.

electricity: The energy produced from the flow of electrons through a conductor (wire).

electric spark (flame) igniter: A device that is located at the burners that produces a spark to ignite fuel.

electric steam boiler: A steam boiler that uses steam produced by resistance heating elements to generate heat.

electromechanical control system: A control system that uses electricity in combination with a mechanical component to operate.

electromechanical relay: A switching device that has sets of contacts that are closed by a magnetic force.

electron: A negatively charged particle that makes up the cloud of an atom.

electronic air cleaner: An air filtering system that passes an electrical charge through a wire grid inside the filter to charge particles and then attracts those particles to plates charged with the opposite charge, where they are trapped.

electronic control system: A type of control system that uses solid-state devices to control refrigeration and HVAC systems.

electronic expansion valve (EEV): An expansion device that uses a stepper motor to open and close a valve orifice to control the amount of refrigerant entering an evaporator.

electronic leak detector: A leak detector that detects the presence of halogen gas.

electrostatic air filter: An air filtering system in which particles pass through electrostatic air filter media where friction gives them an electrical charge and then they are attracted to another layer of media that acts like a magnet.

electrostatic discharge (ESD): The movement of electrons from a source to an object.

element: A substance that cannot be broken down into a simpler form of matter.

energy audit: A comprehensive review of the energy use in a building.

energy cost index (ECI): A ratio of the dollars spent annually on energy used in a building divided by the floor area of the building.

energy recovery ventilator (ERV): A device used to introduce fresh outside air into a building and exhaust stale indoor air while minimizing the energy lost during the process.

energy utilization index (EUI): A ratio of the energy used in a building annually divided by the floor area of the building.

enthalpy: The total amount of heat contained in a substance.

enthalpy economizer system: An economizer system that uses the temperature and humidity levels of the outdoor air to control HVAC operation.

Environmental Protection Agency (EPA): A federal government regulatory agency established in 1970 to control and abate air, water, and land pollution in the form of solid waste, radiation, pesticides, and various other toxic substances.

environmental tobacco smoke (ETS): A mixture of smoke from the burning end of a cigarette, pipe, or cigar and smoke exhaled by a smoker.

equal friction chart: A chart that shows the relationship between the airflow rate (in cfm), static pressure drop (in in. WC/100′), duct size (in in. DIA), and air velocity (in fpm).

equal friction method: A method of sizing ductwork that sets static pressure approximately equal at each branch takeoff in a distribution system.

equal velocity method: A method of sizing ducts that causes each duct section to have the same air velocity.

equipment and reclaimer certification: A credential required by the EPA that contractors and/or technicians must acquire for their particular recovery and/or recycling equipment before use.

equivalent temperature difference: The design temperature difference adjusted for solar gain.

error signal: *See* feedback.

evacuated-tube collector: A solar thermal collection device that consists of rows of parallel, transparent glass tubes from which air has been removed.

evaporation: The process that occurs when a liquid changes to a vapor by absorbing heat.

evaporator: A heat exchanger that transfers heat from a product or from room air into a low-temperature, low-pressure liquid-gas refrigerant.

evaporator pressure regulator (EPR): A valve that is designed to maintain a minimum evaporator pressure at all times. Also known as a holdback valve.

existing building commissioning: *See* retrocommissioning.

expansion device: A device (typically a valve) that causes a pressure drop of high-pressure liquid refrigerant.

expansion tank: A vessel that allows the water in a hydronic heating system to expand without raising the system water pressure to dangerous levels.

explosion warning symbol and signal word: A combination signal and symbol that indicates locations and conditions where exploding parts may cause death or serious personal injury if proper precautions and procedures are not followed.

exposed surfaces: Building surfaces that are exposed to outdoor temperatures.

extended plenum duct system: A supply duct system

consisting of a plenum box attached to the discharge end of a heating unit with a main duct extending across the length of a building.

F

face shield: An eye and face protection device that covers the entire face with a plastic shield and is used for protection from flying objects.

fan: A mechanical device with rotating blades that moves air.

feedback: The measurement of the results of a controller action by a sensor or switch. Also known as an error signal.

feedwater: Boiler-ready water that is a combination of steam condensate and makeup water.

feedwater check valve (backflow preventer): A valve that controls the direction of city water flow by allowing flow in only one direction through the valve.

feedwater heater: A light commercial, commercial, and industrial steam boiler accessory that heats feedwater before the feedwater enters the boiler.

feedwater pump: A steam boiler accessory used to pump feedwater into a steam boiler.

feedwater regulator: A steam boiler accessory that maintains the normal operating water level (NOWL) in a boiler.

feedwater system accessory: Equipment that is not directly attached to a steam boiler but controls the amount, pressure, and/or temperature of water supplied to the boiler.

feedwater valve: A valve that controls the flow of makeup water into a steam boiler to compensate for water that is lost.

fiberglass: A material composed of thin strands of glass.

field-effect transistor (FET): A type of transistor that controls the flow of current through the drain (D) and source (S) when the gate (G) is properly biased.

file: An abrasive hand tool with single or double rows of fine teeth cut into the surface.

filter: An air cleaning device consisting of fine-fiber fabric that is used to remove particles from the airstream.

filter airflow resistance: The pressure drop across a filter at a given velocity.

filter-drier: A device that removes particulate matter, moisture, and acids from the refrigerant in a system.

fin comb: A hand tool used to straighten the edges of thin metal fins that extend from condenser or evaporator coils.

finned tube: A copper pipe with aluminum fins.

finned-tube evaporator: A type of air-medium evaporator that consists of two heat exchange surfaces, which are the tubing and fins.

fire point: The temperature at which a substance, such as oil, sustains ignition.

firetube boiler: A boiler that heats the water surrounding fire tubes as hot gases of combustion pass through the tubes.

fish tape: A retractable tape, usually with a rectangular cross section, that is pushed through an inaccessible space, such as a conduit or partition, in order to draw in electrical conductors.

fitting: A device directly attached to a boiler that is required for proper operation of the boiler and system.

fixed bore pulley: A one-piece pulley that has a machined bore.

fixed leak detector: A stationary leak detector system with sensors and controllers to detect one specific type of refrigerant.

fixed-plate orifice expansion device: An object that has a small fixed nozzle, or orifice, to create a predetermined resistance to flow (pressure drop).

flame detector: An electronic combustion safety control device that consists of a light-sensitive component that detects a flame.

flame sensor: An electronic combustion safety control device that determines when a flame is present by using the burner flame to conduct electricity.

flammability: The ability of a substance to support ignition.

flapper valve: A valve that opens to let an air-fuel mixture in but closes when a firing cycle begins because of backpressure.

flaring tool: A tool applied to the end of tubing used to flare the end of the tubing.

flash point: The temperature at which oil vapors ignite without the oil itself burning.

flat file: A file that is used to file flat surfaces as well as for other operations that require a fast-cutting file.

flat-plate collector: A solar thermal collection device that consists of copper-pipe flow tubes attached to a dark-colored absorber plate in an insulated box underneath a hard glass or polymer cover.

flexible coupling: A coupling containing a resilient member that allows two misaligned shafts to be joined together.

float valve: A valve consisting of a hollow ball that floats on the liquid in an internal reservoir to restrict flow through the valve when the liquid reaches a certain level.

flooded evaporator: A heat exchanger in which a float or other device ensures that the entire evaporator is full of liquid refrigerant.

flow coefficient: The water flow rate through a valve or any control mechanism in a piping system that causes a pressure drop of 1 psi through the valve or mechanism.

flow control valve: A valve that regulates the flow of water (in gpm) in a hydronic heating system.

flow hood: A device that is placed over a grill or diffuser to measure the rate of airflow.

flow rate: The rate at which a furnace burns fuel.

flue gas: The carbon dioxide and other vapors that exit a furnace or boiler to the atmosphere using a flue, which is a pipe or channel for conveying exhaust gases.

fluorescent leak detector: A leak detector that uses a UV light to detect fluorescent dye that is added to a system.

forced-air heating system: A heating system that uses a fan to force air across a heat source to heat a building.

forced convection heating: A form of heating in which convective currents are mechanically moved over a steam coil.

forced draft: A type of mechanical draft created by air being pushed through a combustion chamber by a fan located on the outside of a furnace or boiler.

forced-draft cooling tower: A cooling tower in which fans are located in the airstream, typically at the top of the tower, in order to force the ambient air through.

forced steam cabinet convector: A self-contained forced convection heater terminal device that contains a blower, steam coil, filter, and number of controls in one cabinet.

forced steam cabinet-ventilator convector: A forced convection heater terminal device that contains filters and controls as well as a blower, steam coil, and a cabinet with an opening for outdoor ventilation.

forced steam unit convector: A self-contained forced convection heater terminal device that contains a fan, steam coil, and number of controls in one cabinet and is suspended from ceiling framework.

formaldehyde: A colorless gas with a strong, pungent odor used as a preservative.

fossil fuel: A fuel derived from organic matter that has decayed over millions of years beneath the earth's surface.

fractionation: A problem that occurs when one refrigerant of a blend leaks out of a refrigeration system at a different rate from the other refrigerants of the blend due to the varying temperature-pressure characteristics of the refrigerants.

free area: The total area of a duct outlet opening (hole size) minus the area that is used by the register or diffuser slots, vanes, and frame located within the outlet opening.

freezing point: The temperature at which a substance changes state from a liquid to a solid.

friction head: The effect of friction in a pipe.

friction loss: The decrease in air pressure due to the friction created when air moves through a duct.

fuel: Any substance that is burned to produce heat.

fuel oil: A crude oil distillate that is burned in a furnace or boiler to create heat or used by engines to create power.

fuel oil heater: A device in a fuel oil supply system that is used to heat cold heavy fuel oil to assist in pumping and allow for efficient burning.

fuel oil nozzle: A gun-type fuel oil burner component that atomizes fuel oil by converting it from a solid stream into very small droplets (mist).

fuel oil pump: A gun-type fuel oil burner component that forces fuel oil into the combustion chamber of furnaces and boilers under pressure.

fuel oil strainer: A device that removes solid foreign matter from a fuel oil system.

fuel oil supply system: A system of tanks, valves, piping, strainers, and pumps that transport and condition fuel oil for combustion.

full-face respirator mask: A type of respiratory protection that covers the entire face, including the eyes, and has filter cartridges that protect against dusts, fumes, and particulates.

full-load amperage (FLA): The amperage drawn when a motor is at full load.

full-wave rectifier: A circuit containing two diodes and a center-tapped transformer that permits both halves of an input AC sine wave to pass.

furnace: A self-contained heating unit that includes a blower, at least one burner, a heat exchanger, and a number of controls.

furnace blower drive system: A type of furnace connection used between the electric motor and the blower wheel, or motor-to-wheel connection, of a furnace.

furnace component: A major piece of equipment that makes up a furnace.

fuse: An overcurrent protection device with a fusible link that melts and opens a circuit when an overload condition or short circuit occurs.

fusible plug: A warning device used to indicate the overheating of a steam boiler from a low water condition.

G

gas butterfly valve: A type of metering valve that controls the volume of gas that flows to a burner.

gas control valve: A 100% shutoff safety valve that controls the flow of fuel to a main burner and a pilot burner, if used.

gas-fired atmospheric burner: A burner that mixes ambient air with a gas fuel to create a flame.

gas-fired power burner: A type of burner that uses natural or LP gas and contains a fan or blower.

gas pressure regulator: A type of valve that adjusts the supplied gas pressure to a burner.

gas pressure switch: A switch used to monitor the gas pressure in a gas line.

gas shutoff cock: A type of valve used on a natural-gas fuel system to manually isolate the system or parts of the system.

gas vent valve: A valve that has two main gas valves and/or two pilot light valves and is used as a safety device on high-pressure gas systems.

gate valve: A valve that has an internal gate that slides over an opening through which water flows.

gauge: The number of sheet metal sheets that, when stacked together, equals 1″.

gauge glass: A tubular glass column that indicates water level in a steam boiler.

gauge manifold: A device that has two gauges, a manifold with valves, and connecting hoses to control the transfer of a specific refrigerant.

gauge pressure: A pressure measurement above atmospheric pressure that is used to express pressure derived from the operation of a closed system.

geothermal ground loop: Piping that is used to transfer heat either to or from the earth.

geothermal system: A water system that uses the temperature of the earth to heat or cool the water loop of a building.

ghost voltage: A voltage that appears on a meter that is not connected to a circuit.

global data: Data that is needed by all controllers in a network.

globe valve: A valve that has a disc (globe) that rises or lowers over a seat through which fluid flows.

goggles: A type of eye protection with a flexible frame that is secured on the face with an elastic headband.

good IAQ: Indoor air that is free of harmful particle levels and has a comfortable temperature and humidity level.

government regulatory agency: A federal, state, or local government organization that establishes rules and regulations related to safety, health, equipment installation, and equipment operation.

graph: A diagram that shows a variable in comparison to other variables.

grill: A device that covers the opening of return air ductwork.

gross unit output: The heat output of a hydronic or steam boiler when it is fired continuously.

gross wall area: The total area of a wall including the areas of windows, doors, and other openings.

grounding: The connection of all exposed non-current-carrying metal parts to the earth.

ground loop: An unwanted circuit that has more than one point connected to ground with a voltage potential difference between the ground points high enough to produce a circulating current in the ground system.

ground resistance tester: An instrument used to measure ground resistance for grounding rod installations.

group: A designation used to classify gases and dusts.

guideline: A completely voluntary recommendation that is related to a particular trade or environment.

gun-type fuel oil burner: A fuel oil burner that provides atomized fuel oil to a combustion chamber to increase mixability with combustion air for maximum combustion efficiency.

H

hacksaw: A metal-cutting hand tool with an adjustable steel frame for holding various lengths and types of blades.

half-face respirator mask: A type of respiratory protection that covers the nose and mouth and has filter cartridges that protect against dusts, fumes, and particulates.

half-wave rectifier: A circuit containing one diode that allows only half of an input AC sine wave to pass.

hammer: A striking or splitting hand tool with a hardened head fastened perpendicular to a handle.

hand brake: A hand tool used to make accurate bends in sheet metal.

hand level: A hand tool used to establish accurate horizontal and vertical orientations.

hand protection: A set of hand coverings (gloves) that is worn to prevent cuts or injuries to hands caused by chemicals.

hand tool: A tool that uses no external power other than that supplied by the individual using the tool.

hard hat: *See* protective helmet.

harmonics: Power frequencies that are whole integer multiples of the base frequency (60 Hz).

hasp: A multiple lockout/tagout device.

heat: The energy of a substance resulting from the internal motion of its molecular structure.

heat exchanger: Any device that transfers heat from one substance to another without allowing the substances to mix.

heating load: The amount of heat lost by a building.

heating seasonal performance factor (HSPF): A ratio of heat output (in Btu) to the electricity used (in Wh) in the efficiency measurement of air-source heat pumps.

heating value: The amount of heat produced per unit of fuel per hour; expressed in Btu/hr.

heat of compression: The thermal energy equivalent to the mechanical energy expended by the motor that turns the compressor and the heat created by the compression of the refrigeration molecules.

heat of rejection: The amount of heat rejected by a condenser.

heat pump: A mechanical compression refrigeration system that reverses the flow of refrigerant through a system to move heat from one area to another.

heat pump thermostat: A control component that incorporates a system switch, heating thermostat, and cooling thermostat into one housing.

heat rejection rate: The rate at which heat is transferred from refrigerant in a condenser to a condensing medium.

heat sink: A device with a relatively warm or cool surface that is capable of absorbing heat from a substance such as air or water.

heat transfer factor: A conduction factor multiplied by a design temperature difference.

Hertz (Hz): The international unit of frequency equal to one cycle per second.

high-efficiency particulate air (HEPA) filter: An air filter that removes at least 99.97% of airborne particles 0.3μ in diameter.

high/low signal select: A DDC feature in which a building automation system selects among the highest or lowest values from multiple inputs.

high-pressure appliance: A refrigeration or air conditioning system that has a refrigerant charge over 5 lb, has a boiling point between –58°F and 50°F at atmospheric pressure, and is not categorized as low-pressure.

high-pressure control (HPC): A pressure-activated electromechanical switch used to shut off a compressor when pressure in a system becomes too high.

high-pressure gas switch: A normally closed switch that remains closed and opens when gas pressure becomes too high.

high-pressure gas system: A fuel system that mixes gas with air at pressures up to 5 psi.

high temperature limit switch: An electric switch that shuts off a furnace when it begins to overheat.

high voltage: Voltage over 600 VAC.

holdback valve: *See* evaporator pressure regulator (EPR).

horizontal closed loop system: A type of geothermal ground loop system with piping parallel and close to the surface of the ground that circulates water to exchange thermal energy.

horizontal furnace: A furnace that heats air and distributes the heated air horizontally through ductwork.

horsepower: A unit of power equal to 550 ft-lb/sec or 745.7 W of electricity.

hot-gas discharge line: The refrigerant line that connects a compressor to a condenser.

hot pulldown: A condition where the evaporator pressure, temperature, and superheat amounts are higher than normal because of application conditions.

hot surface igniter: A device located at burners that uses a small piece of silicon carbide, which glows when electric current passes through it, to produce extremely high temperatures.

human comfort: A condition that occurs when people cannot sense a difference between themselves and the surrounding air.

humidifier: A device that adds moisture to air by causing water to evaporate into the air.

humidity: The amount of moisture, or water vapor, in air.

humidity ratio: A measurement, by weight, of moisture in the air.

HVAC Excellence: A not-for-profit organization that provides improved competency through validation of technical education for the HVAC and refrigeration industry.

HVAC wiring diagram: A diagram that shows the wiring connections between HVAC control panels and system components such as sensors and actuators.

hydrocarbon: A compound that contains hydrogen and carbon.

hydrochlorofluorocarbon (HCFC) refrigerant: A type of refrigerant that consists of hydrogen, chlorine, fluorine, and carbon.

hydroelectric-power generation plant: A power plant that changes energy from moving water into electrical energy.

hydrofluorocarbon (HFC) refrigerant: A type of refrigerant that consists of hydrogen, fluorine, and carbon.

hydronic design static pressure drop: The pressure drop per unit length of piping for a given size of piping at a given flow rate.

hydronic heating system: A heating system that uses water or other fluid to carry heat from the point of generation to the point of use.

hydronic heating system Btu: A unit that represents the amount of energy it takes to raise 1 pound of water 1 degree Fahrenheit.

hydronic loop control system: A grouping of electrical devices that control water temperature and flow in a hydronic loop and geothermal ground loop.

hydronic radiant heater: A heater that has a radiant surface that is heated by hot water to a temperature high enough to radiate energy.

hygrometer: A humidity analyzer that measures the physical or electrical changes that occur in various materials as they absorb or release moisture.

hygroscopic element: A material that changes its physical size or electrical characteristics as the humidity changes.

I

ignition electrode: A gun-type fuel oil burner component that is used to create a spark, which ignites a mixture of fuel oil and air.

ignition temperature: The heat required to start a chemical reaction between oxygen and the elements in a fuel.

ignition transformer: A step up transformer on a gun-type fuel oil burner that steps up voltage from 120 VAC on the primary side to 10,000 VAC to 17,500 VAC on the secondary side.

immersion heater: A series of copper rods enclosed in waterproof tubes that are installed in a boiler and surrounded by water to provide heat by electricity.

impeller: A plate with vanes that radiate from a central hub (eye).

improper voltage level: A voltage level in a circuit that is at least 10% higher or lower than the designed level.

inches water column (in. WC): The pressure exerted by a column of water one inch in height.

inclined-tube manometer: A very precise manometer designed so a long inclined section of glass or plastic tubing has more graduations per inch than a U-tube manometer, allowing for greater accuracy.

incomplete combustion: A type of combustion that occurs when not enough oxygen is supplied by combustion air to combine with all the elements in a fuel.

individual duct section: Part of a distribution system that is between two fittings or transitions.

indoor air quality (IAQ): The condition of indoor air as it relates to contaminant levels and air properties that affect the health and comfort of building occupants.

indoor design temperature: The temperature selected for the inside of a building.

indoor unit: A component that contains a heat exchanger and a blower.

induced draft: A type of mechanical draft created by air being pulled through a furnace or boiler combustion chamber by a fan located in the breeching.

inductance: The production of current flow due to a magnetic field on a coil.

infiltration: The leaking of outdoor air into a building.

infiltration air: Air that flows into a building when exterior doors or windows are open or when air leaks in through cracks around doors, windows, or other openings.

infinity: An unlimited number or amount.

infrared temperature meter: A meter that measures heat energy by measuring the infrared energy emitted by a material and displays the temperature as a numerical value.

inlet vane: An adjustable damper that controls airflow to a blower.

inner mantle: The center section of a flame that is produced as hydrogen reacts with oxygen.

input rating: The amount of heat produced by a furnace or boiler as fuel is burned; expressed in Btu/hr.

inside micrometer: A micrometer used to measure linear dimensions between two inside points or parallel surfaces.

installation manual: A general guide used by qualified technicians to install residential and light commercial HVAC equipment.

insulation class rating: A rating that describes how hot a motor can safely operate in a specific ambient environment.

insulator: A material that resists the flow of electrons.

integrated circuit: An electronic device in which all devices, such as transistors, diodes, and resistors, are contained in a single package or chip as components.

integration control algorithm: A control algorithm that positions the controlled device in response to the sum of the offset over a certain length of time.

intermediate tap: *See* plug tap.

International Organization for Standardization (ISO): A nongovernmental, international organization that is comprised of national standards institutions from over 160 countries and includes over 2700 technical committees.

International Union of Operating Engineers (IUOE): A diversified trade union established in 1896 that primarily represents stationary engineers, operating engineers, and members in the service industry.

interpersonal skill: A character trait that allows an individual to communicate effectively with others in a variety of situations.

irritant: A contaminant that can cause redness and/or inflammation of the skin, eyes, nose, throat, or upper-respiratory tract.

J

Joint Commission: A major nonprofit organization that provides accreditation services for health care organizations and sets most of the standards by which health care quality is measured in the U.S.

jumper: A conductor used to connect pins on a controller or device.

K

K factor: A number that represents the actual open area and design characteristics of a specific grill or diffuser.

kinetic energy: Energy that is in motion and, therefore, doing work.

knee pad: A rubber, leather, or plastic pad strapped onto the knees for protection.

knife file: A file with a blade cross-section tapering from a square to a pointed edge.

L

lanyard: A rope or webbing device that connects a harness or body belt to a lifeline.

latent heat: The heat applied to or removed from a substance that results in a change of state.

layout drawing: A drawing that shows the walls, partitions, windows, doors, fixtures, and other details of a building that affect the location of ductwork, registers, and grills.

lead: A toxic metallic element.

Leadership in Energy and Environmental Design® (LEED®): A green building certification system that awards points based on specific energy and environmental construction types and activities.

leakage current: Current that flows through insulation.

leak detector: A device that is used to detect refrigerant or other gas leaks in pressurized air conditioning or refrigeration systems.

leather protector gloves: Gloves worn over rubber insulating gloves to prevent penetration of the rubber insulating gloves and provide added protection against electrical shock.

lever bender: A hand tool with a hinge and lever used to bend 1/4″ and 3/8″ annealed copper tubing.

lifeline: A rope or webbing that is attached to a worker and a tie-off device to prevent the worker from hitting the ground or other object during a fall.

light-emitting diode (LED): A type of diode designed to produce light when forward-biased.

linear scale: A scale that is divided into equally spaced segments.

line voltage: Voltage at 115 VAC to 600 VAC.

liquefied petroleum (LP) gas: A fuel refined from crude oil or natural gas.

liquid-in-glass thermometer: A thermal expansion thermometer consisting of a sealed, narrow-bore glass tube with a bulb at the bottom filled with a liquid.

liquid line: The refrigerant line that connects a condenser to an expansion device.

liquid receiver: A liquid-line storage tank for excess liquid refrigerant.

load: A device that performs work.

locked-rotor amperage (LRA): The amount of current drawn on a motor if a rotor is unable to turn or is locked. Also known as startup amperage.

lockout: The process of removing the source of electrical power from equipment and installing a lock that prevents the power from being turned on.

low excess air burner: A burner that uses only the amount of air necessary for complete combustion in a furnace.

low-loss fitting: A fitting used to prevent the release of refrigerant from a system to the atmosphere and to prevent air from entering the system.

low-pressure appliance: A system that uses a refrigerant that boils above 50°F at atmospheric pressure and operates at low pressures.

low-pressure control (LPC): A pressure-activated electromechanical switch used to shut off a compressor when system pressure becomes too low.

low-pressure gas switch: A normally open switch that is held closed by gas pressure and opens when gas pressure is too low.

low-pressure gas system: A type of fuel system that mixes gas with air in a mixing chamber at approximately atmospheric pressure (0 psi) before the gas enters a burner.

low-pressure hot water heating boiler: A hydronic boiler in which water is heated for the purpose of supplying heat at pressures not to exceed 160 psi and temperatures not to exceed 250°F.

low-pressure steam heating boiler: A boiler operated at pressures not to exceed 15 psi of steam.

low-temperature limit control: A temperature-actuated electric or pneumatic switch that can energize a damper actuator to close the damper and open a water or steam valve if ventilation air drops below 35°F.

low voltage: Voltage of 50 VAC or less (typically 24 VAC).

low water cutoff (LWCO): A sensor and switch or probe that opens the combustion circuit of a boiler in the event of a low water condition.

lube oil pressure safety control: An electromechanical safety switch that is designed to open a compressor circuit when oil pressure falls below a specific setpoint.

M

Magnehelic® pressure gauge: A mechanical pressure gauge consisting of two pressure-tight compartments, a flexible diaphragm, a biasing spring, and a magnetic pointer.

magnetic overload: An external motor protection device that measures the magnetic field accompanying current flow from a starter to the motor power leads and opens the starter circuit if needed.

main gas valve: A type of electrically operated valve that either allows gas to flow or stops the flow of gas to a burner.

main steam stop valve: A valve that is used to isolate a steam boiler from the main distribution header.

major duct section: An independent part of a forced-air distribution system through which all or part of the fan air supply flows.

makeup air: Air that is used to replace air that is discharged as exhaust.

manifold: A pipe that has outlets for connecting other pipes.

manometer: A device used for taking precise pressure measurements using a portable, electrically powered device or using a liquid-filled tube.

manual flow control valve: A globe valve that manually controls the flow of water in a hydronic heating system.

manual zone valve: A valve that is set by hand to regulate the flow of water in a piping loop.

manufactured gas: Any gas produced from coal, oil, liquefied petroleum, or natural gas.

mass: The amount of matter a substance contains.

mass flow rate: The rate of refrigerant flow in pounds per minute.

mass psychogenic illness: A condition that occurs when groups of people start feeling sick at the same time even though there is no proven physical or environmental reason.

matter: Any substance that has weight and occupies space.

maximum continuous rating: The amount per hour of evaporation that can be maintained by a steam boiler for 24 hr.

megohmmeter: A meter that detects insulation deterioration by measuring high resistance values under high-voltage test conditions.

mercury barometer: A test instrument used to measure atmospheric pressure.

metal snip: A hand tool resembling pliers that is specifically used for cutting or shearing sheet metal.

microbial contaminant: A contaminant invisible to the naked eye that is either a living organism or derived from a living organism and can cause adverse health effects.

micrometer: A precision measuring instrument used to take high-accuracy measurements.

mill file: An all-purpose, single-cut file especially adapted for finish filing.

minute release: The unavoidable release of a very small amount of refrigerant during the service process while using low-loss fittings.

miscibility: The ability of a substance to mix with another substance.

mixed-air system: An HVAC system that brings a controlled amount of return air into contact with outdoor air for building use.

mixing valve: A three-way valve that has two inlets and one outlet and is used to mix two water supplies into one desired flow to a terminal device.

mixture: A substance that contains various combinations of solids, liquids, and/or gases.

modular forced-air heating system: A heating system that uses multiple heating units to produce heat for a building.

modulating gas valve: A gas control valve that regulates the high-low flow of gas to the burners of a boiler and typically has an attached ignition control system.

modulating valve: An infinite-position valve.

moisture indicator: A chemical patch inside a sight glass that changes color when moisture is present in refrigerant.

mold: Fungal growth produced on damp or decaying matter.

Mollier diagram: *See* pressure-enthalpy diagram.

motor nameplate: A plate attached to a motor that lists pertinent motor and manufacturer information.

motor starter: A heavy-duty switching device that includes motor overload protection and typically uses a low-voltage coil to energize or de-energize the high-voltage motor connected to it.

movable-vane compressor: A type of rotary compressor that has a rotor, with sliding vanes, that is mounted off-center inside a cam ring.

muffler: A device that reduces transmitted vibrations and noise.

multibulb thermostat: A thermostat that contains more than one mercury bulb switch.

multidrop network: A network of communications modules connected to a main trunk wire that runs throughout a building.

multimeter: A meter that is capable of measuring two or more electrical quantities.

multiple chemical sensitivity (MCS): An allergy-like illness resulting from contact with, or proximity to, a spectrum of substances, including airborne agents.

multistage thermostat: A thermostat that contains several mercury bulb switches that make and break contacts in stages.

multizone system: An air distribution system that serves a relatively small number of zones from a single, central AHU.

N

National Fire Protection Association (NFPA): A national organization that provides guidance in assessing hazards of the products of combustion.

natural draft: **1.** Draft produced by the natural action resulting from temperature and density differences between the cool atmospheric air entering a combustion chamber and the hot gases of combustion going up the chimney of a furnace or boiler. **2.** A draft produced by the airflow resulting from temperature and density differences between the ambient air and the heated air.

natural-draft cooling tower: A cooling tower that uses natural draft to circulate the ambient airstream through the tower.

natural gas: A colorless and odorless gas fuel.

negligence: A legal issue that occurs when there is an injury due to a lack of performance by building owners.

net unit output: The gross output of a water or steam boiler multiplied by the percentage of loss due to pickup.

net wall area: The area of a wall after the areas of windows, doors, and other openings have been subtracted.

network address: A unique number given to each controller on an RS-485 communication trunk line.

network communication module (NCM): A controller that coordinates communication between controllers in a network and provides locations for operator interface.

neurotoxin: A contaminant that can damage the central nervous system.

neutron: An uncharged particle in the nucleus of an atom.

new-building commissioning: A commissioning process applied to new systems and buildings during construction.

new unit adapter: A section of duct that has the new unit discharge size on one side and the old plenum size opening on the other.

nitrogen dioxide (NO_2): A toxic gas and corrosive that is a highly reactive oxidant.

noise: Sine wave distortion from motors and electrical devices that enters a power distribution system directly on the wires/grounds or through the magnetic coupling of adjacent wires.

nominal diameter: The approximate distance from outside edge to outside edge of pipes or inside edge to inside edge of tubes.

noncondensable: Any gas that will not condense under the temperatures and pressures present during the normal operation of an air conditioning system.

noncontact temperature probe: A device used for taking temperature measurements on energized circuits or on moving parts.

nonelectrical lockout devices: Lightweight enclosures that allow the lockout of standard control devices.

nonflammable substance: A substance that is not easily set on fire or is incapable of burning.

nonlinear scale: A scale that is divided into unequally spaced segments.

nonparallel misalignment: A condition in which shafts are not parallel with one another.

nonrenewable energy: Energy that is generated from fossil fuels and cannot be generated over and over as needed.

N-type material: Material created by doping a region of a crystal with atoms from an element that has more electrons in its outer shell than the base material.

nucleus: The heavy, dense center of an atom.

nut driver: A hand tool with a tip designed to fit over a hex nut and used for turning or holding.

O

O&M program: A program that integrates specific O&M plans and general O&M procedures to ensure that control devices and equipment are properly operated and maintained.

O&M systems manual: A document used to describe the operations and maintenance information and the limitations of individual systems.

O&M training plan: A written course of action with specific tasks related to the training needs of operating and maintenance personnel and users of systems and buildings.

O&M training program: A program that identifies specific training objectives, the knowledge needed for staff training, and the methods for delivering the information.

Occupational Safety and Health Administration (OSHA): A federal government regulatory agency that requires all employers to provide a safe environment for their employees.

off-gassing: The release of a chemical compound from a material into the air.

offset misalignment: A condition in which coupling hubs or pulleys are not on the same axis.

oil-fired atmospheric burner: A type of burner that consists of an open pot into which fuel oil flows at a controlled rate.

oil-fired power burner: A type of power burner that atomizes fuel oil with the air.

oil separator: A container that allows hot discharge refrigerant vapor to enter the container and then allows the refrigerant to discharge from the container to the condenser, while the oil is piped to the compressor crankcase.

100% outdoor-air system: An HVAC system that does not recirculate any return air for building use.

one-pipe primary-secondary piping system: A piping system that circulates boiler water through a primary loop.

one-pipe series piping system: A piping system that circulates water through one pipe, looped through each terminal device.

ongoing commissioning: A commissioning process applied to systems and buildings on a periodic or continuous basis to verify that the systems and building continue to meet stated objectives.

open-loop control: A control system in which feedback does not occur between the controller, sensor, and controlled device.

open loop system: A type of geothermal ground loop system that draws groundwater from a well directly into a building where it is used by heat pumps to heat or cool the building spaces and returns the water to the earth through a second well.

open pot burner: A burner used in kerosene heaters.

operating control: A device used to cycle on and off or regulate the various components of a refrigeration unit and/or air conditioning system to meet specific conditions for temperature, pressure, and humidity.

operation and maintenance (service) manual: An equipment-specific guide used by qualified technicians to service residential and light commercial HVAC systems.

operations and maintenance (O&M) plan: A written course of action with specific tasks related to the scheduling of repairs to equipment, production systems, and facilities.

operator interface: A device that allows individuals to access and respond to BAS information.

optoisolation: A communications method in which controllers use photonic (light-sensitive) components to prevent communication problems, typically between transmitters and receivers.

orifice: A small precisely sized hole through which gas fuel flows.

outdoor design temperature: The expected outdoor temperature that a heating or cooling load must balance.

outdoor design temperature table: A table of data developed from records of temperatures taken in a geographic area over a period of years.

outdoor unit: A component that contains a heat exchanger and a fan.

outer mantle: The light blue area of a flame that surrounds the inner mantle and is produced as carbon reacts with oxygen.

output rating: The total amount of heat a furnace or boiler will produce in one hour; expressed in Btu/hr.

outside micrometer: A micrometer used for measuring the outside diameter and thickness of parts.

overcurrent protection device: A device that shuts off the power supply to a circuit when current flow is excessive.

owner manual: A booklet that typically offers the best way to use equipment for an application and that contains specific preventive maintenance recommendations for the equipment.

ozone: An unstable gas with a characteristic pungent odor.

P

panel detail drawing: A detail drawing that provides specific information about the mountings and connections found in an HVAC control panel.

panel enclosure detail: A detail drawing that provides an elevation view of the front and side of an HVAC control system panel.

panel layout detail: A detail drawing that provides an elevation view of the interior of an HVAC control panel.

panel mounting detail: A detail drawing that provides an elevation view of HVAC control panels and mounting support.

parallel circuit: A circuit that has two or more components connected so that there is more than one path for current to flow.

parallel device: A device that transmits multiple bits of information (0s or 1s) at the same time.

parallel flow cooling tower: A cooling tower in which the water and ambient air are introduced at the top of the tower and flow downward through the fill.

partial short: A short circuit that occurs only in sections of a machine.

particle: A solid or liquid that is light enough to be suspended in air.

pathogen: A microorganism that can cause fever or disease.

perfluorocarbon (PFC) refrigerant: A type of refrigerant that consists of only fluorine and carbon.

perimeter loop duct system: A supply duct system consisting of a single duct loop that connects all outlets (registers) together and includes supply ducts.

peripheral device: A device that is connected to a personal computer or BAS controller to perform a specific function.

personal protective equipment (PPE): Gear worn by technicians to reduce the possibility of injury in the work area.

personnel lifts: Mechanized devices that lift a technician beyond the typical reach of a ladder.

pesticide: A product used to kill household pests and insects.

phase sequence indicator: A device used to determine phase sequence and open phases.

photodarlington transistor: A type of transistor that consists of a phototransistor and a standard NPN transistor in a single package.

photodiode: A type of diode that changes resistance or switches on when exposed to light.

phototransistor: A type of NPN transistor that has a large, thin base region that is switched ON when exposed to a light source.

physiological function: The natural physical and chemical function of an organism.

pickup: The additional heat that is needed to warm the water in a hydronic or steam heating system once a period of off-time, such as overnight, has passed.

piercing valve: A temporary valve that is attached to and taps into a refrigerant line for service.

pilot burner: A small burner located near burner tubes that ignites the fuel and air mixture released into the burner tubes.

pilot safety control: A safety control that determines when a pilot light is burning.

pipe section: A length of pipe that runs from one fitting to another fitting.

piston stroke: A measurement of the distance a piston moves up or down in a cylinder.

pitot tube: A device that senses static pressure and total pressure in ductwork.

pliers: A hand tool with opposing jaws for gripping and/or cutting.

plug tap: A tap used for through tapping and as the second tap in bottom tapping. Also known as an intermediate tap.

poor IAQ: Indoor air that contains harmful particle levels and/or has an uncomfortable temperature or humidity level.

positive pressure: Pressure greater than atmospheric pressure.

potential energy: Stored energy that is not in motion.

pot-type fuel oil burner: A fuel oil burner that vaporizes fuel oil by heating it from below.

pour point: The lowest temperature at which a liquid, such as gas, can flow.

power: The rate at which work (in ft-lb) is produced per unit of time (in sec).

power burner: A type of burner that uses a fan or blower to supply and control combustion air.

power control: A control system that controls the flow of electricity to equipment, such as a furnace.

power drill: A power-driven rotary tool used with a drill bit with cutting edges for boring holes in materials such as wood, metal, and plastic.

power factor (PF): The ratio of true power to apparent power in an alternating current (AC) circuit and describes the displacement of voltage and current waveforms in AC circuits.

power tool: A hand tool that uses externally supplied power from an electrical outlet or battery.

precision: The level of accuracy or mechanical exactness.

pressure: The force, in pounds, exerted over a surface divided by the area of the surface.

pressure atomizing burner: A burner that sprays oil into a combustion chamber through an atomizing nozzle.

pressure drop: A loss of energy.

pressure-drop chart: A chart used by a designer to properly size each line section of a refrigeration system to produce the required pressure drop.

pressure-enthalpy diagram: A graphic representation of the pressure/enthalpy relationship of a refrigerant. Also known as a Mollier diagram.

pressure gauge: A boiler fitting that displays the amount of pressure in pounds per square inch (psi) inside a boiler.

pressure-reducing valve: A valve that reduces the pressure of makeup water from a water supply line.

pressure switch: An electromechanical combustion safety control device that detects the pressure in a flue created by the flow of air from a draft induction fan.

pressure-temperature gauge (tridicator): A boiler gauge that measures water pressure and temperature at the point where the gauge is located.

pressure vessel: A tank or container that operates at a pressure greater than atmospheric pressure.

primary air: Air that mixes with fuel, such as in a boiler, to produce a combustible air-fuel mixture that controls the rate of combustion.

primary control unit: A special control device for gun-type fuel oil burners that allows a low-voltage (24 VAC) thermostat to control a burner and provides a safety feature when no flame is detected.

primary division: A division with a listed value.

primary piping loop: A piping system that distributes the main supply of chilled water to the building and the main return of water to the chiller.

programmable thermostat: A stand-alone thermostat that has the capability to be programmed by building occupants.

programmer: A control device that functions as the mastermind of a control system.

propane: A gas fuel composed of 90% to 95% propane gas and 5% to 10% propylene (C_3H_6).

propane-butane mixture: A fuel mixture of propane and butane that is used when the properties of each gas alone are not sufficient for an application.

propeller fan: A type of air moving device that consists of angled blades mounted on a central hub.

proportional control algorithm: A control algorithm that positions the controlled device in direct response to the amount of offset in a BAS.

proportional thermostat: A thermostat containing a potentiometer that sends out an electric signal that varies as the temperature varies.

protective clothing: Clothing made of durable material, such as denim, that provides protection from contact with sharp objects, hot equipment, and harmful materials.

protective helmet: A specially designed hat used to prevent injury from impacts to the head, falling and flying objects, and electrical shock in the workplace. Also known as a hard hat.

proton: A positively charged particle in the nucleus of an atom.

psychrometer: A humidity analyzer that uses two thermometers with the bulb of one thermometer kept moist (wet bulb) and the other bulb kept dry (dry bulb).

psychrometric chart: A graph that defines the properties of air at various conditions.

psychrometrics: The mathematical and graphical study of the properties of air.

P-type material: Material created by doping a region of a crystal with atoms from an element that has fewer electrons in its outer shell than the base material.

pulley (sheave): A device used to connect a belt to a shaft.

pulsation chamber: A short pipe in the discharge line of a compressor that is larger in diameter than the largest suction piping to the compressor.

pulse burner: A type of low excess air burner that introduces an air-fuel mixture to a burner face in small amounts or pulses.

Q

qualified person: A person who has special knowledge, training, and experience in the installation, programming, maintenance, and troubleshooting of specialized equipment.

quick-opening valve: A lever-operated boiler fitting valve that functions as a sealing valve.

R

radiant floor heater: A heating system that uses the interior surface of a floor as a radiant surface.

radiant panel: A factory-built panel with a radiant surface that is heated by a piping coil or grid built into it.

radiant surface: The heated surface from which radiant energy waves are generated.

radiation: A method of heat transfer in which invisible heat rays move heat.

radiator: A terminal device that consists of hollow metal coils or tubes.

radon: An odorless, cancer-causing gas that is a form of ionizing radiation.

ratchet bender: A hand tool with gears used to bend ½″ to ¾″ annealed copper tubing.

ratchet clause: A penalty charged for the electricity demanded during an entire 12-month period based on the highest amount measured during a single month.

receptacle tester: An instrument that is plugged into a standard receptacle to determine when the receptacle is properly wired and energized.

reciprocating saw: A multipurpose power tool with a blade that reciprocates (quickly moves back and forth) to create a cutting action.

recirculating water system: A type of water-medium condenser water system in which water is sent into the condenser, moved to a location where the heat can be rejected, and then pumped back to the condenser.

recommissioning: A commissioning process applied to systems or buildings that were previously commissioned and typically involves changes in their use or operation.

recording psychrometer: A psychrometer that has the wet and dry bulb thermometers connected to a recorder.

rectifier: A device that changes AC voltage into DC voltage.

reducing plenum duct system: A supply duct system consisting of a plenum box attached to the discharge end of a heating unit with a plenum that gets progressively smaller and ducts that branch to individual rooms extending along the length of the main duct.

refinery: A plant where crude oil is processed into petroleum products.

refrigerant: A substance that moves heat by changing state from a liquid to a gas when absorbing heat and from a gas to a liquid when rejecting heat.

refrigerant control valve: A combination expansion device and check valve.

refrigerant flow rate: A measurement of the amount of refrigerant flowing through a system or component in one minute.

refrigerant reclaiming: A process that allows used refrigerant to be conditioned to new product specifications by means that may include distillation.

refrigerant recovery: A process that allows refrigerant to be removed from a system in any condition and stored in an external container without necessarily testing or processing it in any way.

refrigerant recycling: A process that allows refrigerant to be cleaned by separation with single- or multiple-pass devices, such as replaceable-core filter-driers, which reduce moisture, acidity, and particulate matter.

refrigerant velocity: The speed at which refrigerant flows through the piping of a refrigeration system and is expressed in feet per minute (fpm).

register: A device that covers the opening of supply air ductwork and consists of a panel with vanes, a frame for mounting, and a damper for controlling airflow.

register free area: The face area of a register minus the area blocked by the frame or vanes.

regulation: A rule that has been made mandatory by a federal, state, or local government.

reheat: Heat added to supply air, at the point of use, when the air traveling through a duct system has dropped in temperature.

reheat system: An air distribution system that permits zone or space control for areas of different exposure, provides heating or cooling of perimeter areas of unequal exposure, or provides process or comfort control where close temperature control is required.

relative humidity: The amount of moisture in air compared to the amount of moisture the air would hold at the same temperature when the air is saturated, or full of moisture.

relay: A device that uses a low voltage coil to close one or more sets of contacts to switch high-voltage loads on and off.

renewable energy: Energy that is generated from rivers, the sun, oceans, and from beneath the surface of the planet to generate energy over and over as needed.

reset control: A DDC feature in which a primary setpoint is automatically reset to another value as the reset variable changes.

resistance: The amount of opposition a material has to the flow of electrons.

resistance heating element: A device consisting of a grid of electrical resistance wires that are attached to a support frame with ceramic insulators.

retrocommissioning: A commissioning process applied to existing systems or buildings that were not previously commissioned. Also known as existing building commissioning.

reversing valve: A four-way directional valve that reverses the flow of refrigerant through a heat pump system.

revolutions per minute (rpm): The rated motor rotation speed per minute at full-load condition.

rigging: Securing equipment in preparation for lifting.

rigid coupling: An inflexible coupling that is used to join two precisely aligned shafts.

rising-stem outside stem and yoke (os&y) valve: A gate valve that indicates whether it is open or closed by the position of the stem.

rooftop unit: A packaged HVAC unit mounted on the roof of a light commercial building.

rope grab: A device that clamps securely to a rope and contains a ring to which a lifeline can be attached.

rotary-type fuel oil burner: A fuel oil burner that sprays fuel oil by rotating a nozzle assembly at high speed in a circular pattern.

rotor: The moving part of a motor that spins in response to current flow.

RS-485: A differential data transmission standard that allows communication from 32 transmitters to 32 receivers at data rates up to 9600 kb/s and distances up to 4000 ft.

RS-232: A single-ended data transmission standard that allows communication from one transmitter to one receiver at data rates up to 20 kb/s and distances up to 50 ft.

rubber insulating gloves: Gloves made of latex rubber and are used to provide maximum insulation from electrical shock.

rule: A semiprecision measuring tool used for measuring length.

S

safety control: A control that monitors the operation of a furnace.

safety data sheet (SDS): A printed document used to relay hazardous material information from the manufacturer, importer, or distributor to the technician.

safety glasses: An eye protection device with special impact-resistant glass or plastic lenses, reinforced frames, and side shields.

safety label: A label that indicates areas or tasks that can pose a hazard to personnel and/or HVAC equipment.

safety net: A net made of rope or webbing designed to catch and protect a falling worker.

safety relief valve: A valve used only on a hot water boiler that prevents excessive pressure from exceeding the maximum allowable working pressure (MAWP) of the boiler during a malfunction.

safety valve: A steam boiler fitting that prevents a boiler from exceeding its maximum allowable working pressure (MAWP).

saturated air: A mixture of water vapor and air where the relative humidity is 100%.

saturated liquid: Liquid at a certain temperature and pressure that will vaporize when pressure is decreased or temperature is increased.

saturated vapor: Vapor at a certain temperature and pressure that will condense when the pressure is increased or temperature is decreased.

saturation pressure: The pressure at which a substance, such as a refrigerant, changes state.

saturation temperature: The temperature at which a substance, such as a refrigerant, changes state.

Schrader (spring-loaded) pressure gauge: A mechanical pressure gauge that has a small, single-acting spring return cylinder that acts as a pressure-sensing element.

screwdriver: A hand tool with a tip designed to fit into a screw head for fastening operations.

scroll: A sheet-metal enclosure that surrounds a blower wheel.

seasonal energy efficiency ratio: The cooling performance rating of a refrigeration system, such as an air conditioner, which operates under normal conditions over a period of time.

secondary air: Air that is added to a combusting air-fuel mixture, such as in a boiler, to ensure complete combustion.

secondary division: A division that divides primary divisions in halves, thirds, fourths, fifths, etc.

secondary piping loop: A piping system connected to the main supply and return piping loop used to circulate chilled water to building terminal units/cooling coils.

seismograph: A device that measures and records vibrations in the earth.

semiconductor: A material in which the electrical conductivity is between that of a conductor (high conductivity) and that of an insulator (low conductivity).

sensible heat: The heat applied to or removed from a substance that results in a temperature change.

serial device: A device that transmits one bit of information (0 or 1) at a time.

series circuit: A circuit that has two or more components connected so that there is one path for current to flow.

series-parallel circuit: A combination of series- and parallel-connected components in a circuit.

service factor (SF): A number that represents the amount of extra load that can be placed temporarily on a motor without causing damage.

service practice: A procedure created by the EPA and used by technicians who service and dispose of refrigeration and air conditioning equipment for the reduction of refrigerant emissions.

service valve: A type of valving system that allows technicians to attach manifold gauges to the system to record system pressures, charge the system, or recover refrigerant.

setback setpoint: The setpoint used during unoccupied heating of a building.

setpoint: The desired value of a controlled variable to be maintained by a system.

setpoint schedule: A description of the change in the primary setpoint based on the reset variable.

setup setpoint: The setpoint used during unoccupied cooling of a building.

shaded-pole motor: A 1ϕ motor that develops low starting torque and is used in applications of 1⁄20 HP or less.

shell-and-coil condenser: A type of heat exchanger that consists of a tubing coil placed inside a shell.

short circuit: A condition in which extremely high current leaves its normal current-carrying path and goes around a load and back to the power source or to ground.

sick building syndrome (SBS): A situation in which common symptoms experienced by building occupants can be associated with their presence in a specific building but cannot be linked to a specific illness or cause.

sight glass: A special pipe fitting that contains a small glass window that provides a visual indication of a refrigerant's condition.

silicon-controlled rectifier (SCR): A type of thyristor that is capable of only switching direct current.

single-line drawing: A drawing in which the walls and partitions are shown by a single line and that does not indicate actual wall thickness.

single-zone system: An air distribution system that serves a single temperature control zone and is the simplest form of air-handling system.

siphon: A bent tube installed between a steam boiler and a boiler fitting that provides a water seal to prevent the fitting from coming into direct contact with steam.

sleeve coupling: A hollow cylinder that is machined to match the size of the shafts.

sling psychrometer: A psychrometer consisting of two glass thermometers attached to an assembly that permits the two thermometers to be rotated through the air.

smoke: A collection of unburned carbon particles that is carried away by the convection currents generated by a flame.

snubber: A pipe-coupling-like device that protects pressure gauges from surges.

solar gain: Heat gain caused by radiant energy from the sun striking opaque objects.

soldering: A copper tubing joining method that uses nonferrous filler metal with a melting temperature less than 840°F.

solenoid: An electric output device that converts electrical energy into linear mechanical force.

solenoid-operated pilot valve: A valve connected to a hot gas refrigerant line that contains a solenoid and is used to control the main valve.

solid-state relay: A switching device that uses semiconductor materials to change on and off characteristics.

soot: A collection of unburned carbon particles on heating and combustion equipment that forms when the surface temperature is lower than the ignition temperature of the carbon.

source control: A method of reducing air pollution by reducing the origin of pollutants.

special clothing: A coverall or bodysuit made of a specific material that is treated to be impervious to gases, chemicals, or specific materials.

specific gravity: The density of a solid, liquid, or gas based on the density of water.

specific volume: The volume that one pound of a gas substance occupies at a particular temperature and pressure condition; expressed in cubic feet per pound (ft^3/lb).

split-phase motor: A 1ϕ motor that has separate start windings and run windings.

spontaneous combustion: Combustion, or fire, caused by self-ignition through chemical action.

spot heating: A heating method in which heat is provided for a particular part of a building space.

spread: The left to right pattern and distance from a unit where air velocity decreases below design fpm.

spring bender: A hand tool used to free-form bend annealed copper tubing from ¼″ to ¾″ in diameter.

stability: A condition that occurs when a substance can withstand changing conditions without breaking down.

stack switch: A mechanical combustion safety control device containing a bimetallic element that measures flue-gas temperature and converts it to mechanical motion to trip a set of contacts.

stagnant air: Air that is not circulating, typically contains an excess of impurities such as carbon dioxide, and lacks sufficient levels of oxygen required to provide comfort.

stand-alone control: The ability of a microprocessor-based controller to function on its own.

standard: An accepted reference or practice.

standards organization: A nongovernmental organization that develops standards based on an accumulation of knowledge and experience with materials, methods, and practices.

standing radiator: A radiator that has a section of vertical tubes that are connected with headers at each end.

standing steam radiator: A type of steam radiator with vertical tubes that are connected by headers all in an upright vertical position with legs and is installed on outside walls.

star network: A network of communication trunk wires extending to the remote controllers located in a building.

startup amperage: *See* locked-rotor amperage (LRA).

static electricity: An electrical charge at rest.

static head: The weight of water in a vertical column above a datum line.

static pressure: The force that nonmoving air exerts against surfaces such as ducts; measured in inches of water column (in. WC).

static regain method: A method of duct sizing in which each duct section is sized so that the static pressure increase at each takeoff offsets the friction loss in the preceding duct section.

stationary-vane compressor: A type of rotary compressor that has a rotor to which the vanes are attached.

stator: The stationary part of a motor that produces a rotating magnetic field.

steam baseboard convector: A type of convection heater terminal device that is enclosed in a 12″ high or lower cabinet and is installed at floor level along the outside wall of a building.

steam baseboard radiator: A section of horizontal tubes that are connected by headers at each end; installed at the base of outside walls.

steam boiler: A pressure vessel that transfers large amounts of heat to water to generate steam.

steam boiler accessory: A component not directly attached to a steam boiler but required for the proper operation of the heating system and/or boiler.

steam boiler fitting: Any device directly attached to a steam boiler and required for the proper operation of the steam boiler and system.

steam boiler vent: *See* steam vent valve.

steam cabinet convector: A type of convection heater terminal device that has its steam coil and controls enclosed in a metal cabinet.

steam convection heating: The heat transfer that occurs when currents circulate between the regions warmed by steam and the cool regions of a room.

steam header (main steam line): A special distribution pipe that supplies steam to the various branch lines of a steam heating system.

steam heating system: A heating system that uses steam to carry heat from the point of generation to the point of use.

steam pressure gauge: A steam boiler fitting that indicates the amount of pressure in a steam boiler; expressed in pounds per square inch (psi).

steam radiant heating: The heating that occurs when a surface is heated by steam and gives off heat in the form of radiant energy waves.

steam radiant panel: A type of radiant heating system terminal device that is factory-built using a panel with a radiant surface that is heated by steam through a piping coil or grid built into it.

steam radiator: A type of radiant heating system terminal device that consists of hollow metal coils or tubes through which steam flows.

steam strainer: A commercial and industrial steam distribution system accessory that is located in steam lines before steam traps to remove foreign matter from the steam that could cause a steam trap or valve to malfunction.

steam supply system: A system of pipes, valves, traps, and strainers that provide a path for steam to flow from a boiler to the point of use.

steam surface radiant system: A type of radiant heating system terminal device that uses steam to heat the interior surfaces of a room for use as radiant surfaces.

steam terminal device: A heating system component that transfers heat from the steam in a heating system to the air in building spaces and includes controls.

steam terminal device control: A steam distribution component that controls the air temperature in a building space by regulating the flow of air and steam through a terminal device in the space.

steam trap: A commercial and industrial steam distribution system accessory that removes air and condensate from steam lines and terminal units and returns the condensate back to the boiler for reuse as feedwater.

steam vent valve: A steam boiler fitting used to vent air from inside a boiler. Also known as an air cock or steam boiler vent.

stoichiometric combustion: A theoretical condition of combustion where the air-fuel ratio is perfect and there is complete combustion with 100% efficiency.

stop valve: A gate valve or globe valve that is used to shut off steam flow.

strobe light tachometer: A tachometer that contains a strobe light that flashes a beam of light at specified time intervals.

subcooling: The process of removing heat from a refrigerant after it has changed state from a vapor to a liquid.

subdivision: A division that divides secondary divisions in halves, thirds, fourths, fifths, etc.

suction accumulator: A device used to keep liquid refrigerant from entering a compressor.

suction stroke: The downward movement of a piston in a cylinder.

summer dry bulb temperature: The warmest dry bulb temperature expected to occur in a geographic area excluding the highest temperature that occurs in 1% to 5% of the total hours in the three hottest months of the year.

summer wet bulb temperature: The wet bulb temperature that occurs concurrently with the summer dry bulb temperature.

superheat: Sensible heat added to a refrigerant after it has changed state from a liquid to a gas.

surface blowdown valve: A valve located on the surface blowdown line of a steam boiler that is used to release water and surface impurities from the boiler.

surface-water closed loop system: A type of geothermal loop system that circulates water through a loop heat exchanger in a lake or pond in order to exchange thermal energy.

system glide: A temperature-pressure change in a system due to fractionation.

system graph: A small four-sided graphic representation of the operation of a refrigeration system on a pressure-enthalpy diagram.

system pressure: The pounds per square inch represented by the resistance to flow of refrigerant moving through a system.

system pressure drop: The total static pressure drop in a distribution system.

T

tachometer: An instrument used to measure the speed of rotation of a shaft or wheel.

tagout: The process of placing a danger tag on the source of electrical power, which indicates that the equipment may not be operated until the danger tag is removed.

tap: A hand tool used to cut internal threads in a predrilled hole or clean threads that have been disfigured by damage or improper operation.

tapered bore pulley: A two-piece pulley consisting of a tapered outside diameter hub.

taper tap: A tap used to start a true and straight thread and when a tap is required completely through a workpiece.

tape rule: A hand tool used for measuring lengths and that consists of a long, continuous strip of steel that is graduated in regular increments.

temperature: The measurement of the intensity of heat.

temperature-actuated defrost control: A control device that consists of a remote bulb thermostat and an electric control switch.

temperature control expansion device: An object that operates by sensing the temperature at a given point in the system (evaporator outlet) to control flow into the evaporator.

temperature probe: The part of a temperature measuring instrument that measures the temperature of liquids, gases, surfaces, and pipes.

temperature stratification: The variation of air temperature in a building space that occurs when warm air rises to the ceiling and cold air drops to the floor.

tension cord: The load-carrying element of a V-belt.

terminal device: A component at the end of a heating system that transfers heat from the hot water in a hydronic heating system to the air in building spaces.

terminal device control: A thermostat and valve combination that controls the temperature of the air in a building space by regulating the air and water flow through a terminal device.

ternary blend: A blend of three refrigerants.

testing: A procedure used to critically examine, observe, or evaluate the condition or operation of HVAC equipment.

test procedure: A description of the test methods, testing equipment, and expectations for tests conducted on components, equipment, systems, and buildings.

therm: A measurement of heat that equals 100,000 Btu.

thermal imager: A device that detects heat levels from infrared energy emitted without making direct contact with the target.

thermal overload (heater): An external motor protection device that uses the high temperatures caused by overcurrent conditions to open a motor circuit.

thermal protection: A temperature safety method that protects a motor from overheating.

thermal radiation: The heat transfer that occurs as radiant energy waves transport energy from one object to another.

thermistor: A solid-state sensor that changes resistance as temperature changes.

thermoanemometer: A permanently installed air-handling system device used to measure air velocity in a duct.

thermocouple: A device that produces electricity when two different metals that are twisted together are heated.

thermocouple and standing pilot gas valve: A gas control valve that is used to control the burners of a boiler and uses the power generated by the thermocouple.

thermodynamics: The scientific study of the behavior of heat.

thermopile: A device that generates a milliamp current that is used with a special thermostat designed to operate with the low current levels.

thermostatic expansion valve (TXV): An expansion device that uses pressure and temperature to control the amount of refrigerant entering the evaporator.

three-cornered file: *See* three-square file.

three-phase (3ϕ) motor: A motor that uses three phases of electricity to create a rotating magnetic field.

three-square file: A file that has angles of 60° and is used for filing internal angles and clearing out corners and threads. Also known as a three-cornered file.

throw: The distance discharged air flows from a register or diffuser.

thyristor: A solid-state switching device that switches current ON by a quick pulse of control current.

ton of cooling: The amount of heat required to melt one ton of ice (2000 lb) over 24 hr.

total head: The sum of friction head and static head.

total pressure: The sum of the static and velocity pressures in a duct system.

toxicity: A condition that occurs when a substance contains harmful components that may cause health problems.

trade association: An organization that represents the manufacturers or contractors that make, install, and service specific products.

transducer: A device that changes one form of analog signal into another form.

transient voltage: A temporary, unwanted voltage that ranges from a few volts to several thousand volts and lasts from a few microseconds to a few milliseconds.

transistor: A three-terminal semiconductor, switch-like device that controls current according to the amount of voltage applied to the base.

triac: A solid-state switching device used to switch alternating current.

try cock: A valve located on the water column of a boiler; used to determine boiler water level when the gauge glass is not functional.

tube-in-tube condenser: A type of heat exchanger that consists of one small diameter metal tube concentrically placed inside a larger tube.

tube-in-tube evaporator: A type of water-medium heat exchanger that consists of a small tube inside a large tube.

tubing cutter: A hand tool that consists of a cutting wheel and guide wheels attached to a rigid frame used to make square cuts on tubing.

two-pipe, direct-return piping system: A piping system that circulates supply water in the opposite direction of return water circulation.

two-pipe, reverse-return piping system: A piping system that circulates return water in the same direction as supply water.

TXV hunting: A condition where the thermostatic expansion valve opens and closes often, causing widely varying evaporator conditions.

U

ultrasonic leak detector: A leak detector that detects the sounds created by a leak.

ultraviolet (UV) air cleaner: An air cleaner that kills biological contaminants using a specific light wavelength.

ultraviolet (UV) light treatment: A treatment that involves using ultraviolet light to kill the microorganisms and biological growth that exist in the presence of moisture.

Underwriters Laboratories Inc. (UL): An independent organization that tests equipment and products to verify conformance to national codes and standards.

unitary controller: A controller designed for basic zone control using a standard wall-mount temperature sensor.

United Association of Journeymen and Apprentices of the Plumbing and Pipe Fitting Industry of the United States and Canada (UA): A multicraft union.

unit heater: A self-contained forced convection heater that is not connected to any type of ductwork and contains a fan, a hot water coil, and controls in one cabinet.

unit ventilator: A forced convection heater that has a blower, a hot water coil, filters, controls, and a cabinet that has an opening for outdoor ventilation.

universal input-output controller (UIOC): A general controller designed to control most HVAC equipment.

upflow (highboy) furnace: A furnace that heats air and distributes the heated air upward through ductwork.

U-tube manometer: A U-shaped section of glass or plastic tubing that is marked in increments and partially filled with water or another liquid used to measure pressure.

V

variable-air-volume (VAV) box controller: A controller that modulates the damper inside a VAV terminal box to maintain a specific building space temperature.

variable-air-volume (VAV) system: An air distribution

system that provides air at a constant temperature but varies the amount of supplied air in order to maintain the desired setpoints within the building space.

variable-frequency drive (VFD): An electronic device that controls the direction, speed, torque, and other operating functions of an electric motor in addition to providing motor protection and monitoring functions.

variable-volume damper: A damper that controls the amount of airflow at a terminal device in a VAV system.

V-belt: A power transmission device with a trapezoidal cross section.

velocity pressure: The pressure created by air as it travels, such as through a duct.

velocity reduction method: A method of sizing ducts that causes air velocity to be reduced at each branch takeoff.

vent damper: An automatic flue damper that shuts off a flue pipe to dramatically reduce heat loss when the burner of a heating unit is not operating.

ventilation: The process of reducing air pollution by introducing enough fresh outdoor air into a building to dilute indoor pollutants.

ventilation air: Outdoor air that is brought into a building to improve indoor air quality.

venting system: A continuous open passageway from a vent collar to the outside.

vent valve: A boiler fitting used to vent air from inside a boiler. Also known as an air cock or boiler vent.

vernier caliper: A caliper with a main scale and a parallel sliding vernier scale.

vernier scale: A short auxiliary scale placed along the main scale of an instrument to provide accurate fractional readings of the smallest division on the main scale.

vertical closed loop system: A type of geothermal ground loop system in which U-tube ground loops that circulate water are inserted into holes drilled perpendicular to the surface in order to exchange thermal energy.

vibration eliminator: A piece of flexible material placed between sheet metal collars used to isolate rigid ductwork from the vibrations of an HVAC unit.

viscosity: The measurement of a liquid's resistance to flow (thickness).

volatile organic compound (VOC): A carbon-containing compound that easily evaporates at room temperature.

voltage: The amount of electrical pressure in a conductor.

voltage rating: The voltage that a motor uses to operate properly.

voltage tester: An instrument that indicates approximate voltage level and type (AC or DC) through the movement and vibration of a pointer on a scale.

volume control expansion device: An expansion device that controls the amount, or volume, of refrigerant that is able to pass through to the evaporator.

volumetric flow rate: The rate of refrigerant flow expressed in cubic feet per minute.

W

wall unit: A packaged HVAC unit that heats a building space by using a fan and heat exchanger mounted in a metal enclosure, which is installed in or against an outside wall.

warning signal word: A word that indicates a potentially hazardous situation that, if not prevented, could result in death or serious injury.

water column: A steam boiler fitting that reduces the movement of boiler water in a gauge glass to provide an accurate boiler water level reading.

water hammer: A banging noise caused when water in the steam lines of a boiler moves rapidly and hits obstructions such as elbows and valves.

water-medium condenser: A type of condenser that uses water as the heat-exchange medium.

water-medium evaporator: An evaporator that uses a pump to circulate water across its coil to give up heat to the refrigerant. Also known as a cooler.

water pressure drop: The decrease in water pressure caused by friction between water and the inside surface of piping as the water moves through the piping.

water-source heat pump control: A thermostat that controls system fans, reversing valves, and expansion devices when the hydronic loop controls are activated.

water strainer: A device in a hydronic heating system that removes solid particles from the water inside the piping system.

water-to-air installation and maintenance manual: A general guide used by qualified technicians to install water-source heat pump systems.

water-to-water installation and maintenance manual: A general guide used by qualified technicians to indicate the best way to use water-to-water heat pump equipment for an application and that contains specific PM recommendations for the equipment.

watertube boiler: A boiler that passes hot gases of combustion over the outside surfaces of water-filled tubes to heat the water.

weight: A measurement of gravity acting on the mass of a substance.

wet base boiler: A boiler that has water on all sides and below a firebox.

wet-bulb temperature: The temperature of air if it was cooled to 100% relative humidity.

wet cooling tower: A cooling tower with open water loops, allowing direct contact between circulating water and outside air.

wet leg boiler: A boiler that has water along the sides of a firebox.

winter dry bulb temperature: The coldest temperature expected to occur in a geographic area excluding the lowest temperatures that occur in 1% to 2½% of the total hours in the three coldest months of the year.

wire stripper/crimper/cutter: A hand tool used for the removal of insulation from small-diameter wire.

wiring detail drawing: A detail drawing that provides wiring termination information for sensors, actuators, or pieces of equipment that have several termination possibilities or that connect to multiple voltage sources.

wood smoke: A smoke mixture of carbon monoxide, cancer-causing polycyclic aromatic hydrocarbons, and various small particles derived from burning wood.

work: The amount of energy transferred by a force (in lb) over a specific distance (in ft).

workers' compensation: Compensation offered to an employee who is temporarily unable to work because of a job-related injury.

wraparound bar graph: A bar graph that displays a fraction of the full range on the graph at one time.

wrench: A hand tool with jaws at one or both ends that is designed to turn bolts, nuts, or pipes.

Z

zener diode: A type of diode designed to operate in a reverse-biased mode without being damaged.

zeotropic refrigerant: A refrigerant that is formed when refrigerants are blended together but remain in their separate component states.

Index

Page numbers in italic refer to figures.

A

B

D

E

F

G

H

I

M

N

O

P

S

T

heating mode scenarios
DDC controller problems, 905–906, *906*
defective transformer, 904, *905*
water loop temperature problems, *906*, 906
wrong thermostat, *905*, 905

U

V

W

Y

Z